CAMBRIDGE LIBRARY COLLECTION

Books of enduring scholarly value

History

The books reissued in this series include accounts of historical events and movements by eye-witnesses and contemporaries, as well as landmark studies that assembled significant source materials or developed new historiographical methods. The series includes work in social, political and military history on a wide range of periods and regions, giving modern scholars ready access to influential publications of the past.

A New Universal Dictionary of the Marine

First published in 1813, this comprehensive dictionary of maritime terminology in the eighteenth and early nineteenth centuries was designed as a reference work for shipbuilders and navigators. Compiled by the naval historian William Burney, it is a revised and expanded version of the classic *Universal Marine Dictionary* originally published in 1769 by the poet and lexicographer William Falconer (1732–89). It provides over 800 pages of technical data on shipbuilding, navigation, the operation of ships, weaponry and provisions, as well as historical, legal and medical information, and even French vocabulary lists. It contains extensive tables and illustrations, and longer articles including, for instance, lists of clothing, books and equipment required by a young midshipman. Both Burney and Falconer were leaders in their field, and the book remains a key historical source on seafaring during the age of Nelson and the French wars, for scholars and enthusiasts alike.

Cambridge University Press has long been a pioneer in the reissuing of out-of-print titles from its own backlist, producing digital reprints of books that are still sought after by scholars and students but could not be reprinted economically using traditional technology. The Cambridge Library Collection extends this activity to a wider range of books which are still of importance to researchers and professionals, either for the source material they contain, or as landmarks in the history of their academic discipline.

Drawing from the world-renowned collections in the Cambridge University Library, and guided by the advice of experts in each subject area, Cambridge University Press is using state-of-the-art scanning machines in its own Printing House to capture the content of each book selected for inclusion. The files are processed to give a consistently clear, crisp image, and the books finished to the high quality standard for which the Press is recognised around the world. The latest print-on-demand technology ensures that the books will remain available indefinitely, and that orders for single or multiple copies can quickly be supplied.

The Cambridge Library Collection will bring back to life books of enduring scholarly value (including out-of-copyright works originally issued by other publishers) across a wide range of disciplines in the humanities and social sciences and in science and technology.

A New Universal Dictionary of the Marine

William Falconer
Edited by William Burney

Cambridge University Press

CAMBRIDGE UNIVERSITY PRESS

Cambridge, New York, Melbourne, Madrid, Cape Town,
Singapore, São Paolo, Delhi, Tokyo, Mexico City

Published in the United States of America by Cambridge University Press, New York

www.cambridge.org
Information on this title: www.cambridge.org/9781108040037

This edition first published 1815
This digitally printed version 2011

ISBN 978-1-108-04003-7 Paperback

A NEW UNIVERSAL DICTIONARY OF THE MARINE;

BEING, A COPIOUS

EXPLANATION OF THE TECHNICAL TERMS AND PHRASES

USUALLY EMPLOYED IN THE

CONSTRUCTION, EQUIPMENT, MACHINERY, MOVEMENTS, AND MILITARY, AS WELL AS NAVAL, OPERATIONS OF SHIPS:

WITH SUCH PARTS OF ASTRONOMY, AND NAVIGATION, AS WILL BE FOUND USEFUL TO

PRACTICAL NAVIGATORS.

Illustrated with a Variety of Modern Designs of Shipping, &c.

TOGETHER WITH SEPARATE VIEWS OF THE

MASTS, YARDS, SAILS, AND RIGGING.

TO WHICH IS ANNEXED,

A VOCABULARY OF FRENCH SEA-PHRASES AND TERMS OF ART,

COLLECTED FROM THE BEST AUTHORITIES.

ORIGINALLY COMPILED BY WILLIAM FALCONER,

AUTHOR OF THE SHIPWRECK, &c.

NOW MODERNIZED AND MUCH ENLARGED,

BY WILLIAM BURNEY, LL.D.

MASTER OF THE NAVAL ACADEMY, GOSPORT.

LONDON:

PRINTED FOR T. CADELL AND W. DAVIES, IN THE STRAND; AND J. MURRAY, BOOKSELLER TO THE ADMIRALTY, 50, ALBEMARLE-STREET.

1815.

PRINTED BY JOYCE GOLD, 103, SHOE-LANE, LONDON.

TO

THE RIGHT HONOURABLE

ROBERT LORD VISCOUNT MELVILLE,

FIRST LORD COMMISSIONER OF THE ADMIRALTY, &c. &c.

THIS IMPROVED AND ENLARGED EDITION OF

FALCONER'S MARINE DICTIONARY

IS MOST HUMBLY DEDICATED BY THE EDITOR;

AS WELL

FROM MOTIVES OF GRATITUDE,

AS

FROM A DESIRE TO PAY A PUBLIC TRIBUTE OF SINCERE RESPECT,

JUSTLY DUE TO HIS LORDSHIP,

FOR

SO ABLY DISCHARGING

THE ARDUOUS DUTIES OF HIS HIGH AND IMPORTANT STATION,

(PARTICULARLY IN RECOMMENDING

THE INCREASE OF PAY AND EXTENSIVE PROMOTION RECENTLY CONFERRED
ON THE OFFICERS OF THE NAVY,

AND FOR HIS ZEAL IN OBTAINING

AN ADDITION TO THE PENSIONS OF THE WIDOWS,)

IN WHICH HE HAS DONE SO MUCH

HONOUR TO HIMSELF

AND

BENEFIT TO HIS COUNTRY.

PREFACE.

IN a country like England, where the principal sources of the wealth and power of the state are derived from the superiority of our Marine, and the skill, as well as undaunted courage of our seamen, an improved and enlarged Edition of FALCONER's celebrated Marine Dictionary, which has now been out of print upwards of thirty years, will undoubtedly be found of the greatest utility; especially at this period, when the nations of Europe are again likely to be involved in war. No work can be more conducive to the conveyance of instruction than such a compilation; being always useful, as a book of reference, and serving in a certain degree to fill up progressively, without the toil of application, the deficiencies of acquirement, or of memory, which a man may occasionally discover in his own mind. The Editor's object has been to collect from the best writers, a mass of useful information relative to the profession of young Sea-Officers; and he has endeavoured to bring into one point of view many scattered remarks, and much general knowledge, drawn from various sources which are not always accessible to the generality of sea-faring men. He has availed himself of some valuable notes and observations expressly drawn up for this work, and furnished him by his intimate friend the late Mr. William Bayly, many years Head Master of the Royal Naval Academy at Portsmouth, who, in conjunction with another of the Editor's friends, John M'Arthur, Esq. (the distinguished author of the "Treatise on Naval Courts Martial," and other valuable publications on nautical subjects), had turned his thoughts to a new Edition of FALCONER's Dictionary when first it became scarce. No men could be better calculated, from their general knowledge and experience, to do justice to such an undertaking; but at the commencement of the late war, Mr. M'Arthur was called into active service, as Secretary to Lord Hood, when the idea was laid aside: the Editor, however, is greatly indebted to him for many original articles, with which he has enriched the present work. He is also much obliged to his friend Mark Beaufoy, Esq. F.R.S. of Hackney Wick, for several valuable contributions.

In the course of his labours he has likewise received much information, and assistance from naval friends and other scientific gentlemen, to all of whom he begs to return his most grateful acknowledgments. He has also derived considerable help from STEEL's Elements and Practice of Rigging and Seamanship; DARCY LEVER's Young Sea-Officer's Sheet Anchor; and from NORIE's Mariner's Encyclopædia.

The terms of art now used in Ship-building, Mast-making, Sail-making, Anchor-making, Block-making, Rope-making, Rigging, &c. have been selected from a variety of sources: they are concisely defined under their respective heads, with a description of most of the implements and materials used in the different departments.

The articles which treat of Naval Architecture and the practical part of Ship-building, have been revised and considerably enlarged. Those on Gunnery are also improved, and some original matter will be found under the heads of CHARGE, GUN, GUNNERY, MORTAR, RANGE, SHELL, SHOT, WINDAGE, &c. The Editor has made several extracts from the Official Regulations and Instructions which relate to his Majesty's service at sea, for the guidance of commissioned and warrant-officers; they are placed under their respective heads, and a reference to that valuable book has been invariably given at the end of the articles: indeed, he has generally annexed his authorities throughout the work to the subjects he has introduced, the value of which he trusts will be duly appreciated, and they will serve to point out the various researches that have been made for suitable information.

The astronomical articles that are interspersed throughout the work for the benefit of practical seamen, have been selected from the best modern authors; these, with the description and use of astronomical instruments, the Editor hopes will be found useful, since they are closely connected with nautical Astronomy and practical Navigation, the theory of which has been briefly touched upon under the different Sailings.

The drawing and engraving of the Plates have been executed by respectable artists; they consist of a variety of Sections of Ships and Boats; Anchors, Buoys, Capstans, Cleats, and Cannon with their Carriages, &c.; a Chain-pump, Compasses, Hitches, Bends, and Knots; Pointing, Splicing, and Snaking of Ropes; new Filtering Apparatus for the use of ships; Masts, Yards, Booms, Sails, Rigging, and a variety of Tackle; Substitute for a rudder; Method of restoring Masts when wounded; and of making and repairing large Yards when sprung; Mortars with their Beds, Shells and Shot; Representations of fleets in the line of battle; Signals and Telegraphic Signals;

Naval Tactics, and the Theory of working ships; Hadley's, Gunter's, and the Gunner's Quadrant; a Sextant, and Troughton's Reflecting Circle; a first-rate Ship at Anchor; the Frame or Disposition of a second-rate; a 74-Gun Ship and a variety of small Vessels under sail, &c.

The plan of FALCONER's original work, so far as relates to sea-phrases and terms of art, has been strictly followed; but the Editor has with some diffidence, altered much of that respectable author's language, where it seemed absolutely necessary from the improved state of the Marine Arts, and the great change that has taken place since his time in our naval Service. Should it therefore be thought that he has been at all successful, he shall most gratefully attribute it to the solid foundation on which this compilation has been built.

Although the present work embraces more than double the quantity of matter in the original, yet, the Editor has been prevented from enlarging on many subjects so extensively as he could have wished, from the proprietors being desirous that it should not exceed one volume quarto. A grand national work, comprehending every thing relative to the Marine of this country, is a desideratum in our literature; but an undertaking of this sort, to be properly executed, should not only have a combination of talents, but likewise the indulgence of time, leisure, and convenience; from which the Editor has been in a great degree excluded, by the nature of his professional engagements: he must, therefore, rely on the indulgence of a discerning public for whatever deficiencies they may discover till another edition is called for.

On account of some alterations and improvements having taken place in the service, since the work was first sent to the press, from some articles being overlooked and a few not correctly defined, an APPENDIX was deemed necessary, to which the reader is requested to have recourse for such words and phrases as are not to be found in the body of the work.

Our naval officers have always found it much to their advantage in time of war, not only to have a general knowledge of the French language, but also a ready acquaintance with the technical terms used in the French Marine; and, in time of peace, FALCONER particularly states in his Preface the obvious reason "of understanding their pilots, when we have occasion for their assistance," as that first induced him to add a Vocabulary of the French sea-phrases and terms of art. This part of the work has therefore been carefully revised and considerably enlarged from the best editions of the *Vocabulaire des Termes de Marine of* LESCALLIER; a small *Vocabulaire de Phrases Marines, et de Termes de l'Art*, published by a Captain of the British Navy; and from articles with which the Editor was furnished by his late friend Captain Matthew Flinders, of the Royal Navy, who amused himself in correcting the Vocabulary in an edition of FALCONER's Marine Dictionary, which he happened to have in his possession during his long and unjust captivity in the Isle of France. In a letter to the Editor soon after his return to this country, he states, "by frequently conversing with some intelligent French officers, I discovered FALCONER's Vocabulary of the French sea-terms and phrases to be very inaccurate, and partly grown obsolete, from the improved state of the French navy."

The Editor has also derived some information from M. DU HAMEL's *Elémens de l'Architecture Navale, ou Traité Pratique de la Construction des Vaisseaux;* and from other French treatises on the same subject of a later date; and in order to render the Vocabulary as complete as possible, the whole has been submitted to the inspection of Mr. Cherpilloud, late Professor of languages at the Academy, whose knowledge in the various departments of the French Marine, has enabled him to supply many deficiencies. The English explanation of each article in the Vocabulary has been drawn up as concise as possible; being more copiously elucidated under the respective terms and phrases in the body of the work.

The Editor expects no praise for his labours, beyond that which must arise from an honest ambition to have deserved well of the British Navy, which, by his humble efforts, he has, during the last twenty-four years, endeavoured to improve; having trained up in the paths of virtue and science, probably, a greater number of young officers for the sea-service than any other individual in the kingdom; and should this work be found useful to them and the navy in general, he shall think himself amply rewarded, in having bestowed his leisure and attention to so gratifying a purpose.

He cannot conclude without again expressing his gratitude to the many learned and liberal friends who have honoured him with their assistance in the progress of the work, and the zeal which they manifested on the occasion encourages him to solicit a continuance of their communications, with a view to a future edition. For he is well aware that a Marine Dictionary must require constant corrections and additions, in order to make it keep pace with the improvements which are so frequently taking place in every department of our naval service.

GOSPORT, *March 25th,* 1815.

BIOGRAPHICAL MEMOIR

OF

WILLIAM FALCONER.

THE materials from which the greater part of the following Memoir has been drawn, were first presented to the public in the Royal and Imperial Octavo edition of the THE SHIPWRECK, 1804, edited by J. S. Clarke, F. R. S. Vicar of Preston; Chaplain of the Household to the Prince; and author of "The Progress of Maritime Discovery, &c." "I had long sought in vain," says Mr. Clarke, "to procure any authentic materials, however scanty, for a Biographical Memoir of poor FALCONER, when I fortunately met a shipmate of his, Governor Hunter, at the house of John M'Arthur, Esq. of York-Place, a literary and naval character of considerable eminence. With the natural cordiality of a seaman, the Governor communicated to me all the information he could remember, and this was also increased by frequent conversations which I enjoyed with that gallant veteran his brother, Lieutenant Hunter of Greenwich Hospital: but for these gentlemen, the little that has survived respecting FALCONER, would have perished; and even this, owing to the years that have elapsed since the loss of the Aurora frigate, in which he perished, must necessarily be scanty."

MR. WILLIAM FALCONER was born about the year 1730, and was the son of a poor but industrious barber at Edinburgh; who like Fielding's celebrated *Partridge*, possessed considerable talents and humour, and maintained a large family by his industry. It is remarkable that all his children, with the sole exception of William, were either deaf or dumb: FALCONER himself mentioned this singular circumstance to Captain Hunter, when they were shipmates together; and, "I had afterwards," adds that officer, "an opportunity of being convinced of its truth; when, long after the commencement of my acquaintance with him, I met two of his family labouring under their infirmities in the Poor-house at Edinburgh, where they continued until their death."

Poverty in Scotland is never depressed by ignorance, or a want of religious principles: FALCONER, therefore, though poor, had a proper bias given to his mind, from the first dawn of its intellectual powers; and it was this bias, impelled by the energy of his genius, which enabled him, without any further education, than what is acquired at the public schools established in every parish in that country, to reach the goal of literary fame.

At an early age, he entered on board a merchant vessel at Leith, and therein served his apprenticeship; we soon afterwards find him on board a man of war in the situation of a servant to Campbell, the well-known author of Lexiphanes, and a purser in the navy. This officer, according to Dr. Currie, the ingenious Editor of Burns' works, delighted in improving the mind of our young seaman; and afterwards, when he had attained celebrity, this early patron felt a pride in boasting of his scholar. Under such a kind master, and able critic, he, no doubt, profited considerably; and probably acquired that turn for satire, which he afterwards displayed in some of his poetical pieces. He began to exercise his genius at an early age, and published, at Edinburgh, "A Poem, sacred to the memory of Frederick Prince of Wales," 8vo. 1751; and although the confined nature of this loyal effusion offered him but little opportunity of displaying his descriptive and imaginary powers, yet there was that harmonious versification in it, which gave sufficient demonstration of a rising genius. But it is justly remarked, that the complimentary efforts of inexperienced and obscure bards are seldom noticed; and such, we believe, was the fate of this poem: however, at the recommendation of some kind friend, probably his old master, Campbell he was promoted to the quarter-deck about 1757, yet still continued to struggle with the hardships of his profession; and it appears his difficulties were likewise attended with misfortunes; as it is recorded in the 5th volume of the Naval Chronicle, under the biographical memoir of Commodore Locker, that FALCONER, while serving as midshipman on board the Ramillies, with Captain Taylor, was wrecked in that ship on the 15th of February, 1760. On the 5th of February, Admiral Boscawen, in the Royal William, sailed from Plymouth Sound with six sail of the line, to take the command of the fleet in Quiberon-Bay. The wind soon after shifted to the westward, and increased to a violent gale, which dispersed the squadron. The Ramillies was so much shattered, that Captain Taylor bore away for Plymouth; on the 15th, it being extremely thick and hazy, in coming up Channel, he discovered the Bolt-head, but mistaking it for the Ram-head, stood on until the ship was so entangled with the shore, that it was impossible to weather it. Captain Taylor ordered the masts to be cut away, and came to an anchor; but the storm raged with such fury, that the cables parted; the ship was in consequence driven among the breakers, and dashed to pieces. Out of 734 men, twenty-five only of the crew and FALCONER the midshipman were saved, by jumping from the stern to the rocks. In some lines, afterwards addressed to his patron the Duke of YORK, he styles himself, "a hapless youth, whose vital page was one sad lengthened tale of woe." It appears he occasionally relieved and strengthened his mind by literary occupation, and contributed several fugitive pieces to the Gentleman's Magazine; among others, lines *On the uncommon Scarcity of Poetry in that Publication,* signed J. W. A Sailor, March, 1756; *The Chaplain's Petition to the Lieutenants of the Ward-Room,* 1758; and *Verses addressed to a Lady,* dated, Ramillies, in the Bay of Biscay, Nov. 25th, 1758. *The Description of a Ninety-gun Ship,* 1759; and *The Loss of the Ramillies,* 1760. He also wrote *The Midshipman,* an excellent little Poem, descriptive of the abode and sentiments of that class of officers; which name it has since lost in the more technical one of *Orlop,* or the deck immediately over the hold; where, far removed from the light of day, and at a considerable distance below the surface of the water, the cabins or births of the midshipmen are placed.

> "In canvassed birth, profoundly deep in thought,
> His busy mind with Sines and Tangents fraught,
> A MID reclines!—in calculation lost!
> His efforts still by some intruder crost:
> * * * * * * *
> * * * * * * *
> "But think not meanly of this humble seat,
> Whence sprung the Guardians of THE BRITISH FLEET:
> Revere the sacred spot, however low,
> Which formed to martial acts—an HAWKE! an HOWE!"*

The well-known song of *Cease, Rude Boreas;* and several other old sea-songs, written about that time, are with great probability also ascribed to his pen.

Soon after being wrecked in the Ramillies, it appears FALCONER left the Royal Navy, and again entered the merchant-service, as mate on board the Britannia, a ship employed in the Levant trade,

* Captain Howe commanded the Alcide, in 1755, in a memorable action with the Lys. He also led the van in the Magnanime, 1757, under Admiral Knowles, in the attack on Aix; was afterwards created Earl Howe, and made First Lord of the Admiralty.

in which he also was wrecked, near Cape Colonna; from this melancholy event our poet afterwards drew the outline and characters of one of the finest poems in our language, entitled, "THE SHIPWRECK, in three Cantos by a SAILOR," 4to. published by Millar, in 1762, in which he feelingly described the dangers he actually experienced. This work he inscribed to Edward Duke of York, brother to his present Majesty; who then had hoisted his flag as Rear-admiral of the Blue, on board the Princess Amelia, of 80 guns, attached to the fleet under Sir Edward Hawke. For the illustration of several passages in it, FALCONER very judiciously prefixed a chart of the ship's way, and a section of the ship itself. To arrest the attention of his reader, and to render his poem doubly interesting, his well-chosen motto,

> " .... Quæ ipse miserrima vidi,
> Et quorum pars magna fui."

declares it to be a tale of facts, not fiction, for that he himself had not only witnessed, but had been a chief sharer in those misfortunes. "The main subject of the poem is the loss of the ship Britannia, a merchantman, bound from Alexandria to Venice, which touched at the Island of Candia; whence, proceeding on her voyage, she met with a violent storm that drove her on the coast of Greece, where she suffered shipwreck near Cape Colonna; and three only of the crew, were saved.

"The ship putting to sea from the port of Candia, the poet takes an opportunity of making several beautiful marine descriptions; such as the prospect of the shore; a shoal of dolphins; a water-spout; the method of taking an azimuth; and of working the ship.

"In the second Canto, the ship having cleared the land, the storm begins; and with it the consultation of the pilots, and operations of the seamen; all which the poet has described with an amazing minuteness, and has found means to reduce the several technical terms of the marine into smooth and harmonious numbers: so that if Homer has been admired by some for reducing a catalogue of ships into tolerable flowing verse, what praise must be due to FALCONER, who has versified his own sea-language with equal skill and propriety?

"Nor is the poet's talents confined to the description of inanimate scenes: he relates, and bewails, the untimely fate of his companions in the most animated and pathetic strains. The close of the master's address to the seamen, in the time of their greatest danger, is noble and philosophical. It is impossible to read the circumstantial account of the unfortunate end of the ship's crew, without being deeply affected by the tale, and charmed with the manner of the relation."* The novelty and interest of this poem not only procured a favourable reception from the public, but established the author's fame as a poet. Its versification is varied and melodious; its description, being drawn from reality, strong, glowing, and often original; and the characters so well chosen and pourtrayed, that, though remote the scene, it is pregnant with domestic sorrow.

The poem of THE SHIPWRECK is of inestimable value to this country, since it contains within itself the rudiments of navigation: if not sufficient to form a complete seaman, it may certainly be considered as the grammar of his professional science. Many experienced officers have declared, that the rules and maxims delivered in this poem, for the conduct of a ship in the most perilous emergency, form the best, indeed the only, opinions which a skilful mariner should adopt. We possess, therefore, a poem, not only eminent for its sublimity and pathos, but for an harmonious poetic assemblage of technical terms and maxims used in navigation, which a young sailor may easily commit to memory; and also, with these, such scientific principles as will enable him to lay a solid foundation for his future professional skill and judgment.

The Royal DUKE, anxious to honour FALCONER with some mark of his favour, recommended him to quit the merchant service again for the royal navy, where he might be able to serve him: accordingly, before the summer had elapsed, he was rated a midshipman on board Sir Edward Hawke's ship, the Royal George. In this ship Governor Hunter, then a midshipman, commenced an acquaintance with FALCONER, which continued until his death: being both of them from the same part of Scotland, their friendship and intimacy soon increased. At the peace of 1763, the Royal George was paid off,

* Monthly Review, vol. XXVII. p. 197.

when FALCONER was introduced to the gallant brother of his friend, Mr. William Hunter, then midshipman on board the Sutherland. Previous to the peace, the Duke of YORK had embarked on board the Centurion, with Commodore Harrison, for the Mediterranean; on which occasion FALCONER published an ode, entitled, *On the Duke of York's Second Departure from England as Rear-Admiral.* As FALCONER had not completed that time of service which is necessary to qualify a midshipman for a lieutenant's commission, his friends advised him to exchange the military, for the civil line in the royal navy; and, accordingly, in the course of the said year 1763, he, through the interest of his patron, was appointed purser of the Glory frigate, 32 guns, afterwards called the Apollo. About this time it is supposed he wrote *The Fond Lover,* a ballad, inserted in the edition of the "British Poets," 1790; and an *Address to Miranda,* a ballad, first published in Dr. Stuart's "Edinburgh Magazine and Review," for 1773. He soon after this married a young lady of the name of Hicks, daughter of the surgeon of Sheerness yard. Mrs. FALCONER is said to have displayed keen abilities; and that it was the lustre of her mind, rather than of her person, which attracted and confirmed the affection of her husband. In 1764 he published a new edition of THE SHIPWRECK, in 8vo. with considerable additions, comprehending several new descriptions, characters, and episodes, amounting in the whole to upwards of a thousand lines.

In the following year he enlisted in the field of political controversy; and being by gratitude, as well as principle, attached to the party of the King's friends, he wrote a satyrical poem, entitled "THE DEMAGOGUE," in which Mr. Pitt, afterwards Earl of Chatham, Wilkes, Churchill, and the opposition in general, were treated with great severity. The sentiments are, for the most part noble and manly, the satire poignantly severe, the expression strong and nervous; but the tendency of the poem cannot be commended, and the obloquy thrown on Mr. Pitt is totally inexcusable.

FALCONER's principal amusement seems always to have consisted in literary pursuits; and when the Glory was laid up in ordinary at Chatham, Commissioner Hanway, brother to the celebrated Jonas Hanway, became delighted with the genius of her purser. The captain's cabin was ordered to be fitted up with a stove, and with every addition of comfort that could be procured; in order that FALCONER might thus be enabled to enjoy his favourite propensity, without either molestation or expense. In this retirement he finished his celebrated Universal Dictionary of the Marine, 4to; a work that had engaged his utmost application for some years, and to which most of our modern Cyclopædias have been indebted. This alone would have immortalized his name had not his poetical reputation stood foremost. The undertaking was first suggested to him by George Lewis Scott, Esq. and its great utility was acknowledged by Sir Edward Hawke, and other professional men in the navy. In a letter which FALCONER received from the celebrated Du Hamel, his opinion of it was thus decidedly given: *Ce livre manquoit absolument; celui qui a été imprimé en Hollande, et qui a eu un débit considérable, est très-imparfait; celui de M. Saverien est encore plus mauvais.* The death of the gallant Duke of YORK at Monacoa, on the 17th of September, 1767, though felt by all the nation, was more particularly a severe loss to FALCONER, whose welfare, owing to the melancholy event, became again precarious. His literary fame, however, was now established; some few friends, among whom the Hunters took the lead, still remained; and he accordingly endeavoured to dry the tear, which the memory of his royal patron frequently called forth, by indulging in the vision of hope that was still prolonged: nor did Providence in this emergency forsake him. From the Glory, FALCONER was appointed to the Swiftsure,* at the close of that year.

Soon after this he was involved in difficulties, which obliged him to take up his residence in the metropolis, where he is supposed to have struggled for a short time against the *res angustæ domus.* Among other resources he is said to have derived some small emolument from writing in the Critical Review and other periodical publications; and in October, 1768, he received proposals from the late Mr. John

* Governor Hunter was doubtful whether it was the Swiftsure, or the Warspight; but from MSS. which Mr. Clarke had seen, he prefers the former ship.

Murray, of Fleet-street, to admit him as a partner in the line of business which that respectable bookseller afterwards established. This, however, does not appear to have been accepted by FALCONER, who, most probably, had then some prospect of being again employed on his own element.

The Marine Dictionary was not finally printed till the beginning of 1769; although it appears to have been compiled during his naval retreat at Chatham. Immediately after it was published, he was appointed purser to the Aurora, Captain Lee, when a third edition of THE SHIPWRECK was loudly called for: considerable improvements and additions had been made to it by FALCONER; but Mr. Clarke is induced to think, that amidst the agitation of his mind on being appointed purser to that frigate, which was then ordered to carry out to India Henry Vansittart, Esq. Luke Scrofton, Esq. and Colonel F. Forde, officers in the Honourable East India Company's service; that FALCONER, who also had the promise of being their private secretary, from the joy of obtaining so lucrative a situation, neglected this edition, and left the last alterations to some friend, as the inferiority of many passages is strikingly evident. Mr. Clarke, however, has endeavoured, with the assistance of the first and second editions, to make the author correct himself, and has thus restored the purity of the original text, which had become strangely impaired. The joy which this appointment gave to the friends of FALCONER may easily be imagined; but this, alas! was of short duration: the Aurora sailed from England on the 30th of September, 1769; and after touching at the Cape of Good Hope, which she left on the 27th of December following, was lost during the course of the voyage. Thus perished poor FALCONER in the bosom of the waves, the occasional fury of which, and consequent disasters, he has so forcibly and elegantly described: thus leaving behind him a work to perpetuate his name, more durable, and far more honourable, than any monument which the artist's hand could erect: a work which affords ample proof of nautical ability, as well as poetical talents. In an interesting little work, entitled, "*The Journal of* PENROSE, *a Seaman;*" written in 1755, and published by Murray, in 1815, is the following tribute to FALCONER's early poetical abilities:—"How often," says he, "have I wished to have the associate of my youth, BILL FALCONER, with me, to explore these beauties, and to record them in his sweet poetry; but, alas! I parted with him in Old England, never, perhaps, to meet more in this world. His may be a happier lot. Led by a gentler star, he may pass through this busy scene with more ease and tranquillity than has been the portion of his humble friend Penrose."*

It appears Captain Lee, though a stranger to the navigation of the Mosambique Channel, would not be dissuaded from attempting it; which so much displeased Mr. Vansittart, that if an outward-bound East Indiaman had been at the Cape, it is said he would have quitted the Aurora.†

To this may be added, that, on the 19th of November, 1773, a Black was examined before the East India Directors, who affirmed, "that he was one of five persons who had been saved from the wreck of the Aurora; that the said frigate had been cast away on a reef of rocks off Mocoa; and that he was two years upon an island after he had escaped; and was at length miraculously preserved by a country ship happening to touch at that island."

Such are the principal events respecting FALCONER, which the Editor has been able to collect. In his person he was about five feet seven inches in height; of a thin light make, with a dark weather-beaten complexion, and rather what is termed hard-featured, being considerably marked with the small-pox: his hair was of a brownish hue. In point of address, his manner was blunt, awkward, and forbidding; but he spoke with great fluency. Though FALCONER possessed a warm and friendly disposition, he was fond of controversy, and inclined to satire. In his natural temper he was cheerful, and frequently used to amuse his messmates by composing acrostics on their favourites; in which he particularly excelled. As a professional man, he was a thorough seaman; and like most of that profession, was kind, generous and benevolent. He often assured Governor Hunter, that his education had been confined merely to reading English, writing, arithmetic, and a little Latin; but notwithstanding, he had by industry and perseverance acquired a tolerable knowledge of most of the modern languages.

* *Penrose's Journal*, vol. ii. p. 94.

† Gentleman's Magazine, vol. XLI. p. 237.

AN

UNIVERSAL DICTIONARY

OF THE

MARINE.

ABACK, (*coiffé*, Fr.) in naval tactics, is the situation of the sails of a ship when they are pressed against the masts by the force of the wind.

Taken ABACK, the sails are said to be taken aback when they are brought into the above situation, either by a sudden change of wind, or by an alteration in the ship's course purposely intended. or through the carelessness of the person stationed at the helm.

Laid ABACK, the sails are said to be laid aback to effect an immediate retreat; or to give the ship sternway, in order to avoid some danger discovered before her in a narrow channel; or when she has advanced beyond her station in the line of battle.

The sails are placed in this position by slackening their lee-braces, and hauling in the weather ones; so that the whole effort of the wind is exerted on the forepart of their surface, which readily pushes the ship astern, unless she is restrained by some counteracting force. See the article BACKING, and BRACING.

It is also usual to spread some sail aback near the stern, as the mizen-topsail, when a ship rides with a single anchor in a road, in order to prevent her from approaching it so as to entangle the flooks of it with her slackened cable, and thereby loosen it from the ground. See the article ANCHOR.

Fig. 1. plate II. discovers the plan of a ship, *a b*, with her main-topsail, *c d*, aback; in which the curved dotted line expresses the cavity of it, as blown back by the wind on each side of the mast. The fore-topsail, which is full, is exhibited by the line *e f*. Fig. 2. represents a perspective view of the ship in the same situation; and the dart shews the direction of the wind upon both.

Lay all flat ABACK! (*masque tout—brasse tout à culer*, Fr.) the order to arrange all the sails in that situation.

ABAFT, (*arrière*, Fr. *abaftan*, Sax. behind) the hinder part of a ship, or all those parts which lie towards the stern, and it is frequently used to signify *further aft*, or *nearer the stern;* as the barricade stands *abaft* the main-mast, that is, nearer to the stern. The stern strictly speaking is only the outside: abaft includes both inside and outside.

ABAFT *the beam*. This term implies that the relative situation of the object spoken of, is, in some part of the horizon, contained between a line drawn at right angles to the keel, and the point to which the ship's stern is directed.

ABERRATION, in astronomy, is a change in the position of the fixed stars, arising from the progressive motion of light, combined with the annual motion of the earth, by means of which they sometimes appear twenty seconds distant from their true position.

Formerly it was supposed that the motion of light was instantaneous, but, from a series of observations made by Roemer, the Danish astronomer, on the satellites of Jupiter, it has been proved that the motion of light is progressive, like all other bodies, and passes from the sun to the earth, a distance of no less than 95,000,000 miles, in the space of eight minutes and seven seconds. Hence it has been deduced, that if light was propagated in time, the apparent place of a fixed object would not be the same when the eye of an observer is at rest, as when it is moving in any other direction than that of the line passing through the eye of the object; and that when the eye is moving in different directions, the apparent place of the object would be different.

This apparent motion of the heavenly bodies was detected, in 1725, by our celebrated countryman,

Dr. Bradley, and is one of the most brilliant discoveries which have enriched the science of astronomy.

It is a remarkable circumstance, that, from the theory of the first satellite of Jupiter, M. La Place has found for the entire aberration, 37″ 5, a result exactly the same as that which Bradley deduced from a great number of accurate observations on the fixed stars. From this curious coincidence, La Place concludes, that the velocity of light, in all the space comprehended by the earth's orbit, is the same as the circumference of this orbit, and that the same thing holds good with regard to the orbit of Jupiter. See Preface to his *Mechanique Celeste*, tom. iv. p. 11.

Aberration *of a Planet*, in longitude, latitude, right ascension, and declination, is its geocentric motion in longitude, latitude, &c. during the time that light travels from the planet to the earth.

The time for example, in which light moves from the sun to the earth, is 8′ 8″; and, during this time, the geocentric motion of the sun in longitude is 20″, consequently 20″ is the sun's aberration in longitude.

The greatest aberration of the several planets when in perihelion is, Mercury 59″ 5, Venus 43″ 5, Mars 36″, Jupiter 29″, Saturn 27″, Georgium Sidus 25″, Moon 0″ 40‴.

ABJURATION, in law, is an oath to be taken by officers, civil and military, when admitted into office; and it expressly renounces and *abjures* any claim of the descendants of the late Pretender to the throne of these kingdoms.

This oath is required to be taken by all persons in any office, trust, or employment, upon divers occasions, under certain pains and penalties by many statutes, particularly 1 W. and M. 13 W. III. 1 Anne, 1 Geo. I. and may be tendered by two justices of the peace to any person whom they shall suspect of disaffection.

ABLE Seaman (*marin expert*, Fr.) implies one who is not only able to work, but who is also well acquainted with his duty as a seaman.

The upper class into which the crews of ships of war are rated on the ship's books, are termed *able seamen*. The other classes are rated ordinary and landmen.

ABOARD (*à bord*, Fr. *abordo*, Ital.) the inside of a ship: hence, any person who enters a ship is said to go *aboard*: but when an enemy enters in the time of battle, he is said to *board*—a phrase which always implies hostility. See the article Boarding.

To fall Aboard *of*, (*aborder*, Fr.) is to strike against another ship while one or both are in motion, whether by design, or by the force of the wind or current.

Aboard-*main-tack!* (*amure la grande voile!* Fr.) denotes the order to draw one of the lower corners of the main-sail down to the chess-tree. See the article Chesstree.

ABOUT, (*reviré*, Fr. *abutan*, Sax.) is the situation of a ship immediately after she has *tacked*, or changed her course, by going about, and standing on the other tack. See the article Tacking.

About-*ship!* (*adieu-va!* Fr.) the order to the ship's crew to prepare for tacking, or going about.

ABREAST, (*par le travers*, Fr. of *breost*, Sax.) side by side, or opposite to; the situation of two or more ships with their sides parallel to each other, and their heads equally advanced; also of any object in a line with the beam of a ship.

This term more particularly regards the line of battle at sea, where, on the different occasions of attack, retreat, or pursuit, the several divisions of a fleet are obliged to vary their dispositions, and yet maintain a proper regularity by sailing in *right* or *curved* lines. When the line is formed *abreast*, the whole squadron advances uniformly, the ships being equally distant from, and parallel to each other, so that the length of each ship forms a right angle with the extent of the squadron or line *abreast*. The commander in chief is always stationed in the centre, and the second and third in command in the centres of their respective divisions. See this further illustrated in the article Line.

Line Abreast. See the article Line.

Abreast *of a place*, means directly opposite to it. Exam. "*We discovered a fleet* Abreast *of Beechy Head*," *i. e.* off, or directly opposite to it.

Abreast, within the ship, implies on a parallel line with the beam, or at right angles with the keel; as abreast the main hatch-way, in opposition to *afore* or *abaft* the hatch-way.

ABSENT. To be absent without leave. (*s'absenter sans congé, ou permission*, Fr.)

ABSTRACT, is applied to a compendious state of stores kept by the respective warrant-officers in the navy, in order to satisfy themselves in balancing their supplies, expenses, and remains.

Abstracts *of Acts of Parliament*, for the payment of seamen's wages, prize-money, &c. together with the articles of war, are ordered to be printed and hung up in the most public place in every ship in the royal navy, and to be read over once in every month to the ship's company; which is to be attested by the captain, or commander, and signing officers, at the foot of the muster-books. See Appendix.

ABURTON, a small tackle formed by two blocks or pullies; is generally used to set up or tighten the top-mast shrouds, and for many other purposes: whence it is frequently termed a Top Burton-tackle: also a term applied to such casks as are stowed athwart ships, or on a line with the beam.

ACADEMY, is a kind of collegiate seminary, either of a public or private nature, where youth are generally boarded and instructed in the arts and sciences.

The Royal Military Academy *at Woolwich*, is a public institution, where young men are taught fortification, gunnery, mining, and such branches of the mathematics as are necessary to qualify them for the service of the artillery and the office of engineers. It was instituted by his late Majesty King George II. by warrants, dated April the 30th and November the 18th, 1741; and is under the direction of the Master-general and Board of Ordnance, for the time being: and, at first, the lectures of the masters in the academy were attended by the practitioner-engineers, with the officers, serjeants, corporals, and private men of the artillery, besides the cadets. The gentlemen educated at this academy, amounting to 188, are the sons of noblemen, gentlemen, and military officers: they are called gentlemen cadets; and are not admitted under 14, nor above 16, years of age. By a late regulation, every young gentleman, on passing his examination for admission into the academy, is kept there; not at first as a cadet, but for twelvemonths probation, during which time, should his abilities and attention to his studies, joined to orderly behaviour, promise the acquirement of those qualifications necessary for the station he is intended to fill, he will, at the expiration of this period, be admitted as an established cadet. The cadets are under the direction of a lieutenant-governor. Each company has a captain and two subalterns, as military directors; and an inspector, who superintends the studies of the cadets. The academy has, at present, about twenty masters; viz. a professor of mathematics, and eight mathematical masters; a professor of fortification, and two masters; five drawing masters; two French masters; with masters for fencing, dancing, and a professor of chemistry.

This institution is of the greatest consequence to the state; and it is hardly credible that so important an object should be accomplished at so trifling an expense. The pay of each cadet is only 2s. 6d. per day; and, as this is sufficient for his maintenance, his parents are at no expense while he continues at the academy. In consequence of the unhealthy and confined situation of the old buildings, government determined to erect new ones on the side of Shooter's-hill. The foundation-stone was accordingly laid by the Duke of York, on the 27th of May, 1803; and the academy was removed to them on the 12th of August, 1806. There is now an infirmary attached to this institution, for the reception of cadets only, which is under the care of one of the first artillery surgeons, with a careful nurse, and is supplied with every requisite for the comfort of those who may have occasion to be admitted. The cadets' barracks are a very handsome building, and have a noble appearance, from the number of field-pieces arranged before it. The academy is an elegant building, in the Gothic style. For further particulars respecting military establishments, see MILITARY COLLEGE at MARLOW.

The Naval ACADEMY *at Gosport*, is a private institution for the instruction of youth intended for the royal navy, and was established in the year 1791; where 80 young gentlemen are boarded, lodged, and instructed in writing, arithmetic, algebra, mathematics, navigation, nautical astronomy, marine surveying, French, drawing, fencing, and dancing. The terms are from 40 to 50 guineas a year, which include all the common expenses. The situation is admirably well calculated for a naval seminary, as the play-ground is adjoining Portsmouth-harbour, and directly opposite the dock-yard. There is a regular course of mathematical and nautical education taught at the academy, which generally requires from two to three years to accomplish; so that youth should be sent there about 11 or 12 years of age, that they may have a chance of getting through it before they go to sea; which most naval officers think ought not to be later than at 14, in order to make good seamen. The plan has been very generally approved of, and the establishment has been constantly full for the last 20 years. For further particulars respecting public naval establishments, see NAVAL COLLEGE, at PORTSMOUTH.

ACCELERATION, in mechanics (*acceleratio*, Lat.), is the act of quickening motion, or giving it a continual impulse, so as to increase, whether falling freely by its own weight, descending an inclined plane, or oscillating round a fixed centre.

ACCELERATION *of a Planet*. A planet is said to be accelerated in its motion, when its real diurnal motion exceeds its mean. And, on the other hand, the planet is said to be retarded in its motion, when the mean diurnal motion exceeds the real.

This inequality arises from the change in the distance of the planet from the sun, which is continually varying; the planet moving always quicker in its orbit when nearer the sun, and slower when farther off.

ACCELERATION *of the Moon*, is a term used to express the increase of the moon's mean motion from the sun, compared with the diurnal motion of the earth; by which it appears that, from some uncertain cause, it is now a little quicker than it was formerly.

According to La Place this acceleration arises from a diminution in the eccentricity of the earth's orbit, produced by the action of the sun. When this eccentricity, therefore, becomes a minimum, the acceleration will increase; and when the eccentricity increases, the acceleration will be converted into a retardation. Dr. Halley was the first who discovered this diminution in the periodical revolution of the moon, by a comparison of ancient and modern eclipses. He could not, however, ascertain the quantity of acceleration, as the longitudes of the places where these eclipses were observed, were not accurately determined. This, however, was afterwards done by Mr. Dunthorne, who found, that the acceleration amounted to 10″ in 100 years. By a comparison of three eclipses, observed near Cairo, in 977, 978, 979, by Ibn-Junis, La Lande has found the acceleration to be 9″ 886 in 100 years. See *Astronomie par* LA LANDE, tom. 2, art. 1483; and VINCE'S *Astronomy*, vol. I. p. 206.

ACCOMMODATION LADDER, (*échelle de commandement*, Fr.) a convenient light staircase fixed occasionally on the gangway. See the article LADDER.

ACHERNER, in astronomy, a star of the first magnitude, in the southern extremity of the constellation Eridanus, known to mariners by the name of the Spring of the River Erida, but invisible in our latitude. For the beginning of the present year 1812, its

Right Ascension is . . .	22°.	40′.	35″,4
Annual Variation in A. R. .			33,7
Declination	58°.	19.	15, S.
And annual Varia. in Decl. .		—	18,5

ACHROMATIC TELESCOPE, a singular species of refracting telescope, said to be invented by the late Mr. John Dolland, and since improved by his son, Mr. Peter Dolland, and others. The term appears to have been first used by M. de la Lande, in his astronomy, to denote telescopes contrived to remedy aberrations and colours.

ACORN, (*pomme de giroüette*, Fr.) a little ornamental piece of wood, fashioned like a cone, and fixed on the uppermost point of the spindle, above the vane, on the mast-head. It is used to keep the vane from being blown off from the spindle in a whirlwind, or when the ship leans much to one side under sail. See plate I. fig. 9. where *a* represents the acorn, *b* the vane and stock, *c* the spindle, and *d* the mast-head.

ACRONICAL, (from ακρονύκος, compounded of ἄκρος, *the extremity*, and νὺξ, *night*). The rising and setting of a star or planet is said to be *acronical* when this phenomenon happens just as the sun is descending below the horizon.

The acronical is one of the three Greek poetical risings and settings of the stars; and stands distinguished from Cosmical and Heliacal: and by means of which, for want of instruments and accurate observations, they regulated the length of their year.

ACTION, is used in the same sense as battle or engagement. See the article BATTLE.

ACTS OF PARLIAMENT, the principal relating to the marine of Great Britain. See APPENDIX.

ADJOURNMENT, is applied to the putting off the meeting, or re-assembling of a Court Martial, until another day. A Court Martial may adjourn from day to day (sundays excepted) until sentence is passed.

ADMIRAL, (*amiral*, Fr.) an officer of the first rank and command in the fleet, and who is distinguished by a flag displayed at his main-top-gallant mast-head. Also an officer who superintends the naval forces of a nation, and who is authorized to determine in all maritime causes.

The origin and denomination of this important office, which seems to have been established in most countries that border on the sea, have given rise to a great variety of opinions. Neither the antiquity nor the early history of their office is well ascertained. Some writers seem to think that the name is of eastern origin, and borrow it from the Greek; others from the Arabic; and some, with a greater degree of probability, derive both the title and dignity from the Saracens.

Du Cange assures us that the Sicilians were the first by whom the title of Admiral was adopted; and says, they took it from the eastern nations, who often visited them: and he farther gives the authority of Matthew Paris, to shew, that the Genoese were the next who applied the word to the commander of a squadron.

This appears to have been done in the time of the crusaders, and about the year 1244.

History is silent with regard to the exact time when the name of this officer was introduced into the English service. Spelman, a learned antiquary, is of opinion that it was not before the reign of Henry the Third, as it is not mentioned in the laws of Oleron, which were made in the year 1266, nor by Bracton, who was Lord Chief Justice about that time, in his Treatise "De Legibus et Consuetudinibus Angliæ."

There were anciently three or four Admirals of the English seas, who held their offices during the pleasure of the king: besides these, there were Admirals of the Cinque Ports; one of whom (William Latimer) is described as Admiralis quinque Portuum, in the reign of Edward the Third. The title of Admiralis Angliæ, or Lord High Admiral of England, was first given to the Earl of Arundel and Surry, by a patent of Richard the Second, in the year 1387; but was not much used till the reign of his successor Henry the Fourth. It has now been extinct for many years, and seems to be sunk in that of the Lords Commissioners of the Admiralty.

The ADMIRAL, or commander in chief of a squadron, being frequently invested with a great charge, on which the fate of a kingdom may depend, ought certainly to be possessed of abilities equal to so important a station and so extensive a command. His squadron is unavoidably exposed to a variety of perplexing situations in a precarious element. A train of dangerous incidents necessarily arise from those situations. The health, order, and discipline of his people are not less the objects of his consideration, than the condition and qualities of his ships. A sudden change of climate, an infectious air, a scarcity, or unwholesomeness of provisions, may be as pernicious to the former, as tempestuous weather or dangerous navigation to the latter. A lee-shore, an injudicious engagement with an enemy greatly superior, may be equally fatal to both. He ought to have sufficient experience to anticipate all the probable events that may happen to his squadron during an expedition or cruise, and, by consequence, to provide against them. His skill should be able to counteract the various disasters which his squadron may suffer from different causes. His vigilance and presence of mind are necessary to seize every favourable opportunity, that his situation may offer, to prosecute his principal design; to extricate himself from any difficulty or distress; to check unfortunate events in the beginning, and retard the progress of any great calamity. He should be endued with resolution, and fortitude to animate his officers by the force of example, and promote a sense of emulation in those who are under his command, as well to improve any advantage, as to frustrate or defeat the efforts of his enemies.

The most essential part of his duty, however, appears to be military conduct. As soon as the squadron under his command shall put to sea, he is to form it into the proper order of battle, called the LINE. In this arrangement he is to make a judicious distribution of strength from the van to the rear, throwing the principal force into the centre, to resist the impression of the enemy's fleet; which might otherwise, at some favourable opportunity, break through his line, and throw the van and rear into confusion.

It is also necessary that he should have a competent knowledge of the seas, weather, and reigning winds, of the coast or region where he is stationed, which will not only greatly facilitate his plans on the enemy, but likewise enable him to avoid being improperly embayed, where he might be surprised in a disadvantageous situation; and to judge whether it will be most expedient to attack his adversary, or lie prepared to receive his assault. When his squadron is forced by stress of weather, or otherwise, to take shelter in a road or bay, it will likewise suggest the necessary conduct of keeping a sufficient number of cruisers at sea, to bring him early intelligence, that they may be ready to cut or slip the cables when they shall be too much hurried to weigh their anchors. It also behoves the Admiral, or commander in chief, as the forming a complete, strong, and uniform line is a very material article in naval war, frequently to arrange his squadron into this order, that the inferior officers may observe to bring their ships, with greater dexterity and alertness, into their several stations, and maintain the regularity of the line when they tack, veer, or sail abreast. See the article LINE.

When the admiral intends a descent on an enemy's coast, or other attack which may be attended with complicated and unforeseen incidents, his orders should be delivered or drawn up with the greatest accuracy and precision; they should be simple, perspicuous, direct and comprehensive; they should collect a number of objects into one point of view, and foreseeing the effects of success or defeat, appoint the proper measures to be adopted in either event. History and experience confirm the necessity of this observation, and present us with a variety of disasters that have happened on such occasions, merely by a deficiency in this material article.

When an admiral shall conquer in battle, he should endeavour to improve his victory, by pushing the acquired advantages as far as prudence directs; a conduct that merits his attention as much as any in the action. When he shall be defeated, he ought to embrace every opportunity of saving as many of his ships as possible, and endeavour, principally, to assist those which have been disabled. In short, it is his duty to avail himself of every practicable expedient rather than sink under his misfortune, and suffer himself to become an easy prey to an enemy.

He should be sufficiently acquainted with civil law, to judge with propriety of the proceedings of courts-martial, and to correct the errors, and restrain the abuses, which may happen therein by mistake, ignorance or inattention.

As secret treaties, propositions, or schemes of the enemy, may occasionally be submitted to his inspection, or fall into his possession by capture; and which it might be improper to discover to any person near him, he ought to have a competent knowledge of the modern languages, or, at least, those of the countries against whom his military operations are directed, to be able to comprehend with facility the full scope and purport of such papers.

Moreover, he should be versed in geometry, so as to be capable of ordering proper and correct surveys of unknown coasts, roads, or harbours to be made, and to judge of their accuracy, and detect their errors. To ascertain the latitude and longitude of different places, he should be also sufficiently skilled in astronomy and the method of taking lunar observations, which, indeed, is essentially necessary to the profession of a sea-officer, although too much neglected.

By his instructions from the board, the admiral should assist at all councils of war that relate to naval affairs; he should visit, as often as convenient, the other ships of his squadron; he should inquire particularly into their condition, and observe the men mustered, seeing that every man is rated and stationed as he ought to be, taking care that no supernumeraries are born on the books. He is directed to acquaint the secretary of the admiralty with all his proceedings relative to the service, for the information of the lords commissioners of the admiralty; and to attend them on his return home, with an account of his voyage or expedition, and to deliver a copy of his journal to their secretary.

Such, and much more, are the necessary qualifications of an admiral; so that his office and duty require greater skill and more comprehensive abilities than are generally supposed necessary to the command of a naval armament. It is further necessary that he should be duly qualified, at least in this kingdom, to assist at the councils of his sovereign, and enter into the enlarged system of protecting his country from an invasion, or of meditating a descent on an enemy's coast; as well as to improve navigation, and open new channels of commerce. For further particulars of his charge, see the articles ENGAGEMENT, LINE, SQUADRON.

ADMIRAL *of the Fleet*, (*l'amiral d'angleterre*, Fr.) the highest officer under the admiralty of Great Britain; when he embarks on any expedition, he is distinguished by the union flag at the main-top-gallant-mast-head. His Royal Highness the Duke of Clarence has lately been promoted to that honour.

Vice-ADMIRAL, (*vice-amiral*, Fr.) the officer next in rank and command to the admiral; his flag is displayed at the fore-top-gallant-mast-head: also a civil officer appointed by the lords-commissioners of the admiralty. There are several of these officers established in different parts of Great Britain, with judges and marshals under them, for executing jurisdiction within their respective districts. Their decisions, however, are not final, an appeal lying to the court of admiralty in London.

Rear-ADMIRAL, (*contre-amiral*, *lieutenant-general des armées navales*, Fr.) the officer next in rank and command to the vice-admiral, and who carries his flag at the mizen-top-gallant-mast-head.

There are at present in England, exclusive of the admiral of the fleet, 21 admirals of the red squadron, 20 admirals of the white, 20 admirals of the blue, 22 vice-admirals of the red, 19 vice-admirals of the white, 24 vice admirals of the blue, 19 rear-admirals of the red, 17 rear-admirals of the white, and 24 rear-admirals of the blue, exclusive of 31 rear-admirals who have never hoisted their flag, but are superannuated on half-pay. 1st January 1812. For further distinction of admirals see the article FLAG.

ADMIRALTY, (*amirauté*, Fr.) the office of lord-high-admiral, whether discharged by one single person, or by joint-commissioners, called Lords of the Admiralty, who are generally seven in number. This important and high office has seldom been entrusted to any single person, except princes of the blood; or to some nobleman meriting such distinction for his eminent services.

ADMIRALTY, *court of*, a supreme court, held by the Lord High Admiral, or the Lords Commissioners of the Admiralty, for the trial of maritime causes, whether of a civil or of a criminal nature. It was instituted by Edward the Third, and takes cognizance by judge and jury, of all criminal offences upon the high seas, or on the shores of any country under the dominion of Great Britain; but, in civil cases, the discussions follow the practice of the civil law. It is enacted, that, in all criminal cases, felons shall be tried in the presence of a jury by commissioners, whom the Lord Chancellor has the right of appointing; and those commissioners are the judges of the Court of Admiralty: but, besides these, there are always three or four others, two of whom must be taken from the number of the Supreme Judges. An appeal to the sovereign in Chancery lies (8 Eliz. c. 5.) from the sentence of the Court of Admiralty. There are likewise courts of an inferior nature held by the vice-admirals, deriving their appointments from the Supreme Court, within their respective jurisdictions, and an appeal lies from their decisions to the Supreme Court: but, in the case of prize vessels, taken in any port of the world, the appeal lies with certain commissioners of appeals, who are chiefly of the privy council, and not with judges delegate. And this, by virtue of divers treaties with foreign nations, by which particular courts are established in all the maritime countries in Europe for the decision of this question—Whether lawful prize or not? For this being a question between subjects of different states, it belongs entirely to the law of nations, and not to the municipal laws of either country, to determine it.

ADRIFT, (from *a* and *drift*, Sax.) the state of a ship or vessel broke loose from her moorings, and driven without controul at the mercy of the wind, seas, or current, or all of them together.

ADVANCE UPON SEAMEN'S WAGES. Every volunteer in his Majesty's service, whether rated able, ordinary, or landman, is entitled, by Act of Parliament, to an *advance* of two months' wages before the ship proceeds to sea. See APPENDIX.

ADVICE-BOAT, (*barque d'avis*, *aviso*, Fr.) a small vessel employed to carry expresses or orders with all possible dispatch.

ADZE, a carpenter's tool in very common use, it is applied to the purpose of chipping, or rendering smooth, timber, in work of less weight than on occasions where the axe is generally employed.

AFLOAT, (*à flot*, Fr.) floating on the surface of the water: a ship is said to be *afloat* when there is a volume of water under her bottom of sufficient depth to buoy her up from the ground.

AFORE, (*avant*, Fr. *fore*, Sax.) all that part of a ship which lies forward, or near the stem.

AFORE, as a preposition, likewise implies *further forward*, or nearer the prow; as, the manger stands *afore* the fore-mast, *i. e.* further forward, or nearer the stem. In both these senses *afore* is used in contradistinction to *abaft*. See the article ABAFT.

AFT, (*arrière*, Fr. from *æfter*, Sax. or *abaft*) behind, or near the stern of the ship; being opposed to *fore*; as, run out the guns *fore and aft!* *i. e.* from one end of the ship to the other.

AFT, an abbreviation of abaft: thus, right aft when applied to any distant object implies that it is in a direct line with the stern.

AFT is also applied to some of the sheets or ropes fastened to the corners of the sails; as, for instance, haul aft the fore-sheet, or main-sheet, is an order to pull the corner of those sails more towards the stern.

AFTER, (*de l'arrière*, Fr. *æfter*, Sax.) a phrase applied to any object situated in the hinder part of the ship; as, the *after*-hatchway, the *after*-capstan, the *after*-sails, &c.

AFTER-*guard*, in the royal navy, the seamen who are stationed on the poop and quarter deck of the vessel, to attend and work the after-sails, &c.

The AFTER-*sails* usually comprehend all those which are extended on the mizen-mast, and on the stays between the mizen and main-masts. They are opposed to the head-sails, which include all those that are spread on the fore-mast and bowsprit, and both, by their mutual operation on the opposite ends of the ship, duly balance her when under sail. See the article TRIM.

AGENT-VICTUALLER, (*avitailleur*, Fr.) an officer stationed at a royal port, to regulate the victualling of the king's ships, under the direction of the commissioners for victualling the navy. He receives all the provisions from the victualling-office in London, and distributes them to the ships in the harbour. He also receives into his store-houses such as may be returned by ships after the expiration of their cruise or voyage, and renders an account thereof to the said commissioners. He also pays the pursers of his Majesty's ships and vessels the sums of money due for providing them with necessaries, such as fuel, lights, turnery ware, lanterns, &c. commonly called Necessary Money, on his producing a certificate, from the captain and master, of the ship being provided for the time victualled.—He, besides, supplies the purser with money to pay the men for whatever arrears of provisions may be due to them, either by their own savings out of their daily rations, or short allowance through necessity, on producing attested vouchers, that the same is due.

AGENT, NAVY, is a person on shore employed by officers and seamen of the navy to manage their pecuniary and other concerns, particularly to receive their pay, prize-money, &c. and apply it to their use, according to their directions.

By an Act of Parliament passed in the 49th year of his present Majesty, entitled an Act to explain and amend an Act made in the 45th year of his present Majesty, for the encouragement of seamen, and the more effectually manning his Majesty's navy during the war, all *Agents* who act for petty officers, seamen, and others, serving in any of his Majesty's ships, to receive their *pay*, *wages*, *prize*, *and bounty money*, are obliged to take out a license from the Navy Pay-office; which license is liable to be immediately withdrawn, on any grounds being produced that they have abused the trust reposed in them.

AGROUND, (*échoué*, Fr. from *a* and *ground*, Sax.) the situation of a ship whose bottom, or any part of it, hangs or rests upon the ground, so as to render her immoveable till a greater quantity of water shall float her off; or till she shall be drawn out into the stream by the application of mechanical powers.

AHEAD, (*avant*, *de l'avant*, Fr. from *a* and *head*, Sax.) further onward than the ship, or at any distance before her, lying immediately on that point of the compass to which her stem is directed. It is used in opposition to *astern*, which expresses the situation of any object behind the ship. See the article ASTERN.

To run AHEAD *of one's reckoning*, (*depasser*, Fr.) to sail beyond the place erroneously estimated in the dead reckoning as the ship's station.

Line AHEAD. See the article LINE.

A-HULL, (*à sec*, *à mâts*, *& à cordes*, Fr. from *a* and *hull*) the situation of a ship when all her sails are furled, and her helm lashed on the lee-side; she then lies nearly with her side to the wind and sea, her head somewhat turned towards the direction of the wind. See this further explained in the article TRYING.

AIM, (*pointer*, Fr.) the direction of a cannon, or other fire-arm, to its object, or the point to which it is directed.

To take Aim, (*prendre sa mire*, Fr.) is to point a gun to its object according to the point-blank range. See the article Cannon and Range.

AIR is the fine, expansive, transparent fluid which we breathe. It constitutes the vast mass of atmosphere that encompasses our globe, and performs a capital part in the œconomy of nature. It is the vehicle of sound, of smell, and even of heat; it exhales moisture from the land and the ocean; and, mingling its contrasted currents in the higher regions, it again precipitates its watery stores in the form of rain or dew, of hail or snow. By the activity, the character, and extent, of its motions, it tempers the unequal energy of the solar beams, and diffuses a more uniform warmth over the surface of our globe; it not only supplies the breath of life, but supports inflammation, and contributes essentially to the germination and growth of plants; and, presenting the more elevated portions of atmosphere to the illumination of the sun's rays, it scatters these, by repeated, though enfeebling, reflexions, along the concave of heaven; and thus prolongs the cheering influence of light, and prevents the inhabitants of this earth from being ever left in total darkness.

The air is justly reckoned among the number of fluids, because its parts yield to any force impressed, and are very easily moved one amongst another; it presses in proportion to its height, and the pressure is every way equal.

In common with water, the air we breath possesses gravity, but the difference between the specific gravity of air and water, when the barometer is at 30 inches, and the thermometer at the mean temperature of 55°, is estimated as 1 to $833\frac{1}{3}$, that is, as $\frac{2500}{3}$ to 1, or as 2500 to 3. By its gravity the air presses with great force upon all bodies, according to the extent of their surface; and it has been computed to be upwards of 15 pounds weight upon every square inch of surface.

A square column of quicksilver, $29\frac{1}{2}$ inches high, and one inch thick, weighs just 15 pounds, which is equal to the pressure of air upon every square inch of the earth's surface, and 144 times as much, or 2160 pounds, upon every square foot; because a square foot contains 144 square inches. At this rate a middle sized man, whose surface may be about 14 square feet, sustains a pressure of 30,240 pounds, when the air is of a mean gravity; a pressure which would be insupportable, and even fatal to us, were it not equal on every part, and counterbalanced by the spring of the air within us, which is diffused through the whole body, and re-acts with an equal force against the outward pressure. For further particulars, see the article Atmosphere.

ALDEBARAN, or Palilicium, the Arabian name of a fixed star of the first magnitude, situated in the south eye of the constellation Taurus, and hence it is called the Bull's-eye. For the beginning of the year 1812, its

Right Ascension is	66°	17	04",98
Annual Variation in A. R. .			51,31
Declination	16°	7	31 N.
And annual Variation in Decl.			+ 8,16

The distance of the moon's centre from this star is set down in the Nautical Ephemeris for every three hours, adapted to apparent time, at Greenwich; and, from its magnitude and brightness, is a good star for the mariner's observing its distance when either east or west of the moon, in order to ascertain the longitude of the ship.

ALEE, (*sous le vent*, Fr. from *a* and *lee*) the situation of the helm when pushed close down to the lee-side of the ship, in order to put the ship about, or to lay her head to the windward.

ALGEBRA. The science of algebra, like that of arithmetic, is employed in reasoning concerning the relations which subsist among different quantities, in respect of their magnitudes. In both, the quantities treated of are expressed by symbols. In arithmetic, symbols are confined to ten numeral characters; but, in algebra, quantities may be expressed by any characters whatever: the letters of the alphabet are, however, at present, always used; and these are employed to denote, not only the known or given quantities, to which alone the arithmetical notation will apply, but also the unknown quantities, or such as are required to be found.

The two sciences thus agreeing in the objects of which they treat, and, to a certain extent, in their reasonings being carried on by means of symbols, they, of course, depend upon the same first principles. Indeed such is their connexion, that algebra has been denominated *universal arithmetic*; a name which, in some respects, is proper enough, although it does not completely express the distinction between them; the essential character of algebra consisting in this, that, when all the quantities concerned in any inquiry, to which it is applied, are denoted by general symbols, the results of its operations do not, like those of arithmetic and geometry, give the individual values of the quantities sought, but only shew what are the arithmetical, or geometrical operations, which ought to be performed on the original given quantities, in order to determine these values. Simpson, Saunderson, and Emerson, have furnished many improvements on this science in England; Euler, in Germany; M. de la Caille, and M. La Place, in France.

ALGOL, or Medusa's Head, a fixed star of variable magnitude, situated in the constellation Perseus, and marked β in our catalogues. Montanari was the first who observed the variations of this star; and, according to Mr. Goodricke, of York, the first astronomer who discovered the period of these variations; he found it to be repeated every 2 days, 20 hours, 48 minutes, and 56 seconds; and La Lande makes it 2 days, 20 hours, 49 minutes, and 2 seconds.

ALIGNMENT, (*alignement*, Fr.) a supposed line drawn, in order to preserve a fleet, or part thereof, in its just and perfect direction.

ALL hands hoay! (*tout le monde en haut!* Fr.) the order by which all the ship's company are summoned upon deck by the boatswain.

All *hands to quarters hoay!* (*tout le monde a son poste!* Fr.) a term given by some collectors of nautical terms to express the order to the crew for preparation for battle. It is only used in small ships where there are no drums and fifes, such as gun-brigs, cutters, &c. The command is now given by the boatswain piping down the hatchway.

All *in the wind*, the state of a ship's sails when they are parallel to the direction of the wind, so as to shake and shiver, by turning the ship's head to windward, either by design, or neglect of the helmsman: therefore hard-a-lee, and luff-a-lee, are the orders given to the helmsman to put the helm down to the lee-side: and helms-a-lee implies the order to the crew, to cause the head-sails to shake in the wind, for the purpose of bringing the vessel about.

All's *well!* (*bon quart*, Fr.) an acclamation of safety or security pronounced by each sentinel every half hour (when the bell is struck) during the night watches.

ALLOTTING, or ALLOTMENT of Goods, (*repartition de marchandises*, Fr.) is when a ship's barge is divided into several parts bought by different persons, whose names are written on as many pieces of paper, which an indifferent person applies to the several lots or parcels, by which the goods are impartially divided and allotted to the proper persons.

Allotting, or Allotment, is also allowing half the pay of carpenters, boatswains, gunners, petty officers, and all other below, also non-commissioned officers, and privates of the royal marines, to be paid monthly, by the clerk of the cheque, collector of customs, or some other public person appointed by Government, nearest the residence of the person to whom the allotment is made, which, by Act of 35 Geo. III. can extend only to wife, children, and mother.

ALLOWANCE, (*ration*, Fr.) a vulgar term given by seamen to the ration or quantum of provision served out to them on board.

Allowance, *short*, (*ration reduite*, Fr.) is when necessity obliges a curtailment of the usual quantity.

Allowance, *two-thirds*, (*deux tiers de ration*, Fr.) is when necessity obliges an allowance of two-thirds of the usual quantity.

Allowance, *to stop the*, (*arrêter la ration*, Fr.) is the last resource when the provisons are nearly exhausted.

ALMANAC, nautical, (*almanac nautique*, Fr. *almon-at*, Sax.) or Astronomical Ephemeris, is published annually, by anticipation, under the direction of the commissioners of longitude, and may be had for several years before-hand, for the convenience of ships going out on long voyages; and, of late years, there are few British mariners who do not avail themselves of this useful performance.

It is well known that the *Nautical Almanac* has contributed more essentially to the improvement of navigation and of practical astronomy in general, than any work of the kind ever published. It was begun in the year 1767, and has been continued up to 1816 inclusive, making in the whole 50 volumes.

This most important work was originally planned, and has been constantly conducted by the Rev. Dr. Maskelyne, the late astronomer royal; whose name must for ever stand high in the annals of science, both as a profound mathematician and a most accurate and able astronomer; and particularly as the founder of the lunar observations, by which the longitude at sea is now accurately determined; and for this great national purpose the *Nautical Almanac* was established. Beside every thing essential to general use that is to be found in any almanac hitherto published, it contains many new and interesting particulars; more especially, the distances of the moon's centre from the sun and fixed stars, east and west of her, for every three hours of apparent time, adapted to the meridian of the Royal Observatory, at Greenwich, by comparing which with the distances carefully observed at sea, the mariner may, readily, and with little danger of mistake, infer his longitude to a degree of exactness, that may be thought sufficient for most nautical purposes. All the calculations of the solar and lunar eclipses; the eclipses of Jupiter's satellites; the sun's longitude, right ascension, declination, and equation of time; the planets heliocentric and geocentric longitudes and latitudes, declination, and passage over the meridian; the moon's longitude and latitude for every noon and midnight passage over the meridian, her right ascension, declination, semi-diameter, and horizontal parallax, are also made, according to apparent time, by the meridian of Greenwich. The uses and application of these tables are copiously explained and illustrated by examples at the end of the *Nautical Almanac*, and in the tables requisite to be used with it, published, likewise, under the direction of the commissioners of longitude. And the publication of it is chiefly designed to facilitate the use of the different lunar tables that are published, by superseding the necessity of intricate calculations, in determining the longitude at sea. The astronomers of France have since modelled their *Connoissance des Temps* on the plan of this publication; and they have been suspected of copying many of its most valuable and laborious calculations, although they state their articles to be from original computation.

A recent discovery, however, seems to remove all doubt on the subject. The Nautical Almanac has been always conducted with such accuracy that there is no instance of any error of importance having been discovered in it. A trivial exception, however, has lately occurred. It has been observed, that in the volume of the present year, 1812, the obliquity of the ecliptic is assumed about nine seconds too little; which has, probably, originated in making a double correction, in reducing the mean obliquity to the true; which inaccuracy pervades all the columns of the sun's right ascension and declination.

This error was discovered by Dr. Kelly, of Finsbury-square, in September, 1811, who took an early opportunity of shewing it to Mr. Pond, the astronomer royal; and that gentleman agreed as to the existence of the error, and the propriety of having it corrected in the second edition, which may be daily expected. This error will not be attended with any inconvenience to seamen, as it is too delicate to affect their calculations, and is only perceptible to astronomers at land, and especially in observatories; and they may be supposed able to correct for themselves.

It seems, however, a little curious, that the American astronomers, who have printed this volume with additions and improvements, should have left the error in question, with all its consequences, wholly uncorrected. But the most extraordinary circumstance relating to this mistake is, that the French have actually copied it into their *Connoissance des Temps*. They have, it is true, given the obliquity of the ecliptic correctly in their fifth page; but in the columns of right ascension and declination of about nine months, they have inserted our inaccuracies (only making their usual allowance for the difference of meridians) ; and even at the solstitial points, where the mistake is most obvious, they have copied it exactly, and have thus made the sun's greatest declination nine seconds less than the obliquity of the ecliptic in the fifth page, though both should be the same.

If any thing could add to the fame of Dr. Maskelyne, it is the entire confidence thus placed in his calculations by the great astronomers of France ; and of our Nautical Almanac, that great astronomer M. De la Lande writes thus : "*On a fait à Bologne, à Vienne, à Berlin, à Milan ; mais le Nautical Almanac de Londres est, l'ephemeride la plus parfaite qu'il y ait jamais eu.*" It is, besides, highly honourable to his memory, that, in the Nautical Almanacs of half a century, only one error should be found, and even that too small to be felt in nautical practice. It is a curiosity in science; and likewise worthy of being recorded as a most gratifying instance of the powers and persevering energies of the human mind.

ALMUCANTARS, in astronomy, is an Arabic word, employed to denote the imaginary circles parallel to the horizon, which are supposed to pass through every degree of the meridian : they are of much use to shew the height of the sun, moon, or stars, &c. : they are the same as the parallels of latitude, and frequently called ALMACANTARS.

ALOFT, (*en haut*, Fr. *loffter*, to lift up, Dan.) up in the tops, at the mast-heads, or any where about the higher yards or rigging.

ALONG-SIDE, (*bord à bord*, *flanc à flanc*, Fr.) side by side, or parallel to a ship, wharf, &c.

To lay ALONG-*side*, (*élonger*, Fr.) to place a ship by the side of another.

ALONG-*shore*, (*le long de la côte*, Fr.) along the coast; this phrase is applied to coasting navigation, or to a course which is in sight of, and nearly parallel to the shore.

ALONG-*shore owner*, (*armateur de cabotage*, Fr.) a term to signify those people who send their ships to sea in want of stores, provisions, &c.

Lying ALONG, (*à la bande*, *au long*, Fr.) the state of being pressed down sideways, by a weight of sail in a fresh wind that crosses the ship's course either directly or obliquely.

ALOOF, (*lof*, Fr.) at a distance.

To keep ALOOF, commonly called *keep the luff*, is the command given by the pilot or officer to the helmsman, to direct the ship's course nearer the wind, or nearer that point of the compass which the wind blows from. This phrase probably regards the dangers of a lee-shore, from which the pilot might order the helmsman to keep aloof. See the article LUFF.

ALTITUDE, in astronomy, is an arch of a verticle circle, measuring the height of the sun, moon, or star above the horizon.

ALTITUDE, *meridian*, is an arch of the meridian, measured from the horizon to any celestial object then upon the meridian. This altitude is either true or apparent, according as it is reckoned from the rational or sensible horizon ; and the difference between these is what is called the parallax of altitude; and, it is evident, that the parallax is greatest when the object is seen in the horizon, and vanishes when in the zenith ; and also that the apparent place of the object is below its true place: therefore, when the altitude of the lower limb of any object is observed, its semidiameter is to be added thereto, in order to obtain the central altitude; but, if the upper limb be observed, the semidiameter is to be subtracted therefrom. If the altitude be taken by a back observation, the contrary rule is to be applied. The dip of the horizon is to be subtracted from, or added to, the observed altitude, according as the fore or back observation is used. The refraction is always to be subtracted from, and the parallax added to, the observed altitude: hence, the true altitude of the object's centre will be obtained. See LYSONS's rules for computing the sun's altitude in the *Nautical Almanac* for 1771; and for the moon or a star's, see *Nautical Almanac* for 1778.

ALTITUDE, *or elevation, of the Pole*, is the angle or arch of the meridian, intercepted between the horizon and the pole. This is equal to the latitude of the place ; and it may be found by observing the meridian altitude of the pole star, when it is both above and below the pole, and taking half the sum, when corrected, for refraction. Or the same may be found by the declination and meridian altitude of the sun.

ALTITUDE, *or elevation, of the Equator*, is the angle or arch of the meridian, between the horizon and the equator ; and it is equal to the co-latitude of the place.

ALTITUDE, *refraction of*, is an arch of a vertical circle, by which the true altitude of the moon, or a star, or other object, is increased by means of the refraction; and is different at different altitudes, being nothing in the zenith, and greatest at the horizon, where it is about 33′.

ALTITUDE, *parallax of*, is an arch of a vertical circle, by which the true altitude, observed at the centre of the earth, exceeds that which is observed on the surface; or the difference between the angles ☾ L M and ☾ I K of altitude there; and it is equal to the angle I ☾ L, formed at the moon, or other body, and subtended by the radius I L of the earth.

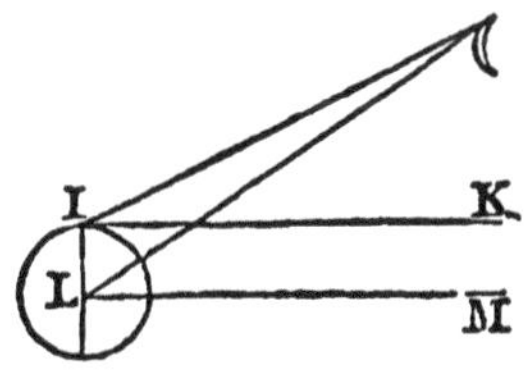

AMAIN, (*cale tout*, Fr. from *main*, or *maigne*, old French) at once, suddenly; as, let go *amain!* *i. e.* let it run at once. This phrase is generally applied to any thing that is hoisted or lowered by a tackle, or complication of pullies.

To lower AMAIN, to lower at once, or let go the fall of the tackle.

To strike AMAIN, to lower or let fall the topsail.

To wave AMAIN, to make a sign to another vessel by waving a bright sword, or something else as a demand for striking its topsails.

AMIDSHIPS, the middle of the ship, either with regard to her length or breadth, as ; the enemy boarded us *amidships*, *i. e.* in the middle, between the stem and stern. Put the helm *amidships*, *i. e.* in the middle, between the two sides.

AMMUNITION, (*munition*, Fr.) military stores, under this title are comprised not only all kinds of cannon, mortars, and all that is necessary for them and the service, as shots, cartridges, wads, grenades, rope, but all sorts of offensive and defensive weapons, such as muskets, swords, powder, match, and every thing that may add to the destruction of the enemy, and your own preservation.

AMPLITUDE, in astronomy, is an arch of the horizon intercepted between the true east or west point thereof, and the centre of the sun, star, or planet, at its rising and setting, and so is either north or south; and is useful in navigation to find the variation of the compass, or magnetic needle. To find the amplitude by trigonometry; say, as the co-sine of the latitude is to radius: so is the sine of the sun or star's declination to the sine of the true amplitude from the east or west point to the north or south, as the declination may be either north or south.

AMPLITUDE, *magnetical*, is an arch of the horizon, contained between the sun or a star, at the time of its rising or setting, and the magnetic east or west point of the horizon pointed out by the compass: the difference between this and the true amplitude is the variation of the compass, the true amplitude being an arch of the horizon included beween the east or west point and the centre of the sun or star, at its rising or setting.

If the magnetic amplitude be found to be 61°55
at the time when it is computed as above to be 39°25

then the difference is the variation westward, 22°30.

AMPLITUDE, in gunnery, implies the complement of the range of the shot, or the horizontal right line drawn from the mouth of the cannon to the spot where the shot finally rests.

ANCHOR, (*ancre*, Fr. *anchora*, Lat. from ἄγκυρα, Greek) is a large heavy iron instrument, with a double hook at one end, and a ring at the other, by which it is fastened to a cable, and dropped from a ship into the bottom of the sea, where, taking its hold, it keeps her in a proper situation from being driven away by the wind or tide.

An anchor is so artfully calculated, by the construction of its parts, as both to sink into the ground when it reaches the bottom, and to bear a very heavy strain before it can be loosened or dislodged from its situation.

Plate 1. Fig. 10. It is composed of a shank, a stock, a ring, and two arms with their flooks or palms. The stock serves to guide the flooks in a direction perpendicular to the surface of the ground; so that one of them sinks into it by its own weight, as soon as it falls, and is still preserved steadily in that position by the stock, which, together with the shank, lies flat on the bottom.

That the form of so very useful an instrument may
Fig. 1. 3. 5. be more clearly understood, we shall describe it as composed of four parts, the *ring*, the *shank*, the *arms*, and the *stock*.

The ring, to which the cable is attached, is fixed on one extremity of the shank, and the arms on the other extremity.

These arms consist of *palms* or *flooks*, and the *bill*.

The *palms* are the broad plates, of a triangular form, at nearly the extremity of the arm; and the *bills* are the sharp points at the very extremity of the arms.

The *throat* of the arms is the rounded angular point where the arm is joined to the shank. A distance equal to that between the throat of one arm and its bill, is marked on the shank from the place where it joins the arms, and is called the *trend;* the *crown* is the place where the arms are joined to the shank; the *small round* is the diameter of the shank where it is smallest, which is near the square part, where the stock is fixed; the *stock* is a beam of wood fixed on the extremity of the shank, at right angles to the plane passing through the arms.

The shank of anchors consists of four square bars of iron, encircled by a number of lesser bars of the same lengths, which are again surrounded by other bars, like the arch-stones of a bridge. The method of building the shank will be understood from the section of it, which is represented in figure 4. The exterior bars Fig. 4.
are then brought to a welding heat, and are welded by the blows of a large mass of iron called a *Hercules*.

This piece of iron, which weighs between four and five hundred weight, is elevated about nine or ten feet above the shank placed upon an anvil, and is then allowed to fall upon the shank. This machine is wrought by men, in the same way as the engine called a ringing-bell for driving piles. The blows of the hercules and of the hammer are continued, till the bars which surround the interior cylinder appear to be one solid mass on the outside, and till every interstice or chink between them is completely closed up.

The shank tapers from the extremity where the arms are fixed, to the other extremity which carries the stock, so that the difference between these diameters is from $1\frac{1}{2}$-inches in small ones, to 3-inches in large ones. The extremity to which the stock is fixed, is of a square form, of the same size as the trend, and haunches into the small round, one sixth the length of the shank. Through this square part is punched a hole for the ring, at the distance of one and a half times the thickness of the ring from the upper extremity of the square part. Between this hole and the lower extremity of the square part, are two small pro- Fig. 3.
minences, *a b*, called *nuts*, raised from the solid iron, for the purpose of securing the stock in its proper position.

At the lower extremity of the shank is formed a flatted surface or scarf, with a shoulder on each side, for the purpose of *shutting on*, or joining the arms to the shank.

The arms are made in the same way as the shank. They are rounded and flatted on the different sides, so as to resemble the shank, and are of the same diameter as the shank at the throat and small round.

The extremities of the arms are scarfed in the same way as that part of the shank to which they are united.

The arms are rounded between the throat and the palms, which are nearly in the middle of the arms; and from this it has a square, called the *blade*, tapering to the bill on the flatted side, and a square nut is made on the inner rounded side for receiving the palm, that the palm when shut on may projectaits thickness at the base or inner part, the exterior part forming a straight surface with the peak or bill.

The outer side of the arm is made straight from the rounded part or haunch to the snape or bill, and there kept to half the substance of the inner part. The snape, which is nearly *one-fourth* the breadth of the palm resembles the bill of a duck. The interior part of the arm is made nearly a straight line from the throat to the bill. It is reckoned stronger from having a small angle in its length, the vertex of the angle being in the middle of the arm.

A line drawn from the trend to the point of the bill forms an angle of nearly 60 degrees with the shank.

The palms or flooks are two thick plates of iron, nearly resembling an isosceles triangle, and formed of different pieces, well wrought and perfectly sound; otherwise, in the process of welding them on the arm, they are likely to separate at the parts where they are joined together. The sides of the triangular palms, which are a little curved, are from *one* and a *fourth* to
Fig. 2. *one* and a *half* inches longer than their base, and the highest points of the curves about the same distance, from a straight line drawn from the extremity of the base to the vertex of the triangle. The base is straight; the inner flat surface is a little curved in its breadth, but is a straight line in its length.

The stock of anchors consists of two beams of oak firmly united by iron bolts and hoops, two hoops being on each side of the middle, and one near each extremity. The length of the stock is equal to the length of the shank, and half the diameter of the ring. Its breadth and thickness in the middle, are each *one-twelfth* of its length At the extremities of the stock, the breadth and thickness is *one twenty-fourth* of its length. The upper surface of the stock next the ring
Fig. 5. is always straight, or forms one continued plane. The under surface is composed of three planes. The central one extends to half the breadth of the stock, on each side of the middle, and from that place the surface descends to each extremity. Lest the wood should shrink, an opening of an inch and a half is generally left between the two beams of oak, that the hooks which bind them together may be driven nearer the middle.

When the various parts of an anchor are thus formed, the next process is, to combine them together. The square which receives the stock is welded on the extremity of the shank; and the ring, whose thickness is half the diameter of the small round, after being forged, is introduced through the opening in the square, and its extremities well joined together. The arms are then welded to the shank by means of the hercules, and the palms by the same means are joined to the arms. When the arms are welded to the shank, the whole crown of the anchor is, by repeated welding heats, made perfectly sound, and worked to its proper proportion. It is then finished by working it with hammers till it is quite cold. This part of the process adds much to the appearance of the anchor, and hardens the surface, so as in some measure to retard its corroding.

As the process of making anchors is a very severe trial to the iron, it should be of the very best quality. While the anchor is heating, the external bars get into a welding state much sooner than the internal ones, and, therefore, they are much more liable to be injured by this part of the process.

By examining the broken part of large anchors, which have been made of iron of an inferior quality, it is very manifest that their strength has been much diminished by being brought to a greater heat than what was necessary. When this is the case, the anchor breaks very short, and none of the internal bars appear, in consequence of their having been consolidated, by over heating, into one mass. But when anchors that are made of the best iron, and not over heated, are exposed to the strain which they cannot resist, they will in many instances bend, and crack on the surface of the shank or arms. When this happens, the centre bars, both in large and small anchors, are always distinctly seen, in consequence of not having been consolidated by an excess of heat. Hence it appears, that if an anchor, large enough for men of war, or East India ships, is perfectly sound, or consolidated in the shanks or arms, from $2\frac{1}{4}$ to $3\frac{1}{2}$ inches from the surface, it is less liable to break, than when it is made one solid body throughout, as this is more likely to be effected by over heating than by the process of hammering.

The greatest care, however, should be taken that the scarfs of the shank and arms and every part of the anchor be perfectly sound; for if any chinks are left, the salt water will find its way through them, and rust the internal bars. The part which contains the water will also swell, and, by producing a separation of the surface, will render the anchor unfit for service.

Fig. 1, 2, 3, of Plate 1, are drawn by a scale of 4-10ths of an inch to a foot, from an anchor of 80-cwt. Fig. 4. is a section of the bars, as laid together in the smallest hoop of the shank for an anchor of 80-cwt. and is drawn by a scale of one-inch to a foot. Fig. 5. is drawn to a different scale, merely to shew the form of the stock.

M. Bouguer, in his Traité de Navire, directs to take the length of the shank in inches, and to divide the cube of it by 1160 for the weight. The reason is obvious, because the quotient of the cube of 201 inches, which is the length of an anchor weighing 7000lb. divided by the weight is 1160; and, therefore, by the rule of three, this will be a common division for the cube of any length, and a single operation will suffice. The same author gives the following dimensions of the several parts of an anchor. The two arms generally form the arch of a circle, whose centre is three-eighths of the shank from the vertex, or point where it is fixed to the shank; and each arm is equal to the same length, or the radius; so that the two arms together make an arch of one hundred and twenty degrees: the flooks are half the length of the arms, and their breadth two-fifths of the said length. With respect to the thickness, the circumference at the throat, or vertex of the shank, is generally made about a fifth part of its length, and the small end two-thirds of the throat; the small end of the arms of the flooks, three-fourths of the circumference of the shank at the throat.

All anchors, says Mr. Bowyer, ought to be the weight in proportion to the cubes of their length.

The following tables contain the dimensions of anchors of various weights, as now made in his Majesty's dock-yards, and the value and number of anchors for ships of each class in the British navy:

TABLE of the dimensions of Anchors, as made in his Majesty's dock-yards, from 1cwt. to 90, which is the weight of the largest Anchors used in the first rates.

Veight.	Length of the shank.		Length of the flooks.		Breadth of the palms.		Thickness of the palms.		Size of the trend.		Size of the small round.		Outer diamet. of the ring.		Thickness of the ring.	
cwt.	feet.	in.	feet.	in.	feet.	in.	feet.	in.	feet.	in.	feet.	in.	feet.	in.	feet.	in.
1	5	8	1	10	0	9	0	0 1/4	0	2 1/2	0	2	0	9	0	1
2	6	6	2	2	0	11	0	0 1/4	0	2 1/4	0	2 1/4	0	11	0	1 1/8
3	7	0	2	4	1	0	0	0 7/8	0	3	0	2 1/2	1	0	0	1 1/4
4	7	6	2	6	1	1	0	0 7/8	0	3 1/4	0	2 3/4	1	1	0	1 3/8
5	8	0	2	8	1	2	0	1	0	3 1/2	0	3	1	2	0	1 1/2
6	8	6	2	10	1	3	0	1	0	3 3/4	0	3 1/4	1	3	0	1 5/8
7	9	0	3	0	1	4	0	1	0	4	0	3 1/2	1	4	0	1 3/4
8	9	6	3	2	1	5	0	1 1/8	0	4 1/4	0	3 3/4	1	5	0	1 7/8
9	10	0	3	4	1	6	0	1 1/8	0	4 3/8	0	3 7/8	1	6	0	1 15/16
10	10	4	3	5	1	7	0	1 1/8	0	4 1/2	0	4	1	7	0	2
11	0	8	3	7	1	8	0	1 1/4	0	4 5/8	0	4 1/8	1	8	0	2 1/16
12	11	0	3	8	1	8 1/4	0	1 1/4	0	4 5/8	0	4 1/8	1	8 1/4	0	2 1/16
13	11	4	3	10	1	8 1/2	0	1 1/4	0	4 3/4	0	4 1/4	1	8 1/2	0	2 1/8
14	11	8	3	11	1	8 3/4	0	1 1/4	0	4 7/8	0	4 3/8	1	8 3/4	0	2 3/16
15	12	0	4	0	1	9	0	1 3/8	0	5	0	4 1/2	1	9	0	2 1/4
16	12	3	4	1	1	9 1/8	0	1 3/8	0	5 1/8	0	4 5/8	1	9 1/8	0	2 5/16
17	12	6	4	2	1	9 1/4	0	1 3/8	0	5 1/4	0	4 3/4	1	9 1/4	0	2 3/8
18	12	8	4	3	1	9 1/2	0	1 3/8	0	5 3/8	0	4 7/8	1	9 1/2	0	2 7/16
19	12	10	4	4	1	9 5/8	0	1 1/2	0	5 1/2	0	5	1	9 5/8	0	2 1/2
20	13	0	4	4 1/2	1	9 3/4	0	1 1/2	0	5 5/8	0	5 1/8	1	9 3/4	0	2 9/16
21	13	2	4	5	1	9 3/4	0	1 1/2	0	5 7/8	0	5 1/8	1	9 3/4	0	2 9/16
22	13	4	4	5 3/8	1	10	0	1 1/2	0	6 1/8	0	5 3/8	1	10	0	2 11/16
23	13	6	4	6	1	10 1/4	0	1 1/2	0	6 1/4	0	5 1/2	1	10 1/4	0	2 3/4
24	13	8	4	6 1/2	1	10 1/2	0	1 1/2	0	6 3/8	0	5 5/8	1	10 1/2	0	2 13/16
25	13	10	4	7	1	11	0	1 5/8	0	6 1/2	0	5 3/4	1	11	0	2 7/8
26	14	0	4	8	1	11 1/2	0	1 5/8	0	6 5/8	0	5 7/8	1	11 1/2	0	2 15/16
27	14	2	4	8 1/2	1	11 3/4	0	1 5/8	0	6 5/8	0	5 7/8	1	11 3/4	0	2 15/16
28	14	4	4	9 1/4	1	11 3/4	0	1 5/8	0	6 5/8	0	5 7/8	1	11 3/4	0	2 15/16
29	14	6	4	10	2	0	0	1 3/4	0	6 5/8	0	5 7/8	2	0	0	2 15/16
30	14	7 1/2	4	10 1/2	2	0 1/2	0	1 3/4	0	6 3/4	0	6	2	0 1/2	0	3
31	14	9	4	11	2	1	0	1 7/8	0	6 3/4	0	6	2	1	0	3
32	14	10	4	11 1/2	2	1 1/2	0	1 7/8	0	6 7/8	0	6 1/8	2	1 1/2	0	3 1/16
33	15	0	5	0	2	1 3/4	0	1 7/8	0	7	0	6 1/4	2	1 3/4	0	3 1/8
34	15	1	5	0 1/4	2	2	0	2	0	7 1/4	0	6 1/2	2	2	0	3 1/4
35	15	2	5	0 3/8	2	2 1/2	0	2	0	7 3/8	0	6 5/8	2	2 1/2	0	3 5/16
36	15	4	5	1 1/4	2	2 3/4	0	2	0	7 1/2	0	6 3/4	2	2 3/4	0	3 3/8
37	15	6	5	2	2	3	0	2	0	7 1/2	0	6 3/4	2	3	0	3 3/8
38	15	7	5	2 1/4	2	3 1/2	0	2 1/8	0	7 5/8	0	6 7/8	2	3 1/2	0	3 7/16
39	15	9	5	3	2	3 3/4	0	2 1/4	0	7 5/8	0	6 7/8	2	3 3/4	0	3 7/16
40	15	10	5	3 1/2	2	4	0	2 3/8	0	7 3/4	0	7	2	4	0	3 1/2
41	16	0	5	4	2	4 1/4	0	2 1/2	0	7 3/4	0	7	2	4 1/4	0	3 1/2
42	16	1	5	4 1/4	2	4 1/2	0	2 1/2	0	7 3/4	0	7	2	4 1/2	0	3 1/2
43	16	2	5	4 1/2	2	5	0	2 1/2	0	7 7/8	0	7 1/8	2	5	0	3 9/16
44	16	3	5	5	2	5 1/4	0	2 1/2	0	7 7/8	0	7 1/8	2	5 1/4	0	3 9/16
45	16	4	5	5 1/4	2	5 1/2	0	2 1/2	0	8	0	7 1/4	2	5 1/2	0	3 5/8
46	16	5	5	5 1/2	2	5 3/4	0	2 1/2	0	8	0	7 1/4	2	5 3/4	0	3 5/8
47	16	6	5	6	2	6	0	2 1/2	0	8 1/8	0	7 3/8	2	6	0	3 11/16
48	16	7	5	6 1/4	2	6 1/4	0	2 1/2	0	8 1/8	0	7 3/8	2	6 1/4	0	3 11/16
49	16	8	5	6 1/2	2	6 1/2	0	2 5/8	0	8	0	7 3/8	2	6 1/2	0	3 13/16
50	16	9	5	6 3/4	2	7	0	2 5/8	0	8 1/4	0	7 1/2	2	7	0	3 3/4

Continuation of Table.

Weight.	Length of the shank.		Length of the flooks.		Breadth of the palms.		Thickness of palms.		Size of the trend.		Size of small round.		Outer diamet. of the ring.		Thickness of the ring.	
cwt.	feet.	in.	feet.	in.	feet.	in.	feet.	in.	feet.	in.	feet.	in.	feet.	in.	feet.	in.
51	16	10	5	7	2	7½	0	2⅝	0	8¼	0	7½	2	7½	0	3¾
52	16	11	5	7½	2	8	0	2⅝	0	8¼	0	7½	2	8	0	3¾
53	17	0	5	7¾	2	8¼	0	2⅝	0	8⅜	0	7⅝	2	8¼	0	3 13/16
54	17	1	5	8¼	2	8⅝	0	2⅝	0	8⅜	0	7⅝	2	8⅝	0	3 13/16
55	17	2	5	8½	2	9	0	2¾	0	8½	0	7¾	2	9	0	3⅞
56	17	3	5	9	2	9¼	0	2¾	0	8½	0	7¾	2	9¼	0	3⅞
57	17	4	5	9½	2	9½	0	2¾	0	8½	0	7¾	2	9½	0	3⅞
58	17	5	5	9¾	2	10	0	2¾	0	8¾	0	7⅞	2	10	0	3 15/16
59	17	6	5	10	2	10¼	0	2¾	0	8⅝	0	7⅞	2	10¼	0	3 15/16
60	17	7	5	10¼	2	10½	0	2⅞	0	8¾	0	7⅞	2	10½	0	3 15/16
61	17	8	5	10½	2	11	0	3	0	8¾	0	8	2	11	0	4
62	17	9	5	11	2	11½	0	3	0	8¾	0	8	2	11½	0	4
63	17	10	5	11¼	2	11¾	0	3	0	8⅞	0	8	2	11¾	0	4
64	17	11	5	11¼	3	0	0	3	0	8⅞	0	8	3	0	0	4
65	18	0	6	0	3	0	0	3⅛	0	9	0	8	3	0	0	4
66	18	1	6	0¼	3	0	0	3⅛	0	9	0	8	3	0	0	4
67	18	2	6	0½	3	1	0	3⅛	0	9	0	8	3	1	0	4
68	18	3	6	1	3	1	0	3⅛	0	9⅛	0	8	3	1	0	4
69	18	3½	6	1⅛	3	1	0	3⅛	0	9⅛	0	8	3	1	0	4
70	18	4	6	1¼	3	1	0	3¼	0	9¼	0	8⅛	3	1	0	4 1/16
71	18	5	6	1½	3	1½	0	3¼	0	9¼	0	8⅛	3	1½	0	4 1/16
72	18	6	6	2	3	1½	0	3¼	0	9¼	0	8⅛	3	1½	0	4 1/16
73	18	8	6	2½	3	1½	0	3¼	0	9⅜	0	8⅛	3	1½	0	4 1/16
74	18	9½	6	3	3	1½	0	3⅜	0	9⅜	0	8⅛	3	1½	0	4 1/16
75	18	11	6	3½	3	1½	0	3⅜	0	9⅜	0	8⅛	3	1½	0	4 1/16
76	19	0⅜	6	4⅛	3	2	0	3⅜	0	9⅜	0	8⅛	3	2	0	4 1/16
77	19	1¾	6	4⅝	3	2	0	3½	0	9½	0	8 3/16	3	2	0	4⅛
78	19	3⅜	6	5	3	2	0	3½	0	9¼	0	8 3/16	3	2	0	4⅛
79	19	4½	6	5½	3	2	0	3½	0	9¾	0	8¼	3	2	0	4⅛
80	19	6	6	6	3	2	0	3½	0	9¾	0	8¼	3	2	0	4⅛
81	19	8	6	6⅝	3	2½	0	3⅝	0	10	0	8¼	3	2½	0	4⅛
82	19	8	6	6⅝	3	2½	0	3⅝	0	10	0	8¼	3	2½	0	4⅛
83	19	8	6	6⅝	3	2¾	0	3⅝	0	10	0	8⅜	3	3½	0	4¼
84	19	9	6	7	3	3	0	3⅝	0	10⅛	0	8⅜	3	3	0	4¼
85	19	9	6	7	3	3¼	0	3⅝	0	10⅛	0	8⅜	3	3¼	0	4¼
86	19	9	6	7	3	3⅜	0	3⅝	0	10⅛	0	8⅜	3	3⅜	0	4¼
87	19	10	6	7¼	3	3⅜	0	3¾	0	10⅛	0	8½	3	3⅜	0	4⅜
88	19	10	6	7¼	3	3½	0	3¾	0	10¼	0	8½	3	3½	0	4⅜
89	19	10	6	7¼	3	3½	0	3¾	0	10¼	0	8½	3	3½	0	4⅜
90	19	11	6	7⅝	3	3⅝	0	3¾	0	10¼	0	8½	3	3⅝	0	4⅜

ESTABLISHMENT and VALUE of ANCHORS for Ships of each Class in the British Navy, 1809.

Anchors.	No. allowed.	Weight of each	Total Weight.	Rate per Cwt.	Value.	No. allowed.	Weight of each.	Total Weight.	Rate per Cwt.	Value.	No. allowed.	Weight of each.	Total Weight.	Rate per Cwt.	Value.
	100 Guns, and upwards.					100 Guns.					98 or 90, 2050 Tons or upwards.				
		Cwt.	Cwt.	£ s. d.	£ s. d.		Cwt.	Cwt	£ s. d.	£ s. d.		Cwt.	Cwt.	£ s. d.	£ s. d.
Bower ..	4	90	360	4 12 4½	1662 15 0	4	84	336	4 12 4½	1551 18 0	4	84	336	4 12 4½	1551 18 0
Stream ..	1	21		2 8 3½	50 14 1½	1	21		2 8 3½	50 14 1½	1	21		2 8 3½	50 14 1½
Kedge ..	1	10½				1	10½				1	10½			
Ditto ..	1	5		2 7 3	36 12 4½	1	5		2 7 3	36 12 4½	1	5		2 7 3	36 12 4½
	98 or 90, under 2050 Tons.					80 Guns.					74 Guns, 1799 Tons and upwards.				
Bower ..	4	76	304	4 12 4½	1450 5 9	4	76	304	4 12 4	1450 5 9	4	73	292	4 4 6½	1234 6 2½
Stream ..	1	18				1	18				1	17½			
Kedge ..	1	9		2 7 3	74 8 4	1	9		2 7 3	74 8 4½	1	8½		2 7 3	61 8 6
Ditto. ...	1	4½				1	4½								
	74 Guns, under 1799 Tons.					74 Guns, 1740 to 1798 Tons.					64 Guns.				
Bower ..	4	70	280	4 1 6	1141 0 0	4	67	268	4 1 6	1092 2 0	4	57	228	3 8 6	780 18 0
Stream ..	1	17½				1	16				1	15			
Kedge ..	1	8½		2 7 3	61 8 6	1	8		2 7 3	56 14 0	1	7½		2 7 3	53 3 1½
	50 Guns, of 1100 Tons.					50 Guns, under 1100 Tons.					44 Guns.				
Bower ..	4	52	208	3 4 9¼	706 0 0½	4	49	196	3 1 5	601 17 8	4	44	176	2 17 2½	503 8 8
Stream ..	1	11				1	11				1	10			
Kedge ..	1	5½		2 7 3	38 19 7	1	5½		2 7 3	38 19 7½	1	5		2 7 3	35 8 9
	40 Guns.					38 Guns.					36 Guns.				
Bower ..	4	46	184	3 1 5	565 0 8	4	44	176	2 17 2½	503 8 8	4	40	160	2 17 2½	457 13 4
Stream ..	1	10				1	10				1	10			
Kedge ..	1	5		2 7 3	35 8 9	1	5		2 7 3	35 8 9	1	5		2 7 3	35 8 9
	32 Guns, large, from 910 to 972					32, Smaller class, that carry 18-pounders.					32 Guns, common class.				
Bower ..	4	40	160	2 17 2½	457 13 4	4	35	140	2 13 6½	374 15 10	4	32	128	2 10 6¾	323 12 0
Stream ..	1	8½				1	8½				1	8½			
Kedge ..	1	4		2 7 3	29 10 7	1	4		2 7 3	29 10 7	1	4		2 7 3	29 10 7
	28 Guns.					24 Guns.					20 Guns.				
Bower ..	4	31	124	2 10 6½	313 9 9	4	29½	118	2 8 3½	284 18 5	4	25	100	2 8 3½	241 9 2
Stream ..	1	8				1	7½				1	7½			
Kedge ..	1	4		2 7 3	28 7 0	1	3½		2 7 3	25 19 9	1	3½		2 7 3	25 19 9
	Sloops of 481 Tons.					Sloops of 422 Tons.					Sloops of 361 to 422 Tons.				
Bower ..	3	25	75	2 8 3½	181 1 10½	3	23	69	2 8 3½	166 12 1¾	3	21	63	2 8 3½	152 2 4½
Stream ..	1	7½				1	7½				1	7½			
Kedge ..	1	3½		2 7 3	25 19 9	1	3½		2 7 3	25 19 9	1	3½		2 7 3	25 19 9
	Sloops 268 to 340 Tons.					Brig Sloops, 370 Tons and upwards.					Brig Sloops, about 365 Tons.				
Bower ..	3	20	60	2 7 3	141 15 0	3	20	60	2 7 3	141 15 0	3	16	48		
Stream ..	1	7½				1	7				1	7		2 7 3	138 4 1½
Kedge ..	1	3½		2 7 3	25 19 9	1	3½		2 7 3	24 16 1½	1	3½			
	Brigs, about 313 Tons, and 282 Tons.					Brigs 250, to Gun Brigs of 187 Tons.									
Bower ..	3	15	45			3	14	42							
Stream ..	1	7		2 7 3	131 2 4½	1	5		2 7 3	115 5 3					
Kedge ..	1	3½				1	2								

The names of the anchors at present in use are,

The Sheet ANCHOR, (*maitresse-ancre*, Fr.) which is the largest and strongest, being that which the seamen call their last hope, and never to be used but in great extremity.

The best Bower ANCHOR, (*la seconde ancre*, Fr.) and *the small Bower* ANCHOR, (*ancre d'affourche*, Fr.) are a little different in size, and take their names from the situation in which they are placed, being carried on the bows of the ship.

The Stream ANCHOR, (*ancre de touée*, Fr.) is less than the preceding: and

The Kedge ANCHOR, (*petite ancre de touée*, Fr.) is the smallest of all, except

The Pilot's ANCHOR, (*ancre de pilote*, Fr.) a small anchor between these two, and used mostly by pilots for dropping a vessel in a stream, or tide's way.

With respect to the situation when in the ground, they are sometimes denominated:—

The Flood ANCHOR, (*ancre de flot*, Fr.) or that by which the ship rides during the flood tide;

The Ebb ANCHOR, (*ancre de jusant*, Fr.) or that by which the ship rides during the ebb tide.

The Sea ANCHOR, (*ancre du large*, Fr.) is that which lies towards the offing.

The Shore ANCHOR, (*ancre de terre*, Fr.) is that which is between the ship and the shore.

The Floating ANCHOR, (*ancre flottante*, Fr.) This is a simple implement, which is sunk below the swell of the sea, where there is no other anchorage, for the purpose of preventing a vessel from drifting. Dr. Franklin proposed that it should consist of a wooden cross, from the ends of which a sail should be stretched, covering the cross, and even all the extremities. It has also been used in the figure of an umbrella. The following plan is now adopted.

Plate 1. fig. 7, 8.

Two flat iron bars, each half the length of the midship beam, are rivetted together with a saucer-headed bolt, clenched at the points, that the bars may be swung together for more convenient stowage. At the four extremities of the bars is a hole in each, for a rope to pass through, which must be drawn tight, so as to extend them at right angles. To this rope is sewed a double or fourfold canvas cloth, of a quadrangular shape, so that when used it is on the side of the bars next the vessel. In each bar are two holes, at equal distances from the centre, and to these holes the ends of two pieces of rope are fixed. The ropes are fastened together in the middle, so as to form a crow foot, having an eye in the centre, which is well secured with spun yarn; and to this is ent a cable, which, when the anchor is used, brings it into an inclined position in the water. An iron ring is fitted to the end of one of the bars, to which a buoy is fastened by a rope of about twelve fathoms long, to prevent the anchor from sinking to the bottom.

A single Flook ANCHOR, (*ancre a une patte*, Fr.) has been recently proposed by Mr. James Stuard. An anchor with a single flook will hold the ship as steadily as those with two, and have the advantage of being more light, and less expensive. They have not, however, the same chance of catching the ground, and are more difficult to prepare for service, as the anchor must be lowered by the crown, in order to keep the flook downwards. As the uppermost flook of the common anchor is sometimes dangerous to ships which happen to ground by it, and often occasions the anchor to be tripped by the cable taking hold of it on the ship's swing, consequently one flook only is employed in this anchor. The shank is made very short, and the bars which compose the shank and arms are all of one piece, without any joining. The palm is made of cast iron, or of a cast iron shell, filled with lead, and the stock is composed of a wrought iron bolt, covered with cast iron. In fixing the stock to the anchor, it has been very lately proposed by Captain Ball, of the Royal Navy, to increase the size of the nut, so that it may extend on each side of the shank to two and an half times the diameter of the shank, for the purpose of receiving two bolts through each projection; these bolts keep together the two beams which form the stock, and completely prevent the stock from slipping off the shank, an accident which very frequently happens in hot climates. In fig. 3, is represented at *m*, in dotted lines only one of the projections. Plate 1.

The ANCHOR *comes home*, (*le vaisseau chasse sur son ancre*, Fr.) implies that the anchor is dislodged from its bed in the ground by the violence of the wind, sea, or current, or all united.

The ANCHOR *drags*, (*chasser sur son ancre*, Fr.) implies the effort of making the anchor come home, so as to drag it along the ground.

The ANCHOR'S *foul*, (*ancre dont le cable a fait un tour autour de la patte superieure*, Fr.) is so called when it either hooks some other anchor, wreck, or cable under the water, or when, by any accident, the ship entangles her slack cable about the stock or upper flook of it.

The ANCHOR *is a cock bill*, (*l'ancre est à la veille*, Fr.) implies that the anchor is suspended perpendicularly from the cat-head, ready to be let go at a moment's warning.

The ANCHOR *is a-peek*, (*l'ancre est à pic*, Fr.) denotes the cable has been drawn in so tight as to bring the ship directly over it.

The ANCHOR *is a-trip*, or *a weigh*, (*l'ancre est dérapée*, Fr.) the state of the anchor when it is just drawn out of the ground in a perpendicular direction, either by the cable or the buoy-rope.

Coming to an ANCHOR (*venir à l'ancre*, Fr.). If the land be made near the destined port, the ship being close hauled (suppose on the larboard tack), and she can fetch a place of safety to remain in till the wind comes round, there being no tide to check her, the top-gallant sails, courses, jib and staysails are taken in; then, with the fore-top-mast staysail, mizen, and three topsails, she stands on till near her birth: the buoy is streamed (having hands stationed at the stopper and shank-painter, and every thing clear of the range of the cable), the fore-topmast staysail hauled down, and the helm is put down to leeward; when she is in the wind, the main and fore-topsail halyards are let fly, the weather braces are hauled in, the sheets started, and the sails clewed up: the mizen-topsail is hove a-back, and when she gets sternway the helm is righted, and the anchor let go: the proper scope of cable is veered away according to the strength of the wind, the stopper is clapped on forwards, the bight of the cable thrown over the bits, and the stoppers are clapped on aft. The hands are then sent on the yards to furl the sails.

If there be so little wind that it is necessary to continue the mizen topsail set, to keep her a-stern of her anchor, the short service only is veered out; because the ship not having power to keep it taught, the cable may be injured by being dragged on the ground. If the wind be pretty strong when she is brought up, the mizen-topsail may be taken in with the others; the mizen will bring her head to wind, she will get sternway, and the anchor may be let go, veering away the proper scope of cable.

When a vessel comes to an anchor, (particularly in a tideway) it is always prudent to take three reefs in the topsails before they are handed, as they will be ready should a sudden gale arise, if there be a necessity for running out to sea.

If there be a tide, the ship is put on that tack by which she can stem it. Thus if the tide run from east, and the wind be at S. S. E. it is plain by the compass that she must be got on the starboard tack; when being close hauled, she just stems it: whereas were she on the larboard tack, she would only lie S. W. as per compass; so that she would drive out against with the tide, her course being only four points from west, its absolute direction.

When a ship, therefore, intends to anchor, and stems the tide, she gets under an easy sail, taking in the top-gallant sails and courses, &c. according to the strength of the wind; and when near enough (having streamed the buoy, &c. as before), clews up the topsails; the tide then checking her way, she lets go the anchor. When she has sufficient cable, and the proper service in the hawse, she rides by the tide with the wind almost across.

Thus in coming to an anchor the ship's head must be always put to the stream. If it comes from leeward, there is nothing to do but to shorten sail, and when the topsails are clewed up, let go the anchor: if from windward, she must be luffed up, hauling out the mizen, and when she meets the stream, and her headway is stopped, the anchor is let go.

Riding at ANCHOR (*être mouillé, ou, à l'ancre*, Fr.). The state of a vessel on the surface of the water, when moored and secured in her proper station. When a fleet of many ships is moored in a port or road, care must be taken to preserve a considerable distance between the vessels, not only for the purpose of keeping them clear of each other, but to prevent them from running foul, when getting under way.

To ANCHOR, or *cast* ANCHOR, (*ancrer, mouiller, &c.* Fr.) to let go the anchor that the ship may ride thereby.

To lie at ANCHOR, (*être à l'ancre*, Fr.) the situation of a ship which rides by her anchor.

To back the ANCHOR, (*empenneller l'ancre*, Fr.) is, to lay down a small anchor ahead of the large one by which the ship rides, the cable of the former being fastened to the crown of the latter, in order to prevent its coming home.

To cat the ANCHOR, (*caponer l'ancre*, Fr.) is when the anchor is drawn perpendicularly up to the cat-head by a sort of strong tackle called a cat.

To fish the ANCHOR, (*traverser l'ancre*, Fr.) to hoist and draw up the flooks of a ship's anchor towards the top of the bow, by a machine called a fish, in order to stow it after it has been catted.

To sheer the ship to her ANCHOR, (*gouverner sur l'ancre*, Fr) is to steer the ship's head to the place where the anchor lies when they are heaving the cable into the ship.

To sweep the ANCHOR (*draguer l'ancre*, Fr.). This is an operation which consists in dragging at the bottom of the sea for an anchor that has been lost.

It is performed by fixing the two ends of a rope to two boats a little asunder, and sinking the middle of the rope, by means of a shot or some other weight: so that, on conducting it along the bottom, it may catch the anchor.

To shoe the ANCHOR, (*couvrir les pattes de l'ancre avec des planches pour la faire mieux tenir dans un fond de vase molle*, Fr.) is to cover the flooks with a broad triangular piece of thick plank, whose area is greater than that of the flooks, in order to give the anchor a stronger hold in soft ground.

To weigh the ANCHOR, (*lever l'ancre*, Fr.) to heave the anchor out of the ground by its cable.

To weigh the ANCHOR *by the long boat*, (*lever l'ancre avec la chaloupe*, Fr.) is performed by applying mechanical powers, fixed in the boat, to the buoy-rope, instead of the cable, and thereby pulling it up to the boat.

ANCHOR-*ground*, (*fond de bonne tenüe*, Fr.) is a bottom which is neither too deep, too shallow, nor rocky: as in the first, the cable bears too nearly perpendicular, and is thereby apt to jerk the anchor out of the ground: in the second, the ship's bottom is apt to strike at low water, or when the sea runs high, by which she is exposed to the danger of sinking: and in the third, the anchor is liable to hook the broken and pointed ends of rocks, and tear away its flooks; whilst the cable, from the same cause, is constantly in danger of being cut through as it rubs on their edges.

In cases of necessity, where the sheet-anchor or best bower has been left behind, in consequence of the cable parting, a substitute has been tried and found to answer very well, by lashing one of the ship's guns to a small anchor, or between the two smallest, if one is not thought sufficient. This was done with good effect by Lord Anson, on board the Centurion, and, afterwards, by Captain Collins of the Terrible, of 74 guns, one of the squadron employed in the year 1758, on the expedition against Quebec.

With respect to anchors, a great improvement might be made in them by the introduction of stocks of iron, tapering from five inches in the middle to four at the edge. The two parts of the stock might be bolted with four bolts and forelocks; and with this sharp form, angulated every way, the stock would sink, in loose ground, deep into the earth, and make a prodigious resistance; by means of a proper scope of cable, it would continue this resistance till the flook of the anchor was almost so much elevated as to lose its hold, and at the same time, help to retard its elevation, as well as preserve its hold.

The action of an anchor can only be comprehended by experiment, and, by the kindness of Providence it seems to be governed by laws peculiar to itself, and adapted to its uses. Its manner of resisting is so favourable, that the retainer annexed, may, without making a very acute angle with the earth, have all the benefit of its powers. The cable being an hypothenuse of twice and a half the length of the perpendicular height from the hawse to the ground, is not, perhaps, sufficient. If the cable veered be three times the length from the ground to the hawse, and it does not appear that any advantage can be gained by additional extent, ships of war may work their cables on any deck they please, and obtain firm anchorage in forty fathoms water, or deeper. If this point is true, as it will most assuredly be found to be, veering a vast length of cable, two or three on one anchor, as has commonly been done in twelve or fewer fathom water, is not only unnecessary, but injurious, since great lengths of rope become weaker in a ratio proportioned to that length.

ANCHORAGE, a duty levied from vessels coming to a particular road, or anchoring ground.

In the first instance, it is due to the crown only, no subjects having rights within the sea-mark, without special grant. In some countries, this duty is regulated more by the size of the vessel, than the value of the cargo; and, in certain ports, it is levied from ships in ballast. Should a vessel be driven in by stress of weather, it is generally admitted, that anchorage is not due, provided no part of the cargo has been discharged. Similar to anchorage duty, are *shore dues*, a right frequently found in the hands of subjects having a grant of a port or haven.

There are instances of an inconsiderable inland borough, destitute of trade, being entitled to levy *shore dues* from merchants coming into a port at many miles distance. Anchorage is also the set of anchors belonging to a ship.

ANCHORING, (*mouillant*, Fr.) If the wind be large, or right aft, and the stream from windward, or what is called a leeward tide, then all the sails are handed, except the fore-topsail, and the cable is bitted: the ship is hauled up sooner, the helm put a-lee, the fore-topsail clewed up, the mizen hauled out, the mizen-staysail sheet aft, to bring her to the wind; and when she comes head to wind, she loses her headway, at which time the anchor is let go, and, according to the strength of the wind and tide, a long scope of cable is veered out.

When the wind is right out of a river, and the ship is to go up, she waits till there is water enough, and the flood sufficiently strong to drive by against the wind: she then drifts in, either stern foremost, or broadside to it, having sufficient sail set to determine the rate of driving. When there is room enough, the first of these is preferred, as she will answer her helm by the tide acting against the sides of the rudder, as if she were going a-head.

ANCHOR STOCK, a method of working planks, whereby the butts of one plank comes nearly over the middle of another plank, and the planks being broadest in the middle, and tapered to the butts, they thereby appear in shape like an anchor stock.

ANDROMEDA, in astronomy, the name of a constellation of the northern hemisphere, representing the figure of a woman almost naked, her feet at a distance from each other, and her arms extended and chained: it is situated to the north of the constellation Aries and Pisces, near Cassiopeia and Perseus.

The body of Andromeda is marked by three brilliant stars, in the form of an arch, and almost at equal distances from one another. This constellation contains about 27 stars that are visible to the naked eye; of which the principal are, α *Andromeda's Head;* β in the girdle, called mirach, or mizar; and γ on the south foot, named alamak, and sometimes alhames.

In the Britannic catalogue the stars of this constellation amount to 66, and to 73 in the Berlin catalogue.

In the head of Andromeda, is a star of the second magnitude, which may be observed for ascertaining the latitude, having an opera-glass fixed to the quadrant.

For the beginning of the year 1812, its

Right Ascension is . . .	359°	40′	16,″95
Annual Variation in A. R.			46,0
Declination	28	3	44 N.
Annual Varia. in Declina.		+	20,25

ANEMOMETER, (of ἄνεμος, wind, and μετρον, measure) an instrument invented for the purpose of ascertaining the strength and velocity of the wind. A great improvement has lately been made in this useful machine, by Professor Poschman, a Russian: the instrument itself is, however, but of recent invention. The properties and form of this machine are thus described: "With the anemometer not only the strength of the wind may be exactly measured, but, by means of other instruments, which are easily adjusted to it, such as barometer, hydrometer, and thermometer, a variety of other physical experiments may be conducted with the greatest convenience. This instrument stands on a hollow pillar, of a pyramidical form, on one side of which are placed over each two dial-plates, with indexes or hands, the one of which belongs to the hydrometer, and the other to the anemometer. It is fixed on an oblong quadrangular board, supported on four feet, which again rest on a pedestal. Through the middle of the pillar a rod is passed, to the upper end of which a weathercock is fixed, and to the lower end an index, standing on a wind-dial, which is furnished below with a magnetic needle, and points out the direction of the wind with the greatest exactness. On one side of the pedestal, in the middle of the outer extremity, a pillar stands with two rollers, between which a roller is passed, on which the wind-board is fixed. This roller is at the further end provided with teeth, or notches, by means of which a notched wheel, which is attached to the axle of the index of the anemometer, is moved. Between the wind-board and the pyramidical pillar a spring passes, which acts constantly on the roller in such a manner that the wind-board, being moved by the current of the wind, strikes against the spring, which in its turn moves the roller; when, by means of the notches in the ends of the roller, which act upon the wheel, the index communicating with it is made to point out the force of the wind. Between the pillars there are also four brass bars attached, which pass through notches in the extremity of the board, and are provided with two small rollers, which serve to give uniformity to their pulses. On the opposite side of the wind-board, a board is attached to the pedestal, and on it the barometer and thermometer are placed. The advantage of this newly invented instrument consists chiefly in this, that the impulse by means of the board above-mentioned, is rendered at all times uniform and incapable of any variation; while the spring, which passes between the board and smaller pillar, prevents the variations of heats and cold from producing any perceptible difference. This spring is also contrived so, as never to exhibit any perceptible change of elasticity, but always to act with the same impulse in proportion to the force of the wind."

ANEMOSCOPE, (from ἄνεμος, and σκέπτομαι, I consider,) a machine invented to tell the changes of the wind. It should consist of an index moving about a circular plate, like the dial of a clock, on which the 32 points of a compass are drawn, instead of hours.

The index, pointing to the divisions in the dial, is turned by an horizontal axis, having an handle head at its outward extremity. This handle-head is moved by a cog-wheel on a perpendicular axis, on the top of which is fixed a vane that moves with the course of the wind, and gives motion to the whole machine. The contrivance is simple; the number of cogs in the wheel, and rounds in the trundle-head must be equal, because it is necessary, that when the vane moves entirely round, the index of the dial should also make a complete revolution.

A different anemoscope is described in the *Phil. Trans.* vol 43, part 2; and one is described in MARTIN's *Phil. Brit.* vol. 2d.

AN-END, (*debout*, Fr.) the situation of any mast or boom, when in a perpendicular situation to the plane of the deck, tops, &c. The top-masts are also said to be an end when they are hoisted up to their usual station at the head of the lower masts.

ANGLE (*angulus*, Lat.) in geometry, the opening or mutual inclination of two lines, or two planes meeting in a point, called the angular point.

Such as the *angles* formed by, or between the two lines A B and A C, at the *vertex*, or angular point A. Also the two lines A B and A C are called the *legs* or *sides* of the angle.

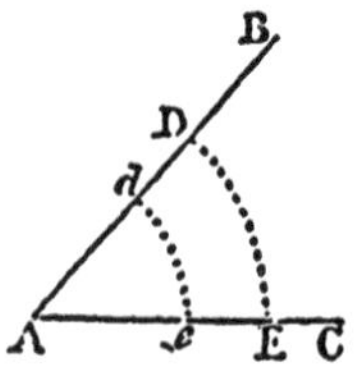

Angles are denoted, or named by a single letter, placing always that of the vertex in the middle.

The measure of an angle, by which its quantity or magnitude is expressed, is an arch, as D E, described from the centre A, with any radius at pleasure, and contained between its legs, A B and A C.—Hence angles are compared and distinguished by the ratio of the arcs which subtend them, to the whole circumference of the circle; or by the number of degrees contained in the arc D E, by which they are measured, to 360, the number of degrees in the whole circumference of the circle. And thus an angle is said to be of so many degrees, minutes, &c. as are contained in the are D E.

It matters not with what radius the arc is described, by which an *angle* is measured, whether great or small, as A D, or A *d*, or any other: for the arcs D E, *d e* being similar, have the same ratio as their respective radii, or circumferences, and therefore they contain the same number of degrees. Hence it follows that the quantity or magnitude of the *angle* remains still the same, though the legs be ever so much increased or diminished.

Acute ANGLE (opposed to obtuse), is that which is less than a right angle, or which does not extend to 90 degrees.

Acute ANGLED *Triangle*, is a triangle whose three angles are all acute.

Obtuse ANGLE (opposed to acute), is greater than a right angle, or whose measure exceeds 90 degrees.

ANOMALY, in astronomy, (from ἀνώμαλος, inequal, Gr. *anomalia*, Lat.) is an irregularity in the mean motion of the planets, whereby they deviate from the aphelion, or apogee; or it is the angular distance of the primary planets from their aphelion, or of the sun and moon from their apogee; and which inequality is either mean, eccentric, or true.

Mean or simple ANOMALY, is where a planet describes an ellipsis about the sun, placed in one of the foci, and is the time it takes up in moving from its aphelion to the mean place or point of its orbit, and is equal to the angle contained between the line of the apsides, or transverse diameter, and a line drawn from the sun to the centre of the planet.

Eccentric ANOMALY, *or of the centre*, is an arch of the eccentric circle cut off by a right line drawn perpendicular to the transverse diameter of the orbit.

True or equated ANOMALY, is the distance of the sun from his apogee, or a planet from its aphelion, seen from the sun; that is, the distance of a planet in signs, degrees, minutes, &c. from that point of its orbit which is farthest from the sun. It is a very difficult, and, at the same time, a very useful problem, to find directly the *true anomaly* from the mean one given, the motion of the earth being in an elliptical orbit, its true anomaly will differ from the mean; this difference is called the equation of the centre, and is the correction to be applied to reduce the mean motions to the true ones. See HUTTON's *Mat.* and *Phil. Dict.* Vol. I. page 120.

To ANSWER, (*repondre*, Fr.) Ex. The chase has *answered* the private signal. That ship does not *answer* her helm. This spar will not *answer* for a main-top-gallant-mast.

ANTARCTIC, (from ἀντὶ, opposite, and ἄρκτος, a bear), a name given to the southern pole of the world; to a circle parallel to the equator; and 23° 28′ distant from that pole. Thus we say the antarctic pole, the antarctic circle. The epithet antarctic is properly derived from the circumstance of the pole and circle, to which it is applied, being exactly opposite to the arctic, or north pole, and the artic circle.

ANTARES, in astronomy, the name of a star of the first magnitude, situated in the heart of the scorpion. This being one of the stars in the moon's path, its distance from that planet is calculated for every three hours of apparent time at Greenwich, and set down in the *Nautical Ephemeris*. Its right ascension in time, adapted to the beginning of the year 1812, is 244° 28′ 21″,15. and annual variation in right ascension 51″,68. Its declination 26° 0′ 5″, South, and annual variation in declination + 8″,4.

ANTINOUS, in astronomy, a part of the constellation *Aquila*, or the Eagle. See the article AQUILA.

ANVIL, (*enclume*, Fr.) a mass of iron, on which the work is forged or hammered.

APEEK, (*à pic*, Fr.) implies when the cable is drawn so tight as to bring the ship directly over the anchor, which is then said to be apeek.

APHELION, in astronomy, is that part of a planet's orbit which is at the greatest distance from the sun, being that end of the greater axis of the elliptic orbit of the planet, most remote from the focus wherein the sun is placed; and is, therefore, opposed to Perihelion.

The times of the aphelia of the primary planets may be known by their apparent diameter appearing the smallest, and also by their moving with the least velocity, in a given time. They may likewise be found by calculation, which De Lande, De la Caille, and other astronomical writers treat of at large.

APOGEE, in astronomy, when applied to the sun, is that part of the earth's orbit, which is at the greatest distance from the sun; consequently, the sun's *apogee* and the earth's aphelion are identically the same point.

Apogee *of the Moon*, is that point of the moon's orbit which is at the greatest distance from the earth.

APPARENT, (from *appareo*, to appear) among mathematicians and astronomers, denotes things as they appear visible, or evident to the eye, or understanding, in contradistinction from their real or true situation: thus we say, the apparent diameter, distance, magnitude, place, &c. of the heavenly bodies.

Apparent *conjunction of the Planets*, is when a right line, supposed to be drawn through their centres, passes through the eye of the spectator, and not through the centre of the earth.—And, in general, the apparent conjunction of any objects, is when they appear or are placed in the same right line with the eye.

Apparent *Diameter*, or *Magnitude of the heavenly bodies*, is not the real length of the diameter, but the angle which they subtend at the eye, or under which they appear: and, the angle, or apparent extent diminishes with the distance of the object. But the apparent magnitude varies not only by the distance, but also by the position of it, as the sun and moon appear much larger when seen in the horizon, than when they are raised higher in the heavens.

Apparent *Distance*, is that distance which we judge an object to be from us, when seen afar off. This is commonly very different from the true distance; because we are apt to think, that very remote objects, whose parts cannot well be distinguished, and which have no other visible objects near them, are at the same distance from us; though, perhaps, they may be thousands or millions of miles off, as in the case of the sun and moon. The apparent distances of objects are also greatly altered, by the refraction of the medium through which they are seen.

Clearing the distance, or finding the true distance, as it is commonly termed, is one of the most intricate calculations in navigation. There are several different methods of clearing the distance—Dunthorne's improved method, and that derived from M. Borda, abridged by the tables of logarithmic differences; both require six places of figures. The late Mr. Witchell has given a rule that only requires four places of figures: but a late method, invented by Mendoza Rios, has the peculiar advantage of not requiring any distinction of cases, an inconvenience that prevails in every other approximate method; and, therefore, this last is most likely to come into general use when sufficiently known. For the rules, &c. see Robertson's *Navigation*, Mackay on the *Longitude*, and Mendoza's *Tables*.

Apparent *Motion*, is either that motion which we perceive in a distant body, the eye at the same time being either in motion or at rest; or that motion which an object at rest seems to have, while the eye itself only is in motion. And the truth of all this is experienced by persons sitting in a boat moving on a river, or in a ship when sailing fast, and viewing houses and mountains, &c. on the shore, or side of the road, or other boats or ships also in motion.

Apparent *Place of a Planet*, &c. in astronomy, is that point in the surface of the sphere of the world, where the centre of the luminary appears from the surface of the earth.

Appearance, or making of a coast. (*apparence d'une côte*, Fr.)

APPOINTMENT of a ship, signifies her equipment, furniture, &c.

APPULSE, in astronomy, the approach of any planet towards a star or the sun. According to some authors, the word *appulse* is employed when two luminaries are apparently in contact; while others, extending its signification, use it as synonimous with *occultation*, when the one body is completely obscured or hid by the other.

APRON, (*platine de canon*, Fr.) in gunnery, a piece of sheet lead which covers the touch-hole of a cannon.

The dimensions of aprons are as follow, viz. for 42, 32 and 24 pounders, 15 inches by 13; for 18, 12 and 2 pounders, 12 inches by 10; and for cannon of less calibre, 10 inches by 8. They are tied by two strings of white marline.

Apron, (from *a* and *foran*, Sax.) of a dock, the platform, or flooring of plank, raised at the entrance of a dock, a little higher than the bottom, against which the gates are shut. See the article Dock.

Apron, (*contre étrave*, Fr.) in ship-building, a piece of curved timber, fixed behind the lower part of the stem, just above the foremost end of the keel.

The apron conforms exactly to the shape of the stem, so that when the convexity of the former is applied to the concavity of the latter, it forms one solid piece, which serves to fortify the stem, and give it a firmer connexion with the keel.

As the apron is composed of two pieces scarfed together, and used to support the scarf of the stem, it is necessary that the scarf thereof should be at some distance from that of the stem. It is formed of the same thickness with the heel of the stem; but its thickness is equal throughout. Sometimes the piece immediately under the apron forms a curve, of which the horizontal part covers the dead wood, whilst the vertical part corresponds with the inside of the stem, to which it is fayed, making the commencement of the apron.

APSIS, in astronomy, is applied to either of the two points in the orbits of planets, where they are at their greatest and least distance from the sun or earth. The *higher apsis* is that which is nearest the sun or earth, and is synonimous with perihelion and perigee; the *lower apsis* is that which is farthest from the sun or earth; and corresponds with the aphelion, or apogee. The line which joins these two points is called the *line of the apses, or apsides.*

APUS, *Avis Indica*, in astronomy, a constellation of the southern hemisphere, placed near the pole, between the *triangulum australe*, and the cameleon, and supposed to represent the bird of paradise.

The number of stars contained in this constellation is 11 in the British catalogue; in Bayer's Maps 12; the largest is only of the fifth magnitude.

AQUARIUS, in astronomy, is one of the celestial constellations, being the eleventh sign in the zodiac, reckoning from Aries, and is marked by the character ♒, representing part of a stream of water, issuing from the vessel of Aquarius, or the *Water-bearer.* This sign also gives name to the eleventh part of the ecliptic, through which the sun moves in part of the months of January and February. It contains 108 stars, in the Britannic catalogue, and 119 in the catalogue published in the *Tables de Berlin.*

AQUEDUCT, or AQUÆDUCT, (*q. d. ductus aquæ, a conduit of water*) is a construction of stone, or timber, built on uneven ground, to preserve the level of water, and convey it, by a canal from one place to another. Some aqueducts are under ground, being conducted through hills, &c. and others are raised above ground, and supported on arches, to conduct the water over vallies, &c.

AQUILA, or the Eagle, in astronomy, is a constellation of the northern hemisphere, generally united with Antinous. In the Britannic catalogue it contains 71 stars, and 76 in the Berlin.

ARA, the Altar, in astronomy, a southern constellation, containing only 7 stars in Ptolemy's catalogue; and, according to the British catalogue, with Sharp's appendix, of 9 stars; none of which exceed the fourth magnitude.

ARCH, or ARC, in geometry, any part of the circumference of a circle, or curved line, lying from one point to another, by which the quantity of the whole circle or line, or some other thing sought after, may be gathered.

All angles are measured by arcs. For this purpose an arc is described having its centre in the point or vertex of the angle: and as every circle is supposed to be divided into 360°, an arc is estimated according to the number of degrees which it contains. Thus an arc is said to be of 30, 50, or 100 degrees, &c.

Equal Arcs, are such arcs as contain the same number of degrees, and whose radii are equal to each other.

Similar Arcs, are those which contain the same number of degrees, and whose radii are unequal.

Diurnal Arch, in astronomy, is that part of a circle which is described by an heavenly object between its rising and setting.

Nocturnal Arch, is that part of a circle described by an heavenly object, between its setting and its rising.

The diurnal and nocturnal arches are always equal to a whole circle.

ARCHITECT, (*architectus*, Lat.) a professor of the art of building.

ARCHITECTURE, (*architectura*, Lat.) the art or science of planning and building, or erecting any edifice, so as properly to answer the end proposed, for solidity, conveniency, and beauty; whether houses, temples, churches, bridges, halls, theatres, ships, &c. Architecture is divided into *civil, military*, and *naval* or *marine.* See each article under its respective head.

ARCTIC Circle, is a lesser circle of the sphere, parallel to the equator, and passing through the north pole of the ecliptic, or distant from the north or arctic pole, by a quantity equal to the obliquity of the ecliptic, which was formerly estimated at 23° 30′, but its mean quantity is now 23° 28′ nearly. This circle is opposed to the *antarctic circle*, which is equally distant from the south pole. These two circles may be conceived to be described by the motion of the poles of the ecliptic round the poles of the equator, or of the world.

Arctic *Pole*, the north pole of the world, and so called from ἄρκτος, *ursa*, the *bear*, from its proximity to the constellation of that name.

ARCTURUS, in astronomy, a fixed star of the first magnitude, between the thighs of the constellation Bootes; so called from ἄρκτος, a *bear*, and οὐρα, a *tail;* as being near the tail of the bear. This star is twice mentioned in the Book of Job, viz. IX. 9, and XXXVIII. 32, by the name of Aish, if the translation be right. Dr. Hornsby concludes that *Arcturus* is the nearest star to our system visible in the northern hemisphere, because the variation of its place, in consequence of a proper motion of its own towards the south, is more remarkable than that of any other of the stars; and, by comparing a variety of observations respecting both the quantity and direction of the motion of this star, he infers, that the obliquity of the ecliptic decreases at the rate of 58″ in a century; a quantity which nearly corresponds to the mean of the computation framed by Euler and De la Lande, upon the principles of attraction.—*Phil. Trans.* vol. 63, p. 1. No. 14.

This star's right ascension, adapted to the beginning of the year 1812, is . . . . 211° 46′ 12″,76

Annual Variation A. R. .			40,83
Declination	20	9	56 N.
Annual Variation in Decl.			— 19,10

ARGO NAVIS, or the ship, in astronomy is a constellation of the fixed stars, in the southern hemisphere, whose stars in the Britannic catalogue and Sharp's appendix are 64.

The Greeks tell us that this was the famous ship in which the Argonauts performed that celebrated expedition so famous in their history.

ARIES, or the Ram, in astronomy, one of the constellations of the northern hemisphere, which is the first sign of the Zodiac, from which a twelfth part of the ecliptic takes its denomination. It is marked ♈, in imitation of a ram's head. The sun enters this sign about the 20th of March, at a time when the fields and hills are overspread with flocks: hence it is reasonable to conjecture that the ancient astronomers, observing the sun to enter this sign at so profitable a time, thought fit to signalize it by the principal of the flock, the *ram.* The stars of this constellation, in the Britannic catalogue are 66; and, in the Berlin catalogue, it contains 67; but they are most of them very small, one only being of the second magnitude, two of the third, and all the rest smaller.

ARITHMETIC, (ἀριθμος, number, and μετρέω, to measure) the art and science of numbers; or, that part of mathematics which considers their powers and properties, and teaches how to compute or calculate truly, and with ease and expedition.

Arithmetic consists chiefly in the four principal rules or operations of *Addition*, *Subtraction*, *Multiplication*, and *Division;* to which may, perhaps, be added *Involution* and *Evolution*, or raising of powers, and extraction of roots, together with the doctrine of the curious and abstract properties of numbers.

Arithmetic, *theoretical*, is the science of the properties, relations, &c. of numbers, considered abstractedly, with the reasons and demonstrations of the several rules. Euclid furnishes a theoretical arithmetic, in the seventh, eighth, and ninth books of his elements.

Arithmetic, *practical*, is the art of numbering or computing, that is, from certain numbers given, of finding certain others whose relation to the former is known. Dr. Hutton, Bonnycastle, and Keith, have furnished ample treatises for practice.

The French arithmeticians divide numbers into periods of three figures, and assign to each of them an appropriate denomination; whereas the English, instead of proceeding to billions, after hundreds of millions, say thousands of millions, ten thousands of millions, billions. Hence the English billions are a thousand times greater than the French; but since the very high numbers which these denominations express, almost never occur in practice, at least in a verbal form, the difference of method is not likely to be often a source of error.

Arithmetic *complement of a logarithm*, the sum or number which a logarithm wants of 10,000000: thus the arithmetical complement of the logarithm 8.154032 is 1.845968.

ARM, each extremity of a bibb, or bracket. See the article Bibb.

Arm also implies that part of the anchor from the crown, on which the palm is shut.

Arm. An arm of the sea. (*Un bras de mer*, Fr.)

ARMED Ship, (*vaisseau armé en guerre*, Fr.) a vessel occasionally taken into the service of the government in time of war, and employed to guard some particular coast, or attend on a fleet. She is therefore armed and equipped, in all respects, like a ship of war, and commanded by an officer of the navy, who has the rank of master and commander. All ships of this sort are upon the establishment of the king's sloops, having a lieutenant, master, purser, surgeon, &c.

ARMILLARY Sphere, (from *armilla*, Lat. a bracelet, or ring) in astronomy, is applied to an artificial sphere, composed of a number of metallic circles, which represent the several circles of the terrestrial sphere, put together in their natural order, to ease and assist the imagination, in conceiving the construction of the heavens, and the motions of the celestial bodies.

The Armillary-*sphere* of glass, in Pembroke-hall, Cambridge, constructed about 34 years since, by Dr. Long, is eighteen feet in diameter; and will conveniently contain 30 persons, who may sit within it, to view, as from a centre, the representation of the celestial sphere. The lower part of the sphere, which, to the inhabitants of these kingdoms, never rises above the horizon, is cut off; and the whole apparatus is so contrived, that, when you are within and in order, it may be turned round with as little labour as it takes to wind up a jack. For a further description and use of the armillary-sphere, see Ferguson's *Lectures*, vol. I. p. 366.

Top-ARMOUR. See the article Top.

ARMOURER of a Ship, (*maître-armurier*, Fr.) an officer appointed by warrant to clean and keep in repair the muskets, pistols, cutlasses, and to do all kinds of smith-work of a ship of war; having a mate to assist him in those duties.

Armourer's *Mate*, (*aide-armurier*, Fr.) the person employed in assisting the armourer in his several duties.

ARMS. (*armes à feu*, Fr.) If the subject of this article require any definition, it may be comprised under "implements to injure by aggression, or to aid the human strength in repelling a foe." The arms used on board ships are, muskets, pistols, cutlasses, and boarding-pikes. Nothing can be more prejudicial to the state of arms than rust, and the sea air is particularly productive of it. Many inventions and projects have been proposed by different persons in different countries, for the preservation of arms from rust. The French, who boast loudly of their experiments and skill in matters of this nature, say they have invented a species of varnish, which will completely effect the desired purpose, and which they direct being made in the following manner.

Take some of that varnish which the French call *vernis gras à l'huile*, of which the basis is gum copal, the whitest that can be found, is most proper; let the essence of turpentine, well rectified, be mixed with this in the proportion of from one-half to four-fifths, according to the degree in which it is intended that the species to be preserved should retain their metallic brilliancy. The metal is, in the first place, to be wiped with a piece of cloth, or leather, strongly impregnated with alkali, then washed with pure water, and afterwards rubbed dry with a linen cloth. Take a bit of fine sponge, let it be washed in water, and afterwards in essence of turpentine, to extract the water; put a little of the above varnish in a vessel, and then place the sponge in the middle of it, till the whole is imbibed. Let the sponge be squeezed between the fingers, till it has discharged nearly all the varnish contained in it; then let it be passed lightly over the barrel of a gun, or any other piece of metal, taking care not to repeat this when the essence has evaporated, which would render the varnish rough and of unequal tint; let the metal then be dried in a place secure from dust.

ARSENAL, a royal public magazine, or place appointed for making and keeping of arms, necessary either for defence or assault. We have excellent *naval arsenals* at Deptford, Portsmouth, and Plymouth; and military arsenals at the Tower, Woolwich, &c. There are several naval and military arsenals on the continent; viz. at Paris, Marseilles, Toulon, Rochefort, and Brest.

ARTICLES of War, with sundry abstracts and acts of parliament, contain certain express rules and orders relating to the government of his majesty's ships, vessels, and forces, by sea. See Articles and Abstracts in Appendix.

In these Articles almost every possible offence that can be committed in the royal navy is set down in explicit terms, and the punishment thereof annexed. In this respect our seamen have considerably the advantage over the soldiers in the army, whose articles of war are not enacted by Act of Parliament, but framed, from time to time, at the pleasure of the crown. "Yet from whence this distinction arose," as that elegant writer, Blackstone, observes, in his *Com.* vol. I. p. 420, "and why the executive power, which is limited so properly with regard to the navy, should be so extensive with regard to the army, it is hard to assign a reason; unless it proceeds from the perpetual establishment of the navy, which rendered a permanent law for their regulation expedient; and the temporary duration of the army, which subsisted only from year to year, and might, therefore, with less danger, be subject to a discretionary government. But, whatever was apprehended at the first formation of the Mutiny Act, the regular renewal of our standing force at the entrance of every year, has made this distinction useless. For if, from past experience, we may judge of future events, the army is now lastingly ingrafted into the British constitution, with this singularly fortunate circumstance, that any branch of the legislature may annually put an end to its legal existence, by refusing to concur in its continuance."

ARTIFICIAL Lines, on a sector, or scale, are so contrived as to represent the logarithmic sines and tangents; which, by the help of the line of numbers, will solve, with tolerable exactness, most questions in *Trigonometry*, *Navigation*, &c.

Artificial *Numbers*, *Sines*, *Tangents*, &c. are the same as the logarithms of the natural numbers, sines, tangents, &c.

ARTILLERY, (*artillerie*, Fr.) in military language, weapons of war. In a more enlarged sense, the word denotes a magazine of all sorts of arms, such as cannon, mortars, bombs, balls, petards, grenades, muskets, and engines of all sorts, ancient and modern, for discharging bullets, darts, shot, shells, stones, &c. by the expansive force of inflamed gunpowder; being what is otherwise called *ordnance*, as used in the land service Also small balls, powder, match, hand-tools, planks, boards, rope, coals, tallow, pitch, rosin, sulphur, saltpetre, quickmatch, fireworks, pontons, &c.

ASCENSION, (from *ascensio*, Lat. a rising upwards) is either right or oblique. Right ascension is an arch of the equator, intercepted between the equinoctial point, or the first of Aries, and the hour circle which passes through any planet or star; or, since the meridian is perpendicular to the equator, it is that point of the equinoctial which comes to the meridian along with any planet or star. Oblique ascension is that point of the equinoctial which comes to the horizon with any of the planets, or fixed stars. The ascensional difference is the difference between the right and oblique ascension of any point in the heavens, or it is the space of time which any of the planets rises or sets before or after the sixth hour from the time of their coming to the meridian.

To find the right ascension of the sun, stars, &c. by trigonometry, say,

As radius
Is to the cosine of the sun's greatest declination, or obliquity of the ecliptic;
So is the tangent of the sun's or star's longitude
To the tangent of the right ascension.

To find the ascensional difference, you must have the latitude of the place, and the sun's declination given: then say,

As radius
Is to the tangent of the latitude;
So is the tangent of the sun's declination
To the sine of the ascensional difference sought.

This, converted into time, shows how much he rises before, or sets after, six o'clock; and by subtracting of which from the right ascension, when the sun is in the northern signs, and adding it, when he is in the southern ones, you will find the oblique ascension.

ASHORE, (*à terre*, Fr. from *a* and *shore*) on the shore, or land, as opposed to aboard.

A ship is said to be Ashore, (*échoué*, Fr.) when she has run upon the ground, or on the sea-coast, either by design or accident.

ASSISTANCE. (*secours*, Fr.) The signal for assistance. (*le signal pour demander du secours.*)

ASSURANCE. See the article Insurance.

ASTERN, (*en arrière du vaisseau*, Fr. from *a* and *steorn*, Sax.) any distance behind a ship, as opposed to *a-head*, which is before her. Thus, when south is *a-head*, or on the line to which the stem is directed, north will be *astern*.

ASTROLABE, (*astrolabium*, Lat.) an instrument for taking the altitude of the sun or stars at sea, being a large brass ring, the limb of which, or a convenient part thereof, is divided into degrees and minutes, with a moveable index, which turns upon the centre, and turns two sights: at the zenith is a ring to hang it by in time of observation, when you need only turn the index to the sun, that the rays may pass freely through both sights, and the edge of the index cuts the altitude upon the divided limb. This instrument, though not much in use now, if well made, and of great weight, that it may hang the steadier, is as good as most instruments that are used at sea for taking altitudes, especially between the tropics, when the sun comes near the zenith, and in calm weather.

ASTRONOMY, (*astronomie*, Fr. *astronomia*, Lat. of ἀστρονομία, of ἄστρον, a star, and νόμος, Gr. the law, rule, &c.) is that branch of natural philosophy which treats of the heavenly bodies, their motions, periods, eclipses, magnitudes, distances, and other phenomena. The determination of their magnitudes, distances, and the orbits which they describe, is called practical astronomy; and the investigation of the causes of their motions, is called physical astronomy. The former is determined from observations on their apparent magnitudes and motions by astronomical instruments and calculations; and the latter from analogy, by applying those principles and laws of motion and gravitation, by which bodies on and near the earth are governed, to the other bodies in the system.

The invention of astronomy has been variously given, and ascribed, to several persons, several nations, and several ages.

Indeed, it is probable that mankind never existed

long without some knowledge of astronomy amongst them. For, besides the motives of mere curiosity, which are sufficient of themselves to have excited men to a contemplation of the glorious and varying celestial canopy, it is obvious, that some parts of the science answer such essential purposes to mankind, as to make the cultivation of it a matter of indispensible necessity. Accordingly, we find traces of it, in different degrees of improvement, among all nations.

For historical accounts of the principal writings and authors on astronomy, see WEIDLER'S, BAILLY'S, and DE LA LANDE'S *Astronomy*; also the more modern books of FERGUSON, LONG, EMERSON, and VINCE, on the same subject.

ASTRONOMICAL CALENDAR, an instrument engraved upon copper-plates, printed on paper, and pasted on a board, with a brass slider carrying a hair: it shews, by inspection, the sun's meridian altitude, right ascension, declination, rising, setting, amplitude, &c. to a greater degree of exactness than the common globes.

ASTRONOMICAL *Clock*. Clocks for astronomical purposes, in which extraordinary nicety in exact measurement of time is necessary, have compensation pendulums, detached escapements with jewelled pallets, and other improvements, with a contrivance first invented by Le Roy, to continue their movement, while the weight is winding up, to prevent the loss of time. They are chiefly used in fixed observatories, and are generally made to show sidereal time.

The late Mr. Ferguson contrived a clock, to shew the apparent diurnal motions of the sun and moon; the age and phases of the moon, with the time of her coming to the meridian; also the time of high and low water; and all this by having only two wheels and a pinion added to the common movement. In this clock the figure of the sun serves as an hour index, by going round the dial-plate in 24 hours; and a figure of the moon goes round in 24 hours, 50½ minutes, the time of her going round the heavens from any meridian to the same meridian again. He has also given a description and drawing of an astronomical clock, in his *Select Exercises*, shewing the apparent daily motions of the sun, moon, and stars, with the time of their rising, southing, and setting; the places of the sun and moon in the ecliptic; and the age and phases of the moon for every day in the year. For a more particular description, see FERGUSON'S *Select Exercises*.

ASTRONOMICAL *Hours*, are such as are accounted from noon or midnight of one natural day, to noon or midnight of another.

ASTRONOMICAL *Observations*. Of these there are records, or mentions, in almost all ages. It is said that the Chinese have observations for a course of many thousand years. But of these, as well as those of the Indians, we have never yet had any benefit. But the astronomical observations of most of the other ancients, are carefully preserved by Ptolemy in his *Almagest*. All the modern observations, both Foreign and English, are handed down to us in the *Historia Cœlestis*, first published by order of the Emperor Ferdinand, 1672; and continued, by our own countryman, Flamsteed, to 1712.

Mr. Flamsteed was appointed the first Astronomer Royal at Greenwich, in the year 1675, and continued to observe the sun, moon, planets, and fixed stars for 44 years, and has given us an improved theory of the whole.

In 1719, Dr. Halley succeeded Mr. Flamsteed, and was the first who discovered the acceleration of the moon, and has given a very ingenious method of finding the parallax. He also composed tables of the sun, moon, and all the planets; and recommended the method of determining the longitude by the moon's distance from the sun and certain fixed stars, which has since been carried into execution, by the late Dr. Maskelyne.

On the death of Dr. Halley, in 1742, he was succeeded by Dr. Bradley. The accuracy of his observations enabled him to detect the smallest inequalities in the motions of the planets and fixed stars, and the consequence was, the discovery of the aberration of light, the nutation of the earth's axis, and a much greater degree of perfection in the lunar tables. He also made an immense number of observations for settling the places of all the stars in the British catalogue.

Dr. Bradley died in 1762, and was succeeded in his office, by Mr. Bliss, who, being in a declining state of health, died in 1765, and was succeeded by the learned Nevil Maskelyne, D.D. the late Astronomer Royal, who continued to discharge the duties of that office with the greatest honour to himself, and benefit to the science, till the beginning of the year, 1811. He has also established the Newtonian doctrine of universal attraction, upon the firmest foundation, by his experiments upon Schehallien. His regular observations of the sun, moon, planets, and fixed stars, which are published annually, are allowed to possess an unrivalled degree of accuracy; and we may consider them as the basis of future improvements of the Tables of the planetary motions. M. De la Lande, in his *Astronomy* (Vol. II. p. 121, last edition) speaking of astronomical observations, says, " Le recueil le plus moderne et le plus précieux de tous est celui de M. Maskelyne, Astronome Royal d'Angleterre, qui commence à 1765, et qui forme déja deux volumes in folio jusqu'à 1786. La precision de ces observations est si grande, qu'on trouve souvent la même seconde pour l'ascension droite d'une planete déduite de différentes étoiles, quoiqu'on y emploie la mesure du temps." His catalogue of fundamental stars is an invaluable treasure. These and his other various important improvements in this science entitle him to the most distinguished rank amongst astronomers, and will render his name illustrious, as long as the science of astronomy shall continue to be cultivated. When he died he was succeeded by Mr. Pond, the present Astronomer Royal, a gentleman hitherto only known as an amateur of that science.

ASTRONOMICAL *place of a Planet or Star*, is the longitude or place in the ecliptic, reckoned from the beginning of Aries, according to the natural order of the signs.

ASTRONOMICAL, or *Mural*, *Quadrant*, is an instrument framed and fitted with telescopes, &c. for observing meridian and other altitudes of the celestial bodies.

This quadrant is in the form of a quarter of a circle, contained under two radii at right angles to one another, and an arch equal to one-fourth part of the circumference of the circle. It is one of the most useful of all astronomical instruments; and, as it is sometimes fixed to the side of a wall, and the plane of it erected exactly in the plane of the meridian, it in this case receives the name of mural quadrant. Tycho Brahe was the first person who applied this arch to a wall; and Flamsteed, the first in England, who fixed one up of 8 feet radius made by the late ingenious Mr. Graham, in the Royal Observatory at Greenwich.

Astronomical, or *Equatorial, Sector*, a very useful mathematical instrument, made by the late ingenious Mr. Graham, for finding the difference in right ascension and declination between two objects, the distance of which is too great to be observed by the micrometer. He constructed the Zenith Sector of 24 feet radius at the Royal Observatory at Greenwich, with which Dr. Bradley discovered the aberration of the fixed stars.

ASYLUM. See the Royal Naval Asylum at Greenwich.

ATAIR, or Altair, the brightest star in the constellation *Aquila*, and of the first magnitude; and its meridian altitude is frequently observed by mariners, for ascertaining the latitude. Its right ascension in time, adapted to the beginning of the year 1812, is 295° 24′ 0″,30; annual variation in right ascension, 43″,78. Its declination, 8° 22′ 40″ N.; annual variation in declination, + 9″,1.

ATHWART, (*par le travers*, Fr. from *a* and *twert*, Dan. transverse) across the line of the ship's course; as, "we discovered a fleet standing *athwart* us," *i.e.* steering across our way.

Athwart-*Hawse*, the situation of a ship when she is driven by the wind, tide, or other accident, across the stem of another, whether they bear against, or are at a small distance from each other; the transverse position of the former with respect to the latter being principally understood.

Athwart *the fore foot*, is generally applied to the flight of a cannon-ball, as fired from one ship across the line of another's course, but a-head of her, as a signal for the latter to bring to.

Athwart-*Ships*, reaching across the ship, from one side to the other, or in that direction.

ATLAS, a large book containing maps and charts, of the principal coasts, harbours, &c. of the known world; as if they were viewed from the top of that celebrated mountain, which the ancients esteemed the highest in the world; or rather on account of their holding the whole world like *Atlas*.

ATMOSPHERE, (of ἀτμὸς, a vapour, and σφαῖρα, a sphere) that invisible elastic fluid which surrounds the earth, and encloses it on all sides.

It receives its name from the Greeks, in consequence of the vapours which are continually mixing with it.

The ancients considered it as one of the four elements of which all things were composed; and some of them seem to have thought that it enters as a constituent principle into other bodies, or at least that air and other bodies are mutually convertible into each other.

No experiments on its nature could well be made by the ancients, as they were unprovided with every instrument fitted for such investigations, and unacquainted with the principles upon which their construction depended. But it has occupied a great deal of the attention of modern philosophers, and has given birth to some of the most brilliant discoveries that grace the annals of science. Its weight was first ascertained by Galileo, and applied by Torricelli to explain the rise of water in pumps, and of mercury in barometrical tubes, and by Paschal to the mensuration of the height of mountains.

Its elasticity was accurately determined by Boyle. Halley and Newton explained the effects produced on it by moisture. Hooke, Newton, and Boyle, pointed out its relation to light, to sound, and to electricity. Its effect upon combustibles and animals was investigated by Hooke, Priestley, and Lavoisier. Its constituents were detected and measured by the experiments of Priestley, Lavoisier, and Cavendish. The effect of heat on it was determined by Shuckburgh, Dalton, and Gay Lussac.

From the experiments of Sir George Shuckburgh Evelyn, (*Phil. Trans.* 1777 and 1798) made with a degree of precision and patient industry, which perhaps have never been surpassed, it appears, that at the temperature of 60°, when the barometer stands at 30 inches, the specific gravity of atmospherical air is 0.001208, that of water being 1.000, or its weight is to that of water as 1 to 828. Hence 100 cubic inches of it under that pressure and at that temperature weigh 30.5 grains: for a cubic inch of pure water at that temperature weighs 252 506 grains, according to experiments of Shuckburgh corrected by Mr. Fletcher. (Nicholson's *Journal*, IV. 35.)

The air when weighed, is supposed to be in its usual state of dryness; when very moist, its specific gravity is diminished. An exact knowledge of the weight of a given bulk of air is of great importance, because it enables us with much facility to ascertain the weight of all other aerial bodies: for it is easy to determine the relative weight of any elastic fluid to that of air.

If the atmosphere were of uniform density, it would be easy to ascertain, with the utmost accuracy, the height to which it extends; for the height of the atmosphere would obviously be to the height of the mercury in the barometer, as the specific gravity of common air is to the specific gravity of mercury. By making the calculation on this supposition, it will be found that the height of the atmosphere is a little more than five miles. As the air, however, gradually diminishes in density, the atmosphere must reach to a much greater distance from the earth than five miles.

It appears from the duration of twilight, that, at the height of 44½ miles, the atmosphere is sufficiently dense to intercept the light of the sun, and reflect it to the earth: it should therefore appear, that it extends to a much greater height.

When a ray of light enters the atmosphere, it is bent from its course by the same cause which refracts the rays of light when they pass through any dense medium, such as glass or water. The refraction sustained by light at its first entrance into the atmosphere must be very small, from the extreme rarity of the air. The deviation, however, will gradually increase as it penetrates the denser strata, and the ray will describe a path increasing in curvature as it approaches the earth. From this property of the atmosphere, the apparent altitude of the sun, moon, and stars, is greater than their real elevation, and they appear to be raised above the horizon when they are actually below it.

"If we consider all the causes," says La Place, "which disturb the equilibrium of the atmosphere; its greatest mobility arising from its fluidity and elasticity; the influence of heat and cold on its elasticity; the great mass of vapour that it alternately absorbs and deposes; and lastly, the changes which the rotation of the earth produces in the relative velocities of its particles, which for this reason are displaced in the direction of the meridians; we should not be surprised at the inconstancy and variety of its motions, which it would be very difficult to subject to any fixed and certain laws."

ATMOSPHERICAL Clock, the name of a machine proposed by Dr. Brewster, for measuring the mean temperature of the atmosphere during any given interval.

This machine records every variation of temperature that takes place during a given period, and the result indicated on the dial-plate is the exact average of all the heights of the mercury in the thermometer. The variations of heat and cold affect the pendulum, which may be either of the tabular or gridiron kind; and which is so constructed, as to render sensible, in the motion of the clock, the alternate contractions and dilatations which it undergoes. See the article Clock.

ATRIP, (*prepor*. Fr. *trippen*, Dutch) is applied to the anchor and the sails: for the first, see the article *The* Anchor *is a-trip*. The topsails are said to be *atrip*, when they are just started from the cap.

ATTACK, (*attaque* Fr.) an assault Also the manner and disposition of a fleet or ship in action.

To Attack, (*attaquer*, Fr.) to assault. Ex. We attacked the enemy's van.

To ATTEMPT. To attempt a passage. (*tenter un passage*, Fr.)

To ATTEND. To attend the signals. (*Veiller aux signaux*, Fr.)

AUGER, (*augure*, *tarrière*, Fr.) a wimble, carpenter's tool for boring.

AURIGA, the Waggoner, in astronomy, the name of one of the constellations in the northern hemisphere, containing 26 stars, according to Tycho Brahe; 40, according to Hevelius; and 66 in the Britannic catalogue.

This is one of the 48 asterisms mentioned by all the ancient astronomers; and represented by the figure of an old man, in a kind of sitting posture, with a goat and her kids in his left hand, and a bridle in his right.

AURORA BOREALIS, an extraordinary luminous appearance or meteor, shewing itself in the night-time, in northern latitudes, whence it has got its name of *northern lights*, or *northern dawn*. It is also known among the vulgar by the name of *streamers*, or *merry-dancers*.

The aurora borealis may, with propriety, be distinguished into two kinds, the *tranquil* and the *varying*.

The tranquil shines with a mild steady light, very much resembling the clearness of twilight; and preserves, for a considerable time, the form in which it first appears, with little or no variation.

The *varying* aurora is still more remarkable in its appearance, and occasionally exhibits the most brilliant and rapidly diversified forms. The duration of the aurora is very various. Sometimes it is formed and disappears in the course of a few minutes. At other times it lasts during the whole night, or even for two or three days together.

In high northern latitudes, as those of Sweden, Lapland, and Siberia, the auroræ boreales are singularly resplendent, and even terrific. They frequently occupy the whole of the heavens; and, according to the testimony of Maupertuis, Middleton, Krafft, and others, eclipse the splendour of the stars, planets, and moon, and sometimes even of the sun himself.

The most accountable of all the circumstances respecting the aurora borealis, is, that it is not much more than a century since this phenomenon has been observed to occur with any degree of frequency in our latitudes; for it cannot be supposed that so beautiful and striking a phenomenon would have passed unnoticed, and unrecorded, during the two preceding centuries, while men of science, and particularly astronomers, were so busily employed in examining every remarkable appearance of the heavens; or that the philosophers of Greece and Rome would have remained silent concerning so beautiful a meteor, had it been in any degree familiarly kn wn to them.

It is in vain to account for their silence by saying, that they inhabited latitudes which are scarcely ever visited by this appearance; for the Romans not only visited, but long resided in, the north of Germany, and in Britain, where the aurora is now frequently seen in great splendour.

The following ingenious theory has been proposed, with a view to resolve this difficulty: There is a very remarkable analogy between the phenomena of electricity and those of magnetism, and, apparently, an intimate dependence of the one upon the other. There are two species of electricity, a positive and a negative; and two species of magnetic polarity, a north and a south. A body positively electrified repels another body positively electrified, and attracts one that is electrified negatively; while the north pole of one magnet repels the north pole, and attracts the south pole of another. The electric shock will deprive a magnetic needle of its power, or communicate it to it again, according to the direction in which it is laid; and during a thunder-storm, the magnetic needle is observed to be powerfully agitated. Thus the intimate connexion between electricity and magnetism seems to be satisfactorily established. Again, that imaginary line, or circle, which traverses the earth irregularly from the north towards the south, and is called the *line of no variation*, because the magnetic needle, when placed upon it, points truly to the poles, is observed to have a gradual, and pretty regular revolution round the earth, performed in about 1000 years; so that, when 1000 years have elapsed, the line of no variation will have reached the same situation which it occupied at the beginning of that period.

This line seems to have a sort of controul over the coruscations of the aurora, which are observed to follow its direction in the heavens; and, as it were, to be attracted towards it, and regulated by its influence in their course. But it appears, by calculation, that during the scientific age of Rome, the only period at which accurate observations of the phenomena of the heavens could be made and recorded, the line of no variation runs across the continent of Europe; and, consequently, the aurora borealis, or stream of electric matter that passes from the north-pole towards the equator, would find a ready vent from the earth into the sky, through the mountains, and other pointed bodies which are so plentifully scattered over the land. By thus flowing constantly, and so greatly subdivided, it could not give rise to any striking atmospheric phenomena.

At present, the direction of this line is through the Atlantic ocean; and, of consequence, the electricity of the polar regions is not imperceptibly transmitted to the atmosphere, but proceeds in such masses, and at such irregular intervals, as to exhibit itself in the beautiful phenomena of the northern lights. A thousand years ago, the line of no variation, no doubt, occupied the same situation as it does at present; and the aurora then shone forth in all its splendour; but, at that period, the nations of Europe were sunk in ignorance and barbarism; and whatever phenomena the heavens presented were lost to posterity, from the rudeness and want of knowledge of the people of that age.

Thus, if this theory be true, when the line of no variation shall again pass over the land, the aurora borealis will become invisible for a time; and when this line reverts to its present situation over the ocean, the aurora will once more shine forth with its wonted lustre.

Maupertuis, Cavallo, Halley, Euler, Hawkesbee, Canton, Franklin, Kirwan, and Libes, have all endeavoured to account for this beautiful phenomena by a great variety of theories. Mr. Dalton has employed a new one in his *Meteorological Observations and Essays*, 1793; and another has lately been employed by M. Monge. See *Leçons de Physique*, par Pujoulz, 1805.

AVAST, (*tiens bon là*, Fr.) the order to stop, or pause, in any exercise; as, a vast heaving! that is to say, desist, or stop from drawing in the cable or hawser, by means of the capstan, &c.

AVERAGE, in commerce, (*avarie*, Fr. *averagium*, Lat.) the accidents and misfortunes which happen to ships and their cargoes, from the time of their loading till their unlading. It is divided into three kinds. 1. The simple or particular *average*, which consists in the extraordinary expenses incurred for the ship alone, or for the merchandise alone; such as the loss of anchors, masts, and rigging, occasioned by the common accidents at sea; the damages which happen to merchandises by storms, capture, shipwreck, wet, or rotting; all which must be borne and paid by the thing that suffered the damage. 2. The large and common *average*, being those expenses incurred, and damages sustained, for the common good and security, both of the merchandise and vessel, consequently to be borne by the ship and cargo, and to be regulated upon the whole. Of this number are the goods or money given for the ransom of the ship and cargo; things thrown overboard for the safety of the ship; the expenses of unlading, or entering into a river or harbour, and the provisions and hire of the sailors when the ship is put under embargo. 3. The small *averages*, which are expences for towing and piloting the ship out of, or into harbours, creeks, or rivers; one-third of which must be charged to the ship, and two-thirds to the cargo.

Average is more particularly used for a certain contribution that merchants make proportionably towards their losses. It also signifies a small duty which those merchants, who send goods in another man's ship, pay to the master, for his care of them, over and above the freight. Hence it is expressed in the bills of lading, paying so much freight for the said goods, with primage and average accustomed.

A-WEATHER, (*au vent*, Fr.) signifies the situation of the helm when pushed to the weather side of the ship, in contradistinction to a-lee; as hard a-weather, &c.

AWEIGH, (*l'ancre à quitté*, Fr. of *a* and *weigh*) the state of the anchor when it is drawn out of the ground in a perpendicular direction, by the application of mechanical powers, as a capstan or windlass, to the cable within the ship; so that aweigh is synonimous to *atrip*.

AWNING, (*tendelet*, from *aulne*, Fr.) a canopy of canvas extending over the decks of a ship or boat in hot weather, for the convenience of the officers and crew, and to preserve the decks from being cracked or split, *ebaroui*, by the heat of the sun. The awning is supported by a range of light posts, called stanchions, which are erected along the ship's side on the right and left; it is also suspended in the middle by a complication of small cords, called a crow-foot.

That part of the poop-deck which is continued forward beyond the bulk-head of the cabin, is also called the awning. See the article Crowfoot.

AXE, (*grande hache ou coignée*, Fr.) a well known carpenter's tool, both in respect to its shape and size. It is applied to the purpose of trimming off, and reducing to its proper size, every species of timber or plank.

AXIS, is either a real or supposed line passing through the centre of any cylindrical or globular body; in short, the centre of any body, when such centre can be determined so accurately, that the body, if required, may turn on the same.

Axis *of the World*, in astronomy, is an imaginary right line, conceived to pass through the centre of the earth, and to terminate at each end in the surface. It always remains parallel to itself, and at right angles with the equator. About this line, as an axis, the earth performs its diurnal rotation, from west to east.

Axis *of a Planet*, is a right line drawn through its centre, about which the planet revolves. The Sun, Earth, Moon, Jupiter, Mars, and Venus, are known, by observation, to move about their several axis; and the like motion may be easily inferred of Mercury, Saturn, and the Georgium Sidus.

AY, ad. (perhaps from *aio*, Lat.) yes, and is always used in lieu thereof at sea, with a repetition, 'ay, ay, Sir.'

AZIMUTH, of the sun, or star, &c. is an Arabic word, used in astronomy, to denote an arch of the horizon, intercepted between the meridian of the place and the azimuth or vertical circle passing through the sun or star; and is equal to the angle at the zenith formed by the said meridian and vertical circle: or it is the complement to the eastern or western amplitude, and is thus found by trigonometry:

As radius is to the tangent of the latitude,
So is the tangent of the altitude of the sun or star,
To the cosine of the azimuth from the south, at
the time of the equinox:—

And at all other times by the following rule in navigation:—

Find the sum of the polar distance and altitude of the sun, and the latitude of the ship. Take the difference between half this sum and the polar distance. To the log. secant of the altitude add the log. secant of the latitude, the log. cosine of the half sum, and the log. cosine of the difference; then half the sum of these will be the log. sine of half the sun's true azimuth, to be reckoned from the south in north latitude; but from the north in south latitude, towards the east, in the morning, and towards the west in the afternoon. Then if the true and observed azimuths be both on the east, or both on the west side of the meridian, their difference is the variation; but if one be on the east, and the other on the west side of the meridian, their sum is the variation; and to know if it be east or west, suppose the observer looking towards that point of the compass representing the magnetic azimuth: then, if the true azimuth be to the right of the magnetic, the variation is east; but if the true be to the left of the magnetic, the variation is west

Azimuth, *magnetical*, an arch of the horizon contained between the magnetical meridian, and the azimuth or vertical circle of the object; or its apparent distance from the north or south point of the compass This is found by observing the sun, or star, &c. with an azimuth compass, when it is ten or fifteen degrees high, either before or after noon.

Azimuth *Compass*, (*compas de variation*, Fr) an instrument adapted to find, in a more accurate manner than by the common sea-compass, the sun's or stars' magnetical amplitude, or azimuth.

It is also used to take the bearings of headlands, ships, and other objects at a distance.

An account of it was first published in the year 1736, by Edward Hanxley, of Wapping, in his book of navigation; but he does not say who was the inventor of it. Mr. Walker has greatly improved this instrument of late, so as to make it show the variation without any *computation*, as also the time from noon, and the sun's altitude. This improved *meridional azimuth compass* is now very generally used by the masters in the navy, and is composed of a common azimuth compass, on which is mounted an universal equatorial sun-dial, that finds its own meridian.

The sun-dial is composed of three parts, viz. an equatorial semicircle, a semicircle of latitude, and a flat piece of brass, called a bridge.

The azimuth card is divided in the usual manner with points, quarter-points, and degrees; and also the brass circle round the card is divided into degrees, &c. It has a stop in the usual way to fix the card.

The equatorial semicircle of the dial part is divided from the ends to its centre, which is marked 90°, or 12 hours of time.

The latitude semicircle is also divided into twice 90°, beginning at its centre, both ways up to 90°.

The bridge is a flat piece of brass, fixed over the equatorial semicircle at right angles to it: it has a slit or opening along its middle from end to end; in this opening is fitted a small piece of brass or slider, which has a small glass lens fixed in its centre, whose focus is equal to the radius of the equatorial semicircle, and collects the sun's image to a point on the divisions in observing. On the bridge is laid down the sun's declination on one side of it; and on the other side are the months and days answering to the declination, so that the middle of the slider may be set to either.

The parts of this instrument are adjusted to each other, and fixed by the maker, except balancing the compass, which must be examined at times; in order to which, set the 90° on the latitude semicircle to the *Zero*, and the slider on the bridge to the middle or Æ; set the bridge to face the sun nearly, turn the instrument round on its centre in the box, and note the degree cut by the sun's image, or light spot on the equatorial semicircle; then turn it half round, and see if the spot of light cuts the same division on the other part of the equatorial semicircle; if not, the compass is not truly balanced, and must be adjusted, so that the bright spot shall cut an equal division both ways.—This adjustment is made by means of a lead weight fixed under the bottom of the compass box, which may be moved at pleasure.

To find the Variation direct. The instrument being adjusted as before directed, set the slider to the sun's declination, or day of the month, (it matters not which, as one answers to the other) the latitude semicircle to the latitude of the place at the time, as near as you can by your log account; set the bridge to face the sun, and turn the instrument round on its centre, till the point of light from the lens rest on the graduated divisions on the equatorial semicircle, then the bridge, and zero's corresponding, will be due north and south, or in the true meridian; therefore, whatever degree the north point of the compass card stands to the right or left of the true north point on the box will be the *variation* east or west, and the time of day will be shown on the equatorial semicircle. This instrument may be used as an azimuth compass. When applied to this purpose, set the slider to the middle of the bridge, and the latitude semicircle to 90, and ship the sights on the bridge, and observe the sun's bearing, and stop the card in the usual way; then turn the instrument round till the point of light cuts the equatorial semicircle, and the degree it falls on will be the altitude of the sun's centre, which with the latitude and declination compute the true azimuth, and find the variation in the usual way.

The *Magnetical Azimuth* may be taken after the sun's altitude is observed, or any other person may read it off without stopping the card.

No great accuracy can be expected either in time or sun's altitude from this instrument, as the sun's image covers a degree or more on the equatorial semicircle: it is, however, near enough in general for the business of variation; but it is not sufficiently accurate for the longitude, either by chronometers or lunar observations. See the article Compass.

Azimuth, or *Verticle Circles*, are great circles of the sphere intersecting each other in the zenith and nadir, and cutting the horizon at right angles. These azimuths are represented by the rhumbs on common sea-charts; and on the globe by the quadrant of altitude, when screwed in the zenith On these azimuths is reckoned the height of the sun or stars, &c. when out of the meridian.

B.

BACK OF THE STERN-POST, *(contre-étambord extérieur*, Fr.) an additional piece behind the stern-post. The difficulty of procuring a stern-post of sufficient breadth in one piece has introduced this practice. it is strongly bolted thereto, and the hinges, which support the rudder, are fixed to it, and it is tenanted into the keel. See the article STERN-POST.

To BACK *an anchor, (empenneller une ancre*, Fr.) to carry out a small anchor, as the stream or kedge, ahead of the large one, by which the ship usually rides, in order to support it, and prevent it from loosening, or *coming home*, in bad ground. In this situation the latter is confined by the former, in the same manner that the ship is restrained by the latter.

To BACK *astern*, in rowing, *(scier à culer*, Fr.) is to manage the oars in a direction contrary to the usual method, so as that the boat or vessel impressed by their force, shall retreat or move with her stern foremost.

BACK *the starboard oars! (scie tribord! avec les avirons*, Fr.) the command to confine the above management to the oars on the right hand side of the boat only, in order to turn her round more speedily to that direction.

To BACK *and fill, (coiffer et faire servir les voiles*, Fr.) is an operation generally performed in narrow rivers, when a snip has the tide in her favour, and the wind is against her—Exam. " We were obliged to back and fill occasionally to get up the Thames."

To BACK *the sails, (mettre à scier*, Fr.) is to arrange them in a situation that will occasion the ship to retreat or move astern, in consequence of the tide or current in her favour, and the wind contrary, but light. This operation is particularly necessary in narrow channels, when a ship is carried along sideways by the strength of that tide or current, and it becomes requisite to avoid any object that may intercept her course, as shoals, or vessels under sail, or at anchor: it is also necessary, in a naval engagement, to bring a ship back, so as to lie opposite to her adversary, when she is too far advanced in the line; and also in fleets under convoy, where ships are too much crowded, by the above operation they may be preserved from falling aboard each other. See the article ABACK.

BACK *the main-topsail! (brasse le grand hunier sur le mat!* Fr.), the command to brace that sail in such a manner that the wind may exert its force against the fore-part of the sail, and by thus laying it aback materially retard the ship's course

BACK-BOARD, *(le dossier d'un bateau*, Fr.) a piece of board of a semi-circular figure, placed transversely in the after-part of a boat, like the back of a chair, and serving the passengers to recline against whilst sitting in the stern-sheets. See the article BOAT.

BACK-FRAME WHEEL, for laying cordage, from a six-thread ratline, to a two-inch rope, is about four or five feet in diameter, and is hung between two uprights, fixed by tenons, on a truck, and supported by a knee of wood. Over its top is a semicircular frame, called the head, to contain three whirls (that run on the brasses) with iron spindles, secured by a hasp and pin. They are worked by means of a leather band encircling the whirls and wheel. Three of the whirls are turned when hardening the strand, and only one when closing the rope, the strands being hung together on it. The truck, on which the *back-frame wheel* is fixed, runs on four wheels, and is made of three-inch oak-plank, about nine feet long, and thirteen inches broad at one end, and eleven inches broad at the other.

BACKSTAFF, an instrument formerly used for taking the sun's altitude at sea; being so called because the back of the observer is turned towards the sun when he makes the observation. It was sometimes called Davis's quadrant, from its inventor, Captain John Davis, a Welchman, and a celebrated navigator, who produced it about the year 1590.

This instrument consists of two concentric arches of box wood, and three vanes; the arch of the longer radius is of 30 degrees, and the other of 60 degrees, making between them 90 degrees, or a quadrant. It is now seldom used, and is called the English quadrant by the French.

BACK-STAYS, *(galhaubans*, Fr.) from *back* and *stay*, long ropes extending from the topmast-heads to the starboard and larboard sides of the ship, where they are farther extended to the channels; they are used to second the efforts of the shrouds, in supporting the masts, when strained by a weight of sail in a resh wind.

They are usually distinguished into breast-back-stays and after back-stays: the intent of the first being to sustain the mast when the ship sails upon a wind; or, in other terms, when the wind acts upon the ship sideways; the second is to enable her to carry sail when the wind is further aft; and the third kind take that name from being shifted or changed from one side to the other, as occasion requires. See the article TRAVELLING BACK STAYS.

There are also back-stays for the top-gallant-masts, in large ships, which are fixed in the same manner with those of the top-masts.

A pair of back-stays is usually formed of one rope, which is doubled in the middle, and fastened there so as to form an eye, which passes over the mast-head, from whence the two ends hang down, and are stretched to the channels, by dead-eyes and lanyards. See the article DEAD-EYES, &c.

BACK-STAY STOOL, a short piece of plank, fitted for the security of the dead-eyes, and chains for the back-stays, though sometimes the channels are left long enough at the after end, for the back-stays to be fitted thereto.

BADGE, (*bouteille, fausse galerie*, Fr.) in naval architecture, is a sort of ornament, placed on the outside of small ships, very near the stern, containing either a window, for the convenience of the cabin, or the representation of it: it is commonly decorated with marine figures, martial instruments, or such like emblems. See the article QUARTER.

BAFFLING, or BAFFLING-WINDS, (*brises folles*, Fr.) are those winds that wave and frequently shift from one point to another.

To BAGPIPE *the mizen*, (*border l'artimon au vent*, Fr.) is to lay it *aback*, by bringing the sheet to the mizen shrouds.

BAG-REEF, (*bas ris*, Fr.) a fourth or lower-reef, sometimes used in the royal navy.

To BALANCE, (*balancer*, Fr.) to contract a sail into a narrower compass, by retrenching or folding up a part of it at one corner. It is peculiar to the mizen of a ship, and the main-sail of those vessels, wherein it is extended by a boom. See the article BOOM.

The operation of balancing the mizen is performed by lowering the mizen-yard, or gaff, a little, then rolling up a small portion of the sail at the *peek*, or upper corner, and lashing it about one fifth down towards the mast. See the article MIZEN.

A boom-mainsail is balanced, after all its reefs are taken in, by rolling up a portion of the clew, or lower aftmost corner, and fastening it strongly to the boom. It is requisite, in both cases, to wrap a piece of old canvas round the sail, under the lashing, to prevent its being fretted by the latter.

BALANCE-REEF, (*ruban qui sert à étrangler la voile*, Fr.) a reef-band, that crosses a sail diagonally, and is used to contract it in a storm.

BALCONY, (*galerie de poupe*, Fr.) the projecting gallery in the stern of old large ships, either for ornament or convenience of the captain's cabin. In modern built ships it is always thrown into the cabin.

BALK, (*balk*, Dut. and Germ. *fourans ou épontilles de apin*, Fr.) a piece of short fir timber; also a beam, such as is used in ship-building.

BALL, (*balle*, Fr.) in the military art, comprehends all sorts of bullets for fire-arms, from the cannon to the pistol; also a composition of divers ingredients, generally of the combustible kinds, serving to burn, give light, smoke, stench or the like; as fire-balls, light-balls, smoke-balls, stink-balls, land-balls, &c. Cannon-balls are made of iron, musket-balls, pistol-balls, &c. are of lead. The experiment has been tried of iron balls for pistols and fusees, but they are justly rejected, not only on account of their lightness, which prevents them from flying straight, but because they are apt to furrow the barrel of the pistol, &c.

Cannon-balls are always distinguished by their respective calibres: thus,

For a	42	pound ball the diameter is	6.68
	32		6.10
	24		5.54
	18		5.04
	12		4.40
	9		4.00
	6		3.49
	3		2.77
	2		2.42
	1		1.92

A new description of inflammable balls, applicable for besieging a town, and peculiar for its small weight, by which means it may be thrown to a great distance, and takes fire on a very curious plan, has lately been invented by Captain Thomas Dundas, of the royal navy. It spreads a flame in three distinct openings, which is so strong that the fire extends a full yard in length from the ball itself, and is so powerful that any thing under, over, or near, cannot escape its effects. See the article SHOT.

BALL *and Socket*, an instrument made of brass, with a perpetual screw, so as to move horizontally, vertically, and obliquely; and is generally used for the managing of surveying and astronomical instruments.

BALLAST, (*lest*, Fr. *ballaste*, Dut. *ballastro*, Span.) a certain portion of stone, iron, gravel, or such like materials, deposited in a ship's hold, when she has either no cargo, or too little to bring her sufficiently low in the water. It is used to counter-balance the effort of the wind upon the masts, and give the ship a proper stability, that she may be enabled to carry sail without danger of upsetting.

There is often great difference in the proportion of ballast required to prepare ships of equal burden for a voyage; the quantity being always more or less, according to the sharpness or flatness of the ship's bottom, which seamen call the *floor*.

The knowledge of ballasting a ship with propriety is certainly an article that deserves the attention of the skilful mariner; for, although it is known that ships in general will not carry a sufficient quantity of sail, till they are laden so deep that the surface of the water will nearly glance on the extreme breadth amidships, yet there is more than this general knowledge required; since, if she has a great weight of heavy ballast, as lead, iron, &c. in the bottom, it will place the centre of gravity too low in the hold; and, although this will enable her to carry a great sail, she will nevertheless sail very heavily, and run the risk of being dismasted by her violent rolling.

To ballast a ship, therefore, is the art of disposing those materials, so that she may be duly poised, and maintain a proper equilibrium on the water, so as neither to be too *stiff*, nor too *crank*, qualities equally pernicious: as in the first, although the ship may be fitted to carry a great sail, yet her velocity will not be proportionably increased; whilst her masts are more endangered by her sudden jerks and excessive labouring: and, in the last, she will be incapable of carrying sail without the risk of upsetting

Stiffness in ballasting is occasioned by disposing a great quantity of heavy ballast, as lead, iron, &c. in the bottom, which naturally places the centre of gravity very near the keel; and that being the centre about which the vibrations are made, the lower it is placed, the more violent will be the motion of rolling.

Crankness, on the other hand, is occasioned by having too little ballast, or by disposing the ship's lading so as to raise the centre of gravity too high, which also endangers the mast in carrying sail when it blows hard: for when the masts lose their perpendicular height, they strain on the shrouds in the nature of a lever, which increases as the sine of their obliquity; and a ship that loses her masts is in great danger of being lost.

The whole art of ballasting, therefore, consists in placing the centre of gravity to correspond with the trim and shape of the vessel, so as neither to be too high, nor too low, too far forward, nor too far aft; and to lade the ship so deep, that the surface of the water may nearly rise to the extreme breadth amidships; and thus she will be enabled to carry a good sail, incline but little, and ply well to the windward.

In the royal navy, the iron ballast is first stowed fore and aft, from bulkhead to bulkhead in the main hold, next to fir cants nailed on the limber-strakes on each side the kelson, five or more inches clear of the limber-boards; and is winged up three or more pigs above the floor-heads in the midships, or bearing part of the ship, and there are two tiers or pigs in the wake of the main hatchway and well-wings. Ships built with a very clear run aft, seldom have any iron ballast stowed abaft the pump-well or after-hold. Ships that have floor and futtock riders, have the iron ballast stowed either lengthways or athwart ships, agreeable to the length of the chambers, which are the clear spaces between the riders.

The shingle ballast is next spread and levelled over the iron ballast, on which is stowed the ground tier of water, bung up and bilge free from the sides, either chine and chine, or bouge and chine, beginning at the coal-room bulkhead, that being the foremost, and making the breakage, if any, at the main hatch. The midship tiers, fore and aft, are the first laid down, and the casks are sunk about one quarter of their diameter into the shingle; the sides are filled in with wingers of small casks, as half hogsheads, gang casks, or breakers; observing not to raise the wingers above the level of the tier, to cause a breakage in the next tier above, which is stowed in the cuntline of the ground tier, bung up and bilge free, and so on, for as many tiers as can be stowed sufficiently clear of the beams.

In the after-hold, between the aft side of the pump-well and fish-room bulkhead, are stowed the provision, above the ground tier; between the casks, billet, or other wood, and shingle ballast.

In the fish room are stowed some of the spirits, or wine, and sometimes coals; and in the spirit-room are stowed the wine and spirits for the ship's use.

In the merchant-service, the stowage consists, besides the ballast, of casks, cases, bales, boxes, &c. which are all carefully wedged off from the bottom, sides, pump-well, &c. and great attention paid that the most weighty materials are stowed nearest to the centre of gravity or bearing of the ship; and higher or lower in the hold agreeable to the form of the vessel. A full low-built vessel requires them to be stowed high up, that the centre of gravity may be raised, to keep her from rolling away her masts, and from being too stiff and laboursome; as, on the contrary, a narrow high built vessel requires the most weighty materials to be stowed low down, nearest the kelson, that the centre of gravity may be kept low, to enable her to carry sail, and to prevent her upsetting. See the article Trim.

Ships are said to be in ballast when they have no other loading. Masters of vessels are obliged to declare the quantity of ballast they bear, and to unload it at certain places. They are prohibited unloading their ballast in havens, roads, &c. the neglect of which prohibition has ruined many excellent ports.

All ships and vessels taking in ballast on the river Thames, are bound to pay the corporation of the Trinity-house, for every ton carried to any ship in the coal trade, 1*s.* and for every other British ship 1*s.* 3*d.* for every ton carried to any foreign ship, 1*s.* 7*d.*

The Trinity-house employ men for that purpose, and regulate them; and their lighters are all marked.

Ballast allowed to the following Ships.

Guns.	Tonnage	Iron Tons.	Shingle Tons.	Guns.	Tonnage.	Iron Tons.	Shingle Tons.
110	2290	180	370	38	930	70	170
100	2090	180	370	36	870	65	160
98	2110	160	350	32	700	65	140
90	1870	160	350	28	600	60	100
80	1620	140	300	24	500	50	80
74	1700	80	270	22	450	50	70
64	1370	70	260	20	400	50	60
50	1100	65	170	Sloop	300	50	40
44	900	65	160	Brig	160	30	15

Ballast, *pea*, (*lest en gravier*, Fr.) is a peculiar kind of fresh water sand, dried by the sun, and used by the ships from China, for stowing their teas with.

Ballast, *shingle*, (*lest de pierre*, Fr.) implies coarse gravel.

BALLATOONS, are large heavy barks, or luggage-boats, used for carrying wood, by the river, from Astracan, and the Caspian sea, to Muscow; They are of an extraordinary length and breadth, their prows and sterns rounding alike to a point; and they carry from 150 to 200 ton, and from 150 to 200 men, to row and tow them along the rivers.

BALUSTERS, (*balustres*, Fr.) small columns, or turned pillars of wood, placed round the balcony in the stern, and likewise in the quarter-gallery of large ships.

BAMBOO Habit, a Chinese contrivance, by which a person, who does not know how to swim, may easily keep himself above water. It is thus constructed: four bamboos, two before, and two behind, their bodies, are placed horizontally, and project about twenty-eight inches. They are crossed on each side by two others, and the whole properly secured, leaving a space for their body; it is put over their heads, may be tied securely in two minutes, packed up in a small compass, and put into a chest.

BANDS, (*bandes*, Fr.) slips of canvas strongly sewed across a sail, to strengthen it, as the reef-bands, &c.

Bands to encircle the wheels and whirls of leather, large line, or catgut.

BANIAN-Days, (*jours maigres*, Fr.) a cant term among sailors, denoting those days on which they have no flesh meat served out to them: it seems to be derived from the practice of a nation amongst the eastern Indians, called Banians, who never eat flesh.

BANK, (*banc de sable*, Fr. *banc*, Sax.) an elevation of the bottom of the sea, some of which are so high as to prevent a ship from floating over them; and, in this sense, amounts nearly to the same as shallows, flats, &c. The shelves that abound with rocks under water, are distinguished by other names, as reefs, ridges, keys, &c.

Banks, on the sea-coast, (*bancs*, Fr.) are usually marked by beacons or buoys, and, in charts, they are distinguished by little dots, as ridges of rocks are by crosses. The principal banks in the Western Ocean, are those of Newfoundland, and the Bahama-bank: the most remarkable one in Newfoundland is called the Grand-bank, which is of a vast extent, being nearly two hundred miles in length, and stretching north and south: its usual depth is from twenty to eighty fathoms: and this is the great scene of the cod-fishery, which is so material an article in European commerce.

An exact knowledge of the banks, their extent, and the depth of water in which they lie, constitutes a very essential part of the science of a pilot, and master of a ship. If the vessel be large, and draws much water, great attention will be necessary to avoid the banks; on the contrary, if it be small, the same banks afford a sure asylum where it may brave the largest and stoutest vessel, which dares not follow it to so dangerous a retreat. By means of this barrier, many small craft have escaped their enemies.

Bank *of oars*, (*banc de rameurs*, Fr.) a seat or bench of rowers in a galley: these are properly called the *thought* by seamen. The common galleys have 25 banks on each side, with one oar to each bank, and four men to each oar. The galeasses have 32 banks on each side, and six or seven rowers to each bank.

BANKER, (*terre neuvier*, Fr.) a vessel employed in the cod-fishery, on the banks of Newfoundland.

BAR of a Port, Harbour, or Haven, (*barre d'une rivière, ou d'un port*, Fr.) a shoal or bank of sand, gravel, &c. thrown up by the surge of the sea, to the mouth of a river or harbour, so as to endanger, and sometimes totally prevent, the navigation into it.

The difficulties and dangers which rarely fail to attend ships, particularly in blowing or stormy weather, in passing these natural impediments to navigation, are, perhaps, among the most formidable that attend a sea-faring life.

BARCA-LONGA, a large Spanish coasting vessel, navigated with lug-sails, and having two or three masts: these are very common in the Mediterranean. See the article Vessel.

BARE-POLES, (*à mâts et à cordes*, Fr.) the term bare-poles implies a ship lying-to, without any sail set whatever, in a gale of wind, generally speaking, in consequence of being on a lee shore.

Bare-poles, *a ship under*, (*un vaisseau à sec*, Fr.) is applied to a ship having no sail set when out at sea.

BARGE, (*bargie*, Dut. *grand canot, ou canot de parade*, Fr.) a boat of state and pleasure, furnished with elegant apartments, canopies, and cushions; equipped with a band of rowers, and adorned with flags and streamers. They are generally used for processions on the water, by noblemen, officers of state, or magistrates of great cities: of this sort we may reasonably suppose the *royal barge*, the *barges of companies*, &c.

There are also barges of a smaller kind, for the use of admirals and captains of men of war. These are of a lighter frame, and may be easily hoisted into and out of the ships to which they occasionally belong. See the article Boat.

Barge, (*cabotiere*, Fr.) is also the name of a flat-bottomed vessel of burden, used on rivers, for conveying goods from one place to another, for loading and unloading ships; and has various names, as a *wear barge*, a *west country barge*, a *sand barge*, a *row barge*, &c.

BARK, (*barque*, Fr.) a general name given to small ships: it is, however, peculiarly appropriated, by seamen, to those which carry three masts without a mizen-topsail. Our mariners, who are trained in the coal trade, apply this distinction to a broad-sterned ship, without a figure head.

Bark, *armed*, (*barque armée*, Fr.) is a kind of fireship filled with soldiers, used for making sallies to attack galleries, and bar the passage over them.

Barks *of Bombay*, called Dingass, are vessels used at Bombay, and places adjacent; and are navigated some times by rowing with paddles. They have one mast, one third the length from the stem, which rakes much forward. On the mast is hoisted a sail, bent to a long yard, resembling a settee-sail. The tack is made fast to the head of the stem, and the sheet to the heel of the mast. These vessels never tack, but wear, in doing which they peek the yard against the mast to shift the sail, at the same time they pass the sheet before the mast. Their rigging consists of a pair of halyards, a bow-line, and brace. Their keels are very much hollowed upwards, to avoid wholly grounding on sand-banks.

Barks *of Cracaloa and Straits of Sunda*, are vessels with flush decks, high sheer, and sharp forward. They have one mast, and the sail is similar to the Caracores, being long and narrow. These vessels are kept from upsetting by a sort of beams crossing the vessel, and bending downwards at the ends, which fasten to a long, round, or flat piece of timber.

Barks, *Japanese*, are vessels similar to the junks, 80 or 90 feet long on one deck, but which have only one mast, that carries a square sail, and forward, one or two jibs made of cotton. They only use sails when the wind is large.

Bark, *long*, (*barque longue*, Fr.) is a little vessel that has no deck, and is longer and lower than the common barks; being sharp afore, and generally going back with sails and oars. It is built like a sloop, and called a double sloop in many places.

Barks, *water*, small vessels used in Holland, for the carriage of fresh water, to places where it is wanting; and also for fetching sea water to make salts of. They have a deck, up to which they are filled with water.

BARNACLE, (*cravan, coquillage qui s'attache au fond des vaisseaux*, Fr.) a species of shell-fish, often found sticking to the bottoms of ships, rocks, &c.

BAROMETER, (from βάρος, weight, and μετρον, a measure) an instrument for measuring the weight or pressure of the atmosphere; and, by that means, the variations in the state of the air, foretelling the changes in the weather, and measuring heights or depths, &c. This instrument commonly consists of a glass tube, open at one end, which, being first filled with quicksilver, and then inverted, with the open end downwards, into a basin of the same, the mercury descends in the tube till it remains at about the height of twenty nine or thirty inches, according to the weight or pressure of the atmosphere at the time, which is just equal to the weight of that column of the quicksilver. Hence it follows, that if, by any means, the pressure of the air be altered, it will be indicated by the rising or falling of the mercury in the tube; or, if the barometer be carried to a higher station, the quicksilver will descend lower in the tube; but, when carried to a lower place, it will rise higher in the tube, according to the difference in elevation between the two places. They are made in various forms, and are named accordingly; viz. the horizontal or rectangular, the diagonal, the wheel, the pendant, and the portable.

BAROMETER, *marine*, (*baromètre*, Fr.) when the barometer is to be used at sea, some contrivance is necessary to prevent the oscillations of the mercury. Two methods are employed for this purpose: either to prevent, by mechanical means, the vibrations of the instrument itself, or to check the motions of the mercury in the tube, by some peculiarity of form. Both methods may also be conjoined.

One of the earliest marine barometers with which we are acquainted, was suggested by Dr. Hooke.

His contrivance consisted of a double thermometer, or a couple of tubes, half filled with spirit of wine; the one sealed at both ends, with a quantity of air included; the other, sealed at one end only The former of these is affected only by the warmth of the air; but the other is affected both by the external warmth, and by the variable pressure of the atmosphere. Hence, considering the spirit thermometer as a standard, the excess of the rise or fall of the other above it will shew the increase or decrease of the pressure of the atmosphere. This instrument is said to be of great use in giving notice of bad weather at sea, and is described by Dr Halley, in *Philos. Trans.* 1700-1, No. 269, p. 791, where he says of it, "I had one of these barometers with me, in my late southern voyage, and it never failed to prognosticate and give early notice of all the bad weather we had, so that I depended thereon, and made provision accordingly; and, from my own experience, I conclude that a more useful contrivance hath not, for this long time, been offered for the benefit of navigation."

Mr. Nairne, an ingenious artist, in London, has also invented a new kind of barometer, which differs from the common barometer, by having the lower part of the tube, for about two feet long, made very small, to check the vibrations of the mercury, which is apt to arise from the motions of the ship. This is also assisted by being hung in gimbals, by a part which subjects it to be the least affected by such motions.

Another sort of marine barometer has also been invented by M. Passement, an ingenious artist, at Paris. This contrivance consists in twisting the middle of the tube into a spiral of two convolutions; by which the impulses the mercury receives from the ship's motions are destroyed, by acting in opposite directions.

One of the most improved kind of marine barometers has lately been constructed by Mr. Troughton, of Fleet-street; the tube consists of two parts, joined together above five inches below the top; the bore in the upper part being about $\frac{4}{10}$ of an inch, and in the lower part only $\frac{2}{100}$. By this construction, partly from the difference of the bores, and partly from the greater friction in the lower end, the motion of the mercury is so much retarded, that any impulse given by the ship having a tendency to raise it, will scarcely have produced a sensible effect, before an opposite impulse will be given having a tendency to depress it. To counteract more effectually the effects of the ship's motions, the instrument is suspended on gimbals, a representation of which is given at fig. 3. plate 2. The whole is attached to the side of the cabin by two tubes of brass, which slide one within the other, and render the instrument capable of being suspended at different distances from the place of support, that the bottom of it may not strike the side of the cabin during any heavy rolling of the vessel. (See fig. 4.) The inner tube carries the gimbals. The external frame of the barometer is a cylindrical tube of wood, on which slides a brass socket; and, in this, is inserted the innermost pair of pivots of the gimbals, or universal joints, which furnishes the instrument with a moveable point of suspension. The top is terminated with a knob of brass, of a weight nearly equal to that of the mercury, &c. at the lower end; and, by this ingenious contrivance of Mr. Troughton, the centre of gravity of the whole will be about the middle; and, if the instrument were of the same specific gravity throughout, the point of suspension that would produce the smallest oscillations in the mercury would be about $\frac{1}{3}$ of the length of the instrument from the top, considering the lower part as a fixed point. But as this is not strictly the case, the point of suspension is best ascertained by experiment. The graduation is on two scales of ivory, about four inches long, for the reception of which two opposite quarters of the cylindrical frame are *crossed out* through that length, their planes pointing towards the centre of the tube. The index is a very light one, and slides upon the glass tube, without touching any other part. At the bottom is the usual screw, which presses up the leather bag, and prevents the mercury from moving when the instrument is carried from one place to another.

The following observations upon the movements and state of the mercury in the marine barometer, were made by Matthew Flinders, Esq. commander of his majesty's ship Investigator, during his examination of the *Coasts of New Holland* and *New South Wales*, the *Terra Australis* of the earlier charts, in the years of 1801, 1802, and 1803. *Phil. Tran.* 1806, Part 2.

The marine barometer, with which these observations were made, was constructed by Nairne and Blunt, and had been used in Captain Cook's Voyages.

The height of the mercury was taken regularly at day-break, at noon, and at eight o'clock in the evening. The temperature of the thermometer was also registered at the same periods. The circumstances that led Captain Flinders to think his observations worth attention were, the coincidence that took place between the rising and falling of the mercury, and the setting in of wind that blew from the sea and from off the land, to which there seemed to be at least as much reference as to the strength of the wind, or the state of the atmosphere.

Among the examples selected from the captain's journals, are nine, that relate to the *south* coast; from these, it appears, generally, that a change of wind from the northern, to any point in the southern half of the compass, caused the mercury to rise, and a contrary change, to fall; and that the mercury stood considerably higher when the wind was from the south side of east and west, than it did in similar weather, when the wind came from the north side. The cause of this appears to be, that the first proportion of air brought in from the sea, is impelled upwards by the land, which it encounters, and along the inclined surface of the land, in a sloping direction: the next portion is, in the same manner, stopped and forced upwards; but it has a shorter space to pass through, because the former portion goes along two of the sides, and the latter along the third side of an obtuse-angled triangle: thus, the succeeding portions of air meet the summit of land before the first portions, and cause the latter to eddy and stagnate; while the stream blowing above this portion, compresses it, and augments its density, whence the increased height of the barometer.

"The barometer," says Captain Flinders, "was of great service to me in the investigation in this dangerous part of the east coast, where the ship was commonly surrounded with rocks, shoals, islands, or coral reefs. Near the main land, if the sea-breeze was dying off at night, and the mercury descending, I made no scruple of anchoring near the shore, knowing that it would either be a calm, or a wind would come off from the land; but if the mercury kept up, I stretched off, in the expectation that it would freshen up again in a few hours. Amongst the barrier-reefs, when the wind was dying away, the barometer told me, almost certainly, from what quarter it would next spring up. If the mercury stood at 30.15, or near it, and was rising, I expected the proper trade wind; and if higher, that it would be well from the southward, or would blow fresh; and, if it was up to 30.30, both. The falling of the mercury to 30.10 was an indication of a breeze from the north-eastward; and its descent below 30 inches, that it would spring up, or shift round to the westward." Hence, it appears, that this skilful commander navigated his vessel in those dangerous parts of the eastern coast that are between the latitudes of 23° and 17°, pursuant to a confident deduction from his own theory; and it is sincerely to be hoped that other navigators will follow the example of this truly ingenious gentleman, in carefully registering their meteorological observations in different latitudes and climates; that some master-hand may, at length, strike out a theory which may satisfactorily connect and elucidate the various results.

Variations of the Barometer.

Instead of introducing any unsatisfactory theory to account for the variations of the barometer, it will be most proper here merely to mention the results of observations upon this instrument, by various philosophers, in different parts of the globe; being sensible that these and many more must be arranged and examined, in order to the establishment of any plausible hypothesis.

Several members of a German meteorological society have registered many observations upon the barometer. The most noted of those observers are, Steilehner, Planer, Chiminello, and Hemmer.

The first of these gentlemen says, that he found, by several comparative observations, that the greatest fall of the barometer does not happen in very remote places at the same time; but that it is earlier towards the west, and later towards the east; and that the difference of the time is nearly equal to the difference of the meridians of the places; an assertion which deserves to be accurately examined.

M. Planer observed the barometer for a whole year, six times every day, viz. at 2, 6, and 10 o'clock in the morning, and at the same hours in the afternoon; and found, in general, that the barometer, between 10 in the morning and 2 in the afternoon, and between 10 at night and 2 in the morning, was less in its rising, and greater in its fall; and that the contrary was the case between the hours of 6 and 10 in the evening and morning.

Chiminello observed the barometer 22 times a day, for three years, but he left a chasm in the night, which he supplied *by calculation*. The principal positions which he then deduced are, that the barometer falls towards noon, as well as towards midnight.

Hemmer deduced the three following general rules from a great number of accurate observations:—

1st. When the sun passes the meridian, the barometer, if in the act of falling, continues to fall, and the falling is accelerated.

2d. When the sun passes the meridian, the barometer, if in the act of rising, falls, or becomes stationary, or rises more slowly.

3d. When the sun passes the meridian, the barometer, which is stationary, falls, if it has not risen before or after being stationary; in which case it usually becomes stationary during the sun's passage.

From a register kept by a Mr. Dunbar, near the banks of the Mississipi, in N. lat. 31° 28′, we find that, for the space of about four days before, and six days after, the summer solstice, the barometer regularly rises from about nine P.M. to about six A.M. then falls till the return of the former hour in the evening, then rises again as before, &c. in alternate periods. In the first four days the direction is *ascending*, and the elevation of a line drawn through the mean is about $\frac{6}{100}$ of an inch. In the latter six days the mean line is perfectly horizontal, the elevation each night amounting to $\frac{2}{100}$, and the depression each day to the same, but occupying double time.

The celebrated Humboldt made some interesting observations at Caraccas, in South America, near the equator. There are, he says, four atmospherical tides every 24 hours, which depend only on the attraction of the sun. The mercury falls from 9 in the morning to 4 in the evening: it rises from 4 to 11 o'clock: it falls from 11 o'clock till past 4 in the morning: and it re-ascends from that time till 9 o'clock. *Neither winds, storms, nor earthquakes, have any influence on this motion.*

Horsburgh, in his last voyage to Bombay, employed two marine barometers, one made by Troughton, and the other by Ramsden; with which he made very minute observations; which were published in the second part of the *Phil. Trans.* for 1805; and in the *Hist. Roy. Soc. Edin.* of the same year, we have a comparison of the diurnal variations of the barometer, made in Peyrouse's voyage round the world, with those made at Calcutta, by Dr. Balfour.

The agreement between these observations seems very remarkable. Dr. Balfour found that, during the whole lunation, in which he observed the barometer from half hour to half hour, the mercury constantly fell from 10 at night to 6 in the morning; from 6 to 10 in the morning it rose; from 10 in the morning to 6 at night it fell again; and, lastly, rose from 6 to 10 at night. The *maximum* height is, therefore, at 10 at night and 10 in the morning; and the *minimum* at 6 at night and 6 in the morning. The only difference is, that in M. Lamanon's observations, the *minimum* is stated to have happened at about 4 instead of 6. This, however, will not seem a very material difference, when it is remembered, that the instant when any quantity attains either its greatest or its least state is not easily ascertained with precision. From the observations, as detailed by M. Lamanon, the time of the *minimum* seems to answer fully as well to 5 as to 4; so that the difference of the results is in every view inconsiderable, and their coincidence on the whole not a little singular. The variation in Dr. Balfour's barometer between the nearest *maximum* and *minimum* is sometimes about $\frac{1}{10}$ of an inch, though, in general, considerably less.

BARRELS FOR POWDER, (*baril à poudre*, Fr.) have each four hoops of copper, and the spaces between them filled with hazel hoops.

Their dimensions are as follow:—

	Whole Barrels.		Half Barrels.		Quarter Barrels.	
	Ft.	In.	Ft.	In.	Ft.	In.
Length	1	9,61	1	5,13	1	2,25
Diameter at top and bottom ..	1	3,61	1	0,37	0	9,35
——— at bulge ..	1	5,36	1	2,00	0	10,71

The whole barrels are made to contain 100 pounds, and the half barrels 50 pounds, and quarter barrels 25 pounds, of powder; but, of late, only 90 pounds have been put into the barrels, 45 into half-barrels, and 22½ into quarter barrels; which, by leaving the powder room to be shifted, preserves it the better.

A Mr. Crawshays, and a Lieutenant Smith, of the royal navy, had each constructed copper powder barrels, some years ago, for the purpose of conveying gun-powder from place to place with more safety.

Mr. Walker has also lately constructed copper powder-barrels, in the form of a cylinder, for the use of the navy; but how far any of these inventions may answer the intended purpose, remains to be ascertained by experience.

BARRELS, *Budge*, are made in the same form as half powder-barrels, with copper and hazel-hoops, but rather less; they have only one head, the other end having a leathern hose at the mouth, nailed round it with copper nails, and has a reeving-string, to reeve it close together; they are used for the purpose of collecting any loose powder that may have been spilt in the act of drawing the guns, &c. Their length is about 10½ inches, and one foot in diameter.

BARRELS for small arm ammunition, &c. are the same as half and quarter powder barrels, with this exception, that they are made lighter, and have no copper hoops.

The half barrel holds 1000, and the quarter-barrel 500 musket-ball cartridges, or double that number of pistol-ball cartridges.

BARRELS, *Water-tight*, have six copper hoops, without any hazel hoops; and are made with very thick staves, to contain 80 pounds each, and the half-barrels 40 pounds each. They are intended for the ground tier in ships' magazines.

To prevent any accident on board, in taking off, or putting hoops on, powder barrels, a copper vice and screw, a driver, and an adze, are allowed to every ship in the royal navy.

BARRICADE, (*barricade*, Fr. *barricada*, Span.) a strong wooden rail, supported by stanchions, extending across the foremost part of the quarter-deck: the upper part, which contains a double rope-netting, above the rail, is stuffed with full hammocks, to intercept the motion, and prevent the execution of small shot, in the time of battle.

BARS, HATCH, (*barres d'écoutilles*, Fr.) are bars which lock over the hatches, for the purpose of securing the hold from depredations.

BARS, *capstan*, (*barres du cabestan*, Fr.) are large thick bars, the end of which is fitted to the hold of the drum-head of the capstan, by which it is turned round the spindle.

BASIN OF A DOCK, (*bassin à radouber les vaisseaux*, Fr.) a place where the water is confined by double flood-gates, or a caisson, and thereby prevented from running out at ebb tide. The use of it is to contain ships either before they enter, or after they come out of, the dock in which they are repaired.

BASIN, (*paradis*, Fr.) also implies some part of a haven, which opens from a narrow channel into a spacious receptacle for shipping.

BATTENS, (*lattes*, Fr.) signify thin pieces of oak or fir, nailed to the mast-heads, and to the midship part of the yards.

BATTENS *of the hatches*, (*lattes de caillebotis*, Fr.) a sort of long narrow laths, scantlings of wooden stuff, or straightened hoops of casks. They serve, by the help of nailing, to confine the edges of the tarpaulings close down to the sides of the hatchways, to prevent the water from penetrating the lower apartments of a ship in a storm.

Tracing BATTENS, are pieces of wood, about three inches thick, nailed to the beams of the ship, instead of cleats, to sling the seamen's hammocks to.

BATTERIES, Floating. Several machines adapted to this purpose have been proposed and constructed in different parts of the kingdom, by various ingenious persons, for the defence of the coast; some of them built new, and bearing the simple form of rafts; others built new also, and somewhat nearer approaching the form of vessels, but totally unfit to bear the sea, or face the danger of a voyage. Others have been formed of old ships, considered, in a great measure, unfit for active service, which had been properly strengthened with pieces of timber and balks, as well on the score of safety from the perils of the waves, as that of better security to the crew in case of hostile attack. Among others, a *mortar battery*, for the bombardment of the enemy's ports, has been invented by Lieutenant-colonel Congreve, equerry to His Royal Highness the Prince Regent, which is proof both against shells and red-hot balls. It is said to be so contrived that, though provided with masts and sails for any voyage, yet they can be securely disposed of in less than a quarter of an hour; so that the battery then presents nothing but a mere hull, with sloping sides, upon the water, which is rowed by forty men, under cover of the bomb-proof, and may, by the peculiar construction of the masts and rigging, be brought under sail again as expeditiously as dismantled. The rudder and moorings are entirely under water, and protected by the bomb-proof, so that no disappointment as to them can possibly arise. The battery is armed with four large *mortars*, for bombardment, and four *42-pounder carronades*, for self-defence; though, from being covered with plates and bars of iron, she can neither be set fire to, nor be carried by boarding. Four such vessels, though they are no more than 250 tons burden each, and draw less than twelve feet water, would throw upwards of 500 shells into any place in one tide, and with the greatest effect and precision; both because from their construction, they have nothing to apprehend from approaching the enemy's *batteries*, and because, from the peculiar contrivance of the *mortar-beds*, the elevation of the mortars is not affected by the rolling or pitching of the vessel. Several of our most eminent naval men have seen and approved of the contrivance; and it is said that *ministers* have attended to this gentleman's *plans*, and intend to institute vigorous and regular bombardments of such of the enemy's ports in which they may collect any considerable flotilla for the invasion of this country.

BATTERY, (*batterie*, Fr.) a parapet wall of stone, earth, faggots, &c. thrown up to cover the men employed about the guns from the enemy's shot.

BATTLE, Naval (*combat naval*, Fr.) denotes an engagement between two fleets, squadrons, or even single ships.

Before a naval battle, every squadron usually subdivides itself into three equal divisions, with a reserve of certain ships out of every squadron, to bring up their rear. Every one of these observing a due birth and distance, are, in the battle, to second one another; and the better to avoid confusion and falling foul of each other, to charge, discharge, and fall off by threes or fives, more or less, as the fleet is greater or smaller. The ships of reserve are instructed either to succour and relieve those that are any way in danger, or to supply and put themselves in the place of those that shall be rendered unserviceable. See the articles Engagement and *Naval* Tactics.

BAVINS, for Fireships, (*bourrees*, Fr.) are small bundles of any sort of brush-wood, which is tough and easily kindled. They are generally two or three feet in length; have all their bush ends lying one way, the other ends being tied together with small cords, and are then dipped in a kettle of melted composition at the bush ends, whose branches are afterwards confined by the hand, to prevent them from breaking off by moving about; and also to make them burn more fiercely.

BAY, (*baie*, Fr.) an inlet of the sea-coast, between two capes or promontories, where shipping frequently ride at anchor.

A Bay is such a gulf or opening into the land as does not run very deep into it, whether large or small; such is the Bay of Biscay. But smaller bays are frequently denominated *creeks*, *havens*, or *roads*; of which description is Milford Haven, and others of inferior note. It may be observed, in general, that a bay has proportionably a wider entrance than either a gulf or haven; and that a *creek* has usually a small inlet, and is always much less than a bay. What is called a *road*, is a place upon any coast where there is anchorage, and a certain degree of protection and shelter from winds.

Bay, in large ships of war, (*gatte*, Fr.) is that part on each side between decks, which lies before the bits.

BAYONET, (*bayonnette*, Fr. from Bayonne, where this weapon is said to have been first invented) is a short broad dagger, made lancet-fashion, and having, instead of a hilt, which it formerly had, an iron handle and ring, to fix to the end of a musket, so as not to hinder its firing or charging. In the reign of Louis XIV. the French, when at war with Spain and Savoy, used bayonets. The bayonet then was merely a dagger, with a round wooden handle, to fix in the mouth of the musket; but the method of fixing it on the outside of the barrel was soon afterwards invented.

BEACH, (*rivage*, *plage*, Fr.) the sea shore, or margin of the sea, particularly that part that is dashed by the waves; the strand.

BEACON, (*balise*, Fr. *beacon*, Sax.) a post or stake erected over a shoal or sand-bank, as a warning to seamen to keep at a distance; also a signal-post on the top of hills, &c.

BEACONAGE, (*branche de ciprès*, Fr.) a small duty paid by shipping in France, for keeping beacons in repair.

BEAK-HEAD, (*coltis*, Fr.) a small platform at the fore part of the upper-deck, in large ships; generally placed the same height from the deck as the port-sills, occasionally for the use of a gun, but chiefly for the conveniency of the men. At the aft-part of the beak-head is a bulk-head, called the *beak-head bulkhead*, which encloses the forecastle.

This name is appropriated to ships whose forecastle is square or oblong; a circumstance common to all vessels of war which have two or more tiers of guns. In smaller ships the forecastle is generally shaped like a parabola, the vertex of which lies immediately above the stem.

The strong, projecting, pointed beaks, used by the ancients in time of battle, are entirely disused since the invention of gunpowder.

BEAMS, (*baux*, Fr. *beam*, Sax. a tree) are strong thick pieces of timber, stretching across the ship from side to side, to support the decks, and retain the sides at their proper distance, with which they are firmly connected by means of strong knees, and sometimes by standards.

The BEAMS of ships of war are usually formed of three pieces scarfed together, and are sustained at each end by thick planks in the ship's side, called clamps, upon which they rest.

Mr. George Williams, master carver, at Chatham dock-yard, has lately invented a method of securing the beams of ships' decks to their sides, without wooden knees, made of one piece; which is as follows:—1. Triangular braces of wrought iron are to be screwed to the upper side of the beam, and also to the ship's side: these braces resemble a Roman capital A, and are let into the beam, and firmly fixed by three bolts; one near the angular point, and two others in the cross part. The ends of these braces are turned up, and bolted to the timbers of the ship's side, by two bolts passing through each end, and through the timbers and outside planks, which secures the beams from lateral motion.

2. In order to secure the beams in a vertical direction, a block is fitted under each, and mortised, both into the beam and the side of the ship. Two iron straps are bolted on each block; one end of each of these straps is fixed to the deck beams by the same bolts as the upper brace: the other ends are bolted against the inside planking: and an oblique bolt goes through the middle of each strap, the block, and the ship's side.

"In this method," says Mr. Williams, "less iron in weight, and fewer bolts are necessary than in the iron knees before in use; there is also less strain upon the bolts, and much more room is gained between decks, for stowage and working the guns, and even a port hole may be made under the beam."

For Mr. Williams's calculation relative to the saving that will be effected by this method in a 74 gun ship, or East Indiaman, and a plan of the invention, see the *Trans. of the Society of Arts*, Vol. 27.

Beak-head BEAM, (*bau de coltis*, Fr.) is the broadest beam in the ship, but generally is made in two breadths, dowelled together.

Midship BEAM, (*maitre-bau*, Fr.) is the longest beam in a ship, being lodged in the midship-frame, or between the widest frame of timbers. At about two-thirds of the height from the keel to the lower-deck, are laid a range of beams, to fortify the hold, and support a platform called the orlop, which contains the cables and stores of the ship.

Orlop BEAMS, (*baux du faux pont*, Fr.) are those beams which support the orlop-deck, but are chiefly intended to fortify the hold.

There are usually twenty-four beams on the lower deck of a ship of 74 guns, and to the other decks additional ones in proportion, as the ship lengthens above.

It is necessary that the beams should have a greater height in the middle than at the two ends, to carry the water more readily off from the decks, and to diminish the recoil of the guns, which will thereby more easily return to their places.

On the BEAM, (*droit par le travers du vaisseau*, Fr.) implies any distance from the ship, on a line with the beams, or at right angles with the keel: thus, if the ship steers or points northward, any object lying east or west, is said to be on her starboard or larboard *beam*.

On the Weather BEAM, (*au vent*, Fr.) implies on the weather-side of the ship.

Before the BEAM, (*un peu par l'avant*, Fr.) is an arch of the horizon comprehended between the line that crosses her length at right angles, and some object at a distance before it; or between the line of the beam and that point of the compass which she stems. Thus, if a ship steering west, discovers an island on the right, three points *before the beam*, the island must bear N. W. by N. from the ship. See the article BEARING.

Abaft the BEAM, (*un peu par l'arrière*, Fr.) See the article ABAFT.

BEAM *ends*, (*vaisseau sur le cotè*, Fr.) a ship is said to be on her beam ends when she inclines very much on one side, so that her beams approach to a vertical position; hence, also a person lying down is said to be on his beam ends.

BEAN-COD, (*bateau pécheur*, Fr.) a small fishing-vessel, or pilot-boat, common on the sea-coasts, and in the rivers of Portugal. It is extremely sharp forward, having its stem bent inward above, in a considerable curve: the stem is also plated on the fore-side with iron, into which a number of bolts are driven, to fortify it, and resist the stroke of another vessel, which may fall athwart-hawse. It is commonly navigated with a large lateen sail, which extends over the whole length of the deck, and sometimes of an outrigger over the stern; and is, accordingly, well fitted to ply to windward.

BEAR, in astronomy, a name given to two constellations, called the Greater and Lesser Bear, or URSA major and minor. The pole star is said to be in the tail of the little bear, and is within less than two degrees of the north pole of the world. See the articles URSA *major* and *minor*.

BEAR, in a ship, is composed of a square piece of wood, on which are fastened some pigs of iron ballast, and used to clean a ship's deck, when a holy-stone cannot be had. See the article HOLYSTONE.

To BEAR, (*rester*, Fr.) Exam. "The land's end bore E. N. E." viz. it was seen from the ship in a line with the E. N. E. point of the compass.

To BEAR *down upon the Enemy*, (*arriver sur l'ennemi*, Fr.) to have the advantage of the wind, or being to windward, so as to approach the enemy by sailing large, or from the wind.

To BEAR *in with the Land*, (*courir vers la terre*, Fr.) implies, to steer a ship towards the land.

To BEAR *off from the Land*, (*s'éloigner de la côte*, Fr.) to steer a ship from the land, lest she should accidentally run *aground*, when under sail.

To BEAR *off*, (*défier, ou parer du bord*, Fr.) also implies to keep any weight, which is being hoisted up, clear from rubbing against the sides; as, "bear off the boat!" "bear off that cask!" &c.

To BEAR *up*, or BEAR *away*, (*arriver faire porter*, Fr.) is to change the course of a ship, in order to make her run before the wind, after she had sailed some time with a side wind, or close hauled; and seems to have been derived from the motion of the helm, by which this is partly produced, as the helm is then borne up to windward, or weather side of the ship. Hence, bear up, seems to have reference to the helm only, as "bear up the helm a-weather!" With respect to any other thing, it is said, "bear away!" or "bear down!" thus, "we bore away for Torbay, in consequence of a heavy gale;" "conceiving a ship to be of fair force, we bore down upon her, and engaged her."

Bear-*a-hand!* (*dépêche!* Fr.) implies make haste, despatch, quick, &c.

BEARDING, (*amoindrer*, Fr.) is the operation of chipping, planking, or otherwise diminishing any piece of timber or plank from a given line, or curve, such as the bearding-lines of the deadwood, bearding of clamps, planksheers, fiferails, &c.

BEARING, (*gissements de la côte*, Fr.) an arch of the horizon intercepted between the nearest meridian and any distinct object, either discovered by the eye, and referred to a point on the compass, or resulting from the sinical proportion; as in the first case, at 4 P.M. Cape Spado, in the isle of Candia, bore S. W. by the compass.

In the second, the longitudes and latitudes of any two places being given, and, consequently, the difference of latitude and longitude between them, the bearing from one to the other is discovered by the following analogy:

As the meridional difference of latitude
Is to the difference of longitude:·
So is radius, to the tangent bearing.

Bearing (*relevement*, Fr.) is also the situation of any distant object, estimated with regard to the ship's position. In this sense an object so discovered, must be either ahead, astern, abreast, on the bow, or on the quarter; if a ship sails with a side wind, a distant object is said to be to leeward, or to windward, on the lee quarter, or bow, or on the weather quarter, or bow.

BEATING, (*louvoyer*, Fr.) the operation of making a progress at sea against the direction of the wind, in a zig-zag line, or traverse: beating, however, is generally understood to be, turning to windward in a storm, or fresh wind.

To BECALM, (*derober*, *abrier*, Fr. from *calme*, Dut.) to intercept the current of the wind, in its passage to a ship, by means of any contiguous object; as, a high shore, or some other ship to windward. At this time the sails remain in a state of rest, and are, consequently, deprived of their power to govern the motion of the ship.

BECALMED, implies also, that, from the weather being calm, and no wind blowing, the sails hang loose against the mast.

BECKETS, (*billes*, Fr.) any thing used to confine loose ropes, tackles, oars, or spars, in a convenient place: hence, beckets are either large hooks, or short pieces of rope, with a knot on one end and an eye in the other, or formed like a circular wreath; or they are wooden brackets; and, probably, from a corruption and misapplication of this last term, arose the word becket, which seems often to be confounded with bracket.

Put the tacks and sheets in the Beckets! the order to hang up the weather main and fore sheet, and the lee main and fore tack, to a small knot and eye-becket on the foremost main and fore-shrouds, when the ship is close hauled, to prevent them from hanging in the water.

BED, (*lit*, Fr.) a slop supplied from the store-keepers of the different dock-yards, in charge of the purser; and, by him, issued to the ship's company; which is charged against the men's wages, and deducted at the payment thereof. The price of a bed and blanket, as charged to the purser, is, at present, 1*l.* 4*s.* 3*d.*; and single beds are 15*s.* 9*d.* For the care of bedding and other slops served out, the purser is allowed a shilling in the pound.

Bed, *for the Casks*, a flat thick piece of timber, lodged under the quarters of casks containing any liquid, and stowed in a ship's hold. The use of the bed is, to support the casks, and keep the bulge, or middle-part of them, from bearing against the ship's floor, or against the body upon which they rest; lest the staves should give way and break in the place where they are weakest: or lie in a wet place, so as to rot in the course of the voyage.

Bed also signifies the place of the greatest diameter in bowsprits; also the main piece of barrel screws through which the puppets work.

Bed *of a River*, (*lit d'une rivière*, Fr.) the bottom of the channel in which the stream or current usually flows.

Beds, *for Cannon*, (*coussins de canon*, Fr.) are pieces of wood, having a stool at one end, which rests on the hind axle-tree; the other end on the bolt through the brackets, or sides of the carriage, for the breach of the cannon to rest on, to form a plane for the coin to move in, in order to elevate or depress the cannon at pleasure.

Beds, *for Mortars*, (*coussins de mortiers*, Fr.) are solid pieces of oak, fixed in the midship of bomb-vessels, to support the mortars. The mortar is hung by trunnions, the breech of which rests in the hollow of the bed, and may be moved round by the hand or tackle.

BEES, in ship-building, are pieces of elm plank bolted to the upper end of the bowsprit.

To BELAY, (*amarrer*, Fr. from *beleygen*, Belg.) to fasten a rope by winding it several times round a cleat, belaying-pin, or kevel: this term is chiefly applied to the running-rigging, there being several other expressions used for large ropes, as bitting, bending, making fast, stoppering, &c. which articles are explained in their due place.

BELAYING Pins, (*chevillots*, Fr.) are wooden pins, turned in a lathe, made of ash, 16 inches long, one inch and three eighths in diameter at the upper end, which is turned as a handle, are three sevenths the length, then shouldered into one inch and an eighth in diameter. Belaying pins are also made of iron, of various sizes.

BELFRY, (*montans de cloche*, Fr.) the shelter under which the ship's bell is suspended.

Strike the BELL! (*pique la cloche!* Fr.) the order to strike the clapper against the bell as many times as there are half hours of the watch elapsed; hence we say, it is two bells, three bells, &c. meaning there are two or three half hours passed.

BELLATRIX, in astronomy, a ruddy glittering star, of the second magnitude, in the left shoulder of Orion. It takes its name from the Latin *bellum*, as being anciently supposed to have great influence in kindling wars and forming warriors. Its right ascension in time adapted to the beginning of the year 1812, is 78° 45 45″,36, annual variation in right ascension 48″,13. Its declination 6° 10′ 11″ N. and annual variation in declination + 4″,0.

BELLY, the hollow part of compass timber; the round part of which is called the back.

BELTS, *Fasciæ*, in astronomy, two zones or girdles surrounding Jupiter's body, more lucid than the rest, and terminated by parallel lines; being sometimes broader and sometimes narrower, and not always occupying exactly the same part of the disc. Dark spots have frequently been observed on the *belts* of Jupiter; and M. Cassini observed a permanent one in the northern part of the most southern belt, by which he has determined the length of Jupiter's days, or the time of his revolution on his axis, which is nine hours and fifty-six minutes. Some astronomers imagine the *belts* to be seas, which alternately cover and leave bare large tracts of the planet's surface; and that the spots are gulfs in those seas, which are sometimes full and sometimes dry. But M. Azout rather conceives that the spots are protuberances of the belts; and others again are of opinion that the transparent and moveable spots are the shadows of Jupiter's satellites. The *belts* of Jupiter were first observed and described by Huygens, in his *Syst. Saturn*, p. 7.

Cassini also speaks of the *belts* of Saturn; being three dark, straight, parallel bands, or *fasciæ*, on the disc of that planet. But it does not appear that Saturn's *belts* adhere to his body, as those of Jupiter do; but rather that they are large dark rings surrounding the planet at a distance. Some imagine that they are clouds in the atmosphere of Saturn, though it would seem that the middlemost is the shadow of his ring.

Belts, in navigation, denotes certain streights near the sound, through which ships pass from the Baltic to the German ocean. The Belts belong to the king of Denmark, who exacts a toll, varying in its amount, and in some circumstances attending the collection of it, from all ships that pass through them, excepting those of Sweden, which are exempted.

BEND, (*avuste*, Fr. probably from *bindan*, Sax. to bind) the knot by which one rope is fastened to another, or to an anchor; hence,

To Bend (*nouer*, Fr.) is to fasten one rope to another, &c.

BENDING the Cable, (*étalinguer le cable*, Fr.) the operation of clinching, or tying the cable to the ring of its anchor.

Bending *a sail*, (*enverguer une voile*, Fr.) fastening it to its yard or stay. See the articles Sail, Stay, and Yard.

BENDS are of various kinds and denominations; such as the carrick-bend, the fisherman's-bend, the sheet-bend, &c.

Fisherman's Bend. To make a fisherman's-bend—With the end part of a rope take two turns around a
Plate II. Fig. 5. spar; a half-hitch round the standing part *b*, and under the turns *a*; then another half-hitch around the standing part *b*. This is used for bending the studding-sail halyards to the yard.

Fig. 6. *Carrick* Bend. To make a carrick-bend—Lay the ends of a rope *a* and *b*, then pass it through the bight, under *d*, and up through the bight *c*.

Sheet Bend. To make a sheet bend—Pass the end of a rope *a* through the bight of another rope *b*, then Fig. 7.
round both parts of the rope *c d*, and down through its own bight.

Bends are also the small ropes used to confine the clinch of a cable.

Bends, in ship-building (*préceintes*, Fr.) are the thickest and strongest planks in a ship's side, on which men set their feet in climbing up; they are more properly called wales, or wails. They are reckoned from the water, and are distinguished by the titles of the first, second, or third, bend. They are the chief strength of a ship's sides, and have the beams, knees, and foot-hooks, bolted to them.

BETELGUESE, or BETALGEUS, in astronomy, a fixed star of the first magnitude, in the right shoulder of Orion. Its right ascension in time, adapted to the beginning of the year 1812, is 86°, 14′ 52″; annual variation in right ascension 48″,6. Its declination 7° 21′ 34,″ north; and annual variation in declination, + 1″5.

BETWEEN-DECKS, (*l'entre-pont*, Fr.) the space contained between any two whole decks of a ship.

BEVELLING, (*enfoncé*, Fr.) in ship-building, the art of hewing a timber with a proper and regular curve, according to a mould which is laid on one side of its surface. Mr. Murray, in his Treatise on Ship-building, gives the following directions respecting bevelling:—

" In order to hew any piece of timber to its proper bevel, it will be necessary first, to make one side *fair* and out of winding—a term used to signify that the side of a timber should be a plane. If this side be uppermost, and placed horizontally, or upon a level, it is plain, if the timber is to be hewed square, it may be done by a plummet and line; but if the timber is not hewed square, the line will not touch both the upper and lower edge of the piece; or if a square be applied to it, there will be wood wanting either at the upper or lower side. This is called within or without a square. When the wood is deficient at the under-side, it is called *under-bevelling*; and when it is deficient in the upper side, it is called *standing-bevelling*; and this deficiency will be more or less, according to the depth of the piece, so that before the proper bevellings of the timbers are found, it will sometimes be very convenient to assign the breadth of the timbers; nay, in most cases, it will be absolutely necessary, especially afore and abaft: though the breadth of two timbers, or the timber and room, which includes the two timbers and the space between them, may be taken without any sensible error, as far as the square body goes. For, as one line represents the moulding-side of two timbers, the fore-side of the one being supposed to unite with the aft-side of the other, the two may be considered as one entire piece of timber."

By under-bevelling is also meant an acute angle; and by standing-bevelling, one that is obtuse.

BEVELLINGS, (*equerrage*, Fr.) are taken by an instrument peculiarly adapted to the purpose, called a *bevel*; it, in some measure, resembles a square, except that the tongue thereof is moveable at pleasure, so that it may be fixed at the angle or bevel at pleasure. To transfer it, or set it on the plank, the same operation is necessary as in the use of the square.

BIBBS, are brackets made of elm plank, and bolted to the hounds of the masts, for the purpose of supporting the trestle-trees.

BIGHT, (*balant d'un cordage*, Fr. *bygan*, Sax. to bend) the double part of a rope when it is folded, in contradistinction to the end: as, "her anchor hooked the *bight* of our cable," *i. e.* caught any part of it between the ends. "The *bight* of his cable has swept our anchor;" that is, the double part of the cable of another ship, as she ranged about, has entangled itself about our anchor.

Bight, (*anse*, Fr.) is also a small bay between two points of land.

BILANDER, (*bilandre*, Fr.) a small merchant vessel, with two masts. It is particularly distinguished from other vessels of two masts, by the form of her mainsail, which is bent to the whole length of a yard hanging fore and aft, and inclined to the horizon in an angle of about 45 degrees, and hanging immediately over the stern, while the fore end slopes downward, and comes as far forward as the middle of the ship; the foremost lower corner, called the tack, being secured to a ring-bolt in the deck, and the aftmost, or sheet, to another in the taffrail. At present, there are few vessels of this description.

BILBOES, (*fers de prisonniers, ou barres de prisonniers*, Fr.) long bars or bolts of iron, with shackles sliding on them, and a lock at the end, used, on some occasions, to confine the feet of prisoners, in a manner similar to the confinement of the hands in handcuffs. The offender is laid in irons, which are more or less ponderous, according to the nature of the offence of which he is guilty.

BILGE, or BILDGE, (*le plafond d'un vaisseau*, Fr. supposed from *bilik*, Sax. a storm) more properly written bulge, that part of a floor in a ship, on either side of the keel, which approaches nearer to an horizontal than to a perpendicular direction, and on which the ship would rest if laid on the ground; or more particularly, those parts of the bottom which are opposite to the heads of the floor-timbers amidships on each side of the keel: hence, when a ship receives a fracture in this place, she is said to be bilged.

Bilge, or Bulge, is also the largest circumference of a cask, or that which extends round by the bung-hole.

Bilge-*Water*, (*eaux qui se trouvent en dessous du vaigrage d'un bâtiment*, Fr.) the rain or sea-water, which occasionally enters the lower apartments of a ship, whence running down to the floor, it remains in the bilge of the ship till pumped out, by reason of her flat-bottom, which prevents it from going to the well of the pump, and is always (if the ship does not leak) of a dirty colour, and disagreeable smell.

BILL, (*bec d'un ancre*, Fr.) the point or extremity of the flook of an anchor: also the ends of compass, or knee-timber.

Bill *of Exchange*, in commerce and law, is an order or request in writing, addressed by one person to another, desiring him to pay a sum of money named in it to a third person, on his account, or to any other to whom that third person shall order it to be paid; or it may be made payable to bearer. In common speech, a bill of exchange is frequently called a *draft;* but the former is the more legal, as well as mercantile, expression.

If a bill is payable to bearer, it is assignable by delivery only; but if it is payable to order, it must be transferred by indorsement and delivery. The person who makes or draws the bill is called the *drawer*, and he to whom it is addressed is denominated the *drawee;* who, when he has undertaken to pay the amount, is termed the *acceptor*. The person in whose favour the bill is drawn, is the *payee;* but if he appoints another person to receive the money, he is then termed the *indorser*, and the person so appointed the *indorsee;* and any one who happens for the time to be in possession of the bill, is called the *holder* of it.

Bills of exchange are either *foreign* or *inland: foreign* when drawn by a merchant residing abroad upon his correspondent in England, or *vice versa;* and *inland*, when both the drawer and the drawee reside in the kingdom. By 9 and 10 W. III. c. 17. and 3 and 4 Anne, c. 9, all distinctions between foreign and inland bills are removed, so that what was the law and custom of merchants, with regard to the one, is now, in most respects, the established law of the country, with regard to the other.

A *Promissory Note*, or note of hand, is a plain and direct engagement in writing, to pay a sum specified at the time therein limited, to a person therein named, or sometimes to his order, or to the bearer at large: this is also made assignable, and indorsable like a bill of exchange. Any persons capable of binding themselves by a contract, may be parties to a bill of exchange, or other negotiable instrument, or be in any manner concerned in negociating either of them.

Where a bill is payable at so many days after sight, or from the date, the day of presentment, or of the date, is excluded. Thus, where a bill payable ten days after sight is presented on the first day of a month, the ten days expire on the eleventh, &c. Ld. Raym. 281. Stra. 829.

It is a custom among merchants, that a person to whom a bill is addressed, shall be allowed a few days for payment, beyond the time mentioned in the bill, called *days of grace*. In Great Britain and Ireland, three days are allowed; in other places, more. If the last of these three days happens to be on Sunday, the bill is to be paid on Saturday; but these days of grace are not allowed on bills payable at sight. If bills become due on Sunday, or on such holidays, when the law forbids business to be done, payment must be demanded, or protest made for non-payment on the preceding day.

Bills of exchange and promissory notes must now be drawn on stamped paper; and the stamp is proportioned under stat. 31 Geo. III. c. 25. 37 Geo. III. and 41 Geo. III. c. 10, to the amount of the bill from sixpence to three shillings for such as are payable on demand; and for those payable after date from one shilling to four shillings. If foreign bills are drawn here, the whole set must be stamped; but bills drawn abroad are not liable to any stamp duty.

A *Protest* is an act of a notary public, stating that a bill has been presented for acceptance, or for payment, and refused, and declaring that the acceptor, indorsers, &c. shall be liable for damages, &c. and to this protestation all foreign courts give entire credit.

In the first place, the notary marks or notes the minute of refusal on the bill itself, (this, in common language, is termed the *noting* of the bill) and afterwards the instrument is drawn out and attested under his hand and seal. If no such notary be resident in the place where the bill is negotiated, protest may be made by any substantial inhabitant, in the presence of two credible witnesses, and should in general be made at the place where payment is refused.

Bill *of lading*, (*connoissement*, Fr.) an acknowledgment signed by the master of a ship, and given to a merchant, containing an account of the goods, which the former has received from the latter, with a promise to deliver them at the intended place for a certain sum of money. Each bill of lading must be treble; one for the merchant who ships the goods, another to be sent to the person to whom they are consigned, and the third to remain in the hands of the master of the said ship. It must, however, be observed, that a bill of lading is only used when the goods sent on board are but part of the cargo; for when a merchant loads a vessel entirely on his own account, the deed passed between him and the master of the ship is called *charter-party*: in this case, the bills of lading are generally signed. See Charter Party.

Bill *of store*, a license granted at the custom-house to merchants, by which they have liberty to carry, custom-free, all such stores and provisions as they may have occasion for during their voyage.

Bill *of sufferance*, a license granted at the custom-house to a merchant, suffering him to trade from one English port to another, without paying custom.

BINDING, (*fer de cap de mouton*, Fr.) the iron wrought round the dead-eyes.

Binding-*Strakes*, (*virures*, Fr.) are two strakes of oak in the deck, generally wrought near the coamings, and worked all fore and aft: they are about one inch or one inch and a half thicker than the rest of the deck, and let down between the beams and ledges, so that the upper side shall be even with the rest of the deck. The design of them is to strengthen and bind the deck so well together, as to prevent its drawing.

BINNACLE, formerly called Bittacle, (*habitacle*, Fr.) a wooden case or box, which contains the compasses, log-glasses, watch-glasses, and lights to shew the compass at night.

Plate II fig. 8. The binnacle (plate II. fig.8.) is furnished with three apartments, with sliding shutters: the two side ones, *a b*, have always a compass in each, *d*, to direct the ship's way, while the middle division, *c*, has a lamp or candle, with a pane of glass on either side, to throw a light upon the compass in the night, whereby the man who steers may observe it in the darkest weather, as it stands immediately before the helm on the quarter-deck.

There are always two binnacles on the deck of a ship of war, one being designed for the man who steers, and the other for the person who superintends the steerage, whose office is called *conning*.

For improvements in this article, see Compass.

BIRTH, (*evitée*, Fr.) the station in which a ship rides at anchor, either alone or in a fleet; as, *she lies in a good birth*, i. e. in a good anchoring ground, well sheltered from the wind and sea, and at a proper distance from the shore and other vessels.

Birth, (*poste*, Fr.) also signifies the room or apartment where any number of the officers or ship's company mess and reside. In a ship of war, there is commonly one of these between every two guns.

To Birth *a ship's company*, (*billeter l'équipage*, Fr.) is to allot to each man the space in which his hammock is to be slung to the beam, or tracing-batten.

BIRTHING, the working a top-side, bulk-head, &c.

BISCUIT, Sea, (*biscuit*, Fr.) is a sort of bread much dried, to make it keep for the use of the navy, and is good for a whole year after it is baked.

The process of biscuit-baking for the navy is simple and ingenious, and is nearly as follows. A large lump of dough, consisting merely of flower and water, is mixed up together, and placed exactly in the centre of a raised platform, where a man sits upon a machine, called a horse, and literally rides up and down throughout its whole circular direction, till the dough is equally indented, and this is repeated till the dough is sufficiently kneaded.

In this state it is handed over to a second workman, who, with a large knife, puts it in a proper state for the use of those bakers who more immediately attend the oven. They are five in number; and their different departments are well calculated for expedition and exactness.

The first man on the farthest side of a large table moulds the dough, till it has the appearance of muffins, and which he does two together, with each hand; and then delivers them over to the man on the other side of the table, who stamps them on both sides with a mark, and throws them on a smaller table, where stands the third workman, whose business is merely to separate the different pieces into two, and place them under the hand of him who supplies the oven, whose work of throwing or chucking the biscuits on the peel must be performed with the greatest exactness and regularity. The fifth arranges them in the oven, and is so expert, that though the different biscuits are thrown to him at the rate of seventy in a minute, the peel is always disengaged in time to receive them separately.

So much critical exactness and neat activity occur in the exercise of this labour, that it is difficult to decide whether the palm of excellence is due to the moulder, the maker, the splitter, the chucker, or the depositor; all of them, like the wheels of a machine, seeming to be actuated by the same principle. The business is to deposit in the oven seventy biscuits in a minute; and this is accomplished with the regularity of a clock; the clack of the peel, during its motion in the oven, operating like the pendulum. The biscuits thus baked are kept in repositories, which receive warmth from being placed in drying lofts over the ovens, till they are sufficiently dry to be packed into bags, without danger of getting mouldy; and when in such a state, they are then packed into bags, of an hundred weight each, and removed into store-houses for immediate use.

At Deptford the bake-house belonging to the victualling-office has twelve ovens; each of which bakes twenty shoots daily; the quantity of flour used for each shoot is two bushels, or 112 pounds; which baked, produce 102 pounds of biscuit. Ten pounds are regularly allowed on each shoot for shrinkage, &c. The allowance of biscuit in the navy is, one pound for each man per day; so that, at Deptford alone, they can furnish bread, daily, for 24,480 men, independent of Portsmouth and Plymouth.

To BITE, (*mordre*, Fr.) is said of the anchor when it holds fast in the ground.

BITS, or BITTS, (*bittes*, Fr. *bitol*, Sax.) a frame composed of two strong pieces of timber, fixed perpendicularly in the fore-part of a ship, whereon to fasten the cables as she rides at anchor.

These pieces, being let down through square mortises, cut in the decks above and below, are bolted and fore-locked to the ship's beams. There are several bits in a ship, the principal of which are those for the cables: their upper ends commonly reach about four or five feet above the lower deck, over which the cable passes. They are supported on the fore part by strong standards; one arm of which is bolted to the deck, and the other to the bits: and on the after-part is fixed a strong beam of timber, parallel to the deck, and at right angles with the bits, to which it is bolted and fore locked. The ends of this beam, which is called the cross-piece, reach about two or three feet beyond the bits, whose upper-ends are nearly two feet above the *cross-piece*. The cable being passed once round about these bits, may be gradually slackened at pleasure; without which, it would be impossible to prevent it from running out with the utmost rapidity when the ship rides a great strain, which is always the case in a storm, or an impetuous tide. In ships of war there are usually two pair of cabl bits, and when they are both used at once the cable is said to be double-bitted. There are also the pawl-bits, carrick-bits &c.

To Bit *the Cable*, (*bitter le cable*, Fr.) is to put it round the bits, in order to fasten it, or slacken it gradually, which last is called *veering away*.

The other bits are of a smaller kind, but constructed nearly in the same manner, and are used to fasten the top-sail-sheets, or the ropes by which the lower corners of the top sails are extended.

BITTER, (*bitture*, Fr.) implies the turn of the cable that is round the bits, in order to its being veered out by little and little at pleasure. A ship stopped by her cable, is said to be brought up to a bitter.

Bitter-*end*, is that part of the cable which is abaft the bits, and therefore within board when the ship rides at anchor. They say, "bend to the bitter-end!" when they would have that end bent to the anchor.

BLACK-STRAKES, (*bandes*, Fr.) a range of planks immediately above the wales in a ship's side; they are always covered with a mixture of tar and lamp-black, which preserves the plank itself, and forms an agreeable variety with the white bottom beneath, and the scraped planks of the side, covered with melted turpentine, or varnish of pine, above. All the yards are likewise daubed with this mixture, which not only preserves them from the heat of the sun and the weather, but gives them a fine gloss, which makes a good appearance, contrasted with the white varnish on the masts.

BLACKSTRAP, a fictitious name, given by our sailors, to that kind of Mediterranean wine with which the ships are supplied on that station; and which, after the grog and wine usually served, they cannot, for a while, relish: hence, to be driven above Gibraltar, is, as they call it, to be *black-strapped*.

BLADE, that part of the arm of the anchor on which the palm is shut.

Blade *of an Oar*, (*plat ou pelle d'un aviron*, Fr.) is the flat part of it, which they plunge into the water in rowing. The force and effect, in a great measure, depend on the length of this part.

BLINK of the Ice, (*blancheur de l'horison occasionnee par les glaces*, Fr.) is that dazzling whiteness about the horizon, which is occasioned by the reflection of light from fields of ice. This is mostly used as a term amongst Greenland ships.

BLOCK, (*poulie*, Fr.) in mechanics, is the name given to a pulley, or system of pullies, mounted in a frame or shell, but considering them as detached from the ropes which run through them. When speaking of the block with its rope, the seaman uses the phrase of a tackle of single or double blocks; hence the term block is applied to the pulley or pullies, with its frame or shell, and its band or strap. Blocks are used for various purposes in a ship, either to increase the mechanical power of the ropes employed in contracting, dilating, or traversing the sails, or to arrange the ends of them in certain places upon the deck, that they may be readily found when wanted. The blocks, which are, for these purposes, disposed in various places upon the masts, yards, and sails, and amongst the rigging, are of various sizes, shapes, powers, and denominations, according to the effect they are calculated to produce; and, consequently, applicable to a multitude of different purposes. Every block is composed of three, and, generally, four parts, viz.

1. The shell, or outside.
2. The sheave, or wheel, on which the rope runs.
3. The pin, or axle, on which the sheave turns.
4. The strap, or part by which the block is made fast to any particular station, and is generally made either of rope or iron.

The single block contains only one sheave, or wheel; a double block, as may be naturally inferred, has two sheaves, one below the other; and so on.

The shells of very large blocks are made of separate pieces of wood, as the cheeks of the shell, its partitions, &c. These are called *made blocks*. The shell is formed of several pieces of elm plank, suited to the thickness of the cheeks, sheave-holes, and partitions, and is strongly bolted together by three bolts at each end, driven through and rivetted, with a washer, at the points.

Blocks are again divided into thick and thin blocks; the former being intended to receive large ropes, and the latter small ones. The following may serve as a general idea of the mode of making any of the common kind of blocks in the old way, before the introduction of the block-machinery in our dock-yards:—

The shells being sawn to their length, breadth, and thickness, the corners or angles are sawn off. The workman then gauges, or marks out, the size of the sheave-hole in the middle, one-sixteenth larger than the thickness of the sheave, and once the thickness longer than the diameter, for a single-sheaved block. In blocks of two sheaves, the partition is kept in the middle, and is one-sixth less than the sheave-hole; each sheave-hole is gauged on the two opposite sides, and in the same manner for blocks with a great number of sheaves. The blocks are then jambed up edge- Plate II. Fig. 9.

wise, with wedges, in a clave, or frame, and the sheave-holes are made in this manner: the length and breadth are first gauged out, and holes are bored, half way through the block, along the part gauged out, with an auger of the size of the sheave-hole; then the sheave-hole is gauged, and bored on the opposite side, in the same manner, so as to meet the opposite holes. Blocks from ten inches and upwards have one hole bored at each end, and cut through with a chisel, and the wood is sawed out with a rib-saw. All blocks have the sheave-holes cleared out by chisels; and by burrs, at the corners. Blocks that are to have iron straps should have the straps fitted on before the wood is cut out of the middle. The hole for the pin is bored through the middle of the block, one-tenth less than the diameter of the pin. The outsides and edges of the shell are next rounded off by the stock-shave, and neatly finished by the spoke-shave. In the royal navy, blocks are left thick upon the edges of the cheeks, but, in merchant-ships, the edges are sometimes thinned off to a small square, and somewhat rounded off.

The scores, which are the grooves to receive the strap, are gauged out along the outside of the cheeks, and tapered in depth, from nothing at the pin, to half the thickness of the strap, at the ends of the block, for a single score; and the same, on each side of the pin, for double scores; which are made when the block is to have double straps. The scores are gauged down across the breast of the block, to half the size of the strap, in order to allow for the serving. After the score is cut, the sheaves are fitted; they are one-tenth thicker than the diameter of the rope intended for running on them, and five times that thickness in diameter. The hole for the pin should be bored through the centre of them by a bit, fixed in the mandrel of a turning lathe, or with a stock and bit and opened out with an auger, one sixteenth larger than the pin, that it may easily turn. They are then put in a lathe, and turned smooth, and the outer circumference hollowed one-third of its thickness, that the rope may embrace it closely.

The diameter of the pin is the thickness of the sheave, and is turned in a lathe, except its head, which is left octagonal, to prevent its turning in the block; and the pin is driven through the holes in the block and sheaves. After the sheaves are fitted, the inside of the sheave-hole, at one end of the block, is gauged hollow, to admit the rope, and correspond with the sheaves; and a small neat chamfer is taken off the edges. The proportions for single, double, treble, fourfold, and other blocks, are, the length is eight times the breadth of the sheave-hole, which is one-sixteenth of an inch more than the thickness of the sheave; the thickness of the sheave is one-tenth more than the diameter of the rope it is intended for; and the diameter of the sheav is five times the thickness. The breadth of the block to be six times the thickness of the sheave, and the thickness to be one half the length, or nearly so. Flat thin blocks must be three-eighths of the length thick; but all blocks, having more than one sheave, increase their thickness more, in the above proportion, by the additional number of sheave-holes, and middle parts, or partitions; the thickness of each partition to be one-sixth less than the breadth of the sheave-hole.

These are the general dimensions, but they sometimes vary, according to the use they are intended for.

In the shell is fixed the Sheave, or Sheaves, which are circular pieces of wood (usually lignum vitæ), or sometimes brass, or cast metal, with a groove, turned on its edge, for the reception of the rope; and, in the best blocks, called coaked sheaves, the sheave has a brass bush fitted in the centre, with a round hole through it, to receive the pin. Plate II. Fig. 11.

The pin, or axis, on which the sheave revolves, is made of lignum vitæ, cocus, or a West Indian wood called green-heart; but the best blocks have iron pins. The pin is supported by passing through the sides of the shell of the block, which is made of elm, ash, or other tough wood, with a hole mortised through it, to receive the sheave, and confine it to revolve steadily, though freely, upon its pin, and, at the same time, keep the rope from getting off the groove in the edge of the sheave. When the block contains two or more sheaves, as many mortises are made. Sometimes the same pin serves all the sheaves; and, at other times, the sheaves are placed one above the other, having, of course, separate pins. Fig. 10.

The Strap is a rope, or, in some cases, a band of iron, encompassing the shell of the block, in a notch, or scoring, cut round the block to receive it. The strap terminates in an eye of rope, or hook of iron, by means of which one of the blocks of a tackle is attached to the object upon which it is to act as a mechanical power, while the other block is suspended from some fixed support. The former is called the running-block; and the latter, the standing block. Fig. 12.

The following articles will explain some of the different kinds of blocks used in shipping:—

The single Block, (*poulie simple*, Fr.) appears like Fig. 13, Plate II. Fig. 13.

The double Block, (*poulie double*, Fr.) like Fig. 14. Fig. 14.

The treble Block, (*poulie à trois rouets*, Fr.) like Fig. 15. Fig. 15.

The fourfold Block, (*poulie à quatre rouets*, Fr.) like Fig. 16. Fig. 16.

Bee Blocks are generally made of elm; they are seven-ninths the length of the *bee*, and two inches in depth, for every foot in length, and seven-eighths that depth in the thickness. They are trimmed square, chamfered on the outside edges, and fitted with a sheave in one end: a hole is cut in the other end, to be fitted with a sheave, should the other fail. The sheave-hole is two-sevenths the length of the block, and one-fourth the length of the sheave-hole in breadth, and half the length of the sheave-hole within the end. Bee-blocks are bolted to the bow-sprit at the outer end under the bees, and the bolts serve as the axis, or pin, for the sheave to work on: the fore-topmast-stay reeves through the sheave-hole at the foremost end of the starboard bee-block; and the fore-topmast-preventer, or spring-stay, through the sheave-hole at the after end of the larboard bee-block. Fig. 17.

Bull's-eye, is a kind of wooden thimble, with a hole in the centre, and a groove in the circumference. Fig. 18.

Cut-Block, (*poulie de capon*, Fr.) is two or three-fold, with an iron strap and a large hook to it, and is employed to draw the anchor up to the cathead. Plate III. Fig. 1.

Cheek Blocks, are half shells, bolted against the masthead; the chief bolt serves for the pin of the sheave; they receive the halyards and stays of their respective masts.

Clew-garnet Blocks, (*poulies de cargue-point*, Fr.) have single sheaves, and are suspended from the yards by a strap with two eyes; a lashing surrounds the yard, and passes through the eyes, so as to suspend the block beneath the yard: these blocks receive the clew-garnets, or ropes, which haul up the clews of the sail; this is applied to the main and fore-yard.

Clew-line Blocks, (*poulies de cargue-point*, Fr.) are for the same purpose as the preceding, but applied to the top-sails, top-gallant, and sprit-sails. A great improvement has lately been made in these blocks, by Mr. Brunell, inventor of the block machinery at Portsmouth. The old clew-line and clew-garnet blocks (for they are the same, except in size), were a single sheave block, strapped with two eyes, as above; a knot was made in the end of the clew-line, or garnet, just at the place where it was attached to the clew of the sail. to prevent the corner thereof being drawn into the block. This was not effective, and frequent inconvenience arose; for the sail being so constantly in motion, the rope had a great tendency to get entangled with the sail, and drawn over the sheave. The improved block in question is shewn in Plate III. Fig. 2. The two holes at *a a*, are where the rope goes in and out again. The sheave is situated in the centre of the block, so as to be wholly inclosed, except a mortise at *b*, where the sheave is put in. The strap surrounds the lower part of the block; then both ends pass through a hole in the upper part about *c*, crossing each other. They are then formed into an eye, by which the block is suspended from the yard. By this means no accident can happen, as the garnet, or rope, is so inclosed in the block, that it cannot be deranged by any violence, nor the sail be drawn into the block. (Plate III. Fig. 2.)

D Blocks are lumps of oak, in the form of the letter D, from 12 to 16 inches long, and eight or ten feet wide, they are bolted to the ship's side in the channels, to receive the lifts, &c.

Fig. 3. *Dead-eye*, is a large circular piece of wood, having a groove in its circumference, for a shroud to lie in. The three holes, or eyes, are for a lanyard to reeve through.

Deep sea-line Block, is a small wooden snatch-block, about from nine to eleven inches long.

Fig. 4. *An Euphroe*, is a long piece of wood, having a number of holes through which the crow-foot, for the awning, &c. is reeved.

Fish Block, (*poulie de candel tte*, Fr.) is hung in a notch at the end of the davit, and serves to haul up the flooks of the anchor to the ship's bow.

Girt-line Blocks, in rigging the fore-mast, main, and mizen-mast, are lashed round the mast-head, above the top of the cap; one to hang on each side. The girt lines, that reeve through them, lead down upon deck, for hoisting the rigging, tops, and cross-trees, and the persons employed to place the rigging over the mast-head.

Fig. 5. *A Heart*, (*grosse moque à un trou*, Fr.) is a block of wood, with a large hole in the centre, at the bottom of which are four or five scores round the outside, a groove is cut, to admit of a rope, called a *stay*, *&c.*

A Heart for a Collar, is sometimes open at the lower end, opposite to which the lanyard is passed. It has a groove on each side *k*, for the seizing to lie in. Fig. 6.

A Heart, is often used in the merchant service, with a round hole *l*, for the heel of the jib-boom to rest in, which is bevelled for that purpose. The bottom of the heart *m*, is also bevelled according to the steeves of the bow-sprit. Fig. 7.

Iron strapped Blocks, frequently have the hook working in a swivel, in order to turn it, that the several parts of the rope, of which the tackle is composed, may not be twisted round each other, which would greatly diminish the mechanical power.

Jear Blocks, (*poulies de drisse à caliorne*, Fr.) are twofold or treble; applied to hoist, or lower, the main, or fore-yards.

Jewel Blocks, two small blocks, which are suspended at the extremity of the main and fore-topsail-yards, by means of an eye-bolt, driven from without, into the middle of the yard-arm, parallel to its axis. The use of these blocks is, to retain the upper part of the top-mast studding-sails beyond the sheets of the top-sails, so that each of these sails may have its full force of action, which would be diminished by the encroachment of the other over its surface. The halyards, by which those studding-sails are hoisted, are passed through the *jewel-block*, whence, communicating with a block on the top-mast-head, they lead downward to the top, or decks, where they may be conveniently hoisted. See the article Sail.

Long-tackle Block, has two single sheaves placed one above the other, in the same shell. The lower sheave is only two-thirds the size of the other; it is used in combination with a common single block, to form the long tackle, for loading, or any other purchase. In the royal navy and East India service they are used as yard tackles. The rope is reeved through it in the same manner as it would be through a common double block; but it is preferred, where it is convenient, because the strap being in the centre of the resistance, it hangs more steadily than when the sheaves are on one pin. Fig. 8.

Main-sheet Block, is used for the sheet tackle of the main-sail-booms of small vessels. The pin projects from each side of the block, being, in all, the same length as the block; the fall or rope of the tackle is belayed or twisted round this pin, to stop it. This block is either single or double, and has a hole through the end, to receive its strap.

Monkey Blocks, are sometimes used on the lower yards of small merchant ships, to lead (into the mast, or down upon the deck) the running rigging belonging to the sails. Some are only small single blocks, attached, by a strap and iron swivel, to iron straps, which embrace and nail to the yard, the block turning to lead the small running ropes in any direction; others are nearly eight square, with a roller working in the middle, and a wooden saddle beneath, to fit and nail to the yard.

Nine-pin Block, is shaped something like the pin from which it derives its name, though flattened on the sides, and used to lead the running ropes in an horizontal direction. Their lengths are generally Fig. 9.

confined to the place in which they are fixed; and this is, for the most part, under the cross pieces of the fore-castle and quarter-deck bits. The breadth of the block, sheave, &c. is governed by the rope, and taper at the ends, to three-eighths of the breadth of the middle; the pins at each end, serving as a vertical axis, is two-thirds of the size of the end, and the thickness is five-eighths of the breadth. These blocks may be turned in a lathe, and flattened after with a spoke-shave.

Rack Blocks, are a range of small single blocks, made from one solid piece of wood, by the same proportion as single blocks, with ends in form of a dove's tail, for the lashing, by which they are fastened athwart the bowsprit, to lead in the running-ropes. They are seldom used.

Plate III. Fig. 10. *Shoe* Blocks, are two single blocks, cut in a solid piece, transversely to each other, so that the sheave of the upper one lies in a contrary direction to that of the lower one; they serve for legs and falls of the bunt-lines, but are seldom used.

Fig. 11. *Shoulder* Block, is a large single block, left nearly square at the upper end, and cut sloping in the direction of the sheave. These blocks are used on the lower yard-arms, to lead in the topsail sheets; and, on topsail-yards, to lead in the top-gallant-sheets; and, by means of the shoulder, are kept upright, and prevent the sheets from jambing between the block and the yard; they are also used at the outer end of the boomkins, to lead in the fore-tackle.

Fig. 12. *Sister* Blocks, are similar to two single blocks, and formed out of a solid piece, about 20 inches long, one above the other. Between the blocks is a scoring for a middle seizing; a round head is turned at each end, and hollowed underneath, to contain the end seizings; along the sides, through which the pins are driven, is a groove or scoring, large enough to receive part of the topmast shrouds, in which it is seized. These blocks receive the lifts and reef-tackle pendants of the topsail-yards.

Fig. 13. *Snatch* Block (*poulie à dent*, Fr.) has a single sheave, with a notch cut through one of its cheeks, to admit the rope, or fall, to be lifted in and out of the block, without putting its end through first. See a figure of this in Plate III. Fig. 13. The strap does not, in this, surround the block, but is put through a hole, bored through the divided end. The figure is represented with two tails, which may be made up for a hook, a thimble, or eye, according to the situation where it is to be used, which is, generally, for the main, or fore-sheet, blocks of square-rigged vessels.

The snatch blocks are iron bound, terminating at the notched end of the block, with a swivel-hook, or an eye-bolt, large enough to receive several turns of lashing, which fastens the block to its fixed support. That part of the strap over the notch, in the side, lifts up with a hinge, and is confined down, when the rope is in the block, by a small pin, put across, through the end of the pin of the sheave, which projects up from the block sufficiently to pass through an eye made in the hinge part of the strap. The strap on the other part of the block is let into the block, and confined by the pin and some nails. These blocks are used for heavy purchases, where a warp, or hawser, is brought to the capstan.

Fig. 14. See Plate III. Fig. 14.

Spring Block (*poulie à ressort*, Fr.) a new kind of block, invented by Francis Hopkinson, Esq. of Philadelphia, and designed to assist a vessel in sailing, by increasing the acting spring of her rigging. It is proposed to apply to all such parts of the rigging as will admit of it with safety and convenience, and where its operation will be most advantageous; but particularly to the sheet-ropes; and, if practicable, to the dead-eyes, in lieu of what are called the chains. A (Plate III. Fig. 15) is a block made in the usual Fig. 15. manner, having a ring, or eye, B, at one end; *c* is a spiral spring, linked at one end, to the hook D E, and, at the other, to the ring F, which is to be annexed to an eye-bolt at the timber-head, or by some other means, to the place where it is to be applied. The spring *c* must be of well tempered steel, and proportioned in strength to the service it is to perform. When used, two of these blocks are employed, one attached to the timber-head, and the other to the sail. Within the cavity, or pipe, formed by the spiral spring, there must be a chain of suitable strength, called a check-chain (represented in the centre of the spring) connected by the links to the hook D E, and ring F. When the spring is not in action, this chain is slack; but when the spiral spring is extended by the force of the wind, as far as it can be without danger of injury, the check chain must then begin to bear to prevent its further extension, and, if strong enough, will be an effectual security against failure. The inventor of this machine apprehends, on good grounds, that a vessel, thus furnished, will be less liable to heel; and that she will receive the impulses of the wind to better advantage, and sail with a more lively and equable motion, than if rigged in the common way.

It is, however, to be feared, that the weight of a spring, sufficiently strong to have any effect on a large sail, would be very unmanageable, particularly aloft, from its weight, and would also be very expensive.

There is no doubt, if this, or a similar contrivance, could be applied to a sail, it would have a good effect: for instance, when a ship is sailing with a certain rate if a sudden blast of wind comes, but does not continue, it will not advance the vessel at all, because it does not continue long enough to communicate an increased momentum to so large a mass as a ship, though, at the same time, it may make her heel or pitch violently. If her rigging is fitted with these springs, she would receive the impulse of the same blast, in a regular and progressive manner, which would tend to increase her velocity, instead of causing her to heel

Strap-bound Blocks, are single blocks, with a shoulder left on each side, at the upper part, to admit the strap through, a little above the pin. These blocks are used at the clews of the quarter-sails, for the clew-garnets, or clew-lines; and, under the yards, the shoulder preserves the strap from chafing.

Thick and thin, or, *Quarter* Block, is a double block with one sheave thicker than the other, and is used to lead down the topsail-sheets and clew-lines. Although these are used for topsail-sheets, and intended for the clew-lines, a single block would be cheaper and better, as the thin sheave is seldom used for the clew lines, it being found rather to impede than to facilitate.

Top Block, (*poulie de guinderesse*, Fr.) is a large single iron-bound block, with a hook, by which it

Plate III. Fig. 16. is hung to an eye-bolt in the cap, for reeving the top rope pendant through, when swaying up, or lowering down, the topmasts.

Fig. 17. *Voil*, or *Voyal* Block, (*poulie de tournevire*, Fr.) is a single sheave block. The length is ten times the thickness of the sheave-hole, which is three-eighths more than the thickness of the sheave; the thickness of the sheave is one-tenth more than the diameter of the voil; and the diameter of the sheave is seven times the thickness. The breadth of the block should be eight times the thickness of the sheave; and the thickness two-sevenths of the length. This block is double-scored, the sheave is coaked with brass, and the pin is iron, and nearly as thick as the sheave. It is used in heaving up the anchor. The voil passes round the jear capstan, and through the block, which is lashed to the main-mast, and the cable is fastened, in a temporary manner, to the voil in several places. It is seldom used, except in the largest ships of the royal navy.

The various blocks lashed to the ship's principal yards, are as follow:—

To the *lower yards*.—The *jear* block, *bunt-line* blocks, *leech-line* block, *lift* blocks, and *topsail-sheet* blocks, strapped together; *quarter* and *slab-line* blocks, strapped together; *clew-garnet* blocks, *tricing* blocks, *preventer-brace* blocks; *pendant* blocks, *studding-sail-halyards* blocks.

To the *topsail-yards*.—*Buntline* and *tye* blocks, strapped together; *top-gallant-sheet* block, and *lift* block, strapped together; *jewel* block, and *brace-pendant* blocks, *clew-line* blocks, and block to lead down the *top-gallant sheets*.

To the *top-gallant-yards*.—*Jewel*, *clew-line*, and *brace-pendant* blocks.

To the *mizen-yard*.—*Jear* block, *derrick* block, *signal-halyard* block, *throat-brail*, *middle-brail*, and *coak-brail* blocks.

To the *cross-jack-yard*.—*Quarter* blocks, *jear* blocks, and *lift* and *topsail-sheet* blocks, strapped together.

To the *bowsprit*.—The *bee* block, bolted to the bowsprit at the outer end, under the bees; *fore-bowline* blocks, lashed on each side the fore-stay-collar; *fore-topsail-bowline* block, lashed to an eye-bolt in the bowsprit-cap.

The late Mr. Walter Taylor, of Southampton, obtained a patent in the year 1781, for an improvement he made in the construction of sheaves or pullies for ship blocks, by which they are much better adapted for the rigging of ships and other purposes, than any hitherto in use: and the patentee, for many years afterwards, enjoyed the contract for supplying the royal navy with blocks, which he manufactured at Wood-Mills, on the river Itchen, near Southampton, where he had an extensive manufactory, the machinery of which was wrought by water.

It has been an old standing maxim with many, that large thin sheaved, and small running ropes, well housed in the blocks, are of the greatest use, though there may be occasion for renewing them more frequently; and running cordage, having a moderate quantity of tar in the yarn, is the means of keeping them soft and limber in a northern or cold climate, which causes them to bend and run through the blocks with greater facility.

Block Machinery.

The immense number of blocks employed in the navy, and the great importance of having them accurately and substantially made, in order to insure their ready performance in every unfavourable situation which the rigging of a ship is exposed to in bad weather, induced government, in the year 1802, on the recommendation of General Sir Samuel Bentham, to attend to the suggestion of Mr. Brunell, an ingenious mechanic, who had, at that time, invented a complete set of machines, and who made an offer to government of superintending their direction and application for a given time, at a per centage on the profits, for manufacturing every part of ship blocks; consequently, in the year 1803, he employed Mr. Maudslay to erect, from his designs, an extensive suit of machinery, in the dock-yard, at Portsmouth, for the fabrication of these articles.

These machines were set to work in 1804, and have been in constant use ever since; and they make all the blocks now used in the navy, which may be said to be not less than 200 sorts and sizes. They consist of 44 machines, which form three sets, that is, three blocks of different sizes, may be proceeding in all their stages at the same time, though, in some of these stages, one machine may operate upon two, or even ten, blocks at the same time. This machinery performs the several operations from the rough timber to the perfect block, in the completest manner possible, and the whole is set in motion by means of a capital steam-engine, of 32 horses' power, erected by Messrs. Boulton and Watts. The manual labour now required is, simply to supply the wood as it is wanted, and to remove the blocks from one part of the machinery to another, till they are perfectly finished, so that a common labourer may be made to act in this business with very little instruction. The block-house is an extensive building, divided into three parts. The steam-engine occupies the centre. One of the wings contain several sawing machines, for converting the timber into scantling, of the proper size, to form the different blocks. The opposite wing contains 37 small engines for forming, polishing, and finishing the blocks and their sheaves.

It will give some idea of the expedition of these works, to state the number of block shells of different sizes, made by each set of machines in a day.

The first set of machines make blocks from 4 to 7 inches in length, at the rate of 700 per day. These have wooden pins.

The second set from 8 to 10 inches in length, at the rate of 520 per day. These have iron pins.

The third set from 11 to 18 inches in length, at the rate of 200 per day, with iron pins. So that all the machines will make 1420 per day.

This is found to be such an immense saving to government, that, it is said, they have offered to give the inventor a considerable sum, over and above the original contract, and to allow him a handsome salary for occasionally superintending the machinery in the dock-yard.

Blocks also signify pieces of fir timber, placed under a large mast while making, in order to raise it from the ground: they are from 7 to 9 feet long, and about 10 inches square: also solid pieces of timber placed under the keel of a ship, the upper pieces are generally free from knots, especially the foremost ones, these being the *splitting blocks* for launching a ship, and should therefore be as clear as possible.

Mr. Seppings, Builder, of Chatham-yard, has lately invented an ingenious and useful contrivance for driving out blocks from under a ship's keel, when in dock, in order to shift the main keel, put on a false one, instead of lifting the ships as heretofore, at a great expence of money, time, and labour, and which is now adopted in all our dock-yards.

Block *and* Block, is the situation of a tackle when the two opposite blocks are drawn close together, so that the mechanical power becomes destroyed, till the tackle is again *over-hauled*, by drawing the blocks asunder.

BLOCKADE, (*blocus*, Fr.) in the art of war, is the blockading up a place by posting troops at all the avenues leading to it, to keep supplies of men and provisions from getting into it; and by these means proposing to starve it out, without making any regular attacks.

To Blockade *a Sea-port*, is to place ships to cruise before the entrance, so as to prevent any vessels from going in or coming out.

During the present war with France, a regular system of *blockading* all the enemy's ports has been practised with great success by our navy, which not only prevents their ships from getting out and forming a large fleet, but also hinders them from training up seamen and marines; and may therefore be considered, in a great measure, the preservation of the British empire.

To raise a Blockade, is to force the troops or ships, that keep the place blocked up, from their respective stations.

BLUFF, (*morne*, Fr.) an high land, projecting almost perpendicularly into the sea.

Bluff-*bow*, (*avant joufflu*, Fr.) A vessel is said to be bluff-bowed, that has broad and flat bows.

Bluff-*headed*, (*vaisseau dont l'étrave a peu d'élancement*, Fr.) is when a ship has but a small rake forward; or, being built with her stem too straight up.

Bluff-*headed* ships, are opposed to those that are sharp-headed. They are shorter, and less masted.

BOARD, (*bordée*, Fr.) the space comprehended between any two places where the ship changes her course by tacking; or it is the line over which she runs between tack and tack, when turning to windward, or sailing against the direction of the wind. See the articles Beating and Tacking.

Board, or Aboard, (*à bord*, Fr.) is used in speaking of things within a ship, or other vessel; hence, to go aboard, signifies, to go into a ship: to heave over-board, is to throw any thing out of a vessel into the sea; to slip by the board, is to slip down by the ship's side; board and board, is when two ships come so near as to touch each other, or when they lie side by side.

By the Board, over the ship's side.

To make a good Board, or *stretch*, (*faire une bordée favourable*, Fr.) to sail in a straight line, when close hauled, without deviating to leeward. See the article Close Hauled.

To make short Boards, (*se tenir bord sur bord*, Fr.) is to tack frequently before the ship has run any great length.

To make a stern Board, (*faire culer*, Fr.) is when, by a current, or any other accident, the vessel has fallen back from the point she has gained on the last tack, instead of having advanced beyond it.

Weather-Board, (*le coté du vent*, Fr.) is that side of a ship which is to windward.

To Board *a Ship*, is to enter an enemy's ship in a fight, in order to take possession. See the article Boarding.

To make a Board, or, as it is otherwise expressed, to board it up to a place, is to turn to windward, and to beat, sometimes upon one tack, and sometimes upon another; in which it is to be noted, that the farther you stand off to one point of the compass, the better board you will make; and that it is better to make long boards than short ones, if you have sea-room.

A long Board, is when you stand a good way off before you tack, or turn; a short board is when you stand off a little; a good board is when you have got up much to windward, or when a ship advances much at one tack, and sails upon a straight line.

To leave the Land on Back-Board, is to leave it a-stern or behind; the back-board being that which, in boats or ships, we lean our backs against.

Board, implies all timber sawed to a less thickness than plank, viz. such as is under one inch, and may be distinguished thus:—Beech, elm, fir, oak, sheathing, and wainscot.

BOARDERS, (*matelots destinés à l'abordage*, Fr.) sailors appointed to make an attack, by boarding, or to repel such attempt by the enemy.

BOARDING-PIKE, a defensive weapon used by sailors in boarding an enemy's ship. See the article Half-Pike.

BOARDING a Ship, (*abordage*, Fr) an assault made by one ship upon another, by entering her, in time of battle, with a detachment of armed men; and is chiefly practised by privateers upon merchant-ships, who are not so well provided with men. This stratagem is seldom made use of in ships of war, the battle being generally decided by the vigorous execution of a close cannonade.

An officer should maturely consider the danger of boarding a ship of war before he attempts it, and be well assured that his adversary is weakly manned; as, perhaps, he may wish to be boarded; if so, a great slaughter will necessarily follow.

The swell of the sea ought also to be considered, because it may run so high as to expose both the ships to the danger of sinking.

There is, perhaps, very little prudence in boarding a ship of equal force; and, when it is attempted, it may be either to windward or to leeward, according to the comparative force or situation of the ships. If there be any swell at sea, it may be more adviseable to lay the enemy aboard on the lee-side, as the water is there the smoothest; besides, if the boarder is repulsed in that situation, he may more easily withdraw his men, and stand off from his adversary. But, as the weather-ship can generally fall to leeward at any time, it is, perhaps, more eligible to keep to windward, by which she will be enabled to rake her antagonist, or fire the broadside into her stern, as she crosses it in passing to leeward, which will do great execution amongst her men, by scouring the whole length of the deck.

Boarding may be performed in different places of the ship, according to the circumstances, preparation, and position of both: the assailant having previously selected a number of prime men, armed with pistols, cutlasses, or boarding-pikes. A number of powder-flasks, charged with gun-powder and fitted with a fuse, are also provided to be thrown upon the enemy's deck immediately before the assault. Besides this, the boarder is generally furnished with an earthen shell, called a stink-pot; which, on that occasion, is suspended from his yard-arms, or bowsprit-end. This machine is also charged with powder, mixed with other inflammable and suffocating materials, with a lighted fuse at the aperture. Thus prepared for the action, and having grappled his adversary, the boarder displays his signal to begin the assault. The fuses of the stink-pot and powder-flasks being lighted, they are immediately thrown upon the deck of the enemy, where they burst and catch fire, producing an intolerable stench and smoke, and filling the deck with tumult and distraction. Amidst the confusion occasioned by this infernal apparatus, the detachment provided rush aboard, sword in hand, under cover of the smoke, on their antagonist, who is in the same predicament with a citadel stormed by the besiegers, and generally overpowered, unless he is furnished with extraordinary means of defence, or equipped with places of retreat, furnished with small arms, &c. which may be fired at any time, upon the boarders, and frequently with success.

In addition to what has been said on this article, it may be necessary to observe, that, in boarding a ship, first making a good observation on the force of your enemy, with other favourable advantages of the wind and current, it is best to bear up directly with him, and to cause all the ports to leeward to be beat open, then bring as many guns from the weather-side as there are ports for; and, laying the enemy's ship on board, loof for loof, order the tops and yards to be manned and furnished with necessaries, and let all the small shot be in readiness; then charge at once with both small and great; and, at the same time, enter the men, under cover of the smoke, either on the bow of the enemy's ship, or bring the midship close up with her quarter, and so enter the men by the shrouds: or, if it be found more convenient to use the ordnance, it is best to board the enemy's ship athwart her hawse; for, in that case, most of the great guns may be used, and the enemy can only use those of her prow. Let some of the men endeavour to cut down the enemy's yards and tackle, whilst others clear the decks, and beat the enemy from aloft. Then let the scuttles and hatches be broke open with all possible speed, to avoid trains, and the danger of being blown up by gunpowder, placed under the decks.

BOAT, (*bœt*, Sax. *boot*, Belg. *embarcation*, Fr.) a small open vessel, conducted on the water by rowing or sailing. The construction, machinery, and the names of boats, are very different, according to the various purposes for which they are calculated, and the services on which they are to be employed.

Thus they are occasionally slight or strong; sharp or flat-bottomed; open or decked; plain or ornamented; as they may be designed for swiftness or burden; for deep or shallow water; for sailing in a harbour or at sea; and for convenience or pleasure.

The Long-Boat, (*chaloupe d'un vaisseau de guerre* Fr.) the largest boat that usually accompanies a ship, is generally furnished with a mast and sails, and may be armed and equipped, for cruising short distances against merchant-ships of the enemy, or smugglers, or for impressing seamen, &c. her principal employ, however, is to bring heavy stores or provisions on board, and also to go up small rivers to fetch water, wood, &c.

The Launch, (*chaloupe*, Fr.) is a boat which of late years has greatly superseded the use of the long boat; it is longer, more flat bottomed, and by rowing a greater number of oars, is better adapted for going up narrow and shallow rivers. See the article Launch.

The Barge, (*canot de parade*, Fr.) is a long, narrow, light boat, employed to carry the principal sea-officers, as admirals, and captains of ships of war on shore, and are very unfit for sea. See the article Barge.

A Pinnace, (*pinnace, ou canot des officiers*, Fr.) resembles a barge, but is smaller, never rowing more than eight oars; whereas, a barge never rows less than ten. The pinnace is for the accomodation of the lieutenants, &c.

Cutters *of a Ship*, (*bateaux à l'usage des vaisseaux de guerre*, Fr.) are broader, deeper, and shorter than the barge or pinnace; they are fitter for sailing, and commonly employed in carrying light stores, provisions, passengers, &c. to and from the ships. In the structure of this sort of boats, the lower edge of every plank in the side over-lays the upper edge of the plank below it, which is called clinch-work. They are generally rowed with six oars, sometimes only with four; in which case they are termed jolly-boats.

Yawls, (*canots*, *yoles*, Fr.) are something less than cutters, nearly of the same form, and used for similar services.

Of all the small boats, a Norway yawl seems to be the best calculated for a high sea, as it will often venture out to a great distance from the coast of that country, when a stout ship can hardly carry any sail.

A Gig, is a long narrow boat, used for expedition, generally rowed with six or eight oars, and is mostly the private property of the captain, or commander.

Jolly Boat, (*petit canot*, Fr.) a smaller boat than a yawl, kept on board ships, for going on shore, and other light work.

The above boats more particularly belong to ships of war; as merchant-ships seldom have more than two, viz. a long boat and yawl: when they have a third, it is generally calculated for the countries to which they trade, and varies in its construction accordingly.

A Wherry, (*bateau de passage*, Fr.) is a light sharp boat, used in a river or harbour for carrying passengers from place to place. The wherries allowed to ply about London, are either scullers, which means a single person with two oars, or two persons rowing with each an oar.

The Portsmouth wherries are built broader and

larger than those on the Thames, and allowed to be a very safe description of boats, and will often keep the sea when those of a larger class are obliged to run into port: they have frequently been known to cross the Channel to France with safety, at different seasons of the year, and seldom meet with any accident.

A PUNT, (*ras de carene, pont volant,* Fr.) is a sort of oblong flat-bottomed boat, nearly resembling a floating-stage used by shipwrights and calkers, for breaming, calking, or repairing a ship's bottom, and is chiefly used for one person to go on shore in from small vessels. It is also frequently used to go after wild-fowl during the winter season, when they hover about the coasts. The boatman is furnished with a long musket, fixed in the bow of the boat, and generally steers with a scull in the stern, while he lays on his back in the bottom of the boat; he takes the advantage of the tide to go down on the birds, and fires, without taking any particular aim, as the flock are rising out of the water, by which means he frequently brings down a great many.

A MOSES, (*bateau plat,* Fr.) is a flat-bottomed boat, used in the West Indies, for bringing off hogsheads of sugar from the sea-beach to the shipping, which are anchored in the roads, and is termed single or double, according to its size.

A FELUCCA, (*felouque,* Fr.) is a large and strong passage-boat, used in the Mediterranean, having from ten to sixteen banks of oars. The natives of Barbary often employ boats of this sort as cruisers.

For the larger sort of boats, see the articles CRAFT, CUTTER, PERIAGUA, and SHALLOP.

The Society of Arts voted the gold medal in the year 1807, to Mr. Christopher Wilson, *Richard-street, Commercial-road, London, for a secure sailing-boat,* which is thus described in Vol. XXV. of the Society's Transactions: Mr. Wilson's *Neutral-built-self-balanced-Boat.*

By the term neutral built is meant the two modes of building, *clincher* and *carvel,* both united, viz. clincher in the inside and carvel on the outside, which particularizes both the two into a third; and as every thing has a distinguishing name, Mr. WILSON has thought proper to introduce it to the public, under the name of a neutral boat.

The two modes of clincher and carvel-built have each their separate advantages and disadvantages in regard to each other; but in this neutral system, both their advantages are said to be united, and both their disadvantages got clear of.

Mr. Wilson claims two distinct improvements, viz. the neutral mode of building, and the application of the hollow sides or balance bodies. The first of these improvements relates to boats, barges, &c. in general: the other to such boats only as are designed for dispatch, safety, or pleasure.

But with respect to these claims of Mr. Wilson, it is but justice to state, that he has been anticipated in them both. Mr. Boswell, to whom the navy is much indebted for many useful hints, had a boat built in June 1803, the internal framing of which was precisely the same as in that of Mr. Wilson's: and the only difference was, that, instead of having slips laid on the inside of the seams with blair or other calking, they were rivetted on without, and the seams calked.

This boat was exhibited in London, and a description of it published in No. LXII. of the *Repertory of Arts, New Series, &c.* The second improvement claimed by Mr. Wilson, is essentially the same as that for which Mr. Lionel Lukin obtained a patent in the year 1785, the specification of which was published in the *Repertory of Arts, Vol. III. First Series.*

The difficulties and danger of passing a bar in stormy weather, and of landing troops on a beach, when there is much surf, or a great swell, have given rise to various experiments, to prevent boats being swamped or stove; and, among others, that of applying *air trunks* to boats built in the common way, agreeable to a plan lately proposed to the Admiralty by Vice Admiral Hunter, who, after forty years experience in different parts of the world, strongly recommends it as the very best possible method yet thought of, to prevent boats sinking in case of being struck by a heavy sea, or filled in going through a surf.

This plan is simple and *practicable* for preventing the loss of lives when a boat is sent from her ship in bad weather; or when troops are to be carried in boats through broken water, which may occasion their shipping enough to fill or swamp the boat in deep water; and the Admiral is of opinion, that, if every ship, whether in his Majesty's or merchant's service, were to have at all times a set of these trunks ready to apply when they have occasion to use their boats in bad weather, many valuable lives would be saved.

To those acquainted with the force of a heavy sea, it must appear much more difficult to prevent a boat from being stove, than from being sunk.

An ingenious mechanic has suggested, that a much greater degree of stability might be given to our common boats or small vessels, and that they might be made to resist the shock of the waves, by diminishing their centre of gravity at the instant such a measure was found necessary, which would greatly augment their general gravity and resistance. He proposes, that in the vertical plane of the boat's centre of gravity, there be placed a rod of iron, formed of several fillets of the same metal; at the extremity of this rod let there be fixed a weight of iron or lead, the specific gravity of which may be in proportion to the effect required to result from it.

When there is no occasion for employing this weight, it might be contained in a place made in the bottom, or keel, of the vessel, which part might be easily appropriated for that purpose, without the weight occasioning any embarrassment or friction.

The rod might be secured by an apparatus very easily made; and it might be provided with teeth, by which it could be manœuvred by a crane, by means of a single handle; the weight might be made to descend to the depth of eighteen, twenty, or twenty-four inches, beneath the lower plane of the keel; and it will appear evident, that the descent of this weight must cause a difference in the power of the boat's resistance against the efforts of the waves. See the article LIFE BOAT.

Trim the BOAT! (*barque droite!* Fr.) the order to sit in the boat in such a manner as that she shall float upright in the water, without leaning to either side.

To bale the BOAT, (*vuider l'eau d'un canot;* Fr.) is to throw or scoop out the water that may have got in her bottom, by leakage or otherwise.

Moor the BOAT! (*amarre le canot!* Fr.) the order to fasten a boat with two ropes, so as that the one shall counteract the other, and keep her in a steady position.

BOAT'S *Crew*, (*équipage de canot*, Fr.) the men appointed to man any particular boat, as the barge's crew, cutter's crew, &c.

Train of BOATS, are small vessels fastened to each other, ascending up the Loire, in France, by sails, when the wind serves, or else towed by men, sometimes to the number of 70 or 80 to a rope.

For a representation of some of the principal boats
Plate IV. of a ship of war, see plate IV. where fig. 1. exhibits
Fig. 1. the elevation, or side view, of a ten-oared barge; *a a*, its keel; *b*, the stern-post; *c*, the stem; *b c*, the water-line, which separates what is under the surface of the water from what is above it; *e*, the row-locks, which contain the oars between them; *f*, the top of the stern; *g*, the back-board; *f g*, the place where the coxswain stands or sits while steering the boat; *l*, the rudder; and *m*, the tiller, which is framed of iron.

Fig. 2. represents the plan of the same barge, where *d* is the thwarts, or seats where the rowers sit to manage their oars; *f*, *i*, *h*, the stern-sheets; *i k*, the benches whereon the passengers sit in the stern-sheets: the rest is explained in fig. 1.

Fig. 3. is a stern view of the same barge, with the projection of all the timbers in the after body; and fig. 4. a head view, with the curves of all the timbers in the fore-body.

Having thus explained the different views of the barge, the reader will easily comprehend the several corresponding parts in the other boats; where fig. 5. is the plan, and fig. 6. the elevation of a twelve-oared cutter that rows double banked: which, although seldom employed unless in capital ships, because requiring twelve rowers, is, nevertheless, a very excellent boat, both for rowing and sailing. Fig. 7. and 8. are the head and stern-view of this boat.

Fig. 9. is the plan of a long-boat, of which fig. 10. is the elevation, 11. the stern-view, and 12. the head-view.

BOAT-*hook*, (*gaffe*, Fr.) an iron hook, with a sharp point on the hinder part thereof. It is fixed upon a long pole or shaft, *n*, (pl. IV. fig. 13.) by the help of which, a person in the boat may either hook any thing to confine the boat in a particular place, or push her off by the sharp point attached to the back of the hook.

BOATING, was a severe punishment inflicted by the ancient Persians on capital offenders, in the following manner: the condemned person being laid on his back in a boat, and having his hands stretched out and tied fast on each side of it, had another boat put over him, his head being left out through a place fit for it. In this posture they fed him, till the worms which were bred in the excrements he voided as he thus lay, ate out his bowels, and so caused his death, which was usually twenty days in effecting, the criminal lying all this while between the boats, in most exquisite torments.

BOATSKIDS, are long square pieces of fir, extending across the ship from the gang-boards, and on which the boats, spare masts, &c. are stowed.

BOATSWAIN, (*maître d'équipage*, Fr.) the officer who has the boats, sails, rigging, colours, anchors, cables, and cordage, committed to his charge.

It is the duty of the boatswain, under the master, particularly to direct whatever relates to the rigging of a ship, after she is equipped from a royal dock-yard. Thus he is to observe, that the masts are properly supported by their shrouds, stays, and back-stays, so that each of those ropes may sustain a proportional effort when the mast is strained by the violence of the wind, or the agitation of the ship. He ought also to take care that the blocks and running-ropes are regularly placed, so as to answer the purposes for which they are intended; and that the sails are properly fitted to their yards and stays, and well furled or reefed when occasion requires.

It is likewise his office to summon the crew to their duty, to assist with his mates in the necessary business of the ship; and to relieve the watch when it expires. He ought frequently to examine the condition of the masts, sails, and rigging, and remove whatever may be judged unfit for service, or supply what is deficient: and he is ordered by his instructions to perform his duty *with as little noise as possible*.

BOATSWAIN'S *Mate*, (*contre maître*, Fr.) is an assistant to the boatswain, and does his duty under an officer where only part of the ship's company is employed; and generally inflicts all punishments, under the direction of the captain or commander.

BOB-STAY, (*sous-barbe de beaupré*, Fr.) a rope used to confine the bowsprit downward to the stem, or cut-water. It is fixed by thrusting one of its ends through a hole bored in the fore-part of the cut-water for this purpose, then splicing both ends together so as to make it two-fold, or like the link of a chain: a *dead-eye* is then seized into it, and a *lanyard* passing through this, and communicating with another dead-eye upon the bowsprit, is drawn extremely tight by the help of mechanical powers. See the article BOWSPRIT.

The use of the bob-stay is, to draw down the bowsprit, and keep it steady; and to counteract the force of the stays of the fore-mast, which draw it upwards. The bowsprit is also fortified by shrouds from the bows on each side, which are all very necessary, as the fore-mast and the upper part of the main-mast are stayed and greatly supported by the bowsprit. For this reason, the bob-stay is the first part of a ship's rigging which is drawn tight to support the masts. To perform this task more effectually, it is usual to suspend a boat, anchor, or other weighty body, at the bowsprit-end, to press it downwards during this operation.

BOB-STAY-*holes*, (*trous de la sous-barbe*, Fr.) are those in the fore part of the knee of the head, for the security of the bob-stay.

A BOLD SHORE, (*côte écore*, Fr.) an epithet applied to the sea-coast, signifying steep, and abrupt, as to admit the approach of shipping, without exposing them to the danger of being run aground, or stranded.

BOLLARDS, are large posts set in the ground on each side of a dock. On docking or undocking ships, large blocks are lashed to them; and through these blocks are reeved the transporting hawsers to be brought to the capstans.

BOLLARD-TIMBERS, or KNIGHT-HEADS, (*les apôtres*, Fr.) are two pieces of timber rising just within the stem, one to each side of the bow-sprit to secure its inner end.

BOLLOCK BLOCKS, (*poulies d'itague des drisses*

des huniers, Fr.) are blocks secured on the middle of the topsail yards, and receive the topsail ties through them, in order to increase the mechanical power used in hoisting them up.

BOLSTER, a piece of timber placed upon the upper or lower cheek, worked up about half the depth of the hawse-holes, and cut away for the easement of the cable, and to prevent its rubbing the cheek: likewise the solid piece of timber that is bolted to the ship's side, on which the stantions for the lining of the anchors are placed.

BOLSTERS, (*coussins de fourrure et paillers*, Fr.) small cushions or bags, filled with tarred canvas, laid between the collars of the stays and the edge of some piece of wood on which they lie: they are used to preserve the stays from being chafed, or worn by the motion of the masts, as the ship pitches at sea. They are also pieces of wood, rounded on the upper part, and covered with several layers of tarred canvas, to prevent the rigging being chafed by the motion of the ship or masts.

BOLSTERS, are also cylindrical pieces of iron, with a hole through the middle, used when holes are to be punched, or opened with pins.

BOLT, (*cheville de fer*, Fr.) a cylindrical pin of iron, of which there are various sorts, (see plate IV. fig. 14, 15, 16, and 17) used for sundry occasions in ship-building.

The bolts are principally employed either to unite several members of a ship's frame into one solid piece, or to fasten any moveable body on a particular occasion. Those which are calculated for the former purpose have commonly small round heads, somewhat flatted, called saucer heads, with a mortise on the opposite end or point, which are used to fasten moveable pieces to those that are fixed: on the contrary, the bolts which are intended for the latter use, have either a large round head, as those of the chains, fig. 14. or an eye, with or without a ring in the same place, fig. 15 and 16, as those which are designed to secure the great guns, the *jears* of the mainsail and foresail, the stoppers of the cables, &c. They are also used for lashing or hooking blocks, &c. and are driven into mast-heads, yards, caps, &c.

The bolts are short or long, according to the thickness of the timber wherein they are to be lodged: they penetrate either quite through the pieces into which they are driven, or to a certain determinate depth. The last of these, called a rag-bolt, is retained in its situation by means of several barbs, fig. 17, which, fastening into the timbers, prevent the bolt from loosening from its station by the working of the ship. The first, after being driven through the pieces it is intended to unite, is confined by a flat iron wedge, called the forelock, which is thrust through a narrow hole in the small end of the bolt, where it is hardened home by a hammer; and to prevent the forelock from cutting the wood-work in this position, a little iron ring is fixed over the end of the bolt, between the forelock and the timber.

Those bolts, which have the largest of the round-heads, are called fender-bolts, being driven into the wales, stem, or sides of some small vessels of burden, as lighters, beancods, prames, &c. to defend their timber-work from the shock of any other vessels which may fall aboard by accident.

Some bolts have a square part left at the back of the eye, that they may not be driven on the eye, and consequently are not in danger of splitting.

BOLT-*rope*, (*ralingue*, Fr.) a rope to which the edges or skirts of the sails are sewed, to strengthen and prevent them from rending. That part of the bolt-rope, which is on the perpendicular parts of the sail, is called the leech-rope; that at the bottom, the foot-rope; and that on the top, the head-rope. Stay-sails, whose heads are formed like an acute angle, have no head-rope. To different parts of the bolt-rope are fastened all the ropes employed to contract or dilate the sails.

BOMB, (*bombe*, Fr.) in artillery, a shell or hollow ball of cast-iron, having a large vent, by which it is filled with gunpowder, and which is fitted with a fuse or hollow plug to give fire by when thrown out of a mortar, &c.: about the time when the shell arrives at the intended place, the composition in the pipe of the fuse sets fire to the powder in the shell, which blows it all in pieces, to the great annoyance of the enemy, by killing the people, or firing the houses, &c. They are now commonly called shells, simply, in the English artillery.

These shells or bombs are of various sizes, that of 13 inches diameter being the highest size now employed: the weight, dimensions, and other circumstances of them, are as in the following table:

Diameter of the shell.	Weight of the Shell.	Powder to fill them.		Powder to burst them in most pieces.	
	lbs.	lb.	oz.	lbs.	oz.
13 inch	195	9	$4\frac{1}{2}$	7	8
10	89	4	$14\frac{1}{2}$	3	4
8	46	2	$3\frac{1}{2}$	2	0

The following proportion is given for all shells. Dividing the diameter of the mortar into 30 equal parts, then the other dimensions in 30ths of that diameter will be thus:

Diameter of the bore or mortar 30
Diameter of the shell.................... $29\frac{1}{2}$
Diameter of the hollow sphere........... 21
Thickness of metal at the fuse hole..... $3\frac{1}{2}$
Thickness at the opposite part 5
Diameter of the fuse hole 4
Weight of shell empty................. $\frac{10}{117}d$
Weight of powder to fill it............ $\frac{2}{473}d$

Where d denotes the cube of the diameter of the bore in inches. But shells have also lately been made with the metal all of the same thickness quite around.

In general, the windage or difference between the diameter of the shell and mortar is $\frac{1}{20}$th of the latter; also the diameter of the hollow part of the shell is $\frac{7}{10}$th of the same. Bombs are thrown out of mortars or howitzers; but they may also be thrown out of cannon; and a very small sort are thrown by the hand, which are called granades. See the article MORTARS.

BOMB-*vessel*, (*bombarde*, Fr.) a small vessel particularly calculated to throw bombs into a fortress. They are generally rigged as ketches, and are said to be invented by M. Reyneau, and to have been first put in action at the bombardment of Algiers.

Till then it had been judged impracticable to bombard a place from the sea.

Bomb-*ketch*, (*galiote à bombes*, Fr.) a vessel built and strengthened with large beams, and equipped with two masts, viz. the main and mizen-mast, usually from 100 to 250 tons burden, for the use of mortars at sea. The bomb-ketch is, therefore, furnished with all the apparatus necessary for a vigorous bombardment.

Bomb-ketches are built remarkably strong, as being fitted with a greater number of riders than any other vessel of war; see Plate IV. fig. 18; and indeed this reinforcement is absolutely necessary to sustain the violent shock produced by the discharge of their mortars, which would otherwise, in a very short time, shatter them to pieces.

Plate IV. Fig. 18.

The modern bomb-vessels carry two 10-inch mortars, four 68-pounders, and six 18-pounders carronades: and the mortars may be fired at as low an angle as 20 degrees; though these mortars are not intended to be used at sea, but on very particular occasions, their principal intention, at these low angles being, to cover the landing of troops, and protect our coast and harbours.

A bomb-ketch is generally from sixty to seventy feet long from stem to stern, and draws eight or nine feet water. The tender is generally a brig, on board of which the party of artillery remain, till their services are required on board the bomb-vessel.

Instructions for their management and security in action:

1. A Dutch pump, filled with water, must be placed in each round-top, one upon the fore-castle, one on the main-deck, and one on the quarter-deck; and furnished with leather buckets, for a fresh supply of water.

2. The booms must be wetted by the pumps before the tarpaulins and mortar hatches are taken off; and a wooden skreen, five feet square, is to be hung under the booms, over each mortar, to receive the fire from the vents.

3. The embrazures being fixed and properly secured, the port must be let down low enough to be covered by the sole of the embrazure. Previous to its being let down, a spar must be lashed across it, to which the tackles for raising it again must be fixed: this spar serves to project the tackles clear of the explosion.

4. The mortars must not be fired through the embrazures at a lower angle than twenty degrees, nor with a greater charge than five pounds of powder.

5. Previous to firing, the doors of the bulk-head, under the quarter-deck, must be shut, to prevent the cabin being injured by the explosion.

6. The bed must be wedged in the circular curb, as soon as the mortar is pointed, to prevent re-action; the first wedge being driven tight before the rear ones are fixed, in order to give the full bearing on the table, as well as the rear of the bed. The holes for dog-bolts must be corked up, to prevent the sparks falling into them.

7. When any shells are to be used on board the bomb, they must be fixed on board the tender, and brought from thence, in boxes, in her long boat; and kept alongside the bomb-ship till wanted, carefully covered up.

In modern bomb-vessels, with embrazures, only the boats need be hoisted out; after which the mortars may be prepared for action in ten minutes.

To BOMBARD, (*bombarder*, Fr.) to fall upon a city or fortress, with bombs, rockets, &c.

BOMBARDIER, (*bombardier*, Fr.) a person employed about a mortar. His business is to drive the fusee, fix the shell, load and fire the mortar, and to work with the fire-workers on all sorts of fire-works, whether for war or recreation.

BOMBARDMENT, (*bombardement*, Fr.) is the act of assaulting a city or fortress by throwing shells, rockets, &c. into it, in order to set it on fire, or otherwise demolish it. As one of the effects of the bomb results from its weight, it is never discharged as a ball from a cannon, that is, by pointing it at a certain object: the mortars in England are generally fixed at an elevation of 45°.

BONNET, (*bonnette*, Fr.) an additional part made to fasten with latchings to the foot of the sails of small vessels with one mast, in moderate winds. It is exactly similar to the foot of the sail it is intended for. They are commonly one-third of the depth of the sails they belong to.

BOOM of a harbour, (from *boom*, a tree, Dutch, *chaîne d'un port*, Fr.) in marine fortification, is a strong iron chain, fastened to a number of spars or poles, and extending athwart the mouth of a harbour or river, to prevent the enemy's ships of war or other vessels from entering. It may be occasionally sunk, or drawn up to the surface of the water, by capstans, and other mechanical powers.

Fire Boom, (*boute-hors pour défendre l'approche des brûlots*, Fr.) a strong pole, occasionally thrust out from a ship's side, &c. to prevent the approach of fire-ships, fire-stages, or vessels accidentally on fire.

The booms imply a space where the spare booms and top-masts are stowed, their ends being supported by the gallows, and affording a receptacle for the barge between them.

Booming, (*boute-hors*, Fr.) among sailors, denotes the application of a boom to the sails. Booming of the sails is never used but in quarter winds, or before a wind. When a ship is said to come booming towards us, it signifies that she comes with all the sail she can make.

Boom-*iron*, (*cercle de boute-hors*, Fr.) is composed of two iron rings, formed into one piece, so as nearly to resemble the figure of 8. It is employed to connect two cylindrical pieces of wood together, when the one is used as a continuation of the other; such is the jib-boom to the bowsprit; and such are the *studding-sail* booms to the respective yards from whose extremities they are prolonged. The rims, or circles of the *boom-irons*, are broad and flat; and one of them, which is firmly driven upon the main, or fore-yard-arm, is somewhat larger than the other, Fig. 19.
as exhibited in Plate IV. fig. 19. The studding-sail-boom usually rests in the small ring, through which it is occasionally thrust outwards from the yard-arm, when the studding-sail is to be set. Every boom of this kind has, or ought to have, two boom-irons, one of which is fixed on the extremity of the yard, and the other further inward. The former of these is frequently framed of one ring only, which projects from the end of the yard, where it is fastened

by a strong iron bar, opening into a sort of fork or crotch that slides upon the yard lengthways, where it is fastened by nails driven from above and below.

Booms, (*boute dehors*, Fr.) are long poles run out from different places in the ship, to extend the bottoms of particular sails. Of these there are several sorts; as the jib-boom, flying jib-boom, studding-sail-booms, ring-tail-boom, driver or spanker-boom, main-boom, and square-sail-boom; the two last, however, are only appropriated to small ships of one or two masts. Booms somewhat differ in their shapes to each other, some taper regularly from the middle to the end like the yards; others from one third the length to the extremity. See the article Jib, &c.

BOOTES, in astronomy, the name of a constellation in the northern hemisphere, containing 54 stars in Flamstead's catalogue, and 64 in the catalogue published in the *Tables de Berlin;* one of which, between his thighs, is of the first magnitude, and called *Arcturus*, which article see.

Bootes is represented on the globe as a man in the posture of walking; his right-hand grasping a club, and his left extended, holding the cord of the two dogs, which seem to be barking at the Great Bear.

BOOT-TOPPING, (*demi-bande, ou demi carène*, Fr.) is the operation of scraping off the grass, slime, shells, &c. which adhere to a ship's bottom, or that part which lies immediately under the surface of the water, and daubing it over with a mixture of tallow, sulphur, rosin, &c.

Boot-topping is chiefly performed where there is no dock, or other commodious situation for breaming or careening; or when the hurry of a voyage renders it inconvenient to have the whole bottom properly trimmed and cleansed from the filth which gathers to it in the course of a voyage. It is executed by making the ship lean to one side, as much as they can with safety, and then scraping off the filth, &c. on the other side. Having thus finished one side, they change the position of the vessel, and perform the same operation on the other side, which not only preserves the bottom from the worm, but makes the ship slide smoothly through the water. See the articles Careen and Dock.

BORE of a cannon, (*calibre d'un canon ou âme*, Fr.) among engineers, denotes the diameter of the barrel of a gun or cannon, or rather its whole cavity.

BOTH SHEETS AFT, (*entre deux écoutes*, Fr.) the situation of a ship that sails right before the wind, or with the wind right astern.

BOTTOM, (*fond-carène d'un vaisseau*, Fr. *botm.* Sax. *bodem*, Belg.) is used to denote either the bottom of a ship, or that of the water: thus, in the former sense we say, a clean or a foul bottom; a British, French, or Dutch bottom: and in the latter sense, a rocky, stony, sandy, coral, muddy, or oozy bottom. By statute, certain commodities imported in foreign bottoms, pay a duty called petty customs, over and above what they are liable to, if imported in British bottoms, &c.

The bottom of a ship comprehends all that part which is under water when the ship is laden; the figure of it must, therefore, be determined by the qualities required in the ship, and the purposes for which she is designed.

It has been remarked, that a ship of war should carry her lowest tier of cannon sufficiently above the surface of the water to be used when necessary. If this quality be neglected, a small ship will have the advantage of a large one, inasmuch as the latter cannot open her lower battery in a fresh side-wind, without being exposed to extreme danger, by receiving a great quantity of water in at her ports between decks.

A ship should be duly poised, so as not to dive or pitch heavily, but go smoothly and easily through the water, rising to the waves when they run high, or when the vessel has reduced her sail to the storm. If she is deficient in this article, the seas will frequently burst aboard, and strain the decks, or carry away the boats. The masts are also greatly endangered from the same cause.

A ship should sail well when large, or before the wind; but particularly when *close-hauled*, or sailing with a side-wind. She should also be enabled in the latter situation to keep her wind, without deviating much to leeward; to work and tack easily, and lie in a turbulent sea without straining violently.

Many of our shipwrights have considered it extremely difficult, if not impracticable, to make a ship carry her cannon well, bear a competent sail, and advance swiftly through the water; because a very full bottom is necessary to acquire the two first qualities; whereas, a sharp floor is better fitted to procure the latter. But when it is remembered, that a full ship will carry a much greater force of sail than a sharp one, a good artist may form the body so as to unite all these three qualities with the additional one of steering easily, by paying a proper attention to the following general rules.

To make a ship carry a good sail. A flat floor-timber somewhat long, or the lower-futtocks pretty round, a straight upper-futtock, the top-timber to throw out the breadth aloft; at any rate to carry the main-breadth as high as the lower-deck. Now if the rigging be well adapted to such a body, and the upper-works lightened as much as possible, so that the whole contributes to lower the centre of gravity, there will be no reason to doubt of the ship's carrying a good sail.

To make a ship steer well, and answer the helm readily. If the fashion-pieces be well formed, the tuck, or spreading-parts under the stern, carried pretty high; the midship frame well forward; a considerable additional depth in the draught of water abaft more than forward; a great rake forward and none abaft; a snug quarter-deck and forecastle: all these will greatly facilitate the steerage; and a ship that sails well will always steer easily.

To make a ship carry her guns well out of the water. A long floor-timber, and not of great rising; a very full midship-frame, and low tuck, with light upper-works.

To make a ship go smoothly through the water, and prevent her from pitching heavily. A long keel; a long floor; not to rise too high afore and abaft; but the area, or space contained in the fore-body, according to the respective weight it is destined to carry: all these are necessary to make a ship pass easily through the sea.

To make a ship keep a good wind, and drive little

to leeward. A good length by the keel; not too broad, but pretty deep in the hold, which will occasion her to have a short floor-timber and a very great rising. As such a ship will meet with great resistance in driving sideways, and feel very little in advancing or going ahead, so she will fall very little to leeward.

Being thus furnished with the methods to qualify a ship for the different purposes of navigation, the only difficulty remains to apply them properly in the construction, which must, in a great measure, be left to the judgment of the artist. The whole art then is, evidently, to form the body in such a manner, as that none of these qualities shall be entirely destroyed; and in giving the preference to that which is principally required in the service for which the ship is destined.

All ships (says a celebrated Writer on Practical Seamanship), whether full for burden, or sharp for sailing, should be built with the full parts of their bottoms to form an arch downwards, instead of having long straight floors, as it is well known when a round cannon-ball is shot slanting into the water, the re-action makes it rise into the air several times, before it immerses and sinks, though it is about eight times heavier than its bulk of water, which will naturally act with the same principles upon ships which have their bottoms somewhat circular all the way, as they are forced forward through the water, the entrance and fore body of the ship in the water, have the passage to open, by dividing and sinking the particles of water, till they have passed the main-frame, or lowest flooring; the re-action of the water upwards against this shaped bottom, will naturally tend to lighten or lift the ship from plunging into the sea, all the way that there is any resistance, which is reckoned to cease at the lowest main-frame, for there the water begins to rise and close again to come to its level, which acts upon the tapered after-body, and runs, if well formed, so as to help the ship forward in her progressive motion. And she will naturally act and turn upon this lowest part of the bottom, because within this main or midship-frame must be the centre of the greatest cavity, or hollow part of the ship's hold; and thereabout will be the centre of gravity of the ship and all that she contains, and will thereby become the centre of motion: the water's pressure upwards being greatest against this lowest part of the bottom, all will contribute naturally to make this the proper centre for the ship to turn upon, to make her more manageable in steering, staying, or wearing: and this last is a very valuable property, for it is most commonly wanted when ships are in the greatest danger, and expeditiously performed may be the saving of the whole. This seems to be pointed out by natural judgment, and what may be seen in the mechanism of nature; and the curve of a ship's bends and upper works do not offend, but please the eye, and all straight flat parts about a ship disgust. The swiftest fish seen in motion at sea, as well as those fowl which swim and dive in water, are all formed with their bodies rounding, as is above recommended.

This form of rounding ship's bottoms, will also recommend itself to take the ground with the least damage; for it is well known, that a ship strains and receives damage, less or more, in proportion as she sits upright, or lays along on one side; and what makes a ship lie along so much upon level ground, but the present practice of having a deep keel under the main body of the ship, which cants her over so far, that the ends of the floorings or lower futtocks bear the stress upon the ground, by which they are often broke, and if the ship has a long straight floor, it is liable to be bent upwards, which causes the whole frame to strain the more, by being thus unnaturally constructed.

Where ships are liable, or are designed to take the ground, as in dry tide harbours, the rounding bottoms will certainly answer that purpose best, as they may be built either full for carrying, or sharp for sailing, as occasion may require; with little or no depth of keel, under the main body, so as not to make them lie along, but as they may sit quite upright, upon level ground; which will make the principal weight and stress tend towards the middle of the lowest floorings, under the main or midship-frame; where the centre of gravity of the whole will be accumulated and supported, as upon the crown of an inverted arch, which form must be allowed, to be capable of bearing the most pressure, with the least damage, and to lay the greatest strain where it ought to be, upon the middle, or strongest part of the floor-timbers.

BOTTOMRY, (*bomerie*, Fr. from *bottom*,) a contract for borrowing money upon the keel or bottom of a ship; that is to say, the master of a ship binds the ship herself, that if the money be not paid at the time appointed, the creditors shall have the ship.

Bottomry is also lending money to a merchant or adventurer who wants it in traffic, and the lender is to be paid a much greater sum at the return of the ship, standing to the hazard of the voyage: on which account, though the interest be greater than what the law commonly allows, yet it is not usury, because the money being furnished at the lender's risk, if the ship perishes, he shares in the loss.

BOUND is used in the following terms: when we say a vessel is wind-bound, it implies that she is prevented from sailing, on account of the wind being contrary: (*vaisseau retenu dans un port par les vents contraires*, Fr.) when ice-bound, that it is totally surrounded with ice, and thereby rendered incapable of proceeding on her voyage: (*être fermé par les glaces*, Fr.) Where are you bound to? (*où allez vous?* Fr.) which is the usual term when you hail a ship, and signifies, to what port are you going?

BOUNTY, (*une pension accordée aux veuves d'officiers*, Fr.) to widows of officers, &c. in the royal navy, was first instituted by Queen Anne, in the year 1704, after a severe storm, which happened in *November* 1703, in which were lost six ships of the line, three frigates, a bomb-vessel, and an advice-boat, on our own coasts; besides several other men of war that were driven on shore, dismasted and otherwise damaged. Her Majesty issued a proclamation, ordering that all the widows and families of such officers and seamen as had perished by the storm, in her Majesty's service, be entitled to her *bounty*, in the same manner as if they had actually been killed in battle.

Bounty, in the royal navy, is a premium or reward offered, upon any emergency of the state, to able and ordinary seamen, as well as landmen, that may voluntarily enter his Majesty's navy, and is continued by proclamation for a limited time.

BOW, (*épaule, avant d'un vaisseau*, Fr.) is the rounding part of a ship's side forward, beginning where the planks arch inwards, and terminating where they close, at the stem or prow. See the article Head, where the bow of a ship is represented at large. It has been proved by a variety of experiments, that a ship with a narrow bow is much better calculated for sailing swiftly, than one with a broad bow; but is not so well fitted for a high sea, into which she always *pitches*, or plunges, her fore-part very deep, for want of sufficient breadth to repel the volume of water, which so easily divides in her fall. The former of these is called by seamen a *lean*, and the latter a *bluff* bow.

A French author observes, that the bow which meets with the least resistance in a direct course, not only meets with the least resistance in oblique courses, but also has the additional property of driving the least to leeward, which is a double advantage gained by forming the bow so as to give it that figure which will be least opposed in moving through any medium.

Doubling of the Bow. The vertical thick stuff placed there to prevent the anchor-bill from tearing it, in passing up and down.

On the Bow, (*plus en avant que le travers du vaisseau*, Fr.) in navigation, an arch of the horizon, (not exceeding 45 degrees) comprehended between some distant object and that point of the compass which is right ahead, or to which the ship's stem is directed. This phrase is equally applicable, when the object is beheld from the ship, or discovered by trigonometrical calculation: as, "we saw a fleet at daybreak bearing three points *on the starboard bow*;" (*par le bossoir de tribord*, Fr.) that is, three points from that part of the horizon which is right ahead, towards the right hand. See also the article Bearing.

BOWER. See the article Anchor.

BOWGRACE, a frame of old rope or junk, laid round the bows, stems, and sides of ships, to prevent them from being injured by flakes of ice.

BOWLINE, (*bouline*, Fr.) a rope fastened near the middle of the leech, or perpendicular edge of the square sails, by three or four subordinate parts, called bridles. It is only used when the wind is so unfavourable that the sails must be all braced sideways, or *close-hauled* to the wind: in this situation the bowlines are employed to keep the weather, or windward, edges of the principal sails tight forward and steady, without which they would be always shivering, and rendered incapable of service. See the articles Bridle, Close-hauling, and Sail.

To check the-Bowline, (*choquer la bouline*, Fr.) is to slacken it, when the wind becomes large or free.

BOWLING-BRIDLES, (*pattes de bouline*, Fr.) are the ropes by which the bow-line is fastened to the leech of the sail.

BOWMAN of a Boat, (*brigadier de canot*, Fr.) is the man who rows the foremost oar in a boat.

BOWSE, (*haler sur un cordage*, Fr.) is chiefly used by the gunners when they haul upon their tackles to thrust a piece out of port, in which case they cry "Bowse hoa!" *i.e.* pull more upon the tackle; also when there is occasion to pull more on the tackle than otherwise, they say "Bowse upon the tackle!"

To Bowse, (*palanquer*, Fr.) to pull upon any body with a tackle, or complication of pullies, in order to remove it, or otherwise alter its state or situation: this is chiefly practised, when such alteration or removal cannot be conveniently effected without the application of mechanical powers. This term is pronounced *bowse*, as when they would have the men pull altogether, they cry "Bowse away!"

BOWSPRIT, (*beaupré*, Fr. from *bow* and *sprit*) a large boom or mast, which projects over the stem, to carry sail forward, in order to govern the fore-part of a ship, and counteract the force of the after-sails, or those extended behind. It is otherwise of great use, as being the principal support of the fore-mast, by confining the *stays* whereby it is secured and enabled to carry sail: these are great ropes stretching from the mast-head to the middle of the bowsprit, where they are drawn tight. See the articles Stay and Dead Eye.

The bowsprit is, like the masts, rounded, except at the outer end, which is square. The rule for its rise is an angle of nearly thirty-six degrees above the horizon.

The bowsprit should be two-thirds the length of the main-mast, and its thickness equal to the mizen-mast: when it is twelve fathoms five feet long, its yards must be eight fathoms two feet long, and the topmast of the bowsprit three fathoms and one foot.

BOX of a rib-saw, two thin iron-plates fixed to a wooden handle; in one of the iron plates is an opening to receive a wedge, by which it is fixed to the saw.

Boxes *of the Pump*, (*pistons de pompe*, Fr.) the materials requisite for repairing the pump.

BOX-HAULING, (*manière prompte de virer vent arrière*, Fr.) a particular method of veering a ship, when the swell of the sea renders tacking impracticable. It is performed by keeping the helm *hard-a-lee*, and hauling off all; bracing about the head as well as the after-sails, hauling close forward the lee, fore and fore-top-bowlines, and up mizen and down after-stay-sails at the same time; the wind will then act upon the sails thus *aback*, and the water upon the lee-side of the rudder, by her stern way, will box the ship short round upon her keel, with her stern up to the wind, far enough aft for the after-sails to draw full the right way to act with the helm, which must be shifted *hard-a-weather*, when the stern-way ceases, so that the head-way with the wind so far aft, will readily bring the ship round on the other tack. The main and fore-tacks are easily got down when the wind is upon the quarter, and shivers the sails, the main-sheet is easily hauled aft, and the after-sails braced up and trimmed sharp, as the ship brings the wind more aft, which helps her round the faster, till the wind comes on the other quarter, that the mizen and mizen-stay-sail may be set to take the right way to bring her to the wind, while you tend and trim the head-sails as she comes home.

Boxhauling may be proved to be the surest and readiest method to get a ship under command of the helm and sails, to answer many occasions in little sea-room, as well as wear, and bring her from one tack to the other, with the least loss of ground to leeward when a ship refuses stays.

Nice managers of sloop rigged vessels, turning to windward in narrow channels, when they want but little to weather a point, rather than make another tack, have a practice of running up in the wind till the headway ceases, then they fill again upon the same tack: this they call making a half tack. Thus a ship in boxhauling may be said to make two half boards, first running with her head, then with her stern up in the wind, by which two motions a ship, if well managed, rather gains to windward, and brings the wind almost aft, by the time the sternway ceases, so that she is under the command of the weather-helm, and the after-sails, to bring her short round on the other tack, with the first of the headway, in which time only it is that the ship goes any way to leeward worth notice in boxhauling; therefore it should always be put in practice on those occasions, by putting the ship in stays, though it is known she will not come about with her head to windward; and in a gale of wind and high waves, or when there are not people enough to manage and haul the head-sails, the after-sails only may be hauled, and the fore-sheet hauled close aft again when it is perceived that the ship has done coming to.

Boxhauling is generally performed when the ship is too near the shore to have room for veering in the usual way. See the article VEERING.

BOXING OFF, (*action de mettre les voiles d'avant sur le mat*, Fr.) an operation in sailing, somewhat similar to box-hauling. It is performed by laying the head-sails aback, to receive the greatest force of the wind in a line perpendicular to their surfaces, in order to throw the ship's head into the line of her course, after she had inclined to windward of it, by the neglect of the helmsman, or otherwise.

BOXING, or BOXING *of the Stem*, (*écart de l'étrave avec le brion*, Fr.) implies also the protection left on the hawse-pieces in the wake of the hawse-holes, when the planks do not run through; but this is nearly laid aside in all ships, and the plank is worked through the stem, both inside and outside, which is a very great improvement; because timber, younger and smaller, will answer the same purpose, and be more durable; and by working the planks through, the bow is much stronger.

BOXING *the Compass*, (*répéter le compas*, Fr.) is a phrase among the sailors for rehearsing the several points of the compass in their proper order.

BRACE, (*bras de vergues*, Fr.) a rope employed to wheel, or traverse the sails upon the mast, in a direction with the horizon, when it is necessary to shift the sails, that they may correspond with the direction of the wind and the course of the ship. Braces are, for this purpose, fastened to the extremities of the yard, which are called the *yard-arms*.

All the braces of the yards are double, except those of the top-gallant and spritsail-topsail-yards. The mizen-yard is furnished with *fangs*, or vangs, in the room of braces. See the article MIZEN.

BRACE, is also a name given to pieces of iron, as supports to various machines in a ship, such as the poop, lanterns, &c.

To BRACE *the Yards*, (*brasser*, Fr.) is to move the yards by means of braces, to any direction required.

To BRACE *about*, (*brasser à virer*, Fr.) is to turn the yards round for the contrary tack.

To BRACE *the Yards sharp up*, (*brasser les vergues au plus près du vent*, Fr.) is to brace the yards to a position, in which they will make the smallest possible angle with the keel; for the ship to have head-way.

To BRACE-*to*, (*brasse au vent*, Fr.) is to ease off the lee-braces, and round in the weather-braces, to assist the motion of the ship's head in tacking.

BRACES, that security for the rudder which is fixed to the stern-post, and to the bottom of the ship

BRACKETS, (*courbâtons des herpes*, Fr.) short crooked timbers resembling knees, fixed in the frame of a ship's head, to support the gratings: they also serve to support the gallery.

BRACKETS, ornaments; the hair bracket is the boundary of the aft-part of the figure of the head, the lower part of which finishes with the fore part of the upper cheek. The console-bracket is a light piece of ornament at the fore part of the quarter-gallery, sometimes called a *canting livre*.

BRAILS, (*cargues*, Fr.) are ropes passing through pullies on the mizen-mast and yard, and fastened, in different places, to the aftermost leech of the sail, to truss it close up, as occasion requires. See the article MIZEN.

BRAILS, is also a general name given to all the ropes which are employed to *haul up*, or collect to their yards, the bottoms, lower corners, and skirts of the other great sails, for the more ready *furling* them whenever it may be found necessary. The operation of thus drawing them together, is called brailing them up, or hauling them up in the brails.

BRAILS are likewise used to the mizen-stay-sail and main-top-sail, either for the purpose of furling them, or preventing them from being chafed by flapping about from side to side, when going before the wind.

To BRAIL *up*, (*carguer*, Fr.) to haul up a sail by means of the brails, for the more readily furnishing it when necessary; as "brail up the main-topmast-stay-sail!" (*cargue la voile d'étai de hune*, Fr.)

BRAKE, (*brimbale de pompe*, Fr.) the handle, or lever, by which a common ship-pump is usually worked: it operates by means of two iron bolts thrust through the inner end of it, one of which rolling across two cheeks or ears, in the upper end of the pump, serves as a fulcrum for the brake, supporting it between the cheeks. The other bolt connects the extremity of the brake to the pump-spear, which draws up the *box*, or piston, charged with the water in the tube. See the article PUMP.

BRASSES, let into the heads of laying or spinning wheels, are about four inches long, and two broad. In the middle of the upper end is a hole for the spindles of the wheels to work in.

BREACH, in fortification, (*brèche*, Fr.) a gap made in any part of the works of a town by the cannon of the besiegers, in order to make an attack upon the place. To make the attack more difficult, the besieged sow the breach with crow-feet, or stop it with *chevaux de frize*.

The batteries to make a breach should commence by marking out, as near as possible, the extent of the breach intended to be made; first, by a horizontal line within a fathom of the bottom of the revetement in a dry ditch, and close to the water's edge in a wet one; and then by lines perpendicular to this line, at short distances from each other, as high as the cordon; then, by continuing to deepen all these cuts, the wall will give way in a body. The guns to produce the greatest effect, should be fired in as near as possible in salvos or vollies. The breach should be one-third the length of the face, from the centre towards the flanked angle. When the wall has given way, the firing must be continued, to make the slope of the breach practicable.

BREAD, (from βρωτὸν, Gr. *food*, *pain*, Fr.) is a light porous, spongy substance, prepared from the flour of certain farinaceous seeds, especially wheat; kneaded with yeast, lukewarm water, and a small portion of salt, and then properly baked.

BREAD, for the navy. See the article BISCUIT.

BREAD-*room*, (*soute à pain*, Fr.) in a ship, that part in the hold adjoining the after-magazine, destined to hold the bread or biscuit. The bread-room is feather-edged, well lined with board, and sometimes with tin plates. It is also proper to warm it well with hanging stoves used for that purpose, for several days before the biscuit is put into it; since nothing is more injurious to bread than moisture.

BREADTH, (*largeur*, Fr.) the measure of a ship from side to side in any particular place: it is usually distinguished into extreme breadth, *ligne du fort*, main-breadth, and top-timber-breadth.

As the sides of the ship are formed by a variety of ribs, called timbers, and the areas of those timbers being of different breadths above and below, it is necessary to distinguish them in the construction, in order to form their several curves, and fix the corresponding pieces with more accuracy and precision. The part of every timber which encloses the greatest space from the middle-line of the ship's length, is therefore called the *main*-breadth; and the distance between the upper-part of the same timber and the middle-line of the ship's length, is called the *top-timber*-breadth.

As the ship is also broader at the midship-frame than in any other point of her length, the distance between her sides in the main-breadth of that timber, is called the *extreme* breadth of the ship.

BREADTH-*sweep*, the radius of the arch which forms part of the curve of a ship's timber.

BREAK OF A BLOCK, the opposite end to that which reeves the fall.

BREAK *of a Deck*, is that part where it terminates, and the descent on the next deck below it commences.

BREAKERS, (*brisans*, Fr.) a name given by sailors to those billows that break violently over rocks lying under the surface of the sea. They are distinguished both by their appearance and sound, as they cover that part of the sea with a perpetual foam, and produce a hoarse and terrible roaring, very different from what the waves usually have in a deeper bottom. When a ship is unhappily driven amongst breakers, it is hardly possible to save her, as every billow that heaves her upwards serves to dash her down with additional force, when it breaks over the rocks or sands beneath it.

To BREAK *ground*, (*point de départ*, Fr.) the first weighing anchor and quitting a place. West Indiamen are obliged to break ground at an established time, in order to save insurance. In the army, it signifies the first opening of trenches against a place.

BREAKING-*bulk*, (*commencer la décharge d'un vaisseau*, Fr.) the act of beginning to unload a ship, or of discharging the first part of the cargo.

To BREAK *sheer*, (*vaisseau à l'ancre forcé hors de sa position*, Fr.) when a ship at anchor is laid in a proper position to keep clear of her anchor, but is forced by the wind or current out of that position, she is said to break her sheer.

To BREAK-*up*, (*démolir un vaisseau*, Fr.) to rip off the planks of a ship, and take her to pieces, when she becomes old and unserviceable.

BREAK-*water*, (*vieux vaisseau placé à l'entrée d'un port pour rompre la mer*, Fr.) the hull of some old ship or vessel, sunk at the entrance of a small harbour, to break off, and diminish the force of the waves, as they advance towards the vessels moored within.

BREAK-*water* is also a sort of small buoy, fastened to a large one in the water, when the buoy-rope of the latter is not long enough to reach from the anchor, lying at the bottom, to the surface of the water. The use of this break-water is therefore to shew where the large buoy swims. See the article BUOY.

The Right Hon. Charles Yorke, late first Lord of the Admiralty, did, on the 16th day of March, 1812, obtain a grant in the House of Commons for a sum of 80,000*l*. for the purpose of purchasing quarries, and defraying other expenses incident to making a *breakwater* in Plymouth Sound, which when completed, will, according to the Hon. Gentleman's statement, make a sufficient roadstead for 36 ships of the line to ride in; and that every person the least acquainted with the naval concerns of this country, knew that port for the last twenty years had not been considered safe, and our fleets were obliged in all cases of danger to have recourse to Torbay and Cawsand Bay.

There is one consideration which sufficiently shews that it is not an idle speculation, as a similar plan was projected by the celebrated Vauban, for the improvement of the harbour of Cherbourg, and carried into effect under the administration of the Duc de Castries, and no good reason can be assigned why the improvement of Plymouth Sound should not be equally practicable.

BREAMING, (*action de chauffer un vaisseau; chauffage*, Fr.) burning off the filth, such as grass, ooze, shells, or sea-weed, from a ship's bottom, that has gathered to it in a voyage, or by lying long in a harbour. It is performed by holding kindled furze, fagots, or reeds to the bottom, which, by melting the pitch, sulphur, &c. that had formerly covered it, loosens whatever filth may have adhered to the planks, which is both scraped and brushed off: the bottom is then covered anew with a composition of sulphur, tallow, &c. which not only makes it smooth and slippery, so as to divide the fluid more readily, but also poisons and destroys those worms which eat through the planks in the course of a voyage. This operation may be performed either by laying the ship aground after the tide has ebbed from her, or by *docking* or *careening*, which articles see; as also COAT and STUFF.

BREAST-FAST, (*amarre qui tient un vaisseau par le travers*, Fr.) a large rope, employed to confine a ship side-ways to a wharf or quay, or to some other ship; as the head-fast confines her forward, and the stern-fast, abaft.

BREAST-HOOKS, (*guirlandes pour fortifier l'avant-d'un vaisseau*, Fr. from *breast* and *hook*) are thick pieces of timber, incurvated into the form of knees, and used to strengthen the fore-part of the ship, where they are placed, at different heights, directly across the stem, so as to unite it with the bows on each side.

The breast-hooks are strongly connected to the stem and hawse-pieces, by tree-nails, and by bolts driven from without, through the planks and hawse-pieces, and the whole thickness of the breast-hooks, upon whose inside those bolts are forelocked, or clinched, upon rings. They are usually about one-third thicker, and twice longer than the knees of the decks which they support.

There are generally four or five of these pieces in the hold, between the kelson and the lower-deck, upon the uppermost of which the planks of that deck are rabetted. There are two placed between the lower and the second decks, one of which is immediately beneath the hawse-holes, and the other under the second deck, whose planks are inlaid thereon, and upon which the inner-end of the bowsprit frequently rests.

The fore-side of the breast-hook, which is convex, is formed so as to correspond with the place in which it is stationed, that is to say, it conforms exactly to the interior figure of that part of the bow where it ought to be fayed: accordingly, the branches, or arms, of the breast-hooks, make a greater angle, as they are more elevated above the keel, whilst the lower ones are more incurvated, and are almost figured like the crotches.

It not being necessary that the inner, or concave, side of these pieces should retain a regular form, the artificers frequently let them remain as thick as possible, to give additional support to the ship's fore-part, where she sustains the whole shock of resistance in dividing the fluid, or in plunging into it.

It is evident that the connexion and solidity of the ship in this place will be reinforced in proportion to the strength and extent of the breast-hooks, so that they may cover a greater number of the head-timbers across the stem, to strengthen the fore-part of the ship, and unite the bows on each side.

BREAST-RAIL, (*lisses du fronteau*, Fr.) the upper rail of the balcony, or of the breast-work, on the quarter-deck.

BREAST-WORK, (*fronteau*, Fr.) a sort of balustrade of rails, or mouldings, which terminates the quarter-deck and poop at the fore-ends, and also encloses the forecastle both before and behind.

BREATH, (*une fraicheur—un souffle de vent*, Fr. a small breath of wind.

BREECH OF A GUN, (*la culasse*, Fr.) the distance from the hind part of the base ring to the beginning of the bore, and is always equal to the thickness of the metal at the vent.

BREECH, in ship-building, the angle of knee-timber, the inside of which is called the *throat*.

BREECHING, (*brague d'affût*, Fr. from *breech*) a strong rope, used to secure the cannon, an prevent them from recoiling too much in the time of battle.

It is fixed by reeving it through a thimble, strapped upon the cascabel, or pumelion of the gun, and the two ends of it are then clinched on each side of the port, to ringbolts, in the ship's sides.

The breeching is of sufficient length to let the muzzle of the cannon come within the ship's side, to be charged, or to lie housed.

BREEZE, (*brise*, Fr.) a shifting wind, that blows from sea and land, alternately, during certain hours of the day or night, and only sensibly felt near the coasts; it commonly rises in the morning, about nine o'clock, proceeding slowly in a fine small black curl on the water, towards the shore; it increases gradually till twelve, and dies away about five. Upon its ceasing, the land breeze commences, which increases till twelve at night; and is succeeded, in the morning, by the sea breeze again.

Mr. Dalrymple, late hydrographer to the Right Honourable the Lords Commissioners of the Admiralty and East India Company, scientifically arranged the wind in the following order:—

1. Faint Air.	7. Moderate Gale.
2. Light Air.	8. Brisk Gale.
3. Light Breeze.	9. Fresh Gale.
4. Gentle Breeze.	10. Strong Gale.
5. Fresh Breeze.	11. Hard Gale.
6. Gentle Gale.	12. Storm.

Land-BREEZE, (*brise de terre*, Fr.) a current of air, which, in many parts within the tropics, regularly sets, from the land, towards the sea during the night; and this, even on opposite points of the coast, it is very dangerous sleeping in the open air, during the time it lasts, particularly in the East or West Indies.

Sea BREEZE, (*brise du large*, Fr.) the current of air, which blows during the day, from the sea, upon the shore, in hot climates. The sea-breezes do not all come from the same point of the compass, but from different points, as the land lies.

BREWING, A SQUALL BREWING, (*un grain qui se forme*, Fr.) the appearance of black tempestuous clouds, gradually collecting, in a particular part of the hemisphere, as the fore-runner of a storm.

BRIDGE, (*planche placée à travers de la chaudière*, Fr.) in rope-making an oak plank, thirty-two inches broad, and two thick, fixed across the top of the kettle in which the yarn is boiled, with a mortise through the middle to admit the step, and a hole at the end, for the yarn to pass through to the nipper.

BRIDLES, the upper-part of the moorings, laid in the king's harbours, to ride ships or vessels of war. See the article MOORINGS.

BRIDLES *of the Bowline*, (*pattes de bouline*, Fr.) short legs, or pieces of rope, running through iron thimbles, by which the bowline is fastened to different places on the edge, or skirt, of a large sail.

As the current of air enters the cavity of the sail in a direction nearly parallel to its surface, it follows that the ridge of the sail must necessarily be shaken by the wind, unless it is kept tight forward; but, as a single rope has not been found sufficient to confine the whole skirt of the sail, inasmuch as it only draws upon one part thereof, it became necessary to apply bridles, or legs, spreading out from the bowline.

BRIG, or BRIGANTINE, (*brigantin ou brig*, Fr.) a small merchant-ship, with two masts. Amongst English seamen this vessel is distinguished by having her main-sail set nearly in the plane of her keel; whereas the main-sails of larger ships are hung athwart, or at right angles, with the ship's length, and fastened to a yard, which hangs parallel to the deck: but, in a brig, the foremost edge of the main-sail is fastened, in different places, to hoops, which encircle the main-mast, and slide up and down it, as the sail is hoisted or lowered; it is extended by a gaff above, and by a boom below.

This term is not universally confined to vessels of a particular construction, or which are masted and rigged in a method different from all others. It is variously applied, by the mariners of different European nations, to a peculiar sort of vessel of their own marine.

To BRING BY THE LEE, (*faire chapelle*, Fr.) to incline so rapidly to leeward of the course, when the ship sails large, as to bring the lee-side unexpectedly to windward; and, by laying all the sails aback, expose her to the danger of upsetting. See the article *To* BROACH-TO.

To BRING-TO, (*mettre en panne*, Fr.) to check the course of a ship, by arranging the sails in such a manner as that they shall counteract each other, and keep her nearly stationary, when she is said to lie-by, or lie-to, having, according to the sea-phrase, some of her sails *aback*, to oppose the force of those which are *full*; or having them otherwise shortened, by being *furled*, or *hauled up in the brails*.

To BRING-TO, is also used in applying a rope to the capstan; as, "to bring-to the messenger," (*frapper le tournevire sur le cable*, Fr.)

To BRING *to Action*, (*forcer au combat*, Fr.) Exam. "We brought the enemy's rear to close action."

To BRING-*up*, (*jeter l'ancre ou mouiller*, Fr.) to cast anchor, is a provincial phrase peculiar to the seamen in the coal trade.

BRINGING-TO, (*mettre en travers*, Fr.) is the act of putting a ship in a situation, so that she shall wait the approach of some other, that may be advancing towards her, in order to her being examined, &c. It also implies, to retard a ship's course, occasionally, near any port, in the course of a voyage.

To BROACH-TO, (*faire chapelle*, Fr.) differs from to bring by the lee, in that it is a rapid inclination to windward of the ship's course; in other respects the effects and danger are nearly the same: these accidents may happen by the negligence or incapacity of the helmsman, by the force of the sea, by carrying away some of the sails, or by disabling the rudder or its appendages.

It is easy to conceive that a ship will carry much more sail before the wind than when she makes a progress with her side to its direction; because, when the current of wind acts nearly endways on her hull, the pressure of it on the masts must be considerably diminished, as she yields to its impulse, and flies before it; and that if she carries a great sail at this time, it can only press her fore-part lower down in the water. But if, when she carries a great extension of sail, her side is suddenly brought to the wind, it may be attended with the most fatal consequences, as the whole force of it then pours, like a torrent, into the cavities of the sails. The masts, therefore, unavoidably yield to this strong impression, acting like levers on the ship sideways, so as nearly to upset her, unless she is relieved by some other event, which may be also extremely dangerous, such as the sails rending to pieces, or the masts being carried away.

BROADSIDE, (*bordée*, Fr.) in a naval engagement the whole discharge of the artillery on one side of a ship of war, above and below; as,

"We poured a broadside into the enemy's ship," *i. e.* discharged all the ship's cannon on one side upon her.

"She brought her broadside to bear on the castle;" that is, the ship was so disposed, as to point all her cannon to it within point-blank range.

"A squall of wind laid the ship on her broadside;" that is, pressed her down in the water, so as nearly to upset her.

BROADSIDE also implies all that part of a ship above the water, which is situated between the bow and quarter, and in a position nearly perpendicular to the horizon.

BROKEN-BACKED, (*arqué*, Fr) the state of a ship so loosened in her frame, either by age, weakness, or some great strain, as to droop at each end.

This circumstance is more common amongst the French, than the English or Dutch ships, owing partly to their great length, and to the sharpness of the floor, whose breadth is not sufficiently carried from the middle towards each end; and partly from being frequently obliged to have a great weight on both ends, when they are empty in the middle, at the time of discharging one cargo and taking in another. See the article CAMBERED.

BUCANIERS, (*flibustiers, ou, boucaniers*, Fr.) a name given to certain piratical rovers of various European nations, who formerly infested the coasts of Spanish America; and, under pretence of traffic with the inhabitants, frequently seized their treasure, plundered their houses, and committed many other depredations. They were originally inoffensive settlers in Hispaniola, but were inhumanly driven from their habitations by the jealous policy of the Spaniards—whence originated their implacable hatred to that nation.

BUCKETS, (*seaux*, Fr.) are made either of canvas, leather, or of wood; the latter are used, principally, for washing the decks, and, therefore, answer the purpose of pails.

BUCKLERS, (*deux pièces de bois accouplées pour boucher les ecùbiers*, Fr.) two pieces of wood fitted together, to stop the hawse holes, leaving only sufficient space between them for the cable to pass, and thereby preventing the ship taking in much water in a heavy sea.

BUGALET, (*bugalet*, Fr.) a small vessel with two masts, used on the coast of Brittany. The fore-mast is very short; and, on each mast is carried a square sail, and, sometimes, a topsail over the main-sail. They have a bowsprit, and set one or two jibs.

BUILDINGS, Marine, (*édifices de marine*, Fr.) are those constructed for making or preserving ships, and other vessels, as docks, arsenals, store-houses, &c.

BUILT, (*construit, bâti*, Fr.) the particular form or construction of a ship, by which she is distinguished from others of a different class, construction, or nation. Thus we say, a ship is said to be frigate-built, galley-built, carvel-work-built, clincher-work-built; or, English-built, French-built, American-built, &c.

BULGE, it was originally written bilge, the part of a ship which bulges out at the floorheads, to assist the ship when taking the ground.

BULGEWAY, a large piece or pieces of timber bolted together, making one solid piece, which is placed under the bulge of a ship, to support her when launching. The support for the bulgeways to lie on, is called *ways*, which sometimes are placed straight, and sometimes camber; but if they do camber, it should be truly circular; though sometimes the curve is quicker at the lower-part, but this is liable to strain the sheer of the ship. Their extreme distance is generally about one-third the breadth of the ship, but this must depend on the form of the midship-bend.

BULK of a Ship, implies the whole cargo stowed in the hold.

Bulk-*heads*, (*cloisons qui traversent le vaisseau*, Fr.) are partitions built up in several parts of a ship between two decks, either lengthways or across, to form and separate the various apartments. Some of those which are built across the ship, are remarkably strong. Those which inclose the magazine are oak-plank and rabbetted, as are those of the spirituous liquor-room; others in the hold are generally oak-plank, the edges are cyphered, to keep the gravel, or dirt, of the ballast in the hold from getting into the store-rooms, or into the well to damage the pumps.

The Bulk-*head afore*, is the partition between the fore-castle and gratings in the head, and in which are the chase ports.

BULLET, (*une balle*, Fr.) a leaden ball, or shot, with which small-arms are loaded.

Bullets, balls, or shot, have different denominations, according to the use that is made of them, viz.

Hollow Bullets, or Shells, are of a cylindrical shape, and have an opening and a fuse at the end, by which fire is communicated to combustibles within, when the shell bursts with a destructive explosion.

Branch Bullets, are two balls, connected with an iron bar.

Two Headed Bullets, or Angles, are two halves of a bullet connected by a chain or bar.

The diameter of a common leaden bullet is found by dividing 1.6706 by the cube root of the number contained in one pound.

The diameter of musket bullets differs but one-fiftieth part from that of the musket bore; for if the shot but just rolls into the barrel, it will be sufficient.

BULL'S-EYE, (*margouillet, ou cosse*, Fr.) a sort of small pulley in the form of a ring, having a rope spliced round the outer edge of it, (which is hollowed to admit of the rope) and a large hole in the middle for another rope to slide in. See plate II. fig. 18. It is seldom used by the English seamen, and then only for the main and fore bowline bridles. It is spliced in the outer end of the bowline, and sliding along the bridle, to rest in the most apposite place, draws it tight above and below.

Bull's-eye, a term given to the patent reflectors let into the ports or deck.

BUM-BOAT, (*bateau de provision*, Fr.) a sort of wherry, used in and about harbours to carry provisions, &c. for sale to ships lying at a distance from the shore, and are frequently called shore-boats, to distinguish them from those belonging to ships.

BUMKIN, or Boomkin, (*boute-lof*, Fr.) a short boom or beam of timber, projecting from each *bow* of a ship, to extend the clew, or lower edge of the foresail to windward; for which purpose there is a large block fixed on its outer end, through which the *tack* is passed, which being drawn tight down, brings the corner of the sail close to the block, then the *tack* is said to be aboard.

The bumkin is fitted above the main-rail in the head, and secured by a strong rope, or iron brace, which confines it downward to the ship's bow, to counteract the strain it bears from the fore-sail above, dragging it upwards.

Bumkin *of a Boat*, (*boute-hors de tappe-cul*, Fr.) is a small outrigger over the stern, usually serving to extend the mizen.

BUNT, (*les fonds, ou la cavité des voiles*, Fr.) the middle part, or cavity of the square sails, as the main-sail, fore-sail, top-sails, and top-gallant-sails. If one of those sails is supposed to be divided into four equal parts, from one side to the other, then may the two middle divisions, which comprehend half of the sail, be properly called the limits of the bunt.

BUNTINE, (*étamine*, Fr.) a thin woollen stuff, of which the colours, or flags, and signals of a ship are usually formed.

BUNTLINE-CLOTH, the lining sewed up the sail in the direction of the buntline, to prevent the sail being chafed.

BUNTLINES, (*cargues fonds*, Fr.) are ropes fastened to cringles on the bottoms of the square sails, to draw them up to their yards: they are inserted through certain blocks above, or on the upper part of the yard, whence passing downwards on the fore-part of the sail, they are fastened below to the lower edge, in several places of the *bolt-rope*.

BUOY, (*bouée*, Fr.) a sort of close cask, or block of wood, fastened by a rope to the anchor, to point out its situation, that the ship may not come too near it, to entangle her cable about the stock, or the flooks of it.

Buoys are of various kinds; as,

Cable-Buoys, (*barrils ordinaires qui servent de bouée au cable*, Fr.) are common casks employed to buoy up the cables in rocky anchorage, to prevent their rubbing against the rocks. In the harbour of Alexandria, in Egypt, every ship is moored with at least three cables, and has three or four of these buoys on each cable for this purpose.

Can-Buoys, (*bouée en forme de cone*, Fr.) these are in the form of a cone, and of this construction are all the buoys which are floated over sands, banks, and shallows, as marks for ships to avoid them. They are made extremely large, that they may be seen at a distance, where there are several near each other: Plate IV. Fig. 20.

they are fastened by strong chains to the anchors which are sunk for this purpose at such places, and are distinguished by the colour, as black, red, and white.

Some are of opinion, that can-buoys to swim the buoy-rope, and to watch in a tide-way, are the most buoyant, but very cumbersome to get into a boat or launch, or to haul up a ship's side: therefore, they give the preference to a nun-buoy, as being handier to get into a launch, or haul up a ship's side, or to carry a rope to, to haul a ship round in a calm, and to take the turns out of the hawser, when a ship is moored, or in transporting ships.

Floating Buoy. Mr. Peache, of Lambeth, has taken out a patent for a *Floating Buoy* on a new construction, dated April, 1807.

It is formed by pieces of wood longitudinally applied to each other, after the manner of staves; but, instead of making the joints to consist of flat surfaces, evenly or squarely applied to or against each other, they are rabbetted together, by making the half thickness of each piece to overlap the other.

The faces are to be tarred, and applied to each other, and upon one of them is spread a covering of hair, &c. and upon this is applied, in a melted state, a coat of pitch and tar, after which the joints are put together, and secured by screws, pins, bolts, and by hoops enveloping the whole. The heads of the buoys are put in by rabbetting the same in with the staves: the fitting is made good by several steps or successive faces in the joints; between which, pitch, tar, hair, &c. are applied, and, for greater security, there are sometimes two, or even more additional heads, fixed within the external heads. In the buoy is fixed a trunk, or perforated piece, so made as not to allow the water to enter the said buoy without; but to afford a passage for a chain, rope, &c. for mooring. The shape which Mr. Peache recommends as the best for his newly invented buoys is cylindrical, or prismical, or conical pyramidal figures, and when the cross section of the figure is not circular, the number of sides should be eight. Buoys constructed on this plan, it is generally thought, will be found to answer much better than those in common use, as they will last longer in a perfect state.

Life-Buoy, (*bouée de sauvetage*, Fr.) is generally of the Can kind, though sometimes it is made of cork. It is furnished with a small flag on the top, and is used to throw overboard for a person, who has fallen into the sea, to lay hold of: while the flag serves to direct a boat to the spot, and thereby frequently saves the life of a fellow-creature.

Nun-Buoys, (*bouée en baril, en forme de deux cônes*, Fr.) are shaped like the middle frustum of two cones, abutting upon one common base, (plate

Plate V. Fig. 1. V. fig. 1) being casks, which are large in the middle, and tapering nearly to a point, at each end.

Wooden-Buoy, (*bouée de bois*, Fr.) a solid piece of light timber, sometimes in the shape of a cylinder, and sometimes of a nun-buoy, having one or two holes through the ends, in which is fixed a short piece of rope, whose two ends being spliced together, make a sort of circle or ring called the strap.

Buoy-*rope*, (*orin de l'ancre*, Fr.) the rope which

Plate V. Fig. 2. fastens the buoy to the anchor: it should be a little more than equal in length to the depth of the water where the anchor lies, as it is intended to float near, or immediately above the bed of it, that the pilot may at all times know the situation thereof.

See Plate V. fig. 2; *b* is the anchor, *c* the buoy-rope, and *d* the buoy floating on the surface of the water.

The buoy-rope is often extremely useful in drawing up the anchor when the cable is broke; it should therefore be always of sufficient strength for this purpose, else the anchor may be lost through negligence.

The Slings of the Buoy, (*garniture de la bouée*, Fr.) are the ropes which are fastened about it, and by which it is hung: they are curiously spliced round the buoy, something resembling the braces of a drum.

To stream the Buoy, (*mettre la bouée à l'eau*, Fr.) is to let it fall from the ship's side into the water, preparatory to letting go the anchor, that it may not be retarded by the buoy-rope as it sinks to the bottom.

Buoys, (*bouées*, Fr.) for general purposes, were first laid down at the entrance of the river Thames in 1538. They were then nearly of the same construction that they are at present, a very faulty construction, as in stormy weather they are frequently removed, which occasions the loss of many lives and vessels. During a period of upwards of 270 years, very little improvement has taken place in the system of buoyage, though much has been done for every other branch of navigation.

A few premiums for buoys, on improved principles, might be attended with very happy consequences to navigation.

Philanthropy and the general interest of humanity have induced a great number of persons, in different countries, to exert the best abilities of their mind to effect this very desirable purpose, though not strictly to be denominated Life-Buoys, yet as closely connected with the same salutary and valuable purpose: the most important of these inventions are, with more propriety, perhaps, to be classed under this head than any other.

At Elsinore, a place about eight English miles from Helsingburg, in Sweden, a remarkable experiment was tried, which is expected to be of the utmost consequence to society. Professor Pelt had in October 1802, notwithstanding the severity of the season, produced in the presence of some well known mechanics, many experiments; among which was a newly invented nautical machine, founded upon the hydrostatic principle for saving persons from drowning. During his late residence there, he repeated the experiment of his new machine in the Sound, in order, to try its effect against the currents of the North Sea and the Baltic.

After a sea-wanderer (a young waggoner, of the name of Isaac Peterson) had equipped himself in this machine, he sprang into the sea, in the presence of numerous spectators, many of whom accompanied him in sailing-boats.

At his departure he fired a pistol twice; and on his passage, he ate, drank, and smoked a pipe, to shew he had the entire use of his arms. Being benumbed with cold, he got into a boat to warm himself by rowing; but soon sprang into the water again and happily reached the Swedish coast, after a passage of one hour and three quarters. On his arrival, he saluted the spectators with a pistol-shot, and waved a flag which he drew from the sea. Having remained a short time at Helsingburg, he returned to the Danish coast, where he arrived amidst the shouts of the curious who lined the shore. According to the opinion of experienced seamen, this machine far surpasses any former invention of the kind, both in regard to its solidity and lightness (weighing but 7 lbs. English), and on account of the simplicity of its structure. It is fastened round the waist, and the person wearing it enters the water up to the middle of his body.

Another invention for the improvement of buoys is thus described: a three inch plank, eighteen inches wide at bottom, and nine inches wide at top, let through the centre of a piece of timber one foot square, and six feet long, the lower part of the plank to be under this piece. The length of this plank and its chain to be proportioned to the depth and rise of water, and the cross piece to be always four feet under the low water line.

A vane may be fixed on the top of the buoy, which should consist of a balk or spar of six inches diameter, lined on two sides with thin strips of iron, and a piece of timber, to be called the float, bored through like a water-pipe, six feet long by one foot thick, two or three feet wide at one end, and one foot wide at the other, tapered off from the thickness to the width. The bore to be large enough for the spar to pass through it freely, so that the float cannot fail to rise and fall with the water. There should be a bolt, or other contrivance, to prevent the possibility of the floats slipping off the spar.

The length of spar and chain is also to be proportioned to the depth and rise of water.

A stout iron rod, with an eye at each end, may be used instead of the chain.

Several other contrivances have been suggested, as tending to the preservation of sea-faring persons, whenever the dreadful accidents sometimes occurring take place. Among these are the common expedients of Cork Jackets, Seaman's Friend, &c. variously constructed, according to the whim or fancy of the inventor. To say nothing of the difficulty attending the immediate use of them, the possession of such articles of supposed preservation may, perhaps, be extremely pregnant of mischief in the hour of danger, and produce a relaxation of those energetic exertions, which, if properly made, might tend to the preservation of all parties.

Among other schemes, though not strictly to be called visionary, a very singular experiment made on the river Thames must not be passed over in silence. The invention is described in the following terms:—Six persons who had previously had the machine girthed on, leaped out of their boats above Westminster-bridge, and afforded a very curious sight to the spectators, the body being in a perpendicular position, the head and upper part of the shoulders only being above the water. In this state their arms, legs, and all their limbs being at perfect liberty, they went down with the tide through Blackfriars-bridge, surrounded by a great number of boats; they went through the middle of the stream, until they passed Strand-lane, where they came closer on the shore: they then came abreast of the Temple, where they again put off, which they did, with no farther trouble than a trifling steerage with their hands.

The machine is very simple in contrivance, being composed of eight divisions, each made of strong sheet copper, firmly soldered and japanned, much on the principle of balls used to float on cisterns; they are connected by straps, and when the machine is put on, the lower girth fastens round the chest; from thence it is rolled up over the breast, but not higher than the arm-pits, from the upper part straps go over the shoulders, and are secured to the girth round the chest. When on, it has much of the appearance of a horse's-collar, the large end downwards. It is very safe, because, if even two or three of the divisions should by any accident, but which is hardly possible, lose their wind and take water, there will still be buoy enough to keep up the body.

An invention somewhat similar to the foregoing, was tried off Greenwich Hospital a few years before, as appears by the following certificate, containing a description of the apparatus, its mode of application, and effect:—

"We whose names are hereunto subscribed, do "certify and declare, that, at the solicitation of "Lawrence Collin, Brazier and Tinman, we did "this day attend in a boat on the river Thames, "off the Royal College, Greenwich, to see a fair "trial made of an invention of the said Lawrence "Collin, for saving persons (though not swimmers) "from drowning, at the same time allowing them the "free use of their arms, of which the following is "the result.

"A copper tube, in the form of a crescent, with "which the experiment was made, weighs nearly "five pounds, and is, we understand, divided into "several cells or compartments lined with block tin, "so as to prevent the natural air pent up from "escaping, and in case of injury or accident to any "one or two of the internal divisions of the machine, "it would still have sufficient power to float bodies. "This *bent* tube is formed to fit the body under the "arms, and any person may fit it upon himself with "the straps in about half a minute.

"With this instrument James Butters, the partner "of Collin, a man of five feet ten inches in stature, "and of extraordinary weight, plunged into the "water and floated in an erect posture, as well as "on his sides, back, and belly, with his hands and "feet at rest; and it appeared to us that he could, "when he pleased, take a sitting posture with his "arms resting upon his knees, having his head and "shoulders perfectly afloat; he also rolled himself "about, and turned with great facility in every "direction.

"Witness our hands this 15th day of Oct. 1801.

(Signed)

"JOHN HUNTER, *Capt. R.N. and late Governor of New South Wales.*

"JOHN BOURCHIER, *Capt. R.N. Greenwich Hospital.*

"WILLIAM HUNTER, *Lieut. R.N. Greenwich Hospital.*

"GEORGE KNIGHT, *Clerk of the Works, Greenwich Hospital.*

"CHARLES VENNER, *Barrister at Law, Stone-buildings, Lincoln's-inn.*

"JOHN M'ARTHUR, *York-place, Portman-square.*"

A variety of other schemes have also been suggested by different persons, for the conveyance of hawsers or ropes on shore from a stranded vessel, and also from the shore to the vessel. See the article SHIPWRECK.

None of these appear fit to be depended on in the hour of danger; and the best hope, or foundation for hope that the crew of any vessel in danger of being wrecked can or may have, ought to spring from their own perseverance and activity: not that these preservatives or experiments ought to be condemned, but recommended to be only used where they can, without interfering with other duties. Under that idea, this list of contrivances, &c. has been given.

BURDEN, or BURTHEN, (*port en tonneaux d'un bâtiment*, Fr. *byrthen*, Sax.) the weight or measure of any species of merchandise that a ship will carry when fit for sea.

The burden of a ship is commonly determined thus, by builders: multiply the length of the keel, taken within board, by the breadth of the ship within board, taken from the midship-beam from plank to plank, and multiply the product by the depth of the hold, taken from the plank below the kelson, to the under part of the upper-deck plank, and divide the last product by 94, then the quotient is the content of the tonnage required. By this result, builders are paid a certain price per ton for building vessels of different sizes: but for a more particular account, see the article TONNAGE.

BURGOO, a sea-faring dish, made by gradually adding two quarts of water to one of oatmeal, so that the whole may mix smoothly; then boiling it for a quarter of an hour, stirring it constantly; after which, a little salt, butter, and sugar, is generally added. It is a cheap and strengthening diet, held by Cockburn very proper to correct that viscidity of humours and costiveness to which the other diet of sailors so much disposes them.

BURNING-GLASS, or BURNING-MIRROR, a machine by which the sun's rays are collected into a point, and thus their force and effect considerably increased, so as to consume objects within its reach.

There are two kinds of burning glasses, namely, *convex* and *concave*. Those of the convex form are lenses, which, by acting according to the laws of refraction, incline the rays of light towards the axis, and unite them in a point, or focus. The concave ones are mirrors, or reflectors, whether made of polished metal, or silvered glass; which by the laws of refraction, throw back the rays into a point before the glass.

These instruments are, undoubtedly, of very ancient origin: the most celebrated were those of Archimedes and Proclus; by the former of which, the fleet of Marcellus was destroyed or reduced to ashes, at the distance of a bow-shot.

In modern times, there have been several inventions of this kind, remarkable for their large diameter and powerful effects; the principal of which are those of Magine, of 20 inches diameter; of Sepatala of Milan, near 42 inches diameter; of Settala of Villette, of Tschirnhausen, of Trudaine, of Parker, and of Buffon; the latter of whom made one that consisted of 400 mirrors, which reflected all their rays to one point, and with this he could melt lead and tin, at the distance of 140 feet.

Sir Isaac Newton presented a burning-glass to the Royal Society, which consisted of seven concave glasses, so placed that all their foci join in one physical point. This glass vitrifies brick or tile in one second, and melts gold in half a minute.

A large burning lens, for fusing and vitrifying such substances as resist the fires of furnaces, and especially for the application of heat *in vacuo*, and in certain other circumstances in which heat cannot be applied by other means, has long been a *desideratum* with persons concerned in philosophical experiments: and this it appears is now in a great measure accomplished by Mr. Parker, an ingenious glass-manufacturer in *Fleet-street, London*. His lens is made of flint glass, and is 3 feet in diameter, but when fixed in its frame, exposes a surface of 32 inches in the clear; the length of the focus is 6 feet 8 inches, and its diameter one inch. The rays from this large lens are received and transmitted through a smaller, of 13 inches diameter in the clear within the frame, its focal length 29 inches, and diameter of its focus three-eighths of an inch: so that this second lens increases the power of the former more than seven times, or as the square of 8 to the square of 3.

From a great number of experiments made with this lens, it is found to fuse substances in much less time than any yet invented.

BURR, (*ciseau à mortaise*, Fr.) a triangular hollow chisel, used to clear the corner of mortises.

BURTON, (*palanquin*, Fr.) a small tackle, formed by two blocks or pullies.

It is generally used to set up or tighten the top-masts shrouds, whence it is frequently termed a TOP-BURTON-TACKLE: but it is equally useful to move or draw along any weighty body in the *hold*, or on the *deck*, as anchors, bales of goods, large casks, &c.

BUSH, a circular piece of metal, usually of brass, let into the lignum vitæ sheaves of such blocks as have iron pins, and thereby preventing the sheave from wearing, without adding much to its weight.

BUSS, (*bûche*, Fr. *busse*, Germ.) a ship of two masts, used by the English and Dutch in their herring fisheries. It is generally from fifty to seventy tons burden; being furnished with two small sheds or cabins, one at the prow and the other at the stern; the former of which is employed as a kitchen.

BUTT, (*about*, Fr.) the end of a plank in a ship's side or bottom, uniting with the end of another.

BUTT also signifies the lower end of any part of made masts, and the ends of haunches, cokes, &c.

To start or spring a BUTT, is to loosen the end of a plank by the ship's weakness or labouring.

BUTT *and* BUTT, (*bout à bout*, Fr.) a term denoting that the butt ends of two planks come together, but do not over-lay each other.

BUTTOCK, (*fesses d'un bâtiment*, Fr.) the convexity of a ship abaft, under the stern; it is terminated by the counter above, by the after part of the bilge below, by the rudder in the middle, and by the quarter on the side.

BUTTON AND LOOP, a short piece of rope, having at one end a walnut knot crowned, and at the other end an eye. It is used as a becket to confine ropes in.

BUTTONS, or BUTTONS OF A BONNET, (*boutons d'une bonnette maillée*, Fr.) are sometimes used in fastening a bonnet upon a sail.

C.

CABIN, (*cabane*, Fr.) a room or apartment in a ship, where any of the officers usually reside.

In large ships there are many cabins; the principal of which, distinguished by the name of great cabin, is designed for the captain, or chief commander. In ships of the line, this chamber is furnished with an open gallery in the ship's stern, as also a little gallery on each quarter. The apartments where the inferior officers or sailors sleep and mess, are usually called births; which see.

Cabin is also a name given to the bed-places built up at the ship's side between decks in merchantmen, for the officers, passengers, and sailors.

Cabin *Boy*, (*mousse de la chambre*, Fr.) a boy whose duty it is to attend and serve the officers and passengers in the cabin.

CABLE, (*câble*, Fr.) a thick large, strong rope, of a considerable length, to which the anchor is fastened, and used to retain a ship at anchor in a road, bay, or haven.

Cables are of various sorts and sizes. In Europe they are usually manufactured of hemp; in Africa they are more frequently composed of bass, which is a sort of long straw or rushes; and in Asia of a peculiar sort of Indian grass.

Every cable, of what thickness soever it be, is generally formed of three ropes twisted together, which are then called *strands:* each of these is composed of three smaller strands; and those last of a certain number of rope-yarns. This number is therefore greater or smaller, in proportion to the size of the cable required.

There are some cables, however, manufactured of four strands; which are chiefly the production of Italy and Provence.

Every ship ought to be furnished with at least three good cables, viz. the *chief cable*, or cable of the sheet-anchor, and the two *bowers*, being a common cable, and a smaller one.

All cables ought to be 120 fathoms in length; for which purpose the threads or yarns must be 180 fathoms; inasmuch as they are diminished one-third in length by twisting. Besides this length, it is necessary to splice at least two cables together, in order to double the length when a ship is obliged to anchor in deep water. For although it is not common to anchor in a greater depth than forty fathoms, yet if there is only one cable, and the ship rides in a storm and tempestuous sea, the anchor will of necessity sustain the whole weight and violent jerking of the ship, in a direction too nearly perpendicular. By this effort it will unavoidably be loosened from its hold, and dragged by the ship, which, thus driven from her station, is in immediate danger of being wrecked on the nearest rocks or shallows; whereas it is evident, that if the cable, by its great length, were to draw more horizontally on the anchor, it would bear a much greater force. See the article Anchor.

The long cable is not so apt to break as the short one; because it will bear a great deal more stretching before it comes to the greatest strain: it, therefore, resembles a sort of spring, which may be very easily extended, and afterwards recovers its first state, as soon as the force which extended it is removed.

A long cable rides a ship better than a short one.

1st. By its weight, which corrects the motion like a spring, while the curve it forms acts on the anchor in a line more horizontal, and thereby holds it faster.

2d. By increasing the radius of motion, the arch grows nearer to a right line, and, consequently, the change of the place is made less sensible.

3d. The elasticity of the rope is increased with its length, which helps the first cause.

4th. The line of traction is more horizontal, and therefore the anchor is less liable to be lifted out of its place.

Besides all this, a ship will ride much smoother with a long cable, and be less apt to *pitch*, or plunge deep in the water with her fore-part.

From what has been said on this subject, it may be justly conceived how very necessary it is to furnish a ship with a sufficiency of cables, or what is called ground-tackle; for owing to a deficiency of this necessary article, many excellent vessels have been lost; it is, therefore, an inconsiderate policy in merchants to expose their vessels to such evident dangers, for the want of them.

A cable ought neither to be twisted too much nor too little; as in the former state it will be extremely stiff, and difficult to manage; and, in the latter, it will be considerably diminished in its strength.

All cables are to each other as the cubes of their diameters.

In the manufacture of cables, after the ropes are made, they use sticks, which they pass first between the ropes of which they make the strands, and afterwards between the strands of which they make the cable, to the end that they may all twist the better, and be more regularly wound together; and also to prevent them from twining or entangling: they hang at the end of each strand and of each rope a considerable weight. For further particulars, see the article Cordage.

The number of threads each cable is composed of, is always proportioned to its length and thickness; and it is by this number of threads that its weight and value are ascertained. Thus a cable of three inches circumference, or one inch diameter, ought to consist of forty-eight ordinary threads, and weigh 192 pounds; and on this foundation is calculated the following table, very useful for all persons engaged in marine commerce, who equip merchant ships on their own account, or freight them for the account of others.

A Table of the Number of Threads and Weight of Cables of different Circumferences.

Circumference in inches.		*Threads or rope yarns.*		*Weight in pounds.*
3	——	48	——	192
4	——	77	——	308
5	——	121	——	484
6	——	174	——	696
7	——	238	——	952
8	——	311	——	1244
9	——	393	——	1572
10	——	485	——	1940
11	——	598	——	2392
12	——	699	——	2796
13	——	821	——	3284
14	——	952	——	3808
15	——	1093	——	4372
16	——	1244	——	4976
17	——	1404	——	5616
18	——	1574	——	6296
19	——	1754	——	7016
20	——	1943	——	7772

Sheet-Anchor-Cable, (*maître cable*, Fr.) is the greatest cable belonging to a ship.

Stream-Cable, (*cable de touée*, Fr.) a hawser, or rope, smaller than the bower-cable, and used with the stream-anchor, to moor the ship in a river, or haven, sheltered from the wind and sea, &c.

To bit the Cable, (*bitter le cable*, Fr.) See the article Bits.

To serve or plait the Cable, (*fourrer le cable*, Fr.) is to bind it round with ropes, canvas, leather, &c. to prevent it from being galled, or worn in the hawse by friction.

Heave in the slack of the Cable! (*abraque le mou du cable!* Fr.) the order to draw it into the ship, by means of the capstan or windlass.

Pay out the Cable! *or veer away the* Cable! (*file du cable!* Fr.) slacken it, that it may run out of the ship. Pay cheap the cable! hand it out apace!

Cable's *length*, (*une encablure*, Fr.) is a measure of 120 fathoms, by which the distances of ships in a fleet are frequently estimated.

To shoot the Cable, is to splice two parts of a cable together.

To slip the Cable, (*filer le cable par le bout*, Fr.) is to let it run quite out, when there is no time to weigh anchor: this method is preferable to cutting it. Before it is either cut or slipped, a spare buoy-rope should be passed through the hawse-hole, and fastened near the end, with a rolling-hitch, that it and the anchor may be easily regained.

Cablet, (*cablot, ou, grelin*, Fr.) a small cable.

Cable-*Tier*, (*la fosse aux cables*, Fr.) is that place on the orlop-deck, where the cables are coiled up and stowed away.

A Table of the Number, Size, and Weight of Cables for a Ship of each Class.

Rate of Ship.	No. of Guns.	No. of Cables.	Size in Inches.	Length in Fathoms.	Weight of each Cable. Cwt.	qr.	lb
1st.	100	9	23½	120	138	0	0
		1	14	120	49	0	0
2d.	90	9	22½	120	126	2	0
		1	13½	120	45	2	0
3d.	80	8	21½	120	115	2	0
		1	13	120	42	1	0
	70	7	20½	120	105	0	0
		1	12½	120	39	0	0
4th.	60	7	19½	120	95	0	0
		1	11½	120	33	0	0
	50	7	18½	120	85	2	0
		1	11	120	30	1	0
5th.	40	7	17	120	70	1	0
		1	10	120	25	0	0
6th.	24	6	15	120	56	1	0
		1	8½	120	18	0	0
Sloops.		5	13	120	42	1	0
		1	8	120	16	0	0

CAISSON, in the military art, (*caisson*, Fr.) a wooden chest, into which several bombs are put, and sometimes only filled with gunpowder. This is buried under some work, whereof the enemy intend to possess themselves; and when they are masters of it, is fired, in order to blow them up.

It is also the name given to a wooden frame, sunk to keep off the water while laying the foundations of the piers of bridges, &c.

Caissons, in the form of a boat, are also used in our dock-yards, instead of flood-gates, in getting ships out or into the basins and slips to be repaired.

The following is a brief description of a wrought iron moveable Caisson, with a rudder, for docking a ship, while riding at her moorings, in any depth of water, leaving her keel dry in three hours, without removing her stores or masts.

This caisson or floating dock is made of wrought iron, half an inch thick, 220 feet long, 64 feet wide, and 30 feet deep, and will weigh about 400 tons, with a stanch six feet wide on the top, for the workmen to stand upon, and also to strengthen the caisson.

The weight of this caisson, when immersed in water, is nearly 350 tons, but, for reasons mentioned below, it is rendered nearly buoyant, being surrounded by an air receptacle capable of suspending the whole weight with great exactness, and which is rivetted to it in such a manner, as also to strengthen the caisson, and support the principal shoars from the ship.

While light, this caisson will draw nine feet of water: when taken to the ship intended to be docked, the water is to be let into it at an opening or plug-hole in the bottom, and it is to be suffered to sink until the upper part is even with the surface of the water, the air tube still keeping it buoyant.

A small quantity of air is then to be discharged,

by opening a plug-hole in the air receptacle, until a quantity of water is let in just sufficient to sink the caisson below the ship's bottom. This being effected, the caisson (nearly buoyant) is then to be raised to the surface of the water, by ropes made fast from the caisson to each quarter of the ship. A pump, plac d within the caisson, is then to be worked by a steam engine of 12 horses' power, placed in a barge alongside, which will empty it in three hours, and reduce the draught eight feet of water; that is, from 26 to 18 feet; when she may be carried up into shoal water, if required, or alongside wharfs, or jetty-heads of the dock-yards.

The ship's sides and bottom tending to fall outwards, by their own weight, and the sides and bottom of the caisson tending to be forced inwards, by the external pressure of the water, it is obvious, that by placing props, or shoars between, both will be supported, while the ship will ride with all her stores on board, and masts standing, nearly as easy as when in water.

Should inconvenience be apprehended at any time from blowing weather, the caisson may be cast off, and let fall to the bottom, where it cannot be injured, and whence it may be raised to the ship's bottom again at pleasure, with as little labour as weighing an anchor.

The caisson will be 12 feet above water, when there is a first rate ship in it; this is a sufficient height to prevent the sea breaking over.

By this plan a ship may have her bottom examined, and be out of dock again in six hours, without coming above the Nore, and without undergoing the tedious process of unshipping and reshipping her stores, waiting for spring tides, or fair wind, to enable her to reach, or to return from the dock; which, on an average, now requires three months, accompanied with an expense of nearly 10,000*l.* a month in wages, subsistence, &c.

This plan may be practised in all countries, where there are no dry-docks, or flowing of tide.

Ships on many foreign stations, when required to be docked, are now obliged to be sent home, at a great expense of money, and waste of time, others being sent to replace them.

This may be avoided in future. Docks made in England may be sent out in pieces of five or six tons, with the necessary rivets and bolts, and ready to be put together where they may be wanted.

From estimation, it has been ascertained, that a caisson capable of docking a first rate ship, will not cost above 19 or 20,000*l.* (for merchantmen and smaller ships, the size and cost will be proportionably less), and, judging from the duration of wrought iron salt pans, will last 20 years without repair. When worn out, it will break up and sell for one-third of its original cost.

By constructing this caisson, adapted to the local circumstances, ships of war and merchant-ships, with all their stores and cargoes on board, can be carried up rivers to wharfs and storehouses, where the depth of water is not above one half the ship's draft.

CALENDAR. See the article ASTRONOMICAL CALENDAR.

CALIBER, (*calibre*, Fr.) is the thickness or diameter of any round body, particularly the bore or width of a piece of ordnance, or that of its ball.

CALIBER *Compasses*, (*compas à mesurer les bois ronds*, Fr.) a sort of compasses made with bowed legs, and used by gunners for measuring the diameter of bombs, shot, shell , &c. as also the cylinder of cannon, mortars, and howitzers. They resemble other compasses, except in their legs, which are formed of wood and arched, so that the points may touch the extremities of the object. These legs are so broad, as to contain a variety of scales, tables, proportions, &c. necessary to be known to gunners, and of which a description is generally given with the instrument.

To find the true diameter of a circle, they have a quadrant fastened to one leg, and passing through the other, marked with inches and parts, to express the diameter required: the length of each rule or leg is usually between the limits of six inches and a foot.

To CALK, or CAULK, (*calfater*, probably from *calage*, Fr. hemp,) to drive a quantity of oakum, or old ropes untwisted and pulled asunder, into the seams of the planks, in the ship's decks or sides, in order to prevent the entrance of water. After the oakum is driven very hard into these seams, it is covered with hot melted pitch or rosin, to keep it from rotting.

CALKERS, (*calfats*, Fr.) are first mentioned in the reign of Charles the First, as employed and victualled, as occasion required, in our dock-yards.

CALKING IRONS, (*fers à calfat*, Fr.) are iron chisels for that purpose. Some of these irons are broad, some round, and others grooved.

CALL, (*sifflet du maître d'équipage*, &c. Fr.) a sort of whistle, or pipe, of silver or brass, used by the boatswain and his mates to summon the sailors to their duty, and direct them in the different employments of the ship.

It is sounded to various strains, adapted to the different exercises, as hoisting, heaving, lowering, veering away, belaying, letting-go a tackle, &c. and the *piping* of it is as attentively observed by sailors, as the beat of the drum to march, retreat, rally, charge, &c. is obeyed by soldiers.

CALL, *a gold call*, with a chain, was formerly the ensign of the office of an admiral. See the *Memoirs of Sir Edward Howard.*

CALLIPERS, an instrument made with bowed legs, and used by carpenters and shipwrights, for taking the diameter of timber, masts, yards, and bowsprits.

CALM, (*calme*, Fr.) the state of rest which appears in the air and sea when there is no wind stirring.

CALM *Latitudes*, the tract of the Atlantic Ocean, situated between the tropic of Cancer and the latitude of 29° N.; or the space between the *trade* and the variable winds: this being frequently subject to calms of very long continuance, has, therefore, justly acquired the name of the *Calm Latitudes*.

A dead CALM, *stark* CALM, *flat* CALM, (*calme plat*, Fr.) are used to denote the greatest possible calm.

A long calm is often more fatal to a ship than the severest tempest, for if tight and in good condition, she may sustain the latter without much injury; whereas in a long calm, the provision and water may be entirely consumed, without any opportunity of obtaining a fresh supply.

Calms are more dreaded by seafaring men than storms; for, under the line, excessive heat sometimes produces such *dead-calms*, that ships are obliged to stay several weeks, without being able to stir. Two opposite winds will produce a calm. This is frequently observed in the gulf of Mexico, at no great distance from the shore, where the land-wind will so poise the general easterly wind, as to produce a *perfect* calm.

Calms are never so great on the ocean as on the Mediterranean, because the flux and reflux of the former keep the water in a continual agitation, even where there is no wind; whereas, there being no tides in the latter, the calm is sometimes so dead, that the surface of the water is as clear as a looking-glass; but such calms are almost constant presages of an approaching storm. On the coasts about Smyrna, a long calm is generally considered as the fore-runner of an earthquake.

It is not uncommon for vessels to be becalmed in the road of the constant Levantine-winds, in places where they ride near the land. Thus, between the two capes of Cartooch toward the main, and Cape Antonia in Cuba, the sea is narrow, and there is often a calm produced by the land-wind that poises the Levantine-wind, and renders the whole perfectly still for two or three days. In this case, the current that runs here is of use to the vessels, if it sets right; when it sets easterly, a ship will have a passage in three or four days to Havannah; but, if otherwise, it is often a fortnight or three weeks' sail, the ship being embayed in the gulf of Mexico.

CAMBER, in ship-building, a term for any thing that rounds, but chiefly expressed to camber the *ways* for launching the ship.

CAMBERED DECK, (*pont arqué*, Fr.) the deck or flooring of a ship is said to be cambered, or to lie cambering, when it is higher in the middle of the length, than towards the stem and stern. Also, when it lies irregular; a circumstance which renders the ship very unfit for war. See the article BROKEN BACKED.

CAMEL, (*chameau*, Fr.) a machine used in Holland for raising or lifting ships, in order to bring them over the Pampus, which is at the mouth of the river Y, where the shallowness of the water hinders large ships from passing.

It is also used in other places, particularly at the dock of Petersburg, the vessels built there being in their passage to Cronstadt lifted over the bar by means of camels.

A camel is composed of two separate parts, whose outsides are perpendicular, and whose insides are concave, shaped so as to embrace the hull of a ship on both sides. Each part has a small cabin, with 16 pumps, and 10 plugs, and contain 20 men. They are braced to a ship underneath, by means of cables, and entirely enclose its sides and bottom; being then towed to the bar, the plugs are opened, and the water admitted until the camel sinks with the ship, and runs aground. Then, the water being pumped out, the camel rises, lifts up the vessel, and the whole is towed over the bar. This machine can rise the ship 11 feet; or, in other words, make it draw 11 feet less water. Professor Beckman says, that the Dutch first contrived to carry their numerous fleet through the Pampus, by means of large empty chests fastened to the bottom of each ship; and this contrivance gave rise to the invention of the camel.

CAMELEON, in astronomy, one of the constellations of the southern hemisphere, near the south pole, and invisible in our latitude. There are ten stars in this constellation, according to Sharp's catalogue; but they are very small, being only of the fifth and sixth magnitude.

CAMELOPARDALUS, in astronomy, a new constellation of the Northern hemisphere, consisting of 32 stars, first observed by Hevelius. It is situated between Cepheus, Cassiopeia, Perseus, the two Bears, and Draco; and, in the British Catalogue, it contains 58 stars, the brightest being only of the fourth magnitude.

CAMFERING, in ship-building, the operation of taking off any angle, or edge of a piece of timber.

CAN, (*bidon*, Fr.) a well known vessel, used by sailors to drink grog out of.

CANCER, the Crab, in astronomy, one of the twelve signs of the Zodiac, drawn on globes in the form of a crab, and thus marked (♋) in books of astronomy. It is the fourth constellation in the starry zodiac.

The reason generally assigned for its name, as well as figure, is a supposed resemblance which the sun's motion in this sign bears to the crab-fish. As the latter is said to walk backwards, so the former, in this part of his course, begins to go backwards, or recede from us. The number of stars in this sign, in the Britannic Catalogue, are 83.

CANCER, *Tropic of*, in astronomy, a lesser circle of the sphere parallel to the equator, and passing through the beginning of the sign Cancer.

CANES VENATICI, the Greyhounds, in astronomy, two constellations of the Northern hemisphere, first established by Hevelius, between the tail of the Great Bear and the arm of Bootes, above the Coma Berenices. The first is called Asterion, being that next the Bear's tail; the other, Chara.

They comprehend 25 stars, according to the British Catalogue; two of which were first observed by Tycho Brahe.

CAN-HOOKS, (*élingues à pattes*, Fr.) an instrument used to sling a cask by the ends of its staves: Plate V.
it is formed by fixing a broad and flat copper hook Fig. 3.
at each end of a short rope, and the tackle, by which the cask so slung may be hoisted or lowered, is hooked to the middle of the rope.

The can-hooks commonly used ashore by brewers, &c. are all iron, the middle part being fitted with a chain in the place of a rope.

CANIS MAJOR, in astronomy, the Great Dog, a constellation of the Southern hemisphere, a little to the westward of Orion's feet. The stars in this constellation, according to the Britannic Catalogue, are 31.

CANIS *Minor*, in astronomy, the Little Dog, a constellation of the Northern hemisphere, just below Gemini; called also by the Greeks *Procyon*. The stars in this constellation are, in the British Catalogue, 14.

CANNON, (*canon*, Fr.) a well known piece of artillery, mounted in battery on the decks of a ship, and used in all naval engagements. It may be defined a long round hollow engine, made of iron or brass, &c. for throwing balls, &c. by means of gunpowder.

They were first made of several bars of iron adapted to each other lengthways; and hooped together with strong iron rings; and appear to have been used in England about the year 1335. They are now made of cast iron or brass, and the tube bored out of the middle of the solid metal. Mezeray, a French writer, acknowledges that King Edward the IIIrd. of England, struck terror into the French army, by five or six pieces of cannon, at the battle of Cressi, in the year 1346, which appears to have been the first used in France. But it is difficult to ascertain by whom they were first used on ship-board; and the earliest authentic record we have, is by Father Daniel, who mentions that cannon were used on board the combined fleet of France and Spain, under Henry, king of Castile, against the English fleet off Rochelle, under the orders of the Earl of Pembroke, in the year 1372.

The metal of which brass cannon is made is, in a manner, kept a secret by the founders; yet, with all their art, they have not hitherto found out a composition that will stand a hot engagement without melting, or being rendered useless. Those cast at Woolwich, bid fairest towards this amendment. The respective qualities which should enter into this composition, is a point not decided; every founder has his own proportions, which are peculiar to himself. The most common proportions of the ingredients used in this country, are the following, viz. To 240lb. of metal fit for casting, they put 68lb. of copper, 52lb. of brass, and 12lb. of tin. To 4000lb. of metal fit for casting, the Germans put $3687\frac{3}{4}\frac{3}{1}$lb. of copper, $201\frac{1}{4}\frac{3}{1}$lb. of brass, and $307\frac{3}{4}\frac{6}{1}$lb. of tin. The French again use 100lb. of copper, 6lb. of brass, and 9lb. of tin; and others 100lb. of copper, 10lb. of brass, and 15lb. of tin.

Cannon were at first made of great length, which rendered them exceedingly heavy, and inconvenient for use. But experience soon taught, that little was added to the force of the ball by a great length of cannon; therefore they were soon reduced, and rendered much more manageable, both by sea and land.

It has been found by experience, that a ball moves with a greater impetus through a less space than a greater, and that of two cannons of an equal bore, but, of different lengths, the longest requires a greater charge of powder than the shortest. The ordinary charge of cannon is equal to one-third the weight of the ball, but that quantity is reduced in time of action, when the guns get warm, as the commanding-officer may judge proper.

All pieces of artillery were formerly distinguished by many names, now totally discontinued, but they now take their names from the weight of ball they carry; as a 6-pounder, a 12-pounder, a 24-pounder, or a 32-pounder, which is the largest now used in the royal navy, the 42-pounders having been laid aside, as being inconvenient, since the year 1790. The general rule for their length seems to be, that the whole charge of powder be on fire, before the ball quit the mouth of the piece.

The principal parts of a sea-cannon, as represented in plate V. fig. 4, are, 1st. The breech, A C, and its button, or cascabel, A h, called by seamen the pomelion. The breech is generally understood to be the solid metal from the bottom of the concave cylinder to the cascabel, which is the extremity of the cannon opposite to its muzzle.

2d. The trunnions, T, which project on each side like arms, and serve to support the cannon near the middle of its length: on these it may be poised, and held almost in *equilibrio*. As the metal is thicker at the breech than towards the mouth, the trunnions are placed nearer to that end than the other.

3d. The bore or calibre, is comprehended between the dotted lines, and particularly expressed in the longitudinal section of a 32-pounder, fig. 5. This represents the interior, or concave cylinder, wherein the powder and shot are lodged with which the cannon is charged: the entrance of the bore is called the mouth.

Names of the other parts, included in Plate V. *Fig.* 4.

A B, the length of the cannon.
A E, the first reinforce.
E F, the second reinforce.
F B, the chace.
H B, the muzzle.
A o, the cascabel, or pomelion.
A C, the breech.
C D, the vent-field.
F I, the chace-girdle.
r s, the base-ring and ogee.
t, the vent-astragal and fillets.
p q, the first reinforce-ring and ogee.
v w, the second reinforce-ring and ogee.
x, the chace-astragal and fillets.
z, the muzzle-astragal and fillets.
n, the muzzle-mouldings.
m, the swelling of the muzzle.
A i, the breech-mouldings.

The use of these machines is to discharge upon the enemy balls of iron, which are of various sizes, in proportion to the calibre of the cannon. The diameter of the ball is always somewhat less than the bore of the piece, that it may be discharged with greater ease, and not damage the piece by rubbing it too forcibly in its passage; and the difference between these diameters is called the windage of the cannon.

The length of any cannon is always reckoned from the hind part of the base ring, or beginning of the cascabel, to the extremity of the muzzle. The second reinforce begins at the same circle where the first terminates; and the chace at the same circle where the second reinforce ends.

The first reinforce, therefore, includes the base-ring; the ogee nearest thereto; the vent-field; the vent-astragal; and first reinforce-ring. The second reinforce contains the ogee next to the first reinforce-ring and the second reinforce-ring. The chace comprehends the ogee nearest to the second reinforce-ring; the chace-girdle and astragal; and the muzzle and astragal. The trunnions are always placed on the second reinforce, so as that the breech-part of the cannon may weigh something more than the muzzle-part, to prevent the piece from starting up behind when it is fired.

A variety of experiments, made with great care and accuracy, prove that powder, when on fire possesses at least 4000 times more space than when in grains. Therefore, if we suppose that the quantity of powder with which a cannon is charged, possesses one-fourth of a cubical foot in grains, it will, when on fire, occupy the space of about 1000 cubical feet. The same experiments evince also that the powder, when inflamed, is dilated equally round its centre. One grain of powder fired in the centre of different concentric circles, round which grains of powder are placed, shall, therefore, set fire to all those grains at once.

From this principle it necessarily follows, that powder, when fired in a cannon, makes at the same instant an equal effort on every part of the inside of the piece, in order to expand itself about its centre every way. But, as the resistance from the sides of the piece turns the action of the powder, so as to follow the direction of the bore of the cannon, when it presses upon the ball, so as to force it outwards, it presses also on the breech of the cannon; and this gives the piece a motion backwards, that is called its *recoil*, which, as we have already observed, is restrained by the *breeching* and the convexity of the decks. The recoil, in some degree, diminishes the action of the powder upon the shot, but this cannot be avoided; for, if the carriages were fixed so as not to give way to this motion, the action of the powder, or the effort that causes the recoil, would tear them to pieces in a very short time.

The metal of cannon is not equally thick in all parts; but is, in some measure, proportioned to the force of the powder which it is to resist. At the breech, where the effort is strongest, the thickness of the metal is equal to the diameter of the corresponding shot. At the first reinforce, where this begins to slacken, the thickness is somewhat less than at the breech: at the second, where the force is still further diminished, the thickness is more reduced than at the first: and, by the same rule, the chace has less thickness than the second reinforce. The thickness of the chace gradually diminishes from the trunnions to the mouth of the piece; so that if a cannon was without cascabel, trunnion, and mouldings, it would exactly resemble the frustum of a cone, or a cone deprived of the small end.

In a vessel of war, cannon are placed on a sort of wheeled sledge, called the *carriage*, of which fig. 6, plate V. is the plan, and fig. 2, plate VI. the elevation. This carriage is composed of two large pieces of plank, called sides or cheeks, connected together by means of cross-pieces, which are either bolts, axle-trees, or transoms. The two axle-trees are fixed across under the fore and hinder parts of the carriage, being supported at their extremities by solid wooden wheels, called trucks. The transom is placed directly over the fore axle-tree, and exactly in the middle of the height of the cheeks or side-pieces. The height of the transom is equal to two diameters of the shot, and the breadth to one diameter.

The following reference will explain the iron-work, and different parts of a sea-carriage, as exhibited in the plan and elevation of a 32-pounder, Plate V. *Fig.* 6, *and Plate* VI. *Fig.* 2.

a. The cap-squares, commonly called clamps in the sea-service.

b. Eye-bolts, by which one end of the clamp is fixed to the carriage.

c. Joint-bolts, upon which the other end of the clamp is fixed over the trunnions; after which it is fore-locked, to prevent the cannon from starting out of its carriage, when fired.

b g. The cheeks or sides of the carriage.

d. The transom-bolt.

e. The bed-bolt, upon which the bed rests, to support the breech of the cannon. This bolt is expressed by the letter A.

f. Hind axle-tree bolts.

g. Breeching-bolts, with rings, through which the breechings pass.

h. Loops, or eye-bolts, to which the gun-tackles are hooked.

i. The fore axle-tree, with its trucks, k.

l. The hind axle-tree, with its trucks, k.

The wheels are firmly retained upon their axle-trees, by means of iron bolts passing through the latter without the wheels: these bolts are called linch-pins.

The breadth of the wheels is always equal to that of the cheeks; but the height of the cheeks and diameter of the trucks must conform to the height of the gun-ports above the deck. The carriages of the lower tiers should, therefore, be so formed, that when the breech of the cannon lies upon the hind axle-tree, the muzzle of the piece should touch above the port, as expressed in plate VI. fig. 4, which represents a cannon secured by its tackles and breechings, to prevent it from straining the ship as she rolls in a stormy sea.

Cannon are charged by putting down into the bottom, first, a quantity of powder, one-third the weight of the ball. This is done with an instrument, fig. 5, termed a *ladle*, which is a kind of cylindrical spoon, generally made of copper, and fixed to the end of a staff, called its handle. Upon the powder is put in, a wad made of rope-yarn, formed like a ball, which is pressed down upon the powder with the instrument expressed by fig. 6, called a *rammer.* Upon this wad is put the ball or shot; and to secure it in its place, another wad is firmly pressed down upon it, which operation is called *ramming home* the wad and shot. The piece is then primed by filling the vent or touch-hole with powder, or by putting a tube in the vent, and is now ready to be fired either by match or a lock screwed thereto. See the article Exercise.

The powder being kindled, immediately expands, so as to occupy a much greater space than when in grains, and thus dilated it makes an effort on every side to force itself out. The ball making less resistance than the sides of the piece, upon which the powder presses at the same time, is driven out by its whole effort, and acquires that violent motion which is well known to the world.

Besides the before-mentioned instruments for

charging cannons, there is the sponge, *fig.* 7, which is used to clean the piece after firing, and to extinguish any sparks that may remain behind. In the land-service, the handle of the sponge is nothing else than a long wooden staff; but in ships of war this handle, that usually contains the rammer at its other end, is a piece of rope well stiffened by *spun-yarn*, which is for this purpose firmly wound about it.

By this convenience the rammer becomes flexible, so that the piece is charged within the ship, as the person who loads it may bend and accommodate the length of the rammer to the distance between the muzzle and the ship's side; being, at the same time, sheltered from the enemy's musketry, to which he would be exposed when using a wooden rammer without the ship. To sponge a piece, therefore, is to introduce this instrument into the bore, and thrusting it home to the farthest end thereof, to clean the whole cavity. The figures 7 and 8 represent sponges of a different kind; one of which is formed of sheep-skin, and the other of coarse wool wove, and fixed on the head of the sponge.

The *worm*, of which there are also different kinds, fig. 8 and 9, is used to draw the charge, when necessary.

The *bit*, or priming-iron, is a kind of large needle, whose lower end is formed into a gimblet, serving to clear the inside of the touch-hole, and render it fit to receive the prime.

The lint-stock is a kind of staff about three feet long, to the end of which a match is occasionally fastened to fire the piece, but not used at sea.

The fluctuating motion of the sea renders it necessary to secure and confine the artillery in vessels of war, by several ropes and pullies, which are called the *gun-tackles* and *breechings*, without which they could never be managed in a naval engagement. The breeching has been already explained, as employed to restrain the recoil. The tackles, fig. 10, are hooked to ring-bolts in the sides of the carriage, and to other ring-bolts in the side of the ship, near the edges of the gun-ports, and are used to draw the piece out into its place, after it is loaded. Besides these, there is another tackle, hooked to the rear or train of the carriage, to prevent the cannon from rolling into its place till it is charged: this is called the *train-tackle*, and is exhibited in fig. 2.

In ships of war, the cannon of the lower-decks are usually drawn into the ship during the course of an expedition at sea, unless when they are used in battle. They are secured by lowering the breech, so as that the muzzle shall bear against the upper-edge of the port, after which the two parts of the breeching are firmly braced together by a rope which crosses them between the front of the carriage and the port; which operation is called *frapping* the breeching. The tackles are then securely fastened about it, with several turns of the rope extended from the tackle and breeching, over the chace of the cannon, as represented in fig. 4.

The service of the artillery, or the method of employing it in a naval action, is explained in the articles Engagement and Exercise. The manner of pointing, or directing them to different objects; the effects of different quantities of powder upon the cannon-ball; and the different lines described by its flights, are also treated at large in the article Range.

The following table contains the Length, Weight, and Diameter of Cannon, or *Long Iron Guns*, with the Diameter of their Shot, and Proportion of Powder, as now used with the same for service in the royal navy.

Pounders.	Length in Ft.	In.	Weight. Cwt.	qrs.	lbs.	Diameter in In.	th.	Diam. of Shot. In.	th.	Prop. of Powder. lbs.	oz.
32..	9	6	55	2	0	6	$\frac{4}{10}$	6	$\frac{2}{10}$	10	10
24	9	6	50	2	0	6	$\frac{8}{10}$	5	$\frac{6}{10}$	8	0
	9	0	47	3	0						
18	9	0	42	2	0	5	$\frac{3}{10}$	5	$\frac{1}{10}$	6	0
	8	0	37	3	0						
12	9	0	34	3	0	4	$\frac{7}{10}$	4	$\frac{5}{10}$	4	0
	8	6	33	1	0						
	7	6	29	1	0						
	7	0	21	0	0						
9	9	0	31	0	0	4	$\frac{2}{10}$	4		3	0
	8	6	29	2	0						
	7	6	26	2	0						
	7	0	25	1	0						
6	8	6	22	1	0	3	$\frac{7}{10}$	3	$\frac{5}{10}$	2	0
	8	0	21	2	0						
	7	6	20	1	0						
	7	0	19	1	0						
	6	6	8	2	0						
	6	0	17	2	0						
4..	5	6	11	3	0	3	$\frac{2}{10}$	3	$\frac{1}{10}$	1	5
3..	4	6	7	1	0	2	$\frac{9}{10}$	2	$\frac{8}{10}$	1	0

For an account of the particular number of men appointed to manage the different degrees of cannon, and the arrangement or distribution of the cannon, according to the several classes of ships, see the articles Quarters and Rate. And for a table of the number, species, length, and mean weight of different iron guns used in each class of ships in the royal navy, see the article Gun.

Swivel Cannon, (*pierriers*, Fr.) a small piece of artillery, carrying a shot of a pound or half a pound, and fixed in a socket on different parts of the ship. The trunnions of this piece are contained in a sort of iron crotch, whose lower end terminates in a cylindrical pivot resting in the socket, so as to support the weight of the cannon.

The socket is bored in a strong piece of oak, reinforced with iron hoops, to enable it to sustain the recoil. By means of this frame, which is called the *swivel*, and an iron handle on its cascabel, the gun may be directed by hand to any object. It is generally used in the tops of ships of war, and frequently loaded with musket-balls, to fire down on the upper-decks of the adversary in time of action.

CANNONADE, (*canonnade*, Fr.) in marine affairs, is the application of artillery to the purposes of naval war, or the direction of its efforts against some distant object intended to be seized or destroyed, as a ship, battery, fortress, &c.

CANNONADING, (*canonner*, Fr.) is therefore used in a vessel of war to take, sink, or burn the ships of an enemy, or to drive them from their defences ashore, and to batter and ruin their fortifications.

As a large ship of war may be considered as a

combination of floating batteries, it is evident that the efforts of her artillery must in general be greatly superior to those of a fortress on the sea-coast; but this is not always the case, for on some particular occasions, her situation may be extremely dangerous, and her cannonading ineffectual. There are several circumstances, in which her superiority consists, viz. the power of bringing her different batteries to converge to one point; of shifting the line of her attack, so as to do the greatest possible execution against the enemy; or to lie where she will be the least exposed to his shot; and chiefly because, by employing a much greater number of cannon against a fort than it can possibly return, the impression of her artillery against stone-walls soon becomes decisive and irresistible. Besides these advantages in the attack, she is also greatly superior in point of defence; because the cannon shot, passing with rapidity through her sides, seldom do any execution out of the line of their flight, or occasion much mischief by their splinters; whereas they very soon shatter and destroy the faces of a parapet, and produce incredible havock amongst the men, by the fragments of the stones, &c. A ship may also retreat when she finds it too dangerous to remain longer exposed to the enemy's fire, or when her own fire cannot produce the desired effect. Finally, the fluctuating situation of a ship, and of the element on which she rests, renders the efforts of shells very uncertain, and altogether destroys the effect of the *ricochet*, or rolling and bounding shot, whose execution is so pernicious and destructive in a fortress or land-engagement; both of which, however, a ship of war may apply with great success. See the article RANGE.

On the contrary, the chief inconvenience to which she is exposed, is, that the low-laid cannon in a fort near the brink of the sea, may strike her repeatedly, on or under the surface of the water, so as to sink her before her cannonade can have any considerable efficacy.

CANOE, (*pirogue, ou, canot des Indiens*, Fr.)· a sort of Indian boat or vessel, formed of the trunk of a hollowed tree, and sometimes of several pieces of the bark fastened together. They are used on several occasions, as fishing, passage, trade, &c. and are of various sizes, according to the uses for which they may be designed, or the countries wherein they are formed. The largest are made of the cotton tree, some of which will carry between twenty and thirty hogsheads of sugar or molasses. Some are made to carry sail, and for this purpose are steeped in water till they become pliant, after which their sides are extended, and strong beams placed between them, on which a deck is afterwards laid that serves to support their sides. The other sorts very rarely carry sail, unless when going before the wind: their sails are made of a sort of silk grass or rushes: they seldom have any rudder, the want of which is supplied by a dexterous management of the hind oar. They are commonly rowed with paddles, instead of oars, which are pieces of light wood nearly resembling a corn-shovel; and instead of rowing with it horizontally, like an oar, they manage it perpendicularly. The small canoes are very narrow, having only room for one person in breadth, and eight or ten lengthways. The Indians who navigate them, are very expert in rowing uniformly, and in balancing them properly with their bodies, which would be difficult for a stranger to do, how well accustomed soever to the conducting of our boats, on account of the extreme lightness of the canoes, and their aptness to be upset.

The American Indians, when they are under the necessity of landing to avoid a water-fall, or of crossing the land from one river to another, carry their canoes on their heads or shoulders, till they arrive at a place where they may again be launched.

The canoe of the Esquemaux Indians, in Labrador, has a light wooden frame, and the shell, instead of plank, is made with seal-skins sewed together, which are not only extended round the bottom and sides, but likewise over the top; forming a complete deck, and having only one opening, conveniently framed and situated to admit the Indian into his seat. A flat hoop is fitted to this hole, rising about four inches, to which the surrounding skin is sewed. The Indian's seal-skin jacket, being of a proper length, he can occasionally bind the skirt of it round the outside of this hoop; by which means he keeps the canoe free from water, and is enabled to pursue his game far from land, or in stormy seas. His paddle is about ten feet long, light, and flat at each end, with which he both rows and steers with great velocity and exactness.

Mr. Crantz, in his History of Greenland, informs us, that the young men in their exercise are taught to upset their canoes, and when the bottom is upward, to recover, by the dexterous management of their paddle, their former upright position, the men rising again either on the side by which they went down, or on the contrary, as they please. The construction of this extraordinary little vessel, so admirably well adapted to the purposes of its owner, does the greatest credit to the ingenuity of this savage people. Though natives of the extensive country of Labrador, they inhabit only the sea-coasts, particularly the islands, the interior parts being no less barren, and possessed by other wandering tribes, their perpetual enemies and superiors at land; so that they are reduced to almost an entire dependence upon the sea for the common necessaries of life. Seals-flesh and oil are amongst the chief articles of their food; and with the skins they make tents, canoes, and apparel. Those islands on which the sea-fowl breed, they visit for their eggs and young; and kill birds in the water with their darts. We are surprised, that provided thus, they should do so much execution amongst these creatures; but when we behold a party of savages, each in his canoe, with only his harpoon and his lance, pursue, attack, and kill the largest whale, how justly are we filled with admiration. The whales flesh and oil they eat; and the tough substance of the gills, commonly called whalebone, they apply very ingeniously to a great variety of uses; trafficking with the overplus for such European goods as they want. In their language, the canoe is called *kaiak*, or man's-boat, to distinguish it from *umiak*, the woman's boat. The latter is a large boat, managed by the women for transporting their families and possessions, when they shift their encampment from place to place, as most convenient for the particular hunting of the season. A kind of wolf-dog, natural to that country, is the only animal they breed for food. The same canoes, language, customs, and way of life, being common to the Greenlanders with the Eskimaux, it is evident they have been originally one people.

There is a Greenland canoe in the repository of the Royal Society, covered with seal-skins, and exactly conformable to the above description.

CANOPUS, in astronomy, a bright star of the first magnitude, situated in the rudder of the ship Argo, by which mariners sometimes observe the latitude. For the beginning of the year 1812, its

Right ascension is	94° 56′ 46″, 0		
Annual Variation in A. R.....	20, 0		
Declination	52	35	48 S.
Annual Variation in Declination	+ 1,7		

CANT, (*oblique, ou dévoié*, Fr.) is a term used to express the position of any piece of timber that does not stand square, and then it is said to be on the *cant*.

CANTING, the act of turning plank or timber to see the opposite side.

CANT-*Pieces*, are pieces of timber or wood, inserted and annexed to the angles of fishes and side-trees, so as to supply any part that may prove sappy or rotten.

CANT-*Timbers*, (*couples dévoyés*, Fr.) those timbers which are situated at the two ends of a ship. They derive their name from being *canted*, or raised obliquely, from the keel, in contradistinction to those whose planes are perpendicular to it. The upper-ends of those on the *bow*, or fore-part of the ship, are inclined to the stem; as those in the *after*, or hind-part, incline to the stern-post above. See the article TIMBER.

CANVAS, (*canevas, toile à voile, ou, à prélat*, Fr.) a piece of strong cloth with which the sails are made, and are distinguished into degrees of fineness by the first eight digits, of which No. 1. is the coarsest and strongest. The numbers go on progressively, increasing in fineness as they advance.

CAP, (*chouquet*, Fr.) a strong, thick block of elm, having two large holes through it, perpendicular to its length and breadth, and parallel to its thickness; one of these is square, the other round; the former being fixed upon the upper end of the lower mast, whilst the latter receives the top-mast employed to lengthen it, and secures it in this position.

The principal caps of a ship are those of the lower masts, which are fitted with a strong eye-bolt on each side, wherein to hook the block by which the top-mast is drawn up through the cap; the process of which is explained in the article MAST.

The breadth of all caps is equal to twice the diameter of the top-mast, and the length to twice the breadth. The thickness of the main and four-caps is half the diameter of their breadths; the mizen-cap three-sevenths, and the top-mast caps two-fifths of their respective breadths.

In the same manner as the top-mast slides up through the cap of the lower-mast, the *top-gallant-mast* slides up through the cap of the top-mast. The cap is represented by fig. 11, plate VI.

CAP *of a Block*, a semicircular projection from the sides, and round the end of a block above the pin; there are two holes bored through it obliquely from the sides which meet and form an angle at the end; through these holes the strap is passed, to prevent its being chafed.

CAP *of a Gun*. See the article APRON.

To CAP *a Rope*, is to cover the end of it with tarred canvas, which is whipt in the twine or spun yarn.

CAPE, a promontory or head-land, which projects into the sea farther than the rest of the coast.

CAPELLA, in astronomy, a bright fixed star, of the first magnitude, in the left shoulder of the constellation Auriga. It is, in the Britannic catalogue, the fourteenth in order of that constellation.

Its right ascension in time, adapted to the beginning of the year 1812, is 75° 42′ 16″,80, annual variation in right ascension, 66″,21. Its declination' 45° 47′ 51″ N. and annual variation in declination, + 5″,09.

CAPITULATION, (*capitulum*, Lat.) yielding or surrendering on certain conditions.

CAPPANUS, the worm which adheres to, and gnaws the bottom of a ship.

The *cappanus* is extremely pernicious to ships, particularly in the East and West Indies; to prevent which, all ships in the royal navy, and many others, are now sheathed with copper: the first trial of which was made on his Majesty's frigate Alarm, in the year 1758.

CAPRICORN, or the Goat, in astronomy, is one of the twelve signs of the Zodiac, represented on globes in the form of a goat and characterised in astronomical writings by this mark ♑.

It is the tenth sign in order, and contains, in the Britannic Catalogue, 51 stars.

Tropic of CAPRICORN, a lesser circle of the sphere, which is parallel to the equinoctial, and at 23° 30′ distance from it southwards.

CAPSIZE THE COIL OF ROPE! is the order to turn it over.

To CAPSIZE, (*chavirer une embarcation*. Fr.) is to upset or overturn a ship, boat, or any other body.

CAP-SQUARES, (*plates-bandes d'affût*, Fr.) in gunnery, are strong plates of iron which come over the trunnions of a gun, and keep it in the carriage.

CAPSTAN, or MAIN CAPSTAN, (*cabestan*, Fr.) is a strong massy column of timber, formed like a truncated cone, and having its upper extremity formed into several squares with holes in them, to receive the bars or levers. It is let down perpendicularly through the decks of the ship, and placed in such a manner, that the men, by turning it horizontally with their bars, may perform any work which requires an extraordinary effort, as heaving in the cable, and thereby bringing the ship's head nearly perpendicular to the anchor, in order to dislodge it from its bed in the ground: it is also used to wind up any weighty body, as the masts, artillery, &c. In merchant ships it is likewise used to take in, or to discharge the cargo, particularly when it consists of weighty materials, that require great mechanical powers to remove them.

The capstan appears to have been first used in Queen Elizabeth's reign. Sir Walter Raleigh, in giving an account of the improvements in the navy, in Queen Elizabeth's or King James's reign, says, "the shape of our English ships has been greatly bettered of late. We have contrived the striking of the top-mast, added the chain-pump, devised studding-sails, top-gallant-sails, sprit-sails, and top-sails. We have also lengthened our cables, and contrived weighing of anchors by the capstan."

There are two capstans in a ship of the line, namely, the *fore* and the *main*; the former is placed in the fore part of the ship abaft the fore-mast, and the latter abaft the main-mast. Fig. 1. plate VII. represents a first rate's double capstan, whose parts are as follow:

Plate VII. Fig. 1.

A is the step in which the spindle works.
B B the beams under the lower and middle-decks.
C C the planks of those decks.
D the iron-pawl-rim.
E the pawl-rim.
F the lifter for raising the messenger.
G G the chocks.
H H the whelps.
I the trundle-head.
K the barrel.
L the partners.
M the drum-head.
N N the pawls.
O the spindle.

Fig. 2. Fig. 2. represents a brig's capstan with its spindle in the step: this is a single capstan, and is made smaller than those used in line of battle-ships.

The drum-head is a broad cylindrical piece of wood, resembling a mill-stone, and fixed on the top of the barrel and whelps. There are a number of square holes cut in it to receive the bars.

The whelps rise out from the main body of the capstan like buttresses, to enlarge the sweep; so that a greater portion of the messenger may be wound about it at one turn, without adding much to the weight of the capstan.

The pawls are two short bars of iron, bolted at one end through the deck to the beams close to the lower part of the whelps; and agreeable to modern practice, are bolted to the pawl-rim; the other end, which occasionally turns round on the iron pawl-rim, being fixed on the deck, as the capstan turns, prevents it from recoiling or turning back by any sudden jerk of the cable as the ship rises on the sea, which might greatly endanger the men who heave.

Should an accident happen at any time by shot or otherwise to the main-capstan, so as to render it unfit for immediate use, the fore-capstan may be shifted in its stead in forty minutes.

The spindle is shod with iron, and turns round in the saucer, which is a sort of iron socket let into a wooden standard, called the step, resting on, and bolted to the beams.

The bars are long pieces of wood, or arms, thrust into a number of square holes in the drum-head, in which they are as the radii of a circle, or the spokes to the nave of a wheel. They are used to heave the capstan round, which is done by the men setting their breasts against them and walking about, like the machinery of a horse-mill, till the operation is finished.

There are small iron pins, or bolts, thrust perpendicularly through the holes of the drum-head, and through a correspondent hole in the ends of the bars, made to receive them when the bars are fixed. They are used to confine the bars, and prevent them from working out as the men heave, or when the ship labours. Every pin is fastened to the drum-head with a small iron chain; and, that the bars may exactly fit their respective holes, they are all numbered.

The Society of Arts voted the gold medal, in the year 1807, to J. W. Boswell, Esq. of No. 15, Clifford's-inn, *for his Model of a Capstan, which works without requiring the messenger or cable coiled round it, to be ever surged.*

Sir Samuel Standridge, Mr. Plucknet, Captain Hamilton, and several other gentlemen, have, at different times, contrived various methods to prevent the necessity of surging, which always causes considerable delay; and when the messenger chances to slip in changing its position, no small danger is incurred by those who are employed about the capstan. Mr. Boswell's method seems much superior to any yet used, and consists in the simple addition of a second smaller barrel, or capstan of less dimensions to the large one; besides which it is to be placed in a similar manner, and which need not in general exceed the size of a half-barrel cask. The coils of the messenger are to be passed alternately round the large capstan and this small barrel, but with their direction reversed on the different barrels, so that they may cross each other in the interval between the barrels, in order that they may have the more extensive contact with, and better gripe on each barrel.

To keep the coils distinct, and prevent their touching each other in passing from one barrel to the other, projecting rings are fastened round each barrel, at a distance from each other equal to about two diameters of the messenger and the thickness of the ring. Those rings should be so fixed on the two barrels, that those on one barrel should be exactly opposite the middle of the intervals between those on the other: and this is the only circumstance which requires any particular attention in the construction of this capstan. The rings should project about as much as the cable or messenger from the barrels, which may be formed with whelps, and in every other respect, not before mentioned, in the usual manner for capstan barrels, only it is recommended that the whelps should be formed without any inclination inwards at the top, and to stand upright all round, so as to form the body of the capstan in the shape of a polygonal prism, if the intervals between the whelps are filled up, in order that the coils may have equal tension at the top and bottom of the barrels, and that the defect which conical barrels cause in this respect may be avoided.

The small barrels should be furnished with falling pawls as well as the large ones; a fixed iron spindle ascending from the deck will be the best for it, as it will take up less room. This spindle may be secured below the deck, so as to bear any strain, as the small barrel needs not be much above half the height of the large barrel; the capstan bars can easily pass over it in heaving round, when it is thought fit to use capstan bars on the same deck with the small barrel. As two turns of the messenger round both barrels will be at least equivalent to three turns round the common capstan, it will hardly ever be necessary to use more than four turns round the two barrels.

The circumstance which prevents the lateral friction of the messenger in this double capstan, is, that in it each coil is kept distinct from the rest, and must pass on to the second barrel, before it can gain the next elevation on the first, by which no one coil can have any influence in raising or depressing another; and what each separate coil descends in a single revolution, it regains as much as is necessary in its passage between the barrels, and, free from all contact with any part of the apparatus, it attains a higher elevation without a possibility of friction or wear. Its construction is so very simple, that it is no more liable to derangement or injury than the capstan itself. Its cost can be but small, and every part of it can be made by a common ship-carpenter, and be repaired by him at sea, if damaged by shot. It will take up but little room, only that of a half-barrel cask; and, it is of a nature so analagous to that kind of machinery to which sailors are accustomed, that it can be readily understood and managed by them.

A Represents the larger, or common capstan used on board ships.

Plate VII. Fig. 4.

B Another capstan of less dimensions, placed in a similar situation.

C The coils of the messenger, passing alternately round the large and small capstans, but with their directions reversed on the different barrels, so that they may cross each other in the interval between them.

D D D D Projecting rings round each capstan or barrel, so fixed on the two barrels, that those on one barrel should be exactly opposite the middle of the intervals between those on the other.

To man the CAPSTAN, (*mettre du monde au cabestan*, Fr.) is to place the sailors at it, in readiness to heave.

To rig the CAPSTAN, (*armer, ou garnir le cabestan*, Fr.) is to fix the bars in their respective holes, thrust in the pins in order to confine them, and reeve the swifter through the ends.

To surge the CAPSTAN, (*choquer au cabestan*, Fr.) is to slacken the rope wound round upon it, of which there are generally two and a half, and sometimes three turns about the barrel at once.

To heave in at the CAPSTAN, (*virer au cabestan*, Fr.) is to go round with it by pushing with the breast against the bars, and drawing in any rope of which the purchase is created.

To come up the CAPSTAN, (*dévirer*, Fr.) is to turn the capstan the contrary way, thereby letting out some of the rope upon which they had been heaving.

To pawl the CAPSTAN, (*mettre les linguets au cabestan*, Fr.) is to fix the pawls to prevent it from recoiling during any pause of heaving.

CAPTAIN OF A SHIP OF WAR, OR POST CAPTAIN, (*capitaine de haut bord*, Fr.) is an officer qualified to command any ship of war from a first rate down to a ship-rigged sloop. Formerly, a 20-gun ship was the smallest that gave post-rank, but by a late regulation of the Lords Commissioners of the Admiralty, the largest class of ship sloops has been added to the list of post-ships, and post-captains under three years standing are now appointed to them, unless they happen to be selected as flag-captains to admirals' ships: after being three years posted, they are appointed to frigates, which they may continue to command till they are of ten years standing, when they are generally removed to 50, or 64-gun ships, preparatory to their taking the command of ships of the line. The charge of a captain in his Majesty's navy is very comprehensive, inasmuch as he is not only answerable for any bad conduct in the military government, navigation, and equipment of the ship he commands; but also for any neglect of duty, or ill management in his inferior officers, whose several charges he is appointed to superintend and regulate.

On his first receiving information of the condition and quality of the ship he is appointed to command, he must attend her constantly, and hasten the necessary preparations to fit her for sea. So strict, indeed, are the injunctions laid on him by the Lords Commissioners of the Admiralty, that he is forbid to sleep out of his ship, from his arrival on board, till the day of his discharge, unless by particular leave from the admiralty, or his commander in chief.

He is enjoined to show a laudable example of honour and virtue to the officers and men, and to discountenance all dissolute, immoral, and disorderly practices, and such as are contrary to the rules of discipline and subordination, as well as to correct those who are guilty of such offences as are punishable according to the usage of the sea.

He is ordered particularly to survey all the military stores which are sent on board, and to return whatever is deemed unfit for service. His diligence and application are required to procure his complement of men; observing carefully to enter only such as are fit for the necessary duty, that the government may not be put to improper expense. When his ship is fully manned, he is expected to keep the established number of men complete, and superintend the muster himself, if there is no clerk of the check at the port.

When his ship is employed on a cruising station, he is expected to keep the sea the whole length of time previously appointed; but if he is compelled, by some unexpected accident, to return to port sooner than the limited time, he ought to be very cautious in the choice of a good situation for anchoring, ordering the master, or other careful officers, to sound, and discover the depths of water, and dangers of the coast.

Previous to any possibility of engagement with an enemy, he is to quarter the officers and men to the necessary stations, according to their office or abilities, and to exercise them in the management of the artillery, that they may be more expert in the time of battle.

His station in an engagement is on the quarter-deck; at which time he is expected to take all opportunities of annoying his enemy, and improving every advantage over him; to exhibit an example of courage and fortitude to his officers and crew; and to place his ship opposite to his adversary, in such a position, as that every cannon shall do effectual execution. See the article ENGAGEMENT.

At the time of his arrival in port, after his return from abroad, he is to assemble his officers, and draw up a detail of the observations that have been made during the voyage, of the qualities of the ship, as to her trim, ballast, stowage, and manner of sailing, for the information and direction of those who may succeed in command: and this account is to be signed by himself and officers, and to be returned to the resident commissioner of the navy at the port where the ship is discharged.

The number of captains allowed to take post in the

navy, on the 1st of January, 1812, was 777; and the number of superannuated and retired captains, 29.

CAPTAIN *of a Merchant Ship*, (*capitaine marchand*, Fr.) is he who has the direction of the ship's crew and cargo. In small ships and short voyages, he is more commonly called *Master*.

Captain is also a title applied in the royal navy to the chief sailor of particular gangs of men, as captain of the after-guard, captain of the fore-castle, captain of the hold, captain of a gun, captain of the main-top, &c.

CAPTAINS *of Port*, (*capitaine de port*, Fr.) among the French, are officers established in some considerable sea-ports where there are arsenals, as at Brest, Toulon, Dunkirk, &c.

CAPTAIN'S *Clerk*, (*commis, ou secrétaire du capitaine chargé de tenir ses comptes et ses écritures*, Fr.) a person employed by the captain, to keep all the books necessary for passing his accounts.—These are as follow:—

One *Copy* of the *Ship's Log-Book*, which is sent away half yearly; containing every word, letter, and figure, that are in the original.

Remarks on *Coasts*, *Roads*, *&c.* or a certificate that none were made, signed by the captain and master.

Muster-Books, properly signed, to be regularly sent every two months, and the signatures of all the lieutenants, written by them on a page thereof.

Slop-Book, with the issue totalled, invoice of supplies, and account of dead men's clothes sold.

Tickets, dead and remove, to be sent with the muster-books.

Sick-Book, to be signed by the captain and purser.

Copy of the *Captain's* commission.

Receipt for a *Complete-Book*, left with his successor.

Gunner's, *Boatswain's*, and *Carpenter's*, *Expense*, supplies, and returns.

Certificate of no *Backstays* shifted or top-mast lost.

Certificate of no *Survey-Book* supplied.

Account of receipt and expence of *Paper*, of sorts, and of tickets.

Vouchers for *Stores* purchased.

Sailing Qualities of the ship or vessel, if paid off.

General Accounts of *Provisions*, &c. received and returned, as per form delivered from the Victualling-office.

Quarterly Accounts of ditto, to be sent to the commissioners of the victualling, together with a certificate of the quantity of spirits, wine, cocoa, and tea, issued.

Ditto of the receipt and distribution of fresh beef.

An Account of *Cattle* received and killed.

Certificate of *Complement*, and title-page.

Certificate from *Purser* to Captain.

Orders and *Reports* of *Surveys* on *Provisions* and *Stores*.

Vouchers for *Savings* by the ship's company.

Vouchers of *Provisions* purchased.

An Account of *Casks* shaken; or a

Certificate that not any casks were shaken.

Letters to the *Commissioners* of the *Navy*, *Victualling*, and *Transport Board*; to grant certificates of no imprest outstanding.

All which *accounts*, it is absolutely necessary should be kept up to the very day, as also the complete-book, and description-book; for the production of them may be required in a complete state, at a short notice.

Besides the above-mentioned, the captain should have an Order Book, and a Letter Book. In the Order Book should be entered copies of all orders given and received, which the captain may, at any time, give to junior captains, or other officers, with particular care as to the wording, and dates of such orders; and in the Letter Book should be entered copies of all letters which the captain may, at any time, send upon his Majesty's service.

Great punctuality is to be used in keeping the account of the receipt and expense of tickets and certificates; and, if the captain is removed from the ship, the remains of them must be left on board, and a specific receipt taken for them of the captain who supersedes.

An open list must also be kept by the captain's clerk, from which he must take the muster for his muster-book, and it may be of use for references.

For a description and use of the before-mentioned books, see LIDDEL'S *Vade Mecum*, 5th edition.

It is the duty of the captain's clerk to attend at the issuing of slops, and to see that the men do not exceed the stipulated sum.

Whenever the ship comes into port where there is a clerk of the check, the captain's clerk is to have a complete open list, signed by the captain and other signing officers, ready to deliver to the muster-clerk afloat, with whom he is to attend, for the purpose of mustering the ship's company. He must also furnish the muster-clerk with an account of the provisions on board.

A state and condition of the ship must be made out weekly, when in port, which is called a *weekly account*, and is sent to the admiral, or, if none, to the secretary of the Admiralty: the same to be observed on joining an admiral or senior officer. This account must be signed by the captain.

When a ship is fitting in time of war, and bounty is allowed to seamen, the bounty list must be made out, and given to the clerk of the check, and must contain their qualities.

Men sick on shore to be noted in the general muster-book by Ss, instead of their muster letters, putting over the mark the day they went sick.

CAPTURE, (*capture*, Fr.) a prize taken by a ship of war at sea. See the article PRIZE.

CARABINE, or CARBINE. (*carabine*, Fr.) a sort of musketoon, the barrel of which is rifled spirally from the breech, so that when the ball, which is forced into it, is driven out by the strength of the powder, it is lengthened about the breadth of a finger, and marked with the rifle of the bore. This piece has an iron rammer; the barrel, including the stock, is three feet long: it has a much greater *range* than the fusil or musket, because the rifle of the barrel impedes the ball, which thereby makes the greater resistance at the first inflammation of the powder, and giving time for the whole charge to take fire before it goes out of the bore, it is at length thrown out with greater force than from the common musket.

CARACORES, are light vessels, used by the natives of Borneo and the islands adjacent, and by the Dutch as *guarda-costas* in those latitudes. They are high at each end, and chiefly navigated with paddles, to use which, they sit both within and without board, on narrow *platforms* of reeds, supported by bars rigged out across the vessel, and one at the outer end on each side, which serve as balances to prevent its being upset. By placing three or four ranks of rowers on the platform of reeds outside, and some within, they can multiply their number so as to produce a very great velocity. They have triple-sheers of bamboo for a mast, supported by shrouds, on which is hoisted an oblong sail, bent to a sort of bamboo yard at the head, and a boom at the foot. The sail is hauled aft by a sheet, the yard has a bowline to keep to windward, and a brace or vang that leads aft. The sail rolls up or furls by a winch at the end of the boom.

CARAVEL, (*caravela*, Span.) a light round old-fashioned ship, with a square poop, formerly used in Spain and Portugal.

Caravel is also a small vessel used on the coast of France, in fishing for herrings on the banks. They are commonly from 25 to 30 tons burden. Those designed for the same fishery in the British Channel, are called by the French, *trinquarts;* which are from 12 to 15 tons burden.

CARCASS, (from *carcasse*, Fr.) the circles which pass from one ring, or plate, to another, seeming to represent the ribs of a human carcass. It is an iron case, or hollow capacity, about the bigness of a bomb, of an oval figure, made of ribs of iron, and filled with a composition of combustible substances made up in proper dimensions, to be placed, instead of a shell, in a mortar.

The design of it is to be thrown out of a mortar into towns, in order to set fire to the buildings, &c.

It also serves by its light, while burning, as a direction in throwing the shells, and to deter persons from approaching to put it out: it is equipped with several pistol barrels loaded with powder and balls to the muzzles, which explode at various times as the composition burns down to them.

The carcasses for sea-service differ from a shell only in the composition, and the four holes from which the fire issues when let off. The composition is, pitch, 2 parts; saltpetre, 4; sulphur, 1; and corned powder, 3. The pitch is melted in an earthen pot; and, when the pot is taken off the fire, the ingredients, first well mixed, are put in; then the carcass is filled with as much as can be pressed in.

They are primed with fuse composition, and pitched over until used.

Another composition for carcasses is,

Saltpetre . . . . 50 parts.
Sulphur 25
Antimony . . . 5
Resin 8
Pitch 5

The Valenciennes composition, so called, from its having been used by the Austrians at the siege of that place, has the effect of making shells answer the purpose of carcasses after they burst. This is formed of

Saltpetre . . . . . 50 parts
Sulphur 28
Antimony 18
Resin, or Swedish Pitch 6

The composition is cast in copper cylindric moulds, 6 inches long, and of different diameters, according to the shell in which it is to be used. It is put in along with the bursting powder, in pieces as large as the shell will admit, without preventing the fusé's being driven down.

The following is a Table of the Dimensions and Weight of Carcasses, as established in 1796.

NATURE.		WEIGHT. Empty.			Of Composition.			Complete.			Time each will burn.
Round, for	In.	lbs.	oz.	drs.	lbs.	oz.	drs.	lbs.	oz.	drs.	Mins.
Mortars and Howitzers.	13	194	10	11	18	14	0	213	8	16	11
	10	89	13	11	7	8	11	97	6	11	8½
	8	44	9	5	4	4	11	48	14	0	5½
For Guns.	42	27	3	0	2	7	11	29	10	11	5
	32	20	13	5	1	14	5	22	11	11	4½
	24	14	12	0	1	9	11	16	5	11	4
	18	11	13	11	1	1	5	12	15	0	4
For Carronades.	68										
	42	26	0	0	2	7	0	28	7	0	4½
	32	21	10	0	1	13	0	23	7	0	4
	24	14	5	0	2	5	0	16	10	0	3½
	18	10	4	0	1	2	0	11	6	0	3
Oblong, for											
Mortars & Howitzers.	10	37	7	5	35	10	0	72	1	5	12
	8	16	5	5	18	2	0	34	7	5	10
	5½	1	12	2	6	15	0	8	11	3	6
	4⅔	1	0	6	3	11	7	4	11	13	4

It being found, at the siege of Quebec, that the quantity of powder requisite for throwing the carcasses into the town, always destroyed them, the method of filling the interval between the powder and carcass with turf was therefore adopted; and found to preserve the carcass, and to produce every desired effect.

CARD of a Compass, (*rose de vents*, Fr.) is applied to the circular paper on which the different points of the compass are projected over the mariner's needles. See the article Compass.

CARDINAL Points, (*points cardinaux*, Fr.) in geography, are the east, west, north, and south points of the horizon.

Cardinal *Signs*, in astronomy, are those at the four quarters, or the equinoxes and solstices, viz. the signs Aries, Libra, Cancer, and Capricorn.

Cardinal *Winds*, are those that blow from the north, south, east, or west points of the horizon.

CAREENING, (*faire abattre, carêner*, Fr.) the operation of heaving the ship down on one side, by the application of a strong purchase to her masts, which are properly supported for the occasion, to prevent their breaking with so great a strain; by which means, one side of the bottom being elevated above the surface of the water, (see plate VII. fig. 5.) may be cleansed from any filth which adheres to it, by Breaming; which article see.

When a ship is laid on a careen, every thing is taken out of her: but this operation is now nearly superseded by sheathing the ships with copper, whereby they keep a clean bottom for several years.

A ship is, also, said to careen when she inclines to one side at sea, by a press of sail.

Careening is a practice very rarely adopted in the British navy, never, indeed, unless there is an absolute necessity, from the want of a dock. It is not only an extremely troublesome, but a very dangerous, operation, as will appear from an account of the process pursued in *heaving down* the Courageux, in the harbour of Toulon.

The French do not heave their ships down to a wharf, but to a hulk constructed solely for that purpose, which is generally a small ship of the line cut down, and the hold provided with a great quantity of iron and single ballast. In the mid-ships of the spar-deck of this hulk, are placed the capstans to which the purchase falls are brought, and so contrived they lead down between decks, through scuttles cut for that purpose; which place being appropriated to that purpose, will cause little confusion on the capstan-deck, where the men lay in at the bars for heaving.

The ship, when perfectly cleared, was warped and lashed to this hulk, to be hove down on her larboard side. The inside and outside of the sides of the ship, wales, and topsides, were well calked, the scupper and port rope holes were plugged up and pitched over, as were all the bolt-heads and other places that appeared any way suspicious of leaking.

Broad pieces of oak of half a foot thick, were placed abreast of the main-mast on the quarter-deck; this was intended for a bedding, on which the heels of the spar shores were to rest and be secured, after the heads of them had been properly placed against the mast.

Beddings of large oak pieces were laid on the upper and lower deck, abreast of the mast; on them were placed a double tier of shores, wedged and secured with brace pieces and cleats, to enable the decks to support the heavy pressure of the mast shores, when in the act of heaving down.

Beds and deck shores were placed for the foremast, in the same manner as the main.

The masts were brought close over to the partners, and the starboard rigging well set up.

The upper deck ports were closed up with double pieces of deal, calked and pitched over.

It was thought advisable not to calk in the lower deck port-leads, probably on the account that it may have occasioned their leaking afterwards.

The ports were closed in, in the same manner as those of the upper-deck.

The gun-room ports and hawse-holes were left open; part of the quarter-galleries taken down, the entrances of which were filled in and calked.

From the Courageux being French built, and consequently wall sided, two breadths of thick deal were annexed to the gangways, to remedy the supposed inconvenience of water getting into the waist, when the ship was hove keel out

Two spar shores, one of fifty French feet, were placed against the main-mast, the heads of which were fitted on a mat doubly lashed and wedged, to prevent them from slipping. The other spar shore measured forty-three French feet; the heels of them rested on the larboard bedding on the quarter-deck, well lashed and secured.

The fore-mast had two spar shores, one of forty-three French feet, and the other of thirty-eight, fitted in the same manner as those on the main-mast.

The ship's rudder was unhung.

In the enclosed part of the manger, many shot were placed as a weight, to bring the ship more by the head, and to facilitate the process of heaving down.

The fifth, sixth, eighth, and ninth ports on the lower-deck were left open; they were doubly bolstered, and secured with strong cleats on the outside, batted under the lower part of the channel-wale.

To add to the security of the shores and rigging, and to take off from a great part of the heavy strain, that the decks and larboard side of the ship must inevitably be subject to, strong pendants (of the dimension of the ship's stream cable) were brought in to the above mentioned bolstered lower-deck ports, after the eyes had been placed over the mast-heads, the ends of which were set up with double purchase tackle, brought round the lower-deck beams (the deck being scuttled for that purpose), and spanned into each other. For the fore-mast, the second, third port, and the hawse-hole (bolstered as before), strong pendants were brought into, set up, and belayed round the bit-heads.

The main-mast had four-fold purchase blocks, two of which were placed and lashed to the mast-head, the other two about ten feet below the hounds. The purchase-tackles were six-inch, and not hawser laid. The fore-mast had three of the same, one to the mast-head, and two seven feet below the hounds. The mizen-mast was not used in this process, but the rigging well set up.

As soon as the people began to heave away in the hulk, and the ship's side far enough to admit of a floating stage, proper people attended to every suspicion of complaint as she went down, provided with pitch, tallow, oakum, boards, and saw-dust; the latter was introduced in a net fixed to a long pole, under water, and applied to particular parts, in case of any leaking.

The greatest strain was observed on the purchase-falls, when the upper-deck ports were touching the water. Four pumps were constantly used at the time of heaving down, two of which were stepped in tubs, on the footling, over the floor-head, and worked on the upper-deck. Two more were fixed on the orlop-deck, which worked on the lower. The first two acted until such a time as the ship was keel out, when the floor-pumps ceased, and those of the orlop came into play, and kept the ship entirely free during the time of repair.

The time of heaving the purchase-falls taught, to their being secured, and the ship keel out, was one hour and forty minutes.

The ship, when hove down, was two feet four inches by the stern. It was observed, in heaving down, that the main-purchase tackle acted almost independently of the fore ones, which sufficiently accounts for the ship being keel out of the water aft, almost as soon as forward.

When the new keel was put on, in order to calk the seam next to the water, the French had a machine like a punt, about ten feet long and four wide, being three feet six inches deep; at each end was a partition, about one quarter the length, filled with shot, to bring this machine down within six inches of the water, being well calked inside and out; the vacancy between the end partitions are where the men stand who are to calk, which they do with much ease and despatch, having people, with ropes, to haul the machine under the ship's bottom, &c.

Eighteen days did the ship remain keel out, on her larboard side, without being righted. The ship being righted at the expiration of the above time, the process for heaving the ship down on her starboard side, commenced.

The shores were removed to that side, and the larboard rigging set up. The lower-deck parts were closed up, and the opposite ones opened to receive the pendants, and the purchases shifted.

The preparation and process were exactly the same as performed on the other side of the ship.

The starboard side of the ship was brought to the same side of the hulk which she was before hove down to, and the process commenced. The ship hove down easier and better than before, and the greatest strain on the purchase-falls were observed to be when the ship was down about a foot higher than the uppermost part of the lower-deck ports. The ship remained not quite two days in this situation; when she was righted, she appeared to have a small list to port.

Much danger may be supposed, when a ship of the line is suffered to remain keel out of the water for the space of three weeks, as the different falls, consequently give more or less, and do not bear the same strain with each other when hove down for that time without being righted: but this cannot exactly be applied to the Courageux, that ship being hove down by two masts, in a basin of water, not subject to a tide.

There are many circumstances in the repair of a ship that may not admit of her being righted when it shall be thought necessary so to do; therefore, judgment cannot at all times be exercised on this point: to remedy this inconvenience, the falls should be particularly attended to; and if there should not be the smallest tide in the place, a current will sometimes make a rise and fall of water, and this observation should be the more attended to when ships are hove down to a wharf; and which, in this latter instance, will require, before right, that some inches of the different falls should either be hove on or come up at discretion.

CARGO (*cargaison, ou chargement*, Fr.) the lading, or whole quantity of whatever species of merchandise a ship is freighted with. It also denotes an invoice of the goods put on board.

CARLINGS (*entremises, ou traversins des baux*, Fr.) short pieces of timber ranging fore and aft, from one of the deck beams to another, into which their ends are scored: they are used to sustain and fortify the smaller beams of the ship, and are exhibited in the deck. They are pieces of oak timber, about eight feet long and eight inches square, or more, used for the purpose of framing the partners.

CARPENTER OF A SHIP (*maitre charpentier d'un vaisseau*, Fr.) an officer appointed to examine and keep in order the frame of the ship, together with her masts, yards, boats, and all other wooden machinery, and stores committed to him by indenture from the surveyor of the dock-yard.

It is the carpenter's duty in particular to keep the ship *tight*; for which purpose he ought frequently to review the decks and sides, and to calk them when it is found necessary. In the time of battle he is to examine up and down, with all possible attention, in the lower apartments of the ship, to stop any holes that may have been made in the sides by shot, with wooden plugs provided, of several sizes, for that purpose. For further particulars, see NAVAL INSTRUCTIONS.

So far back as the year 1638, by the Lord High Admiral's warrant, dated the 25th of April, carpenters were not to be appointed to ships until they had been examined and licensed at Shipwright's-hall.

CARPENTER's-*Mate* (*aide charpentier*, Fr.) the person appointed to assist the carpenter in his duty.

CARPENTER's-*Crew*, (*gens du charpentier*, Fr.) a set of men employed under the carpenter to repair what is necessary to be done for the security of the ship.

CARRIAGE OF A GUN, (*affût d'un canon*, Fr.) is a strong frame of wood fixed on four solid wheels, or trucks, on which the gun is placed. For a description of the parts of a sea-carriage, see the article CANNON; and for improvements, see Lieut.-Col. CONGREVE's *Treatise on the Mounting of Naval Ordnance*.

CARRICK-BEND. See the article BEND.

CARRICK-*Bits*, are the bits which support the windlass.

CARRONADE, (*carronade*, Fr.) a short piece of ordnance invented by Mr. Gascoine, being much lighter than the common cannon, and having a chamber for the powder like a mortar: from their peculiar construction, they are capable of carrying large balls, and are carried on the upper works, as the poop and forecastle: they are much used in close engagements at sea, mostly in frigates, sloops, and small vessels.

The carronade derives its name from Carron in Scotland, the town in which it was first made; and which, in June, 1779, was instituted by the King and Council, a standard gun in the navy.

The Length, Diameter of the Bores, and Weight of Carronade.

Nature.	Diameter of the Bore in Inches.	Length in Ft.	Length in In.	Length in Calibres.	Weight. Cwt.	Weight. qrs.	Weight. lbs.	Proportion between the Shot and carronades
68 Pounders.	8.05	5	2	7.702	36	0	0	59 to 1
		4	0	5.962	29	0	0	
42	6.84	4	3½	7.518	22	1	0	58 to 1
32	6.35	4	0½	7.679	17	0	14	62 to 1
24	5.68	3	7½	7.656	13	0	0	56 to 1
		3	0	6.336	11	2	25	
18	5.16	3	3	7.587	9	0	0	56 to 1
		2	4	5.447	8	1	25	
12	4.52	2	2	5.778	5	3	10	56 to 1

N.B. Carronades have not so much windage as guns.

Notwithstanding the many obvious advantages of carronades, during the course of last war, they got in disrepute, and did not meet with the general approbation of our naval commanders. First, they were so insecurely filled, that, when heated and fired in an oblique direction, they would tear away their fastenings, and upset. And secondly, from their shortness, and scarcely clearing the sill of the ports of line of battle ships, they were liable to communicate fire to the rigging that might lead near them.

It is, however, generally admitted, that they are better adapted for frigates than ships of the line, as they may have the option of distance in single combat; and when within the reach of their point blank, those of a large bore do wonderful execution, and require but few hands to manage them.

An improvement has been made, by Admiral Bertie, in the carriages of the 42 pounder carronades, which consisted simply in depressing the stole of the carriage two inches. This not only imparts to it the good property of being worked, and run out, with a smaller number of men, but it also checks the recoil, and necessarily adds to the force of the shot. The 42 pounders are now fought, and exercised, with only four men each, without any exertion or difficulty.

The following references will serve to explain the different parts of the carronade, as exhibited in plate VI. fig. 3.

A. Solid metal, instead of the button of a gun, through which a screw passes, to serve as a levelling-screw.

B. Metal to confine the rope or hawser.—C, *d*, *d*. Small knobs which mark the *line of metal*, and by which the carronade is levelled.

The parts of the common carronade carriage, as used on deck, in firing over the side of a ship, are exhibited in plate VI. at the letter B, where A represents the side of the ship, *b* the carriage over the side traversing on the centre; and B B the nails for fixing the platform to the deck.

To CARRY AWAY (*emporter*, Fr.) to break; as "that ship has carried away her fore-topmast," i.e. has broken it off.

CARTEL, (*cartel, bâtiment parlementaire,* Fr.) a ship commissioned, in time of war, to exchange the prisoners of any two hostile powers; or to carry a proposal, &c. from one to another: for this reason the officer who commands her is particularly ordered to carry no cargo, ammunition, or implements of war, except a single gun, for the purpose of firing signals.

CARTRIDGE, (*gargousse*, Fr.) is a case of paper or flannel, or partly of both, fitted to the bore of a gun, and containing exactly its proper charge. Cartridges for heavy guns are now made partly of cured paper only, and partly of cured paper, with flannel bottoms. Those for field ordnance and carronades are all of flannel, and suited to the bore or chamber of the pieces for which they are intended. The ball-cartridges, for wall-pieces, muskets, carabines, and pistols, are made of whited-brown paper, on formers of wood.

Flannel cartridges are preferred to paper, because they do not retain the fire, and are, therefore, less liable to accidents in loading. The best way of making flannel cartridges is, to boil the flannel in size, which will prevent the powder-dust from passing through, and render it more stiff and manageable.

CARTRIDGE-*Box*, (*garde-feu*, Fr.) a circular wooden box, with a lid sliding upon the two parts of a small rope. It contains one cartridge, and its use is to preserve it from bursting and from fire.

CARVED-WORK (*la sculpture d'un vaisseau*, Fr.) the ornaments of a ship, which are wrought by the carver.

CARVEL-WORK (*maniere ordinaire de border les vaisseaux*, Fr.), in contradistinction to *clincher-work*, the common method of planking vessels; it is performed by laying the edges close to each other, and calking them to make them water-tight.

CASES FOR CARTRIDGES, are vehicles made of wood, to the exact size of the different species of cartridges, for the purpose of carrying them, with safety, from the magazines, to the different guns on board.

CASE-SHOT, or CANISTER-SHOT (*charge à mitraille*, Fr.), a species of shell, containing a number of small balls, which are inclosed in a round tin case, or canister, and so prepared as to be shot out of great guns, to the great annoyance and destruction of the enemy. These have been superseded, and been substituted for grape-shot, formerly used. The tin case, which contains the shot, is cylindrical, and, in diameter, a little less than the calibre of the gun or howitzer, in which it is to be used. It is filled with iron balls, so as to make up the weight of the shot. These balls are seldom less than one ounce and a quarter in weight.

CASSIOPEIA, in astronomy, a constellation of the northern hemisphere, situated next to Cepheus, and opposite to the Great Bear, on the other side of the pole. The stars in this constellation are, according to the Britannic catalogue, 55.

In the year 1572, a remarkable new star appeared in this constellation, which, at first, surpassed Jupiter himself in brightness and magnitude; but, in a month, it began to diminish in lustre; and, in eighteen months entirely disappeared.

CAST AWAY (*faire naufrage*, Fr.), the state of a ship, which is lost or wrecked on a lee-shore, bank, rock, &c.

To CAST *off*, or, *to* CAST *loose the Guns* (*démarrer les canons*, Fr.) is to untie them.

CAST *of the Lead* (*coup de sonde*, Fr.) the act of once heaving the lead into the sea, to find what depth of water there is.

CASTING, (*abattre*, Fr.) in navigation, the motion of falling off, so as to bring the direction of the wind on one side of the ship, which before was right a-head.

This term is particularly applied to a ship riding head to wind, when her anchor first loosens from the ground.

CASTOR, in astronomy, a moiety of the constellation Gemini; called also Apollo.

CASTOR, is also the name of a bright fixed star of the first magnitude, in the head of the foremost twin of Gemini. For the beginning of the year 1812, its

Right Ascension is . . .	110° 38′ 45″,75
Annual Variation in AR .	57,83
Declination	32 17 22 N.
Annual Variation in Declin.	— 7,1

CASTOR *and* POLLUX. See the article GEMINI.

CASTOR *and* POLLUX, in meteorology. See the article CORPOSANT.

CAT, (*chat*, Fr.) a ship usually employed in the

coal trade, and formed from the Norwegian model. It is distinguished by a narrow stern, projecting *quarters*, a deep *waist*, and by having no ornamental figure on the prows.

These vessels are generally built remarkably strong, and carry from four to six hundred tons; or, in the language of their own mariners, from twenty to thirty *keels* of coals.

Cat, (*capon*, Fr.) is also a sort of strong tackle, used to draw the anchor perpendicularly up to the *cat-head*. The use of this machine is represented in plate VII. fig. 6.

CATAMARAN, (*radeau servant de bac*, Fr.) in its original state, is composed of three pieces of wood lashed together; one of which serves as a keel, the other two as sides: the rower stands or kneels in it, according as he is employed, and paddles with a piece of bamboo: in this state it is now used on the coast of Coromandel, and particularly at Madras, when, by means of a letter being put in a leather cap, messages, &c. are conveyed to the shipping in the roads, when no other communication could possibly take place, owing to the heavy surf. Many lives are also saved, by means of the catamaran-men, when an accident happens to a Mussula-boat, which carries passengers, &c through the surf at that place.

On this original plan, various improvements have been made, by enlargening the *raft*, as it may very properly at first be termed, and laying boards across it, whereby all sorts of goods, and even ordnance, are easily shipped or landed.

An improved catamaran, or coffer, has lately been introduced in the British navy, and an attempt was made, to destroy the enemy's flotilla at Boulogne, which, although it was unsuccessful, owing to adverse circumstances, has given sufficient proofs, that under judicious directions, it may still be of some advantage, yet, ought never to be resorted to, but for the accomplishment of some great and important purpose, and when all other means shall be considered inadequate; for the primary object of war is not to slaughter our fellow-creatures; and, when the energies and resources of the enemy can be impaired, by any other mode, humanity requires that the sacrifice of blood shall be spared, as those machines are understood to be employed in the darkness and silence of night, against a helpless and unsuspecting enemy.

Cat-*Fall*, (*garant de capon*, Fr.) the rope that forms the tackle for heaving up the anchor from the water's edge to the bow. It reeves through the sheaves at the outer end of the *cat-head*, and through the sheaves of the *cat-block* alternately.

Cat-*Harpings*, (*trélingage des haubans sous la hune*, Fr.) small ropes serving to brace in the *shrouds* of the lower masts behind their respective yards, for the double purpose of making the shrouds more tight, and of affording room to draw the yards in more obliquely, to *trim* the sails for a side-wind, when they are said to be close-hauled.

Cat-*Heads*, (*bossoirs*, Fr.) two strong short beams of timber, projecting almost horizontally over the ship's bows, on each side of the bow-sprit, being like two radii which extend from a centre taken in the direction of the bow-sprit.

That part of the cat-head which rests upon the forecastle, is securely bolted to the beams; the other projecting part carries in its extremity two or three small wheels, or *sheaves*, of brass, or strong wood, about which a rope called the *cat-fall* passes, and communicates with the cat-block, which also contains three sheaves.

The cat-head also serves to suspend the anchor clear of the bow, when it is necessary to let it go: it is supported by a sort of knee, which is generally ornamented with sculpture.

Cat-*Holes*, are over the parts as right with the capstan as they can be: their use is to heave the ship astern, upon occasion, by a cable, or a hawse, called *sternfast*.

Cat-*Hook*, is a strong hook fitted to the cat-block, to hook the ring of the anchor when it is to be drawn up or catted. See a representation of this article, plate VII. fig. 6.

Cat-*o'nine-tails*, an instrument of punishment used on board ships in the navy: it is composed of nine pieces of line or cord, about half a yard long, fixed upon a piece of thick rope for a handle, and having three knots on each at small intervals, nearest one end: with this the seamen, who transgress the articles of war and rules of the service, are *flogged* upon the bare back.

Thieves' Cat, a cat o'nine tails, having larger and harder knots upon it than those generally employed, and is only used for the punishment of *theft*.

CATS-Paw, (*fraicheur, ou petit vent sur l'eau*, Fr.) a light air of wind perceived at a distance in a calm, by the impression made on the surface of the sea, which it sweeps very lightly, and then decays. Catspaw is also a name given to a particular turn made in the bight of a rope, in order to hook a tackle on it.

CAULK. See the article Calk.

CEILING, (*vaigres, ou vaigrage*, Fr.) the inside planks of a ship. See the article Foot-Waleing.

CENTAURUS, the Centaur, in astronomy, a constellation of the southern hemisphere, represented on globes in the form of half man and half horse, usually joined with the wolf, and called *centaurus cum lupo*.

The stars of this constellation are, in the Britannic catalogue, 35.

The brightest fixed star in this constellation is in the right hoof, and of the first magnitude.

Its right ascension in time, adapted to the beginning of the year 1812, is 216° 51′ 3″,20, annual variation in right ascension 66″,7. Its declination 60° 3′ 50″ South, and annual variation in declination +16″,1.

CENTRE, in geometry, a point equally distant from the extremities of a line, figure, or body.

Centre *of a Fleet*, (*centre, ou corps de bataille d'une armée navale*, Fr.) the middle of the line, which is always the station of the admiral or commander in chief, and ought to be the strongest proportionably, as it reaches from the van and rear. See the article Line of Battle.

Centre *of Gravity*, in mechanics, is that point about which all the parts of a body do, in any situation, exactly balance each other. Hence, 1. If a body be suspended by this point as the centre of motion, it will remain at rest in any position indifferently. 2. If a body be suspended in any other point, it can rest only in two positions, viz. when the centre of gravity is exactly above or below the point of suspension. 3. When the centre of gravity is supported, the whole body is kept from falling. 4. Because this point has a constant endeavour to descend to the centre of the earth; therefore, 5. When the point is at liberty to descend, the whole body must also descend, either by sliding, rolling, or tumbling down. 6. The centre of gravity in regular uniform and homogeneal bodies, as squares, circles, &c. is the middle point in a line connecting any two opposite points or angles: wherefore, if such a line be bisected, the point of section will be the centre of gravity.

CENTRY, or SENTINEL, (*centinelle*, Fr.) in nautical language, is a *private marine*, from the guard party, posted in some responsible part of the ship, such as the magazine, captain's cabin, &c. to stand and watch carefully, to prevent any surprise, and to see that all is quiet in the night, and no improper conduct carried on.

He is obliged, when the ship is at anchor, to repeat aloud, "All's well!" in that part of the ship where he is posted.

All sentinels are required to be very vigilant on their posts; neither are they allowed to whistle, sing, smoke, or suffer any noise to be made near them; they are not to sit down, lay their arms out of their hands, or to go to sleep; but to keep moving about on their posts.

CEPHEUS, in astronomy, a constellation of the northern hemisphere; whose stars are, in the Britannic catalogue, 35.

CERBERUS, in astronomy, a small northern constellation near Hercules, formed by Hevelius out of the unformed stars. It contains only four stars of the fourth and fifth magnitude, which are enumerated under Hercules in the Britannic catalogue.

CESSATION, (*cessatio*, Lat. *trêve*, Fr.) *of Arms*, an armistice, or occasional truce. When a governor of a place besieged, finds himself reduced to such extremities, so that he must either surrender, or sacrifice the garrison and inhabitants to the mercy of the enemy, he plants a white flag on the breach, or beats the chamade: on which a *cessation of arms* and hostilities, takes place, to give opportunity for capitulation.

CETUS, the Whale, in astronomy, a large constellation of the southern hemisphere, under Pisces, and next the water of Aquarius.

The stars in this constellation, in the British catalogue, are 97.

In the neck of Cetus is a remarkable star, which appears and disappears periodically, passing through the several degrees of magnitude, both increasing and diminishing in about 333 days.

To CHAFE, (*érailler*, *raguer*, Fr.) to rub or fret the surface of a cable, mast, yard, &c. by the motion of the ship, or otherwise.

CHAIN-Boat, a large boat fitted with a davit over its stern, and two windlasses, one forward and the other aft, in the inside. It is used for getting up mooring chains, anchors, &c.

CHAIN-*Bolts*, those bolts which are drove through the upper end of the preventer-plates, and the toe-link of the chains.

CHAIN-*Plates*, are iron plates bolted to the ship's side, and to which the *chain* and *dead eyes* that support the masts by the shrouds are connected.

CHAIN-*Pump*. See the article PUMP.

CHAIN-*Shot*. See the article SHOT.

CHAINS, (*chaînes de haubans*, Fr.) strong links or plates of iron, the lower ends of which are bolted through the ship's side to the timbers.

They are placed at short distances from each other on the ship's outside, as being used to contain the blocks, called *dead eyes*, by which the *shrouds* of the masts are extended. The chains, and that of their *channels*, are represented by plate VII. fig. 7.

Top-CHAINS, (*chaînes des basses vergues*, Fr.) are chains to sling the lower yards in time of battle, to prevent them from falling down when the ropes, by which they are hung, are shot away, or rendered incapable of service.

CHAMBER, OF A MORTAR, is a cell, or cavity at the bottom of the bore, to receive the charge of powder.

CHAMPANS, are small flat-bottomed vessels used by the Chinese and Japanese. They have one mast rigged the same as the main-mast of a junk, with a single sail made of caw. They seldom exceed 80 tons burden; are constructed without iron or nails; and are unfit for rough weather.

CHANGE THE MIZEN! (*change l'artimon!* Fr.) is the order to bring the mizen yard over the other side of the mast.

CHANNEL, (*manche*, Fr.) in hydrography, the deepest part of a river, harbour, or straight, which is most convenient for the track of shipping: also an arm of the sea running between an island and the main or continent, as the British Channel, &c.

CHANNELS, or CHAIN-WALES, (*porte-haubans*, Fr.) broad thick planks, projecting horizontally from the ship's outside, beginning abreast of, and continuing somewhat abaft each mast. See plate VII. fig. 7.

They are formed to extend the shrouds from each other, and from the axis or middle line of the ship, so as to give a greater security and support to the masts, and to prevent the shrouds from rubbing against the gun-wale.

Every mast has its chain-wales, which are either built above or below the second deck-ports in a ship of the line: they are strongly connected to the side by knees, bolts, and standards, besides being confined thereto by the chains, whose upper ends pass through notches on the outer edge of the chain-wales, so as to unite with the shrouds above.

CHAPELLING A SHIP, (*coiffer*, *faire chapelle*, Fr.) the act of turning her round in a light breeze of wind when she is close-hauled, so as that she will lie the same way she did before. This is commonly occasioned by the negligence of the steersman, or by a sudden change of the wind.

CHAPLAIN, NAVAL, (*aumonier*, Fr.) a clergyman of the established church, appointed to perform divine service on board ships in the royal navy.

The introduction of chaplains to the establishment of our ships of war, seems to have been first adopted in the year 1626, as appears by a letter, dated York-house, July 29th, from George Duke of Buckingham to the University of Cambridge, in behalf of Daniel

Ambrose, A.M. and fellow of one of the colleges, who was appointed *chaplain* to one of the King's ships, stating, that his Majesty expected the University would not suffer him to experience any loss or inconvenience in the college to which he belonged, from being so employed in the navy; and it was accordingly ordered, that Mr. Ambrose should have the benefit of his fellowship during the whole period of his service at sea.

By an order in council, dated March 4th, 1812, made by his Royal Highness the Prince Regent, in the name and on behalf of his Majesty, the following regulations, in respect to half-pay or pension to be allowed to chaplains of his Majesty's fleet, after long and meritorious service, as well as the pay and allowances during the period of such servitude, shall be established in future; viz.

1.—" That every ship in his Majesty's navy, from a first to a fifth rate inclusive, shall be allowed a chaplain on her establishment.

2.—" That every chaplain, after eight years of actual sea service, (or, if in a guard-ship, ten years) during which period he shall not have been absent from his duty six weeks at any one time, except by special leave from the board of Admiralty, who shall produce certificates of his good conduct and moral behaviour from the captains he may have served under, shall be entitled to a pension in the nature of half-pay; but no chaplain to be entitled to half-pay or pension, if he shall accept any preferment, with cure of souls, during the term of his required service at sea.

3.—" That the amount of this pension, or half-pay, to each chaplain, shall be 5*s*. per day.

4.—" That if, in the event of a peace, or from other circumstances not depending on himself, a chaplain shall not have been able to complete the full servitude of eight years, a pension may be conferred by the board of Admiralty, according to a scale to be determined on, proportionate to the term of service: in no case, however, to be granted for less than three years service at sea.

5.—" That on the expiration of the period of eight (or ten) years, as the case may be, every chaplain shall give notice in writing, whether it is his intention to continue to serve or not: in the first case, every year's additional service shall entitle him, when placed on the pension list, to receive at the rate of 6*d*. a day, in addition to the pension which his eight (or ten) years servitude may have entitled him to, until the aggregate rate amounts to 10*s*. a day, being the greatest pension which any length of service can confer: in the second case, where no further service is intended, the pension shall remain at 5*s*. a day, as above-mentioned.

6.—" That the length of service and meritorious conduct shall render chaplains eligible to all the chaplaincies of all naval establishments whatever; the disposal of which shall, or may be left, to the consideration of the board of Admiralty; and that no other clergyman shall be eligible to any of those pieces of preferment, than a navy chaplain; and that the presentation to any of those pieces of preferment, whose emoluments may amount to 400*l*. per annum, shall cause the half-pay of the respective chaplains to cease, as is provided by law in the case of the divided living of Simonbourn.

7.—" That the pay of a chaplain, while in actual service, shall be according to the following rates; viz.—

" One hundred and fifty pounds per annum, and the established compensation of 11*l*. 8*s*. a year, for a servant, in each rate, and to have a cabin allotted for him in the ward-room or gun-room, where he is to mess with the lieutenants, and to be rated for victuals; and when the chaplain shall be willing to act as schoolmaster, he shall be entitled to the bounty of 20*l*. a year, granted by her late Majesty Queen Anne, by her order in council of the 21st of April, 1702, provided he shall pass an examination before the Lieutenant-governor, Professor, and Preceptor of the Royal Naval College at Portsmouth, instead of at the Trinity-house, as required by the above-mentioned order in council; and he shall be further entitled to 5*l*. per annum, to be paid to him by every young midshipman and volunteer of the first class, as a remuneration for his education; the same to be stopped out of the said young gentleman's pay.

8.—" That a Chaplain-general shall be appointed, with such emoluments as may be deemed proper by the board of Admiralty, to whom all applications for appointments shall be made, or will be referred, and all regulations entrusted, relating to the establishment of chaplains for the royal navy, in the same manner as is practised with regard to army chaplains.

9.—" That no warrant will be granted by the board of Admiralty to any candidate for an appointment, unless recommended by the Archbishops of Canterbury and York, and the Bishop of London, through the Chaplain-general, as in every way properly qualified.

10.—" That chaplains now serving in the navy, who may, upon due examination, be deemed proper to be continued, shall be allowed the time they may have served, as part (not exceeding one half) of that required to entitle them to the pension or half-pay."

For the duty of a chaplain, see NAVAL INSTRUCTIONS.

CHARACTER, in a general sense, denotes any mark whatever, serving to represent either things or ideas, whether for the sake of brevity, or perspicuity, in their operations: thus, letters are characters, types, or marks of certain sounds, words, of ideas, &c.

Astronomical characters are as follow:

Of the Planets.

☉ The Sun.	♂ Mars.
☾ The Moon.	♃ Jupiter.
⊕ The Earth.	♄ Saturn.
☿ Mercury.	♅ Herschel, or the Gorgian Planet.
♀ Venus.	

Of the Twelve Signs or Constellations of the Zodiac.

♈ Aries, the Ram.	♏ Scorpio, the Scorpion.
♉ Taurus, the Bull.	♐ Sagittary, the Archer.
♊ Gemini, the Twins.	♑ Capricorn, the Goat.
♋ Cancer, the Crab.	♒ Aquarius, the Water-Bearer.
♌ Leo, the Lion.	♓ Pisces, the Fishes.
♍ Virgo, the Maid.	
♎ Libra, the Balance.	

Of the Aspects.

☌ or S. Conjunction.	△ Trine.
SS. Semisextile.	Bq. Biquintile.
* Sextile.	Vc. Quincunx.
Q Quintile.	☍ Opposition.
□ Quartile.	☊ Dragon's head.
Td Tredecile.	☋ Dragon's tail.

Of Time, Motion, &c.

° Degrees.
′ Minutes or Primes.
″ Seconds, &c.
A.M. *ante meridian*, or m. morn.
P.M. *post meridian*, or a. aftern.
h. m. s. Hours, minutes, seconds.

CHARGE, (*charge*, Fr.) is the quantity of powder, ball, &c. put into a gun to be fired at one time: the weight of the powder for service is usually one-third the weight of the ball; but the allowance for 32 pounders is seven sixteenths of the same weight. The charge for battering guns is one-third of the weight; for case-shot one-fourth; that for carronades is usually one twelfth of the weight of the shot; the highest charge is one-eighth, and the lowest one-sixteenth. See the article CANNON.

In charging a gun care must be taken, that the powder, in ramming, be not bruised, because that weakens its effect; that a wad made of rope-yarn, or the like, be rammed over it, and that the ball or shot be intruded. If the ball be red hot, a tampion, or trencher of green wood is to be driven in before it.

The charge for all natures of guns upon service, at present consists of a proportion equal to one-third the weight of the shot when one shot is fired: and when two shot shall be fired, the charge is to consist of a quantity equal to one-fourth the weight of one shot.

To confirm the propriety of fixing the charges at the proportions above specified, the following result of experiments made by a committee of colonels, commandant colonels, and field officers of the royal regiment of artillery, in the year 1796, with different natures of sea service ordnance is adduced; viz.

That with a charge of one-third of powder, and with a single shot at two degrees of elevation, shall at its first fall or graze, strike an object, at the distance of - - - - - 1200 Yards and upwards.

And that with a charge of one-fourth of powder, and a single shot, shall at its first fall or graze, strike an object at the distance of - - - 1000 Yards and upwards.

Further, that when it may be judged advantageous to render the effect of the fire of the guns more formidable and destructive to an enemy, by firing two shots at each round, it was proved that both shots fired with a charge of powder equal to one-fourth the weight of one shot, will strike an object, or a ship, before their first fall or graze upon the water, at the distance of - 500 Yards and upwards.

An experiment for ascertaining the quantity of a charge necessary for certain sea-guns was also tried by Captain Frazier, of the Horse Artillery, on the 21st of November, 1810, with a 24 pounder 6½ feet long, on a butt of wood made of sleepers of nine inches square, firmly connected by iron bars passing from front to rear, and strongly shoared up behind by thick deal planks, and two unserviceable 18 pounder garrison carriages, run close to the butt with their hind trucks taken off.

The 24 pounder at 100 yards distance, with the usual breech tackle (and the carriage made fast in all respects as on board of ships), was fired double shotted, and at point blank twenty one times at this butt with a charge of 4lbs. The effect of the gun's recoil on the breeching was, in every respect, equal and moderate, and without any strain on the tackle. On examining the butt, it appeared much broken, and ten of the balls had gone quite through it to the average distance of 60 yards.

The average penetration of such balls as did not go through the butt was about 4 feet, that is, one foot eight inches more than the thickness of a 74 gun ship near her lower deck ports.

Three rounds double shotted were then fired with a charge of 6lbs. from the same 24 pounder, but the effect of the recoil on the breeching was so violent, as to prove its being unsafe to fire with so great a charge between decks.

Five rounds double shotted were next fired with a charge of 5lbs. from two 18 pounders of eight and nine feet long: their recoil on the breeching was very equal and moderate, though the shot went through the butt and totally ruined the gun-carriages behind it; one of these shot split after passing through the butt, and two others before they struck it. The guns were fired as fast as could be without hurry.

It is recommended that the use of two shots for each round should be restrained as much as possible, it having appeared, from various experiments which have been made with different natures of sea service ordnance, that after firing twenty rounds repeatedly at only the interval of five minutes between each round, the gun has become so violently heated, as to render it extremely hazardous to continue the firing at that rate.

The charges of cylinder gunpowder for the carronades used in the navy are to be arranged conformably to the following scale; viz. one-twelfth the weight of the ball for ships of the 1st, 2d, and 3d class, and one-sixteenth the weight of the ball for gun-vessels. The allowance for carronades in gun-brigs, &c. is only 1 oz. to the pound with respect to the weight of the ball.

But it is suggested that the rule, which the captains of men of war have hitherto practised in reducing the charges of powder for carronades, as well as for cannon when the pieces become warm in a long engagement, may be rigidly attended to.

CHART, or SEA CHART, (*carte marine*, Fr.) a hydrographical or sea-map, for the use of navigators; being a projection of some part of the sea *in plano*, shewing the sea coasts, rocks, sands, bearings, &c. Fournier ascribes the invention of sea-charts to Henry, son of John King of Portugal. These charts are of various kinds, the Plane chart, Mercator's or Wright's chart, the Globular chart, &c.

In the construction of charts, great care should be taken that the several parts of them preserve their position to one another, in the same order as on the earth; and it is probable that the finding out of proper methods to do this gave rise to various modes of projection. There are many ways of constructing maps and charts; but they depend chiefly on two principles. First, by considering the earth as a large extended flat surface; and the charts made on this supposition, are usually called plane charts.

Secondly, by considering the earth as a sphere; and the charts made on this principle are sometimes called Mercator's charts, Globular charts, Hydrographical charts, &c.

Plane charts have the meridians, as well as the parallels of latitude, drawn parallel to each other, and the degrees of longitude and latitude every where equal to those at the equator. And, therefore, such charts must be deficient in several respects.

For, 1st, since in reality all the meridians meet in the poles, it is absurd to represent them, especially in large charts, by parallel right lines. 2dly, As plane charts shew the degrees of the several parallels as equal to those of the equator; therefore, the distance of places lying east and west, must be represented much larger than they really are. And 3dly, in a plane chart, while the same rhumb is kept, the vessel appears to sail on a great circle, which is not really the case. Yet plane charts made for a small extent, as a few degrees in length and breadth, may be tolerably exact, especially for any part within the torrid zone; and even a plane chart made for the whole of this zone will differ but little from truth.

Construction of the plane chart.—The number of degrees of latitude which the chart is intended to contain, and the extent from east to west being fixed upon; a line is to be drawn near the side or end of a sheet of paper, in length equal to the whole length of the chart from north to south; and this line is to be divided into degrees, and numbered accordingly. From each end of this line perpendiculars are to be drawn, and made equal to the intended extent of the chart from east to west, and their extremities are to be joined by a straight line.

If the chart is to commence at or near the equator, and to extend only a few degrees of latitude, the divisions of the parallels may be equal to those of the meridian: but if the chart begins at any considerable distance from the equator, it will conduce to accuracy, to make the length of each degree of the parallel equal to the co-sine of the mean latitude, the radius being 60 minutes; or the extreme parallels may be divided according to the above proportion, and in that case it will become a reduced chart. Meridians and parallels are there to be drawn at convenient distances.

A scale is now to be made of stiff paper, or pasteboard, equal in length to the extent of the chart from east to west, and divided and numbered accordingly.

By this scale, the positions of those places contained within the limits of the chart are very easily laid down, by placing the divided edge of the scale over the latitude of the given place; and under the given longitude, a mark being made will represent the position of the place on the chart. A compass is to be inserted in any convenient place of the chart, an arrow shewing the direction of the flood tide or current.

The times of high water at full and change are to be marked in their proper places, expressed in Roman characters; sounding and quality of the ground at bottom, the leading marks to avoid dangers, &c.

Mercator's chart, like the plane charts, has the meridians represented by parallel right lines, and the degrees of the parallels or longitude, every where equal to those at the equator, so that they are increased more and more, above their natural size, as they approach towards the pole; but then the degrees of the meridians, or of latitude, are increased in the same proportion at the same part; so that the same proportion is preserved between them as on the globe itself. This chart has its name from Girard Mercator, who first proposed it for use in the year 1556, and made the first charts of this kind; though they were not altogether on true or exact principles, nor does it appear that he perfectly understood them; nor was it perfected till Mr. Wright first demonstrated it about the year 1590, and shewed a ready way of constructing it, by enlarging the meridian line by the continual addition of the secants.

Construction of a Mercator's chart.—The limits of the proposed chart is first to be determined, that is, the number of degrees of latitude and longitude which it is to contain, and the degree of latitude and longitude of its commencement. Find the meridional parts answering to each degree of latitude, within the intended limits of the chart, and take the difference between each, and that corresponding to the least degree of latitude in the chart; and reduce these differences to degrees, by dividing by 60.

A parallel, representing that of the least latitude, is to be drawn; upon which the number of degrees in the proposed difference of longitude, from a scale of equal parts, is to be laid off, and divided into degrees, and smaller portions, if convenient, and numbered at each fifth or tenth degree. From each end of this parallel, a perpendicular is to be drawn, and made equal to the difference of the meridional parts of the extreme latitudes taken from the divided parallel; and the ends of these meridians are to be joined by a straight line, which will represent the other extreme parallel, and which is to be divided and numbered in the same manner as the first drawn parallel; the meridians are then to be divided into degrees, and numbered at every fifth or tenth degree. Take the meridional difference of latitude between the beginning of the chart, and the next fifth or tenth degree of latitude from the divided parallel, and lay it off from the first parallel on each of the scale meridians, and join these points by a straight line. In like manner, the meridional difference of latitude answering to each successive interval of five or ten degrees, is to be taken from the first drawn parallel and laid off, and the corresponding parallels are to be drawn and numbered accordingly, and the intermediate spaces are to be subdivided. If the chart is upon a large scale, the meridional difference of latitude answering to each degree, is to be laid off from the least parallel, and equi-multiples of the intervals are to be taken, such as will answer to the proposed extent of the chart. A slip of strong paper is to be divided, and numbered in the same manner as the first drawn parallel. Now each place with the limits of the chart is to be laid down, by placing the slip of paper, so that its extreme points of division may be at the latitude of the given place on each meridian; then, under the longitude of the place a mark is to be made, which will represent the position of that place. In like manner all the places on the coast are to be laid down, and connected by observations made on the coast; or if no sketch had been previously made, the contour of the coast is to be drawn agreeable to the best charts. Meridians and parallels are to be drawn through every fifth or tenth degree of latitude and longitude, and extended to the coast.

A compass is to be inserted in some convenient part of the chart, and the points extended to the land: an anchor is to be drawn where there is good anchoring ground; and in places where it is safe only to stop a tide, an anchor without a stock is to be laid down. The soundings, the quality of the ground, the times of high water at full and change, &c. are to be marked in their proper places.

Globular chart, is a projection so called from the conformity it bears to the globe itself; and was proposed by Messrs. Senex, Wilson, and Harris. This is a meridional projection, in which the parallels are equidistant circles, having the pole for their common centre, and the meridians curvilinear and inclined, so as all to meet in the pole, or common centre of the parallels. By which means, the several parts of the earth have their proper proportion of magnitude, distance, and situation, nearly the same as on the globe itself; which renders it a good method for geographical maps.

Hydrographical charts, are sheets of large paper, on which several parts of the land and sea are described, with their respective coasts, harbours, sounds, flats, rocks, shelves, sands, &c. also the points of the compass, and the latitudes and longitudes of the places.

The manner of using charts.—The principal use of a chart is, to find the course and distance between any two places within its limits, and to lay down the place of a ship on it, so that the position of the ship with respect to the intended port, the adjacent land, islands, &c. may be readily perceived.

CHARTER-PARTY, (*charte-partie*, Fr.) is an instrument of freightage, or articles of agreement made between merchants and sea-faring men, concerning their merchandise and maritime affairs.

A charter-party of affreightment settles the agreement in relation to the freight and cargo of a ship between the merchant and master, or commander of the vessel. It binds the master to deliver the cargo in good condition at the place where his ship is to be discharged, &c.

In those charter-parties, if the dangers of the sea are excepted, it has been adjudged that such exception extends as well to any danger upon sea from ships of war or pirates, as to common hazards of shipwreck, tempests, &c.

A charter-party is the same in civil law, as an indenture at common law, and is distinguished from a bill of lading, inasmuch as the former adjusts the term of the freight, and the latter ascertains the contents of the cargo.

CHASE, (*vaisseau ou bâtiment chassé*, Fr.) is a term applied to a vessel pursued by some other.

Chase *of a Gun*, is its bore or cylinder.

Bow Chasers, (*canons de chasse*, Fr.) the cannon situated in the fore part of a ship, to fire upon any object ahead of her.

Stern Chasers, (*canons de retraite*, Fr.) are those placed in the after part of a ship, pointing astern, and intended to annoy any vessel in pursuit of her.

Chaser, (*vaisseau en chasse*, Fr.) implies the vessel pursuing any other ship or vessel.

Chasing, (*chasser*, Fr.) the act of pursuing a ship or fleet, supposed or known to be hostile, as, "The admiral displayed the signal for a general chase," *i.e.* gave the alarm to the whole fleet or squadron to pursue some other fleet in sight.

CHECK, Clerk of the, (*officier d'administration dans les arsenaux de marine*, Fr.) an officer in the royal dock-yards, who goes on board his Majesty's ships and vessels at the port where he is settled, to muster the ship's company, of whom he keeps a register.

To Check, (*choquer*, *mollir*, *ou larguer*, Fr.) is to ease off a little of a rope which is found to be too stiffly extended: it is also used in a contrary sense when applied to the cable running out, and then implies to stopper the cable.

CHEEKS, are knee pieces of timber, fastened to the ship's bow, and to the knee of the head.

Cheeks *of a Block*, are the two sides of its shell.

Cheeks *of the Mast*, (*jottereaux*, Fr.) the faces or projecting parts on each side of the masts, used to sustain the trestle-trees upon which the frame of the top, together with the top-mast, immediately rests: also the head of the masts, being that part above the steps.

To CHEER, (*animer*, *saluer par acclamations*, Fr.) to salute a ship *en passant*, by the people all coming upon deck and huzzaing three times, called three cheers: it also implies to encourage or animate.

Cheerly, (*hardi!* Fr.) implies heartily, cheerfully, or quickly; as, "row cheerly in the boats!" "lower away cheerly!" that is, pull heartily, lower speedily, &c.

CHESS-TREES, (*pouliots d'amure de grande voile*, Fr.) two pieces of wood bolted perpendicularly, one on each side of the ship: they are used to confine the clews of the main-sail, for which purpose there is a hole in the upper part, through which the tack passes that extends the clew of the sail to windward. See the article Tack.

The *chess-trees* are commonly placed as far before the main-mast as the length of the main-beam.

CHEST at Chatham, now called *the* Chest *at Greenwich*, (in consequence of its removal by an Act of Parliament in the year 1803,) was first advised by Sir Francis Drake and Sir John Hawkins, for the relief of English mariners wounded and superannuated in their country's service.

The original written constitution of the *chest* is supposed to have been lost during the usurpation of Cromwell; but it appears, from the record of a decree, dated the 1st of May, 1617, made in consequence of an inquisition held at the castle of Rochester, by virtue of an Order in Council, dated the 22d of July in the preceding year, that in 1590, the masters, mariners, shipwrights, and sea-faring men, employed in the service of Queen Elizabeth, did, with the consent and persuasion of the Lord High Admiral of England, and the principal officers of the royal navy, voluntarily give and bestow, and consent to have stopped out of their wages, the following sums; viz.

Out of the wages of every mariner, seaman, and shipwright, receiving ten shillings or more per month, sixpence per month.

Out of the wages of every grommett, receiving seven shillings and sixpence, fourpence; and

Out of the wages of every boy receiving five shillings per month, two-pence.

For the perpetual relief of such mariners, shipwrights, and sea-faring men, as by reason of hurts and maims received in the service, were driven into great distress and want, it was also consented to, and established by the said parties, that the like defalcations should, from time to time, for ever, be made for the relief of such hurt and maimed mariners, shipwrights, and seamen, as then did, or should hereafter, serve in the royal navy, or in any other ship or vessel in the royal service, either at sea or in harbour; the money to be distributed at the discretion of the masters attendant, master shipwrights, boatswains, and pursers, of the royal navy, for the time being, or some of every of the said several sorts and ranks, from time to time, as cause shall require, and by the oversight of the principal officers of the navy for the time being, or some of them; that the overplus of the money should be preserved in a strong chest with five locks, of which the keys were to be kept separately; one by a principal officer of the navy, another by a master attendant, another by a master shipwright, another by a boatswain, and the other by a purser, and so to remain for one whole year; and at the end of every year, to be delivered by each of the said several officers to one other officer of the same rank, and so yearly in succession.

To the amount of the defalcations before mentioned. King Charles the IId gave to the chest twelve acres of marsh-land, situated near Rochester, called Delce; and it is ascertained by the records, that, previous to 1672, the fund was increased by the fourpences and two-pences deducted from the monthly wages of all seamen, for the pay of chaplains and surgeons of the navy in such ships where none are borne; and, in 1688, the fines and mulcts imposed by naval courts martial were added, by a grant of King James the IId.

Upon this basis, the business of the chest has been carried on, without any material alteration, to the present time; but the shipwrights no longer contributing, are excluded from receiving any benefit from the fund.

The affairs of the chest are managed by four supervisors and seven directors; the former are the first commissioner of the admiralty, comptroller of the navy, governor of the royal hospital, and auditor of the same for the time being; which said supervisors are one body politic and corporate, and have one common seal; and the latter, with the addition of two gunners, are chosen in the manner pointed out by the decree.

The allotment and payment of pensions, the administration of the funds, and all other ordinary business, is conducted by the governors: but no change of system, or general regulations, can be carried into effect without the concurrence of the supervisors; and the particular accounts of the chest are annually laid before Parliament.

The following statement of the receipts and disbursements for the year 1802, will shew the importance and magnitude of the concerns of the institution.

General Abstract of the Receipts and Disbursements of the Chest at Chatham, for the Year 1802.

RECEIPTS.

	£	s.	d.
Balance remaining in the Chest brought forward	10,056	7	11
Received from the Treasurer of the Navy, on Account of Defalcations from the Wages of Seamen	50,000	0	0
——— Abatement on Naval Officers Bills	288	6	6
——— Interest on Money in the Funds	6,660	0	0
——— Rents and Fines of Estates	1,376	2	0
——— Gift, being the Produce of 10,000*l.* *3 per cent.* Consols, given by a Person unknown, who presented a like sum to Greenwich Hospital	6,849	11	9
	£75,230	8	2

DISBURSEMENTS.

	£	s.	d.
The Number of Pensioners on the Books in the Year 1802 was 8,094; those actually paid were 5,205, whose Pensions amounted to	31,463	9	8
For present Relief	3,577	3	4
To Men whose Cases were not deemed pensionable	3,139	1	0
To Men discharged well, and to Greenwich Hospital	383	18	10
For the purchase of Stock	28,786	13	6
Expenses of the Establishment	1,259	17	8¼
Repairs, &c. of Estates	2,159	1	6½
	£70,769	5	7

N.B. The actual defalcations taken from the ships' books in this year, amounted to 78,863*l.* 18*s.* 5½*d.* of which 50,000*l.* only was received on account.

It appears from the general account of receipts and disbursements, that previous to 1794, large sums of money arising from defalcations, remained in the hands of the Treasurer of the Navy; but from that time the surplus has been annually vested in the 3 *per cent.* consolidated Bank annuities, by which, on the 9th of February 1803, a fund had been created, of 281,500*l.* stock, standing in the names of Sir Andrew Snape Hamond, Bart. Comptroller of the Navy; Francis John Hartwell, Esq. one of the Commissioners of the Navy; Mr. John Madgshon, Master-attendant of Chatham-yard; and Mr. Edward Sison, formerly master-shipwright of that yard.

There is also a further sum in the 3 *per cent.* consolidated annuities, of 2,208*l.* 18*s.* 10*d.* in the name of the Deputy Remembrancer of the Exchequer, as the value of a certain part of the estate of Godsight, taken by government for the purpose of increasing the fortifications at Chatham. The net average profits of the real estates, including the houses built thereon, for the last twenty years, ending in 1802, appear to have been 657*l.* 7*s.* 5*d.* and, no doubt, are much increased since that period.

The funds of the chest are considerably increased during war, the defalcations alone very much exceed the expenditure; but, in time of peace, the reverse is the case, as the paying off the navy gives opportunity to all who have been wounded or hurt to apply for relief; by which the expenditure is increased, whilst the revenue is diminished by the reduced amount of the defalcations arising from a peace establishment.

The process by which a person wounded or hurt in his Majesty's naval service, is enabled to claim relief from the chest, is as follows:—

The surgeon, with the approbation of the captain, fills up a certificate, or smart ticket, (printed forms of which he is furnished with) by inserting the name and age of the party, the nature of the wound or hurt, when received, and the service he was employed on at the time; which is signed by the captain, surgeon, and other officers, and delivered to the party previous to his being discharged, or sent to any of his Majesty's hospitals; the surgeon keeps an account of such smart tickets as may be granted, which is sent to the Navy Office, and forwarded from thence to the governors of the chest for their guidance.

Upon the parties producing the smart ticket, or in case of its being lost, if it should appear, upon reference to the surgeon's list, that one had been granted, the governors and surgeon proceed to examine him; and if the hurt received should not have deprived the party of the power to gain a livelihood, a sum of money is given in full satisfaction for the injury received, called "Smart Money;" but, in case the wounds or hurts are severe, a yearly pension is granted, and an allowance, not exceeding the amount of the pension, is made to the party, for the purpose of defraying the expense of his journey to his home, which is denominated "Present Relief."

At the time of granting pensions, tickets, or certificates, are given to the parties, expressing the amount of their pensions, and their duration; being for one, three, five, or seven years, according to the nature of the injury received. The object in granting pensions for a limited time is not only to prevent frauds, by ascertaining the actual existence of the parties at stated periods, but that the governors may have opportunity of proportioning the pensions to the ability of the parties to maintain themselves. On their re-appearance at the Chest, they are examined by the governors and surgeon, and if their wounds or hurts continue in the same state, their pensions are renewed; if better, they are reduced; if worse, augmented; and, if the party is perfectly recovered, his pension is entirely discontinued.

In cases of total blindness, the loss of a limb, or very severe wounds, the governors have granted pensions for life; and such pensioners only be required at any time hereafter to appear at the chest, the nature of whose wounds may afford a reasonable probability of amendment.

The business of the trust is conducted by the governors, at meetings held monthly, beginning on the first Tuesday in every month; and at an annual meeting early in June.

At the former, the governors meet for the purpose of examining such persons as appear with smart tickets, as also such whose pensions may have ceased from the termination of their tickets; and at the latter, for the general payment of the annual pensions, which become due on the 25th of March.

To relieve the pensioners from the extortionate commission charged by the agents, and to enable them by habits of industry in their respective parishes, to contribute to the support of their wives and families, their pensions are remitted to them half yearly, and made payable at their respective homes, by means of the officers of the customs and excise.

If the pensioner's place of abode be at any distance, a remittance bill is made out, and forwarded to him agreeable to his direction; at the same time, a counterpart is sent to the collector of the customs or excise in the town nearest the place of the pensioner's home, and upon his producing such bill, properly vouched as to the identity of his person, to the collector, and shewing his ticket, the money is paid him without any deduction.

A bill is sent to the collector, and another to the pensioner, whereby any mistake can more easily be rectified in case of miscarriage, &c.

If a pensioner loses his ticket, he must make an affidavit to its loss before he can receive his pension.

Ammunition Chests, (*coffres d'ammunition*, Fr.) are the chests placed in the tops of ships of war, to contain the ammunition of the swivels, &c.

Arm Chests, (*coffres d'armes*, Fr.) are the chests placed in various parts of the ship, to afford a ready supply of muskets, pistols, cutlasses, hand-spikes, &c.

Colour Chests, (*coffres de signaux*, Fr.) are those appropriated for the reception of flags for making signals, and generally kept on the poop.

CHINE, that part of the water-way which is left above the deck, that the lower beam of spirketting may more conveniently be calked.

To CHINSE, (*remplir d'étoupe une couture*, Fr.) is to thrust oakum into a seam, or chink, with the point of a knife, or chisel. This is chiefly used as a temporary expedient when calking cannot be safely or conveniently performed.

CHOCK, (*cale, acore*, Fr.) a sort of wedge used to confine a cask, or other weighty body, in a certain place, and to prevent it from fetching way when the ship is in motion, &c. Also pieces introduced for the purpose of fastening some part of the vessel, or supplying some defect, or to fix between the head of a lower-mast, or the head of a top-mast.

Chock-*a*-*Block*, (*à joindre*, Fr.) is the same with block a-block; which article see.

Chocks *of the Rudder*, are large pieces of timber kept in readiness to stop the motion of the rudder in case of any accident, and while a new tiller is shipped, &c.

CHOKING the Luff, the act of placing the bight of the leading part, or fall of a tackle close up between the nest part and jaw of the block.

To CHOP ABOUT, (*sauter*, *en parlant du vent*, Fr.) is applied to the wind when it varies frequently and suddenly.

CHRONOMETER, (from χρόνος, time, and μετρον, Gr. measure, *chronomêtre*, Fr.) an instrument or machine, for measuring time.

The word is more particularly used by navigators to denote a watch, in which, by the nature of the escapement, and compensations for heat and cold, mean time is, or ought to be kept, to determine the longitude at sea. As the relation between time and longitude will be fully explained under the latter word, it will be sufficient here to observe that the rotation of the earth upon its axis brings the several places upon its surface, in succession, opposite the sun, causing day and night; so that the absolute instant of noon must be earlier at a place which lies to the eastward of another, with which that place may be compared.

Upon the same principle it is that clocks in a large town ought not to indicate the same time. Thus the clocks at St. Paul's, St. Clement's, St. Martin's, and St. George's Hanover-square, in London, ought to strike each four seconds after the other; and this difference, it may be added, would nearly vanish, if heard from any of the western stations; on account of the time employed for the passage of sound; and, for the same reason, it would be nearly doubled in the opposite direction.

And from this general fact it follows, that, allowing twenty-four hours for the whole rotation of the earth, and proportionally for every smaller part, we may determine what is the difference of longitude between them. For example, suppose a chronometer, set to the time at Greenwich, were to be carried to Petersburgh, in Russia, it would indicate time two hours later than Petersburgh clocks, that is to say, it would shew when it was noon at Greenwich, instead of when it was noon at Petersburgh. The obvious conclusion is, that the sun arrives at the meridian of Petersburgh earlier, and, consequently, that this town lies more easterly than Greenwich; and as two hours are in proportion to 24, so is 360°, the earth's circumference, to 30°, the longitude of Petersburgh, reckoned from Greenwich. This is supposing the chronometer to have gone at mean time, or, in other words, that its hands have passed over twenty-four hours on the dial plate from mean noon on one day to mean noon on the next. It is then said to go at mean time, or to have no rate. But this is not necessary to its use; for provided its rate or daily error be known, the time at Greenwich, or the first meridian, may always be found by applying that rate to the time shown by the chronometer.

The rate of a chronometer is that number of seconds which its hands indicate more or less than twenty four hours in a mean day, and is called *plus* when more is indicated than twenty-four hours in a day, and *minus* when less.

For example, suppose the chronometer when set to the time at Greenwich, instead of going at mean time, had been ascertained to have a gaining rate of +7",5 a day, and that it was forty days on its passage to Petersburgh; it is evident that, on its arrival, it would have been forty times seven seconds and a half, or five minutes too fast at Greenwich, and must, therefore, have five minutes deducted from the time it indicated to give that at Greenwich; and so much added if the rate had been losing.

The rate should be well ascertained by the makers of chronometers before they are delivered, and be verified as frequently as possible afterwards; for which purpose, public observatories at the sea-ports would be highly beneficial.

But although an observatory with a good regulator is unquestionably the best means of finding the rate of a chronometer, this advantage is but seldom attainable; and a rate may be very well ascertained in any part of the world, by any person acquainted with the practice of nautical observations, by the following methods:

Suppose a ship to be stationary for a short period, the time should be taken as soon as possible, after her coming to anchor, to ascertain the error of the chronometer, and again on her departure, then the encreased error, divided by the number of days that have elapsed between the two observations, will give the daily error or rate; and, if intermediate observations can conveniently be made, it will be desirable, as the results of the number, will be a corroboration or correction for the rate deduced from the whole period.

Again, suppose a chronometer to be received on board without a rate, and the ship sail too soon for a rate to be settled by the above method, then, if the error be ascertained by observation before her departure, and again on touching at the first land, and the whole error divided by the number of days elapsed, a rate may be obtained for the use of the remainder of the voyage.

Either, or both of the above methods, practised as often as convenient, will be highly serviceable, as well for verifying and correcting a rate originally given, as for finding a new rate.

The invention of chronometers, and the improvements in them, have been variously claimed; but no doubt seems to exist of their having derived great perfection from the researches of French artists. Peter Leroy constructed them with fluid thermometers on the balance, and afterwards invented the present expansion balance of brass and steel united, for correcting the effects of heat and cold on those machines, which is now brought to the greatest perfection by Arnold and others; and to which, with the isochronism of the vibrations of the balance produced by the proper adjustment of the pendulum spring, chronometers principally owe their present degree of accuracy.

Chronometers made upon these principles have measured time to a wonderful degree of perfection. Most captains of men of war have one allowed them by government, when they apply for it on going on a foreign station, and many of them have one of their own private property.

The best makers at present are allowed to be Arnold, Earnshaw, Barraud, Hatton, Brockbanks, and Pennington.

CIRCLE, (*cercle*, Fr.) in geometry, a plane figure, comprehended by a single curve line, called its circumference; in which right lines drawn from a point in the middle, called the centre, are equal to each other.

The area of the circle is found by multiplying the circumference by the fourth part of the diameter, or half the circumference by half the diameter: for every circle may be conceived to be a polygon of an infinite number of sides, and the semi-diameter must be equal to the perpendicular of such a polygon, and the circumference of the circle equal to the periphery of the polygon; therefore, half the circumference multiplied by half the diameter, gives the area of the circle.

Circles, and similar figures inscribed in them, are always as the squares of the diameters; so that they are in a duplicate ratio of their diameters, and, consequently, of their radii.

A circle is equal to a triangle, the base of which is equal to the periphery, and its altitudes to its radius: circles, therefore, are in a ratio, compounded of the peripheries and the radii.

Great CIRCLE *of the Sphere*, is that which divides it into two equal parts, or hemispheres, having the same centre and diameter with it; such are the equator, ecliptic, horizon, the colures, and the azimuths, &c.

Lesser CIRCLE *of the Sphere*, is that which having its centre in the axis of the sphere, divides it into two unequal parts; these are usually denominated the tropic polar circles, and parallels of latitude, all of which are parallel to the equator.

Reflecting CIRCLE, an astronomical instrument for observing the distances of celestial objects. It has long been known but seldom used, owing to the complicated manner of observing to the right and left successively.

This instrument was invented by the late celebrated astronomer, Professor Mayer, and has since been greatly improved by Chevalier de Borda, M. de Magellan; and is now much simplified by Captain Mendoza Rios, and that ingenious artist Mr. Troughton. The circular instrument, or, as it is sometimes called, the *multiplying circle*, is evidently preferable to a quadrant or sextant, inasmuch as the observation of the distance between any two objects may be repeated round the whole of the limb, and the errors of the instrument thereby diminished, or rendered almost insensible.

The sum of all the measured distances may be found with the same simplicity as observing with the sextant, and when the observations are repeated, the difference between the reading after the last observation and the first reading, divided by the number of observations, will give a quotient, from which, deducting the error of the stop, the result will be the mean of the observed distances.

The adjustments of this instrument are, simply to make the mirrors perpendicular to its plane, and the line of cullimation parallel thereto, to prevent any false light from entering the telescope, &c. These being once made, will continue perfect for a considerable time, provided the instrument is taken care of.

There are three different methods of observing the angular distance between two objects with this instrument. The first is called an *observation to the right*, and is that wherein the object, whose image is to be reflected, and the central mirror are on the same side of the telescope; the second an *observation to the left*, when the object to be reflected and the central mirror are on opposite sides of the telescope, which, in both cases, is supposed to be directed to the other object; and the third, a *cross observation*, which is the combination of two successive observations, the one being to the right and the other to the left.

For a more particular account of the adjustments and general precepts for using the instrument, see MACKAY *on the Longitude*.

CIRCLES *of Declination*, on the globe, are great circles, intersecting each other in the poles of the world; or, with some authors, the meridians on which the declination, or distance, of any star from the equinoctial is measured.

Diurnal CIRCLES are parallel to the equinoctial, supposed to be described by the several stars and other points of the heavens, in their apparent diurnal rotation round the earth, or rather, in the rotation of the earth round its axis.

CIRCLES *of Latitude*, or *Secondaries of the Ecliptic*, are great circles perpendicular to the plane of the ecliptic, passing through the poles of it, and through every star and planet. They serve to measure the latitude of the stars, which is an arch of one of those circles intercepted between the star and the ecliptic.

CIRCLES *of Longitude*, are several lesser circles parallel to the ecliptic, diminishing in proportion as they recede from it, or as they approach the pole of that circle; on these the longitude of the stars is reckoned.

Polar CIRCLES, are parallel to the equator, and at the same distance from the poles that the tropics are from the equator. See the article ARCTIC.

CIRCULAR SAILING, is that performed in the arch of a great circle. It is chiefly on account of the shortest distance that this method of sailing has been proposed; and, for the most part it is advantageous for a ship to reach her port by the shortest course.

CIRCULAR *Velocity*, in astronomy, signifies the velocity of any planet, or revolving body, which is measured by the arch of a circle.

CIRCUMFERENCE, (*circonference*, Fr.) the curve line which bounds a circle, and otherwise called a periphery, the boundary of a right-lined figure being expressed by the term perimeter. Any part of the circumference is called an arch, and a right line drawn from one extreme of the arch to the other is called a chord. The circumference of every circle is supposed to be divided into 360 degrees. The angle at the circumference of a circle is double that at the centre.

CIRCUM POLAR STARS, are stars which, by reason of their vicinity to the pole, move round it without setting; such as those in the constellations Camelopardalus, Ursa Major, Ursa Minor, Draco, Cepheus, and Cassiopea.

CISTERN, (*citerne*, Fr.) a large wooden trough, placed in the well, just below the orlop, and having a lead pipe, which goes through the ship's side, whereby it is occasionally filled with sea-water, which is thence pumped up to wash the decks, &c.

CIVIL ARCHITECTURE, is the art of defending and erecting edifices of every kind, for the uses of civil life in every capacity; as churches, palaces, private houses, &c. and it has been divided into five orders, or manners of building, under the names of the *Tuscan*, *Doric*, *Ionic*, *Corinthian*, and *Composite*. There were many authors on architecture among the

Greeks and Romans before Vitruvius, but he is the first whose work is entire and extant. He lived in the reigns of Julius Cæsar and Augustus, and composed a complete system of architecture, in ten books, which he dedicated to this prince. The principal authors on architecture since Vitruvius are, Philander, Barbarus, Salmasius, Baldus, Alberti, Gauricus, Demoniosus, Perrault, De l'Orme, Rivius, Wotton, Serlio, Palladio, Strada Vignola, Scamozzi, Dieussart, Catanei, Freard, De Cambray, Blondel, Goldman, Sturmy, Wolfius, De Rosi, Desgodetz, Baratteri, Mayer, Gulielmus, Ware, Sir William Chambers, &c. &c.

CLAMPS, (*bauquieres*, Fr.) thick planks on the inner part of a ship's side, used to sustain the ends of the beams, and extending from stem to stern, including the whole interior range of the side. They are placed close under each deck, so as to be securely fayed to all the timbers, to which they are fastened by nails driven through the clamp, and penetrating two-thirds of the thickness of the timbers.

The clamps of the lower and second decks ought to be equal in thickness to half the corresponding timbers in that part, and as broad as can be procured. In their disposition it is essentially necessary to avoid their being wounded by the ports, as the strength and firmness of a ship greatly depend on the substance and solidity of those pieces which lie horizontally in her frame.

Clamps are also small crooked plates of iron, forelocked upon the trunnions of the cannon, to keep them fast upon their carriages: these, however, are more properly termed cap-squares. See the article Carriage.

Clamps of the latter sort are likewise frequently used to fasten the masts or bowsprits of small vessels or boats.

Clamp-*Nails*, are those used to fasten on clamps in the building and repairing of ships.

To CLAP on, (*frapper*, *établir*, *&c.* Fr.) is to fasten or lay hold of, as, "*clap on* the stoppers before the bits," *i. e.* fasten the stoppers; "*clap on* the cat-fall," *i. e.* lay hold of the cat-fall.

CLAVE, a stool 14 inches high, made of elm, and supported by four legs; the top six feet long, two or three feet wide, and eight inches thick at each end, and only four inches thick in the middle, on which the shells are set up with wedges for making the sheave-holes.

CLAWING, or CLAWING-off, (*chicaner*, *s'éloigner d'une côte*, *gagner le vent au plus près*, Fr.) the act of *beating*, or turning, to windward from a lee-shore, so as to acquire a sufficient distance from it, to escape the dangers of shipwreck, which often attend so hazardous a situation.

CLEAN, a term given to the sharp part of a ship under water, both forward and aft.

CLEAR, (*clair*, *dégagé*, *&c.* Fr.) as a naval term, is variously applied to the weather, the sea-coasts, cordage, navigation, &c. The weather is said to be clear when it is fair and open, as opposed to cloudy or foggy.

The sea-coast is called clear (*saine*, Fr.) when the navigation is not interrupted, or rendered dangerous by rocks, sands, or breakers, &c.

Clear is expressed of cordage, cables, &c. when they are disentangled, so as to be ready for immediate service. It is usually opposed to *foul* in all these cases.

To Clear *the Anchor*, (*dégager l'ancre*, Fr.) is to get the cable off the flooks, and disencumber it of ropes, ready for dropping.

To Clear *the Hawse*, (*défaire la croix ou dépasser le tour des cables*, Fr.) is to untwist the cables when they are entangled by having either a *cross*, an *elbow*, or a *round-turn*.

To Clear *out a ship*, in the merchant service, is to obtain leave for sailing, or filling the cargo, by paying the customs.

CLEARER, a tool on which the hemp is always finished for lines and twines, for sail-makers, &c.

CLEATS, (*taquets de manœuvres*, Fr.) pieces of wood of different shapes, used, occasionally, in a ship, to fasten ropes upon; some have one, and some two arms; others have no arms, being hollowed in the middle, and are nailed to the deck, or sides, to fasten any thing to.

They are used for different purposes; those intended for the purpose of stopping shores are principally made of elm, not unlike wedges, except that they taper from one side only; those made for the purpose of stopping or fastening rigging are haunched on the back for about one-third the length, the thin end shaped like a duck's bill; they are, in general, made of oak, but those used at the mast-head of elm.

Belaying Cleat, plate VII. fig. 8, is formed with two arms, one on each side the centre or middle part, and nailed or bolted to the side, for the purpose of belaying the running-rigging to.

Comb Cleat, fig. 9, is straight on the inner edge, and round on the back, having a hollow cavity in the middle. It is merely used for leading a rope through, or for keeping it in its place.

Mast Cleat, fig. 10, is made with a *score*, to admit a *seizing*, a long hole in the centre, for an under seizing, and two round holes, by which the seizing may be crossed.

Shroud Cleat, fig. 11, is formed like the belaying cleats, having two arms, the remaining part being straight and grooved on the edge. It has scores cut towards the extremity for the seizings to lie in, which are naked, and a groove in the part where the shroud lies.

Single Cleat, fig. 12, is formed with one arm only, and nailed, on each side of the *slings*, on the yards.

Stop Cleat, fig. 13, is nailed on the bowsprit for the gammoning collars, &c. and sometimes on the yard-arms.

Thumb Cleat, fig. 14, is not unlike a *sling cleat*, with this exception, that it is smaller, and is made for the purpose of hanging any thing on. The number of thumb cleats is indefinite, and are variously fixed in point of strength, according to the strain they have to resist.

CLEW of a Sail, (*point d'une voile*, Fr.) the lower corner of square sails, but the aftmost only of stay-sails, &c. the other lower corner being called the tack.

Clew-*Garnets*, (*cargue-points de basses voiles*, Fr.)

are a sort of tackle fastened to the clews of the main and fore-sail, to truss them up to the yard, which is termed clewing up those sails.

Clews *of a Hammock*, (*points d'un hamac*, Fr.) the combination of small lines by which it is suspended, being formed of knittles, grommets, and lanyards, and are termed double or single clews, according as they are one or two at each end.

Clew-*Lines*, (*cargue-points des voiles quarrées*, Fr.) are for the same purpose as clew-garnets, only that the latter term is solely appropriated to the courses, while the word clew-lines is applied to those ropes on all the other square-sails. See those ropes as represented in the article Sail.

From Clew *to Earing*, a phrase, implying from the bottom to the top, or synonimous with, from "top to toe."

A Spanish Clew, is formed by fixing the knittles at equal distances upon a piece of rope, instead of a grommet, which having an eye spliced and a lanyard placed at each end, extends the hammock similar to a double clew.

To Clew *up*, (*carguer une voile*, Fr.) to haul up the clews of a sail to its yard, by means of the clew-lines.

CLICKS, are small pieces of iron, falling into notched wheels attached to the winches in cutters, &c. and thereby serving the office of pawls.

To CLINCH, (*étalinguer un cable*, Fr.) a particular method of fastening large ropes by a kind of knot, and seizings instead of splicing, and is chiefly used to fasten the cable to the ring of the anchor, and the breechings of guns to the ring-bolts in the ship's side.

CLINCHER-Built, (*bordé à clins*, Fr.) is applied to a vessel made of clincher-work.

Clincher-*Work*, the disposition of the planks in the side of any boat or vessel, when the lower edge of every plank overlays the next under it, like the slates on the roof of a house.

CLINCHING, the operation of fastening the point of a bolt, or nail, on a ring, or rove of iron, by hammering it so as to make it spread.

Clinching, a kind of slight calking, used about the ports, on a prospect of foul weather: it is done by driving a little oakum into their seams, that the water may not come in at them.

CLOCK, in horology, is a machine constructed in such a manner, and so regulated by the uniform motion of a pendulum, as to measure, time, and all its divisions, with exactness, that gives it a just preference over every other method used for the same purpose.

The sphere of Archimedes, mentioned by Claudian, to have been made two hundred years before the birth of Christ, is usually considered as the first attempt at the formation of a clock; it had, indeed, a maintaining power, but no regulator, its motion originating from some hidden weights, or springs, with wheels or pullies, or some such clock-work principle; therefore could only measure time, as a planetarium exhibits the motion of the stars with relative, but not with positive, precision.

There are many authentic records in history to prove the existence of clocks with wheels and weights at the beginning of the 14th century; consequently, there is more reason for assigning this period to the invention than any other; yet the first inventor is not certainly known.

Clock-makers were first introduced into England in the year 1368, when Edward the 3d granted a licence for three artists to come over from Delft, in Holland, and practise their occupation in this country.

The invention of pendulum clocks is owing to the happy industry of the 16th century; and the person to whom mankind is really indebted for bringing this important discovery into universal notice, is the celebrated Christian Huygens, of Zuylichem, who, in his excellent treatise "*De Horologio Oscillatorio*," has described the construction of a pendulum clock, and proved that he made one before the year 1658.

Becher, *De Nova Temporis dimetiendi Theoria*, A.D. 1680, contends that Galileo had claims to the priority of the invention of the mode of applying the pendulum to clock-work, and his son Vincentio Galilei is reported (*Exper. del Acad. del Cimento*) to have made a pendulum clock so early as in 1649, at Venice, suggested by his father's discovery.

The next improvement of consequence on clocks, after the pendulum, was the escapement, performed with anchor pallets, which Berthoud states to have been the invention of Clement, a London clock-maker, in the year 1680. The escapement used by Huygens, and still continued in many chamber clocks, and all the wooden clocks, is that made by two flat pallets attached to an horizontal arbor, acting at opposite sides of the upper part of a horizontal crown wheel; the anchor pallets on the contrary, act on a vertical swing wheel, and move in the plane of the wheel.

The chief advantage of the anchor pallets is, that they will permit the escape to take place with a small angle of vibration, so as to prevent the maintaining power from acting on the pallets a long time by a direct push, as was the case with the crown wheel escapement. The advantage gained by the anchor escapement may be considered, in reality, an approximation to a detached escapement; a farther step was made towards this improvement about the year 1715, by the celebrated George Graham, in the contrivance of a dead-beat escapement, which is principally distinguished from the anchor escapement by having no recoil.

The detached escapement completed the improvement of this part of clock-work. Its object is to make the pendulum perform the greatest part of each vibration entirely free from contact, or connexion, with any part of the train. To effect this, a catch, or locking-piece, restrains all the motion of the swing wheel till the instant when the pallet is to be impelled by it, when it raises the catch, sets the wheel free, and is driven forward by its impulse; immediately after which the catch again falls into its place. A great variety of escapements have been contrived on this principle, by various ingenious men; those in which springs are used in the locking-pieces instead of pivots, invented by Arnold, seem now most preferable. The detached escapement was applied first to chronometers, or time-pieces, but is now used for astronomical clocks.

Amongst the various sorts of pendulums, one suggested by Mr. Troughton, deserves particular notice, and seems worthy of trial. He proposes that its rod should be made of baked potter's earth, of the same composition of Wedgewood's thermometer, and furnished with a metallic cap, by which it should be sustained by the knife-edge suspension, which the celebrated Berthoud affirms has less friction than the spring suspension. Mr. Barraud, of London, has also invented a tubular pendulum, filled with quicksilver, which is now much approved of for astronomical clocks.

The chief advantages which tubular pendulums have over those of the gridiron construction are, that they admit of being much lighter above the bob, with equal strength; that they experience less resistance from the air in their vibrations; and that they are less liable to those shakes and irregular motions in their expansions which the others experience: on the other hand, as the outside tube alone in them comes in contact with the air, through which it passes in its vibrations, the inner tubes can receive much less of its influence as to temperature, which arises from this motion, and which Cummings has shewn to be of considerable consequence.

It appears that the first pendulum clock made in England was in the year 1662, by one Fromantil, a Dutchman.

For an account of the progressive improvements made in pendulums by Graham, Harrison, Elliot, Chandler, Cummings, Troughton, and others, see NICHOLSON'S *British Encyclopædia.*

CLOCK-*Work*, properly so called, is that part of the movement which strikes the hours, &c. on a bell, in contradistinction to that part of the movement of a clock, or watch, which is designed to measure and exhibit the time on a dial-plate, and which is termed watch-work.

CLOCK, *Atmospherical*, see the article ATMOSPHERICAL CLOCK, and Dr. D. BREWSTER'S *Treatise on Astronomical Instruments.*

CLOCK, *Astronomical*, see the article ASTRONOMICAL CLOCK.

CLOSE-FIGHTS, are *bulk-heads*, erected fore and aft in the ship, for the men to stand behind in close-engagement, and to fire on the enemy; or, if the ships be boarded, to scour the decks.

CLOSE-HAULED, (*au plus près*, Fr.) the arrangement, or trim, of a ship's sails, when she endeavours to make a progress in the nearest direction possible towards that point of the compass from which the wind blows: in this manner of sailing the keel of square-rigged vessels commonly makes an angle of six-points with the line of the wind; but cutters, luggers, and other fore-and-aft rigged vessels, will sail much nearer. All vessels, indeed, are supposed to make nearly a point of lee-way when close-hauled, even when they have the advantage of a good sailing-breeze and smooth water. The angle of lee-way, however, enlarges in proportion to the increase of the wind and sea.

In this disposition of the sails, they are all extended sideways on the ship, so that the wind, as it crosses the ship obliquely towards the stern from forwards, may fill their cavities. But, as the current of wind also unties the cavities of the sails, in an oblique direction, the effort of it, to make the ship advance, is considerably diminished: she will, therefore, make but little progress when sailing in this manner.

The ship is said to be close-hauled, because at this time her *tacks*, or lower corners of the principal sails, are drawn close down to her side, to windward; the sheets hauled close aft; and all the bowlines drawn to their greatest extension, in order to keep the sails steady.

CLOSE-QUARTERS, (*cloisons servant de retranchement encas d'abordage*, Fr.) are certain strong barriers of wood stretching across a merchant ship in several places: they are used as a place of retreat when a ship is boarded by her adversary, and are therefore fitted with loop-holes, through which to fire the small arms, whereby the ship's crew may defend themselves, and annoy the enemy; they are likewise furnished with caissons, or powder-chests, fixed upon the deck, and filled with powder, old nails, &c. which may be fired at any time from the close-quarters, upon the boarders.

An English merchant-ship, of 16 guns, being properly fitted with close-quarters, has been known lately to defeat the united efforts of three French privateers who boarded her, after having engaged at some distance nearly two days, with very few intervals of rest. Two of the cruisers were equipped with 12 guns each, and the other with eight. The French sailors were, after boarding, so much exposed to the continued fire of musketry, and coehorns charged with grenades, that a dreadful scene of carnage ensued, in which the decks were soon covered with the dead bodies of the enemy, several of which the boarders, in their hurry to escape, had left behind.

CLOTHED. A mast is said to be clothed when the sail is so long as to reach down to the gratings of the hatches, so that no wind can blow below the sail.

CLOTHING OF THE BOLSTERS. Laying several thicknesses of worn canvas, well tarred, over them, to make an easy bed for the shrouds.

CLOTHS IN A SAIL, (*cüeilles qui composent une voile*, Fr.) are the breadths of canvas in its whole width.

CLOVE-HITCH, (*deux demi-clés*, Fr.) a knot or noose by which a rope is fastened to another. See the article HITCH.

To CLUB HAUL A SHIP (*manière particulière de virer vent devant*, Fr.) is a method of tacking, by letting go the lee-anchor as soon as the wind is out of the sails, which brings her head to wind, and as soon as she pays off, the cable is cut, and the sails trimmed; this is never had recourse to but in perilous situations, and when it is expected the ship will miss stays.

COACH, or COUCH, a sort of chamber or apartment in a large ship of war near the *stern.* The floor of it is formed by the aftmost part of the quarter-deck, and the roof of it by the poop: it is generally the habitation of the captain.

COAKING, the operation of uniting two or more pieces together in the centre, by means of small tabular projections formed by cutting away the solid of one piece into a hollow, so as exactly to make a projection on the other in such manner that they may correctly fit, and the butts prevent the pieces from drawing asunder lengthways. Different methods of coaking are pursued.

COAK *and Plane*, signifies when a coak is formed

and a plane surface follows between, that and the next running coaks are those which are continued the whole length along the middle: they, however, answer the above purpose, inasmuch as the butts of each coak spread one third their breadth within and without each other alternately.

COAKS, are the metal holes in a sheave through which the pin runs.

COAKS, are also certain oblong ridges left on the surfaces of main-masts, by cutting away the wood round them: the intermediate part is called the plane.

Chain COAKS, are formed one on the end of the other, on the opposite sides of the centre line.

COAL-HOLE, (*charbonnière* Fr.) is the part of a ship's hold lying near to the after magazine, and appropriated for the reception of coals, wood, &c.

COAMINGS OF THE HATCHES, (*chambranles, ou vassoles des écoutilles*, Fr.) certain raised borders about the edge of the hatches of a ship, to prevent the water on the deck from running down into the lower apartments. They are represented in the deck, as enclosing their respective hatchways.

COAST, (*côte*, Fr.) the sea shore, or the country adjoining the edge of the sea.

COASTER, (*caboteur*, Fr.) a vessel employed in going from one port to another on the same coast, and, therefore, seldom loses sight of land during the whole voyage.

COASTING, (*aller terre à terre*, Fr.) in navigation, the act of making a progress along the sea-coast of any country, for which purpose it is necessary to observe the time and direction of the tide, to know the reigning winds, the roads and havens, the different depths of the water, and the qualities of the ground, &c.

COASTING-*Pilot*, (*pilote côtier*, Fr.) a pilot, who by long experience has become sufficiently acquainted with the nature of any particular coast, and the requisites mentioned in the preceding article, to conduct a ship or fleet from one part of it to another.

COAT, (*braye*, Fr.) a piece of tarred canvas nailed round that part of the masts and bowsprit which joins to the deck, or lies over the stem. Its use is to prevent the water from running down between decks.

There is also a coat for the rudder nailed round the hole where the rudder traverses in the ship's counter. This hole is represented at the upper part of the stern post.

COAT, (*suage*, Fr.) also implies the *stuff* with which the ship's sides or masts are varnished, to preserve them from the sun and weather, as turpentine, tar, &c. In this sense we say, "Give her a good coat of tar."

COBBING, (*sorte de punition pour les matelôts*, Fr.) a punishment sometimes inflicted at sea. It is performed by striking the offender a certain number of blows on the breech, with a flat piece of wood, called the cobbing-board.

COBBING-BOARD, (*bâton de justice*, Fr.) the board with which offenders on board ships are *cobbed*.

COBOOSE, (*fogone*, Fr. *kambuis*, Dutch,) a sort of box or house to cover the chimney of some merchant-ships. It somewhat resembles a centry-box, and generally stands against the barricade on the fore-part of the quarter-deck. It is the place where victuals are cooked on board merchant-ships.

COCK-BOAT, a small boat used on rivers, or near the shore, which is of no service at sea, because too tender, weak, and small. In ancient days, a cock was the general name of a yawl.

COCKPIT OF A SHIP OF WAR, (*poste du chirurgien*, Fr.) is near the apartments of the surgeon and his mates, being the place where the wounded men are dressed. It is situated near the after hatchway, and under the lower gun-deck.

Fore COCKPIT, (*emplacement qui environne l'échelle de la fosse aux lions*, Fr.) a place leading to the magazine-passage, and the boatswain's, gunner's, and carpenter's store-rooms: in large ships, and during war time, the boatswain and carpenter generally have their cabins in the fore-cockpit, instead of being under the fore-castle.

COCKS FOR BLOCKS, are little square pieces of brass, with holes in them, and put into wooden sheaves, to keep them from splitting and galling by the pins of the blocks in which they move.

COCKSWAIN, or COXEN, (*patron de chaloupe*, Fr.) the officer who steers a boat, and has the command of the boat's crew, and all things belonging to it. He has a whistle to call and encourage his men, and must be ready with his crew to man the boat on all occasions: he sits at the stern of the boat, and steers.

COD-FISHER, (*caplanier*, Fr.) is the name of a vessel employed to cure cod; also of the man appointed to that service.

COIL, (*cueillir*, Fr.) the manner in which all ropes are disposed aboard ships for the convenience of stowage.

Flemish COIL, is a rope coiled up in a spiral manner, forming but one tier, and lying flat on the deck, the end being in the middle of it.

COILING, (*rouer un cordage*, Fr.) is a sort of serpentine winding the ropes, by which they occupy a small space, and are not liable to be entangled amongst one another in working the sails. Each winding of this sort in a cable is called a *fake*, and one range of fakes is called a *tier*; there are generally from five to seven fakes in a tier; and three or four tiers in a cable's length: the small ropes are frequently coiled by hand, and hung upon *cleats*, to prevent their being entangled amongst one another in traversing, contracting, or extending the sails.

COIN, or QUOIN, is the wedge laid under the breech of a gun, for the purpose of elevating or depressing it at pleasure.

COLLAR, (*collier d'étai*, Fr.) the lower part of any of the principal stays of the masts, or the part by which the stay is confined at its lower end. Thus the collar of the main-stay connects the lower end of the stay to the ship's stem. See the article STAY.

COLLAR, a rope formed into a wreath, with a heart or dead-eye seized in the bight, to which the stay is confined at the lower part.

THE ROYAL NAVAL COLLEGE AT PORTSMOUTH, formerly called the *Royal Naval Academy*, is a collegiate seminary founded by George II. in the year 1729, and was approved of by his present Majesty, King George III. whose order in council, dated the

8th October, 1773, directed an academy to be established in the dock-yard, with a view to the training up, constantly, 40 young gentlemen in the principles and practice of navigation, marine-surveying, gunnery, fortification, mechanics, drawing, and various other studies, preparatory to their entrance into actual sea service, of which number 15 were proposed to be sons of commissioned officers of the royal navy.

It having, however, been lately found, that the plan on which this seminary had been conducted, since its establishment, was very inadequate to the increased naval force; and that the importance of establishing a national institution upon a more liberal and extensive scale, would be found indispensably necessary to the state, for the education of officers for the sea service; a memorial from the lords commissioners of the Admiralty to that effect was therefore proposed, and read at the court at the Queen's palace, the 1st February, 1806.

His Majesty having taken the said memorial into consideration, was pleased, by and with the advice of his privy council, to approve of the plan therein recommended, and thereby authorized the lords commissioners of the Admiralty to carry the said plan into execution, subject, however, to such further rules and orders for the internal management of the proposed institution, as they should, from time to time, find expedient for the government of the same. under the name of the *Royal Naval College* at Portsmouth.

The building of the academy was accordingly altered and enlarged, in proportion to the increased establishment, viz. from 40 to 70 young gentlemen, of which 40 scholarships are invariably to be reserved for the sons of commissioned officers; the estimated expense of which, with additional teachers, amounted to 4886*l.* 2*s.* 6*d.*

The establishment of officers for conducting the affairs of the college, consists of the following persons; viz. A governor, lieutenant-governor and inspector.

A professor, preceptor, mathematical master, and housekeeper. A writing, drawing, French, fencing, and dancing master. A surgeon, the master-attendant of the dock-yard, the master shipwright of the yard, and a gunner of marine artillery.

The establishment is under the direction of the lords commissioners of the Admiralty, and the first lord for the time being, is considered as governor of the college.

The qualifications and other requisites for the admission of students into the college, are as follow:

1. Every candidate for admission, is expected to know perfectly the first four rules of arithmetic, reduction, and the rule of three.

2. He will be expected to write English from dictation, and to construct English sentences.

3. And it is recommended that every candidate should have in his memory the definitions, axioms, and postulates, in the beginning of Euclid's Elements, by R. Simpson.

4. Every candidate for admission must transmit to the secretary of the Admiralty, a certificate of his age, which is not to be under 13, nor to exceed 16; and a certificate of his good moral conduct, from the master or person under whom he was last instructed: or if he has used the sea, from the captain under whom he last served.

5. The time of his remaining in the college is not to exceed three years, whether he shall have completed the plan of education or not, which, by a youth of moderate talent, properly qualified on entering, it is calculated may be completed in little more than two years. But whether the time he requires be more or less than two years, it shall be reckoned only as two of the six years servitude, which is indispensably required to qualify a young man to receive the commission of a lieutenant; the other four to be passed actually at sea in a ship or ships in commission, not being a guard-ship at any of the ports in the united kingdom.

6. A young man who has used the sea before his admission, is likely to derive a greater advantage from the established plan of education, than one wholly ignorant of sea service: and it is intended that such candidates for admission as may have actually been at sea from one to three years, and whose ages come within the time limited, shall be considered as entitled to the preference to fill the vacancies that may occur in the college.

The friends of every student admitted into the college, will be required to lodge with the solicitor of the Admiralty, security to the amount of 200*l.* that such student, after finishing his education, shall enter and continue his Majesty's naval service, until he shall have completed the time required to qualify him for the commission of a lieutenant; unless he shall have used the sea two years before his admission, in which case no security will be required.

No student will be permitted to carry with him to the college, on his entrance, or return from the vacations, more than one guinea in his pocket; and as one shilling a-week is allowed out of the public fund to every student, as pocket-money, it is particularly desired that their friends do not, on any account, increase this allowance.

The general vacations take place at Christmas and Midsummer: the former for four weeks, and the latter for five; when all the students are expected to go home, or their friends will be required to pay the housekeeper one guinea per week during that time; but it is strongly recommended that they should all leave the college at the vacations, as in all probability the masters will then be absent.

Each student's pay, during his residence at the college, is 4*s.* per diem; out of which, the professor and housekeeper receive a certain sum for his instruction and board, and the balance is appropriated to defray the contingent expenses.

If a student should be expelled from the college, he can never after be admitted as a commissioned officer in his Majesty's navy.

The Royal Military College *at Marlow* was founded by his Royal Highness the Duke of York, commander in chief of his Majesty's land forces, and is under his control and superintendence.

The college consists of two departments, the *senior* and the *junior*.

The *senior department* was established in the year 1799, and consists of 30 students holding commissions in the regular army, who are under the direction of a commandant, Sir Howard Douglas.

This department is established, for the purpose of instructing officers in the scientific parts of the profession, with a view to enable them the better to discharge their duty, when acting in the command of regiments; and, at the same time, qualify them for being employed in quarter-master general's, and adjutant-general's department.

The studies pursued in this department are, mathematics, fortification, castrametation, military drawing, and surveying; the reconnoitring of ground; the disposition and movements of troops under all the various circumstances of offensive and defensive war; the ascertaining the military resources of a country; the German, Spanish, and French languages; all of which they are supposed to complete within a period of two years and a half from their first entrance.

Before an officer can be admitted, he must have completed his 21st year, and have been actually serving as a commissioned officer for three years abroad, or four years at home; and must produce testimonials of his being well grounded in the duties of the particular branches of the service to which he belongs: he must also undergo an examination. The students of this department pay an annual subscription of thirty guineas, which serves to lighten the expense of instruction. When they have gone through the regular course, they are examined; and, if properly qualified for quitting the college, they receive certificates testifying their fitness for staff appointments.

No student has the liberty of withdrawing from the college, without previous permission from the commander in chief, obtained through the governor of the institution.

The *junior department* was established in 1802, and is appropriated to the instruction of those, who, from early life, are exclusively intended to embrace the military profession; and who, by this means, may be grounded in science previously to their obtaining commissions in the army.

This department is also intended to afford a provision for the sons of meritorious officers who have fallen, or been disabled in the service of their country; and the means of education to the sons of those officers who belong to the regular service.

At present, it consists of three companies, each containing 103 cadets. These are under the direction of the lieutenant-governor. There is also a major, and each company has a captain; and in order to give the whole establishment a military character, these are placed on the establishment of the army.

The daily pay allowed for each gentleman cadet is 2*s.* 6*d.* which is applied towards his instruction, clothing, and maintenance. But there are other funds applied to the same purpose, arising from regulated sums annually paid on account of the cadets, by their parents.

Every candidate must bring a certificate of good conduct from his last schoolmaster, and cannot be admitted under 13, nor above 15 years of age; and he must be free from any mental or bodily defect. Every candidate undergoes an examination to qualify him for admission, and cannot be received until he is perfectly acquainted with the rudiments of arithmetic, and of English and Latin grammar: and he must likewise write a very good hand.

The cadets are admitted into the college on three different establishments.

1. The legitimate orphan sons of subaltern officers, and also the sons of such officers of whatever rank in the regular army, and of masters and commanders, and officers of superior ranks in the navy as have died in the service, and left their families in pecuniary distress. These receive their education, board, and clothing, free from expense to their connexions, for a period of four years, or until they have attained a certain progress in education to render them eligible to the holding a commission as cornet or ensign.

2. This class comprises the sons of officers of the navy or army actually in the service, and not under the rank of a captain, who each contribute a certain yearly subscription appropriated to the rank, &c. of their respective fathers.

The sons of officers dying in the service, but who do not leave their families in circumstances of pecuniary difficulty, are admitted on this establishment with a reference to the rank of their fathers, held at the time of their decease, estimating the sums to be paid on the scale of the half-pay of the rank in which their fathers died.

3. The remaining class of gentlemen cadets, are the sons of noblemen and gentlemen, not of a military profession. These pay a sum annually, which is settled by the board of commissioners, and which at present is 100*l.* per annum.

The gentlemen cadets are instructed under the most able and scientific professors and masters in mathematics, fortification, military and landscape drawing, geography, history, and such as have got any knowledge of the classics continue to read them a certain number of days in the week; the French and German languages, fencing, &c. The whole body of cadets wear uniforms, and are distributed into companies, under careful and competent officers; they are thoroughly instructed in the perfect use of arms, and in the military evolutions practised in the army.

The affairs of the college are managed by a board of commissioners, comprised of the secretary at war, the master-general of the Ordnance, the quarter-master-general, the adjutant-general, the governor and lieutenant-governor of the college; all for the time being: also the Duke of Kent, Sir D. Dundas, the Earl of Harrington, the Duke of Cumberland, the Duke of Cambridge, the Duke of Gloucester, and Lord Cathcart. The immediate command of the college is committed to the governor (Gen. the Hon. A. Hope), and the lieutenant-governor, Colonel Butler.

The staff is composed of a chaplain, agent, secretary to the board of commissioners, paymaster, adjutant to the junior department, quarter-master, surgeon, and assistant-surgeon.

There is a collegiate board that meets at the college, and takes under its cognizance all matters relating to the interior regulation and economy of both departments. This consists of the governor, the lieutenant-governor, and the commandant of the senior department.

No cadet can be withdrawn from the college, without permission from the governor.

It is presumed, that nothing but a want of talent can hinder *any* young man, however friendless, from becoming sufficiently advanced in education, to entitle him to a gratuitous commission within the prescribed period, provided the state of the public service admits of this mode of disposing of commissions.

It is understood that orphan cadets, and the sons of certain officers of the army, are liberally permitted to remain during the vacations, to be supported and instructed at the expense of the institution.

The Royal Military College will be removed to Sandhurst, on Bagshot-heath, in the course of the summer of the present year, (1812,) where a building is erected for its reception, in a style suited to the importance of the establishment and the national character, when it is intended that the establishment shall be increased to four companies.

COLLIERS, (*charbonniers, vaisseaux employés au commerce du charbon*, Fr.) are vessels employed to carry coals from one port to another, chiefly from the northern parts of England to the capital, and more southern parts, as well as to foreign markets. This trade is known to be an excellent nursery for seamen, although they are often found, from the constitution of their climate, to be not so well calculated for southern navigation.

COLOURS, (*pavillon distinctif de chaque nation*, Fr.) the flags or banners which distinguish the ships of different nations. See the articles ENSIGN, JACK, and PENDENT.

COLUMBA NOACHI, Noah's Dove, a small constellation in the southern hemisphere, consisting of 10 stars.

COLUMN OF A FLEET, (*ligne d'une flotte*, Fr.) is a long row of ships following one another, in an established order: it is general to divide a fleet into two or three columns.

COLURES, in astronomy and geography, are two great circles supposed to intersect each other at right angles in the poles of the world, and to pass through the solstitial and equinoctial points of the ecliptic.

That which passes through the two equinoctial points, is called the equinoctial colure, and determines the equinoxes; and the other which passes through the poles of the ecliptic is called the solstitial colure, because it determines the solstices. By thus dividing the ecliptic into four equal parts, they mark the four seasons, or quarters of the year.

COMA BERENICES, *Berenice's Hair*, in astronomy, a modern constellation of the northern hemisphere, composed of unformed stars between the Lion's tail and Bootes.

The stars in this constellation are, in the British catalogue, 43.

To COME HOME, (*l'ancre est dérapée*, Fr.) The anchor is said to come home when it loosens from the ground by the effort of the cable, and approaches the place where the ship floated, at the length of her moorings.

COME *no near!* (*pas au vent!* Fr.) the order to the helmsman, not to steer the ship so close to the wind.

COME *up the Capstan!* (*dévire au cabestan!* Fr.) the order to turn it the contrary way to that which it was heaving, so as to slacken or let out some of the rope which is about it.

COME *up the Tackle-fall!* (*largue le palan!* Fr.) the order to slacken it gently.

To COME *up with*, (*joindre, en parlant d'un vaisseau, qu' on chasse*, Fr.) to overtake.

COMET, (*cometa*, Lat.) in astronomy, a heavenly body which is only occasionally seen in the planetary region, appearing suddenly, and again disappearing; and during the time of its appearance moving in a proper, though very eccentric orbit, like a planet.

Comets have no visible disc, and shine with a faint and nebulous light, accompanied with a beard or tail of luminous matter, which is sometimes of astonishing length.

They are divided into three kinds, *bearded, tailed,* and *haired comets;* though the division rather relates to the different circumstances of the same comet, than to the phenomena of the several. Thus, when the comet is eastward of the sun, and moves from it, it is said to be bearded, because the light goes before it. When the light is westward of the sun, the comet is said to be tailed, because the train follows it. And when the comet and the sun are diametrically opposite, the earth being between them, the train is hid behind the body of the comet, excepting a little that appears around it, in form of a border of hair.

When a comet is examined with a good telescope, it appears like a mass of vapours surrounding a dark nucleus, of different degrees of opacity in different comets. As the comet approaches the sun, its pale and cloudy light becomes more brilliant; and when it reaches its perihelion, it is often brighter than the planets.

Tycho Brahe and Cassini were the first among the modern astronomers, who assigned these bodies a place in our system; but they appear to be unacquainted, both with their motion round the sun, and the true figure of their orbit. These particulars were left for the determination of Newton, who has discovered the paths they describe, and the laws to which they are subject.

Their revolutions are now known to be performed in very long ellipses, whose lower focus is very near the sun; being governed throughout by the same law, of describing equal areas in equal times, which is known to regulate the motions of all the other bodies in the system. By observations of the parallax, it is also found, that at their first appearance they are nearer to us than Jupiter; whence it is concluded, that they are in general less than that planet; for if they were as large, they would be seen at as great a distance.

In their motions round the sun, they are also subject to the same irregularities as the planets; but as their orbits are extremely eccentric, those variations are much more considerable. When they are near the sun their motion is very rapid, and in the more distant parts of their orbits extremely slow; so that their vicissitudes, in this respect, are as much in the extreme, as what they must undergo from heat and cold.

The celebrated Dr. Halley, from the theory of Newton, has calculated tables for determining the orbits of the comets, which, in several instances, have been found agreeable to observation.

Amongst those contained in Halley's tables, we have two only whose perihelion is farther removed from the sun than the mean distance of the earth: all the others have it between the earth and the sun: two of these last pass between the Earth and Venus, eleven between Venus and Mercury, and six between Mercury and the Sun.

The comets which descend below Mercury are almost all visible, either as they approach their perihelion, or as they ascend towards their aphelion. The only case in which they may conceal themselves, is, when the extended part of their orbit lies beyond the Sun, relatively to the Earth, and when they have, at the same time, a southern latitude in the situations where otherwise they might be visible: then they are scarcely to be observed, except on the other side of the equator. In our climates, they skulk under the horizon at the time we should have the best chance of observing them.

That remarkable comet which was observed in the year 1680, is shewn to be the same with one that appeared in 1106, its period being 574 years and upwards. The distance of this comet from the sun when in its perihelion, was to the distance of the earth from the sun, as 6 to 1000; its heat, therefore, at that time, was to the heat of the summer's sun with us, in the ratio of about 28,000 to 1.

The comet of 1759 did not make any considerable appearance, by reason of the unfavourable situation of the earth, all the time its tail might otherwise have been conspicuous; the comet being then too near the sun to be seen by us; but deserves our particular consideration, as it was the first that ever had its return foretold. It performs its revolution in about 75 or 76 years, and may be expected to return in the year 1835.

A comet was observed in the eastern part of Europe, in its descent to the sun about the middle of March 1811, and in the West Indies in the month of June following, when it had passed its descending node, and was coming to its perihelion. It was observed at Paris, after its ascending node, with a right ascension of 147° 18′; and a north declination of 32° 30′ in the constellation of Leo Minor, and was very conspicuous under the square of Ursa Major; whence it seemed to pass in a direction through the square, the tail of Draco, and the body of Ursa Minor near the tail. It was first observed in England on the 21st August, when it had the appearance of a large nebula nearly circular, and of about one degree in extent, with a central light like that of Andromeda; its right ascension then was 150°, and north declination nearly 36°.

On the 7th September, its right ascension was about 262°, and north declination about 42°, still advancing towards the square.

The train then viewed with a good four foot and and a half refractor of Dolland, and a very good night glass, adapted for use by Nairne and Blunt, was evidently divided, by a darkish shade near the farthest extremity, and was about 6 degrees in length, and nearly 4 at its farthest extremity. This comet has fully confirmed the Newtonian and Halleian theory, having been seen so long before and after its perihelion.

Of all the celestial bodies, comets have given rise to the greatest number of speculations and conjectures; and their appearance has always been a source of curiosity to the learned, and of terror to the ignorant: and, notwithstanding all that has been done and written on the subject, from the time of Aristotle and Pythagoras to the present moment, we are forced to confess, that our knowledge respecting these wandering bodies is still very imperfect. There is, perhaps, no class of bodies in the whole material universe, respecting which the reins of judgment and philosophic investigation have been so much abandoned to the power of imagination, as in what relates to comets; and the result has been, what it must always be in similar circumstances, a multiplicity of wild, and, in many instances, absurd conjectures. It is, therefore, a department of science which demands all the vigilance and skill of modern astronomers; and the celebrated Lalande says, in his History of Astronomy, for 1801, "the knowledge of comets is alone wanting, to complete the science of astronomy."

For a table containing the elements of 97 comets, which have been observed and calculated till the year 1810, see the Edinburgh Encyclopædia. And for farther information on comets, see M. PINGRE, *Cometographie*, 2 vols. 4to, *Paris*, 1784; or Sir H. ENGLEFIELD'S *Determination of the Orbits of Comets*, 4to. *London*, 1793; a very valuable work, in which the ingenious Author has explained, with great clearness and accuracy, the manner of computing the orbits of comets, according to the methods of BOSCOVICH and M. de la PLACE.

COMING-TO, or COMING-UP, (*embardée au vent*, Fr.) that part where a vessel stops in approaching the wind.

COMMANDANT, (*commandant ad interim*, Fr.) is that officer whom the command devolves on in the absence of the commander in chief, to which it is usual to prefix his rank, as lieutenant-commandant, captain-commandant.

COMMANDER, (*capitaine de corvette*, Fr.) in the royal navy, is an officer next in rank to a post-captain, and has the command of a ship of war under 18 guns, a sloop of war, armed ship, or bomb-vessel. He is entitled *master* and *commander*, and ranks with a major in the army. The number of commanders in the royal navy on the 1st January, 1812, was 586.

COMMANDER, (*maillet*, Fr.) is also the name of a large wooden mallet, used on sundry occasions in a ship.

COMMANDER *in Chief*, (*commandant en chef*, Fr.) is the chief admiral in any port, or on any station, appointed to hold command over all other admirals within that jurisdiction. See the article ADMIRAL.

COMMISSION IN THE NAVY, (*commission, brevet*, Fr.) is a warrant of office granted from the Admiralty, to officers of certain rank, to enable them to receive their personal pay. It is conferred on them when promoted from one step to another in the service; and is signed by, at least, three of the Lords of the Admiralty.

Lieutenants are the lowest class of commissioned officers; commanders next; then post captains; and, lastly, admirals.

Officers are said to be in *commission* when they are appointed to ships, as they then receive another commission from the Admiralty, expressly for the ships they are appointed to command; and admirals the same, when appointed to a squadron, or station; which enables them not only to receive full pay according to the rate of their ship and the rank and station they hold, but also prize-money.

Officers in the navy take rank in their respective classes according to the seniority of their *commissions*; or, if they have commissions of the same date, according to the order in which they stand on the list of the navy; but when doing duty, or in any service with the army, they take rank agreeable to the King's order in council.

When a captain, or other commissioned officer, is appointed to a ship already in commission, his *commission* is read to the officers and ship's company publicly.

When a captain and officers are appointed to a ship not in commission, that ship is in *commission* as soon as the *pendant* is hoisted, but not before, though the officers receive full pay from the date of their commissions, but not provisions, until they make their appearance and the ship is commissioned.

The *commissions* are not read on the first commissioning of a ship; but officers joining afterwards, when part of the company is formed, have then their commissions read.

COMMISSIONERS OF THE NAVY, (*commissaires de la marine*, Fr.) are certain officers appointed to superintend the affairs of the marine, under the direction of the lords commissioners of the Admiralty.

Their duty is more immediately concerned in the building, docking, repairing, and cleaning of ships in the dock-yards. In consideration of this, all ships of war are commissioned from a report of their qualities presented to the Admiralty by the Navy-board.

They have also the appointment of some of the warrant-officers, as surgeons, masters of ships, &c.

The principal officers and commissioners residing at the board, are,

1. The comptroller. 2 and 3. Two surveyors, who are shipwrights. 4. Clerk of the acts. 5. Comptroller of the treasurer's accounts. 6. Comptroller of the victualling accompts. 7. Comptroller of the storekeeper's accounts., 8. An extraordinary commissioner. Besides these, there are three resident commissioners, who manage the affairs of the three *dock-yards* at Chatham, Portsmouth, and Plymouth, under the direction of the Navy-board in London.

COMMISSIONERS *of the Victualling*, (*commissaires nommés pour veiller à l'approvisionement des vaisseaux*, Fr.) are officers appointed to superintend the supply of provisions for the royal navy.

COMMODORE, (*commandant provisoire d'une escadre*, Fr.) is a general officer in the British navy, invested with the command of a detachment of ships of war, destined on any particular enterprise; during which time he bears the rank of brigadier-general in the army, and is distinguished from the inferior ships of his squadron by a broad red pendent tapering towards the outer end, and sometimes forked.

COMMODORE, (*titre donné à un capitaine qui a plus d'un vaisseau sous ses ordres*, Fr.) is also a title given by courtesy to the senior captain, where three or more ships of war are cruising in company.

COMMODORE also denotes the convoy ship in a fleet of merchantmen, who carries a light in his top, to conduct the rest, and keep them together.

COMMODORE cannot, with propriety, be used, unless he is invested; for to usurp, is to possess without right.

COMPANION, (*capot d'échelle dans les bâtimens marchands*, Fr.) a sort of wooden porch placed over the entrance or stair-case of the master's cabin in a merchant-ship.

COMPANION-*Ladder*, (*échelle de commandement*, Fr.) in ships of war, denotes the ladder by which the officers ascend to, and descend from, the quarter-deck.

COMPANY, (*équipage*, Fr.) the whole crew of any ship, including her officers.

COMPANY *of Ships*, a fleet of merchantmen, who make a charter-party among themselves, the principal conditions whereof usually are, that certain vessels shall be acknowledged admiral, vice-admiral, and rear-admiral; that such and such signals shall be observed; that those which bear no guns shall pay so much per cent. of their cargo; and in case they be attacked, that what damages are sustained, shall be reimbursed by the company in general.

In the Mediterranean, such companies are called *conserves*.

COMPASS, or MARINER'S COMPASS, (*compas de mer, ou boussole*, Fr.) is an instrument used to point out the way or course of a ship at sea. It consists of three parts, namely, the box, the card or fly, and the needle.

The *box*, which contains the card and needle, is a circular brass box hung within a square wooden one, by two concentric rings, called gimbals, so fixed by cross centres to the box, that the inner one, or compass box shall retain an horizontal position in all motions of the ship.

The *card*, or *fly*, is a circle of stiff varnished paper, representing the horizon, and is divided into 32 equal parts, by lines drawn from the centre to the circumference, called points or rhumbs: the intervals between the points are also subdivided into halves and quarters; and also the whole circumference into equal parts called degrees, 360 of which complete the circle; and, consequently, the distance or angle comprehended between any two rhumbs, is equal to eleven and a quarter degrees.

The four principal points are called *cardinal points*; two of which, opposite to each other, are called *north* and *south* points: that which is towards the right hand as we look north is termed *east*, and its opposite the *west* point. The names of all the inferior ones are compounded of these, according to their situation.

The *needle* is a small bar of steel made magnetical, which has the property of pointing one of its ends, towards the north pole of the world. It is fixed to the under side of the card, and in the centre is placed a conical socket, which is put on an upright pointed pin, fixed in the bottom of the box; so that the card, hanging on the pin, turns freely round its centre; and one of the points, by the property of the needle, will always be directed towards the north pole.

The top of the box is covered with a glass pane, to prevent the wind from disturbing the motion of the card.

This apparatus is called the *Compass.*

The whole is enclosed in another box of wood, where it is suspended by brass hoops or gimbals, to keep the card in an horizontal position during the motions of the ship. The whole is to be so placed in the ship, that the middle section of the box, parallel to its sides, may be parallel to the middle section of the ship along its keel.

The Invention of the Mariner's Compass.

Most authors fix on the year 1302, for the date of the incomparable invention or discovery of the *mariner's compass*, or magnetic needle, for the direction of ships at sea. The inventor was Flavio John de Gioja, or Gioga, a native of Amalfi, an ancient commercial city in the kingdom of Naples; and hence it is, that the territory of Principato Citra (that part of the kingdom where he was born) has a compass for its arms.

In commemoration of his discovery, the following verse of one Anthony, of Palermo, is recorded by the Neapolitan historians, viz.

"Prima dedit nautis usum magnetis Amalphi."

By which it is understood, that as the poles of the magnet, or loadstone, answered to the poles of the world, it could also communicate that wonderful property to an iron needle, placed on a chart, marking the points of the world.

The power of the magnet to attract iron was known to the ancients, and is mentioned by Plato, Aristotle, and Pliny; but its directive power, to cause a piece of iron touched with it to point north and south, is, undoubtedly, of later date.

Flavio John de Gioja, according to Abraham Ortelius and others, divided his compass only into eight points, till it was improved by the people of Burges, to thirty-two points: nor, indeed, does this excellent invention seem to have been generally known and used even long after Flavio's time, as appears too plainly, from the Portuguese creeping along the shores, even so late as their first discoveries on the western coasts of Africa, in the fifteenth century.

The poet, Dryden, has happily described the infancy of navigation in the following lines:

"Rude as their ships was navigated then,
No useful *compass*, or meridian known;
Coasting they kept the land within their ken,
And knew no north but when the pole star shone."

Although the English pretend not to the invention of the compass, yet they are said to be the contrivers of the most convenient method of suspending the box which contains the magnetic needle, so as to keep it always horizontal. The variation of the needle, or its declination from the true north point, was discovered by an Italian, the well-known Sebastian Cabot, in the year 1497; and the inclination, or dipping, of that needle, when hung so as to play vertically to a point beneath the horizon, was first discovered by Robert Norman, an Englishman, in 1576.

Lastly, the variation, or the different declinations of the needle at different times in the same place, was first discovered by our countryman Gillibrand, about the year 1634; though some attribute the merit of this discovery to Gassendi, a French mathematician. See the articles Dipping Needle and Variation, in their respective places.

Many are the encomiums justly bestowed by all maritime nations on the *mariner's compass*, for the benefit of navigation and commerce; for by the help of this noble instrument it was, that the Spaniards made their discoveries of a new western world; the Portuguese the way by sea to India and China; and the English and Dutch the several useful discoveries towards the North Pole; all which, but for the compass, would have probably still remained unknown; and all the wealth acquired from such discoveries, and most of the knowledge obtained in consequence thereof, would never have been possessed.

The compass being at all times of the utmost importance to the purposes of navigation, it is reasonable to expect that the greatest attention should be used in its construction, and every attempt to improve it carefully examined, and adopted. Great errors and irregularities, however, have been found incident to the construction of common compasses, arising from the shape of their needles, by which they have not only turned from the true direction, but from that of each other.

The wires, of which the needle has hitherto been generally composed, were only hardened at their ends; and if those ends are not equally hard, or if one end be hardened up higher than the other, when they come to be put together, in fixing them to the card, that end which is hardest will destroy much of the virtue of the other; by which means, the hardest end will have the greatest power in directing the card, and consequently make it vary towards its own direction; and, as the wires are disposed in the form of a lozenge, these cards can have but little force; so that they will often, when drawn aside, stand at the distance of several degrees on either side of the point from whence they are drawn; for all magnetical bodies receive an additional strength by being placed in the direction of the earth's magnetism, and act proportionably less vigorously when turned out of it. Therefore when these kind of needles are drawn aside from their true point, two of the parallel sides of the lozenge will conspire more directly than before with the earth's magnetism, and the other two will be less in that direction: by these means the two former sides will very much impede its return, and the two latter will have that impediment to overcome, as well as the friction, by their own force alone.

The compass, although the most useful instrument in a ship, has, certainly, hitherto, been very much neglected.

In the royal navy it is committed to the care of the boatswain, who cannot be expected to know much of its merits, and who, of course, pays no more attention to it than he does to his other stores; and, even when returned to our dock-yards, they are stowed away like other stores, without any sort of regard being paid to preserve their attraction, or to place them in such manner that one might counteract the other, which might easily be done, with very little more trouble, and without any additional expense.

The mechanical parts of our compasses, too, are, generally, very indifferently executed; indeed, the low price allowed by the government for supplying the navy with that instrument will not admit of their being made by the best workmen and in the most correct manner. It is really very much to be regretted that so little attention has been paid to that most valuable instrument in the sea-service, as, no doubt, many serious losses must have happened from the incorrectness of it.

The several imperfections attending the construction of the common compass, have been happily removed by the labours of Dr. Gowen Knight, who discovered the useful practice of making artificial magnets, by which, and the farther emendations of Mr. Smeaton and Mr. M'Colloch, the needles are larger and stronger than formerly, and, instead of swinging in gimbals, the compass is supported in its very centre upon a prop, and the centres of motion, gravity, and magnetism, are brought almost all to the same point.

The compass thus improved by Dr. Knight is generally used in the royal navy.

The principal, and, indeed, the only circumstance in which Knight's compasses are superior to those which were formerly used, is, that their needles being tempered much higher than usual, are thereby enabled to contain a much greater quantity of magnetical stream, which is, certainly, a real advantage.

Needles of this construction, after vibrating a long time, will always point exactly in the same direction; and, if drawn, ever so little, on one side, will return to it again, without any sensible difference.

In order to illustrate the above description, a view of the several parts of the compass is exhibited in plate VIII. where fig. 1 is the *card*, with the needle N. S. and its cap fixed upon it; fig. 2, is the pedestal that supports the card, containing a sewing needle, fixed in two small grooves to receive it, by means of a collet C and at D, the stem is filed into an octagon, that it may be the more easily unscrewed.

Plate VIII. Fig. 1.

A B represents the outer wooden box; C and D are two milled nuts, by means of which the axis of the inner box G H, and ring E F, by which it is supported, are lifted from their edges on which they move, in order to increase the friction, and prevent the too great vibrations occasioned by the motion of the ship in a high sea; I is one of the axes of the inner box, by which it is suspended on the ring E F; K L is the magnet, or needle, of a regular parallelopiped shape; and M, a small beam of ivory, best seen in fig. 4, that confines the cap to its place. The cap is ivory, so turned as to be capable of receiving a small bit of agate, which is ground concave, and polished on that side where it forms the apex of the hollow cone in the cap.

N, O, P, Q, is the index bar, with its two stiles, and cat-gut threads, which being taken off from the top of the box, when the index is not wanted for use, is placed in two pieces, R and S, properly notched to receive it; T represents the handle of the box.

The instrument thus constructed serves the purpose of an azimuth and amplitude *compass*.

The compass is retained in the binuacle at sea, as exhibited in plate II. fig. 8.

For the description and use of Walker's Meridional Azimuth Compass, see the article AZIMUTH; and, for its figure, see plate VIII. fig. 5.

A patent for improvements in the action and other parts of sea compasses, was granted in September 1810, to Mr. George Stebbing, of Portsmouth, who states, in his specification—

1st. That, in the different compasses, as commonly made with all their parts, according to the different purposes for which they are intended to be used, except the centre, cap and needle, he substitutes, instead of metal or steel for centre or point, that supports or carries the card or needle, *a ruby jewel, precious or any other hard stone or composition*, made into a circular or conical form, but not brought to so sharp a point as the common centre, or point; and that this ruby, stone, or composition, is to be set in metal, or any materials fit for the purpose, long enough to bring the card, or needle, as near the centre of gravity as possible.

2d. That he substitutes instead of agate for the top, or by some, called the bottom, of the cap, *hard steel*, or *hard metal*, either simple or allayed, or *hard stone*; in the centre of which is sunk the section of a small circle, or sphere, well polished, to receive the head of the ruby, stone, or composition, of the centre, that supports or carries the card or needle, instead of the socket in the agate, as before used.

3d. That he makes the magnet or needle of the compass in the usual way, except the ends, which he forms into a square, by reducing the width of the needle to its thickness, whereby the angles are equal, and the magnetic fluid cannot draw the magnet, or needle, from its poles, so much as at present in the compasses now in use.

4th. That he also, in addition, uses for a compass, in time of action or very bad weather, *silk thread*, or any proper material fit for the purpose, to suspend the card and needle about one quarter of an inch off the centre, hung from the glass by means of a swivel, or hook, or any other fit means, to the top of the cap, or from a piece of metal put across the kettle or bowl, instead of through the glass; and, at the bottom of the cap under the card, puts a piece of tube, about an inch long, in a conical shape, to prevent the card, when suspended, touching the side of the kettle or bowl, or the card being thrown off the centre when the guns are firing or ship rolling, or as occasion may require, suspends the compass with *elastic springs*.

That active and ingenious officer, Sir Home Popham, in hopes of turning the attention of our naval officers to this subject, has lately submitted a binnacle and improved compass-card to the Admiralty. It is lighted from under; the card is transparent, and the bottom of the compass-box is of glass. He has adopted Mr. Stebbing's plan of traversing the card on a ruby, to prevent the friction which all other compasses are liable to; and, as there appears to be but little difference of opinion as to the superior merits of the compass, both for night and day, it is expected, shortly, to be brought into general use in the navy; and hopes are entertained that our ships will be supplied, in future, with compasses of a very superior construction. For further particulars see the article VARIATION.

The use of the sea-compass is as follows: the course a ship is to sail in, being known by the chart, and the compass so placed, as that the two parallel sides of the square box be disposed according to the length of the ship, *i. e.* parallel to a line drawn from the head to the stern; the rudder is to be directed accordingly: if the course be found on the chart between the S. W. and S. S. W. *i. e.* S. W. $\frac{1}{4}$ to the S.; turn the stern so as that a line from the S. W. $\frac{1}{4}$ S. exactly answer the mark on the middle of the side of the box. This is all that is required.

The compass is sometimes disturbed by the electricity of its glass cover; the remedy for this inconvenience is, to moisten the surface of the glass with a wet finger, which removes it immediately and effectually.

The *mariner's compass*, with a chart, is much less dangerously removed than the common compass with a bare needle; and the deeper, or farther distant, the needle hangs below the glass, the less disturbance it is likely to receive. But, notwithstanding the various contrivances that have been made to prevent the card from being much affected by the motions of the ship, they have generally been found too delicate to encounter the shocks of a tempestuous sea.

COMPASS, *Application for taking bearings on a Chart.* This instrument was constructed by N. D. Starck, Esq. and consists of an inner and outer brass concentric circle; the latter of which, when in use, is to be applied to a chart, so that its cardinal points may agree with those of the draught, and its central (metallic) point to be directly over the ship's place. The inner circle is to be set to the variation; and the thread from the centre being laid, will shew either bearings by compass, or true bearings, according to the circle upon which they are read. This instrument may also be used in delineating, plotting, and many other necessary operations.

COMPASSES, *Beam*, consist of a long branch, or beam, made of brass, or wood, carrying two brass cursors, or boxes, the one fixed at one end, the other sliding along the beam, with a screw to fasten it on, occasionally. To the cursors may be screwed points of any kind, whether steel, for pencils, or the like. It is used for taking off and transferring divisions from a diagonal scale, likewise for drawing large circles to great extents, &c.

To the fixed cursor, or box, is sometimes applied an adjusting or micrometer-screw, by which an extent is obtained with extreme nicety and regularity.

COMPASSES, *Caliber*. See the article CALIBER.

Hanging COMPASS, *(compas renversé*, Fr.) an instrument resembling the sea-compass, except that the point which supports the card, is fixed in the centre of the glass, and the gimbals are attached to a beam over the observer's head. There is usually one hung in the cabin, that, by looking up to it, the ship's course may be observed without the trouble of going upon deck.

COMPASSES, (*compas a pointes*, Fr.) are mathematical instruments well known for describing circles, measuring figures, &c. They consist of two sharp pointed branches, or legs of iron, brass, steel, or other metal, joined at the top by a rivet, whereon they move, as on a centre: those having the joints made of different metals are esteemed the best, as the difference of texture of the two metals causes the parts to adhere less, and diminishes the wear; thereby promoting equality and uniformity in the motion of the joints.

COMPASSING, or COMPASS TIMBER, (*devers*, Fr.) a name given by shipwrights to such pieces of timber as are incurvated or arched.

COMPENSATION, (*compensation*, Fr.) is an allowance made to flag officers, in addition to their pay, for servants, as follows:—

	£	s.	d.	
The Admiral of a Fleet	£56	0	0	Per month, subject to the usual deductions.
An Admiral . . .	33	0	0	
A Vice-admiral . . .	21	10	0	
A Rear-admiral . . . } A Commodore . . . } A Captain of a Fleet . }	15	10	0	

Captains and Commodores, as a *compensation* for the servants' formerly allowed them, are paid at the rate of 45*l.* 10*s.* per annum, for each hundred men of the ship's complement.

A *compensation*, in lieu of servants, is also made to lieutenants and warrant officers, equal to the net wages of the number of servants to which they were respectively entitled, at the rate of 11*l.* 8*s.* 2*d.* per annum for each servant.

COMPLEMENT, the limited number of men employed in any ship, either for navigation or battle.

COMPLEMENT, in astronomy, the distance of a star from the zenith; or the arc comprehended between the place of the star above the horizon and the zenith.

COMPLEMENT, in geometry, is what remains of a quadrant of a circle, or of 90°, after any certain arc has been taken away from it. Thus, if the arc taken away be 40°, its complement is 50, because $50+40=90$. The sine of the complement of an arc is called the co-sine, and that of the tangent, the co-tangent, &c.

COMPLEMENT *of the Course*, in navigation, is the quantity which the course wants of 90°, or 8 points, viz. a quarter of the compass.

COMPLEMENTS OF OR IN A PARALLELOGRAM, are the two smaller parallelograms made by drawing two right lines parallel to each side of the given parallelogram, through the same point on the diagonal.

COMPOSITION, is the compound ingredients of which gun-powder, fire-works, &c. are made. It also particularly signifies a mixture of bees-wax, with pitch and tar, used in making up fuses and shells.

All dry compositions must be well mixed; first by the hands, and then passed several times through fine hair sieves, that the ingredients may be thoroughly incorporated.

In mixing compositions which require fire, the greatest precautions are necessary; particularly in those where gunpowder enters. The dry part of the composition may, in general, be mixed together first, and put, by degrees, into the caldron, where the other ingredients are fluid, being well stirred all the time of putting in. When the dry ingredients are inflammable, the caldron must not only be taken off the fire, but the bottom must be dipped in water, to prevent the possibility of accidents while mixing them.

Receipts for a Variety of artificial Fire-Works.

Fire and Light Balls are composed of, and filled with, an illuminating or inflammable composition.

The *Fire Balls* are used to set fire to buildings, &c. their composition is, mealed powder 10 parts, saltpetre 2, sulphur 4, and resin 1: or, it may rather be mealed powder 48; saltpetre 32; sulphur 16;

resin 4; steel or iron filings 2; saw-dust of fir, boiled in a ley of saltpetre, 2; charcoal of birch-wood, 1.

The *Light Ball* is used for giving a strong light on distant objects, and is made with mealed powder 2 parts; sulphur 1; resin 1; turpentine 2½; and saltpetre 1½; tow is then taken and dipped and mixed in this composition, till of a proper size, when a coat of mealed powder is applied. A stone may be made use of, covered with several coats of the composition, until it has acquired the size intended: the last coat should be of grained powder.

The best sort of light-ball is, however, said to be made of thick brown paper, to the size of the mortar, filled with a composition, consisting of an equal quantity of sulphur, resin, and mealed powder; which, being well mixed, and put in warm, will produce a clear fire, and burn for some time.

With either of these compositions a thick bag or sack is to be filled, ramming it, if possible, as hard as a stone; putting in the opening a fuse, and, about the same, an iron ring, one-fifth of the ball's diameter wide; and, on the opposite end, another ring, of one-sixth of the ball's diameter; then, with a strong cord, of a quarter of an inch in diameter, lace round the hoops, or rings, from one end of the ball to the other, as often as it is requisite, this is called the *ribbed coat*; then lace it again, on the contrary way, which is called the *check coat*. Between each square cord, iron barrels are driven in, one third of which are filled with powder, and a bullet; at the end of each a small vent is made, so that the composition may inflame the powder, and drive the balls out on every side, which not only destroys whatever lies within the extent of its explosion, but prevents any one from extinguishing it.

The balls, when finished, are dipped in a composition of melted pitch, resin, and turpentine oil, which fastens the whole together.

Smoke Balls are prepared similar to the above described, but with five parts to one of pitch, resin, and saw-dust; this is put into shells made for the purpose, having four holes to let out the smoke. They are thrown from mortars, and continue to smoke from 25 to 30 minutes. They serve to conceal a design or project, answer as signals, &c.

Stink Balls, for the annoyance of an enemy, are similar to the former, but prepared by a composition of mealed powder, resin, saltpetre, pitch, sulphur, rasped horses and asses hoofs burnt in the fire, asafœtida, and other ingredients of a similar nature. These balls may be formed of two half globes, made of coarse paper, glued together until half an inch thick, then filled and covered over with a net.

The composition for *Bengal Lights* is, saltpetre 7 lbs. sulphur 1 lb. 12 oz. and red orpiment 8 oz. or, saltpetre 2 lbs. 4 oz. sulphur 8 oz. antimony 4 oz. and orpiment 1½ oz.

Blue Lights, a well-known species of useful artificial fire-works. The composition is as follows:—Saltpetre 1 lb. 4 oz. sulpher 5 oz. antimony 7 oz. mealed powder 1 oz.

To make blue lights, provide, in the first instance, a piece of trunnel, about 10 inches long, boring a hole at one end, about three inches deep and three quarters of an inch in diameter, then woold it with a rope-yarn while loading it with the composition, putting in a tea-spoonful at a time, and, with an iron rammer to fit the hole, striking it hard with a wooden mallet, five or six times, till it is filled up. At the top put a small quantity of mealed powder, as priming, tying a piece of paper over it. Thus it will be fit for use, as signals, by night.

For *Chinese*, or *White Lights*, the composition is, nitre from 50 to 60 parts, sulphur 16 to 20, antimony 5, orpiment 8 to 10.

The *Valenciennes Composition*, so called from its having been used by the Austrians at the siege of that place, has the effect of making shells answer the purpose of carcasses after they burst.

This is formed of saltpetre 50 parts; sulphur 28; antimony 18; resin, or Swedish pitch, 6. This composition is cast in copper cylindrical moulds, six inches long, and of different diameters, according to the shell in which it is to be used. It is put in along with the bursting powder, in pieces as large as the shell will admit, without preventing the fusés being driven down. For a few general remarks on the composition of artificial fire-works, see the article Fireworks.

COMPTROLLER of the Navy, *(contrôleur de la Marine*, Fr.) one of the commissioners of the Navy Board, at which he presides, to direct the inferior and civil department of the marine, as the Admiralty superintends the superior and military operations of it.

CONCAVE, an appellation used in speaking of the inner surface of hollow bodies, but more especially of spherical or circular ones.

Concave *Glasses*, are such as are ground hollow, and usually of a spherical figure, though they may be of any other, as parabolic, &c. All objects seen through concave glasses appear erect and diminished.

Marine CONDUCTOR, *(paratonnerre*, Fr.) in electricity, is a thick metal wire, generally of copper, extending from above the maintop-gallant truck, downwards, into the water. Its use is to defend the ship from the effects of lightning.

For the use and advantage of conductors on board ships and magazines, see the Naval Chronicle for 1799, page 197; being an abridged translation of a memoir, by Jean Baptiste Le Roy, in the *Histoire de l'Academie des Sciences*, for 1790, of a tour made to the different parts of France, to erect conductors on the different magazines, and to cause them to be placed on ships.

Chains have been used for the preservation of ships; but, as the electric matter does not pass readily through the links of them, copper wires, a little thicker than a goose's quill, have been preferred, and are now generally used. They should reach two or three feet above the highest mast, and be continued down in any convenient direction, so as always to touch the sea-water. These were first recommended to Lord Anson by Dr. Watson, in 1762. *Phil. Trans*. Vol. 52, part 1. p. 633. See Franklin's *Letters*, &c. ed. 1769, p. 65, 124, 479, &c. and Cavallo's *Elect.* chap. 9.

CONJUNCTION, in astronomy, the meeting of two stars or planets in the same degree of the zodiac; and is either true or apparent.

True Conjunction, is when a right line, drawn through the centre of two stars, passes also through the centre of the earth.

Apparent CONJUNCTION, is when that line does not pass through the earth's centre. See the article APPARENT *Conjunction.*

CONNING, or CUNNING, *(faire gouverner,* Fr.) is the art of directing the steersman to guide the ship in her proper course: the officer who performs this duty is either the pilot or quarter-master.

CONSORT, implies any vessel keeping company with another.

CONSTELLATION, in astronomy, is a system of several stars that are seen in the heavens near to each other. This assemblage is also sometimes called asterisms. The ancient astronomers not only marked out the stars, but, that they might more readily bring them into order, they distinguished them by their situation and position in respect to one another, and therefore distributed them into asterisms, or constellations, allowing several stars to make up one *constellation*; and, for the better distinguishing and observing them, they reduced them to the forms of animals, as men, bulls, bears, &c. or to the images of some well-known things, as of a crown, a harp, a balance, &c. or gave them the names of those whose memories, in consideration of some notable exploit, they had a mind to transmit to future ages. The division of the heavens into constellations is of great antiquity; being known to the most early authors; for, in the ancient book of Job is this sublime expostulation—*Canst thou restrain the sweet influence of the* Pleiades, *or loosen the bands of* Orion? And the names of many of the constellations are met with in the writings of the earliest poets, Homer, Hesiod, &c.

The ancients, in their division of the firmament, took in only so much as came under their notice, as visible to the naked eye; this they distributed into forty eight constellations. Modern astronomers have thought proper to retain all the ancient representations of the constellations, and have added a few others; they have also divided the whole starry firmament into three distinct regions:—1. The Zodiac, which includes twelve constellations, commonly called the signs of the Zodiac, viz. Aries, Taurus, Gemini, Cancer, Leo, Virgo, Libra, Scorpio, Sagittarius, Capricornus, Aquarius, and Pisces. 2. All the region of the heavens that lies on the north side of the Zodiac, which contains twenty-one constellations, viz. Ursa Minor, and Ursa Major, Draco, Cepheus, Bootes, Corona Borealis or Septentrionalis, Hercules, Lyra, Cygnus, Cassiopeia, Perseus, Auriga, Ophiucus or Serpentarius, Serpens, Sagitta, Aquila, Delphinus, Equuleus, Pegasus, Andromeda, and the Triangulum; to which were added, afterwards, by Tycho Brahe, two others, viz. Coma Berenices, near the Lion's Tail; and Antinous, near the Eagle; both of which he made of the stars not included in any image. 3. The region on the southern side of the Zodiac, which contains fifteen constellations, known to the ancients, viz. Cetus, the Eridanus, Orion, Lepus, Canis Major, Canis Minor, Argo, Hydra, Crater, Corvus, Centaurus, Lupus, Ara, Corona Australis, or Meridionalis, and Piscis Australis: to these are lately added twelve more constellations, which were formed by Bayer out of the unformed stars, and which are not to be seen by us, who inhabit the northern regions, because of the convexity of the earth, but in the southern parts they are very conspicuous; these are the Pavo, Toucan, Grus, Phœnix, Dorado, Piscis Volans, Hydrus, Cameleon, Musca Indica, Apus, or Apous (Avis Indica), Triangulum Australe, and the Indus. The Galaxy, or Milky-way, is also to be reckoned among the constellations.

See each CONSTELLATION, and the stars contained in it, under its proper head.

Besides the above, Bayer has added eleven other constellations, which he formed out of the unformed stars; five of these are to the north, and are called the Giraffe, the River Jordan, the River Tigris, the Sceptre, and the Flower-de-luce; and six to the south; called the Great Cloud, the Little Cloud, the Cross, the Dove, the Rhomboide, and the Unicorn.

Hevelius has also collected together several stars of the unformed class, into eleven new constellations. These are, the Camelopardalus described by Bartschius, the Cerberus, the Sobieski's Crown, the Dogs, the Fox and Goose, the Little Lion, the Lizard, the Lynx, the Sextant of Urania, the Little Triangle, and the Unicorn; to which Gregory has added the Ring and Armilla.

CONSUL, *(consul,* Fr.) is an officer established by virtue of a commission from the king and other princes, in all foreign countries of any considerable trade, to facilitate and dispatch business, and protect the merchants of the nation. The *consuls* are empowered to keep up a correspondence with the ministers of England, residing in the courts whereon their consulate depends.

They are to support the commerce and interest of the nation; to dispose of the sums given, and the presents made, to the lords and principals of places; to obtain their protection, and prevent the insults of the natives on the merchants of the nation.

CONTINENT, *(continent,* Fr.) in Geography, a great extent of land not interrupted by seas, in contradistinction to island, peninsula, &c. According to what relations we have of the disposition of the globe from late navigators, we may count four *continents*; of which there are but two well known. The first, called the antient continent, comprehends Europe, Asia, and Africa; the second is the new continent, called America; the third, which is called the northern, or arctic, continent, comprehends Greenland, the lands of Spitzburg, Nova Zembla, and the lands of Jesso; and the fourth comprehends New Guinea, New Zealand, New Holland, and several others hitherto little known. Some authors think the two first continents are, in reality, only one, imagining the northern parts of Tartary to join with those of North America.

CONVEX, an appellation given to the exterior surface of gibbous, or globular bodies, in opposition to the hollow inner surface of such bodies, which is called concave: thus we say, a *convex* lens, mirror, superficies, &c.

CONVOY, *(convoi,* Fr.) a fleet of merchant-ships bound on a voyage to some particular part, and protected by an armed force.

To CONVOY, *(escorter,* Fr.) also implies one or more ships of war appointed to accompany and defend them on their passage from being insulted by pirates, or the enemies of the state in time of war.

COOK, (*le coq d' un navire*, Fr.) a person appointed to dress victuals for the ship's company, and take charge of the boilers: besides which there is the admiral's, or captain's, cook.

Cook-*Room*, the galley, or coboose, where victuals are dressed.

COPERNICAN System, is that system of the world in which the sun is supposed at rest in the centre, and the planets, with the earth, to move in the ellipses round him. The sun and stars are here supposed at rest, and that diurnal motion, which they appear to have from east to west, is imputed to the earth's motion from west to east, round its axis. This system was maintained by many of the ancients, particularly Archimedes; but, after him it became neglected, and even forgotten for many ages, till it was revived by Copernicus, about the year 1500; and from him named the *Copernican System*. According to this hypothesis, the sun is supposed very near the centre of gravity of the whole system, and in the common focus of every one of the planetary orbits: next to him Mercury performs his revolution around him; next to Mercury is the orbit of Venus; then the earth, with its attendant, or secondary, the moon, performing a joint course, and, in their revolution, measuring out the annual period. Next the earth is Mars, the first of the superior planets; next to him Jupiter; then Saturn; and, lastly, the Herschel planet. Between Mars and Jupiter have been discovered four very small bodies, called *Asteroides*.

These and the comets are the constituent parts of the solar system, which is now universally received by all philosophers, as the only true one.

COPPERED, or a Copper-bottomed Vessel, is a vessel sheathed with thin sheets of copper, which prevents the worms eating into the planks, or filth, accumulating on the bottom, whereby a ship is made to sail heavily. The Alarm frigate was the first ship ever coppered in the British navy, in 1758; and, in the course of the war with America, which ended in 1783, ships of every class were coppered. See the article Sheathing.

Copper-*fastened*: the bolts, and other metal work in the exterior of the ship's bottom, made of copper instead of iron; the advantage of which is, that the vessel may afterwards be coppered without danger of the sheathing corroding the heads of the bolts, which it is found to do if they are made of iron. It was ordered, in November 1783, that, in future, ships of all classes should be *copper-fastened* under the load draught of water.

COR CAROLI, in astronomy, an extra-constellated star of the third magnitude, in the northern hemisphere, situated between the Coma Berenices and Ursa Major, so called by Dr. Halley, in honour of King Charles, This star's right ascension, in time, adapted to the beginning of the year 1812, is 191° 49′ 2″,40; annual variation in right ascension 42″,8. Its declination 39° 20′ 12″ N. and annual variation in declination — 19″,6.

CORDAGE, (*cordages*, Fr.) is a general term for the running rigging of a ship, or all that part of her rigging which is employed to extend, contract, or traverse the sails; as also for the rope which is kept in reserve, to supply the place of such as may be rendered unserviceable. See the article Rigging.

Cable-laid Cordage, is made with nine strands, that is to say, the first three strands are laid slack, and then three of them being closed together, make a cable or cablet. All cordage manufactured in Britain, or imported into Britain, for the use of the British navy, is, or ought to be, made of the best hemp

In the South Seas several substitutes are used, the growth of hemp being unknown to them, the fibres of the cocoa-nut, with many other plants and vegetable productions, have been applied to this purpose, and with tolerable success.

In Britain many improvements have, of late years, been suggested, relative to the mode of manufacturing ropes and cordage of different natures. Some of these have been found to answer very well, and others still continue in the state of trial; for, on a subject of such magnitude, and of so much importance to the safety of shipping, as also to the lives of seafaring persons, too much circumspection cannot be used.

Sir Samuel Standidge observes, in an answer written by him to a letter to Lord Barham, that, as it is still a doubt whether cable or shroud laid cordage is the best for standing-rigging, and what is the most efficacious mode of preserving it from the weather without injuring the materials. During his long practice in the use of cordage, he says, that he has often examined and taken particular notice of the wear of standing rigging, and has used hawser-laid, and four strand shroud-laid, as also three strand shroud-laid rope for standing-rigging. He has found the cable-laid rope so full of cavities that it admitted of a great deal of water, which has frequently occasioned a decay; and he is of opinion that it has not the strength of the three strand rope, owing to its having so many small strands twisted into a hawser.

As to the four strand shroud-laid rope for standing-rigging, it has, generally, one eighth part of the weight and substance in the middle of the rope, called a goke, in order to make it round, which, either by stretching or want of air, or of being made of inferior hemp, when the four strand rope is opened he has always found it to be rotten, and to have decayed, in some measure, the four strands.

The three strand shroud-laid rope, well tarred with Swedes, or Stockholm tar, and laid short, so as the tar will sprout out to fill up the secret cavities, and, after stretched, when in use, to keep the cavities filled with Stockholm tar, is the best means of keeping out the weather and not injuring the materials.

He has had sets of standing-rigging, he observes (after oft examining, once fresh parcelling, and serving round the mast-heads and foot-hook staves, and fresh seizing in the dead-eye), of three strand shroud-laid rope in constant use fourteen years; and, in his opinion, the three strand shroud-laid rope is the strongest and best for standing rigging.

Mr. William Chapman, of Newcastle-upon-Tyne, obtained a patent, in the year 1801, for the application of certain substances, either separately or combined, as a preservation for cordage; which substances being insoluble in water, or very imperfectly soluble, will, when used in the manufacture of cordage, be the means of making it more durable than it has hitherto been found.

It being well known that cordage, exposed to great

heat, as it necessarily must be in hot climates, frequently becomes useless in two or three years, and, sometimes, less, from a species of destruction, called, by seamen *the dry-rot*; because the rope, looking well to the eye, and no way injured by moisture, is found to have its yarns quite rotten.

It is also known that ropes exposed to water soon lose the tar with which they are covered; which circumstance quickly takes place in warm climates and shallow water; under which predicament that element becomes considerably more heated than it is in the ocean in the same latitude, and its effect is quickly shown upon a new cable, as the part lying under water is soon covered with a slime arising from the solution of the tar.

The first of these disorders, the dry-rot, Mr. Chapman attributes to a superabundant acid, which is disengaged from the tar when exposed to heat.

The other deficiency in ropes prepared with common tar, clearly arises from that substance containing a mucilage, or gum, along with its resinous matter, which latter is insoluble in water.

The remedy, the patentee proposes for both these evils, is very simple, and is effected during the necessary inspissation of tar, to that state in which it is used in the manufacture of ropes. The common process is by boiling the tar until it has thrown off so much of its essential oil as to come to the state in which it is wanted. Instead of which, Mr. Chapman proposes to boil the tar in water two or three times, until it nearly attains the necessary degree of inspissation. In the first boiling, the same as in distillations, it throws off, with the steam, its super-abundant acid, and parts with much of its mucilage to the water; and, by a second process, it is rendered fit for use. Seamen, accustomed to hot climates, experience that tarred ropes, although sufficiently pliable whilst there, become rigid on their return to cold countries; which sometimes occasions the loss of sails, by the breaking of the ropes, which form their borders.

These ropes, which are called bolt-ropes, the patentee proposes to prepare with inspissated gross oils, mixed, if required, with a small quantity of resin. He also prepares crane ropes in the same manner, so as to prevent their receiving injury from the weather, and yet keep them always pliable.

The cause of the rigidity of tarred ropes, after exposure to heat is, the dissipation of the essential oil of the tar. To prevent this inconvenience, Mr. Chapman proposes that his purified tar should be more inspissated than if used by itself, and that a due proportion of tallow, suet, whale-oil, rape-oil, or other fixed oil, be mixed with the tar, and that the oils should previously be deprived of their mucilage after the same process as the tar. The utility of this invention to the navy and merchant ships of these kingdoms must evidently be very considerable.

Mr. Joseph Huddart obtained a patent for a new method of *spinning yarn*. Dated September, 1804.

Mr. Huddart's method of spinning, consists of one the reverse of that used by rope-makers; in theirs, the twisting machinery is stationary, and the man who gives out the hemp moves; in Mr. Huddart's, the twisting machinery moves on, as the thread is formed, and the man who gives out the hemp or flax is stationary.

For this purpose, it is directed that a large band should communicate with a roller in the moveable frame, after passing over stationary wheels at each end of the building, or spinning ground; one of these wheels, turned by men or horses, will give motion, by the band, to the said roller, which turns the spindles that twist the yarn, and also turns two of the wheels on which the frame is supported, so as to give it progressive motion. The hemp or flax is directed to be laid out on a table, furnished with an instrument to hold it, the same as a flax-dresser's hackle, by which the patentee conceives great advantage will be obtained in delivering the hemp even and regular, without any crossing of its fibres, but what is occasioned by the twist.

From a strict examination of this method of spinning yarn, it is imagined that the only part of the invention which seems of much utility, is the mode of delivering the hemp from the hackle.

But with respect to the mode of delivering the hemp or flax, it is conceived a light frame might be easily contrived to hold the flax in some manner over a hackle, and either be carried by the spinner, or drawn by him, as he moved along in the usual manner, which would have every advantage of the other method, without the additional labour and complication.

Another patent for improvements in the manufacture of *cables* and *cordage*, was granted to Mr. Joseph Huddart, in October 1805. *Repert. of Arts*, No. 46, New Series.

In the method of manufacturing cables and cordage, recited in Mr. Huddart's specification, the strands are wound on reels, placed in frames which turn round on axes, between two circular planes connected to a large spindle, or arbor, which receives its motion from the prime mover; the part of this arbor on which it revolves is formed hollow, to permit the strands from the reels to pass through, in which particular, it resembles the spindle of the common spinning-wheel for flaxen thread; the frames of the reels receive a circular motion quicker than the circular planes in which their axes are fixed, and in a contrary direction to their motion, by being each connected with a fixed wheel, by a smaller intermediate wheel, which operates between it and a pinion fixed to the axis of the frame of the reel. The strands are drawn out from the reels by grooved rollers or *whirls*, which receive their motion from a fixed wheel on one of the circular planes in a similar manner to that described, by which the frames of the reels receive their motion: the reels are prevented from moving too freely, by a contrivance which presses on their axes, by which the friction is increased at pleasure, by turning a screw; by this means, the strands can only come off the reels as they are drawn by the grooved rollers: after the cable, or cord, has passed the cavity of the large arbor, it is moved onward between grooved rollers, or *whirls*, of a large size, by which it is drawn off regularly, as it is completed, to the place where it is to be coiled up.

The principal advantage of this method of making cables or ropes, is, that it admits of the work being performed in a small space, and thereby renders needless the great extent of ground, in length at least, necessary in the usual method of making ropes: but in this respect, it has no advantage over the invention of Mr. William Chapman for the same purpose, for which he obtained a patent in January 1798, (*Vide Repertory of Arts*, Vol. IX.) The chief difference between Mr. Huddart's plan and Mr. Chapman's is, that in Mr. Huddart's, the twist of the cable is given by the whole machinery, or frame-work, which holds the strands, turning round. In Mr. Chapman's, the twist is given by the coiled cable turning round, while the frames which contain the revolving reels of the strands remain stationary.

The effect which this difference will occasion is, that, in Mr. Huddart's plan, the labour of turning the machinery will be greatest at first (when all the reels are fully loaded with the weight of the strands), and become gradually less as the weight of the strands on the reels is diminished by their being drawn off to form the cable.

In Mr. Chapman's, the labour will be nearly equal through the whole process, as the mass in motion is increased in one part of the machinery in proportion as it is diminished in another part: this consideration, at first view, would seem to give an advantage to Mr. Huddart's plan; but it will not be found any in practice, for the same force will be required to turn the machinery of each in the beginning, and therefore an equal moving power must be provided for each; on which account, the expense must be equal in this respect for both.

The saving in both plans, from their admitting of the work being performed in a small space, will be ultimately an advantage to the public, as well as to the proprietors of them; for by the heavy rents of extensive rope-yards being removed from the manufacture, the articles in time can be afforded at a cheaper rate. Ropes made in this manner, it is thought, will be twisted more equally in every part, as in the common method of rope-making, in which the twist is communicated through a great length of rope perpetually altering in its extent, an inequality of twist must necessarily ensue.

A patent has also been obtained by Mr. Ralph Walker, for making *ropes* and *cordage*. Dated August, 1806. *Rep. of Arts*, &c. No. 60, New Series.

In Mr. Walker's engine for making ropes for every strand contained in them, an oblong vertical frame is made to revolve on its centre by machinery, which frame supports as many upright bobbins along its sides as there are yarns in each strand of the rope; in ropes for his Majesty's service, an additional bobbin is put at the bottom of each frame to hold the central coloured yarn, called the *royal yarn*, which marks the ropes.

The frames (which support the bobbins), called *flyers* by Mr. Walker, are sustained by another frame, consisting of a vertical shaft and two cross pieces, in which latter the pivots of the flyers turn; at the bottom of the flyers are placed toothed wheels, which interlock with a large fixed wheel, or periphery of a wheel, whose teeth are at the inside, which when the shaft revolves, causes them to twine round; the upper pivots, both of the flyers and shaft, are hollow; through the first the united yarns of each strand pass, and then move over revolving rollers to the hollow pivot of the shaft, through which all the strands pass at once to form the rope, and are twisted together by the motion of the shaft; the rope afterwards passes between grooved rollers, put in motion by shafts and wheels communicating with the wheel that causes the main shaft to revolve, and these rollers draw it forward to the place where it is to be coiled up; or, when of a small size, where it is to be rolled upon a reel, moved likewise by the general machinery.

On the main shaft, three or more toothed wheels of different sizes are placed, to any one of which the wheel which gives it motion may be applied, so that its speed may be increased or diminished as thought fit: there is a contrivance for the same purpose annexed to the rollers, for drawing out forward the rope, which is very novel and ingenious; a shaft extends from the roller to the wheel which gives it motion; near the roller this shaft has an universal joint, at its farther end, where it is to receive its impulse from the revolving wheel; there is annexed to it a segment of a hollow sphere, whose radius is the length of the shaft from the joint to the wheel; on this spherical segment a number of toothed peripheries are fixed, whose centres lie in the line of the shaft, and which increase in size as they are farther removed from the apex of the segment; by this arrangement of parts, any of the toothed peripheries can be brought into contact with the revolving wheel, by merely raising or lowering the end of the shaft: the spherical form of the part on which they are placed, causing them all to be at equal distances from their centre of motion, or the universal joint.

A method of tarring the yarns is inserted in the specification of this patent, in which the tar is kept hot by steam entering into the interval between the boiler and an external case, placed at a due distance from it all round, and secured properly to it at top, so as to be air-tight; and an oblong roller is placed horizontally in the boiler, beneath which the yarns are drawn in their passage through the tar, from bobbins arranged at one side of the boiler, by others at the other side, which are turned by bands, and which are filled equally by guides, that are made to raise or fall by a heart wheel movement; the roller that lies in the tar boiler is placed so that it may be raised or lowered according to the quantity of tar in the boiler.

Twelve well executed figures accompany this specification; the three first are different views of the machinery for making the cordage, described; the fourth and fifth represent machinery on the same principle for making single strands for large cables; the sixth shews the engines for making large ropes from the single strands; the seventh and eighth figures shew the method of laying small ropes; and the ninth, tenth, eleventh, and twelfth figures, exhibit the apparatus described for tarring the yarn.

By this mode of making ropes, the patentee states, that the yarns will all bear an equal proportion of the strain, and the strands be laid uniformly in the rope, and every strand receive an equal degree of twist, by which the rope will be rendered stronger, and of a uniform degree of strength throughout.

Several other patents have been obtained for making cordage by machinery, viz. Mr. Edward Cartwright procured one in May, 1792; another was granted to

Mr. Daniel Belfour, of Elsineur; and, in the same year, one was also obtained by Mr. Joseph Huddart, of Islington, who procured a second in September, 1804; one was also granted in January 1798, to Mr. William Chapman, of Newcastle; and two or three more have been obtained by other persons.

Mr. Huddart's last patent principally regards the method of laying cables from strands previously prepared, as does Mr. Cartwright's also: Mr. Belfour's patent, and Mr. Huddart's first and second patents, relate to the preparation of the yarn for the formation of the strands. Mr. Chapman s patent, as well as this of Mr. Walker's, includes both the formation of the strands from the yarn, and of cordage from the strands.

The chief difference between Mr. Walker's contrivance and Mr. Chapman's, for the formation of strands from the yarn, is, that in Mr. Chapman's, the reels, or bobbins, which contain the yarn, are stationary as to place, and the twist is given by the revolution of the strand; in Mr. Walker's, on the contrary, the frames containing the reels or bobbins of yarn move round, and that containing the strand is stationary.

Mr. Walker's machine for the formation of the cable or cord from the strands, appears in all essential parts to be the same as that for which Mr. Huddart obtained his last patent in October, 1805. The engine for which Mr. Huddart's first patent was taken out, was principally for making strands from the yarn, which were contrived to be laid more regularly by it. This differed from Mr. Walker's contrivance in the same manner that Mr. Chapman's did, in having the reels which held the yarn stationary.

The sections of the frames for supporting the bobbins, annexed to Mr. Walker's patent, have a remarkable neat appearance, and shew a very ingenious arrangement of the parts; the accompanying machinery seems also contrived in a masterly manner, and contains a mechanical novelty which may be useful in other engines, in the contrivance of the shaft with the universal joint, and spherical termination, for regulating the speed of the rotation of certain parts of the engine to the degree required.

Among other extraordinary proposals as a substitute for hemp in the manufacture of standing rigging, is one offered to the public by a Mr. John Slater, of Huddersfield, for forming the *rigging* and *cables* of ships of *metal chains;* the property of which invention was secured to him by patent, in January, 1804.

"My invention," says he, "consists in the substitution of metals in lieu of hemp in the fabrication of cables, shrouds, stays, and other articles for the rigging of ships, which is to be applied in the form of chain-work, and which every workman in chain-work knows how to apply without further instructions."

He then adds, that he claims no discovery in the construction of chains, but in the application of them to the rigging of ships, of whatever form, metal, or metallic substance they may be constructed.

This invention is intended chiefly for standing rigging, and the links of the chain are made of a short oval form, thick in proportion to their size, and every two nearly touching in the centre of the third, in which they are inserted. The weight of the different articles to be the same as that of the cordage now in use. The mast-heads are to be surrounded with a plate of rolled iron, or copper, to keep the shrouds from chafing. One leg of a pair of shrouds is to be received through a large link in the others, just at the place where the present kind are seized; and they are to be set up with dead-eyes, and lanyards, as in the present manner. The fore-stays are to be hooked where they are now seized, and the back-stays fitted in the same manner as the shrouds.

The advantages are, 1st, the duration of one suit of rigging would exceed that of any ship made of the best materials. 2d. As there would be no sensible stretching, it would obviate a serious inconvenience, by which many a ship has lost her masts, and ultimately both cargo and hull; as it is often impossible to set the shrouds up in a heavy gale, or even to secure them in any manner that will preserve the masts. 3d. This substitute for cordage will never require any mechanical stretching, and will be always pliant, and easy to handle. 4th. In time of action, this kind of rigging will be much less liable to be cut with shot from its great cohesion, elasticity, and slipperiness; and in case of its being cut, may be much more easily repaired with string, which will last during the action, and afterwards by a new link or links.

5th. This kind of rigging has a much lighter and more beautiful appearance to the eye, and when blacked with lamp black and pitch, is effectually preserved from rust. The first cost is not greater than that now in use, and when worn out, it can be re-manufactured; of course, the saving to this country would be immense, if it were adopted in the British navy.

Such are the advantages which the inventor describes resulting from the application of chain-work to the rigging of ships. He is, however, aware it may be objected, that the metal of which the chains are composed, will have a tendency to attract the electric fluid. In answering this objection, he observes, and justly, that when the fluid does strike the metal, it will follow every turn it makes till it reaches the surface of the earth or water, which is a strong reason for the adoption of the metallic rigging; as in the case of a ship being struck with lightning, the shrouds having small appendages to let down, it would be easily conducted into the water, without injury to the crew, ship, or cargo, except a man happened to be on the shrouds leading from where it fell. And it is presumed, that accidents of this kind would more seldom happen than they ever do at present, as the men upon decks, in the tops, and on the yards, would be almost entirely secure; and in climates where the surrounding atmosphere is pregnant with the electric fluid, no accumulation could take place, as the rigging would be constantly conducting it harmlessly to the surface, in the same manner as a chain suspended from the conductor of an electrical machine, and resting on the ground, would prevent a person touching it from receiving a shock.

The chain work may be made either of iron or of copper.

Mr. Slater's design is, no doubt, a good one; some experienced captains of ships have acknowledged it as such, long before the patent was obtained.

The chief objection is, that the chain is not sufficiently elastic for standing rigging; but it is very possible to obviate this, by the addition of properly constructed springs; and some degree of elasticity may be given to the chain itself, by forming it with spiral links. The danger from lightning might be easily removed by proper conductors, as Mr. Slater observes.

It is the opinion of an experienced captain, that iron chain shrouds should be served with old canvas well greased, and spun-yarn; this would keep them from rusting, and make them more convenient to the men to handle.

West-India ships often have their lower yards slung with chain; naval ships have always had chain in addition to the hempen slings; and, no doubt, other ships also.

A West-India captain has slung the three lower yards of his ship with chain, in a new method, by which they were much more moveable, and would go about farther, than with hempen slings; and used only two fathom of chain, which cost but eleven shillings.

The same captain also used chain hawsers in the West-Indies, several years ago, and speaks much in favour of them. Many more instances might be mentioned of the good effect of the use of chains.

There is, however, no instance of chains having been actually used in this part of the world for standing rigging, though it had been proposed by the captain above mentioned, previous to the year 1804.

For further information on chain-work, see the article MOORING CHAINS.

COR HYDRÆ, a fixed star of the second magnitude in the heart of the constellation Hydra. Its right ascension in time adapted to the beginning of the year 1812, is 139° 34′ 41,″0, annual variation in right ascension 44″2. Its declination 7° 50′ 56″ South, and annual variation in declination, + 15,″1.

COR *Leonis*, or *Regulus*, in astronomy, a fixed star of the first magnitude, in the constellation Leo. See the article REGULUS.

CORK-JACKET, (*chemise de liège*, Fr.) a machine made somewhat in form of a seaman's jacket, lined with a particular kind of select cork in pieces, so artfully shaped and disposed, as to give it the strongest buoyancy, and also to preserve an easy degree of flexibility, so that the activity of the wearer is not impeded.

Dr. Wilkinson devised, and in his "*Testamen Nauticum*; or, Seaman's Preservation," recommended this machine, or apparatus, to all sea-faring people, as an easy and indubitable means to escape drowning in shipwreck; and adds, a seaman who commits himself to the sea, without having first so provided himself against drowning in shipwreck, seems to merit the same title to prudence, as one who sets sail without either anchor or compass on board.

The use of these jackets in the navy has been sanctioned by the Admiralty, and the Society of Arts have most strenuously recommended them to all who go to sea; hence, it is supposed, the self-evident security afforded by this simple invention, is sufficient to operate an universal conviction in its favour.

A cork jacket or spencer may be constructed as follows: pieces of cork about three inches long, by two wide, and the usual thickness of the bark are enclosed between two pieces of strong cloth or canvas, and formed like a jacket without sleeves; the pieces of cloth are quilted round each piece of cork, to keep them in their proper situations; the lower part of the jacket is made like the same part of women's stays, to give freedom to the thighs in swimming: it is made sufficiently large to fit a robust man, and is secured to the body by two or three strong tapes sewed far back on each side, and tied before; the strings are thus placed, to enable any wearer to tighten it to his own convenience. For further contrivances for saving the lives of seamen, see the article SHIPWRECK.

CORNETTE, (*pavillon à queue d'hirondelle*, Fr.) a swallow tailed flag.

CORONA AUSTRALIS, or MERIDIONALIS, the *Southern Crown*, a constellation in the southern hemisphere, containing 12 stars, according to the British catalogue.

CORONA *Borealis*, or *Septentrionalis*, the *Northern Crown*, or Garland, in astronomy, a constellation of the northern hemisphere, whose stars, in the Britannic Catalogue, are 21.

CORPORAL OF A SHIP OF WAR, (*caporal*, Fr.) is a person under the master at arms, employed to teach the seamen the use of small arms, or musketry; and to attend at the gang-way, or entering-ports, and observe that no spirituous liquors are brought into the ship, unless by express leave from the officers. He is also to extinguish the fire and candles at eight o'clock in winter, and nine in summer, when the evening gun is fired; to walk frequently down on the lower decks in his watch, to see that there are no lights, but such as are under the charge of proper sentinels; and to see that no disorderly conduct, as drunkenness, &c. is practised. See the article MASTER AT ARMS.

CORPOSANT, (*feu St. Elme*, Fr. *corpo santo*, Ital.) a name given to the volatile meteor, or *ignis fatuus*, often beheld in a dark tempestuous night, about the decks or rigging of a ship, but particularly at the extremities, as the mast-heads, and yard-arms, and is most frequent in heavy rain, accompanied with lightning. This appearance, which is nothing more than the electric fluid, passing silently from the clouds to the water, or the contrary, by means of the humidity on the masts and rigging; was, in the dark ages of superstition, esteemed by some a good omen, and by others an evil one; but modern philosophy has so happily explored its cause, that none but the most ignorant are now intimidated by it. "They usually wander, with uncertain motion, from place to place, sometimes appearing to cleave close to the sails and masts; but they frequently leap up and down with intermission, affording an obscure flame, like that of a candle burning faintly. They are produced by some sulphureous and bituminous matter, which being beat down by the motion of the air above, and gathering together, is kindled by the agitation of the air, as butter is gathered together by the agitation of the cream. And from this appearance we infer, that storms come from sulphureous spirits that rarify the air, and fuel into a motion." VARENIUS.

CORSAIR, (*pirate barbaresque*, Fr.) a name commonly given to the piratical cruisers of Barbary, who frequently plunder the merchant ships of European nations with whom they are at peace.

CORVUS, in antiquity, a machine invented by the Romans, at the time of their wars in Sicily, when they first engaged the Carthaginian fleet.

According to Polybius, the *corvus* was framed after this manner:

On the prow of their ships they erected a round piece of timber, about a foot and a half in diameter, and twelve feet in length; on the top of which, was a block, or pulley: round this piece of timber was a platform of boards, four feet broad, and about eighteen feet long, which was well framed, and fastened with iron.

The entrance was longways, and it moved about the aforesaid upright piece of timber, as on a spindle, and could be hoisted up within six feet of the top: about this was a sort of parapet, knee high, which was defended with upright bars of iron, sharp at the ends; towards the top there was a ring, by the help of which and a pulley, or tackle, it was hoisted or lowered at pleasure; with this movable gallery they boarded the enemy's ships, sometimes on their bow, and sometimes on the after-part of the ship, as occasion best served. When they had grappled the enemy with the iron spikes, if they happened to swing broadside to broadside, then they entered from all parts; but in case they attacked them on the bow, they entered two and two, by the help of this machine, the foremost defending the fore-part, and those that followed the flanks, keeping the boss of their bucklers level with the top of the parapet.

CORVUS, the *Raven*, in astronomy, a constellation of the southern hemisphere, wherein, according to the Britannic Catalogue, are 9 stars.

CO-SECANT, in geometry, is the secant of an arc, which is the complement of another to 90 degrees; *co* being, in this case, a contraction of the word complement, and was first introduced by Gunter.

CO-SINE, in trigonometry, is the right sine of an arc, which is the complement of another to 90 degrees.

COSMICAL ASPECT, among astrologers, is the aspect of a planet with respect to the earth.

COSMICAL, in astronomy, a term used to express one of the poetical risings of a star: thus, a star is said to rise cosmically, when it rises with the sun, or with that point of the ecliptic in which the sun is at that time: and the cosmical setting, is when a star sets in the west at the same time that the sun rises in the east.

COT, (*cadre*, Fr.) a particular sort of bed frame, suspended from the beams of a ship, for the officers to sleep in. It is made of canvas sewed in the form of a chest, about six feet long, one foot deep, and from two to three feet wide, and is extended by a square wooden frame with a canvas bottom, on which the matrass is laid. It is reckoned much more comfortable at sea, than either the hammocks or fixed bed-places in passage vessels.

CO-TANGENT, in trigonometry, is the tangent of an arc, which is the complement of another to 90 degrees.

COVE, (*crique, ou petite anse*, Fr.) a small creek or inlet; also a harbour for shipping: as the Cove of Cork.

To **COVER,** (*couvrir, protéger*, Fr.) to protect; as 'our ship covered the landing.'

CO-VERSED SINE, in geometry, is the remaining part of the diameter of a circle, after the versed sine is taken from it.

To **COUCH** WELL, is to lay close and even.

COUNCIL OF WAR, (*conseil de guerre*, Fr.) in the royal navy, an assembly of the admirals and principal officers of a fleet, occasionally called by the commander in chief to concert measures for the conduct of the fleet, with respect to engaging, retreating, besieging, &c. whenever any emergency occurs, that renders him doubtful of his own judgment, or unwilling to risk his own reputation, by taking the whole responsibility upon himself.

When there is a sufficient number of flag-officers together, all *councils of war* are to be held by them only.

The first captain to the admiral of the fleet is, on such occasions, to be esteemed as a rear-admiral, and to take place at all councils of war, and also at courts-martial, next to the junior rear-admiral.

COUNTER, (*contre arcasse*, Fr.) an arc or vault, whose upper-part is terminated by the bottom of the *stern*, and the lower part by the wing-transom and buttock, being referred to from the article QUARTER.

There is also another counter above, parallel to this, but not vaulted, and extends from the top part of the lower counter, to the bottom moulding of the cabin or ward-room windows.

COUNTER-*Brace*, the lee-brace of the fore-top-sail-yard: it is only distinguished by this name at the time of the ship's going about, called *tacking*; at which time, when the sail begins to shiver in the wind, this brace is hauled in, to flatten the sail against the lee-side of the top-mast, and increase the effort of the wind in forcing her to turn round. See the article TACKING.

COUNTER-*Timbers*, are those short timbers in the stern, put in only for the purpose of strengthening the counter.

To **COUNTERSIGN,** is to sign an order, warrant, commission, &c. in quality of a secretary, to render the thing more authentic.

COUNTERSIGN is also the name of a word and given out with the parole, and made use of in a similar manner.

COUNTERSUNK, a hollow, cut by a bit, round the edge of a hole.

COUNTERSUNK-*Bit*, a bit having two cutting edges at the end, reversed to each other, which form an angle from the point.

COUPED-PENDANTS, are those pendants that are alternately divided, but in a vertical direction: it is taken from couped, a term in heraldry, and derivation *coupé*. A pendant may be couped blue, red, and yellow, but alternate is an indefinite term, for a pendant may be alternately divided in an horizontal direction, and then should be called not alternate, but striped, blue, red, and yellow.

COURSE, (*route*, Fr.) in navigation, is that point of the compass or horison which a ship steers on; or the angle which the rhumb-line on which it sails makes with the meridian; being sometimes reckoned in degrees, and sometimes in points of the compass.

When a ship sails either due north or south, she sails on a meridian, makes no departure, and her distance and difference of latitude are the same.

When she sails due east or west, her course makes right angles with the meridian, and she sails either upon the equator or parallel to it; in which case she makes no difference of latitude, but her distance and departure are the same.

But when the ship sails between the cardinal points, on a course making always the same oblique angle with the meridians, her path is then a *spiral curve* cutting all the meridians in the same angle, and terminating in the pole.

As the course is the angle which a ship's track makes with the meridian, it is generally expressed either in points, or degrees. Thus, when a ship sails in a north east direction, we say her course is four points, or 45 degrees; and then the difference of latitude and departure are equal.

When the difference of latitude is greater than the departure, the course is less than four points; but when the departure is greater than the difference of latitude, the course is more than four points.

The course is found by the following canon, when the difference of latitude and departure are known:—

As the difference of latitude is to radius:
So is the departure to tangent course.

Oblique Course, (*loxodromie*, Fr.) the course which crosses the meridian at equal and oblique angles.

COURSES, (*basses voiles*, Fr.) a name by which the principal sails of a ship are usually distinguished, viz. the main-sail, fore-sail, and mizen: the mizen-stay-sail and fore-stay-sail are also sometimes comprehended under this denomination, as are the main-stay-sails of all brigs and schooners. See the article Sail. When a ship sails under the main-sail and fore-sail only, without lacing on any bonnets, she is then said to go under a pair of courses.

COURT-MARTIAL, in the royal navy, is a court composed of admirals, captains, and commanders, for the trial of offences against the articles of war; the powers of which are regulated by the articles and orders contained in the Act of Parliament made in the 22d year of the reign of King George the IId. entitled, "An Act for amending, explaining, and reducing into one Act of Parliament, the Laws relating to the Government of his Majesty's Ships, Vessels, and Forces by Sea;" and also, an act made in the 19th year of the reign of his present Majesty, entitled, "An Act to explain and amend the foregoing Act."

In the articles of war, almost every possible offence that can be committed in the navy is set down, and the punishment thereof annexed. See Articles of War.

The *courts-martial* are to be assembled in the forenoon, and held in the most convenient and public place of the ship, where all, who will, may be present; and the captains of all his Majesty's ships in company, which take post, have a right to assist there. It is required to set from day to day, (*Sunday* always excepted) until the sentence be given; and no member shall absent himself from the court during the whole course of the trial, except in cases of sickness, &c.

All complaints are to be made in writing to the commander in chief, unless where he shall see cause of himself to assemble the same; in which are to be set forth the facts, time, and place, and the manner how they were committed. And if any captain, entitled by his rank to sit in the court, be personally concerned in the matter to be tried, he is not to be permitted to be of the number of members, by whom the court shall be composed.

When a court martial is appointed to be held, the officer who is to preside thereat, is to take care that a copy of the charge, or complaint, be delivered to the person accused, as soon as may be after he shall have received the order to hold such court martial, and not less than twenty-four hours before the trial.

When the court is assembled, the person to be tried, as also the accuser, if any, are to be brought before them; the judge-advocate is to read the warrant, or power, authorising the court to assemble; and, immediately after, to administer to the members, and to take himself, the oath prescribed by the act first mentioned; he is next to read the charge, or complaint against the person to be tried, and thereupon the court may proceed to call witnesses; but they are not to question the said witnesses until an oath hath been administered by the judge-advocate, in the words following, which the witness is to repeat:—

"I, A. B. do most solemnly swear, that, in the evidence I shall give before the court, respecting the present trial, I will, whether demanded of me by question or not, and whether favourable or unfavourable to the prisoner, declare the truth, the whole truth, and nothing but the truth. So help me God."

The judge-advocate is to take down, in writing, the evidence given by each witness, and read the same to the court, in his hearing, that, in case of mistake, it may be corrected: he is also to deliver his opinion in any doubts or difficulties that may arise in the course of the trial.

In taking the opinion of the court, upon all questions, the youngest officer shall vote first, proceeding in order, up to the president.

When the evidence is closed, the accused person is to be removed, and the accuser and standers-by are to withdraw; the court is then to consider the matter in evidence before it, and the judge-advocate, by the direction of the court, is to draw up such questions as shall be agreed upon, whereon to form a determination in regard to the innocence or guilt of the person upon trial. If the party be found guilty of a breach of any of the Articles of War, established by law, the court is to consider and determine on the punishment proper to be inflicted conformably thereto. The judge-advocate is to draw up the sentence accordingly, being careful to specify therein the charge, or substance, of it; and the same to be signed by every member of the court, by way of attestation, notwithstanding any difference of opinion there may have been among them.

After the sentence is drawn up and signed, all persons are to be re-admitted; and, the party accused being also present, the judge-advocate is, by direction of the court, to pronounce the same.

The judge-advocate is to send the original sentence, and an attested copy of the minutes of the evidence and proceedings of the court, to the secretary of the Admiralty, by the first opportunity.

The sentences of naval courts martial and those of the army differ in these particulars—In the navy they are final, and pronounced *instanter*, in open court; but the members of army courts martial are sworn not to divulge the sentence of the court until it shall be approved of by his Majesty, or by some person authorised by him.

In carrying the sentences of courts-martial into execution for corporal punishment, the admiral or commanding officer is to issue orders to the captain of the flag, or other particular ship, to make the signal for the boats of the squadron to assemble, manned and armed on the day appointed, to attend the punishment, and likewise orders to the other captains to send a lieutenant with a boat, manned and armed, from their respective ships to attend and assist thereat.

An order is, at the same time, issued to the captain or commander of the ship to which the prisoner belongs, (accompanied with a copy of the sentence) directing him to cause the punishment to be inflicted along-side of the different ships, in the manner, and in such proportions, as are therein specified. In the order, it is usual to include directions to the captain to cause the surgeon of his ship to attend in the boat with the lieutenant, as well as one of his mates in the long-boat with the prisoner, for the purpose of judging of the prisoner's ability to bear all his punishment, which the surgeon may put a stop to, when he conceives the prisoner is not able to bear any more with safety.

The provost-martial is ordered to attend the punishment, and to read publicly the sentence of the court-martial along-side each ship respectively.

When a sentence of death is approved of by the king, the warrant for execution is transmitted by the Admiralty to the officer commanding the ships and vessels at the place for the time being, who issue the necessary orders, agreeably to the forms of the service, and preparations are made accordingly.

On the morning destined for the execution, the signal of death is displayed, and the boats of the squadron, manned and armed, surround the ship appointed for the execution. The crews of the respective ships are arranged on deck, and are acquainted with the crime for which the punishment is inflicted; after which a gun is fired; and, at the same time, the unhappy victim, who has violated the laws of his country, is run up, by the neck, to the yard-arm, a terrible example to the surrounding spectators.

For further particulars, see the *Naval Instructions* relating to his Majesty's Service at Sea: also Mr. M'Arthur's *Treatise on Naval Courts Martial.*

CRAB, (*petit cabestan*, Fr.) a sort of wooden pillar, whose lower end, being let down through a ship's decks, rests upon a socket like the capstan; and having in its upper end three or four holes, at different heights; through the middle of it, above one another, into which long bars are thrust, whose length is nearly equal to the breadth of the deck: It is employed to wind in the cable, or to purchase any other weighty matter which requires a great mechanical power; but not being so convenient, is now generally laid aside, except in rope walks, &c. This differs from a capstan, as not being furnished with a *drum-head*, and by having the bars to go intirely through it, reaching from one side of the deck to the other; whereas those of the capstan, which are superior in number, reach only about eight inches or a foot into the drum-head, according to the size thereof. This machine is represented in plate VII. by fig. 3, and plate VIII. fig. 6.

Crab, or *Capstan*, a machine fixed in the ground at the lower end of rope-walks, (see plate VIII. fig. 6,) and is used in stretching the yarn to its fullest extent, before it is worked into strands, by means of tackle falls, led from the sledge to the capstan, they being about eighteen yards distant from each other.

Crab is also an engine of wood, with three claws, placed on the ground like a capstan, and used at launching, or heaving ships into the dock, or off the quay.

CRADLE, (*berceau; servant à lancer un vaisseau à la mer*, Fr.) is a timber frame made along the outside of a ship, in order to conduct her smoothly and steadily into the water when she is to be launched; at which time it supports her weight while she slides down the descent, or sloping passage, called the *ways*, which are for this purpose daubed with soap or tallow. This frame is exhibited by plate VIII. fig. 7.

Cradles are also standing bedsteads made up for the wounded seamen, that they may be more comfortable than it is possible to be in a hammock.

CRAFT, a general name for all sorts of vessels employed to load or discharge merchant ships, or to carry alongside, or return the guns, stores, or provisions of a man of war: such are lighters, hoys, barges, &c.

Small Craft, is a term used to denote the small vessels of war attendant on a fleet, such as cutters, schooners, gun-boats, &c. which are generally commanded by lieutenants.

CRANE, (*grue*, Fr.) a machine used on wharfs, and in warehouses, for raising and lowering huge stones, ponderous weights, packages &c.

The crane is of two kinds; in the first kind, called the rat-tailed crane, the whole machine, with the load, turns upon a strong axis: in the second kind, the gibbet alone moves on its axis.

For a particular account of the different cranes, and of the gradual improvements they have received, see Desaguliers's *Exper. Philos.* See also Ferguson's *Lectures*, or *Philos. Trans.* vol. 54, p. 24, for a description of a new and safe crane, with four different powers adapted to different weights.

Crane *Lines*, are lines going from the upper end of the spritsail-topmast to the middle of the fore-stays. They serve to keep the spritsail-topmast upright and steady in its place, and to strengthen it.

CRANK, or CRANK-Sided, (*coté foible*, Fr.) the quality of a ship, which, for want of a sufficient quantity of *ballast* or cargo, is rendered incapable of carrying sail, without being exposed to the danger of upsetting. See the articles Ballast and Trim.

Crank *by the Ground*, is also the quality of a ship, whose floor is so narrow, that she cannot be brought on the ground without danger.

Crank *Wheel*, in rope-making, for spinning of lines, boxcord, &c. is a machine fixed on an iron spindle, or axis, with a handle to turn it by: it hangs between two posts; the after one is six feet high, one foot broad, and five inches thick; in its upper part, above the wheel, is let in a semicircular board, two feet six inches long, two feet broad, and five inches thick, to receive three sets of whirl-bolts, with whirls on them, for the spinners to hang their threads on; at the front side of the wheel is a short post, supported by a knee of oak for the spindle to rest on.

Cranks are also iron braces, which support the lanterns on the poop-quarters, &c.

CRATER, or the Cup, in astronomy, a constellation of the southern hemisphere, whose stars, in the Britannic Catalogue, are 31.

CRAWL, (*bordigue*, Fr.) a sort of pen, or place of confinement, formed by a barrier of stakes and hurdles on the sea-coast, to contain any sort of fish.

CREEK, (*crique*, Fr.) the port of the haven where any thing is landed from the sea. It is also defined by some to be a shore or bank, on which the water beats, running in a small channel from any part of the sea.

CREEPER, (*croc à quatre branches*, Fr.) is an instrument of iron resembling a grapnel, having a *shank* and four hooks or claws, as represented in
Plate IX. plate IX. fig. 1. It is used to drag along a river or
Fig. 1. harbour, with a rope fastened to it, to hook and draw up any thing from the bottom that may have been lost overboard.

CREPUSCULUM, in astronomy, twilight; the time from the first dawn, or appearance of the morning to the rising of the sun: and again, between the setting of the sun, and the last remains of day. The *crepusculum*, or twilight, it is supposed, usually begins and ends when the sun is about 18 degrees below the horizon; for then the stars of the sixth magnitude disappear in the morning, and appear in the evening. It is of longer duration at the solstices than in the equinoxes; because the sun, by the obliquity of his path, is longer in ascending through 18 degrees of altitude.

Twilight is occasioned by the sun's rays refracted in our atmosphere, and reflected from the particles of it to the eye.

The chief differences of twilight are owing to the different situations of places upon the earth, or to the difference of the sun's place in the heavens. Thus, the twilight is longest in a parallel sphere, and shortest in a right sphere, and longer to places in an oblique sphere in proportion as they are nearer to one of the poles; a circumstance which affords relief to the inhabitants of the more northern countries in their long winter nights.

And the twilights are longest in all places of north latitude, when the sun is in the tropic of Cancer; and to those in south latitude, when he is in the tropic of Capricorn. The time of the shortest twilight is also different in different latitudes; in England, it is about the beginning of October and March, when the sun is in the signs ♎ and ♓. For the method of determining it by trigonometry, see Gregor. *Astron.* lib. 2, prob. 41. See also Robertson's *Navigation*, book 5, prob. 12.—Hence, when the difference between the sun's declination and the depth of the equator is less than 18°, so that the sun does not descend more than 18° below the horizon; the *crepusculum* will continue nearly the whole night, as is the case in England from about the 22d of May to the 22d of July.

CREW, the company of sailors belonging to a ship, boat, or other vessel. The number of sailors who are to work and manage a merchant-ship is regulated by the number of lasts she may carry, each last making two tons. The crew of a Dutch ship from 40 to 50 lasts, are seven sailors, and a swabber; from 50 to 60 lasts the crew consists of eight men and a boy, and thus increases at the rate of one man for every ten lasts. English and French crews are usually stronger than Dutch, but always in about the same proportion. In a man of war, the ship's company is divided into several particular crews, or gangs, as the boatswain's crew, the gunner's crew, the carpenter's crew, &c.

The Crew *of a Ship*, (*l'équipage*, Fr.) comprehends the officers, sailors, seamen, marines, ordinary men, servants, and boys; but exclusive of the captain and lieutenants, in the French service.

CRINGLE, (*herseau de boulines, et autres herseaux*, Fr.) is a small hole formed in the *bolt-rope* of a sail, by intertwisting the strand of a rope alternately round itself, and through the *strands* of the bolt-rope, till it becomes three-fold, and assumes the shape of a ring. See plate IX. fig. 2, where *a*, *b*, Fig. 2.
represent part of the bolt-rope of a sail; and *c*, the cringle.

The use of the cringles is generally to receive the ends of ropes, which are fastened to them for the purpose of drawing up the sail to its yard, or of extending the leech by the *bowline bridles*, &c.

Iron Cringles, or *Hanks*, (*anneaux de voiles d'étai, dans certains bâtimens*, Fr.) are open at the lower part, to go over the stays; they are made broader in the middle than at the ends, which causes them to spring to their shape; the ends are drawn out fine, and turned into a small ring, for securing the head of the sail by the seizing. These hanks are mostly used to the foresails of vessels with one mast.

To CRIPPLE, (*désemparer*, Fr.) implies to damage a ship in her masts, yards, &c.

CROSIER, in astronomy, four fixed stars in the southern hemisphere, in the form of a cross, and situated under the constellation Centaurus; by the help of which, those who sail in southern latitudes may readily find the antarctic pole.

The brightest of those stars is of the first magnitude, and at the bottom of the cross: its right ascension in time adapted to the beginning of the year 1812, is 184° 3′ 27″, 60, annual variation in right ascension 48″ 6. Its declination is 62° 4′ 06″ S. and annual variation in declination + 20″,0.

CROSS-Bars, are round bars of iron, bent at each end, and used as levers to turn the shank of the anchor. They are from three to five feet long, and about one inch and a quarter in diameter.

Cross *Chocks*, (*espèce d'accotars*, Fr.) are pieces of timber fayed across the dead-wood in midships, to make good the deficiency of the heels of the lower futtocks.

Cross *in the Hawse*, (*croix dans les cables*, Fr.) is when a ship moored with two anchors from her bows, has swung the wrong way once, whereby the two cables lie across each other. See the article Hawse.

Cross-*Jack*, pronounced *crojeck*, (*vergue sèche*, Fr.) the lower yard on the *mizen*-mast, which is hence called the *cross-jack-yard*, to the arms of which the clews of the mizen topsail are extended.

Cross-*Jack-sail*, (*voile de fortune*, Fr.) is a sail bent to that yard. This sail has been found of little service, and is therefore very seldom used.

Cross *Pawls*, are those pieces of fir or deal timber, which keep the ship together whilst in her frames.

Cross *Piece*, (*râteau, ou ratelier*, Fr.) a rail of timber extended over the *windlass* of a merchant-ship from the knight-heads to the belfry. It is furnished with wooden pins to fasten the running-rigging to, as occasion requires. See the article Windlass.

Cross-*Spales*, in ship-building, are pieces of timber placed across the ship, and nailed to the frames, securing both sides of the ship together till the knees are bolted.

Cross-*Staff*. See the article Forestaff.

Cross-*Trees*, (*barres traversières des hunes*, Fr.) certain pieces of oak timber, supported by the *cheeks* and trestle-trees, at the upper ends of the lower and top-masts, athwart which they are laid, to sustain the frame of the *tops* on the one, and extend the top-gallant shrouds on the other. They are let in and bolted to the trestle-trees.

CROTCHES, (*fourcats*, Fr. *croccia*, Ital.) the crooked timbers that are placed upon the keel in the fore and hind parts of a ship, upon which the frame of her hull grows narrower below, as it approaches the stem afore, and the stern-post abaft.

Crotches, (*cornes*, Fr.) are also pieces of wood or iron, whose upper part opens into two horns, or arms, like a half-moon. They are fixed in different parts of the ship, according to the uses for which they are designed, which is usually to support booms, spare top-masts, yards, &c. The iron crotches are
Fig. 3. exhibited in plate IX. fig. 3.

CROW, (*pince*, Fr.) an iron lever, well known in mechanics, furnished with a sharp point at one
Fig. 4. end, and two claws at the other, as appears in fig. 4, plate IX.

It is used for various purposes, by shipwrights and mariners, to remove weighty bodies, as pieces of timber, to draw spike-nails, &c.; also to direct and manage the great guns, by moving them into their ports, and levelling or pointing them to any particular object.

To CROWD, (*forcer de voiles*, Fr. *cruth*, Sax.) to carry an extraordinary force of sail upon a ship, in order to accelerate her course on some important occasion, as in pursuit of, or flight from, an enemy; to escape immediate danger, &c.

CROW-FOOT, (*araignée*, Fr.) a complication of small cords spreading out from a long block, like the smaller parts which extend from the back bone of a herring. See plate IX. fig. 5. It is used to Fig. 5
suspend the *awnings*, or to keep the *top-sails* from getting under the top rim.

CROWN, in astronomy, a name given to two constellations, the southern and northern.

Crown *of an anchor*, (*le collet d une ancre*, Fr.) the lowest end of the shank of an anchor, where the arms are united.

Crown *of a Cable*, the bights which are formed by the several turns.

CROWNING, (*cul-de-porc*, Fr.) is the finishing part of a knot made on the end of a rope. It is performed by interweaving the ends of the different strands artfully among each other, so that they may not become loosened or untwisted. The design of these knots is, to keep the end of the rope fast in some place assigned for it; they are more particularly useful in all kinds of stoppers.

CRUISE, (*compagne de croisière*, Fr.) a voyage or expedition in quest of vessels or fleets of an enemy, which may be expected to sail through any particular tract of the sea at a certain season of the year. It is performed by traversing that particular tract which is called the cruising latitude.

CRUISER, (*vaisseau croiseur*, Fr.) a vessel employed in cruising.

They are small men of war, employed to sail to and fro in the Channel, and elsewhere, to secure our merchant ships and vessels from the enemy's small frigates and privateers. They are generally such as sail well, and are commonly well manned: indeed, the safety of the trade in the Channel, and up and down the Soundings and other places, absolutely require the constant keeping out of ships at sea. When the ships employed for this purpose, have arrived at their destined station, they traverse the sea backward and forward under an easy sail, and within a limited space, conjectured to be nearly in the track of their expected adversaries.

CRUTCH, (*corne de gui*, Fr.) a support for the main-boom of a sloop, brig, or cutter, &c. and for the driver-boom of a ship, when their respective sails are furled.

Crutches, in ship-building, are pieces of knee-timber, placed within side the ship, for the security of the heels of the cant-timbers abaft.

CUBE, in solid geometry, is a regular solid body, consisting of six equal square sides or faces, and is otherwise called a Hexædron.

CUDDY, (*cuisine, ou foyer dans certaines barques*, Fr.) a sort of cabin or cook-room, in the fore-part, or near the stern, of a lighter, or barge of burden.

CULMINATION, in astronomy, the passage of any heavenly body over the meridian, or its greatest altitude for that day. Hence a star is said to culminate, when it passes the meridian

CURRENT, (*courant*, Fr. *currens*, Lat.) in navigation, is a certain progressive motion of the water of the sea, in several places, by which all bodies floating therein are compelled to alter their course, or velocity, or both, according to the direction of the *current*.

The setting of the current is that point of the

compass towards which the waters run; and the drift of a current is the rate it runs in an hour.

Currents in the sea, are either natural and general, as arising from the diurnal rotation of the earth on its axis, or the tides, &c. or accidental and particular, caused by the waters being driven against promontories, or into gulfs and streights; from whence they are forced back, and thus disturb the natural flux of the sea.

The following observations are made by Varenius:—

"Currents are various, and directed towards different parts of the ocean, of which some are constant, and others periodical. The most extraordinary current of the sea is that by which part of the Atlantic or African ocean moves about Guinea from Cape Verd towards the curvature or bay of Africa, which they call Fernando Poo, viz. from west to east, contrary to the general motion. And such is the force of this current, that when ships approach too near the shore, it carries them violently towards that bay, and deceives the mariners in their reckoning.

"There is a great variety of shifting currents, which do not last, but return at certain periods; and these do, most of them, depend upon, and follow the anniversary winds or monsoons, which by blowing in one place, may cause a current in another."

At Java, in the Streights of Sunda, when the monsoons blow from the west, viz. in the month of May, the currents set to the eastward, contrary to the general motion.

Also between the islands of Celebes and Madura, when the western monsoons set in, viz. in December, January and February, when the winds blow from the N.W. or between the north and west, the currents set to the S.E. or between the south and east.

At Ceylon, from the middle of March to October, the currents set to the southward, and in the other parts of the year to the northward; because at this time the southern monsoons blow, and at the other, the northern.

Between Cochin China and Malacca, when the western monsoons blow, viz. from April to August, the currents set eastward against the general motion, but the rest of the year set westward; the monsoon conspiring with the general motion. They run so strongly in these seas, that unexperienced sailors mistake them for waves that beat upon the rocks, known by the name of breakers.

So for some months after the 15th of February the currents set from the Maldives towards India on the east, against the general motion of the sea.

On the shore of China and Cambodia, in the months of October, November, and December, the currents set to the N.W. and from January to the S.W. when they run with such a rapidity of motion about the shoals of Parcel, as to seem swifter than the flight of an arrow.

At Pulo Condore, upon the coast of Cambodia, though the monsoons are shifting, yet the currents set strongly towards the east, even when they blow to a contrary point.

Along the coasts of the Bay of Bengal, as far as the Cape Romania, at the extreme point of Malacca, the current runs southward in November and December.

When the monsoons blow from China to Malacca, the sea runs swiftly from Pulo Cambi to Pulo Condore, on the coast of Cambodia.

In the Bay of Sans Bras, not far from the Cape of Good Hope, there is a current particularly remarkable, where the sea runs from east to west to the landward; and this more vehemently as it becomes opposed by the winds from a contrary direction: the cause is undoubtedly owing to some adjacent shore, which is higher than this.

These currents constantly follow the winds, and set to the same point with the monsoon, or trade-wind, at sea. See the article Monsoon.

Currents, as they relate to navigation, may carry a ship forward more swiftly, or retard her course, or way, according to the direction or setting in of the current.

Under the equator, where the motion of the earth is greatest, the currents are so violent, that they carry vessels very speedily from Africa to America; but absolutely prevent their return the same way: so that ships are forced to run as far as the 40th or 45th degree of north latitude, to fall into the return of the current, to find a passage to Europe.

It is shewn by Governor Pownal, that this current performs a continual circulation, setting out from the Guinea coast in Africa, for example, from thence crossing straight over the Atlantic ocean, and so setting into the gulf of Mexico by the south side of it; then sweeping round by the bottom of the gulf, it issues out by the north side of it, and thence takes a direction north-easterly along the coast of North America, till it arrives near Newfoundland, where it is turned by a roundling motion backward across the Atlantic again, upon the coasts of Europe, and from thence southward again to the coast of Africa, from whence it set out.

In the Streights of Gibraltar, the currents set in by the south side, sweep along the coast of Africa to Egypt, by Palestine, and return by the northern side of the Streights.

The great violence and danger of the sea in the Streights of Magellan, is attributed to two contrary currents setting in, one from the south, the other from the north, sea.

The currents in the Streights of Gibraltar almost constantly drive to the eastward, and carry ships into the Mediterranean: they are also usually found to drive the same way into St. George's-channel.

The setting, or progressive motion of the current, may be either quite down to the bottom, or to a certain determinate depth.

As the knowledge of the direction and velocity of currents is a very material article in navigation; it is highly necessary to discover both, in order to ascertain the ship's situation and course, with as much accuracy as possible. This some do by the ripplings of the water, and by the driving of the froth along the shore, when in sight of it; but the most successful method which has been hitherto attempted by mariners, is as follows:—A common iron pot, which may contain four or five gallons, is suspended by a small rope fastened to its ears or handles, so as to hang directly upright, as when placed upon the fire. This rope, which may be from 80 to 100 fathoms in length, being prepared for the experiment, is coiled in the boat, which is hoisted out of the ship at a proper opportunity, when there is little or no wind to ruffle the surface of the sea. The pot being then thrown overboard into the water, and immediately sinking, the line is slackened till about 70 or 80 fathoms run out, after which the line is fastened to the boat's stem, by which she is accordingly restrained, and rides as at anchor. The velocity of the current is then easily tried by the *log* and half minute glass, the usual method of discovering the rate of a ship's sailing at sea. (See the article CALM:) The course of the stream is next obtained by means of the compass provided for this operation.

This shews whether there be any current or none; and if any, which way it sets, and at what rate it drives: observing, however, to add something to the drift, for the boat's drift, for though she appear to stand still, yet, in reality, she is found to move. This addition, experience has thus determined; if the line she rides by be 60 fathoms, a third part of the drift is to be added, if 80 fathoms a fourth, if 100 a fifth.

If a ship sail along the direction of a current, it is evident that the velocity of the current must be added to that of the vessel: but if her course be directly against the current, it must be subtracted: and if she sail athwart it, her motion will be compounded, and her velocity augmented or retarded, according to the angle of her direction, with that of the current; that is, she will proceed in the diagonal of the parallelogram formed according to the two lines of direction, and will describe or pass over that diagonal in the same time, in which she would have described either of the sides, by the separate velocities.

For suppose A B D C be a parallelogram, the diagonal of which is A D. Now if the wind alone would drive the ship from A to B, in the same time as the current alone would drive it from A to C: then, as the wind neither helps nor hinders the ship from coming towards the line C D, the current will bring it there in the same time as if the wind did not act:—And as the current neither helps nor hinders the ship from coming towards the line B D, the wind will bring it there in the same time as if the current did not act; therefore, the ship must, at the end of that time, be found in both those lines, that is, in their meeting D: consequently the ship must have passed from A to D in the diagonal A D.

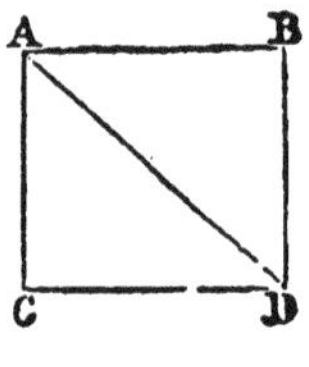

Hence it is plain, when the ship's course is directly against the current, 1. If the velocity of the current be less than that of the ship, then the ship will advance so much as is the difference of these velocities. 2. If the velocity of the current be more than that of the ship, then will the ship fall as much *astern* as is the difference of these velocities. 3. If the velocity of the current be equal to that of the ship, then will the ship remain stationary, the one velocity destroying the other.

If the current thwarts the course of a ship, it not only diminishes or increases her velocity, but gives her a new direction compounded of the course she steers, and the setting of the current.

It is a circumstance well known to seamen, that ships, in coming from the Atlantic, and steering a course for the British Channel, in a parallel somewhat to the south of the Scilly islands; do, notwithstanding, often find themselves to the north of those islands: or, in other words, in the mouth of St. George's, or of the Bristol Channel. This extraordinary error has passed for the effects, either of bad steerage, bad observation of latitude, or the in-draught of the Bristol Channel: but none of these account for it satisfactorily; because, admitting that at times there may be an in-draught, it cannot be supposed to extend to Scilly; and the case has happened in weather the most favourable for navigation, and for taking observations.

The consequences of this deviation from the intended track have very often been fatal; particularly in the loss of the Nancy packet, in our own times; and that of Sir Cloudesley Shovel, and others of his fleet, at the beginning of the eighteenth century. Numbers of cases, equally melancholy, but of less celebrity, have occurred; and many others, in which the danger has been imminent but not fatal, have scarcely reached the public ear.

All of these have been referred to accident; and, therefore, no attempt seems to have been made to investigate the cause of them. Major Rennel is, however, of opinion, that they are to be imputed solely to a current: and he, therefore, endeavours to investigate both that, and its effects, in order that seamen may be apprized of the times, when they are particularly to expect it, in any considerable degree of strength, for then only it is likely to occasion mischief; the current that prevails at ordinary times being, probably, too weak to produce an error in the reckoning, equal to the difference of parallel, between the south part of Scilly, and the track in which a commander, prudent in his measures, but unsuspicious of a current, would choose to sail. The original cause of this current, he says, is, the prevalence of westerly winds in the Atlantic, which impel the waters along the north coast of Spain, and accumulate them in the Bay of Biscay; whence they are projected along the north coast of France, in a direction north-west by west, to the west of Scilly and Ireland.

The Major assigns strong reasons for the existence of this current between Ushant and Ireland, in a chart of the tracts of the Hector and Atlas, East India ships, in 1778 and 1787. His observations on the subject in general, require a continued inspection of his chart; and, consequently, cannot be here inserted, He observes, 1st. If a ship crosses it obliquely, that is, in an east by south, or more southerly direction, she will continue much longer in it, and of course be more affected by it, than if she crossed it more directly. The same consequence will happen, if she crosses it with light winds. 2dly. A good observation of latitude, at noon, would be thought a sufficient warrant for running eastward, during a long night; yet as it may be possible to remain in the current long enough to be carried from a parallel, which may be deemed a very safe one, to that of the rocks of Scilly, it would appear prudent, after experiencing a continuance of strong westerly winds in the Atlantic, and approaching the Channel with light southerly winds, either to make Ushant in time of peace, or at all events to keep in the parallel of 48° 45′, at the highest. 3dly, Ships bound to the westward, from the mouth of the Channel, with the wind in the south-west quarter, should prefer the larboard-tack.

4thly. He recommends the sending a vessel with time-keepers on board, to examine the soundings between the parallels of Scilly and Ushant, from the meridian of the Lizard-point as far west as the moderate depths extend.

Under CURRENTS, (*contre courants*, Fr.) are distinct from the upper or apparent, and in different places set or drive a contrary way. Dr. Thomas Smith makes it highly probable, that in the Downs, in the Streights of Gibraltar, &c. there is an *under-current*, whereby as much water is carried out, as is brought in by the upper-current. This was confirmed by an experiment made in the Baltic Sound, by the seamen on board one of the King's frigates: they went with the pinnace into the midstream, and were carried violently by the current. Soon after that, they sunk a basket with a large cannon bullet, to a certain depth of water, which gave check to the boat's motion; and sinking it still lower and lower, the boat was driven a-head to the windward, against the upper-current, the current aloft not being above four or five fathoms deep, and the lower the basket was let down, the stronger the *under-current* was found.

From this principle, it is easy to account for that continual in-draught of water out of the Atlantic into the Mediterranean, through the Streights of Gibraltar; a passage about twenty miles broad: yet, without any sensible rising of the water along the coasts of Barbary, &c. or any overflowing of the lands, which there lie very low.

Dr. Halley solves the currents setting in at the Streights without overflowing the banks, by the great evaporation, without supposing any *under-current*.

CURSOR, in mathematical instruments, is any small piece that slides, as the brass pointed screw on beam-compasses, which may be moved along the beam for the striking of great or less circles, &c.

CURTATE DISTANCE, in astronomy, is the distance of a planet from the sun to that point where a perpendicular let fall from the planet meets with the ecliptic.

CUSP, in astronomy, is used to express the points or horns of the moon, or other luminary.

CUSTOM, (*droits de la douane*, Fr.) is the tribute or toll paid by the merchants to the King, for goods exported or imported: they are otherwise called duties.

It was enacted anno 6 Edw. III. that no new custom could be levied, nor old ones increased, but by authority of parliament, which was afterwards confirmed by 16 Car. I. c. 8.

The duty of tonnage and poundage was granted to Charles II. for his life, and to his two immediate successors; but now by three several statutes, 9 Ann. c. 6. 1 Geo. I. c. 12, and 3 Geo. I. c. 7, it is made perpetual, and mortgaged for the debt of the public. The customs imposed by parliament are chiefly contained in two books of rates, to which many subsequent additions have been made.

Aliens pay a larger proportion than natural subjects.

In case goods and merchandise are brought to a port, and part of the goods are sold there, but never landed, they must pay the customs. Ships outward bound, and coming from beyond sea, having goods and merchandise on board, are to be entered at the custom-house, and the customs paid or agreed for under penalties or forfeiture of the goods, one moiety to the King, and another to the seizer, &c. 13 and 14 Car. II. Officers of the customs may search ships; and having writ of assistance search houses. By other statutes, foreign goods taken in at sea by a coasting vessel, shall be forfeited and treble value: and for prevention of clandestine running of goods, if any foreign brandy, &c. is imported in vessels under 40 tons, the importer shall forfeit the vessel and brandy. Run goods concealed or offered for sale, are liable to forfeiture and treble value, 8 and 11 Geo. I. When three persons are assembled and armed with fire-arms, &c. to be assisting in the running of goods, they shall be adjudged guilty of felony.

And two or more found in company within five miles of the sea coast, with any horses, carts, &c. on which are put above six pounds of tea, or five gallons of brandy, or other foreign goods of 30*l*. value, landed without entry, and not having permits, who shall carry any offensive weapons, &c. or assault any officer of the customs, shall be deemed runners of goods, and treated as felons, and the goods shall be seized and forfeited. If any person offers tea, brandy, &c. to sale without a permit, the persons to whom it is offered, may seize and carry it to the next warehouse belonging to the customs or excise, and be entitled to the third part of the produce, on condemnation.

And persons offering any bribe to officers of the customs, to connive at the running of goods, are liable to a forfeiture of 50*l*. Obstructing such officers in entering and searching ships, incurs a forfeiture of 100*l*. and if the officers are wounded, or beaten on board any ship, the offenders incur the penalty of transportation, &c. 9 Geo. II.

CUSTOM-*House*, (*douane*, Fr.) an office established on the frontiers of a state, or in some chief city or port, for the reception of customs and duties of importation and exportation, imposed on merchandises, by the authority of the sovereign, and regulated by writs, or books of rates. There are several *custom-houses* in the different ports of England: the most considerable is that of London.

It is under the direction of commissioners, appointed by patent, who have the charge and manage ment of the customs in the several ports of England.

Other officers are, a secretary, solicitor, receiver general, comptroller-general, surveyor-general, &c. all holding their places by patents, with other inferior officers, appointed by warrant from the board of the treasury.

These officers shall have no ship of their own, nor use merchandise, factorage, &c. under penalty of 40*l*.

They are also prohibited to trade in any exciseable liquor, on pain of 50*l*. and forfeiture of office: for taking a bribe, they shall forfeit 100*l*. and 50*l*. for making collusive seizures.

The new custom-house of Dublin is reckoned a most convenient, as well as elegant building.

CUT A FEATHER, a term used by sailors on a ship's having too broad a bow, when they say she will not *cut a feather*, meaning that she will not pass through the water so swift, as to make it foam or froth, which a ship with a narrow bow would do.

To CUT *and Run*, is to cut the cable and make sail instantly, without waiting to weigh anchor.

If a ship shall *cut*, or slip, or part a cable, the captain is to use his endeavours to recover the anchor and cable that are lost; but if the ship put to sea, the senior officer left in the road is to use his utmost endeavours to recover them; but neither he, nor the captain of the ship to which such anchor belongs, is to hire any vessel or boat for the purpose of recovering it, if it can be done by any of his Majesty's ships present.

If an anchor, &c. be left in a roadstead or harbour, the captain is to report the same to the commander in chief, or to the Navy Board, mentioning the bearings and distances of the nearest points of land from the spot where the anchor may lie. See the article *To slip the* CABLE.

To CUT *the Sail*, is to unfurl it, and to let it fall down.

CUTTER, (*cutter*, Fr.) a small vessel commonly navigated in the channel of England, furnished with one mast and a straight running bowsprit that can be run in on the deck occasionally; except which, and the largeness of the sails, they are rigged much like sloops. Many of these vessels are used on an illicit trade, and others employed by the government to seize them; the latter of which are either under the direction of the Admiralty or Custom-house. See a representation of a cutter of this sort, in a plate referred to from the article VESSEL.

CUTTER is also a small boat used by ships of war. See the article BOAT.

CUTTING-DOWN LINE, is a curved line used by shipwrights in the delineation of ships; it determines the thickness of all the floor-timbers, and likewise the height of the *dead-wood afore* and *abaft*. It is limited in the middle of the ship by the thickness of the floor-timber, and abaft by the breadth of the *kelson;* and must be carried up so high upon the stern, as to leave sufficient substance for the brecches of the rising timbers.

CUT-WATER, or KNEE OF THE HEAD, (*guibre*, Fr.) is the foremost part of a ship's prow, formed of an assemblage of several pieces of timber, to render it broad at the upper-part, where it projects forward from the stem to open the column of water as the ship sails along, and also to make her keep to windward better, when she is close-hauled. See the article STEM.

CYCLE, in chronology, a certain period or series of years, which regularly proceed from the first to the last, and then return again to the first, and circulate perpetually.

CYCLE *of the Moon*, or *Lunar Cycle*, called also the *Golden Number*, from its excellent use in the calendar, is a period of nineteen years, after which the new and full moons return on the same days of the months, only one hour and twenty-eight minutes sooner; so that, on whatever days the new and full moons fall this year, they will happen nineteen years hence on the same days of the months, except when a centesimal common year falls within the cycle, which will move the new and full moons a day later in the calendar than otherwise they would have fallen; insomuch that a new moon which fell, before the centesimal year, suppose on March 19, will fall nineteen years afterwards on March 20.

The number of years elapsed in this cycle is called the prime, from its use in pointing out the day of the new moon, *primum lunæ;* and the golden number, as deserving to be written in letters of gold.

The golden numbers are those placed in the first column of the calendar, between March 21 and April 18, both inclusive, to denote the days upon which those full moons fall, which happen upon, or next after March 21, in those years of which they are respectively the golden numbers.

For finding the golden number, add one to the current year of our Lord, because one year of this cycle was elapsed before the christian æra began, and divide by nineteen, the remainder is the current year of this cycle, or the golden number; but if nothing remains, it shews that it is the last year of the cycle, and consequently the golden number is 19.

CYCLE *of the Sun*, or *Solar Cycle*, is a period or revolution of 28 years, which contain all the possible combinations of the dominical letters, in respect to their successive order, as pointing out the common years and leap-years; so that after the expivation of the cycle, the days of the month return in the same order to the same days of the week, throughout the next cycle; except that upon every centesimal year which is not a leap-year, the letters must always be removed one place forward, to make them answer to the years of the cycle. For instance, if the year 1800 was a leap-year, as every centesimal year is in the Julian account, the dominical letters would be E D, and C would be the dominical letter of the next year; but as it is a common year in the Gregorian account, D is the dominical letter of 1801, which answers to the eighteenth of the cycle, C to the nineteenth, &c. until the next centesimal year.

To find the cycle of the sun for any given year of the christian æra, add 9 to the current year of Christ, because the cycle commenced nine years before the christian æra, and divide the sum by 28, the quotient will shew the number of cycles which have revolved since the beginning of that in which the christian æra commenced; and the remainder, if any, shews the current year of the cycle: but if there be no remainder, it will be the twenty-eighth year of the cycle.

CYCLE of the Roman indication, is a period of 15 years, in use among the Romans, commencing from the third year before Christ. This cycle has no connexion with the celestial motions; but was instituted, according to Baronius, by Constantine, who, having reduced the time which the Romans were obliged to serve to fifteen years, was consequently obliged, every fifteen years, to impose an extraordinary tax for the payment of those who were discharged; and hence arose this cycle.

CYGNUS, the Swan, in astronomy, a northern constellation, between Lyra and Cepheus, and, according to the British catalogue, contains 81 stars.

CYLINDER, in geometry, a solid body having two equal circular ends, and every plane section parallel to the ends a circle equal to them also. The cylinder may be supposed to be generated by the rotation of a parallelogram.

D.

DAGGER, in ship-building, is a piece of timber that crosses all the puppets of the bulgeways to keep them together; the plank that secures the heads of the puppets, is called the *dagger-plank*.

DAGGER-*Knees*, *(courbes un peu obliques*, Fr.) sometimes called *Lodging-Knees*, in ship-building, are certain pieces, whose side-arms cast down and bolt through the clamp. See the article KNEE.

DAILY, in astronomy. See the article DIURNAL.

DALE. See the article PUMP-DALE.

DAM, *(batardeau*, Fr.) a mole or bank where water is confined.

Plate IX. Fig. 6. DAVIT, *(davier minot*, Fr.) is a piece of timber, represented by *a a*, plate IX. fig. 6, and used as a crane to hoist the flooks of the anchor to the top of the *bow*, without injuring the planks of the ship's side as it ascends; an operation which, by seamen, is called fishing the anchor. The anchors being situated on both the bows, the davit may be occasionally shifted to either side of the ship, according to the position of that anchor on which it is to be employed. The inner end of the davit is fixed in the fore channel, *b*, and secured by cants, to prevent the heel from sliding out of its place. The bolt, exhibited by fig. 7, is the fish-tackle eye-bolt, which is driven in the sheer-streak, abaft the step of the side. Upon the outer-end of the davit is hung a large block *c*, through which a strong rope traverses, called the fish-pendent, *d*, to whose foremost end is fitted a large iron hook, *e*, and to its after end a tackle or complication of pulleys *f*; the former of which is called the fish-hook, and the latter the fish-tackle. (Fig. 7.)

The anchor being first *catted*, the fish-hook is fastened on its flooks; and is, by means of the fish-pendent and tackle drawn up sufficiently high upon the bow to be made fast by the *shank-painter*. Thus the davit, according to the sea-phrase, is employed to fish the anchor.

There is also a davit of a smaller kind, occasionally fixed in the long-boat, and, with the assistance of a small windlass, used to weigh the anchor by the rope, &c.

DAY, the time which elapses from the rising to the setting of the sun; the interval of time which elapses from the setting to the rising of the sun, or the time he continues below the horison, is called the night.

DAY, *Natural*, implies the time of a whole apparent revolution of the sun round the earth; or rather the time in which the earth makes a rotation on its axis.

DAY, *Astronomical*, is the interval of time which passes from 12 o'clock at noon till the next succeeding noon, and begins when the sun's centre is on the meridian of that place.

It is divided into 24 hours; the first twelve are distinguished by mariners by the mark P.M. signifying *post meridiem*, or afternoon; and the latter twelve are marked A.M. signifying *(ante meridiem*, or before noon.

But astronomers generally reckon through the 24 hours, from noon to noon; and they reckon their morning hours in succession from 12, or midnight, to 24 hours.

Thus, 9 o'clock in the morning of April 0th, is, by astronomers called April the 29th, at 21 hours.

An astronomical day is somewhat greater than a complete revolution of the heavens, which forms a sidereal day. A sidereal day is less than the solar day, for it is measured by 360°; whereas the mean solar day is measured by 360° 59′ 8″ nearly.

DAY'S-*Work*, *(le point, ou la route estimée d'un vaisseau d'un midi à l'autre*, Fr.) in navigation, the reckoning or account of a ship's course, and distance run, during 24 hours, or from noon to noon, according to the rules of trigonometry. See the article DEAD-RECKONING.

DEAD DOORS, *(portes de rechange pour la galerie*, Fr.) in ship-building, are those fitted to the outside of the quarter-gallery-doors, in case the quarter-gallery should be carried away.

DEAD-*Eye*, (*cap de mouton*, Fr.) a sort of round flattish wooden block, see fig. 3, plate III. It is usually encircled with a rope, or with an iron band, plate IX. fig. 8, *g*, and pierced with three holes Fig. 8.
through the flat part, in order to receive a rope called the *lanyard h*, which corresponding with three holes in another dead-eye *i*, creates a purchase employed for various uses, but chiefly to extend the *shrouds* and *stays*, otherwise called the standing-rigging.

In order to form this purchase, one of the dead-eyes is fastened in the upper-link of each *chain* on the ship's side, which is made round to receive and encompass the hollowed outer-edge of the dead-eye. After this the lanyard is passed alternately through the holes in the upper and lower-dead-eyes, till it becomes six-fold; and is then drawn tight by the application of mechanical powers.

The dead-eye used for the stays, *(moques*, Fr.) has only one hole; which, however, is large enough to receive ten or twelve turns of the lanyard: it is generally termed a *heart*, and is expressed by fig. 5, plate III.

Crow-feet DEAD-*Eyes*, (*moques de trelingage*, Fr.) are long cylindrical blocks, with a number of small holes in them, to receive the legs or lines of which the crow-foot, fig. 5, plate IX. is composed. It is generally termed an *euphroe*, and represented by fig. 4, plate III.

DEAD-*Flat*, in ship-building, is the name of a midship-bend.

DEAD-*Lights*, (*faux sabords*, Fr.) are strong wooden ports, made exactly to fit the cabin-windows, in which they are fixed on the approach of a storm, the glass-frames being taken out, which would otherwise be shattered by the surges, and suffer great quantities of water to pour into the ship.

DEAD-*Reckoning*, (*route estimée*, Fr.) in navigation, the judgment or estimation which is made of the place where a ship is situated, without any observation of the heavenly bodies. It is discovered by keeping an account of the distance she has run by the *log*, and of her course steered by the *compass*; and by rectifying these data by the usual allowances for *drift*, *lee-way*, &c. according to the ship's known trim. This reckoning, however, is always to be corrected, as often as any good observation of the sun can be obtained.

DEAD-*Rising*, or RISING-*Line of the Floor*, (*la lisse des facons*, Fr.) those parts of a ship's floor, or bottom, throughout her whole length, where the floor-timber is terminated upon the lower futtock.

DEAD-*Ropes*, are those which do not run in any block.

DEAD-*Water*, (*remoux du sillage*, Fr.) the eddy of water, which appears like little whirl-pools, closing in with the ship's stern as she sails through it.

DEAD-*Wind*, (*vent de bout*, Fr.) the wind right against the ship, or that blowing from the very point to which she wants to go.

DEAD-*Wood*, (*courbes de remplissage*, Fr.) implies certain blocks of timber laid upon the keel, particularly at the extremities afore and abaft, where these pieces are placed upon each other to a considerable height, because the ship is there so narrow as not to admit of the two half timbers, which are therefore scored into this *dead-wood*, where the angle of the floor-timbers gradually diminishes, as approaching the stem and stern-post.

In the fore-part of the ship, the *dead-wood* generally extends from the stemson, upon which it is scarfed to the loof-frame; and in the after-end from the stern-post, where it is confined by the knee, to the after-balance-frame. It is connected to the keel by strong spike-nails.

The *dead-wood* afore and abaft is equal in depth to two-thirds of the depth of the keel, and as broad as can be procured, so as not to exceed the breadth of the keel.

DEAD-*Works*, (*œuvres mortes*, Fr.) a name given to all that part of a ship which is above the water when she is laden. See the article UPPER-WORK.

To DEADEN A SHIP'S WAY, is to impede her velocity through the water.

DEALS, (*bordages, ou planches de sapin*, Fr.) are pieces of timber in form and thickness exactly similar to plank; but the term is applied to fir timber only.

DECKER relates to the rate of a ship of force, as a two or three decker, (*vaisseau à deux ou à trois ponts*, Fr.) *i. e.* carrying two or three entire tiers or ranges of cannon.

DECKS, (*ponts d'un vaisseau*, Fr. *decken*, Dan. to cover) the planked floors of a ship, which connect the sides together, and serve as different platforms to support the artillery and lodge the men, as also to preserve the cargo from the sea and rain.

As all ships are broader at the lower-deck than on the next above it, and, as the cannon thereof are always heaviest, it is necessary that the frame of it should be much stronger than that of the others; and, for the same reason, the second or middle-deck ought to be stronger than the upper-deck or fore-castle.

Ships of the first and second rates are furnished with three whole decks, reaching from the stem to the stern, besides a fore-castle and a quarter-deck; the former extending aft from the stem to the belfry, and the latter forward from the stern to the gang-way, a vacancy being left in the middle, which opens to the upper-deck, and forms what is called the *waist*.

The inferior ships of the line of battle are equipped with two decks and a half; and frigates, sloops, &c. with one gun-deck and half, with a spar-deck below, to lodge the crew.

The decks are sustained by the beams, the clamps, the water-ways, the carlings, the ledges, the knees, and two rows of stanchions, &c. See those articles.

That the figure of a deck, together with its corresponding parts, may be more clearly understood, we have exhibited a plan of the lower-deck of a 74-gun-ship in plate IX. fig. 9. And, as both sides of Fig. 9
the deck are exactly similar, the pieces by which it is supported appear on one side, and on the other side the planks or floor of which it is composed, as laid upon those pieces.

Explanation of the principal Figures represented in the Deck.

A, the principal, or main hatch-way.

B, the stern-post.

C, the stem.

D, the beams, composed of three pieces, as exhibited by D, in one of which the dotted lines shew the arrangement of one of the beams under the other side of the deck.

E E, the plate knee, which is represented in fig. 10. Fig. 10.

F F, the horizontal ears with bolts, which fasten the beam to the side, as shewn in fig. Z.

G, the carlings, ranging fore and aft, from one beam to another.

H, the ledges let into the carlings.

I, the pump-dale, being a large wooden tube which returns the water from the pumps into the sea.

K, the prong or forked beams; being curved pieces of timber serving as half-beams to support the decks, where a whole beam cannot be placed on account of the hatch-ways.

L, the wing-transom, which is bolted by the middle to the stern-post, and the ends let into fashion-pieces.

M, the manger-boards, which enclose the manger, and prevent the water that enters at the hawse-holes from running aft between decks.

N N, the fore hatch-way.
O O, the after hatch-way.
P, the step for the jear capstan.
P *p*, the step for the main capstan.
Q, the wing-transom-knee.
R, the chock under the beam, fig. 10.
S, the lower-deck hook.
T T, the cistern for the chain-pumps.
V, the breadth and thickness of the timbers at the height of the gun-deck.
U, the scuttles leading to the gunner's store-room.
u, the scuttle leading to the bread-room.
W, the station of the fore-mast.
X, the station of the main-mast.
Y, the mizen step.
Z, the ring-bolts of the decks, used to retain the cannon whilst charging.
a a, the ring-bolts of the sides, whereon the tackles are hooked that secure the cannon at sea.
c a a d, the water-ways, through which the scupper-holes are pierced, to carry the water off from the deck into the sea.
b b, plan of the foremost and aftmost cable-bits, with their cross-pieces *g g*, and their standards *e e*.

Thus we have represented, on one side, all the pieces which sustain the deck with its cannon; and, on the other side, the deck itself, with a tier of 32-pounders planted in battery thereon. In order also to shew the use of the breeching and train-tackle, one of the guns is drawn in, as ready for charging. See the articles Breeching and Cannon.

The number of beams by which the decks of ships are supported, is often very different, according to the practice of different countries; the strength of the timber of which the beams are framed; and the service for which the ships are calculated.

The deck which contains the train of a fire-ship is furnished with an equipage peculiar to itself; a description of which will be found under the article Fire-Ship.

Lower Gun-Deck, *(le premier pont dans les vaisseaux de guerre*, Fr.) in first and second rate ships, is denominated the first deck, whose frame is made broader and stronger than either of the other decks: it is laid next the orlop, and sustains the heaviest tiers of guns, as 32-pounders.

Middle-Deck, *(le second pont, dans les vaisseaux à trois ponts*, Fr.) is the second deck, and lays between the lower and upper-deck: it is slighter in its construction than the lower-deck, and sustains the second tiers of guns, as 18-pounders.

Upper-Deck, or *Main*-Deck, *(le troisième pont dans les vaisseaux à trois ponts*, Fr.) is the third deck, and sustains the third tiers of guns, as 12-pounders: it is constructed much slighter than the middle-deck, and, in third-rate ships, is called the second-deck.

Quarter-Deck, *(le gaillard d'arrière*, Fr.) is that above the upper-deck, reaching forward from the stern to the gangway, and supports the carronades, &c.

Gun-Deck, *(le pont dans les frégates, et le premier pont dans les vaisseaux de guerre*, Fr.) in frigates, sloops of war, brigantines, gun-brigs, and cutters, is the main or upper-deck, whereon the guns are placed in battery.

Flush-Deck, or Deck-*Flush fore and aft*, *(pont entier, sans ravalement, ni interruptions*, Fr.) in corvettes, &c. implies a continued floor laid from stem to stern, upon one line, without any stops or intervals.

Half-Deck, *(espace compris entre le mât d'artimon et le grand mât dessous le guillard*, Fr.) is the under part of the quarter-deck of a ship of war, contained between the foremost bulk head of the cabin or ward-room, and the break of the quarter-deck. In the colliers of Northumberland, the steerage itself is called the half-deck, and is usually the habitation of the ship's crew.

Spar-Deck, *(faux-pont, dans les frégates*, Fr.) in frigates and men of war converted into troop ships, is that continued in a straight line from the quarter-deck to the forecastle, and appropriated for the reception of spars, &c. under which the hammocks are stowed away, and the crew sleep.

For Cambered-Deck, Fore-Castle, Orlop, and Poop, see each article under its respective head.

DECLINATION, in astronomy, is the distance of any celestial object from the equinoctial, either northward or southward. It is either true or apparent, according as the real or apparent place of the object is considered.

The declination of any point S. is an arc of the meridian S.E. contained between the given point, and the equinoctial E Q.

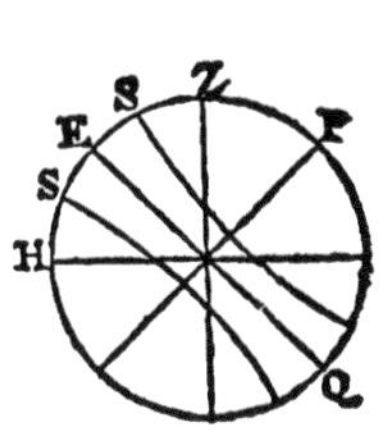

The declination of a star, &c. is found by knowing or observing the latitude of the place, *i. e.* the height of the pole, and then the meridian altitude of the star, &c. hence the difference between the co-latitude, and the altitude of the star, &c. is the declination, viz. the difference between E H the co-latitude, and S H the altitude, is E S the declination.

To find the Sun's Declination at any time: having his place in the ecliptic given, the rule is, as radius is to the sine of the sun's longitude, so is the sine of the greatest declination, or obliquity of the ecliptic, to the sine and his present declination sought.

By comparing ancient observations with the modern, it appears that the declination of the fixed stars is variable; and that differently in different stars; for in some it increases, and in others decreases, and that in different quantities.

Declination *of the Compass*, or *Needle*, is its variation from the true meridian. See the article Variation.

DECOY, *(leurre, piége*, Fr.) a stratagem employed by a small ship of war to betray a vessel of inferior force into an incautious pursuit, till she has drawn her within the range of her cannon, or what is called within gun-shot.

It is usually performed by painting the stern and sides in such a manner as to disguise the ship, and represent her either much smaller, and of inferior force, or as a friend to the hostile vessel, which she endeavours to ensnare, by assuming the emblems and ornaments of the nation to which the stranger is supposed to belong. When she has thus provoked the adversary to chase, in hopes of acquiring a prize, (in the former case) she continues the *decoy* by spreading a great sail, as endeavouring to escape, at the same time that her course is considerably retarded by an artful alteration of her *trim* till the enemy approaches.

Sometimes the art of decoying is used by a single ship, to induce an enemy's squadron to follow her into the view of her own fleet.

It is also performed to elude the chace of a ship of superior force in a dark night, by throwing out a lighted cask of pitch into the sea, which will burn for a considerable time, and misguide the enemy. Immediately after the cask is thrown out, the ship changes her course, and may easily escape if at any tolerable distance from the foe.

DEEP, (*profond*, Fr.) on the coast of Germany to the northward of Friesland, is of the same import as Gulf on the coasts of France, Spain, Italy, &c.

Deep *Sea-Lead.* See the article Lead.

Deep *Sea-Line.* See the article Line.

Deep *Sea-Line Block.* See the article Block.

Deep-*Waisted*, (*haut accastillé*, Fr.) the distinguishing fabric of a ship's decks, when the quarter-deck and fore-castle are elevated from four to six feet above the level of the upper or main-deck, so as to leave a vacant space, called the waist, on the middle of the upper-deck.

DEGREE, (*degré*, Fr.) in geometry, is a division of a circle, including a three hundred and sixtieth part of its circumference. Every circle is supposed to be divided into 360 parts, called degrees, and each degree divided into 60 other parts, called minutes; each of these minutes being again divided into sixty seconds, each second into thirds, and each third into fourths, and so on.

By this means no more degrees or parts are reckoned in the greatest circle than in the least; and, therefore, if the same angle at the centre be subtended by two concentrical arcs, as many degrees are counted in the less as in the greater, as they have the same proportion to their whole peripheries.

Degree *of Latitude.* See the article Latitude.

Degree *of Longitude.* See the article Longitude.

DELPHINUS, the Dolphin, in astronomy, a constellation of the northern hemisphere, whose stars in the Britannic catalogue are 18.

DEMURRAGE, (*starie*, Fr.) is an allowance made to the master of a ship by the merchants, for having detained him longer in port than the time previously appointed and agreed for his departure. The rate of this allowance is generally settled in the charter-party. It is now firmly established, that the claim of demurrage ceases as soon as the ship is cleared out, and ready for sailing. Jameson *v.* Lawrie, House of Lords, Nov. 10, 1796.

DENEB, is an Arabic term, signifying tail, used by astronomers to denote several fixed stars.

Deneb *Adigege*, in astronomy, is a bright fixed star of the second magnitude, situated in the Swan's tail. Its right ascension adapted to the beginning of the year 1812, is 308° 45′ 22″20; annual variation in right ascension 30″.51. Its declination 44° 36′ 50″ North, and annual variation in declination + 12″.6.

Deneb *Elecet*, or Denebola, in astronomy, is a bright fixed star of the second magnitude in the Lion's tail.

For the beginning of the year 1812, its

Right Ascension is . . .	174°	51°	55″,35
Annual Variation in A.R.			45,93
Declination	15	37	26 N.
And Annual Variation in Decl.			— 22,0

DEPARTURE, in navigation, (*point de départ* Fr.) is the easting or westing of a ship in respect of the meridian she departed or sailed from, and is reckoned on a parallel of latitude: so that if a ship sail due north or south, she makes no departure, and her distance and difference of latitude are the same: but, if she sails due east or west, she makes no difference of latitude, and her departure and distance is the same: when a ship sails in any other direction, she makes both difference of latitude and departure, and these, with the distance, form a right-angled triangle, the hypothenuse of which is the distance sailed, the perpendicular is the difference of latitude; the base the departure, the angle opposite the base is the course, and that opposite the perpendicular the complement of the course: hence any two of these parts being given, the rest may be found by plane trigonometry: thus,

As radius, is to the distance,
So is sine of the course,
To the departure from the meridian.

DEPRESSION of the Pole. When a person sails or travels towards the equator, he is said to depress the pole; because as many degrees as he approaches nearer the equator, so many degrees will the pole be nearer the horizon. This phenomenon arises from the spherical figure of the earth.

Depression *of a Star*, or *of the Sun*, is its distance below the horizon; and is measured by an arc of a vertical circle, intercepted between the horizon and the place of the star.

Depression *of the Visible Horizon*, or *Dip of the Horizon*, (*depression de l'horizon*, Fr.) denotes its sinking or dipping below the true horizontal plane, by the observer's eye being raised above the surface of the sea; in consequence of which, the observed altitude of an object is by so much too great.

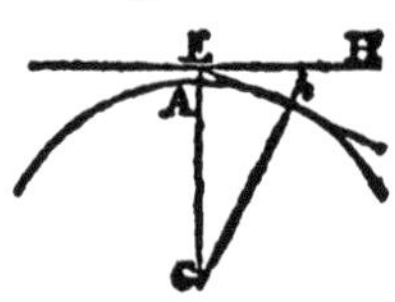

Thus, the eye being at E, the height A E above the surface of the earth, whose centre is C; then E H is the real horizon, and E h the visible one, below the former by the angle H E h, by reason of the elevation A E of the eye.

To compute the Depression or Dip of the Horizon.—

In the right-angled triangle C E h, are given C h, the earth's radius=21000000 feet, and the hypothenuse C E= the radius increased by the height A E of the eye; to find the angle C, which is = the angle H E h, or depression sought; viz.

As C h : C E : : radius : sec. $\angle$ C,
Or As C E : C h : : radius : cosin. $\angle$ C.

By either of these canons are computed the numbers in the following table: which shews the depression or dip of the horizon at sea for different heights of the eye, from 1 foot to 100 feet.

Height of the Eye.	Dip of the Horizon.		Height of the Eye.	Dip of the Horizon.		Height of the Eye.	Dip of the Horizon.	
Feet.	′	″	Feet.	′	″	Feet.	′	″
1	0	57	13	3	26	26	4	52
2	1	21	14	3	34	28	5	3
3	1	39	15	3	42	30	5	14
4	1	55	16	3	49	35	5	39
5	2	8	17	3	56	40	6	2
6	2	20	18	4	3	45	6	24
7	2	31	19	4	10	50	6	44
8	2	42	20	4	16	60	7	23
9	2	52	21	4	22	70	7	59
10	3	1	22	4	28	80	8	32
11	3	10	23	4	34	90	9	3
12	3	18	24	4	40	100	9	33

DEPTH of a Sail, (*chute d'une voile*, Fr.) the extent of the square sails from the head-rope to the foot-rope; or the length of the after-leech of a stay-sail or boom-sail; in other words, it is the extent of the longest cloth of canvas in any sail. See the article Sail.

DERRICK, a contrivance by way of temporary crane, for the purpose of hoisting goods or provisions in or out of a ship. Considerable improvements have been made in this simple apparatus that composes it, and which is singularly deserving commendation, on account of its combining simplicity with utility.

One lately contrived by Mr. Kentish is described in the following manner:—

The principal parts of this machinery are a purchase wheel, properly fastened; a roller to take the rope off that wheel, a break-wheel, and a lever to press on this to lower the weight down; a racket-wheel and pawl to stop the roller from going back when hoisting up the weight; a ship's mast or a post in the ground on shore with a guy, a sheave to keep the rope extended, and a hook on which the weight is fixed.

By the cranes of almost any construction there is little difficulty in raising a weight, but accidents too frequently happen in lowering the lever. The common method of delivering cargo from ships is, by a large luff tackle, fixed to the head of a *derrick*, the end of the face being brought to the capstan, by which four men can with difficulty raise a hogshead of sugar; but it requires at least five to lower one down with safety, and obliges them to drop from the bars.

To prevent such accidents, Mr. Kentish has invented the roller with two inclined planes to his *derrick*; three turns or more with great weight of rope, from the purchase wheel at the head, are to turn round the centre of the roller, which is taken off as the weight ascends by a small roller, with two inclined planes, placed a little above the large roller, a man holding or drawing the rope off until the weight is raised as high as wanted; then the lever is pressed on the break-wheel, by a weight at the end, sufficient to check the roller, and make it perfectly easy to the man or boy to ease the rope away as fast or as slow as he pleases. The weight being checked by three powers; viz. the boy holding on the rope; the handle to the roller; and the lever pressing on the break-wheel.

The derrick may be applied to various uses on shore, on board his Majesty's ships in ordinary, hospital and prison ships, or to get in guns and stores of a man of war, by a few hands before the proper officers and ship's company are ordered on board.

The wheel and roller may be applied to hoist up water from a deep well, and it will bring up a large quantity with less strength and labour than are now required by a common wheel and pinion; to hoisting up on board men of war, or merchant ships, with more ease and expedition than by the common tackle; and to lowering them with safety, as she must go down in a horizontal position into the water, and is disengaged in a moment; which prevents accidents, particularly when the ship has sternway, the boat is often in danger under the counter, before the tackle can be unhooked, four men will be sufficient to deliver a sugar ship, which, with a capstan, requires eight.

With this *derrick*, the patentee asserts, that there is a saving of time and of rope; and the machine itself is not liable to get out of order. It will last as long as the ship, and if an accident occurs, it is so simple, that any mechanic can repair it.

Derrick, (*martinet d'artimon*, Fr.) a tackle used at the outer quarter of a mizen-yard, consisting of a double and single block, connected by a fall: also a diagonal shore, as a support to sheers.

DESCENSION, in astronomy, is either right or oblique.

Right Descension, is an arc of the equinoctial, intercepted between the next equinoctial point and the intersection of the meridian, passing through the centre of the object, at its setting, in a right sphere.

Oblique Descension, is an arc of the equinoctial, intercepted between the next equinoctial point and the horizon, passing through the centre of the object, at its setting, in an oblique sphere.

Descensional *Difference*, is the difference between the right and oblique descension of the same star, or point of the heavens.

DESERTION, in the royal navy, the act of forsaking a ship or boat, or running from sick quarters without leave of absence. If any man shall absent himself three successive musters, he is marked as a *runaway* on the ship's books, from the day on which he left his duty.

If a man so absenting himself shall be discovered and brought on board, the captain is to give a certificate of the same to the person by whom he is apprehended, stating therein that he has charged three pounds against the man's wages on the ship's books; and, should this charge not be made thereon, the captain must be responsible for it.

By the sixteenth article of war, persons deserting, or receiving deserters, or enticing persons to desert, " shall suffer death, or such other punishment as the circumstances of the offence shall deserve and a court-martial shall think fit; and, if any commanding officer of any of his Majesty's ships or vessels of war shall receive or entertain a deserter from any other of his Majesty's ships or vessels, after discovering him to be such deserter, and shall not, with all convenient speed give notice to the captain of the ship to which he belongs; or, if the ships are at a considerable distance from each other, to the secretary of the admiralty, or to the commander in chief; every person so offending, and being convicted thereof by the sentence of the court martial, shall be cashiered."

The numerous abuses arising from the *desertion* of seamen in the West India Trade, induced the legislature to pass an act in the 37th year of his present Majesty, entitled, "An Act for preventing the Desertion of Seamen from British Merchant Ships trading to his Majesty's Colonies and Plantations in the West Indies."

"From the 1st of July 1797, every seaman, mariner, and other person, who shall desert at any time during the voyage, either out or home, from any British merchant ship trading to or from the West India colonies or plantations, shall, over and above all punishments, penalties, and forfeitures, to which he is now by law subject, forfeit all the wages he may have agreed for or be entitled to, during the voyage, from the master or owner of the ship on board of which he shall enter, immediately after such *desertion.* Every master or commander of any British merchant ship, who shall hire and engage to serve on board his ship or vessel any seaman, &c. who shall, to the knowledge of such master, have deserted from any other ship or vessel, shall forfeit and pay 100*l.*" See MAXWELL'S *Spirit of Marine Law.*

DETACHMENT OF A FLEET OR SQUADRON, (*escadre*, Fr.) a certain number of ships chosen by an admiral or commodore from the rest of the fleet, and charged to execute some particular service.

DIAMETER, in astronomy. The diameters of the planets are either *apparent* or *real:* the real diameters are such as they are in themselves: the apparent diameters are such as they appear to the eye; and being measured by a micrometer, are found different in different circumstances, and parts of their orbit. See the article APPARENT DIAMETER.

DIFFERENCE, ASCENSIONAL. See ASCENSION.

DIFFERENCE *of Latitude*, in navigation, is the difference between any two places lying on the same meridian; or the distance between the parallels of latitude of any two places, expressed in miles on the equator, and is found by subtraction when the latitudes are of the same name, and by addition when they are of contrary names.

DIFFERENCE *of Longitude*, of two places, is an arc of the equator contained between the meridians of those two places, or the measure of the angle formed by their meridians, and is found by subtraction when the longitudes are of the same name; and by addition when they are of contrary names.

DIGIT, in astronomy, is the twelfth part of the diameter of the sun or moon, and used to express the quantity of an eclipse. Thus an eclipse is said to be of six digits, when six of these parts are hidden.

DIGITS, in arithmetic, signify any integer under 10, as 1, 2, 3, 4, 5, 6, 7, 8, 9.

DINNAGE. See the article DUNNAGE.

DIP OF THE HORIZON. See DEPRESSION.

DIP *of the Magnetic Needle, (inclinaison de l'aiguille aimantée*, Fr.) a certain property which all needles possess when rubbed with the loadstone of inclining the north end below the level of the horizon: this property is found to increase in going northward.

DIPPING-*Needle*, or *Inclinatory* NEEDLE, is a magnetic needle, so hung, as that, instead of playing horizontally, like the common compass, and pointing out north and south, one end dips or inclines to the horizon, and the other points to a certain degree of elevation above it. Or, according to Mr. Whiston, a *dipping-needle* may be defined to be a long straight piece of steel, every way equally poised on its centre, and afterwards touched with a loadstone; but so continued as not to play on the point of a pin, as does the common horizontal needle, but to swing in a vertical plane, about an axis parallel to the horizon; and this, in order to discover the exact tendency of the power of magnetism.

The inventor of the dipping needle, Mr. Whiston observes, was, without all question, an Englishman, Robert Norman by name, a compass maker at Wapping, about the year 1576; this is not only testified by his own account, in his 'New Attractive,' but was allowed by Dr. Gilbert and other writers of that time. The occasion of the discovery, he himself relates; viz. that it being his custom to finish and hang the needles of his compasses before he touched them, he always found that, immediately after the touch, the north point would bend, or decline downward, under the horizon, insomuch that, to balance the needle again, he was always forced to put a piece of wax on the south end, as a counter-poise.

The constancy of this effect, led him at length to observe the precise quantity of the dip, or to measure the greatest angle which the needle would make with the horizon. This, in the above-mentioned year, (1576) he found at London to be 71° 50′. Mr. Whiston being furnished with the further observations of Colonel Windham, Dr. Halley, Mr. Pound, Mr. Cunningham, Pere Noel, Pere Feuille, and his own, has improved very much on the doctrine and use of the *dipping-needle*; brought it to more certain rules, and endeavoured in good earnest to find the longitude thereby. In order to this, he observes:

1st. That the true tendency of the north or south end of every magnetic needle, is not that point of the horizon, to which the horizontal needle points, but towards another, directly under it, in the same vertical, and in different degrees under it, in different ages, and at different places.

2dly. That the power by which an horizontal needle is governed, and all our navigation ordinarily directed, is proved to be one quarter of the power by which the *dipping-needle* is moved; which should render the latter far the more effectual and accurate instrument.

3dly. That a *dipping-needle*, a foot long, will plainly shew an alteration of the angle of inclination in these parts of the world in half a quarter of a degree, or seven and a half geographical miles; *i.e.* supposing that distance taken along or near a meridian; and a needle of four feet, in two or three miles.

4thly. A *dipping-needle*, four feet long, in these parts of the world, will shew an equal alteration along a parallel, as one of a foot long will shew along a meridian; *i.e.* this will, with equal exactness, shew the longitude, as that will the latitude.

This depends on the position of the lines of equal dip in these parts of the world, which are found to lie about 14 or 15 degrees from the parallels. Hence he argues, that as we can have needles of five, six, seven, eight, or more, feet long, which will move with strength sufficient for exact observation; and since microscopes may be applied for viewing the smallest divisions of degrees on the limb of the instrument, it is evident that the longitude at land may be found thereby to be less than four miles.

To find the Longitude or Latitude by the Dipping Needle.

In order to find the longitude or latitude by the *dipping-needle*. If the lines of equal dip below the horizon, be drawn on maps, or sea charts, from good observations, it will be easy, from the longitude known, to find the latitude; and from the latitude known, to find the longitude. Suppose, for example, a person travelling or sailing along the meridian of London, should find that the angle of dip, with a needle of one foot, was 75°; the chart will shew that this meridian, and the line of dip, meet in the latitude of 53° 11′; which is therefore the latitude sought. Or if he be travelling or sailing along the parallel of London, *i.e.* in 51° 32′ north latitude, and find the angle of dip to be 74°; then this parallel, and the line of this dip, will meet on the map in 1° 46′ of east longitude from London; which therefore is the longitude sought.

The dip or depression of the needle is found to be different in different latitudes, and it increases in going northward, and decreases in going southward.

It appears from a table of observations made with a marine *dipping-needle*, invented by Mr. Nairne, for the Board of Longitude, which was used in a voyage to the north pole, in 1773, that

In latitude 60° 18′ the dip was 75° 0′,
In latitude 70 45′ the dip was 77 52′,
In latitude 80 12′ the dip was 81 52′,
In latitude 80 27′ the dip was 82 2′½.

This needle was twelve inches long, and its axis, the ends of which were made of gold alloyed with copper, rested on friction wheels of four inches diameter, each end on two friction wheels, which were balanced with great care. The ends of the axis of the friction wheels, were likewise of gold alloyed with copper, and moved in small holes made in bell-metal; and opposite to the ends of the axis of the needle and the friction wheels, were flat agates, finely polished. The magnetic needle vibrated within a circle of bell-metal, divided into degrees and half degrees; and a line, passing through the middle of the needle to the ends pointed to the divisions. The needle of this instrument was balanced before it was made magnetical; but by means of a cross, the ends of which (contrived by the Rev. Mr. Mitchell) were fixed on the axis of the needle, on the arms of which were cut very fine screws to receive small buttons, that might be screwed nearer to or farther from the axis, the needle could be adjusted both ways to a great nicety, after being made magnetical by reversing the poles, and changing the sides of the needle. When this needle is constructed for sea, it is suspended by an universal joint on a triangular stand, and adjusted vertically by a plumb-line and button, above the divided circle and the dove-tail work: and the divisions on the circle are adjusted so as to be perpendicular to the horizon, by the same plumb-line, and an adjoining screw; and when it is adjusted, a pointer annexed to a screw is fixed, which serves to move the divided circle. Whenever the instrument is used to find the dip, it must be so placed that the needle may vibrate exactly in the magnetic meridian. See *Phil. Trans.* Vol. LXII. Phipps's *Voyage*, &c. Appendix.

The *dipping-needle*, though of late much improved, is, however, still far from perfection. The general mode of constructing it is, to pass an axis quite through the needle, to let the extremities of this axis, like those of the beam of a balance, rest upon its supports, so that the needle may move itself vertically round; and, when situated in the magnetic-meridian, it may place itself in the magnetic-line. The degrees of inclination are shewn upon a divided circle, in the centre of which a needle is suspended. A *dipping-needle* of the simplest construction is represented by fig. 11, Fig. 11.
plate IX, where A B is the needle, the axis of which, F E, rests upon the middle of two lateral bars, C D, C D, which are made fast to the frame that contains the divided circle, A I B K.

This machine is fixed on a stand, G; but, when used at sea, is suspended by a ring, H, so as to hang perpendicularly.

DIRK, (*sorte de poignard*, Fr.) is a kind of dagger, or short sword, worn more for ornament than for use, by officers in the royal navy. It is hung at the left side, by means of a belt going round the waist, and buckling in front.

DISABLED, (*désemparé*, Fr.) the state of a ship when, by the loss of her masts, sails, yards, or rigging; by springing a leak, or receiving some fracture in the hull, or other disaster, she is rendered incapable of prosecuting her voyage, without great difficulty and danger.

DISC, in astronomy, is the body and face of the sun or moon, such as it appears to us on the earth; or the body or face of the earth, such as appears to a spectator in the moon, &c.

The *disc*, in eclipses, is supposed to be divided into twelve equal parts, called digits. In a total eclipse of any luminary, the whole *disc* is obscured; but in a partial one, only a part thereof.

To DISCHARGE a Ship, (*décharger, ou désarmer un vaisseau*, Fr.) signifies to unlade her, or take out her stores, ammunition, artillery, &c. When expressed of the officers or crew, it implies to disband them from immediate service. When spoken of cannon, it means when it is fired off.

To DISEMBARK, (*débarquer*, Fr.) implies to descend from, or leave a ship for the land: for exam. "the troop was disembarked from a store-ship."

DISCIPLINARIAN, (*officier qui maintient la discipline avec rigueur et exactitude*, Fr.) an officer who maintains strict and rigorous discipline, now commonly called a Martinet. The great difficulty is, however, in drawing the line between strict discipline and tyranny.

DISCIPLINE, Marine, (*discipline navale*, Fr.) signifies the training up soldiers for the sea-service, in such exercises and manœuvres as can be performed on board ships of war at sea.

To DISMANTLE, (*dégréer*, Fr.) is to unrig a ship, and take out all her stores, guns, &c. in readiness for being laid up in ordinary, or for any other purpose.

DISMASTED, (*démâté*, Fr.) the state of a ship

deprived of her masts, whether by design or accident.

DISMOUNTED, the state of a cannon taken off its carriage, or when, by the enemy's shot, the carriage is so broken, as that the gun is rendered unmanageable.

DISPART, (*marquer le point de mire d'un canon*, Fr.) a term in gunnery, used for a mark set upon the muzzle ring of a piece of ordnance, of such height, that a sight-line taken from the top of the base-ring near the vent or touch-hole, to the top of the dispart, near the muzzle, may be parallel to the axis of the concave cylinder. For the method of disparting, see the article GUN.

DISTANCE, properly speaking, denotes the shortest line between any two points, or objects, &c.

DISTANCE, in astronomy. The distance of the sun, planets, and comets, is only found from their parallax, as it cannot be found either by eclipses, or their different phases: for from the theory of the motions of the earth and planets we know, at any time, the proportion of the distances of the sun and planets from us; and the horizontal parallaxes are in a reciprocal proportion to these distances. See the article PARALLAX.

DISTANCE *of the Sun from the Moon's node*, or *apogee*, is an arc of the ecliptic, intercepted between the sun's true place and the moon's node, or apogee.

DISTANCE, in navigation, is the number of miles or leagues a ship has sailed from any one given place to another, and the best rules for reducing the apparent distance of the moon from the sun or a fixed star, to the true distance, may be found in Mendoza Rios's Tables; and in Mackay on the Longitude. The different methods of clearing the distances are also given in the Requisite Tables and in the Nautical Almanac.

DIVING, the art or practice of descending under water, to considerable depths, and abiding there a competent time. The uses of diving are in many respects very considerable, particularly in recovering property out of wrecks of ships; in fishing for pearls, corals, sponges, &c.

Various engines have been contrived to render the business of diving safe and easy; the great point is, to furnish the diver with fresh air, without which he must either make but a short stay, or perish.

Those who dive for sponges in the Mediterranean, carry down sponges dipt in oil in their mouths, but considering the small quantity of air that can be contained in the pores of a sponge, and how much that little will be contracted by the pressure of the incumbent water, such a supply cannot subsist a diver long, since a gallon of air is not fit for respiration above a minute. Dr. Halley assures us, a naked diver cannot subsist above two minutes under water, without a sponge: besides, if the depth be considerable, the pressure of the water on the breast and organs is so great, that it makes the eyes blood-shotten, and occasions a spitting of blood, which if persisted in, would quickly prove fatal, and often has done so.

DIVING-*Bell*, a machine contrived to remedy the inconveniences in diving, and is most conveniently made in the form of a truncated cone, the smaller end being closed, and the larger opened. It is to be poised with lead, and so suspended, that the vessel may sink full of air, and with its open base downward, and as nearly as may be in a situation parallel to the horizon, so as to close with the surface of the water all at once.

The diver sitting under this, sinks down with the included air to the depth desired; and if the cavity of the vessel can contain a tun of water, a single man may remain a full hour without much inconvenience at five or six fathoms deep; but the lower he goes, the included air contracts itself according to the weight of the water which compresses it, so that at thirty-three feet deep the bell becomes half full of water, the pressure of the incumbent water being then equal to that of the atmosphere, and at all other depths the space occupied by the compressed air in the upper part of the bell will be to the under part of its capacity filled with water, as thirty-three feet to the surface of the water in the bell below the common surface, and this condensed air being taken in with the breath, soon accommodates itself to the existing circumstances, so as to have no ill effect, provided the bell is admitted to descend slowly; but the greatest inconvenience of this engine is, that the water entering it, contracts the bulk of air into so small a compass, that it soon heats and becomes unfit for respiration, so that there is a necessity for its being drawn up to recruit it, besides the uncomfortable situation of the diver, who must be almost covered with water.

To obviate these difficulties of the *diving-bell*, Dr. Halley, to whom we owe the preceding account, contrived a farther apparatus, whereby, not only to recruit the air from time to time, but also to keep the water out of the machine at any depth. This bell was made of wood, containing about sixty cubic feet in its concavity, and was the form of a truncated cone, whose diameter at the top was three feet, and at the bottom five. It was so coated externally with lead, that it would go down in a perpendicular direction, and no other.

In the top was a window to let in light, and likewise a cock to let out the hot air that had been breathed; and below, about a yard under the bell, was a stage suspended by three ropes, each of which was charged with about one hundred weight to keep it steady.

To supply air, the bell had a couple of barrels so cased with lead as to sink when empty, each having a bung-hole in its lowest part to let in the water as the air in them condensed on their descent, and to let it out again when they were drawn up full from below. To a hole in the uppermost part of these was fixed a leathern trunk or hose, long enough to fall below the bung-hole, and kept down by a weight in such a way, that the air in the upper part of the barrels could not escape, unless the lower ends of these hose were first lifted up. These air barrels were made to rise and fall like two buckets in a well; by means of these barrels fresh air was continually supplied from above, and it was done with so much ease, that two men with less than half their strength could perform all the labour required. By an additional contrivance it was found practicable, for a diver to go out of the engine, to some distance from it, the air being conveyed to him in a continuable stream by small flexible pipes.

A great improvement has been made in the *diving-bell* by the late Mr. Spalding, of Edinburgh, and many successful trials have recently been made to recover property wrecked, particularly from the Abergavenny East Indiaman, wrecked near Weymouth.

For the manner in which these inconveniences are remedied by Mr. Spalding's new contrivance, we shall refer the reader to GREGORY'S *Cyclopedia*.

DIVING *Bladder*, a term used by Borelli, for a machine, which he contrived for diving under the water to great depths, with great facility, and which is preferred to the common diving-bell.

The vesica, or bladder, usually so called, is to be of brass or copper, and about two feet in diameter.

This is to contain the diver's head, and is to be fixed to a goat-skin habit, exactly fitted to the shape of the body of the person. Within this vesica there are pipes, by means of which a circulation of air is contrived, and the person carries an air pump by his side, by means of which he may make himself heavier or lighter, as the fishes do, by contracting or dilating their air bladder.

DIVISION, (*division*, Fr.) a select number of ships in a fleet or squadron of men of war, distinguished by a particular flag pendant or vane, and usually commanded by a general officer. A squadron is commonly ranged into three divisions, and the commanding officer is stationed in the centre.

In a large fleet, the admiral divides it into three squadrons, each of which is commanded by an admiral, and is again divided into three divisions; each squadron has its proper colours, according to the rank of the admiral who commands it; and each division has its proper mast. The private ships carry pendants of the same colour with their respective squadrons at the masts of their particular divisions, so that the ships in the last division of the blue squadron carry a blue pendant at their mizen-top-gallant-mast-head. These distinctions of divisions are not, however, constantly practised.

The general officers or commanders of divisions, place themselves in the centre of the divisions: the three commanding admirals excepted, who, in a sailing position, lead their respective squadrons.

DIURNAL, in astronomy, something relating to the day, in opposition to nocturnal, which regards the night.

DIURNAL *Arc*, the arc or number of degrees that the sun, moon, or stars describe between their rising and setting.

DIURNAL *Circle*, is the apparent circle described by the sun, moon, or stars, in consequence of the rotation of the earth.

DIURNAL *Motion of a Planet*, is so many degrees and minutes as any planet moves in twenty-four hours. Hence the motion of the earth about its axis, is called its *diurnal* motion.

DOCK, (*chantier de construction*, Fr. imagined of δοχεῖον) in maritime affairs, a broad and deep trench, formed on the side of a harbour, or on the banks of a river; and commodiously fitted either to build ships, or receive them to be repaired and *breamed*.

Dry-DOCK, (*bassin de radoub*, Fr.) has generally strong flood-gates or caissons, to prevent the flux of the tide from entering it while the ship is built or repaired, when the gates are opened, and the water let in to float or launch her.

Wet-DOCK, (*bassin d'un port, ou darce*, Fr.) a place where the ship may be hauled into, out of the tide's way, and so dock herself, or sink herself a place to lie in. This sort of dock has no flood-gates.

There are also wet-docks of another description for loading and unloading ships, as the docks in the Isle of Dogs for the West-India trade, the docks at Wapping, Hull, Liverpool, and Bristol.

Of late years plans have been brought forward and executed for improving the London-docks; making wet-docks for the West India shipping only; and others for the East India shipping only: these have proved very advantageous to the commerce of the port of London, and to the companies who raised the money for accomplishing them.

London DOCK *Company*. This company was established by an act of the 39th and 40th Geo. III. passed the 20th of June 1800, by which they were empowered to raise a capital stock of 1,200,000*l.* This has since been increased; so that the total capital stock which the company are authorized to raise, if necessary, is 2,200,000*l.* The company is under the management of twenty-four directors, who are elected annually. Two general courts of proprietors are held every year, at which all persons are entitled to vote who possess 500*l.* stock or upward. The dividends on their stock are restricted to 10 per cent. per annum, and are paid on the 1st of January and the 1st of July. The present dividend is 5½ per cent. and the company pay the property tax thereon. The dividends are paid, and transfers made on any day except holidays. The company was required to complete the docks within seven years; and, so early as the close of January 1805, the dock was opened for public use.

West India DOCK *Company*. This company was established by an act of 39 Geo. III. passed the 12th July 1799, and was empowered to raise a joint stock of 500,000*l.* with liberty to increase the same to 600,000*l.* which has since been increased to 1,200,000*l.* The company is under the management of twenty-one directors, eight of whom must be members of the corporation of London. They are elected annually, five going out every year for three years, and six the fourth year. Two general meetings of proprietors are to be held every year, in January and July, at which all persons are entitled to vote who hold 500*l.* stock, or upwards. Votes may be given by power of attorney. The present dividend is 10 per cent. per annum, and becomes due the 1st of January and the 1st of July. The dividends are payable every day, and the company pay the property tax thereon. The dock was opened in August 1802.

East India DOCK *Company*. This company was established by act of 43 Geo. III. passed the 27th of July 1803, and was authorized to raise a capital of 200,000*l.* divided into shares of 100*l.* each. They have since been empowered to increase their capital to 400,000*l.* nearly the whole of which has been raised. They were to pay 5 per cent. interest on the money advanced; and, after the docks should be completed, to make dividends not exceeding 10 per cent. per annum. The company is under the management of thirteen directors, who must be holders of at least twenty shares of the company's stock, and four of them must be directors of the East India Company. They are elected annually in July, and three go out every year. Two general meetings are held yearly, in January and July, at which proprietors of five shares and upwards are entitled to vote. They must also be directors of the East India Company, or have been so within two years, or be an agent, husband, consignee, or owner to the value of 500*l.* or upwards, of East India shipping. The present dividend is at the rate of 5 per cent. per annum, and is payable at Lady Day and Michaelmas, except Fridays, between the hours of eleven and two o'clock. The dock was opened in August 1806.

To DOCK *a Ship*, (*caréner un vaisseau dans le bassin*, Fr.) is to draw her into dock, in order to give her a proper repair, cleanse the bottom, and cover it anew with a preparation of stuff. See the article BREAMING.

DOCK-*Yards*, (*arsenaux de marine*, Fr.) arsenals containing all sorts of naval stores, and timber for ship-building. In England, the royal dock-yards are at Chatham, Portsmouth, Plymouth, Deptford, Woolwich, and Sheerness, where his Majesty's ships and vessels of war are generally moored, during the time of peace; and such as want repairing are taken into the docks, examined, and refitted for service. See the article REPAIR.

The three first of these yards are governed by a commissioner resident at the port, who superintends all the musters of the officers, artificers, and labourers, employed in the dock-yard, and ordinary: he also controls their payment therein, examines their accounts; contracts, and draws bills on the Navy office to supply the deficiency of stores; and, finally, regulates whatever belongs to the dock-yard, maintaining due order in the respective offices.

These yards are generally supplied from the northern crowns with hemp, pitch, tar, rosin, canvas, oak plank, and several other species of stores. With regard to the masts, particularly those of the largest size, they are usually imported from New England.

DOG, (*renard à embarquer les bois*, Fr.) a sort of iron hook, or bar, with a sharp fang at one end, so formed as to be easily driven into a piece of timber: it is used to drag it along by means of a rope fastened to it, upon which several men pull, and so draw the plank towards the place where it is to be stowed. It is also used for the same purpose in unlading the ship.

DOG, in astronomy. See the articles CANIS MAJOR and MINOR.

DOGGER, (*dogre, sorte de bâtiment de pêche hollandais*, Fr.) a Dutch fishing-vessel navigated in the German ocean: it is equipped with two masts; a main-mast, and a mizen-mast, and somewhat resembles a *ketch.* It is principally used for fishing on the Dogger Bank.

DOGGER-*Men*, are fishermen belonging to doggers.

DOG-*Vane*, (*penon*, Fr.) a small vane composed of thread, cork, and feathers, fastened on the end of a half-pike, and placed on the weather gun-wale, to steer the ship by when sailing on a wind.

DOG-*Watch.* See the article WATCH.

DOLPHIN OF THE MAST, (*baderne des mâts majeurs*, Fr.) a kind of wreath, formed of plaited cordage, to be fastened occasionally round the masts, as a support to the *puddening;* the use of which is, to sustain the weight of the fore and main-yards by the jears, in case the rigging, or chains, by which those yards are suspended, should be shot away in the time of battle. See the article PUDDENING.

DORADO, in astronomy, a southern constellation, not visible in our latitude; it is also called *Xiphias*, or the *Sword-fish.* The stars of this constellation, in Sharp's catalogue, are six.

DOUBLE-BANKED OARS. See the article OAR.

To DOUBLE *a Cape*, (*doubler un cap*, Fr.) in navigation, the act of sailing round or passing beyond it, so as that the point of land separates the ship from her former situation, or lies between her and any distant observer.

DOUBLE-*Headed Shot.* See the article SHOT.

To DOUBLE *upon a Fleet*, (*doubler les vaisseaux ennemis*, Fr.) in a naval engagement, the act of enclosing any part of a hostile fleet between two fires, or of cannonading it on both sides.

It is usually performed by the van or rear of that fleet which is superior in number, taking the advantage of the wind, or of its situation and circumstances, and tacking or running round the van or rear of the enemy, who will thereby be exposed to great danger, and can scarcely avoid being thrown into a general confusion.

DOUBLING-NAILS, (*clous de différentes dimensions, servant à clouer les planches et bordages*, Fr.) amongst shipwrights, are the nails commonly used to fasten the lining of the gun-ports, &c.

To DOUSE, (*molir, ou larguer*, Fr.) to lower or slacken suddenly; expressed of a sail in a squall of wind, an extended hawser, &c.

DOUSING-CHOCKS, in ship-building, are pieces fayed across the apron, and lapped on the knight-heads, or inside stuff above the upper-deck.

DOWELLING, among shipwrights, is a method of coaking performed by letting pieces into the solid, or by uniting two pieces together by tenons.

DOWN ALL CHESTS! (*à la cale tous les coffres!* Fr.) an order to get all the officers and seamens chests down below from off the gun-decks, when clearing the ship for an engagement.

DOWN *Foresail!* (*laisse tomber la misaine!* Fr.) the command to set the foresail.

DOWN-*Haul*, (*hale-bas des focs et voiles d'étai*, Fr.) a rope passing up along a stay through the cringles of the stay-sail or jib, and tied to the upper corner of the sail, to pull it down, when *shortening* sail.

DOWN-*Haul-Tackle*, (*cargue-bas des vergues de hune*, Fr.) a complication of pullies employed to pull down the main or fore-top-sail-yard in a tempest, in order to reef the sail, because the violence of the wind prevents the weight of the yard from having its natural effect, of descending, when the ropes by which it is suspended are slackened.

DOWN *Jib and Stay-Sails!* (*hale-bas le grand foc et les voiles d'étai!* Fr.) an order to take in those sails. It is also applied to the studding-sails.

DOWNS, a bank or elevation of sand, which the sea gathers and forms along its shores, and which serves as a barrier.

DOWNS, (*dunes*, Fr.) the name given to a famous

road for ships along the eastern coast of Kent, from Dover to the North Foreland. This road has excellent anchorage, and is well defended by the castles of Sandwich, Deal, and Dover. The English fleets usually meet there.

DRABLER, (*sorte de bonnette dans les sloops et goëlettes*, Fr.) an additional part of a sail, sometimes laced to the bottom of the *bonnet* on a *square-sail*, in sloops and schooners.

DRACO, THE DRAGON, in astronomy, a constellation of the northern hemisphere, whose stars, according to Flamstead, are 80.

DRAG, (*drague*, Fr.) a machine consisting of a sharp square frame of iron, encircled with a net, and commonly used to rake the mud off from the platform or bottom of the docks, or to clean rivers: also for the purpose of catching oysters sometimes, improperly called a Drudge. See plate IX. fig. 12.

Fig. 12. *To* DRAG *the Anchor*, (*chasser sur son ancre*, Fr.) is to trail it along the bottom, after it is loosened from the ground, by the effort of the wind or current upon the ship. See the article ANCHOR.

DRAGON'S HEAD and TAIL, are the nodes of the planets, but more particularly of the moon, being the two points wherein the ecliptic is intersected by her orbit, making with it angles of 5° 18′ One of these points looks northward. the moon beginning then to have north latitude; and the other southward, where she commences south latitude.

Thus her deviation from the ecliptic seems, according to the fancy of some, to make a figure like that of a dragon, whose belly is where she has the greatest latitude; the intersection representing the head and tail, from which resemblance the denomination arises. But it must be observed that the points do not always abide in one place, but have a motion of their own in the zodiac and retrogradewise, at the rate of 3′ 11″ per day, completing their circle in 18 years 225 days: so that the moon can be but twice in the ecliptic during her monthly period, but at all other times she will have a latitude or declination from the ecliptic.

It is about these points of intersection that all eclipses happen. They are usually denoted by these characters, ☊ dragon's head, ☋ dragon's tail.

DRAGS, in rope making, are sometimes formed like the after part of the sledge, to which they are fastened by ropes, and are lined with a board on the upper side; they contain weight, as a press, when the rope requires more than the sledge can carry to keep the strands of a proper stretch, and prevent their kinking, as they get hard, and as the rope is brought to its intended size.

DRAUGHT, (*tirant d'eau*, Fr.) the depth of a body of water necessary to float a ship; hence a ship is said to draw so many feet of water, when she is borne up by a column of water of that particular depth: thus, if it requires a body of water, whose depth is equal to twelve feet, to float or buoy up a ship on its surface, she is said to draw twelve feet water; and that this draught may be more readily known, the feet are marked on the stem and stern-post from the keel upwards.

DRAUGHT also implies the drawing, or design by which the ship is to be built, which is generally by a scale of one fourth of an inch to a foot.

To DRAW UPON A SHIP, (*gagner sur un vaisseau*, Fr.) is to gain upon a vessel, when in pursuit of her.

DRAWING, (*porter*, Fr.) the state of a sail when it is inflated by the wind, so as to advance the vessel in her course.

DRAWING-*Knife*, a long narrow edged tool, with a handle at each end.

DREDGE, (*drague pour prendre les huitres*, Fr.) a kind of drag used with a long rope, to catch oysters in deep water.

To DRESS A SHIP, (*pavoiser un vaisseau*, Fr.) is to ornament her with a variety of colours; as ensigns, flags, pendents, &c. of various nations, displayed from different parts of her masts and rigging, &c. on a day of festivity.

DRIFT, (*dérive d'un vaisseau à la cape*, Fr. from *drive*,) in navigation, denotes the angle which the line of a ship's motion makes with the nearest meridian, when she drives with her side to the wind and waves, and is not governed by the power of the helm; and also the distance which the ship drives on that line, so called only in a storm.

A ship's way is only called *drift* in a storm, and then, when it blows so vehemently as to prevent her from carrying any sail, or at least restrain her to such a portion of sail as may be necessary to keep her sufficiently inclined to one side, that she may not be dismasted by her violent labouring, produced by the turbulence of the sea.

DRIFT also signifies the difference between the size of a bolt, and the hole into which it is to be driven; for instance, if a bolt be one eighth of an inch larger than the hole, the bolt is said to have one-eighth *drift*, the same as is applicable to a hoop which is to be driven on a mast.

DRIFT *of a Current*, (*la force d'un courant*, Fr.) is its angle and velocity. See the article CURRENT.

DRIFTS *in the Sheer-Draught*, are those pieces where the rails are cut off; they are ended with scores, and called *drift-pieces*.

To DRIVE, (*dériver, aller en dérive*, Fr. *drifan*, Sax.) is to carry at random along the surface of the water, as impelled by a storm, or impetuous current. Driving is generally expressed of a ship, when accidentally broke loose from her anchors or moorings.

DRIVER, (*tappe-cul*, Fr.) a large sail occasionally set upon the mizen-yard or gaff, the foot thereof being extended by a boom considerably over the stern, in the manner of a cutter's main-sail. It is sometimes fitted to hoist with a half-yard to the peek, like a lower studding-sail, and the fore-leech is laced down the mizen-mast. Driver also signifies the foremost spur in the bulgeways, the heel of which is fayed to the fore side of the foremost puppet, and the sides placed to look fore and aft.

DRIVER-*Boom*, (*bonte-hors de paille-en-cul*, Fr.) the boom on which the preceding article is extended.

DROP, (*chûte*, *en parlant des principales voiles quarrées*, Fr.) a name sometimes given to the depth of the square sails; as, "her main-top-sail *drops* seventeen yards."

To DROP *Anchor*, (*mouiller, ou laisser tomber l'ancre*, Fr.) is used synonimously with to anchor.

To DROP *Astern*, (*culer*, Fr.) implies the retrograde motion of a ship.

DRUM, (*tambour*, Fr.) an instrument of military music used on board ships, as a signal for different purposes agreeable to the beat, or as a cautionary signal to other ships in foggy weather.

DRUXEY, the name given to timber in a state of decay, with white spongy veins.

DRY-ROT, a disease incident to timber for building, such as flooring-boards, joists, wainscoting, &c. which is produced by the putrefactive fermentation of the component parts of the wood, in connexion with moisture. Wood felled in the spring is most liable to this disease.

The *dry-rot* may, in all cases, be infallibly prevented where it is practicable to cover the surface of the wood, properly dried, with a varnish which is impenetrable and not distructible by water. Dr. Darwin is of opinion, that the *dry-rot* may be entirely prevented, by soaking the timber first in lime-water, till it has absorbed as much of it as possible, and, after it has become dry, immersing it in a weak solution of vitriolic acid in water, which he supposes will not only preserve it from decay for many centuries if it be kept dry, but also render it less inflammable; a circumstance that merits considerable attention in constructing houses, ships, &c.

Under this head it may not be improper to observe, that the premature decay of our wooden bulwarks has frequently been considered, but more with a view to complaint, than to the demonstration of remedies. The remedies, however, appear to have been consulted with some confidence of success, in a pamphlet just published by Richard Pering, Esq. clerk of the cheque in his Majesty's yard at Plymouth dock.

After very properly recommending that the shipwright should be regularly educated in naval mechanics, before he is permitted to undertake the performance of any duties in the dock-yard, the author proceeds to attribute the premature decay of our ships (the general duration of which he limits to eight years, unless they undergo a thorough repair) to the use of unfit materials, and improper methods of building, fastening, and calking. To prolong this duration, he then proposes that all our ships should be framed, planked, and calked, under cover, and remain so to season; that they should be fastened with copper, instead of the tree-nails now used; that screws should be employed to heave up the bolts, instead of clenches; that foreign timber should be entirely expelled from our dock-yards, it being much more susceptible of the dry-rot than timber of English growth, except fir, and oak from some parts of America, the latter of which he considers nearly equal in goodness to our own; that the calking-iron should be the last thing applied to a ship before she is launched; that the several docks in a yard, except those which admit vessels with their masts in, should be covered or housed over, as well as the ships on the stocks; and that no ship should ever have any material repairs given to her, the expenses of which may exceed one-third of her original cost!

By carrying his plans into execution, the author conceives that our men of war would be better able to combat the dangers to which they are exposed; that forty millions of money, besides interest, would be saved to the nation every twenty-four years; that the consumption of timber would be diminished, as far as its use extends to ship-building, two-thirds; that thousands of lives would be preserved; and that the *dry-rot* and formation of fungus, which now prevail in our navy, to such an alarming degree, would be annihilated. "These," says the author, "are objects of no trivial importance to a kingdom that possesses the empire of the seas; and if they can be effected," which he pledges himself to prove, whenever called on, "most assuredly have imperious and irresistible claims on the immediate attention of Government." See PERING's *Enquiry into the Cause of the Dry-rot, &c. in our wooden Bulwarks*, lately published.

It has been remarked by a gentleman, (a friend to the navy) that the premises of the writer are pretty correct, and most of his arguments conclusive. The circumstances of Mr. Pering's having been in his Majesty's service nearly thirty years, and very frequently seeing the construction, as also the premature decay of our ships of war caused by the *dry-rot*, adds great weight to what he alleges, as during this long period, he must, necessarily, have acquired considerable experience: at all events, whether his plans are fraught with those benefits which he imagines them to be, or not, they will merit consideration: consideration may give birth to discussion, and discussion may, eventually, lead to something useful.

To DUB, to work with the adze.

DUBBING, the act of reducing or diminishing a piece of timber, by means of an adze.

DUBHE, a bright fixed star of the second magnitude, situated in the back of the constellation Ursa Major or the Great Bear, and marked α in our catalogues. For the beginning of the year 1812, its

Right ascension is	162°	36′	25″
Annual variation in A.R.			57.7
Declination	62	45	57 N.
And annual variation in decl.			—19,2

DUCK, or RUSSIA DUCK, (*toile de Russie*, Fr.) a name given to the finest canvas for sails.

DUCK-*up!* (*cargue à vue!* Fr.) a term used by the steersman, when the main-sail, fore-sail, or sprit-sail, hinders his seeing to steer by a land-mark, upon which he calls out, "Duck-up the clew-lines of those sails!" that is, haul the sails out of the way! also when a shot is made by a chase-piece, if the clew of the sprit-sail hinders the sight, they call out, "Duck up! &c."

DUCKING, (*la cale: punition*, Fr.) a sort of marine punishment unknown, except by name, in the British navy, formerly much used by the French.

It is also a penalty which veteran sailors inflict on those, who, for the first time, pass the tropic of Cancer, or the Equator. See the article *Crossing the* EQUATOR.

Dry-DUCKING, (*cale sèche*, Fr.) a punishment used in some nations, which consists in letting a man drop from the yard-arm by means of a rope, close to the surface of the water, but so as not to plunge him into it.

DUNNAGE, (*fardage*, Fr.) a quantity of fagots, boughs of trees, or other loose wood, laid in the bottom of a ship, either to raise the heavy goods which might make her too stiff, or to keep the cargo sufficiently above the bottom, that it may sustain no damage from the water, if the ship is leaky.

E.

EARINGS, (*rubans de pointure*, Fr.) in a ship, are certain small ropes, employed to fasten the upper corners of a sail to its respective yard; for which purpose one end of the earing is spliced to the *cringle*, fixed in that part of the sail, and the other end is passed six or seven times round the yard-arm and through the cringle, thereby fastening the latter to the former; the two first turns, which are intended to stretch the head of the sail tight along the yard, are passed beyond the lift and rigging on the yard-arm, and are called *outer* turns, while the rest which draw it close up to the yard, and are passed within the lift, &c. are called *inner* turns. Every reef on a yard has its respective *earings*, which are passed in the same manner.

Reef-EARINGS. The only difference between earings and *reef-earings* is, the former is *spliced* to the cringle, and the latter is *rove* through the cringle, having an eye spliced in it for the more easy reeving and unreeving it, in case of its being chafed, that another may easily replace it for fear of splitting the sail.

EARTH, in astronomy and geography, is one of the primary planets, being the terraqueous globe which we inhabit: its astronomical character is ⊕.

Figure of the EARTH.—Notwithstanding the various hypotheses of the ancients of the earth's being cylindrical, flat, &c. it is thought manifest, from the following observations, that its exterior figure is very nearly globular.

1st. A spectator placed on any part of the earth, sees around him a limited portion of its surface, which is called the visible horizon of that place. If the spectator advances twenty or thirty miles from his first position, either to the east, west, north, or south, he will have a new visible horizon, which does not contain any one of the objects which were comprehended in the horizon of his first position.

By advancing still farther, he will have another visible horizon filled with new objects, and bounded by a different portion of the earth. Now this perpetual change in the visible horizon, cannot arise from the inequalities on the earth's surface, for even from the top of the highest hill in one of the horizons we cannot see the objects contained in another: it follows, therefore, that the surface of the earth is convex, and since this change in the visible horizon takes place in every part of the globe that has been visited, it is natural to conclude that the earth is spherical.

2d. If the visible horizon is composed wholly or partly of sea, we obtain an ocular demonstration of the earth's convexity: for in the case of a ship heaving in sight, her masts first appear, while the hull and the lower parts are invisible; as she approaches, more of the mast and rigging become visible, till the whole of the vessel gets above, as it were, the convexity of the earth.

3d. The globular form of the earth is still more satisfactorily proved by the eclipses of the moon; for all eclipses of that planet are caused by the shadow of the earth falling upon the moon; this shadow is always circular, and no other body but a globe, can, in all situations, cast a circular shadow.

It is true, the surface of the earth is not an exact geometrical globe, but then the inequalities are so inconsiderable, that the highest mountain bears no greater proportion to the bulk of the earth, than a grain of dust does to a common globe. The figure of the earth was formerly considered, by mathematicians and geographers, as perfectly spherical, excepting the small inequalities on its surface, of mountains and valleys; till an accident engaged the attention of Sir Isaac Newton and Mr. Huygens, who demonstrated from the laws of hydrostatics, and the revolution of the earth about its axis, that its figure was not a true sphere, but an oblate spheroid flattened towards the poles.

Several degrees of the meridian have been measured in different parts of the earth, and the result of these measurements is, that the earth is about 7,957¼ English miles in diameter; that a degree is longer at the poles than at the equator; the proportion between the polar and equatorial diameters of the earth, being very nearly as 300 to 301. The difference, however, in the results which have been obtained by measuring degrees in various parts of the earth, has been sufficient to induce La Place to maintain, that the earth is not a solid of revolution, but that the terrestrial meridian is a curve of double curvature Professor Playfair, with more justice, ascribes these irregularities to the unequal density of the materials near the surface of the earth where the degrees have been measured, by means of which the direction of gravity is disturbed.

The compression of the earth's poles, and the accumulation of matter at the equator, evidently arise from its diurnal rotation. The gravity of the equatorial parts is diminished by the centrifugal force arising from their rapid motion, while the gravity at the poles suffers no diminution. If the earth, therefore, were fluid, it would obviously assume a spheroidal form; and though there is no authority for saying, that it ever was in such a soft state as to yield to the effects of the centrifugal force, yet it is manifest, that even if it were originally spherical, and in its present consolidated state, it must have been constantly approximating to a spheroidal form. See LA LANDE'S *Astronomie*, and PLAYFAIR'S *Observations on the Earth*, in the *Phil. Trans.* of Edin.

Magnitude of the EARTH. This has been variously determined by different authors, both ancient and modern. Professor Snell, of Leyden, about the year 1620, measured considerable distances between two parallels, and found one degree amount to 19 Dutch miles, and the whole circumference to 6,840 miles.

Richard Norwood measured the space between London and York with a chain; and, on the 11th of June, 1635, old style, he took the sun's altitude at the meridian, with a sextant of five feet radius, and found a degree of 69½ miles, and 14 poles, whence he inferred that the diameter of the earth was about 7,966 miles, and the circuit 25,036 miles. This measurement, though far superior to those of the ancients, was tried by several French mathematicians, who suspected some slight errors, by the King's command, with a quadrant of 3⅓ French feet radius; by which they found a degree contained 542,360 French feet.

M. Cassini, jun. in the year 1700, by the King's command also, renewed the experiment with a quadrant of 10 feet radius, for taking the latitude, and another of 3⅛ feet for taking the angles of the triangles; and found a degree from his calculation, contained 27,292 toises, or nearly 69½ English miles.

From these, and various other attempts since, of a similar nature, to obtain the length of one degree of the meridian, which is to be multiplied by 360, the following dimensions may be taken as near the truth:

The Earth's circumference......	25,000 miles.
The diameter	7,957¾ miles.
The superficies.........	198,944,206 square miles.
The solid	26,393,000,000 cubic miles.

It is besides conjectured, from the measurement of the most approved maps, that the unexplored portions of the earth and seas contain 160,522,026 square miles, the inhabited part of the former 38,922,180, thus divided, Europe contains 4,456,065; Asia, 10,768,823; Africa, 9,654,807; and America, 14,110,874.

The Quantity of Matter, Density, and Attractive Power of the EARTH. Although the relative densities of the earth and most of the other planets have been known a considerable time, it is but very lately that we have come to the knowledge of the absolute gravity or density of the whole mass of the earth. This has been calculated and deduced from the observations made by Dr. Maskelyne, late astronomer royal, at the mountain Schehallien, in Perthshire, in the years 1774, 5, & 6. The attraction of that mountain on a plummet, was observed on both sides of it, and its mass computed from a number of sections in all directions, and consisting of stone; these data being then compared with the known attraction and magnitude of the earth, gave, by proportion, its mean density, which is to that of water as 9 to 2, and to common stone as 9 to 5: from which very considerable mean density, it may be presumed that the internal parts contain some great quantities of metals.

From the density now found, its quantity of matter becomes known, being equal to the product of its density by its magnitude. From various experiments, too, we know that its attractive force, at the surface, is such, that bodies fall there through a space of $16\frac{1}{12}$ feet in the first second of time: whence the force at any other place, either within or without it, becomes known; for the force at any part within it, is directly as its distance from the centre; but the force of any part without it, reciprocally as the square of its distance from the centre.

The Motion of the EARTH. It is now generally granted that, besides the small motion of the earth which causes the precession of the equinoxes, the earth has two great and independant motions; viz. the one by which it turns round its own axis, in the space of 24 hours nearly, and causing the continual succession of day and night; and the other, an absolute motion of its whole mass in a large orbit about the sun, having that luminary for its centre, in such manner that its axis keeps always parallel to itself, inclined in the same angle to its path, and, by that means, causing the vicissitudes of seasons; spring, summer, autumn, and winter.

This annual motion of the earth round the sun is performed between the orbits of Venus and Mars, with the former and that of Mercury within its own, or between it and the sun in the centre, and those of Mars, Jupiter, Saturn, &c. above it or without, which are called from this circumstance superior planets, and the others inferior. The time occupied in performing this revolution, is 365 days, 6 hours, and 49 minutes, or a tropical year, calculated from an equinox or solstice to the same again; the sidereal year, as computed from any fixed star to the same point, and seen from the sun, makes the revolving of the earth to occupy 365 days, 6 hours, and 9 minutes: the figure of the orbit is elliptical with the sun in one focus; the supposed distance is 95 millions of miles, admitting the sun's parallax to be $8\frac{2}{3}''$, or the angles beneath which the semi-diameter of the earth would appear from the sun; and the eccentricity of the orbit, or distance of the sun in the focus from the centre of this elliptic orbit, is about $\frac{1}{60}$th of the mean distance.

EASE THE SHIP! (*lof à la risée!* Fr.) the command given by the pilot to the steersman, to put the helm close to the lee-side; or, in the sea-phrase, *hard-a-lee*, when the ship is expected to *pitch* or plunge her fore-part deep in the water, while close-hauled. The reason usually given for this practice is, that the sudden movement of the helm prevents the ship's head from falling with so much weight and rapidity into the hollow of the sea, as it would do otherwise; which is presuming that the slow and uncertain effect of the helm is sufficient to retard the certain and violent action of gravity: a position that necessarily infers a very singular theory of mechanics. We shall not endeavour to advance any argument in favour of this practice, only remark, that it is most strictly observed, both in merchant ships and his Majesty's navy.

To EASE *off*, or EASE *away*, (*filer en douceur, mollir, ou larguer peu-à-peu un cordage*, Fr.) to slacken, gradually, any rope formed into a tackle.

EAST, (*est, levant; l'un des points cardinaux de la boussole*, Fr.) one of the cardinal points of the compass, or of the world; being that point of the horizon where the sun is seen to rise when in the equinoctial.

EBB, *(reflux, ou jusant,* Fr.) the reflux of the tide, or the return of it after the highest of the flood, usually termed full sea, or high-water. See the article Tide.

ECCENTRIC, or Excentric, in geometry, is a term applied to circles and spheres which have not the same centre, and consequently are not parallel; in opposition to concentric, where they are parallel, having one common centre.

Eccentric, or *Eccentric Circle,* in astronomy, is the circle described from the centre of the orbit of a planet, with half the greatest axis as a radius; or it is the circle that circumscribes the elliptic orbit of the planet, as the circle A Q B.

Eccentric *Anomaly,* or *Anomaly of the Centre,* is an arc A Q of the eccentric circle, intercepted between the aphelion A, and the right line Q H, drawn through the centre P of the planet, perpendicular to the line of the apses A B.

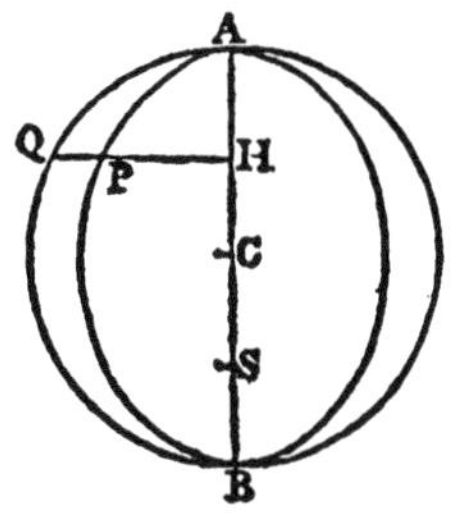

Eccentric *Place of a Planet,* is the point of the orbit, where the circle of inclination coming from the place of a planet in its orbit, falls thereon with right angles.

ECCENTRICITY, in astronomy, is the distance C S (see the above diagram) between the sun S and the centre C of a planet's orbit; or the distance of the centre from the focus of the elliptic orbit, called also the *Simple,* or *Single Eccentricity.*

ECLIPSE, in astronomy, the deprivation of the light of the sun, or of some heavenly body, by the interposition of another heavenly body between our sight and it.

Partial Eclipse, is when only a part of the luminary is eclipsed.

Total Eclipse, is that in which the whole luminary is darkened.

Eclipse *of the Moon,* is a privation of the light of the moon, occasioned by an interposition of the body of the earth directly between the sun and moon, and so intercepting the sun's rays that they cannot arrive at the moon, to illuminate her. Or, the obscuration of the moon may be considered as a section of the earth's conical shadow, by the moon passing through some part of it.

The manner of this eclipse is represented in plate
Fig. 13. IX. fig. 13, where S is the sun, E the earth, and M or M the moon.

Lunar eclipses can only happen at the time of full moon, because it is only then the earth is between the sun and moon; nor do they happen every full moon, because of the obliquity of the moon's path with respect to the sun's, but only in such full moons as happen either at the intersection of those two paths, called the moon's nodes, or very near them; viz. when the moon's latitude, or distance between the centres of the earth and moon, is less than the sum of the apparent semi-diameters of the moon and the earth's shadow.

The chief circumstances in lunar eclipses are the following:—

1. All lunar eclipses are universal, or visible in all parts of the earth which have the moon above their horizon; and are every where of the same magnitude, with the same beginning and end. 2. In all lunar eclipses, the eastern side is what first immerges and emerges again, *i.e.* the left-hand side of the moon as we look towards her, from the north; for the proper motion of the moon being swifter than that of the earth's shadow, the moon approaches it from the west, overtakes and passes through it with the moon's east side foremost, leaving the shadow behind, or to the westward. 3. Total eclipses, and those of the longest duration, happen in the very nodes of the ecliptic; because the section of the earth's shadow, then falling on the moon, is considerably larger than her disc. There may, however, be total eclipses within a small distance of the nodes; but their duration is the less as they are farther from it, till they become only partial ones, and at last none at all. 4. The moon, even in the middle of an eclipse, has usually a faint appearance of light, resembling tarnished copper, which astronomers attribute to the light of the sun, refracted by the earth's atmosphere, and so transmitted thither. Lastly, She grows sensibly paler and dimmer, before entering into the real shadow, owing to a penumbra which surrounds that shadow to some distance.

Eclipse *of the Sun,* is an occultation of the sun's body, occasioned by an interposition of the moon between the earth and sun. On which account, it is by some considered as an eclipse of the earth, since the light of the sun is hid from the earth by the moon, whose shadow involves a part of the earth.

The manner of a solar eclipse is represented at fig.
14, plate IX, where S is the sun, *m* the moon, and Fig. 14
C D the earth, *r m s o* the moon's conical shadow, travelling over a part of the earth C *o* D, and making a complete eclipse to all the inhabitants residing in that tract, but no where else; excepting that for a large space around it there is a fainter shade, included within all the space *r* C D *s,* which is called the *Penumbra.*

Hence, solar eclipses happen when the moon is in conjunction with the sun, or at the new moon, and also in the nodes or near them, the limit being about 17 degrees on each side of it; and such eclipses only happen when the latitude of the moon, viewed from the earth, is less than the sum of the apparent semi-diameters of the sun and moon; because the moon's way is oblique to the ecliptic, or sun's path, making an angle of nearly 5° 35′ with it.

In the nodes, when the moon has no visible latitude, the occultation is total; and with some continuance, when the disc of the moon in perigee appears greater than that of the sun in apogee, and its shadow is extended beyond the surface of the earth; and without continuance at moderate distances when the cusp, or point of the moon's shadow, barely touches the earth. Lastly, out of the nodes, but near them, the eclipses are partial.

The other circumstances of solar eclipses are, 1. That none of them are universal; that is, none of them are seen throughout the whole hemisphere which the sun is then above; the moon's disc being much too little, and much too near the earth, to hide the sun from the whole disc of the earth.

Commonly, the moon's dark shadow covers only a spot on the earth's surface, about 180 miles broad when the sun's distance is greatest, and the moon's least. But her partial shadow, or penumbra, may then cover a circular space of 4900 miles in diameter, within which the sun is more or less eclipsed, as the places are nearer to or farther from the centre of the penumbra. In this case, the axis of the shade passes through the centre of the earth, or the new moon happens exactly in the node, and then it is evident that the section of the shadow is circular; but in every other case the conical shadow is cut obliquely by the surface of the earth, and the section will be an oval, and very nearly a true ellipsis.

2. Nor does the eclipse appear the same in all parts of the earth, where it is seen; but when in one place it is total, in another it is only partial. Moreover, when the moon appears much less than the sun, as is chiefly the case when she is in apogee and he in perigee, the vertex of the lunar shadow is then too short to reach the earth, and though she be in a central conjunction with the sun, is yet not large enough to cover his whole disc, but lets his limb appear like a lucid ring or bracelet, and so causes an *annular eclipse*. According to M. du Sejour, an eclipse can never be annular longer than 12′ 24″, nor total longer than 7′ 58″.

3. A solar eclipse does not happen at the same time, in all places where it is seen; but appears more early to the western parts, and later to the eastern; as the motion of the moon, and consequently of her shadow, is from west to east.

4. In most solar eclipses, the moon's disc is covered with a faint light, which is attributed to the reflexion of the light from the illuminated parts of the earth.

Lastly, in total eclipses of the sun, the moon's limb is seen surrounded by a pale circle of light; which some astronomers consider as an indication of a lunar atmosphere; but others as the atmosphere of the sun, because it has been observed to move equally with the sun, and not with the moon; and besides, it is generally believed that the moon is without any atmosphere, unless it be one that is very small, and very rare.

An eclipse of the sun, or rather of the earth, with respect to any particular place, may be calculated exactly in the same manner as an eclipse of the moon; that is, the times when the moon's umbra or penumbra first touches and leaves the earth: but to find the times of the beginning, middle, and end, at any particular place, the apparent place of the moon, as seen from thence, must be determined, and consequently its parallax in latitude and longitude must be computed.

ECLIPTIC, in astronomy, a great circle of the sphere, supposed to be drawn through the middle of the zodiac, making an angle with the equinoctial of about 23° 30′, which is the sun's greatest declination; or, more strictly speaking, it is that path or way among the fixed stars, that the earth appears to describe, to an eye placed in the sun.

Obliquity of the ECLIPTIC, is the angle which its plane makes with that of the equinoctial. The inclination of the equator to the ecliptic is measured by the arc of a great circle intercepted between their poles, which was taken with very great accuracy by the late Dr. Maskelyne, in the year 1769, and found to be 23° 28′ 10″, or 23° 46944. It was formerly found by Dr. Bradley to be 23° 28′ 30″, who supposed that there was a gradual approach of the ecliptic to the equinoctial at the rate of 1′ in 100 years. The mean obliquity of the ecliptic, is augmented by 9″, when the moon's ascending node is in the vernal equinox. It is, on the contrary, diminished 9″, when the node is in the autumnal equinox, and it is equal to the mean when the node is in the colure of the solstices. This change of the inclination of the earth's axis to the plane of the ecliptic, was called the nutation of the axis by Sir Isaac Newton.

It is now well known, that this change in the obliquity of the ecliptic is wholly owing to the actions of the planets upon the earth, and especially the planets Venus and Jupiter.

According to La Grange, who proceeds upon theory, the annual change of obliquity is variable, and has its limits; about 2000 years, he thinks it was after the rate of about 38″ in 100 years; that it is now, and will be for 400 years to come, 56″ per century; but 2000 years hence 49″ per century. According to Cassini, who computes from observations of the obliquity between the years 1739 and 1778, the annual change then was 60″ or 1′ in 100 years. But according to La Lande, the diminution is at the rate of 88″ per century; while the late Dr. Maskelyne makes it only 50″ in the same time.

Besides the regular diminution of the obliquity of the ecliptic, at the rate of near 50 seconds in a century, or half a second a year, which arises from a change of the ecliptic itself, it is subject to two periodical inequalities, the one produced by the unequal force of the sun in causing the precession of the equinoxes, and the other depending on the nutation of the earth's axis. See the Explanation and Use of the late Dr. Maskelyne's Tables and Observations, p. 6, where we are shewn how to calculate those inequalities, and where he shews, from his own observations, the mean obliquity of the ecliptic.

To find the obliquity of the ecliptic, or the greatest declination of the sun; about the time of the summer solstice, observe very carefully the sun's zenith distance for several days together, then the difference between this distance and the latitude of the place, will be the obliquity sought, when the sun and equator are both on one side of the place of observation; but their sum will be the obliquity when they are on different sides of it. Or, it may be found by observing the meridian altitude or zenith distance of the sun's centre, on the days of the summer and winter solstice; then the difference of the two will be the distance between the tropics, the half of which will be the obliquity sought.

From eighteen observations made at the summer, and thirteen at the winter solstices, reduced to January 1807, the following is the result:

	True	Zen.	Dist.
Sum. sols. .	28°	0′	8″,68
Winter do. .	74	55	56,02
	102	56	4,70
Latitude	51	28	2,35

The mean obliquity of the ecliptic for the same period will be

74°	55′	56″,02
28	0	8,68
46	55	47,34
23	27	53,67

The new solar tables state the obliquity of the ecliptic for 1800, at 23° 27′ 57″, and taking the decrease at half a second per annum, shews the above result to be very near the truth.

EDDY-WATER, (*remoux*, Fr. *ed*, backward, again, and *ea*, water, Sax.) the water that, by some interruption in its course, runs contrary to the direction of the tide or current, and appears like the motion of a whirlpool.

To EDGE AWAY FROM A COAST, (*s'éloigner, ou s'écarter en dépendant d'une côte, &c.* Fr.) to decline gradually from the shore, or from the line of the course which the ship formerly steered; it is particularly applied when a ship changes her course, by sailing *larger*, or more *afore* the wind than she had done before.

To EDGE *in with a Ship*, (*joindre un vaisseau en dépendant*, Fr.) to advance gradually towards the shore, or any other object.

EIGHTS, a former customary, but otherwise unauthorized, deduction of pursers, in the issuing of provisions.

EIKING, in ship-building, a piece fitted to make good a deficiency in length, as the lower part of the supporter under the cat-head, &c likewise the piece of carved-work under the lower end of the quarter-piece, at the aft part of the quarter-gallery.

ELBOW IN THE HAWSE, (*demi-tour dans les cables*, Fr.) a particular twist in the cables by which a ship rides at anchor. In this situation each of the cables, after crossing the other before the stem, is directed outwards on the same *bow* from which it issued: that is to say, the starboard cable *grows* out on the starboard bow, and the larboard cable on the larboard bow. See the article HAWSE.

ELEVATION, the same with altitude or height.

Angle of ELEVATION, in gunnery, that comprehended between the horizon and the line of direction of a cannon or mortar; or it is that which the chase of a piece, or the axis of its hollow cylinder, makes with the plane of the horizon.

ELM, (*orme*, Fr.) is a timber of peculiar property and singular use in situations where it is continually exposed either to the wet, or to be kept dry. Its grain being rough and curly, causes it to be not liable to split. If felled between the months of November and February, it is found to have no sap, and is of peculiar service in mast-making, for the caps, bibbs, bees, and part of the tops.

ST. ELMOS FIRE, or JACK WITH A LANTERN. See the article CORPOSANT.

ELONGATION, in astronomy, the digression or recession of a planet from the sun, with respect to an eye placed on our earth The term is chiefly used in speaking of Venus and Mercury, the arc of a great circle intercepted between either of these planets and the sun, being called the *elongation* of that planet from the sun. But here it is to be observed, that it is only a circle which has the sun for its centre; that the greatest elongation is in a line touching the planet's orbit. For in an elliptic orbit it may be, that the elongation from the sun may grow still greater, even after it has left the place where the line joining the earth and planet touches its orbit. For after that, the true distance of the planet from the sun may increase, whilst the distance of the sun and planet from the earth does not increase, but rather decrease.

But because the orbits of the planets are nearly circular, such small differences are frequently neglected in astronomy.

The greatest elongation of Venus is found by observation to be about 48°, and of Mercury only about 28°, and on that account this planet is rarely seen with the naked eye, being lost in the light of the sun.

Angle of ELONGATION, is an angle contained under lines supposed to be drawn from the centres of the sun and planet to the centre of the earth.

EMBARGO, (*embargo*, Fr. *embargar*, Span.) an arrest laid on ships or merchandise by public authority, or a prohibition of state, sometimes general, to prevent all ships departing, and sometimes partial or particular, as upon foreign ships only, or to prevent their coming in.

EMBARKATION, (*embarquement*, Fr.) the act of shipping, or putting on board troops and stores of all kinds, intended for the public service. As this operation is extremely important, and rendered much less difficult on being properly regulated and attended to, the following instructions will, perhaps, prove useful:—

The first thing necessary to the embarkation of ordnance and stores, is that of preparing a list of all the articles to be embarked, with the weight of each. This list must have a large column for remarks. The tonnage required for bulky articles will be, generally, one-third more than their actual weight; but the tonnage of ordnance, shot, shell, &c. will be equal to their weight.

If vessels be paid according to the tonnage they carry, the masters will, of course, stow away as much as the ships will hold; but if by the voyage, they will be averse to loading their ships too much; a naval officer should, therefore, always attend to see that the ships are properly stowed.

Ordnance and stores may be embarked, either for the purpose of merely transporting them to another situation, or for a military expedition.

In the first case, each ship must be stowed with as much as it will carry, and every article that relates to one particular species of service, or ordnance, must be put on board the same ship: that, in case one ship be lost, the others may remain in themselves complete. This principle must, of course, be likewise attended to in an embarkation for an expedition; but a more particular distribution must take place of the stores when on board. With each piece of ordnance must be placed every thing necessary for its service; its side arms, carriage, timber, ammunition, &c. so as to be readily come at, when required to be disembarked.

If it be an embarkation of ordnance, &c. for a siege, not only every thing necessary for the service of the pieces of ordnance should be arranged with them, but also every thing necessary for the construction of the battery on which they are mounted. It will be advisable, in this case, to put different natures of ordnance in the same ship, in proportion, according to the service required of them. In general it will be best to put the heavy articles in first, and every thing that is light, easy to be moved, or likely to be first wanted. on the top. Previous to embarkation, the guns, carriages, waggons, &c. must be dismounted, but first numbered as follows, and the number of each article marked in the list, in the column of remarks: give each piece of ordnance and its carriage the same number; give the ammunition and other carriages different numbers from the ordnance carriages. Then give every limber, whether of ordnance carriage, ammunition carriage, or waggon, the number of its respective carriage. If, for a simple transport, arrange the small stores, side-arms, &c. according to their several natures; but if for an expedition, every thing belonging to each particular piece of ordnance must be collected together, and the cases or chests in which they are put, marked with the number of the piece of ordnance to which they belong, their natures and description. If there be any doubt of the different parts of the carriages being made with that uniformity so essentially necessary, every part which is separated must bear the number of its carriage. This precaution, at any rate, may be a good one, if the same vessel contain different natures of ordnance or carriages. The axletrees need not be taken off the carriages if the vessels be of a sufficient size to admit them when fixed, as they are not easily replaced without workmen and a tedious operation.

When a carriage is dismounted, all the small articles, such as elevating-screws, linch-pins, drag washers, cap squares, &c. must be carefully collected and secured in a box, marked with the description of stores, and number of the carriage to which they belong.

All carriages or waggons embarked with their axletrees fixed, must be arranged in the ship, side by side, and alternately front and rear, that their axletrees may not interfere with each other, and take too much room. Every transport or other vessel employed in carrying troops or stores for an expedition, should be numbered on the quarters and on the bows, with figures as large as two or three feet, and on the sails, that they may be known at a distance. The number of the ship, her name and tonnage, and the master's name, must be entered in the list of the stores which she carries.

In disembarking ordnance and stores, they must be landed exactly in order, the reverse of what they were shipped. The carriages and waggons must be mounted as soon as possible, and every nature must be arranged as far from the shore as possible to prevent confusion.

If the disembarkation take place in the presence of an enemy, the vessel of course must be loaded accordingly; and the field ordnance, with their carriages, ammunition, &c. must be so arranged as to be first landed, and with the greatest ease possible. In this case, the entrenching tools must also be kept in the greatest readiness.

Embarkation of Troops. All transports taken into his Majesty's service, are under the direction of the Transport Board, and of their agents at the different ports at home and abroad. No troops, or other persons, can be put on board them, or victualled, but by an order from this Board, or one of its agents. Troops embarked on board transports or ships of war (excepting marines) are only allowed two-thirds of a seaman's allowance of provisions. It is, therefore, necessary to divide the men into messes of six each.

Six women to one hundred men embarked on foreign service, are allowed provisions; and ten women to one hundred men on home service. The births on board transports, are usually made six feet square, and each admits four men at a time; but agreeable to the regulations for the army for the year 1799, one-third of the men are always to be on deck; therefore, six men (or one mess) are told off to each birth, one-third of whom are always on watch. The commanding officer of the troops on board a transport, has a right to peruse the charter-party of the ship, which points out every different article, as firing, candles, boats, utensils, &c. which the ship is engaged to find for the use of the troops on board. It likewise expresses the part of the ship allotted to the officers, to the master, the mate, and the agent, should there be one on board.

EMBAYED, from *bay*, (*affalé sur la côte*, Fr.) the situation of a ship when she is enclosed between two capes or promontories: it is particularly applied when the wind, by blowing strongly into any bay or gulf, makes it extremely difficult, and perhaps impracticable for the vessel, thus enclosed, to *draw* off from the shore, so as to weather the capes and gain the offing.

EMERSION, in astronomy, is when any planet that is eclipsed begins to emerge or get out of the shadow of the eclipsing body. It is also used when a star, before hidden by the sun as being too near him, begins to re-appear or emerge out of his rays.

END-FOR-END, (*bout-pour-bout*, Fr.) A reversal of the disposition of any thing is turning it *end-for-end*. It is applied also to a rope that has run quite out of the block in which it was reeved; or to a cable which has all run out of the ship.

END-*On*, (*de bout-au-corps*, Fr.) When a ship advances to a shore, rock, &c. without an apparent possibility of preventing her, she is said to go *end-on* for the shore, &c.

ENGAGEMENT, (*combat naval*, Fr.) in a naval sense, implies a battle at sea; or an action of hostility between single ships, squadrons, or fleets of men of war.

The reader who wishes to be thoroughly acquainted with this subject, must consult all the articles which have any tendency thereto; viz. CANNON, DIVISION, EXERCISE, FLEET, and LINE OF BATTLE.

The *engagements* of the ancients were usually carried on in two different manners. Advanced by the force of their oars, the gallies ran violently *aboard* of each other, and by the mutual encounter of their beaks and prows, and sometimes of their sterns, endeavoured to dash in pieces, or sink their enemies.

For this purpose the prow was commonly armed with a brazen point or trident, nearly as low as the surface of the water. Some of the gallies were furnished with large turrets, and other accessions of building, either for attack or defence. The soldiers also annoyed their enemies with darts and slings, and, on their nearer approach, with swords and javelins;

and, in order that their missive weapons might be directed with greater force and certainty, the ships were equipped with several platforms, or elevations, above the level of the deck. The sides of the ship were fortified with a thick fence of hides, which served to repel the darts of their adversaries, and to cover their own soldiers, who thereby annoyed the enemy with greater security.

As the boring and sinking the enemy's ships with the rostra was the chief manner of sea engagements among the ancients, high and bulky ships had, accordingly, a great advantage over their adversaries, by the force of the stroke of a large ship. The height was likewise no small convenience in boarding and throwing of missile weapons, so that it was much more true among them than among us, that a little ship durst not lay her side to a great one; and though great ships were commonly bad sea vessels, they had a superior force in a sea engagement, the shock of them being sometimes so violent that it would throw a crew on the upper deck of lesser ships overboard. This occasioned the ancients gradually to increase the bulk of their ships, till they came, at last, to an enormous size.

Several of the machines which were employed by the ancients in their naval engagements are now unknown: the following are a few recorded by the ancient writers:—

The Dolphin, which was a large and massy piece of lead or iron, cast in the form of a dolphin; and, being suspended by blocks at their mast-heads or yard-arms ready for a proper occasion, was let down violently from thence into the adverse ships, and either penetrated through the bottom, and opened a passage for the entering waters, or, by its weight, immediately sunk the vessel.

The Drepreiian was an engine of iron crooked like a sickle, aud fixed on the top of a long pole; the use of which was, to cut asunder the *slings* of the sail-yards, and thereby letting the sails fall down, to disable the vessel from escaping, and incommode her greatly during the action. Similar to this was another instrument, armed at the head with a broad two-edged blade of iron, wherewith they usually cut away the ropes that fastened the rudder in the vessel.

They had also a sort of spears or maces, of an extraordinary length, sometimes exceeding twenty cubits: also certain machines used to throw large stones into the enemies ships.

Vegetius mentions another engine, which was suspended to the main-mast, and resembled a battering-ram; for it consisted of a long beam, and a head of iron, and was, with great violence, pushed against the sides of the enemies galleys.

They had also a grappling-iron, which was usually thrown into the adverse ships by means of an engine; this instrument facilitated the entrance of the soldiers appointed to *board*, which was done by means of wooden bridges, that were generally kept ready for this purpose in the fore-part of the vessel.

The arms used by the ancients rendered the disposition of their fleets very different, according to the time, place, and circumstances. They generally considered it an advantage to be to windward, and to have the sun shining directly on the front of their enemy. The order of battle chiefly depended on their power of managing the ships, or of drawing them readily into form; and on the schemes which their officers had concerted. The fleet being composed of rowing vessels, they lowered their sails previous to the action; they presented their prows to the enemy, and advanced against each other by the force of their oars. Before they joined battle, the admirals went from ship to ship, and exhorted their soldiers to behave gallantly. All things being in readiness, the signal was displayed by hanging out of the admiral's galley a gilded shield, or a red garment or banner. During the elevation of this the action continued, and, by its depression or inclination towards the right or left, the rest of the ships were directed how to attack or retreat from their enemies. To this was added, the sound of trumpets, which began in the admiral's galley, and continued round the whole navy. The fight was also begun by the admiral's galley, by grappling, boarding, and endeavouring to upset, sink, or destroy the adversary. Sometimes, for want of grappling-irons, they fixed their oars in such a manner as to hinder the enemy from retreating.

The first warlike preparations which the Romans made by sea, and the order which they observed in drawing up their fleet for battle, are recorded by Polybius:—Soon after the expulsion of Tarquin, Brutus and Horatius being consuls, the Romans were restrained, by a treaty with Carthage, from sailing be yond the Fair Promontory, &c. and indeed they had then such ordinary vessels, that when they resolved to contend for the dominion of the sea with the Carthaginians, who had held it uncontested from their ancestors, they began anew, by building a whole fleet, after the model of one of their enemy's gallies, that was stranded on their coast; and, as they never wanted expedients in their military concerns, they placed banks of rowers on board, in the same form as those of the gallies, and instructed their men to strike and recover their oars by a proper signal, till they were so perfect in the exercise, and so expert in their discipline and management of their fleet.(which was improved with the Corvus, for the purpose of boarding, as already described under that article) that they soon defeated their enemies.

If they could not manage their oars as dexterously as their antagonist, or fall alongside so as to board him, they penetrated his vessel with the brazen prow. The vessels approached each other as well as their circumstances would permit, and the soldiers were obliged to fight hand to hand, till the battle was decided: nor, indeed, could they fight otherwise with any certainty, since the shortest distance rendered their slings and arrows, and almost all their offensive weapons, ineffectual, if not useless. The squadrons were sometimes ranged in two or three right lines, parallel to each other; being seldom drawn up in one line, unless when formed into a crescent. This order, indeed, appears to be the most convenient for rowing vessels, that engage by advancing with their prows towards the enemy. At the battle of Ecnomus, between the Romans and the Carthaginians, the fleet of the former was ranged into a triangle, or a sort of wedge in front, and towards the middle of its depth, of two right parallel lines. That of the latter was formed into a rectangle, or two sides of a square, of which one branch extended behind, and, as the opening of the other prosecuted the attack was ready to fall upon the flank of such of the Roman gallies as should attempt to break their line. Ancient history has preserved many of these orders, of which some have been followed in later times. Thus in a battle in the year 1340, the English fleet was formed in two lines, the first of which contained the larger ships, and the second consisted of all the smaller vessels, used as a reserve to support the former whenever necessary. In 1545, the French fleet under the command of the Mareschal d'Annebault, in an engagement with the English in the Channel, was arranged in the form of a crescent. The whole of it was divided into three bodies, the centre being composed of thirty-six ships, and each of the wings of thirty He had also many gallies; but these fell not into the line, being designed to attack the enemy occasionally. This last disposition was continued down to the reigns of James I. and Louis XIII.

The following curious description of the method of attack and defence, as practised in the royal navy about the twelfth century, is translated from an ancient treatise, called *Speculum Regale.*

"In a naval engagement provide yourself with two spears, which you must not lose in throwing; let one of them be a long one, capable of reaching out of one vessel to another, the other with a shorter handle, so that you may be able to use it conveniently on boarding an enemy. Divers darts are to be used in a sea-fight, as well as heavy spears fenced with iron as the lighter ones, and heated like a dart. On ship-board you must fight with an undaunted mind and wary strokes; nor must your arms be cast away in an inconsiderate manner. In sea-fights, scythes firmly fixed to very long spears, axes, with broad blades, fixed to long handles, boathooks, slings fixed to a staff, catajac and others of that kind, stones, the bow, and the rest of the missive weapons, are indispensible; but of these, bituminous sea-coal mixed with sulphur, holds the first place. Towers are also essential; from which you may use the arms above-mentioned: also a leaver or crow, its four sides being armed with iron nails."

The invention of gunpowder took place in 1330, and the use of fire-arms was gradually introduced into naval war, without finally superseding the ancient method of engagement. The Spaniards were armed with cannon in a sea-fight against the English and the people of Poitou abreast of Rochelle, in 1372; and this battle is the first wherein mention is made of artillery in our navies. Many years elapsed before the marine armaments were sufficiently provided with fire-arms. So great a revolution in the manner of fighting, and which necessarily introduced a total change in the construction of ships, could not be suddenly effected. In short, the squadrons of men of war are no longer formed of rowing-vessels, or composed of gallies and ships of the line, but intirely of the latter, which engage under sail, and discharge the whole force of their artillery from their sides. Accordingly, they are now disposed in a right line parallel to the enemy; every ship keeping *close-hauled* upon a wind on the same tack. Indeed, the difference between the force and manner of fighting of ships and gallies rendered their service in the same line incompatible. When we consider the change introduced, both in the construction and working of ships, occasioned by the use of cannon, it necessarily follows, that squadrons of men of war must appear in the order that is now generally adopted. Finally, the ships ought to present their broadsides to the enemy; and to sail close upon a wind in the wake of each other; as well to retain their own uniformity, as to preserve or acquire the advantage which the weather-gage gives them over their adversary.

As a general engagement of fleets or squadrons of ships of war is nothing else than a variety of particular actions of single ships with each other, in a line of battle; it appears necessary, according to the plan of this work, to begin by describing the latter, and then proceed to represent the usual manner of conducting the former.

The whole economy of a naval engagement may be arranged under the following heads; viz. 1. the preparation; 2. the action; and 3. the repair, or refitting for the purposes of navigation.

The preparation is begun by issuing an order to clear the ship for action, which is repeated by the boatswain and his mates at all the *hatchways*, leading to the different batteries. In a ship of war the management of the artillery requires a considerable number of men, and the officers and sailors are, consequently, restrained to a narrow space in their usual habitations, in order to preserve the internal regularity of the ship. Accordingly, the *hammocks*, or hanging-beds, are crowded together as close as possible between the decks, each of them being limited to the breadth of fourteen inches, and are hung parallel to each other, in rows stretching from one side of the ship to the other, nearly throughout her whole length, so as to admit of no passage but by stooping under them. While in this situation it would be impossible to work the cannon, and therefore they must be removed with the greatest expedition. At the summons of the boatswain, who cries, "*Up all hammocks!*" every sailor repairs to his own, and, having stowed his bedding properly, he cords it firmly with a *lashing*, or line, provided for that purpose, and carries it to the quarter-deck, poop, or forecastle, or wherever it may be necessary. As each side of the quarter-deck and poop is furnished, generally, with a double net-work, supported by iron cranes fixed immediately above the *gunwale*, or top of the ship's side, the hammocks, thus corded, are firmly stowed by the quarter-masters between the two parts of the netting, so as to form an excellent barrier. The tops, waist, and forecastle are then fenced in the same manner. During the performance of these offices below, the boatswain and his mates are employed in securing the sails and yards, to prevent them from falling down when the ship is cannonaded, as she might thereby be disabled, and rendered incapable of attack, retreat, or pursuit. The yards are likewise secured by strong chains, or ropes, in addition to those by which they are usually suspended. The boatswain also provides the necessary materials to repair the rigging, wherever it may be damaged by the shot of the enemy; and to supply whatever parts of it may be entirely destroyed. The carpenter and his crew, in the mean time, prepare their shot-plugs and mauls, to close up any dangerous breaches that may be made near the surface of the water; and provide the iron-work necessary to refit the chain pumps, in case their machinery should be injured in the engagement. The gunner with his mates and quarter-gunners are busied in examining the cannon of the different batteries, to see that their charges are thoroughly dry and fit for execution: to have every thing ready for furnishing the great guns and small arms with powder, as soon as the action begins: and to keep a sufficient number of cartridges continually filled, to supply the place of those expended in battle. The master and his mates are attentive to have the sails properly trimmed, according to the situation of the ship; and to reduce or multiply them, as occasion requires, with all possible expedition. The lieutenants visit the different decks, to see that they are effectually cleared of all incumbrance, so that nothing may retard the execution of the artillery: and to enjoin the other officers to diligence and alertness, in making the necessary dispositions for the expected engagement, so that every thing may be in readiness at a moment's warning.

When the hostile ships have approached each other to a competent distance, the drums beat to arms, and the boatswain and his mates pipe "*All hands to quarters!*" at every hatchway. The persons appointed to manage the great guns immediately repair to their respective stations, and crows, handspikes, rammers, sponges, powder-horns, matches, train-tackles, &c. are placed in order by the side of every cannon. The hatches are laid, to prevent any one from deserting his post by escaping into the lower apartments. The marines are drawn up in rank and file, on the quarter-deck, poop, and fore-castle. The lashings of the great guns are let loose, and the tomp ons withdrawn. The whole artillery, above and below, is run out at the ports, and levelled to the point-blank range, ready for firing.

When the necessary preparations are completed, and the officers and crew are ready at their respective stations, to obey every occasional order, the commencement of the action is determined by the mutual distance and situation of the adverse ships, or by the signal from the commander in chief of the fleet or squadron. The cannon being levelled in parallel rows, projecting from the ship's side, the most natural order of battle is evidently to range the ships abreast of each other, especially if the engagement is general. The most convenient distance is properly within the point-blank range of a musket, so that all the artillery may do effectual execution. The action usually begins by a vigorous cannonade, accompanied by the united efforts of all the swivel guns and small arms. As the method of firing platoons, or vollies of cannon at once, is generally found injurious in the sea service, it should seldom or never be attempted, unless in the battering of a fortification: for the sides and decks of the ship, although sufficiently strong for all the purposes of war, would be too much shaken by so violent an explosion and recoil. The general practice on this occasion throughout the ship is to load, fire, and sponge, the guns with all possible expedition, yet without confusion or precipitation. The captain of each gun is particularly enjoined to fire only when the piece is properly directed to its object, that the shot may not be fruitlessly expended. The lieutenants, who command the different batteries, traverse the deck, to see that the battle is prosecuted with vigour; and to exhort and animate the men in their duty. The midshipmen second these injunctions, and give assistance where it is required, at the guns committed to their charge. The gunner takes care that all the artillery is sufficiently supplied with powder, and that the cartridges are carefully conveyed along the decks in covered boxes. The havoc produced by a continuation of this mutual assault can be more easily conceived than described; battering, penetrating, and splintering the sides and decks; shattering or dismounting the cannon; mangling and destroying the rigging; cutting asunder or carrying away the masts and yards; piercing and tearing the sails, so as to render them useless; and wounding, disabling, or killing the ship's company! The comparative vigour and resolution of the assailants to effect these dreadful consequences, generally determine their success or defeat; but sometimes the fate of the combat may be decided by an unforeseen incident, which may be as fortunate for the one as it is fatal to the other. The ship that has been defeated acknowledges the victory by striking her colours, and is immediately taken possession of by the conqueror, who secures her officers and crew as prisoners in his own ship, and invests two principal officers with the command of the prize, until a captain is appointed by the commander in chief.

When the engagement is concluded, they begin the repair, or refitting, for the purposes of navigation. Accordingly, the cannon are secured by their breechings and tackles, with all convenient expedition. Whatever sails have been rendered unserviceable are unbent; and the wounded masts and yards struck upon the deck, and *fished*, or replaced by others: the standing rigging is *knotted*, and the running rigging spliced wherever this is found necessary. Proper sails are bent in the room of those which have been removed as useless. The carpenter and his crew are employed in repairing the breaches made in the ship's hull, by shot plugs, pieces of plank, and sheet lead. The gunner and his assistants are busied in replenishing the allotted number of charged cartridges, to supply the place of those which have been expended, and in refitting whatever furniture of the cannon may have been damaged by the late action.

Such are the usual process and consequences of an engagement between two ships of war, which may be considered as an epitome of a general battle between fleets or squadrons. The latter, however, involves a greater variety of incidents, and necessarily requires more comprehensive skill and judgment in the commanding officer.

When the commander in chief or admiral of a naval armament has discovered an enemy's fleet, his principal concern is, usually, to approach it, and endeavour to come to action as soon as possible. Every inferior consideration must be sacrificed to this important object; and every rule of action should tend to hasten and prepare for so material an event. The state of the wind and the situation of his adversary, will, in some measure, dictate the conduct necessary to be pursued with regard to the disposition of his ships on this occasion. To facilitate the execution of the admiral's orders, the whole fleet is ranged into three squadrons, each of which is classed into three divisions, under the command of different officers.

Before the action begins, the adverse fleets are commonly drawn up in two lines, parallel to each other, and close-hauled. As soon as the admiral displays the signal for the line of battle, the several divisions separate from the columns, in which they were disposed in the usual order of sailing, and every ship crowds into its station in the *wake* of the next a-head; and a proper distance from each other (which is generally about fifty fathoms) is regularly observed from the van to the rear. The admiral, however, will occasionally contract or extend his line, so as to conform to the length of that of his adversary, whose neglect or inferior skill in this respect he will naturally convert to his own advantage, as well as to prevent his own line from being *doubled* upon; a circumstance which might cause great confusion among his van and rear.

When the adverse fleets approach each other, the *courses* are commonly hauled up in the brails, and the top gallant sails and stay-sails furled. The movement of each ship is chiefly regulated by the main and fore-top-sails, and the jib; the mizen-top-sail being reserved to hasten or retard the course of the ship, and, in fine, by *filling* or *backing*, *hoisting* or *lowering* it, to determine her velocity.

The frigates, tenders, and fire-ships, being also hauled upon a wind, lie at some distance, ready to execute the admiral's orders, or those of his seconds, leaving the line of battle between them and the enemy. If there are any transports or store-ships attendant on the fleet, these are disposed at a still further distance from the scene of action. If the fleet is superior in number to that of the enemy, the admiral usually selects a body of reserve from the different squadrons, which will always be of use to cover the fire-ships, bomb-vessels, &c. and may fall into the line in any case of necessity: these also are stationed at a convenient distance from the line, and should evidently be opposite to the weakest parts thereof.

Monsieur de Morogues, a French author, observes, "that order and discipline give additional strength and activity to a fleet. If thus a double advantage is acquired by every fleet, it is certainly more favourable to the inferior, which may thereby change its disposition with greater facility and dispatch than one more numerous, yet without being separated. When courage is equal to both, good order is then the only resource of the smaller number. Hence we may infer that a smaller squadron of ships of war, whose officers are perfectly disciplined in working their ships, may, by its superior dexterity, vanquish a more powerful one, even at the commencement of the engagement; because the latter being less expert in the order of battle, will, by its separation, suffer many of the ships to remain useless or not sufficiently near to protect each other."

The signal for a general engagement is usually displayed when the opposite fleets are sufficiently within the *range* of point blank-shot, so that they may level the artillery with certainty of execution, which is near enough for a line of battle. The action is begun and carried on throughout the fleet in the manner as described between single ships, at which time the admiral carries little sail, observing, however, to regulate his own motions by those of the enemy. The ships of the line, meanwhile, keep close in their stations, none of which should hesitate to advance in their order, although interrupted by the situation of some ship a-head, which has negligently fallen astern of her station.

Such is now the practice of naval war, that the necessary order of battle and the fabric of our ships very seldom permit the assault of *boarding*, unless in single actions. No captain ought, therefore, to abandon his station in the line, under any pretence whatsoever, unless his ship is too much disabled to continue the combat. The small quantity of sail carried on this occasion will permit the bulk of the fleet, although somewhat impaired, to continue their cannonade a long time without quitting the line.

No captain should be induced to break the line through a false ambition to distinguish himself, or with the hope of achieving any distant enterprise, however flattering the prospect may be. He ought to wait the signal of the admiral, or his commanding officer; because it is more essential to preserve the regularity of a close line, which constitutes the principal force of the fleet, than to prosecute a particular action, which, although brilliant in itself, has seldom any material consequences, unless its object is to seize a flag ship, and even this can only be justified by success.

The various exigences of the engagement call forth the skill and resources of the admiral, to keep his line as complete as possible, notwithstanding unequal attacks and damages. He must order ships from those in reserve, to supply the place of others which may have been rendered unqualified by the action: he must direct his fire-ships, at a convenient time, to fall aboard the enemy; and he must attack ships from one part of the line, or wing which is stronger, to another which is greatly pressed by superior force, and requires assistance. His vigilance is ever necessary to review the situation of the enemy from van to rear, every motion of whom he should, if possible, anticipate and frustrate. He should seize the favourable moments of occasion, which are rapid in their progress, and never return: an opportunity lost may lose the victory. Far from being disconcerted by any unforeseen incident, however distressing it may be, he should endeavour to overcome all difficulties, and make them, if possible, subservient to his designs. His experience and reflection will naturally furnish him with every method of intelligence to discover the state of his different squadrons and divisions. Signals of enquiry and answers; of request and assent; of command and obedience, must be displayed and repeated on this occasion. Tenders and boats must also continually be detached between the admiral and the commanders of the several squadrons or divisions.

When danger presses he should be fortified by resolution and presence of mind, because the whole fleet is committed to his charge, and the conduct of his officers may, in a great degree, be influenced by his intrepidity and perseverance. In short, his renown or infamy may depend on the fate of the day.

If he proves victorious he should prosecute the victory as much as possible, by seizing, burning, or otherwise destroying the enemy's ships. If he is defeated, he should endeavour, by every resource his experience can suggest, to save as many of his fleet as possible, by employing his tenders, &c. to take out the wounded and put fresh men in their places; by towing the disabled ships to a competent distance, and by preventing the execution of the enemy's fire-ships. In order to retreat with more security, he may judge it expedient to range his fleet into the form of an half-moon, or crescent, placing himself in the centre. By this disposition the enemy's ships which attempt to fall upon his rear will at once expose themselves to the fire of the admiral and his seconds, in an advantageous situation; a circumstance which will serve to facilitate the escape of his own ships, and retard the pursuit of those of his adversary.

Should his fleet be too much extended by this arrangement, the wings or quarters are easily closed, and the half-moon rendered more complete; in the midst of which may be placed his store-ships, tenders, &c. In retreating, the uncertainty of the weather is to be considered: it may become calm, or the wind may shift in his favour. His schemes may be assisted by the approach of night, or the proximity of the land; and he ought rather to run the ships ashore, if practicable, than suffer them to be taken afloat, and thereby transfer additional strength to the enemy. In short, nothing should be neglected that may contribute to the preservation of his fleet, or prevent any part of it from falling into the hands of the conqueror. Upon the whole, it appears, the real force or superiority of a fleet consists less in the number of vessels and the vivacity of the action, than in good order, dexterity in working the ships, presence of mind, and skilful conduct in the admiral and captains.

ENGINE. See the article FIRE-ENGINE.

ENSIGN, (*enseigne ou pavillon de poupe*, Fr.) a large flag, or banner, hoisted on a long pole erected over the poop, and called the ensign-staff.

The ensign is used to distinguish the ships of different nations from each other, as also to characterize the different squadrons of the navy.

The British ensign, in ships of war, is known by a double cross, viz. that of St. George and St. Andrew, formed into an union, upon a field which is either red, white or blue.

To ENTER (*s'enrôler*, *s'engager*, Fr.) to enrol or engage for service.

ENTERING-PORT, (*port d'entrée des vaisseaux, à trois ponts*, Fr.) a large port cut down on the middle gun-deck of three-deckers, to serve as a door for persons going in or coming out of the ship.

ENTERING-*Ropes*, or *Man-Ropes*, (*tire-veilles*, Fr.) are two ropes hanging from the upper-part of a ship's side, on the right and left of the ladder, or steps leading up to the gang-way. See the article GANGWAY.

ENTRANCE, (*avant du vaisseaux*, Fr.) a name frequently given to the foremost part of a ship under the surface of the sea.

EPAULETTE, (*épaulette*, Fr.) a shoulder-knot. It is a mark of distinction, made either of silver or gold lace. For the navy they are made of gold, and worn on one or both shoulders, according to the rank of the wearer.

The following are the gradations of rank, as distinguished by the *epaulettes* worn by officers in the royal navy, agreeable to an order from the Admiralty, dated March 23d, 1812:—Captains above three years post have two epaulettes, with a silver crown over a silver anchor; and post captains under three years have two, with a silver anchor without the crown: commanders have two plain epaulettes; and lieutenants one plain one on the right shoulder: rear-admirals have two epaulettes, with one star on the strap of each: vice-admirals two stars; and admirals, three stars. The marines used to wear two silver epaulettes, but, in consequence of their good conduct during the mutiny, they are now styled royal, and wear the gold epaulettes, according to their respective ranks.

EPHEMERIS. See the article ALMANAC, *Nautical.*

EQUATOR, (*équateur*, Fr.) in geography, a great circle of the terrestrial globe, equi-distant from its poles, and dividing it into two equal hemispheres, the one called north and the other south.

Crossing the EQUATOR, a term generally understood by mariners to mean sailing across the line supposed to divide the earth into northern and southern hemispheres, when it has been a common custom, from time immemorial with our sailors, both in the navy and in the merchant-service, to perform a certain ceremony upon those who have never crossed it, or who cannot give a satisfactory account of their having done so; which is as follows:—

Two of the most suitable persons are selected by the superior officers to preside over the ceremony, and to see that it is executed according to ancient custom. One is distinguished by the name of *Neptune*, who is dressed and painted in the character of that heathen god, having a trident in his right hand and a sort of crown on his head; the other is named *Amphitrite*, the supposed wife of Neptune, whose dress is also made suitable to the character. Next a gang is formed from the men who have before crossed the equator, and they give notice to the ship's company and passengers, if any, of the approach of their expected visitors. They then order the largest tubs in the ship to be brought upon deck, and to be filled with salt-water; across each tub a spar or handspike is placed, as a seat, whereon those whom Neptune finds have not crossed the equator are set, whilst they are lathered with a composition of tar, grease, &c. and *shaved with a piece of old iron hoop*, whose edge is well notched. This being done, each person attempts to rise from his hazardous seat, when it is generally so contrived that he falls heels over head into the tub of water; and, after having gone through this ceremony, he becomes *free of the equator.*

All persons solicitous of purchasing their freedom of old Neptune, bribe the superintendents of the ceremony with a gallon of rum and a few pounds of sugar; but if they neglect to do this, they must expect to go through the whole of the operation.

EQUINOCTIAL, *la ligne équinoctiale*, Fr.) in astronomy, a great circle in the heavens, under which the equator moves in its diurnal motion.

The poles of this circle are the poles of the world. It divides the sphere into two equal parts, the northern and southern. It cuts the horizon of any place in the east and west points; and, at the meridian, its elevation above the horizon is equal to the co-latitude of the place. The equinoctial has also various other properties; as,

1. Whenever the sun comes to this circle, he makes equal days and nights all round the globe; because he then rises due east, and sets due west. Hence it has the name *equinoctial.*

All stars directly under this circle, have no declination, and always rise due east and set full west.

2. All people living under this, or upon the equator, or line, have their days and nights at all times equal to each other.

3. From this circle, on the globe, are counted, upon the meridian, the declination in the heavens and the latitude on the earth.

4. Upon the equinoctial, or equator, is counted the longitude, making in all 360° quite round, or 180° east and 180° west.

5. And, as the natural day is measured by one revolution of the equator, it follows that one hour answers to $\frac{360}{24} = 15$ degrees: hence, one degree of the equator will contain four minutes of time; 15 minutes of a degree will make a minute of an hour; and, consequently, four seconds answer to one minute of a degree.

And 6. The shadows of those people who live under this circle are cast to the southward of them for one half of the year, and to the northward during the other half; and twice a year, viz. at the time of the equinoxes the sun, at noon, casts no shadow, being exactly in their zenith.

EQUINOX, (*équinoxe*, Fr.) the time when the sun enters either of the equinoctial points, where the ecliptic intersects the equinoctial. It is so called because, when the sun is in these points the days and nights are of an equal length all the world over. As the sun is in one of them in the spring, viz. about the 20th of March, it is called the vernal equinox; and in the other in autumn, viz. about the 23d of September, it is called the autumnal equinox.

As the sun's motion is unequal, being sometimes quicker and sometimes slower, it happens that there are about eight days more from the vernal to the autumnal equinox, or while the sun is on the northern side of the equator, than while he is moving through the southern signs from the autumnal to the vernal equinox, or on the southern side of the equator. According to the observations of M. Cassini, the

Sun is	186^d	14^h	53^m	in the northern signs,
and only	178	14	56	in the southern signs.
So that	7	23	57	is the difference of them, or nearly eight days.

EQUINOXES, *Precession of the*, is a very slow motion of them, by which they change their place going from east to west, or contrary to the order of the signs; and from the improvements now made in astronomy, it plainly appears that the poles, the solstices, the equinoxes, and all the other points of the ecliptic, have a retrograde motion, and are constantly moving from east to west, or from *Aries* towards *Pisces*, &c. Hence, as the stars remain immovable, and the equinoxes go backward, the stars will seem to move more eastward with respect to them; for which reason the longitudes of all the stars being reckoned from the first point of Aries, or the vernal equinox, are continually increasing, and the constellations seem all to have changed the places assigned to them by the ancient astronomers, who have been accustomed, ever since the days of Hipparchus, to speak of the *vernal equinox*, and the *summer solstice*, as answering to the first degree of Aries and Cancer; yet the fact is, that the equinoctial and solstitial colures at the above-mentioned seasons, are now in the first degree of *Pisces* and *Gemini.*

The tropics of *Cancer* and *Capricorn* are now really become the tropics of *Gemini* and *Sagittarius:* and it has been clearly demonstrated, by comparing the observations of the ancient with the modern astronomers, that the equinoxes recede about 50''¼ annually, or a sign in the ecliptic in the space of 2150 years.

Diodorus Siculus mentions a *golden circle*, of 365 cubits, answering to the days of the year (placed on the tomb of Osymandias, in Thebes), being carried away by Cambyses, which proves that the Egyptians were acquainted with the solar year from a very remote period of time. Mr. Hamilton, a late traveller, in his Egyptiaca, mentions, with admiration, the great Temple of Esné or Latapolis and describes the signs of the zodiac represented on the cieling, which must have been constructed when the sun at the summer solstice was in the middle of *Leo*, which is there placed as the last of the ascending signs, and which must assign an antiquity to it of more than 5300 years.

He also visited the site of the ancient Tentyra, now called Dendera, and in the great temple there, which appears to have been repaired about the reign of Tiberius, he found the two zodiacs, which have been copied by Denon, a celebrated French traveller, who states that they must have been constructed when the sun at the summer solstice was about the twenty-fourth degree of *Cancer;* therefore cannot be referred to a much later period than 3800 years ago: thus, by a medium of a great number of comparisons, the quantity of the annual change has now been fixed at 50''¼, according to which rate it will require 25,791 years for the equinoxes to make their revolution westward quite round the circle, and return to the same point again.

Professor Robertson, of Oxford in the Phil. Trans. for 1807, after observing, that, although the investigations of Newton respecting this subject, evince the same transcendant abilities as a e displayed in other parts of his immortal works, yet it has been justly asserted that he made a mistake in his process, which rendered his conclusions erroneous; and since this discovery, some of the most eminent mathematicians in Europe have attempted solutions of the problem with various success.

He proceeds to observe, that most of their errors and obscurities have arisen from the uncultivated state of the doctrine of compound rotatory motion: he then enters at large into the solution of the problem, and arrives, by a method not materially different from that of Simpson and La Lande, although, perhaps, in some respects more perspicuously and demonstratively stated, at a conclusion precisely the same as that which these astronomers have already obtained.

To EQUIP, (*équiper, ou armer un vaisseau*, Fr.) a term frequently applied to the business of fitting a ship for sea, or arming her for war. See the article FITTING.

EQUULEUS, the *Horse's Head*, in astronomy, a constellation of the northern hemisphere, whose stars in Flamstead's catalogue are ten.

ERIDANUS, the *River*, in astronomy, a constellation of the southern hemisphere, containing, in the British catalogue, 84 stars.

ESCUTCHEON, (*écusson*, Fr.) the compartment in the middle of the ship's stern, where her name is written, and is more peculiar to the French and other foreigners, than to English-built vessels.

EVEN-KEEL, implies an even position of a ship on the water: thus, a ship is said to swim upon an *even-keel* when she draws the same draught of water forward as abaft.

EXCHANGE, (*bourse*, Fr.) a place of resort for merchants, mariners, &c. in a commercial sea-port.

EXERCISE, (*exercise*, Fr.) is the preparatory practice of managing the artillery and small arms, in order to make the ship's crew perfectly skilled therein, so as to direct its execution successfully in the time of battle.

The exercise of the great guns in our navy, is as follows:

Upon beating to arms (every person having immediately repaired to his quarters), the midshipman, commanding a number of guns, is to see that they are not without every necessary article, as (at every gun) a sponge, powder-horn, with its priming wires, and a sufficient quantity of powder, shot, crow, handspike, bed, quoin, train-tackle, &c. sending, without delay, for a supply of any thing that may be missing; and, for the greater certainty of not overlooking any deficiency, he is to give strict orders to every captain under him, to make the like examination at his respective gun, and to take care that every requisite is in a serviceable condition, which he is to report accordingly. And (besides the other advantages of this regulation) for the still more certain and speedy account being taken upon these occasions, the midshipman is to give each man his charge at quarters, (as expressed in the form of the monthly report) who is to search for his particular implements, and, not finding them, is immediately to acquaint his captain, that, upon his report to the midshipman, they may be replaced.

The man who takes care of the powder, is to place himself on the opposite side of the deck from that where we engage, except when fighting both sides at once, when he is to be amid-ships. He is not to suffer any other man to take a cartridge from him, but he who is appointed to serve the gun with that article, either in time of a real engagement, or at exercise.

Lanterns are not to be brought to quarters in the night, until the midshipman gives his orders for so doing to the person he charges with that article. Every thing being in its place, and not the least lumber in the way of the guns, the exercise begins with,

1st. Silence.—At this word, every one is to observe a silent attention to the officers.

2d. Cast loose your guns.—The muzzle lashing is to be taken off from the guns, and (being coiled up in a small compass) is to be made fast to the eye-bolt above the port. The lashing-tackles at the same time to be cast loose, and the middle of the breeching seized to the thimble of the pomelion. The sponge to be taken down, and, with the crow, handspike, &c. laid upon the deck by the gun.

When prepared for engaging an enemy, the seizing within the clinch of the breeching is to be cut, that the gun may come sufficiently within-board for loading, and that the force of the recoil may be more spent before it acts upon the breeching.

3d. Level your guns.—The breech of your metal is to be raised, so as to admit the foot of the bed's being placed upon the axle-tree of the carriage, with the quoin upon the bed, both their ends being even one with the other.

When levelled for firing, the bed is to be lashed to the bolt which supports the inner end of it, that it may not be thrown out of its place by the violence of the gun's motion, when hot with frequent discharges. See fig. 2, plate VI.

4th. Take out your tompions.—The tompion is to be taken out of the gun's mouth, and left hanging by its lanyard.

5th. Run out your guns.—With the tackles hooked to the upper-bolts of the carriage, the gun is to be bowsed out as close as possible, without the assistance of crows or handspikes, taking care, at the same time, to keep the breeching clear of the trucks, by hauling it through the rings; it is then to be bent so as to run clear when the gun is fired. When the gun is run out, the tackle-falls are to be laid alongside the carriages in neat fakes, that when the gun, by recoiling, overhauls them, they may not be subject to get foul, as they would do if in a common coil.

6th. Prime.—If the cartridge is to be pierced with the priming wire, and the vent filled with powder, the pan is also to be filled; and the flat space, having a score through it at the end of the pan, is to be covered, and this part of the priming is to be bruised with the round part of the horn. The apron is to be laid over, and the horn hung up out of danger from the flash of the priming.

7th. Point your guns.—At this command the gun is, in the first place, to be elevated to the height of the object, by means of the side-sights; and then the person pointing is to direct his fire by the upper-sight, having a crow on one side and a handspike on the other, to heave the gun, by his direction, till he catches the object.

The men who heave the gun for pointing are to stand between the ship's side and their crows or handspikes, to escape the injury they might otherwise receive from their being struck against them, or splintered by a shot; and the man who attends the captain with a match is to bring it at the word, "Point your guns," and kneeling upon one knee opposite the train-truck of the carriage, and at such a distance as to be able to touch the priming, is to turn his head from the gun, and keep blowing gently upon the lighted match to keep it clear from ashes. And, as the missing of an enemy in action, by neglect or want of coolness, is most inexcusable, it is particularly recommended to have the people thoroughly instructed in pointing well, and taught to know the ill consequences of not taking proper means to hit their mark; wherefore they should be made to elevate their guns to the utmost nicety, and then to point with the same exactness; and having caught the object through the upper-sight, at the word,

8th. Fire.—The match is instantly to be put to the bruised part of the priming: and, when the gun is discharged the vent is to be closed, in order to smother any spark of fire that may remain in the chamber of the gun; and the man who sponges is immediately to place himself by the muzzle of the gun in readiness.

9th. Worm and Sponge your guns.—The sponge is to be rammed down to the bottom of the chamber, and then twisted round, to extinguish effectually any remains of fire; and when drawn out, to be struck against the outside of the muzzle, to shake off any sparks or scraps of the cartridge that may have come out with it; and next its end is to be shifted ready for loading; and while this is doing, the man appointed to provide a cartridge is to go to the box, and by the time the sponge is out of the gun, he is to have it ready.

10th. Load with cartridge.—The cartridge (with the bottom end first, seam downwards, and a wad after it) is to be put into the gun, and thrust a little way within the mouth, when the rammer is to be entered; the cartridge is then to be forcibly rammed down, and the captain, at the same time, is to keep his priming-wire in the vent, and, feeling the cartridge, is to give the word *home*, when the rammer is to be drawn, and not before. While this is doing, the man appointed to provide a shot is to provide one (or two, according to the order at that time) ready at the muzzle, with a wad likewise; and when the rammer is drawn, at the words,

11th. Load with shot, and wad to your shot.—The shot and wad upon it are to be put into the gun, and thrust a little way down, when the rammer is to be entered as before.

12th. Ram home shot and wad.—The shot and wad are to be rammed down to the cartridge, and there have a couple of forcible strokes, when the rammer is to be drawn, and laid out of the way of the guns and tackles, if the exercise or action is continued; but if it is over, the sponge is to be secured in the place it is at all times kept in.

13th. Put in your tompions.—The tompions to be put into the muzzle of the cannon.

14th. House your guns.—The seizing is to be put on again upon the clinched end of the breeching, leaving it no slacker than to admit of the gun's being housed with ease. The quoin is to be taken from under the breech of the gun, and the bed, still resting upon the bolt, within the carriage, thrust under, till the foot of it falls off the axle-tree, leaving it to rest upon the end which projects out from the foot. The metal is to be let down upon this. The gun is to be placed exactly square, and the muzzle is to be close to the wood, in its proper place for passing the muzzle lashings. See Cannon, and plate VI. fig. 4.

15th. Secure your guns.—The muzzle lashings must first be made secure, and then with one tackle, having all its parts equally taught with the breeching, the gun is to be lashed. The other tackle is to be bowsed taught, and by itself made fast, that it may be ready to cast off for lashing a second breeching.

Care must be taken to hook the first tackle to the upper bolt of the carriage, that it may not otherwise obstruct the reeving of the second breeching, and to give the greater length to the end part of the fall.

No pains should be spared in bowsing the lashing very taught, that the gun may have the least play that is possible, as their being loose may be productive of very dangerous consequences.

The quoin, crow, and handspike, are to be put under the gun, and the powder-horn hung up in its place, &c.

Being engaged at any time when there is a large swell, a rough sea, or in squally weather, &c. as the ship may be liable to be suddenly much heeled, the port-tackle fall is to be kept clear, and, whenever the working of the gun will admit of it, the man charged with that office is to keep it in his hand; at the same time the muzzle lashing is to be kept fast to the ring of the port, and being hauled taught, is to be fastened to the eye-bolt over the port-hole, so as to be out of the gun's way in firing, in order to haul it in at any time of danger.

This precaution is not to be omitted, when engaging to the windward, any more than when to the leeward, those situations being very subject to alter at too short a warning.

A train-tackle is always to be made use of with the lee-guns, and the man stationed to attend it is to be very careful in preventing the gun's running out at an improper time.

The words of command in the short exercise are as follow:—1. Silence; 2. The guns are now loaded; 3. Cast loose the guns; 4. Prime; 5. Point your guns; 6. Fire; 7. Sponge and load; 8. Run out your guns.

Exercise may also be applied, with propriety, to the forming our fleets into the order of sailing, line of battle, &c. an art which the French have termed *evolutions*, or *tactiques*. In this sense, exercise may be defined, the execution of the movements which the different orders and dispositions of fleets occasionally require, and which the several ships are directed to perform by means of signals.

EYE of a Block-strap, is that part by which it is fastened, or suspended, to any particular place upon the sails, masts, or rigging, the eye whereof is represented by fig. 1, plate X. The eye is sometimes formed by making two eye-splices on the ends of the strap, and then fastening them together with a small line, so as to bind round a mast, yard, or boom, as occasion requires. See fig. 2, of the same plate. Plate X. Fig. 1.

Eye *of a Stay*, (*oeillet*, Fr.) is that part of a stay which is formed into a sort of collar to go round a mast-head.

Eye-*Bolt*, a long bar of iron with an eye in one end of it, represented by fig. 16, plate IV. It is formed to be driven into the decks or sides of a ship, for the purpose of fastening ropes or hooking tackles to.

Flemish Eye, is a phrase particularly applied to the eye of a stay, which is either formed at the making of the rope, or by dividing the yarns into two equal parts, knotting each pair separately, and pointing the whole over.

Eye-*Let-Holes*. See the article Sails.

Eye *of a Shroud*, the upper part which is formed of a sort of collar to go over the mast-heads.

Eye-*Splice*. See the article Splice.

EYES of a Ship, (*oeils*, Fr.) a name frequently given to those parts which lie near the hawse-holes, particularly in the lower apartments within the vessel.

F.

FACE-PIECE, in ship-building, a piece of wood wrought on the fore-part of the knee of the head, to assist the conversion of the main-piece, and to shorten the upper bolts of the knee of the head.

FACING, is the art of letting in one piece of timber into another, with a rabbet.

FACTOR, (*facteur*, Fr.) in commerce, an agent, or correspondent, residing beyond the seas, or in some remote part, and commissioned by merchants to buy or sell goods on their account, or assist them to carry on their trade: hence factory is a place where a considerable number of factors reside, to negotiate for their masters, or employers; as the factories of Lisbon, of Leghorn, of Calcutta, &c.

FACULÆ, in astronomy, are certain bright and shining parts, which the modern astronomers have, by means of telescopes, observed upon or about the surface of the sun; they are but very seldom seen.

FAG END OF A ROPE, the end which is become untwisted and loosened by frequent use. To prevent this effect, the ends of ropes are wound round with a piece of twine, or packthread, which operation is *whipping*.

To FAG *out*, (*se raguer*, Fr.) when speaking of a rope, is to become untwisted and loosened.

FAIR, (*bon, dégagé*, Fr.) a general term for the disposition of the wind, when it is favourable to a ship's course, in opposition to that which is contrary, or *foul*.

This term, when applied to the wind, is much more comprehensive than *large*, since the former seems to include about eighteen points of the compass, or at least sixteen; whereas *large* is confined to the beam or quarter, that is, to a wind which crosses the keel at right angles, or obliquely from the stern, but never to one right astern. See the articles LARGE and SCANT.

FAIR, is also applied to ropes as suffering the least friction in a pulley, when they are said to lead fair.

FAIR-*Curve*, in delineating ships, is a winding-line, whose shape is varied according to the part of the ship it is intended to describe: this curve is not answerable to any of the figures of conic sections, although it occasionally partakes of them all.

FAIR-*Way of a Channel*, (*passe, ou milieu d'un chenal*, Fr.) is the path of a narrow bay, river, or harbour, in which ships usually advance in their passage up and down; so that if any vessels are anchored therein, they are said to lie in the *fair-way*.

FAKE, (*pli, cercle d'un cordage*, Fr.) one of the circles, or windings, of a cable, or hawser, as it lies disposed in the coil. See the article COILING.

The fakes are greater or smaller in proportion to the extent of space which a cable is allowed to occupy where it lies.

FALCATED, something in the form of a sickle, or reaping-hook: thus the moon is said to be *falcated* when she appears horned, which happens while she moves from the third quarter to the conjunction, and so on from thence to the first quarter; the bright part appearing then like a crescent, viz. during the first and last quarters: but during the second and third quarters, the light part appears gibbous, and the dark part *falcated*.

FALCONET, a short cannon, about six feet long: this sort of ordnance was formerly used at sea.

FALL OF A TACKLE, (*bout du garant d'un palan*, Fr.) the loose end of a tackle; or that part upon which it is pulled, to produce the desired effect. See the article TACKLE.

To FALL *aboard*, (*aborder un vaisseau*, Fr.) See the article ABOARD.

To FALL *astern*, (*culer*, Fr.) to be driven backwards; to retreat with the stern foremost: it is also spoken of two ships falling together, if one goes faster than the other, the latter is said to *fall astern*.

To FALL *calm*, (*calmer, en parlant du tems*, Fr.) when speaking of the weather, implies a state of rest by a total cessation of the wind.

Cat-FALL. See the article CAT.

To FALL *down*, (*descendre une rivière*, Fr.) to sail, or be conducted from any part of a river, to some other part nearer to its mouth or opening.

To FALL *in with a Ship*, (*rencontrer un vaisseau en mer*, Fr.) is to meet her, or to discover her when off the land.

FALL *not off! (n'arrive pas, lof!)*. The command of the steersman to keep the ship near the wind.

FALLING-HOME, amongst shipwrights, is a term generally applied to the upper part of the top side, when it is very much within a perpendicular.

FALLING-*off*, (*abattée*, Fr.) the movement or direction of the ship's head to leeward of the point whither it was lately directed, particularly when she sails near the wind, or lies by. When a ship is under sail, and keeps not so near the wind as she should do, it is said she *falls off*.

FALLING-*off*, is also the angle contained between her nearest approach to the direction of the wind, and her farthest declination from it, when TRYING. See that article.

FALSE FIRE, a composition of combustibles for the purpose of deceiving an enemy.

FALSE *Keel of a Ship*, (*fausse quille*, Fr.) is composed of several pieces, and fitted under the main keel, to preserve it from friction, and to make the ship hold a better wind; they are generally formed of elm. In such ships as are not intended to be frequently in harbours, where they ground, the *false keel* is slenderly secured, that if, by accident, the ship should take the ground, it may, by coming off, be an assistant in saving her.

FALSE *Lights*, are those which may be used as signals of deception. See the article COMPOSITION.

FALSE *Post*, a piece of timber fixed on the aft part of the stern post, to make good a deficiency therein.

The *false post* is generally tabled to the main post, but seldom sufficiently secured to the keel; if it were lapped about two thirds down the side of the keel, with a hook in the middle, like the boxing of the fore foot, it would be a greater security to the main-keel than the dove-tail plates usually put for that purpose.

FALLS, when a ship is not flush, is a term given by seamen to those risings of some parts of her decks more than others.

FASCIÆ, in astronomy, are certain parts of Jupiter's body, resembling swathes or belts. They are more lucid than the rest of that planet's disc, and are terminated by parallel lines, sometimes broader and sometimes narrower.

FASHION-PIECES, (*cornières, et estains*, Fr.) in ship-building, are the aftmost or hindmost timbers of a ship, which terminate the breadth, and form the shape of the stern: they are united to the stern-post and to the extremity of the wing-transom by a rabbet and a number of strong nails or spikes driven from without. See their connexion with the stern-post and transom, as explained under the article STERN.

To FASTEN, or MAKE FAST, (*reprendre, amarrer, &c.* Fr.) is to tie a rope, or make it secure when it is absolutely necessary to be done.

FATHOM, (*brasse, mesure de six pieds anglois*, Fr.) a measure of six feet, used to regulate the length of the cables and rigging, and to divide the lead and sounding lines, &c.

To FAVOUR, (*ménager, favoriser*, Fr.) to be careful of any thing, as, "We must favour the mast."

To FAY, (*affleurer*, Fr.) in ship building, to fit any two pieces of wood so as to join close together. The plank is said to *fay* to the timbers when it bears or lies close to them all.

It is performed by moulds, bevellings, or by laying one piece on the other, and setting them as close together as possible; then, with compasses, take the distance they may be asunder, and let that distance or spoiling be set off from the surface or edge of the fairest piece on the sides of the other, or as many places as may be necessary for lining. If the pieces coak or table together, the thickness of the coaks or tabling must be added to the former gage or spoiling.

FEAR-NOUGHT, (*frise*, Fr) a peculiar sort of thick woollen stuff, much used in ships for the purpose of lining the port-holes, &c. It is generally hung on a skreen without the magazine door in time of action, to prevent any sparks from communicating to the powder; thence it derives the appellation *fear-nought*.

FEATHER-EDGED, a name given to planks or boards which have one side thicker than the other.

FELUCCA, (*felouque*, Fr.) a small vessel, navigated in the Mediterranean, with from ten to sixteen oars and lateen sails.

FEND, in sea language, imports the same as defend; as, "*Fend* the boat!" that is, to prevent its striking against any thing that might endanger it.

To FEND *off*, (*parer, défier du bord*, Fr.) is to prevent a boat or vessel running foul of another, or against a wharf, &c with too much violence, which is effected by the application of a boat-hook, &c.

FENDERS, (*défenses*, Fr.) certain pieces of old cable, timber, or other materials, hung over the side of a vessel, to prevent it from striking or rubbing against a wharf or quay; as also to preserve a smaller vessel from being damaged by a larger one.

FERRULE, a small iron hook fixed on the extremities of the yards, booms, &c.

To FETCH, (*chercher, gagner*, Fr.) to attain; as, "We shall *fetch* to windward of the light-house this tack."

To FETCH *Way*, to be shaken or agitated from one side to another. It is usually applied to a mast, bowsprit, &c. when it is not sufficiently wedged, being loose in the partners; it is also said of a cask, box, or such body which moves by the rocking of the ship at sea, as not being well secured and enclosed.

To FETCH *the Pump* (*engréner la pompe*, Fr.) is to pour a can of water into the upper part of it, in order to expel the air which is contained between the lower box and the piston, and the lower end of the pump that rests upon the ship's floor; and, accordingly, to make the water poured into the chamber communicate with that in the bottom of the pump-well, so as to be thrown out above by *striking* with the brake or handle. See the article PUMP.

FID, or MAST-FID, (*clef des mâts de hune*, Fr.) a square bar of wood, or iron, with a shoulder at one end, as represented in plate X, fig. 3. It is used to Fig. 3.
support the weight of the top-mast, when erected at the head of the lower-mast, by passing through a mortise or hole at the lower end of the former, and resting its ends on the trestle-trees, which are sustained by the head of the latter: the *fid*, therefore, must be withdrawn every time the top-mast is lowered. The top-gallant-mast is retained at the head of the top-mast in the same manner. See the article MAST.

Splicing FID, (*epissoire*, Fr. *fitta*, Ital.) is also a large pin, made of lignum vitæ, tapering to a point, and used for splicing of cables, or large cordage: they are from eighteen to twenty inches long. See plate X, fig. 4. There are some of a smaller dimen- Fig. 4.
sion for splicing bolt-ropes of sails, &c. When the *fid* is made of iron, it has an eye in the upper end, like fig. 5.

FIFE-RAILS. See the article RAILS.

SEA-FIGHT. See the articles BATTLE, ENGAGEMENT, and EXERCISE.

FIGURE, (*figure*, Fr.) the principal piece of carved-work at the head of a ship. See the article HEAD.

To **FILL**, (*faire servir*, Fr.) in navigation, to brace the sails in such a manner, as that the wind, entering their cavities from behind, dilates them so as to advance the ship in her course, after the sails had for some time been shivering, or braced aback.

It may appear something remarkable, but it is certainly true, that a ship may be forced backward or forward, or may remain in her place, with any wind, while her stern is all the while directed to one part of the horizon; and these different states of motion or rest, may be communicated by backing, filling, or shivering the sails, by drawing the braces on one side and loosening them on the other. See the articles *To* BACK *the Sails*, BRACE, and SHIVERING.

FILLING A SHIP'S BOTTOM, (*mailleter un vaisseau*, Fr.) the act of driving a number of nails with broad heads into her, so as to give her a sheathing of iron, to prevent the worms from getting into the wood: this operation is now much disused, the sheathing with sheets of copper being found superior to it. See the article COPPER-FASTENED.

FILLINGS, in ship-building, are small pieces of timber used for the purpose of making the curve fair for the mouldings, between the edges of the fish-front, and the sides of the mast

FILTER, (*couloir à filtrer*, Fr.) in chemistry, a strainer commonly made of bibulous or filtering paper in the form of a funnel, through which any fluid is passed, in order to separate the gross particles from it, and render it limpid.

There are several filters made of flannel and linen cloth.

Mr. James Peacock, of Finsbury-square, London, obtained a patent, dated December 23, 1791, for a new filtering machine for purifying water by ascent.

This machine, which has been fully proved to be highly useful on shipboard, particularly in long voyages; and which has received the sanction and approbation of some of the most scientific and learned men in the navy, has the following peculiar advantages over all others.

1. It accomplishes the purposes of complete filtration, by ascent of the fluid to be purified through a medium (probably that by which nature performs the same operation) of pure gravel, of progressive degrees of fineness, to the minutest particles of sand; by which means the foulest water or other fluid becomes perfectly freed from all its impurities, without acquiring any noxious mineral quality, which the pumice, or common filtering stones are justly suspected to communicate.

2. If, from continued use, its operation is impeded in the smallest degree, it may be completely cleansed in one minute with the utmost facility; an advantage not known to any species of filtration by descent.

3. Its operation is perpetual; and without occupying more room than a drip-stone with its apparatus, it yields a pure and common stream of more than 300 gallons in every 24 hours. This circumstance evidently puts it far above all comparison with any filter hitherto attempted.

The advantage of cleansing fluids, particularly water, needs not be enlarged upon. Clean linen, clean food, pure beverage, and the consequent comfort and health of inhabitants of great cities, are materially concerned; but the seaman, who is confined in distant and hot climates, to the constant use of water abounding with insects, animalcules, and other impurities which render it disgusting, and frequently noxious, must acquire by this means an invaluable benefit.

It is with confidence derived from long experience and repeated trials of its efficacy, that the inventor respectfully offers this machine to the public, convinced that it is powerfully and effectually applicable to promote the health, comfort, and convenience of mankind.

This machine is exhibited at the Surveyor's office, Guildhall, London, where every information respecting it may be had.

A patent was also granted to Mr. Joshua Collier, of Southwark, for a most ingenious method of filtering and sweetening water, oil, and every other liquid.

The principle of the improved filtering machines consists in combining hydrostatic pressure with the mode of filtering *per ascensum*, which procures the peculiar advantage of causing the fluid and its sediment to take opposite directions. The filtering surface remains the same, while the dimensions of the chamber in which the sediment is received may be varied.

If water is to be freed from noxious particles, it must be made to pass through an iron-box in its way to the filtering chamber, and the box must contain charcoal finely powdered, or any other antiseptic substance insoluble in water; the water being forced into it by hydrostatic pressure, through a tube of any sufficient height. This box has two apertures to receive and deliver the fluid, and these are opened and closed by cocks, or screws, or any other method used for such purposes, and being affixed to the machine by other screws, may be easily detached from the same. Thus, whenever the charcoal begins to lose its antiseptic properties, the box is removed and heated till it is red hot; by which means the foreign matter escapes through the small apertures, after which the box is cooled, and the charcoal becomes sweet, pure, and equally fit for use as at first, though the process be ever so often repeated.

Professor Parrot, jun. of Paris, has also invented a very ingenious and portable filtering machine, as represented in plate X, fig. 6. Fig. 6

The reservoir G may be of any form or dimension which is convenient; the principal part of the machinery consisting of a square vessel, bent in the form of an inverted syphon. The curve may be circular, elliptic, or in any other direction, and made of block-tin. This vessel is filled with fine sand, till nearly the height of the dotted line $x\,y$, which denotes the ascent of the water to D, whence it flows into the receiver. To the part marked A B, which must always be above this line (according to the size of the filtering machine), a woollen bag is attached, open at the top, the lower part of which touches the sand: this collects the coursest impurities, and preserves the sand for a longer time from becoming foul. The bag, therefore, may occasionally be removed, and rinsed in clean water. It is evident, that the water flows at A, through the bag into the filter, and rises at the place marked D, which is considerably lower than the former. It affords a pleasing sight to observe the most limpid fluid penetrating the uppermost stratum of sand, perfectly similar to that oozing from the purest natural spring.

In large machines a water-tight trap-door may be made at F, for the purpose of removing the sand when it is overcharged with impurities. The small diameter of the machine from which the drawing was taken, was eight Paris inches from B to E; the perpendicular height of the lower side, from C, its basis, to the rim D, whence the water issues, four inches and one-twelfth; the opposite height of the mouth, A, B, eight inches and three-fourths; and the height of the sand on the side marked D, three inches and one-sixth.

Although, in experiments of this nature, much depends on the relative size and purity of the sand, which necessarily afford different results, yet Professor Parrot has, after repeated trials, deduced the following conclusions, which appear to be well founded and satisfactory: 1. That the difference of *niveau*, or water-level, has an essential influence on the quantity of the purified water thus obtained. 2. That a prolongation of the stratum of sand does not considerably diminish the product of the filter, but remarkably contributes to the purity of the fluid. 3. That if the water be forced to pass through the sand with increased velocity, it will be less pure than by allowing it a proper time for its passage. And 4. That a machine of the dimensions above described, will furnish about three quarts of water in an hour, or eighteen gallons in twenty-four hours.

This quantity, however, being too large in proportion to the size of the machine, it is advisable either to lessen the difference of the water-fall; or which is still better, to prolong the stratum of sand, in order to reduce the filtration of the water to half the quantity above stated, and to obtain it in greater purity.

Thus, a filtering apparatus eighteen inches long from A to D, two inches thick, and four broad, would afford every hour six pints of very pure water.

If, therefore, so small a machine, containing a very moderate stratum of sand, and requiring only a difference of two or three inches in the height of the water, furnishes a clear and pure fluid, it follows that an apparatus on a larger scale, provided with a bed of sand from five to six feet long, and admitting of a difference from twelve to eighteen inches in the fall of the water, might be usefully employed in public wells, hydraulic machines, and even in camps, for the supply of pure and wholesome water for an army.

Filtering-*Stone*, (*pierre à filtrer*, Fr.) a soft white stone, fixed in a stand and hollowed at the top, for the reception of water, which, after passing gradually through its pores, is received in a vessel placed underneath, when it is freed from all its impurities. They are very necessary in ships at sea when the water becomes foul, and on that account each ship in the royal navy is furnished with one for the use of the captain or commander.

The common *filtering-stone*, which has been used with so much advantage at Paris for the last thirty-five years, is a yellow free stone, of a middle sized grain, soft enough to be cut with a toothed saw, and admitting its grain to be rubbed out by the fingers.

An easy and effectual method of purifying water, is of such constant utility, as to render it a matter of greater importance than it may at first appear.

The Society of Arts, &c. voted their silver medal in the session of 1810, to Mr. William Moult, No. 37, Bedford-square, London, for his new method of applying the *filtering-stone*, for purifying water.

The method of using this stone is that of placing it in the water to be purified; by which means the water presses against the outside of the filter, and oozing through its pores, fills the stone, from which it is to be conveyed into a proper receptacle. The stone is suspended in the cistern by a ring round the inside of it, upon which a projecting part round the top of the stone rests.

The apparatus are exhibited by fig. 7, plate X, where A A is the cistern containing the water to be filtered; B is the *filtering-stone*, suspended in the cistern by a ring round the inside of it, which catches the projecting part of the stone; the water in the cistern filters through into the stone. Fig. 7.

C is a syphon which conveys the filtered water from the inside of the stone into a cistern D, which is the reservoir for clean water. *e* a cock to draw it off as it is wanted. "By this mode of filtration, the impurities of the water are deposited in the bottom of the cistern A, instead of being left in the bottom of the stone as in the usual way." Mr. Moult also states, that double the quantity of pure water is procured by this method in the same time; and that he had used an apparatus of this kind with great success, for more than three years.

The slighest reflection will, undoubtedly, convince any one, that this mode is highly preferable to that in common use, in which the filtration is performed in the contrary direction; for by this valuable process, the principal objections to drip-stones used both in ships of war and on shore, are removed, viz. the constant labour they require to keep them clean, by means of brushes, without eventually producing the intended effect, and without preventing their being finally rendered useless.

FINISHING, in ship-building, the carved ornaments of the quarter-gallery; those below the second counter are called the *lower-finishing*, and those above the upper lights the *upper-finishing*.

FIR-TIMBER, (*sapin*, Fr.) in mast-making, is the most useful wood for length, size, and lightness. It forms the body of the mast, the spindle, side-trees, fishes, cheeks, fillings, and cant-pieces. Masts are also made of single trees, as are yards, booms, topmasts, &c.

Fir *Trees* of Riga and Gottenburg exceed in strength about one-tenth those of Norway, and one-sixth those of New England; and the fir-trees of Scotland those of Norway.

Though the Riga and Gottenburg trees exceed those of New England in strength, the latter have the advantage in size, as a large mast may be made of a single tree; but masts made of Gottenburg, Riga, or Norway trees, are composed of the most substantial parts of several.

FIRE, (*feu*, Fr.) is that subtile invisible cause by which bodies are made hot to the touch, and expanded or enlarged in bulk; by which fluids are rarefied into vapour; or solid bodies become fluid, and at last either dissipated and carried off in vapour, or else melted into glass. It seems also to be the chief agent in nature on which animal and vegetable life have an immediate dependence.

Among all the various disasters to which ships at

sea are subject, none is more destructive in its effects than fire. Various inventions have been offered to the public, many of them extremely ingenious, and likely to be useful for the purpose of remedying or palliating this dreadful disaster: nor has ingenuity been confined solely to mechanical contrivances.

Various chemical mixtures have been tried with good effect: among the rest, one offered by a Mr. Driuzzi, and though its uses are but partial, it would be improper to totally omit the invention, especially as the account of it is comprised in a very few words. It consists of two parts of common soda dissolved in seven pints of water, which is reduced to two-thirds by ebullition, and strained through a piece of cloth. The liquor has no more effect on wood, however, than common water, but it is very useful in extinguishing fires produced by oily fat, and bituminous substances.

Professor Palmer, of Brunswick, has invented a powder for extinguishing fire; from the use of which, advantage is expected to arise, especially in winter, when the water is frozen. This powder is composed of equal parts of sulphur and ochre, mixed with six times their weight of vitriol. These ingredients are mixed, and the mass afterwards pulverized.

The powder is to be scattered over the places on fire; two ounces are sufficient for a surface of a foot square. When it is not possible to approach the flames, cartridges may be made of it, and shot with a cross-bow against such parts of the building where the fire rages with the greatest violence.

In order to preserve timber from fire, the Professor directs, to rub it over with common carpenter's glue, and then sprinkle the powder over it, repeating the operation three or four times, as the preceding layer becomes dry.

If you wish to preserve cloth, paper, ropes, cables, &c. against fire, use water instead of glue, in applying the powder.

To extinguish accidental Fires in Ships.

Mr. Alexander Tilloch, the learned Editor of the *Phil. Mag.* observes, that fire is more supplied from the oxygen of the air than from the combustible body, and that water may itself act as an inciter to fire, instead of extinguishing it, when exposed to a sufficient temperature, in circumstances which shall decompose it into its constituent gases, hydrogen, a combustible, and oxygen, the only supporter of flame.

He also states, that water acts as an extinguisher, *merely* by interposing itself between the flame and the air, and excluding the latter, thus acting "as a wall of separation between the burning materials and the atmosphere:" for this reason he recommends mixing sand, mould, or clay, with the water employed to extinguish fire, to render its effect in this respect more powerful.

The superior efficacy of cutting off all communication between the burning bodies and the atmosphere, he strongly states as the most certain means of stopping conflagration: wherefore, when this takes place in a ship, the hatches should be closed, and all hands employed in stopping every crevice, through which the air could pass to the burning body, with oakum, or any other matter: by which means the ship would become "one large extinguisher," and the fire go out when it had consumed the oxygen of the confined air.

By this method, if the fire forced its way through the lower deck, it might be stopped in the second deck, by closing the ports, hatches, and all other apertures belonging to it in every direction. To hasten the consumption of oxygen of the confined air, Mr. Tilloch recommends, before closing the hatches, to light as many other fires below, in pots, pans, and other vessels, as time will permit, and also to light up as many candles as possible, each of which consuming a gallon of air in a minute, will assist in sooner rendering the enclosed air unfit for combustion.

It is also advised to separate the part of the ship where fires are most likely to originate, from the rest, by air-tight partitions, or bulk-heads, so that when a fire happens there, it would be only necessary to close and calk up the entrance to this part, to put it out.

Observations.—As every thing which may assist to subdue so dreadful an enemy as fire at sea should be known, and as even discussion on this subject may be of use, the benevolent views of the author should be made as public as possible.

Water certainly operates to extinguish fire, by excluding the air; but then it is not the only way in which it acts for this purpose: it puts out fire *by cooling the burning body* also, or, to use the mode of expression adopted in the author's paper, by abstracting caloric from it, in passing to the state of vapour, which is thought, by most, to be the chief way in which water puts out fire; therefore, it would be useful to throw water, not only on the burning wood, but on that which was growing hot.

The closing of all passages, from the burning bodies to the atmosphere, cannot be too much enforced: this, sailors would agree to readily, and, in a great measure, already are aware of.

If time would permit, and if the fire-vessels could be properly secured from doing mischief, it would also be useful to light more fires below, before closing up; but confusion occasioned by the fire, would probably make this a dangerous attempt, and it would be such as sailors would not readily come into.

The dividing particular parts of the ship from the rest, which has often been recommended by others before, would be of excellent service, not only for facilitating the extinction of fire, but in preventing danger from leaks, and strengthening the ship: but yet, such is the aversion of seamen to all divisions of the hold, that it is to be feared no plan of this kind will ever be adopted.

When a fire took place in the lower part of a ship, the method of extinguishing, by excluding the air, might probably be effectual; but, if this happened on the gun-decks of a vessel, it is feared that the number of ports, and other passages for air, to be closed, would render the attempt mostly unsuccessful. It should, however, be tried at all events, as the least chance of relief is of consequence in such desperate circumstances.

FIRE! in the art of war, (*feu! commandement aux cannoniers de tirer*, Fr.) the order to put the match to the priming, or pull the trigger of a cannon or other fire-arm, so as to discharge it. See the article Exercise.

Fire-*Arrow*, (*dard à feu*, Fr.) a small iron or steel dart, formerly used by privateers and pirates to fire the sails of an enemy in battle. It was furnished with springs and bars, together with a match, impregnated with powder and sulphur round about its shaft, and then fired from a musketoon, or swivel-gun. The match, being kindled by the explosion, communicates the flame to the sail against which it is directed, where the arrow is fastened by means of its bars and springs. As this is peculiar to hot climates, particularly the West Indies, the sails being extremely dry, are instantly inflamed, and, of course, convey the fire to the masts and rigging, and, finally, to the vessel itself.

Fire-*Ball*. See the article Composition.

Fire-*Barrels*, a sort of small barrels used for fire-ships, of a cylindrical form, as the best adapted to contain the *reeds* with which they are filled, and more convenient for stowing them between the troughs in the fire-room. Their inside diameters should not be less than 21 inches, and 30 inches are sufficient for their length. The bottom parts are, first, well stowed with short double-dipped reeds, placed upright, and the remaining vacancy is filled with fire-brand composition, well mixed and melted, and then poured over them. The composition used for this purpose is a mass of sulphur, pitch, tar, and tallow.

The barrels should be made very strong, not only to support the weight of the composition before firing, when they are moved or carried from place to place, but to keep them together whilst burning; for, if the staves are too light and thin, so as to burn very soon, the remaining composition will tumble out and be dissipated, and the intention of the barrels to carry the flame aloft, will accordingly be frustrated. The curtain is a piece of canvas, nearly a yard in breadth and length, thickened with melted composition, and covered with saw-dust on both sides.

Fire-*Engine*, a machine for extinguishing accidental fires by means of a stream or jet of water. The common squirting fire-engine consists of a lifting-pump placed in a vessel of water, and wrought by two leavers, that act alternately. During the stroke, the water raised by the piston of the pump spouts forcibly through a pipe joined to the pump-barrel, and made capable of any degree of elevation by means of a yielding leather pipe, or by a ball and socket turning every way, screwed on the top of the pump. The vessel containing the water is covered with a strainer, to prevent the gravel, dirt, &c. which is often poured in with the water, from choking the pump-work.

The melancholy and destructive consequences of fires are such, particularly when they happen in the night, that the utmost ingenuity of man, in almost all countries, has, of late been most laudably exerted in the invention and improvement of fire-engines, and other contrivances, for counteracting the extension and rapidity of the flames. Not the least valuable among these inventions, particularly on the score of simplicity and the ready accommodation which it offers to almost every housekeeper, is an apparatus that might be added to any common pump, and by which it immediately answers the purpose of a fire-engine, either on land or on board of ships.

This contrivance owes its origin to Mr. Benjamin Dearborn, who first published it in the *Transactions of the American Academy of Arts for* 1794.

Mr. Dearborn had raised a tube of thirty feet on his pump, by means of which any single person could throw water on the top of a neighbouring building, full thirty-seven feet from the pump, and between thirty and forty feet high.

A further contrivance for converting a common pump into a fire-engine, particularly on board of ships, has been recently invented by M. Robertson Buchanan, of Rothsay, in Scotland; and for which he obtained the King's Letters Patent, dated March 8th, 1796. We shall give the particulars, as set forth by himself; viz.—This pump, like the common pump, acts by the pressure of the atmosphere, but possesses the following superior advantages: First, the water is discharged from this pump through two valves; the one called the inner valve, situated on the bore of the suction-piece, through which the water, on raising up the piston, first passes, and thence is communicated, by means of an aperture in the side of the pipe, to the second or outer valve, which is placed at the bottom of the cistern, whence it passes into the spout. Secondly, the outer valve is not only within the command of a person's hand, and may be at all times cleared, but is made so as to afford a communication to clear the inner valve. Thirdly, these valves not being confined to any particular dimensions, are made large enough to admit any body or substance through them that is capable of passing through the suction-piece, and, being placed near to each other and in an inclined position, gravel, dirt, or other things, cannot easily lodge in the pump, and the water is thus resisted or wire-drawn as little as possible, in passing through the valves; and the pump is thus also made not liable to choke, or to have any unnecessary wear of its parts; but the valves may also be placed horizontally, or in any other position, when circumstances require it. Fourthly, from the size and situation of the valves, they may be at all times easily taken out, so that they may, without difficulty, be readily repaired or renewed; and hence this pump may be effectually used as an engine for extinguishing fire.

With respect to the common fire-engine, some very valuable additions and improvements have been made to it by Mr. Joseph Bramah, of Piccadilly, London, and for which he obtained the King's Letters Patent, dated April 18th, 1793.

Lieutenant Jekyll, of the royal navy, has invented a contrivance for *turning a ship's common hand-pump into a powerful fire-engine*; which was ordered, a few months ago, to be fitted on board the Venerable and Tiger, in Portsmouth harbour; and to be added to every ship as they came into port for repair. The invention is simple, and does not interfere with the common use of the ship's pump. It is so powerful, that a strong column of water may be thrown over a 20-gun ship's top-gallant-yard. It was worked on board the Royal William, by seven men, and found to throw water 76 feet perpendicular, and 108 feet in a diagonal line: when the water of three pumps was united by a receiver, with one discharging pipe it was propelled much higher, and with an amazing force.

Captains Paterson, Hall, and Austen, lately attended an experiment of its efficacy, on board the Royal William, at Spithead; and from their report it has been adopted. The result will, no doubt, prove beneficial to the service.

Fire-*Men*, (*matelots destinés à éteindre le feu dans le combat*, Fr.) are persons employed to extinguish fire that may happen in any part of the ship during an engagement or otherwise.

Fire-*Rail*, in ship building, a rail fixed above the plank-sheer on the forecastle and quarter-deck, worked similar to the plank-sheer.

Fire-*Ship*, (*brûlot*, Fr.) is generally an old vessel filled with combustible materials, and fitted with sheer-hooks to the yard-arms, and grapling-irons to hook and set fire to the enemy's ships.

Some English vessels, filled with combustible matter, and sent among the Spanish ships composing the Invincible Armada in the year 1588, are said to have given rise to the terrible invention of fire-ships. However, Livy informs us that the Rhodians had invented a kind of fire-ships, which were used in junction with the Roman fleet, in their engagements with the Syrians, in the 190th year before Christ.

Caldrons of combustible and burning materials were hung out at their prows, so that none of the enemy's ships durst approach them; these fell on the enemy's gallies, struck their beaks into them, and, at the same time, set them on fire.

As there is nothing peculiar in the construction of a modern fire-ship, except the apparatus by which the fire is instantly conveyed from one part to another, and from thence to the enemy, it will be sufficient to describe the fire-room, in which the combustibles are enclosed, together with the instruments necessary to grapple the ship intended to be destroyed.

The fire-room is built between decks, and extends from the bulk-head at the forecastle to a *bulk-head*, L, raised behind the main mast, as represented by
Fig. 8. fig. 8, plate X. The train enclosed in this apartment is contained in a number of wooden troughs, D, G, which intersect each other in different parts of the ship's length; being supported at proper distances by cross-pieces and stanchions. On each side of the ship are six or seven ports, H, about eighteen inches broad, and fifteen inches high, and having their lids to open downward, contrary to the usual method.

Against every port is placed an iron chamber, ten inches long, and 3, 5 in diameter. They are breeched against a piece of wood fixed across the ports, and let into another a little higher. When loaded, they are almost filled with corn-powder, and have a wooden tompion well driven into their muzzles. They are primed with a small piece of *quick match*, thrust through their vents into the powder, with a part of it hanging out. When the ports are blown open by means of the iron chambers, the port-lids either fall downward, or are carried away by the explosion. At the time of firing the ship, the iron chambers blow out the port-lids, and open a passage for the flame.

The ports, funnels, and scuttles, not only communicate the flames to the outside and upper-works of the ship and her rigging, but likewise open a passage for the inward air, confined in the fire-room, which is thereby expanded so as to force impetuously through those out-lets, and prevent the blowing up of the decks, which must, of necessity, happen, from such a sudden and violent rarefaction of the air as will then be produced.

On each side of the bulk-head behind, is cut a hole, L, of sufficient size to admit a trough of the same dimensions as the others. A leading trough, L I, whose foremost end communicates with another trough within the fire-room, is laid close to this opening, from whence it extends obliquely to a sally port, I, cut through the ship's side. The decks and troughs are well covered with melted resin. At the time of firing either of the leading troughs, the flame is immediately conveyed to the opposite side of the ship, whereby both sides burn together.

The spaces N, O, behind the fire-room, represent the cabins of the lieutenant and master; one of which is on the *starboard*, and the other on the *larboard* side. The captain's cabin, which is separated from these by a bulk-head, is exhibited also by P.

The proportion of combustible stores for a fire-ship of 150 tons, is as follows:

8	Fire-barrels, filled with composition
12	Iron chambers, to blow open the ports
250	Bavins, single dipped
24	Port-fires
3½	Priming composition barrels
1	Quick match ditto
48	Dipped curtains
150	Long reeds, single dipped
75	Short reeds, single dipped
75	Short reeds, double dipped
60	Hand Grenades.

The quantity of composition for preparing the stores of a fire-ship, is exhibited in the following table:—

	Saltpetre. lb.	Sulphur. lb.	Corn Pr. lb.	Pitch. lb.	Resin lb.	Tallow. lb.	Tar. lb.	Oil. pts.
For 8 barrels.								
	0	0	960	480	0	80	0	0
For 3½ barrels priming composition.								
	175	140	350	0	21	0	0	11
For the curtains, bavins, and reeds, for the ship, and sulphur for salting them.								
	0	200	0	350	175	50	25	0
Total	175	340	1310	830	196	130	25	11

For reeds for barrels 160lb. being one-fifth of the whole of the last article.

The following instructions are given in the last regulations for a fire-ship of 150 tons burden:—

The fire-barrels are to be two feet four inches high, and one foot six inches diameter. Each barrel must have four holes of about six inches square, cut in its sides; and these holes must have a square piece of canvas nailed over them quite close. They are then filled with the same composition as for carcasses, and four plugs, of about one inch diameter and three inches long, and well greased, are thrust into the top, and then left to dry. When dry these plugs are taken out, and the holes filled with fuse-composition, and quick-match at the top; which goes from one hole to the other: after this the top is smeared over with mealed powder mixed up with spirits of wine. When dry again, a sheet or two of brown paper is laid over the top, and then one of the canvas covers, which is made secure by the upper hoop of the barrel.

The composition for dipping reeds, bavins, and curtains, is

Resin	120lbs.
Coarse sulphur	90
Swedish pitch	60
Tallow	6
Mealed powder	12

This proportion will dip 100 reeds and 25 bavins. Each curtain contains one square yard of barras. Each cover for fire-barrels, one square yard of sacking. Immediately that the curtains, covers, &c. are dipped, they are to be strewed over with fine brimstone, before the composition grows cold.

The iron chambers for blowing open the ports hold from nine to eleven ounces of powder: they are fixed in such a manner as to prevent their recoil, and to ensure the ports being blown open. The vents are generally corked up, and covered with a piece of barras, till required to be primed.

To fit out a Fire-ship.

The whole breadth of the fire-room is to be divided into nine parts, and troughs laid the whole length of the room. Cross troughs of communication are laid between them, about twenty in each row, perpendicular to the long troughs. These troughs are usually four inches wide, and four deep. There are two fire-trunks and two fire-scuttles on each side, under which the eight fire-barrels are to be placed.

The reeds and bavins are to be tied down in the troughs. The curtains are to be nailed up to the beams, equally through the fire-room. The ship is not to be primed when fitted out, but only when intended to be fired.

To Prime.—Composition for priming:—

Saltpetre pulverised	22lbs.	8oz.
Resin	2	11
Sulphur	18	
Mealed powder	45	
Linseed oil	1 pint.	

All the reeds and bavins are to be taken up, and a little of the above composition sprinkled in the bottom of the troughs; the reeds, &c. to be then gently tied down again. Quick-match, of six or eight threads doubled must be laid along on the tops of all the reeds, &c. and the priming composition strewed over it, and over all the fire-rooms. The covers of all the fire-barrels must be cut open, and made to hang down on the sides of the barrels. Leaders of strong quick-match must be laid from the reeds to the barrels and to the chambers, and must be tied down to the vents to ensure its not falling off. Strong leaders of quick-match, four or five times doubled, must be laid from the reeds to the sally-ports; and the sally-ports must be connected by quick-match, that the whole may take fire at once.

The following method is now adopted of producing an external fire, in addition to the internal fire before gained by the fire-room:—

Fire boxes, filled with the carcass composition, are distributed in the following manner, in a ship of three masts:—

	No.
One suspended from each of the catheads and davits, on each side of the bow	4
Eight slung across the bowsprit	8
Four across each of the out-riggers abaft	8
Two from the graplins of each of the lower yard-arms	12
One from the dead-eyes on each side of the three round tops	6
One from the middle of the inside of the main, fore, and mizen shrouds	6
	44

The boxes are suspended by chains and hooks, and those flung across the bowsprit and out-riggers are fixed by staples. The two inner ones are laid with leaders of quick-match, which fire instantly, or with port-fires, which burn a given time; they communicate with the outer ones by reeds, which are tied down on the bowsprit and out-riggers. The boxes hanging from the dead-eyes and shrouds are fired by curtains suspended from the shrouds; the lower one of which hangs immediately over one of the large fire-barrels. The two boxes on each yard-arm are hung one over the other; the upper one having a leader of quick-match carried along the yard from the shrouds, and, in burning, will, no doubt, fire the lower one.

Besides the boxes there are fire barrels arranged as follows —Two half barrels on the forecastle; two abaft the main-deck, and four on the main-deck; two in each round-top, placed against the masts; and four large fire-barrels, under fire-trunks, to convey fire to the curtains on the shrouds. All these fire-barrels and boxes are to be fired by separate leaders of quick-match, or port-fire, in order that any part of the ship may be fired, to cover its approach by the smoke; and the remaining part instantaneously upon quitting the ship. It has been found by experiment, that two men, with lighted port-fires, can set fire to the whole of the leaders on the deck, bowsprit, cat-heads, out-riggers, &c. in less than a minute; therefore, the risk of trusting to one main leader to the whole may be avoided. The leaders are laid in painted canvas hose, made for the purpose. See ADYE's *Bombardier*.

Four of the eight fire-barrels are placed under the four fire trunks: and the other four between them, two on each side the fire-skuttles, where they are securely *cleated* to the deck. The longest reeds are put into the fore and aft troughs, and tied down: the shortest reeds are laid in the troughs athwart, and tied down also. The bavins, dipped at one end, are tied fast to the troughs over the reeds, and the curtains are nailed up to the beams, in equal quantities, on each side of the fire-room.

The remainder of the reeds are placed in a position nearly upright, at all the angles of every square in the fire-room, and there tied down. If any reeds are left, they are to be put round the fire-barrels, and other vacant places, and there tied fast.

The port-fires used for firing the ship, burn about twelve minutes. Great care must, therefore, be taken to have no powder on board when the ship is fired.

The sheer-hooks represented by fig. 9, plate X, are Fig. 9.

fitted so as to fasten on the yard-arms of the fire-ship, where they hook the enemy's rigging. The fire-
Fig. 10. graplings, fig. 10, are either fixed on the yard-arms, or thrown by hand, having a chain to confine the ships together, or fasten those instruments wherever it may be found necessary.

When the commanding officer of a fleet displays the signal to prepare for action, the fire-ships fix their sheer-hooks, and dispose their graplings in readiness. The battle being begun, they proceed immediately to prime and prepare their fire-works. When they are ready for grapling, they inform the admiral thereof by a particular signal.

To avoid being disabled by the enemy's cannon during a general engagement, the fire-ships continue sufficiently distant from their line of battle either to windward or to leeward.

They cautiously shun the openings or intervals of the line, where they would be directly exposed to the enemy's fire, from which they are covered by lying on the opposite side of their own ships. They are attentively to observe the signals of the admiral, or his seconds, in order to put their designs immediately into execution.

Although no ship of the line should be previously appointed to protect any fire-ship, except a few of the smallest particularly destined to this service, yet the ship before whom she passes in order to approach the enemy, should escort her thither, and assist her with an armed boat, or whatever succour may be necessary in her situation.

The captain of the fire-ship should himself be particularly attentive that the above instructions are punctually executed, and that the yards may be so braced, when he falls alongside of the ship intended to be destroyed, that the sheer-hooks and graplings fastened to the yard-arms, &c. may effectually hook the enemy. He is expected to be the last person who quits the vessel, and being furnished with every necessary assistance and support, his reputation will greatly depend on the success of his enterprise.

The following is a description of a machine, called the Infernal, which was used at the bombardment of St. Malo, in the year 1693.

The vessel was a new galliot of about 300 tons. In the bottom of the hold were placed above a hundred barrels of gunpowder, covered with pitch, tar, brimstone, resin, tow, straw, and fagots. Over these was a row of thick planks and beams, with holes pierced through them, in order to communicate the fire from above, and upon them were placed three hundred and forty carcasses, or rather chests, or mortars, filled with grenades, cannon-balls, iron chains, fire-arms loaded with ball, large pieces of metal wrapped in tarpaulings, and other combustible matters. They were open in six parts, like six mouths, to let out the flames, so fierce as to consume the hardest substances, and not to be quenched but by hot water, as was, upon trial, observed.

The design was to have fastened it to the town wall, and it was not doubted that if it had taken its effect the town would soon have been in one heap of ruins. It was sent in before the wind, and was at the very foot of the wall where it was to be fastened, when a gust of wind blew it off and forced it upon a rock, where it struck, near the place where it was to have blown up. The engineer, percei ing it begin to burst, had time to set fire to it before he retired; and it did, indeed, blow up soon after: but the carcasses, which were to have done the most execution, being wet with sea-water, did not take fire; yet the shock was so terrible, though at some distance, that it threw down part of the town wall, shook every house in the town, and overthrew the roofs of above three hundred, which were the nearest. The capstan of the vessel, which weighed above a ton, was thrown over the wall on the top of a house, which it beat down.

Among the most formidable contrivances ever used, either as a fire-ship or explosion-vessel, is that which was used to destroy the bridge of boats at the siege of Antwerp, in the year 1585; which an author of that period states to have been a ship strongly timbered, containing a vaulted arch of stone or mortar, filled with 200 barrels of gunpowder, over which were placed large stones of all forms, cannon-shot, iron-chains, &c. sufficient to destroy a whole city, that were exploded by a secret fusee, contrived so as not to set fire to the charge till the vessel came in contact with the bridges, which it blew to atoms.

The use of fire-ships, and the mode of applying them is very imperfectly understood by the French: a circumstance somewhat singular, when we consider the habits of that nation. All the attempts made by that country to use them, from their very first invention, have almost uniformly failed of success. The last attempt they made was in the year 1758, at the siege of Quebec, when they sent three fire-ships down the river St. Lawrence, with the hopes of destroying the British squadron there, under the command of Admiral Sir Charles Saunders; but the gallant commander no sooner saw the enemy's fire-ships coming down with a fair wind and tide, than he sent boats to board those vessels of destruction, who bravely navigated them through our squadron, and then quitting them, saw them all blow up about a mile below, accompanied by the universal cheers of the British fleet.

Fire-*Works*, (*feu d'artifice*, Fr.) otherwise called Pyrotechnia, are artificial fires, made of gunpowder, sulphur, and other inflammable and combustible ingredients, used on occasion of public rejoicings, and other solemnities. The principal of these are rockets, serpents, stars, hail, mines, bombs, garlands, letters, and other devices.

The invention of fire-works is attributed, by M. Mahudel, to the Florentines and people of Sienna; who found out, likewise, the method of adding to them decorations of statues, with fire issuing from their eyes and mouths.

In the composition of fire-works, generally, it is to be observed—1. That *mealed powder* does not contribute most effectually to *expedite* effect. 2. *Nitre*, or saltpetre, used for these purposes, should be as pure as possible: this article causes a slow fire. 3. *Sulphur* tends to prolong a fire. 4. *Charcoal*, if well pulverised, is of great use, and maintains the fire. 5. *Tar*, *pitch*, and *colophony*, or *black resin*, kindle easily, and render a composition slower in its operation. 6. *Antimony* renders a composition quicker, and contributes to produce a greater explosion. 7. *Camphor* causes a beautiful white flame. 8. *Iron filings* make a sparkling fire. 9. *Oil* and *fat* feed a fire, and render its operation slower. 10. *Spirits of wine*, *brandy*, and other spirits, are used to moisten a composition when mixing it. 11. *Gums* serve to glue or stiffen a composition when mixing it.

Several other articles, as *mercury*, *potass*, *saffron*, *acids*, *&c.* are also used in making artificial fireworks.

In all cases it is to be observed, that the strength of the composition does not more depend on the ingredients, than on the effectual manner in which they are blended, and the mode in which they are made up for service. For the composition of artificial fireworks, see the articles COMPOSITION, ROCKETS, &c.

FIRMAMENT, (*firmament*, Fr.) in the Ptolemaic astronomy, the eighth heaven, or sphere, with respect to the seven spheres of the planets which it surrounds.

It is supposed to have two motions, a diurnal motion, given to it by the primum mobile, from east to west about the poles of the ecliptic; and another opposite motion from west to east, which last it finishes, according to Tycho Brahe, in 25,412 years; according to Ptolemy, in 36,000; and, according to Copernicus, in 25,800; in which time the fixed stars return to the same points in which they were at the beginning This period is commonly called the Platonic year, or the great year. See the article *Precession of the* EQUINOXES.

FISH, (*candelette*, Fr.) a machine employed to hoist and draw up the flooks of the ship's anchor towards the top of the bow, in order to stow it, after it has been catted. It is composed of four parts; viz. the pendent, the block, the hook, and the tackle; which, with their several uses, are described under the article DAVIT.

FISH-*Front*, or PAUNCH, (*jumelle*, Fr.) is a long piece of oak or fir timber, convex on one side and concave on the other, used to strengthen the lower masts or the yards, when they are sprung, or have received some damage in battle or in tempestuous weather, &c. to effect which they are well secured by stout rope, called woolding.

FISH *Gig*, (*foëne, instrument de pêche*, Fr.) an instrument used to strike fish at sea, particularly dolphins. It consists of a staff, with three, four, or more, barbed prongs of steel, and a line fastened to the end on which the prongs are fixed: to the other end is fitted a piece of lead, which gives additional force to the stroke when the weapon flies, and causes the points to turn upward after the fish is penetrated.

FISH-*Room*, (*espace compris entre la cale d'arrière et la cale au vin*, Fr.) in a ship, is that place between the after-hold and the spirit room.

FISHES-SIDE, two long pieces of fir, coaked on the opposite side of a made-mast, to give it the diameter required.

To FIT OUT A SHIP, (*armer un vaisseau*, Fr.) is to provide her with a sufficient number of men, to navigate and arm her for attack or defence; also to furnish her with proper masts, sails, yards, ammunition, artillery, cordage, anchors, and other naval furniture, together with sufficient provisions for the ship's company.

FIXED STARS, are such as constantly retain the same position and distance with respect to each other; by which they are contra-distinguished from erratic or wandering stars, which are continually varying their situation and distance. The *fixed stars* only are properly and absolutely called stars: the rest having their peculiar denomination of planet or comet.

FLAG, (*pavillon*, Fr. *flag*, Dut.) a certain banner or standard, by which an admiral is distinguished at sea from the inferior ships of his squadron; also the colours by which one nation is distinguished from another.

In the British navy, flags are either red, white, or blue, and are displayed from the top of the main-mast, fore-mast, or mizen-mast, according to the rank of the admiral.

The first flag in Great Britain is the royal standard, which is only to be hoisted when the king or queen are on board the vessel: the second is that of the anchor of hope, which characterizes the lord high admiral, or lords commissioners of the Admiralty: and the third is the union flag, in which the crosses of St. George, St. Andrew, and St. Patrick, are blended. This flag is appropriated to the admiral of the fleet, who is the first military officer under the lord high admiral. The Navy Board, Custom House, &c. have each their respective flags.

When a flag is displayed from the main-top-gallant-mast-head, the officer distinguished thereby is known to be an admiral; when from the fore-top-gallant-mast-head, a vice-admiral; and when from the mizen-top-gallant-mast-head, a rear-admiral.

The next flag after the union is that of the white squadron, at the main-mast-head; and the last, which characterizes an admiral, is the blue, at the same mast-head.

For a vice-admiral, the first flag is the red; the second, the white; the third, the blue; at the fore-top-gallant-mast-head.

The same order is observed with regard to the rear-admirals, whose flags are displayed at the main-top-gallant-mast-head; the lowest flag in our navy is, accordingly, blue at the mizen.

All the white flags have a red St. George's cross in them, in order the more readily to be distinguished from the French white flag with a white cross. Besides the national flag, merchant ships frequently bear lesser on the mizen-masts, with the arms of the city where the master ordinarily resides; and on the fore-mast, with the arms of the place where the person who freights them lives.

When a council of war is held at sea, if it be on board the admiral, they hang a flag on the main-shrouds; if in the vice-admiral, in the fore-shrouds; and if in the rear-admiral, in the mizen-shrouds. The flags borne on the mizen are particularly called gallants. See the article MAST.

To heave out the FLAG, is to put out, or hang abroad the flag.

To hang out the White FLAG, is to call for quarter; or it shews when a vessel arrives on a coast that it has no hostile intention, but comes to trade, or the like.

To hang out the Red FLAG, is to give a signal of defiance and battle.

To lower or strike the FLAG, (*amener le pavillon*, Fr.) is to pull it down upon the cap, or to take it in, out of the respect or submission due from all ships or fleets inferior to those that are any way justly their superiors.

To lower or strike the Flag, in an engagement, is a sign of yielding, or submission.

The method of leading a ship in triumph is, to tie the flags to the shrouds or the gallery, in the hind part of the ship, and let them hang down towards the water, and tow the vessel by the stern. Livy relates that this was the way the Romans used to do to those of Carthage.

Flags are also used particularly for signals at a certain distance by day, and are of different colours and shapes; those in general used are either pierced, quartered, chequed, fessed, paled, gironed, traversed, party per chief, or party per pale; which last our sailors call the *first*, half one colour half the other; the *last*, such a colour with such a fly.

Flag-*Officers*, (*officiers généraux de la marine*, Fr.) are those who command the several squadrons of a fleet, such are the admirals, vice-admirals, and rear-admirals. The *flag-officers* in our sea-pay are the admiral, vice-admiral, and rear-admiral, of the white, red, and blue.

Flag-*Ship*, (*vaisseau portant le pavillon d'un officier général*, Fr.) a ship commanded by a general or flag-officer, who has a right to carry a flag, in contra distinction to the secondary vessels under the command thereof.

Flag-*Staff*, (*bâton de pavillon, ou bâton d'enseigne*, Fr.) is generally a continuation of the top-gallant-mast above the top-gallant-rigging; but is, sometimes, especially in guard-ships, a spar, occupying the place of the top-gallant-mast, and is only of use to display the flag or pendant; when it is a continuation of the top-gallant-mast, it is frequently termed the royal mast.

FLAKE, (*échaffaud*, Fr.) a sort of platform made of hurdles, and supported by stanchions, and used for drying cod-fish in Newfoundland. These flakes are usually placed near the shores of fishing harbours. It is also a small stage hung over a ship's side, to calk or repair any breach.

To FLARE, or To FLARE Over, implies to hang over, or that a vessel is let out broader aloft than the due proportion will allow: this happens when the ship is housed-in near the water, so that the work hangs over a little too much, when seamen say that the work doth *flare* over.

FLARING, (*élancé ou dévoyé*, Fr.) in contradistinction to falling home, is when a ship's side forward falls out from a perpendicular, then they say she has a *flaring* bow.

FLAT, (*bas-fond, ou batture*, Fr.) a level ground, lying at a small depth under the surface of the sea; otherwise called a shoal, or shallow.

Flat *aft*, is the situation of the sails when their surfaces are pressed aft against the mast by the force of the wind. See the article Aback, &c.

Flat-*bottomed Boats*, (*bateaux plats*, Fr.) are such as are built to swim in shallow water, and to carry a great number of troops, artillery, ammunition, &c. They are sometimes constructed with a twelve-pounder bow-chase, and an eighteen-pounder stern-chase; their keel is from ninety to one hundred feet in length, and from twelve to twenty-four feet beam: they have one mast, a large square main-sail, and a jib-sail; are rowed by eighteen or twenty oars; and carry four hundred men each. The gun takes up one bow, and a bridge the other, over which the troops are to march. Those that carry horses have the fore-part of the boat made to open, when the men are to mount and ride over a bridge.

To Flat-*in the Sail*, (*traverser la voile*, Fr.) is to draw in the aftmost clew of a sail towards the middle of the ship, to give the sail the greater power of turning the vessel: thus, if the mizen, or after-sails are flatted-in, it is evident that the intention is to carry the stern to leeward, and turn the head nearer to the direction of the wind: and if the head-sails are flatted in, the intention is, accordingly, to make the ship *fall off*, when, by design or accident she has come so near the wind as to make the sails shiver. Hence,

Flat-*in forward!* (*traverse les focs et la misaine!* Fr.) is the order to draw in the jib and fore-top-mast-stay-sail-sheets, towards the middle of the ship. This operation, however, is seldom performed, except in light breezes of wind, when the helm has not sufficient government of the ship.

Flats, in ship-building, the name given to all the timbers in midships.

FLAW, (*risée de vent*, Fr.) a sudden breeze, or gust of wind.

FLEET, (*escadre, armée navale, ou flotte de bâtimens marchands*, Fr.) a general name given to the British navy, or to any number of ships, whether designed for war or commerce, when keeping company together.

The admirals of his Majesty's fleet are classed into three squadrons; viz. the red, white, and blue. When any of these officers are invested with the command of a squadron or detachment of ships of war, the particular ships are distinguished by the colours of their respective squadron: that is to say, the ships of the red squadron wear an ensign, whose union is displayed on a red field; the ensigns of the white squadron have a white field; and those of the blue squadron, a blue field; the union being common to all three. The ships of war, therefore, are occasionally annexed to any of the three squadrons, or shifted from one to another.

Of whatever number a fleet of ships of war is composed, it is usually divided into three squadrons; and these, if numerous, are again separated into divisions. The admiral, or principal officer, commands the centre; the vice-admiral, or second in command, superintends the van-guard; and the operations of the rear are directed by the rear-admiral, or the officer next in rank. See the article Division.

The disposition of a fleet, while proceeding on a voyage, will, in some measure, depend on particular circumstances; as the difficulty of navigation; the necessity of dispatch, according to the urgency or importance of the expedition; or the expectation of an enemy in the passage. The most convenient order is, probably, to arrange it into three lines or columns, each of which is parallel to a line close-hauled, according to the tack on which the line of battle is designed to be formed. This arrangement is more used than any, because it contains the advantages of every other form, without their inconveniences. The fleet being thus more enclosed, will more readily observe the signals, and with greater facility form itself into the line of battle; a circumstance which should be observed in every order of sailing. See the article ENGAGEMENT.

Merchant fleets generally take their denomination from the place they are bound to; as, the Turkey fleet, the East India fleet, &c. These, in time of peace, go in fleets for their mutual aid and assistance: in time of war, besides this security, they likewise procure convoys of men of war, either to escort them to the places whither they are bound, or only a part of the way, to a certain place or latitude, beyond which they are judged out of danger of privateers, &c.

To FLEET, (*affaler, ou reprendre un palan, &c.* Fr.) is to change the situation of a tackle, when the blocks are drawn together; or, what is called *block-and-block* by sailors: also to change the position of the dead-eyes, when the shrouds are become too long, which is done by shortening the shroud and turning in the dead-eye again, higher up. The use of fleeting is, accordingly, to replace the mechanical powers into a state of action; the force by which they operated before being destroyed by the meeting of the blocks or dead-eyes.

Fleeting, therefore, is nearly similar to the winding up of a watch or clock. See the article TACKLE.

FLINT, (*pierre à fusil*, Fr.) a well-known hard and semi-pellucid stone, of the chrystal kind, made use of to strike fire with steel. They are usually packed for service in half-barrels.

			qrs.	lb.
One half barrel contains, of	Musket,	2000,	2	14
	Carbine,	3000,	2	10
	Pistol,	4000,	2	15

The most transparent and free from veins are esteemed the best.

	cwt.	qrs.	lb.
In tonnage, 28 kegs of musket-flints take	18	0	0
And 10 kegs of pistol ditto	3	2	0

FLITCH, the name of a piece of small timber, supplied to ships for the purpose of sawing up into boat timber; so called, perhaps, from its small parts resembling a flitch of bacon.

FLOAT, (*radeau de mâts, ou de bois de construction*, Fr.) a raft, or a quantity of pieces of timber fastened together with rafters athwart, to be driven along a river with the tide or current; and even, sometimes, to carry burdens down a river with the stream.

To FLOAT, (*flotter, être à flot*, Fr.) to be borne up or wafted along with the tide on the surface of the water, the theory of which is explained in the article TRIM.

FLOATING-BATTERY. See the article BATTERY.

FLOATING BODIES, are those which swim on the surface of a fluid; the most interesting of which are ships and vessels employed in war and commerce. It is known to every seaman of what vast moment it is to ascertain the stability of such vessels, and the positions they assume when they float freely on the surface of the water. To be able to accomplish this, it is necessary to understand the principles on which that stability and these positions depend. A *floating body* is pressed downwards by its own weight in a vertical line passing through its centre of gravity; and it is supported by the upward pressure of a fluid, which acts in a vertical line that passes through the centre of gravity of the part which is under the water; and, without a coincidence between these two lines, in such a manner as that both centres of gravity may be in the same vertical line, the solid will turn on an axis, till it gains a position in which the equilibrium of floating will be permanent.

From this it is obviously necessary to find what proportion the part immersed bears to the whole; to do which, the specific gravity of the *floating body* must be known; after which, it must be found, by geometrical method, in what positions the solid can be placed on the surface of the fluid, so that both centres of gravity may be in the same vertical line when any given part of the solid is immersed under the surface.

Colonel Beaufoy, of Hackney Wick, who is allowed to be one of our most accurate astronomers, and also well acquainted with maritime affairs, published in 1805, a number of *Experiments for the Purpose of ascertaining the stability of Bodies floating on the Surface of the Water;* which may serve as a standard for comparing the theorems given by writers on this subject. He justly observes that, "of all the qualities necessary for a vessel to possess, none is of greater consequence than stability; and without this most essential property, the advantage gained by giving ships that form best adapted for dividing the fluid in which they float must be unattended by any practical benefit." And, considering the form of a man of war, he thinks it probable, "that, in certain cases, the middle part has a tendency to overset, which is prevented by the triangular form given to the two extremities; and, supposing the idea to be well founded, the labouring, and consequently straining, of a ship when in motion, may be attributed to this circumstance." He also states "the metacentre of a seventy-four gun ship is situated about two feet below the surface of the water, and the centre of gravity four feet." See his *Experiments, &c.* published by Mr. J. Gold, Shoe-lane.

Captain James Burney, an experienced naval officer, has published *Experiments on Bodies floating in a Stream;* which were read before the Royal Society in February 1809, in which he states that, having frequently observed that laden barges, &c. on the river Thames, unimpelled by wind, oars, or towing, moved faster than the stream, or the tide on its surface, he was induced to inquire respecting both the fact and the cause. The several bargemen, whom he questioned respecting them, all agreed in the fact, as being familiar to them. They added also, that a laden barge made a greater progress than a light one, and assigned as a cause that the laden barge had more hold of the tide: whence it may be inferred, that they supposed the tide ran slower at the surface than at some depth beneath it. But Captain Burney, from an experiment which he made with a stick and a riband attached to it, concludes that not to be the case. The captain considers the surface of a stream or current of water, as not being horizontal, but an inclined plane, and the inclination of the surface as the cause of the current, and then has recourse to the effect of pressure, in order to account for the circumstance. He observes, "an oblique pressure, not strong enough to submerge the body, affects it in two directions; one downwards in the manner of weight, to which the body yields to a certain and definite extent; the other horizontal, in which direction the body continually gives way to the pressure. Almost every person has experienced the readiness of a boat to glide from under him, on putting his first foot in her. These two effects of inclined pressure are separately in proportion to the whole pressure, one as the sine, the other as the co-sine, of the angle of incidence is to radius.

"If to a floating body on a horizontal surface, a pressure is applied in a direction making with the horizon an angle of 89° 59′, the proportion of the pressure which would act horizontally, is to the whole pressure, as the sine of 1′ is to radius: and this proportion is $\frac{29}{100000}$ of the whole pressure. In like manner, if the surface inclines 1′ from the true horizontal level, weight applied to a body floating on that surface will give an impulse towards the declining part of the surface equal to $\frac{29}{1000000}$ of the weight applied. Consequently, a barge, having in her 100 tons weight, floating with the stream where the declivity of the surface is 1′, will receive an impulse towards the declining part of the surface, nearly equal to 65lb. which is little short of what is estimated to be the average pull of a horse."

The paper concludes with the following remarks on the progress made by bodies floating in a stream, and descending faster than the stream itself; and endeavours to shew that this progress was the effect of perpendicular pressure, producing impulse towards the declining part of the surface. "The same cause," says the captain, "evidently applies to the production of the stream itself; consequently, the surface, and whatsoever floats on it, are, in this respect, on an equal footing, and the whole agree in pressing forward and in opposing resistance to whatsoever endeavours to overtake them. Without some auxiliary cause, therefore, a floating body cannot overtake the stream. The different shapes of bodies, and likewise the directions in which they are placed with respect to the direction of the impelling power, expose them to more or less resistance. A barge floating crossways to the stream receives the progressive impulse with the least advantage, her whole length acting in resistance to her overtaking the stream. The same barge, when endways with the stream, is acted upon by the same quantity of impelling power, and her progress is opposed by less resistance. With increase of weight, both the impelling power and the resistance are increased: but when the barge is lengthways with the line of the stream, weight added will increase the impelling power in a greater proportion than the resistance is increased. Hence, the heavy barge in a calm will overtake the light barge." See *Captain* BURNEY's *Experiments* (lately published) *for ascertaining the Direction of Currents.*

FLOATING-*Light*, a hollow vessel of tinned iron plate, invented by Mr. W. Shipley, for saving the lives of persons who have the misfortune to fall overboard in the night from any ship. It is made in the form of a boat, and is about 27 inches long, 13 broad in the middle, and 12 deep, the joints of which are carefully soldered, so as to keep out the water: from the gunwale of this boat a handle projects on each side, intended for the man who has fallen into the water to lay hold of, and thus support himself; the buoyancy of the boat being amply sufficient to keep a man from sinking. A frame-work and gimbals are affixed above the middle part of the deck of the boat, for the reception of a swinging lantern, which is to contain a lamp, and which, by means of the gimbals, will always adjust itself to a position nearly vertical. It is recommended that this *float* be placed every night under the care of the officers on watch; that its lamp be frequently trimmed and supplied with fresh oil, and its wick moistened with oil of turpentine, in order that it may take fire with the least touch of a lamp or candle.

When a man falls into the water the lamp is immediately lighted, and the float which carries it is to be let down into the water by a small cord attached to an iron reel; the remainder of the cord is then made fast to the reel, and both thrown into the sea, so that the weight of the reel causes the cord to hang vertically, and thus be in no danger of entangling the legs of the unfortunate man. The light will direct him to the float; and, as soon as he has reached it (for it is supposed he can swim), he is to take hold of the handles, and may move it very fast which way he pleases, only by striking his legs in the same manner he does when he swims. The ship, on tacking about, is directed by the light to the float; and, having come near it, the man is rescued from the watery element by the following process:—A lantern, with a rope ladder, is to be let down by a cord, from the ship, till a cross-bar, below the lantern, but not at the bottom of the ladder, touches the water, which the mariners in the ship may see by means of the light from the bottom of the lantern; and thus the man in the water may lay hold of the cross-bar, and fix his feet on one of the steps of the rope-ladder, when he may lay hold of the iron handle of the float with one hand, and hang it on a hook projecting from the ladder above the cross-bar; which, being done; the man and the float may be both safely raised into the ship.

This contrivance may be useful, either by night or by day, in moderate calm weather; and, on that account, deserves to be introduced on board our commercial vessels and ships of war. For further particulars see *Trans. Soc. Arts*, Vol. 25.

FLOOD, (*le flot*, *le flux*, Fr.) the flux of the tide, or the time the water continues rising. When the water begins to rise it is called a *young flood*; after which it is *quarter flood*, *half flood*, and *high flood*.

FLOOD-*Mark*, the mark which the sea makes on the shore, at flowing water and the highest tide; it is also called *high-water-mark*.

FLOOKS, or FLUKES OF THE ANCHOR. See the article ANCHOR.

FLOOR, (*le fond*, *ou plat-fond d'un vaisseau*, Fr.) the bottom of a ship, or all that part on each side of the keel which approaches nearer to an horizontal than a perpendicular situation, and whereon she rests when aground. Thus we say, "a sharp floor, a flat floor, a long floor, &c."

Hollow FLOOR, an elliptical mould for the hollow of the floor-timbers and lower-futtocks.

FLOOR-*Timbers*, (*varangues*, Fr.) are those parts of the ship's timbers which are placed immediately across the keel, and upon which the bottom of the ship is framed; to these the upper parts of the *timbers* are united, being only a continuation of floor-timbers upwards.

FLOTA, a Spanish fleet. See the article GALLEON.

FLOWING, or FLOWING-SHEETS, (*écoutes largues*, Fr.) the position of the *sheets*, or lower corners of the principal sails, when they are loosened to the wind, so as to receive it into their cavities in a direction more nearly perpendicular than when they are *close-hauled*, although more obliquely than when the vessel is sailing before the wind.

A ship is, therefore, said to have a *flowing sheet* when the wind crosses the line of her course nearly at right angles: that is to say, a ship steering due north, with the wind at the east, or directly on her side, will have a flowing sheet; whereas if the sheets were extended close aft, she would sail two points nearer the wind; viz. N. N. E. See the articles CLOSE-HAULED, LARGE, and TRIM.

FLURRY OF WIND, (*risée, raffale, ou grain de vent*, Fr.) a light breeze of wind shifting to different places, and causing a little ruffling on the calm surface of the sea.

FLUSH-DECK. See the article DECK.

FLY OF THE COMPASS, (*le carton de la boussole*, Fr.) is that part of the compass on which the thirty-two points or winds are drawn, and to which the needle is fastened underneath.

FLY *of an Ensign, Pendant, &c.* (*battant d'un pavillon, &c.* Fr.) the breadth or extent from the staff to the extreme edge or end that flutters loose in the wind.

FLY-*Boat*, (*flûte Hollandoise, ou flibot*, Fr.) a large flat-bottomed Dutch vessel, whose burden is generally from four to six hundred tons. It is distinguished by a stern remarkably high, resembling a Gothic turret, and by very broad buttocks below.

Let FLY *the Sheets!* (*largue les écoutes!* Fr.) is the order to let them go suddenly, lest the ship should upset, or spend her top-sails and masts; which is prevented by letting the sheet go *a-main*, that it may hold no wind.

FLYING OF SAILS, is the act of setting them in a loose manner, as royal-sails without lifts or clews being lashed: as small top-gallant-sails, jibs without stays; and as studding-sails without booms.

FOG, or MIST, (*brume, brouillard*, Fr.) a meteor, consisting of condensed vapours, floating near the surface of the earth, or sea.

FOG-*Bank*, (*banc de brume*, Fr.) an appearance in hazy weather, which frequently resembles land at a distance, but which vanishes as you approach it.

FOMAHAUT, in astronomy, a fixed star of the first magnitude in the constellation Aquarius, or in the mouth of the Southern Fish. For the beginning of the year 1812, its

Right Ascension is	341°	48′	23″,00
Annual Variation in A.R. . .			49,95
Declination	30	37	30 S.
Annual Variation in Decl. . .			— 19,01

THE FOOT OF A MAST, (*le pied d'un mât*, Fr.) is the lower-end, or that which goes into the step.

FOOT *of a Sail*, (*fond d'une voile*, Fr.) denotes the lower edge or bottom.

To FOOT, to push with the feet; as, "foot the top-sail out of the top!"

FOOT-*Rope*, (*ralingue de fond*, Fr.) the rope to which the lower-edge of the sail is sewed. See the article BOLT-ROPE.

FOOT-*Ropes* are also the same with HORSES OF THE YARDS, which article see.

FOOTSPACE-RAIL, in ship-building, is that rail in the balcony in which the ballusters step.

FOOT-WALING, (*vaigrage*, Fr.) the whole inside planks or lining of a ship, used to prevent any part of the ballast or cargo from falling between the floor-timbers. See the article MIDSHIP-FRAME.

FORE, (*avant*, Fr.) the distinguishing character of all that part of a ship's frame and machinery which lies near the stem.

FORE *and Aft*, (*de l'avant à l'arrière*, Fr.) implies throughout the ship's whole length, or from stem to stern. It also implies in a line with the keel.

FORE-*Bowline*, (*bouline de misaine*, Fr.) the bowline of the fore-sail. See the article BOWLINE.

FORE-*Braces*, (*bras de misaine*, Fr.) are ropes applied to the fore-yard-arms, to change the position of the fore-sail occasionally.

FORE-*Castle*, (*gaillard d'avant*, Fr.) a short deck placed in the fore part of the ship, above the upper-deck: it is usually terminated, both before and behind, by a breast-work in vessels of war; the foremost part forming the top of the *beak-head*, and the hind-part reaching to the after-part of the fore-chains.

FORE-*Castle Men*, (*matelots du gaillard d'avant*, Fr.) are those stationed on the fore-castle for the purpose of doing duty, and are generally prime seamen.

FORE-*Cat-Harpings*, (*trélingage des haubans de misaine*, Fr.) a complication of ropes used to brace-in the upper-part of the fore-shrouds. See the article CAT-HARPINGS.

FORE-*Cockpit*. See the article COCKPIT.

FORE-*Foot*, (*brion*, Fr.) a piece of timber which terminates the keel at the fore-end. It is connected by a scarf to the extremity of the keel, of which it makes a part; and the other end of it, which is incurvated upwards into a sort of knee, is attached to the lower end of the stem: of which it also makes a part, being also called the *gripe*.

As the lower arm of the *fore-foot* lies on the same level with the keel, so the upper one coincides with the middle line of the stem: its breadth and thickness, therefore, correspond with the dimensions of those pieces, and the heel of the cut-water is scarfed to its upper end.

FORE-FOOT also signifies, one ship lying or sailing across another's way: as if two ships being under sail, and in ken one of another, one of them lying in her course with her stern so much a-weather of the other that, holding on their several ways, neither of them altering their courses, therefore the windward ship will run a-head of the other; then it is said, such a ship lies with others *fore-foot*.

FORE-*Ganger*, a short piece of rope, one quarter of an inch in circumference (larger than a whale-line), to fix on the harpoon when it is intended to strike a whale.

FORE-*Hooks*, the same with BREAST-HOOKS, which article see.

Fore-*Land*, (*cap, ou pointe*, Fr.) a cape or promontory, projecting into the sea; as the North or South Forelands.

Fore-*Lock*, (*goupille*, Fr.) a little flat-pointed wedge of iron, used to drive through a hole in the end of a bolt, to retain it firmly in its place, and prevent its drawing.

Fore-*Lock Hooks*, in rope making, a sort of hooks made of iron, with a long neck and handle; they have an eye at the end of the neck, for the fore-lock to go through, and are used to hang the yarns on, to harden and close ropes from two and a quarter inches and upwards.

Spring Fore-*Locks*, are made of iron, and having heads, are shouldered underneath, to prevent their going through the mortises; they have also thin tongues at the points, as springs, to keep them from coming out: these are used in bolts that often require shifting.

Fore-*Mast*. See the article Mast.

Fore-*Mast Men*, are those employed on board a ship to take in the top-sails, sling the yards, furl the sails, bowse, trice, and take their turn at the helm, &c.

Fore-*Jears*.	See the articles	Jears.
Fore-*Mast*.		Masts.
Fore-*Sail*.		Sail.
Fore-*Shrouds*.		Shrouds.
Fore-*Stay*.		Stay.
Fore-*Top*.		Top.
Fore-*Top-Mast*.		Top-*Mast*.
Fore-*Top-Gallant-Mast*.		Top-*Gallant-Mast*.
Fore-*Tye*.		Tye.
Fore-*Yard*.		Yard, &c.

To Fore-*Reach upon a Ship*, (*dépasser un vaisseau*, Fr.) is to advance upon or gain ground of some ship or ships in company.

Fore-*Runner of the Log-line*, a small piece of red buntin, laid into that line at a certain distance from the log; the space between them being called the stray-line, which is usually from twelve to fifteen fathoms, and is an allowance for the log to be entirely out of the ship's dead-water before the ship's velocity is begun to be estimated; consequently, the knots begin from that point. See the article Log.

Fore-*Staff*, (*flèche, ou arbalestrille*, Fr.) an instrument formerly used at sea, for taking the altitudes of heavenly bodies.

The *fore-staff*, called also cross-staff, takes its denomination hence, that the observer, in using it, turns his face towards the object, in contra-distinction to the back-staff, with which he turns his back to the object.

Fore-*Tackle*, (*palan de la caliorne de misaine*, Fr.) a tackle on the fore-mast, similar to the Main-tackle, which article see. It is used for similar purposes, and also in stowing the anchor, &c.

Fore-*Top Men*, are those stationed on the *fore-top* in readiness to set or take in the small sails, and to keep the upper rigging in order.

FOREIGN, signifies any person or thing not of this country.

If foreign seamen serve two years on board British ships, whether of war, trade, or privateers, during the time of war, they shall be deemed natural born subjects. But the persons so naturalized are not capable of being of the privy council or members of Parliament, or of taking any place of trust, civil or military, or to have any grant from the crown of lands.—Stat. 13, Geo. II. chap. 3, sect. 2.

To FORGE over, is to force a ship violently over a shoal, by the effort of a great quantity of sail.

FORMERS, (*gargoussieres, ou moules à cartouches*, Fr.) are small cylindrical pieces of wood, fitted to the diameter of the bore of a gun, whereon the cartridge, paper, or cotton, is rolled before it is sewed or fastened.

To FORM the Order of Sailing. See the article Sailing.

FORMING the Line. See the article Line.

FORT, (*fort*, Fr.) a small fortified place, environed on all sides with a moat, rampart, and parapet.

FORTHWITH, (an adverb, from *forth* and *with*) a term used in written orders, signifying immediately, without delay, at once, straight.

FORWARD, (*en avant*, Fr.) implies the fore-part of the ship. See the article Afore.

To FOTHER, or FODDER, (*aveugler une voie d'eau*, Fr.) is to endeavour to stop a leak in the bottom of a ship while she is afloat, either under sail or at anchor. It has been usually performed in the following manner:—A basket is filled with ashes, cinders, and chopped rope-yarns, and loosely covered with a piece of canvas; to this is fastened a long pole, by which it is plunged repeatedly in the water, as close as possible to the place where the leak is conjectured to lie. The oakum, or chopped rope-yarns, being thus gradually shaken through the twigs, or over the top of the basket, are frequently sucked into the hole along with the water, so that the leak becomes immediately choked, and the future entrance of the water is thereby prevented.

A superior method of fothering is now practised, which is performed in the following manner; that is, by fastening a sail at the four corners, letting it down under the ship's bottom, and then putting a quantity of chopped rope-yarns, oakum, wool, cotton, &c. between it and the ship's side; by repeating the latter part of this operation several times, the leak generally sucks in a portion of the loose stuff, and thereby becomes partly, and sometimes wholly stopped. Some persons prefer thrumming the sail, instead of letting down the loose stuff; but in this mode the sail is soon chafed through by the hole if the leak is considerable, without affording sufficient substance to stop it.

FOUL, (*contraire, &c.*) is generally used in opposition to clear, and implies entangled, embarrassed, or contrary to; as, "a ship ran *foul* of us in the river," *i. e.* entangled herself amongst our rigging.

This happens sometimes by the violence of the wind, and sometimes by the carelessness of the people on board, to ships in the same convoy, and to ships in port, by means of others coming in. The damages occasioned by running *foul* of a vessel are of the nature of those in which both parties must bear a part. They are usually made half to fall upon the sufferer, and half upon the vessel which did the injury; but in cases where it is evidently the fault of the master of the vessel, he alone is to bear the damage.

Foul *Anchor*, (*ancre dont le cable a fait un tour*, Fr.) implies, that the cable is twisted round the stock, or one of the flooks, and thereby endangers the ship's drifting.

Foul-*Bottom*, (*fond, chargé de mousses et de coquillages, &c.* Fr.) denotes the bottom of a ship very dirty; as being covered with grass, sea-weeds, barnacles, shells, or other filth, which gathers and adheres thereto, under water.

Foul-*Ground of a Road, Bay, Sea-coast*, or *Harbour*, (*fond de mauvaise tenue*, Fr.) signifies that which is rocky, or abounding with shallows, or otherwise dangerous.

Foul-*Hawse*, (*tour des cables*, Fr.) means that the cables are turned round each other by the winding or turning about of the ship while she rides at anchor. See the articles Elbow and Hawse.

Foul-*Rope*, (*corde embarassée, ou engagée*, Fr.) a rope entangled in itself or hindered by another, so that it cannot run, or be overhauled.

Foul-*Water*, (*eau salie, ou troublée par le vaisseau qui a touché le fond*, Fr.). A ship is said to make *foul-water* when, being under sail, she comes into such shoal or low water that, though her keel does not touch the ground, yet it comes so near it that the motion of the water under raises the mud from the bottom, and so fouls the water.

Foul-*Wind*, (*vent contraire*, Fr.) is used to express that which is unfavourable, or contrary to the ship's course, as opposed to *large* or *fair*.

To FOUNDER, (*couler bas, ou couler à fond*, Fr.) to sink or go down.

A ship is said to *founder* when, by an extraordinary leak, or by a great sea breaking in upon her, she is so filled with water that she cannot be freed of it; so that she can neither veer nor steer, but lie like a log, not being able to swim along, and will, eventually, sink.

FOX, (*tresse de vieux cordages*, Fr.) a sort of *strand*, formed by twisting several rope-yarns together, and used as a *seizing*, or to weave a *mat* or *paunch*, &c.

Spanish Fox, a single rope-yarn untwisted and then twisted up the contrary way, and rubbed smooth. It is used for small seizings.

FRAME. See the article Timber.

To FRAP a Ship, (*ceintrer un vaisseau*, Fr.) is to pass four or five turns of a large cable-laid rope round the hull or frame of a ship, in the middle. to support her in a great storm or otherwise, when it is apprehended that she is not strong enough to resist the violent efforts of the sea. This expedient, however, is rarely put in practice, unless in very old ships, which their owners are willing to venture to sea as long as possible, by ensuring them deeply.

To Frap *a Tackle*, (*éguilleter un palan, &c.* Fr.) to cross and draw together the several parts or complication of ropes, which had been already strained to their utmost extent; in this sense it exactly resembles the operation of bracing up a drum, &c. The frapping increases the tension, and of course adds to the security acquired by the purchase. Hence the cat-harpings are no other than frappings to the shrouds.

To FREE, (*affranchir, en parlant de la pompe*, Fr.) to pump or otherwise throw out the water which has leaked into a ship's bottom at sea. &c.

To FREEZE. See the article Frost.

FREEZING, or FREEZE, (*frize, sorte d'ornement en sculpture*, Fr.) is a sort of ornamental painting on the upper part of a ship's *quarter, stern*, or *bow*. It generally consists of armour, instruments of war, marine emblems, &c.

FREIGHT, or FREIGHT of a Ship, (*affrètement*, Fr.) the hire, or a part thereof, usually paid for the carriage and conveyance of goods from one part or place to another; or the sum agreed on between the owner and the merchant for the hire and use of a vessel.

The freight of a vessel is usually agreed on either at the rate of so much for the voyage, or by the month, or per ton.

To Freight, or *Let out Vessels on Freight or Hire*, (*fréter, louer*, Fr.) is one of the principal articles in the trade of the Hollanders: they are the carriers of all the nations of Europe, and their purveyors, notwithstanding their country produces little or nothing, and, therefore, they are forced to have every thing necessary for the building of their vessels from other countries.

The principal laws and rules relating to freighting are as follow:—if a whole vessel be hired, and the merchant or person who hires it does not give it its full load or burden, the master of the vessel cannot, without his consent, take in any goods without accounting to him for freight.

That, though the merchant does not load the full quantity of goods agreed on in the charter-party, yet he shall pay the whole freight: and if he load more he shall pay for the excess.

If a time be appointed by charter-party, and either the ship be not ready to take in, or the merchant prepared to put on board, the parties are at liberty, with remedy, by action, for the detriment.

If part be on board, and some intervening accident prevent the merchant from shipping the whole in time, the master may contract with another, and have freight as damage for the time these goods were on board, longer than limited. On the other hand, if the vessel be not ready to proceed on the voyage at the time appointed, the merchant may ship the remainder of the goods aboard another, discharging the first, and recover damages against the first master or owners; therefore, by the marine law, chance, or other notorious necessity, will excuse the master, but he loses his freight till he *breaks ground*. But if the merchant be in fault, he must answer the damage, or be liable to maintain the crew ten days; and if, after that, the full freight; if damage afterwards, it is the merchant's risk: but, by the common law, while the goods are on board, the master must see them forth-coming. If goods are fully laded, and the ship hath broken ground, but the merchant afterwards declines the adventure, and unlades again, by the marine law the freight is due.

If a set time be agreed on between the merchant and master to begin and end the voyage, it may not be altered by the supercargo, without special commission: and if a master shall fail on his voyage, after the time agreed on for his departure, and damage happens afterwards, he shall make it good. If a ship be freighted from one port to another, thence to a third, &c. and so home to the port whence she first sailed, (commonly called a trading voyage) the whole is one and the same voyage, if performed according to the charter-party. If the ship be freighted out and in, no freight is due till the voyage is performed; if, therefore, the ship perish coming home, the whole freight is lost.

The master may set ashore such goods as he finds in his vessel, which were not notified to him; or take them at a higher rate than was agreed on for the rest. But if the master freight his ship, and afterwards secretly take in other goods, he loses his freight; and if any of the freighter's goods should, for the ship's safety, be cast overboard, the rest shall not be subject to average, but the master must make it good. If a ship be stopped or detained in its course, either through the master's or merchant's default, the delinquent shall be accountable to the other. Thus, if the freighter load the ship with prohibited goods, he shall answer the freight contracted; but if the ship put into any other port than she is freighted to, the master shall answer damage to the merchant; but if forced in by storm, enemy, or pirates, he must then sail to the stipulated port at his own costs.

If the master be obliged to refit his vessel during his voyage, the merchant shall wait, or else pay the whole freight; if the vessel could not be refitted, the master is obliged to hire another immediately, otherwise only to be paid his freight in proportion to the part of the voyage performed; though, in case the merchant proves that the vessel, at the time it set sail, was not capable of the voyage, the master must lose his freight, and account for damages to the merchant.

Freight shall be paid for merchandises which the master was obliged to sell for victuals, or refitting, or other necessary occasions, paying for the goods at the rate the rest were sold at, where they were landed.

If a ship in her voyage happens to be taken by an enemy, and afterwards is retaken by another ship in amity, and restitution is made, and she proceeds on her voyage, the contract is not determined, though the taking by the enemy divested the property out of the owners; yet by the law of war that possession was defeasible, and being recovered in battle afterwards, the owners became re-invested; so the contract, by fiction of law, became as if she had never been taken, and the entire freight becomes due.

It was covenanted by a charter-party, that a ship should return by a certain time within the river Thames (the danger of the sea excepted), and afterwards in the voyage, and within the time of the return, she was taken upon the sea by enemies unknown to the covenanter; and being detained by them, could not return within the time mentioned by the covenant. *Resolved*—This impediment was within the exception; for these words intend as well any danger upon the sea by pirates, or men of war, as dangers of the sea by shipwreck, tempest, &c.

In case of a prohibition of commerce with the country whither the vessel is bound, so that it is obliged to be brought back again, the master shall only be paid freight for going.

And if a ship be stopped or detained in its voyage, by an embargo by order of the prince, there shall neither be any freight paid for the time of the detention, in case it be hired per month: nor shall the freight be increased if hired for the voyage; but the pay and the victuals of the sailors, during the detention, shall be deemed average. See the article AVERAGE.

If, before the departure of the ship, there should happen an embargo, occasioned by war, reprisals, or otherwise, with the country to which the ship is bound, so that she cannot proceed on her voyage, the charter-party shall be dissolved, without damages or charges to either party; and the merchant shall pay the charges of unlading the goods; but if the restraint arises from a difference between the parties themselves, the charter party shall still remain valid in all points.

If the ports be only shut, and the vessels stopped for a time, the charter-party shall still be valid, and the master shall be reciprocally obliged to wait the opening of the ports and the liberty of the ships without any pretensions for damages on either side.

The merchant, however, at his own charges, may unlade his goods during the shutting up of the ports, upon condition either to relade them, or indemnify the master. For further particulars, see MAXWELL's *Spirit of Marine Law.*

FREIGHT also implies the burden or lading of a ship, or the cargo of goods, &c. which she has on board.

FRESH-SHOT, (*courant d'eau douce à l'embouchure d'une grande rivière*, Fr.) signifies the falling down of any great river into the sea, by means whereof the sea hath fresh water a good way from the mouth of the river. As this is more or less, it is called a great or small *fresh-shot.*

FRESH-*Spell*, implies a fresh gang to relieve the rowers in the long-boat.

FRESH-*Water*, (*eau douce*, Fr.) a name given to water fit to drink, in opposition to sea or salt-water.

FRESH-*Way* (*beaucoup d'air*, Fr.). When a ship increases her velocity she is said to get a *fresh-way.*

FRESH-*Wind*, (*vent frais*, Fr.) signifies, that the wind is strong, but not violent or dangerous; hence, when the gale increases, it is said to freshen.

To FRESHEN THE HAWSE, (*raffraichir le cable*, Fr.) to relieve that part of the cable which has for some time been exposed to the friction in one of the *hawse-holes*, when a ship rocks and pitches at anchor in a high sea.

While a ship remains in such a situation, it is always necessary to wrap some old canvas, mat, leather, or such like material, round that part of the cable which rubs against the *stem*, &c. The matter used for this purpose is called *service:* but, as the violent agitation of the ship, produced by the tempest or sea, as she rides in an open roadstead, must communicate a great friction to the cable, the service will, consequently, be soon worn through: it is necessary, therefore, to have it frequently renewed, by a fresh application of the like materials, behind the former, for the preservation of the cable, on which every thing depends; and this renewal of service is called *freshening the hawse*, a circumstance which cannot be too vigilantly observed.

FRESHES, (*eaux sauvages*, Fr.) imply the impetuosity of an ebb-tide, increased by heavy rains and flowing out into the sea, which it often discolours to a considerable distance from the shore, insomuch that the line which divides the two colours, may be perceived distinctly for a great length along the coast.

FRIGATE, (*frégate*, Fr.) in the navy, a light nimble ship, built for the purpose of sailing swiftly. These vessels mount from 20 to 50 guns, and are esteemed excellent cruisers.

FRIGATE-*Built*, (*frégaté—construit à la manière des frégates*, Fr.) implies the disposition of the decks of such merchant-ships as have a descent of four or five steps from the *quarter-deck* and *forecastle* into the *waist*, in contra-distinction to those whose decks are on a continued line for the whole length of the ship, which are called *galley-built*. See the article DECKS.

Formerly, the name of frigate was only known in the Mediterranean, and applied to a kind of long vessel, navigated in that sea with sails and oars. The English were the first who appeared on the ocean with those ships, and equipped them for war as well as for commerce.

FROST, (*gelée*, Fr.) such a state of the atmosphere as causes the congelation or freezing of water, or other fluids, into ice. In the more northern parts of the wo ld even solid bodies are affected by frost, though this is only or chiefly in consequence of the moisture they contain, which being frozen into ice, and so expanding, as water is known to do when frozen, it bursts and rends any thing in which it is contained, as plants, trees, stones, and large rocks. Many fluids expand by frost, as water, which expands about the one tenth part; for which reason ice floats in water; but others again contract, as quicksilver, and thence frozen quicksilver sinks in the fluid metal.

Frost, being derived from the atmosphere, naturally proceeds from the upper parts of bodies downwards, as the water and the earth: so, the longer a frost is continued, the thicker the ice becomes upon the water in ponds, and the deeper in the earth the ground is frozen. In about 16 or 17 days frost, Mr. Boyle found it had penetrated 14 inches deep into the ground.

At Moscow, in a hard season, the frost will penetrate two feet in the ground; and Captain James found it penetrate 10 feet deep in Charlton Island; and the water in the same island was frozen to the depth of six feet. Sheffer assures us, that in Sweden the frost pierces two cubits, or Swedish ells, into the earth, and turns what moisture is found there into a whitish substance, like ice; and standing water, to three ells or more. The same author also mentions sudden cracks or rifts in the ice of the lakes of Sweden nine or ten feet deep, and many leagues long; the rupture being made with a noise not less loud than if many guns were discharged together. By such means, however, the fishes are furnished with air; so that they are rarely found dead.

The close of the year 1708, and the beginning of 1709, were remarkable, throughout the greatest part of Europe, for a severe frost. Dr. Derham says it was the greatest in degree, if not the most universal, in the memory of man; extending through most parts of Europe, though scarcely felt in Scotland or Ireland. In 1788, we had a severe frost for many weeks, insomuch that the Thames were frozen below the bridges, and booths built thereon.

FULL AND BY, (*près et plein*, Fr.) a term used by seamen, implying the situation of a ship with regard to the wind, when she is close-hauled, and sailing in such a manner as neither to steer too nigh the direction of the wind, nor to deviate to leeward; both of which movements are unfavourable to her course, as in the former her sails will shiver, and render the effort of the wind precarious and ineffectual; and, in the latter, she will advance in a direction widely distant from her real course. Hence, keep her full! (*plein la voile!* Fr.) is the order from the pilot, or other officer, to the helmsman, not to incline too much to windward, and thereby shake the sails so as to retard the ship's velocity.

To FUMIGATE, to medicate or heal by vapours; to clear or purify the air of any infected place, by smoke, &c.

The frequent fumigation of ships is highly necessary, in order to prevent diseases produced by confined or infected air. The materials may be brimstone with saw-dust, or brimstone thrown over hot coals.

Nitre, to which a little vitriolic acid is added; or common salt, with the same addition of vitriolic acid. Gunpowder wetted, or the heated loggerhead in a pitch-pot. The operation should always be performed under the eye of a medical officer, to prevent an improper expenditure of the respective articles applied.

FUMIGATION LAMPS, a recent invention for the purpose of expelling foul air from the holds and other confined places of ships, and particularly useful for cleansing and keeping sweet and wholesome all receptacles for the sick.

Where many persons breathe in a confined place, the air is deprived of its oxygen, or animating principle. And as this destructive effect is still more increased by flame, used to illume the darkness of places where persons are necessarily confined, these lamps are provided with bevel tubes, to prevent this inconvenience to those persons who are on the orlop-deck, which is nearly dark and beneath the surface of the water. And to prevent the necessity of a number of lights, which would thus still more deprive the air of its oxygen, or vital principle, Sir Jerome Fitzpatrick has introduced reflectors into the lamps, by which means, the light is only considerably increased with the respirable air being diminished. And by well placing the lamp flame oven the pan, containing oil of vitriol and nitre, a proportion of oxygen is thus produced, sufficient to supply what the air loses, by the forementioned spoliation, and corrective of its malignant principle. The utility and convenience of these lamps, must, where light and salubrious air are so essential as in hospital ships, be incalculable.

The lamps, it must be observed, are so contrived as to prevent the men from endangering the ship by lighting their pipes clandestinely.

A very ingenious engineer having lately contrived the following method of supplying deep mines with fresh air, and extracting the foul by means of fire, the principle might be usefully applied on board our ships of war, by means of a furnace in the galley, and fresh air supplied with facility to the remotest corner of a ship.

Description of the Method of supplying deep Mines with fresh Air by means of Fire; and, upon the same Principles, the remotest Part of a Ship may be supplied with fresh Air.

It frequently happens, that in sinking deep pits, and also in cutting extensive horizontal mines, that the air either fails entirely, or is so contaminated with fiery particles, or other noxious vapours, as to become extremely pernicious to the healths, and too often instantly fatal to the lives of the people employed in these works.

To remedy this inconvenience, many ingenious inventions have been put into practice; the most effectual of which is, that of extracting foul air by means of fire. It is well known that fire consumes a vast quantity of air; and provided you can admit of a sufficiency to your kindled fewel, it will make no difference whether it is pure air or poisonous vapour. This being established, it appears obvious, that if you can place a fire in a close furnace, and have a tube from the said fire to the bottom of your pit, the fire having no other communication with the air, except through the said tube, it will draw off the foul air with great force, whose place will be as quickly supplied by fresh atmospheric air.

The method of destroying contagion, in crowded places, by fumigation, was first brought into practice by Dr. James Carmichael Smyth, who, having given some striking proofs of its efficacy, received a reward from parliament.

When this fumigation is undertaken on board ships, the ports and scuttles are closed, a number of pipkins, containing hot sand, are procured, and into each is plunged a gallipot, containing half an ounce of concentrated acid.

As soon as the acid is properly heated, an equal quantity of pulverized nitre is gradually added, and the mixture stirred with a glass spatula. The vapour resulting from the decomposition of nitre ascends, and is by the nurses conducted to every part of the apartment, which not only abates the malignity of the fever, but effectually stops the progress of infection.

FUNNEL. See the article Fire-Ship.

To FURL, (*ferler, ou serrer les voiles*, Fr.) to wrap or roll a sail close up to the *yard*, *stay*, or *mast* to which it belongs, and wind a gasket or cord about it to fasten it thereto.

FURLING in a Body, is a particular method of rolling up a top-sail only practised in harbours, and is performed by gathering all the loose part of the sail into the bunt, about the top-mast, whereby the yard appears much thinner and lighter than when the sail is furled over all at sea.

Furling-*Line*, (*raban de ferlage*, Fr.) denotes a cord employed, in the operation of furling: those which are used for the larger sails are generally flat, and are known by the name of *gaskets*.

FURRENS, in ship-building, are pieces to supply the deficiency of the timber in the moulding way.

FUSES, in artificial fire-works, are wooden tubes, used to fire the charge of a shell. They are filled with combustible composition, and furnished at the end with a quick-match. The composition is, saltpetre, 3lbs. 4 oz.; sulphur, 1lb.; and mealed powder, 2lbs.

Diameter of Fuses and Composition.

Species.	Diameter of Fuse.			Composition.			Drove by one man in one day.
	Below the Cup.	At the bottom.	At the cup.	Diameter.	Length.	Time it burns.	
Inches.	Inch.	Inches.	Inches	Inches.	Inches	Seconds	No.
13	2.1	1.575	2.49	.5	8.4	35	25
10	1.8	1.35	2.13	.438	7.2	33	25
8	1.3	1.25	1.78	.375	6.37	29	30
5½	1.1	.825	1.3	.275	4.4	18	50
4⅔	1.0	.75	1.18	.25	3.5	15	700
Grenades	0.8	.6	.9	.2	2.25		1000

The diamĕter inside the cup is three diameters of the bore.

Depth of the cup .... 1½ ditto.

Thickness of the wood at the bottom of the bore, 2 diameters.

The 13 and 10-inch *fuses* of the same length burn so nearly equal, that one common length answers both, as do the 8-inch, 5½, and 4⅔.—Therefore, to confine the length of fuse for any range, multiply the time of flight by 22 for the 13 and 10-inch, and 24 for the 8, 5½, and 4⅔; which is the decimal part of an inch a fuse burns in a second.

Fuses are thought to keep better by being painted; and, for field service are often marked off with black lines into seconds and half seconds.

FUTTOCKS, the middle division of a ship's timbers; or those parts which are situated between the *floor* and the top-timbers: those next the keel are called *ground-futtocks*, and the rest *upper-futtocks*. See the article Timbers.

The epithet *hooked* is frequently applied in common language to any thing bent or incurvated, as the *breast-hooks*, *fore-hooks*, *after-hooks*, &c.: this term is evidently derived from the lowest part or *foot* of the timber, and from the shape of the piece.

Futtock-*Plates*, are certain iron plates, the upper part whereof is open like a ring, to fix the dead-eyes in, and round holes are punched at the lower end of these plates for the futtock-shrouds to hook in, or bolts to be driven through when used for the lower shrouds.

Futtock-*Shrouds*. See the article Shrouds.

Futtock, or Foothook-*Stave*, a short piece of rope served over with spun-yarn, to which the shrouds are confined at the catharpings.

The *futtock-staves* are seized along the lower shrouds horizontally, as much below the upper-side of the trestle-trees as the cap is above; a spar is then lashed to the outside of the shrouds, about six feet below the *futtock-stave*, and a single block lashed round each shroud and spar (except the fore and after shrouds), so that all come in together: the swifting-line is then reeved through each block, from side to side, beginning in the middle, one end leading aft, the other forward; it then crosses, and reeves through two leading blocks, one on each side the deck, and is bowsed taught till the shrouds come into the length of the catharping-legs, which are seized at each end, round the *futtock-stave* and shroud.

G.

GAFF, (*pic, ou vergue à corne*, Fr.) in small ships, a sort of boom, used to extend the upper edge of the mizen, and always employed for the same purpose on those sails whose foremost edges are joined to the masts by hoops or lacings, and which are usually extended by a *boom* below: such are the main-sails of all sloops, brigs, and schooners.

The foremost end of the *gaff* is furnished with two cheeks forming a semi-circle, which enclose the after part of the mast, so as to confine the *gaff* close to its respective mast whilst the sail is hoisting or lowering. It is further secured in this situation by a rope passing from one of the cheeks to the other on the fore-side of the mast; and to prevent the friction of this rope upon the mast, by hoisting or lowering, several little wooden balls, called *trucks*, are hung upon it, in the same manner as beads are hung upon a catholic's rosary.

GAFF *Halyards*. See the article HALYARDS.

GAFF *Top-sail*, (*flèche-en-cul*, Fr.) is a light quadrilateral sail, the head being extended on a small *gaff*, which hoists on the top-mast, and the foot spreading from the throat to the peek, or the extent of the lower *gaff*.

Additional GAFFS. A gentleman of the name of Matthew, who, speaking of his own invention with the utmost diffidence, has proposed the introduction of *additional gaffs* on board ships, even of the line, proposes these should be constructed in such a manner, as to be rigged for service in a very short space of time; and, when no longer needed, to be unshipped, and stowed away with the other spars.

The advantage he thinks of obtaining by this measure is, the prevention of the ship from accident, in case of her missing stays, and is thus described:—

"A gaff should be fixed on the foreside of the foremast, just above the fore-yard; the same on the main and mizen-masts, each of them supporting a schooner-like sail; and the same on the fore-side of the top-masts. If these sails, when the ship was falling off, after missing stays, could have been immediately spread, all the other sails clewed up, and furled as fast as possible, the yards braced in such a manner as to have the least power in pressing the ship to leeward; the rudder placed in a direct line with the keel, and the ship had been provided with a light, yet wide, rudder hung to the stern, so contrived as to have been turned back against it"—had the ship been thus equipped, she might then, in his opinion, have sailed back stern foremost, have cleared the rocks, and have been brought to safe anchorage in the middle of the bay. "I am aware," says he, "that it may be thought, the draught of water abaft being so much superior to that forward, the vessel would not answer the helm at the bows; yet, by a judicious management of the sails, the ship might have been kept on the wind, stern foremost, for some time, and probably the rudder at the stern-post would have assisted in doing it; by being turned in a contrary direction to which it is usually moved, to answer the same purposes.

"Ships may be provided with sails in the manner above described, with another of a jib-like shape, from the taffrail and from the mizen-mast-head; and when they are embayed, and it becomes necessary to work to windward, and tack often, these gaffs and sails might be got up, before they get under weigh, and ready to be used in case of necessity.

"Suppose a ship thus provided, and in the situation before mentioned, probably if the sea ran high, and there was reason to fear the ship might miss stays, it would be the best way not to attempt it, but throw all the sails *aback*, and instantly set the schooner sails, &c. and the moment the ship's headway is stopped, to trace round the yards, and furl the sails. By adopting this method she might make less lee-way, before her motion stern first, commenced, than if she failed in stays: but both should be tried, and that mode abided by, in extreme peril, which causes the ship to drive least to leeward; for it might so happen, that a few feet to windward might save a fine ship, and the lives of many hundred brave men. The lower stays may, in all likelihood, be made useful sails in going stern first, by carrying the sheets to the water gunwale, but the leech-rope must be very strong.

"Experience," says this gentleman, "will soon make sailors masters of this kind of rigging."

GAGE. See the article WEATHER-GAGE.

GAGE, is also a measurement taken by a rule, or a pair of compasses.

GAGGING, a mode of punishment used in the navy, to prevent a continuation of insolent language during confinement for drunkenness or other misconduct. The gag, however, is never resorted to in the service, but under the most aggravating circumstances.

To GAIN THE WIND OF A SHIP, (*gagner le vent d'un vaisseau*, Fr.) in navigation, to arrive on the weather-side, or to windward of some other vessel in sight, when both are plying to windward, or sailing as near the wind as possible.

GALAXY, in astronomy, the *Via Lactea*, or *milky way* in the heavens. This is a broad circle, sometimes double, but for the most part of its course runs in a single stream, and surrounds the whole celestial concave. It is composed of an infinite number of minute stars, too remote from the earth to be distinguished by the naked eye; but are discovered in all parts of it, in great numbers, by the assistance of a good telescope.

Dr. Herschel first began to survey the via lactea with a powerful telescope, and found that it completely resolved the whitish appearance into stars, which the telescope he formerly used had not power to do. The portion he first observed, was that about the hand and club of Orion; and found therein an astonishing multitude of stars, whose number he endeavoured to estimate, by counting many fields, and computing from a mean of these how many might be contained in a given portion of the milky-way.

In the most vacant place to be met with in that neighbourhood, he found sixty-three stars; other six fields contained 110, 60, 70, 90, 70, and 74 stars; a mean of all which gave 79 for the number of stars to each field; and thus he found, that by allowing fifteen minutes for the diameter of his field of view, a belt of fifteen degrees long and two broad, which he had often seen pass before his telescope in an hour's time, could not contain less than 50,000 stars, large enough to be distinctly numbered: besides which, he suspected twice as many more, which could be seen only now and then by faint glimpses, for want of sufficient light.

The success he had within the milky-way soon induced him to turn his telescope to the nebulous parts of the heavens, of which an accurate list had been published in the "Connoissance des Temps, for 1783 and 1784." Most of these yielded to a Newtonian reflector, of 20 feet focal distance, and 12 inches aperture, which plainly discovered them to be composed of stars, or at least to contain stars, and to show every other indication of its consisting of them entirely.

GALE OF WIND, (*coup de vent*, Fr.) a phrase used by sailors, to signify a storm or tempest. It is more particularly termed a hard gale, or strong gale, &c. See the article BREEZE.

A Hard GALE, (*vent grand frais*, Fr.) implies the diminutive of the preceding article.

A Fresh GALE, (*vent frais, ou bon frais*, Fr.) is still a greater diminutive, and may be considered as not too strong for a ship to carry single reefed-top-sails in it when close-hauled.

A Top-Gallant GALE, (*tems à perroquet*, Fr.) implies that sort of wind in which a ship may carry her top-gallant-sails.

Equinoctial GALES, are those storms which are observed to take place about the time of the sun's crossing the equator, or equinoctial line, at which time there is equal day and night.

GALLED, (*éraillé, usé par le frottement*, Fr.) the state of a mast, yard, cable, or other rope, when it is deprived of its surface by friction. To preserve those articles from being damaged by this effect, it is therefore usual to cover them with skins, mats, canvas, or such materials, in the places where they are the most exposed to it. See the article TO SERVE A ROPE.

GALLEON, (*galion*, Fr.) a name formerly given to ships of war, furnished with three or four batteries of cannon. It is now retained only by the Spaniards, and applied to the largest size of their merchant ships, employed in West Indian voyages, and usually furnished with four decks. They likewise bestow the same name on those vessels, whether great or small, which proceed annually to Vera Cruz. The Portuguese also have several ships which they send to India and the Brazils, nearly resembling the *galleons*, and by them called *caragues*.

The Spaniards send every year two fleets; the one for Mexico, which they call the flota; and the other for Peru, which they call the galleons.

The galleons are loaded at Cadiz, from whence they may put out at any time.

They were formerly appointed to sail from Cadiz, in January, that they might arrive at Porto Bello about the middle of April, where the fair being over, they might take aboard the plate, and be at Havannah with it about the middle of June, where they were joined by the flota, that they might return to Spain with the greater safety. For this purpose, the viceroy of Peru was to take care that the plate should be at Panama by the middle of March. The plate is fifteen days removing from Potosi to Africa, eight days generally from thence by sea to Callao, and from that place to Panama twenty days, taking in by the way the plate at Paita and Truxillo. It has, however, been found by experience, that the month of September is the fittest for the fleet to sail: they are about two years in the whole voyage. However often or seldom the galleons go out, the next fleet never go out till the last are returned. When the galleons and flota put out together, they separate about the Antilles Islands; the galleons for Carthagena and Porto Bello, and the flota for Vera Cruz. At their return, they rejoin at Havannah, in the Isle of Cuba.

The loading of the galleons is always the richest: an estimate of the yearly returns of cargoes, both of the flota and galleons, was formerly as follows:—

Of gold, the galleons bring yearly about two or three millions of crowns, and the flota about one. Of silver, the galleons bring eighteen or twenty thousand crowns, and the flota ten or twelve. Of precious stones, the galleons bring two hundred thousand crowns worth of pearls, two or three hundred thousand crowns worth of emeralds, and twenty or thirty thousand crowns worth of bezoar, amethysts, and other stones of less value; the flota brings none at all.

Of wools, the galleons bring forty or fifty thousand crowns, the flota none. Of quinquina, the galleons bring the value of twenty thousand crowns, the flota none. Of skins and leather, the galleons bring seventy thousand crowns worth; the flota as much. Of Campeachy wood, the galleons bring sixty thousand crowns worth; the flota none. Of skins and leather from Buenos Ayres, the register-ships may bring to about two hundred thousand crowns; of cochineal, about a million of crowns; and of indigo, about six hundred thousand.

By a general regulation made in Spain, it has been established, that there should be twelve men of war, and five tenders, annually fitted out for the armada or galleons; eighty ships of six hundred tons burden each, and three tenders, one of an hundred tons for Margaretta Island, and two of eighty each, to follow the armada: for the New Spain fleet, two ships of six hundred tons each, and two tenders of eighty each; and for the Honduras fleet, two ships of five hundred tons each: and, in case no fleet happens to sail any year, three galleons and a tender should be sent to New Spain for the plate.

But the number of galleons has been different at

different times; it has increased in time of war, and diminished in time of peace.

GALLERY, (*galerie de poupe, et plus souvent les bouteilles d'un vaisseau*, Fr.) a balcony projecting from the *stern* or *quarter* of a ship of war, or of a large merchantman. The stern-gallery is wholly at the stern of the ship, and is usually decorated with a balustrade, extending from one side of the ship to the other; the fore-part is limited by a partition called the skreen-bulk head, in which are framed the cabin windows; and the roof of it is formed by a sort of vault, termed the *cove*, which is frequently ornamented with sculpture. See the article STERN.

The quarter-gallery of a ship of 74 guns is represented in plate XI. fig. 9.

It is that part which projects on each quarter, and is generally fitted up as a water-closet.

Ships of twenty guns and upwards, on one deck, have quarter galleries, but no stern gallery; two and three deckers have quarter galleries, with their proper conveniences, and one or two stern galleries.

GALLEY, (*galère*, Fr.) a kind of low flat-built vessel, furnished with one deck, and navigated with sails and oars, particularly in the Mediterranean.

The largest sort of these vessels, (*galeasse*, Fr.) were formerly employed only by the Venetians. They were about 162 feet long above, and 133 feet by the keel; 32 feet wide, with 23 feet length of stern-post. They were furnished with three masts, and thirty-two banks of oars; each bank containing two oars, and every oar being managed by six or seven slaves, who were usually chained thereto. In the fore-part they had three small batteries of cannon, viz. two 36-pounders, two 24-pounders, and two 2-pounders: they had also three 18-pounders on each quarter, and carried from 1000 to 12,000 men. They were found, however, by experience, to be of but little utility, except in fine weather; notwithstanding they were deemed extremely convenient for bombarding or making a descent upon an enemy's coast, as drawing but little water; and having by their oars frequently the advantage of a ship of war, in light winds or calms, by cannonading the latter near the surface of the water, by scouring her whole length with their shot, and at the same time keeping on her quarter or bow, so as to be out of the direction of her cannon.

The *galleys* next in size to these are called half-galleys, and are from 120 to 130 feet long, 18 feet broad, and 9 or 10 feet deep. They have two masts, which may be struck at pleasure, and are furnished with two large lateen sails, and five pieces of cannon. They have commonly 25 banks of oars, as described above. A size still less than these are called quarter galleys, carrying from twelve to sixteen banks of oars. They generally keep close under the shore, but sometimes venture out to sea to perform a summer cruise.

Half and quarter galleys are rigged and navigated the same as galleys, and take this denomination from their being much shorter.

Bombay galleys are like the former, and mostly used by Corsairs on the coast of Barbary.

All the galleys, both ancient and modern, are of a finer and slenderer make than ships. Formerly they made divers kinds; at present the galleys are all alike, the only difference being in respect to size, but nothing as to figure.

In France are forty galleys for the use of the Mediterranean, the arsenal thereof being at Marseilles. The general of the galley bears a double anchor placed in pale behind the escutcheon of his arms, as a mark of his office.

The captain galley is the principal galley of a state, commanded by the captain-general of the galleys.

In France, before the revolution, the royal-galley was the first.

GALLEY, *or Cook-room of a Ship*, (*la cuisine d'un vaisseau*, Fr.) the name of the kitchen of a ship of war, or the place where the grates are put up, fires lighted, and the victuals generally boiled or roasted.

In East-India ships, it is generally termed the cook-room, and on board of merchantmen it is called the coboose.

Row-GALLEY, (*galère à rames*, Fr.) the name given to an open boat, rowing six or eight oars, and used on the river Thames by custom-house officers, press-gangs, and also for pleasure; hence the appellation of custom-house-galley, press-galley, &c.

GALLEY-*Slave*, (*forçat, ou galerien*, Fr.) a person condemned to work at the oar on board a galley, being chained to the deck.

Condemnation to the galleys is a punishment particularly in France, whereby criminals and delinquents are adjudged to serve as slaves on board the galleys, either during life, or for a limited time.

A man condemned for perpetuity is dead in a civil sense; he cannot dispose of his effects; cannot inherit; and if he be married, his marriage is null, nor can his widow have any of her dower out of his goods, which, with his lands, are hereby confiscated.

GALLIOT, (*galiote Hollandoise, sorte de bâtiment à deux mâts avec une vergue à corne*, Fr.) a Dutch vessel, carrying a main and a mizen-mast, and a large gaff-main-sail.

The mast is supported by four or five pair of shrouds, and a stay which sets up to the stem: over this stay is another, that leads to the bowsprit-end. Abaft the mast there is a large gaff or spritsail, and over it a flying topsail, a staysail upon the main-stay, and one or two jibs on the bowsprit. Sometimes a small mast is stepped quite aft, to which is set a sail, hauled out to a boom at the foot.

GALLOWS-BITS, (*potences, pour porter les mâts de hune et vergues de rechange*, Fr.) on flush-decks, the name of a strong frame of oak about eight inches square, made in the form of a gallows, and fixed at the fore and main hatchway, to support the spare top-masts, yards, booms, boats, &c.

GAMMONING THE BOWSPRIT, (*liûre de beaupré*, Fr.) the act of binding the inner quarter of the bowsprit close down to the ship's stem, in order to enable it the better to support the stays of the fore-mast, and carry sail in the fore part of the vessel. This is performed by passing seven or eight turns of a rope over the bowsprit, and through a large hole called the *gammoning-hole*, in the stem or knee of the head alternately: afterwards all the turns are drawn as firm as possible, the opposite ones are braced together under the bowsprit by a *frapping*, as exhibited in plate XI. fig. 6.

GANG, (*détachement de matelots*, Fr.) a select

number of a ship's crew appointed on any particular service, and commanded by an officer suitable to the occasion. See the article PRESS-GANG.

GANG BOARD, (*planche à débarquer*, Fr.) a board or plank with several cleats or steps nailed to it for the convenience of walking into or out of a boat upon the shore, where the water is not deep enough to float the boat close to the landing-place.

GANGWAY, (*passe-avant*, Fr.) a narrow platform, or range of planks, laid horizontally along the upper part of a ship's side, from the quarter-deck to the forecastle, and is peculiar to ships that are *deep-waisted*, for the convenience of walking more expeditiously *fore and aft*, than by descending into the waist. It is fenced on the outside by iron stanchions, and ropes or rails, extended from one to the other; and in vessels of war, with a netting, in which part of the hammocks are stowed. This is frequently called the gang-board in merchant vessels.

GANGWAY, (*échelle*, Fr.) is also that part of a ship's side, both within and without, by which persons enter and depart. It is provided with a sufficient number of steps, or *cleats*, nailed upon the ship's side, nearly as low as the surface of the water; and sometimes furnished with a railed accommodation-ladder, resembling a flight of stairs, projecting from the ship's side, and secured by iron braces, so as to render the ascent and descent extremely convenient.

GANGWAY, is likewise used to signify a narrow passage left in the hold, when a ship is laden, in order to enter at any particular place, as occasion may require; whether to examine the situation of the provisions or cargo, to discover and stop a leak, or to bring out any article required for service, &c. Finally, a *gangway* implies a thoroughfare, or narrow passage of any kind.

To bring to the GANGWAY, implies, to punish a seaman, by tying him up and flogging him with a cat-o'nine-tails.

GANTLOPE, (*la bouline punition*, Fr.) vulgarly pronounced GANTLET, a race which a criminal is sentenced to run in a vessel of war, as a punishment for felony, or some other heinous offence.

To Run the GANTLET, (*courir la bouline*, Fr.) is to go through the above punishment, which is executed in the following manner:—the whole ship's crew is disposed in two rows, standing face to face on both sides of the deck, so as to form a lane, whereby to go *forward* on one side, and return *aft* on the other; each person being furnished with a small twisted cord, called a knittle, having two or three knots upon it; the delinquent is then stripped naked above the waist, and ordered to pass forward between the two rows of men on one side, and aft on the other side, a certain number of times, rarely exceeding three; during which every person gives him stripes as he runs along; he is preceded by the master-at-arms, having a drawn cutlass under his arm, with the point behind him. In his passage through this painful ordeal he is sometimes tripped up and very severely handled while incapable of proceeding. This punishment is seldom inflicted, except for such crimes as will naturally excite a general antipathy amongst the seamen.

GARBOARD-STREAK, (*virure de gabord*, Fr.) in ship building, the first range or *streak* of planks laid upon a ship's bottom next to the keel, throughout the whole length of the floor. The edge of this plank is let into a groove or channel in the side of the keel, which is called the rabbet of the *garboard-streak*.

GARLAND, (*sorte de filet*, Fr.) a sort of net, whose opening is extended by a wooden hoop of sufficient size to admit a bowl or platter within it. It is accordingly used by the sailors as a locker or cupboard to contain their provisions, being hung up to the beams within the *birth*, where they commonly mess between decks.

Shot-GARLAND, (*petit parquet pour les boulets*, Fr.) a piece of timber nailed horizontally along the ship's side, from one gun-port to another, and filled with several hemispherical cavities, to contain the round-shot, ready for charging the great guns in time of an engagement.

GARNET, (*bredindin, sorte de palan*, Fr.) a sort of tackle fixed to the main-stay of a merchant ship, and is used to hoist the cargo in and out at the time of landing and delivering her.

Clew-GARNET. See the article CLEW, &c.

GASKET, (*garcette*, Fr.) a sort of platted cord fastened to the sail-yards of a ship, and used to *furl* or tie up the sail firmly to the yard, by wrapping it round both six or seven times, the turns being at a competent distance from each other.

Bunt-GASKET, (*maîtresse garcette, ou garcette de fond*, Fr.) is that which supports or ties up the bunt of the sail, and should consequently be the strongest, as having the greatest weight to support; it is sometimes made in a peculiar manner.

Quarter-GASKET, is used only for large sails, and fastened about half way out upon the yard, which part is called the quarter.

The Yard-arm GASKET, (*la garcette de bout de vergues*, Fr.) is made fast to the yard-arm, and serves to bind the sail as far as the quarter-gasket on large yards, but extends quite into the bunt of small sails.

GATES OF A DOCK, (*portes d'un bassin*, Fr.) is a kind of thick double door, hung at the entrance, and made to open and shut, so as to confine or exclude the water on docking or undocking a ship.

GATHER AFT THE JIB-SHEET! (*reprends l'écoute du grand foc!* Fr.) the order to pull it tight in.

To GATHER *Way*, (*prendre de l'air*, Fr.) to increase the velocity.

GATT, the same as channel, and is a term constantly used on the Flemish coast and in the Baltic for that purpose.

GAUT, a term made use of in the East Indies, to denote a passage or road from the coast to the mountains or upland country.

GEARS. See the article JEARS, which is the more general way of writing it.

GEMINI, the twins, in astronomy, one of the twelve signs of the zodiac, the third in order, beginning with Aries, and denoted by this mark (♊) in astronomical books. The stars in this constellation, according to the Britannic catalogue, are 85. It is represented on celestial globes by the figure of two twin children, looking each other affectionately in the face, and supposed to be Castor and Pollux.

GEOCENTRIC, in astronomy, is applied to a planet, or its orbit to denote it concentric with the earth, or as having the earth for its centre, or the same centre with the earth.

GEOCENTRIC *Latitude of a Planet*, is its distance from the ecliptic, as it is seen from the earth, which, even though the planet be in the same point of her orbit, is not constantly the same, but alters according to the position of the earth, in respect to the planet.

GEOCENTRIC *Longitude of a Planet*, is the distance measured on the ecliptic, in the order of the signs, between the *geocentric* place, and the first point of Aries.

GEOCENTRIC *Place of a Planet*, is the place in which it appears to us from the earth, supposing the eye there fixed; or it is a point in the ecliptic to which a planet seen from the earth is referred.

GEOGRAPHY, implies a description of the terrestrial globe, or the science which exhibits the results of our investigation respecting it. It is sometimes contrasted with hydrography, which signifies a description of the water; viz. of seas, lakes, rivers, &c. including marine charts. Anciently both were considered in connexion with astronomy, as parts of cosmography, which attempted to delineate the universe. Geography is more properly contrasted with chorography, which illustrates a country or province; and still more with topography, which describes a particular place, or smaller district.

What is called *general* or *universal* geography embraces a wide view of the subject; regarding the earth astronomically as a planet, the grand divisions of land and water, the winds, tides, meteorology, and may extend to what is called mechanical geography, including directions for the construction of globes, maps, and charts. Geography, popularly considered, is occupied in the description of the various regions of this globe, chiefly as being divided among various nations, and improved by human art and industry.

GIMBALS, (*balanciers de boussoles, &c.* Fr.) the brass rings by which a sea-compass is suspended in its box so as to counteract the effects of the ship's motion, and keep the card horizontal.

GIMBLETING, (*action de l'ancre qui se déplante en tournant sur sa verge; ce qui arrive par le trop de tors du cable*, Fr.) a term applied to the anchor, to denote the action of turning it round by the stock, so that the motion of the stock appears similar to that of the handle of a gimblet, when it is employed to turn the wire.

GIRT, (*vaisseau ceintré par ces amarres, de façon à ne pouvoir éviter au vent, ou à la marée*, Fr.) the situation of a ship which is moored so taught by her cables, extending from the *hawse* to two distant anchors, as to be prevented from swinging or turning about, according to any change of the wind or tide, to the current of which her head would otherwise be directed.

The cables are extended in this manner, by a strong application of mechanical powers within the ship; so that as she veers or endeavours to swing about, her side bears upon one of the cables, which catches on her heel, and interrupts her in the act of traversing. In this position she must ride with her broadside or stern to the wind or current, till one or both of the cables are slackened so as to sink under the keel; after which the ship will readily yield to the effort of the wind or current, and turn her head thither. See the article RIDING.

GIRT-LINE, (*cartaheu*, Fr.) a rope passing through a single block, on the head of the lower masts, to hoist up the rigging thereof; and the persons employed to place the rigging and cross-trees upon the mast-heads. The *girt-line* is therefore the first rope employed to rig a ship; after which operation it is removed till the ship is to be unrigged.

GIRTH, is the circumference of any thing, as a piece of timber, &c.

GIVE WAY! (*nage!* Fr.) the order to a boat's crew to row, after having ceased for a short time, or to increase their exertions if they were before rowing.

GIVE *Way together!* (*nage ensemble!* Fr.) is, also, the order to keep time together in rowing, so as that the oars shall all dip and rise together, whereby their several forces are exerted as one.

To GIVE *Chase*, (*donner la chasse*, Fr.) is to pursue a ship or fleet.

GLASS, (*lunette, longue-vue*, Fr.) the common term used to express the instrument called a telescope.

Half-hour GLASS, (*sablier d'une demi heure*, Fr.) frequently called the watch glass, is used to measure the time which each watch has to stay upon deck.

Hour GLASS, (*sabliere d'un heure*, Fr.) a popular kind of chronometer, which serves to measure time by the running of sand from one vessel into another. Glasses of this kind, for half and quarter hours, &c. are much used at sea.

Half-minute and *Quarter-minute* GLASSES, are used to ascertain the rate of a ship's velocity, measured by the log; these glasses should be frequently compared with a good stop watch, to determine exactly how many seconds they run.

Night-GLASS, (*lunette de nuit*, Fr.) a telescope for viewing objects at night.

Sand GLASS, (*sabliere, orloge*, Fr.) is also used in the plural, to denote the duration of any action, as, "We fought yard-arm and yard-arm three glasses;" that is, three half-hours, or an hour and a half.

To flog or sweat the GLASS, (*manger du sable*, Fr.) is to turn it before the sand has quite run out, and thereby gaining a few minutes in each half hour, which make the watch too short.

GLOBE, in geometry, is a round or spherical body, more usually called a sphere.

GLOBES are of two kinds, celestial and terrestrial; each of very considerable use, the former in astronomy, and the latter in geography, for performing many of the operations in an easy obvious manner, so as to be conceived without any knowledge of the mathematical grounds of those sciences.

Description of the Globes.

The fundamental parts, common to both globes, are an axis, representing that of the world, and a spherical shell or cover, which makes the body of the globe, on the external surface of which the representation is drawn. The globe consists of several parts; as (1) The two poles, being the ends of the axis on which the globe turns, representing the diurnal motion of the earth, and the apparent diurnal motion of the heavens. (2) The brazen meridian, on which are marked the degrees from the equator to the poles. (3) The wooden horizon or frame that supports the whole globe, on which are drawn several circles, as that on which the signs of the zodiac are marked; that containing the calendar or days of the month which correspond to the several degrees of the zodiac; and that on which are delineated the thirty-two points of the compass. (4) The hour circle, divided into twelve powers, representing the division of time for the day and the night. And (5) A thin slip of brass, called the quadrant of altitude, useful in several problems to measure the distances and directions of the places from each other.

There are ten principal circles represented upon globes; that is, six greater and four lesser ones. The greater circles are, the horizon, the brazen meridian, already described; the equinoctial, as it is called on the celestial, and equator on the terrestrial globe; the ecliptic drawn along the middle of the zodiac; and the two colures.

The lesser circles, of principal use, are the two tropics and the two polar circles. Of these circles some are fixed, and always obtain the same position; others moveable, according to the position of the observer. The fixed circles are, the equator and ecliptic, with their parallels and secondaries; which are usually delineated upon the surface of the globes. The moveable circles are the horizon, with its parallels and secondaries.

The horizon is that great and broad wooden circle surrounding the globe, and dividing it into two equal parts, called the upper and lower hemispheres. It has two notches, to let the brazen meridian slip up and down, according to the different heights of the pole. On the flat side of this circle are described the twelve signs, the months of the year, the points of the compass, &c. The brazen meridian is an annulus or ring of brass, divided into degrees. It divides the globe into two equal parts, called eastern and western hemispheres.

The quadrant of altitude is a thin pliable plate of brass, answering exactly to a quadrant of the meridian. It is divided into 90 degrees, and has a notch, nut, and screw, to fix to the brazen meridian in the zenith of any place, where it turns round a pivot, and supplies the room of vertical circles. The hour circle is a flat ring of brass, divided into twenty-four equal parts, or hour distances; and in the pole of the globe is fixed an index, that turns round with the globe and points out the hours upon the hour circle. Lastly, there are now generally added a magnetical compass and needle, under the pillar and claw of the frame, for the purpose of placing the globe to represent the apparent situation of the heavens or the earth.

The surface of the celestial globe may be esteemed a just representation of the concave expanse of the heavens, notwithstanding its convexity; for it is easy to conceive the eye placed in the centre of the globe, and viewing the stars on its surface; supposing it made of glass, as some globes are: also, that if holes were made in the centre of each star, the eye in the centre of the globe, properly placed, would view through each of the holes the very stars in the heavens represented by them.

As it would be impossible to have any distant notion of the stars, in respect to their number, order, and distances, without arranging them in certain forms, called constellations; this the first observers of the heavens took care to do; and these, like kingdoms and countries upon the terrestrial globe, serve to distinguish the different parts of the superfices of the celestial globe. The stars, therefore, are all disposed in constellations under the forms of various animals, whose names and figures are represented on the celestial globe, which were first invented by the ancient astronomers and poets, and are still retained for the better distinction of these luminaries.

The globes commonly used are composed of plaister and paper, or other matter.

For the use of the celestial and terrestrial globes, see NICHOLSON'S *Encyclopedia*, and ADAMS and KEITH, *on the Use of the Globes.*

GLOBULAR CHART. See the article CHART.

GLOBULAR *Sailing*, in navigation, is the method of resolving the cases of sailing upon the principles deduced from the spherical figure of the earth, such as Mercator's sailing, or Great-circle sailing. See the article SAILING.

GOLDEN NUMBER. See the article CYCLE OF THE MOON, &c.

GONDOLA, (*gondole*, Fr.) a sort of *barge*, curiously ornamented, and navigated on the canals of Venice; also a passage-boat with six or eight oars, used on other parts of the coast of Italy: hence the rowers are called Gondoliers.

The middle-sized gondolas are upwards of thirty feet long and four broad; they always terminate at each end in a very sharp point, which is raised perpendicularly to the full height of a man.

GONDOLIER, (*gondolier*, Fr.) a man who works or navigates a gondola.

GONGA, from whence comes Ganges, a general name for a river.

GOODGEONS, or BRACES, (*femelots de gouvernail*, Fr.) certain clamps of iron, or other metal, bolted on the stern-post of a ship, whereon to hang the rudder; for which purpose there is a hole in each of them, to receive a correspondent pintle, bolted on the back of the rudder, which turns thereby as upon hinges. There are generally four, five, or six *goodgeons* on a ship's stern-post and rudder, according to her size, and upon these the rudder is supported, and traverses from side to side as upon an axis. See the article HELM.

GOOSE-NECK, (*crochet de fer fixê au bout inférieur d'un gui*, Fr.) a sort of iron hook fitted on the inner end of a boom, and introduced into a clamp of iron, or eye-bolt, which encircles the mast, or is fitted to some other place in the ship, so that it may be unhooked at pleasure.

GOOSE-WINGS OF A SAIL, (*pointes, ou angles d'une voile dont le fond est cargue*, Fr.) the clews or lower corners of a ship's main-sail or fore-sail, when the middle part is furled or tied up to the yard.

The *goose-wings* are only used in a storm, to scud before the wind, when the sail, even diminished by a *reef*, would be two great a pressure on the ship, in that situation.

GORES, (*pointes de toile*, Fr.) in sail-making, the angles cut slopewise, at one or both ends of such cloth as widen or increase the depth of a sail.

GORING, or GORING-CLOTH, (*pointe, ou toile de pointe*, Fr.) that part of the skirts of a sail, where it gradually widens from the upper part or head, towards the bottom or foot: the *goring-cloths* are, therefore, those which are cut obliquely, and added to the breadth. See the article SAIL.

GOUGE, a long, sharp, circular edged tool, similar to chissels. Gouges are flat tools used in turning;

they are stouter and longer than those used with a mallet.

GOUGE-*Bit*, a bit smaller than a centre-bit, with a hollow edge at its end like a *gouge*.

Plate XI. Fig. 1. GRAFTING A ROPE, the act of unlaying the two ends of it, placing the strands one within the other, as for splicing, plate XI. fig. 1, and stopping them at the joining. The yarns are then opened out, split, and made into knittles for pointing.

GRAPE-SHOT, (*charge en grappe*, Fr.) a combination of small shot, enclosed in a thick canvas bag, and corded strongly together, in a cylindrical form, equal in diameter to that of a ball adapted to a cannon. The bag is made exactly to hold a bottom which is put into it; as many shot are then thrown in as the grape is to contain. The whole is next quilted over with strong pack-thread, to keep the shot from moving. When finished, the bags are put into boxes for the purpose of being conveniently carried.

The number of shot in a grape varies according to the size of the guns; in sea service nine is always the number; but for land it is increased to any number or size, from an ounce and a quarter in weight to four pounds. It forms the thickest fire that can be produced in the same space; and can be exceeded in effect only by the CASE-SHOT, which article see.

The following statement exhibits the number and sorts of shot contained in the grape, for guns of different species:—

Guns.	Species of Shot in the Grape.	Number in each.	Number in each Box.
42 pounders	4 lb.	9	4
32 ———	3	9	4
24 ———		9	6
18 ———	1½	9	8
12 ———	1	9	10
9 ———	13 oz.	9	12
6 ———	8	9	20
4 ———	6	9	20

Fig. 2. GRAPLING, or GRAPNEL, (*grapin*, Fr.) a sort of small anchor, fitted with four or five flooks or claws, plate XI. fig. 2, and commonly used to fasten boats or other small vessels.

Fire-GRAPLING, (*grapin d' abordage*, Fr.) an instrument nearly resembling the former, but differing in the construction of its flooks, which are furnished with strong barbs on their points, fig. 10, plate X. These machines are usually fixed by a chain on the yard-arms of a ship, in order to grapple any adversary whom she intends to board: they are particularly useful in *fire-ships*.

GRATINGS, (*caillebotis*, Fr.) a sort of open covers, for the hatches, resembling lattice-work, serving to give light to the lower apartments, and to permit a circulation of air; both of which are particularly necessary, when, from the turbulence of the sea, the ports between decks are obliged to be shut. In very bad weather the *gratings* are covered with tarpaulings, to prevent the water from washing down the hatchways.

GRATUITY, in the royal navy, is a recompense or royal bounty made by his Majesty to the widows, orphans, and mothers of sea and marine officers slain in fight with the enemy.

The gratuities in compassion to the distressed condition of the relations of officers, &c. slain in fight, are as follow:—

1. The allowance to the widow is a full year's pay, according to the post her husband served in at the time of his death. Widows of pilots have the same allowance as the widows of warrant officers under the like circumstances.

2. Orphans, each one third, in proportion to the allowance to the widow; and posthumous children are esteemed orphans.

3. Orphans married are not entitled to the bounty.

4. If there be no widow, a mother (if a widow and above fifty years of age) is entitled to a widow's share.

5. The relations of officers of fire-ships are entitled to the same bounty as those officers of like rank, slain in ships of the fourth rate.

6. Captains of his Majesty's ships are to set down the names of the slain at the end of the muster-books, and on what occasion; the omission of which will be made an objection to the captain's receiving his pay.

7. This bounty extends to those who are slain in tenders, in boats, or on shore, as well as to those slain on board the ships; also to those who are slain in fight with pirates, or in any encounter with the ships of friends by mistake; also those who die of their wounds after battle, are all esteemed as persons slain.

Gratuities to Officers wounded in Fight with the Enemy, and to Seamen hurt in the Service.

1. If any commission, marine, or warrant officer, serving in the navy, receives a wound that shall occasion the loss of an eye or a limb, or the total loss of the use of a limb, or that (upon certificate from the Master, Wardens, and Assistants of the Royal College of Surgeons in London) shall, in effect be of equal prejudice to the habit of the body with the loss of a limb, he shall receive one full year's pay, and shall be allowed such expenses relating to the cure (if it is not performed at his Majesty's expense) as shall be certified reasonable by the said college; he shall also continue in pay during the time of his cure.

2. With respect to the year's gratuity, ships under the third rate are to be esteemed as such.

3. Volunteers are also entitled to the same bounty as lieutenants.

4. If the wounds received do not amount to the loss of a limb, nor are of equal prejudice to the body, the charges of cure are only allowed, but the continuance of pay is not to extend beyond eighteen months from the time the party was wounded.

5. Under the term "Wounded in Fight with the Enemy," is comprehended all the several cases relating to those slain in fight; and those so wounded are paid the gratuities and charges allowed them, on application to the Navy Board.

6. Inferior officers or seamen hurt, maimed, or disabled in the service, by certificate signed by the captain, lieutenant, master, surgeon, and one or more warrant officers of the ship, expressing the nature of the hurt, &c. are entitled to the benefit of the Chest at Greenwich, according to the rules and constitutions thereof. See the articles CHEST AT GREENWICH, and PENSIONS to superannuated officers.

To GRAVE, (*suiver*, *ou espalmer un bâtiment*, Fr.) to clean a ship's bottom, when she is laid aground during the recess of the tide. See the article BREAMING.

GRAVITATION, is the exercise of gravity, or the pressure that one body exerts on another by the power of gravity.

Gravity, in physics, is the natural tendency of bodies towards a centre, or towards each other. The most familiar effect of this power is the weight of bodies, or their tendency towards the centre of the earth.

Gravity equally affects all bodies, without regard to bulk, figure, or matter; so that, abstracting from the resistance of the medium, as air, the most compact, and the most loose, the greatest and the smallest bodies would all descend through an equal space in the same time, as appears from the quick descent of every light body in an exhausted receiver. The space which bodies fall through *in vacuo* is $16\frac{1}{12}$ feet in the first second of time, in the latitude of London.

This power is the greatest at the earth's surface; whence it decreases both upwards and downwards: but not both ways in the same proportion; for upwards, the force of gravity is less, or decreases as the square of the distance from the centre increases; so that, at a double distance from the centre, above the surface, the force will be only one fourth of what it is at the surface; but, below the surface, the power decreases in the direct ratio of the distance from the centre.

Specific Gravity, is the relative, comparative, or apparent gravity in any body, in respect of that of an equal bulk, or magnitude of another body; denoting that gravity or weight which is peculiar to each species or kind of body, and by which it is distinguished from all other kinds.

The comparative gravity of different bodies being necessary for the naval architect to possess the knowledge of, the following table may be found serviceable:—

A Table of the comparative Gravity between different solid and fluid Bodies.

	Comp. Wt.		Comp. Wt.
Platina	23400	Common earth ...	1984
Fine gold	19640	Nitre............	1900
Standard gold....	18888	Ivory	1825
Quicksilver......	11325	Brimstone	1810
Fine silver	11091	Solid gunpowder ..	1745
Standard silver ..	10535	Sand	1520
Copper	9000	Coal	1250
Copper halfpence	8915	Box-wood........	1030
Gun metal	8784	Sea-water	1030
Cast brass	8000	Common water....	1000
Steel	7850	Oak	925
Iron	7645	Gunpowder close stacken	937
Cast Iron	7425		
Tin	7320	Ditto in loose heap	836
Crystal glass	3150	Ash	800
Marble	2700	Maple	755
Common green glass	2600	Elm	600
		Fir	548
Flint	2570	Charcoal	
Common stone...	2520	Cork............	240
Clay	2160	Common air	1.232
Brick	2000	or very nearly $1\frac{7}{30}$	

The several sorts of wood are supposed to be dry.

These numbers being the weight of a cubic foot, or 1728 cubic inches, of each body in avoirdupois ounces; by proportion, the quantity in any other weight, or the weight of any other quantity, may be readily known: Hence result the following rules:—

Problem. 1. To find the magnitude of any body from its weight.

Rule. As the tabular specific gravity of the body
Is to its weight in avoirdupois ounces;
So is one cubic foot, or 1728 cubic inches,
To its content in feet or inches, respectively.

Problem. 2. To find the weight of a body from its magnitude.

Rule. As one cubic foot, or 1728 cubic inches,
Is to the content of the body;
So is the tabular specific gravity
To the weight of the body.

GREEN-Heart, a wood imported from the West Indies, and used for the pins of blocks.

GREENWICH Hospital. The Royal Hospital at Greenwich was founded in the year 1694, by the royal donations of King William and Queen Mary, increased by many private subscriptions and bequests, for the reception of *seamen* and *marines* who have served in his Majesty's navy; and who, from *old age*, *wounds*, or *infirmities*, are incapable of farther active service. They are chosen by the Admiralty, and are lodged, clothed, and victualled, at the expense of the hospital. In addition to this each boatswain is allowed 2*s*. 6*d*. each mate 1*s*. 6*d*. and every private man 1*s*. a week pocket money.

In the early part of last century it received a grant from the crown, of the estate forfeited by the Earl of Derwentwater, and it has been assisted, occasionally, by votes of money from parliament, to the amount of 19,000*l*. the last sum thus voted was 4,000*l*. in 1778. An extraordinary donation of 10,000*l*. was, a few years ago, ordered to be given to the hospital by a gentleman, who signed himself a *Friend to the Navy*, and a like sum to the Chest at Greenwich, which is in the three per cent. consolidated fund.

The number of seamen provided for by the hospital not being fixed, has gradually increased. In 1708 there were only 300 seamen on the establishment; in 1728, there were 450; in 1738, 1000; in 1751, there were 1300; in 1782, 2300. The establishment of in-pensioners, on the 31st of December 1805, was 2410; and the annual expense as estimated for each person, including the expense of provisions, weekly allowance, clothing, linen, bedding, and washing, 27*l*. 10*s*. $9\frac{3}{4}$*d*.

	£.	s.	d.
Amounting in the whole to	66,372	18	$1\frac{1}{2}$
And 144 nurses, at the estimated expense of 29*l*. 15*s*. $8\frac{1}{4}$*d*. each	4,289	5	0
	£70,662	3	$1\frac{1}{2}$

The number of out-pensioners, at the same time, was 3,234, which, at the rate of 7*l*. a year each, as settled in the year 1763, amounts to 22,638*l*.

The nurses are the widows of seamen, and must be under 45 years of age to be admitted into the hospital; they are requested to take out certificates of their husband's service in the navy in the same mode as the pensioners, and produce certificates of their age and marriage to the Board of Admiralty, by whom they are appointed.

Many of the pensioners in the hospital are permitted to have the option of taking up their provisions *in kind*, or of *receiving money* in lieu of the whole, or certain parts, of their daily allowances. This measure was adopted in August 1725, with a view, as it appears from the minutes of the board of directors and the council at that time, of remedying the inconveniences that resulted from the men taking their provisions out of the hospital for sale, which occasioned "a resort of loose, vagabond, and disorderly people."

The following is a Table of the Allowance of Provisions to a Pensioner for every Day in the Week.

	Bread. A Loaf of 16 oz.	Beer.	Beef.	Mutton.	Butter.	Cheese.	Pease.
		Qrs.	lb.	lb.	oz.	oz.	Pint.
Sunday	1	2	—	1	—	4	—
Monday	1	2	1	—	—	4	—
Tuesday	1	2	—	1	—	4	—
Wednesday ..	1	2	—	—	1	8	$\frac{1}{2}$
Thursday ...	1	2	1	—	—	4	—
Friday.....	1	2	—	—	1	8	$\frac{1}{2}$
Saturday	1	2	1	—	—	4	—

The soup is made of the pease, and broth of the meat.

Provisions, and all other articles of considerable consumption, are obtained by contracts entered into in consequence of public advertisements.

The meat is delivered by the contractor on the morning of every meat day, and the weight ascertained by the clerks.

The persons who receive money instead of different articles of provision, may be divided into several classes; for the particulars of which see the *14th Report of the Commissioners of Naval Enquiry, June* 1806.

The revenues of this magnificent national establishment have been recently very much improved in consequence of the investigations and reports of the Commissioners of Naval Enquiry. They arise from sixpence per man per month, paid by *merchant seamen* and *seamen* and *marines* in his Majesty's navy, which is collected at the Sixpenny Receiver's Office, and which is a branch of Greenwich Hospital establishment, by a grant from Parliament, and acts under the direction of the Admiralty Board. Also the duties arising from the North and South Foreland lighthouses; the half-pay and allowance of certain officers of the hospital; the forfeited and unclaimed shares of prize and bounty money, granted by Act of Parliament to the hospital; the rents of the market of Greenwich, and of houses there and in London; the rents and profits of the Derwentwater estate, including lead-mines, and of other estates purchased in the north, and the interest of money invested in the public funds.

By the reform introduced into the Sixpenny Office, founded on the ***Third Naval Report***, the six-pences per month have been more certainly and regularly collected. The functions of that office are, to receive this duty; to appoint deputy-receivers at the outports of England, Scotland, and Ireland, as also at several islands and dominions in America and the West Indies, for the purpose of collecting the said duty; to receive all forfeitures payable to Greenwich hospital; to receive payment of run men's wages employed in the African trade; and to receive, by a late Act of Parliament, the wages of seamen employed in the West India trade, who die on their voyage, for the use of their executors, &c.

In February 1802, the amount of stock of Greenwich Hospital invested in the 3 per cent. consols appears to be 250,300*l.*

Establishment of Officers, &c for 1805.

A Master and Governor	A Surgeon
A Lieutenant Governor	A Clerk of the Cheque
Four Captains	A Surveyor
Eight Lieutenants	A Clerk of the Works
A Treasurer and Receiver-general	A Dispenser
	Three Matrons
A Secretary	A Schoolmaster
An Auditor	An Organist
Two Chaplains	A Butler
A Physician	Several Clerks, under Officers, &c.
A Steward	

The governors have power to grant out-pensions to seamen worn out or become infirm or decrepit.

Mode of applying for Admission.

Persons desirous of being admitted pensioners are to apply at the Admiralty Office at *least ten days* before the day of examination (the days at present appointed for that purpose are, the *first Thursday* in *January*, *April*, *July*, and *October*) where they receive letters directed to the proper officer at the navy office, for certificates of their time of service in the navy; which certificates are sent to the Admiralty before the day of examination, when the candidates are examined by the board (the surgeon of the hospital attending at the same time), and those who are found to be proper objects are minuted to be sent to the hospital; and, as vacancies happen, are sent accordingly.

It will, undoubtedly, afford general satisfaction to the public that the permanent revenues of the hospital are more than adequate to meet the expenditure.

The following is an abstract of the revenue and expenditure of the Royal Hospital at Greenwich, as stated in the *14th Report of the Commissioners of Naval Enquiry for the Year* 1805:—

Revenue.

	£.	s.	d.
Sixpence per man per month, paid by seamen in the merchant service, on an average of the last ten years....	15,666	9	$3\frac{1}{4}$
Ditto from seamen in the king's service, on an average of ten years, to 1804, the account for 1805 not being complete..................	33,173	17	$5\frac{1}{4}$
Income from real property according to the present rents	46,279	6	$1\frac{1}{4}$

	£.	s.	d.
Income from the woods estimated at 1,000	47,279	6	1¼
Interest on 1,168,926*l.* 17*s.* 2*d.* in the funds	35,067	16	1
Duties from the North and South Foreland Lighthouses, on an average of the last ten years.........	5,877	6	4½
Half pay and emoluments of certain officers of the hospital, on an average of the last ten years	1,503	13	10
	£138,568	9	2

EXPENDITURE.

	£.	s.	d.
Establishment, consisting of salaries to officers and servants, charges of the building, &c. according to the actual expenditure of 1805.......	21,837	12	11½
In-pensioners and nurses, according to the numbers on the establishment, estimated at the expense of each person, in 1805	70,662	3	1½
Out-pensioners, according to the number on the establishment, at the present rate of 7*l.* a year	22,638	0	0
Balance in favour of the hospital ...	23,430	13	1
	£138,568	9	2

In the consideration of the revenue, from the property actually in possession of the hospital, an increase may be fairly admitted, arising from the interest on prize-money to be hereafter received by the hospital, a subject mentioned in a former part of this report, which the said commissioners could not, under any probable state of things, estimate at less than 10,000*l.* per annum; this increased, will, within a few years, make the surplus revenue of the hospital above 33,000*l.* besides the advance on the rents which was expected to take place on the expiration of the leases, which determine in 1809.

In addition to the above-mentioned national establishment, there are, in England, many other naval hospitals for the relief of sick and wounded sailors; viz. the Royal Hospital at Haslar, near Gosport: for an account of which, see that article; the Royal Hospital at Plymouth; the Royal Hospital at Deal; and the Royal Hospital at Great Yarmouth.

GREENWICH *School.* Annexed to the establishment of the royal hospital at Greenwich, is a school for the education and maintenance of poor boys, the sons of seamen, who are appointed by 24 directors of the hospital in turn, each director having about *three* appointments in the space of a year.

The number at first was very small, and has increased, from time to time, to 200 boys. They are attended by the best masters; by whom they are instructed in reading, writing, cyphering, and navigation, &c. which are requisite to make them expert seamen, either for the royal navy or the merchant service, and remain at the school between three and four years.

The instructions for candidates are as follow; viz. Every candidate for this school must be *the son of a seaman*, who has served in the navy or merchant service, or *marine*, who has served in one of his Majesty's ships. He must be 11 and under 13 years of age. Must be able to *read a chapter in the Bible*, and be without impediment of speech, or any other bodily infirmity that might incapacitate him from serving in the sea service. At the time of his admission he must produce *a certificate of his age*, stating *the time of his birth*; a certificate of his being *an object of charity;* and *a certificate of his father's servitude at sea.*

If the father has served in the navy, the certificate of his servitude must be obtained *from the Navy-office, by letters signed by one of the directors;* and if in the merchant service, *the certificate must be signed by the person who collects the six-penny duty for the hospital, at the port from which the father has usually sailed;* stating that the duty has been regularly paid by or from him. At the time of his admission he must produce *two respectable housekeepers*, to *be jointly* bound in the sum of *fifty pounds*, that he *shall not run away from the school*, nor *embezzle any of the clothes, books, or instruments*, that may be committed to his care; and that, whenever the directors of the hospital think proper, *he shall consent to be bound an apprentice to the sea service, and to be continued therein for the term of seven years.*

The revenues for the support of the school are derived from interest of money in the funds; the savings on the payment of money to the pensioners in lieu of provisions; the amount of all checks and mulcts imposed on the pensioners; the sums arising from shewing the painted hall and chapel, and from the sale of stores, &c.

In 1805, the receipts amounted to 9,002*l.* 13*s.* 6*d.* and the expenditure to 6218*l.* 17*s.* 11¼*d.* leaving a surplus of 2784*l.* 5*s.* 6¾*d.* of which 2629*l.* 7*s.* 6*d.* were invested in the funds.

It has been suggested by a gentleman that this laudable establishment for the boys might be considerably extended, and an adequate revenue obtained, on a similar application of the profits arising on board his Majesty's ships, from the seamen and marines frequently not taking up their whole allowance, and which is now usually paid to them in money by the pursers, at about *two thirds* of the value paid to them by government, on passing their accounts, and which is nearly another third less than the real value paid by the Victualling-office to the several contractors.

The late increased allowance of provisions to the seamen in the navy, which is greater than what is given by any other maritime state in Europe, would render a fund of this nature extremely productive, and might be applied to other purposes as well as that just mentioned; more especially, as the whole allowance now given cannot be fairly consumed. With the former allowance even, it was no uncommon thing for five men in a mess to take up the allowance of four only, and be paid in money, at the end of the month, the *odd man's allowance;* which proves the redundancy of the provisions. See M'ARTHUR'S *Financial and Political Facts of the present Century.*

Besides the school there is an INFIRMARY attached to the Royal Hospital, which is a capacious building, divided into two principal parts; one for the patients under the care of the physician, and the other for those whose cases require the attendance of a surgeon. It contains a double row of rooms, being altogether in number 64, calculated to hold 256 patients. In the upper story is a small chapel, where prayers are read by the chaplains twice a week, for the benefit of the patients. Other parts of this noble structure comprise the dispensary and surgery, and apartments for the physician, for the surgeon, and dispenser, with their respective assistants, and for the matron; and adjacent, within the walls, are hot and cold baths.

Such are the provisions which the legislature, the genuine organ of a country grateful to her brave defenders, has established for their comfort and accommodation! And this superb and hospitable structure bears splendid testimony to surrounding nations that Great Britain is not more pre-eminently distinguished for her naval triumphs, than for a grateful attention to those intrepid seamen by whom those triumphs have been achieved, as they are not only liberally provided for while in actual service, but when disabled or superannuated, food, clothing, medicine, and every necessary accommodation and comfort, including even the education and maintenance of their children, are here provided by the government.

But not to the government alone is this kind and grateful attention exclusively confined; the country at large, with laudable and generous emulation, rivals the legislature; and, independent of the ample bounties paid upon entering the service, immense subscriptions have been, at various times, received, for the relief of the wounded and the relatives of the killed.

GRENADE, or GRANADO, (*grenade*, Fr.) is a kind of small bomb or shell, of the same diameter as a four-pound bullet; it weighs about two pounds, being charged with four or five ounces of powder, and is thrown from the tops, &c. by the hands of seamen, whence they are most generally styled *hand-grenades*. They have a touch-hole in the same manner as a shell, and a fusee of the same composition. The sailors fire the fusee with a match, and throw the grenades as they are directed: the powder being inflamed, the shell instantly bursts into splinters, that kill or maim whomsoever it may reach on the decks of the enemy.

As this instrument cannot be thrown by hand above fifteen or sixteen fathoms, the ship must be rather near to render it useful in an engagement. The best way to secure a man's body from the effects of a grenade is, to lie flat down on the ground before it bursts.

GRIPE, the same with FORE-FOOT. See that article.

GRIPE *of a Ship*, (*partie inférieure du taille-mer*, Fr.) is the compass or sharpness of her stern under water, chiefly towards the bottom of her stern. The design of shaping her so is to make her *gripe* the more, or keep a good wind, for which purpose, sometimes a false stern is put on upon the true one.

GRIPES, (*risses de chaloupe, ou de canot*, Fr.) a machine formed by an assemblage of ropes, hooks, and *dead-eyes*, and used to secure the boats upon the deck of a ship at sea, and prevent them from being shaken by the labouring of the vessel. The hooks, which are fastened at their ends, are fixed to ring-bolts in the deck on each side of the boat; whence, passing over her middle and extremities, they are extended by means of the dead-eyes, so as to render the boats firm and secure.

To GRIPE, (*être ardent*, Fr.) a ship is said to gripe when she is inclined to run to windward of her course, particularly when she sails with the wind on her beam or quarter: this effect is partly occasioned by the shock of the waves that strike the ship perpetually on the weather-quarter, and force the stern to leeward; but chiefly by the arrangement of the sails, which disposes the ship continually to edge to windward, while in this situation of sailing: in such case they say, "she gripes," or is griping.

GROG, (*boisson faite avec du rum, ou de l'eau de vie et de l'eau*, Fr.) a general name for any spirituous liquor and water mixed together, but is more particularly applied to rum and water cold, without sugar. This mode of mixing the spirit with water is for the purpose of preventing intoxication among the seamen; and the measure is said to have been first introduced into the royal navy by Admiral Vernon: the rum is served out to them merely diluted by water without any other addition, and is used on foreign voyages and stations in lieu of wine or beer (as well as for the purpose of saving stowage) because the latter will not keep for any length of time.

To stop a Sailor's GROG, (*retrancher à un matelot sa portion de cette boisson*, Fr.) is to retrench the portion of grog allowed him, which may happen from several causes, as from selling his grog, for drunkenness, or from sickness; in which latter case it is made up to him again.

GROMMET, (*anneau de corde, pour les voiles d'étai*, Fr.) a sort of ring, or small wreath, formed of a *strand* of rope, laid in three times round, and used to fasten the upper edge of a sail to its respective stay, in different places, by means of which the sail is accordingly hoisted or lowered. See also the article HANKS.

To GROUND A SHIP, (*tirer à terre un bâtiment pour le radouber ou le carèner*, Fr.) is to lay her on shore, in order to bream or repair her. It is also applied to running aground accidentally when under sail, or driving in a tempest.

GROUND-*Tackle*, (*garniture des ancres, &c.* Fr.) a general name given to all sorts of ropes and furniture which belong to the anchors, or which are employed in securing a ship in a road or harbour; as cables, hawsers, tow-lines, warps, and buoy-ropes.

GROUND-*Tier*. See the article TIER.

GROUND-*Tow*, (*second brin de chanvre*, Fr.) in rope-making the loose hemp that comes from the sides of the hatchellers and spinners.

GROUND-*Ways*, the name of large pieces of timber laid across the slip or dock to place the blocks upon.

GROWING, implies the direction of the cable from the ship towards the anchors; as, "The cable grows on the starboard-bow," (*le cable est tendu à tribord du vaisseau*, Fr.) *i. e.* stretches out towards the starboard or right side.

GRUS, the Crane, in astronomy, one of the new constellations in the southern hemisphere; containing, according to Mr. Sharp's catalogue, 13 stars.

GUARD-BOAT, (*canot de ronde*, Fr.) a boat appointed to row the rounds amongst the ships of war in any harbour, &c. to observe that their officers keep a good look-out, calling to the *guard-boat* as she passes, and not suffering her crew to come aboard, without having previously communicated the watch-word of the night.

GUARD-*Irons*, (*barres de fer courbes qui entourent la figure de la poulaine*, Fr.) are curved bars of iron placed over the ornamental figures, on a ship's head or quarter, to defend them from injury.

GUARD-*Not*, a term used on board frigates and other small vessels, to impede the enemy in boarding: also the name of a sort of thick netting, which being extended on ropes all round the vessel, at the height of about six feet, prevents the enemy's boarding.

GUARD-*Ship*, (*vaisseau de garde*, Fr.) a vessel of war appointed to superintend the marine affairs in a harbour or river, and to see that the ships which are not commissioned have their proper watch duly kept, by sending her guard-boats around them every night: she is also to receive seamen who are impressed in time of war: she generally has an admiral's flag at one of her mast-heads.

GUDGEONS. See the article GOODGEONS.

GUESS-ROPE, or GUEST-ROPE OF A BOAT, (*cablot servant à remorquer, ou à amarrer une embarcation*, Fr.) a rope used to tow or to make fast a boat.

GULF (*golfe*, Fr. *golfo*, Ital.) a broad and capacious bay, comprehended between two promontories, and sometimes taking the name of a sea, when it is very extensive, but particularly when it only communicates with the ocean by means of a streight: such are the Euxine, or Black Sea, otherwise called the gulf of Constantinople; the Adriatic Sea, called also the gulf of Venice; the gulf of Sidra, near Barbary; and the gulf of Lions, near France: all these gulfs are in the Mediterranean. There are, besides, the gulf of Mexico, the gulf of St. Lawrence, and the gulf of California, which are in North America. There are also the gulf of Persia, otherwise called the Red Sea, between Persia and Arabia; the gulf of Bengal, in India; and the gulfs of Cochin-China and Kamtschatca, near the countries of the same name.

Some writers define a gulf as running into the land through a straight and narrow passage, and, except at the communication of the sea, to be surrounded by the land, as the gulf of Corinth, the gulf of Lepanto, &c. A very large gulf (such as the above-mentioned) being, in their opinion, an inland sea.

A gulf is strictly distinguished from a sea in being smaller, and from a bay in being larger. It is observed that the sea is always most dangerous near gulfs, by reason of the currents being pent up by the shores.

GUN, (*canon*, Fr.) a fire-arm, or weapon of defence which forcibly discharges a ball, shot, or other offensive matter, through a cylindrical barrel, by means of inflamed gunpowder.

GUN, is a general name, under which are included divers, or even most species of fire arms; mortars and pistols being almost the only kind excepted from this denomination.

In the navy, *gun* is generally considered as synonimous with CANNON (which article see), and distinguished by the epithet *great gun*, from the small guns, as fire-locks, muskets, carbines, blunderbusses, fowling-pieces, &c. Under this single denomination is comprised all ordnance, from a 42-pounder, the heaviest, down to a swivel.

A Table of the Calibre of English and Foreign Guns, expressed in English Inches.

English.		French.		Spanish.		Dutch.		Russian.		Portuguese.	
Pr	Inc.	Pr.	Inc.	Pr.	Inc.	Pr.	Inc.	Pr.	Inc.	Pr.	Inc.
42	7.018	—	—	—	—	—	—	36	6.86	48	7.49
32	6.41	36	6.9	36	6.84	32	6.4	30	6.47	36	6.8
24	5.823	24	6.03	24	6.03	24	5.92	24	6.	24	5.93
18	5.292	16	5.26	18	5.52	18	5.45	18	5.45	18	5.4
12	4.623	12	4.78	12	4.8	12	4.76	12	4.76	12	4.7
9	4.200	8	4.18	9	4.2	8	4.13	8	4.17	9	4.3
6	3.668	—	—	—	—	6	3.78	6	3.78	6	3.75
4	3.204	4	3.315	—	—	—	—	—	—	—	—
3	3.913	—	—	—	—	—	—	—	—	—	—
1	2.019	—	—	—	—	—	—	—	—	—	—

A truly fortified great iron gun ought to measure eleven diameters of the bore at the circumference of the base ring, nine diameters at the trunnions, and seven at the circumference of the muzzle ring.

A truly fortified great brass gun should measure two diameters less at each place of measurement than the iron gun; that is to say, nine diameters of the bore at the circumference of the base ring, seven at the trunnions, and five at the muzzle ring.

In order to discover when a gun quadrates, or hangs well in the carriage, it ought to measure in length seven times her own diameter at the neat; the trunnions ought to be placed at the distance of three diameters from the base-ring; then there will remain four diameters in distance from the muzzle.

In order to discover whether the carriage is proper and of due length for the gun, it ought to be five-eighths the length of the gun, and then the eye will easily discover if it be wide enough, and high enough, or too high.

To dispart a GUN, in order to take proper aim at the given object, insert a priming-wire into the vent, and let it touch the lower part of the metal of the bore; mark the wire close to the vent, take it out and rest it on the lower metal of the rose at the muzzle, and the distance between the muzzle-ring and marked part of the wire is the height of the *dispart*.

In order to find the thickness of the metal at the vent, trunnions, and muzzle, take the diameter of the gun at the vent, and lay it down thus |——|, which will express the diameter; then insert a priming wire into the vent, and let it rest on the lower metal; mark it close to the vent, and, taking it out, lay the mark on the line of the diameter, thus: |——|——|. Crook then the end of the wire a little, that it may enter the vent, and, inserting it a second time, turn it round till it catches the upper metal of the bore; then mark it again close to the vent, set off the distance on the same line of the diameter, and mark how far it reaches from the end of the line, thus:

A B A
|——|——|——|. Then will A and A represent the thickness of the metal, and B, the bore of the gun; and if the portions A A of the line are equal to each other, the thickness of the metal is equal, and,

of course, the gun centrally bored. Girt then the gun at the trunnions with waxed twine, and if it measures nine diameters of the bore the gun is so far truly fortified. Observing the same operation at the muzzle, where it is to measure seven diameters, the process is complete.

In order to discover whether a gun is truly bored, take a spare sponge-staff, and fix on it a rammer-head, strike a chalk-line on it, from one end to the other, and put it into the gun as far as it will go, keeping the chalk-line uppermost, and exactly in the centre; then prick down the vent with a priming wire, and if you find, on taking out the rammer, you have pricked into the chalk-line, you may reasonably conclude the gun is truly bored; but, if you miss the chalk-line, that it is not.

In order to discover when a gun is honey-combed, take a spring searcher with five prongs and a reliever; muzzle the searcher, and ram it home in the gun; take off the reliever, and keep turning the searcher backwards and forwards; you will easily discover whether it catches; when it does, mark the staff close to the muzzle; then turn the searcher again as before, and, whenever it catches again, mark the staff as in the former instance; so that, by laying the staff, when drawn out on the outside of the gun, you may nearly judge where the honey-combs are.

In order to discover the depth of the honey-combs, take a searcher with one prong and a reliever: arm the end of the prong with wax, then ram it home in the gun; take off the reliever, and turn the searcher till it catches; then will the impression made in the wax shew the shape and depth of the honey-comb.

If the honey-comb on either side, or on the lower metal between the breech and the reinforce ring is three-tenths of an inch deep, the gun is to be condemned; if on the upper metal, four tenths; if on any part without, or beyond the reinforce-ring, five-tenths are sufficient.

A most ingenious instrument, invented by the late General Desaguliers, and since brought to the greatest perfection, has totally superseded the use of this contrivance.

All guns intended for sea service are now previously examined, by proper officers belonging to the ordnance board; who, by means of this instrument, being able to ascertain with the greatest precision the state and defects of any gun, after a very short examination, of course, reject all those which, either from natural defect, or subsequent injury, appear unfit for his Majesty's service.

To discover whether a gun is sound or cracked, strike a smart blow on it with a hammer; if it rings clear it may be concluded the gun is sound; if it jars or emits a hoarse sound, it is most probable the gun is cracked. Or the following method may be taken:—stop the vent, and take a piece of touchwood, put it into the gun, and stop the muzzle securely; let the touchwood remain in the gun four or five minutes; if the gun is cracked the touchwood will burn out; if the gun is sound it will be extinguished.

In fitting a shot to a gun, divide the diameter of the bore into twenty equal parts, and the diameters of the shot ought to be nineteen of those parts. With respect to the proper proportion of powder, eighteen-pounders, and all inferior calibres, require half the weight of the shot; for all above there are certain rules to find the proper proportion by.

In order to secure a gun if it break loose, cut down the hammocks, trip the gun, and lash it to the ring-bolts of the side till fine weather. To clear it when a bit is broke in it, draw the gun, and sprinkle powder with a ladle from the breech to the muzzle; this done, drive a tight tompion with a small score in it, and blow the gun off. If a shot has fetched way in the gun, in order to secure it, damp the powder, or split the tompion; then insert a rope-sponge of a small size, and drive the wad home.

If, in loading the gun, the shot sticks by the way, and if, in firing it, it splits, and you cannot draw the gun, in order to free it, the powder must be damped, and, while that is soaking, some powder must be mealed, and the gun primed, getting as much powder down the touch-hole as possible, when the gun must be fired off.

When a ship is going to sea immediately, the articles which should be ready for action are, the powder filled, the powder-horns and cartridge or grape-shot between the guns, hammered shot in the buckets, crows and hand-crows, levers at the guns; nets and cheeses of wads fore and aft; the match-tubs in their proper places, the matches ready, the lockers full of shot, the spare tackles and breechings ready, wet swabs at the door of the magazine and heads of the ladders; the boxes of hand-grenades ready for the tops.

The thickness of the metal of a gun at the vent should be one diameter and a quarter of the bore; and, in an engagement, there should be one man to every five hundred weight of metal.

The pointing of a gun, so as to strike distant objects, depends on two things; viz. 1st. Tracing on the outside of the piece a visual line, parallel to the axis, which is called *disparting*, and is performed by taking half the difference of the diameters of the muzzle and base-ring, and setting it perpendicularly on the muzzle-ring, directly over the centre; for then a line which passes from that point in the base-ring, will, when the piece is truly bored, be parallel to its axis. 2d. The other operation is the determining the allowance to be made in distant shot for the incurvation of the flight of the bullet; this is greater or less (*cœteris paribus*), according to the different charges of powder made use of.

The Great Gun, at Dover castle, is a large brass gun, twenty-four feet long without, and twenty-two feet in the bore, beautifully adorned with flowers and emblematical figures, in relievo, and several inscriptions. This curious gun, vulgarly called Queen Elizabeth's pocket pistol, was a present from the Emperor Charles Vth. to Henry VIIIth. while they were engaged together in a war with France.

The charge of this gun is fifteen pounds of powder, and it is said to carry a ball to a distance of seven or eight miles.

The following Table contains the Number of Guns and Carronades used in each Class of every Rate in the Royal Navy, together with the Species, Length, and Mean Weight of each Gun, &c.

N.B.—The length of a gun or carronade is reckoned from the face of the piece to the back part of the base-ring.

NAMES OF THE DECKS.	110 GUNS.					100 GUNS.					98 GUNS.				
	Number.	Species.	Length.	Mean weight of each gun.	Average weight of all the metal.	Number.	Species.	Length.	Mean weight of each gun.	Average weight of all the metal.	Number.	Species.	Length.	Mean weight of each gun.	Average weight of all the metal.
	No.	Pr.	Ft.	Cwt.	Tons.	No.	Pr.	Ft.	Cwt.	Tons.	No.	Pr.	Ft.	Cwt.	Tons.
Guns on the Gun-deck	30	32	9½	55½		28	32	9½	55½		28	32	9½	55½	
Guns on the Middle-deck	30	24	9½	50½		26	24	9½	50½		30	18	9	43	
Guns on the Upper-deck	32	18	9	43	255½	30	12	9	31	221	30	12	9	31	208½
Guns on the Quarter-deck	14	12	7½	29½		4	12	8½	32		0	0	0	0	
Guns on the Forecastle	4	12	8½	32		10	12	7½	29½		10	12	7½	29½	
Total of guns	110				255½	100				221	98				208½
Carronades on the Forecastle	2	32	4	17½		2	32	4	17½		2	32	4	17½	
Carronades on the Round-house	7	24	3½	13	6½	6	24	3½	13	5½	6	18	3½	10	4½
Carronades on the Quarter-deck	0	0	0	0		0	0	0	0		0	0	0	0	
Total number and weight of guns and carronades	119				261½	108				226½	106				213
Total value of metal, at 18*l*. per ton	*L*. 4707 0 0					*L*. 4077 0 0					*L*. 3834 0 0				

NAMES OF THE DECKS.	90 GUNS.					80 GUNS.					74 GUNS, *(large.)*				
	No.	Pr.	Ft.	Cwt.	Tons.	No.	Pr.	Ft.	Cwt.	Tons.	No.	Pr.	Ft.	Cwt.	Tons.
Guns on the Gun-deck	28	32	9½	55½		26	32	9½	55½		30	32	9½	55½	
Guns on the Middle-deck	30	18	9	43		26	18	9	43		0	0	0	0	
Guns on the Upper-deck	30	12	9	34	208½	24	9	9	29	167½	30	18	9	43	160
Guns on the Quarter-deck	8	12	7½	29½		0	0	0	0		4	18	8	38	
Guns on the Forecastle	2	12	8½	32		4	6	7½	20½		2	18	9	43	
Total of guns	98				208½	80				167½	66				160
Carronades on the Forecastle	2	32	4	17½		2	32	4	7½		2	32	4	17½	
Carronades on the Round-house	6	18	3½	10	4½	6	18	3½	10	4½	6	18	3½	10	13½
Carronades on the Quarter-deck	0	0	0	0		0	0	0	0		10	32	4	17½	
Total number and weight of guns and carronades	106				213½	68				172	84				173½
Total value of metal, at 18*l*. per ton	*L*. 3938 10 0					*L*. 3096 0 0					*L*. 3118 10 0				

NAMES OF THE DECKS.	74 GUNS, *(small)*					64 GUNS.					50 GUNS.				
	No.	Pr.	Ft.	Cwt.	Tons.	No.	Pr.	Ft.	Cwt.	Tons.	No.	Pr.	Ft.	Cwt.	Tons.
Guns on the Gun-deck	28	32	9½	55½		26	24	9½	50½		22	24	9	48	
Guns on the Middle-deck	0	0	0	0		0	0	0	0		0	0	0	0	
Guns on the Upper-deck	28	18	9	43	152	26	18	9	43	137½	22	12	8½	32	94½
Guns on the Quarter-deck	4	9	8½	27½		2	9	8½	27½		2	6	8	21½	
Guns on the Forecastle	14	9	7½	26		10	9	7½	26		4	6	7½	20½	
Total of guns	74				152	64				137½	50				94½
Carronades on the Forecastle	2	32	4	17½		2	24	3½	13		2	24	3½	13	
Carronades on the Round-house	6	18	3½	10	4½	6	18	3½	10	4½	6	12	2½	6½	5½
Carronades on the Quarter-deck	0	0	0	0		0	0	0	0		4	24	3½	13	
Total number and weight of guns and carronades	82				156½	72				141½	62				100
Total value of metal, at 18*l*. per ton	*L*. 2821 10 0					*L*. 2547 0 0					*L*. 1800 0 0				

NAMES OF THE DECKS.	44 GUNS.					38 GUNS.					36 GUNS.				
	No.	Pr.	Ft.	Cwt.	Tons.	No.	Pr.	Ft.	Cwt.	Tons.	No.	Pr.	Ft.	Cwt.	Tons.
Guns on the Gun-deck	20	18	9	43		0	0	0	0		0	0	0	0	
Guns on the Middle-deck	0	0	0	0		0	0	0	0		0	0	0	0	
Guns on the Upper-deck	22	12	7½	29½	80½	28	18	8	36	65½	26	18	9	43	68
Guns on the Quarter-deck	4	6	6	17		8	9	7	24½		10	9	7	24½	
Guns on the Forecastle	2	6	6½	18		2	9	7½	26		0	0	0	0	
Total of guns	48				80½	38				65½	36				68
Carronades on the Forecastle	2	24	3½	3		0	0	0	0		2	32	4	17½	
Carronades on the Round-house	0	0	0	0	5½	0	0	0	0	7	0	0	0	0	7
Carronades on the Quarter-deck	6	24	3½	13		8	32	4	17½		6	32	4	17½	
Total number and weight of guns and carronades	56				85½	46				72½	44				75
Total value of metal, at 18*l*. per ton.	*L*. 1543 10 0					*L*. 1305 0 0					*L*. 1250 0 0				

NAMES OF THE DECKS.	32 GUNS.					28 GUNS.					24 GUNS.				
	No.	Pr.	Ft.	Cwt.	Tons.	No.	Pr.	Ft.	Cwt.	Tons.	No.	Pr.	Ft.	Cwt.	Tons.
Guns on the Gun-deck	26	18	8	38		24	9	7	24½		22	9	7	24½	
Guns on the Quarter-deck	6	6	6	17	54½	4	6	6	17	32½	2	6	6	17	27½
Guns on the Forecastle	0	0	0	0		0	0	0	0		0	0	0	0	
Total of guns	32				54½	28				32½	24				27½
Carronades on the Forecastle	2	24	3½	13		2	24	3½	13		2	24	3½	13	
Carronades on the Round-house	0	0	0	0	4	0	0	0	0	4	0	0	0	0	4½
Carronades on the Quarter-deck	4	24	3½	13		4	24	3½	13		6	18	3½	10	
Total number and weight of guns and carronades	38				58½	34				36½	32				31½
Total value of metal, at 18*l*. per ton	*L*. 1053 0 0					*L*. 661 10 0					*L*. 571 10 0				

NAMES OF THE DECKS.	20 GUNS.					18 GUNS.					16 GUNS.				
	No.	Pr.	Ft.	Cwt.	Tons.	No.	Pr.	Ft.	Cwt.	Tons.	No.	Pr.	Ft.	Cwt.	Tons.
Guns on the Gun-deck	20	9	7½	26		18	6	6	17		16	6	6	17	
Guns on the Quarter-deck	4	6	6	17	29½	0	0	0	0	15½	0	0	0	0	13½
Guns on the Forecastle	0	0	0	0		0	0	0	0		0	0	0	0	
Total of guns	24				29½	18				15½	16				13½
Carronades on the Forecastle	2	12	2½	6½		2	12	2½	6½		0	0	0	0	
Carronades on the Round-house	0	0	0	0	½	0	0	0	0	2½	0	0	0	0	
Carronades on the Quarter-deck	0	0	0	0		6	12	2½	6½		0	0	0	0	
Total number and weight of guns and carronades	26				30	26				17½					13½
Total value of metal, at 18*l*. per ton	*L*. 548 5 0					*L*. 319 10 0					*L*. 243 0 0				

Sloops, gun-brigs, cutters, &c. have their establishment according to their size, or the service they are intended for. And, although the preceding table contains the general establishment for the navy, yet there have been many ships built, and others captured from the enemy, belonging to the several classes that carry greater proportions of guns and carronades, and vary considerably from it: a few of the most remarkable that have, of late years, been in the service, are mentioned as under; but the real establishment for each ship cannot be ascertained for any length of time, as they are so often altered by the different commanding officers applying to the Admiralty to have the guns changed for carronades, and *vice versa*.

Names of Ships of different Rates.		Species of Guns. Pounders. 32	24	18	12	9	6	Species of Carronades. Pounders. 68	42	32	24	18	12
1st Rates.	Commerce de Marseille. F.	34	34	—	52	—	—	—	—	2	8	—	—
	Ville de Paris .... E.	30	30	32	18	—	—	—	—	—	—	—	—
	Caledonia............. E.	32	34	34	—	—	—	—	—	12	—	—	10
	San Josef............. S.	30	32	30	12	—	—	—	—	6	—	2	2
3d Rates.	Sans Pareil .... F.	30	36	—	—	—	—	2	—	16	8	—	—
	Gibraltar .... S	—	62	—	—	—	—	—	—	—	20	—	—
	Cæsar............... E.	30	32	—	18	—	—	—	—	—	6	—	—
	Christian the VIIth... D.	30	—	32	6	—	—	—	—	10	—	—	2
	Centaur.............. E	28	30	—	—	—	—	—	—	18	6	—	—
	Ardent .... E.	—	26	—	—	—	—	—	28	—	14	6	—
5th Rates.	Cambrian............. E.	—	28	—	12	—	—	—	—	8	—	—	—
	Lavinia............... E.	—	—	—	30	—	14	—	—	8	—	—	—
	Indefatigable .... E.	—	26	—	14	—	—	—	6	—	—	—	—
	Galatea .... E.	—	—	26	—	—	6	—	—	—	—	8	—

N.B.—F. signifies French, E. English, S. Spanish, and D. Danish.

The Morning GUN, (*le coup de canon de diane*, Fr.) is the gun fired by an admiral or commodore, at day-break every morning.

The Evening GUN, (*le coup de canon de retraite*, Fr.) is that fired by the above at nine o'clock in the summer, and eight o'clock in winter every night.

GUN-*Boat*, (*chaloupe cannonière*, Fr.) a boat fitted to carry one or more cannon in the bow, so as to cannonade an enemy while she is end on, or advancing towards him; they are principally useful in fine weather, smooth water, and shallow ground, to cover the landing of troops.

The following is an accurate description of the French *gun-boats*, constructed to form the flotilla collected a few years ago at Bologna for the invasion of Great Britain:—They measured 80 feet in length upon deck, nearly 19 in the extremest width, 17 in the broadest part of the bow, and 12 in the stern. In the midship they were four feet from the water edge to the gunwale, five feet at the bow and in the stern, eight feet to the taffrail, or highest part of it.

Their draft of water was three feet ten inches at the bow, and four feet ten inches abaft; having bilge pieces on their bottoms, thereby enabling them to lay on a beach as though they were flat-bottomed; and thus were better constructed either for sailing or rowing than if they were entirely flat; but they were not coppered.

Their measurement was 120 tons; and it is imagined they would each convey from 150 to 200 men. They carry twenty brass swivels, and are also each armed with one long eighteen-pounder in the stern, and two long twenty-four pounders for the bow; which last are stowed longways in midship. And, besides being fired at the bow-ports, may also be fired abaft the fore-rigging, in an oblique direction, or by the broadside: they have no carronades.

These vessels were brig-rigged, taunt, and square; each were provided with forty oars, which were stowed on the quarters. The steering was effected by yokes, fixed to the head of the rudder, on the outside, in order to give the stern-gun room to fire over it. In three months after they were first laid on the stocks they were completely rigged, stored, and launched. Excepting the heavy guns, every article contained in them was new and of good quality, particularly the sails and cordage: they were abundantly supplied with boarding-pikes, tomahawks, cutlasses, and muskets, the latter being longer and lighter than ours. The bow-ports were on each side so close to the bowsprit that it appeared probable, that if the bow-guns were there frequently fired, they would either be disabled, or would damage the bowsprit; nor can the guns be in that place,trained to right or left. It is also the opinion of naval characters, that, as a covering battery, they would be less effective if run aground than afloat, as the weight of the guns are too great for the strength of the vessel in that situation. With respect to the sailing, it is thought they are much too long for their width: that they could not beat up to windward; and, premising a brisk gale, and armed cutter would, at sea, work round and annoy them greatly; a stout frigate would run them down; and a 74 gun ship make dreadful havoc among them.

Of their power of rowing in a calm, it is conjectured that, having their full complement of troops, &c. on board, they could hardly advance more than three knots an hour; and, if collected in any considerable number, their progress would be slower.

Having approached our shores, if they run aground for the purpose of landing troops (which they cannot effect unless they do), their efficacy, as before observed, would be greatly diminished as a covering battery; and, having no platforms for the purpose of disembarking, that operation would be difficult, more especially in any surf: so that, on the whole, it may be fairly presumed, that, with all the ingenuity of their construction, wind and tide must greatly favour their operations to render them effective, aided by the good fortune of escaping our cruisers; and, should any squadron of them, notwithstanding these hazards, succeed in getting their troops on shore, there would be little probability of their ever returning for a second freight, or of remaining long enough undestroyed, even to receive the shattered remains of their first disembarkation: but the construction of these vessels plainly points out the object of attack; namely, the flat coast of this country.

Gun-*Metal*, for brass guns, is composed of eight or ten pounds of tin to one hundred pounds of copper. The largest proportion of tin is used for mortars.

GUNNER of a Ship of War, or Master Gunner, (*maître canonnier d'un vaisseau de guerre*, Fr.) an officer appointed to take charge of all the ordnance, with the ordnance stores and ammunition on board, to observe that they be properly placed and sufficiently supplied with sponges, ladles, and rammers; that, in foul weather, they be housed within board, especially those of the lower tier, and that the ports be shut and calked up; and that, at all times, they may be well lashed and made fast, lest any of them break loose, to the imminent danger of foundering the ship. In the time of an engagement, the gunner is to see that every piece be sufficiently manned: he is answerable to give an account of all his charge upon demand, and to teach the sailors the exercise of the cannon. For further particulars see Naval Instructions.

Gunner's-*Mate*, (*second canonnier*, Fr.) a petty officer appointed to assist the gunner.

Gunner's-*Crew*, (*gens du canonnier*, Fr.) in any ship of war, a certain number of men, under the gunner, appointed to manage the ordnance, keep the tackles and breechings in good repair, and to fire the guns in time of action, &c.

It has been lately remarked, by an intelligent officer, that the gunner's crew appears to be a class much neglected; for, instead of making them skilful artillery-men, and reserving them principally for that service, they are too often suffered to remain totally ignorant of it, and only perform that part of the duty of seamen which is attached to their station. As gunners, it was certainly intended, from the appellation they bear, that they should be acquainted with some part of the art of gunnery; and, if this were the case, they ought first to be well acquainted with it themselves, that it may become their duty to instruct the raw hands in the knowledge of it afterwards.

GUNNERY, (*l'art de l'artillerie* Fr.) is the art, of determining the course and directing the motion of bodies shot from artillery, or other warlike engines. To the art of gunnery belongs the knowledge of the force and effect of gunpowder, the dimensions of pieces, and the proportions of the powder and ball they carry, with the methods of managing, charging, pointing, sponging, &c.

The first application of gunpowder to military affairs, it seems, was made soon after the year 1300, for which the proposal of Friar Bacon, about the year 1280, for applying its enormous explosion to the destruction of armies, might give the first hint; and Schwartz, to whom the invention of gunpowder has been erroneously ascribed, might have been the first who actually applied it in this way.

The first species of artillery, however, which were charged with gunpowder and stone bullets of a prodigious size, were of very clumsy and inconvenient structure and weight. Thus, when Mahomet II. besieged Constantinople, in 1453, he battered the walls with stones of this kind, and with pieces of the calibre of 1200 pounds; which could not be fired more than four times a day.

For the last two hundred years, the formation of cannon has been very little improved: the battering cannon now approved are those that were formerly called demi-cannon, carrying a ball of thirty-two pounds weight; this weight having been found fully sufficient.

The Italians were the first people that made any attempts to ascertain the theory of projectiles, which they did about the beginning of the 16th century. It was then determined that the greatest range of a shot, was, when discharged at an elevation of 45 degrees; and that no part of the path described by a ball is a right line. See the articles Projectiles and Range.

Mr. Robins informs us, in the preface to his *New Principles of Gunnery*, that he had met with no more than four authors who had treated experimentally on this subject. The first of these is Callado, in 1642, who has given the ranges of a falconet, carrying a three pound shot, to every point of the gunner's quadrant, each point being the 12th part, or seven degrees and a half. But, from his numbers, it is manifest that the piece was not charged with its usual allotment of powder.

It is extraordinary that, before Mr. Robins, there were but four authors who had treated experimentally on gunnery, and those to little purpose; but the treatise of Mr. Robins is still regarded as the foundation of the science. The first thing considered by him, and which is, indeed, the foundation of all other particulars relating to gunnery is, the explosive force of gunpowder. The intensity of this force Mr. Robins ascertains in different ways. One of these is, by firing the powder in the air thus: a small quantity of the powder is placed in the upper part of a glass tube, and the lower part of the tube is immersed in water, the water being made to rise so near the top that only a small portion of air is left in that part where the powder is placed; in this situation the communication between the upper part of the tube, and the external air being closed, the powder is fired by means of a burning glass, or otherwise; the water descends upon the explosion, and stands lower in the tube than before, by a space proportioned to the quantity of powder fired. Another was, by firing the powder in vacuo; viz. in an exhausted receiver, by dropping the grains of powder upon a hot iron, excluded in the receiver.

Having determined the force of the gunpowder, or intensity of the agent by which the projectile is to be urged, Mr. Robins next determined the effects it will produce, or the velocity with which it will impel a shot, of a given weight, from a piece of ordnance of given dimensions; which is a problem strictly limited and perfectly soluble by mathematical rules, and is, in general, this: given the first force, and the law of its variation, to determine the velocity with which it will impel a given body in passing through a given space, which is the length of the bore of the gun.

In the solution of this problem, Mr. Robins assumes these two postulates; viz. 1. That the action of the powder on the bullet ceases as soon as the bullet is out of the piece; and 2. That all the powder of the charge is fired and converted into elastic fluid before the bullet is sensibly moved from its place: assumptions which, for good reasons, are found to be, in many cases, very near the truth. It is to be noted, also, that the law by which the force of the elastic fluid varies, is this; viz. that its intensity is directly as its density, or reciprocally proportional to the space it occupies, being so much the stronger as the space is less; a principle well known, and common to all elastic fluids. Upon these principles, then, Mr. Robins resolves this problem, by means of the 39th proposition of NEWTON's *Principia*, in a direct way, and the result is equivalent to this theorem, where the quantities are expressed by algebraic symbols; viz. the velocity of the ball

$$v = 27130\sqrt{\frac{10a}{cd} \times \log.\frac{b}{a}}$$

$$\text{or} = 100\sqrt{\frac{23ad^2}{w} \times \log.\frac{b}{a}}; \text{ whence}$$

v is the velocity of the ball;
a the length of the charge of powder;
b the whole length of the bore;
c the specific gravity of the ball, or weight of a cubic foot of the same matter in ounces;
d the diameter of the bore;
w the weight of the ball in ounces.

For example, suppose $a = 2\frac{5}{8}$ inches; $b = 45$ inches; $c = 11345$ oz. for a ball of lead, and $d = \frac{3}{4}$ inches; then $v = 27130\sqrt{\frac{7}{2269} \times \log.\frac{120}{7}} = 1674$ feet per second, the velocity of the ball.

Or, if the weight of the bullet be $w = 1\frac{9}{20}$ oz. $= \frac{29}{20}$ oz.

Then $v = 100\sqrt{\frac{1115 \times 189}{29 \times 32} \times \log.\frac{120}{7}} = 1674$ feet, as before.

"Having in this proposition," says Mr. Robins, "shewn how the velocity which any bullet acquires from the force of powder, may be computed, upon the principles of the theory laid down in the preceding propositions; we shall next shew, that the actual velocities with which bullets of different magnitudes are impelled from different pieces, with different quantities of powder, are really the same with the velocities assigned by these computations; and, consequently, that this theory of the force of powder, here delivered, does unquestionably ascertain the true action and modification of this enormous powder.

"But, in order to compare the velocities communicated to bullets by the explosion, with the velocities resulting from the theory by computation, it is necessary that the actual velocities with which bullets move, should be capable of being discovered, which yet is impossible to be done by any methods hitherto made public. The only means hitherto practised by others for that purpose, have been either by observing the time of the flight of the shot through a given space; or by measuring the range of the shot at a given elevation, and thence computing, on the parabolic hypothesis, what velocity would produce this range. The first method labours under this insurmountable difficulty, that the velocities of these bodies are often so swift, and, consequently, the time observed is so short, that an imperceptible error in that time may occasion an error in the velocity thus found, of 2, 3, 4, 5, or 600 feet in a second. The other method is so fallacious, from the resistance of the air (to which inequality the first is also liable), that the velocities thus assigned may not be, perhaps, the tenth part of the actual velocities sought.

"To remedy, then, these inconveniences, I have invented a new method of finding the real velocities of bullets of all kinds; and this to such a degree of exactness (which may be augmented, too, at pleasure), that, in a bullet moving with a velocity of 1700 feet in 1″, the error in the estimation of it need never amount to its 500th part; and this without any extraordinary nicety in the construction of the machine."

Mr. Robins then gives an account of the machine, by which he measures the velocities of the balls, which is simply this; viz. a pendulous block of wood suspended freely by a horizontal axis; against which block are to be fired the balls whose velocities are to be determined.

"This instrument thus fitted, if the weight of the pendulum is known, and likewise the respective distances of its centre of gravity, and of its centre of oscillation from its axis of suspension, it will thence be known what motion will be communicated to this pendulum by the percussion of a body of a known weight moving with a known degree of celerity, and striking it in a given point; that is, if the pendulum is supposed at rest before the percussion, it will be known what vibration it ought to make in consequence of such a determined blow; and, on the contrary, if the pendulum, being at rest, is struck by a body of a known weight, and the vibration which the pendulum makes after the blow is known, the velocity of the striking body may thence be determined.

"Hence, then, if a bullet of a known weight strikes the pendulum, and the vibration which the pendulum makes in consequence of the stroke is ascertained, the velocity with which the ball moved is thence to be known."

Mr. Robins then explains his method of computing velocities from experiments with this machine; which is rather troublesome and perplexed, as well as the rules of Euler and Antoni, who followed him in this business: but a much simpler rule is given in Dr. HUTTON's *Tracts*, vol. I. p. 119, where the experiments are explained at full length, and this rule is expressed by either of the two following formulas:—

$$v = 5{,}6727\, c\, g \times \frac{p+b}{bir} \sqrt{o} = 614.58\, c\, g \times \frac{p+h}{birn},$$

the velocity; where v denotes the velocity of the ball when it strikes the pendulum, p the weight of the pendulum, b the weight of the ball, c the chord of the arc described by the vibration to the radius r, g the distance below the axis of motion to the centre of gravity, o the distance to the centre of oscillation, i the distance to the point of impact, and n, the number of oscillations the pendulum will perform in one minute, when made to oscillate in small arcs. The latter of these two theorems is much the easiest, both because it is free of radicals, and because the value of the radical $\sqrt{o}$, in the former, is to be first computed from the number n, or number of oscillations the pendulum is observed to make.

Soon after the first publication of ROBINS's *New Principles of Gunnery*, in 1742, the learned in several other nations, treading in his steps, repeated and farther extended the same subject, sometimes varying and enlarging the machinery; particularly Euler, in Germany; D'Antoni, in Italy; and Messrs. D'Arcy and Le Roy, in France: but most of these, like Mr. Robins, with small fire-arms, such as muskets and fusils.

But, in the year 1775, in conjunction with several able officers of the royal artillery, and other ingenious gentlemen, Dr. Hutton undertook a course of experiments with the ballistic pendulum, in which the machinery was extended to cannon-shot of 1, 2, and 3 pounds weight; for which the Royal Society honoured the doctor with the prizes of the gold medal. For an account of these experiments, see *Phil. Trans.* for 1778.

GUNPOWDER, *(poudre à canon*, Fr.) a composition of nitre, sulphur, and charcoal, well mixed together, and granulated, which readily takes fire; and, when fired, it rarifies or expands with great vehemence, by means of its elastic force. It is to this powder that we owe all the effects and action of guns and ordnance of all sorts, so that fortification, with the modern military art, &c. in a great measure depends upon it.

Although it has been remarked, under the article ENGAGEMENT, that the invention of gunpowder took place in the fourteenth century; yet there are good grounds for ascribing it to a period much more remote. Possibly it came from the east: it can hardly be supposed the explicit terms used to denote it, setting aside those of Roger Bacon, can apply to any other substance.

Some eminent authors have judged the invention of this dreadful combustible as of real utility to mankind, not in aiding their hostile operations, but in lessening slaughter and destruction. They affirm that it is more consonant with humanity that war should be conducted at a distance, where superiority soon becomes evident, than that troops should come to close encounter. They justly observe, too, that personal prowess has a smaller share in characterising a modern hero; and that, when men engaged hand to hand of old, the victor could only assure himself of success by finding no more of his enemy to kill. Now, on the other hand, the very terror inspired by the formidable apparatus of a siege, the thunder of cannon, and preparations for assault, frequently determine the contest, before the combatants have had time to spill each other's blood.

The invention of cannon followed that of gunpowder, and then different modifications of them were devised in smaller weapons.

Proportion of the different Ingredients for making Gunpowder, by different Powers in Europe.

	England.	France.	Sweden.	Poland.	Italy.	Russia.
	lbs.	lbs.	lbs.	lbs.	lbs.	lbs.
Saltpetre	75	75	75	80	76	70
Sulphur	10	$9\frac{1}{2}$	9	8	12	$11\frac{1}{2}$
Charcoal	15	$15\frac{1}{2}$	16	12	12	$18\frac{1}{2}$
Pounds..	100	100	100	100	100	100

To make gunpowder duly, regard is to be had to the purity or goodness of the ingredients, as well as the proportions of them, for the strength of the powder depends much on that circumstance, and also on the due working or mixing of them together.

These three ingredients, in their purest state, being procured, they are then to be mixed together, in the proportion of six parts nitre, one part sulphur, and one part charcoal; the latter containing one-third in weight more than the sulphur.

But it is not the due proportion of the materials only, which is necessary to the making of good powder; another circumstance, not less essential, is the mixing them well together; if this be not effectually done, some parts of the composition will have too much nitre in them, and others too little; and, in either case there will be a defect of strength in the powder.

After the materials have been reduced to fine dust, they are mixed together, and moistened occasionally with water, or vinegar, or urine, or spirit of wine, &c. and then beaten for some time in a stone mortar with a wooden pestle, so as to form the whole into an uniform paste, which is afterwards reduced to grains, by breaking it into small pieces, and passing it through a sieve of copper wire; by which means the grains may have any size given them, according to the nature of the sieve employed, either finer or coarser; and thus, also, the dust is separated from the grains, and again mixed with other manufacturing powder, or worked up into cakes again. Mills are employed for manufacturing great quantities, by which more work may be done in one day than a man can do in a hundred.

To prove GUNPOWDER. The first examination of powder in the king's mills, is by rubbing it in the hands, to find whether it contains any irregular hard lumps. The second is, by blasting two drachms of each sort on a copper-plate, and, in this, comparing it with an approved powder; in this proof it should not emit any sparks, nor leave any beads of foulness on the copper. It is then compared with an improved powder, in projecting an iron ball of 64lbs. from an eight-inch mortar, with a charge of two ounces. The best cylinder powder generally gives about 180 feet range, and pit 150; but the weakest powder, or powder that has been redried, &c. only from 107 to 117 feet.

The merchant's powder, before it is received into the king's service, is tried against powder of the same kind made at the king's mills; and it is received if it gives a range of $\frac{1}{16}$ less than the king's powder with which it is compared. In this comparison both sorts are tried on the same day, and at the same time, and under exactly the same circumstances. The proof of fine grained, or musket-powder is, with a charge of four drachms from a musket-barrel, to perforate, with a steel ball, a certain number of half-inch wet elm boards, placed three-quarters of an inch asunder, and the first 39 feet 10 inches from the barrel: the king's powder generally passes through fifteen or sixteen, and restored powder from nine to twelve. The last trial of powder is, by exposing about one pound of each sort, accurately weighed, to the atmosphere for seventeen or eighteen days; during which time, if the materials are pure, it will not increase any thing material in weight, by attracting moisture from the atmosphere. In this exposure 100lbs. of good gunpowder should not absorb more than twelve ounces, or somewhat less than one per cent.

The present proportion of gunpowder delivered to his Majesty's navy consists of the following quantities and descriptions:—

For foreign service	6-7ths of red L. G. 1-7ths of white or blue L. G.
For Channel service	5-7ths of red L. G. 2-7ths of white or blue L. G.

It is to be understood that the red L. G. gunpowder is to be used for distant engagements, and the white or blue L. G. gunpowder for close engagements; and that the two latter only be used for scaling or salutes.

N.B. The letters L G signify *large* grain; S G, or F G, *small* or *fine* grain.

The above-mentioned proportions consist of several sorts, and the barrels are marked with letters in *red*, *blue*, and *white*, for the sake of distinction.

The *red* letters signify that the powder is made with charcoal burnt in a cylinder, and therefore called cylinder powder, which is the strongest and keeps best.

The *blue* is made with pit-charcoal.

The *white* is restored powder, that has been redried, and mixed with new strong powder, to bring it to a sufficient strength for service.

There are more marks used to distinguish other sorts of powder, but they are not in general use in the navy. See the article MARKS.

The powder marked *blue* and *white* are considered as equal in strength, and the quantity of each of them that is supplied is intended to be first used; and for the purposes of signals, &c. to save as much as possible the cylinder powder for future and actual service.

When powder is returned from ships, it is sent to stoves, that are erected at the different ports, to be dried (under the superintendance of proper officers appointed for that purpose by the Board of Ordnance); then proved, and, if found to have lost any of its strength, it is mixed with new powder, to make it sufficiently strong for service; it is then marked with white letters, as explained above.

French GUNPOWDER. The French proof ball is of brass, and weighs 60lbs. French; the diameter of the mortar 7 inches 9 points, or $\frac{3}{4}$ of a line, and has one line of windage. The chamber holds exactly three ounces; and their best powder must give a range of ninety toises, and their restored powder a range of eighty toises, to be received in the service. But the powder they now make, when new, will give a range of one hundred and one hundred and twenty toises; and Mr. Lombard calculates all his ablest from experiments made with powder, giving one hundred and twenty-five toises with the eprovette.

The above dimensions and weights are all of French standard.

Proportions of POWDER *for the Supply of the different Classes of Ships in his Majesty's Navy; for Foreign and Channel Service.*

Rates of Ships.		Total Quantity of Barrels. L G for general Service. Whole Barrels 90lb. each.	Total Quantity of Barrels. L G for general Service. Half Barrels. 45lb. each.	Total Quantity of Barrels. F G. Half Barrels for priming. 45lb. each.
100	Foreign	479	......	5
	Channel	431	......	5
98	Foreign	402	......	5
	Channel	356	......	5
90	Foreign	355	......	5
	Channel	320	......	5
74	Foreign	330	......	4
	Channel	295	......	4
64	Foreign	256	......	3
	Channel	232	......	3
50	Foreign	194	......	3
	Channel	174	......	3
38	Foreign	147	......	2
	Channel	128	......	2
36	Foreign	138	......	2
	Channel	120	......	2
32	Foreign	....	182	2
	Channel	....	154	2
28	Foreign	....	140	1
	Channel	....	122	1
14	Foreign	....	55	1
	Channel	....	49	1

There are *water-tight barrels*, that do not contain so much powder as mentioned in the above table; viz. whole barrels 80lbs. each, and half barrels 40lbs. each; these are intended for the ground tier only: when any of these are supplied, the number of barrels are increased, so as to give the full proportion of powder.

To restore damaged Gunpowder *to its proper Strength*:—If powder be kept long in a damp place, it will become weak, and be formed partly into hard lumps, which is a sure sign of its being damaged. When powder is thus found, some saltpetre may be perceived at the bottom of the barrel, which, by being wet, will separate from the sulphur and coal, and fall to the bottom of the vessel, settling there in the form of white downy matter. The only method to prevent this is, to move the barrels as often as convenient, and place them on their opposite sides or ends; but, though the greatest care be taken, length of time will greatly lessen the primitive strength.

If it be supposed that the powder has received but little damage spread it on canvas or dry boards, and expose it to the sun; then add to it an equal quantity of good powder, and mix them well; and, when thoroughly dry, barrel, and put it in a dry place. But, if the powder be found entirely bad, first ascertain what it weighed when good; then, by weighing it again, it will be found how much it has lost by the separation and evaporation of the saltpetre; then add to it as much refined saltpetre as it has wasted; but, as a large quantity of this would be difficult to mix, it will be best to put a proportion of nitre to every 20lb. of powder; when done, put one of these proportions into a mealing table, and grind it till it is brought to an impalpable powder, then searce it with a fine sieve; but, if any remain in the sieve that will not pass, return it to the table and grind it again, until it is made all fine enough to go through; being thus well ground and sifted, it must be granulated thus:—first, procure some copper wire sieves, made to suit the required size of the grains (these are called corning sieves or grainers), which are to be filled with the composition; then shake them about, and the powder will pass through, formed into grains. Having thus corned it, set it to dry in the sun; and, when quite dry, searce it with a fine hair-sieve, to separate the dust from the grains.

This dust may be worked with another mixture, so that none may be wasted. Sometimes it may happen that the weight when good cannot be known; in this case, add to each pound one ounce, or one ounce and a half of saltpetre, according to the decay of the powder; then grind, sift, and granulate it, as before directed.

If a large quantity of gunpowder appear to be quite spoiled, the best way is to extract the saltpetre from it, as it can scarcely be worth while to attempt its recovery. See the article Saltpetre.

Laws concerning Gunpowder *and Combustibles*. No person shall make gunpowder but in the regular manufactories established at the time of making the stat. 12 Geo. III. c. 61, or licensed by the sessions, pursuant to certain provisions, under forfeiture of the gunpowder, and 2*s*. per lb. nor are pestle-mills to be used under a similar penalty.

Only 40lbs. of powder to be made at one time, under one pair of stones, except Battle-powder, made at Battle, and elsewhere, in Sussex.

Not more than 40 cwt. to be dried at one time in one stove; and the quantity only required for immediate use to be kept in or near the place of making, except in brick or stone magazines, 50 yards at least from the mill.

Not more than 25 barrels to be carried in any land carriage, nor more than 200 barrels by water, unless going by sea, or coastwise; each barrel to be strongly joined and hooped without iron, and not to contain more than 100lbs.

No dealer to keep more than 200lbs. of powder; nor any person, not a dealer, more than 50lbs. in the cities of London and Westminster, or within three miles thereof, or within any other city, borough, or market town, or one mile thereof; or within two miles of the king's palaces or magazines, or half a mile of any parish church, on pain of forfeiture, and 2*s*. per lb except in licensed mills, or to the amount of 300lbs. for the use of collieries, within 200 yards of them.

By an Act of Parliament made in the 51st year of the reign of King *George* III. entitled "An Act for the better security of his Majesty's Naval Arsenals in the River *Medway*, and *Portsmouth*, and *Hamoaze* Harbours, and of his Majesty's Ships and Vessels lying at and resorting to the same," the masters of all vessels, not belonging to his Majesty, shall, before entering the above ports or harbours, clear or unload their guns, and deposit all gunpowder on board above 5lbs. weight, in such warehouses as shall be appointed by the Admiralty. All masters or owners neglecting to do so, by the above-mentioned Act, shall forfeit and pay any sum not exceeding ten pounds, nor less than five.

GUN-Room, (*sainte barbe*, Fr.) an apartment on the after end of the lower gun-deck of a ship of war, partly occupied by the gunner in large ships, but in frigates and small vessels, where it is below, it is used by the lieutenants as a dining room, &c.

Gun-*Shot*, implies the distance of the point-blank range of a cannon-shot; a ship is, therefore, said to be within gun-shot when she is within that distance.

Gun-*Tackle*, are the ropes, blocks, &c. affixed to each side of the carriage; their use is to run the guns out of the ports, and to secure them to the ship's side in bad weather.

GUNTER'S-Line, called also the Line of Lines and Line of Numbers, is a graduated line, usually placed upon scales, rules, sectors, &c. and is of great use in navigation.

This line is no other than a logarithmic scale of proportionals, wherein the distance between each division is equal to the number of mean proportionals contained between the two terms, in such parts as the distance between 1 and 10 is 10,000, &c. wherefore, if the distance between 1 and 10 upon the scale be made equal to 10,000, &c. equal parts, and 954, &c. the logarithm of 9 of the same parts be set off from 1 to 9, it will give the division standing against the number 9. In like manner, if .903, .845, .778, which are the logarithms of 8, 7, and 6, of the same equal parts, be set off from 1, to 8, 7, and 6, they will give the divisions answering to the numbers 8, 7, 6, upon the line. And, after the same manner, may the whole line be divided.

This line has been contrived various ways for the advantage of having it as long as possible. It was first placed by its inventor, Mr. Edmund Gunter, on a two-foot scale, and called Gunter's Scale. After this Wingate doubled the line in order to render it susceptible of working right on or across.

Then the learned Oughtred projected it in a circle, and also made it to slide; Mackay also improved it; and it was projected in a kind of spiral by Brown.

But the method of using or applying it in all is not very different. In Gunter's and Wingate's projection, common compasses are used; in Oughtred's and Brown's, flat compasses, or an opening index; and, in the sliding-rule, no compasses at all, the slider supplying the place of the compasses. See Mackay *on the Sliding Gunter*.

Gunter's-*Line*, is also usually divided into an hundred parts, every tenth whereof is numbered, beginning with one, and ending with ten. So that if the first great division, 1, represent one-tenth of any integer, or whole number; the next, 2, will represent two-tenths; 3, three-tenths, &c. and the intermediate divisions so many 100th parts of the same integer. Or, if these sub-divisions represent 10 integers, then each of the larger divisions will represent 100, and the whole line will be 1000.

In like manner it may be extended to 1000, by making each sub-division 100. Hence, it is easy to conceive, that any number whatsoever may be found upon the rule, by increasing or decreasing the large divisions; and, consequently, this single line will represent the whole table of logarithms. For the use of Gunter's-line, see Robertson's *Description of Mathematical Instruments*.

Gunter's-*Quadrant*, is a quadrant made of wood or brass, or some other substance, being a kind of stereographic projection of the sphere, on the plane of the equinoctial, the eye being supposed to be placed in one of the poles; so that the tropic, ecliptic, and horizon, form the arcs of circles, but the hour circles all curves, drawn by means of several altitudes of the sun, for some particular altitude, every day in the year.

The use of this instrument is, to find the hour of the day, the sun's azimuth, and other common problems of the globe; as also to take the altitude of an object in degrees. It has been considerably improved by others. See the article Quadrant.

Gunter's-*Scale*, usually called by seamen the Gunter, is a large plain scale, having various lines thereon, of great use in working the cases or questions in navigation. This scale is usually two feet long, and about an inch and a half broad, with various lines upon it, both natural and logarithmic, relating to trigonometry, navigation, &c. On the one side are the natural lines, and on the other the artificial or logarithmic ones. The former side is first divided into inches and tenths, and numbered from one to twenty-four inches, running the whole length near one edge. One half the length of this side consists of two plain diagonal scales, for taking off dimensions to three places of figures. On the other half or foot of this side, are contained various lines relating to trigonometry, in the natural numbers. See Nicholson's *Encyclopedia.*

GUNWALE, or Gunnel of a Ship, (*plat-bord*, Fr.) is that piece of timber which reaches on either side of the ship, from the half-deck to the forecastle, being the uppermost bend, which finishes the upper works of the hull in that part in which are put the stanchions which support the waist-trees. This is called the *gunwale*, whether there be guns in the ship or not. The lower part of any port, where any ordnance are, is also termed the *gunwale.*

GUST, (*bouffée de vent, ou risée*, Fr.) a sudden and violent squall of wind, bursting from the hills upon the sea, so as to endanger the shipping near the shore. These are peculiar to some coasts, as those of South Barbary and Guinea.

GUTT, in the West India Islands, particularly in the island of St. Christopher's or St. Kitts, is a term for the opening of a river or brook, such river or brook also being frequently so called.

GUTTER-Ledge, (*traversier d'écoutille, ou traversin*, Fr.) a cross-bar laid along the middle of a large hatchway in some vessels, to support the covers and enable them the better to sustain any weighty body which may be laid thereon.

GUY, (*corde de retenue pour tenir en respect un fardeau que l'on hisse*, Fr.) a rope used to keep steady any weighty body from bearing or falling against the ship's side, while it is hoisting or lowering, particularly when the ship is shaken by a tempestuous sea.

Guy, is also the name of a tackle, used to confine a boom forward when a vessel is going large, and to prevent the sail from gybing, by any accidental change of the wind or course, which would endanger the springing of the boom, or, perhaps, the upsetting of the vessel.

Guy, is likewise a large slack rope, extending from the head of the main-mast to the head of the fore-mast, and having two or three large blocks fastened to the middle of it. It is employed to sustain a tackle to load or unload a ship with, and is accordingly removed as soon as that operation is finished.

To GYBE, (*trélucher, ou changer, en parlant des voiles auriques*, Fr.) to shift any boom-sail from one side of the mast to the other.

In order to understand this operation more clearly, it is necessary to remark, that, by a boom-sail is meant any sail whose bottom is extended by a *boom*, the fore-end of which is hooked to its respective mast, so as to swing occasionally on either side of the vessel, describing an arc, of which the mast will be the centre. As the wind or the course changes, it also becomes frequently necessary to change the position of the boom, together with its sail, which is accordingly shifted to the other side of the vessel, as a door turns upon its hinges. The boom is pushed out by the effort of the wind upon the sail, and is restrained in a proper situation by a strong *tackle*, communicating with the vessel's stern, and called the *sheet*. It is also confined on the fore-part by another tackle, called the Guy, which article see.

H.

HADLEY'S QUADRANT, an excellent instrument so called from its inventor, John Hadley, Esq. and used at sea for taking the sun's meridian altitude in finding the latitude. For a description of this instrument see the article QUADRANT.

HADLEY'S *Sextant*, an instrument used at sea for taking distances between the sun, moon, and stars, in determining the longitude. For a description of it see the article SEXTANT.

HAG'S-TEETH, or HAKE'S-TEETH, (*défauts, et inégalités dans une tresse, fourrure, &c.* Fr.) are those parts of a *matting*, *pointing*, &c. which are interwoven with the rest, in an erroneous and irregular manner, so as to spoil the general uniformity of the work. See the article POINTING, &c.

HAKE'S-*Teeth*, is also a phrase used to describe some parts of the soundings in the British channel.

HAIL, (*grêle*, Fr.) in natural history, a meteor generally defined frozen rain, but differing from it, in that the hailstones are not formed of single pieces of ice, but of many little spherules agglutinated together; neither are these spherules all of the same consistence; some of them being hard and solid, like perfect ice; others soft, and mostly like snow hardened by a severe frost. Sometimes the hailstone has a kind of core of this soft matter; but more frequently the core is solid and hard, while the outside is formed of a softer matter. Hailstones assume various figures, being sometimes round, at other times pyramidal, crenated, angular, thin, and flat, and sometimes stellated with six radii, like the small chrystals of snow. Natural historians furnish us with various accounts of surprising showers of hail, in which the hailstones were of extraordinary magnitude. Mezeray, speaking of the war of Louis XII. in Italy, in the year 1510, relates that there was, for some time, a horrible darkness, thicker than that of night; after which the clouds broke into thunder and lightning, and there fell a shower of hailstones, or rather (as he calls them) pebble-stones, which destroyed all the fish, birds, and beasts of the country. It was attended with a strong smell of sulphur; and the stones were of a bluish colour, some of them weighing a hundred pounds. *Hist. de France*, tom. ii. p. 339.

At Lisle, in Flanders, in 1686, fell hailstones of a very large size; some of which contained, in the middle, a dark brown matter, which, thrown on the fire, gave a very great report. *Philosophical Transactions*, No. 203.

Dr. Halley and others also relate, that, in Cheshire, Lancashire, &c. April 29, 1697, a thick black cloud coming from Carnarvonshire, disposed the vapours to congeal in such a manner that, for about the breadth of two miles, which was the limit of the cloud, in its progress, for the space of 60 miles, it did inconceivable damage; not only killing all sorts of fowls and small animals, but splitting trees, knocking down horses and men, and even ploughing up the earth; so that the hailstones buried themselves under ground an inch, or an inch and a half deep. The hailstones, many of which weighed five ounces, and some half a pound, and being five or six inches about, were of various figures; some round, others half-round; some smooth, others embossed and crenated: the icy substance of them was very transparent and hard, but there was a snowy kernel in the middle of them.

In Herefordshire, May 4, the same year, after a severe storm of thunder and lightning, a shower of hail succeeded, which far exceeded the former: some persons were killed by it, and their bodies beaten all black and blue; vast oaks were split, and fields of rye cut down as with a scythe. The stones measured from 10 to 13 or 14 inches about. Their figures were various, some oval, others picked, and some flat. *Phil. Trans.* No. 229.

Methods have lately been proposed in England and France, to draw off the lightning and dissipate hailstorms. See *Monthly Magazine for July* 1806.

To HAIL, (*héler un vaisseau*, Fr.) to salute or accost a ship at a distance, either at sea or in a harbour, which is generally performed with a speaking trumpet. The usual expression is, "Hoa, the ship ahoay!" To which she answers, "Holloa!" then follow the requisite questions, as "Whence came ye?" "Where are ye bound?" &c.

HAIR-*Bracket*, in ship building, the moulding which comes at the back of the figure, and breaks in with the upper cheek.

HALBARD, or HALBERT, (*hallebarde*, Fr.) in the art of war, a well known weapon, carried by all military and marine serjeants in time of duty. It is a sort of spear, the shaft of which is about seven feet long, and made of ash; its head is armed with a steel point, edged on both sides, not unlike the point of a two-edged sword: besides this sharp point, which is in a line with the shaft, there is a cross piece of steel, with a small round knob at each end; but generally with a cutting edge at one extremity, and a bent sharp point at the other, so that it serves equally to cut down or push withal. It is also useful in determining the ground between the ranks, and in adjusting the files of a battalion.

HALO, in physiology, a meteor in the form of a luminous ring or circle, of various colours, appearing round the bodies of the sun, moon, or stars, but more particularly the moon. This has been almost universally ascribed to the rays of light issuing from those bodies passing through a frozen medium of hail or snow; as also through the fine rare vesiculæ of a thin vapour towards the top of the atmosphere. And it is probable that the rays of the sun breaking through an uniform dense cloud, nearly exhausted by rain falling from it, may produce a similar effect on moisture in a fluid state, and this is demonstrated frequently, by the sun appearing through such clouds.

The *halos* are sometimes white and sometimes coloured, like the rainbow; but their colours are more dilute than those of the rainbow, and they are in a different order, according to their size. Sometimes only one is visible, and sometimes several concentric *halos* or *coronas* appear at the same time. They are very various in their size; those which have been seen about Sirius and Jupiter were never more than three four, or five degrees in diameter; those, also, which surround the moon are sometimes no more than four or five degrees; but these, as well as those which surround the sun, are of very different magnitudes, some of them being 90° or even larger, and their diameters vary during the time of observation. For further information on the subject of this article, see NEWTON's *Optics*, *p.* 288. PRIESTLEY's *Hist. of Vision*, and SMITH's *Optics.*

To HALE A SHIP, signifies to pull her on shore.

To Over-HALE *a Rope*, is to hale it too stiff, or to hale it the contrary way.

HALF-*Pike*, (*demi-pique*, Fr.) a defensive weapon, composed of an iron spike, fixed on an ashen staff; its use is to repel the assault of boarders in a manner similar to the defence of the charged bayonet among infantry; hence it is frequently termed a *boarding-pike*, which takes the epithet of *half* from its having a much shorter staff than the whole pike.

HALF-*Ports*, (*faux sabords*, Fr.) are shutters made of slit deal, to fit the ports, with a hole cut for the muzzle of a gun to go through.

HALF-*Timbers*, in ship-building, those timbers in the cant-bodies which are answerable to the lower futtocks in the square body.

HALYARDS, (*drisses des huniers et autres voiles supérieures*, Fr.) the ropes or tackles usually employed to hoist or lower any sail upon its respective masts or stay, except the cross-jack and sprit-sail-yards, which are always slung; but in small craft the sprit-sail has halyards. See the article JEARS.

HAMMERS, (*marteaux*, Fr.) in anchor-making, are of five sizes, and only differ in weight and length of handle. They weigh from three to twenty-six pounds; and the handles are, in length, from one foot to four feet; and are used as the size of the work may require.

HAMMOCK, (*branle*, *ou hamac*, Fr.) a piece of hempen cloth, six feet long, and three feet wide, gathered together at the two ends by means of a clew, and slung horizontally under the deck, forming a receptacle for a bed for the sailors to sleep in. There are usually from fourteen to twenty inches in breadth allowed between decks for every hammock in a ship of war: this space however must in some measure depend on the number of the crew, &c. in proportion to the room of the vessel.

In preparing for battle, the hammocks together with their contents, are all firmly corded, taken upon deck, and fixed in the nettings, or wherever the men are too much exposed to the fire of the enemy, so as to form a barricade against small shot. See the article ENGAGEMENT.

Down all HAMMOCKS! (*en bas les hamacs!* Fr.) an order to carry the *hammocks* down between decks, and sling them up in their respective births in readiness to go to bed.

Up all HAMMOCKS! (*branle-bas!* Fr.) an order to unsling the *hammocks*, and to stow them away in their respective places.

HAND, (*on se sert de ce mot pour désigner un homme*, Fr.) is a phrase sometimes used for the word man, as a hand to the lead; clap on more hands, &c.

To HAND *the Sails*, (*ferler*, *ou serrer les voiles*, Fr.) is synonimous with FURLING them, which article see.

HAND-*Grenade*, (*grenade*, Fr.) a small cast-iron shell, filled with powder, and equipped with a touch-hole and fuse, which being set fire to, is thrown by hand from the tops upon the enemy's deck, where, by its bursting, it causes dreadful carnage; it is, however, now much less used than formerly. See the article GRENADE.

HAND-*Lead.* See the article LEAD.

HAND-*over-Hand!* (*main avant!* Fr.) the order to the men, who pull upon any rope, to pass their hands alternately one before the other, or one above the other, if they are hoisting, for the sake of expedition.

A sailor is said to go *hand-over-hand*, when, by the dexterity of throwing one hand above the other, he ascends a single rope, as a shroud or back-stay, without the help of the *ratlings*.

HAND-*over-Hand*, (*main-sur-main*, Fr.) also implies rapidly, as "we are coming up with the chase *hand-over-hand*."

HAND-*Screw*, (*cric*, Fr.) see the article SCREW.

HANDSOMELY, (*en douceur, en parlant d'un palan, ou autre manœuvre, qu'on veut larguer*, Fr.) signifies moderately, as "lower away *handsomely*"!

HAND-*Spike*, (*anspec, et barre de vindas*, Fr.) a wooden bar used as a lever to heave about the windlass, in order to draw up the anchor from the bottom, particularly in merchant ships: for this purpose the handle or small end is round and tapering; and the other end is square, in order to conform to the shape of the holes in the windlass. It is also employed as a lever on many other occasions, as stowing the anchors, or provisions, or cargo, in the ship's hold.

Gunner's HAND-*Crow-Lever*, (*anspec pour le canonnage*, Fr.) an hand-spike shorter and flatter than the above, and armed with iron claws, for the purpose of managing the artillery.

HAND-*Tight*, implies a moderate degree of tension on a rope, so as to make it nearly straight.

To HANG, when speaking of a mast, implies to incline.

To HANG-FIRE, is said of a gun when the flame does not instantly communicate from the pan to the charge; which defect may arise from the powder's being damp, or the vent foul.

To HANG *on a Rope or Tackle-fall*, is to hold it fast without belaying; also to pull forcibly.

HANGING, in ship-building, the name of a plank with a hollow edge, and the hanging of the sheers and decks.

HANK FOR HANK, (*bord sur bord*, Fr.) a phrase expressed of two ships which *tack* and make a progress to windward together; as, the Vulture and Mercury turned up the river *hank for hank*, without being able to get to windward of each other.

HANKS, (*anneaux de bois pour les voiles d'étai*, Fr.) are wooden rings fixed upon the stays to confine the stay-sails thereto at different distances. They are used in lieu of *grommets*, being of a later invention and much more convenient, and are framed by the bending of a tough piece of wood into the form of a wreath, and fastened at the two ends by means of notches, thereby retaining their circular figure and elasticity; whereas the grommets, which are formed of rope, are apt to relax in warm weather and adhere to the stays, so as to prevent the sails from being readily hoisted or lowered. See the articles GROMMET and CRINGLE.

HARBOUR, (*port de mer; havre*, Fr.) a general name given to any sea-port or haven; as also to any place convenient for mooring shipping, although at some distance from the sea. The qualities requisite in a good harbour are, that the bottom be entirely free from rocks or shallows; that the opening be of sufficient extent to admit the entrance or departure of large ships, without difficulty; that it should have good anchoring ground, and be easy of access; that it should be well defended from the violence of the wind and sea; that it should have room and convenience to receive the shipping of different nations, and those which are laden with different merchandises; that it be furnished with a good light-house, and have a variety of proper rings, posts, moorings, &c. in order to remove or secure the vessels contained therein: and finally, that it have at command plenty of wood, and other materials for firing; besides hemp, iron, mariners, &c. See the article ROAD.

HARBOUR-*Master*, (*capitaine de port*, Fr.) an officer appointed to inspect the moorings, and to see that the regulations of the harbour are strictly attended to by the ships which anchor therein.

HARD-A-LEE! (*la barre toute sous le vent; lof tout!* Fr.) the order to the helmsman to put the helm close to the *lee* side of the ship, either to *tack* or keep her head to the wind, when lying to or *trying;* also the situation of the helm in this position.

HARD-*a-Port!* (*babord tout!* Fr.) the order to put the helm close to the larboard, or left side of the ship.

HARD-*a-Starboard!* (*tribord tout!* Fr.) the order to put the helm close to the right side of the ship.

HARD-*a-Weather*, or HARD-*up!* (*arrive tout!* Fr.) the order to put the helm close to the weather or windward side of the ship, so as to bear away. It is likewise the position of the helm, in consequence of that order; being in both senses opposed to *hard-a-lee.*

HARPINGS, (*la partie de l'avant des préceintes d'un vaisseau*, Fr.) in ship-building, the fore-parts of the wales which encompass the bow of a ship, and are fastened to the stem, being thicker than the after part of the wales, in order to reinforce the ship in this place, where she sustains the greatest shock in plunging into the sea, or dividing it, under a great pressure of sail.

Cat-HARPINGS. See the article CAT-HARPINGS.

HARPOON, or HARPING-IRON, (*harpon, à pecher les baleines*, Fr.) a sort of spear used to strike the whales in the Greenland and South Sea fisheries.

The harpoon is sometimes furnished with a long shank, and has at one end a broad and flat triangular head, sharpened at both edges, so as to penetrate the whale with facility; to the other end of this weapon is fastened a long cord, called the whale-line, which lies carefully *coiled* in the boat, in such a manner, as to run out without being interrupted or entangled. As soon as the boat has rowed within a competent distance of the whale, the harpooner launches his instrument (on the upper end of which, near the ring, his name is generally engraved), and the fish, being wounded, immediately descends under the ice with amazing rapidity, carrying the harpoon along with him, and a considerable length of the line, which is purposely let down to give him room to dive. Being soon exhausted with the fatigue and loss of blood, he re-ascends in order to breathe, where he presently expires, and floats upon the surface of the water, when they approach the carcase by drawing in the whale-line.

This line is sixty to seventy fathoms long, and made of the finest and softest hemp, that it may slip the easier; if not well watered, by its friction against the boat, it would be soon set on fire, and if not sufficiently long, it would be soon overset, as it frequently is. With the harpoon they also catch other large fish, as sturgeons, &c.

Gun-HARPOON, (*harpon qui se darde dans un mousqeton*, Fr.) a weapon used for the same purpose as the above, but is fired out of a gun, instead of being thrown by hand. It is made of steel, and has a chain attached to it, to which the line is fastened.

HARPOONER, (*harponneur*, Fr.) a person employed in a whale-boat, whose duty it is to throw or fire the harpoon.

HASLAR HOSPITAL. The Royal Naval Hospital at Haslar, near Gosport, for the reception and cure of sick and wounded seamen and marines, was begun in the year 1746, at the earnest recommendation of the Earl of Sandwich, who was then one of the Lords of the Admiralty, and completed in 1762. It is a regular brick building, four story high, besides cellars, situated on a fine dry gravelly soil in the middle of a field, on the west shore, and near the entrance of Portsmouth harbour, and commands a fine view of Spithead, Portsmouth, and the Isle of Wight.

It is surrounded with an airing ground of 33 acres, formerly part of an estate belonging to a respectable family of the name of Eames, on which was a copse of wood containing much hazle, near the site of an old castle, called *Hazle-worth Castle.*

At the approach to the hospital, near the entrance, is the guard house, a neat military pavilion; where a constant guard is kept, to prevent tumult or desertion.

To the left as you enter the outer court by handsome iron gates, is the lieutenant's office; and farther on the left the wash-house and drying ground: and, in the other angle, between that and the building stand two houses, originally intended for the physician and surgeon: to the right of the entrance is the porter's lodge; and at the northern extremity of the front wall, the fumigating-house and store for the clothes of all patients admitted, which, having undergone the necessary process of cleaning, are stowed away upon racks alphabetically arranged. On the same side between the fumigating-house, &c. and the building, are two houses for the steward and agent: and nearly opposite the centre of the north wing is the dispensary.

The hospital forms three sides of a square, and is composed of a double range of building with wide spaces between them. On the fourth side, a little retired on the brow of the hill, is a handsome chapel, 72 feet long and 36 feet broad, which has been repaired and ornamented within these three years. The front being entire, contains the surgery patients with apartments for the matrons, and other offices at each extremity, with the principal entrance in the centre. The pediment, designed by Mr. Pearce, is a neat piece of sculpture in Portland-stone, which represents his Majesty's arms; Navigation leaning on a rudder, pouring balm into the wounds of a sailor; and Commerce, with their appropriate decorations. This front is 567 feet in length, and the wings 553 feet, and contains 114 wards.

The side-wings are separated near their centre by arcades, beyond which is another double range of wards, intended for patients labouring under infectious disorders, whilst those near the front of the building, are occupied by other classes of medical diseases. The inner court forms a fine quadrangle, surrounded with arcades, 24 feet broad, where the patients take exercise. To this the approach is through a large entrance, with three ranges of massive arches, supporting a large hall, being 100 feet long and 50 feet wide, originally a dining-room, but for many years past has been used occasionally as a ward for surgical patients. Out of this hall the operation-room is taken, and in another part have been lately erected vapour-baths, on the plan of the Hon. Basil Cochrane: the centre-room above is appropriated for the reception of agent's stores, and the upper room likewise appertains to the surgeon's department. Both these rooms, as well as six lobbies containing from six to eight beds each, are comprised in the 114 wards. The other wards are all uniform, 60 feet long and 24 feet broad, each containing 20 beds. The water conveyed from the tower, at the N. W. angle, outside the great enclosure, is by no means good, and is only used for cleaning the wards and utensils; but most excellent water is procured from two wells which are sunk in the airing-ground. The airing-ground is nearly a mile round, and enclosed with a brick wall about 12 feet high.

Since the year 1794, there has been built a row of houses at the extremity of the airing-ground, west of the chapel, for the additional establishment of the governor and officers; behind this building is an enclosed burial-ground of about 12 acres. There are also 32 cabins, containing two beds each, for the accommodation of the attendants, and occasionally for the captains and wounded officers.

Each ward has a water-closet communicating with drains, through which the sea is made to flow at every tide.

There is a bathing-house, for washing patients having infectious disorders, who are supplied with clothes and linen by the house.

The Council formerly consisted of a resident commissioner, with a physician, surgeon, steward, and agent; the physician presiding in the commissioner's absence.

Establishment of the Royal Hospital at Haslar, in October, 1812.

1 Governor	1 Assistant matron
3 Lieutenants	14 Hospital mates
2 Physicians	1 Nurse to every 7 patients
3 Surgeons	1 Butler
1 Chaplain	1 Butcher
1 Steward	1 Barber
1 Agent	1 Plumber
1 Dispenser	1 Cook
9 Clerks	2 Overseers of labourers
2 Matrons	3 Porters

The hospital contains about 1800 patients, exclusive of attendants.

Seamen and marines only were formerly admitted into the hospital, but since the return of the army from Corunna, in the year 1809, soldiers have been admitted.

The number of seamen, marines, soldiers, and every other description of patients *victualled* at the hospital for the last seven years, ending with 1811, is 1,048,238; the average being 410 a day; and the number *admitted* during that period, is 23,540; being upon an average 9 per day.

But when we were obliged to keep our navy in very large fleets to cope with our enemies, this hospital was scarcely found sufficiently large to accommodate all the sick; and in Admiral Keppel's time, when the scurvy was very prevalent in the navy, many with that disorder were encamped in a field near the hospital; but since the time of our celebrated circum-navigator, Captain Cook, who, in his different voyages, shewed, by the internal regulations of his ship and his attention to the health of his seamen, that this disease was to be avoided, and, by the improvements since made in medicine, and the attention now paid by our naval officers to the comforts of the seamen, that disorder very seldom makes its appearance on board any of our ships of war. Our navy being now kept in small squadrons, for the purpose of blockading the few ships that remain in our enemies ports, the sick and hurt are distributed into the different hospitals at the respective ports, which accounts for the small number of seamen received here of late years, and the reason why the military have been admitted on emergencies, as on the return of the late expedition to Walcheren.

HATCH, or HATCHWAY, (*écoutille*, Fr.) a square or oblong opening in the deck of a ship, of which there are several, forming the passages from one deck to another, and into the *hold*, or lower apartments. See the article DECK, plate IX. where A represents the main-hatchway of the lower deck; N N, the fore-hatchway; and O O, the after-hatchway.

The After-HATCHWAY, (*écoutille d'arrière*, Fr.) is placed between the main and mizen-masts.

The Fore-HATCHWAY (*écoutille d'avant*, Fr.) stands a little abaft the fore-mast, or in large vessels, at the break of the fore-castle.

The Main-Hatchway, (*grande écoutille*, Fr.) is placed just before the main-mast, and is the largest in the ship.

Hatches, in merchant vessels, are likewise a sort of trap-doors in the midship, or between the main-mast and fore-mast, through which goods of bulk are let down into the hold; and *hatchway*, properly speaking, is that place where the hatches are.

The hatches of a smaller kind, are distinguished by the name of scuttles. See U U in the deck, plate IX, as also the article Scuttle.

Hatches, (*panneaux d'écoutilles*, Fr.) is also a name improperly applied by sailors to the covers or lids of the hatches.

Hatches are also denominated flood-gates set in a river, &c. to stop the current of water.

Hatch-Bars, (*barres d'écoutilles*, Fr.) are bars of wood or iron, used to lay fore and aft over the hatches, being fitted with a padlock at one end, by which the contents of the hold may be secured from plunder.

HATCHEL or HITCHEL, (*peigne de corderie*, Fr.) a tool with which flax or hemp are combed into fine hairs. It consists of forty sharp-pointed iron pins, or teeth, one foot long, regularly set in a piece of board. Of these there are several sorts, some with finer and shorter teeth, and others with coarser and longer.

HAVEN, a seaport or harbour for ships, as Milford Haven. See the article Harbour.

HAUL-ABOUT, in rope-making, a term used in making a short cable-laid rope, when one strand is made long enough to make three; or for a four-strand rope, long enough to make two, and form an eye at the lower end for a stay.

To Haul, (*haler, ou tirer sur un cordage à force de bras*, Fr.) an expression peculiar to seamen, implying to pull a single rope, without the assistance of blocks, or other mechanical powers upon it; as, "haul in; haul down; haul aft; haul together." See also the articles Bowse, To Hoist, and To Rouse.

To Haul *the Wind*, (*serrer le vent*, Fr.) to direct the ship's course nearer to that point of the compass from which the wind arises. If a ship sail south-west, with the wind northerly, and some particular occasion renders it necessary to *haul the wind* farther to the westward; to perform this operation it is necessary to arrange the sails more obliquely with her keel; to brace the yards more forward, by slackening the starboard and pulling in the larboard braces, and to haul the lower *sheets* further aft: and finally, to put the helm a-port, *i. e.* over the larboard side of the vessel. When her head is turned directly to the westward, and her sails are trimmed accordingly, she is said to have hauled the wind four points, that is to say, from S. W. to W. She may still go two points nearer to the direction of the wind, by disposing her sails according to their greatest obliquity; or, in the sea-phrase, by *trimming all sharp:* and in this situation she is said to be close-hauled, as sailing W. N. W. See the articles Close-hauled and Sailing.

A Haul *of Yarn*, in rope-making, is about four hundred threads, when warped off the winches, with a slight turn in it, to be tarred.

HAUNCH, signifies a sudden decrease in the size of a piece of timber, &c.

HAWSE, (*situation des cables au sortir des écubiers, lorsqu' un vaisseau est affourché*, Fr.) is generally understood to imply the situation of the cables before the ship's stem, when she is moored with two anchors out from the bows, viz. one on the starboard, and the other on the larboard *bow*. Hence it is usual to say, she has a clear hawse, or a foul hawse. It also denotes any small distance *a-head* of a ship, or between her head and the anchors by which she rides; as, "he has anchored in our hawse; "the brig fell athwart our hawse," &c.

A Clear or Open Hawse, (*c'est lorsque les deux cables se dirigent chacun vers leurs ancres, sans se croiser*, Fr.) implies that the cables are directed to their anchors, without lying athwart the stem; or crossing, or being twisted round each other, by the ship's winding about.

A Foul Hawse, (*tour, ou croix dans les cables*, Fr.) on the contrary, implies that the cables lie across the stem, or bear upon each other, so as to be rubbed or chafed by the motion of the vessel.

The hawse may be foul, by having either a cross, an elbow, or a round turn, thus:

Cross in the Hawse, implies one cable lying over another; or that the larboard cable, lying across the stem, points out on the starboard side, while the starboard cable at the same time grows out on the larboard side, fig. 3. plate XI. Fig. 3.

An Elbow in the Hawse, is produced, after having a cross by the ship riding round, but stopping in the middle of that revolution: or, in other words, by riding with her head northward with a clear hawse, and afterwards turning quite round so as to direct her head northward again, fig. 4. Fig. 4.

A Round Turn in the Hawse, is produced, after having a cross, by the ship continuing to wind about the same way, so as to perform an entire revolution, when each of the cables will be twisted round the other, and directed out from the opposite bow, fig. 5. Fig. 5.

If, after this, it again crosses, it is called a Round Turn and Elbow. See the articles Athwart, Cross, and Elbow.

To Clear the Hawse *when Foul*, is to lash a block round the bowsprit, through which a hawser or towline is reeved, having a large hook, called a *fish-hook*, bent to its end. A strong lashing is then passed round both cables below the turns; and the fish-hook placed under the larboard, or riding cable, before the lashing, which is hove well up by the hawser. The end of the *larboard* fore-top bowline, is taken through the starboard hawse-hole, being passed *over* and *under* the cables (because the starboard cable, which is the clearing one, is the undermost), and bent to the starboard cable within board, about two or three fathoms. The *starboard* fore-top-bowline, must be then over-hauled, led through the hawse-hole on its own side, made fast to the end of the cable, and stopped to it with spun-yarn in different places: a rope within board, called a Hawse-Rope, is also made fast to the end of the cable. The *starboard* fore-top-bowline is then hauled upon, the stops being cut as they come out of the hawse-hole, and as the bight comes out, it is triced up by the *larboard* bowline, which is belayed. The hawse-rope is then made fast, the *starboard* bowline cast off, taken over and under the cables (as the larboard one was), and again made fast near the end. The hawse-rope is let go, the starboard bowline being now bowsed upon, hauls the end of the clearing cable over the other. The hawse-rope, being shifted over the cables, is taken into the hawse-hole again, the cable hauled in by it (the bowline being let go and cast off), and bitted. When there is more cable than can be hauled out in one bight, it must be hung by a slip rope to the bowsprit, the bowline cast off and sent in for another bight, proceeding as before. If it blow fresh, it will not be safe to trust to the lashing alone; but a hawser should be made to the *starboard* or clearing cable, with a MIDSHIPMAN'S HITCH, below the lashing or turns. This precaution of clearing the hawse is, however, too much neglected, by which, if nothing worse happen, great injury is done to the cables. It is not uncommon to see large ships with two round turns in the hawse; which must be a disgusting sight to an active seaman.

There are some terms in the sea-language that have also an immediate relation to the hawses; as, "a bold hawse," which signifies the holes are high above the water. "Veer out more cable!" is the order, when part of the cable that lies in the hawse is fretted or chafed, to veer out more cable; so that another part of it may rest in the hawses. "Fresh the Hawse!" is also an order to lay new pieces upon the cable in the hawses to preserve it from fretting. "Burning in the hawses," is when the cables endure a violent stress. "Clearing the hawses," is the act of disentangling two cables that come through different hawses. "To ride hawse-full," is when, in stress of weather, the ship falls with her head deep in the sea, so that the water runs in at the hawses.

HAWSE-*Bags*, (*sacs de voile remplis d'étoupe, pour faire l'office de tampons d'écubier*, Fr.) are canvas bags filled with oakum; they are used in a heavy sea to stop the hawse-holes, and thereby prevent the admission of the water.

HAWSE-*Holes*, (*écubiers*, Fr.) are certain cylindrical holes cut through the bows of a ship on each side of the stem, through which the cables pass in order to be drawn into, or let out of the vessel, as occasion requires. They are represented by *d d*, in fig. 8. plate XI. being fortified on each side by the

HAWSE-*Pieces*, (*alonges d'écubiers*, Fr.) a name given to the foremost timbers of a ship, whose lower ends rest upon the knuckle timber, or the foremost of the cant-timbers. They are generally parallel to the stem, having their upper ends sometimes terminated by the lower part of the beak-head; and otherwise, particularly in small ships, by the top of the bow.

HAWSE-*Plugs*, (*tampons d'écubiers*, Fr.) are certain plugs to stop the hawses, to prevent the water from washing into the manger.

HAWSER, (*haussiere, grelin ou cablot*, Fr.) a large rope, or a kind of small cable, consisting of three and sometimes four strands, serving for various uses on board a ship; as to fasten the main and fore shrouds, to warp a ship as she lies at anchor, and wind her up to it by a capstan, &c. It holds the middle degree between the *cable* and *tow-line*, in any ship whereto it belongs, being a size smaller than the former, and as much larger than the latter. The *hawser* of a man of war may serve for a cable to the sheet-anchor of a small ship. See the articles CORDAGE and ROPES.

HAZY, (*tems embrumé*, Fr.) is that state of the weather which appears much like a fog, yet differs from it, as in foggy weather the atmosphere is dense and moist; but in *hazy* weather the air is more dry.

HEAD, (*éperon, ou poulaine d'un vaisseau*, Fr.) an ornamental figure erected on the continuation of a ship's stem, as being expressive of her name, and emblematical of war, navigation, commerce, &c.

The heads which have any affinity to war or navigation, are, in general, either historical, as referring to some of the deities or heroes of antiquity; or allegorical, as alluding to some of the natural consequences of battle, or the virtues most essential to a life exposed to perpetual danger. Thus, in the former sense, they represent a Neptune, an Alcides, a Mars, an Achilles, a Minerva, or a Jason; and in the latter they produce a *Magnanime*, an Intrepid, a Revenge, or a Victory.

The head of a ship, however, has not always an immediate relation to her name, at least in the British navy. Various instances might be produced to shew, that our artists, formerly as it suited their convenience or judgment, dispensed with this supposed idea of propriety. Hence we sometimes observe the place of a Jason supplied by a Medea; or a beast of prey made the representative of an illustrious lady. The same liberty of design may therefore, with equal propriety, be allowed to symbolize the successes of our arms, by a group of heterogeneous figures, of sundry shapes and sizes, according to the artist's opinion of their superiority or subordination. Their attitude and situation, as well as their size, must, accordingly, depend, in a great measure, on the space into which they are to be crowded; for although the figures may be of equal importance in themselves, yet as there is not room for them all, as large as the life, on a ship's head, it becomes expedient to diminish a few, in order to give place to others. The emblems by which allegorical figures are usually characterized in painting, poetry, and sculpture, are not always thought necessary in a work of this kind, nor even the postures in which these figures are exhibited. And indeed, if we reflect with how much labour and application the workman has endeavoured to fill up every vacancy with some little figure of a convenient form and size, we ought rather to admire his ingenuity than censure him for a violation of those general rules of art, by which it is supposed necessary, on such occasions, to relieve the eye from a scene of perplexity and confusion.

The heads of many of our old ships of war have undoubtedly great beauty and propriety; and candour must acknowledge that some of them have been borrowed from the French designs, which are never left to the invention of illiterate mechanics. A multitude of ornaments appears rather unnecessary in any building calculated for the purposes of war, and if there be any general rule to determine the subjects, and the quantity of sculpture employed in ship-building, it seems to be connected with the ideas of dignity and simplicity. These too are the genuine characteristics of the Grecian and Roman orders of architecture, as opposed to that perplexity, and rage for embellishment, which peculiarly distinguish the Gothic.

Image heads are those founded on practical fiction, and should be bold, warlike, and classical, such as Hercules brandishing his club over the head of Cerberus; Jupiter riding on his eagle, armed with his thunder-bolts, &c.

Emblematical heads consist of appropriate figures, such as an Eagle, denoting dignity, force, and velocity; a Dragon, denoting power, vigilance, &c.

Formerly there was a master-carver, and a considerable number of mechanics in our different dock-yards constantly employed under him in carving ornaments for the heads and sterns of ships; but this establishment is much reduced, and our modern ship's heads are of the plainest order, and more immediately relate to the names they are to bear.

Fig. 6 & 7. Fig. 6. and 7. in plate XI. represent heads of modern ships of the line.

Head, (*l'avant d'un bâtiment*, Fr.) is also used, in a more enlarged sense, to signify the whole front or fore part of a ship, including the bows on each side: the head, therefore, opens the column of water through which the ship passes when advancing. Hence we say, head-sails, head-sea, head-way, &c.

Fig. 8. Thus fig. 8, plate XI. represents one side of the fore part or head of a 74-gun ship, together with part of the bow, keel, and gunnel. The names of the several pieces exhibited therein are as follow:—

A A Fore part of the keel, with *a a* the two false-keels beneath it.

C The stem.

a a The cat-head.

b b The supporter of the cat-head.

c c The knight-head, or bollard timber, of which there is one on each side, to secure the inner end of the bowsprit.

d d The hawse holes.

e The naval whoods, *i. e.* thick pieces of plank laid upon the bow to strengthen the edges of the hawse-holes.

g The beak-head, which terminates the fore-castle on the fore-side, being called the beak-head bulk-head by shipwrights.

H H The gun-ports of the lower deck.

h The gun-ports of the upper deck and forecastle.

I The channels, with their dead-eyes and chain-plates.

i The gripe, or fore-foot, which unites the keel with the stem, forming a part of either.

k k These dotted lines represent the thickness and descent of the different decks from the fore part of the ship at the middle line. The lowest of the three dotted lines, *l*, expresses the convexity of the beams, or the difference between the height of the deck in the middle of its breadth, and at the ship's side. This is also exhibited more clearly in the midship-section, where the real curve of the beam is delineated. See plate VI. fig. 1.

These lines must be always parallel to the lines which terminate the gun-ports above and below.

m m The timbers of the head.

X The rails of the head, which lie across the timbers.

Z The main-wale.

X The channel-wale.

C The load water-line.

See also the continuation of a ship throughout her whole length, as referred to from the article Naval Architecture.

It is evident that the fore-part of a ship is called its head, from the affinity of motion and position it bears to a fish, and in general to the horizontal situation of all animals while swimming.

Head, in a confined sense, also signifies that part on each side of the stem, which is appropriated to the private use of the sailors

By the Head, (*sur l'avant*, Fr.) signifies the state of a ship which is laden deeper forward than aft.

Head *of a Mast*, (*tête d'un mat*, Fr.) the upper part of any mast, or that whereon the caps or trucks are fitted.

Head-*Fast*, (*amarre de l'avant*, Fr.) a rope employed to fasten the head of a ship or boat to a wharf, chain, or buoy, or to some other vessel along-side.

Head-*Land*, (*cap, ou promontoire*, Fr.) a name frequently given to a high cape, or promontory.

Head-*Ledges*, in ship building, are the thwartship-pieces that frame the hatchways or ladder-ways.

Head-*Lines*, imply the ropes of all sails which are next to the yards, and by which the sails are made fast to the yards.

Headmost, (*la tête d'une escadre*, Fr.) the situation of any ship or ships which are the most advanced in a fleet or line of battle.

Head-*Rails*, (*lisses de herpes*, Fr.) in ship-building, are the elliptic rails at the head of the ship.

Head-*Rope*, (*ralingue de têtière*, Fr.) that part of the bolt-rope which terminates any of the principal sails on the upper-edge, and to which it is accordingly sewed. See the article Bolt-Rope.

Head-*Sails*, (*voiles de l'avant*, Fr.) a general name for all those sails which are extended on the foremast and bowsprit, and employed to command the fore-part of the ship: such are the fore-sail, fore-top-sail, fore-top-gallant-sail, and royal, the jib, fore-top-mast, and fore stay-sails, and the sprit-sail with its top-sail. This term is used in opposition to *after-sails.*

Head-*Sea*, (*mer de l'avant*, Fr.) a phrase denoting that the waves meet the head of a ship in her course.

Head-*Stick*, a short round stick, with a hole at each end, through which the head-rope of some triangular sails is thrust, before it is sewed on. Its use is to prevent the head of the sail from twisting.

The Wind Heads *us*, (*le vent nous refuse*, Fr.) that is, veers round to the direction of the ship's course, so as to become more contrary.

Head-*to-Wind*, (*debout au vent*, Fr.) the situation of a ship or boat, when her head is turned to the direction of the wind.

Head-*Way*, (*chemin de l'avant*, Fr.) the motion of advancing at sea. It is thus termed when a ship first begins to advance; or in calm weather, when it is doubtful whether she is in a state of rest or motion. It is in both senses opposed to retreating, or moving with the stern foremost. See the article Stern-Way.

To give a Ship Head-*Way*, in a calm, has hitherto been a subject of much inquiry amongst mariners;

and some effectual contrivance to effect it has often been wished for.

Captain Clem. Sneyd, of the royal navy, has contrived a plan for this purpose, which he communicated to the editor of the *Naval Chronicle*, in 1810, in the following manner:—

"Having been becalmed in his Majesty's ship Sir Francis Drake, in the East Indies, for several days, about three degrees to the southward of the line, on our return from a cruise, and getting very short of provisions, &c. we found it necessary to try some scheme to give her *head-way*. Several methods were tried, but all were found to have some objections. At last I suggested the following plan, which, on putting to trial, I was happy to find succeeded beyond my expectations.

"Two spars, about eight feet longer than the topmast studding-sail-booms, and stouter in proportion, were pointed out of the ports, abreast of the main-mast; holes were cut in the heels, for lashing them in the ports, so as to allow them to swivel fore and aft. At the outer end of each was nailed on a wash of two inch plank three feet wide and six long; without this was rigged on a topping-lift. that went up to the main-mast-head, a smaller hawser that led forward to the cat-head, and a stouter hawser that led aft, close to the taffrail; so that, in working it, the men stationed at the topping-lift topped it up a few feet above the surface of the water; it was then hauled forward, and when far enough so, the men stationed at the after hawser ran aft as fast as they could. The faster it is worked the faster the ship must move through the water. Two of these sweeps may be worked abreast of the main-mast, and two at the fore."

HEART, (*grosse moque à un trou*, Fr.) a peculiar sort of dead-eye, somewhat resembling the shape of a heart; it is only furnished with one large hole in the middle, fig. 5, plate III. whereas a common dead-eye has always three holes; and is principally used to the stays, as the dead-eyes are to the shrouds.

HEAT, in opposition to cold, is a term denoting the property of fire, or of those bodies we denominate hot; being in us a sensation excited by the action of fire.

The diversity of the heat of climates and seasons arises chiefly from the different angles under which the sun's rays strike upon the surface of the earth. Dr. Halley has given a mathematical computation of the effect of the sun under the different seasons and climates. The different degrees of heat and cold in different places, depend, in a very great measure, upon the accidents of situation, with regard to mountains and valleys and the soil. The first greatly helps to chill the air by the winds which come over them, and which blow in eddies through the levels beyond; and mountains, sometimes turning a concave side to the sun, have the effects of a burning mirror upon the subject plain; and the like effect is sometimes had from the convex parts of clouds, either by refraction or reflection. As to soils, a stony, sandy, or chalky, earth, it is known, reflects most of the sun's rays into the air again, and retains but few, by which means a considerable accession of heat is derived to the air; as, on the contrary, black loose soils absorb most of the rays, and return few into the air, so that the ground is much the hotter.

The following table of the heat of different climates is computed for every tenth degree of latitude, to the equinoctial and tropical sun; by which an estimate may be made of the intermediate degrees:

Lat.	Sun in ♈ or ♎	Sun in ♋	Sun in ♑
0	20000	18341	18341
10	19696	20290	15854
20	18797	21737	13166
30	17321	22651	10124
40	15321	23048	6944
50	12855	22991	3798
60	10000	22773	1075
70	6840	23543	000
80	3473	24673	000
90	0000	25055	000

Hence are deducible the following corollaries:—

1. That the equinoctial heat, when the sun becomes vertical, is as twice the square of the radius, which may be proposed as a standard to compare with in all other cases. 2. That, under the equinoctial, the heat is as the sine of the sun's declination. 3. That, in the frigid zones, when the sun sets not, the heat is as the circumference of a circle into the sine of the altitude at 6; and consequently, that, in the same latitude these aggregates of warmth are as the sine of the sun's declination; and, at the same declination of the sun, they are as the sines of the latitudes into the sines of declination. 4. That the equinoctial day's heat is every where as the co-sine of the latitude. 5. In all places where the sun sets, the difference between the summer and winter heats, when the declinations are contrary, is equal to a circle into the sine of the altitude at 6, in the summer parallel; and, consequently, these differences are as the sine of the latitude into, or multiplied by, the sines of declination. 6. From the foregoing table, it appears that the tropical sun under the equinoctial, has of all others the least force. Under the pole it is greater than any other day's heat whatever; being to that of the equinoctial as 5 to 4.

To HEAVE ANY THING OVERBOARD, (*jeter quelque chose à la mer*, Fr.) is to throw any thing from the ship into the water.

To HEAVE *a Flag aboard*, is to hang it out.

HEAVE *and a-way!* an order importing that the next effort shall dislodge and weigh the anchor.

HEAVE *and Pawl!* (*vire pour mettre les linguets!* Fr.) is the order to turn the capstan or windlass till the pawl may be put in, by which it is prevented from coming up, and is sometimes similar to *belay*, when applied to a running-rope.

HEAVE *and Rally!* a cheering order given to the men at the capstan to heave quickly and with spirit.

To HEAVE *at the Capstan or Windlass*, (*virer au cabestan, ou vindas*, Fr.) is to turn it about by means of bars or handspikes.

To HEAVE *the Lead.* See the articles LEAD and SOUNDING.

To HEAVE *a-head*, (*courir en avant sur son ancre*, Fr.) is to draw a ship by the cable, or other rope, fastened to an anchor at some distance before her.

To HEAVE *a-stern*, (*courir en arrière sur son ancre*, Fr.) is to cause her to recede, or go backwards by the same operation.

To Heave *a-strain*, (*virer de force*, Fr.) is to work at the windlass or capstan with more than usual exertion.

To Heave *down a Ship*. See the article Careening.

To Heave *Keel out*, (*abattre en quille*, Fr.) is to raise the keel out of the water by careening, in order to repair or clean it.

To Heave *in Stays*, (*donner vent devant*, Fr.) is to tack or put about.

To Heave *out the Stay-sails*, (*larguer les voiles d'étai* Fr.) is to unfurl and throw them loose from the place where they had been rolled and fastened; which is accordingly done, either to *set* or *dry* them.

To Heave *short*, (*virer à pic*, Fr.) is to draw so much of the cable into the ship, by means of the capstan or windlass, as that by advancing, she is almost perpendicularly above the anchor, and in a proper situation to set sail.

To Heave *taught*, (*roidir le cable*, *en virant au cabestan*, Fr.) is to heave about the capstan, &c. till the cable or rope applied thereto becomes straight and ready for action.

HEAVEN, an azure transparent orb, investing our earth, where the celestial bodies perform their motions. It is of divers denominations, as the highest or empyrean heaven, the ethereal or starry heaven, the planetary heaven, &c.

Heaven, taken in a general sense, or the whole expanse between our earth and the remotest regions of the fixed stars, may be divided into two very unequal parts, according to the matter occupying them; viz. the atmosphere, or aerial heaven, possessed by air; and the ethereal heaven, possessed by a thin and unresisting medium, called ether.

Heaven, in astronomy, is more particularly used for an orb, or circular region, of the ethereal heaven.

HEAVER, (*levier de bois, ou anspec*, Fr.) a wooden staff, employed as a lever on many occasions; particularly in setting up the top-mast-shrouds, frapping the top-masts, strapping the large blocks, seizing the standing-rigging, &c.

HEAVING-Mallet. See the article Mallet.

HEAVY-Metal, implies guns of a large calibre, as, "that ship carries heavy metal," (*ce vaisseau porte une forte artillerie.*)

A Heavy *Sea*, (*grosse mer*, Fr.) imports very strong and high waves.

HEEL, (*talon de la quille tenon de l'étambord*, Fr.) a name usually given to the after-end of a ship's keel; as also to the lower end of the stern-post, to which it is firmly connected.

Heel *of a Mast*, (*pied d'un mât*, Fr.) the lower end, which fits into the step attached to the ship's keel.

Heel *of a Top-Mast*, (*pied du mât de hune*, Fr.) the lower end, which is sustained by the fid upon the *trestle-trees*. See the article Mast.

To Heel, (*plier*, *ou donner à la bande*, Fr.) to stoop or incline to either side; as "the ship heels to starboard." It is usually applied to a ship when she is forced into this position by the wind acting upon her sails, while braced obliquely across her; or by being ballasted more on one side than the other. See the articles Crank, Stiff, and Trim.

When it shall be absolutely necessary to *heel* the ship in order to get at leaks, or repair the copper sheathing, the captain is to superintend that operation himself, and to see that every possible precaution is taken to prevent its being attended with any accident; he is to direct the carpenter to see that all the pumps are in good order, and to appoint careful men to observe the state of the well, who are to report to the officers immediately on their discovering any material increase of water. In three deckers, the lower-deck ports of the side to which the ship is heeled are to be barred in, and the scuppers plugged; and when the lower-deck guns of two-deckers are run out, a sufficient number of men shall be stationed near them, to run them in if necessary; and, during the whole of the time that a ship continues *on the heel*, all the officers and men are to be stationed so as to be able to right her on the shortest notice, if the shifting of the tide, a leak, or a gust of wind should suddenly make it necessary. See Naval Instructions.

HEELING, in mast making, the square part left at the lower end of a top-mast and top-gallant-mast.

HELIACAL, in astronomy, a term applied to the rising or setting of the stars, or, more strictly speaking, to their emersion out of and immersion into the rays and superior splendor of the sun.

A star is said to rise heliacally, when, after having been in conjunction with the sun, and on that account invisible, it comes to be at such a distance from him, as to be seen in the morning before sun-rising; the sun, by his apparent motion, receding from the star towards the east: on the contrary, the heliacal setting is when the sun approaches so near a star as to hide it with his beams, which prevent the fainter light of the star from being perceived, so that the terms apparition and occultation would be more proper than rising and setting.

HELIOCENTRIC Latitude of a Planet, the inclination of a line drawn between the centre of the sun and the centre of a planet, to the plane of the ecliptic.

Heliocentric *Place of a Planet*, in astronomy, the place of the ecliptic wherein the planet would appear to a spectator placed at the centre of the sun.

HELM, (*gouvernail*, *ou timon*, Fr. *helma*, Sax.) a long flat piece of timber, or an assemblage of several pieces, suspended to the hind part of a ship's stern-post by means of pintles and braces, which operate as hinges, and serve to direct the course of the vessel, as the tail of a fish guides the body.

The helm is usually composed of three parts; viz. the rudder, the tiller, and the wheel, except in small vessels, where the wheel is unnecessary.

The length and breadth of the rudder are represented in plate XI. where it is evident that it becomes Fig. 2.
gradually broader in proportion to its distance from the top, or to its depth under the water. The *back*, or inner part of it, which joins to the stern-post, is diminished into the form of a wedge throughout its whole length, so as that it may be more easily turned from one side to the other, where it makes an obtuse angle with the keel. The hinges upon which it is supported are also expressed in this figure. Those which are bolted round the stern-post to the after extremity of the ship, are called braces, and are furnished with a large hole on the after-part of the stern-post. The other parts of the hinges, which are bolted to the back of the rudder, are called pintles, being strong cylindrical pins, which enter into the braces,

and rest upon them. The length and thickness of the rudder are nearly equal to that of the stern-post.

The tiller is a long bar of timber, fixed horizontally in the upper end of the rudder, within the vessel; the movements of the tiller to the right and left, accordingly, direct the efforts of the rudder to the government of the ship's course, as she advances, which is called steering. The operations of the tiller are guided and assisted by a sort of tackle, communicating with the ship's side, called the tiller-rope, which is usually composed of untarred rope-yarns, for the purpose of traversing more readily through the blocks or pullies.

In order to facilitate the management of the helm, the tiller-rope, in all large vessels, is wound about a wheel, which acts upon it with the powers of a windlass. The rope employed in this service being conveyed from the fore end of the tiller to a single block on each side of the ship, forms a communication with the wheel, by means of two blocks suspended near the mizen-mast, and two holes immediately above, leading up to the wheel, which is fixed upon an axis, on the quarter-deck, almost perpendicularly over the fore end of the tiller. Five turns of the tiller-rope are usually wound about the barrel of the wheel, and, when the helm is amidship, the middle turn is nailed to the top of the barrel, with a mark, by which the helmsman readily discovers the situation of the helm, as the wheel turns it from the starboard to the larboard side. The spokes of the wheel generally reach about eight inches beyond the rim or circumference, serving as handles to the person who steers the vessel. As the effect of a lever increases in proportion to the length of its arm, it is evident that the power of the helmsman to turn the wheel will be increased according to the length of the spokes beyond the circumference of the barrel.

When the helm, instead of lying in a right line with the keel, is turned to one side or the other, as in B D,

Fig. 10. fig. 10, plate XI. it receives an immediate shock from the water, which glides along the ship's bottom in running *aft* from A to B; and this fluid turns it towards the opposite side, while it continues in that position; so that the stern, to which the rudder is confined, receives the same movement, and the ship receives an impulse sideways, her stern turns accordingly from B to *b*, or any point whatever *c*, while her head passes from A to *a*. It must be observed, that the water strikes the rudder obliquely, and only with that part of its motion which acts according to the sine of incidence in impelling it in the direction N P, with a force which not only depends on the velocity of the ship's course, but also upon the extent of the sine of incidence. This force is therefore composed of the square of the velocity with which the ship advances, and the square of the sine of incidence, which will necessarily be greater or smaller, according to circumstances; so that if the vessel runs three or four times more swiftly, the absolute shock of the water upon the rudder will be nine or sixteen times stronger under the same angle of incidence: and will be augmented in a greater proportion if the sine of incidence be increased. This impression, or, what is the same thing, the power of the helm is always very feeble, when compared with the weight of the vessel; but it acts with a very long arm of a lever, which occasions it to work very advantageously in turning the ship. For the helm being applied to a great distance from the centre of gravity, G, or from the point about which the vessel turns horizontally, if the direction P N of the impression of the water upon the rudder be prolonged, it is evident that it will pass perpendicularly at the point, R, widely distant from the centre of gravity G; therefore the absolute effort of the water is very powerful. It is not, therefore, surprising that this machine impresses the ship with a considerable circular movement, by forcing the stern from B to *b*, and the head from A to *a*; and even much farther, when the velocity of the ship is preserved, because the effect of the helm always keeps pace with the rapidity of the ship's way.

Amongst all the obliquities which may be given to the rudder, there is one position more favourable than any of the others, to make it more readily produce, with more rapidity, the effect of turning the ship, in order to change her course. To ascertain this, it must be considered, that if the obliquity of the rudder with the keel is greater than the obtuse angle A B D, so as to diminish that angle, the action of the water upon the rudder will increase, and at the same time oppose the course of the ship in a greater degree; because the angle of incidence will be more open, so as to present a greater surface to the shock of the water, by opposing its passage more perpendicularly: but then the direction N P of the effort of the helm upon the ship will pass, with a smaller distance from the centre of gravity G towards R, and less approach the perpendicular N L, according to which it is evident that the power applied should act with a greater effect to turn the vessel. Thus it is evident, that if the obtuse angle A B D is too much inclosed, the greatest impulse of the water will not counterbalance the loss sustained by the distance of the direction N P from N L; or by the great obliquity, which is given to the same direction N P of the absolute effort of the helm with the keel A B. If, on the other hand, the angle A B D is too much opened, the direction N P of the force of action of the helm will become more advantageous to turn the vessel, because it will approach nearer the perpendicular N L; so that the line prolonged from N P will increase the line G R, by removing R to a greater distance from the centre of gravity G; but then the helm will receive the impression of the water too obliquely, for the angle of incidence will be more acute; so that it will only present a small portion of its breadth to the shock of the water, and will, of consequence, only receive a feeble effort. By this principle, it is easy to conceive, that the greatest distance G R from the centre of gravity G, is not sufficient to repair the diminution of force occasioned by the too great obliquity of the shock of the water. Hence we may conclude, that when the water strikes the helm too obliquely, or too perpendicularly, it loses a great deal of the effect it ought to produce; therefore, between the two extremes, there is a mean position, which must be the most favourable to its operations.

The diagonal N P of the rectangle I L represents the absolute direction of the effort of the water upon the helm: N I expresses the portion of this effort, which opposes the ship's headway, or which forces her astern, in the direction of the keel. It is easily perceived that this part N I of the whole power of the helm contributes little to turn the vessel; for if I N is prolonged, it appears that its direction approaches to a very small distance G V from the centre of gravity G, and that the arm of the lever B N=G V, to which the force is applied, is not in the whole more than equal to half the breadth of the rudder: but the relative force N L, which acts perpendicular to the keel, is extremely different. If the first N I is almost useless, and even injurious, by retarding the velocity; the second N L is capable of a very great effect, because it operates at a considerable distance from the centre of gravity G of the ship, and acts upon the arm of a lever G E, which is very long. Thus it appears, that between the two effects N L and N I, which result from the absolute effort N P, there is one which always opposes the ship's head-way, and contributes little to her motion of turning; whilst the other produces only this movement of evolution without retarding her velocity.

Geometricians have determined the most advantageous angle made by the helm with a line prolonged from the keel, and fixed it at 54° 44′, presuming that the ship is as narrow at her floating-line, or at the line described by the surface of the water round her bottom, as at the keel. But as this supposition is absolutely false, since all vessels augment their breadth from the keel upward to the extreme breadth, where the floating-line or the highest water-line is terminated; it follows that this angle is too large by a certain number of degrees. For the rudder is impressed by the water, at the height of the floating-line, more directly than at the keel, because the fluid exactly follows the horizontal outlines of the bottom; so that a particular position of the helm might be supposed necessary for each different incidence which it encounters from the keel upwards. But as a middle position may be taken between all these points, it will be sufficient to consider the angle formed by the sides of the ship, and her *axis*, or the middle-line of her length, at the surface of the water, in order to determine afterwards the mean point, and the mean angle of incidence.

It appears from M. Bouguer's *Traite de la Manœuvre*, sect. 1, liv. II. that in most ships, the angle of the rudder with the prolonged line of the keel should be made to be 46° 40′. Without following the calculations of that able geometrician, we shall, perhaps, be able to explain what he has discussed in a more abstruse manner. When it is required to turn the ship by means of the rudder, and, at the same time, keep the head-way as much as possible, it is evident that the angle 54° 44′, which some have determined to be the most favourable with the line of the keel prolonged, is, in that case, too open; because the water acts upon the rudder with two great a sine of incidence, and which is equal to that of the angle which it makes with the line prolonged from the keel below: but above, the shock of the water is almost perpendicular to the rudder, on account of the width of the ship's sides, as has been shewn before. If then the rudder opposes the fluid, by making only with the line prolonged from the keel, an angle of 45° 1′, the impression, by becoming weaker, will be less opposed to the ship's head-way; and the direction N P (fig. 10, plate XI.) of the absolute effort of the water upon the helm drawing nearer to the lateral perpendicular N L, will be placed more advantageously; since the prolongation of the absolute effort passes at a greater distance, G R, from the centre of gravity G. On the other hand, experience daily testifies, that ships steer well when they do not make the angle D B E more than 35°.

It has been already remarked, that the effect of moving the wheel to govern the helm increases in proportion to the length of the spokes; and so great is the power of the wheel, that if the helmsman employs a force upon its spokes equivalent to 30 lbs. weight, it will produce an effect of 90 or 120 lbs. by the disposition of the wheel alone. On the contrary, the impulse of the water is collected into the middle of the breadth of the rudder, which is very narrow, compared with the length of the tiller; so the effort of the water is very little distant from the fulcrum B, upon which it turns; whereas the tiller forms the arm of a lever ten or fifteen times longer, which also increases the power of the helmsman in the same proportion that the tiller bears to the lever upon which the impulse of the water is directed. This force is, therefore, ten or fifteen times stronger, and the effort of 30 lbs. which at first gave the helmsman a power of 90 or 120 lbs. becomes accumulated to one of 900 or 1800 lbs. upon the rudder. This advantage proceeds from the water's acting on a very short arm of a lever, while the helmsman works on one very powerful, in comparison; and because this lever is moved by a wheel, which multiplies its force. This demonstration ought to remove the surprise with which the prodigious effect of the helm is sometimes considered, from an inattention to its mechanism: for we need only to observe the pressure of the water, which acts at a great distance from the centre of gravity G, of the ship, as well as from the point C, upon which she is supposed to turn, and we shall easily perceive the difference there is between the effort of the water against the helmsman, and the effect of the same impulse against the vessel. With regard to the helmsman, the water acts only with the arm of a very short lever N B, of which B is the fulcrum: on the contrary, with regard to the ship, the force of the water is impressed in the direction N P, which passes perpendicularly at a very great distance from the centre of gravity G, and acts upon a very long lever E G, which renders the action of the rudder extremely powerful in turning the vessel; so that, in a large ship, the rudder receives a shock from the water of 2,700 or 2,800 lbs. which is frequently the case, when she sails at the rate of nine or twelve knots an hour; and this force being applied in E, perhaps 100 or 110 feet distant from the centre of gravity G, will operate upon the ship, to turn her about, with 270,000 or 308,000 lbs.; while the helmsman need not act with a greater power than 30 lbs. upon the spokes of the wheel.

After what has been said respecting the helm, it is easy to judge, that the more a ship increases her velocity with regard to the sea, the more powerful will be the effect of the rudder, because it acts against the water with a force which increases as the square of the velocity of the fluid, whether the ship has head-way or stern-way; with this distinction, that in these two circumstances, the effects will be contrary. For if the vessel goes astern, the helm will be impressed from I to N (fig. 10, plate XI.) and instead of being pushed from N to P, it will receive the effort of the water from N to R; so that the stern will be transported according to the same movement, and the head turned in a contrary direction, and move towards the same side as the tiller B F.

When the helm operates by itself, the centre of rotation of the ship and her movement are determined by estimating the force of this machine; that is to say, by multiplying the surface of the rudder by the square of the ship's velocity.

The frequent accidents to which the rudders of ships are liable in gales of wind, have given birth to many contrivances and inventions of substitutes for this necessary appendage in the hour of distress. In general they vary but immaterially from each other. The form, as originally designed, is always adhered to in the substitute, as closely as possible; but, according to different circumstances, it has been made from topmasts or spars, boards or planking, or even cables. The mode of attaching it also varies but little in one contrivance from another. The first, or preliminary point, always necessary to be gained is, to secure it sufficiently at the top, so that it may work without difficulty, and, at the same time, may, with certainty, be prevented from being torn adrift.

The under part of it is steadied, either by cables or chains, and brought in upon the quarter-deck, as also through the gun-ports; and by this method the new rudder is secured in the best manner possible to the stern of the vessel. This ingenious contrivance, arranged in the hours of the greatest distress, has, at different times, preserved many capital ships from destruction.

From all that has been said of the rudder, it ought to be employed as little as possible, that is to say, the ship and her sails ought to be disposed, that the smallest motion of this machine may bring her to her course, if she deviates from it, or make her perform any evolution which may be thought proper. See the articles RUDDER, SAILING, TO STEER, TRIM, and TO WORK.

There are several phrases relating to the helm, particularly

Down with the HELM! (*la barre dessous!* Fr.) the order to push it down to the lee side of the ship, in order to put the ship about, or lay her to the windward.

Ease, or bear up the HELM! (*mollis la barre!* Fr.) the order to let the ship go more large before the wind.

Put the HELM *a-midships!* (*droit la barre!* Fr.) the order to keep it even with the middle of the ship.

Port the HELM! (*babord la barre!* Fr.) the order to put it over the left side of the ship.

Starbourd the HELM! (*tribord la barre!* Fr.) the order to put it on the right side of the ship.

HELM *Port*, in ship building, the hole in the counter, through which the head of the rudder passes.

HELM *Port Transom*, in ship building, the piece of timber placed across the lower counter, within side, at the height of the helm-port, and bolted through every timber for the security of that port.

HELMSMAN, (*timonier*, Fr.) the man who is entrusted with the management of the helm.

A bad HELMSMAN, (*un mauvais timonier*, Fr.) instead of endeavouring to confine the ship to her course, by moving the helm each way, commonly lett the helm lie, until he sees the ship has got on one side of her course, then moves it so far as to bring it to her course again before he offers to stop her, or meet her with it, till she gets on the other side of her course, so as to require a great deal of helm both ways, by which the ship is steered but little right forward, but is kept yawing about from one side of her course to the other; which shortens the distance gone, and makes both course and distance uncertain, and works the helmsman from side to side; and if the waves run high when carrying a press of sail large, by such bad steering there is great danger of broaching the ship to: therefore, none but the best helmsman should be permitted to steer at such times.

A good HELMSMAN, (*un bon timonier*, Fr.) when a ship is difficult to be steered, at taking the helm, first observes how it lies, then looks with a sharp eye which way the ship is inclined to go from her course, and moves the helm, with a brisk motion, far enough to stop her that way, and feels, by the stress upon the rudder ceasing (which feel ought always to be noticed, as well as any alteration to the eye), when it is a proper time to ease the helm, to prevent her from going on the other side of her course; for the ship is no sooner stopped by the helm from going to one hand, than she will incline to the other, if the helm is let lie in that place: therefore, he keeps moving the helm with a brisk motion, as far as is found necessary to confine her to the course, and by feeling the marked spokes of the wheel, as they come into his hands, soon perceives how much helm, and how far she requires it each way, to command and steer her steadily along, with the least helm, and least trouble to himself.

HEMISPHERE, in geometry, the half of a globe or sphere, when it is supposed to be cut through its centre, in the plane of one of its great circles. Thus the equator divides the terrestrial globe into northern and southern hemispheres: in the same manner the meridian divides it into the eastern and western; and the horizon into upper and lower.

HEMISPHERE is also used to denote a projection of half the terrestrial globe, or half the celestial sphere, on a plane, and frequently called planisphere.

HERCULES, in astronomy, a constellation of the northern hemisphere, in which are contained, according to the Britannic catalogue, 113 stars.

HERCULES, in anchor-making, an instrument used for setting straight the shank of an anchor, welding the palms to the arms, and the arms to the shank of a large anchor.

It consists of a weight about 400 lbs. faced with steel, and a long iron shank. It is suspended perpendicularly over the work by a rope, encircling a small iron wheel, fixed in the mortise cut in the beam. It performs its duty by being drawn up about seven feet high, by men at the tails of the rope; and increasing in weight by the velocity of its fall, it is found to produce more effect than the most powerful exertions with the hammer, or strength of man. That the weight may fall on the spot required, a large iron rod, with a hook, is occasionally fixed to the shank above the weight, for the workmen to direct the stroke.

HERSCHEL, or the GEORGIUM SIDUS, in astronomy, a new primary planet (marked thus ♅), discovered by Dr. Herschel, at Bath, and called by the latter name in compliment to its royal patron; though, on the continent, it is better known by the name of *Herschel*, or *Uranus*. It has six satellites, and is the remotest of all those that are yet known, though not the largest, being, in point of magnitude, less than Saturn and Jupiter.

The cause of this discovery is, in some measure, attributed to the following conjecture of some astronomers; that, from the inequalities in the motion of Jupiter and Saturn, which could not be accounted for from the mutual action of these planets, there existed beyond the orbit of Saturn another planet, by whose action these irregularities were produced. This happy conjecture was confirmed by Dr. Herschel, on the 13th of March 1781; and before that period this new planet had been observed as a small star by Flamstead, Mayer, and Le Monnier, and introduced into their catalogues of the fixed stars. It is situated beyond the orbit of Saturn, at the distance of 1,800,000,000 miles from the centre of the system, and performs its sidereal revolution round the sun in 83 years, 150 days, and 18 hours. Its diameter is about $4\frac{1}{2}$ times larger than that of the earth, being nearly 35,112 English miles. When seen from the earth, its apparent diameter, or the angle which it subtends at the eye, is 3″ 32‴, and its mean diameter as seen from the sun, is 4″.

As the distance of the *Georgium Sidus* from the sun is twice as great as that of Saturn, it can scarcely be distinguished by the naked eye. When the sky, however is serene, it appears like a fixed star of the sixth magnitude, with a bluish white light, and a brilliancy between that of Venus and the moon; but with a power of 2 or 300, its disc is visible and well defined.

The want of light, arising from the great distance of this planet from the sun, is supplied by its six satellites. The second and fourth of these satellites were discovered by Dr. Herschel, on the 11th of January 1787; the other four in 1790 and 1794, but their distances and periodical times have not been so accurately ascertained as the other two.

It is a remarkable circumstance, that all the six satellites move in a retrograde direction, and in orbits lying in the same plane, and almost perpendicular to the ecliptic.

Dr. Herschel remarks, that we shall have eclipses of them in 1818, when they will appear to ascend through the shadow of the planet in a direction almost perpendicular to the ecliptic. See *Phil. Trans.* for the years 1781, 1783, and 1788.

HEWING, the operation of cutting off the rough parts of timber with an axe.

HIDES, (*peaux*, Fr.) the skin of animals, either raw or dressed; the first are allowed ships for service, the second for the gunner's department.

HIGH AND DRY, (*vaisseau à sec sur le rivage*, Fr.) a phrase implying the situation of a ship which is run aground, so as to be seen dry upon the beech when the tide ebbs on her.

HIGH-WATER, (*haute-mer, ou marée haute*, Fr.) is that state of the flood tides when they have flowed to their greatest height, or have ceased to flow or rise.

At *high-water* the motion commonly ceases for about twenty minutes, before it begins to ebb again. The times at high-water of every day of the moon's age, is usually computed from that which is observed on the day of the full or change; viz. by taking 4-5ths of the moon's age on any day of the month, and adding it to the time of high-water on the day of the full or change; then is the sum nearly equal to the time of high-water on the day of the month proposed. See rules and tables for more accurately ascertaining the exact time of high-water in MACKAY and NORIE's *Navigation*.

HIGH-*Water Mark*, (*ligne de démarcation que la mer, à sa plus grande hauteur, laisse sur le rivage*, Fr.) the line made by the tide upon the shore when at its greatest height, at the full and change of the moon. All below this mark is considered under the jurisdiction of the Admiralty.

HIGH-*Sea*, the same as a HEAVY-SEA, which article see.

HINGES, (*pentures, ou couplets*, Fr.) are iron joints, used for sundry purposes, in and about a ship; of which there are various sorts, as the butt and dove-tail hinges, scuttle hinges, locker-joint hinges, port-side hinges, &c.

HITCH, (*clé, sorte de nœud*, Fr.) a sort of knot or noose, by which one rope is fastened to another, or to some other object, as a post, ring, timber-head, mast, &c. They are distinguished by several names, as a half-inch *demi-clef*, a clove-hitch, a rolling-hitch, &c.

To HITCH, (*amarrer, ou saisir un cordage*, Fr.) is to make fast a rope, &c. Thus, when a boat is said to be hoisted in, they say, "hitch the tackles into the rings of the boat!" and when about to weigh anchor, "hitch the fish hook to the flook of the anchor!"

To HITCH *a Rope*, is to pass its end *a*, plate XII. fig. 1, round the standing part, then bring it up through the bight, and seize it to the standing part at *b*. This is called a *half-hitch*; and two of these, one above the other, fig. 2, is called a *clove-hitch*. Plate XII. Fig. 1. Fig. 2.

Blackwall HITCH, is made by putting a bight of a rope over the hook of a tackle, as represented by fig. 3, and letting the part *a* rest upon it, and the part *b* be jambed by the standing part at the cross. This is used with a lanyard, when setting up the shrouds. Fig. 3.

Magnus HITCH, is made by passing two round turns with the end of a rope over a spar *a*, then bringing it before the standing part *b*, passing it again under the spar, and up through the bight which it made (fig. 4), the end part being jambed by the bight *c*. Fig. 4.

A Midshipman's HITCH is made by taking a half-hitch with the end of a rope *a* (fig. 5), round the standing part *b*, then taking another turn through the same bight, jambing it between the parts of the *hitch*: when hauled taught the end may be taken round the standing part at c, or stopped to it. It is in this manner a tail-tackle is clapped on a rope or fall, to augment the purchase. Fig. 5.

A rolling HITCH, (*amarrage en fouet*, Fr.) is made in the following manner:—With the end of a rope, as represented at *a*, fig. 6, take two round turns over the spar, &c. at *c*, then pass two half-hitches (see fig. 1), round the standing part *b*, and it is finished: the end may be stopped to the standing part. Fig. 6.

A Timber HITCH, (*nœud d'anguille*, Fr.) is made by taking the end of a rope *a*, round the spar, or timber head, fig. 7, leading it under and over the standing part *b*, and passing several turns round its own part *c*. See also the articles BEND and KNOT. Fig. 7.

HOAY, a word frequently added to an exclamation, bespeaking attention, as "Main-top, *hoay!*" (*ho de la grande hune, en parlant aux gabiers!*) and is chiefly used to persons aloft, or without the ship.

HOG, (*goret*, Fr.) a sort of flat scrubbing-broom, serving to scrape off the filth from a ship's bottom, under water, particularly in the act of BOOT-TOPPING, which article see.

This instrument is formed by enclosing a multitude of short twigs of birch, or such wood, between two pieces of plank, which are firmly attached to each other, after which the ends of the twigs or branches are cut off even, so as to form a sort of brush of considerable strength and extent. To this machine is fitted a long staff, together with two ropes, the former of which is used to thrust the hog under the ship's bottom, and the latter to guide, and pull it up again close to the planks thereof, so as to rub off the filth effectually. This exercise is usually performed in the ship's boat; which is, accordingly, confined as close as possible to the vessel's side during the operation, and shifted from one part of the side to another, till the whole is completed. Since the invention of coppering a ship's bottom is become so general, the hog is not much used.

To HOG *a Ship*, implies, to scrub it with the instrument as above described. Also a ship is said to be hogged when she has been on shore, and the two ends of her deck seem to lie lower than the middle. See the article BROKEN-BACKED.

HOIST OF A FLAG OR SAIL, (*guindant d'un pavillon, ou d'une voile*, Fr.) is its perpendicular height; in the former it is opposed to the *fly*, which implies its breadth from the staff to the outer edge.

HOIST, also implies the foremost leeches of stay-sails and mast-leeches of boom-sails.

To HOIST, (*hisser, ou guinder, arborer*, Fr.) is to draw up any body by the assistance of tackles. It is also invariably applied to the drawing up the sails along the masts or stays, and displaying of flags and pendants, though by the help of a single block only.

HOISTERS OF THE SWIFTER. See the article SWIFTERS.

HOLD, (*cale du vaisseau*, Fr.) the whole interior cavity or belly of a ship, or all that part of her inside, which is comprehended between the floor and the lower-deck, throughout her whole length.

This capacious apartment usually contains the ballast, provisions, and stores of a ship of war, and the principal part of the cargo in a merchantman. The several store-rooms in a ship of war are separated from each other by *bulk-heads*, and are denominated according to the articles they contain, as the sail-room, the bread-room, the fish-room, the spirit-room, &c. See the article STOWAGE.

After-HOLD, (*cale d'arrière*, Fr.) denotes all that part of the hold which lies abaft the main-mast.

Fore-HOLD, (*cale d'avant*, Fr.) is that part of the hold which is situated in the fore part of the ship, or before the main hatch-way. It is usually in continuation with the main-hold, and serves the same purpose.

Main-HOLD, (*grand cale, ou cale à eau*, Fr.) is that part which is just before the main-mast, and which generally contains the fresh water and beer, for the use of the ship's company.

To rummage the HOLD, (*changer l'arrimage*, Fr.) is to examine its contents.

To Trim or Stow the HOLD, (*arrimer*, Fr.) is to arrange its contents in the most secure and commodious manner possible. See the article TRIM.

HOLD, in navigation, is generally understood to signify a particular situation of a ship, with regard to the shore or to another ship. Hence we say, "Keep a good hold of the land!" or, "Keep the shore well aboard!" which are synonymous phrases, implying to keep near or in sight of the land: when applied to a ship, we say, "She holds her own;" viz. goes as fast as the other ship.

To HOLD *on*, (*retenir le tournevire, ou tel autre cordage qu'on manœuvre au cabestan*, Fr.) is to pull back the hind part of any cable, or other rope, which is heaved round by the capstan or windlass, or drawn in by the purchase of a tackle.

There are very seldom more than three turns of any rope passed about the barrel of the capstan at once, when it is employed in heaving; because a great number of turns of a large rope would soon cover the whole barrel, and utterly destroy the effect of this motion, till those turns could be removed; a circumstance which might be attended with very bad consequences. There being but three or four turns at most upon the barrel at once, a quantity of the rope must necessarily come off, equal to that which is wound on; and this is accordingly pulled back as strongly as possible, to prevent it from *surging* or jerking round the barrel, by being held too loosely. Holding on may therefore be defined the act of retaining any quantity of rope, acquired by the effort of a capstan, windlass, or tackle, as being employed in hoisting as well as heaving.

To HOLD *Water*, (*scier; tenir les avirons dans l'eau en sens contraire, pour arrêter la marche d'un canot*, Fr.) is to stop a boat in her course, which is performed by holding the oars in the water and bearing the blade, or flat part, strongly against the current made *along-side*, by her passing swiftly through the water. See the articles To BACK-ASTERN, OAR, and To ROW.

HOLDERS, (*gens de la cale*, Fr.) the people employed in stowing the hold of a ship.

HOLDFAST, in block making, an iron instrument from 15 to 30 inches long, and from half an inch to one inch and a half in diameter; used to confine the work on the holdfast bench: this iron is bent at one third the length, and forms nearly a square; the longest part is round, and is called the foot; the shortest part is flattened in the inside, and is called the head, as represented at *a b*, fig. 8, plate XII. Fig. 8.

HOLDFAST *Bench*, in block making, a bench of elm plank, whose top is 4 feet long, 12 inches wide, and 3 inches thick, with four legs, 20 inches long (see fig. 8) 4 inches of one end are too upright pieces, 3 inches within square, fixed in mortises; and along the middle of the bench are two rows of holes, in which the *holdfast* is jambed, to confine the work that is to be sawed or trimmed.

HOLES IN THE SAILS, in sail making, are made with an instrument called a stabber, or pegging-awl, and are fenced round by stitching the edge to a small grommet; such are the holes on the head of a sail for the rope-bands, or lacing of square sails, and for seizings on sails that bend to hoops and hanks. Holes are also made across the sails in the reef-bands; at the clews, for marling on the clew-rope; and at the top brims of top-sails, for marling on the foot-rope. The holes, when finished, should be stretched up with a pricker, or marline-spike.

HOLLOA! (*hola!* Fr.) an exclamation of answer to any person, who calls to another to ask some question, or to give a particular order, to signify that they hear him and are ready. Thus, if the master intends to give any order to the people in the main-top, he previously calls "Main-top, hoay!" to which they answer, "Holloa!" It is also the first answer in hailing a ship at a distance. See the articles To Hail and Hoay.

HOLSOM, is a term applied to a ship that rides without rolling or labouring.

HOLY-STONE, a soft porous stone, used in most ships, for the purpose of rubbing and scouring the decks with sand every morning, soon after daylight.

HOME, in a naval sense, implies the situation of some object, either where it retains its full force of action; or where it is properly lodged for convenience or security. In the former sense it is applied to the sails; and in the latter, it usually refers to the stowage of the hold, to the anchors, cartridge, &c.

When it is expressed of the sails, it denotes that their *clews*, or lower corners, are close to the blocks upon the yard-arm, immediately beneath them; it is, therefore, understood only of the loftier sails, as the top-sails, top-gallant-sails, and the studding-sails thereto belonging: hence,

Haul Home *the Top-sail Sheets!* (*borde l'écoute du grand hunier à toucher!* Fr.) is the order to extend the bottom of the top-sail to the lower-yard-arms, by means of the sheets.

Sheet Home *the Top-gallant-sails!* the order to extend the clews of those sails to the top-sail yard-arms.

In the stowage of the hold, a cask, bale, or case, is said to be *home* when it bears against, or lies close to some other object; and, indeed, the security or firmness of the stowage, greatly depends on this circumstance.

Home, when spoken of the anchor, implies the station of the ship, with regard to her anchor; which is accordingly said to *come home*, when it loosens from the ground, by the effort of the wind or current, &c. See the article *The* Anchor *comes Home.*

The Cartridge is Home, (*la gargousse est rendue*, Fr.) implies that it is rammed close down, so that the priming-wire may pierce it through the touch-hole.

HOMEWARD Bound Ship, (*vaisseau sur le retour*, Fr.) is spoken of a vessel when returning from a voyage to the place from whence she was fitted out.

HONEY-COMB, (*chambre, ou défectuosité dans une piece d'artillerie*, Fr.) a flaw or cavity in the cylinder of a piece of ordnance, when it is ill cast and spongeous. The Board of Ordnance rejects all guns having an *honey-comb* of one ninth of an inch deep, as being unfit for service. See the article Gun.

HOOD, (*capuchon*, Fr.) a sort of low wooden porch, resembling the *companion* of the master's cabin (which see), and is placed over the ladder which leads to the steerage in a merchant-ship. Its use is to admit the air and light, and, at the same time, prevent the rain from falling into the steerage.

Hood, in ship building, is also a name given to the upper part of the galley chimney, which being in the shape of the letter L reversed, is trimmed or turned round, according to the various directions of the wind, that the smoke may always fly to leeward.

Hood *of a Pump*, (*capot qui recouvre la roue d'une pompe à chapelet*, Fr.) a short semi-cylindrical frame of wood, serving to cover the upper wheel of a chain-pump.

Naval Whoods, or Hoods. See this article under the letter N.

HOODING-ENDS, in ship building, are those ends of the planks which fit into the rabbets of the stem and stern-post.

HOOK, (*croc, ou crochet*, Fr.) a crooked piece of iron, of which there are several kinds of different shapes and sizes used at sea, as boat-hooks, breast-hooks, can-hooks, cat-hooks, foot-hooks, &c. See these articles. Foot-hooks are the same as Futtocks, which article see.

To Hook *on*, (*crocher*, Fr.) is to apply the hook of a tackle to any weighty body which is to be hoisted in or out; as a boat, a cask, &c.

Hook *and Butt*, (*écart double*, Fr.) the scarfing or laying the two ends of timbers over each other. See the article Butt.

Back-Hooks, in rope making, are large strong iron hooks, to close ropes and cables, fixed in the breast-board of the sledge.

Draught-Hooks, are large hooks of iron, fixed on the cheeks of a gun-carriage, two on each side; one near the trunnion-hole, and the other at the train.

Laying-Hooks, (*manivelle de corderie*, Fr.) in rope-making, the hooks on which the strands are hung together ready for laying or closing.

Loof-Hooks, a tackle with two hooks, one to hitch into a cringle of the main or fore-sail in the bolt-rope, at the leech of the sail by the clew, and the other is to hitch into a strap which is spliced to the chess-tree.

Their use is to pull down the sail, and succour the tackles in a large sail and stiff gale, that all the stress may not bear upon the tack. It is also used when the tack is to be seized more secure, and to take off or put on a bonnet or drabler.

Clasp-HOOPS, for masts, are hoops with a hinge in the centre, and are fastened together in the fore-locks through mortises cut at the ends.

Iron-Hoops, (*cercles de fer*, Fr.) flat pieces of iron formed into a circle, and woolded together at the extremities. They are used for the purpose of driving on masts and yards.

Mould-Hoops, are such as form an ellipsis, and not a circle, and are used chiefly for strengthening masts.

HORARY, something relating to hours, as

Horary, or *Hour Circles of the Globe*, is a brazen circle, fixed upon the brazen meridian, and divided into 24 hours, having an index moveable round the axis of the globe; which, upon turning the globe fifteen degrees, will shew what places have the sun an hour before or after us.

By means of the horary circle many useful problems on the globes are performed.

HORARY *Motion of the Earth*, the arc it describes in the space of an hour, which is nearly fifteen degrees, though not accurately so as the earth moves with different velocities, according to its greater or lesser distance from the sun. Hence we perceive the method of reducing motion into time, and *vice versa;* since $15° = 60'$, or one hour, $1° = 4'$: therefore the clocks at places 15° east of London are one hour faster than those at London; and the clocks at places 15° west of London are one hour later than those at London, and so in proportion. See the article CHRONOMETER.

HORIZON, (*horizon*, Fr.) in astronomy and geography, the line which bounds our vision at sea; or it is that great circle which divides the heavens and the earth into two equal parts, or hemispheres, distinguishing the upper from the lower.

The *horizon* is either rational or sensible: the rational, or true horizon, is a great circle of the apparent celestial sphere, dividing it into two equal hemispheres, and serving as the limits of elevation or depression of celestial objects. It determines the rising and setting of the sun, moon, or stars, in any particular latitude; for, when any of these appear just at the eastern part of the horizon, we say, it rises; and, when it does so at the western part, we say it sets. And from hence, also, the altitude of the sun or stars is accounted, which is their height above the horizon. The poles of the horizon are the zenith and nadir: and the innumerable circles supposed to be drawn through these poles to the horizon, are called the azimuths, or vertical circles.

The sensible, or visible horizon, is that circle which seems to bound our view, or limit our prospect, either at sea or on land.

If a man, six feet high, stand upon an extended plain, or the surface of the sea, he cannot see quite three miles round. To find the distance to which a person can see, at any given height of the eye, or the extent of the visible horizon, is a problem of some utility, particularly to mariners: the rule is, "multiply the square root of the height of the eye in feet, by 1.225, and the product is the distance in miles to which we can see from that height:" thus a sailor, standing at the top-mast of a ship, 120 feet high, can see a distance in miles $= \sqrt{120} \times 1.225 = 13.45 =$ to 13 miles and a half nearly.

HORIZONTAL, (*horizontal*, Fr.) something relating to the horizon, or that is taken in, or on a level with the horizon: thus we say, an horizontal plane, &c. One of the greatest inconveniences mariners have to struggle with, is the frequent want of an horizon; for, though the atmosphere may, at the height of ten or twelve degrees and upwards, be clear enough to give a view of the sun and other objects, yet all below that height is often so hazy as to hinder a distinct sight of the horizon, and, consequently, to prevent observations from having the requisite correctness. This inconvenience is removed by an

HORIZONTAL *Speculum*, invented by Mr. Serson, who was lost in the old Victory man of war; in which ship he was sent out to make trial of his machine.

This curious and useful instrument, as since improved by Mr. Smeaton, consists of a well polished metal speculum, about three inches and a half in diameter, inclosed within a circular rim of brass; so fitted that the centre of gravity of the whole shall fall near the point whereon it spins. This is the end of a steel axis running through the centre of the speculum, above which it finishes in a square, for the convenience of fitting a roller on it, by which it is set in motion by means of a piece of tape wound round the roller. The cup in which it turns is made of agate, flint, or other hard substance; and a pyramidal cover may be made to the whole, composed of glass panes; by this means an observation may be made with it as well covered as open, and it will thereby be prevented from tarnishing by the moist air and spray of the sea.

These specula are as useful by night as by day; for, as the images of the smaller stars may be seen in the speculum, consequently, any object that can be seen reflected from the glasses of the quadrants may be observed by the speculum, and these are all the stars of the first magnitude, the planets Venus, Mars, Jupiter, Saturn, and the moon; so that, by having the declinations of these bodies in the *Nautical Almanac*, or in the common books of navigation, they may be used in observations as well as the sun.

HORIZONTAL *Range of a Piece of Ordnance*, is the distance at which it falls on or strikes the horizon, or on a horizontal plane, whatever is the angle of elevation or direction of the piece. When the piece is pointed parallel to the horizon, the range is then called the point-blank-range.

The greatest horizontal range, in the parabolic theory, or in a vacuum, is that made with the piece elevated to 45 degrees, and is equal to double the height from which a heavy body must freely fall to acquire the velocity with which the shot is discharged. See the article RANGE.

POWDER-HORN, (*corne d'amorce*, Fr.) a polished cow's-horn, used to hold the priming powder, during the time of an engagement, or otherwise, and is issued with the gunner's stores.

HORNS, (*cornes de vergue*, Fr.) the jaws, or semi-circular inner ends of booms and gaffs.

HORSE, (*marche-pied de vergues*, Fr.) a rope reaching from the middle of a yard to its arms or extremity, and hanging about two or three feet under the yard, for the sailors to tread upon whilst they are loosing, reefing, or furling the sails, rigging out the studding-sail booms, &c. In order to keep the horse more parallel with the yard, it is usually suspended thereto, at proper distances, by certain ropes called *stirrups*, which hang about two feet under the yard, having an iron thimble spliced into their lower ends through which the horse passes. See the article RIGGING.

HORSE, is also a thick rope, extended in a perpendicular direction near the *fore* or *after*-side of a mast, for the purpose of hoisting some yard, or extending a sail upon it: when fixed before the mast, it is used for the *square-sail*, whose yard is attached to the horse, by means of a *traveller*, or *bull's-eye*, which slides up and down occasionally: when placed *abaft* the mast, it is intended for the *try-sail* of a snow, but is seldom used in this position, except in those sloops of war which occasionally assume the appearance of snows, in order to deceive the enemy. It also signifies a thick iron rod, fastened at the ends to the inside of the stern of vessels that carry a fore and aft main-sail for the main-sheet to travel on, and also across the deck before the mast, for the sheet of the fore-sail.

Horse, (*baudet*, Fr.) the name of a sawyer's frame or trestle.

Flemish Horse, is a smaller kind of horse, placed at the top-sail-yard-arms, on which the man who passes the earing usually stands.

Iron Horse, (*batayoles de la poulaine*, Fr.) in ship building, the name given to a large round bar of iron, fixed in the heads of ships, with stanchions and netting.

Leather HOSE, (*manche de toile*, Fr.) a long flexible tube, formed of leather or tarred canvas, but chiefly of the latter, and employed to conduct water from the main-deck into the casks in the hold, or by the assistance of a pump, from one cask to another. For the latter use, one of the ends or openings of the hose is fixed in the empty cask, whilst the other is applied to the pump that extracts the water out of the full one. This exercise is, on some occasions, necessary to alter or preserve the trim of the vessel, without disturbing her stowage.

HOSPITAL, Naval, (*naval hôpital*, Fr.) a receptacle for the cure of seamen and marines, sick or wounded, in his Majesty's service. See the articles Greenwich, Haslah, and Plymouth Hospitals.

Hospital-*Ship*. See the article Ship.

HOUARIOS, are small vessels with two masts and a bowsprit, sometimes used as coasters or pleasure-boats, in inlets and rivers of the Mediterranean. Abaft the masts are set lateen-sails and sliding top-masts. The lower part of the sail is bent to hoops that encircle the lower or standing mast; and the upper part of the sail is laced to the top-mast, which slides up and down the lower-mast by grommets, or iron rings, fastened to the heel of the mast. The sail is fastened at the lower part, with a tack to the mast, and, at the peek, with a small earing. The sail is hoisted by a halyard, one end fastened to the heel of the top-mast, and the other end reeved through a sheave-hole in the lower mast-head; it leads down towards the heel of the mast, and there belays. The sail is extended by a sheet, fastened to the clew, and led aft. To the heel of the top-mast is sometimes fastened a down haul rope, that leads down towards the heel of the mast.

These sails furl in a close manner to the lower-mast, by lowering the top-mast and confining the sail in folds by a furling-line. On the bowsprit is spread a jib, which assists the vessel in going about. These sails are called sliding gunters, and used in the navy's pinnaces and barges.

HOUNDS, (*jottereaux, ou flasques des mâts*, Fr.) a name given to those parts of a mast-head which gradually project on the right and left side, beyond the cylindrical or conical surface, which it preserves from the *partners* upwards. The hounds, whose upper parts are also called *cheeks*, are used as shoulders to support the frame of the top and trestle-trees, together with the top-mast and the rigging of the lower-mast. See the article Mast.

To HOUSE the Guns, (*mettre les canons à la serre*, Fr.) is first to run them in upon the decks, and, by taking the quoins from under them, the muzzles rest against the sides above the ports; then securing them by their tackle, muzzle-lashings, and breechings. A gun is sometimes housed fore and aft, to make room, as in the cabin, &c.

HOUSING, or House-Line, (*luzin*, Fr.) a small line, formed of three fine strands, smaller than rope-yarn, and is chiefly used for seizings of block's straps, fastening sails to their bolt-ropes, &c. See the article Bolt-Rope.

HOWITZER, or HOWITZ, (*obusier*, Fr.) a kind of small mortar, or short gun, mounted upon a field-carriage like a gun; but differing from the latter by having the trunnions at the middle instead of the end. It has been superseded in the sea-service by the carronades.

HOWKER, (*hourque, sorte de bâtiment*, Fr.) a Dutch vessel, commonly navigated with two masts; viz. a main and a mizen-mast, and being from sixty to two hundred tons burden. It is also the name of a fishing-boat, used on the southern coasts of Ireland, and carrying only one mast.

HOWLE, among ship carpenters, is said of a ship whose futtocks are scarfed and bolted into the ground-timbers, and the plank laid on them to the orlop.

HOY, (*bateau de passage*, Fr.) a small vessel, usually rigged as a sloop, and employed in carrying passengers and goods from one place to another, particularly on the sea-coast, where the ordinary lighters cannot be managed with safety or convenience.

It would be very difficult to describe, precisely, the marks of distinction between this vessel and some others of the same size, which are also rigged in the same manner; because what is called a *hoy* in one place, would assume the name of a *sloop* or *smack* in another; and even the people who navigate these vessels have, upon examination, very vague ideas of the marks by which they are distinguished from the above-mentioned. In Holland, the hoy has two masts; in England it has but one, where the main-sail is sometimes extended by a boom, and sometimes without it.

To HUG, (*serrer, pincer, &c.*) when spoken of a ship, is to keep close in with the shore; hence

To Hug *the Land*, (*serrer la terre*, Fr.) is to sail as near it as possible.

To Hug *the Wind*, (*pincer le vent*, Fr.) to keep the ship close-hauled.

HULK, a name given to any old vessel laid by, as unfit for further sea service: it is probably derived from the ὁλκαδες, or vessels of burden of the ancient Grecians. In the royal ports they are used for the accommodation of a ship's company, while their own vessel is in dock under repair.

HULL of a Ship, (*corps d'un vaisseau*, Fr.) is her frame, or main body, exclusive of the masts, yards, sails, and rigging: it is usually expressed of a ship, either before she is furnished with masts, &c. or after she is stripped and dismantled. Thus, to strike a-hull in a storm is to take in her sails, and to lash the helm on the lee-side of the ship; and to hull, or lie a-hull, is said of a ship whose sails are thus taken in, and the helm lashed a-lee. See the article A-Hull.

To Hull *a Ship*, (*canonner un vaisseau dans son bois*, Fr.) is to fire cannon-balls into her hull within the point-blank range.

Hull-*Down*, is said of a ship when she is at such a distance as that only her masts and sails are to be seen.

Hull-*to*, (*à la cape*, Fr.) the situation of a ship when she is *trying a-hull*, or with all her sails furled; as in trying. See the article Trying.

HULLOCK of a Sail, is when, in a great storm,

some part is cut and left loose. It is chiefly used in the middle sail, to keep the ship's head to the sea; then all the rest of the sail is made up, except a little at the mizen-yard-arm.

HUMMOCK, (*mamelon, tertre, ou éminence détachée sur les côtes de la mer*, Fr.) a name given by mariners to a hillock, or small eminence of land resembling the figure of a cone, and appearing on the sea-coast of any country.

HURRICANE, (*ouragan, ou tempête*, Fr. *huracan*, Span.) a violent and prodigious tempest, accompanied with lightning, and is occasioned by the collection and opposition of several winds, that sometimes blow from almost every point of the compass, causing a dangerous agitation in the sea, where the waves break and dash against each other with astonishing fury.

On the approach of a hurricane the sea and air become perfectly calm and motionless, without a breath of wind stirring either. Soon after this the sky is darkened, the clouds accumulate, and the light of the day is replaced by terrible flashes of lightning. During the continuance of this general calamity, the vessels which were anchored in the roads frequently cut their cables and put to sea, where they are driven at the mercy of the winds and waves, after having struck their ards and top-masts.

Hurricanes are most frequently between the tropics, where they sometimes produce the greatest devastation. They generally take place about the time of the sun's passing the equinox; *i. e.* 21st March and 21st September. Hurricanes are frequently in the West Indies, where they make prodigious ravages, by rooting up trees, destroying houses, shipping, and the like.

The natives, it is said, can foretel hurricanes by the following prognostics:—

1. All hurricanes happen either on the day of the full, change, or quarter of the moon.

2. From the unusual redness of the sun, great stillness, and, at the same time, turbulence of the skies, swelling of the sea, &c. happening at the change of the moon, the Caribbees conclude there will be a hurricane next full moon; and, if the same signs be observed on the full moon, they may expect one next new moon.

Now, as the wind betwixt the tropics is generally easterly, and upon the sun's going back from the northern tropic, the western winds pour down with violence upon those parts, the opposition of these contrary winds cannot fail to produce a hurricane. Hurricanes do not shift through all the points of the compass, but begin always with a north wind, veer to the south-east, and then cease; and their shifting between these two points is so sudden and violent, that it is impossible for any ship to veer with it; whence it happens that the sails are carried away, yards and all, and sometimes the masts themselves wreathed round like an osier.

Swisserland is subject to very violent hurricanes, which do great mischief, and that in a very singular manner.

Thunder and lightning are frequent with them in winter as well as in summer; and the more violent storms of these are sometimes attended with whirlwinds and hurricanes, which has been seen to raise the water of some lakes, in form of a thick pillar, up to the cloud, and, carrying on before the wind, this vast body of water will sometimes fall in other places, on dry land, and drown the houses and gardens where it chances to fall.

These are all pheuomena of electrical origin, and depend on the same general causes. See the articles LIGHTNING and WATER-SPOUT.

From a number of observations made in different parts of America, on the rate with which hurricanes travel, their velocity has been found to be one hundred miles per hour; the effect in this case cannot be less than fifty pounds avoirdupoise on a square foot. Hence it ceases to be a matter of surprise that houses are overthrown, trees torn up by the roots, ships sunk and dismasted.

A common paper kite will determine the velocity of the wind with great accuracy, by letting it loose, and measuring, with a stop watch, how long it is running out a given quantity of line. The rapidity has been found to be at the rate of seventy-two statute miles per hour; which was not the case at the earth's surface, owing to the impediments which the wind meets with, neither is the current so irregular aloft as below. See the article WIND.

SHIP'S HUSBAND, among merchants, the person who takes the direction and management of a ship's concerns upon himself, the owners paying him a commission for his trouble.

HYDRA, in astronomy, a southern constellation imagined to represent a water-serpent. The number of stars in this constellation is, in the Britannic Catalogue, 68.

HYDROGRAPHER, a person skilled in the art of measuring and describing the sea, rivers, lakes, and canals.

HYDROGRAPHY, is that part of geography which considers the sea, principally, as it is navigable; and gives an account of the tides, counter-tides, soundings, bays, gulfs, creeks, &c. as also of the rocks, shelves, sands, shallows, promontories, harbours, the distance and bearing of one port from another; with every thing that is remarkable, whether out at sea or on the coast.

Some of our best writers use the term in a more extensive sense, so as to denote the same with navigation.

In this sense *hydrography* includes the doctrine of sailing; the art of making sea charts, with the uses thereof; and every thing necessary to be known, in order to the safest and most expeditious performance of a voyage.

The Admiralty have very properly taken into their consideration the very incorrect state of many of our sea charts, and have now appointed men of eminent abilities to assist their hydrographer in examining and reporting upon them; and it is now ordered that all king's ships shall be furnished with a set of the most correct charts, to be accounted for by the commanding officer, as any other necessary store.

In France they have public professors of *hydrography* established in all their sea-ports, who are to instruct youth intended for the sea in all parts of navigation, sailing, steering, &c. with the several branches of mathematics necessary thereto.

They teach gratis, having salaries allowed by the government, and are also charged with the examination of pilots.

J.

JACK, *(pavillon de beaupré*, Fr.) a sort of flag or colours, displayed from a staff erected on the outer end of a ship's bowsprit. In the British navy, the jack is a small union flag; composed of the intersection of the red and white crosses; but in merchant-ships this union is bordered with a red field. See the article FLAG.

JACK, or *Sawyer's* JACK, *(chevalet de scieur de long*, Fr.) an engine of the saw kind, much used about guns and mortars, and always carried with the artillery.

JACK *in the Box*, *(cric à vis*, Fr.) a large wooden male screw, turning in a female one, which forms the upper part of a strong wooden box, shaped like the frustum of a pyramid. It is used by means of levers passing through holes in it, as a press in packing, and for other purposes.

JACK *of the Bread-room*, *(valet de la soute au pain*, Fr.) an assistant to the purser or ship's steward.

JACK *Block*, *(poulie où passe la manœuvre servant à amener*, *&c.* Fr.) a block occasionally attached to the top-gallant-tie, and through which the top-gallant top-rope, is reeved to sway up or to strike the yard.

JACOB'S-STAFF, *(astrolabe*, Fr.) sometimes called a CROSS-STAFF, a mathematical instrument to take altitudes at sea, consisting of a brass circle, divided into four equal parts, by two lines cutting each other in the centre: at each extremity of either line is fixed a sight perpendicular over the lines, with holes below each slit, for the better discovery of distant objects.

The cross is of course mounted on a stand for use. Sometimes instead of four sights there are eight.

To JAMB, *(gêner*, *assujettir*, *acorer*, Fr.) is to inclose any object between two bodies, so as to render it immoveable, whilst they continue in the same position. This expression is usually applied to the situation of a running-rope, when it is so compressed by other bodies as to be incapable of traversing in the blocks, till it is released from this confinement. In this sense jambing is opposed to RENDER, which article see.

A cask, box, &c. is also said to be jambed, when it cannot be dislodged from its confinement without great difficulty.

JAWS, sometimes called HORNS, at the ends of gaffs or booms, are made semi-circular, for the purpose of confining them to the mast, which they effectually do by their forms, which embrace half the circumference of the masts. See the article HORNS.

ICE, *(glace*, Fr.) a brittle transparent body, formed of some fluid, as water frozen or fixed by cold. When water is exposed to a diminished temperature, it assumes the solid state, by shooting into chrystals which cross each other in angles of about 60 degrees. During this process of solidity, the temperature remains constant.

The specific gravity of *ice* to water is various, according to the nature and circumstances of the water, degree of cold, &c. Dr. Irving found the densest *ice* he could meet with in Phipps's voyage to the north pole, about a 14th part lighter than water. M. de Mairan found it, at different trials, 1-14th, 18th, or 19th lighter than water; and, when the water was previously purged of air, only a 22d part.

The northern *ice* extends generally about 9 degrees around the pole; the southern 18 or 20°, in some parts even 30°; and floating ice has occasionally been found in both hemispheres as far as 40° from the poles, and sometimes, as it has been said, even in latitude 41° or 42°. Between 54° and 60° south latitude the snow lies on the ground, at the sea-side, throughout the summer. The line of perpetual congelation is three miles above the surface of the equator, where the mean heat is 84°; at Teneriffe, in latitude 28°, two miles; in the latitude of London, a little more than a mile; and in latitude 80° north, only 1,200 feet. At the pole, according to the analogy deduced by Mr. Kirwan, from a comparison of various observations, the mean temperature should be 31°. In London the mean temperature is 50°; at Rome and at Montpelier, a little more than 60°; in the island of Madeira, 70°; and in Jamaica, 80°.

To be ICE-*Bound.* See the article BOUND.

Fields of ICE, *(bancs de glace*, Fr.) are large bodies of ice, drifted by the course of the current, at sea.

IDLER, *(on nomme ainsi*, *sur les vaisseaux de guerre Anglois*, *tout individu non assujetti à faire le quart*, Fr.) a general name given to all those on board a ship of war, who, from being liable to constant day duty, are not subjected to keep the night watch; but must, nevertheless, go upon deck if all hands are called during the night.

JEARS, *(drisses des basses vergues*, Fr.) an assemblage of tackles, by which the lower yards of a ship are hoisted up along the mast to their usual station, or lowered from thence, as occasion requires; the former of which operations is called *swaying*, and the latter *striking*. See the articles To SWAY and To STRIKE.

In a ship of war the jears are usually composed of two strong tackles, each of which has two blocks;

viz. one fastened to the lower-mast head, and the other to the middle of the yard. The two blocks which are *lashed* to the *slings* of the yard, are retained in this situation by means of two cleats, nailed on each side, whose arms enclose the ropes by which the blocks are fastened to the yard. The two ropes, which communicate with these tackles, lead down to the deck on the opposite side of the mast, according to the situation of the upper jear-blocks.

Within the last few years one *hanging jear-block*, with a long and short leg round the mast-head, has been substituted in large ships in the navy, instead of the two that were formerly used with lashings, which is not only more convenient to the mechanical powers, but also relieves the mast-head from much unnecessary weight.

The jears, in merchant ships, have usually two large single blocks on the opposite side of the mast-head, and another of the same size in the middle of the yard. The rope, which communicates with these, passes through one of the blocks hanging at the mast-head, then through the block on the yard, and afterwards through the other hanging-block upon the mast. To the two lower ends of this rope, on the opposite sides of the mast are fixed two tackles, each of which is formed of two double blocks, the lower one being hooked to a ring-bolt in the deck, and the upper one spliced or seized into the lower end of the great rope above, which is called the tye. By this contrivance the mechanical power of the tackle below is transmitted to the tye, which, communicating with blocks on the yard, readily *sways-up*, or lowers it, either by the effort of both jears at once, on the opposite sides of the mast, or by each of them separately, one after the other.

JETTY-HEAD, (*jettée d'un port, ou môle*, Fr.) a name given, in the royal dock-yards, to that part of a wharf which projects beyond the rest; but more particularly the front of a wharf, whose side forms one of the cheeks of a wet or dry dock.

JEWEL-BLOCKS, (*poulies de drisses des bonnettes, frappées aux bouts des vergues basses*, Fr.) See the article Block.

JIB, (*grand foc*, Fr.) the foremost sail of a ship, being a large stay-sail extended from the outer end of the bowsprit, prolonged by the jib-boom, towards the fore-top-mast-head. In cutters and sloops the *jib* is on the bowsprit, and extended on the lower-mast-head. See the article Sail.

The *jib* is a sail of great command with any side wind, but especially when the ship is *close-hauled*, or has the wind upon her beam; and its effort in *casting* the ship, or turning her head to leeward is very powerful, and of great utility, particularly when the ship is *working* through a narrow channel. See the article Sailing.

Clear away the Jib! (*largue le grand foc!* Fr.) the order to loose it preparatory to its being set.

Flying Jib, (*clin foc*, Fr.) a sail sometimes set upon a boom, rigged out beyond the Jib-Boom, which article see.

Middle Jib, (*second foc*, Fr.) a similar sail, sometimes set before the two preceding, being extended from the end of the jib-boom, while the inner jib-tack is near half way down, or on the boom.

Jib-Boom, (*bâton de foc*, Fr.) is a continuation of the bowsprit forward, being run out from the extremity in a similar manner to a top-mast on a lower-mast, and serving to extend the boom of the jibs and the stay of the fore-top-gallant-mast. It is usually attached to the bowsprit by means of two large boom irons (see the article Iron-Work) or by one boom-iron, and a *cap* on the outer-end of the bowsprit; or by a cap without, and a strong lashing within, instead of a boom iron, which is generally the method of securing it in small merchant-ships. It may be drawn in upon the bowsprit as occasion requires, which is usually practised when a ship enters a harbour, where it might very soon be broke or carried away, by the vessels which are moored therein, or passing by under sail.

Flying Jib-Boom, (*bâton de clin foc*, Fr.) is a boom extended beyond the preceding by means of two boom-irons, and to the foremost end of which the tack of the flying-jib is hauled out.

To JIBE. See the article To Gybe.

JIGGER, (*palan servant à retenir le cable, &c.* Fr.) a machine, consisting of a piece of rope about five feet long, with a block at one end and a sheave at the other, and used to *hold on* the cable when it is heaved into the ship by the revolution of the *windlass*. See the article To Hold on.

The jigger is particularly useful when the cable is either slippery with mud or ooze, or when it is stiff and unwieldy; in both of which cases it is very difficult to stretch it back from the windlass by hand, which, however, is done with facility and expedition, by means of the jigger, in the following manner: the end of the rope, to which the sheave is fastened by a knot, is passed round the cable close to the windlass, and the hind part of the rope, coming over the sheave, is stretched aft by means of another rope passing through the jigger-block. As soon as the last rope is extended, the turn of the former about the cable is firmly retained in its position, by the compression of its hind part under the sheave, acting upon what may be called the neck of the jigger.

Fleet Jigger, a term used by the man who holds on the jigger, when, by its distance from the windlass, it becomes necessary to fleet or replace it in a proper state of action, for as the cable continues to be heaved into the ship, it is evident that the jigger, which is fastened on a particular part thereof, stretching it back, will be removed further aft by every turn of the windlass, and the effort of the jigger will be lessened in proportion to its distance from the windlass; accordingly, when the man gives the above notice, another at the windlass immediately fixes his hand-spike between the deck and the cable, so as to jamb the latter to the windlass, and prevent it from running out till the jigger is replaced near the windlass.

Jigger-*Tackle*, (*palan léger dont les marins se servent en plusieurs occasions*, Fr.) a light small tackle, consisting of a double and single block, and used by seamen on sundry occasions. See the article Tackle.

IMMERSION, the act of plunging any thing into water, or some other fluid.

Immersion, in astronomy, is when a star or planet is so near the sun with regard to our observations, that we cannot see it, being, as it were, enveloped and hidden in the rays of that luminary. It also denotes the beginning of an eclipse of the moon, or that moment when the moon begins to be darkened, and to enter into the shadow of the earth; and the same term is also used with regard to an eclipse of the sun, when the disc of the moon begins to cover it.

Immersion is frequently applied to the satellites of Jupiter, and especially to the first satellite; the observation whereof is of so much use for determining the longitude. The immersion of that satellite is the moment in which it appears to enter within the disc of Jupiter, and its emersion the moment when it appears to come out.

IMPORTATION, (*importation*, Fr.) in commerce, the act of importing or bringing merchandise into any country from foreign parts.

Goods may not be put on board ships of war, except for the sole use of the ship; but, after being put on board, to be subject to the same regulations as other merchandise in merchant-ships.

Goods of the growth, production, or manufacture of Africa, America, or Asia, can only be imported from thence into Great Britain, or from the parts where they are, usually have been, or can be, first shipped for transportation, in British built ships, legally navigated, of which the master and three-fourths of the mariners, at the least, are British subjects; or in ships built of the countries of which the goods are the growth, production, or manufacture, navigated by the master and three-fourths of the mariners, at the least, subjects of the states from which the same are imported.

These goods can only be so imported from each of those quarters of the globe directly; the growth, production, or manufacture of Africa cannot be imported from America, Asia. or Europe; and *vice versa*; nor can any growth or production of Africa, America, or Asia, manufactured in foreign parts, except in the country or place of which they are particularly specified; and those of the Spanish or Portuguese dominions, or plantations, which may be imported from Spain, Portugal, the Western Islands called the Azores, Madeira, or the Canaries, respectively, in British ships, but the goods of the dominions or plantations of either of those two sovereigns cannot be imported from any port or place belonging to any other sovereign.

In case it shall be found necessary to declare war against any foreign power, his Majesty may, if he shall judge it requisite, by proclamation permit trading ships and privateers to be navigated by three-fourths foreign seamen, one-fourth to be his Majesty's native and naturalized subjects of Great Britain (sudden death, hazard, and casualties of war and the seas excepted).

The production of Guernsey, Jersey, Sark, and Alderney, may be imported by the inhabitants of these places into Great Britain, free of duty, upon the oath of the growth, &c. before the magistrate of the said islands respectively, and certificates from the governors, lieutenant-governors, or deputy governors, or commanders in chief; but salt imported from any of those islands is subject to the same duty as any foreign salt imported.

Cotton wool of any sort may be imported from any port or place whatsoever in British-built ships, legally navigated, free of duty; being legally imported in any other ship, is subject to a duty of one penny for every pound weight.

Goods wrecked, or in danger of being wrecked, are excepted from all the restrictions of importation and exportation.

All goods and merchandise imported into Great Britain pay duties by gauge, measure, tale, or weight; the quantities being ascertained by the proper officers at the water side; and, as the exact quantity of any goods liable to the duty, by the gauge, measure, or weight, cannot be accurately ascertained before they are landed, it is uniformly the practice to make a fair entry of a quantity short of the whole estimated to be imported, upon which the duty is received, and after the goods have been examined, and the whole quantity charged by the officers, the duties are to be paid upon the remaining quantity: these two acts of the merchant produce from the collector, comptroller, and examiner, certificates of the payment of those duties which authorize the examining officers to deliver the goods to the merchant, and are called *prime entries* and *post entries*. All goods or merchandise which pay duty by tale, the merchant is supposed to know the exact quantity of; therefore no second, or *post entry* is permitted for any such goods after they have been unshipped with intent to land; and if, by examination, the quantity is found to be greater than the duties have been paid for before they were unshipped, the surplus is, by law and long established custom, forfeited. But the commissioners of the customs and excise have a power to relieve in all cases where it shall be made appear to their satisfaction that there was no intention of fraud.

In those instances, where the merchant is doubtful of the quality or quantity of the goods, which frequently occurs, the law has permitted him to take out *a bill of view*, or *sight*, and the officers then ascertain the quantity of the different sorts of goods, which they make a return of to the collector, upon the bill of sight, and the merchant is to declare of what country they are the growth, production, or manufacture. See the article CUSTOM. And for farther information on the laws of importation and exportation, see MAXWELL's *Spirit of Marine Law*, 1804.

With the exception of the prohibition of some foreign manufactures, the import trade of this country is probably as free as the regulations necessary to secure the payment of heavy duties on almost every article of trade will admit.

To IMPRESS, (*forcer des matelots à servir l'état sur les vaisseaux de guerre*, Fr.) in the royal navy, is to compel, by force, men who have been accustomed to the sea, or to work on any river, as also landmen, upon any emergency of the state, to serve in his Majesty's navy. For this purpose a rendezvous, under a captain and one or more lieutenants, is established in such parts of the United Kingdom where seamen chiefly resort; at which stations volunteers enter, impressed men are received, and deserters from the naval service apprehended.

The practice of thus manning his Majesty's navy is not only sanctioned by custom, but is also established by law; and, where no other adequate mode can be substituted, the law of imperious necessity must be complied with.

The very high bounties, however, which have been

given, and the Act of Parliament of the 35th Geo. III. c. 19, appointing certain proportions of men to serve in the navy, to be levied in the several counties of England and Wales, has been found to answer, without having recourse often to impressing; which, notwithstanding, must, in cases of urgency, be still resorted to, as the most certain, when every other method is found to be ineffectual. See the article Bounty.

Laws concerning the Impress Service.

By an act of the 2d and 3d Philip and Mary, c. 16, s. 3, entitled, "An Act touching watermen and bargemen upon the river Thames," it is enacted, that persons using and exercising the occupation of rowing betwixt Gravesend and Windsor, who shall willingly, voluntarily, and obstinately, withdraw, hide, or convey themselves, in the time of pressing, into secret places, and return again when such time of pressing is overpassed, to row between Gravesend and Windsor, the person so offending shall suffer imprisonment for two weeks, and be disabled from rowing on the Thames for a year and a day following.

13 Geo. II. c. 7, s. 1. Persons under eighteen, or above fifty-five years of age, and foreigners serving in merchant-ships or privateers, are exempt from being pressed.

Sect. 2. Landmen, using the sea, are exempted from being pressed for two years; to be computed from the time of their first going to sea.

Persons not having used the sea, binding themselves apprentices to the sea service, to be exempt from being pressed for three years.

Sect. 3. The Admiralty to grant protections accordingly, without fee.

13 Geo. II. c. 17, s. 2. No persons of the age of eighteen years, who shall have been in any sea service before they bound themselves apprentices, shall be exempt from his Majesty's service.

2 & 3 Anne, c. 6, s. 17. When such apprentices are impressed, or have entered voluntarily into the king's service, the owners or masters of such apprentices shall be entitled to able seamen's wages for such of them as shall be found duly qualified for the same, notwithstanding their indentures of apprenticeship.

4 Anne, c. 19, s. 18. Watermen on the Thames, summoned by the rulers of their company to serve on board the fleet, and not appearing, to be imprisoned for one month, and disabled from working on the Thames for two years.

6 Anne, c. 31, s. 2. Watermen on the Thames, belonging to insurance companies, are protected from being impressed; not exceeding thirty for each office, and their names being registered at the Admiralty-office.

9 Anne, c. 26, s. 1. Fishermen on the Thames, summoned by their court of assistants to serve in the navy, and neglecting to appear, shall be sent on board, and disabled from fishing on the Thames for two years.

5 Eliz. c. 5, s. 43. No fisherman to be taken to serve at sea, unless the commission be first brought to two justices near, who are to choose out the number of men therein specified.

Every ship in the coal trade has the following number of persons protected; viz. *two* able seamen (such as the master shall nominate) for every ship of 100 tons burden; and *one* for every 50 tons for every ship of 100 tons and upwards; and any officer who presumes to impress any of the above, shall forfeit to the master or owner of such vessel 10*l.* for every man so impressed; and such officer shall be incapable of holding any place, office, or employment, in any of his Majesty's ships of war. (6 and 7 Will. III. c. 11, s. 12.) (19 Geo. II. c. 30, s. 7.)

In cases of threatened or actual invasion of a colony, or other unforeseen necessity, commanders of ships of war there may, with consent of the governor and council of the colony, press seamen, to make up their usual complement.

Such seamen to be discharged, if they require it, when the service is performed. Sect. 7

Captains refusing their discharge to pay the penalty of 50*l.*

Persons described in the Act 2 Geo. III. c. 15, to be protected by the Admiralty.

1. Masters of fishing vessels or boats, who either themselves or their owners have, or, within six months before applying for a protection shall have had, one apprentice, or more, under sixteen years of age, bound for five years, and employed in fishing on the coasts, or in any of the rivers.

2. All such apprentices, not exceeding four to every master or owner of any fishing vessel, of 30 tons burden or upwards, and two to every vessel or boat under 30 tons, during the time of their apprenticeship, and until they are of the age of twenty years, they continuing for the time in the business of fishing only.

3. One mariner, besides the master and apprentices, to every fishing vessel of ten tons or upwards, employed on the sea coast, during his continuance in such service of fishing.

4. Any landman, entering and employed on board such vessel, for two years from his first going to sea, and to the end of the voyage in which he is then engaged, if he continue so long in such service.

No master or owner of such fishing vessels or boats above described, is, knowingly, to harbour any deserter from his Majesty's service, under penalty of 20*l.*

No harpooner, line manager, or boat-steerer, belonging to any vessel fitted out for the Southern Whale Fishery, shall be impressed from the said service, so long as he belongs to any ship or vessel employed in the same.

It was held by the late Mr. Justice Buller, that a licenced ferryman is likewise exempt from being impressed. 5 *Term Rep.* 277.

IN, (*serré ou ferlé, en parlant des voiles*, Fr.) the state of any of a ship's sails when they are furled or stowed. It is used in opposition to *out*, which implies that they are *set*, or extended to assist the ship's course.

In, is also used as an order to shorten sail, where the word take is understood, and is usually applied to the square upper-sails; as "In top-gallant-sails!" (*serre les perroquets!* Fr.)

In-and-Out, in ship building, is a term sometimes used for the scantling of the timbers, the moulding way, but particularly for those bolts in the hanging and lodging knees that are drove through the ship's side, which are called *in and out* bolts, &c.

INBOARD, in ship buildi g, a term used to signify any thing that is within the ship; as the inboard works, &c.

INCH, or Innis, a general name for an island.

INCLINATION, is a word frequently used by mathematicians, and signifies the mutual approach, tendency, or leaning, of two lines or planes towards each other, so as to make an angle.

Inclination *of the Axis of the Earth*, is the angle which it makes with the plane of the ecliptic; or the angle contained between the planes of the equator and ecliptic.

Inclination *of a Planet*, is an arc of the circle of inclination, comprehended between the ecliptic and the plane of a planet in its orbit. The greatest inclination of Saturn, according to Kepler, is 2° 32′; of Jupiter, 1° 20′; of Mars, 1° 50′ 30″; of Venus, 3° 22′; of Mercury, 6° 54′.

INDEX of the Globe, the little style or gnomon which, being fixed on the pole of the globe, and turning round with it, points out the hours upon the hour-circle; or on globes, where the horary-circle is represented on the equator, the *index* is made to slide on a brass wire, fixed directly over the equator. See Adams's *Globes*.

INGRESS, in astronomy, implies the sun's entering the first scruple of one of the four cardinal signs, especially Aries.

IN-HAULER, a rope employed to haul in the jib-boom.

INNER-POST, in ship building, is a piece brought on at the fore-side of the main-post, and generally continued as high as the wing-transom, to seat the other transoms upon. The *inner-post* is a great security for the ends of the planks, and, as the main-post is seldom sufficient before the rabbet for that purpose, it likewise greatly strengthens the ship in that part; for if the ship should take ground, especially by swinging, so as to break the stern-post, it might be expected that much of the after dead-wood would be carried away, because the joints are nearly in the same direction as the planks of the bottom.

INSURANCE, (*assurance*, Fr.) in marine affairs, is a contract of Indemnity, whereby the party, in consideration of a stipulated sum, undertakes to indemnify the other against certain specific perils or risks to which he is exposed, or against the occurrence of such events. The party who takes on himself the risk, is called the *insurer;* the party protected by the insurance is called the *insured*; the sum paid to the insurer, as the price of his risk, is called the *premium;* and the written instrument, in which the contract is set forth and reduced into form, is called *a policy of insurance.*

Marine Insurance, is made for the protection of persons having an interest in ships or goods on board, from the loss or damage which may happen from the perils of the sea, during a certain voyage, or for a fixed period of time.

In this country all persons, whether British subjects or aliens, may, in general, be insured; the only exception is in the case of an alien enemy. He cannot maintain an action on a policy on goods, though they were shipped before the war commenced; nor can an agent of such insured maintain the action, though he be a creditor of the insured, for more than the sum insured.

The statute 6 George I, c. 18, anno 1720, authorised the king to grant charters to two distinct companies or corporations, called the Royal Exchange Assurance, and London Assurance, for the insurance of ships, goods, and merchandises at sea, or going to sea, and for lending money on bottomry. They are invested with all the powers usually granted to corporations, and the privilege of purchasing lands to the amount of one thousand pounds per annum each, to provide a sufficient capital to insure all demands on their policies. All other companies are restrained from insuring ships and goods at sea, or lending money on bottomry. And all policies made by any other corporation and any copartnerships shall be void, and the sums under-written forfeited, and all bottomry bonds deemed usurious: but the right of individual insurers continues as before the act. Contracts made in derogation of the rights of the insurance companies are illegal and void.

Ships, freight, goods, and merchandises, &c. are the proper subjects of marine insurance, and there are certain articles which, from motives of public policy, cannot be legally insured in this country, and others, which can only be insured under partial restrictions.

Insurance being a contract of indemnity from loss or damage arising upon an uncertain event, there cannot be an indemnity without a loss, nor a loss without an interest; a policy, therefore, without interest, is not an insurance, but a mere wager. Different persons, having each a qualified property in goods, may insure them to the full value. A reasonable expectation of profit, on a well-founded expectation of a future interest in the thing insured, is an insurable interest. Every fact and circumstance relating to the contract of insurance must be stated, with the most scrupulous regard to truth. The voyage insured must, therefore, be truly and accurately described in the policy; namely, the time and place at which the risk is to begin; the place of the ship's departure, the place of her destination and the time when the risk shall end; whether on goods or on the ship. If a blank be left for the place either of the ship's departure or destination, the policy will be void for the uncertainty.

There are five things essential to a sailing with convoy:—

1. It must be with the regular convoy appointed by government.
2. It must be from the place of rendezvous appointed by government.
3. It must be a convoy for the voyage.
4. The ship insured must have sailing instructions.
5. She must depart and continue with the convoy till the end of the voyage, unless separated by necessity.

In every insurance, whether of ship or goods, there is an implied warrantry of the sea-worthiness of the ship, that is to say, that she shall be tight, staunch, and strong, properly manned, and provided with all necessary stores, and in every other respect fit for the voyage.

Losses by perils of the sea are generally understood to be such accidents or misfortunes as proceed from stress of weather, winds, lightning, tempest, rocks, &c.

If a ship be not heard of for a reasonable time, she shall be presumed to have foundered at sea, and the insured has a right to recover as such, from the underwriter.

A loss by fire, which is merely accidental, and not imputable to the master or mariners, is undoubtedly within the policy. If a ship be burnt by order of the state where she happens to be, to prevent infection, this also has been held a loss within the policy.

If a ship be attacked by an enemy, and the captain, unable to defend her, leave and set fire to her to prevent her from falling into the enemy's hands, the insurer is said to be liable.

To INVERT, (*changer bout pour bout, &c.* Fr. *inverto*, Lat.) to place the last first.

To JOIN, (*joindre*, Fr.) to unite any two pieces of timber into one.

JOINT, in ship building, a term signifying the place where any two pieces are joined, but is more particularly meant to express those lines which are laid down in the mould loft, for the purpose of making the moulds for the timbers, as those lines are the shape of the body between every two timbers, which consequently is the joint.

JOLLY-BOAT. See the article Boat.

JONK, or Junk, (*jonque, sorte de bâtiment Chinois*, Fr.) a kind of small ship, very common in the East Indies: these vessels are about the bigness of our fly-boats, and differ in the form of their building, according to the different methods of naval architecture used by the nations to which they belong. Their sails are frequently made of mats, and their anchors are made of wood. The Chinese junk is a large flat-bottomed vessel, from 100 to 500 tons burden. It has three masts, and a short bowsprit placed on the starboard-bow. The masts are supported by two or three shrouds, which, at times, are all carried on the windward-side.

On the fore and main-mast is hoisted a sort of lug-sail made of cane or bamboo. These sails are confined by iron travellers that encircle the mast, and fix to bamboos at several divisions on the sail. The sail is kept to the wind by two ropes, fastened to wood stirrups, fixed to the foot of the sail, and lead to the mast-head. The lee-part of the sail is hauled aft, by a rope that branches into short legs, that are made fast to each fold of the sail. On the mizen-mast is a gaff-sail, made of coarse cotton; a top-sail, made of the same, is carried on the main-mast; also a jib and sprit-sail, that are set on the bowsprit.

JOURNAL, (*journal nautique*, Fr.) in navigation, a daily register of the ship's course and distance, kept by the pilot and others, wherein notice is taken of every thing that happens to the ship from day to day, with regard to the winds, the rhumbs, the rake, the soundings, &c. in order to enable him to adjust the reckoning, and determine the place where the ship is, together with a general account of whatever is material to be remarked in the period of a ship's voyage, such as the shifting, reducing, or enlarging the quantity of sail; the condition of the ship and her crew; together with the discovery of other ships or fleets, land, shoals, breakers, soundings, &c.

In sea journals, the day, or twenty-four hours, terminates at noon, because the errors of the dead-reckoning are at that period generally corrected by a solar observation. The first twelve hours, from noon to midnight, are marked with P.M. signifying after mid-day; and the second twelve hours from midnight to noon, are marked with A.M. signifying after midnight; so that the ship account is twelve hours earlier than the shore account of time. There are various ways of keeping journals, according to the different notions of mariners concerning the articles that are to be entered. Some keep such a journal as is only an abstract of each day's transactions, specifying the weather, what ships or lands were seen, accidents on board, the latitude, longitude, the meridional distance, course, and run. These particulars are to be drawn from the ship's log-book, or from that kept by the pilot himself. Others keep only one account, including the log-book, and all the work of each day, with the deductions drawn from it.

IRON-GARTERS. See the article Bilboes.

Iron-Jack, in rope-making, a machine sometimes used instead of the table-wheel, or back-frame wheel. It differs from the latter by having an iron wheel with cogs, which work in the whirls; it has iron cogs likewise.

Iron-Sick, (*se dit d'un vieux vaisseau dont les ferrures sont usées par la rouille, et jouent*, Fr.) a term used to signify the state of old vessels when the iron work becomes loose in the timbers or planking.

Iron-Work, (*ferrures*, Fr.) a general name for all the pieces of iron, of whatever figure or size, which are used in the construction and equipment of a ship: as bolts, boom-irons, nails, spikes, chains, and chain-plates, block-straps, cranks, pintles, braces, &c.; which articles see.

ISLAND, or Isle, (*île*, Fr.) is a quantity of land entirely surrounded with water.

Several naturalists are of opinion that islands were formed at the deluge: others think they have been rent and separated from the continent by violent storms, inundations, and earthquakes. These last have observed that the East and West Indies, which abound in islands more than any other part of the world, are likewise more annoyed with earthquakes, tempests, lightnings, volcanoes, &c. than any other part.

Island *of Ice*, (*banc de glace*, Fr.) a name given to a great quantity of ice collected into one huge solid mass, and floating about upon the seas, near or within the polar circle.

Many of these fluctuating islands are met with on the coasts of Spitzbergen, to the great danger of the shipping employed in the Greenland fishery.

JUNK, (*bouts de cables*, Fr.) any remnants or pieces of old cable, which is usually cut into small portions for the purpose of making points, mats, gaskets, sennit, &c. See the article Points, &c.

Junk. See the article Jonk.

JUNO, in astronomy, a new planet, situated between the orbits of Mars and Jupiter; it was discovered by Mr. Harding, at the observatory of Lilienthal, near Bremen, on the evening of the 1st of September 1804.

It is of a reddish colour, and free from that nebulosity which surrounds Pallas. Its diameter and mean distance are less than those of the other new planets, and it is distinguished from all other planets, by the great eccentricity of its orbit; and the effect of this is so extremely sensible, that it passes over that half of its orbit which is bisected by its perihelion, in half the time that it employs in describing the other half, which is farther from the sun. From the same cause, its greatest distance from the sun is double the least distance, the difference between the two distances being about 127 millions of miles.

Though there is no nebulous appearance around the planet Juno, yet it appears from the observations of Schroeter, that it must have an atmosphere more dense than that of the old planets of the system. A very remarkable variation in the brilliancy of this planet has been observed by this astronomer. He attributes it chiefly to changes that are going on in its atmosphere, though he thinks it not improbable that these changes may arise from a diurnal rotation performed in 27 hours.

The following elements were calculated by Burckhardt:—

	Years.	Days.
Annual revolution	4	128
Mean longitude, 31st December, 1804, noon	$1^s\ 12° \ 17' \ 23''$	
Place of ascending node . . .	$5^s\ 21° \ 6' \ 0''$	
Place of perihelion in 1805 . .	$1^s\ 29° \ 49' \ 33''$	
Eccentricity		0.25096
Inclination of orbit	$21° \ 0'$	
Mean distance from the sun in English miles		253,000,000
Diameter in English miles, according to Schroeter . . .		1425
Apparent mean diameter, as seen from the earth, according to Schroeter		3″057

For farther particulars on the subject of this article see SCHROETER, *Phil. Trans.* for the year 1807; LA LANDE, *Journal de Physique*, *Brum.* An. 12.

JUPITER, ♃, in astronomy, one of the superior planets, remarkable for its brightness, being the brightest of all, except Venus. Jupiter is situated between Mars and Saturn, being the fifth in order of the primary planets from the sun. His diameter is 89,170 miles. He turns round his axis in 9 hours and 56 minutes, and performs his periodical revolution in 11 years, 315 days, 14 hours, and 39 minutes, at the distance of 490 millions of miles.

The figure of Jupiter is evidently an oblate spheroid, the equatorial being to the polar diameter as 14 to 13. This result was obtained from the accurate observations of Dr. Herschel; and it is a remarkable coincidence between theory and observation, that, from the influence of the equatorial parts of Jupiter upon the motion of the nodes of his satellites, La Place has found the proportion between his equatorial and polar diameters to be as 10,000,000 to 9286992; a result which differs only a little from the ratio of 14 to 13, and which should be regarded as the more accurate of the two. According to Cassini, the difference of Jupiter's diameters is $\frac{1}{13}$; Pound made it $\frac{1}{13.37}$; Short $\frac{1}{14}$; and Newton $\frac{1}{16}$.

When we look at Jupiter through a good telescope, we perceive several belts or bands extending across his disc, in lines parallel to his equator. These appearances were first observed by two Jesuits, Zuppi and Bartoli. They were afterwards examined in 1633, by Fontana, Rheita, Riccioli, Grimaldi, and Campani; the latter of whom, on the 1st of July 1664, perceived four dark belts, and two white ones. These belts are variable, both in number, distance, and position. Sometimes seven or eight belts have been observed; and, on the 28th of May 1780, Dr. Herschel perceived the whole disc of Jupiter covered with small curved belts, or rather lines, that were not continuous across his disc. The parallel belts, however, are most common, and, in clear weather, may be seen by a good achromatic telescope, with a magnifying power of 40.

For farther particulars respecting the cause of the belts and spots of Jupiter, see the article BELT.

On the Satellites of Jupiter.

Jupiter has four satellites, which constantly accompany him, and which were discovered by Galileo. Their relative situation is continually changing; yet they may be seen with a telescope which magnifies 30 times. The third and fourth, indeed, have been sometimes seen with the naked eye, but it is only when the air is uncommonly pure that we can expect to be gratified with such a sight. Though a century and a half has scarcely elapsed since their discovery, yet, from the extreme shortness of their revolutions, they present to an observer those great and interesting changes, which are not effected in the course of many centuries in the planetary system.

These small bodies have been observed by astronomers with great assiduity during the last century, and the tables of their motions have been brought to a degree of perfection, which the most sanguine expectations of astronomers could never have anticipated.

They oscillate on both sides of Jupiter, and their relative rank is determined by the length of these oscillations. The one in which the oscillation is shortest is called the first satellite, and so on.

These satellites are analogous to our moon. They are all supposed to move in ellipses; though the eccentricities of all of them are too small to be measured, excepting that of the fourth; and even this amounts to no more than 0.007 of its mean distance from the primary.

The orbits of these planets were thought, by Galileo, to be in the same plane with that of their primary; but M. Cassini has found that their orbits make a small angle with it; and, as he did not find any difference in the place of their nodes, he concluded that they were all in the same place, and that their ascending nodes were in the middle of Aquarius.

After observing them for more than thirty-six years, he found their greatest latitude, or deviation from the plane of Jupiter's orbit, to be 2° 55″.

The *first* of these satellites revolves at the distance of 5.697 of Jupiter's semi-diameters, or 1′ 51″, as measured by proper instruments; its periodical time is $1^d\ 18^h\ 27'\ 34''$. The *second* satellite revolves at the distance of 9.017 semi-diameters, or 2′ 56″, in $3^d\ 13^h\ 13'\ 43''$; the *third*, at the distance of 14.384 semi diameters, or 4′ 42″, in $7^d\ 3^h\ 42'\ 36''$; and the *fourth*, at the distance of 25.266, or 8′ 16″, in $16^d\ 16^h\ 32'\ 09''$. Since the time of Cassini it has been found that the nodes of Jupiter's satellites are not in the same place; and from the different points of view in which we have an opportunity of observing them from the earth, we see them sometimes apparently moving in straight lines, and at other times in elliptic curves. All of them, by reason of their immense distance, seem to keep near their primary, and their apparent motion is a kind of oscillation, going alternately from their greatest distance on one side to the greatest distance on the other, sometimes in a straight line, and sometimes in an elliptic curve.

The occultations of these satellites must frequently happen by their going behind their primary, or by coming in betwixt us and it. The former takes place when they proceed towards the middle of their upper semi-circle; the latter, when they pass through the same part of their inferior semi-circle. Occultations of the former kind happen to the first and second satellites; at every revolution the third very rarely escapes an occultation; but the fourth more frequently, by reason of its greater distance.

The beginnings and endings of the eclipses are easily seen by a telescope, when the earth is in a proper situation with regard to Jupiter and the sun; but when this or any other planet is in conjunction with the sun, the superior brightness of that luminary renders both it and the satellites invisible. From the time of the first appearing after a conjunction until near the opposition, only the immersions of the satellites into his shadow, or the beginnings of the eclipses, are visible; at the opposition, only the occultations of the satellites, by going behind or coming before their primary, are observable; and from the opposition to the conjunction, only the immersions, or end of the eclipses are to be seen.

When Jupiter is deprived of the light of the first and second satellites, or of the first and third, the remaining one of the three first cannot possibly be eclipsed at the same time, but is in such a point of its orbit as to give considerable light to the planet. The simultaneous eclipse of the second and third satellites forms an exception to this remark, for, at the same instant, the first satellite has its dark side turned to the planet.

The position of Jupiter's satellites, with respect to the planet, and the figure of their orbits, as seen from the earth, may be easily found, for any given time, from tables that are published in the *Tables de Berlin*. Their relative position is given in the *Nautical Almanac* for every day when Jupiter is at such a distance from the sun that the satellites can be readily seen; or it may be easily found, by an instrument called a *jovilable*, which was first constructed by Flamstead, and afterwards by Cassini. See HERSCHEL, *Phil. Trans* 1797; NICHOLSON'S *British Encyclopedia*; and the EDINBURGH *Encyclopedia*.

JURY-MAST, (*mat de fortune*, *ou mât de rechange*, Fr.) a temporary or occasional mast, erected in a ship, in the place of one which has been lost in a storm, or otherwise, and to which a smaller yard, ropes, and sails are fixed.

A *jury-mast* is sometimes erected in a new ship, to navigate her down a river or to a neighbouring port, where her proper masts are prepared for her.

When masts have been damaged or broken, either in action or by stress of weather, a ready method of supplying the deficiency in such a manner as to enable ships to prosecute their voyage with the usual quantity of sail, is certainly an object of considerable importance to a nation like this, which reigns the undoubted queen of the ocean.

An improvement on *jury-masts* to this effect was proposed by Captain William Bolton, of the royal navy, in a letter to Dr. Taylor, secretary of the Society of Arts, dated October 31, 1807, for which improvement the said society awarded to him their silver medal, in May 1808.

In the captain's method of fitting up *jury-masts*, "the main-mast being broken about one-third of its length above the deck, proper partners are secured on the deck, in which a hand-mast and spare main-top-mast are fixed on each side of the broken main-mast, and secured thereto by two spare caps, mortised on a square made in its centre.

A strengthening cap, moveable on these additional masts, connects them; and the upper parts of these masts are secured firmly by trestle-trees in the main-top. The square fore-top-mast passes through a cap made of strong plank, into the square holes of which the heads of the two temporary masts above-mentioned are inserted, and the heel of the top-mast is fidded on the trestle-trees, or top, as in common, and the mast rigged as usually. The object of the strengthening cap is to steady the spars, and also serve to fid the top-mast on. The fore- top-mast being thus adjusted, the cross-trees and top-gallant-mast are mounted upon it, in the usual manner.' '

The captain also observes, that the two caps are the only things necessary to be made expressly for the purpose, as the other articles are usually ready on board.

Some doubts having arisen relative to the practicability of rigging the jury-masts invented by Captain Bolton, he therefore transmitted to the Society of Arts, in May 1809, a model, completely rigged, in order to remove these doubts, and for which they also voted to him their gold medal. From this it appears, in consequence of the angle being more obtuse, that the rigging so placed will give greater security, supersede the necessity of cathur pinning, and suffer the lower yards to be braced sharp up.

Reference to the Drawing of Captain Bolton's rigged Jury-Mast.

Plate XII. fig. 9, where A A represent the partners or pieces of timber which are bolted to the quarter-deck, for the mast to rest upon. B is the stump of the lower-mast, which is cut square at the top, and of the same size as the head of the mast originally was; upon this square the main and spare lower caps *a a* are fixed; two mortises must be cut in the partners A A, to receive squares made at the lower ends of the two temporary masts D D, which are supported by the caps *a a*, one of them is a spare main-top-mast, the other a hand-mast; these two support the main-top E, additional squares being made on the trestle-trees to receive each of them. *b* is a cap, shown in fig. 10, made of four-inch plank, doubled for the purpose, and fitted upon the heads of the masts D D, for a fore top-mast F F, the heel of which rests in a mortise made in the stump of the lower mast; it is also steadied by a double cap G, separately shown in fig. 11, on which it fids finally on the top. The top-gallant-mast H, is fixed to the mast F, by the top and cap in the usual manner. The figures 10 and 11 show the caps separated from the masts, and are the only things necessary to be made for the purpose; and the object of the cap, fig. 10, is to steady and prevent any wringing of the lower *jury-mast*, and to fid the top-mast whenever it is reefed.

Fig. 9. Fig. 10. Fig. 11.

JUTTY-HEAD. See the article JETTY-HEAD.

K.

KECKLING, (*fourrure de cable*, Fr.) the art of winding old rope, &c. about a cable, to preserve its surface from being fretted, when it rubs against the ship's bow, or *fore-foot*; but more particularly it implies the winding of iron chains round the cable to defend it from the friction of a rocky bottom, or from ice, &c.

KEDGE, or KEDGE ANCHOR, (*petite ancre de touée*, Fr.) a small anchor used to keep a ship steady and clear from her bower anchor, while she rides in a harbour or river, particularly at the turn of the tide, when she might otherwise drive over her principal anchor, and entangle the stock or flooks with her slack cable, so as to loosen it from the ground.

The *kedge anchors* are also used to transport a ship, or remove her from one part of a harbour to another, being carried out from her in the long boat, and let go by means of ropes fastened to these anchors. They are also generally furnished with an iron stock, which is easily displaced for the convenience of stowing them. See the articles ANCHOR and WARP.

Kedges, or small anchors, are made in proportion to the large anchors; their stocks were formerly of iron, but are now mostly of wood; if iron, they are to be the same in length as if wood, and two thirds of the trend in diameter in the middle. Near the middle of the stock is a shoulder, that bears against the hole on one side the square of the shank, with a forelock hole in the stock on the other side of the square of the shank, for a forelock to go through to confine the stock in its place. It is tapered from the middle each way, and when put through the shank, one end is bent, to prevent its falling out, and to keep it close, when brought to the side of the shank, for less stowage.

To KEDGE, (*cajoler*, Fr.) to bring a ship up or down a narrow river by means of the tide, though the wind be contrary. To do this, the fore-course, or fore-top-sail and mizen are set, that they may flat her about; and if she happen to come too near the shore, the kedge-anchor is let fall, with a hawser fastened to it from the ship, in order to turn her head about, which work is called kedging.

KEDGE-*Rope*, a rope which belongs to the kedge-anchor, and employed to restrain the vessel from approaching her bower-anchor.

KEEL, (*quille*, Fr.) the principal piece of timber in a ship, which is usually first laid on the blocks in building.

By comparing the carcase of a ship to the skeleton of a human body, the keel appears as the back-bone, and the timbers as the ribs. Accordingly, the keel supports and unites the whole fabric, since the stem and stern-post, which are elevated on its ends, are, in some measure, a continuation of the keel, and serve to connect and enclose the extremities of the sides by transoms; as the keel forms and unites the bottom by timbers.

The keel is generally composed of several thick pieces placed lengthways, which, after being scarfed together, are bolted, and clinched upon the side.

False KEEL, (*fausse quille*, Fr.) a strong thick piece of timber bolted to the bottom of the keel, which is also very useful in preserving the lower-side. The *false-keel* is provided for the purpose of preserving the under part of the main keel.

In large ships of war the *false-keel* is composed of two pieces, called the upper and lower false-keels. See MIDSHIP-SECTION, plate VI. fig. 1.

The lowest plank in a ship's bottom, called the *garboard-streak*, has its inner edge let into a groove, or channel, cut longitudinally on the side of the keel; the depth of this channel is therefore regulated by the thickness of the garboard-streak.

An invention of a keel has lately been given to the world by a gentleman of the name of Steuard, and he gives the following account of it:—The use of this invention is, to preserve life and property, by making a ship much stronger than on the common construction: so that if she, by stress of weather, be driven on shore, she will keep upright, and be better able to resist danger, which is effected by laying a keel on each side the middle line, distant from it, including the thickness thereof, one sixth of the average breadth at loading draft of water, or a little more or less, according to the fulness and sharpness of the ship.

The patentee observes, that a ship thus secured from straining is more likely not to work her plank loose, which often occasions foundering; and if, by stress of weather, she is unavoidably driven on shore, she has advantage of ships on the common construction, as they, lying on the edge, easily roll by the agitation of the sea, and strike against the ground, and when left by the tide, are often bilged by the overhanging weight; but ships of this construction having one-third of their breadth to stand on, will neither roll, nor be likely to bilge; and if, from the declivity of the ground another ship should upset, this will not, having two-thirds of the weight to counteract it. Therefore, as the preservation of lives and property at sea are of great importance, the above construction of ships is recommended to all it may concern. The patentee requires only a small compensation per ton to allow any ship-builder the benefit of his invention.

Keel, (*sorte de bateau à fond plat, servant à transporter le charbon*, Fr.) is also a name given to a low flat-bottomed vessel, used in the river Tyne to bring the coals down from Newcastle, for loading the colliers for transportation: hence a collier is said to carry so many *keels* of coal.

Upon an Even Keel, (*sans différence, égal tirant d'eau*, Fr.) the position of a ship when her keel is parallel to the plane of the horizon, so that she is equally deep in the water at head and stern.

Keel-*Hauling*, (*la cale*, Fr.) a punishment inflicted for various offences in the Dutch navy: hence,

To Keel-*Haul*, (*donner la cale*, Fr.) is to suspend the culprit by a rope from one yard-arm, with a weight of lead or iron upon his legs, to sink him to a competent depth, and having another rope fastened to him, leading under the ship's bottom, and through a block at its opposite yard-arm; he is then repeatedly and suddenly let fall into the sea, where, passing under the ship's bottom, he is hoisted up, on the opposite side of the vessel, to the other yard-arm. As this extraordinary sentence is executed with a serenity of temper peculiar to the Dutch, the culprit is allowed sufficient intervals to recover the sense of pain, of which, indeed, he is frequently deprived during the operation. In truth, a temporary insensibility to his sufferings ought by no means to be construed into a disrespect for his judges, when we consider that this punishment is supposed to have peculiar propriety in the depth of winter, whilst the flakes of ice are floating on the stream; and that it is continued till the culprit is almost suffocated for want of air, benumbed with the cold of the water, or stunned with the blows his head receives by striking the ship's bottom.

Keel-*Rope*, a hair rope, running between the kelson and the keel of a ship, to clear the limber-holes, when they are choked up with ballast, &c.

KEELERS, (*barrils, ou caisses à étoupes*, Fr.) are small tubs, which hold stuff for the calking of ships.

KEELSON, or Kelson, (*carlingue du fond du vaisseau*, Fr.) in ship-building, a piece of timber, forming the interior, or counter-part of the keel, as it is laid upon the middle of the floor-timbers, immediately over the keel, and, like it, composed of several pieces scarfed together. The scarfs are, if possible, disposed clear of the main and fore-mast, and likewise the main-hatch, as the scarf may be injured by accident in lowering goods and heavy matters. As every floor-timber is bolted through the keel, and every other through the kelson and keel together, it requires that the middle of the scarfs of the kelson be disposed over a floor-timber, that is designed to be bolted through the kelson.

In order to fit with more security upon the floor-timbers and crotches, it is notched about an inch and a half deep, opposite to each of those pieces, and thereby scored down upon them to that depth, where it is secured by copper-bolts. The pieces of which it is formed are about two-thirds of the breadth and thickness of those of the keel.

To KEEP, (*tenir*, Fr.) a term used on several occasions in navigation; as,

To Keep *the Land Aboard*, (*serrer la terre*, Fr.) is to keep within sight of land as much as possible.

To Keep *the Luff*, or *To* Keep *the Wind*, (*tenir le plus près*, Fr.) to continue close to the wind, *i. e.* to sail with a course inclined to the direction of the wind, as much as possible, without deviating to leeward. See the article Close-hauled.

To Keep *off from the Shore*, (*alarguer*, Fr.) to keep at a proper distance from the shore, ship, &c. See the article Offing.

KEEPER of a Boat, (*canotier qui garde un canot*, Fr.) is one of the boat's crew who remains as a sentinel, in his turn, to take care of the boat and her contents, when she lies by the shore, or along-side of a ship, or when she is towed astern of her.

KENTLEDGE, (*saumons de fer pour lest*, Fr.) a term used to signify pigs of iron for ballast, which are laid upon the floor, near the kelson, fore and aft.

KERF, (*trait de scie dans une pièce de bois*, Fr.) the sawn-away slit in any piece of timber.

KERSEY, (*serge croisée*, Fr.) a course stuff, used on many occasions in a ship, such as in *boxing of the stem*, and lining the ports, for the purpose of excluding the water, also to cover the man-ropes, &c.

KETCH. See the article Bomb-Ketch.

Ketches, are used as bomb-vessels, and are, therefore, furnished with all the apparatus necessary for a vigorous bombardment.

Ketches are likewise used as yachts, for conveying princes of the blood, ambassadors, or other great personages, from one place to another. See the article Yacht.

KEVELS, or Ranges, (*taquets à cœur pour amarrer les manœuvres contre le bord*, Fr.) a frame composed of two pieces of timber, whose lower end rests in a sort of step or foot, nailed to the ship's side, from whence the upper ends branch outward into arms or horns, serving to belay the tacks and sheets or great ropes by which the bottoms of the main-sail and fore-sail are extended. These are represented by fig. 12, plate XII. Fig. 12.

Kevel-*Heads*, (*bittons, ou taquets, formés des bouts des alonges de revers du vaisseau, pour amarrer les manœuvres*, Fr.) the ends of the top timbers, which, rising above the gun-wale, serve to belay the ropes, or take a round turn to hold on.

KEY, or Quay, (*quai, espace bordé de maisons le long d'une rivière*, Fr.) a long wharf by the side of a harbour or river, usually built of stone, and having several store-houses for the convenience of lading and discharging merchant-ships. It is, accordingly, furnished with posts and rings, whereby ships may be secured; as also with cranes, capstans, and other engines, to load the vessels which lie alongside.

Keys, (*cayes, roches sous l'eau*, Fr.) are also certain sunken rocks, lying near the surface of the water, particularly in the West Indies.

Keys *with Chains and Staples*, are those fixed on the side-pieces of a carriage or mortar-bed, to fasten the cap-squares, by passing through the eyes of the eye-bolts.

Keys *of Gun-Carriages*, are those denominated Fore-Locks and Spring Fore-Locks; which articles see.

KINK, (*coque dans les cordages*, Fr.) a twist or turn in any cable or other rope, occasioned by its being very stiff or close-laid, or by being drawn too hastily out of the roll or tier in which it lay coiled.

KIT, a composition, made of resin 9lb. pitch 6lb. and tallow 1 lb. used for the last covering of carcasses. To apply it, first break it into small pieces, and put it into an iron pot over a fire, where it must be kept in agitation until it be thoroughly dissolved. When very hot and completely liquid it is fit for use.

KNAPSACK, (*havresac*, Fr. from *snapsack*, Sax.) a rough leather or canvas bag, which is strapped to an infantry soldier's or marine's back when he marches, and which contains his necessaries. Square painted *knapsacks* are considered to be most convenient. They are generally made with a division to hold the shoes, black-balls, and brushes, apart from the linen.

Knapsack is said to have been originally so called from the circumstance of a soldier making use of a sack which had been full of corn, &c. In those days there were no roads, and every thing was carried on packhorses. When the soldiers reposed, they hung up the empty sacks and slept on them. The word should be napsack, from napping, &c. to slumber. The army was supplied by packhorses, and all things were in sacks, so that every soldier had his sack. Such is the account given by a very respectable author; but others will have it that knapsack comes from the Saxon word *snapsack*, a bag to carry food. See JAMES's *Military Dictionary*.

KNAVE-LINE. See the article NAVEL-LINE.

KNEES, (*courbes*, Fr.) are crooked pieces of timber, having two branches or arms, and generally used to connect the beams of a ship with her sides or timbers.

The branches of the knees form an angle of greater or smaller extent, according to the mutual situation of the pieces which they are designed to unite. One branch is securely bolted to one of the deck-beams, and the other is, in the same manner, strongly attached to a corresponding timber in the ship's side.

Besides the great utility of knees in connecting the beams and timbers into one compact frame, they contribute greatly to the strength and solidity of the ship, in the different parts of her frame to which they are bolted, and thereby enable her, with greater firmness, to resist the effects of a turbulent sea.

Dagger-KNEES, (*courbes un peu obliques*, Fr.) are those which are fixed rather obliquely, to avoid, as before-mentioned, an adjacent gun-port, or where, from the vicinity of the next beam, there is not space for the arms of two lodging-knees.

KNEE *of the Head*, or *Cut-Water*, (*la guibre d'un vaisseau, comprenant la gorgère, la flèche et le taille-mer*, Fr.) a large flat piece of timber, fixed edgeways, upon the foremost part of a ship's stem, and supporting the ornamental figure or image, placed under the bowsprit. See the article HEAD.

The knee of the head, which may properly be defined a continuation of the stem, as being prolonged from the stem forwards, is extremely broad at the upper-part. It is fayed and bolted to the stem, let into the head, and secured to the ship's *bows* by strong knees fixed horizontally upon both, and called the *cheeks of the head*. The heel of it is scarfed to the upper end of the fore-foot, and it is fastened to the stem above by a lacing.

Besides supporting the figure of the head, this piece is otherwise useful, as serving to secure the boom, or *bumkin*, by which the fore-tack is extended to windward; and, by its great breadth, preventing the ship from falling to leeward, when *close-hauled*, so much as she would otherwise do. It also affords a greater security to the bowsprit, by increasing the angle of the bob-stay, so as to make it act more perpendicularly on the bowsprit.

The knee of the head is a phrase peculiar to shipwrights, but by seamen it is called the CUT-WATER, which article see.

Iron KNEES, (*courbes de fer*, Fr.) are frequently used in all the various applications instead of wooden ones, on account of the scarcity of timber fit for the purpose.

Iron-Plate KNEES, (*courbes verticales, ou obliques*, Fr.) are of iron; and, from their vertical situation under a deck, appear to support the beams: they are represented by E E, fig. 10, plate IX. and by *l l*, in the MIDSHIP-SECTION, fig. 1, plate VI.

Mr. Thomas Roberts, of the Navy-office, has lately succeeded in the invention of a method of securing the beams of ships to their sides, and of superseding the necessity of standards, top and breadth riders, and nearly of wooden knees altogether. For this invention he was rewarded with a premium of 800*l*. by the King's Order in Council of the 8th October 1807, in consequence of a recommendation from the Lords of the Admiralty; and he has since been honoured with the silver medal of the Society of Arts.

Upwards of thirty ships of different rates have already been built, for his Majesty's service, the beams of which are secured by *iron-plate knees*, agreeable to Mr. Roberts's plan. For farther particulars, and an explanatory engraving of the invention, see *Trans. of the Society of Arts*, Vol. XXVI.

Lodging-KNEES, (*courbes horizontales*, Fr.) are fixed horizontally in the ship's frame, having one arm bolted to the beam, and the other across two or three of the timbers, as represented by F F, fig. Z, plate IX.

Standard-KNEES, are vertical curves employed generally with the orlop-beams; they are fastened on one side against the vertical side of the beam, and on the other against the ship's side: thus they differ from those which are simply called standards. See the article STANDARD.

Transom-KNEES, (*courbes d'arcasse*, Fr.) are those which support the highest transoms of the ship's quarter, and are bolted thereto, and to the after-timbers.

KNIGHT-HEADS, or BOLLARD-TIMBERS, (*les apôtres*, Fr.) are those timbers on each side of the ship nearest to the stem, and continued high enough to secure the bowsprit.

KNIGHT-HEADS, also denote in a merchant-ship, two strong frames of timber, fixed on the opposite sides of the main-deck, a little behind the foremast, which support and inclose the ends of the windlass, that is turned therein as upon an axis: as each of these is formed of two pieces, they may be occasionally separated, in order to take off the turns of the cable from the windlass, or replace them upon it. They are sometimes called the *bits*, and then their upper parts only are denominated knight-heads, which being formerly embellished with a figure, designed to resemble a human head, gave rise to the name they have ever since retained. See the article WINDLASS.

KNIGHT-HEADS, (*seps de drisse*, Fr.) was formerly a name given to the lower jear blocks, which were then no other than bits, containing several sheaves, and nearly resembling our present top-sail-sheet bits.

KNITTING, in rope-making, the act of tying together certain quantities of yarn, when warping into hauls to be tarred.

KNITTLES, (*rabans*, *éguillettes*, Fr. from *knit*) are small lines, composed of two or three rope-yarns, either plaited or twisted, and used for various purposes at sea, particularly to fasten the *service* on the cable, to reef the sails by the bottom, and to sling the sailors hammocks between decks. Knittle is also a name given to the loops or buttons of a bonnet.

KNOCK OFF THERE! (*cesse l'ouvrage!* Fr.) the order to cease from any work.

KNOT, (*nœud et bouton fait au bout d'un cordage*, Fr.) a large knob, formed on the extremity of a rope, generally by untwisting the ends thereof, and interweaving them regularly amongst each other: there are several sorts of knots, which differ in their form, size, and name, according to the uses for which they are designed; as,

Bowline KNOT, (*nœud de bouline*, Fr.) is made
Fig. 13. by laying the end of a rope, *a* (fig. 13, plate XII.) over the standing part *b*, and turning a bight over the standing part; then leading the end round the standing part, through the bight again. When this knot is drawn close, it makes a loop; and, when fastened to the cringles of the sails, they must break or the sails split before it will slip.

Buoy-Rope KNOT, (*nœud de l'orin de l'ancre*, is made by unlaying the strands of the cable-laid rope, and also one of the small strands out of each large one; laying the large ones again as before, and leaving the small ones out; then *single* and *double wall* the small strands (as directed for the stopper-knot,
Fig. 14. fig. 4, plate XIII.) round the rope, fig. 14, worm them along the divisions, and stop their ends with spun-yarn, *a*.

Diamond KNOT, (*nœud de diamant*, Fr.) is made by unlaying the end of a hawser-laid rope for a con-
Fig. 15. siderable length, fig. 15, and with the strands from three bights down its side, holding them fast. Put the end of the strand *a*, over the strand *b*, and through the bight of the strand *c*, as represented in the figure: then put the strand *b* over the strand *c*, and through the bight formed by the strand *a*; and the end *c* over *a* and through the bight of *b*. Haul these taught, lay the rope up again, and the knot will appear like
Fig. 16. fig. 16. This knot is used for the side-ropes, jib-guys, bell-ropes, &c.

Double Diamond KNOT, (*double nœud de diamant*, Fr.) is made with the strands opened out again, following the lead of the single knot through two single bights, the ends coming out at the top of the knot, and leading the last strand through two double bights: then, by laying the rope up again as before to where the next knot is to be made, it will appear like fig. Plate XIII.
1, plate XIII. Fig. 1.

Over-Hand, or *Figure of Eight* KNOT, (*nœud à plein poing*, Fr.) is made by taking the end of a rope *a* (fig. 2.) round the standing part *b*, under its own Fig. 2.
part *d*, and through the bight *c*.

Reef KNOT, (*nœud plat, ou nœud de garcette*, Fr.) is made by first forming a rope into an over-hand knot (see fig. 2.) round a yard or spar; bringing the end in the right hand over to the left, and the left to the right; then take the right hand end round the left, and draw them taught, and it will appear like fig. 3. Fig. 3.

Sheep-Shank KNOT, serves to shorten a rope without cutting it, which may be presently loosened, and the rope not the worse for it.

Stopper KNOT, is made by single walling and double walling, without crowning, a three stranded rope against the lay, and stopping the ends together as represented by fig. 4. If the ends are very short they Fig. 4.
should be wipped, without being stopped.

Wale KNOT, is so made with the lays of a rope that it cannot slip, and serves for sheets, tacks, and stoppers.

Single Wall KNOT, (*cul de-porc simple*, Fr.) is made by unlaying the end of a rope, fig. 5, and with Fig. 5.
the strand *a* form a bight, holding it down on the side of the rope at *b*; passing the end of the next *c* round the strand *a*; the end of the strand *d* round the strand *c*, and through the bight which was made at first by the strand *a*; then haul them rather taught, and the knot will appear like fig. 6. Fig. 6.

To Crown the Single Wall KNOT, is to lay one of the ends over the top of the knot, fig. 7, which call Fig. 7.
the first *a*, lay the second *b* over it, and the third *c* over *b* and through the bight *a*; haul them taught, and the knot with the crown will appear like fig. 8, Fig. 8.
which is drawn open, in order to render it more clear. This is called a *single wall*, and *single crown*, from which the *double wall* and *double crown knots* are easily made.

The knots are generally used to act as a button, in preventing the end of a rope from slipping through an eye, or through the turns of a lanyard, by which they are sometimes made fast to other ropes.

To KNOT, signifies to tie two ropes together, or the end of a rope to a bight in the same. See the articles BEND and HITCH.

KNOT, (*nœud*, Fr.) also implies a division of the log-line, which answers to half a minute, as a mile does to an hour; *i.e.* it is the $\frac{1}{120}$ part of a mile; hence we say, the ship was going eight knots, which signifies eight miles per hour.

KNUCKLE-TIMBERS. See the article TIMBER.

KOFF, a Dutch vessel of burden, with a main and foremast, and a large spritsail set abaft each. Thus they sail very close to the wind; but when the wind is aft, they carry flying topsails and a square-sail upon the fore-mast, and upon the bowsprit two or three jibs.

L.

TO LABOUR, (*fatiguer*, Fr.) implies to roll or pitch heavily in a turbulent sea; an effect, by which the masts and hull of the ship are greatly endangered; because, by the rolling motion, the masts strain upon their shrouds with an effort, which increases as the sine of their obliquity: and the continual agitation of the vessel gradually loosens her joints, and often makes her extremely leaky.

LACERTA, *the Lizard*, in astronomy, one of the new constellations of the northern hemisphere, added by Hevelius to the 48 old ones, near Cepheus and Cassiopeia.

LACING, the rope or line used to confine the heads of sails to their yards or gaffs: Also a piece of compass or knee-timber, fayed to the back of the figure, and the knee of the head, and bolted to each.

LADDER, (*échelle*, Fr.) a well-known convenience, of which there are a great number in a ship, serving as stairs, whereby to ascend or descend from one deck to another.

The ladders are distinguished by epithets according to the several hatchways, or other parts of a ship, wherein they are situated.

Accommodation-LADDER, (*échelle de commandement*, Fr.) is a sort of light stair-case, occasionally fixed on the gangway of the admiral, or commander in chief, of a fleet. It is furnished with rails and entering-ropes, covered with red-bays, and the lower-end of it is retained at a competent distance from the ship's side by iron bars, or braces, to render the passage for entrance or departure more convenient. See the article GANGWAY.

Stern-LADDERS, (*échelles de corde d'arrière*, Fr.) two ladders of rope, suspended from the right and left side of a ship's stern, whereby to descend into the boats which are moored astern, in order to bring them up along-side of the ship; or to use them for any other occasion.

LADDER-*Ways*, are apertures made in the decks of ships where ladders are fixed.

LADEN, (*chargé*, Fr.) the state of a ship when she is charged with a weight or quantity of any sort of materials, equal to her tonnage or burden. If the goods with which she is laden be extremely heavy, her burden is determined by the weight thereof; but if light, she carries as much as she can *stow* to be fit for the purposes of navigation. As a ton in measure is generally estimated at 2000lbs. in weight, a vessel of 200 tons ought accordingly to carry a weight equal to 400,000 lbs.; therefore, when the matter of which the cargo is composed is specifically heavier than the water in which she floats; or, in other words, when the cargo is so heavy that she cannot float high enough, with so great a quantity of it, as her hold will contain, a diminution thereof becomes absolutely necessary.

LADEN *in bulk*, (*chargé en grenier*, Fr.) the state of being loaded with a cargo which is neither in casks, boxes, bales, or cases, but lies loose in the hold; being defended from the moisture, or wet of the hold, by a number of mats and a quantity of *dunnage*. Such are usually the cargoes of corn, salt, or such materials.

LADLE FOR A GUN, (*cuiller à canon*, Fr.) a copper instrument used to draw the charge of a great gun: it is sometimes in the form of an apple scoop, and fixed on a long staff, see plate VI. fig. 5.

Paying LADLE, (*cuiller à bec, pour goudronner les coutures des ponts*, Fr.) an iron ladle with a long spout opposite to the handle; it is used to pour melted pitch into the seams after they are calked.

LAGOON, (*lagune*, Fr.) a name given to large ponds in some parts of America and the West Indies.

LAID-UP, (*désarmé*, Fr.) the situation of a ship when she is unrigged or dismantled during the winter, or laid by for want of employment; or when from age, &c. she is rendered incapable of further service.

LAKE, (*lac*, Fr.) a large collection of water entirely surrounded by land; and stands in geography, opposed to island.

A lake, strictly considered, has no visible communication with the sea: but this is not always the case; for many of the loughs of Ireland and the northern parts of Scotland are exceptions to this rule, and partake of the nature of havens or gulfs.

LAND-BREEZE. See the article BREEZE.

LAND-FALL, (*atterage*, Fr.) the first land discovered after a sea-voyage. A good land-fall implies a discovery of land, at or near the place to which the course was directed, and a bad land-fall implies the contrary.

LAND-LOCKED, (*fermé entre les terres*, Fr.) is said of a harbour which is environed by the land on all sides, so as to exclude the prospect of the sea, unless over some intervening land.

LAND-MARK, (*amaye, ou amer*, Fr.) any mountain, rock, staple, or the like, near the sea-side, which serves to direct ships passing by how to steer, so as to avoid certain dangers, rocks, shoals, whirlpools, &c. For other particulars relative to land, See the articles LAYING, SET, SHUT-IN, HEAD-LAND, &c.

Land-to, (*à la vue de terre*, Fr.) is when a ship lies so far from the shore that she can but just ken land.

Land-Turn, (*brise de terre*, Fr.) is a wind that blows in the night, at certain times in most hot countries.

LANDING. See the article To Disembark.

Landing-*Place*, (*débarcadere*, Fr.) a convenient spot at the edge of a river or harbour, where boats may easily run in and put any one on shore.

LANDMAN, (*homme qui n est pas marin*, Fr.) the distinctive appellation of any person on board a ship who has never before been at sea.

LANGREL, or Langrage, (*mitraille*, Fr.) a particular kind of shot, formed of bolts, nails, bars, or other pieces of iron tied together, and forming a sort of cylinder, which corresponds with the bore of the cannon, from which it is discharged, in order to wound or carry away the masts, or tear the sails and rigging of the adversary, so as to disable him from flight or pursuit. It is seldom used but by privateers and merchantmen.

LANTERN, (*fanal*, Fr.) a well-known machine, of which there are many used in a ship, such are the poop, quarter, and top-lanterns, signal lanterns, store-room lanterns, powder-room lanterns, &c.

A patent has lately been granted to Mr. Brine, of Queen-street, Portsea, for the invention of a lantern, intended as a signal lantern, and for top-lights; which, for its brilliant light, certainty of burning, with its simple construction and durability, is said to surpass any hitherto made for those purposes.

It is shaped like the middle frustum of two cones, abutting upon one common base, and conveys its light from six equidistant magnifying points, as represented at *a a a*, fig. 9, plate XIII. Its properties are as follow:— Fig. 9.

No weather will affect its light; no attendance is necessary after it is once lighted; it will burn twelve successive hours; is not liable to entangle in the rigging; and its light always preserves its central position, let the lantern be in any direction whatever. It is not likely to want repair; the expense of burning is very trivial; and is fully adequate to all the purposes of night signal.

Several of our ships of war, going on foreign stations, have already been fitted with sets of this patent lantern, as also the Royal William, at Spithead.

LANYARDS, (*rides, et autres menus cordages*, Fr.) are short pieces of rope or line, fastened to several machines in a ship, and serving to secure them in a particular place, or to manage them more conveniently; such are the lanyards of the gun-ports, the lanyard of the buoy, the lanyard of the cat-hook, the lanyards of the stoppers, &c.

The principal lanyards used in a ship, are those employed to extend the shrouds and stays of the masts, by their communication with the dead-eyes and hearts, so as to form a sort of mechanical power, resembling that of a tackle. See the article Dead-Eyes and Heart.

These lanyards, are fixed in the dead-eyes as follows: one end of the lanyard is thrust through one of the holes in the upper dead-eye, and then knotted, to prevent it from drawing out; the other end is then passed through one of the holes in the lower dead-eye, whence, returning upward, it is inserted through the second hole in the upper dead-eye, and next through the second in the lower dead-eye, and finally through the third holes in both dead-eyes. The end of the lanyard, being then directed upwards from the lowest dead-eye, is stretched as stiff as possible by the application of tackles; and that the several parts of it may slide with more facility through the holes in the dead-eyes, it is well smeared with hog's lard or tallow, so that the strain is immediately communicated to all the turns at once.

To LAP, (*couvrir*, Fr.) to cover one piece with part of another.

A Lap-*Sided Ship*, (*vaisseau qui a un faux côté*, Fr.) a ship built in such a manner as to have one side heavier than the other, and thereby retains a constant *heel*, or inclination towards the heaviest side; unless when she is brought upright, by placing a greater quantity of the cargo or ballast on the other side.

LARBOARD, (*babord, ou côte gauche du vaisseau*, Fr.) a name given by seamen to the left side of a ship, when looking forward from the stem, wherein the right and left are apparently determined by the analogy of a ship's position, on the water, to that of a fish.

Larboard-*Tack*, (*amures à babord*, Fr.) is when a ship is close hauled, with the wind on her larboard side.

Larboard-*Watch*, (*le quart de babord*, Fr.) a division of a ship's company on duty, while the other is relieved from it. See the article Watch.

LARBOWLINES, a cant term used by the boatswain's mate, implying the larboard watch.

LARGE, (*largue, en parlant du vent*, Fr.) a phrase applied to the wind, when it crosses the line of a ship's course in a favourable direction, particularly on the *beam* or *quarter*: for instance, if a ship is steering west, the wind, in any point of the compass to the eastward of the south or north, may be called *large*, unless it is directly east, and then it is said to be right aft.

To Sail Large, (*aller ou courir vent largue*, Fr.) is therefore to advance with a large wind, so as that the *sheets* are slackened and *flowing*, and the *bowlines* entirely disused. This phrase is generally opposed to sailing *close-hauled*, or with a *scant* wind, in which situation the sheets and bowlines are extended as much as possible.

To LASH, (*amarrer ou lier*, Fr.) to bind or make fast with ropes.

LASHERS, are properly those ropes only which bind fast the tackles, and breeches of the ordnance, when they are haled or made fast within board.

LASHING, (*éguillettes ou lignes d'amarrage*, Fr.) denotes a piece of rope used to fasten or secure any moveable body in a ship, or about her masts, sails, and rigging, and is chiefly used for binding up the ship's side muskets, butts of water or beer, or pieces of timber to make spare topmasts.

Lashing, is also the act of fastening or securing one thing to another with several turns of a rope.

LASKETS, (*lacets d'une bonnette maillée*, Fr.) are small lines like hoops, sewed to the bonnets and drablers of a ship, to lash or lace the bonnets to the courses, or the drablers to the bonnets.

To Go LASKING, (*aller au largue*, Fr.) is much the same with going larger or veering, that is, going with a quarterly wind. See the article VEER.

LASTAGE, (*lest*, Fr.) signifies the ballast or lading of a ship.

LATCHINGS, are loops formed on the line that is sewed to the head of a bonnet to connect it with the foot of a sail; these loops are six inches asunder, except the two middle ones, which are twelve inches long, to fasten off with. The loops are alternately reeved through holes in the foot of the sail, and through each other, and fastened by the two long loops in the middle with two half hitches, by loosing of which they unreeve themselves.

LATEEN-SAILS, (*voiles latines*, Fr.) are triangular sails, frequently used by xebecs, polacres, settees, and other vessels navigated in the Mediterranean sea.

LATEEN-YARD, (*antenne*, Fr.) a long yard, used to extend the lateen-sail upon, is slung about one quarter from the lower end, which is brought down as the tack, while the upper end is raised in the air, in an angle of about 45 degrees. See the article YARD.

LATHE, (*tour*, Fr.) in block-making, an engine used in turning wood, brass, iron, and other materials.

LATITUDE, (*latitude*, Fr.) in navigation, is the distance of a place from the equator, measured by an arc of a meridian contained between the equator and the given place; it is called either north or south, as the given place is situate in the northern or southern hemisphere: thus, London being situate in the northern hemisphere is said to be in north latitude, and is distant from the equator 51° 32′, because degrees of latitude are reckoned from the equator towards the poles, and can never exceed ninety.

Circles parallel to the equator are called *parallels of latitude*, because they shew the latitude of places by their intersection with the meridian; and every meridian is an imaginary circle passing through the poles of the world and cutting the equator at right angles, so that every place on the earth's surface has its proper meridian on which its degrees of latitude are measured.

The latitude of a place, and the elevation of the pole of that place above the horizon, are terms used indifferently for each other, because the latitude and elevation of the pole are always equal.

Since the elevation of the pole is always equal to the latitude of the place, it is therefore best found by observing the poles height; but as the pole is only a mathematical point, and not to be observed by our senses, its height cannot be determined in the same manner as that of the sun or stars; for which reason the latitude of a place is found much more expeditiously by having the true meridian altitude of a celestial object and its declination. Then, if the zenith distance (or the complement of the true meridian altitude) and declination be both north or both south, add them together; but if one be north and the other south, subtract the less from the greater, and the sum or difference will be the latitude of the same name with the greater, as may be seen by the following example:—

Given the sun's true meridional altitude 43°. 2′. 42″ above the north point of the horizon, and declination 16°. 40′. 48″ N.

Take the sum or difference of the declination and the true zenith distance, and the particular operation may be known by constructing a figure as follows:—

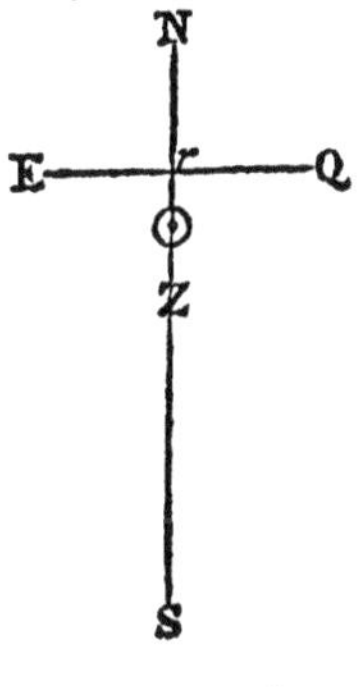

Let the line, N Z S, represent the meridian, N and S, the north and south points of the horizon, and Z the zenith.

Assume ⊙ for the sun's place at the transit, and draw E *r* Q for the equator to the north or south of ⊙, according as the body has the north or south declination; then line Z *r* will represent the *latitude*, which will be north or south, according as Z is to the north or south of E Q.

Hence the true zenith distance	47°	58′	16″	
Declination.................	15	30	8	N.
Lat.	32	28	8	S.

because Z is to the south of E Q.

But since various circumstances frequently occur which render it impossible to observe the meridian altitude of a celestial object, and the latitude of the place being of the utmost consequence in navigation, a variety of useful and ingenious methods have been invented to find the latitude independent of the meridian altitude of any celestial object, which see in ROBERTSON'S *Elements of Navigation*, Vol. I.; also in MACKAY'S and NORRIE'S *Practical Navigation*.

LATITUDE, *By Account*, (*latitude estimée*, Fr.) is the distance from the equator, north or south, estimated by the log-board.

LATITUDE, *By Observation*, (*latitude observée*, Fr.) is the latitude determined by an observation of the sun or a star.

When the latitude by observation, and the latitude by account differ, the longitude may be corrected by the following rule:—

1. Augment the distance by two thirds of the difference betwixt the latitudes by observation and dead reckoning, when the observed latitude is before that by account, and diminish it in the same proportion, when the latitude by observation is behind the dead reckoning.

2. With the augmented or diminished distance, and proper difference of latitude by observation, find a course and departure.

3. With this departure and co. middle latitude, find the difference of longitude; or with the course and meridional difference of latitude find the difference of longitude.

LATITUDE, in astronomy, the distance of a star or planet from the ecliptic, in degrees, minutes, and seconds, measured on a circle of latitude drawn through that star or planet, being either north or south, as the object is situated either on the north or south side of the ecliptic.

LATITUDE *of the Moon, North and South ascending*, is when she proceeds from the ascending node towards her northern or southern limits, or greatest elongation: and her north or south descending is, when she returns from her northern or southern limits towards the descending node. The same holds good of the other planets.

Middle LATITUDE, is the middle point between any two latitudes; and is found by taking half the sum of the two latitudes. See the article MIDDLE LATITUDE *Sailing*.

LAUNCH, (*chaloupe*, Fr.) a peculiar sort of long boat. See the article BOAT.

The principal superiority of the launch to the long-boat, consists in being, by its construction, much fitter to under-run the cable, which is a very necessary employment in the harbours of the Levant sea, where the cables of different ships are fastened across each other, and frequently render this exercise extremely necessary.

LAUNCH, is also the movement by which a ship or boat descends from the shore, either when she is at first built, or at any time afterwards; hence

To LAUNCH, (*lancer à l'eau*, Fr.) is to put out: thus they say, launch the ship, viz. put her out of the dock, &c. To facilitate this operation and prevent any interruption therein, the ship is supported by two strong platforms, laid with a gradual inclination to the water, on the opposite sides of her keel, to which they are parallel. Upon the surface of this declivity are placed two corresponding ranges of planks, which compose the base of a frame called the *cradle*, whose upper part envelopes the ship's bottom, whereto it is securely attached. Thus the lower surface of the cradle, conforming exactly to that of the frame below, lies flat upon it, lengthways, under the opposite sides of the ship's bottom; and as the former is intended to slide downwards upon the latter, carrying the ship along with it, the planes or faces of both are well daubed with soap and tallow.

The necessary preparations for the launch being made, all the blocks and wedges, by which the ship was formerly supported, are driven out from under her keel, till her whole weight gradually subsides upon the platforms above described, which are accordingly called the *ways*. The *shores* and stanchions, by which she is retained upon the stocks till the period approaches for launching, are at length cut away, and the screws applied to move her, if necessary. The motion usually begins on the instant when the shores are cut, and the ship slides downward along the ways, which are generally prolonged under the surface of the water, to a sufficient depth, to float her as soon as she arrives at the farthest end thereof.

When a ship is to be launched, the ensign, jack, and pendant, are always hoisted, the last being displayed from a staff erected in the middle of the ship.
Fig. 10. Plate XIII. fig. 10. represents a ship of war ready to be launched from the stocks.

The largest ship that ever was launched in England, is the Caledonia, of 120 guns, and 2602 tons, built at Plymouth. Ships of the first rate are commonly constructed in dry docks, and afterwards floated out,' by throwing open the flood-gates, and suffering the tide to enter, as soon as they are finished.

To LAUNCH, also, is to convey the ship from the slip into the water.

LAUNCH-*Ho!* (*laisse aller!* Fr.) the order to let go the *top-rope*, after any top-mast is *fidded*.

LAUNCHING PLANKS, are planks fitted in the slip, on each side, for the purpose of launching the ship.

To LAY, to bury or inter; as, "we interred our brave captain in a watery grave."

To LAY *the Land*, (*noyer la terre, c'est-à-dire, s'éloigner des côtes, ou perdre la terre de vue*, Fr.) in navigation, is to increase the distance of a ship from the coast, so as to make it appear lower and smaller; a circumstance which evidently arises from the intervening convexity of the surface of the sea. It is used in contradistinction to *raising* the land, which is produced by the opposite motion of approach towards it.

To LAY-*in off a Yard*, (*rentrer de dessus une vergue*, Fr.) is to come from the yard-arms towards the mast, so as to quit it at the rigging.

To LAY *out upon a Yard*, (*se ranger sur une vergue*, Fr.) to go out towards the yard-arms.

LAYING, in rope-making, the art of closing the strands together, to compose the rope.

LAYING-*Hook*. See the article HOOK.

LAZARETTO, (*lazaret*, Fr.) a building, or a vessel, fitted up and appointed for the performance of quarantine, in which all persons are confined who are suspected to have come from places infected with the plague.

In some merchant-ships, *lazaretto* is the fore part of the lower-deck, parted off for the reception of provisions and stores.

LEAD, or HAND LEAD, (*petit plomb de sonde*, Fr.) an instrument for discovering the depth of water: it is composed of a large piece of lead, from seven to ten pounds weight, and attached, by means of a strap, to a long line called *the lead-line*. It is shaped like fig. 11. plate XIII. having a hole in the Fig. 11.
upper extremity, through which is reeved a gromet (a), being well served over to keep it from chafing; and the lower extremity is hollowed for the purpose of being armed (filled with tallow), to ascertain what kind of ground you strike soundings on. In the end of the lead-line there is a long eye spliced, which is also served over. The eye is reeved through the gromet and taken over the lead, being thus secured. This line is about twenty fathoms in length, and particularly marked as follows: at the distance of two and three fathoms, a piece of black leather is thrust through the strands—at five fathoms, a white rag—at seven fathoms a red one—at ten, and thirteen, black leather; and at seventeen, a red rag.

Deep-Sea LEAD, (*grand plomb de sonde*, Fr.) a lead of a larger size, being from twenty-five to thirty pounds weight, is attached to a much longer line, and shaped like the former.

To Heave the LEAD, (*sonder, ou jeter le plomb*, Fr.) is to throw it into the sea in a manner calculated to produce the desired effect.

To LEAD A ROPE, to guide or pass it, to shew the way.

LEADING PART, is that part of a tackle which is hauled upon.

LEADING *Ships of a Fleet*, (*vaisseaux de tête d'une escadre*, Fr.) are those that are first in the order of battle, or of sailing.

LEADING *Wind*, (*vent largue*, Fr.) a free or fair wind, and is used in contradistinction to a scant wind. See the article WIND.

LEADSMAN, (*homme de la sonde*, Fr.) the man who heaves the lead.

LEAGUE, (*lieue*, Fr.) a measure of three miles in length, generally used in estimating sea distances. A nautical league, or three nautical miles, is the 20th part of a degree of a great circle. The French league sometimes contains 3,000 geometrical paces, or three English miles, and in some parts of France it consists of 3,500 paces: the mean or common league consists of 2,400 paces, and the little league of 2,000. Seventeen Spanish leagues make a degree, or twenty French leagues, or sixty-nine and an half English statute miles.

LEAK, (*voie d'eau*, Fr.) a chink or breach in the deck, sides, or bottom of a ship, through which the water passes into her hull. When a leak first commences, a vessel is said to have sprung a leak.

To stop a LEAK, (*boucher une voie d'eau*, Fr.) is to fill it with a plug wrapt in oakum and well tarred; or putting in a tar-pawling cloth, to keep the water out, or nailing a piece of sheet-lead upon the place.

The present method of securing the magazines of powder, and the stowing of such other dangerous and precious articles on board ships of war, transports, and Indiamen, not guarding sufficiently against the detrimental and destructive consequences of fire and water, of which so many melancholy instances almost every day occur; it therefore becomes the indispensable duty of every man to offer his well-founded sentiments on the most probable means of preventing, not only the melancholy consequences which attend such disasters, but the dreadful apprehensions occasioned by the least alarm of *fire*, or of *the leaks gaining on the pumps*.

An expedient has been lately offered for combining these advantages, which is as follows:—

The ship's hold to be divided longitudinally into three parts, by a *trunk*, extending nearly from the fore to the after part of the ship below, and from the floor of the ship to the under deck, occupying the centre of the hold, and one-third its diameter, with *transverse partitions* within, as well to stiffen the trunk as to divide and accommodate its various contents; the sides thereof to be composed of plank proportioned to the purpose, which are to be two-fold, or double, with a space of about three or four inches between the outer and inner, which may be filled with common putty, made of linseed oil and whiting: the planks throughout to be closely and firmly jointed, and the outer surface thereof to be plated over with thin sheet-iron, and likewise the under surface of the lower deck; from either side of the trunk two or more metal tubes are to extend to the ship's sides, and passing through the same, either close by the floor of the hold, or close by the under side of the deck, they will be out of the way of giving any interruption in either situation.

These tubes or pipes are to be furnished and secured at the ends next to, and passing between the ship's timbers, by close and strong *stop-cocks*, to be opened and shut by *winches* or *keys*, which must extend from the cocks to the upper side of the lower or middle deck, that neither fire, smoke, water, or any matter in the hold may any way obstruct or prevent their being speedily applied to, so that the trunk, or such division thereof, as the powder and other combustible and precious matters are lodged in, may be filled with water on the first appearance of danger from fire; the casks which contain the powder must be water-tight, as, indeed, all casks should be which contain perishable matters, and into which only should such be stowed.

The entrance into the trunk for putting down or taking up what is required, must be from the upper side of the lower-deck, by close hatchways; a pump is to be joined to the trunk, to clear it from the water when the danger is over, and also to keep it free from any water that might insinuate itself, should the ship at any time become leaky.

The trunk will be as convenient and every way as commodious for stowage as any other part of the ship; and, about midway of the trunk's length, there may be a break or hatch, which will admit a free and open passage from either side of the hold, which, also corresponding with the main hatchway, will obviate any inconvenience; but this may be deemed unnecessary, as the trunk is not to extend within a few feet of either extreme of the ship's hold, and, of course, there will be a communication in the hold at those parts.

The inconvenience, if any, which may arise from this measure, must be left to those who disapprove it, to point out. The advantages which appear evidently to result from its adoption are as follow:—

1st. The trunk will be as a partition, dividing the cargo into three distinct parts, which will be attended with much convenience.

2d. Neither the lading nor the ballast can possibly shift in the hold in the worst weather.

3d. When a ship is to sail in ballast, or has a light cargo, no more need be done than to open the cocks and fill the trunk, or any component divisions thereof, with water; which will be as steady and compact therein as if it were congealed, and will prove by far the best sailing ballast, as it will every way lay nearest to the centre of motion, and may be removed or taken in without expence or trouble.

4th. One portion of the trunk may be applied to the ship's store water; which may be taken up by a small pump, as occasion requires.

5th. The hold being thus subdivided, the contents of either portion may be come at with ease and expedition, and the labour of rummaging a confused cargo, for a remote and necessary article, avoided.

6th. The magazine or spirit stores cannot blow up, though the ship were on fire in every other part.

7th. The ship will be in no danger of foundering, although filled with water, as the trunk would then be as a steady, tight, and capacious empty cask in the ship's hold, preserving more than a sufficient area therein to keep her for ever afloat, were she loaded with any thing besides minerals.

8th. All matters in the trunk will be preserved from being spoiled by bilge or salt water, in any case of leakage.

9th. The trunk extending along the centre of the ship's hold, and below the level of the water, cannot be injured by gun-shot, or affected by any external accident the ship may meet with.

10th. The magazine may be come at with facility, without the necessity and danger of taking down a light.

11th. The ship's hold will be so substantially braced and fortified by the trunk, that she can neither strain, bilge, or break her back, and thereby the ship and her whole contents will be preserved, when otherwise all might perish.

Lastly, the owner will be under no apprehension of his ship, thus provided against the worst that may happen; the merchant will not be in fear for his goods, nor the seamen for their lives, as they need not dread fire, foundering, nor blowing up, and, of course, will not be induced to quit their ship, which too frequently happens from such apprehensions.

The learned and ingenious Dr. Franklin has taken no inconsiderable pains in attempting to tranquillize the minds of seamen who have the misfortune to be present in any ship meeting with disaster.

He states, and with great truth and accuracy, that no inconsiderable degree of imaginary fear exists on those occasions; that the cases are sometimes by no means so desperate as they frequently appear to the distracted crew; who, abandoning themselves to despair, not unfrequently quit their ship and perish, when, by continuing with her, both their own lives would have been saved, and the vessel itself preserved. Proofs of this fact, even now, frequently occur, and no one more strong need be adduced than that of his Majesty's ship Guardian; in fact, the commencement of the disaster is rendered additionally frightful by two circumstances: first, the vacuity of the hold and the weight of the surrounding water causes it to rush in at first with almost unnatural velocity, until that velocity shall become weakened by the circumstance of the water within board rising nearly to a level with that without. A second cause exists in the shape of the hold, which being narrow at the bottom, or near the keel, fills at first with much greater speed than is afterwards the case, when the frame expands and the space to be occupied is extended. Thus the crew will be more terrified with finding at first three or four feet water in the hold; when, as it rises in the vessel, it would require, perhaps, as much in quantity to increase its depth twice as many inches. In many vessels the cargo is of so light and buoyant a nature as absolutely to prevent the hull from sinking, let the water within board rise as high as it may. This buoyancy might be materially assisted, first, by throwing overboard all heavy articles, and, secondly, by emptying such of the water-tight casks, and effectually securing them from the re-admission of water, so that, on being returned into the hold, they may act as a species of float.

The Doctor suggests an improvement in the internal construction of a ship, which, he considers would prove an effectual preventative to all danger of foundering in consequence of *springing a leak*. "While," says the Doctor, "we are on this subject of sinking, one cannot help recollecting the well-known practice of the Chinese, to divide the hold of a great ship into a number of separate chambers, by partitions tightly calked, so that if a leak should spring in one of them, the others are not affected by it; and, though that chamber should fill to a level with the sea, it would not be sufficient to sink the vessel. We have not imitated this practice. Some little disadvantage it might occasion in the stowage, is, perhaps, one reason; though that might be more than compensated by an abatement in the insurance, that would be reasonable, and by a higher price taken of passengers, who would rather prefer going in such a vessel. But our sea-faring people are brave, despise danger, and reject such precautions of safety, being cowards only in one sense—that of *fearing* to be thought afraid."

LEAKAGE, (*coulage, en parlant des futailles*, Fr.) is the quantity which runs out of a cask through a leak.

Leakage, in commerce, is an allowance of 12 per cent. in the customs, allowed to importers of wines, for the waste and damage it is supposed to have received in the passage: an allowance of two barrels in twenty-two is also made to the brewers of ale and beer by the Excise-office.

A LEAKY Ship, (*vaisseau qui a fait des voies d'eau*, Fr.) the state of a ship when abounding with leaks: also of a cask, which suffers the liquor within it to run out.

LEAP-YEAR, or Bissextile, (*l'année bissextile*, Fr.) a year consisting of 366 days, and happening every fourth year, by the addition of a day in the month of February, which makes that month consist of 29 days. And this is done to recover the six hours which the sun takes up nearly in his course more than the 365 days commonly allowed for it in other years.

The day thus added was, by Julius Cæsar, appointed to be the day before the 24th of February, which, among the Romans, was the sixth of the calends, and which, on this occasion, was reckoned twice; whence it was called the bissextile.

By the statute *De anno Bissextile*, 21 Hen. III. to prevent misunderstandings, the intercalary day and that next before it are to be accounted as one day. See the article New Style.

Every centesimal, or hundredth year, is a leap-year, according to the Julian account; but, according to the Gregorian, it is always a common year, except when the number of centuries can be divided by four without a remainder, for then it is a leap-year; but the intermediate centesimal years are common ones: hence, to know if it be leap-year, the rule is—if the year consists of complete centuries, and can be divided by 4, it is a leap-year: as it is also, when the intermediate years can be divided by 4: thus the year 1812, is a leap-year; for 12 may be divided by 4, without a remainder.

If the intermediate years cannot be divided by 4, the remainder shews the number of years over leap-year.

The Gregorian computation has been received in most foreign countries ever since the reformation of the calendar, in the year 1582, excepting some northern countries, as Russia, &c. And, by Act of Parliament passed in 1751, it commenced in all the dominions under the crown of Great Britain in the year following; it being ordered by that act that the natural day next following the 2d of September should be accounted the 14th; omitting the intermediate eleven days of the common calendar. The supernumerary day in leap-years being added at the end of the month February, and called the 29th of that month.

LEATHERING, a name given to prepared or tanned leather, fitted close round the circular holes of caps, and the sweeps of jaws belonging to booms and gaffs. It is nailed on the upper and under sides.

LEDGE, (*ressif, ou chaîne de rochers*, Fr.) a long ridge of rocks, near the surface of the sea.

LEDGES, (*sortes de traversins placés entre les baux*, Fr.) are small pieces of timber placed *athwart-ships*, under the decks of a ship, in the intervals between the beams, as exhibited by H, in the representation of the deck, plate IX.

LEE, (*côté de dessous le vent*, Fr.) an epithet used by seamen to distinguish that part of the hemisphere to which the wind is directed, from the other part, whence it arises; which latter is accordingly called *to-windward*. This expression is chiefly used when the wind crosses the line of a ship's course, so that all on one side of her is called to windward, and all on the opposite side *to-leeward*: and hence,

Under the LEE, (*sous le vent*, Fr.) implies farther from that part of the horizon from whence the wind blows; as,

Under the LEE *of the Land*, (*sous le vent de la côte*, Fr.) implies at a short distance from the shore which lies in the direction of the wind. This phrase is commonly understood to express the situation of a vessel anchored, or sailing under the weather shore, where there is always smoother water and less danger of heavy seas, than at a great distance from it.

To lay a Ship by the LEE, (*mettre un vaisseau vent dessus, ou coiffer un vaisseau*, Fr.) is to bring the ship so, that all her sails may lie flat against the masts and shrouds, and that the wind may come right upon her broadside.

A-LEE. See the letter A.

LEE-*Boards*, (*semelle*, Fr.) are strong frames of plank affixed to the sides of flat-bottomed vessels, such as river barges, &c. which draw but little water; these, by being let down into the water when the vessel is close-hauled, prevent her from falling to leeward.

The LEE-*Gage*, (*le dessous du vent*, Fr.) implies farther from the point whence the wind blows than another vessel.

Take care of the LEE-*Hatch!* an order given to the helmsman to take care that the ship do not go to the leeward of her course.

LEE-*Lurches*, (*embardées d'un vaisseau sous le vent ou arrivées*, Fr.) the sudden and violent rolls which a ship often takes to the leeward in a high-sea, particularly when a large wave strikes her on the weather-side.

A LEE-*Shore*, (*terre, ou côte sous le vent du vaisseau*, Fr.) a ship is said to be on a lee-shore when she is near the land, with the wind blowing right upon it.

LEE-*Side*, (*côté de dessous le vent*, Fr.) all that part of a ship or boat which lies between the mast and the side farthest from the direction of the wind; or that half of a ship which is pressed down towards the water by the effort of the sails, as separated from the other half, by a line drawn through the middle of her length. That part of the ship which lies to windward of this line is accordingly called the *weather-side*.

Thus, if a ship sail southward, with the wind at east, then is her starboard, or right side, the *lee-side*; and the larboard, or left, the *weather-side*.

LEE-*Way*, or LEEWARD-*Way*, (*dérive—angle de la dérive*, Fr.) is the lateral movement of a ship to leeward of her course, or the angle which the line of her way makes with the keel when she is *close-hauled*.

In moderate weather and smooth water, when the course which a vessel steers is made good, then the vessel is only actuated by forces, which drive her directly forward, and her head-way becomes her true way. But, in more stormy weather and a rough sea, the violence with which the wind acts, and the impetuosity with which the sea strikes upon the vessel, have a tendency to drive her directly to leeward.

The lee-way is neither increased nor diminished by a less or greater head-way while the state of the weather remains the same, although the angle which indicates the difference between the *course steered* and the *course made good*, becomes on that account greater or less.

As the causes for lee-way are two; viz. the violence of the wind and impetuosity of the sea, it is necessary to consider the direction which these two combined forces would drive a vessel, provided she was left on the water entirely destitute of head-way. This direction is readily ascertained by the eye of an old seaman, and is generally towards the opposite point from whence the wind blows; for, as the irregular motions of the sea are occasioned by the wind, so the powers by which they act on a vessel are usually in the same direction. But this is not always the case, for a sudden change of wind leaves the sea acting in the same direction as before, while the wind is acting in another. The variation, however, in the direction in which the lee-way takes place, is not so great, or of such long continuance, as to make any material alteration in the reckoning.

A vessel, therefore, may be considered to be impelled by two forces, the one driving her directly a-head, and the other driving her directly to leeward; and the doctrine of the composition and resolution of forces may be properly applied to this branch of navigation.

If the violence of the wind and impetuosity of the sea continue the same, the causes for lee-way continue the same; and the effect must consequently remain the same, whatever may be the vessel's head-way. But, although the vessel may drift as much bodily to leeward one time as another, yet if the head-way be increased, the angle, which is the difference between the *course steered* and the *course made good*, becomes less, but if diminished, greater than before. See COLE's *Stereogoniometry*.

The lee-way is produced by the mutual effort of wind and sea upon the ship's side, forcing her to leeward of the line upon which she appears to sail; and in this situation her course is necessarily a compound of the two motions by which she is impelled, of which the one presses forward, according to the line of her keel, from H to K, fig. 1, plate XIV. whilst the other, acting in the line B A, pushes her to leeward of the course from B towards A, with a motion which is usually in proportion to the force of the wind and the rate of her velocity, as appears by the following theory:— Plate XIV. Fig. 1.

When a ship is close-hauled, and the head-sails are in perfect equilibrio with those abaft, the resistance of the water from A to B, (fig. 1.) is equal to the impulse of the sails, whether it is impressed upon the centre of gravity H of the ship, or any other point of her length before or abaft. In this situation, the ship will as readily bear away as come nearer to the wind, with regard to the resistance of the water upon her bottom on one side, and the impulsion of the wind upon the sails on the other. But it must be observed, that the united effort of the sails acts upon the ship according to a direction B A, perpendicular to their surfaces, and commencing its action in some point H, being the mean D between the different effects C G, of the sails *afore* and abaft, which should exactly correspond with the resistance of the water from A towards B; so that the vessel is pushed to leeward of the course I K, which she steers in the direction B A of the effort of the sails. But the resistance of the water, acting upon the lee-side of her bottom, counterbalances this effort, and becomes stronger in proportion to the greater facility with which she divides the fluid with her stem: so that she will really advance in the course N R, which lies nearer the line of her keel than B A. Thus the angle K H R of the lee-way is proportional to the greater or less resistance the ship meets with from the fluid upon her lee-side, respectively with her greater or less facility of dividing it with her fore-part; so that the lee-way is very inconsiderable, except when the ship is close-hauled, and is, accordingly, disregarded whenever the wind is large.

This demonstration might be pushed farther by a fact, founded on daily experience, which proves that not only the lee-way depends on the form of the vessel, but also the degree of velocity with which she advances; and perhaps never, intirely, upon the greater or less obliquity of the sails with the keel, as some authors have pretended. For when a swift-sailing ship is *close-hauled*, with all her sails out, in a very light wind, and scarcely having *steerage-way*, the lee-way is considerable, even in smooth water. This is occasioned by the tardy motion of the vessel, which being feebly pushed forward cannot impress the water with a forcible effect, and, consequently, feels no resistance from it, but is, accordingly, carried with facility by her sails, in the direction of their effort B A; and if we consider the situation of the ship's side, which presents a great surface of sail above the water, it appears that the lee-way will become yet more perpendicular to the keel But when the wind makes a forcible impression, the velocity of moving forward is considerably augmented; the ship strikes the fluid with a force, expressed by the square of two or three leagues of swiftness, from B towards A, in the space of an hour, whilst the water repels her effort in a contrary direction. The resistance of the water is then in the ratio of this square to the square of her first velocity, or head-way; and in this state will not readily yield to its effort. The lee-way immediately decreases, and will be still farther diminished, if the ship's course is accelerated. If, then, at the moment when the ship advances with great rapidity, she *bears away* 12 or 15 degrees, or even two points, without altering the general arrangement of her sails, their obliquity remains the same; the ship, therefore, ought to have the same lee-way, according to the opinion of those who have written on the theory of sailing. The velocity is augmented, because the sails then receive the wind by a greater sine of incidence, and thereby acquire a more powerful effort, whilst the ship's *head* is always struck by the water in the same parts, and by the same sine of incidence; so that the lee-way is also diminished, because the water resists more, in consequence of the accelerated swiftness, and because the resistance is more exerted on the ship's side than on her head, which is less opposed to its impulsion. Hence, we may conclude, that the lee-way of a ship does not entirely depend on the disposition of her sails; that it is different in different vessels, because they are neither formed alike, nor are their sails equally *trimmed* in the same oblique courses: and, finally, because they have always a different velocity, at the same time, and under the same sail. Thus, it is evident, that the lee-way is always composed of the ship's comparative velocity; of her form, which gives more or less proportional resistance upon the side than on the fore-part; and of the disposition of her sails, as forming a greater or smaller obliquity with the keel. See also the articles CLOSE-HAULED, DRIFT, and SAILING.

As all ships are apt to make some lee-way, in casting up the log-book something must be allowed for it. But the lee-way made by different ships, under the same circumstances, will be different: and even the same ship, with different lading, and having more or less sail on board, will make more or less lee-way. However, the common allowances made for lee-way are these:—

1st. If the ship be close-hauled, has all her sails set, the water smooth, and a moderate gale of wind, she is supposed to make little or no lee-way.

2d. If it blow so fresh as to cause the small sails to be handed, it is usual to allow one point.

3d. If it blow so hard that the top-sail must be close reefed, the ship then makes about two points lee-way.

4th. If one top-sail must be handed, it is common to allow two and three quarters, or three points lee-way.

5th. When both top-sails must be handed, they allow about four points lee-way.

6th. When it blows so hard as to occasion the fore-course to be handed, the allowance is between five and a half and six points.

7th. When both main and fore-courses must be handed, then six, or six and a half points must be allowed for her lee-way.

8th. When the mizen is handed, and the ship is trying a-hull, she then makes her way good about one point before the beam; that is, about seven points lee-way.

Though these rules are such as are generally used, yet, as the lee-way depends much upon the mould and trim of the ship, it may be necessary here to give the method of ascertaining it by observation. Thus, let the ship's wake be set by a compass in the poop, and the opposite rhumb is the true course made good by the ship; then the difference between this and the course given by the compass in the binnacle, is the lee-way required. If the ship be within sight of land, the lee-way may be exactly found by observing a point on the land which continues to bear the same way; for the distance between the point of the compass it lies on and the point the ship capes at will be the lee-way.

The lee-way once known, and the weather continuing the same, the angles of correction for every rate the vessel goes are to be calculated; and thereby the *courses made good* become known.

If the lee-way and vessel's rate by the log be given, the angle of correction may be found by the following rule:—

As the rate the vessel sails by the log,
Is to the sine of the angle, which determines the direction of her lee-way;
So is her lee-way to the sine of the angle of correction.

If the angle of correction be known, the lee-way may be discovered, when the rate of the vessel is given.

The rule for finding the lee-way is as follows:—

As the sine of the angle, which determines the direction of lee-way,
Is to the rate the vessel goes by the log;
So is the sine of the angle of correction, to lee-way.

LEECHES OF A SAIL, (*bords des voiles*, Fr.) the borders or edges of a sail, which are either sloping or perpendicular; those of the square sails, that is, the sails whose tops and bottoms are parallel to the deck, or at right angles with the mast, are denominated from the ship's side; as the *starboard* leech of the main-sail, the *lee* leech of the fore-top-sail; but the sails which are fixed obliquely on the masts, have their leeches named from their situation with respect to the ship's length; as the fore-leech of the mizen, the after-leech of the jib, or fore-stay-sail, &c.

LEECH-LINES, (*cargue-boulines*, Fr.) are ropes fastened to the middle of the leeches of the main-sail and fore-sail, and communicating with blocks under the opposite sides of the top, whence they pass downwards to the deck, serving to truss those sails up to the yards, as occasion requires. See the article BRAILS.

Harbour LEECH-LINES, are ropes made fast at the middl of the top-sail-yards, then passing round the leeches of the top-sails, and through blocks upon the top-sail-tye, serving to truss the sails very close up to the yard, previous to their being *furled in a body*.

LEECH-ROPE, (*ralingue de chûte*, Fr.) a name given to that part of the bolt-rope to which the border, or edge of a sail is sewed. In all sails, whose opposite leeches are of the same length, it is terminated above by the earing, and below by the clew. See the articles BOLT-ROPE, CLEW, and EARING.

LEEFANGE, (*barre de fer sur laquelle coule le palan d'une voile aurique, &c.* Fr.) a bar of iron fastened at its extremities to the deck, and upon which the sheets of some fore and aft sails traverse, when tacking, particularly the fore-sail of a sloop, hoy, and cutter.

LEEFANGES, are also ropes reeved into the cringles of the courses, to haul in the bottom of a yacht's or hoy's sails, that the bonnets may be laced on, or the sails taken in.

To LEEWARD, (*sous le vent*, Fr.) denotes towards that part of the horizon which lies under the lee, or whither the wind bloweth: thus, "We saw a fleet under the lee," and "We saw a fleet to leeward," are synonymous expressions.

A LEEWARD *Ship*, (*vaisseau, mauvais boulinier*, Fr.) is a vessel that does not sail so near the wind, nor make so good way as she should; or which is much to leeward of her course when sailing *close-hauled*.

A LEEWARD *Tide*, (*marée qui porte du même côté que le vent*, Fr.) is a tide running in the same direction that the wind blows, and is directly contrary to a tide under the lee, which implies a stream in an opposite direction to the wind.

LEGS, a name given to short ropes branching out into two or more parts, as the bow-line legs, bunt-line legs, crow foot legs, &c.

LEITH, a sea-port in Scotland; but, on the coast of Sweden, particularly, the passage round the point of Landfoort to the city of Stockholm, means a channel, and is used as a general appellation for that purpose; but we have not met with any other instance of such an application of the word.

To LENGTHEN A SHIP, (*alonger un bâtiment*, Fr.) is to cut her down across the middle, and to add a certain portion to her length. This is performed by sawing her planks asunder, in different places of her length, on each side of the midship frame, to prevent her from being weakened too much in one place. The two ends are then drawn apart, to a limited distance, which must be equal to the proposed addition of length. An intermediate piece of timber is next added to the keel, upon which a sufficient number of timbers are erected, to fill up the vacancy produced by the separation. The two parts of the kelson are afterwards united by an additional piece, which is scored down upon the floor-timbers; and as many beams as may be necessary are fixed across the ship in the new interval. Finally, the planks of the side are prolonged, so as to unite with each other, and those of the ceiling refitted in the same manner; by which the whole process is completed. Several ships in the British navy have been lengthened agreeable to this process, and found to answer very well.

LEO, *the Lion*, in astronomy, a constellation of the northern hemisphere, being the fifth sign of the zodiac, and marked thus ♌, as a rude sketch of the animal. The number of stars in this constellation are, according to the Britannic catalogue, 95. For the star called the Lion's-heart, or Cor Leonis, see the article REGULUS.

LEO *Minor, the Little Lion*, in astronomy, a constellation of the northern hemisphere, and one of the new ones that were formed out of the unformed stars and added to the 48 old ones. It contains 53 stars in Flamstead's catalogue.

LEPUS, *the Hare*, in astronomy, a constellation of the southern hemisphere, containing, according to the Britannic catalogue, 19 stars.

LET FALL! (*laisse tomber!* Fr.) the order for putting out a sail when the yards are aloft and the sail is to come down from the yard; but when the yards are stricken down, then the sail is loosed below before they hoist the yard.

To LET *In*, (*entrer les bouts des bordages*, Fr.) amongst shipwrights, is to fix a diminished part of one plank, or piece of timber, into a vacancy, formed in another for this purpose. See the article RABBET.

To LET *Out a Reef*, (*larguer un ris*, Fr.) is to increase the dimensions of a sail, by untying the points of a reef in it.

LETTER OF MARQUE, (*lettre de marque*, Fr.) an extraordinary commission, granted by the Lords Commissioners of the Admiralty, or by the vice-admiral of any distant province, to captains or commanders of merchant ships, or privateers, to cruise against and make prizes of the enemy's ships and vessels, either at sea or in their harbours, in order to make a reparation for those damages they have sustained, or the goods they have been deprived of by strangers at sea: and *reprisals* is only the retaking or taking of one thing for another. In the prosecution of these letters there must be,

1. The oath of the person injured, or other sufficient proof touching the injury sustained.
2. A proof of due prosecution for satisfaction in a legal way.
3. The deferring or denial of justice.
4. A complaint to his own prince or state.
5. A requisition of justice made to the supreme head of the state. After all which, *letters of reprisal* under certain restrictions, are issued; but if the supreme power think these letters of reprisal may affect the peace of the state, they are put off till a more convenient time. The ships so commissioned are also called *letters of marque*.

LEVANTER, (*vent de Médine*, Fr.) a name given to an easterly wind up the Mediterranean.

LEVEL YOUR GUNS! See the article EXERCISE.

Artillery Foot-LEVEL, an instrument in form of a square, having its two legs or branches of an equal length, at the junction of which is a small hole, whence hangs a thread and plummet, playing on a perpendicular line in the middle of a quadrant, which is divided both ways from that point, into 45 degrees. This instrument may be used on other occasions, by placing the ends of its two branches on a plane; for, when the thread plays perpendicularly over the middle division of the quadrant, the plane is then level.

To use it in gunnery, place the two ends on the piece of artillery, which may be raised to any proposed height by means of the plummet, whose thread will cut the degree above the level. See the article GUNNER'S QUADRANT.

Gunner's LEVEL, for levelling cannon and mortars, consists of a triangular brass plate of a radius of about four inches, at the bottom of which is a portion of a circle, divided into forty-five degrees; which angle is sufficient for the highest elevation of cannon and mortars, and for giving shot the greatest range: on the centre of this segment of a circle is screwed a piece of brass, by means of which it may be fixed or screwed at pleasure; the end of this piece of brass is made so as to serve for a plummet and index, in order to show the different degree of elevation. This instrument has also a brass foot, to set upon the piece, so that when it is horizontal, the instrument will be perpendicular. The foot of this instrument is to be placed on the piece to be elevated, in such a manner as that the point of the plummet may fall on the proper degree: this is what is called levelling the piece.

An improved and curious instrument of this description has been invented by Colonel Congreve, of the royal artillery, containing numerous particulars not shewn on the common level, with various useful tables, &c.

LEVELLED OUT, implies any line continued out from a given spot in a level or horizontal direction.

LIBRA, *the Balance*, in astronomy, one of the twelve signs of the zodiac, the seventh in order, and marked like a part of a pair of scales, thus ♎ It is so called, because when the sun enters it, the days and nights are equal, as if weighed in a balance.

The stars in this sign are, according to Flamstead, 51.

LIBRATION, in astronomy, is an apparent irregularity of the moon's motion, whereby she seems to librate about her axis, sometimes from the east to the west, and now and then from the west to the east, so that the parts in the western limb or margin of the moon sometimes recede from the centre of the disc, and sometimes move towards it, by which means they become alternately visible and invisible to the inhabitants of the earth. This *libration* is owing to her equable rotation round her own axis, once in a month, in conjunction with her unequal motion in the perimeter of her orbit round the earth.

LIBRATION *of the Earth*, is sometimes used to denote the parallelism of the earth's axis in every part of its orbit round the sun.

To LIE ALONG, or LIE OVER. See the article ALONG.

To LIE *to*, or *To* LIE *by*, (*être en panne*, Fr.) is to retard a ship in her course, by arranging the sails in such a manner as to counteract each other with nearly an equal effort, and render the ship almost stationary, with respect to her progressive motion, or *head-way*. A ship is usually *brought-to* by laying either her main-top-sail, or fore-top-sail *aback*, the helm being put close down to leeward. This is particularly practised in a general engagement, when the hostile fleets are drawn up in two lines of battle opposite each other. See also the article LYING-TO.

To LIE-*to in a Storm*. See the article TRYING.

LIEUTENANT, (from *lieu*, a place, and *tenir*, to hold) an officer who supplies the place and discharges the office of a superior in his absence. Of these some are civil, as the lords-lieutenants of kingdoms and counties; others military, as the lieutenant-general, lieutenant-colonel, &c.

LIEUTENANT *of a Man of War*, (*lieutenant d'un vaisseau de guerre*, Fr.) the officer next in rank and power to the captain; of these there are several in a large ship, who take precedence according to the dates of their first commissions. The first *lieutenant*, during the absence of the captain, is charged with the command of the ship, as also the execution of whatever orders he may have received from the commander relating to the King's service.

The first lieutenant is to prevent all profane swearing and abusive language; all disturbance, noise, and confusion; to enforce a strict obedience to orders, a proper respect to all superiors, and an observance of discipline and good order; and he is to report to the captain all those whose misconduct he shall think deserving of reprehension or punishment.

The *lieutenant* who commands the watch at sea, keeps a list of all the officers and men thereto belonging, in order to muster them when he judges it expedient, and report to the captain the names of those who are absent from their duty. During the night-watch he occasionally sends a careful officer, to see that the proper sentinels are at their duty, and that there is no disorderly conduct amongst the men; no tobacco smoked between decks, nor any fire or candles burning there, except the lights which are in lanterns, under the care of a proper watch, for particular purposes. He is expected to be always upon deck in his watch, as well to give the necessary orders with regard to *trimming* the sails and superintending the navigation; but he is never to change the ship's course without the captain's directions, unless to avoid an immediate danger.

He is also to see that the master's-mates and midshipmen are attentive in performing their duty with regard to the sub-divisions placed under their inspection.

In the absence of the captain, the senior lieutenant on board the ship is to be responsible for every thing done on board: he is to see every part of the duty as punctually performed as it could be if the captain were present.

In time of action, he is to see that all the men under his command are at their quarters, and that they do their duty with spirit and alacrity. He is to be particularly attentive to prevent the men from loading the guns improperly, from firing them before they are well pointed, and from wetting them after they have been fired; and he is very carefully to prevent their making any improper accumulation of powder in any part of his quarters.

When he has the watch, he is to be constantly on deck until he is relieved by the officer who is to succeed him.

He is to see that the men are alert and attentive to their duty, that every necessary precaution is taken to prevent accident from squalls or sudden gusts of wind; and that the ship is as perfectly prepared for battle as circumstances will admit. He is to be particularly careful that the ship be properly steered, and that a correct account be kept of her way, by the log being duly hove, and the lee-way for each hour marked on the log-board.

He is to keep a log-book, according to the form (No. 25) in the Appendix to the Naval Instructions, the edition of 1808, carefully inserting the information required in each column, as far as the situation of the ship will admit, with whatever additional information, useful to navigation, or to his Majesty's service, his observations may have enabled him to obtain; and, at the end of the voyage, or when he wishes to receive his pay, he is to send it, signed by himself, to the Navy Board.

He is constantly to ascertain the latitude by observation, at noon, or by double altitudes, as circumstances may require; and to keep an account of the ship's way, specifying the course steered and the distance run for each twenty-four hours, with the latitude and longitude she is in, and the bearings and distance of some principal head-land from which she sailed, or towards which she is going with any other particulars, and in any form that the captain shall direct: which account he is to deliver to the captain every day, as soon after noon as the other duties of the ship will allow.

He is to keep men at the mast heads during the day, and in proper stations during the night, to look out, that no ship may surprise or pass near without being noticed, and by him reported to the captain; he is often to remind them of their duty, and to relieve them more or less frequently, according to the state of the weather and the degree of their attention.

During a fog, he is to be particularly attentive to the guns fired by the admiral, that, by observing any alteration which may take place, either in the direction or strength of the report, he may take such steps as may be necessary to prevent the ships being separated from the fleet. He is to be very careful to order the drum to be beaten and the bell to be sounded, according to the tack the ship may be on, for the information of ships that may be near.

The lieutenant may put under arrest any officer whose conduct he shall think so reprehensible as to require it; and he may confine such men as he shall think deserving of punishment; but neither he nor any other lieutenant who may become commanding officer, is to release an officer from his arrest, nor to release nor punish any man who has been confined; which is to be done by the captain only; unless he be absent from the ship with leave from the Admiralty, or from the commander in chief, in which case it is to be done only by the senior lieutenant commanding the ship in his absence. See the last edition of the Naval Instructions.

The number of *lieutenants* in the royal navy, in September 1812, was 3,327 of whom 223 were noted as unfit for sea service; in addition to these there were about 54 superannuated with the rank of commanders.

LIFE-BOAT, (*bateau, ou canot de sauvetage*, Fr.) a modern invention of the utmost consequence to the lives and safety of sea faring persons.

The construction of a boat for the preservation of lives from ships driven on the shores of this kingdom, was originally suggested by the subscribers to the News Room at the Law House, South Shields, in 1789; who, from situation, were the more immediate spectators of the destruction inevitably attending vessels and their crews coming on the sand at the south entrance of Tynemouth haven.

In September 1789, the ship Adventure, of Newcastle, was stranded on the Herd sands, on the south side of Tynemouth haven, in the midst of the most tremendous breakers; and all the crew dropped from the rigging, one by one, in the presence of thousands of spectators; not one of whom could be prevailed on by any reward to venture out to her assistance in any boat of the common construction. On this melancholy occasion the gentlemen of South Shields called a meeting of the inhabitants, and premiums were instantly offered for plans of a boat which should be the best calculated to brave the dangers of the sea, particularly of broken water. Many proposals were accordingly offered; but the preference was given unanimously to Mr. Greathead's, who was immediately directed to build a boat at the expence of the committee.

This boat went off on the 30th of January 1790; and so well has it answered, and even exceeded every expectation, in the most tremendous sea, that during the last 21 years not less than 300 lives have been saved at the entrance of the Tyne alone, which otherwise must have been lost.

The principle of this boat appears to have been

suggested to Mr. Greathead by the following simple fact:—Take a spheroid, and divide it into quarters; each quarter is elliptical, and nearly resembles the half of a wooden bowl, having a curvature with projecting ends; this, thrown into the sea or broken water, cannot be upset or lie with the bottom upwards. The length of the boat is thirty feet; the breadth ten feet; the depth, from the top of the gunwale to the lower part of the keel in midships, three feet three inches; from the gunwale to the platform (within), two feet four inches; from the top of the stems (both ends being similar) to the horizontal line of the bottom of the keel, five feet nine inches. The keel is a plank of three inches thick, of a proportionate breadth in midships, narrowing gradually towards the ends, to the breadth of the stems at the bottom, and forming a great convexity downwards.

The stems are segments of a circle, with considerable *rakes*. The bottom section, to the floor-heads is a curve fore and aft, with the sweep of the keel.

The floor timber has a small rise curving from the keel to the floor-heads. A bilge plank is wrought in on each side, next the floor-heads, with a double rabbet or groove, of a similar thickness with the keel; and on the outside of this are fixed two bilge-trees, corresponding nearly with the level of the keel.

The ends of the bottom section form that fine kind of entrance observable in the lower part of the bow of the fishing-boat called a *coble*, much used in the north. From this part to the top of the stem it is more elliptical, forming a considerable projection.

The sides, from the floor-heads to the top of the gunwale, flaunch off on each side, in proportion to above half the breadth of the floor. The breadth is continued far forwards towards the ends, leaving a sufficient length of straight side at the top. The sheer is regular along the straight side, and more elevated towards the ends. The gunwale fixed to the outside is three inches thick. The sides, from the under part of the gunwale, along the whole length of the regular sheer, extending twenty-one feet six inches, are cased with layers of cork to the depth of sixteen inches downwards; and the thickness of this casing of cork being four inches, it projects at the top a little without the gunwale. The cork on the outside is secured with thin plate, or slips of copper, and the boat is fastened with copper nails.

The thwarts, or seats, are five in number, *double banked*; consequently, the boat may be rowed with ten oars.

The thwarts are firmly stanchioned. The side oars are short, with iron tholes and rope gromets, so that the rower can pull either way The boat is steered with an oar at each end; and the steering oar is one-third longer than the rowing oar.

The platform placed at the bottom, within the boat, is horizontal, the length of the midships, and elevated at the ends for the convenience of the steersman, to give him a greater power with the oar. The internal part of the boat, next the sides, from the under part of the thwarts down to the platform, is cased with cork; the whole quantity of which, affixed to the life-boat, is nearly seven hundred weight. The cork indisputably contributes much to the buoyancy of the boat, is a good defence in going along-side a vessel, and is of principal use in keeping the boat in an erect position in the sea, or rather for giving her a very lively and quick disposition to recover from any sudden *cant* or *lurch*, which she may receive from the stroke of a heavy wave. But, exclusively of the cork, the admirable construction of this boat gives it a decided pre-eminence. The ends being similar, the boat can be rowed either way; and this peculiarity of form assists her in rising over the waves. The curvature of the keel and bottom facilitates her movement in turning, and contributes to the ease of the steerage, as a single stroke of the steering oar has an immediate effect, the boat moving as upon a centre. The fine entrance below is of use in dividing the waves, when rowing against them; and, combined with the convexity of the bottom, and the elliptical form of the stem, admits her to rise with wonderful buoyancy in a high sea, and to launch forward with rapidity, without shipping any water, when a common boat would be in danger of being filled. The flaunching or spreading form of the boat, from her floor-heads to the gunwale, gives her a considerable bearing; and the continuation of the breadth well forward, is a great support to her in the sea; and it has been found, by experience, that boats of this construction are the best sea-boats for rowing against turbulent waves.

The internal shallowness of the boat from the gunwale down to the platform, the convexity of the form, and the bulk of cork within, leave a very diminished space for the water to occupy; so that the life-boat, when filled with water, contains a considerable less quantity than the common boat, and is in no danger either of sinking or upsetting.

It may be presumed by some, that in cases of high wind, agitated sea, and broken waves, a boat of such bulk could not prevail against them by the force of oars; but the life-boat, from her peculiar form, may be rowed a-head, when the attempt in other boats would fail. Boats of the common form, adapted for speed, are of course, put in motion with a small power; but, for want of buoyancy and bearing, are over-run by the waves and sunk, when impelled against them; and boats constructed for burden meet with too much resistance from the wind and sea when opposed to them, and cannot, in such cases, be rowed from the shore to a ship in distress.

Mr. Greathead gives the following instructions for the management of the life-boat:—

The boats in general of this description, are painted white on the outside; this colour more immediately engaging the eye of the spectator, when rising from the hollow of the sea, than any other. The bottom of the boat is at first varnished (which will take paint afterwards), for the more minute inspection of purchasers. The oars she is equipped with are made of fir of the best quality; having found, by experience, that a rove ash oar, that will dress clean and light, is too pliant among the breakers; and, when strong and heavy, from rowing double-banked, the purchase being short, sooner exhausts the rower; which renders the fir oar, when made stiff, preferable.

In the management of the boat she requires twelve men to work her; that is, five men on each side, rowing double-banked, with an oar slung over an iron thole, with a gromet (as provided), so as to enable the rower to pull either way, and one man at each end to steer her, and to be ready at the opposite end to take the steer oar when wanted. As, from the construction of the boat, she is always in a position to be rowed either way, without turning the boat: when manned, the person who steers her should be well acquainted with the course of the tides, in order to take every possible advantage: the best method, if the direction will admit of it, is to head the sea.

The steersman should keep his eye fixed upon the wave or breaker, and encourage the rowers to give way as the boat rises to it, being then aided by the force of the oars, she launches over it with vast rapidity, without shipping any water. It is necessary to observe, that there is often a strong reflux of sea, occasioned by the stranded wrecks, which requires both dispatch and care in the people employed, that the boat be not damaged. When the wreck is reached, if the wind blows to the land, the boat will come in shore without any other effort than steering.

The particular construction of this boat will be best understood by referring to plate XIV. where fig. 2. represents *a cross section of the life-boat.*

Fig. 2.

F F, the outside coatings of cork.

G G, the inside cork filling.

H H, the outside planks of the boat.

I, one of the stems of the boat.

K, the keel.

N N, the timber-heads.

P, the thwarts, or rowers' seats.

R, one of the stanchions under the thwarts, each being thus firmly supported.

S, a section of the gang-board, which crosses the thwarts, and forms the passage from one end of the boat to the other.

T, the floor-heads, or platform for the rowers feet.

U U, the two bilge pieces, nearly level with the keel.

W W, the gunwales.

X, a ring-bolt for the head-fast, there being another also at the other end.

Y, platform for the steersman.

Fig. 3. *A Longitudinal Section of the Life Boat.*

E E E, the sheer or curve of the boat.

I I, the two sterns or ends.

K, the keel.

L L, the aprons, to strengthen the stems.

M M, The sheets, or place for passengers.

N N, timber-heads or boat-fastenings.

O O O O, the tholes, on which the oars are slung by gromets.

T, the flooring under the rowers' feet.

Fig. 4. *Plan of a Truck or Carriage, with four Wheels, to convey the Boat to and from the Sea.*

a, an oblong frame of wood, consisting of two long pieces, hollowed a little, to admit the body of the boat, and secured by the cross-pieces *b b.*

c c c c, four low wheels, each sunk or hollowed in the middle, to run better upon a rail-way or timber-road.

d d, two indents made in the side-timbers, that the bottom of the boat may be firm therein.

e e, two small rollers, moveable in the cross-timbers, for the keel of the boat to slide upon.

f f, two long rollers, one at each end of the frame, to assist in raising the boat upon, or sliding it off, the truck or carriage.

Fig. 5. A, the copper bolt, with one end entered into the wood previous to fixing the tube.

B, a piece of timber or ship's side, into which the bolt is intended to be driven.

Fig. 6. C C C C, the parts of the iron tube fastened together, ready to be put on the bolt A.

D D D D, iron or brass rings, with thumb-screws placed over the joints of the tube, to hold them firm together.

E E E E, the thumb-screws, which keep the rings and tube firm in their proper places.

F, two points formed on the lower ring; they are to stick into the timber, and to enable the tube to be held firm in its place.

Fig. 7. Shews the separation of the parts of the tube, which is effected by slackening the thumb-screws and rings. To put them together, you slide the rings over the joints, placed as close as possible, then, by tightening the thumb-screws, you will have them firm together, and may continue the tube to any length, from one foot to whatever number is required.

Fig. 8. G H, two steel punches or drifts, to be placed on the head of the copper bolt, within the tube, whilst driving. The blow given upon the punch drives forward the bolt. The shortest of them should be used first, and, when driven nearly to its head, should be taken out of the tube and the longer punch applied in its place.

The life-boat is generally kept in a boat-house close to the beach, where it rests on four low wheels, or trucks, concave, for rolling on oars or spars laid on the sand, so as to run out at a moment's notice. But, where the way over which she must be dragged is rough, and the safety of her frame would be endangered, another expedient is adopted. This consists of two wheels, twelve feet in diameter, with a moveable arched axis, to which a pole is fixed for a lever. The boat is suspended between the wheels, under the axis; towards each extremity of which is an iron pin. When the pole is perpendicularly elevated, the upper part of the axis becomes depressed, and a pair of rope slings encompassing the boat, being fixed to the iron pins, she is raised with great facility by means of the pole, which is then fastened down to her stern. There are commonly two crews, each consisting of twelve men, employed to navigate the life-boat, to whom rewards are distributed according to the success of their enterprise.

On the first alarm of a vessel in danger, the life-boat immediately puts to sea; when some experienced steady person takes the command. Her head should be kept to the sea, and she must possess an accelerated velocity to meet the wave. Great caution is to be observed in approaching a wreck, where the reflux of the waves is often productive of danger, and it is considered safest to go to the lee quarter. This, however, depends on circumstances. See GREATHEAD'S *Report of the Evidence respecting the Invention of the Life-Boat.—Transactions of the Society of Arts*, Vol. X. XX. and XXV.

After the value of the invention had been acknowledged, by the presentation of a gold medallion to Mr. Greathead, by the Society of Arts, as also one by the Royal Humane Society, and various gratuities in money, Parliament, on the 9th of June 1802, unanimously voted him 1200*l.* The committee of Underwriters, likewise, at Lloyd's Coffee-house in London, having voted Mr. Greathead 100 guineas, appropriated 2000*l.* of their funds for the purpose of encouraging the building of life-boats on different parts of the coasts of the kingdom. Life-boats have been sent from Britain, on the order of the Emperor of Russia, who signified his approbation by presenting the inventor with a diamond ring, and by the Kings of Prussia, Denmark, and those of other states.

Previous to Mr. Greathead's invention, a patent was granted to Mr. Lionel Lukin, a coach-maker in London, for "an improvement in the construction of boats and small vessels, which will neither sink nor upset."

The essence of the invention rests on the propriety and practice of out riggers, well known to the natives of the South Sea Islands. Projecting gunwales are built to vessels of the ordinary construction, sloping from the top of the common gunwale towards the water, so as not to interrupt the oars in rowing; and from the extreme projection returning to the side in a faint curve, at a suitable distance above the water-line. These projecting gunwales are very small at the stem and stern, and gradually increase to the requisite dimensions: and they may either be solid, consisting of light substances, of cork, or hollow. In the inside of the vessel at stem and stern, and at the sides where projecting gunwales are unnecessary, as also under the seats and thwarts, are to be inclosures or bulk-heads, water-tight, or filled with substances specifically lighter than water. "By this means," the inventor observes, "the boat or vessel will be so much lighter than the body of the water it must displace in sinking, that it will with safety carry more than its common burden, though the remaining space should by any accident be filled with water." To give stability to the vessel, the inventor further proposes to affix a false keel of cast iron or other metal along the centre of the real one. The patent granted to Mr. Lukin is dated in 1785. It has not been discovered that he obtained any honorary reward for his invention; but, in the year 1807, a gold medal was voted by the Society for the Encouragement of Arts, to Mr. Christopher Wilson, for a "secure sailing-boat, or life-boat," which is balanced exactly according to Mr. Lukin's device, by empty projecting gunwales. For a particular description of this boat, which is called the *neutral-built self-balanced boat*, see the article BOAT.

Captain G. W. Manby, of the royal navy, has observed in his *Essay on the Preservation of shipwrecked Persons*, that the simplest method of giving the properties of preservation to any boat, is to well lash empty casks securely in it, so as to give it buoyancy; and, for the advantage of keeping it in an upright position, when launching from a flat shore or beaching, it should have bilge boards of equal depth with the keel, which likewise resists its upsetting, and a stout projecting rope, with swellings upon it, surrounding its gunwale, would serve as a fender to prevent its being stove. See the article SHIPWRECK.

Metallic LIFE-BOAT. A metallic life-boat, on pneumatic and hydrostatic principles, that will neither sink nor upset, yet serve all the ordinary purposes of ships' boats, either for rowing or sailing, was lately tried at London-bridge, on the ebb-tide, during the time of the greatest fall, with her crew on board, and filled with water; when she passed through with the greatest safety, and discharged a considerable portion of the water purposely put into her. These life-boats are described as being made of malleable iron, lead, and tin; one of twenty-feet long and six feet wide, drawing only ten inches water, with twenty-five persons; and possessing valves that, without pumping or personal aid, discharge all the water from them; which valves act occasionally as pneumatic or air valves: they are hydrostatically ballasted with confined water, taken in or put out at pleasure; and are remarkably buoyant and lively in agitated water.

LIFE-PRESERVER, an apparatus lately contrived by Mr. F. C. Daniel, of Wapping, to secure persons from sinking in water, when shipwrecked.

This machine resembles a broad belt, or circular girdle, composed of two folds of pliable leather attached together, and perfectly impervious to water: it is large enough to admit its encircling the body of the wearer, whose head is to pass between two fixed straps, which rest upon his shoulders: the arms of the wearer pass through the spaces on the outside of the straps; one on each side, admitting the machine under them to encircle the body like a large hollow belt.

There is a strap on the lower part of the machine, attached to the back of it, which is to be passed between the thighs of the wearer, and buckled before, so as to hold the machine sufficiently firm to his body, without too much pressure under the arms. The machine, being thus fixed, is inflated with air from the lungs, by the application of the mouth to a stop-cock attached to a flexible tube; when the whole is sufficiently filled, the turning of the cock retains the air in the machine, which is made capacious enough to hold such a quantity of air as will render buoyant upon water *four* persons suspended by the machine.

Mr. Daniel recommends his life-preservers to be prepared as follows:—viz. To select sound German horse-hides, and to cut a piece 6 feet long and 2 feet 6 inches wide, free from blemish or shell; it is first to be curried, and then rendered water-proof by Mollerstein's patent varnish, of Osborn-street, Whitechapel, which keeps the leather supple, and admits it to be easier inflated than any other water-proof leather. The leather is nailed on a board, and the varnish applied upon it; it is then to be passed into an oven several times, the varnish being each time repeated, till the leather is completely covered; then it is cut in the form of a jacket or waistcoat as above described, and firmly stitched, the seams being afterwards perfectly secured by the following black elastic varnish:—

℞. Gum asphaltum, 2 pounds; amber, half a pound; gum benzoin, 6 ounces; linseed-oil, 2 lbs.; spirits of turpentine, 8 lbs.; and lamp-black half a pound; united together in an earthen vessel with a gentle heat.

Mr. Daniel's description in the 25th vol. of the Transactions of the Society of Arts, is accompanied by testimonials from different persons proving the great utility of the life-preserver.

The gold medal of the Society of Arts was voted to Mr. Daniel for this contrivance in the Session of 1806. *See Trans. of the Society of Arts*, for that year.

TOPPING-LIFT. See the article TOPPING.

LIFTING, or SETTING-UP *a Ship in a Dock for Repair*, the act of resting her keel, as the water is gradually let out, on a number of wooden blocks, which are first fixed on the bottom of the dock to the height of about two or three feet; then setting up a variety of sustaining shores on both sides, with wedges, whereon she remains suspended after the blocks are removed: this being done, the calkers and shipwrights are enabled to perform the necessary repairs of the ship, such as calking the garboard seams, shifting the main keel, or adding an additional false keel, &c. When the repairs are accomplished, all the shores, except those at the breast, are knocked down; the water is then let into the dock again, high enough for the ship to ride afloat, when she is taken out for service.

This operation had been found extremely expensive and laborious, till Mr. Robert Seppings, master shipwright of Chatham-yard, invented a method a few years ago to obviate the necessity of lifting ships. His method of suspending, instead of lifting, ships, for the purpose of clearing them from their blocks, affords a very great saving to the public, and abridges two-thirds of the time formerly used in this operation. From the saving of time another very important advantage is derived, namely, that of enabling large ships to be taken into dock, suspended, and undocked, the same spring tides. Without enumerating the inconveniences arising, and, perhaps, injuries, which ships are liable to sustain from the former practice of *lifting* them, and which are removed by the present plan, that which relates to manual labour deserves particular attention; twenty men being sufficient to suspend a first rate, whereas it would require upwards of 500 to lift her.

The blocks constructed by Mr. Seppings, upon which the ship rests, are so contrived that the facility in removing them is proportionate to the quantity of pressure; and this circumstance is always absolutely under command, by increasing or diminishing the angle of three wedges, which constitute one of the blocks; two of which are horizontal, and one vertical. By enlarging the angle of the horizontal wedges, the vertical wedge becomes, of consequence, more acute; and its power may be so increased, that it shall have a great tendency to displace the horizontal wedges.

Mr. Seppings caused three blocks to be made of hard wood, agreeable to his invention, and the wedges of various angles. The horizontal wedges of the first block were nine degrees; of the second, seven; and of the third, five; of course, the angle of the vertical wedge of the first block was 162 degrees; of the second, 166; and of the third, 170. These blocks or wedges were well executed, and rubbed over with soft soap, for the purpose of experiment. They were then placed in a dock in his Majesty's yard at Plymouth, in which a sloop of war was to be docked; on examining them, after the vessel was in and the water let out, they were all found to have kept their situations, as placed before the ship rested upon them. Shores in their wake were then erected to sustain the ship, prior to the said blocks being taken from under the keel. The process of clearing them was, by applying the power of battering-rams to the sides of the outer ends of the horizontal wedges; alternate blows being given fore and aft; by which means they immediately receded, and the vertical wedges were disengaged. It was observed, even in this small ship, that the block which was formed of horizontal wedges of nine degrees, came away much easier than those of seven; and the one of seven, than that of five. In removing the aforesaid blocks by the power of the battering-rams, which were suspended in the hands of the men employed, by their holding ropes passed through holes for that purpose; it was remarked by Mr. Seppings, that the operation was very laborious to the people, they having to support the weight of the battering-rams, as well as to set them in motion. He then conceived an idea of affixing wheels near the extremity of that part of the rams, which strikes the wedges. This was done before the blocks were again placed; and it has since been found fully to answer the purpose intended, particularly in returning the horizontal wedges to their original situations, when the work is performed for which they were displaced; the wheels also giving a great increase of power to the rams, and decrease of labour to the artificers, besides which, the blows are given with much more exactness. This and other experiments being communicated to the Navy Board, Mr. Seppings was directed to attend them, and explain the principle of his invention; which explanation, further corroborated by the testimonials of his then superior officers, was so satisfactory that a dock was ordered to be fitted at Plymouth, under his immediate directions. The horizontal wedges in this and in other docks that were afterwards fitted by him are of cast iron, with an angle of about five degrees and a half, which, from repeated trials, are found equal to any pressure, having in no instance receded; and, when required, were easily removed.

The vertical wedge is of wood, lined with a plate of wrought iron, half an inch thick. On the bottom of the dock, in the wake of each block is a plate of iron, three quarters of an inch thick, so that iron at all times acts in contact with iron.

The dock being prepared at Plymouth in August 1801, the Canopus, a large French 80-gun ship, was taken in, and rested upon the blocks; and the complete success of the experiment was such, that other docks were ordered to be fitted at Sheerness and Portsmouth dock-yards, under Mr. Seppings's directions. At the former place a frigate, and at the latter a three decked ship, were suspended in like manner. This happened in December 1802, and January 1803; and the reports were so favourable as to cause directions to be given for the general adoption of these blocks in his Majesty's yards. This invention being thought of national consequence, with respect to ships, but particularly those of the navy, government has been pleased to notice and reward Mr. Seppings for it.

The time required to disengage each block, is from one to three minutes after the shores are placed; and a first rate fits on about fifty blocks.

The following is the process of obviating the necessity of *lifting* ships, when there may be occasion to put an additional false keel to them, or to make good the imperfections of those already on; also when it may be necessary to calk the garboard seams, scarples of the keel, &c.

The blocks are first laid down, and a sufficient number of shores placed under the ship to sustain her weight, and set taught, stationed as near the keel as the working of the battering-rams fore and aft will admit. Care, however, must be taken to avoid placing any opposite the blocks, as they would, in that case, hinder the return of the wedges with the battering-rams. A blow must then be given forward, on the outer end of the iron wedges, with the battering-rams, in a fore and aft direction, which will cause them to slide aft. The battering-rams abaft then return the blow, and the wedges again come forward; by the repetition of this operation, the wedges will be with great ease cleared, and the angular block on the top will drop down. When the work is performed, the block must be replaced under the keel, and the wedges driven back by working the rams athwart ships.

In returning the iron wedges, to avoid straining the angular blocks, it is proposed to leave a few of them out forward and aft, and stop the ship up, by laying one iron wedge on the other. To facilitate the business, blocks may be cleared forward and aft at the same time, sufficient to get in place one length of false keel. If the false keel should want repairing, it may be done without any additional shores, by clearing one block at a time; and, when the keel is repaired in the wake of that block, return the wedges, as above directed, and clear the next, &c. For an engraving, &c. of this invention, see Nicholson's *Phil. Journal*, Vol. XI.

LIFTS, (*balancines*, Fr.) certain ropes, descending from the cap and mast-head, to the opposite extremities of the yard, immediately under; where, passing through a block or pulley, they become double. They are used to keep the yard in equilibrio; or to pull one of its extremities higher than the other, if occasion requires; but particularly to support the weight of it, when a number of seamen are employed thereon, to furl or *reef* the sail.

In some merchant vessels, the lifts of the top-sail-yards, called the top-sail-lifts, are also used as *sheets*, to extend the clews of the top-gallant-sail.

The yards are said to be squared by the lifts, when they hang at right angles with the mast; that is to say, parallel with the horizon, when the vessel is upright in the water.

LIGHT, (*lumière*, Fr.) is that principle or thing by which objects are made perceptible to our sense of seeing; or the sensation occasioned in the mind by the view of luminous objects. The nature of light has been subject of speculation from the first dawnings of philosophy; but the arguments of philosophers on this subject are too numerous for insertion here.

The Newtonians, however, maintain that light is not a fluid, but consists of a great number of very small particles, thrown off from the luminous body by a repulsive power, with an immense velocity, and in all directions. And these particles, it is also held, are emitted in right lines; which rectilinear motion they preserve till they are turned out of their path by some of the following causes; viz. by the attraction of some other body near which they pass, which is called inflection, or by passing obliquely through a medium of different density, which is called refraction; or by being turned aside by the opposition of some intervening body, which is called reflection; or lastly, by being totally stopped by some substance into which they penetrate, and which is called their extinction. A succession of these particles following one another in an exact straight line, is called a ray of light; and this ray, in whatever manner its direction may be changed, whether by refraction, reflection, or inflection, always preserves a rectilinear course, till it be again changed; neither is it possible to make it move in the arc of a circle, ellipsis, or other curve.

The velocity of the rays of light is truly astonishing, amounting to nearly two hundred thousand miles in a second of time, which is about a million times greater than the velocity of a cannon ball. And this amazing motion of light has been manifested in various ways, and first from the eclipses of Jupiter's satellites. See the article Aberration; and for farther particulars on the subject of light, see Hutton's, *Mat. and Phil. Dictionary*.

Light, (*léger*, *ou lége*, Fr.) is used in contradistinction to laden; a ship is, therefore, said to be light when she has no cargo, or when she is not sufficiently ballasted.

To Light, (*faire recourir une manœuvre*, Fr.) is a term sometimes used by sailors instead of to help; as, " Light along that rope?"

Light *Composition*, consists of eight parts of saltpetre, four parts of sulphur, and one part of antimony; or sometimes of other proportions, together with some mealed and grained powder. See the article Composition.

LIGHTER, (*barque*, *ou gabare*, Fr.) a large open flat-bottomed vessel, generally managed with oars, and employed to carry goods to or from a ship when she is to be laden with, or delivered of, her stores, &c.

Ballast Lighter, (*lesteur*, *ou bâtiment destiné à porter le lest*, Fr.) a vessel fitted up to heave ballast from the bottom of a harbour or river, and to carry it to and from ships.

Covered or *Close* Lighter, is furnished with a deck throughout her whole length, in order to contain those merchandises which would be damaged by rainy weather, as also to prevent pillage.

LIGHT-HOUSE, (*fanal*, *phare*, *ou tour à feu*, Fr.) in the marine, a sort of tower, or high building, erected upon a headland or promontory, on the seacoast, or upon some rock in the sea, and having a great fire, or light from oil with reflectors, upon its top, in the night time, which is constantly attended by some careful person, so as to be seen at a great distance from the land. Its use is to direct the shipping on the coast, that might otherwise run ashore, or steer an improper course, when the darkness of the night, and the uncertainty of currents, &c. might render their situation, with regard to the shore, extremely doubtful.

As many valuable lives and great property often depend upon the certainty of seeing light-houses, at a sufficient and proper distance, no pains or expense should be spared to make them as perfect as possible, to answer their designed purpose, especially where there is a sufficient fund allowed to support them; or where the increase of trade or shipping make large profits arise from them.

It certainly deserves the attention of the public, to get our light-houses improved upon the best plan, for the greater safety to shipping, and the consequent advantages to all concerned therein.

It is well known, from reason as well as experience, that an open coal-fire light, exposed to all winds and weathers, cannot be made to burn and show a constant steady blaze, to be seen at a sufficient distance with any certainty; for, in storms of wind, when lights are most wanted, these open fires are made to burn furiously and very soon away, so as to melt the very iron work about the grate; and, in cold weather, when it snows, hails, or rains hard, the keepers of the lights, not caring to expose themselves to the bad weather, are apt to neglect the fire till it is too low, and then throw on a large quantity of coals at once, which darkens the lights for a time, and till the fire burns up again; and in some weathers it must be difficult to make them burn with any brightness. And when they are enclosed in a glazed close light-house, they are apt to smoke the windows greatly; nor do they afford so constant a blaze as oil lamps, with reflectors behind them; a method now generally adopted.

The following is an account of the most remarkable lights and light-houses around the British coast, with directions to evade the most dangerous rocks, shoals, &c.

In consequence of the number of vessels being wrecked within the last twelve years, on the *Inch Cape*, or *Bell Rock*, at the entrance of the Friths of Forth and Tay, in north latitude 56° 29′, and west longitude 2° 22′, a light-house of considerable magnitude has lately been built thereon, by an Act of Parliament, upon principles similar to those of Edd̄ystone light-house.

The building is 116 feet in height; its diameter, at the base, is 42 feet; the solid contents of the whole being upwards of 3000 tons. The first 30 feet of the light-house is entirely solid, and the stones and courses of this part are connected after Mr. Smeaton's method at the Eddystone. The ascent to the door, which is at the top of the solid, is by a kind of rope-ladder. The staircase occupies 14 feet of the height; and the remaining 56 feet is laid out in apartments for the light-keepers, and in rooms for the stores, which communicate by means of wooden ladders. The building terminates with the light-room, and is covered in by its roof.

The foundation is on a level with low water of spring tides; the sea, accordingly, by the influence of the tide, seldom rises more than 16 feet upon the building; yet, in July 1810, when the light-house was built 70 feet high, the workmen were actually beat off the walls by the sea spray.

The light in this stupendous edifice was, by order of the commissioners of the northern light-houses, first exhibited in the night of Friday, the 1st of February 1811, and is exhibited every night, from the departure of day-light in the evening, till the return of it in the morning. The light is from oil, with reflectors, at the height of about 115 feet from low water of spring tides. To distinguish it from others on the coast, it is made to revolve horizontally, and to exhibit from all points of the compass a bright light, and a light of a red colour, alternately, both shewing themselves in the space of three minutes; so that, in each revolution of three minutes, there will be seen a brilliant light, appearing at a distance like a star of the first magnitude, which, after attaining full strength, is gradually eclipsed; and, after a short interval of darkness, is succeeded by a light of a red colour, which, in like manner, increases to full strength, diminishes, and disappears. The coloured light, being less powerful, may not be seen when the bright light is first noticed; but the periodical revolution of the bright light is, however, sufficiently distinguishable. In thick foggy weather, a bell is tolled by machinery, night and day, at intervals of half a minute. The floating-light, which was moored two miles and a half N.W. half N. from the Bell Rock, is, consequently, discontinued, and the vessel removed from her station.

The accommodation of this important national work has already been felt, and no doubt can be entertained that the future benefits of the light-house, arising to the trade and navigation of the kingdom will be great; that it will, unquestionably, be the means of preventing many wrecks, and of preserving the lives and property of thousands of brave seamen in the North Sea; as ships, instead of avoiding it, as they have heretofore done, from the perilous situation of the Bell Rock, will, in thick weather, be glad of finding it, wishing to ascertain their situation, from not having before made the land, or being out of their reckoning, or from the influence of the tides.

This rock is situated at the opening of the bay, formed by the Readhead, in Forfarshire, on the north side, and Fifeness on the south; and is about five miles on the outside of a line drawn between these two points; it lies in a N.E. and S.W. direction, and is dangerous for about three quarters of a mile; and for about two-thirds of that length, is dry at low water. The N.E. end, which is the highest and broadest part, continues dry till half flood, and there is very deep water close to it all round, except at the S.W. point, where it is a flat or shallow, which extends about a quarter of a mile, and in that direction the danger from shallow water extends further. The breadth of the rock is about 150 yards; its length, including the S.W. point, half a mile; and its surface is very rugged.

Casket Lights. About four miles N.W. by W. from Ortac, and seven miles, in the same direction, from the westernmost point of Alderney, lie the Caskets, a cluster of rocks, some above the water, and many below it. On the largest of those above the water, stand three light houses, in a triangular form, which may be seen, in clear weather, three or four leagues off. The lights exhibited at the Caskets are composed of a number of Argand lamps, fixed on a ring, moving in a circular revolution, and presenting, alternately, a bright body of light in every direction.

The Casket lights are in latitude 49° 45′, longitude 2° 25′ W

Cromer Light-house. This light-house is situate

on the sea-coast of Norfolk, and lies in latitude 52° 55′, longitude 1° 20′ E. Cromer was once a much larger town, but a considerable part of it has gradually been swallowed up by the sea.

Cromer Outer Bank. N. W. by W. 10 miles from the north end of the Leman, 14 miles N. by E. ½ E. from the north end of Hasborough-sand, and 19 miles N. E. ½ E. from Foul-ness light, lies Cromer Outer-bank; which extends N. W. and S. E. one mile and a half, and three-quarters of a mile broad. On its south-east end are 2½ fathoms, and, on the north-west end, 3½ fathoms, at low water.

Cromer Inner Bank. W. S. W. 7 miles from Cromer Outer-bank, N. by W. ½ W. 9 miles from the north end of Hasborough-sand, and 11 miles N. E. ½ N. from Foul-ness light, lies Cromer Inner-bank; which extends one mile and three-quarters N. W. by N. and S. E. by S. and is a mile broad. On the middle of this bank are 4 fathoms, 6 on its north end, and 7 on the south end, at low water. Between this bank and the outer one are 17 fathoms: between the inner bank and Hasborough-sand are 12, 18, and 20 fathoms; and between it and Foul-ness are 16, 18, 16, 12, and 9 fathoms, at low-water.

Description of the Shoals, and Directions for sailing between Foul-ness and Flamborough-head. To the northward of Hasborough-gat, or rather between Foul-ness and the Spurn, there are so many shoals and shifting overfalls, as to render this part of the North Sea unsafe for line of battle ships, except in neap-tides and very moderate weather. It will, therefore, be necessary to give a description of the principal banks and overfalls which impede this navigation.

Leman and Ower. The body of these dangerous shoals, which have been growing up for a long time past, is now, in some places, nearly dry at low water. The body of them lies E. by N. 9 leagues from Foulness; S. E. by E. 11 leagues from the Dudgeon-floating-light; and S. E. 22 leagues from the Spurn. The Leman extends about five miles and a half nearly S. ½ E. and N. ½ W. and is more than a mile broad. The Ower lies nearly parallel with the Leman, and is about four miles long, and a mile broad. Between the sands is a channel of from 10 to 18 fathoms. Near the southern part of the shoals are 9 fathoms; and, near the northern, 15 fathoms. Near the Ower, on the east side, are 17 fathoms, and near the Leman, on the west, 20 fathoms, at low water. On the east side of the Ower the tide runs in various directions; the first of the flood runs E. by S. and afterwards S. E. for the greater part of the tide: near high water it runs S. W. by S. and the ebb-tide in contrary directions. The tide flows here on full and change days, at 7 o'clock, and the stream runs till half past 10 o'clock.

Haddock Bank. N. ½ E. 15 miles from the north end of the Ower, and 35 miles N.E. from Foul-ness, lies, the south end of the Haddock-bank, extending 6½ miles N. N. W. ½ W. one mile and three-quarters broad. On the middle of this bank are 5 fathoms; and towards each end are 6 and 7 fathoms, at low water: near it, on each side, are 14 fathoms. Between this bank and the Ower are from 14 to 24 fathoms. N. by W. 12 miles from the north end of the Haddock-bank, and N. N. E. ½ E. 50 miles from Foul-ness, lies the N. N. E. hole, in which are 40 fathoms. Between this hole and the Haddock-bank are from 13 to 16 fathoms water.

Overfalls. Parallel to, and about five miles from, the shore, extending N. W. by W. ½ W. lie two narrow banks, called Sheringham and Blakeney Overfalls. The east end of Sheringham Overfall, on which are 4 fathoms, bears N. from Cromer; it thence extends about 5 miles. The shoalest water on it, 2¼ fathoms, is on the middle of the sand: on the west end are 5 fathoms; and thence to the east end of Blakeney Overfall, nearly a league distant, are from 6 to 7½ fathoms. On the east end of Blakeney Overfall are 5 fathoms; on the west end 3 fathoms; and, with Blakeney-church bearing S. but 10 feet at low water. In the middle of the channel, between these overfalls and the shore, and N. W. by N. 10 miles from Foul-ness, is a small shoal, called the Pollard, on which are but 3½ fathoms at low water. N. W. 4½ miles from Blakeney, and 2 miles N. of Stuky, is also a danger, more than a mile in length, called Stuky Overfalls, on which are but 9 feet at low water. The depth between this shoal in mid-channel and the shore, is 4 fathoms.

Dudgeon Light. This floating light lies 24 miles N. of Foul-ness light, and rides a little to the westward of the shoal called the Dudgeon. The shoal extends N. N.W. and S. S. E. is about three miles long, and one broad, with 10 feet on its shoalest part, near the light, at low water. On the southern part are 5 and 6 fathoms. About a mile N. E. from the north-east part of the shoal, lies a small knowl of 3 fathoms water: and, 2¼ miles N. ½ W. from the north-west part of the shoal, is another small knowl of 2 fathoms. Near these knowls and Dudgeon's shoal and bank, on all sides, are from 7 to 10 fathoms at low water. To the S. W. of Dudgeon is a shoal, of which, from a cursory survey made of it, by order of the corporation of Trinity-house, in 1804, we have the following particulars: — "From the shoal the Dudgeon-light bears N. E. about six miles; Blakeney-church S. ½ W. Cromer or Foul-ness light S. by E. ½ E. and Holliham-church S. W. ½ W. the north end is nearest the light, but that is not the shoalest part; the south end has about 3 fathoms at low neap-tide; but, at a low spring, not more than 14 feet: it is very narrow, and ranges about N. and S. in length, from 7 fathoms at one end to 7 fathoms at the other, about 3 miles." The Dudgeon-light is in latitude 53° 17′, longitude 1° 2′ E.

Eddystone Light-House. The Trinity-house corporation, of London, has recently caused new lights to be made at the Eddystone light-house, which are upon the improved principle, with *Argand* (otherwise *Quinquet*) lamps, and reflectors; producing a light of great brilliancy, and visible to a much greater distance than the old light with candles.

Numerous spectators assembled on the Hoe, at seven o'clock, in the evening of Monday, the 10th of September 1810, to see the light exhibited the first time; when the effect was instantaneous, and exceedingly brilliant; and the night being very fine before the moon rose, the lights had a very pleasing appearance. This is a great improvement in this elegant structure.

Plymouth Hoe is exactly 15 miles from the light-house; and the light-house is in latitude 50° 11', longitude 4° 15' W.

Galloper. Nearly S.S.E. ¼ E. seven leagues from Orfordness, lies the north end of a shoal called the Galloper, which thence extends S.S.W. about six miles. It is about one mile across in the broadest part, and has, where shoalest, only nine feet over it at low water. The channel between it and the long-sand is six leagues and a half wide, and has from 15 to 24 fathoms.

Near the sand, on the east side, are 20 fathoms; and 2½ miles from it, on the same side, are 27 fathoms, coarse ground, with small black stones. Near the south end are 16 fathoms, stony ground; about three-quarters of a mile without the north end are 14 fathoms, with coarse stones; and close to this end are 12 fathoms. On the north end are 6 fathoms at low water.

Pursuant to the directions of the Lords Commissioners of the Admiralty, in the year 1804, a vessel has been moored at the S.W. end of the Galloper, in which two lights are exhibited from sun-set to sun-rise, for the benefit of his Majesty's cruisers; and, in order to prevent any mistake of this for the other lights in the vicinity, it is to be observed, that the vessel has two masts, with a light at each mast-head, that it may be readily distinguished from the Sunk and Goodwin lights; the former of which is exhibited with one light only, and the latter with three.

The vessel was originally moored within less than half a mile of the sand, in 13 fathoms; but it now lies in 15 fathoms, about two miles S.W. from the shoalest part of the sand. N.E. by N. three miles from the light, there are six fathoms. With the light N.W. by W vessels may cross the sand in five fathoms. Within the light-vessel a buoy is placed, that the usual situation of the light-vessel may be known, should it drift or be taken from its place.

Lizard-Point. The Lizard is a bold land, and several rocks lie near it, called the Stags; some of them, about a quarter of a mile without the point, appear at half-tide. The mark to clear them is, Black-head-point kept just open of the Lizard-point, until the two light-houses on the Lizard appear in one, or until Godolphin-hill appears above the land, to the westward of the Lizard.

The Lizard-point may be seen seven or eight leagues off; it lies 39 miles W. ½ N. from the Eddystone. From the Lizard to Pealea-point, the course is N.N.W ¼ W. 14 miles; from the Lizard the Rundel-stone lies 18 miles N.W. ½ W. and the Wolf-rock, N.W. by W. ¼ W. northerly, distance 28 miles.

The Trinity-house corporation has also, in compliance with the request of the owners and masters of ships navigating the Channel, caused two lights at the Lizard Point, which have hitherto been shewn by coal fires, and found very defective, to be altered to oil, upon the improved principle, with air lamps and reflectors, producing lights of superior brilliancy, that will be visible to a greater distance in every direction where requisite for the guidance of navigation. This alteration took place on the 16th of January 1812.

The Lizard flag-staff is in latitude 49° 57' 55",8''' N. longitude, 5° 11' 17",7 W. from Greenwich; the difference of time being 20° 45',2, according to the *Requisite Tables* (1802).

In the table of geographical positions, published by the French, *Bureau des Longitudes*, 1808, this headland is placed in the same latitude, without the fraction (8'''); but its longitude is made 15",7 more east.

Longships. About three miles N.N.W. ½ W. from the S.W. point of the Land's-end, and one mile W.N.W. from the westernmost point, lie the high rocks, called the Longships; they extend about half a mile north and south.

A light-house stands upon the largest of the Longships, the lantern of which is so lighted with Argand-lamps and reflectors, as to be clearly seen from Point Tol-Pedan-Penwith to Cape Cornwall. The light bears from Point Tol-Pedan-Penwith N.N.W. ½ W. distant 3½ miles; from Cape Cornwall, S.W. ¼ W. 4 miles; from the Brissons, S.W. ¼ S. 3¼ miles; from the Rundel-stone, N.N.W. northerly, near 4 miles; from the Wolf-rock, N.E. northerly, 7½ miles; and from the light-house of St. Agnes, southerly, 25 miles.

Ships sailing down the British Channel, and bound round the Land's-end, cannot make the light till it bears N.N.W. ½ W. or open of Point Tol-Pedan-Penwith, on account of the high land which covers it from this point to Cape Cornwall; but, having seen it, and brought it to bear N. 21° ½ W. will have the Rundel-stone in the directions of the light; and, by bringing the light to bear N. by W. or N. may steer safely for the light clear of the Rundel-stone and all danger, and may give the light any convenient berth, as the westernmost rock of the Longships lies only about half a cable's length from the light-house.

Ships bound from the south-westward of the Light to the northward, must keep this light clear of a N.E. direction, on account of the Wolf-rock; but, by keeping the light a point or two to the eastward or northward of this direction, till they have passed the Wolf, may with certainty avoid it. The same observations will hold good, if bound to the southward, for both these rocks. Ships bound either to the northward or southward, when to the northward of the light, ought to keep it a point or two to the southward of S.W. ¼ S. in order to go to the westward of the Brissons, which are two high and bold rocks, or small islands; but there is no safe passage between them and Cape Cornwall.

Lowestoff Lights. Lowestoff is a town of Suffolk, seated on the sea-shore, and is 10 miles S. of Yarmouth.

The coast is very dangerous for strangers. The light-house is situate just open of the north end of the town.

In sailing from Orfordness to Lowestoff, go no nearer to the shore than 8 or 9 fathoms; this depth of water will carry you without the Onion, Sizewell-bank, and the Barnard-sand. The soundings towards and in the offing are regular.

Between Sizewell-bank and the Barnard lies Sole-bay, or Southwold-bay, where is good riding with off-shore winds, in 8 or 9 fathoms. The south end of the Barnard lies E. and about three-quarters of a mile from the shore, off Covehithe church. This sand extends a mile and a half N.E. by N. and is a quarter of a mile broad, having on it two feet at low water. Between the south end and the shore are 9 feet water; and, towards the north end, 6, 4, and 3 fathoms. To keep clear of the Barnard, avoid bringing Southwold-church on with Easton-houses, until Lowestoff church comes open of Pakefield-cliff. Go no nearer to this sand than 9 fathoms, as, close to it, on the east side, are 8 fathoms.

The course from off the south end of the Barnard to a black buoy on the Holm-sand-head, is N.N.E. ½ E. 5 miles. Between are 8, 7, 6, 4, and 3 fathoms at low water.

In sailing from Lowestoff to Yarmouth.—Having run into the Sandford, and passed the outer buoy, steer N.N.E. until you have passed the N.E. buoy of the Newcome; thence in a fair-way along shore, to abreast of Yarmouth-jetty, the course is N. by E. and distance 6 miles. In working you may stand towards the shore in 4 or 5 fathoms, and towards the Corton to 7 fathoms. In all the space within the Corton there is good anchorage, but the best is with Corton-church N. by W. ½ W. and Lowestoff high light S. W. by W. in 7 and 8 fathoms. Off the south part of Corton-road is also good riding with southerly winds, in 6 fathoms; with easterly winds you may ride well, near the sand, in 8 fathoms; but, with the wind northerly, you will ride best to the northward of Yarmouth-jetty.

Lowestoff light-house is in latitude 52° 30′, longitude 1° 47′ E.

The Needles Lights and Passage. Two light-houses are erected for the benefit of ships *going in* at the Needles in the night; one on the point to the eastward of the Needles, and the other on the beach to the southward of Hurst-castle. As the north-west part of Hurst light-house is darkened, so as to prevent the light from being seen by ships that are farther to the westward from it than S.W. by W. ¼ W. ships, when coming *from the westward*, for this passage in the night, should bring the light open of the north-west darkener in time; it bears, when just open, about N.E. by E. ¼ E; and, as this line of direction runs close along the east side of the Shingles, it will be necessary in ebb-tides, which set strongly over them, to bring the light N.E. by N. and, with a flood tide, to keep it barely open, until they draw near Hurst-point, provided the wind be free; if otherwise, you must be directed by judgment.

If bound out from Yarmouth-roads through the Needles, first steer for Sconce-point, which is bold; you may approach it to about half a cable's length. FromSconce-point steer towardsHurst-castle. In mid-channel is from 15 to 25 fathoms, the deepest water nearest the main. When abreast of the castle, steer for the Needles-point, in from 16 to 5 fathoms. To run through this channel, bring Hurst-castle N.E. by E. ¼ E. it will then be in one with a cupola-house, to the eastward of Lymington-creek, called Pilewell-bath. This mark will carry you over the bridge, clear out to sea. In standing off from Hurst-castle for the Needles-point, be careful to keep the vanes of the windmill, which stands on the Isle of Wight, in sight above the land; or, keeping the Needles light-house open of Hetherwood-point, is a better mark to avoid the Warden-ledge.

The thwart mark for Warden-ledge is, the cupola on Hurst-castle, in one with the flag-staff; but, the best mark is a salt-house that stands on the beach, just within Hurst-castle, in one with Milton-tower. This ledge of rocks runs off about a quarter of a mile from Warden-point; on its extremity is a buoy, lying with the windmill about S.S.E. About W.N.W. from the Needles-point is the bar-shoal, called the Bridge; on which there is no more than 3½ fathoms at low water spring tides. This shoal lies with the Needles all on with each other; and it is so very narrow, that a ship's going at the rate of four miles per hour, will prevent the leadsman from getting a second cast upon it.

The Needle-rocks in one, when on the Bridge.—The course from Hurst-castle is S.W. by W. ¼ W

To avoid the Shingles, in working through the Needles, with the wind from the south-westward, take care to keep Cowes-point shut in with Sconce-point. Should the wind baffle, which is often the case in this narrow passage, take particular care not to open West Cowes-point with Sconce-point. Should the weather be hazy, do not fear bordering on the island side; but stand no farther into Totland-bay than 7½ or 8 fathoms. When clear, and both sides can be distinctly seen, your eye is the best pilot for the island side.

Ships, in order to be ready to pass through the passage at the Needles, may anchor under Hurst-beach, in from 8 to 10 fathoms water. The best mark for anchoring there is Chisel's-door (which is very remarkable from its standing a little to the west and southward of a black house) and a light-house in one.

The mark for the north-east point of the Shingles is, the windmill, which stands upon the island, on with the wood to the westward of the hill, off which Warden-ledge lies: there are two woods, but the easternmost one is that which is now referred to. In the fair-way, Sconce-point will be on with the trees on Portsdown-hill. Always be careful to keep the vanes of the windwill, which stands on the island, in sight, which will keep you clear of Warden-ledge. Do not approach nearer the Shingles than 6 fathoms, but you may stand towards the island to about a cable's length, as it has a bold shore. After passing the Warden-ledge, back and fill, with your ship's head towards the island, observing to keep within a cable's length of it, and the tide will set you through the Needles.

As the tides run with great violence, especially springs, this channel is but seldom attempted, unless with a fresh gale and leading wind. The ebb-tide sets strongly over the Shingles, and the flood towards the Needles-point.

Between Hurst-castle and the Shingles is a small narrow channel of 4½ and 3½ fathoms at low water. It is but seldom used by any, except small coasters. The leading mark through it is, Sconce-point just touching Hurst-castle; which will carry you to the northward of the Shingles, into Christ-church bay, in 5½ and 6 fathoms water.

The Needles lights are in latitude 50° 40′, longitude 1° 34′ W.

The Lords Commissioners of the Admiralty having given directions for a vessel, fitted as a floating light, to be moored off the point of Bembridge, at the east end of the Isle of Wight, as a guide to shipping bound in or out of St. Helen's-road, by night or by day; notice was, on the 20th of July 1812, given by the corporation of the Trinity-house at London, that the said floating-light should be completed and moored at the station above-mentioned, by the 29th of September following: it was accordingly completed, and the lights are exhibited thereon every night, and continued constantly, from sun-set to sunrise, for the benefit of his Majesty's ships and navigation in general.

And masters and pilots are to observe, that two distinct lights are shewn from this vessel, in two lanterns, suspended from two separate masts, at 43 feet distance asunder, and hung at different heights; the one lantern at 25 feet, the other at 18 feet height above the vessel's deck; by which this floating-light may be readily distinguishable from the Ower's floating light, and other lights in that vicinity.

Bembridge-ledge is in latitude 50° 41′, longitude 1° 2¼′ W

Portland Lights. The lower light-house, from the east end of the Shambles, bears nearly N.W. by W. ¼ W. and the west end N.N.W. ½ W.

Wyke church is about a mile to the westward of Weymouth, and may be easily known.

In Portland-road is excellent anchoring, with westerly and southerly winds. In going in, pass either to the eastward or to westward of the Shambles, according as is most convenient; to go to the westward of them, the mark is, the north-east point of Portland on with Wyke-Regis church, or the north-east point, bearing N. by W. ¾ W. You may edge towards the north-east point of Portland, in 5 fathoms; when within this point, you may haul closer to the south shore. From St. Alban's head, the east end of the Shambles lies about 11½ miles, W. by N. and 4½ miles E.S.E. from the Bill of Portland; thence they extend W. by N. 2¼ miles. This is a dangerous shoal of coarse sand and shingles, and steep all round, which always shews itself by a break or ripple. On the east end are from 3 to 5 fathoms; and, on the west end, from 6 to 7 fathoms; but, on the shoalest part, nearly the middle of it, are only 12 feet at low water, spring-tides.

The Race of Portland is a great rippling, about S.S.W. from the Bill, or the southernmost point of Portland; it is about one mile and three quarters in length from east to west, extending about one mile to the eastward of the point, and three quarters of a mile to the westward of it. The Race is caused by the unevenness of the ground over which they pass.

The leading marks for sailing between the Race and the Shambles are, the light-houses in one, or N.N.W ¼ W. by compass, or the east part of Portland N.N.E. ½ E. which will lead ships clear between them. This passage should not be attempted at flood-tide with little wind, as the tide sets right on the Shambles at new and full moon, at the rate of 6½ knots an hour. The time of high-water, at the light-house point, is 9¼ hours, at full and change. Portland lights are in latitude 50° 31′, longitude 2° 27′ W.

Scilly Islands and St. Agnes' Light-house.—A light-house is erected on St. Agnes' Island, in which are exhibited a number of Argand lamps, fixed in a ring, moving in a circular revolution; and presenting a bright and conspicuous light, in every direction, once in a minute.

Scilly is remarkable for a cluster of islands and rocks; which lie almost 10 leagues to the W. of the Land's-end, in Cornwall, and are easily discerned from it.

The chief of these islands is St. Mary's; which has a good port.

The Scilly rocks have been destructive to numbers of ships entering the Channel. One of the most disasterous events of this kind happened in 1707, when three men of war, under Sir Cloudesley Shovel, were lost, with all their crews.

Directions for sailing into the Scilly Islands. In coming from the southward, you will descry Scilly, in clear weather, 6 or 7 leagues off; and have 60 fathoms, grey sand with broken shells; it may be seen, also, on the south side, in 55 fathoms water, stony ground, with some red shells. But, coming from the north, at 7 leagues distance, you will have sand and oaze mixed together.

The island of St. Mary is nearly W. N. W. ¼ W. 24 miles from the Land's-end, and W. N.W. 6 leagues from the Wolf-rock.

The Seven Stones, dangerous rocks, over which the sea breaks, lie 5½ leagues N. W. by W. from the Land's-end. They extend near two miles N. W. and S. E. and are about a mile broad. The breakers, in clear weather, may be seen 3 or 4 leagues off. On the east and west side of these rocks are 52 fathoms; on the north from 30 to 40 fathoms; and on the south side 50 fathoms. Between the Land's-end and the Seven Stones, and between the Seven Stones and Scilly, the tide runs 9 hours to the northward, and but 3 hours to the southward.

There are four principal channels or sounds to go into Scilly; viz. St. Mary's Sound, Crow Sound, Smith's Sound, and Broad Sound; which are divided into N. W. and S. W. Channels. There are, besides, Old and New Grimsby, St. Helen's-gap, Tean Sound, &c. These islands are surrounded by a vast number of dangers; to avoid which the utmost care is required, and directions given in the *Coaster's Companion* for 1812.

St. Agnes' light-house is in latitude 49° 54′, longitude 6° 19′ W.

Besides the light-houses before-mentioned, there are many other lights on the British coast, such as

	Lat. D.	Lat. M.	Long. D.	Long. M.
St. Anne's Lights	51	45	5	7 W.
S. Foreland Light	51	8	1	22 E.
Hook's Light-house (Ireland)	52	4½	7	0 W.
Isle of May Light	56	11½	2	37 W.
Kennaird's-head Light	57	42	2	1 W.
N. Ronaldsha Light	59	25	2	36 W.
Small's Light-house	51	47	5	35 W.
Spurn Lights	53	37	0	11 E.
Staple's Lights	55	38	1	43 W.

For farther particulars see the *New Seaman's Guide and Coaster's Companion* for 1812.

LIGHTNING, (*éclair*, Fr.) in physiology, is a large bright flame, shooting swiftly through the atmosphere, of momentary or very short duration, and commonly attended with thunder.

It is now universally allowed, that *lightning* is really an electrical explosion or phenomenon. Philosophers had not proceeded far in their experiments and inquiries on this subject, before they perceived the obvious analogy between lightning and electricity, and they produced many arguments to evince their similarity. But the method of proving this hypothesis beyond a doubt, was first proposed by Dr. Franklin; who, about the close of the year 1749, conceived the practicability of drawing lightning down from the clouds. Various circumstances of resemblance between lightning and electricity were remarked by this philosopher, and have been abundantly confirmed by later discoveries; such as the following:—Flashes of lightning are usually seen crooked and waving in the air; so the electric spark drawn from an irregular body at some distance, and when it is drawn by an irregular body, or through a space in which the best conductors are disposed in an irregular manner, always exhibits the same appearance.

Lightning strikes the highest and most pointed objects in its course, in preference to others, as hills, trees, spires, masts of ships, &c. so all pointed conductors receive and throw off the electric fluid more readily than those that are terminated by flat surfaces. See the article MARINE CONDUCTOR.

Lightning is observed to take and follow the readiest and best conductor; and the same is the case with electricity in the discharge of the Leyden phial; from whence the Doctor infers, that in a thunder-storm it would be safer to have one's clothes wet than dry.

Lightning burns, dissolves metals, rends some bodies, sometimes strikes persons blind, destroys animal life, deprives magnets of their virtue, or reverses their poles; and all these are well-known properties of electricity.

Lord Mahon observes, that damage may be done by lightning, not only by the main stroke and lateral explosion, but also by what he calls the returning stroke, by which is meant the sudden and violent return of that part of the natural share of electricity which had been gradually expelled from some body or bodies, by the superinduced elastic, electrical pressure of the electrical atmosphere of a thunder-cloud.

With regard to places of safety in times of thunder and lightning, Dr. Franklin's advice is, to sit in the middle of a room, provided it be not under a metal lustre suspended by a chain, sitting on one chair, and laying the feet on another; but the safest place is in a hammock hung by silken cords, at an equal distance from all the sides of a room. See HUTTON'S *Mat. and Phil. Dictionary.*

Artificial LIGHTNING, an imitation of real or natural lightning, by gunpowder, aurum fulminans, phosphorus, &c. but especially the last; between which and lightning there is much more resemblance than the others.

LIGHT-ROOM, (*soute vitrée, où l'on tient des fanaux pour donner du jour aux canonniers dans les soutes aux poudres*, Fr.) a small apartment, having double glass windows, near the magazine. It is used to contain the lights, by which the gunner and his assistants are enabled to fill the cartridges with powder, to be ready for action.

Large ships of war generally have two light-rooms; viz.

After LIGHT-ROOM, (*soute vitrée de l'arrière*, Fr.) that apartment which is attached to the after magazine; and

Fore LIGHT-ROOM, (*soute vitrée de l'avant*, Fr.) is that which gives light to the fore or great magazine.

LIMB, (*bord*, Fr.) in astronomy, implies the edge of the sun or moon; as, the upper limb or edge, the lower limb, the sun and moon's nearest limbs, &c. It also implies the border, or graduated edge of a quadrant, and other mathematical instruments.

LIMBERS, or LIMBER-HOLES, (*anguillères*, Fr.) in merchant-ships, are square holes cut through the lower-part of the floor-timbers, very near the keel, forming a channel for water, and communicating with the pump-well, throughout the whole length of the floor. Every floor-timber has two such holes cut through it, one on each side of the *kelson.*

LIMBER-BOARDS, (*bordages des anguillères*, Fr.) are short pieces of plank, which form a part of the lining of a ship's floor, close to the kelson, and immediately above the limbers. They are occasionally removed, to clear the limbers of any filth, sand, chips, or gravel, by which they may be clogged, so as to interrupt the passage of the water to the pump-well.

LIMBER-KENTLEDGE, (*saumons de fer faits pour entrer dans les vuides des anguillères*, Fr.) in merchant ships, are pigs of iron or lead, cast to fit in between the floor-timbers, or in the limbers.

LIMBER-PASSAGE, a passage formed on each side by the limber-strake and the kelson, for the purpose of the waters having a free communication with the pumps.

LIMBER-ROPE, (*corde des anguillères*, Fr.) a long rope, frequently retained in the limber-holes of a ship, in order to clear them by pulling the rope backwards and forwards, so as to loosen any dirt by which they may occasionally be choked.

LIMBER-STREAK, the strake of the foot-waling nearest the kelson, from the upper side of which the depth in the hold is measured.

LINCH-PINS, or LINSPIN, (*aisses*, Fr.) are small iron pins, which pass through the ends of the axle-trees, to keep the trucks or wheels of a cannon from slipping off.

LINE, (*ligne*, Fr.) a general name given to the arrangement or order in which a fleet of ships of war are disposed to engage an enemy.

This disposition, which is the best calculated for the operations of naval war, is formed by drawing up the ships in a long file, or right line, prolonged from the keel of the hindmost to that of the foremost, and passing longitudinally through the keels of all the others, from the van to the rear; so that they are, according to the sea-phrase, in the *wake* of each other.

In the line, or order of battle, all the ships of which it is composed sail one point free when upon a wind, on the starboard or larboard tack, and about 100 fathoms distant from each other. See plate XV. Plate XV.
fig. 1, where *a b* represents the elevation, and A B Fig. 1.
the plan of this order, upon the starboard tack; the

direction of the wind in both being expressed by the arrow in the latter.

A fleet is more particularly drawn up in the line when in presence of an enemy. It ought to be formed in such a manner as that the ships should mutually sustain and reinforce each other, and yet preserve a sufficient space in their stations, to *work* or direct their movements with facility during the action. Thus they will be enabled effectually to cannonade the enemy, without incommoding the ships of their own squadron.

The line close-hauled is peculiarly chosen as the order of battle, because if the fleet, which is to windward, were arranged in any other line, the enemy might soon gain the *weather-gage* of it; and even if he thinks it expedient to decline that advantage, it will yet be in his power to determine the distance between the adverse fleets, in an engagement, and to compel the other to action. The fleet to leeward, being in a line close-hauled, parallel to the enemy, can more readily avail itself of a change of the wind, or of the neglect of its adversary, by which it may, by a dextrous management, get to windward of him; or, should it fail in this attempt, it will, nevertheless, be enabled, by the favourable state of the wind, to avoid coming to action, if the enemy is greatly superior; or to prevent him from escaping, if he should attempt it.

Besides these advantages, this order of battle is singularly convenient and proper in other respects. The sails of each ship are disposed in such a manner as to counteract each other, so that the ships in general neither advance nor retreat during the action. By this circumstance they are enabled to retain their stations with greater stability, and to prosecute the battle with vigour and resolution, yet without perplexity and disorder. The uniformity of the line will be preserved, so that the admiral's orders may be readily communicated by signals from the van to the rear. The distress of any particular ship, that is disabled and rendered incapable to continue the action, will be presently discovered, and her place accordingly supplied by one of the ships in reserve. The circumstances and situation of the enemy's line will be ever open to the view of the commander in chief, so that he may be enabled to convert any disaster that may happen therein to his own advantage.

It may be alleged, indeed, that the same reasons hold good with regard to the enemy, to whom this arrangement will be equally beneficial. It may also be observed, that particular occasions have rendered it necessary to break the order of the line; and that sometimes this expedient has been practised with equal judgment and success. To the first of these allegations it may be answered, that in war as well as politics, there are certain general rules absolutely necessary to be observed by the hostile powers: rules which are founded on mutual convenience, and authorized by the invariable example of all ages! Whatever tends to facilitate the designs of the adverse parties on each other, or whatever operates to shorten the period of war, and render it less destructive and fatal, are objects which ought never to be disregarded. Disorder has not only a tendency to protract the war, but to make it more bloody and ruinous, and to aggravate all the calamities with which it is inseparably attended. Perhaps this observation is particularly applicable to our present purpose, unless the consequences of disorder in a sea-fight, as hereafter-mentioned, should rather be considered as the creation of fancy, than a recital of facts, naturally resulting from known causes. Although peculiar circumstances have sometimes, by their success, justified the measure of engaging an enemy's fleet, without forming the line; or, after the line has been separated: there is, nevertheless, very few operations in war that require greater delicacy and vigilance, if the hostile fleets are very near to each other. Perhaps no military enterprise can be attended with greater hazard, or with fewer hopes of success. The incessant fire of so large an assembly of ships in a very short time covers the scene of action with a cloud of smoke, which is constantly accumulating. The winds that enabled the two fleets to approach each other are soon become extremely feeble, or, perhaps, perfectly lulled, by the explosions of a vigorous cannonade; they are, of course, incapable any longer to dissipate the smoke which then darkens the air, and is almost impenetrable to the eye. If in this situation the hostile ships are promiscuously scattered amongst each other, it is easy to foretel the mischief, perplexity, and distraction, to which the whole will be inevitably exposed. Not only is the most comprehensive skill of the commander in chief rendered useless; the smaller ships, abandoned to their ill fortune, may be torn to pieces by superior force, without relief or succour: and, what is infinitely worse than all, the ships of the same fleet may cannonade each other, with all the resolution and spirit which they exert against their enemies! If the design of war is conquest, and not massacre, it is thus totally perverted! The battle, instead of being brought to a speedy issue, and decided by a victory and defeat, is unhappily protracted into a scene of slaughter and ruin, equally fatal and undecisive to both parties.

If then, disorder and confusion are fraught with such dangerous consequences in a naval armament, it is no less certain that the principal sinews of its strength are discipline, regularity, vigilance, and activity. It has been already remarked, that the ships of the line should be sufficiently close to sustain each other: for, if they are farther apart than those of the enemy's line, many single ships will suffer the fire of two at once. Hence the fleet is rendered inferior to that of the enemy, at the onset of battle; a circumstance which evinces the superiority of larger ships, accompanied with weightier metal! the enemy is defeated by the efforts of a more numerous and more powerful artillery.

Besides these advantages, the larger ships are in other respects highly preferable in a line of battle. They overlook those of an inferior rate, which are, accordingly, laid open to the fire of their musketry. In a high sea they can more safely employ the artillery of their lower deck than a smaller ship; and, if both are obliged to shut their lower deck ports, the advantage of the three-decked ships, with regard to their cannon, will yet be considerable: they have three tier against two, and two against one. The same superiority subsists in case they are dismasted, when the upper deck is encumbered with the ruins.

The large ships, being higher *between-decks*, are less incommoded with the smoke; and their cannon is managed with greater facility.

The large ships, having greater solidity of frame, are better calculated to resist the effects of battle and tempest. In general, also, they sail better than the small ones, except in fine weather; for, in a fresh wind, when the sea becomes agitated, they have always the superiority.

The fire-ships do not succeed so well against large ships as the smaller ones; the artillery will sink them, or oblige them sooner to relinquish their design; and they are easily *towed* away by the great long-boats.

The line of a fleet, which has abundance of capital ships, need not be so much inclosed as that of an enemy who has fewer. The former may be also less numerous, without being weaker.

An open line will, on many occasions, work more easily than one which is more inclosed; and if it is less numerous, the movements thereof are more expeditious; the signals better attended to; the general order more exactly preserved; and the ships less liable to be separated. Hence it will be less embarrassed by a change of wind, and the order will be sooner re-established.

A less numerous line will more readily approach or escape from an enemy, or a hostile shore; and, finally, when cruising in a smaller space, it will not be so much contracted.

From the preceding reflections it results, that the line which contains more capital ships will be stronger than one more numerous, if composed of smaller ships. This reflection, however, does not exclude a certain number of the third and fourth *rates*, which are necessary in all naval armaments.

As the hostile fleets are drawn up in two opposite lines, with their sides to the wind, it is evident that one must be to the leeward of the other, as appears

Fig. 2. in fig. 2, plate XV. Both situations, however, have their defects as well as advantages.

In a line of battle, the weather-most fleet, or that, which, in sea language, has the weather-gage, is generally allowed to have the advantage, although there are several arguments, on the other hand, in favour of the lee-side; accordingly, we shall endeavour to state the mutual advantages and disadvantages.

Advantages of the Weather-Gage.

1. The weather-gage is the sooner clear of smoke; and, of course, that line can better observe the signals which are spread, than the ships to leeward can, which must have the continuance of both its own and of the enemy longer.

2. If the weather-ships are more in number than the enemy's, they can detach some from their squadron; which, bearing down upon the rear of the enemy, must infallibly throw them into disorder.

3. The fire-ships of the weather-line can, when they are ordered, more easily bear down upon the enemy than those of the lee can ply to windward, which can never be done against a line in action; but the weather fire-ships can bear down against all the resistance that can be made by the enemy.

Advantages of the Lee-Gage.

1. If one, two, or more of the ships to windward should be disabled, they must inevitably drive to leeward, and become a prey to the enemy.

2. The ships of the lee-line can more readily bear away before the wind, and have their places supplied by ships from the *corps-de-reserve*, in case of being disabled or meeting with any disaster.

3. The line to leeward can keep their ports longer open in a strong wind with a high sea, when those to windward, in all probability, may be obliged to shut the ports of their lower tier of guns, to prevent the water from rushing in between decks, which may be attended with the most fatal consequences.

4. The lee-line can more easily observe the men on the decks of the ships to windward, as they heel, and when the smoke does not interrupt their sight; at which time the marines and top-men may easily take aim at and destroy them with muskets and carabins.

The disadvantages of the Weather-Line are,

1. If the sea is rough and the wind boisterous, it cannot readily fight with the lower-deck guns.

2. The weather-line cannot decline the action, without the dangerous expedient of forcing through the enemy's line; and if it *keeps the wind*, the lee-line may inclose and totally destroy it, especially if it is inferior in number to the latter, or if the ships thereof are in bad condition; for it then can find no other resource but in the dexterity of its manœuvres, unless it is favoured by the wind, or any oversight of the enemy.

3. The disabled ships of the weather-line must tack, to avoid falling into the enemy's fleet; and if they are much shattered, they may be altogether separated from their own fleet, particularly if they are in the rear of the line.

The disadvantages of the Lee-Line are,

1. It cannot decide the time and distance of the battle, which may commence before it is sufficiently formed; and it will, perhaps, be attacked by an enemy, who bears away upon it in regular order.

2. It suffers much inconvenience from the fire and smoke of the weather-line, as remarked in the advantages of the weather-line (1).

3. It cannot easily break the enemy's line with its fire-ships, which formerly were very slowly and with great difficulty conveyed to windward. On the contrary, the fire-ships of the weather-line have a considerable advantage (3).

It must be remarked, that the admiral's ship attentively preserves her station in the centre of the line; for if the commander in chief should give way to the caprice or inattention of any of those under his direction, it would introduce an endless confusion into the squadron.

To illustrate this article, and enable the reader to form a clearer idea of the line, we have, in plate XV represented several distinct views, according to the different situations which it occasionally assumes.

Fig. 3, exhibits a perspective view of the line of battle, on the starboard-tack, A B being the plan thereof. Fig. 3.

Fig. 2, *a* represents the profile of the same line on

the starboard tack, as brought to action by the opposite line *b*. The plan of these squadrons, A B, appears immediately below.

Fig. 4. It is necessary to remark here, that a fleet frequently retains the order of the same tack, occasionally, when the whole fleet goes about at once, as expressed by *a*, fig. 4, of which A is the plan. Or it goes about gradually, the headmost ship having *tacked* first, and the next tacked as soon as she arrived in her wake; the rest following the same example. See *c*, fig. 3, and C in the plan of the same figure.

Fig. 5. It also frequently preserves the order of the line close-hauled, although steering with a large wind, either in pursuit of a flying enemy or proceeding in a particular course. Thus the fleet *b*, fig. 5, although ranged so as to be in a line upon the larboard-tack, if close to the wind, is chasing the fleet *a* to leeward, which is either parallel to the former and preserving the same order, or sails on a line abreast, as expressed by the plan C. See also the article ABREAST.

Fig. 6. Fig. 6, exhibits a fleet formed into a line, on the larboard-tack, bearing away upon the continuation of the same line astern. Thus, supposing them to be formed on the starboard-tack, and sailing due north, in a line a-head; it is evident that every ship, at one and the same time, bears away and steers south, the whole fleet will again be upon a line a-head, with the wind upon the larboard-quarter, as expressed in this figure, and in the plan under it.

Fig. 7. Fig. 7, represents a fleet bearing away, in a crescent before the wind, having half of the ships ranged on the starboard-tack, and the other half on the larboard-tack, so as to form the two sides of the angle *b c a*, of which the commander in chief, *a*, makes the central point. This disposition is sometimes used to force through a passage which is guarded by an enemy. See also the plan thereof, A B C below, where it is evident that the admiral is the foremost ship, whilst bearing away, although she would be the last in both lines if they were close-hauled.

Fig. 8. Fig. 8, expresses the order of retreat, which is frequently practised by the French, and is directly the reverse of this; because the angular point is farthest to leeward in the former, whereas it is to windward of both lines in the latter; being also the headmost of both, when close-hauled, although the sternmost ship while they are bearing away.

In an engagement, the ships are generally *brought to*, with the main top-sails laid aback, and their fore-top-sails full, for the purpose of bearing away more readily, when occasion requires. This disposition of the sails is represented in fig. 2, plate II. See also To LYE-TO.

The line is said to be formed abreast when the ship's sides are all parallel to each other, on a line which crosses their keels at right angles. This is more frequently used in pursuing or retreating, with the wind right aft, so that the line forms a perpendicular with the direction of the wind, as exhibited by the ships C, in the plan annexed to fig. 5.

For farther particulars, see M. DE MOROGUE's, and CLERK's *Naval Tactics*, and the Admiralty's *Signal Book:* also the article *Naval* TACTICS.

LINE, in geography and navigation, is emphatically used for the EQUATOR, or EQUINOCTIAL-LINE; which articles see.

LINE *(ligne, ou cordelette,* Fr.) is also the general appellation of a number of small ropes in a ship, formed of two or more fine strands of hemp; as,

Cod-LINE, a small line of 18 threads, used on the banks of Newfoundland. The thread is spun 100 yards, having six threads to a strand, three strands to the line, and standing 40 fathoms, which weigh 3 lbs.

There are also *cod-lines* of a smaller size, as those of 15, 12, and 9 threads.

Concluding LINE, *(petite ligne du milieu des echelles de corde, qui pendent en arriere des vaisseaux,* Fr.) a small rope, which is hitched to the middle of every step of a stern ladder.

Cutting-Down LINE. See the article CUTTING-DOWN.

Deep-Sea LINE, (*ligne de sonde,* Fr.) a long line, marked at every five fathoms with small strands of line, knotted. It is used with the deep-sea lead.

Fancy LINE, a small rope, used in some vessels, to haul up the throat and peek-brails of the mizen, that they may not chafe the lee-side of the sail. The fancy-line reeves through blocks lashed at the peek end, and each end bends to the thimble in the bight of the span on each side.

Fishing LINE, *(ligne de peche,* Fr.) a particular kind of line, generally used for fishing.

Hand LINE, *(petite ligne de sonde,* Fr.) a line about twenty fathoms long, marked with black leather, white rag, and red buntine at different distances. It is made fast to the *hand-lead*, (fig. 10, plate XIII.) and used to determine the depth of water in going in or out of a harbour, river, channel, &c.

Hauling LINE, (*cartaheu,* Fr.) the name of any small rope let down out of a top, &c. to haul up some light body by hand.

Life LINE, (*corde servant aux matelots pour se tenir sur les vergues,* Fr.) a rope occasionally extended in several situations in a ship, for persons to lay hold of, to prevent their falling.

LINE *of Numbers*, or LINE *of Lines*. See the article GUNTER's LINE.

Navel LINE, *(cravatte,* Fr.) a small rope that passes from the tops to the trusses, to keep them up in their place when overhauled, for the purpose of bracing the lower-yards sharp up for sailing by the wind, &c.

A Ship of the LINE, (*vaisseau de ligne,* Fr.) a vessel large enough to be drawn up in the line, and to have a place in an engagement; viz. from 120 to 64 guns.

Spilling LINE, (*corde, pour deventer une voile quarrée,* Fr.) a rope fixed, occasionally, to the square sails, particularly the main and fore courses of a ship in tempestuous weather, for reefing or furling them more conveniently. These lines are reeved through blocks upon the yard, whence leading round the sail, they are fastened behind to the yard, so that the sail is, by their effort, very closely confined.

Tarred LINE, (*ligne noire, ou ligne goudronnee,* Fr.) is that which has been dipped in the tar, and is used on many occasions.

White LINE, (*ligne blanche,* Fr.) implies that which has not been tarred, in contradistinction to tarred line.

The description and uses of the other lines are arranged under their proper heads.

LINING, is the act of marking the length, breadth, or depth of any piece of timber, according to instruction and design, by a cord, rubbed with red or white chalk, fastened at the extremities, and forcibly pulled up in the centre, or towards one end, and suffered to fall perpendicularly, if the cut in the timber is intended to be straight, or thrown sideways, so as to form a curve. It requires considerable practice to perform the latter with exactness.

LINSTOCK, (*boute-feu, ou baton a meche*, Fr.) a staff about three feet long, having a sharp point at one end, and a sort of fork or crotch on the other, the latter serves to contain a lighted match, and, by the former, the linstock is occasionally stuck in the deck, in an upright position. It is frequently used in small vessels, in an engagement, where there is commonly one fixed between every two guns, by which the match is always kept dry and ready for firing.

LIST, (*faux cote*, Fr.) implies an inclination to one side; as, "Our ship has a *list* to starboard;" viz. is depressed more in the water on that side.

LIZARD, an iron thimble, spliced into the main bowlines, and pointed over, to hook a tackle to.

LOADED-SHELL, an ingenious contrivance for saving the lives of people in imminent danger on board a ship, when the vessel is either stranded, or in danger of being so, near the coast. It is a bomb-shell, filled with lead, wherein a staple or ring is fixed, to which a rope is fastened; and the shell, thus loaded, being thrown on shore, fixes itself in the ground, and the other end of the rope being fastened to the ship, the persons on board are thereby enabled to haul themselves ashore. It was invented by John Bell, serjeant (now lieutenant) of the regiment of artillery, who obtained a premium for it of fifty guineas, in 1792, from the Society for the Encouragement of Arts, &c. after exhibiting its utility and practicability before a committee of the society.

On this occasion a loaded shell, weighing about seventy pounds, was thrown on shore, from a small mortar, fixed in a boat moored in the Thames, about two hundred yards from the coast. The shell falling about a hundred yards within land, buried itself eighteen inches in the gravel, when Bell and his assistant, on a raft, floated by casks, properly ballasted, hauled themselves ashore in a few minutes, by the rope affixed to the shell.

For farther information on the subject of *Lieutenant* Bell's invention, see *Transactions of the Society of Arts, &c.* Vol. X. and for the particulars of his method, with an engraving of the apparatus, see Vol. XXV. of the same work.

LOADING of a Gun, (*charger un canon*, Fr.) is the act of charging it, or the charge itself. See the article Charge.

Loading *of a Ship*. See the articles Cargo and Lading.

LOADMANAGE, in maritime affairs, is the hire which the pilot of a ship receives of the master or captain, for conducting the ship up the river, or into port. See Mortimer's *General Commercial Dictionary*.

LOADSMAN, or Lodesman, (*pilote côtier*, Fr.) a kind of pilot, established for the safe conduct of ships and vessels in and out of harbours, or up and down navigable rivers. See the article Pilot.

LOADSTONE, or Lodestone, the stone on which the mariner's compass needle is touched, to give it a direction north and south. See the articles Dipping Needle and Magnetism.

LOBBY, in a ship, is a small apartment adjoining the fore part of the bread-room, and appropriated to the use of the surgeon.

LOB-LOLLY, (*bouillie de gruau*, Fr.) a seafaring dish, otherwise called Burgoo, which article see.

Lob-Lolly *Boy*, (*homme assigne au chirurgien-major et a ses aides, pour infirmier, ou garde malade*, Fr.) a name given to the man who attends the surgeon and his mates, to summon the sick, and to perform any service in their attendance on them.

LOCK, or Weir, (*bonde d un canal*, Fr.) the general name for all those works of wood or stone, made to confine and raise the water of a river; the banks, also, which are made to divert the course of a river, are called by these names in some places. But the term *lock* is more particularly appropriated to express a kind of canal inclosed between two gates; the upper, called by workmen the *sluice-gate*, and the lower, called the *flood-gate*. These serve, in artificial navigation, to confine the water, and render the passage of boats easy in passing up and down the stream.

LOCKER, (*caisson*, Fr.) a kind of box or chest, made along the side of a ship, to put or stow any thing in.

Shot-Locker, (*parc aux boulets*, Fr.) a strong frame of plank, near the pump-well in the hold, in which the shot are kept till they are wanted for service: hence comes the phrase among seamen, of having a shot in the locker, or a guinea in their pocket.

LOG, (*loc*, Fr.) a machine used to measure the rate of a ship's velocity through the water. For this purpose there are several various inventions, but the one most generally used is the following, called the *common log*. It is a piece of board, about a quarter of an inch thick, and five or six inches from the angular point *a*, to the circumference *b*, fig. 9, plate XIV. and balanced by a small plate of lead, nailed on the circular part, so as to swim perpendicularly in the water, with the greater part immersed.

Mr. Edward Massey, of Hanley, in Staffordshire, has obtained a patent for a new sea log. The principal objects of this log are, a set of wheel-work, which operates upon the fingers of the several indices; and a register, which registers the distance sailed. It has also a rotator, or hollow cylinder, made air-tight, and so nearly of the same specific gravity as water, as to float when drawn forwards with the velocity of mere steerage way.

On this rotator are fixed four vanes, placed obliquely. It is then fastened to the register by a cord, about six feet long; to the loop-hole at the other end of the register is secured another line, of sufficient length to extend beyond the eddy of the vessel's wake.

The finger on the index 1, revolves once while the log moves forward one mile; that on the index 2, moves once round in going ten miles; that on the index 3, makes one revolution when the distance sailed is one hundred miles. When the machine is to be used, all the fingers of the indices are set to 0, and both register and rotator committed to the water. As the vessel moves forward, the log must follow; and, from the obliquity of the vanes, it is evident the rotator must revolve quicker or slower, correspondent to the ship's velocity. This rotatory motion is communicated by the cord to the universal joint, connected with the wheels, which, consequently, revolve with the rotator and cord; and thus the actual space passed through, in any given time, is registered on the indices.

General Properties of this Log.

1. Every occasional or momentary acceleration or retardation of the vessel, from irregularity of wind, or other causes, which are either altogether passed over or very vaguely guessed at, in general, are accurately registered on this machine; which not only gives the actual rate of sailing, but the actual distance sailed, since the last inspection.

2. It will give the true distance sailed, from steerage way to any velocity with which the swiftest sailing vessel can move.

3. It is attended with less trouble than the common log, and no mistakes can possibly arise from the result it gives.

A very little reflection will convince any observer of the great superiority of this machine over all others which have been hitherto introduced. For farther particulars, and an engraving of this machine, see NICHOLSON'S *Philosophical Journal*, for 1808.

LOG AND LINE, or the LOG-LINE, (*ligne de loc*, Fr.) a small cord or line, about 150 fathoms long, fastened to the log by means of two legs, *a b*, fig. 10, plate XIV. one of which passes through a hole at the corner, and is knotted on the opposite side, while the other leg is attached to the arc by a pin fixed into another hole, so as to draw out occasionally. By these legs the log is hung in equilibrio; and the line thus annexed to it is wound round a reel, fixed for that purpose, in the gallery of the ship. This line, from the distance of about ten, twelve, or fifteen fathoms off the log, has certain knots or divisions, which ought to be at least 50 feet from each other; though it was the common practice at sea not to have them above forty-two feet asunder.

The length of each knot ought to be the same part of a sea mile as half a minute is of an hour; and, admitting the measurement of Mr. Norwood, who makes a degree on a great circle of the earth to contain 367,200 English feet, or $69\frac{1}{2}$ English statute miles, and, therefore, $\frac{1}{60}$th part of it, or a nautical mile, will be 6120 feet; $\frac{1}{120}$th of 6120, or 51 feet, should be the length of each knot. But, because it is safer to have the reckoning rather before the ship than after it, therefore 50 feet may be taken as the proper length of each knot. The knots are sometimes made to consist only of 42 feet each, even in the present practice; and this method of dividing the *log-line* was founded on the supposition that 60 miles, each of 5000 English feet, made a degree; for $\frac{1}{120}$ of 5000 is $41\frac{2}{3}$, or, in round numbers, 42 feet. Mariners, rather than quit the old way, though known to be erroneous, use glasses for half-minute ones, that run but 24 or 25 seconds. They have also used a line of 45 feet to 30 seconds, or a glass of 28 seconds to 42 feet. When this is the case, the distance between the knots should be corrected by the following proportion:—as 30 is to 50, so is the number of seconds of the glass to the distance between the knots upon the line.

The heat or moisture of the weather has often a considerable effect upon the glass, so as to make it run slower or faster; it should, therefore, be frequently tried by the pendulum in the following manner:- On a round nail hang a string that has a musket-ball fixed to one end, carefully measuring betwixt the centre of the ball and the string's loop over the peg $39\frac{1}{8}$ inches, being the length of a second pendulum; then swing it, and count one for every time it passes under the peg, beginning at the second time it passes; and the number of swings made during the time the glass is running out shews the seconds it contains. The line also is liable to relax and shrink, and should, therefore, be occasionally measured.

The use of the log and line is to keep account and make an estimate of the ship's way or distance run; which is done by observing the length of the line unwound in half a minute's time, told by a half-minute glass; for so many knots as is run out in that time, so many miles the ship sails in an hour. Thus, if there be four knots veered out in half a minute, the ship is computed to run four miles an hour.

No mention of this device for measuring the ship's way occurs till 1607, in an East India voyage, published by Purchas; but from that time its name occurs in other voyages among his collections; and henceforward it became famous, being taken notice of by all the succeeding writers on navigation of every country.

To heave the LOG, (*jeter le loc*, Fr.) is to throw it into the water on the lee-side, and let it run till it comes without the eddy of the ship's wake, before they begin to count; and, for the ready discovery of this point of commencement, there is commonly fastened to the line a red rag: then one, holding a half-minute glass, turns it up just as the first knot, or the mark from which the knots begin to be reckoned, turns off the reel, fig. 11, plate XIV. or passes over the stern. As soon as the sand in the glass is out, the reel is stopped, and the knots run off are told, and their parts estimated.

The log, however, is a very precarious way of computing, and must always be corrected by experience and good sense; there being a great deal of uncertainty in the yawing of the ship going with the wind aft, or upon the quarter, in the heaving of it, by its coming home, or being drawn after the ship; on account of the friction of the reel and lightness of the log in the course of the current, and in the strength of the wind, which seldom keeps the same tenor for two hours together.

In ships of war and East Indiamen, it is usual to heave the log once every hour, and in all other vessels once in two hours; and if, at any time of the watch, the wind has increased or abated in the intervals, so as to affect the ship's velocity, the officer generally makes a suitable allowance for it, at the close of the watch.

To correct the LOG. Suppose Beachy Head bore, by the compass, N.W. at 6 hours A.M. distance 6 miles: required the ship's latitude and longitude.

Now, to an observer at Beachy Head, it is evident the ship would bear S.E. (which is opposite to N.W.):

therefore, with the course S.E. and the distance 6 miles the ship's latitude and longitude may be corrected.

Log-Board, (*table de loc*, Fr.) are two boards shutting together like a book, and divided into several columns, containing the hours of the day and night, the direction of the winds, the course of the ship, with all the material occurrences that happen during the twenty-four hours, or from noon to noon; together with the latitude by observation. From this table the officers work the ship's way and compile their journals. The whole being written with chalk, is rubbed out every day at noon.

Log-Book, (*journal de navigation*, Fr.) a book ruled in columns like the log-board, into which the account of the log-board is transcribed every day at noon, together with every circumstance deserving notice, that may happen to the ship, or within her cognizance, either at sea or in a harbour, &c. from whence, after it is corrected, &c. it is entered into the Journal, which article see.

The intermediate divisions or watches of a log-book, containing four hours each, are usually signed by the commanding officer thereof, in ships of war or East Indiamen.

Log-Reel. See the article Reel.

LOGARITHMS, (from the Greek λογος, ratio, and αριθμος, number, *logarithmes*, Fr.) are numbers so contrived and adapted to other numbers, that the sums and differences of the former shall correspond to, and shew the products and quotients of the latter. Or, more generally speaking, logarithms are a series of numbers in arithmetical progression, answering to another series of numbers in geometrical progression, Thus,

0, 1, 2, 3, 4, 5, indices, or logarithms.
1, 2, 4, 8, 16, 32, geometrical progression.

Or,

0, 1, 2, 3, 4, 5, indices, or logarithms.
1, 3, 9, 27, 81, 243, geometrical progression.

Or,

0, 1, 2, 3, 4, indices, or logarithms.
1, 10, 100, 1000, 10000, geometrical progression.

Where it is evident, that the same indices serve equally for any geometric series; and, consequently, there may be an endless variety of systems of logarithms to the same common numbers, by only changing the second term, 2, 3, or 10, &c. of the geometrical series.

It is also apparent, from the nature of these series, that if any two indices be added together, their sum will be the index of that number which is equal to the product of the two terms, in the geometric progression, to which those indices belong. Thus, the indices 2 and 3, being added together, are = 5; and the numbers 4 and 8, or the terms corresponding with those indices, being multiplied together, are = 32, which is the number answering to the index 5. And, in like manner, if any one index be subtracted from another, the difference will be the index of that number, which is equal to the quotient of the two terms to which those indices belong. Thus, the index 6, minus the index 4, is = 2; and the terms corresponding to those indices are 64 and 16, whose quotient is = 4; which is the number answering to the index 2. For the same reason, if the logarithm of any number is multiplied by the index of its power, the product will be equal to the logarithm of that power. Thus, the index or logarithm of 4, in the above series, is 2; and if this number is multiplied by 3, the product will be = 6; which is the logarithm of 64, or the third power of 4.

And, if the logarithm of any number is divided by the index of its root, the quotient will be equal to the logarithm of that root. Thus, the index or logarithm of 64 is 6; and, if this number is divided by 2, the quotient will be = 3; which is the logarithm of 8, or the square root of 64. Here it may be observed that these properties are general in all cases, as will appear evident from a geometrical series, according to the algebraic method of notation; thus r^0, r^1, r^2, r^3, r^4, r^5, r^6, &c. where the ratio of the geometrical series is denoted by r, and the arithmetical series is supplied by the numeral exponents of the terms. Hence it follows, that the properties, here ascribed to any geometrical, and its corresponding arithmetical series, depend on two well-known propositions in algebra; namely, that the sum of the exponents of any two powers of an algebraic quantity is equal to the exponent of their product; and that the difference of their exponents is equal to the exponent of their quotient. Hence, in all cases when the terms of an arithmetical progression are adapted to those of a geometrical progression, as in the foregoing examples; the terms of the arithmetical series are called the logarithms of the corresponding terms of the geometrical series.

The logarithms most convenient for practice are such as are adapted to a geometrical series increasing in a tenfold proportion, as in the last of the above forms, and are those which may be found at present in most of the common tables upon this subject. The distinguishing mark of this system of logarithms is, that the index or logarithm of 1 is 0; that of 10, 1; that of 100, 2; that of 1000, 3, &c. And in decimals the logarithm of .1 is — 1; that of .01, — 2; that of .001, — 3, &c. From whence it follows, that the logarithm of any number between 1 and 10 must be 0 and some fractional parts; and that of a number between 10 and 100 will be 1 and some fractional parts; and so on for any other number whatever. And, since the integral part of a logarithm is always thus readily found, it is usually called the index, or characteristic, and is commonly omitted in the tables, being left to be supplied by the operator himself, as occasion requires.

History of Logarithms.—The properties of the geometrical series, which constitute the foundation of the doctrine of logarithms, appear to have been known as far back as the days of Archimedes; for that celebrated mathematician makes use of them in his work, entitled "*Arenarius*;" or, "*Treatise on the Number of Sands.*" The same properties are also mentioned in the writings of Stifelius, a German mathematician, who lived in the 16th century.

The learned calculators, about the close of the 16th and beginning of the 17th century, finding the operations of multiplication and division, by very long numbers of 7 or 8 places of figures, which they had frequently occasion to perform, in resolving problems relating to geography and astronomy, to be exceedingly troublesome, set themselves to consider whether it was not possible to find some method of lessening this labour, by substituting other easier operations in their stead. In pursuit of this object, they reflected, that since, in every multiplication by a whole number, the ratio or proportion of the product to the multiplicand is the same as the ratio of the multiplier to unity, it will follow that the ratio of the product to unity (which, according to Euclid's definition of compound ratios, is compounded of the ratios of the said product to the multiplicand, and of the multiplicand to unity), must be equal to the sum of the two ratios of the multiplier to unity, and of the multiplicand to unity. Consequently, if they could find a set of artificial numbers, that should be the representatives of, or should be proportional to, the ratios of all sorts of numbers to unity, the addition of the two artificial numbers that should represent the ratios of any multiplier and multiplicand to unity, would answer to the multiplication of the said multiplicand by the said multiplier, or the sum arising from the addition of the said representative numbers, would be the representative number of the ratio of the product to unity; and, consequently, the natural number to which it should be found, in the table of the said artificial or representative numbers, that the said sum belonged, would be the product of the said multiplicand and multiplier. Having settled this principle, as the foundation of their wished-for method of abridging the labour of calculations, they resolved to compose a table of such artificial numbers, or numbers that should be representatives of, or proportional to, the ratios of all the common or natural numbers to unity.

The first observation that naturally occurred to them in the pursuit of this scheme was, that whatever artificial numbers should be chosen to represent the ratios of tneir whole numbers to unity, the ratio of equality, or of unity to unity, must be represented by 0; because *that* ratio is properly no magnitude, since, when it is added to, or subtracted from, any other ratio, it neither increases nor diminishes it.

The second observation that occurred to them was, that any number whatever might be chosen at pleasure, for the representative of the ratio of any given natural number to unity, but that, when once such choice was made, all the other representative numbers would be thereby determined, because they must be greater or less than that first representative number, in the same proportions in which the ratios represented by them, or the ratios of the corresponding natural numbers to unity, were greater or less than the ratio of the said given natural number to unity.

The third observation that occurred to them was, that, as these artificial numbers were representatives of, or proportional to, ratios of the natural numbers to unity, they must be expressions of the numbers of some smaller equal ratios, that are contained in the said ratios. And, therefore, they thought these artificial numbers, which thus represent, or are proportional to, the magnitudes of the ratios of the natural numbers to unity, might not improperly, be called the LOGARITHMS of those ratios, since they express the numbers of smaller ratios, of which they are composed And then, for the sake of brevity, they called them the *logarithms of the said natural numbers themselves*, which are the antecedents of the said ratios to unity, of which they are in truth the representatives.

The foregoing method of considering this property leads to much the same conclusions as the other way, in which the relations between a geometrical series of terms and their exponents, or the terms of an arithmetical series, are contemplated. In this latter way it readily occurred that the addition of the terms of the arithmetical series corresponded to the multiplication of the terms of the geometrical series; and that the arithmeticals would therefore form a set of artificial numbers, which, when arranged in tables, with their geometricals, would answer the purposes desired, as has been explained above.

It does not, however, appear, that any person perceived all the advantages that might be derived from these properties till about the beginning of the 17th century; when their utility was rendered evident by the happy invention of logarithms. Notwithstanding whatever may have been said or conjectured by others, it is certain that this discovery, one of the most valuable that ever was made in mathematics, is undoubtedly due to John Napier, baron of Merchiston, in Scotland; who published it to the world, in 1614, in a work which he called "*Mirifici Logarithmorum Canonis Descriptio*," and which contained a large table of logarithms, together with their description and uses.

Next to the discovery of logarithms, is the improvement which they received in their form from Henry Briggs, who was, at the time of the publication of Napier's logarithms, professor of geometry in Gresham College, in London, and afterwards Savilian professor of geometry at Oxford.

On the first publication of Napier's logarithms, Briggs immediately applied himself to the study and improvement of them; and presently saw that it would be of advantage to change the scale; so that the logarithm of 1 being 0, as in Napier's form, the logarithm of 10 might be 1, that of 100, 2; of 1000, 3; and so on, as in Briggs's form; whereas the logarithms of the same numbers, according to Napier's construction, were 2.302585, 4.605168, 6.907753, &c. This improvement Briggs communicated to the public in the preface to his "*Arithmetica Logarithmica*;" so that it appears he was the inventor of the present scale of logarithms; and also lived to complete a table of logarithmic sines and tangents, for the 100th part of every degree of the quadrant, to 14 places of figures, besides the index; together with a table of natural sines, to 15 places, and the tangents and secants to 10 places; with the construction of the whole.

These tables were printed at Gouda, in Holland, under the care of Adrian Valcq, who had now completed the logarithms of numbers from 1 to 100,000; and nearly finished them, when the death of the author prevented him from completing the application and uses of them. However, this office he recommended, on his death bed, to his friend Henry Gellibrand, then professor of astronomy in Gresham College; who added a preface, and the application of the logarithms to plane and spherical trigonometry, &c. This work was published in 1633, under the title of "*Trigonometria Britannica*."

Many other authors have written on this subject, and several subsequent improvements have been made, which cannot be inserted here. It may, however, be proper to mention that the arrangement of the logarithms, in the tables, was first reduced to the most convenient form by John Newton, in his "*Trigonometria Britannica*;" published at London, in 1658.

In 1706, Sherwin published a complete collection of *Mathematical Tables*, in 8vo. containing, besides the logarithms of all numbers to 101,000, the sines, tangents, secants, and versed sines, both natural and logarithmic, to every minute of the quadrant. Gardiner revised these tables, and published an edition in 1742, which is esteemed the most correct of any. In the same year he published by subscription a 4to. volume, of "*Tables of Logarithms for all Numbers from* 1 *to* 102,100, *and for the Sines and Tangents to every Ten Seconds of a Degree in the Quadrant*; *as also, for the Sines of the first* 72 *Minutes to every single Second; with other useful and necessary Tables*:" for finding the logarithms of numbers to 20 places of figures. This was esteemed a very correct and useful collection.

Among the tables of logarithms that have been published of late years in this country, there are two works most deservedly in repute, both for accuracy and convenience of arrangement; these are, Dr. Hutton's "*Mathematical Tables, in 8vo. containing common Hyperbolic and Logistic Logarithms, &c.*" and "Taylor's *Tables of Logarithms, in* 4*to. of all Numbers, from* 1 *to* 101,000, *each to* 7 *places of figures; and of the Sines and Tangents, to every Second of the Quadrant.*"

On this, as well as on many other occasions, it seems but justice to remark the extraordinary spirit and elegance with which the learned men and the artisans of the French nation undertake and execute works of merit. "*Tables Portatives de Logarithmes, publiees a Londres par Gardiner,*" is a work most beautifully printed, in a neat portable 8vo. volume, and contains all the tables in Gardiner's 4to. volume, with some additions and improvements by Callet, and printed with a considerable degree of accuracy, at Paris, by Didot, in 1793. Callet's second edition of the same work, very neatly printed, at Paris, with stereotypes of solid pages, in 8vo. by Didot, in 1795, contain the logarithms of numbers from 1 to 101,000; and the logarithmic sines and tangents for every second in the first five degrees, and for every 10 seconds of the remaining degrees of the quadrant; and also for every 10,000th part of the arc, according to the new centesimal division of the quadrant: the logarithms are to 7 decimal places.

The Board of Longitude in France have long had under their consideration another very extensive collection of mathematical and astronomical tables, which they have now published under the direction of M. Prony, in which they have extended the table of natural sines and tangents to 22 decimal places, with 5 orders of differences; the tables of logarithmic sines and tangents to every 100,000th part of the quadrant to 12 places of decimals and 3 orders of differences; and the tables of logarithms of numbers from 1 to 200,000, to 12 places of decimals, and 3 orders of differences. They have added astronomical tables from calculations and observations, agreeable to the new centesimal divisions.

For farther information of the uses of logarithms in numerical calculations, with the shortest method of constructing them, see the *Historical Introduction to Dr.* Hutton's *Logarithms.*

LOGGERHEAT, (*boulet de fer avec un manche, qu on fait rougir au feu pour le tremper dans le goudron, pour le faire fondre,* Fr.) a spherical mass of iron with a long handle, as represented by fig. 9, plate XV. Its use is to heat tar, by making the iron hot in the fire, and then plunging it into tar. Fig.

LONG-BOAT. See the article Boat.

LONGITUDE of a Star, in astronomy, is an arc of the ecliptic, intercepted between the beginning of Aries and the point of the ecliptic cut by the star's circle of longitude, and is found on the celestial globe by the following operation;—Fix the centre of the quadrant of altitude on the pole of the ecliptic, and its graduated edge on the given star: then the degree of the ecliptic, cut by the quadrant, reckoned among the signs, is the longitude required. Thus Arcturus's longitude will be 23° of ♎.

Longitude *of a Place*, in geography, is an arc of the equator, intercepted between the first meridian and the meridian passing through the proposed place; which is always equal to the angle at the pole, formed by the first meridian and the meridian of the place.

The first meridian may be placed at pleasure, passing through any place, as London, Paris, Teneriffe, Ferro Islands, &c. but with Britons, it is generally fixed at Greenwich, where the Royal Observatory is placed; and the longitudes counted from it will be either east or west, according as they lie on the east or west side of that meridian. The difference of longitude between two places upon the earth, is an arc of the equator comprehended between the two meridians of these places; and the greatest possible is 180 degrees, when the two places lie on opposite meridians.

Since the parallels of latitude always decrease, the nearer they approach the pole; it is plain a degree upon any of them must be less than a degree upon the equator, in the ratio of the co-sine of the latitude to the radius. Hence, as the radius is to the co-sine of any latitude, so is the minutes of difference of longitude between two meridians, or their difference in miles upon the equator, to the distance of these two meridians on the parallel of that latitude, in miles. And, by this theorem, is constructed the table shewing how many miles answer to a degree of longitude, at every degree of latitude, which may be found in the common books of geography and navigation.

The longitude of a place on land may be found with considerable exactness, by observing the passage of the moon over the meridian, and comparing it with the passage observed in some fixed observatory. A much greater accuracy will be obtained by this method, if several successive transits of the moon be taken at either place of observation, as then the motion of the moon in right ascension will be obtained without the aid of calculation: but it will be requisite to attend to the equation of second differences, and even then the irregularity of the moon's motion in 24 hours is so great, that a very sensible error may still remain uncorrected.

Several writers, in explaining this method, appear to have fallen into a misconception of the subject, by confounding together the retardation of the moon in

24 hours, with the real retardation observed between two successive transits: and which latter should evidently be used in calculating the proportional retardation corresponding to a given difference in longitude.

Longitude, in navigation, the distance of a ship or place, east or west, from another in any known meridian, reckoned in degrees on an arc of the equator; and is, therefore, the same as longitude in geography.

The early navigators had no means of estimating their longitude but by the computed run of the ship; and the dangers they incurred by this inaccurate method, were sufficient to convince every enlightened government, particularly of maritime states, of the importance of encouraging, to the utmost effort of human ingenuity, what could be directed to the improvement of this defective state of navigation.

The early speculations of astronomers were of but little practical utility to the navigation of those times. In the 16th century, eclipses of the moon were strongly recommended; but they happened very seldom, and were too inaccurately computed to be of any great use. Perhaps, now and then, the approximate longitude of an almost unknown country, where a mariner might accidentally be on shore, was computed by this method, but to determine the place of a ship it was perfectly inadequate.

To find the longitude of a ship at sea is a problem to which the attention of navigators and mathematicians has been drawn, ever since navigation began to be improved. The importance of this problem soon became so well known, that, in 1598, Philip III. of Spain, offered a reward of 1000 crowns for the solution; and his example was soon followed by the States General, who offered 10,000 florins. By an act, passed in the British Parliament in 1714, a reward was also offered for the discovery of the longitude: viz.—the sum of 10,000*l.* if the method determined the longitude to one degree of a great circle, or to sixty geographical miles; of 15,000*l.* if it determined it to forty miles; and of 20,000*l.* if it determined it to thirty miles: with this proviso, that if any such method extend no farther than thirty miles adjoining to the coast, the proposer should have no more than half the rewards. The act also appoints the first Lord of the Admiralty, the Speaker of the House of Commons, the first Commissioner of Trade, the Admirals of the Red, White, and Blue Squadrons, the Master of the Trinity-House, the President of the Royal Society, the Royal Astronomer at Greenwich, the two Savilian Professors at Oxford, and the Lucasian, Plumian, and Lowndian Professors at Cambridge, with several other persons, as commissioners for the Board of Longitude.

After this Act of Parliament, several other acts passed in the reigns of George II. and III. for the encouragement of finding the longitude. At last, in 1774, an act passed, repealing all other acts, and offering separate rewards to any person who should discover the longitude, either by the watch keeping true time within certain limits, or by the lunar method, or by any other means. The act proposes, as a reward for a time-keeper, the sum of 5000*l.* if it determine the longitude to one degree, or sixty geographical miles; the sum of 7500*l.* if it determine it to forty miles; and the sum of 10,000*l.* if it determine it to thirty miles, after proper trials specified in the act. If the method is by improved solar and lunar tables, constructed upon Sir Isaac Newton's theory of gravitation, the author shall be entitled to 5000*l.* if such tables shall shew the distance of the moon from the sun and stars, within fifteen seconds of a degree, answering to about seven minutes of longitude, after allowing half a degree for the errors of observation. And for any other method, the same rewards are offered as those for time-keepers, provided it gives the longitude true within the same limits, and is practicable at sea.

The commissioners have also a power to give smaller rewards, as they shall judge proper, to any one who small make any discovery for finding the longitude at sea, though not within the above limits; provided, however, that if such person or persons shall afterwards make any further discovery, as to come within the above-mentioned limits, such sum or sums as they may have received shall be considered as part of such greater reward, and deducted therefrom accordingly.

For the improvements made in Chronometers or Time-Keepers, for the purpose of ascertaining the longitude, see these respective articles.

Various astronomical methods for finding the longitude have been found out by Lyon, Dunthorn, Witchell, Mackay, Mendoza, &c. These methods, which chiefly depend on having an ephemeris or almanac, suited to the meridian of some place, as Greenwich, to which the *Nautical Almanac* is adapted, which shall contain for every day computations of the times of all remarkable celestial motions and appearances, as adapted to that meridian. So that if the hour and minute are known when any of the same phenomena are observed in any other place whose longitude is desired, the difference between this time and that to which the time of the said phenomenon was calculated and set down in the almanac, will be known, and consequently the difference of longitude also becomes known between tnat place and Greenwich, allowing at the rate of fifteen degrees to an hour. Now it is easy to find the time at any place, by means of the altitude or azimuth of the sun or stars, but the great difficulty in these methods seems to lie in the fewness of p oper phenomena, capable of being thus observed, for all slow motions, such as belong to the planet Saturn for instance, are quite excluded, as affording too small a difference, in a considerable space of time, to be properly observed; and it appears that there are no phenomena in the heavens proper for this purpose, except the eclipses or motions of Jupiter's satellites, and the eclipses or motions of the moon, viz. such as her distance from the sun or certain fixed stars lying near her path, or her longitude or place in the zodiac, &c. Now, of these methods—

1st. That by the eclipses of the moon is very easy, and sufficiently accurate, if they did but happen often For, at the moment when the beginning, or middle, or end of an eclipse is observed by a telescope, there is no more to be done but to determine the time, by observing the altitude or azimuth of some known star; which time being compared with that in the tables, set down for the happening of the same phenomenon at Greenwich, gives the difference in time and consequently of longitude sought. But, as the beginning or end of an eclipse of the moon cannot generally be observed nearer than one minute, and sometimes two or three; therefore, the longitude cannot certainly be ascertained by this method.

2d. The moon's place in the zodiac is a phenomenon more frequent than her eclipses; but then the observation of it is difficult, and the calculus perplexed and intricate, by reason of two parallaxes; so that it is hardly practicable to any tolerable degree of accuracy.

3d. But the moon's distance from the sun or certain fixed stars, are phenomena to be observed many times in almost every night, and afford a good practical method of determining the longitude of a ship at almost any time; either by computing from thence the moon's true place, to compare with the same in the almanac; or by comparing her observed distance itself with the same as set down in the ephemeris; by the help of which, and the requisite tables, or those published by Mendoza Rios, which are also provided by the Board of Longitude, the calculations relating to the longitude may now be completed with great ease and accuracy in half an hour.

For farther information see to the instructions that accompany MENDOZA's and MACKAY's *Tables*, and also those that are annexed to the tables requisite to be used with the Astronomical and Nautical Ephemeris.

4th. The phenomena of Jupiter's satellites have commonly been preferred to those of the moon, for finding the longitude; because they are less liable to parallaxes than these are, and besides they afford a very commodious observation whenever the planet is above the horizon. Their motion is very swift, and must be calculated for every hour.

Now, to find the longitude by these satellites, with a good telescope, observe some of their phenomena, as the conjunction of two of them, or of one of them with Jupiter, &c. and at the same time find the hour and minute, from the altitudes of the stars, or by means of a clock or watch, previously regulated for the place of observation; then, consulting tables of the satellites, observe the time when the same appearance happens in the meridian of the place for which the tables are calculated; and the difference of time, as before, will give the longitude.

The eclipses of the first and second of Jupiter's satellites are the most proper for this purpose; and, as they happen almost daily, they afford a ready means of determining the longitude of places at land, having, indeed, contributed much to the modern improvements in geography; and, if it were possible to observe them with proper telescopes, in a ship under sail, they would be of great service in ascertaining its longitude from time to time.

LONGITUDE, *Board of*, a board formed of commissioners, who are appointed for the purpose of examining, trying, and judging of proposals, experiments, and improvements, relating to the longitude; and generally selected from among the most respectable scientific men in the country. See an Act of Parliament for 1714, for appointing a Board of Longitude.

LONGITUDE *by Account*, (*longitude estimée*, Fr.) is the longitude east or west, as computed daily from the ship's course and distance run.

LONGITUDE *by Lunar Observations*, (*longitude observée*, Fr.) is the above calculated from observing the moon's distance from the sun or a fixed star.

The requisite data for determining the longitude at sea, by the lunar method, are the apparent distance of the centre of the moon from the centre of the sun or star, and the apparent altitude of the centres of each at the moment of observation. Hence, three observers are usually employed: one observes the distance between the sun and moon, one the altitude of the sun, and the other the altitude of the moon. When this cannot be done, the place of the other two may be supplied by computation.

By means of lunar tables, the exact distance of the moon from the sun or star is computed for every three hours, for the meridian of Greenwich. We are not, however, to suppose that these distances are such as the moon and sun would *appear* to have at Greenwich; but such as they would appear to an observer at the centre of the earth. It is for *Greenwich time* only that they are computed; a circumstance not sufficiently insisted upon by elementary writers on this subject. From these tables (of the Nautical Almanac) it is easy to infer the distance for any intermediate interval: a simple proportion will be sufficient for this purpose. We may, therefore, consider ourselves as in possession of an instantaneous phenomenon, answering to every instant of time at Greenwich; since the distance of the sun and the moon is never the same for two successive instants of time.

Now, if we consider the converse of this proposition, it is equally evident, that if we have given the distance of the moon from the sun, as seen from the centre of the earth; we can, by the same tables, infer the exact time at Greenwich corresponding to this distance. Now, the object of a lunar observation is, to determine this distance at a given moment of actual time, to ascertain the apparent time at this moment for the meridian of the observer, and to compare it with the moment of Greenwich time, which is to be inferred from the given distance.

Now, the difficulty of the process arises from this circumstance, that since, to an observer on the surface of the earth, the moon appears always depressed by the effect of parallax, and the sun elevated by the effect of refraction, the angular distance observed with a sextant, or any other instrument, is not the same as the distance seen from the centre of the earth, and for which alone the nautical tables are calculated: hence, a spherical computation becomes necessary. Two cases of oblique spherical triangles must be computed, before the observed distance can be corrected, and the true distance ascertained.

To shorten the solution of this problem, and to reduce it within the compass of a mariner's ordinary powers of computation, has been an object with the first geometricians in Europe. It would lead us much beyond our limits to give a history of the numerous solutions that have been proposed. The French mathematicians, probably not having a great facility of constructing tables, have directed their attention chiefly to such methods as require only the common tables of logarithms. In our own country, where the Board of Longitude is always ready to publish any useful tables that may be approved, those methods and formulæ have been in general preferred, which admitted of the shortest solution by means of tables. In this respect, a progressive series of improvement has taken place since the first introduction of this method; and a skilful mariner will now compute the true distance from the apparent in five minutes, when, formerly, as many hours were required.

Besides the methods founded on a direct trigonometrical solution, there are many (such as Lyon's and Dr. Maskelyne's) which are founded on rather a different principle. The small triangle ☽ *m m'* is computed as if a plane one, ☽ *m* being the effect of the total depression of the moon in changing the distance: a similar triangle is formed for computing the effect of refraction for the sun or star. Various formulæ have been deduced from each of these principles, for the investigation of which the reader may consult Cagnole's *Trigonometry*, and various volumes in the *Connoissance des Temps*. A very clear and scientific investigation of all these methods was given by Captain Mendoza, in the *Philosophical Transactions* for 1797. See the article Lunar *Observations*, and also Rees's *Cyclopædia*, new edition.

Longitude *by Time-Keeper, (longitude par l'horloge marine,* Fr.) is estimated by the difference between the time at the place, and the time indicated by one of those improved watches called time-keepers.

The sun appears to move round the earth from east to west, or to describe 360° in 24 hours, and therefore he appears to move 15° in an hour.

If, therefore, the meridians to two places make an angle of 15° with each other; or, if the two places differ 15° in longitude, the sun will come to the eastern meridian one hour before he comes to the western, and, therefore, when it is twelve o'clock at the former place, it is only eleven at the latter; and, in general the difference between the times by the clock at any two places, will be the difference of their longitudes converted into time at the rate of 15 degrees for an hour, the time at the eastern place being the forwardest. If, therefore, we can tell what o'clock it is at any two places at the same instant of time, we can find the difference of their longitudes, by allowing 15 degrees for every hour that the clocks differ.

Let, therefore, the time-keeper be well regulated and set to the time at Greenwich, that being the place from which we reckon our longitude; then, if the watch neither gains nor loses, it will always shew the time at Greenwich, at wherever place it may be. Now, to find the time by the clock at any other place, take the sun's altitude, and thence find the time; now the time thus found is apparent time, or that found by the sun, which differs from the time shown by the clock by the equation of time. We must, therefore, apply the equation of time to the time found by the sun, and we shall get the time by the clock; and the difference between the time by the clock so found, and the time by the time-keeper, or the time at Greenwich, converted into degrees at the rate of 15° for an hour, gives the longitude of the place from Greenwich. For example: let the time by the time-keeper, when the sun's altitude was taken, be $6^h\ 19$, and let the time deducted from the sun's altitude be $9^h\ 27'$, and suppose at that time the equation of time be 7', shewing how much the sun is that day behind the clock, then the time by the clock is $9^h\ 34$, the difference between which and $6^h\ 19$, is $3^h\ 15$, and this converted into degrees, at the rate of 15° for one hour, gives 48° 45', the longitude of the place from Greenwich; and as the time is forwarder than that at Greenwich, the place lies to the east of Greenwich. Thus the longitude would be very easily determined if the time-keeper could be depended upon. But, as a time-keeper is liable to gain or lose, before it is sent out, its gaining or losing every day for some time, should be ascertained to find its true rate of going. See the article Chronometer and Time-Keeper.

To find the Longitude of a Ship at Sea by a Time-Keeper.

Rule.—Let several altitudes of the sun, or of any fixed star, be observed, and correct the mean altitude as usual, with which, the ship's latitude, and object's declination, compute the apparent time of observation, to which apply the equation of time, reduced to the time and place of observation, according to its title in the Nautical Almanac, and hence the mean time of observation will be known.

To the mean of the times of observation, as shown by the time-keeper, apply its error and accumulated rate. Hence the mean time, under the meridian of the place where the error and rate were established will be known; to which apply the difference of longitude in time between that place and Greenwich, and the mean time of observation under the meridian of Greenwich will be obtained. Now, the difference between the time at the place of observation and that at Greenwich will be the longitude of the place in time; and which is east or west, according as the time by observation is later or earlier than the Greenwich time. See Mackay *on the Longitude.*

LONGITUDINAL, in general, denotes something placed lengthways: thus, some of the fibres of the vessels in the human body are placed longitudinally, others transversely, or across.

LONG-TACKLE Block. See the article Block.

LONG-TIMBERS, in ship-building, are those timbers in the cant-bodies, which reach from the dead-wood to the head of the second futtock.

LOOF, *(partie de l'avant du vaisseau, appellee aussi en Francois, le lof du vaisseau,* Fr.) the after part of a ship's bow; or that part of her side forward where the planks begin to be incurvated as they approach the *stem:* hence, the guns which lie here are called loof-pieces.

LOOK-OUT, *(découverte, ou vigie,* Fr.) a watchful attention to some important object or event, which is expected to arise from the present situation of a ship, &c.

There is always a look-out kept on a ship's forecastle at sea, to watch for any dangerous objects lying near her track, or for any strange sail heaving in sight, &c. The officer of the watch, accordingly, calls often from the quarter-deck, to the persons appointed for this service, "Look well out afore there!"

A LOOM Gale, *(petit vent frais, ou vent sur l'eau,* Fr.) is a gentle easy gale of wind, in which a ship can carry her topsails *a-trip.*

Loom *of an Oar, (bras d'un aviron,* Fr.). See the article Oar.

LOOMING, *(mirage, vue indistincte d'un objet,* Fr.) an indistinct appearance of any distant object, as ships, mountains, &c. as "That ship looms large afore the wind."

Looming *of the Land, (mirage de la terre.* Fr.) implies its being high above the water. &c.

LOOP, (*noeud coulant*, Fr.) a noose made in a rope.

LOOP-HOLES, (*meurtrieres*, Fr.) are small apertures, formed in the *bulk-heads* and other parts of a merchant-ship, through which the small arms are fired on an enemy who attempts to board her.

LOOPS, (*anneaux*, Fr.) in a gun-carriage, are iron rings fastened, one on the front of the fore axle-tree, and two on each side, through which the rope or tackle pass, whereby the guns are moved backwards and forwards.

To LOOSE A SAIL, (*deployer, ou larguer une voile*, Fr.) is to unfurl or cast it loose, in order to its being *set*, or dried, after rainy weather.

To LOOSEN A ROPE, (*larguer une manœuvre*, Fr.) is to cast it off, or let it go.

Fig. 10. LOPER, in rope making, a machine used to lay lines with, see fig. 10, plate XV. It has two iron swivel-hooks (that run round in a brass or iron box) at each end, for the line to hang on and work. by the power of the fore turn, from the wheel at the upper end.

LOSING COMPANY AT SEA. is the separating of one or more ships from a convoy bound to a certain place, or one ship from another on a cruising station, either from the effects of a hard gale of wind, or any other disaster that might be met with during a voyage.

The necessity of keeping ships together on a cruising station, and the important benefits resulting from any contrivance that can tend to prevent their separation, reflects not only the highest honour on the ingenuity of the inventor, but promises to be of the highest national advantage.

A gentleman of the name of Hardacre, an officer in the royal navy, presented a plan for this purpose, a few years ago, to the Lords of the Admiralty, which seemed to merit some notice; it was highly spoken of in all the periodical works, and an idea entertained of the success of the project.

From the same source it is to be collected cursorily that the machine is likewise of great utility to prevent a ship from falling into the brow of the sea; and, in a gale of wind, to keep her from the danger of being driven upon a lee shore, where there is no anchorage.

It will also be found of great importance in case a ship should lose her sails, carry away her masts, lose a rudder, get becalmed after a strong gale in a heavy sea, or become otherwise in a distressed situation.

The invention is rendered invaluable, on account of its being on so plain and easy a principle, that it may be comprehended in an instant; and two or three seamen can, at any time, construct the whole machine in a few minutes, out of a part of the ship's materials, without any loss or expense whatever; as the machine, after it is done with, can easily be taken to pieces: and those pieces which compose it used again for their ordinary purposes.

LOST, (*perdu, ou naufrage*, Fr.) the state of being foundered or cast away; expressed of a ship when she has either sunk by means of some disaster, or struck upon a rock, shelf, or lee-shore, where she has been beaten to pieces by the violence of the sea.

LOUGH, or LOCH, the former is the Irish, and the latter the Scotch term for LAKE, which article see.

LOW-WATER, (*mer basse*, Fr.) the lowest point to which the tide ebbs. See the article TIDE.

To LOWER, (*amener, &c.* Fr.) is to ease down gradually, expressed of some weighty body suspended by tackles, or other ropes, which being slackened, suffer the said body to descend as slowly or expeditiously as occasion requires: hence,

LOWER *Cheerly!* (*amène tout plat! ou en bande!* Fr.) the order to lower expeditiously.

LOWER *Handsomely!* (*amene en douceur*, Fr.) the order to lower gradually.

LOXODROMICS, the art of oblique sailing, by the rhomb, which always makes an equal angle with every meridian; that is, when you sail neither directly under the equator, nor under one and the same meridian, but across them.

LUBBER, (*marin d'eau douce*, Fr.) a contemptuous name given by sailors to those who are unacquainted with the duty of a seaman.

LUBBER'S-*Hole*, (*trou du chat*, Fr) is the vacant space between the head of the lower-mast and the edge of the top; it is so termed from a supposition that a *lubber*, not caring to trust himself up the futtock-shrouds will prefer that way of getting into the top.

LUFF! (*lof!* Fr.) the order to the helmsman to put the tiller towards the *lee*-side of the ship, in order to make the ship sail nearer to the direction of the wind.

LUFF, also signifies the roundest part of the bow of the ship.

LUFF *of a Sail*, is the fore, or weather, part thereof.

LUFF *Round!* (*lof tout!* Fr.) is the order to throw the ship's head up in the wind, in order to tack her, &c.

To LUFF *into a Harbour*, (*gagner un port en tenant le plus prés*, Fr.) is to sail into it close by the wind.

To Spring a LUFF, (*faire une aulofée*, Fr.) is to yield to the effort of the helm, by sailing nearer to the direction of the wind, than the ship had done before.

Keep your LUFF! (*serre le vent!* Fr.) the order to the steersman to keep nearer to the wind.

LUFF-TACKLE, (*palan qui sert a divers usages*, Fr.) a name given to any large tackle that is not destined for a particular place, but may be variously employed, as occasion requires. It is, generally, somewhat larger than the *jigger-tackle*, although smaller than those which serve to hoist the heavier materials into and out of the vessel; which latter are the main and fore tackles, the stay and quarter tackles, &c.

LUGGER, (*lougre*, Fr.) a small vessel carrying either two or three masts, with a running bowsprit, upon which she sets lug-sails, and sometimes has top-sails adapted to them.

On the bowsprit are set two or three jibs; and the lug-sails hang obliquely to the masts, their yards being slung at one-third their length, one on each lower-mast and topmast: the topmast fixes abaft the mast-heads, as those of schooners. Luggers sail well close-hauled, and very near the wind.

The rigging is very light and simple: the masts are supported by shrouds and stays; and the yards have halyards, lifts, and braces. To the lee-clew of the sail is a sheet, and to the windward-clew a tack, which is occasionally shifted as the vessel goes about.

Some luggers are mere open boats, and have only a small mast and a ring-sail set to it over the stern, and the foot spread by a small boom. The French employ luggers as privateers; but there are not many vessels of this description in our service.

LUG-SAIL, (*voile a la façon de celles d'un lougre*, Fr.) a quadrilateral sail, bent upon a yard which hangs obliquely to the mast, at one third of its length. These are more particularly used in the barca longas, navigated by the Spaniards in the Mediterranean.

LUG-SAIL *Boat*, (*canot gree en lougre*, Fr.) a boat carrying sails of the preceding description.

LUMBER, in trading ships, denotes stowage-wood and small timber, such as spars, joists, boards, planks, hoops, staves, anchor-stocks, handspikes, &c.

LUMINARIES, a term used for the sun and moon, by way of eminence for their extraordinary lustre and the great quantity of light they give us.

LUMPERS, (*gabariers, employes a charger et decharger un navire marchand*, Fr.) the appellation given to labourers employed to load and unload a merchant-ship when in harbour.

LUNAR, something relating to the moon; thus we say, lunar month, lunar year, lunar eclipse, &c.

LUNAR *Cycle*. See the article CYCLE *of the Moon*.

LUNAR *Distance*, is the apparent distance of the moon from the sun or a fixed star, at the time of making an observation for finding the longitude; the reduction of which, from the effect of parallax and refraction, to the *true distance*, requires some tedious calculations. See the article APPARENT *Distance*.

Few problems have ever been more investigated or studied than that of clearing the *lunar distance*, and many ingenious methods have been devised for contracting the operation. These methods are founded on some of the following principles:—The first general principle is spheric trigonometry; the second is the doctrine of proportional errors, by which the effects that the errors in the altitudes produce on the distance are solved by fluxions, or by the *differential calculus*. and a third principle has been lately discovered, which is founded on the properties of a quadrangle inscribed in a circle, as explained and exemplified by the inventor, Dr. Andrew, in his Astronomical and Nautical Tables.

LUNAR *Observations*, or *the* LUNAR *Method*, (*observation de distance*, Fr.) is a method of observing the angular distance between the moon and the sun, or a fixed star, and comparing it with the computed distance in the Nautical Ephemeris, for the purpose of ascertaining the longitude at sea.

This method of finding the longitude is the greatest modern improvement in navigation: the idea, however, is not modern, but it has not been applied with any success until within the last fifty years. M. de la Lande mentions certain astronomers, who, above two hundred years ago, proposed this method, and contended for the honour of the discovery; but its present state of improved and universal practice he very justly ascribes to Dr. Maskelyne. The discovery, indeed, seems to claim less honour than its subsequent improvements; it is one of those things which are obvious in theory, but difficult in practice. The most ancient method of finding the longitude was by the lunar eclipses; and that of finding it by the lunar distances is perfectly analogous: it is, therefore, highly probable that the method was thought of at a very early period, but the want of proper tables and apparatus prevented its being reduced to practice.

The late Astronomer Royal was selected by the Board of Longitude, in the year 1761, to observe the transit of Venus over the sun's disc, at the Island of St. Helena: this afforded him an opportunity of taking lunar observations with great effect, which had long been contemplated as a desideratum in navigation. Hadley's quadrant, which had been recently invented, and Professor Mayer's lunar tables contributed much to his success in working the lunar observations; and it is to him this country is indebted for some of the greatest improvements that have been made in navigation. It was he who first proposed and superintended the construction of the Nautical Almanac, which has now been published for fifty years, and which relieves the calculator from all the very laborious part of the process; and the remaining part of the computation has been so simplified by successive improvements, both in the formulæ and construction of the requisite tables, that, at present, the necessary observations may be both made and computed by any mariner, who has received a tolerably good nautical education.

As the practical methods of making and computing a lunar observation are given at great length in every nautical book, we shall confine ourselves to explaining the general nature and object of the problem, and refer the reader to professed treatises on navigation for farther information. In MACKAY'S *Treatise on the Longitude*, the reader will find some excellent methods of solving both this and a variety of other nautical problems, accompanied by very useful tables. MENDOZA'S *Tables* contain his own valuable method of computing a lunar observation, besides general tables for every nautical purpose. The REQUISITE TABLES (which article see) are well known, and are in the hands of every navigator.

It may be observed, that, in the most practical methods of finding the longitude at sea by celestial observations, the moon is the chief guide or instrument; for the quickness of her motion renders her peculiarly well adapted for measuring small portions of correspondent time. Now, as she is seen in the same part of the heavens nearly at the same instant of absolute time, from all parts of the earth where she is visible, and as she is continually and sensibly changing her place, it is evident, that, if two correspondent observers were to note the precise moment of their respective times, when she is seen in any particular part of the heavens, *the difference between those times would shew the difference of longitude.*

In every method of finding the longitude by the moon, the first object is, to be able to ascertain the part of the heavens where she is: this is easily seen at the time of her eclipses, or at the occultation of a fixed star; and these were naturally the first methods resorted to, but they occur too seldom for general use; the moon's place, however, may be marked with equal precision, by taking her distance from some fixed object, whose latitude and longitude are known; and stars in or near the zodiac are preferred, as the nearer such objects are to the moon's orbit, the greater will be her motion with respect to them: and, though her motion is not uniform; yet, during the short space of time that she is near any star, she may be considered as moving uniformly.

The Nautical Almanac may be considered a perpetual observer, that communicates universally and instantaneously, certain celestial appearances, as they take place at Greenwich observatory. Here the distances are given between the moon and the sun, and certain remarkable stars in or near the zodiac, for every three hours; and any intermediate distance or time may be thence found by the rule of proportion, with sufficient accuracy. If, therefore, under any meridian, a lunar distance be observed, *the difference between the time of observation and the time in the Almanac, when the same distance was to take place at Greenwich, will shew the longitude.*

This method, so apparently simple, would, no doubt, have been adopted long ago, had it not been for the want of proper instruments for making observations, and correct lunar tables; for the moon, though so near to the earth, has always perplexed astronomers more than any other celestial body: quadrants, sextants, and reflecting circles, have now removed the former difficulty, and Professor Mayer's lunar tables, as improved by Mason and Maskelyne, the latter; as from them those tables in the Nautical Almanac, of the lunar distances, have been calculated under the direction of Dr. Maskelyne for many years. See the article LUNAR *Tables*.

Notwithstanding the assistance of the Nautical Almanac and improved Requisite Tables, this calculation is still tedious; which arises from the circumstance of the *observed distances* between the heavenly bodies not being the *true distances;* for the altitudes of these bodies are more or less affected both by refraction and parallax; and, though these effects only operate in a vertical direction, yet that which changes the altitude of two bodies must also change their distance asunder. This is evident from the consideration that the altitude of a celestial object is an arc of an azimuth circle intercepted between the object and the horizon; and, as all azimuth circles incline gradually to each other, from the horizon to the zenith, where they meet, it is plain that the more two bodies are apparently raised, the less will be their apparent distance asunder, and the contrary. It is well known that the heavenly bodies are raised by refraction, and depressed by parallax; and that these effects are greatest in the horizon, and gradually diminish to the zenith, where they become nothing. Refraction depends upon altitude alone; but parallax depends upon both altitude and distance. All celestial objects, except the moon, are more affected by refraction than by parallax, and therefore appear above their true places; but the moon is always seen, excepting in the zenith, below her true place, being more affected by parallax than refraction, on account of her proximity to the earth.

These effects of parallax and refraction, though counteracting each other, seldom do it so equally as to render all correction unnecessary. Sometimes the *apparent* distance is nearly a whole degree more or less than the *true* distance; and the principal cause of so great a difference is the moon's parallax: for this body, which is the chief guide to the longitude, is also the great cause of error in the distances, and is therefore the principal object of correction.

The following remarks, made by the late Reuben Burrow, well known as an eminent astronomer and an able mathematician, are intended to point out an error in the usual practice of making the lunar observations, and another in the method of computation.

"It is well known," he observes, "that a little before and after the conjunction, the whole hemisphere of the moon is visible, and the enlightened crescent seems to extend some distance beyond the dusky part Now, having determined the longitude of a place from the eclipses of Jupiter's satellites, I took several sets of distances of the moon's limb from a star near the time of conjunction, both from the bright and dusky part of the circumference; and having calculated the results, I found that those taken from the dusky part were much nearer the truth than the others. The nature of the error evidently showed that the star had really been at some distance from the limb when it appeared to be in contact with it; and, as the error was a considerable part of a degree, I saw it would be of consequence to discover the cause of it; which, however, was obvious enough from Newton's principles, and may be explained as follows:—

"Let A D be the diameter of the moon, and A the centre of a star in contact with the moon's limb: now, as the enlightened part of the moon evidently appears to extend beyond the dusky part, let the concentric circle B C represent the moon's limb, thus apparently magnified: then, with the centre A, and the distance D C, describe a circle, which, consequently, will touch the moon's apparent circumference inwardly: now, as this last is a consequence of supposing the centre of the star to touch the circumference of the moon, exclusive of the deception, it follows, that *the proper method of taking the distance is to make the star appear to touch the moon inwardly.*

"But all the writers on this subject have particularly directed that the star be made to touch *outwardly.*

"Let B, therefore, be the point of contact, and *a* the centre, the error then is *A a*, or the sum of the apparent increase of the moon's radius, and the apparent radius of the star; this quantity, it is evident, will make a considerable error in the result; and errors arising from this source are more to be attended to as they are not of a kind to be lessened by increasing the number of observations. The same reason is applicable to the sun and moon, with very little alteration.

"The distance of the moon from the sun or a star, at each three hours, is given in the Nautical Ephemeris; and the method of inferring the time for any intermediate distance, is by simple proportion. this would be just, if the moon's motion was uniform; but, as this is not the case, the velocity should be taken into the account, as well as the space, in determining the time taken by the moon to move any given distance; and the proper measure of the velocity is such a quantity as has the same ratio to the space described, as three hours have to the time that has been actually taken to move the given distance. To find this quantity correctly, would require interpolation; but it will be sufficient in practice to find the time first by the common method, and then to correct the interval for three hours to that time, by taking a proportional part of the second difference of the moon's distance at the beginning of each three hours, supposing the first differences to answer to the middle of each interval.

" The last correction, though not so considerable as the first, will often bring the result nearer to the truth by three, four, five, or six miles, and sometimes more; which, in geographical determinations, is of consequence; and, by paying attention to those and some other causes of error, the results in general will be much nearer to the truth than is usually imagined."

The instruments used for these lunar observations are sextants and reflecting circles. Under CIRCLE is already described the reflecting circle of Mr. Troughton's construction; which we conceive to be vastly superior to any sextant for obtaining with accuracy the angle subtended by the moon and star. As each observation has three readings, two observations (one on each side) of zero, give six results. This instrument, in the hands of a careful observer, will not give a greater error than 20″, or about ten geographical miles at the equator. The error of the lunar tables may amount to about as much more, and an error in the altitudes and other data, about the same quantity. Should all these conspire to produce a total error in the same direction, which very rarely happens, this total error would amount to 60 , or 30 miles.

Some persons still prefer sextants, from an opinion that they derive some advantage from length of radius; but they are subject to errors which have no tendency to correct each other; and should only be used for the altitudes, and the circle to be taken in preference for the observations of the distance. When circumstances do not admit of three observers, the altitudes of the sun and moon may be computed, and we are disposed to think that this would always be the more preferable method, where the observer is sufficiently skilful to make the additional computation without fear of mistake. The altitudes found in this case, by computing the horary angles are the true altitudes, and must be corrected by applying the refraction and parallax *inversely*, but for doing this accurately, tables should be computed for reducing *true distances* to *apparent*. However, a little attention to this circumstance will render the whole operation sufficiently easy; and, if the computations are well made, the accuracy of this method will, probably, exceed that of direct observation. The manner of adjusting the instruments, and of making the observations, is best taught by practice; and for written instructions, see Dr. MASKELYNE's *Mariner's Guide*, Dr. MACKAY *on the Longitude*, and PROFESSOR VINCE's *Practical Astronomy*.

Various methods of working the lunar observations have been devised chiefly by Halley, Euler, Mayer, Maskelyne, Lyons, Witchell, Burrow, Borda, Gerrard, Wales, Mackay, Kelly, Andrew, and Mendoza. The methods of the three last authors appear the most concise, but all are sufficiently correct; and seamen generally prefer such as they have first learned.

No method has hitherto obtained an exclusive preference over the rest, nor does it appear possible to reduce the calculation to a conciseness to answer the general purposes or wishes of seamen; and hence, other modes have been devised, of obtaining approximate solutions by projection or graphic operation.

The first graphic method for clearing the distances was that by La Caille, called the *Chassis de reduction*, which has since been copied by La Lande, Mackay, and others. It is an orthographic projection, consisting of a great number of lines accurately drawn, and various scales for obtaining the different corrections.

Another graphic operation, of a different description, was executed by the late George Margetts, and published in 1790. It consists of seventy large plates, containing numerous lines drawn from the solutions of lunar distances in Dr. Shepherd's Tables, published at Cambridge, By Margett's Longitude Tables, the solution of a lunar observation may be obtained in about one-fourth of the time required by calculation; and the answer, though not perfectly accurate, is sufficiently correct for the purposes of navigation.

Another orthographic projection, founded on the fluxional analogies of spheric triangles, has been devised by Dr. Kelly, of Finsbury-square, and published in his excellent Introduction to Spherics and Nautical Astronomy; where an investigation of its principles is given with a demonstration, shewing, that in all proper altitudes it will give the true distance within a few seconds.

The simplicity of this projection is extremely curious, as giving an approximate solution of a complicated problem, by drawing four right lines only from the scale of chords, and it must therefore be very useful where great expedition is required.

Mr. Donne, of Bristol, has lately formed and delineated what he calls a *Corrector*, on the principles of Dr. Kelly's Orthographic Projection, which he seems to think will become very useful to practical seamen.

The late Mr. Ferguson invented a Parallactic Rotula, some years ago, for the purpose of clearing the distance.

M. Bouguer contrived a method for clearing the distance with a common scale and compass, by a graphical operation, and invented an instrument called a Heliometre, for measuring the distance between the moon and a star; which M. de la Lande describes as an accurate and useful instrument for that purpose.

M. de Charnieres also constructed an instrument, which he called a *Megametre*, upon the principles of the object glass micrometers, for the purpose of measuring distances less than ten degrees.

The Chevalier de Borda's ingenious contrivance, called a *Cercle de Reflexion*, for clearing the distance is also well known on the continent.

Dr. Andrew, of the East India Academy at Croydon, has also published a new method of clearing observed lunar distances, which, he conceives, has several advantages over the common methods, as the answer will in every possible case, be determinate; the diverse affections of the sides of the spherical triangles commonly used never occasioning by his method any embarrassment to the operator, and as there never happens any change of signs from *plus* to *minus*, or from *minus* to *plus*, in the calculation, it is plain the operation must be clear from ambiguity.

It will also be found, on trial, he observes, that the nature of the numbers employed in this mode of calculation tends very much to preclude small errors, or to render their effects of no consequence. A table of logarithms of natural numbers, a table of the squares of chords, and table the ninth, with its corrections from the Requisite Tables, he als observes, are all that are necessary for this method: and, as they require to be opened only about seven times, the whole operation might be easily performed in less than as many minutes.

Mr. Turner, a teacher of navigation, at Portsea, has also lately published a new practical set of lunar tables and rules, whereby he states, the *apparent* distance may be reduced to the *true* by a very short process.

But it may be justly observed, that operations which appear the most concise are not always the most expeditiously performed, as much depends on the number and variety of tables required, and the manner of applying them.

Lunar *Periodical Month*. See the article Month.

Lunar *Tables*, are a series of numbers calculated from the theory of the Moon, to be used with the lunar observations, in finding the longitude at sea.

Mr. Flamstead, the astronomer royal, complained, soon after his appointment in 1675, that he found the best lunar tables extant erring sometimes 12 minutes or more in the moon's apparent place, which would cause a mistake of half an hour, or 7½ degrees in the longitude, by comparing her place in the heavens with that given in the tables.

In 1732, Mr. Robert Wright, of Winwick, in Lancashire, published new lunar tables, according to the Newtonian theory; but the best are those originally calculated by Professor Mayer, of Gottingen, in 1755, for which he was rewarded by the English Board of Longitude, with 3,000*l*. An edition of them was published in 1770, by Dr. Maskelyne, under the title of "*Tabulæ motuum Solis et Lunæ, &c.*" in which were added several tracts and tables of his own, with explanations in Latin and English.

The late Mr. Charles Mason, who had been employed to observe the transit of Venus in 1761, at Bencoolen, and who was afterwards an assistant to the late Dr. Mask lyne, at the royal observatory at Greenwich, having had many opportunities of comparing them accurately with observations, published an improved edition in 1780, in which he introduced eight more equations, which Mayer had given in his Theory, but which he thought of too little consequence, or too uncertain, to introduce in his tables; he also made use of an *empirical* equation of about 20″, when a *maximum*, which was not founded on theory, but employed because it made the tables agree better with observation: this, however, not appearing to be derived from principle was often rejected by other astronomers, till La Place, in his *Mechanique Céleste*, found that it proceeded from the compression of the earth at the poles, which prevents the gravitation to the earth from decreasing precisely as the squares of the distances increase, and by that means produces this small irregularity, which Mr. Mason, although an accurate observer and calculator, perhaps was not sufficiently skilled in mathematics to deduce from principle.

M. Burg, an accurate astronomer and able mathematician at Berlin, has also proved the necessity of the equations introduced by Mason; and published, in France, an edition of these tables, in 1806, in which he added six other equations, which were entirely new. Besides these, the secular equations of the anomaly and node, and the equation which has a period of 185, discovered by La Place, have been introduced; they are now to be found in the 3d vol. of Vince's Astronomy, and exceed in accuracy any tables yet known. M. Burg was rewarded by the National In titute of France in 1800, after their mathematicians had examined the tables with much care, and compared th m with many of Dr. Maskelyne's observations, as well as their own, and found the error in few cases amount to 10″, a space which the moon describes in her orbit in less than 20″ of time; the longitude may, therefore, now be found by lunar observation, to an arc that corresponds to the third part of a minute, or to five nautical miles; a degree of accuracy fully sufficient for all the purposes of naviga ion.

Lunar *Year*, consists of 354 days, or 12 synodical months, of 29½ days each. See the article Year.

In the early ages, the lunar year was used by all nations; the variety, of course, being more frequent and conspicuous in this planet, and, consequently, better k own to men than those of any other. The Romans regulated their year, in part, by the moon, even till the time of Julius Cæsar; the Jews, too, had their lunar month and year.

LUNATION, the period or time between one new moon and another: it is also called the synodical month, consisting of $29^d\ 12^h\ 44'\ 3''\ 11.3'''$; exceeding the periodical month by $2^d\ 5^h\ 0'\ 55''$.

LUPUS, the Wolf, in astronomy, a southern constellation, joined to the Centaur, containing 19, or, according to Flamstead's catalogue 24 stars.

LURCH, (*embardée*, Fr.) the sudden jerk or rolling of a ship on either side, which is caused by a heavy wave striking either upon the rudder or quarter, and which often strains her principal fastenings. See the article Lee-Lurches.

LYING-TO, is the act of waiting for some other ship, either approaching or expected; or to avoid pursuing a dangerous course, especially in dark or foggy weather, &c. See the article To Lie-to.

LYNX, in astronomy, a constellation of the northern hemisphere, composed by Hevelius out of the unformed stars. In his catalogue it consists of 19, but in the Britannic of 44 stars.

LYRA, the Harp, in astronomy, a constellation of the northern hemisphere, containing, according to the Britannic catalogue, 21 stars.

M.

MACULÆ, in astronomy, are dark spots appearing on the luminous faces of the sun, moon, and even some of the planets; in which sense they stand contradistinguished from faculæ. See the article FACULÆ.

A MADE BLOCK, is a block which has its shell made of several pieces of elm plank, suited to the thickness of the cheeks, sheave-holes, and middle-parts, and is strongly bolted together with three bolts at each end, driven through and clenched on a ring at the points. These blocks have flatter cheeks and squarer edges than other treble and four-fold blocks, for heaving down ships or other heavy purchases.

Smaller *made blocks*, of modern invention, are made of two pieces, joining in the middle, and the pin to work on patent rollers, let into the inside of the checks, which are bolted or rivetted together at the ends. These blocks, however, are thought too complex for the royal navy, and are not so easily repaired in case of failure. See the article BLOCK MACHINERY.

MAGAZINE, (*soute aux poudres*, Fr.) a close room or store-house, built in the fore or after-part of a ship's hold, to contain the gunpowder; it is lighted, occasionally, by means of candles fixed in the *light-room* contiguous to it. This apartment is strongly secured against fire, and no person is suffered to enter it with a lamp or candle. See the article LIGHT-ROOM.

After MAGAZINE, (*soute aux poudres de l'arriere*, Fr.) in a ship of war, is built in the after-part of the ship's hold, and, as it is generally smaller than the fore magazine, it contains only a sufficient supply of cartridges for the after-guns, during an action; whereas the fore magazine (*soute aux poudres de l'avant*, Fr.) contains a quantity of powder sufficient to supply the ship for a length of time.

Powder MAGAZINE, a place in which powder is kept in large quantities, and which, on account of the nature of the substance preserved, should be arched and bomb-proof. According to the plan of Vauban, they are sixty feet long and twenty-five broad in the inside. The foundations are eight or nine feet thick, and about as many feet high, from the foundation to the spring of the arc. As some inconveniences have arisen from this structure, Dr. Hutton proposes to find an arc of equilibration, which he would have constructed to a span of twenty-feet, the pitch being ten feet: the exterior walls at top forming an angle of 113 degrees, and the height of the angular point above the top of the arc to be seven feet.

MAGNET, THE LOADSTONE, (*l'aimant*, Fr.) a kind of ferruginous stone, resembling iron ore in weight and colour, though rather harder and heavier; and it is endued with divers extraordinary properties. attractive, directive, inclinatory, &c.

The *magnet* is usually found in iron mines, and sometimes in very large pieces, half magnet half iron. Its colour is different, as found in different countries. Norman observes, that the best are those brought from China and Bengal, which are of an irony or sanguine colour; those of Arabia are reddish; those of Macedonia, blackish; and those of Hungary, Germany, England, &c. the colour of unwrought iron. Neither its figure nor bulk are constant or determined; being found of all shapes and sizes. The ancients made use of the *magnet* in medicine.—The moderns, more happy, take it to conduct them in their voyages.

The most distinguished properties of the *magnet* are, that it attracts iron, and that it points towards the north pole; and, in other circumstances, also dips or inclines to a point beneath the horizon, directly under the pole: it also communicates these properties, by touch, to iron. By means of which, are obtained the mariners compass needles, both horizontal, and inclinatory or dipping-needles. See the article DIPPING-NEEDLE.

MAGNETIC-SAILING, is the art of navigating a ship on any given point of the compass; but when the needle is subject to variation, the course will require a correction.

Let N S represent the true meridian N *s* the magnetic, at west variation, the angle *s* C *b* the magnetic course and C A the distance run: then it is evident the angle SC*b* is the true course, C B the difference of latitude, and AB the departure. Hence, the course given by the compass may be corrected by the following rule:—

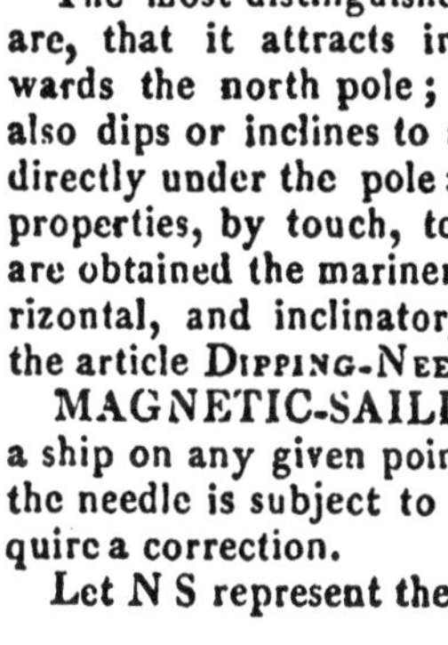

Suppose yourself placed at the centre (C) of the compass, and looking the way the ship sails, then, if the given variation be westerly, allow it to the left hand; but if easterly, allow it to the right hand, and the true course will be obtained.

The following plan for correcting the dead reckoning of ships for the variation of the compass, is proposed for the sake of simplifying the operation.

Rule.—Use all the courses as you otherwise would do, provided there was no variation; and call the difference of latitude and departure so found, the magnetic difference of latitude, and magnetic departure. Then find the course and distance corresponding to this magnetic difference of latitude and departure; the course being corrected, for the variation will give the true course, and the distance will be the true distance.

Example.—A vessel sails, by compass S.S.W. five miles, E. by N. six miles; S.W. nine miles; and E.S.E. four miles; when the variation of the compass is 25° westerly; required her difference of latitude and departure.

By the Traverse Table.

Courses.	Dist.	N.	S.	E.	W.
S. 39°½ W.	5	——	3.86	——	3.18
S. 27½ W.	3	——	2.66	——	1.40
S. 10½ W.	2	——	1.97	——	.36
S. 30 W.	10		8.49		4.94

Now the difference of latitude 11.34, found in the common way, with the departure 1.32, gives a course of 6°½, with a distance of 11½ miles, without allowing for the variation. Because the departure is easterly, and the variation westerly, this course must be added to the variation of the compass, to obtain the true course, which will consequently be 31°½. Therefore the true course is S. 31°½ E. and the true distance is the same as the magnetic distance 11½ miles. The difference of latitude and departure answering to this course and distance, is 9,72 difference of latitude, and 6 departure.

Had the departure and variation been both easterly, or both westerly, the difference of the magnetic course and the variation of the compass would have produced the true course.

The foregoing plan is both more expeditious and more correct than the one usually practised. See Cole's *Stereogoniometry*.

MAGNETISM, (*magnétisme*, Fr.) the quality or constitution of a body, by which it is rendered magnetical, or a magnet, sensibly attracting iron, and giving it a meridional direction. This is a transient power, capable of being produced, destroyed, or restored.

Magnetism is supposed to have been first rendered useful about the beginning of the 13th century, by John de Gioja, a handycraft of Naples, who noticed the peculiar attraction of metals, iron in particular, towards certain masses of rude ore; the touch of which communicated to other substances of a ferruginous nature, especially iron or steel bars, the same property of attraction: these touched bars he observed to have a peculiar and similar tendency towards one particular point; that, when suspended in equilibrio, by means of threads around their centres, they invariably indicated the same point; and that, when placed in a row, however adversely directed, they soon disposed themselves in perfectly parallel order. In this instance he improved upon the property long known to, but not comprehended or applied to use by the ancients. Gioja was not long in further ascertaining the more sensible purposes to which the magnet might be appropriated. He, accordingly, fixed various magnets upon pivots, supporting their centres in such a manner as allowed the bars to traverse freely. Finding that, however situate within the reach of observation and comparison, they all had the same tendency, he naturally concluded them to be governed by some attraction which might be ultimately ascertained and acted upon. He, therefore, removed into various parts of Italy, to satisfy himself whether or not the extraordinary impulse which agitated these bars that had been magnetised by friction, existed only in the vicinity of Naples, or was general. The result of his researches appears to be, that the influence was general, but that the magnets were rendered extremely variable, and fluctuated much, when near large masses of iron.

The whole power of *magnetism* in this country, as it affects needles a foot long, is to that of gravity nearly as 1 to 300, and as it affects needles four feet, as 1 to 600.

The quantity of magnetic power accelerating the same dipping-needle, as it oscillates in different vertical planes, is always as the co-sines of the angles made by those planes with the magnetic meridian, taken on the horizon.

The time of oscillation and vibration, both in dipping and horizontal needles, that are equally good, is as their length directly; and the actual velocities of their points along their arcs, are always unequal. And hence, magnetical needles are, *cæteris paribus*, still better, the longer they are; and that in the same proportion with their lengths.

Of the Causes of Magnetism.—Though many authors have proposed hypotheses, or written concerning the cause of *magnetism*, as Plutarch, Descartes, Boyle, Newton, Gilbert, Hartsoeker, Halley, Whiston, Knight, Beccaria, &c. yet no satisfactory solution of its phenomena has appeared. It is certain, indeed, that both natural and artificial electricity will give polarity to needles, and even reverse their poles; but though from this it may appear probable that the electric fluid is also the cause of *magnetism*, yet, in what manner the fluid acts while producing the magnetic phenomena, seems to be quite unknown.

Dr. Knight, indeed, deduces from several experiments the following propositions, which he offers, not so much to explain the nature of the cause of *magnetism*, as the manner in which it acts; the magnetic matter of a loadstone, he says, moves in a stream from one pole to the other internally, and is then carried back in a curve line externally, till it arrives again at the pole where it first entered, to be again admitted; the immediate cause why two or more magnetical bodies attract each other is, the flux of one and the same stream of magnetical matter through them; and the immediate cause of magnetic repulsion, is the conflux and accumulation of the magnetic matter. See *Phil. Trans.* Vol. 44, p. 665.

For the particulars of Dr. Knight's improvements in artificial magnets, see the article Compass.

Signor Beccaria, from observing that a sudden stroke of lightning gives polarity to magnets, conjectures that a regular and constant circulation of the whole mass of the electric fluid from north to south may be the original cause of *magnetism* in general. This current he would not suppose to arise from one source, but from several, in the northern hemisphere of the earth: the aberration of the common centre of all the currents from the north point, may be the cause of the variation of the needle; the period of this declination of the currents, may be the period of the variation; and the obliquity with which the currents strike into the earth, may be the cause of the dipping of the needle, and also why bars of iron more easily receive the magnetic virtue in one particular direction. *Lettre dell' Elettricismo*, p. 269; or PRIESTLEY'S *Hist. Elec.* Vol. I. See also CAVALLO'S *Treatise on Magnetism.*

MAGNETISM *of Ships.* Captain Matthew Flinders, of the royal navy, during the course of his voyage in his Majesty's sloop Investigator, for completing the discovery of New Holland and New South Wales, remarked that the variation and the bearings of land, taken with an azimuth compass upon the binnacle, were very different when the ship's head was in different directions; and, at length, found that the following circumstances obtained throughout the whole of the observations.

"1. When the head was east, the variation differed from the truth; and always on the same side, while the ship remained in the same hemisphere.

"2. When the head was west, the differences were equally great; but a contrary way.

"3. The head being north or south, made no difference in the variation; and it was then a medium between what was found at east and at west.

"4. At the intermediate points, between the magnetic meridian and east or west, the difference from the true variation bore a proportion to the angle made by the ship's head with the meridian. If the head were on the western side, the difference was of the same nature as that when the head was west; if on the east side, as at east.

"5. The proportion at the intermediate points obeyed the following law.

As radius is, to the difference at east or west (for eight points):

So is the *sine* of the angle between the ship's head and magnetic meridian, to the difference for that angle.

"Or, if the number of points, which the head was to the right or left of the meridian, were taken as a *course*; and the difference for eight points, reduced to minutes, taken as a *distance*; then the difference for the number of points was found in the *departure* column of the traverse table.

"6. These differences were of a direct contrary nature in the south, to what they were in the northern hemisphere. In the English Channel, the compass gave *too much* west variation when the head was west; but, in the southern hemisphere, it always gave too little; and the great west variations were found when the head was east.

"7. The differences did not change suddenly on crossing the equator; but, all the way from England, they diminished gradually; and, to all appearance, as the dip of the needle diminished. When the south end of the needle began to dip, the differences commenced the other way; and increased gradually as we advanced southward; until, having arrived in Bass' Strait, where the south dip is nearly as great as the north is in England, the differences became almost as great as when we first sailed; but, as I said before, of an opposite nature."

The experiments lately made in England, by Captain Flinders, prove that similar differences, obeying the same laws, take place in most, or all ships of war; and perhaps they do in merchant ships also, "For," says he, "I have found them in vessels of 80 and 25 tons. Differences were also found in other parts, often greater than at the binnacles; but these being of less importance, the differences only which were observed at, or above the binnacles, on changing the ships heads *from east to west*, are here inserted.

"Sheerness,	Starling gun-brig, diff. of variation...	7° 29 greater.
	Helder frigate........	13 3 do.
	Raisonable 64, armed *en flute*.....	0 42 do.
Portsmouth,	Loire frigate	2 7 do.
	Devastation bomb	4 2 do.
Plymouth,	Orestes brig..........	Uncertain.
Channel,	Investigator armed ship	8 4 do.

"In Captain Cook's ships, the Endeavour and Resolution, and also in the Discovery, commanded by Captain Vancouver, the differences appear to have been nearly the same as in the Investigator; and also of a contrary nature in the two hemispheres.

The cause of all these changes in a compass on shipboard, and the modes by which they may be obviated, he has endeavoured to explain in the account of the experiments drawn up by order of the Admiralty; and, to obtain further proofs, it is desirable that the differences should be observed in as many ships, and as accurately as possible, in the following manner:—

"The azimuth compass to be used, should be a good one; the card traversing freely; and if possible, the needle should be retouched with magnetic bars, before the observations are made. A low stand or stool must be prepared; so that when the compass is placed upon it, it may be of the same height with that by which the ship is steered. The binnacle being then taken away, substitute the stand and azimuth compass; or if the sun cannot be there seen all round, fix the stand as near to the situation as it can be seen; but *clear of any iron*, and *exactly amidships*. Let azimuths be taken, using both sides of the vane; and this as often as convenient, and with the ship's head in various directions; but more particularly at and near east and west; noting the direction of the head to the nearest quarter of a point.

"These observations should always be made with the same instrument, and in the same place; and be entered in a table of ten columns, under the following heads:—

Time, containing the year, day, and hour.

Latitude, of the place of observation.

Longitude, of ditto.

Dip, of the needle, if an instrument is on board.

Alt. ☉ *cent.* corrected for dip, refraction, and semidiameter.

The *hour angle* may be substituted, where there is a time keeper, and the altitude cannot be observed.

Azim. Obsd. being the mean of three or four sights with each side of the vane.

(If Walker's meridional compass be used, the two last are unnecessary.)

By whom, mentioning the observer's name against each observation.

Ship's head, the mean of what it was at the beginning and end.

Variation.

Circumstances, specifying if under sail or at anchor: also, whether the ship was steady, or had motion.

At the head of the table should be mentioned the number of guns mounted, and whether the place was at the binnacle; or if not, how far before or abaft it. Observations may be made in other parts of the ship; but these should be kept in a separate table.

Captain Flinders concludes by observing, that "a good number of such experiments, carefully made in a variety of ships, and particularly when lying steadily at anchor, would show whether the differences at the binnacle are usually so great as to cause much error; and, consequently, how far the discovery of them is of importance to ordinary navigation. With marine surveyors, and all officers who wish to fix the position of places or of their ships, by compass bearings, the subject must necessarily be one of much interest."

MAIL, in rope making, a kind of steel chain-work, flat, and fastened upon leather, about nine inches long and seven broad. It is used to rub off the loose hemp that remains on white cordage.

MAIN, (*grand, premier ou principal*, Fr.) an epithet applied to whatever is principal, as opposed to what is inferior or secondary. Thus, the main land is used in contradistinction to an island; and the main-mast, the main-wale, the main-keel, and the main-hatchway, are in like manner distinguished from the fore and mizen-masts, the channel-wales, the false-keel, and the fore and after-hatchways, &c.

For the sails, yards, and rigging of the main-mast, see those particular articles.

Main-Breadth, in ship-building, is the broadest part of the ship at any particular frame.

Main-Keel, in ship-building, a term of distinction between it and the false-keel. See the article Keel.

Main-Mast. See the article Mast.

Main-Post, the same with stern-post.

Main-Sheet *Block*. See the article Block.

Main-Tackle, (*palan de grande caliorne*, Fr.) a large strong tackle, hooked occasionally upon the main pendant, and used for various purposes, particularly in securing the mast, by setting up the rigging, stays, &c. See the article Pendant.

To MAKE, (*faire, &c.* Fr.) is variously applied, in the sea language; as,

To Make *a good Board*. See the article Board.

To Make *the Land*, (*découvrir la terre*, Fr.) is to discover it from a distant situation, in consequence of approaching it after a sea voyage: as, "In your passage to cape Tiburon, it will be necessary to make Turk's Island."

To Make *Sail*, (*faire de la voile*, Fr.) is to increase the quantity of sail already extended, either by letting out the *reefs*, and by hoisting an additional number of small sails, or by performing either of those exercises separately.

To Make *Sternway*, (*culer*, Fr.) is to retreat or move with the stern foremost.

To Make *Water*, or *To Leak*, (*faire eau*, Fr.) generally signifies to leak, unless when the epithet *foul* is added. See the article Foul-Water.

MAKING-IRON, (*ciseau de calfat double ou cannele*, Fr.) a sort of iron chisel, used by calkers, for the purpose of driving oakum into the seams of ships. It is about seven or eight inches in length and two in breadth; and has a groove cut along its edge or lower extremity, in order to spread the oakum more securely.

MALL, in mast-making, an iron tool, used for the purpose of driving bolts, &c.

MALLET, (*maillet*, Fr.) a sort of wooden hammer, of which there are several sorts, used for different purposes on ship-board, as the

Calking Mallet, (*maillet de calfat*, Fr.) an instrument chiefly employed to drive the oakum into the *seams* of a ship, where the edges of the planks are joined to each other in the sides, decks, or bottom.

The head of this *mallet* is long and cylindrical, being hooped with iron, to prevent it from splitting in the exercise of calking.

Heaving Mallet, in sail-making, a mallet with a small cylindrical head, used as a lever to strain tight the cross stitches and beat them close, when sewing on the bolt-rope.

Serving Mallet, (*maillet a fourrer*, Fr.) a mallet used in *serving* the rigging, which binds the spun-yarn more firmly about it, than could possibly be done by hand; which is performed in the following manner: two or three turns of the spun-yarn, which has been previously rolled up in a large ball, or clew, are passed about the rope, and about the body of the mallet, which for this purpose is furnished with a round channel in its surface, (see fig. 11, plate XV.) Fig. 1
that conforms to the convexity of the rope intended to be served. The turns of the spun-yarn being strained round the mallet so as to confine it firmly to the rope, which is extended above the deck, one man passes the ball continually about the rope, whilst the other, at the same time, winds on the spun-yarn by means of the mallet, whose handle, acting as a lever, strains every turn about the rope as firm as possible.

MAN, (*batiment, vaisseau*, Fr.) in sea language, is used to imply a ship, as a man of war, a merchantman, a Guineaman, an East Indiaman, a Greenlandman, &c. in all which instances the word ship is implied.

To Man, (*garnir de monde, &c.* Fr.) is to place a sufficient number of men for any particular exercise at the proper station; as,

To Man *a Prize*, (*amariner une prise*, Fr.) is to send a proper number of men on board to navigate her to her destination.

To Man *Ship*, (*ranger du monde sur les vergues pour donner des cris de salut*, Fr.) is to range the seamen on the yards and rigging, in readiness to give three cheers, as a salute.

Man *the Capstan!* (*du monde a garnir les barres du cabestan!* Fr.) the order to place the men to the bars in readiness to heave round the capstan.

Man *the Topsail-Sheets!* (*du monde sur les ecoutes de hune!* Fr.) the order to let the men lay hold of, and be ready to pull up the topsail-sheets.

Man *the Yards!* (*du monde sur les vergues!* Fr.)

the order to a sufficient number of men to go upon the yards to reef or furl the sails

MANDREL, of the turning lathe, in block-making, is an iron spindle, with a screw-cut box at one end, to screw in the chock which confines the work; round the middle of the spindle is a collar with a groove, in which the cord turns that works it. It is used in a lathe that is worked by a wheel instead of a treddle.

MANDRELS, in anchor-making, are circular iron instruments, forming a cone four feet high, on which hoops are driven, to be made perfectly round.

MANGER, (*gatte*, Fr.) a small apartment, extending athwart the lower-deck of a ship of war, immediately within the hawse-holes, and fenced at the after-part by the

MANGER-BOARD, (*cloison de la gatte*, Fr.) a strong bulk-head, fixed on the fore part of the deck, serving to stop the water which sometimes rushes in at the hawse-holes (see the letter M in the deck, plate IX.) and would otherwise run aft in great streams on the deck; the water thus stopped is again returned into the sea, through the SCUPPERS, which article see.

MANIFEST, (*declaration de la cargaison d'un bâtiment*, Fr) in commerce, is a regular list of a ship's cargo, containing the mark and number of each separate package, the names of the persons by whom the different parcels of goods are shipped, and those of the persons to whom they are consigned; a specification of the quality of the goods contained in each package, as rum, sugar, tea, coffee, &c. and also an account of the freight, that the captain is to receive from the consignee of such goods, on his arrival, corresponding with the bills of lading, which he has already signed. The *manifest* is usually signed by the ship-broker who clears the vessel out at the custom-house, and by the captain, and serves as a voucher for the latter, whereby to settle his account with his owners, &c. For Abstracts of the Manifest Act, including Bounties and Drawbacks, see MAXWELL'S *Spirit of Marine Law*; also an Act of the 26 Geo. III. c. 40.

MANNING THE YARDS, (*ranger du monde sur les vergues*, Fr.) the act of ranging the people on the yards, rigging, &c. either in honour of some illustrious visitor, upon hoisting the flag of some favourite admiral, or in commemoration of some splendid victory.

This exhibition has often proved fatal to some of our brave seamen, and therefore ought to be dispensed with, unless upon very particular occasions, as from its very nature it must always be attended with considerable danger; and it is well known that many of our admirals are too firmly convinced of the attachment of their tars to require such a testimony of their zeal or affection, and, much to their credit, have forbidden such a compliment to be paid them.

MANŒUVRE, (*manœuvre*, Fr.) a change of position, &c. as, "The expeditious *manœuvre* of the captain extricated the ship from her perilous situation: also, "The skilful *manœuvre* of the admiral, by changing the position of the line, was the means of obtaining the victory over his enemies."

MAN OF WAR, (*vaisseau de guerre*, Fr.) a term generally applied to ships carrying from 20 to 120 guns, invariably employed in naval engagements. See the article SHIP.

* MAN-ROPES, (*tire-veilles*, Fr.) a general name given to the small set of ropes used for ascending and descending a ship's side, hatchways, &c. they are usually covered with kersey or canvas.

MAP, (*carte*, Fr.) a plane figure, representing the surface of the earth, or a part thereof. See the article CHART.

MARINE, a general name for the navy of a kingdom or state; as also the whole economy of naval affairs, or whatever respects the building, rigging, arming, equipping, navigating, and fighting of ships. It comprehends also the government of naval armaments, and the state of all the persons employed therein, whether civil or military.

MARINE *Barometer*. See the article BAROMETER.

MARINE *Chair*, a machine invented by Dr. Irwin, for viewing the satellites of Jupiter at sea, and thereby facilitating the necessary observations in determining the longitude of their eclipses.

The longitude might easily be found at sea by the eclipses of Jupiter's satellites, provided a telescope, possessed of sufficient magnifying power could be employed for that purpose; but this is found to be extremely difficult, by reason of the perpetual agitation of the sea. To remedy this inconvenience, Dr. Irwin invented a *Marine-Chair*, which Dr. Maskelyne was ordered to try in a voyage to Barbadoes, under the direction of the Commissioners of Longitude, in 1763. The Royal Astronomer reports thus: "I made full trial of the late Dr. Irwin's *Marine Chair*, but could not derive any advantage from the use of it; and, considering the great power requisite in a telescope for making these observations well, and the violence as well as irregularities of the motion of the ship, I am afraid the complete management of a telescope on shipboard will always remain among the desiderata for this discovery." Other methods have since been proposed by M. Kratzenstein, M. Foyt, &c. and are described in RUTLEDGE'S *Theorie Astronomique*, and mentioned in the *Cosmolabe de Jacques Besson*, printed at Paris in 1807.

MARINE *Officer*, (*officier des troupes de la marine*, Fr.) is a military officer, who holds a commission from the Lords of the Admiralty, and employed either on shore or in a ship of war, to command a certain complement of marines in their respective duties.

Marine officers take rank in their respective classes according to the seniority of their commissions, or according to the order in which they are placed on the marine list; but, when doing duty, or in any service with the army, they take rank agreeable to the date of their commissions, which are signed by the King as well as the Lords of the Admiralty, and entered at the War-office.

The marine officers are upon all occasions to be treated, as well by the captain of the ship as by all other officers and people belonging to her, with the decency and regard due to the commissions they bear: and though lieutenants of marines share in prizes only with warrant officers of ships, upon consideration of their different sea-duty, yet it is not intended to degrade their rank; and they are, while they do their duty, to be considered and treated in all respects as a commissioned officer should be.

The commanding officer of marines serving on board any of his Majesty's ships is strictly directed to transmit, immediately after the end of every calendar month, or as soon afterwards as opportunity offers, to the secretary of the Admiralty, a return of the marines under his command; or, if they belong to different divisions a separate return for each division, expressing their state and condition, in order to their being forwarded to the commanding officer at the head-quarters of the division to which they respectively belong, that they may see the number and disposition of the marines serving on ship-board, and the time when last clothed, and be thereby enabled to make their returns to the Admiralty and Navy-board as correct as possible; and the captain of the ship is to attest such returns, that they may agree with the ship's books. See the NAVAL INSTRUCTIONS; and, for the establishment and number of marine officers and men allowed to each class of ships, see the article ROYAL MARINES.

MARINE *Society*. The Marine Society, in Bishopsgate-street, London, was founded in 1756, and originated from the following fact:—Lord Harry Pawlett, since Duke of Bolton, in the spring of 1756, then commanding his Majesty's ship Barfleur, requested John Fielding, Esq. afterwards Sir John Fielding, Knt. a magistrate of Middlesex, to collect a number of poor boys for the use of his ship, desiring that they might be clothed at his expense. Fowler Walker, Esq. late of Lincoln's-inn, happened to meet these boys on their journey to Portsmouth, and being struck with their appearance, his humanity suggested to him that a greater number of such poor boys might be fitted out by a subscription. On his arrival in town he proposed to Mr. Fielding to solicit the public for a subscription for this purpose, offering to open it by a small donation. This worthy magistrate, in his written answer, expressed his doubts of the event, but acquiesced with Mr. Walker in his design, and happily succeeded so far that he collected sufficient to clothe three or four hundred boys.

It was very evident that the war would call for a greater number of seamen than existed; and, therefore, landmen must be called in. No bounty could be so proper as seamen's clothing. The late Mr. Jonas Hanway, a merchant, totally unconnected with the noble lord and both the gentlemen above-mentioned, desired a meeting of the merchants and owners of ships, and proposed to them to form themselves into a society to give clothing to landmen and boys for the sea service. The first part was eagerly embraced, and the design as speedily carried into execution: and many days had not elapsed when the part relating to the boys also fell into the hands of this society. A regular society was formed, and a proper committee and other officers appointed. Mr. Fielding was invited to join as a member, but he did not long continue; Mr. Walker afterwards became a member, and is comprehended in the act of incorporation.

How grateful the design proved, was obvious from the conduct of numbers of our fellow subjects of all ranks and conditions, who, under this association, contributed with a liberal hand. The society found a number of young fellows who were in danger of becoming a prey to vice through idleness, and others who entertained hopes of gain by prizes, as soon as the garb of a seaman was presented to them gratis gladly entered into the king's service.

The motive for clothing of landmen was urgent in proportion to the call for seamen.

The society boys are in general from thirteen to fifteen years of age; but some of them appear to great disadvantage from their growth being checked by the poverty and insufficiency of their diet, and the defect of due warmth, occasioned by the want of such clothing as our climate requires to nourish their limbs: as hunger and nakedness prey on the human frame, good clothing and ship provision soon make a great alteration in their favour. With regard to the proper care of their morals, this society could not be more attentive in recommending them to the commanders were they their own children, as the regulations of the society abundantly testify.

The principles on which this society set out, prompted it to act this part, at the same time esteeming the service done to the navy and his Majesty's officers as an object of common national interest.

Mr. Hickes, a merchant of Hamburgh, seeing the great utility of this design, and not being connected by any near relation, followed the dictates of his affections and christian charity towards his country, which pleaded strongly in his breast for the lowest and most forsaken part of his fellow subjects; and knowing what was further intended by the Marine Society, he, therefore, bequeathed to it, in September 1763, a legacy, amounting to 22,000*l.* although it produced only about 18,000*l.*

Shortly after this the society opened an office at the Royal Exchange, as the most centrical place; and formed a numerous committee, composed chiefly of merchants of consideration, and likewise of owners of ships. Having made this arrangement, poor boys from all parts of London and Westminster, and such as were found fit and proper for sea service, and appeared ready to do their duty as apprentices, were provided for, to the number of six or eight in a week, according as fit and proper masters applied to take them upon such terms as might probably answer the end. Now, they give clothing and chest to the apprentice, the indentures, conveyance, cure of the sick, inoculation to those who have not had the small pox; and, until masters can be provided, lodging and food to such boys as are totally destitute and friendless.

To omit nothing which could give stability to this institution, the next object considered was its *incorporation*; for which purpose an act of the 12 Geo. III. was obtained.

The advantage derived from 18,000*l.* arising from the bequest made by Mr. Hickes, though the interest of this sum alone is but a small object for such an enterprise, was a good foundation for this measure, leaving it to the liberality of our fellow subjects to complete the work.

Fifty-two governors were incorporated, from which number were appointed a president, six vice-presidents, a treasurer, and forty-four assistants; but, in 1808, there were sixty assistants, in whom are vested certain powers and conditions. Having completed the work of a political establishment, nothing appeared more consistent than an attempt to form a local residence more suitable than the office over the Royal Exchange. The metropolis of so vast an empire, devoted to commerce, and frequently involved in war, required a regular rendezvous, where the distressed might apply for employment at sea. And the society's friends in the East Indies having remitted two generous collections, they were encouraged to open a subscription. They contracted with the committee of the Bridge-lands for a convenient spot in Bishopsgate-street, adj ining St. Ethelburga's church, the nearest that could be procured in the vicinity of the Royal Exchange; the ground rent is 82*l.* the buildings on it, exclusive of what the society occupies, at present let for 160*l.* On this spot the society fixed their establishment, and erected the present buildings: contiguous to the leasehold estate they purchased the freehold adjoining to Clark's-alley, of Mr. J. Jordan, the builder of the house.

Quality of Boys.—1. Such boys as are literally in a vagrant state; of whom. some are recommended by magistrates, either as found wandering, or as guilty of some petty offence.

2. Those who live chiefly by begging, or seldom do any work, but appear in filth and rags, and sometimes half naked: of these the smallest number ofler.

3. Some who have occasionally earned their bread by going on errands, or in markets, brick-kilns, feeding hackney coach horses, draw boys, &c.: the number of these is greater than of the other.

4. The sons of poor people who have large families, and, upon enquiry, are in too great a state of indigence to provide clothing for the sea. These being often in a naked condition, and unemployed, are exposed to every temptation which indolence or idleness can create, and apply for employment at this office; of these there seems to be a majority.

5. Boys whose parts have been wrong cast, being so contrary to their genius, that they are more inclined to hazard their necks than to live a sedentary life: of this class there is no inconsiderable number.

Sir Brook Watson, Bart. alderman, who had the interest of the society very much at heart, proposed, at a general court, on the 4th of May 1786, that a stout ship should be procured, of from three to four hundred tons burden, which might be capable of receiving from one hundred and fifty to two hundred boys; that this ship should be moored in some convenient part of the river Thames; that an officer of the navy, of unexceptionable character, should be appointed to command tho ship, and to take upon him the general superintendence of the boys; that under him there should be a boatswain, a boatswain's-mate, a carpenter, and a cook, and an able well-qualified schoolmaster should also be chosen to instruct the boys in reading, morality, and in the principles of the Christian religion. This plan was carried into execution in the following September, and a number of boys were immediately sent on board. Since this regulation the society has had every reason to be more than satisfied with the establishment. The boys are taken from the town, separated from their evil companions, cleansed from their rags and filthiness, initiated into their profession, softened into habits of subordination and obedience, inured to gentle discipline, and made useful and fit persons for the notice and choice of officers and merchants, who want a supply of this kind of boys for servants and apprentices. They have boats, which they are taught to row, and which they are soon enabled to manage with vigour and dexterity: nor is their confinement on board either irksome or dispiriting. Their employment is varied as much as possible in their duties as young seamen; and much of their time taken up with their attendance on the school-master.

Their pay on board ships of war is 7*l.* per year to those from thirteen to fifteen years of age; and 8*l.* from fifteen to seventeen: this is supposed to be sufficient to supply them with clothing till such time as they learn the duty of a seaman, and are big enough to be treated as such, and to receive the wages of seamen in the king's service.

They are allowed good and sufficient provision, being the same allowance as a man's: and, in time of war, they are also entitled to prize-money.

The Officers of this Institution are as follow:—

1 President	1 Treasurer
6 Vice-presidents	1 Secretary
1 Chaplain	1 Schoolmaster
1 Surgeon	1 Messenger
1 Apothecary	1 Housekeeper.

Qualifications of Governors.—A subscription of 2*l.* 2*s.* per annum, a donation of 12*l.* 12*s.* or 2*l.* 2*s.* per annum for a number not less than eight years, is a qualification for a governor for life. Legacies may be bequeathed to the treasurer (for the time being) of the corporation called the Marine Society, London.

	£	s.	d.
The last account published by the Marine Society closed on the 30th of September 1809; by which it appeared, that from the 30th of June to that date; viz. six months' donations, dividends, and annual subscriptions, received in that quarter, amounted to	1554	1	11
Which, added to the balance remaining on the 30th of June..............	1088	4	11
Renders a total sum of.............	2642	6	10

	£	s.	d.	£	s.	d.
From which sum the disbursements for 126 boys sent to sea, as part of the crews of his Majesty's ships, 47 apprenticed in the merchants' service, and 101 landmen volunteers, clothed to serve as seamen in his Majesty's navy, cost..	1273	15	7			
J. Knowles, for cloth....	128	9	0			
Purchase of 1144*l.* 19*s.* consols of 3 per cent. in aid of the widows' fund	780	0	0			
Bridge-masters, for ground rent...............	74	8	11			
				2256	13	6
Balance in the banker's hands, Sept. 30				£385	13	4

Thus the society clothed, during the last year, two hundred and seventy-four men and boys; and, since its commencement, 36,670 men, and 28,074 boys. The saving of so many distressed lads, and making them useful to their country, while the clothing given to the land volunteers, making them fit to associate with their brave comrades, will at once prove the attention that is paid to the public benefit as well as to the private individual. For further particulars of this article see HIGHMORE's *Public Charities* for 1810.

MARINE *Surveyor*, is the name of a machine, contrived by Mr. H. de Saumarez, for measuring the way of a ship at sea. The machine is in the form of the letter Y, and is made of iron or other metal. At each end of the lines which constitute the angle or upper part of the letter, are two pallets, not much unlike the figure of the log; one of which falls in the same proportion as the other rises. The falling or pendant pallet meeting a resistance from the water as the ship moves, has by that means a circular motion under water, which is faster or slower, according as the vessel moves. This motion is communicated to a dial within the ship, by means of a rope fastened to the tail of the Y, and carried to the dial. The motion being thus communicated to the dial, which has a bell in it, it strikes exactly the number of paces, miles, &c. which the ship has run. Thus the ship's distance is ascertained, and the forces of tides and currents may also be discovered by this instrument. See *Phil. Trans.*

MARINER, (*marin*, *matelot*, Fr) a person who gets his living by following the profession of a seaman.

MARINER'S-COMPASS. See the article COMPASS.

THE ROYAL MARINES, (*soldats*, *ou troupes de marine*, Fr.) a body of troops employed in the sea-service, under the direction of the Lords Commissioners of the Admiralty; who were originally raised and embodied with the sole view of being a nursery to man our fleets and ships of war.

Many are the causes which originally suggested the expediency of forming an establishment of marines. Experience has always shewn that landmen were very improper substitutes for the supplies of seamen, under a system of impress. The marines were always quartered in the vicinity of our principal sea-ports, where they were regularly trained to the different ways of ship fighting, and to the various manœuvres of a vessel in which numbers were necessary. Being thus locally placed, their value was early felt by their exertions in equipping the squadrons fitted out, when but little confidence could be placed in the sailor, perhaps, just impressed into the service.

The general principles and regulations that were instituted for the conduct of marine regiments, from their formation to the close of the reign of King William III. evidently shew that they were entirely devoted to naval purposes.

As each individual became properly qualified to act on shipboard as a foremast-man, which was uniformly encouraged, he was discharged from his regiment, entered upon the books as such, and levy-money was granted to the officers of his corps, in order to supply the vacancy of him who was thus transferred.

The first authentic instance of any regiment of this description appears in the Army List of 1684, and from the return of the general review on Putney-heath, on the 1st of October in that year. Neither the exact æra of its establishment, or that of the other maritime forces is clearly ascertained; and it would be absurd to speculate upon dates which can yield no material result.

The return above-mentioned runs thus, and is annexed at large: "The Lord High Admiral of England, his Royal Highness the Duke of York and Albany's *Maritime regiment of foot*, commanded by the Hon. Sir Charles Littleton, called also the Admiral's Regiment."

This regiment consisted of twelve companies, without any grenadiers; the officers of which were,

1 Colonel, 1 Lieutenant-colonel, 1 Major,
12 Captains, 12 Lieutenants, 12 Ensigns,
1 Adjutant, 1 Quarter-master,
1 Chirurgeon, and 1 Chirurgeon's-mate.

They had yellow coats lined with red, and their colours were, a red cross with rays of the sun issuing from each of its angles. They stood the third in seniority in the line of that day; and it may be presumed from its subsequent reduction, that a step was obtained in it by the 4th, then the regiment of Holland, commanded by John, the second Lord Mulgrave, and now called *the Old Buffs*.

Between 1687 and 1698, there were several maritime regiments raised for the purposes, and under the regulations before mentioned. They were Colonels Mordaunt, Colt, Seymour, and Brundenell's; also Sir Cloudesley Shovel's, my Lord Torrington's, and the Marquis of Carmarthen's; all of which were disbanded during the currency of 1697 and 1698.

In the early part of 1756, the French had collected immense forces upon their coasts, with the avowed object of invading England; consequently, measures of precaution and defence were adopted by a regular and annual increase of seamen and marines. The following are the numbers of marine corps voted by Parliament for the service, from 1756 to 1763, both years inclusive:—

Year.	Marine Corps.
For 1756	9,138
—— 1757	11,419
—— 1758	14,845
—— 1759	14,845
—— 1760	18,355
—— 1762	19,C61
—— 1763	4,287

Thus the numbers of marine corps increase and decrease according to the hostile or pacific state of the realm.

It was in the year 1758 that that highly useful body of men the National Militia, was ordered to be raised and organized for domestic security, while the regular armies were employed in retaliating the terrors of invasion, and in carrying the war into the foreign possessions of our enemies.

The naval establishment for the year 1762, consisted of 70,000 seamen, including the 19,061, or 36 companies of marine corps.

The merits and usefulness of the latter became conspicuous by the most zealous and spirited services, on either element, to which their mingled duties called them.

It was not until the year 1771 that the corps of marines felt the fostering influence of a patron and a friend. Earl Sandwich, placed at the head of the Admiralty, was now enabled to realize a system he had long entertained in his thoughts; he beheld them orphans, and adopted them as his children. From him originated the *colonels commandant* of divisions, and many of those interior regulations, which, for their economy and wisdom, do honour to the establishment.

His lordship, by a memorial to the throne, soon after his accession to power, obtained an Order in Council to enlarge the number of field officers. These happy reforms held out a hope to the veteran, from which he had, till now, been precluded, that of being at the head of a division. By a train of considerate attentions the corps was placed on a respectable footing; its discipline became much improved; and his lordship soon saw those objects of his tender care surpassed by no troops in the world, for subordination, intrepidity, and cool valour.

From the æra of 1764 may be dated the fatal American rebellion, and those revolutionary principles which have long continued, and which still menace the quietude and independence of Europe and the world.

After a lapse of ten years, to ascertain the views and opinions of so many provinces, differing in their manners, and in some respects opposite in their interests, a general congress was held at Philadelphia, on the 5th of September 1774, by deputies from the whole. Their proceedings were marked by moderation, yet with firmness.

The terms of conciliation arising from congress having been rejected by the mother country, by whose gratitude those colonists had frequently been delivered from the terrors of a hostile and restless neighbour, his Majesty communicated a message, recommending the augmentation of his forces by sea and land. This being done, hostilities proceeded, and in the resulting dangers and hardships of this war the marine soldiers were remarkably conspicuous.

Total numbers of marine corps voted by Parliament, for the service from 1778 to 1782, both years inclusive:—

Year.	Marine Corps.
For 1778	11,000
—— 1779	17,389
—— 1780	18,779
—— 1781	20,317
—— 1782	21,305

Thus it appears, that in every war, state necessity required an annual increase to the establishment of marines.

At the close of the last year, a general peace was ratified between America and Great Britain, and all her contending powers.

Towards the close of the year 1792, a small addition was granted to the corps of marines, and a much greater took place the following year, in consequence of the intermeddling policy of France, and the war which succeeded it.

The narrow views that had influenced so great a reduction at the peace of 178 , was bitterly felt at the commencement of hostilities.

It is to the faithful marine soldier who knows not the meaning of desertion, that the nation, under such sudden emergencies, is often obliged to have recourse, in equipping as well as manning her fleets; for it is not in human nature to place confidence in that man who has been constrained to serve, and who recognizes neither a voluntary will nor a sacred obligation to discharge his allotted duties.

Several additional companies were now deemed absolutely necessary to the marine corps.—The following are the numbers voted by Parliament for the service, from 1794 to 1801, both years inclusive:—

Year.	Marine Corps.
For 1794	12,115
—— 1795	15,000
—— 1796	18,000
—— 1798	20,000
—— 1800	20,000
—— 1801	30,000

The latter number was kept up till the 27th of March 1802, when a definitive treaty of peace with France was made. On the 29th of April following, his Majesty, in order to mark his royal approbation of the very meritorious and loyal conduct of the marine corps, during the late war, was graciously pleased to direct that, in future, the corps shall be styled, "The Royal Marines."

Besides the many useful reforms adopted by Earl St. Vincent, for the internal economy of the British navy, his lordship also turned his attention to the royal marine corps. On the 18th of March 1803, a new code of instructions was published for their regulation when on shore, which vests in the four senior captains of each division, the management and superintendence of many concerns that had formerly been placed under separate departments.

The pleasing and consolatory work of peace had scarcely been accomplished, when disappointed ambition began to wreak its rancour through the low channels of a mercenary press. Hitherto the organs of knowledge, and sometimes of liberal opinion, the papers of republican France, under the controul of its first consul, became the contemptible interpreters of degrading scurrility and the most inflammatory philippics against the English government. They proclaimed a latent danger, whilst all the public institutes of Buonaparte were levelled at the very foundations of our commercial greatness. A banditti of unauthorized agents, too, set themselves down in our seaports; who, under the specious pretexts of trade, were commissioned to explore our coasts and to contaminate our principles. By such subtle means had the courage and energies of almost every state which had fallen under the French yoke, been previously undermined, and they well merited the instant and retaliating vengeance of a great kingdom. But as war is a serious evil at all times, every expedient to avert it should be tried by a commercial country, ere it has recourse to arms. However, our national security and honour were at stake; and, in the early part of 1803, we again took up arms in defence of our dearest birthrights · when the Lords of the Admiralty judged it necessary to augment the Royal Marine Corps to the war establishment; it having been reduced during the short peace of 1802, to one hundred companies, each containing one hundred men, exclusive of officers.

See Gillespie's *Review of the Royal Marine Corps.*

In August 1804, three companies of artillery (being one to each division) were ordered to be formed for the purpose of serving afloat, with additional pay for officers and men. The officers belonging to this small corps are not separated from the general marine list, neither are they promoted, except in their turn with the others. On attaining the rank of full major they lose their appointment in the artillery.

By a warrant from the Admiralty, dated the 15th of August 1805, another division was established; which is called the fourth, or Woolwich division, consisting of 41 companies, including one of artillery. It has the same number of field officers, pay captains, &c. as the other divisions.

Few alterations of material consequence have taken place from that period to the present time (January 1st, 1813), except that, during the administration of Lord Mulgrave, four additional colonel commandants en second were appointed; one to each division, which now makes the whole number of second commandants eight: and also forage was allowed for one horse for lieutenant-colonels, majors, and adjutants.

The annexed list will shew the numbers of marine companies which are now stationed at the different head-quarters:—

Chatham, 48 *Companies, viz.*

1	19	37	55	73	91	103	121
4	22	40	58	76	94	106	124
7	25	43	61	79	97	109	127
10	28	46	64	82	98	112	130
13	31	49	67	85	99	115	133
16	34	52	70	88	100	118	1st Artillery.

Portsmouth, 49 *Companies, viz.*

2	20	38	56	74	92	110	125	138
5	23	41	59	77	95	113	128	140
8	26	44	62	80	101	116	131	142
11	29	47	65	83	104	119	134	2d Artil-
14	32	50	68	86	107	122	136	lery.
17	35	53	71	89				

Plymouth, 49 *Companies, viz.*

3	21	39	57	75	93	111	126	139
6	24	42	60	78	96	114	129	141
9	27	45	63	81	102	117	132	143
12	30	48	66	84	105	120	135	3d Artil-
15	33	51	69	87	108	123	137	lery.
18	36	54	72	90				

Woolwich, 41 *Companies, viz.*

144	150	156	162	168	174	180
145	151	157	163	169	175	181
146	152	158	164	170	176	182
147	153	159	165	171	177	183
148	154	160	166	172	178	4th Artil-
149	155	161	167	173	179	lery.

Present establishment of Royal Marines, 32,668; including 10 boys to each company, or 1,830.

Number of Officers of his Majesty's Royal Marine Forces, exclusive of the 4 Artillery Companies.

1 General, 1 Lieutenant-general, and 1 Major-general.
4 Colonels.
1 Colonel Commandant-in-chief.
1 Assistant to ditto.
4 Colonels Commandant.
8 Second Colonels Commandant.
19 Lieutenant-colonels.
16 Majors.
191 Captains.
*16 Second Captains.
389 First Lieutenants.
357 Second Lieutenants.

* The Second Captains supply the place of the Resident Pay Captains.

Staff.

8 Adjutants.
4 Quarter-masters.
4 Paymasters of Divisions. } On half-pay.
4 Barrack-masters.
4 Surgeons.
4 Assistant Surgeons.

Number of Officers of the Royal Marine Artillery Companies.

7 Captains. | 8 Second Lieutenants.
23 First Lieutenants. | 1 Acting Adjutant.

Complement of Royal Marines in each Rate.

Rates.	Captains.	1st Lieut.	2d Lieut.	Serjeant.	Corporal.	Drum.	Privates	Total.
First Rate, 100 guns if a flag	1	1	2	4	4	2	156	170
Do. without a flag....	1	—	3	4	4	2	156	170
Second Rate, 98	1	—	3	3	3	2	138	150
Third Rate, 80	1	—	2	3	3	2	130	141
Ditto, 74............	1	—	2	3	2	2	115	125
Ditto, 64............	1	—	2	2	2	1	82	90
Fourth Rate, 50......	1	—	2	2	2	1	52	60
Ditto, 41............	0	1	1	2	2	1	47	54
Fifth Rate, 38	0	1	1	2	2	1	4	48
Ditto, 32............	0	1	0	1	1	1	38	42
Sixth Rate, 28	0	1	0	1	1	1	27	31
Ditto, 20	0	1	0	1	1	1	26	30
Sloops	0	0	0	1	1	0	18	20
Gun-brigs	0	0	0	1	1	0	13	15

Officers retired on Full Pay.

3 Colonels-commandant. | 36 Captains.
1 Lieutenant-colonel. | 11 First Lieutenants.
4 Majors. | 6 Second Lieutenants.

Reduced Officers on Half Pay.

1 Colonel. | 8 Captain Lieutenants.
1 Major. | 139 First Lieutenants.
78 Captains. | 111 Second Lieutenants.

Paymaster and *Inspector*, with 6 Clerks.

MARITIME, (*maritime*, Fr.) something relating to, bounded by, or near the sea: thus, a *maritime* province or country is one bounded by the sea; and a *maritime* kingdom or state is one that makes a considerable figure, or is very powerful at sea.

Maritime-Institution. The London Maritime Institution was established in October 1790, by the ship-owners of the port of London, for the relief of decayed and unfortunate master-mariners and their families, who have become members of this society, by paying one guinea entrance money, and two annual subscriptions of that amount: ten guineas constitute a member for life. A considerable subscription has already been raised for building alms-houses on Bow-common.

The Merchant-seaman's Office is near the Royal Exchange, where the society's affairs are conducted.

Maritime-Powers, (*les puissances maritimes*, Fr.) among the European states, are understood Great Britain and Holland.

MARK, (*marque*, Fr.) in commerce, a certain note which a merchant puts upon his goods, or upon the cask, hogshead, &c. that contains them, in order to distinguish them from others; such as a grape, a crow's foot, a diamond, a cross, an asterisk, &c. Some use one or other of these marks by themselves; others join them with the initial letters of their own name; and others use the letters only.

MARKAB, in astronomy, a bright fixed star, of the second magnitude, in the head of the wing of the constellation Pegasus. It is in the Britannic catalogue the fifty-fourth in order of that constellation, and marked by this character (*a*). Its right ascension in time adapted to the beginning of the year 1812, is 343° 50 54; annual variation in right ascension, 44″,6. Its declination, 14° 11′ 48″ N. and annual variation in declination, + 19″,4.

MARKING-YARN, in rope-making, a white thread, untarred, laid in rope, for the King's or East India Company's mark. That for the King's mark is distinguished from the other by being spun the contrary way. See the article Rogue's-Yarn.

MARKS, in the royal navy, are used to distinguish his Majesty's goods and stores from private property; and are as follow:—The King's mark, commonly called the *broad arrow*, is the most common in use, and is stamped upon all wrought metals belonging to his Majesty, such as copper, lead, iron, staples, nails, locks, &c. used either in the navy or ordnance department.

The cast metals receive this mark from their respective moulds; the timber, wooden-frames, &c. are generally marked in a similar way by a rasing-knife.

The mark in canvas is distinguished by a blue thread being wove in it, and a serpentine line, painted black, down the centre of each breadth.

The marks upon the heads of cask and packages, containing provisions for his Majesty's service, are painted white, with the initials of their contents, as to species and quantity, and the initials of the contractors name; as also a number, the date of warrantry, and the time when, and place where, they were packed.

For the various marks painted upon the heads of the powder barrels, see the article Gunpowder.

To MARL, (*embromer, ou merliner une manœuvre*, Fr.) is to wind any small line, as marline, spun yarn, twine, &c. about a rope, so that every turn is secured by a sort of knot, and remains fixed, in case all the rest should be cut through by friction. This expedient is much preferable to the winding a line spirally about a rope for the same purpose, because as the turns are at some distance from each other, the same quantity of line will serve as well for the one method as the other, with this difference, that if one of the spiral turns are cut through, the whole will be rendered useless; whereas by marling, this is entirely prevented.

Marling is commonly used to fasten slips of canvas, called *parsling*, upon the surface of a rope, to prevent its being galled, or to attach the foot of a sail to its bolt-rope, &c.

MAR-LINE, (*merlin*, Fr.) a particular kind of small line, composed of two strands very little twisted; there is both tarred and white *mar-line*.

MARLING-SPIKE, (*épissoire*, Fr.) an iron pin tapering to a point (see fig. 12, plate XV.), furnished with a large round head, and principally used to penetrate the twists or strands of a rope, in order to introduce the ends of some other through the intervals, in the act of *knotting* or *splicing*. It is also used as a lever about the rigging, particularly in fixing the seizings upon the *shrouds*, *block-straps*, *clews* of the sails, &c. Fig. 12.

To MAROON, (*abandonner des matelots malfaiteurs sur une terre deserte*, Fr.) to put one or more sailors on shore upon a desolate island, under pretence of their having committed some great crime. This detestable expedient has been too often practised by some inhuman commanders of merchant-ships, particularly in the West Indies.

To MARRY, (*entrelacer*, Fr.) in splicing ropes, is to join one rope to another, for the purpose of reeving it, which is performed by placing the end of each close together, and then attaching them by worming.

To Marry *two Ropes*, (*entrelacer les torons, en faisant une épissure*, Fr.) is to knot the yarns together in a kind of splice, so as not to be thicker at the juncture than at any other part.

MARS, in astronomy, one of the seven primary planets, usually denoted by this character , and revolves next beyond the earth in our system; is of a red firy colour; and always gives a much duller light than Venus, though sometimes he equals her in size. He is not subject to the same limitation in his motions as Mercury or Venus; but appears sometimes very near the sun, and, at others, a great distance from him; sometimes rising when the sun sets, or setting when he rises. It is remarkable of this planet, that when he approaches any of the fixed stars, which all the planets frequently do, these stars change their colour, grow dim, and often become totally invisible, though at some little distance from the body of the planet; but Dr. Herschell thinks this has been exaggerated by former astronomers.

Mars is 4189 miles in diameter, and revolves round the sun in $686^{d}\ 22^{h}\ 18^{m}$, at the distance of 144,000,000 of miles. He is remarkable for the red colour of his light, the brightness of his polar regions, and the great number and variety of spots which mark his surface. The atmosphere of this planet, which astronomers have long considered as of an extraordinary size and density, is the cause of the singular redness of its light. When a beam of white light passes through any dense medium, its colour approaches to red, in proportion to the density of the medium and the space through which it has passed. The momentum of the red, or least refrangible rays, being greater than that of the violet, or most refrangible rays; the former will force their way through the resisting medium, while the latter are either reflected or absorbed The colour of the light, therefore, when it reaches the eye, must partake of the colour of the least refrangible rays; and this colour must increase with the number of violet rays that have been obstructed in their passage. Hence, we see that the morning and evening clouds are beautifully tinged with red; that the sun, moon, and stars appear of the same colour when near the horizon; and that every luminous object seen through a mist is of a ruddy hue.

When we first perceive Mars in the morning, when he begins to separate from the sun, his motion is direct, and the most rapid possible. This rapidity diminishes gradually, and the motion ceases altogether when the planet is about 137 degrees distant from the sun; then his motion becomes retrograde, and increases in rapidity, till he comes into opposition with the sun. It then gradually diminishes again, and becomes nothing when Mars approaches within 137 degrees of the sun.

Then the motion becomes direct, after having been retrograde for seventy-three days; during which interval the planet described an arc of about 16 degrees. Continuing to approach the sun, the planet, at last, is lost in the evening rays of that luminary. All these different phenomena are renewed after every opposition of Mars; but there are considerable differences both in the extent and duration of his retrogradations.

After Galileo had discovered that Mars sometimes appears gibbous, Dr. Hook and Cassini perceived upon the disc of this planet a number of dark spots. Dr. Hooke noticed some trifling changes in their position, which he supposed to arise from a rotation performed once or twice in 24 hours; but Cassini had the merit of determining, from these changes, that the diurnal revolution of the planet was completed in 24^h 40. The same spots were observed at Rome by Campani and Divini. "Besides the permanent spots on the surface of Mars," says Dr. Herschel, "I have often noticed occasional changes of partial bright belts, and also once a darkish one in a pretty high latitude; and these alterations we can hardly ascribe to any other cause than the variable disposition of clouds and vapours floating in the atmosphere of the planet." As the diurnal rotation of Mars has been accurately established by the motions of its spots, it was natural to expect, that, in conformity to the laws of gravity, it should exhibit a spheroidal form. Owing to his gibbous appearance, there is some difficulty in taking accurate measures of the equatorial and polar diameters of Mars. Dr. Herschel, however, has succeeded in the attempt; and found that his figure was an oblate spheroid, whose equatorial diameter is to the polar as 1355 to 1272, or nearly as 16to 15.

From the circumstance of Mars having no satellite, and appearing to require light in the sun's absence, M. Fontenelle has imagined that this planet is phosphorescent, and gives out light during night; the light which it has imbibed in the day.

For further particulars on the subject of this article, see the EDINBURGH *Encyclopædia*, NICHOLSON'S *British Encyclopædia*, and *Phil. Trans.* for 1784.

MARTINGALE, in a ship, a name given to the rope extending downwards from the jib-boom end to a kind of bumkin, and generally fixed perpendicularly under the cap of the bowsprit; its use is, to confine the jib-boom down in the same manner as the bob-stays retain the bowsprit.

MARTNETS, (*cargue-boulines*, Fr.) in a ship are small lines fastened to the leech of a sail, being reeved through a block on the topmast head, and coming down by the mast to the deck. Their use is to bring the leech of the sail close down to the yard to be furled.

MAST, (*mat*, Fr.) a long round piece of timber, generally elevated perpendicularly upon the keel of a ship; to which are attached the yards, sails, and rigging.

A mast, with regard to its length, is either formed of one single piece, which is called a *pole-mast*, or composed of several pieces joined together, each of which retains the name of mast separately. The lowest of these is accordingly named the lower-mast, *a*, fig. 1, plate XVI. The next in height is the top-mast, *b*, which is erected at the head of the former; and the highest is the top-gallant-mast, *c*, which is prolonged from the upper end of the topmast. Thus the two last are no other than a continuation of the first upwards. Plate XVI Fig 1.

Lower MAST, (*mat majeur*, Fr.) is the lowest mast in the ship; the foot or heel of it rests in a block of timber called the step, which is fixed upon the *kelson*. It is fixed in the ship by an apparatus described in the article SHEER-HULK.

Top MAST, (*mat de hune*, Fr.) is that which is raised at the head or top of the lower-mast through a *cap*, and supported by the *trestle-trees*. It is composed of two strong bars of timber, supported by two prominencies, which are as shoulders on the opposite sides of the mast, a little under its upper end: athwart these bars are fixed the *cross-trees*, upon which the frame of the top is supported. Between the lower-mast head, and the foremost of the cross-trees, a square space remains vacant, the sides of which are bounded by the two trestle-trees. Perpendicularly above this is the foremost hole in the cap, whose after-hole is solidly fixed on the head of the lower-mast. The top-mast is erected by a tackle, whose effort is communicated from the head of the lower-mast to the foot of the top-mast; and the upper end of the latter is accordingly guided into, and conveyed up through, the holes between the trestle-trees and the cap, as above-mentioned.

Top-gallant MAST, (*mat de perroquet*, Fr.) a smaller mast than the preceding, and raised and secured to its head in the same manner.

Top-gallant Royal MAST, (*mat de perroquet volant*, Fr.) is sometimes a yet smaller mast, elevated through irons at the head of the top-gallant-mast; but it is more generally a continuation of the latter above the rigging; when it is termed a pole top-gallant-mast, to distinguish it from a stump top-gallant-mast, which terminates just above the rigging.

Main-MAST, (*grand mât*, Fr.) is the largest mast in a ship, standing nearly in midships between the stem and stern.

Fore-Mast, (*mat de misaine*, Fr.) is that which stands near the stem of the ship, and is next in size to the main-mast, and carries the fore-sail and fore-topsail-yard.

Mizen-Mast, (*d'artimon*, Fr.) is the smallest mast, which stands about half-way between the main-mast and the stern. See the article Mizen-Mast.

Armed Mast, a mast that is made of more than one tree.

Hollow-Mast. See the article Ship-Building.

Jury-Mast. See the article Jury.

Made-Mast, (*mât d'assemblage*, Fr.) a mast composed of several united pieces of the soundest trees, in contradistinction to one consisting of a single tree, or piece, and is stronger than when made of single trees, and less liable to spring.

The counter-mizen, in large vessels and galleons, is in the stern.

Rough-Mast, (*mat brut*, Fr.) denotes a spar fit for making a mast. See the articles Bowsprit, Jib-Boom, and Rough-Tree.

To Mast *a Ship*, (*mater un bâtiment*, Fr.) is to hoist her masts into her by means of a Sheer-Hulk, which article see.

To spend a Mast, (*casser un mât*, Fr.) a term used by seamen to signify the breaking of a mast in foul weather.

Mast-Carlings, are large carlings at the side of the mast-rooms, that are left deep enough to receive the cross-chocks.

Mast-Cloth, in sail-making, is the lining in the middle on the aft side of topsails, to prevent the sail being chafed by the mast.

Mast-Coats, are coverings made of well-tarred canvas, to prevent the water from going down the mast-hole.

The present practice of making the largest masts is, by a *spindle*, or *upper-tree*, two *side-trees*, with their *heel-pieces*, *side-fishes*, *cheeks*, *front-fish*, *cant-pieces*, and *fillings*.

The *spindle*, or *upper-tree*, is made of two pieces, the substantial part of two lesser trees, coaked one into the other in the middle, and bolted together with bolts five feet asunder, and it tapers towards the upper and lower ends.

The spindle of some masts is made of a single tree.

The *side-trees* of large masts are made of two pieces, as above-mentioned, coaked one into the other in the middle, and bolted together with bolts ten feet asunder.

Side-trees of smaller masts are made of one tree; and, after they are so made, are coaked into the lower taper, or tongue of the spindle, and below that into each other, and bolted through all with bolts five feet asunder.

The diameter athwart-ships is gained by the side-trees and spindle.

The *heel-pieces* are short pieces, scarfed on the lower-part of the side-trees, to make good their deficient length. They are often worked short, to gain substance in the middle.

The *side-fishes* are long planks of fir, let in and coaked into the flat surface made by the side-trees and spindle, of sufficient thickness to gain the diameter required fore and aft. The mast, thus far completed, is rounded, and secured by strong iron hoops, driven on the outside.

The *cheeks* are large pieces of fir (formerly oak), coaked or tabled into the upper taper or tongue of the spindle, and sufficiently thick to fashion out the head of the mast, and leave a stop for the support of the trestle-trees: they are bolted together and secured by strong iron hoops on the square part or head

The *front-fish*, or *paunch*, is a long plank of fir, hollowed to the convexity of the mast, and fastened on the fore-side of the mast over the iron hoops.

The *cant-pieces* and *fillings* are various pieces of fir; the former make good the angles occasioned by the side-trees and side-fishes; and the latter form a fair round with the side of the mast and the upper part of the front-fish, for the convenience of woolding.

The several pieces are formed and united together, as represented in the section of a lower-mast of this sort, fig. 4, plate XVI. The whole is then strongly woolded together, between every hoop, with twelve or thirteen close turns of rope (see *a*, fig. 1.), each turn secured by woolding-nails, with leather under their heads, to prevent their cutting. Fig. 4

Masts of less diameter than the former are constructed of two trees, called the *upper* and *lower tree*, which are scarfed and coaked together in the middle; the upper part forming the tongue for the reception of the cheeks. The lower tree sometimes has the addition of a *heel-piece*, to make good the length of the upper tree; and the two trees give the diameter the fore and aft way; then, with the addition of two *side-fishes*, the mast is completed for hooping, and the *cheeks* are then let on.

In fitting masts, particular attention should be paid to place the rounding sides of curved or crooked masts forward.

Bowsprits of large ships are made similar to the above; but, when made of a single tree, not sufficiently large to obtain the diameter required, it is common to make good such deficiency by a fish or paunch of fir, extending the whole length; and run, coaked, and fayed close to the surface. made on the upper side for that purpose, nailed fast to the same, and secured by wooldings, as the masts are. This method of strengthening masts, bowsprits, &c. is commonly used at sea, when sprung, or otherwise weakened, by fishes made of fir, oak, plank, anchor-stocks, or such materials as can be had.

The Method of fixing Masts. The machinery by which it is elevated, or, according to the sea-phrase, *swayed up*, is fixed in the following manner: the top-rope *d*, fig. 2, passing through a block *e*, which is hooked on one side of the cap, and afterwards through a hole, furnished with a sheave or pulley *f*, on the lower end of the top-mast, is again brought upwards on the other side of the mast, where it is at length fastened to an eye-bolt in the cap *g*, which is always on the side opposite to the top-block *e*. To the lower end of the top-rope is fixed the top-tackle *h*, the effort of which being transmitted to the top-rope *d*, and thence to the heel of the top-mast *f*, necessarily lifts the latter upwards, parallel to the lower-mast. When the top-mast is raised to its proper height, fig. 3, the lower end of it becomes firmly Fig. 2. Fig. 3.

wedged in the square hole, above described, between the trestle-trees. A bar of wood, or iron called the *fid*, is then thrust through a hole, *i*, in the heel of it, across the trestle-trees, by which the whole weight of the top-mast is supported. See the articles CAP and TRESTLE-TREES.

The principal articles to be considered in equipping a ship with masts are, 1st. the number: 2d. their situation in the vessel; and 3d. their height above the water.

The masts being used to extend the sails by means of their yards it is evident, that, if their numbers were multiplied beyond what is necessary, the yards must be extremely short, that they may not entangle each other in *working* the ship, and, in consequence, their sails will be very narrow, and receive a small portion of wind. If, on the contrary, there is not a sufficient number of masts in the vessel, the yards will be too large and heavy, so as not to be managed without difficulty. There is a mean between these extremes, which experience and the general practice of the sea have determined; by which it appears, that, in large ships, every advantage of sailing is retained by three masts and a bowsprit.

The following rules have been given, as being the result of long practice and experience, for the situation of the masts in the King's service. Perhaps, after all that can be said, there may exist more whim and fancy in the regulation, than real truth. The centre of the foremast to be abaft the aft side of the rabbet of the stem, or inside of the plank on the lower-deck, being one-ninth part of the length of its lower-deck: the centre of the main-mast to be abaft the aft-side of the rabbet ot the stem, also on the lower-deck, taking five-ninths of its length: and the centre of the mizen-mast to be abaft the aft side of the stem on the lower-deck, at seventeen-twentieths of its length. In the merchant service these rules are materially deviated from; and this deviation is occasioned as well by the fancy of the commander, as that of the person employed in constructing or fitting the vessel; the point, however, still remaining positively undetermined, it is evident no certain rule has hitherto been discovered. It is a general practice, however, to fix the main-mast as near as may be in the centre of the vessel. With respect to the fore-mast and mizen-mast, no certainty of opinion is pretended to be held: as to the rake, that is generally given up as a matter for which no rule can be laid down; generally speaking, the practice is, that all masts should rake aft; but, in some ships (the late Eagle, of 64 guns, may be instanced in particular), the raking aft has been totally destructive of every good sailing principle; when the alteration of it has immediately caused an opposite and most advantageous effect. The trim of some ships has been found so nice, that the most trivial alteration of the rake has instantly produced it. Those who consider themselves the most experienced critics and judges, assert, that all single-masted vessels, as cutters, &c. and even those which have two, as brigs and snows, should be fitted o that their main, or principal mast, should rake aft, in order that it may better withstand the impulse and strain occasioned by the violence of the wind. It is also laid down as a rule never to be deviated from, that the masts are never to touch the partners, for that a neglect in this particular renders them extremely liable to spring or to be carried away. The most general opinion with respect to the rake of one and two-masted vessels is, that the main-mast should rake aft, from its head, three-quarters of an inch for every yard in length; the fore-masts the eighth part of an inch for every yard in length, calculated by a perpendicular let-fall from the head of the mast to the keel. In cutters, or vessels having one mast only, the inclination is fixed at one inch and an half for every yard in length, ascertained by the same rule. The bowsprit is to be nearly perpendicular.

The situation of the main-mast in brigs or vessels having two masts, is fixed at two-thirds of the whole length, reckoning from abaft the head of the stem, and that of the fore-mast at three-twentieths of the same length.

The most advantageous position of the masts is undoubtedly that from whence there results an equilibrium between the resistance of the water, on the body of the ship, on one part, and of the direction of their effort on the other. By every other position this equilibrium is destroyed, and the greatest effort of the masts will operate to turn the ship horizontally about its direction; a circumstance which retards her velocity. It is counterbalanced, indeed, by the helm; but the same inconvenience till continues; for the force of the wind, having the resistance of the helm to overcome, is not entirely employed to push the vessel forward. The axis of the resistance of the water should then be previously determined, to discover the place of the *main-mast*, in order to suspend the efforts of the water equally, and place the other masts so as that their particular direction will coincide with that of the main-mast. The whole of this would be capable of a solution if the figure of the vessel were regular, because the point, about which the resistance of the water would be in equilibrium, might be discovered by calculation: but when the real figure of the ship is considered, these flattering ideas will instantly vanish. This observation induced M. Saverien to employ a mechanical method to discover the axis of resistance of the water, which he apprehended might be used with success in the manner following:—

When the vessel is launched before the places of the masts are determined, extend a rope A B, fig. 5, plate XVI. from the head to the stern. To the extremities A and B attach two other ropes, A D, B C, and apply to the other ends of these ropes two mechanical powers, to draw the ship according to the direction B C, parallel to itself. The whole being thus disposed, let a moveable tube Z, fixed upon the rope A B, have another rope Z R, attached to it, whose other end communicates with a mechanical power R, equal to the two powers D and C. This last being applied to the same vessel, in such manner as to take off the effects of the two others, by sliding upon the rope A B, so as to discover some point Z, by the parallelism of the ropes A D, B C, feebly extended with the rope Z R; the line Z R will be the axis of the equilibrium of the water's resistance, and by consequence the main-mast should be planted in the point Z. Fig. 5.

The figures E, E, E, are three windlasses on the shore, by which this experiment is applied.

With regard to the situation of the other masts, it is necessary, in the same manner, to discover two points; so that the direction of the two mechanical powers operating, will be parallel to the axis of resistance R Z already found.

The exact height of the masts, in proportion to the form and size of the ship, remains yet a problem to be determined. The more the masts are elevated above the centre of gravity, the greater will be the surface of sail, which they are enabled to present to the wind; so far an additional height seems to be advantageous. But this advantage is diminished by the circular movement of the mast, which operates to make the vessel stoop to its effort; and this inclination is increased, in proportion to the additional height of the mast; an inconvenience which it is necessary to guard against. Thus what is gained upon one hand is lost upon the other. To reconcile these differences, it is certain, that the height of the mast ought to be determined by the inclination of the vessel, and that the point of her greatest inclination should be the term of this height above the centre of gravity. See the article TRIM.

Proportions for the Lengths and Diameters of Masts in the Royal Navy.

LENGTHS OF MASTS.—The length of the lower-deck and extreme breadth being added together, the half is the length of the *main-mast*, thus, the length of the lower-deck of a 74-gun ship is 176 feet; breadth, extreme, 48 feet 8 inches, added together, they make 224 feet 8 inches. The half, or 112 feet 4 inches, is the length of the main-mast; which being determined, the other masts bear the following proportion:—

Fore-Mast, eight-ninths of the main-mast.

Mizen-Mast, 100 guns, seven-eighths; 90 guns to 50, six-sevenths; 44 guns to 20, five-sixths of the main-mast.

In sloops, the mizen-mast is three-fourths of the main-mast

When the mizen-mast steps on the lower-deck, the depth of the hold is to be deducted from the length here given. This method is now generally practised in the royal navy.

Main-Topmast, three-fifths of the main-mast.

Fore-Topmast, eight-ninths of the main-topmast.

Mizen-Topmast, three-fourths of the main-topmast.

In sloops, the mizen-topmast is five-sevenths of the main-topmast.

Top-gallant-Masts, one-half the length of the topmast.

Bowsprits, of 80-gun ships and upwards, seven-elevenths of their main-masts. Bowsprits of 74-gun ships and under, three-fifths of their main-masts.

Brig's Main-Masts. The depth in the hold is added to the length and breadth, and one-half is the length of the main-mast.

Cutter's Main-Masts. Add together the length of the deck, the breadth extreme, and depth in the hold; and three-quarters of that sum is the length of the main-mast.

DIAMETERS OF MASTS.—*Main* and *Fore-Masts* of ships of 100 to 50 guns inclusive, are one inch in diameter at the partners to every yard in length. Ships of 44 to 32 guns inclusive, nine-tenths of an inch to every yard in length. And ships of 28 guns and under, seven-eighths of an inch to every yard in the length.

Main-Masts of Brigs, one inch to every yard in length; and the *fore-mast* nine-tenths of the diameter of the main-mast.

Masts of Cutters, three-fourths of an inch in diameter to every yard in length.

Mizen-Masts of Ships of 100 to 64 guns inclusive, three-fifths of the diameter of the main-mast. 50-gun ships and under, two-thirds of the diameter of the main-mast.

Main and Fore-Topmasts, one-inch to every yard in the length of the fore-topmast.

Mizen-Topmasts, seven-tenths of the diameter of the main-topmast.

Topgallant Masts, one-inch to every yard in their lengths.

Royal-Masts, two-thirds of the diameter of their topgallant-masts.

Bowsprits of Ships of 100 to 64 guns inclusive, two-inches less than the diameter of the main-mast. In 50-gun ships and under, the same diameter as the main-mast.

Proportions for the Lengths and Diameters of Booms.

LENGTHS OF BOOMS.—Lower-studdingsail-booms, five-ninths of the main-yard.

Top-Studdingsail-Booms, one-half the length of the yards they go on.

Flying Jib-Boom, five-sevenths of the bowsprit.

Driver-Boom, the same length as the main-topsail-yard.

DIAMETERS OF BOOMS.—Studdingsail-Booms, one inch to every five feet in the length.

Flying Jib-Booms, seven-eighths of an inch to every yard in the length.

Driver-Boom, the same as the fore-topsail-yard.

Proportions for the Lengths and Diameters of Masts, &c. for Sloops, Smacks, and Hoys.

LENGTHS OF THE MASTS, &c.

Mast and *topmast* in one, thrice and three-fourths the whole length.

Mast to the rigging-stop or hounds, three-fourths the whole length.

Mast and *topmast* to the stop of the topmast, forty one parts of the whole length.

Topgallant-mast to the rigging stop, four-sevenths of the length of the mast.

Boom, two-thirds of the mast.

Gaff, three-fifths of the boom.

DIAMETER OF THE MASTS, &c.

Mast, one-fourth of an inch to every foot in length.

Topgallant-mast, three-eighths of an inch to every foot in length.

Boom, three-sixteenths of an inch to every foot in in length.

Gaff, one-fourth of an inch to every foot in length.

TABLE of the Dimensions of Masts now used in the Royal Navy

NAMES OF THE MASTS.	110 GUNS. 2164 TONS.		90 GUNS. 1870 TONS.		80 GUNS. 1620 TONS.		74 GUNS. 1800 TONS.		64 GUNS. 1369 TONS.		50 GUNS. 1044 TONS.		44 GUNS. 879 TONS.		38 GUNS. 951 TONS.	
	Length Ft. In.	Dia In.	Length Ft. In.	Dia In.	Length Ft. In.	Dia In.	Length Ft. In.	Dia In.	Length Ft. In.	Dia In.	Length Ft. In.	Dia In.	Leng Ft. In.	Dia In.	Length Ft. In.	Dia In.
Main-mast	117 0	39	112 0	37¼	107 0	35½	111 0	37	101 0	33½	92 0	29	88 0	26½	90 0	27
Main-topmast	70 0	20¾	66 0	19¾	64 0	19¼	66 0	19¾	58 6	17⅛	53 0	16	53 0	15¾	54 0	16⅛
Main-topgallant-mast	35 0	11⅜	33 0	11	32 0	10½	33 0	11¼	29 3	9¼	26 6	8⅞	26 6	8¼	27 0	9
Fore-mast	103 6	34½	100 0	33½	95 9	31⅞	98 6	32¾	89 7	29¾	81 6	26⅝	73 0	23¾	80 0	23½
Fore-topmast	62 10	20¼	59 0	19¾	59 9	19¼	58 8	19¾	53 0	17⅛	48 0	16	47 0	15¾	48 0	16⅛
Fore-topgallant-mast	31 0	10¼	29 0	9⅜	28 6	9½	29 4	9¾	26 6	8¾	24 0	8	23 6	7¾	24 0	8
Mizen-mast	101 4	23	96 6	22½	93 0	21⅛	95 0	22¼	86 0	19¼	78 9	19⅜	74 0	17⅞	75 7	18⅛
Mizen-topmast	52 0	14	49 0	13½	46 0	13¼	49 0	13½	43 6	12¼	40 0	11⅛	40 0	11	41 0	11½
Mizen-topgallant-mast	26 0	8⅜	24 0	8	23 0	7⅜	24 6	8⅛	21 9	7¼	20 0	6⅜	20 0	6⅜	″9 6	6¼
Bowsprit	74 0	37	71 0	35½	68 0	34	67 6	35	60 4	31½	56 0	29	52 6	26⅓	55 0	27

NAMES OF THE MASTS.	36 GUNS. 871 TONS		32 GUNS. 677 TONS.		28 GUNS. 594 TONS		24 GUNS 513 TONS		20 GUNS. 429 TONS.		SLOOPS OF 300 TONS.		BRIGS OF 200 TONS		CUTTERS OF 200 TONS.	
	Length Ft. In	Dia In.	Length Ft. In.	Dia In.	Length Ft. In.	Dia In.	Length Ft. In.	Dia In.	Length Ft. In.	Dia In.	Length Ft. In.	Dia In.	Length Ft. In.	Dia In	Length Ft. In.	Dia In.
Main-mast	89 0	26	85 0	24¾	81 4	23⅜	75 0	22¼	72 0	21⅛	63 0	18⅜	56 0	19	88 0	22
Main-topmast	33 4	15¾	51 0	15⅓	48 9	14⅜	45 0	13¼	43 2	12⅔	37 6	11¼	31 0	10	——	—
Main-topgallant-mast	25 8	8⅝	25 6	8½	24 4	8	22 6	7½	21 7	7¼	18 9	6⅓	23 6	6¼	44 0	9¾
Fore-mast	79 6	23⅛	75 0	22	72 0	20⅞	66 6	19½	64 0	19	56 0	16⅜	49 0	17	——	—
Fore-topmast	47 0	15¼	45 0	15⅛	43 0	14⅛	40 0	13¼	38 4	12¾	33 4	14¼	31 0	10	——	—
Fore-topgallant-mast	22 5	7½	22 6	7½	21 6	7	20 0	6½	19 2	6⅜	16 8	5⅜	23 6	6¼	——	—
Mizen-mast	74 8	17¾	72 0	17	69 0	16¾	64 0	15	61 0	14	48 0	12	——	—	——	—
Mizen-topmast	40 0	11⅛	38 0	10½	36 7	10	33 9	9¼	32 5	9	26 9	7¾	——	—	——	—
Mizen-topgallant-mast	20 0	6½	19 0	6¼	18 3	6	16 10	5½	16 0	5¼	13 0	4¼	——	—	——	—
Bowsprit	54 0	26	52 0	25	48 9	23⅜	45 0	22¼	43 6	21⅛	37 6	18⅜	34 0	17	64 0	20

TABLE of the Dimensions of Masts now used in the Merchant Service.

NAMES OF THE MASTS.	1300 TONS.		1200 TONS.		1100 TONS.		1000 TONS.		900 TONS.		800 TONS.		700 TONS.		600 TONS.	
	Length Ft. In.	Dia In.	Length Ft. In.	Dia In.	Length Ft. In.	Dia In	Length Ft. In.	Dia In.	Length Ft. In.	Dia In.	Length Ft. In.	Dia In.	Length Ft. In.	Dia In.	Length Ft. In	Dia In.
Main-mast	96 0	31½	94 0	31	90 0	30	87 0	29	86 0	28	84 0	27	82 0	26	78 0	24
Main-topmast	56 0	17½	56 0	17½	56 0	17¼	52 0	16½	50 0	16	49 0	16	48 0	15½	48 0	15
Main-topgallant-mast	27 0	9	27 0	9	27 0	9	26 0	9	26 0	9	25 9	9	24 0	8½	23 0	8
Fore-mast	90 0	30	88 0	30	84 0	29½	81 0	29	79 0	28	77 0	27	75 0	26	73 0	23⅓
Fore-topmast	56 0	17½	54 0	17½	52 0	17¼	50 0	16½	48 0	16	47 0	16	46 6	15½	46 0	15
Fore-topgallant-mast	26 0	9	26 0	9	26 0	9	24 0	9	23 0	9	22 0	8½	22 0	8½	22 0	8
Mizen-mast	78 0	21½	78 0	21½	78 0	21	76 0	20	76 0	20	73 0	19	70 0	18	68 0	16
Mizen-topmast	11 0	13	40 0	12½	39 0	12	38 0	12	37 0	12	36 0	12	34 0	11	38 0	10⅔
Mizen-topgallant-mast	21 0	7	21 0	7	20 0	6⅓	19 0	6⅓	19 0	6	19 0	6	18 0	5½	18 0	5½
Bowsprit	60 0	31	59 0	30½	58 0	30	54 0	29	54 0	28	53 0	27	50 0	25	50 0	24⅛

NAMES OF THE MASTS.	500 TONS.		400 TONS.		360 to 350 TONS.		330 to 300 TONS.		280 to 250 TONS.		BRIGS. 150 TONS		KETCHES 170 TONS.		SCHOONERS, 110 TONS.	
	Length Ft. In.	Dia In.	Length Ft. In.	Dia In.	Length Ft. In.	Dia In.	Length Ft. In.	Dia In.	Length Ft. In.	Dia In.	Length Ft. In.	Dia In.	Length Ft. In	Dia In.	Length Ft. In.	Dia In.
Main-mast	72 0	22	70 0	21½	68 0	20½	66 0	19	61 0	18	60 0	16½	72 0	19	75 0	15
Main-topmast	43 0	14	41 0	13	39 0	12½	38 0	12	36 0	11	30 0	10	36 0	10	——	—
Main-topgallant-mast	23 0	8	23 0	7½	20 0	7	19 0	6½	18 0	6	16 6	5½	20 0	4½	32 0	7½
Fore-mast	67 0	22	65 0	21	63 0	20½	61 0	19	58 0	18	54 0	16½	——	—	66 0	15
Fore-topmast	41 0	14	40 0	13	38 0	12½	37 0	12	35 0	11	32 0	10	——	—	——	—
Fore-topgallant-mast	22 0	8	21 0	7½	19 0	7	18 0	6½	17 0	6	16 0	5½	——	—	26 0	6½
Mizen-mast	63 0	15	62 0	15	60 0	14	58 0	13	54 0	12	——	—	49 0	10	——	—
Mizen-topmast	30 0	10	30 0	9½	30 0	9½	28 0	9	26 0	8	——	—	30 0	5½	——	—
Mizen-topgallant-mast	17 0	5	16 0	5	16 0	5	15 0	4½	14 0	5	——	—	——	—	——	—
Bowsprit	48 0	23	47 0	21	14 0	20⅓	12 0	19	38 0	18	32 0	16	44 0	14½	50 0	14½

SLOOPS, 70 TONS.

NAMES OF THE MASTS.	LENGTH.		DIAM.
	Ft.	In.	Inc.
Main-mast	64	0	16
Main-topgallant-mast	30	0	$6\frac{1}{2}$
Bowsprit	40	0	14

The diameters, in the foregoing tables of dimensions, are taken at the *partners* of lower-masts (which are at the middle deck in three-decked ships, and at the upper-deck in all other ships); at the *cap* in top-masts, topgallant-masts, and royal masts; and at the *bed* of bowsprits.

N.B. Masts with hoops are made one inch larger than the given diameter in these tables, two feet each way from the partners, to give the hoops a quicker drift.

See the *Art of making Masts, Yards, &c. &c.*

In order to secure the masts, and counterbalance the strain they receive from the effort of the sails impressed by the wind, and the agitation of the ship at sea, they are sustained by several strong ropes, extended from their upper-ends to the outside of the vessel called *shrouds*, as represented at fig. 5, plate XVI. They are further supported by other ropes, stretched from their heads towards the fore-part of the vessel. See the article RIGGING.

MASTER OF A SHIP OF WAR, (*maitre pilote, officier de vaisseau de guerre anglois, qui est chargé de naviguer le bâtiment, sous les ordres du capitaine*, Fr.) an officer, ranking immediately after the lieutenants, appointed by the commissioners of the navy to assist in fitting, and to take charge of the navigating and conducting a ship from port to port, under the direction of the captain. The management and disposition of the sails, the working of the ship into her station in the order of battle, and the direction of her movements in the time of action, are also particularly under his inspection; and he is moreover charged with the stowage of the hold. He is to be careful that the rigging, sails, and stores, be duly preserved; to see that the log and log-book be regularly and correctly kept; accurately to observe the appearances of coasts, rocks, and shoals, with their depths of water and bearings, noting them in his journal. He is to keep the hawser clear when the ship is at anchor, and to provide himself with proper instruments, maps, and books of navigation. It is likewise his duty to examine the provisions, and accordingly to admit none into the ship but such as are sound, sweet, and wholesome.

Whenever he shall be ordered to survey stores or provisions reported to be decayed or unserviceable, he is to examine the state of each article, with the most scrupulous attention, and, as far as his judgment shall enable him to determine, to point out the cause to which the defective state of such stores or provisions is to be attributed. If he find any articles no longer fit for the service for which they were intended, he is to mention, in his report, any other service for which he shall think them fit.

He is to keep the keys of the spirit-room and after-hold; and, when they are wanted, to deliver them to one of the master's-mates, strictly charging him not to suffer a light to be carried into the spirit-room; to attend himself, without quitting on any account, either the spirit-room or after-hold while it is open; to see it properly secured when the service for which it was opened is executed, and to return the keys to him as soon as he has done so.

He is to be particularly careful to prevent any waste or improper use of water; and never to allow of its being started, nor pumped out in the hold, without particular directions from the captain, nor to suffer more to be hoisted up in a day than the quantity allowed to be used: and he is every day at noon, to report to the captain the quantity of beer and water expended during the last twenty-four hours, and the quantity remaining on board.

When the ship shall be laid up, he is to deposit a copy of the log-book and journal with the commissioners of the navy; and to enable him the better to perform these services, he is allowed several assistants, who are properly termed MATES and QUARTER-MASTERS, which articles see.

Any claim that he may have for pilotage, he is in no case to receive abroad, but is to transmit the certificate of the same to the Navy Board, by whom payment will be ordered. See NAVAL INSTRUCTIONS.

MASTER'S-MATE, (*officier sous le commandement du maître-pilote* (*master*) *pour l'aider dans ses fonctions*, Fr.) an officer under the direction of the master, by whose choice he is generally appointed, to assist him in the several branches of his duty. Accordingly, he is to pay particular attention to the *log*-line and glass by which the ship's course is measured, to keep a log-book according to the form given in the NAVAL INSTRUCTIONS, and to adjust the sails to the wind in the fore-part of the ship. He is also to see that the cables are well *coiled* and kept clean in the *tiers*, and sufficiently *served* when employed to ride the ship. And finally, he is to superintend the stowage of the hold, taking especial care that all the ballast and provisions are properly stowed therein.

MASTER *of a Merchant Ship*, (*capitaine marchand, ou patron*, Fr.) the commanding officer who is appointed by the owners to manage the navigation, and every thing relating to her cargo, voyage, sailors, &c. He is generally denominated captain.

Nothing more materially concerns the master of a merchant-ship than to know what degree of responsibility is attached to his situation, and what privileges it invests him with. The master and owner are so intimately connected together, that many observations which will be found under the article OWNERS OF SHIPS, are equally applicable to the one and the other, and will prove to be necessary knowledge for the master of a ship. The doctrine of marine insurances, too, he should be well acquainted with, in order that he may not, by his conduct, inadvertently prejudice his owner's claim on the under-writers.

Masters of ships responsible for goods committed to their charge.—The law makes no distinction between carriers *by land* and carriers *by water*; and, for whatever losses that arise from the neglect of persons employed under them, they are answerable; whatever cases, therefore, that are contained under this article, relative to carriers by land, must be understood to be equally applicable to carriers by water, such as masters of ships or hoymen.

The masters of vessels laden with coals are directed, by 47 Geo. III. sess. 2, c. 68, s. 24, to deliver, on arrival at or to the westward of Blackwall, Gravesend, or moorings for the delivery in the port of London, a copy of the certificate originally delivered to them by the fitters or venders, as enacted in the 9th year of Queen Anne.

The master of a ship is liable for goods of which the ship is robbed in the river.—The reasons are, 1. Because he is an officer known; 2. Because he receives his salary out of that which is paid for the freight. But the master may reimburse himself out of the mariner's wages for a loss happening by their neglect.

The merchants of London have agreed upon a new form for their bills of lading, the difference be'ween it and the former bills of lading consists in the introduction of an exception, which runs in these words: —"*The act of God, the king's enemies, fire, and all and every other dangers and accidents of the seas, rivers, and navigation, of whatever nature or kind soever, save risk of boats, so far as ships are liable thereto, excepted.*"

It is, however, proper to observe, that, if no bill of lading be signed, then the master or owner is bound by the general law of the land.

Passengers dying on board.—If any passenger die on board, the master is obliged to take an inventory of his effects; and, if no claim be made to them within a year, the master becomes proprietor of the goods, but answerable for them to the deceased's legal representatives. Bedding and furniture become the master's and his mate's; but the clothing must be brought to the mast-head, and there appraised and distributed among the crew.

Captain dying.—If a captain die, leaving money on board, and the mate, becoming captain, shall improve the money, he shall, on allowance for his care, account both for interest and profits.

Master to agree in writing with his seamen.—No master is to proceed on any voyage, for parts beyond the seas, without first coming to an agreement, in writing, with his mariners, for their wages: if he do so, he shall forfeit, for every mariner so taken without written agreement, five pounds. 2 Geo. II. c. 36, s. 1.

Masters of merchant-vessels to bring home seafaring persons.—British governors, ministers, and consuls, residing in foreign parts, or where none such is present, any two British merchants, are required to provide for sea-faring men and boys, subjects of Great Britain, who, by shipwreck, capture, or other unavoidable accident, shall be in foreign parts, or or who shall have been discharged there as unserviceable from the royal navy, and subsist them at 9*d per diem* each; and send them home as soon as conveniently may be, in any ship belonging to the royal navy, or in any merchant-ship. 32 Geo. III. c. 83.

Every master of a merchant-ship homeward-bound, is required to take all such sea-faring persons on board (not exceeding four to each ten tons), and bring them to some port in Great Britain. Upon arrival, if he produce to the Navy Board a certificate from the governors or consuls, &c. of the number of names of the men and boys taken on board by their directions, and his own affidavit specifying the time during which he subsisted them, and whether his own complement was full or not, he will be paid 9*d per diem* for the passage and provisions of each person above the number of his full complement. 32 Geo. III. c. 33.

Masters to keep an account of forfeitures to Greenwich Hospital.—The master of every vessel is required, by the 2 Geo. II. c. 36, to keep a regular account of the penalties and forfeitures due to Greenwich-hospital, by the mariner's disobedience, to deduct the same from their wages, and to pay the amount thereof to the collector of the Greenwich-hospital duty, within three months after such deduction, upon pain of forfeiting treble the value thereof to the use of the said hospital.

Master *at Arms*, (*capitaine d'armes*, Fr.) an officer appointed by warrant from the Board of Admiralty, to teach the officers and crew of a ship of war the exercise of small arms, and, above all other things, to direct their attention to the pointing of their muskets well before they fire them; to plant sentinels over the prisoners, and superintend them during their confinement. He is also to see all the fires and lights extinguished at the time the captain shall direct, except such as shall be permitted by proper authority, or under the inspection of sentinels. It is likewise his duty to attend the *gangway*, to prevent spirituous liquors being conveyed into the ship, unless by permission of the commanding officer. He is to see that the small arms be kept in proper order. He is to visit all vessels coming to or going from the ship, and prevent the crew from going from the ship without leave. He is also to acquaint the officer of the watch with all irregularities in the ship which shall come to his knowledge. When there are no marines on board, the *master at arms* is to place the sentinels on their posts; to instruct them in their duty, and to see that they perform it diligently; he is to see that their arms and the ammunition they are directed to be supplied with are in good order, and perfectly fit for service: he is frequently to visit them, to enquire whether they have properly received and perfectly understood their orders; and to insure their being alert. His assistants in these several duties are called Ship's Corporals, who also relieve the sentinels, and one another, at certain periods. See the article Corporal of a Ship of War.

Master *Attendant* (*officier employé dans les arsenaux de marine d'Angleterre, et dont les fonctions sont à-peu-près semblables à celles du capitaine de port en France*, Fr.) an officer in the royal dock-yards, appointed to assist in the fitting or dismantling, removing or securing vessels of war, &c. at the port where he resides. He is frequently to inspect the *moorings* which are sunk in the harbour, and observe that they are kept in proper repair, to be always ready when occasion requires. It is also his duty to visit all the ships in *ordindry*, and see that they are frequently cleaned and kept in order; and to attend at the general musters in the dock-yards, taking care that all the officers, artificers, and labourers, registered in the navy-books, are present at their duty.

MAT, (*sangle, paillet, ou baderne, &c.* Fr.) a sort of thick web, or texture, formed of spun-yarn, or *foxes*, containing each a greater or lesser number of rope-yarns, in proportion to the thickness of the mat intended to be woven.

Mats are used to fasten upon the outside of such parts of the standing rigging as are exposed to the friction of the yards, bolt-ropes of sails, or other ropes, in extending, shifting, or trussing up the sails, particularly the lower ones. The largest and strongest sort of these mats are called *paunches*.

MATCH, (*mèche*, Fr.) a kind of rope, slightly twisted, and prepared with inflammable ingredients, for the purpose of firing cannon. It burns slowly and dead, so that less danger is to be apprehended from the use of it on ship-board, where powder is too apt to be loosely scattered about in time of battle, than there would from that of port-fires in land-service.

Common or *Slow* MATCH, is made of hemp or coarse flax, spun like cord, but very slack; it is composed of three twists or strands, which are again covered with tow, so that the twists do not appear: lastly, it is boiled in the lees of old wine. This, when once lighted at the end, burns on gradually, without going out, until the whole be consumed.

The *slow match* may also be made by boiling the hemp for 24 hours in a ley of ashes, lime, and water; or merely of beech-ashes and water. A good match ought to burn equally and gradually, and with a good hard point; to light with ease, and feel soft and dry. In order to be always fit for use, they should be carefully kept where they can imbibe no moisture. The *slow match* used by the English is made by contract: one yard of it will burn eight hours. The French slow match is usually made by soaking light-twisted white rope for three days in a strong ley. It burns about three feet in six hours.

Quick MATCH, so called, in contradistinction to slow, is a more rapid combustible than any other species of fire-work or preparation of that kind ever invented. It is made of three cotton or worsted strands, drawn into lengths, and put into a kettle, just covered with white wine vinegar, and a quantity of saltpetre and mealed powder then added, and boiled till well-mixed; or, sometimes, saltpetre only is used; taking the strands out hot, and laying them into a trough with some mealed powder, moistened with spirits of wine, thoroughly wrought into the cotton by rolling it backwards and forwards with the hands; when this is done, they are taken out separately, and dried upon a line.

Quick-match is used for conveying fire to any remote distance, and in as many different directions as may be necessary at the same instant; in short, it is frequently applied to the same purposes for which trains of gunpowder were formerly laid, and were subject to fail in their effects from a variety of causes and accidents.

The cotton match is generally made of such cotton as is put in candles, of several sizes, from one to six threads thick, according to the pipes for which it is designed. The composition is, cotton, 1 lb. 12 oz.; saltpetre, 1 lb. 8 oz.; mealed powder, 10 lb.; spirits of wine, 2 quarts; water, 3 pints; and isinglass, 3 gills. That for the worsted match is, worsted, 10 oz.; mealed powder, 10 lb.; spirits of wine, 3 pints; and white wine vinegar, 3 pints.

The worsted or cotton must be laid evenly in an earthen or other pan, and the different ingredients poured over it, and about half the powder; being left a short time to soak, it is afterwards wound smoothly on a reel, and laid to dry: the remaining half of the powder is then sifted over it, and it is ready for use when dry.

Gunners are strictly ordered not to allow any match to be burnt in the day, nor more than two lengths at the same time in the night, without being ordered so to do by the captain; and, when the match is burning, it is always to be hung over water in tubs, and the gunner's mate of the watch is to attend to it.

MATE OF A MERCHANT-SHIP, (*second d'un vaisseau marchand*, Fr.) the officer who commands in the absence of the master, and shares the duty with him at sea.

The first, second, and third mates of East India ships may indeed, with great propriety, be compared to lieutenants in the royal navy, particularly with regard to their duty: the third takes also the care of the hold, and, in this respect, performs part of the master's duty in the navy

The number of mates allowed to ships of war and merchantmen is always in proportion to the size of the vessel. Thus a first-rate man of war has six mates, and an East Indiaman the same number; a frigate of 20 guns, and a small merchant ship, have only one mate in each; and the intermediate ships have a greater or smaller number, according to their several sizes, or to the services on which they are employed.

MATE, (*second ou aide*, Fr.) generally implies an assistant, as quarter-master's mate, boatswain's mate, gunner's-mate, carpenter's mate, armourer's mate, sail maker's mate, steward's mate, cook's mate, surgeon's first, second, third, fourth, and fifth mate.

MATHEMATICIAN, (*mathématicien*, Fr.) a person skilled in mathematics. The efforts of *mathematicians* to improve the sciences are well known from the days of Euclid and Archimedes to the time of Newton and La Place.

Euler, Delambre, La Grange, La Lande, and La Place, have contributed much on the continent: while Herschel, Hutton, Vince, Playfair, Woodhouse, and Dalby, are actively employed in their respective stations; some in simplifying and smoothing the road, and others in enlarging the boundaries of science in our own country.

A periodical work, called the Ladies' Diary, has now been published upwards of a century, and contains many curious problems, not of the highest order, indeed, but that still require a knowledge far beyond the mere elements of science to resolve; and the number of ingenious men, who take a share in proposing and solving the problems in that and other popular publications of the same kind; such as Leybourn's Mathematical Repository, the Gentleman's Diary, the Mathematical Companion, the Mathematical Miscellany, &c. seems not a little surprising.

The problems contained in these works have tended to awaken curiosity, and the solutions to convey instruction in a much better manner than is always to be found in more splendid publications; and have been the means of no inconsiderable degree of useful mathematical knowledge being more widely diffused in England than in any other country in the world.

MATHEMATICS, (from μαθησις, Gr. *les mathématiques*, Fr.) the science or doctrine which teaches, or contemplates, whatever is capable of being numbered or measured, so far as is computable or measurable, and is so connected with the general scope of this work, that a concise definition seems requisite to be given.

Mathematics are generally distinguished into *speculative* and *practical*, *pure* and *mixed*.

Speculative MATHEMATICS, is that which barely contemplates the properties of things: and

Practical MATHEMATICS, that which applies the knowledge of those properties to some useful purpose in life.

Pure MATHEMATICS, is that branch which considers quantity abstractedly, and so either computable, as ARITHMETIC; or measurable, as GEOMETRY.

Mixed MATHEMATICS, considers quantity, as subsisting in material being; they are very extensive, and distinguished by various names, according to the different subjects they consider, as ASTRONOMY, GEOGRAPHY, NAVIGATION, &c. which articles see.

MAUL, (*masse de fer*, Fr.) a large iron hammer, used for various purposes of driving any bolt or nail.

Top-MAUL, is distinguished from the above, by having an iron handle with an eye at the end, by which it is tied fast to the mast-head, to prevent accidents by its falling out of the top; it is principally used to drive the fid in or out of the top-mast.

MEALED, denotes that which is pulverized, or reduced to fine powder.

MEAN, (*moyen*, Fr.) a middle state between two extremes; thus we have a mean distance, mean motion, &c.

MEAN *Distance of a Planet from the Sun*, in astronomy, is the right line drawn from the sun to the extremity of the conjugate axis of the ellipsis the planet moves in; and this is equal to the semi-transverse axis, and is so called, because it is a mean between the planet's greatest and least distance from the sun.

MEAN *Motion*, in astronomy, that whereby a planet is supposed to move equally in its orbit, and is always proportional to the time.

MEASURE, denotes any given quantity, assumed as one, to which the ratio of other similar quantities may be expressed.

MEASURE *of an Angle*. See the article ANGLE.

MEDICINE-CHEST, (*coffre à médicine*, Fr.) a large chest, containing the medicine for the use of the sick and hurt on board of his Majesty's ships. It is generally kept in the cockpit, under the care of the surgeon.

MEDITERRANEAN PASS, is an indenture of security, under the hands and seal of the Lords Commissioners of the Admiralty, granted to masters of British merchant-ships, for the purpose of enabling them to pass the Barbary states unmolested by their cruisers.

These passes are granted only to British-built ships navigated with a master and three-fourths of the mariners British subjects, or foreign protestants made English denizens. In the Barbary states, it has been agreed, that the subjects of the king of Great Britain should pass the seas unmolested by the cruisers of those states; and, for better ascertaining what ships and vessels belong to British subjects, it is provided, that they shall produce *a pass* from the Lords Commissioners of the Admiralty.

These passes contain but a very few words, and are written on parchment, with ornaments at the top, through which a scolloped indenture is made; the *scolloped tops* are sent to Barbary, and being put in possession of their cruisers, the commanders are instructed to suffer all persons to pass who have passes that will fit these scolloped tops.

The protection afforded by these passes is such, that no ships which traverse the seas frequented by these rovers, ever fail to furnish themselves with them; whether in the trade to Portugal, Spain, Italy, Levant, or any part of the Mediterranean.

The rules and orders under which Mediterranean passes are now granted, were made by the King in Council, on the 14th of June 1722, and on the 28th of August 1776, upon representations made by the Board of Admiralty of the abuses then practised.

It has been found expedient, at the conclusion of a war, and sometimes during a peace, to recall and cancel all passes that have been issued, and to issue others in a new form. This has been done for two reasons: first, that these useful instruments, by various means, either accidental or fraudulent, came into the hands of foreigners, who, under cover of them, carried on in security a trade which otherwise would belong to British subjects, and which had been purchased by the crown at the expense of keeping up this sort of alliance. Secondly, that the Barbary states complained, that, adhering to the rule of fitting the other part of the indenture to the passes, they were obliged to suffer ships to pass that did not belong to British subjects. For these reasons the passes were called in, in the years 1722, 1729, 1750, 1765, 1776, 1783, and, for the last time, in 1802, and new ones issued. By stat. 4 Geo. II. c. 18. it is made felony, without benefit of clergy, to forge, counterfeit, or alter Mediterranean passes; and such offences committed out of the kingdom may be prosecuted in any country. See MORTIMER'S *General Dictionary of Commerce*.

MENKAR, in astronomy, a bright fixed star of the second magnitude, situated in the throat of the constellation Cetus, and marked α in our catalogues. For the beginning of the year 1812, its

Right Ascension is . . .	42° 39′ 4″
Annual Variation in A. R. .	46,7
Declination	3 19 27 N.
Annual Variation in Declin.	+ 14,7

MERCATOR'S CHART, or PROJECTION, is a projection of the surface of the earth in plano, so called from Gerrard Mercator, a Flemish geographer, who first published maps of this sort in the year 1556, though it was Edward Wright who first gave the true principles of such charts, with their application to navigation in 1590. See the article CHART.

MERCATOR'S *Sailing*, or, more properly, WRIGHT'S *Sailing*, is the method of computing the cases of sailing on the principles of Mercator's chart; which principles were laid down by Edward Wright, in the beginning of the 16th century; or, the art of finding on a plane the motion of a ship upon any assigned course, that shall be true as well in longitude and latitude, as distance; the meridians being all parallel, and the parallels of latitude straight lines.

In the right-angled triangle A *b c*, let A *b* be the true difference of latitude between two places, the angle *b* A *c* the angle of the course sailed, and A *c* the true distance sailed; then will *b c* be what is called the departure, as in plane sailing: produce A *b* till A B be equal to the meridional difference of latitude, and draw B C parallel to *b c*; so shall B C be the difference of longitude.

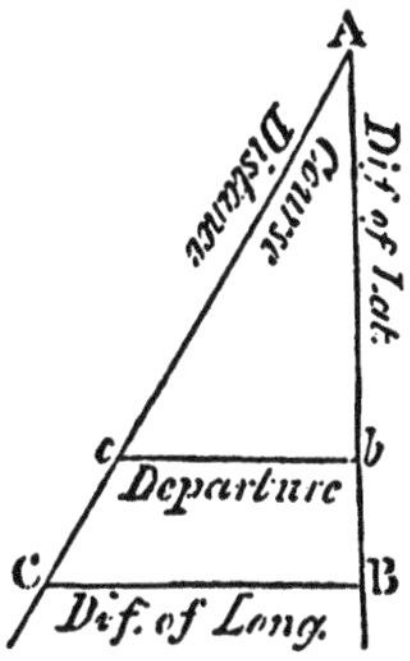

Now, from the similarity of the two triangles A *b c*, A B C, when three of the parts are given, the rest may be found; as in the following analogies: As
Radius : sin. course : : distance : departure;
Radius : cos. course : : distance : dif. lat.;
Radius : tan. course : : merid. dif. lat. dif. longitude.

And, by means of these analogies, may all the cases of Mercator's sailing be resolved.

MERCHANDISE, (*marchandise*, Fr.) implies all sorts of goods and wares exposed to sale in fairs or markets.

MERCHANT, (*négociant*, Fr.) a wholesale dealer in all sorts of merchandise, who exports and imports from different parts of the world, deals in exchange, and buys and sells goods in their original packages, without breaking bulk.

Formerly, every one who was a buyer or seller in the retail way was called a *merchant*, as they are still, both in France and Holland; but, in this country, shopkeepers, or those who retail goods, have long lost that appellation, and the name of English merchant is now highly respected in every part of the world.

Merchant-Company, a society of merchants, consisting of a considerable number of members associated together, in one common interest, by a charter obtained from the sovereign. Of these there are several well-regulated companies of Great Britain for foreign commerce; which are as follow:—

The Hamburgh Company, is the oldest trading company in the kingdom, though not always known by that name, nor restrained to those narrow bounds to which it is now confined. It was first called the Company of Merchants trading to Calais, Holland, Zealand, Brabant, and Flanders: then it acquired the general title of Merchant Adventurers of England, as being composed of all the English merchants who traded to the Low Countries, the Baltic, and the German Ocean. Lastly, it was called by its present name. This company was first incorporated by Edward I. in 1296, and their privileges have been confirmed by many of his successors. The revolutions which happened in the Low Countries towards the end of the 16th century, and which laid the foundation of the republic of Holland, having hindered the company from continuing their commerce with their ancient freedom, they were obliged to turn it almost wholly to the side of Hamburgh and the cities on the German ocean.

The Russian Company was first projected towards the end of the reign of King Edward VI. and executed in the first and second years of Philip and Mary; but had not its perfection till its charter was confirmed by Act of Parliament under Queen Elizabeth, in 1566. It had its rise from certain adventurers who were sent in three vessels, on the discovery of new countries, and to find out a north east passage to China. These, falling into the White Sea, and putting into the port of Archangel, were exceedingly well received by the Muscovites; and, at their return, solicited letters patent, to secure themselves the commerce of Russia, for which they had formed an association.

This company subsisted with reputation, almost a whole century, till the time of the civil wars. It is said, the Czar then reigning, hearing of the death of Charles I. ordered all the English in his states to be expelled; of which, the Dutch taking advantage, settled in their room. After the restoration the remains of the company re-established part of their commerce at Archangel, but never with the same success as before, the Russians being now well acquainted with the Dutch merchants and merchandise. This company subsists still, under the direction of a governor, four consuls, and assistants; and, by the 10th and 11th of William III. c. 6, the fine for admission was reduced to five pounds.

The Eastland Company was incorporated by Queen Elizabeth. Its charter is dated in the year 1579. By the first article, the company is erected into a body politic, under the title of the Company of Merchants of the East, to consist of Englishmen, all real merchants, who have exercised the business thereof, and trafficked through the Sound before the year 1568, into Norway, Sweden, Poland, Livonia, Prussia, Pomerania, &c. excepting Narva, Muscovy, and its dependencies. Most of the remaining articles grant them the usual prerogatives of such companies, as a seal, governor, courts, laws, &c.

The Turkey or Levant Company had its rise under Queen Elizabeth, in 1581. James I. confirmed its charter in 1605, adding new privileges. During the civil wars there happened some innovations in the government of the company, many having been admitted members, not qualified by the charters of Queen Elizabeth and King James, or that did not conform to the regulations prescribed. Charles II. upon his restoration, endeavoured to settle it upon its ancient basis; to which end he gave them a charter, containing not only a confirmation of their old one, but also several new articles of regulation. By this the company is erected in a body politic, capable of making laws, &c under the title of the Company of Merchants of England trading to the Seas of the Levant. The number of members is not limited, but is ordinarily about 300. The principal qualification required is, that the candidate be a freeman of London, and a wholesale merchant, either by family, or by serving an apprenticeship of seven years.

The East India Company. The first, or, as it is called, the Old East India Company, was established by charter from Queen Elizabeth, in 1600; but, for some time, the partners seem to have traded with different stocks, though only in the ships belonging to the whole company.

In 1612, they joined their stocks into one common capital, and though their charter was not as yet confirmed by Act of Parliament, it was looked upon at that early period to be sufficiently valid, and nobody ventured to interfere with their trade. At this time their capital amounted to about 740,000*l.* and the shares were as low as 50*l.* Their trade was, in general, successful, notwithstanding some heavy losses, chiefly sustained through the malice of the Dutch East India Company. In process of time, however, it came to be understood, that a royal charter could not by itself convey an exclusive privilege to traders; and the company was reduced to distress by the multitude of interlopers who carried off their trade. This continued during the latter part of the reign of Charles II. the whole of that of James II. and part of William III. when, in 1698, a proposal was made to Parliament, for advancing the sum of 2,000,000*l.* to government, on condition of erecting the subscribers into a new company, with exclusive privileges. The old company endeavoured to prevent the appearance of so formidable a rival, by offering government 700,000*l.* nearly the amount of their capital at that time; but such were the exigencies of the state, that the larger sum, though at 8 per cent. interest, was preferred to the former, at one half the expense. Thus were two East India companies erected. Through the negligence of those who prepared the Act of Parliament, the new company were not obliged to unite in one joint stock. The consequence was, that a few private traders, whose subscriptions scarcely exceeded 7200*l.* insisted on a right of trading separately, at their own risk. Thus a kind of third company was established; and, by their mutual contentions, all the three were brought to the brink of ruin. Upon a subsequent occasion in 1700, a proposal was made to Parliament for putting the trade under the management of a well regulated company, and thus laying it in some measure open. This, however, was opposed by the company, who represented in strong terms the mischiefs likely to ensue from such a proceeding.

In 1702 the companies were, in some measure, united by an indenture tripartite, to which the Queen was the third party: and, in 1708, they were, by Act of Parliament, perfectly consolidated into one company, by their present name, "The United Company of Merchants trading to the East Indies." In this act it was thought worthy to insert a clause, allowing the separate traders to continue their traffic till 1711; but, at the same time, empowering the directors, upon three years notice, to redeem their capital of 7,200*l.* and thereby convert the capital of the whole company into a joint stock. By the same act, the capital of the company, in consequence of a new loan to government, was augmented from 2,000,000*l.* to 3,200,000*l.*

In the year 1793, when the company applied to government for a renewal of their charter, petitions were sent up from all the principal trading towns against it: but his Majesty's government, after mature deliberation, were, at that time, convinced that the regulated monopoly of the company was essential to the interests of the country, and thought that consideration alone a sufficient answer to all pretences for interference with that monopoly. Now, their charter being again nearly expired, the court of directors have again solicited for a renewal of it; and state, that, unless this takes place, advantages of the highest importance will be lost to the state: they have also distinctly stated to government that, at the beginning of the year 1813, they have

1. A marine of 100 large ships, containing a tonnage of 103,333 tons, employing 1400 officers and 10,000 seamen, of acknowledged skill and bravery, and which has been found ready and able, at all times, to minister to the national convenience, and to augment the national glory.

2. An extensive establishment of yards, which has often rendered large and seasonable assistance to the royal navy, and might, in cases of necessity, be made available to the national defence.

3. The remittance of the fortunes accumulated by above 3,000 officers of the company's military service, and some hundreds of civil servants, which are transferred to the capital, and stimulate the industry of Britain.

That, in addition to these considerations, the liberal disbursements of the company for national enterprises, and the meritorious contribution of their service to the National fund of strength and glory, are entitled to a just and honourable remembrance, in the preparation of any measure which may vitally affect their interests.

The following particulars, from the close of the Fourth Report of the Committee of the House of Commons, appointed in the late Parliament, are confidently appealed to, for a proof, not only that the company have carried on their affairs without any expense to the finances of this country, but that the operations of their system have been manifold ways beneficial to its interests.

1st. That, since 1791, British industry has been encouraged by the employment of 46,000,000*l.*

2d. That the produce and manufactures of India, purchased with this sum, together with those of China sold in England, realized in sale, amount to nearly 140,000,000*l.*

3d. That the purchase of the produce and manufactures of England amounted to 29,200,000*l.*

4th. That the employment of British shipping amounted to 25,000,000*l.*

5th. That the duties on imports, collected through the company at a trifling expense to government, amounted to 39,300,000*l.* and, on exports, to 660,000*l.*—together, 39,960,000*l.*

6th. That the combination of these and other sums, producing 185,960,000*l.* shews that, on an average of the last seventeen years, 10,900,000*l.* have been diffused in various channels, through the whole circulation of the British empire, &c.

But, for proof of the moral and political benefits derived to the population of India from the government of the company, this court appeals to the Fifth Report of the Committee of the House of Commons, which exhibits the gratifying picture of a people raised from the lowest disorder and degradation, to a state of industry, security, and freedom; and that this court confidently hopes, the improvement in the condition of fifty millions of natives will not be interrupted, by the pernicious experiments to which the sanction of Parliament is now so loudly solicited by petitions from almost every quarter.

That the interests of the company are now become so interwoven with the political and commercial system of Britain, that, by their sudden dissolution, the credit and grandeur of this country would receive a shock, which *must* be perilous, and might be fatal.

That, with such a case in favour of the company, the court is assured they may approach Parliament with confidence, and without the smallest alarm from the misrepresentations with which they have been assailed; and they are persuaded, that the wisdom of that enlightened body will never countenance the sacrifice of clear and positive interests of one class of men, to the contingent and uncertain advantage of another; nor demolish a mighty practical system, which has been raised by such immense exertions, in order to place its materials at the disposal of interested speculations.

Our commerce with India, now more than a hundred years, has been exclusively conducted by the said company, and has produced great national advantages, by the vast importation of eastern produce, by the extended exportation of our manufactures, and by yielding an increasing revenue, of certain and easy collection. Since the establishment of this company by Queen Anne, its prosperity has, in many instances, been subsidiary to the necessities of government, and, under its auspices, an empire has been acquired and consolidated, more important for its extent, opulence, and population, than the conquests of any other nation.

The company have established a school and a college at Hertford, where students are properly qualified for the situations of writers and civil officers in the service of the United East India Company of Merchants. They have also established a military school at Addiscombe, near Croydon, in Surrey, for the education of young gentlemen cadets for the artillery, engineer, and infantry corps in India.

The Company of Merchants trading to Africa, was established in 1750. Contrary to the former practice with regard to regulated companies who were reckoned unfit for such sort of service, this company was subjected to the obligation of maintaining forts and garrisons. It was expressly charged at first with the maintenance of all the British forts and garrisons that lie between Cape Blanc and the Cape of Good Hope; and, afterwards, that of those only which lie between Cape Rouge and the Cape of Good Hope. The act which establishes this company (the 23d of Geo. II. c. 31.) seems to have had two distinct objects in view; first, to restrain effectually the monopolizing spirit which is natural to the directors of a regulated company; and, secondly, to force them, as much as possible, to give an attention, which is not natural to them, towards the maintenance of forts and garrisons. For the first of these purposes the fine for admission is limited to forty shillings. The company is prohibited from trading in their corporate capacity, or upon a joint stock; from borrowing money upon a common seal; or, from laying any restraints upon the trade, which may be carried on freely from all places, and by all persons, being British subjects and paying the fine. The government is in a committee of nine persons, who meet in London, but who are chosen annually by the freemen of the company at London, Bristol, and Liverpool; three from each place. No committee-man can be continued in office for more than three years together. Any committee-man might formerly be removed by the Board of Trade and Plantations, now by a Committee of Council, after having been heard in his defence. The company are forbidden to export negroes from Africa, or to import any African goods into Great Britain; but, as they are charged with the maintenance of forts and garrisons, they may, for that purpose, export, from Great Britain to Africa, goods and stores of different kinds. Out of the money which they shall receive of the company, they are allowed a sum not exceeding 800*l.* for the salaries of their clerks and agents at London, Liverpool, and Bristol; the house-rent of their offices in London, and all other expenses of management, commission, and agency in England. What remains of this sum, after defraying those different expenses, they may divide among themselves, as a compensation for their trouble, in what manner they think fit. For the second purpose mentioned (the maintenance of the forts and garrisons), an annual sum has been allotted to them by Parliament, generally about 13,000*l.* For the proper application of this sum, the committee is obliged to account annually to the cursitor baron of the exchequer, which account is afterwards to be laid before Parliament.

Merchant-Man, (*navire marchand*, Fr.) a trading ship employed in importing and exporting goods to and from foreign countries.

The captain of a merchant-ship should know the relation between the money, weights, and measures, of several countries; the goodness and value of every thing requisite for the construction or refitting of vessels; the prices of wood, cordage, masts, anchors, sails, &c. He should also have a knowledge of foreign languages, and the laws, customs, &c. of different countries. For farther particulars see the article Master *of a Merchant Ship.*

Merchant-Service, (*marine marchande*, Fr.) in commerce, a marine service of Great Britain, which is set apart from the royal service, and carried on to a great extent, particularly in time of peace, under the auspices of government. Merchant ships are, in time of hostilities, protected by our ships of war, when sailing to and from foreign countries, for the purpose of trafficking and carrying on their commercial dealings. For farther particulars relative to this article, see Bill of Lading, Charter-Party, Importation, Marine-Insurance, Merchant-Company, Master of a Merchant-Ship, and Pilot.

MERCURY, in astronomy, the smallest of the inferior planets, represented in astronomical books by this character ☿, is about 3224 English miles in diameter, and revolves round the sun at the distance of 37 millions of miles. He emits a brilliant white light, and twinkles like the fixed stars; though, by reason of his always keeping near the sun, he is seldom to be seen; and, when he does make his appearance, his motion toward the sun is so swift, that he can only be discerned for a short time. He appears a little after sun-set, and again a little before sun-rise. Mercury never goes to a greater distance from the sun than about 27° 5′; so that he is never longer in setting after the sun than an hour and fifty minutes; nor does he ever rise sooner than one hour and fifty minutes before that luminary. Very frequently, he goes so near the sun as to be lost altogether in his rays. When he begins to make his appearance in the evening, after sun-set, he can scarcely, at first, be distinguished in the rays of the twilight. But the planet disengages itself more and more, and is seen at a greater distance from the sun every successive evening; and, having got to the distance of about 22° 5′, it begins to return again. During this interval, the motion of Mercury, referred to the stars, is direct; but, when it approaches within 18 degrees of the sun, it appears for some time stationary; and then its motion begins to be retrograde. The dazzling splendour of his rays, the shortness of the interval during which observations can be made upon his disc, and his proximity to the vapours of the horizon, when he is observed, have prevented astronomers from making any interesting discoveries respecting this planet. When the planet passes the sun in its retrograde motion, its diameter is the greatest possible; when it passes the sun in its direct motion, it is the smallest possible. Sometimes, when the planet plunges into the sun's rays in the evening, it may be seen crossing the sun, under the form of a black spot, which describes a chord along the disc of the sun. This black spot is recognized to be the planet, by its position, its apparent diameter, and its retrograde motion.

These transits of Mercury, as they are termed, are real annular eclipses of the sun: they demonstrate that the planet is an opaque body, and that it borrows its light from the sun.

Dr. Herschel has frequently examined Mercury with telescopes magnifying 200 and 300 times, and he always appeared equally luminous in every part of his disc, without any dark spot or ragged edge: but he exhibits the same difference of phases with the moon, being sometimes horned, sometimes gibbous, and sometimes shining almost with a round face, though not entirely full, because his enlightened side is never turned directly towards us. Like the moon, the crescent is always turned towards the sun. These different phases throw considerable light on the orbit of Mercury. By examining the variation in the daily appearance of Mercury's horns, Schroeter found the period of his diurnal rotation about his axis to be 24 days, 5 hours, and 28 minutes. Wallot imagined that Mercury had a horizontal refraction of 276″; but Bugge, when observing the transit of this planet in 1802, could perceive no traces of an atmosphere. See the Edinburgh *Encyclopædia*, and Nicholson's *Encyclopædia.*

MERIDIAN, (*méridien*, Fr.) in astronomy, is a great circle passing through the poles of the world, and both the zenith and nadir crosses the equinoctial at right angles, and divides the sphere into two hemispheres, termed the eastern and western: it has its poles in the east and west points of the horizon.

It is called meridian, from the Latin *meridies*, because, when the sun appears to come to this circle above the horizon, it is then mid-day; and then the sun has his greatest altitude for that day. On this circle the latitude of a place is reckoned, being always equal to the elevation of the pole above the horizon, or to the distance of the zenith from the meridian.

These meridians are imaginary lines, and change according to the longitudes of places; so that they may be said to be infinite in number, for all places from east to west have their several meridians: but different nations have one fixed, which is called the first meridian, the fixing of which is a matter merely arbitrary; and hence different persons, nations, and ages, have fixed it differently.

Geographers and map makers often assume the meridian of a place where they live, or the capital of their country, or its chief observatory, for a first meridian; and from thence reckon the longitudes of places, east and west. Astronomers, in their calculations, usually choose the meridian of the place where their observatories are made, for their first meridian; as the Royal Observatory at Greenwich, and the Observatory at Paris.

Meridian *Altitude of the Sun and Stars*, is their altitude when in the meridian of the place where they are observed. Or it may be defined, an arc of a great circle perpendicular to the horizon, and comprehended between the horizon and the sun or star then in the meridian of the place.

To take the meridian altitude with a quadrant.—If the position of the meridian be known, and the plane of an astronomical quadrant be placed in the meridian line, by means of the plumb-line suspended at the centre, the meridian altitudes of the stars, which are the principal observations whereon the whole art of astronomy is founded, may easily be determined. The meridian altitude of a star may likewise be had by means of a pendulum-clock, if the exact time of the star's passage over the meridian be known. Now, it must be observed, that stars have the same altitude for a minute before and after their passage by their meridian, if they be not in or near the zenith; but if they be, their altitudes must be taken every minute when they are near the meridian, and their greatest altitudes will be the meridian altitudes sought. See the article Altitude, *Meridian.*

Ante-Meridian, or A. M. (*avant midi*, Fr.) signifies before noon.

Magnetical Meridian, is a great circle, passing through the magnetical poles, to which the magnetic needle, or needle of the mariner's compass, conforms itself.

Post-Meridian, or P.M. (*après midi*, Fr.) after noon.

MERIDIONAL Distance, (*lieues mineures*, Fr.) in navigation, is the same with departure, easting or westing, or the difference of longitude between the meridian under which the ship now is, and any other meridian she was before under.

Meridional *Parts*, *Miles*, or *Minutes*, (*latitudes croissantes*, Fr.) in navigation, are the parts by which the meridians in Mr. Wright's chart, (commonly called Mercator's) increase as the parallels of latitude decrease: and, as the co-sine of the latitude of any place is equal to the radius or semi-diameter of that parallel; therefore, in the true sea-chart, or nautical planisphere, this radius being the radius of the equinoctial, or whole sine of 90°, the meridional parts at each degree of latitude must increase, as the secants of the arc, contained between that latitude and the equinoctial, decrease. The tables, therefore, of meridional parts, which may be seen in books of navigation, are made by a continual addition of secants; they are calculated in some books for every degree and minute of latitude; and they will serve either to construct a Mercator's chart, or to work Mercator's sailing. For other uses, see ROBERTSON'S *Elements of Navigation.*

MESHES, a name given to the spaces between the lines of a netting.

MESS, (*ordinaire des officiers gamelle, ou plat des gens de l'équipage*, Fr.) among seamen, implies any company of the officers or crew of a ship, who eat, drink, and associate together.

MESS-MATE, (*matelot, compagnon d'ordinaire, ou matelot de plat*, Fr.) a companion or associate in such a company as above described.

MESSENGER, (*tournevire*, Fr.) in a ship, a large rope, used to unmoor or heave up the anchors, by transmitting the efforts of the capstan to the cable. This operation is performed by fastening one part of the *messenger* to the cable, in several places, by a particular kind of rope, called nippers, and by winding another part thereof three or four times about the capstan, which answers the same purpose as if the cable itself were in that manner wound about the capstan; and the messenger being much lighter and more pliant, it is infinitely more convenient. The messenger has an eye-splice at each end, through which several turns of a strong lashing are passed, forming a continuation, whereby a quantity passes forward on one side equal to what is hove in on the other. See the article VOYAL.

Light forward the MESSENGER! (*allège le tournevire!* Fr.) the order to pull the slack of it towards the hawse-holes, so as to be ready to fasten upon the cable which is being hove in.

METAL OF A SHIP, a term used by seamen for all the guns and carronades on board.

Thus, a large 74-gun ship carries about 173¼ tons of *metal*, besides copper sheets and nails on her bottom.

The number of guns being known, the weight of *metal* in a ship may easily be found, as their species and mean weight are marked on them and their carriages. See the table containing the species, length, and mean weight of all the guns and carronades used in the royal navy, under the article GUN.

To be laid under METAL, among gunners, implies the situation of a gun, when the mouth of it falls below an horizontal position.

METEOR, a term used by some writers, which comprehends all the visible phenomena of meteorology, but it is more generally confined to luminous bodies appearing suddenly at uncertain times, and with more or less of motion in the atmosphere. These may be reduced under three classes; viz. fire-balls, falling or shooting stars, and ignes fatui. See the articles CORPOSANT, WATER-SPOUT, &c.

METE-STICK, (from *meter*, to measure; and *stick*, from *steek*, Dut.) a staff of a certain length, fixed on a broad board at right angles, in order to measure or determine the necessary height of a hold, and to level the ballast: in merchant-ships one ought to be very particular, but in the royal navy a moderate attention is necessary in their use for stowing the ground tier, the others, consequently, arrange themselves.

MICROMETER, (μικρὸς and μέτρον, Gr. *micromèter*, Fr.) an instrument generally fitted to a telescope, in the focus of the object glass, for measuring the angles subtended by objects contained within the field view of the telescope, and is chiefly used to measure the apparent diameter of the sun, moon, or planets. This instrument has been much improved of late, and its use extended so as to be capable of measuring the distance of bodies at a considerable distance apart. There have been various contrivances for this purpose.

The first was invented by M. Huygens, a French mathematician and astronomer: his *micrometer* was composed of a fine wire, stretched across the eye-tube near the principal focus, and two straight plates with fine edges, placed at right angles to the wire, the one of them was fixed, and the other movable by means of a screw; and the whole rendered capable of being turned round into any position, at pleasure: whence, by opening the plates until the object or objects were contained between the inner edges of them, the indices would shew the distance. But to ascertain the distance, or make a scale for such micrometer, first, make the inner edge of the moveable plate to coincide with that of the fixed one, and set the index to zero; then separate them by any even number of revolutions, and observe the time any star is moving over that space when on or near the meridian: then 15′ multiplied by the time, and by the natural co-sine of the star's declination will give the angle corresponding to this distance of the plates in degrees, &c. in decimals; and, from thence a table may be made answering to any part by even proportion. Or, a table may be readily made to give the value of any part of the scale, by setting up an object of a known diameter, or two well defined objects, at a given distance from each other, and from the micrometer; turn the micrometer screw till the plates coincide with the outer edges, or just take in the objects, the number of revolutions, and parts will answer to the angle subtended, which is readily found by computation. Sir Isaac Newton remarked, that the measures, by the measure on the plates, were a little too large; in consequence, light objects seen on dark ones appeared too great, and dark objects on light ones appeared too little than when they were equally bright.

Dr. Bradley applied fine wires instead of plates, and also added a small motion by tooth and pinion, to set the wires accurately in any position; and also placed the wires in various positions, which rendered it more general and extensive. Notwithstanding all that has, or probably can be done, the wire micrometer is subject to some errors, that cannot easily be overcome.

The late Mr. John Dollond, from some hints he got from a Mr. S. Savary, of Exeter, invented an object glass micrometer, by cutting an object glass into two equal parts, and applying it to the eye-end of a Gregorian reflecting telescope; and the present Mr. Peter Dollond has improved it, and applied it to an achromatic telescope; and, in its present improved state, it is generally used by astronomers.

There have been various other micrometers contrived by Dr. Maskelyne, Mr. Ramsden, Dr. Herschel, &c. an account of them may be found in the *Philosophical Transactions.*—And also various methods of determining the difference of right ascension and declension of stars, by means of cross wires in the eye-tubes of telescopes.

By considering the principles on which these instruments are constructed, Dr. Brewster has found them liable to numerous sources of error; and, in order to remove these, and give simplicity to the micrometer, he has proposed to construct the instrument on optical instead of mechanical principles. See the *Edinburgh Encyclopædia*, Vol. II. parts 3 and 4.

MID-CHANNEL, (*milieu d'une passe, ou mi-canal*, Fr.) implies a way across, or in the middle of any channel, river, &c.

MIDDLE LATITUDE SAILING, in navigation, is an easy, though not strictly accurate method of resolving those problems in sailing, in which the course is neither upon a meridian nor a parallel, the earth being supposed to be a sphere. This sailing is a combination of plane and parallel sailing; and is so named, because the difference of longitude is reckoned upon the middle parallel, between the latitudes sailed from and come to.

The line which a ship describes, when sailing upon the same oblique rhumb, makes equal angles with every meridian; and, therefore, since the earth is a sphere, and the meridians meet at the poles, this line will be a curve of the spiral kind, and is called a *loxodromic*, or an oblique course.

The latitude of that parallel whose arc is equal to the departure, is not easily found with accuracy; and, therefore, various methods have been employed to find it nearly, with as little trouble as possible. That which is most generally known and practised, is, by taking half the sum of the latitudes sailed from and come to, for the latitude of the mean parallel. This latitude is, however, always less than the true latitude of the above parallel; yet, when the distance is not very considerable, the difference of longitude inferred therefrom, and the departure, in places near the equator, or when the course is above six or seven points, is very near the truth. In other situations, and when the distance exceeds 5 or 600 miles, this method is not to be depended on. If, however, the latitude of the place sailed from, or of that come to, be given, and also any two of these four; namely, the course, distance, difference of latitude, and departure, then the resulting problems in middle latitude sailing may be resolved with sufficient accuracy, by assuming a series of latitudes between the two extreme latitudes, at intervals of one or two degrees, and computing the difference of longitude answering to each pair of successive latitudes, and the other data, and the sum will be the whole difference of longitude. Other methods, besides that of taking half the sum of the latitudes for the middle latitude, proposed for resolving problems in this sailing, were, by using the latitude, inferred from the arithmetical mean of the co-sines of the two latitudes; by the geometrical mean of the co-sines of the latitudes; by the latitude deduced from half the sum of the meridional parts of the latitudes; and from the geometrical mean of the same parts, &c.

Hence, the various problems in *middle latitude sailing* may be resolved by connecting two right-angled triangles, the departure being a side common to both Thus, let the triangle ABC represent a figure in plane sailing, in which AC is the distance, AB the difference of latitude, BC the departure, and the angle A the course. Again, let DBC be a triangle in parallel sailing, in which DC is the difference of longitude, BC the departure, and the angle BCD the middle latitude. In one of these triangles there will be two terms given, and in the other triangle, one term, to find the required parts: and that triangle is to be resolved first, in which two parts are given; and hence, the unknown parts of the other triangle are easily found. If the departure be neither one of the given, nor required terms, the whole triangle may be considered as an oblique-angled triangle, and resolved as such. Thus, if the course, distance, middle latitude, and difference of longitude, are the terms concerned, any three of which being given, the other may be found by one proportion: for, in the triangle ACD, wherein AC is the distance, the angle A the course, BCD the middle latitude, and consequently ADC the complement of the middle latitude, and DC the difference of longitude; it is scarcely necessary to observe, that the middle latitude is found by taking half the sum of the latitudes of the two places, if they are upon the same side of the equator, or half their difference, if on opposite sides. For resolving the different cases in this sailing, see MACKAY's and other common *Treatises on Navigation.*

MID-HEAVEN, *Medium Cœli*, in astronomy, is that point of the ecliptic which culminates, or is highest, or is in the meridian at any time.

MIDSHIP, (*milieu du vaisseau*, Fr.) a term applied by shipwrights to several pieces of timber, which lie in the broadest part of the vessel; as,

MIDSHIP-BEAM, (*maître-bau*, Fr.) in ship-building, the beam upon which the extreme breadth of a ship is formed, and which is situated in the *midship-frame*, nearly in the middle of her length, serving as a standard from whence the dimensions and proportions of the masts and yards are to be taken.

MIDSHIP-BEND, (*maitre gabarit, ou contour du maitre couple*, Fr.) in ship-building, is the broadest frame in a ship, called *deadflat.*

MIDSHIPMAN, (*garde de la marine, ou aspirant*, Fr.) in the royal navy, is a young gentleman generally appointed by the captain of a man of war, as a junior officer, from the first class of volunteers, to second the orders of the lieutenants and other officers, and assist in carrying on the necessary duty of the ship, either aboard or on shore, at the dock-yards, &c. Formerly, *midshipmen* were entered in the service at a very early age, but, by a late regulation no young gentleman can be borne on any ship's books until he is *thirteen* years of age, unless he is an officer's son, in which case he may be entered at *eleven;* and it is also thought expedient that he should serve at least two years on board as a volunteer, before he can be rated midshipman, except he has been educated at the Naval College, where he is allowed two years' time, and is supposed to have had opportunities afforded him of learning much of the duty expected of a midshipman, and in that case is eligible to the situation on his first going to sea.

The number of midshipmen in a man of war, like that of many other officers, is always in proportion to the size of the ship; viz. a first rate is allowed to have 24; a second rate, 18; a third rate, 12; and so on in proportion.

All the young gentlemen of the first class of volunteers are allowed to walk the quarter-deck, and are considered as looking forward to become rated midshipmen; therefore, are permitted to wear the naval uniform from their first entering into the service. During their servitude as volunteers they are generally styled *youngsters*; and are, in most ships, put under the care of the gunner, who is, generally, a steady officer, and who caters for them, looks after their clothes, and superintends their mess: if they have not finished their nautical studies, they are also put under the care of the school-master, for some hours every morning; but, where there is no school-master, the captain frequently takes that trouble on himself; and the rest of the day the youngsters are exercised in learning the use of the ropes, rigging, &c.

When they are fifteen years of age, and get rated midshipmen, their pay is increased; at this period they become what is generally termed *oldsters* on board, and form a mess among themselves with the mates.

After serving two years as a midshipman, they become eligible to be promoted to the situation of master's-mate, provided they have been well instructed in the theory of navigation, and are become expert in seamanship, as master's-mates have much duty to attend to, are often placed in responsible situations, in charge of boats, and frequently entrusted to navigate prizes into port. For farther particulars respecting their duty, see the article MASTER'S-MATE.

At the age of nineteen, provided they have been constantly at sea for the last six years, and can pass their examination in navigation and seamanship, they become eligible to be made lieutenants, and to hold a commission in the navy, whether they are in the situation of master's-mates or midshipmen. Formerly, after having served six years, they were obliged, at much inconvenience and expense, to repair to the Navy-office in London, to pass their examination; but they are now very properly permitted to be examined before the three senior captains at any of the regular sea-ports, where they may happen to be at the expiration of their six years servitude; and, on the home station, the examinations take place on the first Wednesday of every month. This is certainly a very proper regulation, and not only advantageous to the midshipman, but beneficial to the service in general.

Mates and midshipmen, on going to pass their examination, are expected to be prepared with a *regular certificate*, that they are actually nineteen years of age (as, by an Order in Council, it is stated that they shall forfeit their rank, whatever it may be in the service, should it appear, at any future period, that there was any deception used in this certificate); and an abstract of their time of servitude, taken from the different ship's books at the Navy-office, they may have served on board of. They are also expected to shew their log-books, for, at least, the two last years, fairly written and correctly kept, and are required to produce certificates from the captains with whom they have sailed during the whole time of their servitude, not only expressive of their having been diligent and attentive to the duties of their profession, but that they have at all times been obedient to the commands of their superior officers.

Without these documents are produced they cannot be examined for commissions. They are expected to be well acquainted with every thing relative to navigation and seamanship; are strictly examined in the different sailings, working tides, day's works, and double altitudes; and are expected to give some account of the different methods of finding the longitude by a time-keeper and the lunar observations. In practical seamanship they are expected not only to be able to conduct a ship from one place to another, under every disadvantage of wind, tide, &c. but also to give an account how they would proceed in cases of danger and distress, from the loss of masts, rigging, &c. After this examination is gone through in a satisfactory way, the candidate obtains a certificate from the captains that he is duly qualified to hold a commission as a lieutenant, whenever it may please the Lords Commissioners of the Admiralty to grant him one.

This is a most anxious period with the young naval hero; and the step from a midshipman to a lieutenant is generally considered the greatest in the service. But it often happens that, after going through their regular servitude and examination, even with credit, they are obliged to remain in the situation of midshipmen and mates for years afterwards, for want of opportunities and friends to get them promoted; and, at the beginning of the year 1813, there were nearly 2000 young gentlemen in the service, who had not only served their time, but passed their examination for lieutenancies.

Though this is much to be lamented, yet, it must be considered, that, from the great magnitude of our present naval force, and the numbers of mates and midshipmen employed, many of them must, in consequence, remain a considerable time after they pass their examinations for lieutenants, before they can be promoted, or, indeed, before vacancies can happen, even though they should be fortunate in having friends to second their views, and, eventually many will be obliged to accept of warrants instead of obtaining commissions.

The circumstance of there being so many candidates for promotion in the navy has arisen from the long protracted war, and from the enterprising spirit and brilliant achievements of our gallant naval heroes (who have long excited the astonishment of surrounding nations, and the admiration of a grateful country), which have inspired so many of our youths with an ardent desire to embark in the profession, and enrol their names in the list of fame, in hopes of obtaining a sprig of those laurels that overshadow the tomb of our illustrious Nelson of Trafalgar; yet this will be a very serious consideration to the friends and relatives of those young men who do not get promoted before a peace takes place, and a reduction made in our naval establishments, which must necessarily follow; as, from their early habits and confined education, they cannot be expected to be eligible for many situations on shore; it is, therefore, sincerely to be hoped, that the Admiralty may take some step, when a peace is likely to take place, either to promote or find employment for so numerous and respectable a class of young officers; many of whom, whenever opportunities are afforded them, will, in all probability, do honour to the profession they have chosen, and credit to the country in which they have been brought up.

Midshipmen have many little hardships to struggle with, and difficulties to overcome, on their first entering into the service: these arise partly from the nature of the service they have to perform, and from the description of people they have to associate with. They are necessitated to be much with the common seamen, in order to see them perform their duty and to learn their own. Many of our officers, who have their youngster's welfare at heart, procure school-masters, and see them regularly instructed every day in the theory of navigation, &c. during the two first years of their servitude: and some officers will not take any young gentleman who has not been regularly educated for the service. Were this more strictly adhered to by captains in general, it would tend much to the respectability of the service, and put a stop to that facility with which boys, who become tired of study at school, now get into the service, whenever they can get friends to recommend them to a captain.

No young gentleman ought ever to be received as a volunteer of the first class, unless he has had at least, a good English education, so as to be able to read, and spell correctly, and has gone through the English grammar, and is able to construe a sentence in any common English author. He ought also to be able to write a good plain hand, and be well acquainted with vulgar and decimal arithmetic, as almost all nautical calculations are performed by logarithms, which is a species of decimals: and it is strongly recommended that he should be acquainted with the French grammar, and know something of drawing before he goes to sea.

His friends should also be expected to allow him from thirty to fifty pounds per annum, besides his pay, for his mess, clothes, and other expenses. This sum should always be paid in advance to the captain's agent in London, and then he will have it in his power to furnish the youngster with money, or to order him clothes whenever he sees him in want of them.

While considered youngsters, they ought on no account to have the care of money themselves, as it would, probably, be the means of leading them into every kind of vice and idleness when they get on shore, and can be of little use to them on board: should they not be inclined to spend it themselves, the senior midshipmen will soon induce them to do so, or else contrive, on some pretence, to borrow it of them.

For the information of parents and guardians living remote from sea-ports, it may be necessary to touch upon the expense of fitting out a young gentleman on his first entering into the sea service. Should the ship in which he is about to embark be on the home station, or on Channel service, the expense will generally be from 70 to 80*l.* and, for a foreign station, from 90 to 100*l.* including his entrance to the mess, which generally comes to from eight to twelve pounds, and pocket money to pay for his washing, and other contingent expenses. Afterwards, he should only depend on his pay and what is allowed by his friends, which he ought to receive through the hands of the captain, as, by that means, he will be more likely to avoid extravagance, and learn to set a proper value on money when he receives it through that channel. Captains differ in opinion, both as to the quantity of clothes and the annual allowance for their young gentlemen: some seem to think thirty pounds a year sufficient, others deem forty quite little enough, and many require their friends to allow them fifty pounds a year, who are desirous that they should appear at all times like young gentlemen. Their pay as volunteers is only 9*l.* a year, from which 5*l.* are deducted for the schoolmaster, when he is chaplain of the ship; and they do not receive any part of it till they have been twelvemonths at sea, and then only six months; as, according to the rules of the service, there is always six month's pay kept back from the petty officers and seamen. Almost all captains have a list of what clothes, &c. they wish their youngsters to have, and it will be advisable for their friends to apply to the captain, who generally has some regular tradesmen at most of the sea-ports, that he recommends to fit them out, and who knows exactly what he wishes them to have, and will get every thing ready for them in a few days, without any sort of trouble to their friends.

The following is a general inventory of what is mostly required at first going to sea:—

1 Uniform jacket or coatée suit, superfine cloth.
1 Ditto, ditto, second best cloth.
1 Round jacket suit, ditto.
1 Surtout coat, and a watch-coat.
3 Pair of white jean trowsers and waistcoats.
3 Pair of nankeen trowsers, and 3 kerseymere waistcoats.
2 Round hats, with gold loop and cockade.
1 Glazed hat, hanger and belt.
18 Linen shirts, frilled.
12 Plain calico shirts.
3 Black silk handkerchiefs.
18 Pocket cotton ditto.
12 Pair of brown cotton stockings.
6 Pair of white cotton ditto.
6 Pair of worsted, or lamb's wool ditto.
2 Strong pair of shoes, and 2 light pair of ditto.
6 Towels, and 3 pair of sheets and pillow-cases.
2 Table-cloths, about three yards long.
A mattress, 3 blankets, and a coverlet.
A set of combs and clothes brushes.
A set of tooth-brushes and tooth-powder.
A pewter wash-hand-basin, and a pewter cup.
A set of shoe-brushes, and a dozen cakes of blacking, or half a dozen bottles of ditto.
A strong sea-chest, with a till and two trays in it, and a good lock, with two keys.
A quadrant, and a small day and night glass.
A silver table-spoon and tea-spoon.
A knife and fork, and a pocket-knife and pen knife.
A log book and journal, with paper, pens, and ink

ROBINSON's *Elements of Navigation*; or MACKAY's, NORIE's, or MOORE's *Epitome* of ditto.
The Requisite Tables and Nautical Almanac.
A Bible, Prayer Book, and an Introduction to the Christian Religion.

The above articles and apparel are deemed absolutely necessary on a home station; and, when going on a foreign one, an additional dress-suit must be added, with more light waistcoats and trowsers, a cocked hat, and some additional linen.

The following books are recommended as concise abridgements in one volume each, and well calculated for youth going to sea, who have but little room for larger works:—

The *British Neptune*, or *An Abridgement of Naval History*.
The *Naval Heroes*, or *Lives of Celebrated Seamen*.
ROBINSON's *Abridgement of Modern History*.
ROBINSON's *Abridgement of Ancient History*.
GREGORY's *Polite Education*.
GREGORY's *Scripture History*.
WOODARD's *Narrative of Shipwrecks*, &c.

As parents and guardians are frequently at a loss to know what class of ships to place their sons and wards in, on their first going to sea, it may be proper here to observe, that it is generally allowed a ship of the line is preferable to any other where there is a school-master, and they have also much more room and many more comforts in a line-of-battle ship; and, of course, get sooner reconciled to the nature of the service.

After being properly qualified to do their duty as midshipmen, it is thought desirable by many that they should get removed to frigates and small vessels, in order to learn their duty as seamen, during the last two or three years of their servitude; as, from the complement of midshipmen being small in all those classes of ships, they are always actively employed, and soon acquire a thorough knowledge of seamanship. For farther particulars see NAVAL OFFICER.

MIDSHIPMEN, in the honourable East India Company's service, are the junior officers on board the Company's ships: the situation is similar to that described in the preceding article, as they gradually rise from midshipmen, to the rank of mates and commanders.

MIDSHIP-SECTION, or MIDSHIP-FRAME, (*maître-couple*, Fr.) a name given to that timber, or combination of pieces, formed into one frame, which determines the extreme breadth of the ship, as well as the figure and dimension of all the inferior timbers.

In the article NAVAL ARCHITECTURE, the reader will find a full explanation of what is meant by a frame of timbers. He will also perceive the out-lines of all the principal frames, with their gradual dimensions, from the midship-section, delineated in the plane of projection annexed to that article. As the parts, of which the several sections are composed, have the same relation to each other throughout the vessel; and as all the corresponding pieces, without and within those sections are also nearly alike, and fixed in the same manner, it will be sufficient for our purpose to represent the principal, or midship-section, together with its corresponding parts, which are as follow:—

Explanation of the MIDSHIP-SECTION, plate VI. fig. 1, which exhibits a transverse section of a 74-gun ship, at the broadest part, answering to the same scale by which are delineated the head, quarter, and stern of a ship, of the same size, in plates XI. &c. to which the reader is referred.

A the keel, with *a* the false keel beneath it.
B the kelson.
C one of the beams of the orlop.
D one of the lower-deck beams; with *b*, the beam of the upper-deck.
E the hanging-knees, by which the beams are attached to the timbers.
F a common standard.
G the clamps, which sustain the extremities of the beams.
H the gun-ports of the lower-deck; with *c*, the ports of the upper-deck.
I K the different pieces of *thick stuff*, placed opposite to the several scarfs, or joinings, in the frame of timbers.
L the rising, or dead-wood.
M the planks of the decks.
N the water-ways.
O the planks of the ceiling, between the several ranges of thick stuff.
P the spirkettings.
Q the main-wale, to fortify the ship's side opposite to the lower-deck.
R the channel-wale, opposite to the upper-deck.
S the sheer-strake.
T the string.
V the floor-timbers, which are laid across the keel, and bolted to it.
U the angle-chocks, at the head of the floor-timber and heel of the second futtock.
W the angle-chocks, at the head of the first futtock and heel of the third futtock.
X the angle-chocks, at the head of the second futtock and heel of the fourth futtock.
Y the angle-chocks, at the head of the third futtock and heel of the top-timber.
Z a cross chock, at the heels of the first futtocks.
d the coamings to the hatchways on the different decks.
e the floor-rider.
f the second futtock-rider.
g the shell-pieces under the beam.
h the chock under the shelf.
i the fillings between the futtock-riders, for faying the ship to.
k k the chocks under the beams, in lieu of wood knees.
l l the plate-knees, with ears, which secure the beams to the ship's side.

MILITARY ACADEMY. The Royal Military Academy at Woolwich, for the engineer and artillery department of the army, is fully described under the article ACADEMY.

MILITARY *Architecture*, otherwise more usually called *fortification*, is the art of strengthening and fortifying places, to screen them from the insults or attacks of enemies, and the violence of arms; by erecting forts, castles, and other fortresses, with ramparts, bastions, &c. The authors who have chiefly excelled in this art are, Coehorn, Pagan, Vauban, Scheiter, Blondel, and Montalembert.

MILITARY *Asylum*. The Royal Military Asylum, for the children of soldiers of the regular army, is situated near the Royal Hospital at Chelsea.

The first stone of this elegant structure was laid by His Royal Highness the Duke of York, on the 19th day of June 1801.

The motives which gave rise to this establishment, and the principles upon which it is founded, are alike honourable to the present age, and congenial with the soundest maxims of policy, humanity, and benevolence. The necessity, likewise, of such an institution, will appear obvious to all, when we consider the helpless and forlorn condition of many of these orphan objects of commiseration, who, in this comfortable asylum, are now well clothed, have good wholesome food, acquire a decent education, are taught the principles of Christianity; and, finally, become useful members of society, in whatever situation it may please Providence to place them.

The boys wear red jackets, blue breeches, blue stockings, and black caps; and are taught to read, write, and cast accounts, and are admitted from the age of five to twelve years.

The girls wear red gowns, blue petticoats, straw bonnets, and white aprons; they are taught to read, write, and cast accounts, they are also taught knitting and needle-work of different kinds, and are constantly employed in all manner of household-work, and admitted from five to ten years of age. Children under the age of five years will not be admitted, except when belonging to *regiments ordered on foreign service*, or in the case of orphans, or under other circumstances of *peculiar distress*, which must be *specially stated.*

It was originally intended that the number of boys admitted should not exceed seven hundred, and that the girls should not exceed three hundred; making in all one thousand children.

The affairs of the Royal Military Asylum are regulated by commissioners appointed by the King's sign manual, who hold four quarterly meetings or boards yearly, when the general business transacted is, the appointment of officers, approving of contracts, examining of accounts, and application to Parliament for the sums of money necessary for the support of the institution.

The board also selects the children for admission, who are chosen with the following regulations:—1st. Orphans, or those whose fathers have been killed, or died on foreign stations; or those who have lost their mothers, and whose fathers are ordered on foreign service, or whose parents have other children to maintain. The merit of the father, as to regimental character, is always considered as a principal recommendation. None are admitted but children born in wedlock, of warrant and non-commissioned officers and soldiers of the regular army, whose fathers, if living at the time of application, must either be still in the *regular service*, or *out-pensioners*. Every child admitted must be free from mental or bodily infirmity. The parents or friends who apply for the admission of children, are requested to sign their consent to such children's remaining in the asylum as long as the commissioners may judge proper, and to their being disposed of, when of proper age, as apprentices or servants; or, if boys, to their being placed, with their own free consent, in the regular army as private soldiers.

The contemplation of this noble building is calculated to excite the liveliest sensations of gratitude towards those illustrious personages under whose immediate patronage it was first planned, and finally completed.

To the soldier it must continually afford the most pleasing prospects for the comfort and support of his infant children, and will induce him to fight, if possible, with greater confidence; at least, he will enter the field against the enemies of his king and country, with a full assurance that, if he falls in battle, his family will never feel the pangs of misery, famine, or want.

For farther particulars of this benevolent institution, and the internal regulations and management thereof, see FAULKNER'S *Historical Account of Chelsea.*

MILITARY *College*. See the article COLLEGE. This institution was removed from Marlow to Sandhurst, near Bagshot, in September 1812; where an elegant building has been erected by government for the accommodation of cadets for general service, with an excellent house for the governor, and extensive grounds for military evolutions.

MILITARY *School*, for the engineer, artillery, and infantry service of the honourable East India Company. See the article SCHOOL.

MILKY-WAY, VIA LACTEA, or GALAXY, a white and luminous zone, which encircles the heavens, and which appears to be the nearest of all the nebulæ. This track consists of an immense multitude of stars, seemingly very close together; whose mingled light gives this appearance of whiteness, by Milton beautifully described as a path "powdered with stars." See the article GALAXY.

MILL, in block-making, is a patent invention used by some block-makers; it consists of a large horizontal wheel, turned by horses, which, by means of other wheels, iron cranks, leather bands, &c. gives motion to different turning-lathes and frame-saws, that cut the sheaves, and turn the sheaves and pins.

MINUTE, in geometry, is the sixtieth part of a degree of a circle.

The *minute* of a degree is marked with an acute accent, thus ; the second by two, ″; and the third by three, ‴.

MIRACH, in astronomy, a bright fixed star, of the second magnitude, in the girdle of the constellation Andromeda, marked β in our catalogues. For the beginning of the year 1812, its

Right ascension is . . .	14°	48′	28″
Annual variation in A. R. .			49,9
Declination	34	31	21 N.
Annual variation in declination			+ 19,4

MIRROR, in catoptrics, any polished body, impervious to the rays of light, and which reflects them equally.

Concave MIRRORS *for Ship Lights*, in light-houses on sea-coasts, are generally made of copper, plated with silver, and polished in the curve of the parabola, by which their light neither spreads nor converges, but darts a cylinder of light to a vast distance.

Mirrors for the production of these lights are now generally placed in a circle, which has one in its centre, each being about twenty inches in diameter, with an Argand-lamp in its focus, supplied with oil from behind. The frame in which they are fixed stands perpendicularly to the horizon, on a shaft, united with a machine below, that turns the whole round in the space of one or two minutes, as circumstances require; hence, a cylinder of light, five feet and a half in diameter, sweeps the whole horizon, and by its intermission, cannot be mistaken for any other.

Lights of this nature have lately been exhibited in light-houses on the British coast; viz. at the Bell-rock, Caskets, Eddystone, Isle of Wight, Land's-end, Scilly Islands, Skerries Islands; and have even been copied by foreigners, and shown on the Continent, some stationary and others with motion. See the article Light-House, &c.

The first used of the concave mirrors was about the year 1770, to direct ships at Liverpool and Newcastle; into the first the road is very intricate, through sands and shoals, from Hoylake to the Black-rock. To direct mariners in the night, that ingenious and useful dock-master, Captain William Hutchinson, placed polished copper mirrors in light-houses, facing each turning of the Channel; so that the pilot made way towards the first light he saw, till he saw another towards his right or left; he then turned immediately towards the second light, and pursued that channel till another light broke in upon him, to which he then directed the ship. Thus is the harbour obtained by night as well as by day. At Newcastle the Tyne flows perpendicularly into the German Ocean, and vessels in the night might sail past its mouth, were it not for a copper mirror, that throws out a light into the sea, in the direction of the river; so that, as soon as that light is beginning to be crossed, the ship turns towards it, and is directed by it into the river.

MISSILE, any weapon thrown by hand, or projected, and which strikes at a distance from the moving power, as a dart, &c.

MIZEN, (from *misana*, Ital. *artimon*, Fr.) the aftermost or hindmost of the fixed sails of a ship, extended sometimes by a *gaff*, and sometimes by a *yard*, which crosses the mast obliquely; the fore-end reaching almost down to the deck, and the after-end being peeked up as high above the middle of the yard, which is attached to the mast: the head and fore-leech, or the mizen, are laced upon the gaff (or yard) and mast; and the sheet hauls out near the taffrail.

Mizen-Mast, (*mât d'artimon*, Fr.) the mast which supports all the after-sails.

The explanations of the rigging, yards, and sails of this mast being, in general, applicable also to the same furniture of both the other masts, the reader is therefore referred to the articles Shroud, Stay, Yard, &c. observing only, that the epithet of fore, main, or mizen, is added to each term, to distinguish them from each other.

MOLE, or Mole-Head, (*mole, ou jetée d'un port*, Fr.) a name given, in the Mediterranean, to a long pier, or artificial bulwark of masonry, extending obliquely across the entrance of a harbour, in order to break the force of the sea from the vessels which are anchored within.

Mole, (*mole*, Fr.) is also applied to the harbour or haven which is formed by the bulwark above-described.

MOMENTUM, in mechanics, signifies the same with *impetus*, or the quantity of motion in a moving body; which is always equal to the quantity of matter, multiplied into the velocity; or, which is the same thing, it may be considered as a rectangle, under the quantity of matter and velocity.

MONITION, an instrument issued from an Admiralty court.

MONOCEROS, the Unicorn, in astronomy, a constellation of the northern hemisphere, situated under Canis Minor, and formed by Hevelius out of the Stellæ Informes, or those which were not comprised within the outlines of any of the others. In Hevelius's catalogue this constellation contains 19 stars; but in the Britannic catalogue 31.

MONSOON, (*mousson*, Fr.) a name given to the periodical or trade winds, which blow regularly in certain latitudes of the Indian ocean. They continue five or six months invariably in one direction, and then alter their course, and blow, during an equal space of time, from a different point of the compass with the same uniformity. See the article Wind.

MONTH, (*mois*, Fr.) in chronology, the 12th part of the year, and is so called from the moon, by whose motions it was regulated. There are several denominations of months, such as lunar, solar, civil, and calendar.

Lunar Month, is that space of time which the moon occupies in passing from any determined point of the ecliptic, until she arrives again at the same point, and is either *illuminative*, *periodical*, or *synodical*.

Lunar Illuminative Month, is the space from the first time of the moon's appearance after new moons, to her first appearance after the new moon following. Hence, as the moon appears sometimes sooner after the new moon, and sometimes later; the quantity of the illuminative month is not always the same. By this month the Turks and Arabs reckon.

Lunar Periodical Month, is the time in which the moon runs through the zodiac, or returns to the same point again; the quantity of which is 27 days, 7 hours, 43 minutes, 12 seconds.

Lunar Synodical Month, called also a *Lunation*, is the time between two conjunctions of the moon with the sun, or between two new moons, the quantity of which is 29 days, 12 hours, 44 minutes, 12 seconds. See the article Moon.

The ancient Romans used lunar months, and made them alternately of 29 and 30 days: they marked the days of each month by three terms; viz. Calends, Nones, and Ides.

Solar Month, is the time in which the sun runs through one entire sign of the ecliptic; the mean quantity of which is 30 days, 10 hours, 29 minutes, 5 seconds; being the 12th part of 365 days, 5 hours, 49 minutes, the mean solar year.

Astronomical or *Natural* Month, is that measured by some exact interval corresponding to the motion of the sun or moon. Such are the lunar and solar months above-mentioned.

Civil or *Common* Month, is an interval of a certain number of whole days, approaching nearly to the quantity of some astronomical month, lunar or solar. Civil months are various, according to the astronomical month to which they are accommodated.

MOON, or Luna, ☾, in astronomy, a satellite, or secondary planet, belonging to the earth, about which she revolves in 27 days, 7 hours, 43 minutes, 12 seconds; and turns once round her axis exactly in the time that she goes round the earth, which is the reason that she shews always the same side towards us, and that her day and night, taken together, are just as long as our lunar month.

The moon, from her proximity to our globe, and next to the sun, is the most remarkable body in our system, both on account of its apparent magnitude and diversity of aspect, and also useful for the division of time.

The motion of the moon in its orbit about the earth, is from west to east, and its orbit is found to be inclined to the ecliptic. By observing in the evening the position of the moon with regard to the fixed stars, she will appear about 13 degrees farther east than she was on the preceding evening, and will perform a complete *sidereal revolution* from one fixed star to another, or from one point in absolute space to the same point again, in $27^d\ 7^h\ 43'\ 12''$. As the equinoxes during this period recede about 4″, a space which the moon will pass over in 7″ of time, her *periodic revolution*, with regard to the equinoxes, will be $27^d\ 7^h\ 43'\ 5''$. The nodes move backwards in the heavens $1°\ 26'\ 6''$ in the time of a revolution; and, since the moon passes through this space in $2^h\ 36'\ 49''$, her revolution, with regard to her nodes, will be $27^d\ 6'\ 23''$. The motion of the moon's apogee being about $3°\ 2'\ 38''$ direct in the time of a revolution, she will describe this space in $7^h\ 43'\ 12''$, and will perform her anomalistic revolution in $27^d\ 13^h\ 15'\ 53''$.

The revolution of the moon with regard to the sun, that is, the time which she takes to revolve from one conjunction to the next conjunction with the sun, is called her *synodical revolution*. While the moon is going round the earth, the earth and consequently the sun, advance about $29°\ 6'\ 25''$, a space which the moon passes over in $2^d\ 5^h\ 1^m$. Hence the synodical revolution will be $29^d\ 12^h\ 44'\ 12''$.

When the moon performs her monthly revolution, she presents various appearances to a spectator on the earth. When she is seen near the sun in the western part of the horizon, she appears like a bow or crescent of light. As her distance from the sun increases, the enlightened part of her disc gradually augments, till she presents her fully illuminated disc to the earth. The enlightened portion again diminishes, till she is seen in the morning, a little before sun-rise, in the form of a crescent; and these phases are regularly repeated in the same order during every revolution.

When the moon is new, or is in conjunction with the sun, her dark side is turned to the earth, and she is invisible. When she reaches her first octant, or has gone through one-eighth of her orbit, one-fourth of her enlightened side is turned to the earth, and she appears like a crescent.

When she reaches her first quarter, the half of her illuminated disc is seen at the earth, and she appears a half moon.

In her second quarter more of her enlightened hemisphere comes into view, and she appears gibbous.

When she is in opp sition to the sun, her enlightened side is completely turned to the earth, and she again appears full moon.

In her third quarter a part of the enlightened hemisphere is turned away from the earth, and the moon appears gibbous; and she again appears as a half moon.

In her fourth, or last quarter, she has again the form of a crescent, and disappears, after having completed her synodical revolution round the earth. The moon is said to be in syzigy when she is in conjunction or opposition with the sun.

By attending to the different positions of the earth with regard to the moon, it will appear that the earth exhibits similar phases to an inhabitant of the moon, with this difference, that the phases of the earth are always opposite to those of the moon. When the moon is full, the earth is invisible; when the moon is new, the earth is full; and, when the moon has $\frac{1}{12}$ of her diameter enlightened, the earth has $\frac{11}{12}$ of its disc illuminated. The light, therefore, in the dark part of the moon, arises from the light reflected from the earth, which is nearly full at the time when the secondary light of the moon is visible. The earth's surface being about thirteen times greater than that of the moon, it affords thirteen times more light to the moon than the moon does to us.

If we look at a large map of the moon, which exhibits even a tolerable representation of the lunar surface, we shall find that the eastern limb of the moon is separated from the central parts of her disc by darker regions, and that the luminous portion comprehended between these darker regions and the circular line which bounds her eastern limb, has actually the form of a bow, which is broadest towards her southern limb, and gradually diminishes in breadth towards her northern horn. The immediate cause, therefore, of the lucid bow, is to be sought for in the accidental circumstance of the moon's eastern limb being more luminous than the adjacent regions towards the centre. The central parts of the moon, indeed, are equally luminous with her eastern limb; but their brilliancy is impaired by their proximity to the illuminated portion: and, it is obvious, that this explanation of the phenomenon may be equally just, whether the secondary light of the moon is caused by phosphorescence, or by reflection from the earth.

When we look at the surface of the moon with a good telescope, we find that its appearance is wonderfully diversified. Besides the large dark spots, which are visible to the naked eye, we perceive extensive valleys and long ridges of highly elevated mountains, projecting their shadows on the plains below. Single mountains occasionally rise to a great height, while hollows, more than three miles deep, and almost exactly circular, are excavated in the plains. The margin of these circular cavities is often elevated a little above the general level, and a high eminence rises in the centre of the cavity. When the moon approaches to her opposition with the sun, the elevations and depressions upon her surface in a great measure disappear, while her disc is marked with a number of brilliant points and permanent radiations.

These various appearances have been accurately represented in maps of the moon's surface.

Excellent drawings of the moon were made by the celebrated astronomer Tobias Mayer, and by Mr. Russel; but the most accurate and complete that have yet been published, are those of the celebrated

Schroeter, who has given highly magnified views of several parts of the moon's surface.

Astronomers have not been content with merely inspecting the surface of the moon, they have even attempted to measure the height of the mountains, and the depth of her cavities; and on this point there is a difference of opinion, greater than might have been expected.

Dr. Herschel, however, has measured several of the lunar mountains with great care, and finds that, a few excepted, they, generally, do not much exceed half a mile. Before he measured them, they were reckoned much higher, being generally over-rated. He observes, that it should be examined whether the mountain stands on level ground, which is necessary that the measurement may be exact.

The lunar surface is not only diversified with rocks and cavities, but some parts of it might be clearly distinguished from others by their superior illumination. The dark parts of the moon's disc are always smooth, and apparently level; while the luminous portions are elevated tracts, which either rise into high mountains, or sink into deep and extensive cavities. The dark parts have, by some, been thought immense collections of water, and by others, to be only a great number of caverns and pits; but, when we examine the moon's disc with minute attention, we find that these obscure portions are not exactly level, like a fluid surface. The great irregularity of the line bounding the light and dark parts, on every part of the surface, proves that there can be no very large tracts of water, as such a regular surface would necessarily produce a line, terminating the bright part, perfectly free from all irregularity. Also, if there was much water upon its surface, and an atmosphere, as conjectured by some astronomers, the clouds and vapours might easily be discovered by our telescopes; but no such phenomena have ever been observed.

On April 19, 1787, $10^h\ 36'$ sidereal time, Dr. Herschel discovered three volcanoes in different places of the dark part of the new moon; two of them seemed to be almost extinct, but the third shewed an actual eruption of fire, or luminous matter, resembling a small piece of burning charcoal, covered by a thin coat of white ashes; it had a degree of brightness about it, as strong as that with which such a coal would be seen to glow in faint daylight. The adjacent parts of the volcanic mountain seemed faintly illuminated by the eruption. A similar eruption appeared on May 4, 1783. On March 7, 1794, a few minutes before eight o'clock in the evening, Mr. Wilkins, of Norwich, an eminent architect, observed, with the naked eye, a very bright spot upon the dark part of the moon; it was there when he first looked at the moon and the whole time he saw it, which was about five minutes; it was a fixed steady light, except the moment before it disappeared, when its brightness increased. The same phenomenon was also observed by Mr. T. Stretton, in St. John's-square, Clerkenwell, London; and several astronomers have observed the same phenomenon.

When the sun and moon rise above the horizon, or set below it, they generally appear much larger than they do when seen on the meridian; though it is certain that they subtend a smaller angle in the former case than in the latter, when measured with a micrometer. When we estimate the magnitude of remote objects, our judgment is formed from comparing it with adjacent objects whose magnitude is known. Thus, if we saw a man standing upon a rock in the middle of the sea, we may form a tolerable correct estimate of the size of the rock, by comparing it with the man, whose size may be reckoned about five feet eight inches; but if the rock is seen by itself, it is impossible to form any probable estimate of its size. Hence, a buoy, seen at sea, always appears much less than it really is. The same thing happens with regard to the estimation of distances.

For the same reason, the apparent size of the moon when in the horizon, is estimated by a comparison with intervening objects; but, when she is at a considerable height above the horizon, she diminishes in size like the ball on the top of a spire.

The mean distance of the moon from the earth is about 239,000 miles; and her semi-diameter is nearly three-elevenths of the radius of the earth, or about 1,081. And, as the magnitudes of spherical bodies are as the cubes of their radii, the magnitude of the moon : magnitude of the earth : : $3^3 : 11^3$: : 1 : 49 nearly. See VINCE's *Astronomy*.

While the moon is performing her monthly revolution, she always presents the same side to the earth. This remarkable circumstance arises from her rotation about an axis in $29^d\ 12^h\ 44'\ 12''$, the same time that she revolves round the earth; for, if the moon had no rotation upon an axis, she would exhibit every part of her surface to a spectator upon the earth.

From observations upon the lunar spots, Cassini found that the moon revolves round an axis inclined $88°\ 17'$ to the ecliptic, and that the nodes of the moon's equator have the same position, and the same retrograde motion, as the nodes of the moon's orbit.

The equality between the rotation and revolution of the moon, was ascribed by Newton, to her having an oval form, one side of which was denser than the other. La Grange, however, has shewn, that though the moon ought to be elevated at the equator, by the diminution of the centrifugal force, yet the elevation is four times as great in the direction of the diameter of the equator that is directed to the earth, in the same way as the waters of the earth are always of a spheroidal form, the axis of the spheroid being directed to the moon. In consequence of the attraction of the earth upon this elevated portion, La Grange has demonstrated, that the velocity of the moon's motion is sometimes accelerated and sometimes retarded; and that the tendency of this attraction is to produce an equality between the rotation and revolution of the moon, even if they had been different, and to occasion a coincidence both in the position and motion of the nodes of the lunar equator and the lunar orbit.

La Place supposes that the high mountains of the moon will have an influence upon these phenomena, and that this effect will be greater in proportion to the smallness of the compression at the lunar poles, and the smallness of her mass. See LA GRANGE, *Mem. Berlin*, 1780; LA PLACE, *System du Monde*.

For farther information on the subject of this article, see *Phil. Trans.* for the years 1780, 1787, and 1792; also FERGUSON's *Astronomy*; the EDINBURGH *Encyclopædia*; and NICHOLSON's *Encyclopædia*.

To find the MOON's *Age or Change*.—The following canon, in which the twelve numbers answer to the twelve months, beginning with January, will serve for this purpose:—

Janus 0, 2, 1, 2, 3, 4, 5, 6,
8, 8, 10, 10, these to the epact fix,
The sum, bate 30, to the month's day add,
Or take from 30, age or change is had.

The reason of adding these numbers to the epact in the several months, is because the lunar synodical months fall short of the calendar months; so that the epact, which expresses how much the lunar year falls short of the solar or calendar year, must be considered as continually increasing; and, therefore, to find the new moons, which are the beginnings of the synodical months, an addition must be made to the epact in every month, and more and more as the year advances; which additional numbers are called the menstrual epacts. Only, nothing is to be added to the epact in January, because the annual epact, together with the day of the month, does then express the true age of the moon: but, as January has 31 days, which is near two days more than a synodical month, therefore, the beginning of the lunar month in February will fall two days sooner than it did in January; consequently, 2 is the menstrual epact of February; and then, as February has but 28, or, at most, 29 days, which may be accounted one day less than a synodical month, the next lunar month will begin one day later in March than it did in February; consequently, the menstrual epact of March decreases instead of increasing, and is but 1. If from thence you reckon the lunar months to consist of 30 days and 29 interchangeably, the new moons will fall so much earlier in the following months than the new moon did in January, as is expressed by the menstrual epacts in the canon; *viz.* 2 days in April, 3 in May, &c, until they amount to 11 days at the end of the year, which are then added to the annual epact.

To find the Time of the MOON's *coming to the Meridian, or Southing, as it is commonly called.*— Multiply her age by 4, and divide the product by 5: the quotient gives the hour, and the remainder, multiplied by 12, the minute.

The reason of this rule is, that as the moon at the change comes to the south with the sun, or at twelve o'clock, and, as there are thirty days, nearly, from one new moon to another, and twenty-four hours in a day, therefore, she loses, one day with another, $\frac{24}{30}$ths, or $\frac{4}{5}$ths of an hour in the time of her southing: now, the moon's age, a number of days from the change, being multiplied by 4, the product is so many fifths of an hour as she has lost, which, divided by five, is reduced to hours, and the remainder, if any, multiplied by 12, will be minutes.

Harvest MOON. It is remarkable that the moon, during the week in which she is full in harvest, rises sooner after sun-setting than she does in any other full moon week in the year. By doing so, she affords an immediate supply of light after sun-set, which is very beneficial to the husbandmen, for reaping and gathering in the fruits of the earth.

In these northern latitudes, towards the time of the autumnal equinox, from two to three days before, till as long after the time of full moon, the retardation, or difference between the times of the rising of the moon, in that interval, is only about two hours; whereas, at the opposite time of the year, and in an equal interval, this difference in the time of rising will amount to seven or eight hours. Hence, since the time of harvest is towards the autumnal equinox, this phenomenon has obtained the name of the *Harvest Moon*, and may be clearly explained by a common celestial globe in the following manner:—

Make chalk-marks all round the globe, on the ecliptic, at $12\frac{1}{5}$ degrees from each other (beginning at Capricorn) which is equal to the moon's daily mean motion from the sun: then elevate the north pole of the globe to the latitude of any place in Europe; suppose London.

This done, turn the globe round, westward, and you will see that different parts of the ecliptic make very different angles with the horizon, as they rise above it in the east: and, therefore, in equal times, very unequal portions of the ecliptic will be seen. About Pisces and Aries, seven of these chalk-marks will rise in little more than two hours, as measured by the motion of the index on the horary circle: but about the opposite signs, Virgo and Libra, the index will go over eight hours in the times that seven marks will rise. The intermediate signs will more or less partake of these differences, as they are more or less remote from those above-mentioned.

Hence, it is plain, that when the Moon is in Pisces and Aries, the difference of her rising will be little more than two hours in seven days; but in Virgo and Libra it will be eight hours in seven days: and this happens every month of the year, because the moon goes through all the signs of the ecliptic in a month, or rather, in 27 days, 8 hours.

If the moon's orbit coincided with the ecliptic, and that orbit at the same time invariable in form and position, the interval in the time of rising in two successive nights, under the same circumstances, would take place every year. But the case is otherwise; for the moon's orbit is inclined to the ecliptic in an angle of about 5° 8′ 49″; and, therefore, the interval of time in the rising of the moon, on two successive nights, will vary according to the position of the moon's nodes at that time; that interval being least when the moon's ascending node is in Aries, and greatest when in Libra; it is hence obvious, that the harvest moons are more or less beneficial, according to the position of the moon's nodes. And, since a revolution of the nodes is completed in about 19 years, in that interval, therefore, a period of the most and least beneficial harvest moons will be completed. For a table, containing 50 years forward, in which the harvest moons are most and least beneficial to the inhabitants of the northern hemisphere, see FERGUSON's *Astronomy*, and REES's *Cyclopædia*.

Acceleration of the MOON. See the article ACCELERATION.

To MOOR, (*affourcher, ou amarrer un vaisseau*, Fr.) is to confine or secure a ship in a particular station, by chains or cables, which are either fastened to the adjacent shore, or to anchors in the bottom of a river or harbour.

To MOOR *by the Head*, (*s'affourcher par l'avant*, Fr.) is to secure a ship by her anchors before, without any behind; or her cables may be attached to posts, rings, or moorings, which answer the same purpose.

When a ship is *moored by the head* with her own anchors, they are disposed according to the circumstances of the place where she lies, and the time she is to continue therein. Thus, wherever a tide ebbs and flows, it is usual to carry one anchor out towards the flood, and another towards the ebb, particularly where there is little room to range about; and the anchors are laid in the same manner, if the vessel is moored head and stern in the same place. The situation of the anchors, in a road or bay, is usually opposed to the reigning winds, or those which are most dangerous; so that the ship rides therein with the effort of both her cables. Thus, if she rides in a bay or road which is exposed to a northerly wind and heavy sea from the same quarter, the anchors passing from the opposite *bows* ought to lie east and west from each other: hence both the cables will retain the ship in her station with equal effort against the action of the wind and sea.

When a ship is come to anchor in a situation where she is to remain, she is secured by moorings, and it is of essential importance that she should be well secured, the method of which must depend upon the setting in of the tides, and other circumstances. When the best bower is gone, and about the length of two cables is run out, the small bower is let go; and, when that has taken the ground, you heave in one cable's length upon the best bower, and veer away a cable's length of the small bower; and thus a ship is kept at an equal distance between both, the one lying to the head, and the other to the stern.

In roads where there is much tide, and freshes are expected, ships moor according to the set of the current, one anchor riding to the flood, and the best anchor and cable to the ebb.

In roads where there is little or no tide, ships moor according to the set of the most prevailing wind on the coast. It may so happen, according to circumstances of situation, that the vessel must ride by a greater scope on one cable than the other, but still the manner of letting go the anchors is the same.

Another method of mooring is by carrying the last anchor out in a boat to the place appointed, and then letting it go. Steadying, or mooring with a kedge, is usually done in this manner.

Mooring *for east, west*, &c. is when the seamen observe which way, and on what point of the compass the wind or sea is most likely to endanger the ship, and there laying out an anchor.

Mooring *a fair Birth*, at Sea, is the act of mooring in a place free from any annoyance.

Mooring *a Proviso*, is having an anchor out and a hawser ashore, then the ship is moored with her head ashore; and two cables are the least, and four the best to moor by.

MOORING-BLOCK, a sort of cast iron anchor, about 150 cwt. used in some of his Majesty's ports, for riding ships by, in lieu of anchors.

The silver medal of the Society of Arts was voted to Mr. Samuel Hemman, *of his Majesty's dock-yard at Chatham*, in the year 1810, for his *improved mooring-block for ships*, made of cast iron. This *mooring-block*, from its construction and weight, is decidedly superior to any anchor whatever. The largest anchors appropriated to the moorings for a ship of the line are from 65 to 80 cwt. and even those sizes are found insufficient at times; whereas the weight of a mooring-block is about 150 cwt. which is 60 cwt. heavier than a first-rate's anchor, besides a form better calculated to hold in addition to its increased weight; this double advantage will allow a reduction from the length of the chain, which is very expensive; and, in one pair of moorings for a ship of the line, a saving of about 874*l.* will arise out of 2472*l.* the present expense of a pair of moorings with anchors. There is another advantage attending the use of these mooring-blocks; viz. that there are no anchors appropriated to moorings but might be made serviceable; and, should there be no defective anchors in store, we are necessitated to take sound ones. It is further to be observed, that there is no species of stores in his Majesty's navy that is so difficult to procure as large anchors, owing to the want of anchor-smiths; and, moreover, an anchor of 76 cwt. will take upwards of thirty days to make it; whereas, a mooring-block may be cast in as many hours, in any of the iron-foundries in the kingdom.—These circumstances are of the greatest importance, and moorings of this description will answer for ships of all sizes, provided there is a sufficient depth of water under them. See the *Transactions of the Society of Arts*, Vol. XXVIII. for further particulars, and a figure of the improved mooring block.

Chain Moorings, (*corps morts pour l'amarrage*, Fr.) an assemblage of anchors, chains, and bridles, laid athwart the bottom of a river, or harbour, to ride the shipping contained therein.

The anchors employed on this occasion have generally but one flook, which is sunk and secured in the ground by piles in the river near low-water mark. Two anchors being fixed in this manner, on the opposite sides of the river, are furnished with a chain, extending across from one to the other; in the middle of the chain is a large square link, whose lower end terminates in a swivel, to which are attached the bridles, which are short pieces of cable, well served, whose upper ends are drawn into the ship, and afterwards secured to the bits, &c. By this means the vessel veers round very readily, according to the change of the wind or tide; in some places, however, particularly in rivers, each ship takes in a bridle astern also, by which she becomes moored head and stern.

A great number of moorings of this sort are fixed in the royal ports, or the harbours adjacent to the king's dock-yards, as Deptford, Chatham, Portsmouth, Plymouth, &c.

Swinging Moorings, (*amarrages à tourniquet*, Fr.) These differ from the former in this: the ground-work being the same, to the ring of the chain is fixed one chain-pendent, the end of which is passed through a large buoy; and to that is connected a swivel ring by a shackle; and thus kept constantly from sinking to the ground. This is used for mooring one or two ships to. Near each end of the chain is fixed one more chain-pendant, supported likewise by a buoy, which is used only for the purpose of warping ships from the middle to any intended place.

A ship, when launched from the stocks, is furnished with one stream anchor, and another which is termed a launching-anchor, somewhat heavier than the stream-anchor, hanging by stoppers to the cat-heads; by one or both of which she is brought up, and afterwards warped to her head and stern moorings.

To moor a Ship to Head and Stern MOORINGS; first clench the end of a bridle to the jew's-harp of the chain-pendent, and pass it into the hawse-hole; then heave it up and bit it. Take the after-bridles into the gun-room port, or in upon the quarters; and take a round turn round the beam of the after-hatch, and cross it with a throat and round seizing; or if more convenient, round the jear bits.

To moor a Ship to Swinging MOORINGS; pass out the ends of the bridle, on each side, through the hawse-hole next the stem; then pass them through the swivel ring of the chain-pendent; then return them into the ship through the after hawse-holes, and bit them. In this situation a ship is fitted for sea, by regularly stowing her ballast, and getting on board her rigging, stores, &c.

In respect to temporary moorings, rules cannot be fixed to suit all places and all ships, but different methods may be necessary, according to the road, the size of the ship, her draught of water, and how she may be provided with ground-tackle, &c. Where the tide runs very strong, and there is water enough, it is certainly best to moor *water-shot*, the anchors and cables lying in the same direction with the tide, which can then have but little effect on them, in comparison with what it has when a ship is moored across the tide, where the power of the tide alone, upon the moorings, is of itself sometimes enough to bring the anchors home. Therefore, great ships, that have sheet anchors, and are well manned, to clear the hawse, when obliged to lie where the tide runs strong, moor with the best and small bower anchors, water-shot; but small ships, having only two bower anchors and but few hands to clear the hawse, drawing less water, may moor out of the strength of the tide, with the small bower and stream anchor, and hawser athwart the tide, and keep the best bower anchor in reserve, as occasion may require.

When a ship is to be moored only to wait for a fair wind, the best anchor and open hawse should be towards the foul wind quarter; but, when a ship is to lie with all winds that may blow, the best anchor and open hawse should be towards the worst wind that may blow to raise the waves and give the ship a pitching motion, and must leave no more of the smallest moorings within board than just enough to *freshen the hawse* on occasions; by which means they will hold the longer, by having the longer scope, and the hawse will be the easier cleared.

Where a ship is liable to ground with the ebb, and not to float till the flood tide runs strong, the safest method is to moor across the tide, which lays the ship clear of getting damage upon her own anchors; when it requires to attend and slack the mooring, that her stern may swing over upon the flood, to prevent her girting. And, when moored with a small anchor and hawser, the helm should be always shifted, to sheer her towards the small anchor, to ease the strain upon the small moorings, and to lay the principal stress upon the bower anchor and cable, in order to ride the ship more securely when the tide runs strong.

A successful method was tried some years ago by Captain W. Hutchinson, and which has since been frequently practised, of mooring a ship in a very narrow river, where the shore on each side was covered with nothing but loose shelly stones. The best bower anchor was laid close over to one side of the river, and a hawser, with a long bow-line knot, made fast to the middle of a twelve feet deal-board, on the opposite shore, under the shelly stones, broadside, and its lower edge a little slanting towards the ship, which held her better than an anchor would have done in its place. Another remark of this kind deserves notice, and has been practised where anchors placed one to back another could not be made to hold in sand: a bulky bundle, made of rushes that grow on sand hills, buried under the sand, and trampled upon, has held so firmly, that it would break a cable made fast to it; and no doubt but any thing of the same bulk would answer the same purpose.

Method of applying the weakest Moorings, to help a Ship to ride out a Storm.—In a dangerous situation, all practicable hints that may contribute any thing towards safety, deserve notice; and it may often happen that the small bower cable may be too much worn, or the small moorings known to be too weak to bear the strain of the ship to ride out a storm, when the anchor at the best cable is in danger, or expected to come home; in this case it is advisable to make the weak moorings serve for a backing to the best anchor and cable, contriving a traveller (as it may be properly called) of sufficient rope to go slack round the best cable, without the hawse, and well secured with rolling hitches seized, &c. to the weak moorings that may be veered away, or let go, as occasion may require: or, if short of rope, to make a proper traveller, a stopper may be put on without the hawse, till the end of the small moorings is put round the best cable, with a bowline-knot open enough to slide along the best cable, till it comes to the anchor-stock, which may prove such a sure backing to it as to prevent its coming farther home; so that by this means, there is a much better chance to ride out a storm by the best cable, singly, than to run the risk of either anchor coming home, to bring them both a-head together, by which means one cable is apt to get foul of the other anchor, and to get so entangled as to make the anchors come home, or make it necessary to cut the cable which is foul of the anchor, if the ship has a pitching motion.

But it must be allowed, that this as well as every other uncommon method, requires judgment to contrive and resolution to put the design speedily in practice, as the occasion and necessity may require, to preserve the whole when safety depends upon riding out a storm.

MOOT, in block-making, a piece of hard wood, about nine inches square and two or three inches thick, hooped with iron at each end, and coaked with brass in the centre, similar to a sheave. A mouth is made in the coak, and fitted with a sharp strong steel cutting-knife, resembling a spoke-shave. It is used for turning block-pins by the block-mill. The pin to be turned is fixed by its head in the mandrel, and the *moot* is advanced to it on the sliding puppet.

MOOTER, or TREE-NAIL MOOTER, (*maître-chevilleur*, Fr.) a name given to the person who turns the tree-nails by the assistance of a moot, for fastening the planks and timbers of ships.

MORTAR, (*mortier*, Fr.) a short piece of ordnance, thick and wide, having a chamber less than the size of its bore, and used to discharge bombs or carcasses into a fortified place. The bomb, or shell, is a great hollow ball, filled with powder, which, falling into a fortification, &c. destroys the most substantial buildings by its weight, and bursting asunder, creates the greatest disorder and mischief by its splinters. To prevent the shell from bursting at the first moment of discharge, it is furnished with a fuse, calculated to continue burning during its flight; and to increase the weight of its fall, the mortar is elevated to a considerable angle above the horizon.

The chambers of mortars are extremely different in their figures, and each of those figures is defended by better or worse arguments. Thus they are spherical, cylindrical, conical, bottled, or concave. Indeed, nothing appears to be less determined upon true principles or experiments than the proportions of the several parts of a mortar.

As the sea-mortars, or those which are placed in the bomb-vessels, are generally fixed at a much greater distance from the object than is required on shore, they are made somewhat longer, and much heavier than the land-mortars.

Fig. 7. Plate XVI. fig. 7, represents a sea-mortar, the principal parts of which are, A, the chace; B, the reinforce; C, the breech; and D, the trunnions. The interior part, comprehended between the dotted lines, is called the bore, wherein the bomb is lodged; and the inner part of the bore, which is diminished towards the breech, and contains the powder, is termed the chamber.

Mr. Muller, in his *Treatise of Artillery*, very justly observes, that the breech of our 13 inch sea-mortars is loaded with an unnecessary weight of metal: the chamber thereof contains 32 pounds of powder, and at the same time they are never charged with more than 12 or 15 pounds, by the most expert officers, because the bomb-vessel is unable to bear the violent shock of their full charge. Thus the action of the powder is diminished by the vacancy left in the chamber, which is never above half filled. As a charge of 12 or 15 pounds of powder at most is therefore, sufficient, it is evidently proved, by the theory of powder, that this will produce the greatest effect when discharged from a mortar with a cylindrical chamber, represented by fig. 8. He also proves,
Fig. 8.
by a variety of experiments made by Captain Desaguliers and himself, that the conical chamber, now used, is considerably inferior to the cylindrical one with the last charge of powder.

To facilitate the use of the mortar, it is placed in a solid carriage of timber, called the bed, whose different parts are strongly bolted together. By means of this it is firmly secured in its situation, so that the explosion of the powder may not alter its direction. In the middle of the upper side of this carriage, plate XVI. fig. 9, are two semi-circular notches, to receive the trunnions; over these are fixed two very strong bands of iron, called the cap-squares, *a*, the middle of which is bent into a semicircle, to embrace the trunnions, and keep them fast in the mortar-bed. The cap-squares are confined to the timber-work by strong pins of iron, called the eye-bolts, *b*, into whose upper ends are driven the keys, chained beneath them. On the fore-part of the bed a piece of timber is placed transversely, upon which rests the belly of the mortar, or that part which contains the chamber. The plan of this piece is represented by fig. 12, and the elevation, which is called the bed-bolster, by fig. 13, it is used to elevate and support the mortar whilst firing.

These beds are placed upon very strong frames of timber, which are fixed in the bomb-ketch, and represented in fig. 18 and 21, plate IV. the former of which exhibits the transverse section of a bomb-vessel, with the mortar fixed in its place, at an elevation of forty-five degrees. They are securely attached to the frames by means of a strong bolt of iron, fig. 14, Fig. 14.
plate XVI. called the pintle, passing perpendicularly through both, and afterwards through one of the beams of the vessel. Thus the pintle, which passes through the whole in the centre of the plan, fig. 10, serves as an axis to the bed; so that the mortar may be turned about horizontally as occasion requires.

Plate XVI. fig. 9, represents the elevation of the bed of a 10-inch sea-mortar; fig. 10, is the plan, and 11, the front view thereof; fig. 12, exhibits the plan, and fig. 13, the elevation of the bed-bolster. Fig. 9. Fig. 10, 11, 12, 13.

The shell, as already observed, is a great hollow ball, charged with powder. Fig. 15, is a perspective view of it, and fig. 16, a section, whereby the thickness is exhibited. The parts *a* and *b* of the shell are its handles, by which it is lifted up or removed; and *c* is the fuse-hole, or aperture, through which the powder is poured in to charge it. Fig. 15 and 16.

The lower part of the shell, as appears by fig. 16, is the thickest, by which it becomes heavier on that side, and accordingly falls thereon, and never on the fuse. It is also the better enabled thereby to resist the impression of the powder, by which it is discharged from the mortar. Both of these reasons, however, Mr. Muller conceives to be immaterial, because nothing but an absolute stoppage of the air can exhaust the fuses, as their composition enables them to burn in water, as well as air or earth; and the explosion of the mortar would not, in his opinion, be able to break them, if they are equally thick every where. The most proper quantity of powder to charge a shell is probably two-thirds of the weight which would fill the cavity.

The fuse of it is represented by *c d*, fig. 16. This is generally a conical tube, formed of beech, willow, or some dry wood, and filled with a composition of sulphur, saltpetre, and mealed-powder. The shell being charged, this fuse is inserted in the cavity through the fuse-hole; and when fired, communicates the fire to the powder in the shell.

The fuses are charged with great care, that nothing may prevent them from communicating the fire to the powder in the centre of the bomb. They are driven into it so as that only about an inch and a half comes out beyond the fuse-hole; and then the shell is said to be fixed.

These fuses are also charged long before there is occasion to use them; and that the composition with which they are filled may not fall out, or be damaged by growing damp, the two ends are covered with a composition of tallow, mixed either with pitch or bees'-wax. When the fuse is to be put into the shell, the little end is opened or cut off; but the great end is never opened till the mortar is to be fired.

The proper quantity of gunpowder being put into the chamber, if there be any vacant place, it is generally filled up with hay: some choose a wooden plug; over this they lay a turf; some a tompion fitted to the bore of the piece, and lastly the bomb; taking care that the fire be in the axis thereof, and the orifice be turned from the muzzle of the piece. What space remains is to be filled up with hay, straw, turf, &c. so as that the whole charge may not be exploded without the utmost violence.

The shell being placed upon the wad, as near the middle of the mortar as possible, with the fuse-hole uppermost, another wad is pressed down close upon it, so as to keep the shell firm in its position. The officer then points the mortar, or gives it the inclination necessary to throw the shell to the place designed. When the mortar is thus fixed, the fuse is opened; the priming-iron is also thrust into the touch-hole of the mortar to clear it, after which it is primed with the finest powder. This done, two of the matrosses, or sailors, taking each one of the matches, the first lights the fuse, and the other fires the mortar. The shell thrown out by the explosion of the powder, is thrown to the place intended; and the fuse, which ought to be exhausted at the instant of the shell's falling, inflames the powder contained therein, and bursts it into splinters; which, flying off circularly, occasion incredible mischief wheresoever they reach.

To elevate the mortar, so as its axis may make any given angle with the horizon, they apply the artillery level, or gunner's quadrant: the use whereof see under the articles Level and Quadrant.

An elevation of 70 or 80 degrees, is what is commonly chosen for rendering mortars most serviceable in throwing shells into towns, forts, &c. though the greatest range be at 45 degrees.

All the English mortars are fixed at an angle of 45 degrees, and lashed strongly with ropes at that elevation.

The following are the necessary Orders before a Bombardment by Sea.

1. When any fixed shells are issued from the tenders, the artillery people on board are immediately to fix others in their room, and are always to keep in their tenders the same number they had at first.

2. The shells are to be fixed in the boat appointed to carry them, provided the weather permits; otherwise, in the safest place on deck, and to be *kited*, or lowered down into a spare rack, which must be in each boat for that purpose. While the shells are fixing, the powder-room is to be shut, the hatches laid down and well secured against fire, and the place where they are fixed is to be well watered.

3. The shells being carefully examined, in order that no spike is left therein, by which the fuse may be split, the fuses are to be cut the full length, and to be set home into the shell very strongly.

4. No shells fixed during the service are to be kited; but if any should be left, when the service is over, they are immediately to be kited.

5. The powder in the bomb-vessels is to be used first; and none to be opened or measured out, except in the captain's cabin, the door of which is to be kept shut during the whole time, and covered with tanned hides, to make it as secure as possible.

6. The fixed shells in the boats are to be likewise covered from fire or wet, with hair-cloth and tanned hides with the utmost care.

7. If the service is carried on at night, all the powder is to be ready measured out in cartridges, which may be kept in the powder-magazine and captain's cabin, in the empty powder-barrels and powder-bags; and all the shells requisite are to be ready. The tin tubes, one powder-horn, and the port-fires, also the punches and bits for the vents, are to be kept in the captain's cabin.

8. No fire nor light, except match and port-fires, to be on board either bomb-vessel or tender during the service.

9. The captain's cabin and the passage to it, also the way to the magazine and decks, are to be constantly watered.

10. The sponges for the mortars are to be all examined and tried, and if too large, they are to be cut so as to enter easily.

11. The vents of the mortars are to be examined, and the punches and tubes tried in them.

12. A laboratory-chest is to be on board each bomb-vessel, in the captain's cabin, in which all the small stores are to be kept.

13. Two tubs of water are to be on deck, for the lightest port-fires and match, which must be constantly held in them till ordered to fire.

14. Two careful men are also to be appointed for this service, who are to do nothing else on any account.

15. Two careful men of the artillery are to be left on board each tender, for the filling and fixing of the shells.

16. Application must be made to the admiral for two men of war's boats to attend on each bomb-ketch and tender, for carrying shells and stores. One of these is to be loaded with fixed shells, which, when sent to the bomb-vessel, must remain with her until they are all taken out, which should be only as they are wanted for loading the mortars; it is then to return to the tender. The other boat, meanwhile, will be receiving more fixed shells, and on the signal given from the bomb-ketch for more shells, must immediately repair to her with them.

17. A gang of warrant-officers, and eight seamen, are to be at each mortar, to give whatever assistance may be required.

18. A gang from the navy, with a careful warrant-officer, and a non-commissioned officer of the artillery, are to have the charge between decks on board each bomb-vessel and tender, to get up the fixed shells that are in the rack; and a careful person is to remain constantly at the powder-room door, which must be kept shut as much as possible.

19. When any powder is wanted from the tender for loading the mortar, it should be measured out in the tender, and the proper charge put into paper cartridges, upon which should be written the quantity, and the mortar for which it is allotted.

In shooting with mortars, the following general rules should always be observed:—

1. To measure the distance of the object aimed at.

2. That the bombs be of equal weight, otherwise the shots will vary.

3. That the carriage be on an exact level to prevent its leaping.

4. That the powder with which the piece is charged be always of the same strength and quantity.

5. That the charge be always equally rammed down.

6. That the wads be always of wood, tompions, or oakum.

7. That the fuses be fresh made the days on which they are to be used, and that they be of a composition proportionable to the range of the shot in the air, so that the bomb may break at the very moment of, or soon after, its fall; which composition must be such as shall not be extinguished though it fall in water, but continue burning till the bomb breaks.

If the service of mortars should render it necessary to use pound-shots, 200 of them, with a wooden bottom, are to be put into the 13 inch mortar, and a quantity of powder, not exceeding five pounds; and 100 of the above shot, with $2\frac{1}{2}$ pounds of powder, for the 10 inch mortar, or 3 pounds at most.

One inch of fuse burns 4 seconds and 48 parts.

Weight of the Sea-Mortars and Shells, as also of their full Charges.

Nature of the Mortar.	Powder contained in the Chamber when full.		Weight of the Mortar.			Weight of the Shell when fixed.	Weight of Powder contained in the Shell.	
	lb.	oz.	cwt.	qrs.	lbs.	lb.	lb.	oz.
10 inch howitzer	12	0	31	2	26	....	....	
13 inch mortar..	30	0	81	2	1	195	7	0
10 inch mortar..	12	0	34	2	11	89	....	

In respect to the chambers of mortars in 1787 and 1789, experiments were made at Woolwich with four shifting chambers for the purpose of ascertaining which form gives the greatest range. The chambers were all of the same capacity; that is, 63.7 cubic inches, and contained two pounds of powder. Their forms were as follow:—

1. The common conical chamber, with a circular bottom.

2. The same reversed.

3. The cylinder chamber with a circular bottom.

4. The spheric chamber. The ranges were the medium of six pounds; from them it appears, that when the spheric chamber is filled with powder, it has the advantage in point of range; but when smaller charges are used, its ranges are found to be shorter than those of other forms.

The conical (No. 1.) chamber of the present establishment gives the longest range under other circumstances.

For an account of the elevation of the mortar, and flight of shells according to the different charges of powder, see the article RANGE, and ROBINSON'S *Tracts*.

A description of two brass mortars landed on the 10th of November, 1812, from his Majesty's ship Revenge, being part of the battering train used by the French, at the late siege of Cadiz. and sent home by the Regency of Spain, as a present to his Royal Highness the Prince Regent of England.

Le Grand Napoleon.

Length, 8 feet.
Diameter of the bore, $11\frac{1}{4}$ inches.
Weight, 7,590 lbs. Spanish.
Charge of powder, 38 lbs.
Elevation, 36°.

Cast at Seville, in March 1811.

The external form is like that of a gun, with trunnions and cascable; it is fired either by slow match or a train, and supported by two brass brackets, resembling in shape those of English carriages, with a chock of wood in the inner side of each, and the whole secured together by long iron lever bolts.

The shell weighs from 170 to 175 lbs. Spanish nearly filled with lead.

Quantity of powder for bursting, 8lbs.
Range of shell, 6,200 yards.

Le Duc de Treviso.

Length, 8 feet.
Diameter of the bore, $12\frac{1}{2}$ inches.
Weight, 7,550 lbs. Spanish including the brass bed.
Charge of powder, 38 lbs.
Elevation, 45°.
The shell is in all respects similar to the other.
Range of shell, 6,000 yards.

The form is like that of an English mortar, with the exception of the bed being cast on to it, and placed on another bed made of wood with trucks.

On Fort Napoleon, whence these were taken, there were twelve mortars, six large and six small, all of the same description as the above-mentioned. There were also other batteries with similar mortars, the whole of which were battered and rendered unserviceable from being fired into by a 24-pounder, when the siege was raised.

Coehorn MORTAR, a small kind of mortar, invented by the celebrated engineer, Baron Coehorn, to throw small shells or grenades.

MOTION, in astronomy, is peculiarly applied to the orderly courses of the heavenly bodies. See the articles ANOMALY, DIURNAL, and MEAN.

MOULD, (*gabarit*, Fr.) a thin flexible piece of timber, used by shipwrights, as a pattern whereby to form the different curves of the timbers, and other *compassing* pieces in a ship's frame; of these there are two sorts, namely, the bend mould and hollow mould. The former of these determines the convexity of the timbers; and the latter, the concavity on the outside, where they approach the keel, particularly towards the extremities of the vessel. The figure given to the timbers by this pattern is called their *bevelling*. See the article BEVELLINGS.

MOULD. also signifies the shape of such pieces as are wanted to make good deficiencies; others, for forming hoops, are made to the shape and size of the mast, and nailed together at the corners.

MOULDED, implies the breadth of any thing, or the sides of a piece formed by a mould.

MOULDING, the act of marking out the true shape of any timber from the mould. It is also a name given to any ornamental projection from the side, &c. of a ship.

MOULDINGS *of a Gun or Mortar*, (*moulures*, Fr.)

are the prominent parts which serve for ornament, such as the breech mouldings, rings, &c.

Mould-Loft, (*salle des gabarits*, Fr.) in dock-yards, a spacious loft, wherein ships are laid down from certain draughts, for the purpose of making moulds and taking the bevellings of their different parts.

To MOUNT, (*monter*, Fr.) expressed of a vessel of war, implies, to carry; as, "She mounts twenty guns."

To Mount *a Gun*, (*monter un canon*, Fr.) is to place a piece of ordnance on its carriage, for its management in firing.

MOUSE, (*bouton, ou pomme d'étai et de tournevire*, Fr.) a knot or knob, wrought on the outside of a rope, by means of spun-yarn, parcelling, &c. as the knot wrought on the stay of a ship, which prevents the collar from closing round the mast-head. See the article Puddening.

It is particularly used on the stays of the lower-mast, to prevent the *eye* from slipping up to the mast; a circumstance which would render it extremely difficult to remove the stay from the mast-head, when necessary.

To Mouse *a Hook*, (*amarrer un fil de caret sur un croc*, Fr.) is to pass a yarn or fox round the hook of a block, in order to prevent its clearing itself from any thing it may be hooked to.

To MUFFLE the Oars, (*mettre des paillets aux avirons*, Fr.) is to put some matting, &c. round those parts of the oars which lie on the edge or gunnel of the boat, when rowing, to prevent them from making a noise against the tholes and row-locks.

MUNIONS, in ship-building, the pieces that part the lights in the stern and quarter-gallery.

MUSCA INDICA, the Bee or Fly, in astronomy, a southern constellation, containing four stars.

MUSKET, (*fusil*, Fr.) a fire-arm used in war, and borne on the shoulder of foot soldiers. The length of a musket is fixed at three feet six inches, from the muzzle to the pan, and it carries a ball from sixteen to twenty-nine to the pound.

MUSKETOON, (*mousqueton*, Fr.) a short thick species of musket, whose bore is the thirty-eighth part of its length; it carries five ounces of iron, or seven and a half of lead, with a proportionate quantity of powder. The name is also applied to a fire-arm resembling a horse pistol, of a very wide bore, and sometimes bell-mouthed.

To MUSTER, (*passer la revue des équipages*, Fr.) is to call over the whole ship's company, or any detachment thereof, who are accordingly expected to answer their names personally on the occasion.

All Hands to Muster! (*en-haut tout le monde pour l'appel général!* Fr.) the order to the boatswain to pipe all hands to muster.

Muster-Master, (*commis aux revues*, Fr.) in royal dock-yards, a person appointed by the Navy Board, to call over the lists of all his Majesty's ship's companies in the different ports; as also the artificers working on board, in order to prevent false musters.

MUTINEER, (*mutin*, Fr.) a person wilfully concerned in a *mutiny*.

MUTINY, (*révolte*, Fr.) in a naval or military sense, the act of rising against authority, or revolting and disobeying of orders. Any officer or soldier who shall presume to use traitorous or disrespectful words against the sacred person of his Majesty, or any of the royal family, is guilty of *mutiny*.

If any person in or belonging to any of his Majesty's ships shall make, or endeavour to make, any mutinous assembly, upon any pretence whatsoever, he shall be deemed guilty of *mutiny*; and, being convicted thereof by the sentence of a court-martial, shall suffer death. And, if any person in or belonging to the fleet shall utter any words of sedition or mutiny, he shall suffer such punishment as a court martial shall judge him to deserve.

If any officer, seaman, or marine, in or belonging to the fleet, shall behave himself with contempt to his superior officer, such superior officer being in the execution of his duty, he shall be punished according to the nature of the offence by the judgment of a court martial.

If any person in the fleet shall *conceal* any traitorous or mutinous design, being convicted thereof by the sentence of a court martial, he shall suffer death, or such other punishment as a court martial shall think fit; and if any person shall *conceal* any traitorous or mutinous words spoken by any, to the prejudice of his Majesty or government: or any words, practice, or design tending to the hindrance of the service, and shall not forthwith reveal the same to the commanding officer; or, being present at any mutiny or sedition, shall not use his utmost endeavours to suppress the same, he shall be punished as a court martial shall think he deserves.

If any officer, seaman, or marine, or other person in the fleet, shall *strike* any of his superior officers, or draw, or offer to draw, or lift up any weapon against him, being in the execution of his duty, on any pretence whatsoever, every such person, being convicted of any such offence by the sentence of a court martial, shall suffer death. And if any officer, seaman, or marine, or other person in the fleet, shall presume to *quarrel* with any of his superior officers, being in the execution of his duty, or shall *disobey* any lawful command of any of his superior officers; every such person, being convicted of any such offence by the sentence of a court martial, shall suffer death, or such other punishment as shall, according to the nature and degree of the offence, be inflicted upon him by the sentence of a court martial.

Every person, being in actual service and full pay, and part of the crew in or belonging to any of his Majesty's ships or vessels of war, who shall be guilty of *mutiny*, *desertion*, or *disobedience*, to any lawful command, in any part of his Majesty's dominions on shore, when in actual service relative to the fleet, shall be liable to be tried by a court martial, and suffer the like punishment for every offence, as if the same had been committed at sea, on board any of his Majesty's ships or vessels of war. See the Articles of War; 19th, 20th, 22d, and 31th article.

MUZZLE of a Gun, (*la bouche, d'un canon*, Fr.) the extreme end around the mouth; hence, *muzzle-ring* is that ring which strengthens the mouth or *muzzle*.

Muzzle-Lashings, are two and a half inch ropes, about four or five fathoms in length, used to lash the muzzles of guns, so as to confine them to the upper part of the ports.

N.

NADIR, in astronomy, that point of the heavens which is diametrically opposite to the zenith, or point directly over our heads. The zenith and nadir are the two poles of the horizon, each being 90° distant from it.

NAIL, (*clou*, Fr.) a small spike made either of iron, copper, or mixed metal, which being driven into wood, serves to bind several pieces together, or to fasten something upon them.

The various sorts of nails used in ship-building are as follow:—

Clasp-headed, or Bradded NAILS, (*clous à tête piquée*, Fr.) are used for clasping, when driven into the wood, that their heads shall not be seen, and thereby render the work smooth, so as to admit a plane over it.

Clincher NAILS, (*clous à vis*, Fr.) have square shanks tapering but little, and are used by boat-builders; they are proper to be used in any boarded buildings, as they will drive without splitting the wood, and draw without breaking.

Copper NAILS (*clous à cuivre*, Fr.) are made of mixed metal of various lengths and sizes; they have a flat round head, with a square shank, about one inch and a half long, and are used for fastening the copper sheathing to the bottoms of ships. There are also pure copper nails, used in fastening the planks of the deck near the binnacle, to prevent the attraction of the magnetic needle; also in sheathing boards on the bottoms of ships, and in lining the magazines. There is another description of copper nails, that have round convex heads, and used for fastening lead to the knee of the head.

Single-deck NAILS, (*clous de six pouces*, Fr.) are between five and six inches in length, and used for fastening deck planks to the beams, carlings, and ledges.

Double-deck NAILS, (*clous de sept pouces de longueur*, Fr.) are seven inches in length, and used for doubling of shipping, and fastening planks to the beams.

Flat NAILS, or *Cooper's Flats*, (*clous d'un pouce à pointe platte*, Fr.) have small flat shanks nearly an inch long, and used to fasten tarred paper to the bottoms of ships, before the sheathing is applied; they are used also in fastening together the scarfs of boats.

Filling NAILS, (*clous d un pouce de longueur*, Fr.) have large clout heads, are one inch long, and chiefly used for filling ship's bottoms, as also for bulk-heads in prison ships. There are also filling brads used for the same purpose.

Plate NAILS, (*clous de deux pouces à tête ronde*, Fr.) have round heads, are two inches in length, and used chiefly to fasten bill plates to the bill-boards, and to hang scuttle-hinges to the port-lids, &c.

Port NAILS, (*clous de sabords*, Fr.) are used in fastening hinges to the port-lids, as also for fastening plates to the bill-boards.

Ribbing NAILS (*clous de 9 et 10 pouces*, Fr.) are generally about nine or ten inches in length, have large round heads, with rings or chocks under them, to prevent their heads from drawing through when ripped off, and are used for fastening the ribands and harpings to ship's timbers.

Rudder NAILS, (*clous de ferrure de gouvernail*, Fr.) are about five inches in length, with a full head, and used for fastening pintles to the rudder.

Scupper NAILS, (*clous à mangère*, Fr.) are about an inch in length, have broad heads, and are used for fastening leather and canvas to the scuppers.

Sheathing NAILS, (*clous de tillac*, Fr.) are either of iron or copper, and used for fastening sheathing boards on ship's bottoms.

Two-penny NAILS, (*clous d'un pouce à tête platte*, Fr.) are flat-headed, about an inch long, and used for nailing lead and tin to different parts of a ship.

Four-penny NAILS, (*clous ou dessous de ceux de demi-tillac*, Fr.) are one inch and a half long, have sharp points, and are generally used for lining bulk-heads.

Six-penny NAILS, (*clous de demi-tillac*, Fr.) are one inch and a half long, they have flat points, and are used in making gratings and half-ports.

Ten-penny NAILS, (*clous de deux pouces un quart*, Fr.) are two inches and a quarter in length, and are generally used for the same purpose as twenty-penny-nails.

Tree NAILS. See the article TREE.

Twenty-penny NAILS, (*clous de deux pouces trois quarts*, Fr.) are about three inches in length, and used principally for fastening bulk-heads, and for laying flat the orlop.

Twenty-four-penny NAILS, (*clous de trois pouces et demi*, Fr.) have flattish heads and thick shanks, are about three inches and a half in length, and used in fastening bulk-heads to their respective places, &c.

Thirty-penny NAILS, (*clous de demi caravelle*, Fr.) are shaped like the twenty-four-penny; they are three inches and a half long, and used for similar purposes.

Forty-penny NAILS, (*clous de caravelle*, Fr.) are four inches and a half long, shaped similar to the thirty-penny, and applied to the same purposes.

Weight or Spike NAILS, (*clous au poids*, Fr.) are of various lengths and sizes, and used for fastening bulk-heads to their respective places.

To Nail *or Spike a Gun, (enclouer un canon,* Fr.) is to drive a nail or iron spike into the vent or touch-hole, so as to render it for a time, unserviceable. There are many methods of forcing a nail out, but they have never been found of general use. The best method, when this is impracticable, is to drill a new hole. Another way by which cannon may be rendered useless, is by driving a ball of sufficient calibre to stop the bore tightly.

NARROWS, *(passage étroit entre deux terres,* Fr.) a phrase used among seamen to denote a small passage between two lands.

NAVAL, *(naval, ou de marine,* Fr.) of or belonging to a ship, or to the royal navy; thus we say, naval officers, naval stores, &c.

Naval Architecture, *(architecture navale,* Fr.) or the science of ship-building, comprehends the theory of delineating marine vessels upon a plane; and the art of framing them upon the stocks, according to the proportions exhibited in a regular design.

All edifices, whether civil or military, are known to be erected in consequence of certain established plans, which have been previously altered or improved till they have arrived at the desired point of perfection. The construction of ships appears also to require at least as much correctness and precision as the buildings which are founded upon *terra firma:* it is therefore absolutely necessary that the mechanical skill of the shipwright should be assisted by plans and sections, which have been drawn with all possible exactness, examined by proper calculations, and submitted to the most accurate scrutiny.

History of Naval Architecture. It would be a fruitless attempt to investigate either the first inventor of naval architecture, or even the country or quarter of the world from whence it derived its origin. The remote distance of time renders the attempt useless, as the result of the inquiry cannot produce any decisive determination. It may be conjectured that the inquisitive and active spirit of enterprise, constitutionally, as it were, implanted in our nature, displayed itself at one and the same time, in a variety of quarters and districts; for the primitive ideas of men dispersed over the face of the globe, unconnected with each other, and totally ignorant of each others existence, appear in such perfect unison as to invention, that they well warrant this supposition. It must be allowed that the moderns never could have attained that summit of knowledge now reached, had it not been for the labours of the ancients, which laid the foundation of that structure, which the modern artist has, to give him every merit he claims, only borne a less laborious part of bringing nearer to perfection. But the nations which stand foremost as joint candidates for the honour, appear to have been the Ægyptians and Phœnicians; at least, it is to them the invention is ascribed by authors of the highest antiquity, as well as credit, on the score of affording the most authentic information.

The Nile, which intersected their country, presented to the former a less dangerous opportunity of making the first rude attempts in the art of navigation than the sea itself did; while the enterprising spirit that appears peculiarly to have marked the Phœnician character, as well as the advantageous situation of their two principal cities, Tyre and Sidon, urged them, by gradual steps to bolder enterprises, which raised them to an unrivalled pre-eminence, as navigators, among the nations which surrounded them, and enabled them to engross, almost entirely to themselves, the commerce of the universe.

According to the best authority, the original method used by the Ægyptians in constructing vessels is as follows:—The bark used on the Nile appears to have been formed of small planks cut out of the acantha, or Ægyptian thorn: these were not, as might be naturally supposed, cut into lengths, as planks, but nearly square, measuring about three feet each way; they were lapped over each other like tiles, and fastened together by a proper number of wooden pins, nearly of the same shape with the tree-nails of modern times.

This mode of construction was found sufficiently strong for the purposes to which it was applied, even without the obvious assistance of any internal frame of timber; and, proving equal to the necessities and ambition of the inventors, they for a long time troubled not themselves with attempting any additional improvement.

The hull of the vessel being completed, a competent number of seats or benches, for the accommodation of the rowers were added; and when the joints or seams were carefully calked with the papyrus, so as completely to exclude the water, the floating fabric then became fit for immediate use. We must not, however, forget to mention, that experience very early suggested the necessity of some directing as well as impelling power, in aid of human labour. A mast, formed out of a straight stick of the acantha, and a sail, made of papyrus, supplied the latter; at the same time a rudder, which is said to have passed through the keel, or bottom of the vessel, remedied the defect occasioned by the want of the former.

These vessels being, as well from their construction as equipment, almost incapable of stemming the current of the river, were generally towed up against it by persons on shore, unless the wind fortunately proved sufficiently strong and favourable for the proposed course, to enable the sail to be used as a substitute. On returning with the current, it was customary for the Ægyptians to fasten, with ropes across the prow of the vessel, a hurdle of tamarisk, which being let down into the water, and steadied by ropes, or bands made of twisted reeds, caused it to move forward with increased velocity, in consequence of the stream acting with greater force on the surface of the hurdle which extended beyond the sides, than it would have done on the mere vessel itself, without this ingenious aid. In order to preserve a due balance between the head and stern, which might otherwise have been affected by the action of the water on the hurdle, and in some degree also, by the weight of it, as well as to cause the boat to swim nearly with *an even keel,* a stone of considerable magnitude, pierced through the middle, was suspended by a rope from the stern; a contrivance which was found to answer the purpose so well, that the unskilled navigators were enabled to pass to and fro without either danger or difficulty.

It is a general idea, founded, we believe, on the best information and opinions now to be procured, that the first species of a commercial vessel in most frequent, as well as extensive use, among nations widely separated from each other, was the RAFT, a collection of trees, rudely fastened together with ropes, formed, most probably, from the barks of the very trees which constituted the float, or from some other coarse material which the dawning genius of our early ancestors had discovered to be applicable to that purpose. Experience soon taught the navigators that they were deficient in the power of directing the course of this unwieldy machine, so as to be certain, in spite of the natural opposition of winds and currents, of reaching in safety the precise spot they wished: and, notwithstanding the manifest inconvenience which must have attended the continued use of so imperfect a structure, a very extensive period of time appears to have elapsed before the improvement on it became general. To remedy this inconvenience a simple addition was first devised, which consisted of nothing more than a few thick planks of wood thrust down into the water, to the depth of three or four feet, between the joints of the trees which composed the raft; these being raised or lowered according to the wish of the untutored pilot, were found, by experience, to aid him considerably in the management of his vessel.

Besides the Ægyptians, to whom the original invention has been attributed, the same contrivance appears to have been adopted by the Phœnicians and Ethiopians, the latter of whom are said to have undertaken what, speaking comparatively, might be considered very distant voyages, with no better means of maritime conveyance. Sicily, Corsica, and various other islands in the Mediterranean are said to have been first colonised by navigators who had no other means of transporting themselves.

Instances are not wanting, also, if any dependance can be placed on the terms used by the ancients, of the application of such vessels to the purposes of war.

Floats exactly answering the same description with those in use among the ancients, have been found in the South Seas within the last century. While genius applied itself to the improvement of this rude system in those quarters, where, from the peculiarity of situation, and their approximation to the ocean, it had first gained footing, the cotemporary inhabitants of countries very far distant, struck out boldly at once into a more scientific, though more contracted scale, for a single tree only, artificially hollowed, served them for every purpose which either their necessities or their ambition appeared to require. The invention is supposed by Pliny and others to have originated with the inhabitants of Germany, who being little known, were considered even by the Romans themselves as barbarians and savages. These boats or vessels, varying very little from the modern canoes, found, almost without exception, throughout all the newly discovered islands, were, in many instances, so capacious as to contain thirty persons, and were on that account either extremely formidable or useful, according to the tempers of the navigators. See Pliny, lib. XVI. cap. 31.

The Romans were little inclined to expeditions of mere discovery, and sought not to become acquainted with any country whose remote situation appeared to defy its arms; they hesitated not to bestow the name of barbarians on the inhabitants of those districts whose manners or customs differed from their own. This fact is a sufficient reason why a people, whose judgment, taste, and consummate knowledge, in what were considered the polite arts, should have so little knowledge of the science of ship-building, which might have gratified their ambition to the greatest extent.

The strict analogy the gallies of the Romans and Grecians bore to those possessed by the inhabitants of the Sandwich Islands, in the South Seas, seems to prove that they must have originally gone thither from Athens or from Rome. Although ignorant of the principles of the science, the ancients soon discovered, without much investigation, many of those essential points which, even at the present time, are considered among the most valuable and interesting that relate to it. They found out, without much enquiry, and probably without a knowledge of the cause, that the breadth of a vessel, extended beyond a certain proportion, materially retarded her progress through the water: they discovered that a rotundity of shape caused their gallies to roll; and that while the extension of breadth retarded their motion, too great a diminution of it produced an inconvenience of greater consequence, rendering their vessels so constructed extremely liable to be upset by any sudden shock, either from the wind or any enemy. In short, by comparing the knowledge of the ancients with improvements of the moderns, it plainly appears that not only this but almost all the sciences may be considerably simplified by a strict attention to what has been the practice of times far remote; and the modern artist will find, by looking back, that those ideas which he fancied were his own, had been promulgated to the world long before he was in existence: hence he will find his labours considerably shortened by data established on actual experiment, on which he may raise a superstructure without tormenting himself about rendering the foundation of it secure.

The discovery of ship-building, or rather the invention of it, is attributed by the ancients to a casual observation on the facility with which a split reed (in Latin, canna) floated on the surface of the water, and from that term is derived the Indian word CANOE.

The canoe of the North American, called by him Periague, which serves him as a fishing-boat on the coast, and with which he travels and trades along his rivers, is made of bark. The Aborigines of Canada used the bark of the birch, and sometimes constructed them of a sufficient size to hold four or five persons. For a description of the canoe of the Indian, and the Caribbee, &c. see the article CANOE.

The raft and the canoe at length became inadequate to the wishes of the possessors, and ingenuity of course was stimulated to the contrivance of what was considered as a necessary extension.

The variety of inconveniences to which the primary invention was liable, was afterwards much reduced by that more ingenious piece of mechanism, called by the Romans navis oneraria, by the Grecians Φορτηγός, a ship of burden, built either for the purpose of commerce, or for the conveyance of troops and different warlike and other stores, which the frequent contentions between nations rendered in some measure indispensably necessary.

The use of the sail appears, from the most authentic testimony, to have been very particularly appropriated to this class of vessels.

Previously to our entering farther on this subject, it may be necessary to state here concisely the various additions and progressive improvements made in the construction of a ship.

The hull, of course, consisted of three parts; viz. the prow, or head; the poop, or stern; and the body, or midship-frame: under the bottom, and along the centre of the latter, passed the carina, or keel, which, by dividing the element on which the vessel floated, in an acute angle, it was found to contribute, in a very great degree, towards increasing the celerity with which the hull passed through the water; at the same time it produced another good effect, serving as it does in modern use for the foundation of the ribs or timbers, which formed the body, bearing no slight resemblance to the vertebræ of the back, in the animal and human frame; continuing up the head or bow of the vessel in a curve line, conforming to its shape, it became what is now called the stem: as well adding stability and firmness to the front most opposed to the assaults of the sea, as enabling it to divide and pass through the swell of it, with greater ease and velocity than the adoption of any other form would have permitted.

The keel is reported to have been generally omitted during the early ages of navigation in vessels intended merely for commerce: but as many inconveniences and dangers were soon discovered from this omission, the keel was universally and indiscriminately applied to them as well as to vessels of war.

To this improvement of the keel subsequent experience suggested the addition of what the Greeks call φαλκις, or kelson, which confining the heads of the floor-timbers, then in two parts, joined into and divided by the keel, very materially contributed to strengthen the vessel. Close to the kelson was the well, contrived as a receptable for all that bilge-water which the working of the vessel through a rough sea caused the admission of, added to the impossibility of closing the joints or seams so completely by calking, but that under such circumstances some must find its way.

The part immediately above the kelson was called the κοίλη, or hold, and from thence is derived the English word keel, which forms the bottom of it. Aloft beams were fixed, which naturally strengthened the vessel, and supported that necessary covering, the deck. The frame consisted of such timbers as formed the principal strength of the vessel, which might then have been considered as complete; but if it be fair to give any pre-eminence where two distinct parts mutually contribute to the support and perfection of each other, that which may be deemed the most material, is the ἀποζώμα, the side, or exterior planking of the vessel.

As the frame, especially in midships, rose at right angles from the keel, so was the planking in former times, as now, put on in a line nearly parallel to it, allowing, as was necessary for the curve, or sheer of the frame: it completely enclosed it, being closely as well as firmly attached and fastened to it by means of large nails or bolts, formed of iron, some of which, as necessity required, passing through both, were bent or clinched thereby rendering the whole structure firm and compact. As it was found impossible, particularly in vessels of large dimensions, to procure planks of sufficient length to extend from the stem to the stern, the danger or inconvenience that might otherwise have arisen from the end of either starting, was in a very great measure obviated by the ingenious and useful introduction of a *dovetail*, which connected them so securely, that little danger appeared to remain of their ever separating. The side was, as now, divided into different parts, and distinguished by different appellations; the lowest was termed θάλαμος, or the floor timber; the second, corresponding to that part of the hull now distinguished by the name of second futtock, was styled ζύγος, so termed from the junction of the timber at that part; the upper division, θράνος. signifying a bench or seat; the deck, which was called καταστωμα, was thrown over at the part now termed the top timber. In each of these divisions that particular class of warlike vessels, called Triremes, which, as they were the most frequent in use, so were they most commonly noticed; and from which all deductions or descriptions relative to the ancient marine have been usually drawn, had, as is now commonly thought, a tier or range of oars. The opening in the side of the vessel through which these were worked, was called τράφηξ, probably from τρέπω, to turn, and are now, when adopted in modern use, styled row, or row-lock ports.

There exists some difference of opinion among antiquaries relative to the form and disposition of these ports; some asserting that the aperture was continued through the whole range; while others, with much more reason on their side, attribute a distinct port to each oar. It is evident this supposition is correct, from the term which we find frequently bestowed on them by the ancients themselves, of columbaria, or pigeon-holes, a term so congruous to their form, as to prevent all possibility of doubt or dispute. These different tiers were distinguished in the Trireme by terms analagous to their situation; the lower being called θαλαμια; the second, ζυγια; the upper, θρανια.

Historians and others have been so extremely vague, irregular, and contradictory in the accounts they have afforded, not only of the particular form in which the galley was constructed, but also as to other points not less consequential, that investigation, were they to be implicitly relied on, would be extremely difficult. In aid of this inquiry the curious have therefore had recourse to the very indeterminate information of coins, and such remnants of sculpture as the ravages of time, and the barbarous fury of invaders, have left to treasure up in the cabinets of the curious. The information they afford, though founded, perhaps, on the most respectable evidence now existing, is at best extremely imperfect.

Among the most probable and the most rational explanations that have yet been given, is one by Mons. Lescallier. It solves many of those strange assertions made by the ancients of the magnitude of particular vessels, which, throwing an air of fiction and romance on their descriptions, consequently induce posterity to doubt, if not totally discredit them. "We have, for a long time," says he, "treated as a kind of visionary chimera, the account of three, four, five, and even eight tiers of oars, one above the other, by which the curious, who are unacquainted with naval matters, wish to explain the different appellations bestowed on ancient gallies, called Triremes, Quadriremes, Quinquiremes, and Octoremes. Whoever will give himself the least trouble to reflect on the subject, will very easily perceive the absolute impossibility of any vessel being able to carry even four rows or ranks of oars thus disposed. In modern gallies, which have only one tier, and are in length equal to a ship carrying sixty-four guns; the oars, though the supporting point, or row-lock port, is as near the water-line as possible, are forty-four feet long. Allowing a space of four feet and a half between the lower tier of row-ports, and that immediately above it, the oars of the second must, pursuing this rule, be seventy-seven feet in length; those of the third an hundred and ten; those of the fourth one hundred and forty-three, &c." Where can we, as judiciously remarked by this author, either find wood proper for the formation of these oars, or men powerful enough to use them? Even the third tier could not be managed properly, were not the vessel perfectly straight, or, according to the English term, *wall-sided*. The oars of the lower rank, too, must have been extremely short, so as to act on the surface of the water at a very small distance from the side of the vessel, in which case it must be remarked, that it is very evident they could not be of any service except in a dead calm. As to the quadragintiremes, or vessels usually described as having forty ranks or tiers of oars, we cannot reconcile the report to our understandings, except by supposing them nothing more than gallies fitted with as many oars in each rank. Those who pretend to give any other interpretation may as well attempt to prove, that a modern ship of war, mounting 100 guns, had as many tiers of cannon, one above the other.

"It may probably be interesting," says Lescallier, "to explain this opinion more fully; even should it be deemed erroneous, it will be some consolation to reflect it is not the first error the investigation of this subject has given birth to.

"The Uniremes may be supposed to have been those gallies or vessels which had only one row of oars extending between their masts, or, perhaps, the entire length of the vessel, like the modern feluccas of Barbary; and, consequently, required only one rank of rowers.

"The Biremes had one tier of oars between their masts, and another abaft the main-mast.

"The Triremes appear to have been gallies of a still more formidable description than the preceding, having one tier of oars extending between the masts, a second abaft the main-mast, and a third forward, near the prow or stem, before the fore-mast.

"The Quadriremes had their oars ranged like the Triremes, with the difference of having two tiers of oars one above the other abaft the main-mast.

"The Quinquiremes were also of the same description, with the addition of a second tier of oars forward.

"The Octoremes had two tiers of oars in the midships, and three at the stem and stern, making in the whole eight. It cannot be denied that some vessels had three entire tiers of oars; this is, indeed, established to have been the case from the evidence of a multitude of ancient sculptures; but there is no certain proof of any having been constructed with a greater number. With regard to the Octoremes, they were enormous floating structures, built merely for the purposes of luxury, and to gratify a ridiculous ostentation; so unfit for war, or even navigation, that they could not venture to sea without manifest danger. Of this description was the celebrated galley of Philopater; such also was that constructed by Archimedes, for Hiero, King of Syracuse, and presented to Ptolemy; and lastly, of the same class may that, built in the reign of the Emperor Caligula, be supposed to have been, which foundered in the reign of Claudius, and was irrecoverably lost in the port of Ostia."

The foregoing explanation, which appears perfectly simple and conformable to reason, enlarges our ideas of the marine of the ancients, which has hitherto been very much misrepresented or misunderstood.

The gallies of war, when the custom of naval hostility was introduced, and gradually advanced in general practice, were certainly improved, and perhaps enlarged; but the peculiar exigences of the state and the mode of fighting then practised not requiring an attention to those points which have become necessary since the revival of the science in modern times, the ease with which vessels were at that time built, rendered it possible for a powerful nation to send forth an armament as formidable, in respect to numbers, as it thought proper, or could find persons to navigate and man it. The gradual diminution in the numbers of vessels composing fleets, as the improvement in the construction of those vessels gradually advanced, forms no slight internal evidence in favour of the correctness with which historians have given such numerical statements, as, at first sight, may have been considered as exaggerated and untrue. The same points pervades the chronology of naval war, and serve for, at least circumstantial evidence, as well in the support of the historians of antiquity, as of those in modern Europe, whose accounts have been rejected by many as legendary tales, fit only for the extravagance of romance, or the allowed effusions of poetical fancy.

To conclude, we may venture to assert, that the gallies of the ancients were as long as any modern ships of war, though very narrow, and much less raised from the surface of the water, if we except the Octoremes, vessels with eight ranks, or as some will have it, distinct tiers of oars.

Though in the time of action the success of every manœuvre, and the event of the encounter itself, in a great measure depended on the discipline and strength of the rowers, yet it is not from thence to be concluded, that gallies were not at other times deprived of the use of the mast and sail. When the wind was fair, they constantly made use of the sails, and worked their vessels either with those only, or aided by the oars also, if it was so nearly a calm, as to enable the latter to afford any supplementary assistance: this custom has been continued in the Mediterranean among vessels still retaining the same name up even to the present time. For farther particulars see the article GALLEY.

Some of the ancient vessels were of wonderful magnitude, if we may credit the testimony of authors. In particular, Hiero, King of Syracuse, is said to have possessed one, intended for the sole purpose of carrying merchandise, which was of four thousand tons burden: and the Ægyptians, at a still earlier period, built a ship, which they called the Isis, that was 180 feet in length, 45 in breadth, and 43 in perpendicular height from the upper-deck to the bottom of the pump-well.

The inhabitants of Alexandria were also much noticed, in ages somewhat later, on account of the immense size of the vessels which they constructed for similar service to that last-mentioned.

After making all possible allowance for the extravagant notions of ancient writers, we may fairly conclude that the science of naval architecture very rapidly advanced, in some particular countries, soon after its first discovery; for though we may doubt some particulars, still there will remain behind firm and immoveable facts, fully sufficient to convince us, that it must, in point of strength at least, have been conducted on fixed and determinate principles, established by close attention and considerable experience. The principal proportions, if those of the Isis are to be taken as a pattern, differed not very materially from those even of our own times: but we are much in the dark as to any other of those more minute particulars, the knowledge of which would enable us to form that perfect representation of their vessels, which future ages, no matter how remote, will be able to amuse themselves with, of those built at the present time.

At a period no farther distant than the 15th century, there is some doubt and much obscurity among writers in their descriptions of the form of particular celebrated ships; and historians have also given extravagant accounts of vessels constructed even in these times, which appear to be derived more from fancy than from fact.

King Henry the Seventh, who, from his long residence on the continent, had opportunities of acquiring greater skill in maritime affairs than most of his predecessors, and seems to have been the first king that thought of raising such a naval force as might be at all times sufficient for the services of the state, built a ship called the *Great Harry*, which cost him about 14,000*l.* and which, properly speaking, was the first ship of the royal navy. She was burnt by accident, at Woolwich, in 1553. See DERRICK's *Memoirs of the Rise and Progress of the Royal Navy.*

Henry the Eighth, in the year 1515, built at Erith the Henry Grace-de-Dieu, of 1000 tons burden; she carried 19 brass and 103 iron pieces of ordnance, was manned with 350 soldiers, 300 mariners, and 50 gunners; and is of very different force from that which the drawing preserved in the Pepysian library seems to convey the idea of. Taken in a strict sense, it can no more be considered as an actual portrait of a ship, rudely as vessels might then be constructed, than can the uncouth figures of vessels which are so frequently seen on the reverses of ancient coins, be taken as the actual and correct records of the form of gallies. Not to speak contemptuously of so curious a document, it can only be esteemed as the general resemblance of a ship, such as might be sketched by the most artless hand, upon mere recollection, and at a very remote period from actual inspection.

It is very evident, that, at the period now treated of, there were two distinct fashions in ship-building observed by marine architects, particularly those in Britain. In the tapestry woven to commemorate in this country the destruction of the Spanish Armada, this point, when joined with other confirming evidence, appears established almost beyond controversy. The first of these fashions, derived originally from the Venetians, and transmitted from them to the English, was certainly adopted by the constructor of the ship in question. There are very many in the hangings just mentioned, which bear so close a resemblance, in the principal particulars, to the Pepysian drawing, as may silence those who boldly reject the drawing in question, because some parts of it are incorrect and absurd. On the other hand, there is a second description of vessels, which appear to have been peculiar to the English, and contrived as a very wise improvement on the ridiculous height of the first, which bears so great a similitude to a print published by a person of the name of Allen, in 1756, and professed to be a representation of the Great Harry; that it may also serve to convince us this print is not so bad a representation of the ship, as many persons have considered it. In fact, it seems certain, that such vessels, so differing from each other, were actually cotemporary, and engaged at one and the same time, in the same line of service: therefore, taking the whole of the evidence into consideration, it seems but fair to assert, that the drawing and the print were both of them tolerably correct, that one was the production of a very inferior, and the other of a much more polished, artist.

Our historical accounts of the Spanish Armada, in respect to the vessels of which it consisted, compared to the English fleet that opposed it, represents their magnitude and lofty appearance nearly as far superior as a vessel is to a boat she carries to attend her; yet it will be found, on examination, that there were only four ships in the whole Spanish fleet superior to the Triumph, commanded by Sir Martin Frobisher. Although in the comparative statement of the two armaments, that of England was certainly very far inferior in respect to tonnage, yet the terms made use of to excite wonder and applause were undoubtedly carried beyond the bounds of truth and propriety. Of the same nature, and probably owing to the same cause, are the accounts of the fleets of Darius, Xerxes, and of Anthony.

Another circumstance exactly in point will be found in different portraits of the *Sovereign of the Seas*, which was built at Woolwich, in the year 1637, in the reign of King Charles the First. According to Mr. Thomas Heywood's publication, addressed, at that time, to the King, the Sovereign of the Seas was in length by the keel, 128 feet, within some few inches; her main breadth 48 feet; in length, from the fore-end of the beak-head to the after-end of the stern, *à prora ad puppim*, 232 feet; and in height, from the bottom of her keel to the top of her lantern, 76 feet: bore five lanterns, the largest of which would hold ten persons upright; had three flush decks, a fore-castle, half-deck, quarter-deck, and round-house. Her lower tier had thirty ports for cannon and demi-cannon,

Middle tier,	30 for culverines and demi ditto.
Third tier,	26 for other ordnance.
Forecastle,	12, and two.

Half-decks had 13 or 14 ports more within-board, for murdering-pieces, besides 10 pieces of chase-ordnance forward, and 10 right aft, and many loop-holes in the cabins, for musket-shot. She had eleven anchors, one of 4,400 lbs.; was of 1637 tons burden; and built by Peter Pett, Esq. under the direction of his father, Captain Phineas Pett, one of the principal officers of the navy. She had two galleries besides, all of most curious carved work, and all sides of the ship carved with trophies of artillery and types of honour, as well belonging to sea as land, with symbols appertaining to navigation; also their two sacred majesties badges of honour; arms, with several angels holding their letters in compartments; all which works were gilded over, and no other colour but gold and black. One tree, or oak, made four of the principal beams, which was 44 feet, of strong serviceable timber, in length; three feet in diameter at the top, and ten feet at the bottom.

She was the largest ship that had ever been built in England, and is said to have been designed only for splendor and magnificence: but being taken down a deck lower, she became, according to report, one of the best men of war in the world.

The renowned Sir Walter Raleigh, speaking of the ability and knowledge possessed by the British shipwrights, observes, "To say the truth, a miserable shame and dishonour it were for our shipwrights, if they did not exceed all others in the setting up of our royal ships, the errors of other nations being far more excusable than ours; for the kings of England have for many years been at the charge to build and to furnish a navy of powerful ships for their own defence, and for war only; whereas the French, the Spaniards, the Portuguese, and the Hollanders, till of late, have had no proper fleet belonging to their principal state."

It may be necessary to mention here the different methods that have been practised in the sheathing of ships, from the original adoption of the measure to the present time: the first was a thin covering of deal, or fir plank, into which the worm penetrated; but, which, consequently, preserved the bottom itself till they so far demolished the covering as to acquire an easy passage into the interior part, an injury which took them some time to effect. An attempt was made in the year 1675 to introduce sheet lead for thin deal, as a more lasting substitute; but, after some experiments, the project was totally abandoned, and the original method persevered in.

Some few years afterwards an addition was made to the wood covering, from which much advantage was looked for, and no inconsiderable benefit derived: the bottom of the ship having first received a coat of pitch, the whole was completely covered with brown paper, which, of course, closely adhered to it; a second paying of pitch, mixed with tar, was then laid on the paper; and a fourth coating of short hair carefully attached to the tar; the deal sheathing was then brought over the whole, and being firmly fastened to the bottom by a great number of nails, the whole operation was rendered complete. A number of experiments, with various materials introduced between the wood sheathing and the bottom, among which, that of coating the latter with lime is to be remembered, have been tried at different times; but none of them have proved so effectual or useful as that just described.

In the year 1758, a trial was made on a small British frigate called the Alarm, of the effect which a sheathing of copper would produce in the double purposes of preserving the bottom from the injury occasioned by the worm, and contributing to the swift sailing of the vessel. It was proved, that no adhesion of barnacles, or any other substances, could take place on the copper; so that not only the expense as well as time of graving and cleaning them would be saved, but that they might remain, so long as they continued fit for service in other respects, in the same condition with respect to sailing, as they were the first hour when they were sent to sea. The hope was realised in every respect; and, in the course of the war with America, which terminated in 1783, our ships of every class were coppered; and it was ordered by the government, in November of that year, that, in future, all ships should be copper fastened under the load draught of water.

The custom of sheathing, however, is certainly of very ancient date, which has been fully proved by the discovery and rescue of Trajan's galley from the lake Riccio, where it had remained under water for more than thirteen centuries. Leo Baptisti Alberti, who records the circumstance, states, on his own inspection and knowledge, "that the pine and cypress, of which it was built, had endured, and were then in so sound a state as to be nearly incredible: the bottom was, according to the modern, doubled: the seams had been evidently calked with linen, and the whole of the external part carefully smeared, or payed with a coat of Greek pitch, over which was brought a sheathing, formed of lead, rolled or beaten to a proper thinness, and closely attached to the bottom by a sufficient number of small copper nails." For further particulars see CHARNOCK's *History of Marine Architecture.*

Locke, who has noticed the above circumstance in his History of Navigation, observes, "here we have calking and sheathing together above sixteen hundred years ago; for," adds our author, "I suppose no man can doubt that the sheets of lead nailed over the outside with copper nails was sheathing, and that in great perfection; the copper nails being used rather than iron, which, when once rusted in water with the working of the ship, soon lose their hold, and drop out."

As the limits of this volume will not permit us to proceed farther in the details of this history; nor to recapitulate the many late improvements in the science of naval architecture; we shall, therefore, refer the reader for the state to which it is now arrived, to the different articles relating thereto in this work, and particularly to the following article.

NAVAL *Architecture*, or Ship-building, may be distinguished into three principal parts.

1st. To give the ship such an exterior form as may

be most suitable to the service for which she is designed.

2d. To give the various pieces of a ship their proper figures, and unite them into a firm and compact frame, so that by their combination and disposition they may form a solid fabric, sufficient to answer all the purposes for which it is intended.

3d. To provide convenient accommodations for the officers and crew, and also suitable places of stowage for the cargo, furniture, provisions, artillery, ammunition, &c.

The exterior figure of a ship may be divided into the bottom and upper works.

The figure of the bottom is determined by the qualities which are necessary for the vessel, and conformable to the service for which she is proposed.

The upper works comprehend all that part which is usually above the water when the ship is laden.

The limits of our design will not admit of a minute description and enumeration of all the pieces of timber which enter into the construction of a ship, nor of a particular description of their assemblage and union; or the manner in which they reciprocally contribute to the solidity of those floating citadels. It nevertheless appears necessary to give a general idea of the use, figure, and station of the principal pieces, to those who are entirely unacquainted with the subject. As our definitions will be greatly illustrated also by the proper figures, we have annexed to this article a plate, which comprehends some of the most material draughts, as well as a representation of the principal pieces employed in naval architecture.

It is usual among shipwrights to delineate three several draughts.

1st. The whole length of the ship is represented according to a side view, perpendicular to the keel, and is termed the plane of elevation, or sheer-draught. Plate XVII.

2d. The ship is exhibited according to an end view, and stripped of her planks, so as to present the outlines of the principal timbers; and this is properly termed the plane of projection, or the vertical plane of the timbers, because it shews the projection of their frames relatively to each other.

3d. It is not sufficient to have the vertical curves of the bottom in different places, for a distinct idea of the horizontal curves is also equally necessary and useful: this is obtained by means of *water-lines* traced upon what is called the horizontal plane.

The elevation or sheer draught determines the length and depth of the keel; the length and projection, or *rake* of the stem and stern-post; the position of the midship-frame upon the keel, together with that of the principal frames afore and abaft; the wales, the dimensions and situations of the gun-ports, the projection of the rails of the head and quarter gallery, with the stations of the masts and channels.

This draught, however, conveys no idea of the vertical curve of the ribs or timbers; for, as their projection will be only represented in a plane elevated upon the length of the keel, they will appear in this direction no otherwise than as straight lines. To perceive these curves accurately, they must be regarded in another point of view, which will represent their projection upon a vertical plane, supposed to cut the keel at right angles in the place where the ship is broadest. For, as all ships are broader near the middle of their length than towards the extremities, it is evident that the timbers are more extended in proportion. The most capacious of these represents what is called the *midship-section*; and upon the area of this section is delineated the projection of all the others.

Thus the plane of projection limits the different breadths of a ship in various points of her length, and exhibits the outline of the timbers respectively to each other, as they are erected upon the keel. Accordingly, this draught ought to present a variety of sections of the ship in different places of her length, and always perpendicular to the upper edge of the rabbet of the keel, so that the eye of the observer, when placed in what may be properly termed the *axis* of the ship, may perceive the several sections at one glance; that is to say, when looking full on the stem, from before the ship, discover the fore timbers; and when looking from behind, directly on the stern, he shall perceive the form of the after timbers.

To form a just idea of this plane, therefore, we ought to suppose a ship resting upon the stocks, in the same position as when afloat upon the water. Thus a variety of black vertical lines may be drawn at equal distances upon the bottom, which is white, to form different outlines of the ship corresponding to the timbers within. It is to be observed, that the fashion of the inferior timbers must conform to the figure of the midship-section, which is placed in the fullest part of the ship; and as the planes of all the other timbers diminish in a certain progression as they approach the stem and stern, they are properly delineated on the plane of the midship-section, which also represents the depth of the keel and length of the midship-beam.

As the two sides of a ship ought to be exactly alike, it is judged sufficient to represent the sections of the fore-part of the ship on the right side, and those in the after-part on the left, so as to perceive all the sections, as well afore as abaft, upon one plane. See PROJECTION, plate XVII.

However necessary it may be to understand precisely the vertical curves of the bottom, it is no less requisite to have a just idea of those which are horizontal.

The horizontal, or floor plane, is that upon which the whole frame is erected, and will be more clearly understood by previously describing the water-lines and ribands of which it is composed.

When a ship floats upon the stream, it is evident that her upperworks will appear to be separated from the bottom by the surface of the water, which will accordingly describe an imaginary horizontal line upon the bottom from the stem to the stern-post.

The most elevated of those lines is termed the upper horizontal or water-line; and, in order to prove the fairness of the body, the several horizontal lines are set up at equal distances above the base-line, and distinguished by the water-lines H I J K L, in the PROJECTION. This we would always recommend to be done, although the water-lines of many ships are not parallel with the line of the keel; for, unless they are parallel with the keel, they cannot answer the intended purpose of laying off the ship.

By lightening a ship gradually, and at the same

time preserving the direction of her keel, or the angle which it makes with the surface of the water, a variety of water-lines may be drawn parallel to each other. See a further illustration of these lines in the article WATER-LINE. See also their figure on a ship's bottom, plate VII. fig. 5.

The ribands are likewise of great utility in ship-building; they are narrow and flexible planks placed on the bottom at different heights, so as to form a sort of mould for stationing the inferior timbers between the principal ones. They differ from the water-lines, inasmuch as the latter have only one curve, which is horizontal, whereas the ribands, besides their horizontal one, have vertical curves from the main-breadth upwards. See the article RIBANDS.

We have already observed, that the qualities required in a ship ought to determine the figure of the bottom: a ship of war, therefore, should be able to sail swiftly, and carry her lower tier of guns sufficiently out of the water. A merchant ship ought to contain a large cargo of mercantile goods, and be navigated with few hands; and both should be able to carry sail firmly, steer well, drive little to leeward, and sustain the shocks of the sea without being violently strained.

The first thing to be established in the draught of a ship is her length; and, as a ship of war, according to her rate, is furnished with a certain number of cannon, which are placed in battery on her decks, it is necessary that a sufficient distance should be left between the ports to work the guns with facility, and particularly to leave space enough between the foremost gun and the stem, and between the aftmost gun and the stern-post on each side, on account of the arching, or inward curve of the ship towards her extremities.

When the length of a ship is determined, it is usual to fix her breadth by the dimensions of the midship-beam. On this occasion the shipwrights, for the most part, are conducted by rules founded on their own observation; for, having remarked, that some vessels, which by repeated experience have been found to answer all the purposes of navigation, have a certain breadth in proportion to their length, they have inferred that it would be improper to depart from this proportion: but, as other ships have been constructed with different breadths, which were equally perfect, a variety of different general rules have been adopted by these artists, who are accordingly divided in their opinions about the breadth which ought to be assigned to a ship relatively with her length, whilst each one produces reasons and experience in support of his own standard. Those who would diminish the breadth allege, 1st. That a narrow vessel meets with less resistance in passing through the water; 2dly, That by increasing the length she will drive less to leeward; 3dly, That, according to this principle, the water-lines will be more conveniently formed to divide the fluid; 4thly, That a long and narrow ship will require less sail to advance swiftly; that her masts will be lower, and her rigging lighter; and, by consequence, the seamen less fatigued with managing the sails, &c.

Those, on the contrary, who would enlarge the breadth, pretend, 1st, That this form is better fitted to receive a good battery of guns; 2dly, That there will be more room to work the guns conveniently; 3dly, That by carrying more sail, the ship will be enabled to run faster; or, that this quality will at least overbalance the advantage which the others have of more easily dividing the flood; 4thly, That, being broader at the load-water line, or place where the surface of the water describes a line round the bottom, they will admit of being very narrow on the floor, particularly towards the extremities; and, 5thly, That a broad vessel will more readily rise upon the waves than a narrow one.

From such opposite principles has resulted that variety of standards adopted by different shipwrights; and a servile imitation of these mechanical methods has, to the great reproach of the art, produced all these pretended rules of proportion: for the various models they have hitherto adopted indisputably prove their doubt and uncertainty with regard to their proper standard. Hence these pretended mysteries which are only to be revealed to such as are initiated into the craft! And hence violence of opposition, and mutual contempt amongst the artists! Indeed, nothing appears more effectually to have retarded the progress of naval architecture, than the involving it in mysteries which the professors would gravely insinuate are only intelligible to themselves. This ridiculous affectation is, nevertheless, two generally retained, notwithstanding the example to the contrary of some of the most able shipwrights in Europe, who are real masters of the theory of their art, and do honour to their profession, and who are justly exempted from the censure to which the others are often exposed.

It is not to be expected that an art so complicated and various, comprehending such a diversity of structures, can be treated at large in a work of this sort. To enter into a particular detail of the theory and practice, to explain the different parts with sufficient accuracy and perspicuity, would of itself require a large volume, and, consequently, much exceed the limits of our design. See the article SHIP-BUILDING.

Being thus necessitated to contract our description into a narrow compass, it will be sufficient to give a general idea of the subject, to describe the principal pieces of which a ship is composed, and to explain the principal draughts used in the construction thereof.

As the several lines exhibited in the planes of elevation, projection, &c. will be rendered more intelligible by a previous account of the principal of those pieces, it may be proper to begin with reciting their names, and giving a summary description of their uses and stations. They are, for the most part, represented according to the order of their disposition, in plate XVII. fig. 1, and termed PIECES OF THE HULL. Plate XVII. Fig. 1.

A, the pieces which compose the keel, and which are securely bolted together, and clinched.

B, the stern-post, which is tenoned into the keel, and which unites the sides of the ship abaft.

C, the stem, which is composed of three pieces scarfed together, into which the ship's sides are united forwards.

D, the inner post, into which the transoms are let.

E, the false post, which serves to augment the breadth of the stern-post.

F, the pieces of dead-wood, which form the after-part of the ship.

G, the knee of the stern-post, which unites it to the keel.

H, the apron, which is fayed on the inside of the stem, to support the scarf thereof; for which reason, the scarf of the former must be at some distance from that of the latter.

I, the stemson, which reinforces the scarf of the apron.

K, the wing transom, which is fayed across the stern-post, and bolted to the head of it, having its two ends let into the fashion-pieces.

L, the deck transom, parallel to the wing transom, which is secured in the same manner.

M, the filling transom.

N, the lower transoms.

O, the foremost fashion-piece on one side; the heel of which is connected with the dead-wood, and the head is secured to the wing-transom.

P, the middle fashion piece.

Q, the after fashion piece.

R, the side counter-timber, which steps on the end of the wing-transom.

S, the sternson.

T, the gripe.

U, the lengthening piece on the side counter-timber.

V, the scarf of the after-piece of the keel.

W, the pieces which compose the kelson, and which are scarfed together, and placed over the middle of the floor timbers, upon each of which they are scored about an inch.

Y, the several pieces of the knee of the head; the lower part of which is fayed to the stem; the heel being scarfed to the gripe.

In vessels of war, the general dimensions are established by authority of officers appointed by the government to superintend the building of ships. In the merchant-service, the extreme breadth, length of the keel, depth of the hold, height between decks and in the waist, are agreed on by contract; and from these dimensions the shipwright is to form a draught suitable to the trade for which the ship is designed.

In projecting the draught of a vessel of war, the first article to be considered is her length. As every ship is much longer above than below, it is also necessary to distinguish the precise part of her height, from which her length is taken: this is usually the lower gun-deck, or the load-water-line. It has been already observed, that water-lines are described longitudinally on a ship's bottom by the surface of the water in which she floats, and that the line which determines her depth under the water is usually termed the load-water-line. In this draught it will be particularly necessary to leave sufficient distance between the ports.

The next object is to establish the breadth by the midship-beam. Although there is great difference of opinion about proportioning the breadth to the length, yet it is most usual to conform to the dimensions of ships of the same rate. After the dimensions of the breadth and length are determined, the depth of the hold must be fixed, which is generally half the breadth: but the form of the body should be considered on this occasion; for a flat floor will require less depth in the hold than a sharp one. The distance between the decks must also be settled.

We may then proceed to fix the length of the keel, by which we shall be enabled to judge of the rake of the stem and stern-post. The rake is known to be the projection of the ship at the height of the stem and stern-post, beyond the ends of the keel afore and abaft; or the angle by which the length is increased as the fabric rises. To these we may also add the height of the stem and wing-transom.

As a ship is much broader at the middle than at the extremities, the arms of the floor-timber will form a very obtuse angle at the extreme breadth; but this angle decreases in proportion to the distance of the timbers from the midship-section, so that the foremost and aftmost ones will form a very acute angle.

Shipwrights differ extremely in determining the station of the midship-section; some placing it at the middle of the ship's length, and others farther forward. They who place it before the middle, allege, that if a ship is full forward, she will meet with no resistance after she has opened a column of water; and that the water so displaced will easily unite abaft, and by that means force the ship forward; besides having more power on the rudder, in proportion to its distance from the centre of gravity; this also comes nearer the form of fishes, which should seem the most advantageous for dividing the fluid.

When the rising of the midship-floor-timber is decided, we may then proceed to describe the rising-line of the floor, on the stern-post abaft, and on the stem afore.

The height of the lower-deck is the next thing to be considered. It is determined in the middle by the depth of the hold; and some builders make it no higher than the stem; but they raise it abaft as much above its height in the middle as the load-water-mark, or draught of water abaft, exceeds that afore. With regard to the height between decks, it is altogether arbitrary, and must be determined by the rate of the ship, and the service she is designed for.

It is also necessary to remember the sheer of the wales, and to give them a proper *hanging*; because the beauty and stateliness of a ship greatly depend upon their figure and curve, which, if properly drawn, will make her appear airy and graceful on the water.

We come now to consider the upper-works, and all that is above water, called the dead-work: and here the ship must be narrower, so that all the weight lying above the load-water-line may thereby be brought nearer the middle of the breadth, when of course the ship will be less strained by the working of her guns, &c. But although some advantages are acquired by diminishing the breadth above water, we must be careful not to narrow her too much; as there must be sufficient room left on the upper-deck for the guns to recoil. The security of the masts should likewise be remembered, which requires sufficient breadth to spread the shrouds. A deficiency of this sort may indeed be in some measure supplied by enlarging the breadth of the channels.

With regard to the qualities required in the construction of a ship, to fit her out for the various purposes of navigation, the reader is referred to the article BOTTOM.

We shall now proceed to explain the plane of elevation or sheer-draught of a modern seventy-four gun-ship as represented in plate XVII. fig. 2. Fig. 2.

ELEVATION, OR SHEER-DRAUGHT.

A A, the keel; whose upper edge is prolonged by the dotted line, upon the extremities of which are erected perpendiculars, which determine the height of the wing-transom.

B, the stern-post.

C, the stem.

D D, the quarter-gallery.

E, the quarter-piece, which limits the stern on each side.

F, the taffrail, or upper piece of the stern.

G G, Profile of the stern, with its galleries.

H H H, the gun ports.

I I I, the channels, with their dead-eyes and chain-plates.

K, the tuckrail.

L, the lower finishing.

M, the upper finishing.

N, the fretwork on the upper finishing.

O, the rudder.

P, the top timber line.

Q, the upper edge of the main-wale.

R, the lower edge of the main-wale.

S, the upper edge of the channel-wale.

T, the lower edge of the channel-wale.

U, the water lines.

X X, the rails of the head.

Y, the knee of the head, or cutwater.

Z Z, the cheeks of the head.

a, the cat-head.

b b, the hawse-holes.

The frame timbers are represented by A C E G I K M O Q S U W Y, in the fore-body of the ship, before the midship-frame.

⊕, represents the midship-frame, called *dead-flat.*

The timbers in the after-body are represented by 2, 4, 6, 8, 10, 12, 14, 16, 18, 20, 22, 24, 26, 28, 30, 32, 34, which are erected abaft the midship-frame.

As the eye of a spectator is supposed in this projection to view the ship's side in a line perpendicular to the plane of elevation, it is evident that the convexity will vanish, like that of a cylinder or globe, when viewed at a considerable distance.

It has been already observed, that the plane of projection may be defined a vertical delineation of the curves of the timbers upon the plane of the midship-section, which is perpendicular to that of the elevation. It is necessary to observe here, that the various methods, by which these curves are described, are equally mechanical and arbitrary. In the latter sense, they are calculated to make a ship fuller or narrower, according to the service for which she is designed, and in the former they are drawn according to those rules which the artist has been implicitly taught to follow, or which his fancy or judgment has esteemed the most accurate. They are generally composed of several arcs of a circle, reconciled together by moulds framed for that purpose. The radii of those arcs, therefore, are of different lengths, according to the breadth of the ship in the place where such arcs are swept; and they are expressed on the plane of projection either by horizontal or perpendicular lines; the radii of the breadth-sweeps being always in the former, and the radii of the floor-sweeps in the latter direction. These two arcs are joined by a third, which coincides with both, without intersecting either.

The curve of the top-timber is either formed by a mould which corresponds to the arc of the breadth-sweep, or by another sweep, whose centre and radius are without the plane of projection. The breadth of the ship at every top-timber is limited by an horizontal line drawn on the floor-plane, called the half-breadth of the top-timbers. The extreme breadth is also determined by another horizontal line on the floor-plane; and the lines of half-breadth are thus mutually transferable, from the projection and floor-planes to each other.

The necessary data by which the curves of the timbers are delineated then are, the perpendicular height from the keel, the main or principal breadth; and the top-timber-breadth: for, as a ship is much broader near the middle of her length than towards the ends, so she is broader in the middle of her height than above and below; and this latter difference of breadth is continued throughout every point of her length. The main breadth of each frame of timbers is, therefore, the ship's breadth nearly in the middle of her height in that part: and the top-timber breadth is the line of her breadth near the upper end of each timber. It has been already observed, that as both sides of the ship are alike, the artificers only draw one side, from which both sides of the ship are built; therefore, the timbers before the midship-section are exhibited on one side of the plane of projection, as in plate XVII. fig. 4, and those abaft before it on the other, as in fig. 5. Fig. 4, and 5.

PLANE OF PROJECTION.

A, the keel.

B B, the line which expresses the upper-edge of the keel, from which the height of each timber and height of its different breadths are measured.

C C, the perpendiculars raised on the line B, to limit the ship's extreme breadth and height amidships; or, in other words, to limit the breadth and height of the midship-section.

D, a perpendicular erected from the middle of the keel, to bisect the line of the ship's breadth in two equal parts.

E, the half-breadth of the top-timber line.

F, the upper height of the main-breadth line.

2, 4, 6, 8, 10, 12, 14, 16, 18, 20, 22, 24, 26, 28, 30, 32, 34, the radii of the breadth-sweeps, in the after-body, fig. 5, abaft the midship-section.

⊕, the midship-section, called *dead-flat.*

A C E G I K M O Q S U W Y, are those of the radii of the breadth-sweeps in the forebody, fig. 4, before the midship-section.

2, 4, 6, 8, 10, 12, 14, 16, 18, 20, 22, 24, 26, 28, 30, 32, 34, are the outlines of the timbers abaft the midship-section, in different parts of their height.

⊕, The midship-section, called *dead-flat.*

A C E G I K M O Q S U W Y, are the outlines of the timbers before the midship-section, in different parts of their height.

G G, the crosses on the dicked line in the fore and after bodies, which are the radii of the floor-sweeps.

The horizontal plane, plate XVII. fig. 3, is composed of water-lines and ribands; it also contains the main, and top-timber breadth-lines, or the longitudinal lines by which the main-breadth and top-timber-breadth are limited in every point of the ship's length. Fig. 3

HORIZONTAL PLANE.

A, the diagonal or riband-line.

B, the main half-breadth line.

C, the top-timber half-breadth line.

D E F G, the horizontal, or water-lines.

I, the round aft of the stern at the height of the main-breadth.

J, the round aft of the stern at the top-timber breadth.

K, the stem.

L, the stern-post.

a c e g i k m o q s u w y, are the joints of the frame-timbers before the midship-frame.

⊕, the midship-frame, called *dead-flat*.

2, 4, 6, 8, 10, 12, 14, 16, 18, 20, 22, 24, 26, 28, 30, 32, 34, are the joints of the frame-timbers abaft the midship-frame.

The plan and elevation of this 74-gun ship are drawn to a scale of $\frac{1}{12}$th of an inch to a foot.

Thus we have endeavoured briefly to explain the nature and uses of the principal draughts used in the construction of a ship, which reciprocally correspond with each other in the dimensions of length, breadth, and depth. Thus the plane of elevation is exactly of the same length with the horizontal or floor-plane. The several breadths of the timbers in the floor-plane and that of the projection are mutually transferable; and the real height of the timbers in the projection exactly conforms to their height in the elevation.

In the same manner, the breadths of all the timbers may be laid from the projection to the horizontal plane, and *vice versa*, from that to the projection. Thus, the height of each timber may also be transferred from the elevation to the projection, &c.

The principal utility of these draughts, is, therefore, to exhibit the various curves of the ship's body, and of the pieces of which it is framed, in different points of view, which are either transverse or longitudinal, and will, accordingly, present them in very different directions. Thus, the horizontal curves of the transoms and water-lines are represented on the floor-plane, all of which are nearly straight lines in the elevation and projection; and thus the vertical curves of the timbers are all exhibited on the projection, although they appear as straight lines in the elevation and floor-plane.

Before this article is closed, it may be necessary to remark, that the various PIECES of the HULL, as well as the lines in the draughts which have not been already defined, are copiously explained in their proper places; as it would have been contrary to the plan of this work to have given a more enlarged description of them here.

That the reader, however, might be the better enabled to comprehend the scope of this article, it was judged necessary to give a general sketch of naval architecture itself; to collect into one point of view the most material draughts by which a ship is constructed, and to describe, as concisely as possible, the several parts of which they are composed.

The principal parts of a ship, also, which are here reduced into a narrow compass, will be represented at large in different places of this work, to illustrate those explanations to which it may be necessary to refer, in order to understand the subject more clearly. Thus, the head, the stern, the quarter, the bow, and the midship-section, of a ship of 74 guns, are exhibited on a scale of $\frac{1}{8}$ of an inch to a foot; by which all the subordinate parts may be distinctly viewed, and their combination and arrangement sufficiently understood, For farther particulars see the article SHIP-BUILDING.

The following are the most celebrated works on Naval Architecture, that have been published by the French and Spanish.—

Elémens de l'Architecture Navale, ou Traité Pratique de la Construction des vaisseaux, by M. du Hamel du Monceau, an excellent work, which possesses the merit of having been the first to illustrate in a satisfactory manner, the theory and practice of the art.

Traité du Navire de Bouguer, which dwells on the matter scientifically; but is rather deficient in the practice, and on the laws of fluids.

Théorie complette de la Construction, et de la Manœuvre des vaisseaux, by Euler.

Scientia Navalis a Mathæo Eulero, is a learned and conspicuous work.

Essai Géométrique et Pratique sur l'Architecture Navale, by M. Vial du Clairbois.

Traité Elémentaire de Construction des vaisseaux, by the same.

Manuel de Construction Pratique, by M. de Lironcourt.

Traité de la Construction des vaisseaux, by M. Dumaits de Goimpy.

Examen Maritimo, in Spanish, by Don Jorge Juane. This work contains a most excellent theory, founded on experience and practice.

A Treatise on Naval Architecture, in Swedish, by Chapman.

Naval Architecture is, also, much indebted to Chevalier de Borda, who was the first founder of the French schools of naval architecture, and applied the principles of Euler to the uniform construction of ships; so that all those of the French navy might be similar with respect to sailing.

In 1781, Mr. Marmaduke Stalkartt, of Deptford, published an excellent Treatise on Naval Architecture, or the Rudiments and Rules of Ship-Building, which is exemplified in a series of Draughts and Plans, with observations tending to the further improvements of that important art.

A second edition was published in 1787.

In 1794, Mr. William Southerland, shipwright, &c. published a small book, called the Ship Builder's Assistant; or, Naval Architecture: containing the method of drawing the plans of ships, and moulding their timbers; and also the practical rules necessary to be observed in building the hulls of all sorts of ships. To which is added, the Scantling, or Mensuration of Ship's Timbers, &c.

In 1804, Steel and Co. of Cornhill, London, published the Elements and Practice of Naval Architecture, in one volume, quarto, developing in a clear and familiar manner, the principles of the art, both theoretic and practical, with all the requisite directions for constructing and completing a ship of every class, from the forming of the draught to the launching into the water.

In the work is included a new and extensive set of Tables of the Dimensions of the most approved ships of each class in the British navy, of the different species of merchant vessels, on the most improved construction, and of numerous boats, &c. from the long-boat of a line-of-battle ship to a wherry. The whole is illustrated with a separate volume of engravings, consisting principally in a series of elegant and accurate draughts of different classes of ships and vessels, from a line-of-battle ship to a boat of the smallest description; with another series on the same grand scale, elucidating the principles of laying-off, &c.

This is considered an invaluable work, both for the purpose of imitation and for constant use.

Steel and Co. have also recently published, The Shipwright's Vade Mecum; a clear and familiar introduction to Ship-building; including practical geometry, &c. with four draughts. The Art of making Masts, Yards, &c. with large plates separate.

There are many other writers on this subject, too numerous to mention here.

NAVAL *Asylum.* The Royal Naval Asylum at Greenwich, is due, originally, to the philanthropy of the late Messrs. Benjamin and Abraham Goldsmid, chiefs of the Hebrew nation, aided by the public and professional skill of Admiral Sir Sidney Smith.

This benevolent institution was founded, at Paddington, in the year 1800, and has been transplanted from thence to Greenwich, for the orphans of his Majesty's seamen and marines, and destined for his Majesty's fleet. From the 21st of October, 1805, it was made a royal foundation by command of his Majesty, in commemoration of the glorious victory of TRAFALGAR, and has obtained official organization. It has for its object, the education, clothing, and maintenance, of 1000 children, whose fathers are, or have been, engaged in the naval service of their country. There are, at present, 800 boys and 200 girls The boys are taught reading, writing, and cyphering; and, where their capacities display fitness, are instructed in navigation; and, during the hours of relaxation, the elder boys are taught rope and sail-making; and they are instructed in the rudiments of naval discipline, by regular veteran boatswains. The girls are taught to read and write; are instructed in needle-work, and household industry, and settled as apprentices. This institution is supported by public grants from Parliament, and private bequests of estates, or donations, being exempt from the operation of the Mortmain Act, on account of its great public object, and national utility.

The building fills up the *vista* between both wings of Greenwich college, to which it seems to form an appropriate centre. The contemplation of this noble institution, and the principles upon which it is founded, reflects the highest honour on its founders, and on the country at large.

For further particulars, and a list of persons appointed by the government to manage and regulate the establishment, see NAVAL CHRONICLE, for 1807, 1808, and 1811.

NAVAL *Charitable Society.* This meritorious and excellent institution was established in 1791, for the purpose of affording relief to the indigent orphans and widows of naval officers, and to officers reduced by misfortune.

Many necessitous objects have been relieved, and a fund is established to enable the society to persevere in their laudable design. The *modus* agreed upon for the annual subscriptions is as follows:—

	£	s.	d.		£	s.	d.
Admirals	3	3	0	Lieutenants	0	5	0
Vice ditto	2	2	0	Physicians	0	10	0
Rear ditto	1	1	0	Surgeons and masters	0	5	0
Captains	0	10	6				
				Pursers	0	5	0

The sum given to each applicant, is limited to 20*l.*

There is a president, vice-presidents, and a secretary for conducting the affairs of the Society.

Subscriptions, in aid of this laudable undertaking, are received at the principal banking houses in London, Portsmouth, and Plymouth; and by the treasurers and secretary to the Society, of whom particulars may be known, on application, by letter, or otherwise. The Society meet at the Thatched House Tavern, the first Monday in January, April, July, and October.

A similar charity was established at the Admiralty in 1732.

NAVAL *Discipline,* (*discipline navale,* Fr.) denotes the exercise and performance of the established code of naval laws and instructions, and the internal regulations of a ship of war.

Long experience has proved the efficacy of this system; and we need only look back on our late naval victories, to be perfectly convinced of the propriety and goodness of that code of laws; which forms the ground-work of all our naval successes; notwithstanding the flimsy arguments and absurd statements that have lately appeared in an essay on Naval Discipline, by a disappointed officer in that service.

NAVAL *Officer.* See the article OFFICER.

NAVAL *Station.* See the article STATION.

NAVAL *Stores.* See the article STORES.

NAVAL *Tactics.* See the article TACTICS.

NAVAL HOODS, or HAWSE-BOLSTERS, (*pièces de bois, ou sont percés les écubiers,* Fr.) in ship-building, are large pieces of plank, or thick stuff, wrought above and below the hawse-holes.

NAVAL LINE. See the article LINE.

NAVIGATION, (*navigation, l'art et l'action de naviguer,* Fr.) is the art of conducting a ship at sea, from one port to another by observation of the heavenly bodies, calculation of the distance or way, made daily, and by steering such a course, under guidance of the compass, as may lead, in the most direct manner, from the place quitted to the ship's destination.

This art is of the highest antiquity, having been cultivated by people independent of each other, who inhabited the sea-coasts of different countries, and had occasion, or found it convenient to convey themselves upon the water from place to place, beginning from rafts and logs of wood, and gradually improving the structure and management of their vessels, according to the length of time and extent of their voyages.

History of NAVIGATION. Of all the inventions and improvements the wit and industry of man has discovered and brought to perfection, none seems to be so universally useful, profitable, and necessary, as the art of navigation. Historians ascribe the invention of it to the Ægenites, the Phœnicians, Syrians, and the ancient inhabitants of Britain. Some will have it, that the first hint was taken from the flight of the kite; others, from the motion of fishes in general; and others, to that of the nautilus, a curious shell-fish, in particular; while a learned editor of Virgil's Georgics believes, that an alder-tree, grown hollow with age, and falling into the river on which it was planted, gave the first hint towards navigation.

Scripture refers the origin of so useful an invention to God himself, who gave the first specimen in the ark built by Noah: for the raillery which that good patriarch underwent on account of his enterprise, is a sufficient demonstration that the world was then ignorant of any thing like navigation, and that they even thought it impossible. But, whatever be the origin of this art, and whatever nation may claim the honour of inventing it, or of having rendered it subservient to the noble advantages of commerce, it is certain, that among all the nations of antiquity, the structure of their vessels was extremely rude, and their method of working them no less defective. They were unacquainted with some of the great principles and operations of navigation, which are now considered as the first elements on which that science is founded. Though that property of the magnet, by which it attracts iron, was well known to the ancients, its most amazing and important virtue of pointing to the north pole had escaped their observation. Destitute of this faithful guide, which now conducts the pilot with so much certainty in the unbounded ocean, during the darkness of night, and when the heavens are covered with clouds, the ancients had no other method of regulating their course, than by observing the sun and stars. Their navigation was, of course, uncertain and timid. They durst seldom quit sight of land, but crept along the coast, exposed to all the dangers, and retarded by all the obstructions, unavoidable in holding such an awkward course.

An incredible length of time was requisite for performing voyages, which are now finished in a short space. Even in the mildest climates, and in seas the least tempestuous, it was only during the summer months that the ancients ventured out of their harbours. The remainder of the year was lost in inactivity.

Those who have written more diffusively upon the subject, have taken a survey of the progress of discovery and navigation among the ancients; beginning with the Egyptians, and proceeding successively with the Phœnicians, Jews, Carthaginians, Greeks, and Romans. From this survey, which may be traced from the earliest dawn of historical knowledge, to the full establishment of the Roman empire, the progress of the ancients appears to be wonderfully slow. It seems neither adequate to what we might have expected from the activity and enterprise of the human mind, nor to what might have been performed by the powers of the great empires, that successively governed the world. If we reject accounts that are fabulous and obscure; if we adhere steadily to the light and information of authentic history, without substituting in its place the conjectures of fancy, or the dreams of etymologists, we must conclude, that the knowledge which the ancients had acquired of the habitable globe was extremely confined.

Nevertheless, the discoveries of the Greeks and Romans were still very considerable, when compared to those of remoter times; and, in the second century of the christian æra, geography, enriched by new observations, made a very conspicuous figure, under the auspices of Ptolemy the philosopher. "Of all the arts and professions which have at any time attracted my notice," says Foster in his History of Voyages and Discoveries, "none has ever appeared to me to be more astonishing and marvellous than that of navigation, in the state in which it is at present; an art, which doubtless, affords one of the most certain irrefragable proofs of the amazing powers of the human understanding. This cannot be made more evident, than when taking a retrospective view of the tottering inartificial craft to which navigation owes its origin, we compare it with a noble and majestic first rate, containing a thousand men, together with their provisions, drink, furniture, wearing apparel, and other necessaries for many months, besides one hundred and twenty pieces of heavy ordnance: and bearing all this vast apparatus safely, and, as it were, on the wings of the wind, across immense seas, to the distant shores."

We have every reason to believe by the authorities of both sacred and profane history, that the Phœnicians were the first, and, for a long period of time, the most successful of the ancient navigators. We find the king of Tyre, whose subjects were of that nation, assisting king Solomon with gold and curious materials for building the famous temple at Jerusalem. Though the virtue of the magnetic needle was totally unknown in those days, yet it is morally certain that these bold navigators not only coasted along the neighbouring shores of the Mediterranean, but sailed southward to Africa, and north as far as Britain, trading for tin to the coast of Cornwall, at a time when the existence of this island was not known to the greater part of the nations inhabiting the continent. The Assyrians, Egyptians, and other ancient states, are reported to have had great fleets before the days of David, or even of Moses. The accounts of the naval power of Semiramis are to be suspected as fabulous; that the Egyptians and some other nations have been represented as covering the seas with their fleets, may, probably, have arisen from the number of Phœnician vessels employed in their service. The Greeks, who learned other arts from them, acquired that of navigation among the rest, and almost as soon as they were formed into states, began to think of making themselves respectable for their fleets, with which they repeatedly defeated those of the Persians, and made themselves masters of the eastern coasts of the Mediterranean, while the Phœnicians were employed in trading and planting colonies in various other parts of the world.

The famous Carthage, a colony from Tyre, from small beginnings, rose to high estimation for her naval power, by which means, in a great measure, she was enabled to contend with Rome for the empire of the world. But the indefatigable industry, unwearied application, and boundless thirst of conquest which marked the character of the Romans, caused them at last to triumph. Though at first they were little skilled in maritime affairs, to which their ancient genius had not led them, yet, finding themselves powerfully supported by a people whose great resources were in trade, and whose naval strength contributed chiefly to their support, they resolved on manning great fleets, in which, though at first they were unsuccessful, they were at length enabled to combat their enemies, as it were, upon their own element, and finally gave them such a terrible overthrow as induced them to accept of such a peace as the conquerors were disposed to grant.

The fleets at this time consisted of gallies of various sizes, with several benches of rowers, and were filled with soldiers who fought with their usual weapons of war, to which were added, various engines peculiar to their situation; and some of the gallies had towers, from whence they shot or threw darts, stones, and other missiles. By the help of these, they could make safe approaches to the walls of towns in offensive war, and when attacked in their own vessels, could carry on a defensive with as great security as if they were on shore.

That distinguished power which the Romans thus wrested from the Greeks and Carthaginians, they preserved till the division of the empire, after which it began to decline. The distant provinces revolted. The nations shook off the yoke of the conquerors; barbarians, whose names were almost unknown, poured in upon the various parts of the divided empire. Rome, sinking under her own weight, was at last sacked by the Goths, and Constantinople taken by the Saracens; which event put an end to the eastern empire, in the reign of Michael Palæologus. Hence originated that consequence which the Arabs derived: though, at first, apparently enemies to learning and the arts, yet, as they extended their power, they became encouragers of them, and, while the destruction of the Western empire had involved Europe in ignorance and distraction, these people began to cultivate useful knowledge, and to carry on an extensive trade with divers nations, though in ships of a very slight construction.

Nor did the division of the Arabian empire prove the extinction of this commerce, which long survived the destruction of the Khalifate, and the remains of which were yet visible to the Portuguese when they entered the Indian seas, along whose coasts the pilots, it is said, were found to have a knowledge of sea-charts, and even that of the compass, the discovery of which was then but recently known in Europe.

During this period the rival republics of Genoa and Venice were almost the only powers that attended to trade and navigation in the western world, the crusades abroad, and the feudal system which prevailed amongst the most respectable powers at home, joined to their intestine divisions, proving most unfavourable to the arts, and prolonging that night of ignorance, whose shades began to be dispelled about the beginning of the fourteenth century.

According to Locke, in his History of Navigation, the magnetic needle, or sea compass, appears to have been invented by Flavio John de Gioja, of Amalfi, about the beginning of the thirteenth century, which wonderful discovery gave rise to many improvements in the art of navigation. Locke's observations on the reason of its tending to the pole, or pointing out the north, are of a very interesting and philosophical nature, and may be found in the 9th volume of the octavo edition of his works.

According to Signor Gregorio Grimaldi's Dissertation, published at Rome, in 1742, the mariner's compass was invented in the year 1302.

M. Montucla, in his *Histoire des Mathématiques*, published at Paris in 1758, affirms, that this invention came from the east, and was known at an earlier period in Europe. However that may be, it certainly had not been commonly used in navigation till during the voyages that were made in the years 1420 and 1421, by Henry Duke of Viséo. This learned Prince, brother to Edward King of Portugal, was a great encourager of the arts, and sent for eminent men from foreign parts, to teach navigation, and make instruments and sea-charts, as appears by a history of Spain, published in 1605.

Herrera, in his *Descriptione de las Indias Occidentale*, published at Madrid, in 1601, says, the celebrated Christopher Columbus, before he attempted the discovery of America, consulted Martin de Bohemia, and prevailed on him to instruct the Spaniards in navigation; for the improvement of which art, the Emperor Charles the Fifth afterwards founded a lecture at Seville, as appears by Hackluyt's dedication to his first volume of voyages, printed in 1599.

Columbus discovered the variation of the sea-compass in the year 1492, if we may give credit to his son Ferdinand, who has asserted this in his life of Columbus, written in Spanish, and printed at Venice in the year 1571: but several other respectable writers about that period seems to ascribe the discovery to the celebrated Sebastian Cabot. See Dr. GILBERT'S *History of the Magnet*, published at London, in 1600; and PURCHAS'S *Pilgrim*, vol. I. published in 1625.

Cabot was a Venetian by birth, and first served Henry VII.; then the king of Spain; and, afterwards returning to England, he was appointed grand pilot by Edward the Sixth, with a salary of 160*l.* a year. Of this famous navigator and his expeditions, many writers have made mention, both English and foreign, in their collection of voyages.

John Werner, of Nuremberg, one of the greatest astronomers of the time, in his Annotations on Ptolemy's Geography, printed in 1514, describes that ancient instrument the cross-staff, and recommends it for observing the distance between the moon and some star, in order to determine the longitude. But the art of navigation still remained very imperfect, as a plane chart was the only thing they had to guide them in their distant voyages, which must have often misled them, especially in those distant from the equator; and from the want of books of instruction for seamen.

About the same time proposals were made for finding the longitude by observations of the moon.

Gemma Frisius, in his *De Principiis Astronomiæ et Cosmographiæ*, printed at Antwerp, in 1530, advised the keeping of time by means of watches and clocks, then newly invented: he contrived a new sort of cross-staff, and in a new edition of his work, printed in 1553, he has given a figure of an instrument he calls a nautical quadrant, which he describes as being very useful in navigation.

The following is an account of the principal writers on navigation:—

The Spanish and Portuguese Writers on Navigation.

It appears by Maffeii Histor. Indic. printed at Florence, that Roderic and Joseph, physicians to King John the Second, of Portugal, together with Martin de Bohemia, a native of Fayal, about the year 1485, calculated tables of the sun's declination, for the use of the sailors, and recommended the use of the astrolabe for taking observations at sea.

A celebrated mathematician, Pedro Nunez, or Nonius, so early as 1537, turned his attention to navigation, and published a book, written in the Portuguese language, to explain a difficulty in that art, proposed to him by the commander, Don Martin Alphonso de Susa; which, with the addition of a second book, he published in Latin, about thirty years after, at Basil or Basle; the whole entitled, *De Arte et Ratione Navigandi*, where he explains the errors of the plane-chart, and gives solutions of several curious problems in astronomy and navigation, particularly that of determining the latitude from two observations of the sun's altitude, and the intermediate azimuth being given; and shews how most of the problems in navigation may be performed on the globe. He also shews a new method of dividing the quadrant by concentric circles; this admirable division is still practised, and, in compliment to the inventor, was first called a *Nonius*, by our royal astronomer, Dr. Edmund Halley.

Two treatises were published in Spanish, containing a system of the art of navigation; the first by Pedro de Medina, at Valladolid, in 1545; and the other by Martin Cortes, or Curtis, at Seville, in 1556. These appear to be the oldest writers who have entered fully into that subject; as Medina, in his dedication to Philip Prince of Spain, seems to lament, that ships were then daily lost for want of teachers and books of navigation. This work being the first of the kind, was soon translated into Italian, French, and Flemish, and served for a long time as a guide to navigators of foreign countries. Cortes, in his dedication to the emperor, boasts of being the first who had reduced navigation into a compendium, and his work was translated into English by Richard Eden, and published in London, in the year 1561, on the recommendation of that great navigator Stephen Burrough.

Cortes's system of navigation consisted of an account of the Ptolemaic hypothesis, the circles of the sphere, roundness of the earth, its latitudes, longitudes, climates, eclipses of the luminaries; a calendar; how to find the prime, epact, &c. and by the last the moon's age, and thence the tides; a description of the sea-compass, not forgetting the loadstone, with something about the variation, calling it north-easting and north-westing, for the discovering of which, by night as well as by day, he said an instrument might be easily contrived. It also contained tables of the sun's declination for four years, in order to find the latitude from his meridian altitude; of the course of the sun and moon; the length of the days; of time and its divisions; to find the hour of the day, and by the nocturnal, that of the night; lastly, Cortes clearly explained the errors of the old sea-chart, in which he made several improvements, and described several curious instruments; one to find the place and declination of the sun, with the days and place of the moon; certain dials, the astrolabe and cross-staff, with a complex machine to discover the hour and latitude at once.

In 1628, Antonio de Naiera published, at Lisbon, a treatise, intitled, *Navegacion Especulativa y Practica*; where, though the author rectifies the tables of the declination of the sun and fixed stars, he proceeds no farther in the theory of navigation than what had been advanced by Nonius, and followed by Cespedes: but, in 1712, was published at the same place, *Arte de Navegar por Manuel Pimental*; where is shewn the use of Wright's Chart. He likewise describes Davis's quadrant.

In 1757, a treatise was published at Cadiz, intitled *Compendio de Navegacion para el uso de los Cavelleros Guardias Marinas*, written by that ingenious gentleman, Don Jorge Juan. This is an excellent performance, and comprehends all the modern improvements.

In 1776 was published, in Spain, *Rudimentos de Tactica Naval pr Instruccion de los ofr. Subalternos de Marina, ordenado pr Dr. Jose de Mazarredo y Salazar Teniente de Navio de la Rl Armada.* This work is considered a good one. The present Spanish naval laws were compiled by this author; who died in 1811, at Madrid, with the title of vice-admiral. He was considered as a learned man, and a good admiral.

An excellent treatise on navigation, in 2 volumes, 4to. intitled *Tratato de Navegacion, por Don Josef de Mendoza y Rios*, was published at Madrid, in 1787; in which the method of finding the longitude by the lunar observations is clearly explained. The investigation of five different methods of reducing the apparent to the true distance, are given in this work. There is also a valuable paper, by the same gentleman, intitled *Recherches sur l'Astronomie Nautique*, in the Philosophical Transactions for 1797; which, besides other things relating to navigation, contains the investigations of many different methods of reducing the apparent distance between the moon and the sun, or a fixed star, to the true distance.

In 1800, his *Colleccion de Tables para vareos usos de la Navegacion*, was published at Madrid, and an explanation of the tables is prefixed, with their use exemplified in the solution of the various computations in nautical astronomy. In the following year, his "*Tables for facilitating the Calculations of Nautical Astronomy*," was printed in London, in one volume, 4to. To this work is added an appendix, containing tables by that ingenious gentleman, H. Cavendish, Esq. for reducing the apparent to the true distance between the moon and the sun, or a fixed star, with precepts for using the same. In this year he also published, in the Philosophical Transactions, an account of an *Improved Reflexion Circle*; on which it appeared he was then making farther improvements. Again, in 1805, Captain Mendoza published his "*Complete Collection of Tables for Navigation and Nautical Astronomy*," in one volume, 4to. in which he has greatly simplified the method of reducing the apparent to the true distance, by means of new and extensive tables. To that work an explanation of the tables is annexed; and their use illustrated in a series of problems and examples. A third edition, much enlarged and improved, was published in 1809.

In 1791, *Reflexiones sobre las Maguinas y Maniobras del uso de Abordo, Ordenadas pr Dr. Francisco Siscar Tente de Navio*; which is the best work the Spaniards have on the subject.

In 1793, *Examen Maritimo Teorico Practico ó Tratado de Mecanica Aplicado a la Construccion con Conosimto y Manejo de los Navios y demas Embarcasiones pr el Gefe de Escuadra, aumtada pr Dr. Gabriel Siscar, Capitan de Fragata.* This work is much in use.

In 1796, *Memoria Sobre las Observaciones de Latitud y Longitud en el mar pr Don Dionicio Arcala Galeano, Capitan de Navio.* The Spanish officers make use indifferently of this work or Mendoza's.

In 1801, *Lecciones de Navegacion ó Principios Nesesarios a la Ciencia del Piloto pr Dr. Dionicio Macarte y Dias, Teniente de Fragata Maestro de la Academia de Pilotos del Ferrol.* This work is in much repute, and considered as an excellent guide for masters in the Spanish navy.

In 1803, *Curso de Estudios Elementales de Marina Escrito de Orden de S. M. C. pr Dr. Gabriel de Siscar*, in 4 volumes. This work is made use of by the midshipmen in the royal schools.

Although the Spaniards were the earliest writers on navigation, yet they appear to have been very slow in adopting the modern improvements.

The English Writers on Navigation.

Dr. William Cunningham, in his 'Cosmographical Glass,' printed in 1559, briefly treats of navigation, and shews the use of the nautical quadrant.

Mr. William Bourne published a supplement to Cortes's work in the year 1577, in which he gives a table of the places and declinations of 32 principal stars, in order to find the latitude, and added a large tide table. By considering attentively the irregularities of the moon's motion, he discovered the errors that sailors had made in finding her age by the epact, and also in their determining the hour from observing on what point of the compass the sun and moon appeared.

Bourne published several other tracts, called *Inventions and Devices*, where he describes a method by wheel-work, of measuring the velocity of a ship at sea.

In 1581, Mr. Robert Norman published a discovery he had long before made, of the dipping-needle, in a pamphlet, called the New Attractive, wherein he shews how to determine its quantity, and sets forth the errors that appeared to arise from the common method of making compasses in different countries.

Mr. William Burrough, the famous practical navigator, who had been many years at sea, and afterwards made comptroller of the navy to Queen Elizabeth, published a discourse on the variation of the compass, in 1599, wherein he shews how to determine the variation several ways; demonstrates the falsehood of the rules then commonly used, to find the latitude by the guard-stars; and exposes many errors in the sea-charts, by the neglect of the variation.

The celebrated Gerard Mercator published a pair of globes in 1541, as being very serviceable to mariners; and Mr. Edward Mullineux, in 1592, set forth a much larger pair: on the terrestial one were described many new discovered countries, and the respective voyages round the world by Sir Francis Drake in 1577, and Mr. Thomas Candish in 1586, with the progress Sir Martin Frobisher had made towards the north in 1576, were traced out.

In 1594, Mr. Robert Hues published an elaborate treatise on the globes in Latin; wherein he solves by the globe the problem of determining the latitude by double altitudes with the intermediate time between the observations.

According to Anthony Wood, Hues was a man of letters, had been far at sea, and was a member of the university of Oxford. His book was well received, and has been, indeed, a pattern for such as afterwards treated on the same subject.

Captain John Davis, a celebrated English navigator, who gave name to the streights which he discovered in 1594, published a small treatise, entitled, *The Seaman's Secrets*; the eighth edition of which appears to have been printed in 1657, and seems to have superseded Cortes's. In this work Davis treats of plane sailing, and shews the form of keeping a reckoning at sea. He likewise shews how to sail by the globe, and commends great circular sailing. He has also given, at the end of his book, a figure of a staff of his contrivance, to take a back observation, which seems to have continued in use for some time, and was known among seamen by the name of back-staff.

This instrument was succeeded by one of a more commodious form, now well known by the name of Davis's Quadrant, which was also of his invention.

Contemporary with Davis was Mr. Richard Polter, a master in the royal navy, who published a small book intitled, *The Pathway to Perfect Sailing*, wherein he infers, that different loadstones communicated different degrees of variation to the magnetic needle.

Gerard Mercator seems to have been the first who conceived the method of correcting the errors frequently complained of in the plane-chart, which was much in use at that time, and served tolerably well in voyages near the equator. He began by representing both the meridians and parallels of latitude by parallel straight lines, as in the plane-chart, but gradually augmented the distances between the degrees of latitude in advancing from the equator towards either pole, that the rhumbs also might be extended into straight lines; so that a straight line between any two places, laid down in this chart by their longitudes and latitudes, should make an angle with the meridians, expressing the rhumb leading from one to the other. Although he published his Universal Map, thus constructed, in the year 1569, yet he does not appear to have clearly understood the principles upon which he proceeded, and was ignorant of the genuine method of dividing the meridian. Mercator's scheme was not indeed contrived for representing the parts of a country in a just proportion to each other, but is appropriated to the use of mariners who sail upon rhumbs by the guidance of the compass.

Barlow, in his book called the Navigator's Supply, published in the year 1597, mentions, that in a personal conference with two East Indians, they affirmed, that instead of our compass they used a magnetical needle, six inches in length, upon a pin, in a white China dish, filled with water, in the bottom of which they had two cross lines to represent the four principal winds; the rest of the divisions being left to the skill of their pilots: he also mentions in the same book that the Portuguese, in their first discovery of the East Indies, got a pilot at Mahinde, who brought them from thence in 33 days within sight of Calicut.

Our countryman Mr. Edward Wright, a fellow of Caius College, Cambridge, seems about this time to have discovered a true way of dividing the meridian, which, with a short table for that purpose, and a specimen of a chart so divided, he furnished to his friend Mr. Thomas Blundeville, who published it in 1594, among his *Exercises*, that treat on navigation. In 1599, Mr. Wright himself published his famous treatise, entitled, *The Correction of certain Errors in Navigation*; where he shews the reason of this division, the manner of constructing his table, and its uses in navigation, with other improvements. In 1610 a second edition of his book was published, and dedicated to Prince Henry; the improvements in this edition were the Tables of Latitudes for dividing the Meridian, computed to ten minutes; and an instrument, which he called the sea-rings, by which the variation of the compass, altitude of the sun, and time of the day, may be readily determined at once in any place where the latitude is known—a third edition was published by Moxon in 1657. He was allowed fifty pounds a year by the East India Company, for reading a lecture annually on navigation. His improvements on Mercator's chart soon became known and copied, and some of the writers on the continent even disputed the discovery with him, although the French writers acknowledged he was the author.

In 1614 Mr. Raphe Handson, among his nautical questions, solved very distinctly every case of navigation, by applying arithmetical calculations to Wright's table of latitudes, or meridional parts, as they have been since called.

In 1620, Mr. Edmund Gunter published a table of artificial sines and tangents to every minute of the quadrant, and applied to navigation his admirable *Scale*, on which were inscribed the logarithmic lines for numbers, and for sines, tangents, and secants. He also greatly improved the *Sector* for the same purpose. He first shewed how to take a back observation by the cross-staff, and invented an instrument, called a *Cross-bow*, for taking altitudes of the sun or stars.

In an edition of this work, published in 1623, the method of approximation, by what is called by our sailors, middle latitude, was first introduced and explained; and he there also shews the application of logarithms in various ways to navigation; though the first application of the Logarithmic Table to the cases of sailing, was by Mr. Thomas Addison, in his Arithmetical Navigation, published in 1625.

In 1633, the Rev. William Oughtred projected Gunter's ruler or scale into a circular arc, and applied it to several important points in navigation. It was afterwards converted into a sliding ruler, now called a *sliding Gunter*.

In 1635 Mr. Henry Gellibrand published a Mathematical Discourse on the Variation of the Magnetic Needle.

This extraordinary phenomenon he discovered by comparing the observations found out at different times near the same place by Mr. Burrough, Mr. Gunter, and himself. This discovery was soon known abroad, and several accounts of it shortly published on the continent.

About this time navigation began to be considered as an art, depending in a great measure, on mathematical principles. Mr. Richard Norwood, who had been much at sea, seems first to have shewn the use of Trigonometry in the three principal kinds of sailing; and towards the farther improvement of that art, he undertook a laborious work for examining the division of the log-line. In a treatise, called *The Seaman's Practice*, which he published in 1637, he describes his own excellent method of keeping a sea-reckoning, using a traverse-table, and shews how to rectify the course by the variation of the compass; as also how to discover currents, and to make a proper allowance for them. This treatise, and that of Trigonometry were often reprinted, as the principal books for learning, scientifically, the art of navigation. He also published a small manual for sailors, which has gone through a great number of editions. About the year 1645 Mr. Bond added an improvement to Wright's method in a new edition of Norwood's Epitome, by a property in his meridian line, whereby its divisions are more scientifically assigned, than the author himself was able to effect, which he deduced from this theorem, that these divisions are analogous to the excesses of the logarithmic tangents of half the respective latitudes augmented by 45 degrees above the logarithm of the radius. This theorem was fully demonstrated by Mr. James Gregory of Aberdeen, in his *Exercitationes Geometricæ*, published in 1668; and afterwards more concisely by Dr. Halley, in the Philosophical Transactions for 1695; and still made more simple by Mr. Roger Cotes, in the Philosophical Transactions for 1714. This Mr. Bond, who introduced so useful an improvement in navigation, was a teacher of mathematics in London, and employed to superintend all the publications, or treatises on navigation. In an edition of the *Seaman's Calendar*, he declared he had discovered the longitude, by having found out the true theory of the magnetic variation, and to gain credit to his assertion, he foretold, that, at London, in 1657, there would be no variation of the compass, and from that time it would gradually increase the other way; which happened accordingly He afterwards calculated a table of the variations for 49 years to come, and published it in the Phil. Trans. for 1668, which procured him great reputation, as his treatise, called *The Longitude Found*, was published by the special command of King Charles II. in 1676.

About two years afterward, Mr. Peter Blackborrow published a book, called *The Longitude not Found*; wherein he exposed the hypothesis of Mr. Bond, which induced the celebrated Dr. Halley to undertake the business. His speculations on this subject are published in the Phil. Trans. for 1683, and 1692. Dr. Halley published a dissertation on the monsoons in the Phil. Trans. for 1690, containing many observations very useful to all such as sail to places that are subject to such winds.

In 1700 Mr. Bond published his celebrated Variation Chart, which was received with great applause, and found useful on many occasions for determining the longitude.

In 1686, Newhouse's Whole Art of Navigation appeared in 4to. and was republished with additions and improvements in 1701.

The writers on navigation about this period appear to be many, but the improvements few. Wright, Norwood, Bond, and Halley, had acquired most of the art, and having now settled the true principles of navigation, all the succeeding authors had only to tread in their steps, making some little improvements, which they generally mentioned in the titles of their books.

The first we meet with of any note after this period is Seller's *Practical Navigation*, which, though without demonstrations, has the rules for all the different sailings, with various other matters relative to the art; as also the use of the azimuth compass, the ring-dial, the sea-ring, cross-staff, Davis's quadrant, nocturnal dipping-needle, and globe, together with all the necessary tables; the whole being so well adapted to the use of practical navigators, that the book ran through many editions, the last of which appears to be dated in 1739.

In 1702, we meet with a small tract, entitled, *A New Compendium of the whole Art of Navigation*, by Mr. William Jones, and published by John Buckler, in which are some new and useful observations respecting the allowances for the lee-way of a ship. See Dr. WILSON's *Dissertation on the Rise and Progress of the Art of Navigation*, first published in 1772.

In 1709, Keith's New Theory of the Longitude was published in 4to. and, in 1722, Haselden's Description of Mercator's Chart.

In 1723, Hodgson's Elements of Navigation, which was republished in 4to. in 1738.

In 1725, Rowe's Navigation and Discovery of the Longitude.

In 1729, Collier's Art of Navigation, in 4to. and, in 1730, Harris's Treatise of Navigation, in 4to.

In 1733, Kelly's Navigator's Complete Tutor, in 2 vol. 8vo.

In 1743, Hauxley's Navigation, in 2 vol. 8vo. and, in 1747, Neal's Method of finding the Longitude.

Nor ought we, in this place, to pass over in silence the very useful invention of Dr. Gowin Knight, in making artificial magnets, that are of greater efficacy than the natural ones. Although the doctor did not think proper to reveal the secret, yet the Rev. John Mitchel soon after discovered it, and published an account of artificial magnets in 1750; and Mr. John Canton, having also made the discovery, published it in the Philosophical Transactions for the year ensuing.

In 1750, Barrow's System of Navigation was published in 4to.

In the year 1754, Mr. J. Robertson, then master of the Royal Mathematical School at Christ's Hospital, and afterwards master of the Royal Naval Academy at Portsmouth, published his *Elements of Navigation*, containing both the theory and practice, with the necessary tables; to which is added, a small treatise on Marine Fortification. This was composed for the use of the above-mentioned school and academy. It is an excellent work, and the best adapted for teaching the art of navigation in a scientific manner that has yet been published. A second edition appeared in 1764, and a third in 1772. Mr. Robertson died in 1776. Mr. William Wales, master of the Royal Mathematical School at Christ's Hospital, corrected and improved a new edition of this valuable work in 1780, a second edition in 1786, and a third in 1796. Lieutenant Glynn, who succeeded Mr. Wales at Christ's Hospital, superintended the publication of the seventh and last edition of this work in 1805; wherein he states having corrected the errors that had crept into the former edition, and computed new tables of the sun's declination.

In 1755, Emerson's Navigation was published; and, in the same year, Williams's Essay on the Longitude at Sea, in 4to. and Martin's Geography and Navigation.

In 1758, Atkinson's Epitome of Navigation was published, in 8vo. revised by Mountaine in 1778, and by Adams in 1782.

In 1760, Heath's Royal Astronomer and Navigator was published, in 4to.

In 1762, Patoun published his Navigation, in 8vo. and the second edition was republished in 1770.

The Rev. Dr. Nevil Maskelyne, our late astronomer royal, deserved the thanks, not only of every navigator, but of the public at large, for his persevering industry and great exertions in bringing the method of finding the longitude at sea, by the lunar observations, into common practice. He had been sent out by the Commissioners of the Board of Longitude to St. Helena, in the year 1761, to observe the transit of Venus, which afforded him a good opportunity of working the lunar observations on the voyage out and home; and also of proving Mayer's lunar tables, which have been lately sent over from Gottingen to the Board of Longitude. In 1763 he published his *British Mariner's Guide*, containing new precepts for making the necessary observations and computations. He also added useful rules for computing the effects of parallax and refraction, which he had formerly communicated to the Royal Society. In 1765, he presented a memorial to the Board of Longitude, setting forth the great accuracy of Mayer's lunar tables, which he had experienced both on his late voyage, and also on shore; he also pointed out the great utility of their being published, and concluded by stating it not only to be his own opinion, but that of several of our eminent navigators, that nothing would then be wanting to bring the lunar method to be generally practised at sea, but a Nautical Ephemeris, an assistance which they humbly hoped to receive from this honourable Board. In consequence of this memorial, the Board came to the resolution of printing the tables; they applied to Parliament for power to reward Professor Mayer for this valuable work; and also for power to reward other persons, to be employed in compiling a Nautical Almanac, which seemed necessary, in order to render the lunar tables of general utility. Parliament readily granted the sum of 3,000*l.* to the widow, as the professor died at the beginning of the year 1762; and also the sum of 300*l.* to the celebrated Euler, who, it appeared, had contributed greatly to assist Mayer in correcting and completing the last edition of the lunar tables. They also granted power to the Board of Longitude to appoint eminent calculators to compute a Nautical Ephemeris, which was immediately put in execution, and the first published in the following year, for 1767; and which has been continued annually ever since, by order of the Commissioners of Longitude, under the direction of the royal astronomer. For farther particulars see the article ALMANAC, *Nautical*.

In the year 1766, Dr. Maskelyne published, with the first Nautical Almanac, a set of tables requisite to be used with it, to facilitate the calculations. See the article REQUISITE TABLES.

In 1770, J. H. Moore published a good practical book on navigation; which was copied and published, with some additions, by Bowditch, in America, in 1802, and republished in England by Kerby, in 1804, and a second edition in 1806.

In 1771, Ludlam's use of Hadley's Quadrant appeared, and was found a very useful little tract.

In 1772, the Cambridge Tables were published in folio, compiled by Dr. Shepherd, professor of astronomy. This valuable work, known to every scientific navigator, was published by order of the Commissioners of Longitude, for correcting the apparent distance of the moon and a star from the effects of refraction and parallax.

In 1773, a new edition of Henry Wilson's Epitome of Navigation, much improved, was published at Edinburgh, in 8vo.; and, in 1774, a Treatise on the Longitude, &c. by Waddington, in 4to.; and Kean's Method of finding the Longitude, in 8vo.

In 1778, Mr. Dunn's Method of finding the Longitude was published; and several other tracts of his on the subject of navigation, in the years 1782, 1786, and 1794.

In 1779, Crosby's System of Navigation; and,

In 1782, Martin's Mariner's Mirror, in 8vo.

In 1784, Rymer's Practice of Navigation, in 4to.; and, in 1785, Donne's British Mariner's Guide, in 8vo.

In 1786, Haselden's Seaman's Assistant, in 4to.; and, in 1787, Garrard's Method of finding the Longitude in 8vo. which was reprinted in 1791.

In the same year, Ferguson's Navigation, in 8vo.; and, in 1788, Harrison's Complete Observer.

In 1790, Margett's Longitude Tables appeared in 4to.; and, in the following year, he published a more complete set, in folio.

About this time Malham and Dickson published each a small tract on Navigation.

In 1792, Taylor's Tables of Logarithms were published, being three years after the author's death. For an account of which see the article LOGARITHMS.

In the same year, Nicholson's Treatise on Practical Navigation, in 4to. was published; and a second edition appeared in 1806.

In 1793, Dr. Mackay published his first edition of the Theory and Practice of finding the Longitude at Sea or Land, in two volumes, 8vo.; a second edition appeared in 1801; and a third, much enlarged and improved, in 1810.

In 1794, Mr. William Wales published a small tract on the longitude; and Churchman a Magnetic Atlas, in 4to.; and, in the same year, Walcot's Method of finding the Longitude was published.

In 1796, Margett's Horary Tables were published in folio; and Dr. Kelly's Treatise on Spherics and Nautical Astronomy also appeared in this year.

In 1798, Nicholson's Navigator's Assistant, in 8vo.; and, in 1800, Clarke's Seaman's Desiderata, in a small 4to. vol.

In 1804, Mackay's Complete Practical Navigator, in 8vo. which was republished in 1810.

About this time Dr. Andrew published a set of Tables, in 8vo. for simplifying the method of clearing the lunar distance.

In 1806, Grant's Method of finding the Longitude at Sea, &c. in 4to.; and Norie's Epitome of Practical Navigation was published in 8vo.; a second edition in 1808, and a third in 1810. In the same year Turner's Tables for clearing the Lunar Distance were published.

In 1812, Dr. Kelly's Astronomical Computations, comprising new tables of the sun's longitude, right ascension, and declination for that year, with explanations and calculations of apparent mean and sidereal time; to which is added, an easy and expeditious method of clearing the lunar distance by projection, in order to find the longitude at sea.

The common treatises on navigation that now seem most in use at sea, are, Norie's, Mackay's, and Moore's Epitome, which are merely calculated for practical seamen, who have not studied the theory; and on that account are published in a small convenient size, and at a low price.

In all schools where navigation is taught, scientifically, Robertson's Elements are used, and is allowed to be by far the best book for that purpose of any yet published. For an account of the several instruments used in this art, with the methods for finding the latitude and longitude at sea, tides, variation of the compass, and the different kinds of sailing, see each respective article.

The Dutch Writers on Navigation.

Michael Coignet, a native of Antwerp, published a small treatise, in 1581, as a supplement to Medina's Treatise on Navigation; the errors of which he has clearly pointed out. In treating of the sun's declination, he took notice of the gradual decrease of the obliquity of the ecliptic, a point which seems to have been long disputed, but now settled by the theory of attraction.

He also improved the cross-staff with three transverse pieces, as it was then in common use among the sailors; and invented some instruments for nautical purposes, which have been long since laid aside, except his Nocturnal. He concludes his work by describing a method of sailing on a parallel by means of a ring-dial and a twenty-four hour glass.

The Dutch have been great navigators, and famous for their atlasses and charts: the oldest appear to be those published at Leyden in 1584, called the Mirror of Navigation, by L. J. Waghenair: in their late ones there is described an instrument to be used after the manner of Davis's quadrant, but where, instead of circular arcs, are substituted straight lines.

A very good work, entitled, "A Treatise concerning the Determination of the Longitude at Sea, by the Distance between the Moon and the Sun, or a fixed Star," was published by order of the committee for Marine Affairs, at Amsterdam, in the year 1796; wherein the methods given by Borda, Dunthorne, Lyons, Maskelyne, and Mackay, are explained and demonstrated.

The French Writers on Navigation.

The Jesuits, who had long been considered as a well-informed set of men, and had treated on almost every branch of science, notwithstanding their being reproached by historians with converting learning into superstition, did not deem the art of navigation beneath their consideration, as we find by the three following authors, who were of their society:—

At Paris, in 1633, Father George Fournier, published an *Hydrography*, principally relating to navigation.

John Baptist Riccioli, in his *Geographia et Hydrographia Reformata*, printed at Bologna in 1661, has inserted a treatise of navigation, which he collected from most of the contemporary writers.

Father Millet Dechalles wrote on this subject in a more masterly manner, both in his *Cursus Mathematicus*, printed at Lyons, in 1674, and in a French treatise, entitled, *L'Art de navigeur demontré par principes*, published at Paris in 1677.

These three authors, besides treating of the different kinds of sailing, shew various methods of taking altitudes, finding the variation, and estimating a ship's way, &c.

In 1698, M. John Bouguer composed, by authority, his *Traité complet de la Navigation*, which was well received, as it contained most of the practical methods then known. In 1733, Father Pezenas, a Jesuit, published at Marseilles, a tract, called *Elémens de Pilotage*; and, in 1741, a larger work, entitled, *Pratique du Pilotage*. This author shewed how to find the meridional parts by the artificial tangents, which had long been known in this country from Norwood's Epitome of Navigation.

In 1748, Le Roy's Account of his and Harrison's Attempts for finding the Longitude, &c. was published in 4to.

In 1753, M. Peter Bouguer, published a very elaborate work on this subject, entitled, *Nouveau Traité de Navigation*, which is extremely well written, and the author appears to be an excellent mathematician. He there gives a variation-compass of his own invention, similar to that which had been described by Captain Middleton, in the Phil. Trans. for 1738; and afterwards by Dr. Knight, as improved by the ingenious Mr. John Smeaton, and described in the Phil. Trans. for 1750. M. P. Bouguer is also very particular in determining the lunations more accurately than by the common methods, and in describing the necessary corrections of the dead-reckoning.

Mr. Severien, in his Marine Dictionary, printed at Paris, in 1758, mentions several early French writers on this subject, among others, M. G. Denys, who, in 1666, published several tracts on navigation at Dieppe. The Sieur Blondel St. Aubin, in 1673, published his *Trésor de la Navigation*, in which he treats the art very fully by calculations, and describes an instrument much in use among the French sailors at that time for performing the different operations of navigation mechanically, and which appears to have been a sort of skeleton of Wright's Universal Map. In 1683, M. Dassier, in his *Pilote Expert*, seems to have been the first French writer who shewed the use of Gunter's scales; and in 1693 was published, by order of government, the first part of a well known work, intitled, *Le Neptune François*, consisting of sea-charts on Wright's principles, made from the latest observations, and received by Messrs. Pene, Cassini, and others.

The excellent astronomer M. de la Caille, in 1760, published a new edition of Bouguer's book, which he abridged and much improved.

In 1766 appeared at Paris, a treatise, under the direction of M. le Monnier, intitled, *Abrégé du Pilotage devisé en deux parties, ou l'on traite principalement des Amplitudes, des Loxodromies, dans l'hypothèse de la Sphère et de Sphèroïde des marées, des variations de l'aimant*; the former part of which had been published some years before.

In the same year, Pére Pezenas published his *Astronomie des Marins*, at Avignon; which, besides examples to the problems in the *Astronomie Nautique* of Maupertuis, and other articles, contains various methods of finding the longitude from observations of the moon.

M. de Charniéres published a treatise, entitled *Expériences sur les Longitudes, faites à la mer en 1768; publiées, par ordre du Roi, en 1768*, in which he has shewn the utility of his instrument, called a *Mégamètre*, in observing the distance between the moon and a star. In 1772, he also published his *Théorie et pratique des Longitudes en mer*, which contains a description of the above-mentioned instrument; and another, called a *Héliomètre*, which had lately been improved by the Abbé de Rochon; and also an *Astromètre*, invented by the Abbé, and the methods of making and reducing the observations for the latitude and longitude. The Abbé de la Caille recommends it as the only practical method to be used at sea, in his edition of Bouguer's Navigation, published at Paris, in 1781. This subject is also fully treated in the sixth volume of Bezout's *Cours de Mathématiques*, published at Paris, in the same year, and used in all the national schools. It is also recommended in M. Callet's edition of Gardiner's Logarithmic Tables, printed at Paris, in 1783.

In 1771 appeared, *Mémoire ou l'on propose une nouvelle méthode pour déterminer les Longitudes en mer*, in 8vo.

In 1772, *Exposition des Moyens, par le Monnier.*

In 1773, *Tables et Instructions pour la Théorie & la pratique des Longitudes en mer, par l'Académie Royale de Marine.*

In 1775, *Description des Octans & des Sextans Anglois, par Magellan*, in 4to.

In 1778, *Les Logarithmes des Tems*, was published in 8vo. at Vienna.

In the same year, *Le Guide du Navigateur*, in 8vo. by Lévêque.

In 1779, *Description d'un nouveau Compas d'azimuth*, published in 4to. at Havre.

An excellent tract was published at Paris in 1787, intitled *Description et Usage du Cercle de Reflexion, par le Chevalier de Borda;* in which that instrument is very particularly described, its use shown in nautical observations, and the manner of calculating these observations.

Lecons de Navigation par M. Dulague, Professeur d'Hydrographie au Collège de Rouen, was first published at Paris, in 1790. This is allowed to be a clear and easy epitome, and is generally used in private schools in France.

In 1792, was published *Bouguer par La Caille, augmenté par La Lande*, at Paris, in 8vo.

In 1793, M. De La Lande, published his *Abrégé de Navigation*, at Paris, in 4to, which contains a brief history of navigation, a catalogue of nautical books, with various other particulars; and lastly, the method of finding the longitude by lunar observations: but the principal purpose for which this book was published, is the set of horary tables which it contains, and by which the apparent time at the ship may be readily deducted from the altitude of any heavenly body, whose declination is contained within the limits of the table and the latitude of the place. In this work, M. De La Lande says, "the moon furnishes the most general and certain method of finding the time under the first meridian; and it is the distance between the moon and sun, or stars, that is employed for this purpose."

In 1801, was published at Paris, in 4to. *L'Art du calcul Astronomique du Navigateur, par Dubourquet.* Since that period the French have been so completely enveloped in war, and have had their shipping, colonies, and commerce, so cut up, that they have had no inducement to turn their attention to improvements in navigation.

Navigation is either proper or common.

Navigation, *Common*, usually called coasting, in which the places are at no great distance from one another, and the ship sails generally in sight of land, and mostly within soundings. In this, little else is required besides an acquaintance with the lands, the compass, and sounding-line; each of which articles see under its respective head, as also the article Coasting.

Navigation, *Proper*, is where the voyage is long, and pursued through the main ocean. And here, besides the requisites in the former case, are likewise required the use of Mercator's chart, the azimuth and amplitude compasses, the log-line, and half minute glass, with other instruments for celestial observations; as quadrants, reflecting circles, &c.

Navigation turns chiefly upon four things; two of which being given or known, the rest are thence easily found out. These four things are, the difference of latitude, difference of longitude, the reckoning or distance run, and the course or rhumb sailed on. The latitudes are easily found by observation, and that with sufficient accuracy: the course and distance are had by the log-line, or dead-reckoning, together with the compass. Nor is there any thing wanting to the perfection of navigation, but to determine the longitude, and that may now be easily found with sufficient accuracy for all nautical purposes. See the articles Longitude, Longitude *by Time-Keeper*, Lunar *Distance, Observations, Tables*, and Time-Keeper.

Navigation *Acts*, are those statutes which have been passed at different times by the legislature of this country, for the protection and extension of Navigation.

But the Navigation Act, particularly and emphatically so called, is meant the statute of 12 Charles II. c. 18, entitled, "An Act for the encouraging and increasing of Shipping and Navigation:" all subsequent laws on this subject are but comments on this, modifying and explaining its different regulations, but leaving untouched its general and fundamental principles.

This statute was passed by the Usurpation Parliament, on the 9th of October 1651. In the year preceding, indeed, the same parliament had sketched a faint and rude outline of the Navigation Act; the sugar islands still adhering to Charles II. the statute of 1650 prohibited all ships of foreign nations trading with any English plantations without license from the council of state. The statute of 1651 arose partly from the same cause, and partly from the circumstance that the English merchants used to freight Dutch ships to bring home merchandise from America and the West Indies, as they could be had at a lower rate of freight than native ships. Till this act passed, it appears that notwithstanding the existence of former unrepealed acts for the protection and increase of English shipping, all nations in amity with England actually did import into England what commodities they pleased, and in what ships they pleased; and as the Dutch, at that period, were the principal carriers of Europe, most of the merchandise imported into this country was imported by them.

The law of 1651 enacted, that no commodities (colonial or of any other description) should be imported into England, unless in vessels solely owned, and commanded, and principally manned by English subjects; and, where the commodities were foreign, unless they were entirely conveyed in such vessels from the place where the commodities grew, or to which they were usually, in the first instance, transported by sea. There were some peculiar exceptions, and one grand and general one; *viz.* goods originally produced, or first transported by sea from an European state, might be imported thence into England, in vessels owned by the subjects of that state.

The title of this act was suited to the design which it had in view; it was, "Goods from foreign Parts, by whom to be imported," (Scobell's Acts, 1651, c. 22, p. 132); and the consequence was, that the Dutch were cut off from the carrying trade with our colonies, and their importation of fish into this country was laid under great restrictions and heavy burdens.

Before the passing of this law great jealousy existed, and frequent disputes had taken place between this country and holland; these were greatly increased and embittered by it: acting in conformity with it, our ships of war insisted on their right to search Dutch vessels, and, in several instances, made prizes of them, where they found that they contravened the act. The war, which broke out in the subsequent year, and which was carried on with great rancour and obstinacy, arose, partly, at least, from the Navigation Act of 1651.

The restoration Parliament were so clearly and strongly convinced of the necessity and benefit of the Navigation Act, that they sanctioned it, first in 1660, and afterwards in 1662; the first of these statutes, 12 Charles II. c. 18 (as has been already observed), is that emphatically called the Navigation Act.

The principal dispositions of the Navigation Act are thus summed up by Adam Smith, in his Wealth of Nations:

First, all ships, of which the owners, masters, and three-fourths of the mariners are not British subjects, are prohibited, under pain of forfeiting ship and cargo, from trading to the British settlements and plantations, or from being employed in the coasting trade of Great Britain.

Secondly, the great variety of the most bulky articles of importation can be brought into Great Britain only, either in such ships as are above described, or in ships of the country where those goods are produced, and of which the owners, masters, and three-fourths of the mariners are of that particular country; and when imported even in ships of the latter kind, they are subject to double alien's duty. If imported in ships of any other country, the penalty is forfeiture of ships and goods.

Thirdly, a great variety of the most bulky articles of importation are prohibited from being imported, even in British ships, from any country but that in which they are produced, under pain of forfeiting ship and cargo.

Fourthly, salt fish of all kinds, whale-fins, whale-bone, oil, and blubber, not caught by and cured on board British vessels, when imported into Great Britain, are subject to double alien's duty.

This act differed from that of 1651, in its second enacting clause, which did not permit any but English subjects to be factors or agents in English colonies. Before this time the Dutch were the principal factors or agents in our colonies; and the third enacting clause secured the East India, Guinea, and Levant companies, from foreign shipping.

The act 14 Charles II. c. 11, also differed in some respects from the act 12 Charles II. c. 18; by the latter it was sufficient, with respect to the vessel, if she was either owned or built at home; while, by the former of these acts, both these circumstances were declared indispensable.

The principal objects which the Navigation Act, 12 Charles II. c. 18, had in view, will be best explained by its preamble, which declares it to be passed, "not only for the sake of employing and increasing English shipping and seamen, and securing a vent for woollen and other manufactures; but also to make this kingdom a staple of the commodities of those plantations, as well as of the commodities of other countries, for the supplying them (it being the usage of other nations to keep their plantation trade to themselves); and, farther, if colonial commodities should be taken from any part but the plantations, that the trade of them would thereby, in a great measure, be deserted from hence, and carried elsewhere; his Majesty's customs and other revenues much lessened; the fair trader prejudiced; and this kingdom not continue a staple of plantation commodities, nor that vent, for the future, of the victual and other native commodities of this kingdom."

British built ships are declared to be such as have been built in Great Britain, Ireland, Jersey, Guernsey, or the Isle of Man, or in some colony at the time belonging to Great Britain; any ship whatever, taken and condemned as lawful prize, is to be regarded as a British built ship. Ever since the Navigation Act passed, "Surveyors of the Act of Navigation" have been appointed at the Custom-house, whose duty it is to keep a register of all British-built ships. British seamen are declared to be such as are natural born subjects of the empire; or naturalized by act of Parliament; or made denizens; or become subjects by conquest. And by 33 Geo. III. c. 68. foreign seamen, who shall serve three years in the British navy, during the time of war, if they have not taken the oath of allegiance to a foreign state, may be employed, and shall be considered as British seamen. The same statute also introduced the principal modification which has taken place in the Navigation Act, by enacting, that from the conclusion of the war, no goods shall be *exported* in British vessels, unless the master and three fourths of the crew were British; the Navigation Act not interfering with the exportation of goods from British ports.

Such is the general view of those acts which have been passed for encouraging and encreasing shipping and navigation; and there can be as little doubt that the Navigation Act has been beneficial to Great Britain, considered as a nation dependent for her safety and well-being on her maritime power, as that it has been prejudicial to her, considered as a nation purely commercial. Under the circumstances in which Great Britain is at present placed, it would be highly impolitic and unsafe to alter, essentially, the navigation laws; she stands now more in need of maritime defence than she ever did; and the means of that maritime defence require more than ever to be preserved untouched by the encroachments of foreign nations; since, if the Navigation Act was abrogated, the operation of our taxes is such, that foreign ships, in almost every case, would be able to carry goods at a much lower rate of freight than our own ship-owners could do. But, on the other hand, it may well be questioned, whether, at present, the regulation of the Long Parliament should not be adopted, and full liberty given to purchase ships wherever they could be found cheapest: as has been already remarked, freight would thus become more reasonable, and, consequently, the great object of the navigation laws would be more completely answered, while our merchants would have less reason to complain: if this alteration were made, the loss of the merchants would be entirely for the benefit of the nation, now it is only partly for the benefit of the nation; while, by benefitting at the same time the ship-builders, that benefit is essentially lessened.

NAVIGATOR, (*navigateur*, Fr.) a person who is skilled in the art of navigation, and capable of conducting a ship at sea to any place proposed: this applies as well to mathematical knowledge or theory, as to seamanship or practice. For the necessary qualifications of a good *navigator*, see the articles MASTER OF A SHIP OF WAR, MIDSHIPMAN, and PILOT.

NAVIS, Argo Navis, or the ship Argo, in astronomy, a constellation of the southern hemisphere. See the article Argo.

NAVY, (from *navis*, Lat. *marine—forces maritimes d'un état*, Fr.) implies, in general, any fleet or assembly of ships. It is, however more particularly understood to be the fleet of vessels of war that belong to a prince or state. See the article Fleet.

Although the English were considerable for power at sea in the reign of Henry VII. yet they had not much occasion to assert it till the reign of Henry VIII. who, about the year 1510, by his prerogative, and at his own expense, laid the foundation and settled the constitution of the royal navy. An Admiralty and Navy Board were constituted by him; regular salaries were settled, not only for the admirals and vice-admirals, but for his other officers and seamen; and about this period the sea service became a distinct and regular profession.

The royal navy of England has ever been considered its greatest defence and ornament, the floating bulwarks and "wooden walls" of the island, have, therefore, been assiduously cultivated and improved from the earliest ages (see *Laws of* Oleron). And yet, so vastly inferior were our ancestors, in this respect, to the present age, that even in the maritime reign of Queen Elizabeth, Sir Edward Coke (3 Inst. 50), thinks it natter of boast that the royal navy of England then consisted of *thirty-three* ships. The present condition of our marine is in a great measure owing to the salutary provisions of the statutes called the Navigation Acts, by which the constant increase of British seamen and shipping was not only encouraged, but rendered absolutely necessary. See the article Navigation Acts.

The management of the *royal navy* formerly under the Lord High Admiral of Great Britain, but now vested in a body, styled the Lords of the Admiralty, is entrusted to principal officers and commissioners of the navy, who hold their places by patent. The principal officers of the naval department are—The Treasurer of the Navy, whose business it is to receive money out of the exchequer, and to pay all the charges of the navy, by warrant from the principal officers;—Comptroller of the Navy, who attends and comptrols all payment of wages, is to know the rates of stores, to examine and audit all accounts, &c there is also a deputy-comptroller to assist;—Surveyors of the Navy, who are to know the condition of all ships in the different ports, see wants suppl ed, and to estimate repairs; there are, at present, three joint-surveyors and a sub-surveyor of the navy;—Clerk of the Acts, whose business it is to record all orders, contracts, bills, warrants, &c.

The principal Commissioners of the Navy are five: the first executes that part of the Comptroller's duty which relates to the comptrolling the Victualler's accounts; the second, another part of the said Comptroller's duty, relating to the account of the store-keepers of the yards; the third has the direction of the navy at the port of Portsmouth; the fourth has the same at Chatham; and the fifth at Plymouth.

There are also commissioners for Deptford, Woolwich, and Sheerness; and, in time of war, commissioners are appointed to reside at some of the principal yards abroad, as Gibraltar, Halifax, Nova-Scotia, Malta, Antigua, Jamaica, Cape of Good Hope, Madras, Bombay, &c. and, since the increase of the royal navy, these have several subordinate clerks under them, with salaries allowed by the government.

The navy had formerly been victualled by contract, but the victualling is now under the commissioners, who hold their office at Somerset House, in the Strand, London.

The following is a general statement of the navies of the different countries, as correct as could possibly be obtained from existing circumstances, to the dates hereafter mentioned.

The British Navy. The number of ships and vessels composing his Majesty's navy on the 1st of June 1813, are as follows:—

Rates.	No.	Guns.
1st.....	4 of	120
	2 of	112
	5 of	110
	4 of	100
2d.....	17 of	98
	1 of	90
3d....	1 of	84
	13 of	80
	1 of	78
	172 of	74
	36 of	64
4th.....	1 of	60
	1 of	52
	26 of	50
	1 of	48
5th.....	10 of	44
	13 of	40
	80 of	38
	70 of	36
	35 of	32
6th.....	10 of	28
	7 of	24
	7 of	22
	28 of	20
	161 Sloops and Yachts.	
	13 Bombs and Fireships.	
	191 Brigs.	
	42 Cutters.	
	65 Schooners, Gun-vessels, Luggs. &c.	

There are 723 in commission, 220 in ordinary, and 74 building; making together 1017.

French Navy. The number of ships in the French Navy stationed off Toulon, Rochefort, Flushing, Cherburg, Brest, L'Orient, Naples, Genoa, Triest, and Venice (without taking into the account those of Holland, of Italy, and of other naval powers), at the beginning of the year 1813, was as follows:—

Ships.	Guns.
9 of	120
1 of	90
8 of	80
62 of	74
10 Sail of the Line building.	
72 Frigates.	
12 Ditto building.	

Of these there are 57 sail of the line and 48 frigates ready for sea, and 32 of the line and 36 frigates fitting and building.

Spanish NAVY. The number of ships in the Spanish navy, at the beginning of 1813:—

4 three-deckers. 12 two-deckers. 7 frigates. 9 Sloops. 18 Brigs. 21 Schooners. 14 Polacres, Luggers, &c.	Fit for service.
6 two-deckers. 5 frigates.	In want of repair.
4 two-deckers. 3 frigates.	Laid up, as being unserviceable.

Portuguese NAVY. An account of the navy of Portugal, at the beginning of 1813:—

At Brazil.

Ships.		Guns.
1	of	84
4	of	74
3	of	64
1	of	44
1	of	36
1	of	32
2	of	22
1	of	20
1	of	12

In the Port of Lisbon.

5 in want of repair.
3 in bad condition.
1 only fit for condemnation.

Dutch NAVY. The number of ships in the Dutch navy, at the beginning of 1813:—

1 of 86 guns.
1 of 84 ditto.
2 of 80 ditto.
4 of 74 ditto
1 of 64 ditto
5 sail of the line in the Texel.
21 sail of the line in the Scheldt.

American NAVY. The number of ships in the American navy on the 1st of June 1813:—

1	of	74 guns.	
3	of	44 ditto.	Frigates.
5	of	36 ditto.	Frigates.
2	of	32 ditto.	Frigates.
2	of	20 ditto.	Sloops.
1	of	18 ditto.	Sloops.
3	of	16 ditto.	Brigs.
1	of	14 ditto.	Brigs.
2	of	12 ditto.	Schooners.
4		bomb-vessels,	
1		Prison ship.	

Besides the above, there are several small vessels on the lakes, and at different stations; and it is reported, from 170 to 180 gun-boats.

Turkish NAVY. The state of the Turkish navy, at the beginning of 1810, was as follows:

3 three deckers.
7 two ditto
5 of 74 guns.
3 sail of the line building.
7 frigates.
7 corvettes.
2 bomb-vessels, and 2 brigs.

Russian NAVY. The number of ships in the Russian navy, according to a late report made by the Russian minister of marine.

The great fleet of the Baltic, consisting of 59 vessels, and carrying 2,260 guns.
Baltic galley fleet, 41 sail, 705 guns.
Black Sea fleet, 41 sail, 1,225 guns.
Rowing fleet of the Black Sea, 40 gun-boats, carrying 52 guns and 80 falconets.
The flotilla in the Caspian Sea, 6 vessels, carrying 70 guns.
The flotilla of Ochozk, on the coast of Siberia, 11 vessels carrying 36 guns.
Building, 10 ships of the line.

In this estimate are included imperial ships of every class and condition, from a first-rate to a gun-brig; those that are building, under repair and laid up in ordinary as unserviceable, as well as those that are in commission and fit for immediate service.

The greater part of the Baltic fleet are now in England.

Swedish NAVY. The number of ships in the Swedish navy, at the beginning of 1813, was composed of 12 sail of the line, and 8 frigates, besides cutters, gun-boats, &c. and there are two ships of the line and three frigates building.

Danish NAVY. The number of ships in the Danish navy, at the beginning of 1813, was

4 ships of the line.
2 frigates.
120 gun-boats.

There are two ships of the line and three frigates on the stocks; but, owing to the annihilation of commerce and the embarrassment of the government since the capitulation of Copenhagen, on the 7th of September 1807 (when 18 ships of the line, 15 frigates, 6 brigs, and 25 gun-boats, were delivered up to his Majesty's forces, and brought to England), scarcely any progress has been made in the construction of them.

NAVY denotes, also, the collective body of officers and men employed in his Majesty's sea-service.

Manning the NAVY, or a FLEET, is the providing of it with a sufficient number of seamen for working the ships. Various plans have been adopted by the government at different times for this purpose, and one of the methods commonly resorted to in cases of emergency is, that of impressing seamen, by warrants from the Lords Commissioners of the Admiralty to the captains, which are, by them assigned to their lieutenants, who go on shore with a body of men, and with the sanction of the civil power, pick up all seamen they can find, who are not legally protected; and to render this method more effectual, a rendezvous, under a captain, a lieutenant, and a midshipman, is established in those sea-ports, where seamen generally resort, not only to receive volunteers, who have a bounty allowed them, but to impress any seamen whom they find lurking about in idleness and not legally protected. For farther information, see the articles To IMPRESS, and PRESS-GANG.

As long as the war is kept up on its present footing, whilst the very inferior fleets of our enemies are either blockaded up in harbour, or obliged to confine their operations to a run from one port to another, it is evident, that the advantages for forming seamen are greatly on our side, and that the crews of the privateers, or of the small coasting vessels of the Continent, that sneak from one anchorage to another, cannot form the ground-work of a marine capable of contending with the colossal navy of Great Britain.

The state of our navigation, and the number of our seamen, every one of whom the French empire must dread as the warlike follower of some future Nelson, must ever be serious objects of jealousy and alarm; and, in peace, as well as in war, they will endeavour, by their internal regulations, to depress the one, and to diminish the other.

Of two hostile empires, the power that has, and is likely in future to have most money, in order to purchase materials and naval stores, and to employ most ship-wrights, will beable to build and equip most ships. The power that has the greatest population, can put most men into its ships after they are built; and that which rears most seamen by its ordinary commercial navigation during peace, will be able most speedily to man its fleets with good sailors at the commencement of a war, and most readily replace their loss during its continuance. See MAJOR PASLEY's *Essay on Military Policy, &c.*

NAVY *Bills,* include all bills issued or drawn for the use of the navy, and may be divided into three sorts, viz. officers' bills, for their personal pay, &c. Bills drawn or issued by the Navy Board. Bills drawn or issued by the Victualling Board.

1. *Officers Bills for their personal Pay, &c.*— Commissioned officers, masters, and surgeons, commanding or serving in any of his Majesty's ships, and entitled to the pay of such service, may, at the end of three, six, or twelve lunar months, draw bills in duplicate or triplicate, as the case may require, upon the commissioners of the navy, for the amount of the net personal pay due for the time, which bill shall specify the rate and name of the ship or vessel, the rank of the officer drawing the same, and the full amount of the pay due to him; also the amount of the usual deductions and abatements to which the same shall be liable, and the net residue of the pay to be drawn for. But no commissioned officer, master, or surgeon, shall be permitted to draw for the first three months after he shall have received three months pay in advance.

Form of Bill.

1st rate [Caledonia, Lieut. James Jackson], entered day of

Full pay from day of to
day of ....£
Amount of deductions ....£

Nett pay..£

Gentlemen,

Ten days after sight [my 2d and 3d bill of this tenor and date not being paid], please to pay to [Thomas Moss], or order, the sum of being the nett personal pay due to me as of his Majesty's ship [Caledonia], between day of and day of for value received.

[James Jackson.]

"Approved,"
G. H. Captain of

To the Commissioners of his Majesty's Navy, London.

Assignment.

No. 1.

To be paid by the Treasurer of his Majesty's Navy, out of £ received day of under the head of wages to pay officers.

A. B.
C. D. } Commissioners of the Navy
E. F.

By virtue of the act of 35th George the Third.

N.B. The forging of this bill, or procuring any other person to forge the same, in order to receive any part of the wages, or pay, of any commissioned officer, master, or surgeon in the royal navy, is made felony, without benefit of clergy, by the act of 35th of George the Third.

If a bill be drawn by a lieutenant, master, or surgeon, it must be 'approved' by the captain or commander, in testimony of his knowledge of such pay being justly entitled to be drawn for, and commanding officers are particularly enjoined not to approve any bill, unless the same is previously dated, whence and when drawn, also filled and signed; and are also desired to be careful that there own bills are complete in these particulars previous to their being negociated.

A lieutenant, master, or surg on being sent to sick quarters, who may be desirous of drawing for pay up to the end of the six weeks allowed, provided he remains so long at sick quarters, must be careful to take with him the time he is entitled to pay up to the day of his going sick, certified by his comm nding officer; he must also have annexed thereto a certificate under the hands of the physician or surgeon of the hospital or sick quarters, of his remaining under cure; otherwise no captain is authorized to approve the bill that may be drawn, nor can such bill, if defective in any of these particulars, be accepted and signed for payment.

Captains and commanders removed from one ship or vessel to another, before three months pay shall be due, may draw bills at the expiration of three months from the last payment, which bills shall distinguish the periods they shall have served on board such ships or vessels.

No officer, whose command ceases, nor any lieutenant, master, or surgeon, who is superseded or discharged by reason of a ship's being paid off, or lost, is entitled to draw bills for personal pay up to the day of being put upon half-pay or ceasing to be entitled to pay for the ship on service, but is required forthwith to transmit vouchers or journals for passing his accounts, when the usual certificate being granted of no objection to such officer receiving his wages, his full pay and compensation will be ascertained thereon, (including any additional pay allowed by reason of a flag having been flying on board the ship) the sums that may have been drawn for, will be abated therefrom, and the residue of pay immediately assigned for payment at the office of the Treasurer of the Navy in town, but which may be paid at any of the outports, where the officer may himself be desirous of receiving the same.

All such bills to be drawn upon the commissioners of the navy, shall be payable ten days after sight, and shall be negociable by indorsement in the same manner as bills of exchange: and, when they shall be presented to the commissioners of the navy, together with the extracts from the ship's books and surgeon's certificates, as the case may be, if they shall not see cause against it, they shall assign the same upon the treasurer of the navy for payment.

Flag-officers shall be allowed to include their table-money in any bills they may draw under the authority of an act 35th Geo. III. . All orders, bills, remittance bills, or certificates, paid by the treasurer of the navy, shall be deemed and taken as sufficient vouchers on passing his accounts.

And all orders, bills, remittance bills, or certificates drawn under the authority of this act; and all other bills drawn by the commissioners of the navy, victualling, sick and hurt, and transport, upon the treasurer of the navy, under the authority of any other acts, shall henceforward be free from stamp-duty; nor shall any stamped receipt, whatever, be required for monies paid on the same; but the receipts shall be taken on the back of the instrument by which the money is paid. The forging any bill, certificate, order, or extract, drawn under the authority of this act, shall be punishable with death.

The bills drawn for pay by an officer will remain as a charge (or imprest) against him, till his accounts or journals for the ship are passed, and any officer transmitting vouchers or journals for passing an account to any specified time, is, therefore, to draw bills for pay for a subsequent time only. See an act of the 35th Geo. III. entitled, "An Act for establishing a more easy and expeditious method for the punctual and frequent payment of the wages and pay of certain officers belonging to his Majesty's navy," dated 22d June, 1795.

By his Majesty's Order in Council of the 15th of June, 1796, chaplains of his Majesty's ships may draw bills for their personal pay, and two-thirds of fourpences under the regulations prescribed by the act of parliament.

A captain may draw bills on the commissioners of the navy according to the form (No. 21) given in the Naval Instructions, for all money which may be required to pay the expenses incurred by the careening and repairing of a ship, and the supplying her with the necessary stores for sea. These, and all other bills, which he may have occasion to draw on account of the ships, he is to draw on the most favourable terms for the public, bringing to account the full benefit of any exchange in its favour: the rate of which or of the premium or discount thereon is to be noted on the bill. He is to give to every office as early notice as possible of the bills he draws upon it, with his reasons for drawing the same; and to send with the bills, or as soon after as may be, accounts with vouchers, as until such vouchers be received, the amount of the bills drawn will be charged as an imprest against his pay.

Form of a Bill of Exchange.

Barbadoes, 24th of June, 1813.

First of Exchange,
£ sterling.

Exchange, premium, or discount, at per cent.

Gentlemen,

Thirty days after sight of this my first bill of exchange (second and third of the same tenor and date not being paid) be pleased to pay to or order, the sum of [expressed in words] according to the rate of exchange above stated, being for stores, or repairs, &c. (as the case may be) for the use of his Majesty's ships under my command, as per voucher herewith forwarded, or that will be forwarded the first opportunity (as the case may be).

I am, gentlemen, your humble servant,

Captain of his Majesty's ship.

These are to certify, that the above bill was drawn for the use of his Majesty's ship and that the stores have been supplied and the workmanship performed, &c. (as the case may be).

To be signed by the signing officers of the ship.

We the subscribers, being merchants at this place, do certify that the above bill is not drawn under the present governing rate of exchange.

To be signed by two or three merchants.

} Merchants.

2. *Bills drawn or issued by the Navy Board.*—A commissioned officer, master, or surgeon, may on appointment to a ship from half-pay, apply to the commissioners of his Majesty's navy, for an order on the treasurer of the navy, to pay such officer the arrears of half-pay that may be due to him, up to the day of his commencing full pay, when the commissioners, being satisfied as to the appointment, and that no imprest stands against his wages, will grant an order for the payment of the arrears of half-pay appearing to be due, which the act of Parliament says, shall be paid at the pay-office of the treasurer of the navy.

A commissioned officer, master, or surgeon, may also, on being appointed to a ship from half-pay, apply to the commissioners of the navy for an order from them (or the commissioner at the port where the ship may be) on the treasurer of the navy, to pay such officer the amount of three months net personal pay, which will be accordingly granted, provided no imprest stands against his wages, and he transmits along with his application, a certificate of the day he is entitled to commence full pay, but no officer is authorized to draw a bill for pay in advance, nor can any officer be allowed three months full pay in advance, whose name does not appear on the half-pay list previous to the time of his appointment.

3. *Bills drawn or issued by the Victualling Board.* Victualling Bills are orders for the payment of money, issued by the commissioners of the navy and victualling offices, on the treasurer of the navy, &c. for stores and provisions purchased for the service of the navy.

That the contractors may not be losers by receiving these bills in payment, the current discount, at the time of making them out, is added to the amount of the goods delivered. The inconveniences arising from too great a number of these bills remaining undischarged at any time, have induced government at different periods, to allow the proprietors to exchange them for a certain proportion of redeemable stock; and thus they become possessed of at least an equal security, with this advantage, that any part of it can be disposed of, as occasion may require, instead of being limited in this respect, by the amount of the bills, which is generally considerable; the interest also becomes, by this mode of management, payable at certain periods; whereas the time of payment of the interest on the unfunded bills, previously to the regulation which took place in 1794, was always uncertain; no payment being made till the principal and the interest due upon it were discharged together. In 1794, all navy and victualling bills were made payable on a certain day, to be expressed in each particular bill, which day was not to exceed 15 months from the date of the bill, but the bills now issued from the navy and victualling offices are at 90 days date, bearing interest at *3d per cent. per diem*, commencing from the date of the bill, and payable with it; and these bills being negociated as bills of exchange, of course cannot be bought and sold at a discount. In negociating these bills, the whole property-tax on the 90 days interest is deducted, and the interest for the time they have to run is taken on the net sum and added to the bill.

NAVY-BOARD. (*conseil, ou comité des commissaires de la marine*, Fr.) the commissioners of the navy collectively considered.

NAVY-OFFICE, (*les bureaux de la marine subordonnés aux Lords de l'Amirauté*, Fr.) the place where the chief business of the navy is transacted by the officers and commissioners; and is held at Somerset-house, in the Strand, London. This office was constituted by Henry VIII. in the year 1509.

NAUTICAL, (*nautique*, Fr.) any thing properly relating to sailors or sailing.

It is a duty incumbent on every captain in the royal navy, in whatever part of the world he may be employed, to exert himself to obtain the best possible information on all foreign coasts, with the harbours, bays, and other roadsteads thereon; to ascertain with accuracy their latitudes and longitudes; the variation of the compass; the winds and currents which prevail on them; the dangers with which the approach to them is attended, and the means of avoiding them; and also to obtain the best information on all other subjects which may in any degree tend to the improvement of navigation, or the increase of nautical knowledge. He is to direct all the officers under his command, who may be properly qualified, to determine, with the utmost accuracy in their power, the latitude and longitude of every port he goes to, and of every remarkable promontory, or head-land, with the setting of the tides and currents.

He is to order the master to keep a book, according to the form (No. 26) in the Naval Instructions, and to insert in the several columns the most perfect and circumstantial account he can obtain of the subjects to which they are respectively allotted; and, to enable him to do so, he is to direct him to survey all coasts on which he may be employed, whenever he service will admit of it; and to direct any officers, properly qualified, to assist him in taking such survey, in making charts, and in drawing views of the coasts; and he is to furnish him with boats for that purpose, whenever the other duties of the ship will admit of it. The Captain is himself to examine the accounts given in this book; to correct, improve, or enlarge them as far as his own observations shall enable him, and to sign it with the master. See NAVAL INSTRUCTIONS.

NAUTICAL *Angle*, (*angle nautique*, Fr.) an instrument invented by Mr. Thomas Pattrick, optician, &c. No. 29, King-street, Covent Garden, whereby a ship's departure, meridional difference of latitude, &c. are obtained from inspection, with the greatest accuracy, in plane, oblique, current, and Mercator's sailing, with a sliding scale of meridional parts, graduated to four miles.

NEAPED, OR BENAPED, (from *nepflod* Sax. *Amorti—Situation d'un bâtiment échoué dans le tems des grandes marées*, Fr.) the situation of a ship which is left aground on the height of a spring tide, so that she cannot be floated off till the return of the next spring.

NEAP-TIDES, (*morte marée*, Fr.) are those tides that happen when the moon is in the middle of the second and fourth quarters. The *neap-tides* are low, in contradistinction to spring-tides; and the lowest of them falls about three days after those quarters, or four days before the full and change, when it is called *deep* neap See the article TIDE.

NEAR. See the article NO-NEAR.

NEBULÆ, in astronomy, a term applied to certain fixed stars, less than the sixth magnitude, which shew a dim, hazy, light, and, therefore, only appear like little dusky specks or clouds. There are certain spots in the heavens called nebulæ, some of which consist of clusters of telescopic stars, and others appear like luminous spots of different forms. The most considerable is one in the midway, between the two stars on the blade of Orion's sword, marked θ by Bayer, and was discovered in the year 1656, by Huygens; it contains only seven stars, and the other part is a bright spot upon a dark ground, and appears like an opening into brighter regions beyond. With instruments of high space-penetrating powers, such as Dr. Herschel's 40 feet telescope, nebulæ are the objects that may be perceived at the greatest distance. For farther information on Nebulous Stars, see *Phil Trans* vol. 81: also Dr. Herschel's account of them in the same work, for 1800, 1802, and 1804.

NECESSARY MONEY, is a certain allowance made to pursers in the royal navy, to provide the ships at all times with a sufficiency of coals, wood, turnery ware, candles, lanterns, &c. and is regulated according to the following scheme, agreeably to his Majesty's Order in Council of the 23d of April, 1803.

Complement of the Ship.	Allowance per Man per Lunar Month of 28 Days. s.	d.
From 343 to 265 inclusive	1	5
—— 264 to 156	1	6
—— 155 to 122	1	7
—— 121 to 97	1	9
—— 96 to 87	2	0
—— 86 to 77	2	2
—— 76 to 68	2	4
—— 67 and under	2	6

But in case the complement should exceed 343 men, the purser is allowed a halfpenny per man a day, or one shilling and two pence per man every 28 days for the whole; for the amount whereof money is furnished him by the clerk of the cheque of the victualling department, at the port, through the agent to the commissioners of the victualling, upon his producing a certificate under the hands of the commander and master, of the time to which the ship is victualled, and of her being completed with the usual necessaries for that purpose.

When the ship carries a flag or broad pendant, or shall be employed in conducting a convoy, so as to occasion the carrying of top and poop lights, the purser is allowed credit on his account under the head of money, on producing a certificate, as aforesaid, of his having provided such extra lights.

NEEDLE. See the article COMPASS.

Magnetical NEEDLE, (*aiguille de boussole, ou aiguille aimantée*, Fr.) in navigation, a needle touched with a loadstone, and sustained on a pivot or centre; on which, playing at lib rty, it directs itself to certain points in or under the horizon; whence the magnetical needle is of two kinds, viz. horizontal, and inclinatory.

Horizontal NEEDLES, are those equally balanced on each side the pivot that sustains them: and which, playing horizontally with their two extremes, point out the north and south points of the horizon.

In the construction of the horizontal needle, a piece of pure steel is provided of a length not exceeding six inches, lest its weight should impede its volubility, very thin, to take its verticity the better, and not pierced with any holes or the like, for ornament's sake, which prevent the equable diffusion of the ma netic virtu . A perforation is then made in the middle of its length and a brass cap or head soldered on, whose inner cavity is conical, so as to play freely on a style or pivot, headed with a fine steel point or agate, to prevent injury by friction. See the article COMPASS.

The north paint of the needle, in our hemisphere, is made a little lighter than the southern; the touch always destroying the balance if well adjusted before, and rendering the north end heavier than the south, and thus occasioning the needle to dip.

The needle is not found to point precisely to the north except in very few places, but deviates from it more or less in different places, and that too at different times, which deviation is called its declination or variation from the meridian. See the articles MAGNETISM OF SHIPS, and VARIATION.

Inclinatory or Dipping NEEDLE. See the article DIPPING NEEDLE.

Sail NEEDLE, (*aiguille à voile*, Fr.) a small steel instrument with three sides towards the point. Sail needles are of various sizes, according to the work they are intended for, as large marline, small marline, double bolt-rope, large bolt-rope, small bolt-rope, store, old work, tabling, and flat-seam, *needles*. The needles should be no larger than is necessary to carry the twine, and the edges should be taken off, that the canvas may not be cut.

NESS, (*promontoire, cap, ou pointe*, Fr.) a point or promontory or jetting out of land, such as Dungeness, Orfordness, &c.

NETTING, (*filets de bastingage*, Fr.) a fence made by seizing together the bights of small ropes, leaving uniform spaces or meshes between it.

The *netting*, in merchant-ships, is usually stretched along the upper part of a ship's quarter, to contain some of the seamen's hammocks, and secured in this position by rails and stanchions.

Boarding NETTING, (*filets d'abordage*, Fr.) a netting extended fore and aft, from the gunwale, to a proper height up the rigging, to prevent an enemy from boarding in time of an engagement.

Bowsprit NETTING, (*filets de beaupré*, Fr.) a netting fastened at the outer end of the bowsprit to the horses, or man-ropes, to stow away the fore-topmaststay-sail and jib.

Hatchway NETTINGS, are made of inch rope, and the meshes formed in different ways, according to fancy: they are placed over the hatchways, and firmly secured, in order to save any person from falling into the holds; at the same time they afford a proper circulation of air in the ship.

Head NETTING, in sloops, &c. is that which is fastened to the horses in the head and upper rail, to save the men from slipping overboard.

Over-head NETTING, (*casse-tête*, Fr) is stretched from the main-mast aft to the mizen shrouds, in an horizontal position, about twelve feet above the quarter-deck. Its use is, to secure the officers from accidents of blocks, &c. falling from the mast-heads.

Splinter NETTINGS, (*filets qu'on établit contre le bord pour préserver des éclats*, Fr.) are made of sinnet, and nailed to the inner part of the ship's side, for the purpose of lessening the injury that might otherwise be occasioned by the splinters.

Top NETTING, is fastened to the rail, shrouds, and top, to preserve the men from falling.

NEUTRAL or NEUTRAL-SHIP, (*vaisseau neutre*, Fr.) an appellation given to any ship that belongs to a nation, state, or port, which is in amity with all those powers with whom she may traffick. Vessels of this description are always to be had in time of war, and are of the utmost utility in a commercial point of view, as, without their interference, the current of trade between different parts of the globe, would, in all probability be entirely impeded. Such nations as can boast of the possession of a well-ordered marine, must, in time of war, acquire immense profits, by acting as carriers between hostile powers; for the freight which they obtain is necessarily greater than that allowed to vessels in the service of the belligerent states, and must, of course, increase in proportion to the duration of hostilities. In the present war between France and England, the Americans, who have been justly styled the carriers of the world, have reaped great emoluments from their neutrality, and astonishingly improved the state of their marine: this, however, is now at an end, as they declared war against this country, towards the close of the year 1812.

Many of the northern powers have likewise benefitted in a surprising degree by our differences with the French, but, on the other hand, whatever glory our navy may have acquired, the merchant-service of Great Britain has, no doubt, been very much injured during the war.

"Captains of his Majesty's ships are not to take under their protection the ships of any power which is at war with any other power, with which his Majesty is not at war, nor the ships of a neutral power, unless they shall be ordered so to do, or some very particular circumstances shall occur to render it necessary, of which they are to inform the Secretary of the Admiralty, as early as possible." See NAVAL INSTRUCTIONS.

NEW Moon. See the article Moon.

NIBBED-HOOKS, in rope making, are used to hang the yarn on, to harden the strands, and to lay ropes from two inches and a quarter, to three inches and a quarter.

NIGHT, (*nuit*, Fr.) that part of the natural day during which the sun is below the horizon, or that space wherein it is dusky.

Night was originally divided by the Hebrews, and other eastern nations, into three parts, or watchings, and is still continued so in our navy: the first watch is from 8 to 12 o'clock; the second from 12 to 4; and the third from 4 till 8 in the morning.

The Romans, and afterwards the Jews, from them, divided the night into four parts or watches, the first of which began at sun-set and lasted till nine at night, according to our way of reckoning; the second lasted till midnight; the third till three in the morning; and the fourth ended at sun-rise.

The ancient Gauls and Germans divided their time not by days but by nights; and the people of Iceland and the Arabs do the same at this day. The like was also observed by our Saxon ancestors.

NINE-PIN Block See the article Block.

To NIP the Cable, (*saisir le cable avec les garcettes de tournevire*, Fr.) is to tie or secure it with a Seizing; which article see.

NIPPER, in rope-making, a machine formed of two steel plates, eight inches square, and half an inch thick, with a semi-oval hole in each four inches wide, which, by the upper plate moving, enlarges or contracts as the tarring of the yarn requires.

Nipper-Men, (*matelots postés pour frapper les garcettes, de tournevire sur le cable*, Fr.) are persons employed to bind the nippers about the cable and messenger, and to whom the boys return the nippers when they are taken off.

NIPPERS, (*garcettes de tournevire*, Fr.) certain lengths of the best rope-yarn carelessly fastened together, and used to fasten the cable to the messenger in a ship of war, when the former is drawn into the ship by mechanical powers applied to the latter.

They are usually nine or ten feet in length, according to the size of the cable; and five or six of them are commonly fastened about the cable and voyal at once; those which are farthest aft are always taken off as the cable approaches the main hatchway; and others are, at the same time, fastened on, in the forepart of the ship, to supply their places, the boys of the ship receiving the ends to walk after them, and carrying them forward again when cast off from the cable.

Selvagee Nippers, are used, when, from a very great strain, the common nippers are not found sufficiently secure; selvagees are then put on, and held fast by means of tree-nails. See the articles Selvagee and Tree-Nails.

NITTLES. See the article Knittles.

NOCK, in sail making, the foremost upper corner of boom-sails, and of staysails cut with a square tack.

Nock-Earing, the rope which fastens the nock of the sail.

NOCTURNAL, (*nocturne*, Fr.) something relating to the night in contradistinction to diurnal.

Nocturnal *Arc*, in astronomy, the arc of a circle described by the sun, or a star, in the night.

Semi Nocturnal *Arc of the Sun*, is that portion of a circle which he passes over between the lower part of our meridian and the point of the horizon, wherein he arises; or between the point of the horizon wherein he sets, and the lower part of our meridian.

Nocturnal, or Nocturlabium, (*nocturlabe*, Fr.) an instrument chiefly used at sea, to take the altitude or depression of some stars about the pole, in order to find the latitude and hour of the night. Some *nocturnals* are hemispheres or planispheres, on the plane of the equinoctial. Those commonly in use among seamen are two; the one adapted to the pole-star, and the first of the guards of the Little Bear; the other to the pole-star, and the pointers of the Great Bear. Fig. 6

This instrument consists of two circular plates, (fig. 6. plate XVII.) applied to each other. The greater, which has a handle to hold the instrument, is about two inches and a half in diameter, and is divided into twelve parts, agreeing to twelve months, and each month sub-divided into every fifth day; and so as that the middle of the handle corresponds to that day of the year wherein the star here regarded has the same right ascension with the sun.

If the instrument be fitted for two stars, the handle is moveable. The upper left circle is divided into twenty-four equal parts, for the twenty-four hours of the day, and each hour sub-divided into quarters, as in the figure. These twenty-four hours are noted by twenty-four teeth, to be told in the night. Those at the hour twelve, are distinguished by their length. In the centre of the two circular plates is adjusted a long index A, moveable upon the upper plate. And three pieces, viz. the two circles and an index, are joined by a rivet which is pierced through the centre with a hole, two inches in diameter, through which the star is to be observed.

To use the Nocturnal. Turn the upper plate till the long tooth, marked 12, be against the day of the month on the under plate: then, bringing the instrument near the eye, suspend it by the handle with the plane nearly parallel to the equinoctial; and viewing the pole-star through the hole of the centre, turn the index about till, by the edge coming from the centre, you see the bright star or guard of the Little Bear (if the instrument be fitted to that star): then that tooth of the upper circle, under the edge of the index, is at the hour of the night on the edge of the hour circle; which may be known wi hout a light, by counting the teeth from the longest, which is for the hour 12. See Chambers's *Encyclopædia*.

NODES, in astronomy, the two points wherein the orbit of a planet intersects the ecliptic, whereof the node, where the planet ascends northwards, above the plane of the ecliptic, is called the ascending node, the northern node, and the head of the Dragon, and is marked thus, ☊; the other node, where the planet descends to the south, is called the descending node, the southward node, or the Dragon's tail, marked thus, ☋.

The line wherein the two circles intersect, is called the line of nodes. It appears from observation, that the line of the nodes of all the planets constantly changes its place, and shifts its situation from east to west, contrary to the order of the signs; and that the line of the moon's nodes, by a retrograde motion, finishes its circulation in the period of nineteen years; after which time, either of the nodes having receded from any point of the ecliptic, returns to the same again; and when the moon is in the node, she is also seen in the ecliptic. If the line of nodes were immoveable, or had no other motion than that whereby it is carried round the sun, it would always look to the same point of the ecliptic, or would keep parallel to itself as the axis of the earth does.

NOG, a name given by shipwrights to a tree-nail drove through the heel of each shore that supports the ship on the slip.

NOGGING, the act of securing the heels of the shores with tree-nails.

NO-MAN's-LAND, (*espèce d'échaffaut porté sur les potences entre le gaillard d avant et le gaillard d'arrière*, Fr.) a space in midships, between the after part of the belfrey and the fore part of a ship's boat, when she is stowed upon the booms, as in a deep-waisted vessel. This term is, probably, derived from the situation, as being neither on the starboard nor larboard side of the ship, but being situate in the middle, partakes equally of all those places.

NONAGESIMAL, (*nonagésime*, Fr.) in astronomy, the 90th degree of the ecliptic, reckoned from the eastern term or point. The altitude of the nonagesimal is equal to the angle of the east, and, if continued, passes through the poles of the ecliptic; whence the altitude of the nonagesimal at a given time, under a given elevation of the pole, is easily found. If the altitude of the nonagesimal be subtracted from 90°, the remainder is the distance of the nonagesimal from the vertex.

NO NEAR! or No Nearer, (*pas au vent!* Fr.) the order to the helmsman, to steer the ship no higher to the direction of the wind than the sails will operate to advance the ship in her course; or, when the ship is close-hauled, to put the tiller a little to the windward side, in order to keep the sails full.

NORMAN, (*barre de bois qui se place dans un des trous du vindas dans certains bâtimens marchands pour servir à amarrer les cables*, Fr.) a name given to a short wooden bar, thrust into one of the holes of the capstan, to take a turn for lowering any thing.

NORTH, (*nord*, Fr.) one of the four cardinal points of the horizon, being that intersection of the horizon and meridian which is nearest our pole.

North-East *Passage*. The advocates for the *north east passage* have divided that navigation into three parts; and, by endeavouring to shew that these three parts have been passed at different times, they conclude from them that the whole taken collectively is practicable. These three parts are, 1. From Archangel to the river Lena; 2. From the Lena round Tschukotskoi Noss (or the north-eastern promontory of Asia) to Kamtschatka; and 3. From Kamtschatka to Japan. With respect to the first part, no one has ever asserted that it has been performed in one voyage. From an account of the several voyages that have been made in these seas, it appears that there is a cape between the rivers Chalanja and Piasida, that has never yet been doubled. As to the second division, it has been affirmed that a passage has been effected by several vessels, which have, at different times, sailed round the northern extremity of Asia. But from the Russian accounts, it is inferred, that it has been performed but once, viz. by one Deshneff, who, in 1648, is said to have doubled this formidable cape. Of the third, or remaining part, of this passage, no doubt can be entertained. The connexion between the seas of Kamtschatka and Japan has been established by many voyages.

NORTHING, (*différence de latitude nord*, Fr.) in navigation, the difference of latitude which a ship makes in sailing towards the north pole.

North-Star, or North-Pole-Star, is the last in the tail of the Little Bear. See the article Pole-Star.

North-*North-East*, *North-East-and-by-East*, are the subdivisions of the compass between the north and east.

North-*Pole*. See the article Poles.

North-West *Passage*. A *north-west passage*, by Hudson's-bay into the Pacific Ocean, has been more than once attempted, but hitherto without success. Some greatly doubt of the practicability of such an enterprise, and think the observations made by the Russians give us small hopes of success. It appears (from the Phil. Trans. No. 482), that the Russians have passed between the land of Nova Zembla and the coast of Asia; and, as the Dutch formerly discovered the northern coasts of Nova Zembla, we may be now well assured that that country is really an island.

North Westing. See the article Variation.

NOSE-BIT, in block-making, a bit similar to a gouge-bit, having a cutting edge on one side of its end.

NOTHING-OFF! See the article Off.

NUCLEUS, the Kernel, a term used by Hevelius, and some other astronomers, for the body of a comet, which others call its head, as distinguished from its tail, or beard.

Nucleus, is also a term used by some writers for the central part of the earth, and other planets, which they suppose firmer, and as it were separated from them, considered as a cortex or shell.

NUMERARY-Signals. See the article Signals.

NUT, a small piece of iron, with a screw-hole cut in the middle.

NUTATION, in astronomy, a kind of trepidation, or tremulous motion of the axis of the earth, whereby, its inclination to the plane of the ecliptic is not always the same, but varies backwards and forwards some seconds; and the period of these variations is nine years. This nutation was discovered by Dr. Bradley, who published an account of his discovery in the year 1737.

Sir Isaac Newton observes, that the moon has the like motion, only very small and scarcely sensible.

NUTS of the Anchor, (*tenons de la verge d'une ancre*, Fr.) are two little prominencies, appearing like short square bars of iron, fixed across the upper part of the anchor shank, to secure the stock thereof in its place; for which purpose there is a corresponding notch, or channel, cut in the opposite parts of the stock, of the same dimensions with the nuts. See the article Anchor.

O.

OAK, (*chene*, Fr.) the most useful timber for strength and durability, when exposed to the weather. It is used by Europeans in ship-building; as also in mast-making, for trestle-trees, cross-trees, open-tops, &c.

The English oak claims the precedence as a timber tree for its prodigious height and bulk, and the superior worth of its timber. Every possessor of considerable estates, ought, therefore, to be particularly assiduous in raising woods of them: which is effected by sowing the acorns either in a nursery, and the plants transplanted where they are to remain, or sown at once in the places where they are always to stand.

All the sorts of oak will prosper in any middling soil and open situation, though in a loamy soil they are generally more prosperous; they will thrive tolerably in gravelly, sandy, and clayey land, as may be observed in many parts of this country of the common oak.

But it is to the crown lands that we must look with any degree of confidence for the future supply of naval timber. If 60,000 or 70,000 acres of the royal forests had been enclosed and planted, as recommended by the commissioners of the land revenue in 1792, we should at this time have had a valuable nursery for future use. In the session of 1811, however, bills were passed for planting and enclosing of 1,600 acres in the forest of Alice Holt, and 2,000 acres in that of Woolmar. It is, indeed, more than time to look minutely into the state of our naval timber. The crisis is fast approaching, when the scarcity will be felt in a most alarming degree. It is generally allowed, that, if we go on building ships of the line at the rate in which we have proceeded for the last seven years the whole of our native oak will be exhausted in less than half a century. The expenditure of oak in the royal navy, has been generally computed to be the one eighth part of the whole consumption of the country.

OAKUM, (*étoupe*, Fr.) the substance into which old ropes are reduced, when they are untwisted, loosened, and drawn asunder. It is principally used in calking the seams, tree-nails, and bends of a ship, to prevent leaks. See the article CALKING.

Black or Tarred OAKUM, (*étoupe noire, ou goudronnée*, Fr.) is that which is formed from old tarred ropes.

White OAKUM, (*étoupe blanche*, Fr.) is that which is formed from untarred ropes.

OAKUM-BOY, (*calfatin*, Fr.) the boy who attends calkers to bring them oakum, pitch, &c.

OAR, (*aviron, or rame*, Fr. *are* Sax.) in navigation, a long piece of timber, flat at one end, and round or square at the other, and which being applied to the side of a floating vessel or boat, serves to make it advance upon the water.

The flat part of the oar which is dipped into the water is called the *blade*; and that which is within board, is termed the *loom*, whose extremity, being small enough to be grasped by the rowers, is called the *handle*.

To push the boat or vessel forwards, by means of this instrument, the rowers turn their backs forward, and, dipping the blade of the oar in the water, pull the handle forward, so that the blade, at the same time, may move aft in the water. But since the blade cannot be so moved without striking the water, this impulsion is the same as if the water were to strike the blade from the stern towards the head; the vessel is, therefore, necessarily moved according to this direction. Hence, it follows that she will advance with the greater rapidity, by as much as the oar strikes the water more forcibly. Consequently an oar acts upon the side of a boat or vessel like a lever of the second class, whose fulcrum is the station, upon which the oar rests on the boat's *gunnel*. In large vessels, this station is usually called the *row-port*; but in lighters and boats it is always termed the *row-lock*, as exhibited by *e e*, fig. 1, plate IV.

The oars for ships are generally made out of fir timber; those for barges are made out of New England, or Dantzic rafters; and those for boats, either out of English ash, beech, or fir rafters from Norway.

Boat your OARS! (*désarme les avirons!* Fr.) the order to the men to cease rowing, and to lay the oars in the boat.

Double-banked OARS, (*avirons à couples*, Fr.) imply the situation of the oars of a boat, when two opposite ones are managed by rowers seated on the same bench or thwart. The oars are also said to be *double-banked*, when there are two men rowing at each oar.

To feather the OARS, in rowing, is the art of turning their blades in an horizontal position, that they shall not hold the wind.

Get your OARS *to Pass! (arme les avirons!* Fr.) the order to prepare them for rowing.

Lay on your OARS! the order to the men to cease rowing for a short time, in order to speak, or to pay homage to a superior when passing.

To ship the OARS, (*armer les avirons*, Fr.) is to place them in the row-locks ready for rowing.

To unship the OARS, (*désarmer les avirons*, Fr.) is to heave them out of the row-locks, which is frequently done on ferries, &c. in passing close to a vessel, to prevent them from being broke.

OBJECT-GLASS, of a telescope or microscope, the glass placed at the end of the tube, which is next the object to be viewed.

To prove whether object-glasses be well centred, hold the glass at a due distance from the eye, and observe the two reflected images of a candle, varying the distance till the two images unite, which is the true centre point; then if this fall in the middle or central point of the glass, it is known to be truly centred.

As object-glasses are commonly included in cells that screw upon the end of the tube of a telescope, it may be proved whether they be well centred, by fixing the tube, and observing, while the cell is unscrewed, whether the cross-hairs keep fixed upon the same lines of an object seen through the telescope.

OBLIQUE ANGLE, one that is not a right angle, but is either greater or less than this, being either obtuse or acute.

OBLIQUE *Angled-Triangle*, is that whose angles are all oblique.

OBLIQUE *Ascension*, is that point of the equinoctial, which rises with the centre of the sun, or star, or any other point of the heavens in an oblique sphere.

OBLIQUE *Circle*, in the stereographic projection, is any circle that is oblique to the plane of projection.

OBLIQUE *Descension*, is that point of the equinoctial which sets with the centre of the sun or star, or other point of the heavens in an oblique sphere.

OBLIQUE *Sailing*, is the application of oblique-angled plane triangles, to the solution of problems at sea. In going along shore, this sailing is particularly useful in ascertaining the distance of places from the ship, and from each other; and hence, it is applied in taking a survey of a coast or harbour.

In this sort of sailing, *to set an object*, signifies to observe what rhumb or point of the compass is directed to it, and the *bearing* of an object is the rhumb on which it is seen; also the bearing of one place from another is reckoned by the name of the rhumb passing through those two places; hence the bearing of two places from each other are upon opposite points of the compass; thus, if one place bears E.N.E. from another, the latter will bear W.S.W. from the former, being in the same line, but in opposite directions.

EXAMPLE.

Coasting along shore we observed two capes, the first bore N. by E. and at the same time the second bore N. E. ¾ E.; now, by the chart, these capes bear from each other N. W. ¼ W. and S. E. ¼ E. (by compass) distant 21 miles; required the distance from both places at that time.

BY CONSTRUCTION.

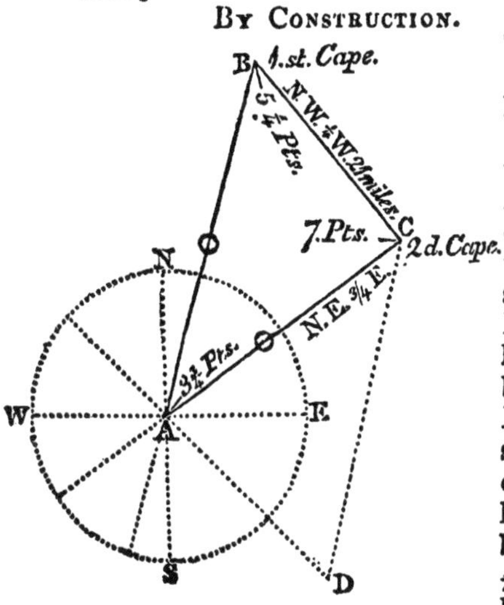

To illustrate this example, let N.W. S. E. represent the compass, A the centre, or place of the ship, observing cape B. (or the line A B) bearing N. by E. and, at the same time, cape C (or the line A C) bearing N.E. ¾E. and since by the chart these two capes bear from each other N.W.¼W. draw the opposite rhumb S.E.¼E. or line A D=21 miles; then from the point D draw the line D C parallel to A B, and the line C B parallel to A D; the former will cut the N.E.¾E rhumb in C, making cape C's distance, or the line A C=27 miles, and the latter the N. by E. rhumb in B, making cape B's distance, or the line A B=31 miles.

BY CALCULATION.

In the triangle A B C are given the angle B A C 3¾ points, the arc between the N. by E. and N. E. ¾ E. lines; the angle A B C 5 ¼ points, the interval between the S. by W and S. E. ¼ E. lines; and the angle A C B 7 points, the interval between the N. W. ¼ W. and S. W ¾ W. lines: and the side B C 21 miles; to find the sides A B and A C.

To find the side A B.

As sine angle A. 3 ¾ points ar. co.	0 17292
Is to the side B C 21	1.32222
So is sine angle C 7 points . . .	9.99157
To the side A B 30. 67	1.48671

To find the side A. C.

As sine angle A 3¾ points ar. co.	0.17292
Is to side B C 21	1.32222
So is sine angle B 5¼ points . . .	9.93335
To the side A C 26. 82	1.42849

Hence, the distance of the ship from the first cape is 31 miles, and from the second cape 27 miles nearly.

OBLIQUE *Sphere*, in geography, is where the pole is elevated any number of degrees less than 90°; in which case, the axis of the world, the equator, and parallels of declination, will cut the horizon obliquely. It is this obliquity that occasions the inequality of days and nights, and the variation of the seasons.

OBLIQUITY OF THE ECLIPTIC, is the angle which the ecliptic makes with the equator. See the article ECLIPTIC.

OBSERVATION, (*observation*, Fr.) among mariners, is observing with an instrument the meridian altitude of some celestial phenomenon, as the sun, moon, and stars, or their distances from each other. The finding a latitude from the meridian altitude observed, they call *working an observation.* For the method of making an *observation*, see the articles LATITUDE and LONGITUDE.

OBSERVATIONS, in astronomy. See the articles CHRONOMETER, *Reflecting* CIRCLE, LUNAR *Observations, &c.*

OBSERVATORY, a building furnished with instruments for observing the heavenly bodies. It was formerly in the shape of a tower, erected on an eminence, and covered with a terrace; but now, firmness of foundation, and a convenient disposition of instruments, are considered of more importance than any particular form of structure, or eminence of situation.

Fixed OBSERVATORIES, are those where instruments are fixed in the meridian; by which, with the aid of astronomical clocks, the right ascensions and declinations of the heavenly bodies are determined, and motion, time, and space, made to measure each other.

Such buildings and apparatus only are called *regular* observatories, although very useful operations are sometimes performed, and important discoveries made, where there are no instruments fixed in the meridian.

As navigation essentially depends on the determinations made in observatories, such establishments have been considered of great national importance, especially in maritime states; and hence they have been liberally endowed by different governments. Even private observatories have been, in many places, erected at a considerable expence, and their number has been greatly increased of late years; a circumstance which, while it marks the progress of science, does honour to the age in which we live.

Besides the important advantages of improving nautical astronomy, observatories are, in themselves, objects of the highest interest to every scientific and liberal mind. Here the inventions and discoveries of the greatest men in all ages are united, and applied to the most grand and useful purposes. Here theory is reduced to practice, and science illustrated by art. In short, an observatory may be considered as affording the most sublime display of the works of God and man.

The Greenwich observatory, or the royal observatory of England, was built and endowed by order of King Charles II. at the instance of Sir Jonas Moore, and Sir Christopher Wren; the former of these gentlemen being surveyor-general of the Ordnance, the office of astronomer royal was placed under that department, in which it has continued ever since.

This building was erected on the site of the ancient moated tower of Duke Humphrey, uncle to Henry VI. and the first stone of it was laid August the 10th, 1675, by Mr. Flamsteed, who had been appointed astronomer royal.

It is situate on the highest eminence of Greenwich park, about 160 feet above low water mark; and stands in latitude 51° 28′ 40″. The soil here is particularly favourable for such an institution, being of a flinty gravel, through which the rain soon passes, and thus the atmosphere is generally dry, which contributes to the preservation of the instruments, as well as to the uniformity of refraction.

It is furnished with several very accurate instruments; particularly two mural quadrants, each of eight feet radius, with telescopic sights; that which observes the southern meridian was made by Bird; and the other, which observes the northern, was constructed by Graham, a celebrated artist, who not only improved clocks and watches to a surprising degree, but also astronomical instruments; he also contrived the equatorial sector for making observations out of the meridian, and constructed a zenith sector, of 24 feet radius for the royal observatory, with which Dr. Bradley discovered the aberration of the fixed stars. The astronomical, or transit clock, was also made by Graham, and has been rendered very accurate by Earnshaw. The transit instrument, which is eight feet long, and the axis three feet, is famous as having been used by Halley, Bradley, and Maskelyne. It was originally made by Bird, and has been successively improved by Dolland and Troughton. There are also some excellent telescopes, particularly a forty-inch achromatic, with a triple object-glass, and a five-feet achromatic, by Dolland; with a six-feet reflector, by Dr. Herschel. Also two equatorial sectors, by Sisson; and a clock, by Arnold. These are chiefly used for observing comets.

A valuable equatorial instrument, constructed by Ramsden, has lately been presented to the royal observatory by the Hon. Mr. Jenkinson: it is intended to be used both in a vertical and inclined position: and as an altitude and azimuth instrument, there can be no doubt of its proving a valuable acquisition.

A grand transit circle of six feet radius, constructed by Troughton, under the direction of the late Dr. Maskelyne, and a clock of great value by Hardy, were also placed in the royal observatory in May 1812, with which Mr. Pond, the present astronomer royal, made his observations on the summer and winter solstice of that year, which, with a catalogue of some of the principal fixed stars, he published in the first volume of the *Philosophical Transactions* for the present year 1813. From the construction of this instrument, no use whatever is made either of the level or plumb line, but the position of the polar point on the instrument is found by assiduously observing the pole-star, and is inferred, by comparing the observed, with the real places of certain numbers of standard stars, previously determined by the instrument itself; by which means its position may always be ascertained to within a tenth of a second, a degree of precision to which no plumb-line or level can ever be expected to arrive.

The first astronomer royal, or the person to whom the province of observing was first committed, was Mr. John Flamsteed, who held the situation 43 years; Dr. Halley succeeded Flamsteed, and discharged the duties of the situation with much credit to himself for 23 years; Bradley, 20 years; Bliss, 2 years; and Maskelyne, 46 years: the latter of whom has been succeeded by John Pond, Esq. F.R.S. who was appointed astronomer royal in February 1812. For farther particulars see the article Astronomical *Observations*, &c.

There is an observatory belonging to the royal naval college in Portsmouth dock-yard, which is furnished with an excellent astronomical clock, a good transit instrument, and a telescope at the public expense.

A series of astronomical observations have been carried on at this observatory for many years by the late Mr William Bayly, head-master of the Royal Academy, who was one of the best calculators and accurate astronomers of the age, having been several years employed as an assistant to that able astronomer Dr. Maskelyne, at the royal observatory at Greenwich, where he calculated several of the early volumes of the Nautical Almanac; he was afterwards appointed, by the Board of Longitude, to accompany Sir Charles Douglas to the North Cape to observe the transit of Venus, and, subsequently, accompanied the celebrated Captain Cook in his voyages round the world. His astronomical observations were published in an 8vo. volume; and his valuable papers and observations were bequeathed to his learned friend and executor, Col. M. Beaufoy, of Hackney Wick, who is fully competent to appreciate their value.

There is, also, a public observatory near Oxford,

belonging to the University, at present under the care of Dr. Robertson ; and another at Cambridge, under the direction of Mr. Vince, professor of astronomy at that University.

There has lately been an Astronomical Institution formed at Edinburgh, with a view to the cultivation of astronomy, which seems to have two objects in view ; viz.

1. A popular observatory, furnished with telescopes, level, sextant, repeating-circle, a small astronomical circle, portable transit, clock, &c.

2. A scientific observatory provided with instruments for the nicest observations, such as a mural circle, transit instrument, reflecting circle, a zenith, sector, astronomical clock, reflecting and achromatic telescopes ; the whole to be under the inspection of the professors of mathematics and astronomy in the University.

The observatory is on the Calton Hill, near the city, was begun some years ago, but never finished, and is now granted by the magistrates to the present institution.

There is also a very good observatory at Glasgow, belonging to the University, and under the direction of Dr. Urie ; it is fitted up in the best manner, with all the modern instruments.

There was, formerly, a good public observatory at Aberdeen, belonging to the University, under the direction of the late Dr. A Mackay, who has long been known to the mathematical world by his writings: he possessed great assiduity and vigour of mind, which were principally directed to the advancement of the sciences connected with navigation. Since his time, however, it has been suffer. d to go to decay.

There is a public observatory about two miles from Dublin, under the care of Dr. Brinkley, where there is a very large astronomical circle, which was begun by the late Mr. Ramsden, and finished by Mr. Berge, in 1807. The vertical circle of this instrument is eight feet in diameter, with which the learned Doctor has made several very minute observations, and has extenuated the apparent annual parallax of Lyræ to be nearly three seconds.

There is, also, another public observatory at Armagh, under the direction of Dr. Hamilton, in which there is an excellent equatorial instrument constructed by Mr. Troughton, with a vertical circle of three feet and a half in diameter, with which the Doctor has made some accurate observations, which were published in the *Philosophical Transactions* for 1806.

Besides the above mentioned, there are many private observatories, of which the following is an alphabetical list.

In England.

Blackheath . . .	Stephen Groombridge, Esq.
Blenheim . . .	The Duke of Marlborough.
Cambridge . . .	Rev. Mr. Catton.
Chislehurst . . .	Rev. Francis Wollaston.
Christ's Hospital .	Governors and Masters
Derby	William Strutt, Esq.
East Sheen . . .	Rev. Wm. Pearson.
Finsbury-square, .	Dr. P. Kelly.
Godwood . . .	The Duke of Richmond.
Gosport	Dr. William Burney.
Hackney Wick . .	Col. Beaufoy.
Hayes	William Walker, Esq.
Highbury . . .	Mr. Bentley,
Highbury Terrace .	Capt. Huddart.
Hoddesdon . . .	Wm. Hodgson, Esq.
Islington	Gavin Lowe, Esq.
Paragon, Southwark .	James Strode Butt, Esq.
Park Lane . . .	Sir Harry Englefield, Bart.
Richmond Gardens .	His Majesty.
Rose Hill, Sussex .	John Fuller, Esq.
Sherburn . . .	Earl of Macclesfield.
Slough	Dr. Herschel.
Somerset House . .	The Sec. to the Royal Soc.
St. Ibbs, Hitchen .	Mr. Professor Lax.
Woolwich . . .	Rev. Lewis Evans.

In Scotland.

General Brisbane, in Renfrewshire, has a very neat observatory fitted up in the best manner, where there is a beautiful two feet circle, which is actually the model of the six-feet transit-circle, made by Troughton for the royal Observatory at Greenwich.

Within these few years, there was a private observatory at Hawk-hill, near Edinburgh, in latitude 55° 57′ 5″; in which astronomical observations used to be made, and meteorological observations registered. And there was also a private observatory at Perth, belonging to the lord privy seal for Scotland.

OBTUSE, (*obtus*, Fr.) signifies blunt, dull, &c. in opposition to acute, sharp, &c. thus, we say, *Obtuse* Angle, *Obtuse* Angled-*Triangle*, which articles see.

OCCIDENT, (*occident*, Fr.) in geography, the westward quarter of the horizon, or that part of the horizon where the ecliptic, or the sun therein descends into the lower hemisphere.

OCCULTATION, in astronomy, the time a star or planet is hidden from our sight, by the interposition of the body of the moon, or some other planet. The occultation of a star by the moon, if observed in a place whose latitude and longitude are well determined, may be applied to the correction of the lunar tables ; but, if observed in a place whose latitude only is well known, may be applied to the determining the longitude of the place.

Circle of Perpetual Occultation, is a parallel in an oblique sphere, as far distant from the depressed pole, as the elevated pole is from the horizon. All the stars between this parallel and the depressed pole, never rise, but lie constantly hidden under the horizon of the place.

OCEAN, (*océan*, Fr.) in geography, is that vast collection of salt and navigable waters, in which the two continents, the first including Europe, Asia, and Africa, and the last America, are inclosed like islands. The ocean is distinguished into three grand divisions : 1. The Atlantic Ocean, which divides Europe and Africa from America, which is generally about 3,000 miles wide. 2. The Pacific Ocean, or South Sea, which divides America from Asia, and is generally about 10,000 miles over; and 3. The Indian Ocean, which separates the East Indies from Africa, which is 3,000 miles over. The other seas, which are called oceans, are only parts or branches of these, and usually receive their names from the countries they border upon.

By taking a good map of the world, and, with a pair of scissars, clipping out all the water from the land, and weighing the two parts separately, it will be found, that the water occupies about two thirds of the whole surface of the globe.

OCHRE, (*ochre*, Fr.) a sort of red chalk, used by shipwrights to mark timber with, in hewing and forming it.

OCTANT, or Octile, in astronomy, the aspect of two planets, wherein they are distant an eighth part of a circle, or 45° from each other.

OFF, (*à la hauteur*, Fr.) is applied to the movement of a ship, when she sails out from the shore towards the distant sea; it also implies abreast of, or near; as, "We were off Cape Finisterre."

Off and On, (*bord à terre, bord au large*, Fr.) when a ship, beating to windward, approaches the shore by one board, and, by the other, sails out to leeward, she is said to stand *off and on*.

Nothing Off! (*sans arriver!* Fr.) the order to the helmsman not to suffer the ship to bear away, or fall off from the wind.

Naval OFFICER, (*officier de la marine*, Fr.) an officer holding a commission from the lords of the Admiralty in the royal navy; the term is applied in opposition to marine or military officer, equally to lieutenants, captains, and admirals; the term is also applied to commissioners in his Majesty's dock-yards, to governors of hospitals, and, frequently, to men in other civil situations at the out ports, who are not commissioned officers. Our naval officers are now considered of much greater consequence in the state than formerly, they, as well as our military officers, being frequently called to the House of Commons or legislative assembly, and are often employed as diplomatic characters at foreign courts; it, therefore, becomes necessary that they should be men of liberal education; otherwise they can never be expected to fill even the higher stations in their own profession with credit to themselves, and honour to their country; and even when they have fought a glorious battle, cannot set down and give a good account of it.

There are three classes of officers in the royal navy, viz. 1. Commission officers, who receive their appointments from the lords of the Admiralty: 2. Warrant officers, who receive them from the Navy Board: and 3. Petty officers, who are appointed by the captains.

Officer *of the Watch*, (*officier de quart*, Fr.) the lieutenant who has the charge of the watch. See the article Watch.

OFFING, (*le large, le côté du large, ou de la pleine mer*, Fr.) implies out at sea, or at a good distance from the shore, where there is deep water, and no need of a pilot to conduct the ship; thus, if a ship from shore be seen sailing out towards the sea, they say, "She stands for the *offing*." And if a ship having the shore near her, have another a good way without her, or towards the sea, they say, "That ship is in the *offing*."

OFFWARD, (*vers le large*, Fr.) the situation of a ship which lies aground, and leans from the shore: Thus, they say, "The ship heels *offward*," when, being aground, she heels towards the water-side; and, "The ship lies with her stern to the *offward*, and the head to the shoreward" when her stern is towards the sea, and head to the shore.

OIL. (*huile*, Fr.) The general character of oils are combustibility, insolubility in water, and fluidity. The well known properties of this article, and the uses to which it is applied need no recapitulation. The effect produced by oil on the surface of the sea, when agitated by the wind, would seem incredible, were not the fact so easily ascertainable as to remove every incredulous doubt. Considerable precaution ought to be used in attending to it when shipped on board in any quantity. It is extremely combustible though a liquid, and, even without any inflammable substance approaching it, it has been known frequently to take fire, and materially endanger the safety of the ship. This was occasioned by one of the vessels, in which the oil was contained, having leaked, and the fermenting oil (if that term might be used) set fire to some light articles which lay over the spot. This anecdote, singular as it may appear, literally took place, some years ago, on board an East Indiaman, called the Ajax.

OKER. See the article Ochre.

Laws of OLERON, are certain laws of the navy or marine, which were framed and drawn up by Richard I. at the Island of Oleron, near the coast of Poictou, the inhabitants of which have been deemed able mariners for these seven hundred years past. These sea-laws, which have been reckoned the most excellent of the kind, are recorded in the Black Book of the Admiralty.

ON the Beam. See the article Beam.

ON the Bow. See the article Bow.

ON the Quarter. See the article Quarter.

OPEN, (*à découvert*, Fr.) the situation of a place which is exposed to the wind and sea.

Open, (*ouvert*, Fr.) is also expressed of any distant object, to which the sight or passage is not intercepted by any thing lying or coming between. Thus, to be open with any place, is to be opposite to it; as the entry of a port, road, or haven.

OPENING, (*ouverture, ou passage entre deux terres*, Fr.) a passage, or streight, between two adjacent coasts or islands.

OPHIUCUS, in astronomy, a constellation of the northern hemisphere; called also Serpentarius, which article see.

OPPOSITION, is that aspect or situation of two planets or stars, when they are diametrically opposite to each other; being 180°, or a semi-circle apart; and marked in astronomical books thus, ☍.

OPTIC-Glasses, are glasses ground either concave or convex; so as either to collect or disperse the rays of light; by which means vision is improved, and the eye strengthened, preserved, &c. Among these, are telescopes, microscopes, &c.

Optic *Inequality*, in astronomy, is an apparent irregularity in the motions of far distant bodies; so called, because it is not really in the moving bodies, but arising from the situation of the observer's eye. For, if the eye were in the centre, it would always see the motions as they really are.

Optic *Place*, of a star, &c. is that point or part of its orbit, which is determined by our sight, when the star is seen there. This is either true or apparent; true, when the observer's eye is supposed to be at the centre of the motion; or apparent, when his eye is at the circumference of the earth.

ORB, a spherical shell, hollow sphere, or space contained between two concentric spherical surfaces.

The ancient astronomers conceived the heavens as consisting of several vast azure transparent orbs or spheres inclosing one another, and including the bodies of the planets.

The Orbis *Magnus, or Great* Orb, is that in which the sun is supposed to revolve: or rather it is that in which the earth makes its annual circuit.

ORBIT, in astronomy, the path of a planet or comet, or the curve it describes in its revolution round its central body; thus the Earth's orbit is the curve which it describes in its annual course, and usually called the ecliptic.

The orbits of all the planets are ellipses having the Sun in their common focus; in which curve they move according to an invariable law. However, the orbit of the Earth is considerably disfigured by the action of the Moon; as is also the orbi of Saturn by the action of Jupiter, when they happ n to be in conjunction. The orbits of the planets are not all in the same plane with the ecliptic, but are variously inclined to it, and to each other; but still the plane of the ecliptic intersects the plane of the orbit of every other planet, in a right line which passes through the Sun, called the line of the nodes, and the points of intersection of the orbits themselves are called the nodes.

ORDER, in astronomy. A planet is said to go according to the order of the signs when it is direct, proceeding from Aries to Taurus, thence to Gemini, &c. It goes contrary to the order of the signs when it is retrograde, or goes back from Pisces to Aquarius.

Order *of Sailing*. See the article Sailing.

ORDERS, *(instructions,* Fr.*)* the injunctions of officers to the men, in respect to the duty of the ship, &c. Every officer is strictly enjoined to refrain from making any remarks on the orders of any of his superior officers, which may tend to bring him into contempt; and most carefully to avoid the doing of any thing, which, if seen or heard by, or reported to, the ship's company, may discourage them, or render them dissatisfied with their condition, or with the service on which they shall then be employed, or with any service on which they may be ordered. And if any officer shall so far forget so essential a part of his duty, as either, by his conduct or observations, to endeavour to lessen the respect due to his superior, or his orders; or to dishearten the ship's company, or any part thereof, or to render them dissatisfied, every officer, and every other person, who shall observe such conduct, or shall hear such observations, is strictly charged to report it immediately to the captain of the ship, or to the commander-in-chief, as circumstances may require; and if any officer or other person, shall observe such conduct, or hear such observations, and shall not report them as herein directed, he shall be considered as accessary to the offence, and shall be punished accordingly.

If an officer shall receive from his superior an order which may be contrary in any respect, to any article in the general Naval Instructions, or to any particular order he may have received from the lords commissioners of the Admiralty, or from any superior officer, he is to represent in writing, such contrariety to the officer from whom he shall have received the order; but if, after such representation, that officer shall direct him to obey the order he has given him, he is to obey it, and report the circumstances to his commander-in-chief, or to the secretary of the Admiralty as may be necessary.

ORDINARY, *(état d'ouvriers et de gens de mer, employés à garder et à entretenir, les vaisseaux désarmés,* Fr.*)* the establishment of the persons employed by the government to take charge of the ships of war, which are laid up in the several harbours adjacent to the royal dock-yards. These are principally composed of the warrant-officers of each ship, as the gunner, boatswain, carpenter, purser, and cook, and their servants. There is besides a crew of labourers, termed extra men, enrolled in the list of the ordinary, who pass from ship to ship occasionally to pump, moor, remove, or clean them, when it is necessary.

Ordinary *Deals*, an inferior sort of deals, from 10 to 12 feet long, supplied to all ships in commission, for repairing the half-ports, hatches, &c.

Ordinary *Seamen, (matelots à la basse paie,* Fr) implies one who can make himself useful on board, but is not an expert or skilful sailor; the latter being termed an Able Seaman, which article see.

Ships in Ordinary, *(vaisseaux désarmés, qui ont seulement un petit état, composé d'officiers mariniers, et de gardiens pour leur entretien, dans l'usage des arsenaux d'Angleterre,* Fr.*)* are those, which, being laid up, are under the direction of the commissioners at the different royal ports.

ORDNANCE, *(artillerie,* Fr.*)* a general name for every species of artillery used in war; as cannon, mortars, howitzers, coehorns, swivels, &c. In time of service, each man is supposed to work 5 cwt. of metal; hence, may be computed the number of men required to attend each piece. For farther particulars, see the articles Cannon, Gun, and Mortar.

Boring of Ordnance. Guns are bored in the following manner; the piece is placed upon two standards, by means of two journeys, turned round by a water-wheel; the breech being introduced into the central line of the wheel, with the muzzle towards the sliding carriage, which is pressed forward by a ratch and weights: upon this sliding carriage is fixed, truly horizontal and central to the gun, the drill-bar, to the end of which is fixed a carp's-tongue-drill, or cutter; which, being pressed forward upon the piece whilst it is turning round, perforates the bore, which is afterwards finished with bars and cutters.

The machinery for boring of ordnance is sometimes put in motion by a steam engine; and, in this way, from 18 to 24 great guns have been boring at the same time; the borer in each piece being brought up to its proper place in the gun by a lever and weights. In this method of bringing up the borer, the pressure may always be made equable, and the motion of the borer regular: but the disadvantage is, that, without due attention, the borer may work up too far towards the breech, and the piece be spoiled.

In the royal arsenal at Woolwich, only one piece is bored at a time in the same mill: the gun to be bored, lies with its axis parallel to the horizon, and in that position is turned round its axis by means of wheel-work, moved by one or more horses. The borer is laid as above described, in the direction of the axis of the gun, and is incapable of motion in any direction except that of its length; and, in this direction, it is constantly moved by means of a small rack-wheel, kept in proper motion by two men, who thus make the point of the borer so to bear against the part of the gun that is boring, as to pierce and cut it. The outside of the gun is smoothed, at the same time, by men with instruments fit for the purpose, whilst it turns round, so that the bore may be exactly in the centre of the metal. See GREGORY's *Mechanics*.

ORDNANCE *Debentures*, are bills issued by the Board of Ordnance, on the treasurer of that office for the payment of stores, &c.

Office of ORDNANCE, (*bureau de l'artillerie*, Fr.) an office kept within the Tower of London, which superintends and disposes of all the arms, instruments, and utensils of war, both by sea and land, in all the magazines, garrisons and forts, in Great Britain. The officers of the Ordnance are, 1. The master-general, from whom are derived all orders and dispatches relating to the same. 2. The lieutenant-general, who receives orders from the master-general, and sees them duly executed; orders the firing of guns on days of rejoicing, and sees the train of artillery fitted out when ordered to the field. 3. The surveyor-general, who has the inspection of the ordnance, stores, and provisions of war, in the custody of the store-keepers; he allows all bills of debt, keeps a check on labourers, &c. 4. The treasurer, through whose hands passes the money of the whole office, as well for payment of salaries as debentures; as also a clerk of the ordnance, and a clerk of the deliveries.

ORIENT, (*orient*, Fr.) the east, or the eastern point of the horizon.

ORIENTAL, (*oriental*, Fr.) situate towards the east, with regard to us, in opposition to occidental or the west.

ORIENTAL *Astronomy*, *Philosophy*, *&c.* used for those of the east, or of the Arabians, Chaldeans, Persians, Indians, &c.

ORION, in astronomy, a constellation of the southern hemisphere, containing eighty stars, according to the Britannic catalogue. The late improved telescopes have discovered several thousand stars in this constellation; of these, there are two of the first magnitude, four of the second, and several of the third and fourth. One of the stars of the first magnitude, is upon the middle of the left foot, and is called *Regel*; the other is on the right shoulder, and called *Betelguese*; of the four of the second magnitude, one is on the left shoulder, and called *Bellatrix*, and the other three are in the belt, lying nearly in a right line, and at equal distances from each other, forming, what is popularly called the *Yardwand*.

ORLOP, (*faux-pont*, Fr. *over loop*, Du.) the lower but tmporary deck in a ship of war, whereon the cablese,s usually coiled, the sails deposited, and the several officers' store-rooms contained. Small ships have a kind of platform in midships, which is also called the *orlop*, and is chiefly for the use of the cables.

ORLOP-*Beams*. See the article BEAMS.

ORRERY, an astronomical machine, for representing the motions and appearances of the heavenly bodies; and hence often called PLANETARIUM, which article see.

The reason of the name Orrery is as follows:— Mr. Rowley, a mathematical instrument maker, having got one from Mr. George Graham, the original inventor, to be sent on board a ship, with some of his own instruments, he copied it, and made the first for the Earl of Orrery; Sir Richard Steel, who knew nothing of Mr. Graham's Invention, thinking to do justice to the first encourager, as well as to the inventor of such a curious instrument, called it an Orrery, and gave Rowley the praise which was due to Mr. Graham. See DESAGULIER's *Experim. Philos.* vol. I. p. 430.

ORTHOGRAPHY, in fortification, is the profile or representation of a work, or a draught so conducted, as that the length, breadth, height, and thickness of the several parts are expressed, such as they would appear if perpendicularly cut from top to bottom.

ORTIVE, or EASTERN AMPLITUDE, in astronomy, is an arc of the horizon intercepted between the point where a star rises, and the east point of the horizon.

TO OVERBLOW, (*surventer*, Fr.) in speaking of the wind, is to blow so hard that the ship can bear no top-sails.

OVERBOARD, (*hors le bord, ou à la mer*, Fr.) the state of being thrown out, or the act of falling from a ship or boat into the water on which she swims; as "the ship sprung a leak, and obliged us to throw the guns *overboard*;" "a heavy sea broke over the deck, and carried two of our men *overboard*."

OVERCAST, (*couvert, en parlant du tems*, Fr.) in speaking of the weather, implies cloudy, dull, &c.

OVERCAST-STAFF, (*trébuchet*, Fr.) a scale, or measure, employed by shipwrights to determine the difference between the curves of those timbers which are placed near the greatest breadth, and those which are situate near the extremities of the keel, where the floor rises and grows narrower.

OVERFALL, (*bas fond*, Fr.) a term used by mariners, for a bank or shoal lying near the surface of the sea, so as to endanger the safety of ships that approach it; as the Blakeny and Sheringham overfalls, the Stuky overfalls, near the coast of Norfolk.

OVERGROWN, (*grosse mer, mer très haute*, Fr.) is said of the sea, when the surges and billows are unusually high; but, when the waves are no more than commonly high, it is called a rough sea.

TO OVERHAUL, (*reprendre, gagner, visiter, &c.* Fr.) implies to examine a ship, person, or thing.

To OVERHAUL *a Tackle*, (*reprendre un palan*, F.) is to open and extend the several parts of a tackle, or other assemblage of ropes, communicating with blocks or dead-eyes, so that they may be again placed in a state of action.

OVERHAULING *a Ship*, (*gagner sur un vaisseau*, Fr.) signifies that a vessel is gaining fast upon her adversary, in a chase, as "we are overhauling that ship very fast."

OVER-LAUNCHING, in ship-building, a term used to signify the splicing or scarphing one piece of timber to another, to make the stronger work.

OVER-MASTED, (*trop de mature*, Fr.) the state of a ship whose masts are too high, or too heavy for the weight of her hull to counterbalance, as "that frigate is overmasted."

To OVER-RAKE, (*se dit d'un vaisseau à l'ancre, qui faisant tête à une grosse mer, y est très exposé, et en recoit à bord des coups fréquents*, Fr.) in speaking of a ship riding at anchor in a head-sea, when the waves thereof frequently break in upon her, they are said to *over-rake* her.

To OVER-SET, or *To* UPSET, (*chavirer*, *faire capot*, *renverser*, Fr.) is to turn any thing upside-down. The Royal George was over-set at Spithead, on the 29th of August, 1782, in consequence of being too much heeled whilst under repairs, with her lower-deck ports open, by which catastrophe upwards of 800 souls on board perished.

OUT, (*dehors; c'est-à-dire, déployé et orienté, en parlant des voiles*, Fr.) an expression frequently used at sea, implying the situation of the sails when they are *set*, or extended, to assist the ship's course; as, "What sail have you out?" This expression is opposed to IN, which signifies that such sails are furled. See the article IN.

OUT-BOARD, a term signifying any thing that is without the ship; as the out-board works, &c.

OUT-FIT, (*armement, or dépenses de l'armement*, Fr.) is generally used to signify the expenses of equipping a ship for a sea-voyage, or of arming her for war, or both together. The term is also applied in providing necessary apparel, &c. for naval officers and seamen before they go a voyage. See the articles *To* FIT *out a Ship*, and MIDSHIPMAN.

OUT-HAULER, a name given to a rope used to haul out the tack of the jib.

OUT-OF-TRIM, (*hors de son assiette, par défaut d'arrimage*, Fr.) the state of a ship when she is not properly balanced for the purposes of navigation; which may be occasioned by a defect in the rigging, or in the stowage of the hold.

OUT-OF-WINDING. See the article WINDING.

OUT-RIGGER, (*aiguille de carène*, Fr.) a strong beam of timber, of which there are several fixed on the side of a ship, and projecting from it, in order to secure the masts in the act of careening, by counteracting the strain they suffer from the effort of the careening tackles, which being applied to the mast-head draw it downwards, so as to act upon the vessel with the power of a lever, whose fulcrum is in her centre of gravity.

OUT-RIGGER, (*boute-hors, ou boute-lof*, Fr.) is also a small boom, occasionally used in the tops, to thrust out the breast back-stays to windward, in order to increase the angle of tension, and thereby give additional security to the topmast.

It is usually furnished with a tackle at its inner-end, communicating with one of the topmast-shrouds; and has notches cut on the outer end, to contain the back-stays, and keep them steady therein. As soon as the back-stays are set up, by means of their tackles in the chains, the out-rigger is applied aloft, which it forces out to windward, beyond the circle of the top, so as to increase the angle which the mast makes with the back-stays, and accordingly enable the latter the better to support the former.

This machine is sometimes applied without any tackle; it is then thrust out to its usual distance beyond the top-rim, where it is securely fastened, after which the back-stays are placed in the notches, and extended below.

To OUT-SAIL *a Ship*, (*dépasser un vaisseau; avoir la marche sur un vaisseau*, Fr.) is to sail faster than another.

OUTWARD, (*qui signifie hors du royaume*, Fr.) implies out of the port or kingdom; as,

OUTWARD-BOUND *Ships*, are those vessels that have received orders to sail to any foreign port, for the purpose of trafficking, &c. in contradistinction to homeward-bound ships.

OWNER, (*proprietaire-armateur, ou marchand chargeur d'un bâtiment*, Fr.) the proprietor of a ship, by whom she is freighted to the merchant for a sea voyage.

By the 21 Geo. III. c. 29, it is enacted, That,

1. If the master of any vessel shall conceal, or suffer to be concealed, foreign spirits, tea, or coffee, above two gallons of spirits for each seaman, above 6 lbs. of tea, or 19 lbs. of coffee; or if the master, &c. of such ship or vessel shall clandestinely import, or suffer to be clandestinely imported therein, any such foreign spiritous liquors, or any other uncustomed goods, whereby the owners become liable to forfeit the ship, he shall forfeit all his wages to the owners, with treble the value of the uncustomed goods on board.

2. If the mates or seamen shall conceal any of the said goods (above the quantity allowed by law), they shall forfeit their wages, and also 10*s*. for every gallon of foreign spirits, and 10*s*. for every pound of tea or coffee.

3. And, if in time of war, the owner may send them to serve three years on board a man of war, except such as are old, disabled, or unfit for service.

4. A printed copy of this act is to be put up on some conspicuous part of every British trading-vessel; and, when any clauses become defaced, the master shall cause the same to be replaced, under the forfeiture of one shilling per day during such omission, to be paid to the owner.

5. Mates or seamen, who shall refuse to pay the penalties, &c. may be committed for three months.

6. Justices may mitigate penalties, but not reduce them under one half.

7. Proceedings of the justices not to be quashed for want of form, and to be final.

8. Commencement of actions is limited to six months; the party sued may plead the general issue; and, on verdict against the plaintiff, recover treble costs.

But, by a discretionary power, invested in the honourable Commissioners, the stores before-mentioned are allowed duty free only for the use of the crew during the delivery of the cargo; as soon as the goods are all out, and the voyage thereby at an end, the stores that shall remain unconsumed when the ship is cleared, are to be entered, and not more left than will exceed 40*s*. in customs and excise, by way of clearing stores.

If a ship arrive in ballast, she is to be considered as fit for clearing, and to be allowed only the clear-stores as above-mentioned.

Where spirits are found on board, no wine is to be allowed as clearing-stores. But where there are no spirits, then two dozen quarts of ordinary wine, for the use of the crew on board, in short voyages, from Holland, Germany, Flanders, &c.

By the 26 Geo. III. c. 86, it is enacted, That,

1. *Owners not answerable for loss of Goods by Master or Mariners, beyond the Value of Ship and Freight.*—No person, who is owner of any ship or vessel, shall be liable to answer or make good any loss or damage, by reason of any robbery, embezzlement, secreting, or making away with, of any gold, silver, diamonds, jewels, precious stones, or other goods or merchandise, which shall be shipped on board any ship or vessel; or for any act, damage, or forfeiture, done or incurred, from the passing of this act, without the knowledge of such owner; farther than the value of the ship or vessel, with all her appurtenances, and the full amount of the freight due, or to grow due, for the voyage wherein such robbery, embezzlement, secreting, or making away with, shall be made, although the master or mariners shall not be concerned in, or privy to, such robbery, embezzlement, secreting, or making away with.

2. *Owners not answerable for loss by Fire.*—No owner of any ship or vessel shall be liable to answer or make good any loss or damage which may happen to any goods or merchandise whatsoever, which shall be shipped and put on board such ship or vessel, by reason or means of any fire happening to, or on board of, the said ship or vessel.

3. *Masters or Owners not answerable for Loss of Goods, unless their Nature, Quality, and Value, are made known upon shipping.*—No master or owner of any ship or vessel shall be subject or liable to answer for, or make good, any loss or damage which may happen to any gold, silver, diamonds, watches, jewels, or precious stones, which shall be shipped on board any such ship or vessel, by reason of any robbery, embezzlement, making away with, or secreting thereof, unless the owner or shipper thereof shall, at the time of shipping the same, insert in his bill of lading, or declare, in writing, to the master or owner of such ship or vessel, the true nature, quality, and value, of such gold, silver, diamonds, &c.

4. *Where Ship and Freight are not equal to all the Losses,*—If several freighters or proprietors of any such gold, silver, &c. shall suffer any loss or damage, by any of the means aforesaid, in the same voyage (fire only excepted), and the value of the ship or vessel, and freight, shall not be sufficient to make full compensation to all of them; then such freighters or proprietors shall receive satisfaction thereout in proportion to their respective losses; and, in every such case, it shall be lawful for such freighters or proprietors, or any of them, on behalf of himself and all other such freighters or proprietors, or for the owners of such ship, to exhibit a bill in any court of equity, for a discovery of the total amount of such losses, and of the value of such ship or vessel, appurtenances, and freight, and for an equal distribution and payment amongst such freighters or proprietors, in proportion to their respective losses, according to the rules of equity. Provided that, if any such bill shall be exhibited on behalf of the part-owners of such ship, the plaintiff, or plaintiffs, shall thereto annex an affidavit, that he or they do not collude with any of the defendants; and shall thereby offer to pay the value of such ship or sassel, and freight, as such court shall direct.

5. *Remedies against the Master and Mariners*—Provided that nothing shall extend to discharge any remedy which any person or persons now hath or have, or may hereafter have, against the master or mariners of such ship, in respect of any embezzlement, secreting, or making away with, any gold, silver, &c. shipped or loaden on board such ship or vessel; or on account of any fraud, abuse, or malversation, of and in such master and mariners respectively; but it shall be lawful for every such person or persons, so injured, to pursue such remedy against the said master and mariners as they might have done before the making of this act.

6. This act shall be considered a public act.

Besides these two acts of parliament, which affect the interest of ship-owners, there are other rules of law to which they are subject; the principal of which are as follow:—

Goods spoiled by Default of a Master.—If goods are spoiled by default of a master of a ship employed by the owners, the owners are liable; but the action must be brought against all the part-owners, who make but one master.

Part-Owners not agreeing about a Voyage.—If several part-owners wish to send a ship on a voyage, but two or three other part-owners refuse their consent, the former may send her on the voyage, but they must enter into a recognizance in the Admiralty for her safe return. See the article PART-OWNERS.

Repairing a Ship.—If a ship be repaired in the river Thames, and fitted out there with new rigging and apparel, the ship itself is not liable, but the owners. If she be repaired abroad, the ship is liable, and the master may hypothecate (or pawn) her for payment of the charges.

The repairer of the ship may sue either the master who employs him, or the owners; but if he undertakes it on a special promise from either, the other is discharged. An action was brought by a shipwright for repairing the defendant's ship in his dock. About three hours before the ship's repairs were finished, a fire happened, and she was burnt. Notwithstanding which, the court held that the owner was liable to pay for the repairs that had been done.

Owners liable for Provisions, &c. bought by the Master.—If the master of a ship buys provisions for her, and has money from the owners to pay for the same, but sails without paying the money, the owners are liable to pay, in proportion to their respective shares in the ship, the master being but a servant to the owners. See 2 VERN. *Rep.* 643.

Masters and owners being intimately connected in many respects, we shall therefore refer the reader to the article MASTER *of a Merchant-Ship.*

OX-EYE, (*œil de bœuf, météore qui se remarque au cap de Bonne Espérance, et qui présage un coup de vent,* Fr.) a name given by mariners to a small cloud or meteor, seen at the Cape of Good Hope, &c. which presages a dreadful storm. It appears at first in the form or size of an ox's eye, but descends with such celerity that it seems suddenly to overspread the whole hemisphere, and at the same time forces the air with such violence, that ships are sometimes scattered several ways, some directly contrary, and many sunk downright.

P.

PACK, or PACKAGE, *(balle de marchandise, futaille, &c.* Fr.) among merchants, is generally defined, a single lot or parcel of any species of goods made up for exportation, &c. such as a barrel or pipe of wine, a bale of cloth, a case of hats, &c.

PACKET, or PACKET-BOAT, *(paquebot,* Fr.) a vessel appointed by the government to carry the mail of letters, packets, and expresses, from one kingdom to another by sea, in the most expeditious manner. Thus the packet-boats, under the direction of the postmaster-general of Great Britain, carry the mails from Dover to Calais, and from Harwich to Helvoetsluys, in time of peace; also from Parkgate to Dublin, and from Falmouth, to Lisbon, Gibraltar, Malta, the West Indies, America, &c.

PADDLE, *(pagaie,* Fr.) a sort of oar sometimes used in men-of-war's boats, as also by the natives of Africa and America, to navigate their canoes. It is much shorter and broader in the blade than the common oar, and is equally employed in rowing and steering. See the article CANOE.

PAINT, *(couleur à l'huile,* Fr.) in the arts, is prepared by grinding, with a proper quantity of oil, the charcoal, or lamp-black, or red or yellow ochre, &c. according to the colour wanted for use, till they are united into a smooth, uniform, thick compound, which is occasionally diluted with more oil, to a due consistence for being worked freely with the brush.

There are sundry sorts of paint used in and about his Majesty's ships, either for ornament or for the preservation of such parts to which they are applied, as black, red, white, and yellow, which are in the gunner's and carpenter's charge.

PAINT-*Cisterns,* in ships of war, are cisterns made of wood, and lined with lead, to contain the different kinds of paint.

PAINTER, *(cablot, ou cordage servant à amarrer une chaloupe, ou autre embarcation,* Fr. probably from *bindar,* Sax. to bind) a rope employed to fasten a boat along-side of a ship, wharf, or key, &c.

Shank-PAINTER. See the article SHANK.

PALILICUM, in astronomy, the same as ALDEBERAN; which article see.

PALLAS, in astronomy, a new planet, situate between the orbits of Mars and Jupiter, and is a little larger in magnitude than Ceres, but of a less ruddy colour.

It was discovered by Dr. Olbers, at Bremen, in Lower Saxony, on the 28th of March 1802, the same active astronomer who discovered Ceres some years after Piazza had first discovered it at Palermo, in Sicily. It is surrounded with a nebulosity of less extent than that of Ceres, and performs its annual revolution in nearly the same period.

The planet Pallas, however, is distinguished in a very remarkable manner from Ceres and all the other primary planets, by the immense inclination of its orbit. While these celestial bodies are revolving round the sun in almost circular paths, rising only a few degrees above the plane of the ecliptic, Pallas ascends above this plane, at an angle of about 35 degrees, which is nearly five times greater than the inclination of Mercury. From the eccentricity of Pallas being greater than that of Ceres, or from a difference of position in the line of their apsides, while their mean distances are nearly equal, the orbits of these two planets mutually intersect each other; a phenomenon which is altogether anomalous in the solar system.

The atmosphere of Pallas, according to the observation of Schroeter, is to that of Ceres as 101 to 146, or nearly as 2 to 3. It undergoes similar changes, but the light of the planet exhibits greater variations On the 1st of April, the atmosphere of Pallas suddenly cleared up, and the solid nucleus or disc of the planet was alone visible. About 24 hours afterwards she appeared pale and surrounded with fog; and this appearance continued during the 3d and 4th of April. Schroeter has shewn, that this phenomenon does not arise from the diurnal rotation of the planet.

The diameter of Pallas has not yet been determined with sufficient accuracy. Dr. Herschel makes it only 80 miles, which is not quite one half the diameter of Ceres, while Schroeter makes it not less than 2099 miles, which is considerably larger than the magnitude that he assigned to Ceres.

For farther information on the subject of this article, see SCHROETER, *Lilienthalische Beobachtungen der neu-entdeckten Planet Pallas, Gottingen,* 1805; *Phil. Trans.* 1807; *Connoissance des Temps,* 1809.

PALLETTING HATCHES, in ship-building, are small apertures, about 20 inches square, formed by the palletting-beams and carlings in the fore-magazine. Drawers are fitted to them for the purpose of holding charcoal, which in a great measure draws the damp that would otherwise communicate with the powder.

PALM, *(paumelle de voilier,* Fr.) in sail-making, &c. an instrument used instead of a thimble in sewing of canvas, sails, &c. It is composed of a flat round piece of iron, an inch in diameter, whose surface is full of cavities, to receive the head of the needle, and is fixed upon a piece of canvas or leather, which encircles the hand, keeping the iron in the palm of the hand, whence it has its name.

PALM, among anchor-smiths, is also the broadest part of the arm of an anchor, which terminates in a point, in order to fix readily into the ground. See the article ANCHOR.

PANCH. See the article PAUNCH.

PARALLACTIC, something relating to the parallax of the heavenly bodies. See the article PARALLAX.

PARALLACTIC *Angle of a Star*, is the difference of the two angles, under which its true and apparent distance from the zenith is seen.

The sines of the parallactic angle, at the same or equal distances from the zenith, are in the reciprocal ratio of the distances from the centre of the earth.

PARALLACTIC *Rotula*, an instrument invented by the late Mr. Fergu on, and afterwards improved by Chevalier de Borda, for clearing the distance, in taking lunar observations.

PARALLAX, (*parallaxe*, Fr.) in astronomy, denotes a change of the apparent place of any heavenly body, caused by being seen from different points of view; or it is the difference between the true and apparent distance of any heavenly body from the zenith.

That part of the heavens in which any celestial object would appear, if viewed from the surface of the earth, is called its *apparent place*; and the point wherein it would be seen at the same instant from the earth's centre, is called its *true place*. The difference between the true and apparent places is called the *parallax in altitude*.

The PARALLAX *of Altitude*, is properly the difference between the true distance from the zenith, and the apparent distance. Hence the parallax diminishes the altitude of a star, or increases its distance from the zenith. See the article ALTITUDE.

The Horizontal PARALLAX *of a Planet*, is the angle which the earth's semi-diameter, when viewed directly, subtends at that place, and is always greatest in the horizon. From hence it decreases all the way to the zenith, where it is nothing; the real and apparent places there coinciding. The horizontal parallax being given, the parallax corresponding to any given altitude may be found by the following rule:—

To secant of the apparent altitude, add the proportional logarithm of the horizontal parallax; the sum, rejecting radius, will be the proportional logarithm of the parallax in altitude.

The parallax always diminishes the altitude of a phenomenon, or makes it appear lower than it would do, if viewed from the centre of the earth; and this change of the altitude may, according to the different situation of the ecliptic and equator in respect of the horizon of the spectator, cause a change of the latitude, longitude, declination, and right ascension of any phenomenon, which is called their parallax. The parallax, therefore, increases the right and oblique ascension; diminishes the descension; diminishes the northern declination and latitude in the eastern part, and increases them in the western; but increases the southern both in the eastern and western part; diminishes the longitude in the western part, and increases it in the eastern. Hence it appears, that the parallax has just opposite effects to refraction. See the article REFRACTION.

PARALLAX *of Right Ascension and Descension*, is an arc of the equinoctial, by which the parallax of altitude increases the ascension and diminishes the descension.

PARALLAX *of Declination*, is an arc of a circle of declination, by which the parallax of altitude increases or diminishes the declination of a star.

PARALLAX *of Latitude*, is an arc of a circle of latitude, by which the parallax of altitude increases or diminishes the latitude.

PARALLAX *of Longitude*, is an arc of the ecliptic, whereby the parallax of altitude increases or diminishes the longitude.

PARALLAX *of the Annual Orbit of the Earth*, is the difference between the heliocentric and geocentric place of a planet, or the angle at any planet, subtended by the distance between the earth and sun.

For the various methods of finding the parallaxes of the celestial bodies, see HUTTON's *Mat. and Phil. Dictionary*, and MACKAY *on the Longitude*.

PARALLEL, (*parallèle*, Fr.) is a word sometimes used by mariners in the same sense as latitude, as, "Our orders were to cruise in the parallel of Teneriffe;" which signifies, "We were ordered to cruise in the latitude of Teneriffe."

PARALLEL *Sailing*, (*route parallèle*, Fr.) is the art of finding what distance a ship should run due east or west, in sailing from the meridian of one place to that of another place, in any parallel of latitude; the method of performing which is by sailing to the parallel of latitude the place is in; keeping a good account, so as to be certain whether the place is then to the eastward or westward; and also, if possible, to know the longitude arrived at, and then to run due east or west, till the ship comes near the longitude of the given place, where she is then sure to make the port required.

This is usually done in the torrid zone, to avoid or take advantage of the TRADE WINDS, or MONSOONS, as they are frequently called: see these articles.

The computations in parallel sailing depend on the following rules:—

As radius,
Is to the difference of longitude;
So is the co-sine of latitude,
To the meridional distance. And,
As the difference of longitude,
Is to radius;
So is the meridional distance,
To the co-sine of latitude.

Hence, the length of a degree in any parallel, is as the co-sine of latitude is to radius; or, which is the same thing, as radius is to the secant of latitude.

PARALLEL *Sphere*, (*sphère parallèle*, Fr.) is that situation of the sphere wherein the equator coincides with the horizon, and the poles with the zenith and nadir. In this sphere all the parallels of the equator become parallels of the horizon, consequently, no stars ever rise or set, but all turn round in circles parallel to the horizon; and the sun, when in the equinoctial, wheels round the horizon the whole day. After his rising to the elevated pole, he never sets for six months; and after his entering again on the other side of the line, never rises for six months longer. This is the position of the sphere to such as live under the poles, and to whom the sun is never higher than 23° 30′.

PARALLELS, or PARALLEL *Circles*, in geography,

called also parallels or circles of latitude, are lesser circles of the sphere, conceived to be drawn from west to east, through all the points of the meridian, commencing from the equator to which they are parallel, and terminating with the poles. They are called parallels of latitude, because all places lying under the same parallel have the same latitude.

PARALLELS *of Latitude*, in astronomy, are lesser circles of the sphere parallel to the ecliptic, imagined to pass through every degree and minute of the colures. They are represented on the globe by the divisions on the quadrant of altitude, in its motion round the globe, when screwed over the pole of the ecliptic.

PARALLELS *of Altitude*, or ALMUCANTARS, are circles parallel to the horizon, imagined to pass through every degree and minute of the meridian between the horizon and zenith, having their poles in the zenith. They are represented on the globe by the divisions on the quadrant of altitude, in its motion about the body of the globe, when screwed to the zenith.

PARALLELS *of Declination*, in astronomy, are the same with parallels of latitude in geography.

PARBUCKLE, (*trévire*, Fr.) is a contrivance to haul up, or lower down a cask, &c. where there is no crane or tackle.

It is formed by passing the middle of a rope round a post or ring, or under a boat's thwart; the two parts of the rope are then passed under the two quarters of the cask, bringing the two ends back again over it, which being both hauled or slackened together, either raise or lower the barrel, &c. as may be required.

To PARCEL A SEAM or A ROPE, (*couvrir une couture, ou un cordage avec des bandes de vieille toile*, Fr.) is to put a sufficient quantity of canvas round it, that is well daubed over with tar, and bound with spunyarn, to keep it from chafing, which is commonly done by the help of a serving-board.

PARCELLING, (*morceaux, ou bandes de toile goudronnée, pour servir de fourrure*, Fr.) a name given to long narrow slips of canvas, daubed with tar, and frequently bound about a rope, in the manner of bandages, previous to its being *served*.

It is laid in spiral turns, as smoothly upon the surface as possible, that the rope may not become uneven and full of ridges. Parcelling is also used to raise a *mouse* on the *stays*, &c. and is firmly fastened by *marling*, from one end to the other.

PARDO, the name of a vessel used in the Chinese seas, both for trade and war; they are not so large as Junks, but similar, except that their sails are slackly laced by one side to the masts, instead of being suspended by a yard.

PARLIAMENT-HEEL, (*demi-bande*, Fr.) a term used to imply the situation of a ship, when she is made to stoop a little to one side, so as to clean the upper part of her bottom on the other side, and cover it with a fresh composition; and afterwards to perform the same office on that part of the bottom which was first immersed. The application of a new composition, or *coat* of stuff, above the wale, is called BOOT-TOPPING, which article see.

PARRAL, (*racage*, Fr.) an assemblage of ribs and trucks, which, with the parral-rope, form a sort of collar (see fig. 1, plate XVIII.), to attach the top-sail yards to the masts, so that the yards may slide easily up or down the masts. Pl. XVIII. Fig. 1.

The rib, fig. 2, is made of ash-board, the diameter of the yard in length, and one-third the length in breadth. It is straight on the inner edge, and divided into two semi-circles on the back, with a hollow cavity between. Through the flat sides are bored two holes, one in the middle of each semi-circle, for the reception of the rope. Fig. 2.

The truck, fig. 3, is made of ash, or harder wood, the breadth of the rib in diameter, and one-third more than the diameter in length: it is turned in a lathe, and, through the centre, lengthways, is bored a hole, to answer those in the rib. Fig. 3.

There are four sorts of parrals; viz.

PARRAL-ROPE, (*bâtard de racage*, Fr.) is formed of a single rope, well served, and furnished with an eye at each end; this being passed round the yard, is seized fast on, the two ends are passed round the after-part of the mast; and one of them being brought under, and the other, over the yard, the two eyes are lashed together with a piece of spun-yarn on the fore-side thereof: this, however, is seldom used but for the top-gallant and smaller yards.

PARRAL *with Ribs and Trucks*, (*bigots de racage à pommes*, Fr.) is formed by passing the two parts of the parral-rope through the two holes in the ribs, observing that between every two ribs is strung a truck on each part of the rope. The ends of the parral-rope are made fast with seizings; these are chiefly used on the topsail-yards.

In order to fasten this machine more conveniently about the mast and yard, so as to attach the latter to the former, the parral-rope is formed of two pieces, each of which is furnished with an eye at one end, and both eyes lie on one side of the mast; that is to say, one piece of the rope passes through the lower part of the parral, and thence under the yard, whilst the other comes through the upper part of the parral and over the yard, till both eyes meet on the fore-side of the yard, where they are joined together. The two other ends of the parral-rope are passed about the yard and the hind part of the parral alternately, till the latter is sufficiently secured to the former.

PARRAL *with Trucks*, (*pommes de racage*, Fr.) is composed of a single rope, passing through a number of trucks, sufficient to embrace the mast; these are calculated for the cheeks of a gaff.

Truss-PARRAL, (*racage simple, ou racage à l'Angloise*, Fr.) is formed by fixing a rope upon the middle of the yard, which, passing at the back of the mast, is reeved through an iron thimble, spliced into another rope (also fastened upon the yard), and communicates with a tackle reaching to the deck, whereby it may be occasionally slackened or straightened; ships of war generally have two of these, one leading from each side, and they are peculiar to the lower-yards.

To PART, (*aller en dérive, en parlant d'un bâtiment qui a rompu son cable*, Fr.) to be driven from the anchors; and is expressed of a ship when she has broken her cable by the violence of the wind, waves, &c.

PARTIAL LOSS, in maritime affairs, implies a damage which a ship may have sustained in the course of her voyage, from any of the perils mentioned in the policy of insurance. When applied to the cargo, it means the damage which goods may have received, without any fault of the master, by storm, capture, stranding, or shipwreck; although the whole or the greater part thereof may arrive in port. These *partial losses* fall upon the owners of the property so damaged, who must be indemnified by the underwriter. But the underwriters of London expressly declare, as appears from a memorandum at the foot of the policy, that they will not answer for partial losses, not amounting to 3 per cent. This clause was intended to prevent the underwriters from being continually harassed by trifling demands. But at the same time that they provide against trifling claims for partial losses, they undertake to indemnify against losses, however inconsiderable, that arise from a general average.

When we speak of the underwriters being liable to pay, whether for total or partial losses, it must always be understood, that they are liable only in proportion to the sums which they have underwritten. Thus, if a man underwrites 100*l.* upon property valued at 500*l.* and a total loss happen, he shall be answerable for 100*l.* and no more, that being the amount of his subscription; if only a partial loss, amounting to 60*l.* or 70*l.* per cent. upon the whole value, he shall pay 60*l.* or 70*l.* being his proportion of the loss. See the article INSURANCE.

PARTNERS, (*étambrais; les trous où passent les mâts sont apellés de même en François, étambrais*, Fr.) in ship-building, are pieces of plank, fastened in certain frames which surround the capstans, masts, pumps, and bowsprit, for the purpose of strengthening them, and preserving them from the effects of pressure: hence,

PARTNERS *of the Bowsprit*, (*étambrai du beaupré*, Fr.) are three pieces of oak timber douled together, and bolted to the upper and lower-deck beams, to receive the heel of the bowsprit. In 74-gun ships they are about twelve inches thick, and three feet nine inches in breadth.

PARTNERS *of the Main-Capstan*, (*étambrais du grand cabestan*, Fr) are several pieces of four-inch oak plank shut in the frame, and fayed to the barrel on the main-deck; they are bolted and calked, but the bolts are not clinched, and sufficient room is always left for the capstan to work easy, and for the admission of an iron hoop, which projects an inch above the partners, to prevent the water from going between decks.

PARTNERS *of the Main-Mast*, (*l'étambrai du grand mât*, Fr.) a frame fitted in the apertures of the decks, for the reception of this mast, and is intended to strengthen the decks in those places, but particularly to secure them from the pressure of the mast.

PARTNERS *of the Pumps*, (*les étambrais des pompes*, Fr.) are pieces of two and a half inch plank fayed close to the pumps, and fastened to the decks: they are calked and coated, to prevent the water from going between decks.

PART-OWNERS, are partners interested and possessed of certain shares in a ship. When there are part-owners of a ship, the majority may fit her out, without the consent of the rest; and if they do, such majority run all hazards, and have a right to partake of the profits.

Actions lie as well against the part-owners of ships for the loss or damage of goods delivered to the masters, as against the masters; for, as the masters of ships are chargeable in respect of their wages, so are the part-owners in respect to the freight: but an action against the part-owners of a ship must be brought against all of them, or the defendants may take advantage of it, by pleading in abatement, &c. SHOW. *Rep.* 30. 105; 3 LEV. 259. See the article OWNER.

To PASS, (*passer*, Fr.) a term used by seamen to give any thing that is portable from one to the other: also to take certain turns of a rope round a yard, &c. as, "Pass the line along!" "Pass the gasket!" "Pass a heel lashing!" "Pass a seizing!" &c.

PASSAGE, (*traversée, ou passage d'un port à un autre; aussi le prix couvenu avec le capitaine pour le passage*, Fr.) a voyage from one place to another by sea; and, according to stat. 4 Edward III. c. 7, is the hire a man pays for being transported over any sea or river.

Homeward-Bound PASSAGE, (*vaisseau sur son retour*, Fr.) See the article HOMEWARD-*Bound Ship*.

Outward-Bound PASSAGE, (*vaisseau sur sa traversée de départ*, Fr.) See the article OUTWARD-*Bound Ships*.

PASSAGE-BOAT, (*bateau de passage*, Fr.) a ferry-boat, or any small vessel employed to carry passengers or luggage by water, from one port to another.

PASSAREE, (*cordage servant, lorsque le vent est largue, à assujettir l'amure de la misaine au boissoir*, Fr.) a rope formerly used to confine the tacks down to the ship's side, when she is going large in light breezes.

PASSENGER, (*passager*, Fr.) a person who pays for his passage in any vessel, and, therefore, is not expected to assist either in working or fighting her.

PASSPORT, or PASS, (*congé, ou passeport, de bâtiment marchand*, Fr.) a license or writing obtained from a prince or governor, granting permission and a safe conduct to pass through his territories: it is also a permission granted by any state to navigate in some particular sea, without hindrance or molestation from it; and contains the name of the vessel, and that of the master, together with her tonnage, and the number of her crew, certifying that she belongs to the subjects of a particular state, and requiring all persons at peace with that state, to suffer her to proceed on her voyage without interruption. See the article MEDITERRANEAN-PASS.

This term is also used for a license granted by a prince, for the importation or exportation of merchandise, moveables, &c. without paying the duties. Merchants procure such passports for certain sorts of commodities.

PASSPORT, is also a license obtained for the importing or exporting of merchandize deemed contraband, and declared such by tariffs, &c. as gold, silver, precious stones, ammunition, horses, corn, wool, &c. upon payment of the duties.

PATRON, a name given to the commanders of small vessels, such as passage-boats, &c. and is particularly applied to the man who steers the ship's long-boat. This term is in most common use in the Mediterranean. See the article COXSWAIN.

PAVO, *the Peacock*, in astronomy, a constellation

in the southern hemisphere, added by the modern astronomers. It contains 14 stars.

PAUNCH, (*natte, baderne, ou tissu de vieux cordages pour fourrer les cables, vergues, &c.* Fr.) a sort of thick and strong mat, formed by interweaving twists of rope-yarn into foxes, as close as possible. It is chiefly used to fasten on the outside of the yards or rigging, to prevent their surfaces from being chafed by the friction of some other contiguous object, and particularly when the vessel is rocked by a tempestuous sea.

PAWL, (*élinguet de cabestan, de vindas, et tous cliquets en général, soit de bois ou de fer*, Fr.) a piece of iron, about two feet in length and four inches square at its largest end, the other end is furnished with a hole, and bolted through the partners of the capstan; in this position it traverses round the bolt, and pawls the capstan, which it prevents from turning back, when it is employed to heave in the cable or hawsers; thus, they say, "Heave a pawl!" that is, heave a little more for the pawls to get hold of the welps. See the article CAPSTAN.

PAWL-BITS, (*cadre de charpente, et potence établie vers le milieu du vindas, où se fixent les élinguets ou cliquets*, Fr.) in merchant-vessels, are pieces of timber fixed perpendicularly before the windlass, near the middle of it, and serving as supports to the pawls which are pinned into them.

PAY, (*appointemens*, Fr.) in the royal navy, is generally applied to the officers' income from government, for their personal services; but the term WAGES is more frequently applied to the pay of seamen.

As the British navy advanced progressively to that meridian of power and glory, which it has now attained, various acts of the legislature were passed for the increase and encouragement of seamen, which, to use the language of the legislature, "had greatly tended to augment the marine force of this realm, upon which the security of these realms, and preservation of their trade and commerce, did immediately depend."

The several acts of King William, Queen Anne, and George I. when the navy might comparatively be said to have been in its cradle, having been found in a great measure ineffectual to answer the benevolent purposes of protection and encouragement for which they were intended; Mr. Grenville, then treasurer of the navy, in the year 1758, brought into parliament a bill, 31 Geo. II. c. 10. which formed the foundation upon which seamen's wages have since been paid.

By this act, seamen, whether at home or on foreign service, were enabled to remit by bill a part of their wages for the support of their wives and families.—Tickets were made out and transmitted to the navy-office, which enabled their relations to receive immediately the wages of those who died abroad, and such modes were prescribed for witnessing seamen's powers, as were expected to put a stop to those frauds and forgeries which had previously been so repeatedly practised.

The objects, however, of this act having been, in many instances, found ineffectual, particularly with respect to seamen's wills and powers, an act was passed (26 Geo. III. c. 6), establishing *specific modes* for executing wills and powers of attorney, under every predicament in which a seaman could be placed, and appointing an officer, under the denomination of *Inspector of Wills*, whose express duty it was, minutely to investigate the authenticity of these instruments.

In the year 1792, other acts were passed by the legislature for the encouragement of seamen, 35 Geo. III. c. 33 and 34, viz.—The first, an act for explaining and amending the 26 Geo. III. and for further extending the benefits thereof to *petty-officers and seamen, non-commissioned officers of marines, and marines.*—The principal objects of these acts were,

1st. To extend to marines serving in the navy, the liberty of remitting a part of their wages to their families; to allow them the privilege of being paid by tickets, and other advantages, which by the 31 Geo. II. were confined to seamen exclusively.

2d. Enabling seamen, who might be removed abroad from one ship to another, and, perhaps, afterwards to a third or fourth, to receive upon their arrival at any port in England where wages are paid, all the pay to which, for their past services, they were intitled—to empower such as were disabled in the service to receive their pay from the revenue-officer nearest to the place where they should arrive or reside, without obliging them to travel, as had been formerly the case, from the remotest part of the kingdom, to London, Portsmouth, Plymouth, or Chatham, and without making them wait till the arrival of different ships, in which they might successively have served, which might often have been kept on foreign stations for many years; enabling them also to receive their pay without waiting for the return of their ships, if left on shore at hospitals.

3d. To stop the practice too common with officers of the navy, of anticipating their pay, by assigning to agents and others, not only such part as was due at the time of the transaction, but also such as they expected to receive for their future services.

4th. To stop the practice of taking seamen out of the service by fraudulent arrests, which was continued, notwithstanding the regulations of former acts: by those acts seamen were exempted from arrests for any sum under 20*l.*; they were trusted, therefore, till their debts exceeded that amount, and then imprisoned; by purposely granting fictitious notes also, and procuring themselves to be arrested thereon, seamen often avoided going on particular duties, to the detriment of the public service.

The second, 36 Geo. III. c. 6, "for explaining and amending the 26th of the same king, and for further extending the benefits thereof to petty officers and seamen, non-commissioned officers of marines, and marines, serving on board any of his Majesty's ships." The objects of this act relate principally to *the wives, the families*, and *relations of seamen*, enabling them in the most remote situations to receive their allowances from the public, without the assistance or interference of agents or attorneys; and by making the money pass immediately from the public purse into the hands of the party entitled thereto, to put a stop to the heavy deductions and abatements which are almost always made *when a third person* is employed. This is done by means of remittances, and is also an extension of a principle introduced by former acts. Under the regulations of this act, the widow, the child, the father, or other relation, the person entitled by will, or even the fair creditor of deceased seamen, who have served in the navy, has only to make application, by letter, to the treasurer of the navy, stating his connexion or relation to the deceased, and the nature of his expectations from his estate; and desiring that all the wages and allowances due may be remitted to him, at whatever place he finds most convenient. No acquaintance with the proceedings or forms of office is necessary. In consequence of the first application the necessary papers and vouchers are sent to be executed; and as soon after as the steps for examination can be taken, and the justness of the claim is admitted, a bill is sent for the clear balance due, payable by the revenue officer most convenient to the party, without fee, deduction, or abatement.

The third act, 37 Geo. III. c. 53, extends the benefits of all the former acts to seamen and their families who reside in Ireland, gives them the same powers of remitting wages and pensions to themselves and families, and puts them in every respect upon the same footing with seamen and their families who reside in Great Britain.

The fourth act, viz. 37 Geo. III. c. 53, is entitled, "An Act for carrying into execution his Majesty's Order in Council of the 3d of May 1797, for an *increase of pay and provisions* to the seamen and marines serving in his Majesty's navy; and to amend so much of the 35th year of the reign of his present Majesty, as enables petty officers and seamen, non-commissioned officers of marines, and marines, *to allot part* of their pay for the maintenance of their wives, children, or mothers."

This act is founded upon his Majesty's order in council, for an *increase in the allowance of wages and provisions* for the petty officers, seamen, landmen, and marines; and for a continuation of the pay of all petty officers, seamen, landmen, and marines, who may hereafter be wounded in action with the enemy, until their wounds shall be healed, or some other provision shall be made for them. By the first clause of this act the following additions are made to the wages—viz. 5*s.* 6*d.* per month in addition to the wages of the petty officers and able seamen; an addition of 4*s.* 6*d.* per month to the wages of landmen; and to the marines, when embarked and serving on board his Majesty's ships, the allowances usually called consolidated allowances made to marines when serving on shore, viz.—to serjeants, corporals, and drummers, at the rate of *two pence*, and to privates at the rate of *two-pence farthing* per day; and with respect to provisions, *the full allowance* is to be issued to the crews of his Majesty's ships, without any deduction whatsoever on account of leakage or waste.

By the 2d clause of this act, petty officers, seamen, landmen, and marines, wounded in action with the enemy, shall receive the full amount of their wages and allowances until their wounds shall be healed; or until, being declared incurable, they shall receive a pension from the chest at Chatham (now the chest at Greenwich), or be admitted into the royal hospital at Greenwich.

The 3d section empowers the commissioners of the navy to direct the treasurer of the navy, or any receiver-general of land tax, collector of customs, or collector of excise, or clerk of the check, to whom an allotment shall have been made pursuant to the act of 35 Geo. III. c. 28, *to increase the allowance* to the wives or mothers of petty officers or seamen, non-commissioned officers, or marines, in a proportion *equal to half the increased* pay provided by this act; and which shall thereupon be paid in the same manner as if such declaration and order of allotment had been originally, according to the rate of half the pay of such petty officers, seamen, and marines, as by this act is provided. This calculation is to be made as nearly equal as may be to *half the pay*, and to be calculated according to a schedule annexed to the act.

In the year 1803, another act was passed; viz.—43 Geo. III. c. 166, "for the encouragement of seamen, and for the better and more effectually manning his Majesty's navy; for regulating the payment of prize money," &c.—This act, however, was repealed by the 45 Geo. III. c. 72, which makes further regulations relative to prize agency, and consolidates all the provisions for the distribution of prize money into one single act; which will be found detailed under the article Prizes.

The following Tables of Officers and Seamen's personal Pay, are calculated, agreeably to an Order in Council, dated August 23, 1806, upon the increased pay in the navy, which commenced May 1st, 1806:—

*Table of Admiral's Pay.**

	Day.	Week.	Month.	Year.	Compens.
	£. s. d.	£. s. d.	£. s d.	£. s. d.	£. s. d
Admiral and Com. in Chief of the Fleet	5 0 0	35 0 0	140 0 0	1825 0 0	729 19 6
Admiral	3 10 0	24 10 0	98 0 0	1277 10 0	450 3 0
Vice-Admiral ..	2 10 0	17 10 0	70 0 0	912 10 0	280 5 0
Rear-Admiral ..	1 15 0	12 5 0	49 0 0	638 15 0	202 0 0

Table of Flag-Officers Pay,† *Compensation, Deductions, &c. for Three, Six, and Nine, Lunar Months.*

Three Months.		Full.	Deduct.	Nett.	6 Months	9 Months.
		£. s. d.	£. s. d.	£. s. d.	£. s. d	£. s. d
Adm. C. of Fleet	Flag Pay	420 0 0	5 5 0	414 15 0	829 10 0	1214 0 0
	Compens.	168 0 0	11 5 0	156 15 0	313 10 0	470 5 0
Admiral	Flag Pay	294 0 0	3 13 6	290 6 6	580 13 0	870 19 0
	Compens.	99 0 0	6 15 0	92 5 0	184 10 0	276 15 0
Vice-Ad.	Flag Pay	210 0 0	2 12 6	207 7 6	414 15 0	622 2 6
	Compens.	64 10 0	4 10 0	60 0 0	120 0 0	180 0 0
Rear-Ad.	Flag Pay	147 0 0	1 16 9	145 3 3	290 6 6	435 9 9
	Compens.	46 10 0	3 7 6	43 2 6	86 5 9	129 7 0
First Cap. to Adm.	Flag Pay	147 0 0	1 16 9	145 3 3	290 6 6	435 9 9
	Compens.	28 10 0	2 5 0	26 5 0	52 10 0	78 15 0

Number of Men allowed to Flag-Officers.

	Men	Retinue 1*l.* 4*s.* per Mon.	Servants 19*s.* per Month.	Borne on Ship Books as Servants to Adm and paid Able.
Commander in Chief	50	34	16	12
Admiral	30	18	12	10
Vice-Admiral	20	10	10	7
Rear-Admiral......	15		10	5
First Captain	10		10	

* Entitled to draw for their personal pay and compensation for *three*, *six*, and *nine* lunar months.

† Exclusive of the allowance of 1*l.* per day table money to the commander in chief, which may be included in his bills.

Flag officers are appointed by the Admiralty Board, and upon appointment to a command from half-pay, are entitled to *three months'* personal pay in advance, to commence from the date of their commission, or order to hoist their flags.

Table of the Number of Officers, &c. in the different Rates, with their full Pay per Lunar Month.

OFFICERS.	No.	RATE	PAY PER MONTH.
			£. s. d.
Captain	1 in all	1 2 3 4 5 6	32 4 0 26 12 0 23 2 0 18 4 0 15 8 0 16 16 0
Lieutenant (*a*)	8 6 4 3 2	1 and 2 3 4 5 6	8 8 0
Master	1 in all	1 2 3 4 5 6	12 12 0 11 11 0 10 10 0 9 9 0 8 8 0 7 7 0
2nd Ditto	1 in	1, 2, 3, only	5 5 0
Boatswain Gunner Purser	1 in all	1 2 3 4 and 5 6	4 16 0 4 6 0 3 16 0 3 6 0 3 1 0
Carpenter (*b*)	1 in all	1 2 3	5 16 0 5 6 0 4 16 0
Master's Mates	6 4 3 2 2	1 2 3 4 5 and 6	3 16 6 3 10 6 3 6 8 2 18 4 2 12 6
Midshipmen (*c*)	24 24 16 10 6 4	1 2 3 4 5 6	2 15 6 2 10 6 2 8 0 2 4 3 2 0 6 2 0 6
Captain's Clerk (*d*)	1 in all	1 2 3 4 5 and 6	4 7 0 4 0 6 3 16 0 3 5 4½ 2 18 6
Schoolmaster Armourer (*e*) Master at Arms	1 in all	3 4 5 and 6	2 8 0 2 4 3 2 0 6
Carpenter's Mates (*f*)	2	1 and 2	2 10 6

OFFICERS.	No.	RATE.	PAY PER MONTH.
			£. s. d.
Calker and Ropemaker (*g* and *h*)	1 in all	1 2 and 3	2 10 6
Quarter Masters (*i*) Boatswain's Mates (*k*)		1 2	2 5 6
Sailmaker and Cooper (*l*)	1 in all	3 4 5 6	2 5 6 2 4 6 2 2 6 2 0 6
Gunner's Mate (*m*) Yeoman of the Powd. Room Armourer's Mates (*n*)		3 4 5	2 2 6 2 0 6 1 18 6
Corporals		1 2 3 and 4	2 5 6 2 2 6 2 0 6
Calker's Mates (*o*)	1 in 1 to 4	1 to 4	2 6 6
Yeoman of Sheets (*p*) Coxswain	1 in all	1 2 3 and 4 5 and 6	2 2 6 2 0 6 1 18 6 1 16 6
Quartermaster's Mates		1 and 2	2 0 6
Captain: Forecastle, Tops, Afterguard, Mast Trumpeter (*q*)	Same No. as Quarter Masters.	3 and 4 5 6	1 18 6 1 16 6 1 14 6
Sailmaker's Mate Cooper's ditto	1 in	all	1 18 6
Quarter Gunner (*r*) Carpenter's Crew (*s*) Cooper's ditto	2 in 1, 2 1 in 1 to 5	1, 2 3, 4, 5, 6,	1 16 6 1 15 6 1 15 6
Sailmaker's Crew (*t*) Gunsmith (*u*) Steward (*x*) Cook		in all 3 6	1 15 6 1 15 6 1 15 6 1 14 6
Steward's Mate	1 in 1 to 4		1 5 2
Chaplain (*y*)	in all	all	11 10 9¼
Able Seamen (*z*) Ordinary Landmen		in all do. do.	1 13 6 1 5 6 1 2 6
Boys (Volunteer) 1st. Class Do. do. 2d. do. Do. do. 3d. do.	The same Pay in all Rates.		Per Year. 9 0 0 8 0 0 7 0 0

(*a*) There are six lieutenants to a 74 gun ship, whose complement is 640; and sometimes four in 5th rates. Sub-Lieutenants of sloops, brigs, &c. have 4*l.* 18*s.* per month.

(*b*) In 4th, 5th, and 6th rates the same pay as boatswain, gunner, and purser. Of hired armed-ships, 2*l.* per month, and found by the owners.

(*c*) None to be *rated* but such as have been two years at sea, and, with master's mates, to produce journals before they can be paid.

(*d*) Paid as above by a late regulation.

(*e*) Must produce a certificate from the storekeeper of the ordnance, that the arms have been received in good order.

(*f*) In all other rates one

(*g* and *h*) Calkers and ropemakers in 4th, 5th, and 6th rates, have 2*l.* 10*s.* 6*d.* per month.

(*i*) In 1st and 2d rate, eight; in 3d, six; in 4th, 4; in 5th, three; and in 6th rates, two.

(*k*) In 1st and 2d rates, four; in 3d and 4th, two; and in 6th rates, one.

(*l*) In 1st and 2d rates, only 2*l.* 5*s.* 6*d.* per month.

(*m*) Four in 1st, two in 2d and 3d rates, and one in the rest.

(*n*) Two in 1st, 2d, 3d, and 4th rates; one armourer's mate in 5th and 6th rates; two corporals in 5th, and 1 in 6th, rates.

(*o*) Only in the four 1st rates.

(*p*) In the 1st rates, 2*l.* 2*s.* 6*d.* per month; in other rates, 2*l.* 0*s.* 6*d.* 1*l.* 18*s.* 6*d.* and 1*l.* 16*s.* 6*d.*

(*q*) In 1st and 2d rates, 1*l.* 18*s.* 6*d.*; in 3d, 4th, and 5th 1*l.* 15*s.* 6*d.*

(*r*) One to every four guns.

(*s*) Ten in the 1st and 2d, eight in the 3d, six in 4th, five in 5th, and four in 6th rates.

(*t*) Two in 1st, 2d, 3d, and 4th rates; in the rest one.

(*u*) One only in 1st and 2nd rates.

(*x*) In 4th rates, 1*l.* 13*s.* 10*d.* per month, 1*s.* 2*d.* per day, and 8*s.* 5*d.* per week; in 5th rates 1*l.* 10*s.* 2*d.* and in 6th rates 1*l.* 9*s.* 6*d.* per month. Purser's stewards, and coopers, cannot be paid unless they produce a certificate at the pay table from the purser.

(*y*) Besides 5*l.* a year, from each midshipman and volunteer of the first class; and a bounty of 20*l.* per annum, if doing the duty of a schoolmaster.

(*z*) None to be rated able but such as have been three years at sea.

Table of Pay to Medical Officers, &c.

MEDICAL OFFICERS.	FULL PAY PER DAY.		
	£.	s.	d.
Physician	1	1	0
——— having served three years	1	11	6
Surgeons not having served six years	0	10	0
——— after six years *	0	11	0
——— after ten years	0	14	0
——— after twenty years	0	18	0
——— of receiving ships, sloops, prison ships, and all other (hospital ships employed only in harbour duty excepted)	0	10	0
——— of hospital ships	0	15	0
——— of hospitals	0	15	0
——— of depôts of prisoners	1	1	0
Hospital Mates employed *at home*	0	6	6
——— *abroad*	0	7	6
Dispensers of Hospitals	0	10	0

ASSISTANTS.	RATES.	No.	FULL PAY PER DAY.		
			£.	s.	d.
Assistants serving on board	First	3	0	6	6
	Second	3	0	6	6
	Third	2			
	Fourth	2			
	Hos. Ship	3			
All other ships according to the accustomed regulation to bear rates	....	1			
1st or 2d mates or assistants, not qualified nor provided with instruments	....	....	0	5	0
3d ditto ditto	....	....	0	4	0

General Observations relative to Officers.

CAPTAINS,

When appointed to ships from half pay, may draw three months in advance, or *three, six*, or *nine months*, when due. Commissioned officers appointed to ships abroad have half-pay to the day of embarkation; and personal pay from that time to the day of joining.

They are not paid by ship's books, but by bill.—Captains not describing the ship's company, not to be paid wages without the board's order.

Compensations in lieu of servants 19*s*. per month.

First captain to a flag officer commanding 10 sail of the line is entitled to the pay of rear-admiral, &c.—Captains pay 3*d per pound* from their personal pay toward officers' widows, and 6*d per month* to Greenwich Hospital.

Senior captains set aside by promotions of flag officers, to be placed on the ordinary estimate of the navy, with premiums equal to half pay.

Captains holding civil employments not to be continued on the half pay list.

Captains producing certificates, agreeably to the tenor of the 4th article of the printed instructions, shall be sufficient to entitle them to flag pay.

Compensation for Servants.

4 to every 100 men of the complement.
4 servants if complement 80 men.
3 ditto ——— 53 ditto.

LIEUTENANTS

In all rates are allowed 6*s*. per day, and at the commencement of a war 6*d*. per mile for travelling expenses to join their ships.—They have a compensation for one servant: have an additional pay of 6*d*. per day when a flag is on board; and pay the same as captains towards Greenwich Hospital. Lieutenants commanding cutters, luggers, &c. registered in the navy estimate, are allowed two servants—entitled to three months' advance, or may draw bills, when due at three, six, or nine months.—Lieutenants commanding vessels not registered in the navy, to be allowed only one servant—are paid by bill, and pay towards the widows fund and Greenwich Hospital.

An additional lieutenant to be appointed to all ships where flags are hoisted, during that time.

Lieutenants in hired cutters and tenders are allowed compensation for a servant to pass journals and send monthly to the navy office the entries and discharges of the men.

Lieutenants are to be paid their half pay and advance, notwithstanding their victualling accounts are not passed, provided the victualling and sick books are delivered.

Lieutenants commanding hired armed-vessels to be paid as other officers, by the cashier paying tickets.

MASTERS.

Masters of king's yachts and tenders, registered at the navy, are allowed 5*l*. per month.

Masters of hired armed ships, 4*l*. per month, and found by the owners.—Hired armed tenders, the same pay.

Second master and pilot of a gun-boat or fire-ship, 4*l*. per month.

Masters of cutters of 55 men are allowed a servant.

Masters may draw for three months' advance, the same as captains and lieutenants—they pay 3*d*. per pound to officers' widows, 1*s*. per month to the chest at Greenwich, and 6*d*. to Greenwich Hospital.

Masters are entitled to one servant each to complement down to 60 men.

A second master to be allowed each line of battle ship at 5*l*. 5*s*. per month; to be included in the complement.

PHYSICIANS.

No person to be appointed as such to a fleet or to an hospital, unless he shall have served at least five years as surgeon.

The widows of physicians and surgeons to be allowed pensions at the discretion of the Admiralty, in proportion to the time of their husband's service.

None of the above officers retiring, without the approbation of the *sick and hurt board*, or refusing to serve when called upon, if judged capable, shall receive half pay, nor shall their names remain on the navy list. Their widows will, consequently, not be entitled to any pension.

No officer to be entitled to the advantages of the above regulations, who shall not have served during the present war, or satisfied the *sick and hurt board* of his inability to serve; but such persons shall remain upon the same establishment as formerly.

SURGEONS OF SHIPS.

Every surgeon in the navy (excepting those serving on board receiving sloops, or convalescent ships, and any other ships than hospital ships, employed only on harbour duty), shall, after *twenty-years'*

* Not more than three years of this time must have been served as hospital mate, or assistant surgeon.

service, on full pay (including not more than *three* years as an hospital mate or assistant surgeon), be allowed 18*s.* per day; and after such length of service, in whatever ships they may have served, shall have a claim to retire on a *half pay* of *six shillings* per day: but if the cause of retirement be ill health contracted in the service, and it shall be so certified to the *sick and hurt board*, the rate of such half pay, after 20 years' service, shall be 10*s.* per day.

Every surgeon in the navy, after 30 years' service, of which not more than three as hospital mate, or assistant surgeon, shall have an unqualified right to retire on *half pay*, at the rate of 15*s.* per day.

Medicines and utensils to be supplied by the *sick and hurt board;* but surgeons shall provide, at their own expense, such surgical instruments as the said board shall judge necessary.

SURGEONS OF HOSPITALS, DOCK-YARDS, INFIRMARIES, and DEPÔTS

Must be selected from the list of naval surgeons; and the time served by surgeons or assistant surgeons in hospitals to be included in the time served on board, for retirement.

Surgeons of *naval hospitals*, *dock yards*, or *royal marine infirmaries*, from completing the terms of 20 or 30 years respectively, shall derive the same advantage as those serving on board his Majesty's ships. The advantages also of this regulation to extend to those surgeons appointed previous to the order in council.

ASSISTANT SURGEONS

Of dock-yards are also, by a late order, selected from the list of naval surgeons; but the time served in that capacity only to be allowed at the discretion of the Admiralty, it not being mentioned in the order in council.

Assistant surgeons, hospital mates, and dispensers of hospitals, depôts, &c. have half-pay allowed them.

WARRANT OFFICERS

Belonging to ships reduced, not to suffer in their pay by the said reducement.

Where ships are lost or taken, wages to be continued up to the day of the court martial; if prisoners, until they arrive in England, or join some king's ship.—Discharged abroad, being superseded, or invalided, to be allowed able pay while on their passage to England.

Boatswains, gunners, and carpenters, may allot half their personal pay for the benefit of their *wives, children*, or *mothers*.

Warrant officers are subject to the same deductions as masters, and commence pay from their appearance.

Boatswain, gunner, and carpenter, where the complement is 96 men, are allowed a compensation for two servants; from 95 to 60, only one servant each. Purser, chaplain, and cook, one servant.

When sent to sick quarters, to be allowed only six weeks if an acting officer is appointed; otherwise to be paid to the time of passing their account.

CHAPLAINS,

On their first appointment, are entitled to three months pay in advance, and are allowed a compensation for one servant. See the articles CHAPLAIN and NAVY *Bills*.

The PAY to the royal marines is as follows:—

MARINES.

Pay on Shore per Lunar Month.

		£.	s.	d.	
General	Admirals.	140	0	0	Not liable to serve afloat.
Lieutenant General		112	0	0	
Major General		84	0	0	
Colonels		80	5	4	

Pay on Shore per Lunar Month.

	£.	s.	d.	
Colonels (Commandant)	66	5	4	Not liable to serve afloat.
Second ditto	39	15	8	
Lieutenant Colonels	23	16	0	
Majors	22	8	0	
Majors (by Brevet)	22	8	0	No increase of pay when serving afloat, but shore allowances cease.
Captains	14	14	0	
Second Captains	14	14	0	
First Lieutenants	10	10	0	
Second Lieutenants	7	7	0	

STAFF.

	£.	s.	d.	
Adjutants	5	12	0	Not liable to serve afloat.
Quarter Masters	6	10	8	
Barrack Masters	14	0	0	
Pay Masters (of Districts)	22	19	8	
Surgeons	28	0	0	
Assistant Surgeons	10	10	0	

	Pay on Shore per Lun. Mon.			Pay on Board Per Lun. Mon.		
	£.	s.	d.	£.	s.	d.
Serjeants	2	11	4	2	0	3
Corporals after 14 years	2	2	0	1	12	1
From 7 to 14 years	1	19	8	1	9	9
Under 7 years	1	17	4	1	7	5
Drummers	1	12	1	1	3	4
Privates, after 14 years	1	12	8	1	3	4
From 7 to 14 years	1	10	4	1	1	0
Under 7 years	1	8	0	0	19	3
Boys of the 1st Class	1	8	0	0	14	0
Ditto 2d Class	1	3	4	0	14	0

MARINE ARTILLERY.

	£.	s.	d.			
Captains	15	8	0	Not liable.		
First Lieuts. under 7 years	9	11	4	Have 5s. per day when serving in bomb-vessels.		
Ditto above 10 ditto	10	19	4			
Second Lieutenants	7	16	4			
Serjeants	3	18	2	3	7	1
Corporals, after 14 years	3	13	6	3	4	9
From 7 to 14 years	3	11	2	3	2	5
Under 7 years	3	8	10	3	0	1
Drummers	1	15	7	1	15	6
Bombardiers, after 14 years	3	8	10	3	0	1
From 7 to 14 years	3	6	6	2	17	9
Under 7 years	3	4	2	2	15	5
Gunners, after 14 years	2	6	6	1	19	8
From 7 to 14 years	2	3	9	1	17	4
Under 7 years	1	15	7	1	15	6

By a late regulation, marine officers, when serving on board ship, are furnished with forms for drawing on the paymaster, without the medium of an agent, at the rate for

Captains........ 8 shillings per day.
1st Lieutenants .. 5 ditto.
2d ditto .... 4 ditto.

The remainder is afterwards paid as arrears, deducting out of it, income tax, fees of commissions, and any other charge that may appear against them

To PAY, (*espalmer, brayer-filer-payer, &c.* Fr.)

as a naval term, implies to daub or anoint the surface of any body, in order to preserve it from the injuries of the water, weather, &c.

To PAY *Off*, *(faire une embardée du côté de dessous le vent*, Fr.) is to let a ship's head fall to leeward of the point whither it was previously directed; particularly when, by neglect of the helmsman, she had inclined to windward of her course, so as to make the head-sails shiver in the wind. See the article FALLING-OFF.

To PAY *a Mast or Yard*, *(enduire de résine, de goudron, de noir de fumée, ou de vernis noir, un mât, ou une vergue, &c.* Fr.) is to anoint with tar, turpentine, rosin, tallow, or varnish; tallow is particularly useful for those masts upon which the sails are frequently hoisted and lowered, such as topmasts, the lower-mast of sloops, schooners, &c.

To PAY *out a Cable or other Rope*, *(filer un cable, ou autre cordage*, Fr.) is to slacken it so as to let it run out of the vessel for some particular purpose.

To PAY *a Seam*, *(brayer une couture*, Fr.) is to pour melted pitch along it, so as to defend the oakum, with which it is calked, from the effects of wet.

To PAY *a Ship*, *(payer l'équipage d'un vaisseau*, Fr.) in the royal navy, commonly implies, to pay the men and petty officers their established wages, under the superintendance of the commissioner afloat, at the port, and, for certain reasons, is generally done a day or two before the ship puts to sea. But, according to the custom of the navy, there are six months' wages kept in reserve from the crew, to cover the expenses incurred by desertion, &c. except when they are paid off.

To PAY *off a Ship*, *(désarmer un vaisseau et en congédier l'équipage*, Fr.) is to pay the wages of the ship's petty officers and crew, and to discharge the ship from actual service, in order to be laid up in ordinary.

To PAY *a Vessel's Bottom*, *(espalmer, un bâtiment, ou donner un suif, ou couroi à un bâtiment*, Fr.) is to cover it with a composition of tallow, sulphur, resin, &c. See the article BREAMING.

PEAK, *(pic*, Fr.) a name given to the upper corner of those sails which are extended by a *gaff*, or by a yard, which crosses the mast obliquely, as the mizen-yard of a ship, the main-yard of a *bilander*, &c. The upper extremity of those yards and gaffs are also denominated the *peak.*

To PEAK, *(apiquer*, Fr.) is to raise a yard or gaff more obliquely to the mast. The term peak is used in various senses, for which see the articles ANCHOR and APEAK.

Fore-PEAK, a place in the fore part of small vessels, where the spare-stores are usually kept. To this there is a scuttle or hatch in the deck.

PEAK-HALYARDS, *(martinet du pic*, Fr.) are the ropes or tackles by which the outer end of a gaff is hoisted, as opposed to the THROAT-HALYARDS, which article see.

PEDERERO, or PETERERO, *(pierrier*, Fr.) a small piece of ordnance formerly used on board some foreign ships, for the discharging of nails, broken iron, or partridge shot, on an enemy attempting to board It is generally open at the breach, and the chamber made to take out to be loaded that way, instead of at the muzzle. This species of ordnance is managed by a swivel, and was formerly much in use among the Spaniards.

PEGASUS, the *Horse*, in astronomy, a constellation of the northern hemisphere, in form of a flying horse; and, according to the Britannic catalogue, contains 93 stars.

PEN, a place enclosed by hurdles, for fishing on the sea-coast.

PENDANT, or PENNANT, *(flamme*, Fr.) a sort of long narrow banner, displayed from the mast-head of a ship of war, and usually terminating in two ends or points, called the swallow's tail, as expressed by *a*, fig. 10, plate XIII. It denotes that a vessel is in commission or actual service.

Broad PENDANT, *(guidon, ou cornette*, Fr.) is a kind of flag, terminating in one or two points, and used to distinguish the chief of a squadron. See the article COMMODORE.

PENDANT, *(pendeur*, Fr.) is also a short piece of rope, fixed on each side, under the shrouds, upon the head of the main and fore masts, from which it depends as low as the *cat-harpings*, having an iron thimble spliced into an eye at the lower end, to receive the hooks of the main and fore tackles.

Fore and Main-stay-tackle PENDANTS, *(pendeur de la caliorne de misaine, et pendeur de palan du grand étui*, Fr.) are ropes fastened to the upper part of the fore and main-stay of a ship; which, by the help of tackles, are employed to hoist up the boats, casks of beer, water, provisions, &c.

There are, besides, many other pendants of the latter kind, which are generally single or double ropes, to whose lower extremity is attached a block, or tackle: such are the fish-pendant, the yard-tackle-pendants, the reef-tackle-pendants, the pendants of boat-davits, &c. all of which are employed to transmit the efforts of their respective tackles to some distant object.

Rudder PENDANTS, *(fausse sauve-garde de gouvernail*, Fr.) are strong ropes spliced in the rings of the rudder-chain, to prevent the loss of the rudder, if by any accident, it should get unshipped, or disentangled from the braces.

PENDULUM, in mechanics, denotes any heavy body so suspended as that it may vibrate or swing backwards and forwards, about some fixed point, by the force of gravity. These alternate ascents and descents of the pendulum, are called its oscillations or vibrations.

PENDULUM-CLOCK, a clock having its motion regulated by the vibration of a pendulum. For the improvements in pendulums, see the article CLOCK.

PENINSULA, *(presqu'île*, Fr.) in geography, is a portion or extent of land which is almost surrounded with water, being joined to the continent only by an isthmus, or narrow neck. Such is Africa, the greatest peninsula in the world, which is joined to Asia by a narrow neck of land at the end of the Red Sea; such also is Pelopennesus, or the Morea, joined to Greece; and Jutland, &c. This term has also been applied lately to Spain and Portugal, since this country has been so much interested in assisting the inhabitants in repelling the French invaders.

PENSION, *(pension*, Fr.) in the royal navy, is an allowance or remuneration granted by his Majesty to naval officers, who, from age and long services, are rendered incapable of discharging their duty.

Commissioned officers disabled in the service, are considered as their cases may deserve, or as his Majesty may think fit; and are generally rewarded according to the rank they hold, the nature of their wounds, and length of servitude.

The following are the regulations of pensions to superannuated warrant officers:—

Boatswains, gunners, carpenters, pursers, and cooks, whose employments are constant, must serve full *fifteen years*; masters and surgeons, whose employments are constant, *eight years* in their respective offices, before they are entitled to superannuation.

Officers applying for superannuation are to be examined by the physicians of the sick and hurt board, in conjunction with the surgeons' company; who are to report their opinion in writing, of their incapacity to serve his majesty.

Officers whose employments are constant, being admitted to superannuation, receive an annual pension equal to their pay in the ship of the highest rate they have served in.

The pensions of masters are equal to half their pay, and surgeons are allowed the same pensions as masters of equal rates.

PENSIONERS, (*pensionnaires;—invalides*, Fr.) in the royal navy, are seamen and marines who have been engaged in his Majesty's naval service, and who, from old age, wounds, or infirmities, are rendered incapable of further service; in consideration whereof they have an annual allowance paid them out of the Chest at Greenwich. The pensioners may be distinguished into two sorts; viz.

In-PENSIONERS, are those seamen, marines, &c. who have been selected by the Admiralty, and received into GREENWICH *Hospital*, according to the mode of admission described under that article: they are clothed, lodged, and victualled, and have a weekly allowance at the expense of the said hospital.

Out-PENSIONERS, are those seamen, marines, &c. who have a yearly allowance paid to them out of the Chest, according to the nature of their hurts and length of their servitude. For the process by which persons wounded or hurt in his Majesty's naval service, are enabled to claim relief from the CHEST *at Greenwich*, see that article.

By an Act of the 32d Geo. III. c. 67, pensions due to the out-pensioners shall be paid by bill in any part of Great Britain or Ireland, upon application being made by the party, in writing, to the governors of Greenwich hospital.

PENUMBRA, in astronomy, a partial shade observed between the perfect shadow and the full light in an eclipse. It arises from the magnitude of the sun's body; for were he only a luminous point, the shadow would be all perfect; but by reason of the diameter of the sun, it happens that a place which is not illuminated by the whole body of the sun, does yet receive rays from a part thereof. See the article ECLIPSE *of the Sun*.

PERIAGUA, (*pirogue*, Fr.) a sort of large canoe, composed of the trunks of two trees, hollowed and united into one fabric; whereas canoes in general are formed of the body of one tree. The periagua is used in the Leeward Islands, South America, and in the Gulf of Mexico. Their masts and sails have a great resemblance to the proa, &c.

There are also double periguas used in Otaheite, &c. which are constructed much longer and wider than that above-mentioned. See the article CANOE.

PERIGEE, in astronomy, is that point of the orbit of the sun or moon, wherein they are at their least distance from the earth, in which sense it stands opposed to apogee.

PERIHELIUM, in astronomy, is that point of a planet's or comet's orbit which is nearest the sun, in which sense it stands opposed to aphelium.

PERILS OF THE SEA. It may, in general, be said that every thing happening to a ship in the course of her voyage, by the intermediate act of God, without the intervention of human agency, is a peril of the sea. Thus every accident happening by the violence of the winds and waves, by thunder or lightning, by driving against rocks, by the stranding of the ship, or by any other violence which human prudence could not foresee, nor human strength resist, may be considered as a loss within the meaning of a policy of insurance, and the assurer must answer for all damages sustained in consequence of such accident. But if a ship be driven by stress of weather on an enemy's coast, and is there captured, it is a loss by capture, and not by *perils of the sea*. GREEN v. ELMSLEY—*Peake*, 212.

In order to charge the underwriter for a loss by perils of the sea, the ship must be equipped with every thing necessary for the voyage; she must be sea-worthy, have a sufficient crew, and a captain and pilot of competent skill.

A practice prevails amongst assurers, that the ship shall be deemed lost, if not heard of in six months after her departure (or after the time of the last intelligence from her) for any port in Europe, and in twelve months, if for a greater distance. If, under this usage, the assurer should pay the money, supposing the ship to be lost, when it really is not the case, he may recover it back by action.

PERIOD, in astronomy, the time taken up by a star or planet in making a revolution round the sun; or the duration of its course till it returns to the same point of its orbit. There is a wonderful harmony between the distances of the planets from the sun, and their periods round him, the greatest law of which is, that the squares of the periodical times of the primary planets are to each other as the cubes of their distances from the sun. This harmony among the planets is one of the greatest confirmations of the Copernican hypothesis.

PERIODICAL, any thing appertaining to period, or going by periods. Thus the periodical motion of the moon, is that of her monthly period or course about the earth, called her periodical month.

PERIODICAL *Month*. See the article MONTH.

PERIPHERY, in geometry, the circumference of a circle, ellipse, or any other regular curvilinear figure.

To PERISH, (*périr*, Fr.) speaking of a ship, is to wreck or to founder; as, "The ship foundered at her anchors, and every soul on board perished."

PERPENDICULAR, in geometry, a line falling directly on another line, so as to make equal angles on each side, which are called right angles.

PERPENDICULAR, in gunnery, a small instrument used for finding the centre line of a piece of ordnance, in the operation of pointing it to a given object.

PERSEUS, in astronomy, a constellation of the northern hemisphere, which, according to the Britannic catalogue, contains 67 stars.

PHASES, in astronomy, the several appearances or quantities of illumination of the Moon, Venus, Mercury, and other planets; or the forms in which they appear illuminated by the sun The phases of the moon are very observable with the naked eye, by which she sometimes increases and sometimes wanes; is now falcated, and again appears as half a circle; at other times she is gibbous, and again has a full circular face. By the help of a good telescope, the like phases may be observed in Venus, Mars, &c. Copernicus, before it was possible to ascertain the fact by means of glasses, foretold that it would, at some future period, be ascertained, that Venus underwent all the changes to which the moon was subject; and Galileo was the first person who, by actual observation, confirmed the truth of his theory.

PHENOMENON, or PHÆNOMENON, an extraordinary appearance in the heavens or on earth; either discovered by observation of the celestial bodies, or by physical experiments: such are meteors, comets, uncommon appearance of the stars and planets, earthquakes, &c. such also are the effects of the magnet, phosphorus, &c.

PHŒNIX, in astronomy, a constellation of the southern hemisphere, unknown to the ancients, and invisible in our northern parts of the globe. It is said to contain 13 stars.

PHYSICIAN OF A FLEET, in the royal navy, is a person appointed by the Admiralty with the medical superintendence of a fleet or squadron of ships employed on any particular station, as the Channel, Mediterranean, West Indies, &c. and is under the immediate orders of the commander in chief: if an hospital ship is with the fleet, his residence is generally on board her; if not, it is usual for him to be in the flag-ship with the commander in chief.

The arrangement of every thing appropriated to the reception of the sick, sent on board the hospital ship for cure shall be under his direction; and the surgeon, and all other persons appointed to attend them shall be under his orders. He is to propose to the commander in chief every thing which he may think likely to be of service to the sick, to increase their comforts, or to accelerate their cure; and, as far as circumstances will admit, he should comply with his proposals.

He is also to visit the different ships of the squadron frequently; to inquire into the health of the ships' companies, and the treatment of the sick; and, if any unusual sickness prevails on board, he is to represent the same to the commander in chief, with the nature of the disease, and the necessary means which he deems requisite for eradicating and putting a stop to the progress of the malady. He is also authorised to examine the journals of the medical officers, to inquire into the practice of the surgeon of any ship he visits, and his manner of treating the diseases of the men under his care, and to give him such directions as he may deem necessary. He is also to inquire into the conduct and abilities of the assistant surgeons, that he may be able to point out to the sick and hurt board, or to the commander in chief, those who may be best qualified for any particular service, or for promotion.

And whenever he shall think it necessary, he is to examine the instruments, medicines, and necessaries, on board any ship, and if he finds them bad in quality or deficient in quantity, he is to report it to the commander in chief, that he may take such measures as circumstances may require. For farther information see the NAVAL INSTRUCTIONS.

PHYSICIAN OF *a Naval Hospital*, is a person appointed by the Admiralty to one of his Majesty's naval hospitals, to superintend and prescribe medicine for the inward complaints of the sick and wounded seamen; and to attend such physical patients every day, or as often as circumstances may require.

PIAZZI, or CERES FERDINANDEA, in astronomy, a new planet, situate between the orbits of Mars and Jupiter: it was discovered at Palermo, in Sicily, on the 1st of January, 1801, by M. Piazzi, an ingenious Italian astronomer, who has since distinguished himself by his numerous observations.

It was then situate in Taurus, and was observed by Piazzi till the 12th of February, when a dangerous illness compelled him to discontinue his observations. It was, however, again discovered by Dr. Olbers, of Bremen, on the 1st of January 1807, nearly in the place where it was expected from the calculations of Baron Zach. The nebulæ with which it was surrounded, gave it the appearance of a comet; and it was, in consequence of the suggestion of Professor Bode, of Berlin, or of Baron Zach, that Piazzi and other astronomers ranked it among the planetary bodies.

Ceres is of a ruddy colour, though not very deep, and appears about the size of a star of the eighth magnitude. It seems to be surrounded with a large dense atmosphere, and plainly exhibits a disc, when examined with a magnifying power of about 200. From a great number of observations, Schroeter found the atmosphere of Ceres to be 675 English miles high, and he perceived that it was subject to numerous changes. The visible hemisphere was sometimes overshadowed, and at other times cleared up; so that he thinks there is little chance of discovering the period of its diurnal rotation.

When Ceres is approaching to the earth, its diameter increases much more rapidly than it ought to do from the diminution of the distance. "This," says Schroeter, "arises from the finer exterior strata of its atmosphere becoming visible while it approaches the earth."

Ceres performs her revolution round the sun in 4 years, 7 months, and 10 days; and her mean distance from that luminary is nearly 260,000,000 of English miles. The eccentricity of her orbit is a little greater than that of Mercury, while its inclination to the ecliptic exceeds that of all the old planets. According to the measurements of Dr. Herschel, the diameter of Ceres does not exceed 163 miles, while the observations of the German astronomer Schroeter makes it 1624 miles. Schroeter accounts for this remarkable difference between his measurements and those of Dr. Herschel, by maintaining, that the projection-micrometer, used by the English astronomer, was placed at too great a distance from the eye, and that he measured only the middle clear part of the nucleus of the planet.

For further particulars on the subject of this article, see HERSCHEL's account in the *Phil. Trans.* for 1802; SCHROETER, *Lilienthalische Beobachtungen der neu-entdeckten Planet Ceres*, Gottingen, 1805.

PIECES, in mast-making, are short pieces of plank used to make good some deficiency in the heels of top-masts, &c. There are also PIECES of oak plank coaked and bolted to the hounds of topmasts, in order to support the trestle-trees, and the rigging attached thereto.

PIECES OF ORDNANCE, include all sorts of great guns and mortars. A soldier's firelock is likewise called his piece.

Battering PIECES, are those used at sieges to make breaches, such as 24-pounders. Field-pieces are 12-pounders, &c.

Shelve PIECES, in a ship of war, are large pieces of oak or fir timber, sometimes placed under the beams, and secured with bolts, to strengthen the ship.

PIER, *(jettée d'un port*, Fr.) a strong mound or fence, usually composed of durable stones, and projecting into the sea, for the convenience of vessels lading and unlading their cargoes, whatever may be the state of the tide; and, in some instances, it is built for breaking off the violence of the waves from the entrance of a harbour.

HALF-PIKE. See the article HALF-*Pike*.

PIGS OF BALLAST, (*saumons de fer*, Fr.) are large masses of cast iron, used for ballasting ships.

PILASTERS, in ship-building, are fluted ornaments, with base and capital fastened to the munions between the lights.

PILE, in artillery, implies a collection or heap of shot or shells, piled up by horizontal courses into either a pyramidal or else a wedge-like form; the base being an equilateral triangle, a square, or a rectangle. In the triangle and square, the pile terminates in a single ball or point, and forms a pyramid; but with the rectangular base, it finishes at top, in a row of balls, or an edge forming a wedge.

In the triangular and square piles, the number of horizontal rows, or courses, or the number counted on one of the angles from the bottom to the top, is always equal to the number counted on one side, in the bottom row. And in rectangular piles, the number of rows or courses is equal to the number of balls in the breadth of the bottom row, or shorter side of the base; also, in this case, the number in the top row or edge is one more than the difference between the length and breadth of the base.

PILLARS, in ship-building, are pieces of wood or iron fitted under the beams of the decks to support them.

PILLOW, (*coussin du mât de beaupré*, Fr.) a block of timber, whereon the inner-end of the bow-sprit is supported. See the article BOWSPRIT.

PILOT, (*pilote*, Fr.) in a general sense, implies a person properly qualified and licensed to conduct ships on or near the sea-coast, or through intricate channels, and into the roads, bays, rivers, &c. within his respective district. Pilots are not constant standing officers on board our ships, but are called in occasionally, on coasts or shores unknown to the master; and having piloted the vessel, they return to the shore where they reside.

After a pilot is taken on board a merchant ship, the master has no longer any command of her till she is safe in harbour; but then the master resumes the government of the same, and is to see her bed and lying; the pilot being no longer liable, though for his own convenience he may still be on board. The same rule holds good if a pilot goes on board only to conduct a ship through some dangerous place, as for instance, Yarmouth-roads: after passing them the master must resume the command, and the pilot is no longer responsible.

The regulations with regard to pilots in the royal navy are as follow:—The commanders of the king's ships, in order to give all reasonable encouragement to so useful a body of men as pilots, and to remove all their objections to his Majesty's service, are strictly charged to treat them with good usage, and an equal respect with warrant officers.

The purser of the ship is always to have a set of bedding provided on board for the pilots, and the captain is to order the boatswain to supply them with hammocks, and a convenient place to lie in, near their duty, and apart from the common men; which bedding and hammocks are to be returned when the pilots leave the ship.

A pilot, when conducting one of his Majesty's ships in pilot-water, shall have the sole charge and command of the ship, and may give orders for steering; setting, trimming, or furling the sails; tacking the ship, or whatever concerns the navigation: and the captain is to take care that all the officers and crew obey his orders. But the captain is diligently to observe the conduct of the pilot, and take particular care that the hand-lead be kept constantly going, whether the pilot thinks this precaution necessary or not; and if he judges him unqualified to conduct the ship, or that he is carrying her into danger, he may remove him from the command and charge of the ship, and take such methods for her preservation as shall be judged necessary; remarking upon the log-book the exact hour and time when the pilot was removed from his office, and the reasons assigned for it.

Captains of his Majesty's ships, which are to sail from the river Thames or the Nore, are to apply to the Commissioners of the navy for pilots, and are to employ those only who shall be sent them by the Trinity-House of Deptford-Stroud; but captains of ships which are to sail from the Downs to the Nore or the Thames, are to employ pilots belonging to the Society of Pilots at Dover.

The captain of a king's ship employing a pilot in a foreign part of his Majesty's dominions, shall give him three certificates of the services he performs (to be marked respectively, 1st, 2d, and 3d) of the same tenor and date; which being produced to the proper naval officer, he shall cause the same to be immediately paid; but if there be no naval officer there, the captain shall pay him, and send the proper vouchers, with his bill, to the navy-board, in order to be paid as bills of exchange.

Captains of his majesty's ships, employing foreign pilots to carry the ships they command, into or out of foreign ports, shall pay them the rates due by the establishment or custom of the country, before they discharge them; whose receipts being duly vouched, and sent with a certificate of the service performed, to the navy-board, they shall cause them to be paid with the same exactness as they do bills of exchange.

The owners of hired ships, cutters, and other armed vessels, are, by their contracts, to provide for them masters who have been accustomed to sail in ships and vessels of this description; it is therefore incumbent upon the latter to take charge of them upon all ordinary cases of cruising or pilotage, where it is not usual for a king's ship to employ a pilot extra; and it has been deemed advisable, in order to prevent any unnecessary expense to the public on this account, to require, as indispensable to the payment of certificates for pilotage of hired vessels, that they be approved by the commander-in-chief or commanding officer on the station; or that his order for the employment of the pilot is produced. For further particulars see NAVAL INSTRUCTIONS, the edition for 1810.

Penalty on Pilots not regularly examined.—If any person shall take upon him to conduct or pilot a ship by or from Dover, Deal, or the Isle of Thanet, to any place on the river Thames or Medway, before he has been examined by the master and wardens of the Society or Fellowship of Pilots of the Trinity-house of Dover, Deal, and the Isle of Thanet, touching his ability; and approved and admitted into the said society, at a court of load-manage, by the lord warden of the cinque ports, or his deputy, and the master and wardens; such person, for the first offence, shall forfeit 10*l* for the second, 20*l.* and for every other offence, 40*l.* &c. 3 Geo. I. c. 13, sec. 1.

This act shall not hinder any person from assisting a ship in distress.—Sec. 7. Nor shall it extend to the taking away any liberties vested in the corporation of the Trinity-house of Deptford-Stroud.—Sec. 9.

Penalty on unlicensed Pilots.—If any person shall take charge of any ship as pilot, down the river of Thames, or through the North Channel, to or by Orfordness, or round the Long-Sand-Head into the Downs, or down the South Channel into the Downs, or from or by Orfordness upon the North Channel, or the river Medway, other than such as should be licensed to act as a pilot, by the master, wardens, and assistants of the Trinity-house of Deptford-Stroud, under the common seal of the corporation; every person so offending, and being convicted before two justices of the peace for the city of London, or the counties of Middlesex, Essex, Kent, or Surrey, shall, for every offence forfeit 20*l.*; provided that nothing in this act shall extend to the obliging any master of any ship in the coal-trade or coasting-trade to employ a pilot. 5 Geo. II. c. 20, sec. 1.

Pilots subject to proper Regulations.—The pilots already admitted by the said corporation, shall be subject to the regulations of the corporation, provided the regulations do not relate to the pilots keeping of turns, or to setting the rates of pilotage; and shall pay the ancient dues, provided the same do not exceed one shilling in the pound, out of their pilotage, for the use of the poor of the corporation, within ten days after the service of such pilots shall be ended.—Sec. 8.

Pilots refusing the Charge of his Majesty's Ships.—In case such pilots shall refuse to take the charge of any of his Majesty's ships when appointed thereto by the said corporation, or shall have misbehaved themselves in the conduct of any ships, or in any other part of their duty; or if pilots shall refuse to obey any summons of the corporation, or such orders as the corporation shall make in the premises, the general court of the said corporation, upon examination thereof, are required to recall the warrants granted to such pilots; and if such persons shall (after notice given by the clerk of the said corporation, to them in person, or left at their place of abode) act as pilots within the limits afore-mentioned, they shall be subject to all the penalties inflicted on unlicensed pilots.—Sec. 9.

This Act not to extend to certain Places.—Nothing in this act shall extend to the impeaching of any privileges enjoyed by the pilots of the Trinity house of Kingston-upon-Hull, or the Trinity-House of Newcastle-upon-Tyne.—Sec. 11. Nor shall this act extend to the impeaching of the franchises, nor to take away the sole right of piloting merchant-ships from or by Dover, Deal, or the Isle of Thanet, upon the river of Thames or Medway, granted to the society and fellowship of the master, wardens, and pilots of the Trinity-house of Dover, Deal, and the Isle of Thanet, by 3 Geo. I. c. 13, sec. 12. See also the article TRINITY-HOUSE COMPANY.

Rates of Pilotage.—For conducting any ship from Dover, Deal, or the Isle of Thanet, to any place on the river Thames and Medway, the following, and no greater prices shall be taken; viz. for every ship drawing less than seven feet water, 5*l.*; seven feet, 5*l.* 5*s.*; seven and a half feet, 5*l.* 12*s.* 6*d.*; eight feet, 6*l.*; eight and a half feet, 6*l.* 7*s.* 6*d.*; nine feet, 6*l.* 15*s.*; nine and a half feet, 7*l.* 2*s.* 6*d.*; ten feet, 7*l.* 10*s.*; ten and a half feet, 7*l.* 17*s.* 6*d.*; eleven feet, 8*l.* 5*s.*; eleven and a half feet, 8*l.* 12*s.* 6*d.*; twelve feet, 9*l.*; and so increasing at the rate of 7*s.* 6*d.* per half foot, until a draught of seventeen feet and a half, when the price is 14*l.* 17*s.* 6*d.*; eighteen feet, 15*l.* 6*s.*; eighteen and a half feet, 15*l.* 14*s.*; nineteen feet, 16*l.* 13*s.*; nineteen and a half feet, 20*l.* 9*s.* 6*d.*; twenty feet, 21*l.* But for conducting a vessel from Gravesend to Long-Reach, the following prices are allowed, in addition to the preceding: ten feet draught of water, 10*s.* 6*d.*; ten and a half to twelve feet, 15*s.*; twelve and a half to 14 feet, 17*s.* 6*d.*; fourteen and a half to sixteen feet, 1*l.* 1*s.*; sixteen and a half to eighteen feet, 1*l.* 11*s.* 6*d.*; eighteen and a half to twenty feet, 2*l.* 2*s.* Additional for piloting a ship from Gravesend to Blackwall and London: ten feet draught of water, 1*l.* 1*s*; ten and a half to twelve feet, 1*l.* 10*s.*; twelve and a half to fourteen feet, 1*l.* 15*s.*; fourteen feet and a half to sixteen feet, 2*l.* 2*s.*; sixteen and a half to eighteen feet, 3*l.* 3*s.*; eighteen and a half to twenty-feet, 4*l.* 4*s.*—47 Geo. III. sess. 2, c. 10, sec. 4.

Negligence of Pilots.—If any pilot shall negligently lose the ship under his care, and be thereof convicted, he shall for ever be incapacitated from acting as a pilot; and the number of such pilots shall not be less than a hundred and twenty, whose names, ages, and places of abode shall, every 25th of March, be affixed in some public place in the custom-houses at London and Dover; and for not returning such list, the master and wardens of the society shall forfeit 10*l.* &c. 3 Geo. I. c. 13, sec. 5.

Branch Pilot, is a pilot duly authorized by the Trinity Board to pilot ships up particular channels and rivers. They have regular wages from the Trinity-house Company.

Coasting Pilot. See the article Coasting.

To Pilot *a Ship, (piloter un vaisseau,* Fr.) is to navigate her into or out of a port, harbour, or through any place of danger, where rocks, shoals, &c. are to be met with.

PIN of a Block, (*essieu de poulie,* Fr.) is the axis on which the sheave revolves, being supported by the shell. See the article Block.

PINCERS, (*pinces,* Fr.) a small iron instrument used by the carpenter's crew, in a ship of war, to draw the forelocks out of the saucer-links of the chain-pump, when under repairs.

PINK, (*pinque,* Fr.) a name given to a ship with a very narrow stern. Those used in the Mediterranean Sea differ from the xebecs only in being more lofty, and not sharp in the bottom: they are vessels of burden, have three masts, and carry lateen sails.

All vessels, however small, whose sterns are very narrow, are called *pink-sterned.*

PINNACE, (*pinnace, sorte d'embarcation mâtée en goëlette,* Fr.) a small vessel, navigated with oars and sails, and having generally two masts, rigged like those of a schooner. They are chiefly used for procuring intelligence, and for landing of men, &c.

Pinnace, (*pinnace, ou chaloupe à huit avirons,* Fr.) is also a boat, usually rowed with eight oars. See the article Boat.

Belaying PINS, (*chevillots de fer, ou de bois, pour amarrer les manœuvres,* Fr.) are pieces of wood, or iron, fixed in a rack for making fast the small running-rigging. See the article Belaying.

Pins *and Chains,* for securing the bars in the capstan, are iron pins connected with small chains, which are fastened to the drum-head of the capstan by means of staples. See their use explained under the article Capstan.

PINTLES, (*éguillots de gouvernail,* Fr.) a sort of small mixed metal bolts, fastened upon the back part of the rudder, with their points downwards, in order to enter into, and rest upon the braces, fixed on the stern-post to hang the rudder. They are represented by I in fig. 9, plate XI.

PIRACY, (*piraterie,* Fr.) the seizing or plundering a vessel on the high seas, without having a proper authority and commission from the government to commit such a depredation.

Pirate, (*pirate, ou forban,* Fr. πειρατης, Gr.) a sea-robber, or an armed ship that roams the seas without any legal commission, and seizes and plunders every vessel she meets indiscriminately, whether friends or enemies.

If any ship or vessel shall be taken, acting as a ship of war or privateer, without having a commission authorising her so to do, her crew are to be considered as pirates, and are to be carefully confined until an opportunity offers of delivering them up, to be tried by the High Court of Admiralty. And if any of his Majesty's subjects be found serving on board an enemy's ship of war or privateer, they are to be closely confined until an opportunity offers for their being tried as traitors.

The colours usually displayed by pirates are said to be a black field, with a death's head, a battle-axe, and hour-glass. The last instrument is generally supposed to determine the time allowed to the prisoners, to consider whether they will join the pirates in their felonious combination, or suffer speedy death, which is often perpetrated in the most cruel manner. See also the article Proa.

Pirate, also implies a person who enriches himself by marine robberies, committed either by force, fraud, or surprise, on merchan s or other traders at sea. Those who rob in rivers, and who are, perhaps, not less numerous though not so open in their depredations as the sea-pirates, are usually distinguished from the latter by the appellation of river-pirates.

PISCES, in astronomy, the 12th sign or constellation of the zodiac, and is represented on the globes in the form of two fishes, tied together by the tails.

The number of stars in Pisces, is, according to the Britannic catalogue, 113.

Pisces *Australis,* the Southern Fish, is a small constellation in the southern hemisphere, containing, according to Flamsteed's catalogue, 24 stars. The star Fomahaut, of the first magnitude, is in the mouth of this fish.

Pisces *Volans,* the Flying Fish, is a small constellation of the southern hemisphere, added by the moderns: it contains eight stars, but is not visible in our latitude.

PISTOLS, (*pistolets,* Fr.) are small fire-arms, with which seamen selected for boarding the enemy are always furnished.—Each boarder is provided with a brace, and a cutlass.

The pistol is said to have taken its name from Pistoya, in Italy; where, as Fauchet tells us, it was first made.

PITCH, (*brai sec,* Fr. *pix.* Lat.) a resinous substance of a darkish brown colour, produced from tar boiled down to dryness: this part of the process is commonly performed in a still, in order to save an essential oil which arises on boiling, and which is called, from the name of the tree which tar is principally prepared from *oleum pini,* and *oleum tædæ.* It is used in calking of ships, to fill the chinks or intervals between the planks of their sides, decks, or bottoms. The want of this article might, in cases of real necessity, be supplied by preparations of paint and other substances.

In the island of Trinidad, there is a lake of asphaltum, or mineral pitch, which furnishes an inexhaustible supply of this article. When this substance exudes from the ground, it is in the state of liquid tar. The Spaniards found it to answer so well when laid on boiling hot, and mixed with tallow or oil, in the proportion of about four pounds to every hundred pounds of pitch, that Admiral Apodaca, in the year 1797, when the island fell into our possession, had received orders to form an establishment for the preparation of this pitch for the use of the navy.

It is said to possess the valuable qualities of resisting the worm which abounds in the Gulf of Paria, and of preserving iron. But there appears to be a prejudice against it, of which we know not the foundation.

There is in India a sort of resin called *dammer*, which, when mixed with sulphur, is an excellent substance for pitch. It is produced from a tree which grows abundantly on Prince of Wales's Island, and in the neighbourhood of the Streights of Malacca. In the upper parts of India the Sulla pine produces such quantities of *kota*, or pure turpentine, that the whole of the consumption of Europe might be supplied from these districts. In fact, there are few materials which we have been in the habit of drawing from other countries, that India does not, or would not afford.

To PITCH *the Seams.* See the article *To* PAY *a Seam.*

PITCH *Spout-Ladle*, (*cuiller à brai*, Fr.) is a small iron ladle, having the spout opposite to the handle, and used to pay the seams of the decks, &c.

PITCH *Mop*, (*torchon à manche*, Fr.) is a mop made of thrums, &c. to pay the seams where the spout-ladle cannot properly be used, as on the exterior parts of the ship.

PITCH *Pot*, (*chaudière à goudron* Fr.) is a large iron pot used on board ships, for the purpose of boiling pitch.

PITCHING, (*tangage*, Fr. *appicciare*, Ital.) may be defined the vertical vibration which the length of a ship makes about her centre of gravity; or the movement, by which she plunges her head and after-part alternately into the hollow of the sea.

This motion may proceed from the waves, which agitate the vessel, or the wind acting upon the sails, which makes her stoop at every blast thereof.

When the wind acts upon the sails the masts yield to its effort, with an inclination which increases in proportion to the length of the mast to the augmentation of the wind, and to the comparative weight and distribution of the ship's lading.

The repulsion of the water, to the effort of gravity, opposes itself to this inclination, or at least sustains it, by as much as the repulsion exceeds the momentum, or absolute effort of the mast, upon which the wind operates. At the end of each blast, when the wind suspends its action, this repulsion lifts the vessel; and these successive inclinations and repulsions produce the movement of *pitching*, which is very inconvenient; and when it is considerable, will greatly retard the course, as well as endanger the mast, and strain the vessel.

PIT-SAW, a large two-handed saw, used to saw timber in pits. It is set rank for coarse stuff, so as to make a kerf of almost a quarter of an inch; but for finer stuff it is set finer.

PLACE, in optics, or *Optical* PLACE, is the point to which the eye refers an object.

Optic PLACE *of a Star*, is a point in the surface of the mundane sphere, in which a spectator sees the centre of the star, &c.

PLACE *of the Sun, Moon, Star*, or *Planet*, in astronomy, simply denotes the sign and degree of the zodiac which the luminary is in; and is usually expressed either by its latitude and longitude, or by its right ascension and declination.

PLAIN, in mast-making, an even surface between two coaks.

PLAN, in general, denotes the representation of something drawn on a plane; as maps, charts, &c.

PLANE, a term used by shipwrights, implying the area, or imaginary surface, contained within any particular outlines. Thus the plane of elevation, plate XVII. exhibits a surface limited by the head before, by the stern abaft, by the keel below, and by the upper part of the ship's side above. Thus the horizontal plane, in the same plate, is comprehended within the lines which describe the ship's greatest breadth and length; and thus also the plane of projection circumscribes the greatest height and breadth of the same vessel.

PLANE, in geometry, denotes a plain surface, or one that lies evenly between its bounding lines: and as a right line is to the shortest extension from one point to another, so a plain surface is the shortest extension from one line to another.

PLANE, a well-known tool to make smooth with.

PLANE *Angle*, is an angle contained under two lines or surfaces; and so called in contradistinction to a solid angle, which is formed by three or more planes.

PLANE *Chart*, in navigation, is a sea-chart, having the meridians and parallels represented by parallel straight lines; and consequently having the degrees of longitude the same in every part. See the article CHART.

PLANE *of Projection*, in the stereographic projection of the sphere, is that in which the projection is made; corresponding to the perspective plane.

PLANE *Sailing*, (*naviguer sur la carte plate*, Fr.) is that which is performed by means of a plane chart; in which case, the meridians are considered as parallel lines, the parallels of latitude are at right angles to the meridians, the lengths of the degrees on the meridians, equator, and parallels of latitude, are every where equal; and the degrees of longitude are reckoned on the parallels of latitude, as well as on the equator.

In plain sailing, the principal terms and circumstances made use of are, COURSE, DISTANCE, DEPARTURE, and DIFFERENCE OF LATITUDE, &c. which articles see.

If a ship sails either due north or south, she sails on a meridian, makes no departure, and her distance and difference of latitude are the same; but where the ship sails either due east or west, she runs on a parallel of latitude, makes no difference of latitude, and her departure and distance are the same.

It may farther be observed, that the distance run, difference of latitude, and departure form a plane right-angled triangle; to illustrate this, let C A represent the distance, C B the difference of latitude, A B the departure, and the angle A C B the course, then will C A B be the complement of the course.

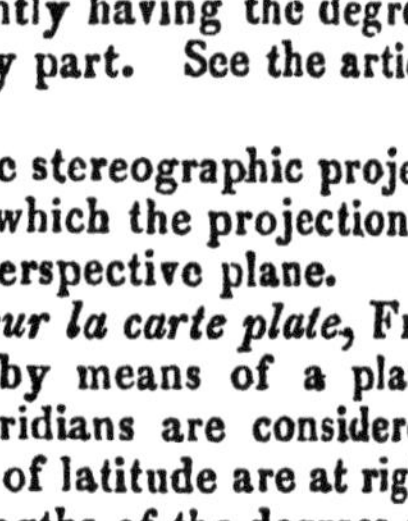

Hence are deduced the following proportions for solving all the cases that occur in the practice of plane sailing.

Case	Given	Required	Solutions.
1	Course & Distance.	Diff. Lat. Depart.	Rad. : Dist. :: Cos. Cou. : Diff. Lat. Rad. : Dist. :: Sin. Cou. : Depart.
2	Course Diff. Lat.	Distance Depart.	Cos. Cou. : Diff. Lat. :: Rad. : Distance. Cos. Cou. : Diff. Lat. :: Sin. Cou. : Depart.
3	Course Depart.	Distance Diff. Lat.	Sin. Cou. : Depart. :: Rad. : Dist. Sin. Cou. : Depart. :: Cos. Cou. : Diff. Lat.
4	Distance Diff. Lat.	Course Depart.	Dist. : Rad. :: Diff. Lat. : Cos. Cou. Rad. : Dist. :: Sin. Cou. : Depart.
5	Distance Depart.	Course Diff. Lat.	Dist. : Rad. :: Depart. : Sin. Cou. Rad. : Dist. :: Cos. Cou. : Diff. Lat.
6	Diff. Lat. Depart.	Course Distance.	Diff. Lat. : Depart. :: Rad. : Tan. Cou. Sin. Cou. : Depart. :: Rad. : Distance.

Plane *Scale*, a thin ruler, upon which are graduated the lines of chords, sines, tangents, secants, leagues, rhumbs, &c. being of much use in most of the practical parts of mathematics, but particularly in Navigation. See the article Scale.

Plane *Triangle*, is a triangle formed by the production of three right lines, in opposition to a spherical and a mixed triangle.

Plane *Trigonometry*, is the doctrine of plane triangles, their measures, proportions, &c. See the article Trigonometry.

Planes are much used in astronomy, conics, &c. for imaginary surfaces, supposed to cut and pass through solid bodies; and on this foundation is the whole doctrine of conic sections built.

PLANET, in astronomy, is a celestial body, revolving round the Sun as a centre, or at least as a focus, and continually changing its position, with respect to the fixed stars; whence the name planet, from πλανητης, signifying wanderer.

The planets are usually distinguished into primary and secondary.

Primary Planets, called also simply, planets, are those which move round the Sun as a centre, or focus of their orbit: such are Mercury, Venus, the Earth, Mars, Jupiter, Saturn, the Georgium Sidus, and Ceres, Juno, Pallas, and Vesta; the four last having been recently discovered on the continent, and are situate between the orbits of Jupiter and Mars.

Secondary Planets, more usually called satellites, or moons, are those which revolve round a primary planet as a centre, in the same manner as the primary ones do about the sun: such are the moon which moves round the earth, as a secondary; and Jupiter, Saturn, and the Georgium Sidus, have each several secondary planets, or moons, moving round them. For an account of the different planets, see these articles under their respective heads.

PLANETARIUM, an astronomical machine, contrived to represent the motions, orbits, &c. of the planets as they really are in nature, or according to the Copernican system. The larger sort of them are called Orreries. See that article.

A very remarkable machine of this sort was invented by Huygens, which is still preserved among the curiosities of the University at Leyden.

The planetariums most commonly in use, do not represent the true times of the celestial motions, but only their proportions; and are not kept in motion by clock-work, but are only turned round occasionally with the hand, to help to give young beginners an idea of the planetary system; or, if constructed with sufficient accuracy, and go by clock-work, to resolve problems relating to the motions of the planets, and of the earth, moon, &c.

Dr. Desaguliers, in his Experimental Philosophy, vol. i. describes a planetarium of his own contrivance, which is considered to be one of the best of the common sort.

Mr. William Jones, mathematical instrument maker, of Holborn, London, also constructed a planetarium, and has paid considerable attention to such machines to bring them to a greater degree of simplicity and perfection.

PLANETARY System, is an assemblage of the planets and comets, moving in their respective orbits, round their common centre, the Sun, which gives light and heat to all.

PLANISPHERE, a projection of the sphere, and its various circles on a plane, in which sense maps, wherein are exhibited the meridians, and other circles, are called Planispheres.

Planisphere, is also the name of an astronomical instrument, used in observing the motions of the heavenly bodies; being a projection of the celestial sphere upon a plane, representing the stars, constellations, &c. in their proper order, distance, &c.; as the astrolabe, which is a common name for all such projections.

PLANK, (*bordages*, Fr.) are strong boards from one to four inches thick, cut from various kinds of wood, especially oak, pine, and fir. They are imported in large quantities from the northern parts of Europe, particularly from the ports of Dantzic, Archangel, Petersburgh, Narva, Revel, Riga, Memel, &c.

To Plank *a Ship*, (*border un vaisseau*, Fr.) is to cover and line the sides with oak planks, which is sometimes called by the artificers, *laying on the skin*. This completes the process of Ship-Building, which article see.

The breadth and thickness of all the planks of a seventy-four gun ship, as also of her *wales* and *thick-stuff*, are exhibited in the Midship-Section, fig. 1. plate VI.

PLAT, (*fourrure de vieux cordages pour les cables*, Fr.) a sort of braided cordage, formed of the *strands* of old rope-yarn, twisted into *foxes*. It is used to wind about that part of the cable which lies in the *hawse-hole*, where it would otherwise be greatly injured by the continual friction produced by the agitation of the ship in stormy weather. See the articles *To* Freshen *the Hawse, and To* Serve.

PLATE, (*bande de fer*, Fr.) a flat piece of iron or tin, either single or double, used on various occasions in a ship, to strengthen the part to which it is fastened.

Plate *Nails*. See the article Nails.

Backstay Plates, (*chaînes de gal-haubans*, Fr.) are pieces of iron used instead of chains to confine the lower dead-eyes of the back-stays.

Puttock or *Foot-hook* Plates, (*bandes de hune*, Fr.) are narrow plates of iron attached to the dead-eyes of the topmast shrouds. See the article Futtock.

Patent Plates, square concave plates of wrought

iron, invented by Messrs. Brodie, and Co. iron-founders, of London, for repairing ships boilers, when an accident happens to them by shot, or otherwise. There is also another sort invented by the same persons called *screw dish-plates*, which are used by the carpenters for the same purpose.

PLATFORM, (*plateforme*, Fr.) a number of planks laid together in form of a floor, in the holds or any other part of the ship, for a temporary or particular purpose.

PLATOON, a small square body of men assembled together for the purpose of being exercised in the use of small-arms.

PLEASURE BOAT, (*bateau de plaisance*, Fr.) a boat fitted up for the reception of company to sail up and down a river, harbour, lake, &c.

PLEIADES, in astronomy, an assemblage of seven stars in the neck of the constellation Taurus, the bull, although there are only six of them visible to the naked eye The largest is of the third magnitude, and named Lucida Pleiadum.

PLUMB, a term used by mast-makers, &c. to denote perpendicular.

PLUMB-LINE, a line to which is suspended a lump of lead that keeps it straight and perpendicular; by it is taken, at the sides of masts, &c. the diametrical dimensions, to correspond with the line or lines struck on the surface of the upper side to which it is sawed.

A mast, or piece to be plumbed, should be placed in such a manner as, when hewed in its direction, may gain the greatest substance clear of sap or rot.

PLUG, (*tampon*, Fr.) a piece of timber, formed like the frustum of a cone, and is used for different purposes in a ship, as

Hawse-PLUGS, (*tampons d'écubiers*, Fr.) are made for the purpose of stopping the hawse-holes when the cables are unbent, or not in them. Their use is to prevent the water from going in when the ship pitches.

Shot-PLUGS, (*tampons de différens calibres, pour boucher les trous faits par le canon*, Fr.) are used to stop the breaches made by cannon-balls, and are formed of various sizes, according to the different sizes of shot which may penetrate the ship's sides or bottom in an engagement, for which purpose they are always ready.

PLUNDER, (*butin, pillage*, Fr.) a name given to the effects of the officers or crew of a prize, when pillaged by the captors.

To PLY, (*louvoyer*, Fr.) is to make a progress against the direction of the wind; hence

PLYER, (*boulinier*, Fr.) is a vessel that makes great advances in this manner of sailing.

PLYMOUTH HOSPITAL. The Royal Naval Hospital at Plymouth, for the reception and cure of sick and hurt seamen and marines; was begun while the Earl of Sandwich was second Lord of the Admiralty, and completed so far as to be fit to receive patients in the year 1762. It is a very spacious and convenient building, surrounded with a large airy ground, situate at Stonehouse, about a mile from Plymouth, and is built of rough marble, with Purbeck rusticated quoins; and though without ornament, acquires great dignity from its bulk and simplicity. The Hospital forms a square, and consists of ten buildings, besides offices, detached from each other, with a view to prevent the spreading of fire or contagious diseases, and for the greater advantage of ventilation. There are six wards in each building, and sufficient room for twenty beds in each ward; so that 1200 patients may be conveniently received into the hospital.

The deficiencies in this building, as pointed out by Howard, have long since been remedied, and, in the conveniencies of a receiving-room, hot and cold baths, with a vapour-bath on the plan of the Hon. Basil Cochrane, the establishment is now on an equal footing with the Royal Naval Hospital at Haslar.

Establishment of the Royal Hospital at Plymouth, for 1812.

1 Governor.	14 Hospital Mates.
3 Lieutenants.	2 Matrons.
2 Physicians.	1 Assistant ditto.
2 Surgeons.	1 Nurse to every 7 patients.
1 Agent.	1 Butler.
1 Steward.	1 Butcher.
1 Dispenser.	1 Barber.
1 Chaplain.	1 Cook.
6 Clerks.	3 Porters.

The following is the number of seamen and marines admitted into the Hospital during the last seven years, ending with 1811.

	Patients.
For the year 1805,.............	3136.
Ditto....1806,.............	2898.
Ditto....1807,.............	1836.
Ditto....1808,.............	2401.
Ditto....1809,.............	2616.
Ditto....1810,.............	2886.
Ditto....1811,.............	2144.

POINT, (*pointe, partie avancée d'une côte peu élevée*, Fr.) a low arm of the shore, which projects into the sea, or into a river, beyond the contiguous part of the beach.

POINT, in astronomy, is a term applied to certain parts or places marked in the heavens, and distinguished by proper epithets.

The four grand points or divisions of the horizon, viz. the east, west, north, and south, are called the Cardinal Points. The points where the ecliptic and equator intersect, are called the Equinoctial Points. In particular, that where the sun ascends towards the north pole is called the Vernal Point; and that where he descends towards the south, the Autumnal Point. The highest and lowest points of the ecliptic are called Solstitial Points: the former of them, the Estival or Summer Point; the latter, the Brumal or Winter Point.

To POINT *a Gun*, (*pointer un canon*, Fr.) in gunnery, is to direct it towards any particular object or point. See the article EXERCISE.

POINT-BLANK, (*de but en blanc*, Fr.) in gunnery, denotes the horizontal level position of the gun, or having its muzzle neither elevated nor depressed. And the point-blank range, is the distance the shot goes before it strikes the level ground, when discharged in the horizontal or point-blank direction: or sometimes this means the distance the ball goes horizontally in a straight-lined direction.

To POINT *a Rope*, &c. (*mettre une queue de rat à un cordage ou manœuvre*, Fr.) is to unlay the end of it, as for splicing; then weave a sort of mat about the diminished part, and stop it, so as to thrust it more easily through any hole, and prevent it from being readily untwisted. This is performed in the following manner:

Take out as many yarns as are necessary, and make knittles, which is done by taking separate parts of the yarns when split, and twisting them: comb the rest down with a knife, like fig. 4. plate XVIII. Make two knittles out of every yarn which is left, and lay half of them down upon the scraped part, and the other, back upon the rope, as represented by fig. 5. Take a length of twine, which is generally termed the *warp*, and pass three turns of it very taught, jambing them with a hitch as at *a*. Then lay the knittles backward and forward, as before, and pass the warp. The ends may be whipped and snaked with twine, or the knittles hitched over the warp, and hauled taught. The upper seizing must also be snaked, which will appear like fig. 6. Thus the end of a *reef-line* is pointed, so that, being stiffer, it may more readily penetrate the eyelet-holes of the *reef;* and the ends of the strands of a cable are occasionally pointed for the greater convenience of *splicing* it to another cable, especially when this task is frequently performed. The extremities of the splice of a cable are also pointed, that it may pass with more facility through the hawse-holes.

Fig. 4. Fig. 5. Fig. 6.

To POINT *a Sail*, (*mettre des garcettes de ris à une voile*, Fr.) is to affix points through the eye-let hole of the reefs. See the article POINTS.

POINT *the Yards to the Wind!* (*mets les voiles en ralingue!*) the order to brace the yards sharp up, when at anchor, in order that they shall not receive the impulse of the wind.

POINTING, (*pointer un canon*, Fr.) in gunnery, is the laying a piece of ordnance in any proposed direction, either horizontal, elevated, or depressed to any angle. This is usually effected by means of the gunner's level or quadrant, which being applied to the muzzle of the piece, shews by a plummet the degree of elevation or depression. See the article *Gunner's* QUADRANT.

POINTING, or *Pricking the Chart*, in navigation, is the act of marking on the chart what point or place the vessel is in, and is commonly practised by the Dutch. This is done by means of the latitude and longitude, after these are known, or found by observation or computation. Thus, draw a line, with a pencil, across the chart, according to the latitude; and another across the other way, according to the longitude; then the intersection of these two lines, is the point or place on the chart where the ship is; which is then marked black with a pen, and the pencil lines rubbed out. From the point or place, thus found, the chart readily shews the direct course and distance run, and also the course and distance to steer to the intended port, &c.

POINTS, (*garcettes de ris*, Fr.) are flat pieces of braided cordage, tapering from the middle towards each end, whose lengths are nearly double the circumference of the yard, and used to reef the courses and top-sails of square-rigged vessels: they are fastened to the sails by passing one through every eyelet-hole in the reef-bands, and making two knots upon it, one on each side of the sail, to prevent its falling out. See the article REEF.

POINTS *of the Compass*, or *Horizon*, &c. in geography and navigation, are the points of division when the whole circle is divided into 32 equal parts. These points are, therefore, at the distance of the 32d part of the circle, or 11° 15′, from each other; hence 5° 37′ 30″ is the distance of the half points, and 2° 48′ 45″ the distance of the quarter points. The principal of these are the four cardinal points, east, west, north, and south. See the article COMPASS.

POKER, or DRIVER, a wrought iron instrument of various lengths and sizes, used for driving hoops on masts. It has a flat foot at one end, and a round knob at the other.

POLACRE, (*polacre*, Fr.) a ship with three masts, usually navigated in the Mediterranean: each of the masts is commonly formed of one piece, so that they have neither tops nor cross-trees; neither have they any *horses* to their yards, for the men stand upon the top-sail yards to loose or furl the top-gallant-sails, and upon the lower yards to reef, loose, or furl the top-sails, the yards being lowered sufficiently down for that purpose. These vessels are generally furnished with square sails upon the main-mast, and lateen sails upon the fore and mizen-masts. Some of them, however, carry square sails upon all the three masts, particularly those of Provence in France. See also the article XEBEC.

POLAR, in general, something relating to the poles of the world, or poles of the artificial globes; thus we meet with polar circles, polar projections, &c.

POLARITY, the quality of a thing considered as having poles, but chiefly used in speaking of the magnet.

A WATERMAN'S POLE, (*croc de batelier*, Fr.) is a sort of small mast used by watermen in wherries and pleasure-boats, and to which is fastened the sail.

POLE-AXE, (*hache d'armes*, Fr.) a sort of hatchet nearly resembling a battle-axe, having an handle about fifteen inches in length, and being furnished with a sharp point, bending downwards from the back of its head. It is principally used to cut away the rigging of any adversary who endeavours to board. They have also been sometimes used in boarding an enemy's ship, whose hull was more lofty than that of the boarders, by driving the points into her side, one above another, so as to form a sort of scaling-ladder; whence they are sometimes called Boarding Axes.

POLE-MAST, (*mât à pible*, Fr.) implies any mast that is formed of one single piece or tree, in contradistinction to one composed of several pieces.

POLE-STAR, in astronomy, is a star of the second magnitude, and the last in the tail of Ursa Minor. Its longitude Mr. Flamsteed makes 24° 14′ 41″; its latitude 66° 4′ 11″ The nearness of this star to the pole, renders it of great service in navigation, &c. for determining the meridian line, the elevation of the pole, and, consequently, the latitude of the place, &c. See the article LATITUDE. For the beginning of 1813, its

Right Ascension is	13° 47′ 4″
Annual Variation in ditto....	3′ 13
Declination, ...	88 18 39 N.
Annual Variation in ditto....	+ 19. 5.

POLES, in astronomy, the extremities of the axis upon which the whole sphere of the world revolves. These two points, each 90° distant from the equinoctial or equator, are by way of eminence called the poles of the world; and the extremities of the axis of the artificial globes, corresponding to these points in the heavens, are termed the poles thereof.

Poles, in geography, are the extremities of the earth's axis, or the points on the surface of the sphere through which the axis passes. Of which, that elevated above the horizon is called the Arctic or North Pole; and the opposite one, the Antarctic or South Pole.

Poles *of the Ecliptic*, are two points on the surface of the sphere, 23° 30′ distant from the poles of the world, and 90° distant from every part of the ecliptic.

Poles, in magnetics, are two points of a loadstone, corresponding to the poles of the world; one pointing to the north, and the other to the south.

Under bare Poles, (*à sec, ou à mâts et à cordes*, Fr.) the situation of a ship at sea when all her sails are furled, particularly in a tempest. See the articles *To* Scud, and Trying.

POLLUX, in astronomy, the hindermost twin, or posterior part of the constellation Gemini.

Pollux, is also the name of a fixed star of the second magnitude, in the constellation Gemini. See the articles Castor and Pollux, also Gemini. For the beginning of 1813, its right ascension is 113° 27′ 48″, annual variation in right ascension, 55″ 3.; declination 28° 23′ 8″ N. annual variation in declination, —7″ 9.

POMELION, (*bouton, d'un canon*, Fr.) a name given by seamen to the cascabel, or hindmost knob of a Cannon, which article see.

PONTOON, (*ponton, ou fosse pour le carénage des vaisseaux*, Fr.) a large low flat vessel, nearly resembling a barge of burden, and furnished with cranes, capstans, tackles, and other machinery necessary for careening ships. These are principally used in the Mediterranean, but very seldom in the northern parts of Europe.

POOP, (*dunette*, Fr. *puppis*, Lat.) the highest and aftmost deck of a ship.

Poop-Royal, (*carosse, ou logemens pratiqués sur la dunette des vaisseaux François*, Fr.) a short deck, or platform, placed over the aftmost part of the poop in the largest of the French and Spanish men of war, and serving as a cabin for their masters and pilots. This is usually called the top-gallant poop by our shipwrights.

To Poop, (*aborder un vaisseau par l'arrière*, Fr.) is to run the head, bowsprit, or jib-boom of one vessel into the stern or poop of another, which sometimes happens accidentally in harbours, &c. but ought most certainly to be avoided as much as possible.

POOPING, or, *A* Pooping *Sea*, (*grosse mer de vent arrière*, Fr.) implies the breaking of the sea over the taffrail on the poop. When the sea is so turbulent as to break in the stern-sashes, &c. it is also called *a pooping sea*

POPPETS, the name of perpendicular pieces of timber which are fixed on the fore and aftmost parts of the bulgeways, to support the ship when launching.

PORT, (*port de mer, ou havre*, Fr.) a harbour or haven on the sea-coast, where ships arrive with their freight, and customs from goods are taken. See the article Harbour.

Port is also a name given, on some occasions, to the larboard, or left side of the ship, as in the following instances:

The ship heels to Port, (*le vaisseau est sur babord*, Fr.) that is, stoops or inclines to the larboard side.

Top the fore-yard to Port! (*pèse sur la balancine de babord de misaine!* Fr.) the order to sway the larboard extremity of that yard higher than the other. See the article *To* Top a Yard.

Port *the helm!* (*babord la barre!* Fr.) the order to put the helm over to the larboard-side of the vessel.

Hard a Port! (*babord tout!* Fr.) the order to put the helm close to the larboard side of the ship.

In all these cases this word appears intended to prevent any mistakes happening from the similarity of sounds in the words starboard and larboard, particularly when they relate to the helm, where a misapprehension might be attended with very dangerous consequences: therefore the word larboard is never used in conning.

Bar of a Port, (*port a barre*, Fr.) implies a rock or sand-bank, lying before the mouth of a port or harbour, and which prevents ships from entering therein, except with the tide.

Close-Port, (*port renfermé dans les terres*, Fr.) is one within the body of a city, as the ports of Rhodes, of Venice, Amsterdam, Rochelle, Bayonne, and St. Jean de Luz.

Free Port, (*port franc*, Fr.) is one open and free for merchants of any nation to load and unload their vessels in, without paying any duty or customs; such are the ports of Genoa and Leghorn.

The term free port is also used for a total exemption and franchise which any set of merchants enjoy, for goods imported into a state, or for those of the growth of the country exported. Such was the privilege the English enjoyed for several years after their discovery of the port of Archangel, and which was taken from them on account of the regicide, in 1648.

Port-*Bars*, (*barres de sabords*, Fr.) are strong pieces of oak fitted in the port-sills; they are about six inches square, and, having two mortices cut near their ends, receive the hooks which hook into the port-shackles. The iron hooks and wedges are attached to these bars, whereby the ports are properly secured.

Port-*Fire*, in gunnery, a paper tube, about 16 inches long, filled with a composition of mealed-powder, sulphur, and nitre, rammed moderately hard, and used instead of match to fire the guns.

Port-fires and lights are compositions that kindle readily and burn clearly, and should not go out even under water.

One that is from 10 to 12 inches in length, and from three to 5-twelfths of an inch thick, will burn from 12 to 15 minutes.

A substitute for port-fires has been made of two ounces of nitre dissolved in a gallon of water, and sheets of soft brown paper dipped therein. These, when dry, were rolled up for use to about the size of common port-fires. Port-fires may be made of any length, but seldom exceed 20 inches. The interior diameter or mould should be 10-sixteenths of an inch, and the diameter of the whole about half an inch. The paper cases should be rolled wet with paste, and one end folded down. The composition for *wet port-fires* is—salt-petre, 6; sulphur, 2; and meal-powder, 1: when it is well mixed and sieved, it is to be moistened with a little linseed oil. The composition for *dry port-fires* is—saltpetre, 4; sulphur, 1; mealed-powder, 2; and antimony, 1.

Port-*Hooks*, (*crocs à sabords*, Fr). are hooks driven through the side of the ship, and clinched, for the purpose of hooking the hinges that are fastened to the port-lids.

Port-*Lids*, (*mantelets de sabords*, Fr.) a sort of hanging-doors, which shut in the ports at sea, they are fastened by hinges to their upper edges; so as to let down when the cannon are drawn into the ship, whereby the water is prevented from entering the lower deck in a turbulent sea. The lower and upper edges of the ports are always parallel to the deck, so that the guns, when levelled in their carriages, are all equally high above the lower extremity of the ports. The ports are exhibited, throughout the ship's whole length, by H. in the Elevation, plate XVII.: they are also represented upon a larger scale in plate XI. figures 8. and 9. See also the articles Cannon, Deck, and Stern.

Port *Nails*. See the article Nails.

Port-*Riggles*, the name of small semi-circular pieces of wood, about two inches thick, nailed to the ship's side over the ports, to prevent the water that may run down the side from going therein.

Port-*Ropes*, or *Pendants*, (*rabans de sabords*, Fr.) are ropes spliced into the rings on the outside of the port-lids: they lead through the lead pipes in the ship's side, and are used to haul up the port-lids by the assistance of blocks.

Port-*Sashes*, (*sabords vitrés*, Fr.) are glass frames put in the gun-room, ward-room, and cabin ports of a ship of war, to give light to those apartments. They are generally fastened with lanyards, and sometimes with hooks and eyes.

Port-*Shackles*, (*boucles de sabords*, Fr.) a sort of iron ring-bolts, driven through the lower-deck port-lids, and clinched on their hinges, for the purpose of securing the port-lids, which is done by means of wedges and hooks that hook into the rings.

Port-*Tuckles*, (*palans de sabords*, Fr.) are those mentioned in the preceding article, which serve to haul up the port-lids.

Man the Port-*Tackle Falls!* (*du monde aux sabords!* Fr.) the order given to the men to prepare themselves to haul up the port-lids.

PORTER, in anchor-making, is a straight bar of iron, two inches square, confined at one end to the end of the shank; it has holes punched through at the other end, for the admission of the cross-bars, which act as levers in turning the shank when making.

PORTLAST, or Portoise, (*plat-bord*, Fr.) is synonymous with gunwale; as, "Lower the yards down a portlast!" that is, down to the gunwale.

To ride a Portoise, (*être au mouillage avec ses mâts de hune et ses basses vergues amenés, dans un coup de vent*, Fr.) is to have the lower yards and top-mast struck, or lowered down, when at anchor in a gale of wind.

PORTS, or Port-holes, (*sabords*, Fr.) the embrasures or openings in the side of a ship of war, wherein the artillery is ranged in battery upon the decks above and below: they are said to have been invented by Descharges, a French builder, at Brest, in the year 1500.

The ports are of a sufficient extent to point and fire the cannon, without injuring the ship's side by the recoil; and the shipwrights have certain established dimensions, by which they are cut in proportion to the size of the cannon.

Gun-Room Ports, (*sabords de Sainte-Barbe*, Fr.) the ports which are in the ship's counter, and used for stern-chases, and also for passing out a small cable or hawser, either to moor head and stern, to tow a ship, or to spring upon the cable.

Half-Ports, (*faux sabords*, Fr.) a kind of shutters, with circular holes in their centre, large enough to go over the muzzles of the guns, and furnished with pieces of canvas, nailed round their edges, to tie upon the guns, whereby the water is prevented from entering in at the ports, although the guns remain run out. They are principally used upon the main-deck, and particularly in ships carrying one tier of guns.

An improvement has been made in the half-ports of ships of war, which is as follows:—The lower part of the half-port to the centre of the gun is a piece of four-inch fir or elm plank, hung with hinges to the side, and rabbeted on the upper edge to receive the upper part thereof, and then secured with bolts; so that the improvement consists in the safety of the lower part of the half-port, and the ease with which it may be knocked out upon coming to action.

Lower-Deck Ports, (*sabords de la première batterie*, Fr.) are those situate on the lower gun-deck.

Main-Deck Ports, (*sabords de la seconde batterie*, Fr.) are those on the middle gun-deck of three-decked ships.

Hook on the Ports *fore and aft!* the order to the carpenter's crew to hook on the port-lids fore and aft the ship, on the appearance of a squall.

Lower the Ports *fore and aft!* (*amène les mantelets de sabords, ou ferme les sabords!* Fr.) the order to let down the port-lids fore and aft the ship, so as to hook them on.

Slope the Ports! the order to raise or lower the port-lids to an inclined position, to prevent the rain from going in between decks.

Square the Ports! the order to the men, after the port-lids are hauled up, to haul them in a position that they shall appear square with each other.

POST-CAPTAIN in the royal navy, is the next in rank to an admiral. See the article Captain.

POUNDER, in artillery, a term used to express the weight of a shot or ball, or how many pounds the proper ball is for any cannon; as a 32-pounder, 24-pounder, 18-pounder, 12-pounder, &c. It is also applied to the pieces of ordnance that carry those balls.

POWDER. See the article Gunpowder.

Powder-*Barrels*. See the article Barrels.

Powder-*Chests*, (*caisses d'artifices*, Fr.) are small

boxes, charged with powder and a quantity of old nails or splinters of iron, and fastened occasionally on the decks or sides of a merchant-ship, when furnished with close-quarters, having a train of powder which communicates with the inner apartments, so as to be fired at pleasure to annoy an enemy who attempts to seize he by boarding.

These chests are usually from 12 to 18 inches in length, and about 8 or 10 in breadth, having their outer or upper part terminating in an edge. They are nailed to several places of the quarter, the quarter-deck, and bulk-head of the waist, having a train of powder which communicates with the inner apartments of the ship.

Powder-*Horn*. See the article Horn.

Powder-*Marks*. See the article Marks.

Powder-*Mill*, (*moulin à poudre*, Fr.) generally implies the whole of the premises whereon gunpowder is manufactured.

The principal powder-mills in England, are those of Hounslow, Dartford, Waltham-Abbey, &c.

PRAM, or PRAME, (*prame, sorte d'allége, en usage dans les ports de la Hollande et de la mer Baltique*, Fr.) a sort of lighter, formerly used in Holland and the ports of the Baltic Sea, for loading and unloading ships.

PRATIC, or Pratique, (*entrée, ou faculté de communiquer dans les ports sujets à quarantaine*, Fr.) a term used in the European ports of the Mediterranean Sea, implying permission to trade and communicate with the natives of the country, after having performed a limited quarantine.

PRECESSION *of the Equinoxes*. This is fully explained under the article Equinoxes.

PRIMARY Planets. See the article Planet.

Press *of Sail*, (*avoir autant de voile dehors qu'un bâtiment en peut porter*, Fr.) denotes as much sail as the then state of the wind, &c. will permit the ship to carry.

To PRESS, or *To* Impress *Seamen*, is to compel those who have been brought up to the sea-service, or accustomed to boats in harbours or on rivers, to serve in his Majesty's navy during the war.

Humanity has often raised a plaintive tone against the practice of impressing seamen; yet this has hitherto been successfully opposed by the voice of ages and the verdict of experience; the prejudice of the people being always in favour of the manner handed down to them by their forefathers, there appears, probably, more risk in wounding their attachment by a change even for the better, than by persisting in the good old way.

It may be farther observed, that persons brought up to the sea are not liable to the militia ballot, and, in common cases, the impressed man has little to complain of more than the man drawn and compelled to serve in the militia, as both are taken for the same period; that is, during the continuance of war. He who voluntarily goes to sea, is fully aware beforehand, that the line of life he has chosen subjects him to the impress, and at the same time frees him from the chance of being ballotted for the militia: therefore, having made his election, he must take his chance of being compelled to serve his country, should his Majesty at any time deem it necessary.

A PRESSED MAN, or Impressed Seaman, (*matelot pressé, ou retenu au service malgré lui*, Fr.) is one who has been pressed into the King's service during war, in contradistinction to a volunteer. A pressed man may, in the course of a fortnight, enter as a volunteer, and receive the bounty; he may also, at any future period, become a volunteer, when he will be entitled to promotion, but not to the bounty. Several of our naval officers were originally pressed into the navy from the merchant-service.

PRESS-GANG, (*escouade armée, ou détachement, qui sous les ordres d'un officier de la marine saisit de force tout matelot, soit de la marine marchande, ou de l'Etat, qui n'est pas muni d'un certificat qui le protège*, Fr.) a detachment of seamen employed on shore, who, under the command of a captain, a lieutenant, and a midshipman of the royal navy, are empowered, by the civil magistrates, in time of war, to impress the seamen whom they find skulking about or unemployed belonging to the merchant-service, &c. and oblige them to serve on board his Majesty's ships. See the article *To* Impress.

PREVENTER, (*fausses manœuvres, ou manœuvres de précaution pour prévenir les accidents*, Fr.) an additional rope, employed to support any other, when the latter suffers an unusual strain, particularly in a hard gale of wind.

Preventer-*Braces*, (*faux-bras*, Fr.) are temporary braces, fixed occasionally to succour the main or fore-yard of a ship, or to supply the place of the usual braces, in the event of their being shot away in action.

Preventer-*Bolts*, are bolts driven in the lower end of the preventer-plates, to assist the strain of the chain-bolts.

Preventer-*Plate*, a broad plate of iron, fixed below the toe-link of the chains to support them against the efforts of the masts and shrouds, having a chain-bolt driven through its upper end, and a preventer-bolt through the lower: hence

Preventer-*Shrouds*, (*haubans de fortune, ou faux-haubans*, Fr.) are applied in the same manner to serve the same purposes. See the articles Shroud and Stay.

Preventer-*Stay*, (*faux-étai*, Fr.) is a small stay fixed above the standing one, and serves to relieve the latter, or to supply its place.

To PRICK the Chart, (*pointer la carte*, Fr.) is to trace a ship's course upon a marine chart, by the help of a scale and compasses, so as to discover her present situation.

To Prick *a Sail*, (*recoudre une voile dans l'entre-deux de chaque couture, lorsqu'elle est vieille*, Fr.) is to stitch two cloths of a sail together along the space comprehended between the two edges or selvages that overlay each other. Or, it is to run a middle-seam between the two seams which unite every cloth of a sail to the next adjoining, and is rarely performed till the sails have been worn for a considerable time.

PRICKER, in sail-making, a small iron instrument like a marline-spike, but straight, to make the holes with.

PRIMAGE, in maritime affairs, is a certain allowance paid by the shipper or consignee of goods, to the mariners and master of a merchant vessel for loading the same. It is mentioned in stat. Henry VIII. c. 14. In some places it is a penny in the pound; in others sixpence for every pack or bale, or otherwise according to the custom of the place.

To PRIME A GUN, (*amorcer un canon*, Fr.) is to lay the train of powder from the opening of the touch-hole along the cavity made for that purpose, in order to fire the piece.

To PRIME *a Fire-ship*, (*apprêter un brulot*, Fr.) is to lay the train, and get her in readiness for being set on fire. See the article FIRE-SHIP.

PRIMING-WIRE, (*épinglette, ou dégorgeoir*, Fr.) a sort of iron needle, employed to penetrate the touch-hole of a cannon, when it is loaded, in order to pierce the cartridge, and as a rammer to charge the touch-hole with powder.

PRIVATEERS, (*corsaires*, Fr.) are vessels of war, armed and equipped by particular merchants, and furnished with commissions from the state, to cruise against and annoy the enemy, by taking, sinking, or burning their shipping. These vessels are generally governed on the same plan with his Majesty's ships.

The commission obtained by the merchants empowers them to appropriate to their own use whatever prize they make, after legal condemnation; and government allows them besides 5*l*. (35. Geo. III. c. 66), for every man on board a man of war or privateer, taken or destroyed, at the beginning of the engagement: and, in case we are at war with more potentates than one, they must have commissions for acting against each of them; otherwise, if a captain, carrying only one against the Danes, should in his course meet with and take a Frenchman, this prize is not good, but would be taken from him by any man of war he met, and could not be condemned (*for him*) in the Admiralty, as many have experienced.

PRIZAGE, is the custom or share that belongs to the king, out of such merchandises as are taken at sea by way of lawful prize. 13 Eliz. c. 5.

PRIZE, (*prise, capture*, Fr.) a vessel taken at sea from the enemies of a state by a ship of war, privateer, or armed merchantman, having a commission for that purpose.

Vessels are looked on as prizes if they fight under any other standard than that of the state from which they have their commission: if they have no charter-party, invoice, or bill of lading on board; if loaded with effects belonging to the king's enemies, or with contraband goods, those of the king's subjects recovered from the enemy, after remaining twenty-four hours in their hands, are deemed lawful prizes.

By stat. 13 Geo. II. c. 4., judges and officers failing of their duty in respect to the condemnation of prizes, forfeit 500*l*. with full costs of suit, one moiety to the king, and the other to the informer.

The following are the Regulations with regard to Prizes in the Royal Navy:

When any ship or vessel is taken from the enemy, the hatches are to be immediately spiked up, and her lading and furniture carefully secured from embezzlement, till sentence is passed upon her in some court of Admiralty empowered to take cognizance of causes of that nature.

The captain is to cause the principal officers, &c. of the prize to be examined: three or more of the company, who can give the best evidence, to be brought to the said court of Admiralty, together with the charter-parties, bills of lading, and other ship's papers found on board.

When prisoners of war are put on shore, the captain of the ship from which they are sent is to deliver to the commanding officer present, to be by him transmitted to the Commissioners of Transports, a list containing their names; the name of the vessel in which they were taken; the day they were first victualled on-board, and the day they are landed. He is also to deliver a similar list to the Agents for Prisoners, at the place where they are put on shore. For further particulars, see the NAVAL INSTRUCTIONS.

All prizes taken by ships and vessels having commissions of letters of marque and reprisals, may be sold and disposed of by the merchants, owners, fitters, and others, to whom such letters of marque and reprisals are granted, for their own use and benefit, after final adjudication, and not before.

The nett produce of all prizes which are or shall be taken by any of our ships or vessels of war, shall be for the entire benefit and encouragement of the flag-officers, captains, commanders, and other commissioned officers, and of the seamen, marines, and soldiers on board of the said ships and vessels at the time of capture: and such prizes may be lawfully sold and disposed of by them and their agents, after the same shall have been finally adjudged lawful prize, and not otherwise.

PRIZE-MONEY, (*part de prise*, Fr.) the profits accruing from the sale of a prize. In ships of war, the prize money is to be divided among the officers, &c. according to the following Proclamation; but among privateers, the division is according to the agreement between the owners.

The following is the Distribution of Prize-Money for Captures from France, Spain, Holland, and the Italian and Ligurian Republics, and Ships bearing the Flags of Prussia and Pappenburgh, pursuant to the King's Proclamations of the 7th of July, 1803, 31st of January, 1805, and 5th of June, 1806.

The distribution shall be made as follows:—the whole of the nett produce being first divided into eight equal parts:

To captains actually on board at the taking of any prize, three eighth parts: but in case any such prize shall be taken by any ships of war under the command of a flag or flags, the flag officer or officers being actually on board, or directing and assisting in the capture shall have one of the three eighth parts; the said one eighth part to be paid to such flag or flag officers in such proportions, and subject to such regulations, as are hereinafter mentioned.

Captains of marines and land-forces, sea-lieutenants, and master on board, shall have one eighth part, to be equally divided amongst them; but that every physician appointed to a fleet or squadron of ships of war, shall, in the distribution of prizes taken by the ships in which he shall serve, or in which such ship's company shall be entitled to share, be classed with the sea-lieutenants with respect to the said one eighth part, and be allowed to share equally with them; provided such physician be actually on board at the time of taking such prizes.

The lieutenants and quarter-masters of marines, and lieutenants, ensigns, and quarter-masters of land forces, secretaries of admirals or commodores, with captains under them, second masters of line of battle ships, boatswains, gunners, pursers, carpenters, master's mates, chirurgeons, pilots, and chaplains, on board, shall have one eighth part, to be equally divided amongst them.

The midshipmen, captain's clerks, master sailmakers, carpenter's mates, boatswain's mates, gunner's mates, masters at arms, corporals, yeomen of the sheets, coxswains, quarter masters, quarter master's mates, chirurgeon's mates, yeomen of the powder-room, serjeants of marines and land forces on board, shall have one eighth part, to be equally divided amongst them.

The trumpeters, quarter gunners, carpenter's crew, steward's cook, armourers, steward's mate, cook's mates, gunsmiths, coopers, swabbers, ordinary trumpeters, barbers, able seamen, ordinary seamen, and marines, and other soldiers, and all other persons doing duty and assisting on board, shall have two eighth parts, to be equally divided amongst them.

Cutters, Schooners, Brigs, &c.

In the case of cutters, schooners, brigs, and other armed vessels, commanded by lieutenants, the share of such lieutenants shall be three eighth parts of the prize, unless such lieutenants shall be under the command of a flag officer or officers, in which case the flag officer or officers shall have one of the said three eighths, to be divided among such flag officer or officers, in the same manner herein directed in the case of captains serving under flag officers.

The share of the sub-lieutenant, master, and pilot, shall be one eighth; the said eighth, if there be all three such persons on board, to be divided into four parts, two parts to be taken by the sub-lieutenant, one part by the master, and one part by the pilot: if there be only two such persons on board, then the eighth to be divided into three parts, of which two thirds shall go to the person second in command, and one third to the other person. If there be only a sub-lieutenant or a master, and no pilot, then the sub-lieutenant or master to take the whole eighth.

The share of the chirurgeon, or chirurgeon's mate (where there is no chirurgeon,) midshipmen, clerk, and steward, shall be one eighth. The share of the boatswains, gunners, and carpenter's mates, yeomen of the sheets, sail-maker, quarter master, and quarter master's mate, shall be one-eighth. And the share of the seamen, marines, and other persons on board, assisting in the capture, shall be two eighth parts.

Shares of Ships of War in Sight when Captures made.

But it is our intention, nevertheless, that the above distribution shall only extend to such captures as shall be made by any cutter, schooner, brig, or armed vessel, without any ships of war being present, or within sight of, and adding to the encouragement of the captors, and terror of the enemy: but in case any ships of war shall be present, or in sight, that then the officers, pilots, petty officers, and men, on board such cutters, schooners, brigs, or armed vessels, shall share in the same proportion as is allowed to persons of the like rank and denomination on board of our ships of war; the sub-lieutenant and master to be considered as warrant officers; and such cutters, schooners, brigs, or armed vessels, shall not, in respect to such captures, convey any interest or share in the flag eighth to the flag officer or officers under whose orders such cutters, schooners, brigs, and armed vessels, may happen to be: and we further order, that the directions herein given, with respect to the shares of sub-lieutenants, masters, and pilots, on board vessels commanded by lieutenants, be applied to captures made from France and the Batavian Republic.

Hired Armed Vessels.

The nett produce of all prizes taken by such hired armed vessels, (except as hereinafter mentioned) shall be for the benefit of our commissioned officers in our pay, and of the masters and crews on board the said hired armed vessels at the time of the capture, and that such prizes may be lawfully sold and disposed of by them and their agents, after the same shall have been to us finally adjudged lawful prize, and not otherwise; the distribution whereof shall be as follows:

The whole of the nett produce being divided into eight equal parts;—our officer commanding any hired armed vessel, who shall be actually on board at the taking of any prize, shall have three eighths; but in case such hired armed vessel shall be under the command of a flag or flags, the flag officer or officers being actually on board, or directing or assisting in the capture, shall have one of the said three eighth parts, the said one eighth part to be paid to such flag or flag officers in such proportions, and subject to such regulations, as are hereinafter mentioned. In case there be acting on board such hired armed vessel, besides our officer commanding the same, one or more commissioned sea-lieutenants in our pay, such lieutenant or lieutenants shall take one eighth.

One eighth shall belong to the master and mate, of which the master shall have two thirds, and the mate one third; but in case there shall be acting on board such hired armed vessel one or more midshipmen, or other person in our pay, of those who are classed with midshipmen in the former part of this our proclamation, in that case the master shall have one half of the eighth, and the other half shall be equally divided between the mate, midshipmen, and such other persons in our pay.

Three eighths shall belong to, and be divided among the other officers and the rest of the crew.

And in the case of prizes taken by any hired armed vessel not commanded by any of our commissioned officers, one eighth shall belong to the flag officers, to be divided as aforesaid, in case such hired armed vessel shall be under the command of a flag. Two eighths shall belong to the master and mate, of which the master shall have two thirds, and the mate one third.

Three eighths shall belong to and be divided among the other officers and the rest of the crew in manner aforesaid. The surplus, the distribution of which is not herein directed, shall remain at our disposal; and if not disposed of within a year, after final adjudication, the same shall belong and be paid to Greenwich Hospital.

Prizes taken jointly.

And in the case of prizes taken jointly, by any of our ships of war, and any hired armed vessel our commissioned officer or officers on board such hired armed vessel shall share with our commissioned officer or officers of the same rank on board our ship of war, being joint captors; the master of such hired armed vessel shall share with the warrant officers; the mate of such hired armed vessel with the petty officers; and the seamen of such hired armed vessel with the seamen on board our said ship or ships of war; save and except, that in case such hired armed vessel shall be commanded by one of our commissioned officers, having the rank of master and commander, and there shall be none of our lieutenants on board, or in case such hired armed vessel shall be commanded by the master, in both those cases the master of such hired armed vessel shall share with the lieutenants of our ships of war, and the mate with the warrant officers.

Admiralty to settle Disputes, in Cases of Difficulty.

And in case any difficulty shall arise in respect to the said distribution, not herein sufficiently provided for, the same shall be referred to our lords commissioners of the Admiralty, whose directions thereupon shall be final, and have the same force and effect as if herein inserted.—Provided, that if any officer being on board any of our ships of war at the time of taking any prize, shall have more commissions or offices than one, such officer shall be entitled only to the share or shares of the prizes, which, according to the above-mentioned distribution, shall belong to his superior commission or office.—Provided also, that in all prizes taken by any of our squadrons, ships, or vessels, while acting in conjunction with any squadron, ship, or vessel of any other power that may be in alliance with us, a share of such prizes shall be set apart, and be at our further disposal, equal to that share which the flag, and other officers and crews of such squadron, ships, or vessels, would have been entitled to, if they had belonged to us.

Regulations respecting Flag Officers' Eighths.

We do hereby further will and direct, that the following regulations shall be observed concerning the one eighth part, herein before mentioned to be granted to the flag officers who shall actually be on board at the taking of any prize, or shall be directing or assisting therein. First, that a captain of a ship shall be deemed to be under the command of a flag, when he shall actually have received some order directly from, or be acting in execution of some order issued by a flag officer; and shall be deemed to continue under the command of such flag, so long as the flag officer by whom the order was issued, or any other flag officer acting upon the same station, shall continue upon such station; or until such captain shall have received some order directly from, or be acting in execution of some order issued by some other flag officer, or the lords commissioners of the Admiralty.

Secondly, that a flag officer, commander in chief, when there is but one flag officer upon service, shall have to his own use the said one eighth part of the prizes taken by ships and vessels under his command.

Thirdly, that a flag officer, sent to command on any station, shall have no right to any share of prizes taken by ships or vessels employed there, before he arrives within the limits of such station, and actually takes upon him the command, by communicating orders to the flag officer previously in command; save only, that he shall be entitled to a share of prizes taken by those particular ships to which he shall actually have given some order, and taken under his command within the limits of such station.

Fourthly, that a commander in chief, or other flag officer, appointed or belonging to any station, and passing through or into any other station, shall not be entitled to share in any prize taken out of the limits of the station to which he belongs, by any ship under the command of a flag officer of any other station, or under Admiralty orders; unless such commander in chief, or flag officer, is expressly authorized by the lords commissioners of the Admiralty to take upon him the command on that station in which the prize is taken, and shall actually have taken upon him such command in manner aforesaid.

Fifthly, that when an inferior flag officer is sent to reinforce a superior flag officer on any station, the superior flag officer shall have no right to any share of prizes taken by the inferior flag officer before the inferior flag officer shall arrive within the limits of the station, and, moreover, shall actually receive some order directly from him, or be acting in execution of some order issued by him.

Sixthly, that a chief flag officer quitting a station, either to return home, or to assume another command, or otherwise, except upon some particular urgent service, with the intention of returning to the station as soon as such service is performed, shall have no share of prizes taken by the ships or vessels left behind, after he shall have passed the limits of the station, or after he shall have surrendered the command to another flag officer, appointed by the Admiralty to be commander in chief upon such station.

Seventhly, that an inferior flag officer quitting a station, except when detached by orders from his commander in chief, out of the limits thereof, upon a special service, with orders to return to such station as soon as such service is performed, shall have no share in prizes taken by the ships remaining on the station, after he shall have passed the limits thereof; and in like manner the flag officers remaining on the station shall have no share of the prizes taken by such inferior flag officer, or by the ships and vessels under his immediate command, after he shall have quitted the limits of his station, except when detached as aforesaid.

Eighthly, that when vessels under the command of a flag, which belong to separate stations, shall happen to be joint captors, the captain of each ship shall pay one third of the share to which he is entitled, to the flag officers of the station to which he belongs. But the captains of vessels under admiralty orders, being joint captors with other vessels under a flag, shall retain the whole of their share.

Ninthly, that if a flag officer is sent to command in the out-ports of this kingdom, he shall have no share of the prizes taken by ships which have sailed, or shall sail, from that port by order from the Admiralty.

Tenthly, that when more flag officers than one serve together, the eighth part of the prizes taken by any ships of the fleet or squadron, shall be divided in the following proportions; viz. if there be but two flag officers, the chief shall have two third parts of the said one eighth, and the other shall have the remaining third part; but if the number of flag officers be more than two, the chief shall have only one half, and the other half shall be equally divided among the other flag officers.

Eleventhly, that commodores, with captains under them, shall be esteemed as flag officers, with respect to the eighth part of prizes taken, whether commanding in chief or serving under command.

Twelfthly, that the first captain to the admiral and commander in chief of our fleet, and also the first captain to our flag officer appointed to command a fleet or squadron of ten ships of the line of battle, or upwards, shall be deemed and taken to be a flag officer, and shall be entitled to a part or share of prizes, as the junior flag officer of such fleet or squadron.

To Prize, (*établir un levier pour mouvoir un fardeau*, Fr.) is, to apply a lever to move any weighty body, as a cask, anchor, cannon, &c.

PROA, or Flying Proa, a name given to a sort of narrow canoe, about 30 feet long and 3 feet wide, used about the Ladrone Islands. The head and stern of the *proa* are exactly alike; but the lee side is flat, and the weather-side round; and to prevent her upsetting, which from her small breadth, and the straight run of her leeward side, would, without this precaution, infallibly happen, there is a frame about 10 feet in length laid out of her from windward, to the end of which is fastened a block of wood, in the shape of a small vessel made hollow. The weight of this frame, commonly called an outrigger, is intended to balance the *proa*, and the small boat or hollow block of wood is, by its buoyancy, being always in the water, to prevent her upsetting. The body of the vessel is made of two pieces joined endwise, and sewed together with bark, for there is no iron used about her; she is about two inches thick at the bottom, which at the gunwale is reduced to less than one.

This construction is so extremely light, that she seems to feel no resistance in her passage through the water, and, with a brisk trade-wind, will sail near 20 miles an hour. They are steered by paddles at each end. Their rigging consists of two stays, that set up at the ends of the *proa*, and four shrouds that set up at the four corners of the frame. The mast-yard and boom, are of bamboo. The sail is made of matting, and shaped like a settee-sail; and the lower end of the yard is confined forward in a shoe-block. In going about, they keep her way so, that the stern becomes the head; and, to shift the sail, the yard is raised, and the lower end taken along the gunwale, and fixed in a shoe-block as before; the boom is shifted at the same, by slackening the sheet, and peaking the boom up along the mast, then by hauling upon another sheet, the end of the boom is brought to the place where the lower yard-arm was before, and is hauled aft at the other end.

PROCYON, in astronomy, a bright fixed star of the second magnitude, in the body of the constellation Canis Minor, or the Little Dog. For the beginning of 1813, its right ascension is 112° 22′ 32″, annual variation in right ascension 47″.1; declination, 5° 41′ 52″ N. and annual variation in declination 8″.5.

PROJECTILE, or Project, in mechanics, is any body which, being put into a violent motion by an external force impressed upon it, is dismissed from the agent, and left to pursue its course; such as a stone thrown out of the hand or a sling; an arrow from a bow, a ball from a gun, &c.

PROJECTILES, the science of the motion, velocity, flight, range, &c. of a projectile put in violent motion by some external cause, as the force of gunpowder, &c. This is the foundation of gunnery, under which article may be found what relates peculiarly to that branch.

All bodies, being indifferent as to motion or rest, will necessarily continue the state they are put into, except so far as they are hindered, and forced to change it by some new cause. Hence, a projectile put in motion, must continue eternally to move on in the same right line, and with the same uniform or constant velocity, were it to meet with no resistance from the medium, nor had any force of gravity to encounter.

In the first case, the theory of projectiles would be very simple indeed; for there would be nothing more to do, than to compute the space passed over in a given time by a given constant velocity: or either of these from the other two being given.

But by the constant action of gravity, the projectile is continually reflected more and more from its right-lined course, and that with an accelerated velocity, which, being combined with its projectile impulse, causes the body to move in a curvilineal path, with a variable motion, which path is the curve of a parabola; and the determination of the range, time of flight, angle of projection, and variable velocity, constitute what is usually meant by the doctrine of projectiles. See Robins's *Tracts*.

What is said above, however, is to be understood of projectiles moving in a non-resisting medium; for when the resistance of the air is also considered, which is enormously great, and which very much impedes the first projectile velocity, the path deviates greatly from the parabola, and the determination of the circumstances of its motion becomes one of the most complex and difficult problems in nature.

The ancient philosophers took the line described by a body projected horizontally, such as a bullet out of a cannon, while the force of the powder greatly exceeded the weight of the bullet, to be a right line. Nicholas Trataglia was the first who perceived the mistake, and maintained that the path of the bullet was a curved line through the whole of its extent. But it was Galileo who first demonstrated what particular curve it is that a projectile describes; shewing that the path of a bullet projected horizontally from an eminence, was a parabola, the vertex of which is the point where the bullet quits the cannon.

Philosophers seemed to be satisfied for a long time with this projection, that the curve is always a parabola, supposing the body moves in a non-resisting medium, deeming the effect of the air's resistance on the path as of no consequence. In process of time, however, they began to suspect that the resistance of the medium might have some effect upon the projectile curve, and they set themselves to consider this subject with some attention.

Huygens, supposing that the resistance of the air was proportional to the velocity of the moving body, concluded that the line described by it would be a kind of logarithmic curve.

But Newton, having clearly proved, that the resistance to the body is not proportional to the velocity itself, but to the square of it, shews, in his PRINCIPIA, that the line a projectile describes, approaches nearer to an hyperbola than a parabola.

Mr. Robins has shewn that, in some cases the resistance to a cannon-ball amounts to more than 20 times the weight of the ball; and Dr. Hutton, having prosecuted this subject far beyond any former example, has sometimes found this resistance amount to near a hundred times the weight of the ball; viz. when it moved with a velocity of 2000 feet per second, which is a rate of almost 23 miles in a minute.

Mr. Robins has not only detected the errors of the parabolic theory of gunnery, which takes no account of the resistance of the air, but shews how to compute the real range of resisted bodies.

There is an odd circumstance which often takes place in the motion of bodies projected with considerable force, which shews the great complication and difficulty of this subject; namely, that bullets in their flight are not only depressed beneath their original direction by the action of gravity, but are also frequently driven to the right or left of that direction by the action of some other force. Now if it was true that bullets varied their direction by the action of gravity only, then it ought to happen that the errors in their flight to the right or left of the mark they were aimed at, should increase in the proportion of the distance of the mark from the piece only. But this is contrary to all experience; the same piece which will carry its bullet within an inch of the intended mark at ten yards distance, cannot be relied on to 10 inches in 100 yards, much less to 30 in 300 yards.

And this inequality can only arise from the track of the bullet being incurvated sideways as well as downwards; for by this means the distance between the incurvated line and the line of direction, will increase in a much greater ratio than that of the distance; these lines coinciding at the mouth of the piece, and afterwards separating in the manner of a curve from its tangent, if the mouth of the piece is considered as the point of contact.

This is put beyond a doubt from the experiments made by Mr. Robins; who found also that the direction of the shot in the perpendicular line was not less uncertain, falling sometimes 200 yards short of what it did at other times, although there was no visible cause of difference in making the experiment. See the article RANGE.

PROJECTION, in mechanics, is the act of giving motion to a body, from thence called projectile.

If the direction of the force by which the projectile is put in motion is perpendicular to the horizon, the projection is said to be perpendicular; if parallel to the apparent horizon, it is said to be an horizontal projection; and if it makes an oblique angle with the horizon, the projection is oblique. In all cases the angle which the line of direction makes with the horizontal line, is called the angle of elevation of the projectile, or of depression when the line of direction points below the horizontal line.

PROJECTION, in perspective, is the appearance or representation of an object on the perspective plane.

PROJECTION *of the Sphere in Plano*, a representation of the several points or places of the surface of the sphere, and of the circles described upon it, upon a transparent plane placed between the eye and the sphere, or such as they appear to the eye placed at a given distance.

The projection of the sphere is commonly divided into Orthographic and Stereographic.

Orthographic PROJECTION, is that in which the surface of the sphere is drawn upon a plane, cutting it in the middle; the eye being placed at an infinite distance vertically to one of the hemispheres. This projection is useful in some parts of astronomy, particularly in determining the transits of planets over the sun's disc, and in the projection of the lunar and solar eclipses.

Stereographic PROJECTION *of the Sphere*, is that in which the surface and circles of the sphere are drawn upon the plane of a great circle, the eye being supposed in the pole of that circle. This projection not only affords the most natural representation of the prominence and convexity of the sphere, but is also performed with more ease and accuracy, and therefore justly preferred to any other both for its simplicity and correctness.

Common maps of the world and of countries are generally drawn upon the stereographic principles.

PROMONTORY, (*promontoire*, Fr.) in geography, is a rock or high point of land projecting out into the sea: the extremity of which towards the sea is commonly called a Cape or Head-land.

PROMOTION, in the royal navy, the advancement of officers to higher situations, which is done at the pleasure of the first lord commissioner of the Admiralty, according to the practice of the navy, either from the personal claims of those who have signalized themselves by acts of bravery, or from long services on foreign stations.

The gradations of promotion, among naval officers, are from midshipmen to master's mates, lieutenants, commanders, post-captains, rear-admirals, vice-admirals, and admirals.

The petty officers are promoted from mates to boatswains, gunners, and carpenters: and the seamen, generally, get rated by the captains, according to their behaviour and expertness in the performance of their duty, from landmen to ordinary and able-bodied seamen, from thence to captains of the forecastle, captains of the tops, captains of the after-guard, captains of the mast, and so on.

PROOF of ARTILLERY and SMALL-ARMS, implies a trial made upon them to ascertain whether they will stand the quantity of powder allotted for that purpose.

The Board of Ordnance allows bullets of lead of 11 in the pound for the proof of muskets, and of 29 in two pounds for service; 17 in the pound for the proof of carbines, and 20 for service; 28 in the pound for the proof of pistols, and 34 for service.

When guns of a new metal, or of lighter construction are proved, they are, besides the common proof, fired 150 or 200 times, as quick as possible, loaded with the common charge given in actual service. Cannon are proved to ascertain their being well cast, their having no cavities or defects in the metal.

In making this proof, the piece is laid upon the ground, supported only by a piece of wood in the middle, of about five or six inches thick, to raise the muzzle a little; and then the piece is fired against a solid butt of earth.

Every species of ordnance undergoes different kinds of proof before it is received into his Majesty's service; viz.

1st. They are gauged as to their several dimensions, internal and external, as to the accuracy of the position of the bore, the chamber, the vent, the trunnions, &c.

2d. They are fired with a regular charge of powder and shot, and afterwards *searched* to discover irregularities or holes produced by the firing.

3d. By means of engines, an endeavour is made to force water through them.

4th. They are examined internally, by means of light reflected from a mirror. See the articles CANNON, CHARGE, GUN, and GUNPOWDER.

PROTECTION, (*protection*, Fr.) is a privilege granted by the Admiralty to different descriptions of persons, from being impressed into his Majesty's sea-service. For the laws relating to the impress service, and the persons described in the Act, 2 Geo. III. c. 15, to be protected by the Admiralty, See the article *To* IMPRESS.

PROTEST, (*procès verbal, fait par le capitaine et l'équipage d'un vaisseau*, Fr.) an instrument, drawn up in writing, and attested before a justice of the peace (or consul or vice-consul in foreign parts), by the master of a merchant ship, and a part of the ship's crew, after the expiration of a voyage, describing the severity of the voyage, whereby the ship has suffered, or may suffer, in her hull, masts, rigging, or cargo. It is chiefly intended to shew, that such damages did not happen through any neglect or misconduct of the master or his officers, &c.

PROTRACTOR, is an instrument used in surveying, for laying down the angles of a field, or other figure, upon paper. The protractor is a small semi-circle of brass, or other solid metal, the limb or circumference of which is nicely divided into 180°: it serves not only to draw angles on paper, or any plane, but also to examine the extent of those already laid down. For this last purpose, let the small point in the centre of the protractor be placed above the angular point, and let the side coincide with one of the sides that contain the angle proposed; then the number of degrees cut off by the other side, computing on the protractor, will show the quantity of the angle that was to be measured.

An improved protractor for measuring angles on paper, or laying them down with great accuracy and to the smallest radius; serving also to perform any problem of trigonometry by construction, has lately been invented by Mr. Hart, and called by him a Goniometer, for which he took out a patent about 7 years since, but never made it public.

This instrument consists of two concentric rings of brass A and B, fig. 7. plate XVIII. whose surfaces are coincident; the inner ring turning round in a groove in the interior circumference of the outer ring: the latter is divided into half-degrees and the former carries an index with a vernier, by which the divisions are subdivided into minutes. To the inner ring is attached a brass bar D E, as a diameter, and at right angles thereto, the semi-diameter F C: the edge of the former is chamfered down to the paper, and may have on it a set of divisions consisting of any number of equal parts, figured both ways from the centre C, in order to serve as a scale for setting off the length of any line from that point; a tangent screw is attached to the outer ring by which the index of the vernier may be set to any given or required angle with great accuracy. This is all that is essential to the protracting of angles on the paper, or measuring them when done, and the method of doing it is evident; for if a right line be drawn from C, along the diameter D E, to any distance within the circle, and the inner ring be turned till the index has gone over the degrees and minutes of the intended angle, then another right line drawn from C to any point within the circle is the required angle, whose vertex is C, and the scale on C E or C D would serve to make the sides of the angle of any measure whatever. Fig. 7.

Along the diameter D E, and semi-diameter C F. there is a groove in which slides a dove-tailed cursor, carrying a bar with a pencil, &c. attached, and moveable at right angles to D E, by which on turning the inner ring of the instrument quite round, any circle may be drawn; or having two dove-tailed cursors sliding one in each of the grooves D E and C F, with a beam attached, the instrument is converted into a trammel for drawing ellipses of any size or form.

By means of these bars sliding in all directions, it is easy to perceive how triangles or any other geometrical figures may be drawn without the help of rules or compasses.

For the description and figure of a new Protractor for laying down and measuring angles upon paper, see BREWSTER's *Treatise on New Philosophical Instruments for* 1813.

PROVING OF POWDER. See the article GUNPOWDER.

To PROVISION, properly to victual, to furnish with provisions. This word is of very modern adoption, and evidently grows out of the French term *approvisioner*.

PROVISIONS, (*provisions*, Fr.) those articles of food and sustenance which are served out daily to each person on board his Majesty's ships, under the superintendance of the purser, and consist of such as are under-mentioned:

Table of the daily Proportion of Provisions allowed to every Person on board his Majesty's Ships.

	Sun.	Mo.	Tu.	We.	Th.	Fr.	Sat.	Tot. pr. Wk.
Bread, lbs.	1	1	1	1	1	1	1	7 lbs.
Beer galls.	1	1	1	1	1	1	1	7 galls.
Beef, lbs.			2				2	4 lbs.
Pork, lbs.	1				1			2 lbs.
Pease, pts.	½			½	½	½		1 qt.
Oatmeal, pt.		½		½		½		1½ pts.
Sugar, oz.		2		2		2		6 oz.
Butter, oz.		2		2		2		6 oz.
Cheese, oz.		4		4		4		12 oz.

together with an allowance of vinegar, not exceeding half a pint to each man per week. There is cocoa with sugar allowed for breakfast, as a substitute for butter or oatmeal, and lime-juice with sugar, as a beverage, when subsisting on salt provisions.

When the beer is expended, a pint of wine, or half a pint of rum, brandy, or other spirits is allowed daily, the spirits being always mixed with water, to each person, instead.

But no wine or spirits, are to be issued while in port, nor at sea until all the beer is expended: and no more is to be served out than the daily allowance; nor any for paying debts, or in lieu of other provisions, or on any other pretence whatever. Officers concerned in sending any allowance of wine *out of the ship* are liable to be dismissed the service.

Captains may shorten this allowance, if necessity requires it, taking due care that the men be paid for the deficiency; nor is any officer to have whole allowance while the company is at short.

In foreign voyages, some of the above species may be changed; that is, half a pint of brandy, rum, or arrack, for a gallon of beer. Four pounds of flour, sometimes three pounds, with a pound of raisins, or half a pound of currants, with half a pound of beef suet pickled, are equal to a four pound piece of beef, or a two pound piece of pork with pease: half a pound of rice, or two ounces of sugar, for a pint of oatmeal.

Prisoners of war, on board any of his Majesty's ships, have only two thirds of the aforesaid allowance of provisions.

Beef is cut into four pound pieces, and pork into two, and every cask is to have its contents marked on the head.

Every fourteen pieces of beef cut for eight pound pieces taken out as they rise, and the salt shaken off, are to weigh one hundred and twelve pounds avoirdupoise; and every twenty-eight pieces of pork cut for four pound pieces, one hundred and twelve pounds avoirdupoise.

By the regulations respecting provisions issued by the Victualling Office, May 18, 1797, if, according to this standard, upon the weighing of a whole cask of beef or pork in the presence of two or more warrant officers of the ship, there shall be found a deficiency of weight, the captain may order the purser to issue to the seamen so much more beef or pork as shall make up the deficiency.

If there be a want of pork, the captain may order three pounds of beef to be given out in lieu of two pounds of pork.

One day in every week there shall be issued out a proportion of flour and suet in lieu of beef, but this is not to extend beyond four months victualling at one time.

Only three months butter and cheese shall be supplied for foreign voyages, the remainder to be made up in olive oil.

Fresh meat is to be allowed every day (when it can be conveniently done), instead of salt meat; three pounds of mutton accounted for a four pound piece of beef, or a two pound piece of pork with pease.

PROVOST-MARTIAL, (*grand prévot*, Fr) an officer appointed by the Admiralty to take charge of prisoners at a court-martial, and to hold them in custody afterwards till the sentence passed by the court (if any) be carried into execution.

PROW, (*proue—berthelot, ou flêche des bâtimens Latins*, Fr.) a name given by seamen to the beak, or pointed cut-water of a galley, polacre, or xebec. The upper part of the prow, in those vessels, is usually furnished with a grating-platform for the convenience of the seamen who walk out to perform whatever is necessary about the sails or rigging in the bowsprit.

PUDDENING of a BOAT'S STEM, (*collier de défense d'un canot*, &c. Fr.) is a quantity of rope-yarn pointed and placed firmly on the stem of the boat as a kind of fender.

PUDDENING *the Ring of an Anchor*, (*embouding nure d'une ancre*. Fr.) is the act of well parcelling it with tarred canvas, and then warping it round with twice-laid-stuff, which is done thus:— A number of lengths are cut, each three times the diameter of the ring: they are laid on the ring, and stopped by a temporary seizing in the middle, as shewn at *a*, fig. 8. plate XVIII. They Fig. 8.
are laid by hand as far as *b*, when a turn or two of ratline stuff is taken round, and the heaver *b*, being put through it, it is hove well round, which stretches all the turns of the pudding or wreath, making them lie taught and even. A seizing is clapped on within the heaver, and snaked, and the heaver is then taken off.

Fig. 9. The parts are then laid and hove in the same manner to *c*, fig. 9, where another seizing is clapped on: the same operation is performed on the other side to *d* and *e*, when it will appear like the figure, the temporary seizing being taken off. The ends of the pudding are then opened out like those represented by *f*, and well tarred.

PUDDING, or PUDDENING *of a Mast*, (*bourrelet d'un mât*, Fr.) a thick wreath, or circle of cordage, tapering from the middle towards the ends, pointed all over, and fastened about the main-mast and fore-mast of a ship, directly between the trusses, to prevent their yards from falling down, when the ropes by which they are usually suspended are shot away in battle.

The pudding is made by splicing an eye at each end of a piece of rope, whose length is about twice the diameter of the mast, then firmly *serving* it over with spun-yarn throughout its whole length, increasing the turns from each end towards the middle, and *pointing* it over for neatness, as represented by
Fig. 10. fig. 10, plate XVIII. It is then fitted with a lanyard, which is spliced into one of the eyes, and fixed about the mast, by passing the lanyard alternately through both eyes on the fore-side of the mast. See the article DOLPHIN.

There are also puddings for the yards made in a similar manner.

To PULL, (*nager*, *ramer*, &c. Fr) implies to row with the oars, as "Pull away!" "Pull the starboard oars!" "Pull together!" &c.

PULLEY, one of the five mechanical powers, consisting of a small circular wheel, either of wood or metal, turning on an axis, and having a channel round its edge or circumference, in which a cord slides, and so raises up weights, &c.

Some pullies are put in blocks, which are fixed; and others in blocks which are moveable, and rise with the weight. The wheel is called the sheave or shiver; the axis on which it turns, the pin; and the fixed piece of wood or iron into which it is put, the block. See the article BLOCK.

PUMP, (*pompe*, Fr.) in hydraulics, a machine used to discharge the water from the ship's hold into the sea. There are several pumps used on board ships, and may be distinguished thus:

Chain-PUMP, (*pompe à chapelet*, Fr.) in a ship of war, is a long chain equipped with a sufficient number of valves at proper distances from each other, working upon a sproket-wheel, and passing downward through the back-case, then returning upward through the round chamber, by which means the water is pumped out of the ship's hold. It is managed by two winches, whereon several persons may be employed at once; and thus it discharges a greater quantity of water than any other pump hitherto invented, and is worked with less inconvenience to the seamen.

It appears to have been first applied to the pump by Mr. Mylne, to exhaust the water from the caissons at Blackfriar's Bridge. It has thence been transferred to the marine by Captain John Bentinck, of the royal navy, after having received some material improvements by Mr. William Cole, engine-maker, under his direction, to answer that service.

The chain of this machine is now more simple and mechanical, and less exposed to danger. It has been very recently improved by Mr. William Collins, engine, chain, and pump-maker, Bridge-road, Lambeth; the improvement consists in the application of brass metal-links, instead of cast-iron ones, as formerly used.

We have exhibited in plate XIX. fig. 1. a section Pl. XIX.
of the chain-pump (drawn to a scale ¼ of an inch to Fig. 1.
a foot), as fixed in a 74 gun ship, where

A A, represents the cistern.
B B, part of the pump-dale.
C C, the ends of the cistern continued down to the deck in lieu of chocks.
D D, the square chamber.
E, the gun-deck beam.
F F, the iron hoop to join the square and round chambers together.
G G, the working chamber, which is 8½ feet long.
H, the cast iron roller and pin, at the bottom of the pump.
I I, the back case.
K K, the poppets.
L L, the planks of the ship's bottom.
M, the kelson.
N N, the limber-streaks.
O, the hood of the cistern, which prevents any filth, &c. from going into the pump.
P P, the metal rhodings of the winch, for the spindle to work in.
Q Q, the spindles of the winch.
R R, the screw hoops.
S S, the sproket-wheel.
T T, the winches.
V, the length of the chain, containing three double, two single, and one saucer-link, with two saucers and two leathers.
U, the double link
W, the saucer-link.
X, the saucer.
Y, the leather.
Z, the main keel.

N.B. The figures represented by V, U, W, X, and Y, are drawn to half an inch to a foot.

It has been found upon a fair trial, that when the winches are well manned, a tun of water can be discharged in a minute by means of the chain-pump; and by thus working it in a ship's well, with water occasionally let in for the purpose, it may reasonably be expected to promote the general health of the men by cleanliness and ventilation; it would, moreover, insure the machine to perform well on every emergency, and render the carpenter expert in the management of it.

The iron-work, particularly the wheel and axis, should be paid over with tar varnish; or, if that is not to be had, with tallow. This treatment will secure it from speedy corrosion, and consequently preserve a sufficient strength for a longer time.

Directions for repairing the Chain of this Pump when a link is broken.

First, try if it can be hooked in the pump; if it cannot, turn the chain gently back until it will go no farther; then you may be sure the end of it is down at the bottom of the pump. Ease off the back-case, by taking out the iron pins, haul out the chain, and put it aside; apply a hook-rod with the hooks toward you, and hook the end of the chain at the bottom of the pump: when you have it fast, let the men above put back the chain gently, to enable you to haul it up to the wheel, where the broken link is to be taken out and a new one put in. Observe always to put the stops or hooks of the new link in the same direction as those on the chain you join them to, overhaul the pump, examine the forelocks, make all safe, put on the back-case, and the pump is ready to work.

It is strongly recommended to all persons who may have charge of a chain-pump, to practise the method of recovering the end of the chain until they are expert in the performance of that operation, and putting in a new link.

In case the chain draws to one side of the wheel in working, raise the rhoding on that side until you find it keeps at equal distance from the sides. To prove what water the pump loses in working, turn the winches with a velocity that will keep the water to the top of the pump, but without flowing over; observe how many revolutions per minute are required to do this, for so many are lost in that time.

The Common or *Hand* Pump, (*pompe à main*, Fr.) is a long wooden tube, whose lower end rests upon the ship's bottom, between the timbers, in an apartment called the *well*, inclosed for this purpose near the middle of the ship's length.

This pump is managed by means of the brake, and the two boxes or pistons. Near the middle of the tube, in the chamber of the pump, is fixed the lower-box, which is furnished with a staple, by which it may at any time be hooked and drawn out, in order to examine it. To the upper-box is fixed a long bar of iron, called the spear, hooked to an eye-bolt in the brake. Another sort of spear is sometimes used to this pump, having two holes in it. There is also an iron stanchion, with two ears, fixed to the pump, which receives the brake; and two bolts passing through both, serve to connect the whole together, by means of forelocks. At a small distance from the bolt, which passes through the spear, the brake is confined by another bolt between two cheeks or ears, fixed perpendicularly on the top of the pump. Thus the brake acts upon the spear as a lever, whose fulcrum is the bolt between the two cheeks, and discharges the water by means of the valves and clappers, fixed on the upper and lower boxes.

A copper basket, cut full of holes, is fitted on the bottom of the hand pump, to prevent any substance from passing that would choke the pump.

A Hand Pump has lately been invented and brought forward, by Captain George Truscott, of the royal navy, for the use of ships of war. It is commonly called *the patent pump*; and is composed of a metal barrel, near three feet long, with two boxes, and two bolts, upon which the spear is secured, and worked by means of two iron handles directly opposite to each other.

The most convenient as well as less expensive place for fixing this pump, has been found by experience, near the fore hatchway, on the orlop; although it has sometimes been fixed near the main-mast on that deck.

It is furnished with two lead pipes that have brass screws at the ends; one at the bottom of the barrel, with branches leading into the holds, which with a leather hose screwed on at the end, and extending by joints into the bung-hole of any cask or tank, serves to draw the water out of it, whilst the other pipe is screwed upon a projection from the barrel, and conveys the water up to an iron tank on the main-deck. At the extremity of this pipe there is fixed a stop-cock, which, when turned, serves to fill the tank; and at the end of the said cock there is a male screw, that receives a leather hose made to screw thereon, and which conducts the water into the iron boilers for the purpose of cooking the provisions.

This pump not only supersedes the necessity of hoisting up and striking down water-casks, which is certainly a dangerous operation when a ship has a violent motion in a gale of wind, and has often created accidents both to the crew and casks, but is found to be of the utmost utility in thus simply conducting the water to those parts where it is wanted for immediate use, and has already been supplied to many of our ships of war.

A Hand Pump, called also a *Stuffing-Box* Pump, has lately been invented for the use of his Majesty's ships.

It has for its object the saving of a cistern and an additional pump, and is not liable to admit the water into the well, as is often the case with the cistern: it, moreover, answers the purpose of pumping the water out of the well and washing the decks. A box called a *dumb-box* is put in to the pump below the lead pipe, which leads through the side of the ship into the pump; to the side thereof is fixed a stop-cock, which may always be kept turned without the trouble of any person attending it.

When the pump is wanted to pump ship the cock must be stopped and the boxes drawn out. This being done, put the boxes in again, which is all that is required to be done.

This stuffing-box pump has already been found very serviceable to several ships in the royal navy that have been fitted with it, and is coming into general use.

Head Pump, (*pompe de l'avant du vaisseau, ou à mettre hors du bord, pour puiser de l'eau*, Fr.) a moveable pump to put over the bows or side of a ship. These were formerly used in the royal navy, to pump water into the ship for washing the decks, &c. but since the invention of the stuffing-box pump it has not been used. See the article Cistern.

Pump-Brake, (*bringuebale*, Fr.) in the hand-pump, is the wooden lever or handle by which the hand-pump is worked.

Pump-Bolts, (*chevilles de pompe*, Fr.) belonging to the hand-pump, are two pieces of iron, each having a knob at one end, and a hole for a pin or forelock in the other; one serves to fasten the pump-spear to the brake, and the other as a fulcrum for the brake to work upon.

Pump-Cistern. See the article Cistern.

Pump-Dale, (*manche en bois d'une pompe*, Fr.) a piece of elm timber, about twelve inches square, extending from the cistern of the chain-pump across the ship; it has a hole bored through its centre, to convey water from the cistern through the side of the ship, and thereby preserves the decks from the effects of wet.

Pump-Gears, *(garniture de pompe*, Fr.) a term used to imply any materials requisite for fitting or repairing the pump, as boxes, leather, &c. See the article Jears.

Pump-Hood. See the article Hood *of a Pump*.

Pump-Hook, a bar of iron, with an eye at one end for the convenience of the hand, and a hook at the other to draw the lower-box out of the hand-pump.

Pump-Print, a description and drawing of the chain-pump, by the late Captain John Bentinck, of the royal navy, with directions for the management thereof, when an accident happens. It is supplied to all carpenters of his Majesty's ships of war.

Pump-Spear, (*verge de pompe*, Fr.) in the hand-pump, is a bar of iron, with a hook at the upper end to receive the eye in the brake. It communicates with the upper box, which is put in motion by the brake.

To Pump *Ship*, (*pomper un vaisseau*, Fr.) is to work at the pump by spells; that is, to relieve the men with fresh ones; when the ship springs a leak, the carpenter and his crew always examine the well, and attend the pumps whilst working.

The Pump *Sucks*, is said of the pump when all the water is drawn out, and there comes up nothing but froth and wind.

To make an Artificial Pump *at Sea*.—Take a spare topmast, or any other spare spar that will work near the size; saw it right down the middle; after which hollow it in both sides with gouges or chisels, taking care to form a chamber for the boxes of a pump to work and to be tight, then put it together with a few fine small drawing spike-nails, clinch it with fine oakum, and coat it with a composition of pitch and tar; after which (not having iron-hoops), bind it round with strong sinnet, which will make it complete for working, and to answer equally the same as when bored. An *artificial pump* on this construction has lately served every purpose during a voyage from Rio Janeiro to Liverpool.

PUNISHMENT, (*punition*, Fr.) among seamen and marines, is the pain inflicted on them either for the neglect of duty, or the commission of crimes against the established code of marine laws. Notwithstanding this has of late become a subject of public conversation and an object of parliamentary inquiry, yet it is found morally impossible to carry on our naval service without occasionally resorting to punishment; nor should this be considered strange when it is recollected that the great body of our seamen and marines are composed of the lowest classes of the community; and that depraved characters of almost every description often take refuge in the navy.

Several modes of punishment are practised on board men of war, but that in general use is the cat-o'-nine-tails, with which the offenders are flogged upon the bare back at the gang-way; and the number of stripes inflicted on them by the boatswain's mate, under the direction of the captain or commander, is always proportioned to the nature of the offence or magnitude of the crime. All the officers and crew in the ship are obliged to attend to see punishment inflicted, which is performed with much solemnity, in order that a public example may have a proper effect on the men.

Punishments, however, are now much less resorted to than formerly, and in many of our ships of war the duty is carried on without often having recourse to flogging, as our seamen are become much more enlightened, and seem to have a more proper sense of their moral and religious duty. Our naval officers are also become as much famed for honour and humanity as they are for courage and bravery.

The nature of crimes and punishments inflicted on offenders are regularly entered in the ship's log-book; and, by an order from the Admiralty, dated April 1811, a quarterly return is made to them of the number of seamen and marines flogged in each of his Majesty's ships, together with their crimes, age, time they were confined, and the quantity of punishment; still very properly leaving it in the power of the captain or commander to inflict punishment whenever he shall deem it absolutely necessary for the good of the service; it being generally allowed, that the fear of *immediate* punishment operates most powerfully to the prompt performance of duty and good order among the crew of a man of war.

PUNT, (*ras de carène, ou pont flattant*, Fr.) a sort of flat-bottomed boat, whose floor resembles the platform of a floating stage. It is used by the naval artificers, either in calking, breaming, or repairing the bottom of a ship. Small punts are also used in shallow rivers and lakes for the purpose of fishing and shooting. See the article Boat.

To PURCHASE, (*lever, à l'aide d'un pouvoir mécanique—acheter*, Fr.) is to raise or remove heavy bodies by means of mechanical powers, such as the tackles, windlasses, capstans, screws, and hand-spikes, which are frequently employed in fixing or extending the rigging of a ship.

PURSER, (*commis aux vivres*, Fr.) in the royal navy, an officer appointed by warrant from the lords commissioners of the Admiralty, to take charge of the provisions and victualling stores of a ship of war, to see that they are carefully distributed to the officers and crew; and upon urgent occasions to purchase and provide, and finally to account for the same, agreeably to the printed Naval Instructions which he has received for that purpose.

He, with two proper and competent persons as his securities for the due discharge of his trust, is required to enter into a bond to his Majesty in the penal sum established for the rate or class of the ship or vessel to which he may belong, as under-mentioned; viz.

1st rate, 1200*l.*; 2d rate, 1000*l.*
3d rate, 800*l.*; 4th and 5th rate, 600*l.*;
6th rate, and sloops, &c. 400*l.*

He is to ascertain from the captain the time for which the ship is ordered to be victualled, and to see the full quantity of provisions, water-casks, necessaries, &c. received on board; to take care that all the provisions be sweet, good, and wholesome, and that the casks appear to be sound, or otherwise to object against their being received into the ship without a survey.

For encouraging him to a zealous and faithful discharge of his duty, he shall, after passing his accounts in the Victualling Office, be paid, in addition to his own and servants wages, for all such quantities of the species of provision hereafter mentioned, as he may appear creditor for, by a bill payable in the general course, or according to the existing practice of the victualling, at the following rates, viz.

	s.	d.		
Bread	0	1¼	per lb.	Into which all such articles as may be supplied in lieu, will be reduced, according to their just proportions, on making up his accounts.
Beer	0	1¼	per gallon.	
Beef	1	8	per piece of 8 lbs.	
Pork	1	3	per piece of 4 lbs.	
Pease	4	0	per bushel.	
Oatmeal	0	6	per gallon.	
Sugar	0	4	per lb.	
Butter	0	6	per lb.	
Cheese	0	3	per lb.	

Also an allowance of ten pounds per centum on the amount of the sums paid to the ship's company for savings of provisions, subject to the regulations contained in the Naval Instructions on that head.

As he is charged with all victualling stores for which he gives receipts or vouchers, there will be allowed on his account for ordinary waste, as under.

Casks	1 Tun	For every 100 men of the complement victualled for a month,
Iron hoops	3	
Bags	4	

together with two bags in addition, for the like number of men, after the ship shall have been absent from a victualling port in England more than twelve months. But all deficiencies, after making such allowances, are to be paid for, in Channel service, at the following rates, viz.

	l.	s.	d.	
Casks	2	0	0	per ton.
Iron-hoops	0	1	0	each.
Bags	0	1	6	each.
Jars	0	10	0	each;

and upon foreign service, the highest prices that shall have been paid by the crown during the period of his account.

He is to provide the ship sufficiently at all times with coals, wood, turnery-ware, candles, lanterns, &c.; see this fully explained under the article NECESSARY *Money*.

The following annual allowance will also be made to him, upon passing his account, viz.

	£	s.	d.
For a complement of 800 men and upwards	25	0	0
From that to 700	22	10	0
—— Ditto to 600	20	0	0
—— Ditto to 550	19	0	0
—— Ditto to 450	16	10	0
—— Ditto to 350	14	0	0
—— Ditto to 50	13	0	0
—— Ditto to 40	12	0	0
—— Ditto to 30	11	0	0
—— Ditto to 25	10	0	0
—— Ditto to 20	9	0	0
—— Ditto to 15	8	0	0
—— Ditto to .... all under	6	0	0

together with a further allowance of sixpence a man per lunar month, for such number of supernumeraries as shall exceed the number of men short of the full complement.

The purser is frequently to look into the stowage, good order, and preservation of the provisions, and to issue out first such part thereof as he shall have perceived most liable to decay, or which shall have the marks of old provisions; informing the master from time to time, that they may be kept at hand.

If at any time whilst on foreign service there should be a necessity for him to receive sea provisions from a contractor, consul, correspondent, or other instrument of the Victualling Office, he is to present to his captain or commander a certificate under his hand of the then exact remains of each species of provision, if any on board: whereupon the captain or commander is to make a demand under his hand upon such contractor, consul, &c. to supply the provisions wanted; which demand, being forthwith entered in the log-book, and a certificate of its being so entered, placed at the foot by the master, are, with duplicates thereof, to be delivered to such contractor, consul, &c.; and after the provisions are received on board, he is, on separate paper from the demand, &c. to give the person applying receipts for the same, accompanied by a certificate from the captain or commander, a lieutenant, and the master, of the provisions having been received on board in kind; in both of which are to be stated the number and denomination of the packages containing them, as per form given in the Naval Instructions; taking care, that proper reference be made in the vouchers given for provisions received, to the date of the demand upon which they are supplied; and that the number of men, and time for which the provisions are intended to be and are completed to, be never omitted in the captain's demand and officer's certificate.

He is not to purchase any provisions in places where there is an agent, contractor, consul, correspondent, or other instrument of the Victualling Office for supplying the same, or when there is a probability of coming to such place, under pain of having the full costs charged against him in addition to the provisions; but in all cases of absolute necessity for provisions, where there is no agent, contractor, consul, &c. of the victualling, he is to certify under his hand, the remains of each species, if any on board, or that there is none of any particular species, and having obtained his captain's or commander's warrant, and the master's certificate as before mentioned, to procure the necessary quantities to complete the victualling generally, or any species particularly, as the case may require, he is to execute the service so committed to his management, with the utmost frugality, observing that the provisions he purchases are good, wholesome, and fit for his Majesty's service; and that condemnations of provisions purchased by him will not be allowed, unless satisfactorily accounted for. In order to pass his accounts for the same, he must produce the certificate of his remains, together with the aforesaid order and certificates of the provisions being received on board; a bill of parcels, expressing, in words at length, the quantity of each species, and (if a substitute) for what particular article, and the price thereof, in the weights or measures, and currency of the place; receipts for payment of the amount, witnessed by two commissioned or warrant officers: and a certificate from the governor or consul, if any, or else from two or three of the most eminent merchants of the place, testifying that the prices charged in the said bills were the current rates at the time when the said purchases were made; and stating the then currency of exchange, and the proportion between the aforesaid weights and measures, and those used in England. If purchases are to be made by him in consequence of orders from a superior officer to the captain, he is to observe the rules before laid down, and the captain's order to him is to be accompanied by a copy of that from his superior officer.

If it shall be necessary for him to draw bills of exchange upon the commissioners of the victualling, either for money taken up abroad to purchase provisions, or for the payment of provisions supplied by merchants, or for other services, the money is, if practicable, to be taken up by public advertisement: the captain or commander is to attest the bills, certifying that they are drawn upon account of the necessary victualling of the ship, and by his order: and the purser is to procure certificates from the governor, consul, or merchant, as before directed, of the currency of the exchange at that time, which he is to transmit home with the bills; and at the same time, or as soon after as possible, to send to the Victualling Office a distinct account, shewing how the said money has been laid out, together with a certificate from the captain or commander, a lieutenant, and the master, of the provisions charged therein to have been bought, having been received on board in kind.

Form for a Bill of Exchange given in payment for Provisions purchased by the Purser.

Exchange for 170*l.* 11*s.* 3*d.* *sterling.*

Palermo, August 10, 1812.

Gentlemen,

At thirty days sight of this my first bill of exchange (my second or third of the same tenor and date not being paid) please to pay to Mr. Anthony Weatherport, or his order, the sum of one hundred and seventy pounds, eleven shillings, and three pence sterling, value received, in bread bought of him, for the use of his Majesty's ship A———, per vouchers to be transmitted by,

Gentlemen,

Your most humble servant,

B. C. Purser.

To the Commissioners for victualling
his Majesty's Navy, London.

Attestation.

I hereby certify that this bill is drawn for the service above mentioned, and by my order,

D. E. Captain.

On the purser's return home, he is to deliver into the Victualling Office an account current of all his disbursements for the service of the ship, setting forth the quantity and price of every species of provision or victualling stores bought, and the particular charge attending the same, reduced into sterling money; at the foot of which account current he is previously to make oath, that the several sums therein mentioned were actually paid as against each article expressed, without profit, benefit, or advantage to himself, or any person on his account, or to any other person or persons whatsoever, except the venders or others properly entitled either by exchange, negociation, sale of bills drawn, or otherwise, that the several and respective quantities of provision mentioned therein, in foreign weights or measures, amounted in English weights and measures to the quantities wherewith he has charged himself, and no more; that they were all received on board in kind; and that he has not procured provisions of any sort at the same time and place on his own account. And when money has been taken up by public advertisement, he has also to deliver the original tenders in writing, and a copy of the publication attested by the captain.

Whenever the captain or commander shall be desirous of passing his accounts, or the captain or purser shall be removed from the ship, the purser is to give the captain a certificate agreeably to the form in the Naval Instructions, unless circumstances shall have occurred to render it improper, in which case the purser is to represent the same by letter to the commissioners of the victualling.

When the ship whereof he is purser, shall return into port, to be laid up, vessels or lighters will, upon due application being made to the Victualling Office, be immediately sent to bring away the remainder of the provisions and victualling stores. He is therefore to send with the said provisions or stores, his steward, or some other careful person, to see the safe delivery of them to the agent or other officer appointed to receive them; and at the same time he is to send to the said agent or officer a particular account, under his hand, of the provisions returned, specifying the times when, persons from whom, and places from whence they were received; and if under warrantry, when the same will expire; observing, that unless such an account be sent with the provisions, no receipt or credit will be given him for them.

He is to present his accounts and papers to his captain in due time, that he may have a fair opportunity of examining them minutely.

All books and papers necessary for passing his victualling accounts, are to be delivered into the proper offices by schedule under his hand, or that of his attorney or agent, at one and the same time, or notice be then given of such as are wanting, with reasons why they are so; without which, no vouchers tending to procure allowances of provisions, stores, or money, will be received or admitted after the general delivery which is to be within the space of four months after he shall be superseded, or the ship be paid off. And in case he shall have received any provisions or stores from contractors, he is to produce certificates from them or their agents of his having signed vouchers for the same. But if he shall be superseded, or the ship be paid off on a foreign station, he is to cause his accounts to be made up in the same space of time, and to send them home by the first safe opportunity, acquainting the commissioners of the victualling when and by what conveyance they were so forwarded.

Besides the affidavits and other proofs directed and required in different parts of the Naval Instructions, the purser is to vouch and substantiate all claims on any part of his accounts, by such further affidavits, or other methods, as circumstances may render necessary, the case admit, or the Victualling Board require.

If he has any cause of complaint against any of the officers of the ship, in relation to the disposition of the provisions or stores belonging to or under his charge, he is to represent the same to the Navy Board, before the payment of the ship, in order to their doing right between his Majesty, them, and him, before the said officers receive their pay.

When a ship shall have been employed on a foreign station more than six months, and the purser shall have an account of twelve months not closed, the captain or commander shall, upon application being made to him by the purser for that purpose, and whenever the service will admit of the provisions being got up, apply to the commander in chief, or the senior naval officer, to grant an order for three of the masters of the squadron, whenever so many ships can be collected, and for one of the lieutenants of the ship, to take a strict and careful survey or account of the number and contents of every cask of provisions which shall remain in the ship;—of the weight or measure of the hullage of any casks which may have been opened—of the actual weight of the bread, and of the number and description of the casks, iron-hoops, and biscuit-bags, which shall actually remain on board at the time of such survey; and to cause an account of the whole to be drawn out and signed by the surveyors in the manner described in the Naval Instructions. But in the event of there being no other ship on the station, the captain or commander is in such case to grant an order to one of the lieutenants, the master, and one of the master's mates, to proceed upon and take the survey in the manner above pointed out. The purser and his steward are to be present at, and to assist in, the taking of such survey; and are to make oath at the foot thereof, that the survey was actually taken in their presence by the officers who have signed the same, and that it contains a just, correct, and true statement of the provisions and victualling stores which really and *bona fide* remained on board the day on which the said survey was taken.

The purser is charged with the clothes and effects of persons who die on board; and he sells the same before the mast to the rest of the crew, and is obliged to give a satisfactory account of the same to the executors or administrators of the deceased. It is his duty also to keep a regular slop-book, wherein he is to make distinct and separate entries of the particulars and value of the slop-clothes, bedding, dead-men's clothes, &c. issued or supplied to the ship's company. For farther particulars, see the articles Provisions, Slops, and Tobacco.

He is to receive bibles and prayer-books from the naval store-keepers, and distribute them to the ship's company, under the direction of the chaplain, as these books are now, by a late order from the Admiralty, to be considered as ship's stores, and to be accounted for accordingly.

He is also to keep a Sick-Book for those sent out of the ship: also a Number and Content-Book, wherein, as far as possible, every separate package of provisions, &c. the marks, numbers, and contents thereof; time when, from whence, and from whom received, is to be correctly entered. See that part of the Naval Instructions relating to the Instructions for Pursers.

A regulation has been lately made, that all candidates for the office of purser in the royal navy, shall, previous to their appointment, be examined as to their qualifications. The first examination took place at Portsmouth, on the 11th of September, 1813, before the senior captain and the three senior pursers at that port.

Purser's *Steward*. See the article Steward.

PYROTECHNY, the science which teaches the management and application of fire in several operations; but in a more limited sense, it refers to the composition, structure, and use of artificial fire-works. The ingredients are, 1. saltpetre, purified for the purpose: 2. sulphur, and 3. charcoal. Gunpowder is likewise used in the composition of fire-works, being first ground, or as it is technically termed, *mealed*. Camphor and gum-benzoin are employed as ingredients in odoriferous fire-works. The proportions of the materials differ very much in different fire-works, and the utmost care and precaution are necessary in the working them to a state fit for use, and then in the mixing. Our limits will not allow us to enter on the subject with a sufficient degree of minuteness to teach the method of making of fire-works, and shall therefore only give a brief account of the proportions of the materials in some of the more common and more interesting articles in use.

The charges for sky-rockets are made of saltpetre, four pounds; brimstone, one pound; and charcoal, one pound and a half; by another direction, saltpetre, four pounds; brimstone, one pound and a half; charcoal, twelve ounces; and mealed-powder, two ounces. These proportions vary again according to the size of the rocket; in rockets of four ounces, mealed-powder, salt-petre, and charcoal, are used in the proportions of 10:2 and 1; but in very large rockets the proportions are saltpetre, four; mealed-powder and sulphur one each. When stars are wanted, camphor, alcohol, antimony, and other ingredients are required, according as the stars are to be blue, white, &c. In some cases gold and silver rain is required; then the brass-dust, steel-dust, saw-dust, &c. enter into the composition; hence the varieties may be almost indefinite.

With respect to colour, sulphur gives a blue, camphor a white or pale colour, saltpetre a clear white yellow, sal-ammoniac a green, antimony a reddish, rosin a copper colour. These materials require preparation before they are fit for use; and before a person can be qualified for the business of fire-work making, he must understand the method of making the moulds, cases, &c. and be acquainted with the instruments used in the art, their dimensions, and materials.

To discuss the several topics connected with the business, would require a space much larger than could be afforded in this work, we shall therefore refer the reader to distinct treatises on the subject: also to the articles Composition, Fire-Works, and Rockets, each in alphabetical order.

Q.

QUADRANGLE, in geometry, the same with a quadrilateral figure, or one consisting of a square having four right angles. To the class of quadrangles belongs the square, parallelogram, trapezium, rhombus, and rhomboides.—A square is a regular quadrangle; a trapezium, an irregular one.

QUADRANT, (*octant, ou quart de cercle*, Fr.) in geometry, is either the quarter or 4th part of a circle, or the 4th part of its circumference, the arc of which therefore contains 90°.

QUADRANT, a mathematical instrument of great use in astronomy and navigation, for taking the altitudes of the sun and stars; as also, for taking angles in surveying heights and distances, &c.; and is variously contrived, and furnished with different apparatus, according to the uses it is intended for.

Astronomical QUADRANT, is one usually made of brass or iron bars, having its limb nicely divided, either diagonally or otherwise, into degrees, minutes, and seconds, and is in the form of a quarter of a circle, and furnished either with two pair of plain sights or two telescopes, one on the side of the quadrant, and the other, moveable about the centre, by means of a screw. The dented wheels serve to direct the instrument to any object or phenomenon.

The application of this useful instrument, in taking observations of the sun, planets, and fixed stars, is obvious; for being turned horizontally upon its axis, by means of a telescope, till the object is seen through the moveable telescope, then the degrees, &c. cut by the index, give the altitude, &c. required. When it is fixed to the side of a wall, it is called a *mural quadrant*. See the article ASTRONOMICAL *Quadrant*.

Davis's QUADRANT, the same as BACKSTAFF, which article see.

Gunner's QUADRANT, sometimes called the *Gunner's Square*, is an instrument made of brass or wood, divided into degrees, and each degree into ten parts, to lay guns or mortars to any degree of elevation.

Fig. 11. The common sort is represented by fig. 11, plate XVIII., and consists of two branches, between which is a quadrantal arc, divided into 90°, and furnished with a thread and plummet, which point out the exact angle made by the chase, or bore.

The longer branch or bar being put into the mouth of the piece, so as to lay in contact evenly with its lower side, and the quadrant part being turned down, immediately beyond the muzzle, the pendulum wire or thread, which is fixed to a small pivot in the right angle exactly at the centre, whence the quadrant was described, will be kept perpendicular by the weight attached thereto; and will thus indicate the exact elevation of the piece. The point of oscillation, that is, the pivot, must, however, be always kept very smooth; that there may not be the least roughness; else the action will be affected, and the index prove erroneous. Sometimes, on the sides of the longer bar, are noted the divisions of diameters and weights of iron balls, as also the bores of pieces. See also the article LEVEL.

Gunter's QUADRANT, so called from its inventor, Edmund Gunter (fig. 12, plate XVIII), besides the apparatus of other quadrants, has a stereographic projection of the sphere on the plane of the equinoctial; and also a calendar of the months next to the divisions of the limb; by which, besides the common purposes of other quadrants, several useful questions in astronomy, &c. are easily resolved. For the use of this quadrant, see GREGORIE's *Dictionary of Arts and Sciences*. Fig. 12.

Hadley's QUADRANT, so called from its inventor, John Hadley, Esq. who first laid the account of this admirable instrument before the Royal Society, May the 13th, 1731.

The form of this instrument (fig. 13, plate XVIII.), according to the present mode of construction, is an octagonal sector of a circle, and, therefore, the arc contains 45°; but because of the double reflection, the limb is divided into 90°; wherefore it is usually called a quadrant. The following parts are those which require the particular attention of the observer. Fig. 13.

1. An octant, consisting of an arc or limb, connected with two radii.
2. An index, moveable round a centre, and accompanied with a dividing scale, to shew the observed altitude, or angular distance.
3. A speculum, or index-glass, placed on the index in such a manner, that its plane is over the centre of motion of the index.
4. Two horizon-glasses, with their adjusters; being the fore-horizon-glass, and the back-horizon-glass.
5. A set of coloured glasses, to prevent the effect of the solar rays on the eye of the observer: and
6. Two sight-vanes.

The *Octant* consists of two radii, A B, A C, a limb, and two braces L M, which last are intended to strengthen the instrument, and prevent it from warping. The arc or limb B C, although only an eighth part of a circle, is nevertheless divided into 90°, on account of the double reflection. These degrees are numbered from right to left, and each is commonly divided into three equal parts. Hence one of these intervals contains twenty minutes; and by means of a scale at the end of the index, divided into twenty equal parts, an observation may easily be read off to the nearest minute. The graduation on the limb continued a few degrees to the right of *o*, is usually called the *arc of excess*, and is found very convenient for several purposes.

The *Index* D, is a flat bar, commonly made of brass, that turns on the centre of the octant, and broader towards the axis of motion; at the other end of the index, a piece of brass turns up behind the limb, having a spring to make the divided scale lie close to the limb, and a screw to fasten it in any position. Some quadrants have an adjusting-screw affixed to the lower part of the index, by which means the index is made to move gently along the limb, without those sudden startings to which it is liable when moved by hand, and which may render the observation uncertain to two or three minutes. It may be observed, that when the index is moved by hand, it should be taken hold of by the lower part, and not by the middle, as is sometimes the practice. That part of the index which moves along the limb, has a small scale attached to one side of a rectangular aperture, called a dividing scale, but most commonly a *Vernier*, or *Nonius*, from the names of the supposed inventors. It usually contains a space equal to 21 or 19 divisions of the limb, and is divided into 20 equal parts; hence, the difference between a division on the limb, and a division on the dividing scale, is one-twentieth of a division on the limb, or one minute, and therefore if any division on the dividing scale is in the same straight line with a division on the limb, then no other division on the dividing scale can coincide with a division on the limb, the extreme divisions excepted.

The old practice of reckoning the divisions on the Vernier from its middle towards the right, and from the left towards the middle, being found inconvenient, a much more commodious method has been lately introduced, wherein the divisions are numbered from right to left.

The *Index Glass*, or *Speculum* E, is a rectangular piece of plane glass, one of whose planes is for the most part completely quicksilvered. The intention of this glass is to receive the image of the sun, or any other observed object, S, and reflect it upon either of the two horizon-glasses F and G, according to the nature of the observation.

This glass is set in a brass frame, and is placed perpendicular to the plane of the instrument, immediately over the centre of motion of the index, by means of two screws in a small plate of brass, connected with the under part of the brass frame, and projecting at right angles from the back of that frame; one of these screws C, takes into, and the other presses against the index.

The index-glass, according to the common method of setting it in its frame, is often liable to be bent; and hence the difference between the true and observed altitudes might be considerable. This source of error, however, is now removed by Mr. Dolland, who contrived the frame so, that the glass lies on three points, which, by means of a spring and screw, press the glass against other three points immediately opposite to the former; and since the position of a plane can be adapted to any three points; therefore, the plane of the glass cannot be bent, though the pressure be considerable.

As every silvered glass, although its planes be perfectly parallel to each other, gives two reflections, one from the fore surface of the glass, being the faintest, and the other from the silvered surface, which is the brightest, it is evident from these double reflections that errors may arise; to prevent which, the late Dr. Maskelyne proposed, that the back of the upper half of the index-glass, or that farthest from the plane of the instrument, instead of being silvered as usual, should be ground rough, and painted black, the line joining the silvered and blacked parts being parallel to the plane of the instrument. Hence, when the object to be observed is bright, its reflection is to be taken from this half of the glass, the reflection from the silvered part being prevented by a screen set before it; but if the object is not sufficiently bright, the screen may be turned down, and its reflection taken from the silvered part as usual.

On the radius A B of the octant, are two small speculums F G. The surface of the upper one is parallel to the index-glass, when the counting division of the index is at *o* on the arc; but the surface of the lower one is perpendicular to the index-glass, when the index is at *o* degrees on the arc; these mirrors receive the reflected rays from the object, and transmit them to the observer.

The *Horizon-Glasses*, one for the fore, and the other for the back observations, are not entirely quicksilvered; the upper one F, is only silvered on its lower part, or that half next the quadrant, the other part being transparent; and the back part of the frame is cut away, that nothing may impede the sight through the unsilvered part of the glass. The edge of the foil of this glass is nearly parallel to the plane of the instrument, and ought to be very sharp, and without a flaw. The lower horizon-glass G, is silvered at both ends, and in the middle there is a transparent slit, through which the direct object is to be viewed. Each of these glasses is set in a brass frame, having an axis passing through the woodwork, and is fitted to a lever on the under side of the quadrant; by this lever the glass may be turned a few degrees on its axis, in order to set it parallel to the index-glass. The lever has a contrivance to turn it slowly, and a button to fix it. To set the glasses perpendicular to the plane of the quadrant, there are two sunk screws, one before and one behind each glass; these screws pass through the plate on which the frame is fixed, into another plate; so that by loosening one and tightening the other of these screws, the direction of the frame with its mirror may be altered, and thus be set perpendicular to the plane of the instrument.

There are three *Coloured Glasses*, two of which are tinged red or dark, and the other green K, by which the eye of the observer is preserved from the effects of the solar rays; and the glare of the moon, in night observations, is taken away. Each of these glasses is set in a separate cell, and so connected, that either one or more may be used at the time of observation. The green glass may be used alone, if the sun is very faint; it is also used for taking the altitude of the moon, and in ascertaining her distance from a fixed star.

When these glasses are used for the fore observation, they are fixed at K; when used for the back observation, they are removed to N.

The *Sight Vanes* H and I, are perforated pieces of brass, standing perpendicular to the plane of the quadrant. That which is at H is used for the back, the other for the fore, observation.

There are two holes in the fore-sight vane, the lower of which, and the upper edge of the silvered part of the fore-horizon-glass, are equidistant from the plane of the quadrant, and the other is opposite to the middle of the transparent part of that glass. The back-sight vane has only one perforation, which is exactly opposite to the middle of the transparent slit in the horizon-glass to which it belongs.

Adjustment of Hadley's QUADRANT. Previous to observation, the quadrant should be carefully adjusted, otherwise it will be impossible to deduce accurate conclusions from the observations. For this purpose, the mirrors must be set perpendicular to the plane of the quadrant, the fore horizon-glass must be set parallel to the speculum, and the planes of the speculum and back horizon-glass produced, must be perpendicular to each other when the index is at *o*.

Various other methods have been proposed by Messrs. Dolland, Blair, and Gilkerson, for effecting the adjustment of the back horizon-glass, some of which are as follow:

In Mr. Dolland's method, an index is applied to the back horizon-glass, by which it may be moved, so as to be parallel to the index-glass, when *o* on the index scale coincides with *o* on the limb. When this is effected, the index of the back horizon-glass is to be moved exactly 90° from its former position, which is known by means of an arc divided for that purpose; and then, the plane of the back horizon-glass will be perpendicular to the plane of the index-glass produced.

In Mr. Blair's method, the under edge of the index-glass is ground and polished, so as to be at right angles to the plane of that glass; and, hence, the back horizon-glass is adjusted, by making the direct and reflected horizons agree, the index being at *o*, exactly in the same manner as the fore horizon-glass is rendered parallel to the speculum.

Mr. Gilkerson, Postern-row, Tower-hill, London, has a new method of adjusting the index-glass, by means of a screw at the upper end of that glass, for which a patent was obtained.

The quadrant is liable to be put out of order by many accidents; for although it generally comes perfectly adjusted from the maker's hand, yet, the heat of the sun, and the spray of the sea, causing it to warp, or expanding and contracting the frame, and sometimes a fall, will materially alter it; it is, therefore, highly necessary that the mariner should be able to verify and re-adjust the several parts of his instrument at sea, whenever any of these circumstances take place.

The different modes of adjusting the quadrant are five; three of these being once made, are not so liable to alteration as the other two, which last should be always examined before every observation: they are as follow:

1. *To set the Index-Glass perpendicular to the Plane of the Instrument.*

Set the index to about 45°, and hold the quadrant so, that its plane may be nearly parallel to the horizon, the limb being from the observer; then look into the index-glass, and observe if the portion of the limb seen by reflection appears in the same plane with that seen directly; if they do, the speculum is perpendicular to the plane of the quadrant; if otherwise, the error is to be taken away, by turning the screws behind the speculum in contrary directions, and both screws must be left equally tight.

2. *To set the fore Horizon-Glass perpendicular to the Plane of the Quadrant.*

Set the index to *o*; hold the plane of the quadrant parallel to the horizon, and apply the eye to the lower hole in the sight-vane; then direct the sight to the horizon, or to any other well defined object; and if the horizons seen through the transparent and in the silvered part of the glass, are apparently in the same straight line, the fore horizon-glass is perpendicular to the plane of the instrument; if not, one of the horizons will appear higher than the other: now, if the horizon seen by reflection is higher than the same seen directly, unscrew the nearest screw in the pedestal of the glass, and screw up that on the farther side of the glass, till the direct and reflected horizons appear to be in the same straight line. But if the reflected is lower than that seen directly, unscrew the farthest, and screw up the nearest screw, till the coincidence of the horizons is perfect; observing to leave both screws equally tight; and the fore horizon-glass will be perpendicular to the plane of the quadrant.

3. *To set the fore Horizon-Glass parallel to the Index-Glass, the Index being at Zero, or the point from which the divisions on the limb are numbered.*

Hold the plane of the quadrant in a vertical position, with the arc lowermost, the index being previously set exactly to *o*: now the eye being applied to the hole in the vane next the quadrant, and the sight directed to a well defined part of the horizon, or any well defined distant object, then, if the horizon seen in the silvered part of the glass coincides with that seen through the transparent part, the horizon-glass is adjusted; but if the horizons do not coincide, unscrew the middle screw in the middle of the lever on the other side of the quadrant, and turn the nut at the end of the lever, till both horizons coincide: then fix the lever in this position, by tightening the milled screw.

If this adjustment be again examined, it will perhaps be found imperfect; in this case, it remains either to repeat the adjustment, or rather, to find the error of adjustment, or, as it is usually called, the *index error*, which may be done thus:

Let the sight be directed through the vane to the horizon, move the index till the reflected horizon coincides with that seen directly, and the difference between *o* on the limb, and *o* on the Vernier, is the index error; which is additive when the beginning of the Vernier is to the right of *o* on the limb, otherwise subtractive.

4. *To set the back Horizon-Glass perpendicular to the Plane of the Instrument.*

Put the index to *o*, hold the plane of the quadrant parallel to the horizon, and direct the sight towards the horizon, through the back sight-vane. Now, if the horizon reflected from the silvered part of the glass is in the same straight line with that seen through the transparent part, the glass is perpendicular to the plane of the instrument.

If the horizons do not unite, turn the screws in the pedestal of the glass, till both appear to make one continued straight line.

5. *To set the back Horizon-Glass perpendicular to the Plane of the Index-Glass, the Index being at Zero.*

Let the index be put as much to the right of *o* as twice the dip of the horizon amounts to; hold the quadrant in a vertical position, and apply the eye to the back horizon vane; then, if the reflected horizon coincides with that seen directly, the glass is adjusted; otherwise, the screw, in the middle of the lever, on the other side of the quadrant, must be unturned, and the nut at its extremity turned, till both horizons coincide. It may be observed, that the reflected horizon will be inverted, that is, the sea will be apparently uppermost, and the sky lowermost.

The Use of Hadley's Quadrant in observing Altitudes at Sea.

The altitude of any object is determined by the position of the index on the limb, when by reflection that object appears to be in the horizon. In observing altitudes, it will be found most convenient to hold the quadrant with the left hand, and move the index with the right: hence, the left hand is to be applied to the farthest radius, towards the end of the curved brace, and the right hand to the lower part of the index.

There are two different methods of taking observations with Hadley's quadrant. In the first of these the face of the observer is directed towards that part of the horizon immediately under the sun, and is, therefore, called the *Fore Observation.* In the other method, the observer's back is to the sun, and it is hence called the *Back Observation.* This last method of observation is to be used only when the horizon under the sun is obscured, or rendered indistinct by fog, or any other impediment.

In observing the altitude of any object, the quadrant must be held perpendicular to the horizon, and so that its plane produced would pass through that object. It frequently happens, that when the object is nearly in contact with the horizon, the observer, by endeavouring to make this contact better, often pushes the index too far, and thereby renders the contact worse than before. In order to remove this inconvenience, the best quadrants are provided with a screw at the lower end of the index, called an *adjusting*, or *tangent screw;* by which a slow motion may be given to the index; and, consequently, those sudden starts, to which it is liable when removed by hand, prevented.

A small magnifying glass will be found of great service in reading off the observed angles; in doing which, the limb may be held towards the observer, and the eye should be in a plane perpendicular to the quadrant, passing through the coinciding divisions.

To take the Sun's Altitude by the Fore Observation.

Turn down either of the coloured glasses, before the horizon-glass, according to the brightness of the sun; and if the sun is bright enough to be seen in the transparent part of the glass, let the eye be placed at the upper hole in the sight-vane, otherwise at the hole next the plane of the quadrant; direct the sight to that part of the horizon which is under the sun, and move the index till the coloured image of the sun appears in the horizon-glass; then give the quadrant a slow vibratory motion about the axis of vision; move the index till the sun's lower or upper limb, at the lowest part of the arc, described by this motion, is in contact with the horizon, and the degrees and minutes, shown by the index on the limb of the quadrant, will be the observed altitude of the sun's upper or lower limb.

To take the Altitude of the Moon by the Fore Observation.

Put the index to *o*, turn down the green-tinged glass, place the eye at the lower hole in the sight-vane, and direct the sight to the moon; which being found in the silvered part of the horizon-glass, move the index gradually, and follow the moon's reflected image, till the enlightened limb is in contact with the horizon, at the lower part of the arc described by the vibratory motion, as in the last article; and the index will shew the observed altitude of that limb of the moon which was brought in contact with the horizon. If the observation is made in the day-time, the coloured glass is unnecessary.

To take the Altitude of a Star or Planet by the Fore Observation.

Put the index to *o*, direct the sight to the star, through the lower hole in the sight-vane and transparent part of the horizon-glass; move the plane of the quadrant a very little to the left, and the image of the star will be seen in the silvered part of the glass; then move the index, and the image of the star will descend; continue this motion of the index till the star is in contact with the horizon, at the lowest part of the arc described; and the degrees and minutes shown by the index on the limb, will be the observed altitude of the star.

When that part of the horizon, under the object whose altitude is intended to be observed, is obscured or rendered indistinct, by a fog or any other impediment, the back observation becomes necessary; we would, however, advise the observer, before he depends entirely on the altitude taken by the back observation, to accustom himself to this mode of observing, and to ascertain the accuracy to which he is capable of attaining, by comparing altitudes taken by this method, with those of the same object, taken at the same time by the fore observation, allowance being previously made for the dip and index error, when required.

To take the Sun's Altitude by the Back Observation.

Put the stem of the coloured glasses into the place made to receive it between the horizon-glasses; turn down either, according to the brightness of the sun, and hold the instrument vertically; then direct the sight through the hole in the back sight-vane, and transparent slit in the horizon-glass, to that part of the horizon which is opposite to the sun; move the index till the sun is in the silvered part of the glass, and by giving the quadrant a vibratory motion, whose axis is that of vision, the image of the sun will describe an arc whose convex side is upwards; now bring the limb of the sun, when in the upper part of the arc, in contact with the horizon; and the degrees and minutes shewn by the index on the limb will be the altitude of the other limb of the sun.

To take the Altitude of the Moon, &c.

The altitude of the moon is observed in the same manner as that of the sun, with this difference only, that the use of the coloured glass is unnecessary, unless the moon is very bright; and that the enlightened limb, whether it be the upper or lower, is to be brought into contact with the horizon. In the day time, the moon's altitude may be observed as directed in the next article.

To take the Altitude of a Star, or Planet, &c.

Look directly to the star, through the vane and transparent slit in the horizon-glass; move the index till the opposite horizon with respect to the star is seen in the silvered part of the glass, and make the contact perfect as formerly. If the altitude of the star is known nearly, the index may be set to that altitude, the sight directed to the opposite horizon, and the observation made as before.

Unless observations are made accurately, the conclusions to be deduced therefrom cannot be depended on; and since a single altitude is not to be so much relied on, as the mean of several, taken at short intervals of time; therefore, four or five altitudes should be taken at nearly equal intervals of time, and their sum, divided by their number, will be the mean altitude which the object would probably have had at the mean interval, or the instant arising from the division of the sum of the times of observation by their number. If the intervals between the successive observations are short, and nearly equal the corresponding differences of altitude will also be either nearly equal, or gradually increasing or decreasing. Hence, an erroneous altitude may easily be discovered; thus, if the altitudes of an object were 19° 58′, 20° 4′, 20° 11′, and 20° 15′, and the corresponding times of observation, 9h. 15′ 1″, 9h. 15′ 45″, 9h. 16′ 36″, and 9h. 17′ 30″, respectively, it may hence be presumed, that the three first altitudes are tolerably accurate, but that the last should have been about 20° 18′, and, therefore, such altitude should be absolutely rejected.

This method of deducing the mean altitude of the same limb of the sun is, however, attended with an error arising from the irradiation, or apparent spreading of its rays beyond the true limb; and, in order to diminish, if not wholly to destroy it, the mean of an equal number of altitudes of each limb, observed at nearly equal intervals of time, should be taken. As this error amounts only to a few seconds, it, therefore, becomes insensible in observations taken at sea. See MACKAY, *on the Longitude*.

As the DIP *of the Horizon*, PARALLAX, and REFRACTION, are necessary corrections to be made in taking the sun's and moon's altitude, see these articles under their respective heads; also the REQUISITE TABLES.

Reflecting QUADRANT, an instrument by Mr. Ezekiel Walker, to measure an angle by the *fore* observation, without very oblique reflections from the mirrors, greater than can be taken by the sextant; and also with greater accuracy when the angle is large, and causes the rays of light to fall so obliquely upon the sextant's index-glass as to occasion some doubt respecting the truth of the observation. For the description and principles of this instrument, see NICHOLSON'S *Philosophical Journal* for 1802.

Besides those already described, there are other quadrants invented by Adams, Cole, Collins, or Sutton, and also the Horoditical, the Sinical, and Surveying quadrants.

QUADRANT *of Altitude*, is an appendix to the artificial globe, consisting of a thin slip of brass, the length of a quarter part of one of the great circles of the globe, and graduated. At the end, where the division terminates, is a nut riveted on, and furnished with a screw, by means of which the instrument is fitted on the meridian, and moveable round upon the rivet to all points of the horizon. Its use is to serve as a scale in measuring of altitudes and distances; and it also shews the amplitudes and azimuths in performing problems on the globes.

QUADRANTAL TRIANGLE, is a spherical triangle, one of whose sides is 90°, or the quarter part of a circle.

To QUADRATE A GUN, is to see it placed truly on its carriage, and that the wheels be of an equal height.

QUADRATURE, in astronomy, that aspect of the moon when she is 90° degrees distant from the sun; or when she is in a middle point of her orbit, between the points of Conjunction and Opposition; that is, in the first and third quarters.

QUADRATURE, in geometry, denotes the squaring or reducing a figure to a square. Thus, the quadrature of a circle, ellipse, &c. is the finding a square which shall contain just as much surface, or area, as those figures.

QUADRILATERAL, or QUADRILATERAL *Figure*, in geometry, is a figure whose perimeter consists of four right lines, making four angles. The quadrilateral figures are either a parallelogram, trapesium, rectangle, square, rhombus, or rhomboides.

QUANTITY, denotes any thing capable of estimation or measurement; or which, being compared with another thing of the same kind, may be said to be greater or less, equal or unequal to it; and is increased by addition, and diminished by subtraction; which are, therefore, two primary operations that relate to quantity.

A quantity that is to be added, is called a positive

quantity; and a quantity to be subtracted, is said to be negative. A quantity consisting of more than one term is called a compound quantity; whereas, that consisting of one term only is denominated a simple quantity.

The quantity of matter in any body, is the product of its density into its bulk; or, a quantity arising from the joint consideration of its magnitude and density; as, if a body be twice as dense, and take up twice as much space as another, it will be four times as great. This quantity of matter is best discoverable by the absolute weight of bodies.

The quantity of motion in any body is the factum of the velocity into the mass, or it is a measure arising from the joint consideration of the quantity of matter, and the velocity of the motion of the body; the motion of any whole being the sum or aggregate of the motion in all its several parts. Hence, in a body twice as great as another, moved with an equal velocity, the quantity of motion is double; if the velocity be double also, the quantity of motion will be quadruple. Hence, the quantity of motion is the same with what we call the momentum or impetus of a moving body.

QUARANTINE, (*quarantaine*, Fr.) in maritime affairs, is a period of time in which all ships going to or coming from foreign stations are compelled to lay at anchor in certain places established by the different governments for that purpose, and without having intercourse with the shore, except for necessary provisions, which are conveyed with the utmost precaution.

Quarantine takes its name from a term of forty days, but is, generally, long or short, according to the places where ships come from and have touched at, the nature of the bills of health they exhibit, or the general state of the health of the crews, &c. during their voyage.

Masters of merchant-ships coming to English ports have liberty to land their goods, crews, and passengers, while they are under quarantine, at places called *lazarets*, which are instituted for that purpose.

At almost every sea-port, both at home and abroad, there is a person denominated *pratique-master*, who is appointed by the governments to relieve ships from quarantine, immediately he receives commissions so to do.

The prevalence of many infectious disorders of late years, in foreign countries, and the consequent increase of the danger of their importation in articles of merchandise, have drawn the attention of the legislature to means of prevention.

To this end it appeared necessary to establish a more extensive system of quarantine; and, accordingly, new regulations, more connected, extensive, and more effectual, are instituted by 45 Geo. III. c. 10, which act

1. Empowers the lords of the treasury to erect a lazaret upon Chetney-hill, in Kent, &c.

2. To erect floating lazarets, until those on shore shall be completed.

3. That from the 5th of April 1805, the owners of ships, *which shall have performed quarantine*, are to pay the following duties of customs; viz.

Having arrived from any part of Turkey, or from any place in Africa, within the Straits of Gibraltar, or in the West Barbary, on the Atlantic Ocean, *with* a clean bill of health, 7*s.* 6*d.* per ton burden. Having arrived from any place whatever (except from any part of Turkey, or from any place in Africa, within the Straits of Gibraltar, or in the West Barbary, on the Atlantic Ocean), *with* a clean bill of health, 3*s.* per ton burden, and having so arrived *without* a clean bill of health, 10*s.* per ton burden. Having arrived in this kingdom, with a cargo which in whole or part shall consist of goods, wares, or merchandise, the growth, produce, or manufacture of Turkey, or of any port or place in Africa, within the Straits of Gibraltar, or in the West Barbary, on the Atlantic Ocean, and which shall have arrived from any port or place whatever, or shall have performed quarantine in any port or place of Great Britain, or as aforesaid, 7*s.* 6*d.* per ton burden: and having so arrived under such circumstances as shall induce his Majesty, by and with the advice of his privy council, to subject them to the like quarantine as ships arriving from Turkey *without* clean bills of health, 15*s.* per ton burden, or as a ship arriving *with* clean bills of health, 7*s.* 6*d.* Every ship or vessel, the cargo of which, or any part thereof, shall have performed quarantine in this kingdom, and which shall enter inwards in the port of London, an additional duty of 1*s.* per ton burden. But no ship or vessel arriving from any place whatever, with a clean bill of health (except from any part of Turkey, or from any place in Africa, within the Straits of Gibraltar, or in the West Barbary, on the Atlantic Ocean, in ballast, or whose cargo shall consist wholly of salt), shall be liable to any duty under this act; and no ship or vessel which shall, together with its cargo, have duly performed quarantine in either of the lazarets of Malta, Ancona, Venice, or Messina, or Leghorn, and shall sail and arrive thence in this kingdom with proper documents attesting the same, shall be liable to or charged with any duty under this act.

4. The treasury may order the duties to be reduced, and afterwards raised again.

5. No vessel, upon which the quarantine duty is imposed, shall be cleared inwards, in any port of Great Britain, the islands of Guernsey, Jersey, Alderney, Sark, or Man, unless it be first paid; and the tonnage is to be ascertained according to the register under 26 Geo. III. c. 60.

6. The owners of vessels, who have paid the quarantine duties, may demand from importers a proportion of those duties.

7. The duties are to be paid into the exchequer, and a separate account to be kept.

8. This section regulates the application of the duties to the lazaret.

9. From April 5, 1805, all former acts, so far as they relate to the performance of quarantine, are repealed.

10 and 11. Any goods or ships that shall be specified by the orders of his Majesty, his heirs, or successors, in council, as arriving from any place whatever under suspicious circumstances of infection, shall be subject and liable to such regulations and restrictions as may be directed respecting the same in the orders of council; as also any ships, vessels, and boats, which may have received any person, merchandise, article, &c. from on board of them.

12. The privy council, on emergencies, can make such regulations as may to them appear requisite.

13. If the plague, or other infectious distemper, shall appear on board any vessel within the Straits of Gibraltar, the master shall immediately proceed to one of the foreign lazarets, and there perform quarantine, until such time as the land lazaret be fit for the due performance of quarantine therein; but if such plague, or other infectious distemper, shall appear on board any vessel without the Straits of Gibraltar, then the master shall (unless such land lazaret shall have been so declared to be fit for the due performance of quarantine therein) immediately proceed to the harbour of St. Helen's Pool, between the islands of St. Helen's Tean and North Withel, being two of the islands of Scilly, or such other place as his Majesty, in council, shall from time to time direct; where, being arrived, he shall immediately acquaint the principal magistrate thereof, and also some custom-house officer of some port of England near thereunto; and the said custom-house officer of such near port in England shall, with all possible speed, send intelligence thereof to the commissioners of the customs at London; and the said principal magistrate shall, in like manner, with all possible speed, send intelligence thereof to one of the secretaries of state, to the end that such measures may be taken for the comfort and support of the crew and passengers on board, and such precautions used to prevent the spreading of the infection, as the case shall require; and the said vessel shall there remain until his Majesty's pleasure be known; nor shall any of the crew or passengers thereof go on shore. But, if the said master or commander shall not be able to make the said island of Scilly, or other place so appointed by his Majesty as aforesaid, or shall be forced to go up either of the channels it shall not be lawful for him to enter into any port, but he shall remain in some open road till he receive directions by some order of the privy council, and he shall use every necessary means in his power, to prevent any of his ship's company or passengers from going out of his ship, and to avoid all intercourse with other vessels, or persons, until such master or commander shall have received such directions; and the said master, or any other person on board, who shall be disobedient herein, shall be adjudged guilty of felony, and suffer death.

14. The commander of any vessel, liable to perform quarantine, shall, from 5th April 1805, at all times when such vessel shall meet with any other vessel at sea, or shall be within four leagues of the coast of Great Britain or Ireland, or of Guernsey, Jersey, Alderney, Sark, or Man, hoist a signal to denote that his vessel is liable to quarantine: which signal shall be, in the day time (if such vessel shall have a clean bill of health) a large yellow flag, of six breadths of buntine at the main-top mast head; and (if such vessel shall not have a clean bill of health) then a yellow flag with a circular mark or ball entirely black thereon, whose diameter shall be equal to two breadths of bunting; and, in the night time, the signal shall, in both cases, be a large signal-lantern, such as is commonly used on board king's ships, with a light at the same mast-head; and such commander shall keep such signals, respectively, hoisted during such time as the vessel shall continue within sight of such other vessel, or within four leagues of the said coasts or islands, and while so in sight or within such distance, until such vessel shall have arrived at the place where it is to perform quarantine, and until it shall have been legally discharged therefrom, on failure whereof such commander shall forfeit for every such offence 200*l.*

15. If the commander of any vessel, not liable to perform quarantine, shall hoist such yellow flag, whereby other ships may be put to the inconvenience of changing their course, such commander shall forfeit 200*l.*

16. The commander, master, &c. of any ship arriving from foreign parts, shall give to the pilot, who shall go on board, a true account of the place and places where such ship shall have loaded, and touched in the course of the voyage; and shall forfeit 200*l.* in case of misrepresentation. And if, after such vessel shall have sailed, any places where she may have touched have been made liable to the performance of quarantine, then the pilot, under the penalty of 50*l.* shall give notice thereof to the commander, who shall thereupon act accordingly.

17. In case any pilot shall conduct any vessel, liable to perform quarantine, into any place, which is not specially appointed for the reception of such vessels (unless compelled by stress of weather, adverse winds, or accidents of the seas), such pilot shall, for every such offence, forfeit 100*l.*

18. When any place shall be infected with the plague, or other infectious disease, or when any order shall be made by his Majesty in council concerning quarantine, and the prevention of infection as aforesaid, then, as often as any vessel shall attempt to enter any place in Great Britain, Guernsey, Jersey, Alderney, Sark, or Man, (whether appointed for the performance of quarantine or not) the principal officer of customs in such place, or such person as shall be properly authorized, shall go off to such vessel, and such officer or other person shall, at a convenient distance from such vessel, demand of the commander thereof, a true answer in writing or otherwise, upon oath or not, according as he shall, by such officer or other person be required, to all such questions as shall be put to him, in pursuance of his Majesty's order in council; and, if such commander shall thereupon refuse to make a true discovery in any of the particulars concerning which he shall be interrogated, he shall for every such offence forfeit 200*l.*

19. If it should appear that such vessel is liable to perform quarantine, and that the port where it so arrives is not the port where it ought to perform quarantine, in such case the officers of any of his Majesty's ships of war, forts or garrisons, and all other his Majesty's officers whom it may concern, upon notice thereof, and any other persons whom they shall call to their aid and assistance, shall oblige such vessel to repair to such places as shall be appointed for performance of quarantine. And, in case any such vessel shall come from any place visited with the plague, or other infectious disease, or shall have any person on board actually infected, and the commander shall conceal the same, he shall be adjudged guilty of felony, and suffer death.

20. Every commander of any vessel ordered to perform quarantine, shall forthwith deliver to the superintendent of quarantine, or other person appointed to see quarantine duly performed there, such bill of health and manifest as he shall have received during his voyage, together with his log-book and journal, under pain of forfeiting 100*l.* if he shall wilfully refuse and neglect so to do.

21. If the commander of any vessel, liable to perform quarantine, having notice that such vessel is so liable, shall himself quit, or shall knowingly suffer any seamen or passengers to quit such vessel by going on shore, or by going on board any other vessel or boat, before such quarantine shall be fully performed (unless in such cases, and by such proper licence as shall be directed concerning quarantine, and the prevention of infection, as aforesaid), or if the commander of such vessel shall not, within a convenient time after due notice given for that purpose, cause such vessel and her lading to be conveyed into the places appointed to perform the quarantine in, then such commander, for every such offence, shall forfeit five hundred pounds. And, if any person shall so quit such vessel, contrary to the true meaning of this act, it shall be lawful for all persons whatsoever, by any kind of necessary force, to compel such person to return on board, and every such person so quitting such vessel shall, for every such offence, be imprisoned for six months, and shall also forfeit 200*l.*

22. When any vessel, which has performed quarantine in a foreign lazaret, shall arrive in Great Britain, Guernsey, Jersey, Alderney, Sark, or Man, with a clean bill of health, no goods or merchandise liable to retain the infection of the plague, or other infectious disease, shall be unshipped or moved in order to be landed; but the commander of such vessel shall, immediately upon arrival, give notice thereof, and of the contents of his manifests as aforesaid, and of the foreign port in which such vessel hath performed quarantine, to the principal officer of the customs at the port of arrival, or at the port nearest thereto; and, if the commander, or any person whatsoever, shall land, unship, unload, or move in order to land, any goods or merchandise out of the said vessel, before an order of privy council shall be made, giving directions therein, or otherwise than shall be directed in the said order, every such person shall, for every offence, forfeit 200*l.*

23. All persons liable to perform quarantine, shall be subject, during the said quarantine, to such orders as they shall receive from the proper officers, and the said officers are to enforce all necessary obedience to the said orders; and, in case of necessity, to call in others to their assistance, and all persons so called in are to assist accordingly; and such officers shall compel all persons obliged to perform quarantine, to repair, and to cause all goods and merchandise, comprised within any such orders, to be conveyed to the lazaret, or other place duly appointed in that behalf, according to such directions as shall be made by order of council; and, if any person, so obliged to perform quarantine, or any person having had intercourse with them, shall neglect to repair within convenient time, after notice, to the said lazaret, or other place appointed, or shall escape, or attempt to escape out of the same before quarantine duly performed, the watchmen, and other persons, appointed to see quarantine performed, are required, by necessary force, to compel every such person to return; and every person so neglecting to repair into the said lazaret or other place, and also every person actually escaping as aforesaid, shall be adjudged guilty of felony, and suffer death.

24. Persons quitting any ship or place liable to perform quarantine shall be open to seizure by any constable, &c. and to be conveyed, by the warrant of any justice of the peace, to the place appointed for the performance of it.

25. Contains an indemnity for orders in council to send afloat, for the performance of quarantine, any person who has had intercourse with a suspected vessey

26. If any officer of customs, or any person whatsoever, to whom it appertains to execute orders concerning quarantine, or the prevention of infection, shall be guilty of any wilful breach or neglect of duty, every such officer and person shall forfeit such office or employment, and become thenceforth incapable to hold the same; and every such officer and person shall forfeit 100*l.*; and if any such officer or person shall desert from duty when employed, or permit any person, ship, vessel, goods, or merchandise, to depart, or be conveyed out of the lazaret, &c. unless by order of council; or, if any person, authorized to give a certificate of a ship having duly performed quarantine, shall knowingly give a false certificate thereof; every such officer and person so offending shall be deemed guilty of felony, and suffer death; and, if any such officer or person shall knowingly and wilfully damage any goods performing quarantine, he shall be liable to pay treble damages and full costs of suit to the owner thereof.

27. If any person, not infected with the plague or other infectious disease, nor liable to perform quarantine, shall enter the said lazaret or other place so appointed, whilst any person being under quarantine shall be therein, and shall return or attempt to return thence, unless in such cases and by such license as shall be directed by virtue of any orders of council, the watchmen, or other persons appointed to guard the said lazaret or other place, shall, by necessary force, compel such person to repair into the said lazaret or other place, there to perform quarantine, and in case such person shall actually escape out of the said lazaret, or other place where he shall be so placed to perform quarantine, he shall be adjudged guilty of felony, and suffer death.

28. After quarantine shall have been duly performed by any vessel or persons; and, upon proof made by the oaths of the commander of such vessel, and of two persons belonging thereto, or upon proof by the oath of two or more credible witnesses, before the collector or principal officer of the customs, or before any justice of the peace living near the place, or (when such quarantine shall have been performed within Guernsey, Jersey, Alderney, Sark, or Man) before any two jurats or magistrates of the said isles, that such vessel and every such person have duly performed quarantine as aforesaid, and that the same are free from infection; and, after producing a certificate to that purpose from the chief officer who superintend the quarantine, then such collector, or principal officer of the customs, or such justice of

the peace, or such jurats or magistrates, are to give a certificate thereof; and thereupon such vessel, and persons so having performed quarantine, shall be liable to no farther restraint.

29. All goods, wares, and merchandise liable to quarantine, shall be opened and aired in such places, and for such time and in such manner, as shall be directed by order in council; and, after such orders shall have been duly complied with, a certificate thereof shall be given by the chief officer appointed to superintend such quarantine, and proof thereof shall be mad by the oaths of two or more credible witnesses before the collector or principal officer of the customs at the port lying next to such places where the same were opened and aired, or before any justices of the peace living near the same, or before any two jurats or magistrates of Guernsey, Jersey, Alderney, Sark, or Man, respectively; and, upon the production of certificates thereof to the commissioners of customs in England or Scotland, or to the chief magistrate of any of the said last-mentioned islands, such goods, wares, and merchandise, shall be forthwith discharged from any restraint or detention upon the same account.

30. Persons forging certificates, &c. shall suffer death.

31. If any person conceal from the officers of quarantine, or clandestinely convey any letters, &c. from any ship performing, or liable to perform quarantine, or from the lazaret, every such person shall be adjudged guilty of felony, and suffer death: and if any person unload any goods, merchandise, &c. from a vessel liable to perform quarantine, and shall receive such articles, every such person shall forfeit a sum not exceeding five (nor less than one) hundred pounds sterling.

32. If it shall at any time happen that any part of Great Britain, Ireland, Guernsey, Jersey, Alderney, Sark, or Man, or France, Spain, Portugal, or the Low Countries, be infected with the plague, or any other infectious disease, his Majesty, by proclamation, may prohibit all small boats and vessels, under twenty tons burden, from sailing out of Great Britain, Guernsey, Jersey, Alderney, Sark, and Man, until security be first given by the master thereof, in the penalty of 300*l.* that such mentioned boat or vessel shall not go to or touch at any place mentioned in such proclamation; and that neither the master of such boat or vessel, nor any mariner or passenger therein, shall, during such time, go on board any other ship or vessel at sea; and that such master shall not suffer any person to come on board such boat or vessel at sea from any other ship or vessel, and shall not, during the time aforesaid, receive any goods or merchandise whatsoever, out of any other ship or vessel; and, in case any boat or vessel, for which such security shall be required by such proclamation, shall sail, before such security be given as aforesaid, every such boat or vessel, together with her tackle, apparel, and furniture, shall be forfeited to his Majesty; and the master, and every mariner thereof, shall severally forfeit 20*l.*

33. Publication in the London Gazette of any order in council to be sufficient notice to all concerned.

34 to 43 inclusive. These sections relate to the application of the penalties, modes of conviction, &c.

44. This act shall take effect from the 5th day of April, 1805.

An Act for making Additional Provisions for the effectual Performance of Quarantine—46 Geo. III. c. 98.

1. Any vessel, on board whereof the plague, or an infectious distemper shall actually be, is required, when meeting with another ship within four leagues of Great Britain, or the islands of Guernsey, Jersey, Alderney, Sark, or Man, to hoist a signal denoting its existence; in the day-time, a yellow and black flag, of eight breadths of bunting at the main-mast-head; and in the night-time, two large signal lanterns (those commonly used), one over the other, at the same mast-head; and such commander, master, &c. shall keep it hoisted until arrival at the port or place where the ship is to perform quarantine: a penalty of 200*l.* is enacted on failure thereof.

2. The commander shall deliver to the pilot coming on board, in addition to the particulars enacted by 45 Geo. III. c. 10, a complete account of the cargo, on penalty of 200*l.* for neglect, refusal, misrepresentation, or omission thereof or therein: whereupon such pilot, under forfeit of 100*l.* shall communicate to the commander information of such of the articles as may be liable to quarantine. Such commander shall hoist a proper signal, and comply in every respect with the aforesaid recited act; and if, unless by stress of weather or other accidents, the pilot conduct the ship to the place not specially appointed for her reception, having omitted to require such account of the cargo, for each offence 100*l.* shall be paid by him.

3. No vessel that has performed quarantine, having been solely rendered liable to it by communication, against the will of the master and crew, with a ship that was originally subject to perform it, shall be charged with the duties imposed upon the latter by the said recited act; provided such circumstance appear to the satisfaction of the commissioners of the customs.

5. So much of the recited act as relates to the certificate and proof of goods having been opened and aired, is repealed; and such proof shall be hereafter made from the oath of the master of the lazaret, and of the quarantine officers already mentioned in the aforesaid act (or of two or more credible witnesses serving in the lazaret or vessel), before the superintendent of quarantine, or his assistant; after the administration of which oath, certificate shall be made by the officers of quarantine; on the production of which to the officers of customs, such articles, goods, &c. shall be liable to no further restraint, either at the port of quarantine, or any other place to which they may be conveyed.

6. His Majesty's privy council may order ships coming from America, or the West Indies, on apprehension of the yellow fever, &c. to come to anchor at certain places appointed from time to time by the commissioners of the customs, in order to ascertain whether or not the performance of quarantine be necessary; which shall not, however, take place unless specially ordered afterwards.

7. All persons, ships, boats, and vessels may be prohibited, by order in council, to go within the limits of any station assigned to ships without clean bills of health, for the performance of quarantine: and every person, for such offence shall forfeit 500*l.*

8. Forging, altering, or uttering, knowingly and wilfully, any certificate of quarantine, shall be adjudged felony, and punished with death, without benefit of clergy.

9. Consuls and vice-consuls can administer oaths respecting quarantine, in the several places and towns where they reside.

10. Persons authorised to take examinations may administer on oath; and those who swear falsely, or procure others so to do, shall be deemed guilty of wilful and corrupt perjury.

In pursuance of the preceding, a proclamation was issued, dated the 5th April 1805, and noticed in subsequent orders, which contains such a variety of regulations relative to the performance of quarantine that we have thought it right to refer for farther information to the act itself, which commenced and took effect on the 1st day of July, 1805, for all ships coming from any port or place in Europe, without the Straits; and from any port or place on the continent of America, on the 1st day of October 1805, and concludes thus:—

The lords commissioners of his Majesty's treasury, the commissioners for executing the office of lord-high-admiral of Great Britain, the lord-warden of the cinque ports, the master-general, and the rest of the principal officers of the ordnance, his Majesty's secretary at war, and the governors and commanders-in-chief for the time being, of the isles of Guernsey, Jersey, Alderney, Sark, and Man, are to give the necessary directions herein as to them may respectively appertain.

QUARTER, (*la hanche, ou l'arrière du vaisseau, depuis les grands porte-haubans jusqu'à la poupe*, Fr.) that part of a ship's side which lies towards the stern, or which is comprehended between the aftmost end of the main-chains and the sides of the stern, where it is terminated by the quarter-pieces.—Or the quarter of a ship is generally considered to be all that part aft, from the middle of the main-channel.

Although the lines by which the quarter and bow of a ship, with respect to her length, are only imaginary, yet experience appears sufficiently to have ascertained their limits; so that if we were to divide the ship's sides into five equal portions, the names of each space would be readily enough expressed. Thus the first, from the stern, would be the quarter; the second, abaft the midships; the third, the midships; the fourth, before the midships; and the fifth, the bow. Whether these divisions, which, in reality, are somewhat arbitrary, are altogether improper, may be readily discovered by referring to the mutual situation or approach of two adjacent vessels. The enemy boarded us on the larboard side! Whereabouts? Abaft the midships, before the midships, &c.

Fig. 9. plate XI. represents a geometrical elevation of the quarter of a modern 74 gun ship, as corresponding with the other figures of a ship of the same rate, delineated upon the same plate. See the articles HEAD, MIDSHIP-SECTION, and STERN.

In the plane of elevation, fig. 2. plate XVII. the quarter is continued into the side, upon a smaller scale.

Reference to, and Explanation of fig. 9, *plate* XI.

A the keel, with *a* the false keel beneath it.

B the stern-post.

C C C rails to the upper, middle, and lower stools to the quarter-galleries.

D D the quarter-gallery, with its munions and windows.

E the quarter-pieces.

F the taffrail.

G the braces for the pintles of the rudder to work in.

H the gun-ports of the lower deck.

h the gun-ports of the upper and quarter-deck.

I the pintles which work in the braces.

K K the breast-rails to the quarter-gallery.

L the lower finishings.

Q the upper finishing.

R the after-part of the channel-wale, parallel to the main-wale.

S the main-wale.

T the rudder.

U U the small rails over the upper finishing.

V the rough-tree rail.

W the after-part of the load water-line.

k k l the curve of the several decks corresponding to those represented in the head.

On the QUARTER, (*par la hanche*, Fr.) implies any point of the compass between the keel and beam, when an observer looks from the centre of the ship towards the stern. See the article BEARING.

To play upon the QUARTER *of a Ship*, (*canonner un vaisseau dans la hanche*, Fr.) is to point the guns in such manner, that the shot shall strike on that part of an enemy's ship in action.

QUARTER, in astronomy, the fourth part of the moon's period: thus, from the new moon to the quadrature is the first quarter; from this to full moon, the second quarter, &c.

QUARTER *of a point*, (*le quart d'un air de vent du compas*, Fr.) in navigation, is the fourth part of the distance between two points of the compass, and is 2° 48′ 45″.

Oak or Fir QUARTER, a term given to any piece of oak or fir timber that is $2\frac{1}{2}$ inches by 3, or 3 inches by 4; and from 10 to 40 feet in length: it is used for various purposes in a ship.

QUARTER-BADGE. See the article BADGE.

QUARTER-BILL, (*rôle de combat*, Fr.) a list, containing the different stations to which the officers and crew are quartered, in the time of action, and the names of the persons appointed to those stations.

QUARTER-CLOTHS, (*prélats du bastingage du gaillard d'arrière et de la dunette*, Fr.) are long pieces of painted canvas, fastened to the outside of the quarter-netting along the rough-tree rail on the quarter-deck, and extending from thence to the plank-sheer, to keep out the spray of the sea.

They contain in general, two whole cloths, and

one gored cloth, which is always placed at the lower part. The seams are an inch broad, and a two or three inch tabling is made all round. They are used principally in small vessels, as sloops, brigs, transports, &c.

QUARTER-DECK. See the article DECK.

QUARTER-GALLERY, (*le haut des bouteilles*, Fr.) a small balcony on the quarter of a ship, generally communicating by doors with that on the stern. See the article GALLERY.

QUARTER-GUNNER, (*aide-canonnier*, Fr.) a petty officer under the direction of the gunner of a ship of war, whom he is to assist in every branch of his duty; as keeping the guns and their carriages in proper order, and duly furnished with whatever is necessary; filling the powder into cartridges; scaling the guns, and keeping them always in a condition for service. The number of quarter-gunners in any ship is always in proportion to the number of her artillery, one quarter-gunner being allowed to every four guns.

QUARTER-MASTERS, (*quartier-maîtres*, Fr.) in a ship of war, are petty officers appointed by the captain to assist in the several duties of the ship, as stowing the ballast and provisions in the hold, coiling the cables on their platforms, overlooking the steerage of the ship, keeping the time by the watch-glasses, and, in turn, over-looking the purser's steward in his delivery of provisions, &c.

QUARTER-MASTERS-MATES, (*bossemans*, Fr.) are persons subordinate to the preceding.

QUARTER-MEN, (*quartier maitre*, Fr.) are officers employed under the master-shipwrights, to superintend companies of shipwrights and calkers in building, calking, and repairing his Majesty's ships in docks and afloat. There are, in time of war, about 200 *quartermen* employed for that purpose at the different dock-yards in England.

QUARTER-NETTING, (*filets du bastingage, du gaillard d'arrière et de la dunette*, Fr.) a sort of net-work, extended along the rails on the upper-part of a ship's quarter. In a ship of war these are always double, and supported by rails and stanchions, placed at proper distances. They contain some of the seamen's hammocks.

QUARTER-PIECES, (*les montans des côtês de la galerie*, Fr.) are pieces of timber at the after-part of the quarter-gallery, near the taffrail.

QUARTER-RAILS, (*filarets de bastingage, ou lisses du bastingage du gaillard d'arrière et de la dunette*, Fr.) are narrow-moulded planks, generally of fir, reaching from the top of the stern to the gangway, and serve as a fence to the quarter-deck, to prevent the men from falling into the sea by the rolling of the ship, particularly in small vessels.

QUARTER-WIND, (*vent grand largue*, Fr.) implies a wind that blows directly on the ship's quarter, or that which comes in abaft the main-mast shrouds, even with the quarter of the ship. It is said to be the best of all winds for filling the sails; as, a wind blowing right astern, is, therefore, in some measure kept off by the interposition of the mizen and main-masts.

QUARTER-TACKLE, (*palan de bout de vergue*, Fr.) a strong tackle, fixed occasionally upon the quarter of the main or fore-yard, to hoist boats and heavy packages into and out of the ship.

QUARTERING, (*grand largue*, Fr.) is said of a ship when sailing at large, that she neither goes by the wind, nor before the wind, but directly between both.

QUARTERING, in gunnery, is when a piece of ordnance is so traversed, that it will shoot on the same line, or on the same point of the compass whereon the ship's quarter has its bearing.

QUARTERLY-BILLS. By an act of 35 Geo. III. flag-officers, captains, lieutenants, masters, surgeons, and assistant-surgeons, can draw upon the commissioners of the navy for their nett *personal pay* every three lunar months, by a draft properly filled up and approved by the captain, called a *quarterly-bill;* printed forms of which are supplied to all his Majesty's ships for that purpose.

QUARTERS, (*postes désignés à chacun pour le combat*, Fr.) in a ship of war, implies the several stations where the officers and crew are posted in time of action.

It may be necessary to observe here, that the strictest attention is required in quartering a ship's company, and that when a ship is first put in commission, the seamen should be often exercised at their respective quarters, that they may be familiarized to the various occurrences which are like to happen in action. When the seamen are perfect in one station of the gun, they should be put to another, in order to acquire a proficiency in the whole.

The men should be quartered by the watch-bill, taking equal proportions from the several watches. This plan will be found more advantageous in small ships; in which, when at sea, all hands are frequently called to their quarters in the night, and, in this case, those whose watch is upon deck will naturally be employed in getting their quarters ready, while the others are bringing up their hammocks.

In fine weather, the hammocks should always be brought up, stowed, and covered well up, in the nettings, which, in the event of falling in with an enemy, would be of incalculable advantage.

The men at the guns should be stationed in the following manner, viz. The whole number divided into two equal parts, under a first and second captain, having their respective boarders, with their belts and cutlasses; sail-trimmers; lantern-men, to fetch fire-lanterns; firemen, with buckets and swabs; winch-men, to pump the ship, if required; port-tackle-fall-men, to lower the portlids; and powder-men, to fetch powder; under the orders and directions of officers specifically nominated to superintend these different duties.

On beating or piping all hands to quarters, the first captain, with his party, is to get the starboard guns ready, and the second captain, with his party, should get the larboard ones ready.

In fighting both sides at once, the seamen should be taught to work at two guns together: thus, while one gun is in, and the men are worming, sponging, and loading it, the other should be run with the spare men from the gun which is in, in addition to those belonging to the gun which is to be run out, and so continue alternately to assist each other. When one side only is engaged, the first party of men stationed at the gun should be at the fore-part of it, and the second party at the after-part. The first captain should prick, prime, and point the gun to the object; the second captain should secure the powder-horn and stop the vent. The boarders should sponge and load the gun; the fire-men should choke the tackle-falls, and handle the crows and handspikes; and the rest should work the gun.

As nothing in action conduces more to keep up a brisk fire than being well supplied with powder, every necessary arrangement should be made for that purpose, by stationing a sufficient number of men to hand it along the passages, and dividing the supply as much as possible, that no confusion or mistake may happen in distributing the cartridges. In three-decked ships it is usual to supply the lower-deck with cartridges from the grand magazine, handed up at the fore-hatch-way; the middle-deck, from the fore-hanging magazine, handed along by the sail-room up the main-hatchway; the main-deck, fore-castle, and quarter-deck, from the after-magazine, with whips at the foot of the ladder coming off the middle-deck. Positive orders should be given that no cartridge be given out to any person who does not deliver an empty cartridge-box.

The number of men appointed to manage the artillery is always in proportion to the nature of the guns, and the number and condition of the ship's crew; and is regulated at the pleasure of the captain or commander.

The number of men to be trained to the use of muskets on board his Majesty's ships and sloops of war, is as follows:—

Rate of the Ship.	No. of men to the small arms.
1st	150
2d	120
3d of 74 guns and upwards	100
— of 64 guns and upwards	80
4th of 50 guns	60
5th	50
6th	40
Sloops of war	30

The lieutenants command the different batteries; the master superintends the movements of the ship, and whatever relates to the sails; the boatswain, and a sufficient number of men, are stationed to repair the damaged rigging; and the gunner and carpenter, wherever necessary, according to their respective offices. See also the articles CANNON, ENGAGEMENT, and EXERCISE.

The marines are generally quartered on the poop and forecastle, or gangway, under the direction of their officers; although, on some occasions, they assist at the great guns, particularly in distant cannonading

All Hands to QUARTERS! (*chacun à son poste!* Fr.) is the order to assemble the whole of the officers and ship's company to their respective stations, when an enemy's ship heaves in sight, to prepare themselves for action. This is done either by beating the drum, or by the boatswain and his mates piping down the hatchways.

In ships of war all hands are beat or piped to quarters every evening before sun-set, to see if they are orderly; and that they, under directions of the officers, arrange properly all implements belonging to the guns, &c. so as to be ready for action at a minute's notice.

Close QUARTERS. See that article.

Sick-QUARTERS, implies, when there is not room in a naval hospital for an officer sent sick thereto, and he provides himself, or is provided with lodgings, for the payment of which, and for subsistence, he is allowed a certain sum, according to the place he is at.

QUARTERS! is also an exclamation to implore mercy from a victorious enemy.

QUARTERS *of a Mast*, a term applied to some of the divisions on a mast, where the different diameters are set off, for lining or marking.

QUARTERS *of the Yards*, (*brasseiage des vergues; partie des vergues comprise entre le grand diamètre et les bouts de vergue*, Fr.) are certain divisions on the surface of the yards, where the diameters are set off from the slings: they are termed first, second, and third quarter, and the outer end.

QUAY, (*quai*, Fr.) a place to land goods upon. See the article KEY.

QUICK-MATCH. See the article MATCH.

QUICK-SAND, (*sable mouvant*, Fr.) a loose sand, into which a ship sinks by her own weight, as soon as the water retreats from her bottom.

QUICK-WORK, (*œuvres vives d'un bâtiment*, Fr.) in ship building, that part of a ship's sides, both within and without board, which is above the channel-wales and decks. It is sometimes performed with deal, which does not require the fastening, nor the time to work it, as the other parts do: hence it is called *quick-work*.

QUILTING, (*action de trélinguer, ou garnir une jarre, &c. avec de vieux cordages; c'est aussi la garniture, ou le trélingage de cette jarre, &c.* Fr.) the operation of weaving a sort of coating, formed of the *strands* of rope, about the outside of any vessel, to contain water, &c. as a jar, bottle, &c. The term is also applied to the coating wove on these vessels.

QUINTILE, in astronomy, an aspect of the planets, when they are 72 degrees, or a fifth part of the zodiac, distant from each other.

QUOIL. See the article COIL.

QUOIN, (*coin*, Fr.) in artillery, is a wedge of wood, placed under the breech of a gun that is mounted on the bed of its carriage, in order to elevate or depress it at pleasure, or point it more truly to the object to be shot at.

QUOINS, (*toutes sortes de coins, cales, ou pièces de bois servant à arrimer et acorer des tonneaux, &c.* Fr.) are also employed to wedge off the casks of wine, oil, spiritous liquors, &c. from each other, that their bilges may not be chafed so as to occasion a leak, by the agitation of the ship at sea.

R.

RABBET, (*rablure*, Fr.) in carpentry, a deep groove or channel, cut in a piece of timber longitudinally, to receive the edge of a plank, or the ends of a number of planks, which are to be securely fastened therein. The depth of this channel is equal to the thickness of the plank, so that when the end of the latter is let into the rabbet, it will be level with the outside of the piece. Thus the ends of the lower planks of a ship's bottom terminate upon the stem afore and the stern-post abaft, with whose sides their surfaces are even. The surface of the garboard-streak, whose edge is let into the keel, is, in the same manner, level with the side of the keel at the extremities of the vessel: hence the rabbet of the keel, of the stem, and of the sternpost.

RACE, (*ras de marée, ou de courant*, Fr.) a name given to a strong rippling tide or current; as Portland Race, which is caused by the projection of the land with the unevenness of the ground over which the tide flows, and which is one mile and three quarters long, in the direction east and west.

RACK, (*rateau de poulies*, or *rateliers*, Fr.) a frame of timber, containing several sheaves, and usually fixed on the opposite sides of a ship's bowsprit, to direct the sailors to the respective ropes passing through it; all of which are attached to the sails on the bowsprit.

To RACK *a Tackle*, (*éguilleter un palan*, &c. Fr.) is to fasten the two opposite parts of it together with a seizing, so that any weighty body suspended thereby shall not fall down, although the rope, which forms the tackle, should be loosened by accident or neglect.

RADIUS, in geometry, is the semi-diameter of a circle, or a right line drawn from the centre to the circumference. In trigonometry, the radius is termed the whole sine, or sine of 90°.

RAFT, (*radeau de bois de construction*, Fr.) a sort of float, formed by an assemblage of various planks, or pieces of timber, fastened together side by side, so as to be conveyed more commodiously to any short distance in a harbour or road, than if they were separate. The timber and plank with which merchant ships are laden, in the different parts of the Baltic Sea, are attached together in this manner, in order to float them off to the shipping. See also the article CATAMARAN.

RAFT-PORT, (*sabord de charge, servant à embarquer les bois, dans les bâtimens marchands ou flûtes*, &c. Fr.) a square hole cut through the buttock, and sometimes in the bow, of some merchant-ships and naval transports, immediately under the counter, to load and unload the planks or pieces of timber for transportation; and which on account of their great length, could not be received on board otherwise.

RAG-BOLT, (*fiche*, Fr.) an iron pin, having several barbs cut on its shank, to retain it in the wood, as explained under the article bolt, and represented by fig. 17. plate IV.

RAILS, (*lisses d'accastillage*, Fr.) are narrow planks, generally of oak and fir, upon which there is a moulding stuck: they are nailed for ornament on several parts of a ship's upper-works, as the drift-rails, fife-rails rough-tree-rails, sheer-rails, tuck-rails, waist-rails, &c. some of which are also intended as a fence to prevent the sailors from falling overboard.

RAILS *of the Head*, (*lisses de herpes*, Fr.) are certain curved pieces of timber, extending from the bows on each side to the continuation of the ship's stem, to support the knee of the head, and the ornamental figure fixed thereon. The former of these rails is represented at large in fig. 8. plate XI., as explained under the article HEAD.

RAIN, (*pluie*, Fr) is the water that descends from the atmosphere in the form of drops of a considerable size.

Rain is apparently a precipitated cloud; as clouds are nothing but vapours raised from moisture, waters, &c. By this circumstance it is distinguished from dew and fog: in the former of which the drops are so small, that they are quite invisible; and in the latter, though their size be larger, yet they seem to have very little more specific gravity than the atmosphere itself, and may therefore be reckoned hollow spherules rather than drops.

It is universally agreed, that rain is produced by the water previously absorbed by the heat of the sun, or otherwise, from the terraqueous globe, into the atmosphere, as vapours, or vesiculæ. These vesiculæ being specifically lighter than the atmosphere, are buoyed up by it, till they arrive at a region where the air is in a just balance with them; and there they float, till by some new agent they are converted into clouds, and thence either into rain, snow, hail, mist, or the like.

But the agent in this formation of the clouds into rain, and even of the vapours into clouds, has been much controverted.

Most philosophers, however, will have it, that the cold, which constantly occupies the superior regions of the air, chills and condenses the vesiculæ, at their arrival from a warmer quarter; congregates them together, and occasions several of them to coalesce into little masses: and thus their quantity of matter increasing in a higher proportion than their surface, they become an overload to the thin air, and so descend in rain.

The late writers on this part of philosophical science have considered rain as an electrical phenomenon. Signior Beccaria reckons rain, hail, and snow, among the effects of a moderate electricity in the atmosphere. Clouds that bring rain, he thinks, are produced in the same manner as thunder clouds, only by a moderate electricity. He describes them at large; and the resemblance which all their phenomena bear to those of thunder-clouds, is very striking. He notes several circumstances attending rain without lightning, which render it probable that it is produced by the same cause as when it is accompanied with lightning.

Light has been seen among the clouds by night in rainy weather; and even by day rainy clouds are sometimes seen to have a brightness evidently independent of the sun. The uniformity with which the clouds are spread, and with which the rain falls, he thinks, are evidences of an uniform cause like that of electricity. The intensity also of electricity in his apparatus, usually corresponded very nearly to the quantity of rain that fell in the same time. Sometimes all the phenomena of thunder, lightning, hail, rain, snow, and wind, have been observed at one time; which shews the connection they all have with some common cause. Signior Beccaria therefore supposes that, previous to rain, a quantity of electric matter escapes out of the earth, in some place where there is a redundancy of it; and its ascent to the higher regions of the air, collects and conducts into its path a great quantity of vapours. The same cause that collects, will condense them more and more; till, in the places of the nearest intervals, they come almost into contact, so as to form small drops, which uniting with others as they fall, come down in the form of rain. The rain will be heavier in proportion as the electricity is more vigorous, and the cloud approaches more nearly to a thunder-cloud, &c. See *Lettere dell Elettricismo*; and Priestley's *History*, &c. *of Electricity*. And for further accounts of the phenomena of rain, &c. see the article *Marine* Barometer. See also *Philosophical Transactions*, Abr. vols. 4 and 10: also the *Meteorological Journal* of the Royal Society, published annually in the Philos. Trans.

RAINBOW, Iris, or simply the *Bow*, (*arc en ciel*, Fr.) is a meteor in form of a party-coloured arc, or semicircle, exhibited in a rainy sky, opposite to the sun, by the refraction and reflection of his rays in the drops of falling rain. There is also a secondary, or fainter bow, usually seen investing the former at some distance. Among naturalists, we also read of lunar rainbows, marine rainbows, &c.

The rainbow, Sir Isaac Newton observes, never appears but where it rains in the sun-shine; and it may be represented artificially, by contriving water to fall in small drops, like rain, through which the sun shining, exhibits a bow to a spectator placed between the sun and the drops, especially if there be disposed beyond the drops some dark body as a black cloth, or such like.

Dr. Barrow, in his Lectiones Opticæ, at Lect. 12, n. 14, says, that a friend of his (by whom we are to understand Mr. Newton) communicated to him a way of determining the angle of the rainbow, which was hinted to Newton by Slucius, without making a table of the refractions, as Des Cartes did. The Doctor shews the method; as also several other matters, at n. 14, 15, 16, relating to the rainbow, worthy the genius of those two eminent men. But the subject was given more perfectly by Newton afterwards, viz. in his optics, prop. 9; where he makes the breadth of the interior bow to be nearly 2° 15′, that of the exterior 3° 40′ their distance 8° 25′, the greatest semi-diameter of the interior bow 42° 17′, and the least of the exterior 50° 42′, when their colours appear strong and perfect.

The doctrine of the rainbow may be illustrated and confirmed by experiment in several different ways.

Thus, by hanging up a glass globe full of water, in the sun-shine, and viewing it in such a posture that the rays which come from the globe to the eye, may include an angle either of 42° or 50° with the sun's rays; for example, if the angle be 42°, the spectator will see a full red colour in that side of the globe opposite to the sun. And by varying the position so as to make that angle gradually less, the other colours, yellow, green, and blue, will appear successively, in the same side of the globe, and that very bright. But if the angle be made about 50°, suppose by raising the globe, there will appear a red colour in that side of the globe towards the sun, though somewhat faint; and if the angle be made greater, as by raising the globe still higher, this red will change successively to the other colours yellow, green, and blue. And the same changes are observed by raising or depressing the eye, while the globe is at rest. See Newton's *Optics*, part 2, prop. 9, prob. 4.

Lunar Rainbow. The moon sometimes also exhibits the phenomenon of an iris, by the refraction of her rays in the drops of rain in the night time.

Aristotle says, he was the first that ever observed it; and adds, it is never seen but at the time of full moon; her light at other times being too faint to affect the sight after two refractions and one reflection.

The lunar iris has all the colours of the solar, very distinct and pleasant; only fainter, both from the different intensity of the rays, and the different disposition of the medium.

Marine Rainbow, or the *Sea Bow*, is a phenomenon sometimes observed in a much agitated sea; when the wind sweeping part of the tops of the waves, carries them aloft; so that the sun's rays, falling upon them, are refracted, &c. as in a common shower, and there paint the colours of the bow. These bows are less distinguishable and bright than the common bow; but then they exceed as to numbers, there being sometimes twenty or thirty seen together. They appear at noon-day, and in a position opposite to that of the common bow, the concave side being turned upwards, as, indeed, it is necessary it should be, from what has been said in accounting for the phenomena of the solar bow.

To RAISE, (*lever*, Fr.) to elevate any distant object at sea by a gradual approach towards it: this effect is produced by the convex surface of the sea, which previously intercepted the view, when directed towards the lower part of the object. This term is opposed to that of *To* LAY *the Land*, which article see.

To RAISE *a Purchase*, (*faire un appareil*, *et en général*, *preparer une force méchanique*, Fr.) is to prepare certain instruments, or machines, so that, by their mutual effects, they may produce a mechanical force sufficient to overcome the weight or resistance of the object to which this machinery is applied.

RAKE, (*quête; élancement*, Fr.) the projection of the upper parts of a ship at the height of the stem and stern beyond the extremities of the keel. Thus if a plummet be hung from the top of a ship's stern, so as to be level with the continuation of the keel, the distance between the after end of the keel and the plummet will be the length of the rake of the stern, or the rake abaft.

RAKE also signifies the inclination of the masts from a perpendicular situation with the keel; so that they cease to form right angles therewith; as those of American schooners, Bermudian sloops, French chassé-marées, &c.

RAKE *of the Rudder*. See the article RUDDER.

To RAKE *a Ship*, (*enfiler un vaisseau*, *en le canonnant de long en long*, Fr.) is to cannonade her on the stern, or head, so as that the balls shall scour the whole length of her decks; which is one of the most dangerous incidents that can happen in a naval action. This is frequently called *raking fore and aft*, and is similar to what is called by egineers *enfilading*.

RAM, in astronomy. See the article ARIES.

To RAM *home the Charge of a Gun*, (*bourrer un canon*, Fr.) is to apply the rammer to the mouth thereof, and then force back the charge as far as possible.

RAMLINE, in mast-making, a long line used to gain a straight middle line on a tree or mast: it is made fast at one end, and pulled upon at the other till it is quite straight, and then made fast.

It is also sometimes used to form the sheer of a ship, and to set the beams of the deck fair.

RAMMER, (*refouloir*, Fr.) in gunnery, is a cylinder of wood nearly fitting the bore of a cannon, and fastened on a wooden staff, or on a rope well stiffened with spun-yarn, see figures 6 and 7, plate VI. It is used to drive the charge of a cannon home, or to the innermost part of it.

RAMPART, (*rempart*, Fr.) in fortification, is an elevation of earth round a place capable of resisting the cannon of an enemy; and formed into bastions, curtins, &c. A parapet is raised upon this bank or elevation which always looks towards the country. It is encompassed with a ditch, and is sometimes lined or fortified on the inside. Sentinels are regularly placed round the ramparts, and pieces of heavy ordnance are planted, at given distances, for the protection of the place.

RAMROD OF A MUSKET OR PISTOL, (*baguette*, Fr.) is the stick, or piece of iron used in charging, to drive home the powder and shot, as also the wad, which prevents the shot from rolling out. The ramrod was formerly called *scourer*.

RAN, in rope-making, a term used to imply twenty cords of twine, wound on a reel, and every cord so parted by a knot as to be easily separated.

RANDOM-SHOT, (*à toute volée de canons*, *ou de mortiers*, Fr.) in gunnery, is a shot made when the muzzle of a gun is raised above the horizontal line, and is not designed to shoot directly or point-blank. The utmost random of any piece is about ten times as far as the bullet will go point-blank. The bullet will go farthest when the piece is mounted to about 45° above the level range. See the articles GUNNERY, PROJECTILES, and RANGE.

RANGE, (*longueur du cable paré sur le pont*, *pour mouiller l'ancre*, Fr.) a sufficient length of the cable drawn in upon the deck, before the anchor is cast loose from the bow, that by sinking to the bottom without being interrupted, the flooks may be forced deeper into the ground; therefore, the range drawn up out of the tier, ought to be equal in length to the depth of the water where the ship anchors.

RANGE, is also a name given to a large cleat, with two arms or branches, bolted in the waist of ships, to belay the tacks and sheets to. See also the articles *Belaying* CLEAT, and KEVELS.

RANGE, (*portée des canons*, *ou mortiers*, Fr.) in gunnery, the path of a bullet, or the distance to which a bomb or cannon ball is thrown from a piece of artillery, by the explosion of gunpowder.

If the piece is in a line parallel to the horizon it is called the *right* or *level range*; but when it is elevated to 45°, it is said to have the *utmost range*: all others between 0° and 45°, are called the *intermediate ranges*.

On account of the resistance of the air, the ranges are altered, and that in different proportions, both for the different sizes of the shot, and their different velocities: so that the greatest range, in practice, always lies below the elevation of 45°, and the more below it as the shot is smaller, and as its velocity is greater: so as that the smallest balls, discharged with the greatest velocity in practice, ranges the farthest with an elevation of 30° or under, while the largest shot, with very small velocities, range farthest with an elevation of nearly 45°; and at all the intermediate degrees in the other cases. See the articles CANNON, GUNNERY, CHARGE, MORTAR, PROJECTILES, RESISTANCE, RICOCHET-FIRING, and ROCKET.

The following Tables, which are copied from SIMMONS'S *Sea-Gunner's Vade-Mecum*, exhibit the ranges of caronnades, cannon, and mortars, as ascertained by actual experiment:

1. *Ranges with Carronades.*—The charge being one-twelfth of the weight of the shot; and with one shell and one wad. The line of fire from six to nine feet above the level of the water.

	Species.	68	42	32	24	18	12
	Charge.	5lb. 8oz.	3lb. 8oz.	2lb. 10oz	2 lbs.	1lb. 8oz.	1 lb.
		Yards.	Yards.	Yards.	Yards.	Yards.	Yards.
Elevation.	Point-blank	450	400	330	300	270	230
	1 Degree,	650	600	560	500	470	400
	2 do.	890	860	830	780	730	690.
	3 do.	1000	980	900	870	800	740
	4 do.	1100	1020	970	920	870	810
	5 do.	1280	1170	1087	1050	1000	870

Ranges, with 8-inch Shells, from 68-pounder Carronades.

Weight of Shell.	Charge	Flight.	Elevation.	First graze.	Extreme range.
				Yards.	Yards.
43lbs 11oz.	3 lbs.	1½″	Point-blank	302	1365
	8....	—	5 Degrees	1140	
	4....	1½	1	358	1843
	.....	5	5	1137	1250
	.....	—	11½	1767	

2. *Ranges with Cannon, 32, 24, and 18-pounders.*

Elevation.	Proportion of Powder.	Species of Shot.	Range.
			Yards.
2°	⅓	With single shot to the first graze	1200
2°	¼	Ditto.	1000
2°	¼	Two shot, ranged close together, to.......	500
4°	⅓	Single shot	1600
4°	¼	Ditto	1500
7°	⅛	Ditto	2150
7°	¼	Ditto	2020
2°	¼	One round shot, and one round of grape, will range with effect together, to	600
4°	¼	One round of grape shot alone, to.........	1000
2°	¼	One double-headed or bar-shot, will range to the first graze......	800

Ranges with 5½-inch Shells from an Iron 24-pounder. Length of Gun, 9½ feet.—Weight, 49 cwt. 26lbs.

Elevation.	2 Pounds.			2 lbs. 8 oz.			3 Pounds.		
	Flight.	Range to First Graze	Range to Extreme	Flight.	Range to First Graze	Range to Extreme	Flight.	Range to First Graze	Range to Extreme
Deg	Sec.	Yards	Yards.	Sec.	Yards	Yards.	Sec.	Yards	Yards
1	1	213	1139	2¼	562	1456	1	277	1424
2	1¾	384	1267	1½	442	1413	1¾	526	1464
3	2¼	565	1413	2½	647	1553	2¼	740	1600
4	2½	750	1479	3¾	896	1639	3½	880	1679
5	3¾	836	1670	4	915	1510	5	1182	1733
6	4	896	1495	5	1140	1657	6¼	1384	1787
7	6¼	1180	1492	6	1205	1481	6¼	1410	1749
8	6¾	1305	1526	6½	1259	1544	7	1520	1744
9	7½	1329	1527	7	1341	1561	7¼	1722	1938
9½	6¾	1229	1453	.....	....	....	8½	1748	1831

3. *Ranges of 13 and 10-inch Sea-Mortars.—The 13-inch, at 45 degrees, 4100 yards: the 10-inch, 3800. The former with 32 lbs., the latter with 12 lb. 8 oz. of powder.*

Ranges with a 10-inch Sea-Mortar, at 21 degrees, on a Horizontal Plane.

Weight of Mortar.	Weight of Shell.	Charge.	Elevation.	Flight.	Range.
cwt. qrs. lbs.	lb. oz.	lb. oz.	Degrees.	Seconds.	Yards.
34 2 14	86 —	5 8	21	14½	2335
	87 —			16	2510

Ranges with Sea-service Iron Mortars, at 45 degrees, upon a Horizontal Plane.

13-Inch.			10 Inch.		
Charge.	Flight.	Range.	Charge.	Flight.	Range.
lbs. oz.	Seconds.	Yards.	lbs. oz.	Seconds.	Yards.
2 —	13	690	1 —	13	680
4 —	18	1400	2 —	18	1340
6 —	21	1900	3 —	21	1900
8 —	24½	2575	4 —	24½	2500
10 —	26½	2975	5 —	26	2800
12 —	29	3500	6 —	27	3200
14 —	29½	3860	7 —	29	3500
16 —	30	3900	8 —	30	3800
18 —	30½	4000	9 —	30½	3900
20 —	31	4200	9 8	30½	4000

As one of the effects of the shell results from its weight, the range of mortars is extremely different from that of cannon, because the former is not pointed at a certain object, like the latter, but inclined to the horizon at a certain angle; so that the shell being thrown up obliquely, much in the same direction as a tennis-ball struck by the racket, may fall upon the place intended. Hence it appears that the mortar has no point-blank range, or at least that no use is made of it.

Though a mortar may be elevated indifferently, either so much above or below 45° as to carry a shell to a given distance, yet when any building is to be destroyed, it should be raised above 45°, because the shell, rising to a greater height when fired according to a greater angle, falls with greater force, and consequently will do more damage to the place on which it is thrown. But when the business is, to fire on a body of men, the mortar must be pointed below 45°, that the shell may not have force enough to enter far into the ground, and that the splinters in the explosion may do more execution.

"It appears," says Dr. Hutton, "that the velocity of the ball increases with the increase of charge only to a certain point, which is peculiar to each gun, where it is greatest; and that by farther increasing the charge, the velocity gradually *diminishes*, till the bore is quite full of powder. That this charge for the greatest velocity is greater as the gun is longer, but yet not greater in so high a proportion as the length of the gun is; so that the part of the bore filled with powder bears a less proportion to the whole bore in the long guns, than it does in the shorter ones; the part which is filled being indeed nearly in the inverse ratio of the square root of the empty part.

"It appears too, that the velocity, with equal charges, always increases as the gun is longer; though the increase in velocity is but very small in comparison with the increase in length; the velocities being in a ratio somewhat less than that of the square roots of the length of the bore, but greater than that of the cube roots of the same, and is indeed nearly in the middle ratio between the two.

"It appears, again, from the table of ranges, that the range increases in a much lower ratio than the velocity, the gun and elevation being the same. And when this is compared with the proportion of the velocity and length of the gun in the last paragraph, it is evident that we gain *extremely little* in the range by a *great* increase in the length of the gun, with the same charge of powder. In fact the range is nearly as the 5th root of the length of the bore, which is so small an increase, as to amount only to about a seventh part more range for a double length of gun. From the same table it also appears, that the time of the ball's flight is nearly as the range; the gun and elevation being the same.

" It has been found, by these experiments, that no difference is caused in the velocity or the range, by varying the weight of the gun, nor by the use of wads, nor by different degrees of ramming, nor by firing the charge of powder in different parts of it. But that a very great difference in the velocity arises from a small degree in the windage: indeed, with the usual established windage only, viz. about 1-twentieth of the calibre, no less than between 1-third and 1-fourth of the powder escapes and is lost; and as the balls are often smaller than the regulated size, it frequently happens that *half* the powder is lost by unnecessary windage.

" It appears too, that the resisting force of wood, to balls fired into it, is not constant; and that the depth penetrated by balls, with different velocities or charges, are nearly as the *logarithms* of the charges, instead of being as the charges themselves, or, which is the same thing, as the square of the velocity.—Lastly, these and most other experiments shew, that balls are greatly deflected from the direction in which they are projected: and that frequently as much as 300 or 400 yards in a range of a mile, or almost 1-fourth of the range." See HUTTON's *Mathematical and Philosophical Tracts*, in 3 vols. 8vo.

Before we close this article, it may be proper to observe, that the range of cannon is greater in the morning and at night, than at noon, and in cold than in hot weather: for, at these times, the air being less heated, gives less way to the dilation of the powder, which being by this means confined as it were to a smaller sphere of action, must have a stronger effect in proportion.

Horizontal RANGE *of a Piece of Ordnance.* See the article HORIZONTAL.

Point-Blank RANGE. *(portée d'un canon tiré de but en blanc,* Fr.) See the article POINT-BLANK.

RANK, (*rang*, Fr.) the relative situation which officers hold with respect to each other, or to naval and military things in general.

The commissioned officers of his Majesty's navy are divided into the following ranks and denominations, viz Flag-officers.—Captains, who command post ships.—Commanders, who command sloops.—Lieutenants.—Sub-Lieutenants.

The flag-officers are divided into three ranks, viz. admirals, vice-admirals, rear-admirals.—Each rank is divided into three squadrons, distinguished by flags of different colours; they are to rank and command in the following order, viz.

Admiral and commander in chief of the fleet.

Admirals of the red, of the white, and of the blue.

Vice-admirals of the red, of the white, and of the blue.

Rear-admirals of the red, of the white, and of the blue.

The admiral of the fleet is to carry an union flag at the main-top-gallant-mast-head; the other admirals are to carry the flag by which the squadron they belong to is distinguished, admirals carrying it at the main, vice-admirals at the fore, and rear-admirals at the mizen-top-gallant-mast-head.

Flag-officers of the same rank are to command according to the priority of their commissions, or, if they have commissions of the same date, according to the order in which they stand on the list of the navy.

There is frequently a temporary rank of commodore, which is distinguished by a broad pendant, red, white, or blue; and commences when the pendant is hoisted, and ceases when it is struck.

If a captain is appointed to command the ship which carries such pendant, the commodore shall be paid as the junior rear-admiral, and rank next to the captain of the fleet, who, if not a flag-officer, is to take rank after the junior rear-admiral.

Captains and commanders are to take precedence, and to command in their respective ranks, according to the dates of their commissions, or having commissions of the same date, according to the order in which they stand on the list of the navy. The same observation holds good with respect to lieutenants, as well in the ships to which they belong, as on any service in which the lieutenants of several ships shall be employed; except the lieutenants of flag-ships, who are in all situations to take precedence, and to command, in such order as the flag-officer shall direct. The same regulations apply to sub-lieutenants.

The order in which officers shall take precedence and command in the ship to which they belong, is as follows:—Captain or commander, lieutenant, sub-lieutenant, master, second master, gunner, boatswain, carpenter, master's mate, and midshipmen.

For farther particulars respecting Naval Rank, &c. See the NAVAL INSTRUCTIONS.

To RANK with, is to hold the same relative situation with regard to others. Thus the corresponding rank between naval and military officers is as follows:

NAVY.	ARMY.
Admiral, or Commander in Chief of his Majesty's fleet	Field-Marshal.
Admirals, with their flags at the main-top-mast head	Generals of horse and foot.
Vice Admirals	Lieutenant Generals.
Rear Admirals	Major Generals.
Commodores, with broad pendants	Brigadier Generals.
Captains commanding post-ships, after three years from the date of their first commisssion	Colonels.
All captains commanding post-ships	Lieutenant Colonels.
Captains not taking post	Majors.
Lieutenants	Captains.

Nothing in this, however, shall give any pretence to military officers to command any of his Majesty's squadrons, nor to naval officers to command on shore; nor shall either have right to demand the military honours due to their respective ranks, unless upon actual service.

Rank is sometimes given to persons holding civil situations in the navy and army, in order to secure to them the privilege and advantages, particularly in case of capture, which military commissioned officers enjoy. Thus pay-masters in the army, and surgeons in the navy, rank as captains youngest of the line. The surgeons in the navy obtained this privilege in the year 1798.

RANSOM, (*rançon*, Fr.) was the sum formerly given by captains or passengers for the redemption of a vessel captured by pirates. This is now prohibited by statute.

RAREFACTION, (*raréfaction*, Fr.) is the extension of the parts of a body, by which it is made to take up more room than it did before. It is essentially connected with gunnery; for in proportion to the rapid combustion and consequent rarefaction of air, produced by the ignition of gunpowder confined in the chamber of a gun, so will be the force of explosion with which the charge is propelled.

RASING, the act of marking, by the edges of moulds, any figure upon timber, &c. with a raising-knife, or with the points of compasses.

Rasing-Knife, a small edged tool, fixed in a handle, and used for rasing particular marks on timber, lead, tin, &c.

RASP, (*râpe*, Fr.) in block-making, a steel implement with short sharp teeth upon it: it is fixed in a wooden handle, and used in rubbing down the edges of the shells of blocks, &c.

RATE, (*rang des vaisseaux de guerre*, Fr.) in the royal navy, is the order or class, degree, or distinction into which the ships of war are divided, according to their force, burden, &c.

The regulation which limits the rates of men of war to the smallest number possible, seems to have been dictated by considerations of political economy, or of that of the simplicity of the service in the royal dock-yards.

The British navy is accordingly distributed into *six* rates, exclusive of the inferior vessels that usually attend on naval armaments; as sloops of war, gun-vessels, armed ships, bomb-ketches, fire-ships, and cutters, or schooners.

In general, the ships of every rate, besides the captain, have a chaplain, master, surgeon, purser, boatswain, gunner, and carpenter; all of whom, except the chaplain, have their mates or assistants, in which are comprehended the sail-maker, the calker, the rope-maker, the master at arms, the armourer, the captain's clerk, the gunsmith, &c.

The number of officers are always in proportion to the rate of the ship. Thus, a first-rate has 8 lieutenants, 6 master's mates, 24 midshipmen, and 3 surgeon's mates, who are considered as gentlemen; besides the following petty officers: quarter-masters, and their mates, 14; boatswains mates and yeomen, 8; gunners mates and assistants, 8; carpenters mates, 2, besides 10 assistants; quarter-gunners, 30; with one steward, and a steward's mate to the purser. For the established number of commissioned, warrant, and petty officers, together with their mates and assistants, allowed to each rate, see the table under the article Pay, and the article Royal Marines.

The first rate ships of 100 guns and upwards, were formerly conceived to be incapable of government, unless in the most favourable weather in the summer; but, on account of the great improvement of the marine art, and the degree of perfection to which seamanship is arrived in this country, they are now managed with the greatest facility.

Ships of the second rate, below 100 guns, which have three decks, carry their sails remarkably well, and labour very little at sea. They are well calculated for a general action, or for cannonading a fortress.

Those of the third rate, carrying from 64 to 74 guns, and having two tiers, are fit for the line of battle, to lead the squadrons of ships of war into action; and, in general, to suit the different exigences of the naval service.

The fourth rates of 50 guns, having two decks and a poop, may be employed on the same occasions as the third rates, and may also be destined to protect our foreign colonies, or on expeditions of great distance, since they are generally excellent sea-going ships. They are also destined to lead the convoys of merchant-ships, to protect our colonies and commerce, to cruise in different stations, to accompany squadrons, or to convey intelligence, and are frequently used as flag-ships on foreign stations, in time of peace.

All vessels of war under the fourth rate are usually comprehended under the denomination of frigates, and never appear in the line of battle, although there are some attached to every squadron to look out, repeat signals, and to assist ships that may have been disabled after an action. They are divided into fifth and sixth rates; the former carry from 32 to 44 guns upon one deck, those carrying 44 guns upon two decks being now discontinued; and the latter, from 20 to 28, both inclusive. Those which carry from 28 to 38 guns upon one deck, with the quarter-deck, are extremely proper for cruising against an enemy in short expeditions, being light, long, and usually excellent sailers.

Bomb-vessels are on the same establishment as sloops; but fire-ships and gun-vessels are on that of fifth rates.

Cutters and schooners are generally commanded by lieutenants; and, being swift sailers, are employed to carry despatches and convey intelligence from one ship or place to another.

If the dimensions of all ships of the same rate were equal, it would be the simplest and most perspicuous method to collect them into one point of view in a table; but as there is no invariable rule for the general dimensions, it must suffice to give those of some particular ships in each rate; for which purpose we have been kindly furnished with a list of some of the latest construction, which is as follows:

TABLE OF DIMENSIONS, &c. *of the different Rates, when and where built; the Number of Men and Guns to each, and the Nature of Metal on each Deck, &c.*

Rate	Guns.	Ships Names.	Dimensions. Length of the GunDeck. Feet In.	Dimensions. Length of the Keel for Tonnage. Feet In.	Dimensions. Breadth Extreme Feet In.	Dimensions. Depth in Hold. Feet In.	Tons.	Built. When	Built. Where.	Light Draught of Water. Afore. Feet. In.	Light Draught of Water. Abaft. Feet. In.	No. of Men.	Gun Deck. No.	Gun Deck. Prs.	Middle Deck. No.	Middle Deck. Prs.	Upper Deck. No.	Upper Deck. Prs.	Quarter Deck. No.	Quarter Deck. Prs.	Fore-castle. No.	Fore-castle. Prs.	By whose Draught, or how came into the Service.
FIRST RATE.	120	Caledonia	205 0	170 9¼	53 8	23 2	2616	1808	Plymouth	15 10	18 2	875	32	32	32	24	34	18	6 a10	12 32	2 a2	12 32	Sir Wm. Rule.
	—	Nelson	205 0	170 10¼	53 6	24 0	2601	Build	ing at Woolwich.			875	32	32	34	24	34	18	6 a10 b6	12 32 18	2 a2	12 32	Surveyors of the Navy.
	110	Hibernia	201 2	167 4¼	53 1	22 4	2508	1804	Plymouth.	16 4	18 7	875	32	32	32	24	34	18	4 a12	18 32	2 a4	18 32	Sir John Henslow.
	100	Queen Charlotte	190 0½	156 1¾	52 6	23 9	2289	1810	Deptford,	16 3	19 9	837	30	32	30	24	30	12	6 a8 b6	12 32 18	2 a2	12 32	Similar to the old Q. Charlotte.
SECOND RATE.	98	Boyne	186 3	153 2¾	51 5	22 1	2155	—	Portsmouth.	14 6	18 1	738	28	32	30	18	30	12	4 a8	12 32	2 a2	12 32	Draught of the Victory.
	—	Impregnable	196 0	163 6¼	51 2	22 1½	2278	—	Chatham.	14 3	17 4	738	28	32	30	18	30	12	6 a8 b6	12 32 18	2 a2	12 32	
THIRD RATE.	80	Talavera	192 0	159 10	49 0	21 0	2041	Build	ing at Portsmouth.			670	30	32			32	18	4 a10	12 32	2 a2	12 32	
	—	Foudroyante	183 8½	151 1¼	50 10	22 6	2062	1798	Plymouth.	14 3	18 7	700	30	32			32	24	14	12	4	12	
	—	Christian VIIth.	187 0	154 9½	51 6¼	20 10	2131	1804	Denmark.			670	30	32			32	18	4 a10	24 32	2 a2	24 32	Captured from the Danes in 1607.
	74	America	176 0	144 11½	47 9	21 0	1758	1810	River Thames.	12 9	17 8	590	28	32			28	18	4 a10 b6	12 32 18	2 a2	12 32	
	—	Centaur	176 0	144 3	49 0	20 0	1842	1797	Woolwich.	13 0	17 9	640	28	32			30	24	14	9	2	24	
	—	Sultan	175 0	144 1½	47 6	20 6	1729¾	1607	Ditto.	13 8	17 7	640	28	32			28	18	4 a10 b6	12 32 18	2 a2	9 32	
	—	Superb	184 1	149 10	49 2	21 10½	1927	1798	Northfleet.	13 6	18 3	640	30	32			30	24	4 2 a10 b6	9 24 32 24	4 2 a2	9 24 32	Similar to the Pompée (Fr.)
	64	Standard	159 6	131 0	44 4	19 0	1370	1782	Deptford.	12 2	17 1	491	26	24			26	18	2 a10 b4	9 32 18	4 a2	9 32	Similar to the Intrepid (Fr.)
FOURTH RATE.	50	Isis	154 0	127 3¾	41 11	17 6	1119	Build	ing at Woolwich.														
	—	Grampus	151 0	124 7½	41 0	17 8	1114	1802	Portsmouth.	11 9	16 7	343	22	24			22	12	4	6	2	6	
FIFTH RATE.	—	Leopard	146 5	120 0¼	40 8	17 6	1056	1790	Sheerness.	10 3	15 8	343	22	24			22	12	4	6	2	6	Similar to the Old Portland.
	44	Ulysses	140 0	115 5½	38 0½	16 4¾	887	1779	Liverpool.	10 11	13 0	294	20	18			22	18	2	18			Roebuck.
	40	Endymion	159 3½	132 3	42 7¼	12 4	1277	1797	River Thames.	10 3	15 5	320					26	24	a2	24	a5	24	Old Pomone.
	38	Lacedemonian	150 1½	125 4¾	39 11	12 9	1063	1813	Portsmouth.			300					28	18	a14	32	a6 2	32 9	Old Hebe.
	—	Horatio	154 3	129 11¾	39 8½	13 7	1090	1807	Bursledon.	10 6	15 1	284					28	18	a14 2	32 9	a2 2	32 9	Lively.
	36	Belvidera	145 0¾	121 1¼	38 2½	13 3	946	1809	Deptford.	10 9	14 0	274					26	18	a14	32	a14 2	32 9	Euryalus.
	—	Pyramus	141 1	117 2½	38 5	12 0	920	1810	Portsmouth	10 3	12 9	264					26	18	a12 2	32 9	a2 2	32 9	Belle Poule.
	32	Medusa	144 0	121 6	37 8½	12 5½	920	1801	Northfleet.	10 5	13 10	254					26	24	a,2 2	32 6	a2 2	32 6	Amphion.
SIXTH RATE.	—	Thames	127 1	107 0¾	34 1	11 9	661	1805	Chatham.	9 10	12 6	215					26	12	a8	24	a2	24	
	28	Aurora	120 6	99 4	33 7	11 0	596	1777	River Thames.	8 2½	12 4½	195					24	9	a8 4	24 6	a4	24	Enterprize.
	22	Cossack	117 11¼	98 4½	33 3½	10 6	546	1806	South Shields.	9 1	12 11	175					2 a20	9 32	2	6	2	6	
	—	Comus	120 10	98 7	31 6¼	10 3	522	—	Yarmouth.	9 0	11 11	155							a4	18	a2 2	18 6	
	20	Camilla	108 1¼	89 10½	30 1	9 8	433	1776	Chatham.	7 4	12 6	155					a22 20	32 9	a6	24			
		Cygnet (Sloop) rigged as a ship	106 0	87 7	28 0	13 9	365	1804	Yarmouth.	7 1½	10 7	121					2 a16	6 32					
		Peruvian (Sloop) rigged as a Brig	99 10	77 0¼	30 7	13 0	383	1808	Bursledon.	6 9	11 1	121					2 a16	6 32					
		Orestes (Brig)	93 0	76 0¼	26 6	12 0	284	1805	Ipswich.	7 8	8 2	95					a14 2	24 6					

N. B. As types could not be procured small enough to give the correct fractional parts arising from the dimensions of the gun-deck and keel for tonnage, the nearest proportional fractions, viz. ¼, ½, and ¾, have, therefore, been substituted. *a* indicates Carronades on the upper and quarter-decks; and *b* on the Round House.

RATE *of Sailing*, (*marche d'un bâtiment*, Fr.) in navigation, denotes the distance a ship has run from one place to another in a given period of time, which may be ascertained pretty correctly by a proper attention to the log-line and half-minute glass. See the article LOG.

Captain James Burney, of the royal navy, has lately proposed a "*New Method of measuring a Ship's Rate of Sailing*;" which is as follows:—

"A line towing astern of a vessel which is passing through the water, will pull against her head-way. As the ship's way increases, the pull of the line will increase; and *vice versa*.

"If this, with a proper scope of line (about 25 fathoms may probably be sufficient) shall be found to be a regulated quantity of pull, corresponding in the same manner, at all times, to the rate of sailing, it will answer the purpose of a log.

"Many experiments have been made upon the same principle; but the most plain and easy one, of towing a measured length of line, has escaped trial; though less liable to give erroneous or variable results than any which can be made near a ship. By it the rate of sailing may be obtained either constantly or occasionally, and can be taken with ease by one person: in which respect it would have great advantage over the common log; the use of which requires three persons.

"By a trial made in a boat with about twenty fathoms of line, rather larger than log-line, towing astern, and fastened to a spring steelyard, the strength of the pull was found to vary with the rate of sailing, which, however, was not ascertained by measurement, but by estimation; the boat's rate of sailing during the trial varied between $2\frac{1}{2}$ knots and 5 knots per hour, and the pull of the line upon the steelyards was observed to vary from 2 lbs. to $5\frac{1}{4}$ lbs. increasing and decreasing with the velocity. So great a variation in the strength of the pull gives all the advantage which can be desired for forming a scale, and will allow of the experiment being made with smaller line.

"If the proposed length of the line is passed through a pulley, so as to go clear out at the stern-port or cabin window, and the inner end is fastened to a loose chain, of weight adapted to the purpose, on the deck under the pulley; or to a number of small weights made consecutive by short intervals of line, the chain or weights will be drawn up more or less, according to the ship's velocity. By a few comparisons of the quantity of weight raised from the deck with the rate of sailing, a scale may be marked.

"In an improved state of the experiment, instead of using weights or a pulley, the inner end of the line (coming direct from the water) can be fastened to a spring, and communicate with an index that shall express the rate of sailing.

"This machine (if so plain a contrivance deserves that name) may be put on constant duty, or dropped occasionally to ascertain the rate.

"The objections which are likely to occur are,

1st. "The line being liable to contraction or expansion, as the temperature of the water varies. But it is scarcely to be supposed that the greatest contraction or expansion of line from its mean state (after it has been properly stretched and seasoned) will occasion an alteration of a hundredth part in the force of the pull.

2d. "That in a fresh wind the part of the line between the ship and the surface of the water, will be liable to some additional pull from being exposed to the wind. To this inconvenience the log-line, in the common way of heaving the log, is likewise exposed, when the wind is much aft. In either case, when the ship is not right before the wind, the remedy is the same; which is, to throw the log or the line over from before the lee gangway, and to give a few fathoms more of stray line; which, however, in the new method proposed, it would be necessary to apply a correction, the quantity of which may be accurately ascertained.

3d. "The motion of a ship in pitching. But this is not to be regarded as an objection; for the rate of sailing is to be estimated only by what the experiment shews when the ship is going steadily; in the same manner as in taking bearings, if the compass swings we wait till it is quiet. Whenever the ship goes steadily for ten seconds together, or even five seconds, the pull of the line will be regulated to the average rate of sailing." See *Captain* BURNEY'S *Experiments* (lately published) *for ascertaining the Directions of Currents, &c.*

Mr. Pitot has contrived an instrument for measuring a ship's velocity in the water, which consists of two tubes, one with a trumpet mouth, turned forward, to face the impulse of the water, the other rising vertically from the bottom; the difference of the height of the water in these two tubes, he supposes, will denote the velocity of the ship. See GREGORY'S *Mechanics*.

Mr. Hamill has made some observations in No. 58 of the Philosophical Journal, on this contrivance; and seems to think the fluctuation of the sea would render the rise of the water in the tubes undulatory and irregular, and to remedy this defect proposes that the orifice of the trumpet-tube, within the ship, shall be on a level with the water, and have a receiver annexed to it to catch the water which will run over from it when the ship is under way; by the quantity of water flowing into the receiver in a given time, and through a given aperture, he supposes the rate of a vessel may be better ascertained than by the method of Mr. Pitot.

In the third volume of the American Philosophical Transactions, a similar instrument is described by a gentleman of the name of Hopkinson, of Philadelphia, but without a trumpet head or secondary tube.

Captain Hamilton, now a commissioner in the royal navy, published an account of a contrivance of this sort in the *Naval Chronicle* for July 1801, which Mr. Pitot also recommended, and is as follows:—

Let a half-inch pipe of stout copper, with the end closed, be run through a hole bored for the purpose, in the bottom of a ship, at nearly the distance of a foot from the keel. The pipe to be about five inches without the hole bored. Near the middle length of the ship, or in the foremost part of the well, is the best and most convenient place for the apparatus.

On the kelson, opposite to the hole, let a tube of wood, square, three inches in the clear, perpendicular to the kelson, and six feet or more above the load water-line of the ship, be fixed; the lower end of the tube to be closed. Let the part of the copper pipe that is within the ship be bent, and the end left open let into the lower part of the wooden tube, a little above its base.

Let a quarter-inch round hole be bored in the fore part of the copper pipe, the eighth of an inch from its closed end, below the ship (to admit the water when afloat into the wooden tube), the plane of the hole perpendicular to the keel. And let every part of the described apparatus be water-tight, except the last-mentioned hole.

The Captain further proposed, the following *Contrivance to determine the quantity sailed in a given time.*

Assume a point on the after-part of the wooden tube, supposed in the plane of the ship's flotation, when the ship is in *lightest* sailing trim; at three or four inches below that point make a mark, and with the distance between deep and light load water-line, make a second and lower-mark. Between the marks attach a small pipe or tube of wood, an inch or an inch and a quarter in the clear, closed at the lower end, and communicating with the larger tube by a small pipe. The length of the attached tube for all ships of war may be three feet. Within this attached tube, through collars of leather, are two small copper pipes, to move up and down, conformably to the ship's plane of flotation.

At the lower end of the least of the copper pipes, is bored a capillary hole of the thirty-second of an inch in diameter, or else a glass end with a capillary bore as above. This size is chosen merely from convenience. The capillary hole should be covered with a screen or cap, full of smaller holes, to prevent its being choked, though the water at the bottom of a ship is usually clear; the small holes should be made only in the sides, and not in the bottom of the screen.

Let a cistern, made cubical, if convenient, containing about thirty gallons, with a guage to indicate the quantity of water within it, and cocks on each side to let the water out. At the head of the small pipe above mentioned, is a spout to discharge water into the cistern. For further particulars and a figure of this contrivance, see the *Naval Chronicle* for July 1801.

It appears that the height of the water in these tubes will depend on that of the external water, and on the *obliquity* of the aperture to the impulse of the water, as well as on the velocity of that impulse. Every wave that the ship passes must, therefore, affect the elevation of the water in the tube, and this very irregularly, as it would require *some time* to convey the effect through the tube.

In Captain Hamilton's tube the velocity of the impulse would be, in a great measure, lost against the opposite side of the tube, by its vertical position. The others (where this defect is removed by the gradual flexure of the pipes from a horizontal to a perpendicular position), as well as his, are merely calculated to operate when sailing before the wind; when sailing with a side wind, the impulse of the water on the external aperture would be *oblique*; and this, according to Venturi's accurate experiments, would produce variations so great, as in some cases even to cause the water to descend in the tubes when those gentlemen would expect it to rise; and we much doubt that an accurate instrument for measuring a ship's way can be made on these principles.

RATING, implies the advancement of deserving seamen to petty officers in his Majesty's ships, which is done at the pleasure of the captain, according to their behaviour and expertness in their duty; as, landmen to ordinary and able-bodied seamen, able-bodied seamen to captains of the fore-castle, tops, after-guard, mast, hold, &c.

Although the sail-maker, calker, rope-maker, armourer, and the ship's cook are considered as petty officers, yet they are not rated in the manner before mentioned, but have their appointment to those situations by warrant from the Navy Board. For farther particulars see the article PROMOTION.

RATION, (*ration*, Fr.) a certain allowance of beef or pork, biscuit, with beer, wine or rum, served out to each person on board his Majesty's ships for his daily subsistence. For the exact daily rations thus distributed see the article PROVISIONS.

RATLINES, or RATLINGS, (*enflêchures; c'est aussi les quarantaniers, menus cordages dont on fait les enflêchures*, Fr.) are small lines, which traverse the shrouds of a ship horizontally, at regular distances from the deck upwards, and forming a variety of ladders, whereby to climb or to descend from any of the mast-heads.

To RATTLE DOWN THE SHROUDS, (*mettre des enflêchures aux haubans*, Fr.) is to fix the ratlines to them, in order to prevent them from slipping down by the weight of the sailors: they are firmly attached by a knot called a clove-hitch, to all the shrouds except the fore-most or aft-most; where one of the ends being fitted with an eye-splice, is previously fastened with twine or packthread.

RAY, in optics, a beam or line of light, propagated from a radiant point, through any medium.

If the parts of a ray of light lie all in a straight line between the radiant point and the eye, the ray is said to be *direct*: the laws and properties of which make the subject of optics. If any of them be turned out of that direction, or bent in their passage, the ray is said to be *refracted*. If it strike on the surface of any body, and be thrown off again, it is said to be *reflected*. In each case, the ray, as it falls either directly on the eye, or on the point of reflection or of refraction, is said to be *incident*.

Again, if several rays be propagated from the radiant object equidistantly from one another, they are called *parallel* rays. If they come inclining towards each other, they are called *converging* rays. And if they go continually receding from each other, they are called *diverging* rays.

It is from the different circumstances of rays that the several kinds of bodies are distinguished in optics. See the articles LIGHT, REFRACTION *of Light*, and REFLECTION *of the Rays of Light*.

RAZEE (*vaisseau rasé*, Fr.) in the royal navy, an appellation given to a two-decked ship, when the round-house, quarter-deck, and forecastle, are cut down forward and aft to the upper-deck sills, and in midships flush with the deck. In this operation the upper-deck is converted into a quarter-deck, forecastle, and gangways, and the lower-deck of course becomes the upper-deck. Two-decked ships thus cut down have great advantages over the enemies frigates, as they carry their guns much higher out of water, and bear a greater weight of metal; they have also a greater height between decks, which is more convenient both to officers and men. They generally carry 28 long 24 pounders on what now becomes the main-deck, and carronades, &c. on the quarter-deck and forecastle; and have a complement of 470 men.

The Indefatigable, Majestic, Magnanime, and Saturn have been cut down for razees; but a larger class of frigates are now laid down in our dock-yards, in order to supersede the necessity of such a measure in future.

REACH, (*l'étendue en ligne droite du courant d'une rivière, comprise entre deux pointes ou contours; au bout de ces étendues & routes en ligne droite un vaisseau est obligé de manoeuvrer, ou de changer de route, ainsi on dit, dans la Tamise, Long-reach, Blackwall-reach, &c. ce terme n'a point d'équivalent en Francois*, Fr.) the line or distance comprehended between any two points or stations on the banks of a river, wherein the current flows in a strait uninterrupted course; as, Woolwich-reach, Blackwall-reach, Long-reach, in the county of Kent.

REACHING-POST, in rope-making, a post sixteen or eighteen inches in diameter, and about four feet high, fixed in the ground at the lower-end of a rope-walk. It is used in stretching the yarn by means of a tackle, one of the blocks of which is hooked to a strap round the post, the other to a pendant at the sledge, and are about eighteen yards distant from each other.

READY ABOUT! (*pare à virer!* Fr.) an order given to the crew, implying that all hands are to be attentive and at their stations, ready for tacking.

REAMING, in block-making, the act of increasing the size of a hole with a large instrument.

REAR, (*arrière-garde*, Fr.) a name given to the last division of a squadron, or the last squadron of a fleet, and which is accordingly commanded by a third officer of the said fleet or squadron. See the article DIVISION.

REAR-ADMIRAL, (*contre-amiral; il porte le pavillon quarré au mât d'artimon*, Fr.) in the royal navy, an officer next in rank to a vice-admiral. See the article ADMIRAL.

RECHANGE, among seamen, a term sometimes used for such tackle and materials as are kept in reserve on board the ship, to serve in case of failure of those already in use.

RECKONING, or, *a Ship's* RECKONING, (*estime*, Fr.) in navigation, is that account by which it may be known where the ship is at any time, and on what course or courses she is to steer, in order to gain her intended port. See the article DEAD-RECKONING.

RECOIL, (*recul*, Fr.) the falling back of any piece of fire arms, when discharged of its contents. This repulsive motion on the breach of a piece of ordnance, which is always subject to a recoil, according to the size and the charge it contains, is occasioned by the rarified air pressing on all sides, in order to expand itself with freedom. Guns, whose vents are a little forward in the chase, recoil most. To lessen the recoil of a gun, the platforms are generally made sloping towards the embrasures of the battery.

To RECOIL, (*reculer*, Fr.) is to run back in consequence of resistance or repercussion.

The Recoil of Sea-Service Iron Guns, on Ship-Carriages, upon a Horizontal Platform, is as follows:—

Charge of Powder and Shot.	Elevation.	32 pounders	24 pounders	18 pounders
	Deg.	Ft. In.	Ft. In.	Ft. In.
$\frac{1}{3}$ of Powder and 1 Shot	2	11 —	11 —	10 6
$\frac{1}{3}$ of Powder and 2 Shot	4	19 6	18 6	18 —
$\frac{1}{4}$ of Powder and 2 Shot	7	11 6	12 —	12 —

For a plan and explanation lately contrived by Vice-admiral John Hunter, for checking the recoil of guns in line of battle ships. See *Naval Chronicle*, Vol. XXVI.

RECONCILER, or TOP-TIMBER-HOLLOW, among ship-builders, is a mould sometimes used to form the hollow in the top-side, called the *reconciling mould*.

RECONCILING, in mast-making, denotes the making one fair surface of two pieces. Thus the workmen are frequently directed to leave a certain substance for *reconciling*.

To RECONNOITRE A FLEET, PORT, or HARBOUR, (*reconnoître*, Fr.) is to view or examine its present force, state, or condition, as accurate as circumstances will allow, in order to make a report thereof: as, "We reconnoitred the enemy's fleet, and saw 20 sail of the line, besides frigates."

To RECRUIT FOR THE NAVY, (*recruter*, Fr.) is to raise men for manning his Majesty's ships and vessels at-sea, which is done by holding out a certain bounty for the services of volunteers: but on any emergency of the state, impressing seamen is resorted to for that purpose. See the article *Manning the* NAVY.

RECTANGLE, in geometry, is a right-angled parallelogram, or a right-angled quadrilateral figure. In arithmetic and algebra, a rectangle signifies the same with factum or product.

RECTANGLED, RECTANGULAR, or RIGHT-ANGLED, are appellations given to figures and solids which have one or more right-angles; thus, a triangle with one right-angle is termed a rectangled-triangle: also parallelograms with right-angles, squares, cubes, &c. are rectangular. Solids, as cones, cylinders, &c. are also said to be rectangular, with respect to their situation, when their axes are perpendicular to the plane of the horizon. The ancient geometricians always called the parabola the rectangular section of a cone.

RECTIFIER, in navigation, an instrument consisting of two parts, which are two circles, either laid one upon, or let into, the other, and so fastened together in their centres, that they represent two compasses, one fixed, the other moveable; each of them divided into thirty-two points of the compass, and three hundred and sixty degrees, and numbered both ways, from the north and south, ending at the east and west, in ninety degrees. The fixed compass represents the horizon in which the north and all other points, are liable to variation. In the centre of the moveable compass is fastened a silk thread, long enough to reach the outside of the fixed compass; but if the instrument be made of wood, an index is used instead of the thread. It is used to determine the variation of the compass, in order to rectify the ship's course, &c.

RECTIFYING OF THE GLOBE OR SPHERE, is a previous adjustment of it, to prepare it for the solution of problems. This usually consists in placing it in the same position as the true sphere of the world is at some certain time proposed; which is done, first, by elevating the pole above the horizon as much as the latitude of the place is, then bringing the sun's place for the given day, found in the ecliptic, to the graduated side of the brass meridian, next move the hour index to the upper hour of 12, so shall the globe be rectified for noon of that day; and if the globe be turned about till the hour index point at any proposed hour, then is the globe in the real position of the earth at that time, if placed due north and south by means of the compass.

RECTILINEAL, or RECTILINEAR, is the quality or nature of figures that are bounded by right lines, or formed by right lines.

RED-HOT SHOT, (*boulets rouges*, Fr.) an appellation given to large iron balls that are heated red-hot in certain furnaces made for that purpose, when used on shore; or in a forge, when used on board ships.

When fired, they are intended not only to dilapidate, but to set on fire the place or the ship in which they lodge, and to cause irreparable damage.

The mode of charging with those shot, without the precaution and use of tin cannisters, is as follows: The powder for firing must be put in strong flannel cartridges, without any holes, lest some grains should remain in the bore, when putting the cartridge home. Over the powder must be rammed a good dry wad, then a damp one, and then the hot shot; and if the gun lies at a depression, there must be a wad over the shot, which may be rammed home. If these precautions be attended to, the gun may be pointed after being loaded, without the smallest danger; as it is well known that the shot will grow cold in the gun without burning more than a few threads of the wads next to it. This was the manner of loading with hot shot during the siege of Gibraltar: but by putting the shot, when heated, into a tin cannister, is certainly the safest way, and an effectual method of preventing accidents.

REEDS FOR FIRESHIPS, (*un roseau, une canne*, Fr.) are made up in small bundles of about a foot in circumference, cut even at both ends, and tied together in two places. They are distinguished into two kinds; viz. the long and short; the former of which are four feet, and the latter two feet five inches in length. One part of them are singly dipped, that is, at one end; the rest are dipped at both ends in a kettle of melted composition, and being immersed about seven or eight inches in this preparation and then drained, they are sprinkled over with pulverised sulphur upon a tanned hide.

REEF, (*ressif*, Fr.) a chain of rocks, lying generally near the surface of the water.

REEF, (*ris des voiles*, Fr. *reef*, Dut.) a certain portion of a sail, comprehended between the top or bottom, and a row of eyelet-holes generally parallel thereto.

The intention of the reef is to reduce the surface of the sail in proportion to the increase of the wind; for which reason there are several reefs parallel to each other in the superior sails, whereby they may be still farther diminished, in order to correspond with the several degrees of the gale. Thus the top-sails of ships are usually furnished with four reefs, as shewn at *l m n*, and the reef of the main-course *s*, plate XX, parallel to the yard; and there are always three or four reefs, parallel to the bottom on those main-sails and fore-sails, which are extended upon booms; a circumstance common to many of the small vessels.

To REEF, (*prendre des ris*, Fr.) is to reduce a sail, by taking in one or more of the reefs, and is either performed with lines, points, or knittles.

All Hands REEF *Topsails!* (*tout le monde en haut à prendre des ris!* Fr.) the order to the seamen to go aloft with all speed, and reef those sails which are next above the courses; viz. the main-sail, fore-sail, and mizen.

The topsails are always, and the courses generally, reefed with points, which are flat braided pieces of cordage, whose lengths are nearly double the circumference of the yard. These being inserted in the eyelet-holes, are fixed in the sail by means of two knots in the middle, one of which is before, and the other behind the reef-band.

In order to reef the top-sails with more facility and expedition, they are lowered down and made to shiver in the wind; the extremities of the reef are then drawn up to the yard arms by the reef-tackle, where they are securely fastened by the earings. The space of sail, comprehended in the reef, is then laid smoothly over the yard, in several folds, and the whole is completed by tying the points about the yard, so as to bind the reef close up to it.

In reefing a course, the after-end of the point should be thrust forward between the head of the sail and the yard; and the fore-leg of the same point should come aft over the head of the sail, and also under the yard, and thus crossed over the head of the sail; the two ends should be tied on the upper side of the yard as light as possible.

When a sail is reefed at the bottom, it is generally done with knittles, in the room of points; or in large sails, such as the main-sails of armed cutters, pieces of line termed reef-hanks, are fixed in the eyelet-holes. For other methods of reducing a sail see the articles BALANCE and GOOSE WING.

REEF-BAND, (*bande des ris*, Fr.) a piece of canvas, sewed across the sail, to strengthen it in the place where the eyelet-holes of the reefs are formed.

Captain Malcolm Cowan, of the royal navy, has, within the last ten years, made an improvement both in the construction and the reefing of sails of ships and vessels, which renders the operation of reefing the courses more easy and expeditious, and with a less number of men than the general method requires. From this improvement, he states, that the courses, in line-of-battle ships, are reefed at the foot by a few men; and it has been proved, by experiment in several ships, that one of the courses can be reefed in two minutes, without hauling it up, or starting tack or sheet, or shaking the sail, which he seems to think may become a valuable acquisition to the naval service, particularly when vessels are on a lee shore in a gale of wind.

Reef-Hanks, are short pieces of small line, sewed at certain distances on the reefs of boom-sails.

Reef-Lines, (*garcettes de ris, dont on faisoit autrefois usage pour prendre des ris aux voiles basses,* Fr.) are small ropes by which the seamen formerly reefed the courses, by passing them spirally through the holes of the reefs, and over the head of the sail, alternately, from the yarms to the slings, and then straining them taught.

Reef-Tackle, (*palan de ris,* Fr.) a rope which passes from the deck through a block at the topmast-head, and thence through a hole in the topsail-yard-arm, and is afterwards attached to a cringle, a little below the lowest reef, as exhibited in plate XX. where *h h* are the reef-tackles, and *i* the pendent thereof. It is used to pull the skirts of the topsails close up to the extremities of the topsail-yards, in order to lighten the sail, the weight of which would otherwise render it very difficult to perform this operation.

Close-Reefed, (*être aux bas ris,* Fr.) denotes the position of the sails when all the reefs are taken in.

REEL, (*tour,* Fr.) a machine turning round on an axis, and serving to wind various lines upon.

Deep-Sea Reel, (*tour de la grande ligne de sonde,* Fr.) is that which contains the deep-sea line.

Hand-Reel, in rope-making, is formed of narrow boards, with three or four holes at each end, in which pegs are fixed to reel marline and other lines on.

Log-Reel, (*tour á loc,* Fr.) the reel appropriated to the use of the log-line, and is represented by fig. 11. plate XIV.

Twine-Reel, in rope-making, is formed, generally, of four small oak bars, about eighteen inches in length, one of which is made to slide, for the convenience of taking off the twine.

Yarn-Reel, a circular board, nailed in the middle to a piece of oak, four inches square and sixteen inches long: through the centre is bored a hole, by which it turns round a bolt, and is employed to wind spun-yarn upon.

Rope Reels, have four ribs fixed at each end in a flat circular piece of wood; round the edges are blades or handles to turn them: one of the circular pieces is called the head, and is made to slide off, in order to take the coil away. They turn on an iron spindle with a round head; are from twelve to thirty-six inches in length, and from twelve to eighteen inches in diameter, and used to reel ropes on, from a six-thread ratline to a two-inch rope.

To REEVE (*passer une manœuvre dans ses poulies, cosses,* &c. Fr.) is to pass the end of a rope through any hole, as the channel of a block, the cavity of a thimble, cleat, ring-bolt, cringle, &c. Hence, to pull a rope out of a block is called unreefing.

To REFIT (*réparer le gréement, &c. d'un vaisseau qui a été désemparé, ou qui a besoin de le changer,* Fr.) is generally understood to repair any damages, which a ship may have sustained in her sails or rigging, by battle or tempestuous weather; but more particularly by the former. See the articles Engagement and Repair.

REFLECTING Circle, (*cercle de réflexion,* Fr.) an astronomical instrument for observing the distances of celestial objects and measuring angles.

The instrument which bears this appellation is of the nature of Hadley's quadrant, and constructed upon the same principles; we shall therefore consider our readers as acquainted with that well-known instrument, a full account of the principles, properties, and use of which, has already been given under its respective head, in a former part of the work.

Although there are few purposes in practical astronomy and surveying, to which this kind of instrument may not be advantageously applied, yet it was originally constructed for, and has been chiefly used in making lunar observations for finding the longitude at sea. The late Professor Mayer of Gottingen, the ingenious and laborious astronomer who first furnished tables for that purpose, seems to have judged well of what kind of instrument would ultimately be used, to effect the best practical solution of a lunar problem. Agreeable to this idea we find, that when he transmitted to England his solar and lunar tables, they were accompanied by drawings, and a description of the first reflecting circle, and were published by the Board of Longitude in the year 1770.

Description of Mayer's *Reflecting Circle.*

Professor Mayer proposed, that his new instrument should be 16 inches in diameter; and for the purpose of relieving the arm of the observer from the weight of so heavy an apparatus, recommended that it should be supported by a staff, which should rest upon the ground or deck. Bird, who probably made the first of those instruments, improved upon that idea. He made the staff communicate with a belt, which buckled round the loins of the observer. The staff was double, being composed of two sliding tubes. These moved one within the other, for the purpose of adjusting the height of the circle to the eye of the observer, and by the help of a ball and socket attached to the centre of the instrument, he was enabled to accommodate its plane to the position of the objects. Bird also applied darkening glasses to the reflecting circle, and gave it every useful appendage which Hadley's quadrant had borne before.

It is a curious circumstance, that Bird divided his circles into 360° instead of 720°, as the nature of all reflecting instruments require; whence it became necessary to double the angle read off in a single observation, and in continued ones, to divide the accumulated angle by only half the number of repetitions. It may be considered an unfortunate circumstance for the credit of the reflecting circle, that Bird was its original maker; for the correctness of his dividing determined Admiral Campbell to neglect the repeating principle altogether, which he deemed a troublesome mode of observing, and to use a single portion of the limb only. He then saw, that the arc of the octant ought to be extended to a sextant, so as to measure 120 or more degrees, and represented to the Board of Longitude, that a sextant, the radius of which should be double that of the circle, would be equally manageable at sea. He was ably seconded by Bird in these speculations; and such was the opinion of astronomers and others, concerning the judgment of these two individuals, that, in this country, the sextant was immediately adopted for the nicer nautical purposes, to the exclusion of the octant and circle which preceded it.

Description of BORDA's *Reflecting Circle.*

About the year 1775, the justly celebrated Chevalier de Borda, turned his thoughts to the improvement of the reflecting circle of Mayer, and, after twelve years application to the subject, gave his labours to the public in an excellent pamphlet, which appeared at Paris in 1787. In this work the chevalier minutely describes the former construction, gives a particular account of his own improvements, as well as ample instructions for using the instrument, and acoompanies these with complete directions for finding the longitude by the lunar method, together with new rules and tables for this and other nautical purposes.

The great objection to Mayer's circle was, that it required two operations in order to obtain one result; namely, one for the parallelism of the glasses, and another for the angular distance of the objects. Borda removed this inconvenience in the neatest manner possible; and in doing this, gave the instrument new and useful properties. Not merely to take a mean of a great number of observations, but, as those observations are made without being read off till we come to the last, the error of reading off is no greater for all the observations put together, than it would have been for one observation only; so that, when divided into as many parts as there have been observations made, the result almost vanishes. The repeating circle, therefore, gives a mean of the errors of observation, and of the division of the instrument: and the error of reading off, it goes near to entirely annihilate.

This instrument has usually been made of 10 inches diameter, and divided to every minute; the whole limb comprehending, like all reflecting instruments, 720°. This is owing to the well-known law in catoptrics, that the angle of reflection is, in all cases, equal to the angle of incidence, and not on account of double reflection, as some writers have erroneously stated.

Description of TROUGHTON's *Reflecting Circle.*

Many of the circles of Borda had been made in London by different artists, chiefly for foreign orders, and for the late Mr. J. H. de Magellan, who made a few trifling alterations in its construction, and who put in an ineffectual claim to the invention of the method of making crossed observation. About the year 1794, that ingenious artist Mr. Troughton turned his attention to an instrument, which hitherto seemed only to have claimed the attention of foreigners; and by dispensing with the repeating principle in his new construction, he found it would afford him the means of making a much more simple and commodious instrument for his customers, which would be greatly superior in precision, and yet in management not very different from those they had been accustomed to use. The following is a representation of both sides of this circle.

Plate XXI. fig. 1, is a plan of the divided side of the instrument, and fig. 2, a perspective view of its upper side AA, in both figures is a circle of brass, with a narrow ring of silver let into a circular groove in it, as is seen in fig. 1, on which silver the divisions are made. BBB are three arms, carrying verniers at their ends; they are all cast in one piece, and screwed to a truly-turned steel axis, fitting into a tube, which is screwed to the centre of the circle; (this tube cannot be seen in the plan, but is denoted in *a* in fig. 2;) the index glass *b*, which is a plane silvered mirror, is fastened to the other end of this axis by three screws, in such a position, that the centre line of the steel axis it is fixed to, if produced, would exactly coincide with the plane of the silvered surface of the mirror, and consequently that the plane of the mirror produced passes through the centre of circle AA, perpendicular to its plane. Pl. XXI Fig. 1.

To the upper end of the tube *a*, fig. 2, a crooked plate of brass, *dd*, is fastened, and connects it with two other tubes, *e* and *f*, whose lower ends are fixed to the cross-bar frame of the circle; one of these *f*, has the mirror, *h*, called the object glass, placed on it; the other has a telescope, *k*, fixed to it, directed to the object-glass, *h*. The instrument is held, when in use, by handles adapted to different occasions, of which there are four; two perpendicular, and two parallel to the plane of the circle; of the letter *e*, is one on the upper side, supported by a small pillar coming from the interception of two of the bars of the frame, and steadied by entering the tube, *a*, the other handle is at the divided side of the circle, and is fastened to the circle at the upper side, in the same manner as *e*; by a crooked hollow tube G, going round the circle; of the other two handles, one *m*, is, above the circle, screwed into a cock fixed to it, so that it is perpendicular to the centre of the circle, it comes over the index-glass, but does not touch it. The other handle is screwed into the handle of the crooked tube G, so as to be in the same line with the upper handle *m*; *o* are three dark glasses, between the index and horizon glasses, turning on a joint, so as to be put out of the way when necessary, or any one or two of them can be turned in the line of the telescope to darken the light, more or less, in observations of the sun; *p*, are three other glasses supported by a small pillar behind the horizon-glass, which can also be turned back as is necessary. The telescope is screwed into a brass ring *r*, this is supported by a square piece of brass, tapped at the corners, so as to form a screw; and by turning *s*, a nut upon the screw, the telescope can be raised or lowered parallel to itself; there is also an adjustment to bring the line of collimation of the telescope to be parallel with the plane of the circle.

The circle is divided on the silver ring shewn in the plan, into 720 parts, each of which answers to a degree, as this instrument measures double the angle shown upon the arc, the same as the Hadley's quadrant; these are subdivided into three, each of which will be twenty minutes. The verniers include fifty-nine of these divisions, and are divided into sixty; the coincidences of these will subdivide each original division of the circle into sixty parts, each equal to twenty seconds. The arm on which the vernier, D, is fixed, has a clamp at its end to fasten it to the circle; and a fine screw, *x*, to move it slowly a small quantity after it is clamped.

The dark glasses, in sets of three, are fixed to the instrument in such a manner, that any or all of them may be used at the same time, as circumstances may require. This gives the circle before us a decided superiority over all the others. Nor is it a matter of small importance, with respect to convenience, that a handle is found both above and below: by one or other of these, in every possible position of the instrument that observations can require, it may be held right or left, as is found most convenient to the observer. It must, however, be acknowledged, that the reflecting circle, under the best circumstances, is obliged to be held two or three inches farther from the observer than the sextant is, and, of course, bears proportionally heavier upon his arm. Mr. Troughton is of opinion, that this inconvenience is the principal, if not the sole cause, that the circle has not yet come into general use; but we would hope, that an inconvenience so trifling, cannot deter a good officer from exercising the best means of obtaining the truth; especially, when we consider how much is entrusted to his care, and that the safety of the whole depends, in a great degree, upon ascertaining, in a long voyage, a few points of longitude and latitude, upon which, with confidence, he can rely. We shall now describe the manner of making an observation, by this instrument, of the angle between two objects, nearly in the same horizontal plane, we suppose all the adjustments of the instrument to be perfect; the observer first holds the instrument in his right hand, by the handle screwed to the lower handle of the tube G; he looks through the telescope, *k*, and unsilvered part of the horizon glass, *h*, and directs it to one of the objects which will be in the dotted line, *kg*; he then turns the index and index glass, *b*, by its arm D, which must be unclamped, until the other object in the line, *yb*, is reflected from *b* to *h*, and from *h*, by the silvered part of the glass, into the line *kg*, in which is placed the observer's eye; he then clamps the arm D, and gently turns the screws *x*, backwards or forwards, until the reflected image of the object in the line *y*, and the other object seen through the telescope, both exactly cover one another. The observation is now half made, and the observer reads off and writes down the degrees, minutes, and seconds, of each vernier; he then inverts the instrument, holding it by the handle, *m*, and directs the telescope to the object, in the line *yb*, and brings the reflected image of the object, in the line *kg*, into view, by turning round the index and index glass the same as before; the observation is then read off, and registered. To determine the angle measured, a mean of the products of both observations must be taken; this is the angle between the lines *yb* and *kg*. A small microscope, M, in the plan, is used to examine the verniers, and it can be applied to either vernier, as required.

The dark glasses *op*, are only wanted in observing the sun or moon. It is evident that by inverting the instrument, as we have described; the index error is of no consequence, as it will be always more in one side and less in the other.

The advantages of this instrument, when compared with the sextant, are chiefly these: the observations for finding the index error are rendered useless, all knowledge of that being put out of the question, by observing both forwards and backwards. By the same means the errors of the dark glasses are also corrected; for, if they increase the angle one way, they must diminish it the other by the same quantity. This also perfectly corrects the errors of the horizon-glass, and those of the index-glass very nearly. But what is still of more consequence, the error of the centre is perfectly corrected, by reading the three branches of the index; while this property, combined with that of serving both ways, probably reduces the errors of dividing to one-sixth part of their simple value. Moreover, angles may be measured as far as one hundred and fifty degrees, consequently the sun's double altitude may be observed when his distance from the zenith is not less than fifteen degrees; at which altitude, the head of the observer begins to intercept the rays of light incident on the artificial horizon; and, of course, if a greater angle could be measured, it would be of no use in this respect.

This instrument, in common with the sextant, requires three adjustments. First, the index-glass perpendicular to the plane of the circle. This being done by the maker, and not liable to alter, has no direct means applied to the purpose: it is known to be right, when, by looking into the index-glass, you see that part of the limb which is next you reflected in contact with the opposite side of the limb, as one continued arc of a circle: on the contrary, when the arc appears broken, where the reflected and direct parts of the limb meet, it is a proof that it wants to be rectified. The second is, to make the horizon-glass perpendicular. This is performed by a capstan-screw, at the lower end of the frame of that glass; and is known to be right, when, by a sweep with the index, the reflected image of any object will pass exactly over, or cover the image of that object seen directly. The third adjustment is, for making the line of collimation parallel to the plane of the circle. This is performed by two small screws which also fasten the collar into which the telescope screws, to the upright stem on which it is mounted: this is known to be right, when the sun and moon, having a distance of one hundred and thirty degrees, or more, their limbs are brought in contact just at the outside of that wire which is next to the circle; and then, examining if it be the same, just at the outside of the other wire: its being so is the proof of adjustment.

When the lunar tables, &c. of Mayer, as improved by Mason, were used for the construction of the nautical almanac, the sextant, as it came from the hands of our best makers, was, in point of exactness, nearly on a level with its columns: but the tables of Gauss, computed from the theorems of La Place, left both far behind. These now, however, form the basis of our almanac, and an instrument of commensurate exactness is also to be found in the reflecting circle.

With the reflecting circle so constructed, the almanac so improved, and with Captain Mendoza's excellent tables, for finding the longitude, either at sea or land, is no longer either a doubtful or a difficult operation.

Particular instructions for adjusting and observing with Troughton's Reflecting Circle, are printed and usually given with the instrument.

Description of MENDOZA'S *Reflecting Circle.*

In a paper published in the *Phil. Trans.* for 1800, Captain Mendoza has described, at considerable length, one construction, and briefly mentioned others, all of which, however, lead only to extend the repeating principle. The general fabric of this circle is, respecting the frame, position of the glasses, and handles, exactly as in Troughton's construction, to which the repeating property has been restored in a very ingenious manner, and so as to produce double the effect that is done in Borda's, by collecting the sum of backward and forward motion of the index into one reading. For this purpose, there have been added two very light circles, one of which he calls a *flying circle*, and the other a *flying nonius*. They are both centred upon the upper end of the socket, which receives the axis of the proper index, and of the great mirror. The motions of these two circles are free and independent of each other: they are situate one above the other, close to the limb of the fixed frame, having no connection whatever with it, except at the centre. It will be remembered, that, in the former construction, the index with three branches had backward and forward motion; it is precisely so in this, only this index has but one branch, namely, that with the screws for quick and slow motion. This index carries no vernier, nor are there any divisions upon the limb of the fixed frame to which it applies. The graduation is made upon the flying circles, the lower of which is divided into degrees, and every 20′; the upper one carries opposite verniers, subdividing the lower down to 20″. There is a clamp attached to the index, which lays hold of the flying nonius, and in an observation forwards carries it along over the face of the flying circle; and there is also a similar clamp, which, laying hold of the flying circle in an observation backwards, carries it back under the flying nonius. The flying circle is, in the former case, clamped to the limb of the fixed frame, and in the latter, the nonius is secured in the same manner.

Now, as in the observation forwards, the nonius advances upon the circle, and in the one backwards, the circle recedes upon the nonius, it is evident that the sum of the two motions will be given at reading off. Instead, however, of reading here, forward and backward observations are to be repeated, until a sufficient set has been procured, and then the whole space that the circle and nonius have been moved respecting each other may be read off, and on being divided by twice the number of forward and backward motions together, will give the required angle.

There are two sextants of coarse divisions made upon the fixed limb, whose zeros are distant from each other by the breadth of the index of the index-glass; contiguous to these divisions slide two pieces, which, being set right and left to the angle about to be measured, will stop that index in the requisite positions for finding the objects nearly together in the field of the telescope. This useful contrivance enables an observer to take a set of sights of any number without looking at the divisions, the first and last excepted.

Mr. Troughton first applied this neat contrivance to the circle of Borda, which, without it, is extremely inconvenient to use.

Notwithstanding the great ingenuity displayed in these contrivances, we think that the reflecting circle last described, can neither be very accurate nor very pleasant in its use: the management of five clamps, which are to be made fast and loose in regular succession, ten distinct operations to every pair of observations, any of which being mistaken or neglected would spoil a whole set, requires a degree of attention not to be found in every common observer. And the liability which there must be, even in the best work, to force the flying circles from their places in clamping, and thereby to cause errors, without any means of detecting them, is a most forbidding circumstance. And should the index of the central mirror have a motion eccentric to those of the flying circles, the latter at every step will feel the effect, and in consequence, the result will be charged with the full error belonging to a single observation.

Professor HASSLER'S *Method of restoring the Repeating Property to Troughton's Reflecting Circle.*

We understand that there has been suggested to Mr. Troughton an unexceptionable method of restoring the repeating property to his reflecting circles, without in the least diminishing the advantages of the fixed frame. In this the circular border of the frame is not employed; it has one flying circle, divided in the usual manner. Two indices are employed; one of them, below the circle, is screwed fast to the frame, the other is above, having, like the circle itself, free and independent motion round the centre; the latter carries at the lower end of the axis the index-glass. Both of the indices have screws for quick and slow-motion, which act upon the flying circle; and they are both furnished with opposite verniers, those of the lower index being brought round from below, in order to count upon the divisions of the circle. When the index and horizon-glasses are parallel, the two indices cross each other at right angles; therefore the upper one has a range of backward and forward motion through the full sextantal angle, without interfering with the lower one.

The effect produced by this arrangement is as follows:—In an observation forwards, the circle being clamped by the fixed index, the moveable index advances upon the circle, and being clamped to it, the contact of objects is made by its proper screw. And in an observation backwards, the circle being released from the fixed index, is carried backwards along with the moveable one, receding upon the former, to which it is again clamped, and then the contact of objects is made as before. These alternate operations must be continued until a set is obtained; and then the upper index will give the amount of forward, and the lower that of backward observations. With respect to repeating the angle, this instrument produces the same effect as Borda's circle; but we cannot here come at the double result, as we did in the arrangement of Mendoza. In this latter, however, the clamping is performed after the contacts are observed, which is its great defect; but the former has no such impediment, the clamps being fixed before the contacts are made; and therefore, in this respect, stands equal to a sextant, or any of the former constructions of the circle.

To such gentlemen as wish for a reflecting circle which repeats the angle, we hesitate not to recommend the one we have just endeavoured to describe; and we have to inform the public, that they are indebted for the idea to Professor Hassler, a native of Switzerland, and a citizen of the United States of America. See EDINBURGH *Encyclopædia*, Vol. VI.

REFLECTION, or REFLEXION, in mechanics, is the return, or regressive motion of a moveable body, occasioned by the resistance of another body, which hinders it from pursuing its former course of direction.

Reflection is conceived, by the latest and best authors, as a motion peculiar to elastic bodies, by which, after striking on others which they cannot remove, they recede, or turn back, or aside, by their elastic power.

On this principle it is asserted, that there may be, and is, a period of rest between the incidence and the reflection; since the reflected motion is not a continuation of the other, but a new motion, arising from a new cause or principle, viz. the power of elasticity.

It is one of the great laws of reflection, that the angle of incidence is equal to the angle of reflection, *i. e.* that the angle which the direction of motion of a striking body makes with the surface of the body struck, is equal to the angle made between the same surface and the direction of motion after the stroke.

REFLECTION *of the Rays of Light*, like that of other bodies, is their motion after being reflected from the surfaces of bodies.

The reflection of the rays of light from the surfaces of bodies, is the means by which those bodies become visible. And the disposition of bodies to reflect this or that kind of rays most copiously, is the cause of their being of this or that colour. Also, the reflection of light, from the surfaces of mirrors, makes the subject of catoptrics.

Newton has shewn, that the reflection of light is not effected by the rays striking on the very parts of the bodies; but by some power of the body equally diffused throughout its whole surface, by which it acts upon the ray, attracting or repelling it without any real immediate contact. This power he also shews is the same by which, in other circumstances, the rays are refracted; and by which they are at first emitted from the lucid body.

REFLECTION, in catoptrics, is the return of a ray of light from the polished surface of a speculum or mirror, as driven thence by some power residing in it.

The ray thus returned is called a *reflected ray*, or a *ray of reflection*; and the point of the speculum where the ray commences, is called the *point of reflection*. Thus, the ray A B, proceeding from the radiant A, and striking on the point of the speculum B, being returned thence to C, B C represents the reflected ray, and B the point of reflection; in respect of which, AB represents the incidental ray, or ray of incidence, and B the point of incidence; also the angle CBE is the angle of reflection, and ABD the angle of incidence; where DE is the reflecting surface, or at least a tangent to it at the point B. Though some count the angle of incidence and of reflection from the perpendicular BF.

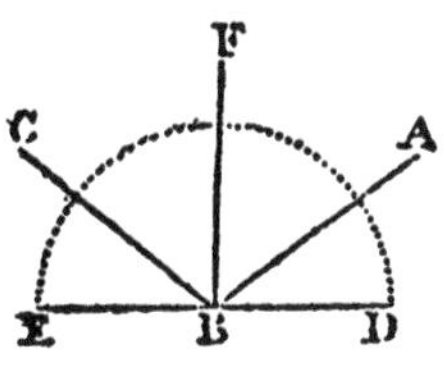

REFRACTION, in mechanics, is the deviation of a moving body from its direct course, by reason of the different density of the medium it moves in; or it is a change of direction, occasioned by a body's falling obliquely out of one medium into another of a different density.

REFRACTION *of Light*, in optics, is an inflection or deviation of the rays from their rectilineal course on passing obliquely out of one medium into another of a different density.

That a body may be refracted, it is necessary that it should fall obliquely on the second medium, as in perpendicular incidence there is no *refraction*.

The ancients confounded refraction with reflection; and it was Newton who first taught the true difference between them. He shews, however, that there is a striking analogy between them, and particularly in the case of light.

By refraction it is, that convex glasses, or lenses, collect the rays, magnify objects, burn, &c.; and hence the foundation of microscopes, telescopes, &c. And by refraction it is, that all remote objects are seen out of their real places; particularly, that the heavenly bodies are apparently higher than they are in reality. The refraction of the air has many times so uncertain an influence on the places of celestial objects, near the horizon, that wherever refraction is concerned, the conclusions deduced from observations that are much affected by it, will always remain doubtful, and sometimes too precarious to be relied on. See Dr. BRADLEY's remarks on this subject in the *Philosophical Transactions*, N°. 485.

It appears from various considerations, and might be shewn by the observations of Mr. Molyneux and Dr. Bradley on the parallax of the fixed stars, that (notwithstanding what has been asserted to the contrary by other writers) their rays are not at all impeded by the rapid motion of the earth's atmosphere, nor by the object-glass of the telescope, through which they pass. And by Newton's theory of refraction, which is grounded on experience only, it appears that light is so far from being resisted and retarded by refraction into any dense medium, that it is swifter there than *in vacuo* in the ratio of the sine of incidence *in vacuo* to the sine of refraction into the dense medium. See PRIESTLEY's *History of Light, &c.*

REFRACTION, in astronomy, or REFRACTION *of the Stars*, is an inflection of the rays of those luminaries, in passing through our atmosphere; by which the apparent altitudes of the heavenly bodies are increased.

This refraction arises from the atmosphere being unequally dense in different stages or regions; rarest of all at the top, and densest of all at the bottom; which inequality in the same medium, makes it equivalent to several unequal mediums, by which the course of a ray of light is continually bent into a continued curve line. See the article ATMOSPHERE.

The effect of this refraction may be thus conceived. Suppose AB a quadrant of a vertical circle described from the centre of the earth C, under which DE a quadrant of a circle on the surface of the earth, and FG a quadrant of the surface of the atmosphere. Then suppose HI a ray of light emitted by a star at H, and falling on the atmosphere at I; this ray coming out of the ethereal medium, which is much rarer than our air, or perhaps out of a perfect vacuum, and falling on the surface of the atmosphere, will be refracted towards the perpendicular, or inclined down more towards the earth; and since the upper air is again rarer than that near the earth, and grows still denser as it approaches the earth's surface, the ray in its progress will be continually refracted, so as to arrive at the eye in the curve line ID. Then supposing the right line DK to be a tangent to the arc at D, the ray will enter the eye at D in the direction of DK; and, therefore, the star will appear in the heavens at L, instead of H, higher or nearer the zenith than it really is.

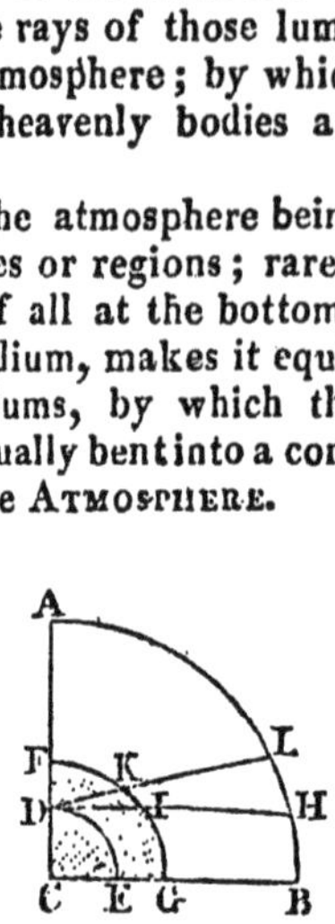

Hence arise the phenomena of the crepusculum or twilight; and hence also it is that the moon is sometimes seen eclipsed, when she is really below the horizon, and the sun above it.

The stars appear higher by refraction than they really are; so that to bring the observed or apparent altitudes to the true ones, the quantity of refraction must be subtracted.

Stars in the zenith are not subject to any refraction; those in the horizon have the greatest of all: from the horizon, the refraction continually decreases to the zenith.

Refraction lengthens the day, and shortens the night, by making the sun appear above the horizon a little before his rising, and a little after his setting. Refraction also makes the moon and stars appear to rise sooner and set later than they really do. The apparent diameter of the sun or moon is about 32′; the horizontal refraction is about 33′; whence the sun and moon appear *wholly* above the horizon when they are entirely below it. Also, from observations it appears that the refractions are greater nearer the pole than at lesser latitudes, causing the sun to appear some days above the horizon, when he is really below it; doubtless from the greater density of the atmosphere, and the greater obliquity of the incidence.

The horizontal refraction, being the greatest, is the cause that the sun and moon appear of an oval form at their rising and setting; for the lower edge of each being more refracted than the upper edge, the perpendicular diameter is shortened, and the under edge appears more flatted also. Hence, if we take with an instrument the distance of two stars when they are in the same vertical line and near the horizon, we shall find it considerably less than if we measure it when they are both at such a height as to suffer little or no refraction; because the lower star is more elevated than the higher.

From observations made on circumpolar stars, it is well known that the greatest and the least zenith-distance of one of those stars gives the co-latitude of the place; and, since this latter is an invariable quantity, it ought to result, whatever be the stars selected, when all due corrections are applied to the observations.

REFRANGIBILITY OF LIGHT, the disposition of the rays to be refracted.—A greater or less refrangibility, is a disposition to be more or less refracted, in passing at equal angles of incidence into the same medium.

Newton has proved, from a great variety of experiments, that the blue rays are more refracted than the red ones, and that there is likewise unequal refraction in the intermediate rays; and upon the whole it appears that the sun's rays have not all the same refrangibility, and consequently are not of the same nature. It is also observed, that those rays which are most refrangible, are also most reflexible.

That the rays of light are differently refrangible, is the foundation of Newton's whole theory of light and colours; and the truth and circumstances of this theory he evinced from his own experiments. See NEWTON's *Optics*, 3d edit.

The difference between refrangibility and reflexibility was first discovered by Sir Isaac Newton about the beginning of the year 1672, and communicated to the Royal Society, in a letter dated February the 6th, of that year, which was published in the Philosophical Transactions, N°. 80, page 3075; and from that time it was vindicated by him, from the objections of several authors; and at length it was more fully laid down, illustrated and confirmed by a great variety of experiments, in his excellent treatise on Optics.

REGION OF THE AIR OR ATMOSPHERE. Authors divide the atmosphere into three stages, called the upper, middle, and lower regions. The lowest region is that in which we breathe, and is bounded by the reflection of the sun's rays, viz. by the height to which they rebound from the earth. The middle region is that in which the clouds reside, and where meteors are formed, &c.; extending from the extremity of the lowest, to the tops of the highest mountains.—The upper region commences from the tops of the mountains, and reaches to the utmost limits of the atmosphere. In this region there probably reigns a perpetual equable calmness, clearness, and serenity.

Ethereal REGION, is the whole extent of the universe, comprising all the heavens, with the orbs of the fixed stars and other celestial bodies.

REGIONS *of the Sea*, are the two parts into which the whole depth of the sea is conceived to be divided. The upper of these extends from the surface of the water, down as low as the rays of the sun can pierce, and extend their influence; and the lower region extends from thence to the bottom of the sea.

REGISTER-SHIPS, in commerce, are such vessels as the government, on certain conditions, admits to all the rights enjoyed by British-built ships, thereby permitting them to trade to and from the ports in amity with or appertaining to Great Britain, and granting them those immunities extended to the British flag, with protection in the aforesaid ports as well as at sea, when circumstances allow of its being afforded.

Register ships, in the Spanish trade, are vessels which obtain a permission either from the king of Spain or the council of the Indies, to traffic in the ports of the Spanish colonies, and are so called from their being registered before they set sail from Cadiz for South America.

Every ship or vessel having a deck, or being of the burden of 15 tons, and belonging to a subject in Great Britain, Guernsey, Jersey, or the Isle of Man, or any colony, plantation, island, or territory, belonging to the king, is to be registered by the person claiming property therein, who is to obtain a certificate of such registry, from the collector and comptroller, if in Great Britain or the Isle of Man; or from the governor, lieutenant-governor, or commander in chief, and the principal officer of the customs, if in Guernsey, Jersey, or any colony or plantation.

Persons applying for registry for any ship built after the 1st August, 1786, are to produce a true account under the builder's hand, of the time and place of building, tonnage, and the name of the first purchaser, and make oath that the ship in question is the same.

If such officer gives false descriptions, or any person makes a false registry, or grants a false certificate, he is to forfeit 100*l*.

A bond is entered into by the master and owners at the time of obtaining the certificate, that the said certificate shall not be sold, lent, or otherwise disposed of; and that in case the ship be lost, taken, or the like, the certificate, if preserved, shall be returned, within one month after the master's arrival in the king's dominions, to the collector or comptroller of some port of Great Britain, or the Isle of Man, or of the plantations, &c.; and if any foreigner shall become entitled to the whole or any share of the interest in the ship, the certificate shall be given up within seven days after such transfer of property, if it shall be within the limits of the king's dominions; and, if at any foreign port, to the British consul resident there; and if at sea, to the consul or British officer at the first port where the ship shall arrive, immediately after arrival: or if the port at which she arrives is in Great Britain, &c. then within fourteen days after arrival.

If a certificate be lost or mislaid, a registry or certificate *de novo*, may be granted according to the regulations of Stat. 15 Geo II. c. 31, provided the above-required security be given; and the oath prescribed by this act be taken, instead of that prescribed by act of 26 Geo. III. c. 60.

If a ship, after registering, shall be altered in form or burden, she must be registered *de novo*, as soon as she returns to the port to which she belongs, or to any other port where she may be legally registered.

As often as the property of a ship is transferred from one British subject to another, in the whole or in part, the certificate of the registry is to be recited in the bill of sale, which is otherwise void. And as often as the master is changed, the certificate is to be delivered by the master or owner at the port where the change is made, to the person authorised there to grant certificates, who is to indorse and subscribe a memorandum of such change, &c.

The owners of prize-ships, before registering, to produce a certificate of the condemnation under the hand and seal of the judge; and also an account of the particulars contained in the before-mentioned certificate, to be made and subscribed by persons appointed to survey the ship.

Certificates to distinguish ships in two ways, either by being intitled, "Certificate of British-Plantation Registry;" or, "Certificate of Foreign Ship's Registry for the European trade, British Property."

No ship to be allowed to clear out as a British ship, or to be entitled in anywise to the privileges of a British ship, unless the owner has obtained a certificate.

The masters of ships are, on demand, to produce their certificates to the principal officer in any port of his Majesty's dominions; or to the British consul or chief officer in any foreign port, under a penalty of 100*l*.

The fees of registry are specially limited by the act

All former acts made for the registry of ships, shall be taken to extend to ships required by this act (26 Geo. III. c. 60.) to be registered.

A Ship's REGISTRY, is a printed instrument, containing the names of the owner and master, the name and exact description of the vessel, the place to which she belongs, when and where built or captured, and, if a prize-ship, the date of condemnation, whether British, foreign or British-plantation built, her precise dimensions, tonnage, and the port at which she was registered.

For the form of the oath and certificate, in order to make a registry, see Stat. 26 Geo. III. c. 60.

REGULATING CAPTAIN, (*capitaine, ou major de vaisseau dont l'emploi consiste à examiner les matelots volontaires et autres, &c.* Fr) is an officer stationed at the different royal ports, in time of war, to examine the seamen intended for the navy, whether pressed or volunteers, and to see that they are fit for his Majesty's service before they are sent on board a man of war. For farther particulars, see the articles To PRESS, and PRESS-GANG.

REGULUS, in astronomy, a bright fixed star of the first magnitude, in the heart of the constellation Leo. For the beginning of the year 1813, its right ascension was, 149° 35′ 55″, annual variation in right ascension 48″. 2; declination, 12° 52′ 38″ N. and annual variation in declination,—17″. 2.

The longitude of Regulus, as fixed by Flamsteed, is 25° 31′ 21″, and its latitude 0° 26′ 38″ N.

REIGNING WINDS, (*vents régnants, ou qui dominent dans chaque parage*, Fr.) a name given to the winds which usually prevail on any particular coast or region, the knowledge of which is essentially necessary to every pilot who is charged with the navigation in those seas. See the article WIND.

REINFORCE, (*renfort*, Fr.) in founding guns, is that part of a gun next to the breech, which is made stronger than the rest of the piece, in order to resist the force of the powder. There are generally two in each piece, called the first and second reinforce, as pointed out by AE and EF, fig. 4, plate V. The second reinforce is smaller than the first, upon the supposition, that, when the powder is inflamed, and occupies a greater space, its force is diminished; but which is not really the case.

REINFORCE *Rings of a Cannon*, (*plate-bande du renfort*, Fr.) are three flat mouldings, like flat iron hoops, placed at the breech end of the first and second reinforce, and projecting beyond the rest of the metal about a quarter of an inch. See the article CANNON.

RELIEVER, in gunnery, is a flat iron ring, with a wooden handle, at right angles to it. It is used, on searching a gun, to relieve one or other of the springs of the searcher that may have hitched into any cavity, which is effected by its passing round the handle, and compressing the springs together.

RELIEVING TACKLES, (*deux forts palans employés lorsqu'on abat un bâtiment en carène, pour faire l'office de ce que les François nomment attrapes*, Fr.) are two strong tackles furnished with guys or pendents, which, passing under the ship's bottom to the opposite side, are attached to the lower gun-ports. The other ends of the tackles are hooked to the wharf, or pontoon, by which the vessel is careened. They are used to prevent a ship from overturning on the careen, and to assist in bringing her upright after that operation is finished. Thus, if the ship is first to be laid down on the larboard side, which is nearest the wharf, the relieving-tackles are passed under her bottom from the said wharf, and attached to the starboard side, by which they will restrain her from falling lower than is necessary. See the articles CAREENING and TO RIGHT.

RELIEVING-*Tackles*, are also those which are occasionally hooked to the tiller in bad weather, or in action, when the wheel or tiller-rope is broken or shot away.

RELIEVING-*Tackle*, is also a name sometimes given to the train-tackle of gun-carriages. See the articles CANNON and EXERCISE.

REMITTANCE, AMONG SEAMEN. By an act of 31 Geo. II. brought forward and extended by the late Lord Melville, 32 Geo. III. when a ship of war shall not be in a port in Great Britain, or on the coast thereof, and shall have twelve calendar months or more wages due, every inferior officer, seaman, or marine, may remit the whole, or any part of his wages then due, except the wages due for the last six months, to his wife, father, mother, grandfather, grand-mother, brother, or sister; and this remittance may be continued every six months, while the ship is abroad, except the last six months, as before mentioned.

Thus, while the British seaman, or marine, is serving his country abroad, well fed and clothed, his wages may be applied to the relief or comfort of those for whom he feels most interested. See also the article ALLOTMENT.

To RENDER, (*couler, en parlant des manœuvres, c'est-à-dire, que rien ne les arrête ou gêne dans leur passage au travers les poulies, cosses, &c.* Fr.) as a sea term, implies to yield or give way, without resistance, to the efforts of some mechanical power. It is usually expressed of a complicated tackle, lanyard, or lashing, when the effect of the power applied is communicated with facility to all the parts, without being interrupted in its passage. It is, therefore, used, in contradistinction to sticking or jamming.

RENDEZVOUS, (*rendez-vous d'une escadre, &c. en cas de séparation.* Fr.) the port or place of destination, where the several ships of a fleet or squadron are appointed to join company, or to rejoin in case of separatio .

RENDEZVOUS, is also a name given to any house where a press-gang resides, and volunteers are invited to enter into the navy. See the articles To IMPRESS, and PRESS-GANG.

REPAIR (*radoub*, Fr.) the operation of repairing any injuries, or supplying any deficiencies which a ship may have suffered from age, battle, storm, &c.

The repair is necessarily greater or smaller, in proportion to the loss which the vessel has sustained. Accordingly, a suitable number of the timbers, beams, or planks, or a sufficient part of either, are removed, and new pieces fixed in their places. The whole is completed by calking, and paying the body with a new composition of stuff See the articles *To* CALK, and to PAY *a Vessel's Bottom.*

To REPEAT SIGNALS, (*répéter des signaux*, Fr.) is to make the same signal with the admiral, in order to its being more readily distinguished at a distance, or through smoke, &c.

To REPEAT *a Signal*, (*répéter un signal*, Fr.) sometimes implies to make a signal over again, on account of its not having been attended to the first time. The repeat is usually accompanied with a gun.

REPEATER, or REPEATING SHIP, (*vaisseau répétiteur*, Fr.) is a vessel, generally a frigate, appointed to attend each admiral in a fleet, and to repeat every signal he makes, with which she immediately sails the whole length of the fleet or squadron, if the signal is general, or to the ship for which it is intended, if particular, and then returns to her station near the admiral's ship, waiting for further instructions.

REPEATING CIRCLE. The principal to which the repeating circle owes its existence, was discovered by Professor Mayer of Gottingen, in 1758, who communicated his invention to the Royal Society of that city, in a memoir upon an instrument which he calls a Goniometer, the principles of which he had mentioned to other astronomers eight years before. The goniometer was intended for observing angular distances between terrestrial objects in land-surveying; and, although certainly not a circle, belongs to the class of repeaters, and was the original step which led to the construction of the repeating instrument, which Mr. Troughton has now brought to such great perfection.

When Hadley's octant became known to the world, Professor Mayer was among those who rightly appreciated its valuable properties, and by extending its arc to a complete circle, for the purpose of reducing the errors of division, he gave it the repeating property of his goniometer, which he had before perhaps only thought of. The professor was fully aware, that the common theodolite, and other surveying instruments, were capable of being used as repeaters; and we can assign no other reason for his giving the preference to the undivided bars, diagonal scale and compasses, than that scales must at that time have been more accurately divided than circles.

It was about the year 1789, when Chevalier de Borda first caused a repeating circle to be constructed for astronomical and geodetic purposes; and one of our motives in being so particular in what is said before, was, to shew exactly how much is due to him. The former repeaters, it has been seen, gave but one result for two operations; but by the use of an additional telescope he procured the means of obtaining double results, but this advantage is not purchased at a less expense than three distinct operations. The astronomers of the continent of Europe, and particularly in France, where the art of dividing, except of kingdoms, is almost unknown, have been lavish of their praises of this instrument; while in this country, where that art is practised with the greatest success upon much more minute objects, a prejudice against it has too long continued. As a proof of this, we believe that the late Mr. Ramsden never made either a reflecting or a repeating circle. This prejudice is now, however, fast wearing off, and this circle begins to take that rank among our instruments to which its merit entitles it.

The repeating circle may be considered either as a geodetical or as an astronomical instrument; and it owes much of its celebrity to its having been employed in this double capacity by the French astronomers, in the great trigonometrical operations lately carried on in that kingdom.

Of the manner in which the repeating circle was used on this occasion, a most ample account has been given by the National Institute, in two quarto volumes, entitled *L'arc du Meridien*. Much useful information may likewise be found in Puissant's *Geodosie*, and in a single quarto volume, published separately by Delambre, entitled, *Methodes analytiques pour la Determination d'un arc du Meridien.*

For a description of Troughton's repeating circle, with its properties, as an astronomical instrument, and use for measuring terrestrial angles, see EDINBURGH *Encyclopædia*, vol. vi.; also the article REFLECTING *Circle*.

REPORT, (*bruit; rapport*, Fr) sound; loud noise, as that made by the discharge of a musket or cannon.

REPORT, (*rapport*, Fr.) is likewise a plain statement of what persons have seen and performed; or a circumstantial detail of facts methodized.

REPORTS *on Surveys upon Provisions*, in the royal navy, are a fair statement drawn up by persons appointed, pursuant to orders from the commander in chief, or the senior officer commanding on the spot, to examine into the state and condition, cause of defect, and quantity defective; also making mention of the time, place where, and from whom received, in order that it may be traced whether the defects have arisen from any neglect on the part of the contractors, agents, or pursers, that the loss sustained by government should be made good by the defaulter.

The report made by officers appointed to survey stores, &c. is to specify by whose order it is taken, and for what purpose; what are the articles ordered to be surveyed; the quantity and quality of those articles remaining on board, or the actual state of any which shall be particularly represented as deficient or defective; the number or quantity is always to be written in words at length; and if any stores complained of be found to be no longer fit for their proper use, the report is to specify whether they be fit for any other, and for what use, or whether they be no longer fit for any purpose whatever.

Any officers wilfully signing a false report of the quantity or condition of the stores or provisions he is ordered to survey, or discovering any fraudulent practices in the management of such stores or provisions, without making proper mention of them in his report can be immediately suspended, and his misconduct reported to the commander in chief, who reports it to the secretary at the Admiralty on home stations; but on foreign stations, the former is invested with full powers to try the offender, and put the sentence in execution immediately afterwards, transmitting to the Admiralty a copy of the whole proceedings and minutes of the trial, which are required by them, that the merits or demerits of the case may be publicly discussed, if it should be found requisite; and that testimonials of their approbation, or *vice versa*, may be sent out to such officers accordingly.

It is, however, to be observed, that no person can give a false account of stores or provisions worn out or defective by which the surveying officers may be deceived, and led to make out an improper report without both parties conniving at it; as it is particularly set forth in the order (which is addressed to three persons, nominated by the commander in chief, &c. to examine into the state, &c. of such stores or provisions represented to be so worn out, and defective) that they are not only to ascertain the quality, but quantity, of provisions, defective, and to see it weighed in their presence. And with regard to sails, masts, yards, &c. they are all of certain dimensions, regulated by government, according to the rate of the ships they are designed for; therefore, as the number of yards of canvas contained in the sails is known; also the length, diameter, &c. of the masts and yards, it is only requisite for the surveyors to state the different descriptions, and the number of each: hence it is evident that no fraudulent practices can be had recourse to, but by general consent. See the article SURVEYS.

REPORTS *on the rowing of the day and night guard* are, the daily information given in respecting the rowing, &c. of that guard; also of prisoners of war, for which forms are to be filled up and abided by, and which may be found in the General Printed Instructions: they are delivered in abstract by the captain or commander of the ship, to the commander in chief at the times given out by him in general orders.

In the army, the accounts rendered of men, stores, or provisions, as to number or quantity, are termed *returns* with the exception of courts-martial or surveys, at the termination of these, the statements must invariably be termed *reports.*

REPORTS *on the Sick*, a statement of the daily state of the sick and hurt on board, which is made by the surgeon (or his assistant in his absence) to the captain or commanding officer, in order that he may inform himself as to the nature of their diseases with respect to the probable time of recovery: also, that when any persons are put on the sick list, their stations may be filled up for the time being, by shifting men from their quarters, &c.

And by late additional instructions to surgeons, they are required to deliver in weekly, or oftener if requested, to the cómmanders in chief, &c. the state and condition of their sick within that period, that the number of effective men in the fleet or squadron may be more readily ascertained. They are also directed by these instructions, when in port, to transmit a weekly report to the commissioners of the sick and hurt, and transport board, to the same effect.

REPRIMAND, (*reprimande*, Fr.) a slight kind of punishment sometimes inflicted on commissioned and warrant officers. It consists in reproving or reprimanding them at a court-martial for misconduct; when the charges preferred against them have been wholly or in part proved, they are reprimanded accordingly.

REPRISALS, (*représailles*, Fr.) in the civil law, a right which princes have to retake from their enemies such things as they illegally detain from them; or other things equivalent thereto. Thus, when places or ships are taken or held from a prince, he seizes others, which is termed making reprisals. See the article LETTER OF MARQUE.

REPRISE, (*reprise*, Fr.) at sea, is a merchant-ship, which, after its being taken by a corsair, privateer, or other enemy is retaken by the opposite party.

If a vessel thus retaken, has been twenty-four hours in the possession of the enemy, she is deemed a lawful prize; but if retaken within that time, she is to be wholly restored to the proprietor, upon his allowing one third of her value for salvage to the recaptors. Also, if the reprise has been abandoned by the enemy, before it has been taken into any port, it is to be restored to the proprietor.

REQUISITE-TABLES, in navigation, a set of tables requisite to be used with the Nautical Ephemeris, published by the late Dr. Maskelyne, in the year 1766, to accompany the first Nautical Almanac for facilitating the calculations, which were judged too difficult for seamen; with these tables he gave two excellent methods of clearing the apparent distance from the effects of parallax and refraction; the first by Mr. Lyons, and the other by Mr. Dunthorne: also two examples are added for ascertaining the longitude from observation.

A second edition of the requisite tables became necessary in 1781, which was new modeled by Dr. Maskelyne, and two other methods of correcting the apparent distance were also given; the one invented by himself, and the other by Mr. George Witchell, master of the Royal Academy at Portsmouth. Several other useful tables were introduced, among which were the Logarithmic Solar Tables for finding the latitude by double altitudes, and a table of latitudes and longitudes of places that had been settled by astronomical observations and geographical surveys, supplied by Captains Cook and Huddart, and also by Messrs. Bailey, Wales, and other scientific navigators. The new tables were constructed and the explanations drawn up by Mr. William Wales, master of the Royal Mathematical School in Christ's Hospital; a person well qualified for this work from his knowledge of the practice, as well as the theory, both of astronomy and navigation.

A third edition being called for, Dr. Maskelyne corrected the errata that had been discovered in the second edition, and introduced several new improvements that are mentioned in the explanation and use of the tables, a part of the work which Mr. Wales found necessary entirely to write anew. Mr. Wales improved Mr. Witchell's method of clearing the distance, and put it in a better form in this edition, which was published in 1802, to which an appendix was added in 1809, containing tables of natural sines and co-sines, natural versed-sines, and an extended table of logarithms to 100,000.

RESISTANCE, or RESISTING-FORCE, in philosophy, denotes, in general, any power which acts in an opposite direction to another, so as to destroy or diminish its effect.

There are many kinds of resistance arising from the various natures and properties of resisting bodies, and governed by a variety of laws; as, the resistance of solids, the resistance of fluids, the resistance of the air, &c.

RESISTANCE *of Solids*, in mechanics, is the force with which the quiescent parts of solid bodies oppose the motion of others contiguous to them.

Of these there are two kinds. The first where the resisting and the resisted parts, *i. e.* the moving and quiescent bodies, are only contiguous, and do not cohere; constituting separate bodies or masses. This resistance is what Liebnitz calls *resistance of the surface*, but which is more properly called *friction.*

The second case of resistance, is where the resisting and resisted parts are not only contiguous, but cohere, being parts of the same continued body or mass. This resistance was first considered by Galileo, and may be properly called *renitency.*

RESISTANCE *of the Air*, in pneumatics, is the force with which the motion of bodies particularly of projectiles, is retarded by the opposition of the air or atmosphere. See the articles GUNNERY, PROJECTILES, &c.

The air being a fluid, the general laws of the resistance of fluids obtain in it; subject only to some variations and irregularities from the different degrees of density in the different stations or regions of the atmosphere.

The resistance of the air is chiefly of use in military projectiles, in order to allow for the differences caused in their flight and range by it. Before the time of Mr. Robins, it was thought that this resistance to the motion of such heavy bodies as iron balls and shells, was too inconsiderable to be regarded, and that the rules and conclusions derived from the common parabolic theory, were sufficiently exact for the common practice of gunnery. But that gentleman shewed, in his valuable work, 'The New Principles of Gunnery,' that so far from being inconsiderable, it is in reality enormously great, and by no means to be rejected without incurring the grossest errors; so much so, that balls or shells which range, at the most, in the air, to the distance of two or three miles, would in a vacuum range to 20 or 30 miles or more.

Chevalier de Borda and others had greatly extended the theory, but principally by means of gratuitous, and as is now known, inaccurate assumptions respecting the resistance of the air. Yet there still remained much to do in order to bring us acquainted with the real nature of the expansive force of gunpowder, the actual velocities of shot at the commencement of their motion or in different points of their path, the laws of the resistance experienced by balls and shells in their motion, and the true nature of the curve they describe. In order, therefore, that this important and intricate department of philosophy might receive some essential improvement, it became desirable, that a person possessing an active and ardent mind with habits of regularity and perseverance, should be so circumstanced as to have both the inclination to enter upon this peculiar investigation, and the means of pursuing it: and this, by a happy coincidence, occurred by the late Duke of Richmond (a man of science and of great public spirit) being master-general of the ordnance just at the period when Dr. Hutton was, with all the zeal and activity of the meridian of life, discharging the duties of the mathematical professorship at the Royal Military Academy at Woolwich.

The mathematical sciences are taught at this institution with a view to their application to military purposes, and particularly to the practice of artillery: and Dr. Hutton was a man not likely to rest satisfied with affecting to teach, what, in truth, there were no data for teaching properly. He knew that if the doctrine of projectiles were ever to be so exalted as to become an integral part of mathematical science, it must rest upon the basis of well-conducted experiments. He, therefore, began a series so early as the year 1775; and afterwards carried on a far more extensive one, under the auspices of the Duke of Richmond (and officially under the direction of General Sir Thomas Blomfield), during the summers of 1783, and of many succeeding years. Many of the results of this extensive series of experiments, are extremely important; but as the limits of this work will not allow us to mention them, we must therefore refer the reader to Dr. HUTTON'S *Mathematical and Philosophical Tracts* themselves, in 3 vols. 8vo. in which will be found a series of extensive and well-conducted experiments upon Robins's whirling machine, to determine the resistance of the air.

These experiments, together with those made by firing balls from artillery, constitute a complete and connected series of resistances to balls, from the slow velocities of 5 or 10 feet per second, to the rapid velocities of 1900 and 2000 feet. It appears from an examination of the results, that, though the resistances are nearly as the squares of the velocities in very slow motions, they are never exactly so. The exponent of the velocity indicating the resistance *always* exceeds 2. At 200 feet per second, that exponent is 2.028: at 500 feet, it is 2.042: at 1000 feet, it is 2.115: from thence it keeps gradually increasing up to the velocity of 1500 or 1600 feet per second, where the exponent is 2.153: and from this velocity the exponent gradually diminishes, being 2.136 at the velocity of 2000 feet, the limit of the experiments.

That the resistance should not be accurately as the square of the velocity, must be evident to every one who attentively reflects upon the subject. But Dr. Hutton has gone farther, and at pp. 221, 222 of the third volume, has very satisfactorily developed the causes of the variable exponent in the ratio of the resistance. He has also investigated three or four theorems for the resistance of balls; of which the following appears to be both accurate and convenient in use. Let v be the velocity in feet with which a ball, whose diameter is d feet, moves in air near the earth's surface, then will the resistance in avoirdupois pounds be expressed by the formula $(.000007565\, v^2 - .00175v)\, d^2$.

Dr. Hutton, having deduced the law of resistances to spherical bodies moving in the air, proceeds in a series of important problems in Tract 37 to apply it to the determination of the most essential particulars in the motion of military projectiles. Here many of the solutions are both elegant and satisfactory. But the grand problem by which the actual trajectory of the projectile may be determined, still remains unsolved, and must do until some philosopher possessing an adequate portion of Dr. Hutton's science and zeal, shall be so favourably circumstanced as to carry through another set of experiments with a special regard to that object.

RESISTANCE *of Fluids*, is the force with which bodies, moving in fluid mediums, are impeded and retarded in their motion.

A body moving in a fluid is resisted from two causes. The first of these is cohesion of the parts of the fluid. For a body, in its motion, separating the parts of a fluid, must overcome the force with which those parts cohere. The second is the inertia, or inactivity of matter, by which a certain force is required to move the particles from their places, in order to let the body pass.

The retardation from the first cause is always the

same in the same space, whatever the velocity may be, the body remaining the same; that is, the resistance is as the space run through in the same time; but the velocity is also in the same ratio of the space run over in the same time; and, therefore, the resistance from this cause, is as the velocity itself.

The resistance from the second cause, when a body moves through the same fluid with different velocities, is as the square of the velocity. For, first, the resistance increases according to the number of particles or quantity of the fluid struck in the same time; which number must be as the space run through in that time, that is, as the velocity: but the resistance also increases in proportion to the force with which the body strikes against every part; which force is also the velocity of the body, so as to be double with a double velocity, and triple with a triple one, &c.; therefore, on both these accounts, the resistance is as the velocity multiplied by the velocity, or as the square of the velocity. Upon the whole, therefore, on account of both causes, viz. the tenacity and inertia of the fluid, the body is resisted partly as the velocity and partly as the square of the velocity.

But when the same body moves through different fluids with the same velocity, the resistance from the second cause follows the proportion of the matter to be removed in the same time, which is as the density of the fluid.

The following experiments for ascertaining the resistance of water on the fore-part or bow of a vessel, as well as the minus or negative pressure on the stern, were communicated for this work by Colonel Mark Beaufoy, of Hackney Wick.

By referring to bodies A B, and C D, it will be seen that these two bodies have the same fore and middle parts, but different stern-ends. The stern-end of A B being 9 times the semi-diameter of the body, and the stern-end of C D only 6 times the semi-diameter. Now it is evident that, as these two bodies have similar fore and middle bodies, the difference of the resistance between them, after the friction is deducted, must arise from the difference of the after-pressure.

TABLE 1. line 1, contains the total resistance of A B.
line 2, contains the friction, which being deducted, leaves in
line 3, the plus and minus pressure.

TABLE 2. line 1, contains the total resistance of C D.
line 2, contains the friction, which being deducted, leaves in
line 3, the plus and minus pressure.
line 4, contains the plus and minus pressure of A B.
line 5, contains the difference between the minus pressure of A B and C D, which is found not to be materially different. The minus pressure of A B may therefore be considered as nothing.

But as more difficulty is experienced in making accurate experiments with right-lined and acute fore bodies than in curved ones, recourse was had to the bodies E F and G H. These bodies, as in the former case have the same shaped (semi-circular) fore and middle bodies, but different shaped sterns; the shape of the after part being like those in the former experiments.

TABLE 3. line 1, contains the total resistance of E F.
line 2, contains the friction, which being deducted, leaves in
line 3, the plus and minus pressure.

TABLE 4. line 1, contains the total resistance of G H.
line 2, contains the friction, which being deducted, leaves in
line 3, the plus and minus pressure.
line 4, contains the plus and minus pressure of E F.
line 5, contains the difference between the plus and minus pressure of E F and G H, or the minus pressure of G H, which, being deducted from line 3, leaves in
line 6, the plus pressure of G H.

The minus pressure of G H is used in the calculations.

The figure I K has the same shaped fore and middle body as figures A B and C D, with the stern part taken away. By proceeding in the manner as in the other tables, the minus pressure was found as is set down in Table 5.

TABLE 5, figure L M is the body G H reversed; that is to say, the stern of G H in this case, is made the bow or fore-part of L M, and the bow of G H the stern of L M.

TABLE 6. line 1, contains the total resistance of L M,
line 2, contains the friction.
line 3, the plus and minus pressure.
line 4, the plus pressure which being deducted, leaves in
line 5, the minus pressure.

For the purpose of determining the accuracy of the above-mentioned results, experiments were made with the body N, the fore-part of N has the same bow as G H, and the after-part is the stern end as L M. It is evident, if to the resistance of the bow G H, we add the minus pressure of L M, the sum should be the resistance of N, after deducting the friction.

TABLE 7. line 1, contains the total resistance of N.
line 2, contains the friction.
line 3, plus and minus pressure.
line 4, plus pressure of G H.
line 5, minus pressure of L M.
line 6, sum of the two pressures, or the plus and minus. The pressure of the body N, as determined by the experiments made with the bodies G H and L M.
line 7, contains the plus and minus pressure of N, being determined by the experiments made with that body.
line 8, contains the difference between the two results, which agree as nearly as could be reasonably expected, considering the nature of the experiments.

HORIZONTAL SECTIONS.

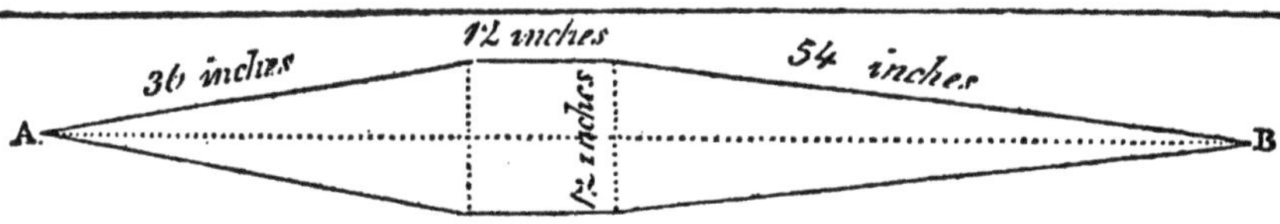

TABLE I. Feet per Second.

	1.	2.	3.	4.	5.	6.	7.	8.	9.	10.	11.	12.
Total resistance in lbs. avoir.	0.2937	1.1616	2.5584	4.442	6.776	9.524	12.657	16.145	19.958	24.075	28.47	33.11
Friction substract	0.0425	0.1634	0·3583	0.625	0.963	1.371	1.846	2.889	2.999	3 675	4.41	5.22
Remains the plus pressure	0.2512	0.9982	2.2001	3.817	5.813	8.153	10.811	13.756	16.959	20.400	24.06	27.89

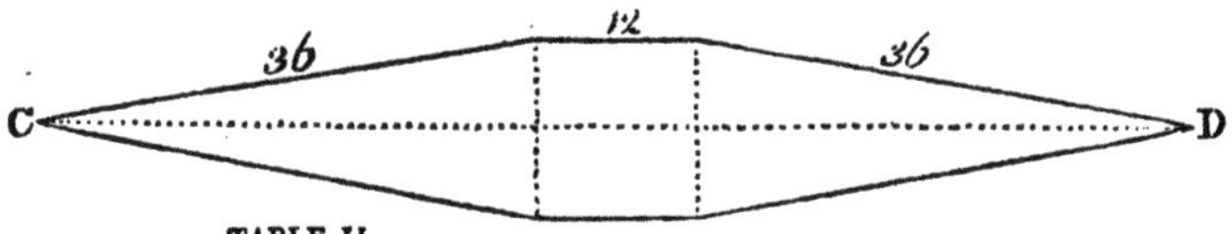

TABLE II.

	1.	2.	3.	4.	5.	6.	7.	8.	9.	10.	11.	12.
Total resistance in lbs. avoir.	0.2517	1.0237	2.2996	4.056	6.273	8.930	12.004	15.479	19.336	23.559	28.13	33.03
Friction substract	0.0368	0.1417	0.3109	0 542	0.836	1.189	1.602	2.073	2.602	3.188	3.83	4.53
Remains the plus and minus pressure	0.2149	0.8820	1.9887	3.514	5.437	7.741	10.402	13.406	16.734	20.371	24.30	28.50
Plus pressure of A B	0.2512	0.9982	2.2001	3.817	5.813	8.153	10.811	13.756	16.959	20.400	24.06	27.89
Difference of minus pressure AB and CD	+0373	+1162	+2114	+303	+376	+412	+ 409	+ 350	+ 225	+ 029	—.24	—.61

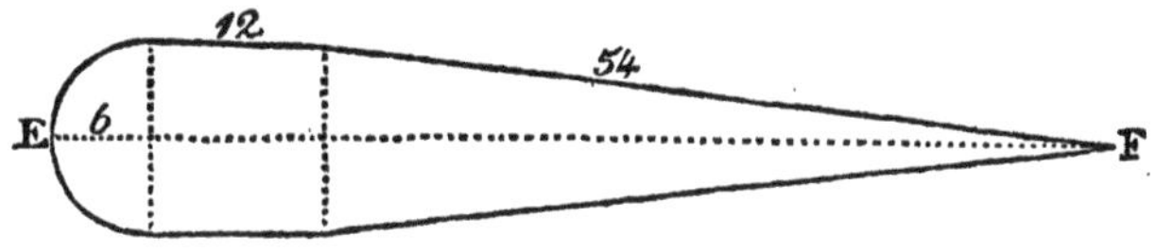

TABLE III.

	1.	2.	3.	4.	5.	6.	7.	8.	9.	10.	11.	12.
Total resistance in lbs. avoir.	0.2706	1.1089	2.5050	4.441	6.898	9.857	13.300	17.210	21.569	26.373	31.59	37.21
Friction substract	0.0344	0.1323	0.2902	0.506	0.780	1.110	1.495	1.935	2.429	2.976	3 58	4.23
Remains the plus pressure	0.2362	0.9766	2.2148	3.935	6.118	8.747	11.805	15.275	19.140	23.397	28.01	32.98

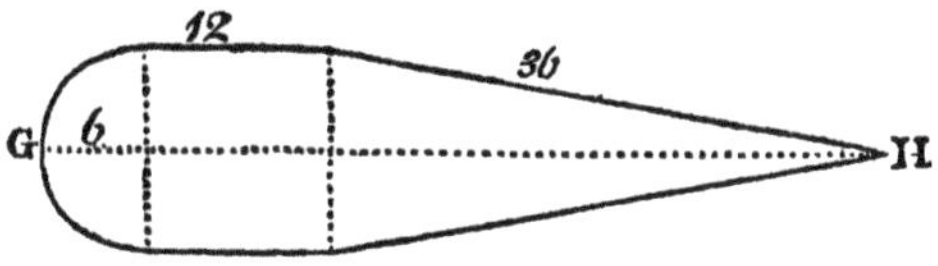

TABLE IV.

	1.	2.	3.	4.	5.	6.	7.	8.	9.	10.	11.	12.
Total resistance in lbs. avoir.	0.2698	1.1069	2.5026	4.440	6.900	9.867	13.321	17.244	21.622	26.442	31.69	37.34
Friction substract	0.0288	0.1107	0.2427	0.423	0.653	0.928	1.250	1.619	2.032	2 489	2.99	3.54
Remains plus and minus pressure	0.2410	0.9962	2.2599	4.017	6.247	8.939	12.071	15.625	19.590	23.953	28.70	33.80
Plus pressure of E F substract	0.2362	0.9766	2.2148	3.935	6.118	8.747	11.805	15.275	19.140	23.397	28.01	32.98
Remains the minus pressure	0.0048	0.0196	0.0451	0 082	0.129	0.192	0.266	0.350	0.450	0.556	0.69	0.82
Which substracted leaves plus pressure of GH	0.2362	0.9766	2.2148	3.935	6.118	8.747	11.805	15.275	19.140	23.397	28.01	32.98

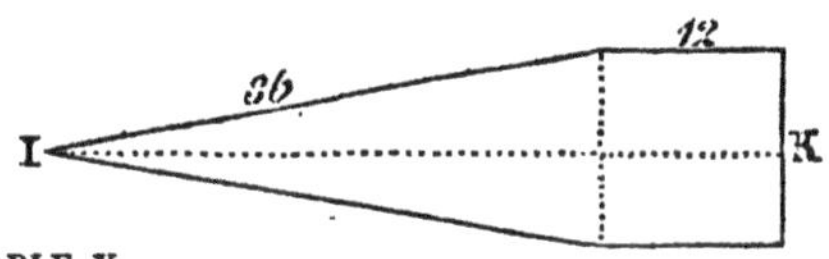

TABLE V.

	1.	2.	3.	4.	5.	6.	7.	8.	9.	10.	11.	12.
Minus pressure of I K	0.1629	0.6278	1.3668	2.361	3.591	5.045	6.707	8.563	10.607	12.820	15.190	17.71

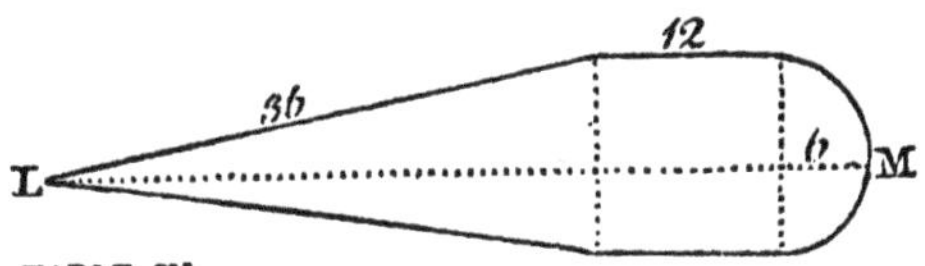

TABLE VI.

	1.	2.	3.	4.	5.	6.	7.	8.	9.	10.	11.	12.
Total resistance in lbs. avoir.	0.3054	1.2544	2.8394	5.043	7.847	11.232	15.180	19.678	24.702	30.25	36 29	42.81
Friction substract	0.0288	0.1107	0.2427	0.423	0.653	0.928	1.250	1.619	2.032	2.49	2.99	3.54
Remains plus and minus pressure	0.2766	1.1437	2.5967	4.620	7.194	10.034	13.930	18.059	22.670	27.760	33.30	39.27
Plus pressure of C D substract	0.2101	0.8624	1.9436	3.432	5.308	7.549	10.136	13.356	16.284	19.815	23.61	27.68
Remains the minus pressure	0.0665	0.2813	0.6531	1.188	1.886	2.755	3.794	4.703	6.386	7.945	9.69	11.59

N. B. The head pressure is found by substracting the minus pressure of GH from the plus and minus pressure of C D, exam. 28.50—82=27.68.

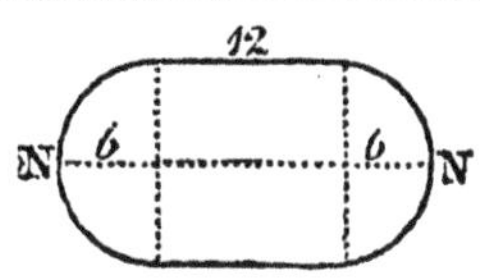

TABLE VII.

	1.	2.	3.	4.	5.	6.	7.	8.	9.	10.	11.	12.
Total resistance in lbs. avoir.	0.3406	1.3965	3.1586	5.608	8.723	12.486	16.874	21.873	27.461	33.63	40.35	47.61
Fraction substract	0.0207	0.0796	0.1746	0.304	0.469	0.668	0.900	1.164	1.461	1.79	2 15	2.54
Remains plus and minus pressure	0.3199	1.3169	2.9840	5.304	8.254	11.818	15.974	20.709	26.000	31.84	38.20	45.07

	1.	2.	3.	4.	5.	6.	7.	8.	9.	10.	11.	12.
To the plus pressure of body G H add	0 2362	0.9766	2.2148	3.935	6 118	8 747	14.805	15.275	19.140	23.397	28.01	32.98
The minus pressure of body L M	0.0665	0.2813	0.6531	1.188	1.886	2.755	3.794	4.703	6.386	7.945	9.69	11.59
Sum is plus minus pressure of N N	0.3027	1.2579	2.8679	5.123	8.004	11.502	15.599	19.978	25.526	31.342	37.70	44.57
Plus and minus pressure of N N	0.3199	1.3169	2.9840	5.304	8.254	11.818	15.974	20.709	26.000	31.840	38.20	45.07
Difference of results by experiments	.0172	.0590	.1161	.181	.250	.316	.375	.731	.474	.498	.50	.50

N. B. The middle part of each body was a cube of 12 inches. The sharp ends are triangles, and the circular ends semi-cylinders; and the centre of each body was immersed six feet below the surface of the water. The minus pressure of A B is considered as nothing.

RESPONSIBLE, (*responsable*, Fr.) answerable, accountable. Thus all commanders in chief, as well as commissioned, warrant, and petty officers, in the royal navy, are more or less responsible for the trust reposed in them.

RESPONSIBILITY, (*responsabilité*, Fr.) is the state of being answerable. All public officers, civil, military, or naval, are in a state of responsibility with respect to the national concerns.

REST, (*repos*, Fr.) the continuance of a body in the same place, or its continual application or contiguity to the same parts of the ambient or contiguous bodies; and therefore is opposed to motion. Sir Isaac Newton defines true or absolute rest to be the continuance of a body in the same part of absolute space; and relative rest to be the continuance of a body in the same part of relative space. Thus in a ship under sail, relative rest is the continuance of a body in the same part of the ship; but absolute rest is its continuance in the same part of universal space, in which the ship itself is contained. It is one of the laws of nature, that matter is indifferent to motion or rest.

RETARDATION, in physics, the act of retarding or diminishing the velocity of a moving body. If bodies of equal bulk, but of different densities, be moved through the same resisting medium, with equal velocity, the medium will act equally on each, so that they will have equal resistances, but their motions will be unequally retarded, in proportion to their densities. Retarded motion from gravity is peculiar to bodies projected upwards, and this in the same manner as a falling body is accelerated; only in the latter, the force of gravity acts in the same direction with the motion of the body; and in the former, in an opposite direction. See the article RESISTANCE.

RETICULA, or RETICULE, (*reticule*, Fr.) in astronomy, is a contrivance for measuring the exact quantity of eclipses. This instrument, introduced by the Academy of Sciences at Paris, is a little frame, consisting of thirteen fine silken threads, equidistant from each other, and parallel, placed in the focus of object-glasses of telescopes; that is, in the place where the image of the luminary is painted in its full extent; consequently, the diameter of the sun or moon is hereby seen divided into twelve equal parts or digits; so that, to find the quantity of the eclipse, there is nothing to do but to number the luminous and the dark parts. As a square reticule is only proper for the diameter, not for the circumference, of the luminary, it is sometimes made circular by drawing six concentric equidistant circles. This represents the phases of the eclipse perfectly.

To RETIRE FROM THE SERVICE, (*se retirer du service*, Fr.) is to quit a naval or military situation or place of trust which one fills, either by commission or warrant.

Commissioned officers, in the royal navy, when they are not employed in actual service, as also when they are no longer fit for sea-service, retire on half pay generally, and in some particular instances on account of servitude, on full pay, according to the rank they hold, provided they have no civil employments.

To RETIRE *upon full Pay*, (*se retirer du service avec la paie entière*, Fr.) is to leave any ship, department, or office under the government, with the full pay or subsistence which is attached to the exercise of its duties or functions. Warrant officers, whose employments are constant, and who, after serving a certain period of time, are admitted to superannuation, may be said to retire upon full pay or pension, equal to the pay in the ship of the highest rate they have served in. See the articles PAY, PENSION, and PENSIONERS.

RETIRED LIST, (*role des officieres retirés du service*, Fr) a list on the ordnance and marine establishment, upon which superannuated or deserving officers whose health or circumstances justify a retirement, are placed.

RETREAT, (*retraite*, Fr.) the order or disposition in which a fleet or squadron declines engagement, or flies from a pursuing enemy.

RETROGRADATION, or RETROGRESSION, in astronomy, is an apparent motion of the planets, by which they seem to an observer placed on the earth, to go backwards in the ecliptic, and to move contrary to the order of the signs, which is from east to west.

When a planet moves according to the order of the signs, as from Aries to Taurus, from Taurus to Gemini, &c. it is said to be *direct.*—When it appears for some days in the same place, or point of the heavens, it is said to be *stationary*; and when it goes contrary to the order of the signs, it is said to be *retrograde.* All these different affections or circumstances may happen in all the planets, except the sun and moon, which are seen to go direct only.

These various circumstances, however, in the motions of the planets are not real, but only apparent, as the inequalities arise from the motion and position of the earth from whence they are viewed; for when they are considered as seen from the sun, their motions appear always uniform and regular.

RETROGRADE, denotes backward, or contrary to the forward or natural direction.

To RETURN, (*retourner*, Fr.) is to replace or give back; as, " We shall return the remainder of our stores." " The admiral returned our salute."

RETURNS, (*comptes*, Fr.) in maritime affairs, principally relate to the delivery of stores or provisions from a ship or ships, into the king's stores or depôts, on behalf of government, by the respective officers, into whose charge they have been entrusted, agreeably to the rules of the navy, and for which they obtain credit on their accounts. Condemned provisions are always returned into store, being so directed in the report made; but the article cheese, after being condemned, may be thrown overboard, by a specific order from the captain, who must give a certificate for that purpose. The butter, when condemned, is sometimes given to the boatswain, to use as grease, &c.

The returns made by boatswains, gunners, and carpenters, in consequence of surveys, where stores condemned are not convertible. One copy, termed report, of the proceedings on that survey, or examination, must accompany the articles sent on shore, to shew the authority for so doing, also to prevent fraud or collusion: a receipt for such returned stores cannot be refused.

In the carpenter's department, and in a few instances of the boatswain's, wherein a mast, yard, &c. is carried away or sprung, &c by the violence of the weather, or even an oar broke, and if the heel or part remaining in possession be returned into store, a new article will be given in lieu thereof, where it is clearly perceived by the fragment, that the case is as has been related; but this accommodation is chiefly allowed in the smaller articles, which form the officers charge. See the article REPORT.

RETURNS, (*remise, cargaison de retour*, Fr.) among merchants, a term implying the value, whether it consists of goods or specie, returned by a consignee to the consigner of the whole or part of a cargo. Thus the returns sent from the West Indies, in exchange for articles of British manufacture, &c. consist of sugar, coffee, cotton, &c.

REVERBERATION, in physics, the act of a body repelling or reflecting another, after its impinging on it.

In the glass-men's furnace the flame reverberates or bends back again to burn the matter on all sides.

Echoes are occasioned by the reverberation of sounds from arched obstacles.

REVOLUTION, in astronomy, is the period of a star, planet, comet, &c.; or its course from any point of its orbit, till it returns to the same again.

The planets have a two-fold revolution: the one about their own axes, usually called their *diurnal rotation*, which constitutes their day; and the other about the sun, called their *annual revolution*, or *period*, constituting their year.

REWARD, (*récompense*, Fr.) a recompense given for good offices performed, as for the apprehension of deserters from his Majesty's ships and vessels, &c. See the article STRAGGLING MONEY.

RHOMBOID, or RHOMBOIDES, in Geometry, a quadrilateral figure, whose opposite sides and angles are equal; but which is neither equilateral, nor equiangular.

RHOMBUS, is an oblique equilateral parallelogram, or a quadrilateral figure, whose sides are equal and parallel, but the four angles, not all equal, two of the opposite ones being obtuse, and the other two opposite ones acute.

The two diagonals of a rhombus intersect at right angles, but not of a rhomboides.

As to the area of the rhombus or rhomboides, it is found, like that of all other parallelograms, by multiplying the length of the base by the perpendicular breadth.

RHUMB, (*rhumb de vent*, Fr.) in navigation, is a vertical circle of any given place, or the intersection of a part of such a circle with the horizon. Rhumbs, therefore, coincide with the points of the world, or of the horizon; and hence the mariners distinguish the rhumbs by the same names as the points and winds.

But we may observe, that the rhumbs are denominated from the points of the compass in a different manner from the winds. Thus, at sea, the N.E. wind is that which blows from the north-east point of the horizon towards the ship in which we are; but we are said to fall upon the north-east rhumb when we go towards the north-east.

Rhumbs are divided and sub-divided like points. Thus the whole rhumb answers to the cardinal point: the half rhumb answers to a collateral point, or makes an angle of 45° with the former: the quarter rhumb makes an angle of 22° 30′ therewith; and the half-quarter rhumb makes an angle of 11° 15′.

Sometimes navigators divide the 32 points into four quarters, and call the rhumb next the east the first rhumb, the next to that the second rhumb, &c.

RHUMB-LINE, (*rhumb de vent, sur la carte*, Fr.) a line prolonged from any point of the compass on a nautical chart, except the four cardinal points: or it is a line which a ship, keeping in the same collateral point or rhumb, describes through its whole course.

The great property of the rhumb-line, or loxodromia, and that from which some authors define it is, that it cuts all the meridians under the same angle. This angle is called the angle of the rhumb, or the loxodromic angle.

The angle which the rhumb-line makes with any parallel to the equator, is called the complement of the rhumb.

An idea of the origin and properties of the rhumb-line, the great foundation of navigation, may be conceived thus: a vessel beginning its course, the wind wherewith it is driven makes a certain angle with the meridian of the place; and as it is supposed the vessel runs exactly in the direction of the wind, it makes the same angle with the meridian which the wind makes. For example: a wind that is north-east, and which of consequence makes an angle of 45° with the meridian, is equally north-east, wherever it blows, and makes the same angle of 45° with all the meridians it meets. A vessel, therefore, driven by the same wind, always makes the same angle with all the meridians it meets with on the surface of the earth.

If the vessel sails north and south, it makes an angle infinitely acute with the meridian, that is, parallel to it, or rather sails in it. If it run east and west, it cuts all the meridians at right angles.

RIBBANDS, or RIBANDS, (*lisses des couples*, Fr. from *rib* and *bend*) in naval architecture, are long narrow flexible pieces of timber, nailed upon the outside of the ribs, near the foremost to the aftmost square-timbers, so as to encompass the ship lengthways.

The ribbands, being judiciously arranged with regard to their height and distance from each other, and forming regular sweeps about the ship's body, will compose a kind of frame, whose interior surface will determine the curve of all the intermediate or filling timbers, which are stationed between the principal ones. As the figure of the ship's bottom approaches to that of a conoid, and the ribbands having a limited breadth, it is apparent that they cannot be applied to this convex surface without forming a double curve, which will be partly vertical and partly horizontal; so that the vertical curve will increase by approaching the stem, and still more by drawing near the stern-post. It is also evident, that by deviating from the middle line of the ship's length, as they approach the extreme breadth at the midship-frame, the ribbands will also form an horizontal curve.

The floor-ribband is terminated by the foremost and aftmost square-timbers.

The first futtock-ribband is placed between the floor and lower futtock-head.

The second is placed between the first and second futtock-head.

The third is placed between the second and third futtock-head.

The fourth is termed the main-breadth ribband.

Those ribbands are connected with the harpings, and reach fore-and-aft the ship in the weak of the square-timbers. See the article HARPINGS.

RIBBING NAILS. See the article NAILS.

RIBS OF A PARRAL. See the article PARRAL.

RIBS *of a Ship*, (*côtes, ou membres d'un bâtiment*, Fr.) a figurative expression for the TIMBERS, which article see.

RICOCHET, (*ricochet*, Fr.) denotes a bound or leap, such as a flat piece of stone makes when thrown obliquely along the surface of water: hence,

RICOCHET-FIRING, in the military art, is a method of firing guns with small charges of powder, and at a low elevation, as from three to six degrees. The ball or shot, thus discharged, goes bounding and rolling along, and killing or destroying every thing in its way.

To RIDE, (*être à l' ancre, &c.* Fr.) when expressed of a ship, signifies to keep her in a particular station, by means of one or more cables with their anchors, which are for this purpose sunk into the bottom of the sea, &c. in order to prevent the vessel from being driven at the mercy of the wind or current. See the article MOORINGS.

To RIDE *Apeak.* See the article PEAK.

To RIDE *Athwart, or between Wind and Tide*, (*être mouillé au courant, et évité en travers du courant et du vent, qui sont opposés l'un à l'autre*, Fr.) is when the wind and tide are in opposition, but so nearly equal in their force that the ship rides with the tide running against one side, and the wind blowing upon the other.

A rope is said to ride, when one of the turns by which it is wound about the capstan or windlass lies over another, so as to interrupt the operation of heaving.

To RIDE *Easy*, (*avoir les mouvemens doux au mouillage par une grosse mer*, Fr.) is said of a ship when she does not labour or feel a great strain upon her cables.

To RIDE *Hard*, (*étant mouillé, avoir les mouvemens durs dans un gros tems*, Fr.) is on the contrary, to pitch violently in the sea, so as to strain her cables, masts, or hull.

To RIDE *Head to Wind*, (*faire tête au vent, étant à l'ancre*, Fr.) is when the wind is so much more powerful than the tide, as to cause the ship to swing till her head is in the direction of the former.

To RIDE *out a Gale*, (*étaler un coup de vent*, Fr.) signifies that the ship does not drive during the storm.

To RIDE *the Head-rope of a Sail*, (*appuyer avec les pieds, sur la ralingue de têtière d'une voile*, Fr.) is to shake and stretch it by treading upon it, while a purchase is employed at the end to extend it.

RIDERS, (*porques*, Fr.) a sort of interior ribs, fixed occasionally in a ship's hold, opposite to some of the principal timbers, to which they are bolted, and reaching from the kelson to the beams of the lower deck, and sometimes higher, in order to strengthen her frame. They are bolted to the other timbers, to support them when it is apprehended the ship is not sufficiently strong in the part where they are fixed, which is generally amidships.

They have also their floor-pieces and futtocks, and sometimes their top-pieces, all of which are scarphed to each other in the same manner as in the timbers.

The riders ought to be stationed so as to lie between two ports of the lower deck, and to correspond with the timbers to which they are attached, in such a manner, as that the scarphs of the riders may be clear of those of the timbers. They are scored upon the kelson, clamps, and thick stuff of the bottom, and are secured by bolts, which are driven from without, so as to penetrate the outside planks, the timbers, the clamps, and the riders; on the inside of which last they are fore-locked.

These pieces are rarely used in merchant ships, on account of the space they occupy in the hold; neither are they always used in vessels of war, at least till the ship is enfeebled by service.

RIDGE, (*écueil, ressif, ou banc de rochers*, Fr.) a long assemblage of rocks, lying near the surface of the sea.

RIFLE-GUN, (*carâbine; arquebuse rayée*, Fr.) in the military art, a fire-arm which has lines formed within its barrel, that run in a vermicular direction; and are more or less numerous, or more indented, according to the fancy of the artificer.

The rifle barrel possesses many advantages over the common one; these advantages are attributed to the threads or rays with which it is indented. The threads are sometimes cut in such a manner, that the line which commences on the right side of the breech, terminates on the left at the muzzle; by which means the ball acquires a rotary movement, revolving once and a half round its own axis before it quits the piece, and then boring through the air with a spiral motion. The rifles are said to carry their balls both farther and truer than the common pieces. For the nature and qualities of them, see ROBINS's *Tracts on Gunnery*, vol. 1. page 328, &c.

RIFLEMEN, (*carabiniers*, Fr.) oa name given to experienced marksmen, armed with *rifles*. The Americans place riflemen in the tops of their ships of war for the destruction of their adversaries.

To RIG, (*gréer, ou garnir, &c.* Fr.) is to fit the shrouds, stays, braces, &c. to their respective masts and yards.

To RIG *in a Boom*, (*faire rentrer un boute-hors*, Fr.) is to draw it in from a situation upon the end of a yard, bowsprit, or another boom, &c. to extend the foot of a sail; as, "Rig in the top-mast studding-sail booms!"

To RIG *out a Boom*, (*pousser en dehors un boute-hors, arboutant, &c.* Fr.) is to run out a pole upon the end of a yard, or bowsprit, to extend the foot of a sail; as, "Rig out the top-gallant studding-sail booms!" These booms are confined in those places by double rings, formed like a figure of 8, one part of which is fastened to the respective yard-arm, or bowsprit end, and the other receives the boom, which is occasionally drawn in tl ough it.

The rings used in this service are termed boom-irons.

RIGEL, in astronomy, a bright fixed star of the first magnitude, in the left foot of the constellation Orion.

For the beginning of the year 1813, its
Right ascension was .... 76° 23′ 18″.3
Annual variation in right ascension 43. 1
Declination .... 8 26 43 S.
Annual variation in declination.. — 5.9

RIGGERS, (*agréeurs*, Fr.) are persons whose business it is to go on board ships to fit the standing and running rigging.

The term is also applied to any party of men sent to his Majesty's rigging-lofts or hulks to prepare the standing-rigging for putting over the mast-heads.

RIGGING, (*manœuvres, agrès et garniture du vaisseau*, Fr.) a general name given to all the ropes employed to support the masts; and to extend or reduce the sails, or arrange them to the disposition of the wind.

In rigging a mast, the first thing usually fixed upon its head is, a circular wreath or rope, called a grommet, or collar, which is firmly beat down upon the top of the hounds. The intent of this is to prevent the shrouds from being fretted or worn by the trestle-trees, or shoulders of the mast; after this are laid on the two pendents, from whose lower ends the main or fore-tackles are suspended; and next, the shrouds of the starboard and larboard side, in pairs, alternately The whole is covered by the stays, which are the largest ropes of the rigging. See the article MAST.

When a yard is to be rigged, a grommet is also driven first on each of its extremities: next to this are fitted on the horses, the braces; and, lastly, the lifts, or top-sail sheet-blocks: all of which are explained in their proper places.

The principal objects to be considered in rigging a ship appear to be strength, convenience, and simplicity, of arranging the whole machinery in the most advantageous manner, to sustain the masts, and facilitate the management of the sails; and of avoiding perplexity, and rejecting whatever is superfluous or unnecessary. The perfection of this art then consists in retaining all those qualities, and in preserving a judicious medium between them.

Within the last thirty years, many useful discoveries have been made and applied in the leading of rigging. For a good plain practical treatise lately published on this subject, the reader is referred to LEVER's *Young Sea Officer's Sheet-Anchor*; or *a Key to the Leading of Rigging and Practical Seamanship*, in which may be found much new and original information, illustrated with elegant engravings, calculated for the private instruction of young gentlemen in the royal navy, as well as for the merchant service.

Lower RIGGING, (*haubans et étais des mâts majeurs*, Fr.) denotes the shrouds, stays, &c. which are set up, and firmly attached to the lower masts.

Standing RIGGING, (*manœuvres dormantes*, Fr.) is that which is employed to sustain the masts, and which remains in a fixed position; as the shrouds, stays, and backstays.

Running RIGGING, (*manœuvres courantes*, Fr.) is that which is fitted for the purpose of arranging the sails, by passing through various blocks in different places about the masts, yards, shrouds, &c. as the braces, sheets, halyards, clew-lines. &c.

Top-mast RIGGING, (*haubans, étais et gal-haubans des mâts de hune*, Fr.) consists of the top-mast-shrouds, stays, and back-stays.

Top-Gallant RIGGING, (*haubans, &c. des mâts de perroquet*, Fr.) that which is fixed to the top-gallant-masts.

The names, with a description and use of the various ropes which comprise the rigging of ships, will be found in this work under their respective heads.

RIGGING-LOFT, (*attelier, de grêement*, Fr.) in dock-yards, a long room wherein the standing-rigging is fitted in readiness for his Majesty's ships, and in which is a master in the navy, as superintendant, to see that it is done in a workmanlike manner.

There is a large tackle fastened either to a post or to a ring-bolt in the floor for bowsing taught the rigging; and, at certain distances, down the middle is a row of large strong posts, firmly fixed, for the purpose of stretching ropes, laying on service, &c.; and on each side are births for the men to prepare the small rigging in.

To RIGHT *a Ship*, (*relever, ou redresser un vaisseau qui a été abattu en carène*, Fr.) is to restore her to an upright position, after she has been laid on a careen, by the mechanical powers usually applied in that operation.

This is generally the natural effect of casting loose the careening pullies by which she had been drawn down. It is, however, necessary sometimes to apply mechanical powers to right the ship in such a situation. The principal of these are the RELIEVING TACKLES, which article see.

A ship is also said to right at sea when she rises, with her masts erected, after having been pressed down on one side by the effort of her sails, or a heavy squall of wind.

To RIGHT *the Helm*, (*dresser la barre*, Fr.) implies to bring it in midships, after having put it out of that position.

RIGHT, in geometry, signifies the same with straight: thus a straight line is called a right one.

RIGHT-*Angle*, is that angle which one line makes with another upon which it stands, so as to incline neither to one side nor the other.

RIGHT-*Angled*, is said of a figure when its sides are at right angles or perpendicular to each other. This sometimes holds good in all the angles of the figure, as in squares and rectangles; sometimes only in part, as in right-angled triangles.

RIGHT-*lined Angle*, one formed by right lines.

RIGHT *Ascension, Descension, Parallax, &c.* See the respective articles in alphabetical order.

RIGHT *Sailing*, is that in which a voyage is performed on some one of the four cardinal points, east, west, north, or south.

If the ship sail on a meridian, that is, on the north or south points, she does not alter her longitude, but only changes the latitude, and that just as much as the number of degrees she has run. But if she sail on the equator, directly east or west, she varies not her latitude, but only changes the longitude, and that just as much as the number of degrees she has run.

And if she sail directly east or west, upon any parallel, she again does not change her latitude, but only the longitude; yet not the same as the number of degrees of a great circle she hath sailed in, as on the equator, but more, according as the parallel is remoter from the equinoctial towards the pole. For the less any parallel is, the greater is the difference of longitude answering to the distance run.

RIM, or Brim, (*bord extérieur, ou listeau des hunes*, Fr.) a name given to a skirting of narrow elm'board round the upper side of any of the tops, as, "Bear the back-stays abaft the top-rim!"

RINDGALL, denotes the injury a tree receives when young, so that the bark or rind grows into the inner substance of the tree.

RING, in anchor-making, is an iron circle shut in the upper part of the shank of an anchor, and to which the cable is bent. See the article Anchor.

Ring, in astronomy and navigation, an instrument in the form of a ring, serving to take altitudes of the sun. It is usually made of brass, about nine inches in diameter, and suspended by a little swivel, at the distance of 45°, from the point of which is a perforation, which is the centre of a quadrant of 90°, divided in the inner concave surface.

To use it, let it be held up by the swivel, and turned round to the sun, till his rays, falling through the hole, make a luminous spot among the degrees, whereby the altitude is found.

This instrument is generally preferred before the astrolabe, because the divisions are here larger than on that instrument; however, it is far from being exact enough to be much depended on in astronomical observations, which are better made by the Quadrant or Reflecting *Circle*, which articles see.

Ring *of Saturn*, is a thin, broad, luminous, circular arc, surrounding the body of that planet, like the wooden horizon of an artificial globe, without touching it, and appearing double, when seen through a good telescope.

This ring was first discovered by Huygens, who, after frequent observation of the planet, perceived two lucid points, like ansæ or handles, arising out from the body in a right line. In subsequent observations he always found the same appearance, and therefore concluded that Saturn was encompassed with a permanent ring; and accordingly produced his 'New System of Saturn,' in the year 1659. However, Galileo first discovered that the figure of Saturn was not round.

Huygens makes a space between the globe of Saturn and the ring, equal to the breadth of the ring, or rather more, being about 22,000 miles broad; and the greatest diameter of the ring, in proportion to that of the globe, as 9 to 4. But Mr. Pound, by an excellent micrometer, applied to the Huygenian glass of 123 feet, determined this proportion, more exactly, to be as 7 to 3. See the article Saturn.

Ring Bolt, (*cheville à boucle*, Fr.) an iron bolt, with an eye at one end, wherein is fitted a circular ring, as represented by figures 15 and 16, plate IV. The ring-bolts are for several uses, but particularly to hook the tackles to, by which the cannon of a ship are managed and secured: accordingly there is one fixed in the deck opposite to every cannon, as represented by Z, fig. 9. plate IX., and they are, for the same purpose, fixed in the edges of the gun-ports, as expressed in the Midship Section, plate VI. They are driven through the plank and the corresponding beam, or timber, and retained in this position by a small pin thrust through a hole in the small end, as appears in fig. 16. plate IV

Ring-Ropes, (*bosses du cable, attachées à des boucles de fer qui sont sur le premier pont*, Fr.) are short pieces of rope, tied occasionally to the ring-bolts in the deck, to stopper or fasten the cable more securely when the ship rides with a heavy strain. They are, however, more particularly necessary in veering away the cable gradually in those circumstances, in order to freshen the hawse; as, without this precaution, it would be extremely difficult to check the cable, which being then charged with a great effort, might be drawn violently out of the ship at random.

Ring-Tail, (*bonnette de baume, ou de brigantine*, Fr.) a small quadrilateral sail, extended on a little mast, which is occasionally erected for that purpose on the top of a ship's taffrail, the lower part being stretched out by a boom, which projects from the stern horizontally. This sail is only used in light and favourable winds.

Ring-Tail *Sail*, (*bonnette de baume, ou de brigantine*, Fr.) is also a name given to a sort of studding-sail, hoisted beyond the after-edge or skirt of those main-sails which are extended by a boom and gaff over the stern; as in all sloops, brigs, and schooners. The two lower corners of this sail are stretched out to a boom, called a

Ring-Tail *Boom*, (*boute-hors de bonnette de baume, ou de brigantine*, Fr.) which rigs in and out upon the main or driver-boom, in the same manner that a studding-sail boom does on the top-sail yards.

Rings *and Starts for Hatches*, (*anneaux des écoutilles*, Fr.) are small iron rings half an inch thick, and 2¼ inches diameter in the clear: they are driven through the hatches, into which their points are turned to prevent them from drawing, and serve to lift them out of their respective places.

Port-Rings, or *Shackles*, (*anneaux des sabords*, Fr.) are iron rings about 7-eighths of an inch thick, and 4 inches diameter in the clear: they are clinched both inside and outside the lower parts of the portlids: the two on the outside of the portlid are used by means of tackles to haul it up; and the other two are for securing it inside by means of the port-fastenings.

RIPPLING, (*bouillonnement des eaux, ou clapotage*, Fr.) a broken and interrupted noise, produced by a current on or near the sea-coast.

RIPPING-IRON, or Chisel, (*ciseau à rompre les bordages, ou à démolir*, Fr.) an iron instrument used by shipwrights to rip the sheathing-boards and copper from off the bottom of ships.

RISING, in astronomy, the appearance of the sun, or a star, or other luminary, above the horizon, which before was hid beneath it.

By reason of the refraction of the atmosphere, the

heavenly bodies always appear to rise before their time; that is, they are seen above the horizon, while they are really below it, by about 33′ of a degree. See the article REFRACTION.

There are three poetical kinds of risings of the stars. See the articles ACRONICAL, COSMICAL, and HELIACAL.

RISING-LINE, (*lisse des façons*, Fr.) a name given by shipwrights to an incurvated line, which is drawn on the plane of elevations or sheer-draughts (see plate XVII. fig. 2.) to determine the height of the ends of all the floor-timbers throughout the ship's length, and which accordingly ascertains the figure of the bottom, with regard to sharpness and flatness.

RIVER, (*rivière*, *fleuve*, Fr.) in geography, a stream or current of fresh water, flowing in a bed or channel, from a source or spring, into the sea.

When a stream is not large enough to bear boats, or small vessels, laden, it is properly called by the diminutive *rivulet* or *brook*; but when it is considerable enough to carry larger vessels, it is called by the general name *river*.

The great, as well as the middle-sized rivers, proceed either from a confluence of brooks and rivulets, or from lakes; but no river of considerable magnitude flows from one spring, or one lake, but is augmented by the succession of others. Thus the Wolga receives above two hundred rivers and brooks before it discharges itself into the Caspian Sea; and the Danube receives no less before it enters the Euxine Sea. Some rivers are much augmented by frequent rains, or melted snow. In the country of Peru and Chili, there are small rivers, that only flow in the day; because they are only fed by the snow upon the mountains of the Andes, which is then melted by the heat of the sun. There are also several rivers upon both sides the extreme parts of Africa, and in India, which, for the same reason, are greater by day than at night. The rivers also in these places are almost dried up in summer, but swell and overflow their banks in winter, or in the wet season. Thus the Wolga in May and June is filled with water, and overflows its shelves and islands, though at other times of the year, it is so shallow, as scarcely to afford a passage for ships laden.

In the months of June and July, 1813, the river Mississippi rose higher than it had been known for these thirty years past. The water had burst the mounds, and inundated the country for many miles on each side, and swept away vast crops, with an immense number of horses, cattle, &c.

The Nile, the Ganges, the Indus, &c. are so much swelled with rain or melted snow, that they overflow their banks, and these deluges happen at different times of the year, because they proceed from various causes. Those that are swelled with rain are generally highest in winter, because it is then usually more frequent than at other times of the year; but if they proceed from snow, which, in some places, is melted in the spring; in others, in summer, or between both, the deluges of the rivers happen accordingly. Again, some rivers hide themselves under ground, and rise up in other places, as if they were new rivers. Thus the Tigris, meeting with Mount Taurus, runs under it, and flows out at the other side of the mountain: also, after it has run through the lake Tospia, it again immerges, and being carried about eighteen miles under ground, breaks out again, &c. The channels of rivers, except such as were formed at the creation, Varenius thinks, are artificial. His reasons are, that, when a new spring breaks out, the water does not make itself a channel, but spreads over the adjacent land; so that men were necessitated to cut a channel for it to secure their grounds. He adds, that a great number of channels of rivers are certainly known from history to have been dug by men.

During the last century, the philosophical world was agitated by a debate concerning the origin of those waters which are necessary for the supply of rivers, &c. One party contended strongly for the existence of a large mass of water within the bowels of the earth, which supplied not only the rivers but the ocean itself; at the head of these we may place the ingenious but fanciful Burnet. The French philosophers, on the contrary, asserted, that the waters of the ocean were conveyed back by some subterraneous passages to the land, and being filtrated in their passage, returned again to the sea in the course of the rivers: but this opinion appears contrary to all the known principles of hydrostatics.

In opposition to these hypotheses, our illustrious countryman, Halley, contended that the process of evaporation, and the immense disposition of water in consequence of it, were fully adequate to the whole supply. If, indeed, we consider the immense quantity of water which is continually carried up into the atmosphere by evaporation, and consider that this is a process which is continually going on, not only from the ocean but from the rivers themselves, and from the whole surface of the earth, we shall see but little reason to doubt of Dr. Halley's hypothesis; but may reasonably conclude, that this kind of circulation is carried on through all nature; and that the sea receives back again through the channel of the rivers, that water which it parts with to the atmosphere.

All rivers have their source either in mountains, or elevated lakes; and it is in their descent from these, that they acquire that velocity which maintains their future current. At first their course is generally rapid and head-long; but it is retarded in its journey by the continual friction against its banks, by the many obstacles it meets to divert its stream, and by the plane's generally becoming more level as it approaches towards the sea.

Rivers are always broadest at the mouth, and narrower towards their source. But what is less known, and probably more deserving curiosity, is, that they run in a more direct channel as they immediately leave their sources; and that their sinuosities and turnings become more numerous as they proceed. It is a certain sign among the savages of North America, that they are near the sea, when they find the rivers winding, and every now and then changing their direction. And this has become an indication to the European travellers themselves, in their journies through those trackless forests. As those sinuosities, therefore, increase as the river approaches the sea, it is not to be wondered at, that they sometimes divide, and thus disembogue by different channels. The Danube disembogues into the Euxine by seven mouths; the Nile, by the same number; and the Wolga, by seventy. For the different rivers in Europe, Asia, Africa, and America, with an account of their nature, situation, and extent, see GREGORY'S *Dictionary of Arts and Sciences*: and for an account of the rivers of Great Britain, see SKRINE'S *History of Rivers*. Also, for an elaborate treatise on rivers and canals, see *Phil. Trans.* vol. 69, by Mr. Maun, who has availed himself of the observations of Gulielmini, Ximenes, and most other writers on this subject.

RIVET, (*rivet*, Fr.) a metal pin clenched at both ends, so as to hold an intermediate substance with more firmness.

RIVETTING-PLATES, in gun-carriages, are small square thin pieces of iron, through which the ends of the bolts pass, and are rivetted upon them.

ROACHING A SAIL, a term used by sail-makers to signify the allowance made for the beauty in the appearance of a sail, as well as allowing for belly, &c.

ROACH-LEECH, a term signifying the curve on the mast-leech of some fore and aft sails, &c.

ROAD, or ROADSTEAD, (*rade*, Fr.) a place of anchorage, at some distance from the shore, on the sea-coast, whither ships or vessels occasionally repair to receive intelligence, orders, or necessary supplies; or to wait for a fair wind, &c.

A good ROADSTEAD, (*bonne rade*; *bon mouillage*, Fr.) is that which is protected from the reigning winds and the swell of the sea, has a good anchoring ground, and is at a competent distance from the shore.

An open ROAD, (*rade foraine*, Fr.) is one that is not sufficiently inclosed from the wind and sea.

ROADER, or ROADSTER, (*vaisseau au mouillage, ou à l'ancre*, Fr) denotes a vessel riding at anchor in a road, bay, or river. If a vessel under sail strike against any roadster, and damage her, the former is obliged by law to make good the damages sustained by the latter.

Roadsters should attentively observe to anchor at a competent distance from each other, so as not to intercept each other's departure.

ROBINS, See the article ROPE-BANDS.

ROCK, (*rocher*, *roche*, Fr. *rocca*, Ital.) a large mass or block of hard stone, firmly fixed in the ground. Rocks are frequently discovered near the surface of the sea, and they greatly endanger ships that accidentally drive foul of them.

A Half-Tide ROCK, (*rocher qui se decouvre à mi-marée*, Fr.) denotes a rock which appears above the water at half ebb.

ROCKET, (*fusée volante*, Fr. *rochetto*, Ital.) in pyrotechny, is an artificial fire-work, contained in a cylindrical case of paper, and filled with the composition under mentioned; which, being tied to a stick, mounts into the air to a considerable height, and there bursts with an explosion resembling stars. In war they are used as signals.

Those for the royal navy generally weigh one pound.—Of these the composition is, saltpetre, 4 lbs.; sulphur, 1 lb.; charcoal-dust, 1 lb. 8 oz.; and mealed powder, 4 oz.

The sulphur, saltpetre, and powder, to be first mealed and mixed well together, so that one shall not appear separated from the other. The charcoal should be mealed by itself, and the whole well mixed together. A port-fire cut in pieces, about half an inch long, will answer for stars.

Dimensions of a one-pound ROCKET.—Stick, 8 feet long; breadth of the top, 2 inches; 1½ inch thick; and tapering to 2-tenths at the end. The number of sheets of thick paper required, is five.

The composition for the stars is, mealed powder, 8 ounces; saltpetre, 8 lbs.; sulphur, 2 lbs.; antimony 2 lbs. isinglass dissolved, 3¾ oz.; spirits of wine, one pint; vinegar, one quart.

General Table of Sky-Rockets.

		2-pound.	1-pound.	½-pound.	¼-pound.
		Inches.	Inches.	Inches.	Inches.
Case	Exterior diameter . . .	2 13	1. 69	1. 34	1. 06
	Interior diameter . . .	1.529	1.214	0,961	0 76½
	Length before driving . .	15. 97	12. 67	13. 05	8. 25
Length of gauge for the choke .		1. 5	1. 25	1. 0	0. 75
Cylinders for heading	Diameter	2. 84	2. 25	1. 79	1. 39
	Length	4. 26	3. 38	2. 68	2. 12
Cones for heading .	Diameter	2. 84	2. 25	1. 79	1. 39
	Height	4. 26	3. 38	2. 68	2. 18
Mallet for driving		1 lb.	8 oz.	2. 1	1. 10
No. of strokes		31	21	18	13

Copper Ladles for filling Sky-Rockets.

Length, 1½ the exterior diameter of the bore.

Diameter, equals the interior diameter of the case.

Circumference, three quarters the interior calibre of the case.

Sky-Rockets are driven with composition up to 4½ exterior diameters of the case from the choke; and 1-fifth of a diameter above the composition with good clay. They are bored and reamed up to 3½ diameters.

Dimensions of Sticks for Rockets.—General Rules.

For rockets from half an ounce to one pound, the stick must be 60 diameters of the rocket in length: for rockets from one pound and upwards, fifty or fifty-two diameters. Their thickness at top about half a diameter, and their breadth very little more. Their square at bottom equal to half the thickness at top.

Species of Rockets.	6-pdr.	4-dr.	2-pdr.	1-pdr.	½ pdr.	¼ pdr.	⅛ pdr.
Distance of poise from the point of the cone . .	Ft. In. 4 1½	Ft. In. 3 9	Ft. In. 2 9	Ft. In. 2 1	Ft. In. 1 10½	Ft. In. 1 8¼	Ft. In. 1 3

Rockets of between three and four inches in diameter have been observed to ascend as high as 1000 or 1200 yards; but the height of common rockets is between 450 and 600 yards; and their flight usually short of seven seconds. See SIMMONS'S *Sea-Gunner's Vade-Mecum*.

Congreve's ROCKETS, so called from the name of the inventor, are of a peculiar internal structure and composition, by which the action of this principle of projectile force is so greatly increased as not only to triple the flight of small rockets so formed, but also to allow of the construction of rockets of such dimensions as on the ordinary principles of combination would not even rise from the ground, and of such powers of flight and burden as have hitherto been considered altogether impracticable.

On the basis of this increase of power, Colonel Congreve has succeeded in making *War Rockets* for various naval and military purposes, and of various descriptions and calibres, either for explosion or conflagration, and armed both with shells and case-shot; of which we have subjoined a table, shewing their different natures, ranges, &c. By this it will be seen, that the 32-pounder rocket carcass, which is the nature hitherto chiefly used for bombardment, will range 3000 yards with the same quantity of combustible matter as that contained in the 10-inch spherical carcass, and 2500 yards with the same quantity as that of the 13-inch spherical carcass. It will be seen, also, that the 12-pounder rocket case-shot, which is so portable that it may be used with the facility of musketry, has a range nearly *double* that of field artillery, carrying as many bullets as the 6-pounder spherical case; and here it ought to be observed, that the projectile force of the rocket is *peculiarly* well calculated for the conveyance of case-shot to great distances, because as it proceeds, its velocity is accelerated instead of being retarded, as happens with every other projectile, while the average velocity of the shell is greater than that of the rocket *only* in the ratio of 9 to 8; independent of which, the case-shot conveyed by the rocket admits of *any desired increase of velocity* in its range by the bursting powder, which cannot be obtained in any other description of case.

Of this description of rocket case-shot, one hundred infantry soldiers will carry into action, in any situation where musketry can act, 300 rounds, and 10 frames, from each of which four rounds may be fired in a minute. And of the same description of shot for the use of cavalry, four horses will carry ninety-six rounds and four frames, from which may be fired sixteen rounds in a minute, each horse not having more than the ordinary burden of a dragoon horse.

The carcass rockets were first used at Boulogne in 1805, in consequence of a demonstration of their powers made at Woolwich by Colonel Congreve, in the presence of Mr. Pitt, and several other cabinet ministers, in the month of September. Sir Sidney Smith was appointed to command this expedition, but from the lateness of the season, it being the end of November before the preparations were completed, nothing was done that year. In 1806, however, Colonel Congreve renewed his proposition for the attack of Boulogne by rockets, and it was ordered, in consequence of Lord Moira, then master-general of the ordnance, and Lord Howick, then first lord of the Admiralty, having attended an experiment at Woolwich, and having satisfied themselves by their own observation of the powers of the weapon. The attack was accordingly made under the command of Commodore Owen, late in October, 1806; having been put off during the summer months in consequence of the negociation for peace at that time pending. From this delay, however, instead of being carried on upon the great scale at first intended, it became a mere desultory attack, in which not more than 200 rockets were fired. The town, however, was set on fire by the first discharge, and continued burning for near two days: it is also believed that some of the shipping were burnt, but the greater part of the rockets certainly went over the basin into the town.

Since this period, the carcass-rockets have been used in almost every expedition, and always under the immediate direction of the inventor. Their reputation was completely established at Copenhagen, where they did incredible execution: after the siege, they were ordered by Lord Chatham, the master-gen. of the ordnance, to be reported upon by a committee of field officers of artillery who had witnessed their effect in that bombardment, and who pronounced them to be "*a powerful auxiliary to the present system of artillery.*" Nor need we go further back for proofs than the late siege of Flushing, where General Monnet, the French commandant, made a formal remonstrance to Lord Chatham respecting the use of them in the bombardment of that place: than which no better proof need be recorded of the effect they must have produced. If such, therefore, be the acknowledged effect of the weapon, when only a handful is used, merely as it were in the way of experiment by the inventor, with not more than twenty or thirty men to assist him, what may not be expected from it when regularly organized in the service, and generally combined with the other implements of bombardment?

Congreve's rockets are formed in metallic cases. The carcass of a 32-pounder, as represented by A H, plate XIX., is 4 feet and a half in length, having a strong iron head, B C, the diameter of which is 6½ inches, and at the smaller end, D E, 4½ inches, filled with a composition as hard and solid as iron itself. The stick, F G, is 18 feet in length, and contrived so as to be fixed firmly and instantaneously, in action, by being first put through the two rings *i k*, and then screwing the end into the ring *l*, which secures it to the rocket. We understand, however, that Colonel Congreve is now employed, and with the most sanguine expectation, of dispensing with the stick altogether.

The great general point of excellence of the rocket system, is, the *facility* with which all the natures of this weapon may be *conveyed* and *applied*.

Its peculiar applicability to naval bombardment hinges on this property, that there is no reaction, nor recoil in firing of the largest rockets; so that by this means carcasses, equal to those projected from the *largest* mortars, may be thrown from the *smallest* boats. And its *peculiar* fitness for land-service, is, that it is a description of extremely powerful ammunition, *without ordnance*, so that the burden of mortars and guns is dispensed with, and all that is to be carried is *actual available missile matter*, capable of the range, and of many of the most important effects of the heaviest artillery. Not that this system has been brought forward with any view to do away any part of the present system of artillery, but merely as a powerful auxiliary to it, *in all cases*, and certainly in some cases, as possessing a facility of application where artillery cannot at all be brought to act: and on that account a *rocket-corps* has been lately established in this country, part of whom were sent out in 1813, under the command of Capt. Bogue, of the artillery, to join the allied armies in the North, and performed the most essential service to the allies, more particularly at the great battle near Leipsic, where that gallant officer fell at the head of his brave corps.

There has been much misapprehension as to the *expense of the rocket system*; and it is therefore proper to explain, that, in fact, it is the cheapest of all ammunition depending on the projectile force of gunpowder; for the proof of which it will be only necessary briefly to state, that the 32-pounder carcass-rocket costs only 1*l*. 11*s*. 0½*d*. complete in every respect for service; whereas its equivalent, the 10-inch spherical carcass, with the charge of powder necessary to convey it 3000 yards, which power is *contained* in the rocket, costs 1*l*. 2*s*. 7*d*., independent of any charge for the *mortar*, *mortar-bed platforms*, *difference of transports*, &c. attached to the spherical carcass and not to the rocket, which actually requires *no apparatus whatever*, to use it in a bombardment, and has therefore no charge attaching to it, beyond the first cost, but that of transport; and a vessel of 300 tons will carry 5000 of them at least. We have indeed seen a calculation by which it appears, that in every 10-inch carcass so thrown, there is an actual saving of 3*l*. In fact, when our previous observation is kept in mind, that "the rocket system *is a system of ammunition without ordnance*," it cannot fail to strike every one reading the above statement, that, in all its applications, the rocket must be the cheapest possible arm. See JAMES's *Military Dictionary*.

Modes of using the Rocket Carcasses in Bombardments.

1*st*.—*From a Slope without any Apparatus.*

Colonel Congreve states that it requires only a slope of a few yards in length to discharge the rockets in considerable quantities, as they may be laid against the slope in notches cut to receive them, and any number may be previously arranged for a volley, according to the length of the bank; laying the rockets about 18 inches apart, supported by small stakes driven into the bank, and the sticks to which they are screwed are placed in holes made to receive them at the bottom of the bank, whilst the rockets communicate with one another by leaders, or they may be fired separately by port-fires with long sticks. Hence it is evident, that the rockets possess the greatest possible facility of being used in large vollies to a great extent, from this capability of discharge, without any apparatus whatever. It is only to be observed, that to give the extreme range, the slope must be so arranged as to ensure at least 55° of elevation to the rocket; and that where any convenient natural slope exists, or where en renchments are to be established, this is the most simple application of the rocket.

2*d*.—*By Means of a portable Stand or Frame.*

Where a rapid operation is intended, and no convenient slope is to be found, and where it is not intended to open any trenches, then the frame is to be used. These frames are very portable, and made on a similar principle to a painter's or gardener's ladder.

When divested of the legs, it may be used from the smallest boat in the navy, the elevation being given it by raising or lowering the front of it by the halyards and the foremast of the boat. With the legs it becomes a convenient frame for the shore, is pointed by means of a plummet, and has a water bucket hung at the top for the convenience of sponging, which must be attended to after every round. A common artillery waggon will carry two of them, and near 100 rounds of 32-pounder carcass-rockets, and in less than five minutes they may be prepared for action in any situation. Their direction and elevation may be varied with great facility, and they are constructed so as to discharge two rockets at a time, and four men to a frame is a sufficient complement.

The rockets are placed by a man going up the ladder, and four rockets may be discharged from each frame in five minutes without confusion or hurry; each rocket is equal to a 10-inch spherical carcass, as to range, and superior to it in quantity of combustible matter.

Exercise for the bombarding Frames.

N°. 1. goes up the ladder to sponge the iron chambers from which the rockets are discharged, which must be done with a very wet sponge after every round. He then, after having primed both the chambers with a quill tube and powder, places a rocket in each, and cocks the two locks after every thing else is done, taking care to lay the lines leading to the lock clear.

N°. 2. receives the rocket from N°. 3, and hands it up the ladder to N°. 1, after having carefully stript off the canvass circle which covers the vent. He also helps N°. 1, in any way that he may require, and fires when the word is given by N° 1, who, having charge of making ready, will give the word when he is sufficiently clear of the frame.

The lines for firing are to be led through the rounds of the ladder into the rear, 15 or 20 feet; at which distance the man firing is not incommoded by the smoke.

N°. 3, has only to bring the rockets from the rear to the foot of the frame, where he delivers them to N°. 2. He may either bring them singly, or two at a time, but must observe not to bring any fresh supply until those in the frame are discharged.

N°. 4, is stationed with the rocket and sticks at a convenient distance in the rear, where he unheads the boxes in which the rockets are packed, six in a box; and screws the sticks firmly to the rocket. As this is an instantaneous process, he should not fix more than he requires from time to time to hand over to N°. 3.

The above distribution makes the work of each extremely simple; and, by this means, it is evident that a very heavy fire may be kept up by 40 men with ten frames. A non-commissioned officer may, with ease, superintend two, or even three frames; and one officer may direct the whole.

Whatever number of frames may be used, they should not be nearer than 10 or 20 yards from each other, that the business of firing, loading, &c. in each, may proceed independently and without interruption from the other.

Of the Use of Rockets in the Field.

There are two descriptions of rocket applicable to field-service against troops; the first is, the 32-pounder shell rocket, which ranges a 9-pounder Sharpnel's shell 3000 yards, a distance at least double that to which it can be used by any other application of artillery:—the second is, a small description of rocket case-shot, equal to a 6-pounder case-shot, with a range of 2000 yards, and which infantry may use by hand.

1.—*The 32-pounder Shell Rocket* is to be discharged by a field carriage, which carries 54 rounds in the limber; and, being merely a slight frame upon wheels, without any piece of ordnance, is, of course, the lightest description of field-carriage, though capable of ranges so much greater than the heaviest field artillery. The same allotment and complement of men are ample for the service of this carriage as for that of the frames for bombardment.

2.—*The Portable Rocket Case-Shot*, are intended to be used from a slight frame, which one man can carry, and the rockets themselves may also be conveyed by hand; one man may carry three or four such rockets; and thus, for any particular service, a company of infantry might, without the incumbrance either of horses or carriage, possess the powers of two or three hundred rounds of 6-pounder case-shot, ranging 2000 yards, which is nearly double the range of ordinary field artillery.

Both these species of rockets, it is conceived, would be a most powerful auxiliary to artillery in the field, both on account of the great length of ranges compared to ordinary field-artillery, and on account of the *extraordinary facility of their use:* for in addition to the carriage and frame, it is conceived that they will give very great recochet ranges without any carriage or frame at all, but merely by laying them on the ground; or at most by making a rest by two small stakes driven into the ground, and crossing at top to receive the head of the rocket, and give it a little elevation.

General Directions for firing Rockets, with respect to the State of the Wind.

A moderate breeze does not at all affect the flight of these large rockets; but if a fresh wind blows across the line of fire from the right, the rocket must be pointed a little to the left of the object.

If across the range from the left, they must be laid a little to the right.

If directly against the flight of the rocket, a little more elevation must be given the frame; thus, with the extreme range, 60° elevation instead of 55°.

If directly with the flight of the rocket, a little less elevation, as for instance, 50° instead of 55°, for the extreme range.

For, in the first two cases, the rocket always inclines towards the wind, from its acting a little more on the stick than on the body of the rocket: from the same cause, in the third case, it tends to depress the elevation of the stick; and, in the fourth, to increase it: the effect, however, is not great, even in a very strong wind, and may, by very little attention, be allowed for. Care also should be taken to screw the stick on as straight as possible to the rocket; and, if particular nicety of direction is required, to reject any sticks which may be warped; the rocket should also *be laid in as true a line* as possible on the frame, the sticks being placed exactly uppermost, and the chambers in which the rockets lie being *fixed perfectly parallel to the sides* of the frames, that the sticks may not be twisted. And if all these directions be properly attended to, Colonel Congreve pledges himself that the rocket will be found to be capable of as much accuracy as the spherical carcass in long ranges.

List *of the different Species of* Rocket Ammunition *at present made by* Colonel Congreve.

Natures of Ammunition.		*Armed with*	*Extreme Range.*	*Elevation for Extreme Range.*
			Yards	Degrees.
42-Pounder Carcass Rockets	*Carcasses.*	large, 18 lbs. of combustible matter; small, 12 lbs.	3,500	60
42-Pounder Shell Rockets	*Shells.*	$5\frac{1}{2}$ inch; 12-Pounder spherical	3,500	60
32 Pounder Carcass Rockets	*Carcasses.*	large, 18 lbs. combustible matter	2,000	60
		medium, 12 lbs. equal 13-inch carcass	2,500	60 to 55
		small, 8 lbs. equal 10 inch ditto	3,000	55
32-Pounder Shell Rockets	*Shells.*	9-pounder spherical	3,000	50
32-Pounder Rocket Case Shot	*Case shot, which receives any increased velocity from the bursting powder*	large, containing 200 carbine balls	2,500	55
		small, 100 ditto	3,000	50
32-Pounder Explosion Rockets	*strong iron cones containing from 5 to 12 lbs. of powder, to burst by fuses*		2,500 to 3,000	55
12-Pounder Rocket Case Shot	*Case Shot*	large, 72 carbine balls	2,000	45
		small, 48 ditto	2,500	45

ROCKET *Light Ball*, also invented by Colonel Congreve, is a species of light ball thrown into the air by means of one of his rockets, where having reached the summit of the rocket's ascent, it is detached from it by an explosion, and remains suspended in the air by a small parachute to which it is connected by a chain. Thus, in lieu of the transient and momentary gleam obtained by the common light ball, a permanent and brilliant light is obtained and suspended in the air for five minutes at least, so as to afford time and light sufficient to observe the motion of an enemy by night, either on shore or at sea, where it is particularly useful in chasing, or for giving more distant and extensive night-signals. It is to be observed, that nothing of this kind can be obtained by the projectile force of either guns or mortars, because the explosion would infallibly destroy any construction that could be made to produce the suspension in the air.

Floating ROCKET *Carcass*. This is another of Colonel Congreve's applications of his rockets, and of the parachute, for the purpose of conveying combustible matter to distances far beyond the range of any known projectile force: at the same time that it is cheap, simple, and portable. The floating carcass, like the light ball, is thrown into the air attached to a rocket, from which being liberated at its greatest altitude, and suspended to a small parachute, it is driven forward by the wind, and will, in a moderate breeze, afford ranges *at least double* those of the common carcass, and may, therefore, for naval purposes, from a blockading squadron, be thrown in great quantities with a fair wind against any fleet or arsenal, without the smallest risk, or without approaching within range either of guns or mortars: thus, in the late blockade of the Russian fleet at Baltic Port, it might have been continually used, at all events, with great prospect of success, *and certainly where no other means of annoyance could be applied*. The rocket containing this carcass, is not longer than the 32-pounder rocket carcass, and the whole expense added to the rocket, does not exceed 5s: nor are the approaches of the carcass itself necessarily visible by night, as it may be arranged not to inflame till some time after it has settled. It is evidently, therefore, capable of becoming a very harassing weapon, with a great chance of doing as much mischief as any other carcass amongst large fleets and flotillas, by lodging unperceived in the rigging, or lighting in extensive arsenals, in situations where no other means of annoyance *whatever* exist.

Indian ROCKETS are used by the native troops in India. These rockets are made of iron, and are lashed to a bamboo cane. The weight is seldom more than two pounds, or less than one. They were used with very destructive effect against the British during the siege of Seringapatam, in 1799.

Signal ROCKETS are seldom more than one pound weight. They are used in the navy for signals, and are fired perpendicularly.

ROCKY, (*plein de rochers*, Fr.) a mass composed or abounding with stone, rock, &c. as distinguished from sandy, muddy, &c.

RODS, or RAMMERS, are instruments either of iron or wood, used to drive home the charges of muskets, carabines, and pistols.

RODS, a name sometimes given to the sticks which are fastened to sky-rockets, to make them rise in a straight line.

ROGUES-YARN, (*fil blanc mis dans un des torons d'un cordage goudronné, et fil goudronné dans les cordages blancs, pour servir à reconnoître les cordages de fabrique des arsenaux royaux d'avec ceux des particuliers*, Fr.) a name given to a rope-yarn which is twisted in a contrary manner to the other part of a rope, and being tarred, if in a white rope, but white if in a tarred rope, is easily discovered. It is placed in the middle of each strand in all cables or cordage made for the king's service, to distinguish them from the merchant's cordage.

To ROLL, (*rouler, ou aller au roulis*, Fr.) when applied to a ship at sea, denotes to reel to and fro, which is caused by the violent agitation of the water,; as, "That ship almost rolls her keel out."

ROLLER, in block-making, a cylindrical pin turning on its own axis, and is used in some blocks instead of a sheave.

The patent rollers are made of two parallel circular plates of brass, about a quarter of an inch thick. Four or more solid brass cylinders are placed at equal distances round these plates, and work upon their own axis, between them, at right angles. Thus any pin working through these plates must touch the rolling surfaces of the solid brass cylinders, by which the friction is considerably lessened.

ROLLER, (*virevaut*, Fr.) a cylindrical piece of timber, fixed either horizontally or perpendicularly to the ship's deck, so as to revolve on an axis. It is used to prevent the cables, hawsers, &c. from being chafed by lessening the friction they would otherwise sustain.

ROLLERS, (*rouleaux servant à conduire et faciliter le transport des pièces de bois et autres objets lourds*, Fr.) are also moveable pieces of wood, of the same figure, which are occasionally placed under boats or long pieces of timber, &c. in order to move them with greater facility either in the dock-yards, or in lading and delivering merchant-ships.

ROLLING, (*roulis; le mouvement du roulis*, Fr.) the motion by which a ship rocks from side to side like a cradle, occasioned by the agitation of the sea.

ROLLING is, therefore, a sort of revolution about an imaginary axis, passing through the centre of gravity of a ship: so that the nearer the centre of gravity is to the keel, the more violent will be the rolling motion; because the centre about which the vibrations are made, is placed so low in the bottom, that the resistance made by the keel to the volume of water which it displaces in rolling, bears very little proportion to the force of the vibration above the centre of gravity, the radius of which extends as high as the mast-heads.

But if the centre of gravity is placed higher above the keel, the radius of vibration will not only be diminished but an additional force to oppose the motion of rolling will be communicated to that part of the ship's bottom, below the centre of gravity.

So far as relates to the effect of rolling, when produced by the quality or stowage of the ballast, and to the manner by which it may be prevented, viz. a change of the quantity or disposition of the ballast, we shall endeavour to explain under the article TRIM. It may, however, be necessary to remark, that the construction of the ship's bottom may also contribute to diminish this movement considerably.

Many fatal disasters have arisen to ships from their violent rolling; as, the loss of the masts, loosening of the cannon, and straining the decks and sides, so as to weaken the ship to a great degree: it is therefore particularly necessary to guard against it as much as possible, not only in the construction of the bottom, but by causing the centre of gravity of the ship to fall as near the load-water line as possible, which can only be effected by a judicious arrangement of the ballast or cargo.

ROLLING-HITCH. See the article HITCH.

ROLLING-TACKLE, (*palan de roulis des vergues*, Fr.) a purchase occasionally fastened to that part of a sail-yard which is to the windward of the mast, in order to confine the yard close down to the leeward when the sail is furled.

It is used to prevent the yard from having a great friction against the mast in a heavy sea, which would be equally pernicious to both.

ROOM, (*soute*, Fr.) a name given to several particular apartments in ships of war, as, the BREAD-ROOM, COOK-ROOM or GALLEY, GUN-ROOM, LIGHT-ROOM, SAIL-ROOM, SLOP-ROOM, SPIRIT-ROOM, STEWARD-ROOM, STORE-ROOM, WARD-ROOM, &c. which articles are explained under their respective heads.

ROOM and *Space*, (*mailles, ou intervalles entre les couples*, Fr.) in ship-building, imply the distance from the joint, or moulding-edge of one floor-timber to the other, which, in all ships that have ports, should be so disposed, that the scantling of the timber on each side the port, and the breadth of the port fore and aft the openings between the timbers of the frames, if any, included, be answerable to the distance of two rooms and spaces.

ROPE, (*corde*, Fr.) an assemblage of several yarns, from thence formed into a certain number of strands, according to the nature and size of the rope to be made, which are twisted together by means of a wheel: it is of various uses, as in binding, staying, drawing, suspending, &c.

ROPE-BANDS, (*rabans de ferlage*, Fr.) pronounced robins, are pieces of small rope, or braided cordage, having an eye at one end; they are used in pairs to the upper edges of the square sails to their respective yards, the long leg passing over the yard two or three times round, and the short leg coming under, is tied to it upon the yard.

ROPE-MAKER, a person appointed by warrant from the commissioners of the navy, to make and superintend the making of such ropes as may be found necessary for the use of the ship.

He is under the immediate direction of the boatswain, from whom he receives whatever quantity of junk and rope-yarns may be requisite for making such rope as shall be required. When directed to make rope of any description, he is to be careful in using such rope-yarns only as shall be fit for the purpose for which that rope may be intended, reserving always the worst yarns for those purposes, where little strength will be required.

If any of the ship's company be directed to assist him, either in making of rope, or spinning of yarn, he is to deliver to them a sufficient quantity of rope-yarns, and to require the quantity of spun-yarn or rope, which that quantity of yarns ought to make. He is to deliver to the boatswain every rope and ball of spun-yarn, as soon as it is completed; and every evening to make a report to the master of all the spun-yarn and rope made in the course of the day.

Before the rope-maker is allowed to receive his wages, he must produce a certificate, signed by the captain, of his having been sober, obedient, and attentive to his duty. See the NAVAL INSTRUCTIONS.

ROPE-MAKING, (*corderie, l'art du cordier*, Fr.) is an art of great importance to a maritime state; and, therefore, deserves the attention of the intelligent observer; as hardly any art can be carried on without the assistance of the rope-maker.

Cordage makes the very sinews and muscles of a ship; and every improvement which can be made in its preparation, either in respect to strength or pliableness, must be of immense service to the mariner, and to the commerce and defence of nations.

The aim of the rope-maker is to unite the strength of a great number of fibres, and the first part of his process is spinning of rope-yarns; that is, twisting the hemp in the first instance. This is done in various ways, and with different machinery, according to the nature of the intended cordage. We shall confine our description to the manufacture of the larger kinds, such as are used for the standing and running-rigging of ships.

An alley or walk is inclosed for the purpose, about two hundred fathoms long, and of a breadth suited to the extent of the manufacture. It is sometimes covered above, and at the upper end is set up the spinning-wheel. The band of the wheel goes over several rollers, called whirls, turning on pivots in brass holes. The pivots at one end come through the frame, and terminate in little hooks, whilst the wheel, being turned by a winch, gives motion in one direction to all those whirls. The spinner has a bundle of dressed hemp round his waist, with the two ends meeting before him. The hemp is laid in this bundle in the same way that women spread the flax on the distaff. There is great variety in this; but the general aim is to lay the fibres in such a manner, that as long as the bundle lasts, there may be an equal number of the ends at the extremity, and that a fibre may never offer itself double, or in a bight. The spinner draws out a proper number of fibres, twists them with his fingers, and having got a sufficient length detached, he fixes it to the hook of a whirl. The wheel is now turned, and the skein is twisted, becoming what is called rope-yarn, and the spinner walks backwards down the rope-walk. The part already twisted draws along with it more fibres out of the bundle. The spinner aids this with his fingers, supplying hemp in due proportion as he walks from the wheel, and taking care that the fibres come in equally from both sides of his bundle, and that they enter always with their ends, and not by the middle, which would double them. He should also endeavour to enter every fibre at the heart of the yarn. This will cause all the fibres to mix equally in making it up, and will make the work smooth, because one end of each fibre is by this means buried among the rest, and the other end only lies outward; and this, in passing through the grasp of the spinner, who presses it tight with his thumb and palm, is also made to lie smooth. A good spinner endeavours always to supply the hemp in the form of a thin flat skein, with his left hand, while his right is employed in grasping firmly the yarn that is twining off, and in holding it tight from the whirl, that it may not run into loops or kinks. It is evident, that both the arrangement of the fibres, and the degree of twisting depend on the skill and dexterity of the spinner, and that he must be instructed, not by a book, but by a master. The degree of twist depends on the rate of the wheel's motion, combined with the retrograde walk of the spinner. We may suppose him arrived at the lower end of the walk, or as far as is necessary for the intended length of his yarn, he then calls out, and another spinner immediately detaches the yarn from the hook of the whirl, gives it to a second spinner, who carries it aside to the reel, and attaches his own hemp to the whirl-hook. In the mean time, the first spinner keeps fast hold of the end of his yarn; for the hemp, being dry, is very elastic, and if he were to let it go out of his hand, it would instantly untwist, and become little better than loose hemp. He waits, therefore, till he sees the reeler begin to turn the reel, and goes slowly up the walk, keeping the yarn of an equal tightness all the way, till he arrives at the wheel, where he waits with his yarn in his hand till another spinner has finished his yarn. The first spinner takes it off the whirl-hook, joins it to his own, that it may follow it on the reel, and begins a new yarn.

The second part of the process is the conversion of the yarns into a rope, cord, or line. That we may have a clear conception of the principle which regulates this part of the process, we shall begin with the simplest possible case, the union of two yarns into one line.

When hemp has been split into very fine fibres by the hatchel, it becomes exceedingly soft and pliant, and after it has lain for some time in the form of fine yarn, it may be unreeled and thrown loose, without losing much of its twist. Two such yarns may be put on the whirl of a spinning-wheel, and thrown like flaxen yarn, so as to make sewing thread. It is in this way, indeed, that the sail-makers sewing-thread is manufactured, and when it has been kept on the reel, balls, or bobbins for some time, it retains its twist as well as its uses require. But this is by no means the case with yarns spun for great cordage. The hemp is so elastic, the number of fibres twisted together is so great, and the diameter of the yarn (which is a sort of lever on which the elasticity of the fibre exerts itself), is so considerable, that no keeping will make the fibres retain this constrained position.

The end of a rope-yarn being thrown loose, it will immediately untwist, and this with considerable force and speed: it would, therefore, be a fruitless attempt to twist two such yarns together; yet the ingenuity of man has contrived to make use of this very tendency to untwist not only to counteract itself, but even to produce another and a permanent twist, which requires force to undo it, and which will recover itself when this force is removed. Every person must recollect, that when he has twisted a packthread very hard with his fingers between his two hands, if he slackens the thread by bringing his hands nearer together, the packthread will immediately curl up, running into loops or kinks, and will even twist itself into a neat and firm cord.

The component parts of a rope are called strands, and the operation of uniting them with a permanent twist is called laying or closing, the latter term being chiefly appropriated to cables and other very large cordage.

The process of laying or closing large cordage is this: the strands of which the rope is composed consist of many yarns, and require a considerable degree of hardening. This cannot be done by a whirl driven by a wheel-band; it requires the power of a crank turned by the hand. The strands, when properly hardened, become very stiff, and when bent round the top, are not able to transmit force enough for laying the heavy and unpliant rope which forms beyond it. The elastic twist of the hardening strands must, therefore, be assisted by an external force. All this requires a different machinery and a different process. At the upper end of the walk is fixed up the tackle-board; this consists of a strong open plank called a breast-board, having three or more holes in it, fitted with brass or iron plates. Into these are put iron cranks, called heavers, which have hooks or forelocks, and keys, on the ends of their spindles. They are placed at such distances from each other, that the workmen do not interfere with each other while turning them round. This breast-board is fixed to the top of strong posts, well secured by struts, braces, facing the lower end of the walk. At the lower end is another breast-board, fixed to the upright post of a sledge, which may be loaded with stones or other weights. Similar cranks are placed in the holes of this breast-board; the whole goes by the name of the sledge. The top necessary for closing large cordage is too heavy to be held in the hand; it therefore has a long staff, which has a truck on the end: this rests on the ground, but even this is not enough in laying great cables. The top must be supported on a carriage, where it must lie very steady, and needs attendance, because the master workman has sufficient employment in attending to the manner in which the strands close behind the top, and in helping them by various methods. The top is therefore fixed to the carriage, by lashing its staff to the two upright posts, when a piece of soft rope, or strap, is attached to the handle of the top by the middle, and its two ends brought back and wrapped several times tight round the rope, in the direction of its twist, and bound down. This greatly assists the laying of the rope by its friction, which both keeps the top from flying too far from the point of union of the strands, and brings the strands more regularly into their places.

The first operation is warping the yarns. At each end of the walk are frames called warping-frames, which carry a great number of reels or winches, filled with rope-yarn. The foreman of the walk takes off a yarn end from each, till he has made up the number necessary for his rope or strand, and bringing the ends together, he passes the whole through an iron ring fixed to the top of a stake driven into the ground, and draws them through: then a knot is tied on the end of the bundle, and a workman pulls it through this ring till the intended length is drawn off the reels. The end is made fast at the bottom of the walk, or at the sledge, and the foreman comes back along the skein of yarns, to see that none are hanging slacker than the rest. He takes up in his hand such as are slack, and draws them tight, keeping them so till he reaches the upper end, where he cuts the yarns to a length, again adjusts their tightness, and joins them altogether in a knot, to which he fixes the hook of a tackle, the other block of which is fixed to a firm post, called the warping post. The skein is well stretched by this tackle, and then separated into its different strands, each being knotted apart at both ends. The knots at their upper ends are made fast to the hooks of the cranks in the tackle-board, and those at the lower ends are fastened to the cranks in the sledge. The sledge itself is kept in its place by a tackle, by which the strands are again stretched in their places, and every thing adjusted, so that the sledge stands square on the walk, and then a proper weight is laid on it. The tackle is now cast off, and the cranks are turned at both ends, in the contrary direction to the twist of the yarns; in some kinds of cordage the cranks are twined the same way with the spinning twist. By this the strands are twisted and hardened up, and as they contract by this operation, the sledge is dragged up the walk. When the foreman thinks the strands sufficiently hardened, which he estimates by the motion of the sledge, he orders the heavers at the cranks to stop. The middle strand at the sledge is taken off from the crank; this crank is taken out, and a stronger one put in its place. The other strands are taken off from their cranks, and all joined on the hook which is now in the middle hole; the top is then placed between the strands, and being pressed home to the point of their union, the carriage is placed under it, and firmly fixed down: some weight is taken off the sledge. The heavers now begin to turn at both ends; those at the tackle-board continue to turn as they did before, but the heavers at the sledge turn in the opposite direction to their former motion, so that the cranks at both ends are now turning one way. By the motion of the sledge-crank, the top is forced away from the knot, and the rope begins to close. The heaving at the upper end restores to the strand the twist which they are constantly losing by the laying of the rope. The workmen judge of this by making a chalk mark on intermediate points of the strands, where they lie on the stakes, which are set up along the walk for their support. If the twist of the strands is diminished by the motion of closing, they will lengthen, and the chalk mark will move away from the tackle-board; but if the twist increases by turning the cranks at the tackle-board, the strands will shorten, and the mark will come nearer to it. As the closing of the rope advances, the whole shortens, and the sledge is dragged up the walk. The top moves faster, and at last reaches the upper end of the walk, the rope being now laid.

In the mean time, the sledge has moved several fathoms from the place where it was when the laying began. These motions of the sledge and top must be exactly adjusted to each other; and the rope must be of a certain length, therefore, the sledge must stop at a certain place. At that moment the rope should be laid; that is, the top should be at the tackle-board. In this consists the address of the foreman; he has his attention directed both ways; he looks at the strands, and when he sees any of them hanging slacker between the stakes than the others, he calls to the heavers at the tackle-board to heave more upon that strand. He finds it more difficult to regulate the motion of the top, as it requires a considerable force to keep it in the angle of the strands, and is always disposed to start forward: to check this, some straps of soft rope are brought round the staff of the top, and then wrapped several times round the rope behind it, and kept firmly down by a lanyard or bandage. This both holds back the top, and greatly assists the laying of the rope, causing the strands to fall into their places, and to keep close to each other, which is sometimes very difficult, especially in ropes composed of more than three strands. It would greatly improve the laying the rope, if the top has a sharp, smooth, tapering pin of hard wood, pointed at the end, projecting so far from the middle of its smaller end, that it gets in between the strands which are closing. This supports them, and makes their closing more gradual and regular. The top, its notches, the pin, and the warp, or strap, which is lapped round the rope, are all smeared with grease or soap, to assist the closing. The foreman judges of the process of closing chiefly by his acquaintance with the walk, knowing that when the sledge is abreast of a certain stake, the top should be abreast of a certain other stake. When he finds the top too far down the walk, he slackens the motion at the tackle-board, and makes the men turn briskly at the sledge. By this the top is forced up the walk, and the laying of the rope accelerates, while the sledge remains in the same place, because the strands are loosing their twist, and are lengthening, while the closed rope is shortening. When, on the other hand, he thinks the top too far advanced, and fears that it will be at the head of the walk before the sledge has got to his proper place, he makes the men heave briskly on the strands, and the heavers at the sledge-crank work softly. This quickens the motion of the sledge by shortening the strands; and by thus compensating what has been over-done, the sledge and top come to their places at once, and the work appears to answer the intention. When the top approaches the tackle-board, the heaving at the sledge could not cause the strands immediately behind the top to close well, without having previously produced an extravagant degree of twist in the intermediate rope. The effort of the crank must therefore be assisted by men stationed along the rope, each furnished with a tool called a woolder, which is a stout oaken stick, about three feet long, having a strap of soft rope-yarn or cordage, fastened on its middle or end. The strap is wrapped round the laid rope, and the workman works with the stick as a lever, twisting the rope round in the direction of the crank's motion. The woolders should keep their eye on the men at the crank, and make their motion correspond with his. Thus they send forward the twist produced by the crank, without either increasing or diminishing it, in that part of the rope which lies between them and the sledge. Such is the general and essential process of rope-making. The fibres of hemp are twisted into yarns, that they may make a line of any length, and stick among each other with a force equal to their own cohesion. The yarns are made into cords of permanent twist by laying them; and that we may have a rope of any degree of strength, many yarns are united in one strand, for the same reason that many fibres were united in one yarn; and in the course of this process it is in our power to give the rope a solidity and hardness which make it less penetrable by water, which would rot it in a short time. Some of these purposes are inconsistent with others; and the skill of a rope-maker lies in making the best compensation; so that the rope may on the whole be the best in point of strengh, pliancy, and duration, that the quantity of hemp in it can produce. The following rule for judging of the weight which a rope will bear, is not far from the truth. It supposes them rather too strong; but it is so easily remembered, that it may be of use. Multiply the circumference in inches by itself, and take the fifth part of the product, it will express the tons which the rope will carry. Thus, if the rope has 6 inches circumference, 6 times 6 is 36, the fifth of which is $7\frac{1}{5}$ tons.

This may suffice for an account of the mechanical part of the manufacture. But we have taken no notice of the operation of tarring; and our reason was, that the methods practised in different rope-works are so exceedingly different, that we could hardly enumerate them, or even give a general account of them. It is evidently proper to tar in the state of twine or yarn, this being the only way that the hemp could be uniformly penetrated. The yarn is made to wind off one reel, and having passed through a vessel containing hot tar, it is wound up on another reel; and the superfluous tar is taken off by passing through a hole surrounded with spongy oakum; or it is tarred in skeins or hauls, which are drawn by a capstan through a tar-kettle, and through a hole formed of two plates of metal, held together by a lever loaded with a weight. See the article Nipper.

It is established beyond a doubt, that a tarred cordage, when new, is weaker than white, and that the difference increases by keeping. The following experiments were made by Mr. Du Hamel, at Rochefort, on cordage of three inches (French) in circumference, made of the best Riga hemp.

	White.	Tarred.
August 8th, 1741.		
Broke with	4500 pounds.	3400 pounds.
	4900	3300
April 25th, 1743.		
	4600	3500
	·5000	3400
Sept. 3d. 1746.		
	4000	2700
	4200	2800

A parcel of white and tarred cordage was taken out of a quantity which had been made February 12th, 1746. It was laid up in the magazines, and comparisons were made from time to time, as follows:

	White bore.	Tarred bore.	Differ.
1746, April 14th,	2645 lbs.	2312 lbs.	333
1747, May 18th,	2762 do.	2155 do.	607
1748, June 19th,	2575 do.	1752 do.	823
1749, Sept. 25th,	2917 do.	1865 do.	1052

Mr. Du Hamel says, that it is decided by experience, 1. That white cordage in continual service is one-third more durable than tarred. 2. That it retains its force much longer while kept in store. 3. That it resists the ordinary injuries of the weather one-fourth longer.

Dr. Gregory has related a remarkable instance to prove this, in his Dictionary of Arts and Sciences, "that in 1758, the shrouds and stays of the sheer-hulk at Portsmouth dock-yard were over-hauled, and when the worming and service were taken off, they were found to be of white cordage. On examining the store-keeper's books, they were found to have been formerly the shrouds and rigging of the Royal William, of 110 guns, built in 1715, and rigged in 1716. She was thought top-heavy, and unfit for sea, and unrigged, and her stores laid up. Some few years afterwards, her shrouds and stays were fitted on the sheer-hulk, where they remained in constant and very hard service for about thirty years, while every tarred rope about her had been repeatedly renewed."

It may be proper to remark here, that the Royal William, which was broken up at Portsmouth in September, 1813, was actually built in the year 1676; and there is an authentic record of her being repaired at Portsmouth, in 1693, as well as in 1719.

Tarred cordage is chiefly serviceable for cables and ground tackle, which must be continually wetted, and even soaked. The result of careful observation is, 1. That white cordage, exposed to be alternately very wet and dry, is weaker than tarred cordage. 2. That cordage which is superficially tarred, is constantly stronger than what is tarred throughout, and it resists better the alternatives of wet and dry. The shrouds of the sheer hulk were well tarred and blacked, so that it was not known that they were of white cordage.

Attempts have been made to increase the strength of cordage by tanning. But, although it remains a constant practice in the manufacture of nets, it does not appear that much addition, either of strength or durability, can be given to cordage by this means. The trial has been made with great care, and by persons fully able to conduct the process with propriety. But it is found that the yarns take so long time in drying, and are so much hurt by drying slowly, that the room required for a considerable rope-work would be immense; and the improvement of the cordage is but trifling. For the improvements which have taken place lately in the machinery and manufacture of cables, cordage, &c. as also for the methods of prolonging their use, see the article *Cable-laid* CORDAGE.

ROPES, (*cordes, cordages et manœuvres*, Fr.) in a general sense, denote all sorts of cordage, above one inch in circumference, used in rigging of ships: they have different appellations, according to their various uses, as CABLE, HAWSER, TOW-LINE, WARP, &c. Ropes are of two sorts, viz.

Cable-laid ROPES, (*cordages deux fois commis*, Fr.) are composed of nine strands, the three greatest strands containing each three small strands, (see the article *Cable-laid* CORDAGE) and

Hawser-laid ROPES, (*cordages une fois commis, ou haussières*, Fr.) are made of three strands, each composed of a certain number of rope-yarns in proportion to its required thickness.

There are, besides those above-mentioned, several ropes, particularly denominated, and their uses described in this work; as, BOLT-ROPE, BUOY-ROPE, ENTERING-ROPES, GUESS-ROPE, GUY-ROPE, PARRAL-ROPE, RING-ROPES, SLIP-ROPE, TILLER-ROPE, TOP-ROPE, &c. which articles see under their respective heads.

To find the Weight of ROPES. Multiply the square of the circumference in inches by the length in fathoms, and divide the product by 480 for the weight in hundred weights.

ROPE-YARN, (*fil de caret*, Fr.) the smallest and simplest part of any rope, being one of the threads of which a strand is composed; so that the size of the latter, and the rope into which it is twisted, is determined by the number of rope-yarns.

ROPE-YARN, denotes also the yarn of any rope untwisted. It commonly consists of cable-ends worn out, and therefore called *junks of the cable*, which serve for many purposes among sailors.

ROTULA. See the ar icle PARALLACTIC *Rotula*.

ROVER, (*forban, ou écumeur de mer*, Fr.) a pirate or free-booter. See the article PIRATE.

ROUGH-TREE, (*lesses de batayoles de bois*, Fr.) a mast, yard, or boom, which remains rough, and in an unfinished state.

ROUGH-TREE *Rail*, a name given in merchant-ships to any long piece of timber, placed as a rail or fence above the ship's side, from the quarter-deck to the fore-castle.

ROUND-HOUSE, (*chambre de conseil dans les vaisseaux de la compagnie des Indes, et dans les gros bâtimens marchands*, Fr.) in merchant ships, a name given to the cabin or apartment built on the after-part of the quarter-deck. In King's ships, it was formerly called the POOP, (see that article) where a number of marines are stationed in time of action, and is where carronades are mounted.

ROUND-HOUSE, (*lieux d'aisance pratiqués auprès de la poulaine pour l'usage des gardes de la marine, &c. sur les vaisseaux de guerre anglois*, Fr.) in King's ships, is also an appellation given to two convenient places built near the head of the ship; the one on the larboard side being appropriated to the private use of the mates, midshipmen, and warrant officers; and the other on the starboard side, to the use of those in the sick-bay.

To ROUND IN, (*haler, en parlant des manœuvres courantes, mais plus particulièrement des bras des vergues*, Fr.) generally implies, to pull upon any slack rope which passes through one or more blocks, in a direction nearly horizontal, and is particularly applied to the braces; as, "Round-in the weather main-top-sail-brace!" &c. It is apparently derived from the circular motion of the rope about the sheave or pulley through which it passes.

To ROUND *Up*, (*haler, en parlant des manœuvres courantes dont la direction approche de la perpendiculaire*, Fr) is used nearly in the same sense, only that it is expressed of a tackle which hangs in a perpendicular direction, without sustaining or hoisting any weighty body, and is opposed to over-hauling, by which operation the blocks are drawn farther asunder.

ROUNDING, (*fourrure de cable, faite avec des quaranteniers, ou autres menus cordages*, Fr.) a name given to old ropes wound firmly and closely about that part of a cable which lies in the hawse, or athwart the stem, &c. It is used to prevent the cable from being chafed in those places. See the articles KECKLING and SERVICE.

ROUNDING, is also a term used in mast-making, when, in large masts, the sixteen-square is divided into four parts, and lined, and those angles dubbed off.

Smaller masts, yards, &c. have the angles of the sixteen-square taken off with a drawing-knife, and then are rounded fair and smooth with planes.

ROUNDING, in rope-making, the act of giving a rope an additional turn, after being closed.

ROUND-ROBIN, a compact or agreement entered into by seamen, when they have cause of complaint against their superior officer, to state their grievances to the Admiralty or commander-in-chief, and to endeavour to obtain redress, without subjecting any one individual more than another to be thought the leader or chief mover. The term appears to be a corruption of *ruban rond*, as their complaints are generally stated in a circular form, and the signatures written all round them, so that none appear first.

ROUND TURN, the situation of the cables of a ship which, when moored, has swung the wrong way, and twisted themselves round each other by the force of the tide or current. See the article HAWSE.

ROUND-TURN, is also the passing a rope once round a timber-head, &c. in order to hold on.

To ROUSE, (*haler ensemble sur un simple cordage, sans l'aide d'aucun palan, ni d'aucune machine pour tirer à soi un fardeau, &c.* Fr.) to pull together upon a cable, hawser, &c. without the assistance of tackles, capstans, or other mechanical powers; as, "Rouse hearty!" It is particularly used in the exercise of removing a ship from one place to another, by means of ropes and anchors. See the article *To* WARP.

To ROW, (*ramer; nager, ou voguer*, Fr) to

impel a boat or vessel along the surface of the water by the force of oars, which are managed in a direction nearly horizontal. See the article OAR.

ROW-DRY! (*nage sec!* Fr.) the order to the men who row, not to splash water into the boat with their oars.

ROW-GALLEY. See the article GALLEY.

ROW-LOCKS, (*tolletières, placées pour les avirons sur le plat-bord d'une embarcation grée à l'Angloise: c'est aussi ce qu'on nomme apostis dans les galères*, Fr.) are those parts of the gunwale, or upper edge of a boat's side, whereon the oar rests in the exercise of rowing.

ROW-PORTS, (*sabords des avirons*, Fr.) little square holes cut in the sides of small vessels of war, nearly parallel to the surface of the water, for the purpose of rowing them in a calm.

ROWED-OFF ALL! (*lève rames!* Fr.) the order for the rowers to cease, and to lay their oars in the boat.

ROWERS, (*rameurs, ou canotiers*, Fr.) an appellation given to persons by whom the oars of a boat are managed.

ROWING-GUARD. See the article GUARD-BOAT.

ROYAL, (*royal*, Fr.) something belonging to a king; kingly.

ROYAL, (*le grand perroquet volant*, Fr.) a name given to the sail which is sometimes spread immediately above the main-top-gallant-sail, to whose yard-arms the lower corners of it are attached. The royal sails are distinguished by three names, viz. fore-top-gallant royal, main-top-gallant royal, and mizen-top-gallant royal; they are only used in light and favourable breezes. See the article SAIL.

ROYAL *Arsenal*. See the article ARSENAL.

ROYAL *Marines*. See the article MARINES.

ROYAL *Military Academy at Woolwich*. See the article ACADEMY.

ROYAL *Military College at Marlow*, and the ROYAL *Naval College at Portsmouth*. See the article COLLEGE.

ROYAL *Naval Hospitals at Greenwich, Haslar,* and *Plymouth*, see these articles.

ROYAL *Navy*. See the article NAVY.

ROYAL *Oak, Robur Carolinum*, in astronomy, is one of the new southern constellations, the stars of which, according to Sharp's catalogue, annexed to the Britannic, are 12.

ROYAL *Standard*. See the article STANDARD.

ROYALS, in artillery, are a sort of small mortars, which carry a shell, whose diameter is five inches and a half. They are mounted on beds the same as other mortars.

ROYAL-SAIL *Sails*, are quadrilateral, and generally made of canvas, N° 8: they are the same as a top-gallant-stay-sail, only with one or two cloths less, and are hoisted next above them.

RUBBER, in sail-making, a small iron implement, fixed in a wooden handle, and used to rub down or flatten the seams of the sails.

RUDDER, (*gouvernail*, Fr.) a long flat piece of timber, hung by pintles to the braces, on the aftmost part of a ship's stern post, being nearly even with the false keel, as represented T, fig. 9, plate XI.

As the rudder is nearly perpendicular and without-side the stern, another piece of timber, called the tiller, is fitted to it at right angles, which comes into the ship, by which it is managed and directed.

A narrow rudder is best for a ship's sailing, provided she can be guided and turned by it: for a broad rudder will hold much water when the tiller is put over to any side; but if a ship has a full quarter, so that the water cannot come quick and strong to her rudder, she will require a broad rudder. See this further explained and illustrated, under the article HELM.

Ships having occasion to use a substitute rudder, have provided themselves, through the means afforded, by securing two spare top-masts and a spare cap, whose square part is generally cut out to fit the stern-post and act as a lower goodgeon or brace. The cap may be secured to the stern-post by one hawser leading forward from the eye-bolt in the cap, and another leading from the keel under the ship's bottom into the hawse-holes, and hove well taught: the rudder, at the same time, being made to sink by pigs of iron, or by an anchor, at the lower part, and removed when it is fixed.

The best invention of this kind was that introduced by Captain Edward Pakenham, of the royal navy; hence called *Pakenham's Substitute Rudder.*

Main-Piece, or *Chocks of the* RUDDER. See the article CHOCKS.

RUDDER-*Coats*, are coverings made of well-tarred canvas, to prevent the water from coming in at the rudder-hole. See the article COAT.

RUDDER-*Irons*, among boat-builders, the irons which are fastened to the rudder, in order to hang it to the stern-post.

RUDDER-*Nails*. See the article NAILS.

RUDDER-*Pendants*. See the article PENDANTS.

Rake of the RUDDER, a term used to signify the fore part of the rudder, which depends entirely on the rake of the stern-post.

RULE, in a general sense, implies government, sway.—In a more confined one, canon, precept, direction: hence

Rules and Regulations to be observed on board his Majesty's ships and vessels of war in commission.

A captain is not to make any alteration in the masts, yards, or sails, &c. of the ship he commands; unless from some peculiar circumstances, of which he is to inform the secretary of the Admiralty, or the commander in chief, by the first opportunity; and he is to allow every officer to possess the cabin allotted him, without any alteration.

He is to order a sentinel to be posted at the hatchway or scuttle leading to the magazine, who is to prevent all persons going down without orders from the commanding-officer of the watch; also to see that a light, if required, be put in a secure lantern, and not to allow wine or spirits to be carried down, unless there be officers cabins in the fore cock-pit.

He is to take every precaution to prevent accidents by fire; and not to allow lights to be used on the orlop or cable tiers, except in good lanterns, nor to allow smoking of tobacco, except in the galley, and strictly to enjoin officers not to read in bed by the light of either lamps or candles, and to forbid the sticking of candles against the beams, or any part of the ship. At the setting of the watch, all fires and lights are to be extinguished by the master-at-arms and ship's corporal, except those expressly permitted by the captain. No spirituous liquors are to be drawn off but upon the upper deck, and by day only; nor moved from one vessel to another, except in cases of emergency.

None are to be allowed to suttle on board, nor are the men to sell, exchange, or dispose of their bedding; neither are debts so contracted to be charged, such as tobacco, or any other articles usually issued to them.

The captain must enforce cleanliness, by causing the upper decks to be washed every morning, and the lower as often as the weather will admit; they are also to be swept twice every day, and he is to obtain for the ship's company dryness and good air, by working the ventilators; the hammocks to be carried upon deck, and the ports to be opened when the weather will permit; fires are also to be made in the stoves when the lower deck-ports cannot be opened, and the ship always to be pumped dry, &c.

He must attend particularly to the cleanliness of the seamen, and never suffer them to sleep in wet clothes or wet beds, if it can possibly be prevented.

When the ship's company are received on board, the captain is, with the assistance of the senior lieutenant, and other officers, impartially to examine and rate them, according to their respective abilities; the petty officers to be rated only from seamen properly qualified, and to order the masters, mates, and midshipmen, to keep log-books or journals, which must be produced on receiving their pay, and on being examined for lieutenants.

He is to quarter the officers and sea-men as early as possible, and the general quarter-bill to be hung up in a public part in the ship; also particular lists over every gun and place where the seamen are quartered, that every one may know his station in time of action.

He is frequently to exercise the ship's company at the guns and musketry, and the times of exercising are to be mentioned in the log-book; he is likewise to allot a part of the day to exercise newly-raised men in bending and unbending sails; to cause a certain number of seamen to be trained to the use of musketry; and when those are expert, a second party is to be trained, &c. he is then to see them taught the use of the half-pike and cutlass, and to exercise them in the various modes of boarding the enemy's ships, and of repelling the attacks of boarders.

He is to direct the officers to see that the gunner is attentive in giving proper instructions to the men under his directions, in pointing the guns, &c. and to supply a certain quantity of powder and shot, and every month, for six months after the guns are first received on-board, as many cartridges of powder and round shot for the upper-deck guns, as shall amount to half of their number; and as many musket cartridges with ball as may furnish one-half of the seamen, and all the marines, with twelve. But, after the first six months, he is to direct the gunner to supply only half that quantity for both guns and muskets.

The captain must, with the assistance of the officers, divide the ship's company, exclusive of marines, into as many equal divisions as there are lieutenants allowed to the ship, and a lieutenant is to command each division. The lieutenants are to attend, and to be responsible for the conduct of the men in their respective divisions. The marines are also to be equally divided, each division to be governed by a subaltern, who is also responsible for their good conduct, and the condition of their arms.

The captain is to send to the secretary of the Admiralty the names of such officers as may be absent when he shall receive orders to sail; and if judged proper, other officers may be substituted by him; and he is on no account to allow women or foreigners to be carried to sea, without orders from his superior officer.

The captain must keep a log-book, according to the form in his instructions, in which he is to insert all circumstances relative to the ship, and to send to the Admiralty a correct copy thereof every six calendar months.

He is to direct the officers to determine with the utmost possible accuracy the latitude and longitude of places; and to direct the master in keeping a book; in the several columns of which are to be inserted the most perfect and circumstantial account he can obtain of the respective subjects to which they are allotted, and to see that he is assisted by the best qualified officers in taking surveys, drawing views of the coasts, making charts, &c.; the accounts given in this book must be examined, corrected, and signed by the captain, who is to make all such observations and remarks as may be beneficial to his Majesty's fleets, or to commerce; and to send one copy of the said book, and of all the observations, charts, and views, every six months to the Admiralty. The duties required in the foregoing article will by no means be dispensed with, except when satisfactory reasons are given of its absolute necessity. The captain, in executing this duty, is strictly enjoined not to give cause of offence to any power in amity with his Majesty, and by every opportunity to send to the secretary of the Admiralty a particular account of his proceedings, and to acknowledge the receipt of all orders from the Admiralty, or any superior officer. He is to make correct returns of the condition of the ship, quantity of stores, &c.; and when at sea, in time of war, always to keep in readiness for battle, as well as when at anchor, and to be prepared to repel any attempts of an enemy to board the ship; also to be careful not to approach a ship of war of any foreign power, even in time of peace, without being prepared for battle.

He is to be particularly attentive that prisoners of war are treated with humanity, their property protected, and that they have their proper allowance of provisions, with every comfort of air and exercise, as circumstances may admit; whatever number there may be, they must be attentively observed, to prevent their cutting the ropes, or doing any damage to the ship; sufficient guards are to be appointed, who are to be particularly watchful when the ship's company are aloft, and every other precaution must be taken to prevent their attempting to rise.

He must not enter any other ports than those to which he is ordered, but keep the sea when ordered to cruise; and if necessitated to put into port, he must send to the secretary of the Admiralty an account of his proceedings, the cause which obliged him to leave his station, and be equipped in readiness, when he may receive fresh orders.

He is to give timely information to masters of merchant vessels when he is about to leave any port, war being expected, and to take under his protection such as are bound the same way.

When he meets with any foreign vessel, he is to send a lieutenant to enquire if there are any British seamen on board, without offering violence or detaining the vessel.

Every captain is to assist the persons employed by the directors of Greenwich Hospital and the Trinity-House, in recovering the duties from foreign ships, but not to employ force or give cause of complaint.

He must place the ship in a safe birth when coming to anchor; if in a port or road to which he is a stranger, and of which he has no accurate chart, the soundings are to be taken three cable lengths round the ship: he is neither to cut nor slip the cable, where there is a possibility of weighing the anchor, except for the safety of the ship, or for chasing the enemy; and he is to be particularly careful in anchoring and weighing anchor, to preserve the copper sheathing.

The captain must hire a vessel to carry any important intelligence that he may think sufficient to justify the expense incurred, and give particular directions, that custom house officers coming on board be not obstructed in their duty, and that they be borne as supernumeraries for victuals only; the boatswain to lend hammocks, and the purser slop-beds, to each of them during their continuance on board.

He is likewise to inform the secretary of the Admiralty of any officer he may discover smuggling or attempting to smuggle liquors, or other articles.

He is particularly enjoined not to make any alteration in the rigging, but to keep the shrouds, stays, &c. as they are fitted in the dock-yard; and must favour the masts as much as possible, never hazarding them by carrying a press of sail, except in chase, or on occasions that may render it absolutely necessary: and he is to direct the carpenter to be attentive in repairing all defects in the ship and boats, so that but *little* assistance may be required from the dock-yards.

He is to be present when it may be necessary to heel the ship, to see that precaution is taken to prevent any accident happening in that operation, and to direct the carpenter to examine and see that the pumps are in good order, and to appoint careful men to observe the state of the well, who are to report to the officers immediately on their discovering any material increase of water. In three-decked ships, the lower-deck ports of the side, to which the ship is heeled, are to be barred in, and the scuppers plugged; and when the lower-deck guns of two-decked ships are run out, a sufficient number of men must be stationed to run them in if necessary: and during the time the ship continues on the heel, all the officers and men are to be stationed so as to right her on the shortest notice, if the shifting of the tide, a leak, or a gust of wind, should suddenly make it necessary. When a ship is careened, if painting be requisite, he may order it to be done.

When a foreign ship of war enters any of his Majesty's ports, the senior captain is to send a lieutenant to the commanding officer, and to shew them such attention and respect as their situation and rank may require, and to afford such assistance to the ship, if wanted, as circumstances will admit, and as a power in amity with his Majesty may reasonably expect.

Captains are particularly to prevent any officer from visiting any ship, until it be ascertained whether or not they are subject to quarantine; and the captains of such vessels as are subject to quarantine, are most strictly enjoined not to conceal, but to make it known by carrying a yellow flag at the mast-head, when they have communicated with any ship at sea subject to Quarantine. See that article.

The captain is to direct births to be prepared in the most airy part of the ship for any sick that may be on board; who are to be carefully attended by the surgeon; the births to be kept clean and ventilated and, if a contagious disease, they must be separated from the rest to prevent its spreading; and to provide warm clothing for men returning to England from a southern climate, to prevent the bad consequences such transition often occasions.

As no ship ought to go to sea without a chaplain, the captain is to see that proper attention and respect, due to his sacred office, be shewn him by all the officers and men under his command.

He must direct the purser and master to report the stores that he may know in what respect the ship he commands is fit for service, previous to his proceeding to sea; likewise at the end of every week to report the quantity expended, and to enjoin the officers to prevent waste or improper expenditure.

He is to enter in a book kept for the purpose, copies of all orders he may either receive or issue, and likewise of all public letters; and to cause the articles of war, being abstracts of acts of parliament, &c. to be hung up in a public part in the ship, and to be read to the ship's company at least once a month.

A captain is not to inflict any punishment on commissioned or warrant officers, but may suspend or put them under arrest till tried by a court-martial, and may apply to the senior officer present, to supply him with a substitute till the commander in chief, or the Admiralty, are informed of such suspension or arrest.

He is to see that strict justice in every respect is done to all officers and seamen under his command, and he alone is to order punishment, according to the enormity of the crime or nature of offence, and always to be present at such punishment, as well as all the officers and ship's company: but the captain of a ship carrying a flag-officer shall never order punishment without the sanction of the flag-officer; and must inform him of all signals made by the commander in chief, and the disposition of the fleet, &c.

A captain, when removed from the command of a ship, must leave to his successor all the orders he has received, which remain unexecuted, and is to give him attested copies of them, except such as are secret, the originals of which he is to deliver to him without allowing any copy to be taken. This shall be a sufficient authority for his successor to execute them, which he is to do in the same manner as if they had been originally addressed to him. He is also to deliver to him all papers of note, signal-books, log-books, muster and slop-books, &c. to be receipted, and the receipts sent to the Navy-Board, when his accounts are passed.

When removed from one ship to another, if they are in the same port, he is allowed to take with him a certain number of seamen, on obtaining an order from the Admiralty.

At his leaving the ship, he is to draw up a correct account of her qualities, from his own observation and experience, which may be of use to his successor; this is to be signed, and lodged at the master shipwright's office at the dock-yard, and a duplicate sent to the Navy-Board.

When a ship is wrecked, the captain is to use every exertion possible to save the lives of the crew, books, stores, &c. of the ship. He is himself to take special care to preserve all secret orders, signals, and instructions, or, if necessary, to destroy them, to prevent their falling into improper hands. He is to place sufficient guards to prevent the stores and provisions, which are saved, from being embezzled. After a court-martial has taken place, he is, with all possible speed to transmit the particulars to the commissioners of the navy; and, to prevent irregularities, he must keep strict discipline. If the ship be wrecked on the coast of any foreign power in amity with his Majesty, the captain must solicit the assistance of the governor, or chief officer, and see that the officers and ship's company conform to the laws and regulations of the country during their stay: and he is further enjoined to be as economical as possible in the expenditure of public stores, and to be extremely attentive to every thing done, particularly by his clerk, as he will be held responsible for every error he may commit in the discharge of his duty. For more minute particulars, see the Naval Instructions.

To RUMMAGE, signifies to clear a ship's hold, in order to examine its contents; or, to remove goods or luggage from one place to another.

RUN, (*les ailes, ou extrêmites de la cale: la partie de la cale, ou du fond du vaisseau qui est vers les façons de l'arrière, et qui va en retrécissant*, Fr.) the aftmost part of a ship's bottom, where it grows extremely narrow, as the floor approaches the stern-post.

Run, or *Course of a Ship*, (*sillage, ou route d'un bâtiment*, Fr.) implies the distance she has sailed in a certain time.

Run, among sailors, implies the agreement they make with the master or owner of a vessel to work a single passage; as, from Lisbon to England, &c.

To Run *down a Coast*, (*descendre une côte*, Fr.) is to sail along by it.

To Run *down Latitude*, or *Longitude*. See the articles Latitude and Longitude.

To Run *down a Vessel*, (*passer par dessus un bâtiment, en l'abordant de bout au corps*, Fr.) is to pass over her, by running against her end-on, so as to sink her.

To Run *the Gantlope*, or *Gantlet*. See the article Gantlope.

Run *out the Guns!* (*les pièces en batterie!* Fr.) the order to force their muzzles out at the port-holes by means of the tackles.

To Run *out a Warp*, (*porter une touée au dehors du bâtiment*, Fr.) is to carry the end of a hawser out from the ship in a boat, and fasten it to some distant place to remove the ship towards that place, or to keep her steady whilst her anchors are lifted, &c.

To let Run *a Rope*, (*larguer un cordage tout à fait*, Fr.) is to let it quite loose.

A Run-Man, (*matelot déserteur*, Fr.) implies a deserter from a ship of war.

RUNG-HEADS, (*fleurs d'un vaisseau*, Fr.) a name sometimes given by shipwrights to the upper ends of the floor-timbers, which are otherwise more properly called floor-heads. See the article Floor-*Timbers*.

RUNNER, (*itague d'un palan*, &c. Fr.) a thick rope used to increase the mechanical powers of a Tackle, which article see.

The runner passes through a large block, and has usually a hook attached to one of its ends, and one of the tackle blocks to the other; and, in applying it, the hook, as well as the lower block of the corresponding tackle, is fixed to the object intended to be removed.

RUNNING-Fight, (*cannonade en chasse; combat en chasse*, Fr.) an action or battle in which the enemy endeavours to escape, whilst the opponent continues to pursue within gun-shot.

Running *of Goods*, is a clandestine landing of goods, without paying the legal customs or duties for the same. See the article Smuggling.

Running *Rigging*, (*manœuvres courantes*, Fr.) is all that part of a ship's rigging which passes through blocks, &c. and is used in contradistinction to standing rigging. See the article Rigging.

Running *Part of a Tackle*, is synonimous with the fall, and is that part on which the power is applied to produce the intended effect.

RUSSIA Company. See the article Merchant Companies.

S.

SADDLE, (*taquet en forme de croissant, placé au bout des vergues, pour le passage des boute-hors de bonnettes*, Fr.) a small cleat, or block of wood, hollowed on the upper and lower sides, and nailed upon the lower yard-arms, to retain the studding-sail-booms in a firm and steady position: for this purpose, the cavity on the lower part of the saddle conforms to the cylindrical surface of the yard to which it is attached; and, in like manner, the hollow on the upper side answers to the figure of the boom, and serves as a channel whereby it may be run out or in, along the yard, as occasion requires.

SADDLE *of the Bowsprit*, (*taquet de beaupré*, Fr.) a name, also given to a semicircular piece of wood, which is shaped and fastened to the upper part of the bowsprit, boom, &c.

SADDLE is also the name of a piece of elm timber fitted on the upper end of the lacing.

To SAG TO LEEWARD, (*dériver, ou tomber sous le vent, en naviguant au plus près*, Fr.) is when a ship makes a considerable lee-way, or driven too far to leeward of the course whereon she apparently sails. It is generally expressed of heavy-sailing vessels, as opposed to keeping well to windward, or, in the sea-phrase, *holding a good wind.*

SAGITTA, *the Arrow, or Dart*, in astronomy, a constellation of the northern hemisphere, near the eagle, consisting, according to Flamsteed's Catalogue, of no less than twenty-three stars.

SAGGITTARIUS, the Archer, in astronomy, the ninth sign of the Zodiac, marked thus ♐. The stars in this constellation are, in the Britannic Catalogue, fifty-two.

SAIK, (*saique*. Fr.) a Turkish vessel, very common in the Levant, used for the conveyance of merchandise. It has only one mast, which, together with its top-mast, is extremely high.

SAIL, (*voile*, Fr. *segl, Sax. seyl, Dutch*,) an assemblage of several breadths of canvas, or other texture, sewed together, and extended on, or between the masts, to receive the wind, and impel the vessel through the water.

The edges of the cloths, or pieces of which a sail is composed, are generally sewed together with a double seam; and the whole is skirted round at the edges with a cord, called the bolt-rope.

Although the form of sails is extremely different, they are, nevertheless, either triangular or quadrilateral; or, in other words, their surfaces are contained either between three or four sides.

The former of these are sometimes spread by a yard, as lateen-sails; or by a stay, as stay-sails; or by a mast, as shoulder-of-mutton-sails; in all which cases, the foremost leech or edge is attached to the said yard, mast, or stay, throughout its whole length. The latter, or those which are four-sided, are either extended by yards, as the principal sails of a ship, or by yards and booms, as the studding-sails, drivers, ring-tails, and all those sails which are set occasionally; or by gaffs and booms, as the main-sails of sloops, brigs, &c.

The principal sails of a ship (Plate XX.) are the courses or lower sails, *aa*, the top-sails, *bb*, are next in order above the courses; the top-gallant-sails, *cc*, are expanded above the top-sails; the top-gallant-royal-sails, AA, are expanded above the top-gallant-sails; and sometimes another sail, called a *sky-scraper*, is expanded above the top-gallant-royal-sails. Plate XX.

The courses are, the main-sail, fore-sail, and mizen; main-stay-sail, fore-stay-sail, and mizen-stay-sail; but more particularly the three first.

The main-stay-sail is rarely used except in small vessels.

In all quadrilateral sails, the upper edge is called the head; the sides or skirts are called leeches; and the bottom or lower edge is termed the foot: if the head is parallel to the foot, the two lower corners are denominated clews, and the upper corners earings.

In all triangular sails, and in those quadrilateral sails wherein the head is not parallel to the foot, the foremost corner at the foot is called the tack, and the after lower corner the clew; the foremost edge is called the fore leech, and the hindmost the after leech.

The heads of most quadrilateral sails, and the fore-leeches of lateen sails, are attached to their respective yard or gaff by a number of small cords called rope-bands, and the upper extremities are tied to the yard-arms, or to the peak of the gaff, by earings.

The stay-sails are extended upon stays between the masts, whereon they are drawn up or down occasionally, as a curtain slides on its rod, and their lower parts are stretched out by a tack and sheet. The clews of a top-sail are drawn out to the extremities of the lower yard by two large ropes, called the top-sail sheets; and the clews of the top-gallant-sails are in like manner extended upon the top-sail yard-arms, as exhibited in Plate XX.

The studding-sails are set beyond the leeches or skirts of the main-sail and fore-sail, or of the top-sails or top-gallant-sails of a ship. The lower edges of the lower studding-sails, represented by *oo*, are extended by a boom, hooked into an eye, and fixed in the main channel for that purpose.

The topsail-studding-sails, *pp*, are extended by studding-sail booms, run out beyond the extremities of the main-yard.

The top-gallant-studding-sails, *qq*, are also extended by studding-sail booms, run out from the top-sail-yard.

Those sails, however, are only set in favourable winds and moderate weather.

All sails derive their names from the mast, yard, or stay, upon which they are extended. Thus, the principal sail extended upon the main-mast, is called the
PlateXXII. Fig. 1. main-sail, or main course, *d*, fig. 1. Plate XXII.; the next above, which stands upon the main-top-mast, is termed the main-top-sail, *e*; and that which is spread across the main-top-gallant-mast, is named the main-top-gallant-sail, *f*; above which, there is the main-top-gallant royal.

In the same manner there are, the fore-sail, or fore-course, *g*; the fore-top-sail, *h*; the fore-top-gallant-sail, *i*, the fore-top-gallant royal, 1; the driver, *k*; the mizen top-sail, *l*; the mizen top-gallant-sail, *m*, and the mizen top-gallant royal, 2. Thus, also, there are the main-stay-sail, *o*, main-top-mast stay-sail, *p*, an dmain-top-gallant stay-sail, *q*, with a middle stay-sail which stands between the two last, and a royal stay-sail, 3: all these stay-sails are between main and fore-masts.

The stay-sails between the main and mizen-masts are, the mizen stay-sail, *r*, and the mizen top-mast the stay-sail, *s*.

The sails between the foremast and the bowsprit are, the fore-top-mast stay-sail, *t*, the jib, and sometimes a flying jib, *u*. There are, besides, two square sails extended by yards under the bowsprit and jib-boom; the first of which is called the sprit-sail, *y*, and the second the sprit-sail-top-sail, *z*.

The studding-sails being extended upon the different yards of the main-mast and fore-mast, are likewise named according to their stations, the lower, top-mast, or top-gallant studding-sails.

A storm main-topsail has been lately introduced into the service without reefs, and of such a depth as to allow the sail to clear the main-cap; it is placed to the main-topsail-yard, and sheets home to the main-yard, like the main-topsail. This sail is used only when a ship is lying-to in a gale of wind: being shallower, and of course much lighter, than the main-topsail, it therefore causes *less pressure* at the main-mast-head, which has been found very advantageous.

For the other sails of a ship, such as the GAFF-TOP-SAIL, RING-TAIL, SKY-*Scraper*, &c. See these articles.

The ropes by which the lower yards of a ship are hoisted up to their proper height on the masts, are called the jears. In all other cases, the ropes employed for this purpose are called halyards; hence the sails are expanded by halyards, tacks, sheets, and bow-lines; and are drawn up together, or trussed up, by bunt-lines, clew-lines, *dd*; leech-lines, *ee*; reef-tackles, *ff*; and lift-tackles, *gg*; which are represented in plate XX. As the bunt-lines and leech-lines pass on the other side of the sail, they are expressed by dotted lines in the same plate. The higher studding-sails, and, in general, all the stay-sails, are drawn down, so as to be furled, or taken in, by down-hauls. The courses, top-sails, and top-gallant-sails, are wheeled about the mast, so as to suit the various directions of the wind, by braces.

In addition to what has already been said we shall briefly mention Captain Malcolm Cowan's improvement in the construction of his patent sails of ships and vessels. It appears, that many trials have been made on those sails, and the mode of reefing them, under the superintendance of some of our naval officers, who have sanctioned their utility; but the old construction of sails is still abided by, both in the Royal Navy and merchant-service.

"The sails of ships and vessels," says Captain Cowan, "from being made with the cloths and seams in a vertical instead of a horizontal direction, are more liable to split up and down, and to be *blown to pieces*, either when carrying a *press of sail* in a gale of wind, on a lee-shore, or from the *shaking of the sails*; and a ship and crew might be *lost* from a few inches of a vertical seam giving way, when there might not be *time* to take the sail in to repair it.

"From the experiment that has been made Capt. Cowan asserts, the difference in the duration of the sails made with horizontal cloths, and those of the old make, has been proved to be as *eighteen months to eleven*, making a difference of *seven months wear in favour of those with the horizontal cloths*; and they are every way stronger, more effective, and stand nearer to the wind."

The Captain also observes, that "when sails are made with horizontal cloths and seams, the sails will stand better, particularly in a gale of wind, as the strongest direction of the cloth and seams are opposed to the greatest force of the wind, which acts horizontally; and should the sail split in that direction, it would still remain full, and be less liable to blow away altogether: which is generally the case when a sail splits in a vertical direction.

"Storm stay-sails set purposely with the cloths horizontally, have proved this beyond a doubt."

"The common sails require to be hauled up, to be reefed, at the risk of splitting them, at a time, perhaps, when the ship is in imminent danger, from the want of sea room; and the best seamen of the crew must be sent on the yards, when they possibly may be much wanted on deck. But the facility with which sails that reef at the foot can at all times be managed, would obviate the difficulty and danger of carrying sail, and enable ships to make quicker voyages; and prevent them often, when weakly manned, from detaining fleets; it would also enable merchant-ships to be navigated with fewer hands, which would be a considerable saving of expense, and a great advantage in time of war in particular, when men are so scarce.

"The origin of the invention of the patent sails was," as Captain Cowan states, "in a gale of wind on a lee-shore, in which a line of battle ship was exposed for many hours to the danger of being totally lost, with all her crew, by the splitting of her sails, and the utter impossibility of reefing them. No human means, then known, were left untried to save the ship; and when, to all appearance, every hope had vanished, and silent horror and resignation prevailed throughout, the obvious defects, and the erroneous construction of the sails for centuries past, became apparent." See *Capt.* MALCOLM COWAN's *Essay on the Construction of Sails of Ships and Vessels, with Plans*, &c.

After SAILS, (*les voiles de l'arrière*, Fr.) are those that belong to the main-mast and mizen. They keep the ship to windward; on which account ships sailing on a quarter wind require a head-sail and an after-sail, one to counteract the other. See the articles AFTER and HEAD-*Sails*.

Shoulder of Mutton-SAILS, (*voiles auriques*, Fr.)

are triangular, and similar to the lateen sails; but are attached to masts instead of the yards.

Sliding-Gunter-Sail, a triangular sail used in boats; it is bent at its foremost leech to hoops or grommets that slide on the lower mast: the peak or head is attached to a small top-mast, that slides up in the direction of the lower-mast, through two hoops fixed at its head, about three feet asunder. When the top-mast is lowered, the sail furls close up to the lower-mast.

Laws on Sails.—Every master of a vessel belonging to any of his Majesty's subjects, navigated with, or having on board, any foreign made sails, is, at the time of reporting his ship, to make an entry and report, on oath, of every such sail, and, before the ship is cleared, to pay duty for them, under penalty of forfeiture of sails, and a fine of 50*l.* for every offence, by the master; and such sails are to be stamped at the port of entry in the same manner as foreign sail-cloth. 19 Geo. II. c. 27.; 26 Geo. II. c. 32. But if the master, before the ship is cleared, declares his intention of not paying the duty, and delivers up the sails to the officers of the customs, the sails only are to be forfeited. *Idem.*

But if the ship come from the East Indies, the captain is not liable to the effects of the above regulations. *Idem.*

Sail, (*voile, on employe ce terme également pour signifier un navire, ou bâtiment quelconque*, Fr.) is also a name applied to any vessel seen at a distance under sail; as, "We saw five *sail* off Cape Finisterre."

To Loose a Sail. See the article To Loose.

To Make Sail, See the article To Make.

To Set Sail, (*mettre a la voile*, Fr.) is to expand the sails upon their respective yards and stays, in order to commence sailing.

To Shorten Sail, (*diminuer de voiles*, Fr.) is to reduce or take in part of the sails, with an intention to diminish the ship's velocity.

To Strike Sail, (*amener les voiles pour saluer des vaisseaux de guerre*, Fr.) is to lower it suddenly, which is particularly used in saluting or doing homage to a superior force, or to one whom the law of nations acknowledges as superior in certain regions. Thus all foreign vessels strike to an English man of war in the British seas. See the article Salute.

Sail Cloth, (*toile à voile*, Fr.) a cloth of strong texture, made of hemp, for the purpose of supplying ships with sails. Although considerable quantities of this valuable article are annually manufactured in Britain, yet as they are not adequate to the demands of the navy, but being subject to mildew, are consequently less durable than the sail-cloth imported from North America; we shall state the following expedient, adopted in that country, by which the cloth may be considerably improved:—it simply consists in moistening the warp in the loom with a decoction, or gelatinous substance, prepared from the refuse of neat's-feet (after the oil is extracted), which is boiled in water till it is converted into a kind of glue: the weavers of this article in Britain employ a paste made of flour and water, which renders the cloth brittle; whereas, by using the animal-preparation above-mentioned, the sail-cloth will not only be rendered more durable, but the expense may be lessened: and an article that is at present thrown away, may thus be usefully employed.

British sail-cloth may be exported duty-free, and all new ships are to have a suit of sails of British cloth.

Every maker of British sail-cloth shall stamp his name and place of abode, in words at length, on every piece, on pain of forfeiting 10*l.*

Sail-Hook, in sail-making, a small iron hook, with an eye at one end, to which a cord is spliced; it is used to confine the work, while sewing, by hooking on the canvass, the cord being fastened to some convenient place.

Sail-Loft, (*voilerie*, Fr.) in dock-yards, a large room or apartment wherein the sails are cut out, made, repaired, and kept, in readiness for his Majesty's ships and vessels.

Sail-Maker, (*voilier*, Fr.) a person appointed by warrant from the Commissioners of the navy, to repair the sails that may at any time be damaged in action, or otherwise.

He has a mate to assist him in his several duties.

He is to examine the sails when they are received on board; and, if they are defective in quality, or deficient in number, to report the same to the boatswain immediately.

He is to see that all the sails are correctly tallied, and so disposed of in the sail-rooms, as to enable him to find directly any that may be wanted.

He is also to inspect frequently the condition of the sails in the sail-rooms, to see that they are not injured by leaks or vermine; and to report to the boatswain, whenever it shall be necessary, to have them taken upon deck to be dried: he is to repair them whenever they require it, and to use his best endeavours to keep them always fit for service.

The sail-maker, before he shall be allowed to receive his wages, is required to produce a certificate, signed by the captain, of his sobriety, obedience, and attention to his duty. See the Naval Instructions.

Sail-Rooms, (*soutes aux voiles*, Fr.) are places on the orlop-deck, inclosed for the reception of the sails; they are distinguished according to their relative situation: as, the fore sail-room, the after sail-room, &c.

Sail-Yard. See the article Yard.

SAILER, (*voilier*, Fr.) a term indiscriminately applied to vessels having masts, yards, sails, &c. hence,

A Heavy Sailer, (*mauvais voilier*, Fr.) a vessel which can advance but slowly on the water.

A Prime Sailer, (*fin voilier*, Fr.) is one that is capable of attaining a great velocity by the impulse of the wind upon her sails, &c.

SAILING, (*action de naviguer, d'être à la voile. Cingler, ou faire route*, Fr.) denotes the movement by which a vessel is impelled through the water, by the action of the wind upon her sails; being otherwise expressed by the more significant term navigating.

The effect of sailing is produced by a judicious arrangement of the sails to the direction of the wind; accordingly, the various modes of sailing are derived from the different degrees and situations of the wind with regard to the course of the vessel.

To illustrate this observation by examples, the plas

of a number of ships proceeding on various courses are represented by fig. 3, plate XXI, which exhibits the thirty-two points of the compass, of which C is the centre: the direction of the wind, which is northerly, being expressed by the arrow.

It has been observed in the article Close-hauled, that a ship in that situation will sail nearly within six points of the wind. Thus the ships B and *y* are close-hauled, the former being on the larboard tack, steering E. N. E. and the latter on the starboard tack sailing W. N. W. with their yards *ab* braced obliquely, as suitable to that manner of sailing. The line of battle on the larboard tack would accordingly be expressed by C B, and on the starboard by C*y*.

When a ship is neither close-hauled, nor steering afore the wind, she is generally said to be sailing large. The relation of the wind to her course, is precisely determined by the number of points between the latter and the course close-hauled. Thus the ships *c* and *x* have the wind one point large, the former steering E. b N. and the latter W. b N. The yards remain almost in the same position as in B and *y*: the bow-lines and sheets of the sails being only a little slackened.

The ships *d* and *u* have the wind two points large, the one steering east and the other west. In this manner of sailing, however, the wind is more particularly said to be upon the beam, as being at right angles with the keel, and coinciding with the position of the ship's beams. The yards are now more across the ship, the bow-lines are cast off, and the sheets more relaxed; so that the effort of the wind being applied nearer to the line of the ship's course, her velocity is greatly augmented.

In *e* and *t*, the ships have the wind three points large, or one point abaft the beam: the course of the former being E. b S. and that of the latter W. b S. The sheets are still more flowing, the angle which the yards make with the keel are further diminished, and the course accelerated in proportion.

The ships *f* and *s*, the first of which steers E. S. E. and the second W. S. W. have the wind four points large, or two points abaft the beam. In *g* and *r*, the wind is five points large, or three points abaft the beam, the former sailing S E. b E. and the latter S. W. b W In both these situations the sheets are still further slackened, and the yards laid yet more athwart the ship's length, in proportion as the wind approaches the quarter.

The ships *h* and *q*, steering S. E. and S. W. have the wind six points large, or more properly on the quarter; which is considered as the most favourable manner of sailing, because all the sails co-operate to increase the ship's velocity: whereas, when the wind is right aft, as in the ship *m*, it is evident that the wind, in its passage to the foremost sails, will be intercepted by those which are further aft. When the wind is on the quarter, the fore tack is brought to the cat-head; and the main-tack being cast off, the weather-clew of the main-sail is hoisted up to the yard, in order to let the wind pass freely to the fore-sail; and the yards are disposed so as to make an angle of about two points, or nearly 22°, with the keel.

The ships *i* and *p*, of which the former sails S. E. b S. and the latter S. W. b S. are said to have the wind three points on the larboard or starboard quarter: and those expressed by *k* and *o*, two points; as steering S. S. E. and S. S. W. in both which positions the yards make nearly an angle of 16°, or about a point and an half, with the ship's length.

When the wind is one point on the quarter, as in the ships *l* and *n*, whose courses are S. b E. and S. b W. the situation of the yards and sails is very little different from the last mentioned; the angle which they make with the keel being somewhat less than a point, and the stay-sails being rendered of very little service. The ship *m* sails right afore the wind, or with the wind right aft. In this position the yards are laid at right angles with the ship's length: the stay-sails being entirely useless, are hauled down, and the main-sail is drawn up in the brails that the fore-sail may operate; a measure which considerably facilitates the steerage, or effort of the helm. As the wind is then intercepted, by the main-top-sail and main-top-gallant-sail, in its passage to the fore top-sail and fore top-gallant-sail, these latter are, by consequence, entirely becalmed, and might therefore be furled to prevent their being fretted by flapping against the mast, but that their effort contributes greatly to prevent the ship from broaching-to, when she deviates from her course to the right or left thereof.

Thus all the different methods of sailing may be divided into four, viz. close-hauled, large, quartering, and afore the wind; all which relate to the direction of the wind with regard to the ship's course, and the arrangement of the sails. See also the articles Drift and Lee-Way.

Sailing is also used for the art or act of navigating, or of determining the cases of a ship's motion by means of sea-charts: these cases are differently distinguished, according to the principles upon which the computations are founded, as Current, Mercator's Middle-Latitude, Oblique, Parallel, Plane, Traverse, and Windward *Sailing*, which articles see.

Sailing-Trim, (*bâtiment bien dans son assiette et dont les voiles sont orientées de manière à lui procurer la marche la plus avantageuse*, Fr.) a term applied to a ship when she is in the best state for sailing.

Order of Sailing, (*ordre de marche*, Fr.) the general disposition of a fleet of ships when proceeding on a voyage or expedition.

It is generally found most convenient in the order of sailing, for ships of war to be formed in three parallel lines, abreast of each other; because lines can more readily assume the order of battle, or retreat. The disposition of this order rests with the commander in chief; and the lines are as near to each other as regularity, and a regard for their common security, will admit. The distance, which ought to be carefully observed in tacking, may be regulated by the movements of some of the ships in the line farthest to windward, which would accordingly govern the operations of the whole squadron.

In forming this order, the leading ships of the starboard and larboard lines are to place themselves abreast of the leading ship of the centre line, at the distance as assigned by signal, and the following ships of each line are to take their respective stations astern of each other, at the distance as expressed by the commodore.

From the Order of Sailing to form the Order of Battle.

The line of battle is complete, when the fleets are arranged upon a wind in a line a-head of each other at equal distances. To attain this position from the order of sailing without confusion, the commodore will make the signal to form in the order of battle on the same tack the fleet then happens to be sailing upon. For example; suppose the ships were in the order of sailing with the wind on their larboard side, as represented by fig. 1. plate XXIII, then the starboard or lee-line of the order is to heave to, as shewn by their yards being braced a-back, while the centre and weather-lines bear up and continue a course along the dotted lines, so as to arrange themselves a-head of the lee-line. The sternmost ship of the centre line having arrived at its position a-head of the lee-line, her line is to heave to; and when the sternmost ship of the weather-line shall have arrived at its position, the whole fleet is to make sail and follow the leading ship. Had the ships been upon the starboard tack, the lee or larboard line is to heave to, and the centre or weather-lines are to form a-head of it as represented by fig. 2.

Pla. XXIII. Fig. 1.

Fig. 2.

Should the fleet have been close hauled, as represented, when the signal was made to form the line, they naturally will be in the order of battle when arrived in a line a-head of the lee line; but should the fleet have been sailing from the wind at the time, the line a-head thus formed must approach the wind before it assumes the proper bearings of the order of battle, which is six points from the wind. In this case, that the line may approach the wind with regularity, the leading ship must haul the wind, and the rest follow in succession, as they arrive in the wake of their next a-head; which evolution being performed, the fleet will then be in the order of battle.

To change the Order of Battle from one Tack to the other.

On the signal being made to change the order of battle, the leading ship is to tack, and the rest are to follow tacking in succession, in the wake of the ship next a-head, which has already gone about.

Any vacancies which may happen in the line of battle, are to be made good by the rear closing up.

All vessels, not included in the line of battle, are so place themselves on the opposite side of the line to the enemy, and give all the assistance in their power to tow away disabled ships.

Order of Retreat.

In retreating from an enemy, the ships of the line will preserve themselves abreast of each other, in near order astern of the convoy, the better to enable them to assume the line of battle on either tack, should the enemy press them too closely. See the articles *Naval* SIGNALS, and *To* TACK.

Rate of SAILING. See the article RATE.

SAILOR, (*matelot, marin,* Fr.) a name indiscriminately given to all persons on board ship, who are trained in the exercise of fixing the machinery of a ship, and managing her either at sea or in harbour.

The principal qualities required in a common sailor to entitle him to the full wages, are, that he should be able to sound, steer, and manage the sails, that is, to set, reef, and furl them; he is then called an ABLE-*Seaman*, which article see.

Fresh-Water SAILOR, (*marin d'eau douce,* Fr.) an epithet of derision applied to those who have never been at sea, at their first coming on board.

SALLY-PORT, a large port on each quarter of a fire-ship, out of which the officers and crew make their escape into the boats, as soon as the train is fired. It is also a term given to the places where men of war boats land from the ships, and where wherries and merchantmen's boats are not allowed to land.

SALT, (*sal,* Lat. *sel,* Fr.) a mineral body, readily soluble in water, not inflammable, and tasting sharp or pungent upon the tongue. Alum, nitre, the different kinds of alkali, &c. are all salts, but we shall only treat of alimentary salt, or *muriat of soda,* in a concise manner.

Alimentary salt is found under a variety of forms in its different states, but is immediately distinguished by applying it to the tongue: and always assumes a cubic, pyramidal, or parallelopiped figure, after regular crystalization. It is sustained in vast quantities in a liquid form, in sea-water, and in salt springs, and is also found solid in the bowels of the earth.

Bay-SALT, (*sel de mer,* Fr.) is that obtained from the water of the sea by evaporation. It had this name from being first made in the bay of St. Ubes, in Portugal. This evaporation is in some places performed by the heat of the sun, the water being let into shallow trenches in order to expose as large a surface as possible. This method is practised in the southern provinces of France, and on a very large scale in Portugal.

Rock-SALT, (*sel minéral,* Fr.) a mineral substance, taken from the bowels of the earth. The rocks of salt, are generally hewn in the form of great cylinders: the workmen using hammers, pick-axes, and chisels, as in stone quarries, to separate the several banks of stone.

The mines near Wieliezka, in Poland, and those near Eperies, in Hungary, are particularly famous; their depth and capacity are surprising; and within them exists a kind of subterraneous republic.

As soon as the massive pieces are got out of the quarry they break them into fragments fit to be thrown into the mill, where they are ground and reduced to coarse flour, which answers all the purposes of sea-salt.

The rock-salt exported from Liverpool, constitutes a considerable branch of commerce, and goes principally to the Baltic.

Refined Rock-SALT, (*sel raffiné,* Fr.) is that obtained by dissolving the article, in its crude state, in salt or fresh-water, and afterwards boiling the solution, &c.

The neighbourhood of Northwich, in Cheshire, is extremely abundant in rock-salt. The Northwich rock-salt is never used at our tables in its crude state; and its application to pickling or curing of flesh or fish, or of preserving any provisions, without its being previously refined into white-salt, that is, without its being dissolved in water, and boiled down in what is called white-salt, is prohibited under a penalty of 40*s.* for every pound of rock-salt so applied.

The pure transparent masses, however, of rock-salt, might probably be used by us with our food, without any sort of danger or inconvenience; at least, we know that rock-salt is so used, without being refined, both in Poland and Spain.

One very important fact ought to be mentioned, with respect to the internal structure of the Cheshire rock-salt, viz. that no organic impressions or remains have ever been met with in any of the beds of the mineral, which have been worked in this district; yet some authors seem to think, that the beds of this mineral have been formed by deposition from the waters of the sea. The nature and form of the plain in which the salt-mines appear, confirm this supposition, which, nevertheless, is attended with some difficulties: as the non-existence of marine exuviæ, either in rock-salt, or in the adjacent strata of clay, makes it very difficult to connect with the idea of a deposition from the waters of the sea.

It must be acknowledged also, that it is difficult to give a satisfactory account of the consolidation of the beds of salt; these circumstances remain for the consideration of future geologists.

White-Salt, (*sel blanc*, Fr.) is distinguished from the bay-salt, by receiving its consistence from the action of fire. See the article *Bay*-Salt. For the process of making white-salt, see Mortimer's *Dictionary of Commerce.*

Laws relative to Salt. Ships laden with salt, hovering on the coast, may be brought into port by the proper officers, who may continue on board till the salt is unladen, or the ship depart on her intended voyage; if neither be done within twenty days, the salt is forfeited, and double the value recovered on the master. 1 Ann, *c.* 21. 5 Geo. II. *c.* 6.

Salt imported in ships, whose burden is under 40 tons, or otherwise than in bulk, except for ship's provisions, is forfeited, and double the value recovered, &c. by the above-mentioned act.

No British manufactured salt to be re-imported from foreign parts. 2 and 3 Ann. *c.* 14. 5 Ann. *c.* 8.

Rock or white salt not to be shipped till weighed by the officers of the excise. 11 Will. III. *c.* 22.

SALTPETRE, or Nitre, (*salpètre*, Fr.) a species of salt, found abundantly on the surface of the earth in India, South America, Southern Africa, and some parts of Europe. It is used in the composition of gunpowder after its third concoction of boiling.

This substance is soluble in seven times its weight of water, at the temperature of 60 degrees, and in rather less than its own weight of boiling water. When exposed to a strong heat it melts, and congeals by cooling into an opaque mass, which has been called mineral crystal.

SALT-PITS, reservoirs made on a coast to contain sea-water, for the purpose of making salt.

Salt, or Sea-Water, (*eau de mer*, Fr.) that which is strongly impregnated with salt, and is very sensible to the taste.

The saltness of the sea, lakes, &c. is a thing that has long puzzled and perplexed philosophers to account for. The honourable Mr. Boyle believes it to be supplied, not only from rocks and masses of salt, which at the beginning were, or in some countries may yet be found, either at the bottom of the sea, or at the sides, where the water can reach them, but also from the salt which the rivers, rain, and other waters, dissolve in their passage through divers parts of the earth, and, at length, carry with them into the sea. Buffon, and most modern philosophers, acquiesce in this opinion.

On comparing the waters of the northern seas with those of a warmer latitude, we find in the former a considerably less proportion of salt. The Baltic has a still smaller share of saline ingredients than the ocean in the same latitude; and, by a chemical analysis, it has been found, that the proportion of them in this inland-sea scarcely exceeds the half of their proportion in the water of the German Ocean. In certain places, where the influx of river-water is great, the saline taste is still farther diminished; and it has been found by experiments made in the Sound, that the deep water is not so fresh as that of the surface.

Sea-water is a solution of common salt, *sal catharticus amarus*, a selenitic substance, and a compound of muriatic acid with magnesia, mixed together in various proportions. It may be freshened by simple distillation without any addition; and thus it has sometimes been useful in long voyages at sea.

Sea-water by itself has a purgative quality, owing to the salts it contains, and has been greatly recommended in scrophulous disorders.

Sea-water is about three parts in 100 heavier than common water, and its temperature at great depths is from 34 to 40 degrees; but near the surface it follows more nearly the temperature of the air.

Method of converting Salt-Water into fresh.

Many attempts have been made to render salt-water fresh and potable; but they have, in general, been little else than different modes of distillation, though many of the authors of them, either from ignorance of the real nature of sea-water, or from a design of rendering their process mysterious, have mixed different ingredients with the water, either before or after its distillation.

Sir R. Hawkins, in the reign of Queen Elizabeth, obtained fresh-water by distillation.

Dr. Hales informs us, that Sir Theophilus Oglethorpe, in the reign of Charles II. applied lime as an ingredient to improve distilled sea-water.

Dr. Lister proposes the freshening of sea-water by means of sea-plants; these vegetables, like other plants, continually exhaling a large quantity of water in form of vapour, and that being all fresh, he proposes on this basis a distillation without fire. The body of a large still was filled three parts with sea-water, and several fresh sea-plants in their growing posture put into it: the head was then fitted on, and a receiver adapted to the nose; then the continual ascent of vapour from the plant condensed into water in the head, and conveyed into the receiver sweet and clear water fit for drinking; the quantity thus produced, though considerable, is only small in proportion to the wants of a ship's company.

Mr. Hauton, therefore, attempted a more general method of doing it in quantities; for this purpose a great quantity of sea-water is to be taken up in a proper vessel, and as much oil of tartar, *per deliquium*, is to be added to this as will render it turbid, and cause a large precipitation. After this, the water is to be distilled, the furnace of the still being so contrived as to take up very little room, and consume only a small quantity of fuel; the sea itself may serve as a worm tub, the worm passing out of the ship at one place, and in again at another. When the water is thus distilled, it is mixed with an alkaline earth, and, after stirring it thoroughly about, the earth is to be suffered to subside to the bottom, and the water poured off; it is then said to be perfectly sweet, and no way distinguishable from river water.

The process of Mr. Appleby, published in 1734 by order of the lords commissioners of the Admiralty, consisted in mixing with twenty gallons of sea-water six ounces of a fixed alkali, prepared with quicklime, as strong as *lapis infernalis*, and six ounces of bones calcined to whiteness and finely powdered, and drawing off in a common still, with a slow fire, fifteen gallons. See *Phil. Trans.* Vol. XLVIII. p. 69.

Dr. Butler, instead of the *lapis infernalis* and calcined bones, proposed the use of soap-leys.

Dr. Hales, in his Philosophical Experiments, suggested a method by keeping the sea-water close shut up till putrified and has again become sweet, and then distilling the water; upon which he found that three-fourths of the sea-water was free from bitterness and the acid. In another place he recommends the use of powdered chalk and ventilation, by blowing up through the distilling water with a double pair of bellows. See *Phil. Trans.* Vol. XLIX. p. 312. &c.

Captain Chapman, in 1757, obtained fresh water by distilling sea-water with wood-ashes.

Dr. Lind, in 1761, discovered that sea-water, distilled without the addition of any ingredients, yielded pure fresh-water; and, in 1762, an account of this was read to the Royal Society, and soon after published by authority of the lords commissioners of the Admiralty.

In 1763, M. Poissonnier invented a still for this purpose, and Lord Mulgrave, in his voyage towards the North Pole in 1773, has done equal justice to its practicability.

Dr. Irving, in 1770, by whom distillation was generally introduced into the British navy, obtained a parliamentary reward of £5000; his invention consisted of a tea-kettle made without a spout, with a hole in the lid in the place of a knob; when filled with sea-water, the fresh vapour will arise as it boils and issue through the hole in the lid; into which the mouth of a tobacco-pipe is fitted, letting the stem incline a little downwards, and the vapour of fresh-water will take its course through the stem of the tube, and may be collected by fitting a proper vessel to its end. He also adapted a tin, iron, or copper tube, of suitable dimensions, to the lid of the common kettle used for boiling the provisions on ship-board; the fresh vapour, arising from boiling the sea-water in the kettle, passes through this tube into a hogshead, which serves as a receiver; and that the vapour may be readily condensed, the tube is kept cool by constantly wetting it with a mop dipped in a tub of cold water.

It appears from the testimony which was delivered to the lords commissioners of the Admiralty, by many respectable officers who witnessed an experiment made at Spithead, in January 1771, on board the Arrogant, that eighty gallons of sea-water did, in twenty-five minutes, after being put into the copper, and a fire made, distil in the proportion of twenty-five gallons per hour into fresh water, perfectly well tasted, and of less specific gravity than the best spring water in that neighbourhood; and the said officers gave it as their opinion, that five hundred gallons of fresh water might be distilled in the course of twenty-four hours, with the same quantity of fuel, in proportion to the time, as it required in the ordinary business of the ship. Every ship's kettle is divided into two parts, by a partition in the middle; one of these is only used when peas or oatmeal are dressed; but water is, at the same time, kept in the other to preserve its bottom. Dr. Irving has availed himself of this circumstance; and, by filling the spare part of the copper with sea-water, and fitting on the lid and tube, he has shewn that sixty gallons of fresh-water may be drawn off during the boiling of the above-mentioned provisions, without any additional fuel. He recommends also the preserving of the water, which may be distilled from the coppers in which peas, oatmeal, &c. are dressed, as salutary for the scorbutic, and the most proper kind of water for boiling salt provisions. He likewise particularly remarks, that only three fourths of the sea-water should be distilled, as the water distilled from the remaining concentrated brine is found to have a disagreeable taste; and the farther continuation may be injurious to the vessel.

For an account of several experiments made on some of the best distilled water, prepared by Dr. Irving from sea-water, by Dr. Watson, see his *Chem. Ess.* vol. 2, p. 168, &c.

Dr. Priestley has suggested a proposal to give to this distilled water the briskness and spirit of fresh spring water, and, at the same time, of rendering it, perhaps, a remedy or preventive against the scurvy, by impregnating it with the gas or fluid called fixed air, obtained by mixing chalk with oil of vitriol.

A patent has lately been taken out by a Mr. Archbold, for a method of converting salt-water into fresh, both on land and on-board ship, which is produced by distillation to pure fresh water, on a principle of filtration. For this purpose, stills of a new construction are used, each of which has an outward case of metal; between the interior sides and the bottom, and the exterior bottom and sides of the still, a place is left vacant; but the still is inserted in the case in such a manner, that there shall be no egress for the steam from the case, except by a safety-valve. The head and neck are fixed to the still; thus the water in the cases, not having the pressure of the atmosphere, will rise much beyond the boiling heat, and make the stills which are inserted in them boil also; and there being no egress for the steam from the case, except by the safety-valve; a small fire will suffice to keep up this degree of heat. From that part of the case which comes in immediate contact with the fire, a flue may be inserted, which, making some horizontal revolutions along the bottom, may pass out into the chimney. The back also of the fire-place can be a narrow boiler, which may communicate with the cases of the still. When performed on-board ship, a reservoir of salt-water is placed upon the deck of the vessel, through which the chimney of the fire may pass and impart its heat; and from thence pipes, having a cock attached to each, lead into the cases and stills for the purpose of their supply. From the necks of the stills pipes are brought, conducting the steam into vessels for cooking provisions. The range has two metal doors in front, each of which is attached by hinges to iron bolts: these fit into staples fixed in the side of the range, so that when the fire is not wanted for cooking, it can be enclosed by these doors; but when required, the doors can be drawn out the length of the bolts, fitting into the staples at the side of the range, and form a screen, between which and the fire the meat can be roasted. For the purpose of condensing steam on-board of ships, the tube containing it may pass through the ship, and along any part of the outside of it which lies immediately in the water, and again entering the ship, it discharges the condensed water into the water designed for its reception: after the sea-water has been distilled in this manner, it is passed through a filterer, consisting of a small cylindrical case made of tin or other metal, and being filled with pounded charcoal, each end is stopped by a circular cover perforated with holes, fine enough to prevent the charcoal from passing through. One end of this case is inserted into a cask, also partly filled with pounded charcoal, and the water being poured into the cask, filters out through the case. See the article FILTERING MACHINE.

A patent has also been taken out by Mr. Lamb, for the invention of an apparatus for rendering salt-water perfectly fresh and pure by distillation. The apparatus is attached to a new-constructed ship's fire-hearth; and it is said, that while a hearth, fitted on this principle, proportioned to the size of a fifty-gun ship, produces 25 gallons of fresh water per hour, without the slightest interruption to cooking, it effects a saving of one-fourth the consumption of fuel.

In consequence of certain experiments which have been made on-board his Majesty's ship Trusty, the invention has received the approbation of the lords commissioners of the Admiralty. Captain Hodgson, of the Trusty, before whom the experiments were made, states, that the fire-hearth fully answers all its intended purposes; that he feels no hesitation in saying, that the invention will be attended with the greatest utility to the navy in general; that it consumes considerably less fuel, and that the coppers boil in one-third less time than by any mode before employed; that it produces fresh-water from sea-water at the rate of 20 to 25 gallons per hour; and that the water is as good, if not better, than any he ever drank at sea.

SALVAGE, (*sauvetage*, *sauvement*, Fr.) an allowance made for the recovery of a ship or goods from the enemy, after having remained in his possession twenty-four hours; or of any thing dragged up from the bottom of the sea. It is paid by the first proprietors to the persons who have so recovered it, or else detained legally by the latter. For the laws relating to salvage, see 12 Ann. Stat. 2. *c.* 18.

SALVAGE-*Money*, (*récompense de sauvetage*, Fr.) is a reward allowed by the civil and statute law for the saving of ships or goods from the dangers of the sea, pirates, or enemies. Where any ship is in danger of being stranded or driven on shore, justices of the peace are to command the constables to assemble as many persons as are necessary to preserve her; and, being preserved by their means, the persons assisting therein shall, in thirty days after, be paid a reasonable reward for the salvage, otherwise the ship or goods shall remain in the custody of the officers of the customs as a security for the same.

SALUTE, (*salut*, Fr. from *saluto*, Lat.) a testimony of respect or homage, rendered by the ships of one nation to those of another, or by ships of the same nation to a superior or equal.

This ceremony is variously performed, according to the circumstances, rank, or situation of the parties. It consists in firing a certain number of cannon, or vollies of small arms; in striking the colours or topsails, or in three general shouts of the whole ship's crew, mounted on the yards and rigging for that purpose.

Regulations relative to SALUTES are: 1. The admiral and commander-in-chief of the fleet is to be saluted with seventeen guns; he is to return fifteen to all flag-officers, and thirteen to captains.

2. Admirals are to be saluted with fifteen guns, and vice and rear-admirals with thirteen.

3. Every flag-officer is to return the salute of another flag-officer, of the same rank, with the same number of guns he is saluted with.

4. An admiral is to return two guns less to a vice or rear admiral, and four guns less to a captain. A vice-admiral is to return two guns less to a rear-admiral, and four less to a captain. A rear-admiral is to return two guns less to a captain. Commodores are to salute, and to be saluted, as rear-admirals.

5. Ships meeting a squadron, in which there are more than one flag-officer, are to salute the commander only of such squadron, who is not to return the salute until all the ships saluting have ceased firing; he is then to make one general return, by firing the number of guns with which an officer of his rank is saluted.

6. When two squadrons meet, only the officers who command them are to salute.

7. A flag-officer, appointed to command in chief, shall be saluted, on hoisting his flag, by all the ships under his command, unless a flag-officer senior to him be present; in which case they are to salute him so soon as he shall be separated from such senior officer.

8. Flag-officers are not to be saluted by other flag-officers, nor by captains, who shall not have been separated from them six calendar months.

9. The captain of one of his Majesty's ships is not to salute the captain of another of his Majesty's ships in any part of the world.

10. Flag-ships are to return the salutes of foreign ships of war, in the same manner as they return the salutes of his Majesty's ships. If a captain be saluted by a foreign ship of war, he is to return it with an equal number of guns.

11. Merchant ships, whether belonging to his Majesty's subjects, or to those of any other nation, are to receive in return to their salutes, six guns less from the admiral and commander-in-chief of the fleet, four less from all other flag-officers, and two less from captains; but, when several merchant ships salute at the same time, the officer saluted may return such a number as he shall think proper.

12. When his Majesty, or any of the royal family, shall go on board any of his Majesty's ships, the standard is to be hoisted at the main-top-gallant-mast head of that ship, and they are to be saluted with twenty-one guns on their going on-board, and on their leaving the ship, and as much oftener as, from

circumstances, the commanding officer shall think proper.

13. When a nobleman, or any other person in a high public station, shall embark on-board any of his Majesty's ships, he may be saluted on his going on-board, and on his quitting the ship, with the following number of guns, viz.

A duke or ambassador with......	15
All other noblemen and envoys..	13

14. If any nobleman shall visit any of his Majesty's ships, he may be saluted, on his leaving the ship, with the following number of guns, viz.

A duke with..................	15
All other noblemen............	13

But, if he visit several ships of any squadron, or in the same port, he shall be saluted by only one of them.

15. If the ship of a flag-officer anchor in any foreign port or road, he is to inform himself what salutes have been usually given or received by flag-officers of his rank, of other nations or by those of his Majesty, and he is to insist on receiving the same marks of respect.

16. If a ship, not carrying a flag, anchor in any foreign port or road, the captain may salute the fort with such a number of guns as may have been customary, on his receiving an assurance from the governor, that the fort will return an equal number; but without such assurance he is never to salute.

17. If any foreign nobleman, flag-officer, or general officer, shall go on-board any of his Majesty's ships, he may be saluted on his leaving the ship with such a number of guns, as, from his rank and quality, may be proper. And the captain of a foreign ship of war may be saluted, on his visiting one of his Majesty's ships, if such a compliment shall have been paid to the captain of any of his Majesty's ships on visiting the ships of his nation.

18. His Majesty's consul or a British factory may be saluted with eleven guns on their leaving any of his Majesty's ships which they shall visit; but this may be done only once on the arrival of a ship in any foreign port, and once before her departure from it.

19. None of the before-mentioned salutes are to be fired without the approbation of the commanding officer present.

20. The anniversary days of the birth, accession, and coronation of the king; of the birth of the Queen, of the restoration of King Charles the Second, and of the Gunpowder Treason, shall be solemnized by such of his Majesty's ships as are in port, with such a number of guns as the commanding officer present shall direct, not exceeding twenty-one in each ship.

21. All salutes are to be fired from the guns of the upper deck.

22. His Majesty's ships are not, on any account, to salute any of his Majesty's forts or castles in the United Kingdom.

23. If a ship, which anchors in any foreign port or road, shall find there a flag-officer of the nation to which the fort belongs, the captain may salute such flag-officer with as many guns as he would salute a British flag-officer of the same rank, on his being assured that his salute shall be returned, in the same manner as it would be returned to a ship of that nation.

24. Within his Majesty's seas his ships are not, on any account, to strike their top-sails, nor take in their flags, nor in any way to salute any foreign ship whatever; nor are they, in any other seas, to strike their top-sails, or take in their flags to any foreign ships, unless such foreign ship shall have first struck, or shall at the same time strike, their flags and top-sails to his Majesty's ships.

25. If any of his Majesty's subjects shall so far forget their duty as to attempt to pass any of his Majesty's ships without striking their top-sails, the names of the ship and the master, the port to which they belong, the place from which they came, and that to which they are bound, together with affidavits of the fact, are to be sent to the secretary of the Admiralty, in order to their being proceeded against in the Admiralty court.

26. If any flag-officer shall die when on actual service, his flag shall be lowered half-mast, and shall continue so until he is buried; and at his funeral the commanding officer present shall direct such a number of minute-guns, not exceeding twenty-five, as he may think proper, to be fired by every ship, to begin when the corpse is put into the sea, or when it is put off from the ship to be carried on-shore.

27. If the captain of one of his Majesty's ships shall die, his pendant shall be lowered as directed in the case of the flag; and at his funeral the commanding officer present shall order such a number of minute-guns, not exceeding twenty, as he shall think proper, to be fired from the ship he commanded, beginning as in the preceding article. If the ship be alone, the officer succeeding to the command is to order this to be done.

28. If the lieutenant of one of his Majesty's ships shall die, the commanding officer present shall order three volleys of musketry to be fired at his funeral, from the ship to which he belonged. See the NAVAL INSTRUCTIONS.

SALUTE *on joining Company*. If a salute be required on joining company with an admiral at sea, irons should be ready heated in the fire, and the guns loaded and primed; as irons afford the most approved method of firing a salute.

SALUTING DAYS, are those on which the King and Queen's birth days, &c. are celebrated. The number of guns fired on the holidays, called Saluting Days in the Royal Navy, are as follows:—

	No. of Guns.
Queen's Birth-day, January 18th	21
King Charles's Restoration, May 29th	17
King's Birth-day, June 4th..........	21
King's Coronation, September 22d....	21
King's Accession, October 25th.....	17
Gunpowder-Plot, November 5th.....	17

SAMPANE, (*champan, sorte de bâtiment chinois*, Fr.) a kind of vessel used by the Chinese.

SAMSON'S POST, (*épontille des écoutilles, servant d'échelle pour descendre dans la cale*, Fr.) a sort of pillar erected in a ship's hold, between the lower deck and the kelson, under the edge of a hatchway, and furnished with several notches, that serve as steps to ascend or descend.

This post, being firmly driven into its place, not only serves to support the beam and fortify the vessel in that place, but also to prevent the cargo or materials, contained in the hold, from shifting to the opposite side, by the rolling of the ship in a turbulent and heavy sea.

SAMSON'S-POST is also the name of a strong piece of timber, which is made to ship and unship, used on board ships of war, which being placed in a sloping position from deck to deck, with the upper end resting against a beam, serves, by means of a single block lashed near its middle, to form a return for a tackle-fall, and thereby affords space for a greater number of hands to clap on, and run away into a more clear part of the ship.

SASHES FOR PORTS. See the article PORT-*Sashes.*

SATELLITES, in astronomy, are certain secondary planets, moving round the other planets, as the Moon does round the earth; they are so called because always found attending them, from rising to setting, and making the tour of the Sun together with them.

The words moon and satellite are sometimes used indifferently: thus we say, either Jupiter's moons or Jupiter's satellites, but usually we distinguish, restraining the term moon to the Earth's attendant and applying the term satellite to the little moons more recently discovered about Jupiter, Saturn, and the Georgium planet, by the assistance of the telescopes, which is necessary to render them visible.

The satellites move round their primary planets as their centres, by the same laws as the primary ones do round their centre the Sun; viz. in such a manner, that, in the satellites of the same planet, the squares of the periodic times are proportional to the cubes of their distances from the primary planet.

SATELLITES *of Jupiter* are four little moons, or secondary planets, performing their revolutions about Jupiter, as that Planet does about the Sun.

Simon Marius, mathematician to the Elector of Brandenburg, in November 1609, discovered three little stars moving round Jupiter's body, and proceeding along with him; and in January 1610, he found a fourth. On January the 8th, 1610, Galileo also observed the same in Italy, and in the same year published his observations. These satellites were also observed in January 1610, by James Harriot, an English mathematician, who made constant observations on them from that time till the 26th of Feb. 1612. Galileo, however, appears to be the first who applied them to the discovery of the longitude; and, for this purpose, dedicated a considerable part of his life in making observations to form a theory of their motions. For further particulars of these satellites, see the article JUPITER. And for an account of the satellites of SATURN, see that article.

The Method of finding the Longitude of a Place by the Eclipses of Jupiter's Satellites.

The observation necessary to find the longitude of a place by this method is, the apparent time of an immersion, or emersion, of either of the satellites, but particularly that of the first, on account of its periodic time being shorter than that of either of the other satellites; and because its elliptic equation, and other inequalities, except that whose period is 437 days, are insensible. The instruments proper for making the above observation are, a clock or watch beating seconds, an achromatic telescope of three or four feet focal length, and a quadrant, or some other instrument proper for determining the error of the watch.

An *immersion* of a satellite is, the instant of its entrance into the shadow of Jupiter; and an *emersion* is, that of its re-appearance out of the shadow.

When the time of Jupiter's passage over the meridian, as expressed in the Nautical Almanac, is less than XII hours, the eclipses of the satellites happen on the west side of the planet; and when that time exceeds XII hours, these observations happen on the east side of Jupiter. In the first case, the emersions of the first and second satellites are visible; and, in the latter, the immersions only may be observed Both the immersion and emersion of the other two satellites are generally visible, except when Jupiter is too near the opposition or conjunction.

In order to observe an immersion or emersion of any of the satellites, the observer must apply to his telescope, a magnifying power suitable to the state of the weather, and adjust it to a distinct vision previous to the time of observation. The watch must be contiguous to him, so that he may be able to catch the second as soon as the observation of the satellite is completed. An assistant clock will, however, be found more convenient. The observer is then to place himself at the telescope about three or four minutes before the expected time of an eclipse of the first satellite, twice as much before that of the second or third satellite, and a quarter of an hour before that of the fourth: the longitude of the place of observation being pretty well known. If its situation is uncertain, let four times as many minutes as there are degrees of uncertainty be added to the above times. If the observation to be made is that of an immersion, the observer must keep his eye direct to the satellite till it disappears; but if it is an emersion, he must keep his sight directed to that part of the shadow from which he expects the satellite to emerge; and in the instant of the last appearance of the satellite, if the observation is that of an immersion, or that of its first appearance, if an emersion, must be wrote down, and the observation is completed.

If an astronomical eye-glass is used, it must be remembered, that it inverts objects, making that which is the right side apparently the left, *et contra.*

The eclipses of Jupiter's satellites are given in the Nautical Almanac, in page 3 of the month. Those visible at Greenwich are marked with an asterisk. The configurations, or apparent relative positions of the satellites, are contained in the last page of the month.

To find the Longitude of a Place from an Observation of an Immersion, or Emersion, of one of the Satellites of Jupiter.

RULE.—To the observed time of the eclipse, apply the error of the watch for apparent time, deduced from an observation of the altitude of the Sun, or a fixed star, taken as near the time of the eclipse as possible: hence, the apparent time of observation will be known. The difference between this time, and that given in the Nautical Almanac, page 3 of the month, will be the longitude of the place in time; being east or west, according as the time by observation is later or earlier than the Greenwich time.

There are now added to the Nautical Almanac, by way of appendix, since the year 1800 inclusive, tables containing the eclipses of the satellites of Jupiter, expressed in mean time, calculated from the tables of M de Lambre, inserted in the tables annexed to the first volume of the third edition of M. de la Lande's Astronomy. And since these tables are far superior to any of the former, agreeing with the observations in a manner truly surprising, it, therefore, becomes necessary to use them. For this purpose, the apparent time of observation must be reduced to mean time, by applying thereto the equation of time from the Nautical Almanac, according to its title.

SATURN, in astronomy, one of the primary planets, being the sixth in order or distance from the Sun, and is marked in astronomical books with this character ♄.

This is a very conspicuous planet, though not so bright as Jupiter, and is perhaps one of the most engaging objects that astronomy offers to our view. His diameter is 79,042 miles, and the period of his revolution round the Sun is 10,759 days, 1 hour, and 51 minutes, at the distance of 900 millions of miles. He is distinguished from all the other planets by a large luminous ring surrounding his body, which was discovered by Huygens. He moves from west to east, nearly in the plane of the ecliptic, and exhibits irregularities similar to those of Jupiter and Mars. He becomes retrograde both before and after his opposition, when at the distance of about 109° from the Sun. His retrograde motion continues about 139 days; and, during its continuance, he describes an arc of about 6°.

The double ring, which was first observed by Mr. Pound, consists of two concentric rings detached from each other, and from the body of the planet: the innermost of which is nearly thrice as broad as the outermost.

The ring of Saturn casts a deep shadow upon the planet. It is sharply defined, both in its inner and outer edge, and appears to be more luminous than Saturn himself. Hence Dr. Herschel has concluded, that it is not any shining fluid, or aurora borealis, as some have imagined, but a solid body, equal in density to the planet. The Doctor is also of opinion, that the edge of the ring is not flat, but of a spherical, or rather spheroidal form. The ring, as seen from Saturn, appears like a large luminous arc in the heavens, as if it did not belong to the plant.

The rotation of the ring, as settled by Dr. Herschel, from observations made in the year 1789, on the apparent motions of the five luminous and protuberant points on the ansæ of the ring, is 10h. 32′. 15″·4. The Doctor also observes, that the ring is so extremely thin, that it seemed not to be the third part of the diameter of the third satellite, which was less than one second; and that the seventh satellite, when in the plane of the ring, appeared in the shape of a bead upon a thread, projecting on both sides of the same arm.

When we look with a good telescope at the body of Saturn he appears, like most of the other planets, to be of a spheroidal form, arising from a rapid rotation about his axis.

On the 14th of September, 1789, Dr. Herschel measured his diameter, and found that the equatorial diameter was 22″·8, and the polar diameter 20″ 6, which gives the proportion of nearly 10 to 11. It appears, however, from more recent observations made by the same astronomer, that the polar is to the equatorial diameter as 32 to 35, or as 11 to 12 nearly. Till the year 1805, Dr. Herschel had always regarded Saturn as an accurate spheroid; but, on the 12th of April of that year, he was struck with a very singular appearance exhibited by the planet, which resembled a square, or rather a parallelogram, with the four corners rounded off deeply, but not so much as to bring it to a spheroid. After examining Saturn with his telescopes, and comparing it with the form of Jupiter, Dr. Herschel concluded that this was the real form of the ring. The following are the proportional dimensions of Saturn's disc:—

Diameter of the greatest curvature ..	36″
Equatorial diameter................	35″
Polar diameter....................	32″
Latitude of the longest diameter	43°20′

The surface of Saturn is diversified, like that of some of the other planets, with dark spots and belts. Huygens observed five belts, which were nearly parallel to the equator. They were also perceived by Cassini and Messier. Dr. Herschel has also observed several belts, which, in general, are parallel with the ring. On the 11th of November, 1793, immediately south of the shadow of the ring upon Saturn, he noticed a bright, uniform, and broad belt; and close to it a broad and darker belt, divided by too narrow white streaks; so that he saw five belts, three of which were dark, and two bright. The dark belt had a yellowish tinge. These belts generally cover a larger zone of the disc of Saturn, than the belts of Jupiter occupy upon his surface.

Dr. Herschel has likewise perceived dark spots upon Saturn's disc; and, by the changes in their position, has determined the daily rotation of the planet to be performed in 10h. 16′ 0″·44, round an axis perpendicular to the plane of the ring. La Place had formerly found, from theory, that the interior ring ought to perform its revolution in 10 hours. The ring of Saturn, therefore, revolves in the same time nearly as the planet, and round the same axis.

Satellites of Saturn. The planet Saturn is surrounded with seven satellites, which supply him with light during the absence of the Sun. The fourth of these satellites was first discovered by Huygens on the 25th of March, 1655. Cassini discovered the fifth in October 1671, the third on the 23d of December 1672, and the first and second in the month of March 1684. The sixth and seventh satellites, which were discovered by Dr. Herschel in 1789, are nearer to Saturn than any of the rest, though, to avoid confusion, they are named in the order of their discoverer. These satellites are all so small, and placed at such a distance from the Earth, that they cannot be seen unless with excellent telescopes. Wargentin saw the five old satellites with an achromatic telescope of ten feet; and, on the 19th of December, 1793, Dr. Herschel saw them distinctly with a power of 60 applied to his ten feet reflector. The sixth and seventh are the smallest of the whole; the first and second are the next smallest; the third is greater than the first and second; and the fourth is the largest of them all. The fifth satellite surpasses all of them but the fouth in brightness, when it is at its western elongation from Saturn; but, at other times, it is extremely small, and entirely disappears at its eastern elongation.

This phenomenon, which was at first discovered by Cassini, appears to arise from one part of the satellite being less luminous than the rest. In consequence of the rotation of the satellite, and its axis, this obscure part of its disc is turned, like the Earth's, when it is in the part of its orbit east of Saturn; and the luminous part of its surface becomes visible while it enters into the western part of its orbit. Dr. Herschel observed this satellite through all the variations of its light, and concluded, that, like our Moon, and the satellites of Jupiter, it turned round its axis in the same time that it performed its revolution round the primary planet.

The *first* satellite of Saturn revolves at the distance of 4.893 semi-diameters of the planet, in 1d 21h 18′ 26″; the *second* at 6.268 semi-diameters of Saturn, in 2d 17h 44′ 51″; the *third* at 8.754 semi-diameters of Saturn, in 4d 12h 25′ 11″; the *fourth* at 20.297 semi-diameters of Saturn, in 15d 22h 41′ 13″; the *fifth* at 59.154 semi-diameters of Saturn, in 79d 7h 53′ 43″; the *sixth* at 3.080, in 23h 37′ 23″; and the *seventh* at 3.952 semi-diameters of Saturn, in 1d 8h 53′ 9″

The position of the satellites of Saturn, and the figure of their orbits, may be easily found for any particular time by tables of their motions, calculated by Cassini, and given in the *Tables de Berlin*; and from tables calculated by Dr. Herschel, and published in the *Philosophical Transactions* for 1790. Their configurations may also be found by a very simple instrument called a *Saturnilabe*, which is described in La Lande's Astronomy, vol. 3. p. 203.

The theory of the satellites of this planet is less perfect than that of the satellites of Jupiter. The difficulty of observing their eclipses, and of measuring their elongations from Saturn, have prevented astronomers from determining, with their usual precision, the mean distances and the revolutions of the secondary planets.

The orbits of the satellites of Saturn move like those of the moon, and the satellites of Jupiter upon fixed planes which pass constantly by nodes of the equator and the orbit of Saturn, between their two last planes. Their orbits preserve their inclination almost invariable, and their nodes have a retrograde motion nearly uniform.

A complete tabular view of the elements of the satellites of Saturn may be found among the Tables of Vince's Astronomy. See Herschel, *Philos. Trans.* for 1790, 1792, 1794, and 1805. La Place, *Mem. Acad. Paris*, 1787 *Méchanique Céleste*, tom. iv. p. 173, 185.

SATURNILABE, an astronomical instrument for finding the configurations of the satellites of the planet Saturn, and is fully described in La Lande's Astronomy, vol. 3. page 203.

SAVEALL, the name of a sail which is sometimes placed to catch the wind that might escape beneath the foot of any sail.

SAVINGS or Provisions, implies the bread, wine, spirits, beef, pork, butter, and cheese, which have been saved by the different persons or messes in any of his Majesty's ships, from the established allowance of those species; and for which the purser pays them at the expiration of one, two, or three months at furthest, agreeable to the credit prices stated in his instructions.

The purser, at the expiration of every one, two, or three months, as before-mentioned, is to make out three accounts, according to the form given in the general printed instructions, specifying in distinct columns the number of each mess, the number of persons in each mess, the name of the person appointed to receive for the mess, the number on the ship's books of the person so receiving, the quantities of provisions saved by each mess, the value thereof at the credit prices before referred to, the time when paid, the signatures of the persons to whom paid, and the signatures of the persons witnessing the payment; which accounts, after being carefully examined by the purser and his steward, shall be signed by them, attesting them to be just and true, and that they are ready to make oath thereto whenever required.

When the ship shall be at a regular established victualling port, the purser, instead of drawing bills upon the commissioners of the victualling for the amount of the savings (as in the case of the ship being at a port where there is no agent, or other proper instrument of the victualling), is to procure the sum necessary for the payment thereof from the agent for the victualling; who will furnish him therewith, upon his delivering to him an account of the quantities, rates, and amount of the provisions saved; with a certificate subjoined or annexed thereto, signed by the captain or commander, and the master, stating that the money therein inserted is actually due to the ship's company, and granting also to the agent a proper receipt for the same.

The captain or commander, immediately after the bill is drawn, or the money received, is to cause the payment to be made publicly in the presence of the master and himself; the purser taking particular care that each of the three accounts before-mentioned be properly signed by the persons to whom the money is paid, and that each payment be witnessed by a commissioned or warrant officer, who is to sign his name and description in the column for that purpose.

All money paid to seamen in foreign coin, on the score of savings of provisions, shall be issued to them according to the sterling value, and they shall have the benefit of the exchange.

All officers, and others, are strictly forbidden to concern themselves, directly or indirectly, in buying the seamen's savings; and the officer who has the payment of the same, is charged to pay the parties themselves, without any regard to notes or other obligation whatsoever.

Two of the three accounts, before-mentioned, are to be immediately delivered to the captain, who is to sign a certificate at the foot of the one, which the purser is to keep for passing his accounts; setting forth, that he had received two accounts of the same tenor and date; one of which the captain is directed to send to the commissioners of the victualling, and to note on the other when, and by what conveyance, it had been so forwarded. The remaining account the captain is to reserve, and deliver into the victualling office on the passing of his accounts; and he is to take notice, that the amount of all the bills attested by him for savings of provisions, will necessarily be charged as imprests against him, and so remain, until one of the said accounts shall, on his part, be delivered into the victualling office. See the NAVAL INSTRUCTIONS.

SAUCER OF A CAPSTAN, (*écuelle de cabestan*, Fr.) is a socket of iron let into a wooden stock or standard, called the step, resting upon, and bolted to, the beams. Its use is to receive the spindle or foot, upon which the capstan rests and turns round.

SAUCER-HEADED *Bolt*. See the article BOLT.

SAW, (*scie*, Fr.) an indented instrument, which serves to cut or divide into pieces various solid bodies, as wood, stone, marble, &c.

The best saws are made of steel, ground bright and smooth; those of iron are only hammer-hardened; whence, the first are not only stiffer, but smoother than the last.

Saws are distinguished into different sorts, viz. the pit-saw, whip-saw, hand-saw, tenon-saw, and compass-saw.

Pit-SAW. See the article PIT.

SAWING, (*sciage*, Fr.) the application of the saw in dividing of timber into boards, &c.

SCALE, a mathematical instrument, consisting of certain lines drawn on ivory, wood, &c. divided into various parts, either equal or unequal. It is of great use in laying down distances in proportion, or in measuring distances already laid down. There are scales of various kinds, according to the several uses to which they are applied, viz. GUNTER'S *Scale*, PLANE *Scale*, &c. which articles see.

To SCALE *the Guns*, (*souffler les canons*, Fr.) is to clean the inside of a ship's cannon, by the explosion of a small quantity of powder; which effectually blows out any dirt or scales of iron which may adhere to the interior surface.

SCALENE TRIANGLE, in geometry, is a triangle whose sides and angles are all unequal. A cylinder or cone, whose axis is oblique, or inclined to its base, is also said to be scalenous, though more frequently it is called oblique.

SCANT, (*vent qui refuse*, Fr.) a term applied to the wind when it becomes unfavourable to a ship's course after having been fair or large. It is distinguished from foul wind, as in the former a ship is still enabled to sail in her course, although her progress is considerably retarded; but in the latter she is obliged to deviate from it.

SCANTLING, (*échantillon des pièces de bois*, Fr.) in ship-building, a name given to any piece of timber, with regard to its breadth and thickness, when reduced to the standard size.

To SCARPH, (*écart, ou empâture, ou assemblage quelconque de deux pièces de bois*, Fr.) is to unite any two pieces of timber together by the extremities, so that the end of one goes over the end of the other, being tapered that the one may be let into the other and become even, as the keel-pieces, clamps, &c. But when the ends of the two pieces are cut square and put together, they are said to butt to one another: and when another piece is laid on and fastened to both, as is the case in all the frame-timbers, this is called scarphing the timbers; and half the piece which fastens the two timbers together, is reckoned the length of the scarph.

SCHEME, a draught or representation of any geometrical or astronomical subject, and is otherwise termed a diagram.

SCHOOL, (*schola*, Lat.) a public place, wherein the languages, arts, and sciences, are taught.

According to Du-Cange, it was anciently used for all places where several persons met together, either to study or converse. Now, we say, a grammar-*school*, a naval and military *school*.

Naval SCHOOLS. See the articles *Royal Naval* COLLEGE, *at Portsmouth*; *Naval* ACADEMY, *at Gosport*; GREENWICH *School*, for the education and maintenance of the sons of warrant and petty officers in his Majesty's navy; and NAVAL *Asylum at Greenwich*, for the clothing, maintenance, and education of the orphans of his Majesty's seamen and marines.

Military SCHOOLS. See the articles *Royal Military* ACADEMY, *at Woolwich*, for the artillery and engineer departments.

The Military ACADEMY, *at Addiscombe*, near Croydon, for cadets, for the engineer, artillery, and infantry service of the honourable East India Company, was originally attached to the military institution at Woolwich, but removed to the present situation a few years ago, and considerably enlarged; there are now one hundred Cadets admitted, seventy for the engineers and artillery departments, and thirty for the infantry. The system of education is similar to that pursued at Woolwich, with the addition of the oriental languages, which are particularly attended to.

Royal Military COLLEGE, *at Marlow*, the senior department of which, for commissioned officers, is now removed to Farnham: and the junior, for Cadets, to Sandhurst, near Bagshot.

MILITARY *Asylum*, *at Chelsea*, for the education, maintenance, and clothing, of the children of soldiers in the regular army.

SCHOOL-MASTER, (*maître d'école, embarqué sur les vaisseaux de guerre Anglois pour l'instruction des Gardes de la marine, &c.* Fr.) in the Royal Navy, a person of approved abilities in mathematics, and the theory and practice of navigation, by a previous examination before the master, warden, and assistants of the Trinity House in London, from whom he must obtain a certificate of his qualifications, in order to receive his warrant from the lords commissioners of the Admiralty.

If a candidate for the appointment of schoolmaster be at Portsmouth, and the ship to which he shall be desirous of being appointed, be likely to sail in so short a time as not to admit of his going to London to be examined as above directed, he shall be examined by the professor and preceptor of the Royal Naval College there, whose certificate of his being properly qualified shall, in that case, be sufficient. He must also produce a certificate from two or more persons of good repute, of the sobriety of his life, and the goodness of his morals.

The heads of the examination of a schoolmaster for the Royal Navy, are as follows:—

1. Arithmetic, with vulgar and decimal fractions, and the square and cube roots.

2. The usual method of finding the Golden Number, Epact, Moon's age, her southing, and the time of high water in any port.

3. The principles of geometry in the construction of such problems as are necessary in plane and spherical trigonometry in the solution of all the various cases of rectangular and oblique triangles.

4. The use of the terrestrial globe in finding the latitude, longitude, angle of position, and distance between any two places.

5. The use of the celestial globe in finding the amplitudes, and times of rising and setting of the sun, or any of the known fixed stars.

6. Spherical trigonometry, applied to the solution of all such problems in astronomy as are commonly necessary in navigation; such as amplitudes, azimuths, and the hour of the day or night.

7. Plane sailing, viz. the working of traverses, the solution of all plane sailing questions, with the application thereof to sailing in currents, and turning to windward.

8. Globular sailing, viz. Middle Latitude and Mercator's, and the solution of all questions relating to sailing on a great circle of the sphere.

9. The use of instruments for observing the altitudes and azimuths of the Sun and stars; such as the Quadrant, Sextant, and amplitude or azimuth Compass, with the use of observations in finding the latitude and longitude of a ship, and variation of the compass.

10. The construction and use of plane and Mercator's Charts.

11. The use and application of the preceding articles in actual working days-work, as by a journal kept at sea.

12. The method of finding the latitude by double altitudes of the Sun.

13. The lunar method of finding the longitude at sea by tables, and the help of an Ephemeris; and,

14. That a plan of instructions be produced in a book, shewing the manner in which the person, desiring to be examined, instructs his pupils, or has been taught himself.

After the Schoolmaster's appointment to one of his Majesty's ships, he is to be extremely attentive to the education of the young gentlemen who may be put under his care, diligently instructing them in such branches of the mathematics as may be necessary for them to know; encouraging the diligent by every means in his power, and reporting the misconduct of the idle to the captain, that he may take such measures with them as he shall think necessary.

He is not only to instruct them in mathematics, but is also to watch over their general conduct, and attend to their morals; and if he should observe any disposition to immorality, debauchery, or any conduct unbecoming an officer and a gentleman, he is to represent it to the captain, that it may be immediately corrected as it shall deserve.

If there be any boys in the ship desirous of learning navigation, he is, with the permission of the captain, to give them such instruction as he shall find them qualified to receive.

It is his duty, also, to assist any officer who may require him in any astronomical observations or calculations he may be desirous of making, provided it does not in any degree interfere with the time appointed for the instruction of his pupils.

He will be required, before he receives his pay, to produce a certificate from the captain of his sobriety and good conduct, and of his ability and diligence in the discharge of his duty.

When he leaves the ship, he is to obtain a similar certificate, which he is to send to the secretary of the Admiralty, without which he cannot again be appointed to any of his Majesty's ships. See the Naval Instructions.

SCHOONER, (*goëlette* Fr.) a small vessel with two masts, whose main and fore-sails are suspended by gaffs, reaching from the mast towards the stern; and stretched out below by booms, whose foremost ends are hooked to an iron, which clasps the mast so as to turn therein as upon an axis, when the after-ends are swung from one side of the vessel to the other.

This vessel is generally a fast sailer, and principally employed in trade by those who make speculations where dispatch is requisite.

SCIATIC Stay, a strong rope fixed from the main to the foremast head in merchant ships: when loading or unloading, it serves to sustain a tackle, which, travelling upon it, may be shifted over the main or fore-hatchways as occasion requires.

SCOOP, (*escop à main*, Fr.) a little hollowed piece of wood, used to throw water out of a boat, which operation is usually called baling the boat.

SCORE, (*entaille*, Fr.) a notch or hollow cut by a saw, gouge, or chisel, out of any piece of wood, to admit another projecting in a similar shape.

Score *of a Dead-Eye*, (*trou de cap de mouton*, Fr.) is the hole through which the rope passes.

SCORPIO, the Scorpion; in astronomy, the eighth sign of the Zodiac, denoted by this character ♏. The number of stars in this sign, according to Flamsteed's Catalogue, are 44.

To SCOUR the Coast, (*purger la côte*, Fr.) a term frequently used by detached parties of seamen to express the act of firing a quick and heavy discharge of ordnance or musketry, for the purpose of dislodging an enemy.

SCRAPER, (*gratte, ou racle*, Fr.) an iron instrument, having two or three sharp edges (see fig. 3, plate XXIII.), and used to scrape off the dirty surface of the planks on a ship's side or decks, or to clean the topmasts, particularly after a voyage.

When the sides of a ship are thus sufficiently scraped, they are varnished over with turpentine, or a mixture of tar and oil, &c., which preserves the planks from being rent or split by the heat of the sun, &c., and gives the ship a more splendid appearance on the water.

SCREEN BULK-HEAD, in ship-building, the aftmost bulk-head under the round-house.

SCREENS, (*entourages de toile, &c.* Fr.) in a ship of war, are partitions made of canvas, which, being nailed up in lieu of the bulk-heads that are taken down in readiness for action, serve for the accommodation of the captain, and the officers in the ward-room See the article BIRTH.

Fire-SCREENS, are pieces of fear-nought sewed together, and hooked round the magazine passages, and also round the hatchways, where it is necessary to pass the powder; they are always wetted previous to commencing an action, in order to prevent their taking fire.

SCREW, one of the mechanical powers, or rather a combination of two of them, the inclined plane and the lever, principally used in pressing bodies together, or in lifting great weights.

It is defined a right cylinder, cut into a furrowed spiral, every where making the same angle with the length of it: and receiving its motion, not from any stroke, but, as Wilkins calls it, "a *vectis* at one end of it."

When the furrowed spiral is cut upon the exterior of a cylinder, it is called a *male* screw: but if the spiral be cut along the interior surface of a hollow cylinder, it is said to be *female*; the latter is also sometimes called the *box* or *nut*.

When motion is to be given to something, the male and female screws are necessarily conjoined, that is, whenever the screw is to be used as a simple engine, or mechanical power: but when joined with an axis in peritrochio, there is no occasion for a female, but in that case it becomes part of a compound engine.

Bed or Barrel SCREW, for raising the heads of large masts, and fixing the trestle-trees, are made of elm and consist of two puppets, a bed, and a sole; the puppets are four feet nine inches long; have their lower parts round, like a cylinder, and are cut with a screw; their upper parts, or heads, are larger, and either octagonal or round, with iron hoops driven on to prevent their splitting; through these heads are holes, on the different sides, through which long iron pins are thrust, which put them in motion by pressing against them, and acting as levers. The bed is a broad thick piece of elm, three times the diameter of the screw in breadth, once its diameter in thickness, and about six feet in length, with a hole at each end of the bed for the screws or puppets to work in; six bolts are driven through the beds, and clenched on a ring to prevent its splitting; two are in the middle, and two at each end. In the sides and ends, mortises are cut for hand-holes; the sole is the same length and breadth as the bed, and half the thickness, with a hole hollowed out at each end for the puppets to work in; sometimes the sole is thicker; and, instead of holes, has an iron plate with a hole let into each end, wherein works an iron spring that is driven into the heel of the puppet with a shoulder.

Hand SCREW, a box of elm containing cogged iron wheels of increasing powers; the outer one, which moves the rest, is put in motion by a winch on the outside, and is called either single or double, according to its increasing force: this is sometimes, though falsely, called a jack.

To SCREW *down*, (*estiver le chargement d'un bâtiment*, Fr.) is to secure a vessel's cargo by means of screws.

SCREWS, in gunnery, are those fastened to the cascable of light guns and howitzers, by means of an iron bolt, which goes through a socket fixed upon the centre transom, to elevate or depress the piece with, instead of wedges.

Lock SCREWS, small screws which are attached to the lock of a musket.

SCROWL, in ship-building, a name given to the two pieces of fir timber which are bolted to the knee of the head, and serve in lieu of a figure.

To SCRUB THE DECKS, (*laver les ponts*, Fr.) is to clean them by means of a holy-stone and sand, or with sea-water and long-handled scrubbing-brushes: some captains prefer the former method, and others the latter.

Washing the decks is now very generally exclaimed against by the medical officers in the navy, as being prejudicial to the health of the seamen: yet an intelligent medical officer, who appears to have spent the best part of his life on the ocean, speaks, in his Naval Hygiene, just published, decidedly in favour of the old practice, and adds, that he has witnessed it in different ships, not only on the home stations, but beyond the arctic circle; on the coast of Jutland; in the Southern Ocean off the Cape of Good Hope; in the Chinese Seas; on the Banks of Newfoundland: in short, in every sea from Iceland to Borneo, without the crew's experiencing any bad effects from the humidity: he farther observes, that dry holy-stoning the decks in northern climates, between the autumnal and vernal equinoxes may be very proper, but between the tropics, washing the decks is not only harmless, but highly salutary, and procures more attention to ventilation than the opposite system. See JOHNSON'S *Essay on the Influence of Tropical Climates on European Constitutions.*

To SCRUB *the Hammocks*, (*laver les hamacs*, Fr.) is to clean them, by laying them flat on the deck and rubbing them with a brush previously dipped in water.

SCUD, a name given by seamen to the low and thin clouds which are most swiftly wafted along by the winds.

To SCUD, (*fuir à la lame, dans un gros tems.* Fr.) to be carried precipitantly before a tempest, and is either performed with a sail extended on the ship's fore-mast, or, if the storm is excessive, without any sail, which is then called scudding under bare poles. In sloops, schooners, and other small vessels, the sail employed for this purpose is called the squaresail. In large ships, it is either the foresail, at large, reefed, or with its goose wings extended, according to the degree of the tempest; or, it is the fore-top-sail close reefed, and lowered on the cap: which last is particularly used when the sea runs so high as to becalm the foresail occasionally; a circumstance which exposes the ship to the danger of broaching-to.

As a ship flies with amazing rapidity through the water whenever this expedient is put in practice, it is never attempted in a contrary wind, unless when her condition renders her incapable of sustaining the mutual effort of the wind and waves any longer on her side, without being exposed to the most imminent danger.

The principal hazards incident to scudding are, generally, a pooping sea; the difficulty of steering, to prevent broaching-to; and the want of sufficient sea-room.

A sea striking the ship violently on the stern, may dash it inwards, by which she must inevitably founder. In broaching-to suddenly, she is threatened with being immediately overset; and, for want of sea-room, she is endangered by shipwreck on a lee-shore.

SCULKER, or Skulker, (*paresseux*, Fr.) on-board ship, is one who, from idleness, or laziness, absents himself, and endeavours to avoid his duty.

SCULL, (*aviron de canot*, Fr.) a kind of short oar, the loom of which is only equal in length to half the breadth of the boat, whereby two may be managed by one man, one on each side.

To Scull, (*gabarer*, Fr.) is to make a boat advance on the water by putting an oar in the notch, which is generally cut in the top of the stern, and then forcing it from the larboard to the starboard side.

SCULLER, (*embarcation maniée par un seul homme*, Fr.) a term used to denote a boat rowed on the river Thames by one man with two sculls, which is used in contradistinction to oars. It is, however, generally applied to a person using only one oar in the stern of a boat, to impel it forward.

SCUPPERS, (*dalots*, Fr. *schoepen*, Dutch, *to draw off* , certain channels cut through the water-ways and sides of a ship, at proper distances, and lined with sheet lead, in order to carry the water off the deck into the sea.

Scupper-*Hose*, (*cuir, ou manche des dalots*, Fr.) a leather pipe or tube, nailed round the outside of the scuppers of the lower decks, and which, by hanging down, prevents the water from entering when the ship inclines under a press of sail.

Scupper-*Nails*. See the article Nails.

Scupper-*Plugs*, (*tampons de dalots*, Fr.) are used to stop the scuppers occasionally.

SCURVY, (*scorbul*, Fr.) a well-known disease to which seamen are peculiarly exposed, from being necessitated to eat salted provisions, and sometimes to drink unwholesome water, &c. in long voyages.

This disease was formerly very prevalent among seamen, but now little known in the British navy, from the improvements in medicine, the great attention of our medical officers, and the wholesome regulations introduced by the naval officers in our ships of war.

A proposition has been submitted to the French government by Citizen Rivoire, of the marine, to contribute to the health of their seamen, by diminishing the quantity of salted meats used at sea, and rescuing the men in a certain degree from the ravages of the scurvy. The government had so far given sanction to the plan, which the inventor asserts to be of tried efficacy, as to insert it in the official paper, the *Redacteur*, of the 10th of August, 1799.

"Salted meats," says Citizen Rivoire, "form the general food of persons at sea, and, however carefully they may be prepared, the use of them is at all times pernicious. When vessels are a long time at sea without putting into any port, the scurvy is found to rage among the crews, and in a manner the more dangerous, as the means of cure are slow and uncertain. The following is the process which Citizen Rivoire employed, which he says is the result of many experiments, with which he had every reason to be satisfied:

"It will be necessary in the naval slaughter-houses, to select the most fleshy parts, the four quarters, for instance, of the beeves which are killed for their daily consumption, which of course must then be doubled. The muscles being cleared of the surrounding fat, are to be cut in large slices perpendicularly, according to the direction of the fibres. These slices are to be exposed to the heat of the ovens in which the bread and biscuit are baked for the use of the navy, when they are at the heat of fifty degrees by the the mometer of Reaumur, and to remain there until the ovens be cold: an opening should be left for the evaporation of the moisture.

"The same operation is to be repeated, taking care to turn the larger slices until they are entirely dried, and of an horny but friable, consistence.

"On the other hand, it will be necessary to melt the fat which has been taken away with the suet of the cattle, and to purify the whole by means of salt and boiling water; but, in the fat thus purified, care must be taken that no moisture remain; and it would also be of advantage, if some aromatic herbs were employed in this part of the process. Every thing being thus prepared, a number of casks should be ready: those which have served for the conveyance of sweet oil are most suited for the purpose; the slices of dried meat are in these to be placed in layers, and the melted fat, not being too hot, is to be poured upon them. The casks are then to be headed, taking care that no vacant spaces are left, and both ends are immediately to be calked."

This process is very simple, and by no means expensive, being executed with precision, has the advantage of securing an agreeable food, of preserving the health of the sailors, and of containing a double quantity of provisions in the same space that would be occupied by meat salted in the ordinary manner.

The provisions being reduced into a small compass, a smaller quantity may of course be given

The fat of the beef being purified with care, may serve to season the roots and greens, and to enrich the soup of the crews.

The process is to be exactly the same with respect to mutton; and where this is not dearer than beef, such a quantity may be laid in as to preclude entirely the necessity of using salted provisions. See Naval Chronicle, Vol. II.

SCUTTLE, (*écoutillon*, Fr.) a small hatchway or hole, cut for some particular purpose through a ship's deck or sides, or through the coverings of her hatchways and ladder-ways, and furnished with a lid which firmly incloses it when necessary. See the articles Deck and Hatchway.

To Scuttle, (*saborder un vaisseau, pour le couler, ou pour en retirer les effets*, Fr.) is to cut large holes through the bottom, sides, or decks of a ship, particularly when she is stranded or overset, and continues to float on the surface. The design of this expedient is usually to take out the whole, or a part of the cargo, provisions, stores, &c. with all possible expedition.

Scuttle-Butt, (*charnier*, Fr.) is a cask having a square piece sawn out of its bilge, and lashed upon deck. It is used to contain the fresh water for daily use, whence it is taken out with a leaden can.

SEA, (*mer*, Fr. *sæ*, Sax. *zee*, Dutch), in geography, is frequently used for that vast tract of water encompassing the whole earth, more properly called Ocean, which article see.

Sea, is more properly used for a particular part of the ocean, almost surrounded by land, and is denominated from the countries it washes, or from other circumstances. Thus we say, "The Irish Sea, the Mediterranean Sea, the Baltic Sea, the Red Sea," &c.

Sea, is variously applied by sailors; to a single wave; to the agitation produced by a multitude of waves in a tempest; or to their particular progress or direction. Thus they say, "We shipped a heavy sea; there is a great sea in the offing; the sea sets to the southward." Hence a ship is said to head the sea, when her course is opposed to the setting or direction of the surges.

What proportion the superfices of the sea bears to that of the land, is not precisely known, though it is said to be somewhat more than two thirds. As the waters of the earth must necessarily rise to the surface thereof, as being specifically lighter than the earth, it was necessary there should be large cavities therein, as receptacles to contain them, otherwise they would have overspread all the superfices of the earth, and so have rendered it utterly uninhabitable for terrestrial animals; for the centre of the earth being the common centre of gravity, and the nature of fluids being such, that they equally yield to equal powers; and the power of attraction being every where equal, at equal distances from the centre, it follows, that the superficial parts of the water will every where conform themselves to an equidistant situation from the centre, and, consequently, will form the surface of a sphere, so far as they extend. Hence, that the sea seems higher than the earth or land, results from the fallacy of vision, whereby all objects and the parts of land, as well as sea, the further they are off from us, the higher they appear; the reason of all which is plain from optics; for it is well known, that the denser any medium is, through which we behold objects, the greater is the refraction: or the more their images appear above the horizontal level; also, the greater quantity of the medium the rays pass through, the more will they be bent from their first direction; on both these accounts, the appearances of things remote, and on the sea, will be somewhat above the horizon, and the more so as they are the more remote.

Depth of the Sea. With regard to the depth or profundity of the sea, Varenius affirms, that it is in some places unfathomable, and in other places very various, being in certain places $\frac{1}{20}$, $\frac{1}{3}$, $\frac{45}{100}$, $1\frac{1}{10}$, $2\frac{2}{10}$, $4\frac{1}{2}$ English miles, in other places deeper, and much less in bays than in oceans. In general, the depths of the sea bear a great analogy to the height of mountains on the land, so far as is hitherto discovered; and it is a general rule among sailors, and is found to hold true in many instances, that the more the shores of any place are steep and high, forming perpendicular cliffs, the deeper the sea is below; and that, on the contrary, level shores denote shallow seas. Thus, the deepest part of the Mediterranean is generally allowed to be under the height of Malta. And the observation of the strata of earth and other fossils, on and near the shores, may serve to form a good judgment as to the materials to be found in its bottom. For the veins of salt and of bitumen, doubtless, run on the same, and in the same order, as we see them at land; and the strata of rocks that serve to support the earth of hills and elevated places on shore, serve also, in the same continued chain, to support the immense quantity of water in the bason of the sea. The coral fisheries have given occasion to observe, that there are many, and those very large, caverns, or hollows, in the bottom of the sea, especially where it is rocky; and that the like caverns are sometimes found in the perpendicular rocks which form the steep sides of those fisheries.

These caverns are often of great depth, as well as extent, and have sometimes wide mouths, and sometimes only narrow entrances into large and spacious hollows. The bottom of the sea is covered with a variety of matters, such as could not be imagined by any but by those who have examined into it, especially in deep water, where the surface only is disturbed by tides and storms; the lower part, and consequently its bed at the bottom, remaining for ages perhaps undisturbed. The soundings, when the plummet first touches the ground, on approaching the shores, gives some idea of this. The bottom of the plummet is hollowed, and in that hollow there is placed a lump of tallow; which, being the part that first touches the ground, the soft nature of the fat receives into it some part of those substances which it meets with at the bottom; the matter, thus brought up, is sometimes pure sand, and sometimes a kind of sand made of the fragment of shells, beaten to a sort of powder; sometimes it is made of a like powder of the several sorts of corals, and sometimes it is composed of fragments of rock; but besides these appearances, which are natural enough, and are what might well be expected, it brings up substances which are of the most beautiful colours. Dr. Donati, in an Italian work containing an essay towards a natural history of the Adriatic Sea, has related many curious observations on this subject; having carefully examined the soil and productions of the various countries that surround the Adriatic Sea, and compared them with those which he took up from the bottom of the sea, he found that there is very little difference between the former and the latter.

At the bottom of the water there are mountains, plains, vallies, and caverns, similar to those upon land. The soil consists of different strata, placed one upon another, and mostly parallel and correspondent to those of the rocks, islands, and neighbouring continents. They contain stones of different sorts; minerals, metals, various putrified bodies, pumice stones, and lavas, formed by volcanos. One of the subjects which most excited his attention, was a crust, which he discovered under the water, composed of crustaceous and testaceous bodies, and beds of polypes of different kinds, confusedly blended with earth, sand, and gravel; the different marine bodies which form this crust, are found at the depth of a foot or more, entirely petrified and reduced into marble; these, he supposes, are naturally placed under the sea, when it covers them, and not by means of volcanos and earthquakes as some have conjectured. On this account he imagines, that the bottom of the sea is constantly rising higher and higher, with which other obvious causes of increase concur; and from this rising of the bottom of the sea, that of its level or surface naturally results; in proof of which, this writer recites a great number of facts.

La Place, from his observations at Brest, concludes, that the mean depth of the sea is four leagues.

Increase of the SEA. There are two principal reasons why the sea does not increase by means of rivers, &c. falling every where into it. The first is, because waters return from the sea by subterranean cavities and aqueducts, through various parts of the earth. Secondly, because the quantity of vapours raised from the seas, and, falling on the land, only cause a circulation, but no increase of water. It has been found, by calculation, that in a summer's day there may be raised in vapour, from the Mediterranean Sea, 5,280,000,000 tuns of water; and yet this sea receiveth not, from all its nine great rivers, above 1,827,000,000 tuns per day, which is but a third part of what is exhausted in vapours.

The ascent of the sea-water, for the formation of springs, by a subterranean circulation of its water to their sources, has been a great objection with many against the system of their being formed of the sea; but Dr. Plot has observed, that there are many ways by which the water may ascend above its own level; 1. By the means of subterranean heats. 2. By filtration. 3. By the unequal heights of several seas. 4. By the distance of the centre of magnitude from the centre of gravity in the terraqueous globe: the superfices of the Pacific Sea is said to be further from the centre of gravity, than the top of the highest hill on the adverse part of the globe. And, 5. By the help of storms. The sea water actually ascends above its own level, coming into wells whose bottoms lie higher than the surface of the sea at high-water mark.

Luminousness of the SEA. The luminousness of the sea is a phenomenon that has been noticed by many nautical and philosophical writers. Mr. Boyle ascribes it to some cosmical law or custom of the terrestrial globe, or at least of the planetary vortex. L'Abbé Nollet was long of opinion, that the light of the sea proceeded from electricity; and others have had recourse to the same principle and shown, that the luminous points in the surface of the sea are produced merely by friction. There are, however, two other hypotheses, which have more generally divided between them the solution of this phenomenon; the one of these ascribes it to the shining of luminous insects or animalcules, and the other to the light proceeding from the putrefaction of animal substances. L'Abbe Nollet, who at first considered this luminousness as an electrical phenomenon, having had an opportunity of observing the circumstances of it when he was at Venice in 1749, relinquished his former opinion, and concluded that it was occasioned either by the luminous aspect, or by some liquor or effluvia of an insect which he particularly describes, though he does not altogether exclude other causes, and especially the spawn or fry of fish. A similar conjecture is proposed by a correspondent of Dr. Franklin's, in a letter read at the Royal Society in 1756; the writer of which apprehends, that this appearance may be caused by a great number of little animals floating on the surface of the sea. And Mr. Forster, in his account of a voyage round the world with Captain Cook, describes this phenomenon as a kind of blaze of the sea; and having attentively examined some of the shining water, expresses his conviction that the appearance was occasioned by innumerable minute animals of a round shape, moving through the water in all directions, which show separately as so many lumiuous sparks when taken up by the hand; he imagines that these small gelatinous luminous specks may be the young fry of a certain species of some medusae or blubber. And M. Dagelat and M. Rigaud observed several times, and in different parts of the ocean, such luminous appearances by vast masses of different animalcules; and a few days after, the sea was covered near the coast with whole banks of small fish in innumerable multitudes, which they supposed had proceeded from the shining animalcules. But M. le Roi, after giving much attention to this phenomenon, concludes, that it is not occasioned by any shining insects, especially as, after carefully examining with a microscope some of the luminous points, he found them to have no appearance of an animal; and he also found, that the mixture of a little spirits of wine, with water just drawn from the sea, would give the appearance of a great number of little sparks, which would continue visible longer than those in the ocean: the same effect was produced by all the acids, and various other liquors. M. le Roi is far from asserting that there are no luminous insects in the sea, for he allows that several gentlemen have found them; but he is satisfied that the sea is luminous chiefly on some other account, though he does not so much as offer a conjecture with respect to the true cause. Other authors, equally dissatisfied with the hypothesis of luminous insects for explaining the phenomenon, which is the subject of this article, have ascribed it to some substance of the phosphoric kind, arising from putrefaction. See *Phil. Trans.* vol. 59. and FRANKLIN'S *Experiments* and *Observations*, page 274.

Motion of the SEA. M. Dassie has been at great pains to prove, that the sea has a general motion, independently of winds and tides, and of more consequence in navigation than is generally supposed. He affirms, that this motion is from east to west, inclining towards the north, when the sun has passed the equinoctial northward, and that during the time the sun is in the northern signs; but the contrary way after the sun has passed the said equinoctial southward: adding, that when this general motion is changed, the diurnal flux is changed also; whence it happens, that in several places the tides come in, during one part of the year, and go out during the other, as on the coasts of Norway, in the Indies at Goa, Cochinchina, &c. where, while the sun is in the summer signs, the sea runs to the shore; and when in the winter signs, runs from it. On the most southern coasts of Tonquin and China, for the six summer months, the diurnal course runs from the north with the ocean; but the sun having repassed the line towards the south, the course declines also southward. See *Phil. Trans.* N°. 135.

Saltness of the SEA. With regard to the saltness of the sea-water, it is very rationally judged to arise from great multitudes, both of mines and mountains of salt, dispersed here and there in the depths of the sea. The salt being continually diluted and dissolved by the water, the sea becomes impregnated with its particles throughout; and, for this reason, the saltness of the sea can never be diminished. Dr. Halley supposes, that it is probable the greatest part of the sea-salt, and of all salt lakes, as the Caspian Sea, the Dead Sea, the Lake of Mexico, and the Titicaca, in Peru, is derived from the water of the rivers which they receive; and since this sort of lakes has no exit or discharge but by the exhalation of vapours; and, also, since these vapours are entirely fresh, or devoid of such particles, it is certain, that the saltness of the sea and such lakes must, from time to time, increase, and therefore the saltness at this time is greater than at any time heretofore. He further adds, that if, by experiments made in different ages, we would find the different quantity of salt, which the same quantity of water (taken up in the same place, and in all other the same circumstances) would afford, it would be easy from thence, by rules of proportion, to find the age of the world very nearly, or the time wherein it has been acquiring its present saltness.

The Bishop of Landaff has recommended a most simple and easy mode of ascertaining the saltness of sea in any latitude: we insert it in his own words:

"As it is not every person who can make himself expert in the use of the common means of estimating the quantity of salt contained in sea-water, I will mention a method of doing it, which is so easy and simple, that every common sailor may understand and practice it; and which, from the trials I have made of it, seems to be as exact a method as any that has yet been thought of. Take a clean towel, or any other piece of cloth, dry it well in the sun, or before the fire, then weigh it accurately and note down its weight; dip it in the sea-water; and when taken out, wring it a little, till it will not drip when hung up to dry; weigh it in this wet state; then dry it either in the sun or at the fire, and when it is perfectly dry weigh it again. The excess of the weight of the wetted cloth above its original weight, is the weight of the sea-water imbibed by the cloth; and the excess of the weight of the cloth after being dried, above its original weight, is the weight of the salt retained by the cloth; and, by comparing this weight with the weight of the sea-water imbibed by the cloth, we obtain the proportion of salt contained in that species of sea-water.

"Who ever undertakes to ascertain the quantity of salt contained in sea-water, either by this or any other method, would do well to observe the state of the weather preceding the time when the sea-water is taken out of the sea, for the quantity of salt contained in the water near the surface may be influenced both by the antecedent moisture and the antecedent heat of the atmosphere." Whether the sea is salter or not at different depths, has not yet been properly ascertained.

Temperature of the SEA. "With respect to the temperature of the sea," says Bishop Watson, "at different depths, it seems reasonable enough to suppose, that in summer time it will be hotter at the surface than at any considerable depth below it, and that in winter it will be colder."

Mr. Wales describes the instrument he made use of for trying the temperature of the sea, at different depths, in the following terms: 'the apparatus for trying the sea-water, at different depths, consisted of a square wooden tube of about eighteen inches long, and three inches square externally. It was fitted with a valve at the bottom and another at the top, and had a contrivance for suspending the thermometer exactly in the middle of it. When it was used, it was fastened to the deep sea-line just above the lead, so that all the way as it descended, the water had a free passage through it, by means of the valves, which were then both open; but the instant it began to be drawn up, both the valves closed by the pressure of the water, and of course the thermometer was brought up in a body of water of the same temperature with that it was let down to.' With this instrument, which is much the same with one formerly described by Mr. Boyle, in his observations about the saltness of the sea, water was fetched up from different depths, and its temperature accurately noticed in different seasons and latitudes. August 27, 1772, south latitude, 24° 40′. The heat of the air was 72½, of the water at the surface 70, of water from the depth of 80 fathoms 68. December 27, 1772, south latitude 58° 24′. The heat of the air was 31, of the water at the surface 32, of water from the depth of 160 fathoms 33½.

In the voyage to the high northern latitudes before-mentioned, they made use of a bottle to bring up water from the bottom, which is thus described: 'the bottle had a coating of wool three inches thick, which was wrapped up in an oiled skin, and let into a leather purse and the whole inclosed in a well-pitched canvas bag, firmly tied to the mouth of the bottle, so that not a drop of water could penetrate to its surface. A bit of lead shaped like a cone, with its base downwards, and a cord fixed to its small end, was put into the bottle; and a piece of valve leather, with half a dozen slips of thin bladder, were strung on the cord, which, when pulled, effectually corked the bottle on the inside.' We have here put down two of the experiments which were made during that voyage.

"August 4, 1773, north latitude 80° 30′. The heat of the air was 32, of the water at the surface 36, of water fetched up from the depth of 60 fathoms under the ice 39. September 4, 1773, north latitude 65°. The heat of the air was 66½, of the water at the surface 55, of water from the depth of 683 fathoms 40. It appears from all these experiments, that when the atmosphere was hotter than the surface of the sea, the superficial water was hotter than that at a great depth; and when the atmosphere was colder than the surface of the sea, it is evident that the superficial water was somewhat colder than at a considerable distance below it."

A Long SEA, (*mer longue*, Fr.) implies an uniform and steady motion of long and extensive waves; on the contrary,

A Short SEA, (*mer courte*, Fr.) is when they run irregularly, broken, and interrupted, so as frequently to break over a vessel's bow, side, or quarter.

A Good SEA-*Boat*, (*bâtiment qui se comporte bien à la mer*, Fr.) a vessel that bears the sea firmly, without labouring heavily, or straining her masts and rigging. See the article LIFE-BOAT.

SEA-*Breeze*. See the article BREEZE.

SEA-*Chart*. See the article CHART.

SEA-*Coast*, (*côte de mer*, Fr.) the shore of any country, or that part which is washed by the sea.

SEA-*Cloths*, (*habits de bord*, Fr.) jackets trowsers, &c. See the article SLOPS.

SEA-*Fight*, (*combat naval*, Fr.) a contest at sea, in which the ships of two hostile fleets come to action, and endeavour to take, burn, sink, or otherwise destroy one another. To record the many sea-fights in which this country has been engaged, particularly with France, Spain, and Holland, would of itself occupy volumes. It will therefore be sufficient for us to state, that, although the majority of our military operations, till within these last few years, have not been considered equal to the valour of our troops, yet our naval contests have seldom failed to answer the most sanguine expectations of the country.

SEA-*Laws*. The code of marine laws in this country is so extensive and important, that in a work of this nature a reference can only be given to it. These laws, with the Naval Instructions, are furnished by the Admiralty to all captains requiring them for their guidance at sea. There is also a very useful compendium, entitled, "The Spirit of Marine Law," edited by John Irving Maxwell, Esq. of the Honourable Society of the Inner Temple, London, for those concerned with the merchant-service.

SEA-*Legs*, (*pied marin*, Fr.) implies the capacity of walking on a ship's decks when she pitches or rolls about at sea.

SEAMAN, in opposition to landman, a person brought up to the sea-service. See the article ABLE *Seaman*.

SEAMAN'S *Friend, and Bather's Companion*, a name given by Mr. Mallison to his invention for the preservation of shipwrecked mariners.

The power of this simple invention to effect the above desirable purpose having been confirmed by a decision of the House of Commons, on the committee of which were three lords of the Admiralty, seems a sufficient reason for introducing the following particulars:

It consists of from two to four solid pieces of cork, in the shape of a parallelopipedon, or long square, the size being in proportion to the person who is to make use of it; one acts as a breastplate, the other as a back-board, connected over the shoulders on each side the throat by two strong pieces of webbing, and applied to the body by throwing it over the head; the front even with the axilla or arm-pit, the back to correspond. It is fastened on the body by tying the strings on each side, and bringing the long strap between the legs, which fasten in front. When this is done, the person and the invention are so completely identified, that you may perform every action in the water you choose. You swim at will, by making use of those actions of the arms and feet that enable you to go through the water. You turn as well as on a feather-bed, and float. When you take the horizontal position, it gives the least possible resistance in going through the water, and completely leaves the wearer the free use of his arms, legs, and thighs; you may fight in the water nearly as well as you can on land; indeed, so perfectly is the end answered, that in wreck, should faintness be induced through fatigue, or the individual fall a-sleep through weariness, his life will be preserved by the depending gravity of his legs and thighs, keeping the body in an upright position, while his head and throat are continually out of water, it enables him to use his feet when he feels the bottom, his hands when any thing presents itself to assist him, or he wishes to assist others; and the shield it gives his body prevents, in a great measure, the dreadful consequences of being thrown against the rocks or shores. The pieces of cork are covered with different substances, from green baize to elastic cloth or kerseymere, and even with silk. They are sowed on the covering, and then through the cork, by packthread, which renders it nearly impossible to be torn off, whatever force is used; and the pieces of cork, by being solid, prevents its being destroyed.

Reference to the Parts of the Seaman's Friend.

AA, fig. 4, plate XXIII. represent four strong pieces of tape or webbing, connecting the invention together.

BB, two ditto to fasten at the sides.

CC, two straps which fasten in front, and completely identify the person and the invention when used.

The advantage of this invention, over all others, for a similar purpose, consists in its durability, portability, cheapness, and simplicity; with common care, it will last a man his life-time; it enables him to get through the water nearly as well as the best swimmer, and allows a poor man the advantage of possessing them as well as the rich.

Directions for Using the Seaman's Friend.

Enter the water breast-high; lift up your feet, and you will find yourself supported, with your head and throat above the water: you may then fearlessly strike into deep water,—swim and float at your pleasure,—as it is impossible to be drowned. If in a boat, fasten a running noose round one of your arms, and give the end of the rope to an attendant; let yourself gradually into the water, and you will find yourself supported as above; cast off the rope, and you may swim or float at will.

The report of the select committee, appointed by the House of Commons in June 1811, to enquire into the merits of this invention, is as follows:—

Resolved unanimously, that the invention of Mr. Mallison is well deserving public attention, and is likely to prove useful in the British Navy, particularly to boats crews in going from and returning to ships in stormy weather, as the person wearing it is not materially impeded in the use of his limbs in rowing, walking, &c. See MALLISON'S *Seaman's Friend, and Bather's Companion.*

SEA-*Mark*, (*amer*, Fr.) a point or conspicuous object distinguished at sea; they are of various kinds, as, promontories, steeples, ruins, trees, &c. and are very beneficial in directing vessels on the coast, and of pointing out their situation. See the article BEACON.

SEAMEN, or SEA-*Faring Men*, (*marins; hommes de mer*, Fr.) are persons trained to the occupation of MARINERS or SAILORS, which articles see.

The relative duties of sea-faring men being of a laborious and dexterous nature, can only be acquired by industrious application; and commercial encouragement seems best calculated to promote its interests, and secure an excellence in maritime affairs.

Previous to the reign of Edgar, England had little share in foreign traffic, resting contented with her domestic produce. But, by the indefatigable exertions of that great prince, in training and augmenting the number of mariners, England became powerful in the maritime world; and although it dwindled in succeeding reigns, yet the spirit of enthusiasm again revived to supply her with exotic delicacies, and guard her from the ravages of foreign intrusion. At this period, no class of men whatever, taken as a body, can possibly exhibit greater attachment to the interests of their country than the British seamen: when encountering the fury of battle, their minds are always elevated and ennobled, and with zeal they check the progress of insatiable and tyrannical ambition. The honour, courage, and power, in the noble exploits of the British Navy, reflect the brightest eulogium in the annals of history, and by their achievements the British Empire stands unrivalled in the scale of political importance.

Laws respecting SEAMEN. Every seaman, who ships himself on board a merchantship for any voyage, shall be obliged to sign an agreement for wages within three days, 2 Geo. II. c. 36. s. 2. And, by the act of the 31 Geo. III. c. 39, this regulation is extended to the coasting trade.

If any seaman desert, or refuse to proceed on the voyage, after signing such contract, he shall forfeit to the owners all the wages due to him. If any seaman absent himself from such ship after signing, the master may apply to any justice for a warrant to apprehend him; and, if he shall refuse to proceed on the voyage, without sufficient reason, the justice shall commit him to the house of correction, 2 Geo. II. c. 36. s. 3 and 4.

If any seaman shall absent himself from the vessel to which he belongs without leave, he shall, for every day's absence, forfeit two days pay to the use of Greenwich Hospital, 2 Geo. III. c. 36, s. 5.

If any seaman leave the vessel to which he belongs, without receiving his discharge in writing from the master, unless he enters into his Majesty's service, he shall forfeit one month's pay, 2 Geo. II. c. 36. s. 6.

On arrival in Great Britain, the master shall pay the seamen their wages, if demanded, in thirty days after entry at the Custom house, or at the time the seamen shall be discharged (whichever shall happen first), deducting therefrom the above-mentioned penalties, under the forfeiture of 20*s*. to every such seaman unpaid. 2 Geo. II. c. 36, s. 7.

If any mariner or inferior officer of any merchant ship shall decline or refuse to defend the ship, or utter words to discourage others from doing so, every mariner so behaving shall lose all his wages, together with such goods as he may have in the ship, and suffer imprisonment for a time not exceeding six months. And every mariner, who shall have laid violent hands on his commander, to hinder him from fighting in defence of his ship and goods, shall suffer death as a felon, 22 and 23 Car. II. c. 11, s. 7 and 9.

If any master, mariner, &c. shall wilfully cast away, burn, or destroy his ship, or procure the same to be done, he shall suffer death as a felon, 22 and 23 Car. II. c. 11, s. 12.

Masters may reimburse themselves out of the wages of their mariners, for losses happening by their negligence. Raymond's Rep. p. 650.

Such seamen as are reserved to serve the King, or other persons, at sea, are not to depart without license, &c. Seamen fighting, quarrelling, or making any disturbance, may be punished by the commissioners of the navy with fine and imprisonment. Registered seamen are exempted from serving in any parish office, &c. and are allowed bounty money besides their pay. By the law of merchants, the seamen of a vessel are accountable to the master or owners, and the owners to the merchants, for damage sustained, either by negligence or otherwise. When a seaman is hired for a voyage, and he deserts before it is ended, he shall lose his wages; and in case a ship be lost by tempest, or in a storm, the seamen lose their wages, as well as the owners their freight.

By various statutes, sailors having served the King for a limited time, are free to use any trade or profession, in any town of the kingdom.

By 2 Geo. II. c. 36, made perpetual by 2 Geo. III. c. 31, no master of any merchant-ship or vessel shall carry to sea any seaman, his own apprentice excepted, without first entering into an agreement with him for his wages, such agreement to be made in writing, and to declare what wages such seaman is to receive during the whole of the voyage, or for such time as shall be therein agreed upon; and such agreement shall also express the voyage for which such seaman was shipped to perform the same, under a penalty of £5. for such mariner carried to sea without such agreement, to be forfeited by the master to the use of Greenwich Hospital, Sect. 1. This agreement is to be signed by each mariner within three days after entering on board such ship, and is, when executed, binding on all parties, Sect. 2.

Mariners deserting after signing such contract or agreement, to forfeit their wages, Sect. 3.

Upon the arrival of any merchant ship or vessel into Great Britain, from parts beyond the seas, the masters or commanders of such ships are obliged to pay the seamen and mariners on-board their wages, if demanded, in thirty days after the said ships or vessels have entered at the Custom House (except in cases where a covenant shall be entered into to the contrary), or at the time the said seamen and mariners shall be discharged, which shall first happen, if demanded, deducting out of such wages the penalties and forfeitures by this act imposed, under the penalty of paying to each seaman or mariner so unpaid, 20*s.* over and above the wages due, Sect. 7.

All forfeitures incurred and appropriated to Greenwich hospital, to be paid within three months, Sect. 10.

Nothing in this act contained shall extend, or be construed to extend, to debar any seaman or mariner belonging to any merchant ship or vessel, from entering, or being entered, into his Majesty's service; nor shall such seaman or mariner, for such entry, forfeit the wages due to him, during the time of his service in such merchant ship or vessel; nor shall such entry be deemed a desertion, Sect. 13.

For information relative to our marine laws, see a very useful compendium [originally compiled, digested, and arranged, to the year 1792, by John M'Arthur, Esq. to whom the navy is much indebted for many improvements in naval tactics, &c. and much information on naval courts martial] published by John Irving Maxwell, Esq. a work that contains the spirit of the marine laws, sufficiently dilated to be practically useful to all persons in any degree connected with shipping: and, at the same time, it will serve as a copious index to those who may wish to consult any particular act.

SEA-*Port*, (*port de mer*, Fr.) a haven near the sea, as distinguished from one that is situate upon a river. See the articles HARBOUR and PORT.

SEA-*Room*, (*belle dérive*;—*de l'eau à courir*, Fr.) implies a sufficient distance from the land, rocks, or shoals, wherein a ship may drive or scud without danger of shipwreck.

SEA-*Sickness*, (*mal de mer*, Fr.) is a peculiar disorder, to which most passengers, &c. are, in some degree, subjected, whilst at sea. Although this malady is not attended with fatal consequences, and may even, in certain cases, be beneficial to health, yet, during its continuance, few disorders are more distressing and irksome to the patient.

Medical gentlemen being themselves equally exposed to the effects of this disorder with the rest of the passengers, they have not been able to make exact observations on the symptems which precede and accompany it, nor satisfactorily to ascertain the result of their prescriptions. It will not, therefore, it is presumed, be deemed arrogance, to offer a few remarks on the subject, founded upon experience and long acquaintance with the sea.

The seas in which this disorder attacks the passenger with the greatest violence, are those where the waves have a long uninterrupted freedom of action; of course, bays, gulphs, and channels, may be navigated with less inconvenience, as the waves, meeting with more frequent resistance, and the repercussion being considerably stronger, the vessel does not experience that gentle, uniform vacillation, which sickens the stomach and renders the head giddy. By the same argument, a person feels less inconvenience from the disorder in the wide ocean in a small vessel, on which the slightest motion of the waves makes a strong impression. He is likewise still less exposed to it in a very large vessel, as in a ship of the line, or a large merchantman deeply laden, as the waves in this case scarcely affect the vessel. It is in ships of the middling size, and which carry but a light cargo, that the passenger suffers most from sea-sickness. It has been observed, that this disorder affects people in years less than young persons; those of a dark, less than those of a fair, complexion; and that it seldom attacks infants. The duration is not limited to any fixed period of time; with some it lasts only a few days, with others weeks, months, and even during the whole course of a voyage; even the oldest experience a relapse, especially if they have quitted the sea-service for a long term of years.

The following may be recommended as the most efficacious modes of precaution for preventing, or at least mitigating, the effects of the disorder:—

1st. Not to go on-board immediately after eating; and when on board, not to eat any great quantity at any one meal.

2nd. To take strong exercise, with as little intermission as conveniently can be done: as indolent and slothful passengers always suffer most from the disorder.

3d. To keep much upon deck, even in stormy and rainy weather, as the sea-breeze is less liable to affect the stomach than the stagnated air of the cabin, which is frequently rendered infectious for the want of sufficient circulation.

4th. Not to watch the motion of the waves, especially when strongly agitated with tempest.

5th. To avoid carefully all employments which harass the mind, as reading, study, meditation, and gaming; and, on the other hand, to seek every opportunity of mirth and mental relaxation.

6th. To drink occasionally carbonic acids, as the froth of strong fermented beer, or wine mixed with Seltzer water, and fermented with pounded sugar, or a glass of champaign.

7th. It will be found of great service to take the acid of sulphur dulcified, dropped upon lump-sugar, or in peppermint-water; or ten drops of sulphureous ether.

With regard to eating, it is advisable to be very sparing, at least not to eat much at one meal. The proper diet is bread and fresh meat, which should be eaten cold with pepper. All sweet flavoured food should be carefully avoided; and the passenger should refrain from fat, but especially from all meat that is in the least degree tainted. Even the odour of flowers is very pernicious; for which reason, it is not expedient to examine marine productions, as these generally have a nauseating smell. The fumes of vinegar may be inhaled with great benefit. The drink should consist of tart wines, lemonade, or Seltzer water. The passenger would do well to drink little and often.

Where the above preventives have not been employed, or have not succeeded in securing the passenger from sea-sickness, he may, however, experience considerable relief from the following:—

1st. If symptoms of vomiting appear, they may frequently be remedied by the patient prostrating himself in a horizontal position, upon his back or belly, and lying perfectly still. But if the fits of vomiting are too violent to be repressed, in that case it is best to promote them by a strong dose of salt-water, an expedient, however, which must not be too often repeated, as it tends still more to weaken the stomach. When the emetic takes effect, let the patient bend his body, advancing his knee towards his breast, and support his head against a firm and solid resting-place: he must be particnlarly careful to untie his garters and neckcloth; as this precaution will secure him from the risk of a rupture, and from the ill effects of the blood rushing violently towards the head and breast.

After the vomiting has subsided, its return may be guarded against by preserving a state of repose, and even keeping the eyes shut for a considerable time. Let the patient choose a cool ventilated place, remembering to keep himself warm and well clothed, as perspiration is highly salutary. But he must not indulge in too long sleep during the day-time, as this induces torpidness. In the morning, he should constantly take a gargle of sugar dissolved in vinegar. Let him eat often, but sparingly; and if he can content himself with a dish of chocolate, coffee, or strong tea, he will reap still greater benefit. He should never drink water in its pure elementary state, but mix it with brandy, vinegar, or wine. In the morning, instead of brandy, he may take a glass of wine, with an infusion of orange peel or Peruvian bark. A glass of punch taken occasionally will prove of very essential service, as it promotes perspiration; it is proper to add, that warm clothing, flannel-shirts, &c. are efficacious remedies against excessive expectoration, and all other symptoms of this terrible disorder.

Sea-*Weeds*, (*herbes et plantes marines*, Fr.) a sort of herbs or tangles floating on the surface of the sea, or washed upon the sea-coast.

Sea-*Worthy*, (*bâtiment en bon état*, Fr.) implies the state or quality of a ship, which is in every respect fit for a voyage.

Every ship assured must, at the time of the assurance, be able to perform the voyage, unless some external accident should happen; and if she has a latent defect, though wholly unknown to the parties, that will vacate the contract, and the assurers are discharged. This doctrine is founded upon that general principle of assurance law, that the assurers shall not be responsible for any loss arising from the insufficient or defective quality or condition of the thing assured.

But, although the assured ought to know whether the ship was sea-worthy or not at the time she set out upon her voyage, yet he may not be able to know the condition she may be in after she is out a twelvemonth; and, therefore, whenever it can be made appear, the decay to which the loss is attributed did not commence till a period subsequent to the assurance, as she was sea-worthy at the time, the underwriter would be liable.

In the case, Eden *v.* Parkinson, Doug. 708, the same principle was much relied upon. Lord Mansfield said, "By an implied warrantry, every ship assured must be right, staunch, and strong; but it is sufficient if she be so at the time of her sailing. She may cease to be so in twenty-four hours after departure, and yet the underwriter will continue liable." Every case of this kind, it is true, must depend upon its own circumstances; but, when they are once ascertained, the rule of law is clear and decisive.

Sea-worthiness being an implied condition in the contract of assurance, it is not necessary there should be any previous representation of the condition of the ship, unless particularly called for; because, unless she be fit for the performance of the voyage assured, there is no binding contract. Shulbred *v.* Nutt; Sittings at Guildhall after Hilary, 1782.--- Haywood and another *v.* Rodger; Term Rep. Hilary, 44 Geo. III.

SEAMS, (*coutures entre les bordages*, Fr.) the intervals between the edges of the planks in the decks and sides of a ship, or the places where the planks join together. These are always filled with a quantity of oakum, and covered with hot pitch, to prevent the entrance of the water. See the article Calking.

Seams, in mast-making, the openings between the pieces that compose made masts, yards, &c. into which oakum is driven.

Seams *of the Sails*, (*coutures des voiles*, Fr.) in sail-making, the parts where two edges of canvas are laid over each other and sewed down.

SEARCHER, an officer of the customs, whose business it is to search and examine, all ships that are outward-bound, to see whether they have any prohibited goods on-board.

Searcher, (*griffe de chat*, Fr.) in gunnery, an iron instrument hollowed at one end to receive a wooden handle, and at the other there are from four to eight flat springs, of about eight or ten inches long, pointed, and turned outwards at the ends. It is used to search a gun after it has been fired, in order to discover if it be defective or not.

SEASONING of Timber, is the preparing of it for use by a natural process, after being felled, &c. at a proper season of the year, either by piling it in stacks and interposing blocks of wood between, or placing it on-end under sheds, or strewing it horizontally on the ground; by siding or squaring the tree when it is felled, or by leaving the outside, or *sap*, as it is called by workmen, to protect the more central parts from cracking.

No discovery has yet been made, that we know of, of a speedy method of seasoning timber by artificial means, though numerous experiments have been tried with that view, such as charring, burning, stoving, burying in unslaked lime, or sand; coating with paint, tar, and compositions of various kinds, the effects of which have generally been found pernicious, by tearing the fibres asunder; or favourable to the process of decay, by driving the moisture into the interior parts of the wood, and preventing its escape.

Various attempts have also been made to destroy the principle of vegetation, by impregnating timber with chemical preparations, all of which have failed; and the experiments are found to be too tedious and expensive to be applied on so large a scale as is required for timber employed in building ships of war: moreover, all acid impregnations would be destructive of iron, which is but too liable to corrosion by the acid of the timber in its natural state while it lies buried in it.

Experiments, we understand are now making at

Woolwich on the speedy seasoning of timber, by stowing some hundred loads in a close kiln, and introducing, by means of a retort filled with saw-dust, an oleaginous substance, which, by penetrating into the pores of the timber, as the aqueous or sappy particles are driven off by heat, prevents the fibres from being torn asunder, and preserves the logs from cracking. The idea is ingenious, and we wish the desired effects may accrue from the experiments themselves.

It has not, however, been discovered from the closest inspection, that any artificial process was adopted in seasoning the timber of the Royal William, built about the year 1676, and only broke up in the summer of 1813, or any other of our most durable ships; and the sound state of timber found in old cathedrals, churches, and mansions, after many centuries, afford the most ample proof that nothing more is necessary to secure its durability than proper seasoning, wet and damp, and exposure to the common air, excluded.

It is not enough, however, that the timber which enters into the construction of ships be thoroughly seasoned; it should be kept as much as possible free from moisture.

The alternate exposure to water and air, to heat and cold, will sow the seeds of decay in the best oak how well soever it may be seasoned. It would be in vain, therefore, to expect any very protracted period of duration in ships which, while building, have been exposed for four, five, or six years to all the vicissitudes of the weather. Yet this exposure, water-soaked at one season, split by the heat of the sun at another, and the crevices filled with ice at a third, is called seasoning! Would it not be more proper to term it *a mode of destroying oak timber by the common rot in the least possible time?* The fact is well known in our dock-yards, that some of the timbers which have been set up the first year, are found so far gone in the third or fourth as to make it necessary to replace them; therefore, instead of six years exposure on the stocks, with the bottom of the ship water-soaked in the slip, and the upper part splitting in the sun, or cracking with the wind, twelve months would be quite sufficient to complete a ship, when the necessary quantity of dry and duly seasoned timber was collected and prepared on the spot. See the article SHIP-BUILDING.

SEASONS, are certain portions or quarters of the year, distinguished by the signs, which the sun then enters. Upon them depend the different temperatures of the air, different works in tillage, &c.

The year is divided into four seasons, viz. spring, summer, autumn, and winter, which take their beginning when the sun enters the first signs Aries, Cancer, Libra, and Capricorn.

The seasons are illustrated by fig. 1. plate XXIV. where the candle at I represents the sun in the centre, about which the earth moves in the ecliptic ABCD, which cuts the equator *abcd* in the two equinoxes E and G. When the earth is in these two points, it is evident that the sun equally illuminates both the poles, and makes the days and nights equal all over the earth. But while the earth moves from G by C to ♑, the upper or north pole becomes more and more enlightened, the days become longer, and the nights shorter; so that when the earth is at ♑, or the sun at ♋, our days are at the longest, as at midsummer. While the earth moves from ♑ by D to E, our days continually decrease, by the north pole gradually declining from the sun, till at E or autumn they become equal to the night, or 12 hours long. Again, while the earth moves from E by A to F, the north pole becomes always more and more involved in darkness, and the days grow always shorter, till at F or ♋, when it is mid-winter to the inhabitants of the northern hemisphere. Lastly, while the earth moves from ♋ by B to G, the north parts become more and more out of darkness, and the days grow continually longer, till at G the two poles are equally enlightened, and the days equal to the nights again; and so on continually year by year.

SECANT, in geometry, is one line cutting another, whether rectilineal or curved.

SECANT, in trigonometry, is a right line drawn from the centre of a circle through the circumference, till it meets with a tangent to the same circle.

SECRETARY, (*secrétaire*), Fr.) a person, who, by the directions of his employer, writes letters, dispatches, orders, &c. which he renders authentic by obtaining the admiral's signature. In the naval department there are,

	per Annum. £.
1st Secretary to the Admiralty, whose salary is	4000
2d Ditto .... ditto	2000
Ditto to an admiral and commander in chief of the fleet	500
Ditto to an admiral commanding in chief	400
Ditto to a vice or rear-admiral, or to a commodore commanding in chief	300
Ditto to flag-officers not commanding in chief, as commodores and captains of the fleet	150

The situation of a secretary to an admiral and commander in chief, &c. was formerly held by the purser of the flag-ship, but now frequently by persons unconnected with the service. Pursers, whose ships are in ordinary, are, however, still employed as secretaries, but those in sea-going ships are not allowed to act in that capacity.

To enable a secretary to receive his pay, it is necessary for him to produce a certificate from the flag-officer he has served, of the time he was actually employed, and of his diligence, sobriety, and good conduct.

SECTOR, a mathematical instrument, consisting of two rulers or legs made of box, ivory, or brass; it is of great use in geometry, trigonometry, surveying, &c. in measuring, laying down, and finding proportional quantities of the same kind: as between lines and lines, surfaces and surfaces, &c. whence the French call it the compass of proportion.

The great use of the sector above the common scales, &c. is, that it is contrived so as to suit all radii, and all scales. By the lines of chords, sines, &c. on the sector, we have lines of chords, sines, &c. to any radius between the length and breadth of the sector when open.

The sector is founded on the fourth proposition of the sixth book of Euclid; where it is demonstrated that similar triangles have their sides proportional.

SECTOR *of a Circle*, is a portion comprehended between two radii, or semidiameters, and the arc of the circle, making an angle at the centre; and an arc or part of the circumference.

SECTOR *of a Sphere*, is the conic solid, whose vertex ends in the centre of the sphere, and its base is a segment of the same sphere.

Equatorial or *Astronomical* SECTOR, an instrument invented by the late ingenious artist Mr. George Graham, for finding the difference in right ascension and declination between two objects, whose distance is too great for the field of view to be observed through a fixed telescope by means of a micrometer; the first instrument made of this sort is now in the royal observatory at Greenwich. But however well adapted this instrument may be to the purposes for which it was intended, yet it will not conveniently admit of a large telescope; Dr. Maskelyne, therefore, thought of a construction which would admit one of a larger size, and which has, besides, several other advantages in respect to the adjustments, which Mr. Sisson executed for him.

With this instrument adjusted, any object in the heavens may be found, whose right ascension and declination are known, and its path traced out; the telescope may also be directed to a comet, or any other celestial body whose place is known, though invisible to the naked eye; if there be not a known fixed star convenient for the purpose, compare the body with one whose place is not known, and afterwards settle its place by taking its right ascension and declination at its passage over the meridian. The convenience of this instrument is, that we can at any time determine the place of a body without waiting for its coming to the meridian; where, besides the inconvenience of waiting, an observation might be hindered by the badness of the weather, or the body might come to the meridian in the day time when it could not be seen.

When the telescope is directed to any star, it is evident, that if the polar axis be turned about, the telescope will describe the star's parallel of declination, consequently it will follow the star in its apparent diurnal motion; this instrument, therefore, becomes a *Paralactic Instrument*, which is an instrument to direct a telescope at any time to any heavenly body whose right ascension and declination are known, and by turning it about a polar axis to keep the star always in the field of view, all which are performed by this instrument. For the adjustments and a particular description, see Vince's Practical Astronomy.

Zenith SECTOR. The first instrument of this kind was made by Dr. Hook, in order to determine the annual parallax of the fixed stars, as from its construction it is capable of measuring small angles with greater accuracy than the quadrant. It was adapted only to measure small angular distances from the zenith, and from thence it was called a *Zenith Sector.* The length of his instrument was 36 feet, which, from the manner of its construction and the care he took in rectifying it, promised great exactness; notwithstanding which, it appears that it was liable to an error of 30″, and therefore totally unfit for the purpose for which it was intended. This attempt, however, induced Mr. Molyneux to have another instrument constructed, upon nearly the same principles, about $24\frac{1}{4}$ feet long, which was executed by Mr. Graham, to a much greater degree of accuracy. This instrument being originally designed to observe γ *Draconis*, was so constructed as to be capable of measuring only 7 or 8 minutes, and there being but a few stars within half that distance from the zenith large enough to be observed, he could not thoroughly examine how stars differently situated in respect to the equinoctial and solstitial points of the ecliptic would be affected. This consideration induced Dr. Bradley to get another instrument made on the same construction, and to have an arc of $12\frac{1}{2}^{\circ}$ fixed to it, by which means he could measure $6\frac{1}{4}^{\circ}$ on each side of the zenith, which gave him an opportunity of making choice of several stars both in magnitude and situation, there being more than 200 in the British catalogue, which could be observed by it. The arc was extended so far in order to take in *Capella*, the only star of the first magnitude near his zenith; the length of his tube was about $12\frac{1}{2}$ feet, which he judged would be long enough to adjust the instrument to a sufficient degree of accuracy; and he observed, "I have had no reason since to change my opinion; for, from all the trials I have yet made, I am very well satisfied that when it is carefully rectified, its situation may be securely depended upon to half a second." With this instrument Dr. Bradley made his two admirable discoveries, the *Aberration of Light* in the fixed stars, and the *Nutation of the Earth's Axis.* This instrument is now in the Royal Observatory at Greenwich.

SECULAR EQUATIONS, in astronomy, are corrections required to compensate such inequalities in the motion of the heavenly bodies as are found to occur in the course of a century. Thus there are secular inequalities in the motion of the moon, which require for their correction so many distinct secular equations: and the like may be said of other bodies. See the treatises on astronomy by VINCE and WOODHOUSE.

SECURE YOUR GUNS! See the article EXERCISE.

To SEE, (*voir*, Fr.) a term applied by seamen to the discovery of any strange sail, or the land from the mast-head; as, "we *see* a strange sail to windward, and land to leeward."

SEGMENT OF A CIRCLE, that part of a circle cut off by a chord, or a portion comprehended by an arc and its chord; and may be either greater or less than a semicircle.

SEGMENT, in geometry, is a part cut off the top of a figure by a line or plane; and the part remaining at the bottom is called a frustum, or zone.

SEINE, the name of a large fishing-net.

To SEIZE, (*eguilleter-faire un amarrage-frapper un cordage*, Fr.) is to bind or fasten any two ropes, or different parts of one rope together, with a small line or cord.

SEIZING, (*amarrage*), Fr.) a name given by seamen to the cord which fastens any thing together.

SEIZURE, an arrest of some merchandise, moveable, or other matter, either in consequence of some law, or of some express order of the sovereign.

Contraband goods, those fraudulently entered, or landed without entering at all, or at wrong places, are subject to seizure. In seizures, among us, one half goes to the informer, and the other half to the king.

SELVAGEE, (*estrop, ou skirre, servant à rider les haubans et étais*, &c. Fr.) a sort of hank or skein of rope-yarn turned into a circular form, and marled together with spun-yarn. It is used to fasten round any rope, as a shroud or stay, so that a tackle may be hooked in it, to extend the said shroud or stay, in order to set it up.

SEMIDIAMETER, a term synonymous with radius, a right line drawn from the centre of a circle or sphere to its circumference. The distances, diameters, &c. of the heavenly bodies, are usually estimated by astronomers in semidiameters of the earth; and the distance of the secondary planets from their respective primary ones, by semidiameters of the body of the primary planet.

Suppose the semidiameter of the earth to be unity, then the measures of the sun and planets will be as follow:—

	Semidiam.		Semidiam.
The Earth ...	1.	Mars	0.65
The Sun.......	111.25	Jupiter	11.81
The Moon....	0.27	Saturn	9.77
Mercury	0.38	Herschel or GeorgiumSidus	4.32
Venus.........	1.15		

To SEND, (*tanguer*), to pitch precipitately into the hollow or interval between two waves; as, "Every time the vessel sends, the topmasts complain."

SENNIT, (*éguillette, ou tresse de fils de caret*, Fr. from *seven* and *knit*,) a sort of flat braided cordage, used for various purposes, and formed by plaiting five or seven rope-yarns together.

SENTENCE OF A NAVAL COURT-MARTIAL, decision; final judgment. The sentence of a naval court-martial is drawn up by the judge-advocate, under the directions of the court, and signed by the members by way of attestation; it is then countersigned by the judge-advocate, and pronounced by him *instanter*, in open court: but in the army, the members of a court-martial are sworn not to divulge the sentence till it shall be approved by his Majesty, or by the commander in chief. See the articles COURT-MARTIAL and REPRIMAND.

SENTINEL, (*sentio*, Lat. *sentinella*, Ital. *centinelle*, Fr.) a private marine, from the guard-party, posted in various parts of a ship of war to prevent any surprise, &c. and to stop such as would pass without order.

No person is to strike or abuse a sentinel on his post; but when he has committed a crime, he is to be relieved, and then punished according to the rules and articles of war. See also the article CENTRY, under which the principal part of this article is given.

SERPENS, in astronomy, is a constellation in the northern hemisphere, being one of the 48 old constellations mentioned by all the ancients, and is called more particularly Serpens Ophiuchi, being grasped in the hands of the constellation Ophiuchus. The stars in this constellation, in Ptolemy's catalogue are 18, in Tycho's 13, in Havelus's 22, and in the Britannic catalogue 64.

SERPENTARIUS, in astronomy, a constellation of the northern hemisphere, being one of the 48 old constellations mentioned by all the antients. It is called also Ophiuchus, and anciently Æsculapius. It is in the figure of a man grasping a serpent. The stars in this constellation in Ptolemy's catalogue, are 29, in Tycho's 15, in Hevelius's 40, and in the Britannic catalogue 74.

SERVANTS, in the royal navy, are boys generally chosen from the second and third classes on board his Majesty's ships, for the purpose of waiting upon the officers on board, and of accompanying them on shore, when they find it necessary. For the allowance to commissioned and warrant officers, for servants, see the articles COMPENSATION and PAY.

By an order of council, dated April 16th, 1794, a compensation is directed to be made to captains, lieutenants, and warrant-officers, equal to the nett wages of the number of servants to which they were respectively entitled, at the rate of 11*l* 8*s*. 2*d*. *per annum* for each servant: and that no boys should be allowed to be in future borne on the ship's books under the denomination of servants to the captains, lieutenants, or warrant-officers; but instead thereof, a certain number (equal, or as near as may be to three-fourths of what were then allowed as servants) to be borne on separate lists, after the ship's company, and divided into CLASSES, which article see in the APPENDIX.

To SERVE, (*servir*, Fr.) in a naval sense, implies to do duty as an officer or seaman on board ship.

To SERVE *a Rope*, (*fourrer un cordage*, Fr.) is to wind round spun-yarn, &c. by means of a mallet, to prevent it from being rubbed. The materials used for this purpose, are called Service.

SERVICE, (*fourrure, soit de toile, ou de vieux cordages*, &c. Fr.) a term given to all sorts of stuff, whether of old canvas, mat, plat, hide, parcelling, spun-yarn, &c. when put round cables, &c. in order to preserve them from being chafed by the strain of the ship, or otherwise.

The line or spun-yarn being wound up in a ball, two or more turns are taken from it round the rope, confining the end under the turns; the mallet is then placed on the rope, and two or more yarns are passed round the rope and mallet, and round the handle; then, turning the mallet, whilst another person passes the ball round the rope, it leaves the spun-yarn on the rope, and draws it tight.

Mooring-SERVICE, is that which is at the first splice, when a ship is moored, and rides at one cable's length.

SERVICE, (*service*, Fr.) denotes also every species of naval duty which is done by an inferior under the influence and command of a superior. It also means exploit, achievement, and points out the particular profession to which a man belongs; as, land-service, sea-service, and the degrees of knowledge which he may have acquired by practice, viz. "he has seen a great deal of service."

SERVICE, implies also the period during which a

man has done duty, or followed the naval profession in an active manner.

To see SERVICE, is to be in actual contact with an adversary at sea, &c.

To retire from the SERVICE, (*se retirer du service*, Fr.) to quit the navy, or resign with or without the advantage of being benefited by an annual allowance of pay.

No officer can resign his commission, retire from the service, or remove to a foreign country, without having previously obtained permission from the lords commissioners of the Admiralty. See also the article To RETIRE.

Home SERVICE, the naval duty which is carried on within the limits of the United Kingdoms of Great Britain and Ireland.

Foreign SERVICE, in contradistinction to home service, is the duty done at any distant place, or on a foreign station.

SERVICEABLE, capable of performing all necessary naval duty; also fit for use, as serviceable stores, &c.

SERVING-BOARD, in sail-making, a small piece of board, seven or eight inches long, and three inches broad, tapering to one end as a handle: it has a small notch or groove cut in the middle of the broad end, and one or two on the sides, in which the spun-yarn is twisted; its use is the same as the mallet, but for small ropes only.

SERVING-*Mallet.* See the article MALLET.

To SET, (*relever un objet avec la boussole*, Fr.) is to observe the bearings or situation of any distant object by the compass, in order to discover the angle which it makes with the nearest meridian. Thus, to set the land, or the sun, by the compass, is to observe how the land bears on any point of the compass, or on what point of the compass the sun is. Also, when two ships sail in sight of one another, to mark on what point the chased bears, is termed, to set the chase by the compass. See the article BEARING.

To SET, (*signifie aussi la direction du vent, de la marée, d'un courant, de la lame*, &c. Fr.) is also used with regard to the direction of the wind, current, or swell of the sea, but particularly the two latter: as, the tide which sets to the south, is opposed to a swelling sea setting to the north-west.

To SET *the Sails*, (*déployer les voiles, ou les mettre au vent*, Fr.) is to loosen and expand them, so as to force the ship through the water; or to accelerate her velocity when she is moving, and perhaps give a new direction to her motion. It is used in contradistinction to taking the sails in.

To SET *a Main Sail.* Before the sail is loosed, let the double-block of a tackle be made fast to the weather-clew, and the single block be hooked low down upon the chess-tree, and the fall led aft; then man well the main tack and fall at the same time; and, when the sail is loosed, ease away the weather clew-garnet, let go the bunt-lines and leech-lines, bowse down upon the tackle, and take in the main tack; the main tack being down, haul aft the sheet, brace up the yard, and haul the main bow-line.

To SET *a Fore-Sail.* A fore-sail is set after the same manner as a main-sail; but as the fore-tacks generally lead double, they are a sufficient purchase without the aid of a tackle to the weather clew.

To SET *a Top-Sail.* Let a tackle be in readiness to clap on either sheet, as may be required. First, man the lee-sheet; and, the sail being loosed, ease down the bunt-lines and weather clew-line, and haul home the lee sheet; then the weather sheet, hoist the sail, and brace up as required. Should the wind be quartering, the lower and top-sail yards should be braced well into the wind, before the sail is sheeted home.

To SET *up the Rigging*, (*rider les haubans*, &c. Fr.) is to extend the shrouds, stays, and back-stays, more firmly than before, to secure the masts, which is performed by the application of mechanical powers, as tackles, &c. See the articles DEAD-EYE, LANYARD, &c.

SETTEE, (*scitie*, Fr.) a vessel with two masts, equipped with triangular sails, commonly called lateen sails. The least of them are of sixty tons burden, and they serve to transport cannon and provisions for ships of war, and the like. They are peculiar to the Mediterranean Sea, and are generally navigated by Italians, Greeks, or Mahometans.

SETTING, in astronomy, the withdrawing of a star or planet, or its sinking below the horizon. For the ACRONICAL, COSMICAL, and HELIACAL settings, see these articles.

To SETTLE, (*faire son effort, en parlant des bois dans la construction d'un bâtiment;—mollir*, &c. Fr.) To lower, also to sink; as, "the deck has settled, namely, has sunk below its first plan by the wood drying." "The ship has settled abaft since yesterday." "We settled the land." "Settle the main-top-sail halyards!" that is, ease off a small portion of them, so as to lower the yard a little. See the article *To* LAY *the Land.*

SETTS, in mast-making, denotes powers made use of, where force is required to bring or unite two or more pieces together, and is performed by screws, shores, cross-setts, or cleats.

Cross-SETTS, are made by two short pieces of spar, about four or six feet in length; one is laid across on the upper side, and the other on the underside of any two pieces that are to be brought together, and their ends lashed together on each side with several turns of rope taken round each end alternately; wedges are then driven in between the upper cross-piece, and the side or part of the mast.

To SEW, (*amortir*, Fr.) when spoken of a ship, is to rest upon the ground; and, while the depth of water around her is not sufficient to float her, she is said to be sewed by as much as is the difference between the surface of the water and the ship's floating-mark, or water-line.

SEXAGESIMALS, or SEXAGESIMAL FRACTIONS, are fractions whose denominators proceed by the powers of 60, instead of by the powers of 10, as in common decimal fractions; that is, in sexagesimals a prime or first minute, is $\frac{1}{60}$, a second $\frac{1}{60^2}$, or $\frac{1}{3600}$; a third $\frac{1}{60^3}$, or $\frac{1}{216000}$, and so on; but here, as in decimal fractions, the denominators being always the power of the same number, they are not put down in any operations performed on these numbers, their real value being indicated by the places they occupy in the result; thus 26°,9′,31″, 16‴, means $26° + \frac{9°}{60} + \frac{31°}{3600} + \frac{16°}{216000}$, and so on.

Anciently there were no other than sexagesimals used in astronomical operations, for which reason they are sometimes called *astronomical fractions*, and they are still retained in many cases, as in the division of time, and of the circles. See TAYLOR's *Sexagesimal Tables*, which exhibit at sight, in minutes, seconds, and tenths of a second, the fourth term in any proportion, where the first term is 60 minutes; and the second term, any number of minutes under 60 minutes; and the third term, any number of minu es and seconds under 60 minutes. The third term is to be sought at the top of the table, and the minutes constituting the second term on the side, answering to both, in the area of the table, is the fourth term required. It must, however, be allowed, that the decimal division in all these cases would be far preferable, an improvement which has been carried into effect by the French mathematicians, who now divide their quadrant into 100 degrees instead of 90, which degrees are divided into 10ths, 100ths, &c. and thus the measure of any angle is expressed by integers and decimals the same as any other quantity; the same method is also used by these astronomers for the division of time, as likewise for their weights and measures of every kind; and the important improvement thus introduced into the sciences only requires to be duly appreciated, in order to be generally adopted by all nations.

SEXTANS, the sextant, in astronomy, is a new constellation placed across the equator, but on the south side of the ecliptic, and by Hevelius made up of some unformed stars, or such as were not included in any of the 48 old constellations. In the Britannic catalogue it contains 41 stars.

SEXTANT, in geometry, is the sixth part of a circle, or an arc containing sixty degrees.

SEXTANT, (*sextant*, Fr.) is also an astronomical instrument, constructed on the same principles as the quadrant, and may be said to be that instrument extended; as, from the double reflection, it is divided into 120°. As the sextant is particularly intended to measure the distance of the moon from the sun, or a fixed star, in order to find the longitude of a place by lunar observations; and as this distance must be obtained as accurately as possible, it is, therefore, constructed with more care, and provided with some additional appendages that are wanting in the quadrant. This instrument is commonly made of metal; the index and limb are of brass, but the frame is of a harder composition; the only wooden part belonging to it is a handle attached to the back, by which, when observing the distance between two objects, it is to be held with one hand, while the other is regulating the motion of the index. A degree on the limb of this instrument is commonly divided into three equal parts, each of which, therefore, contains 20, and the index scale is divided into six equal parts; hence each contains 10′, and the vernier shews 10″. That end of the index next the limb is furnished with an adjusting, or, as it is sometimes called, a tangent screw, by which the index may be moved slowly and regularly, and, therefore, the contact of the limbs of any two objects may be made as perfect as the eye is able to distinguish, when assisted with one of the telescopes which accompany this instrument. In some sextants, the lower half of the index-glass, or that next the plane of the instrument, is silvered as usual, and the back surface of the other half is ground and painted black. A thin plate of brass, equal in length and breadth to the silvered part of the speculum, and moving on an axis, is to be raised before that part of the index-glass when the sun is very bright; in which case the reflection is to be taken from the polished surface of the upper half of that glass, so that any error which might probably arise from the want of parallelism in the planes of the glass, is by this means avoided. This plate is also painted black. A sextant is not fitted up for the back observation; and in some the horizon-glass is so fixed to the instrument, as to be incapable of adjustment. In instruments of this construction, it becomes absolutely necessary to find the index-error.

The coloured glasses are similar to those applied to a common Hadley's quadrant, and are usually four in number; a sextant is, however, generally provided with three more, to be placed occasionally on the farther side of the horizon-glass. These are particularly useful in observing the sun's altitude by reflection at land; or in finding the index-error by means of the sun.

There are two telescopes belonging to the sextant, one of which shows objects in their natural position, and the other is of the common astronomical construction, and, therefore, shewing objects inverted. A tube, without glasses, also accompanies this instrument; by means of these, the line of sight may be rendered parallel to the plane of the sextant, and the contact of the limbs of any two objects more accurately observed. The tube, or either telescope, is to be screwed into a brass ring, which is connected with another brass ring, by means of two screws, in such a manner, as to raise or lower the telescope, in order that the line of collimation may be directed to a proper part of the horizon glass. The line of collimation is an imaginary straight line, joining the centre of refractions of the object-glass of a telescope, and either the intersection of the cross-wires, or middle between the parallel wires.

A circular head, containing a plate, in which are three coloured glasses, and an aperture without glass, sometimes accompanies the sextant. This head is to be screwed on the eye-end of the tube, or on that of either telescope; the edge of the plate projects a little beyond the head on one side, and is moveable by the finger, so as the perforation or either coloured glass, may be brought opposite to the circular aperture in the head.

Fig. 4. plate XXI. is a plan of the sextant, as described above; fig. 5, is one of the telescopes; fig. 6, is the field view of the telescope, with the two parallel wires; and fig. 7, is the tube.

Adjustments of the SEXTANT. The adjustments of a sextant are, to set the mirrors perpendicular to its plane and parallel to each other, when the index is at zero, and to rectify the position of the line of collimation. The deviation of each of these from its true position might indeed be found, and the resulting error of observation from thence computed; but this by no means ought to be admitted, unless in cases of absolute necessity.

1st. To set the Index-Glass perpendicular to the Plane of the Sextant.

Put the index to about 60°, and hold the plane of the sextant nearly parallel to the horizon, the limb being from the observer: then direct the sight to the speculum, and if the reflected limb of the instrument appears to be exactly in the same plane with that seen directly, the glass is perpendicular to the plane of the instrument. But if the limbs are not in the same plane, turn the screws in the projecting plane behind the speculum, till they are apparently so, and the glass will be adjusted. This adjustment might be effected by means of the adjusting tool.

2d. To set the Horizon-Glass perpendicular to the Plane of the Sextant.

The index-glass being previously adjusted, set the beginning of the divisions on the index to zero on the limb, and hold the plane of the instrument in a horizontal position; then direct the sight to the horizon-glass; and if the reflected horizon is apparently in the same straight line with that seen directly, the glass is perpendicular to the plane of the sextant; otherwise, turn the adjusting-screw of the horizon-glass, at the back of the instrument, till the coincidence of the reflected and direct horizons is perfect.

3d. To set the Horizon-Glass parallel to the Speculum.

Set the first division on the index to zero on the limb, fasten the index in this position, and make the coincidence of these divisions perfect, by means of the adjusting-screw at the end of the index, the eye being assisted with the magnifying lens; screw the telescope in its support, and turn the screw belonging thereto at the back of the instrument, till the field of the telescope is bisected by the line which separates the silvered and transparent parts of the horizon-glass; now hold the sextant vertically, direct the sight of the horizon, and if the reflected and direct horizons do not coincide, release the screw which lies nearly behind the support of the telescope, and turn the nut at the extremity of the lever, till the coincidence of the horizons is perfect; and the horizon-glass will be adjusted. After the screw which contains the lever in its place is fastened, it will be proper to examine this adjustment; if the coincidence of the horizons is not perfect, the adjustment is to be repeated till it is so; but as it is difficult to obtain an exact coincidence by this means, the horizon may be brought to coincide, by turning the adjusting-screw of the index, and the difference between the two zeros is the index-error. The index-error may also be found, by measuring the diameter of the sun or moon twice, with a motion of the index, in contrary directions. If both measures are taken either to the right or left of O on the limb, half their sum will be the index-error, and is additive or subtractive accordingly; but if one of the measures is taken to the right, and the other to the left of O, half their difference is the index-error, being additive when the diameter measured to the right of O exceeds that measured to the left; otherwise, subtractive, since in some sextants the horizon-glass cannot be adjusted, the index-error must, therefore, be found: and in this case it may be considered as a constant quantity, to be applied to all angles measured with the same instrument.

4th. To make the Line of Collimation parallel to the Plane of the Sextant.

Turn the eye-end of the telescope containing the two parallel wires, till the wires are parallel to the plane of the instrument; and let two distant objects be selected, as two stars of the first magnitude, or the sun and moon, whose distance may not be less than 90° or 100°; make the contact of the limbs of these objects as perfect as possible at the wire nearest the plane of the instrument; fix the index in this position; move the sextant till the objects are at the other wire, and if the same limbs are in contact, the axis is adjusted; but if the limbs are either apparently separated, or partly cover each other, correct half the error by the screws in the circular part of the supporter, one of which is above, and the other between the telescope and sextant; turn the adjusting-screw at the end of the index, till the limbs are in contact; then bring the objects to the wire next the instrument, and if the limbs are in contact the axis of vision of the telescope is parallel to the plane of the instrument; if not, proceed as at the other wire, and continue till no error remains. If the moon is one of the objects, this adjustment must be performed as quick as possible, otherwise allowance must be made for the moon's apparent motion in the interval. This adjustment may also be made by looking directly to the sun through the telescope; then, if the shadow of the farther radius exactly covers the nearer radius or middle vertical bar, at the same time that the sun appears in the middle between the wires, the telescope is right; otherwise, it must be adjusted by the wires in the circular part of the supporter. The error of the line of collimation being given, the resulting error of observation may be found by entering table XXII, of Mackay on the Longitude, with the above error at the top, and the observed distance in the side-column. Is is sometimes necessary to know the angular distance between the wires of the telescope; to find which, let these wires be placed perpendicular to the instrument; hold the sextant vertical; direct the sight to the horizon, and move the sextant in its own plane, till the horizon and upper wire coincide; keep the sextant in this position, and move the index till the reflected horizon is covered by the lower wire, and the division shewn by the index on the limb, corrected by the index-error, will be the angular distance between the wires.

Use of the Sextant. The sextant is particularly adapted to measure the angular distance between the moon and the sun, or a fixed star. When the distance between the moon and either of these objects is to be observed, the sextant must be held so that its plane produced may pass through the eye of the observer and both objects; and the reflected image of the most luminous of the two is to be brought in contact with the other seen directly. To effect this, therefore, it is evident, that, when the brightest object is to the right of the other, the face of the sextant must be held upwards; but if to the left, downwards. When the face of the sextant is held upwards, the instrument should be supported with the right hand, and the index moved with the left; but when the face of the sextant is from the observer, it should be held with the left hand, and the motion of the index regulated by the right hand.

To observe the Distance between the Moon and any Celestial Object.

1*st. Between the Sun and Moon.* Put the telescope in its place, and the wires parallel to the plane of the instrument; then, if the index-glass is half silvered and half blackened, and if the sun is very bright, raise the plate before the silvered part of the speculum; direct the telescope either to the transparent part of the horizon-glass, or to the line which separates the silvered and transparent parts of that glass, according to the brightness of the sun, and turn down one of the coloured glasses; then hold the sextant so that its plane produced may pass through the sun and moon, having its face either upwards or downwards, according as the sun is to the right or left of the moon; direct the sight through the telescope to the moon, and move the index till the limb of the sun is nearly in contact with the enlightened limb of the moon; now fasten the index, and by a slow motion of the instrument make the image of the sun move alternately past the moon, and, when in that position where the limbs are nearest each other, make the coincidence of the limbs perfect by means of the adjusting-screw. This being effected, read off the degrees and part of a degree pointed out on the limb by the index, using the magnifying-glass; and thus the angular distance between the limbs of the sun and moon is obtained.

2*d. Between the Moon and Star.* Direct the middle of the field of the telescope to the line of separation of the silvered and transparent parts of the horizon-glass; if the moon is very bright, turn down the lightest coloured glass, and hold the sextant so, that its plane may be parallel to that passing through the eye of the observer and both objects, with its face upwards, if the moon is to the right of the star; but if to the left, the face of the sextant is to be held from the observer. Now direct the sight through the telescope to the star, and move the index till the moon appears by reflection to be nearly in contact with the star; fasten the index, and turn the adjusting screw till the coincidence of the star and enlightened limb of the moon is perfect; and the degrees and parts of a degree shown by the index will be the observed distance between the moon's enlightened limb and the star. The contact of the limbs must always be observed in the middle between the parallel wires.

It is sometimes difficult, for those not much accustomed to observations of this kind, to find the reflected image in the horizon-glass; it will, perhaps, be found more convenient to look directly to the object, and by moving the index to make its image coincide with that seen directly; or, if the distance between the objects be known nearly, the index may be set to this distance; the sight being directed to one of the objects, and the sextant held as formerly directed, and the other object will be seen in the field of the telescope. See MACKAY on the *Longitude.*

SEXTILE, in astronomy, the aspect or position of two planets when they are distant the sixth part of the circle; viz. two signs, or 60 degrees; and it is marked thus ⚹.

SHACKLES, a name given to the rings with which the ports are secured, by hooking the port-bars to them; also a sort of iron rings to hook tackles to. See the articles PORT and RINGS.

SHACKLES, (*les boucles intérieures des sabords; aussi fers, ou menilles de prisonniers*, Fr.) are also semicircular pieces of iron put upon the bilboe-bolts, for confining those who have deserved corporal punishment. See the article BILBOES.

SHAKES, or SHAKEY, (*fente et gerçure dans une pièce de bois quelconque*, Fr.) a name given by shipwrights to planks or timber which are full of clefts, and will not bear calking or fastening.

SHALLOP, (*sorte de barque, ou chaloupe, mattée en goëlette*, Fr.) a sort of large boat with two masts, usually rigged like a schooner.

SHALLOW, (*basse, batture, ou bas fond*, Fr.) a term synonymous with shoal, signifying that part of the water not sufficiently deep to sail over by vessels which are navigated in that place: there is, however, this difference, a shallow is never supposed to be dry, even at the lowest ebb; but shoals are often dry at low water.

SHANK OF AN ANCHOR, (*verge d'une ancre*, Fr). a term given to the beam or shaft of an ANCHOR, which article see.

The shanks of anchors taper in their lengths one inch and a half in small anchors; to three inches in large, keeping their proper size at the trend, and three quarters of an inch to two inches the flatted way.

SHANK-PAINTER, (*serre-bosse, composée d'un cordage et d'une chaîne, servant à brider l'ancre contre le bord*, Fr.) a short rope and chain; the latter is fastened to the topside abaft the cat-head, and used to hang the shank and flooks of the anchor up to the ship's side.

To SHAPE A COURSE, (*commander, ou donner la route; faire route*, Fr.) to direct or appoint the track of a ship, in order to prosecute a voyage.

SHARP-BOTTOM, (*fond fin, en parlant de la construction d'un bâtiment*, Fr.) is synonymous with a sharp floor, and used in contradistinction to a flat floor. See the article BOTTOM.

To Brace up SHARP, (*orienter toutes les voiles au plus près*, Fr.) is to turn the yards to the most oblique direction possible, so as that the ship may lie well up to the wind.

SHEATHING, (*doublage, soit de bois, soit de feuilles de cuivre, &c.* Fr.) a sort of casing or covering nailed all over the outside of a ship's bottom, to protect the planks from the pernicious effects of the worms, particularly in hot climates. It consists of a number of fir boards or deals of fir, or what is far preferable, sheets of copper, which last is always used in the royal navy, as well as in the East India Company's service, and is coming into general use in the merchant-service, as all transports are sheathed with copper. See the articles COPPERED and COPPER FASTENED.

In many of the islands with which the Indian Ocean is covered, the inhabitants have a custom of paying the bottom of their vessels with chunam, a composition peculiar to the country, and which is said to be more useful and serviceable than sheathing with copper or any other article hitherto known or discovered.

The following is an account of the sheets of copper, counter-sunk nails, and paper used in coppering the under-mentioned ships:

Ships Names.	Guns.	No. of Sheets.	Counter-sunk Nails.			Paper.	
			Cwts.	qrs.	lbs.	Rs.	qrs.
Caledonia	120	4429	30	3	0	21	0
Hibernia	110	4123	29	3	4	20	0
Victory	98	3640	28	2	5	20	0
Prince George	98	3650	28	3	21	19	0
Alexander	84	3376	29	0	6	17	6
Talavera	80	3388	25	2	2	17	0
Colossus	74	3325	26	2	10	16	0
Isis	50	2226	18	1	12	14	0
Cambrian	44	2422	19	0	21	12	0
Tribune	36	1918	16	0	0	9	10
Medusa	32	2028	17	0	0	10	0
Mercury (en flute)	28	1597	14	1	11	7	6
Ariadne	20	1252	10	1	17	6	12
Spider	14	800	6	0	20	6	0
Snapper (gun-brig)	12	742	5	2	0	3	0
Griper .. (ditto) ..	12	778	6	0	7	4	0

SHEATHING-*Boards*, (*planches de sapin de doublage*, Fr.) are of fir plank, from ⅜ths to ⅞ths of an inch thick, used principally in covering the bottom of merchant-ships, before the copper-sheathing is applied.

SHEATHING-*Nails*. See the article NAILS.

SHEAVE, or SHIVER, (*rouet de poulie*, Fr.) the wheel on which the rope works in a block; it is generally formed of lignum-vitæ, sometimes of brass, and frequently of both: the interior part, or that which sustains the friction against the pin, being of brass, is let into the exterior, which is made of lignum vitæ, and is then termed a sheave with a brass coak or bush.

SHEAVE-*Hole*, (*clan*, Fr.) the channel cut in a mast, yard, or timber, in which to fix a sheave, answering instead of a block.

SHEEP-SHANK, (*jambe de chien; nœud fait a l'itague d'un palan, &c. pour la raccourcir*, Fr.) a sort of knot made on a rope to shorten it, and is particularly used on runners or ties, to prevent the tackle from coming block and block. By this contrivance the body to which the tackle is applied may be hoisted much higher, or removed much further, in a shorter time.

Thus, if any weighty body is to be hoisted into a ship, and it be found that the blocks of the tackle meet before the object reach the top of the side, it will be necessary to lower it again, or hang it by some other method, till the runner of the tackle is sheep-shanked, by which the blocks will again be separated to a competent distance.

To SHEEP-SHANK, (*raccourcir un cordage*, Fr.) is also a term used to shorten the top-gallant back-stays, &c. (*pro tempore*), when the masts are struck.

SHEER, (*tonture des préceintes; relèvement des ponts*, Fr.) in ship-building, the longitudinal curve of a ship's deck or sides.

SHEER, (*signifie aussi la position d'un bâtiment au mouillage qui change cette position en mollissant son cable*, Fr.) is also the position in which a ship is sometimes kept when at single anchor, in order to to keep her clear of it: hence

To Break SHEER, is to deviate from that position, and thereby risk the fouling of the anchor.

If the side of a ship at anchor is presented to the tide by any means, the water will act upon her two ways; one in the direction of her keel, the other in the direction of her beams; the last will cause her to sheer out to one side of her anchor, which was a-head before. Suppose, for example, the power which presents the ship's side to the tide to be a spring; from the anchor coming in aft on the starboard side, upon heaving in the spring, the ship will sheer over to port, bringing the anchor upon the starboard bow: the more the spring is hove in, the more the ship will go a-head, and over to port, until her side makes an angle with the tide of 45°, as the furthest she can go over from her anchor; for the ship being hove in after this, will cause her to return, and be in the stream of her anchor when she is hove round a-broadside to the tide. Now by putting the helm over to starboard, causing the water to act upon her starboard side, the ship will be forced over to port: but, on the contrary, if the helm had been put to port, the larboard side would have been presented to the action of the water, and the ship would go over to starboard, but the power of the rudder being according to the strength of the tide, its action lessens upon the rudder, as the ship sheers obliquely to the stream, and cannot produce such an effect as the spring.

To SHEER, (*faire des embardées*, Fr.) to deviate or stray from the line of the course, so as to form a crooked and irregular path through the water. It may be occasioned by the ship's being difficult to steer, but it happens very often from the negligence or incapacity of the helmsman.

To quicken the SHEER, among shipwrights, is to shorten the radius which strikes out the curve; and

To straighten the SHEER, is to lengthen the radius, &c.

To SHEER *up Alongside*, (*s'approcher d'un bâtiment bord sur bord*, Fr.) to approach a ship in a parallel direction.

To SHEER-*off*, (*s'éloigner, s'écarter, fuir, ou prendre chasse*, Fr.) to remove to a greater distance.

SHEER-HOOKS, (*grapins de bouts de vergue*, Fr.) an iron instrument with two or three prongs at one end like a fork, and four hooks near the other end, as represented by fig 9. plate X. at the extremities of the yards of fire-ships, to entangle the enemy's rigging, &c.

SHEER-HULK, (*vieux vaisseau servant de machine à mâter dans les ports*, Fr.) an old ship of war, cut down to the gun or lower deck, having a mast fixed in midships, and fitted with an apparatus consisting of sheers, tackles, &c. to heave out or in the lower-masts of his Majesty's ships, particularly in the royal ports, as occasion requires.

The mast of the Neptune *sheer-hulk*, in Portsmouth harbour, as represented by *aa*, fig. 2, plate XXIV, is 3 feet 2 inches in diameter, and 119 feet in length, strengthened with shores *bb*; the shores being 21 inches in diameter, and 86 feet in length.

There are shrouds and stays to secure the mast and sheers, which serve as the arm of a crane, supported by a derrick, *cc*, 138 feet in length and 24 inches in diameter, to hoist in or out the lower-masts of any ship, when conveniently laid alongside. The sheers *dd*, being three in number, are, generally, composed of two pieces, 26 inches in diameter, the middle one being 137 feet, and the fore and after one 143 feet in length; their heels resting upon the side of the hulk, and their heads inclining outward from a perpendicular, to hang over the ship, whose masts are to be fixed or displaced. The tackles *eee*, which extend from the head of the mast to the sheers, are intended to support the latter.

The main purchase for heaving in or out the lower masts, is produced by the effort of two capstans, fixed on the deck for that purpose.

SHEERS. See the articles MASTS and SHEER-HULK.

SHEERS, used for masting vessels in the merchant-service, are two hand-masts, or large spars, erected on the vessel whose masts are to be fixed or displaced; the lower-ends or heels rest on opposite sides of the deck, upon thick planks sufficiently long to extend over two or three beams, shored underneath. The two hand-masts cross each other at the upper end, and are securely lashed.

A tackle is lashed in the centre, and hangs perpendicularly over the station where the mast is to be fixed.

The *sheers* are secured by guys or stays, of proportionable rope, extending fore and aft to the opposite extremities of the vessel.

SHEER-STRAKE, in ship-building, the strake under the gunwale in the top-side; it is generally worked thicker than the rest of the top-side, and scarphed between the drifts.

SHEET, (*écoute*, Fr.) a rope fastened to one or both the lower corners of a sail, to extend and retain it in a particular station.

When a ship sails with a side-wind, the lower corner of the main and fore-sails are fastened by a tack and a sheet, the former being to windward and the latter to leeward; the tack is, however, entirely disused with a stern-wind; whereas the sail is never spread without the assistance of one or both of the sheets.

The stay-sails and studding-sails have only one tack and one sheet each: the stay-sail tacks are always fastened forward, and the sheet drawn aft; but the studding-sail tacks draw the outer clew of the sail to the extremity of the bottom, while the sheet is employed to extend the inmost.

SHEET-ANCHOR, (*maîtresse ancre*, Fr.) the largest anchor used in a ship; and, in stress of weather, or when an uncommon hard gale of wind happens, it is the mariner's last refuge. See the article ANCHOR.

SHELL, (*bombe*, Fr.) in artillery, an hollow cast-iron ball to throw out of mortars or howitzers, having a vent or hole, (see figures 15 and 16, plate XVI.) through which the powder for bursting it is put, by means of a copper funnel; and when filled, the fuse for firing is driven firmly into the hole: the metal at the bottom or part opposite the fuse, is generally made thicker than the rest, that the fuse may always be uppermost, which prevents its being beat out by striking against any hard substance, when the shell falls.

The metal of the shells used for guns and carronades, is usually cast of an equal thickness throughout, and consequently bursts into more pieces.

Shells were first applied in naval warfare for the bombardment of maritime towns, by the French, in the reign of Lewis XIV.

Shrapnell's SHELLS, now called *Spherical Case-Shot*, are shells of a peculiar construction, invented by Colonel Shrapnell, of the royal artillery.

They have been used at several times with surprising effect against the French army, by Marquis Wellington, field-marshal of the British army; and are much used in RICOCHET-FIRING. For a farther description of them, see the article *Spherical Case-Shot*.

Table of the Dimensions and Weight of SHELLS *for Mortars and Howitzers.*

Species.	Weight.				Diameter.	Powder contained in Shells.		Powder for bursting.		Diameter of Fuse-holes.		Thickness of Metal.
										Outside.	Inside	
	Cwt.	qrs.	lbs.	oz.	Inch.	lbs.	oz.	lbs.	oz.	Inch.	Inch.	Inch.
13-Inch.	1	3	2		$12\frac{3}{4}$	10	4	6	12	1.837	1.696	2.05
10 do.	—	3	9		$9\frac{3}{4}$	4	5	2	10	1.57	1.45	1.575
8 do.	—	1	11	8	$7\frac{3}{4}$	2	12	1	14	1.219	1.127	1.2
$5\frac{1}{2}$ do.	—	—	15	4	$5\frac{1}{4}$	1	—	—	12	0.894	.826	0.822
$4\frac{2}{5}$ do.	—	—	8		$4\frac{1}{5}$	—	7	—	5	0.852	.769	0.653
H.Gren.	—	—	3	11	3.49							
	—	—	1	13	2.77	.....		—	$1\frac{1}{2}$			

Dimensions of Shells for Guns and Carronades made with an equal Thickness of Metal.

Species.			42-pr	32.	24.	18.	12.
			Inch.	Inch.	Inch.	Inch.	Inch.
Guns.	Diameter of the Shell	Exterior	6.684	6.105	5.547	5.05	4.4
		Interior	4.404	4.005	3.767	3.4	2.8
	Thickness of Metal ..		1.14	1.05	0.89	0.82	0.8
	Diameter of Fuse-hole	Exterior	0.894	0.894	0.893	0.832	0.832
		Interior	0.826	0.826	0 826	0.76	0.769
	Powder for bursting ..		14 oz.	11 oz.	12 oz.	9 oz.	$5\frac{1}{2}$ oz.
Carronades.	Diameter of the Shell.	Exterior	6.64	6.05	5.48	4.935	4.295
		Interior	4.36	3.95	3.48	3.235	2.695
	Thickness of Metal ..		1.14	1.35	1.	0.8[illegible]	0.98
	Shell's weight .... lbs.			.22		.12	
	Contains Powder.. oz.			$.12\frac{1}{2}$		.9	
	Powder for bursting oz.			.10		.7	

The following shells may also be fired from guns:

Hand-grenades, from ...... 6-pounders.
$4\frac{2}{5}$-Shells ....12 —
$5\frac{1}{4}$-Shells ....24 —
8-inch ....68-pr. carronades.

Shells may likewise be thrown from guns to short distances, in cases of necessity, though the bore be not of a diameter sufficient to admit the shell. For this purpose the gun may be elevated to any degree that will retain the shell upon its muzzle, which may be assisted by a small line going from the lugs of the shell round the neck of the gun. To produce a greater effect, the space between the shell and the charge may be filled with wads or other substances.

To find the weight of a Shell of iron.—Take nine sixty-fourths of the difference of the cubes of the external and internal diameters for the weight of the shell.

To find the size of a Shell to contain a given weight of powder. Multiply the pounds of powder by 57.3, and the cube root of the product will be the diameter in inches.

To find how much powder will fill a Shell.—Divide the cube of the internal diameter of the shell in inches by 57.3, for the pounds of powder.

For farther particulars respecting the shell and its fuse; also the manner of charging and firing it, see the article MORTAR.

SHELL *of a Block*, (*corps d'une poulie*, Fr.) the outer frame or case wherein the sheave or wheel is contained, and traverses about its axis. See the article BLOCK.

To SHELTER, (*abriter*, &c. Fr.) when applied to a ship or vessel, is to anchor in a harbour or roadstead that is screened from the winds, &c.

SHELVES, (*écueils*, *dangers*, *ou ressifs*, Fr.) a general name given to any dangerous shallows, sandbanks, or rocks, lying immediately under the surface of the water.

SHELVING, (*penchant*, Fr.) a gradual inclination of the land towards the sea: hence

A SHELVING *Bottom*, (*fond qui se perd*, Fr.) an irregular surface at the bottom of the sea, which sometimes accumulates to such an height as to endanger ships running foul of it.

To SHIFT, (*changer*, *&c.* Fr.) when expressed of the wind, implies to alter.

To SHIFT *the Helm*, (*changer la barre*, Fr.) is to alter its position, by pushing it towards the opposite side of the ship.

To SHIFT *a Tackle*, (*affaler*, *ou reprendre un palan*, Fr.) is to remove the blocks of a tackle to a greater distance from each other, on the object to which they are applied, in order to give a greater scope or extent to their purchase. This operation is otherwise called fleeting. See the article *To* FLEET.

SHIFT *the Messenger!* (*change le tournevine!* Fr.) the order to change its position on the capstan from the right to the left, and *vice versa*.

SHIFTED, (*desarrimé*, Fr.) the state of a ship's ballast or cargo when it is shaken from one side to the other, either by the violence of her rolling in a turbulent sea, or by an extraordinary inclination to one side when under a great pressure of sail. This accident, however, rarely happens, unless to those cargoes stowed in bulk, as corn, salt, or such materials.

SHIFTER, (*matelot*, *ou mousse adjoint au coq*, Fr.) a person appointed to assist the ship's cook in washing, steeping, and shifting the salt provisions.

SHINGLE. See the article BALLAST.

SHINGLES, shifting sand-banks seen at low-water, near Christ Church Bay, which often prove dangerous to shipping coming to Spithead through the Needles. By endeavouring to avoid the Needle Rocks, which extend some way into the sea at the west end of the Isle of Wight, there is much danger, with a strong current setting in, of getting on the shingles; to avoid which, there is alight-house erected on the beach near Hurst Castle.

SHIP, (*vaisseau*, *navire*, *et bâtiment en général*, Fr.) a general name given to all large vessels navigated on the ocean. In the sea-language, however, it is more particularly applied to a vessel furnished with three masts, each of which is composed of a lower mast, top-mast, top-gallant-mast, and royal mast, with the yards and other machinery thereto belonging.

The plans, elevations, and sections used in the construction of a ship; the principal pieces of which she is composed; and the qualities requisite to answer the several purposes of navigation, are described in or referred to from the article NAVAL *Architecture:* and the application of this theory to practice is treated in the article SHIP-BUILDING.

The machinery and furniture with which she is equipped are variously diffused throughout this work, and naturally spring from one another, like a multitude of branches from one general trunk. See the articles ANCHOR, MAST, SAIL, YARD, RIGGING, &c.

The qualities by which she is enabled to encounter a tempestuous sea, are treated in the articles BALLAST and TRIM; and her several movements therein are explained under the articles DRIFT, LEE-WAY, NAVIGATION, PITCHING, ROLLING, TO TACK, and TO WORK.

Considered as a moveable fortress or citadel, her military operations are copiously described in the articles CANNON, CANNONADE, ENGAGEMENT, EXERCISE, LINE, RANGE, SAILING, SHOT, SIGNALS, &c. and as her efforts are occasionally like those of a mine, or bombardment, the reader is also referred to the articles FIRE-SHIP, and MORTAR.

A SHIP *Cut Down*, (*bâtiment rasé*, Fr.) implies one which has had a deck cut off from her, whereby a three-decker is converted into a two-decker, and a two-decker becomes a frigate. See the article RAZEE.

A SHIP *Raised upon*, (*bâtiment dont les œuvres mortes ont ète élévées par des alonges*, Fr.) is one whose dead works have been heightened by additional timbers.

Admiral's SHIP, (*vaisseau amiral*, Fr.) any ship bearing the admiral's flag.

Armed-SHIP. See the article ARMED-*Ship*.

Fire-SHIP. See the article FIRE-*Ship*.

Guard-SHIP. See the article GUARD-*Ship*.

Hospital-SHIP, (*vaisseau servant d'hôpital*, Fr.) a vessel fitted up to attend a fleet of men of war, and receive their sick or wounded; for which purpose her decks should be made high, and her ports large. The gun-deck is entirely appropriated for the reception of the sick, and is flush without cabins or bulk-heads; except one of deal, or canvas, for separating those in malignant distempers. Two pair of chequered linen sheets are allowed to each bed, and scuttles cut in the sides for inlets of air. The sick are occasionally visited by the physician, and constantly attended by the surgeon, a proportionable number of mates, assistants, &c. The cables ought also to run upon the upper deck, in order that the beds or cradles may be more commodiously placed between decks, and admit a free passage of air, to disperse that which is offensive or corrupted.

The captain of an hospital-ship is to be particularly attentive to see that the ship is kept extremely clean, especially that part which is appropriated to the sick; he is to give the strictest orders that the buckets used by the sick should be frequently emptied and washed, and that the dressing of wounds or sores should be thrown overboard, as soon as they are taken off.

He is to attend to all requisitions and proposals of the physician or surgeon, and, as far as circumstances will admit, to do whatever he may recommend for the comfort and convenience of the sick.

If the provisions supplied for the use of the sick should be expended, he is to direct the purser to supply the agent with such quantity as he may demand, with the approbation of the physician or surgeon; and if there be any difference in quality in any articles of provisions, he is always to appropriate the best to the use of the sick.

He is to be particularly careful in preventing wine or spirituous liquors being carried on board, without his express permission; and he is to give the most positive injunctions to the ship's company never to give the sick any thing to eat or drink, except that which the physician or surgeon shall allow.

He is to be very attentive in keeping the station appointed him in the fleet, particularly when he is directed to attend to a ship, whose crew is sickly, and may frequently require the assistance of the physician.

In addition to the complement of an hospital-ship, there shall be borne as attendants on the sick, on a list of supernumeraries for wages and victuals, a physician, a surgeon, three assistant-surgeons, six landmen as nurses, a baker, four washermen, a servant to the physician, and a clerk and servant to the surgeon.

The captain of the hospital-ship, or the captain of any other ship in which the physician of the fleet shall be embarked, is to furnish him with a proper boat, whenever he shall think it necessary to visit any ship of the fleet, or to carry his returns to the commander in chief.

A Leeward-SHIP. See that article.

A SHIP *of the Line*, (*vaisseau de ligne*, Fr.) is usually applied to all men of war carrying sixty guns and upwards. Of late, however, our fifty-gun ships have been formed sufficiently strong to carry the same metal as those of sixty, and accordingly may fall into the line in cases of necessity in time of action.

The ships of seventy-four guns are generally esteemed the most useful in the line of battle, and indeed in almost every other purpose of war. It has therefore been judged conformable to our design to represent different views and sections of a ship of this class.

Merchant-SHIP, (*bâtiment marchand*, Fr.) a vessel employed in commerce, to carry commodities of various sorts from one port to another.

The largest merchant-ships are those employed by the different companies of merchants who trade to the East Indies. They are in general as large as our 44-gun ships, and are commonly mounted with 20 guns on their upper deck, which are nine-pounders, and six on their quarter-deck, which are six-pounders. They are particularly employed in the Chinese trade, and some of them are said to have brought to Europe cargoes amounting to 2000 tons, the principal part of which was tea. See the articles MERCHANT, MERCHANT-*Company*, and MERCHANT-MAN.

Prison-SHIP, (*vaisseau servant de prison dans un port*, Fr.) a vessel fitted up in a port purposely for the accommodation of prisoners of war, and regularly guarded.

Private SHIP *of War*. See the article PRIVATEER.

Receiving-SHIP, (*vaisseau servant d'entrepôt, pour les gens destinés au service de la marine*, Fr.) a ship stationed at any place to receive volunteers and impressed men, and train them to their duty in readiness for any ship of war which may want hands.

Store-SHIP, (*bâtiment armé en flûte, pour charger des munitions*, &c. Fr.) a vessel employed to carry artillery or naval stores for the use of a fleet, fortress, or garrison.

Transport-SHIP, (*bâtiment de transport*, Fr.) a merchant-ship hired by the Transport-Board for the purpose of conveying ammunition, stores, &c. from one place or port to another.

Troop-SHIP, (*bâtiment de transport pour les troupes*, Fr.) a ship carrying her guns on the upper deck, and employed to convey troops, and on that account is commonly called a troop or transport-ship.

SHIPS *of War*, (*vaisseaux de guerre*, Fr.) are properly equipped with artillery, ammunition, and all the necessary martial weapons and instruments for attack or defence. They are distinguished from each other by their several ranks or classes. See the article RATE.

Besides the different kinds of ships above mentioned, which are denominated from the purposes for which they are employed, vessels have also, in general, been named according to the different manner of rigging them. It would, however, be an endless, and at the same time an unnecessary task, to enumerate all the different kinds of vessels with respect to their rigging: but for some of the principal ones, we refer the reader to the article VESSEL.

Pl. XXV. represents a modern first-rate ship of war, with masts, yards, sails, rigging, &c. at anchor; the several parts of which are as follow: Pl. XXV.

1 BOWSPRIT.
2 Gammoning.
3 Bumkin.
4 Horse.
5 Bob-stays.
6 Martingal.
7 Martingal-stays.
8 Bowsprit shrouds.
9 Jib boom.
10 Jib, stay, and sail,
11 Jib-halyards.
12 Horses.
13 Sprit-sail-yard and course
14 Bowsprit-cap.
15 Jack-staff and flag.
16 Braces.
17 FORE-MAST.
18 Shrouds and ratlines.
19 Stay and lanyard.
20 Preventer-stay and lanyard.
21 Yard and course, with studding-sail-booms.
22 Horse.
23 Top.
24 Yard tackles.
25 Lifts.

26 Braces.
27 Sheets.
28 Tack.
29 Bowlines and bridles.
30 Futtock-shrouds.
31 Cap.
32 Fore Top-Mast.
33 Shrouds and Lanyards.
34 Yard and sail, with studding-sail booms.
35 Stay and sail.
36 Preventer-stay.
37 Backstays.
38 Halyards.
39 Lifts.
40 Braces.
41 Horses.
42 Staysail halyards.
43 Bowlines and bridles.
44 Sheets.
45 Cross-trees.
46 Cap.
47 Fore-Top-Gallant Mast.
48 Shrouds.
49 Yard and sail.
50 Backstays.
51 Stay.
52 Lifts.
53 Braces.
54 Bowlines and bridles.
55 Royal stay.
56 Backstay.
57 Royal yard and sail.
58 Royal braces.
59 Royal lifts.
60 Flag of the lord high admiral.
61 Main-Mast.
62 Shrouds and ratlines.
63 Stay.
64 Preventer-stay.
65 Stay-tackles.
66 Yard-tackles.
67 Lifts.
68 Braces.
69 Horse.
70 Sheets.
71 Tack.
72 Bowlines and bridles.
73 Top.
74 Cap.
75 Yard and course, with studding-sail-booms.
76 Futtock shrouds.
77 Main-Top-Mast.
78 Shrouds and lanyards
79 Yard and sail, with studding-sail-booms.
80 Backstay.
81 Preventer-stay.
82 Stay and sail.
83 Halyards.
84 Lifts.
85 Braces.
86 Horse.
87 Sheets.
88 Bowlines and bridles.
89 Cross-trees.
90 Cap.
91 Main-Top-Gallant-Mast.
92 Shrouds.
93 Yard and sail.
94 Backstay.
95 Stay, halyard, and sail.
96 Lifts.
97 Braces.
98 Bowlines and bridles.
99 Royal stay.
100 Backstay.
101 Royal yard and sail.
102 Royal braces.
103 Royal lifts.
104 Royal standard.
105 Mizen-Mast.
106 Shrouds and ratlines.
107 Cross-jack-yard.
108 Stay.
109 Preventer-stay.
110 Cross-jack-lifts.
111 Ditto ditto braces.
112 Horse.
113 Top.
114 Cap.
115 Mizen-Top-Mast.
116 Shrouds.
117 Stay.
118 Backstay.
119 Yard and sail.
120 Lifts.
121 Braces.
122 Bowlines and bridles.
123 Cross-trees.
124 Cap.
125 Mizen-Top-Gallant-Mast.
126 Shrouds.
127 Stay.
128 Backstay.
129 Yard and sail.
130 Bowlines and bridles.
131 Lifts.
132 Braces.
133 Royal yard and sail.
134 Royal lifts.
135 Royal braces.
136 Royal stay.
137 Royal back-stays.
138 Union jack.
139 Driver-boom.
140 Boom topping-lifts.
141 Boom guy-falls.
142 Gaff and Driver.
143 Derrick-fall.
144 Peak-brails.
145 Peak-halyards.
146 Ensign staff.
147 Ensign.
148 Bower cable.

To Ship, (*embarquer des munitions, des marchandises, ou des hommes, &c.* Fr.) is to embark any person, or put ammunition, stores, &c. aboard ship; also to receive any thing into a ship; as, "we shipped a heavy sea."

To Ship, also implies to fix any thing in its place; as, "ship the oars!" that is, place them in their row-locks.

To Ship *the Tiller*, (*mettre en place le gouvernail*, Fr.) is to place it on the head of the rudder.

Ship-*Shape*, (*arrangé à la manière des vaisseau; c'est-à-dire avec exactitude*, Fr.) according to the fashion of a ship; as, "the mast is not rigged ship-shape;" "Trim your sails ship-shape," &c.

SHIP-BUILDING, (*construction — art de la construction et architecture navale*, Fr.) the art of erecting ships from certain draughts, and is performed by scarphing, rabbeting, tenoning, scoring, facing, and bolting. It is distinguished from naval architecture, the former being the practice of building ships, and the latter the theory of delineating them.

We shall here attempt to describe the various parts of a 74-gun ship, and, at the same time, shew how these parts are progressively applied and connected with each other, according to the modern practice of building, from the laying of the keel to the launching of her into the water.

The timbers being provided, the blocks are laid in the middle of the slip, about five feet distant from each other to receive the keel; the lower being much larger than those above, are, therefore, fastened upon the ground-ways, and the upper ones on the lower with tree-nails: their declivity is generally from five-eighths to three-quarters of an inch to a foot; the upper sides being made straight fore-and-aft and level athwart ships.

The *Keel* is formed of several pieces of elm timber, scarphed perpendicularly and bolted together, and is the basis for the whole structure; it is set straight upon the middle of the blocks, and kept in that position by tree-nails being driven into the blocks along its sides.

The *Dead-wood* is generally of oak of various forms and thicknesses, and fayed on the upper side of the main-keel, the scarphs giving shift to the scarphs of the keel, and fastened thereto by tree-nails.

The *Floors* are also of oak, fixed athwart-ship, and scored into the dead-wood.

The *Stem* is composed of two or more circular pieces of oak timber, which are scarphed and bolted together: its lower end is scarphed or boxed into the fore-end of the main keel, and its upper end extends to the under side of the bowsprit.

The *Apron* is also of oak, and conforms to the shape of the stem; the convexity of the former, fayed to the concavity of the latter, forms one solid piece.

The *Knight-heads* are of oak; their heads are sufficiently above the head of the stem to support the bowsprit: narrow timber, or fillings, have recently been introduced between the knight-heads and the stem, in order to prevent the bowsprit from wounding them. The stem, apron, knight-heads, and the narrow timbers, or fillings, are all bolted together and raised to their proper stations; the scarph or boxing of the stem is then bolted securely to the fore-part of the keel.

The *Stern-post* is of oak, is the principal piece of timber in the stern-frame, and into which the wing, filling, and deck-transoms are scored and bolted. Its lower end is tenoned into the keel, and terminates the ship abaft.

The *Inner-post* is fayed to the fore side of the stern-post, for the purpose of seating the transoms below the deck; its lower end is tenoned into the keel, and its upper end terminates at the under side of the lower-deck-transom.

The *Transoms* are of oak, fixed athwart the stern-post, and bolted thereto, in order to form the buttock of the ship, and to fortify her after-part. They are divided into four sorts, viz. wing, filling, lower-deck, and the others below the latter, are distinguished by N°s 1, 2, 3, 4, 5, 6, and 7.

The *Wing-transom* is the upper transom in the stern frame, in which the heels of the counter-timbers are let in and bolted.

The *Filling-transom* is the intermediate transom, which is placed sufficiently above the deck-transom to admit the plank of the deck being worked between them.

The *Deck-transom* is next below the filling-transom, and is worked sufficiently broad to secure the after ends of the gun-deck plank.

The *Transoms* below the lower-deck-transom, are, with those described above, bolted to the stern-post, as before mentioned.

The *Fashion-pieces* are of oak, so called from their fashioning the after-part of the ship, and have scores taken out of their aft sides, in order to receive the ends of all the transoms. The stern-frame being now raised and secured, forms the basis of the whole stern.

The *Harpings*, called the *floor-surmark*, are pieces of oak, similar to the ribbands; they are trimmed and bevelled to the shape of the body of the ship in the wake of the fore and after cant-bodies, and are shored securely, to keep the timbers in their respective stations.

The *Ribbands*, also called the *floor-surmark*, are of fir-timber, nailed to the timbers of the square-body; their fore and after ends are scarphed to the harpings, which extend from the stem to the stern; they are also shored securely to retain the floor-timbers and the heels of the first futtocks in their proper stations.

The *Frame-timbers* are the bends of timbers forming the body of the ship, each of which is composed of a first, second, third, fourth futtock, and a top-timber, which being united, are raised cross-spalled, to their respective stations, shored and bolted sideways to the floor-timbers.

The *Cross-spalls* are pieces of fir plank, nailed, in a temporary manner, athwart the ship in the wake of the ports, by which the frames are kept well to their breadth.

The whole of the harpings and ribbands are now trimmed and fastened to their respective stations, till the plank is wrought.

The *Filling-timbers* are intermediate timbers placed between the frames; they are trimmed and got into their stations singly, and secured by ribbing-nails through the harpings and ribbands.

The *Side Counter-timbers* are of oak, and partake of the shape of the top-side, their lower ends being placed at the extremity of the wing-transom, and bolted to the foremost fashion-pieces.

The *Midship Counter-timbers*, are also of oak; their lower ends are let into the wing-transom, and bolted thereto: they are fixed at equal distances from each other, in order to form the stern-lights, and all the stern-rails are firmly bolted to them.

The *Hawse-pieces* are oak timbers, whose sides stand nearly fore and aft, and form the bow of the ship.

Short hawse-pieces have lately been placed above and below the hawse-holes, and are the means of preventing the timbers from being weakened by cutting the hawse-holes.

The *Cross-chocks* are pieces of oak timber which cross the dead-wood, and secure the heels of the lower futtocks.

The *Angular-chocks* are fastened to the different heads and heels of the timbers; when they are completed, the outside frame of the ship is reconciled, and the inside dubbed to its scantling ready to receive the plank.

The *Kelson* is composed of long square pieces of oak timber, scarphed together in a horizontal direction, and fixed within board exactly over the keel, and may be considered as the counter-part thereof for strengthening the lower part of the ship: it is fitted to and laid upon the middle of the floor-timbers, and bolted through the floors and the keel, the scarphs giving shift to the scarphs of the keel.

The *Stemson* is a piece of compass-oak timber, worked on the aft-side of the apron, having its lower end scarphed into the kelson, which gives shifts to the scarphs of the apron, and its upper end continued as high as the upper-deck.

The *Postson*, or *Sternson*, is a large knee-piece of oak timber worked upon the after dead-wood; the fore-end is scarphed into the kelson, and the aft-side fayed and bolted to the fore sides or throats of the transoms, and continued as high as the lower-deck transom.

The *Dead-wood bolts* are long copper-bolts driven through the kelson, dead-wood and keel, where they are well clinched upon copper rings for the purpose of securing the ship in that part.

The *Lower-deck ports* are then trimmed out to their proper size, and the heads and heels of the timbers cut off in order to let in the port-sills.

The *Lower-deck port-sills* are pieces of oak timber let in horizontally between the frames, to form the upper and lower sides of the ports.

The *Main Wale*, (*black-strake*, and *diminishing-stuff*,) is the lower-wale, which is wrought, fastened with bolts, and bored off, but the tree-nails are not driven through it, until the inside plank is worked: it is generally placed on the lower-breadth, so that the fastenings of the knees or chocks, to the ends of the beams, may be secured to it by the in and out bolts.

The *Fillings* are pieces of timber driven edgeways between the timbers from the floor-heads downward to the dead-wood, and calked within and without board to prevent the ship from leaking, should the plank of the bottom be rubbed off by any accident, which has frequently happened.

The *Plank of the bottom* is wrought and bored off, and sufficiently fastened with bolts made for that purpose, but the tree-nails are not driven until the inside plank is worked.

The *Orlop Clamps* are thick plank worked withinside the ship, and the ends of the beams are let into them: the tree-nail-holes are then put through from the outside, and the tree-nails driven through and wedged, by which means the inside and outside plank are secured; the butt-bolts are afterwards driven, which completes the fastening of the plank.

The *Orlop Beams* are substantial pieces of oak timber, scarphed, dowelled, and bolted together; they are laid athwart-ships to support the orlop-deck, and to keep the sides of the ship together by means of chocks with their iron plate-knees, or wooden knees.

The *Lower-deck clamps* are also thick plank worked withinside the ship, and fastened in a similar manner to the orlop clamps above-mentioned.

The *Lower-deck beams* are also substantial pieces of oak timber, scarphed, dowelled, and bolted together; they are laid athwart-ships to support that deck, &c. by means of chocks and iron plate-knees, which are bolted both to the chocks and to the ends of the beams; these plate-knees have ears attached to them, and are made to receive two in and out bolts; these ears substitute lodging-knees.

The *Lower deck hook*, is a large piece of compass-timber fixed withinside athwart the bow of the ship, to which it is firmly bolted: its upper side is trimmed to the round of the beam to receive the plank of the deck.

The *Lower-deck spirketting* is thick plank wrought within board; its lower edge is wrought sufficient distance from the beam to admit the waterway, and the upper edge kept well with the upper side of the lower port-sills, and fastened similar to the orlop-clamps, as before mentioned.

The *Transom-knees* are pieces of compass-timber, bolted to the wing-transom and to the sides of the ship, in the direction of the transom; being the principal security to the ends of the wing-transom.

The *Upper-deck ports* are next trimmed out to their proper size, and the heads and heels of the timbers cut off, in order to let in the port-sills.

The *Upper-deck port-sills* are let in as the lower-deck port-sills before-mentioned.

The *Channel Wale* is thick plank worked between the upper and lower-deck ports, the chain and preventer-bolts being driven through it.

The *Upper-deck clamps* are thick plank worked within-side the ship, and fastened similar to the lower-deck clamps.

The *Upper-deck beams* are substantial pieces of oak timber, scarphed, dowelled, and bolted together; they are laid athwart-ships to support that deck, &c. by means of chocks and plate-knees, the latter being bolted to the former and to the ends of the beams, as in the lower-deck.

The *Upper-deck hook* is a piece of compass-timber fixed within-side athwart the bow of the ship, and bolted, &c. as the lower-deck hook.

The Upper-deck transom is a piece of oak timber, whose upper side is trimmed to the round of the beam, and bolted to the counter-timbers to receive the after end of the deck plank: its ends are fastened by knees, which cast under the upper-deck beams, and are securely bolted to the transom and side of the ship.

The *Upper-deck spirketting* is thick plank wrought within board, similar to the lower-deck spirketting above-mentioned.

The *Quarter-deck and forecastle ports* are then trimmed out to their proper size, and the port-sills are let in as the upper-deck port-sills above-mentioned.

The *Sheer-strake* is thick plank wrought on the top-sides, and its upper edge is kept well with the top timber-lines, or top of the side.

The *Forecastle and quarter-deck clamps* are thick plank worked withinside the ship, and fastened as the upper-deck clamps.

The *Forecastle and quarter-deck beams* are substantial pieces of oak timber, scarphed, dowelled, and bolted together, &c. the same as the upper-deck beams.

The *Quarter-deck transom* is trimmed as the round of the beam, and bolted to the stern-timbers to receive the after-ends of the deck-plank; its ends are fastened by knees, which cast under the quarter-deck beams, &c.

The *Forecastle-deck hook* is a piece of compass-timber, fixed withinside athwart the bow of the ship and bolted thereto, as the upper-deck hook before-mentioned, in consequence of the beak-heads being discontinued, and round bows introduced into the service.

The *Quarter-deck and forecastle spirketting* is wrought within board, &c. similar to the upper-deck spirketting; and the short stuff between the ports is worked and sufficiently fastened.

The *Round-house clamps* are worked withinside the ship, and secured like the quarter-deck clamps.

The *Round-house beams* are of small scantling, generally of one piece, placed athwart-ships; their ends are secured by chocks and plates, or hanging and lodging-knees.

The *Round-house transom* is bolted to the stern-timbers, and trimmed to the same round as the beams; a rabbet is taken out of the upper side to receive the deck planks, and its end is secured by iron knees.

The *Limber-streak* are thick oak plank, wrought near the kelson with a rabbet trimmed out, on the midship-side, to receive the limber-boards.

The *Thick-plank*, at the floor and first futtock-heads, are applied to strengthen the ship at the different heads and heels of the timbers.

The *Foot-waling* or *ceiling* is composed of plank which is placed between the several pieces of thick-stuff.

The *Breast-hooks* are compass-pieces of oak timber, fayed and bolted athwart the bow of the ship, where they are the principal security.

The *Crutches* are also compass-pieces of oak timber, fayed and bolted on the foot-waling abaft, for the security of the after-part of the ship.

The *Sleepers* are also compass-pieces of oak-timber, fayed and bolted to the transoms and sides of the ship within board, in a diagonal direction, to strengthen the buttock.

The *Bends, of Riders in the hold*, are interior timbers which combine and strengthen the ship; they are fayed on the foot-waling; and cross the kelson, their upper ends continue as high as the orlop and lower-deck beams, and are fayed and bolted through the sides of the ship; they are distinguished by floor, and first, second, and third futtocks.

The *Limber-boards* are composed of short pieces of oak, having their grain placed fore-and-aft, except in the wake of the hatchways, where it runs up and down; they serve to keep the dirt, &c. from getting into the limber-passage; their lower edges are fitted into the rabbet or the limber-strake, and the upper edges fayed and fitted against the sides of the kelson.

The *Pillars* are straight pieces of oak timber erected perpendicularly under the middle of the beams to support the orlop; their heels are chased into the kelson, and their heads tenoned into the beams.

The *Bulk-heads in the hold*, are various partitions separating one part of the ship from the other; they are built with rabbeted and cyphered-edged planks: those erected in the main hold have their seams battened over; those for the magazines and spirit-room are lined with thin deal, and have a cement applied between them to keep the powder and spirits from accidents by fire, &c.

The *Grand magazine* is composed of racks, battens, drawers, &c. for containing the powder.

The *Light-room* is a place forward, separated from the grand magazine; and in order to give light to the latter, the lanterns are fixed in it between strong oak jambs, with splaw-boards cased with lead and tin, to secure them very firmly in their places; and to prevent any accident happening to the magazine, there are wire-guards fixed before the glass.

The *Coal-hole* is placed abaft the after-hold, for keeping coals, &c. in.

The *Spirit-room* is an apartment abaft the coal-hole, and is appropriated for the reception of spirituous liquors, &c.

The *Powder-room* is a convenient place abaft the spirit-room; it contains racks, &c. in which the filled cartridges are placed ready for action.

The *Bread-room* is close abaft the powder-room, and lined with feather-edged boards, with dunnage battens underneath them, to prevent any water that may spring from a leak in the sides of the ship, from damaging the bread.

The *Orlop* is the deck beneath the lower gun-deck, and is supported by an oak pillar placed under each beam, and framed with carlings and ledges to receive the plank, which are generally of three-inch fir. On this deck are erected the purser's, surgeon's, boatswain's, gunner's, and carpenter's cabin and store-rooms; the midshipmens births, the captain's, and lieutenant's store-rooms, the purser's slop-room, the steward's room, and a place for the marine clothing. The greater part of the said deck is also used for stowing and coiling the cables.

The *Lower-gun deck* is that next the orlop, also supported by an oak pillar under each beam, and framed with carlings and ledges to receive the plank, which is of four-inch oak, and on which the lower tiers of guns, consisting of 32-pounders, are placed.

There are also attached to this deck the cisterns for the pumps, the steps for the capstans, mizen-mast, bowsprit, and the riding-bits with their cross-pieces, to which the cables are secured when the ship rides at anchor.

The cabins for the junior lieutenant and chaplain, are erected abaft upon this deck, as also a gun-room for the midshipmen to mess in.

The *Upper-deck* is next above the lower gun-deck, and is also supported by an oak pillar under each beam, and framed with carlings and ledges to receive the deck plank, which is of three-inch fir, and on which the upper tiers of guns, consisting of 24, or long 18-pounders, are placed. On this deck the ward-room abaft, with the lieutenants, master's, and captain of marines' births are erected with screen bulk-heads, which are readily taken down on coming to action. The cook-room, with the fire-hearth, are placed abaft the fore-mast; and the topsail-sheet and jear-bits are also placed on this deck.

The *quarter-deck* is next above the upper-deck, and is of three-inch fir plank; it is supported by oak pillars, except those that come in the wake of the jear capstan, and they are of iron; they are hung to the beams, and hooked up when the capstan is working. On this deck the captain's cabin abaft, and the steering-wheel, which works the tiller, are erected; also the fore brace-bits for bracing the yards; and the blocks; viz. the main-lifts, the main-top-sail halyards, and cheek-blocks: it also contains 24-pounder and 32-pounder carronades.

The *Forecastle*, which is also above the upper-deck, is of three-inch fir, and supported by iron pillars: on it the blocks belonging to the fore-tacks, cat, top-sail-halyards, and fore-lifts are placed.

The *Skid-beams* are of fir timber, fixed in a range between the quarter-deck and forecastle; they have two fir pillars under each of them, and are for the purpose of stowing away the spars, booms, boats, &c.

The *Round-house*, abaft, is that part above the quarter-deck, forming a covering to the captain's cabin; it is laid with three-inch fir plank, and contains a light for the captain's cabin, also the mizen-top-sail sheet-bits, and six 24-pounder carronades.

The *Knee of the head* is generally of oak; the pieces which compose it are of various forms and thicknesses, tabled or dowelled together: the aft-side is fayed and bolted to the stem of the ship, and its lower end continued downwards to the gripe.

The *Rudder* is that useful machine attached to the stern-post by pintles and braces; it is formed of oak and fir timber, the principal of which is called the main-piece, which reaches sufficiently above the upper-deck to allow a hole to be cut in it for the reception of the spare tiller, in case the working tiller on the lower gun-deck should be rendered unserviceable in action, or otherwise.

The *Cheeks to the knee of the head* are knee-pieces of oak timber supporting the knee; their after-arms are bolted to the bow of the ship, and their fore-arms to the knee of the head.

The *Cat-heads* are strong compass-pieces of oak timber, whose outer ends project from the ship's sides, and the inner ends are secured under the fore-castle beams: each of them has three sheaves at the outer end for the purpose of catting the anchor.

The *Main head-rails* are of two sorts, circular and straight; the former extends from the fore-side of the cat-head, to the back of the figure-head; and the latter has lately been introduced, having been proved a greater security to the knee of the head than the circular rail, and is also a saving both as to workmanship and materials.

The *Channels* are an assemblage of planks fayed and dowelled together edgeways; the fore and main channels are bolted to the side of the ship above the upper-deck ports, and the mizen channel above the quarter-deck ports, for the purpose of spreading the shrouds, in order to support the masts.

The *Figure-head* is the principal piece of carved-work or ornament at the head of the ship; it is scored over the knee of the head and bolted to the lace-piece.

The *Bolsters to the hawse* are large pieces of oak or elm timber, fayed and bolted to the bows of the ship between the cheeks: there are two holes cut through each of them in the direction of the hawse-holes, the foresides of which are made with an easy round to prevent the cables from chafing.

The *Timbers in the head* are narrow pieces of oak, standing nearly perpendicular; their lower ends are fayed and fastened on the upper cheek, and the upper ends bolted to the main-rail; the middle and lower head-rails being let into them and secured.

The *Cross-pieces in the head* are also pieces of oak let in and bolted to the main-rails; they have knees at each end, which secure the head-rails. Between the two main-rails the head is framed with carlings and ledges. There are also seats of ease in the head, for the convenience of the seamen.

The *Bumkins* are pieces of round fir timber, bolted to the middle of the main-rail; their inner ends are secured by cleats against the bows, and the outer ends project from the rails for the purpose of hauling down the fore-tack.

The *Rails of the stern* are of oak, fayed to the *round-aft* of the stern and to the *round up*, and bolted to the timbers: they are distinguished thus; the tuck-rail, which is the lowest; the lower counter-rail, which is above the latter; the upper counter-rail; the foot-rail, breast-rail, and also the cove. The stern is then planked up with two-inch rabbeted plank, which is fastened to the timbers.

The *Taffrail is the upper part of the stern*, and composed of light ornamented carved-work.

The *Munions of the stern* are broad pieces of fir about two inches thick; they are placed equidistant from each other, and fastened to the stern-timbers to receive the sashes.

The *Stools of the quarter-gallery* are thick pieces of plank fayed together edgeways, and bolted to the sides of the ship. Of these there are three sorts, viz. the lower, middle, and upper stools, for building the quarter-gallery.

The *Quarter-pieces* are substantial pieces of fir timber, forming the boundaries of the stern, and connecting the quarter-galleries thereto, and to the taff-rail; their upper ends are bolted to the ship's sides, and the lower ends extend downward to the middle stool, where they are secured by iron knees.

The *Quarter-rails* are of oak, and form the boundaries of the quarter-galleries: the lower rails are termed the rails on the lower stools; the next the breast-rails, foot-rails, and the rails on the upper stools.

The *Munions to the quarter-galleries* are of fir, about two inches thick; they are placed equidistant from each other, and fastened to the stools and rails to receive the sashes.

The *Lower finishings* are solid pieces of ornamental carved-work, fastened to the ship's sides under the quarter-galleries.

The *Upper finishings* are also ornamental carved-work, placed upon the upper stools, and fastened at each end with nails. The dead-eyes and chains are let into the channels, and secured thereto; the chain-plates are then let on, and the chain and preventer-bolts driven and secured.

The *Port-lids to the lower gun-deck* are a sort of shutters, hung by hinges, to enclose the ports in tempestuous weather. These port-lids have scuttles cut through them for the admission of fresh air, and lids hung to them with illuminators in the centre, in order to give light between decks.

The *False-keel* is of elm, and is fastened in a temporary manner, under the main-keel, with nails and copper staples made for that purpose, to prevent the latter from being chafed, should the ship at any time strike the ground, in which case the false-keel gives way, and the main-keel is preserved.

The *Launch of the ship*, or the bilge-ways, by which she is conveyed into the water, require some consideration to ascertain with what declivity the ways must be laid; for large ships they are generally from three-quarters to seven-eighths of an inch to a foot; and for the small classes from one inch to one inch and an eighth. The utmost care should always be taken that the fore-foot be clear of the after ground-way, when launching.

In order to give a more comprehensive idea of the different timbers, we have exhibited the disposition of the frame of a second-rate ship, in fig. 13, plate XXIV. where

A is the main-keel.
B the stern-post.
C the stem.
D the knight-head.
E the knuckle-timber.
F the side counter-timber.
G the midship ditto.
H the wing-transom.
I the filling-transom.
J the deck-transom.
K the seven transoms under the deck-transom.

L the after fashion-piece.
M the middle ditto.
N the foremost ditto.
O O O the floor-timbers, which cross the keel.
P P P the first futtock.
Q Q Q the second ditto.
R R R the third ditto.
S S S the fourth ditto.
T T T the top-timbers.
V V V the lengthening pieces to the fourth futtock.
U U the gun-deck port-sills.
W W the middle-deck ditto.
X X the upper-deck ditto.
Y Y the quarter-deck and forecastle ditto.
a a the round-house ports.
b b the quarter-deck and forecastle ditto.
c c the upper-deck ditto.
d d the middle-deck ditto.
e e the lower-deck ditto.
f the entering-port.
g g g the doorways to the quarter-gallery.
h h h h h the hawse-pieces.
i i the hawse-holes.
k k the double futtocks in the fore and after-bodies.

ⒶⒸⒺB D F H K M O Q, the station of the square-timbers of the frame in the fore body.

s u w y &, the station of the cant-timbers of the frame in ditto.

1, 3, 5, 7, 9, 11, 13, 15, 17, 19, and 21, the station of the square-timbers of the frame in the after-body.

23, 25, 27, 29, and 31, the station of the cant-timbers of the frame in ditto.

It may, however, be necessary to observe, that, as the theory ought always to precede the practice, this article would probably be much better understood by previously perusing that of NAVAL *Architecture*, which may be considered as a proper introduction to it.

It has been asserted in the Third Report of the Commissioners for revising the Civil Affairs of the Navy, that, "where we have built exactly after the form of the best of the French ships that we have taken, thus adding our dexterity in building to their knowledge in theory, the ships, it is generally allowed, have proved the best in our navy; but whenever our builders have been so far misled in the science of naval architecture, as to depart from the model before them in any material degree, and attempt improvements, the true principles on which ships ought to be constructed have been mistaken or counteracted, and the alterations, according to the information given to us, have, in many cases, done harm.

"From the same cause there has been infinite variety in the alterations made, and in the forms which have been adopted. The alterations being founded on no certain principles, no similarity in form of the ships could be expected, and they have the appearance of having been constructed on the chance, that, in the multitude of trials made, some one might be found of superior excellence. While, therefore, our rivals in naval power were employing men of the greatest talents, and most extensive acquirements, to call in the aid of science for improving the construction of ships, we have been rather remiss in quest of such discoveries as chance might bring in our way.

"Nothing certainly can be more surprising than that, in a nation so enlightened as this, and whose power, importance, and even safety, depend on its naval superiority, matters so essential to its preservation should so long have been neglected.

"As a remedy for this great evil, it has been proposed, that the ships of each class or rate should be constructed in every particular according to the form of the best ship in the same class in our navy; of the same length, breadth, and depth, the masts of the same dimensions, and placed in the same parts of the ship, with the same form and size of the sails."

It has been generally reported, that most of the plans of our ships of war have heretofore been determined by the surveyors of the navy in favour of some chance draught which may have succeeded; or by the encomiums lavished on some prize-ship by the officer who may have captured her, &c. All this is perfectly natural, and, if it stopped here, might not be much amiss; but the mischief follows: each, in turn, suggests some *change* in the figure of the ship, by which she is to become a paragon of excellence; for instance, a little more sheer,—a little more breadth of beam,—a little more height between decks, &c. without considering how small a deviation from the original draught will alter the line of flotation, and affect her sailing; change the centre of gravity, and affect her stability; and, instead of improving, destroy every good quality which she before possessed. The disadvantages arising from such a variety of models are of serious importance, particularly when they meet with damage at sea and require immediate repairs, as scarcely any two ships take the same-sized masts, yards, &c. these therefore cannot be kept in store to supply accidental losses.

Until of late, most of the youths intended for shipwrights, in the royal dock-yards, were placed under the eye of a superior workman, and continued five or six years at the practical part of their business. At the expiration of that time, young men of the best character and abilities, were admitted into the mould-loft, in order to acquire the knowledge of the theory of ship-building; namely, drawing, laying off ships, &c. From this situation they were frequently employed as overseers, to superintend the building of ships of war, which were to be built by contract in our merchants yards: hence they gradually imbibed such a knowledge of ship-building as to qualify them for officers in that science. We mention this, in order to point out, by way of contrast, the success that may be expected to result from a superior class of shipwrights apprentices, which, by the King's Order in Council, September the 20th, 1809, has been established at Portsmouth dock-yard. This class consists of twenty-five young men of liberal education, who, before admittance, must be examined by the professor of the Royal Naval College, and the instructor in the theory of naval architecture. Their mornings are occupied in the study of mathematics and mechanics, and in the application of them to naval architecture, in drawing the different parts of ships, and making complete draughts and plans. The remainder of the day is employed under the master shipwright in the mould-loft, and in all the various kinds of manual labour connected with ship-building, as well as in the management and conversion of timber, so as to make them fully acquainted with the duties of a practical shipwright. The last year of their apprenticeship is to be served at sea, to afford them an opportunity of acquiring some practical knowledge in the steering, sailing, trimming, and ballasting of ships, &c. during which, the order directs that, "they shall mess with the officers, and be treated in all respects as gentlemen."

Nothing can be more advantageous to this country than such an establishment; and a number of young men of the highest promise have been already entered, who will undoubtedly excel the French in the science of naval architecture as much as our shipwrights at present surpass theirs in the practice of the art. For farther particulars, see the article SHIPWRIGHTS *Apprentices*.

Discontinuing the ceiling or inside lining of ships, substituting diagonal trusses, and filling up the intermediate spaces between the timbers with old wood not applicable to any other purpose, have been lately adopted in the building of many of his Majesty's ships, particularly the Tremendous, built by Mr. Seppings, which was launched without breaking or *hogging* the tenth part of an inch; she sails better than most ships of her class, is perfectly dry, betrays no signs of weakness, and is in every respect what may be called a crack-ship.

Instead of exposing a ship six years on the stocks, with all the inconveniences and ill consequences arising from so long a period; twelve months would be quite sufficient to complete a ship, when the necessary quantity of dry and duly-seasoned timber was collected and prepared on the spot.

This, however, is not the practice with those built for the royal navy; trees which have been felled six years, and trees which have scarcely left the forest as many months, greet one another in the same ship. These timbers are closed up with plank of many inches thick, which must necessarily be soaked with wet, in order to bend it to the frame of the ship; and, as it becomes dry shrinks from its work. Mr. Pering, therefore, suggests the propriety of building ships under cover. This is not a new idea, as the Swedes build all their ships of war, at Carlscrena, in covered docks. The slips in the naval arsenal at Venice are roofed; the French have a covered dock at Brest; and it was reported even by the Commissioners of the Navy to the Commissioners of the Land Revenue, in 1789, that, "if ships were built under cover, and roofs continued over them while they remain on the slips, it would be a means of rendering them still more lasting."

Short as the duration of our present ships of war is, it becomes considerably abridged in those cases in which North American oak is mixed in any quantity with our own; because it is extremely susceptible of the dry rot, which is forthwith communicated to any wood that comes in contact with it. Nothing, therefore, but necessity, and the apprehension of a deficiency of English oak, should induce government to use that of American growth.

England, though the first naval power in the universe, and equal to all the rest united, yet she is deficient in proper accommodations for her navy. Her dock-yards have risen from small beginnings to their present state, by a succession of expedients and make-shifts. The navy which they have created has altogether outgrown them; they are wanting in extent, and in the conveniencies due to so magnificent a fleet. The expense of a covered dock would be doubly saved in the first line-of-battle ship built in it; and the workmen would perform their work sooner and better, because no kind of weather would interrupt their progress; the ship would be calked, painted, and coppered when dry, to the exclusion of 'partial leaks, suffocating damps, and oozing drip.'*

The consumption of timber in a 74-gun ship amounts to 2000 trees of about two tons each; and the expense of a three-decker, in the hull alone, is nearly 100,000*l.*: these are powerful reasons for attending to any expedient which can produce a saving either of our timber or labour.

The number of Shipwrights necessary for building ships of war within twelve months are as follow:

Men.	Guns.	About Tons.
47	74	1,700
27	36	900
11	18	430
9	Brig	380
7	Gun Vessel	180

Many useful hints relative to the construction and preservation of ships have been given by Richard Pering, Esq. of Plymouth dock-yard, in his pamphlet on the "Causes of premature Decay in our Wooden Bulwarks, &c." published in 1812, who thus briefly sums up his new mode of ship-building: "Convert the timbers, set up the frame, and finish the ship out of the way, without at all caring whether the timbers are green or not; let her stand to season, but by no means let a calking-iron approach her sides for two years at least; no more of her bolts should be driven than may be sufficient to hold her together, as every aperture should be left open for the circulation of air; no tree-nails should be used on any account, but the work should be fastened with copper alone, whenever it is practicable. The advantages are, that the timbers, plank, bulkheads and all other parts of the ship, would be equally and properly seasoned together; and the calking of a ship to be done just before she is put into the water, when her plank has so shrunk as to be likely to shrink no more. Every part of the ship would thus be as dry as possible; no fungus, drip, nor unwholesome damp would arise to endanger the health of the ship's company; the sides of the vessel will be both *wind-tight* and *water-tight*; the plank swelling upon the oakum will unite with it, and form one solid body. When the seasoning is deemed complete, then let the screws on the ends of the bolts

* We are happy to find, since the above was written, that the Commissioners Report, and Mr. Pering's suggestion of building ships under cover, are now generally adopted in his Majesty's dock-yards.

be hove up, so as to bring wood and wood together in the closest contact. The ship would now be as tight as a drum, water-proof, and healthy throughout."

The successes of our navy against almost all the European powers have been so signal, and our maritime ascendancy is so complete, that the public are inclined to imagine that little necessity exists for improvement in this most important department; and it was with no small surprise that the nation received from Mr. Pitt, when he began, in the year 1804, to expose the errors of Lord St. Vincent's management, a notice that our ships of war were then in a condition far inferior to that in which they stood at the comparatively inglorious æra of 1792. Succeeding administrations have devoted ample sums to the construction of new vessels; and if number were all that we require, our situation might be deemed completely satisfactory: but the more we investigate the management of our navy, the more room shall we perceive for the introduction of reform.

In addition to what has been said on this article we shall refer our readers to the 9th volume of the NAVAL CHRONICLE for some useful naval projects, on the *Forms and Properties of Ships*: Also to the 16th volume of the same work for the late Admiral Sir Charles Knowles's *Disquisition on Ship-Building*, combined in X letters to Mr. Story.

SHIP-MASTER, (*capitaine marchand,—armateur ou proprietaire*, Fr.) the captain, commander, or owner of a ship or vessel.

SHIPPING, (*vaisseaux, bâtimens, navires en général*, Fr.) a multitude of vessels lying near to each other, as: "the harbour is crowded with shipping."

SHIPPING *Order*, (*ordre de chargement*, Fr.) is a written mandate directed by a merchant to his lighter-man, ordering him to receive and put on board a certain ship the goods specified in the order. This order is generally delivered to the manufacturer or trader who makes up the goods to be shipped, and who sends it down with them to the water-side.

SHIPWRECK, (*naufrage*, Fr.) the destruction of a vessel that is driven among rocks, or stranded on a lee-shore, with an intermediate space of deep water between it and the land. Who ever has or may discover any mode by which even a partial alleviation of this destructive evil in a commercial country like this, surrounded by the sea, can be effected, may be said not only to have deserved well of his country, but of mankind in general; among this number was Mr. John Bell, of the royal artillery, afterwards promoted to a lieutenancy, who contrived a method of saving persons from stranded ships, in the year 1791, by providing a mortar between five and six hundred weight, with a chamber, &c. capable of containing one pound of powder, and a bore to admit a leaden ball 60 pounds or upwards. When a ship is stranded, the mortar is to be elevated about 45 degrees, and a rope or deep-sea-line is to be fastened by one end to the ball, while the line itself, being coiled round parallel handspikes, may be easily and rapidly unfolded to its full extent. On the discharge of the mortar towards the shore, the ball will carry with it the line or rope, and by burying itself in the earth make that end of the rope fast, while the other end is in the stranded vessel: thus the rope becomes stretched between the vessel and the shore, and a communication thence established by means of rafts. The rafts recommended by Lieutenant Bell are each formed by lashing together with ropes five empty water-casks belonging to the ship, and tying above them a seaman's chest, with holes cut in its sides, to prevent its filling, and to allow the person who rides in it better convenience for taking hold: pullies are attached to this chest, through which the rope is to run: the raft is also to be ballasted underneath, to prevent its upsetting.

The mortar and necessary balls or shells that would be used on such an occassion might form a part of the ballast of the vessel; and whenever a ship is driving or unmanageable near the shore, the mortar might be brought on deck, and the apparatus prepared. In some cases grapnels may be advantageously fired from common ordnance to answer the same purpose.

The following is a relation of some trials made before a committee of the Society of Arts at Woolwich, in August, 1791.

From a boat moored about 250 yards from shore, the shell was thrown 150 yards on shore, with the rope attached to it; the shell was of cast iron, filled with lead; it weighed 75lbs.; its diameter 8 inches; the rope in the trial was a deep-sea line, of which 160 yards weighed 18lbs.; the angle of the mortar, from whence the shell was fired, was 45 degrees. By means of the line, Lieut. Bell and another man worked themselves on shore upon his raft of casks; there were many links in the rope, but they were cleared with ease by Lieut. Bell, with the assistance of his match-blocks.

The second trial was repeated in a similar manner, and with equal success, the shell falling within a few yards of the former place; the gale of wind was brisk, and the water rough. The direction of the shell was nearly from north to south, and the wind blew nearly north-west.

In the third trial, the mortar was elevated to 70 degrees; the rope attached to the shell was an inch and half tarred rope, of which every 50 yards weighed 14½lbs. The shell was of the kind above mentioned; it fell 160 yards from the mortar, and buried itself about two-thirds in the ground; the line or rope run out was about 200 yards, and it required the force of three men to draw the shell out of the ground at that distance.

There can be little doubt that, in many cases when vessels are stranded near the shore, the adoption of the means pointed out by Lieutenant Bell would tend to the preservation of many valuable lives; and since a suitable piece of ordnance, with a block carriage and leaden balls, would not cost above 11 or 12 pounds sterling, the expense furnishes no objection of moment. Indeed, in trading vessels, such a piece might farther answer the purpose of making signals of distress, by filling the chamber with powder, and well wadding it, when the report of the firing would be heard to the distance of some miles: and, in defence of a ship, such a gun would be highly useful, on account of the facility which its shortness gives to its loading and firing.

The Society of Arts, &c. rewarded Mr. Bell, at that time a sergeant in the royal artillery, with fifty guineas for his invention: they published the particulars in the tenth volume of their transactions, and thought it expedient again, in the year 1807, to insert a descriptive engraving that had been omitted at that time, with some farther particulars in the 25th vol. Models and drawings of the whole apparatus are reserved in the Society's repository for the instruction of the public.

To save the lives of shipwrecked persons, and to rescue mariners from a watery grave, seems, even in idea, a contemplation so eminently benignant, and in its execution, to a warlike, commercial, and generous nation like this, an undertaking so truly benevolent, so extensively philanthropic, that we cannot sufficiently applaud the humanity of the inventor of an apparatus to save the lives and property of the shipwrecked mariners in the extremity of danger.

As there are few in this great nautical nation that are not directly, collaterally, or generally connected with *marine affairs* and marine adventures; the consideration, therefore, takes strong possession of the feelings of individuals, and of society at large, by the patriotic means of protecting the lives of a class of the most important members of the latter, and giving the former, that kind of moral security, and, consequently, mental happiness, which must arise from the perusal of a small 8vo. volume, just published by Capt. G. W. Manby, of the royal navy, and an honorary member of the Humane Society.

The Captain informs us, that while he was stationed at North Yarmouth, in the year 1807, a place remarkable for the beautiful features its coasts present in a *calm*, and for the terrific distortion of them in a *storm*, he had been witness to the loss of several vessels; and, after describing the dreadful catastrophe, he very appropriately observes, that,

"The horrors of shipwreck at a distance from land, a scene I had unhappily too often witnessed, suggested to me also the benefit that must result from enabling a life or pilot-boat to go over a flat beach, with facility and certainty, to the relief of sufferers. This observation was corroborated by the opinions of various residents on the coast, whose assurances gave evidence, too conclusive, of the many lives and immense property annually lost for the mere want of the means I have hinted at: the chief difficulty was, that no boat could be forced over a high and raging surf without some powerful artificial aid.

"Among other desiderata that have grown out of my researches in this ardent and important work, there is one I presume to think of much worth.—Boats, in common, where occasion has required their being launched in cases of shipwreck, have often failed, owing to their want of buoyancy, and other properties of the life-boat. I have devised a simple method of giving to every kind of boat these advantages, and at a most inconsiderable expense. Thus every vessel provided with a boat so prepared, would possess within herself the power of preserving the life of any person fallen or washed overboard, in a high wind and heavy sea: and boats around the coast, by the same means, may be converted into life-boats."

From this benevolent idea, this treatise, in which Captain Manby has reduced *theory* to *practice*, has emanated: we are, consequently, much gratified to find, that Parliament has, "in the most liberal manner," been pleased to countenance his undertaking; and, by an ample allowance, afforded to him every encouragement: he, therefore, hopes, "that Great Britain will ere long be guarded with this additional belt of succour for the unfortunate."

"And why", he philanthropically observes, "may not the glorious cause of humanity be fostered on foreign shores? It would be heart-cheering indeed to learn, that the invention of an Englishman had been adopted for the salvation of the shipwrecked in every clime, and on every coast!"

This work contains representations of the apparatus, with directions for using it in the assistance of persons on board stranded vessels on a lee-shore, in the day, as well as in a similar situation at night; and also contains the following Advertisement:

That the benevolent intentions of such persons who view the importance of seamen's lives in a National *light, or are disposed by philanthropy to carry into effect the application for alleviating the distresses of the shipwrecked, may not be discouraged under the apprehension of heavy expense, Captain* Manby *avails himself of this opportunity for explaining the precise cost at which he will engage to furnish the essential part of the apparatus; the other materials will not exceed a few pounds.*

"*A mortar of a six pounder calibre, with shot ready strapped for immediate application, whether for effecting communication from the shore, to be used in a boat for the relief of vessels on a bar, or grounded at a distance from the land, or from a ship; for by an observance of the instructions to be given, it is now rendered complete, and it becomes an object of considerable importance to ship-owners and masters, that each vessel may carry the means of her crew's preservation, and which means the crew may convert into a most formidable engine of defence against the attack of an enemy in the attempt of boarding, or when closely engaged—7l. 17s. 6d.*

"*A large mortar, of twenty-four pound calibre, with barbed shot strapped, &c. capable of hauling any boat from a beach—15l. 15s.*

"*The remainder of the apparatus may be purchased and constructed at any place, as his* Essay on the Preservation of Shipwrecked Persons, *explains the whole process, and (the work is strongly recommended to the attention of the lord lieutenants, magistrates, and inhabitants of maritime counties) accompanies each mortar, from which any person may fit up the entire apparatus, without incurring the expense of carriage of the materials.*

"*Orders for Capt. M.'s apparatus are received at the office of Henry Abbott, Esq. Navy Agent, Clement's-inn, London.*"

In addition to what has been said on this article, we have the pleasure to state that, An Act of Parliament was passed on the 2d July, 1813, which enacts, that there shall be allowed to British Governors, Ministers, Consuls, and Merchants, residing in foreign parts for the maintenance of all sea-faring men and boys, subjects of the United Kingdom of Great Britain and Ireland, that shall be shipwrecked, captured, or by other unavoidable accident, be driven or cast away, or that shall be discharged, as unserviceable, from any of his Majesty's ships or vessels at such foreign ports or places where such Governors, Ministers, Consuls, or Merchants reside; and to every Master or other person, having charge of a ship or vessel, taking on board and subsisting such seafaring men and boys, in the manner expressed in the act 31. Geo. II. c. 10. such sum of money per diem as the lords commissioners of the Admiralty, from time to time, shall, in their discretion direct. See also the articles Buoys, Cork-Jacket, Life-Boat, Life-preserver, Seaman's *Friend and Bather's Companion*, &c.

SHIPWRIGHT, (*charpentier de navire*, Fr.) one who is skilled in the art of building ships from certain draughts; and also in the theory and practice of constructing such draughts for ships of various rates and classes.

The company of shipwrights were instituted in the reign of James the First, and consisted of a master, two wardens, and sixteen assistants. It appears in the 12th volume of the Archæologia, that Mr. Phineas Pett, who had been educated at Emanuel College, Cambridge, and afterwards served his apprenticeship in Deptford dock-yard as a shipwright, was appointed master shipwright at Woolwich dock-yard, in November 1605. He was elected and sworn Master of the shipwrights company in April 1606, and their meetings were then held at the King's-Head, in New Fish-street. That a new charter was granted in 1612 for incorporating the shipwrights of England, when Mr. Pett appears also to have been ordained the first master.

According to Derrick's Memoirs of the Royal Navy, he became an assistant commissioner in February, 1630, and a principal commissioner of the navy in December following. In 1637, by order of Charles the First, he built the famous ship named the *Sovereign of the Seas*, the largest that had ever been built in England; she was afterwards cut down and called the Royal Sovereign. See the article *History of* Naval *Architecture* for farther particulars.

In Pepy's Naval Minutes it is mentioned, "that Shipwrights-Hall did anciently view and approve of the draught of the ships that were to be built for the King;" and by the Lord High Admiral's warrant, dated in April, 1638, carpenters were not to be appointed to ships until they had been examined and licensed at Shipwrights'-Hall. It appears from Mr. Pett's Manuscript in the British Museum, that it was customary, in his time, to hire and victual the shipwrights and calkers on any emergency, and to discharge them when the work for which they were employed was performed. It also appears that there was at that time a small permanent establishment of artificers for ordinary service in each of the dock-yards.

Previous to the year 1691, the master shipwrights and artificers at our naval ports were borne on board one of his Majesty's ships of war, fitted up purposely for their reception. The average number employed in the several dock-yards in the year 1700, were about 1780; and in 1800, the number of shipwrights were 3776, an increase by no means proportionate to the augmentation of the number of ships and tonnage of the navy during that period, which may be accounted for in some degree by the increased quantity of work that is now performed by any given number of artificers in that class, and by the great saving of labour from the introduction of coppering and copper-bolting of ships, as they do not now require docking near so often as formerly, in order to have their bottoms cleaned and their bolts replaced; for the copper bolts do not decay as the iron bolts did.

Under the description of shipwrights, are included the persons employed in the occupations of mast-making and boat-building, which in private concerns are generally considered as separate branches; but in our dock-yards, the execution of such work is under the direction of distinct officers, denominated master mast-maker, and master boat-builder.

The ship-wrights are formed into gangs or companies, over whom officers, called quartermen, are placed; each of those gangs, according to the directions of the Navy Board, should consist of twenty men and six apprentices: but the apprentices depend on the proportion they bear to the working shipwrights. The quartermen have the selection of their respective companies; each, according to his seniority, nominating one man in turn, beginning with those who have apprentices.

The artificers and labourers in our dock-yards, when hurt in his Majesty's service, are attended by the surgeons, and are allowed 2*s.* 1*d.* per day, whatever class they may belong to, for six weeks, unless they are able to resume their labour at a shorter period. There is likewise an establishment of superannuation for the different classes, to which they are admitted, if rendered incapable of labour from hurts received, or after an uninterrupted service of thirty years.

Master Shipwright, a superior officer appointed to each of his Majesty's dock-yards, to superintend and direct the building and repairing of the different classes of ships, &c. He has the direction and superintendence of nearly the whole operative business of the dock-yard to which he is appointed; in the execution of which he is assisted by other subordinate professional officers, termed "Assistants to the Master Shipwright;" one of whom is particularly entrusted with the management and conversion of timber, and is styled "Timber Master," who has also an assistant for measuring all the timber taken into the yard. It is likewise his duty to inspect the quality of all stores received from contractors, which are used in his department, and to attest their fitness for the service; to survey and value all vessels hired or purchased into his Majesty's navy; to keep an account and certify to the Navy Board the quantity of all works performed by contract in the dock-yard; and to keep an account of the earnings of the respective artificers under his superintendence, and to certify the amount from time to time to the clerk of the check. Over the shipwrights and calkers, an assistant is appointed, who conducts and directs the execution of the work, which is also considered under the superintendence of the master shipwright.

Foreman of the SHIPWRIGHTS, a class of officers between the quartermen and shipwright's assistants, whose duty it is to direct and superintend the building and repairing of ships, and the several works in the docks. There are also *foremen afloat*, who are appointed to survey and report to the master shipwright the state and condition of the ships and vessels afloat.

SHIPWRIGHTS' *Apprentices*, in his Majesty's dockyards, are boys bound to certain persons under the direction of government, for a term of seven years, who, in the course of their apprenticeship are instructed by some of the deserving shipwrights, in the art of building, calking, and repairing ships and vessels. They are now divided into two classes, viz. the superior and inferior.

Superior Class. By the King's Order in Council, September 20th, 1809, a superior class of shipwrights apprentices, consisting of twenty-five young men of liberal education, has been established at the royal naval college in Portsmouth dock-yard: and on the 5th of November, 1810, the principal officers and commissioners of his Majesty's navy, gave notice that a plan of education had been established for them.

The number of students in this class was at first limited to twelve, but have gradually increased to twenty-five.

The period of apprenticeship is seven years, but for a student who may have previously served in the inferior class of apprentices in his Majesty's yards for the space of two years, it is only six years; and only five years for a student who may have previously served in the said class for the space of three years.

An instructor in the theory of naval architecture is added to the present establishment of the Royal Naval College, to assist particularly in the instruction of the superior Class of Apprentices.

After admission into this class, the apprentices attend at the Royal Naval College three or four hours every morning except Sundays and the established holidays, where they are instructed in such parts of the following subjects, as appear to the Professor of the College to be best adapted to the object of the Establishment, viz. the Formation of able and useful shipbuilders:

Geometry,	Fluxions,
Algebra,	Theory of Naval Architecture,
Plane Trigonometry,	
Mechanics,	Drawing,
Hydrostatics,	The French Language.

On Mondays, Tuesdays, and Wednesdays, the apprentices are employed in the college during the hours above mentioned, in the study of Mathematics, under the Professor or Instructor; on Thursday and Friday mornings in Drawing, under the Drawing Master, the art of Perspective being most particularly applied in drawing the different parts of a ship; and Saturday mornings are appropriated to the study of such books and exercises in the French Language, under the French Master, as may perfect them therein, and at the same time advance their professional knowledge. This division of time allotted for their studies is, however, subject to such alterations as appear to the Professor best calculated for promoting the object of this Institution.

In the remaining part of the day, when not attending at the College, they are employed under the Master Shipwright or Instructor of Naval Architecture, in the Mould-Loft, in the various kinds of manual labour connected with ship-building; in the acquirement of a complete knowledge of the different parts of a ship, and of the method of constructing and fastening the same, as well as in discriminating her different properties, deducible from theoretical calculations; also in the management and conversion of timber, and generally, in such other manner as the master shipwright considers best adapted to make them fully acquainted with a full detail of the duties of a practical shipwright.

At the end of each year, a Public Examination takes place, officers mentioned, and a general statement of the improvement of the respective apprentices in the theoretical part of their profession is transmitted by the Professor of the College, and of their progress in the practical part by the master shipwright, through the Commissioner of the Dockyard, to the Admiralty Office: to the end that, if it shall appear that an apprentice has particularly distinguished himself in any of the departments of his profession, the Lords Commissioners of the Admiralty may confer on him such mark of their approbation, as their lordships may deem expedient.

The last year of their apprenticeship, or such part thereof as the Lords Commissioners of the Admiralty may think proper, is intended to be served at sea, in such of his Majesty's ships as their lordships shall direct, with a view of obtaining some practical knowledge in the steering, sailing, trimming, and ballasting of ships; in making remarks on the motion and working of a ship, in all her parts, in bad weather, and on a high sea; the standing and straining of the masts and yards under such circumstances; the placing of the fixed blocks, with a view to the proper leading of ropes, &c.; and, generally, in obtaining such information as the captain, carpenter, and other officers of the ship may, by their experience, communicate, and such as their own observations may suggest.

During the time that such an apprentice shall be on board, it will be required of him to keep a journal of his remarks and observations, arranged under distinct heads, noting such improvements in any part of the ship, or any thing connected therewith, as he may conceive, from a minute attention to the circumstances by which a ship is affected at sea; a copy of which journal is to be transmitted to the Admiralty.

A cabin, or birth, is to be allotted to him in the gun-room, or in some other suitable part of the ship: he is to mess with some of the officers, and be treated in every respect as a gentleman.

Salaries to be allowed Apprentices of the Superior Class.

If the Apprentices have not served two Years in the inferior class.	If the Apprentices have served two Years in that class.	If the Apprentices have served three Yrs. in that class.	Sum.
1st Year	....	....	.... £.60
2d Year	1st Year	....	 70
3d Year	2d Year	1st Year	 80
4th Year	3d Year	2d Year	 90
5th Year	4th Year	3d Year	100
6th Year	5th Year	4th Year	110
7th Year	6th Year	5th Year	130

Upon their return from sea, until they can be appointed officers, in consequence of vacancies occurring, they are to be employed as overseers of ship-building: in assisting the master shipwrights in the different yards, and their assistants or foremen in the mold-loft; under the timber-master, or in any other situation in which the master shipwright may have occasion to employ them, with the approbation of the commissioner; during that interval they are to have £.150 per annum; and if such vacancies should not happen for three years after their return, they are then to have £.180 per annum, until they be provided for.

In order that the services of young men, whose education has been obtained at a considerable expense to government, may not be lost at the expiration of their apprenticeship, they are each of them, previous to their entry in the yard, required to find two sureties to enter into bond to the amount of £.500; and in the event of an apprentice being discharged for misconduct, the amount of the bond, or so much of it as may indemnify government for the expense incurred in his education, is to be forfeited.

The object of this institution is, to improve the qualifications of the shipwrights in the dock-yards, and thereby to insure a succession of artificers, from among whom officers of the description of master measurers, foremen, &c. will be selected; and from these, the appointments of the superior officers in his Majesty's dock-yards, and surveyors of the navy, will ultimately be made.

Inferior Class. The inferior class of shipwrights apprentices are taken at 14 years of age, provided they stand 4 feet 10 inches high, and can read and write, and have some knowledge of figures. By a recent order they are taught, in the fourth year, the art of calking, which formerly they learnt the last year of their servitude. During the first four years they are now allowed one half of their earnings, the other half goes to those who are appointed to instruct them; and the last three years they have £.30 per annum, provided that sum exceeds one half of their earnings; if not, they are still only allowed one half of their earnings.

They have also a further allowance called chip-money, which is paid as follows: 4*d.* per day for the first four years, and 6*d.* per day for the last three years. After serving 30 years in the constant employment of government from the expiration of their apprenticeship, they are entitled to superannuation, and have from £.20 to £.24 per annum for life.

SHIVER. See the article SHEAVE.

To SHIVER, (*fasier, en parlant des voiles*, Fr.) when speaking of a sail, to shake or flutter in the wind, as being neither full nor aback, but in a medium between both, as well with regard to its absolute position, as to its relative effect on the vessel; as, "shiver the mizen top-sail!"

SHIVERS, a term given to the foul particles that are taken from the hemp when hatchelling.

SHOAL, (*banc, batture, ou bas-fond*, Fr.) a term synonymous with shallow.

SHOAL-*Water*, (*eau peu profonde*, Fr.) implies a very little depth of water.

To SHOE THE ANCHOR, (*couvrir les pattes de l'ancre avec des planches, pour la faire mieux tenir dans un fond de vase molle*, Fr.) is to cover the flooks with a broad triangular piece of plank, whose area, is greater than that of the flooks. It is intended to give the anchor a stronger and surer hold in very soft or oozy ground.

SHOE *of the Anchor*, (*sabatte de l'ancre*, Fr.) a small block of wood, convex on the back, and having a small hole sufficient to contain the point of the anchor-flook, on the fore-side. It is used to prevent the anchor from tearing the plank on the ship's bow, when ascending or descending; for which purpose the shoe slides up and down along the bow, between the flook of the anchor and the plank, as being pressed close to the latter by the weight of the former.

SHOE-*Block*. See the article BLOCK.

To SHOOT A BRIDGE, (*enfiler l'arcade d'un pont, ou porte le courant*, Fr.) is to go through that part of a bridge through which the main current runs; as, "we shot through London bridge."

To SHOOT *a-head of a Ship*, (*dépasser un vaisseau*, Fr.) is to run a-head of her by superior sailing.

SHORE, (*côte de la mer, rivage, plage*, Fr.) the general name for the sea-coast of any country.

A Bold SHORE. See the article BOLD.

SHORES, (*acores, ou épontilles des vaisseaux sur le chantier*, Fr.) are props or stanchions fixed under a ship's side or bottom, to support her on the stocks, or when laid on the blocks in the slip.

SHORT-ALLOWANCE MONEY, is a pecuniary allowance made to the officers and seamen of any of his Majesty's ships, for the period they have been necessitated to subsist on a diminution of the established allowance, which is deemed so from the scarcity of provisions, and generally arises from the service sent on not being performed in the time limited or allowed for that purpose, either at sea or on a foreign station.

When short-allowance money is due to the ship's company, the captain, with the assistance of the purser, is to cause three alphabetical lists to be prepared according to the form prescribed in the Naval Instructions, containing the names of the officers and seamen belonging to the ship within the time of short allowance; and after comparing the same with the muster and sick books, and being satisfied that they agree together, he is to send them signed by himself, the master, and purser, to the Commissioners of the Victualling, or to their officer appointed to pay the same.

The short-allowance money is to be paid every three months, if abroad; and where there is no agent victualler, the purser, by warrant of the captain, may take up money to discharge the same upon the best exchange, and draw bills upon the Victualling Board for that purpose.

If the ship is alone, the captain is directed to act as a commander in chief, or senior captain.

The purser, by the first opportunity after payment, is to send one list to the commissioners of the victualling, and another by the next safe conveyance.

The ship's company is to be paid for their short allowance in sterling value, and to have the benefit of the exchange. See the Naval Instructions.

From the difficulty that has been experienced of late years, in procuring change of monies to pay each man separately, by reason of its scarcity, several commanders have had recourse to the method prescribed for the payment of savings of provisions, substituting only, on the headings, the words "Provisions short allowed," instead of savings, &c. still adhering in every other particular to that form.

It may be necessary here to point out the facility with which the payments are made by the adoption of this method, which is as follows:

From 6 to 12 men are paid at a time, *as its rear*, except in the small ships or vessels wherein a less number of men than six are messed: each mess in a body, is called in turn before the purser, or his steward, some time previous to the payment, when one person from each mess is appointed to receive for the whole, whose name is inserted in the list made out in a column appropriated for that purpose, which prevents the probability of any dispute arising thereafter.

But should any of the mess, previous to the payment, have been detached from the ship on public leave or service, who were entitled to the whole or any part of money due for provisions short allowed, not then returned, the amount due to them is withheld, and paid into their hands on joining their ship; yet, in the event of their discharge by order to another ship, during such absence, a ticket of their servitude is made out and forwarded to them, stating on the face of it the sum due for the provisions short allowed. This money is received either when the ship, which they are then in and belonging to, is paid, or, if discharged from the service before, by reason of infirmity, it is then paid to them. The vouchers made out, and the disposal of them, are similar to those for Savings of Provisions, which article see.

The buying of short-allowance money, directly or indirectly, is strictly forbidden; and the parties themselves are to be paid, without any regard to notes, or any other obligation whatever.

Short-*Hauls*, in rope-making, are hauls of yarn short of the common length.

Short-*Laid*, in rope-making, implies short-twisted.

To SHORTEN, (*raccourcir, &c.* Fr.) when expressed of a ship's sails, is used in opposition to make. See that article, also Sail.

SHORTS, in rope-making, the toppings and tailings of hemp, which are dressed for bolt-ropes and whale-lines.

Shorts also imply the distinction between the long hemp used in making staple-ropes, and inferior hemp.

SHOT, (*boulet, et en général toutes sortes de charges pour les canons*, Fr.) a name given to all sorts of balls used for artillery and fire-arms, from the pistol to the cannon: those for cannon and carronades being of iron, and those for small-arms of lead.

Cannon shot is distinguished by the nature of the ordnance for which it is intended; as, 42, 32, 24, 18, 12, 9-pounders, &c. and those for small-arms by the names of the pieces; as musket, carbine, pistol, &c.

Shot that is globular is frequently called Ball, which article see.

The shot now used in the royal navy is principally confined to three sorts, namely, round, grape, and case or canister.

Round-Shot, (*boulet rond*, Fr.) is the most simple, and composed of a ball or globe of cast-iron, whose weight is in proportion to the size of the cannon, or to the diameter of its bore.

Grape-Shot, (*charge à la suédoise, ou charge en grappe*, Fr.) is a combination of balls, fig. 6. plate V. put into a canvas bag, and corded strongly together, so as to form a sort of cylinder, whose diameter is equal to that of the ball which is adapted to the cannon.

Case or Canister-Shot, (*charge à mitraille*, Fr.) fig. 7, is formed by putting a quantity of small shot into a case or canister.

The tin case is cylindrical, in diameter a little less than the calibre of the gun or howitzer. It is filled with lead balls, so as to make up the weight of the shot. These balls are seldom less than 1¼ oz. in weight. But little effect is to be expected from firing case-shot beyond 300 yards, from the very great divergency of the balls.

The proportions of shot for the royal navy are,

Round for Ships	Lower-Deck . . .	60 round	each gun	for Chan. Serv.
		80	do.	for Foreign do.
	Middle, Upper, and Quarter-Deck, .	70	do.	for Chan. do.
		100	do.	for Foreign do.
Ditto for Sloops		50	do.	for Chan. do.
		60	do.	for Foreign do.

Grape and case-shot are supplied each in the proportion of from 5 to 12 rounds for every gun, as the nature of the service may require.

There are other sorts of shot formerly used, such as the chain, cross-bar, langrel, &c. but these are now nearly discontinued in the British navy.

Chain-Shot, (*boulets enchaînés*, Fr.) fig. 8, consists of two balls linked together, being principally designed to destroy the masts and rigging, which they are better fitted to perform than the single bullets.

Double-headed or Bar-Shot, (*boulet ramé, ou boulet à deux têtes*, Fr.) fig. 9, is a ball cut into two equal parts, and joined together by a bar of iron. In the French service the middle is sometimes filled with a composition, and the whole covered with linen dipped in brimstone; the cannon in firing also inflames the combustibles or composition of this ball, which sets fire to the sails of the vessel. One of the heads of this ball has a hole to receive a fuse, which communicates with the charge of the cannon.

Langrel-Shot. See the article Langrel.

Spherical Case-Shot, or *Shrapnell's Shell*, so termed from the inventor Colonel Shrapnell, of the royal artillery, is a cast iron shell, one half being solid, the other hollow, to contain a number of small shot and powder to fill up the interstices, into which a fuse is driven, which, when projected from the mortar, takes fire, and communicates to the charge. When the shell arrives at a given distance, it explodes, and the solid part, which always goes foremost, receives an additional impulse to its flight; the explosion, however, does not change the direction of the shot, they, therefore, complete the curve, or tract of the shell.

Although we could enter very fully into this destructive, and hitherto unparalleled, mode of annoying an enemy, we shall abstain from giving any particulars which might lead to a discovery of the invention.

The following explanation of the effects and advantages that might be derived from firing this species of shot, is extracted from a book lately published. We should not, otherwise, have thought ourselves justified in saying so much upon the subject.

1st. The whole charge takes effect on the enemy at any distance. By the present mode of firing the case shot, the greatest part of the charge disperses as soon as it leaves the muzzle of the gun, and cannot be directed.

2d. Grape, or case shot, may be fired with an effect equally close and collected, to any distance within the range of the piece; and the artillery need not advance within musket-shot of the enemy, to make use of this sort of fire with its full effect, and are not so subject to have their guns charged either by cavalry or infantry.

3d. It requires less precision and exactness to point a piece of ordnance charged with spherical case shot than with round shot, because case shot is a wide and dispersed fire, and the difficulty in elevation consequently less.

4th. Its comparative destruction with that of round shot will be, generally, as the number of shot within the shells to one; viz. a three-pounder, twenty-two to one in its favour; a six-pounder, fifty to one, &c. in which calculation is not enumerated any effect from the splinters of the shell.

5th. Small balls cannot be projected to very considerable distances, unless enclosed in heavy spherical cases, which from their form and weight, are not much influenced by the resistance of the air, or diverted from their direction.

6th. The explosion of the shell makes no change in the direction of the shot within it; they consequently complete the shell's tract, or curve, which has sometimes been observed to be 400 yards.

7th. From the unevenness of the ground, such as hillocks, banks, fallow-fields, &c. all shots which graze most commonly lodge; whereas, by using this shell, the whole charge will be carried over these irregularities, and reach the object with its full contents.

Firing these kind of shells from guns is managed with more facility than the ordinary howitzer practice, both as to the length of fuse, and to the elevation required, and may be carried on in the field precisely the same as firing round shot.

The method of making Fuses for Spherical Case-Shot, is as follows:

The fuses, after being turned so as to fit the fuse-holes, are bored, and a deep thread grooved inside, to hold the composition firm: and instead of being turned with cups, they are hollowed conical and roughed with a tool that cuts under, the better to receive the priming.

After they are driven with fuse-composition, one and a half inch, they are sawed across the top about one fifth of an inch down, so as not to touch the composition, and divided into five equal parts, of two-tenths of an inch each; after which a bit of quick-match is placed across, and drawn tight in the same grooves; they are then primed, with mealed powder and spirits of wine, capped and packed for service.

Star-Shot consists of four pieces of iron, whose bases, when separate, form the quadrant of a circle; so that the whole, being joined, forms a cylinder equal to the shot of the cannon. Each of those pieces is furnished with an iron bar, the extremity of which is attached to a sort of link, as keys are strung upon a ring. Being discharged from the gun, the four branches or arms extend every way from the link in the centre. These also are chiefly intended to destroy the sails or rigging, but their flight and execution are very precarious at a tolerable distance.

A Shot of a most destructive nature has been invented by a Mr. Fane, and experiments on its effects were exhibited, in 1811, before the Lords of the Admiralty. It is a four-pounder shot, wrapped round with a prepared cotton, and made very hard, so as to appear like a large cannon-ball—on firing of which, it has the usual effect of a cannon-shot; but the moment it starts from the cannon's mouth, it presents one solid mass of fire; and whatever it may happen to strike, whether rigging or hull of a ship, it will immediately take fire.

To find the weight of an iron shot, whose diameter is given; and the contrary. *Rule.* Double the cube of the diameter in inches, and multiply it by 7; so will the product (rejecting the two last or right hand figures) be the weight in pounds. *Example.* What is the weight of an iron shot of 7 inches diameter? The cube of 7 is 343, which doubled is 686, and this multiplied by 7 produces 4802, which, with the right hand figures rejected, gives 48 pounds, the weight required.

To find the diameter of the shot, when the weight is given. *Rule.* Multiply the cube root of the weight in pounds by 1.923, and the product is the diameter in inches. *Example.* What is the diameter of an iron shot of 52 pounds? The cube root of 52 is 3.732, which multiplied by 1.923, gives 7.177 inches, the diameter required.

Rule, by Logarithms.

To one-third of the logarithm of 52	0.572001
Add the constant logarithm	0.283979
Logarithm of the diameter ...... 7.117 =	0.855980

Table of Diameters of English Round-Shot.

Nature.	68	42	32	24	18	12	9	6	3	1
Inches.	8	6.684	6.105	5.475	5.043	4.403	4.000	3.498	2.775	1.92

Table of Grape-Shot for Sea-Service.

Nature.	Weight of each Shot.		Total Weight of the Grape complete.	
	lbs.	oz.	lbs.	oz.
24 pounder	4	—	46	6
32	3	—	34	1
24	2	—	25	5
18	1	8	19	15½
12	1	—	10	15
9	—	13	7	6
6	—	8	5	8½
4	—	6	3	14½
3	—	4	2	10½
½	—	¾ lead.	—	8¼

Small shells, as $4\frac{2}{5}$-inches, and hand-grenades, were quilted into grape for 13-inch mortars, at Gibraltar. The fuses were turned inwards next the iron tompion, and leaders of quick-match for communicating fire to the fuses were introduced through holes made in the wooden bottom, and placed as near the fuse as possible, in the centre of the grape. These answered very well for short ranges.

Table of English Case-Shot for Sea-Service.

Nature.	Sea Service. Weight of each Shot.	Sea Service. Number in each Case.	Sea Service. Weight of each Case filled.	Carronades. Nature.	Carronades. Weight of each Shot.	Carronades. Number in each Case.	Carronades. Weight of each Case filled.
Prs.	oz.	No.	lbs. oz.	Prs.	oz.	No.	lbs. oz.
32	8	70	33 8				
24	8	42	22 15				
18	6	42	16 8				
12	4	42	11 5	68	8	90	46 2
9	3	44	8 9	42	8	66	32 8
6	2	40	5 2	32	8	40	21 4
4	2	28	4 —	24	8	32	16 1
3	2	20	2 15	18	6	31	12 2
1	1¼	12	1 2¼	12	4	32	8 2

Nature.	For Mortars. Weight of each Shot.	For Mortars. Number of Shot in each.	For Mortars. Weight of Case filled.	Howitzers. Weight of each Shot.	Howitzers. Number of Shot in each.	Howitzers. Weight of Case filled.
Inches.	oz.	No.	lbs. oz.	oz.	No.	lbs. oz.
10	—	170	91 8	—	—	— —
8	6	90	38 4	6	90	38 8
5½	3	55	12 6	3	55	12 8
$4\frac{2}{5}$	2	55	8 1¾	2	55	8 2

If at any time case-shot cannot be procured, or tin for the purpose of making cases, a very good substitute may be made by the following method:

Take a wooden bottom of ¼ of an inch less than the ordinary size, roll strong cartridge paper which has been sponged over two or three times with water, and nearly dry, on a former of the same diameter to the proper dimensions; paste down the exterior edge of the cylindrical case, and nail it on to the bottom with small tacks; when dry, pour in a quantity of hot pitch and rosin, or kitt composition; then a layer of musket balls; contiuing them alternately till the case is filled in length equal to two diameters of a round shot, independent of the wooden bottom; over the last tier put another bottom two inches in thickness, and nail it on.

To Shot *the guns* (*mettre les boulets dans les canons*, Fr.) is to load the pieces of ordnance with the necessary quantity of gunpowder and ball.

Shot-*Boxes* (*caisse à munitions*, Fr.) are boxes in which grape, case, and small-arms' shot, are packed for service.

Shot-*Gauge*, an instrument for measuring the diameter of shot.

Shot-*Locker*. See the article Locker.

Shot-*Pile*. See the article Pile.

Shot-*Racks*, wooden frames bolted to the coamings and head ledges round the hatch-ways on the decks, to contain the different species of shot.

To SHOVE Off, (*pousser au large*, Fr.) a term used by those who have the management of boats, to move a boat clear off the shore, which is done by the help of the boat-hook, &c.; as, "Shove off the boat!"

SHOULDER, in block-making, is a projection made upon the surface of blocks, pins, &c. by reducing one part to a less substance.

Shoulder-*Block*. See the article Block.

Shoulder *of Mutton Sail*. See the article Sail.

SHROUDS, (*haubans*, Fr.) a range of large ropes extended from the mast-heads to the right and left sides of the ship, to support the masts and enable them to carry sail, &c.

The shrouds are always divided into pairs or couples: that is to say, one piece of rope is doubled, fig. 5. plate XXIII., and the two parts fastened together at a small distance from the middle *a*, so as to leave a sort of noose or collar *a b* to fix upon the mast-head. This collar being fixed in its place, viz. close down upon the trestle-trees *k*, fig. 2, plate XVI. a pair of shrouds hang from it, whose lower ends ought to reach down to the deck: the lower ends of these shrouds are set up or extended to the channel I. on the outside of the ship, by the application of mechanical powers, as explained in the articles dead-eye and lanyard.

The shrouds, as well as the sails, &c. are denominated from the masts to which they belong: they are the main, fore, and mizen shrouds; the main-top-mast, fore-top-mast, or mizen-top-mast shrouds; and the main-top-gallant, fore-top-gallant, or mizen-top-gallant shrouds.

The number of shrouds by which a mast is sustained, as well as the size of rope of which they are formed, is always in proportion to the size of the mast, and the weight of sail it is intended to carry.

The two foremost shrouds on the starboard and larboard sides of the ship are always fitted first upon the mast-head; the second on the starboard and larboard, and so on till the whole number is fixed. The intention of this arrangement is to brace the yards with greater facility when the sails are close-hauled, which could not be performed without great difficulty if the foremost shrouds were last fitted on the mast-head, because the angle which they would make with the mast would then be greatly increased.

Bentinck SHROUDS, are strong ropes seized on the futtock-staves of the lower rigging, and extending to the opposite channels, where they are set up by means of dead-eyes and lanyards, in the same manner as the other shrouds; their use is to relieve or support the masts when the ship rolls.

Bowsprit SHROUDS, are shrouds put over the head of the bowsprit, and extended on each side to the ship's bows to support the former.

Bumkin SHROUDS, are strong ropes fixed as stays or supports to the bumkin-ends, to prevent their rising by the efforts of the fore-tacks upon them.

Futtock, or Foothook SHROUDS, are pieces of rope communicating with the futtock-plates above and the catharpings below, and forming ladders, whereby the sailors climb up to the top-brim.

The Top-mast SHROUDS, are extended from the top-mast-heads to the edges of the tops, as expressed by fig. 3. plate XVI. and in plate XX. The lower dead-eye, as represented by *q*, fig. 3. plate XVI. employed for this purpose, is fitted with an iron band, called the foot-hook-plate, which passes through a hole in the edge of the top, and communicates with a rope called the foot-hook shroud, whose lower end is attached to the shrouds of the lower mast, in the station *l*. The upper ends of the foot-hook shrouds are furnished with an iron hook, as shewn at *n*, fig. 2. plate XVI. which enters a hole in the lower end of the foot-hook plate, so that when the top-mast-shrouds are extended to secure the mast, the foot-hook shrouds necessarily acquire an equal tension by means of the foot-hook plate, which, passing through the top, transmits the effort of the mechanical powers to the foot-hook shrouds below.

The Top-gallant SHROUDS, are extended to the cross-trees, as represented by *m*, fig. 3. plate XVI., where passing through holes in their ends, they continue over the futtock-staves of the top-mast rigging, and, descending almost to the top, are set up by lanyards passing through thimbles instead of dead-eyes.

To Set up the SHROUDS. See the article *To* SET.

To SHUT IN, (*fermer*, Fr.) is said of landmarks or points of land, when one is brought to cover the other or intercept the view of it.

SHUTTING, in anchor-making, is the act of joining or welding one piece of iron to another.

SICK-BAY, or SICK-BIRTH, (*poste aux malades*, Fr.) in line of battle ships, is a place immediately under the forecastle, usually fitted up on the starboard side, and appropriated to the reception of the sick and wounded seamen.

It is partitioned off by wooden frames screened with canvas, and may be easily taken down and folded on coming to action. When there is not sufficient room in the sick-bay for the patients, other places are fitted up between decks, &c.

SICK *and Hurt and Transport Board for Prisoners of War*, (*comité d'administration, ou commissaires chargés de pourvoir aux besoins & de veiller aux soins des gens de mer malades ou blessés, dans les divers hopitaux, &c; ils brevêtent les chirurgiens, & leurs aides.—Ils sont aussi chargés du soin des prisonniers de guerre & des objets qui y ont rapport*, Fr.) consist of six commissioners and a secretary, one of whom is always physician, who are called commissioners of that board. Their duty consists in arranging, in the most advantageous manner, the various departments relating to the prisoners of war in the United Kingdom, with respect to receiving, discharging, exchanging, victualling, and clothing them. It also extends to our naval hospital departments so far as victualling, clothing, and supplying medicines, necessaries, &c.; and they have the appointment of all officers thereto, except governors, chaplains, physicians, and lieutenants, who are within the patronage of the Admiralty Board. They appoint 1st and 2d lieutenants, stewards, and clerks to prison-ships; also all medical officers to the navy.

The medical commissioner has the examining of the surgeons and assistant surgeons, intended for the royal navy, in physic, chemistry, &c. after they have passed Surgeons Hall; he has also the examination of all the surgeon's journals. The commissioner's power is united in the appointment of all officers within the limits of the board, which has an immediate intercourse with the Admiralty Board.

Regulations for SICK *and Hurt Seamen on board their own Ships.* In every ship convenient room must be made between decks, for the reception of sick and hurt seamen, who are placed under the immediate care of the surgeon, or assistant where no surgeon is allowed, who visits them twice a day, or oftener, if occasion requires.

The captain is to appoint some of the ship's company, by turns, to serve the sick, night and day, and keep the place clean, to be washed with vinegar and fumigated, if the surgeon shall deem it necessary.

By orders from the captain, the cooper may make buckets out of the old staves and hoops; and the carpenter cradles for the use of the sick, and those who have fractured bones.

When men are taken ill with infectious fevers, they are to be stripped at their entry into the sick birth, and, if practicable, washed with soap and warm water; their clothes are also to be immersed and washed in boiling water, in order to destroy infection, and the strictest attention must be paid to the cleanliness of the patient. The surgeon is also to direct in regulating the diet of the sick, according to their several symptoms and disorders, restricting salted provisions, and such articles as he may judge improper; in cases where wine is required, he is to apply to the captain, that the purser may supply them with the quantity he shall judge necessary.

When fish is caught for the ship's company, he is to give the captain a list of the men who stand most in need of such refreshment, that they may be the first attended to.

Regulations for sending SICK *and Hurt Seamen from their Ships for Cure.* Sick men are never to be sent to hospitals, neither attending the ship, nor on shore, except when their distempers or hurts render it inconvenient to keep them on board their own ships. In that case they must be sent under the charge of an officer, and a surgeon's assistant, as the case may require, with their clothes, bedding, and a sick ticket, expressing their names, entry, and number on the ship's books, with a case from the surgeon particularising the previous symptoms and treatment which he or his assistant has experienced and adopted.

An officer is never to be sent to an hospital except in cases of urgency, without the approbation of the commander in chief, who is to direct the captains to visit the hospitals, and enquire into the conduct of the medical attendants, &c. and to attend to any complaints the patients may have to make. The commander in chief, or senior officer present, must occasionally visit the hospital when other duties will admit of it.

The captain is to send an officer on discharging days, to receive such men as are recovered, and to take an account of such as are dead, discharged, &c.

When any are discharged from sick quarters, and do not return to their ship in twenty-eight days, a D.S.Q. is placed against their names, in order to stop their pay.

Captains are to receive such men as the agent of an hospital shall send to them; and if they think them unserviceable, they are to apply for their being surveyed.

Slop clothes are not to be issued to any man during his continuance in sick quarters, unless when it is absolutely necessary, and then they must be entered in the ship's books to be charged against his wages.

When ships are in a foreign port where there is not a naval hospital, nor any person appointed by the commissioners for sick and wounded seamen; when necessity requires, the commander in chief, senior officer, or captain present, must cause tents to be made for their reception with the old sails of the ship, and appoint the best qualified surgeon in the squadron, &c. to attend them. But if the raising of tents and the attendance of surgeons be impracticable, the commander in chief, &c. may contract with proper persons to supply the sick with lodgings, provisions, &c.; and to appoint some well-qualified medical man or surgeon of the place to attend them, who shall be allowed a specified sum for medicines and attendance.

With the sick men are sent the usual sick-tickets, which the surgeon is never to deliver to them, but is to send them to the captain of the ship into which they may be discharged; and he is to receive from the captain a certificate of the day on which the men are sent on board.

SICK-*List*, (*état des malades*, Fr.) a list containing the names of all those who are unable to do duty in the ship, and is daily sent up by the surgeon to the captain, for his inspection and guidance.

SIDE, (*côté d'un bâtiment*, Fr.) a name given to all that part of a ship which is presented to the view between the stem and stern, in a direction nearly perpendicular to the horizon. It is terminated above by the gunwale, and below by the lower edge of the main-wale, which separates it from the bottom: it is inclosed by the stern abaft, and by the bow forward.

The figure of the side is formed by that of the timbers upon which it is constructed. It is covered with plank, extending from one end of the ship to the other; it is also reinforced in different places by BEAMS, CLAMPS, KNEES, RIDERS, and STANDARDS, which articles see.

SIDE-*Trees*, in mast-making, are the lower main-pieces of a made mast. See the article MAST.

SIDED, among shipwrights, a term used to express the dimensions of any piece of timber, in contradistinction to its being moulded.

Wall-SIDED. See the article WALL.

SIDEREAL, or SIDERIAL, something relating to the stars; as, Sidereal Year, Day, &c. being those marked out by the stars.

SIDEREAL *Day*, is the time in which any star appears to revolve from the meridian to the meridian again, which is 23 hours 56′ 4″ 6‴ of mean solar time, there being 366 Sidereal days in a year, or in the time of 365 diurnal revolutions of the sun; that is exactly, if the equinoctial points were at rest in the heavens; but the equinoctial points go backward with respect to the stars, at the rate of 50″ of a degree in a Julian year, which causeth the stars to have an apparent progressive motion eastward 50″ in that time. And as the sun's mean motion in the ecliptic is only 11 signs 29° 45′ 40″ 15‴ in 365 days, it follows, that at the end of that time he will be 14′ 19″ 45‴ short of that point of the ecliptic from which he sets out at the beginning; and the stars will be advanced 50″ of a degree with respect to that point; consequently, if the sun's centre be on the meridian with any star on any given day of the year, that star will be 14′ 19″ 45+50‴, or 15′ 9″ 45‴ east of the sun's centre, on the 365th day afterwards, when the sun's centre is on the meridian; and therefore that star will not come to the meridian on that day till the sun's centre has passed it by 1′ 0″ 38‴ 57⁗ of mean solar time; for the sun takes so much time to go through an arc of 15′ 9″ 45‴; and then in 365 days 0^h 1′ 0″ 38‴ 57⁗, the star will have just completed its 366th revolution to the meridian.

SIDEREAL *Year*. See the article YEAR.

SIGHT, (*vue*, Fr.) a word used by seamen on discovering land, or a vessel at sea, &c.; as, "We have got sight of the land!" "Keep sight of that vessel!" &c.

SIGHT, (*la mire*, Fr.) a small piece of brass or iron which is fixed near to the muzzle of a musket or pistol, to serve as a point of direction, and to assist the eye in levelling; it also serves to fix the bayonet upon the barrel.

SIGN, in astronomy, a 12th part of the ecliptic, or zodiac; or a portion containing 30 degrees of the same.

The ancients divided the zodiac into 12 segments, called signs, commencing at the point where the ecliptic and equinoctial intersect, and so counting forward from the west to east, according to the course of the sun; these signs they named from the 12 constellations which passed those segments in the time of Hipparchus; but the constellations have since so changed their places by the precession of the equinox, that Aries is now found in the sign called Taurus, and Taurus in that of Gemini, &c.

The names and characters of the 12 signs, and

their order, are as follow: Aries ♈, Taurus ♉, Gemini ♊, Cancer ♋, Leo ♌, Virgo ♍, Libra ♎, Scorpio ♏, Sagittarius ♐, Capricornus ♑, Aquarius ♒, Pisces ♓; each of which, with the number of stars in them, see under its respective head, Aries, Taurus, &c.

Naval SIGNALS, (*signaux*, Fr.) are certain signs agreed upon for suddenly communicating intelligence to distant objects at sea, to which the voice cannot possibly reach: like the words of a language they become arbitrary expressions, to which we have previously annexed particular ideas; and hence they are the general sources of immediate intelligence throughout a naval armament, &c.

Signals are of the utmost importance, being all appointed and determined by the Lords Commissioners of the Admiralty, and communicated in the instructions sent to the commander of every ship of the fleet or squadron, before their putting to sea.

It is absolutely necessary that all signals should be simple, and composed in such a manner as to express the same signification at whatever mast-head or yard-arm they may be displayed from, in order to render them effectual. They should be issued without precipitation, exposed in a conspicuous place so as to be seen at a distance, and sufficient time should be allowed to observe and obey them.

Signals were used in the navy in Queen Elizabeth's time, and probably much earlier. James the Second, when Duke of York and admiral of England, commanding the British fleet in the Dutch war of 1665, was the first who reduced into a system a set of day-signals for directing the principal evolutions and movements of a fleet in the line of battle, &c. These signals were indicated by shewing flags at particular parts of a ship, and were comprehended in the *Sailing and Fighting Instructions* issued to the respective flag-officers and captains of the fleet. From that period, and until the close of the American war, they were practised in the navy; but the system now established on numerary principles, possesses the essential qualities of simplicity, perspicuity, and a variety of combinations easily understood, which before closing this head, we shall have occasion to notice in general terms, that the attention of naval officers may be excited to the cultivation and improvement of so interesting a branch of their profession.

All signals may be reduced into three kinds, *viz.* 1st. *Day Signals.* 2d. *Night Signals.* 3d. *Fog Signals.*

Day Signals, (*signaux de jour*, Fr.) are made by displaying flags, jacks, and pendants, and the firing of guns, either to windward or to leeward, as occasion requires.

The *Night* Signals, (*signaux de nuit*, Fr.) are made, by the combination of guns, lights, and false-fires, or rockets, and these are combined by multiplication and repetition; by which combination of signals, previously known, the admiral conveys orders to his fleet, every *squadron*, every *division* and ship of which has its particular signal.

Fog Signals, (*signaux de brume*, Fr.) are made by firing of guns at quick or slow intervals; and are answered by firing of muskets, beating of drums, and ringing of bells occasionally.

The guns to be fired at quick or slow intervals are usually left blank, and filled up at the discretion of the admiral, and therefore unnecessary to notice; but the following are some of the principal significations:

1. To denote being on the *starboard* tack.
2. To denote being on the *larboard* tack.
3. To anchor.
4. To denote being anchored, half an hour after.
5. To moor, or if before moored, to unmoor.
6. To weigh.
7. To tack.
8. To haul to the wind on the *starboard* tack.
9. To haul to the wind on the *larboard* tack.
10. To *wear*, the sternmost and leewardmost ships first in succession, and continue in the same situation, whether of sailing or lying-to, as before.
11. To alter *course* to *starboard.*
12. To alter *course* to *larboard.*
13. To *lie-to* on the *starboard* tack.
14. To *lie-to* on the *larboard* tack.
15. On discovering *danger.*
16. If in *distress* and in want of *immediate* assistance.
17. On striking and sticking fast on a shoal.

In the Instructions, it is observed, that the signal guns will be all fired to windward, or on the same side during the continuance of a fog.

The admiral, in the execution of his intentions to tack, wear, &c. will not begin to change his situation until ten minutes after the respective signals shall have been made, except upon any extraordinary emergency: meaning thereby to allow a sufficient time for the ships near him to be prepared for the same movement.

In case of a sudden shift of wind, or wind springing up after a calm, and not being then favourable for continuing the course before steered; or that the admiral thinks proper to steer a different course on such occasions, he will make the signals for *lying-to*, or the signals for sailing by the wind on either tack that will best correspond with his further intentions. And if the wind by such alteration, becomes favourable for sailing large on the course he intends to steer; he will soon after signify, by the proper signal, the number of points he means to steer from the wind on either tack accordingly.

During a fog, the admiral will continue the same course on which he was steering before the fog commenced, whilst the wind continues favourable for enabling him to do so: but if it comes a-head when before sailing large, or before the wind, he will keep the wind with a full sail, not to go more than three knots an hour. If the wind decreases, he will make sail in proportion, for continuing his former rate of sailing, except by setting studding-sails

If the fleet is to bring to under the top-sail, the main top-sail will be to the mast. If under courses, it will be with the main-sail, and the fore-yard braced up.

When in the channel, or on a coast, and the signal is made for anchoring, the ships are to anchor immediately. And it is to be remembered that the firing of muskets in vollies, is to denote the being anchored; it should, therefore, be continued some time in each ship, until there is reason to think the whole or greater part of the fleet have anchored.

The repeating frigates are not to repeat signals during a fog.

1. *Of Day Signals.* There are certain signals made by means of the ships, sails, and guns, independent of those with flags, which being few in number, it may be proper first to notice; thus,

SIGNIFICATION.	PLACE OR HOW MADE.
To prepare for sailing	*Fore*-top-sail loose, with 1 gun.
For all *boats*, and every person belonging to the fleet to repair to their respective ships	*Fore*-top-gallant-sail loose and hoisted: the top-sail yard down, with 1 gun.
To unmoor	*Main*-top-sail loose in the top.
To weigh	*Fore* and *Main*-top-sail sheets hauled home, with 1 gun.
To moor	*Mizen*-top-sail hoisted and clewed up.
To *cut* or slip the cable, leaving a buoy thereon	*Main* and *Fore* top-sails loose, with 2 guns.
On striking, and sticking fast on a shoal	Guns—till observed.

Compass SIGNALS, (*compas de signaux*, Fr.) are made with certain flags and pendants, and occasionally used as auxiliaries to the Numerical Day Signals to be hereafter noticed. When any general signal is accompanied with either of the Compass Signals, expressive of a particular point, the direction in which the ships are to proceed to carry the purpose of such signal into execution, is meant to be thereby denoted; as when the signal is made to chase, to alter course, &c.

On strange ships being discovered (and so of any other occasion when the bearing is necessary to be made known), the quarter is to be denoted by the proper Compass Signal; and when answered by the admiral, the Compass Signal is to be hauled down once for each strange ship seen; or the number may be indicated by a Numeral Signal.

John M'Arthur, Esq. LL. D. formerly secretary to Admiral Lord Hood, made various improvements in day and night signals, as well as telegraphic communications.

Early in the year 1790, he laid before the Earl of Chatham, then first lord of the Admiralty, a code of day signals, combining a tabular and numerary plan, on very simple principles, and susceptible of great variety and extent of ideas. He observes in his introduction, " that having been frequently stationed during the American war to observe signals in fleets, and with a view of obviating the difficulties that frequently arose from shewing flags at a particular part of a ship, as well as for facilitating the signification of all signals, it had been the pursuit of his leisure moments since the year 1782; when he first digested a compendium on a plan analogous, though not mature, and presented it to Admiral Digby, then commander in chief of his Majesty's ships and vessels in North America.

" The precise execution," he observes, " of all naval evolutions, is attainable by the good order and discipline of the respective ships, and a harmonious and speedy compliance with the signals. It is this which gives force and agility to the movements of a fleet, uniting or separating the divisions, or the bodies composing it, according to the exigencies of the moment; and the fewer evolutions that are put in execution in presence of an enemy the better, unless with a view of obtaining an advantageous position, to arrange the fleet in line of battle. In the formation or change of positions in all evolutions with large bodies, there must at times occur in the best disciplined fleets, a temporary disunion, separation, or irregularity of some of the ships, in getting in to that compact order necessary.

" The order of sailing, in as compact an order as the weather will permit, is essential for the due observance of signals, and for forming with celerity and precision the line of battle. In fleets from 12 to 20 sail, the order of sailing is generally in three squadrons or divisions, and in fleets consisting of 24 to 36 sail of the line, the order of sailing is usually in six divisions; for it is a known truth in naval as well as in military tactics, that the more compact or collected the order of sailing in divisions, or the order of march of armies in columns, are adhered to in approaching an enemy, the more power and agility they will have in the arrangement for action." As, therefore, the regularity and prompt execution of all naval evolutions depend on the nice observation and speedy understanding of ideas attached to signals, it was Mr. M'Arthur's aim to simplify them, though combining different plans, and to divest them of every ambiguity and misunderstanding. For that purpose, he not only devised the plan of signals alluded to, by partly uniting to the methodical ideas of Locke on his Common-place Book, a numerical moveable index of the flags; but also, had arranged and classed all signals under eight general and relative heads, viz. 1. *Anchoring.* 2. *Calling Officers and Boats.* 3. *Chasing.* 4. *Convoys and Prizes.* 5. *Fighting Evolutions.* 6. *Sailing Evolutions.* 7. *Private Ships Signals to the Admiral.* 8. *Miscellaneous.*

The compartments or leaves of the book appropriated to these heads respectively, were subdivided alphabetically into the emphatical words, or purport of the signals, connected with the general heads; thus serving as an index, incorporated with the table of flags and significations, for the admiral's greater promptitude and facility in making the necessary signals, as well as for private ships, immediately referring to the significations indicated.

The signals are comprised into two tables opening from the centre of the book. The tabular flags used are only twelve in number, which, with the aid of an auxiliary or *substitute* flag, to save the expense of having *duplicate* flags, express 312 literal significations, (independent of a vast variety of signals in a numeral sense to be hereafter noticed), being more than is required for any code of naval signals. The plan being simple and homogeneous, no pendants nor triangular flags are requisite, either for denoting the points of the compass, duty of launches and boats, or for distinguishing signals, denoting particular squadrons, divisions, &c. of the fleet. The whole, as well as private ships signals, are compressed into one code, &c. indicated by *thirteen* flags only, instead of *thirty* flags, besides a variety of pendants and triangular flags, &c. as were formerly, and are still used in the navy.

There is also a consideration of some importance recommended in the adoption of this plan, namely, that the captain of every private ship in the fleet being at all times furnished with the thirteen established signal flags, may on any emergency be immediately detached on a particular service, and thereby as a commanding officer can communicate his ideas, in the language of signals, with the same facility he was accustomed to obey them, while in a subordinate situation.*

Each table has twelve leaves annexed, corresponding to the number of tabular flags, and each page or leaf is subdivided into twelve different compartments, on a line with the flags denominated *inferior*, for the more speedy finding the signification of any signal. Flags of any colour or description may be used, but the twelve tabular flags selected for this code, are not only such as are capable of expressing a second arrangement of distinct signals when inverted† or reversed, but have also two colours contrasted in each flag, as are best seen at a distance, and according to the ascertained observations on the change or vicissitudes of the atmosphere, namely, *red and white*, *blue and white*, *yellow and white*.

In the *proper* order of hoisting the signals, the *red*, *blue*, or *yellow* colours of the flag, is always uppermost, or next the mast; in the *inverted* or *reversed* order, the *white* part of the flag, is *uppermost*, or *next* the mast. In the first table the flags are shewn *proper*, and in the second they are exhibited *inverted* or *reversed*. Hence twelve tabular flags, with a substitute flag, indicate, as already noticed, 312 literal significations; and by way of illustration, the first table is exhibited in Pl. XXVII. comprising 156 numbers referring to signals composed in sentences or distinct significations. The same flags inverted and reversed, form a second table (which being easily understood, it is not necessary to explain by a plate), making with the former 312 significations, besides the numerary and alphabetic plans of which the code is susceptible, indicating upwards of 20,000 additional ideas, all of which are illustrated by examples given in the introduction to these signals.

XXVII. Explanation of pl. XXVII. exhibiting the twelve tabular flags in the proper order,‡ with the substitute flag.

The twelve flags are exhibited horizontally in the upper column, and these identical flags are again shewn vertically in the lateral or left side column, and the substitute is shewn separately at the bottom. The numbers from one to twelve, placed immediately under the upper column of flags, not only denote signals appropriated to single flags, but are indices to refer to the compartments of significations allotted to each of the superior flags, whether hoisted singly or with an inferior flag.

The superior flags, with the index numbers attached, have the intervening leaves cut transversely, in order to exhibit at one view the compartment or leaf, on which the signification corresponding to the signal made is to be found, where casting the eye on the lateral or side column flag, the signification is instantly found.

Each table has twelve leaves annexed; that is, one leaf of paper for each flag, and each page or leaf is divided horizontally by lines of separation for the written signification of signals, into thirteen compartments. The superior flags with index numbers being cut transversely, in the form of a merchant's alphabet to a ledger; and the flags in the lateral columns being painted in their proper colours, on the twelfth leaf back from the opening of table, are, by this contrivance, always in view with the superior flag, both in front and whatever part of the book may be turned to; therefore when any signal is made, suppose with two flags, the observer has only to place his thumb on the upper flag; and turning over to its compartment, and looking on the left for the inferior, he will instantly find the same number as is shewn in the angle of meeting in the table, or opening of the Book, as well as the signification of the signals; while, to prevent mistakes, he has at the same time the superior and inferior flags of the signal constantly in view.

Thus, for example; if flag B were hoisted superior, and flag M inferior; or in other words, if N°. 2, *blue* and *white* flag in the upper column, was placed over N° 1, *red* and *white* in the lateral column, the number 25 will be found in the corresponding angle or square of the table; and by placing the thumb on the superior flag B, you turn over the cut leaves immediately to the page where the literal signification of the signal to the corresponding number 25 is expressed, in a line to the inferior flag M. If the same superior flag B were hoisted with the substitute flag, or which, in other words, implies in this tabular form a duplicate of the superior, the number 26 will be found in the angle of meeting, and the signification is written at length on the page back, on a line facing the duplicate or inferior flag on the side column; hence the substitute is here used merely to stand in the place of a duplicate flag, and giving a value to the tabular numbers as expressed in the angles of meeting or squares of the table. Thus the signals expressing the tabular numbers 13, 26, 39, 52, 65, 78, 91, 104, 117, 130, 143, and 156, have the superior and lateral flags represented as of the same denomination, the substitute in these instances is to be always hoisted inferior, as standing in place of a duplicate of the superior flag; but in the numerary application of the plan to be hereafter noticed, the substitute will express a repetition of the superior flag in corresponding units.

When the substitute flag is hoisted alone, in the tabular plan, it signifies annulling the preceding sig-

* It was usual to furnish captains of private ships with certain flags only for making signals to the admiral, but they were not supplied with the general signal-flags of the commander in chief; consequently no junior officer could be detached on any service without much detriment and inconvenience.

† The term *inverted* is to be understood when a flag is shewn upside down; namely, half red and half white horizontally divided, will be inverted, when the *white* is shewn uppermost.—*Reversed*, is applied when a flag is turned the contrary way to the mast, or where it is hoisted, namely, a red and white divided vertically, is reversed when the white part is next to the mast.

‡ *Red* colour is indicated in the plate of flags by vertical strokes or lines—*blue* by horizontal strokes, and *yellow* by small dots.

nal; and by being hoisted superior to any of the other twelve flags, it may be appropriated to particular signals, independent of the tabular or numeral ones.

The twelve flags, by being painted in duplicates on silk or vellum, may be made to shift or slide on thin box-wood attached to the upper and lateral columns, in order that the order or place of the flags may be changed or transposed to different numbers, when the admiral may think it necessary, so that nothing need be apprehended from the code of signals falling into the hands of an enemy. In this respect the substitute flag may, when hoisted over any other flag, indicate the transposition of that tabular flag to the place of N°. 1, and the other eleven flags are to follow such change of place in the progressive order they stand, whereby the tabular system, with respect to flags indicating numbers, becomes totally changed, while the original significations of the signals remain permanent.*

Numerary and Alphabetic Combinations, with the 13 Flags of the Table in Plate XXVII.

The plan is not confined solely to the simple tabular form already explained, and which would suffice to denote a greater number of signals than has heretofore been practised in the navy; but it is susceptible of a numerary combination capable of expressing 20,000 additional ideas applicable to so many words of a dictionary or vocabulary regularly numbered from 1 to 20,000. The letters of the alphabet can also be easily denoted, by applying as marked in the table the signals first in order from 1 to 26 inclusive. This is so obvious that it requires no further explanation.

For the better understanding of the numerary combination, reference will be first made to the ten flags numbered in the lateral column from one to ten, which, with the aid of the substitute, can express 10,000 signals, by hoisting the flags in the *proper* order as represented; and when these flags are shewn inverted or reversed, 10,000 additional ideas can be denoted.

1. *With one or two flags.* The signal flags from 1 to 9, hoisted singly, will express their corresponding numbers as units, and flag 10 is to be considered the cypher flag. When two flags are shewn at the same time, the one hoisted superior will express *tens*, and the under one *units*; thus, flag 2 over flag 4, would express 24: flag 3 over flag 10 or cypher, would express 30: and flag 4 hoisted over the substitute, would denote 44. The substitute in this combination being always considered a repetition in units of the number immediately above it; the signals, therefore, that can be expressed by one or two flags are 99 in number.

2. *With three flags.* When three flags are shewn at the same time, the superior denotes *hundreds*, the next *tens*, and the lower one *units*. Thus, if 346 were to be expressed, flag 3 would be hoisted superior for *hundreds*, the flag 4 next for *tens*, and flag 6 for *units*; consequently, the number of signals, with the combination of three flags, will be 999 in number.

3. *With four flags.* When four flags are shewn at the most conspicuous parts of a ship, the superior flag will denote *thousands*, and the three others *hundreds*, *tens*, and *units*, as in the preceding article. The number of signals, with the combination of 4 flags, will be 9999 in number.

Additional substitute flags. Having assigned one substitute flag for the first combination with two flags, which for distinction's sake is termed the *integral substitute*, it is proper that a second and third should be assigned to the other combinations, so as to leave no chasm in the progressive numbering of hundreds and thousands; therefore, the flag 11, with the letter W prefixed, is termed the *decimal substitute*, when three flags, including it, are hoisted: and flag 12, with the letter X prefixed, is termed the *centesimal substitute*, when four flags are hoisted.

Example with *three flags*. If 333 were to be expressed, flag 3 would be hoisted uppermost; next to it the *decimal substitute*, and the lowest would be the *integral substitute*.

Example with *four flags*. If 4444 were to be denoted, the flag 4 would be superior, the *centesimal substitute* next, then the *decimal* and *integral substitutes*. If 4044, the cypher-flag would be under flag 4 the superior, then the *decimal* and *integral substitutes*, and so on of the other numbers wherever a substitute flag was necessary, to the last in the combination; namely, 9999, which would be expressed by flag 9 superior to the three substitutes in their proper order.

To complete the series to 10,000, the *cypher-flag* is to be hoisted superior to the *three substitutes* in their natural order.

By inverting or reversing the numerical flags, the value of 10,000 is added to each signal, whereby 10,000 additional ideas on the same principles can be expressed; consequently, the two arrangements, with only 13 flags, will denote 20,000 signals or words of a dictionary.

Example in the inverted order of the tabular flags. If 10,023 were to be expressed, it would only be necessary to hoist flag 2 inverted over flag 3, also inverted. To denote 10,123, the three flags 1, 2, 3, would be inverted. To denote 12,345, the superior flag 2 for so many thousands would be shewn inverted, next flag 3 would be shewn inverted, then the diagonal flags 4 and 5 would be shewn *reversed*; that is, the white part next the mast, and so forth, always bearing in mind that by this order of inverting or reversing the flags, there are 10,000 added to the original numbers which the signals would otherwise express, if the flags were hoisted in the original or *proper* order.

The above plan of tabular and numerary signals was illustrated by plates containing a variety of essential evolutions in the order of sailing and lines of battle. But although it was allowed to possess great ingenuity, and met with a favourable reception from the Admiralty, there appears to have been an insurmountable obstacle to its adoption, principally, it was said, arising from a numerary plan of day

* This suggestion of a transposition-flag was afterwards adopted in the New Arrangement of Lord Howe's Signals, prepared by Mr. M'Arthur in 1792; and a single flag was then appropriated to that purpose, accompanied by separate instructions, which, we understand, is still continued in the present code of signals for the navy.

signals digested by Earl Howe, then in contemplation to have been brought into general practice when an opportunity offered.

Alterations and Improvements to the Code of Day and Night Signals now established in the Navy.

In the Spanish armament of 1790, and Russian armament of 1791, Earl Howe's day signals, printed in quarto, with instructions, also his night signals, were issued to his Majesty's ships under the orders of Admiral Lord Hood, commander in chief, then also one of the lords commissioners of the Admiralty. In the summer of 1792, when a rupture with the French Republic was apprehended, Lord Hood was appointed to the command of a squadron in the Channel, "*for the purpose of exercising the officers and men in the order of battle, and of sailing, and other duties in their several stations, which might best qualify them for service, preparatory to and in battle against an enemy.*"

During this service, in which a squadron of 15 sail, including frigates, was exercised in the various evolutions indicated by Earl Howe's day signals. Mr. M'Arthur, the admiral's secretary, turned his attention to render them more simple and perspicuous, by preparing a new arrangement, in which several important alterations and additions were incorporated, drawn from the code which he had, as already noticed, presented two years antecedently to the Admiralty. From motives of respect and delicacy towards Earl Howe, Sir Hyde Parker, then acting as captain of the fleet under Lord Hood, transmitted to the former, a copy of Mr. M'Arthur's explanatory observations on the proposed new arrangement of day signals and instructions. The whole having been subsequently laid before the lords commissioners of the Admiralty, they were approved of and printed in the latter end of 1792, under Mr. M'Arthur's superintendence, with all the alterations and additions he had suggested, together with a new code of night signals, he had at the same time compiled, and submitted for approval.

From that period, these day signals and instructions, together with the new code of night signals alluded to, and to be hereafter noticed, have been issued to his Majesty's ships, with little or no variation in form or substance. The following being the principal heads of the Explanatory Observations prefixed to the new arrangement of Lord Howe's signals, submitted to the Admiralty, they will suffice to give the reader a competent idea of the improvements suggested, and which were ultimately adopted

"SIGNAL BOOK.

1. The column of purport in Earl Howe's quarto Signal Book is proposed to be done away entirely, and a copious Alphabetical and Numerical Index substituted; the present arrangement, therefore, may be printed in a convenient octavo size, with as large a type as Lord Howe's quarto impression.

2. By leaving a small blank space on the outer margin of the leaves, the index to the signals and instructions, as well as the numbers on each page, may be exhibited in the form cut out in the accompanying signal book; whereby both books become at once arranged either for a flag, or private ship Similar attention has been paid to private ship's signals, to which is added a Numerical Index, whereby signals are made with facility, and their significations found without any further arrangement.

3. Every signal has immediate reference to the article relative to it in the Instruction Book; and that the articles may not be mistaken for the pages, they are always referred to in *Roman numerals.*—The like is attended to in the Indices, and the numbers of the signals are placed in figures on the left hand, so as not to be mistaken for the pages referred to on the right.

4. Instead of the leaves formerly allotted to the *Supplemental flags*, they are now exhibited to view on two pages only, under the title of *Single Flags appropriated to particular Signals independent of the Numerals*; therefore, no number or article is attached to them as formerly, but the purport of each retained, by which it is distinguished when referred to, either in the Signal or Instruction Book; thus, *Affirmative Flag*, *Annulling Flag*, *Rendezvous*, *Preparative*, &c.

5. The single flags appropriated to particular signals being arranged as in the preceding article, the numeral signals are made to commence at N°. 1, instead of N°. 13, as formerly in Lord Howe's code, and from 13 the numbers and significations follow as heretofore.

6. To the end that the numbers and significations may be permanent, and to remove every apprehension of their being useful to the enemy in case of getting possession of them in this regular form; the accompanying model of a signal board, in which the numerical flags, painted in a duplicate order, are made to change or slide, with observations on an additional signal proposed to be inserted in the department allotted. Single flags, namely, by the introduction of what is called a *key* or *transposition flag*, which, when hoisted, denotes that the numeral or integral flag hoisted under it, is to be transposed or changed to the upper place allowed, N°. 1; and the other eight integral flags are to follow such change of place in the progressive order they stand, whereby the numeral system becomes totally changed. This suggestion may, perhaps, on a due consideration, be deemed worthy of adoption.

7. The table of triangular flags, page 6, Signal Book, is proposed to be substituted for the former one, as being more simple in form.

8. In some signals it has been thought necessary to insert a *N. B.* illustrative of them taken from the Instructions; such as signals 4, 24, 25, &c. whereby signals that formerly appeared on a perusal of the Instructions only, are now brought forward to view in the proper departments of the respective signals.

9. Should the new signals 10, 11, and 12, relative to enforcing the General Station Signal N°. 4, not be deemed essential, they may remain blanks, or be appropriated for other significations.

10. If the Signal Book were printed on stout paper, and half bound, it would admit the margin to be cut out in the form submitted, with great regularity, as alphabetical and numerical indices, and be equally adapted for a commander in chief to make signals, as well as private ships to refer immediately

to the significations indicated. Eight plates are annexed of the most essential evolutions, with instructions on the principal movements. These are entirely new, and have been introduced for the reasons to be noticed in the subsequent article.

Instruction Book to the DAY SIGNALS.

All the instructions are arranged and consolidated in one book under general heads, having a copious alphabetical index annexed; and it is humbly submitted, whether it may not be deemed proper to print this Instruction Book in folio, as a size more convenient than octavo for being left open on a table under either of the *general heads or running titles* corresponding to the situation of the fleet; for were it printed octavo size, the same as the Signal Book, it would be too thick and bulky to remain open on a table. On the other hand, the necessity of having the Instruction Book carried on the quarter deck, seldom occurs, but the Signal Book is always necessary to have recourse to, and as such, an octavo is the most portable and convenient form for the latter, more particularly as the signals are now arranged for that purpose.

In order to avoid all ambiguity and confusion, one mode of technical expression has been adhered to, both in the Signal Book and Instruction; viz. *Squadrons*, to express the three component parts of a fleet, either in the order of sailing or in line of battle: *Divisions*, to express such parts into which squadrons may be divided either in the order of sailing by divisions, or in the line of battle: *Column* is applied only to the fleet in the order of sailing in two columns or grand divisions.

The only chain of new matter that has been attempted to be introduced into Earl Howe's General Instructions for Ships of War are the following *instructions* for the conduct of the fleet in the *execution* of the *principal movements* of the *evolutions*, which are illustrated by figures projected on principles consonant to those of the best naval tactitians, and in strict conformity to the ideas of Lord Howe, diffused in his General Instructions. In the practice of evolutions it is obvious, that rules for the execution of the movements of a fleet in order of battle are useful and requisite; and the want of which has been often regretted by flag-officers and captains, more especially, as a due observance of them, will tend to make the *squadrons* or *divisions* move with promptitude and regularity in the order indicated by general signal."

I. *Instructions for the Conduct of the Fleet in the Execution of the principal Movements of the Evolutions, as referred to in the preceding Explanatory observations.*

When the admiral makes general signals for any intended evolution, it is to be understood, that the movements thereby indicated are to be put in execution the instant the preparative flag is hauled down, and the general signal left flying; and the respective commanders of squadrons, instead of repeating such general signals, are (so soon as conveniently may be after answering it) to prepare their squadrons to act together, or in succession, by putting abroad the signal corresponding to the established movements herein contained; and when the admiral hauls down the preparative flag to the evolution, the commanders of squadrons are likewise to haul down the preparative to the movements, that the squadrons, divisions, or ships denoted may forthwith put the same in execution.

To the end that the squadrons or divisions may act with regularity and promptitude, in the formation of the evolution, indicated by general signal, and thereby preserve in all movements the compact order necessary, the subordinate flag-officers, and captains of ships respectively, are enjoined to adhere to the established Instructions to the Movements, and not to deviate from them unless upon some extraordinary emergency rendering the same expedient.

II. *Instructions to the Movements of the principal Positions, in forming from the Order of Sailing in Three Squadrons, to the Order and Line of Battle on the Centre Squadron, Starboard or Larboard Lines of Bearing, as may be indicated by General Signal.*

POSITION I. *Forming on the Centre Squadron, and in the prescribed Order of Battle on the Starboard Tack.*

Execution. See plate XXVIII. fig. 1. Pl.XXVIII Fig. 1.

The centre squadron (C—*c*) to *keep under an easy sail*, while the van (V—*v*) *bear up together*, and preserve relative bearing and distance (R—B), and when they bring the ships of the centre squadron (C—*c*) upon the line of bearing, they are to *haul their wind together*, upon the starboard tack. The ships of the rear squadron (R—*r*) are at the same time to *tack together*, preserving relative bearing and distance (R—B), and when arrived in the wake of the sternmost ship of the centre squadron, they are to *retack* and form the order of battle on the starboard tack.

POSITION II. *The Signal being made for forming the Line of Battle on the Centre Squadron, the Van becoming the Rear, and the Rear the Van.*

Execution. See plate XXVIII. fig. 2. Fig. 2.

The ships of the van squadron (V—*v*) are to *bear up together*, preserving relative bearing and distance (R—B), and when arrived in the wake of the centre squadron (C—*c*) they are to *haul their wind together*, on the starboard tack, according to signal; the centre at the same time *keeping under an easy sail*; while the ships of the rear squadron (R—*r*) *preserve order with a press of sail*; and when advanced so far that the sternmost and leewardmost ship can fetch into her station a-head of the weathermost ship of the centre (*which she is to signify by shewing her pendant*), the rear squadron (R—*r*) then *tack together*, preserving relative bearing and distance (R—B), and stand on until they can bring the centre squadron (C—*c*) upon the line of bearing, then *retack together*, and form the line as indicated by signal.

POSITION III. *The Signal made for forming the Line of Battle, on the Centre Squadron, the Van becoming the Centre, and the Centre the Van of the Line.*

Execution.

The ships of the van squadron are to *bear up together*, and when arrived in the wake of the centre, they are to *haul their wind together*; the centre at the same time preserving order, *with a press of sail*, are to take their stations in the van of the line, while the ships of the rear squadron *tack together*, and when arrived in the wake of the van now becoming the centre, they are to *retack together* and form the line.

III. *Instructions to the Movements of the principal Positions in forming from the Order of Sailing to the Order or Line of Battle on the Weather Squadron, Starboard or Larboard Lines of Bearing, as may be indicated by Signal.*

Position IV. *The Signal being made for forming in the prescribed Order of Battle on the Weather Squadron.*

Fig. 3. Execution. See plate XXVIII. fig. 3.

The ships of the centre squadron (C—*c*) and rear squadron (R—*r*) are to *tack together*, and preserve their relative bearings and distance (R—B), and when the centre (C—*c*) arrive at the wake of the van squadron (V—*v*) then (preserving order and under an easy sail) they are to *retack together*, and the rear squadron (R—*r*) is to observe the like with regard to the centre, and form the line.

N. B. The ships of the rear squadron (R—*r*) having a greater distance to run, before they can *retack together*, it may be found necessary for the commander of the squadron to make the signal for them to make all possible sail, preserving the same order.

Position V. *The Signal made for forming the Line of Battle, on the Weather Squadron, the Van becoming the Rear, and the Rear the Van.*

Pl. XXIX. Execution. See plate XXIX. fig. 4.

Fig. 4. The ships of the van squadron (V—*v*) are to *preserve order, and keep under an easy sail* (or if necessary *to bring to*) while the centre (C—*c*) and rear (R—*r*) *keep on under a press of sail*, and when the sternmost ship $(\frac{c}{10})$ of the centre squadron is so far advanced at $(\frac{B}{10})$ as to be able to fetch a-head of the van (V—*v*) (*which she is to make known by shewing her pendant*), then the centre squadron is to *tack together*, preserving relative bearing and distance (R—B), and when they bring the van upon the line of bearing (V—*v*) *retack together* and form the line.

The rear squadron (R—*r*), is to observe the same movements with regard to the centre.

Position VI. *The Signal made for forming the Line of Battle, on the Weather Squadron, the Van becoming the Centre, and the Centre the Van.*

Execution.

The ships of the van squadron are to *preserve order, and keep under a very easy sail.*

The ships of the centre squadron are to *preserve order, and carry a press of sail*, until the sternmost ship can fetch a-head of the headmost ship of the van squadron (which she is to signify by her pendant), *then tack together*, and when they have brought the van upon the line of bearing, *retack together*, and haul on either line of bearing indicated. At the same time the ships of the rear squadron are to *tack together*, and when arrived in the wake of the van (now becoming the centre) they *retack together*, and form the line.

N. B. It was not thought necessary to illustrate this movement by a figure, as it can be easily understood by referring to that of position V.

IV. *Instructions to the Movements of the principal Positions in forming from the Order of Sailing in Three Squadrons to the Order or Line of Battle on the Lee Squadron, Starboard or Larboard Lines of Bearing, as may be indicated by Signal.*

Position VII. *The Signal made for forming in the prescribed Order of Battle on the Lee Squadron.*

Execution. See plate XXIX. fig. 5. Fig. 5.

The ships of the van and centre squadrons (V—*v*) and (C—*c*) are to *bear up together* about two points, preserving relative bearings (R—B); and when they bring the rear on the line of bearing (R—*r*), they are to *haul up together*, and form the line, the rear (R—*r*) at the same time *preserving order, and keeping under an easy sail.*

Position VIII. *The Signal made for forming the Line of Battle, on the Lee Squadron, the Van becoming the Rear, and the Rear the Van.*

Execution. See plate XXIX. fig. 6. Fig. 6.

The ships of the van squadron (V—*v*) are to *bear up together* about six points, while the ships of the centre (C—*c*) *bear up together* about two points, preserving relative bearing (R—B), the rear (R—*r*) *preserve order under a press of sail*, and so soon as the centre (C—*c*) have got into the wake of the rear (R—*r*), they are to *haul up together* in the line; and the ships of the van are to observe the same movement with regard to the centre.

Position IX. *The Signal being made for forming the Line of Battle, on the Lee Squadron, the Van becoming the Centre, and the Centre the Van of the Line.*

Execution.

The ships of the centre squadron *bear up together about one point, and press sail.* The van *bear up together about three points*; and the rear *preserve order under an easy sail.*

When the ships of the centre squadron have brought the rear upon the line of bearing, they are to *haul up together in the line*, and the van is to observe the same with regard to the centre.

N. B. By referring to figure 6, position VIII. this movement will be easily understood.

Night Signals.

Observations on the Numerary Night Signals established in the Navy.

The code of night-signals now used, had been chiefly arranged from a plan devised by that ornament to his profession the late Admiral Kempenfelt, on numerary principles, by the combination of guns, lights, and false fires. Sir Hyde Parker, captain of the fleet under Admiral Lord Hood, having in his possession Admiral Kempenfelt's ideas in manuscript, relative to this new mode of communicating signals in the night, and which from his lamented death had not been brought to maturity; they were put into the hands of Mr. M'Arthur, secretary to the commander in chief, at the time he was as already noticed preparing in the autumn of 1792, a new arrangement of Earl Howe's Day-Signals and Instructions, to be submitted to the consideration of the Admiralty. From these original sources, Mr. M'Arthur presented to the Lords Commissioners of the Admiralty in a digested form, with some alterations and additions, the Numerary Code of Night Signals alluded to.

Admiral Kempenfelt having left no instructions to his plan, the principal part of Earl Howe's old Night Instructions was very judiciously selected, so far as was applicable to the New Code of Night Signals; but, some additional instructions were deemed necessary to complete the arrangement. These Night Signals and Instructions, thus prepared and comprised, with Indices, in one small octavo volume, were then adopted, and have ever since been used in his Majesty's navy. As it may be deemed improper to give publicity to these Night Signals, or to exhibit in this work a detail of the principles on which they are founded, suffice it to observe, that the utility of the plan, compared to that formerly in practice by shewing lights at particular parts of a ship, or in the form of triangles, squares, or lozenges, where best seen, is too obvious to be dwelt upon; yet, when a fleet is in the presence of an enemy, it may be proper to make signals in the night by lights only.

Admiral Lord Hood, knowing well how to appreciate every improvement in naval tactics, was the first commander in chief of a fleet who issued to the flag-officers of squadrons and captains of ships under his orders at the commencement of the war in 1793, the New Arrangement of Day and Night Signals above noticed; and their utility since that period, has been confirmed by the practice and experience of every successive commander in chief.

From the observations of many intelligent officers, it was found that the code of numerary night signals, although it had advantages in some respects over every other known mode practised, yet there was a desideratum sought after, namely, to give privacy to night signals when in expectation of falling in with an enemy, or approximating the line of an enemy's coast. It is, therefore, the object of the following auxiliary plan proposed by Mr. M'Arthur, to remedy the disadvantages that might eventually result by firing guns, as the component part of the numerary combination to night signals, and which may still be continued as the standing system, when the fleet is not in expectation of falling in with the enemy.

This can be effected by the combinations of four distinct lights, without the aid of guns or false fires; and the mode proposed will have the superadded advantage of being occasionally employed as a nocturnal telegraph, for naval or military purposes.

General principles to be adopted.

1st. An improvement in the construction of signal lanterns; and as only four are required for each ship, they ought to be issued as an article of store by the Navy Office.

2d. The signals are to be made from on board ship where best seen, *forward or abaft*; or if a ship be directly a-head or a-stern of other ships to whom these signals are to be communicated. The signal lanterns may be shewn at the *larboard and starboard yard-arms*, where best seen; but if used on shore, the *left hand* post or signal staff; and the *right hand* post or staff, are to be considered synonimous terms to the application, at sea, of *forward* and *abaft*, or *larboard* and *starboard yard-arms.*

3d. The combination of two, three, or four lights, shewn at sea forward or abaft, or at the larboard and starboard yard-arms; but if on shore, hoisted on the left post or right, together with one or two obscurations of the lights, either in a horizontal or vertical position, will denote more ideas than is required for any code of night signals.

The Lanterns are to be made in a cylindrical form; dimensions about 9 inches diameter, and proportionable height. A lamp with four burners is to be placed in a socket in the centre, and so constructed, that if a lantern should happen to upset, the oil cannot be spilt. In stead of horn or glass to the lanterns, it is proposed to substitute *Talk*, or what is commonly known by the name of *Muscovy glass*, which is equally as transparent as common glass, is much lighter, and will not consume by fire.

Each lantern to have an obscurer of tin or canvas, to denote the combinations to be hereafter explained. These obscurers to be in the form of cylinders, perforated with air-holes at the tops and open at the bottoms, so as to encircle or cover the lanterns, and thereby eclipse the lights when necessary.

The obscurations, or eclipses, of the lights, are thus practised with facility:

Suppose the four lights hoisted vertically, or one over the other, at the usual distance of ten or twelve feet asunder, each lantern to have its obscurer placed from five to six feet over it, and by a line passing through a small block attached to the lantern halyards, any of the lights at the word of command would be obscured. In exposing the light again, it is only necessary to haul the topping line of the obscurer, while the lantern remains in the same permanent situation.

The lanterns and mode of obscuring the lights being premised, the plan is susceptible of the following arrangement: The first will denote, as by preparatory signal, that the combinations are to be applied telegraphically, to the letters of the alphabet.

The second arrangement will denote, by preparatory signal, the first 25 numbers allotted to the significations in the standing Night Signal Book for the navy, and in order to express fifty additional numbers, making in all 75 signals, which is more than the numerary code with guns, lights, and false-fires comprehend, an obscuration of one light for an interval of 20 or 30 seconds will denote the numbers from 26 to 50 inclusive; and the obscuration of 2 lights will denote the numbers from 51 to 75 inclusive, as expressed in the following Table: The open cyphers thus, ○ in the annexed table, denotes the lights shewn, and the shaded ones thus, ● denote the lights obscured or eclipsed.

It is to be observed, that when one or two lights are to be obscured, the lights composing the signal are first shewn for a few minutes; then one or two lights, as indicated in the table, to be obscured for 20 or 30 seconds, and which is to be repeated by shewing the lights again until answered and understood by the ships to which the signal may be addressed.

Table *exhibiting the Letters of the Alphabet, and 75 Numeral Signals by the Combination of Four Lights, shewn and obscured.*

		Lights, where shewn.			Lights shewn, and 1 obscured.			Lights shewn, and 2 obscured.	
Letters	No.	Forward.	Aft.	No.	Forward.	Aft.	No.	Forward.	Aft.
A	1	○ ○		26	● ○		51	● ●	
B	2	○ ○ ○		27	● ○ ○		52	● ● ○	
C	3	○ ○ ○ ○		28	● ○ ○ ○		53	● ● ○ ○	
D	4		○ ○	29		● ○	54		● ●
E	5		○ ○ ○	30		● ○ ○	55		● ● ○
F	6		○ ○ ○ ○	31		● ○ ○ ○	56		● ● ○ ○
G	7	○	○	32	●	○	57	●	●
H	8	○ ○	○	33	● ○	○	58	● ●	○
I. J.	9	○ ○	○ ○	34	● ○	○ ○	59	● ●	○ ○
K	10	○ ○ ○	○	35	● ○ ○	○	60	● ● ○	○
L	11	○	○ ○	36	●	○ ○	61	○	● ●
M	12	○	○ ○ ○	37	●	○ ○ ○	62	○	● ● ○
N	13	○	○ ○	38	●	○ ○	63	○	● ●
O	14	○ ○	○	39	● ○	○	64	● ●	○
P	15	○ ○	○ ○	40	● ○	○ ○	65	● ●	○ ○
Q	16	○ ○	○ ○	41	○ ●	○ ○	66	● ●	○ ○
R	17	○ ○	○ ○	42	● ○	○ ○	67	● ●	○ ○
S	18	○ ○		43	● ○		68	● ●	
T	19		○ ○	44		● ○	69		● ●
U	20	○ ○ ○		45	● ○ ○		70	● ● ○	
V	21		○ ○ ○	46		● ○ ○	71		● ● ○
W	22	○ ○ ○	○	47	● ○ ○	○	72	● ● ○	○
X	23	○	○ ○ ○	48	○	● ○ ○	73	○	● ● ○
Y	24	○ ○ ○ ○		49	○ ○ ○		74	● ● ○ ○	
Z	25		○ ○ ○ ○	50		● ○ ○ ○	75		● ● ○ ○

A Table *of the Number of Guns, Lights, and Blue Lights, employed to express Numbers, which refer to certain Significations, as set forth in the following Table.*

Guns.	Blue-Lights.	Lights.	Numbers.	
1	—	—	0	One gun to call the attention of the fleet.
—	—	1	1	A general acknowledgment that the signal is understood.
—	—	2	2	Stand upon the starboard tack.
—	—	3	3	Stand upon the larboard tack.
—	—	4	4	Annul the foregoing signal.
—	1	—	5	To shew my situation.
—	1	1	6	Heave to upon the starboard tack.
—	1	2	7	Heave to upon the larboard tack.
—	1	3	8	Make sail.
—	1	4	9	Shorten sail.
2	—	—	10	We are overpressed with sail.
2	—	1	11	Headmost ship shorten sail.
2	—	2	12	Request to speak the commodore.
2	—	3	13	Continue as before, although the commodore acts otherwise.
2	—	4	14	We have carried away a top-mast, or top-sail-yard.
2	—	1	15	We see danger.
2	1	1	16	A stranger is in the fleet.
2	1	2	17	Show no light.
2	1	3	18	Haul two points to starboard.
2	1	4	19	Haul two points to port.
3	—	—	20	North.
3	—	1	21	N. E.
3	—	2	22	East.
3	—	3	23	S. E.
3	—	4	24	South.
3	1	—	25	S. W.
3	1	1	26	West.
3	1	2	27	N. W.
3	1	3	28	
3	1	4	29	
4	—	—	30	

Explanation of the Table.

Each lantern-light to four, expresses 1, 2, 3, 4, respectively. One blue-light expresses 5. Two guns express 10, three guns express 20, and four guns 30. To express any number, therefore, as for instance 19, fire two guns, burn one blue-light, and hoist four vertical lights where they can best be seen.

Previous to making any of the first nine signals which do not require guns to express the number, one gun will be fired to draw the attention of the fleet, and in making the remaining signals the necessary guns will be fired to draw the attention of the fleet previous to hoisting the lights, or burning the blue-light. To prevent a confusion of lights, when a signal is made from the commodore, the top light will be covered.

It is recommended to use glass signal-lanterns, as horn admits a very dull light. Such lanterns should be well strengthened down the sides, and be provided with a secure ring at the bottom as well as at the top, for the convenience of bending them to each other; and to the top ring of three of the signal-lanterns, a tack should be fitted, about a fathom long, to admit their being bent to each other with regularity and dispatch.

Preparatory Signal, (*signal préparatoire,* Fr.) a signal given by the admiral to the whole, or any part of his fleet, and is immediatly answered by those to whom it is directed; by shewing the same signal, to testify that they are ready to put his orders in execution. Having observed their answer, he will shew the signal which is to direct their operations: as, to chase, to form the line, to begin the engagement, to board, to double upon the enemy, to rally or return to action, to discontinue the fight, to retreat and save themselves. The dexterity of working the ships in a fleet depends on the precise moment of executing these orders; and on the general harmony of their movements: a circumstance which evinces the utility of a signal of preparation.

As the extent of the line of battle, and the fire and smoke of the action, or other circumstances in navigation, will frequently prevent the admiral's signals from being seen throughout the fleet, they are always repeated by the officers next in command; by ships appointed to repeat signals; and, finally, by the ship or ships for which they are intended.

The ships that repeat the signals, besides the chiefs of squadrons or divisions, are usually frigates lying to windward or to leeward of the line. They should be extremely vigilant to observe and repeat the signals, whether they are to transmit the orders of the commander in chief, or his seconds, to any part of the fleet; or to report the fortunate or distressful situation of any part thereof. By this means all the ships from the van to the rear will, unless disabled, be ready at a moment's warning, to put the admiral's designs in execution.

To preserve order in the repetition of signals, and to favour their communication, without embarrassment, from the commander in chief, to the ship for which they are calculated, the commanders of the squadrons repeat after the admiral; the chiefs of the divisions, according to their order in the line, after the commanders of the squadrons; and the particular ships after the chiefs of the divisions; and those in return, after the particular ships, and *vice versa*, when the object is to convey any intelligence from the latter to the admiral.

To Repeat SIGNALS. See the article *To* REPEAT.

SIGNAL *Flags*. See the article FLAGS.

SIGNAL *Trumpet*. See the article TRUMPET.

SILENCE! (*Silence!* Fr.), an order given to the men to keep quiet and steady in the performance of their duty in the ship; as, "Silence fore and aft!"

To SILENCE *a Battery*, (*faire taire une batterie*, Fr.), is by a vigorous cannonade to annoy the enemy, so as to cause them to cease firing from it.

SILLS OF THE PORTS, or PORT-SILLS, (*seuillets des sabords*, Fr.), in ship-building, are pieces of oak timber, let in horizontally between the frames to form the upper and lower sides of the ports.

SINE, or RIGHT-SINE OF AN ARC, in trigonometry, a right line drawn from one extremity of the arc, perpendicular to the radius drawn to the other extremity of it: or, it is half the chord of double the arc. Thus the line DE is the sine of the arc BE, either because it is drawn from one end E of that arc perpendicular to CB the radius drawn to the other end B of the arc; or, also, because it is half the chord DF of double the arc DBF. For the same reason also DE is the sine of the arc AE, which is the supplement of BE to a semicircle or 180 degrees; that is, every sine is common to two arcs, which are supplements to each other, or whose sum make up a semicircle, or 180 degrees.

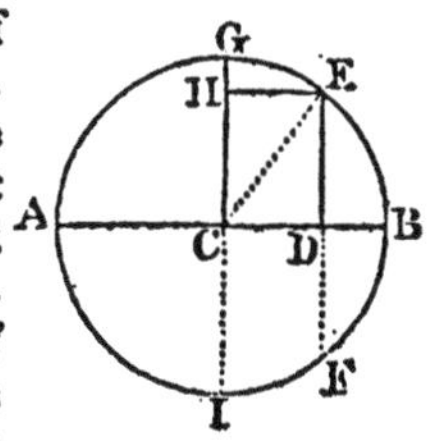

SINE *Complement*, or *Co-Sine*, is the sine of an arc EG, which is the complement of another arc BE, to a quadrant. That is, the line EH is the co-sine of the arc BE; because it is the sine of EG, which is the complement of BE. And for the same reason DE is the co-sine of EG. Hence the sine, co-sine, and radius, of any arc, form a right angled triangle CDE, or DEH, of which the radius CE is the hypothenuse; and, therefore, the square of the radius is equal to the sum of the squares of the sine and co-sine of any arc, that is, $CE^2 = CD^2 + DE^2$ or $CE^2 = CH^2 + EH^2$.

Versed SINE, is that part of the diameter between the sine and the arc. So BD is the versed sine of the arc BE, and AD the versed sine of AE, also GH the versed sine of EG, &c. All versed sines are affirmative, the sum of the versed sine and co-sine, of any arc or angle, is equal to the radius, that is, $BD + CD = AC$. The sine, co-sine, and versed sine, of an arc, are also the same of an angle, or the number of degrees, &c. which it measures.

The sines, &c. of every degree and minute of a quadrant, are calculated to the radius I, and ranged in tables for use. But because operations with these natural sines require much labour in multiplying and dividing by them, the logarithms of them are taken, and ranged in tables also; and these logarithmic sines are commonly used in practice, instead of the natural ones, as they require only additions and subtractions instead of multiplications and divisions.

SINICAL QUADRANT, a quadrant made of wood or metal, with lines drawn from each side intersecting one another, with an index, divided by sines, also with 90 degrees on the limb, and two sights at the edge. Its use is, to take the altitude of the sun. Instead of the sines, it is sometimes divided into equal parts; and then it is used by seamen to solve, by inspection, any problem of plane sailing.

To SINK, (*couler bas, &c.* Fr.) among seamen, is to force a vessel under the surface of the water by means of scuttling, &c.

SIRIUS, the Dog-Star, in astronomy, a very bright star of the first magnitude, in the snout of the constellation Canis Major, or the Great Dog. This is the brightest of all the stars in our firmament, and, therefore, probably, says the late Dr. Maskelyne, in a paper recommending the discovery of its parallax, the nearest to us of all. Philos. Trans. vol. 2, page 889. Some, however, suppose Arcturus to be the nearest. The right ascension of Sirius for the beginning of 1814, was 99° 14′ 15″, annual variation in right ascension, 39″ 8; declination, 16° 27′ 59″ S. annual variation in declination +4″ 2.

SIROCCO, a periodical wind which generally blows in Italy and Dalmatia, every year, about Easter. It blows from the south-east by south, and is attended with heat, but not rain; its ordinary period is twenty days, and it usually ceases at sunset. When the sirocco does not blow in this manner, the summer is almost free from westerly winds, whirlwinds, and storms.

In summer time when the westerly winds cease for a day, it is a sign that the sirocco will blow the day following, which usually begins with a sort of whirlwind.

SISTER-BLOCKS. See the article BLOCK.

SKEET, (*grand escop, ou escop a vaisseau*, Fr.) a sort of long scoop used to wet the decks and sides of a ship, in order to keep them cool, and to prevent them from splitting by the heat of the sun, as represented by fig. 11. plate XIX. It is also employed in small vessels to wet the sails, to render them more efficacious in light breezes: this operation is sometimes performed in large ships by means of the fire-engine.

SKIDS, or SKEEDS (*défenses, pièces de construction dans les vaisseaux*, Fr.) are long compassing pieces of timber, formed so as to answer the vertical curve of a ship's side. They are notched below so as to fit closely upon the wales; and as they are intended to preserve the plank of the sides, when any weighty body is hoisted or lowered against them, they extend from the main-wale to the top of the sides; and are retained in this position by bolts or spike-nails.

SKIFF, (*esquif*, Fr. *scaffa*, Lat.) a small light boat resembling a yawl, also a wherry without masts or sails, usually employed to pass a river.

To SKIN UP A SAIL IN THE BUNT, is to make that part of the canvas which covers the sail when furled, smooth and neat, by turning the sail well up on the yard.

SKIPPER, (*patron de navire*, Fr.) a familiar name derived from the Dutch, and given to the masters of small merchant vessels.

To SKULK, (*se cacher*, Fr.), to hide or absent one's self from duty. See the article SCULKER.

SKY, the blue expanse of the air or atmosphere. The azure colour of the sky is attributed, by Newton, to vapours beginning to condense, having attained consistence enough to reflect the most reflexible rays, viz. the violet ones; but not enough to reflect any of the less reflexible ones.

SKYLARKING, a term used by seamen, to denote wanton play about the rigging, and tops, or in any part of the ship, particularly by the youngsters, and is frequently productive of mischief, and sometimes of very serious accidents.

SKY-ROCKET. See the article ROCKET.

SKY-SCRAPER, (*grecque*, Fr.) a small triangular sail, generally made of canvas, No. 8, and sometimes set above the royal.

SLAB-LINE, (*couillard*, &c. Fr.) a small cord passing up behind a ship's main-sail or fore-sail, and being *reeved* through a block, plate XX. attached to the lower part of the yard, is thence transmitted in two branches to the foot of the sail, to which it is fastened. It is used to truss up the sail, but more particularly for the convenience of the steersman, that he may look forward beneath it, as the ship advances.

SLACK, (*mou,——lâche*, Fr.) implies a decrease in tension or velocity; as, "Slacken the lanyards of our main-stay!" "The tide slackens," &c.

The SLACK *of a Rope*, (*le mou d'un cordage*, Fr.) is that part which hangs loose, as having no strain or stress upon it.

SLACK-*Rigging*, (*grèement qui a molli, ou qui n'est pas tenu*, Fr.) implies that the shrouds, stays, &c. are not so firmly extended as they ought to be.

SLACK-*In-Stays*, (*mou à virer*, Fr.) signifies slow in going about.

SLACK-*Water*, (*mer étale*, Fr.) the interval between the flux and reflux of the tide, or during which the water apparently remains in a state of rest.

SLATCH, (*intervalle, ou durée d'une brise*, Fr.) the period of a transitory breeze of wind, or the length of its duration.

SLAUGHTER, destruction by the sword, cutlass, boarding-pikes, and small-arms, which frequently happens to both parties in the act of boarding, where obstinacy is persisted in.

SLEEPERS, (*vaigres obliques, ou pièces de liaison de l'avant et de l'arrière des vaisseaux de construction Angloise; ces pièces traversent obliquement plusieurs des courbes d'arcasse, et des guirlandes*, Fr.) a name formerly given by shipwrights to the thick-stuff placed longitudinally in a ship's hold, opposite to the several scarphs of the timbers; but now generally applied to the knees, which connect the transoms to the after-timbers on the ship's quarter.

They are fitted within side the ship abaft, one arm lying on the foot-walcing, and the other extended up the transoms; and are particularly used in Greenland ships, to strengthen the bows and stern-frame, to enable them to resist the shocks of the ice.

SLEET, (*pluie et neige ensemble*, Fr.) the drops of rain accompanied with flakes of snow, occasioned by the moisture of the atmosphere, and is prevalent in the winter season.

SLIDING-KEEL, in ship-building, a late invention of Captain John Shanks of the royal navy; an officer who, during the American and late war, gave proofs of his talents for invention on many occasions.

Vessels of this kind have each three keels, made moveable in a trunk or well, so as to be drawn up or let down in shoal water. By means of these keels they sail the faster, steer easier, tack and wear quicker, and in less room: they ride more easy at anchor; take the ground better; and in case of shipwreck, springing a leak, or of fire, they are more safe and more likely to be saved. For fire ships, floating batteries, gun-boats, and flat-bottomed boats for landing troops, they are particularly well adapted. Three vessels with these keels have been built for his Majesty's service, viz. the Trial cutter, Cynthia sloop of war, and the Lady Nelson of 60 tons burden; which latter vessel was fitted out for a voyage of discovery in 1800, under command of lieutenant James Grant, of the Royal Navy, who in her arrived safe at New South Wales the following year. An account of this very interesting voyage, with the origin of sliding keels, and various official documents of their utility, was lately published, to which we refer our readers.

SLIDING-RULE, a mathematical instrument serving to perform computations in gauging, measuring, &c. without the use of compasses, merely by the sliding of the parts of the instrument one by another, the lines and divisions of which give the answer or amount by inspection.

This instrument is variously contrived and applied by different authors, particularly Gunter, Partridge, Hunt, Everard, and Coggeshall; but the most useful are those of Gunter, Everard, and Coggeshall.

To SLING, (*élinguer*), Fr.) is to hoist or lower the boats, casks, ordnance, or any other weighty body, by means of slings for that purpose.

To SLING *the Yards for Action*, (*établir de fausses drisses, &c. aux vergues pour le combat*, Fr.) is to secure them close up by means of iron chains, which are not so liable to be cut through by the enemy's shot, as rope.

SLINGS, (*élingue de corde*, Fr.) a rope fitted to encircle a cask, jar, bale, or case, and suspend it whilst hoisting or lowering. There are various sorts of slings, according to the weight or figure of the object to which they are applied. Those which are most frequently used in lading and delivering ships are represented by fig. 7. plate IV. the tackle being hooked to the middle part *a*, whilst *b* and *c* are placed round the opposite quarters of the cask, &c.

Boat Slings, (*élingues d'embarcation*, Fr.) are strong ropes, furnished with hooks and iron thimbles, whereby to hook the tackles, in order to hoist the boats in or out of the ship, the hooks of the slings are applied to ring-bolts fixed in the keel and extremities of the boat.

Butt-Slings, (*élingues pour les futailles*, Fr.) are those used in lading and delivering ships, and are nearly in the form of a pair of spectacles.

Slings *of a Buoy*. See the article Buoy.

Slings *of a Yard*. (*Suspense d'une vergue, drisses ou manœuvres qu'on établit pour soutenir les vergues dans un combat;—c'est aussi le milieu, ou le grand diamètre d'une vergue*, Fr.) ropes fixed round the middle of the yard, serving to suspend it for the greater ease of working, or for security in an engagement; in the latter case they usually add iron chains to the slings of the lower yards. This term also implies the middle, or that part of the yard on which the slings are placed.

SLIP, (*cale de construction; aussi toute sorte de cales, ou plans inclinés, pour l'embarquement et le débarquement des marchandises et munitions, &c.* Fr.) a place lying with a gradual descent on the banks of a river, or harbour, convenient for ship-building.

To Slip *the Cable*. See the article Cable.

The captain is never to order the cable to be cut nor slipped if there be a possibility of weighing the anchor, except for the purpose of chasing an enemy, or when absolutely obliged to do so for the safety of the ship; in either of which cases, if he does not return to the same port, he is by the first opportunity to inform the commissioner of the port, if there be one, if not, the commissioners of the navy, of the situation of the anchor, that proper means may be taken to recover it. See the Naval Instructions.

Slip-Knot, (*nœud coulant*, Fr.) is one which will not bear any strain, but will either become untied, or will traverse along the other part of the rope.

Slip-Rope, is a rope used to trice the bight of the cable into the head; and is also employed in casting off a vessel, till got in a tide-way, &c.

SLIPPERY-Hitch. See the article Hitch.

SLOOP, (*Sloop, bateau bermudien*, Fr.) a small vessel furnished with one mast, the main-sail of which is attached to a gaff above, to the mast on its foremost edge, and to a boom below; by which it is occasionally shifted to either quarter. It differs from a cutter, by having a fixed steering bowsprit, and a jib-stay; nor are the sails generally so large in proportion to the size of the vessel. See pl. XXXIII.

Sloops *of War*, (*corvettes*, Fr.) are vessels in the navy commanded by officers in a middle rank, between a lieutenant and a post-captain, and styled masters and commanders. These vessels carry from 10 to 18 guns, and are variously rigged as ships, brigs, schooners, and sometimes as cutters.

SLOPS, (*hardes de matelots*, Fr.) a name given to all species of wearing apparel, bedding, &c. which are supplied to his Majesty's ships in commission, by the naval store-keeper, for the out fit of seamen.

The captain or commander of a ship, upon her being fitted for sea, whether for foreign or channel service, or at any time of her being in want of slops or bedding, is to make timely application to the store-keeper at the port where the ship is, for such quantities of slop-clothing as will complete the store, agreeable to the establishment of the 11th of June, 1799; and also such bedding as may be needful for the use of the ship's company, and to get the same on board before he proceeds to sea, to prevent, as much as possible, the purchasing of slops or bedding abroad, or the ship's company suffering for want of them.

This plan has been very beneficial to the seamen, as well as a great saving to government, as pursers formerly had it in their power to make purchases sometimes unnecessarily for private emolument. The only inconvenience complained of by the purser is, the want of store-room, as the place allotted by the Navy Board for the slop-room, has never been enlarged since the commencement of providing these comforts for the crew, and its dimensions are agreeable to the rate or class of the ship, and does not contain any thing like the proportion ordered by the above regulation of the slop-clothes to be received on board; particularly when destined for foreign service; consequently, they are obliged to be placed in other parts of the ship, where they are less secure, and more liable to be injured by damps, and exposed to depredators.

Seamen coming on board, and being destitute of necessaries, may be fitted out with slops so far as two months pay in advance will allow, and every month after that period is elapsed, they may be furnished to the amount of seven shillings to the end of the voyage, if they shall be in want.

Slops are to be issued out publicly; and as the captain is not to suffer those to be supplied who are not really in want, so he is to oblige those who are ragged, or want bedding, to receive such necessaries as they want, not exceeding the limits aforesaid.

The captain is to keep a slop-book, according to the form prescribed, and, before the payment of the ship, or on his removal, to send the same to the Navy Board, signed by himself, the senior lieutenant, master, and purser.

On the discharge of a man by ticket, the value of the clothes he has been supplied with, must be noted upon the same in words at length.

If necessity requires the buying of clothes in foreign parts, the captain must procure them of the kind used in the navy, and at as moderate a rate as possible: he must also, by the first conveyance, send an invoice of the same to the Navy Board. But where there is a store-keeper, neither the captain, nor commander, nor any other person whatever, except the store-keeper, is to purchase or contract for slops or bedding. The purser is to have the charge of the slop-clothes and bedding, for the faithful discharge of which trust he shall enter into bond to the Navy Board on taking up his warrant; and for his care and pains in issuing the same, and keeping the accounts thereof, he shall at the final settlement of this account be paid one shilling in the pound on the amount of the slop-clothes issued by him, except those issued to supernumeraries, for victuals only, discharged to other ships, and those which shall appear to have been purchased by the commander of the said ship.

When a seaman dies on board, his clothes, &c. may be sold by auction, and the produce charged against the buyers on the muster and pay books; a particular account of the same must also be kept on the slop-book.

The purser is to be allowed one shilling in the pound for his trouble, who is also to give the executors, &c. of the deceased, a particular account of the clothes sold, and to whom.

Seamen are not permitted to bid for the clothes of the deceased officers, that are above their wear, nor to bid for effects above their real value, nor purchase more than their wages can answer, agreeable to allowance. See the NAVAL INSTRUCTIONS.

To SLUE, (*tourner sur son axe, ou pivot;—pivoter, ou trévirer*, Fr.) is to turn any cylindrical or conical piece of timber about its axis, without removing it. This term is generally expressed of the movement by which a mast, boom, or spar is turned about, in its cap or boom-iron.

SMACK, (*smaque, sorte de bâtiment, mâté et gréé à peu près comme un cutter, il est bâti plus solidement; on s'en sert principalement à la pêche*, Fr.) a small vessel commonly rigged as a cutter, and used in the coasting or fishing trade; or as a tender in the King's service.

SMALL OF THE ANCHOR, is that part of the shank immediately under the square. See the article ANCHOR.

SMALL-*Arms*, a general name given to CARABINES, MUSKETS, PISTOLS, &c. which articles see.

To Exercise Seamen to the use of SMALL ARMS, is to learn them the use of muskets, &c. either for defending themselves on board ship, or to join troops of the line in service on shore. For this purpose a system of small-arms exercise has lately been published by Mr. Joseph Conolly, and has been inspected and approved of by his Royal Highness the Commander-in-Chief. It comprises all that is necessary for attack and defence, in the most direct and simple manner; and is the exercise of action with the musket.

"All troops of infantry," says Mr. Conolly, "fall in with their arms at the order; standing in that position, they are sized and counted off into divisions, sub-divisions, sections, and wings, and are ready to fix bayonets, prime and load.

"The seaman from the nature of his dress, want of pliability in his hands, and situation of his pouch, finds it necessary to make two motions in shouldering his arms, which requires only one in the line; and he is apt to drop his musket on the toes of his left hand man, and cause confusion.

"The ship's pouch is fastened in front, with a belt round the middle; the bayonet-carriage is on the small belt, to the left of the pouch. The charge is considered part of the platoon, where it presents itself when the ammunition is expended."

It is of great importance that seamen should be skilful in the use of muskets, the captain, is therefore, to order a number of sailors to be trained to the use of them, and, if possible, to be exercised every day under the direction of the junior lieutenant, the master at arms, and the ship's corporal; and when they shall have attained to such perfection as shall fit them for being employed wherever musketry may be required, he is to order another party to be trained; and so on, until the whole of the ship's company become expert. And as it may be necessary occasionally to employ seamen on shore in operations against an enemy, the number of men, as expressed in the following table, are to be trained more particularly for that purpose.

Rate of the Ship.	No. of Men.
1st.	150
2d.	120
3d. of 74 guns and upwards.	100
3d of 64 guns.	80
4th.	60
5th.	50
6th.	40
Sloop.	30

The seamen are also to be taught the cutlass exercise and half-pike, and to be frequently exercised in the various modes of boarding an enemy's ship, and of repelling the attacks of boarders. See the articles BOARDING, CUTLASS, AND HALF-PIKE.

SMART-TICKET, a certificate granted by the captain to any warrant or inferior officer, seaman, or other in sea-pay, when hurt, maimed, or disabled in the service, whether on board or on shore, to the end that he may receive the benefit of the chest at Greenwich.

The smart-ticket is a printed form, filled up by the surgeon, expressing the name and age of the person, the nature of his wound or hurt, when received, and the service on which he was employed at the time; it is signed by the captain, lieutenant, and surgeon, and given by the latter to the person previous to his being discharged, or sent to any of his Majesty's hospitals.

SMOKE, (*fumée*, Fr.) a humid matter exhaled in the form of vapour by the action of heat, either externally or internally.

SMOKE-SAIL, (*masque*, Fr.) a small sail, hoisted against the fore-mast when a ship rides head to wind, to give the smoke of the galley an opportunity of rising, and to prevent its being blown aft on to the quarter-deck.

To SMUGGLE, (*faire la contrebande*, Fr.) to illegally import or export goods; either clandestinely or in an open hostile manner, without paying the accustomed duties thereon. Persons to the number of three or more, assembled and armed, and assisting in running goods, are adjudged guilty of felony, and liable to transportation; and where two or more persons in company are found passing within five miles from the sea-coast, &c. with any horse or cart, wherein are put more than six pounds of tea, or five gallons of brandy, or other foreign goods above £30. value, landed without entry, and not having a permit, or persons carrying any offensive weapons, &c. shall be adjudged guilty of felony, and all the goods shall be forfeited. See the Act of 9. Geo. II. c. 35. s. 10 and 13.

Ships from foreign parts, having on board tea, or brandy in casks under sixty gallons, &c. lying at anchor, hovering near the ports, and not proceeding on their voyages, all such tea, brandy, &c. shall be forfeited, and foreign goods which shall be taken at sea, by any ship or boat, within eight leagues of the coast, without paying the custom duties, shall be liable to forfeiture, and the master of such vessels shall forfeit treble the value. 9 Geo. II. c. 35. s. 22.; and 5 Geo. III. c. 43. s. 5 and 27.

Persons obstructing officers of the customs in entering or searching any coasting vessel shall forfeit £100; and if any officer being on board a ship or boat, be wounded or beaten in the execution of his duty, the offenders shall be transported for seven years, 9 Geo. II. c. 35. s. 28 and 29.

If three or more persons armed be assembled to assist in the illegal transportation, or running of goods, or rescuing the same after seizure, or rescuing any person apprehended for any felony, by any act relating to the customs or excise, or preventing the apprehending of any such persons: or if such person shall use any disguise when passing with such goods, or shall forcibly hinder any officer from seizing such goods, or shall maim or dangerously wound any officer attempting to go on board any vessel, or shoot, or maim, or dangerously wound him when on board; every such offender shall be guilty of a capital felony, without benefit of clergy. 19 Geo. II. c. 34. s. 1.

Persons harbouring any such offenders, to be transported for seven years. Sect. 3.

Any person apprehending an offender to have £500, and, if wounded, £50. extraordinary; and the executors of any person killed in attempting to apprehend offenders, shall have £100 Sect. 10.

One person discovering another, to be acquitted and receive part of the reward; and discovering two or more offenders, to have £50 for each. Sect. 11.

If an officer be killed or wounded, or the goods be rescued, the hundred is to make amends, in the proportion of £100 for every person killed, to be paid to his executors, and not exceeding £40. to every person beat or wounded, and £200 for goods rescued. Sect. 6.

SMUGGLER, (*bâtiment contrebandier*, Fr.) a vessel employed in a contraband trade.

SNAKING, (*serpentage en corde*, Fr.) is the act of winding small ropes spirally round a large one, the former lying in the intervals between the strands of the latter, and is frequently termed Worming, which article see.

Snaking *the Stays, or Ropes on the Quarters, instead of Netting*, is the act of seizing proportioned sized rope, at angles, from one stay or rope to the other, alternately, in a parallel direction along the whole length. Its use to stays is, that one part may remain perfect and independant of the other, should it be shot away.

SNATCH-Block. See the article Block.

SNOTTER, (*collier fixé au mât de certains bâtimens, pour tenir le pied de la livarde, ou baleston de la voile*, Fr.) a short rope spliced together at the ends, and served with spun-yarn, or covered with hide: it is seized to the size of the mast, leaving a bight to fit the lower end of the sprit, which it confines to the mast. See the article Sprit.

SNOW, (*senau, sorte de bâtiment, grée comme un brig, il a un petit mat, près et en arrière du grand mât, pour sa brigantine*, Fr.) a vessel equipped with two masts, resembling the main and fore-masts of a ship, and a third small mast just abaft the main-mast, carrying a sail nearly similar to a ship's mizen. The foot of this mast is fixed in a block of wood, or a kind of step upon the deck, abaft the main-mast; and the head is attached to the after-part of the main-top. The sail, is called the try-sail, and is extended from its mast towards the stern of the vessel. See plate XXXIII.

When the sloops of war are rigged as snows, they are furnished with a strong rope, called a horse, which answers the purpose of the try-sail-mast, the fore-part of the sail being attached by rings to the horse, in different parts of its height.

This is generally the largest of all two masted vessels employed by Europeans, and is reckoned the most convenient for navigation.

Snow, (*neige*, Fr.) a well-known meteor, formed by the freezing of the vapour in the atmosphere. It differs from hail and hoar frost, in being as it were crystallized, which they are not. This appears on examination of a flake of snow by a magnifying glass; when the whole of it appears to be composed of fine shining spicula diverging like rays from a centre. As the flakes descend through the atmosphere, they are continually joined by more of these radiated spicula, and thus increase in bulk like the drops of rain or hailstones, so that it seems as if the whole body of snow were in an infinite mass of icicles irregularly figured.

The lightness of snow, although it is firm ice, is owing to the excess of its surface, in comparison to the matter contained under it, as even gold itself may be extended in surface, till it will float upon the least breath of air. From some late experiments on the quantity of water yielded by snow, it appears that it gives only one tenth part of its bulk in water. In the northern countries, the ground is covered with snow for several months; which proves exceedingly favourable to vegetation, by preserving the plants from those intense frosts which are common in such countries, and which would otherwise certainly destroy them.

The line of perpetual congelation is an imaginary crooked plane, which we conceive in the atmosphere, and the snow situate at a greater elevation than this line, never melts. It has been observed by Mr. Esmark, of Konigsberg. that on the north and north-east declivities of Norway, the line of perpetual snow, is at the height of 3000 feet above the level of the sea, while on the south and west declivities, its height was no less than 7000 feet.

If the curve or the snow-line over the earth's surface could be constructed from accurate rules, its height would probably indicate the temperature of all the places over which it passes, and thus make us acquainted with the laws concerning the change of temperature, hitherto so little known.

SNUBBING, a term used by seamen for the method of checking the sudden jerk of the cable or hawser, after the anchor is let go, or otherwise.

SNYING, among shipwrights, a term used for a circular plank, edgeways, to work in the bows of the ship.

SOCIETY, (*Societas Lat.*) an assemblage or union of several learned persons, for their mutual assistance, improvement, or information, and for the promotion of philosophical or other knowledge.

There are various philosophical societies instituted in different parts of the world, as

The Royal SOCIETY of England, which was instituted by King Charles the II. for promoting natural knowledge. The subjects treated in their transactions are arranged under five heads, viz. natural history, mathematics, mechanical philosophy, chemistry, and miscellaneous articles. Almost every man of science in the British empire, who has flourished since the original establishment of this society, has been enrolled amongst its fellows. The Royal Society of Scotland; Dublin Society; Manchester Literary and Philosophical Society; the Royal Academy of Sciences at Paris; Berlin Society; Brussels Society; American Philosophical Society, &c.

Marine SOCIETY, for educating, clothing, &c. boys and landsmen for the royal navy. See the article MARINE.

Naval Charitable SOCIETY, for affording relief to the indigent orphans and widows of naval officers, &c. See the article NAVAL.

SOLAR, (*solaire*, Fr.) something belonging to the sun; as

SOLAR *Month*. See the article MONTH.

SOLAR *System*, (*Système Solaire*, Fr.) the order and disposition of the heavenly bodies, which revolve round the sun as the centre of their motion, viz. the primary planets, the secondary, and the comets.

The planets are opaque bodies, nearly spherical, only luminous by the reflected light of the sun, and we have reason to believe that they are similar to the earth.

They move round the sun from *west by south, to east*, in orbits nearly circular, and almost in the same plane. The comets move in all manner of directions, in long elliptical orbits, much inclined to one another, and to the orbits of the planets.

Kepler discovered some of the principal laws by which the motions of the planets are governed. He was the first that demonstrated, by calculations equally difficult and laborious, that they must revolve in elliptical, and not in entire circular orbits. He calculated by the observations of Tycho Brahe, the distance of Mars from the sun in different parts of his orbit, and proved that it could not possibly be adjusted to the circumference of a circle.

The second law of Kepler is, that *the square of the times of the planet's revolution, are as the cubes of their distances from the sun:* that is, if we compare the square of the time which any two of the primary planets occupy in completing their orbits, we shall find these two squares the same proportion as between the cubes of the mean distances of these two planets from the sun. Thus if we know the time of the revolution of two planets, we can compute their respective distances from the sun; and when acquainted with the true distance of one, it is easy to find the true distance of the other.

The third law of Kepler is, that *the areas are in proportion to the times:* hence we see that these times are shorter in proportion as the planet is nearer the sun. Newton has proved that this law is a necessary consequence in the projectile force combined with the centripetal or attractive force, which retains the planets in their orbits.

The time in which any planet goes round the sun is the *length of its year*, and the time on which it turns round on its axis, is *the length of its day and night*, taken together.

From these distances, and the angles the planets subtend at those distances, their magnitudes are calculated with tolerable certainty. Their largest apparent diameters, as seen from the earth, are, the sun, 32′ 36″, Mercury, 12″; Venus, 57″; Jupiter, 40″; Saturn, 18″; the ring of Saturn, 42″; and the Georgium Sidus, 4″. Their real diameters, distances from the sun, sidereal revolutions, &c. are expressed in the following table.

	Diameters in English miles.	Distances from the Sun in English miles.	Sidereal revolutions.	Inclination of their orbits.
Sun.	893522		d. h. m. s.	
Mercury.	3261	37,000,000	87 23 15 14	7° 0′ 0″
Venus.	7693	68,000,000	224 16 49 11	3 23 35
Earth.	7920	95,000,000	365 6 9 12	
Moon.	2161		27 7 43 0	
Mars.	5312	145,000,000	686 23 30 36	1 51 0
Jupiter.	90255	493,000,000	4332 14 27 11	1 18 56
Saturn.	80012	906,000,000	10759 1 51 11	2 29 50
GeorgiumSidus	34217	1812,000,000	83yrs. 157ds. 18h.	0 46 20

For a representation of the solar system, see fig. 2, plate XXVI, and for other particulars, see EDINBURGH *Encyclopedia*, Vol. II.

SOLAR *Tables*. See the article TABLES.

MARINE SOLDIER. See the article *Royal* MARINES.

SOLE OF A GUN-PORT, (*sole, ou seuillet de sabord*, Fr.) is the lower part of it, which is more properly called the port-sail.

SOLE *of the Rudder*, (*pièce ajoutée au dessous du gouvernail, pour le mettre de niveau avec la fausse-quille*, Fr.) a piece of timber attached to the lower part of it, to render it nearly level with the false-keel.

SOLID, in geometry, is the third species of magnitude, having three dimensions, viz. length, breadth and thickness, and is frequently used in the same sense with body. A solid may be conceived to be formed by the direct motion, or revolution of any superficies of whatever nature and figure.

A solid is contained, or terminated under one or more planes and surfaces; as a surface is under one or more lines.

Regular SOLIDS, are those terminated by regular and equal planes; under this class comes the *tetrahedron, hexahedron*, or *cube, octahedron, dodecahedron, icosihedron.*

Irregular SOLIDS, are such as do not come under the definition of regular solid: such as the sphere, cylinder, cone, parallelogram, prism, parallelopiped, &c.

SOLIDITY, is the quantity of space which a body fills, or occupies in length, breadth, and thickness; called also the *solid content* and the cube thereof. The space or solidity of a body is ascertained by making the product of the above three dimensions.

Solidity, in physics, a property of matter or body, by which it excludes every other body from that place which is possessed by itself.

SOLSTICE, in astronomy, is the time when the sun is in one of the solstitial points, that is, when he is at the greatest distance from the equator, which is now nearly 23° 28′ on either side of it. It is so called, because the sun then seems to stand still and not to change his place, as to declination, either way.

There are two solstices, in each year, when the sun is at the greatest distance on the north and south sides of the ecliptic; viz. the estival or summer solstice, and the hyemal or winter solstice.

The Summer Solstice is when the sun is in the tropic of cancer; which is about the 21st of June, when he makes the longest day. And

The Winter Solstice, is when he enters the first degree of capricorn; which is about the 22d. day of December, when he makes the shortest day. This is to be understood, as in our northern hemisphere; for in the southern, the sun's entrance into capricorn makes their summer solstice, and that into cancer their winter one. So that it is more precise and determinate, to say the northern and southern solstice.

SOLSTITIAL Colure, is that which passes through the solstitial points.

Solstitial *Points*, are those points of the ecliptic the sun is in at the times of the two solstices, being the first points of cancer and capricorn, which are diametrically opposite to each other.

SOUND, (*détroit entre deux caps, & s'entend en anglois du détroit du Sund à l'entrée de la mer Baltique*, Fr.) in geography, denotes a strait or inlet of the sea, between two capes or head-lands.

The word Sound is used, by way of eminence for that celebrated strait which connects the German sea to the Baltic. It is situate between the island of Zealand and the coast of Schonen. It is about 16 leagues in length, and in general about five in breadth, except near the castle of Cronenburg, where it is but one; so that there is no passage for vessels but under the cannon of the fortress.

Sound, on the coast of Norway, in particular, is used for any opening of a river, or any gulf or deep inlet of the sea in the same sense as Deep on the coast of Germany. In other parts it is more usually understood of a passage between the main land, to which it is contiguous, and some island, which together form a strait or passage within such island.

Sound, in physics, a perception of the mind, communicated by means of the ear; being an effect of the collision of bodies, and their consequent tremulous motion, communicated to the ambient fluid, and so propagated through it to the organs of hearing. To illustrate the cause of sound, it is to be observed, 1st, that a motion is necessary in the sonorous body for the production of sound. 2dly. That this motion exists first in the small and insensible parts of the sonorous bodies, and is excited in them by their mutual collision against each other, which produces the tremulous motion so observable in bodies that have a clear sound, as bells, musical chords, &c. 3dly, That this motion is communicated to, or produces a like motion in the air, or such parts of it as are fit to receive and propagate it. Lastly, That this motion must be communicated to those parts that are the proper and immediate instruments of hearing.

It is established, as well by mathematical reasoning from the nature of an elastic fluid whose compression is as the weight, as from experiment, that all sounds whatever arrive at the ear in equal times, from sounding bodies equally distant. The common velocity is 1142 English feet in a second of time. The knowledge of the velocity of sound is of use in determining the distances of ships or other objects: for instance, a ship fires a gun, the sound of which is heard five seconds after the flash is seen: then 1142 multiplied by 5, gives the distance 5710 feet, or one English mile and 430 feet. It is a common observation, that persons in good health have about 75 pulsations, or beats of the artery at the wrist, in a minute; consequently, in 75 pulsations, sound flies about 13 land miles, or $11\frac{1}{7}$ sea miles, which is about one land mile in 6 pulses, or one sea mile in 7 pulses, or a league in 20 pulses.

And hence the distance of objects may be found by knowing the time employed by sound in moving from those objects to an observer. For example, on seeing the flash of a gun at sea, if 54 beats of the pulse at the wrist were counted before the report was heard, the distance of the gun will easily be found by dividing 54 by 20, which gives 2.7 leagues, or about 8 miles.

Water has been little noticed with respect to its conducting sound; but it will be found to be of the greatest consequence. On the water we could distinctly hear a person read at the distance of 140 feet, on land at 76. It is to be observed, that on land no noise intervened; but on the river some noise was occasioned by the flowing of the water against the boat. Little doubt can remain of the influence water has on sound; and it conducts sound more than any other body whatever. After water, stone may be reckoned the best conductor of sound.

When the aerial waves meet with an obstacle which is hard, and of a regular surface, they are reflected; and consequently, an ear placed in the course of these reflected waves, will perceive a sound similar to the original sound, but which will seem to proceed from a body situated in like position and distance behind the plane of reflection, as the real sounding body is before it. This reflected sound is called an echo.

The waves of sound being thus reflexible, nearly the same in effect as the rays of light, may be deflected or magnified by much the same contrivances as are used in optics. From this property of reflection it happens, that sounds uttered in one focus of an elliptical cavity, are heard much magnified in the other focus; instances which are found in several domes and vaults, particularly the gallery of St. Paul's cathedral, in London, where a whisper uttered at one side of the dome is reflected to the other, and may be heard very distinctly. On this principle also are constructed the *speaking trumpet*, and the *hearing trumpet*, which either are, or ought to be, hollow parabolic conoids, having a perforation at the vertex, to which the mouth is to be applied in speaking, or the ear in hearing.

To Sound, (*sonder, jeter le plomb*, Fr.) is to try the depth of the water, and the quality of the ground, by means of a plummet sunk from a ship to the bottom.

There are two plummets used for sounding, one of which is called the hand-lead, weighing about 8 or 9 pounds, as represented by fig. 11, plate XIII.; the other the deep-sea-lead, which weighs from 25 to 30 pounds, and both are shaped like the frustum of a cone or pyramid. The former is used in shallow waters, and the latter at a great distance from the shore; particularly on approaching the land, after a sea voyage. Accordingly the lines employed for this purpose are called the deep-sea-lead-line, and the hand-lead-line.

The hand-lead-line, which is usually 20 fathoms in length, is marked at every 2 or 3 fathoms; so that the depth of the water may be ascertained either by day or night. At the depth of 2, and 3 fathoms, there are marks of black leather; at 5 fathoms, there is a white rag; at 7, a red rag; at 10, black leather; at 13, black leather; at 15, a white rag; and at 17, a red rag.

Sounding with the hand-lead, which is by seamen, called heaving the lead, is generally performed by a man who stands in the main-chains to windward. Having the line all ready to run out, without interruption, he holds it nearly at the distance of a fathom from the plummet, and having swung the latter backwards and forwards three or four times, in order to acquire the greater velocity, he swings it round his head, and thence, as far forward as is necessary; so that, by the lead's sinking whilst the ship advances, the line may be almost perpendicular when it reaches the bottom. The person sounding then proclaims the depth of the water in a kind of song resembling the cries of hawkers in a city. Thus, if the mark of 5 fathoms is close to the surface of the water, he calls "By the mark five!" and as there is no mark at 4, 6, 8, &c. he estimates those numbers, and calls, "By the dip four," &c. If he judges it to be a quarter or a half more than any particular number, he calls, "And a quarter five! and a half four!" &c. If he conceives the depth to be 3 quarters more than a particular number, he calls it a quarter less than the next: thus, at four fathoms and three quarters, he calls, "A quarter less five!" and so on.

The deep-sea-lead is marked with two knots at 20 fathoms, 3 at 40, 4 at 50, and so on to the end. It is also marked with a single knot in the middle of each interval, as at 25, 35, 45 fathoms, &c. To use this lead more effectually at sea, or in deep water on the sea-coast, it is usual previously to *bring-to* the ship, in order to retard her course; the lead is then thrown as far as possible from the ship on the line of her drift, so that, as it sinks, the ship drives more perpendicularly over it. The pilot, feeling the lead strike the bottom, readily discovers the depth of the water by the mark on the line nearest its surface. See the articles Lead and Line.

The difficulty of getting true soundings in a gale of wind, increases in proportion to the depth of water, and the violence of the wind and waves: The best method that navigators have experienced is to pass the lead from the weather quarter, round the stern to leeward, without all, to the forepart of the quarter deck, and a man to carry the bight of the line to the lee main yard-arm, which is to be laid square when the helm is put a-lee, to bring the ship to; and when the ship begins to shoot to windward by her head-way, then the lead is to be swung right to leeward from the deck, whilst the man at the yard-arm, with a tight line, swings it as far as possible to leeward of the ship, which gains to windward of the lead, whilst the head-way continues, and then runs her stern up to the wind by the steerway, backs near to where the lead was hove, by which means a great deal of line may be run out, nearly up and down; and the lead being armed with tallow, makes it certain whether it has struck the bottom or not, if it has, it shows what sort of bottom there is.

When and how often it may be necessary to try for soundings, must be left to the discretion of the commanding officer, and will be more or less so, according to the nature of the coast, place, or situation of the ship; but the want of its being done in due time, is well known to have occasioned many fatal losses; therefore, they who neglect heaving the lead will always be blamed in proportion to the degree of danger and the loss their negligence may occasion.

To Sound *the Well*, (*sonder la pompe*, Fr.) is to ascertain the depth of water in it by means of the sounding rod.

Self-Acting SOUNDER, a contrivance consisting of a glass tube, having the one end open and the other shut, with a valve at the open end, which opens inwards by the pressure of the water, thereby allowing the air to be compressed in the upper part of the tube into a less space; while it is made to sink below the surface of the water, the air which it contains becomes gradually compressed, and the tube receives a quantity of water proportioned to the depth to which it has descended. When the sounder is drawn up out of the water, the index shews the depth to which it has descended by the valve at its mouth closing, and retaining the water in the tube at the same height that it was when at the bottom of the sea.

This sounder shews, with sufficient accuracy, the perpendicular depth of water, by the degree of space into which the air is compressed at the upper end of the glass tube, and not from the quantity of line given out to it, no attention whatever of the seaman being necessary, but only to give it length enough: neither has the diagonal direction which the current gives the line the smallest effect on the index, which always shews the exact perpendicular depth of water at the time of leaving the bottom of the sea.

SOUNDING-Rod, (*sonde de pompe*, Fr.) in ships of war, an iron instrument with feet and inches marked thereon, having a small rope fastened at its upper end, and is used to ascertain the depth of water in the well.

Soundings, (*marques du fond que rapporte la sonde*, Fr.) a name given to the specimen of tha ground; a piece of tallow being stuck upon the base of the deep-sea-lead, brings up distinguishing marks of the bottom; as sand, shells, ooze, &c. which adhere to it.

The SOUNDINGS, (*fond;—brassiage*, Fr.) *i. e.* the depth of the water and the nature of the ground which are carefully marked in the log-book, as well to determine the distance of the place from the shore, as to correct the observations of former pilots. See the articles COASTING and NAVIGATION.

In SOUNDINGS, (*avoir du fond*, Fr.) implies the being so near the land as that a deep-sea-lead will attain the bottom, which is not practicable in the ocean.

SOUTH, (*sud*, Fr). one of the four cardinal points of the compass, being that directly opposite the north.

SOUTHERN HEMISPHERE, LATITUDES, SIGNS, &c. are those on the south side of the equator.

SOUTHING, in navigation, the difference of latitude made by a ship in sailing to the southward.

SOUTHING *of the Moon*, (*passage de la lune au méridien*, Fr.) the time at which the moon passes the meridian of any particular place. See the article MOON.

SPAN, (*pendeur à deux branches, fixé à un mât, ou à un étai, pour y passer les bras, ou les boulines de quelque voile*, Fr.) a small line or cord, the middle of which is usually attached to a stay, whence the two ends branch outward to the right and left, having either a block or thimble attached to their extremities. Its use is, to confine some rope which passes through the corresponding blocks or thimbles, as well to increase the effort of the said rope, as to prevent it from swinging at too great a distance from the centre of its action in stormy weather. Such are the spans occasionally used for the top-gallant braces, or the fore-top-gallant bowlines, &c.

To SPAN *the Booms*, is to confine them by ropes.

To SPAN *in the Rigging*, (*brider les haubans, pour en frapper le trélingage*, Fr.) is to draw the upper parts of the shrouds together, by tackles, in order to seize on the catharping-legs.

To SPAN *the Runners*, is to take several turns with small rope round both runners abaft the mast, and to frap the turns.

SPAN-SHACKLE, (*cheville & boucle de fer servant d'appui au davied dans les vaisseaux anglois*, Fr.) a large bolt driven through the forecastle and forelocked under the forecastle beam, both under and upon the upper-deck-beam; on the forecastle it has a large square ring for the end of the davit to fix in.

SPANKER, (*voile de baume, ou de brigantine*, Fr.) a name sometimes given to a ship's DRIVER, which article see.

SPANKER-BOOM. See the article BOOMS.

SPARE, (*de rechange*, Fr.) an epithet applied to any part of a ship's equipage or furniture, that lies in reserve, to supply the place of such as may be lost or rendered incapable of service. Hence we say, spare tiller, spare top-masts, spare sails, spare rigging, &c.

SPARS, (*mâtereaux, ou esparres*, Fr.) large round pieces of timber, fit for making top-masts, &c.

To SPEAK WITH A VESSEL, (*parler à un bâtiment*, Fr.) implies that, on the meeting of ships or vessels at sea, one has communication with the other by going on board; but if that should be found impracticable from the state of the weather, the speaking-trumpet is employed by the homeward-bound ship to ask the necessary questions. See the article *To* HAIL.

SPEAKING-TRUMPET. See the article TRUMPET.

SPEAR OF A PUMP. See the article PUMP.

SPECIFIC, in philosophy, that which is proper and peculiar to any thing; or characterises and distinguishes it from every other thing. Thus the attraction of iron is specific to the loadstone, or is a specific property of it.

SPECIFIC *Gravity*, in hydrostatics, is the relative proportion of the weight of bodies of the same bulk. See the article GRAVITY.

SPECIFIC *Gravity of living Men.* Mr. John Robertson, late librarian to the Royal Society, in order to determine the specific gravity of men, prepared a cistern 78 inches long, 30 inches wide, and 30 deep; and having procured 10 men for this purpose, the height and weight of each were taken, they then plunged successively into the cistern. A ruler or scale, graduated to inches and decimal parts, was fixed to one end of the cistern, and the height of the water shewn by it was noted before each man went in, and to what height it rose when he immersed himself under its surface. The following table contains the several results of his experiments:

No. of Men.	Height. Feet.	In.	Weight. lbs.	Water raised Inches.	Solidity. Feet.	Weight of Water. lbs.	Specific Gravity.
1	6	2	161	1.90	2.573	160.8	1.001
2	5	$10\frac{3}{8}$	147	1.91	2.586	161.6	0.901
3	5	$9\frac{1}{2}$	156	1.85	2.505	156.6	0.991
4	5	$6\frac{3}{4}$	140	2.04	2.763	172.6	0.801
5	5	$5\frac{7}{8}$	158	2.08	2.817	176.0	0.900
6	5	$5\frac{1}{2}$	158	2.17	2.939	183.7	0.849
7	5	$4\frac{3}{8}$	140	2.01	2.722	170.1	0.823
8	5	$4\frac{1}{8}$	121	1.79	2.424	151.5	0.800
9	5	$3\frac{1}{4}$	146	1.73	2.343	146.4	0.997
10	5	$3\frac{1}{8}$	132	1.85	2.505	156.6	0.843
Medium of all.	5	$6\frac{2}{3}$	146	1.933	2.618	163.6	0.891

One of the reasons, Mr. Robertson says, that induced him to make those experiments was, a desire of knowing what quantity of timber would be sufficient to keep a man afloat in water, thinking that most men were specifically heavier than river or common fresh water; but the contrary appears from the trials above recited; for except the first, every man was lighter than an equal bulk of fresh water; so that if persons who fall into water had presence of mind enough to avoid the fright usual on such occasions, many might be preserved from drowning; and a piece of wood not larger than an oar, would buoy a man partly above water as long as he had strength or spirits to keep his hold. *Phil. Trans.* vol. L. art. 5.

N. B. From the last line of the table appears the medium of all the circumstances of height, weight, &c.; particularly the mean specific gravity, 0.891, which is about $\frac{1}{9}$ less than common water.

SPECULUM, or MIRROR, in optics, any polished body impervious to the rays of light; such as polished metals, and glasses lined with quicksilver, or any other opake matter, popularly called looking-glasses, or even the surface of mercury or of water, &c.

SPELL, (*tems réglé pendant lequel les matelots font un service quelconque, à la pompe, &c., après lequel tems ils se relèvent alternativement*, Fr.) the period wherein one or more sailors are employed in a particular exercise, from which they are relieved as soon as the limited time expires: such are the spells to the hand-lead in sounding; to the pump; to look out on the mast-head, &c. and to steer the ship; which last, however, is generally called the *trick*.

To SPELL, (*relever, &c.* Fr.) also implies to relieve, or return to those services: thus, we say, spell the pump, spell the lead, &c.

To SPELL *the Mizen*, (*laisser tomber la misaine et l'apiquer*, Fr.) is to let go the sheet and peak it up.

SPELL *the Pump!* (*relève les hommes de la pompe!* Fr.) the order given to the men to work the pump for a limited time, when they are relieved by fresh hands.

MARINE SPENCER, a contrivance by Mr. Knight Spencer, for the preservation of shipwrecked mariners, &c. and for which he was rewarded by the Royal Humane Society with the honorary silver medal, in 1803.

This invention consists of a girdle, whose diameter is intended to fit the body, and is composed of about 800 old tavern corks, strung upon a strong twine, well lashed together with laycord, covered with canvas, and painted in oil, so as to render it waterproof: two cords are fastened to the back of the girdle, with loops at their ends; another cord about three feet comes from the same place, in the middle of which a few corks are strung, covered with canvas and painted as above mentioned; a pin of hard wood, three inches long, and half an inch in diameter, is fastened to the front of the girdle by a tape or cord about two inches long, the same at the bottom: when it is used, slide it from the feet close up to the arms, bring the cords one over each shoulder, and fasten them by the loop to the pin in front, bring the other cord between the legs, which likewise fasten by the other pin. See also the article CORK-JACKET.

To SPEND A MAST. See the article MAST.

SPENT-BALL, (*boulet mort, balle morte*, Fr.) A cannon or musket ball is said to be spent, when it reaches an object without sufficient force to pass through it, or otherwise wound, than by a contusion. Spent-balls, however, are very frequently fatal in their effects, especially when they hit living objects.

SPHERE, in astronomy, that concave orb or expanse which invests our globe, and in which the heavenly bodies, the sun, moon, stars, planets, and comets, appear to be fixed at an equal distance from the eye. This is also called the sphere of the world, and it is the subject of spherical astronomy.

This sphere, as it includes the fixed stars, from whence it is sometimes called the *sphere of the fixed stars*, is immensely great; so much so, that the diameter of the earth's orbit is vastly small in respect of it; and consequently the centre of the sphere is not sensibly changed by any alteration of the spectator's place in the several parts of the orbit: but still, in all points on the earth's surface, and at all times, the inhabitants have the same appearance of the sphere; that is, the fixed stars seem to possess the same points in the surface of the sphere; for our way of judging of the places, &c. of the stars, is to conceive right lines drawn from the eye or from the centre of the earth, through the centres of the stars, and thence continued till they cut the sphere: and the points where these lines so meet the sphere, are the apparent places of those stars.

The better to determine the places of the heavenly bodies in the sphere, several circles are conceived to be drawn on the surface of it, which are called circles of the sphere; and it is generally divided into three parts, the ZODIAC, the NORTHERN and SOUTHERN HEMISPHERES.

Projection of the SPHERE, is a representation of all its great and small circles, poles, &c. upon a plane surface; such as they would appear to the eye, viewing them from some certain distance, or in some certain position. In thus projecting it, only one half of its surface is drawn, for the great and small circles of both hemispheres are alike: this may be illustrated by setting a globe in any position, and the circles above the horizon will be similar to those below it.

There are various projections of the sphere, which depend upon the two following considerations: first, the distance or place of the eye; secondly, the position of the sphere as presented to the eye. The distance or place of the eye is distinguished by three kinds of projection, which are called *Gnomonic*, *Orthographic*, and *Stereographic*. See the article PROJECTION.

SPHERE, in geography, &c. denotes a certain disposition of the circles on the surface of the earth, with regard to one another, which varies in the different parts of it.

The circles originally conceived on the surface of the sphere of the world are almost all transferred, by analogy, to the surface of the earth, where they are conceived to be drawn directly underneath those of the sphere of the stars, they would coincide with the respective circles on it. Thus we have a horizon, meridian or equator in the heavens, which divides the sphere into two equal parts, the one north and the other south; so does the equator on the surface of the earth divide its globe in the same manner. And as the meridians in the heavens pass through the poles of the equinoctial, so do those on the earth, &c. With regard to the position of some of these circles in respect of others, we have a *right*, an *oblique*, and a *parallel* SPHERE.

A Right SPHERE, is that which has the poles of the world in its horizon, and the equator in the zenith and nadir. The inhabitants of this sphere live exactly at the equator of the earth, or under the line. They have, therefore, no latitude nor elevation of the pole. They can see both poles of the world; all the stars do rise, culminate, and set to them; and the sun always rises at right angles to their horizon, making their days and nights always of equal length, because the horizon bisects the circle of the diurnal revolution.

An Oblique SPHERE, is that in which the equator, as also the axis, cuts the horizon obliquely. In this sphere one pole is above the horizon, and the other below it; and therefore the inhabitants of it see always the former pole, but never the latter; the sun and stars, &c. all rise and set obliquely; and the days and nights are always varying, and growing alternately longer and shorter.

A Parallel SPHERE, is that which has the equator in or parallel to the horizon, as well as all the sun's parallels of declination. Hence the poles are in the zenith and nadir; the sun and stars always appear quite around and parallel to the horizon; the inhabitants, if any, being just at the two poles, having 6 months continual day, and 6 months night, in each year; and the greatest height to which the sun rises to them is 23° 28′, or equal to his greatest declination.

SPHERE, in geometry, a solid body contained under one single uniform surface, every point of which is equally distant from a point in the middle called its centre.

A sphere may be considered as made up of an infinite number of pyramids, whose common altitude is equal to the radius of the sphere, and all their bases from the surface of the sphere. And therefore the solid content of the sphere is equal to that of a pyramid whose altitude is the radius, and its base is equal to the surface of the sphere, that is, the solid content is equal to $\frac{1}{3}$ of the product of its radius and surface.

SPHERICAL, something relating to the sphere: hence,

SPHERICAL *Angles*, are formed by the inclination of two great circles on the surface of the sphere meeting in a point, called the angular point.

The measure of a spherical angle is an arc of a great circle intercepted between the legs of the angle, 90 degrees distant from the angular point. Thus, the angle which the equinoctial makes with the ecliptic on the celestial globe, is measured by that arc of the celestial colure which is intercepted between those circles, and is 90 degrees distant from the angular point. Hence, the angular point is the pole of the great circle which measures the angle; and any small circle contains the same number of degrees as a great circle.

SPHERICAL *Triangles*, are formed by the intersection of three great circles on the surface of the sphere. Both the sides and angles of spherical triangles are computed by degrees, minutes, and seconds, in the same manner as angles, only in plane triangles. The three angles of a spherical triangle are always more than the three angles of a plane triangle, owing to the convexity of the spherical surface. The three angles of a spherical triangle are always less than six right angles; for one angle cannot be 180°, and therefore the three angles must be less than 540°.

Thus, the angles of a spherical triangle may vary from something more than 180°, to something less than 540°; and therefore, when two angles are given, the third is not thence known, as in a plane triangle; but when the three angles of a spherical triangle are given, the sides may be thence found, which cannot be done in a plane triangle. Two sides of a spherical triangle, like those of a plane one, are always greater than the third; for the sides of a spherical triangle are the nearest distance between the angular points on the sphere, as the sides of a plane triangle are the nearest distance between the angular points upon a plane.

The three sides of a spherical triangle are always less than two semicircles, or 360 degrees; for if any two sides be continued, they will intersect in opposite points of the sphere, and form two semicircles, it is evident that the third side of the triangle, with the produced sides, will make a new triangle; and as these two new sides are greater than the third, the two semicircles must be greater than the three sides of the triangle. In this, spherical triangles differ essentially from plane ones, the sides of the latter being unlimited; but in both kinds of triangles the sides may be of the smallest possible dimensions. For farther particulars on this subject, see KELLY's *Spherics.*

SPICA VIRGINIS, in astronomy, a bright fixed star of the first magnitude, in the ear of corn of the constellation Virgo. For the beginning of the year 1814, its

Right ascension was............	198°	51	0″	
Annual variation in right ascension,			47.2	
Declination	10.	11.	5	S.
Annual variation in declination ..			+ 18.8.	

SPIKE. See the article HAND-SPIKE.

To SPIKE *a Gun*, (*enclouer un canon*, Fr.) is to choke up the touch-hole of a piece of ordnance, by driving a nail forcibly into it, so as to render the piece unserviceable. See also *To* NAIL or SPIKE *a Gun.*

To SPIKE *a Gun*, also implies, to fasten a quoin with spikes to the deck, close to the breech of the gun-carriage, so that the gun may keep firmly and closely to the sides of the ship, and not break loose when the ship rolls.

To SPILL A SAIL, (*déventer une voile pour la serrer, ou pour y prendre un ris*, Fr.) to discharge the wind out of its cavity or belly, in order to furl or reef it. This is either performed by collecting the sail together, or by bracing its edge to the wind, so as to shiver it.

SPILLING-LINES, (*espèces de dégorgeoirs placés dans un gros tems pour aider à carguer le fond des basses voiles*, Fr.) are certain ropes fixed occasionally to the main and fore-sails of a ship, in tempestuous weather, for reefing or furling them more conveniently. See the article LINE.

SPINDLE, in mast-making, also implies the upper main-piece of a made mast.

SPINDLE *of a Capstan*, (*pivot d'un cabestan*, Fr.) is that part of a single capstan between decks; its lower end or foot, which is shod with iron, becomes the pivot or axis upon which it turns in the saucer. See the article CAPSTERN.

SPINDLE *of a Vane*, (*fer de girouette*, Fr.) a sort of iron pin tapering at the upper end to a point. It is fixed into the upper end of the top-gallant-mast, so as to carry a vane, which, turning threon horizontally, shews the direction of the wind. It is usually crowned with a globular or conical piece of wood called the acorn, which prevents the vane from being blown off. See the article ACORN

SPINNING-WHEEL, in rope-making, for 12 spinners to spin yarn at the same time, is about five feet in diameter, and is hung between two posts fixed in the ground; over its top is fixed a semicircular frame called the head, which contains twelve whirls that turn on iron spindles, with hooks on their front ends to hang the hemp on, and are worked by means of a leather band, which encircles its wheels and whirls.

Spinning *of Yarn*, in rope-making, is the act of combining the fibres of hemp, after it has been properly hatchelled, into certain lengths, by the help of a spinning-wheel, which should be turned regularly, that the yarn may be spun with a regular twist; for if slack-spun, it will break in straining up.

Particular attention should be paid by the spinner that the yarn be spun even, solid, and round; to accomplish which, he must spin with a strong even grip of the hand, taking care not to make his yarn larger in one place than in another, nor make a practice of spinning too much in a hurry, as in that case the yarn will lose its principal support, which is its proper turn, and will be little stronger than a parcel of straight hemp laid together. To every twelve spinners there are allowed two hatchellers, one wheel-turner, and one wheel-tender; the wheel-tender's business is to splice the threads, and wind them on winches. See the article Rope-Making.

SPIRIT-Room, (*cale au vin*, Fr.) a space in the after-part of a ship's hold, set a-part for the brandy, wine, rum, &c.

SPIRKETTING, (*feuilles bretonnes*, Fr.) in ship-building, the strake wrought on the ends of the beams; or, where there are ports, it is the two strakes worked up to the port-sills; in which case the middle of the planks should not be reduced, unless it occasioned the butts to be less than six inches.

To SPLICE, (*épisser*, Fr.) is to join the two ends of a rope together, or to unite the end of a rope to any part thereof, by interweaving the strands in a regular manner.

There are several methods of making a splice, according to the services for which it is intended; all of which are distinguished by particular epithets, as

Short Splice, (*épissure courte*, Fr.) is made by untwisting the ends of two ropes, or the ends of one rope, and, having placed each of the strands of one opposite to and in the interval between two strands of the other, to draw them close together; and then interweave the strands of one into the alternate strands of the other, by penetrating the latter with a fid or marline-spike, parallel to the axis or length of the rope. This splice is used on the cables, slings, block-straps, and in general all ropes which are not intended to run through blocks, or where the splice is not in danger of being loosened or separated.

Long-Splice, (*épissure longue*, Fr.) occupies a greater extent of the rope; but, by the three joinings being fixed at a distance from each other, the increase of bulk is also divided; hence it is much neater and smoother than the short-splice, and better adapted to run through the channel of a block, &c. for which use it is generally calculated.

Eye-Splice, (*épissure d'estrop, ou de gance*, Fr.) forms a sort of eye or circle at the end of a rope, and is used for splicing in thimbles, bull's-eyes, &c. and sometimes on the ends of block-straps: the strands are, therefore, untwisted, and their extremities thrust through the three strands in that part of the rope whereon the splice is to be formed, and thence passing over the surface of the second strand, they are again thrust through the third, which completes the operation.

Cunt-Splice, (*épissure en portière de vache*, Fr.) is constructed in the same manner as the eye-splice, but for different purposes: it is made by splicing the ends of two lines at a short distance from each other, the extremities of either being interwoven into the bight of the other, so that the line becomes double in the extent of the splice. This is chiefly used in lead-lines, log-lines, and fishing-lines, where the short-splice would be liable to separation, as being frequently loosened by the water.

Drawing-Splice, (*épissure de cable*, Fr.) used on cables, is made by unlaying about three fathoms of the ends; then placing them together, and making a short splice; then leave about one fathom, and from thence reduce each strand to a long taper, by gradually cutting away as many yarns as necessary, and neatly point over the taper; then lay the ends up the cunt-lines, and clap on a quarter-seizing at each end of the splice, an end seizing at the beginning of the pointing, and a stop at the ends of the tails. This is the best splice for cables, as it may be readily taken asunder. Another good method of making a Cable *Splice* is, to put the ends in twice each way; then to pick out the strands, worm part of them round the cable, and taper away the rest, which should be marked close down; then clap on a good throat, and two end-seizing of ratline.

Tapered- Splice, (*épissure de cable*, Fr.) mostly used on cables, is made by unlaying a certain length of each cable; then placing them together, and interplacing the strands, as in the short-splice, twice each way, and hauled taught each time; then inlay the strands or ends successively, and reduce them by cutting away one strand; then interpose the two remaining strands, and reduce them to a single strand, which is again thrust through and cut off. The splice is then served over with spun-yarn, something more than the whole length.

To Splice *the Main-brace*, (*boire du grog*, Fr.) is to give each person on board a glass of grog, when the captain deems it necessary.

SPLINTERS, (*éclats de bois*, Fr.) the pieces of a ship's sides, masts, decks, &c. which, being knocked off by a shot, acquire great velocity, and frequently do more damage among the men than the shot itself.

Splinter-*Nettings*. See the article Nettings.

To SPLIT a Sail, (*défoncer, ou crever une voile*, Fr.) is to rent it asunder, which sometimes happens by the violence of a tempest, or by sustaining a greater effort on one part of it's surface than the rest.

Split, (*vaisseau naufragé*, Fr.) when applied to a ship, implies the state of being bilged on a rock or shore.

SPOILING, in mast-making, is the act of taking

the greatest distance of the inequalities between any two pieces of timber to be fayed together.

SPOKE-SHAVE, in block-making, a kind of plane to smooth and finish the shell with: it is a piece of steel four or more inches long, and one inch and a half broad, sharp at one edge as a knife, and one fourth of an inch thick at the back; at each end is left about two inches of narrow steel bent up as pegs, by which it is regulated and secured in a piece of wood ten inches long, and narrowed at each end for handles.

SPONGE OF A GUN, (*écouvillon*, Fr.) in gunnery, a cylindrical piece of wood, whose diameter, when covered with lamb-skin or coarse wool, is equal to that of the shot; see figure 6, plate VI. In the sea-service, it is fixed on a stiff rope, which serves as a handle, and is used to clean the gun when fired.

To SPONGE, (*écouvillonner*, Fr.) is to clean a gun out with the sponge; and should be constantly repeated after every explosion.

SPONGE *your Guns!* (*écouvillonez les canons!* Fr.) the order to clean them out after discharging their contents. See the Article EXERCISE.

SPOON-DRIFT, (*poussière d'eau de mer, qui couvre la surface des eaux, dans un coup de vent,* Fr.) a sort of showery sprinkling of the sea-water, swept from the surface of the waves in a tempest, and flying, according to the direction of the wind, like a vapour

SPOUT. See the article WATER-SPOUT.

SPRAY, (*éclaboussure d'eau de mer, chassée par le vent du sommet d'une vague,* Fr.) the sprinkling of the sea, which is driven from the top of a wave in stormy weather. It differs from spoon-drift, as being only blown occasionally from the broken surface of a high wave, whereas the latter continues to fly horizontally along the sea, without intermission, during the excess of a tempest or hurricane.

SPRIG, a name given to a small eye-bolt which is ragged at the point.

SPRING, (*fente oblique, ou transversale, dans une pièce de mâture, qui la rend douteuse,* Fr.) a crack running transversely or obliquely through any part of a mast or yard, so as to render it unsafe to carry the usual quantity of sail thereon.

SPRING, (*embossure, et en général toute amarre passant par l'arriere, qui a son point d'appui en avant, ou par le travers d'un bâtiment,* Fr.) is also a rope passed out of a ship's stern and attached to a cable proceeding from the bow, when she lies at anchor. It is usually performed to bring the ship's broad-side, or battery of cannon, to bear upon some distant object; as another ship, or a fortress on the coast, &c. When a ship rides by anchors which are only carried out of one end, she will swing upon the surface of the water like a weather-cock, according to the direction of the wind; unless when the wind is opposed by a current. Now, if a rope be extended from the other end to the same anchor, it is evident that by slackening one of the ropes, and keeping fast the other, her side will be presented to the wind; and that, by slackening one of those ropes, and keeping fast the other, her side will lie more or less obliquely to the wind, so as to be opposed to any distant object to the right or left. Thus, if a ship rides with her head northerly, and it is required to cannonade a fortress lying on the south, or south-east, a hawser is run out from the stern, and being carried forward, without her side, is attached to the cable, at a competent distance a-head of the ship; the hawser is then tightened by the capstan or tackles, and the cable being slackened, the ship immediately turns her side towards the object intended to be battered.

SPRING is likewise a rope reaching diagonally from the stern of a ship to the head of another which lies a-breast of her, at a short distance, and is generally performed to make one of the ships sheer off, to a greater distance from the other. Springs of this sort are occasionally applied from a ship to a wharf or pier for the same purposes.

SPRING, in astronomy and cosmography, denotes one of the seasons of the year, commencing in the northern parts of the earth on the day the sun enters the first degree of *Aries*, which is about the 21st of March, and ending when the sun enters *Cancer*, at the summer solstice, about the 21st of June, spring ending when the summer begins. Or, more strictly and generally, for any part of the earth, or on either side of the equator, the spring season begins when the meridian altitude of the sun, being on the increase, is at a medium between the greatest and least; and ends when the meridian altitude is at the greatest. Or the spring is the season or time, from the moment of the sun's crossing the equator till he rises to the greatest height above it.

SPRING-BLOCK. See the article BLOCK.

To SPRING *a Butt*, (*tête de bordage qui s'est lâchée*, Fr.) See the article BUTT.

To SPRING *a Leak.* (*faire une voie d'eau*, Fr.) A ship is said to spring a leak when by any accident the water passes through a breach in her sides or bottom into the hull. See the article LEAK.

To SPRING *the Luff.* See the article LUFF.

To SPRING *a Mast, Yard, &c.* See the article SPRING.

SPRING-SEARCHER. See the articles GUN and SEARCHER.

SPRING-TIDE, (*grande marée, ou maline*, Fr.) the periodical excess of the elevation and depression of the TIDE, which article see.

SPRIT, (*livarde, ou baleston d'une voile de certaines embarcations, qui la traverse diagonalement,* Fr.) a small boom or pole which crosses the sail of a boat diagonally, from the mast to the upper aftmost corner, which it is used to extend and elevate: the lower end of the sprit rests in a sort of wreath, called the snotter, which encircles the mast at that place. These sort of sails are accordingly called sprit-sails.

SPRIT-SAIL, (*civadière dans les vaisseaux*, Fr.) a sail attached to a yard which hangs under the bowsprit, as represented by fig. 3, plate XXXII. It is furnished with a large hole at each of its lower corners, to evacuate the water with which the cavity or belly of it is frequently filled by the surge of the sea, when the ship pitches.

SPRIT-SAIL *Topsail*, (*contre-civadière*, Fr.) a sail extended above the former by a yard which hangs under the jib-boom, and is represented by z in the same figure. The lower corners of this sail are hauled home to the spritsail yard-arms; after which the sail is drawn out towards the extremity of the boom, in light winds, as any other topsail-yard is hoisted upon its mast.

Formerly the spritsail topsails were set on a mast, which was erected perpendicularly on the end of the bowsprit: but this method has of late been justly rejected, as being inconvenient and dangerous to the bowsprit, although serviceable in light breezes.

SPRIT-SAIL *Top-Gallant-Sail*, is a quadrilateral sail cut square on the head, and set upon the flying jib-boom, in the same manner that the sprit-sail top-sail is set upon the inner jib-boom: this sail is, however, very rarely used.

SPRIT-SAIL *Yard*. See the article YARD.

SPRIT-SAIL *Top-sail Yard*. See the article YARD.

SPUNGE. See the article SPONGE *of a Gun*.

SPUN-YARN, (*bitord*, Fr.) a small line or cord formed of two or three rope-yarns twisted together by a winch. The yarns are usually drawn out of the strands of old cables, and knotted together and tarred. It is employed for several purposes; particularly to fasten one rope to another, to seize block-straps to the shrouds, and to serve ropes which are liable to be chafed, &c.

SPURLING-LINE, (*ligne qui fait mouvoir l'axiomètre*, Fr.) the line which forms the communication between the WHEEL and the TELL-TALE. See those articles.

SPURS, are pieces of timber fixed on the bulgeways, their upper ends being bolted to the ship's side above water, for the security of the bulgeways.

SPURS *of the Beams*, (*pièces de construction qui dans les vaisseaux Anglois tiennent lieu des demi-baux à l'endroit des écoutilles dans la construction Françoise*, Fr.) in ship-building, are curved pieces of timber, serving as half-beams to support the decks where a whole beam cannot be placed on account of the hatchways. See the article DECK.

SPURS *of the Bits*, (*courbes des bites, ou arboutans qui dans la construction Anglaise tiennent lieu des courbes de bites dans les vaisseaux François*, Fr.) the same with STANDARDS, which article see.

SQUADRON, (*escadre*, Fr. *squadrone*, Ital.) implies a detachment of ships employed on any particular expedition, or the third part of a naval armament. See the articles FLAG, CENTRE, FLEET, DIVISION, &c.

SQUALL, (*raffale, ou grain*, Fr.) a sudden and violent gust of wind, usually occasioned by the interruption and reverberation of the wind from high mountains. These are very frequent in the Mediterranean; particularly that part of it which is known by the name of the Levant, as produced by the repulsion and new direction which the wind meets with in its passage between the various islands of the Archipelago.

A squall or brisk wind commencing suddenly after a calm, is supposed to be occasioned by descending winds, produced by the impulse of falling rain: they are generally experienced to have their motion in a horizontal direction, when the impulse is perceived on the sail of a ship; but it is probable that the current of wind may descend until near the surface of the sea, when calm, and then be deflected in a horizontal direction on approaching it.

A Black SQUALL, (*grain noir*, Fr.) one attended with a dark cloud diminishing the usual quantity of light.

A White SQUALL, (*grain blanc*, Fr.) in contradistinction to black, produces no diminution of light.

A Thick SQUALL, (*grain bien nourri*, Fr.) is accompanied with hail, sleet, &c.

SQUARE, (*quarré, se dit aussi d'un vaisseau qui a beaucoup d'envergure*, Fr.) a term peculiarly appropriated to the yards and their sails, either implying that they are at right angles with the mast or keel; or that they are of greater extent than usual.

Thus, when the yards hang at right angles with the mast, they are said to be square by the lifts; when they hang perpendicular to the ship's length, they are called square by the braces: but when they lie in a direction perpendicular to the plane of the keel, they are square by the lifts and braces; or, in other words, they hang directly across the ship, and parallel to the horizon.

The yards are said to be very square, when they are of extraordinary length; and the same epithet is then applied to their sails with respect to their breadth.

To SQUARE *the Yards by the Lifts*. See the article LIFTS.

SQUARE, an iron or wooden instrument, with a perpendicular and base, used as a guide to set off by.

SQUARE, in anchor-making, is the upper-part of the shank of an ANCHOR, which article see.

SQUARE, in geometry, a quadrilateral figure, whose angles are right, and sides equal; or it is an equilateral rectangle, or an equilateral rectangular parallelogram.

To find the Area of a SQUARE. Multiply the side by itself, and the product is the area.

An Eight SQUARE, in mast-making, is formed by reducing a four-square nearer to a round, by taking off the angles, which are lined or marked within the edge, or from the middle line on each side. If the former, $\frac{7}{24}$ths of the diameter is set within the edge at as many places as may be necessary for lining; or $\frac{5}{24}$ths the diameter on each side of the middle line, for the latter, and a line struck to range fair through those spots, and the angles dubbed or sawed away to those lines.

A Sixteen SQUARE, in mast-making, is made by reducing a mast, or any other piece nearer to a round, by taking the angles of the eight square in the following manner; viz. divide the eight-square into four equal parts, and strike a line to range through the divisions next the edges or angles, which are dubbed off to these lines, similar to the former.

SQUARE *Foot*, a square whose sides are equal to a foot, or 12 inches; and the area, or square foot, is equal to 144 square inches.

Geometrical SQUARE, a compartment often added on the face of a quadrant, called also Line of SHADOWS.

SQUARE *Number*, is the product arising from a number multiplied by itself; thus, 9 is the square of 3, and 25 the square of 5.

SQUARE-*Rigged*, (*batiment mâté à quarré, ou qui a le gréement d'un bâtiment à trois mats*, Fr.) is a term used in contradistinction to all vessels whose sails are extended by stays, lateen, or lug-sail yards; or by booms and gaffs; the usual situation of which is nearly in a plane with the keel.

Square-*Root*, a number considered as the root of a second power or square number: or a number which multiplied by itself, produces the given number,; as, the square root of 25 is 5, and that of 9 is 3.

Square-*Sail*, (*voile de fortune, ou de tréou, dans une goëlette ou autre bâtiment latin*, Fr.) is any sail extended to a yard, which hangs parallel to the horizon, as distinguished from the other sails which are extended obliquely.

Square-*Sail*, (*voile quarrée*, Fr.) is also the name of a sloop's or cutter's sail which hauls out to the lower yard called the square-sail-yard. This sail is only used in fair winds, or to scud in a tempest. In the former case, it is furnished with a large additional part called the bonnet, which is then attached to its bottom, and removed when it is necessary to scud.

Square-*Sail-Boom* (*vergue de tréou, ou vergue de fortune*, Fr.) a boom lashed across the deck of a vessel with one mast, and used to spread the foot of the square-sail.

Square-*Sterned*, (*vaisseau à poupe quarrée, comme celle des vaisseaux de guerre*, Fr.) implies a stern like a British-built ship of war, in opposition to the Dutch and other northern nations, who build their vessels with *round* sterns.

Square-*Tuck*. See the article Tuck.

SQUARING, in mast-making, the act of making one or more sides from the plumbed sides, which may be performed by a plumb or square, the tongue of which is a perpendicular to its stock or base.

STABBER, in sail-making, an instrument, similar to a pricker, only it is triangular instead of square.

STABILITY of Bodies, (*stabilitas Lat.*) is their firmness and strength to resist any substance that may come in contact with them.

The stability or firmness of bodies, in contradistinction to fluidity, consists in this, that the particles which compose them do so rest, or are entangled, that there is among them a mutual cohesion.

For a series of experiments lately made by Colonel M. Beaufoy, of Hackney Wick, for the purpose of ascertaining the stability of bodies floating on the surface of the water; and Captain Burney's experiments on bodies floating in a stream, see the article Floating Bodies. Also the article Resistance *of Fluids*.

STAFF, Back, Cross, Fore. See these articles.

Staff, (*bâton*, Fr.) a light pole erected in different parts of a ship, whereon to hoist and display the colours.

Ensign-Staff, (*bâton de pavillon*, Fr.) a staff reared immediately over the stern of a ship to display the ensign. See the article Ensign.

Flag-Staff, (*bâton d'enseigne, ou de commandement*, Fr.) is that erected at each of the mast-heads, or formed by their upper ends to support the flag or pendant of the respective squadron or division to which the ship belongs.

Jack-Staff, (*bâton de pavillon de beaupré*, Fr.) is that fixed at the end of the bowsprit, to extend the jack. See the article Jack.

Staff *for the Nipper*, in rope-making, is a piece of oak timber, seven feet long, and four inches square; one end is fixed in an iron strap on the upper plate of the nipper; and on the other end a weight is suspended sufficient to press the tar out of the yarn.

Staffs, in mast-making, are short pieces by which the sets are made; also long narrow pieces, with divisions of ten or twenty feet marked thereon, and again divided into halves: they are used to set off long lengths.

Staffs *for Tops*, in rope-making, are round, from six to eight feet long, and from two and an half to five inches diameter, which go through a hole in the top, or are confined under it by a bolt and nail: they run on a truck-wheel at the lower-end, as the rope closes.

STAGE, (*échaffaud*, Fr.) a machine composed of plank, let over the sides of a ship, by ropes, whereon the people may stand when repairing, calking, or paying the sides, wales, &c.

A Floating Stage, (*ras de carene*, Fr.) is one which needs not the support of ropes, being sufficiently large and firm to bear upon the water.

STAKE-Heads, in rope-making, are about four feet long, and four by three inches square, with four wooden pins to keep the strand asunder, for lines they are about two feet long, and three by two inches square, and have six pins.

Stake-Posts, in rope-making are pieces of oak, about four feet high, and twelve inches diameter, with a mortise hole in each for the stake heads to go in and out, to keep the rope from the ground, they are placed about ten yards distant from each other along the whole length of the walk.

STANCIHON, (*épontille*, Fr.) a sort of small pillar of wood or iron used for various purposes in a ship: as to support the deck, the quarter-rails, the nettings, the awnings, &c.

STANCHIONS, (*epontilles fixes*, Fr.) a name also given to the upright pieces of timber in a bulkhead, breast-work, &c. of a ship.

Stanchions, *that ship and unship*, (*épontilles volantes*, Fr.) are made of iron, and placed on the bow, quarter, and waist for stowing the hammocks on; those on the latter being much larger than the others.

Stanchions *of the Awning*, (*montans de tente*, Fr.) are pieces of fir timber placed perpendicular on the upper decks of ships in ordinary, for the purpose of supporting the awning.

Stanchions *of the Nettings*, (*chandeliers, &c.* Fr.) are either slender bars of iron, whose lower ends are fixed in iron sockets at proper distances; or square wooden pillars let into the upper part of the ship's sides. See the article Quarter-netting.

To STAND, (*faire route vers un objet, en parlant d'un bâtiment à la voile*, Fr.) when expressed of a ship, implies the movement by which she advances towards a certain object, or departs from it; as, "the enemy stands in-shore," "the English fleet is standing off." "At day-break we discovered three sail standing to the northward, &c.

To Stand *by a Rope, &c.* (*veiller à une manœuvre, &c.* Fr.) is to attend to it; as, "Stand by the top-sail-halyards!"

To Stand *On*, (*continuer sa route*, Fr.) is to continue the course on which a ship sails.

STAND *from Under!* (*garre dessous!* Fr.) an order given to those below to keep out of the way of any thing being lowered down or let fall from above.

To STAND *off, or out to Sea,* (*porter au large, ou avoir le cap au large,* Fr.) when speaking of a vessel, is to keep at a competent distance from the shore, so as to be clear of danger.

To STAND *off and On,* (*se tenir bord sur bord,* Fr.) is to keep alternately near to the shore and out to sea; as, "That brig is standing on and off."

To STAND *under easy Sail,* (*faire route à petites voiles,* Fr.) is to advance in any proposed course with a moderate impulse of wind upon the sails.

To STAND *upon the Starboard or Larboard Tack,* (*courir tribord, ou babord amure,* Fr.) is to advance by tacking either to the right or left side of the ship.

THE ROYAL STANDARD, (*étendard royal qui s'arbore au grand mât, lorsque le Roi d'Angleterre, ou quelque prince du sang sont à bord,* Fr.) the state flag of Great Britain and Ireland, as represented by c, plate XXV. This flag, which is never hoisted except on occasions of the first ceremony, is of course the highest in rank of any used, and is never displayed on ship-board, except when the king himself or some personage belonging to the royal family is actually present, or on days of particular rejoicing, such as their Majesties birth days, the restoration of King Charles the second, with some other festivals of the same distinction, or the celebration of some recent victory; at such memorable periods it has been, and still continues the custom for the commander-in-chief of a fleet, to hoist the royal standard at his top-gallant-mast-head.

It displays the arms of the united kingdom quarterly, first and fourth, gules, three lions passant gardant in pale, or, for England; second, a lion rampant, gules within a double tressure flory, counterflory of the last for Scotland. Third, azure, an harp, or, stringed argent for Ireland, on an escutcheon of pretenee, ensigned with the electoral bonnet, and divided per pale, and per cheveren, enarched in three compartments; the arms of his Majesty's dominions in Germany gules, two lions passant gardant in pale, or for Brunswick, or seme of hearts proper, a lion rampant azure Lunenburg, third, gules, a horse current argent for Saxony, in the centre on an escutcheon gules, the crown of Charlemagne proper, being the badge of the office or arch-treasurer to the holy Roman Empire.

STANDARD-*Knees.* See the article KNEES.

STANDARDS, (*courbes verticales des ponts, dont une branche se cheville sur le pont, et l'autre sur le coté du vaisseau entre les sabords. On met souvent à la place de ces pièces des aiguillettes de porques, et quelquefois des courbes de fer,* Fr.) in ship-building, are inverted knees placed above the deck instead of beneath it, and having their verticle branch pointed upwards from those which lie horizontally. Such also are the standards of the BITS and CHANNELS, which articles see.

THE STANDING PART OF A ROPE OR TACKLE, (*dormant d'un eordage, ou d'un garant de palan,* Fr.) in the formation of knots, is the principal part of a rope which is made fast to the mast, deck or block; in contradistinction to that pulled upon, which is called the fall, or running part.

The STANDING *part of a Sheet,* is that part which is made fast to a ring at the ship's quarter.

STANDING-*Rigging.* See the article RIGGING.

STANDING-WATER, (*eaux dormantes,* Fr.) the water in which there is no current or tide.

STAPLE-ROPES, in rope-making, a term used for ropes made of hemp that is not inferior to clean Petersburgh.

STAR, (*stella, Lat. étoile,* Fr.) in astronomy, a general name for all the heavenly bodies which are dispersed throughout the whole heavens. The stars are distinguished, from the phenomena, &c. into *fixed* and *erratic* or *wandering.*

Erratic or Wandering STARS, are those which are continually changing their places, and distances, with regard to each other. These are what are properly called planets. Though to the same class may likewise be referred comets or blazing stars.

Fixed STARS, called also barely STARS by way of eminence, are those which have usually been observed to keep the same distance with regard to each other. Some stars appear to be of a sensible magnitude to the naked eye, but when viewed through a telescope, seem only as lucid points, without any apparent diameter; hence their immense distance from the solar system is inferred, and consequently they emit their own light, otherwise they would be invisible. It is, therefore, reasonable to suppose them to be so many suns, diffusing light and heat to planets revolving round them.

The number of fixed stars appears very considerable to a person who looks inattentively at them; but, from a due contemplation, it will be found that, without the assistance of a telescope there can very seldom be more than a thousand seen above the horizon at the same time: however, when the stars are viewed with a good telescope, their number appears almost infinite. Between the equinoctial and the parallel of 45° north declination, M. de la Lande observed the places of 8000 stars, with an achromatic telescope of 8 feet. In a space 8° in length and 3° in breadth, Dr. Herschel reckoned 44,000 small stars; and in the year 1792, August 22, he found that in 41 minutes of time, no less than 258,000 stars had passed through the field of view of his telescope. See Phil. Transactions, 1795, p. 70. The stars, with respect to their apparent splendour, are divided into orders called *Magnitudes.* The brightest are called stars of the *First Magnitude:* the next to these in splendour, stars of the *Second Magnitude,* and so on to those which are just perceptible to the naked eye, and which are called stars of the *Sixth Magnitude.* Those which cannot be discerned without the assistance of a telescope, are called *Telescopic Stars,* and are divided into orders of the seventh, eighth, ninth, &c. magnitudes accordingly. For farther particulars, see the article SUN.

The stars, for the purpose of finding any one more readily, are divided into parcels called *Constellations.* These, in order to assist the imagination, are supposed to be circumscribed with some known figure, as that of a *man, woman, ship,* &c. and those stars which lie between constellations are called *Unformed Stars.* As it would be an endless task to give a proper name to each star, it has, therefore, been customary to mark the stars of each constellation with the letters of the *Greek* alphabet, in such a manner, that the first letter is prefixed to the brightest star, the second letter to the next in brightness, and so on. Many of the brightest of the fixed stars have also proper names: thus α, Bootæ, is also called Arcturus, &c.

The number of stars of the first magnitude, in the zodiac, and in both hemispheres, do not amount to twenty. Since the equator changes its position with respect to the ecliptic, its axis will also be changeable; and its poles, during the *annus magnus*, will complete a period in a circular, or rather spiral orbit, therefore, on this account, the longitude, right ascension, and declination of every star will be variable; and consequently, the pole of the equator cannot always be directed to the same star, or which is the same, no star can remain at or near the pole of the equinoctial, for any considerable space of time. By this motion of the fixed stars it follows, that the zodiacal constellations have not retained their places with respect to the vernal equinox, but seemed to have moved *in consequentia*, or towards the east, almost a whole sign, since the time of Hipparchus; so that the constellation Aries is now where Taurus then was, Taurus where Gemini, and so on. See the article *Precession of the* EQUINOXES. Besides the motion arising from the recession of the equinoctial points, the stars have two apparent periodical motions, called the *Aberration* and *Nutation;* this latter is also sometimes called the *Deviation* of the fixed stars. The first of these motions arises from the velocity of light combined with that of the earth in its orbit, and hence, each of the stars apparently describes an ellipse about its mean place, the longest diameter being 40″, and period a year. See the article ABERRATION. The second of these motions, or the Nutation, arises from the attraction of the moon upon the equatorial parts of the earth; by which the pole of the equator describes an ellipse about its mean place as a centre, the ratio of the axes of the ellipse, being as 19″.1 to 14″.2: a period is completed in 18 years. See the article NUTATION. Many of the principal fixed stars are observed to change their position with respect to the adjacent stars, and this change of place is called the proper motion of the stars. As a necessary result of the proper motion of the stars, Dr. Herschel infers the motion of the whole solar system in absolute space; which in a great measure, seems to be confirmed from his comparison of the apparent relative motions of many of the stars.

Cary's map of the stars, or plans of the constellations, will be of great service to the navigator, who wishes more information in this particular department; as will also Doctor Wollaston's, Fusciculus, &c.

The Seven STARS, a common name given to the cluster of stars in the neck of the sign Taurus, properly called the *Pleiades*. They are so called from their number seven which appear to the naked eye, though some eyes can discern only six of them; but by the help of a telescope there appears to be a greater number.

STARBOARD, (*tribord, ou côté droit du vaisseau, en allant de l'arrière a l'avant,* Fr.) the right side of the ship when the eye of a spectator is directed forward or towards the head.

STARBOARD *the Helm!* See the article HELM.

To START, (*vuider, en parlant de ce qui est liquide, et déplanter en parlant de l'ancre,* Fr.) when applied to liquids, is to empty; but to any weight, as the anchor, &c. it signifies to move.

STARTING, a vulgar term denoting a summary mode of punishment formerly used on board ships, which was inflicted on the seamen by the boatswain's mate with a rope's end, by order of the commanding officer, for laziness at their duty, and frequently resorted to for want of alacrity in hoisting the top-sails to the mast-head; and to quicken their efforts in getting boats in and out, also in hoisting in beer and water, and in performing such like duties: it is now very seldom practised, and was never sanctioned in the ervice by law, only by custom.

NAVAL STATION, (*poste, station navale,* Fr) implies a safe and commodious shelter or harbour for the warlike and commercial fleets of a nation, where there is a dock-yard, and every thing requisite for the repair of ships, as well as a convenient place of arms for the assembly of its military expeditions.

The naval stations belonging to the British empire are as follow:

Downs.
North Sea and Baltic.
English Channel and Coast of France.
Jersey and Guernsey.
Irish Station.
Spain, Portugal, and Gibraltar.
Mediterranean.
Coast of Africa.
Halifax and Newfoundland.
West Indies.
South America.
Cape of Good Hope and Southward.
East Indies.

STATION *Bill,* (*rôle de postes,* Fr.) a list containing the appointed posts of the ship's company, when navigating the ship.

To STAVE, (*defoncer, &c.* Fr.) is to break a hole in any vessel; as, "Our launch is stove."

STAY, (*étai,* Fr.) a large strong rope employed to support the mast on the fore part, by extending from its upper end towards the fore part of the ship, as the shrouds are extended to the right and left, and behind it. See the articles MAST, RIGGING, and SHROUD.

Fore STAY, (*étai de misaine,* Fr.) is that which reaches from the fore-mast head towards the bowsprit end.

Main STAY, (*grand étai,* Fr.) is that which extends over the forecastle to the ship's stem.

Mizen STAY, (*étai d'artimon,* Fr.) is that which is stretched to a collar on the main-mast, immediately above the quarter-deck.

Fore-Topmast STAY, (*étai du petit mât de hune,* Fr.) is that which comes to the end of the bowsprit, a little beyond the forestay.

Main Topmast STAY, (*étai du grand mât de hune,* Fr.) is attached to the head or hounds of the foremast.

Mizen Topmast STAY, (*étai du mât de perroquet de fougue,* Fr.) is that which comes to the hounds of the main-mast.

Fore-Top-gallant STAY, (*étai du petit mât de*

perroquet, Fr.) is that which comes to the outer end of the jib-boom.

Main-Top-gallant STAY, (*étai du grand mat de perroquet*, Fr.) is that which is extended to the head of the fore-top-mast.

Mizen-Top-gallant STAY, (*étai du mat de perruche*, Fr.) is that which is attached to the head of the main-mast.

Royal STAYS, (*étais de perroquets volants*, Fr.) are those which extend to the jib-boom end, or to the heads of the top or top-gallant-masts next before them. The whole of these stays are nearly in the directions of the upper edges of the several stay-sails, which derive their names from them. See the article SAIL.

Spring STAYS, (*faux étais*, Fr.) a kind of assistant stays extending in a direction nearly parallel to the principal stays; they are much thinner than the others, and only used to the lower-masts and top-masts.

STAY-HOLES, in sail-making, the holes made through stay sails at certain distances along the top, through which they are seized to the hanks of the stay.

STAY-SAILS, (*voiles d'étai*; *aussi les focs, dont chacun est distingué par la dénomination particulière, ou de sa draille, ou de son étai respectif*, Fr.) a sort of triangular sails extended upon a stay. See the article SAIL.

STAY-SAIL STAYS, (*étais des voiles d'étai*, Fr.) are ropes used solely to extend and support the stay-sails, as the middle stay-sails.

STAY-TACKLE, (*palan d'étai*, Fr.) a large tackle, attached by means of a pendant to the main stay. It is used to hoist heavy bodies, such as the boats, or butts of water, beer, &c. in or out of the ship and out of the holds; for which purpose there are generally two; the one over the fore-hatchway, the other perpendicular to the main-hatchway; and they are accordingly distinguished by the epithets main and fore-stay-tackles, though both are upon the main-stay.

STAY-SAIL-*Tackle*. See the article TACKLES.

To STAY *a Ship*, (*donner vent devant, ou faire prendre vent devant à un vaisseau*, Fr.) is to arrange the sails and move the rudder, so as to bring the ship's head to the direction of the wind, in order to get her on the other tack.

To Heave in STAYS. See the article HEAVE.

To Miss STAYS, (*manquer à virer*, Fr.) A ship is said to miss stays, when her head will not come up in the direction of the wind, in order to get her on the other tack: a failure in this operation is often attended with dangerous consequences, from the situation of the ship.

STEADY! (*gouverne là!* Fr.) in sailing large, the command given by the quarter-master or man at the cun, to keep the ship in her course, without deviating to the right or left, to which the helmsman answers, 'steady'; to shew his attention to the order.

STEAM, (*vapeur*, Fr.) the smoke or vapour arising from water, or any other liquid or moist body, when considerably heated.

The steam raised from hot water is an elastic fluid, which, like elastic air, has its elasticity proportional to its density when the heat is the same, or proportional to the heat when the density is the same.

STEAM-BOAT, a name given to boats or vessels navigated without a sail, against the wind and tide, by the force of steam only. The Americans claim the honour of this invention, and assert that they have navigated a vessel whose keel was 160 feet in length, to go by the force of steam six knots an hour. The idea, however, has often been practically tried in England; and it is believed, that the principal merit of the discovery, is owing to a native of Scotland, born at the Calton-Hill, Edinburgh; where a number of ingenious mechanics reside, the son of one of whom, after living at Glasgow as an engineer, went some years ago to America, and having a mechanical turn, completed, with the assistance of an American gentleman this important invention. *Steam Boats*, are already established on the Hudson, where the tide runs at the rate of six miles an hour; and in the Delaware, where it runs four miles; and it is said to be extended to the Ohio and the Mississipi. In the New York steam boats there are four cabins in which they can dine and lodge 100 people, and passengers travel with the same ease, and receive as much, and as good accommodation, as can be obtained on the best roads and in the best regulated inns, in Europe. They can go two or three hundred miles in all sorts of weather, within an hour of the regulated time.

As this discovery has already been brought to such perfection in America, it is to be wished that a plan and description could be obtained; for they might be of use in our navigable rivers and canals, and on the Scotch and Irish lakes: in particular, they might obviate some objections which have been made to the navigation of Lochness, as a part of the Caledonial canal. They might be the means of taking ships out of harbours, and would render navigation more certain.

Even in a military point of view, advantage might be derived from the invention; for it might enable our troops to attack places, which otherwise, owing to the wind and tide, they could not approach. On the Ganges, and other large rivers in the East, such boats might be of singular utility. See *Major James's Military Dictionary*, vol. 2.

STEAM-ENGINE, an engine for raising water by the force of steam produced from boiling-water; and often called the *Fire Engines*, on account of the fire employed in boiling the water to produce the steam. This is one of the most curious and useful machines, which modern art can boast, for raising water from ponds, wells, or pits, for draining mines, &c. Were it not for the use of this important invention, it is probable we should not now have the benefit of coal fires in England; as our forefathers had, before the present century, excavated all the mines of coals as deep as it could be worked without the benefit of this engine to draw the water from great depths.

The principle of this machine is as follows:—There is a forcing pump with its rod fixed to one end of a lever, which is worked by the weight or pressure of the atmosphere upon a piston at the other end, a temporary vacuum being made below it, by suddenly condensing the steam, that had been let into the cylinder in which this piston works, by a jet of cold water thrown into it. A partial vacuum being thus made, the weight of the atmosphere presses down the piston, and raises the other end of the straight lever with water, &c. from the mine. Then immediately a hole is uncovered in the bottom of the cylinder, by which a fresh supply of hot steam rushes in from the boiler, which acts as a counterbalance for the atmosphere above the piston, and the weight of the pump-rods at the other end of the lever carries that end down, and of course raises the piston of the steam cylinder. The orifice for the emission of the steam is immediately shut, and the cock opened for injecting the cold water into the cylinder: this condenses it to water and another vacuum is made below the piston, which is now again forced down by the weight of the atmosphere, and thus the work is continued so long as water and fuel are supplied. It was originally invented by the Marquis of Worcester in the reign of Charles II. This nobleman, who appears to have been possessed of much knowledge, with a fertile imagination, published in 1663, a small book, called "*A Century of Inventions*," giving an account of one hundred discoveries or contrivances of his own; but the descriptions of many of them are so obscure, that they are altogether unintelligible.

Among them is an account of his invention of raising water by the force of steam, which, now that we are possessed of the engine, appears to agree very well with its construction. But as there is no plate to accompany his description, we are entirely unacquainted with the particular mode in which he applied the power of steam. It does not appear, however, that he met with sufficient encouragement; and this useful discovery was long neglected.

Towards the end of the century, Captain Savary, a person of great ingenuity, having probably seen the account of the Marquis of Worcester's invention, was convinced of its practicability, and succeeded in constructing a machine of this kind. He obtained a patent for the invention, and erected several steam-engines, which he published in a book entitled, "The Miner's Friend." published in 1696.

Mr. Savary succeeded in raising water to small heights, and erected several engines in different parts of England; but he could make nothing of deep mines. Many attempts have been made to correct these defects, but hitherto without much success.

In the beginning of the eighteenth century, Newcomen, an ironmonger or smith, and Cauly, a glazier, at Dartmouth, in Devonshire, first conceived the project of applying a piston with a lever, and other machinery. They were contented to share the profits of the invention with Savary, who procured a patent for it in 1705, in which they were all three joined.

Newcomen's engine was first offered to the public in 1705. But many difficulties occurred in the execution of it, which were removed one by one: and it was not till 1712, that the engine seemed to give confidence in its efficacy; at last Mr. Beighton, a very ingenious and well-informed artist, simplified the whole of these subordinate movements, and otherwise very much improved the machine.

The greatest improvement that has since been made on Newcomen's engine, has been in the manner of placing the boiler; instead of placing it underneath the cylinder, it is built at some distance from it, and sometimes in a separate building.

About 1762, Mr. Watt began to turn his attention to this machine, which he has since brought to so great a degree of perfection.

His improvements are founded upon these, and some other collateral observations; he preserves an uniform heat in the cylinder of his engines, by suffering no cold water to touch it, and by protecting it from the air or other cold bodies, by a surrounding case filled with the steam, or with hot air or water and by coating it over with substances that transmit heat slowly. He makes his vacuum to approach nearly to that of the barometer, by condensing the steam in a separate vessel, called the condenser; which may be cooled at pleasure without cooling the cylinder, either by injection of cold water, or by surrounding the cylinder with it; and generally by both. He extracts the injection-water and detached air from the cylinder or condenser, by pumps, which are wrought by the engine itself; or he blows it out by the steam.

As the inside of the cylinder was in the old engine exposed to the air at every stroke when the piston descended, and was considerably cooled thereby, he incloses the top of the cylinder by a metal plate, having a hole in it, through which the piston rod works in a collar of leather, and instead of employing the pressure of the atmosphere to force down the piston, he introduced the steam above the piston, when the vacuum is formed underneath, and employs it to produce this effect: thus making the direct pressure of the steam the moving power, as in the original construction of the engine.

The last great improvement made by Mr. Watt, was his giving an impulse to the piston by the steam, both in descending and ascending, instead of being impelled, as in the old engine during descent of the piston only.

It would far exceed the limits of this work, to enter into an examination of all the steam-engines invented by different persons. It is sufficient to mention, that no engine of this kind has been found, upon careful trial, to be superior to Mr. Watt's. It may not be improper here to state the actual performance of some of these engines, as they have been ascertained by experiment.

A steam engine of the best construction, with a thirty inch cylinder, acts with the force of 40 horses; and since it acts without intermission, will perform the work of 120 horses, or 600 men; each square inch of the piston being nearly equivalent to a labourer. The consumption of about 84 lbs. of good pit coal, will raise 48,000 cubical feet of water 10 feet high, which is equivalent to the daily labour of 8 men: the value of this quantity of coal is seldom so much as that of the work of a single labourer for a day; but the expense of the machinery generally renders a steam engine somewhat more than half as expensive as the number of horses for which it is substituted. See WEBSTER'S *Edition* of IMISON'S *Elements*, and NICHOLSON'S *British Encyclopedia*.

STEELER, in ship-building, the foremost or aftmost plank in a strake, which is dropped short of the stem or stern-post.

A SHORE STEEP-TO, (*côte de fer*; *côte acore*, on *côte à pic*, Fr.) is said of a shore when it descends almost perpendicularly into the water.

To STEER, (*gouverner, &c. Fr.*) is to direct the ship's way by the movements of the helm; or of applying its efforts to regulate her course when she advances.

The perfection of steering consists in a vigilant attention to the motion of the ship's head, so as to check every deviation from the line of her course in the first instant of its motion; and in applying as little of the power of the helm as possible. By this she will run more uniformly in a straight path, as declining less to the right and left; whereas, if a greater effort of the helm is employed, it will produce a greater declination from the course, and not only increase the difficulty of steering, but also make a crooked and irregular track through the water. See the article HELM.

The phrases used in steering a ship vary according to the relation of the wind to her course. Thus, if the wind is fair, or large, the phrases used by the pilot, or officer who superintends the steerage, are port, starboard, and steady. The first is intended to direct the ship's course farther to the right; the second is to guide her farther to the left; and the last, is designed to keep her exactly in the line, whereon she advances, according to her prescribed course. The excess of the first and second movement is called hard-a-port, and hard-a-starboard; the former of which gives her the greatest possible inclination to the right, and the latter an equal tendency to the left. See the articles PORT, STARBOARD, STEADY, HARD-*a-Port*.

If, on the contrary, the wind is foul or scant, the phrases then used are luff, thus, and no near; the first of which is the order to keep her close to the wind; the second, to retain her in her present situation; and the third, to keep her sails full. See the articles LUFF, THUS, NO NEAR, FULL and BY, and CONNING.

In a ship of war, the duties of conning and steering are divided among the quarter-masters, their mates, and the most expert seamen, who attend the helm in turns. The steerage is constantly superintended by the quarter-masters. In merchant-ships every seaman takes his turn in steering, being directed therein by the mate of the watch, or some other officer.

As the safety of a ship, and all contained therein, depend, in a great measure, on the steerage or effects of the helm, the apparatus by which it is managed should often be diligently examined by the proper officers.

Indeed, when the fatal effects which may result from negligence in this important duty are duly considered, such inattention must be pronounced unpardonable.

STEERAGE, (*c'est dans les vaisseaux de guerre Anglois, une chambre en avant de la cloison de la grande chambre, et qui lui sert d'antichambre. Dans les vaisseaux marchands, c'est le logement des matelots, placé ordinairement en avant de la chambre, Fr.*) an apartment without the great cabin of a ship, from which it is separated by a partition or bulk-head. In large ships of war it serves only as a hall or antichamber to the great cabin; but in merchant-ships, it is generally the habitation of the inferior officers and ship's crew.

STEERAGE, (*c'est aussi l'effet du gouvernail, Fr.*) is also used to express the effort of the helm; and hence

STEERAGE-WAY, (*sillage, ou chemin du vaisseau, Fr.*) is that degree of progressive motion communicated to a ship, by which she becomes susceptible of the effects of the helm to govern her course.

STEERING-WHEEL, (*roue de gouvernail, Fr.*) in a ship of war, a wheel to which the tiller-rope is conveyed, for the convenience of steering the ship.

STEERSMAN, (*timonnier, Fr.*) the helmsman, or the person employed at the helm to regulate the ship's course. He is reckoned the best steers-man who uses the least motion in putting the helm over to and again, and who keeps the ship best from making yaws, that is, from running in and out. See the article YAW. For this purpose, he should diligently watch the movements of the head by the land, clouds, moon, or stars; because, although the course is in general regulated by the compass, the vibrations of the needle are not so quickly perceived, as the salies of the ship's head to the right or left, which, if not immediately restrained, will require additional velocity in every instant of their motion, and demand a more powerful impulse of the helm to reduce them; the application of which will operate to turn her head, as far on the contrary side of her course.

To STEEVE, (*donner l'élévation du beaupré, du bâton de foc, &c. Fr.*) in ship-building, is to give the bowsprit a certain angle of elevation with the horizon, which is generally from 26 to 30°; but was formerly from 31 to 36°. The bowsprits of French men of war steeve more than those of English.

STEM, (*étrave, Fr.*) a circular piece of timber, into which the two sides of a ship are united at the fore end; the lower end of it is scarfed to the keel, and the bowsprit rests upon its upper end.

The stem is formed of one or two pieces, according to the size of the vessel; and as it terminates the ship forward, the ends of the wales and plank of the sides and bottom are let into a groove or channel, in the middle of its surface, from the top to the bottom: which operation is called rabbetting.

The outside of the stem is usually marked with a scale of feet, according to its perpendicular height from the keel. Its use is to ascertain the draught of water at the fore part, when the ship is in preparation for a sea-voyage, &c.

The stem at its lower end is of equal breadth and thickness, with the keel, but it grows proportionally broader and thicker towards its upper extremity. See the articles NAVAL *Architecture* and SHIP-BUILDING.

False STEM, (*fausse étrave, Fr.*) is that fixed before the right one. When a ship's stem is too flat, so that she cannot keep the wind well, they put a false stem above, which makes her rid more way, and bear a better sail.

To STEM *the Tide*, (*refouler la marée, Fr.*) is to acquire a velocity in sailing against the tide equal to the force of the current.

From STEM *to Stern*, (*de tête en tête, Fr.*) a term implying from one end of the ship to the other.

STEMSON, (*marsouin d'avant, Fr.*) a compass piece of timber fixed within the apron to reinforce the scarf thereof, in the same manner as the apron supports the scarf of the stem. In large ships it is usually formed of two pieces, as represented by I. plate XVII. in the pieces of the knee of the head, &c.

STEP or Tongue, for the tar-kettle, in rope-making, is made of three inch oak plank, five feet long and thirteen inches broad, which tapers to nine inches at the bottom, and is put into the kettle through a mortise in the bridge. Within four inches of the lower-end of the step is a round hole five inches in diameter, for the yarn to pass through. The step is suspended and regulated by a tackle.

Step *of the Capstan*, (*carlingue de cabestan*, Fr.) a large block of oak let down between, and bolted to the beams of the lower-gun-deck, as represented by A, fig. 1. plate VII.; it has a hole in its upper side, to receive the spindle of the capstan, which works therein.

Steps *of the Main and Fore-Masts*, (*carlingue du grand mât et du mât de misaine*, Fr.) are also large pieces of oak crossing the kelson, as appears in figures 2 and 3, plate XVI.; that of the main-mast is fixed only, but that of the fore-mast is bolted securely, and they receive the heels of these masts.

Step *of the Mizen-Mast*, (*carlingue du mât d'artimon*, Fr.) is also a large piece of oak bolted to the beams of the lower-gun-deck, to receive the heel of that mast.

Steps *of the Sides*, (*échelons, ou taquets d'échelles*, Fr.) are moulded pieces of beech or oak timber, nailed to the sides of the ship, as represented by c, fig. 2, plate XVII., for the convenience of getting on board.

Step *the Masts!* (*mâte le canot!* Fr.) the order to fix the masts in their proper stations. For the method of fixing them, see the article Mast.

To Step *a Boat's Mast* (*mâter un canot*, Fr.) is to erect and secure it in readiness for setting sail.

STERN, (*la poupe d'un vaisseau, ou l'arrière*, Fr.) the posterior part of a ship; or that which is presented to the view of a spectator, *placed* on the continuation of the keel behind, as exhibited in figures 1 and 2, plate XXIV.

These sterns are terminated above by the taffrails, and below by the lower counters: they are limited on the sides by the quarter-pieces, and the intermediate space comprehends the galleries and windows of the different cabins.

Explanation of figure 1, which exhibits the stern of a seventy-four gun-ship.

A. the stern-post.
B. the lower counter-rail, which terminates the height of the counter.
C the upper counter-rail.
D D the quarter galleries.
E E the quarter-pieces.
F the taffrail.
G G the gun-ports.
H the foot-rail.
I the breast-rail.
K K the-tuck-rail.
L the lower finishing.
M M the rails to represent a cove.
N N the side counter-timbers.

Figure 2, represents a modern stern of a first rate; accordingly it is furnished with a middle apartment between the ward-room and the captain's cabin. This apartment is also furnished with galleries on the stern and quarter. The other parts of it are described in the explanation of figure 1. See also the article Quarter.

By the Stern, (*sur l'arrière, ou sur cul*, Fr.) a term denoting the condition of a vessel which is more deeply laden abaft than forward.

Stern-Board. See the article Board.

To make a Stern-Board, see the article Board.

Stern Chases. See the article Chases.

Stern-Fast, (*croupière, amarre qui tient un vaisseau par l'arrière*, Fr.) a rope used to confine the stern of a ship, lighter, or boat, to any wharf, jetty-head, &c.

Stern-Frame, (*arcasse, terme de construction*, Fr.) in ship-building, is that frame of timber which is composed of the stern-post-transoms and fashion-pieces.

Stern-Ladder. See the article Ladder.

Sternmost, (*la queue d'une escadre*, Fr.) implies any ship or ships which are in the rear, or farthest a-stern, as opposed to head-most.

Stern-Post, (*étambord*, Fr.) a long straight piece of timber erected on the extremity of the keel, to sustain the rudder, and terminate the ship behind.

This piece, which is expressed by B in the pieces of the hull, fig. 1, plate XVII., ought to be well secured and supported; because the ends of all the lower planks of the ship's bottom are fixed in a channel cut on its surface, and the whole weight of the rudder is sustained by it.

The dimensions of the stern-post, or the proportional breadth and thickness, in the different parts of its height, are geometrically delineated in the quarter of a 74 gun-ship, fig. 9, plate XI., being expressed by B. It is usually marked like the stem, with a scale of feet from the keel upwards, in order to ascertain the draught of water abaft.

The difficulty of procuring a stern-post of sufficient breadth in one piece, has introduced the practice of fixing an additional piece behind it, which is strongly bolted to the former: the hinges, which support the rudder, are accordingly fixed to this latter, which is also tenoned into the keel, and denominated the back of the post. It is half the breadth of the stern post at the heel, but diminishes gradually towards the upper end, where it is one-third narrower.

The stern-post is strongly attached to the keel by a knee, G, in the pieces of the hull, fig. 1, plate XVII., of which one branch extends along the keel, being scarfed and bolted to the dead-wood, and clinched under the main-keel; whilst the other branch inclines upwards, and corresponds with the inside, or fore part of the stern post, to which it is also bolted in the same manner.

Stern-Sheets *of a Boat*, (*chambre d'un canot*, Fr.) that part contained between the stern and the aftmost seat of the rowers. It is generally furnished with seats to accommodate the passengers. See the article Boat.

Stern-Way, (*chemin que fait un bâtiment par l'arrière lorsqu'il cule*, Fr.) the movement by which a ship retreats, or goes backwards with her stern foremost.

Pink STERNED. See the article Pink.

Square Sterned. See the article Square.

STEWARD, a man appointed to manage the affairs of another. In a ship of war, there are three descriptions of stewards, viz. the captain's, purser's, and ward-room stewards.

Captain's Steward, (*maître d'hôtel du capitaine*, Fr.) a person nominated by the captain, having an optional rating on the ship's books, to take charge of his sea-stock, &c.

Purser's Steward, (*maître valet*, Fr.) a person appointed by the purser to distribute daily the different species of provisions, to the officers and crew, for which purpose he is assisted by a mate, who is rated steward's mate, and commonly called Jack of the bread-room, and sometimes by the cooper of the ship.

He is generally rated purser's steward on the ship's books, and has, besides his wages, an annual allowance from the purser, as a remuneration for his servitude under him.

Ward-Room or *Gun-Room* Steward, (*maître d'hôtel, de l'état major*, Fr.) a person nominated by the first lieutenant and the mess, to take charge of the sea-stock, &c. belonging to the gentlemen in the ward-room, and is generally under the inspection of the caterer of the mess.

Steward-Room, (*cambuse*, Fr.) an apartment adjoining the bread-room, usually on the orlop deck, where the steward weighs, measures, and serves out the provisions to the ship's company.

STICK out the Cable! the order to slacken and push it out at the hawse-hole, when the anchor is hauling up to the cat-head, &c.

STIFF (*vaisseau dur à abattre*, Fr.) when applied to a ship, the quality by which she is enabled to carry a sufficient quantity of sail, without oversetting.

STINK-POT, (*pot à feu*, Fr.) an earthen jar, or shell, charged with powder, grenades, and other materials of an offensive and suffocating smell. It is sometimes used by privateers, to annoy an enemy whom they design to board; for which purpose it is furnished with a lighted fuse, at the opening, or touch-hole. See the article Boarding.

STIRRUPS, (*étriers de marche-pied*, Fr.) short ropes having their upper ends plaited and nailed round the yards, and eyes made in their lower ends, through which the horses are reeved, to keep them parallel to the yards. See the article Horse.

STIVING, among mast-makers, a term used for elevating any thing, so as to make an angle with the horizon. See the article To Steeve.

STOCK, (*provision*, Fr.) generally implies provisions procured by individuals, for the particular accommodation of themselves or messmates; hence we say, *fresh stock, sea stock, live stock*, &c.

Stock, in block making, a wooden instrument, having a bit fixed in one end, and a pin with a round head in the other, the pin and the bit serving as an axis to turn it on; and is used to bore holes with.

Stock *of an Anchor*, (*jât d'ancre*, Fr.) two long pieces of oak, tapering from the middle, and fastened together with iron hoops and tree-nails, and fixed on the shank of an anchor traversely to the arms. See the article Anchor.

To Stock *an Anchor*, (*enjâler une ancre*, Fr.) is to fix the end of it firmly into the stock, as mentioned in the last article.

To Stock-*to the Anchor*, (*serrer le jât de l'ancre contre le bord au moyen d'un palan*, Fr.) is to stow it by means of a tackle upon the upper end of the stock, which is intended to bowse it into a perpendicular direction.

Stock-Shave, in block-making, a large sharp-edged cutting knife, with a handle at one end, and a hook at the other, by which it hooks into an iron staple that is driven into an elm block; it is used to pare off the rough wood from the shells of blocks, &c.

Stock-Tackle, (*palan servant à brider le jât de l'ancre contre le bord*, Fr.) a tackle composed of a double and single block, strapped, with a hook and thimble, and made fast round the upper part of the stock of the anchor, with a tail-block and strap round the main-stay, and is used to keep the stock clear of the ship's side when fishing the anchor.

Stocks, (*chantier pour la construction des vaisseaux, ou chantier de construction*, Fr.) a frame erected on the shore of a river, or harbour, whereon to build shipping. It generally consists of a number of wooden blocks, ranged parallel to each other, at convenient distances and with a gradual declivity towards the water. See the article *To* Launch and Slip.

STOOLS, in ship-building, are small channels fixed to the ship's sides, to contain the dead-eyes for the back stays. See the article Back-Stay Stool.

Stools, are also pieces of plank fastened to the ship's side to receive the birthing of the gallery.

To STOP a Leak. See the article Leak.

STOPPERS, (*bosses*, Fr.) are short pieces of rope, which are usually knotted at one, or both ends, according to the purpose for which they are intended. They are either used to suspend any weighty body, or to retain a cable, shroud, &c. in a fixed position.

STOPPER-Bolts, (*chevilles à bouche pour les bosses du cable*, Fr.) in ship-building, large ring-bolts, which are driven into the deck, as expressed by Z, fig. 9, plate IX., before the main-hatch, &c. for the use of the stoppers.

Stopper *of the Anchor*, (*bosse de bout*, Fr.) a strong rope attached to the cat-head, which, passing through the anchor-ring, is afterwards fastened to the timber head, and thereby securing the anchor to the bow.

Stoppers *of the Cables*, (*bosses des cables*, Fr.) have a large knot, and a lanyard at one end, and are fastened to the stopper-bolts, as represented by *h*, in the deck, fig. 9, plate IX., by the other. They are attached to the cable by the lanyard, which is fastened securely round both, by several turns passed behind the knot, or about the neck of the stopper; by which means the cable is retained from running out of the ship, when she rides at anchor.

Stoppers *for the Rigging*, (*bosses à fouet*, Fr.) have a knot and a lanyard at each end. They are used only when the shroud-stays or back-stays are cut asunder in battle, or disabled by tempestuous weather; at which time they are lashed, in the same manner as those of the cables, to the separated parts of the shrouds, &c. which are thereby reunited, so as to be fit for immediate service. This, however, is only a temporary expedient, applied when there is not time or opportunity to refit them, by a more complete operation.

STOPPER *the Cable!* (*bosse le cable!* Fr.) the order to put stoppers on the cable, to prevent it from running out of the ship when riding at anchor.

Deck-STOPPERS, (*bosses à bouton*, Fr.) are those used to retain the cable when the ship is riding at anchor.

Dog-STOPPERS, (*maîtresse bosses*, Fr.) are used as additional securities when the ship is riding in heavy gales of wind, or in bringing up a ship with much stern-way, to prevent the cable from snapping at the bits, and to ease the deck-stoppers.

Fore-Tack,- and Sheet, STOPPERS, are used to secure the tacks and sheets, till they are belayed.

STOPS, in mast-making, are square projections, or shoulders, left on the outside of the cheeks, at the upper part of the hounds of lower-masts; also on top-masts, and top-gallant-masts, and at the outer end of jib-booms.

STORE-HOUSES, in dock-yards, are large buildings for the purpose of containing naval stores, such as cables, cordage, blocks, &c. for fitting out and repairing the ships and vessels of his Majesty's navy. They are of different forms and sizes, according to the purpose for which they are intended.

STORE-KEEPER, (*garde magasin*, Fr.) an officer in the royal dock-yards, invested with the charge of the principal naval stores, as the sails, anchors, cordage, &c.; it is likewise his duty, in rotation with the clerk of the check, and clerk of the survey, to attend the survey and receipt of all stores from contractors, and the delivery of lots of old stores to the purchasers; and to check the certificates made out by the clerk of the check, for stores received upon contract.

STORE-ROOM, (*soute*, Fr.) an apartment, or place of reserve, of which there are several in a ship, to contain the provisions, or stores, together with those of her officers, during a sea-voyage; as, the captain's, lieutenant's, boatswain's, gunner's, and carpenter's store-rooms.

STORE-SHIP. See the article SHIP.

NAVAL STORES, (*munitions pour la marine*, Fr.) comprehend all those particulars made use of, not only in the royal navy, but in every other kind of navigation; as anchors, cables, cordage, hemp, masts, yards and spars, oars, sail-cloth, pig-iron, bar-iron, resin, pitch, tar, turpentine, oakum, ochre, shot, shells, sheet-copper, gunpowder, with other ship-chandlery wares, ordnance, and fire-arms of every description.

It appears so early as in 1618, when commissioners were appointed to enquire into the state of the navy, &c. that they caused due proportions of stores and rigging for a ship of each class, for the King's approval, and first formed one uniform establishment of sea stores for a ship of each rate.

Two excellent regulations were adopted by the navy board, after the war in 1784.

The first for stores set apart for ships in ordinary. The second was that of an establishment of stores for the general magazines at each of the dock-yards, and also at the several naval stations, both at home and abroad. These were original ideas of Sir Charles Middleton, then comptroller of the navy (afterwards Lord Barham, and First Lord of the Admiralty), which were suggested, no doubt, from the difficulties the Board had experienced in procuring certain articles, and the high prices paid for others, during the war. In conformity to this plan, the establishments consist of specific quantities of all the principal, and many of the inferior, articles of naval stores, at the several dock-yards and naval stations, so far as the nature of the service requires.

They are kept up by means of the annual or occasional contracts, and many good effects have also been produced by it, one, for instance, in 1800, when the confederacy between the Northern powers, and the embargo in the Russian ports, would have been attended with very unpleasant circumstances, had the magazines not been well supplied.

Disposal or Arrangement of STORES. Simmons, in his *Sea-Gunner's Vade-Mecum*, observes, that there are many things kept below in an English ship of war, which, with arrangement and taste, might be disposed of to much more advantage, both to set off their appearance, and to be in readiness at the quarters the moment they are wanted. For example, the match-tubs, wads, salt-boxes, powder-horns, pikes, tomahawks, and tube-boxes, which are absolutely requisite to be at hand; but, particularly, in a cruising ship: and, at the same time, it furnishes the executive officer with an opportunity of displaying his abilities in disposing of them to advantage. The fire-buckets are generally well painted, and hung in a row round the fife-rail, and the lower part of the breast-work of the poop, having their lanyards flemished close down inside, which disposes of the whole that is allowed. Some ships that do not stow hammocks in the breast-work of the poop, have a breast-work made of pikes, which in general appears heavy, and without taste. If there should not be any breast-work, the best substitute is a half circle of tomahawks on each side, and in large ships two, which are light, and look well. The spikes, as they have no beauty in appearance, but are instruments necessary to be at hand, may be kept out of sight (the handles either scraped very clean, or painted white, and the pike black), under the half-deck. The tomahawks crossed, and nailed up over each gun, with a powder-horn between them. The wads made up in cheeses, and covered with canvas, having an emblematical painting on it. The salt-boxes painted in the same manner, and fixed over the gun. The tube-boxes ranged in a dry place, painted and numbered, for the guns; they are commonly put under the half-deck, against the bulk-head of the cabin, under the charge of the sentinel at the cabin-door. The locks are, in some ships, kept constantly on the guns, with covers cut out of the solid blocks of wood, or cast lead; these covers may keep wet from them, but cannot prevent the damp, which will, of course, render them unfit for use, and shews the impropriety of keeping them on. It appears better to place them in a dry situation, after having been fitted with the number of the gun to which they belong, affixed over them. One man at each gun should be taught to fix them on, who should be stationed to perform this particular service when the drummer beats to quarters.

Laws relative to Naval STORES. If any person, who has the charge or custody of any of the king's armour, ordnance, ammunition, shot, powder, or habiliments of war, or of any victuals for victualling the navy, shall, to hinder his Majesty's service, embezzle, purloin, or convey away, the same to the value of 20 shillings, or shall steal or embezzle any of his Majesty's sails, cordage, or any other of his naval stores, to the value of 20 shillings, he shall be adjudged guilty of felony, without the benefit of clergy. 22. Car II. c. 5.

None but the contractors with the commissioners of the navy, shall mark any stores of war, naval or military, with the mark commonly used to his Majesty s stores, upon pain of forfeiting 200*l.*; and persons in whose custody they are found, are liable to the same penalty. 9 and 10 Will. III. c. 41.

The treasurer, comptroller, surveyor, clerk of the acts, or any commissioner of the navy, may act as justices, in causing the offender to be apprehended, committed, and prosecuted for the same. 9 Geo. III. c. 30.

If any person shall wilfully and maliciously set on fire, burn, or destroy, any of his Majesty's military, naval, or victualling stores, or other ammunitions of war, or any place where such stores or ammunition shall be kept, he and his abettors shall be guilty of felony, without the benefit of clergy. 12 Geo. III. c. 24.

Every officer is strictly enjoined to be as economical as possible in the expenditure of stores, and not to admit of their being converted to any other than their proper use; except in such manner, and into such articles as are allowed on the establishment of the navy; also to avoid all unnecessary public expenses; and as far as may depend on him, to prevent it in others.

When any warrant officer receives stores on board, he is to be very particular in ascertaining that they are good in quality, and that he receives the full quantity specified in the note sent with them. If he discovers any defect or deficiency in them, he is immediately to report the same to the captain. When supplied with stores by any other officer, he is to charge himself therewith, and to mention his having done so in the receipt he gives for them. See the NAVAL INSTRUCTIONS.

STORM, (*tempête, ou fort coup de vent*, Fr.) in a general sense, denotes a violent agitation of the wind, and is usually accompanied with thunder, lightning, and heavy rain whose effects are sometimes very disastrous to sh ps at sea. They proceed from the contact of clouds, which are charged with more or less electricity; the electric fluid successively darting from one to another, to obtain an equilibrium. In a storm or tempest the seamen frequently lie-to until it is abated.

A remarkable storm happened in the year 1703; it began about the middle of November, and was at its greatest height on the 27th of that month. In this storm the Eddystone light-house was blown down; the Vanguard, a second rate, lost in the harbour at Chatham; the Resolution, a third rate, on the coast of Sussex; the Northumberland, Stirling Castle, Restoration, third rates, and a gun brig, on the Goodwin Sands; the Reserve at Yarmouth, the York at Harwich, and Newcastle at Spithead; and several other men of war were driven on shore, dismasted, and very much damaged. In consideration of these disasters, Queen Anne immediately issued a proclamation, ordering that all the families of such officers and seamen as had perished by the storm in her Majesty's service, should be entitled to her bounty, in the same manner as if they had actually been killed in battle.

On the evening of the 19th of November, 1813, a most tremendous storm, or rather hurricane, from the south-eastern board, rushed up the harbour of Halifax, in America, with such destructive violence, as has not been witnessed since the tornado in September, 1798. In little more than an hour, it left but few vessels at their anchors, of which there was scarcely one that did not sustain material injury. Its utmost fury was felt about dead low-water: less damage was sustained by the wharfs and stores than might otherwise have been expected; but several shallops, &c. were sunk, and many others torn and totally wrecked by the sea, and more than one hundred ships and vessels suffered materially from its violence.

To Manage in a STORM. When a ship rides leeward-tide, and the wind increases, give her more cable, otherwise the anchor may start by not doing so in time, and the ship will not easily be brought up again; this is the more necessary when riding in the hawse of another ship. Previously to giving a long service, it is usual to take a weather-bit, or a turn of the cable round the windlass-end, that, in veering away, the ship may not overpower you; then grease the services, to prevent its chafing in the hawse.

If the gale increases, the top-masts should be timely struck, but the fore-yard seldom, if ever, should be lowered down, that, in case of parting, the fore-sail may be always ready. At these times, let there be more people on deck than the usual anchor-watch, that no accident may happen through inattention.

STORM *Mizen*, a triangular sail, similar to a fore-top-mast stay-sail, made of canvas, No. 2, or 3, by bends on the forepart, to those abaft, and parallel to the mizen-mast; the foot is extended towards the taffrail by a sheet.

STOVE, (*fourneau*, Fr.) an iron apparatus, for containing fire, from which heat is communicated to the apartment in which it is placed. Stoves on board ship are principally used in the admiral's or captain's cabins, gun room, &c. for the purpose of airing the apartments between decks.

Hanging STOVE, a stove about a foot square, open at the top, and invented by Messrs. Brodie & Co. of London, for airing the different apartments in a ship: it is attached to an iron stand, which receives the ashes and sparks of fire that run through the grate or bottom of the stove.

There are two small bars or handles, about 2 feet in length, which are fastened to the sides; their upper ends terminate in hooks, which are hooked to small eye-splices in the lower ends of ropes, whose upper ends are spliced similar to the lower, and hung to hooks driven into battens under the deck.

One person, under the direction of the master at arms, always attends each stove when hung in the above manner with fire in them, to prevent any accidents.

The sum of fifteen guineas was voted by the Society of Arts, &c. in the year 1807, to Mr. Joseph Collier, No. 11, Crown-street, Soho, London, for an improved ship's stove: its front is partly covered by a circular slide, and moves from the back by a brass handle, with a moveable plate attached to the side, and supported by the latch catching a pin, by which means it acts as a blower, and causes the fire to burn more briskly; but which slides down also, to shut the fire up; there is another plate that hangs on the latch, but that can be let down to shut up the ash-pit, which may be drawn out when the side facings are pulled up; there is likewise a circular plate or cap, which slides so as to shut the chimney up close. For an engraving of the invention, see the *Transactions of the* Society of Arts, *&c, vol.* 25.

STOVING, in rope-making, the act of placing white rope in an iron stove, to which heat is communicated, by means of a flue, which makes the rope more limber, and pliant to receive the tar.

To STOW the Hold. See the articles Hold and Trim.

STOWAGE, (*arrimage*, Fr.) the general disposition of the several materials contained in a ship's hold, with regard to their figure, magnitude, or solidity.

In the stowage of different articles, as ballast, casks, cases, bales, and boxes, there are several general rules to be observed, according to the circumstances or qualities of those materials. The casks which contain any liquid, are, according to the sea phrase, to be bung up and bilge free, *i. e.* closely wedged up, in an horizontal position, and resting on their quarters; so that the bilges, where they are thickest, being entirely free, cannot rub against each other, or the ship's side, by the motion of the vessel. Dry goods, or such as may be damaged by the water, are to be carefully inclosed in casks, bales, cases, or wrappers, and wedged off from the bottom and sides of the ship, as well as from the bows, masts, pump-well, &c. Due attention must likewise be had to their disposition, with regard to each other, and to the trim and centre of gravity of the ship; so that the heaviest may always be nearest the keel, and the lightest gradually above them. See the articles Ballast, Trim, and Rolling.

If a ship be newly commissioned, the master is to obtain the most correct information he can, of the manner in which the hold was stowed, when she was last in commission, and what were then her qualities, that the stowage may be altered if there be reason to suppose it may be done with advantage.

If the ship shall not have been at sea, the master is to consult the master-shipwright of the dock-yard on what may be the best manner of stowing her. But if he finds the hold already stowed, he is to inform himself how it has been done; and he is attentively to examine her qualities at sea, that he may suggest such alterations in her stowage, as he may think likely to improve them.

He is to be present at the stowing of the hold, to see that the ship has the proper quantity of iron and shingle ballast, and to admit the stowage of water and provisions, without risk of damage to the casks.

When any provisions are pointed out to him, by the purser, as being older than the rest, he is to see that they are stowed uppermost, to admit of their being first hoisted up; and, when the stowage is completed, he is to enter in the log-book a particular account of the manner in which it was performed. See the Naval Instructions.

Mr. Robert Dickenson, of 58, Great Queen-street, London, has obtained a patent for some ingenious inventions of iron tanks, casks, and buoys, presenting an improved system of watering and stowing ships, preserving provisions, merchandise, and liquors from waste, leaking, vermin, &c. both at sea and land.

The patentee observes in his Treatise on Stowage, that, "the injury sustained by cargoes and stores, in *wood*, and the waste of room from the spaces necessarily left unoccupied by cask-stowage, are evils of greater magnitude than can be conceived, except after minute and accurate calculations: the disadvantages chiefly arise from the incompetency of the material to withstand and resist the elements, insects, and vicissitudes of climates; but the necessary thickness of the staves, the solidity of the ends, and the depth to which these ends are necessarily sunk within the body of the casks, are circumstances which suggest very powerful objections to their use, where saving of room is an object of importance, independent of the damage of stores.

"The *wood* alone, in a provision cask of Mr. Mellish's (containing 41 gallons) occupies a space equal to 19 gallons. A water cask of one tun, or 252 gallons, is 35 inches diameter at the ends, 43 inches diameter at the middle, and 61 inches long, from out to out, $1\frac{1}{2}$ inch thick, and 2 inches space at each chime, making 76 gallons, for *wood* and *lost space*. An iron tank, containing the same quantity, occupies the space of only 7 gallons, giving an advantage of 69 gallons of room saved on every tun of water, or 26 per cent. exclusive of the space occupied by the fathom-wood, and the difference between squares and circles in stowage.

Wood Casks occupy

	Gallons
For the interstices between the circles -	63
Occupied by wood and loss at the chimes -	76
Ditto by fathom-wood stowed between the casks	35
Total	174
An *iron* vessel occupies only - - -	7
Space lost	167

Thus, on a *one* ton cask made of *wood*, 167 gallons might be saved by being stowed in *iron*.

"The weight of the wood casks and the iron tanks is nearly the same; but it requires half the weight of the wood cask in fathom-wood, to stow them, which increases the weight of the *wood* cask 50 per cent. above the weight of *iron.*

"With a view to remedy the disadvantages resulting from the use of *wooden* casks, an arrangement in point of stowage has been contrived, which cannot fail to excite the attention of every person connected with the naval, military, mercantile, and shipping interests of Great Britain.

"This improvement consists in the adaptation to the holds of vessels of a system of iron cases, of a square, polygonal, or any other form, by which they may adapt themselves to each other, without leaving any interstices.

"These iron vessels are secured, when necessary, from being injured by rust, by means of a varnish, which effectually guards them against all corrosion; but for many purposes this is not necessary, and in every case they would last ten times as long as iron-bound casks, and in some cases for a much longer time."

Having thus stated, generally, the nature of this invention, the patentee proceeds, in his essay just published, to give a clear idea of the manifold advantages that might result from its adoption, both in the navy and merchant-service, by comparative statements, shewing the additional room and convenience obtained by it in our ships of war, and the additional profit likely to accrue to the owners of such merchant-ships as may adopt this improved method of stowage.

STRAGGLERS, (*déserteurs*, Fr.) are seamen who desert and abscond from his Majesty's ships and vessels, contrary to the articles of war. See the article DESERTION.

STRAGGLING-MONEY, a compensation allowed to persons on their apprehending and bringing on board any stragglers, or deserters, from his Majesty's ships or vessels. The sum allowed on this occasion, is £3, payable by the clerk of the check, at the port where the ship may be at, on producing a certificate from the commanding officer, that such service has been duly performed; and that the sum is so charged on the ship's books, against the growing wages of the stragglers or deserters thus apprehended.

STRAIT, (*détroit*, Fr.) in geography, is a narrow channel, or arm of the sea, contained between two opposite shores, affording a passage out of one great sea into another; as the straits of Gibraltar, of Sunda, of Dover, of Magellan, &c.

STRAND, (*toron d'un cordage*, Fr.) one of the twists, or divisions, of which a rope is composed. See the articles ROPE and CABLE.

STRAND, (*est aussi le rivage de la mer, ou la plage*, Fr.) also implies the shore of the sea, or bank of a great river.

STRANDED, (*navire échoué*, Fr.) is said of a ship that is driven on shore by a tempest, or run aground through ill steerage, and so perishes. Where any vessel is stranded, the justices of the peace are empowered to command the constables near the coast, to call assistance, in order to preserve the ship, if possible. See the article SHIPWRECK.

In the event of a vessel being stranded with a cargo packed in Mr. Dickenson's patent *iron stowages*, if circumstances admit of boats approaching the wreck, the contents, whether dry or moist goods, may be recovered to the owners, *perfectly secured from danger.*

STRANDED, (*cable, ou cordage dont un des torons a manqué*, Fr.) speaking of a cable or rope, signifies that one of its strands is broken.

STRANGE SAIL, (*voile inconnue*, Fr.) denotes an unknown or enemy's ship within view; as "We see a strange sail to leeward of our course."

Seamen, under the directions of the lieutenant, are kept at the mast-heads during the day, and in proper stations during the night, to look out; and if, in time of war, they should see a strange sail in the night, the lieutenant is to send a midshipman to inform the captain thereof, and is himself to get the ship ready for action: he is to keep out of gun-shot of the strange ship until every thing is ready; but, in doing this, he is to be careful not to remove to such a distance as to risk the losing sight of her. See the NAVAL INSTRUCTIONS.

STRAP, (*estrop de cordages, ou poulies*, Fr.) a piece of rope generally spliced into a circular wreath (see fig. 12, plate II.) and used to surround the body of a block; so that the latter may be hung to any particular station about the masts, yards, or rigging. See the articles BLOCK and EYE.

Iron STRAP, (*estrop de fer en général*, Fr.) is a hoop of iron, in lieu of rope, fixed round the shell of a block, and is only used to some particular blocks.

STRAP-*Bound Blocks.* See the article BLOCK.

STRAPS, are also used occasionally, to fasten upon any large rope, for the purpose of hooking a tackle to the eye, or double part of the strap, in order to extend, or pull with redoubled effort, upon the same rope; as in setting up the rigging, where one hook of the tackle is fixed in a strap applied to the particular shroud, and the other to its lanyard. See the article LANYARD.

STRATAGEM, (στρατήγημα, Gr. *stratagème*, Fr.) an artifice in war; a trick whereby an enemy is deceived.

STREAKS, (*virures-files, ou cours de bordages*, Fr.) the uniform ranges of planks on the bottom or sides of a ship; or the continuation of planks joined to the end of each other, and reaching from the stem, which limits the vessel forward, to the stern-post and fashion-pieces, which terminate her length abaft. The lowest of these, which is called the GARBOARD-STREAK, is let into the keel below, and into the STEM and STERN-POST. See those articles.

STREAM, (*courant*, Fr.) a name given to any running water. See the article CURRENT.

STREAM *Anchor.* See the article ANCHOR.

To STREAM *the Buoy.* See the article BUOY.

STREAM *Cable.* See the article CABLE.

STRESS OF WEATHER, (*tems forcé*, Fr.) denotes that of a tempestuous nature, and unfavourable to navigation, in which ships are driven at the mercy of the wind and sea; and, if no port or harbour be near, they are frequently obliged to *lie-to* during the violence of the storm.

STRETCH, (*bordée*, Fr.) is generally understood as a sea phrase, to imply the progression of a ship under a great surface of sail, when close-hauled. The difference between this phrase and standing, is apparently in the quantity of sail, which, in the latter, may be very moderate, but in stretching, generally signifies excess; as, "We were standing in shore, (under easy sail) when we discovered the enemy stretching to the southward, i. e. under a crowd of sail."

To STRETCH *along a brace, &c. (parer une manœuvre sur le pont,* Fr.) is to lay it along the decks in readiness for a number of men to lay hold of.

STRETCH *Out fore and aft! (avant par tout!* Fr.) in rowing, the order to the boat's crew to bend and fall back to the utmost, in order to take longer hold of the water with the oar, and thereby give the boat the greater velocity.

STRETCHERS, *(traversins placés dans un canot pour les pieds des rameurs,* Fr.) a sort of staves fixed athwart the bottom of a boat, for the rowers to place their feet against, in order to communicate a greater effort to their oars.

STRIKE! *(amène!* Fr.) an order to lower the colours, in token of surrender, to a victorious enemy. See the article FLAG.

To STRIKE, *(amener,* Fr.) also implies to lower or let down any thing; as the ensign, or topsail, in saluting; or, as the yards, topmasts, and top-gallant-masts, in tempestuous weather.

Within his Majesty's seas, his ships are not on any account to strike their topsails, nor take in their flags, nor in any way to salute any foreign ships whatever; neither are they, in any other seas, to strike their topsails, nor take in their flags, to any foreign ships, unless they shall have first struck, or shall at the same time strike their topsails, flags, &c. to his Majesty's ships.

If any merchantmen shall so far forget their duty, as to attempt to pass any of his Majesty's ships without striking their topsails, the name of the ship and of the master, the port to which they belong, the place from whence they came, and that to which they are bound, together with the affidavits of the fact, are to be sent to the secretary of the Admiralty, in order to their being proceeded against in the Admiralty Court. See the NAVAL INSTRUCTIONS.

To STRIKE, *(caler,—toucher de la quille,* Fr.) to run ashore, or to beat upon the ground in passing over a bank or shallow.

To STRIKE *the Bell.* See the article BELL.

To STRIKE *Soundings, (prendre la sonde,* Fr.) to ascertain the depth of water with the hand-lead, &c.

STRING, *(bauquière des gaillards, à l'endroit des passé-avants, c'est-à-dire, proprement le bordage qui est intérieurement placé contre le bord, pour former la jonction des deux bauquières,* Fr.) in ship building, the highest range of plank in a ship's ceiling; or that which lies between the gunnel and the upper edge of the upper deck-ports, as expressed by T, in the midship-section, figure 1, plate VI.

To STRIP THE MASTS, *(décapeler les haubans; dégréer, ou dégarnir les mâts,* Fr.) is to unrig a ship, or deprive the masts of their machinery and furniture; an exercise which is otherwise called dismantling.

STROKE, *(coup de rame,* Fr.) a single sweep of the oars in rowing. Hence they say, "Row a long stroke!" which is intended to push the vessel forward more steadily.

STROKESMAN, *(vogue-avant, ou le rameur le plus en arrière qui donne le coup de rame pour regler la vogue,* Fr.) the person who rows the aftmost oar in a boat, and gives the stroke, which the rest are to follow; so that all the oars may operate togethor.

STUDDING-SAILS, *(bonnettes,* Fr.) certain light sails extended, in moderate and steady breezes, beyond the skirts of the principal sails, where they appear as wings to the yard-arms.

Lower STUDDING-SAILS, *(bonnettes basses,* Fr.) are spread beyond the leeches of the main-sail and fore-sail, and fixed nearly in the same manner; only that the booms, which extend the feet, are generally hooked to the chains by means of a goose-neck; or else swings off with the sail, to which they are suspended; being kept steady abaft by a rope called the GUY, which article see.

Topmast and Top-gallant STUDDING-SAILS, *(bonnëttes de hune,* Fr.) are those which are set on the outside of the top-sails and top-gallant-sails: they are spread at the foot by booms, which, sliding out from the extremities of the main and fore yards, their heads or upper edges are attached to small yards, which are hoisted up to the topsail and top-gallant-yard-arms. See the articles SADDLE and BOOM-IRON.

STUDDING-SAIL *Booms, (boute-hors de bonnettes,* Fr.) are long poles sliding through boom-irons at the extremities of the yards, and from the vessel's sides, used to spread the studding sails. See also the article BOOM.

STUDDING-SAIL *Yard.* See the article YARD.

STUFF, *(ce mot signifie en general tout couroi, suif, brai, ou resine dont on enduit le fond, ou carène d'un bâtiment, ses œuvres mortes, &c.* Fr.) any composition or melted mass, used to smear or pay the sides or bottom of a ship. That which is chiefly used for the lower masts is simply turpentine, resin, or varnish of pine: for the topmasts, tallow or butter; for the sides, turpentine, varnish of pine, tar and oil, or tar mixed with oil and red ochre; and for the bottom, a mixture of tallow, sulphur, and resin, or tar, whale oil and broken glass, or any part of these ingredients; and this application is called giving a new coat of stuff to the masts, sides, &c.

STYLE, in chronology, a particular manner of counting time; as the *Old Style,* the *New Style.* See the article CALENDAR in the APPENDIX.

Old STYLE, is the Julian manner of computing, as instituted by Julius Cæsar, in which the mean year consists of 365¼ days.

New STYLE, is the Gregorian manner of computation, instituted by Pope Gregory the 13th, in the year 1582, and is used by most catholic countries, and many other states of Europe. For farther particulars, see the article LEAP-YEAR.

SUB-LIEUTENANT, *(sous lieutenant, ou enseigne,* Fr.) a rank introduced between the full lieutenant and the midshipman, whose pay is inferior to that of a lieutenant; but at the time of adoption by Earl St. Vincent, in 1802, who was then First Lord of the Admiralty, it was intended as a temporary relief to the very many deserving young gentlemen, who had served their time and passed for lieutenants. It is, however, nearly done away in the navy, there being but few *subs* now remaining.

Sub-Marine *Explosion*. See the article Torpedo.

Sub-Marine *Navigation*, the art of sailing under the surface of the water in vessels peculiarly constructed for that purpose; of which several inventions have appeared in the different countries of Europe and America. See the article Diving-*Bell*.

Sub-Marine *Nautilus*. In the French Archives of Discoveries and Inventions for the year 1810 and 1811, an account of an apparatus, called a *Sub-marine Nautilus*, is described at full length, and is said to have been approved by some of the learned members of the institute. This nautilus is a vessel, which, as its name imports, is to sail under the water, and to be capable of containing a considerable number of persons. It appears to be of an oval form, and divided into three apartments; the centre for the reception of the passengers, and the two ends to contain the apparatus for raising or sinking the machine, and for supplying a change of air. A particular account is given of the method employed for directing the machine, and for causing it to rise or sink; but we can scarcely form an opinion of its practicability. In all such contrivances, the greatest difficulty is, to afford a supply of respirable air, and in this respect the nautilus is evidently defective. After an account of two mechanical contrivances, one of which is a system of flexible tubes, intended to reach to the surface of the water, the author adds, "the navigators provide for the imperious necessity of breathing, by an ample stock of compressed oxygen, which is kept in reserve, and which they use with the economy imposed on them by the interest of their own preservation."

SUBORDINATION, (*subordination*, Fr.) in the naval service, is a perfect submission to the orders of superiors; a dependence which is regulated by the rights and duties of every seaman, &c.; subordination should shew the spirit of the chief in all the members, and this single idea, which is manifest to the dullest apprehension, suffices to shew its importance. Without subordination it would be absolutely impossible that good order and harmony could be kept up in the navy, where there are so many different characters to deal with. Indeed, it is subordination that gives soul and harmony to the service: it adds strength to authority, and merit to obedience: and while it secures the efficacy of command, reflects honour upon its execution. See the article Naval *Discipline*.

SUBSTITUTE, (*substitut*, Fr.) one placed by another to act with delegate power.

Substitute *for a Rudder*. See the article Rudder.

To Substitute, (*substituer*, Fr.) is to put one thing in the place of another; as substitutes for the regular Provisions, which article see.

SUCCESSION of Rank, the relative gradation according to the dates of commissions. See the articles Commission and Rank.

Succession *of Signs*, in astronomy, is the order in which they are reckoned, or follow one another, and according to which the sun enters them; called also *consequentia*. As *Aries*, *Taurus*, *Gemini*, *Cancer*, &c. When a planet goes according to the order and succession of the signs, or *in consequentia*, it is said to be direct; but retrograde when contrary to the succession of the signs, or *in antecedentia*, as from *Gemini* to *Taurus*, then to *Aries*, &c.

SUIT of Sails, (*jeu de voiles*, Fr.) implies any set of sails that may be appropriated to the use of a ship, or vessel.

SUMMER, (*été*, Fr.) the name of one of the seasons of the year, beginning one of the quarters when the year is divided into four quarters, or one half when the year is divided only into two, summer and winter. In the former case, summer is the quarter during which, in northern climates, the sun is passing the three signs *Cancer*, *Leo*, *Virgo*, or from the time of the greatest declination, till the sun comes to the equinoctial again, or has no declination; which is from about the 21st of June, till about the 22d of September. In the latter case, summer contains the six warmer months, while the sun is on one side of the equinoctial; and winter the other six months, when the sun is on the other side of it.

It is generally said, that a frosty winter produces a dry summer, and a mild winter a wet summer. See *Philos. Trans.* N°. 458. Sect. 10.

Summer *Solstice*, (*solstice d'été*, Fr.) the time or point when the sun comes to its greatest declination, and nearest the zenith of the place. See the article Solstice.

SUN, (*sol*, Lat. *soleil*, Fr.) in astronomy, the most conspicuous of the heavenly bodies, occupying the centre of the system, which comprehends the earth, the primary and secondary planets, and comets, marked thus, ☉. The sun is that magnificent luminary which enlightens these worlds, and by its presence constitutes day. We have referred to this article from the fixed Stars, because the sun agrees with them in several particulars, as in the property of emitting light continually, and in retaining constantly its relative situation with but little variation: they may have probably many other properties in common. The sun is, therefore, justly considered as a fixed star comparatively near us; and the stars as suns at immense distances from our earth; and we reasonably infer, from the same analogy, that the stars are possessed of gravitation and of the other general properties of matter; they are supposed to emit heat as well as light; and it has been conjectured that they serve to cherish the inhabitants of a multitude of planetary bodies revolving round them. In a paper on the "Construction of the Heavens," Dr. Herschel says, it is very probable, that the great stratum called the milky way is that in which the sun is placed, though perhaps not in the centre of its thickness, but not far from the place where some smaller stratum branches from it. Such a supposition will satisfactorily, and with great simplicity, account for all the phenomena of the milky-way, which according to this hypothesis, is no other than the appearance of the projection of the stars contained in this stratum, and its secondary branch. See the article Galaxy. In another paper on the same subject, he says, that the milky way is a most extensive stratum of stars of various sizes admits no longer of the least doubt; and that our sun is actually one of the heavenly bodies belonging to it is as evident.

The sun, like many other stars, has probably a progressive motion, directed towards the constellation Hercules. Dr. Herschel, on this subject, observes that the apparent proper motions of 44 stars out of 56, are nearly in the direction which would be the result of such a real motion of the solar system; and that the bright stars Arcturus and Sirius, which are probably the nearest to us, have, as they ought, according to this theory, the greatest apparent motions. Again, the star Castor, appears when viewed with a telescope, to consist of two stars, of nearly equal magnitude; and though they have both an apparent motion, yet have never been found to change their distance with respect to one another a single second, a circumstance which is easily understood, if both their apparent motions are supposed to arise from the real motion of the sun. The sun revolves on his axis in 25d. 10h. with respect to the fixed stars; this axis is directed towards a point about half way between the pole star and Lyra, the plane of the rotation being inclined a little more than 7° to that in which the earth revolves. The direction of this motion is from west to east. All the rotations of the different bodies which compose the solar system as far as they have been ascertained, are in the same direction, and likewise all their revolutions, excepting those of some of the comets, and those of some of the satellites of the Herschel planet. The time and the direction of the sun's rotation are ascertained by the change of situation of the spots, which are usually visible on his disc, and which some astronomers supposed to be elevations, and others to be excavations in the luminous matter covering the sun's surface. These spots are frequently observed to appear and disappear, and they are in the mean time liable to great variations, though they are generally found about the same points of the sun's surface. M. Lalande supposes them to be parts of the solid body of the sun, which by some agitations of the luminous ocean, with which he conceives the sun to be surrounded, are left nearly or entirely bare. Dr. Wilson and Dr. Herschel are disposed to consider this ocean as consisting rather of a flame than of a liquid substance; and Dr. Herschel, in an ingenious paper, attributes the spots to the emission of an aeriform fluid, not yet in combustion, which displaces the general luminous atmosphere, and which is afterwards to serve as fuel for supporting the process; hence he supposes the appearance of copious spots to be indicative of the approach of warm seasons on the surface of the earth, a theory which he has attempted to maintain by historical evidence. The exterior luminous atmosphere has an appearance somewhat mottled, some parts of it, appearing brighter than others, have been called faculæ, but Dr. Herschel distinguishes them by the names of ridges and nodules. The spots are usually surrounded by margins less dark than themselves, which are called shallows, and which are considered as parts of an inferior stratum, consisting of opaque clouds, capable of protecting the immediate surface of the sun from the excessive heat produced by combustion in the superior stratum, and perhaps rendering it habitable to animated beings. To which Dr. Young replies, if we inquire into the intensity of the heat, which must necessarily exist wherever this combustion is performed, we shall soon be convinced that no cloud, however dense, could impede its rapid transmission to the parts below. Besides the diameter of the sun is 111 times as great as that of the earth; and at its surface, a heavy body would fall through no less than 450 feet in a single second; so that if every other circumstance permitted human beings to reside on it, their own weight would present an insuperable difficulty, since it would become thirty times as great as upon the surface of the earth, and a man of moderate size would weigh above two tons.

It is also observed, that from the effects produced at the distance of 95,000,000 miles, we may infer, that every thing must be scorched up at its surface. Yet we must admit, that there are many facts in natural philosophy which shew that heat is produced by the sun's rays only when they act on a calorific medium: they are the cause of the production of heat by uniting with the matter of fire which is contained in the substances that are heated; as the collision of the flint and steel will inflame a magazine of gunpowder, by putting all the latent fire which it contains into action. On the tops of mountains of sufficient height, at the altitude where clouds can seldom reach to shelter them from the direct rays of the sun, we always find regions of ice and snow. Now, if the solar rays themselves conveyed all the heat we find on this globe, it ought to be hottest where their course is the least interrupted: Again, our aëronauts all confirm the coldness of the upper regions of the atmosphere; and since, therefore, even on our earth the heat of the situation depends upon the readiness of the medium to yield to the impression of the solar rays, we have only to admit, that on the sun itself, the elastic fluids composing its atmosphere, and the matter on its surface, are of such a nature as not to be capable of any extensive affection of its own rays; and this seems to be proved by the copious emission of them; for if the elastic fluids of the atmosphere, or of the matter contained on the surface of the sun, were of such a nature as to admit of an easy chemical combination with its rays, their emission would be very much impeded. Another well known fact is, that the solar focus of the largest lens thrown into the air, will occasion no sensible heat in the place where it has been kept for a considerable time, although its power of exciting combustion, when proper bodies are exposed, should be sufficient to fuse the most refractory substances.

Apparent diurnal Motion of the Sun.—The phenomenon which first arrests the attention of an inhabitant of the earth, is the apparent motion of the sun through the heavens. After ascending above the horizon, his height gradually increases; and when he has reached the highest point of his path, his altitude begins to diminish, till he sinks below the horizon, and withdraws the cheering influence which he had shed over nature. After continuing invi ible for some time, the sun re-appears near the same place where he formerly rose to our view. He again traverses the heavens, and strikes below the verge of the distant landscape. Since the sun is stationary in the course of the planetary system, as will afterwards be proved, his apparent motion through the heavens must arise from the motion of the earth about its axis, in the same way as a piece of matter, whirled about in the air, turns round an axis, or a line joining two opposite points of it, which are at rest, the time which elapses between the arrival of the sun at the meridian of any place, and his subsequent arrival at the same meridian, is called a day, and is obviously equal to the time in which the earth turns round its axis. This interval is the universal standard adopted by all nations for the measure of time. It is divided into 24 equal parts, called *hours*, each of which are subdivided into 60 equal parts, called *minutes*, and each minute into 60 equal parts, called *seconds*.

Apparent Annual Motion of the SUN.—When we examine the motions of the sun with a little more attention, we shall find that he has another motion in the heavens, opposite to his apparent daily motion from east to west. This phenomenon may be distinctly observed, by marking on any clear evening, a bright star which is first visible after sun-set near the place where the sun sinks below the horizon. On the following evening the star will cease to become visible, on account of the approach of the sun; and all the stars to the east of it will be successively eclipsed by the rays of the sun, till he has made a complete revolution in the heavens. The motion of the sun from west to east, or the path which he describes, may be more accurately ascertained, by observing every day the point of the heavens which comes to the meridian with the sun, and its altitude above the horizon. By this means a series of points will be obtained, which will mark the apparent path of the sun. The line which is thus traced among the fixed stars, and which is called the Ecliptic, will be found to be inclined to the equator, at an angle of 23° 28′, cutting it in two opposite points. The apparent motion of the sun from west to east, evidently arises from the motion of the earth in the same direction, and in the same plane. If the earth is in any point of its orbit, the sun will always be seen in the opposite point of the heavens. When the earth moves one degree to the east, the sun will also appear to move one degree in the opposite direction; and when the earth has completed one revolution in its orbit, the sun will appear to have completed a revolution in the heavens. Hence it follows, that the ecliptic, or the apparent path of the sun in the heavens, is the path or orbit, in which the earth revolves round the sun; and that the earth's equator is inclined 23° 28′ to its annual orbit, or, what is the same thing, that the earth's axis is inclined 66° 32′ to its path round the sun.

The change of Seasons occasioned by the SUN.—While the sun appears to have a daily motion from east to west, and an annual motion in the opposite direction, his height, when he comes to the meridian of any place, and his continuance above the horizon are perpetually changing. On the 21st of March he rises at six in the morning, and reaches the western horizon, after having continued for twelve hours to shed his light over the earth, and to cheer it with the genial temperature of spring. During the remaining days of March, the whole of May, and the greater part of June, his meridian altitude gradually increases: he lingers above the horizon during a longer period, and man is obliged to shelter himself from the fervour of his heat, and the dazzling brilliancy of his rays. On the 21st of June, his meridian altitude, having reached its maximum, begins to diminish till the 21st of September, when it is exactly the same as it was on the 21st of March, and when he remains above the horizon during the same period. The length of the day continues to diminish, with the diminishing altitude of the sun, till the 21st of December, when he continues only about eight hours above the horizon, and when the obliquity of his rays deprives them of their usual heat, and brings on the dulness and rigour of winter. After the 21st of December, the meridian altitude of the sun, and the length of the day, gradually increase; and the same phenomena are repeated, in a similar order, during every revolution of the sun.

We shall conclude this article, with some particulars respecting the sun, by Sir Isaac Newton. 1. That the density of the sun's heat, which is proportional to his light, is 7 times as great in Mercury as with us, and that water there would be all carried off in the shape of steam, for, he found, by experiments with the thermometer, that a heat seven times greater than that of the sun's beams in summer will serve to make water boil. 2. That the quantity of matter in the sun is to that in Jupiter nearly as 1100 to 1, and that the distance of that planet from the sun is in the same ratio to the sun's semidiameter; consequently, that the centre of gravity of the sun and Jupiter is nearly in the superficies of the sun. 3. That the quantity of matter in the sun is to that in Saturn as 2360 to 1, and that the distance of Saturn from the sun is in a ratio but little less than that of the sun's semidiameter. And hence the common centre of gravity of Saturn and the sun is a little within the sun. 4. By the same method of calculation it will be found, that the common centre of gravity of all the planets cannot be more than the length of the solar diameter distant from the centre of the sun. 5. the sun's diameter is upwards of 100 diameters of the earth, and therefore its magnitude must exceed that of the earth more than one million of times. 6. If 360 degrees (the whole ecliptic) be divided by the quantity of the solar year, it will give 59′ 8″, which therefore is the medium quantity of the sun's apparent daily motion; hence his horary motion is equal to 2′ 27″. By this method the tables of the sun's mean motion are constructed. See the *Solar Tables* in Vol. II. part 3 and 4, of the EDINBURGH *Encyclopædia*.

SUN-*Rise*, (*lever du soleil*, Fr.) is when the sun makes his appearance above the horizon.

SUN-*Set*, (*coucher du soleil*, Fr.) is when he withdraws, or sinks below the horizon.

SUPERANNUATED, (*invalide*, Fr.) in a naval sense implies that an officer is unfit to serve, either from age, or infirmity, and is generally allowed to retire from the service upon two-thirds of his pay.

Such masters as have not served during the present war, or who are now considered to be unfit for farther duty at sea, are to be superannuated on the half-pay to which from their services and qualifications they may be respectively entitled by the *old establishment*, provided unquestionable testimony be produced that they have not at any time declined to serve when capable of performing their duty; and such as may not have it in their power to produce such testimonials (except in cases where by misconduct they have forfeited all claims to remuneration for past services) will be also superannuated at the former established allowance of superannuation. Gunners, boatswains, carpenters, pursers, and cooks, whose employments in the navy are constant, must have served full fifteen years before they are entitled to apply for superannuation, then, if their claims are admitted, they receive an annual pension equal to their pay in the ship of the highest rate they have served in. For farther particulars, see the articles GRATUITY and PENSION.

SUPERCARGO, (*supercargue, ou écrivain, de vaisseau marchand*, Fr.) an officer charged with the accounts of the cargo, and all other commercial affairs in a merchant-ship.

SUPERFICIES, or SURFACE, in geometry, the outside or exterior face of any body. This is considered as having the two dimensions of length and breadth only, but not thickness; and therefore it makes no part of the substance or solid content or matter of the body.

SUPERIOR, (*superieur*, Fr.) is one above another in rank, power, authority, strength, &c. In the navy, as well as in every other branch of society, the individual who holds an inferior post, or place, is bound to pay due obedience and respect (as far as regards the exercise of his duties or functions) to his superior in rank.

SUPERIOR OFFICER, (*officier supérieur*, Fr.) is an officer in a higher rank than another.

In the royal navy, all flag officers, captains, and commanders, and every other officer in his station, is to conduct himself in the discharge of his duty, with the most perfect respect to his *superiors*, and the most implicit obedience to their orders; and every officer is, on all occasions to shew to all other officers, every degree of respect and attention, which their respective situations may entitle them to expect.

Every officer is to observe attentively the conduct of all the officers under his command, and of the ship's company in general; urging them, individually and collectively, to a zealous and perfect performance of their duty; commending all those whose good conduct shall merit commendation; and censuring, punishing, or reporting to his superior, as circumstances may require, those whose misconduct shall deserve it. See the article TYRANNY and OPPRESSION.

SUPERNUMERARY, (*surnuméraire*, Fr.) beyond a fixed or stated number.

The commander-in-chief is directed by his instructions not to order any captain or commander of any ship or vessel under his command to bear supernumeraries, unless he can shew reason for his doing so, which is to be specified in the order he gives for that purpose. The supernumeraries for wages, and those for victuals only; as also the volunteer boys of the first and second classes are to be entered on a separate list from the seamen and marines, which list shall specify whether at whole, or at what other allowance, and by whose order they are borne on the ship's books.

Notwithstanding any general order to the captain to bear supernumeraries, yet he is strictly enjoined never to proceed to sea with more than the complement of the ship, without an express order for so doing; except when he shall sail from a port in which there is no other ship to receive them.

Supernumeraries of every description borne on board his Majesty's receiving ships, and also those borne on board other ships for victuals only, are not to have any tobacco issued to them from the public stock in the purser's charge.

All officers, artificers, or other persons belonging to his Majesty's dock-yards or in ordinary, who may have been victualled on board any ship or vessel during her continuance in port, are to be borne as supernumeraries, and a list of their names, specifying their several entries and discharges, is to be sent under the captain's or commander's direction to the commissioner, that he may order the clerk of the check to check all of them that are borne for victuals on board such ship or vessel. See the NAVAL INSTRUCTIONS.

SUPERSEDING, in the royal navy, implies the succession of one officer to the identical situation of another by special appointment.

SUPERSEDING *Signal*, a signal hoisted by order of a superior officer on board a ship, giving notice that some individual has been deprived of his authority.

SUPPLEMENT OF AN ARC, in geometry, or trigonometry, is the number of degrees that it wants of being an entire semicircle; as a complement signifies what an arc wants of being a quadrant.

SUPPLY, (*remplacement de vivres, de rafraîchissemens, ou munitions*, Fr.) a fresh recruit of provisions or stores sent to a ship or fleet.

SUPPORTERS, in ship building, a name given to the knee-pieces of oak-timber under the cat-heads.

SURF, (*ressac de la mer contre une côte, ou des rochers*, Fr.) the swell of the sea which breaks upon the shore, or any rock lying near the surface of the sea.

SURGE, (*lame, ou vague*, Fr.) a swelling sea; or a great wave rolling above the general surface of the water. See the article WAVE.

To SURGE, (*choquer*, Fr.) is to let go a portion of a rope suddenly; as, "Surge the messenger!"

SURGEON OF MAN OF WAR, (*chirurgien-major d'un vaisseau de guerre*, Fr.) an officer of approved abilities in surgery and physic, by a previous examination at Surgeon's Hall for the former, and at the Office of the Commissioners for Sick and Hurt and Transport Board, for the latter, by which board he is appointed, by warrant, to one of his Majesty's ships or vessels.

A naval surgeon of abilities and circumspection is generally the most independent officer in the ship, as his line of duty is unconnected with the others.

He has the entire charge and management of the sick and hurt seamen on board his ship; is to perform such surgical operations on the wounded as he may deem absolutely necessary to the safety of their lives; and to see that the medicines and necessaries with which he is supplied from the said board, or their agents, are good in kind, and administered faithfully to the sick patients under his care.

He will be furnished on applying to the sick

and hurt board, if on a home station, or to their agent, if on a foreign station, with medicines, and necessary utensils, agreeable to the rules of the navy; and he will be required to provide, at his own expense, such surgical instruments as the said board shall judge necessary, a list of which will be put on the back of the warrants given by that board to every surgeon and assistant.

When any supplies of necessaries shall be received on board, they are to be inspected by the surgeon, master, and purser, who are to certify whether the quantity and quality agree with the established allowance, but if they do not, the defective articles shall be returned, and all deficiencies made good.

He is to provide himself at all times with a competent number of smart-tickets which will be delivered to him on application to the Commissioners for sick or wounded seamen, or to their agents at the out-ports: to inspect and take care of the necessaries sent on board for the use of the sick, and see that they are duly served out for their relief.

He is also to visit the men under his care twice a day, or oftener, if circumstances require it; he must see that his assistants do their duty, so that none want due attendance or relief. In cases that are difficult, he is to consult with the physician of the fleet, and afterwards to follow his directions in such cases; and report to the captain daily, the state of his patients.

When the sick are ordered to the hospital, he is to send with them to the surgeon an account of the time and manner of their being taken ill, and how they have been treated. But none are to be sent to sick quarters, unless their distempers, or the number of sick on board, are such, that they cannot be taken due care of; and this the surgeon is to certify under his hand, before removal.

He must be ready with his mates and assistants in an engagement, and have all things at hand necessary for dressing the wounded men, if any.

He is to keep a journal of his practice, according to the form delivered to him on his appointment, noting the disease and symptoms, with the medicines daily prescribed, which journal is to be sent to the commissioners for sick and wounded seamen, with the other papers necessary for passing his accounts at the expiration of each year, and not oftener, unless in case of his quitting the ship. He is also to give a general history of the prevailing complaints in the ship, during the above periods, and if any malignant or infectious diseases shall have existed, to trace them to their source, and to account for their introduction, stating the means used to destroy the infection and prevent its re-appearance. He is also to note therein the particulars of the cases in which he may have demanded wine of the purser.

When the ship shall be in any port of Great Britain, he is not to fail to transmit regularly to the commissioners for sick and wounded seamen, a weekly return of the state of the sick, signed by himself, with remarks and observations, according to the form in the NAVAL INSTRUCTIONS, which see for farther particulars.

SURGEON'S-MATES OR ASSISTANTS, (*aides-chirurgiens*, Fr.) are medical gentlemen appointed, after being duly examined at Surgeon's Hall, to assist the surgeons in their respective duties in his Majesty's ships, &c.

The following hints to young naval surgeons are lately published by an old veteran in the service, and may be useful even to those who have had some experience.

1. In order to reap the greatest advantage from the field on which he is entering, he should set out on a fixed plan, as to what he is to pursue.

He ought to dedicate a certain sum annually for the purchase of well selected professional books, and that he may peruse them with advantage, he ought minutely to examine the nature of each complaint that occurs within the sphere of his observation, and then consult every author in his possession on that particular topic, noting how their sentiments agree with the actual phenomena. By this means he will not only have many resources at hand in the treatment of diseases, but their symptoms will become technically rivetted, as it were, on his mind, by the comparison; while his notes will be a ready reference to sources of information, to meet accidental emergencies.

2. He should systematise his studies; allotting certain days for medical, and others for general literature, taking care that the former should always preponderate, and that he should early accustom himself to digest what he has read, and to *think* at least three times as much as he *reads*, in order to become equal to original and independent exertion.

3. He should preserve a dignified deportment in respect to temperance and morality; as a mild and steady discountenance of vice and debauchery will often have more effect on his brother officers than the frowns or lectures of the chaplain, who is too often very illiberally suspected of censuring vice merely *ex officio*.

4. Although mental relaxation be as necessary as corporeal rest, yet he should be exceedingly cautious how he gives way to amusements commonly resorted to by civilians on board ship, as there will be a danger of estranging him from professional pursuits, which will be apt to appear dull and insipid, when contrasted with those more fascinating entertainments. If, however, his intellectual enjoyments are well regulated, he will not often require these adventitious aids in passing his time at sea.

5. His conduct to the men should ever be characterised by humanity and tenderness, always keeping in mind the *uncertainty* of his art, and the danger of suddenly deciding on what he might conceive to be imposition; as it is better that we should be ten times duped by fictitious complaints, and thus protect a *sculker*, than once reject a real case of pain or disease, however appearances might justify us at the moment.

6. He should keep an accurate record of every circumstance touching the health of the ship's company, or in any manner connected with his profession, or the general science of philosophy, so that the most important facts and observations might be transmitted to the medical board from time to time; and if the medical commissioner were to examine with any degree of minuteness the journals that flow in from the different quarters of the globe, an accurate estimate might soon be found of ability, from such an extensive source of information as would thus

be established; and a powerful stimulus to the exertion of talent and industry would be diffused among the meritorious orders of the profession, without any expense to the nation! See JOHNSON's *Essay on the Influence of Tropical Climates on European Constitutions, &c.*

SURMARKS, in ship-building, the stations of the ribands and harpings, which are marked on the timbers.

SURRENDER, (*se rendre*, Fr.) to give up to a victorious adversary; as, " the fort surrendered to our squadron."

SURVEY, (*inspection, ou visite des officiers dans un port*, Fr.) in a general sense, implies a strict and impartial examination, made by several naval officers, into the state or condition of any materials, &c. belonging to a ship or fleet of men of war: hence

SURVEY *on Masts, Yards, Sails, and other Stores*, are taken when they appear to be worn out, and rendered unserviceable by use: and on provisions, which, when received on board, appear to be deficient in quantity, or defective in quality.

The officers employed on surveys are, one master, and two officers of the class of him whose stores are to be surveyed; they are all, if possible, to belong to other ships than that to which the stores to be surveyed belong; but the master of that ship, and the officer who has charge of those stores, are to be present to give what information may be required, and to prevent partiality or injustice, or to represent it to the captain if they perceive without being able to prevent it. But if there be a necessity for an immediate survey when there are not a sufficient number of ships present to furnish, or when the sickness of their officers prevents other ships from furnishing, the number of officers required, the master of the ship may be ordered to assist at such survey; but if the ship be alone, such survey is to be taken by one of the lieutenants, the master, and one of the master's mates; but in this case the survey is to be taken again, if other ships shall join company before the surveyed stores, &c. have been disposed of.

Surveys may be applied for when the parties who are ordered to examine into the articles, cannot conscientiously condemn, by reason of finding, on minute investigation, that the said articles, if masts, yards, sails, &c., are not sufficiently worn, but capable of being longer made use of in their present service; in which case, a reprimand from the captain of the ship, or commander-in-chief, &c. frequently follows, for the trouble and hinderance to the service.

SURVEY *on Provisions, (inspection des provisions*, Fr.) The purser applying for a survey on the provisions, is not required to state any particular quantity defective, but to specify the quantities of such and such species complained of, and mention is made of the cause, in his opinion, of such defect; nor is the number even of casks or bags of bread or biscuit required to be specified, as the bread is picked, and the defective set apart from the other; also all other species are selected in like manner, and when the whole that appears unquestionably of a quality unwholesome and not fit for men to eat is ascertained, it is weighed in their presence, and the totals of each species stated in their report with their opinions as to the cause, to which they are ready to make oath if required.

SURVEY *on Officers, (inspection des officiers malades*, Fr.) When an officer is incapable of performing his duty on board from ill health, &c. and it is deemed necessary by the surgeon that he should be sent to an hospital, or sick-quarters, it is usual for a survey to be held on him by three captains and three surgeons afloat, to report to the admiral, or senior officer, whether, in their opinion, he should be sent to the hospital or sick-quarters. When the officer is to be sent to any one of the royal hospitals and his ship is in port, the survey is to be held at the hospital by the physician and surgeon thereof, who report his state to the admiral; but in case the officer should come into harbour dangerously ill, or wounded, such survey is dispensed with. The admiral or senior officer should always sign the officer's sick ticket.

Surveys are also held on men at sea, who, from the surgeon's report, may be deemed unfit for service; but when the ship is in harbour, the survey takes place at the naval hospital, if there is one.

SURVEY *on Surgeon's Instruments, Medicines, &c.* is to be held by the physician of the fleet and two surgeons, or by three surgeons, as the commander-in-chief shall direct; who are to specify the quantity, quality, and condition of each of them. Should any of the medicines be found unfit to be administered, they are reported unserviceable, and to be returned into store. If there shall be a necessity for a survey, when three surgeons cannot be obtained, the commanding officer present may order a surgeon's assistant to attend as one of the surveying officers. See also the article REPORTS.

Clerk of the SURVEY, an officer in his Majesty's dock-yards, warranted by the Admiralty, who, with the master shipwright and master attendant, directs the issue of all stores to the warrant officers of ships, and for the use of the yard, and keeps a charge against the persons to whom they are issued: it is his duty also to survey all articles returned into store, and with the clerk of the check and store keeper, to attend the survey and receipt of all stores from contractors, and the delivery of lots of old stores to the purchasers; and to examine the certificates made out by the clerk of the check, of the stores received, by which the contractors obtain payment. He also surveys all remains of stores of boatswains and carpenters, at the returns of ships from sea, and passes their accounts.

To SURVEY *a Harbour*, is to take the plan of it, similar to the plotting of land on shore.

SURVEYING, (*arpentage*, Fr.) in military mathematics, is the art or act of measuring lands; that is, of taking the dimensions of any tract of ground, laying down the same in a map or drawing, and finding the content or area thereof.

Surveying consists of three parts: the first is the taking of the necessary measures, and making the most necessary observations, on the ground itself; the second is, the laying down of these measures and observations on paper: and the third, the finding the area, or quantity of ground there laid down. The first is what we properly call *surveying;* the second *plotting*, or *mapping*, and the third *casting up*.

Marine SURVEYING, (*construction d'un plan ou d'une carte marine*, Fr.) is the art of properly representing on a map or chart a correct view of a coast, or harbour, carefully shewing all the various projections, and bearings of the headlands, as well as the extent of sand-banks, shoals, rocks, &c. in a bay or harbour. It depends upon the same principles, and is chiefly conducted like that of military surveying, except that the operation is more complicated and laborious, as it is necessary to erect signals and to mark a variety of objects along the coast, with their different bearings from one another, and the several parts of the harbours, and likewise to measure a great number of angles, at the different stations.

The utility of making accurate surveys of coasts and harbours is of the greatest importance to navigators, we shall therefore give a few general directions, shewing how they may be easily surveyed with such instruments only as are commonly used at sea.

In surveying of sea coasts, islands, bays, harbours, &c. there are two principal things necessary; first that a base line of a sufficient length be measured or ascertained with a degree of accuracy sufficient for the intended survey, which base should be at least $\frac{1}{3}$ or $\frac{1}{4}$ of the greatest distance of the objects to be laid down on the map from it where it is possible to be obtained; but the longer it can be procured the better, so that it does not exceed the length of the survey; and distinct vision is preserved of all the objects to be laid down from each end of it. Secondly, it is necessary that the angles formed by the said base line and lines drawn from each end of it to the different objects to be laid down should be determined as accurate as possible, and where more observations than two of the same object can be taken, its true position will be determined with more accuracy. Where it is possible, the observation should be so taken, that the lines drawn from each end of the base to the same object may cut each other, at right angles as near as possible, by which means any uncertainty in their crossing will be removed. There are various methods of measuring a base line, according as circumstances may require; first in running along a coast in a ship, or boat, the log must be used, and in order to be somewhat certain it should be hove 3 or 4 times every hour, and sometimes oftener when among tides, or currents, &c.: moreover the currents or drift of the tides must be determined both as to direction, and strength, or velocity.

To Measure a Base Line.---There are various methods of measuring a base line according as the situation will best admit of. When the distance is great, the motion of sound may perhaps be the most convenient of any, as it may be performed in any situation either on land or at sea. It being generally agreed on by mathematicians that sound moves at the rate of 1142 feet per second, and at the rate of 571 feet per half second; and that uniformly so: therefore, with a watch that has a second hand, count the number of seconds between your seeing the flash of the gun, and hearing its report, which multiplied by 1142 will give the distance in English feet; which divide by 5280, the quotient will be English miles, or divide by 6116, the quotient will be nautical miles, or sound moves over a statute mile in 4''. 62 seconds of time, or over a nautical mile in 5''. 35 seconds of time: therefore, divide the time observed between the seeing of the flash, and hearing of the sound, by 5. 35, the quotient will be the distance in nautical miles. If a second watch is not at hand, a half second pendulum may be readily made; for pendulums being in length as the squares of their times of vibrating; and it being found that a pendulum 39. 2 inches long will vibrate seconds: therefore say, as $1^2 : 39.2 :: 5^2 : 9.8$ inches, the length of an half second pendulum: consequently, if a plummet, or a musket-ball is suspended by a string, or small wire that is 9. 8 inches, from the point of suspension to the centre of the ball, set it going and it will swing half seconds, make a mark where it hangs when at rest, and for every time it crosses that mark count half a second, then the number of half seconds or vibrations between the seeing of the flash and hearing of the report multiplied by 571 will give the distance in English feet required. The foregoing method of measuring distances by the motion of sound is better adapted to distances of a considerable extent as of 3, 4, or more miles, either at sea or on shore; for in general a person will not observe the interval of time much nearer than a quarter of a second, therefore will be liable to err 2 or 300 feet by any single observation; consequently, a number of observations should be made, and a mean of all the results taken; which will probably give the distance pretty near the truth.

It may sometimes happen that a base line is wanted on shore, at or near the sea side; which base should be measured with the greatest care and accuracy possible; as the truth of the whole survey will depend on it.

For this purpose, a good steel chain should be provided, the wire of which should be strong and soldered together at each limb and ring; also a standard brass rod or scale of 6 feet long, divided into feet, inches, and tenths, with a diagonal scale at one end subdividing to hundredths of an inch; also 3 or 4 straight grained white fir rods, 2 or $2\frac{1}{2}$ inches square, and ten feet long, tipt with brass at each end; so that the ends may be easily brought in contact with each other, and the brass at one end of each rod shouldbe capable of being adjusted for expansion and contraction by heat and cold; also two wooden levels should be provided of about 20 or 22 feet each, like those used by masons with the addition of moveable legs: with such an apparatus a base line may be measured with great accuracy, though the ground may be very uneven. When the line has a considerable declivity, or one end higher than the other, the difference of heights must be determined either by a water level, or it must be found by observation and the base computed by trigonometry. But a coast may be surveyed without measuring a base line on the shore; for example,

Make choice of two hills, inland, or remarkable headlands, that can be seen from most parts of the sea coast, and bring them to bear due east or west, and observe the ship's latitude, which will be the latitude of the hills or headlands; this may be done by bringing the ship to an anchor, or standing forward in the same parallel with the hills, till an observation can be made, then run on till the two hills or headlands are in a line, and set them by the compass; as you run along the coast, observe by your compass when most of the remarkable points of the coast are in a line with the hills, as also the mutual bearings from each other; whence the true situation of the whole will be readily determined, and a true chart constructed.

To measure a Base line on the surface of the water.—Provide two or three cords, each 100 fathoms long, with corks to float them, in the same manner as the line of a fishing net is floated; these lines must be stretched, and then wetted, before they are measured, and kept in a wet state during the time of the measurement; chains, with floats of cork would be more exact than lines, being less liable to alter their lengths.

Take three small boats, with three or four men in each boat; let one boat anchor at one end of the line proposed to be measured, where let one end of the chain or line be held; let a second boat move forward with the other end, until it is stretched on the surface of the water, and there anchor, and hold the ends of the first and second chains together, while the third boat moves forward with the other end of the second chain, until it is on a stretch, and then anchor; when the first boat quits her station, and moves forward with the third chain, leaving the first chain for the second boat, she delivers one end of the third chain to the third boat, and rows on until it is on a stretch, and then anchor, when the second boats moves forward with the first chain, and so on alternately, till the line is measured.

The chains or lines used, must be measured immediately before, and after the line is measured, to see that it has not altered its length: this must be performed on shore, on some level place. If care be taken, a base line may be measured this way with a considerable degree of accuracy.

In surveying bays or harbours, the general way is, to go to some point of it from which the wind blows, and sail across the bay or harbour, noting the time in sailing across: during which the log must be hove once, or oftener; from whence the distance run, or width of the bay or harbour will be determined, which will serve as a base line; a base line measured thus, will be sufficiently accurate for small surveys.

To measure Angles.—The azimuth compass is generally used for determining the angle between the base line, and lines drawn from each end of it to the object to be laid down. But Hadley's sextant is much preferable for many reasons; it is well known that compasses will differ greatly from each other, and even from themselves, to the amount of several degrees; they therefore should be made use of as seldom as possible: besides, when the ship has any considerable motion, the card will seldom be steady enough to observe with any tolerable degree of accuracy; and moreover, while you are waiting for the card to be steady, the lands will alter their relative situation, if the ship is under way. For these and many other obvious reasons, the sextant should have the preference, and will be found much more convenient than any other instrument in marine surveying. When among islands and shoals, &c. it may be used in the main or mizen tops to good advantage, where the parts may be seen better, and the angles taken with equal facility and exactness: or if one or two can be set by the compass, or compared with some determinate object, the true position of the whole will be determined. By holding the plane of the sextant in a plane passing through any two objects, and moving the index till the one seen by reflection on the quick-silvered part of the small speculum coincides with the other seen through the clear part of it, the degrees, &c. on the arc will be the angle between them measured on the plane, but the angle so measured may differ considerably from the horizontal angle subtended between the objects.

The relative distance of islands, rocks, shoals, &c. may be determined very near the truth, by the following method: where an eminence is near from whence all may be seen, set one of them by your compass, and take the horizontal angles between them with your sextants, also take the angles of depression of each within the visible horizon of the sea; then, having determined the height of your eye above the water, this may be performed by a good barometer, or by a spirit level, or geometrically; from whence the position and distance of all the objects may be readily computed, and a chart made of them. See *Philos. Trans. vol.* 55, also Mc. Kenzie's *Maritime Surveying.*

SURVEYORS of the Navy, (*inspecteurs de la marine d'angleterre*, Fr.) officers, who sit at the navy board, being invested with the charge of building and repairing his Majesty's ships, at the different dock-yards of the kingdom: for which purpose they are selected from those who have been regularly trained to the theory and practice of ship building.

To SUSPEND an Officer, (*suspendre un officier de ses fonctions*, in a naval sense, implies the putting under arrest any officer for misconduct.

The commander in chief, where there is no commissioner resident, may suspend any officer from his duty, whose misconduct has made it necessary for him to do so, and appoint another, provisionally in his stead. He is to send to the secretary of the Admiralty, by the first opportunity, an account of such misconduct, and of his reasons for suspending him; and if he be an officer having the charge of stores, he is to order a survey of them to be taken, and is to give them in charge to the officer whom he appoints to act in his stead, with directions to open a new account of the receipts, returns, conversions, and issues, of all stores, from the period of closing the said survey, in order to ensure the means of investigating the preceding officer's accounts.

A captain is not to inflict any punishment on a commissioned or warrant officer, but may suspend and put under arrest, any officer who misbehaves, and may continue such arrest and suspension until an opportunity offers of trying such officer by a court martial; if the service requite it. he may apply to the senior officer present to appoint some person to act in his stead, but must not appoint one himself; and must inform the commander in chief, or the Lords Commissioners of the Admiralty, of such arrest and suspension.

SUSPENDED, (*officier suspendu de ses fonctions*, Fr.) deprived of the pay and emolument of a situation, and rendered incapable of exercising the duties attached to it, during the pleasure of the commander in chief.

SUSPENSION, (*suspension*, Fr.) a temporary arrest from performing the duties of a situation.

SWAB, (*faubert*, Fr.) a sort of mop formed of a large bunch of old rope yarns, and used to clean the decks and cabins of a ship.

Hand SWAB, (*faubert à manche en bois*, Fr.) a smaller kind, used for wiping dry the stern-sheets of a boat, washing of plates and dishes, &c.

SWAB-*Wringers*, (*matelots du gaillard d'avant tenus à égouter les fauberts*, Fr.) people appointed to wash the swabs and keep them dry and ready for use.

SWABBER, (*balayeur de vaisseau*, Fr.) a man appointed to use the swabs in drying up the decks. He is sometimes called the ship's sweeper, but more commonly the captain's swabber.

SWAN. See the article CYGNUS.

To SWAY, (*hisser*, Fr.) implies to hoist; and is particularly applied to the lower yards, top-masts, and top-gallant-masts and yards; as, "Sway up the lower yards, topmasts!" &c.

To SWEEP, (*balayer, draguer*, Fr.) is to drag the bight, or loose part of a small rope, along the surface of the ground, in a harbour, or road, in order to hook and recover some anchor, wreck, or other material, sunk at the bottom. It is performed by fastening the two ends of the rope to the sides of two boats, abreast of each other, at some distance. To the middle of the rope is suspended a weight, to sink it to the ground; so that as the boats advance by rowing a-head, the rope drags along the bottom, in order to hook any anchor, &c. for which they are searching.

SWEEP *of the Tiller*, (*tamise, ou tamissaille*, Fr.) a circular plank fitted to support the foremost end of the tiller, serving to convey the tiller-rope round it, and to keep it always taught.

SWEEPER OF THE SKY, (*nom que donnent les marins anglois au vent du N. O. en Amerique*, Fr.) a name given by sailors to the N.W. winds of America.

SWEEPERS, (*balayeurs*, Fr.) persons appointed, sometimes by way of punishment, to sweep the decks with brooms.

Captain-SWEEPER, (*balayeur en chef*, Fr.) a man who has the command over the preceding gang. See the article SWABBER.

SWEEPS, (*avirons de vaisseau*, Fr.) a name given to the large oars used on board ships of war, in a calm, either to assist the rudder in turning them round, or to increase the ship's velocity in a chase.

SWELL, (*houle; agitation sourde des eaux de la mer*, Fr.) generally denotes a heavy and continued agitation of the waves, rolling in any particular direction; as "There is a great swell setting into the bay." It is, however, more particularly applied to the fluctuating motion of the sea, which remains after a storm; as also to that which breaks on the sea-shore, or upon rocks, or shallows.

SWIFTER, (*garde-corps, ou tire-veille du cabestan, raban qui tient au bout de toutes ses barres lorsqu'il est armé, pour les assujettir, et aider à virer*, Fr.) a rope used to confine the bars of the capstan in their sockets, whilst the men are turning it round; for which purpose is is passed through holes in the extremities of the bars, so as to strain them firmly together, like the spokes of a wheel; which operation is called swifting. See the article CAPSTAN.

SWIFTER, (*ceinture d'un bâtiment, &c.* Fr.) is also a strong rope, sometimes used to encircle a boat lengthways, as well to strengthen and defend her sides, as to enable her the better to resist the impression of other boats which may run against her. It is usually fixed about ten inches below the boat's upper edge, or gun-wale.

SWIFTERS, (*c'est encore les haubans de fortune, ou faux-haubans, et plus proprement les haubans impairs*, Fr.) are likewise two shrouds fixed on the starboard and larboard sides of the lower masts, above all the other shrouds, as an additional security to the masts, and are never confined, like them, to the catharpings.

SWIFTER *of the Shrouds*, (*bride des haubans, ou cordage servant à les brider*, Fr.) a rope used to stretch the shrouds, to prevent their future extension.

SWIGGING OFF, the act of pulling upon the middle of a tight rope, which is made fast at both ends.

SWIMMING, (*nager*, Fr.) the act of supporting the body towards the surface of the water, so as to float and to advance by means of combined and regular strokes, performed by the hands and feet.

Man is the only animal that learns to swim, the brute creation having an aptitude more or less by instinct: the best example is exhibited by the motion of a frog, on the surface of the water. When a man falls into deep water, he will rise to the surface by floatage, and will continue there if he do not elevate his hands. If he move his hands in any manner he pleases, his head will rise so high as to allow him free liberty to breathe; and if he move his legs, as in the action of walking, or rather walking up stairs, his shoulders will rise above the water; so that he may use less exertion with his hands or apply them to other purposes. By these plain directions, there is not the least doubt, but that any person would swim tolerably well: and when once the ability to support one's-self in the water is acquired, other changes of position and action are performed with facility. Doctor Franklin observes in his Experiments and Observations on this useful art,

1st. That the limbs and head of any human body are heavier than fresh water; the trunk is lighter merely because the lungs are inflated.

2d. That the head is heavier than sea-water; but the limbs and trunk are lighter.

3d. That a person throwing himself on his back in sea-water, will float with his face clear so as to breathe with ease.

4th. That in fresh water the legs will gradually sink; and the body will float in an erect posture.

5th. That in this posture, if the head be held in its natural position, the surface of the water will reach just above the nostrils.

6th. That if the head be leaned quite back, the

nose and mouth will be always above the water and free for breathing, and the body will rise and fall on every such inspiration and expiration.

The rules for swimming well, in our method, that is to say, swiftly, and with little fatigue are few, the body must lie as near the surface, and the head as low as conveniently may be; the knees must be kept wide asunder, in order that the obliquity of action in one leg may counteract that of the other, instead of their joint action producing a libratory motion of the body; and the stroke or impulse must be given with much more velocity than that employed in drawing the legs up again. It is not easy to make any mistake of importance in using the arms by imitating other swimmers. To swim on the back is so easy, that it is frequently learnt at the first trial: even by those who have had no previous knowledge of supporting themselves in the water. The first thing requisite is fortitude to remain in a horizontal position on the surface of the water, without fear of sinking; then gently to move the hands in the way of paddling or sculling, which though not absolutely necessary, would insure the support of the head, and prevent the feet from sinking. And lastly, the legs should be drawn up very gently and then struck out, at the same time bringing the chin towards the breast, to prevent the water flowing over the face.

To SWING, (*éviter au changement de vent ou de la marée, en parlant d'un bâtiment à l'ancre,* Fr.) to turn a ship round the anchors, or moorings, at the change of the wind, or tide; and is only applicable when moored by the head, or riding at a single anchor.

SWIVEL, (*pierrier,* Fr.) a small piece of artillery, carrying a shot of half a pound, and fixed in a socket on the top of a ship's side, stern, or bow, and also in the tops; the trunnions of this piece are contained in a sort of iron crotch, whose lower end terminates in a cylindrical pivot resting in the socket so as to support the weight of the cannon. By means of this swivel (which gives name to the piece of artillery, and an iron handle, on its cascabel, the gun may be directed by hand to any object. See the article ENGAGEMENT.

SWIVEL, (*tourniquet de fer,* Fr.) is also a strong link of iron used in mooring-chains, &c. which permits the bridles to be turned repeatedly round, as occasion requires.

SWIVEL-HOOK, (*croc à tourniquet,* Fr.) a hook that turns in the end of an iron block-strap, and readily takes the turns out of a tackle.

SWORD, (*épée,* Fr.) a weapon worn at the side, and serving either to cut or stab. Its parts are the handle, guard, and blade; to which may be added the bow, scabbard, pummel, &c. Fencing-masters, however, divide the sword into three parts, viz. the upper, middle, and lower; or the fort, middle, and foible.

Broad SWORD, (*Sabre* Fr.) an original weapon of Scotland, having but one edge: it is basket-handled, and three-feet two inches in length.

Small Broad SWORD, a weapon of the same construction as the common broad-sword, but less, and much lighter.

Regulation SWORD, (*épée d'uniforme,* Fr.) a sword which is ordered to be worn by all British officers: it has a spring shell and embossed blade, and may be properly called a long cut and thrust. It is an imitation of the Austrian sword, and was introduced last war.

The admirals and captains in the royal navy wear the cut and thrust blade, the same size as the infantry, with a stirrup gilt-hilt lion's head, back piece and ivory gripe. The lieutenants and midshipmen wear the same sort of blade, only not the end gilt, and the hilt differs in having plain back-piece instead of a lion's head, and the gripe either ebony or black fish-skin instead of ivory. Midshipmen, however, generally wear a hanger or dirk. The cutlass, used on board ships, is a broad cut and thrust with a common black japan-hilt.

SWORD OR CUTLASS *Exercise,* the art of attack and defence with the sword or cutlass.

Mr. Angelo, junior, of Boulton-row, Berkeley-square, London, has reduced the sword exercise to certain principles, and published a plate containing all the positions by which it may be easily studied and practised. He has lately been directed by the Admiralty to instruct a certain number of officers and men from his Majesty's ships at the different ports in the cutlass exercise for the purpose of better defending themselves in the act of boarding an enemy's ship, and they are ordered to instruct the others on board their respective ships: this exercise is now intended to form a part of the education of our naval and military officers.

TO BE SWORN IN, OR TO TAKE THE OATHS, a ceremony which every officer in the British navy of and above the rank of lieutenant, is obliged to go through before he receives his commission. These oaths are administered at the Admiralty, and consist of a total abjuration of the Roman catholic doctrine, and the consequent admission of the power, spiritual, as well as temporal, of the king, and are called oaths of allegiance and supremacy.

LUNAR SYNODICAL MONTH. See the article MONTH.

SYPHERING, in ship-building the act of lapping one edge of a plank over the edge of another for bulk-heads, making the edges of the plank, and also the sides of the bulk-heads, plain surfaces.

SYSTEM, in a general sense, denotes an assemblage or chain of principles and conclusions: or the whole of any doctrine, the several parts of which are bound together, and follow or depend on each other. As a System of Astronomy, a System of Planets, a System of Philosophy, a System of Motion, &c.

SYSTEM, in astronomy, denotes an hypothesis, or a supposition of a certain order or arrangement of the several parts of the universe; by which astronomers explain all the phenomena or appearances of the heavenly bodies, their motions, changes, &c.

This is more particularly called the System of the World, or the SOLAR *System,* which article see: also LAPLACE's *System of the World.*

T.

TABLE MONEY, (*addition aux appointemens des amiraux pour pourvoir aux dépenses de leur table à bord*, Fr.) an allowance to flag officers in addition to their pay, drawn therewith, as a compensation for the necessary expenses they are put to in furnishing their tables.

All flag officers and all commodores having captains under them, who have commissions as commanders in chief, shall be allowed thirty shillings per day for their table, on whatever station or service they may be employed.

The allowance for table money to the admiral and commander in chief of the fleet, was established in the reign of James II. and commences from the date of their appointment to command in chief, and continues until they strike their flags in consequence of being ordered to do so; or until they are informed by directions from the Lords Commissioners of the Admiralty, that their appointment as commander in chief is at an end.

If a commander in chief at any port, or of any fleet or squadron, has leave to be absent from his duty, and shall strike his flag and leave the command to another; the officer being a flag officer, who shall command during his absence, shall receive the table money until the commander in chief return to the squadron, and resume the command. See the NAVAL INSTRUCTIONS.

TABLE-SHORE, (*terre en forme de table*, Fr.) denotes a low level shore.

TABLE WHEEL, in rope-making, is a wheel to lay ropes, from a six-thread ratline, to a two and a half-inch rope, and is fixed in the wheel-house, at the upper end of the rope walk, in a frame fixed in the ground, with two sliding cheeks, and bands to work the whirls, to go separately over each whirl, and round the turning wheel. Some have six sets of whirls, of different sizes, with iron spindles and nibbed, or fore-locked hooks at the outer end. A tackle-board, twelve inches broad, and three inches thick, with six holes for the hooks to go through, is fixed above the cheeks upon cleats.

TABLES, in mathematics, are systems or series of numbers, calculated to be ready at hand for expediting any sort of calculations in the various branches of mathematics.

Astronomical TABLES, are computations of the motions, places, and other phenomena of the planets, both primary and secondary.

The oldest astronomical tables now extant, are those of *Ptolemy*, found in his *Almagest*; these, however, are not now of much use, as they no longer agree with the motions of the heavens.

Many attempts were made by astronomers in different countries, to perfect a set of tables; but with little success, as appears by the historical account drawn up by that eminent mathematician, Dr. HUTTON, in his *Mathematical and Philosophical Dictionary*, till about the beginning of the 18th century, when De la Hire, constructed wholly from his own observations, and without the assistance of any hypothesis, the *Ludovician* tables, which, before the invention of the micrometer, telescope, and the pendulum clock, was held impossible. 'Soon after this, Dr. Halley laboured hard to perfect another set, which were printed in 1719, but not published till 1752.

M. Monnier, in 1746, published, in his *Institutions Astronomiques*, tables of the motions of the sun and moon, with the satellites, as also of refractions and places of the fixed stars. De la Hire also published tables of the planets, and La Caille tables of the sun: Gael Morris published tables of the sun and moon, and Mayer constructed tables of the moon, which were published by the Board of Longitude. Tables of the same kind have also been computed by Mr. Charles Mason, from the principles of the Newtonian philosophy, which are found to be very accurate, and are employed in computing the nautical ephemeris. Many other sets of astronomical tables have also been published by various persons and academies; and divers sets of them may be found in the modern books on astronomy, navigation, &c., of which those are esteemed the best and most complete that are printed in LALANDE'S and VINCE'S *Astronomy*.

For an account of several, and especially of those published annually under the direction of the Commissioners of Longitude. See the article *Nautical* ALMANACK, LONGITUDE, LUNAR *Observations*, and REQUISITE *Tables*.

TABLES *of the Sun*. The solar tables are those from which are computed the sun's semi-diameter, horizontal parallax, hourly motion in latitude, longitude, right ascension, and declination; by which we convert apparent into mean time, and make all calculations for mean time, such as finding the sun's place, &c. M. Monnier published tables of the motions of the sun in his Institutions Astronomiques; afterwards La Caille, Gael Morris, and the celebrated Mayer, also computed solar tables, which have since been improved by Mason, from the principles of the Newtonian philosophy, and are now found very correct: but the most complete set of astronomical tables are those constructed by M. De Lambre, from the observations of Dr. Maskelyne, and the theory of La Place. See *Les Memoires de l'Académie de Berlin*, for 1784 and 1785, published in VINCE's *Astronomy*, vol. 3d.

TABLES *of Mercury*, contain the epochs of the mean motion in longitude reckoned from the mean equinox for years, days, hours, minutes and seconds, with the equation of the orbit of Mercury for any degree of mean anomaly, with their differences, and also the differences of the logarithms of the distance of Mercury from the sun.

TABLES *of Venus*, are in every respect similar to those of Mercury, and the use of the tables are the same.

TABLES *of the Moon*. See the article LUNAR *Tables*.

TABLES *of Mars*, contain the epochs of the mean longitude of the aphelion and node of Mars, with the mean motion for years, days, hours, minutes, and seconds, and were formed on the same principle as those of the sun, and adapted to the situations of Mars and Jupiter, at the beginning of each respective year.

TABLES *of Jupiter*. All the equations of the motion of Jupiter, that of the orbit excepted, arise from the attraction of Saturn, and depend on the relative situations of Jupiter and Saturn. The theory of these equations was given by M. De La Place, in the Memoirs for 1786, from the theory of gravity, and were computed by M. De Lambre, and may be found in VINCE's *Astronomy*, vol. 3d.

TABLES *for the Satellites of Jupiter*. The best tables for computing the eclipses of Jupiter's satellites, are those computed by M. De Lambre from the theory of the mutual attraction of the satellites, by M. De La Place, in the *Memoires de l'Académie*, for 1784 and 1788; the forms of the equations were there given, but their values were found from observation.

TABLES *of Saturn*, are in all respects similar to those of Jupiter, the theory of the equations was given by M. De La Place, in 1785 and 1786, and were computed by M. De Lambre.

Planetary TABLES. If the planets moved uniformly in circles, about the sun in the centre: after ascertaining their periodical times, their angular velocities in any given time would be known; for the periodical time is to any given time, as 360° is to the angle described in that time: hence if tables were constructed for each planet shewing its angular motion, or what is called simply its motion, for years, days, hours, minutes and seconds; then if their places were once determined for any given instant; for instance, at the beginning of some one year, called an epoch, their places at any other time might be immediately found, by applying to their places at that epoch, their motions from that time, to the given time: but as their motions are not uniform, tables are constructed to shew the difference between their true places, and the places where they would have been if their motions had been uniform, or their mean places: if therefore we first calculate the mean place, and then apply that difference; we get the true place. This difference is made up of several parts called equations, and these arise from the elliptical term of the orbits, and the mutual attraction of the heavenly bodies.

TABLING, in mast-making, the act of uniting pieces of timber together, in a manner similar to the chain-coak, but broader, and comes quite to or very near the edge, from the line in the middle.

TABLING *of the Beams*, (*entailles et adents faits dans l'assemblage des baux*, Fr.) in ship building, the act of letting one beam-piece into another, after the scarphs are trimmed out for that purpose.

The common way of putting beams together, is to divide the tablings in the middle of the beam, and that part which is taken out at the upper side to be left at the lower-side: when done in this manner, especially when kersey or flannel is put in the scarphs, the water is liable to lay in the scarphs, and must be the means of rotting the beams.

The method practised in our dock-yards is, to unite the beam-pieces together by dowelling, and putting tar paper between them, which hardens the wood, and the water occasioned by the leaking of the decks has a free passage, and the beam drys again. This method is quite equal to the other in point of strength, and is sufficient for the service required.

TABLING *of the Sails*, (*tablier et renfort des voiles*, Fr.) the act of making broad hems on the skirts and bottoms of a ship's sails, to strengthen them in that part which is attached to the bolt-rope.

TACK, (*amure*, Fr.) a rope used to confine the foremost lower corners of the courses and stay-stails in a fixed position, when the wind crosses the ship's course obliquely. The same name is also given to the rope employed to pull out the lower corner of a studding-sail, or driver, to the extremity of its boom.

The main-sail and fore-sail of a ship are furnished with a tack on each side, which is formed of a thick rope tapering to the end, and having a knot wrought upon the largest end, by which it is firmly retained in the clew of the sail: the tack, therefore, extends the sail to windward, whilst the sheet extends it to leeward. See the article CHESS-TREE.

TACK *of a flag*, a line spliced into the eye at the bottom of the tabling, for securing the flag to the halyard.

TACK *of a Sail*, (*amure d'une voile*, Fr.) is also applied, by analogy, to that part of any sail to which the tack is fastened.

A ship is said to be on the starboard or larboard tack, when she is close-hauled, with the wind on the starboard or larboard side: and in this sense the distance which she sails in that position is considered as the length of the tack; although this is more frequently called a BOARD, which article see.

To TACK, (*virer vent devant*, Fr.) to change the course from one board to another, or to turn the ship about from the starboard to the larboard tack, or *vice versa*, in a contrary wind. Thus the ship A, being close-hauled on the larboard tack, and turning her prow suddenly to windward, receives the impression of the wind on her head-sails *a*, by which she falls off upon the line of the starboard tack *a*. Tacking is also used, in a more enlarged sense, to denote that manœuvre by which a ship makes an oblique progression to windward, in a zigzag direction. This, however, is more usually called BEATING or TURNING *to windward*, which articles see.

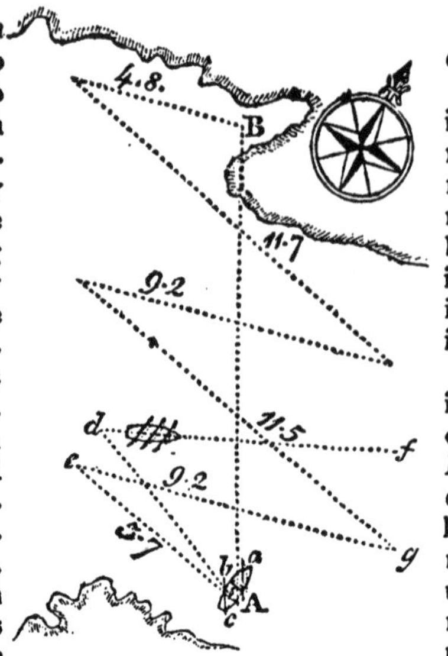

Thus, suppose a ship A, bound to a port B, lying to windward, with the wind northerly, as expressed by the arrow. The sails *a*, *b*, *c*, being braced obliquely with the keel, the wind also falls upon their surfaces in an oblique direction, by which the ship is pushed to leeward, as explained in the article LEE-*Way*. Hence, although she apparently sails W.N.W. upon the larboard tack, as expressed in the dotted line A *d*, and E.N.E. upon the other, *df*, yet if the lee-way is only one point (and indeed it is seldom less in the smoothest water), the course will accordingly be W. by N. upon one tack, and E. by N. upon the other, as represented by the lines A *e*, and *eg*.

If the port A were directly to windward of the ship, it is evident that both tacks ought to be of equal length: or, in other words, that she ought to run the same distance upon each tack: but as the place of her destination lies obliquely to windward, she must run a greater distance upon one tack than the other; because the extremities of both boards should be equally distant from the line of her true course BA; so the larboard tack A *e*, crossing the course more obliquely than the other, *e g*, will necessarily be much longer.

As the true course, or the direct distance from B to A, is only 12 leagues, it is evident, that with a favourable wind she could reach it in a few hours. On the contrary, her distance is considerably increased by the length of her boards, in a contrary wind; which, by its obliquity with her sails, operates also to retard her velocity. Thus, her first board, A *e*, on a W. by N. course, is equal to 5. 7 leagues. The second tack, *eg*, is 9. 2 leagues E. by N.: the third tack, parallel to A *e*, is 11. 5: the fourth, parallel to *eg*, is 9. 2: and the fifth, parallel to the first, 11. 7 leagues. Finally, the sixth board is 4. 8 leagues parallel to the second, which brings her to the port B. By this scheme it appears that she has run more than four times the extent of the line A B, her primitive distance; and this in the most favourable circumstances of a contrary wind, viz. when the sea is smooth, and when she may carry her full top-sails. For if the wind blows stronger, to render it necessary to reef the topsails, she will soon make two points of leeway, and accordingly run east on one board and west on the other. In this situation she will neither approach nor recede from the place of her destination: but if the wind increases, the sea will also be enlarged; a circumstance that still further augments the leeway. Hence the vessel will gradually fall off from the port, in proportion to the augmentation of the wind and sea, which occasion a proportional increase of lee-way.

In order to explain the theory of tacking a ship, it may be necessary to premise a known axiom in natural philosophy, 'That every body will remain in a state of rest, or move uniformly in a right line, unless it be compelled to change its state by forces impressed; and that the change of motion is proportional to the moving force impressed, and is made according to the right line in which that force is exerted.'

By this principle it is easy to conceive how a ship is compelled to turn into any direction, by the force of the wind acting upon her sails in horizontal lines. For the sails may be so arranged as to receive the current of air, either directly, or more or less obliquely: hence the motion communicated to the sails must of necessity conspire with that of the wind upon their surface. To make the ship tack, or turn round with her head to windward, it is therefore necessary, after she has received the first impression from the helm, that the head-sails should be so disposed as to diminish the effort of the wind, in the first instant of her motion, and that the whole force of the wind should be exerted on the after-sails, which, operating on the ship's stern, carries it round like a weather-cock. But since the action of the after-sails, to turn the ship, will unavoidably cease when her head points to windward, it then becomes necessary to use the head-sails, to prevent her from falling-off, and returning to her former situation. These are accordingly laid aback on the lee-side, to push the vessel's fore-part towards the opposite side, till she has fallen into the line of her course thereon, and fixed her sails to conform with that situation.

It has been observed above, that the first effort to turn the ship in tacking, is communicated by the helm, which is then put to the lee-side. This circumstance being announced by the commanding officer, who then calls out, Helm's a-lee! the head-sails are immediately made to shiver in the wind, by casting loose their sheets or bowlines. The officer then calls, "Raise up tacks and sheets!" which is executed by loosening all the ropes which confine the corners of the lower sails, in order that they may be more readily shifted to the other side. When the ship has turned her head directly to windward, as in *d*, the order is given to turn about the sails on the main and mizen-masts, by the exclamation, "Haul main-sail, haul!" the bowlines and braces are then instantly cast off on one side, and as expeditiously drawn in on the other side, so as to wheel the yards about their masts: the lower corner of the main-sail is, by means of its tack, pulled down to its station at the chess-tree; and all the after sails are, at the same time, adjusted to stand upon the other board. Finally, when the ship has fallen off five or six points, the commanding officer calls, "Haul off all!" or, "Let go, and haul!" then the sails on the fore-mast are wheeled about by their braces: and as the ship has then a tendency to fall off, she is checked by the effort of the helm, which for that purpose is put hard-a-lee. The fore-tack, or the lower corner of the fore-sail, being fixed in its place, the bowlines are hauled, and the other sails are properly arranged to the wind; which exercise is called *trimming all sharp*. See the articles Lee-*way* and Sailing.

Tack-Tackle. See the article Tackle.

TACKLE, (*palan*, Fr.) pronounced *taicle*, a machine formed by the communication of a rope with an assemblage of blocks, and known in mechanics by the name of pulley.

Tackles are used in a ship to raise, remove, or secure weighty bodies; to support the masts, or to extend the sails and rigging. They are either moveable, as communicating with a runner, or fixed, as being hooked in an immoveable station; and they are more or less complicated, in proportion to the effects which they are intended to produce.

Pl. XXIII. Fig. 3. If *a b d e*, fig. 3, plate XXIII, be a single block, upon which are suspended the weights *f g*, then since the nearest distance of the ropes *f g*, from the centre of motion *c*, and *a c* equal to *d c*, the block will be reduced to the lever or balance *a d* with respect to its power; since *a c* is then equal to *d c*, it is apparent that *f g* will always be *in equilibrio*. As no advantage therefore can be acquired, in raising a weight by an immoveable single block, it is only rendered useful by changing the direction of the moving power. This circumstance is extremely convenient to the labourers, and often absolutely necessary, particularly in raising bodies to a higher station, as from the hold to the upper decks, or from the deck to the masts or yards, &c. which would otherwise be difficult or impracticable to perform. See also the articles Block and Whip.

Fig. 4. When a single block is moveable along with the body to which it is attached, fig. 4, as the blocks of the brace-pendents, reef-tackle, pendents, jiggers, &c. the *momentum* of the power is doubled; because it moves twice as fast as the weight, or body to which it is attached. For in the same time that any part of the rope *f*, moves upward from *f* to *g*, equal in length to the two equal ropes *d* and *c*, the block, and consequently the weight annexed, will be drawn through the space *e h*, whose length is equal to one of the ropes only.

Fig. 5, 6, and 7. When a tackle consists of two or more fixed and moveable blocks, wherein one rope communicates with the whole; if one end of the rope be fixed, as in figures 5, 6, and 7, in order to proportion the weight to the resistance, the power applied must be to the weight, as one, to twice the number of sheaves in the moveable blocks: because, in the efforts of a tackle, the velocity of the moving power is to the velocity of the rising or moving body, as twice the number of moveable sheaves to unity, as appears in fig. 5, which consists of one fixed block *a*, and another moveable as *e*. For since one rope operates on all the sheaves from *g* to *f*, fig. 3, the part at *f*, lying beyond the fixed block, and called the fall, cannot be drawn down and lengthened, unless the two parts *d* and *c*, on each side of the moveable block, be at the same time equally drawn up and shortened. Hence it is evident, that the part *a f* will be lengthened twice as much as either *d* or *e* is shortened, because whatever is taken from each of those parts is added to the length of *a f*; but the point *f*, to which the power is applied, descends as fast as *a f* is lengthened; and the point *e*, to which the weight is fastened, ascends as fast as *d* or *c* is shortened. If, therefore, a weight suspended at *f*, be to a weight suspended at *c*, as one to two, they will balance each other, as being in the reciprocal ratio of their velocities.

Whatever has been observed with regard to the tackles above-mentioned, is equally applicable to all others, and is in the same manner demonstrable, viz. that the velocity with which the mechanical force moves, in raising a weight, is to the velocity wherewith the weight rises, as twice the number of moveable sheaves to unity.

A tackle wherein one of the blocks is moveable, is represented by fig. 8. That part of the tackle which is fixed to one of the blocks, is called the standing part; all the rest are called running parts; and that which is applied to the capstan to be hove upon, is called the fall. It is used in sheer-hulks for masting ships. The application of the tackle to mechanical purposes is termed hoisting or bowsing. See those articles. Fig. 8.

Boom Tackles, are tackles composed of double and single blocks, strapped, with tails, and used in getting the studding-sail-booms in or out.

Bowline Tackle, is composed of a long tackle, and a single block, strapped, with a hook and thimble, and used to bowse up the main bowline when the ship is upon a wind.

Burton Tackle. See the article Burton.

Fish Tackle, (*palan de candelette*, Fr.) is composed of a long tackle, and a single block, strapped, with eyes, and used with a pendant, to fish the anchor, and get it into its place. See the articles Fish and Davit.

Garnet Tackle, (*bredindin*, Fr.) is composed of a double and single block, strapped, with a hook and thimble: it is hooked to the skiatic-stay in merchant-ships, and used to hoist goods in or out.

Jigger Tackles, are composed, of double and single blocks, strapped, with tails, and are used for topping the main and fore-yards by the lifts, &c. See the article Jigger.

Luff Tackle. See the article Luff.

Out-hauler Tackle, is composed of two single blocks, strapped, with tails, and used to bowse out the jib-boom.

Port Tackle. See the article Port.

Quarter Tackles, are composed of double and single blocks, strapped, with eyes, and the lower blocks with a hook and thimble; they are used to hoist up water, provisions, &c.

Reef Tackles, are composed of two double, or two single, blocks; one of which is spliced into a pendant, and the other is strapped with an eye; they are used to draw the extremities of the reefs close up to the yard-arms, for reefing the sail.

Relieving Tackles. See the article Relieving.

Ridge Tackle, is composed of a double and a single block, strapped, with an eye, and used to suspend the awning in the middle.

Rolling Tackles. See the article Rolling.

Rudder Tackles, are composed of long tackle blocks and single blocks, strapped, with hooks and thimbles; they are used to save or direct the rudder when any accident happens to the tiller.

Runner Tackles, are composed of double and single blocks, and a pendant; the lower blocks are strapped, with a hook and thimble; they are used to set up the shrouds, and to get the mast-heads forward, for staying the masts.

Main and Fore-Stay Tackles, are composed of double and single blocks, strapped, with hooks and thimbles, except the blocks spliced into a pendant.

The pendant formerly travelled on the stay, by iron thimbles; but this is now discontinued in the royal navy, as they much injured the stay by friction. See the article Stay-Tackle.

Stay-Sail Tackles, are composed of double and single blocks; the lower blocks are strapped, with a hook and thimble; they are used to set up the jib, and other stay-sail-stays.

Shifting Back-stay Tackles, are composed of double and single blocks, strapped, with a hook and thimble, and used to set up the shifting back-stays where they may be wanted.

Tack Tackle, (*palan d'amure*, Fr.) is composed of a double and single block, strapped, with hooks and thimbles; it is used for bowsing down the tacks of the principal sails to their respective stations, and is particularly attached to the tacks of the main-sails of brigs, sloops, cutters, and schooners, for the same purpose.

Topmast-stay, and Preventer-Stay Tackles, (*palans des mats de hune et des faux étai*, Fr.) are composed of long tackle blocks, and single blocks; the lower blocks are strapped, with a hook and thimble; they are used to set up the top-mast and preventer stays.

Fore-Topgallant-Stay Tackle, is composed of a double and a single block, and is used to set up the fore-top-gallant-stay.

Stock Tackle. See the article Stock.

Top Tackle. See the article Top.

Train Tackle. See the article Train.

Winding Tackle, is composed of a fourfold and a treble block, or a treble and a double block, strapped, with eyes; it is attached to the winding-tackle-pendant, and chiefly used to get in and out the guns. See also the article Winding.

Yard Tackles, are composed of double and single blocks; the double blocks are spliced into the lower ends of pendants, and the single blocks are strapped with hooks and thimbles; they are used to hoist the boats in or out. See also the article Yard.

Tackle-*Fall*, that end of a tackle which is bowsed on, or the rope which connects the blocks together. The whole assemblage is called a tackle, and is used for stretching the yarn, &c. See the article Fall.

Naval TACTICS, (*la tactique navale*, Fr.) in the art of war, is the method of arranging squadrons or fleets in the order of battle, and of regulating their movements and operations, as also the necessary signals on every occasion, to the greatest possible advantage. In a more enlarged sense, *tactics* denote those manœuvres, stratagems, and deceptions, employed by the commander of a fleet, for the purpose of gaining a weather-gage, cutting off any part of a line, or attacking any particular portion thereof, in such manner as may either defeat the views of a hostile fleet, or subject it to loss and discomfiture.

The old system of *tactics* was burdened with ceremonies, and received opinions, which were held to be inviolable: the difference of one or two ships in favour of the enemy, was considered a sufficient excuse for a variety of precautions, generally amounting to forbearance from engaging the superior power; and, although we certainly can recollect a number of gallant exploits performed by our fleets when somewhat inferior to the enemy, yet it has been reserved for latter times to exhibit what could be done by the British navy, even when opposed to nearly double their own force.

For a new and ingenious system of naval tactics, we are indebted to John Clerk, Esq. of Eldin, Fellow of the Society of Scotish Antiquaries, and of the Royal Society of Edinburgh; who, in the preface of his work, explains a circumstance which has excited the surprise of every reader; namely, how he, a gentleman not bred to the sea, and who never performed even a single voyage, was conducted to discoveries in naval tactics, which had escaped the observation of professional men, and to which the nation is so deeply indebted for its naval victories.

As it will be impossible, from any detached quotation, and unassisted by the plates, to comprehend the whole detail of a system which is explained with mathematical precision, in a series of demonstrations, we shall endeavour to give an accurate analysis of the principles upon which this system is founded, and of the chief illustrations by which it is developed and explained.

Mr. Clerk begins, with observing, that during the greater part of the American war, and during the whole of the two preceding wars, British sailors, in the rencounter of single ships, or of small squadrons, never failed to exhibit the most distinguished seamanship, intrepidity, and perseverance, attended with the most uninterrupted success: yet, wherever large fleets, of ten, twenty, or thirty ships were assembled and *formed in a line of battle*, nothing memorable had ever been achieved, and not a single ship had been lost or won. The defeat of Conflans, and other fortunate occurrences during the same war, form no exception to this general observation; as on most of these occasions, the English possessed a decisive superiority in numbers; and, in the defeat of Conflans, the French, though nearly of equal force, betrayed the utmost irresolution, and fairly run away, without even attempting to sustain an engagement. From the amazing exertions and uniform success of our seamen in lesser conflicts, it cannot be ascribed to any diminution, but to the improper direction of their characteristical intrepidity, that during the whole of that war, our fleets 'were invariably baffled; nay, were worsted, without having ever lost a single ship, or almost a man.' As little can it be attributed to the superior construction or sailing of the enemy's vessels, when we find that they were so frequently overtaken and captured in single engagements. Mr. Clerk, therefore, concludes that the French must have adopted a new system of tactics, which we had not then discovered; and that the method hitherto pursued on our part since it was always unsuccessful, must have been radically wrong.

Mr. Clerk proceeds, in the first part of his work,

to a series of demonstrations on the mode of attack from windward, of which, for the reasons already assigned, it is impossible to give any more than the general result. A single vessel to windward in pursuit of another to leeward, will never bear down, endwise, in a direct line, on the broadside of the other; because she is then exposed to be raked from end to end by the fire of the enemy, and to be disabled from pursuit. She will either bear down a-stern of the other, and continue the pursuit in a parallel line, till she get alongside; or, having shot a-head, she will bear down athwart the other, to intercept her in her course. But a *fleet* to windward has invariably born down in a perpendicular, or at least in a slanting line on another to leeward. each ship in a line of battle abreast of the other, till they brought up, within a proper distance, for a close and general engagement from van to rear. A fleet to leeward, formed in a line of battle a-head of each other, and desirous to avoid a general engagement, had therefore full leisure to disable the other during its approach. And when the latter had assumed a situation for close encounter, the former might bear away at intervals, while enveloped with smoke; or, by making more sail, might shoot a-head, and pour its whole fire into the opposite van, as it passed and wore in succession, to form a new line to leeward, on the opposite tack, where, if the enemy were not already disabled, the same manœuvre might be repeated with the same success. Some idea may be formed of the effect of a raking fire, from the supposition of a column of infantry exposed to the fire of a battery through the whole line. But a ship is infinitely more disabled by the damage sustained in a few shrouds or principal stays, in a yard or topmast, than by the loss of men; and the area which the hull and rigging together present to the shot, appears from our author's calculations to be twenty times larger than the area formed by the decks alone where the men are exposed.

Mr. Clerk, in concluding this part of his demonstrations, says, " after a proper examination of the late sea engagements, or rencounters, it will be found, that the French, have never once shewn a willingness to risk the making of the attack, but have invariably made choice of a leeward position: and, when extended in line of battle, in that position, they have disabled the British fleets in coming down to the attack. Upon seeing the British fleet disabled, they have made sail, and demolished the van in passing: and upon feeling the effect of the British fire, they have also withdrawn at pleasure, either a part, or the whole of their fleet, and have formed a new line of battle to leeward. The French have repeatedly done this upon every occasion: and, on the other hand, it will be found that the British, from an irresistible desire of making the attack, have as constantly and uniformly courted the windward position, and have repeatedly had their ships so disabled and separated, by making the attack, that they have not once been able to bring them to close with, to follow up, or even to detain one ship of the enemy; therefore, there was every reason to believe, that the French had adop ed, and put in execution, some system which the British either had not discovered, or had not yet profited by the discovery "

That such was the new system adopted by the French, to preserve their own ships, while they disabled ours, and such the mode of attack to which the English uniformly adhered, is illustrated by the details of a variety of naval engagements, from Admiral Byng's in the Mediterranean, 1756, to Admiral Graves's rencounter off the Chesapeak, 5th of September, 1781. In Byng's unfortunate engagement, the British having weathered the French fleet, edged down in a slanting or oblique line to bring the latter to close action from van to rear. The headmost ships suffered a raking fire, and received three broadsides, before they could reach their stations to return a shot. The sixth ship in the line was disabled by the loss of a topmast; and from the interruption which she occasioned to the line, the van was separated from the centre and rear. The van of the French fleet bore away at intervals, amidst the mistaken shouts of our seamen, as soon as it felt the effects of our fire. The centre and rear, by making more sail, poured their whole fire with impunity into our five headmost ships, as each vessel ranged along our van: and as they bore away in succession, they formed a new line three miles to leeward; while our van which had sustained the whole action, was too much disabled for the fleet to renew such a disadvantageous attack.

In Pocock's engagement in the East Indies, two years afterwards, a similar mode of attack, and the same system of defence were employed upon each side, and with the same success. So early had the French adopted a defensive plan, which preserved their own fleet, while it disabled ours, and which, in Byng's engagement, prevented the relief of Fort St. Philip.

Admiral Byron's engagement off Granada, on the 6th of July, 1779, is described as similar to Byng's, in almost every respect. Our fleet bore down from windward, in the same oblique line; but as the enemy kept bearing away, we were unable either to bring their rear into action, or to produce a close engagement in the van. Our headmost ships were either disabled in making the attack, as they received the whole fire of the enemy's line, as each ship of the latter passed and wore in succession, in order to form to leeward upon the opposite tack. The French adhered so closely to this defensive system that, to avoid all danger of a general engagement, they forbore even to intercept our disabled ships which had necessarily fallen to leeward; and their caution was rewarded with the capture of Granada.

Admiral Arbuthnot's engagement off the Chesapeak, exhibits a repetition of the same manœuvres, with this remarkable addition, that the French fleet, which had the weather-gage, being apprehensive of an engagement in that situation, run down and formed to leeward of the British line. Accordingly, our head most ships were so much disabled in bearing down to engage, that when the enemy wore as usual, and formed again to leeward, our fleet was unable to renew the attack. In Admiral Graves's engagement off the Chesapeak, 5th September, 1781, the same manœuvre was practised with equal success. While the enemy's van bore away, their centre bore up in passing, not only to protect their own van, but to pour their whole fire successively into ours.

The last instance of an unsuccessful attack from

windward, which our author has produced, is Lord Rodney's engagement off Martinico, on the 17th of April, 1780. His lordship's first design was to attack the rear of the enemy with his whole force. The French Admiral, however, discovering the meaning of the signal, wore and formed on the opposite tack; and the manœuvre, though it was still practicable, was exchanged for a general attack upon the whole line. Notwithstanding the personal gallantry of Lord Rodney, and the example of close action given by the Sandwich, the French fleet bore alternately away and escaped; while the English, from the damage sustained in the hulls and rigging, were unable to continue the pursuit that night.

Mr. Clerk next proceeds to shew, that wherever the French, in opposition to their usual practice, had kept to windward, as if aware of the peculiar danger of their situation, they were careful never to make the attack themselves. Their anxiety to preserve a secure distance, is illustrated by Rodney's two engagements on the 15th and 19th of May, 1780, to windward of Martinico; by Sir Samuel Hood's engagement on the 17th of April 1781, off Martinico; and by Admiral Keppel's in 1778, off Ushant; in each of which, a smart cannonade was maintained, while the two adverse fleets passed each other upon opposite tacks. In the last of these engagements, the French fleet, as in the subsequent engagement with Arbuthnot, off the Chesapeak, run down and formed to leeward, after having passed the fire of the British line.

Such was our naval situation in the beginning of the year 1782, when the first part of this work appeared. During the whole war, our fleets had invariably been baffled, disabled, and worsted; in fact, in every general engagement, *without the loss of a single ship on either side, or almost of a man.* Our admirals adhered invariably to the established mode of attack, and endeavoured to obtain a windward position before they began to engage. Each ship steered directly upon her opponent in the adverse line, and brought up in order to produce a general engagement from van to rear. In this situation, our admirals could not avail themselves of the superior skill, perseverance, and spirit of our seamen; nor, in bearing down, could our ships retaliate upon the enemy with a single shot. The French, relying upon our want of penetration to discover, or of skill to counteract this new system of defence, never fail to accomplish the object of their expedition, and to disable our ships, while they preserved their own. Dispirited by the failure of our arms in the American war, we beheld ourselves uniformly baffled in our own element; and we began to apprehend a decay of spirit in our officers and seamen, when we reflected upon the victories obtained at Messina and La Hogue, upon our obstinate battles with the Dutch in the preceding century, and on the glorious annals of Queen Elizabeth's reign.

Mr. Clerk's merit would have been sufficiently conspicuous, had he merely discovered this concealed system of French tactics, which, during three successive wars, had escaped the penetration, or at least had eluded the skill of our naval commanders. But he has also taught us how to counteract this system, and to bring the enemy, in every situation, to a close engagement, in which the superiority of our seamen might be successfully exerted. Instead of the former erroneous mode of attack, the system which he has substituted, possesses this peculiar merit, that it cannot avail our enemies even when divulged. That innate spirit and habitual skill, which have rendered our seamen superior in every close encounter, must first be communicated to our enemies, before the two modes of attack which he proposes, and which have been successfully adopted, from windward and from leeward, can be employed against us even upon equal terms. We will proceed to explain concisely the principles of each mode of attack, without attempting to follow the minute and accurate demonstrations which the author has given, of the various situations to which they may extend.

The mode of attack proposed, whether from windward or from leeward, is founded upon the same principle which a skilful general adopts in engagements by land; that is, to exert the chief force against the weakest or most vulnerable part of the opposite line. The impossibility of carrying the whole fleet by a general attack, has been sufficiently proved. Instead, therefore, of bearing directly down to stop the van, of which the result has been generally unsuccessful, and always indecisive, our author demonstrates that a fleet to windward, arranged in three divisions as the service may require, should continue the pursuit, like a single ship, in nearly a parallel line of approach; and should confine the attack to as many vessels as it can reach and cut off in the centre or rear. Admitting even the superiority of the enemy's sailing, the swiftest vessels to windward will necessarily outsail and intercept the heaviest in the fleet to leeward, which last must either abandon its rear, or return to hazard a general and close engagement, which it has endeavoured to avoid. But, in whatever manner the enemy shall attempt to return, whether the whole fleet shall double round in succession, or each vessel shall tack or wear in the line, the time and course required for the execution of the movement, will increase the distance between the van and centre, and that part of the line which is intersected and attacked. If they tack or wear in the line, the ships are exposed to a raking fire, while the windward divisions of the opposite fleet are ready to interpose, and to prevent their junction with the rear. If, anticipating the intended attack upon their rear, as in Rodney's first engagement off Martinico, they would endeavour to avoid it by wearing round, in order to pass on the opposite tack, the fleet to windward may either bear down athwart their course, and bring the headmost ships to a close action, or may renew the attack with the same success against their former van, which is now their rear. Or, if they should endeavour to escape before the wind, their retreat must soon degenerate into an absolute flight, in which the slowest vessels must always be overtaken by the fleet in pursuit.

The mode of attack, from leeward occupies the second part of the work. As the course of a fleet in working to windward is in angular lines, if the advantages upon each side were equal, the distance between two fleets working to windward would continue always the same. But the least disaster to the masts or rigging, even of a single vessel, retards the progress of a fleet to windward, lest that vessel should fall to leeward into the hands of the enemy. Every alteration in the wind enables the fleet to leeward to weather the other; and as these accidents are unavoidable in cruising, Mr. Clerk concludes, that, in the course of a few days, a fleet to leeward must fetch some part at least of the opposite line. Were they to meet on the same tack, the engagement might continue while they held the same course; but the fleet to windward, when desirous to avoid an engagement, has invariably passed the other on an opposite tack. Our fleet, instead of continuing its former course, invariably bore away, when to leeward, in order to engage, as it ranged along the opposite line; but the two fleets, as their courses were different, were necessarily separated after a short cannonade. As the two fleets in Keppel's engagement off Ushant, passed each other at the rate of five miles an hour, it appears that each ship was engaged for no more than a quarter of a minute with each opponent, and was exposed but for six minutes and a half to the whole fire of the adverse fleet.

"Let any one imagine," says Mr. Clerk in his observations upon this engagement, "a rencounter of horsemen, where the parties on coming to the ground appointed, had pushed their horses at full speed, exchanging a few pistol shot as they passed one another in opposite directions, at the distance of forty or fifty yards, and then some idea may be formed of the effect of rencounters, where adverse fleets are brought to pass each other on contrary tacks." He concludes that "by such investigations only can it be explained, how two adverse fleets, amounting to thirty ships of the line each, carrying above 36,000 men, after having been brought in opposition of battle, and sustaining a furious cannonade from 4000 guns, besides musketry, how, I say, they have been brought to be separated again without effect, without the smallest apparent decision, that is, without the loss of a ship on either side, and sometimes without the loss of a man, although the rencounter has often been said to have been within pistol shot." p. 108.

Mr. Clerk's mode of attack from leeward, seems to have been first suggested by the examination of this unsuccessful engagement. If our fleet, instead of bearing away when it reached the enemy's line, had continued close to the wind, in its former course, the opposite line must have been *cut in twain* and intersected by ours; and either the rear, when thus separated and obstructed in its course, must have been abandoned by the van, or a close and general engagement must have ensued. He demonstrates, by a variety of examples, that whether the attack be directed against the centre or rear of the enemy's fleet, whether it be made by the van or centre of the fleet to leeward, the portion of the line which is then intersected, and cut off from the rest, must be forced to leeward, and taken or destroyed. When the attack is made by the centre, the headmost ships bear away as usual and engage to leeward; those of the centre pierce the opposite line, and stretching to windward, place the enemy between two fires; the rear remains interposed to obstruct the course of the enemy's rear, or the return of their van. When the line is thus intersected and broken, the van and rear present two distinct objects of attack; but as these objects are not both to be accomplished, the attack of the rear, which is already far advanced, and where the success is certain, is shewn to be far preferable to that of the van, which requires a long pursuit. The trials of Mathews, Byng, Keppel, and Palliser, had been instituted to punish the authors, if not to discover the causes, of our miscarriages by sea. During the whole of these trials, there was no idea, nor indeed the most distant surmise entertained, that the established modes of attack were in themselves injudicious, and incapable of being made effectual. Our officers were eminently distinguished by their gallantry and seamanship; but they had hitherto bestowed no adequate degree of attention upon naval tactics. The truth undoubtedly is, that the first idea of cutting the line originated with Mr. Clerk. The outlines of his mode of attack from leeward are contained in his observations upon Keppel's engagement, which were written soon after the 27th of July, 1778, and communicated to a few friends.

In January 1780, when he was in London, being fully impressed with the importance of the naval ideas, which long had been working in his imagination, and in consequence of the strictures on Lord Keppel's engagement which had been sent to the Admiralty the year before, some appointments for the purpose of farther communication on this subject were made by his friends, among whom was an officer of distinguished merit, afterwards Sir Charles Douglas, then first captain to Sir George Rodney, with Sir George Rodney, who was in London, and about to set out to take the command of the fleet in the West Indies. At this meeting the whole of his acquisitions on the subject of naval tactics, for many years back, was discussed. Lord Rodney himself at all times acknowledged the communication; and having from the first approved of his system, declared, before he left London, that he would strictly adhere to it in fighting the enemy.

In the engagement off Martinico, April 17th, 1780, it was evidently his first intention, to execute our author's mode of attack from windward, against the rear of the enemy; but when that was exchanged for an attack upon the whole line, his fleet was baffled as usual, and disabled from pursuit.

In the subsequent engagements of the 15th and 19th of May, his lordship still hesitated to execute a new mode of attack; and, during a short and distant cannonade upon opposite tacks, he passed to leeward of the French fleet, which had weathered his van. Upon the 12th of April 1782, the British fleet was still to leeward, and the two fleets were on opposite tacks. The van bore away along the opposite line; and had it been followed by the centre, nothing memorable would have been achieved that day. But the Formidable, the Admiral's ship in the centre, kept close to the wind. On perceiving an opening near the centre of the enemy, Rodney broke through at the head of the rear division, and gave the first example of cutting the line. All the consequences predicted by our author immediately ensued. The rear of the French fleet was driven to leeward in the utmost confusion, and torn to pieces by a raking fire. The van and centre, instead of attempting to rejoin their rear, fled in different directions, under a press of sail. The most unbounded praise is undoubtedly due to the gallant Rodney; nor can it detract in the least from his merit, that he was unable to avail himself to the utmost of a mode of attack never practised before. His van, unprepared, perhaps, to improve the attack, continued to stretch and to tack to windward, leaving an opening through which the rear of the enemy was permitted to escape. Abandoning the proper object of attack, namely, the enemy's rear, the proximity of which would have rendered its destruction unavoidable, he directed the pursuit of his whole fleet against the distant van; and after a chase of five leagues, which continued till sunset, five ships of the line were captured, but the rest were preserved by the approach of night.

From this first execution of our author's system, a new æra has been fixed in the history of our naval transactions. During three successive wars, no decisive engagement had almost ever happened, till Rodney, in the execution of our author's system, gave the first example of cutting the enemy's line. Since that period, no engagement has ever proved indecisive; and, with the exception of the battle of the Nile, where the French fleet was at anchor, the same manœuvre has been uniformly practised with the same success. Three days before Lord Howe's distinguished victory on the 1st of June 1793, the signal was twice given (the first, we believe, that was ever given) for the British fleet to leeward to tack in succession, to cut and pass through the opposite line; but from the misconduct of the Cæsar, in neglecting to keep to the wind, the Admiral's ship in the centre, with her two seconds only, cut the line, while the rest of the fleet passed to leeward, having tacked before they were sufficiently advanced. On the 1st of June, the Queen Charlotte, Lord's Howe's ship, cut the French fleet in the centre, between the Admiral's ship and her second; and when the two fleets were thus intermixed together, the superiority of our British seamen, decided the event of that important day. So sensible were the French of the cause of our victory, that, if we recollect right, the convention, upon the report of Jean Bon St. André, passed a decree of death against that captain who should suffer the line to be cut.

In Lord St. Vincent's engagement on the 14th of February 1797, to which the perpendicular attack, or attack at right angles, (p. 194) seems to allude, the Spanish fleet, amounting to twenty-seven sail of the line, was discovered at day-break, extending from windward. By carrying a press of sail, his lordship, with fifteen ships of the line, "disregarding the regular system," intersected and cut off the division to windward, of which four were taken, before the remainder of the fleet to leeward could return to their relief.

In Lord Duncan's victory off Camperdown, his lordship, on discovering the Dutch fleet to leeward, bore down, not in a line of battle abreast, but in two divisions or perpendicular lines; and his attack was directed, not as formerly, against the enemy's van, but, according to Mr. Clerk's system, against the centre and rear. His own division cut the line between the eighth and ninth ships. Onslow's division passed between the fourteenth and fifteenth ships from the van; and while the former, or perhaps the Venerable, and her two seconds kept the van at bay, the rest engaged with the centre and rear. By departing from the customary line of battle, and confining the attack, or the chief force, to a portion of the enemy's line, six ships of the van escaped; but the centre and rear, with the exception of a single ship, were overpowered and taken.

Mr. Clerk's Essay on Naval Tactics was the first original scientific treatise published on that subject in this kingdom: all the other essays that appeared in Great Britain prior to it, being either translations from the French, or remarks upon French authors.

"Our affairs at sea," says Mr. Clerk, "soon after spring, 1782, took a different turn; and I have since had the great satisfaction to see, by the adoption of my system, a decided and permanent superiority given to our fleets. I shall say nothing, in this place, of the brilliant enterprise of Lord Hood, for the relief of St. Christopher's, the account of which arrived about this time. The public joy on this glorious occasion had not subsided, when intelligence came of the memorable and glorious victory gained by Lord Rodney, upon the 12th of April, 1782; a victory far more decisive and important than any which had been gained by our fleets during the last century. The general exultation was excessive; and I flatter myself I could distinctly perceive, even from the first accounts of the engagement, that the victory was owing to the adoption of my system; and especially to the manœuvre of cutting the enemy's line in attacking from leeward.

"Sir George Rodney himself, when he arrived in Britain, made no scruple to acknowledge, that I had suggested the manœuvres by which he had gained the victory of the 12th of April, 1782; and the system on which it proceeded might have remained unknown and unexplained, and perhaps it would not have been followed in other instances, had not my Essay attracted the notice of the navy: for the manœuvre was so new and uncommon, and so little agreeable to the former practice, that its adoption by Sir George Rodney, as well as its consequences in that instance, must naturally have been ascribed to accident or good fortune, more especially as Sir George had not, on former occasions, departed from the old rules." In contemplating the beautiful simplicity and unquestioned efficacy of Mr. Clerk's system, it is peculiarly pleasing to reflect, that it is constructed upon principles not less congenial than honourable to the character of the nation for whose use it was intended. This system does not consist in any trick or manœuvre, by which courage may be rendered unnecessary, or gallantry decoyed to its destruction; on the contrary, it proceeds upon the proud presumption, that we shall certainly beat our enemies, if we can only get near enough to grapple with them: and its sole object is to give the valiant an opportunity of fair fighting, to counteract the shifting policy of a wily adversary, and to ensure a fair field for the display of courage, discipline, and perseverance in strenuous and decisive contest.

The nation at large is indebted to his genius, and should be proud and forward to acknowledge and to discharge the obligation. His situation in life, we understand renders him independent of patronage, and his character leads him to disdain any honour that is not offered to his acceptance.

Since the year 1782, Admiral Sir John Jervis, by a skilful manœuvre, cut off a large portion of a Spanish fleet, near Cape St. Vincent (whence the peerage bestowed on him received its designation); but the late Lord Nelson appears most conspicuous in that mode of attack which, in general, secured a victory. The battle of the Nile was doubtless a master-piece of tactical science; it merits notice from its simplicity, and, if we may be allowed to use the term, its infallibility. The manœuvre he used was, to throw two of his ships upon every one of the weathermost of the enemy's line, by causing his fleet to divide as it approached them; consequently including each French ship between two of ours. The residue which were moored in a line a-head, fully expected to see ours range up their whole length, and oppose ship to ship. They saw their error when it was too late; being to leeward, it was impossible for them to render efficient aid, and they fell in detail; with the exception of a very small portion, which escaped by putting out to sea, whither we were not in a condition to follow with any hopes of overtaking them.

In the famous battle of Trafalgar, in which the immortal Nelson quitted his earthly frame, the combined fleets were drawn up in the form of a crescent, and awaited our attack, which was made in a double column, apparently bearing down on their centre. This novel mode of coming into action kept the enemy completely in suspense: it threatened every part of their line. If our two columns had turned the same way, they would have been able to do infinite damage in that quarter, before the other wing of the enemy could come up to succour their overpowered friends: if the two columns should cut through the centre, they must destroy it, and effectually separate the two wings, so as to leave them ignorant of each other's fate. Such was the fact: the enemy, though superior in numbers, lost no less than nineteen sail of the line.

Perhaps nothing can place a fleet in a more dangerous state, and render it less able to resist an attack, than making sail before the wind, in a line of battle a-head, to avoid a pursuing enemy. In such a case, whenever the rear of that line can be brought to action, it becomes subject to an accumulating force, in consequence of the pursuing fleet thickening upon it; while the van of its own line, being to leeward, must make many tacks, or at least two long ones, before it can succour its rear. The disadvantage must be very great, even if all the ships, on both sides, sail upon an exact equality; but as that is never the case, many of the flying ships will probably be driven completely to leeward, and never be able to afford the smallest assistance. Yet British seamen, even when compelled to retire before a very superior force, generally manage by some well-contrived device, to intimidate their pursuers, or to put on so good a face, as to convince them of the dear price at which the victory is to be bought. Of this we cannot quote a more appropriate instance than the escape of five sail of our ships, under the command of Admiral Cornwallis, from no less than nineteen sail of French ships of the line; an escape resulting entirely from the manœuvres of the British Admiral, whereby he fully convinced the French that a large force was at hand.

In a late French publication entitled "Essais Historique." by an old officer of the royal navy, speaking of naval tactics he mentions, that numerous as are the French publications on the practice of seamanship, none of them treat on the management of a fleet in action. The incidents of battle are indeed so various as to make it a matter of great difficulty to apply general rules; but this circumstance would only confer the greater utility on a work containing detailed accounts of former actions. The author appears to incline more to the English plan of directing a broadside against the hull of a ship than to the customary way among Frenchmen of firing at the rigging when vessels are in close action; because it is very difficult to render a ship unmanageable, and the mischief received in the interval from the enemy's guns is likely to be of the most serious nature. He also comments on the advantage of the method invented by Mr. Clerk, of cutting off a portion of the enemy's fleet, so as to fight the remainder separately, but does not enlarge on the plan to the degree which the frequency of its repetition, and success attending its adoption by our naval officers, during the late wars, appeared to require.

The present unparalleled state of discipline, throughout our navy, would, of itself, give us the command of the ocean; but we are greatly indebted, at the same time, to an excellent code of signals, both for the day and the night, whereby every operation and manœuvre may be directed with readiness and perspicuity. The day signals are, for the most part, made by flags, jacks, and pendants; and the firing of guns, either to windward or to leeward; the night signals are made, by the combination of guns, lights, and false fires, or rockets. When fleets are large, or their duty extensive, especially in cruising in the day-time to intercept a convoy of merchant vessels, &c. there are repeating frigates, which display the several signals made by the commander; so that they may be communicated to all the vessels; every signal being kept flying until answered by all ships to which they may relate. See the article *Naval* SIGNALS.

TAFFRAIL, (*le tableau et le couronnement de la poupe des bâtimens*, Fr.) in ship-building, the carved-work at the upper part of the ship's-stern, as expressed by F. in fig. 2, plate XVII.

TAIL, the long end of a block-strap; hence

TAIL-BLOCK, (*poulie simple estropée à fouet pour divers services*, Fr.) a single block, having a short piece of rope attached to it, by which it may be fastened to any object at pleasure; either for convenience, or to increase the force applied to the said object, as explained in the first part of the article tackle.

TAIL-BOARDS, in ship-building, the carved-work between the cheeks, that is fastened to the knee of the head.

TAIL *of a Gale*, (*fin d'un coup de vent, lorsqu'il a beaucoup diminué*, Fr.) a name given by sailors to the latter part of a storm, wherein the violence is considerably abated.

To TAKE A-BACK. See the article A-BACK.

To TAKE *in a Reef*. See the article REEF.

To TAKE-*In the Sails*, (*serrer les voiles*, Fr.) is to brail up and furl them at sea, particularly when the wind increases. It is generally used in opposition to setting. See the articles To FURL and To SHORTEN.

To TAKE-*In a Course*, (*serrer une basse voile*, Fr.) Man well the weather clew-garnet, ease off the tack and bow-line, and run it up; then man the lee-clew-garnet, bunt-lines, leech-lines, and weather-brace: and, being all-ready, ease away the sheet, haul up the clew-garnet, bunt-lines, and leech-lines, and round-in the weather-brace, till the yard is pointed to the wind. Then haul tight the trusses, braces, lifts, and rolling-tackle, and let the hands furl the sail.

To TAKE-*In the Fore-Sail* in the time of Veering. When the ship begins to veer, the yard being kept braced sharp up, let go the tack and bow-line, and haul up the weather clew-garnet; when the ship is nearly before the wind, the bunt and leech-lines, and the other clew-garnet may be hauled up; and, if the situation admits of it, and occasion requires, the ship may be steered with the wind on the quarter, till the sail is secured.

To TAKE-*In a Topsail.* There are many opinions about the best mode of performing this; some approve of clewing up to windward first, and others to leeward. If the weather-side is to be clewed up first, the weather-brace must be rounded well in, and the yard got close down upon the lifts, otherwise the lee-rigging will be in danger of being carried away by the great pressure of the lee yard-arm. If the weather-brace can be rounded well in, and the yard be got close down, it will be best to clew up to windward first, for thus the sail may be taken in without a shake; but, if the weather-brace cannot be hauled in, to ease the yard of the lee-rigging, recourse must be had to clewing up to leeward first; in this case, it will be best, if hands can be spared, to man both the clew-lines, bunt-lines, and weather-brace, at the same time; thus when the lee-sheet is eased off, the weather-brace may be hauled in with ease, and the yard laid to the wind; and, when the lee clew-line is half up, ease off the weather sheet, and run up the weather clew-line; then haul tight the lee-brace, bowse tight the rolling-tackle, and furl the sail.

To TAKE-*In a Top-gallant Sail.* The lee-sheet must be started first, for if the weather-sheet is first ased off, the yard will fly fore and aft.

To TAKE-*In a Jib.* Man well the down-haul, let go the halyards, ease off the sheet and haul down briskly, when the sail is close down, ease away the out-haul, and haul the sail into the bowsprit-cap; then let it be stowed away in the forestay-sail-netting.

To TALLY AFT THE SHEETS, (*border les basses voiles*, Fr.) a phrase used by sailors, implying to pull aft the sheets or lower corners of the mainsail and fore-sail.

TANGENT, (*tangens, Lat.*) in geometry, is a line that touches a circle or curve, that is, meets it in a point without cutting it, though it be produced both ways *ad infinitum*; as the tangent E I F, of the circle A L I K. The point I, where the tangent touches the curve is called the point of contact, and the tangent E I F, is perpendicular to the radius C I.

A tangent touches a curve only in a geometrical point, but at that point the direction of the curve is the same as that of the tangent.

TANGENT *of an Arc*, in trigonometry, is a right line drawn touching one extremity of the arc, and limited by a secant or line drawn through the centre and the other extremity of the arc: as A G is the tangent of the arc A B I, and A H is the tangent of the arc A I K.

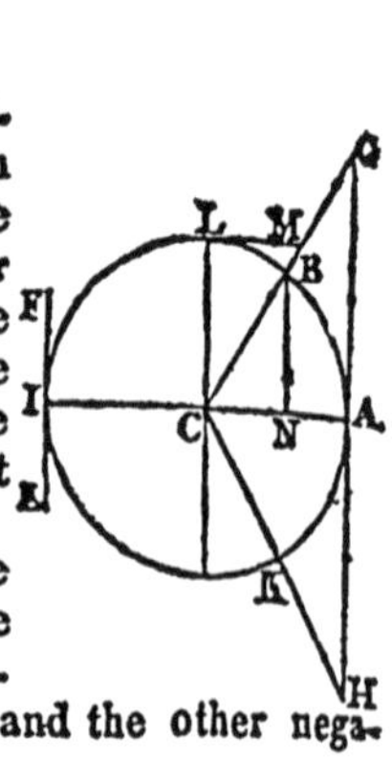

The tangent of an arc and the tangent of its supplement are equal, but of contrary affections, the one being positive, and the other negative.

The tangent of an arc is a fourth proportional to the cosine, the sine, and the radius; that is, C N : N B :: C A : A G. Hence, a canon of sines being made or given, the canon of tangents is easily constructed from them.

Co-TANGENT, contracted from complement tangent, is the tangent of the complement of the arc or angle, or of what it wants of a quadrant or 90°. So L M is the cotangent of its complement B I.

The tangent is reciprocally as the cotangent; or the tangent and cotangent are reciprocally with the radius. That is, tangent is as $\frac{1}{\text{cotangent}}$, or tangent : radius :: radius : cotangent; and the rectangle of the tangent and cotangent is equal to the square of the radius; that is, tangent + cotangent = radius².

Artificial TANGENTS, or logarithmic tangents, are the logarithms of the tangents of arcs, so called, in contradistinction from the natural tangents, or the tangents expressed by natural numbers.

Line of TANGENTS, is a line usually placed on the Sector and Gunter's scale, the description and uses of which, see under the article SECTOR.

Sub-TANGENT, a line lying beneath the tangent being part of the axis intercepted by the tangent and the ordinate to the point of contact.

Method of TANGENTS, is determining the quantity of the tangent and subtangent of any algebraic curve; the equation of the curve being given. This method is one of the greatest results of the doctrine of fluxions. It is of great use in geometry; because that in determining the tangents of curves, we determine at the same time the quadrature of the curvilinear spaces.

IRON TANKS, are square iron vessels lately invented by Mr. Robert Dickinson, of Great Queen-street, Lincoln's Inn Fields, London, for the stowage of ship's water, and other stores, cargoes, &c. for which he has obtained his Majesty's Royal Letters Patent. They may be made of any height, which is regulated by the depth in the hold. The patent iron tank will hold near two tons, if four feet square, or five hundred gallons; however, it can be made to any size, pattern, or sweep, and to fit the sides of the ship, and will last as long as the best built ship.

They will preserve the ship's water perfectly sweet and clear, during the longest voyage, and a double quantity of all ship's cask stores, a double cargo of oil, rum, sugars, and every description of cargo heretofore in casks, may be stowed. The quantity, however, depends on the depth, which is arbitrary; for a first rate the tanks ought to be 8 feet deep and four feet square, in which case she would very well stow 400 or 500 tons of water, and leave plenty of room for dry stores.

The Right Honourable the Lords of the Admiralty, on hearing of the advantage that would accrue from the adoption of the iron tanks, and that water might be preserved perfectly sweet in them, for any length of time, immediately ordered some of them to be tried on board vessels sent on voyages to different quarters of the globe: "the result," says Mr. Dickinson, "was creditable, and accorded with their judgment, and now iron tanks are coming into general use throughout his Majesty's navy.

"The iron tanks are easily filled without the trouble of removing them, and their contents may be pumped out by the common means employed to pump beer from the cellar; consequently, the casualties, ruptures, and deaths, formerly occasioned in the hold by rousing about the wooden casks, and getting up the water every morning watch, and other labour now superseded, numbers of valuable lives are annually saved to this country." See also the articles STOWAGE, *Iron* BUOY and CASK, in the APPENDIX.

TAR, (*goudron*, Fr.) a sort of liquid gum of a blackish hue, which is procured from pines or fir-trees, and is used to pay the sides of ships and boats, and their rigging and yards, in order to preserve them from the effects of the weather, by which they would otherwise soon become cracked, split, or rotten.

It is a known fact that tar is always found to leak, as also oil, black-varnish, &c. above all other things, as if it was from the nature and subtilty of these bodies, coupled with the inability of the barrels, and that they were not to be wholly contained. "This however," says Mr. Dickinson, inventor and patentee of the wrought iron buoys, casks, and tanks, "is not the fact, which will be proved from the following observations.

"Tar is put into barrels at Archangel, generally, of a temperature of about 40 degrees: it is then put into the holds of ships, and, at some seasons of the year, by the time it arrives at London, is at 60°, and sometimes considerably higher: the barrels are then shipped off in vessels to the East and West Indies, where the temperature is generally 96°. That in a change of temperature from 40° to 90°, the tar expands more than four per cent, and to this expansion is attributed the leakage. By taking a cast iron shell, and filling it with tar, then closing the aperture with a screw, and heating it in an open place, the effect of expansion will be readily perceived."

In order to convince the Honourable the Commissioners of the navy of the expansion of tar, Mr. Dickinson was lately induced to make the following experiment before them. He took a glass cylinder 18 inches in height and 4 in diameter: on it he marked a quadrated scale, divided into 20 parts, then filled it to 90°, and placed it in the heat of the sun, till the thermometer, which was placed in the cylinder, rose to 90°, and the tar had expanded from 90° to 94°. To render the thing more evident he took a lump of putty, placed it on the tar, and tied it tight down with a cloth, then perforated and passed through it the end of a long glass tube half an inch in diameter. Thus furnished, he took it to the Navy Office in London, placed it on the sill of their window in the sun, and in a few minutes the tar rose in the tube more than two feet three inches. They were perfectly satisfied with the experiment, and immediately ordered that, at the different dock-yards, a windage of 5 per cent. should in all cases be allowed.

From these experiments one natural conclusion may be drawn, namely, if more than 32 gallons of tar be put into a 34 gallon cask, it is liable to burst. Not that it is necessary for 32 gallons to expand to 34 to produce this effect; but in expanding to $33\frac{1}{2}$, it will compress the two gallons of windage air into half a gallon, and thereby compressing *four* atmospheres into one, (15lbs. being an inch of atmosphere), a pressure of 45lbs. to the inch is inflated, and on the whole internal surface there will be a pressure of 95,760 lbs. as will appear from the following calculation. Suppose the length of the band to be 27 inches, in circumference 60 inches, ends 18 inches each, area of both 508. & $27 \times 60 + 508 \times 45 = 95,760$lbs. the internal pressure, which no cask can bear.

A TAR, or JACK-TAR, (*un matelot, par dérision, comme on dit en François cul goudronné*, Fr.) is a figurative expression for a sailor.

TAR-KETTLE, in rope-making, is made of copper, and holds from ten to twenty barrels of tar. It is set in strong brick-work, and over it is fastened, from side to side, in the direction of the nipper, a bridge made of three-inch oak-plank, thirteen inches broad, through the middle of which is a mortise for the step to go through to keep the yarn down when drawing through the kettle. On that side of the kettle next the capstan is an upright post, twelve inches square, in which is fixed a nipper, to press the tar out of the yarn; and a staff, with a weight suspended at the end, is fixed on the side of the nipper, to keep it down, that the yarn may have no more tar than is necessary.

TAR-PAULING, (*prélat goudronné*, Fr.) a broad piece of canvas payed over with tar, and used to cover the hatchways of a ship at sea, to prevent the penetration of the rain, or sea-water, which may at times rush over the decks; also to cover the blocks at the sheer-heads of hulks, &c. See the article BATTENS.

TARRING OF YARN, in rope-making, is a process which requires particular attention. The person employed in this operation should be extremely careful that the tar does not boil too fast, nor too slow; if too fast, the tar will not stay in; if not fast enough, it will not have sufficient power to penetrate the yarn; therefore, a strict medium must be observed by the kettle-heater, as also, to keep the horse which draws the capstan round, going at a gentle, steady, pace, thereby giving the yarn a proper time to imbibe its necessary proportion of tar; but, at the same time not suffering it to keep in the boiling tar too long, which is apt to make the yarn very tender. Particular attention should also be directed in paying, or coiling, the yarn into the kettle, that too great a length is not payed at once; if so done, the yarn will of course touch the bottom of the kettle, which should be avoided, as it will imbibe the dregs and discolour the yarn. When tarred, it should always be overcast, or removed the same day; if omitted, it will heat and tender the yarn, and must be carefully avoided: in oversetting the yarn, it should be well shook and opened, as it admits the air, separates and hardens the yarns, and contributes much to the strength of the cordage.

TARTAN, (*tartane, sorte de bâtiment de la Mediterranée*, Fr.) a small coasting vessel navigated in the Mediterranean Sea, and having only one mast and a bowsprit, the principal sail, which is extremely large, being extended by a lateen-yard. When the wind is aft, a square-sail is generally hoisted like a cross-jack.

TAUGHT, (*roide, ou tendu, &c.* Fr.) the state of being extended or stretched out, and is usually applied to a rope or sail, in opposition to its being slack.

TAUGHT *Sail*, (*porter beaucoup de voile*, Fr.) implies a great quantity of sail set.

TAUNT, or TAUNT-MASTED VESSEL, (*bâtiment qui à une haute mâture, ou une mâture élévée*, Fr.) an epithet signifying that it is high or tall. It is also expressed of an extraordinary length, as square is applied to the yards on the same occasion.

TAURUS, *the Bull*, in astronomy, one of the twelve signs in the zodiac, and the second in order, marked thus ♉. In this constellation there are some remarkable stars that have names; as Aldebaran in the south, or right eye of the bull, the cluster called the Pleiades in the neck, and the cluster called Hyades in the face.

The stars in the constellation Taurus, in the Britannic catalogue, are 141.

TEAK, a hard and durable East India wood, used for masts and hound-pieces of top-masts.

In the year 1803, a 36 and a 74 gun-ship were ordered to be built in the East India Company's docks at Bombay, with teak timber; since which several other ships have been built there.

Among the returns laid before the House of Commons, in the Session of 1813, is one of all the ships built in India, and admitted to registry, from the 1st of January, 1794, to the 5th of April, 1813; from which it appears that the total number is 76, and the measurement 48,438 tons.

The superior quality and abundance of the teak timber in India, and the naval establishments in that country, are available resources for keeping up our naval strength; and if the supply which the forests of India yield, is not made use of in ship-building, it must, like many other resources of India, be lost to the nation. For farther particulars on the quality and durability of teak, see the article TIMBER.

TELEGRAPH, (from τηλε, distance, γραφω, I write Gr. *télégraphe*, Fr.) a word derived from the Greek, and descriptive of a contrivance by which information is almost instantaneously conveyed to a considerable distance.

The *telegraph*, though it has been generally known and used by the moderns only for a few years, is by no means a modern invention. It is generally regarded as a well-authenticated fact, that the burning of Troy was known in Greece a very short time after it happened, and long before it was possible for any person to have arrived there from the scene of conflagration. One of the Greek plays opens with a scene in which a watchman descends from a tower in Greece, with the news that Troy was taken. "I have been looking out these ten years, (says he) to see when that would happen, and this night it is done:" a proof that the ancients possessed some mode of conveying information very quickly to a great distance; but what that mode was, it is not easy to determine.

When the Chinese courtiers travel to distant parts of the empire, signals are made by means of fire, from one day's journey to another, in order that preparations may be made for their reception. In most barbarous nations it was customary to give the alarm of war by fires lighted on the tops of hills; and Polybius styles the instruments used by the ancients for communicating intelligence in this manner, *pyrsiæ*, because the signals were made by means of fire. A new method of accomplishing this object, was invented by Cleoxenus (some say by Democritus), and much improved by Polybius. He divided the Greek alphabet into five parts, and expressed the letters on boards in five columns. The order in which these letters were to be taken, was signified by torches held up in certain positions, previously agreed upon by the parties.

Neither this last-mentioned method, nor any other known to the ancients, however, appears to have been carried into general use; nor have we any account of any modern invention for the same purpose, previous to the Marquis of Worcester's, contained in his "*Century of Inventions*," which was published in 1663. In this work, he affirms that he had discovered "a method by which, at a window, as far as the eye can discover black from white, a man may hold discourse with his correspondent, without noise made or notice taken; being according to occasion given, or means afforded, *ex re nata*, and no need of provision before-hand; though much better if foreseen, and course taken by mutual consent of parties." This means of correspondence he also asserts to have been rendered so perfect, that it could be put in practice "by night as well as by day, though as dark as pitch is black."

William Amontons, an ingenious French philosopher, about the year 1702, invented a method of conveying information from one place to another by signals; and it is probable, without any knowledge of what had been previously advanced on the same subject, by the Marquis of Worcester and Dr. Hooke, as his method seems to be much inferior to that pointed out by this last gentleman.

None of these inventions, however, were applied to purposes of utility before the French revolution, when M. Chappe, towards the end of 1793, invented an instrument of a similar nature, and applied to it the name of *telegraph*: whether or not he knew any thing of the inventions of Dr. Hooke and Amontons, it is impossible to say. The first chain of telegraphs formed a communication between the Louvre, at Paris, and Lisle, a distance of about thirty-five leagues, near which town the French army then was. The first description of this instrument is said to have been brought from Paris to Frankfort, on the Maine, by a former Member of Parliament of Bourdeaux, who had seen that which was erected at the Mountain of Belville, near Paris. Two working models of it were made at Frankfort, and sent by Mr. W. Playfair to the Duke of York: and hence the plan and alphabet of the machine came into England. M. Chappe's contrivance consisted chiefly of an upright post bearing a transverse bar about 10 or 12 feet in length, and 9 inches broad, moveable about a pivot in its centre; to each extremity of this bar, an arm of about 3 feet long, was attached by a moveable joint: by means of these three, the combination of movement was very extensive, simple, and easy to be performed. A number of arbitrary positions were fixed to denote the letters of the alphabet; and these being always made precisely in the same manner, by means of the mechanism employed to work the arms, intelligence was conveyed with astonishing accuracy and rapidity.

In consequence of this information, various experiments were made in this country, and a communication by means of a chain of signals was soon established from the Admiralty-office to the sea-coast. The telegraph adopted for this purpose, consisted of six octagonal boards, each moveable about an axis, and capable of being placed either vertically or horizontally, so as to be either visible or invisible at the nearest station at pleasure. These six boards are capable of exhibiting thirty-six changes by the most simple and easy mode of working them; and by these the letters of the alphabet and the numeral cyphers are denoted. This telegraph, however, was by no means considered as being free from objections and inconveniences, and various contrivances for remedying these have been invented since that period. To give an account of all these, would greatly exceed our limits; we shall, therefore, mention a few of the principal of them.

Mr. John Garnet proposed a very simple and ingenious instrument, which consisted merely of a bar or plank, moveable about a centre, round which a circle was described, having the letters of the alphabet and figures depicted round its circumference. An index which corresponded with the bar, enabled the operator to set it to any letter or figure he wished to communicate. A wire was fixed across the tube of the telescope used by the observer, which was to be brought to coincide with, or to be parallel to the bar; then the letter or figure to be communicated was denoted by a small index at the eye end of the telescope, which contained a circle similar to that encompassing the centre of motion of the bar.

None of the preceding telegraphs, however, appear to exceed, either in simplicity, cheapness, or facility of working, that described by J. N. Esq. in Vol. I. of the *Repertory of Arts*, *Old Series*. For a nocturnal telegraph, it consists of four large patent reflectors, situated in the same plane, and parallel to the horizon, but sufficiently elevated above it. Each of these reflectors is susceptible of being elevated or depressed at pleasure, by means of winches, and the letters of the alphabet are denoted by the changes. When this telegraph is used in the day, gilt balls, or any other conspicuous objects, are to be used instead of the reflectors.

Captain Pasley, of the Royal Engineers, also invented two telegraphs, one of them to be used by day and the other by night. Chevalier A. N. Edelchantz, a Member of the Society of Arts, &c. has invented a telegraph, which is composed of three vertical rows of flat vanes, or indices, similar to those used in the English telegraphs. The middle row contains four, and each of the others, three of these vanes, or shutters; the fourth in the middle row being above the rest. Each index has a crank fixed to it, from which a wire or small rope, comes down to the machine, by which it is worked. The principal part of this machine is a barrel with a winch handle, by the revolution of which a bar of wood, having an iron plate fixed to the top of it, is drawn down; and then wires, or strings, connected with the signal-boards, being attached to this at pleasure, the signals are brought into an upright position. When the wires, or strings, are relieved from their connection with the plate, a weight suspended from each signal-board brings it to its original position. The business of connecting these strings with the plate, is facilitated by means of ten keys, each marked with a figure denoting the signal board, the string of which is opposite to it, and which it causes to slide into its proper notch in the plate when required. In working this telegraph, six signals are usually made in a minute and 1024 changes, or signals, may be clearly shewn. A much greater number might also be exhibited, but it would require an attention to too many circumstances. Words which frequently occur in correspondence are represented by such signals as are most easily recollected: these are termed fixed signals, and may be readily known by examining the numbers of a table previously constructed, and referred to by the signals. "The great advantage of this contrivance is, that, by the powerful leverage afforded by the winch and barrel, one man is enabled to move any number, or even the whole of the shutters at once, and to retain them in that situation, be the weather ever so boisterous, until the signal is answered; which is the more necessary in this telegraph, on account of its greater number of vanes, or indices; but which cannot be done in the government telegraphs in England, even with their more limited number of vanes, without employing several persons at each telegraph." The inventor of this telegraph states, that one man is sufficient to manage it at the two extreme stations, and two men at the intermediate ones.

Major Charles Le Hardy has also invented a telegraph, which is intended to express numbers, to which words may be referred at pleasure, and which consists of nine bars or radii, answering to the nine digits; and of four polygonal or concentrical bars, which intersect the radii, and beginning with the outside one, represent units, tens, hundreds, and thousands. Over each of these bars an index traverses which marks the number of thousands, of hundreds, of tens, and of units, as far as ten thousand.

The signals are worked by means of a wheel and rack, put in motion by a winch, except the two at the corners, which are fixed on iron bars, and made to revolve a quarter round by levers connected with cross bars, by means of ropes or wires. To prevent the necessity of inspecting the signals, an index, which revolves with the signal-board, points out the number on a dial-plate that it designates; so that the telegraph may be worked correctly within doors. The signal-boards are secured at any particular place by means of a bolt adapted to the purpose.

It has been generally remarked, that Chevalier Edelcrantz's telegraph is superior, in some respects, to those employed by the Government of this country; but inferior in point of general utility to Major Le Hardy's. Each, however, received the same honorary reward from the Society to which they transmitted their improvements. See the *Trans. of the Society of Arts*, &c. Vol. 26.

The use of telegraphs has hitherto been confined to naval and military purposes; but it has been suggested that they might be used with great advantage in commercial and other concerns; that they might be erected at a trifling expense, and a correspondence carried on with them to the most distant parts of the kingdom.

The following plan for facilitating telegraphic signals by day, and at the same time for establishing on simple principles, telegraphic correspondence by night, was communicated, in December 1797, to the Lords Commissioners of the Admiralty, by John M'Arthur, Esq. LL.D. whose improvements in the present system of naval signals by day and night were duly appreciated at the commencement of the war in 1793, as already noticed under the article Signals.

In the construction of the proposed telegraph, it is intended to remedy the defects to which others are liable, namely, their being too heavy and clumsy to admit of being raised to any considerable height above the building where erected; since it is obvious that high posts are essentially necessary for denoting night signals, from the great distance that the lights must be placed asunder, so as the eclipses or obscurations, indicated by the following plan, may be distinctly seen, under the visual angle of the observers, at the respective stations. At the same time, simplicity, cheapness, and facility in working the telegraph, is attempted to be united with distinctness in observation; it may be so contrived as to change its direction, and consequently might be seen from any particular point or station.

Construction of the Telegraph. — Two posts, or poles, are to be erected, about 30 ft. above the officer's cabin, or tent, A B, and nearly the same distance asunder as the posts now used, and to be securely fixed or stepped in either a square or circular bottom frame, or flooring of the cabin, having concentric rollers, and to stand on a solid circle platform, moveable upon a pivot, so as to turn the whole to any direction, and be thereby enabled to correspond with the line of the several stations marked on the platform.

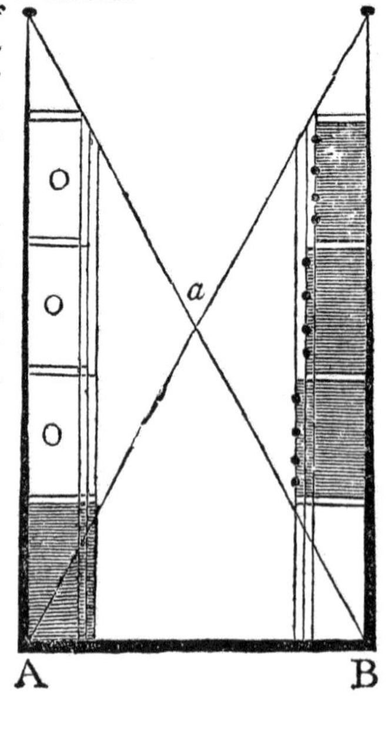

Each pole to be secured by a rope, in imitation of a stay-rope, passing from the top of the one, and fixed to ring-bolts, near the bottom of the other, on the roof of the cabin; which will remove the weight of the additional wood-work used in the construction of other telegraphs. From each stay-rope would fall (as in the annexed diagram, with the three lights on the right obscured) three parallel cords, or ropes, at a given distance from each post, to be also fixed to ring-bolts on the officer's cabin, viz. suppose the first cord at the distance of two feet; the second at the distance of two feet three inches; and the third, or inner one, at the distance of two feet six inches; that is, each cord three inches apart; on which three strong frames, with black painted canvas, might be made to slide, having the edges next the post to run in parallel grooves, and the edges next the parallel cords, to embrace them respectively with spring rings, as denoted by the four black dots, thus , on each of the three screens hoisted to obscure the three lights on the right.

The upper screen would slide on the first cord, and in its corresponding groove; the centre screen on the second cord; and the lower screen on the last and third cord, and in their corresponding grooves or channels; and by either giving the stay-ropes, or the parallel cords, a small deviation, the screens would easily slide, and not interfere with each other. The blank space or distance left on the right hand between the roof of the cabin and the aperture allotted the lower screen, is adapted for any of the screens, when hauled down, whilst a light is displayed in working the telegraph; that on the left exhibits all the screens down, while the three lights are exposed.

Hence, the difference in the dimensions of the screens shewn in working, even by day, would be advantageous for distinctly denoting the signals, and might be worked with equal celerity as the present mode with boards. The angular point *a*, where the cords called stays meet or cross each other, may be termed the centrical point of the telegraph, or the centre of the field of vision of the observer's telescope; and at any time, when the telescope is required to be adjusted, a ball placed at that point by day, or a light by night, would regulate the axis of the instrument, by bringing the ball, or light, to bear exactly on the centre of the telescope, which would be easily practised, by means of the vertical and horizontal wires, proposed also to be used in the object and eye-glasses, for the purposes hereafter mentioned.

Three patent reflectors, or lamps, are proposed to be fixed on each post, at least 12 feet asunder; and their proper diameters and focus, so as to produce the greatest effect, at given distances, might be easily ascertained.

A telegraph, on the above construction, might be adapted for day as well as night correspondence, and be so easily moved from place to place, that it might be equally employed for the operations of an army, by day or night, and could with facility be erected in a tent: limited to the following night signals only, it might perhaps be deemed more simple to obscure the lights, by means of three cylinders, vertically placed, and moving on brass rods in horizontal directions from each pole, worked by pullies or small winches; or the obscurations might be made by means of a simple apparatus attached to the lamp.

Telescope.—The next object, as a desideratum to the practical simplicity of the plan proposed, is a good refracting telescope, properly fixed in the officer's cabin, with its axis directed to the centre of the observed telegraph, whose field of vision should be confined, and take in little more than the radius of the telegraph observed; whereby the distinctness required in night observations would be better produced; and such telescope should have placed, in the common focus of the object and eye glasses, two vertical silver or brass wires, parallel to each other; also two horizontal wires, parallel, dividing them in equal parts, fixed by pins or screws to a small brass circle, as represented in the annexed diagram, and similar to the mode used in the glasses of transit instruments. This would render the obscurations of the light, on the right or left of the operator, to be more easily and accurately defined by the observer.

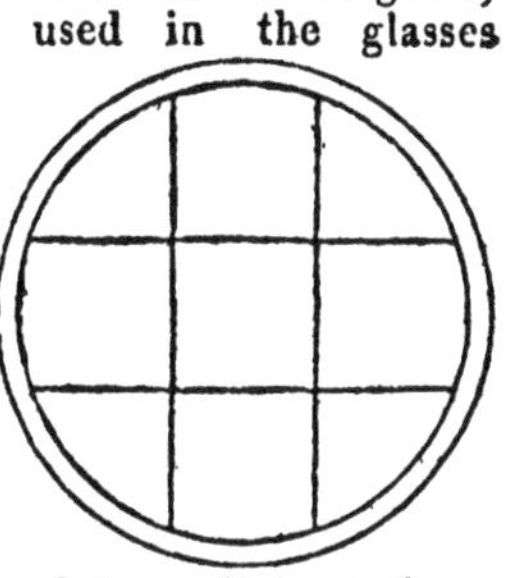

At the first and last stations there would only be occasion for a set of simple reflectors or lamps; but in the intermediate stations, each reflector or lamp should be double, so as to face both the preceding and subsequent observatories; and each observatory should be furnished with two telescopes having wires, as above mentioned.

Manner of denoting the Twenty-six Letters of the Alphabet; also Numbers, to any extent.

The six vowels to be denoted, as hereafter specified, on the general principle of obscuring one light on the left, and another on the right of the operator.

N.B. The black cyphers thus ●, denote obscurations of the lights, and the open cyphers thus ○, denote the lights exposed.

The twenty consonants to be arranged in two divisions; viz. the first ten, in the order of the alphabet, to be denoted by a majority of lights obscured on the left: the second division, containing the last ten consonants, by a majority of lights obscured on the right of the operator: hence, each letter of the alphabet can be at once marked by a single operation, without any auxiliary signal for denoting a division of the alphabet.

Thus, *for denoting the vowels*,

a		*e*		*i*		*o*		*u*		*y*	
●	●	●	○	●	○	○	○	○	○	○	○
○	○	○	●	○	○	●	●	○	○	●	○
○	○	○	○	○	●	○	○	●	●	○	●

a to be denoted by an obscuration of the upper light on the left hand, and an obscuration of the upper light on the right; and so forth, as expressed, to each vowel in the above table.

N.B. When left, or right, is mentioned in the explanations, the left, or right, of the operator is understood by it.

For denoting the first ten consonants, b, c, d, f, g, h, j, k, l, m.

General Principle.—Never more than three or four lights obscured at a time; and the majority always obscured on the left hand of the operator.

b		*c*		*d*		*f*		*g*		*h*		*j*		*k*		*l*		*m*	
●	●	●	○	●	○	○	●	○	○	○	○	●	○	●	●	●	○	●	○
●	○	●	●	●	○	●	○	●	●	●	○	○	●	●	○	●	●	●	○
○	○	○	○	○	●	●	○	●	○	●	●	●	○	●	○	●	○	●	●

Example. *b* is denoted by obsc ing the upper and middle lights on the left, and the upper light on the right. *c*, by obscuring the same lights on the left, and the middle light on the right, and so forth, as expressed under each consonant in the above table

For denoting the last ten consonants, n, p, q, r, s, t, v, w, x, z.

General Principle.—Never more than three or four lights obscured at a time; and the majority always obscured on the right hand of the operator; hence this is the converse of the foregoing.

n		*p*		*q*		*r*		*s*		*t*		*v*		*w*		*x*		*z*	
●	●	○	●	○	●	●	○	○	○	○	○	○	●	●	●	○	●	○	●
○	●	●	●	○	●	○	●	●	●	○	●	●	○	○	●	●	●	○	●
○	○	○	○	●	○	○	●	○	●	●	●	○	●	○	●	○	●	●	●

Example. *n*, is expressed by obscuring the upper and middle lights on the right, and obscuring the upper light on the left; and so forth, as expressed under each consonant in the above table.

For denoting Numbers.

General Principle.—One light only obscured at a time, on either hand, or two lights obscured at a time, on either hand.

1		2		3		4		5		6		7		8		9		0	
●	○	○	○	○	○	○	●	○	○	○	○	●	○	○	○	○	●	○	○
○	○	●	○	○	○	○	○	○	●	○	○	●	○	●	○	○	●	○	●
○	○	○	○	●	○	○	○	○	○	○	●	○	○	●	○	○	○	○	●

Example. One, is denoted by an obscuration of

the upper light on the left; Two, by an obscuration of the middle light on the left; Three, &c. as expressed in the above table, under each figure, including the cypher.

As the signal for the termination of a word, or number, may be made by obscuring the upper and centre lights on the left, and the upper and centre lights on the right, it is easy to conceive how numbers may be expressed to hundreds of millions. Thus, 3650: the numeral sign 3, denoted by the obscuration of the lower light, on the right; then 5, by the obscuration of the centre light on the right; and the 0, or cypher, by obscuring the middle and lower lights, on the right; after which the termination signal would be made. If *tens of thousands* were to be expressed, then the five signals, corresponding to the numbers, would be successively denoted; to close with the termination signal.

The preparatory signal to be made, when the telegraph is to be worked, is by displaying the six lights in their proper places; and several auxiliary signals, for particular purposes, may in words, or sentences, be made by the combination of four or five lights obscured: such as, to *annul the preceding signal made—the quarters of the compass—interrogatory—affirmative and negative answers—Ships, under sailing orders, to put to sea—ships to remain till further orders, &c.*

The principle of the foregoing plan of night signals is so simple and self-evident, that it may be practised at a window, by allotting three panes on the left, and three on the right, for the six lights; and every evolution and combination, indicated by the obscurations, may be satisfactorily demonstrated: and a correspondence might easily be carried on with a person in a house situated as far opposite as lights could be distinguished.

Nocturnal telegraphs, on similar principles, it is presumed, might, on a future consideration, more especially in times of peace, be used by Government, as an object of revenue, in conveying commercial intelligence to any part of the kingdom, much cheaper than an express could travel. Should a merchant or banker, after the ordinary post hours, wish to communicate any important commercial information to his correspondent, at a distant port or station, he would only have to deliver it to the inspector of the first telegraph to be conveyed; paying in the ratio of the words employed, and the distance of the place, in such communication. A considerable revenue might by this means be derived, after defraying all the expenses of such telegraphs.

A new Plan or Arrangement for communicating the Words of a Dictionary, by adhering to the Combinations of the Day Telegraph now in Use; and which may be equally applied to the Day and Night Telegraph proposed in the above article

There being 63 mutations or movements of our Telegraph as at present established, it is proposed to arrange them as in the following table, in seven series or ranks; by which, were it necessary, 9,999,999 words could be denoted.

The first 26 mutations not only denote numbers from 1 to 26 inclusive, but also the letters of the alphabet in the usual manner, and the telegraph can be worked accordingly. But in applying (as by preparatory movement, to be hereafter noticed) the mutations to the words of a Dictionary numbered from 1 to 99,000 inclusive, the first five series of numbers will be sufficient for the purpose; namely, the first series will indicate units, the second series tens, the third hundreds, the fourth thousands, and the fifth tens of thousands, as noted in the table of movements.

All the words, therefore, under the letters A, B, C, D, and E of a Dictionary, will fall to be numbered from 1 to 9,900 odd, and the remaining words of the Dictionary will extend from 10,000 to nearly 99,000. The numeral signal made will refer to the number of the word, consequently, some words will require one or two movements, others three, and some four; and in general, the movements for conveying ideas in this manner will be one half fewer than the usual mode, and many sentences will be composed in one third of the time sooner than by letters, and with greater certainty and distinctness.

In the following table, the six squares represent the six shutters or spaces of the telegraph; and where there is a mark thus , it denotes the movement by shutting the board in the day, or obscuring the light in the night.

When the telegraph is to be worked by night, the obscurations of the lamps placed in the open spaces of the shutters, are effected by a simple apparatus having a semi-globular lens and reflector attached to each burner, which has the peculiar advantage of transmitting the rays of light, on the line of bearing from one telegraphic station to another.

The apparatus consists of a square tin case placed over the lamp, being perforated with holes at the bottom, and having in the upper part a funnel with small lateral tubes to admit a sufficient quantity of air; the side of this tin case behind the burner has a polished steel reflector, and the front or opposite side has a tin tube of six inches in diameter, and twelve inches in length, which slides into a short tube fixed to the case in a direct line with the burner; at the end of the sliding tube nearest the light, there is placed a semi-globular lens of four or five inches in diameter, which concentrates the rays of light, and increases its force and brilliancy. At the outer end of this tube there is a square plate of tin which slides into a groove; and having a small cord over a pulley, the operator may either open or shut it, so as to shew or obscure the lights at pleasure in making the combinations indicated in the table of signals.

From the experiments made with an apparatus of this kind applied to a common table-lamp with Argand's burner, a light may be distinctly seen at the distance of nine or ten miles, provided the atmosphere be not cloudy. Should it be required to make the lights visible at a greater distance, it can be effected by enlarging the diameters of the lenses and reflectors, whereby the strength and brilliancy of the lights may be so increased, as to be visible from 20 to 25 miles; but in such case the telegraphs ought to be erected on very elevated situations, and the lamps should be placed from 18 to 20 feet asunder, so as the lights might be clearly defined, and the obscurations accurately distinguished.

The annexed table exhibits the different combinations.

TABLE OF THE TELEGRAPHIC MOVEMENTS,

FOR WORDS OF A DICTIONARY, SENTENCES, OR LETTERS OF THE ALPHABET, &c.

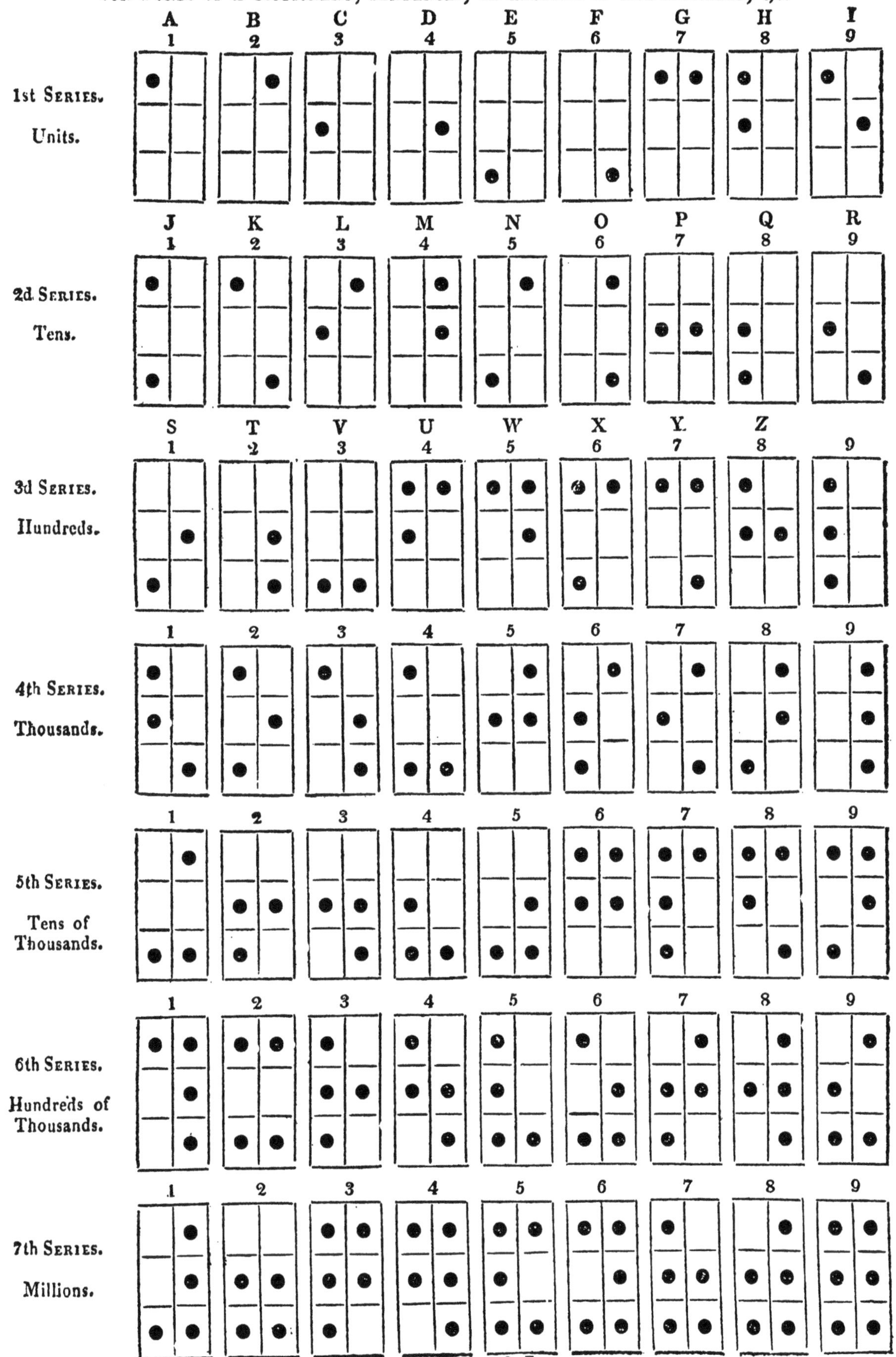

Illustration of the annexed Table, and Examples.

1st. To denote the word in a dictionary numbered 90, it would only require the movement 9 in the second series of tens; but in denoting 97, the movement 9 in the second series would be first made, then followed by the movement 7 in the first series for units.

2d. Suppose a word was numbered 5,860, it would be denoted first by the movement 5 in the fourth series for thousands; secondly, by the movement 8 for hundreds in the third series; and lastly, by the movement 6 in the second series.

3d. Suppose a word was numbered 70,806, it would be denoted by three movements: first, the movement 7 in the fifth series; next the movement 8 in the third series; and lastly, the movement 6 in the first series. Indeed, were it necessary to express words or ideas in the rank of millions, many of them could be done by two or three movements: Thus, 2,000,300 would be denoted by two movements: first, the movement 2 in the seventh series, and the movement 3 in the third series, indicate *two millions* and *three hundred.*

But as the 18 principal movements in the sixth and seventh series of the table for expressing hundreds of thousands and millions, are not required for the purposes of a dictionary, they can be allotted to express other particular signals.

Preparatory and other Signals.

The first movement in the sixth series may be made the preparatory signal, to denote that the subsequent movements of the telegraph are to be applied in expressing the numbers attached to words of a dictionary. After the close of a sentence referring to objects, the next movement would imply the number in a literal sense; then the names or qualities might be denoted. Thus the movements expressing the words "Ships have arrived," being made, the next movement would indicate the *number of ships;* and then the subsequent movements might indicate the names of the several Ships of War, according to their numbers, on a separate list. In like manner, after the sentence announcing the arrival of troops was made, their number, rank, and file, might be denoted, and then the number of the respective regiments or corps.

The movement 2 in the sixth series, is preparatory to the telegraph's working by the sentences or significations numbered in a naval or military signal book.

The movement 3 in the sixth series, will be interrogatory of sentences. Thus in the naval signal book, suppose No. 30, had this signification, "ships are ready to proceed on the service ordered," the interrogatory would be first made, and afterwards the movement 3 in the second series, indicating the sentence in the sense of, "Are *the ships ready to proceed on the service ordered?*" or, "*Are the troops ordered for service, ready to be embarked?*" &c.

The movement 4 in the sixth series will be interrogatory as to the subsequent movements denoting words; and in this sense implies, how-many? What number? &c.; namely, *ships, seamen, troops.*

The movement 5 in the sixth series will be an *affirmative* answer to any signal.

The movement 6 in the sixth series will be the *negative* answer.

The movement 7 in the sixth series will denote *inability* to comply with a particular order or signal.

The movement 8 in the sixth series will denote *annulling* the preceding signal or sentence.

The movement 9 in the sixth series will denote the peremptory or imperative sense, by enforcing the signal or sentence previously or subsequently made.

The movement 1 in the 7th series will denote a colon, or the first branch of a sentence.

The movement 2 in the seventh series will denote the termination of the second branch of a sentence.

The movement 3 in the seventh series will denote a stop, or the sentence terminated.

The remaining six movements in the seventh series; viz. 4, 5, 6, 7, 8, and 9, may be preparatory to signals for calling officers to town, of the first, second, or third post, courts-martial, rendezvous, &c.

TELEGRAPHIC SIGNALS, on a new plan, adapted to military and naval services, in separate or conjunct operations, invented by John M'Arthur, Esq. LL. D. and first published in the Royal Military Panorama, for April, 1814. They are made by three soldiers armed with fire-locks, or three sailors furnished with oars; also by the symbols of six flags in arithmetical progression of numbers.

The principles on which these signals are conducted, are illustrated by the following diagrams and tables.

Diagrams of the Six Different Positions of Three Signal Men.

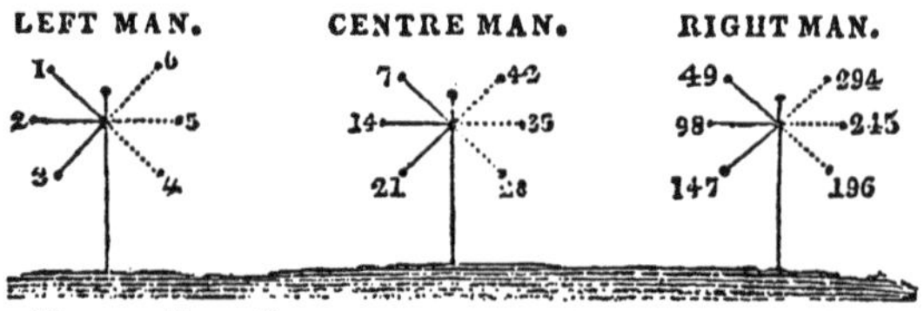

TABLE I. *Combinations to the above positions.*

Apply to the Corresponding No. of the Right Man's positions, that of the Centre Man's positions, and add the No. of the Left Man's positions, as denoted in the upper line.							
Right Man's positions and Correspondent Numbers as underneath.	Left Man's No.	1	2	3	4	5	6
	Centre Man's No.	7	14	21	28	35	42
Position 1st	49	56	63	70	77	84	91
2d	98	105	112	119	126	133	140
3d	147	154	161	168	175	182	189
4th	196	203	210	217	221	231	238
5th	245	252	259	266	273	280	287
6th	294	301	308	315	322	329	336

General Observations on the Positions.

The mode proposed in this telegraphic plan of signals, is peculiarly adapted for the military operations of a large army, either in making forward movements against an enemy, or in retreating from a superior force.

Instead of having a telegraphic post erected with three moveable arms, one over the other, and all the apparatus necessary for what the French call a *Semaphore*, the simultaneous principles are applied to the movements of three soldiers armed with firelocks, who by two or three lessons can be correctly trained in the execution of the six positions. There is an advantage in this mode of making signals in the line of an extensive army; namely, that they can at all times be readily communicated in a line of posts or stations, by placing the signal men on the most elevated grounds, and within such limited distances as not to be seen by an enemy.

Explanation and Examples.—The three signal men, distinguished *left man*, *centre man*, and *right man*, stand in a line about twelve feet asunder, facing the station to which the communication is to be made (that is corresponding to the reader's position), and to each are allotted the six similar positions or movements as the following, applicable to the left signal man's six positions with firelock and correspondent numbers from 1 to 6 inclusive (See *Diagram*). The first position, to *the left raise arms*, is made from the poize or *present arms*, by throwing the firelock across the chest to the left, elevated to the angle of 45 degrees The second position, to *the left square arms*, is executed by bringing the firelock in a horizontal lineto the left. Third position, to *the left lower arms*, is executed by depressing the firelock in an angle of 45 degrees on the left side. Fourth position, to *the right lower arms*, is executed in a similar manner on the right side of signal man. Fifth position, to *the right square arms*, is executed by bringing the firelock horizontally on the right side. And the sixth position, to *the right raise arms*, is executed by elevating the firelock in an angle of 45 degrees to the right.

These positions of the left-hand man have the correspondent numbers (as prefixed to Diagram), 1, 2, 3, 4, 5, 6, as well as in the upper horizontal line of the table of combinations.

The centre signal man's positions are executed in the same manner, and have the following correspondent numbers, as marked on the Centre Diagram, as well as shewn in the second horizontal line of the table of combinations; namely, to the first position of centre man is the correspondent number, 7; to the second, 14; to the third, 21; to the fourth, 28; to the fifth, 35; and to the sixth position, 42.

The *right signal man* has also six positions executed in a similar manner, and to which are the following correspondent numbers, as denoted in the Diagram on the right hand, and shewn in the vertical column of the table; viz. First position has the correspondent number, 49; the second position, 98; the third position, 147; the fourth position, 196; the fifth position, 245; and the sixth position, 294.

Examples of the Single and Combined Positions.

The single and combined positions of the arms, as above noticed, of three signal men, are capable of denoting 342 signals or sentences. Thus, the left-hand man represents, by his six different positions, numbers from 1 to 6 inclusive.

2d. The centre man's positions denote correspondent numbers, from 7 to 48 inclusive, by having occasionally the correspondent numbers of the left man's positions added, as denoted in the first horizontal column of table. Thus, 7 is represented by the centre man's first position of *arms raised to the left.* No. 8 would be denoted by the left man being at the same time in the first position: but if 9 were to be expressed, the left man would place arms in second position, and so forth to 13, by the combination in this line. If 17 were to be expressed, the centre man would have arms in position 2d, indicating its correspondent number 14, while the left man would be in position 3d; and so forth, making, with the combined positions of two signal men, 48 signals, as represented in the two upper columns in the table.

The words of command, expressive of any combined positions, might be briefly delivered by the officer in the rear of signal men, so as to be performed at the same time together, on pronouncing the last word of command. Thus the above signal 17 would be ordered in the following words: "*the left man's 3d position with the centre man's 2d position*—Execute."

3d. The right man's positions singly and combined with the others, will denote 312 signals.—Examples combined with the centre man's positions. If 56 were to be represented, the right and centre men would be respectively in the first position; the former denoting its correspondent number 49, and the latter 7, and the sum 56 will be found in the angle of meeting of positions in the table. If 57 were to be represented, the left man's first position would be made and added to the others. If 58, the left man's second position would be made at the same time; and if 59, the left man's third position would be made at the same time, and its correspondent number 3, added to the sum of the right and centre men's number, as above.

Another example of the combined positions of the three signal men shall be given, which, it is trusted, will render the nature and extent of this plan of arithmetical progression sufficiently clear and comprehensive. Suppose it were required to make the signal numbered 303? The following would be the positions of the three signal men made together. The right man to throw arms in position 6th, representing 294; the centre man in first position represents 7; and the left-man's second position would represent correspondent number 2; which added together make 303.

The arms of the three diagrams shew the correspondent numbers to the six different positions, and the table I. annexed, exhibit the combinations of numbers to all the positions from 1 to 336, and by adding the six numbers of the upper horizontal line, the combination and extent of the table, as demonstrated, will be 342.

To prevent an enemy's discovering the signals, and in order to change or transpose the order of the system, the right man's positions in the annexed table may be made a moveable index. And suppose another key to that represented in the table were to be used; for instance, the 2d position of moveable index to be transposed to the correspondent numbers of first, would change the correspondent numbers of the system, for the 2d position would then represent and stand against 49 instead of 98, as in the table. The third position would represent 98; the 4th position, 147; the 5th position, 196; the 6th position, 245; and the 1st position, 294; and so forth, for any changes that might be desirable to be made.

A second arrangement of the figures in the table may be made applicable to 342 of the most essential words falling under each of the 26 alphabetic letters of a vocabulary or dictionary. This could be effected by numbering the words in the margin, from 1 to 342, in every letter of the dictionary, which, multiplied by 26, would give 8,892 words. But in this arrangement two signals must necessarily be made, the first to indicate the letter of the alphabet, and the second the number of the word under such letter. The positions in the table, representing figures from one to 26 inclusive, would, in alphabetical order, be termed A to Z. and the second signal would refer to the number of the word under any of the respective letters. Table II. will not only answer the purpose of representing the letters of the alphabet for spelling words, but, will also refer to the letters of alphabet, when words of a vocabulary or dictionary are meant to be represented.

The principles of the above plan may be applied to boat service in the navy, and may at times be important in conjunct operations of navy and army, either in embarking or disembarking troops, &c. The table of positions and correspondent numbers is easily adapted to ship or boat service, by considering the left signal man as synonimous to the forward man of a vessel, or bow man of a boat. The centre man requires no additional explanation, and the right man may be considered the same as the after man, or oar, of a vessel or boat.

In *boat service*, the positions to the correspondent numbers of table I. will be made by the *bow man*, the *centre man*, and *after man*, standing erect in the boat, having the boat's side parallel to the station to which the signal is to be communicated, and the position of the oars to the left or right, as indicated in the table of the three soldiers' positions with firelocks, will indicate the correspondent numbers.

TABLE II. *For Representing Letters of the Alphabet, and for communicating Words of a Vocabulary or Dictionary by the combination of the Numeral Table I.*

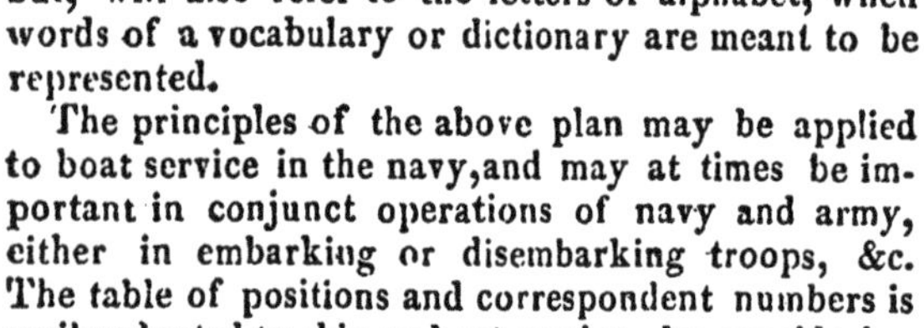

II. Table of Positions and Combinations.

Left Man's positions & letters represented		1st A	2d B	3d C	4th D	5th E	6th F
Right Man's Position 1st	G	H	I	J	K	L	M
Position 2d	N	O	P	Q	R	S	T
Position 3d	V	U	W	X	Y	Z	&
Position 4th	Termination of the word spelt.						
Position 5th	Termination of the Sentence.						
Position 6th	Annulling Signal, or Negative Answer to any signal made.						

EXPLANATION.—The left man's positions, as arranged in the upper column, represent the first six letters of the alphabet, by making one position at a time.

The right man's positions, singly or combined with those of the left, represent the remaining letters of the alphabet; viz. the right man's 1st position represents G, but when this position is combined with the left man's 1st position it will represent H. If combined with the left man's 2d position it will represent I. If the 3d position, J; and so forth, as in the second column of letters.

The right man's *second* position, singly, represents letter N, but if this 2d position is combined with the left man s 1st position, it will represent letter O, and so forth for the other letters, as in the 3d column from the top in table II.

The right man's 3d position, singly, represents letter V. and when combined with the left man's 1st position, it will represent letter U, and so forth for the other letters, as in the fourth or last column of the table.

As two men, by their combined positions of firelocks, can represent 48 distinct signals, and that there are only 30 applied in the preceding table to express all the letters of the alphabet, and three general significations, there are still 18 distinct signals or combinations that may be made with the 4th, 5th, and 6th positions of the right-hand man, combined with those of the left, which may be appropriated to preparative or distinctive signals for divisions or brigades of the army, or to so many separate sentences or significations in the conjunct operations of army and navy.

Combinations of Tables I. and II. for referring to Words of a Vocabulary.

It having been explained that table I. represents numbers by the positions of three men, from 1 to 342 inclusive, and that table II. represents all the letters of the alphabet by the combined positions of two men; if, therefore, each of the 26 letters of a Vocabulary were to have the most essential words for military or naval purposes numbered on the margin, from 1 to 342 inclusive, two signals would refer to any word; namely, the first signal in table II. would refer to the letter, and the second, as in table I. to the number of the word, consequently 8,892 words would be denoted. This is demonstrated by multiplying 342, the extent of numbers in table I. by 26, the letters of the alphabet in table II. Instead of employing three signal men for military purposes, if three posts, of about eight feet in length, were erected twelve feet asunder, and each having an arm of four feet moveable as a radius, the numbers could with equal facility be represented, and this mode would have the advantage of having the positions visible at a greater distance. All the posts necessary could, without much inconvenience, be transported with an army, and easily put up in suitable stations. Suppose six stations in the line of an army, and each distant from the other three miles, and at each station three posts erected, any signal might be communicated from one extremity to the other, being a distance of 15 miles, in less time than one minute.

Table III.—*Telegraphic Signals by Si xFlags on the Semaphoric Plan, referring, 1st, to sentences, numbered from 1 to 342 inclusive; 2dly, to Letters of the Alphabet; and 3dly, to Words of a Vocabulary.*

	Colour of Flags.		 A			 B		
	Where hoisted.		Fore or Left.	Main or Centre	Mizen or Right.	Fore or Left.	Main or Centre	Mizen or Right.
	Nos. represented.		1	2	3	4	5	6
	Colour of Flags.		 C			 D		
	Nos. represented.		7	14	21	28	35	42
 E	Fore or Left.	49	56	63	70	77	84	91
	Main or Centre	98	105	112	119	126	133	140
	Mizen or Right.	147	154	161	168	175	182	189
 F	Fore or Left.	196	203	210	217	221	231	238
	Main or Centre	245	252	259	266	273	280	287
	Mizen or Right.	294	301	308	315	322	329	336

N. B. The tinctures or colours of the flags are distinguished as follows; viz. *red* by perpendicular lines thus  *blue* by horizontal lines thus  and *yellow* by dots thus, 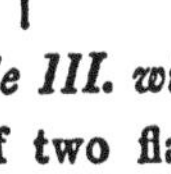

Explanation of Table III. with Six Flags.

By the combination of two flags, as in the two upper horizontal columns, 48 signals may be expressed, and by the combination of three flags at the different parts, as expressed in the vertical column on the left, 342 signals may be represented on similar principles of arithmetical progression, as explained in the Semaphoric positions of three signal men, in table I.

The flags in table III. are to be shewn at the fore main or mizen masts of a ship, or, which implies the same thing, *forward*, *centre*, and *aft*. of a vessel with one or two masts; or larboard yard-arm, mast-head, and starboard yard-arm, when it may be necessary to communicate signals from the shore, with a mast and yard across; or if from one military station to another, the flags may be shewn at three posts, twelve or fifteen feet asunder, distinguished *left*, *centre*, and *right* hand posts.

In the annexed table there are six different flags, each of two colours, and each flag separately or conjointly hoisted, represents the numbers expressed in the table. The flag distinguished by letter A (*red and blue* horizontally divided), shewn at the fore, or left-hand post, represents 1; at the main or centre 2; and at the mizen or right, 3. The flag B (*red and yellow* horizontally divided), shewn at the fore or left, represents 4; at the centre, 5; and at the mizen or right, 6.

The second horizontal line contains also two flags; viz. the flag *blue and yellow* horizontally divided, distinguished by letter C; and the flag *red and blue* quartered, called D, which flags denote, in arithmetical progression of numbers, 7, 14, 21, 28, 35, and 42, as represented under the parts where hoisted.

There are two other distinct flags; viz. *red and yellow* quartered, and *blue and yellow* quartered, termed E and F, placed in the vertical column to the left, with their correspondent numbers prefixed when hoisted singly, but if hoisted with any of the flags in the second horizontal column, the number will be found in the table expressed at the angle of meeting of both flags, and the intermediate numbers, from 1 to 6 inclusive, will be represented by adding the correspondent number of the upper horizontal column to those already denoted. Consequently by the combination of these six flags, 342 signals or sentences so numbered, can, by this arrangement, be made.

Examples.—*With one flag at a time.* The flags A and B represent, as expressed under them, when shewn at the parts specified, 1, 2, 3, 4, 5, and 6.—Secondly, *with two flags at a time.* To denote or make the signal 56, the flag E would be shewn at the fore, representing 49, and the flag C hoisted over it at the fore, representing 7, making together, as expressed at the angle of meeting in the table, the signal 56. If flag C were to accompany the former, but shewn at the main or centre, representing 14, it would represent 63, and so forth with any two flags.—Thirdly, *with three flags.* The flag E, shewn at the main, representing 98; the flag C, at the fore, representing 7; and the flag A over it at the fore, representing 1, make together 106. To denote 312 the flag F would be shewn at the mizen or right, denoting 294, with the flag C at the main, denoting 14, which gives, at the angle of meeting in table, 308. And at the same time hoisting B at the fore, representing 4, will give 312, the number required, and so forth from 1 to 342 inclusive, as represented in the figures of the table.

As each flag is composed of two distinct colours, and so arranged as to make different signals when inverted, another table, if required, in the inverted order, may be made, so as to represent 342 additional correspondent numbers, making in all, with six flags, 684 signals; but if only 342 are required, the six flags may be of any colours or forms.

Preparative signals for the different arrangements of which the table is susceptible, can be made without interfering with the tabular numbers; viz. When flag A is hoisted over flag B, where best seen, it will denote preparative to the signals going to be made, having references to significations or sentences, numbered from 1 to 342. When flag B is hoisted over flag A, where best seen, it will be preparative to the alphabetic arrangement in table IV. for spelling of words. When flag C is hoisted over flag D, at the most conspicuous place, it will denote the preparative of the 3d arrangement of the tables to words of a vocabulary or dictionary, each letter, to have the most essential words numbered on the margin from 1 to 342 inclusive. Several other independent or preparative signals may be made.

TABLE IV.—*Arrangement, representing the Letters of the Alphabet by the combination of the First Three Flags of the preceding Table, and only One or Two of them shewn at a time.*

	Colour of Flags.		 I.			 II.		
	Where hoisted.		Fore or Left.	Main or Centre	Mizen or Right.	Fore or Left.	Main or Centre	Mizen or Right.
	Letters represented.		A	B	C	D	E	F
 III.	Fore or Left.	G	H	I	J	K	L	M
	Main or Centre	N	O	P	Q	R	S	T
	Mizen or Right.	V	U	W	X	Y	Z	&
Red and Blue quartered	Fore or Left.	Termination of the word spelt.						
	Main or Centre	Termination of the sentence or words of signal.						
	Mizen or Right.	Annulling a preceding signal, or negative answer.						

It is to be observed that this table, representing the letters of the alphabet by flags, is on the principles analogous to the Semaphoric plan by the positions of two signal men in table II.

The two flags in the upper column, hoisted singly at the parts specified, represent letters A to F inclusive. The flag III. hoisted at the fore, or left post, represents G, but when combined with flag I. hoisted over it at the fore, letter H will be represented, as expressed at the angle of meeting. If flag III. at the fore be accompanied by flag I. at the main, it will represent letter I, and so forth for the other letters in that horizontal column as expressed at the respective angles of meeting. If flag III. be shewn at the main or centre post it will represent N, but when accompanied with flag I. at the fore it will represent O, and so forth for the other letters in this column, and the remaining column in the table to the end of the alphabet.

Third arrangement for representing the Words of a Vocabulary.

This is effected by the combination of the two tables, consequently two separate signals for each word must be made; the first will refer to the letter in table IV. and the second to the number in table III. correspondent to the word.—Example. Suppose a word under letter L had the number 70 prefixed to it, the alphabetic signal in table IV. for L would be made, then the numeral signal in table III. for 70 would be made: thus the several words necessary to compose a sentence in a concise manner, would be communicated to a great distance, in a very short time.

As the flags of two colours, selected for table III. are (as noticed page 541) capable of representing 342 additional signals when hoisted in the inverted order, another table on similar principles may be constructed in continuation of the numbers from 343 to 684. This last multiplied by 26, the letters of alphabet, will give 17,784, the number of words which can be denoted by the combination of six flags.

Telegraphic Signals, *by the Combination of Six Flags only, and a Substitute Pendant, adapted to Naval, Military, Commercial, and Political Correspondence, to a vast variety and extent, arranged on the Arithmetical Progression of Numbers, with Integral, Decimal, and Centesimal Powers. By* John M'Arthur, *Esq. LL.D. first published in the Naval Chronicle, Vol. XXXI.*

Table I. *Representing Flags and Numbers.*

 Substitute Pendant.	Flags 1st Series.	 A.			 B.			 C.		
	Where hoisted	Fore or Left.	Main or Centre.	Mizen or Right.	Fore or Left.	Main or Centre.	Mizen or Right.	Fore or Left.	Main or Centre.	Mizen or Right.
	Units.	1	2	3	4	5	6	7	8	9
The flags of the first Series are shewn inverted as in the vertical column underneath.	Flags 2d Series.	 D.			 E.			 F.		
	Where hoisted	Fore or Left.	Main or Centre.	Mizen or Right.	Fore or Left.	Main or Centre	Mizen or Right.	Fore or Left.	Main or Centre	Mizen or Right.
	Tens.	10	20	30	40	50	60	70	80	90
 A. Fore.	100	110	120	130	140	150	160	170	180	190
A. Main.	200	210	220	230	240	250	260	270	280	290
A. Mizen.	300	310	320	330	340	350	360	370	380	390
 B. Fore.	400	410	420	430	440	450	460	470	480	490
B. Main.	500	510	520	530	540	550	560	570	580	590
B. Mizen.	600	610	620	630	640	650	660	670	680	690
 C. Fore.	700	710	720	730	740	750	760	770	780	790
C. Main.	800	810	820	830	840	850	860	870	880	890
C. Mizen.	900	910	920	930	940	950	960	970	980	990
Flags D, E, F, of the 2d Series being inverted, as in the lower line will represent Thousands.		 D.			 E.			 F.		
		Fore.	Main.	Mizen.	Fore.	Main.	Mizen.	Fore.	Main.	Mizen
		1000	2000	3000	4000	5000	6000	7000	8000	9000

General Observations on the Nature, Extent, and Advantages of the annexed Table and Plan.

1st. By the arithmetical combination of numbers, made with six different flags and a substitute pendant, in the series of units, tens, and hundreds, a greater number of signals or ideas may be expressed, than by either the numerary or telegraphic code of signals, as practised with 26 flags and several pendants, in the navy.

2d. The plan now submitted, is so simplified in principles, that it is equally adapted for the navy and army, either in separate or conjunct operations; and is peculiarly useful as a code for the military, commercial, and political correspondence of the East India Company, to a very great variety and extent.

3d. So few flags (only six in number) being required, the plan for cheapness and facility is not only universally adapted to ships or hired transports, in the merchant service, but is also convenient for communicating signals between ships and boats detached on particular service; also between ships of war and the signal stations on the coast, as well as between one military station to another on an extended line of an army. In a military point of view the plan may be considered of some importance, from the variety and extent of signals that may with celerity be communicated from one wing of an army to another, either in making forward movements in front of an enemy, or in retreating from a superior force, and that in situations where the flags may be observed in most occasions at $1\frac{1}{2}$ or two miles distance, from one station to another, unless the atmosphere be very dense with fog, or that the wind should blow the flags in the direction of the observer at the respective stations.

4th. With respect to the advantages to be derived by the mercantile body, insurers, and underwriters, from so cheap and easy a mode of communication, it may suffice to mention, that every ship or vessel furnished with six signal flags and a pendant, could at all times, and in every situation communicate ideas of importance to their correspondents or agents, involving either the safety of the vessel, crew, and cargo, or accelerating the benefit of the concern, by landing and receiving of the cargo, the supplies of necessary stores, provisions, &c. 1st. To effect this, as will be more fully explained in the examples to be given, one signal at a time can express preconcerted sentences applicable to numbers from 1 to 999 inclusive.

A second arrangement can be made, by representing letters of the alphabet, for the purpose of spelling words, composing sentences.

A third arrangement, if necessary, by two signals, may be made, applicable to the words of a vocabulary or dictionary, to the extent of no less than 25,974 words, most common in use, as will be explained at the conclusion of examples.

Explanation of the Table I. prefixed, and Combination of Numbers.

1st.—*Description of Flags.*

The table of flags, six in number, and a substitute pendant, having correspondent numbers annexed, is capable of representing in the first order, 999 distinct signals or ideas, and by an additional substitute pendant, 9,999 signals may be communicated, by inverting the flags D E and F, as exhibited under the double line at the bottom of table I. But as 999 signals may be deemed fully adequate for either naval or military operations, separately or conjunctly, as well as for all commercial purposes, the correspondent numbers to that extent are only exhibited in table I. But, to those persons who may be curious to ascertain, or apply the combinations, with an additional pendant, some observations and examples of this power, from 1,000 to 9,999, shall be given at the end, which probably some readers may consider rather a subject of curiosity than of public utility.

The six flags in the annexed table I. for the sake of distinction, and that they may not be confounded with the numbers or figures correspondent to their positions, as well as to facilitate the references to the explanations and examples, are defined by the letters A, B, C, D, E, F, and to each flag, correspondent marks or labels of distinction should be attached, for the facility of practice.

The first flag, marked A, is red and blue horizontally divided.

The 2d flag, B, is red and yellow, do. do.

The 3d flag, C, is blue and yellow, do. do.

The 4th flag, D, is red and blue quartered.

The 5th flag, E, is red and yellow quartered.

The 6th flag, F, is blue and yellow quartered.

Here it may be necessary to observe, that three tinctures or colours are only employed, and that each of the six flags has two distinct colours, so as to represent different signals, when the flag is inverted. The tincture or colour of each flag in the annexed table I. is distinguished as follows, according to the rules of heraldry; *viz.* red, by perpendicular lines thus,  blue by horizontal lines thus  and yellow by dots thus 

2d.—*Method of the shewing the Flags afloat, or on Shore.*

1. The flags may be shewn at the fore, main, or mizen-top-masts of a ship, or which implies the same thing, *forward, centre,* or *aft,* if made by vessels with one mast, or by ships' boats; or if made from three distinct posts on shore, the term *left, centre,* or *right* post will make the distinction.

2. They may be shewn by ships or vessels at the most conspicuous mast-head and yard across; and in like manner may be answered or communicated from signal stations on the coast, where a mast and yard are only erected; and in such case the term larboard, or left, when applied to yard-arm, is to be considered as a synonymous term to the position of fore-mast or forward, and the term centre synonymous to main-mast, and starboard or right synonymous to the position of mizen or left.

3d.—*Order of the Flags and Correspondent Numbers represented in Table I.*

1st. The three flags, A, B, C, arranged as the first series or order in the upper horizontal column, shewn at the different parts specified, represent units as expressed.

2d. The second series or horizontal order of flags, *viz.* D, E, F, being hoisted at three different parts, will represent tens or decimals, as expressed by the numbers under them respectively.

3d. The vertical column of three flags, being the first series or order, A, B, C, hoisted inverted, will represent hundreds, as expressed by the figures prefixed to their different positions.

N. B. If thousands were necessary to be represented, the flags of the 2d order, D, E, F, would be inverted as in the horizontal column at the bottom of table I.

Examples of the Combinations representing Numbers.

1. The order in the first horizontal column is explained by the integral numbers, from 1 to 9 inclusive; namely, the flag A, (red and blue), hoisted at the fore, or left, represents 1; if hoisted at the main or centre, 2; if at the mizen or right, 3. The flag B, (red and yellow), at the fore or left, is 4; at the main or centre, 5; and at the mizen or right, 6. The flag C, (blue and yellow), hoisted at the fore or left, is 7; at the main or centre, 8; and at the mizen or right, 9.

2d. The flags D, E, F, in the second order, represent tens, and when accompanied by the integral or first order of flags, A, B, and C, the correspondent units are to be added. Thus the decimal flag D, hoisted at the fore or left, will represent 10, and A at the fore or left hoisted above it, will represent 1; making together 11, the number represented by the combination of both flags, and so forth, to 19 inclusive, by hoisting with the flag D at the fore, any of the other unit flags, at the part or parts indicated. By a similar combination of the other decimal flags, E and F, with the unit flags, 99 signals or ideas can be expressed, as in the annexed table I.

3d. The flags A, B, and C, being inverted, as exhibited in the vertical column, on the left, they will, as previously mentioned, represent the series of hundreds, and the blue pendant will then become a substitute to the respective flags, in the first order of units. Thus, the flag A, (which in the inverted order will be shewn half blue and red divided horizontally) at the main or centre will represent 200, as prefixed. The flag D, of the second or decimal order, shewn over it, at the main or centre will represent 20, and the substitute pendant over this last at the main, will represent in the order of units No. 2, making, when added together, 222. If these two flags were hoisted, as already expressed, and the substitute pendant at the mizen or right, the number represented would be 223. But, if 224 were to be represented, the aforesaid two flags would be hoisted at the main, and the flag B, (red and yellow) at the fore, making together 224.

If the flag A, in the inverted order, as in the vertical column, were hoisted at the mizen or right, it would represent 300, as expressed against it; and if accompanied by the flag D at the mizen representing 30, and flag C at the main representing 8. these numbers added together represent 338, and so forth, from 1 to 999 inclusive.

It is to be observed, that when two or three flags representing a number are shewn at the same place, their correspondent numbers are to be taken in the order represented in the table; that is, the lower flag represents hundreds, the middle or centre flag tens, the superior flag, or if the substitute pendant be occasionally used in its stead, it will represent units, as in the foregoing examples.

The table is so constructed, by arranging the three first flags in the upper horizontal column, or series, as units, the three flags in the second horizontal column or series, as *tens*, in arithmetical progression; and the third series, being the flags of the upper horizontal column inverted, and shewn in the vertical column, represent hundreds; so that the eye can immediately perceive the number of any signal represented by one, two, or three flags.

Example by the Combination of the Flags in the Vertical Column, representing Hundreds, with those in the second Horizontal Series of Tens.

Suppose the flag B, (yellow and red), inverted as in vertical column, were hoisted at the main, with the flag D, (red and blue, quartered), at the mizen, casting the eye on the angle of meeting of the position of vertical and horizontal flags in table I. the correspondent number represented will be found to be 530; and if any of the flags of the first series were at the same time hoisted with the two flags named, it would add the correspondent number of units, to the number so found as above in the angle of meeting of the centesimal and decimal flags.

It having been stated in the outset, that the table of six flags exhibited, representing the combination of numbers from 1 to 999, is susceptible of representing, by the aid of an additional pendant, no less than 9,999 distinct numbers or signals, it is thought proper to give the following explanations and examples.

It is to be observed, that the three flags in table I. representing the decimal powers from 10 to 90 inclusive; *viz.* flags D, E, F, quartered red and blue, &c. may be allotted to represent the series of *thousands*, when hoisted in the inverted order, at the parts specified, that is, from 1000 to 9000 inclusive; and the second substitute pendant to be used will represent the correspondent decimal number, or tens of this flag, so inverted.

Example, if 1010 were to be represented, the flag D would be shewn inverted at the fore (making in this order *blue* and *red* quartered) representing 1000; and the second substitute pendant for decimals would be shewn at the same time over the said flag at the fore, representing its place 10, making together 1010. If 2110 were to be represented, the flag D would be hoisted in the inverted order at the main, denoting 2000, with the flag A also inverted at the fore, representing 100, accompanied by the decimal substitute pendant at the fore, representing 10, making, when added together, 2110, and so forth, from 1000 to 9,999 signals inclusive.

Preparative signals to the different arrangements of which the table is susceptible, can be made without interfering with the tabular numbers; *viz.* flag A hoisted over flag B at the most conspicuous part, would denote the numbers of the table to be applied to the significations or sentences so numbered. 2dly. Flag B hoisted over flag A at the most conspicuous part, would denote the alphabetic table II. for spelling of words. 3dly. Flag C hoisted over flag D at the most conspicuous part, would be the preparative for the subsequent signals and numbers to be applied to the words of a vocabulary or dictionary. Several other preparative signals may be made, without interfering with the tabular numbers.

TABLE II.—Representing the letters of the alphabet by the six flags in the preceding Table I. having only one flag hoisted at a time, at the parts specified.

N. B. The six flags are distinguished by Roman numerals.

Colour of Flags.	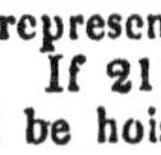 I.			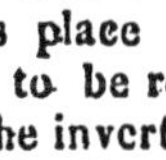 II.			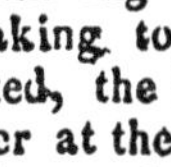 III.		
Where hoisted.	Fore or Left.	Main or Centre.	Mizen or Right.	Fore or Left.	Main or Centre.	Mizen or Right.	Fore or Left.	Main or Centre.	Mizen or Right.
Letters represented.	A.	B.	C.	D.	E.	F.	G.	H.	I.
Colour of Flags.	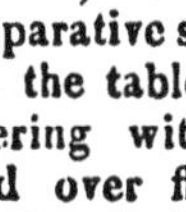 IV.			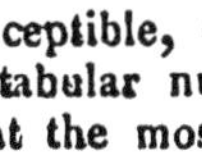 V.			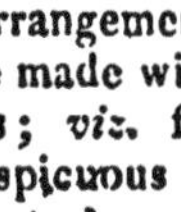 VI.		
Where hoisted.	Fore or Left.	Main or Centre.	Mizen or Right.	Fore or Left.	Main or Centre.	Mizen or Right.	Fore or Left.	Main or Centre.	Mizen or Right
Letters represented.	J.	K.	L.	M.	N.	O.	P.	Q.	R.
Colour of Flags.	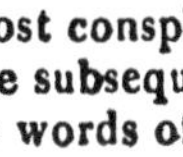 I.			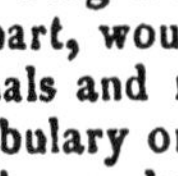 II.			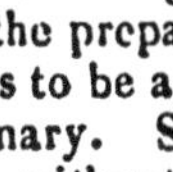 III.		
Where hoisted.	Fore or Left.	Main or Centre.	Mizen or Right.	Fore or Left.	Main or Centre.	Mizen or Right	Fore or Left.	Main or Centre.	Mizen or Right.
Letters represented.	S.	T.	V.	U.	W.	X.	Y.	Z.	&c.

The annexed table, representing letters of the alphabet, requires little or no explanation, as the flags I. II. and III. express the first nine letters; namely, from A. to I. inclusive, under the parts where they are to be respectively hoisted, and so forth of the other flags as represented in the table, each flag indicating a letter of the alphabet.

The termination denoting a word being spelt, may be represented by hoisting flag IV. inverted. The termination of a sentence by flag V. inverted; and the annulling or negative signal by flag VI. inverted, or by any other flag or combination of flags that are not in the alphabetic table.

Third arrangement to represent words of a vocabulary or dictionary, by resorting to the alphabetic table of letters last explained, and table I. representing numbers from 1 to 999 inclusive.

By having the most essential words, or those in common use for naval or military purposes, falling under the 26 letters of the alphabet, numbered from 1 to 999 inclusive, affixed on the margin of a vocabulary or dictionary, it would only be necessary in referring to the correspondent number prefixed to the word, to make in the first instance the alphabetic signal or letter under which the word was numbered; and 2dly, the signal in table I. correspondent to the number of the word; consequently, by making two signals for a word, no less than 25,974 words may be indicated, which is demonstrated by multiplying the 999 numbers of table I. by the 26 letters of table II.

N. B. As several letters of the dictionary do not contain 999 words, the blanks might be filled up with sentences, names of ships, &c.

To prevent discovery of the signals, the order of the flags may be transposed, while the numbers and letters in the tables would remain permanent.

TELESCOPE, (τηλε, far, and σκοπεω, I observe, Gr.) an optical instrument which serves for discovering and viewing distant objects, either directly by glasses, or by reflection by means of specula, or mirrors. Accordingly, telescopes are either refracting or reflecting; the former, consisting of different lenses through which the objects are seen by rays refracted through them to the eye; and the latter of specula, from which the rays are reflected and passed to the eye.

The lens or glass turned towards the object, is called the *object-glass;* and that next the eye, the *eye-glass;* and when the telescope consists of more than two lenses, all but that next the object are called *eye-glasses.* The telescope is one of the noblest and most useful inventions these ages have to boast of: by means of it, the wonders of the heavens are discovered to us, and astronomy is brought to a degree of perfection which former ages could have no idea of. The discovery was owing rather to chance than design, since the theory it depends upon was not then known. In 1560, John Baptista Porta, it appears, was the first who contrived a telescope, by joining concave glasses together, for the purpose of assisting some friends, who, either saw remote things dimly or near ones confusedly, but not understanding clearly his own invention, he neither improved, nor applied it, to celestial observation. About 30 years afterwards, viz. in 1590, a telescope 16 inches long, was made and presented to prince Maurice of Nassau, by Zachariah Jansen, a spectacle-maker of Middleburg, in Zealand. It appears Fontana, of Naples, put in his claim to the invention, without knowing of Jansen's discovery: but none of them had yet made any above eighteen inches in length. Simon Marius in Germany, and Galileo in Italy, Dr. Hutton asserts in his Mathematical and Philosophical Dictionary, were the first who made long ones fit for celestial observation, and seems surprised that the world was so long without that admirable invention, as the principles of optics are contained in Euclid's Elements, and were well-known to the ancient geometricians. Mr. W. Molyneux, in his Dioptrica Nova, declares his opinion, that our own celebrated countryman friar Bacon, who died in 1294, did perfectly well understand all sorts of optical glasses, and knew likewise the way of combining them, so as to compose an instrument like a telescope: and his son Samuel Molyneux asserts more positively, that the invention of telescopes, in it first original, was certainly put in practice by, friar Bacon, although its first application to astronomical purposes, may probably be ascribed to Galileo. The passage to which Mr. Molyneux refers, in support of Bacon's claims, occurs in his *Opus Majus*, p. 348 and 357 of Jebb's edition, in 1773; and Dr. Jebb himself, in his dedication of the *Opus Majus*, produces a passage from a manuscript to shew that Bacon actually applied telescopes to astronomical purposes.

There is also a passage in Thomas Digges's Stratioticos, p. 359, where he affirms that his father, Leonard Digges, had a method of discovering by perspective glasses set at due angles, all objects pretty far distant, that the sun shone upon, which lay in the country round about; and that this was by the help of a manuscript book, of Roger Bacon, of Oxford, whom he conceived to be the only person besides his father, who knew it since Archimedes. This is more remarkable, as the Stratioticos was printed in 1579, more than 30 years before Metius or Galileo made their discovery of those glasses; therefore it would appear that Roger Bacon was the first inventor of telescopes, and Leonard Digges the next reviver of them in this country; but to whomsoever we ascribe the honour of first inventing the telescope, certain it is that the rationale of this admirable instrument, depending on the refraction of light in passing through mediums of different forms, was first explained by the celebrated Kepler, who also pointed out methods of constructing others of superior powers, and more commodious applications than those first used. The principal effects of telescopes depend upon this plain maxim, viz. that objects appear larger in proportion to the angles they subtend at the eye; and the effect is the same whether the pencils of rays, by which objects are visible to us, come directly from the objects themselves, or from any place nearer to the eye, where they may have been united, so as to form an image of the object; because they issue again from those points in certain directions, in the same manner as they did from the corresponding points in the objects themselves. In fact all that is effected by a telescope is first to make such an image of a distant object, by means of a lens or mirror, and then to give the eye some assistance for viewing that image as near as possible; so that the angle which it shall subtend at the eye, may be very large compared with the angle which the object itself would subtend in the same situation. This is done by means of an eye-glass, which so refracts the pencils of rays as that they may afterwards be brought to their several foci, by the natural humours of the eye. But if the eye had been so formed as to be able to see the image, with sufficient distinctness, at the same distance, without an eye-glass, it would appear to him as much magnified, as it does to another person who makes use of a glass for that purpose, though he would not in all cases have so large a field of view.

Although no image be actually formed by the foci of the pencil without the eye, yet if, by the help of an eye-glass, the pencils of rays shall enter the pupil, just as they would have done from any place without the eye, the visual angle will be the same as if an image had been actually formed in that place. See PRIESTLEY's *History of Light*. SMITH's *Optics*, &c.

Astronomical TELESCOPE, is one that consists of an object-glass, and an eye-glass, both convex; and is so called from being wholly used in astronomical observations.

It was Kepler who first suggested the idea of this telescope; having explained the rationale, and pointed out the advantages of it in his Catoptrics, in 1611. But the first person who actually made an instrument of this construction was father Scheiner, who has given a description of it in his *Rosa Ursina*, published in 1630. To this purpose, he says, "if you insert two similar convex lenses in a tube, and place your eye at a convenient distance, you will see all terrestrial objects, inverted, but magnified and very distinct, with a considerable extent of view." He afterwards subjoined an account of a telescope of a different construction, with two convex eye-glasses which again reverses the images, and makes them appear in their natural position. Father Reita, however, soon after, proposed a better construction, using three eye-glasses instead of two.

As the astronomical telescope exhibits objects inverted, it serves commodiously enough for observing the stars, as it is not material, whether they be seen erect or inverted; but for terrestrial objects it is much less proper, as the inverting often prevents them being known. But if a plane well polished metal speculum, of an oval figure, and about an inch long, and inclined to the axis in angle of 45°, be placed behind the eye-glass; then the eye, conveniently placed, will see the image, hence reflected, in the same magnitude as before, but in an erect situation; and, therefore, by the addition of such a speculum, the astronomical telescope is thus rendered fit for terrestrial objects.

Achromatic TELESCOPE, a name given to the refracting telescope, invented by the late Mr. John Dollond, and so contrived as to remedy the aberration arising from colours, or the different refrangibility of the rays of light. The improvement made by Mr. Dollond, in his telescopes, by making two object glasses of crown glass, and one of flint, which was tried with success when concave eye-glasses were used, was completed by his son Mr. Peter Dollond, who, conceiving that the same method might be practised with success with convex eye-glasses, found, after a few trials, that it would answer the purpose. But besides the obligation we are under to Mr. Dollond, for correcting the aberration of the rays of light in the focus of object-glasses, arising from their different refrangibility, he made another improvement in telescopes, by correcting in a great measure, both this kind of aberration, and also that which arises from the spherical form of lenses, by an expedient of a very different nature, viz. increasing the number of eye-glasses. Whence it appears, that, by thus increasing the number of eye-glasses, the indistinctness near the border of the field of a telescope, may be very much diminished, though not entirely taken away. These considerations first set Mr. Dollond upon contriving how to enlarge the field, by increasing the number of eye-glasses, without any hinderance to the distinctness or brightness of the image; and though others had been about the same work before, yet observing that the five-glass telescope commonly used would admit of farther improvement, he endeavoured to construct one with the same number of glasses in a better manner, which so far answered his expectations, as to be allowed by the best judges to be a considerable improvement on the former. Encouraged by this success, he resolved to try if he could not make some farther enlargement of the field, by the addition of another glass, and by placing and proportioning the glasses in such a manner, as to correct the aberrations as much as possible, without any detriment to the distinctness; and at last he obtained as large a field as is convenient or necessary, and that even in the longest telescope that can be made.

Of the achromatic telescopes, invented by Mr. Dollond, there are several different sizes, from one foot to eight feet in length, made and sold by his sons, P. and J. Dollond, St. Paul's Church-yard, London.

Reflecting TELESCOPE, is one that may be enlarged to almost any power. Instead of lenses, it consists chiefly of metallic mirrors, and remote objects are exhibited by reflection instead of refraction. This noble invention has been ascribed to the ingenious Mr. James Gregory, a native of Aberdeen, and professor of mathematics at St. Andrews, and is described in his "*Optica Promota*," published in 1663; but, according to his own account, he possessed no mechanical skill, and after some fruitless trials to find a workman capable of realising his invention, he was obliged to abandon the thoughts of bringing into use this kind of telescope. Mr. Gregory died in 1675, aged only 36 years, and deserves a niche between Des Cartes, and the immortal Newton, who brought the invention of his precursor forward to maturity, and the reflecting telescope in time became the chief source of modern discoveries in astronomy. Dr. Birch, in his *History of the Royal Society*, *Vol.* 3, mentions a reflecting telescope constructed by Dr. Hooke, in which the great mirror was perforated, so that the spectator looked directly towards the object; a full description of which may be seen in HOOKE's *Experiments*, by DERHAM.

Mr. John Hadley in the year 1723, presented to the Royal Society a telescope, which he had constructed upon Newton's plan. The two telescopes which Newton had made were but 6 inches long, held in the hand for viewing objects, and in power they were compared to a 6 feet refractor; but the radius of the sphere, to which the principal speculum of Hadley's was ground, was 10 feet 5¼ inches, and consequently its focal length was 62⅝ inches. In the *Philos. Trans. Abr. Vol.* VI. *p.* 133, may be seen a drawing and description of this telescope, and also of a very ingenious but complex apparatus, by which it was managed. Mr. Hadley, after finishing two telescopes of the Newtonian construction, applied himself to make them in the Gregorian form, in which the large mirror is perforated. This scheme he completed in the year 1726.

Dr. Smith prefers the Newtonian construction to that of Gregory; but if long experience be admitted as a final judge in such matters, the superiority must be adjudged to the Gregorian, as it is now, and has been for many years past, the principal instrument in request.

The magnifying power of any telescope may be easily found by experiment, viz. by looking with one eye through the telescope upon an object of known dimensions, and at a given distance, and throwing the image upon another object, seen with the naked eye. Dr. Smith has given a particular account of the process, but the easiest method of all is, to measure the diameter of the aperture of the object-glass, and that of the little image of it which is formed at the place of the eye; for the proportion between these gives the ratio of the magnifying power, provided no part of the original pencil be intercepted by the bad construction of the telescope; for in all cases, the magnifying power of telescopes, or microscopes, is measured by the proportion of the diameter of the original pencil, to that of the pencil which enters the eye. See PRIESTLEY's *History of Light, p.* 747.

But the most considerable, and indeed truly astonishing magnifying powers, that have ever been used, are those of Dr. Herschel's reflecting telescopes. Some account of these, and of the discoveries made by them, may be seen in several of the late volumes of the *Philosophical Transactions.*

Dr. Herschel observes, that though optician s have proved that two eye-glasses will give a more correct image than one, he has always, from experience, persisted in refusing the assistance of a second glass, which is sure to introduce errors greater than those it would correct.

Mr. Green, of Deptford, has lately added, both to the reflecting and refracting telescopes, an apparatus, which fits it for the purpose of surveying, levelling, measuring angles and distances, &c. See his description and use of the improved reflecting and refracting telescopes, and scale of surveying, &c. published in 1778. Mr. Ramsden too, has lately adapted telescopes to the like purpose, of measuring distances from one station to another, &c.

Double Image TELESCOPE, *and Coming up Glass for measuring distances.* An ingenious instrument of this kind, invented by the celebrated Mr. Ramsden, has been, for several years used in the British navy. It consists of a common telescope, having one of the eye-glasses divided into two equal segments, the centres of which are separated from each other by a micrometer screw. By this means the centres of the two images of the object formed by the segments of the eye-glass are separated till they come in contact. The distance between the centres of the glasses is shewn in revolutions, and parts of a revolution of the finger screw, by which they are separated. The whole revolutions are indicated by a scale on a small slip of brass fixed to the eye-tube, while each of these units is subdivided into hundredth parts, by a circular head fixed on the finger screw. Another instrument for the same purpose, invented by M. Rochon, of the National Institute, and director of the naval observatory at Brest, has been used in the French navy, and found of very great utility. This instrument, founded on the double refraction of Iceland crystal, consists of two prisms of that substance, moveable along the axis of the telescope.

Two images are of course formed, and the angle subtended by the line joining their centres, is measured by the distance between the prisms and the object-glass. This instrument does great credit to the ingenuity of its inventor; but its utility is extremely limited by the small extent of the scale, which, without a particular construction of the prisms, cannot measure angles above 20 minutes, and also by the great difficulty of forming the prisms with sufficient accuracy. "I was desirous," says Mr. Rochon, "of extending the effect of the double refraction, so as to measure the diameter of the sun; and I accomplished this, which appeared to exceed the known power of the double refraction of rock crystal, which does not go beyond 20 minutes, when it is cut in the most advantageous prismatic shape. For this I employed two equal prisms, cut in the direction most suitable for my purpose, and on placing them in opposite directions, I found that the double refraction was not perceptible, but on reversing their directions, the double refraction of each prism was nearly doubled, so that I obtained two images, separated by the interval of 40 minutes.

"I ought not to omit, that in this new construction, there are difficulties of execution not easy to surmount; and I know only one person, citizen Narci, who is capable of giving rock crystal the prismatic form, in the proper direction for obtaining the double refraction necessary to the goodness of the micrometer."

It appears, that the instrument of M. Rochon will measure no farther than 20 minutes when it is made in the ordinary way; and no farther than 40 minutes when the new construction is adopted. But we have often seen, that in measuring distances, any error in the instrument, or in the observation, produces an error in the result, which increases with the smallness of the angle, and therefore in the Iceland crystal micrometer the result must be very inaccurate.

Dr. Brewster, editor of the Edinburgh Encyclopædia, in his treatise on new philosophical instruments, published in 1813, has described a double image telescope and coming-up-glass, as possessing the same property of the micrometer instrument, in requiring no correction for the angle, in consequence of the aberration of focal length, while it has the additional advantage of requiring no stand or support. In the double image telescope, however, the two images always keep close together, so that their contact may be observed with the utmost accuracy, not only when it is exposed to a slight tremor in the hand of the observer, but even during the motion of a vessel at sea, or during the gentle agitation of a carriage.

This valuable property renders the double image telescope of great utility in situations where there is neither time nor opportunity for steady observation. As a naval telescope, therefore, for measuring distances in general at sea, but particularly as a *coming-up-glass*, for ascertaining whether a ship is approaching to, or receding from, the observer, it is entitled to particular notice. This instrument which he has described, may have an extent of scale above twelve times greater than that of M. Rochon's micrometer, and may measure angles even of ten and twelve degrees. Its construction, too, is extremely simple, and it can scarcely be put out of order, unless it is broken in pieces.

The method of employing it as a coming-up-glass, may be very shortly explained. When a vessel is seen at a distance, it subtends a certain angle at the eye of the observer. If its distance increases, this angle will of course be proportionally diminished; and if it approaches to the observer, the angle will be increased. Now if this vessel be viewed through the double image telescope, and if the semi-lenses are separated from the object glass, till the two images, either of the whole vessel, or of any part of it, are in contact, we obtain from the scale the angle which that part subtends. If the images continue in contact, the distance of the vessel must have remained the same; but if the images have separated, from each other, the vessel must be receding from the observer: and the proportion of the two distances may be found, by again bringing the images in contact, and finding the angle anew. If the first angle was 50′, and the second 40′, then the first distance must have been to the second, as 40 to 50, that is in the inverse ratio of the angles. If the images, on the contrary, had overlapped each other at the second observation, the distance between the two vessels must have been diminishing, and the degree of diminution would have been found, by bringing the images into contact and again observing the angle. If the angle were now 60′, then the ratio of the distances must have been as 50 to 60.

Neither the length of the ship, nor any part of it lying in a horizontal direction, or in a direction inclined less than 90° to a horizontal line, should be chosen for the subtense of the angle, as the angle which it forms at the eye of the observer may vary merely from the turning of the ship, while its distance suffers no deviation. The length of the masts, or any part of them, or of any vertical line in the vessel, ought always to be taken for this purpose.

A Coming-up Glass, or Telescope, has also been made by Mr. Thomas Jones, and is constructed so as to enable a person on board a chasing ship, to ascertain whether she lessens or increases her distance from the vessel she is pursuing. To convert a refracting telescope into what the sailors call a *coming-up glass*, the third glass from the eye, in the drawer of the telescope, is divided in two, by cutting it across its centre at right angles to its surface; and in using it, these two semi-lenses are separated from each other in the direction of their line of separation.

By turning a micrometer finger-screw on one side of the eye-tube, each semi-lens forms an image of the same object; and these images will be more or less separated in proportion of the distance of the centres of the semi-lenses from each other, which distance is shewn in the revolutions, and parts of a revolution, of the finger-screw that separates them. For this purpose, a circular head is fixed on the finger-screw, the edge of which is divided into 100 parts; and in order to know the number of revolutions, a small slip of brass that passes over the graduated surface of the head (and serves as its index for shewing the centesimal parts of a revolution), is fixed to the eye-tube, and has its chamfered edge also divided, each division being equal to one entire revolution of the screw.

The adjustment of this telescope will be obvious enough to those who are acquainted with the application of a micrometer. As to its use; having directed the telescope to the vessel chased, turn the finger-screw till the two images of some well-defined part of the vessel appear to have their extreme edges in contact with each other; then read off the number of revolutions of the screw, shewn on the chamfered edge, also the parts of a revolution shewn on the edge of the lead: then, after some time, to know whether distance has been gained or lost in the chase, again bring the edges of the images of the same object in contact as before. If the revolutions of the screw be the same as before, there has been neither gain nor loss in the chase. But if the revolutions be less, the distance from the chased vessel will be greater in proportion to the difference of these numbers to the former; on the contrary, if the number of revolutions be greater, a gain has been made on the vessel, in proportion to the difference of these numbers to the number of revolutions and parts of the first observation.

Luminous Image Telescope. Dr. Brewster has given a description of an ingenious instrument of this kind, for measuring angles and distances at night, in his treatise on new philosophical instruments, in which he proves that the images of two luminous points formed in the focus of an object-glass, may be expanded into circular images of light, till they touch each other, merely by pushing in the eye-piece of the telescope, and has demonstrated that the angle formed by the line joining the centres of those images when in contact, varies directly as the distance of the intersection of the cones of rays from the focus of the object-glass, and inversely as the distance of the same part from the object-glass itself; he has likewise shewn, that, in bringing the luminous images in contact, the eye-piece should be pushed in, so that we may perceive the sections of the cones of rays between the object-glass and its principal focus, though the position of the scale upon the tube will in reality be without the focal point. The sections of the cones of light are, in general, better defined, when they are taken between the object-glass and its principal focus; but in good telescopes, the difference is not so great as to prevent the scale from being made on either side, or on both sides of the principal focus. Aided by these principles, the construction of the instrument becomes remarkably simple. It is of the greatest importance that the apertures be exactly circular; and when the form of the luminous points can be altered by the observer, they should likewise be made as round as possible. As the accuracy of observation must in a great measure depend on the distinctness of the line, by which the circular images are bounded, it is of the utmost importance that an excellent achromatic object-glass should be selected for the instrument. The objects being always luminous, there is no occasion for much light, and a great magnifying power being equally unnecessary, the diameter of the object-glass may be made much smaller than in common telescopes.

When the instrument is constructed according to these principles, its application to the measurement of distances is precisely the same as that of the micrometrical and the double image telescope: the use of this instrument, however, is in some respects limited, from the necessity of having the angle always subtended by luminous points. But there are numerous cases, and these too, of great importance, where it may be successfully employed. The distance of a vessel from the coast, or the rate at which it approaches to or recedes from the shore, may frequently be determined by taking the angle subtended by accidental or by fixed lights on the land.

Simple TELESCOPES, are used to observe eclipses of the sun and moon; immersions and emersions of Jupiter's satellites; occultations of fixed stars by the moon; the external and internal contact of Mercury and Venus, with the sun's limb in their transit over his disc, and by the latter to determine the sun's parallax; the variation of the apparent magnitude of some of the fixed stars; and, in short, in all cases where the measure of space is not required.

An eclipse of the moon, if the shadow of the earth were well terminated, would afford an accurate and ready method of finding the longitude; but as the shadow is not well defined, it is subject to a small degree of inaccuracy. The time of the beginning of an eclipse is rendered uncertain, because the darkness of the penumbra keeps continually increasing till it comes to the umbra, so that until the umbra actually comes upon the moon it is not discovered, and therefore you miss the accurate time of the beginning.

The beginning and end of a lunar eclipse cannot be generally observed nearer than 1′ of time, and frequently not nearer than 2′ or 3′; therefore the longitude cannot be depended upon, from a single observation, to nearer than one degree. The mean, however, of a great number of observations will reduce it to much smaller limits. Now, as the phases of the moon are universal, or appear at the same instant to every observer, it is manifest that the difference of the times at different places, at which any one phase is observed, or the difference of the times of observation at one place, and the computed time for any other, converted into degrees at the rate of 15° for an hour, will give the difference of the meridians. The telescope made use of to observe a lunar eclipse, should be one that has a great deal of light and does not magnify much. Those called night glasses may be conveniently used for this purpose, in which case Dr. Maskelyne advises the observer to place a circular aperture at the object-end, about one-half or one-third of the size of the common aperture. The shadow comes upon the moon on the east side, and goes off on the west.

The beginning of a solar eclipse is also liable to some uncertainty; for as you cannot see the body of the moon approaching the sun, its contact cannot be observed till it has sensibly altered the appearance of the sun's limb where it enters. The observer, therefore, should be apprized of that part of the sun's limb where the moon enters, and keep his eye about that point, and whenever he suspects the moon is entering, he should note the time, which he may strike out if he finds that he was deceived. The best way to find the instant of the contact is, to take the distance of the horns (with a micrometer) at some instant after the commencement, and from thence calculate how far the moon is entered upon the sun's disc; from whence, knowing the velocity with which the moon approaches the sun, the time of the beginning will be known. The same difficulty, although in a smaller degree, will attend the determination of the time of the end; for as the observer is here apprized of the event, and actually sees the moon leaving the sun, he will be able to judge more accurately of the end than he could of the beginning. As the eclipses of the sun are not universal, or do not begin at different places at the same instant, the difference of meridians is not found immediately by comparing the times of the same phase at two places. The reader will find the method of reducing the difference of longitude, in this case, in most books of astronomy. If the telescope be furnished with a micrometer, the quantity of the eclipse may also be found by measuring the distance from the middle, terminating the dark part, to the opposite part of the sun's limb. If an eclipse be observed in a place whose latitude and longitude is known, it furnishes data for correcting the lunar tables, by comparing the calculations with the observations.

The eclipses of Jupiter's satellites afford the readiest method of determining the longitude of places at land. It was also hoped that some method might be invented to observe them at sea, and Mr. Irwin made a chair to swing for that purpose, for the observer to sit in; but Dr. Maskelyne, in a voyage to Barbadoes, under the direction of the Commissioners of Longitude, found it totally impracticable to derive any advantage from it, and he observes that, "considering the great power requisite in a telescope, for making these observations well, and the violence as well as irregularities of the motion of a ship, I am afraid the complete management of a telescope on ship board will always remain among the desiderata for finding the longitude."

The telescope proper for making these observations are common refracting ones, from 15 to 20 feet, reflecting ones of 18 inches or 2 feet, or the 46 inch achromatic with three object-glasses, which were first made by Mr. Dollond. On account of the uncertainty of the theory of the satellites, Dr. Maskelyne advises the observer to be settled at his telescope three minutes before the expected time of an immersion of the first satellites, six or eight minutes before that of the second or third, and at least a quarter of an hour before that of the fourth. And if the longitude of the place be also uncertain, he must look out proportionably sooner: thus, if the longitude be uncertain to 2° answering to 8′ of time, he must begin to look out 8 minutes sooner than is mentioned above. However, when he has observed one eclipse, and found the error of the tables, he may allow the same correction to the calculations of the ephemeris for several months, which will advertise him very nearly of the time of expecting the eclipses of the same satellite, and dispense with his attending so long. As it is of the utmost importance for the observer to know at what distance from Jupiter an eclipse happens, particularly the emersion, in order that his eye may be directed to the proper place when he is waiting for the satellite's emerging from the shadow,

The first page of each month in the Nautical Almanack will inform the astronomer what phenomena it may be proper for him to observe. And previous to any observation, the instrument should be ready, and examined; the position of the object should also be attended to, so that if there be any local obstacles in the way, they may be removed. For want of attending to these circumstances, observations are frequently lost. Dr. Maskelyne has given a correct table and particular directions in the *British Mariner's Guide*. See also VINCE's *Practical Astronomy*.

A Naval and Military TELESCOPE, for measuring distances and the extensions of objects at sight, by means of a new micrometrical adjustment, for which a patent is obtained, has been invented by Mr. Cater Rand, of Lewes, Sussex, and now sold by Mr. J. Nanier, No. 70, St. James's-street, London.

This instrument consists of a micrometer of a new and improved construction, applied to the eye-tube of a telescope, in order to measure small angles subtended by objects, and thence to compute their distances and extensions by means of a table engraved upon the telescope.

Besides its use to the army and navy for the above purposes, it has all the properties in common with other telescopes for viewing objects on the horizon and in the heavens; for measuring the apparent diameters of the sun, moon, and planets; the distances of the satellites from their respective planets, &c. It may also be used as a microscope for viewing and measuring the linear extensions of the smallest objects.

Figure 8, plate XXI, shews, *in plano*, the micrometer, which is situated in the focus of the first eye-glass of the telescope; A is the nut or screw-head that regulates the mechanism of the parts, and gives motion to the several parallel wires, *a*, *b*, *c*, *d*; *a*, *b* being two wires seen in one, which open parallel to, and to the extent of *c* and *d* on turning the screw at A; at the same time, and by the same motion of the screw, *c* and *d* move parallel to themselves to the extent *e*, *f*, always keeping at an equal distance from the wires *a*, *b*, respectively. The wires *a*, *b*, or *c*, *d*, thus opened include an angle, of which the eye of the observer is the vertex; this angle may be accurately measured in the following manner, by the motion of the graduated scale B.

This scale is attached to the plate which carries the two wires *a*, *c*, and consequently moves up and down with them by the turning of the screw A; another plate carrying the other two wires, *b* and *d*, moves in a contrary direction, and thus the opening of the wires *a* and *b*, and *c* and *d*, is produced. A scale, C, is fixed to the telescope by the side of the scale B, and has on it an index, serving to mark the number of divisions on the scale B, which the latter passes over while the wires are opening; but as the wire *a* moves upwards while the wire *b* moves downwards, and that the scale B alone moves, while C is fixed, it is evident that the distance between the wires *a* and *b* is always equal to double the space passed over by the scale B: hence arises a necessity of doubling the number of divisions which are read on B, to get the value of the angle subtended between the wires *a* and *b* when opened. The graduations on the scale B are hundredths of an inch, each of which in these telescopes is equal to an arc of one minute of a degree, but each division being doubled for the reason above given, its value is in reality equal to two minutes of a degree. On the scale C, a vernier or nonius is placed, subdividing the graduations of the scale B into tenth-parts, each of which is consequently equal to six seconds, but when doubled, to 12 seconds.

The value, in parts of a degree, of the divisions on the scale B, depends upon the magnifying power of the telescope; if, therefore the micrometer be applied to any telescopes which have not the same power as those to which the inventor has hitherto applied them, it will be necessary to have the value of its divisions calculated expressly for every telescope of different power, which should be engraven on some part of the telescope. The value of the divisions of the vernier depends of course upon that of the micrometer scale. For reading off the fine divisions of the scales B and C, a small lens is placed before them.

The use of the exterior parallel wires *c*, *d*, is, to give the micrometer a means of measuring larger angles than the wires *a*, *b*, would allow; in this case the angle being subtended between the wires *c*, *d*, the value of the angles contained between the wires *a*, *b*, must be added to the value of the angles contained between the wires *a*, *c*, and *b*, *d*, respectively, which latter angles in each telescope are constant quantities, and their values are engraved on the eye-tubes, to their true amount, they are therefore simply to be added to the angle between *a* and *b*, without being doubled, which this must always be.

The following example will be sufficient to explain the use of the instrument in taking angles, and the manner of reading them on the scale.

Applying the telescope to a sentinel on a distant fort, and opening the parallel wires *a*, *b*, by means of the screw head, A, till the whole man, from the top of his hat to the bottom of his feet, is neatly defined between the two wires, then reading the divisions of the scales, it was found that the index stood a little below the second division on the micrometer scale, and the fourth division of the vernier coincided with a line of the micrometer scale, thus making $2\frac{4}{10}$ or 2′ 24″, which doubled, give 4′ 48″, the apparent angle subtended by the sentinel.

Note. As in this instrument the vernier is fixed and the scale moveable, in order to find the number of centesms of the micrometer divisions, the divisions on the vernier must be read from the bottom upwards to the line of coincidence.

Having found the angle subtended at the eye of the observer by a distant object, the value of that angle must be sought in a table, which is engraved on the telescope, shewing how often the size of an object seen under certain angles is contained in its distance from the observer, and by means of the number in such table, corresponding to the given angle, the real distance of any object may be computed as in the two following problems.

PROBLEM I.

To find the distance of an object from the place of observation, its size being known.

Multiply the tabular number corresponding to the angle under which the object is seen, by the size of the object, the product is the distance required; and *vice versa* if the distance be known and the size required, divide the distance by the tabular number corresponding to the angle, the quotient will be the size required.

The truth of this rule is evident, because the table always gives the number of times that the size of an object is contained in its distance.

Example. Laying off an enemy's battery at sea, the height of whose flag-staff had been previously ascertained to be 27 feet, required its distance from the ship.

The angle subtended by the flag-staff, being found by the foregoing method to be 26′ 24″. Seek 26′ 24″ in the table of distances, which will be found 130.24 (taking a proportional part for the 4″); this number multiplied by 27, the height of the staff gives 3516½ feet nearly, the distance between the ship and battery.

If no object, whose size is known, should present itself by which to ascertain the distance, the angle subtended by a man whose height, including his hat and shoes, may be assumed at 6 feet, will give the distance near enough for military or nautical purposes, such an object as a man being almost always to be seen, both in a fortress and on the sea-shore. Experience has taught the most able artillerist that nothing in the practice of gunnery is more difficult than to judge truly of distances by the eye, as it is called; and the more especially where objects are viewed across a deep ravin or valley: but with this instrument, the distance of an object may almost instantaneously be found, if its height be known, even to a foot, while without it, much ammunition may be injudiciously and fruitlessly expended, without the least annoyance to the enemy.

PROBLEM II.

To find the distance and size of an object, when both are unknown.

Here *two* subtended angles of the same object must be taken, at any convenient distance from each other, in a right line, approaching to or receding from the object, as may be most convenient: then, to find the *distance*, multiply the larger angle into the distance between the *two* places of observation, and divide the product by the difference between the *two* angles, which will give the distance of the object from the farther place of observation: and to find the *size* of the object, divide the distance at either place of observation by the tabular number answering to the subtended angle at that place.

Example. Let the distance and size of the object D E, fig. 9, plate XXI. from either place of observation, as at A or C, be required. Apply the telescope to the object, beginning at either station, suppose A, and note the angle subtended, which let be 17′; pace, or measure any distance, from A to C, in a right line with the object, which suppose 45 feet; at C, apply the telescope again, and suppose the angle subtended by the object, from this station, to be 20′: then by the rule $\frac{20' \quad 45}{20'-17'} = 300$ feet, the distance from A, consequently 255 feet is the distance from C. To find the *size* of the object D E, seek the tabular number opposed to 17′, which will be 202 by dividing 300 feet, the distance from A, the quotient will be 1.485 feet, or 18 inches nearly, the size of the object; and the same result would have been obtained if 255, the distance from C, had been divided by 172, the tabular number opposed to 20′.

To find the *distance* and *height* of an object at SEA is attended with a little difficulty, owing to several causes; but principally to the irregular and almost constant *motion* of a SHIP, and the difficulty of measuring the *true* distance between *two* places of observation; but even here, if due care be taken, every thing may be ascertained with wonderful accuracy; it is, however, recommended to make use of *three*, instead of *two* observations, to guard against any material error, where very great accuracy is required. It is also adviseable to take the mean of two or three observations from the *size* or *height* of a *known* object, where the distance only is required, which may be done in a few seconds, so that the distance of a ship's way between the times of taking the observations can be but a few yards, which may be accounted for.

The more perpendicular the object whose distance or height is required, is to the axis of the telescope, the more exact will the result be; and the distance and extension of an object upon any very considerable eminence cannot be so truly taken by the foregoing rules as if it were upon a *true* horizontal level, unless it be perpendicular to the axis of the telescope: but in military and nautical observations this will seldom be of any consequence.

In trigonometrical surveys, it will be proper to make an allowance for the diminishing angle of objects, in proportion to their elevation above the plane of the *horizon*, where they are not perpendicular to the axis of the telescope, according to the following analogy, viz.

As co-sine of the elevation of the object above the horizon, is to radius, so is the observed angle subtended by the object, to the true angle. With which angle operating according to the circumstances of the case as in the preceding problems, the distance of the object may be very correctly found.

TELL-TALE, (*axiometre*, Fr.) a small piece of wood, traversing in a groove across the front of the poop-deck, and which, by communicating with a small barrel on the axis of the steering-wheel, indicates the situation of the helm.

TEMPEST, (*tempête*, Fr.) implies the utmost violence of the wind. See the article STORM.

To TEND, (*éviter au changement de la marée en parlant d'un vaisseau à l'ancre*, Fr.) is to turn or swing a ship round when at single anchor, or moored by the head in a tide-way, at the beginning of the flood or ebb. Thus, if the flood sets northerly, it is evident that the ship, unless when moored head and stern, will fall into the line of the current, turning her head to the southward. But as the reflux will, for the same reason. set to the southward, the ship will of necessity turn about at the change of the tide, and carry her head to the northward; and the transition from one situation to the other is called tending or swinging.

To TEND *a Ship for a Weather-Tide.* A ship riding at anchor upon a lee-tide, with the wind in the direction of the tide, requires, upon the tide's setting to windward, to tend clear of her anchor: for this purpose, when the weather-tide sets, and brings the wind broad upon either bow, hoist the jib and fore-stay-sail, and brace full the yards to shoot the ship a tight cable from her anchor: then put the helm a-lee, and wait until the buoy comes upon the lee-side; this done, brace-to the head-yard, and keep the after-yards full to assist the helm. If the buoy bears nearly a-beam, the jib and fore-stay-sail may be hauled down; but, if the wind is fresh, and shoots the ship nearly end-on with the cable, bringing the buoy upon the quarter, keep up the fore-stay-sail; as, in this situation, the ship will be in danger of breaking her sheer against the helm; and the fore-stay- ail will be ready to catch her before she can fall to windward of her anchor. As the weather-tide slacks, the ship will gradually fall wind-road; then haul down the fore-stay-sail, and right the helm: should it blow fresh, let the yards be braced forward, and give the ship sufficient cable.

To TEND *a Ship with the Wind a few points across the Tide.* If a ship riding at anchor upon a lee-tide, with the wind two or three points upon the bow, is to be cast for a weather-tide; when the tide is done, the ship will become wind-road, and of course must cast with her head to the weather-shore. As the lee-tide makes and brings the wind on either side, put the helm a-lee, hoist the jib and fore-stay-sail with the sheets to windward, brace a-back the head-yards, and fill the after-yards. When the ship has sheered tight to windward of her anchor, haul down the jib and fore-stay-sail; and, as the wind is broad upon the quarter, she will lie quiet the remainder of the tide. As the weather-tide slacks, the ship will shoot end-on with the cable, bringing the buoy upon the weather quarter. When the wind is a-beam, if it is thought necessary, the jib and fore-stay-sail should be hoisted, to force the ship a tight cable from her anchor, and hauled down when they shiver. Should it blow fresh, give the ship sufficient cable before the lee-tide makes, and point the yards to the wind.

To TEND *a Ship with the Wind across the Tide.* The simplest way of tending a ship is, to keep each tide to leeward of her anchor. At each slack water the ship will become wind-road; and as she tends and brings the wind on either side, put the helm a-weather, and hoist the fore-stay-sail, with the sheet to windward, to force the ship a tight cable from her anchor. When the tide is set, and the ship upon a proper sheer to leeward of her anchor, the fore-stay-sail may be hauled down.

TENDER, (*patache, sorte de bâtiment servant de cayenne, de corvette, &c.* Fr.) a small vessel employed to attend a larger one, to supply her with stores, to carry intelligence, &c. Vessels appointed to receive volunteers and impressed men, and convey them to receiving ships, &c. are also called tenders.

TENON, in carpentry, the end of a mast, bowsprit, or any piece of timber, cut smaller to enter into a mortise.

TERM, in ship-building, a piece of carved-work, placed under each end of the taffrail, at the side timber of the stern, and extended down as low as the foot-rail of the balcony.

TERMINATOR, in asfronomy, a name sometimes given to the circle of illumination, from its property of terminating the boundaries of light and darkness

TERRA. See the article EARTH.

TERRA-FIRMA, in geography, is sometimes used for a continent, in contradistinction to island. Thus, Asia, the Indies, and South America, are usually distinguished into terra-firmas and islands.

TERRAQUEOUS, (*terra, aqua* Lat) in geography, an epithet given to our globe or earth, considered as a mass of land and water.

TERRESTRIAL, something relating to the earth. As the terrestrial globe, terrestrial line, &c.

To TERTIATE, is to examine the thickness of the metal of a piece of ordnance, in order to ascertain its strength; or, in other words, to find whether it has its due thickness at the vent, trunnions, and neck; if the trunnions and neck are in due order, and the chase straight, &c. This is usually done with a pair of calliper compasses.

TETRAEDRON, or TETRAHEDRON, (τετρα, & ἑδρα, Gr.) in geometry, is one of the five Platonic or regular bodies, being solids, comprehended under four equilateral and equal triangles. Or it is a triangular pyramid of four equal and equilateral faces.

TETRAGON, (τετραγωνος, Gr.) in geometry, a quadrangle, or a figure having four angles. Such as a square, a parallelogram, a rhombus, and a trapezium. It sometimes also means peculiarly a square.

TETRAGONIAS, a meteor, whose head is of a quadrangular figure, and its tail or train long, thick, and uniform. It does not differ much from the meteor called *trabs* or beam.

TETRAGONISM, (τετραγωνισμος, Gr.) a term which some authors use to express the quadrature of the circle, because the quadrature is the finding a square equal to it.

THEODOLITE, a mathematical instrument generally useful, and particularly so to engineers and artillerists, in surveying and taking heights and distances.

THERMOMETER, (from θερμος, *heat*, and μετρον, *measure*, Gr.) an instrument for measuring the temperature of the air, &c. as to heat and cold.

The thermometer and thermoscope are generally accounted the same thing. But Wolsius makes a difference, and also shews that what we call thermometers, are really no more than thermoscopes.

The invention of the thermometer is attributed to several persons by different authors, viz. to Sanctorio, Galileo, Father Paul, and to Drebbel. Thus, it is ascribed to Cornelius Drebbel of Alcmar, about the beginning of the seventeenth century, by his countryman Boerhaave. But whoever was the first inventor of this instrument, it was at first very rude and imperfect; and as the various degrees of heat were indicated by the different contraction or expansion of air, it was afterwards found to be an uncertain, and sometimes a deceiving measure of heat, because the bulk of the air was affected, not only by the difference of heat, but also by the variable weight of the atmosphere.

The Air THERMOMETER depends on the rarefaction of the air, and consists of a glass tube connected at one end with a large glass ball, and at the other end immersed in an open vessel, or terminating in a ball with a narrow orifice, which vessel or ball contains any coloured liquor that will not easily freeze. *Aquafortis*, tinged of a fine blue colour, with solution of vitriol or copper, or spirit of wine, tinged with cochineal, will answer this purpose. But the ball must be first moderately warmed, so that a part of the air contained in it may be expelled through the orifice; and the liquor pressed by the weight of the atmosphere will enter the ball, and rise to the middle of the tube, at a mean temperature of the weather; and in this state, the liquor by its weight, and the air included in the ball and tube by its elasticity, will counterbalance the weight of the atmosphere. As the surrounding air becomes warmer, the air in the ball and the upper part of the tube, expanding by heat, will drive the liquor into the lower ball, and consequently its surface will ascend; on the contrary, as the ambient air becomes colder, that in the ball is condensed, and the liquor, pressed by the weight of the atmosphere, will descend; so that the liquor in the tube will ascend or descend more or less, according to the state of the air contiguous to the instrument. To the tub is affixed a scale of the same length, divided upwards and downwards, from the middle, into 100 equal parts, by means of which may be observed the ascent and descent of the liquor in the tube, and consequently the variations also in the temperature of the atmosphere.

Mercurial THERMOMETER. It is a most important circumstance in the construction of thermometers, to procure a fluid that measures equal variations of heat by corresponding equal variations in its own bulk: and the fluid which possesses this essential requisite in the most perfect degree, is mercury: the variations in its bulk approaching nearer to a proportion with the corresponding variations of its heat, than any other fluid. Besides, it is the most easy to purge of its air, and is also the most proper for measuring very considerable variations of heat and cold, as it will bear more cold before freezing, and more heat before boiling than any other fluid. Mercury is also more sensible than any other fluid, air excepted, or conforms more speedily to the several variations of heat. Moreover, as mercury is a homogeneous fluid, it will, in every thermometer, exhibit the same dilatation or condensation by the same variations of heat.

Dr. Halley, though apprized only of some of the remarkable properties of mercury above recited, seems to have been the first who suggested the application of this fluid to the construction of thermometers. See *Philos. Trans. Abr. Vol.* II. p. 34.

Boerhaave says, these mercurial thermometers were first contrived by Olaus Roemer; but the claims of Fahrenheit of Amsterdam, who gave an account of his invention to the Royal Society, in 1724, has been generally allowed. And though Prius and others, in England, Holland, France, and other countries, have made this instrument, as well as Fahrenheit, most of the mercurial thermometers are graduated according to his scale, and are called *Fahrenheit's Thermometers*. See *Philos. Trans. Num.* 381, or *Abr. Vol.* VII. *p.* 49.

To construct Fahrenheit's thermometer, a small ball is blown on the end of a glass tube, of an uniform width throughout. The ball and part of the tube are then to be filled with quicksilver, which has been previously boiled to expel the air. The open end of the tube is then to be hermetically sealed. The next object is to construct the scale. It is found, by experiment, that melting snow, or freezing water, are always at the same temperature. A thermometer, therefore, immersed in the one or the other, will uniformly stand at the same point. It has likewise been observed, that water boils under the same pressure of the atmosphere at the same temperature. A thermometer, therefore, immersed in boiling water, will uniformly stand at the same point. Here then are two fixed points, from which a scale may be constructed, by dividing the intermediate space into equal parts, and carrying the same divisions as far above and below the two fixed points as may be wanted: thermometers constructed in this way may be compared together, for if they are accurately made, and placed in the same temperature, they will always point to the same degree on the scale.

The freezing point of Fahrenheit's thermometer is marked 32°, and the reason of this is said to have been, that this artist thought that he had produced the greatest degree of cold, by a mixture of snow and salt; and the point at which the thermometer then stood, in this temperature, was marked zero. The boiling point, in this thermometer, is 212°, and the intermediate space between the boiling and freezing points, is therefore divided into 180°. This is the thermometer commonly used in Great Britain.

There are other thermometers used in different countries of Europe, which differ from each other in the number of degrees between the freezing and boiling points. Reaumur's thermometer was generally used in France before the revolution, and is still employed in different countries on the continent. The freezing point in this thermometer is marked zero, and the boiling point 80°. To convert the degrees of Reaumur's thermometer to those of Fahrenheit, is, by this *formula*, Reaumur, $\frac{\times 9}{4}+32$ = Fahrenheit, that is, multiply the degrees of Reaumur by 9, divide by 4, and add 32. This gives the corresponding degrees on Fahrenheit's scale. The thermometer of Celsus has the space between the freezing and boiling points divided into 100°. The boiling point is 100°, and the freezing point zero.

This thermometer is used in Sweden. "The *thermometre centigrade*," now used in France, has the scale divided in the same way. To convert the degrees of this thermometer into those of Fahrenheit, Celsus, $\frac{\times 9}{5} + 32 =$ Fahrenheit. In De lisle's thermometer, which is used in Russia, the space between the boiling and freezing points is divided into 150°; but the degrees are reckoned downwards. The boiling point is marked zero, and the freezing point 150°. To reduce the degrees of this thermometer under the boiling point to those of Fahrenheit, Delisle $\frac{\times 6}{5} - 212 =$ Fahrenheit. And above the boiling point Del. $\frac{\times 6}{5} + 212 =$ Fah.

Such are the principles and mode of construction of the thermometers; an instrument which has been of the utmost importance in enabling us to discover many of the properties of caloric, as by it only we can ascertain, with accuracy, the relative temperatures.

The common contrivance for a self-registering thermometer, now sold in most of the London shops, consists chiefly of two thermometers, one mercurial, and the other of alcohol having their stems horizontal; the former has for its index a small bit of magnetical steel wire, and the latter a minute thread of glass, having its two ends formed into small knobs, by fusion in the flame of a candle.

The magnetical bit of wire lies in the vacant space of the mercurial thermometer, and is pushed forward by the mercury whenever the temperature rises, and pushes the fluid against it; but when the temperature falls, and the fluid retires, this index is left behind, and consequently shews the maximum. The other index, or bit of glass, lies in the tube of the spirit thermometer, immersed in the alcohol: and when the spirit retires, by depression of temperature, the index is carried along with it, in apparent contact with its interior surface; but, on increase of temperature, the spirit goes forward and leaves the index, which therefore shews the minimum of temperature since it was set. As these indices merely lie in the tubes, their resistance to motion is altogether inconsiderable. The steel index is brought to the mercury by applying a magnet on the outside of the tube, and the other is duly placed at the end of the column of alcohol, by inclining the whole instrument. See NICHOLSON'S *Encyclopedia*.

The thermometrical cold of the frost in Russia, in November 1812, being at from 10 to 15 degrees of Reaumur's thermometer, must of course mean below the freezing point, and that so early in the winter is prodigious; as it is equivalent to 27 and 37 below the freezing point upon Fahrenheit's scale, that degree would be nearly equal to the most severe cold ever felt in this country. At St. Petersburgh, in the month of December, 1812, Reaumur's thermometer stood two days at 19, two at 20, and one at 21 degrees. On the 1st of January, 1813, it was at 18; on the 2d at 20; it was four times at 21, and twice at 23, and twice at 24; on the 18th it was at 25, and the next day at 28. At Doubossan, in the temperate climate of the government of Cherson, the thermometer was at 22 degrees in the month of December.

At Chester, on the evening of the 16th of January 1814, the thermometer stood at 7 degrees, that is 25 degrees below the freezing point. It has not been so low for 19 years, when it was one night at 5 degrees.

At Leeds, on the 22d of January, 1814, at half past 11, p.m. Fahrenheit's thermometer, which was suspended in a garden, stood at 6¼ degrees, and during the night at 5 degrees. On the 27th of December, 1795, at 11 p.m. the thermometer stood at 4 degrees.

THICK and THIN, or QUARTER-BLOCKS. See the article BLOCK.

THICK-STUFF, (*vaigres plus épaisses qui les vaigres ordinaires, qui se placent dans la construction des vaisseaux anglois vis-à-vis les jonctions des alonges des couples*, Fr.) in ship-building, a term for all plank exceeding four inches in thickness. See the articles MIDSHIP-SECTION, and SHIP-BUILDING.

THIMBLE, (*cosse de fer*, Fr.) in sail-making, an iron ring, whose outer surface is hollowed throughout its whole circumference, in order to contain, in the channel or cavity, a rope which is spliced about it, and by which it may be hung in any particular station: see fig. 9, plate VIII. It is used to defend the eye of a rope which surrounds it from being injured by another rope passing through it, or by the hook of a tackle which is hung upon it.

THOLES, (*toulets*, Fr.) are small pins driven perpendicularly into the gunwale of a boat. In the exercise of rowing, the oar is contained between the two tholes, in the space which is called the row-lock. Sometimes there is only one pin to each oar, as in the boats navigated in the Mediterranean sea. In that case the oar is fastened upon the pin by means of a strap or rope; and indeed this method is much more ancient than the former. See the article To Row.

THROAT, (*corne d'une vergue supérieure de brigantine, ou partie du pic voisine du mât*, Fr.) a name given to the end of a gaff which is next the mast, and is opposed to peak, which implies the outer extremity. Hence the ropes employed to hoist up, and lower a gaff, being applied to those parts of it, are called the throat and peak halyards.

THROAT, in anchor-making, the inner part of the arms of an anchor, joining the shank. See the article ANCHOR.

THROAT-BRAILS, (*cargues*, Fr.) are those which are attached to the gaff close to the mast.

THROAT-HALYARDS, (*drisses d'en dedans des voiles auriques*, Fr.) are ropes or tackles applied to hoist the inner part of the gaff and its appendant portion of the sail.

THROAT *of a Knee*, (*collet d'une courbe*, Fr.) in ship-building, implies the hollow part of a piece of knee-timber.

To THRUM, (*larder*, *piquer*, Fr.) is to insert in a sail, mat, &c. through small holes made by a bolt-rope-needle, or a marline-spike, a number of short pieces of rope-yarn or spun-yarn.

THRUST, (*botte*, Fr.) in fencing, hostile attack with any pointed weapon. When one party makes a push with his sword, to wound his adversary with the point, it is called a thrust.

To THRUST *Home*, (*tirer à fond*), in fencing, denotes a thrust after any parade of tierce, or of half circle, when the adversary does not throw in a rèposte, or is late in doing so.

THUMB-STALL, in sail-making, a ferrule made of iron, horn, or leather, with the edges turned up, to receive the thread in sewing. It is worn on the thumb, to tighten the stitches.

THUNDER, (*tonitru*, Lat. *tonnerre*, Fr.) is that noise in the lower region of the air, excited by a sudden explosion of electrical clouds; which are therefore called thunder clouds.

The *phenomena* of thunder is variously accounted for. Seneca, Rohaut, and some other authors, both ancient and modern, account for it, by supposing two clouds impending over one another, the upper and rarer of which, becoming condensed by a fresh accession of air raised by warmth, from the lower parts of the atmosphere, or driven upon it by the wind, immediately falls forcibly down upon the lower and denser cloud; by which fall, the air interposed between the two being compressed, that next the extremities of the two clouds is squeezed out, and leaves room for the extremity of the upper cloud to close tight upon the under; thus a great quantity of the air is enclosed, which at length escaping through some winding irregular vent or passage occasions the noise called thunder.

But this lame device could only reach at most to the case of thunder heard without lightning; and, therefore, recourse has been had to other modes of solution. Thus, it has been said that thunder is not occasioned by the falling of clouds, but by the kindling of sulphurous exhalations, in the same manner as the noise of the *aurum fulminans*. "There are sulphurous exhalations, says Sir Isaac Newton, always ascending into the air when the earth is dry; there they ferment with the nitrous acids, and, sometimes taking fire, generate thunder, lightning, &c."

The effects of thunder are so like those of fired gunpowder, that Dr. Wallis thinks we need not scruple, to ascribe them to the same cause; and the principal ingredients in gunpowder, we know, are nitre and sulphur; charcoal only serving to keep the parts separate, for their better kindling. Hence, if we conceive in the air a convenient mixture of nitrous and sulphurous particles; and those, by any cause, to be set on fire, such explosion may well follow, and with such noise and light as attend the firing of gunpowder, and being once kindled, it will run from place to place, different ways, as the exhalations happen to lead it; much as is found in a train of gunpowder.

But a third, and most probable opinion is, that thunder is the report or noise produced by an electrical explosion in the clouds. Ever since the year 1752, in which the identity of the matter of lightning, and of the electrical fluid has been ascertained; philosophers have generally agreed in considering thunder as a concussion produced in the air by an explosion of electricity. See the article Lightning.

THUNDERBOLT, (*foudre*, Fr.) is when lightning acts with extraordinary violence, and breaks or shatters any thing, it is called a thunderbolt, which the vulgar, to fit it for such effects, suppose it to be a hard body, and even a stone. But that we need not have recourse to a hard solid body to account for the effects commonly attributed to the thunderbolt, will be evident to any one, who considers those of the *pulvis fulminans*, and of gunpowder; but more especially the astonishing power of electricity, when only collected and employed by human art, and much more when directed and exercised in the course of nature. As stones and bricks struck by lightning are often found in a vitrified state, we may naturally suppose, with Beccaria, that some stones in the earth having been struck in this manner, gave occasion to the vulgar opinion of the thunderbolt.

Thunder-Cloud, in physiology, are those clouds which are in a state fit for producing lightning and thunder.

THUNDERING-Barrels, (*barriques à feu*, Fr.) are casks which contain the fire-pots in a fire-ship.

THUS! (*comme çà!* Fr.) an order to the helmsman to keep the ship in her present situation when sailing with a scant wind; so that she may not approach too near the direction of the wind, and thereby shiver her sails, nor fall to leeward, and run farther out of her course. See the article To Steer.

THWARTS, (*bancs de rameurs*, Fr.) the seats or benches of a boat whereon the rowers sit to manage the oars.

Thwart-Ships, (*en travers du vaisseau*, Fr.) across the ship. See the article Athwart.

TIDE, (*marée*, Fr.) a regular periodical current of the water setting alternately in a flux, and reflux, and is produced by the mutual attrac ion of the sun and moon, but chiefly that of the latter. This attraction cannot alter the shape of the solid parts of the earth, but it has a great effect on the water, and causes it to assume a spheroidical figure, the longest axis being in the direction of the moon. It is the highest tide at the place which is perpendicularly under the moon, or where the moon crosses the meridian: not when the moon is on the meridian, but some time afterwards, because the force by which the moon raises the tide continues to act for some time after it has passed the meridian. The oval figure of the waters keeps pace, with the moon in her monthly journey round the earth, by whose daily rotation upon its axis, it presents each part of its surface to the action of the moon. In great oceans this rising and falling, in other words, the flux and reflux of the sea, takes place twice a day; that is, about every six hours the waters of the ocean extend themselves over its shores: this is called the flux or flood; in this state they remain a short space of time, after which they retire or fall back, and this is called the reflux or ebb tide.

During the flux, the waters of rivers communicating with the ocean are stopped by the advance of the sea-water, and after reflux they resume their usual courses.

Where the motion of the waters is not retarded by capes, islands, straits, or other similar obstacles, three periods are remarkable in the tides, the daily monthly, and the annual periods.

The mean daily period is 24 hours, 49 minutes, during which there are two flood and two ebb tides. This interval is the time in which the moon performs her mean apparent daily revolution round the earth. During this diurnal period we observe,

1st. That the high tide reaches the eastern harbours and roads, sooner than those to the west.

2dly, That between the tropics, the tides always seem to proceed from east to west.

3dly, That in the torrid zone, unless there be some particular obstacle, the flood tides come regularly at the same time to all places under the same meridian. On the contrary, in the temperate zones, it comes sooner to a lower than to a higher meridian of latitude; but beyond 65° of latitude the tide is not sensible.

The monthly period is distinguished, 1st, by this circumstance, that, at the new and full moons the tides rise much higher than at other periods, called spring tides, and when the moon is in the quarters, the tides are lowest, and called neap tides. The new and full moons are called the zyzigies, the quarters the quadrature: the tides go on increasing from the quadrature to the syzigies, and decreasing from the syzigies to the quadratures.

2dly, When the moon is in the syzigies or quadratures, the tide is at the highest three hours after the moon has passed the meridian. When the moon is going from the syzigies to the quadratures, the time of high water is rather sooner than these three hours. The contrary happens when the moon passes from the quadratures to the syzigies.

3dly, Whether the moon is in the southern or the northern hemisphere, the time of high tide does not happen any later in northern climates.

The annual period is distinguished by these circumstances: 1st, that at the time of the equinoxes, the spring tides are higher than at any other season of the year, and the neap tides the lowest, because at these periods the sun and moon are in the equator. At the solstices, the spring tides are not so high as in other lunations; and the neap tides are not so low as at other periods. The tides are also higher at the winter than at the summer solstice.

2dly, The tides are higher in proportion as the moon is near the earth, that is, when she is in her perigee. They are also higher when the moon is near the equator, and has of course less declination.

In general then it may be said, the highest tides are when the moon is at once near the equator, in perigee, and in the syzigies.

3dly, In northern climates the spring tides are higher in the evening during winter; and in the summer, they are higher in the morning. It is evident from the detail of these phenomena, that the tides have a marked connection with the motions of the moon; and that they are also in some degree governed by those of the sun. Wherefore we may fairly conclude that these luminaries, and particularly the former, are the natural causes of the phenomena of the tides.

Kepler conjectured that the gravitation of the earth towards the sun and moon was the cause of the tides. "If the earth ceased," said he, "to attract the waters of the ocean, they would be elevated towards the moon; for the moon's sphere of attraction extends to our earth and evidently acts upon the waters." What was mere conjecture in this great astronomer, was reduced to certainty by the superior genius of Sir Isaac Newton.

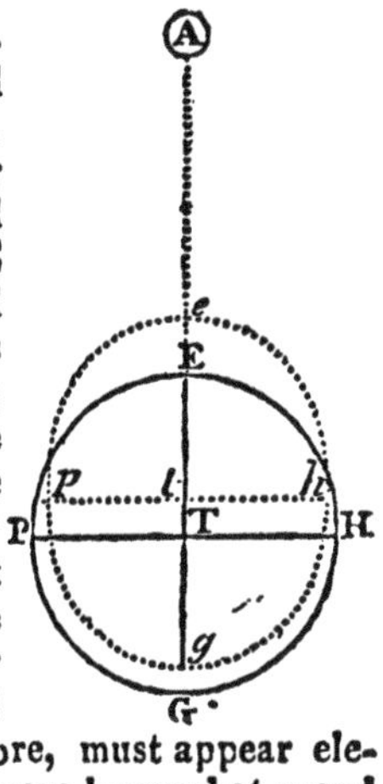

He imagines the moon A acting upon the terrestrial spheroid by the force of her attraction, draws the centre T towards her to the extent of 20 feet and brings it to t; that the part E being nearer to the moon, and still more forcibly attracted, is carried to the extent of 30 feet; and that the point G being more distant from the moon, and more feebly than the centre T, is only drawn as far as g to the extent of 10 feet; it is evident that the radii $t\,e$ and $t\,g$ must be longer by ten feet than the radii T E and T G. The waters, therefore, must appear elevated to that extent, while they are lowered at p and h. Thus the fluid (as appears evidently by the diagram) will be elevated at two opposite points in the line A G, in which line are the centres of the earth and the moon. If the attraction of the sun be added to that of the moon, the effect will be proportionably greater; but if these two attractions are placed counterpoise to each other, the effect will be proportionably less.

The motion of the waters of the sea (at least that of which we are sensible, and which is not common to them with the whole mass of the terrestrial globe) is not the effect of the entire action of the sun and moon, but of the difference between the action of these luminaries upon the centre of the earth, and upon the fluid with which it is covered, as well on the upper as the lower surface. It is this difference which we call action, force, or attraction, solar or lunar. The lunar action is much more powerful than that of the sun. We have seen that the waters of the ocean must rise at the same time both at that part of the ocean immediately under the moon and at the opposite point. Consequently, at ninety degrees from these points on each side, the water must be lowered. In the same manner the solar action must elevate the waters in that part which is immediately under the sun, and at the part diametrically opposite. Combining the two actions, we shall find that the elevation of the water at the same place must be subject to some variations both with respect to quantity, and time, according as the solar and lunar action are combined; or according as these forces act differently, or against each other.

In general, in conjunctions and oppositions of the sun and moon, their forces are combined. In conjunction these bodies act on the same meridian; and in opposition, they still act in the same line, and each raises the water on that side which is immediately under it.

On the contrary, in the quadratures, the water which is elevated by the sun, is depressed by the moon's attraction, for she is then ninety degrees from the sun. This is the time of the lowest or neap tides: and the highest or spring tides happen at new and full moon, when the two luminaries are in conjunction or opposition.

In the course of every natural day there are two tides, which depend upon the action of the sun, as in every lunar day there are two which depend on that of the moon; all follow the same law. The effect of the sun, as has been stated, is, however, much feebler than that of the moon, because the mass of the sun though much greater than the earth and moon together, yet its immense distance vastly lessens its force. The relative force of the sun and moon are estimated by Newton in the proportion of one to four and an half. In general, the nearer the moon happens to be to the earth, the greater is its attraction, and the same may be said of the sun.

The highest tides would be at the moment when the moon passed the meridian, if the waters had not, like all bodies in motion, a *vis inertiæ* by which they are inclined to retain the impression they have received But this produces two effects; it retards the time of high water and diminishes the height of the tides. Both after the flux and reflux, the ocean continues some time quiescent, neither disposed to rise nor fall, because the waters have a tendency to preserve the state of rest and equilibrium in which they are at the flood and ebb tides; and because the motion of the earth, displacing the waters with relation to the moon, lessens the intensity of the action of that luminary. These two efforts counterbalance each other for some moments. It must be observed, that the attraction of the particles of the fluid to each other, and obstacles of different kinds retarding their motion prevent them from passing all at once from a state of flood to that of ebb.

The moon passes above the eastern parts of the globe before the western. The flood tide, therefore, always proceeds in this direction. But the general motion of the sea between the tropics from east to west is more difficult to explain. This motion is evinced by the direction in which all floating bodies proceed there. It is observed also, that all other things being equal, it is much easier to navigate towards the west than in the contrary direction. M. D'Alembert has demonstrated, in his enquiry into the causes of winds, that the action of the sun and moon must cause a motion in the waters under the equator from east to west: this action must, according to the same writer, equally affect the air, and is one of the principal causes of the trade winds.

If the moon remained always in the equator, it is evident she would then be always ninety degrees distant from the poles, and that there could be there neither flux nor reflux, for the waters at the poles would always be low. Though the moon, however, is not always in the equator, she is never more distant from it than twenty degrees. It is no wonder, therefore, that, near the poles, and even at the latitude of sixty-five degrees, the tide is not perceptible.

As it only happens twice in a month, that the sun and the moon, are in the same line or direction, (that is when they are in conjunction or opposition), the elevation of the water is nei her immediately under the sun nor the moon, but in a point between the two. When the moon passes from the syzigies to the quadratures, the highest elevation of the waters ought to take place at the setting of the moon; and the contrary happens when the moon passes from the quadratures to the syzigies. In the first case, the time of high water ought to precede the three lunar hours: for on one side the *vis inertiæ* of the waters produces the elevation three hours after the moon passes the meridian; and on the other, the relative situation of the sun and moon affects this elevation before the moon passes the meridian, and for similar reasons in the second case, the time of high water must happen rather after the three hours.

As there is some retardation of the tide by the *vis inertiæ* of the waters, and their tendency to preserve an equilibrium, the highest tides do not take place exactly at the time of the oppositions and conjunctions of the sun and moon, but two or three tides after. In the same manner, the lowest neap tides happen a little after the quadratures. Since in winter the sun is a little nearer the earth than in the summer, it is observed that, when all other circumstances are equal, the tides about the winter solstice are rather higher than those of the summer solstice, such would be the regular phenomena of the tides, if the sea was in all parts of the same depth.

All the defects in the Newtonian theory of the tides are amply supplied by that eminent philosopher La Place, who, in the years 1775, 1779, and 1790, communicated to the Academy of Science at Paris, a series of memoirs on this subject; without considering the waters of the ocean in a state of equilibrium, or to be brought by the action of gravitation, towards the earth, and toward the sun and moon, into the figure of an aqueous spheroid, and maintaining the particles of water by the action of these different forces to be at rest: he, on the contrary, asserts, that the rotation of the earth does not allow time to this spheroid ever to be accurately formed; and long before the three attractions are able to produce their full effect, they are changed relatively to one another, and disposed to produce a different effect; instead of the formation of an aqueous spheroid, the tendency to it produces a continual oscillation in the waters of the ocean, which are thus preserved in perpetual movement, and can never attain a state of equilibrium and of rest. To determine the nature of these oscillations is a matter of extreme difficulty, and is a problem which neither Newton, nor any of the three geometers who pursued his tract, were able to resolve in the state of mechanical and mathematical science which then existed.

La Place considers each particle of water as acted upon by three forces, its gravitation to the earth, to the sun, and to the moon, and also as impelled by the rotation of the earth; he enquires into the nature of the oscillations that will be excited in the fluid, and finds that the oscillations thus arising, may be divided into three classes.

The first do not depend on the rotation of the earth, but only on the motion of the sun or moon in their respective orbits, and on the place of the moon's nodes. These oscillations vary periodically, but slowly; so that they do not return in the same order till after a very long interval of time. The oscillations of the second class depend principally on the rotation of the earth, and return in the same order, after the interval of a day nearly. The oscillations of the third class depend on an angle that is double the angular rotation of the earth; so that they return after the interval nearly of half a day. Each of these classes of oscillations proceeds just as if the other two had no existence; a circumstance that tends very much to simplify the investigation into their combined effect.

The oscillations of the first kind are proved to be almost entirely destroyed by the resistance of any motion the sea must necessarily meet with; and they amount nearly to the same as if the sea were reduced at every instant to an equilibrium under the attracting body. Those of the second class involve, in the expression of them, the rotation of the earth; and they are also affected by the depth of the sea. The difference of the two tides in the same day, depend chiefly on these oscillations, and it is from thence that La Place determines the mean depth of the sea to be about four leagues. The oscillations of the third kind are calculated in the same manner; and from the combination of all these circumstances, the height of the tides in different latitudes, in different situations of the sun and moon, the difference between the consecutive; the difference between the time of high water and the times when the sun and moon come to the meridian, all these circumstances are better explained in this method, than they have ever been by any other theory. La Place has instituted a very elaborate comparison between his theory and observations on the tides, made during a successsion of years at Brest.

1st. Between the laws by which the tides diminish from their maximum at the full and change, to their minimum at the first and third quarters, and by which they increase again from the minimum to the maximum, as deduced from the observations at Brest, and as determined by the theory of gravitation, there is an exact coincidence.

2d. According to theory, the height of the tides, at their maximum, near the equinoxes, is to their height in similar circumstances at the solstices, nearly as the square of the radius to the square of the co-sine of the declination of the sun at the solstice; and this is found to agree nearly with observation.

3d. The influence of the moon on the tides increase as the cube of her parallax; and this agrees so well with observation, that the law might have been deduced from observation alone.

4th. The retardation of the tides from one day to another is but half as great at the syzigies, as at the quadratures. This is the conclusion from theory; and it agrees well with observation which makes the daily retardation of the tide 27′ in the one case, and 55′ in the other.

La Place has treated a subject connected with the tides that has not been touched on by any other author before him. This is, the stability of the equilibrium of the sea. A fluid surrounding a solid nucleus may either be so attracted to that nucleus, that, when any motion is communicated to it, it will oscillate backwards and forwards till its motion is destroyed by the resistance it meets with, when it will again settle into rest; or it may be in such a state, that when any motion is communicated to it, its vibrations may increase, and become of enormous magnitude. Whether the sea may not, by such means, have risen above the tops of the highest mountains deserves to be considered; as that hypothesis, were it found to be consistent with the laws of nature, would serve to explain many of the phenomena of natural history. La Place with this view, has inquired into the nature of the equilibrium of the sea, or into the possibility of such vast undulations being propagated through it. The result is, that the equilibrium of the sea must be stable, and its oscillations continually tending to diminish, if the density of its waters be less than the mean density of the earth; and that its equilibrium does not admit of subversion, unless the mean density of the earth was equal to that of water, or less. As we know from the experiments made on the attraction of mountains, as well as from other facts, that the sea is more than four times less dense than the materials which compose the solid nucleus of the globe are at a medium, the possibility of these great undulations is entirely excluded; and therefore says La Place, if, as cannot well be questioned, the sea has formerly covered continents that are now much elevated above its level, the cause must be sought for elsewhere than in the instability of its equilibrium.

With the questions of the figure of the earth, and of the flux and reflux of the sea, that of the precession of the equinoxes is closely connected; this motion though slow, being always in the same direction, and, therefore, continually accumulating, has early been remarked, and was the first of the celestial appearances that suggested the idea of an *annus magnus*, one of those great astronomical periods by which so many days and years are circumscribed. As it affects the whole heavens, and as the changes it produces are spread out over the vast extent of 25,000 years, it has proved a valuable guide amid the darkness of antiquity, and has enabled the astronomer to steer his course with tolerable certainty, and here and there to discover a truth in the midst of the traditions and fables of the heroic ages.

The ebb and flow of tides are most perceptible on the shores of the ocean, where the influence of the sun and the moon is not obstructed by local impediments. In narrow seas, or in seas like the Baltic and the Mediterranean (which, though wide in themselves, are connected with the ocean by narrow straits), the ebb and flow of the tides are scarcely visible. So light are they in the Mediterranean, that the ancients never became properly aware of them till they passed the Straits of Gibraltar: and they imagined, on seeing the water leave the shore, that the gods were about to withdraw their protection from them. In the Cattegat, the tides in many parts are feeble and irregular; in the Sound and the Belts, this is still more the case; and, throughout the upper part of the Baltic, the alternation, if it exists, is imperceptible. Not that this sea is without its fluctuations from other causes. After the lapse of certain intervals, the waters of the Baltic are found to swell in such a manner as to rise three or four feet above their ordinary level; and these extraordinary risings occur in all seasons, but particularly in autumn, when the atmosphere is obscured by clouds, and loaded with humidity. When they take place in winter, they have the effect of lifting up the icy surface of the sea along the coast, of giving it a convex shape, and often of making it burst asunder. These swoln waters regain their level sometimes at the expiration of a few days; and, at other times, not for several weeks. They cause great agitation in the straits and gulfs; and it not unfrequently happens that they produce an inundation throughout the adjoining lands, which gives a salt taste to the fresh-water lakes that communicate with the sea: in the Mœlar, the great lake near Stockholm, this taste is so strong as to render the water unfit for domestic purposes. Several causes have been assigned for this remarkable phænomenon. Some persons have attributed it to the winds, which, by impelling the waters from the middle of the sea, augment their height in the gulfs and along the shores: but this explanation is neither compatible with the perpetual tendency of water to a level, nor is it conformable to experience, the rise of the Baltic generally preceding a tempest, and disappearing before a change to calm weather. It would be equally fruitless to seek for a solution in the supposed influx of water from the ocean in consequence of westerly winds; since five days of continued influx would not produce a rising of two feet; a change which is often seen effected in the Baltic in the space of twenty-four hours. "The most probable cause," says M. *Catteau*, "is to be sought in the state of the atmosphere. M. *Schulten*, in a memoir delivered to the Academy of Sciences of Stockholm, has remarked the striking correspondence between the state of the barometer and the swelling of the water. He found that the barometer regularly fell, shortly before the rise of the waters, and that it as constantly rose a little time before their fall. He inferred that the cause was to be sought in the unequal pressure of the atmosphere on different parts of this inland-sea. As the barometer varies in the north of Europe from twenty-four inches to twenty-six and a half, the difference between these extreme points will be found to correspond to a difference of three-feet, or three-feet and a half in the elevation of water. If to this circumstance we add local allowances for the swelling of rivers, and the effect of westerly winds, we shall, says M. *Schulten*, be able to account for the inequality of the rise in different parts of the shore of the Baltic; and, if farther observations are found to confirm these, the conclusions resulting from them will explain similar phænomena in other quarters.

The Caspian is known to rise and fall perceptibly, and similar variations have been remarked in the lake of Geneva; variations ascribed by *Saussure* to the same cause as by *Schulten*. Let us hope that a farther progress in such inquiries will lead to discoveries of great utility, and enable us to anticipate many of those changes of weather which at present elude our sagacity.

The fishermen in the Baltic and the Cattegat have acquired, from the observation of the rising of the waters, the habit of predicting the approach of rain and storms with surprising accuracy; a knowledge which proves extremely useful to them in their maritime occupations."

Atmospheric Tides, are certain periodical changes in the atmosphere, similar, in some respects, to those of the ocean, and produced in a great measure by the same causes.

A Leeward Tide. See the article Leeward.

Neap Tides. See the article Neap.

Slack-Tide. See the article Slack-Water.

Spring-Tide. See the article Spring.

Weather-Tide. See the article Weather.

To-Tide. (*étaler la marée; remonter à la faveur de la marée*, Fr.) to work in or out of a river, harbour, or channel, by favour of the tide, and anchoring whenever it becomes adverse.

To Tide It, (*étaler la marée*, Fr.) is to pursue a ship's course by means of the tide, with a foul wind, anchoring during the intervals of the contrary tide.

Tide-Gate, (*ras de marée, passage étroit ou la marée se porte avec violence*, Fr.) a place in which the tide runs with great velocity.

Tide-Road, (*situation d'un bâtiment qui, étant mouillé par un vent et une marée opposés l'un à l'autre, est évité de bout au courant*, Fr.) the situation of a vessel, which, being at anchor when the tide and wind are opposed to each other, has her head towards the current.

Tide *Tables*, are those which set forth the times of high water at sundry places, as they fall on the days of the full and change of the moon, and may be seen in most almanacs.

TIDESMAN, (*commis de la douane, qui veille à bord des bâtimens marchands, jusqu'à ce que le droit sur les marchandises soit payé*. Fr.) an officer appointed by the custom-house, to reside on board of merchant ships, while they have any customable goods on board.

Tide-Waiter, (*commis qui veille aux débarquement des marchandises à la douane*, Fr) a custom-house officer, whose duty it is to attend on the quays, to see goods of merchants and others landed.

Tide-Way, (*lit de marée*, Fr.) that part of a channel or river, in which the tide ebbs and flows strongly.

TIER, (*plan en parlant des pièces d'arrimage*, Fr.) also implies a range of casks or packages in the hold; hence we say, the

Ground Tier, (*premier plan*, Fr.) or that which is next above the kelson, the second tier and upper tier.

Tier, (*rang de canons, ou la batterie d'un vaisseau, d'un coté*, Fr.) the range of cannon mounted on one side of a ship's deck. See the articles Deck and Cannon.

Tier *of the Cable*, (*rang de cable*, Fr.) is a range of the fakes or windings of the cable, which are laid within one another in a horizontal position, so that the last becomes the innermost. See the article Coiling.

Cable Tier, (*la fosse aux cables*, Fr.) is the space in the midst of a cable, when it is coiled; also the place in which it is coiled.

TIGHT, (*étanche, ou qui tient l'eau*, Fr.) in opposition to leaky, the quality whereby a vessel resists the penetration of any fluid, whether compressing its surface, or contained within it: hence a ship is said to be tight, when her planks are so compact and solid as to prevent the entrance of the water in which she is immersed: and a cask is called tight, when the staves are so close that none of the liquor contained therein can issue through or between them. In both senses it is opposed to leaky.

TILLER, (*barre du gouvernail*, Fr.) a piece of timber fitted into the head of the rudder, to steer the ship with. See the article Helm.

Tiller-Rope, (*drosse de gouvernail*, Fr.) the rope which forms a communication between the fore-end of the tiller and the wheel, and is usually made of untarred rope.

TILT, (*tendelet de canot*, Fr.) a small canopy or awning of canvas, or other cloth, extended over the stern sheets of a boat, and supported by iron-work, or broad laths of flexible wood incurvated. It is principally used to keep off rain, as an awning is intended to defend from the sun's heat.

TIMBER, (*bois de charpente, bois à bâtir*, Fr.) a general name for all sorts of felled and seasoned wood.

Of all the different kinds of European timber, the oak is undoubtedly esteemed the best for ship-building. The India teak timber is on many accounts peculiarly well adapted for ship-building; 1st, its durability far exceeds that of the best oak; 2dly, it requires little seasoning and never shrinks; 3dly, it does not splinter when struck with shot; 4thly, it bears exposure to all the heat of the torrid zone, and to the extreme cold of the frozen ocean, without injury; and is, therefore, perhaps the only timber in the world that can stand the changes of climate to which ships are so much subjected, when employed in long and distant voyages. But what renders it more peculiarly adapted for this purpose, is its quality of destroying the worm, and preserving iron, in consequence of its containing a considerable quantity of oil, and no ligneous acid; while the gallic acid of the oak corrodes the iron, and appears to be particularly grateful to the taste of the worm. Fir timber is, however, more generally useful in other respects; it is used for flooring, wainscotting, and the ornamental parts of buildings within-doors. Elm is the next in use, especially in England and France; it is very tough and pliable, and therefore easily worked; it does not readily split, and it bears the driving of bolts and nails better than any other wood.

Ash is chiefly used by wheelwrights and coach-makers, for shafts, naves, &c. Beech is also used for many purposes; it is very tough and white when young, and of great strength, but liable to warp very much when exposed to the weather, and to be worm-eaten when used within-doors; its greatest use is for planks, bedsteads, chairs, and other household goods. Wild chesnut timber is by many esteemed to be as good as oak, and seems to have been much used in old buildings; but whether these trees are more scarce at present than formerly, or have been found not to answer so well as was imagined, this timber is now but little used. Walnut-tree is excellent for the joiner's use, it being of a more curious brown colour than beech, and not so subject to the worms, and is now become scarce from its being much used in the ordnance department for gun-stocks, &c. The poplar, abel, and aspen trees, which are very little different from each other, are sometimes used instead of fir, but mostly by turners, &c. The goodness of timber not only depends on the soil and situation in which it stands, but likewise on the season wherein it is felled.

Converting of TIMBER, (*préparation des bois de construction*, Fr.) the act of bringing it into a fit shape for ship-building, by sawing and hewing it to the form required.

Defects in TIMBER, (*défauts dans les bois*, Fr.) The defects in timber are various, and chiefly owing to the barrenness of the soil, as in loose and broken ground, timber is generally found to be shakey, which is very injurious.

Lopping of timber, or suffering cattle to browse upon it, often occasions it to rot and decay. But the greatest enemies to the growth of young trees, are rabbits; for where there is a number of those animals, spontaneous shoots can no sooner appear above the ground, than they are destroyed by them.

Sound timber is generally produced in places where the earth consists of strong clay, wherefore, timber of English growth is found far preferable to that of most other countries; and although some of the best sort of east-country plank is very flexible, and consequently useful for many purposes; yet it is often found very unserviceable, being either shakey, foxey, druxey, worm-eaten, or full of rotten knots; therefore, timber of English growth has certainly the preference; and even after it has stood so long, and past the time allowed for growth, it is allowed to be as durable as any other in its full strength.

Felling of TIMBER, (*coupe des bois*, Fr.) is the act of cutting it down for different purposes. In this, people disagree very much: some recommend felling it as soon as its fruit is ripe, others in the spring, and many in the autumn. But as the sap and moisture of timber are certainly the causes of its perishing sooner than it otherwise would do, it seems evident that timber should be felled when there is the least sap in it, viz. from the time that the leaves begin to fall, till the trees begin to bud. In England, this work usually commences about the end of April, because the bark then rises most freely; for where a quantity of timber is to be felled, the statute requires it to be done then, for the advantage of tanning.

Doctor Plot, in his discourse on the most seasonable time for felling timber, published in the Philosophical Transactions for 1691, observes, that trees felled in the spring are full of sap, which remains in the pores and putrifies, leaving the wood full of cavities, which make the timber weak, subjects it to rifts or gapes, and causes it to shrink: that in Staffordshire, they, therefore, bark the trees in the spring, and leave them standing till winter, and fell them when the sap has subsided:" and he quotes the concurring testimony of Vitruvius and Pliny, that if the sap be let out of the oak while yet living, it will acquire almost a perpetuity of duration.

There can be no question that winter-felled timber is far superior to that which is cut down in the spring.

The Montague, 74-gun ship, launched at Chatham, in 1779, was wholly built of winter-felled timber; but the advantages of stripping off the bark while standing, are not clearly established. Some think that timber acquires considerable durability by it. The only experiment in this country, we believe, is that of the Hawke sloop of war, which was built at Deptford, in 1793, one half with timber stripped of its bark while standing, the other half with timber felled in the ordinary way. Ten years afterwards she was broken up, when no visible difference was apparent in the state of decay of the two kinds of timber.

The interesting experiments lately made by Captain Layman of the Royal Navy, on the means discovered by him of extracting from green timber all the corruptive juices, so as to apply it at once to the purpose intended, as well as to increase the strength and render the sapwood superior to the heart in its natural state, was stated in the House of Lords, as deserving the perusal of all those who were desirous of information on a subject so important to this country, particularly as furnishing the means to prevent the rapid decay of our ships of war. See the article SEASONING *of Timber*.

TIMBER-HEADS. See the article KEVEL-HEADS.

TIMBER AND ROOM, (*mailles, ou intervalle entre les couples*, Fr.) the distance betwixt the moulding edges of two adjoining timbers, which always contain the breadth of two timbers, and sometimes two or three inches between them. It must be observed that one mould serves for two timbers; the fore-side of the one being supposed to unite with the after side of the other, and so make only one line; which is actually the case in all the frames, which in some ships are every third, and in others every fourth timber. The frames are first put up, and fastened to the ribbands, and afterwards the others are put up. See also the article ROOM and SPACE.

TIMBERS *of a Ship*, (*couples, ou membres d'un vaisseau, et toutes les pièces dont ils sont composés*, Fr.) the ribs, or the incurvated pieces of wood, branching outward from the keel in a vertical direction, so as to give strength, figure, and solidity to the whole fabric.

One timber being composed of several pieces united into one frame, is accordingly called a frame of timbers by the artificers. These different pieces are exhibited in fig. 1, plate XVII, Pieces of the Knee of the head, &c. The head of the floor-timber is cut square, to join the heel of the next above it. To support the connection of the timber in that place, another assemblage of pieces is formed, and joined in the same manner; so that when both the sets are fastened together, the joinings in one set will be nearly opposite to the middle of the pieces in the other. Hence it is evident, that the mould which serves for the lowest piece will conform to the under part of the corresponding piece above it: and thus the mould, appropriated to every division of a timber, will determine, or answer to the figure of the next adjoining thereto.

The outlines, or bends, of the principal timbers of the ship, are geometrically delineated in the plane of projection, plate XVII, and their particular stations in the ship's length are represented in the HORIZONTAL PLANE, and that of the ELEVATION, plate XVII.

CANT-TIMBERS. See the article CANT.

FILLING-TIMBERS, (*remplissage des couples*, Fr.) are those which are put up between the frames.

FLOOR-TIMBERS. See the article FLOOR.

KNUCKLE-TIMBERS, (*couples de coltis*, Fr.) are the foremost cant timbers on a ship's bow; the hindmost on the quarter are termed fashion-pieces.

SQUARE-TIMBERS, (*couples perpendiculaires à la quille, ou placés a l'equerre, par opposition à ceux qui sont dévoyés*, Fr.) are those whose planes are perpendicular to the keel

TOP-TIMBERS, (*alonges de revers*, Fr.) in ship-building, are the uppermost timbers: the first general tier of timbers that reaches the top of the side, are, or ought to be, called top-timbers: those which scarph on the heads of the upper-futtocks, are called short top-timbers.

TIME, (*tempus*, Lat. *tems*, Fr.) may be defined a portion of infinite duration, and is generally measured by motion; the heavenly bodies have, in every age, been selected for that purpose, and the ancients distinguished their different seasons by the appearance of particular constellations. The rising of the Pleiades they made denote the commencement of summer, and that of the Dog-star its declination, while subsequent knowledge enabled mankind to arrive at a more exact method of tracing the course of time, and to distinguish it by epochs, cycles, centuries, years, months, weeks, days, hours, minutes, and seconds. The principal divisions of time are the year and the day, which are measured by the sun's apparent revolutions, both annual and diurnal. Time is generally distinguished into *apparent*, *mean*, and *sidereal*, according to the manner of measuring the day.

Apparent TIME, also called *True Solar* and *Astronomical* TIME, (*tems vrai*, Fr.) is regulated by the apparent motions of the sun.

Mean, or *Mean Solar* TIME, also called *Equated* TIME, (*tems moyen*, Fr.) is a mean or average of apparent time.

Sidereal TIME, is shewn by the diurnal revolutions of the fixed stars.

An *Apparent Day*, is the interval between two successive transits of the sun's centre over the same meridian, which interval is two continual variations, owing to the eccentricity of the earth's orbit, and the obliquity of the ecliptic to the equator. These variations are computed in a table called the Equation of Time, which is given in the Nautical Almanac, with rules for its use. The equation of time may also be seen in the difference between a *good clock* and a *perfect sun-dial*, and is computed to be the difference between the sun's mean longitude and true right ascension, corrected by the equation of the equinoxes in right ascension. For a clear explanation of the equation of time, see WOODHOUSE'S *Astronomy*, lately published.

A *Mean Day*, is the interval that would be observed between two successive transits of the sun's centre over the same meridian, supposing the earth's orbit was circular, and the sun always in the equinoctial; and is such as is shewn by a clock that goes exactly 24 hours in a day, and 365 days, 5 hours, 49 minutes, in a year, which is said to be adjusted to mean time.

A *Sidereal Day*, is the interval between two successive transits of a star over the same meridian, which interval is uniform, because all the fixed stars make their revolutions in equal times, owing to the uniformity of the earth's diurnal rotation on its axis.

The sidereal day is shorter than the *Mean Solar Day*, by 3′, 55″.55 sidereal time. This difference arises from the sun's apparent annual motion from west to east, which leaves the star, as it were, behind. Thus, if the sun and a star be observed on any one day to pass the meridian at the same instant, the next day, when the star returns to the meridian, the sun will have advanced about a degree easterly; and as the earth's diurnal rotation on its axis is from west to east, the star will come to the meridian before the sun, so much, that at the end of the year, it will have gained a day on the sun, or have passed the meridian 366 times, while the sun will have passed it only 365 times. It therefore appears, that the earth describes about its axis an arc of 360° 59′ 8.″3 in a mean solar day, and an arc of 360° in a sidereal day; therefore, as 360° 59′ 8″.3 : 360° : : 24 hours : 23^h 56′ 4″.09, the length of a sidereal day in mean solar time, or the interval between the two successive transits of a star over the same meridian. Hence the following general rule for converting sidereal time into mean time, and the contrary:

As 24^h : 23^h 56′ 4″.09 : : any portion of sidereal time to its equivalent in mean time. And,

As 23^h 56′ 4″.09 : 24^h : : any portion of mean time to its equivalent in sidereal time.

From this it is evident, that apparent and mean time are the same with respect to the length of the hour, minute, and second of each; as well as of the year. It is only their days that differ, and this variation is marked by the time of commencement. Thus the apparent day always begins when the sun's centre is on the meridian, but the mean day commences sometimes sooner and sometimes later, as computed by the tables of the equation of time.

A sidereal day, being the interval between two successive transits of a star over the same meridian, may commence with any star; but the usual method is to begin the day when the first point of Aries is on the meridian; and a sidereal clock thus set will shew the right ascension of all the heavenly bodies as they pass the meridian. The difference, therefore, between the observed time, found by a meridian transit, and the computed time in the tables, shews the error of the clock. See Dr. Kelly's *Astronomical Computations* lately published.

Time-Keeper, in a general sense, denotes an instrument adapted for measuring time; but, in a more confined sense, it is a term now used in common with chronometer, to denote watches made on a particular construction, for the purpose of determining the longitude at sea.

We find that soon after the invention of watches, they were applied to the discovery of the longitude, and so early as 1530, Gemma Frisius projected a method of finding the longitude by means of watches; and another attempt was made by Metius in the year 1631. In 1662, Dr. Hook made some trials of a pendulum clock, which vibrated half seconds at sea.

Among the manuscripts in the Sloane collection at the British Museum, is one containing an account of the going of two watches at sea, from April 29th, 1663, to the 4th of September: in the journal kept of their rates, the watches are distinguished from each other, by the letters A and B. Between the 13th of August, and September the 5th, the time by the watch A was frequently compared with the apparent or solar time, and the difference compared with the longitude according to the dead-reckoning.

In 1664, Mr. Huygens made very great improvements in watch and clock making, by the introduction of pendulums and balance-springs, which Dr. Hook had invented a few years before.

The next person who applied his thoughts this way, was Henry Sully, who left England in 1713, and resided some time at Vienna, where he published a small treatise with the following title, *Régle Artificielle du Tems, par* H. S. de Londres, *imprimé à Vienne en Aûtriche*, 1714. Soon after this, he quitted Germany, for France, where he spent the rest of his life, chiefly in improving time-keepers for the discovery of the longitude; and to him that country is principally indebted for the great improvements that were afterwards made in that art. In the year 1716, he presented a watch of his own making to the Academy of Sciences, which was highly approved of; first, on account of the considerable diminution of friction; secondly, because what remained, was by a singular circumstance made uniform: and thirdly, because the equality of the parts of the watch shewed great ingenuity in the inventor. In 1726, he went to Bourdeaux, for the purpose of trying some experiments on his marine watches, where he was seized with a disorder in his breast, which carried him off in a few days. He had communicated many of his ingenious contrivances to his pupil, the celebrated M. Julien le Roy, who afterwards perfected many of his inventions. However, the first who succeeded to any considerable degree in making time-keepers useful for finding the longitude, was Mr. John Harrison, who, in 1726, produced a watch which measured time with a greater degree of accuracy, than any that had at that time been produced. In 1736, it was tried in a voyage to Lisbon and back, on board one of his Majesty's ships, during which time it corrected an error in the ship's dead-reckoning of almost a degree and a half.

M. Julien le Roy, who had studied the art of watch-making under the celebrated Sully, made considerable advances towards the improvement of time-keepers in France, by means of introducing a new escapement; and, in 1748, placed an instrument in the royal observatory at Paris, of the performance of which, Cassini gave a most favourable account.

In 1754, his son presented a paper to the Academy of Sciences at Paris, containing a description of a new time-keeper, proper to be used at sea for finding the longitude.

In 1763, he presented a time-keeper, three feet in length, to the Academy of Sciences; and in the following year, another of half the length of the former: and in 1766, he presented one to the King, which, after being compared with observations on land for the space of three months, by M. Le Monnier, and afterwards in a voyage undertaken for the purpose, the celebrated Cassini pronounced it to be the most perfect of its kind.

M. Berthoud published in 1761, his *Art de Conduire et Regler les Pendules et les Montres*, and soon after constructed two time-keepers, that were tried for upwards of a year, on board the French frigate Isis, by order of the King, which were found to give the longitude, upon an average, within one-fourth of a degree; and in consequence of this artist's abilities, he was appointed watch-maker to the King, with a pension of 3000 livres.

In England, Mr. Harrison had applied himself with indefatigable industry in improving his time-keepers, and by order of the Commissioners of Longitude, in 1739, he finished a second watch, more perfect than the first; and the result of the experiments made upon it was, that the motion was sufficiently regular and exact for finding the longitude of a ship within the limits proposed by Parliament. Being thus encouraged, he engaged in a third time-keeper, which he conceived would be still more correct, and brought it to such a state of forwardness, that after being reported on by twelve members of the Royal Society, they honoured him, in 1749, with the gold medal, which is given annually for the most important and ingenious discoveries.

In 1758, Mr. Harrison finished his third time-keeper, which performed so much to his own satisfaction, that he applied to the Commissioners of Longitude to have this instrument tried in a voyage to some port in the West-Indies, according to the directions of the statute of the 12th of Anne. Consequently, Mr. William Harrison, son of the inventor, embarked in the Deptford, at Portsmouth, November 18th, 1761, on a voyage for Jamaica, with his father's fourth time-keeper; after having ascertained its error and rate by the master of the Royal Academy at Portsmouth, in the presence of the commissioner and the captain of the ship, who attested the same, and sent it to the Admiralty. On his arrival there, on the 19th of January, 1762, the longitude, as shewn by the time-keeper, differed but one geographical mile and a quarter from the true longitude, deducted from astronomical observations. The same gentleman left Jamaica on the 28th of the same month, with the time-keeper, and arrived at Portsmouth on the 26th of March following, when he found that it had erred in the four months, having passed through a variety of climates, and undergone violent agitations at sea, no more than 1′ 53″.5, or 28½ minutes of longitude in time, which, in the parallel of Portsmouth, does not amount to 18 nautical miles; whereas, the act requires no greater exactnesss than 30 geographical miles, or minutes of a great circle, in such a voyage. Mr. Harrison now claimed the whole reward of 20,000*l.* offered by an act passed in the British Parliament, in the year 1714; but some doubts arising in the minds of the Commissioners, concerning the true situation of the island of Jamaica, and the manner in which the time at that place had been found, as well as at Portsmouth; they only awarded him 5000*l.* and he was ordered to make a second trial, in a voyage to the island of Barbadoes, in which precautions were taken to obviate as many of the former objections as possible. Accordingly the Commissioners previously sent out Dr. Maskelyne and Mr. Green, to make astronomical observations on that island, which, when compared with other corresponding ones made in England, would determine, beyond a doubt, its true situation; and Mr. William Harrison again set out with his father's time-keeper, on the 28th of March, 1764, the watch having been compared with equal altitudes, both in London and at Portsmouth, before he set out, and reported to the Admiralty, and he arrived at Barbadoes on the 13th of May following; where, on comparing it again by equal altitudes of the sun, it was found to shew the difference of longitude between Portsmouth and Barbadoes, to be 3h. 55m. 3s. the true difference of longitude between these places, by astronomical observations made by Messrs. Maskelyne and Green, the commissioners, being 3h. 54m. 20s.; so that the error of the watch was 43s., or 10′ 45″ of longitude. Mr. Harrison sailed from Barbadoes on the 4th of June, and arrived in London on the 18th of July, where the time-keeper was compared with the regulator it had been compared with before he set out, and it was found to have gained only 54″ in 156 days, the interval between its leaving London and return.

In consequence of this and the former trials, Mr. Harrison received another moiety of the reward offered, after explaining the principles on which his watch was constructed, and delivering this, as well as the three former, to the Commissioners of the Longitude, for the use of the public; and he was promised the remainder of the reward when other time-keepers should be made on the same principles, either by himself or others, performing equally well with that which he had last made. In the mean-time, this last time-keeper was sent down to the Royal Observatory at Greenwich, to be tried there under the direction of the Rev. Dr Maskelyne, the late astronomer-royal. But it did not appear, during this trial, that the watch went with the regularity that was expected; from which it was apprehended that the performance even of the same watch was not at all times equal; and consequently, that little certainty could be expected in the performance of different ones.

Soon after this trial, the Commissioners of Longitude agreed with Mr. Kendal, one of the watchmakers appointed by them to receive Mr. Harrison's discoveries, to make another watch on the same construction with this, to determine whether such watches could be made from the account which Mr. Harrison had given, by other persons as well as himself. The event proved the affirmative; for the watch produced by Mr. Kendal, in consequence of this agreement, went considerably better than Mr. Harrison's did. Mr. Kendal's watch was sent out, under the care of Mr. Wales, with Captain Cook, in his second voyage towards the south pole, and round the globe, in the years 1772, 1773, 1774, and 1775; when the only fault found in the watch was, that its rate of going was continually accelerated; though in this trial of three years and a half, it never amounted to 14″½ a day. The consequence was, that the House of Commons, in 1775, to whom an appeal had been made, were pleased to order the remaining moiety to be given to Mr. Harrison, agreeably to the act of Parliament. Mr. Harrison had also at different times received other sums of money, as encouragements to him to continue his endeavours, from the Board of Longitude, and from the East India Company, as well as from other individuals, amounting in all to upwards of 24,000*l.*

Of late years, several improvements have been made in Chronometers, particularly that of a new balance spring, and an improved thermometer-piece, or a compensation for the effects of heat and cold in the balance, by Mr. John Arnold, for which he claimed one of the new rewards offered for improvements in discovering the longitude. In 1780, he published an account of the going of a pocket chronometer, kept at the royal observatory, wherein he tates that its greatest error in the period of thirteen months, the time it had been under trial at the observatory, was only 2′ 32″.2, which, in the Channel would amount to about 25 geographical miles; but, that during the last eight months, the extreme difference did not amount to one minute of time, which would be about 10 miles in the Channel. He concludes by observing, that the chronometer is better adapted for determining the relative longitude of places than any other method. For this improvement Mr. Arnold received 1300*l.* from the Board of Longitude; and in December 1805, his son received the farther sum of 1700*l.* making up, with what his late father had received, the sum of 3000*l.* which had been voted by the Board of Longitude as a reward for the excellence of the principle and performance of his time-keeper.

Mr. Thomas Mudge has also obtained a part of the parliamentary reward for time-keepers. His first was received at the royal observatory on the 11th of November, 1776, for trial, and at a Board of the Commissioners, held on the 1st of March, 1777, Dr. Maskelyne made a report of the going of the watch from the time it was admitted, a period of 109 days, during which time it had gained only 1′ 19″. In consequence of the above favourable report, and being informed that he meant to persevere in making others still more perfect, they agreed to give him 500*l.* to indemnify him in case he did not succeed in obtaining the great reward still held out by act of Parliament. Mr. Mudge in consequence finished two more time-keepers, which were tried at the royal observatory three several times, viz. in 1779, 1783, and in 1789. On the 4th of June, 1790, the Board of Longitude determined unanimously, that, as Mr. Mudge's time-keepers had not gone, upon the last twelvemonth's trial, with the exactness required by the act of Parliament, the Board were not authorised to order any farther trial of them. In the following year, he presented a memorial to the Board, and solicited a reward on the ground of his time-keepers being superior to any hitherto invented, and being constructed on such principles as will render them permanently useful. Mr. Mudge now offered to make these principles known, on condition of receiving 2000*l*, for the discovery. The Board did not chuse at that time to comply with his request; however, in June 1793, he received 1500*l.* from Parliament for his improvements.

Messrs. Earnshaw, Brockbanks, Barraud, Hatton, and several others, have constructed time-keepers, each artist introducing some improvement or alteration which he thinks for the better; so that it is to be hoped they will soon be brought to a very considerable degree of perfection, and afforded at a much cheaper rate than hitherto, the common price now being from 75 to 100 guineas.

The following remarks on time-keepers were drawn up by an old experienced officer, who had been many years employed in marine surveying, particularly on the N.W. coast of America, and was communicated in Norie's Mariner's Encyclopædia, some years ago.

"When a time-keeper is received on board a ship, the greatest care should be taken to have it immediately secured in some convenient place in the cabin, where it may be the least liable to be removed during the voyage: it should never be touched but at the time of winding up, which ought to be at noon, and then with the greatest care, particularly avoiding circular motion, as time-keepers have frequently been known to alter their rates in carrying them from the ship to the observatory, and in returning again to the ship; when there was every reason to suppose that had they remained quiet on board, there would have been no alteration in their rates.

"When a time-keeper is received on board, its rate on the number of seconds it gains or loses in twenty-four hours, with the time which it was fast or slow of mean time at Greenwich, on some certain day, ought to be given from the observatory with it, and which should be noted carefully every day at noon, in a book kept for that purpose, so that it may be known at any time, without any farther calculation, what it is fast or slow of the mean time at Greenwich. If the time-piece is slow of mean time at Greenwich, and losing, add its daily rate to its error every day. What is here meant by error is, the quantity the time-piece is fast or slow, of the meridian reckoned from, and not the error that is occasioned by any irregularity in the going of the machine, and which, to the extent of it, would occasion a wrong determination of the longitude. This sum must be added to the time it shews, to give Greenwich mean time; if it is slow, and gaining, subtract its daily rate, and the remainder to the time shewn by it for the Greenwich mean time. If the time-keeper is fast, and gaining, add the daily rate, and subtract the sum; or if it is fast and losing, subtract the daily rate, and the remainder from the time shewn by the time-keeper, to obtain the Greenwich mean time. It is evident, that when the time-piece is fast, and losing, its error will decrease till it becomes slow; and then the error will increase; also if it is slow, and gaining, its error will decrease, till it becomes fast, and then will increase. When by this correction of the time-piece, the mean time at Greenwich is known, all that is wanted is the mean time at the ship. The difference between them is the longitude in time, which is west, when the Greenwich mean time is the greater of the two, and east when it is the less.

"To prevent the moving of a time-piece, the observer should be provided with a pocket-watch, having a second's hand, This should be compared with the time-piece, noting the difference, both immediately before, and after taking altitudes. If the difference be not the same before and after, correct it in the proportion of the elapsed times, before you apply it to the time shewn by the pocket watch, to give the time shewn by the time-piece. The rate of a time-piece should often be tried, and if it is found to have altered, a new rate should be given. The alteration may be discovered, and a new rate given as follows:—In any harbour, which has been regularly surveyed, and the latitudes and longitudes well ascertained, the bearing of any two points will determine the ship's position on the chart: thus the ship's latitude is known. Let altitudes be taken, and from them find the longitude by the time-piece: if this differs materially from the ship's longitude, it is fair to conclude that the time-keeper has altered its rate.—If the ship stays a week or ten days in the same place, a new rate may be given it, by taking altitudes as often as is convenient until the ship sails; noting the difference of the longitudes given by the time-piece the first and last day: then turning this difference into time, and dividing that time by the number of days from the time of taking the first altitudes, to the time of taking the last, the quotient will be an increase, or decrease, to be applied to the old rate.

"By this mode, the rate of a time-piece may be ascertained on board; and, indeed, this method is generally recommended to those who have the charge of navigating a ship. Yet the best and surest mode of finding the rate of time-keepers is, by a well-regulated clock on shore, or by equal altitudes taken by an astronomical quadrant. As neither of these can be used on board a ship, a substitute is required," and, he observes, "I know of no better than a quicksilver horizon; which, with care, can never lead to any error: it is therefore far preferable to the sea horizon, which is always liable to a considerable error. When I have been in an observatory, finding the rate of time-pieces by a well adjusted astronomical quadrant, I have frequently, for experiment's sake, tried the quicksilver horizon, and brought out the rate the same as by equal altitude. I have found the quicksilver horizon capable of fixing the latitude of a place, nearly as accurately as an astronomical quadrant. It may be made small enough to be carried in the pocket, and with a sextant, called a portable observatory; and a person may with these two instruments, regulate his time-pieces wherever he may chance to stop. The quicksilver horizon shews twice the true altitudes, *i. e.* suppose the sun's altitude as taken by a quicksilver horizon, measures on the sextant 90°, half that sum, or 45° will be the altitude of the sun's lower limb; which is to be treated as a common altitude, except that there is no correction for the dip of the horizon.

"Let us now suppose a time-piece to be fixed in a ship; the observer to be provided with a quicksilver horizon, a sextant, and a pocket-watch, with a second's hand, which goes tolerably well, and that he wishes to find the rate of his time-piece: let him set his pocket-watch to the same second with the time-piece, if it is a stop-watch; otherwise let him compare it, noting the difference; then let him go on shore with his quicksilver horizon, and take altitudes: let him immediately return on board, and compare again, and if the pocket-watch has altered its difference from the time piece, let him make the proper proportion for the error, according to the elapsed times, as above directed, to find the time which the time-piece shewed when the altitudes were taken: with this, the altitudes may be used as if the time-piece had been taken on shore; and in this way the rate may be found.

"It should also be remarked, that when the observation is made, the sun's altitude should not be less than 8°, as late as 9h, or half-past 9h, if it is in the morning, or about the same distance from noon, namely, half-past two in the afternoon. This method is much superior to that where the sea horizon is used: indeed, it is very little inferior to that by the astronomical quadrant. If a person is near an observatory, a rate may be given to a time-piece without moving it, merely by setting the pocket-watch with the time-piece, and comparing it with the clock at the observatory several times a week.

"The various uses of time-pieces can only be known to those who are in the habits of working and studying them. The first, and most essential is with the latitude, to ascertain the exact place of the ship, any day, and several times a day, if necessary, in any part of the ocean, when the sun is visible. Secondly, it has been recommended before to take altitudes about 9h, or half-past 9h, in the morning, for finding the longitude. Having these, you will always be sure of the first of double altitudes, should there be any appearance of losing the meridional altitude by clouds, or mist, arising near noon. Thus, it may frequently happen that the navigator may have both longitude and latitude, when, without the use of a time-keeper, he might be disappointed of both; as Lord Hugh Seymour, in his voyage to the West-Indies, in 1796, had been at sea nearly six weeks, before the first lunar observation was obtained, in which period the longitude had been ascertained thirty different times by his time-keepers. La Perouse, in his voyage round the world, was also at one time 31 days without being able to make a single lunar observation; but, during that period, he had determined the longitude twenty different times by means of his chronometers. Thirdly, it very often happens, particularly in high latitudes, that in a cloudy day, the sun will shew itself on or about passing the meridian. Having a time-piece, you can tell, within a minute, when the sun is on the meridian; if an altitude can be caught about that time, even so far as ten minutes before or after the sun passes the meridian, that altitude will give the latitude nearly: for in such latitude, particularly when the sun's declination and the latitude are of different names, the sun's altitude alters very little in ten minutes on either side of noon. Fourthly, time-keepers are of great use, in carrying on lunar distances, from one day to another, so that a number of observations may be brought to one point. Altitudes being taken at the proper time by the time-piece, the apparent time thus obtained will answer for any lunar distances taken in the course of the succeeding twenty-four hours. The method of carrying on or reducing the time of taking the distances, to the time of taking altitudes for the longitude, may be illustrated as follows:

"Suppose that altitudes are taken for the longitude in the morning, and that several sets of lunar distances are observed in the afternoon. Take the elapsed time, from observing the altitudes, to the time of each set of distances, and when the Greenwich time of the distances is brought out, subtract this elapsed time from the Greenwich time of each set; each remainder will be a Greenwich time when the altitudes were taken, and will produce an observed longitude for that time; not for the time when the distances were observed. On the contrary, if lunar distances are observed, before altitudes are taken for the apparent time, the elapsed time of each set must be added to the Greenwich time, to bring each set up to the time when the altitudes were taken. Observations of lunar distances also may be carried on from one time to another, so as to enable the observer, after having gained a number of sets on each side of the moon, to bring them all to one point by the time-piece, and get a mean of all the observations, which will give nearly the true longitude, if the observations are good." See the articles LONGITUDE *by Time-Keeper*, and LUNAR *Observations*.

In addition to the observations already made on time-pieces, it may be necessary to advert to the very great importance of these instruments to officers charged with fleets or convoys, with respect to making short passages from one part of the world to another.—A person who has a good time-piece will be enabled to steer straight courses. This will shorten the distance much, and by that very circumstance will frequently enable him to get a fleet into port, before a shift of wind, which might otherwise keep him out a long time. It will likewise give an officer a good chance of escaping the vigilance of a superior enemy; as he need not, according to the common system of navigation, run into the latitude of his port, in which his enemy may be cruising a few degrees of longitude to windward for him. He may run his latitude down thirty or forty leagues north or south of the port, as it may happen, till he is nearly or exactly in the longitude of it and then steer north, or south, to his destination, as the case requires. This mode of navigation might prevent the capture of many a valuable ship, and would certainly, by shortening the passage, in many instances, cause a great saving, not only to the merchant, but also to the government.

Mr. Ezekiel Walker, of Lynn, submitted a method to the Board of Longitude, in 1783, of determining the longitude, in which he observes, that, "although the best time-keeper, be too imperfect a machine to be depended upon for that purpose in long voyages, yet, if five or six be taken to sea in the same ship, the longitude computed by each separately, and the mean of their results taken, it will come exceedingly near the truth, even at the end of three or four months."

Another great advantage pointed out in this method is, that should any one of the time-keepers stop, it may be set a-going again by the rest; or should any one of them be found to go incorrectly, its rate may be rejected, and the longitude still determined by the others; in short, he makes it appear that this method is almost entirely free from all those accidents to which a single chronometer is liable.

To shew the perfection of this method more clearly, Mr. Walker has produced two examples in the 8th volume of NICHOLSON'S *Philosophical Journal* for 1804, which were taken from the rates of chronometers that had been determined in fixed observatories.

In the first example, made with three of Arnold's chronometers, which contains a period of thirty days, there are only two instances in which this method differs so much as a single second from mean time.

In the second example, made with two of Mudge's and one of Arnold's chronometers, containing a period of six weeks, the greatest error on any one day, during that time, is only 5″.9, and the total error at the end of that period, is only 1″.9. "From these examples," says Mr. Walker, "we may see the nature of fortuitous events, how regularity rises out of irregularity; the error of one time-keeper correcting that of another in a most wonderful manner; and as the number of time-keepers are increased, the error will be decreased, until it be almost annihilated. Consequently the method of finding the longitude at sea need no longer be doubtful, as it may, by this method, be found to any degree of precision that may be useful in navigation. The only objection seems to be the expense: but two or three hundred pounds would purchase a sufficient number to secure any ship from that danger which might arise from the want of knowing the longitude; a sum which certainly bears no sort of proportion to the value of a British squadron, or even to that of a single East-Indiaman."

TIMENOGUY, a rope fastened at one end to the fore-shrouds, and nailed at the other end to the anchor-stock, on the bow, to prevent the fore-sheet from entangling.

TIMONEER, (*timonnier*, Fr.) the helmsman, or person who manages the helm to direct the ship's course. See the article *To* STEER.

TOBACCO, (*tabac*, Fr.) a plant, whose leaves are in such general request in most parts of the world, that it forms a most extensive and important branch of commerce. It is said to have derived its name from the island of Tobago, where the Spaniards first found it, after their discovery of America; and was first imported about the year 1560.

By an order of Admiralty, December 27th, 1798, tobacco was ordered to be provided and supplied by and under the direction of the Victualling Board; for the use of the companies of His Majesty's ships and vessels, and proper deposits thereof are now made in the victualling stores at Portsmouth, Plymouth, Deptford, Chatham, Dover, and Yarmouth, and also at Hull, and at Leith, under the care of contractors, &c.

No greater quantity of tobacco is to be allowed to any of his Majesty's ships or vessels, than two pounds weight avoirdupois, per lunar month, to each of the seamen, or others serving on board such ships or vessels; nor is any greater quantity to be permitted to be shipped at one and the same time, than will be sufficient to serve the crews of such ships or vessels for the period to which they shall be severally victualled.

All tobacco to be charged to each person respectively, on the ship's books, at the rate of 1*s*. 7*d*. per pound avoirdupois weight.

Supernumeraries of every description, borne on board receiving or other ships for victualling only, are not to have any tobacco issued to them from the public stock in the purser's charge.

Pursers to be allowed a commission, and an allowance for wastage, upon all tobacco issued after the above mentioned rate. See the NAVAL INSTRUCTIONS.

TOGETHER! (*ensemble!* Fr.) the order given to the men in the exercises of heaving, rowing, hoisting, &c. to act all in concert, or at the same instant.

TOGGEL, (*chevillot*, ou *quinçonneau*, Fr.) a small wooden pin, from four to six inches in length, and usually tapering from the middle towards the extremities. It is used to fix transversely in the lower part of a tackle, in which it serves as an hook, whereby to attach the tackle to a strap, slings, or any body whereon the effort of the tackle is to be employed.

There are also toggels of another kind, employed to fasten the top-gallant sheets to the span, which is knotted round the cap at the top-mast-head. For as the lifts of the topsail-yard are out of use when the topsail is hoisted, they are always converted into top-gallant-sheets, to render the rigging at the mast-heads as light and simple as possible. Before the topsail-yards can be lowered so as to be sustained by their lifts, it therefore becomes necessary to transfer that part of the lift to the top-mast-head, that so the whole weight of the yard may be sustained by its mast-head, and no part thereof by the top-gallant-yard, which would otherwise be the case. This is performed by fixing the double part, or bight of the lift, within the eye of the span above-mentioned, and inserting the toggel through the former, so as to confine it to the latter, which operation, amongst sailors, is called putting the sheets in the beckets.

Toggel-Bolt, in block-making, a small iron bolt, having a flat square head, with a hole or eye punched through it.

TOMPION, (*tampon, ou tappe de canon*, Fr.) a sort of bung or cork used to stop the mouth of a cannon. At sea the tompions are carefully encircled with tallow or putty, to prevent the penetration of the water into the bore, whereby the powder contained in the chamber might be damaged, or rendered unserviceable.

Tompions, (*plateaux de bois de la charge à mitraille*, Fr.) also signify iron bottoms to which grape-shot are fixed; the diameters of which are as follow:

42 pounders....$6\frac{6}{10}$ inches	9 pounders....$3\frac{9}{10}$ inches
32 ....ditto....6	6 ....ditto....$3\frac{3}{4}$
24 ....ditto....$5\frac{4}{10}$	4 ....ditto....$2\frac{9}{10}$
18 ....ditto....$4\frac{9}{10}$	$1\frac{1}{2}$ ...ditto....$2\frac{1}{10}$
12 ....ditto....$4\frac{1}{4}$	$\frac{1}{2}$ ...ditto....$1\frac{3}{10}$

TONGUE, in mast-making, the taper part of the lower end of a spindle, or of a scarph.

Tongue, a short piece of rope spliced into the upper part of standing-backstays, &c. to the size of the topmast-head: it is served over with spun-yarn, and is used to keep them open to the size of the mast-head.

TONNAGE, (*port, en tonneaux, d'un bâtiment*, Fr.) a custom or impost due for merchandise brought or carried in tons from or to other nations, after a certain rate in every ton. Various ways for determining the length of the *keel for tonnage*, as part of the data by which to ascertain the tonnage of a ship have been adopted at different periods.

"The old mode of calculating the tonnage of men of war," says Mr. Willett, "was by multiplying the length of the keel by the extreme breadth, and the product by the depth in hold, and then dividing by 96." See *Archaeologia, Vol.* XI.

The next method of determining the length of the keel for tonnage, and rule for casting it, were settled by the Lords Commissioners of the Admiralty, in the year 1719.

On a straight line with the lower part of the rabbet of the keel, erect a perpendicular or square line to the upper edge of the wing-transom, at the after part of the plank, and at the stem to the fore part of the plank at $\frac{5}{6}$ parts of the height of the wing-transom; the length between the said perpendiculars added to $\frac{1}{34}$ of the extreme breadth (allowing for the stem and stern posts without the rabbets) from which subtract $\frac{6}{25}$ of the height of the wing-transom for the rake abaft, and also $\frac{3}{5}$ of the main breadth for the rake afore, leaves the length of the keel for tonnage: which multiply by the breadth, and the product by half the breadth, and divide by 94, gives the tonnage.

The following method for ascertaining the tonnage of ships of war, transports, cutters, &c. was settled by an order of the Navy Board, in 1781. and is now invariably adopted in his Majesty's naval service.

The length to be taken from the fore part of the stem at the height of the upper deck to the aft part of the main-post at the height of the wing-transom. In three decked ships, the length to be taken at the height of the middle-deck to the main-post at the wing-transom; from the length thus taken, subtract $\frac{3}{5}$ of the breadth extreme to the plank of the bottom for the rake forward, and $2\frac{1}{2}$ inches for every foot the wing-transom is high above the lower part of the rabbet of the keel for the rake abaft; the remainder is the length of the keel for tonnage. But cutters, or any other vessels, whose posts rake more than $2\frac{1}{2}$ inches in a foot, they are to be taken from the extreme fore part of the stem, at any height that may chance to be to the back of the main post on a range with the lower part of the rabbet of the keel, the breadth to be taken from outside to outside the plank or thick-stuff in the broadest place, either above or below the wale; then to get the thickness of the stuff at the breadth, and the thickness of the bottom; whatever the thick-stuff at the breadth exceeds the thickness of the plank of the bottom, that must be deducted from the breadth taken from the outside of the thick-stuff; the remainder is the breadth extreme; then multiply the length of the keel for tonnage by the breadth extreme, and the product by the half breadth, and dividing the whole by 94, the quotient will be the tonnage.

Calculation of the burden of a 74-gun ship, according to the last-mentioned rule.

			Ft.	In.
Length from the fore side of the stem, at the height of the upper-deck, to the aft-side of the main-post at the height of the wing-transom			185	6
$\frac{3}{5}$ of the extreme breadth is..........	29	$9\frac{3}{5}$		
The height of the wing-transom is 28 feet 3 inches, which produces for every $2\frac{1}{2}$ inches..................	5	$10\frac{5}{8}$		
		Total	35	$8\frac{9}{40}$
Length of the keel for tonnage			149	10
Multiplied by the extreme breadth...........			49	8
		Product	7441	8.8″
Multiplied by half the extreme breadth.......			24	10
Divided by 94			184802	9.2″.8′
Burden in tons			1964	$\frac{66}{94}$

The method of finding the tonnage of any ship among merchants and ship-owners, is by the following rule:—Multiply the length of the keel by the breadth f the beam, and that product by half the breadth of the beam, and divide the last product by 94, and the quotient will be the tonnage. Example, Suppose the ship's keel 72 feet, breadth of the beam 24 feet, then $\frac{72 \times 24 \times 12}{94} = 220.6$.

The tonnage of goods is sometimes taken by weight, and sometimes by measurement. The method which allows most is allowed to a vessel. In weight, twenty hundred make one ton, but by measurement, forty cubic feet are equal to one ton. All carriages, or other stores, to be measured by tonnage, are taken to pieces, and packed in the best manner which will occupy least room on board ship. All ordnance, whether brass or iron, is taken in tonnage by its actual weight. Musket-cartridges, in barrels or boxes, all ammunition in boxes, and other articles of great weight, are taken in tonnage according to their actual weight.

The following is the tonnage allowed to the military officers of the ordnance embarked for foreign service, for their camp-equipage and baggage:

For a field officer..............5 tons.
For a captain ....3 do.
For a subaltern ....$1\frac{1}{2}$ do.

There are few subjects which require more investigation than the tonnage of ships; for at present the mode of finding the tonnage of a ship is so replete with erorr, that a new method is required; and it would be worth the attention of the public to bestow some great reward on the person who gave the simplest and best method of calculating the tonnage of ships and vessels of various classes. It appears from the general construction of merchant ships, that more attention is paid to evade the tax on tonnage, than to their sailing well with the wind in different directions; and that if the real tonnage of ships were taken, an alteration would soon be made in the construction for the better.

Men of war have less tonnage than they measure, as appears from the Victory of 100 guns, whose supposed tonnage, by the established measurement, is 2143 tons, but whose actual tonnage is only found to be 1839 tons. The Hindostan, East-Indiaman, measures 1248 tons, and actually carries $1890\frac{1}{2}$ tons, which makes her larger than the Victory by $51\frac{1}{2}$ tons freightage. See *Naval Chronicle*, Vol. II. p. 311

TOP (*hune*, Fr.) a sort of platform, surrounding the lower mast-head, from which it projects on all sides like a scaffold.

The principal intention of the top is, to extend the top-mast-shrouds, so as to form a greater angle with the mast, and thereby give additional support to the latter. It is sustained by certain timbers fixed across the hounds or cheeks of the mast, and called the trestle-trees and cross-trees, the former of which are expressed by *k*, *k*, fig. 1, plate XVI. and the latter by *l*, *l*, fig. 2.

The plan of the top is now generally made in two parts, for the convenience of getting it in its place, and is represented by fig. 6, where *a*, *a*, represent the two span-pieces, which are bolted through the top to keep its parts together: *b*, *b*, the cross-trees upon the trestle-trees, which are also bolted through the top to secure it in its place; *c*, *c*, the knees; and *d*, *d*, the rim.

The top is extremely convenient to contain the materials necessary for extending the small sails, and for fixing or repairing the rigging and machinery, with more facility and expedition. In ships of war, it is used as a kind of redoubt, and is accordingly fortified for attack or defence, being furnished with swivels, musketry, and other fire-arms, and guarded with a thick fence of corded hammocks: finally, it is employed as a place for looking out, either in the day or night.

The frame of the top is either close-planked like a platform, or open like a grating: the latter kind is much better than the former in tempestuous weather, as presenting a smaller surface to the wind when the ship leans over to one side, and by consequence being less exposed to its efforts.

The dimensions of tops in the royal navy are as follow:—The breadth of the top athwart-ships, is one-third the length of its corresponding top-mast; the length fore and aft, three-fourths their breadth. The size of the square hole is, athwart-ships, two-fifths the breadth of the top; fore and aft, thirteen-fourteenths the breadth; and its aft-side is one-fifth the length of the top from the aft-side of the platform; and it is placed in the middle athwart-ships. The trestle-trees and cross-trees extend nearly to the edge of the tops.

Top-Armour, (*pavois des hunes*, Fr.) a rail about three feet high, extending the width of the top on the afterside, supported by stanchions, and equipped with a netting, and sometimes with painted canvass. By this name it seems to have been considered as a sort of blind, behind which the men may conceal themselves from the aims of the enemy's fire-arms, in time of action, whilst they are charging their own muskets, carabines, or swivels.

Top-Block. See the articles Block and Mast.

Top-Brim, in sail-making, a space in the middle of the foot of a top-sail, containing one-fifth of the number of its cloths. It is so called from its situation, being near the fore-part of the top, or platform on the mast, when the sail is extended.

Top and Butt, in ship-building, a general method of working the English plank (except in the topside) to make good work and conversion, which is done by disposing of the top-end of every plank, within six feet of the butt-end of the plank above or below it, leaving all the planks to work as broad as possible, so that every other seam is fair.

Top-Chains. See the article Chains.

Top-Cloth, a large piece of canvas, used to cover the hammocks which are lashed in the top when prepared for action.

Top-Lantern, (*fanal de hune*, Fr.) a large lantern placed in the after part of the top in any ship where an admiral or commodore is personally aboard. It is supported on each side by iron braces.

Laying Top, (*toupin*, *ou couchoir*, Fr.) in rope-making, a cylindrical piece of wood, having three or four scores or notches on its surface, and varying in its size according to the thickness of the rope, for which it is intended.

Top-Lining, in sail-making, the lining sewed to the aft-side of top-sails, to preserve the sail from chafing the top.

Top-Mast. See the article Mast.

Top-Gallant-Mast. See the article Mast.

Fore, Main, and Mizen Top-Men, (*gabiers*, Fr.) are persons stationed in the several tops to attend the taking in or setting of the small sails, and to keep the upper rigging in order; they are generally chosen from the most young and active of the seamen, the principal of whom is termed the captain.

Top-Rope, (*guinderesse*, Fr.) a rope employed to sway-up a top-mast or top-gallant-mast, in order to fix it in its place, or to lower it in tempestuous weather, or when it is no longer necessary. The rope used on this occasion for the top-masts is, on account of their great weight, furnished with an assemblage of pullies, at its lower end, called the top-tackle, to hoist or lower the mast with greater facility. The whole of this is particularly explained in the article mast, and the plate therein referred to.

Top-Sails, (*huniers*, Fr.) are large sails extended across the top-masts, by the top-sail-yards above, and by the yards beneath; being fastened to the former by ear-rings, rope-bands, and to the latter by topsail-sheets, which pass through two great blocks fixed on its extremities; and from thence to two other blocks fixed on the inner part of the yard close by the mast, lead downwards to the deck, where they may be slackened or extended at pleasure.

Top-Gallant-Sails, (*perroquets*, Fr.) are expanded above the top-sail-yards, in the same manner as the top-sails are extended above the lower yards. See the article Sail.

Top-Side. See the article *Top*-Timbers.

Top-Tackle, (*palan de guinderesse, fixé dans les navires Anglois au bas des guinderesse de hune*, Fr.) a large tackle hooked to the lower end of the top-mast, top-rope, and to the deck, in order to increase the mechanical power in hoisting the top-mast. It is composed of two strong iron-bound double or triple blocks, the hooks of which work on a swivel.

To Top *a Yard*, (*apiquer une vergue, ou peser sur ses balancines*, Fr.) is to draw one of the extremities of it higher than the other, by slackening one lift, and pulling upon the opposite one, so as to place the yard at a greater or less obliquity with the mast.

TOPPING-LIFT, (*balancine de gui*, Fr.) a large and strong tackle, used to suspend or top the outer end of a gaff, or of the boom of a cutter's, brig's, sloop's, or schooner's main-sail.

Topping and Tailing, in rope-making, the act of clearing both ends of the hemp with the hatchel.

Toppings, a term used by hatchellers for that which comes from the hemp in the act of hatchelling.

TORNADO, (*tourbillon de vent*, &c. Fr.) a violent squall or gust of wind rising suddenly from the shore, and afterwards veering round the compass like a hurricane. They are most frequent on the coasts of Guinea and South Barbary.

TORPEDO (*torpedo*, Fr.) an infernal machine, said to have been the invention of Mr. David Bushnell, of Saybrook, in Connecticut, in the late American war, for the purpose of destroying the British ships, by sub-marine explosion, and was then called the *American Turtle*.

It consisted of a small decked boat, capable of containing one or more persons, and furnished with apparatus enabling them to remain thirty minutes under water. The method of providing them with fresh air for respiration was, by having a reservoir of air, of suitable dimensions, to the size of the boat, and the number of persons in it. By means of a condensing pump, the air in this reservoir, was condensed about 400 times, and, by a spring, let out at intervals, as circumstances required; the impure air being rectified by carbonic acid, neutralized with chalk. Within the boat were flaps, like those of a rundle, by which it was moved; also two rudders one vertical, the other horizontal, to steer it by; a compass marked with phosphorus, and a pump to empty the hold, or air reservoir. The persons within could elevate the boat to the surface of the water, or depress it at pleasure. To this vessel was attached a large powder magazine, containing 150 pounds of powder, intended to be fastened to the bottom of an enemy's ship with a driving screw, in such a way, that the same stroke, which disengaged it from the machine, should put the internal clock-work in motion; this being done, the ordinary operations of a gun-lock, at the distance of half an hour, or any time that should be deemed sufficient for the conductor to escape, would cause the powder to explode, and leave the effects to the common laws of nature.

Mr. Bushnell's experiments, however, did not succeed to his expectations, some unforeseen accident, or mismanagement, always prevented their success, sometimes the operator in boring the vessel destined for destruction, was stopped in his progress, by encountering a piece of iron; and, in removing to a more convenient place for his purpose, has absolutely been unable to discover the vessel a second time, also the motion of the tides and currents, tended much to his disappointment.

Mr. Bushnell invented several other curious machines for the annoyance of the British shipping, which only partially succeeded; he destroyed a vessel in the charge of Commodore Symmonds; the explosion of one of his *kegs*, also demolished a vessel on the Long Island shore. In December, 1777, he committed to the Delawar river, a number of kegs, or turtles, destined to fall among the British fleet at Philadelphia; but this squadron of kegs being separated and retarded by the ice, demolished only a single boat. He boasted, however, as his countrymen generally do, that this catastrophe created an alarm, unprecedented in its nature and degree.

This invention has been generally attributed to a Mr. Robert Fulton, an American, who, in 1804, (under the name of Mr. Francis), was patronised to a certain extent by the British government: but after his unsuccessful experiments against the enemy's flotilla, then assembled at Boulogne, and from the principles, and policy of its measures, the Fultonian, projects were discountenanced in this country.

Soon after this, Mr. Fulton proffered his services successively to the French, and Americans, by whom he has also been encouraged and rewarded.

The Americans in August, 1813, attempted to destroy the British ship Plantagenet, in Lynnhaven Bay, by the explosion of a combustible machine, invented by a Mr. Mix of the American navy. It was found to be a case, containing about six barrels of gunpowder, to which a lock was affixed, and a line attached to the lock, reaching to the boat, or the person who had the execution of the design. It was next suspended to a stage of planks, at each end of which were about 50 fathoms of small line, with a buoy at both ends.

The machine is thus put together and let into the water, the combustible case sinks about 12 feet, and being kept at that depth from the surface of the stage, it has many chances of success in close anchorage, the buoys being extended by the line 100 fathoms, may convey the line by the help of the tide across the cable of any ship at anchor, which the moment it touches will cause the machine to swing round to the side, or the bottom of the ship, and the person using it, finding by the line that it has stopt, and judging it to have reached the intended object, pulls the trigger of the lock by the string, and the explosion takes place. Should it, thus situated, not blow the ship up; if it start a butt-end of one of her planks, she would inevitably founder.

It fortunately did not succeed upon the Plantagenet, the explosion taking place half a cable's length from the ship, the immense column of flame and water thrown up, excited a temporary alarm, but nothing farther. However, our blockading ships on the coast of America have kept a sharp look-out in their guard-boats since this infernal attempt was made, and it is hoped the infamous designs of the enemy will again be baffled with disappointment.

TORRENT, (*torrent, cours rapide*, Fr.) in hydrography, a temporary stream of water, falling suddenly from mountains, &c. where there have been great rains, or an extraordinary thaw of snow; sometimes making great ravages in the plains.

TORRID ZONE, (*Zône torride*, Fr.) is that tract of the earth lying on each side of the equator as far as the two tropics, or 23° 30′ of north and south latitude.

To TOSS THE OARS UP, (*mettre les avirons debout*, Fr.) is to put them in a perpendicular direction, ready to let them all fall at once into the water, and is intended as a compliment to the passengers in the boat.

To TOUCH, (*voiles qui commencent à faséyer*, Fr.) speaking of a ship's sails, is when they first begin to shiver, with their edges in the direction of the wind, and is either occasioned by an alteration in the ship's course, or by a change of the wind. See the articles FULL AND BY.

To TOUCH *and Go*, (*toucher et parer*, Fr.) is spoken of a ship, when, under sail, she rubs against the ground with her keel, without much diminution of her velocity.

To TOUCH *at any Port or Bay*, &c. (*relâcher dans quelque port, ou baie*, &c. Fr.) implies to stop or anchor occasionally, at some intermediate port, or bay, in the course of the voyage.

TOUCH-HOLE, (*lumière d'une arme à feu*, Fr.) the vent through which the fire is conveyed to the powder in the chamber of a gun.

To TOW, (*touer, remorquer, &c.* Fr. *teon, teohan*, Sax.) to draw a ship or boat forward in the water, by means of a rope attached to another vessel or boat, which advances by the effort of rowing or sailing.

Towing is either practised when a ship is disabled, and rendered incapable of carrying sail at sea; or when her sails are not fixed upon the masts, as in a harbour; or when they are deprived of their force of action by a cessation of the wind.

When a ship of war is dismasted, or otherwise disabled from carrying sail at sea, she is usually towed by a cable reaching from her bow to another ship a-head. In a harbour, towing is practised by one or more boats, wherein all the force of the oars is exerted to make her advance.

Mr. Dickinson, the ingenious inventor and patentee of the cast iron buoys, barrels, and tanks, has suggested to the Editor, that, the mode of towing ships and vessels up and down rivers, into and out of action, or in a calm, has hitherto been very faulty and inconsiderate. "For, when we see a small boat towing a vessel," says Mr. Dickinson, "The rope, being run out the whole length, becomes taught, with a jerk, and then flies back; and when the towers take two or three more strokes with their oars, the jerk is repeated, and so it continues. Now, the effect produced on the vessel by those jerks, is much like that which would be produced by a blow on the stern with a sledge hammer. A ship in a state of rest, like all heavy bodies, is not to be put in motion instantaneously, and yet sailors pursue this ridiculous system." To remedy this evil he advises that the towing-line be passed through a block at the cat-head, or suspended from, and fastened to the end of the bowsprit, with a weight at its end, by which means the line, instead of its coming with a jerk, would be always taught, and the weight pulling up or dropping down: *i. e.* in the interval of not pulling, and between the strokes the weight will drop and pull the boat, and the taughtness of the rope, and the pressure will always be the same.

If a boat be manned with four men, tie a weight of about 24lbs. more or less, to the end of the towing-line, and with this 6lbs. a piece besides the line, they will find work enough to keep that weight constantly out of the water, and thus tow the vessel steadily along without incurring those sudden jerks, which facilitate their progress but little, but for want of something better, it has been submitted to time immemorial.

To take a Vessel in TOW, (*prendre un bâtiment, à la remorque*, Fr.) is sometimes a figurative expression for taking care of her.

TOW-BOAT, (*bateau servant à touer*, Fr) implies any boat employed in towing a ship or vessel out of a harbour, &c.

TOW-LINE, (*haussière, ou touée*, Fr.) a small hawser generally used to remove a ship from one part of a harbour or road to another, by means of anchors, capstans, &c. as explained in the article *To* WARP. It is also employed occasionally to moor a small vessel in a harbour, conveniently sheltered from the wind and sea.

TOW-ROPE, (*grelin, ou haussière à touer*, Fr.) a name given to any cable or other rope used in the exercise of towing.

TOWING OVERBOARD, (*prendre à la remorque*, Fr.)

the act of drawing any thing after a ship or boat, when she is sailing or rowing, which has previously been on board that ship or boat.

TRACK OF A SHIP. See the article WAKE.

To TRACK, (*tirer à la cordelle une embarcation, &c. le long d'une rivière, ou d'un canal, à force de bras, ou avec des chevaux, &c.* Fr.) to pull any vessel or floating body along the stream of a canal or river, by means of a rope extended from the vessel, &c. to the adjacent shore, and drawn along the banks of the river, by men or horses.

TRACK-SCHUYT, (*bateau de Hollande et des pays voisins de la mer Baltique, que l'on hale le long des canaux et rivières,* Fr.) a vessel employed to carry goods or passengers up and down the rivers or canals in Holland, and the countries bordering on the Baltic sea. It is usually tracted by a horse, who trots along the margin to a limited distance, after which he is relieved by another.

TRADE, (*commerce, trafić, &c.* Fr.) is the exchanging of one commodity for an other, or for some representative of value, with which other commodities may be procured. At first, trade consisted in exchanging the surplus of some particular article of produce, which an individual possessed beyond what he found necessary for his own use, for the similar surplus of some other article possessed by another individual, which kind of traffic is now generally called barter. By the introduction of metallic and paper money, the operations of trade were much facilitated, while the increase of population rendered them of greater magnitude; improvements in the art of navigation extended their boundaries; the advancement of science and increase of luxury multiplied the objects of commerce; and the various duties and regulations established by different governments rendered the subject more complicated to individuals, but, at the same time, of increased importance to the state. The advantages of commerce began to engage the attention of mankind at a very early period. About 1300 years after the flood, the commerce of the Phœnicians had attained to a considerable extent; they possessed manufactures, entered into commercial partnerships, made long voyages, had resident agents in foreign countries, and had adopted the practice of lending money on bottomry. The Greeks and Romans, particularly the latter, carried on an extensive trade, considering the imperfect state of navigation. The reign of Augustus was particularly favourable to commerce, as the general peace, which then prevailed, enabled the merchants to pursue their concerns unmolested. At this period, Malta was famous for its number of workmen and its various manufactures, particularly of fine cloths. The inhabitants were the descendants of the ancient Phœnicians, and, like them, pursued commerce with avidity. Under Tiberius, we find the Romans extending their protection to the north; and they founded the town of Havern, the most ancient in Friesland. The cultivation of vines in France, Spain, and Portugal, seems to have increased about this time, and from the inhabitants neglecting to cultivate corn, and turning much of their arable land into vineyards, we may presume the latter afforded considerable profit. About the year 50, the capital of England was first described as a place famous for merchandise. In the 9th century, the Venetians carried on a very beneficial commerce with the Levant, whence they brought spices, silks, drugs, and fruits from the east in great abundance, and supplied the greater part of Europe with these commodities. The Crusades contributed materially to the revival of commerce during the 12th and 13th centuries, by introducing an acquaintance with the luxuries of the east, which multiplied the objects of trade. About this time, the commercial towns bordering on the Baltic, finding their trade much interrupted by pirates, entered into an agreement for mutual support and assistance. This union, which became very celebrated under the title of the Hanseatic League, at first consisted only of twelve towns, but finally, of upwards of seventy, including all the principal commercial towns of Europe, and embracing nearly all the foreign trade then carried on. The invention of the mariner's compass, which enabled trading vessels to make much longer voyages, and led to a discovery of a passage to the East Indies by the Cape of Good Hope, with the settlements made on the coast of Africa, in Arabia, and in India, affected very materially the commerce of the Mediterranean and of the Hanse Towns, and threw the principal parts of the foreign trade into the hands of the Portuguese. The discovery of America opened a new and extensive field for commercial adventure, while the establishment of manufactures produced new articles of trade. The persecutions of the Spaniards drove great numbers of industrious workmen from the Netherlands into England and other countries, who introduced several manufactures, and greatly improved others which were already established. While the different branches of British manufactures were thus improving, its colonial possessions in the East and West Indies were greatly increased, and England was enabled to rival the Dutch, not only in manufactured goods, but also in the different articles of colonial produce. During the last century, the trade of Holland has rapidly declined; and since the invasion of that country by the French, has been almost annihilated; while the decided naval superiority maintained by Great Britain, has extended and secured its commerce, and raised it to a magnitude unparalleled in the history of the world.

The policy of the times was by no means calculated to encourage trade: our merchants were very jealous of foreigners participating in their commercial profits, though it was to such persons they were indebted for the little knowledge of trade they possessed; and the government being at times still more jealous in this respect than individuals, many injudicious restrictive laws were made: amongst others, an act was passed in 1429, prohibiting persons from selling goods to aliens except for ready money. It was, however, soon found necessary to repeal this prohibition, and permission was then granted, to sell to aliens, upon giving not more than six months credit. Charles II. imposed a duty of 5 per cent, on goods exported as well as imported, on domestic manufactures, as well as foreign merchandise, and double taxes on all goods when exported by aliens. In 1672, however, the duties payable by aliens on the exportation of all goods of British manufacture were repealed. This salutary principle was still more extended in 1700, by removing the imposts previously laid on every kind of woollen goods exported, and on corn, meal, and bread. It was, however, by the law, "for the further encouragement of manufactures," passed in 1722, that every one was allowed to export, duty free, all kinds of merchandise, the produce of Great Britain, except a few particular articles: while drugs and other commodities used for dying, were also to be permitted to be exported duty-free. This period forms a memorable epoch in commercial policy; and the subsequent increase of trade is the best proof of the wisdom of the regulations then adopted. The various articles of foreign commerce being subject to great fluctuation in their value, a book of rates was adopted at the custom-house, in 1696, for the purpose of valuing uniformly the several articles imported and exported.

The American war produced a memorable effect on the foreign trade of this country. Previous to that event our manufactures were not much in the habit of exporting on their own account. The prohibition in America of importation from this country compelled them to seek new markets, and to improve those already known; and from this date commenced that rapid increase of export to the continent of Europe which has chiefly enabled us to sustain the burden of that war, and of those in which we have been since engaged. From the peace of 1783, our manufacturers have almost universally acted as merchants, and shipped their goods on their own account. They have gained possession of the foreign markets in part from the superiority of their skill, but far more from the superiority of their capital, which has enabled them to give a credit considerably superior to that which the exporters of other countries can allow.

During the last twenty-four years the amount of the imports has more than doubled, and the amount of the exports has still more increased in proportion.

Of the total exports of Great Britain, about two thirds of the corrected official value consists of British produce and manufactures, and one-third of foreign merchandise; of the former, little more than one fourth is exported to the continent of Europe; about the same amount to the United States of America, and the remainder to the East and West Indies, &c.

In the commercial intercourse between Great Britain and Ireland, a very unjust and oppressive species of policy was long exercised, by which the latter country was debarred from the enjoyment and use of its own resources, to make it entirely subservient to the interests and opulence of the former. It was not suffered to share in the bounties of nature, or to derive emolument from the industry of its citizens, but was shut up from every species of commerce, and restrained from sending the produce of its soil to foreign markets. Under such circumstances it cannot be supposed that its commerce, even with the country that monopolized it, could be very great.

Although Ireland now receives a considerable amount of foreign merchandise, as well as manufactured goods from Great Britain, the balance of trade is considerably in her favour, owing principally to the increased importation by Great Britain of Irish linens, which has amounted of late, to upwards of 36,600,000 yards annually. But this unfavourable balance of trade with respect to Great Britain, is probably counteracted by other circumstances.

The principal articles of import from Russia are iron, hemp, flax, tallow, pot-ashes, coarse linens, deals, and lathwood, hog's-bristles, &c. The exports consist of broad-cloth and woollen-stuffs, refined sugar, cotton, lead, tin, iron and steel ware, earthen ware, glass, alum, salt, coals, horses, porter, and various articles of less importance. It appears that the balance of trade is much in favour of Russia, which balance received in cash from Great Britain, enables it to import the wines of France, Spain, and Portugal, brandy, oil, indigo, and many other articles of foreign produce.

The trade between Great Britain and Sweden, has continued to the same extent nearly for the last 20 years. The imports consist principally of iron of a superior quality to that of any other country, pitch, tar, deal-boards, and sail-cloth. The trade with Prussia is carried on principally in foreign vessels.

The produce and manufactures of Great Britain exported to Germany, with the exception of consignments to particular mercantile houses, are sold at the great fairs of Frankfort, Leipsic, Brunswick, and Nuremberg. The Easter fair at Leipsic is greatly celebrated for the immense quantities of British manufactures sold there.

The trade with Holland was formerly the principal branch of our European commerce, it was at its greatest height about 1745. On the French taking possession of Holland during the late war, the trade with this country was of course in a great measure suspended particularly during the years 1795 and 1796, some little relaxation of the restrictions which had been imposed was afterwards found necessary; and on the peace, its commercial connection with this country immediately revived, till it was checked again by war: and at this period Holland has again resumed its commercial intercourse with this country.

The export to Flanders was formerly considerable, but of late years the trade with that part of the continent has been carried on through other ports.

The commercial intercourse with France has been so much deranged for some years past, that it is needless to give any account of the little that remains, which, during the time of war, must be entirely carried on by the vessels of neutral powers.

The trade with Portugal was formerly a very beneficial branch of our commerce, but has decayed of late; having suffered much from the ravages of war.

The Mediterranean trade has suffered much interruption since the year 1793: Great Britain, however, still retains a considerable share of it. The goods usually imported from Venice are currants loaded at Zante, and Cephalonia, or some port in the Morea, silks, cream of tartar, glass beads for the East and West Indies, and African trade, whisks of rice-weeds, sponge, some Levant drugs, cotton, oil, and vallonea.

The exports to Italy and Venice are pilchards and red-herrings, of which the consumption is considerable, a small quantity of salmon, tin, lead, and litharge, alum, woollens, and hardware; besides sugars, cocoa, pepper, and other foreign articles.

The trade with Turkey has been less disturbed than that of any other part of the Mediterranean, particularly since Malta has been in the hands of the English. The exports consist chiefly of lead in pigs, lead-shot, tin, wrought and cast iron, hardware, cloths and stuffs, glass and earthen-ware, clocks and watches, West India coffee, sugar, pepper, &c. The imports are cotton-wool, cotton-yarn, silk, mohair, madder, galls, figs, raisins, &c.

The West India trade is in some respects the most important branch of our commerce. The capital engaged in it has been estimated to be no less than 70,000,000*l.* The value of the goods exported from Great Britain, on an average of four years, ending the 5th of January, 1796, was estimated, according to their real value, at 3,900,000*l.*; but besides the goods sent directly hence to the West Indies, merchandise to a considerable amount used to be exported to the coast of Africa, for the purchase of slaves (prior to the abolition of the slave trade). Adding, therefore, the value of the slaves annually imported and retained in the islands, the whole export to the West Indies may be estimated at least at 4,672,000*l.* The value of the imports is considerably greater; and the annual balance is to be considered in part as a remittance of the income or property of persons who possess estates in the West Indies, and who are wholly or occasionally resident here; and of persons who have lent money on mortgage or otherwise in the West Indies, and receive the interest through the medium of these remittances: a further part defrays the freight and other charges in this country of West India produce.

The British trade with the United States of America, previous to the commencement of hostilities in 1812, was of considerable importance to both countries. The imports from America, consisted chiefly of unmanufactured American produce, such as corn, flax-seed, rice, tar, timber, turpentine, &c. The exports from Britain to America were chiefly British manufactures.

The exports to Canada, and the remaining British possessions in North America in 1792, amounted to 1,120,091*l.* of which 908,105*l.* were the value of British manufactures, and 211,886*l.* that of foreign merchandise; it has since been rather under that amount. The imports consist of skins and furs of all kinds.

The East India trade is a concern of vast magnitude, if considered merely in a commercial point of view. The annual sales of the imports of the company for 16 years, preceding 1757, amounted to about 2,055,000*l.*, on the average; during the succeeding ten years, their sales of imports increased to 2,150,000*l.* per annum; in the following ten years they rose to 3,330,000*l.* per annum. During the late American war, the sales of imports fell off about 200,000*l.* per annum; but from that period they increased again, and amounted on an average to 4,768,242*l.* and in 1804 to 5, 866,073*l.* The exports to India re chiefly woollens, metals, naval and military stores, for the purchase of teas in China, a considerable amount of bullion is annually exported. The investments in India are formed from the proceeds of the company's export, from the proceeds of part of the territorial revenues invested in merchandise, and from the fortunes of private individuals, and the profits of their trade remitted to this country, which are paid into the company's exchequer, and remitted by the company in merchandise from India.

The African trade, lately abolished, was chiefly for the purchase of slaves to supply the West India islands. The total number of vessels which entered inwards from Africa, in the year 1804, was 38; the number which cleared outwards was 176.

The number of trading vessels built and registered in the different ports of the British empire, in the year ending the 5th of January, 1804, was 1402, amount of tonnage 135,349 tons.

Trade-Winds, (*vents alisés*, Fr.) are certain regular winds blowing within or near the tropics, and being either periodical or perpetual. Thus, in the Indian Ocean, they blow alternately from two opposite points of the compass, during a limited season; and, in the Atlantic Ocean, continue almost without intermission in one direction. They are accordingly called trade-winds, from their great utility in navigation and commerce. See the articles Monsoon and Wind.

TRAIL, in gunnery, the end of a travelling carriage, opposite to the wheels, and upon which the carriage slides when unlimbered, or upon the battery.

To Trail, is to draw along the ground.

Trail-Board, (*frise entre les jottereaux, ou frise de l'éperon*, Fr.) in ship-building, the carved-work between the cheeks, that is fastened to the knee of the head.

TRAIN, (*traînée de poudre*, Fr.) the hinder part of a gun-carriage; also a line of gunpowder, or other combustible materials, forming a communication with any body intended to be set on fire or exploded.

Train-Tackle, (*palan de rétraite, ou de recul des canons*, Fr.) a combination of pulleys, which is, during action, hooked to an eye-bolt, in the train of the carriage, and to a ring-bolt in the deck, as represented by fig. 2, plate VI. Its use is, to prevent the gun from running out of the port whilst loading.

TRANSIT, (*transitus*, Lat.) in astronomy, denotes the passage of any planet, just before or over another planet or star; or the passing of a star or planet over the meridian, or before an astronomical instrument.

Venus and Mercury in their transits over the sun's disc have the appearance of dark spots.

If the orbits of these planets were in the plane of the ecliptic, this phenomenon would happen at every inferior conjunction, and the planets would pass exactly across the centre of the solar disc. As their orbits, however, are inclined to the ecliptic, Venus and Mercury are seldom seen upon the sun, excepting when the line of their nodes coincide nearly with a line joining the earth, the inferior planet, and the sun. But at all other times Venus and Mercury, at the time of their inferior conjunction, pass either above or below the disc of the sun.

The transits of Mercury and Venus were first predicted by the celebrated Kepler. He announced a transit of Mercury for 1631, and two of Venus for 1631, and 1761. The transit of Mercury was observed by Gassendi; but as the transit of

Venus in the same year must have been over before the rising of the sun, Gassendi, who had made preparations to observe it, could not fail to be disappointed. Another transit of Venus, which Kepler had not predicted, was observed at Hool, near Liverpool, in 1639, by our countryman Mr. Horrox, the first person who had the singular felicity of observing Venus upon the disc of the sun.

These phenomena return at stated periods: for if a transit happens at any given time at either of the nodes of Mercury or Venus, it is obvious, that before another transit can happen, the earth must perform a certain number of revolutions round the sun, in the same time that the planet performs a greater number of revolutions in its orbit; the number of revolutions performed by the earth, and the inferior planet, must be such, that they are brought into nearly the same relative position to the sun which they had at the first transit, and will evidently be in the inverse ratio of the periodic times of the earth and the planet. In order to determine their periods, therefore, we must find whole numbers for Mercury in the ratio of 87.969 : 365.256. These numbers are 7 : 29; 13 : 54; 33 : 137; 46: 191; 217 : 901; 26 : 1092; consequently in seven years Mercury will have performed nearly twenty-nine revolutions, and so on with the rest. The whole numbers expressing the periods of Venus's transit, being in the ratio of 224.701 to 365.256, are 8 : 13; 235 : 382; 243 : 395; 713 : 1159.

The transits of Venus over the sun's disc are among the most interesting phenomena in astronomy, not only from the rarity of their occurrence, but from the important determinations to which they lead. Dr. Halley pointed out the advantages which astronomers would receive from accurate observations of the transits of Venus, in enabling them to determine the distance of the sun from the earth; and, accordingly, the transits in 1761, and 1769, were observed with the utmost perseverance and zeal in every quarter of the globe. Dr. Maskelyne being sent by the commissioners of the Board of Longitude, to St. Helena, to observe the transit of Venus in 1761; and Mr. William Bayly was also sent by the Board to the North Cape, to accompany Sir Charles Douglas, in observing the transit of Venus, in 1769.

Dr. Halley computed the time of a number of these visible transits, for the two last centuries, and published in the Philosophical Transactions, number 193.

Those of Mercury happen in April and October, those of Venus in May and November, both old style, and by adding 11 days, gives the time new-style.

IN APRIL.

Years.	Times of Mercury's Conjunction.			Distances from the sun's centre.	
	days	hours	min.		
1615	22	21	38*	7′	20″N.
1628	25	5	15*	9	35 S.
1661	23	4	52*	4	27 N.
1674	26	12	29	12	28 S.
1707	24	12	6	1	34 N.
1720	26	19	43*	15	21 S.
1740	21	11	43	15	36 N.
1758	24	19	20*	1	19 S.
1786	22	18	57*	12	43 N.
1799	26	2	34*	4	12 S.

IN OCTOBER.

Years.	Times of Mercury's Conjunction.			Distances from the sun's centre.	
	days	hours	min.		
1730	22	5	28	16′	45″S.
1736	30	22	53**	13	5 N.
1743	24	23	2**	8	42 S.
1756	26	16	36	0	38 S.
1769	29	10	10	7	24 N.
1776	22	10	19	15	23 S.
	November.				
1782	1	3	44*	15	27 N.
	October.				
1789	25	3	53*	7	20 S.

* This mark denotes that the transit is partly visible at London, and these **, that it is entirely visible.

It must be observed, that at the ascending node of Mercury in the month of October, the diameter of the sun takes up 32′ 34″, and therefore, the greatest duration of a central transit, is $5^h\ 29^m$. But in the month of April, the diameter of the sun is 31′ 54″, whence by reason of the slower motion of the planet, there arises the greatest duration 8h. 1m. Now, if Mercury approaches obliquely, these durations become shorter on account of the distance from the centre of the sun. Dr. Halley then exhibits the principles of calculating these transits, from whence he infers that,

"After 18 years, Venus returns to the sun taking away $2^d\ 10^h\ 52\frac{1}{2}^m$, from the moment of the foregoing transit; and the planet proceeds in a path 24′ 41″ more to the south than the former.

"After 235 years, adding $2^d\ 10^h\ 9^m$, Venus may again enter the sun, but in a more northern path by 11′ 33″. But if the foregoing year is bissextile $3^d\ 10^h\ 9^m$ must be added.

"After 243 years Venus may also pass the sun, only taking away 43^m from the time of the former, but proceeds more southerly by 13′ 8″. Now, if the foregoing year be bissextile, add $23^h\ 17^m$.

"And in all these appulses of Venus to the sun, in the month of November, the angle of her path with the ecliptic is 9° 5′, and her horary motion within the sun 4′ 7″. And since the semidiameter of the sun is 16′ 21″, the greatest duration of the transit of the centre of Venus comes out $7^h\ 56^m$.

Then let the sun and Venus be in conjunction at the descending node in the month of May; and by the same numbers the same intervals may be computed.

After 8 years let there be taken away $2^d\ 6^h\ 55''$, and Venus will make her transit in a more northern path by 19′ 58″.

"After 235 years, add $2^d\ 8^h\ 18^m$, or, if the foregoing year be bissextile $3^d\ 8^h\ 18^m$, and you will have Venus more to the south by 9′ 21′

"Lastly, after 243 years, add 1^h 23^m, or, if the foregoing year be bissextile 1^d 1^h 23^m, and Venus will be found in conjunction with the sun, but in a more northerly path by 10′ 37″.

"In every transit within the sun at this node, the angle of Venus's path with the ecliptic, is 8° 28′, and her horary motion is 4′; and the semidiameter of the sun, subtending 15′ 51″, the greatest duration of the central transit comes out also 7^h 56^m, exactly the same as at the other node.

"As to the epochs, from that only ingress which Horrox observed, the sun being just then ready to set, it is concluded, that Venus was in conjunction with the sun at London, in the year 1639, November 24^d 6^h 37^m, and that she declined towards the south 8′ 30″. But in the month of May no mortal has seen her as yet within the sun. It appears, that Venus for the next time will enter the sun in 1761, May 25^d 17^h 55^m, that being the middle of the eclipse, and then will be distant from his centre 4′ 15″, towards the south. Hence and from the foregoing revolutions all the phenomena of this kind will be easily exhibited for a whole millenium as computed in the following table.

IN NOVEMBER.

Year.	Times of Venus's Conjunction.			Distances from the Sun's centre	
	days	hours	min.		
918	20	21	53	6′	12″N.
1161	20	21	10	6	55 S.
1396	23	7	20	4	38 N.
1631	26	17	29	16	11 N.
1639	24	6	37	8	30 S.
1874	26	16	46	3	3 N.
2109	29	2	57	14	36 N.
2117	26	16	3	10	5 S.

IN MAY.

Year.	Times of Conjunction.			Distances from the Sun's centre.	
	days	hours	min.		
1283	23	8	14	5′	31″S.
1291	25	15	9	14	27 N.
1518	25	16	32	14	52 S.
1526	23	9	37	5	6 N.
1761	25	17	55	4	15 S.
1769	23	11	0	15	43 N
1996	28	2	13	13	36 S.
2004	25	19	18	6	22 N

"As for the duration of these eclipses of Venus they may be computed after the same manner as those of Mercury in respect of the centre. But since Venus's diameter is pretty large, and since the parallaxes also may bring a very notable difference as to time, a particular calculation must necessarily be made for every place.

"Now the diameter of Venus is so great, that while she adheres to the sun's limb almost 20 minutes of time will be elapsed, that is, when she applies directly to the sun. But when she is incident obliquely, she continues longer in the limb. Now that diameter according to Horrox's observation, takes up 1′ 18″, when she is in conjunction with the sun at the ascending node and 1′ 12″ at the other node.

"Lastly, the chief use of these conjunctions, as we formerly observed, is to determine the sun's distance from the earth, or his parallax, which astronomers have in vain attempted to find by various other methods: for the minuteness of the angles required easily eludes the nicest instruments. But in observing the ingress of Venus into the sun, and her egress from the same, the space of time between the moments of the internal contacts, observed to a second of time, that is, to $\frac{1}{15}$ of a second or 4‴ of an arc, may be obtained by the assistance of a moderate telescope and a pendulum clock, that is consistent with itself exactly for 6 or 8 hours. From two such observations rightly made in proper places, the distance of the sun within a 500th part may be certainly concluded." See the article PARALLAX. Also VINCE's and LA LANDE's *Astronomy*,

TRANSIT *Instrument*, in astronomy, is a telescope moveable about a horizontal axis, and so adjusted as to make its line of collimation describe a great circle passing through the pole and zenith, or the meridian of the place. Its use is to take the right ascension of the heavenly bodies, and to correct the going of the clock; for which purpose it has the system of wires placed in the principal focus of the object glass. It has only one eye-glass, which is convex, and consequently the objects are inverted. Directing, therefore, the telescope towards the south, the bodies enter the field of view on the west side. The position of the telescope may also be verified by observing whether the interval of the transit of two known stars which differ considerably in declination, be equal to their difference of right ascension in time, and if not, alter the position till they are equal. The position may also be examined by observing equal altitudes of the stars, for if the interval of time be bisected by their transits, the instrument is adjusted. As the operation of fixing a transit telescope accurately in the meridian is tedious; the following directions will facilitate the business:

Bring the telescope as nearly into the meridian as you can by the sun at noon, and by taking the transit of the fixed stars find the length of a sidereal day by the clock. Take the altitude of a star, when on or near the prime vertical, and note the time by the clock; then from the latitude, declination, and altitude, compute its distance from the meridian, from which you will know the time by the clock when it passes, a little before which, bring, by the parallel screw, the middle wire of the telescope; then bisect the star, and by that screw keep the star bisected till you observe the time by the clock when it passes the meridian, and you will then have the telescope very near in the meridian. The adjustments being all made, that is, the axis of the telescope being adjusted parallel to the horizon, the middle wire adjusted perpendicular to the horizon, the line of collimation perpendicular to the axis, and brought to the meridian mark, by the *parallel* screw, the middle wire will always, as the telescope is turned about its axis, coincide with the meridian, and, therefore, every celestial body will be upon the meridian, when it passes over that wire. Observe the interval of the passage of any fixed star over the middle wire, and also over the side wires in order to multiply the observations of the passage over the middle wire; then if that interval be 23^h, $56'$. $4''$ it is adjusted to *solar* time, if 24^h to *sidereal* time; if not, lengthen or shorten the pendulum till you find the interval be either the one or the other as you may want, and the clock is adjusted. But as all clocks will either gain or lose something, it is necessary that the rate of its going should be ascertained: for that purpose observe the intervals continually, and you will get the rate of its going, that is, how much it gains or loses in a day. And to be more exact, take the passage of several stars, and the mean rate of going of them all. *Secondly* by the *sun*, and suppose you want to adjust the clock to mean solar time: take the time of the transit of the sun's centre over the middle wire. Now the equation of time applied by the nautical almanac to 12 o'clock, gives the true time of the transit, which, if it agrees with the time observed, shows that the clock points out the true time, if not, you get the error; and thus by repeating the observation every day, you get the rate of the clock's going, and how much it is too fast or slow, and consequently you will always get the true time. To determine the right ascension of the heavenly bodies, we may observe, that, as the earth revolves uniformly about her axis, the right ascension of stars passing the meridian at different times will differ in proportion to the difference of their times of passing, and as the clock is supposed to go uniformly, we have the following rule:—As the interval of the times of the passage of any fixed star over the meridian: the interval of the passage of any two stars :: 360°: their difference of right ascension. By the same method we find the difference between the right ascensions of the sun at noon, and any star by taking the interval of their transits, and, therefore, if that of the star be known, that of the sun will also; which will be rendered more exact, if we compare it with several stars and take the mean. The uses of the right ascensions are, 1st, *To find the time a star or planet comes to the meridian.* This is necessary to be known, that the astronomer may be ready to make his observation, and, therefore, mathematical accuracy is not necessary. 2dly, *To find at any given time, how long it will be before any body comes to the meridian.* This is found from the last, by finding the time at which it comes to the meridian, and subtracting the given time from it, borrowing 24 hours, if necessary. 3dly, *To determine in what order the heavenly bodies pass the meridian.* This will happen according to the order of their right ascensions, beginning at the least. 4thly, *To make a Catalogue of the fixed stars.* The difference of longitudes of two places may also be found from the times of the moon's passage over their meridians compared with the times of the passage of a fixed star, as proposed by Dr. Maskelyne in the nautical ephemeris for 1769. Find from the nautical ephemeris, or from computation, the increase of the moon's right ascension in 12 hours; observe also the interval of time between the passage of a given star, and of the moon's enlightened limb over the meridian of each place, and the difference of these intervals will show the increase of the moon's right ascension in the time of the star's passage from one meridian to the other; hence, *as the increase of the moon's* A.R. *in 12 hours*: 180° :: *the above intervals*: *difference of longitude.*

TRANSOMS, (*barres d'arcasse*, Fr.) are certain beams or timbers of oak fixed across the stern-post of a ship, and bolted thereto, to fortify her after-part, and give it the figure most suitable to the service for which she is calculated.

The deck-transom is the aftmost or hindmost beam of the lower deck, whereon all the deck-planks are rabbetted: and all the transoms are fixed athwart the stern-post, in the same manner as the floor-timbers are laid upon the keel. As the floor-timbers also, with regard to their general form and arrangement, have a rising, by which the bottom becomes narrower as it ascends towards the extremities; so the arms of the transoms, being gradually closer in proportion to their distance from the wing transom downwards, give a similar figure to that part of the ship, which accordingly becomes extremely narrow, from the counter towards the keel; and this general figure or curve is called the flight of the transoms.

Although these pieces are therefore extremely different in their figures, according to the extent of the angles formed by their branches or horns, each of them has nevertheless a double curve, which is partly vertical, and partly horizontal, with regard to its situation in the ship. The former of these is called, by the artificers, the round-up, and the latter the round-aft.

The vertical direction of the arms or angles of the transoms, with regard to the ship's length, are expressed in the plane of Elevation; and their horizontal curves are also delineated on the plane of Projection, both in plate XVII. and described in the general explanation of the planes in the article Naval *Architecture.*

As the transoms fill up the whole space comprehended between the head of the stern-post above, and the aftmost floor-timbers below, it is necessary to distinguish them by particular names.

Helm Port Transom, (*barre au bout de l'etambord*, Fr.) that which is at the head of the stern-post, and forms the upper part of the gun-room-ports.

Wing Transom, (*lisse de hourdi*, Fr.) the next below, which forms the lower part.

Deck Transom, (*barre du premier pont*, Fr.) that whereon all the lower deck planks are rabbetted.

The 1st, 2d, 3d, Transoms, &c. (*première, seconde, troisième, &c. barre d'arcasse*, Fr) are respectively below the preceding. See the article Ship-Building.

Transom-Knees. See the article Knees.

TRANSPORT. See the article Ship.

To Transport *Goods*, (*transporter des marchandises, ou effects*, Fr.) is to convey them over-seas to foreign countries.

To Transport *a Ship*, (*changer un vaisseau de place dans un port*, Fr.) is to navigate her from one place to another, by means of anchors, ropes, &c.

Transport-*Board*, (*certain bureau de la marine, dont les commissairès, qui le composent frètent les bâtimens de transport en brevètent les officiers; ils sont aussi chargés du soin et de l'entretien des prisonniers de guerre*, Fr.) a board instituted in July, 1794, for the purpose of conducting the transport service, viz. to transport troops and stores; also to provide accommodation and provision for all prisoners of war, and to regulate their exchange by cartel, &c.

It is subordinate to the secretary of state for the home department, has six commissioners, and one secretary, whose office is in Dorset-square, Westminster, London, and consists of one chief clerk, one assistant, three extra clerks, one clerk of the minutes, one clerk for keeping account of the appropriation and service of transports, one assistant, ditto, and one extra clerk.

The province of this board is to negotiate all contracts with the ship-owners, to cause surveys to be made of the vessels, finally to adjust the accounts and pay the balance.

With respect to the terms of this contract, these must vary according to the particular service for which the merchant vessels may be destined. The masters of the vessels thus hired, are furnished by the board with the necessary directions for their guidance, together with blank forms for keeping their accounts of hammocks, provisions, &c. supplied for the troops; and where the service is of a particular nature, the masters of the vessels, upon resorting to the agent for the Transport Board, are furnished with ample and precise instructions for their government.

The general form of a charter-party of affreightment for stores and troops for the service of government, varies but little from charter-parties entered into by ship-owners. For farther particulars, see the article SICK *and Hurt and Transport Board, &c.*

TRANSPORT-*Service*, a department under government, conducted by the commissioners of the TRANSPORT *Board*, which article see.

TRAVELLER, (*racambeau*, Fr.) a large iron thimble, whose diameter is much longer, in proportion to the breadth of its surface, than the common ones, as represented by fig. 10, plate VIII. It is furnished with a tail formed of a piece of rope, about three feet in length, one end of which encircles the ring, to which it is spliced. These machines are principally intended to facilitate the hoisting or lowering of the top-gallant-yards at sea; for which purpose two of them are fixed on each backstay, whereon they slide upwards and downwards, like the ring of a curtain upon its rod: being thus attached to the extremities of the top-gallant-yard, they prevent it from swinging backwards and forwards, by the agitation of the ship, whilst the yard is hoisting or lowering at sea.

Jib TRAVELLER, is a circular iron hoop, with a hook and shackle, used to haul out the tack of the jib.

TRAVELLING BACKSTAYS, are so denominated from their having a traveller upon the topmast, which slides up or down according to the reefs in the topsail, thereby confining the principal support of the backstay to that part of the mast immediately above the top-sail-yard.

TRAVELLING *Martingale*, a similar contrivance adapted to a martingale, to support the jib-boom in that particular part where the jib-tack is fixed. See the article MARTINGALE.

TRAVERSE, or TRANSVERSE, in general, denotes one thing that goes athwart another; that is, crosses and cuts it obliquely.

TRAVERSE, in gunnery, is the turning a piece of ordnance about, as upon a centre, to make it point in any particular direction.

TRAVERSE, in fortification, denotes a trench with a little parapet, sometimes two, one on each side, to serve as a cover from the enemy that might come in flank.

TRAVERSE, (*route oblique*, Fr.) in navigation, is the variation or alteration of a ship's course, occasioned by the shifting of the winds, or currents, &c.; or a traverse is a compound course, consisting of several different courses and distances. The traverses formed by a ship, when making an oblique progression against the direction of the wind, are explained in the article *To* TACK.

TRAVERSE-BOARD, (*renard servant aux pilotes, sur les vaisseaux de guerre François*, Fr.) a thin circular piece of board, marked with all the points of the compass, and having eight holes bored in each, and eight small pegs hanging from the centre of the board. It is hung up in the steerage, and used to determine the different courses run by a ship during the period of the watch; and is performed by putting a peg in that point of the compass whereon the ship has run each half hour. Thus, if the wind is northerly at the beginning of the watch, the ship, being close-hauled on the larboard tack, will steer W.N.W. If, after the first half hour, the wind changes to N. by W. the ship will fall off to W. by N. both of these courses are marked by the helmsman upon the traverse-board, by putting in one peg for every half hour on which she steers the same course; as one peg into W.N.W. and two pegs into W. by N. if she sails an hour on the latter course, and so on. This implement is particularly useful in light and variable winds.

TRAVERSE-SAILING, (*route oblique*, Fr.) is the method of working or calculating traverses or compound courses, so as to bring them into one, &c.

Traverse sailing is used when a ship, having sailed from one port towards another, whose course and distance from the former is known, and is by reason of contrary winds, or other accidents, forced to shift and sail upon several courses, which are to be brought into one course, to learn, after so many turnings and windings, the true course and distance made good, or the true point the ship is arrived at; and so to know what must be the new course and distance to the intended port.

To reduce a compound course to a single one, 1. Make a table of six columns, marked course, distance, N. S. E. W. beginning at the left hand, and write the given courses and distances in their proper columns. 2. Seek the given course and distances in the traverse, and let the corresponding differences of latitude and departure be written in their proper columns, in the table made for the question. 3. Add up the columns of northing, southing, easting, and westing; then the difference between the sums of northing and southing gives the whole difference of latitude, which is of the same name with the greater; and the difference between the sums of easting and westing will be the whole departure, which is likewise of the same name with the greater. 4. The whole difference, latitude, and departure to the compound course being found, the direct course and distance is found by plane-sailing.

To construct a traverse; with the chord of 60°, describe a circle to represent the compass, and divide it into four quadrants, N. S. E. W.

Lay off each course on the circumference, reckoned from its proper meridian, and from the centre to each point, draw so many blank radii, and mark these with the number of the course, 1, 2, 3, 4.

On the first radius, lay the first distance, mark its extremity; through which, and parallel to the second radius, draw the second distance of its proper length; through the extremity of the second distance and parallel to the third radius, draw the third distance of its proper length: and thus proceed till all the distances are drawn.

A line drawn from the extremity of the last distance to the centre of the circle, will represent the distance made good; and a line drawn from the same point, perpendicular to the meridian, will shew the departure, which is illustrated thus: A ship sails from an island, in latitude 69° 36′ N. and has run the following courses and distances, viz. S.W. by S. 45 miles, E.S.E. 50 miles, S.W. 30 miles, S.E. by E. 60 miles, and S.W. by S. ¼ W. 72 miles; required her present latitude, with the direct course and distance between the place sailed from, and the place come to.

Latitude sailed from 69° 36′N.
Diff. Lat. 149 ÷ 60 = 2 29 S.

Latitude come to = 67 7 N.
Course due south, and distance, 149.2 miles.

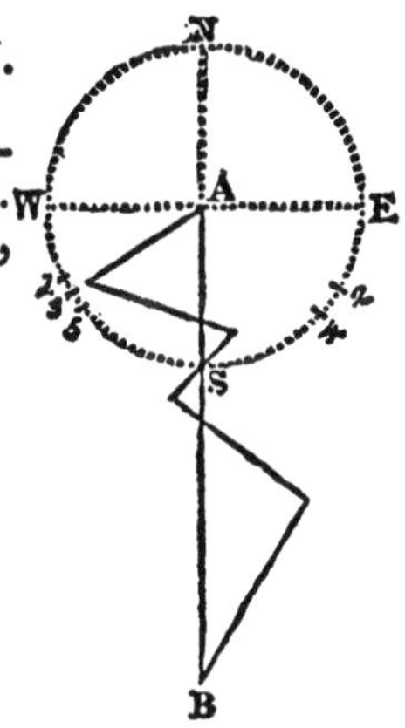

Courses.	Dist.	N.	S.	E.	W.
S.W. by W.	45	..	25.0		37.4
E.S.E.	50	..	19.1	46.2	
S.W.	30	..	21.2		21.2
S.E. by E.	60	..	33.3	49.9	
S.W. by S. ¼ W.	72	..	50.6		37.5
		..	149.2	96.1	96.1

TRAVERSE-TABLE, (*table, ou journal du loc*, Fr.) the same with a table of difference of latitude and departure, being only the difference of latitude and departure, ready calculated to every degree, point, half point and quarter point of the quadrant; and for any instance under one hundred miles, though it may conveniently serve for greater distances, by taking their halves, thirds, fourths, &c. and doubling, tripling, quadrupling, &c. the difference of latitude and departure found to those parts of the distance.

This table is one of the most necessary things a navigator has occasion for; for by it he can readily deduce all his courses and distances run in the space of twenty-four hours, into one course and distance; whence the latitude he is in, and departure from the meridian may be found.

TREADDLE, in block-making, is composed of two thin battens, 4 feet long, laid transversely under the lathe; to the lower batten the cord is fastened that connects it to the lathe, and the work is put in motion by the pressure of the workman's foot: the upper batten is fixed at one end to the ground with a thick piece of leather, resembling a hinge, and at the other end across the under one by an iron pin, for which purpose eight or ten holes are bored along the middle to regulate it to the work.

TREE-NAILS, (*gournables*, Fr.) are long cylindrical pins of cleft oak, from an inch to an inch and a half in diameter, and from one foot to three feet six inches in length, used for fastening the inside and outside plank of a ship to the upright timbers. They have recently been made of American pitch pine, a wood which is said to be more liable to dry-rot and decay than oak, and consequently very improper for the service.

The tree-nails are justly esteemed superior to spike nails or bolts, which are liable to rust, and loosen, as well as to rot the timber; but it is necessary that the oak of which they are formed should be solid, close, and replete with gum, to prevent them from breaking and rotting in the ship's frame. They ought also to be well dried, so as to fill their holes when they are swelled with moisture. They have usually one inch in thickness to 100 feet in the vessel's length; so that the tree-nails of a ship of 100 feet long, are one inch in diameter; and one inch and a half for a ship of 150 feet.

TREE-NAILS, are also cylindrical wooden pins, used by riggers for levers, or heavers.

Rough TREE. See the article ROUGH.

TRENCHES, in fortification, are ditches which the besiegers cut to approach more securely to the place attacked; whence they are called lines of approach. Their breadth is eight or ten feet, and depth six or seven.

TREND, in anchor-making, is that part of the shank of an anchor from which the size is taken. See the article ANCHOR.

To TREND, (*s'incliner*, Fr.) to incline; speaking of a coast, as "the land trends to the south-west."

TRESTLE, in rope-making, a frame with four legs, braced together with stout pins, and used at the upper end of a rope, or put under ropes of a short length, when the strands cannot be put on the stake-heads.

TRESTLE-TREES, (*barres maîtresses des hunes*, Fr.) in mast-making, two strong bars of oak timber fixed horizontally on the opposite sides of the lower mast-head, to support the frame of the top, and the weight of the top-mast. See the articles MAST and TOP.

TRIANGLE, (*triangulum*, Lat.) in geometry, a figure bounded or contained by three lines or sides, and which consequently has three-angles, from whence the figure takes its name.

TRIANGLE, in astronomy, a name common to two constellations; the one in the northern hemisphere, called *triangulum cœleste*, contains, according to the British catalogue, 16 stars; the other, in the southern hemisphere, called *triangulum australe*, contains five stars, according to Mr. Sharpe's catalogue.

TRIANGULAR *Canon*, tables relating to trigonometry; as of sines, tangents, secants, &c.

TRIANGULAR *Compasses*, are such as have three legs or feet, by which any triangle, or three points, may be taken off at once. Those are very useful in the construction of maps, globes, &c.

TRIANGULAR *Numbers*, are a kind of polygonal numbers; being the sums of arithmetical progressions, which have 1 for the common difference of their terms.

TRIANGULAR *Quadrant*, is a sector furnished with a loose piece, by which it forms an equilateral triangle; upon it is graduated and marked the calendar, with the sun's place, and other useful lines; and by the help of a string and plummet, with the divisions graduated on the loose piece, it may be made to serve for a quadrant.

To TRICE, (*hisser*, &c. Fr.) is to haul or tie up by means of a small rope or line.

TRICING-LINE, (*aiguilette*, Fr.) a small cord generally passing through a block or thimble, and used to hoist up any object to a higher station, in order to render it less inconvenient, such as the tricing lines of the yard, tackle, &c. The inner tricing-line hoists the block, and the outer one the parts of the tackle.

TRICK, the same as spell with regard to steering a ship. See the article SPELL.

TRIDENT, with mathematicians, is a kind of parabola, by which *Des Cartes* constructed equations of six dimensions.

To Hold the TRIDENT *of the Seas*, a familiar word used among the British, to express their dominion at sea.

TRIGGER, (*détente*, Fr.) an iron catch, which being pulled with the finger, disengages the cock of a gun-lock, that it may strike fire.

TRIGON, (τριγωνος, Gr.) in astronomy, denotes an aspect of two planets when they are 120° distant from each other; called also a trine, being the third part of 360 degrees.

TRIGONOMETRY, (from τριγωνος, and μετρον, Gr. *trigonométrie*, Fr.) is the art of measuring the sides and angles of triangles, either plane or spherical, and is accordingly called either plane or spherical trigonometry.

Trigonometry dates its origin from the time of Hipparchus, to whom the geography and astronomy of ancient Greece were greatly indebted; to remedy the defects in the construction of practical geometry, he introduced calculation founded on the famous property of the right-angled triangle, said to have been discovered by Pythagoras, and now forming the 47th proposition of Euclid's Elements. It is this theorem which is the connecting principle of arithmetic and geometry, or which renders the relation between the position of lines and their magnitude expressible by number. The merit of perceiving this at so early a period, was certainly great; and it enabled Hipparchus to lay the foundation of trigonometrical calculation. As the work which Hipparchus composed is now lost, we know nothing of the steps by which trigonometrical calculations advanced, till we find it, some centuries later, in the spherics of Menelaus, and in the Almagest of Ptolemy. Regiomontanus, the inventor of decimal fractions, introduced them into the tables and calculations of trigonometry: he added also a great number of new geometrical theorems, and deduced from them nearly the same rules which are now in use.

In proportion as the sciences advanced, greater accuracy of calculation was required; and the difficulty, as well as the time consumed, increased in the same proportion. What contributed most to facilitate mathematical calculations, was the noble invention of logarithms, by Baron Napier, and the improvements made by this illustrious person, in spherical trigonometry, is a lasting monument of his penetration and judgment; the state to which he brought trigonometry, continued with hardly any change, except perhaps a better arrangement of its rules, and a more concise demonstration of its principles, till about the middle of the last century, when Euler introduced the arithmetic of the sines, or, as it may more properly be termed, the application of algebra to trigonometry. By this means, the art of geometrical investigation is enriched with a new branch, properly called *Trigonometrical Analysis*. In this form, the sciences have been cultivated in France and Germany, ever since the change made in it by Euler. In England it has been confined, till within these few years, to its first and original occupation. The methods of the foreign geometers, however, have come gradually into notice, and our trigonometrical treatises within the last ten years, have generally contained some of the fundamental theorems and operations, in the arithmetic of the sines, and have followed the notation of Euler.

By trigonometry we determine the magnitudes of the earth, and planets; and the position of the fixed stars with respect to each other, by which we are enabled to depict the appearance of the heavens in a small compass; also the distances of the planets from the sun, their motions, eclipses, and other phenomena, are calculated by trigonometry, as are likewise the distances and positions of places on the earth, with their latitudes and longitudes; it may therefore justly be considered the basis of astronomy and geography. Navigation, with all its modern improvements, depends entirely on trigonometry; it is also the foundation of maritime surveying, and almost every branch in practical mathematics. For the progressive improvements by the different writers on this subject, see Introduction to Dr. HUTTON's *Logarithms*.

The business of this important science is to find

the angles when the sides are given; and the sides of their respective ratios when the angles are given; and to find sides and angles, when sides and angles are partly given. To effect this, it is necessary not only that the peripheries of circles, but also certain lines in and about circles, be supposed to be divided into a certain number of parts.

Every triangle has six parts, three sides and three angles, and it is necessary that three of these parts be given to find the other three. In plane trigonometry, it is necessary that one of the three parts at least be a side, since from three angles can only be found the proportions of the sides, but not the real quantities of them. But in spherical trigonometry, the three parts that are given, may be of any kind, either all sides or all angles, or part of both.

There are usually three methods of resolving triangles, or the cases of trigonometry; viz. geometrical construction, arithmetical computation, and instrumental operation. In the first method, the triangle is constructed by drawing and laying down the several parts of their magnitudes given, viz. the sides from a scale of equal parts, and the angles from a scale of chords, or other instrument, then the unknown parts are measured, by the same scales, and thus become known.

In the second method, having stated the terms, the sides being proportional to the sines of their opposite angles,

Thus, 1st. To find an angle.

: As one side given : : To the sine of its opposite angle,
: So is another side given : To the sine of its opposite angle.

2d. To find a side.

: As the sine of any angle given : : To its opposite side,
: So is the sine of another angle given : To its opposite side.

This proportion is resolved like all others, in which a fourth term is to be found from three given terms, by multiplying the second and third together, and dividing the product by the first. Or in working with logarithms, add the logarithms of the second and third terms together, and from that sum subtract the logarithm of the first term, then the number answering to the remainder, is the fourth term sought.

Thirdly, to work a case instrumentally, extend the compasses from the first term to the second, or third, which happens to be of the same kind with it, then that extent will reach from the other term to the fourth. In this operation, the line of numbers is used for the sides of triangles, and the lines of sines or tangents, for the angles, according as the proportion respects sines or tangents.

For the solution of spherical triangles, there are various theorems and proportions, which are similar to those in plane trigonometry, by substituting the sines of sides instead of the sides themselves, when the proportion respects sines, or tangents of the sides, for the sides, when the proportion respects tangents, &c. See WOODHOUSE'S *Treatise on Plane and Spherical Trigonometry, &c.*

TRIM, (*assiette, arrangement convenable d'un bâtiment, de son arrimage, de ses mâts, de son grèement, &c.* Fr.) the state or disposition of the ballast, cargo, masts, &c. by which a ship is best calculated for the several purposes of navigation.

The TRIM *of the Hold, (l'arrimage de la cale, le plus convenable, à la marche, et à la stabilité du bâtiment,* Fr.) denotes the most convenient and proper arrangement of the various materials contained therein, relatively to the ship's motion or stability at sea, by which a ship is either brought by the head or stern, or depressed in the water to a sufficient depth to carry sail well, and advance rapidly.

As the stowage of the hold, or the disposition of the several articles of the cargo, considerably affects the ship's motion and stability, it will be necessary to give a general idea of the action of a heavy body upon the fluid that supports it, and the re-action of the fluid on the floating body.

The whole weight of any body, then, may be considered as united in its centre of gravity; so that, if it were suspended by a line fastened to this centre, the line would hang in a perpendicular position, as directed through the centre of gravity to the centre of the earth. A body which floats in a fluid, is not, however, supported by its centre of gravity, but by the compression of the surrounding filaments of water: and each of these, being considered as infinitely small, will act upon a very minute portion of the surface of the floating body, with regard to the specific gravity, and conform to a principle applicable to all fluids, in proportion to the height of these filaments, viz. That the weight of a column of any fluid will be in proportion to a specific gravity of the fluid and the height of the column multiplied by its base.

But as heavy bodies endeavour, by their gravity, to approach the centre of the earth, in a vertical line passing through their centres; so the pressure of fluids endeavours to carry bodies in a vertical, tending from the centre of the earth towards their surface, and passing through the centre of gravity of the submerged part, which forces them towards the surface. So, in any submerged body at rest, these two opposite forces coincide in the same vertical, acting in a direction quite contrary to each other.

From this theory it results, that the stability or trim of a ship chiefly depends upon her construction, as considering the bottom to be homogeneous. This, however, can only happen when her cargo consists of the same materials throughout, as with corn, salt, or any species stowed in bulk, and when her hold is entirely filled. For if the ship has not sufficient breadth to resist the effort of the wind upon her sails; or if she is built too high, or too sharp in the floor, her centre of gravity will be too high, and she will be very crank, *i. e.* apt to overturn.

But as the stiffness of a ship, or quality to carry sail without danger of overturning, depends very much on the stowage of the hold, the centre of gravity may thereby be considerably lowered, by which her stability will be increased in proportion. It is a general maxim amongst mariners, that a ship will not carry sufficient sail till she is laden so deep that the surface of the water may glance on her extreme breadth amidships. She must therefore have a great deal of weight, as ballast, &c. to bring her to this situation, which is called a good sailing trim.

Several circumstances are also to be particularly

considered with regard to the quality, weight, and stowage of the ballast. If the centre of gravity be placed too high, the ship will be rendered incapable of carrying a sufficient quantity of sail; and by having it too low, she will be in danger of rolling away her masts. When it is placed too far forward, the ship will pitch, and labour heavily; and when too far aft, she will occasionally be exposed to the dangerous circumstance of a pooping sea. These extremes being carefully avoided, it remains to proportion the contents of every part of the hold to its capacity, and to place the lightest materials uppermost. See the article STOWAGE.

To TRIM *or Stow the Hold.* See the article HOLD, &c.

TRIM *of the Masts*, (*juste position des mâts*, Fr.) denotes their position with regard to the ship and to each other. Thus, in the latter sense, they should neither be too near nor too far apart; and, in the former, they should not be too far forward or aft; and, according to the situation or quality which communicates a greater velocity to the vessel, they should either be upright, or inclining aft, or forward.

TRIM *of the Sails*, (*orièntement des voiles*;—*voiles bièn orièntées*, Fr.) denotes the general arrangement which is best calculated to accelerate the ship's course, according to the direction of the wind. See the articles CLOSE-HAULED, LARGE, SAILING, and *To* TACK, &c.

If the ship were always to sail before the wind, it would be a very simple operation to trim the sails; because nothing else could be required than to dispose them so as to receive the greatest possible effort of the wind, which is evidently performed by arranging them at right angles with its direction. But when the current of wind acts more directly upon the ship's side, it necessarily falls more obliquely on the surface of the sails, so as to diminish their effort to push the ship forward; and to augment their tendency to make her incline to one side. Hence we may conclude, that an increase of the wind, when accompanied with a variation unfavourable to the ship's course, will by no means augment her velocity; because the force, previously employed to push her forward, will afterwards operate to overturn her; and because this impression renders it necessary to reduce the quantity of sail, the effort of which is further diminished by the obliquity of the action of the wind upon its surface.

By this theory it appears, that the effect of the wind to advance the ship, decreases in proportion to its obliquity with any sail upon which it operates.

TRIM *the Boat!* See the article BOAT.

Sailing TRIM. See the article SAILING.

SHARP-TRIMMED, (*orientè au plus près*, Fr.) the situation of a ship's sails in a scant wind.

TRIMMING, in mast-making, &c. the act of working any piece of timber into the form and shape designed by an axe or adze.

TRINITY-HOUSE COMPANY. The Society of the Trinity-House was founded by Sir Thomas Spert, comptroller of the navy to Henry VIII. at Deptford, in Kent, and incorporated on the 20th of May, 1515, by the name of "The Master, Wardens, and Assistants, of the Guild or Fraternity of the most glorious and undivided Trinity, and of St. Clement, in the parish of Deptford Stroud, in the county of Kent." He was also the first master of the Society. He died in 1541, and was buried in the chancel of Stepney church; to whose memory the company, in 1622, erected a monument there, eighty one years after his decease.

The corporation is governed by a master, four wardens, eight assistants, and eighteen elder brethren, the other members are denominated younger brethren, and are of unlimited number, as every master or mate, expert in navigation, may be admitted as such; and from them the vacancies of elder brethren are supplied. The corporation and court of assistants are invested with the following powers:

1. The examining of the boys studying mathematics at Christ's Hospital.

2. The examining of the masters of his Majesty's ships; the appointing pilots to conduct ships in and out of the river Thames, and the amercing all such as shall presume to act as a master of a ship of war, or pilot, without their approbation, in a pecuniary mulct of 20*l.*

3. The settling the several rates of pilotage, and erecting, ordering, and maintaining light-houses, buoys, beacons, and other sea-marks upon the several coasts of the kingdom, for the security of navigation, particularly those of Scilly and Dungeness, to which all ships pay a duty.

4. The preventing of aliens from serving on board English ships, without their license, upon penalty of 5*l.* for each offence.

5. The punishing of seamen for desertion or mutiny in the merchants' service.

6. The hearing and determining the complaints of officers and seamen in the merchants' service, but subject to an appeal to the Lords' Commissioners, or Judge of the Court of Admiralty.

7. The granting of licenses to poor seamen (non-freemen) to row on the river Thames for their support in the intervals of sea-service, or at an advanced age.

To this company belongs the ballast-office, for clearing and deepening the river Thames, by taking from thence a sufficient quantity of ballast, for the supply of all ships that sail out of the river, in which service sixty barges, of the burden of 30 tons, and two men each, are continually employed. And in consideration of the great increase of the poor of this fraternity, they are by their charter empowered to purchase, in mortmain, lands, tenements, &c. to the amount of 500*l.* per annum, and also to receive charitable benefactions of well-disposed persons.

The revenue of the corporation, which arises from tonnage, ballastage, beaconage, &c. and from contingent benefactions, is applied, after defraying the charges of carrying on the great purposes of the foundation, towards the relief of decayed seamen, their widows, and orphans, of whom they annually receive a large number by pensions; in addition to the maintenance of their alms-houses at Deptford and Mile-end.

The ancient hall at Deptford, where some of their general meetings were held, was taken down in 1787, and a more commodious and elegant structure was raised on the north side of Tower-hill in London, for that purpose; a situation far more convenient for the court of assistants. The building was opened for the dispatch of business in 1795. It is of Portland stone, consisting of a basement story, and one above it, with columns of the Ionic order, ornamented with allegorical and appropriate reliefs; the first stone was laid on the 12th of September, 1793.

Their usual meetings are held on Wednesdays and Saturdays, at this house, but their courts are not fixed to a certain time, as that depends on the business necessarily requiring it.

There are two hospitals at Deptford, belonging to this corporation. The old hospital, of which there is a view engraved by Gribelin, in 1701, was built in the reign of Henry VIII. and consisted originally of twenty-one apartments, but being taken down and rebuilt in 1788, the number was increased to twenty-five; it adjoins the church-yard of St. Nicholas. The other, which is in Church-Street, was built about the time of the Revolution; Sir Richard Browne, in 1672, gave the ground, after the expiration of a short term; and Captain William Maples, in 1680, gave 1300*l.* towards the building. This house consists of fifty-six apartments, forming a spacious quadrangle, in the centre of which is placed a statue of him. On the east side, opposite the entrance, is a plain building, which serves both for a chapel and hall. Here the brethren of the Trinity-House meet annually on Trinity Monday, by ancient custom, to choose a master, and afterwards to go to St. Nicholas church, where they hear divine service, and a sermon from their chaplain.

The pensioners of both hospitals consist of decayed pilots and masters of ships, or their widows. The single men and widows receive about 18*l.* per annum, the married men about 28*l.*

The alms-houses of the Trinity Company, on the north side of Mile-end-road, were founded in the year 1695, by the corporation, on a piece of ground given to them for that purpose, by Captain Henry Mudd, who was an elder brother. They are twenty-eight in number, and devoted to the residence of decayed commanders of ships, or mates or pilots, and their wives or widows, whose pensions are 18*l.* per annum, and one chaldron and a half of coals. In the centre of the quadrangle is the statue of Captain Saunders, who died in 1721, having bequeathed 100*l.* and the reversion of an estate in Lincolnshire of 147*l.* per annum to these alms-houses; this reversion fell in about the year 1746, when this statue was erected. On the north side of the square is a very neat chapel, in the windows of which are some coats of arms in stained glass, which were removed from the old hall of the Trinity-House at Deptford, when it was taken down in 1787, and were set up in the chapel in 1793.

The officers of this corporation are, a master, a deputy-master, twenty-eight elder brethren, a secretary, and a solicitor.

The experience and acknowledged skill of the elder brethren, render their opinion and judgment in all nautical concerns of great moment; and at this period, when our navigation and commerce have been extended to every part of the globe, and most mercantile transactions depend upon shipping, the great questions, that continually arise therein, require the best talents and judgment; and therefore the courts of judicature frequently derive considerable assistance from the opinions of the elder brethren; and in other cases, where the parties have referred to their award, they have generally been satisfied of its justice. See Highmore's *Public Charities.*

TRIONES, in astronomy, is a sort of constellation or assemblage of seven stars in the *Ursa Major*, popularly called *Charles's wain*, From the *Septem Triones*, the north pole takes the denomination *Septentrio.*

TRIP, (*voyage dans les pays étrangers, ou campagne; ce mot signifiè aussi, bordée en louvoyant,* Fr.) a term denoting an outward-bound voyage, particularly in the coasting navigation. It also signifies a single board in plying to windward.

To Trip *the Anchor*, (*faire dèraper l'ancre, ou lui faire quitter le fond,* Fr.) is to loosen it from the bottom by its cable or buoy-rope.

TRIPPING-Line, (*cartaheu servant à dégréer le bout inférieur des vergues de perroquet,* Fr.) a small rope serving to unring the lower top-gallant-yard-arm, when in the act of striking or lowering it down upon deck.

TRISECTION, in geometry, is the dividing any thing into three equal parts.

TROPICS, (*tropiques*, Fr.) in astronomy, two fixed circles of the sphere drawn parallel to the equator, through the solstitial points, or at such distances from the equator, as is equal to the sun's greatest recess or declination.

Tropics, in geography, two lesser circles of the globe, drawn parallel to the equator, through the beginnings of Cancer and Capricorn, being in the planes of the celestial tropics, and consequently at 23° 28' distance either way, from the equator.

TROUGH of the Sea, (*entre-deux les lames,* Fr.) a name given to the hollow, or interval between two waves, which resembles a broad and deep trench perpetually fluctuating. As the setting of the sea is always produced by the wind, it is evident that the waves, and consequently the trough or hollow space between them, will be at right angles with the direction of the wind. Hence a ship rolls heaviest when she lies in the trough of the sea.

Filling Troughs, in the magazines of ships of war, are made of wood, and used to hold the powder that is started out of the barrels, for filling the cartridges.

TROWSERS, (*pantallons*, Fr.) a sort of loose long breeches generally worn by sailors.

TRUCE, (*trève*, Fr.) a suspension of arms, or a cessation of hostilities between two armies, in order to settle articles of peace, bury the dead, &c.

TRUCK-BARROW, in rope-making, a sort of barrow with three wheels, and used to take hauls of yarn from the yarn-house, and remnants of yarn, coils of rope, &c. from the ground to the rope house.

TRUCKS, (*ce mot signifie plusieurs petites pièces de bois,* Fr.) pieces of wood, which are either conical, cylindrical, spherical, or spheroidal, and used for different purposes, as the intention of these is to guide the sailors to the particular rope, which might otherwise be easily mistaken for some other of the same size, especially in the night.

TRUCKS *of a Gun-Carriage.* See the article CARRIAGE.

TRUCKS *of the Mast-Heads,* (*pommes de girouette à la tête des mâts,* Fr.) are circular pieces of wood fixed as a cap on the top-gallant-mast-head; they are generally furnished with two or more small pullies, through which the pendant and signal-halyards are reeved. They are also called halyards.

TRUCKS *of the Parrals.* See the article PARRAL.

TRUCKS *of the Shrouds,* (*cosses de bois, ou margouillets fixés aux haubans pour la conduite des manœuvres,* Fr.) are short cylindrical pieces of elm, &c. having a hole through the middle, lengthways, a groove down the side, of the size of the shrouds, and a score round the middle to admit a seizing: they are seized to the shrouds, to lead ropes through, that they may be more readily found.

TRUCKS *of the Flag-Staffs,* (*pommes des bâtons de pavillon,* Fr.) are flat and round pieces of elm, having a small sheave on each side: they are fixed by a square mortise-hole cut in the middle, upon the upper end of flag-staffs, and are used to reeve the halyards.

SIGNAL TRUMPET, an instrument invented by Mr. Rickman, Bookseller, Upper Mary-le-bone-Street, London, the intention of which is to enable persons to produce a very loud report in circumstances where cannon would be inconvenient, and thus answer the end of signal guns. The invention is perfectly simple and ingenious, being nothing more than affixing a pistol to a common signal-trumpet instead of the mouth-piece. The report thus produced equals that of a nine-pounder, as was satisfactorily proved by experiments made at Woolwich. The common speaking trumpet used at sea, is used by the patentee, and the pistol-barrel, as well as the common mouth-piece, are made to screw on at pleasure. For a still louder report, a swivel, or other small piece of cannon may be used instead of the pistol, and then the trumpet has a parabolic form given to it, that it may not suffer by the great concussion.

The various uses, both civil and military, to which this invention will apply, may be easily imagined; but we may observe, that one great use of the trumpet, that of sending articulate sounds to a greater distance than the mere voice can effect, is here entirely lost; and it becomes merely an alarm signal, the purport of which can be only inferred from collateral circumstances.

Speaking TRUMPET, (*porte voix,* Fr.) a tube formed to collect the impulses of sound in speaking, and convey them forward to a distance, they are used in time of action, and for the purpose of hailing ships at sea.

TRUMPETER, in ships of war, a person who sounds the trumpet or bugle-horn every morning and evening after drum-beating. The person rated as trumpeter was originally employed to pass the word of command from the man at the gun through the hoses which encircle the tiller-ropes, to those at the tackles, when steering the ship by the tiller in the gun-room, after the tiller ropes had been carried away, or the steering-wheel disabled by shot, or otherwise.

FIRE-TRUNKS, (*coffres à feu, qui entrent dans la disposition d'un brulot,* Fr.) are wooden funnels fixed in fire ships under the shrouds, to convey the flames to the masts, rigging, and sails.

TRUNNIONS, (*tourillons d'un canon,* Fr.) the two knobs or short cylinders of iron which project from the opposite sides of a piece of ordnance, and serve to support it in the carriage.

TRUNNION *Ring,* the ring about a cannon, next before the trunnions.

TRUSS, (*drosse de racage, ou servant de racage aux basses vergues,* Fr.) a machine employed to pull a yard home to its respective mast, and retain it firmly in that position. See the article PARRAL.

As the truss is generally used instead of a parral, it is rarely employed, except in flying top-gallant-sails, which are never furnished with parrals. It is no other than a ring or traveller, which encircles the mast, and has a rope fastened to its after-part, leading downward to the top or decks; by means of which the truss may be straightened or slackened at pleasure. The halyards of the top-gallant-sails being passed through this ring, and the sail being hoisted up to its utmost extent, it is evident that the yard will be drawn close to the mast, by pulling down the truss close to the upper part of the sail; for without the truss, the sail and its yard would be blown from the mast, so as to swing about, by the action of the wind, and the rolling of the vessel; unless the yard were hoisted close up to the pulley wherein the halyards run; which is seldom the case in flying top-gallant-sails, because they are usually much shallower than those which are fixed or standing.

TRUSS, in ship-building, short pieces of carved-work, mostly in small ships, fitted under the taff-rail, in the same manner as the terms.

TRUSS-*Pendant.* See the article PENDANT.

TRUSS-*Tackle.* See the article TACKLE.

To TRY A CURRENT. See the article CURRENT.

TRYING, (*action de capeyer, ou d'être à la cape,* Fr.) the situation in which a ship lies nearly in the trough or hollow of the sea in a tempest, particularly when it blows contrary to her course; or it is the act of lying-to in a storm, which may be performed under any of the courses reefed, if requisite, or even under bare-poles, the helm being lashed a-lee.

In trying, as well as in scudding, the sails are always reduced in proportion to the increase of the storm. Thus, in the former state, a ship may lie by the wind under a whole main-sail, a whole fore-sail, or a whole mizen, or under any of those sails, when diminished by the reef or balance. As the least possible quantity of sail used in scudding are the goose-wings of the fore-sail; so in trying, the smallest portion is generally the mizen-stay-sail or main-stay-sail: and in either state, if the storm is excessive, she may lie with all the sails furled, or under bare poles.

The intent of spreading a sail at this time is to keep the ship more steady, and, by pressing her side down in the water, to prevent her from rolling violently; and also to turn her bow towards the direction of the wind, so that the shock of the waves may fall more obliquely on her flank, than when she lies along the trough of the sea. While she remains in this situation, the helm is fastened close to the lee-side, for, hard-a-lee, to prevent her as much as possible from falling-off. But as the ship is not then kept in equilibrium by the effort of her sails, which at other times counterbalance each other at the head and stern, she is moved by a slow but continual vibration, which turns her head alternately to windward and to leeward, forming an angle of three or four points in the interval. That part where she stops, in approaching the direction of the wind, is called her coming-to, and the contrary excess of the angle to leeward is termed her falling-off.

Thus, suppose the wind northerly, and a ship trying with her starboard side to windward; if in turning her head towards the source of the wind, she arrives at N.W.¼N. or N. 39° W. and then declines to the leeward as far W.½S. or S. 84° W. the former will be called her coming-up, and the latter her falling-off.

TRY-SAIL, (*voile de senaut, ou voile de cape des goëlettes, cutters, lougres, &c.* Fr.) a small sail used by cutters, luggers, sloops, &c. in lieu of their main-sail, during a storm.

GROG-TUB, a half cask, set apart for mixing the daily allowance of spirits with water, prior to its being served out to the ship's company.

Match-TUB, (*baille de combat*, Fr.) the half of a cask, having notches sawn in its edge, wherein the lighted matches are placed during action, the bottom being covered with water or sand to extinguish any sparks which may fall from the match.

Top-Sail-Halyard-TUBS, (*baille pour les drisses des huniers*, Fr.) are similar to half casks, in which the top-sail-halyards are coiled.

TUBES, for firing guns and mortars, are of tin or quill. They must pass through a guage of $\frac{7}{10}$ of an inch in diameter. The composition is mealed powder, mixed up stiff with spirits of wine. They are made up in bundles of 100 each.

Length of tin tubes.

Length.	Sea Mortars.
In.	In.
12.2	13
7.75	10

N.B. If tin tubes get damaged by wet, the composition may be cleared out of them, and they may be fresh filled If spirits of wine cannot be had, good rum or brandy will answer the purpose.

TUCK, (*la partie inférieure de la poupe comprise entre la lisse d'hourdi et les estains; c'est ce qu'on appelle aussi, les fesses d un bâtiment*, Fr) that part of the ship where the ends of the bottom-planks are collected together immediately under the stern or counter.

When this part, instead of being incurvated, and forming a convex surface, assumes the shape of a vertical or oblique plane, it is said to be square.

A Square TUCK, (*un cul quarré*, Fr.) is terminated above by the wing-transom, and below and on each side by the fashion-pieces.

TUMBLING-HOME, (*rentrée des œuvres mortes d'un navire*, Fr.) that part of a ship's side which falls inward above the extreme breadth, so as to make the ship gradually narrower from the lower deck upwards. This angle is represented in general throughout all the timbers in the plane of projection, plate XVII. It is also more particularly expressed by Q T, in the midship-section, plate VI. where it is evident, that the ship grows narrower from Q towards T. In all our old sea books, this narrowing of a ship from the extreme breadth upwards is called housing-in. See the article UPPER-WORKS.

To TURN. See the article *To* TACK.

A Round TURN *in the Hawse.* See the article HAWSE.

To TURN *In*, (*se coucher*, Fr.) among seamen, denotes to get in their hammocks or cots; and

To TURN *Out*, (*se lever*, Fr.) implies, on the contrary, to leave them.

To TURN *Over Men*, (*donner du monde à un bâtiment*, Fr.) is to remove them from one ship or vessel to another.

To TURN *Time into Longitude.* See the article LONGITUDE *by Time-Keeper.*

TURNING, in mechanics, an ingenious and useful art, by which articles in metals, wood, stone, ivory, &c. are manufactured; and the instrument employed for this purpose, is called a lathe.

TURNING *Lathe*, in block making, a well known machine for turning sheaves, pins, &c. called cheeks, about 8 feet long and 5 inches square, and 3 inches asunder, between which slide two pieces of wood, called puppets, that are made to fasten at any distance the work may require, by a wedge driven through them under the cheeks. Near the upper part of one is fixed a strong sharp spike of tempered steel, and opposite to it in the other is a sharp pointed iron screw; between which the work is sustained and kept in a circular motion, by means of a cord turned round it, and fastened above to a pliable pole, and underneath to a treddle moved by the foot: against the sides of the puppets is confined a batten with a pin through one end, called a rest, for the workman to keep his tool in a steady position. When the work is too heavy for the treddle to perform, a wheel is substituted, which is turned by one or two handles, and is connected by means of a cord, the ends of which are spliced together, and fixed in a groove round the wheel, then crossed and put round a groove in the collar of a mandrel, to be turned with a swift and regular motion.

TURNING TO WINDWARD, (*louvoyer*, Fr.) denotes that operation in sailing wherein a ship endeavours to advance against the direction of the wind, by a compound course, inclined to the place of her destination. This is otherwise called plying or beating to windward. See the articles BEATING and *To* TACK.

TURNING *to Windward in very narrow Channels.* At weighing, if the wind is partly across the tide, it will cast the ship with her head towards the weather-shore, which she may be kept clear of, by driving with the sails a-back till the anchor is up and stowed; and, as the tack towards the weather-shore is the shortest, it is prudent to back as near the lee-side as possible, in order to make the first board the longer; to get the three top-sails, jib, stay-sail, and mizen, properly set; and to get all ready in time for tacking. Make as bold as possible with the weather-shore, because on that side a ship is always sured in coming about; and, in case of missing stays, a ship may be backed off from the weather-shore, till she has room to fill and set the sails, and get sufficient headway to try her in stays again without danger. But, when the ship is got about, and standing towards the lee-shore, it may be necessary to put her in stays in good time; because she does not so certainly stay when going slanting with the tide, as when going across it.

By staying her thus in good time, if she even miss stays, there may be room enough to fill and try her the second time, or to use such means as may prevent her from going on shore.

But, when the wind is right against the tide, which begins to make to windward, be cautious not to weigh the anchor till the ship swings end-on to the tide, and brings the wind so far aft, that she may be steered right against the tide till the anchor is up and stowed, and the sails, with which the ship is to work, are all ready.

Haul the wind and get ready for tacking when you are close over to one side, to gain the whole breadth of the channel for getting under-way. For this purpose, let the first trip be made as short as possible, till it is found how the ship works upon both tacks, and then make longer or shorter boards accordingly; but take care not to stand into an eddy tide on either side, which has often occasioned ships to miss stays and go on shore. If a ship will not stay, she must be veered, box-hauled, or club-hauled.

TURPENTINE, (*turpentina*, Lat. *térébenthine*, Fr.) a resinous juice, extracted from several species of trees.

The true turpentine-tree is found in Spain, and the southern parts of France, as well as in the island of Chio, and in the Indies.

The common turpentine is prepared from different sorts of the pine, and is quite thick, white, and opaque.

Venice turpentine is a mixture of eight parts of common yellow or black rosin, with five parts of oil of turpentine.

The Strasburg turpentine is extracted from the silver fir, is commonly of a yellowish brown colour, and imported from Germany.

TWILIGHT, in astronomy, is that faint light which is perceived before sun-rising, and after sun-setting: it is occasioned by the earth's atmosphere refracting the rays of the sun, and reflecting them among its particles. The depression of the sun below the horizon, at the beginning of the morning, and end of the evening twilight, is on a medium 18°. See also the article CREPUSCULUM.

TWINE, (*fil*, Fr.) in sail-making, a sort of strong twisted thread. There are two sorts of twine used in sail-making, viz. extra and ordinary; the extra is for seaming, and runs 360 fathoms to the pound; the ordinary is used to sew on the bolt-rope, and runs 200 fathoms to the pound. The twine for the navy is composed of three threads.

TWINING-IRONS, in anchor-making, are square bars, with an ess-hook at one end, which grasps the porter, or the shank, to turn it over.

TWINKLING OF THE STARS, denotes that tremulous motion which is observed in the light proceeding from the fixed stars, a phenomenon which has called forth hypotheses and theories, which it would be useless to enumerate in this place, particularly as the principal cause is now acknowledged to arise from the unequal refraction of light, in consequence of inequalities and undulations in the atmosphere.

This twinkling must be carefully distinguished from the irradiations of remote luminous objects seen under a small angle; the latter being very different in appearance, and attributable to a very different cause.

Besides a variation in the quantity of light, it may be added, that a momentary change of colour has likewise been observed in some of the fixed stars; so that one looking steadfastly at Sirius, or any bright star, not much elevated above the horizon, its colours seem not to be constantly white, but appears tinctured, at every twinkling, with red and blue; for these colours being much fewer than those of the intermediate ones, and therefore much more liable to inequalities from the common effect of chance, a small excess or defect in either of them will make a very sensible difference in the colour of the stars.

TYE, (*itague*, Fr.) a sort of runner or thick rope, used to transmit the effort of a tackle to any yard or gaff, which extends the upper part of a sail.

The tye is either passed through a block fixed to the mast-head, and afterwards through another block attached to the yard or gaff, and turns to the mast-head; or the end of it is simply passed over the sheave in the mast-head, and then fastened to the said yard or gaff.

TYRANNY AND OPPRESSION, is the misery experienced from any unjust treatment and severity which may be displayed by strict disciplinarians over their inferior officers and men; which, notwithstanding its illegality, and being strictly forbidden in the code of Naval Instructions, has sometimes been practised on board of his Majesty's ships and vessels at sea.

Yet if an officer shall observe any misconduct in his superior, or shall suffer any personal oppression, injustice, or other ill-treatment, he is not, on that account, to fail, in any degree, in the respect due to such superior officer; but he is to represent such misconduct to the captain of the ship to which he belongs, or to the flag-officer commanding the squadron in which he serves; or to the commander in chief, as circumstances may require.

And although the order of a superior officer must be complied with by an inferior who receives it, yet he is by no means prevented from making a proper and respectful representation of any bad effects which he shall think may result to his Majesty's service from its being obeyed; nor is any person prevented from representing, either to the officer by whom the order was issued, or to his superior, any injustice or oppression he may suffer, or may have reason to apprehend he shall suffer by such order being carried into execution; nor is an officer, or any person whatever, prevented from taking such measures, as the custom of the service allows, to obtain redress for any injustice or injury done to him from tyranny or any oppression he may suffer by the conduct or orders of his *superiors*. See the NAVAL INSTRUCTIONS.

V.

VACUUM, in physics, denotes a space empty, or devoid of all matter or body.

Whether there be any such thing in nature as an absolute vacuum; or whether the universe be completely full, and there be an absolute plenum; is a question that has been agitated by the philosophers of all ages.

It has been the opinion of some philosophers, particularly the Cartesians, that nature admits of no vacuum, but that the universe is entirely full of matter: in consequence of which opinion they were obliged to assert, that if every thing contained in a vessel could be taken out or annihilated, the sides of that vessel, however strong, would come together; but this is contrary to experience, for the greatest part of the air may be drawn out of a vessel by means of the air-pump, notwithstanding which it will remain whole if its sides are strong enough to support the weight of the incumbent atmosphere. Should it be objected here, that it is impossible to extract all the air out of a vessel, and that there will not be a vacuum on that account; the answer is, that since a very great part of the air that was in the vessel may be drawn out, as appears by the more quick descent of light bodies in a receiver when exhausted of its air, there must be some vacuities between the parts of the remaining air; which is sufficient to constitute a vacuum. Indeed, to this it may be objected by a Cartesian, that those vacuities are filled with *materia subtilis*, that passes freely through the sides of the vessel, and gives no resistance to the falling bodies: but as the existence of this *materia subtilis* can never be proved, we are not obliged to allow the objection, especially since Sir Isaac Newton has found that all matter affords a resistance nearly in proportion to its density. There are many other arguments to prove this, particularly the motion of the comets through the heavenly regions, without any sensible resistance; the different weights of bodies of the same bulk, &c. All the parts of space says Sir Isaac Newton are not equally full; for if they were, the specific gravity of the fluid which would fill the region of the air, could not, by reason of the exceeding great density of its matter, give way to the specific gravity of quicksilver, gold, or any body, how dense soever: whence neither gold, nor any other body, could descend in the air; for no body can descend in a fluid, unless it be specifically heavier than it. But, if a quantity of matter may, by rarefaction, be diminished in a given space, why may it not diminish *ad infinitum?* And if all the particles of bodies are of the same density, and cannot be rarefied, without leaving pores, there must be a vacuum. See CHAMBERS'S *Encyclopædia*.

VALENCIENNES COMPOSITION. See the article COMPOSITION.

VALOUR (*valeur*, Fr.) courage, bravery, intrepidity. Courage is that grandeur of soul, which prompts us to sacrifice all personal advantages, and even the preservation of our beings, to a love of doing our duty. The exercise of this determined courage in the hour of battle, is called valour.

Valour is composed of bravery, reason, and force: by bravery, we understand that lively ardour which fires us for the combat; reason points out to us the method of conducting it with justice and prudence; and force is necessary for the execution of duty. It is bravery which animates the heart, reason springs from the soul, and force depends upon the body; without bravery we fear obstacles, danger, and death, without reason, courage would have no legitimate view; and without force it would be useless: the concurrence of these three qualities, is the basis of true heroic valour.

Coolness is the effect of courage, which knows its danger, but makes no other use of that knowledge, than to give direction with greater certainty; courage is always master of itself, provided against all accidents, and regulated by existing circumstances; never confounded by any danger so as to lose sight of the motions of the enemy, or of the means by which he may be most effectually opposed. See JAMES'S Military Dictionary, vol. 2.

VALVE, in hydraulics, pneumatics, &c. is a kind of lid or cover to a tube or vessel, contrived to open one way, and by which, the more forcible it is pressed the other way, the closer it shuts the aperture; so that it either admits the entrance of a fluid into the tube, or vessel, and prevents its return; or permits its escape, and prevents its re-entrance.

Valves are of great use in the air-pump, and other wind machines; in which they are usually made of pieces of bladder. In hydraulic engines, as the emboli of pumps, they are mostly of strong leather, of a round figure, and fitted to shut the apertures, of the barrels or pipes. Sometimes they are made of two round pieces of leather enclosed between two others of brass; having divers perforations, which are covered with another piece of brass, moveable upwards and downwards, on a kind of axis, which goes through the middle of them all. Sometimes they are made of brass, covered over with leather, and furnished with a fine spring, which gives way upon a force being applied against it; but upon the ceasing of that, returns the valve over the aperture.

VAN, (*avant-garde*, Fr.) the foremost division of a naval armament, or that part which leads the way to battle; or advances first in the order of sailing. See the articles CENTRE, FLEET, and REAR.

VANE, (*girouette*, Fr.) a slip of bunting placed at the mast-head, or some other conspicuous place in the ship, as represented by *b*, fig. 9, plate I. It is commonly sewed upon a wooden frame called the stock, which contains two holes whereby to slip over the spindle, upon which it veers about by the wind to shew its direction or course.

Distinguishing-VANE, (*girouette de distinction, ou girouette que les bâtimens d'une escadre portent pour se faire reconnoître*, Fr.) serves, by its situation, to denote what squadron a ship belongs to, and by its colour or combination of colours, to point out the particular ship which bears it.

Dog-VANE, (*penon*, Fr.) a small light vane, formed of thin slices of cork, stuck full of feathers, and strung upon a piece of twine. It is usually fastened to the top of a half-pike, and placed on the weather-side of the quarter-deck, in order to shew the helmsman the direction of the wind, particularly in a dark night, or when the wind is feeble.

VANGS, (*palans de retenue, ou bras d'un pic, ou d'une vergue d'artimon*, Fr.) a sort of braces to support and keep steady the mizen-gaff: they are fixed on the outer-end or peak, and reach downwards to the aftmost part of the ship's quarters, where they are hooked and drawn tight, so as to be slackened when the wind is fair; and drawn in to windward when it becomes unfavourable to the ship's course.

VARIATION, in astronomy. The variation of the moon called by Bulliald the reflection of her light, is the third inequality observed in the moon's motion; by which, when out of the quadratures, her true place differs from her place twice equated.

Newton makes the moon's variation to arise partly from the form of her orbit, which is an ellipsis; and partly from the inequality of the spaces which the moon describes in equal times, by a radius drawn to the earth.

To find the greatest VARIATION. Observe the moon's longitude in the octants; and to the time of observation compute the moon's place twice equated; then the difference between the computed and observed place is the greatest variation.

Tycho makes the greatest variation 40′ 30″, and Kepler 51′ 49.″ But Newton makes the greatest variation, at a mean distance between the sun and the earth, to be 35′ 10″; at the other distances, the greatest variation is in a ratio of the times of the moon's synodical revolution directly, and the triplicate ratio of the distance of the sun from the earth inversely. And, therefore, in the sun's apogee, the greatest variation is 33′ 14″, and in his perigee 37′ 11″; provided that the eccentricity of the sun is to the transverse semidiameter of the *orbis magnus*, as $16\frac{7}{8}$ to 1000. Or taking the mean motions of the moon from the sun, as they are stated in Dr. Halley's tables, then the greatest variation at the mean distance of the earth from the sun will be 35′ 7″, in the apogee of the sun 33′ 27″, and in his perigee 36′ 51." See VINCE'S *Astronomy*.

VARIATION *of the Compass, or Declination of the Magnetic Needle*, (*variation de la boussole.—Déclinaison de l'aiguille aimantée*, Fr.) is the deviation of the needle or compass, being either east or west from the true north point, or the angle which the magnetic needle makes with the meridian line on a horizontal plane, when suspended at liberty.

In sea language, the variation is usually called *north-easting* or *north-westing*.

All magnetic bodies are found to range themselves, in some sort, according to the meridian; but they seldom agree precisely with it: in one place they decline, from the north toward the east, in another toward the west; and that differently, at different times.

The variation of the compass could not long remain a secret, after the invention of the compass itself: accordingly Ferdinand, the son of Columbus, in his life written in Spanish, and printed in Italian at Venice, in 1571, asserts, that his father on the 14th of September, 1492, observed it: though others seem to attribute the discovery to Sebastian Cabot, a Venetian, in the service of King Henry VII, about the year 1500. It appears, however, that this variation or declination of the needle was known even some centuries earlier, though it does not appear that the use of the needle itself in navigation was then known; for it seems there is in the library of the university of Leyden, a small manuscript tract on the magnet, in Latin, written by one Peter Adsiger, bearing date the 8th of August, 1269; in which the declination of the needle is particularly mentioned. Mr. Cavallo has printed the chief part of this letter in the supplement to his treatise on magnetism, with a translation. The curiosity of this, says Mr. Cavallo, consists in its containing almost all that is at present known of the subject, at least the most remarkable parts of it, mixed, however, with a good deal of absurdity. In it the laws of magnetic attraction, and the communication of that power to iron, the directive property of the natural magnet, as well as of the iron that has been touched by it, and even the declination of the magnetic needle, are clearly and unequivocally mentioned in it.

Various are the hypotheses that have been framed to account for this extraordinary phenomenon: but we shall only mention some of the later and more probable.

The first is that of Gilbert, which is followed by Cabeus, &c.

This idea was, that it is the earth or land that draws the needle out of its meridian direction; and hence they agree that the needle varied more or less as it was more or less distant from any great continent; consequently, that if it were placed in the middle of an ocean, equally distant from equal tracts of land on each side, eastward and westward, it would neither decline to the one nor the other, but point directly north and south. Thus they say, in the Azores Islands, being equally distant from Africa on the east, and America on the west, there is in effect, found no variation; but as you sail from the Azores towards Africa, the needle begins to decline from the north to the east, and that still more till you reach the shore. If you still proceed eastward, the declination gradually diminishes again by reason of the land left behind on the west, which continues to draw the needle. The same holds till you arrive at a place where there are equal tracts of land on each side, and there again the variation will be nothing.

The observations of our mariners, in their East India voyages, seem to confirm this system. As they proceed towards the Cape of Good Hope, the variation is still eastward: at length arriving at the Cape de los Aqüillas, the meridian line then dividing Africa into two equal parts, there is no variation at all, but as they proceed and leave the African coast on the West, the variation becomes westward.

But the misfortune is, the law does not hold universally; in effect, a great number of observations of the variations, in various parts, made and collected by Dr. Halley, overturns the whole theory.

Others, therefore, have recourse to the frame and compages of the earth, considered as interwoven with rocks and shelves, which being generally found to run towards the poles, the needle comes to have a general tendency that way; but it seldom happens that their direction is exactly in the meridian, and the needle has consequently, for the most part some variation.

Others hold various parts of the earth, to have various parts of the magnetic virtue, as some are more intermixed with heterogeneous matters, which prevent the free action or effect thereof, than others. Some again ascribe all to magnetic rocks and iron mines, which, affording more of the magnetic matter than other parts, draw the needle more.

Lastly, others imagine earthquakes, or high tides, to have disturbed and dislocated several considerable parts of the earth, and so changed the magnetic axis of the globe, which originally was the same with the axis of the globe itself.

Dr. Hook communicated to the Royal Society, in 1675, a theory of the variation, the substance of which is, that the magnet has its peculiar pole, distant ten degrees from the pole of the earth, about which it moves, so as to make a revolution in 370 years: whence the variation he adds, has altered of late about 10 or 11 minutes every year, and will probably continue to do so for some time, when it will begin to proceed slower and slower, till at length it becomes stationary and retrograde, and so return back again.

Dr. Halley has given us another system, the result of a great number of observations, and even of a number of voyages made at the public expense on this account. The light which this author has thrown on this obscure part of natural history, is very great, and the consequences thereof in navigation, &c. are very considerable. From these observations, the learned author deduces the following theory.—

He considers the whole globe of the earth as one great magnet, having four magnetical poles, or points of attraction; near each pole of the equator two; and that in those parts of the world which lie nearly adjacent to any one of these poles, the needle is governed by it, the nearest pole being always predominant over the more remote.

The pole which at present is nearest to us he conjectures to lie in or near the meridian of the land's end of England, and not above 7° from the north pole: by this pole, the variation in all Europe, Tartary, and the North Sea, are chiefly governed; though still with some regard to the other northern pole, whose situation is in the meridian passing about the middle of California, and about 15° from the north pole of the world, to which the needle has chiefly a tendency in all North America, and in the two oceans on either side of it, from the Azores westward to Japan, and farther.

The two southern magnetic poles, he imagines, are rather more distant from the south pole of the world; the one being about 16° from it, on a meridian 20° to the westward of the Magellanic Streights, or 95° west from London: this pole commands the needle in all South America, in the Pacific Ocean, and the greatest part of the Ethiopic Ocean. The other magnetic pole seems to have the greatest power, and the largest dominion of all, as it is the most remote from the pole of the world, being little less than 20° distant from it, in the meridian which passes through New Holland, and the island Celebes, about 120° east from London: this pole is predominant in the south part of Africa, Arabia, the Red Sea, Persia, India, and its islands, and all over the Indian sea, from the Cape of Good Hope eastward, to the middle of the Great South Sea that divides Asia from America.

Such, he observes, seems to be the present disposition of the magnetic virtue throughout the whole globe of the earth. It is then shewn how this hypothesis accounts for all the variations that have been observed of late, and from the whole it appears, that the direction of the needle, in the temperate and frigid zones, depends chiefly upon the counterpoise of the forces of two magnetic poles of the same nature: as also why, under the same meridian, the variation should be in one place 29½ degrees west, and in another, 20½ degrees east.

In the torrid zone, and particularly about the equator, respect must be had to all the four poles, and their positions must be well considered, otherwise it will not be easy to determine what the variation should be, the nearest poles being always strongest; yet it is sometimes counterbalanced by the united forces of two more remote ones. Thus in sailing from St. Helena, by the isle of Ascension to the equator, on the north-west course, the variation is very little easterly, and unalterable in that whole tract, because the South American pole, which is nearest in the aforesaid places, requiring a great easterly variation, is counterpoised by the contrary attraction of the North American, and the Asiatic south poles; each of which singly is, in these parts, weaker than the American south pole; and upon the north-west course the distance from this latter is very little varied; and as you recede from the Asiatic south-pole, the balance is still preserved by an access towards the North American Pole. In this case no notice is taken of the European north-pole; its meridian being a little removed from those, of these places, and of itself requiring the same variations which are here found.

After the same manner the variations in other places about the equator are accounted for, upon Dr. Halley's Hypothesis.

To find the Variation of the Compass. Suspend a thread and plummet over the compass, till the shadow pass through the centre of the card; then observe the rhumb, or point of the compass which the shadow touches when it is the shortest. For the shadow is then a meridian line, and consequently the variation is shewn. Or thus: obsérve the point of the compass upon which the sun, or star, rises and sets; bisect the are intercepted between the rising and setting, and the line of bisection will be the meridian line, consequently the variation is had as before. The same may also be obtained from two equal altitudes of the same star, observed either by day or night. Or thus: observe the rhumb upon which the sun or star rises and sets; and from the latitude of the place, find the eastern or western amplitude: for the difference between the amplitude, and the distance of the rhumb observed, from the eastern rhumb of the card, is the variation sought.

Or thus: observe the altitude of the sun, or some star whose declination is known; and note the rhumb in the compass to which it then corresponds, then the difference between the azimuth and the observed distance of the rhumb from the south, is the variation sought. For the method of finding the variation by the Amplitude and Azimuth. See these Articles.

The use of the variation is to correct the courses of a ship as steered by the compass, which must always be done before they are calculated.

Variation of the *Variation*, is a gradual and continual change in the variation, observed in any place by which the quantity of the variation is found to be different at different times.

This continual change in the variation is gradual and universal, as appears by the following observations made in London and Paris at different periods.

At London.

In 1580	11°	5′E.	In 1722	14°	2′W.		
— 1622	6	0 E.	— 1747	17	40 W.		
— 1657	due north.		— 1780	22	41 W.		
— 1692	6	0 W.	— 1811	24	14	2″W.	

At Paris.

— 1550	8°	0 E.	— 1681	2°	2′W.
— 1640	3	0 E.	— 1759	18	10 W.
— 1660	due north.		— 1760	18	20 W.

According to Dr. Halley's theory this change in the variation of the compass, is supposed to be owing to the difference of velocity in the motions of the internal and external parts of the globe. From the observations that have been cited, it seems to follow that all the magnetic poles have a motion westward, but yet not exactly round the axis of the earth, for then the variations would continue the same in the same parallel of latitude, contrary to experience.

From the disagreement of such a supposition with experiments therefore, the learned author invented the following hypothesis to solve his theory; the external parts of the globe he considers as a shell, and the internal as a nucleus, or inner globe; and between the two he conceives a fluid medium. That inner earth having the same common centre or axis of diurnal rotation, may revolve with our earth every 24 hours: only the outer sphere having its turbinating motion somewhat swifter or slower than the internal ball; and a very minute difference in length of time, by many repetitions, becoming sensible; the internal parts will gradually recede from the external, and they will appear to move, either eastward or westward, by the difference of their motions.

Now, supposing such an internal sphere, to have such a motion, the two great difficulties in the former hypothesis are easily solved; for if this exterior shell of earth be a magnet, having its pole at a distance from the poles of a diurnal rotation; and if the internal nucleus be likewise a magnet, having its poles in two other places, distant also from the axis; and these latter, by a slow gradual motion, change their place in respect of the external, a reasonable account may then be given of the four magnetic poles before mentioned, and also the changes of the needle's variation.

The author thinks that two of these poles are fixed, and the other two moveable; viz. that the fixed poles are the poles of the external cortex or shell of the earth, and the other, the poles of the magnetic nucleus, included and moveable with the former. From these observations he infers, that the motion is westward, and consequently that the nucleus has not precisely attained the same velocity with the exterior parts in their diurnal rotation; but so very nearly equals it, that in 365 revolutions the difference is scarcely sensible.

The difference of this kind, arises from hence, that the impulse by which the diurnal motion was impressed on the earth, was given to the external parts, and from thence it, in time, communicated to the internal; but so as not yet perfectly to equal the velocity of the first motion impressed on the superficial parts of the globe, and still preserved by them.

Mr. Whiston in his New Laws of Magnetism raises several objections against this theory.

M. Euler, the son of the celebrated mathematician of that name has likewise, controverted and censured Dr. Halley's theory. He thinks, that two magnetic poles placed on the surface of the earth, will sufficiently account for the variation, and he then endeavours to shew, how we may determine the declination of the needle, at any and on every part of the globe. For particulars of his hypothesis, see *Histoire de l'Academie des Sciences and Belles Letters of Berlin* for 1757; also Mr. Cavallo's *Treatise on* Magnetism, p. 117.

Variation *of the Needle by Heat and Cold.* This amounts to a few seconds of a degree in the same place at different hours of the same day, and is only discoverable by nice observations. Mr. George Graham made several observations of this kind in the year 1722 and 1723. See *Philos. Trans.* No. 383, or *Abr.* vol. 7, p. 290.

About the year 1750, Mr. Wargentin, secretary of the Swedish Academy of Sciences, took notice both of the regular diurnal variation of the needle, and also of its being disturbed by the *aurora borealis*, as recorded in the *Philos. Trans.* vol. 47, p. 126.

Mr. Canton, about the year 1756, commenced a series of observations, amounting to near 4000, with an excellent variation compass, of about 9 inches diameter. The number of days on which these observations were made, was 603, and the diurnal variation on 574 of them was regular, so as that the absolute variation of the needle westward was increasing from about 8 or 9 o'clock in the morning, till about 1 or 2 in the afternoon, when the needle becomes stationary for some time; after that, the absolute variation westward was decreasing, and the needle came back again to its former situation, or nearly so, in the night, or by the next morning. The diurnal variation is irregular when the needle moves slowly eastward in the latter part of the morning, or westward in the latter part of the afternoon; also when it moves much either way after night, or suddenly both ways in a short time. These irregularities seldom happen more than once or twice in a month, and are always accompanied, as far as Mr. Canton observed, with an *aurora borealis*. Mr. Canton lays down and evinces, by experiment, that the attractive power of the magnet, whether natural or artificial, will decrease while the magnet is heating, and increase while it is cooling. He then proceeds to account for both the regular and irregular variation. It is evident, he says, that the magnetic parts of the earth in the north, on the east-side, and on the west-side of the magnetic meridian, equally attract the north-end of the needle. If the eastern magnetic parts be heated more by the sun in the morning, than the western parts; the needle will move westward, and the absolute variation will increase: when the attracting parts of the earth on each side of the magnetic meridian have their heat increasing equally, the needle will be stationary, and the absolute variation will then be greatest; but when the western magnetic parts are either heating faster, or cooling slower, than the eastern, the needle will move eastward, or the absolute variation will decrease; and when the eastern, and western magnetic parts are cooling equally fast, the needle will again be stationary, and the absolute variation will then be least. By this theory the diurnal variation in the summer ought to exceed that in winter, and accordingly it is found by observation, that the diurnal variation in the month of June and July, is almost double that in December and January.

The irregular variation must arise from some other cause than that of heat communicated by the sun; and here Mr. Canton has recourse to subterranean heat, which is generated without any regularity as to time, and which will, when it happens in the north, affect the attractive power of the magnetic parts of the earth, on the north end of the needle. That the air nearest the earth will be most warmed by the heat of it, is obvious; and this has been often noticed in the morning, before day, by means of thermometers at different distances from the ground. See *Philos. Trans.* vol. 48, p. 526.

Mr. Canton has annexed to his paper on this subject, a complete year's observations; from which it appears, that the diurnal variation increases from January to June, and decreases from June to December. See *Philos. Trans.* vol. 51, p. 398.

The following observations on the diurnal maximum and minimum variations of the magnetic needle, were made with an improved compass, &c. by Col. M. Beaufoy, of Hackney Wick, in latitude 51° 32′ 40″ north, longitude west in time $6''\frac{82}{100}$.

1813. Month.		Morning Observations. Hour.	Morning Observations. Variation.	Noon Observations. Hour.	Noon Observations. Variation.	Evening Observations. Hour.	Evening Observations. Variation.
July	17	8h 30′	24° 13′ 34″	1h 55′	24° 23′ 38″	7h 15′	24° 16′ 22″
	18	8 30	24 15 10	1 53	24 22 25	7 42	24 16 52
August	1	8 40	24 27 12	1 50	24 24 16	7 15	24 16 40
	17	8 35	24 14 15	1 55	24 23 46	7 00	24 13 27
January 1814,	16	8 55	24 13 10	1 55	24 18 14	Not	observed
	17	8 55	24 15 08	1 55	24 21 10		
February	18	8 45	24 14 03	1 50	24 20 31		
	19	8 40	24 14 35	1 50	24 23 47		
Mean of observations in July, 1813.		8 37	24 14 32	1 50	24 23 04	7 08	24 13 56
Do. in August,		8 44	24 15 58	2 02	24 23 32	7 05	24 16 08
Do. in Jan. 1814		8 52	24 15 05	1 53	24 19 03	Not	observed
Do. in February,		8 47	24 14 50	1 52	24 20 58		

The variation was observed with two needles, one of them being of the usual shape, the other a cylinder: seven observations were made with one needle, which was then drawn aside, and when settled, seven more were read off; the needle was then changed, and the same operation repeated, and the mean of these 28 readings off was considered the true variation.

It has also been observed, that different needles, especially if touched with different loadstones, will differ a few minutes in their variation. See Poleni *Epist. Philos. Trans.* No. 421.

Dr. Lorimer, in his supplement to Cavallo's Magnetism, adduces some ingenious observations on this subject. It must be allowed, says he, according to the observations of several ingenious gentlemen, that the collective magnetism of this earth, arises from the magnetism of all the ferruginous bodies contained in it, and that the magnetic poles should therefore be considered as the centres of the powers of those magnetic substances. These poles must therefore change their places according as the magnetism of such substances is affected; and if with Mr. Canton we allow, that the general cause of the diurnal variation arises from the sun's heat in the forenoon and afternoon of the same day, it will naturally occur, that the same cause, being continued, may be sufficient to produce the general variation of the magnetic needle for any number of years. For we must consider, that ever since any attentive observations have been made on this subject, the natural direction of the needle in Europe has been constantly moving, from west to east, and that in other parts of the world, it has continued its motion with equal constancy.

As we must therefore admit, says Dr. Lorimer, that the heat in the different seasons depends chiefly on the sun, and that the months of July and August are commonly the hottest, while January and February are the coldest months in the year; and that the temperature of the other months falls into the respective intermediate degrees, so we must consider the influence of heat upon magnetism to operate in the like manner, viz. that in a short time it scarcely manifests itself; yet in the course of a century, the constancy and regularity of it become sufficiently apparent. It would, therefore, be idle to suppose that, such an influence could be derived from an uncertain or fortuitous cause. But if it be allowed to depend upon the constancy of the sun's motion, and this appears to be a cause sufficient to explain the phenomena, we should according to Newton's first law of philosophizing, look no farther.

As we therefore consider, says he, the magnetic

powers of the earth to be concentrated in the magnetic poles, and that there is a diurnal variation in the magnetic needle, these poles must perform a small diurnal revolution proportional to such variation, and return again to the same point nearly. Suppose then the sun in his diurnal revolution passes along the northern tropic, or along any parallel of latitude between it and the equator, when he comes to that meridian in which the magnetic pole is situated, he will be much nearer to it than in any other; and in the opposite meridian he will be of course the farthest from it.

As the influence of the sun's heat will therefore act most powerfully at the least, and less forcibly at the greatest distance, the magnetic pole will consequently describe a figure something of the elliptical kind; and as it is well known that the greatest heat of the day is some time after the sun has passed the meridian, the longest axis of this elliptical figure will lie north-easterly, in the northern, and south-easterly, in the southern hemispheres. Again, as the influence of the sun's heat will not from those quarters have so much power, the magnetic poles cannot be moved back to the very same point, from which they set out; but to one which will be a little more northerly and easterly, or more southerly and easterly, according to the hemispheres in which they are situated. The figures they describe, may more properly be termed elliptoidal spirals.

In this manner the variation of the magnetic needle in the northern hemisphere may be accounted for. But with respect to the southern hemisphere we must recollect, that though the lines of decl nation in the northern hemisphere have constantly moved from west to east, yet in the southern hemisphere, it is equally certain that they have moved from east to west, ever since any observations have been made on the subject. Hence then the lines of magnetic declination, or Halleyan curves, as they are now commonly called, appear to have a contrary motion in the southern hemisphere, to what they have in the northern, though both the magnetic poles of the earth move in the same direction, that is from west to east.

In the northern hemisphere there was a line of no variation, which had east variation on its eastern side, and west variation on its western side. This line evidently moved from west to east during the two last centuries; the lines of east variation moving before it, while the lines of west variation followed it with a proportional pace. These lines first passed the Azores or Western Islands, then the meridian of London, and after a certain number of years still later, they passed the meridian of Paris. But in the southern hemisphere there was another line of no variation, which had east variation on its western side, and west variation on its eastern; the lines of east variation moving before it, while those of the west variation follow it. This line of no variation first passed the Cape des Aiguilles, and the Cape of Good Hope; the lines of 5°, 10°, 15°, and 20° west variation following it, the same as was the case in the northern hemisphere, but in the contrary direction.

Dr. Gowin Knight's opinion was, that this earth had originally received its magnetism, or rather that its magnetical powers had been brought into action, by a shock, which entered near the southern tropic, and passed out at the northern one; his meaning appears to have been, that this was the course of the magnetic fluid, and that the magnetic poles were at first diametrically opposite to each other. Though according to Mr. Canton's doctrine, they would not have long continued so; for from the intense heat of the sun in the torrid zone, according to the principles already laid down, the north pole must have soon retired to the north-eastward, and the south pole to the south-eastward. It is also curious to observe, that on account of the southern hemisphere being colder upon the whole than the northern, the magnetic poles would have moved with unequal pace: that is, the north-magnetic pole would have moved farther in any given time to the north-east, than the south magnetic pole could have moved to the south-east. And, according to the opinions of the most ingenious authors on this subject, it is generally allowed, that at this time the north magnetic pole is considerably nearer to the north than the south magnetic pole is to the south pole of the earth.

Several ingenious sea-officers are of opinion, that in the western parts of the English channel, the variation of the magnetic needle has now nearly reached its western limit; having in no part of it ever amounted to 25°.

There are, however, others who assert that the variation is still increasing in the channel, and as far westward as the 15th degree of longitude, and 51° of latitude, at which place they say that it amounts to about 30°, which was its greatest declination eastward; when it became retrograde.

But there is every reason to believe it to be near its western extreme, and it would tend much to the benefit of science, and would throw considerable light upon the subject, if it could be ascertained that this change begins to take place in all parts of the earth's surface at the same time; this, therefore, is the period to draw the attention of philosophers to this curious point. Those situated at some considerable distance from London will be able to make the most decisive observations. North America, Sweden and Russia, are well situated for the purpose. We hope that this opportunity, which is not likely to recur again these 300 years, will not be allowed to escape, without adding to the little that we do know, about magnetism.

Variation *Chart*. Dr. Halley having collected a multitude of observations made on the variation of the needle in many parts of the world, was hence enabled to draw, on a Mercator's Chart, certain lines, shewing the variation of the compass in all those places over which they passed, in the year 1700, when he published the first chart of this kind called the *Variation Chart*.

From the construction of this chart it appears, that the longitude of any of those places may be found by it, when the latitude and the variation in that place are known. Thus, having found the variation of the compass, draw a parallel of latitude on the chart through the latitude, found by observation; and the point where it cuts the curved line, whose variation is the same with that observed, will be the ship's place. A similar project of thus finding the longitude, from the known latitude and inclination or dip of the needle, was before proposed by Mr. Henry Bond, in his treatise intitled the longitude found, printed in 1676.

This method, however, is attended with two considerable inconveniences: 1st, that wherever the variation lines run east and west, or nearly so, this way of finding the longitude becomes imperfect, as their intersection with the parallel of latitude must be very indefinite: and among all the trading parts of the world, this imperfection is at present found chiefly on the western coasts of Europe, between the latitudes of 45° and 53°; and on the eastern shores of North America, with some parts of the Western Ocean and Hudson's Bay, lying between the said shores; but for the other parts of the world, a variation chart may be consulted with considerable advantage. However, the variation curves, when they run east and west, may sometimes be applied to good purposes in correcting the latitude, when meridian observations cannot be had, as it often happens on the northern coast of America, in the Western Ocean, and about Newfoundland; for if the variation can be found exactly, then the east and west curve, answering to the variation in the chart, will shew the latitude.

2ndly, As the deviation of the magnetical meridian, from the true one, is subject to continual alteration, therefore, a chart to which the variation lines are fitted for any year, must in time become useless, unless new lines, shewing the state of the variation at that time, be drawn on the chart: but as the change in the variation is very slow, therefore new variation charts published every 7 or 8 years, will answer the purpose tolerably well. And thus it has happened that Halley's Variation Chart has become useless, for want of encouragement to renew it from time to time.

However, in the year 1744, Mr. William Mountaine, and Mr. James Dodson, published a new variation chart, adapted for that year, which was well received; and several instances of its great utility having been communicated to them, they fitted the variation lines anew for the year 1756, and in the following year published the 3d variation chart, and also presented to the Royal Society, a curious paper concerning the variation of the magnetic needle, with a set of tables annexed, containing the result of upwards of 50 thousand observations, in six periodical reviews, from the year 1700 to 1756 inclusive, and adapted to every five degrees of latitude and longitude in the more frequented oceans; which paper and tables were printed in the Transactions for the year 1757.

From these tables of observations, such extraordinary and whimsical irregularities occur in the variation, that we cannot think it wholly under the direction of one general and uniform law; but rather conclude with Dr. Gowin, in the 87th proposition of his treatise upon attraction and repulsion, that it is influenced by various and different magnetic attractions, perhaps occasioned by the heterogeneous composition of that great magnet the earth. Many other observations on the variation of the magnetic needle, are to be found in several volumes of the *Philos. Trans.* See particularly vol. 48, p. 874; vol. 50 p. 329; vol. 56, p. 220; and vol. 61, p. 422.

VECTIS, in mechanics, one of the simple mechanical powers, more usually called the Lever.

VECTOR, or Radius Vector, in astronomy, is a line supposed to be drawn from any planet moving round a centre, or the focus of an ellipse to that centre or focus. It is so called, because it is that line by which the planet seems to be carried round its centre; and with which it describes areas proportional to the times.

To VEER, (*virer vent arrière*, Fr.) is to cause a ship to change her course from one board to the other by turning her stern to windward. Hence it is used in the same sense as wearing, and in opposition to tacking, wherein the head is turned to the wind, and the stern to the leeward.

Thus the ship A (see the diagram under the article To Tack), having made the necessary dispositions to veer, bears away gradually before the wind, till it blows obliquely upon the opposite side, which was formerly to leeward, as at *a*; and as the stern necessarily yields to this impression of the wind, assisted by the force of the helm, and the action of the waves upon the same quarter, the side which was formerly to leeward soon becomes to windward, as in the point *a*.

Since, by this movement, a ship loses ground considerably more than by tacking, it is rarely practised except in cases of necessity or delay: as, when the violence of the wind and sea renders tacking impracticable; or when her course is slackened to wait for a pilot, or some other ship in company, &c.

It has been observed in the article To Tack, that the change of motion in any body will be in proportion to the moving force impressed, and made according to the right line in which that force operates. Hence it is evident that veering as well as tacking is a necessary consequence of the same invariable principle; for as, in the latter, almost the whole force of the wind and of the helm are exerted on the hind part of the ship, to turn the prow to windward; so in the former, the same impression, assisted by the efforts of the helm, falls upon the prow, to push it to leeward; and the motion communicated to the ship must in both cases necessarily conspire with the action of the wind.

Thus, when it becomes necessary to veer the ship, the sails towards the stern are either furled or brailed up, and made to shiver in the wind; whilst those near the head are spread abroad, so as to collect the whole current of air which their surfaces can contain. Hence, while the whole force of the wind is exerted on the fore-part of the ship to turn her about, its effect is considerably diminished, or altogether destroyed, on the surfaces of the after-sails. The fore part accordingly yields to the above impulse, and is put in motion; and this movement, conspiring with that of the wind, pushes the ship about as much as is necessary to produce the effect required. When she is turned so that the wind will act upon that quarter which was formerly to leeward, as at the point *a* (see the diagram as before-mentioned), her circular motion will be accelerated by extending some of the sails near the stern, as the mizen, and by placing those at the prow more obliquely, which will wheel the vessel round with her bow to the windward; in the same situation, with regard to the wind, as when close-hauled, or tacking.

When the tempest is so violent as to prevent the use of sails, the effort of the wind operates almost equally on the opposite ends of the ship, so that the masts and yards situated at the head and stern, counterbalance each other. The effect of the helm is also considerably diminished, because the head-way, which gives life and vigour to all its operations, is at this time feeble and ineffectual. Hence it is necessary to destroy this equilibrium which subsists between the mast and yards afore and abaft, and to throw the balance forward, in order to prepare for veering. This is accordingly performed by bracing the foremost yards across the direction of the wind, and arranging those on the main-mast and mizen-mast directly in the line of the wind. If this expedient prove unsuccessful, and it is absolutely necessary to veer, in order to save the ship from destruction, by oversetting or running ashore, the mizen-mast must instantly be cut away, and even the main-mast, if she yet remains incapable of answering the helm by bearing away before the wind.

To VEER *away*, (*filer*, Fr.) is to let go a rope gently.

To VEER *away the Cable*. See the article CABLE.

To VEER *and Haul*, (*haler un cordage par secousses pour lui donner un balancement, et augmenter la force*, Fr.) is to pull a rope tight, by drawing it in and slackening it alternately, till the body to which it is applied acquires an additional motion, like the increased vibrations of a pendulum, so that the rope is straightened to a greater tension with more facility and dispatch. This method is particularly used in hauling the bow-lines.

The Wind VEERS *and Hauls*, (*le vent adonne et refuse*, Fr.) implies that it is altering its direction, and becoming more or less fair. Thus it is said, to veer aft and to haul forward.

To VEER *under Bare-Poles*. This should be performed, if circumstances will allow, by hoisting the fore-staysail: if, however, that cannot be done conveniently, the head-yards are to be braced up as sharp as possible, and those abaft pointed to the wind. Then, if the ship veers, she will steer under the masts and ropes only. A number of seamen, sent up and placed close to each other in the weather fore-shrouds, will be found also of very great service.

VEGA or WEGA, in astronomy, a bright fixed star of the first magnitude, in the top of the beak of the constellation Lyra. For the beginning of the year 1814, its

Right ascension was	227° 39′ 33″
Annual variation in right ascension	30.4
Declination	38 36 51 N.
Annual variation in declination	+ 2.9

VELOCITY or SWIFTNESS, in mechanics, is that affection of motion, by which a moving body passes over a certain space in a certain time. It is also called celerity; and always proportional to the space moved over in a given time, when the velocity is uniform, or always the same during that time.

VELOCITY is either uniform or variable. Uniform, or equal velocity, is that with which a body passes always over equal spaces in equal times. And it is variable and unequal, when the spaces passed over in equal times are unequal; in which case it is either accelerated or retarded velocity; and this acceleration, or retardation, may also be equal or unequal, *i. e.* uniform or variable.

VELOCITY is also either absolute or relative. Absolute velocity is that in which the velocity of a body is considered simply in itself, or as passing over a certain space in a certain time. But relative or respective velocity, is that with which bodies approach to or recede from each other, whether they both move, or one of them be at rest. Thus, if one body move with the absolute velocity of two feet per second, and another with that of six feet per second; then, if they move directly towards each other, the relative velocity with which they approach is that of eight feet per second; but if they move both the same way, so that the latter overtake the former, then the relative velocity with which that overtakes it, is that only of four feet per second, or only half of the former; and consequently it will take double the time of the former before they come in contact together.

Initial VELOCITY, in gunnery, denotes the velocity with which military projectiles issue from the mouth of the piece by which they are discharged.

To measure the velocity of a cannon ball, is a problem of no small difficulty, on account of the magnitude of the thing to be measured, and the constant variation to which it is necessarily subject. A body moving with a velocity from a thousand to two thousand feet per second, must describe a line of small extent in a portion of time that cannot be measured; and if the extent is considerable, the velocity is constantly varying at a rate not well ascertained, and the line passed over is not itself susceptible of accurate measurement. To remedy these inconveniences, Mr. Robins contrived to reduce the velocity of the ball to one that should be less in a great ratio, and in one that could be easily ascertained. This he effected, by making a ball of a given weight strike a large but moveable block of wood, also of a given weight, into which it penetrated; so that the two went on together after the impulse, with a velocity as much less than that of the ball, as the block and ball together were heavier than the ball alone. The velocity thus reduced, was so moderate as to be easily measured. This appears very simple as well as ingenious; but, when the experiment is to be made, the conclusion is not so readily obtained as might be supposed from this general statement. The block of wood, in order to be perfectly moveable, must be suspended; and therefore cannot describe a straight line when it is put in motion, but a circular line, in which it must vibrate backwards and forwards; and it is from the extent of these vibrations that the velocity with which it began to move must be inferred. The properties of compound pendulums, and of the centre of oscillation, must therefore be taken into the account. This was done by Mr. Robins; and the block of wood when properly suspended for measuring the initial velocity of shot, has been called the *Balistic Pendulum.*

A series of very valuable experiments in gunnery was begun by Dr. Hutton, in conjunction with several officers of the royal regiment of artillery, in the year 1775, and carried on at the expense of the Board of Ordnance. An account of these was presented to the Royal Society, and rewarded with the gold medal in 1778. From these experiments, which were conducted according to the method explained by Robins in his *New Principles of Gunnery*, many very important conclusions were deduced. It was made evident by them, that the powder fires almost instantaneously, and that the velocities which it communicates are as the square roots of the quantities of powder divided by the square roots of the weights of the shot. This conclusion is peculiarly valuable; and it deserves to be remarked, that it is conformable to the inference that would be drawn from the abstract principles of dynamics, leaving out many physical circumstances which might be expected very materially to affect the conclusion. This, therefore, is an instance which but rarely occurs where the simplifications made by the mathematician, chiefly with a view to facilitate his calculus, do not occasion any material difference between the conclusions of theory, and the result of experiment.

The same experiments pointed out the importance of pursuing them farther, and of instituting a new series on a plan still more accurate and extensive. This was begun in 1783, by the order, and under the direction of the late Duke of Richmond, who was at that time master-general of the ordnance, and by his abilities, his zeal for the public service, and his indefatigable exertions, was eminently qualified for that important office. The experiments were continued every summer with little interruption till 1791. The subjects in view were various; to ascertain the velocities produced by different charges with the same weight of shot, and the same weight and calibre of gun, supposing the only variation to be in its length; the greatest velocity due to the different lengths of guns; the penetration of balls into blocks of wood, the diminution of velocity, by the resistance of the air; the ranges and times of flight of balls fired with different velocities, and at different elevations, &c.

The necessary theorems are investigated anew; and the conclusions deduced are the same which Euler, in his remarks on Robins, had given, and also General D'Antoni, in his treatise on gunpowder. Dr. Hutton's formula agree also with Robins's theorem; not as he originally gave it, but as he corrected it in the *Philos. Trans.* for 1743. Of this correction it is material to remark (as Dr. Hutton has done), that the editor of Robins's works in 1761 has neglected to avail himself. Dr. Hutton has reduced the rule to greater simplicity than any of the authors just mentioned, and has added a correction for the resistance which the air opposes to the motion of the pendulum, and for some other circumstances that had hitherto escaped notice; in consequence of which his theorem is the most accurate that has yet been given.

Another method of ascertaining the initial velocity of shot, was employed in these experiments, depending on the recoil of the gun, when suspended and free to vibrate like the pendulum. This also requires a variety of circumstances to be taken into account; and it might be expected that this method would be preferable to the other, being the most direct possible, and giving the velocity of the ball immediately as it issued from the mouth of the piece; whereas the pendulum only gives it after it has passed through 20 or 30 feet of air, and consequently suffered a sensible diminution. On this account, one would expect the velocity deduced from the recoil, to exceed that found from the balistic pendulum; the contrary of which, however, happens whenever the velocity is great; and it is only in the case of small velocities that the results from the first sort of experiments, exceed those from the second.

One advantage possessed by these experiments, above all others yet published, consists in the greater size of the guns. They were 1-pounder guns, five very fine brass pieces of that calibre having been cast at Woolwich for these experiments, and bored as true as possible, the bore being 2.02 inches. They were of different lengths; viz. of 15, 20, 30, 40 calibres. The lightest was 290lbs. the heaviest 502; when they were used in the experiments for the recoil, they were made by additional weights to weigh 917lbs.

The following are a few of the general results, extracted from the beginning of the 37th tract on the *Theory* and *Practice* of *Gunnery*, by Dr. Hutton.

1. If c be the charge of powder, and b the weight of the ball, v the velocity with which it leaves the gun, may be always found from this formula, $v = 1600\sqrt{\frac{2c}{b}}$, or which is nearly the same, $v = 2240\sqrt{\frac{c}{b}}$. v is here expressed in feet. If the weight of the ball, and the velocity v are given, the charge of powder may be found from this equation $c = \frac{b\,v^2}{1280000}$.

2d. By increasing the quantity of charges for each gun, it was found, that the velocities continued to increase till they arrived at a certain degree, different in each gun; after which they constantly decreased till the bore was quite filled with the charge.

The quantities of powder, when the velocities arrived at their *maxima*, were different according to the lengths of the guns. In the shortest gun, which was 15 of its calibres in length, the 12 ounces of powder gave the greatest velocity, and the charge occupied $\frac{3}{16}$ of the bore. The gun of 20 calibres, with the same weight of ball, 1 lb. had the maximum of velocity when the charge was 14 ounces, occupying $\frac{2}{9}$ of the bore. With the longest gun of 40 calibres, the maximum velocity was with a charge of 18 ounces, which occupied $\frac{3}{20}$ of the bore. The proportion of the cavity filled by the charge for the maximum velocity diminished therefore from $\frac{1}{3}$ to $\frac{1}{4}$ and $\frac{1}{7}$ the length, as the gun increased. When the charges are equal in guns of different lengths, the velocity is greatest from the longest gun; but the increase is small in comparison to the increase of length, varying in a ratio between that of the square roots and the cube roots of the lengths.

With the same elevation the range increases in a much less ratio than the initial velocity. At an elevation of 15°, with a velocity of 864 feet per second, the range was 4,130 feet; at the same elevation, with a velocity of 1,676, almost double of the former, the range was only 6,700 feet.

No difference is caused in the velocity or range by varying the weight of the gun, by the use of wads, by different degrees of ramming, nor by the situation of the vent: a great difference, however, arises from a small difference in the degree of windage, or excess of the diameter of the bore above that of the shot. If the windage is $\frac{1}{20}$ of the calibre, which is the usual size, no less than $\frac{1}{3}$ or $\frac{1}{4}$ of the powder escapes, and is lost. As the balls are often smaller than the regulated size, it frequently happens that half the powder is lost by unnecessary windage.

The conclusion which Dr. Hutton has deduced concerning the resistance of the air, and the law which it observes in respect of the velocity of the moving body, are of great importance; and indeed the experiments themselves are probably the most accurate and extensive that ever have been made on this subject. They were made partly by the help of the *balistic pendulum*, partly by help of a *whirling machine*, by which a body may be made to move in the circumference of a circle with an uniform velocity; the accelerating force applied to a barrel with which the body is connected, becoming just equal to the resistance of the air, so that the motion is reduced to uniformity. This was the contrivance of *Robins* whose machine made by *Ellicot*, Dr. Hutton, after repairing and improving it, applied to determine the resistance to small velocities, while the *balistic pendulum* did the same for great velocities.

Comparison of the Velocities by the Gun and Pendulum.

Gun.	2 Ounces.			4 Ounces.			8 Ounces.			16 Ounces.		
No.	Velocity by		Diff.	Velocity by		Diff.	Velocity by		Diff.	Velocity by		Diff.
	Gun	Pend.		Gun	Pend.		Gun	Pend.		Gun	Pend.	
1	830	780	50	1135	1100	35	1445	1430	15	1345	1377	—32
2	863	835	28	1203	1180	23	1521	1580	—59	1485	1656	—171
3	919	920	—1	1294	1300	—6	1631	1790	—159	1680	1998	—318
4	929	970	41	1317	1370	—53	1669	1940	—271	1730	2106	—376

In this table, the first column shews the number of the gun, as they were of different lengths; viz. the length of number 1 was $30\frac{1}{3}$ inches, number 2 was $40\frac{1}{3}$ inches, number 3 was 60 inches, and number 4 was 83 inches, nearly. The rest of the table is divided into four spaces, for the four charges 2, 4, 8, 16 ounces of powder: and each of these is divided into three columns: in the first of the three is the velocity of the ball as determined from the vibration of the gun; in the second is the velocity of the ball as determined from the vibration of the pendulum; and in the third is the difference between the two, being so many feet per second, which is marked with the negative sign, or —, when the former velocity is too little, otherwise its + or positive. From the comparison contained in this table, it appears, in general, that the velocities determined by the two different ways, do not agree together; and that therefore the method of determining the velocity of the ball from the recoil of the gun, is not generally true, although Mr. Robins and Mr. Thompson had suspected it to be so: and consequently that the effect of the powder on the recoil of the gun, is not exactly the same when it is fired without a ball, as when it is fired with one. It also appears, that this difference is not always regular, neither is the different guns with the same charge of powder, nor in the same gun with different charges: that with very small charges, the velocity of the gun is greater than that by the pendulum; but that the latter always gains upon the former, as the charge is increased, and soon becomes equal to it; and afterwards goes on to exceed it more and more: the particular charge, at which the two velocities become equal, is different in the different guns; and that this charge is less, or the equality sooner takes place, as the gun is longer. And all this, whether we use the actual velocity with which the ball strikes the pendulum, or the same increased by the velocity lost by the resistance of the air in its flight from the gun to the pendulum. See HUTTON's *Mathematical and Philosophical Tracts.*

VENT, (*lumière d'un canon*, Fr.) in artillery, or, as it is vulgarly called, the touch-hole, is the opening through which the fire is conveyed to the powder that composes the charge. The vents of English guns are all $\frac{2}{10}$ an inch in diameter.

As the placing the vents in mortars, howitzers, and guns, in the best manner, is a delicate point, about which, both authors and practitioners differ, the most common method is, to place the vent about a quarter of an inch from the bottom of the chamber, or bore, others half an inch, and some even an inch from the bottom.

It has been imagined, that if the vent were to come out at the middle of the charge, the powder would be inflamed in less time than in any other case, and consequently produce the greatest range, because, if a tube be filled with powder, and lighted in the centre, the powder will be burnt in half the time it would otherwise be, were it lighted at one end. This gave a grounded supposition, that the greater the quantity of powder which burnt before the shot or shell was sensibly moved from its place, the greater force it would receive. To determine this, the King of Prussia, in 1765, caused a 3-pounder to be cast with three shifting vents, one at the centre of the charge, another at the bottom, and the third at an equal distance from the bottom and centre one; so that when one was used, the others were effectually stopped.

The gun weighed 2 cwt. 1 qr. 20 lb.; its length was 3 feet 3 inches, and the bottom of the bore quite flat. It was loaded each time with one-fourth of the shot's weight of powder; and it was found, that when the lowest or bottom vent was used, the shot went farthest, and the ranges of the others diminished in proportion as they were distant from the bottom. The piece was elevated to 1 degree 30 minutes.

In 1766, the same monarch caused experiments to be tried with three small mortars of equal size and dimensions, but of different forms in their chambers; each of which held seven ounces and a half of powder. From these experiments it appeared, that the concave chamber produced the greatest ranges, and that the bottom of the chamber is the best place for vents having in that place the greatest effect. See JAMES's *Military Dictionary*.

VENT-*Astragal*, is that part of a gun or howitzer which determines the vent-field.

VENT-*Field*, is the part of a gun or howitzer between the breech mouldings and the astragal.

VENTILATOR, (*ventilateur*, Fr.) the name of various machines by which the noxious air of any close places, as, the store-rooms and holds of ships, hospitals, chambers, &c. may be expelled, and fresh air introduced in its stead. The noxious qualities of bad air have been long known; and Dr. Hales and others have taken great pains to point out the mischiefs arising from foul air, and to prevent or remedy them by the use of his ventilators; the account of which was read before the Royal Society in May, 1741. In the November following, M. Triewald, military architect to the King of Sweden, informed Dr. Mortimer, secretary to the Royal Society, that he had, in the preceding spring, invented a machine for the use of his Majesty's men of war, in order to draw out the bad air from under their decks, the least of which exhausted 36,172 cubic feet of air in an hour, or at the rate of 21,732 tons in twenty-four hours. In 1742, he sent one of them, formed for a sixty-gun ship, to France; which was approved of by the Royal Academy of Science at Paris; and the King of France ordered all the men of war to be furnished with the like ventilators.

The ventilators invented by Dr. Hales consist of a square box of any size; in the middle of one of its sides a broad partition or midriff is fixed by hinges, and it moves up and down by means of an iron rod fixed at a proper distance from the other end of the midriff, and passing through a small hole in the cover of the box. Two boxes of this kind may be employed at once, and the two iron rods may be fixed to a lever moving on a fixed centre: so that by the alternate raising and pressing down of the lever, the midriffs are also alternately raised and depressed, whereby the double bellows are, at the same time, both drawing in air, and pouring it out through apertures with valves made on the same side with, and placed both above and below the hinges of the midriffs. See *Description of Ventilators, by* STEPHEN HALES, D. D. Lond. 1743.

The Doctor has shewn the use of his ventilators very fully. As to ships, in particular, he observes, that the wind-sail, made use of at sea to introduce fresh air between decks, is far from being sufficient for that purpose; nor can it be used with equal safety to the sick, and those who are sleeping, by means of the strength of the wind which conveys the air with too much violence. But when the foul air is carried off by means of ventilators, notwithstanding the great velocity with which they throw out the air, which they may do at the rate of sixty tons in a minute, yet the motion of it downwards into the hold, to supply what is carried off, is so very gentle, that it cannot be perceived; because the sum of all the open passages for it through the deck exceeds the opening of the trunk of the ventilator, in so great a proportion as 100 to 1, or more. Besides, in a calm, the wind-sail can do little or no good; nor when the ship is under sail, at which time the wind-sail is not used. And it is to be observed, that it is not the ventilating of a ship now and then with a wind-sail, when wind and weather serve, that will suffice; it ought to be done daily, if due regard be had to the health of the ship's crew. The great quantity of rancid noxious vapours, which are incessantly exhaling from a number of living human bodies, and from the hot, stagnant, putrid air in the hold, makes it very advisable to refresh so bad an air continually, either with the wind-sail, when that can be properly used, or else with ventilators, which are intended to supply the defects of the wind-sail.

Ventilators must also be of particular service in new ships, which are observed to be more unhealthy, on account of a greater quantity of sappy wreck which arises from new timber, and makes the confined air the more unwholesome.

They will also be an effectual preservative of horses in transports, where they are sometimes suffocated, when in a storm there is a necessity to shut down the hatches.

These ventilators will also drive out of the hold of a ship that dangerous vapour which arises from corn, which is so noxious, that sometimes they dare not venture into the hold, till after the hatches have been opened for some time.

Ventilators will not only be of service to preserve several kinds of goods, but also the timbers and planks of the hold itself, when laid up in ordinary, as well as when in use, and will make the air in the hold less noxious, though it will still be offensive to the smell, by reason of the bilge-water. But this may be made less offensive, by often letting in sweet water from the sea, and then pumping it out, which good practice ought to be continued, notwithstanding the use of the ventilators.

What is here said of foul air of ships may be applied to that of barracks, hospitals, mines, &c.

The larger kind of ventilators, used by the doctor, are ten feet long, five feet broad, and two feet high, in the clear within. Those he used by way of experiment on board the Captain, a seventy-gun ship, were ten feet long, four feet three inches wide in the clear within, and thirteen inches deep; one inch of which being occupied by the midriff, there remained a foot deep for it to rise and fall in. A ventilator of these dimensions will, through a trunk of a foot square, drive the air at the rate of twenty-five miles in an hour, which is double of what Marriot assigns for the velocity of a pretty strong wind.

But besides these ventilators, the doctor made a smaller sort, four feet in length, sixteen in breadth and thirteen inches deep, all in the clear within; these serve to purify, most easily and effectually the bad air of a ship's well, when there is occasion for persons to go down into it, by blowing air through a trunk, reaching within a yard of the bottom of the well, both for some time before and during their stay there.

When the hold is to be ventilated from one end to the other, the three doors of the gang-way into the gunner's store-room must be opened, and all the gratings on the gun-deck be covered with tarpaulings, leaving all doors open whose rooms want ventilation, on the orlop and the steward-room hatch. But when it is thought proper to ventilate between decks, then the doors of the gang-way into the gunner's store-room must be shut, and the scuttle in the headmost trunk or pipe upon the gun-deck must be opened, and all the gratings of the middle deck, if the ship be a three-decker, or of the upper deck, if it be a two-decked ship, to be laid with tarpaulings; and, if possible, one of the stern-posts opened, or the aftermost hatchway, or a scuttle on purpose, through the deck, as near the stern as possible. See Dr. HALE's *Treatise*.

A method of drawing off air from ships by means of fire-pipes, which some have preferred to ventilators, was published by Sir Robert Moray, in the *Phil. Trans.* for 1665. These are metal pipes, about two inches and a half in diameter, one of which reaches from the fire-place to the well of the ship; the other three branches go to other parts of the ship; the stove-hole and ash-hole being closed up, the fire is supplied with air through these pipes.

VENUS, in astronomy, one of the inferior planets, but the brightest, and to appearance the largest of all the planets; and is designed by the mark ♀, supposed to be a rude representation of a female figure, with her trailing robe.

Venus is easily distinguished from all the other planets, by her whiteness and brightness, in which she exceeds all the rest, even Jupiter himself, and which is so considerable, that in a dusky place she causes an object to project a sensible shadow, and she is often visible in the day time. Her place in the system is the second from the sun, viz. between Mercury and the Earth, and in magnitude is about equal to the earth, or rather a little larger, according to Dr. Herschel's observations.

As Venus moves round the sun, in a circle beneath that of the earth, she is never seen in opposition to him, nor indeed very far from him; but seems to move backward and forward, passing him from side to side, to the distance of about 47 or 48 degrees, both ways, which is her greatest elongation.

When she appears west of the sun, which is from her inferior conjunction to her superior, she rises before him, or is a morning star, called also Phosphorus, or Lucifer; and when she is eastward from the sun, which is from her superior conjunction to her inferior; she sets after him, and is called Hesperus, or Vesper, or the evening star: being each of these in its turn for 290 days.

The real diameter of Venus is nearly equal to that of the earth, being about 8648 miles; her apparent mean diameter seen from the earth, 59″, seen from the sun, or her horizontal parallax, 30″; but as seen from the earth 18″.79, according to Dr. Herschel: her distance from the sun 78 millions of miles; her eccentricity $\frac{7}{1000}$ths of the same, or 490,000 miles; the inclination of her orbit to the plane of the ecliptic 3° 23′; the points of their intersection or nodes are 14° of ♊ and ♐; the place of the aphelion ♒ 4° 20′; her axis inclined to her orbit 75°; her periodical course round the sun 224 days 17 hours; the diurnal rotation round her axis is very uncertain, being, according to Cassini, only 23^h 20′; but according to the observations of Bianchini, it is in 24 days 8 hours; though Dr. Herschel thinks it cannot be so much as the latter.

Venus, when viewed through a telescope, is rarely seen to shine with a full face, but has phases and changes just like those of the moon, being increasing, decreasing, horned, gibbous, &c.: her illuminated part being constantly turned toward the sun, or directed toward the east when she is a morning star.

The different phases of Venus were first discovered by Galileo; who thus fulfilled the prediction of Copernicus: for when this excellent astronomer revived the ancient Pythagorean system, asserting that the earth and planets move round the sun, it was objected that in such a case the phases of Venus should resemble those of the moon; to which Copernicus replied, that some time or other that resemblance would be found out.

Galileo sent an account of the first discovery of these phases in a letter, written from Florence in 1611, to William de Medici, the duke of Tuscany's ambassador, at Prague; desiring him to communicate it to Kepler. The letter is extant in the preface to Kepler's Dioptrics, and a translation of it in Smith's optics, p. 416; having recited the observations he had made, he adds, "We have hence the most certain, sensible decision and demonstration of two grand questions, which to this day have been doubtful, and disputed among the greatest masters of reason in the world. One is that the planets in their own nature are opaque bodies, attributing to Mercury what we have seen in Venus: and the other is, that Venus necessarily moves round the sun; as also Mercury and the other planets, a thing well believed, indeed by Pythagoras, Copernicus, Kepler, and myself, but never yet proved, as now it is, by ocular inspection upon Venus."

Cassini and Campani, in the years 1665 and 1666, discovered spots in the face of Venus: from the appearances of which the former ascertained her motion round her axis; concluding that this revolution was performed in less than a day; or at least that the bright spot which he observed, finished its period either by revolution or libration in about 23 hours. And De la Hire, in 1700, through a telescope of 16 feet, discovered spots in Venus, which he found to be larger than those in the moon.

The next observations of the same kind that occur, are those of Signior Bianchini at Rome, in 1726, 1727, and 1728, who, with Campani's glasses, discovered several dark spots in the disc of Venus, of which he gave an account and a representation in his book, entitled HESPERI *et Phosphori Nova Phenomena*, published at Rome in 1728. From several successive observations Bianchini concludes, that a rotation of Venus about her axis was not completed in 23 hours, as Cassini imagined, but in $24\frac{1}{3}$ days; that the north pole of this rotation faced the 20th degree of Aquarius, and was elevated 15° above the plane of the ecliptic, and that the axis kept parallel to itself during the planet's revolution about the sun. Cassini the son, though he admits the accuracy of Bianchini's observations, disputes the conclusion drawn from them, and finally observes, that if we suppose the period of the rotation of Venus to be 23h. 20m. it agrees equally well with the observations both of his father and Bianchini; but if she revolve in $24\frac{1}{3}$ days, then his father's observations must be rejected, as of no consequence.

In the *Philos. Trans.* 1792, are published the results of a course of observations on the planet Venus, begun in the year 1780, by Mr. Schroeter, of Lilienthal Bremen. From these observations, the author infers, that Venus has an atmosphere in some respects similar to that of the earth, but far exceeding that of the moon in density, or power to weaken the power of the sun: that the diurnal period of this planet is probably much longer than that of other planets: that the moon also has an atmosphere, though less dense and high than that of Venus; and that the mountains of this planet are five or six times as high as those on the earth.

Dr. Herschel, between the years 1777 and 1793, made a series of observations on this planet, accounts of which are given in the *Philos. Trans.* for 1793. The results of these observations are, that the planet revolves about its axis, but the time of it is uncertain: that the position of its axis is also very uncertain: that the planet's atmosphere is very considerable: and that he never could perceive any inequalities in the terminator of Venus, nor any phenomena which indicated the existence of mountains, though he had no doubt that such inequalities existed.

Schroeter computed the dense part of Venus's atmosphere to be about 16,020 feet high; that it rises far above the highest mountains; that it is more opaque than that of the moon; breadth of its twilight is 2″.45 broad, and that the density of the atmosphere is the reason why we do not discover in the surface of Venus those superficial shades and that variety of appearance which are seen on the other planets.

Venus exhibits the same appearances to us every eight years, except when transits happen; her conjunctions, elongations, and times of rising and setting are very nearly the same on the same days.

In the year 1686, Cassini supposed that he perceived a satellite near Venus, having the same phases as the planet. Short, Baudouin, and Montaigne, imagined that they saw the same phenomenon; but this illusion was completely dissipated by the transit of Venus over the sun, when she appeared unaccompanied by a secondary planet. These astronomers, however, seem to have been deceived by an optical illusion; for the imaginary satellite was probably a secondary image of Venus, formed by a double reflection. M. Wargentin had in his possession a good achromatic telescope, which always shewed Venus with a satellite, but the deception was discovered by turning the telescope about its axis.

M. Lambert has given a theory of the satellite of Venus, deduced from the observations of Cassini, Short, Montaigne, and other astronomers, and he has actually computed tables for finding its position in the heavens. There is a strange consistency among some of the observations of these astronomers, which could scarcely have been expected.

Lambert maintains, from his tables, that the satellite, if it did exist, might not have passed over the sun's disc at the time of the transit of Venus, but he expected that it might be seen alone on the sun when Venus passed near that luminary. See Mason, *Philos. Trans.* 1770, p. 454; Herschel, *Philos. Trans.* 1793. p. 201; also the article Transit.

VERNIER, is a scale, or division, well adapted for the graduation of mathematical instruments, so called from its inventor, Peter Vernier, a gentleman of Franche Comté, who communicated the discovery to the world in a small tract, printed at Brussels, in 1631.

This was an improvement on the method of division proposed by Jacobus Curtius, printed by Tycho, in Clavius's Astrolabe, in 1593. Vernier's method of division, or dividing plate, has been very commonly, though erroneously, called by the name of Nonius; the method of Nonius being very different from that of Vernier, and much less convenient.

When the relative unit of any line is so divided into many small equal parts, those parts may be too numerous to be introduced, or if introduced, they may be too close to one another to be readily counted or estimated; for which reason, there have been various methods contrived for estimating the aliquot parts of the small divisions, into which the relative unit of a line may be commodiously divided; and among those methods, Vernier's has been most justly preferred to all others. For the history of this, and other inventions of a similar nature, see Robins's *Math. Tracts, Vol. II. p* 265, &c.

Vernier's scale is a small moveable arc, or scale, sliding along the limb of a quadrant, or any other graduated scale, and divided into equal parts, that are one less in number than the division of the portion of the limb corresponding to it. So if we want to subdivide the graduations on any scale, into, for example, 10 equal parts, we must make the Vernier equal in length to 11 of those graduations of the scale, but dividing the same length of the Vernier itself only into 10 equal parts; for then it is evident that each division on the Vernier will be $\frac{1}{10}$th part longer than the gradations on the instrument, or that the division of the former is equal to $\frac{11}{10}$ of the degree on the latter, as that gains 1 in 10 upon this.

If the scale were not inches and 10ths, but degrees of a quadrant, &c. then the 8 would be $\frac{8}{10}$ of a degree, or 48′; or if every division on the scale be 10 minutes, then the Vernier will subdivide it into single minutes, and the 8 will then be 8 minutes.

By altering the number of divisions, either in the degrees or in the Vernier, or in both, an angle can be observed to many different degrees of accuracy.

For the method of applying a Vernier to a quadrant, see the article Quadrant: and for the application of it to a telescope, and the principles of its construction, see SMITH's *Optics*, *Book III. Sect.* 861, *&c.*

VERSED-SINE. See the article SINE.

VERTEX, in astronomy, is the same with vertex, or zenith. Hence a star is said to be vertical when it happens to be in that point which is perpendicularly over any place.

VERTEX *of an Angle*, is the angular point, or the point where the legs or sides of the angle meet.

VERTEX *of a Curve*, is the extremity of the axis or diameter, or it is the point where the diameter meets the curve; which is also the vertex of the diameter.

VERTEX *of a Figure*, is the uppermost point, or the vertex of an angle opposite the base.

VERTICAL, implies any thing placed in a direction perpendicular to the horizon.

VERTICAL *Circle*, is a great circle of the sphere, passing through the zenith and nadir of a place. The vertical circles are also called *azimuths*. The meridian of any place is a vertical circle, viz. that particular one which passes through the north or south point of the horizon. All the vertical circles intersect one another in the zenith and nadir.

The use of the vertical circles is to estimate or measure the height of the stars, &c. and their distances from the zenith, which is reckoned on these circles; and to find their eastern and western amplitude, by observing how many degrees the vertical, in which the star rises or sets, is distant from the meridian.

Prime VERTICAL, is that vertical circle, or azimuth, which passes through the equinoctial points.

VERTICAL or *Opposite Angles*, in geometry, are such as have their legs or sides continuations of each other, and which consequently have the same vertex or angular point.

VERTICITY, is that property of the magnet or loadstone, or of a needle, &c. touched with it, by which it turns or directs itself to some peculiar point, as to its poles. The attraction of the magnet was known before its verticity.

VESPER, in astronomy, called also *Hesperus*, and the *Evening Star*, is the planet Venus, when she is eastward of the sun, and consequently sets after him, and shines as an evening star.

VESPERTINE, in astronomy, is when a planet is descending to the west after sun-set, or shines as an evening star.

VESSEL, (*bâtiment de mer en général*, Fr.) a general name given to the different sorts of ships which are navigated on the ocean, or in canals and rivers. It is, however, more particularly applied to those of the smaller kind, furnished with one or two masts.

The first vessel ever known to have floated on the waters, was the ark made by God's appointment, in which Noah and his three sons were saved from the universal deluge. "To the immediate interposition of God," says Burchett in his Naval History, "are we to attribute the invention of shipping, as we are to his concurring providence, hose improvements which have been since made therein, and the perfection it is arrived to at this time." We can neither discern the necessity of controverting the fact, nor perceive the smallest reason to doubt its truth. This ark, ship, or whatever else it may be called, had neither oars, sails, masts, yards, rudder, nor any sort of rigging whatsoever; being guided by Divine Providence, and having no particular port, or coast to arrive at, it was only formed to float upon the waters, and when these were dried up, it rested on the mountains of Ararat, as we read in Gen. viii. 4.

The length of it appears to have been 450 feet, the breadth 75, and the height 45. Its form was rectangular; and the wood with which it was built, and which in Scripture is styled *Gopher-wood*, was from the cypress tree, as being, according to Vitruvius, the least subject to decay; the sap which pervades every part of that wood, is so very offensive, that no worm, or other corrod ng animal, will touch it. The structure of the ark was certainly adapted to the burden it was to carry, and the weather it was to endure; and in these respects may be considered as the most complete and perfect model that ever was devised.

Not many years after the flood, it became necessary for the descendants of Noah to employ themselves in the science of constructing ships, to convey themselves and families to their respective allotments of territory. The Isles of the Gentiles, which included the continent of Europe, the northern parts of Asia, and Asia Minor, with all the islands of the Mediterranean and Ægean sea, were assigned to the sons of Japhet.

Kittim, a grandson of Japhet, according to Josephus, settled in Cyprus; and from his name, not only all islands in general, but most maritime places, were, in the Hebrew language, termed Kittim.

We have, in the XLIXth chap. of Genesis, an authentic testimony of the antiquity of shipping. This likewise shews the great antiquity of the Phœnician commerce and navigation; being prior by almost 200 years, to the coming of Danaus, surnamed Armais, from Egypt to Argos, with the first ship that had ever been seen by the Greeks; who, until that period, had only used floats and rafts. See ANDERSON's *History of Commerce*: and for farther particulars, see the articles *History of* NAVAL *Architecture*, and the *History of* NAVIGATION.

The views of utility, which ought always to be considered in a work of this kind, limit our general account of shipping to those which are most frequently employed in European navigation. We have therefore collected the principal of these in the under-mentioned plates, so that the reader, who is unacquainted with marine affairs, may the more easily perceive their distinguishing characters, which are also more particularly described under the respective articles.

Thus, in plate XXXI. is a lugger chasing a schooner. Pl. XXXI

Fig. 1, plate XXXII, is a brig or brigantine under-sail; fig. 2, a cutter; fig. 4, a xebec; and fig. 5, a dogger. Pl. XXXII

Plate XXXIII, exhibits a ketch, sloop, bilander, pleasure-boat, galliot, and a yacht, all of which are under sail. Pl. XXXIII

The ketch, whose sails are furled, is furnished with a try-sail; and it has a fore-sail, fore-stay-sail, and jib, nearly similar to those of a sloop; but the sails on the main-mast and mizen-mast are like those of a ship. The main-sail and main-top-sail of the brig are like those of the schooner; and the fore-mast is rigged and equipped with sails in the same manner as the ship and snow. The sails, masts, and yards of the XEBEC being extremely different from these are described at large under that article. In the schooner, both the main-sail and fore-sail are extended by a boom and gaff, as likewise is the sloop's main-sail; the sails of the dogger and galliot are sufficiently expressed in the plate.

A new invented vessel, named the *Constellation*, which was intended to sail against wind and tide, arrived above Blackfriars-bridge, on the 16th of November, 1811, from Bristol. This vessel was about 50 feet in length, with only one mast, made of iron, and an upright windlass affixed to it; there were twelve horizontal sails, similar to the shape of window-shutters, which were extended or shortened in an instant: on any occasion the mast, with all its appendages, could also be as quickly struck. She had neither blocks nor any running rigging, except a fore and aft stay and cable; her guns, which were of curious mechanism, kept their own elevation.

Hired Armed VESSELS, are vessels employed as cruisers against the enemy, hired of merchants, and of which the masters and crews are paid by the merchants, although some of them are commanded by officers commissioned or receiving pay in the navy.

VESTA, in astronomy, a new planet, which was discovered, after much labour, in the constellation Virgo, on the 29th of March, 1807, by Dr. Olbers, of Bremen, who gave it this appellation.

As soon as the discovery was made known in England, the planet was observed at Blackheath, on the 26th of April, 1807, by S. Groombridge, Esq. an ingenious and active astronomer, who has successfully devoted his leisure and his fortune to the advancement of astronomy. He continued to observe it with his excellent astronomical circle, till the 20th of May, when, from its having ceased to become visible on the meridian, he had recourse to equatorial instruments.

On the 11th of August, Mr. Groombridge resumed his meridional observations, from which he has computed part of the elements of its orbit; and he observed the ecliptic opposition of the planet, on the 8th of September, 1808, at $7^h\ 30'$, in longitude $11^s\ 15°\ 54'\ 26''$. His observations were continued till the beginning of November, 1808, and he expected to have found the planet again at its opposition in February, 1810; but, from a continuance of cloudy weather, and probably from errors in the elements he did not succeed.

The planet Vesta is of the 5th or 6th magnitude, and may be seen in a clear evening by the naked eye. Its light is more intense, pure, and white than either that of Ceres, Juno, or Pallas; and it is very similar in appearance to the Georgium Sidus. It is not surrounded with any nebulosity; and even with a power of 636, Dr. Herschel could not perceive its real disc. The orbit of Vesta cuts the orbit of Pallas, but not in the same place where it is cut by that of Ceres. According to the observations of Schroeter, the apparent diameter of Vesta is only 0.488 of a second, one half of what he found to be the apparent diameter of the fourth satellite of Saturn; and yet it is very remarkable, that its light is so intense, that Schroeter saw it several times with his naked eye.

M. Burckhardt is of opinion, that Le Monnier had observed this planet as a fixed star, since a small star situated in the same place, and noticed by that astronomer, has since disappeared.

The following are the elements of the orbit of Vesta, computed by Mr. Groombridge, from his own observations:

	Years.	Days.	Hours.
Revolution	3	66	4
Place of perihelion	8^s	$13°$	$0'\ 0''$
Place of ascending node	3^s	$14°$	$38'\ 0''$
Inclination of orbit		$7°$	$8'\ 20''$
Mean distance			2.163
Eccentricity in part of the earth's rad.			0.0953

The elements given by Burckhardt, in the *Connoissance des Temps*, for 1809, from the most recent observations on the continent, do not differ materially from those of Mr. Groombridge.

The orbits of the four new planets appear to intersect each other in various planes; and the points of intersection must be perpetually shifting, according to the changes in the aphelia of the planets. The mean distances employed are,

Ceres	2.765
Pallas	2.791
Juno	2.657
Vesta	2.373

For farther particulars on the subject of this article, see *Edinburgh Encyclopædia*, Vol. II. part the 3d and 4th; GROOMBRIDGE'S *Phil. Mag.* Vol. XXVII. and XXXI.; *Connoissance des Temps*, for 1809.

VIA LACTEA, in astronomy, the milky way, or GALAXY. See that article.

VIA SOLIS, or *Sun's Way*, is used among astronomers, for the ecliptic, or path in which the sun seems always to move.

VIBRATION, in mechanics, is a regular reciprocal motion of a body, as, for example, a pendulum, which being freely suspended, swings or vibrates from side to side.

Mechanical authors, instead of vibration, often use the term *oscillation*, especially when speaking of a body that thus swings by means of its own gravity or weight.

The vibrations of the same pendulum are all isochronal; that is, they are performed in an equal time, at least in the same latitude; for in lower latitudes they are found to be slower than in higher ones. The vibrations of a longer pendulum take up more time than those of a shorter one, and that in the sub-duplicate ratio of the lengths, or the ratio of the square roots of the lengths. Thus, if one pendulum be 40 inches long, and another 10 inches long, the former will be double the time of the latter in performing a vibration; for $\sqrt{40} : \sqrt{10} :: \sqrt{4} : \sqrt{1}$, that is, as 2 to 1. And because the number of vibrations, made in any given time, is reciprocally as the duration of one vibration, therefore the number of such vibrations is in the reciprocal sub-duplicate ratio of the length of the pendulums. For farther particulars on this article, see HUTTON's *Philos. Dict.*

VICE ADMIRAL. See the article ADMIRAL.

VICTUALLER. See the article AGENT.

VICTUALS. See the article PROVISIONS.

VINDEMIATRIX, or VINDEMIATOR, a fixed star of the third magnitude, in the northern wing of the constellation Virgo.

VINEGAR, (*vinaigre*, Fr.) an agreeable, acid, penetrating liquor, prepared from wine, beer, ale, cyder, &c. and was known many ages before the discovery of any other acid, except those ready formed in vegetables.

In long voyages, such as those of discovery, &c. no means has been found to contribute so strongly to the health of the crew, as repeatedly washing all parts that could be got at between decks, particularly those that are less exposed to the air, with vinegar. It is frequently used in the artillery, to cool pieces of ordnance after firing them. Two pints of vinegar to four pints of water is the usual proportion.

VIOL. See the article VOYOL.

VIRGO, in astronomy, one of the signs or constellations of the zodiac, which the sun enters about the 21st or 22d of August; being one of the 48 old constellations, and is mentioned by the astronomers of all ages and nations, whose works have reached us: anciently the figure was that of a girl, almost naked, with an ear of corn in her hand, evidently to denote the time of harvest among the people who invented the sign.

The stars in this constellation in Ptolemy's catalogue, are 32; in Tycho's 33; in Hevelius's 50; and in the Britannic 110.

VISIBLE, something that is an object of vision or sight, or the property of a thing seen.

The Cartesians say that light alone is the proper object of vision. But, according to Newton, colour alone is the proper object of sight; colour being that property of light by which the light itself is visible, and by which the images of opaque bodies are painted on the retina.

VISION, is the act of seeing, or of perceiving external objects by the organs of sight.

When an object is so disposed, that the rays of light, coming from all parts of it, enter the pupil of the eye, and present its image on the retina, that object is then seen.

VISUAL, relating to sight, or seeing.

VISUAL *Angle*, is the angle under which an object is seen, or which it subtends.

VISUAL *Point*, in perspective, is a point in the horizontal line, where all the ocular rays unite. Thus a person standing in a straight, long gallery, and looking forward, the sides, floor, and ceiling seem to meet and touch one another in this point or common centre.

VISUAL *Rays*, are lines of light, conceived to come from an object to the eye.

UNARMED, (*non armé;—désarmé*, Fr.) the state of being without armour or weapons.

To UNBALLAST, (*délester*, Fr.) to discharge or get the ballast out of a ship.

To UNBEND, (*dégréer*. Fr.) implies to take off the sails from their yards and stays; to cast loose the cables from their anchors, or to untie one rope from another.

To UNBEND *a Course*, (*dégréer une basse voile*, Fr.) First furl the sail, then cast off the rope-bands, and make them fast round the sail, clear off the gaskets. When the rope-bands are all off, ease off the lee earing, and lower down the sail; and, when the people upon deck have got hold of the lee part of the sail, ease away the weather earing.

To UNBEND *a Topsail*, (*désen verguer un hunier*, Fr.) First cast off the points of the reefs, keeping fast the earings; then furl the sail, and cast off the rope-bands, which make fast round the sail, clear of the gaskets. After this, cast off the lee earings, and haul the lee-side of the sail into the top; then haul in the weather-side; next unbend the reef-tackle pendants, bunt-lines, and bow-lines; bight the sail snugly up together, and send it down by the clew-lines to windward or leeward, as most convenient.

To UNBIT, (*débitter*, Fr.) is to remove the turns of a cable from off the bits. See the articles BITS and CABLE.

UNDER-CURRENT, (*courant sous l'eau, qui ne se ressent point à la surface*, Fr.) a stream running below the surface of the water. See the article CURRENTS.

UNDER-FOOT, (*qui s'applique à une ancre qu'on laisse tomber à-pic*, Fr.) is said of an anchor when first let go.

UNDER *the* LEE *of the* LAND. See the article LEE.

UNDER-MASTED, or a *Low-Masted Ship*, is such whose masts are too short, or too light, in which case she cannot bear so great a sail as should give her true way.

To UNDER-RUN *a Cable*, (*pomoyer, ou passer sous un cable, &c.* Fr.) is to pass under it in a boat, in order to examine if any part of it is damaged or entangled.

It is usual to under-run the cables in particular harbours, as well to cleanse them with brooms and brushes from any filth, ooze, shells, &c. collected in the stream; as to examine whether they have sustained any injury under the surface of the water; as, from rocky ground, or by the friction against other cables or anchors.

To UNDER-RUN *a Tackle*, (*défaire les tours d'un palan*, Fr.) is to separate the several parts of which it is composed, and range them in order, from one block to the other; so that the general effort may not be interrupted, when it is put in motion.

UNDER-SAIL, (*à la voile*, Fr.) the state of a ship when she is loosened from her moorings, and under the government of her sails and rudder. See the articles HELM and SAIL.

UNDER-SET, (*mouvement du fond des eaux de la mer, qui porte au vent dans une baie, où il bat en pleine côte par un gros tems*, Fr.) a motion of the water beneath the surface, produced by the wind impelling the upper part directly upon the shore of a bay, whereby the water, having no other means of regaining its level, necessarily takes a direction contrary to the wind, in that situation where the wind has less power, namely, below the surface.

To Get UNDER-WAY *from River Moorings.* If

it be in a tide-way, and with a leading wind, so that the ship can stem the tide; let it be a rule, when the tide serves to get under-way, and sail against the flood; which gives time to clear a ship of her moorings, and affords a more powerful effect to the helm, to steer the vessel clear of other ships, and any particular danger.

UNDERWRITER, (*assureur*, Fr.) an insurer, who underwrites, or subscribes his name to policies of insurance on ships, merchandise, &c. for a certain sum, for which he receives a premium, and consequently takes the risk on himself so far as the sum he insures. See the article INSURANCE.

NAVAL UNIFORM, (*uniforme de la marine*, Fr.) the established clothing of officers in the Royal Navy, according to their rank.

By order of his Royal Highness the Prince Regent in 1812, all flag-officers, captains, commanders, lieutenants, master's-mates, and midshipmen of his Majesty's Royal navy shall be altered as follows.

The Admiral of the Fleet's Full Dress. A coat of blue cloth, blue cloth collar, white cloth lapells and cuffs, with five laces round the cuffs; laced as formerly; epaulettes and buttons as formerly, with the addition of a crown over the anchor.

His Undress. A blue cloth coat, blue cloth collar, white lapells, and cuffs with five laces; laced round the collar and lapells to the end of the skirts; flap and frame, hips and back-skirts laced; twist button-holes in lapells and flaps as formerly; epaulettes and buttons, the same as in the dress uniform.

An Admiral's Full Dress, is the same as the admiral of the fleet, with only four laces on the cuffs.

Vice Admiral's, the same, with only three laces on the cuffs.

Rear Admiral's, the same, with only two laces on the cuffs. The epaulettes, with the respective distinctions of three, two, and one star, the same as formerly. Buttons as formerly, with the addition of a crown over the anchor.

The undress or frock uniform of flag-officers, except the admiral of the fleet, to be the same as formerly, with the alteration of the button only.

The captain to the admiral of the fleet, and first captains to commanders in chief, if not flag-officers, to wear, while so employed, the undress or frock uniform of rear-admirals.

Captains and commanders of his Majesty's fleet, are to wear uniforms of the same pattern.

The full dress to be similar to that formerly in use, excepting that the lappells and cuffs are to be white, laced as formerly, with the crown over the anchor on the button. Captains and commanders are both to wear two epaulettes of the same pattern as formerly, with only the following distinctions.

The epaulettes of captains three years post, are to have an addition of a silver crown over a silver anchor. The epaulettes of captains under three years post to have the silver anchor without the crown.

The epaulettes of commanders to be plain.

Lieutenants of his Majesty's fleet are to wear a dress uniform of the same pattern as captains and commanders, but without any lace, and with one plain epaulette, similar to that formerly worn by captains and commanders, on the right shoulder; buttons of the same pattern as for captains.

The undress or frock uniform of captains, commanders, and lieutenants respectively, are to be the same as formerly, with the addition of the epaulettes and button, which are to be worn as in the full dress.

The whole of the commissioned officers of his Majesty's fleet, are to have the linings of their dress uniforms white. The flag-officers are only to have the linings of their dress uniforms white silk.

Masters-mates, and midshipmen, are to wear the same uniform, as formerly, with the alteration of the button only, which is to be of the same pattern as that of the captains and lieutenants.

UNION FLAG. See the articles FLAG and JACK.

To UNLACE, (*dégreer une bonnette*, Fr.) is to loosen and take off the bonnet of a sail from its principal part.

To UNLOAD A GUN, or a MUSKET, (*décharger un canon, ou un mousquet*, Fr.) is to take the powder and ball out of a piece of ordnance, or musket.

To UNMOOR, (*désaffourcher*, Fr.) is to reduce a ship to the state of riding by a single anchor and cable, after she has been moored or fastened by two or more cables. See the articles ANCHOR and MOORING.

All Hands UNMOOR *Ship! (tout le monde à désaffourcher!* Fr.) the order to the ship's company to perform that duty: it is used in contradistinction to that of To MOOR; which article see.

To UNREEVE. (*dépasser une manœuvre de ses poulies, &c.* Fr.) is to withdraw or take out a rope from any block, thimble, dead-eye, &c. through which it had formerly passed. See the article To REEVE.

To UNRIG A SHIP, (*dégréer, ou dégarnir, un bâtiment, un mât, ou une vergue, &c.* Fr.) is to deprive her of the standing and running rigging, &c.

To UNSHIP, (*démonter, &c.* Fr.) is to remove any piece of timber or wood from the place in which it was fitted, as, "Unship the capstan bars!" "Unship your oars!" Unship the tiller."

To UNSLING, (*ôter des élingues*, Fr.) is to take off the slings from boats, butts, buoys, yards, &c.

VOLLEY OF SMALL ARMS, (*salve de mousqueterie*, Fr.) the discharge of a great many muskets at the same time.

VOLUNTEER, (*volontaire*, Fr.) a man who enters willingly to serve on board a ship of war, in contradistinction to an impressed man. It also implies a man, who, being in the King's service, freely offers himself to serve in any particular expedition of danger or fatigue.

VORTEX, a whirlwind, or sudden violent motion of the air, in gyres or circles. With seamen it implies an eddy or whirlpool, as the vortex or whirlpool of Maelstrom, or Moskoestrom, on the coast of Norway, it is heard at the distance of many leagues, and forms a vortex of great extent and depth, so violent, that if a ship comes near it, it is immediately drawn irresistibly into the whirl, and there disappears, being absorbed and carried down in a moment, where, at the bottom, it is dashed to piece against the rocks, and when the water becomes still which it does at the turn of the ebbing and flowing on the tide, the shattered fragments rise, scarcely known to be the parts of a ship. In the time of its greates violence, the danger of its influence is said to extenc to the distance of twelve English miles from its centre Whales, and other animals which happen to be caught by this dreadful whirlpool exhibit signs of their sensibility of destruction by their hideous yells, and desperate, but ineffectual efforts to escape.

VOYAGE, (*voyage*, Fr.) in a general sense implies a space passed over by sea in a ship or vessel, from one port to another; but more particularly to some foreign clime, as to the East or West Indies, &c.

Towards the close of the 15th century, the spirit of adventure on maritime discoveries, seems to have actuated the different European nations, to venture beyond their native boundaries in quest of foreign regions, hitherto unknown.

Venice and Genoa, then the only powers in Europe who owed their support to commerce, engrossed the whole traffic of India, which had hitherto been carried on through the inland parts of Asia, or by the way of Egypt and the Red Sea. In this state of affairs, Christopher Columbus, a native of Genoa, from his knowledge of the true figure of the earth, conceived a project of sailing thither by the west, and of opening to his country a new source of opulence and power.

To defray the expense of this expedition he first solicited the assistance of the Genoese, next, the Portuguese, and afterwards the courts of England and France: but his proposals were rejected as chimerical and absurd; Spain was now his only resource, by which court, after eight years solicitation, he was furnished with all necessaries, and equipped with three ships; and on the 23d of August, 1492, he set sail from Palos de Moguer, on the coast of Andalusia, directing his course to the Canaries; where he refitted, and on the 6th of Sept. put again to sea, steering to the westward; here his difficulties began to increase, the most formidable was the variation of the compass, then first observed, which seemed to indicate a change in the laws of nature in an unknown ocean; his sailors became mutinous, threatening his destruction; but his intent firmness, and the discovery of land on the third of October, put an end to all commotion.

Columbus first landed on one of the Bahama islands, which he named St. Salvador, nine hundred and fifty leagues west of the Canaries; steering still to the southward, he discovered the island of St. Domingo, abounding in all the necessaries of life, gold mines, cotton, &c. this he considered the centre of his discoveries, and after acquiring a general knowledge of all the islands in the great sea between North and South America, and with what gold he had procured in trafficking with the Indians, on the 15th of March, 1493, he returned to Spain, amidst the acclamations of the people, and was received at the Spanish court with all possible demonstrations of honour; thus were the West Indies discovered by seeking a passage to the East, and even after the discovery, it was considered a part of the eastern hemisphere.

While Columbus in the west was making his progress in discovery, the Portuguese navigator, Bartholomew Diaz, venturing along the western coast of Africa, discovered the Cape of Good Hope, now so beneficial to our East India commerce.

The success of Columbus now rendered the court of Spain as eager to forward his designs, as they had been formerly dilatory. A fleet of seventeen sail was immediately fitted out, with all the necessaries for conquest or discovery, and fifteen hundred men, among whom were many of high rank and fortune. On the 15th of September, he sailed from Seville, then steering to the south of his former course, on the 3d of November, discovered the island of Dominica, he then discovered the Caribbee islands, and the island of Porto Rico. November the 22d, he arrived at St. Domingo, where he planted his colony, and sent to Spain twelve ships, with some gold, and a full account of his voyage. On the 24th of April, 1494, steering westward, he came upon the island of Cuba, and in coasting along the southern shore, on the 18th of May, discovered 160 small islands, which he called the Queen's Garden. On the 22d of July, he discovered the island of Jamaica, then returned to St. Domingo, and afterwards to Spain, in order to justify his conduct at court, from certain malicious aspersions that were thrown upon him.

On the 20th of November, 1497, the Portuguese navigator, Vasco de Gama, doubled the Cape of Good Hope, and landed at Calicut in India, on the 22d of May, 1498.

Sebastian Cabot, a Venetian, residing in England, was employed by King Henry the VIIth, to discover a north west-passage to the East-Indies, and with two ships sailed from Bristol, in the summer of 1497, in this voyage he discovered the island of Newfoundland, and all the north-east coast of the British Empire in North America, and without obtaining the object of his pursuit, he returned to England, in the spring of 1498.

On the 13th of May, 1498, Columbus set out with six ships, on a third expedition, more conspicuous than any he had hitherto undertaken: sailing towards the equinoctial, and then to the west, on the first of August, fell in with the island of Trinidad; passing this, and two other small islands in the mouth of the river Oronoco, after which, he discovered the Continent. Columbus landed at several places, and traded with the inhabitants for gold and pearl. He then set sail, and arrived at St. Domingo, on the 2d of August. The news of the discoveries Columbus had made on the Continent, induced many private individuals to adventure. On the 20th of May, 1499, four ships sailed from Port St. Mary, and sailing to the south-west, on the 6th of June came in sight of the Continent; among those who landed was Amerigo Vespuccio, a merchant of Florence, and being a man of address, had the honour of giving his name to this half of the globe.

On the 26th of January, 1500, Vincent Yanez Pinzon, who accompanied Columbus in his first voyage, with four ships, at his own charge, landed on the coast of Brazil, and took possession for the King of Spain. In this voyage, he discovered six hundred leagues to the south-east of Paria.

The same year Emanuel, King of Portugal, fitted out a fleet of thirteen sail for the East-Indies, which sailed from Lisbon in March; but being carried to the westward, on the 24th of April, landed on the coast of Brazil, which became one of their most valuable possessions.

The enemies of Columbus now became inveterate, he was treated as a traitor, and carried over to Europe in irons; but after fully vindicating himself at court, he was restored to favour. On the 9th of May, 1502, he undertook another voyage, in which he discovered the island of Guanaja, and Cape Casinas, in the province of Honduras.

After suffering great fatigues, he returned to Spain, and died at Valladolid, in 1506, in the 59th year of his age, after having by his voyages made known the whole Continent of America, from Cape Honduras, in 18 degrees north-latitude, to Port Seguro on the coast of Brazil, in 17 degrees south-latitude.

After the death of Columbus, the succeeding governors of St. Domingo and Cuba, endeavoured to purchase the same advantages by the blood of the natives, which he had obtained by good sense and humanity. The Spaniards setting no bounds to their cruelty and brutality, in a few years they depopulated the island of St. Domingo, which contained three millions of inhabitants; and Cuba which contained six hundred thousand.

In February, 1519, Ferdinand Cortes, was dispatched from Cuba, with eleven ships, carrying six hundred men, and a small number of field pieces; with this inconsiderable force, he subdued the vast empire of Mexico, the most powerful state on the Continent of America.

He extended his conquest over the adjoining provinces to the South-sea; being above two hundred leagues in length, and one hundred and fifty in breadth, he discovered the rich mines of silver, which, with the gold mines of Peru, have ever since enriched the Universe.

In 1524, John Varrazona, a Florentine, was employed by Francis I. King of France, to make discoveries to the north-west, he sailed from Diep, with four ships, in which voyage he discovered the northern part of the island of Florida, and sailed along the eastern coast of the Continent, from 34, to 50 degrees north-latitude, giving it the name of New France.

The island of St. Bartholomew was discovered by the Spaniards, on the 13th of September, 1525.

On the 20th of May, 1527, two ships sailed from the Thames, sent by King Henry VIII. to discover the north-west passage, entering between the island of Newfoundland and the Continent, one of them was cast away, the other sailed to Cape Briton, and to the coast of Arambec, and returned to England in the month of October.

In 1530, Francis Pizarro, with one hundred and eighty-five Spaniards sailed from Panama, he landed at the bay of St. Matthew, and marching along the shore, fell upon the town of Quapel, and took a vast quantity of gold, silver, and emeralds; he then sailed to the island of Puna, in latitude 4 degrees south, then took possession of Tumbez, and moving still to the southward, he founded the city of St. Michael, where he settled a colony. Pizarro then marched into the country where he had several engagements with the Indians, took the city of Caxamalca; and Curco near two hundred leagues to the south-east of the former; moving from thence to the sea he founded the city of Lima, in 18 degrees south latitude, and subdued all the empire of Peru.

In 1532, Nunho de Guzman, was dispatched from Mexico by Cortes, to reduce the province of Mechoacan; he discovered and subdued the provinces of Culiacan and Cinaloa, opposite the south-end of California.

In 1534, Simon d'Alcazova, was sent by the King of Spain to discover to the southward of Peru, and in the month of January, with two ships and 250 men, entered the straits of Magellan, where Alcazova was murdered; one of the ships was cast away on the coast of Brazil, the other in much distress arrived at the island of St. Domingo.

The same year James Cartier, was dispatched from port St. Mary, by order of Francis I. king of France, to discover the north part of America, and on the 10th of May he landed at Port St. Catherine, in Newfoundland; after sailing from Cape Biz, to Cape de Grace, he arrived at St. Malo, on the 5th of September.

May the 19th, 1535, Cartier sailed again to the coast of Newfoundland; sailing thence, on the 15th of August, he came upon the islands of Assumption and Orleans.

The same year D. Pedro de Mendoza, with eleven ships, and eight hundred men, sailed from St. Lucar; he landed near the river Plate, where he settled the colony of Buenos Ayres.

1536, two ships with 120 men, sailed from London, for North America, but landed at Newfoundland, where they were reduced by extreme want, and a few only returned to England in a French ship, which they surprised, and took on the coast.

On the 7th of March, 1539, F. Mark de Niza, with some Indians, set out from Culiacan, and discovered to the north of California on the south seashore, to the country of Cibola, in 39 degrees north-latitude.

On the 16th of April, 1542, John Francis de la Roche with three ships, and 200 persons, sailed from Rochelle, and landed at Quebec, where he erected a fort and settled a colony.

On the first of November, 1543, Ruy Lopez de Villalobos, sailed from New Spain, and discovered the islands, St. Thoma, Anublada, and ten other small islands which he called the Gardens; also the islands Cæsarea Coroli, Manilla, and Gilolo, near the Moluccas.

In April, 1562, John Ribault with two French ships arrived on the coast of Florida; he then sailed to the river Port Royal, near which he built Charles Fort, and settled a colony.

The same year Mr. Hawkins, made a voyage from England to Guinea, and brought from thence 300 Indians, which he sold at St. Domingo, which seems to have been the commencement of that diabolical traffic the slave trade. On the 22d of April, 1564, Captain Loudonniere, with three French ships, sailed for Florida; he arrived in the river May, near to which, he erected Fort Caroline. In 1565, it was taken by the Spaniards, and retaken by the French in 1567.

On the 12th of June 1576, Mr. Martin Forbisher sailed from Gravesend, to discover a passage to China, and Cathay, by the north-west; in this voyage, he entered a strait in 63 degrees north-latitude, which he called Forbisher's Strait, and touched at the islands St. Gabriel, Butcher's island, &c. and returned to Harwich on the first of October.

He set out on a second voyage in 1577, and touched at the Orkney isles and Friesland; after running thirty leagues up Forbisher's Straits, he returned to England.

In 1578, fifteen ships sailed from Harwich, to make farther discoveries in the north, and being driven to the southward of Forbisher's Straits, they ran 60 leagues up another, returning to the Countess of Warwick's Sound, loaded abundance of ore, and steered for England.

Francis d'Ovalle, in 1582, sailed from Acapulco, in Mexico; running westward to the islands, de Engano, Tandaye, and Manilla, he then reached the port of Macao on the coast of China; after being furnished with necessaries, he returned to port Acapulco.

In June 1583, Sir Humphrey Gilbert sailed from the West of England, with five vessels and 150 men, to plant a colony in North America; after touching at Newfoundland, he met with many disasters at sea, and returned to Dartmouth with only one vessel.

On the 27th of April, 1584, Philip Amadas, and Arthur Barlow, sailed from the west of England in two barks, and landed on the coast of Virginia, and returned in the month of September following.

Sir Richard Greenvil, with seven ships, sailed from Plymouth, in June, 1585, and settled a colony on the island of Wokoken, in Virginia; the planters made many discoveries in the country, but being set upon by the inhabitants, they were reduced, by great distress, and taken on board by Sir Francis Drake, on his return from the expedition to the Spanish plantations.

In 1585-6-7, Mr. John Davies made three voyages to discover the north-west passage to China, but was always unsuccessful.

In 1587, Sir Walter Raleigh sailed from England with three vessels and 150 men, and settled a colony on the island Roanoak.

In 1595, Avaro da Mendana, set out from Peru, for the islands of Solomon, with four ships, and all the necessaries, to settle a colony: but three of his ships were cast away, and in the other he arrived at Manilla.

On the 6th of March, 1602, Captain Gosnals sailed from Falmouth, and on the 24th of April, discovered Elizabeth's Island, in 41 degrees 10 minutes north latitude.

In 1600, four ships sailed from Peru, for the Philippine Islands, they were driven to the south of Solomon s islands, and made several discoveries.

Captain Samuel Champlain sailed from port Honfleur, in Normandy, the 15th of March, 1603, after running a great way up the river of Canada, and getting intelligence of the country, he returned to France.

The same year, two vessels from Bristol, and one from London, sailed to Virginia.

In 1604, M. de Monts, sailed from France with two ships well manned to Acadia and Canada; and afterwards settled a colony at Port Royal, in Newfoundland.

On the 31st of March, 1605, Captain George Weymouth, with one ship, sailed from the Downs, and landed at St. George Island, on the coast of Virginia, at a port which he called Pentecost harbour.

Captain Newport, in 1606, sailed with three ships from London to Virginia, where he founded St. James Town, and settled a colony.

Mr. Hudson, in 1610, again undertook the discovery of the north-west passage, and proceeded one hundred leagues farther than any before him had hitherto done.

The same year, Sir Henry Middleton, with three ships, made a voyage to the Red Sea and Sumatra: he died at Machin, the 24th of May, 1613.

In 1611, Sir Thomas Button was sent by Prince Henry, on the north-west discovery; he passed Hudson's Strait, and sailed two hundred leagues to the south-westward, where he discovered New Wales, and the land called Cary's Swansnest.

In April, 1612, Richard Moore, with one ship and 60 men, was sent to plant the islands of Bermuda, (before discovered by the Spaniards); here he settled a colony, who built several forts.

In 1612, James Hall, and William Baffin, returned to England, after having discovered Cockin's Sound, in 65 degrees, two minutes, north-latitude.

In 1621, a British colony was settled at Plymouth, in New England, North America.

On the 20th of October, 1628, Captain Francis Pelsart, with eleven Dutch vessels, set sail for India; he came to several small islands, about 38 degrees south-latitude, and sailing along the coast of Asia, in 1629, he came among the Indian Islands, and from thence to Batavia, in the island of Java.

Captain James, in 1631, sailed from England to the north-west, and discovered New South Wales.

In 1642, Abel Jansen Tasman, with a yacht, and a fly-boat, set sail from Batavia, and touched at the following places, viz. the island Mauritius; Anthony Van Diemen's Lands; Frederick Hendrick's Bay; the island of New Zealand; Three Kings' Islands; islands Amsterdam and Zealand; Prince William's Islands; a cluster of islands, called Onthong Java; New Guinea; Islands of Mark, Green Island, and the Burning Island; he then returned to Batavia, having performed his voyage in 10 months.

In 1643, a Dutch ship, sailing to the northward of Japan, discovered Staten Island, the Company's Land Island; and the land of Eso or Yedso.

Zachariah Gillam, in 1667, built Fort Charles, in latitude 51 degrees north.

In 1669, Captain Narbrough, afterwards Sir John Narbrough, sailed through the Straits of Magellan, to the South Sea, on discoveries, which he made public in his voyage.

In 1673, F. Marquette, a Jesuit, with six Frenchmen, set out from *Lac des Puans*, in the province of Canada, and came to the great river Mississipi, and discovered all the inland part of North America, along the river, from 38 to 34 degrees north-latitude, lying on the back of Canada, Virginia, &c. down to Florida.

In 1680-1, Captain Sharp, having sailed to the South Sea, and missing the Strait of Magellan, was driven to 60 degrees south-latitude, and named an Island which he discovered, the Duke of York's Island.

On the 14th of January, 1698, Captain Dampier, in his Majesty's ship the Roe-Buck, sailed from the Downs to make discoveries; after passing the Canaries, and Cape Verd, he anchored on the coast of Brazil, running 114 degrees, he came to the coast of New Holland, in 26 degrees south-latitude; sailing thence to the coast of New Guinea, he discovered an Island which he called New Britain. In 1699, Dampier returned by Timor and Java to the Cape of Good Hope, and the island of St. Helena; at the island of Ascension his ship foundered, but his crew were saved, and returned with him to England, on board the East India ship called the Canterbury.

On the 8th of January, 1705, Peter Koben, A.M. sailed from the Texel for the Cape of Good Hope, and continued till 1713, exploring the interior of Africa, and observing the manners and customs of the Hottentots, which he afterwards published.

Commodore Roggewein, a gentleman of Zealand, sailed from the Texel, on the 21st of August, 1721, in search of a southern Continent; but returned in 1723, without obtaining this object.

On the 26th of May, 1735, Don George Juan, and Don Antonio De Ulloa, sailed from Cadiz, in order to determine the true figure of the earth, by which, measuring a degree of the meridian near the equator, it was ascertained to a considerable degree of advantage. In returning, the Deliverance, with Don Ulloa, was taken by an English ship; but he was afterwards admitted to the Royal Society, then liberated, and with all his papers, returned to Madrid, on the 26th of July, 1746.

In March, 1750, Mr. Grose sailed to the East-Indies, and wrote a history of the manners, customs, &c. of the Indians, which was afterwards published.

On the 3d of June, 1773, the Honourable Captain Phipps, afterwards Lord Mulgrave, sailed from Sheerness, with a view to discover a north-east passage to China and Japan; after sailing to 80 degrees 47 minutes north-latitude, he returned to England, finding the passage impracticable.

Voyages *round the World.* On the 20th of September, 1519, Ferdinand Magellan, with 5 ships and 250 men, was sent by the Emperor Charles V. king of Spain, on a voyage round the world; he sailed from St. Lucar de Barameda, and on the 21st of October, 1520, discovered Virgin's Cape, and afterwards the strait, called the Strait of Magellan, through which he sailed to the South Sea, being about 100 leagues in length; sailing W.N.W. three months and twenty days, he touched at the Phillippine Islands, then to the island of Matan, where he was killed by the natives. John Sebastian Cano then took the command; he sailed to the island Bohol, and discovered the island of Borneo, and came to the Molucca islands on the 7th of September, 1522. Steering south-west, he came to the islands Malva and Timor: then proceeding southward, doubled the Cape of Good Hope, and arrived at St. Lucar the 7th of September, 1522, being the first voyage round the world.

On the 13th of September, 1577, Sir Francis Drake, with five ships sailed from Plymouth, on a second voyage round the world. He entered the Straits of Magellan on the 21st of August, 1578, and entered the South Sea on the 6th of September; sailing thence, came to the Molucca Islands the 14th of November. On the 8th of February, 1579, he fell in with the island Barateve, from this to Java and the Cape of Good Hope; and arrived in England on the 3d of November, 1580.

On the 21st of July, 1586, Sir Thomas Candish undertook a third voyage round the world; he sailed from Plymouth with three small vessels; on the 3d of January, 1587, he anchored at the mouth of the Straits of Magellan, and entered the South Sea on the 14th of February, after losing two of his vessels and meeting with many disasters: on the 8th of February, 1588, he discovered the island Batochina, and coming to Java, he sailed for the Cape of Good Hope, touched at St. Helena, and on the 9th of September arrived at Plymouth.

The Dutch fitted out four ships under the command of Oliver Noort, for a voyage round the world; he set sail on the 19th of July, 1598; on the 10th of December came to Prince's Island on the coast of Congo, and in February 1600, came into the South Sea, through the Straits of Magellan; he discovered the island Luzon or Manilla, one of the Philippines; and by the way of Java, the Cape of Good Hope, and St. Helena, he returned to Amsterdam the 26th of August, 1601.

August the 8th, 1614, George Spilbergen sailed from the Texel with five Dutch ships; he passed through the straits of Magellan, touched at the Molucca Islands, and arrived at Zealand in July 1617, by the islands of Java and St. Helena.

On the 16th of June, 1615, Isaac Le Maire, of Amsterdam, and William Cornelison Schouten, of Horn, sailed from the Texel with two ships on the discovery of a new way to the East Indies, by sailing to the south of the Straits of Magellan; on the 9th of December, they entered Port Desire, on the coast of America, where one of their ships was burnt; the 13th of January, 1616, sailing southward, discovered the island Staten, and the point of Tierra del Fuego, between which they found a passage to the South Sea, and called it the Straits of Le Maire; sailing 1600 leagues westward of Chili, then northward, they anchored near the coast of New Guinea and sailed from thence to Batavia, where the president of the Dutch East India Company seized the ship and goods; finding ships homeward bound, they arrived at Amsterdam, in the month of July, 1617.

In 1683, John Cook, with one ship, and one Cowley, as master, sailed from Virginia, he passed the coast of Guinea, and entered the South Sea in 60 degrees south latitude; they next anchored at Canton in China; from thence to the island Borneo, where Cowley the master, with nineteen others, in a boat, came to Batavia, and he, with three others, sailed to Europe in a Dutch vessel, and published an account of the voyage; for the particulars of which, see Hack's *Voyages*, 1699.

On the 1st of August, 1708, Captain Wood Rodgers, and Captains Edward Cooke, Stephen Courtney, with two ships, the Duke and Duchess, and 231 men, sailed from King-Road on a voyage round the world, proceeding by Cape Verd; they reached the coast of Brazil the 18th of November; from thence doubled Cape Horn, and entered the South Sea, and touched at the islands Juan Fernandez, Labos, and Puna; then came to Porto Seguro, in California; from which they sailed on the 10th of January, 1710, to Serpana and Guam; afterwards they touched at Sampo and Batavia: on the 18th of December, they reached the Cape of Good Hope, and remained there till the 6th of May, 1711, when they sailed for the Texel under Dutch convoy, and arrived in the Downs on the 2d of October.

February 13th, 1719, Captain Clipperton in the Success, with 180 men, and Captain Shelvock, in the Speedwell, with 106 men, sailed from Plymouth; the ships parted in a storm: Clipperton entered the South Sea by the Straits of Le Maire; arriving on the coast of Mexico, he met with Captain Shelvock, on board the Jesu Maria, having lost the Speedwell at Juan Fernandez. Clipperton then sailed to Guam, where his ship ran aground, and was fired at by the Spaniards; after getting off, much damaged, she passed the Bashee Islands to Amoy and Macao; here she was surveyed, condemned, and sold. In her he came as a passenger to Batavia, and, in a Dutch East Indiaman, landed at Galway, in Ireland, July 1722: broken down with toils and soured with disappointments, he lived only a few days after his arrival.

We return now to Captain Shelvock, who, after many struggles, arrived at Canton in China, where, being sick, he was deserted by his officers and crew, and divested of all that was valuable, he came to England by an East-Indiaman, having spent three years and seven months in this voyage round the world.

On the 17th of August, 1764, Captain Byron, in quality of a Commodore, with the Dolphin ship of war, and the Tamar frigate, sailed from the Downs on a voyage of discovery in the South Seas; he sailed to Madeira and Palma, thence to Rio de Janeiro, where he was treated with much civility; he weighed anchor and sailed to Cape Blanco, then by the Strait of Magellan into the South Sea, where he discovered some small islands, and returned to England by Batavia and the Cape of Good Hope, and anchored in the Downs the 9th of May, 1766.

On the 19th of August, 1766, Captain Samuel Wallis, with the Dolphin, the Prince Frederick, and the Swallow, sailed from Plymouth Sound on a voyage round the world; sailing by Madeira and Palma, the Swallow parted, but in three days joined them at the Isle of May; passing in sight of Terra del Fuego and the Virgin Island, he entered the Straits of Magellan, where the Swallow, commanded by Captain Carteret, was again left behind; after struggling with storms and currents, he entered the South Sea, and shortly after discovered the following islands: Whitsun, Queen Charlotte's, Egmont, Gloucester, Cumberland, Prince William Henry's, Osnaburg, and the island of Otaheite, where he continued some time; he then prosecuted his voyage, dicovering many more islands, and by Batavia, the Cape of Good Hope, and the island of St. Helena, he arrived in England, and anchored in the Downs the 20th of May, 1768.

Captain Carteret, with the Swallow, pursued a different track, and having sailed round the world with Captain Byron, was well acquainted with the nature of the voyage, he entered the South Sea and steered along the coast of Chili, from thence touched at many islands; he discovered the islands Egmont, Trevanian, Gower's, Simson's, and Carteret's Islands, then proceeded to New Britain, discovered by Dampier; and in sailing to the straits of Macassar, discovered five small islands; he next came to Bonthain where he repaired, and touching at the Straits of Sunda, the Cape of Good Hope, and the island of St. Helena; in March, 1769, he anchored at Spithead, having circumnavigated the globe in a vessel little calculated for the enterprise.

The French nation became animated with the views of signalizing themselves, and began to project voyages for the like purposes.

On the 15th of November, 1766, Bouganville, by an order from the King of France, sailed from Minden with three ships, and the celebrated astronomer M. Vernon, to make observations in order to discover the longitude at sea. M. Bouganville was directed to give up the Falkland Islands to the Spaniards; after being detained on this occasion till the 2d of July, 1767, he sailed to Rio Janeiro, where he met the store-ship Etoile; after repairing, he sailed for Buenos Ayres, and took in provisions; entered the Straits of Magellan, and touching at many islands, came to Otaheite; in proceeding to Port Prasin, he discovered many small islands, and here M. Vernon viewed an eclipse of the sun, and made nautical and astronomical remarks thereon; he next landed at the island Bouro, where he was hospitably treated by the governor and supplied with provisions; after which he sailed by Batavia, the Straits of Sunda, and the Isle of France; and on the 18th of January, 1769, he doubled the Cape of Good Hope. Near St. Helena he met with the Swallow, commanded by Captain Carteret, formerly mentioned, returning from a similar voyage; he then landed in his native country, having displayed great ability and skill in conducting this expedition.

On the 30th of July, 1768, Captain James Cook sailed from Deptford with the Endeavour, in order to observe a transit of Venus over the Sun's disc from the Island of Otaheite; and also to make discoveries in the South Sea. Mr. Charles Green, coadjutor of Dr. Bradley, astronomer royal, was nominated by the Royal Society to assist in the astronomical part of the undertaking; Joseph Banks, Esq. afterwards Sir Joseph Banks, and Dr. Solander, also embarked with them. Sailing by Madeira, Rio de Janeiro, &c. he doubled Cape Horn, and entered Port Royal harbour in the Island of Otaheite, the 13th of April, 1769. Captain Cook traded amicably with the inhabitants, and made all necessary preparations for observations on two conspicuous parts of the island, and on the 3d of June observed the transit of Venus from both stations; after circumnavigating the island, he set sail, and discovered a number of islands, which he named the Society Islands. Captain Cook then sailed to an island which he named East Island, and from thence to another which he gave the name of Mayor Island, where Mr. Green and the other gentlemen landed to observe a transit of Mercury, and named the bay in which the ship lay, Mercury Bay, and discovered a number of islands of different sizes to the north-west, which he called Mercury Islands; sailing thence by New Zealand, he discovered it to be composed of two islands by sailing between them, and came to an island which he called Solander's Island; here it was resolved by a council of officers that they should return to England by the East Indies, and steer by the eastern coast of New Holland. On the 6th of May, 1770, he anchored in a bay which he called Botany Bay; after this he discovered a point of land, and called it Cape Byron; he then discovered York Isles, and that which Captain Cook named New South Wales, from thence he sailed to Saru, Java, Batavia, the Cape of Good Hope, and St. Helena; and on the 12th of June, 1771, anchored in the Downs, having made many brilliant discoveries.

Captain Cook was again dispatched by government with two ships to explore the southern hemisphere, in order to discover a continent; he sailed from Plymouth Sound on the 13th of June, 1772, and traversed the Southern Ocean, so as to leave no possibility of the existence of a contiuent within the reach of navigation.

In his progress he discovered New Caledonia, the Island of Georgia, and an unknown coast, which he named Sandwich Land: he then returned to England, and anchored at Spithead on the 30th of July, 1774.

In the preceding voyage, the question respecting the existence of a southern continent was fully determined; but the practicability of a north-west passage to the Pacific Ocean was still an object of sufficient importance to excite a spirit of investigation.

Captain Cook was again equipped with two ships, the Endeavour and the Discovery; they cleared the Channel on the 14th of July, 1776, and touched at the Cape of Good Hope, the Island of Otaheite, and New Zealand; from thence to Owhyee, one of the Sandwich Islands, where Captain Cook was killed by the natives, and Captain Clerke died soon after; the command then devolved on Captains Gore and King, who prosecuted the voyage with the utmost perseverance and skill, and arrived at the Nore on the 4th of October, 1780. Thus ended a voyage distinguished by the extent and importance of its discoveries; besides other inferior islands it added the Sandwich Islands to the formerly known limits of the terraqueous globe; they fully ascertained the proximity of the two continents Asia and America, and perceived the impracticability of a passage between them, either by an eastern or a western course.

Though Captain Cook's last voyage did not accomplish the original object in view, it furnished philosophy with many new observations, and opened to commerce several extensive prospects; the valuable furs he discovered on the western coast of America, soon induced adventurers to pursue a new traffic so lucrative; and several voyages were undertaken from Canton and Bombay, to Nootka Sound. Captain Nathaniel Portlock and Captain George Dixon, who had both sailed with Captain Cook, were sent out by a company of English merchants, entitled "The King George's Sound Company," for carrying on a fur trade from the western coast of America to China. They sailed from Gravesend on the 29th of August, 1785, with two ships, the King George and the Queen Charlotte; they doubled Cape Horn, and came to the Sandwich Islands, King George's Bay, &c.; and after procuring two cargoes of furs, they sailed to Canton where they sold them to the East India Company; they then loaded with tea and other articles, and both ships arrived in England, the one in August, and the other in September, 1788.

On the 26th of June, 1785, the two French frigates La Boussole and L'Astrolabe, by order of government, sailed from the port of Brest for a voyage of discovery; the former was commanded by J. F. G. De La Pérouse, the latter by De Langle. They sailed by Madeira, Teneriffe, Trinidad, St. Catherine, Conception, Easter, and Sandwich islands; and in July, 1786, discovered Port de François, in latitude 58° 37′ north, and 142° 15′ west longitude; and in November discovered Neckar Island, in latitude 23° 34′ north, and longitude 169° 17′ west; sailing thence by Macao Road, Manilla, and the Island of Kumi, they entered the Japanese sea, and sailed along the coast of China and Tartary to Corea, where some astronomical observations were made. They next discovered the strait which separates Jeso from Oku-Jessu, and sailing through it, came by Staten-Land, &c. to Kamschatka, Avotscha, to the Island Maouna, where M. de Langle was assassinated by the Indians, with 11 of the ship's company. See LOCKE's, BURNEY's, and PINKERTON's *Collection of Voyages*.

In the year 1801, Captain Matthew Flinders sailed from England in his Majesty's ship Investigator, under the direction of the Admiralty, upon a voyage for completing the discovery of Terra Australis, where he had the misfortune to run upon a coral rock, and lose his ship. Out of the wreck he constructed a small schooner, that carried him and the crew to the Mauritius, where, to the disgrace of the French nation, instead of being received with kindness, as is the practice of all civilized nations to nautical discoverers, he was imprisoned by the governor De Caen, and confined nearly seven years, during which time, he suffered much in his bodily health, and contracted diseases in prison that hastened his death. He returned to England, and fortunately for mankind, and his own fame, he survived finishing the printing an account of his voyage, and died in July 1814.

By order of his Imperial Majesty Alexander I. of Russia, Captain A. J. Von Krusenstern sailed with the two ships Nadeshda and Neva, from Cronstadt, in August 1803, on an embassy to the court of Japan, and also for the purpose of increasing their East India commerce. After passing some time at Copenhagen, he proceeded to the Canaries, the coast of Brazil, Cape Horn, and on the 6th of May, 1804, he came in sight of Nukahiwa, one of the Washington islands, which lie towards the middle of the Pacific Ocean, and steered along shore in quest of Port Anna Maria. He had proposed to hold a new course from the Marquesas to Kamtschatka, in hopes of making new discoveries; but the small stock of provisions procured at Nukahiwa, caused him to proceed to the Sandwich Islands, where he stood in for the coast of Owhyhee; at this place he parted for a time from the Neva, whose destination was not to Japan, but to the Aleutic islands. He then proceeded to Kamtschatka, and on the 14th of July, anchored in the harbour of Petropaulosk, or St. Peter and St. Paul. Having repaired his ship, and recruited his stock of provisions, he sailed from the bay of Awatchka, on the 7th of September, for Japan, having on board the ambassador from the Emperor of all the Russias, the Emperor of Japan.

In approaching the south-east cape of Niphon, the principal of the Japanese islands, he searched for some islands, which, according to the early Spanish charts of the Northern Pacific Ocean, are situated near that Cape; but he could not find them, though he sailed close to the situations assigned to them in the charts.

September the 28th, the Nadeshda came in sight of some of the southern of the Japanese islands, and afterwards proceeded to the Gotto islands. On the 8th of October he anchored in the harbour of Nangasaki, the only port of Japan left free for the entrance of foreign traders: here the Nadeshda was boarded by a Japanese officer, and questioned. A first meeting passed in the exchange of compliments; and the second completely terminated the negociation, by a document being delivered into the hands of the ambassador, containing an order that no *Russian ship should again come to Japan*. Having re-embarked the embassy, Captain Krusenstern sailed from Nangasaki on the 17th April, 1805, and in the beginning of May was near the western entrance of the Strait of Sangaar, which separates Jesso from Niphon: thence he proceeded to the northward, and sailed east-ward through La Pérouse's strait. When he had traced the eastern coast of Sachalin from La Pérouse's strait to Cape Patientie, he sailed from Kamtschatka, passing between the Kurile islands, and arrived in the harbour of St. Peter and St. Paul the 5th of June following. On the 5th of July he sailed again for Sachalin, and completed his examination of the east coast. In order to reconnoitre, where he was restrained from making full discovery, he proceeded from the north part of Sachalin to the southward, between it and Tartary, 20 leagues. On the 9th of October he set sail from the Bay of Awatchka, in order to return to Europe by the way of China.

At Macao, the Nadeshda had been rejoined by the Neva, with a rich cargo of furs, from the Aleutic islands, and they sailed together in company for Europe. The Nadeshda anchored at Cronstadt, August the 19th, 1806, having been absent 3 years and 12 days, and lost but one of her crew. The Neva had been equally fortunate.

Since that period the nations of Europe have been so much involved in war, that no voyages for discoveries have lately been made.

VOYOL, (*tournevire*, Fr.) a large rope used to unmoor, or heave up the anchors of a ship, by transmitting the effort of the capstan to the cables.

This is performed by fastening one part of the voyol to the cable in several places, and by winding another part thereof three or four times round the capstan, which answers the same purpose as if the cable itself were in that manner wound about the capstan; and the voyol being much lighter, and more pliant, is infinitely more convenient for the operation. See the articles Capstan and Nipper.

If the cable is drawn into the ship by the main-capstan, the voyol is used without any block: but if the capstan in the fore part of the ship be employed for this purpose, the voyol usually passes through a large block attached to the main-mast; and thence communicates with the jear capstan.

Voyol-Block. See the article Block.

UP Courses! (*cargue ensemble les basses violes*, Fr.) the order to haul them up by means of the clew-garnets, bunt-lines, &c.

Up *with the Helm!* (*la barre au vent!* Fr.) the order to put it to the weather-side of the ship.

Up and Down, (*à-pic*, Fr.) is said of the anchor when the cable is hove in, so as to be perpendicular to the hawse-holes.

UPHROE, an oblong block made of ash, without sheaves, from nine to thirty inches long, and two to five inches in diameter, as represented by fig. 4, plate III. It has several holes bored through the middle at equal distances, and grooved round the outside to receive the rope by which it is suspended. It is used to suspend the awnings, by extending the small ropes through the holes lengthways along the middle of the awning.

UPPER-DECK, (*le pont supérieur; c'est le deuxième pont dans les vaisseaux à deux batteries, et le troisième pont, dans ceux à trois ponts*, Fr.) the highest of those decks which are continued throughout the whole of a ship of war, or merchantman, without any interruption. See the articles Deck and Waist.

UPPER-WORKS, (*œuvres mortes*, Fr.) a general name given to all that part of a ship which is above the surface of the water when she is properly balanced for a sea-voyage: hence it may be considered as separated from the bottom by the main wale, as explained in the article Naval *Architecture*.

UPRIGHT, (*droit*, Fr.) the situation wherein the opposite sides of a ship are equally elevated above the surface of the water, as in fig. 2. plate XVI. or when she neither inclines to the right nor left, with regard to the vertical position of her stem and stern-post.

URSA, in astronomy, the Bear, a name common to two constellations of the northern hemisphere, near the pole, distinguished by *Major* and *Minor*.

Ursa *Major*, or the Great Bear, one of the 48 old constellations.

There are two remarkable stars in this constellation, viz. those in the middle of his body, considered as the two hindermost of the wain, and called the Pointers, because they always point nearly in a direction towards the north pole-star, and so are useful in finding this star out.

The stars in Ursa Major, are, according to Ptolemy's catalogue, 35; in Tycho's 56; in Hevelius's 73; but in the Britannic catalogue 87.

Ursa *Minor*, the *Little Bear*, called also *Arctos Minor*, *Phœnice*, and *Cynosura*, one of the 48 old constellations, and near the north pole, the large star in the tip of its tail being very near to it, and thence called the pole-star. Ptolemy places 8 stars in this constellation, Tycho 7, Hevelius 12, and Flamsteed 24.

USES *and* Customs *of the Sea*, (*les us et coutumes de la Mer*, Fr.) certain general principles which compose the basis of marine jurisprudence, and regulate the affairs of commerce and navigation.

W.

WAD, or WADDING, (*valet à canon*, Fr.) in gunnery, is a quantity of old rope-yarns, or tow, hay, &c. rolled firmly together, into the form of a ball, and used to confine a shot or shell, together with its charge of powder, in the breech of a piece of ordnance.

M. Le Blond observes, in his elements of war, that the wad is necessary to retain the charge closely in the chamber of the cannon, so that it may not, when fired, be dilated around the sides of the ball by its windage as it passes through the chase; a circumstance which would considerably diminish the effort of the powder. But as the wad cannot be fastened to the sides of the bore, it is carried away in the same instant when the charge is inflamed, and that with so little resistance, that it cannot, in any degree, retard the explosion, or give time for the entire inflammation of the powder.

This reasoning may with equal propriety be applied to the wad that covers the bullet; which, nevertheless, is absolutely requisite, to prevent it from rolling out when the piece is fired horizontally or pointed downwards. Both are therefore peculiarly necessary in naval engagements, because without being thus retained in its chamber, the shot would instantly roll out of the chase by the agitation of the vessel.

WAD-HOOK. See the article WORM.

WAFT, or WEFT, (*berne*, Fr.) a signal displayed from the stern of a ship for some particular purpose, by hoisting the ensign, furled up together into a long roll, to the head of its staff. It is particularly used to summon the ships' boats off from the shore; or as a signal for a pilot to repair aboard.

WAGES, (*paie*, Fr.) a certain sum, per month, allowed each class of seamen employed in his Majesty's ships for their respective services.

In the reign of Henry the Eighth, the wages of seamen were only five shillings per month, accounting twenty-eight days to the month; a rule observed in every subsequent increase of their wages, and still practised. In the reign of James the First, their wages were raised to ten shillings per month; but whether by that prince or his predecessor is not ascertained; but it is more probable that it was done by Elizabeth, as Charles the First, in the very beginning of his reign, raised them to fifteen shillings, which it is not likely he would have done if his father had so recently raised them. See also the article PAY.

Abstract of an Act of Parliament for the Payment of Seamen's Wages.

I. Every volunteer entering his name with a commissioned officer, appointed for entering volunteers on board any ship in the royal navy, shall receive a certificate thereof gratis, and be entitled to wages from the date of such certificate, including the day of the date thereof, in case he appears on board within fourteen days, if the place where he enters is not above one hundred miles from the ship; within twenty days if above one hundred miles; or within thirty days if above two hundred miles; and shall be allowed the usual conduct money, and also two months wages advance, at the first fitting out of the ship, and before the ship proceeds to sea.

II. Every supernumerary man serving ten days in any ship, shall be borne for and entitled to his wages, upon the books of such ship: and to all other benefits, as if he was part of the complement of such ship; but men lent from one ship to another, shall continue to be borne for, and entitled to their wages upon the books of the ship from which they were lent, until they shall be regularly discharged from thence, and in no other.

III. Every inferior officer or seaman who shall be turned over from one ship to another (in case the ship into which he is turned over, is then, or shall come into a port of Great Britain where there is a commissioner of the navy) shall be paid by proper pay-lists, all the wages due to him in the ship from which he was so turned over, before the ship into which he shall be turned over proceeds to sea, unless it shall be otherwise directed by special order from the Admiralty, in cases of the greatest exigency only; and if the ship, in pursuance of such order, proceeds to sea before such payment can possibly be made, the wages shall be paid as soon as such ship shall come again into any port of Great Britain, where there is a commissioner of the navy.

IV. Every officer or seaman who shall be turned over from one ship to another, shall not serve or be rated in a worse quality or lower degree than he served in or was rated for in the former ship; and shall have an advance of two months wages before the ship into which he is turned over proceeds to sea, in case he should not have received such advance before.

V. Such sums of money shall, in the first place be appropriated, and shall from time to time be issued and applied out of the supplies granted, or to be granted for any naval services, as shall be sufficient for the regular payment of all tickets made out pursuant to the act, and for the regular discharge of all wages due, or to grow due in manner following; that is to say, as often as any ship shall have been in sea-pay twelve months or more, shall be or arrive in any part of Great Britain, or the coast thereof, the captain or commander shall immediately cause five complete pay-books to be made out for all the time such ships shall have been in pay, except the last six months, and shall forthwith transmit, by the first safe opportunity, such books, together with three alphabets and a slop-book, to the commissioners of the navy at their board; and as soon as such ship shall be or arrive in any port of Great Britain where there is a commissioner of the navy, the said commissioners of the navy shall solicit the necessary sums of money, and shall cause immediate payment to be made of the wages due, deducting the advance-money, and all defalcations; leaving always six months wages unpaid, and no more; and all the wages due upon any ship shall be paid as soon as may be, or within two months at farthest, after the arrival of such ship in the port to be laid up.

VI. The month shall consist of twenty-eight days.

VII. Upon application to the commissioners of the navy at their board, by any inferior officer or seaman, who shall then be in the service, and was absent at the payment of the ship to which he did belong, or from the captain or commander of any ship in which such officer or seaman shall then serve, in case such ship shall be in any port of Great Britain where there is a commissioner of the navy, the commissioners of the navy at their board shall immediately send the pay-books or pay-lists made out from them to such commissioner, who shall forthwith cause the wages to be paid to such officer or seaman.

VIII. The captain or commander shall make out a ticket upon the death of every inferior officer or seaman, and shall transmit the same by the first safe opportunity to the commissioners of the navy at their board, who are to assign the same for payment within one month after the receipt thereof; and the same shall be delivered, and payment thereon made, without fee or reward, to the executors or administrators of such officer or seaman, or to the attorney of such executors or administrators.

IX. The captain or commander shall make out a ticket for every inferior officer or seaman who shall be discharged as unserviceable, pursuant to the directions of the act, and shall send such ticket by the first safe opportunity to the commissioners of the navy at their board. The captain or commander shall not deliver such ticket to such officer or seaman, but shall give him a certificate of such discharge, containing an exact copy of the ticket, and a description of his person. The said commissioners of the navy shall immediately cause the day when such ticket was received to be indorsed thereon, and shall assign the same for payment within one month at farthest, from the day of making such indorsement; and if any inferior officer or seaman shall present such certificate at the Navy Office, the said commissioners of the navy are forthwith to examine such certificate, and the person presenting the same: and being satisfied that the certificate was made out for such person, and that he is rendered unserviceable, they shall testify the same on such certificate. The said ticket being so assigned, shall be immediately delivered, and payment made at the Pay Office of the navy, to such officer or seaman, without fee or reward, and to no person whatsoever. If the ticket shall not have been transmitted to, and received by the commissioners of the navy, the money appearing to be due by the copy of the ticket in the certificate, shall be paid in like manner as if the ticket had been received. Such officer or seaman being desirous to receive his wages at any port in Great Britain, where a commissioner of the navy resides, may produce his certificate to such commissioner of the navy, who, being satisfied that such certificate was made out for such person, and that he is rendered unserviceable, shall sign and transmit the same to the commissioners of the navy at their board; who within four days after the receipt of such certificate, are to send the ticket for such officer or seaman; or if such ticket shall not have been transmitted to and received by them, then the said certificate, containing a copy of the said ticket, instead thereof to the commissioners at such port, who shall cause immediate payment thereon to be made, without fee or reward; and such commissioner shall send such officer or seaman to the nearest hospital, where he is to be received and victualled, from the time of presenting such certificate until payment is made. If any such certificate shall be lost or destroyed, or such officer or seaman shall not present the same in person, or the money due on any such ticket shall not be paid before the general payment of the ship's company, the ticket shall be cancelled, and the wages be payable as if such ticket and certificate had not been made out.

X. When any inferior officer or seaman shall, by order of the captain or commander be sent sick ashore, and be sent into any hospital, or sick quarters, such captain or commander shall make out a sick ticket for the wages due to such officer or seaman, and transmit the same with such officer or seaman to the hospital or sick quarters, and if such officer or seaman shall be regularly discharged from thence as unserviceable, a certificate of his discharge, with the sick ticket annexed thereto, shall be delivered to him; and if he shall present the same to a commissioner at any port in Great Britain, such commissioner being satisfied that such officer or seaman is unserviceable, shall sign the same on the certificate, and forthwith transmit such certificate and sick-ticket to the commissioners of the navy at their board, who within four days after the receipt thereof are required to cause a proper ticket or pay-list to be made out for the wages due to such officer or seaman, and to send such ticket or pay-list, without delay, to the commissioner at such port, who shall cause immediate payment of the wages to be made, without fee or reward, to such officer or seaman, who, notwithstanding such discharge, shall be maintained in such hospital or sick-quarters, from the time he shall present the certificate and sick-ticket, until the payment shall be made.

XI. The payment of tickets, certificates, or pay-lists, shall not be delayed, though the muster or pay-books be not regularly sent to, and received by, the commissioners of the navy; but if any error shall be made in any ticket, certificate or pay-list, the loss shall be made good out of the wages of the captain or commander by whom such ticket, certificate, or pay-list was made out.

XII. As often as any ship which shall not be in any port of Great Britain, or on the coast thereof, shall have twelve months wages due, the captain or commander shall cause the names of all the inferior officers and seamen to be called over, and each to answer to his name; and shall do the same at the end of every six months, when twelve months or more wages shall be due; and if any such officer or seaman shall then declare, or deliver in writing the name and place of abode of his wife, father, or mother, and desire that the whole or any part of his wages then due, except the wages due for the last six months, shall be paid to such wife, father, or mother, by the receiver-general of the land-tax for any county, riding, or city; or collector of the customs for any port; or collector of the excise for any collection in Great Britain; or the clerk of the check at any dock-yard; the captain or commander is strictly required to cause four lists to be made out of the persons so desiring to make such remittances, and shall transmit such lists, by the first safe opportunity, without delay, to the commissioners of the navy at their board; who, on the receipt thereof, shall immediately make out two bills, pursuant to the directions of the act, for the payment of the wages so allotted by each person; one of which bills shall be sent to the persons respectively specified in such lists, and the other to such receiver, collector, or clerk of the cheque; and if the person to whom any such bill is sent, shall, within six months from the date thereof, produce and deliver the same to such receiver, collector, or clerk of the cheque, together with a certificate that such person is the wife, father, or mother of such officer or seaman respectively, under the hands of the minister and churchwardens; or, in Scotland, of the minister and two elders of the parish where such person was married or resides; such receiver, collector, or clerk of the cheque, upon being satisfied of the truth of such certificate, is immediately, without fee or reward, to pay the sum mentioned in such bill, taking a receipt; such bill, together with the duplicate thereof, being produced at the Navy Office, shall be immediately assigned for payment by the commissioners of the navy, and repaid by the treasurer of the navy to such receiver, collector, or clerk of the cheque, or their order respectively. But if payment of the said bill be not demanded of such receiver, collector, cr clerk of the cheque, and the duplicate of the said bill, together with a proper certificate, be not produced and delivered to them respectively, within six months after the date thereof, the bill is to be returned and cancelled, and the sum contained therein is to become payable to such inferior officer or seaman, when the ship shall be paid.

XIII. In like manner, when wages shall be paid at the Pay Office, or any of the out ports, if any inferior officer or seaman shall be desirous to remit the whole, or any part of his wages, to his wife, children, parents, or any other person, and to have a bill drawn for the same upon any such receiver, collector, or clerk of the cheque, two bills are to be made out, one of which is to be delivered to such officer or seaman, and the other to be sent to such receiver, collector, or clerk of the cheque; who is immediately to pay the sum therein mentioned, without any fee or reward, taking a receipt; such bill, together with the duplicate thereof, being produced at the Navy Office, shall be immediately assigned for payment by the commissioners of the navy and be repaid by the treasurer of the navy to such receiver, collector, or clerk of the cheque, or their order respectively. But if payment of the said bill be not demanded of such receiver, collector, or clerk of the cheque; and the duplicate of the said bill shall not be produced and delivered to them respectively within six calendar months from the date thereof, the bill is to be returned and cancelled, and the sum contained therein is to be immediately paid to such officer or seaman or other person duly authorized to receive the same.

XIV. If any such receiver, collector, or clerk of the cheque, shall not have in his hands money to answer any bill tendered by such wife, father, or mother, as aforesaid, and shall refuse or delay the immediate payment thereof, he is immediately to indorse on such bill the cause of his refusal or delay, and appoint for payment some day within two months after such tender; and if, upon complaint to the commissioners respectively appointed to manage the land-tax, customs, or excise, or to the commissioners of the navy respectively, it shall appear that any such receiver, collector, or clerk of the cheque, hath unnecessarily and wilfully refused or delayed payment; or that he, or any person employed by or under him, hath taken any fee, reward, gratuity, discount, or deduction, on account of payment of any such bill, such commissioners may fine such offender in any sum not exceeding fifty pounds.

XV. The wages, pay, and allowances earned by an indentured apprentice, shall be paid to his master, as hath been usual, unless such apprentice was above eighteen years of age when his indentures were executed, or shall be rated as a servant to any officer to whom such apprenticeship is not known; in which case, such officer shall be entitled to the pay and wages of such servant, according to the usual practice of the navy, until such officer shall be informed of such apprenticeship.

XVI. Captains or commanders of ships are to transmit from time to time to the Commissioners of the Navy at their board, complete pay-books, lists, and tickets; and also once in every two months, complete muster-books, under the penalty of forfeiting all their wages to the chest at Greenwich, and of being liable to such further punishment as a court-martial shall inflict, except in cases of necessity, to be made appear to the satisfaction of the Lord High Admiral, or the Commissioners of the Admiralty.

XVII. The tickets, certificates, pay lists, and bills, shall be sufficient vouchers to the Treasurer of the Navy for the payment thereon.

XVIII. Captains or commanders issuing any tickets, other than such as are directed by the Act, shall be subject to a penalty of fifty pounds for every ticket, and shall forfeit all their wages to the chest at Greenwich.

XIX. No captain or commander shall be liable to any penalty for any offence against the Act, before the first of June, One thousand seven hundred and fifty-nine, unless he hath previously received this Abstract.

XX. No letters of attorney made by any inferior officers or seamen, or their executors or administrators, for receiving wages or allowances of money, shall be valid, unless declared therein to be revocable;

and unless the same, if made by any such officer or seaman then in the service, be signed before and attested by the commander, and one other of the signing officers of the ship, or by a clerk of the chequer; and if made by any such officer or seaman after his discharge from the service, unless the same be signed before, and attested by, the mayor, or chief magistrate of the town or place where such officer or seaman shall then reside: or unless the letter of attorney, if made by such executors or administrators, be signed before and attested by the minister and churchwardens; or in Scotland, by the minister and two elders of the parish where such executors or administrators reside.

XXI. All letters of attorney, other than such as are made in manner aforesaid, and all bargains, sales, bills of sale, contracts, agreements, and assignments, concerning wages or money to inferior officers or seamen, shall be null and void.

XXII. No more than one shilling shall be taken by any ecclesiastical court, or other person, for the probate of any will, or letters of administration, granted to the widows, children, fathers, mothers, brothers, or sisters of inferior officers, seamen, or marines, dying in the service, and for the pains, trouble, and expence attending the same, unless the goods and chattels are of the value of twenty pounds, nor more than two shillings, unless such goods and chattels are of the value of forty pounds; nor more than three shillings, unless such goods and chattels are of the value of sixty pounds; nor more than one shilling for issuing commissions to swear such widows, children, fathers, mothers, brothers, or sisters, being executors or administrators to inferior officers, seamen, or marines, and for the pains, trouble, and expense attending the same, unless the goods and chattels are of the value of twenty pounds; nor more than two shillings, unless the goods and chattels are of the value of forty pounds; nor more than three shillings, unless the goods and chattels are of the value of sixty pounds, under the penalty of fifty pounds, to be paid by the offender to the party aggrieved.

XXIII. Whoever willingly and knowingly shall personate or falsely assume, the name or character of, or procure any other to personate or falsely assume the name or character of any officer, seaman, or other person intitled to wages, pay, allowances, or prize-money, for service done on board of any ship in the royal navy; or the executor, administrator, wife, relation, or creditor, of any such officer, seaman, or other person, in order to receive any wages, pay, allowances, or prize-money; or shall forge or counterfeit, or procure to be forged or counterfeited, any letter of attorney, or other power or authority whatsoever, in order to receive any wages, pay, allowances, or prize-money; or shall willingly and knowingly take a false oath, or procure a false oath to be taken, to obtain the probate of a will, or letter of administration, in order to receive any wages, pay, allowances, or prize-money, shall be guilty of felony, and suffer death.

XXIV. When the pay books are closed, tickets shall be made out at the Navy Office, to the seamen who shall not have received their wages, and such tickets shall be paid in course once a month.

XXV British governors, ministers, and consuls, residing at foreign parts, or where no such are present, any two British merchants are required to provide for seafaring men and boys, subjects of Great Britain, who, by shipwreck, capture, or other unavoidable accident, shall be in foreign parts, or who shall be discharged there as unserviceable, from the royal navy, and subsist them at sixpence per diem each, and send them home as soon as conveniently may be, in any ship belonging to the royal navy, or in any merchant ship.

XXVI. Masters of ships shall be allowed sixpence per diem for all such men and boys as shall exceed their complement.

XXVII. Seamen shall not be taken out of the service for any debt under twenty pounds.

XXVIII. But creditors may file a common appearance, so as to entitle them to proceed to judgment and outlawry, and to have an execution thereupon, except against the bodies of such seamen.

XXIX. Receivers of seamen's wages taking more than sixpence in the pound, shall for every offence forfeit fifty pounds; and if any such offender be a clerk, officer, or servant in an office belonging to the navy, he shall also lose his place, and be incapable of holding any place of profit in any such office.

XXX. Clerks, officers, and servants in offices belonging to the navy, taking fees (not allowed by the Act) for doing any thing directed by the Act, shall be subject to the same forfeitures.

XXXI. This Abstract, and also the Articles of War, are to be printed and hung up in the most public place in every ship of the royal navy; and they are also to be read over once in every month to the ship's company, and the reading of them is to be attested by the signing officers at the foot of the muster books; to the end that every seaman in the royal navy may know the punishments he is liable to for any neglect or disobedience, and the encouragements he is entitled to by the performance of his duty; and that upon suffering any injury, he may be enabled to lay his complaint before the Lord High Admiral, or the Commissioners of the Admiralty, who are by this Act directed to enquire into the same, and to grant redress if such complaint be justly founded; and to take care that this Act be punctually carried into execution.

Abstract of an Act of Parliament passed in the Twenty-sixth Year of the Reign of King George the Third, intituled, "An Act for the further preventing Frauds and Abuses attending the Payment of Wages, Prize-Money, and other Allowances due for the Service of Petty Officers and Seamen on Board any of His Majesty's Ships."

From and after the 1st of August, 1786, no letter of attorney made by any petty officer or seaman in his Majesty s service, or by the executors or administrators of any such petty officer or seaman, to empower any person to receive any wages, pay, or other allowances of money, due for such service, shall be valid, unless declared to be revocable by the express words thereof; and that no letter of attorney, or will, made by such petty officer or seaman, whereby any wages, pay, prize-money, or other allowances of money due for such services, is authorized to be received, or bequeathed, shall be valid, unless the same (if made when the party is in his Majesty's service) shall be signed before, and attested by, the captain or commanding officer, and one other of the signing officers of the ship to which he belongs, and shall specify in the body thereof, the name of the ship, and number at which the maker of the will, or letter of attorney, stands upon the ship's books, or by the agent of any of his Majesty's hospitals or quarters appointed to receive sick and wounded seamen, in which the party shall be at the time, and unless such letter of attorney or will, if made by a petty officer or seaman discharged from his Majesty's service, or if such letter of attorney is made by the executors or administrators of any such officer or seaman, and made within the bills of mortality of London and Westminster, is attested by an officer to be appointed by the Treasurer of his Majesty's navy, for the purpose of inspecting such wills and letters of attorney; or if made at any of the ports where seamen's wages are paid, is attested by the Treasurer of the navy's chief or second clerk there; or if made at any other place, is attested by the minister and church-wardens of any parish in England or Ireland; or in Scotland by the minister and two elders of the parish, where such petty officer or seaman, executors or administrators, shall respectively reside.

Every such letter of attorney and will shall contain the ship's name to which the person granting the same last belonged; and also the full description of the residence, profession, or business, of the person to whom the same is made, and also the day of the month and place where executed.

After such letter of attorney or will shall be executed and attested as above, the same shall not be delivered to such party himself, or to any person on his behalf; but if executed abroad, shall be, with all convenient speed, sent by the commander of the ship, or agent of the hospital or sick quarters, at the times when they transmit their returns to the navy, and sick and hurt boards; or if executed in Great Britain or Ireland, shall be sent by the commander of the ship, agent of the hospital or sick quarters, treasurer's clerk, minister of the parish, or whoever of them shall attest such letter of attorney, or will, by the general post, addressed to the Treasurer, or Pay-Master of the navy, at the Navy Pay-Office, London.

The Treasurer or Pay-Master shall immediately deliver over the same to the Inspector, who shall immediately, on the receipt thereof, register the same in a book to be kept for that purpose: and in case it shall appear to him that such letter of attorney or will is not genuine, he shall not pass the same, but shall give notice, by the general post, to the attorney or executor, informing him or them, that the letter of attorney or will is stopt, and the reason thereof: But if it shall appear to the Inspector, that the letter of attorney or will is genuine, he, or a person authorized to officiate for him, shall sign his name thereto, and put a stamp thereon, in token of his approbation: and every such letter of attorney shall be kept as one of the vouchers of the Treasurer of the navy's accounts: and the Inspector shall immediately after such approbation give notice, by the general post, to the attorney, that he has received and approved of the same; and he shall send to such attorney a checque, signed and stamped by the Inspector or his assistant; which checque shall authorize him to demand payment of, and discharge all wages, pay, prize-money, or allowance of money, to which the seaman was entitled for his service on board any of his Majesty's ships. The Inspector shall give like notice to the executors of wills when they are approved of, and shall register the same, and make out the checques in manner above directed, with respect to letters of attorney, and forward such checques in like manner to the executors; upon the authority of which, they, or their attornies, may apply, upon the testator's death, to the Inspector, requesting that the will may be sent by him to a proctor in Doctors' Commons, where they may, on application, obtain probate thereof; which probate, when obtained, shall be lodged with the Inspector; and he or his assistant shall certify upon the checque formerly delivered, that a probate has been granted; and the checque shall then authorize the executors to demand payment of, and discharge all sums due to him, as executor to the party who made the will.

The Inspector shall, in return to all letters of attorney and wills received by him from ministers of parishes, give notice as aforesaid to the said minister who transmitted the same, and not to the grantor thereof, of his having passed and approved of such letter of attorney or will, and send the checque by the general post, made out as above, to such minister, and which notice shall be addressed to the minister of the parish (naming the same) without inserting the minister's name, and be delivered to him at his manse or dwelling house; and every such minister shall deliver such checque to the party who executed such letter of attorney, or will; and all letters and packets addressed to or sent by the Treasurer or Pay-Master of the navy, or Inspector, shall be sent and received free of postage.

All captains and commanders of ships shall, upon their monthly muster books or returns, specify which of the men mentioned in such returns have granted any letter of attorney during that month, or space of time from the preceding returns, by inserting the date thereof opposite to the party's name.

WAGGONER, in astronomy, is the constellation Ursa Major, or the Great Bear, called also vulgarly Charles's Wain.

WAGGONER, is also used for a routier, or book of charts, describing the seas, their coasts, &c.

WAIST, (*le vibord d'un vaisseau, à l'endroit des passe-avants, ou l'entre—deux des gaillards; c'est aussi ce même espace, sur le pont supérieur,* Fr.) that part of a ship which is contained between the quarter-deck and fore-castle, being usually a hollow space, with an ascent of several steps to either of those places.

When the waist of a merchant-ship is only one or two steps of descent from the quarter-deck and fore-castle, she is said to be galley-built; but when it is considerably deeper, as with six or seven steps, she is called frigate-built,. See the articles DECK, DEEP-*Waisted*, and FRIGATE.

WAIST-CLOTHS, (*prêlats des passe-avants,* Fr.) coverings of canvas or tarpauling for the hammocks, which are stowed on the gang-ways, between the quarter-deck and fore-castle.

WAISTERS, (*matelots assignés au service de la grande rue*, Fr.) a name given to the men stationed in the waist in working the ship. As they have little to do besides pulling and hoisting, their business requires only strength, without art or judgment, they are therefore commonly selected from the strongest landsmen and ordinary seamen.

WAKE, (*les eaux du bâtiment*, Fr.) the print or track impressed by the course of a ship on the surface of the water. It is formed by the re-union of the body of water, which was separated by the ship's bottom whilst moving through it; and may be seen to a considerable distance behind the stern, as smoother than the rest of the sea. Hence it is usually observed by the compass, to discover the angle of lee-way.

A ship is said to be in the wake of another, when she follows her upon the same track, or on a line supposed to be formed in the continuation of her keel. Thus the ships *a b*, fig. 3, and A B, fig. 6, plate XV are all in the wake of the foremost, *b*. See the article Line.

Two distant objects observed at sea, are termed in the wake of each other, when the view of the farthest is intercepted by the nearest, so that the observer's eye and the two objects are all placed upon the same right line.

WALES, (*préceintes*, Fr.) an assemblage of strong planks, extending along a ship's side, throughout her whole length, at different heights, and serving to reinforce the decks, and form the curves by which the vessel appears light and graceful upon the water.

As the wales are framed of plank, broader and thicker than the rest, they resemble ranges of hoops encircling the sides and bows. They are usually distinguished into the main-wale and the channel-wale; the breadth and thickness of which are expressed by Q and R in the midship-section, plate VI. and their length is exhibited in the elevation, plate XVII, where Q R is the upper and lower edges of the main-wale, and S T the upper and lower edges of the channel-wale, parallel to the former.

The situation of the wales being ascertained by no invariable rule, is generally submitted to the fancy and judgment of the builder; but the position of the gun-ports and scuppers ought to be particularly considered, that the wales may not be wounded by too many breaches.

WALL KNOT, (*cul-de-porc*, Fr.) a particular sort of large knot raised upon the end of a rope, by untwisting the strands, and interweaving them amongst each other.

Wall-Sided, (*vaisseau qui à le côté droit comme un mur*, Fr.) the figure of a ship's side, when, instead of being incurvated, so as to become gradually narrower towards the upper part, it is nearly perpendicular to the surface of the water, like a wall. This was formerly called *wale-reared*.

WAR, (*bellum*, Lat. *guerre*, Fr.) a contest or difference between princes, states, or large bodies of people; which no being determinable by the ordinary measures of justice and equity, is referred to the decision of the sword.

There are four different kinds of war, viz. the offensive; the defensive; that between equal powers; and the auxiliary, which is carried on out of our own territories, to succour a prince or ally, or to assist a weaker whom a more powerful prince has attacked.

Offensive war is the act of invading an enemy's country or state, or the commencement of hostilities within an enemy's territory.

Defensive war is the act of defending a country, &c. when invaded by an enemy.

Auxiliary war is that in which a prince succours his neighbours, either in consequence of alliances, or engagements entered into with them, or sometimes to prevent their falling under the power of an ambitious prince.

As war, on the one hand, in respect to its effects, is intimately connected with the prosperity and independence of nations; so, on the other, it requires infinite skill, combination, and management, when considered as an art founded on the sciences, and men only of extraordinary talents are found to excel.

As two different elements constitute the theatre of its operations, war is naturally divided into naval and military arrangements.

Naval Warfare, is the art of arranging squadrons or fleets in order of battle, and regulating their movements in such a manner as may be deemed best calculated for attacking, defending, or retreating to the greatest possible advantage, and is termed *Naval* Tactics, which article see.

Our naval warfare has for many years been conducted with superior skill and undaunted courage, unprecedented by any nation in Europe, as will appear by looking back to the memorable victories obtained by our naval heroes over the fleets of our enemies;—the French, at La Hogue, in 1692; off Ushant, in 1747; off Cape François, in 1757; at Brest, in 1758; and in the West Indies, in 1782:—the Spanish fleet, at Cape St. Vincent, in 1797; and the Dutch, at the battle of Camperdown, in 1797; the French, in the battle of the Nile, in 1798; the Danish fleet off Copenhagen, in 1801; and the combined fleets of France and Spain, at Trafalgar, in 1805, &c.; and in short, in every quarter of the globe, till their ships, colonies, and commerce, were almost annihilated.

There is a Naval College established in Portsmouth dock-yard, where professors and masters of the first abilities are appointed to instruct our young naval officers in the principles of their profession; and an extensive private Naval Academy, for the same purpose, at Gosport.

Military Warfare, is the art of arranging armies in the order of battle, and of regulating their movements in such a manner, as may be deemed most proper for possible advantage.

Although war, as a science, has of late years been certainly carried to a great degree of perfection, yet it has varied but little in its principles; for the maximum of the art is to bring troops to attack with the bayonet, in the same manner as they were accustomed to do with the pike some centuries ago; and the British soldiers, by their conduct in Flanders, Egypt, Spain, Portugal, and France, have lately proved, that hand to hand they still preserve their antient reputation; and now, as of old, they stand unequalled at a charge.

As war is to be regulated by certain rules and principles, and that on the knowledge and application of these depends the fate of a campaign, and perhaps of a nation, it necessarily follows, that a general ought to possess extraordinary talents and attainments. Accordingly military seminaries for the instruction of those destined to become officers, have long been established in this country; at Woolwich, for the engineer and artillery department, and at Farnham and Sandhurst for the line; able professors and masters have also been provided, and the arts and sciences connected with war, are now publicly taught. The genius of the nation has also of late years been directed in an eminent degree to military affairs, by the institution of the volunteers; and it seems now to be established as a principle, that an army is as necessary as a fleet, both for offensive and defensive war.

The Duke of Wellington, field-marshal of the allied army, viz. the English, Spanish, and Portuguese troops, has most eminently distinguished himself in conducting the late war on the Continent, which lasted five hard campaigns, during which time it is remarkable that no circumstances happened which could throw a shade once over the military glory of the country, or his own personal character. It was a new feature that this great general had conducted these five campaigns through such extraordinary difficulties in the face of the first armies of Europe, headed by some of the greatest commanders which the French revolution had brought forth; and that in every instance he should have been victorious. The field-marshal had inspired confidence in all the nations whose troops he commanded, and had risen to the highest commands in their service; and so active was he in his exertions, that he received the thanks of Parliament thirteen different times for his distinguished victories, and for those great exploits which have so mainly contributed to the restoration of peace, and the happy deliverance of Europe.

The following are the most remarkable wars in which England has been engaged, since the

War with	Scotland	1068.
Peace with	Ditto	1113.
	France	1113.
War with	France	1116.
Peace with	Ditto	1118.
	Scotland	1139.
War with	France	1161.
Peace with	Ditto	1186.
Successful war with France		1194.
Peace with	Ditto	1195.
Civil war	renewed ....1215,—ended	1216.
	with France ..1224,—ended	1234.
	1262,—ended	1267.
Civil war with	France	1294.
	Scotland	1296.
Peace with	France	1299.
	Scotland	1323.
War with	Scotland	1327.
	ended	1328.
	Scotland	1333.
	France	1339.
Peace with	France, May 8,	1360.
War with	France	1368.
	civil	1400.
	Scotland	1400.
Peace with	France, May 31,	1420.
War with	France	1422.
	civil, between York and Lancaster	1452.
Peace with	France, October	1471.
War	civil	1486.
	with France, October 6	1492.
Peace with	Ditto, Nov. 3	1492.
	Scotland	1502.
War with	France, Feb. 4	1512.
	Scotland	1513.
Peace with	France, August 7	1514.
War with	Ditto	1522.
	Scotland	1522.
Peace with	France	1527.
	Scotland	1542.
War with	Ditto, immediately after.	
Peace with	France and Scotland, June 7	1546.
War with	Scotland	1547.
	France	1549.
Peace with	both, March 6	1550.
Civil war		1553.
War with	France, June 7	1557.
	Scotland	1557.
Peace with	France, April 2	1559.
	Scotland	1560.
War with	France	1562.
Peace with	Ditto	1564.
War with	Scotland	1570.
	Spain	1588.
Peace with	Ditto, August 18	1604.
War with	Spain	1624.
	France	1627.
Peace with	both, April 14	1629.
Civil war		1642.
War with the	Dutch	1651.
Peace with	Ditto, April 5	1654.
War with	Spain	1655.
Peace with	Ditto, Sept. 10	1660.
War with	France, Jan. 26	1666.
	Denmark, Oct. 19	1666.
Peace with the	French, Danes, and Dutch, August 24	1667.
Peace with	Spain, Feb. 13	1668.
War with the	Algerines, Sept. 6	1669.
Peace with	Ditto, Nov. 19	1671.
War with the	Dutch, March	1672.
Peace with	Ditto, Feb. 28	1674.
War with	France, May 7	1689.
Peace, general, of Rhyswick, Sept. 20		1697.
War with	France, May 4	1702.
Peace of	Utrecht, March 13	1713.
War with	Spain, Dec.	1718.
Peace with	Ditto	1721.
War with	Spain	1739.
	France, March 31	1744.
War with	France	1756.
	Spain, Jan. 4	1762.
Peace with	both, Feb. 10	1763.
War with the	Caribbs of St. Vincent, in	1773.
Civil war in	America, commenced July 14	1774.
War with	France, Feb. 6	1778.
	Spain, April 17	1780.
	Holland	1780.
Peace with	France, Spain, Holland, and America, Sept. 3	1783.

War with France; by the English, Prussians, Austrians, and other German powers, in.........1793.
Peace between Prussia and the French Republic 1795.
Peace between Spain and the French Republic..1795.
Peace between the French and the Sardinians, in 1796.
Peace between the French and the Austrians, in 1797.
War between the British and Tippoo Saib, in India......................1797.
War with the French Republic, by the Austrians, Russians, Neapolitans, &c.........................1798.
War with the Turks, and the invasion of Egypt, in.........................1798.
Peace between the French and Russians, in....1799.
Peace between the French and Austrians, in ...1800.
Preliminaries of peace commenced between the French and the Ottoman empire, in consequence of the reduction of Egypt, by the British forces, in................1801.
Preliminaries of peace between France and Great Britain, &c. in.... ....1801.
Peace between France and Great Britain .....1802.
War with France.......................1803.
The Americans declare war against Great Britain, June 18...............1812.
Preliminaries of peace between Great Britain, France, &c. signed May 30 ..1814.

WARDENS of His Majesty's Dock-Yards, are generally old lieutenants in the royal navy, appointed by the Navy Board, to see that no person whatever be admitted without an order from the commissioner, except whom he knows, or those who are employed in the yards.

WARD-ROOM, (*la grande chambre*, Fr.) in ships of war, a room over the gun-room, where the lieutenants, and other principal officers, sleep and mess.

To WARE, or Wear, (*virer vent arrière*, Fr.) is to cause a ship to change her course from one board to another, by turning her stern to the wind: hence it is used in the same sense as To Veer, which article see.

WARP, (*cablot, grelin ou haussière servant à remorquer, ou touer un bâtiment; on dit aussi une touée*, Fr.) a rope or hawser, employed occasionally to remove a ship from one place to another, in a port, road, or river: hence,

To Warp, (*touer*, Fr.) is to change the situation of a ship, by pulling her from one part of a harbour, &c. to some other, by means of warps, which are attached to buoys; to other ships; to anchors sunk in the bottom; or to certain stations on the shore, as posts, rings, trees, &c. The ship is then drawn forward to those stations, either by pulling on the warps by hand, or by the application of some purchase, as a tackle, windlass, or capstan.

When this operation is performed by the ship's lesser anchors, these machines, together with their warps, are carried out in the boats alternately, towards the place where the ship is endeavouring to arrive; so that when she is drawn up close to one another, the other is carried out to a competent distance before her, and being sunk, serves to fix the other warp, by which she may be farther advanced.

Warping is generally used when the sails are unbent, or when they cannot be successfully employed, which may either arise from the unfavourable state of the wind, the opposition of the tide, or the narrow limits of the channel.

To Warp *Yarn*, in rope-making, is to stretch the yarns, previous to their being tarred, all to one given length, which, in full length rope-grounds, is 200 fathoms. The usual method is to warp the yarn, either in whole or half hauls, and is done by putting the number of threads intended to be drawn down at once in a bight, into a block with one sheave, the one end being made fast at the upper end, which being drawn down and fixed over the end of a hook, made fast to a post at 100 fathoms from the warping post, forms, when opened, a length of 200 fathoms, as before-mentioned. The number of half bights, or blocks of yarn, contained in a whole or half haul, is in a great measure governed by the size of the yarn intended to be warped; as for instance, 16 a 19-thread yarn are warped 336 threads in a haul, 20 a 25-thread yarn are warped 400 threads in a haul. In winding the yarns on the winches after it is spun, it is most usual to wind them on in companies of four in a company; but as that method cannot always be pursued, it frequently happens that whole or half hauls of yarn are obliged to be warped in half bights of an odd number.

Warp *of Shrouds*, the first given length, taken from the bolster at the mast-head to the foremost dead-eye.

WARPED into Junks, in rope-making, is expressed of yarn warped into short lengths for spun-yarn.

Warping *Blocks*, in rope-making, are used to warp the yarn into hauls for tarring. The tops and bottoms are made to separate, in order to let in the sheaves and screw down.

Warping-Hook, for hanging the yarn on, when warping into hauls for tarring, is a large iron hook, hung occasionally to the warping-posts.

Warping-Post, a post, 14 or 16 inches in diameter, fixed in the middle of the ground for warping the yarn into hauls.

WARRANT, (*espèce de brevet que l'on donne aux maîtres, &c. de différentes classes sur les vaisseaux de guerre Anglois*, Fr.) the name given to a sort of commission or authority to those officers appointed by the Navy-Board, while the authorities granted by the Admiralty are styled commissions.

Press Warrant, (*brevet que donne l'Amirauté pour autoriser capitaines ou lieutenans de vaisseau à presser les matelots*, Fr.) a warrant issued by the Admiralty, on cases of emergency, authorizing the captain or lieutenant of the impress service at any sea-port, to impress such seamen belonging to the merchant-service, as may be found skulking about, or unemployed.

Warrant-Officers, (*titre donné, sur les vaisseaux de guerre Anglois, au maître-pilote, au chirurgien-major, au commis aux vivres, et aux maîtres des différentes classes*, Fr.) are persons employed in the royal navy, by warrant from the Commissioners of the navy, to take charge of the stores issued to them from his Majesty's dock-yards, and to see that they are not converted into improper uses; also to direct and superintend their crews in the performance of the requisite duties of the ship, and transmit annually to the Navy Board a faithful account of the expenditure of their stores. The gunner, however, obtains his warrant from the Board of Ordnance, to whom he transmits his annual expenditure of stores. For farther particulars of the relative duties of warrant officers, see the articles MASTER, SURGEON, PURSER, BOATSWAIN, GUNNER, CARPENTER, &c.

General Instructions for WARRANT OFFICERS. The warrant officers of his Majesty's ships in ordinary, are frequently to examine the store-rooms, and report their defects to the master-shipwright of the dock-yard, that they may be repaired, in order to receive the stores whenever the ship may be put into commission; and to see that the stores received on board, are good in quality, and the proper quantity; and to report to the captain any defect or deficiency which they may discover in them.

They are to indent for all the stores they receive, and be careful that they do really receive all they indent for.

They are likewise to keep an account of the receipt, expenditure, &c. of stores, according to a form, specifying the place where, and the persons from whom, the stores are received, or the person to whom they are supplied.

Interlineations in the accounts of stores expended are not allowable: but any omission is to be represented to the master, who being satisfied with the truth of such representation, he is to allow it to be inserted in the account of expenses for the next month.

No waste for stores not perishable, will ever be allowed, unless from unavoidable accidents, which must be certified by two of the principal officers present at such accident: but for stores damaged by neglect or misconduct of any officer, their value shall be charged against the wages of such officer.

Every officer must keep his own accounts, and shall be held responsible for the conduct of his yeoman.

Every officer shall be responsible for errors in his accounts, and shall pay out of his wages the full value of all stores not accounted for, or improperly expended.

No officer shall suffer his yeoman to take stores from the store-rooms, without his express order.

When they are supplied with stores by other officers, whether of the same ship or of any other, they are to charge themselves with those stores, and are to mention their having done so in the receipt they give for them: but one officer is not to supply another with stores without an order from the captain in writing, and he is to demand a receipt and to mention by whose order they were supplied.

Officers, when appropriating rope, canvas, or any other article to use, must be attentive to conform to the established length, and other dimensions, of whatever it may be intended to make, and when they convert stores into any other use than that for which they were originally intended, they are to expend them in their accounts, as having been so converted, and are to charge themselves with whatever they convert them into.

They are to be particular in stating the quantity of masts, sails, colours, &c. that may be lost or blown away; and when stores are damaged or worn out, they are to apply to the captain for a survey, and to apply them to whatever use the surveying officers shall appoint; they are to visit their store-rooms frequently, to see that they are kept clean, in order, and well aired.

Officers are never to carry lights into the store-rooms, except in good lanterns.

They are strictly charged, never to put into the magazine, the wings, nor store-rooms, any wine or spirituous liquors, except such as the captain shall expressly permit.

When a ship is to be dismantled, either for the purpose of being refitted, or paid off, they are to be particularly careful in preventing their stores, rigging, &c. from being cut or damaged in sending them to the store-houses.

When a warrant officer is removed from a ship, or wishes to pass his accounts, he is to apply to the captain for a survey on his stores: but when a warrant officer dies, the captain is immediately to apply to the commanding officer present, to order, or if the ship be alone, he is himself to order a survey on the stores remaining on board; one copy of the report of such survey is to be sealed up with the papers of the officer who died, and another copy is to be delivered to his successor, and considered as his first charge.

Warrant officers are charged to survey stores with the strictest attention, so that when called on, they may be able conscientiously to make oath to the correctness of the report they have made. In surveying stores they are to point out, to what other service they may be appropriated.

When ordered to survey stores for the purpose of ascertaining their quantity, whether to enable the officer in whose charge they are placed, to pass his accounts, or to transfer them from one officer to another; they are not to take any account of any part of them, from the officer who has charge of them; but as far as it shall be possible for them to do so, they are themselves to ascertain their real quantity. See the NAVAL INSTRUCTIONS.

WASH, or BLADE OF AN OAR. See the article OAR.

To WASH *the Decks*. See the article *To* SCRUB.

WASH-BOARDS, (*falques ou coulisses qui se montent sur le plat-bord de certains bâtimens à rames quand ils sont à la voile, pour élever leur vibord*, Fr.) are broad thin planks, fixed occasionally on the top of a boat's sides, so as to increase the height thereof, but may be removed at pleasure. They are used to prevent the sea from breaking into the vessel in rough and tempestuous weather.

WATCH, (*quart, ou garde*, Fr.) the division of a ship's company into two parties or guards, one of which is called the starboard, and the other the larboard watch; these epithets allude to the situation of their hammocks when hung up: the two watches are, however, occasionally separated into three or four divisions, as in a road, for an anchor-watch, &c.

Starboard-WATCH, *Hoay! Star-Bowlines, Hoay!* (*tribord au quart!* Fr.) *Larboard*-WATCH, *Hoay! Lar-Bowlines, Hoay!* (*babord au quart!* Fr.) the exclamations used by the boatswain's mates when summoning their respective watches upon deck to relieve each other.

WATCH, the space of time wherein one division of the ship's crew remains upon deck, to perform the necessary services, whilst the rest are relieved from duty.

The ship crews are divided into three watches, in order to lessen, as much as possible, an evil that seems unavoidable in a sea-faring life; and the longest of these, in the British marine, is four hours.

It has been remarked, "that the order of nature was never inverted with impunity; and whatever we detract from the natural period of our sleep, will be deducted in the end from the natural range of our existence. Those whose rest is regularly broken by watch-duty, generally put on an appearance of premature old age, that a person unacquainted with them would be liable to err several years in forming a calculation of their ages from their exterior physiognomy, owing principally to their abridged and broken sleep at night."

In a ship of war, the watch is generally commanded by a lieutenant; and in merchant-ships by one of the mates.

WATCH, is also a word used in throwing the deep-sea-lead, when each man, on letting go the last turn of line in his hand, calls to the next abaft him, watch! to apprize him to be a watch for the soundings; and that the line should have no check in going down.

Anchor-WATCH, (*la garde*, Fr) denotes a small guard kept constantly upon deck, while the ship rides at single anchor in a roadstead, &c.

Dog-WATCHES, (*le quart de quartre heures à six, et celui de six à huit*, Fr.) are the two reliefs which take place between four and eight o'clock P. M. each of which being only two hours on deck. The intent of these watches is to change the turn of the night-watch every twenty-four hours; so that the party watching from eight to twelve in one night, shall watch from midnight till four in the morning on the succeeding one.

The First WATCH, (*le quart de huit à minuit*, Fr.) is from eight o'clock in the evening till twelve at night.

The Middle WATCH, (*le quart de minuit à quartre heures*, Fr.) continues from twelve till four in the morning.

The Morning WATCH, (*le quart de quartre à huit, ou celui de diane*, Fr.) comprehends from four to eight o'clock in the morning.

A buoy is said to watch when it continues floating upon the surface of the water.

To Set the WATCH, (*commencer le quart*, Fr.) is to appoint one division of the crew to enter upon the duty of the watch.

From the 25th of March to the 21st of September, the watch is set at nine o'clock; and from the 21st of September to the 25th of March at eight o'clock in the evening; and is always to be relieved at day break in the morning.

To Relieve the WATCH, (*relever le quart*, Fr.) is to change those who constitute any of the abovementioned watches, when another watch is immediately substituted in their stead.

WATCH-BELL, (*cloche de quart*, Fr.) in ships of war, a large bell hung to the beam of the forecastle, and sometimes to the beam near the main-mast: it is struck when the half-hour-glass is run out, to make known the time or division of the watch: it is also used in foggy weather to point out to other ships on what tack you are sailing; as ships on the starboard tack toll their bell, whilst those on the larboard beat their drum.

WATCH-BILL, (*rôle de quart*, Fr.) a list of the petty officers, seamen, and marines, of a ship of war, which, being hung up in some conspicuous part of the ship, points out the station of each man, and what watch he belongs to.

Copies of the watch-bills are kept by the master's mates and midshipmen, for the purpose of mustering their watches, and seeing that the men are at all times on deck, or in their stations.

WATCH-GLASSES, (*horloges de sable, ou sabliers*, Fr.) a name given to the glasses employed to measure the period of the watch, or to divide it into any number of equal parts, as hours, half-hours, &c. so that the several stations therein may be regularly kept and relieved; as at the helm, pump, look-out, &c.

WATCH-GUN, (*coups de canons de diane et de retraite*, Fr.) the gun which is fired on board ships of war at the setting of the watch in the evening, and relieving it in the morning.

When a flag officer shall be in any port or road in his Majesty's dominions, he is, at the setting of the watch at night, to fire one gun, previous to which the tatoo is to be beaten on board his ship, and all other ships in company.

In all ships not carrying a flag, or broad pendant, the sentinels are to fire their muskets.

At the relieving of the watch in the morning, the senior flag officer is to fire a gun, and the other flag officers or commodore in succession, a volley of musketry, and the sentinels, in all other ships, their muskets; after which the reveillez is to be beaten in every ship present.

If there should not be a flag officer or commodore at any of his Majesty's ports where guard-ships are stationed, the senior captain is to fire the watch gun; but ships at anchor in any other port, without a flag officer or commodore, are not to fire the watch-gun, but are to set and relieve the watch by beating a drum, and the sentinels firing their muskets.

The senior officer of any number of his Majesty's ships, which may anchor in any foreign port or road, is to fire the watch guns, except in those places, in which the regulations of the port do not admit of guns being fired. See the NAVAL INSTRUCTIONS.

WATER, (*aqua*, Lat. *eau*, Fr.) in phisiology, is a transparent fluid, without colour, smell, or scarcely any taste, in a very small degree compressible; it is coagulable into a transparent solid substance, called ice, in a temperature of 30° of Fahrenheit's thermometer, and gaseous at 212°; but volatile and fluid on resuming any degree of heat between these points; and when pure, or freed from heterogeneous particles, is reckoned one of the four elements.

But as there seems to be water in all bodies, and particles of almost all kinds of matter in water, it is hardly found sufficiently pure to be considered as an element. Water, if it could be had alone, and pure, Boerhaave argues, would have all the requisites of an element, and be as simple as fire; but there is no expedient hitherto discovered for procuring it so pure.

Water seems to be diffused every where, and to be present in all space where there is matter. There are hardly any bodies in nature but what will yield water: it is even asserted, that fire itself is not without it. From this consideration, philosophers have been led to hold the opinion, that all things were made of water.

Properties and Effects of WATER.—Water is a very volatile body; it is entirely reduced into vapours, and dissipated when exposed to the fire and unconfined.

Water heated in an open vessel, acquires no more than a certain determinate degree of heat, whatever be the intensity of the fire to which it is exposed; which greatest degree of heat is when it boils violently.

It has been found that the degree of heat necessary to make water boil, is variable, according to the purity of the water and the weight of the atmosphere. The following table shews the degree of heat at which water boils, at various heights of the barometer, being a medium between those resulting from the experiments of Sir George Shuckburgh and M. De Luc.

Height of the Barometer.	Heat of boiling Water.	Height of the Barometer.	Heat of boiling Water.
Inches.	°	Inches.	°
27	206.9	29	210.3
27½	207.7	29½	211.2
28	208.5	30	212.0
28½	209.4	30½	212.8

Weight and Purity of WATER.—Water scarce ever continues two moments exactly of the same weight, by reason of the air and fire contained in it. The expansion of water, in boiling, shews what effect the different degrees of fire have on the gravity of water, in order to settle its degree of purity. However, the purest water we can obtain, according to the experiments of Mr. Hawskbee, is 850 times heavier than air; or, according to the experiments of Mr. Cavendish, the thermometer being at 50° and the barometer at 29¼°, about 800 times as heavy as air; by Sir George Shuckburgh, when the barometer is at 29.27′, and the thermometer at 53°, water is 836 times heavier than air; this general proportion may be accounted a standard; *viz.* that, when the barometer is at 30° and the thermometer at 55°, then water is 820 times heavier than air; however, we have not any very exact standard in air; for water being so much heavier than air, the more water there is contained in the air, the heavier of course must the air be; as indeed a considerable part of the weight of the atmosphere seems to arise from the water that is in it.

Water is found the most penetrative of all bodies, after fire, and the most difficult to confine; passing through leather, bladders, &c. which will confine air, making its way gradually through wood, and is only retainable in glass and metals.

Water, by this penetrative quality alone, may be inferred to enter the composition of all bodies, both vegetable, animal, fossil, and even mineral; with this particular circumstance, that it is easily and with a gentle heat, separable again from bodies it had united with. And yet the same water, as little cohesive as it is, and as easily separated from most bodies, will cohere firmly with some others, and bind them together in the most solid masses; as in the tempering of earth, or ashes, clay, or powdered bones, &c. with water, and then dried and burnt, when the masses become as hard as stones, though without the water they would be mere dust or powder. Indeed it appears wonderful that water, which is otherwise an almost universal dissolvent, should nevertheless be a great coagulator.

To WATER, (*faire de l'eau, ou faire aiguade*, Fr.) among seamen, is to take a fresh supply of water on board when they are in want.

Bilge-WATER. See the article BILGE.

Dead-WATER. See the article DEAD.

Foul-WATER. See the article FOUL.

Hard-WATER, (*eau âpre*, Fr.) is that in which soap does not dissolve completely or uniformly, but is curdled.

The dissolving power of hard water is less than that of soft; and hence its unfitness for washing, boiling kitchen vegetables, &c.

The hardness of water may arise either from salts, or from gas. That which arises from salts, may be discovered and remedied by adding some drops of a solution of fixed alkali; but the latter by boiling, or exposure to the open air.

Spring waters are often hard; but river water soft. Hard waters are remarkably indisposed to corrupt; hence they seem to be best fitted for keeping at sea, especially as they are so easily softened by a little alkaline salt.

High-WATER. See the article HIGH.

Low-WATER. See the article LOW.

Rain-WATER, (*eau de pluie*, Fr.) may be considered as the purest distilled water, but impregnated during its passage through the air with a considerable quantity of phlogistic and putrescent matter; whence it is superior to any other in fertilizing the earth. Hence also it is inferior for domestic purposes to spring or river water, even if it could be readily procured.

River or *Running* WATER, is next in purity to snow or distilled water; and for domestic purposes is superior to both, it having less putrescent matter and more fixed air. That, however, is much the purest that runs over a clean rocky or stony bottom.

River waters generally putrify sooner than those of springs. During the putrefaction, they throw off a part of their heterogeneous matter, and at length become sweet again, and purer than at first; after which they commonly preserve a long time: this is remarkably the case with the Thames water, taken up about London, which is commonly used by seamen in long voyages.

Salt-WATER. See the article SALT or SEA WATER.

Snow-WATER, (*eau de pluie*, Fr.) is the purest of all the common waters; when the snow has been collected pure, and kept in well-stopped bottles, it remains unaltered for several years.

Spring-WATER, (*eau de fontaine*, Fr.) is commonly impregnated with a small portion of imperfect neutral salt, and on account of its coolness is an agreeable drink; but on account of its being usually somewhat hard, is inferior to that which has run for a considerable way in an open channel.

Spring water arises from the rain, and from the mists and moisture in the atmosphere. These falling upon hills and other parts of the earth, soak into the ground, and pass along till they find a vent out again, in the form of a spring.

Motion of WATER, in hydraulics. The theory of the motion of running water is one of the principal objects of hydraulics, and to which many eminent mathematicians have given their attention: yet it is to be wished that their theories were more consistent with each other, and with experience. But notwithstanding their labours, this intricate subject still remains in a great measure obscure and uncertain. Even the simple case of the motion of running water, when it issues from a hole in the bottom of a vessel, has never yet been determined, so as to give universal satisfaction to the learned. On this head, it is now pretty generally allowed, that the velocity of the issuing stream, is equal to that which a heavy body acquires by falling through the height of the fluid above the hole, as may be demonstrated by theory: but in practice, the quantity of the effluent water is much less than what is given by this theory, owing to the obstructions to the motion in the hole, partly from the sides of it, and partly from the different directions of the parts of the water in entering it, which thence obstruct each other's motion. And this obstruction, and the diminution in the quantity of water run out, is still the more in proportion as the hole is the smaller. See HUTTON's *Mat. and Philos. Dictionary*.

Method of conveying Fresh WATER *from the Shore to Ships or Boats at Anchor*. In order to effect this very desirable purpose, Mr. Mark Noble of Battersea, has lately obtained a patent for what he terms a fire-extinguishing Chest Engine, a trial of which took place on the 10th of November, 1813, in Cawsand Bay, in the presence of several naval officers; when it was fully proved, that a vessel, moored more than a quarter of a mile from the shore, might be supplied with six butts of water in thirty minutes, notwithstanding a heavy surf, by four men working the engine on the beach Similar experiments have lately been made at Portsmouth and Deal.

This engine acts in a threefold capacity; viz.—1st. As a ship-engine, in case of fire, but with this difference, that it is more compact, from being much smaller, and discharges double the quantity of water supplied by those which are now in use with the same number of men.

2d. As a portable hand-pump, to be placed in any convenient part of the ship for clearing the vessel of water, in time of danger or distress.

3d. For the purpose of sending fresh water from the shore through a surf, or shoal-water, when it is not possible for boats to fill casks; an event which ships are too frequently subject to in foreign parts.

Every invention, by which labour can be simplified, undoubtedly deserves encouragement: and the fact above related speaks so strongly in favour of itself, as to supersede the necessity of our farther recommendations.

WATERAGE, (*fraix de tranport par eau*, Fr.) money paid for carriage of goods and merchandise, &c. by water.

WATER-BOARDS, or Weather-Boards of a boat, (*plat bord d'un canot*, Fr.) are large boards used to keep out the waves or spray of the sea. See also the articles WASH-BOARDS and WEATHER-BOARDS.

WATER-*Borne*, (*à flot*, Fr.) the state of a ship, with regard to the water surrounding her bottom, when there is barely a sufficient depth of it to float her off the ground; particularly when she had for some time rested thereon.

WATER-*Casks*, (*futailles*, Fr.) casks or vessels which ought to be properly prepared and sweetened, for the purpose of containing salubrious water on board ships of war and transports.

All his Majesty's ships, whether victualled for Channel or foreign service, are to be furnished with as many water-casks as their captains and commanders shall demand; and as the same are to be supplied of such sizes as the captains shall apply for, either leagers, butts, puncheons, hogsheads, barrels, or half hogsheads, all commanders are to have regard thereto, and to apply for such quantities and species of water-casks as they can conveniently stow; observing that leagers are to be demanded and supplied for the ground tiers only.

The very great demand for beer and water-casks at the out-ports, making it extremely difficult, more especially in time of war, to raise and fit them fast enough, and the expense attending the same being very great; and as both the difficulty and expense would be materially lessened if the beer and water-casks on board ships employed in home and Channel service were not to be shaken; all commanders of his Majesty's ships so employed are most strictly required, never to suffer the empty beer or water-casks to be shaken, unless in cases of absolute necessity; and when such necessity shall arise, to give particular orders that the casks be very carefully taken to pieces, and the staves and heading marked and packed in such a manner as shall enable the cooper to set them up again; that the iron hoops be very carefully stowed away in bundles, without being bent or otherwise injured; and that they be either returned when a proper opportunity may offer, into his Majesty's victualling stores, or be again set up, if wanted. And the said commanders are further directed to transmit, quarterly, to the commissioners of the victualling, a particular account of all beer and water-casks, which may have been so shaken, stating whether they have again been set up, or the staves and hoops returned into store, and at what place. As a substitute for water-casks, see the article *Iron* TANKS.

WATER-*Engine*, (*pompe à eau*, Fr.) an engine for extinguishing fires; or any engine to raise water; or an engine moved by the force of water. See the articles FIRE-ENGINE and STEAM-ENGINE.

WATER-*Gage*, (*jauge*, Fr.) an instrument for measuring the depth or quantity of any water. See the article GAGE in the APPENDIX.

WATER-*Lines*, (*lignes d'eau*, Fr.) certain horizontal lines supposed to be drawn about the outside of a ship's bottom, close to the surface of the water on which she floats. They are accordingly higher or lower upon the bottom, in proportion to the depth of the column of water required to float her.

In order to conceive a clearer idea of the curves of those lines when represented on a plane, let us suppose a ship laid upright on a level ground; so that the keel shall lie in the same position, with respect to the horizon, as when she is laden. We may then describe several black horizontal lines about her bottom, which may be whitened for that purpose, as represented by fig. 5, plate VII.

If a spectator is supposed to be placed, at a competent depth, under the middle of her bottom, in a line perpendicular to the plane of the ground; he will then, viewing the bottom upwards, discover the horizontal curves of all the water-lines. These curves are all delineated on a plane, supposed to be formed by an horizontal section of the bottom, at the height of the *load water-line*.

Load-WATER-*Line*, (*ligne d'eau, ou ligne d'eau en chargé*, Fr.) the largest of all the above-mentioned lines, or that which the surface of the water makes on a ship's bottom, when she is loaded, or ready for sea.

WATER-*Logged*, (*bâtiment engagé*, Fr.) the state of a ship when, by receiving a great quantity of water into her hold, by leaking, &c. she has become heavy and inactive upon the sea, so as to yield without resistance to the efforts of every wave rushing over her decks. In this dangerous situation, the centre of gravity is no longer fixed, but fluctuating from place to place, the stability of the ship is utterly lost; she is therefore almost totally deprived of the use of her sails, which would operate to overset her, or press her head under water. Hence there is no resource for the crew, except to free her by the pumps, or abandon her, by getting into the boats as soon as possible.

WATER-*Mark*, (*marque de la marée haute*, Fr.) the mark or limit of a flood: thus we say, high-water mark, low-water mark, &c.

WATER-*Measure*. Salt, sea-coal, &c. while on board vessels in the pool or river, are measured with the corn bushel heaped up; or else five striked pecks are allowed to the bushel. This is called water-measure, and it exceeds Winchester measure by about three gallons in the bushel.

WATER-*Sail*, (*bonnette qui s'établit dans un beau tems, en dessous des bonnettes basses*, Fr.) a small sail spread occasionally under the lower studding-sail, or driver-boom, in a fair wind and smooth sea.

WATER-*Shot*. See this explained under the article *Swinging*-MOORINGS.

WATER-*Spout*, (*trombe*, Fr.) an extraordinary aqueous meteor, most frequently observed at sea. It is a truly formidable phenomenon, and is, indeed, capable of causing great ravages. It generally begins by a cloud, which appears very small, which is called by mariners the squall; this in a little time augments into an enormous cloud of a cylindrical form, or that of a reversed cone, and produces a noise like an agitated sea, sometimes emitting thunder and lightning, and also great quantities of rain or hail, sufficient to inundate large vessels, overset trees and houses, and every thing which opposes its violent impetuosity.

These water-spouts are more frequent at sea than by land; and sailors are so convinced of their dangerous consequences, that when they perceive their approach, they frequently endeavour to break them by firing a cannon before they approach too near the ship. They have also been known to commit great devastation by land: though where there is no water near, they generally assume the harmless form of a whirlwind.

Water-spouts are by the majority of philosophers referred to the same origin as whirlwinds; some have endeavoured to account for them, on electrical principles. They observe that the effluent matter, proceeds from a body actually electrified towards one which is not so; and the affluent matter proceeds from a body not electrified, towards one which is actually so. These two currents occasion two motions analogous to the electrical attraction and repulsion. If the current of the effluent matter is more powerful than the affluent matter, which in this case is composed of particles exhaled from the earth, the particles of vapour which compose the cloud are attracted by this effluent matter, and form the cylindrical column, called the descending water-spout; if, on the contrary, the affluent matter is the strongest, it attracts a sufficient quantity of aqueous particles to form gradually into a cloud, and this is commonly termed the ascending water-spout.

A different theory is, however, adopted by other respectable philosophers; and it is possible, after all, that there may really be two kinds of water-spouts, the one the effect of the electrical attraction, as noticed above; and the other caused by a vacuum, or extreme and sudden rarefaction of the air.

It is well known, that even a common fire produces a kind of circulation of the air in a room, but in a different form. It is therefore not difficult to conceive that when any part of the column of air upon the surface of the earth or water is suddenly rarefied, either by electricity or any other cause, a vacuum at least comparatively to the rest of the air, will immediately take place, and the circumambient air rushing in at once from every quarter to fill the space, a conflict of wind ensues, and consequently a circular motion, by which light bodies will be taken up and turned round with considerable velocity: this violent rushing of the air on all sides into the vacuum, then forms what is commonly called at land a whirlwind.

When this vacuum takes place at sea, from the nature of fluids, the water will rise to a certain height by the pressure of the atmosphere, as in a common pump; but as the vacuum is not quite perfect, the water will be divided into drops; and as that vacuum is generally caused by heat, it will be rarefied when it reaches the upper regions of the atmosphere, and assume the appearance of a cloud.

The point or cone of the water-spout is generally oblique, depending on the force and direction of the wind which drives it along. Dr. Perkins, of Boston, is disposed to adopt a different theory of water-spouts. Captain Melling informed him, that in a voyage from the West Indian Islands to Boston, a water-spout came across the stern of the vessel where he then was; a flood of water fell upon him with such violence as almost to beat him down, and the spout immediately passed off with a roaring noise into the sea. The water from the spout was perfectly fresh. Mariners in general affirm, that water descends from the clouds through the water spouts into the sea; and hence it is inferred, that a whirlwind cannot be the cause of a water-spout, nor can both of these phenomena proceed from the same cause. See the article WHIRLWIND.

WATER-*Tight*, (*bâtiment étanche*, Fr.) the state of a ship when she is not leaky.

WATER-*Ways*, (*gouttières*, Fr.) long pieces of timber serving to connect the sides of a ship to her decks, and form a channel to carry off the water from the latter by means of scuppers.

The convexity of the decks, represented by N M N, in the Midship-Section, plate VI. necessarily carries the water towards the sides, where this piece is fixed, which is principally designed to prevent the water from lodging in the seams, so as to rot the wood and oakum contained therein. The water-ways N N are therefore hollowed in the middle lengthways, so as to form a kind of gutter or channel, one side of which lies almost horizontal, making part of the deck, whilst the other rises upward, and corresponds with the side, of which it likewise makes a part. They are scored down about an inch and a half, or two inches, upon the beams, and rest upon lodging-knees or carlings. They are secured by bolts driven from without through the planks, timbers, and water-ways, and clinched upon rings on the inside of the latter.

The scuppers, which are holes by which the water escapes from off the deck, are accordingly cut through the water-ways.

WATERING-PLACE, (*aiguade*, Fr.) a situation where boats can load with fresh water for the use of ships.

WAVE, (*vague, ou lame*, Fr.) in physics, a volume of water elevated by the action of the wind, &c., upon its surface, into a state of fluctuation, and accompanied by a cavity: the extent from the bottom or lowest point of one cavity, and across the elevation, to the bottom of the next cavity, is the breadth of the wave.

Waves are considered as of two kinds, which may be distinguished from one another by the names of natural and accidental waves. The natural waves are those which are regularly proportioned in size to the strength of the wind which produces them. The accidental waves are those occasioned by the wind's re-acting upon itself by repercussion from hills or high shores, and by the dashing of the waves themselves, otherwise of the natural kind against rocks and shoals; by which means these waves acquire an elevation much above what they can have in their natural state.

Mr. Boyle has proved, by a variety of experiments, that the utmost force of the wind never penetrates deeper than six feet into the water; and it should seem a natural consequence of this, that the water put in motion by it can only be elevated to the same height of six feet from the level of the surface in a calm; and these six feet of elevation being then added to the six of excavation, in the part from whence that water was raised, should give twelve feet for the greatest elevation of a wave, when the height of it is not increased by whirlwinds, or the interruption of rocks or shoals, which always give an additional elevation to the natural swell of the waves.

From this calculation we are not to suppose that no wave of the sea can rise more than six feet above its natural level in open and deep water; for some immensely higher than these are formed in violent tempests, in the great seas. These, however, are not to be accounted waves in their natural state; but they are single waves composed of many others: for in these wide plains of water, when one wave is raised by the wind, and would elevate itself up to the exact height of six feet, and no more, the motion of the water is so great, and the succession of the waves so quick, that during the time wherein this rises, it receives into it several other waves, each of which would have been of the same height with itself. These accordingly run into the first wave, one after another as it rises; by this means its rise is continued much longer than it would naturally have been, and it becomes accumulated to an enormous size. A number of these complicated waves arising together, and being continued in a long succession by the duration of the storm, make the waves, so dangerous to shipping, which the sailors, in their phrase, call mountains high.

Different waves do not disturb one another when they move in different directions. The reason is, that whatever figure the surface of the water has acquired by the motion of the waves, there may be in that an elevation and depression; as also such a motion as is required in the motion of the wave.

Waves are often produced by the motion of a tremulous body, which also expand themselves circularly, though the body goes and returns in a right line; for the water which is raised by the agitation, descending, forms a cavity which is every way surrounded with a rising.

The motion of the WAVES, makes an article in the Newtonian philosophy; Newton having explained their motions, and calculated their velocity from mathematical principles, similar to the motion of a pendulum, and to the reciprocation of water in the legs of a bent and inverted syphon, or tube. See NEWTON's *Principia*.

WAVES *Breaking into Surf*. This seldom happens at sea, unless the wind is very high: but when they approach the shore, they always break sooner or later. The general reason of their breaking appears to be the excess of the velocity with which the upper part of the wave advances above that of the lower part: and this may be derived either from the effect of the wind on the upper part, or from the resistance of the bottom to the motion of the lower part, or from a third cause, which is more general, where the magnitude of the wave is at all considerable in comparison with the depth of the fluid; for in this case, the upper part of the wave must have a natural tendency to advance more rapidly than the lower, on account of the greater depth which determines its velocity. Besides this, the form of the wave itself, where the water is shallow, may be such as to render it incapable of advancing without a change of the direction of its anterior surface into a situation more nearly vertical. See NICHOLSON's *Philosophical Journal*, Vol. IX.

Settling WAVES *by Means of Oil*. This wonderful property, though well known to the ancients, as appears from the writings of Pliny, was for many ages either quite unnoticed, or treated as fabulous by succeeding philosophers. Of late it has, by means of Dr. Franklin, again attracted the attention of the learned; though it appears, from some anecdotes, that scafaring people have always been acquainted with the tendency that oil has for breaking the waves and calming the sea. See Dr. FRANKLIN's *Experiments:* also the article OIL.

WAY, (*chemin, &c.* Fr.) the course or progress which a ship makes on the water under sail. Thus, when she begins her motion, she is said to be under way; and when that motion increases, she is said to have fresh way through the water. Hence also she is said to have head-way or stern-way, to gather way or to lose way, &c.

For farther particulars on the way of a ship, see the articles HEAD-*way*, LEE-*way*, STERN-WAY, &c.

Milky WAY. See the article GALAXY.

To WEAR. See the article To VEER.

WEATHER, (*tems, ou température de l'air*, Fr.) the state or disposition of the atmosphere, with regard to the degree of the wind, to heat or cold, or to dryness and moisture; but particularly to the first. See the articles ATMOSPHERE, FROST, HAIL, HEAT, RAIN, SNOW, &c.

WEATHER, (*signifie aussi le côté du vent, ou tout ce qui est au vent*, Fr.) is also applied to every thing lying to windward of a particular situation: hence a ship is said to have the weather-gage of another, when she is farther to windward. Thus also, when a ship under sail presents either of her sides to the wind, it is then called the weather-side; and all the rigging and furniture situated thereon are distinguished by the same epithet; as, the weather-shrouds, weather-lifts, weather-bow, weather-braces, weather-quarter. See the article LEE.

To WEATHER, (*passer au vent, doubler, &c.* Fr.) is to sail to windward of some ship, bank, or head-land.

To Make a Good WEATHER, implies, when a ship in a gale of wind, or a storm, is easy in a sea, and does not strain herself or rigging, which some ships are more liable to than others, from many circumstances.

To WEATHER *out a Storm*, (*étaler un coup de vent sous voile*, Fr.) speaking of a ship under sail, is to continue on the course during the continuance of the storm, without dropping an anchor.

WEATHER-BEATEN, ((*bâtiment battu de la tempête, ou qui à souffert par le mauvais tems*, Fr.) a term applied to any thing which appears to have born much hard or rough weather.

WEATHER-BIT, (*bitture du cable au côté extérieur du cabestan, ou double tour autour des bittes*, Fr.) a turn of the cable about the end of the windlass, without the knight-heads. It is used to check the cable, in order to slacken it gradually in tempestuous weather, or when the ship rides in a strong current. See also the article RING-ROPE.

WEATHER-BOARDS, (*planches placées en forme de toits, pour garantir de la pluie un navire désarmé, sans empêcher la circulation de l'air*, Fr.) are pieces of plank placed in the ports of a ship, when laid up in ordinary; they are fixed in an inclined position, so as to turn off the rain without preventing the circulation of the air.

WEATHER-CLOTHS, (*prélats de bastingage*, Fr.) are long pieces of canvas or tarpauling, used to preserve the hammocks from injury by weather when stowed: also to defend persons from the wind and spray of the sea.

WEATHER-GAGE, (*avantage, ou dessus du vent*, Fr.) implies the situation of one ship to windward of another, when in action, &c. For the advantages and disadvantages of the weather-gage, see the article LINE.

WEATHER-GLASSES, (*terme général pour exprimér les baromêtres, thermomêtres, &c.* Fr.) are instruments contrived to shew the state of the atmosphere, as to heat, cold, moisture, weight, &c. and so to measure the changes that take place in those respects; by which means we are enabled to predict the alteration of the weather, as to rain, wind, frost, &c.

Under the class of weather-glasses are comprehended barometers, thermometers, hygrometers, manometers, and anemometers. See the articles BAROMETER and THERMOMETER.

WEATHER-HELM, (*signifie litéralement la barre au vent, et s'applique à un navire qui est ardent*, Fr.) a ship is said to carry a weather helm when she is inclined to come too near the wind, and, therefore, requires the helm to be kept constantly a little to windward.

The WEATHER-QUARTER, (*la hanche du vent*, Fr.) is that quarter of the ship which is on the windward side.

WEATHER-ROLLS, (*roulis du bâtiment du côté du vent*, Fr.) those inclinations which a ship makes to windward in a heavy sea; those which she makes to leeward being termed LEE-*Lurches*, which article see.

WEATHER-SHORE, (*côte, ou terre qui est au vent*, Fr.) the shore which lies to windward of a ship.

The WEATHER-SIDE, (*la côte du vent d'un bâtiment*, Fr.) denotes that side of a ship under sail upon which the wind blows, or which is to windward.

WEATHER-TIDE, (*marée qui porte au vent, Fr.*) denotes that which, by setting against a ship's lee-side, while under sail, forces her up to windward.

WEB, in block-making, the thin partition on the inside of the rim, and between the spokes of an iron sheave.

WEDGE, one of the mechanical powers as it is called. The wedge is a triangular prism, whose bases are equilateral acute-angled triangles.

WEDGES. See the article QUOINS.

To WEIGH, (*lever l'ancre, appareiller*, Fr.) is to heave up the anchor of a ship from the ground, in order to prepare her for sailing. See also the articles ANCHOR and AWEIGH.

To WEIGH *an Anchor by the Long-Boat*, is to take the long-boat to the buoy of the anchor, and put the buoy-rope over the davit of the long-boat, and a tackle on the buoy-rope; by which, with the assistance of men on the fall, the anchor is weighed out of the ground. This being accomplished, the cable is hove in on-board; the buoy-rope and tackle being secured in the boat, they approach the ship as the cable is hove in, and the anchor catted and stowed.

Small anchors and grapnels are got up by the davit, hauling upon the cable or grapnel-rope by hand.

To WEIGH *an Anchor by Under-Running*, is to place the cable over the davit-head, and under-run it till it is nearly a-peak, when it is tripped (by means of tackles as before) by the buoy-rope. This method, however, is very troublesome, and is only adopted when the buoy is gone, and a ship cannot get near her anchor for want of water.

WEIGHING THE ANCHOR. Captain Ricketts, of the royal navy, has lately proposed an improvement for weighing anchors. The contrivance consists of *battens*, of a semi-circular shape on the back, to be fixed on the deck around the capstan, so as to ship and unship at pleasure, like the capstan-bars.

The use of these battens, when fixed, is to give the men who work at the capstan a more secure footing, and consequently a greater power to heave round, which will enable them to weigh an anchor, or to heave any weighty body with greater facility than can possibly be acquired by the common mode on a smooth polished deck.

The experiment has, on board merchantmen, been realised, so far as relates to the hoisting in of cargoes; but in such instances the battens were *permanent*, and consequently intruded on the walk of the deck.

To obviate this objection, Captain Ricketts has proposed cutting mortises in the rise on which the capstan stands, and in the planks at the sides, or by nailing pieces of wood to form artificial ones (so that they might be made to slip up and down) by which means a certain number would be readily provided, and the ingenuity of the carpenter would supply the rest.

The idea of this improvement seems to have originated from the common practice of nailing strong battens to the heads of piers, and to platforms, in many places on shore, whereon capstans are erected, against which battens, the people employed at the bars place the soles of their feet, and thereby acquire additional power.

From what has been said on this subject, it is hoped that the easy practicability, and great utility of this simple expedient, will stimulate some of our naval officers to give it a fair trial. The subjoined diagram will serve to elucidate the method of constructing the proposed improvement.

Plan of the capstan, and part of the shifting battens fitted in mortises.

A. Rise on which the capstan stands.
B. Gunwale with mortises.
C. Bits abaft the mast.
D. Battens on the deck.
e, *e*. Mortises.

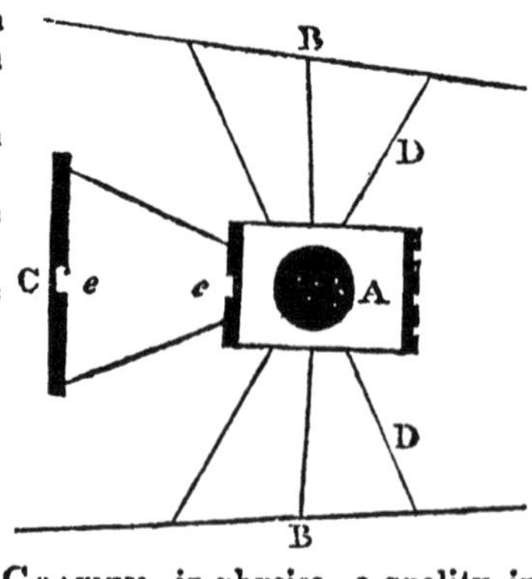

WEIGHT, or GRAVITY, in physics, a quality in natural bodies, by which they tend downwards toward the centre of the earth. See the article GRAVITY.

WELDING, in anchor-making, the act of forging iron when intensely heated.

WELDING-HEAT, is the strongest heat when the iron is most proper to bind.

WELL, (*archipompe, &c.* Fr.) an apartment formed in the middle of a ship's hold, to enclose the pumps, from the bottom to the lower deck. It is used as a barrier to preserve those machines from being damaged by the friction or compression of the materials contained in the hold, and particularly to prevent the entrance of ballast, &c. by which the tubes would presently be choked, and the pumps rendered incapable of service. By means of this inclosure, the artificers may likewise more readily descend into the hold, in order to examine the state of the pumps, and repair them, as occasion requires.

WELL *of a Boat, (sentine d'une embarcation*, Fr.) the place in the bottom where the water lies, between the ceiling and the platform of the stern-sheets, from whence it is thrown out into the sea with a scoop.

WELL *of a Fishing Boat or Smack*, (*réservoir d'un bateau pêcheur, pour conserver le poisson en vie*, Fr.) an apartment in the middle of the hold, which is entirely detached from the rest, being lined with lead on every side, and having the bottom thereof perforated with a number of small holes through the floor, so that the water running into the well is always as fresh as that in the sea, and is nevertheless prevented from communicating with other parts of the hold. Its use is to preserve the fish alive which are put into it.

WEST, (*ouest, couchant, ou occident*, Fr.) one of the cardinal points of the horizon, or of the compass, diametrically opposite to the east, or lying on the left hand when we face the north; or, west is strictly the intersection of the prime vertical with the horizon, on that side where the sun sets.

WESTING, in navigation, denotes the quantity of departure made good to the westward from the meridian.

To WET THE DECKS, (*mouiller les ponts*, Fr.) is to throw water over them, to keep them cool, and prevent them from splitting by the heat of the sun, which is generally done with a skeet or long scoop kept on board for that purpose.

To WET *the Sails of small Vessels, (arroser, ou mouiller les voiles*, Fr.) is to throw water over them with the above-mentioned implement, to render them more efficacious in light breezes. In large ships, this operation is performed by means of an engine.

WHALE. See the article CETUS.

WHARF, (*quai muni de grues, et d'autres machines pour le débarquement des merchandises, &c.* Fr.) a perpendicular building of wood or stone, raised on the shore of a road or harbour, for the convenience of lading or discharging a vessel by means of cranes, tackles, capstans, &c.

A wharf is built stronger or slighter, in proportion to the effort of the tide or sea which it has to resist, and to the weight which it is intended to support.

Gun-WHARF, (*parc d'artillerie*, Fr.) a wharf whereon guns, mortars, shells, shot, and every sort of warlike instrument, are deposited till they are wanted for the service; as the gun-wharfs at Portsmouth, Woolwich, and Chatham.

WHARFAGE, (*droit, ou paye que l'on donne au gardien d'un quai, pour veiller au débarquement et à la sureté des marchandises*, Fr.) When a wharfinger receives a sum of money for wharfage, porterage, and sufferage, according to the custom of trade, he is bound to deliver the goods in safety, on board the destined vessel, and is also responsible for any loss or damage that may be sustained through his negligence.

WHARFINGER, (*gardien d'un quai*, Fr.) the person who has the charge of a wharf, and takes account of all the articles landed thereon, or removed from it into any vessel lying alongside thereof; for which he receives a certain fee called wharfage, as a due to the proprietor for the rent of the quay or wharf, and for the use of his machines and furniture.

WHEEL, (*roue, rouage*, Fr.) one of the six powers of mechanism; and without doubt contributes more to the general convenience of mankind, than any of the other five, by the wonderful variety of purposes wherein it is employed.

Wheel *of the Helm*. See the articles Helm and Steering-Wheel.

WHELPS. See the article Capstan.

WHERRY. See this explained under the article Boats.

Wherry, (*houari; bateau dépêche*, Fr.) is also applied to some decked vessels used in fishing, in different parts of Great Britain and Ireland.

WHIP, (*cartaheu qui passe dans une poulie simple, pour enlever des objets légers*, Fr.) a sort of small tackle, either formed by the communication of a rope with a single immoveable block, as represented by fig. 3, plate XXIII, or with two blocks, one of which is fixed, and the other moveable, as fig. 5; and is generally used to hoist up light bodies, as empty casks, &c.

To Whip, (*surlier, enlever, &c.* Fr.) is also to tie a piece of twine, or spun-yarn, &c. round the end of a rope, to prevent it from being untwisted and loosened.

To Whip-*Up*, (*enlever des tonneaux vuides à l'aide d'un cartaheu passé dans une poulie simple*, Fr.) is to hoist light bodies up into the top, or empty casks out of a ship's hold, &c. by means of a whip.

Whip-*upon-Whip*, (*double cartaheu servant à enlever des objets pesants*, Fr.) the greatest purchase that can be gained by blocks, which is formed by fixing the end of one whip upon another whip-fall. Thus two single blocks will afford the same purchase as a tackle, having a double and a single block, and with much less friction. This purchase should therefore be used whenever the length of the hoist will admit of it. To topsail and top-gallant-yards, that hoist with a single tye, there is sufficient room to apply this purchase as halyards, which will overhaul with great facility.

Whip-Staff, (*manivelle de la barre du gouvernail*, Fr.) a piece of timber in the form of a strong staff, fastened into the helm for the steersman, in small ships, to hold in his hand, in order to move the rudder and direct the ship in her proper course.

WHIRLPOOL, (*gouffre, ou tournant d'eau*, Fr.) an eddy, vortex, or gulf, where the water is continually turning round. In rivers these are very common from various accidents, and are usually very trivial and of little consequence. In the sea they are more rare, but more dangerous.

Sibbald has related the effects of a very remarkable marine whirlpool among the Orcades, which would prove very dangerous to strangers, though it is of no consequence to the people who are used to it. This is not fixed to any particular place, but appears in various parts of the limits of the sea among those islands. Wherever it appears it is very furious, and boats, &c. would inevitably be drawn in and perish thereby; but the people who navigate them are prepared for it, having an empty vessel, a log of wood, or a large bundle of straw, or some such thing, with them, and as soon as they perceive the whirlpool, they toss the article within its vortex, and thus keep themselves out of it. This substance, whatever it may be, is immediately received into the centre, and drawn under water; as soon as this is done, the surface of the place where the whirlpool was, becomes smooth, and they row over it with safety; in about an hour after they see the vortex begin again in some other place, which is usually about a mile distant from the other.

WHIRLS, (*curles, ou molettes*, Fr.) in rope-making, are of beech or ash, five inches long, cylindrically formed, and fixed upon iron spindles in the heads of wheels, with a hook at one end for the spinner to hang his hemp on. They are likewise used to hang the yarn on for hardening, and laying ropes, from a six-thread ratline to a two and a half inch rope. Those for twine and lines are made of box-wood, with a hole through the middle, and two or more grooves round them, one to hold the catgut which encircles the wheel, and the other to hold a small cap, made of catgut, to which the thread is fastened in spinning or laying.

The whirls for large work are 4 inches in diameter, with three or four grooves, and an iron cap.

WHIRLWIND, (*tourbillon de vent*, Fr.) a wind that rises suddenly, and is exceedingly rapid and impetuous, in a whirling direction, and often progressively also, but commonly soon spent.

There are divers sorts of whirlwinds, distinguished by their peculiar names, as, the prester, typho, turbo, exhydria, and ecnephias.

The prester is a violent wind, breaking forth with flashes of lightning. This is rarely observed, and scarce ever without the ecnephias. Seneca says it is a typho, or turbo, re-kindled or ignited in the air.

A typho or vortex, most properly called whirlwind, or hurricane, is an impetuous wind, turning rapidly every way, and sweeping all round the place. It usually descends from on high, and is frequently in the eastern ocean, chiefly about Siam, China, &c. which renders the navigation of those parts very dangerous.

The exhydria is a wind bursting out of a cloud with a great quantity of water. This only seems to differ in degree from the tenephias, which is frequently attended with showers.

The ecnephias is a sudden and impetuous wind, breaking out of some cloud, frequently in the Ethiepic Sea, particularly about the Cape of Good Hope, and is generally called by seamen, a travado.

Dr. Franklin, in his physical and meteorological observations, supposes a water-spout and a whirlwind to proceed from the same cause; their only difference being, that the latter passes over the land, and the former over the water. This opinion is corroborated by M. de la Pryme, in the *Philosophical Transactions*; where he describes two spouts, observed at different times in Yorkshire, whose appearances in the air were exactly like those of the spouts at sea, and their effects the same as those of real whirlwinds.

Whirlwinds have generally a progressive as well as a circular motion; so had what is called the spout, at Topsham, described in the *Transactions;* and this also, by its effects, appears to have been a real whirlwind. Water spouts have also a progressive motion, which is more or less rapid; being in some violent, and in others barely perceptible.

Whirlwinds generally rise after calms and great heats: the same is observed of water-spouts, which are therefore most frequent in the warm latitudes.

The wind blows every way from a large surrounding space to a whirlwind. Three vessels employed in the whale fishery, happening to be becalmed, lay in sight of each other, at about a league distance, and in the form of a triangle. After some time, a water-spout appeared near the middle of the triangle, when a brisk gale arose, and every vessel made sail. It then appeared to them all, by the trimming of their sails, and the course of each vessel, that the spout was to leeward of every one of them; and this observation was farther confirmed by the comparing of accounts, when the different observers afterwards conferred about the subject. Hence whirlwinds and water-spouts agree in this particular likewise.

But if the same meteor, which appears a water-spout at sea, should, in its progressive motion, encounter and pass over land, and there produce all the phenomena and effects of a whirlwind, it would afford a stronger conviction that a whirlwind and a water-spout are the same thing. An ingenious correspondent of Dr. Franklin gives as follows, one instance of this, that fell within his own observation at Antigua, and which convinced him that a water-spout is a whirlwind.

"There appeared, not far from the mouth of the harbour of St. John's, two or three water-spouts, one of which took its course up the harbour. Its progressive motion was slow and unequal, not in a strait line, but as it were by jerks or starts. When just by the wharf, I stood about 100 yards from it. There appeared in the water a circle of about 20 yards diameter, which to me had a dreadful, though pleasing appearance. The water in this circle was violently agitated, being whisked about, and carried up into the air with great rapidity and noise, and reflected a lustre, as if the sun shined bright on that spot, which was more conspicuous, as there appeared a dark circle around it. When it made the shore, it carried up with the same violence, shingles, staves, large pieces of the roofs of houses, &c. and one small wooden house it lifted entirely from the foundation on which it stood, and carried it to the distance of 14 feet, where it settled without breaking or oversetting; and, what is remarkable, though the whirlwind moved from west to east, the house moved from east to west. Two or three negroes and a white woman were killed by the fall of the timber, which it carried up into the air, and dropt again."

A fluid moving from all points horizontally towards a centre, must, at that centre, either mount or descend. If a hole be opened in the middle of the bottom of a tub filled with water, the water will flow from all sides to the centre, and there descend in a whirl. But air flowing on or near the surface of land or water, from all sides towards a centre, must at that centre ascend; because the land or water will hinder its descent.

If these concentrating currents of air be in the upper region, they may indeed descend in the spout or whirlwind; but then, when the united current reached the earth or water, it would spread, and probably blow every way from the centre. There may be whirlwinds of both kinds; but from the effects commonly observed, Dr. Franklin suspects the rising one to be the most frequent: when the upper air descends, it is perhaps in a greater body, extending wider, as in thunder-gusts, and without much whirling; and when air descends in a spout or whirlwind, he conceives that it would rather press the roof of a house inwards, or force in the tiles, shingles, or thatch, and force a boat down into the water, or a piece of timber into the earth, than snatch them upwards, and carry them away.

The whirlwinds and spouts are not always, though most frequently, in the day-time. The terrible whirlwind which damaged a great part of Rome, June 11, 1749, happened in the night; and was supposed to have been previously a water-spout, it being asserted as an undoubted fact, that it gathered in the neighbouring sea, because it could be traced from Ostia to Rome.

This whirlwind is said to have appeared as a very black, long, and lofty cloud, discoverable, notwithstanding the darkness of the night, by its continually lightning, or emitting flashes on all sides, pushing along with a surprising swiftness, and within three or four feet of the ground. Its general effects on houses were, stripping off the roofs, blowing away chimnies, breaking doors and windows, forcing up the floors, and unpaving the rooms (some of these effects seem to agree well with the supposed vacuum in the centre of the whirlwind), and the very rafters of the houses were broke and dispersed, and even hurled against houses at a considerable distance, &c.

Doctor Franklin, in proceeding to explain his conceptions, lays down two or three positions, as a foundation for his hypothesis. 1. That the lower region of air is often more heated, and so more rarefied, than the upper; and by consequence specifically lighter. The coldness of the upper region is manifested by the hail, which sometimes falls from it in warm weather. 2. That heated air may be very moist, and yet the moisture so equally diffused and rarefied as not to be visible till colder air mixes with it, at which time it condenses and becomes visible. Thus our breath, although invisible in summer, becomes visible in winter.

These circumstances being granted, he presupposes a tract of land or sea, of about 60 miles in extent, unsheltered by clouds and unrefreshed by the wind, during a summer's day, or perhaps for several days without intermission, till it becomes violently heated, together with the lower region of the air in contact with it, so that the latter becomes specifically lighter than the superincumbent higher region of the atmosphere, wherein the clouds are usually floated: he supposes also that the air surrounding this tract has not been so much heated during those days, and therefore remains heavier. The consequence of this, he conceives, should be, that the heated lighter air should ascend, and the heavier descend; and as this rising cannot operate throughout the whole tract at once, because that would leave too extensive a vacuum, the rising will begin precisely in that column which happens to be lightest, or most rarefied; and the warm air will flow horizontally from all parts to this column, where the several currents meeting, and joining to rise, a whirl is naturally formed, in the same manner as a whirl is formed in a tub of water, by the descending fluid receding from all sides of the tub towards the hole in the centre.

And as the several currents arrive at this central rising column, with a considerable degree of horizontal motion, they cannot suddenly change it to a vertical motion; therefore, as they gradually, in approaching the whirl, decline from right to curve or circular lines, so, having joined the whirl, they ascend by a spiral motion; in the same manner as the water descends spirally through the hole in the tub before mentioned.

Lastly, as the lower air nearest the surface is more rarified by the heat of the sun, it is more impressed by the current of the surrounding cold and heavy air which is to assume its place, and consequently its motion towards the whirl is swiftest, and so the force of the lower part of the whirl strongest, and the centrifugal force of its particles greatest. Hence the vacuum which encloses the axis of the whirl should be greatest near the earth or sea, and diminish gradually as it approaches the region of the clouds, till it ends in a point.

This circle is of various diameters, sometimes very large.

If the vacuum passes over water, the water may rise in a body or column therein to the height of about 32 feet. This whirl of air may be as invisible as the air itself, though reaching in reality from the water to the region of cool air, in which our low summer thunder-clouds commonly float; but it will soon become visible at its extremities. The agitation of the water under the whirling of the circle, and the swelling and rising of the water in the commencement of the vacuum, renders it visible below. It is perceived above by the warm air being brought up to the cooler region, where its moisture begins to be condensed by the cold into thick vapour; and is then first discovered at the highest part; which being now cooled, condenses what rises behind it, and this latter acts in the same manner on the succeeding body: where, by the contact of the vapours, the cold operates faster in a right line downwards, than the vapours themselves can climb in a spiral line upwards; they climb, however, and as by continual addition they grow denser, and by consequence increase their centrifugal force; and being risen above the concentrating currents that compose the whirl, they fly off, and form a cloud.

It seems easy to conceive, how, by this successive condensation from above, the spout appears to drop or descend from the cloud, although the materials of which it is composed are all the while ascending. The condensation of the moisture contained in so great a quantity of warm air as may be supposed to rise in a short time in this prodigiously rapid whirl, is perhaps sufficient to form a great extent of cloud: and the friction of the whirling air on the sides of the column may detach great quantities of its water, disperse them into drops, and carry them up in the spiral whirl mixed with the air. The heavier drops may indeed fly off, and fall into a shower about the spout; but much of it will be broken into vapour, and yet remain visible.

As the whirl weakens, the tube may apparently separate in the middle; the column of water subsiding, the superior condensed part drawing up to the cloud. The tube or whirl of air may nevertheless remain entire, the middle only becoming invisible, as not containing any visible matter.

Dr. Stuart, in the *Philosophical Transactions*, says, "It was observable of all the spouts he saw, but more perceptible of a large one, that towards the end it began to appear like a hollow canal, only black in the borders, but white in the middle; and though it was at first altogether black and opaque, yet the sea-water could very soon after be perceived to fly up along the middle of this canal, like smoke in a chimney."

When Dr. Stuart's spouts were full charged, that is, when the whirling pipe of the air was filled with quantities of drops and vapour torn off from the column, the whole was rendered so dark that it could not be seen through, nor the spiral ascending motion discovered; but when the quantity ascending lessened, the pipe became more transparent, and the ascending motion visible. The spiral motion of the vapours, whose lines intersect each other on the nearest and farthest side of this transparent part, appeared therefore to Dr. Stuart like smoke ascending in a chimney; for the quantity being still too great in the line of sight through the sides of the tube, the motion could not be discovered there, and so they represented the solid sides of the chimney.

Dr. Franklin concludes, by supposing a whirlwind or spout to be stationary, when the concurring winds are equal; but if unequal, the whirl acquires a progressive motion in the direction of the strongest pressure. When the wind that communicates this progression becomes stronger above than below, or below than above, the spout will be bent or inclined. Hence the horizontal process and obliquity of water-spouts are derived.

WHISTLE. See the article Call.

WHOODINGS. See the article Hooding-Ends.

WIDOWS' Men, (*noms supposès ajoutes au role des equipages Anglois par ordre du gouvernement et dont la paie se retient pour le soulagement des veuves des marins*, Fr.) an appellation given to a certain number of men, according to circumstances, in every hundred of which the complement of the ship shall consist, who are directed by Act of Parliament to be borne on all his Majesty's ship's books as able seamen, the produce of whose wages, and the value of whose provisions, are applied to the relief of poor widows of commissioned and warrant-officers in the royal navy.

For the due performance of this benevolent scheme, every captain, or other officer, commanding any of his Majesty's ships or vessels of war, is directed to enter on the books of the ship or vessel he commands, as part of her complement, one or two (as the clerk of the check shall inform him that circumstances require) fictitious names in every hundred men of her complement, which are to be borne as able seamen, under the appellation of *widows' men*, with numbers 1, 2, 3, &c. set against them; observing, that when the number to be borne in every hundred is two, if the whole complement, or the part exceeding the hundreds of the complement, amounts to twenty-five, and does not exceed seventy-five, one widow's man is to be entered for that number; and if it amounts to seventy-five, two widows' men are to be entered; but, if only one widow's man in each hundred of the complement is to be borne, none are to be entered for any broken number above the hundred, unless it amounts to fifty; and, in small vessels in this case, where the complement consists of fifty, one widow's man is to be borne; these widows' men are to be borne for wages, as if they were men actually serving on board, but not for victuals; and, when pay-books are made out for the payment of the ship's company, the widows' men are to be entered on them in the same manner as the seamen serving on board are entered, but without any deductions whatever being made from their wages.

Widows' *Pension*, *(pensiòn des veuves des marins*, Fr.) an allowance paid quarterly from government to the widows of commissioned and warrant-officers of the royal navy.

On the 1st of January, 1814, the pension of sea-officers' widows was increased to, and established at the following rates, viz.

	per Annum.
The widow of a flag officer of his Majesty's fleet	£120
The widow of a captain superannuated, with the rank of rear admiral	100
The widow of a post captain of three years' standing	90
The widow of a post captain under three years' standing	80
The widow of a commander	70
The widow of a lieutenant, superannuated, with the rank of commander	60
The widow of a lieutenant	50
The widow of a master	40
The widow of a surgeon	40
The widow of a purser	30
The widow of a boatswain	25
The widow of a gunner	25
The widow of a carpenter	25
The widow of a second master of a yacht, or master of a naval vessel warranted by the Navy Board	25

Provided that those widows only shall be reputed proper objects of the charity, who have not pensions on the ordinary estimates of the navy, or by any other grant from Government, and whose annual incomes, arising from their real or personal estates, or otherwise, do not amount to double the sum allowed for their respective pensions.

WILDFIRE, *(feu grégeois*, Fr.) a kind of artificial or factitious fire, which burns even under water, and that with greater violence than out of it. It is composed of sulphur, naphtha, pitch, gum, and bitumen, and is only extinguishable by vinegar mixed with sand and urine, or by covering it with raw hides.

Its motion or tendency is said to be contrary to that of natural fire, and it always follows the direction in which it is thrown, whether it be downwards, sideways, or otherwise.

Several are of opinion, that the ancient Greeks and Romans used this wildfire in their engagements at sea: whether or not that was the case, it was applied against the Saracens in a sea-fight commanded by Constantine Pogonates, near Cyzicus, in the Hellespont, and with such effect, that he burnt the whole fleet therewith, in which there were 30,000 men.

Constantine's successors used it on divers occasions, and with equal advantage; and what is very remarkable, they were so happy as to keep the secret of the composition to themselves, so that no other nation knew it in the year 960.

WINCH, (*virevaut*, Fr.) a cylindrical piece of timber, having an axis, whose extremities rest in two channels placed horizontally or perpendicularly, and furnished with clicks or pawls. It is turned about by means of a handle, resembling that of a grindstone, and is generally employed as a purchase, by which a rope may be more conveniently or more powerfully applied to any object, than when used singly, or without the assistance of mechanical powers. See also the article *Chain* Pump.

Winch, in rope-making, a machine to wind the yarn on as it is spun, and consists of eight spokes, to form the body, and eight blades, four at each end, to contain the spokes, and an iron bolt, with a round head to turn it on.

Pump-Winches. See the article Pump.

WIND, (*vent*, Fr.) a sensible current in the atmosphere, usually blows from one part of the horizon to its opposite part.

The horizon, being divided into 32 points (see the article Compass) the wind which blow from either of them has an assignable name.

Winds are either constant or variable, general or particular. Constant winds are such as blow the same way, at least for one or more days; and variable winds are such as frequently shift within a day. A general or reigning wind is that which blows the same way, over a large track of the earth, almost the whole year. A particular wind is what blows, in any place, sometimes one way, and sometimes another. If the wind blow gently, it is called a breeze; if it blow moderately, it is called a steady breeze; if harder, it is called a gale, or a stiff or taught gale; and if with violence, it is termed a storm or hard gale.

The following observations on the wind have been made by skilful seamen; and particularly the learned Dr. Halley.

1st. Between the limits of 60 degrees, namely,

from 30° of north latitude to 30° of south latitude, there is a constant east wind throughout the year, blowing on the Atlantic and Pacific oceans; and this is called the trade wind.

For as the sun, in moving from east to west, heats the air more immediately under him, and thereby expands it; the air to the eastward is constantly rushing towards the west to restore the equilibrium, or natural state of the atmosphere; and this occasions a perpetual east wind in those limits.

2d. The trade-winds near their northern limits blow between the north and east, and near the southern limits they blow between the south and east.

For as the air is expanded by the heat of the sun near the equator; therefore the air from the northward and southward will both tend towards the equator, to restore the equilibrium. Now these motions from the north and south, joined with the foregoing easterly motion, will produce the motions observed near the said limits between the north and east, and between the south and west.

3d. These general motions of the wind are disturbed on the continents, and near their coasts.

For the nature of the soil may either cause the air to be heated or cooled; and hence will arise motions that may be contrary to the foregoing general one.

4th. In some parts of the Indian Ocean there are periodical winds, which are called monsoons; that is, such as blow half the year one way, and the other half-year the contrary way.

For air that is cool and dense, will force the warm and rarified air in a continual stream upwards, where it must spread itself to preserve the equilibrium: so that the upper course or current of the air shall be contrary to the under current; for the upper air must move from those parts where the greatest heat is; and so, by a kind of circulation, the north-east trade-wind below will be attended with a S.W. above; and a S.E. below with a N.W. above: and this is confirmed by the experience of seamen, who, as soon as they get out of the trade-winds generally find a wind blowing from the opposite quarter.

5th. In the Atlantic ocean, near the coasts of Africa, at about 100 leagues from shore, between the latitude of 28° and 10° north, seamen constantly meet with a fresh gale of wind blowing from the north-east.

6th. Those bound to the Carribee islands, across the Atlantic ocean, find, as they approach the American side, that the said N.E. wind becomes easterly or seldom blows more than a point from the east, either to the northward or southward.

These trade-winds, on the American side, are extended to 30, 31, or even to 32° of north latitude; which is about 4° farther than what they extend to on the African side: also, to the southward of the equator, the trade-winds extend three or four degrees farther towards the coast of Brazil, on the American side, than they do near the Cape of Good Hope on the African side.

7th. Between the latitudes of 4° north and 4° south, the wind always blows between the south and east. On the African side, the winds are nearest the south; and on the American side nearest the east. In these seas Dr. Halley observed, that when the wind was eastward, the weather was gloomy, dark, and rainy, with hard gales of wind; but when the wind veered to the southward, the weather generally became sereno, with gentle breezes, next to a calm

These winds are somewhat changed by the seasons of the year; for when the sun is far northward, the Brazil S.E. wind gets to the south, and the N.E. wind to the east; and when the sun is far south, the S.E. wind gets to the east, and the N.E. winds on this side of the equator veer more to the north.

8th. Along the coast of Guinea, from Sierra Leone to the island of St. Thomas (under the equator) which is about 500 leagues, the southerly and south-west winds blow perpetually: for the S.E. trade-wind having passed the equator, and approaching the Guinea coast within 80 or 100 leagues, inclines towards the shore, and becomes south, then S.E. and by degrees, as it approaches the land, it veers about to south, S.S.W. and when very near the land it is S.W. and sometimes W.S.W. This tract is troubled with frequent calms, and violent sudden gusts of wind, called tornadoes, blowing from all points of the horizon.

The reason of the wind setting in west on the coast of Guinea, is, in all probability, owing to the nature of the coast, which, being greatly heated by the sun, rarefies the air exceedingly, and consequently the cool air from off the sea will keep rushing in to restore the equilibrium.

9th. Between the 4th and 10th degrees of north latitude, and between the longitude of Cape Verd, and the easternmost of the Cape Verd isles, there is a tract of sea which seems to be condemned to perpetual calms, attended with terrible thunder and lightnings, and such frequent rains, that this part of the sea is called the rains. In sailing through these six degrees, ships are said to have been sometimes detained whole months.

The cause of this is, apparently, that the westerly winds setting in on this coast, and meeting the general easterly wind in this track, balance each other, and so produce the calms; and the vapours carried thither by each wind, meeting and condensing, occasion the almost constant rains.

The last three observations shew the reason of two things which mariners experience in sailing from Europe to India, and in the Guinea trade.

And first, The difficulty which ships in going to the southward, especially in the months of July and August, find in passing between the coast of Guinea and Brazil, notwithstanding the width of this sea is more than 500 leagues. This happens, because the S.E. winds at that time of the year commonly extend some degrees beyond the ordinary limits of 4° north latitude; and besides coming so much southerly, as to be sometimes south, sometimes a point or two to the west, it then only remains to ply to windward: and if, on the one side, they steer W.S.W. they get a wind more and more easterly; but then there is danger of falling in with the Brasilian coast, or shoals; and if they steer E.S.E. they fall into the neighbourhood of the coast of Guinea, from whence they cannot depart without running easterly as far as the island of St. Thomas; and this is the constant practice of all the Guinea ships.

Secondly, All ships departing from Guinea for Europe, their direct course is northward; but on this course they cannot proceed, because the coast bending nearly east and west, the land is to the northward. Therefore, as the winds on this coast are generally between the S. and W.S.W. they are obliged to steer S.S.E. or south, and with these courses they run off the shore; but in so doing, they always find the winds more and more contrary; so that when near the shore, they can lie south; but at a greater distance they can make no better than S.E. and afterwards E.S.E. with which courses they commonly fetch the island of St. Thomas and Cape Lopez, where finding the winds to the eastward of the south, they sail westerly with it, till coming to the latitude of 4° south, where they find the S.E. wind blowing perpetually.

On account of these general winds, all those that use the West India trade, and even those bound to Virginia, reckon it their best course to get as soon as they can to the southward, that so they may be certain of a fair and fresh gale to run before it to the westward: and for the same reason those homeward-bound from America endeavour to gain the latitude of 30°, where they first find the winds begin to be variable; though the most ordinary winds in the north Atlantic ocean come from between the south and west.

10th. Between the southern latitudes of 10 and 30 degrees in the Indian ocean, the general trade-wind about the S.E. by S. is found to blow all the year long, in the same manner as in the like latitudes in the Ethiopic ocean: and during the six months from May to December, these winds reach to within two degrees of the equator; but during the other six months, from November to June, a N.W. wind blows in the track lying between the 3d and 10th degrees of southern latitude, in the meridian of the north end of Madagascar; and between the 2d and 12th degrees of south latitude, near the longitude of Sumatra and Java.

11th. In the tract between Sumatra and the African coast, and from three degrees of south latitude quite northward to the Asiatic coasts, including the Arabian sea and the Gulf of Bengal, the monsoons blow from September to April on the N.E. and from March to October on the S.W In the former half-year the wind is more steady and gentle, and the weather clearer, than in the latter six months; and the wind is more strong and steady in the Arabian sea than in the Gulf of Bengal.

12th. Between the island of Madagascar and the coast of Africa, and thence northward as far as the equator, there is a tract, wherein, from April to October, there is a constant fresh S.S.W. wind; which to the northward changes into the W.S.W. wind, blowing at times in the Arabian sea.

13th. To the eastward of Sumatra and Malacca, on the north of the equator, and along the coasts of Cambodia and China, quite through the Philippines as far as Japan, the monsoons blow northerly and southerly; the northern one setting in about October or November, and the southern about May: these winds are not quite so certain as those in the Arabian seas.

14th. Between Sumatra and Java to the west, and New Guinea to the east, the same northerly and southerly winds are observed; but the first half-year monsoon inclines to the N.W. and the latter to the S.E. These winds begin a month or six weeks after those in the Chinese seas set in, and are quite as variable.

15th. These contrary winds do not shift from one point to its opposite all at once; and in some places the time of the change is attended with calms, in others by variable winds: and it often happens on the shores of Coromandel and China, towards the end of the monsoons, that there are most violent storms, greatly resembling the hurricanes in the West Indies, wherein the wind is so excessively strong, that hardly any thing can resist its force.

All navigation in the Indian ocean must necessarily be regulated by these winds; for if the mariners should delay their voyages till the contrary monsoon begins, they must either sail back, or go into harbour, and wait for the return of the trade-wind.

The relative force of the wind upon a ship's sails, and the epithets by which it is distinguished, as fair, large, &c. according to the angle which it makes with her course, are explained in the article SAILING.

Velocity of the WIND. It has been remarked, that our most violent winds take place when neither the heat nor the cold is greatest, or about the 21st of March and the 22d of September, when the sun crosses the equator, at which times the equinoctial gales generally arise, and extend over a great tract of country; they are accompanied by sudden and great falls of the mercury in the barometer. The reason appears to be, that violent winds succeed the precipitation in rain of a large quantity of vapour, which previously constituted a part of the bulk of the atmosphere; and this precipitation cannot take place when the general temperature approaches to either extreme.

"The velocity of wind in a great storm," says Robertson in his Navigation, "is not more than 50 or 60 miles in an hour; and a common brisk gale about 15 miles an hour."

A common paper kite will determine the velocity of the wind with great accuracy, by letting it loose, and measuring with a stop-watch how long it is running out a given quantity of line. Colonel Beaufoy says he has known the rapidity to be at the rate of seventy-two statute miles per hour, which was not the case at the earth's surface, owing to the impediments which the wind there meets with, neither is the current so irregular aloft as below; and that wind has only the 666th part of the effect of water, when moving with equal velocity. He also observes that it frequently happens in violent storms of wind, the current does not reach any considerable altitude; for often, at the height of 1600 feet, there is a perfect calm; on the contrary, it is not uncommon for the wind, at the elevation of three or four miles above the level of the sea, to move with very great celerity, whilst the lower parts of the atmosphere remain in a state of perfect tranquility. Again, it may be illustrated by the motions of dense fluids, which are always impeded in the parts contiguous to the sides and bottom of the vessels; and the same thing happens in tide-rivers, where the boatman, when he wishes to proceed with the tide, commits himself to the middle of the stream; but when he has to strive against it, he keeps close to the shore. It is, therefore, not the upper parts of the atmosphere which are accelerated, but the lower are retarded by friction against the surface of the earth.

Reigning WINDS. See the article REIGNING.

Scant WIND. See the article SCANT.

Trade WINDS. See the article TRADE.

By the WIND, (*être au plus près*, Fr.) is when a ship sails as nearly in the direction of the wind as possible.

The WIND'S *Eye*, (*le lit du vent*, Fr.) denotes the direct point from which the wind blows.

Between WIND *and Water*, (*entre deux eaux*, Fr.) when speaking of a ship, implies that part of her bottom which is frequently brought above the water by her agitation when at sea.

To WIND *a Call*, (*donner un coup de sifflet, commander la manœuvre avec le sifflet*, Fr.) is to blow or pipe upon a boatswain's whistle, so as to communicate the necessary orders of hoisting, heaving, belaying, slackening, veering away, &c. See the article CALL.

To WIND *a Ship, &c.* (*faire présenter l'arrière d'un bâtiment, où il avoit l'avant*, Fr.) is to change her position, by bringing the stern to lie in the situation of the head; or directly opposite to its former situation.

WINDAGE OF A GUN, (*vent d'une pièce d'artillerie, c'est—à dire la différence entre le calibre de l'ame et le diamètre du boulet*, Fr.) is the difference between the diameter of the bore of a gun, and the diameter of the shot or shell corresponding thereto.

The usual windage of guns in the English service, has heretofore been one-twentieth of the caliber.

Dr. Hutton observes, in *Math. and Philos. Dictionary*. "that this allowance of windage is too much, owing perhaps to the first want of roundness in the ball, or to rust, foulness, or irregularities in the bore of the gun. But lately a beginning has been made to diminish the windage, which cannot fail to be of very great advantage, as the shot will both go much truer, and have less room to bounce about from side to side, to the great damage of the gun; and besides, much less powder will serve for the same effect, as in some cases, one-third or one-half the inflamed powder escapes by the windage. The French allowance of windage is $\frac{1}{27}$ of the diameter of the ball."

WIND-BOUND, (*bâtiment retenu au mouillage par des vents contraires*, Fr.) when speaking of a ship or vessel, implies that she is prevented from sailing to her destination, in consequence of the wind being contrary.

WINDING. (*courbure*, Fr.) When the edge or side of a piece of timber or plank is not a direct plain, but twists, it is said to wind; if the contrary, Out-of-winding.

WINDING-TACKLE, (*caliorne à trois rouets, servant à embarquer, ou débarquer l'artillerie du bord, et à d'autres usages*, Fr.) a name usually given to a tackle formed of one fixed triple block, and one double or triple moveable block. It is principally employed to hoist up any weighty materials, such as the cannon into or out of a ship.

WINDLASS, (*cabestan horizontal*, Fr.) a machine used in merchant-ships instead of a capstan, to heave up the anchors from the bottom, &c.

It is a large piece of timber, of an octagonal shape, as represented by *fig.* 21, plate VII, formed on the principles of the axis *in peritrochio*, and is supported at the two ends by two frames of wood, *a, b*, placed on the opposite sides of the deck near the fore-mast, called windlass-bits; it is turned about in this position as upon an axis, by levers, called hand-spikes, which are for this purpose thrust into square holes, cut through the body of the machine. See the article *To* HEAVE.

The lower part of the windlass is usually about 9 inches above the deck. It is like the capstan, furnished with strong pawls, *c, d*, to prevent it from turning backwards by the effort of the cable, when charged with the weight of the anchor, or strained by the violent jerking of the ship in a tempestuous sea. The pawls which are formed of wood or iron, fall into notches, cut in the surface of the windlass, and lined with plates of iron. Each of the pawls, being accordingly hung over a particular part of the windlass, falls eight times into the notches at every revolution of the machine, because there are eight notches placed on its circumference under the pawls. So if the windlass is 20 inches in diameter, and purchases five feet of the cable at every revolution, it will be prevented from turning back, or losing any part thereof, at every seven inches nearly, which is heaved in upon its surface.

As this machine is heaved about in a vertical direction, it is evident that the effort of an equal number of men acting upon it will be much more powerful than on the capstan; because their whole weight and strength are applied more readily to the end of the lever employed to turn it about; whereas, in the horizontal movement of the capstan, the exertion of their force is considerably diminished. It requires, however, some dexterity and address to manage the handspike to the greatest advantage; and to perform this, the sailors must all rise at once upon the windlass, and, fixing their bars therein, give a sudden jerk at the same instant, in which movement they are regulated by a sort of song or howl pronounced by one of their number.

Spanish WINDLASS, (*appareil, composé d'un anspec, &c. au tour du quel on fait prend plusieurs tours d'un cordage qu'on vent roidir*, Fr.) a machine formed of an iron bolt, placed in a hole, which is hove round by a woolder that acts as a lever for turning it round. It is used to stretch small rigging for serving, &c.

WIND-RODE, (*qui se dit d'un bâtiment, qui, étant mouillé par une marée et un vent opposés l'un à l'autre, évite de manière à rester sous le vent de son ancre*, Fr.) a term applied to a ship, which riding where the wind and tide are opposed to each other, is forced by the violence of the former to remain to leeward of her anchor.

WIND-SAIL, (*manche à vent*, Fr.) a sort of wide tube or funnel of canvas, employed to convey a stream of fresh air into the lower apartments of a ship, being let down through the hatches, and kept extended by several wooden hoops in different parts of its height; the upper part is open on the side which is placed to windward, as represented by fig. 9, plate IV, so as to receive the full current of it, which fills the tube, and rushes downward into the lower regions of the ship.

Ships of war, in hot climates, have generally three or four of these windsails, which, together with the ventilators, contribute greatly to the preservation of the health of the crew.

In sultry latitudes, the greatest attention should ever be paid, in keeping the windsails constantly trimmed to the breeze, so that as much pure air as possible may be circulated between decks.

To WINDWARD, (*au vent du butiment et en général le côté du vent*, Fr.) is towards that point of the horizon from whence the wind blows.

Sailing to WINDWARD, (*louvoyer*; *tenir le vent, ou le plus près*, Fr.) is the art of working the ship towards that quarter of the compass from whence the wind blows.

For rightly understanding this part of navigation, it will be necessary to explain the terms that occur in it, though most of them may be seen under their proper heads in this work.

When the wind is directly, or partly against a ship's direct course for the place she is bound to, she reaches her port by a kind of zigzag or z like course, which is made by sailing with the wind first on one side of the ship, and then on the other side.

When a ship sails the same way that the wind blows, she is said to sail or run before the wind; and the wind is said to be right aft, or right astern; and her course is then 16 points, or the farthest possible from the wind, that is, from the point the wind blows from. When the ship sails with the wind blowing directly across her, she is said to have the wind on the beam; and her course is 8 points from the wind. When the wind blows obliquely across the ship, the wind is said to be abaft, or before the beam, according as her course is more or less than 8 points from the wind. When a ship endeavours to sail towards that point of the compass from which the wind blows, she is said to sail on a wind, or to ply to windward. And a vessel sailing as near as she can to the point from which the wind blows, she is said to be close hauled. Most ships will lie within about 6 points of the wind; but sloops, and some other vessels, will lie much nearer. To know how near the wind a ship will lie, observe the course she goes on each tack, when she is close-hauled; then half the number of points between the two courses, will shew how near the wind the ship will lie.

The windward, or weather side, is that side of the ship on which the wind blows; and the other side is called the leeward or lee side. Tacks and sheets are large ropes fastened to the lower corners of the fore and main sails; by which either of these corners is hauled fore or aft. When a ship sails on a wind, the windward tacks are always hauled forward, and leeward sheets aft. The starboard tacks are aboard, when the starboard side is to windward, and the larboard side to leeward: and the larboard tacks are aboard, when the larboard side is to windward, and the starboard to leeward.

The most common cases in turning to windward, may be constructed by the following precepts:

Having drawn a circle with the chord of 60°, for the compass, or the horizon of the place, quarter it by drawing the meridian and parallel of latitude perpendicular to each other, and both through the centre; mark the place of the wind in the circumference; draw the rhumb, passing through the place bound to, and lay on it, from the centre, the distance of that place. On each side of the wind, lay off, in the circumference, the points or degrees shewing how near the wind the ship can lie, and draw these rhumbs. Now the first course will be on one of these rhumbs according to the tack the ship leads with. Draw a line through the place bound to, parallel to the other rhumb, and meeting the first; and this will shew the course and distance on the other tack.

EXAMPLE.—The wind being at north, and a ship bound to a port 25 miles directly to windward; beginning with the starboard-tacks, what must be the course and distance on each of two tacks to reach the ports?

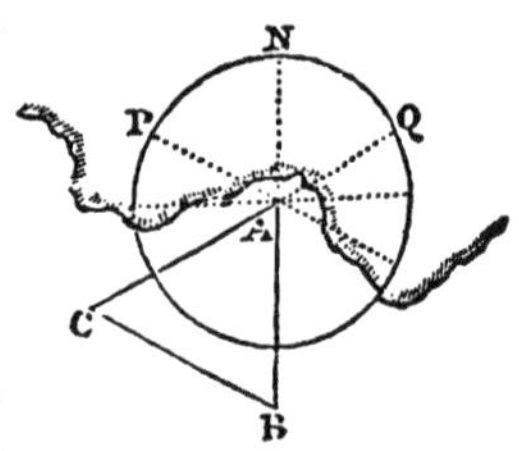

CONSTRUCTION.—Having drawn the circle, &c. as above described, where A is the port, A P and A Q the two rhumbs, each within six points of A N; in N A produced, take A B = 25 miles, then B is the place of the ship; draw B C parallel to A P, and meeting Q A produced in C; so shall B C and C A be the distances on the two tacks; the former being W.N.W and the latter E.N.E.

CALCULATION. — Here ∠ B = N A P = 6 points.
and ∠ A = N A Q = 6 points.
Therefore ∠ C = 4 points.

So that all the angles are given, and the side A B, to find the other two sides A C and B C, which are equal to each other, because the opposite angles A and B are equal. Hence,

As sin. C : A B : : sin. A : B C,
i.e. S. 45° : 25 : : S. 67° 30′ : $32\frac{2}{3}$ = B C or A C, the distance to be run on each tack.

WINDWARD *Tide*, (*marée contraire au vent*, Fr.) a name given to a tide that sets to windward.

WINGERS, (*petites pièces que l'on place sur les ailes d'un arrimage*, Fr.) small casks stowed close to the side in a ship's hold, where the large casks would cause too great a rising in that part of the tier.

WINGS, (*ailes de l'arrimage à l'avant et à l'arrière, ou côtés de la cale d'un bâtiment*, Fr.) a name given to those parts of the hold and orlop-deck which are nearest to the sides.

This term is particularly used in the stowage of the several materials contained in the hold; as, "Stow the large casks amidships, and the smaller barrels in the wings!"

The WINGS *of a Fleet*, (*les ailes d'une armée navale, lorsqu'elle est placée sur une ligne*, Fr.) are the skirts or extremities of a fleet when it is ranged into a line a-breast, or when forming two sides of an angle. Thus the ships *a*, *b*, fig. 5 and 6, plate XV. are in the wings of their fleets or squadron.

It is usual to extend the wings of a fleet in the day-time, in order to discover any enemy which may fall into their tract. To prevent separation, however, they are commonly summoned to draw nearer to the centre of the squadron before night, by a signal from the commander in chief, which is afterwards repeated by ships in the intervals.

WING-SAIL *for Ketches*, a quadrilateral sail, similar to the mizen course of a ship. It is made of canvas, No. 6 or 7, and bends abaft the mainmasts to hoops which encircle the mast. The head is expanded by a gaff.

WING-TRANSOM. See the article TRANSOMS.

WINTER, (*hiver*, Fr.) one of the four seasons or quarters of the year.

Winter properly commences on the day when the sun's distance from the zenith of the place is the greatest, or when his declination is the greatest on the contrary side of the equator; and it ends on the day when that distance is a mean between the greatest and least, or when he next crosses the equinoctial.

At and near the equator, the winter, as well as the other seasons, return twice every year; but all other places have only one winter in the year; which, in the northern hemisphere, begins when the sun is in the tropic of Capricorn, and in the southern hemisphere when he is in the tropic of Cancer: so that all places in the same hemisphere have their winter at the same time.

Notwithstanding the coldness of this season, it is proved in astronomy, that the sun is really nearer to the earth in our winter than in summer: the reason of the defect of heat being owing to the lowness of the sun, or to the obliquity of his rays.

WITHIN BOARD, (*dans le bâtiment*, Fr.) signifies within-side of the ship; and,

WITHOUT-BOARD, (*hors du bord*, Fr.) on the contrary, denotes without-side the ship.

WOODEN-BOTTOMS, in gunnery, a name given to cylindrical pieces of wood, of different lengths and diameters, according to the sizes of guns. They are bottomed at one end to receive the shot, and the flannel cartridge is fastened to the other end, the whole forming one cartridge, which is put into the piece at one motion.

WOODLOCK, a piece of thick-stuff fitted on the rudder to prevent it from unshipping.

WOOF, the twine or thread woven across the knittles in pointing.

To WOOLD, (*rouster, ou faire une rousture*, Fr.) is to wind a piece of rope about a mast or yard, to support it in a place where it may have been fished or scarfed; or when it is composed of several pieces united into one solid.

WOOLDERS, single and double handed, in rope-making, are sticks about 3 feet long, and 4 inches in circumference, with straps of rope-yarn made fast, to fix on the rope and assist the men at the hooks in closing the rope.

WOOLDING, (*rousture*, Fr.) is also the rope employed in this service.

To WORK A SHIP, (*manœuvrer un bâtiment*, Fr.) is to direct her movements, by adapting the sails to the force and direction of the wind.

A ship is also said to work, when she strains and labours heavily in a tempestuous sea, so as to loosen her joints or timbers.

To WORK *Double Tides*, (*faire l'ouvrage de trois jours en deux;—expression usitée dans les arsenaux d'Angleterre*, Fr.) a phrase used in the dock-yards, implying that the people perform the work of three days in two.

To WORK *Up Junk*, (*défaire des torons de vieux cordages, et en faire du bitord, &c.* Fr.) is to draw yarns from old cables, &c. and therewith make foxes, points, gaskets, sennit, or spun-yarn.

To WORK *to Windward*, (*louvoyer*, Fr.) is to make a progress against the direction of the wind. See the articles BEATING, *To* PLY, *To* TACK, and *To* TURN.

THEORY OF WORKING SHIPS.—The theory of working ships is the demonstration, supported by proofs, of the effects of every sail, and of the rudder, separately or altogether considered, both with respect to the points where these machines are placed in the ship, and to the different dispositions which either are given them in the changes of evolutions, or which arise from their various obliquities, when they present, more or less obliquely, their surfaces to the course of the water or the wind.

Of the Action which Water or Wind have, by their Pressure on Surfaces.

Fluids are formed of an infinite number of particles, the minuteness of which is the cause why they communicate, by their shock, but very imperceptible degrees of motion, in the first instant of their action: and such is the weakness of their action, that it requires to be repeated a great many times before they can produce any sensible effect on the bodies they are to move.

It is easy to conceive, that the more specific gravity a body is possessed of, the stronger its impulse must be: therefore water, which weighs nearly eight hundred and fifty times more than air, ought to produce (the velocity being the same) an impulsion eight hundred and fifty times more than air would against a surface of the same size, moved in direction perfectly similar. And when it is known that the impulse of a fluid depends on its specific gravity, it will be easily understood that such an impulse must depend also on the extent of the surface which is struck. For, it is plain that the greater the surface is (the gravity, the velocity, and the direction, of the fluid being the same), the stronger the impulse will be, admitting still the same proportion to be kept between the extent of that fluid's surface and that of any other surface put in comparison with it; because a surface of twelve feet square will always receive twelve times as much impulsion as would a surface of only one foot square. We must observe here, at the same time, that such parts of the fluid as strike find more or less difficulty to recoil after the shock, according as the surface is more or less extensive; because, the greater the surface struck, the longer is the continuance of repulsion from their former directions impressed on the particles, which, by that very act of repulsion, receive a new direction, by which they are made to lose for a while the first movement they had during their primitive ones; whence it follows, that the shock of the subsequent particles must be altered: but this deviation from the direct line of the subsequent particles may be looked upon as almost nothing, since there is very little wanting, indeed, but all impulsions should be in the reciprocal proportions which exist between them and the surfaces on which they strike, allowing always all other circumstances to be alike.

It must be observed, that the rapidity of the fluid contributes doubly to the force of the impulse; for every particle strikes with so much the more strength as it acts with a greater velocity, and is at the same time followed by a greater number of new particles to shock the surface. So that the greater the celerity of the particles, the greater is the number of those which share the action, and the more powerful is the resistance they oppose to their being put out of their direct motion. But, if the fluid is possessed of five or six times more rapidity, it is evident that every particle enjoys likewise five or six times more force to shock the surface which opposes the passage of them all together; as, on the side of the surface, there are five or six times as many particles to encounter in the same space of time: therefore, such a surface, thus exposed to the shock of the fluid, will be struck with twenty-five or thirty-six times more force at one time than at another, since there are five or six times as many particles employed in the act of striking, and supported with five or six times as much rapidity. Whence it may be concluded, that impulsions increase as the squares of velocities; or, rather, that they are between themselves as the squares of their velocities, when all other circumstances are the same.

When a surface is exposed to the course of a fluid, it is indifferent whether we consider that the fluid shocks the surface, or that the surface moves the fluid: or, again, whether we consider the fluid and the surface as having each their respective share of the velocity with which that surface receives the impulse of the fluid.

When the wind has little velocity, its action is observed to be but faint; but, when moving with rapidity, then it becomes capable of producing the greatest effects. This is easy to be conceived; for, if to the action of every particle of air, which is stronger by reason of its increased celerity, be added a greater number of particles striking at the same time, it is evident that its force will increase as the square of its velocity; which has already been demonstrated.

The same may be said of water, the impulse of which is almost like that of a solid when it acts, or is acted upon, with a great rapidity of motion. Whence we must conclude, that if that water meets perpendicularly a body which presents to it a great superficies, such a body must have the greatest solidity to be able to resist it.

Experience confirms this principle. For, a ship which drives to leeward does not divide the fluid with her side in a direct line; there is always some obliquity in the direction she pursues by her act of dividing. This obliquity proceeds from the little resistance she experiences from the fluid either at her stem or at her stern. So that, should she be driven ever so little to leeward, she glides always obliquely on the column of water which opposes her under her lee, following a line more or less close to the direction of her length, than to the perpendicular which may be conceived to be drawn as lateral to her keel.

We have hitherto spoken of the impulse of fluids upon surfaces only, when considered as perpendicular: but, when that impulse becomes oblique, it is clear, that it must receive a great diminution; since the motion of every particle will be discomposed, on account of its acting only by its motion perpendicular to the surface, its impulse will be in proportion to the different angles of incidence, which always express their respective velocities, these being considered in a direction perpendicular to the surface.

If, instead of one particle, we consider the whole surface as exposed to the course of all those which compose a fluid; it will appear evident, from what has been said, that the surface E F (fig. 3. plate XXXIV.) which is oblique to the course of the fluid, presents to that fluid a less surface than it would if it were perpendicular to it, like A B. So that each particle produces a less shock, and the particles which are at the same time contributing to the shock are less in number. Now, as these two causes of diminution follow the same proportion, it results, that the impulsions of fluids are between themselves as the squares of the sines of incidence. Therefore, as soon as the impulse of a fluid, which strikes a surface perpendicularly, is known, that impulse, when it strikes the surface obliquely, is only to be diminished in the same proportion as the sine total I K is to the square of the sine of incidence L K.

Fig. 3. Pl. XXXIV

The surface A B, fig. 3, receives all the direct impulse of the fluid which strikes it perpendicularly, and which is contained between C D: but the same surface, presented obliquely to the fluid in the direction E F, will receive but a part of the impulse, which will be proportional to the sine of incidence K L, compared with the sine total I K, of the direct effort of every particle contained between the parallels E G and F H, which inclose a much less space than the first A C and B D. Whence it is easy to conclude, that the diminution of the impulse of the fluid has diminished on two sides, and has consequently followed the proportion of the square of the sine total I K to the square of the sine of incidence L K; for there is a less number of particles employed in striking the surface, and with a smaller degree of velocity.

It follows, that we ought not to be surprised to see the velocity of a ship diminishing considerably, when, after running with the wind aft or large, the vessel is hauled closer to the wind. For it is evident that all the sails which can possibly be spread in this last direction, will receive but very little impulse, on account of their great obliquity to the wind, with which they cannot make an angle more open than 30 degrees, and sometimes much less, as will be demonstrated hereafter. So that the impulse has diminished in proportion as the square of the sine total is to the sine of incidence of 30 degrees; that is to say, as 4 to 1. Therefore, the sails, receiving but a very faint impulse, can communicate to the ship but a small motion; and that motion is still enfeebled by the resistance of the water on the lee-bow; which resistance increases, on one hand, by the inclination of the ship; and, on the other, by the greater surface which she presents to the water in the direction of her length; to which must be added, the decomposition of the absolute effort of the sails, the lateral part of which is now become much greater than the direct. Hence, we find the rapidity of the ship's way is already diminished from three evident causes, to which another may still be added; and it is this: If the ship has an inclination to the horizon (as this always happens in oblique courses), and if the wind has ever so little force, there will result again from that circumstance a cause of diminution of impulse of the wind on the sails; because, in such a case, the sails follow that particular inclination of the ship called *heeling*; and this diminution of impulse will follow this particular proportion, *viz.* that in such a direction the square of the sine of incidence will be smaller than that of the sine total. Therefore, we see that the absolute sine of incidence diminishes in a two-fold proportion, and receives that diminution from the compound ratio of the proportion which the sine total bears to the two sines of the obliquity of the yard with the wind and of the inclination of the sail with the wind.

The impulse of the wind being continual, must necessarily communicate to the ship degrees of velocity, which, from instant to instant, are increasing, until there happens to be an *equilibrium* between the impulse of the wind on the sails and the resistance of the water on the bows, observing, that, in the courses where the ship sails with the wind abaft the beam, the first moment when the wind strikes the sails is the time when its impulse is greatest, and the resistance of the water the weakest; because, at that instant, the ship does not yet move in the fluid, not having yielded to the power of the wind: but, in a few moments, the velocity of the ship increasing, the resistance of the water on the bows increases also considerably: then the impulse of the wind on the sails is proportionably decreasing, because, the ship, receding, as it were, from the wind, must of course lessen its power on the sails. Thus the accelerating force is incessantly lessening from two causes; first, from the wind striking the sails with less force; and, second, from the greater part of its impulse being destroyed by the resistance of the water on the bows; a resistance which increases in proportion as the ship's way accelerates; for, this opposition of the water is as a deduction from the effort of the wind; since, by its resistance, the water renders part of that effort ineffectual. Hence, the rate of sailing will be the greatest possible when the impulse of the wind upon the sails shall be so diminished, and the resistance of the water on the bows so increased, as that the two forces, acting in contrary directions, are in a perfect equilibrium. Now we must conclude, that the vessel will enjoy a constant and uniform motion; for, the ship advances as if she were not subject to the action of any exterior force, the wind no longer having power to increase her velocity, because the resistance of the water on her bows prevents it; and, on the other hand, the impulse of the wind hinders the water, by its resistance, from retarding her course.

If a ship run on a line perpendicular to the direction of the wind, the impulse on the sails is always the same, because she does not recede from the point from which the wind blows; but, when she sails close-hauled, the impulse must be stronger; because she runs to windward, and draws nearer to that point. So that, if the rate of the ship's sailing be great, the apparent angle of incidence diminishes in proportion to the two velocities, *viz.* that of the wind, and that of the ship.

The moment a surface, which is suspended or a-float, is struck by a fluid, that is the time of the greatest impulsion, if it were not in motion before, and of the greatest resistance of the surface.

If the body were struck at the two points by two mobiles, exerting equal powers with respect to the centre of gravity, it is plain that the whole body will be carried up on parallels, and that the sum of the two powers will act on the centre of gravity, since they are equal in every respect.

Theorem.—*A Sail acts always two ways on a Ship, when it is not perpendicular to her Length.*

Demonstration.—We have only to consider the sail A B, fig. 4, oblique to the ship and the wind, and we must be convinced that it is impelled in the direction C D, with a force expressed by the square of the sine of incidence of the wind upon the sail. Therefore, what is said here for the present case is to be understood as applicable in all others, in which the sail shall not be perpendicular to the length of the ship; for, then, she would go only in the direction of her length from C to E, or from C to F, according as the sail might be full or a-back.

If C D be equal to the impulse of the wind upon the sail, as expressed by the square of the sine of incidence A V, we have only to form the right-angled parallelogram G H, to be convinced that such a direction is composed of the two effects C H and C G, which it produces with respect to the body E F, upon which it acts in impelling it in the direction C D.

The more the yard A B, fig. 4, shall make the angle A C E acute, the more the effect C H will augment, and the other C G diminish; for, the more the angle A C E becomes acute, the more its equal D C H or C D G will be acute also; so that C D, which is perpendicular to the centre of the yard, will approach more to C H, the other perpendicular to the length of the ship E F; which cannot happen without increasing the ship's tendency to fall off in the direction C H, and increasing likewise the cause C D, which follows in that increase the same proportion as the square of the sine of incidence augments. But this increase of the impulse C D is not sufficient to preserve the effect of the sail in the direction of the keel C G. On the contrary, it diminishes in the proportion of the decrease of the sine A C E or C D G; whence it follows, that you never can augment the impulse of the wind by shifting the situation of the sail, when it is properly trimmed, without lessening the rate of sailing, when neither the ship changes her course nor the wind shifts.

It might, in the same manner, be demonstrated, that the more open the angle A C E of the sail A B is with the keel, the more its effect C G, in the direction of the ship's length, will increase in the same proportion as the increase of the sine of that angle, when the impulse of the wind upon the sail is the same; for the sines of the angles are in proportion to their opposite sides in the triangle C D G, of which the angle C D G is equal to the angle A C E.

If the impulse augment also, the two effects, C G and C H, will augment proportionally.

If the sail A B receive the impulse of the wind E on its forward surface, it would still act in two ways on the ship, by forcing her first astern in the direction CF, and then to leeward in the direction H C: to be convinced of this, reverse the parallelogram, by tracing it on the after part of the yard towards F, and use the same reasoning.

THEOREM.—*The Angle* A C F, *fig.* 5, *formed by the Yard* A B *and the Keel* F C *of the Ship, is equal to the Angle* D C H, *comprehended between the Perpendiculars* D C *to the Yard, and* E C *to the Keel, in the like manner as the Angle* C D G.

DEMONSTRATION.—The angle A C D is right, since C D is perpendicular on A B: the other angle F C E is right also; for C H is perpendicular on F I; therefore the arc A D is equal to the arc F H; and if, from these two equal arcs be taken away the common one F D, the remainders A F and H D will be equal: because, when equals are taken from equals, the remainders must be equal.

Secondly, the angle C D G in the parallelogram G E is alternate to the angle D C E: therefore it is equal to it; and also equal to A C E.

It follows, that the angle B C H is equal to the other angle D C F; for, if from the two equal arcs A D and B D be taken the other equal arcs F A and D H, the remainders F D and H B will be equal also.

The few principles of geometry here given, and which will be found of great service in the sequel, ought not to discourage. We make use of them now, only to establish principles as simple as they are sure, and to leave nothing to doubt or conjecture in the following part of this theory, which is in itself very abstruse. We shall however be obliged to use again a few more demonstrations of this kind.

Of the most advantageous Angle of the Sails with the Keel and the Wind, in order to obtain the greatest Rate of Sailing on an oblique Course.

In most ships the sails make with the keel an angle A D R, fig. 6, of 40 degrees, or thereabouts (some more, some less), when close-hauled. We are now to make it appear that this angle is not the most favourable to run with the greatest velocity, in getting to windward. It should be much more oblique; but as it is not possible, in practice, to attain the greatest perfection, we must be contented to approach it as near as possible, in great ships, by reducing the angle A D R to 30 degrees only. This will be so much the more easily done, as in every ship the two foremost shrouds of each lower mast can be suppressed. For it must be observed, that, in the movement of pitching and rolling, the masts always incline forward, in the direction D E of the effort of the sail; so that the shrouds which are abaft, and cat-harpinned in, are sufficient to support the masts, since they act nearly opposite to the effort of the sail. Besides, should there be reason to expect bad weather, preventer-shrouds may easily be fastened to the straps which are always ready hung for that purpose. This practice is so much the better grounded, as the number of those preventers can at any time be increased as circumstances or necessity require.*

We shall, for the future, consider the angle A D R, which is the most oblique in practice, as fixed at 30 degrees, though, in some ships, it may happen to be more acute: a circumstance to which particular attention should be paid.

Among the angles A D R, B D R and H D R, fig. 6, which the sail A Z can make with the keel in the same course D R, it is evident there must be one more advantageous than the rest, to produce the greatest velocity possible in the most oblique course. That angle of the sail and the keel is not what we propose directly to determine, since it is impossible to render it more acute than 30 degrees, the term to which we have fixed it in practice: but it will serve us to determine the most favourable angle of incidence A D V of the wind upon the sail, and which is the most advantageous to run with the greatest rapidity on all oblique courses between close-hauled and the wind abaft.

Before we enter upon the demonstration of the rule which must be followed in practice, the principle which serves to demonstrate its utility must be first established. It may be recollected that impulsions are between themselves as the squares of the sines of incidence. Therefore, to judge if it is advantageous to render the angle of incidence A D S or A D V of the wind upon the sail A Z more or less acute, we must examine if the square of the sine of incidence A F, or the total impulse D E, increases more or less than the squares of the sines of incidence B C and H I, or than their correspondent impulsions D G and D K. proportionally to the diminution or increase of the sines of the angle of obliquity of the sail with the keel A T, B Y, and H L: for, if the square of the sine of incidence H I, or the impulsion D K, does not increase so much proportionally, as the sine of obliquity A T diminishes in becoming equal to L H, it is evident that the position of the sail A D is more favourable than when it is situated in the direction D H: and if the square of the sine of incidence B C, or its correspondent impulsion D G, diminishes more in proportion than the sine A T augments, in becoming equal to the other sine of obliquity B Y, in the other position of the sail; it is an evident proof that its situation A D is still more favourable than if it were in the position B D, and that there is even no better situation than A D, whether the angle of incidence A D V be increased or diminished.

To prove it, we shall consider fig. 6, the absolute impulsions D K, D E, and D G, as correspondent

* This recommendation of M. Bourdé, to suppress the two foremost shrouds of each lower mast, in order to brace up sharper, we are warranted in saying, cannot be followed in the British navy. It was lately determined, upon a consultation of the officers of the King's yards, the question having been referred to them, that the present number and dimensions of the rigging of ships could not be advantageously altered or safely diminished. And, as to occasionally casting them off, it may be rendered, by the inevitable accidents of navigation, highly dangerous; as, for instance, in case of being suddenly taken a-back, this lessening of the support of the mast might be attended with its loss.

with the sines of incidence H I, A F, and B C, and proportional to the squares of these same sines; then, on these diagonal lines, if we draw the rectangles X N and M O, in order to dissolve those total impulsions D E and D G, it will appear evident that the direct effort D X, in the direction of the keel, is the greatest possible, when the tangent A S of the angle of incidence is double the tangent A R of the angle of obliquity of the sail with the keel; for, if the angle of incidence be opened 10 degrees, by placing the sail in the situation H D, it will appear that, though the total impulsion D K is augmented in the ratio of the square of the new sine of incidence H I to the first A F, the partial effort D M, in the direction of the keel, will be nearly by 1-tenth less, in this situation of the sail H D, than in the first A D. The direct impulsions D M, proceeding from the total ones D K and D G, are equal, because these last have augmented or diminished in the same ratio as the sines H L and B Y have lessened or increased in proportion to the square of the sine of incidence A F, and to the sine of obliquity A T. These direct partial impulsions D M and D X are in a compounded ratio of the sines of obliquity H L, A T, B Y, equal to those of the angles D K M, D E X, D G M, and of the total impulsion D K, D E, and D G, for if the total impulsion augment by a movement of the sail, the sine of obliquity diminishes: so that from the total impulsions can at any time be deduced the direct ones for every possible angle of incidence. We might very well verify by calculating this demonstration, which proves that the tangent A S of the angle of incidence must always be double the tangent A R of the angle formed between the sail and the keel, agreeably to the situation of the sail A Z; since if any other position be given to it with respect to the wind V, whether it be in the direction H D ten degrees more open than A Z, or like B D ten degrees more oblique, a result, as D M, in the direction of the keel will ever be found less than D X.

As the vanes always indicate the apparent direction of the wind, on all the courses the ship can sail, the angle which the wind makes with the course, or the keel, cannot fail being easily known if there is no lee way; let that angle with the sails be parted into two others, so that the angle of obliquity of the sails with the course may have its tangent equal to half the tangent of the apparent angle of incidence of the wind upon the sails. On this foundation, it will be easy to form a table which will always shew both the apparent angle of incidence, and that of the obliquity of the sail with the keel or with the course. This table will serve for all oblique courses, provided the after sails cover those forward only in a trifling degree; for, should they becalm them much, they must, for other considerations, be braced up a little more to windward; but always leave the apparent angle of incidence of the wind upon the sail more open than that between the sail and the keel, or the course.

Remarks on sailing by the Wind.

When it is desired to gain to windward as much as possible, without absolutely wishing to sail with the greatest velocity, let the direction of the coast under the lee be supposed to make with the absolute direction of the wind (which must as near as possible be known) an angle of 90 degrees; or, in the sea-phrase, *blowing dead on shore:* let the angle A C E, fig. 7, formed by the sail and the keel, be known to be 30 degrees, let the lee-way be also known to be 10 degrees, the angle E C I between the sail and the course will consequently be 40 degrees, which you must take from the total angle V C L 90 degrees; then there will remain 50 degrees, the half of which, 25 degrees, is to be taken for the absolute angle of incidence V C E, and for its equal I C L; so that the ship A B will go 55 degrees from the wind when she is close-hauled, and will consequently recede as much as possible from the point D on the coast, the direction of which makes an angle of 90 degrees with the absolute direction of the wind V K.

But, if the situation C L, fig. 8, of the point D, from which you wish to move, made an obtuse angle V C L, with the positive direction of the wind V M; then the tangent of the apparent angle of incidence V C E must be made double the angle of obliquity E C I which the sail makes with the course, at the same time that the angle I C L of the course and the coast shall be made equal to the angle V C E, formed by the real direction of the wind V K and of the sail F E: so that two considerations must at once be attended to. For example: the angle A C E, formed by the sail and the keel, is 30 degrees; then, according to the first principle, it will be necessary that the apparent angle of incidence V C E should be 49° 6′; and if the difference between the apparent and real direction of the wind be 10°, there will be 59° 6′ for the angle which the sail E F makes with the real direction of the wind V M: so that the angle L C I, of the course and the object stood from, must be found also to be 59° 6′, and the total angle L C V will then be 148° 12′, adhering to the two principles of sailing with the greatest velocity, and of getting to windward of the point D, as much as possible, at the same time; while the angle L C V, formed by the apparent direction of the wind and that of the coast from which the ship moves, will be only 138° 12. The yawing and the different velocities of the ship render the angle formed by the two directions of the wind (the real and the apparent) more or less open. If the ship has more velocity at the same time, or if the course approaches more to the direction of the wind, it will appear by the vanes that the wind draws forward, and the angle of the two directions of the wind will augment. If the ship falls off, and yet still preserves the same velocity, or if her velocity decreases without altering her course, the wind will seem by the vanes to draw aft, and the angle of the two directions will diminish; so that, whenever the ship shall have velocity or run obliquely to the wind, there will always be a difference between its real and apparent direction. If the ship run exactly before the wind, or have no motion at all, there will be then no other but the real direction of the wind shewn by the vanes: but, in oblique courses, the vanes, being moveable, will fix themselves in a middle direction between the absolute tendency of the wind and the course of the ship; whence we may easily conclude, that the apparent direction of the wind, shewn by the vanes, is a medium between the respective velocities of the ship and of the wind; since that direction necessarily partakes more of the greater velocity than of the less: so that if the ship runs east, with the wind at south, having the fourth part of the velocity of the wind, the vanes will shew S. S. E. 4° 30 S. for the apparent direction of the wind.

Of the Sails which are before the Centre of Gravity of a Ship.

The sails which are before the centre of gravity of a ship are the sprit-sail, sprit-sail top-sail, the jib, the fore-top-mast stay-sail, and the fore stay-sail.

Besides these sails there are, on the fore-mast, the fore-sail, fore-top-sail, fore-top-gallant-sail, and fore-top-gallant-royal-sail, with their respective studding-sails. Now these four last sails may be regarded as only one large sail, wide at the foot and tapering towards the head, and which can be reduced, as occasion requires, either by taking in the royal, or by reefing the fore-top-sail, or even taking it quite in, if necessary, to have the fore-sail only set; or by hauling the fore-sail up, if nothing but the top-sail is wanted. It must notwithstanding be observed, that the different parts of the whole sail may, in certain cases, be worked differently the one from the other; as, for example, in reefing the top-sail, or in taking in either the one or the other. But, when you want to set them to work all together, either for making a course, or performing some evolution, they must all be braced and trimmed in the same form, and with the greatest uniformity possible. Therefore, whatever may be said concerning one of them in any operation, is to be understood to be the same with respect to all the rest.

The main-stay-sail, the main-top-mast stay-sail, the middle-stay-sail, and the main-top-gallant stay-sail, are likewise sails of the fore-part of the ship's centre of gravity.

Of the Effect of the Fore-and-aft Sails, which are on the fore part of the Centre of Gravity of a Ship.

The jib and stay-sails being of a triangular figure, their centre of gravity is easily found; and that point is to be considered as the part, in all these sails, on which the whole effort of the wind is united, when they are exposed to its impulse, in whatever way it strikes them.

The particular effort of each fore-and-aft sail being on the fore part of the centre of gravity of the ship, it follows that the total effort of all these sails must be there too; and that, if the ship was in perfect equilibrium with respect to the wind, before her sails were set, she will lose it immediately after they make the fore part of the ship obey the wind, whenever it strikes them perpendicularly or obliquely. For, it must be observed, that almost all these sails have their tacks in the middle of the ship, and their sheets lead to the sides; so that they make with the keel a very acute angle: whence it is easy to conceive, that the perpendicular which would be raised on the exterior surface of these sails, in the direction of their efforts to leeward, from their centre of gravity, would differ but very little in the lateral direction from a perpendicular to the keel. From this we may therefore conclude, that these sails would have but very little effect to accelerate the rapidity of sailing with respect to their position, if it was not demonstrated that they are very advantageous in going by the wind. They make the ship steer well, and are particularly useful when a ship gripes much: and, when they do not take the wind out of any of the lower sails, they ought to be used, particularly when one is obliged to sail by the wind, or to run not very large. The jib and fore-top-mast stay-sail must be preferred, because they are at all times useful when they can receive the wind; for, by their position, they cannot take the wind out of any of the other sails, and their particular effect in veering is considerable, not only on account of their great surface, but because they act before the point on which the ship turns with a very long arm of a lever. Besides, all sails, which raise a ship's head, assist her progress: for the direction of their effort ascending obliquely towards the horizon, they do not make her plunge in the water, which is an advantage peculiar to these sails. Experience has confirmed their utility on all occasions when they can be employed without taking the wind out of the other sails.

Of the Effect of the Fore-Sail, Fore-Top-Sail, Fore Top-Gallant-Sail, and Sprit-Sail, in their different Situations.

When the sail A B, fig. 9, is trimmed close to the wind which blows from the point V, it is impelled in the direction C D with the force expressed by the square of the sine of incidence, and composed of the two effects C E and E D. But, as the centre of effort of that sail A B is on the fore part of the centre of gravity of the ship H, and as its power C D is always decomposed between those two effects C E and E D, it follows, that the effect of this sail is to cause the ship to bear away; while it keeps up at the same time, and even augments, the rapidity of sailing.

If the fore-sail A B received the impulse of the wind perpendicularly, it would still produce the effects of bearing away, and augmenting the rate of sailing, for the reasons just given above, but more effectually would it do so on account of the increase of the impulse of the wind upon the sail.

It follows, from what has been said, that when the sails upon the fore-mast are full, on the same side they are tacked, being braced obliquely to the keel, there is always one part of their effort, in proportion to the obliquity, which acts to make the ship bear away; while the other part of their effort acts at the same time to accelerate or keep up the rate of her sailing.

Of the Sails which are abaft the Centre of Gravity of a Ship.

The main-sail, main-top-sail, main-top-gallant-sail, and main-top-gallant-royal-sail, and their respective studding-sails; the mizen-stay-sail, mizen-top-mast-stay-sail, the mizen-course, mizen-top-sail, mizen-top-gallant-sail, and mizen-top-gallant-royal-sail, are all the sails which are placed abaft the centre of gravity of a ship, the point round which the total effort of the sails is placed.

Of the Effect of the Fore-and-aft Sails abaft the Centre of Gravity of a Ship.

The centre of effort of these sails being abaft the centre of gravity of the ship, it follows that they always force the after-part of the ship to leeward, and consequently contribute to bring her to the wind, as soon as they receive its impulse; for that movement of the after-part of the ship cannot happen, without the head approaching to the direction of the wind.

The fore-and-aft sails being in general situated very obliquely, it follows, consistently with principles, that they are very advantageous for sailing by the wind. Therefore, we must not neglect augmenting them: observing, at the same time, that they do not take the wind out of one another, nor becalm the principal sails. They are only to fill up the space between the masts fore and aft, in sailing near the wind, in order that no wind may be lost.

Of the Effect of the Square Sails of the Main-Mast, and of the Mizen-Top-Sail, in their different Obliquities.

As we have already demonstrated, that when the sail A B, fig. 10, is trimmed obliquely to the keel, it produces evidently two effects on the ship; it must therefore follow, that, in dissolving its power C D, we shall find its compound effects, the one C F, in the direction of the keel which produces the velocity, and the other C E lateral, which (in forcing the after-part of the ship to leeward, by its action on the point C abaft the centre of gravity G of the ship) occasions her to come to the wind; for that motion of the stern from C to E cannot take place, unless the fore-part H acts contrarily in coming towards the point from which the wind blows, V.

If the sails A B were more or less oblique to the keel, they would still have the same effects of keeping up the ship's velocity, and bringing her to the wind. And if they receive its impulse perpendicularly, it would still be the same thing, producing those two effects, however, with greater efficacy than in any other situation with respect to the wind, because then they receive its greatest possible impulse for the time.

Of the Height of Masts.

The correct height of the masts of ships is still a problem which remains to be solved for the builders. The most skilful of them have not paid attention enough to the solutions and determinations which are contained in the works of the late Mr. Bouguer on that subject. It seems, on the contrary, as if they had endeavoured to deviate as much as possible from the true principles in that respect, by raising the masts a great deal more than they were formerly, although they were already much too high, as the learned author, just mentioned, has asserted. An experience, confirmed by repeated observations, has convinced me of this truth; *viz.* that "as soon as a ship inclines, her velocity diminishes in proportion as her inclination increases." This principle has been verified on different vessels and at different times by several officers, and in various oblique courses. I had no share in those various experiments, and therefore cannot be suspected of partiality: but, as they have always proved, to those who have made them, that the present mode of masting is generally too high, I will not hesitate a moment longer to deliver here an epitome of my own experiments on that subject.*

Having all the sails out, and being hurried on by a strong gale, I have ordered all the top-gallant-sails, the studding and stay-sails, to be taken in, without the ship losing the least perceptible degree of her velocity; nay, I have seen it sometimes to increase by a twentieth, and that at a time when the ship ran already at the rate of nine or twelve knots an hour.

These trials, which I have made with care, and which were performed so quickly, that the wind should not have time to increase or diminish in strength, are sufficient to prove the necessity of lowering the centre of effort of the sails in general, and consequently all the masts. These experiments have been repeated by augmenting the number of sails, sometimes at the risk of fatiguing the masts; and it has always been found that the velocity did not increase, when the ship was more inclined; but that she laboured more and more in all her parts, as her movements became stronger and the concussions of her pitching rougher, although the sea was not more swelled. At other times, when the ship inclined pretty much, though the wind was not quite strong enough to hurt the masts, I have lessened the number of sails; and it happened that the ship, after that suppression of the top-sails, was easier in her movements, steering better, and was, in short, more quiet, though the swells of the sea were still the same; an attention which must not be neglected in these kinds of observations, which should be often repeated before a positive decision. However, we do not recommend any diminution in the surface of the sails, in lessening their height: but it will often happen that we shall rather recommend to increase it upon the whole. For that which is lost in height may be regained by the width. There will even result, from that operation, another advantage: the top-sails, by this reform, being shorter, and, thereby, proportionally wider, than the lower sails, will be more easily cut to their shape; and their sides being formed with lines exactly strait, the sail will be the more tight, by which a much greater effect on the ship will be produced. The mast being shorter, and the sails wider, with less fall, the surface will be the same: but the effort of that surface will, with the same wind, act on shorter levers, the fulcrum of which will not be altered; therefore, it will operate at a shorter distance from that fulcrum; and therefore much less will be the power which makes the ship incline: and the ship, being more upright, will sail with more velocity, because her water-lines will be then more advantageous than when she heels. On the other hand, the sails being less inclined, they present a wider surface to, and receive a stronger impulsion from, the wind; an advantage which must always produce an increase of swiftness and a de-

* We have thought it proper to give the reasoning of M. Bourdé upon this subject, although the practice of high-masting prevails in the British navy. It has been very justly remarked, that high masts have the advantage of giving an option; because they *enable* a ship, without *obliging* her, to carry a greater quantity of sail than ships with shorter masts.

crease of drift. Add to this the real advantages of trimming the sails better, of working them with more ease, of rendering the masting in general more solid, and more capable of resistance in bad weather as well as in battle.

But how must we determine the height for the masts? or, in other words, how much are they to be shortened? The *Treatises on the perfect Working and Masting of Ships*, by Mr. Bouguer, teach us that method. It is from those Treatises I have imbibed the notions of my principles on that subject. But, in order to give a previous idea of that inquiry, and to engage the builders and seamen to bring to perfection this part both of the building and working of ships, upon which, almost as much as from their bottom, their steerage undoubtedly depends, I will subjoin here what M. Brue, a learned and studious officer, made me conceive on that subject.

"That masting," said he, "is absolutely perfect, when the centre of effort of the sails is precisely opposite to, or at the same height as, or parallel with, the *point velique*. What is the *point velique?* It is that point in a perpendicular (raised from the centre of gravity of the horizontal surface of a ship at the floating-line), which is intersected by the direction of the absolute impulse of the sea on the head of the vessel. This is the *point velique* in direct courses."

It is clear. No great effort of imagination is necessary to conceive this principle, which appears so evident, that it may be surprising why it has not yet been made use of. For, this *point* once known, the centre of effort of the sails will be so too; and their right height, as well as that of the masts, will be determined. A little more calculation, and an attention to the plan of the ship, will be necessary, in order to find out that absolute direction of the effort of the impulsion of the water on the bows. But that should not prevent the inquiry. On the contrary, it should be an additional inducement to those, who, building such good vessels as we are now possessed of, and which might still be of a more advantageous form, will be desirous to make them more perfect, by masting them more advantageously. This would undoubtedly be the case; for, several vessels have had their masts cut shorter, and the practice has been attended with decided success. These facts, which could be attested by many able seamen, will always speal highly in favour of this principle; although, when that shortening was made, the sails were not widened in proportion.

"But," continues M. Brue, "in carrying this inquiry farther than it ever was, the intersection of the two above-mentioned lines (*viz.* that of the absolute impulse of the water on the bows, and that of the perpendicular of the centre of gravity of the surface of the floating-line of the ship) cannot take place unless in a direct course; and, as soon as the course becomes oblique, they no longer meet. The centre of gravity of a floating-line's surface of the ship passes to the leeward of the axis, on account of the inclination which always occurs in that sort of course: and the direction of the shock of the fluid, which then takes its origin a little to leeward also of the bow, passes, in its prolongation, to windward, without meeting the perpendicular at the centre of gravity of the floating-line's surface;" (which is easily conceived, if we present to our imagination the horizontal edge of that floating-line's surface but ever so little inclined;) "whence it results that no *point velique* will be found in any course but a direct one. Which is true, unless we could fancy such a ship as would neither drive nor incline in an oblique course. But that is not possible; and hence no perfect mode of masting could be discovered in the last case of the oblique course."

This is true, strictly speaking; for, in each instant of a course, a different point of the bow is struck by the water; which is owing to the pitching of the ship, the continual variations in the strength of the wind, and the greater or smaller inclination produced by the rolling motion of the ship.

"But," says again M. Brue, "the *point velique*, relative to these various circumstances, varies therefore in the proportion of the almost infinite variety of those circumstances, which accompany the course of a ship; that is to say, according to all the degrees of drift, all the degrees of inclination on either board, forward and abaft, as many times, in short, as there are new points of the bow either struck or no longer struck, the *point velique* ascends or descends.

"I pass over the minute examination I could make of each particular cause which contributes to lower that *point* from its utmost height, which is, in the direct course, to its lowest, which takes place in the most oblique course, accompanied with the greatest lateral inclination of the ship. And there is no method to get out of that common road which is pursued in determining the dimensions of the masts, but that of attending to the following considerations; *viz.* Such a ship, being in ended for such a latitude, the wind she is most commonly to expect there will be nearly of such a strength, and generally oblique to her course by so many degrees; so that her most common drift will be nearly so many degrees, and her lateral inclination so many, &c. To give her, therefore, the most suitable masting, making her relatively perfect, we must seek for her *point velique* in what situation we shall think most convenient, and there place the centre of effort of her sails."

All this reasoning tends evidently to the shortening of all the masts, and proves the necessity of doing it, at the same time as it determines their height. The most difficult point in that operation is, to find out the direction of the absolute impulsion of the water on the bows, when the ship steers a course close-hauled and one with the wind on the beam, with such an inclination as the ship could be supposed to have in either of these two courses; when the wind will allow to have four square-sails set, together with the mizen top-sail. Considering these two suppositions of the wind on the beam, and close-hauled, it will be easy to determine the height of the masts proper for that double situation; because, if the gale blows harder, one may lessen the number of sails; if weaker, one may increase it by adding stay-sails, top-gallant-sails, jib, &c. If the wind gets more aft, then the surface of the sails may be increased again by adding the studding and top-gallant royal sails: finally, it is very clear that top-gallant and top-gallant-royal sails will always be of service when the centre of effort of the sails should ascend.

General Observations on the Effect of more or less Surface of Sails exposed, in various Weathers, to the Wind.

I. When a ship, with a certain quantity of sail, has acquired the utmost velocity with the power which then puts her in motion, it is certain, that, if the surface of the sails are either increased or diminished, the rapidity of the headway will likewise augment or lessen in a very complicated ratio. In order to find out the degree of impulsion of the wind upon the sails, multiply their surface by the square of the excess of the velocity of the wind on that of the ship, or which is the same thing, by the square of the apparent velocity of the wind. Then a second multiplication of that product is to be made by the square of the ine of the angle of absolute incidence, or, in the second case, by the square of the sine of apparent incidence. And this second product will give the degree of the absolute impulse of the wind on the sails, in the actual state which we have supposed.

In order to find in what ratio the surface of the sails is to be augmented to make the ship acquire a certain degree of velocity above that which she possessed under a supposed particular quantity of sail, it must first be known by how much the velocity of the wind exceeds that of the ship; then, knowing how many degrees her head-way is wished to be accelerated, the sails must be increased in the ratio of the squares of the two velocities of the ship; *viz.* that which was known before the alteration of the sails, and that which she is afterwards to acquire. But, as the ship recedes so much the more from the action of the impulse of the wind as her velocity increases, it is evident the surface of the sails must be increased also in the ratio of the square of the two excesses, that is, the different excess of the wind over that of the ship both before and after the increase of the sails; then the ship will acquire the wished-for velocity; provided no other cause happens to oppose it, as we have already hinted before, and as we shall have an opportunity to shew more particularly hereafter.

Suppose the wind has 12 degrees of velocity, and the ship under a certain set of sails, has three; the velocity of the wind, in the direct course, will exceed that of the ship only by 9 degrees. If the velocity of the ship is intended to equal the third part of that of the wind and to have therefore 4 degrees for head-way; then the sails are to this effect to be increased in the ratio of the squares of the two velocities 9 to 16, because the resistance of the water on the bows will increase in that very proportion. But, in the first case, the velocity of the wind exceeded that of the ship by nine degrees, while, in the second case, it exceeds it no more than 8. Hence it results, that the impulse of the wind on the sails has diminished in the ratio of the two squares 81 to 64; and, in order to repair that loss in the impulsion of the wind, the expansion of the sails is also to be increased in that last ratio of 64 to 81; then the ship will be able to run with the degree of velocity desired.

II. When the masting is perfect, that is to say, when the ship is masted according to the *point velique*, she will rise from the water parallel to herself by a certain quantity relative to her velocity, and she will rise always more and more in proportion as she acquires new degrees of velocity in her head-way: because she is moved by forces which stand exactly and continually in equilibrium with the action of the water on her bows, the inclination of which forward contributes so much the more to that rising out of the water as it is more remote from the perpendicular. For, then, the vertical impulsion will have more power, since it acts more directly on a very oblique bow than it would on a vertical one. This reasoning may be as exactly applied to the direct impulsion, the absolute effort of which may be decomposed, since it acts less against the velocity of the sailing on an oblique bow than on a vertical one, while the other part of its action joins with the vertical impulse to raise the head of the ship, which shocks the water with very great strength when she is arrived to a very great velocity, and which water opposes her so much the more as it is shocked with violence. So it is easy to conclude, that, in any ship whatever, the more rapid the head-way is, the more parallel to herself she rises above the water, if the centre of effort of her sails is at the same height as the *point velique*: for, the point of the bows, on which may be considered as united the action of the water which opposes its progress, may be taken also as the point of bearing. So that all the sails acting from abaft to forward on different points of the axis of the ship (she being considered as a lever in the direction of her length), they raise the after-part of that point, and place it on a level with the elevation of the bows; which never can happen, if the centre of effort of the sails is above or below the *point velique*. If it is placed above, the power of the sails, acting on too long levers, will raise the after-part of the point of bearing of the bows above the level of the elevation of the ship's head. If it is placed below, the power of the sails acting on two short levers, the after-part of the ship will remain plunged in the water, without being able to rise on a level with the bows. Therefore, in either of these two cases, when the centre of effort of the sails is either above or below the *point velique*, the ship, however well-built, will lose some of the qualities of sailing, either in her readiness to obey the helm, or in her steadiness to carry sail, especially if she is over-masted: for, in this last case, she will gripe, incline easily, and lose much of her head-way, since her bows will plunge in the fluid, or, rather, her stern will rise too much out of it; which will diminish the action of the water on the rudder, and increase it on the bows. In the last case (an inconvenience of which ship-builders seldom, if ever, have been guilty), the ship will be slow to obey, and her head-way will be slackened, because she will never present her most advantageous water-lines to the fluid, nor have a sufficient surface of sails, as, although their width is the same, their height is not so. The point of perfection then is this, *viz.* when the centre of effort of the sails is placed at the height of the *point velique*.

III. The next proposition will appear a paradox to many seamen. But, it is no less self-evident.

There are many cases in which the adding of a few

sails, instead of increasing a ship's velocity, retards it. It is, however, an error into which all seamen almost continually fall, when, in a strong gale, they want either to distance or approach a ship. When their own ship is arrived to a very great velocity (sometimes of twelve or more knots an hour), if they have to do with an adversary the rapidity of which is nearly equal to that of their own ship, they fancy that, by adding more sails to those they have already, at the time when their ship is perhaps best disposed and arrived at its utmost degree of swiftness, they shall increase the rapidity of her head-way; and, accordingly, they hoist up some additional stay or studding-sails, especially if the wind happens to be on the beam, or a little more aft. But, by this their expectations are baulked; for the ship becomes more inclining, her head plunges, and the resistance of her bows increasing in the direction of the keel more than the effort of the sails in the direction of the course, the rapidity decreases in so much as the water acts more powerfully than the sail does. Besides all this, those forward and lateral inclinations of the ship, produced by the effort of the new-added sail, which have caused the centre of effort of the sails to ascend, and the *point velique* to descend (if the new-added sail has been set above it), cause also the ship not to rise from the water parallel to herself; she rises her stern, and plunges her head; whence it results that she gripes from two causes: first, because, as her stern lies less in the water, the rudder is of course exposed to a less shock; and the stern, which always acts as a sail, is more easily mastered by the wind which strikes it then on the beam with a great deal more efficacy than it does her head: on the other hand, as the resistance of the water on the lee-bow has increased by the inclining of the head, plunging thereby that part of the ship which is the most full in its shape, and increasing also the surface on which the water acts, which has both diminished the head-way and increased at the same time the lateral impulse on the side of the bow; so that lateral impulsion forces the ship to windward more at the head, than she is impelled to fall off by the lateral part of the effort of the new-added sail. Whence it follows that the ship becomes still more griping, which is an additional cause of the decrease of her head-way; because, the helm being more a-weather, in order to hold the ship better to her course, more of the rudder itself is presented to the run of the water; and, by the great surface it offers more directly to its shock, retards the velocity of the ship. Whence we are to conclude, that as soon as any more sails are added to a ship which carries already a sufficient quantity of them, she will lose her qualities of steering well and making good head-way, whether those additional sails are set forward or a-stern.

Practice of Working *Ships.*—The useful part of a seaman's duty is the application of his theoretic knowledge to the various evolutions of a ship. We are now about to shew how those various evolutions should be performed, consistently with the principles upon which the science of navigation is founded.

A ship, when launched from the stocks, is furnished with one stream-anchor, and another, which is termed a launching-anchor, somewhat heavier than the stream-anchor, hanging by stoppers to the cat-heads: by one or both of these she is first brought up, and afterwards warped to what are termed *Head and Stern* Moorings; which article see.

Of the Ballast and Lading.

When a ship is loading, it should be considered, that her tendency to pitch or roll depends not alone on her form, but even more upon the greater or less advantageous distribution of the heaviest parts of her cargo.

Particular attention is to be paid to moderate her pitching, as that is what most fatigues a ship and her masts; and it is mostly in one of these motions that masts are seen to break, particularly when the head rises after having pitched. Although the rolling be proportionably a more considerable movement than pitching, seldom any accident is seen to arise from it, as it is always a slow one. It is, however, not less proper to prevent it as much as possible. This will in general be easily obtained, without being any way detrimental to the ship's stiff carrying of sail, if, when the ballast is iron, you stow it up to the floor-heads; because it will recall the ship with less violence after her having inclined, and it will act on a point but little distant from the centre of gravity. To make this clearer, let us consider it by the assistance of a figure.

First, let it be understood, that the *centre of gravity* of a body is that point by which it may be suspended, and the parts remain in perfect equilibrium.

That the *centre of cavity* is the centre of that part of the ship's body which is immersed in the water; and which is also the centre of the vertical force that the water exerts to support the vessel.

That the *centre of motion* is a point upon which a vessel oscillates or rolls when put in motion.

That the *metacentre* is that point above which the centre of gravity must by no means be placed; because, if it were, the vessel would overset.

Various methods have been recommended to find all these different points; some of them are in their nature fixed, others varying. Thus, when a ship is completely loaded, the centre of gravity is fixed, howsoever the vessel may alter her position.

The centre of motion is always in a line with the water's edge, when the centre of gravity is even with or below the surface of the water; but whenever the centre of gravity is above the water's surface, it then becomes the centre of motion. This must be understood of bodies not perfectly circular; for, if circular, the centre of motion will be the centre of the circle.

The centre of cavity varies with every inclination of the ship, because that depends upon the shape of the body immersed.

The metacentre (which has been likewise called the *shifting centre*) depends upon the situation of the centre of cavity; for it is that point, where a vertical line, drawn from the centre of cavity, cuts a line passing through the centre of gravity and being perpendicular to the keel.

Let the segment of a circle 1 2 3, (fig. 11, plate XXXIV.) represent the transverse section of a vessel's bottom; W L, the surface of the water; M, the metacentre, as well as the centre of motion, because this is a circle; C, the centre of cavity; G, the centre of gravity; and the line 2 4, the vertical axis of the vessel which may be turned round the point M, as on a fulcrum supported by the centre of cavity. By thus simply considering the vessel as a lever in the direction of her vertical axis playing round the centre of motion, it is plain that, if the centre of gravity was placed above the point M, being the metacentre too, the vessel would upset; therefore, that the ship may have stability, the centre of gravity must be below this point. And it may be observed, that, the farther G is removed from the metacentre; the greater must be its force, as the gravity then acts with a greater length of lever, considering the fulcrum of that lever to be at the centre of motion; or, if the weight at G be augmented, it will likewise increase the force; therefore the force of G may be expressed by multiplying the balance of weight, beneath the centre of motion, by the distance of the centre of gravity from the centre of motion.

The centres of cavity and motion, in circular bodies, will ever be in a line perpendicular to the horizon, but the centre of gravity may be either on one side or the other of this line.. When such a body is at rest, the centre of gravity will be in this line; but, if in motion, it will be diverted from it. Thus the points M and C will always be perpendicular to W L; but the point G, by the body's rolling, may be on either side; for instance, at *g*. While G is perpendicularly beneath the centre of motion, its action can only tend to *preserve* this circular body in her erect position; but, if it is removed to either side, as to *g*, its action is to *return* it to the erect position; and this action increases as the distance G *g*, which is the sine of the angle of roll *g* M G, the distance of M G being considered as the radius. Thus, to gain the force of gravity with any roll, as *g* M G, let the balance of weight beneath the centre of motion be multiplied by the sine of the angle of roll G *g*.

But the tendency to roll may be also diminished by the shape of the hull. For, let us suppose that the transverse section be allowed more beam and increased by the dotted lines. Now, when this vessel is rolled over, it is plain that the cavity will be augmented towards the side L, of course its centre must remove towards L, say to *c*; and, if from *c* be erected a perpendicular to the horizon, it will cut the vertical axis at *n*, which will, in this case, be the metacentre, above which, if the centre of gravity were placed, it would act in conjunction with the centre of cavity to overset the vessel: but, as the centry of gravity is here below it at *g*, her stability will be increased by the increased distance of G from *n*, the metacentre; and the vessel will roll round the point M as her centre of motion.

When sailing in smooth water, the greater the stability the better; but, if a vessel with a heavy cargo, stowed low in her bottom, be sent out into a rough tempestuous sea, where every wave will throw her from her equilibrium, she will return with such violence as to endanger her masts; and, should she be dismasted, her roll will then be with still greater force, possibly to the destruction of her hull. Was the cargo in this laboursome vessel to be removed higher up towards the centre of motion, so as to lessen her stability, she would be found considerably easier; her roll would be by such deliberate motions, such as to lessen the danger to her masts and hull.

The ballast is placed round and very near the centre of gravity of the ship, because it will prevent the motion of the pitching being so hard as it would, if that weight were distant either afore or abaft that point. Whenever the sea runs a little high, the ship is never carried by a single wave; there are generally two or three always passing under at the same time, unless when the sea is extremely long, the swells coming from a great distance, and in latitudes very remote from land: for then it happens that the largest ships are sometimes carried by one single wave. But in either circumstance, the ballast ought not to be stretched afore or abaft the centre of gravity, as soon as the ship is in the parallel to her draught of water marked for the ballast, which it is absolutely essential to pay attention to. To prove this principle, suppose in either case a long or short surge, and that the water strikes the ship forward, that thereby she may be exposed to the greatest and hardest pitching: for, when the wave takes a ship under the stern, her motions, if she has got a little headway, are not dangerous; because, as she flies before the wave, she recedes in some measure from its impulse; while, in the first case, she increases on the contrary that same impulse in the ratio of the square of all her velocity.

First, the ship, whose extremities are light, or little loaded, being supposed to run with any velocity whatever against the wave which comes to her a-head, shocks that wave with a force expressed by the square of the sum of the two velocities: she divides it and goes through it, at the same instant that she is raised by the vertical impulse of that column of water, which opposes to her a supporting power too considerable for her weight to displace: the wave which follows, produces the same effect in receiving the fall of the ship, because the first is already under the middle of the ship, whence it passes to the stern, which is supported by it, while the second takes its place in the middle, and the third is come to support the head; and this is an uninterrupted succession. This motion continuing thus as long as the sea is agitated, it follows that the ship is never at rest: no sooner has she been raised by a wave, but she falls again when that wave is gone, which falling is proportionably less sharp as her head is les heavy: the shake is then less violent, since she shocks the water with a less mass, which prevents her pitching so deep as she would if she were more heavy; consequently the masting does not suffer, and the head-way is less delayed, as the fullest part of the bows is not so much exposed to the shock of the water.

Secondly, When the ship is carried by one single wave, her fall is still less sharp, if little loaded a-head, than when she is carried only by the middle. She rises, therefore, more easily at the moment the other wave comes to strike her, and the shake is not so violent. Were she to plunge deeper into the fluid, it might happen that the column of water would become higher than her head, and, passing partly over it, would expose her to the danger of foundering.

In the stowing of the cargo, it is proper to place the heaviest part of the stowage as low as possible, taking care to preserve that draught of the ship which is most advantageous for her, whether she be in ballast or when laden. Those points are marked both at the head and stern: in a word, the great art of stowing lies, in endeavouring that each of the vertical parts, in which the extremities of a ship may be supposed to be equally divided, be lighter, when her lading is complete, than the weight of the mass of water they are to displace; observing always, that the vertical parts of the middle admit of being loaded more heavily than the weight of water they are able to displace. For the practice of stowing ballast, see the article BALLAST.

Manner of getting on Board, and Stowing the Anchors and Cables.

Anchors, though they bear different names on board of ships in the royal navy, as sheet, best bower, small bower, and spare, are of the same weight. The stream and kedge anchors are smaller, and grapnels are for boats.

Ships of 110, 100, 98, and 90 guns, have seven anchors; from 80 to 20 guns inclusive, six anchors; ships of 300 tons and sloops have five; and brigs and cutters three anchors.

In fitting out ships, the anchors are brought in craft near the bows, being most convenient to the hawse-holes, through which the cables pass to be clenched.

The bower-anchors are first catted; which is performed by hooking the hook of the cat-block into the ring of the anchor, and bowsing upon the fall that leads in through a snatch-block on the fore-castle; the cat-falls being previously reeved through the sheaves of the cat-head and cat-block, keeping the hook of the cat-block downwards, and its point inwards. They are then fished, by means of the half-davit, pendent, and tackle, thus: The davit is first stepped in the channel on the side wanted, and supported by guys; the masthead-guy goes over the end of the davit, with an eye. The other end fastens round the fore-masthead, with a round turn and two half-hitches. The foremost guy goes over the end of the davit; the other end fastens round the cat-head, with two half-hitches, and is securely stopt. The after guy goes over the end of the davit, and makes fast with two-half-hitches, through an eye bolt in the after part of the fore channel. At the outer end of the davit is hung, by its strap, a large single block; through which is reeved the pendent, with a large iron hook spliced in the lower end, to hook the anchor within the flooks; then to the inner end of the pendent, is made fast a tackle, by thrusting a toggle through an eye in the block-strap: after that has passed through an eye in the pendent, the other block of the tackle is hooked in an eye-bolt in the fore-part of the quarter-deck; the effort of the tackle is communicated to the hook, by means of the pendent, by men's bowsing on the tackle-fall. Thus the flooks of the anchor are raised and placed on the gunwale, where it is made fast by the shank-painter chain. That the flooks may lie level, the shock is bowsed upon by the anchor-stock tackle, the double block of which is hooked to a selvagee, fastened round the stock of the anchor under the first hoop, and connected by its fall to a single block, hooked to a selvagee, fastened round the lanyard of the main stay: the fall leads in upon the fore-castle.

The best bower is then placed forward near the bows on the starboard side, the small bower near the bows on the larboard side, a little abaft their respective cat-heads, and are secured by their stoppers, from the cat-heads and shank-painters. The stopper has one end clenched round the cat-head; the other end passes through the ring of the anchor; returns upwards, and leads over a large thumb-cleat bolted to the cat-head, and is made fast with several turns, and the end hitched round the head-rail and timber-head, on the fore side of the cat-head. The shank-painter hangs the shank and flook of the anchor to the ship's side without-board; and, when stowed the shank-painter is passed under the inner flook round the shank of the anchor, and made fast with two or three turns, and the end stopped round timber-heads on the forecastle. With these two bower-anchors, ships are generally moored when lying in a tide's way or in a fleet.

The sheet and spare anchors are hoisted by runners and tackles, main-stay and yard tackles, and are stowed securely with stock and bill lashings at the after part of the fore-shrouds, before the chess-trees on each side of the ship, with one of their arms resting on a chock, bolted to the gunwale; the stock being bowsed-to by the anchor-stock tackle. The sheet-anchor is stowed on the larboard-side, and is the first resource in a gale of wind, after parting with either of the bowers; for which reason, when in port, the sheet cable is kept bent, and the anchor is over the side, suspended by the stopper and shank-painter, ready for cutting away in case of necessity. The spare anchor is stowed on the starboard-side, and is seldom used, but when one of the others is lost.

The stream-anchor is stowed on the spare anchor; and, when used, it is sent in the long-boat or launch, with its cable bent, and let go at any particular spot, either for steadying the ship, when riding by only one bower, or to assist a ship when in shore, or to warp her, &c.

The kedge is stowed on the stream and spare anchors, and is frequently used to stop a ship for a tide in little winds; but, if the wind is too powerful for the kedge, the stream-anchor is substituted. The kedge is sometimes used, in moderate weather, to warp the ship so as to shift her birth.

The cables also are brought in craft along-side; and, should they be new from the rope-walk, let them be coiled down in the craft, the same way they are to be coiled on board.

It is recommended to merchant-ships, especially of the smaller size, that their cables should be coiled the way they are bit, or the way they run round the windlass; and their tiers should be on the side opposite to that on which they lead. But this practice cannot be approved of for the royal navy, nor is it indeed there adopted; because, when heaving in upon one cable, and veering out upon another, the cables, by being crossed, are apt to foul in the hatchway. The best bower, which is mostly the working-cable, should lead foremost up the hatchway on the starboard side; then the small bower on the larboard-side; and, afore the latter, the sheet; which, being the least used, can be triced close round the fore-part of the hatchway, out of the way.

A cable generally grinds or kinks from more turns being forced into it by the coiling than it had when first made; and the only way to get rid of those grinds or kinks is to coil the cable across the hatchway from side to side, in large fakes, with the sun: then take the upper end through the coil, and coil the cable down in the tier the way required. By this means, as many turns will be taken out of the cable as there are fakes coiled round the hatchway.

It should be a rule in coiling cables, never to lay out near the hatchway, but to keep that part of the tier as low as possible, that the bends may have sufficient room to upset.

Were all store-cables first coiled down from the rope-walk against the sun, they would be better adapted to coil on either side of the ship; for a cable coiled against the sun will more easily reverse, and have less grinds or kinks in it than a cable coiled with the sun.

Anchors (except when at sea) should always have a cable clinched to the ring ready for letting go, and a buoy-rope with a buoy.

To clinch a cable; run it through the hawse-hole and the ring of the anchor three or four fathoms in length; then haul the bight up in the head, and pass the end of the cable over the bight, and through the ring, between it and its own part: then pass the cable-bends, and cross them with strands, well greased, one at the end, and the other about one foot from the end; and be careful not to form the clinch larger than the ring of the anchor.

A ship, when fitted, removes from her moorings, in order to take in her guns and proceed to sea; but, before she finally sets sail, she generally comes to anchor, and is moored once with her own anchors and cables, which will be treated of hereafter.

Preparations necessary to be made for Anchoring.

On approaching an anchorage, the anchor and buoy are got clear, and a range of cable stretched along the deck suitable to the depth of water. Care should be taken, that nothing is in the way to check the cable, or stop its running out: then, at a proper distance, a turn is taken round the bits with the cable, thus: First pass the cable from the anchor underneath the cross-piece, then take up a bight of the cable abaft the bits, and throw it over the bit-head. The end of the cable is clinched round the orlop-beams in the royal navy, and round the main-mast in the merchant-service.

It is necessary to have water near the bits, to prevent its taking fire by the friction. Stoppers and ring-ropes of all kinds should be ready for use.

The stock-lashing being cast off, and nothing but the anchor-stopper and shank-painter retaining the anchor, men are stationed to stand by them, and let go at the moment ordered.

To secure the cable when out, *deck-stoppers* are thus previously prepared: they are turned into the ring-bolts on the deck, round a large iron thimble, and fastened with a throat and end seizing. Each stopper has a lanyard spliced round the head, under the knot, by which several turns are taken round the cable, and the end stopt.

Besides the deck-stoppers, others are used as an additional security to the cable; such are the bit-stoppers, &c.

Bit-Stoppers. Each stopper is reeved through a hole in the standard-knee, against the fore-part of the riding-bits, and is turned-in or spliced. It has a lanyard spliced round the head, under the knot. When used, several turns are taken with the lanyard round the cable, and the end stopt. It is to check the cable in bringing up the ship. Another bit-stopper, much approved of is about four fathoms long, and tailed out like a nipper at one end, and knotted at the other. Let this stopper be rove through the hole in the standard-knee. To pass it, let it be held aft, inside, over the cable, and under the bits, outside the cable; then worm it round the cable before the bits. Then, as the cable runs out, and it is required to check the ship, haul tight the worming; and, by the cables drawing forward, it will tighten the stopper, and bend the cable so close to the bits as effectually to bring the ship up. This stopper is not likely to jamb, therefore is extremely well calculated for bringing a ship up with ease: as, by slackening, and hauling tight the worming, the cable may be suffered to run out, or be checked at pleasure.

In heaving up in a heavy sea, when by a sudden pitch of the ship, the messenger or nippers give way, this kind of stopper will be found extremely serviceable; for, upon these occasions, this stopper may be always passed ready, and the bight triced up abaft the bits with a rope-yarn clear of the cable.

Another bit-stopper, made with a large eye, that it may be thrown over the bit-head, and shifted over from side to side, is also much approved of.

Dog-Stoppers. One end is clenched round the main-mast, and the other end wormed in the cunt-lines of the cable, and stopt in several places; then brought back with several turns over its own part, and the end stopped. It is of little service, unless it be long enough to clap on above the coamings.

Wing-Stoppers. One end is clenched round the orlop-beams in the wings, and the other end is clapt on as the dog-stopper.

Ring-Ropes are occasionally made fast to the ring-bolts in the deck and to the cable, by passing the ends through the ring of the bolt, and through the bight, then clapt on the cable with cross turns, and the ends stopped. Ring-ropes may be better single than double; they are passed with less confusion of turns. To pass a single ring-rope, and have it in readiness to check upon veering away the cable, take also three slack turns through the ring-bolt and round the cable, one before the other, and hold up the parts fair; then take as many slack turns of worming round the cable, before the ring, and they held up fair, leaving sufficient room for the cable to pass through. When the cable is to be checked, haul tight the worming; and, by the cable's running out, it will readily draw the turns tight through the ring, and bind the cable so close to the ring as to prove an excellent stopper. Ring-ropes are similar to the lanyards of stoppers, to check the cable when freshening the hawse, or to add security to the stoppers in a heavy sea.

Lengthening of Cables.—Cables are lengthened by splicing one to the end of the other, thus: the closest and best method is to put the ends in twice each way; then, to pick out the strands, and worm part of them round the cable, and taper away the rest; which should be closely marled down, and a good throat and end-seizing clapped on of six-thread ratline.

The strands of the best bower and stream cable had better be pointed, that these cables may be more quickly spliced and unspliced in cases of necessity.

To prevent Cables from chafing by friction in the hawse, and against the stem or cutwater, they are rounded or served thus: they should be served against the lay. The most expeditious way of clapping on rounding is with a top, where room to work it will admit; otherwise it must be beaten on with mallets; and care taken to stop the service with spun-yarn at every six or eight turns.

Mooring-services are clapped on about fifteen fathoms from the end or cable-splice. Large vessels should have twelve or fourteen fathoms of service, half of it rounded and the rest plaited and keckled Upon the best bower, or working cable, there should be a short service of eight or ten fathoms at the half-cable.

Mats of the width of the cable's circumference, and about three fathoms long, are very convenient to have ready to lace on the cable with expedition, in cases of necessity, in the way of the hawse or cutwater.

The best service to prevent a cable's chafing is cut from a tanned horse-hide, big enough to wrap two or three times round the cable. The method of putting it on is, first to parcel the cable with two or three turns of old canvas, the length of the leather service; which, if too stiff to put on, only requires dipping in water and beating, which makes it soft and pliable: then pass it tight and smooth round the canvas, and stop it on with sennit or three-yarn knittles, well greasing them and the service before veering it into the hawse-hole. Avoid raising the surface of the leather by knittles, &c. underneath the canvass.

Of Coming to Anchor.

A ship ought always, unless under some particular circumstances, to be brought to anchor under an easy sail, such as the three top-sails, jib, or fore-topmast-staysail, and sometimes the mizen, according as the vessel has more or less inclination to fall off or come to the wind.

There are, no doubt, cases when more sail may be required; but they are exceptions to, and do not destroy, the general principle. Nor should an anchor ever be dropped to leeward of the place you mean to bring up in; because that would often occasion a necessity of casting two anchors at once, for fear of dropping still more to leeward.

When the wind is so violent as to bring the anchor home, and make the vessel drive, the cable is veered away; and, in veering away, the turns of the stopper-lanyards are slackened, and a portion of the cable suffered to go out of the hawse, to let the vessel further a-stern of her anchor; in which situation she bears less strain on the flooks, and is less liable to drag the anchor; for, the more cable is out, the flooks become deeper buried, and the ship rides in greater safety.

In letting go an anchor, great care should be taken that the water be not so shoal as to endanger the ship hurting herself upon it, and that the anchor be not fouled by the cable getting about the flook or stock.

Nor should the water be too deep, because the cable, when out, should approach as near as possible to a horizontal direction. Indeed, this principle is so true, that three cables spliced together an-end of each other, are kept bent to the best bower-anchor, to be used in cases of necessity; and it is found, that one good anchor, with a long range of cable, is a safer anchorage than two anchors with short cables. However, when the ship has not room to drive, and if the night be dark, let fall a second anchor under foot, with a range of cable above the deck. At all events, the deep-sea lead should be thrown over the gunwale, and the line frequently handled, to be certain that the ship does not drive.

In hard and rocky bottoms, where anchors cannot have much hold, cables are chafed and cut to pieces. When necessitated to anchor in such places, a chain should be run up the cable from the ring of the anchor to a certain distance, to secure it from danger. When a chain is not to be had (although the top chains may serve), empty casks, well bunged, are good substitutes, slung and fastened to the cable at equal distances, to support and keep it from the bottom.

When ground is soft and oozy, and anchors will not hold securely, but come home with little wind, it is common to cover the flooks with a broad triangular piece of plank, much larger than the flook. Sometimes the anchor is backed, or retained, by carrying out the stream, or kedge, a-head of the anchor the ship usually rides by. In this situation, the bower-anchor is confined by the stream, or kedge, in the same manner as the ship is restrained by the bower-anchor.

In preparing to come to anchor, when the wind is not violent, the topsails ought always to be clewed up at the mast-heads: that is, let go the sheets and haul the clew-lines and bunt-lines close up; lower away the topsails and take in the slack of the braces as the yards come down. In this manner you run less danger of splitting and tearing the sails than by any other method.

In all operations hereafter treated of, it should be observed, that whether the wind is moderate or blows fresh, it makes this only difference, viz. the velocity of the ship's movements, in the latter case, being considerably increased, the sails will require an earlier diminution to stop the head-way; and that, of course, less time in general is taken up in performing every operation.

To Anchor with a Spring, in order to present the Vessel's Side to a Place or Ship you wish to Cannonade.

This is executed when you know that the wind or current will bring your head, when at anchor, towards the object you mean to attack: for, should the wind or tide bring your broadside to bear on the object you mean to cannonade, the spring would only be a precaution to get under way more quickly in case you were obliged to retreat, or in case the wind or tide should shift.

Get a large snatch-block in the aftermost port, on the same side you wish to present to the wind or current, and on the same side with the anchor and cable with which you mean to bring up; then, through the block, reeve a hawser, the end of which is to be clinched to the ring of the anchor you mean to let go; the other part is brought to the capstan, with necessary ranges of the cable and hawser on deck. That done, and the ship being arrived at the birth, you are to deaden her way according to circumstances: you let go the anchor, and veer away enough cable and hawser, now a little more of the one, and then a little more of the other, according as you wish to present more head or stern; which you can do by heaving on the spring, or, which is the same, veering away more cable. Should you find it requisite to shift your position, you have only to veer out more of the hawser.

To come to an Anchor in Roads that are often crowded with Ships, and to leave clear Births for others.

The best anchoring-births in these places are mostly known by marks, and of course are occupied by the first ships.

In a tide or trade-wind roadstead, the next ship that comes should not anchor right a-head or a-stern of the first ship and so as to lie on the other's hawse, but should come-to on the bow and quarter at a sufficient distance to prevent other ships from coming between, and in a slanting direction from the tide or wind. This might contribute to the safety of ships when it blows strong upon a lee-tide or in strong sea-breezes, as each single ship may then veer away what cable may be necessary, and keep clear of the other ship's hawse a-stern; or, in case of driving or casting, they have a better chance of keeping clear of each other.

A good anchoring-birth in a crowded roadstead is obtained by first running down through the middle of the fleet, and taking notice where a good birth is left vacant by some ship that has sailed from the middle of the fleet; then steer out from among the ships, and turn to windward so far, as to give time to take in and furl all the sails, and run down before the wind amongst the ships, without any sail, and let go the anchor at the intended birth.

To come to an Anchor with the Wind across the Tide.

The ship should, if possible, be put upon the tack that stems against the tide when the anchor is let go: and, if it be designed to continue at a single anchor, in order to keep it clear, sheer the ship, and keep her to leeward of the anchor, by keeping the helm a-weather and the fore-topmast staysail set, with the sheet to windward.

Much benefit may result from letting go the anchor stemming against the tide, especially with a rapid tide; for it gives an opportunity to observe at what rate the ship drives a-stern, so as to judge whether it may not be necessary to keep sail set, in order to bring the ship up to ride easy in a rapid tide, and keep her clear of shoals, &c. a-stern.

To come to an Anchor when the Wind is right against the Tide, the Ship driving with the Strength of the Tide against the Wind.

Shoot the ship a-head of her anchor, or sheer her clear of it, upon the same tack as she is meant to shoot upon the next tide, always endeavouring to keep the ship, in swinging with the tide, on one side of the anchor to clear it. Suppose that the ship, driving to windward, has got to an anchoring-birth, or that the tide is so far spent that she will drive no farther to windward, and must come to an anchor on the starboard tack. In letting go the anchor, the ship should be shot a-head of it, and kept a-head with the helm a-weather, the yards braced full with the larboard braces, the fore-topmast stay-sail and mizen set full, till the windward tide is done; then she falls to leeward and rides windroad, with the wind and anchor right a-head; in which position she will lie clear of the anchor till the next windward tide.

To come to an Anchor without Tending.

If it happen that a ship is to be brought up in a place where there is not sufficient room to tend her, reduce her head-way as much as possible, before she comes to her anchoring-birth, so that a less scope of cable will bring her up.

To get up an Anchor, in Ships which have a Main and Jear Capstan.

In large ships which have a main and a jear capstan, and the strain is thought too great for the messenger alone, the voyol is used thus: three or four turns are taken round the jear capstan with one end, so as to leave that side clear on which the cable is coming in; and pass the other end through the voyol-block, which is lashed round the main-mast on the lower deck. It is then carried forward, and passed round the rollers in the manger near the hawse-holes; then brought aft, and spliced to the other end with a short splice, and the ends marled down tight. That side of the voyol on which the cable is coming in is fastened to the cable by nippers; and thus the continued efforts of the capstan are conveyed to the cable, until it is hove in. The nippers are clapt on in the manger, from one to two fathoms asunder; and the voyol is applied to the midships or, inside of the cable. Nippers are clapt on by taking three or four turns round the voyol, four turns round the cable and voyol, and then three or four times round the cable. This method is an exceeding good one, and very suitable to quick heaving; but, when the strain is great, and the cable muddy, the nippers clapt on after this method will not nip sufficiently; and sometimes recourse is had to the following method: throw sand or ashes upon the cable, and take a long dry nipper, which middle, and pass one half aft, racking it in and out round the cable and voyol; then worm its end round the voyol only. After this, pass the other half in the same manner forward, but worm its end round the cable only, and let each end of the nipper be held on. The advantages of this method are, that, as the strain of the cable lies forward, and that of the voyol aft, the nipper will be drawn so tight as effectually to hold the cable till something gives way: also they can never jamb, for both ends are clear for taking off. Another method, when the strain is great, is, to have nippers with an overhand-knot made at one end; and with that end a round turn taken round the cable and voyol, leaving three or four feet of the end; then, with the other end, take three or four racking turns, and expend nearly the remainder with turns round the cable and voyol, laying the knotted end under and over each of the last turns: the end is then held fast. The men who clap on the nippers are attended by boys, who hold the ends of them, and follow the progression of the cable as it is hove in: and, as the nippers arrive near the main-hatchway, they are taken off and carried forward, where they are again clapt on: and so in succession, until the cable is hove in sufficiently to raise the anchor above the water. It is then stoppered round all before the bits: that is, round the cable and voyol. The anchor is then catted, and afterwards fished. To shift the voyol for heaving in a second anchor, it must be unspliced, and the turns round the capstan reversed. When the strain is so great as to require other purchases, the top-tackles may be used thus: the double block is lashed to the main-mast or topsail-sheet bits, the treble block is lashed on the cable, and the fall brought to the capstan. If the top-tackle falls are thought insufficient, any hawser may be used that will reeve through the blocks.

To get up an Anchor in Ships which have not a Jear Capstan.

Ships without a jear capstan have no voyol, but heave in their cables by the messenger, which has an eye spliced in each end; one of which ends is passed with three or four turns round the capstan on the upper deck, and the other end passed forwards round the rollers, at the fore-part of the manger; then brought aft to the other end, and lashed thus: several turns are passed through the eyes crossing each other in the middle, then a half-hitch is taken round the parts, and the ends stopped with spun-yarn. The remainder of the operation is performed as by the voyol, with this exception, the messenger is applied to the outside of the cable; and, when the nippers are insufficient, the messenger may be hitched thus: the bight of the messenger is fastened round the cable at the manger with a rolling-hitch, and the bight seized round the cable before the hitch. This practice is by no means so good as the others.

When getting under way in a sea-gale, the voyol is better than a messenger, as the sending of the ship carries all the strain to the main-capstan, and endangers the men at the bars; whereas, with a voyol, the strain is taken to the voyol-block, and the men at the fore-jeer capstan heave in safety.

To Get up a Second Anchor.

Suppose, by the former methods, that the starboard-anchor is got up, and that the cable of the second anchor enters the larboard hawse-hole, the operation of getting up the second anchor is the same, observing only, that the messenger must be shifted, and the turns on the capstan reversed, to change the disposition and side: and the men, who before held on the larboard-side in the first operation, will hold on the starboard-side now: the motion of the capstan is performed the contrary way, and the cable on the larboard-side is fixed and hove in. See *Chevalier de* Sauseuils *Manœuvrer, or Skilful Seaman*, translated from the French of M. Bourde de Villehuet: also the *Theory* and *Practice* of Seamanship.

WORM of a Gun, (*tire-bourre*, Fr.) an instrument used to pieces of artillery (see figures 8 and 9, plate VI.), to draw the charge, or to take out the bottom of the cartridge, which sometimes remains in the chamber after firing, and which, by being on fire, might explode the next charge while being rammed home, whereby a man would probably lose both his hands; hence it is indispensably requisite that a gun should be wormed, at least, every third discharge.

To Worm *a Cable*, (*peigner un cable*, Fr.) is to wind a rope spirally about a cable, so as to lie close along the interval between every two strands. It is generally designed to support and strengthen the cable, that it may be enabled to sustain a greater effort when the ship rides at anchor; and also to preserve the surface of the cable, where it lies flat upon the ground, near the station of the anchor; particularly in moderate weather.

WORM-EATEN, (*piqué, ou rongé des vers*, Fr.) the state of a plank, or of a ship's bottom, when a number of cavities are made in it by a particular kind of worm, which abounds in the tropical climates.

Link WORMING, (*cable supporté par des chaines*, Fr.) Cables may frequently be preserved by means of chains twisted round them in the following manner: Take three chains of about fifteen fathoms each in length, all of an equal size, and proportioned in thickness to the nature of the case. Let them be wound round the cable, so that they may project sufficiently to receive the greatest part of the friction, one end of them being fastened to the ring of the anchor, and each chain being from thence wormed round its respective hollow or channel of the cable, so as not to check it from stretching. Fasten the other ends of the chains to the cable. It is not necessary that the chains should be very heavy, but extremely adviseable that the links should be short, because they will thereby be more pliant, and worn much easier These chains will completely guard the cable against the chafing of the rocks, and they may be put on in a few minutes, so that it is unnecessary to have them fitted on, but when it is probable they may be really serviceable. A due attention to shortening cables at the slack of the tide, and veering out as it flows, must also be highly important to prevent their being damaged.

But it frequently happens that cables are suffered to kink, as it is called, through inattention to these circumstances, and to avoid a little trouble. If the wind should freshen when they are in this condition, though the cables be ever so good, the first foul sea that comes against them is almost sure to occasion their snapping; and vessels may be driven on shore and much damaged at least, if not totally lost, before they can possibly be brought up to other anchors, that may not always be in a state of readiness. For if there be an inattention and carelessness in one respect, there may be the same probably in others of equal importance.

As the weight of the three chains, even for a large cable will not exceed 500 pounds weight, the only difficulty which remains to be obviated will arise from the smallness of the hawse-holes, but these could be enlarged in vessels that shall deem it of sufficient moment to make the trial, and with the additional resistance of a boy or spare man at the head of the vessel, to cast off the chains as they come up, and hand them in above, all inconvenience will be removed. Admitting, indeed, that it may take up a few minutes more to weigh an anchor so secured than a common one, surely this can be no argument against a measure that tends to preserve and insure the safety of the vessel and ship's company.

It may not be improper to recommend farther, in high latitudes, where ice may cut or damage the cables while riding at anchor, the use of three other chains of the same nature as the preceding, but not more than five or six fathoms in length. These may be wormed round the cables in the same manner at the surface of the water, and will be an excellent means of guarding them from the ice, so as either to prevent its wounding them, or chafing them quite asunder; a circumstance which is not unprecedented in our own harbours or rivers, but which frequently happens on the continent of Europe.

WRECK, (*bâtiment naufragé,—les débris, ou la carcasse d'un bâtiment naufragé*, Fr.) the ruins of a ship which has been stranded or dashed to pieces on a shelf, rock, or lee shore, by tempestuous weather, or by accident. See the article SHIPWRECK.

To WRING A MAST, (*forcer un mât, ou le rendre courbe; en ridant trop les haubans, &c.* Fr.) is to bend or strain it out of its natural position by setting the shrouds up too taught. This phrase is also applied to a capstan, &c. when by too great a strain the component parts of the wood become deranged, and are thereby disunited.

WRING-BOLTS, (*longues chevilles de fer munies d'une boucle à une extrêmité, et goupillées à l'autre; elles servent au charpentier à gêner et assujettir les bordages contre les couples lorsqu'on les met en place*, Fr.) in ship-building, are bolts used to bend and secure the planks against the timbers, till they are properly fastened by bolts, spikes, and tree-nails.

WRING-STAVES, (*gros bâtons ou leviers ayant leur point d'appui à la cheville dont on a fait mention dans l'article précédent*, Fr.) are strong pieces of wood used with the *Wring-Bolts*.

X.

XEBEC, (*chébec, sorte de bâtiment de guerre de la Méditerranée*, Fr.) a small three-masted vessel, navigated in the Mediterranean Sea, also on the coasts of Spain, Portugal, and Barbary, and is distinguished from all other European vessels by the great projection of the prow and stern beyond the cutwater and stern-post respectively. See fig. 4, plate XXXIII.

The sails of the xebec are in general similar to those of the polacre, but the hull is extremely different.

Being generally equipped as a corsair, it is constructed with a narrow floor, to be more swift in pursuit of the enemy; and of a great breadth, to enable her to carry a considerable force of sail for this purpose, without danger of overturning. As these vessels are usually very low built, their decks are formed with a great convexity from the middle of their breadth towards the sides, in order to carry off the water, which falls aboard, more readily by their scuppers. But as this convexity would render it very difficult to walk thereon at sea, particularly when the vessel rocks by the agitation of the waves, there is a plat-form of grating extending along the deck from the sides of the vessel towards the middle, whereon the crew may walk dry-footed, whilst the water is conveyed through the grating to the scuppers.

When a xebec is equipped for war, she is occasionally navigated in three different methods, according to the force or direction of the wind.

Thus, when the wind is fair, and nearly a-stern, it is usual to extend square sails upon the main-mast; and, indeed, frequently on the fore-mast: and as those sails are rarely used in a scant wind, they are of an extraordinary breadth.

When the wind is unfavourable to the course, and yet continues moderate, the square yards and sails are removed from the masts, and laid by, in order to make way for the large lateen yards and sails, which soon after assume their place: but if the foul wind increases to a storm, these latter are also lowered down and displaced; and small lateen yards with proportional sails are extended on all the masts.

The xebecs, which are generally armed as vessels of war by the Algerines, mount from sixteen to twenty-four cannon, and carry from 300 to 450 men, two thirds of whom are generally soldiers.

The method of working these vessels is so exceedingly complicated and difficult, that the labour is very considerable.

Y.

YACHT, (*yacht*, Fr.) a vessel of state, usually employed to convey princes, ambassadors, or other great personages, from one kingdom to another.

As the principal design of a yacht is to accommodate the passengers, it is usually fitted with a variety of convenient apartments, with suitable furniture.

The royal yachts are commonly rigged as ketches, except the principal one reserved for the sovereign, which is equipped and rigged as a ship. They are elegantly furnished, and richly ornamented with sculpture; and always commanded by post-captains.

It appears from Pepy's Naval Minutes, that the name of *yacht* was not known in England until the year 1660, when the Dutch presented one, called the Mary, to Charles the Second.

The smaller yachts, which are rigged as sloops, are generally used by the commissioners of the navy, at the different dock-yards.

Private pleasure-vessels, when sufficiently large for a sea-voyage, are also termed yachts.

YARD, (*vergue*, Fr.) a long piece of timber suspended upon the masts of a ship, to extend the sails to the wind. See the articles Mast and Sail.

All yards are either square or lateen; the former of which are suspended across the masts at right angles, and the latter obliquely.

The square yards, plate XX. are nearly of a cylindrical shape, tapering from the middle, which is called the slings, towards the extremities, which are termed the yard-arms; and the distance between the slings and the yard-arms on each side, is, by the artificers, divided into quarters, which are distinguished into the first, second, third quarters, and yard-arms. The middle quarters are formed into eight squares, and each of the end parts is figured like the frustum of a cone. All the yards of a ship are eight square in the middle.

Cross-jack and sprit-sail-yards are equal to the fore-topsail-yard.

Sprit-topsail-yard is equal to the fore-top-gallant-yard.

The lifts of the main-yard are exhibited in the above plate, by *g g*; the horses and their stirrups, by *h h*, *i i*; and the braces and brace-pendants, by *t u*.

The large lateen-yards derive their names from having been peculiar to the ancient Romans, and are usually composed of several pieces fastened together by wooldings, which also serve as steps whereby the sailors climb to the peak, or upper extremity, in order to furl or cast loose the sail.

The mizen-yard of a ship, the main-yard of a bilander, and the yards of a lugger, though they have not, strictly speaking, lateen sails, are nevertheless hung obliquely on the mast, and the slings are nearer the fore-end than the aftmost-end, or peak.

For the different names and positions of all the yards, see the article Ships *of War*, and the figure of the first rate ship referred to from that article.

Proportions for the Lengths and Diameters of Yards Gaffs, and Staffs in the Royal Navy.

Lengths of Yards.

Main-Yard, eight-ninths of the main-mast.

Fore-Yard, seven-eighths of the main-yard.

Mizen-Yard, six-sevenths of the main-yard.

Main-topsail-Yard, five-sevenths of the main-yard.

Fore-topsail-Yard, seven-eighths of the main-topsail-yard.

Mizen-topsail-Yard, two-thirds of the main-topsail-yard.

Topgallant-Yard of 74-gun ships, two-thirds, all under, three-fifths, of their topsail-yards.

Royal-Yards, one-half of the topsail-yards.

Cross-jack-Yard, the same as the fore-topsail-yard.

Spritsail-topsail-Yard, the same as the fore-topgallant-yard.

Studdingsail-Yards, four-sevenths of their booms.

Driver-Yard, the same as the fore-topgallant-yard.

Diameters of Yards.

Main and *Fore-Yards*, at the slings, seven-tenths of an inch to every yard in the length.

Mizen-Yard, two-thirds of the diameter of the main-yard.

Topsail-Yards, five-eighths of an inch to every yard in the length.

Topgallant-Yards, six-tenths of an inch to every yard in the length.

Royal-Yards, one-half the diameter of their topsail-yards.

Spritsail-Yard, the same diameter as the fore-topsail-yard.

Spritsail-topsail-Yard, the same diameter as the fore-topgallant-yard.

Studdingsail-Yards, one inch in diameter to every five feet in the length.

Cross-jack-Yard, the same diameter as the fore-topsail-yard.

Driver-Yard, the same diameter as the fore-topgallant-yard.

Length and Diameter of Gaffs.

Length of the *Gaffs*, five-eighths of their respective booms.

Diameter of the *Gaffs*, the same as their booms.

Lengths and Diameters of Staffs.

Length of the *Ensign-Staff*, one-third of the main-mast, above the taffrail.

Length of the *Jack-Staff*, one-half the length of the ensign-staff, above the taffrail.

Diameter of the *Ensign-Staff*, half an inch to every yard in the length.

Diameter of the *Jack-Staff*, three-fourths of an inch to every yard in the length.

The following Tables contain the lengths and diameters of all the yards now used both in the Royal Navy and Merchant-Service in the different classes of ships and vessels, from a sloop of 70 tons burden, to a first-rate of 110 guns:

TABLE

Of the Dimensions of Yards now used in the Royal Navy.

Names of the Yards.	110 Guns. 2164 Tons.			90 Guns. 1870 Tons.			80 Guns. 1620 Tons.			74 Guns. 1800 Tons.			64 Guns. 1369 Tons.			50 Guns. 1044 Tons.			44 Guns. 879 Tons.			38 Guns. 951 Tons.		
	Length. Ft.	In.	Dia. In.	Length. Ft.	In.	Dia. In.	Length Ft.	In.	Dia. In.	Length. Ft.	In.	Dia. In.	Length. Ft.	In.	Dia. In.	Length. Ft.	In.	Dia. In.	Length. Ft.	In.	Dia. In.	Length. Ft.	In.	Dia. In.
Main-Yard	102	4	24	98	0	22½	93	0	22	99	0	23	90	4	21	82	0	19¼	80	0	18½	81	9	19
Main-Topsail-Yard	73	0	15½	70	0	14¾	66	6	14⅛	72	0	15	66	0	13⅝	60	9	12½	57	6	11⅞	59	0	12¼
Main-Topgallant-Yard	48	9	10	46	6	9⅜	46	0	9¼	46	6	9½	39	0	7⅞	36	0	7⅜	36	0	7⅜	37	0	7½
Main-Royal-Yard	32	0	7¾	30	0	7⅜	28	0	7	28	0	7½	27	0	6¾	26	0	6¼	25	0	5⅞	25	0	6⅛
Fore-Yard	89	1	21	85	9	20	81	4	19⅜	85	0	20	79	6	18½	72	0	17	70	0	6¼	71	6	16⅜
Fore-Topsail-Yard	64	6	13¾	61	0	12⅞	58	0	12¼	63	0	13	57	6	12⅛	52	6	11	52	0	10¾	53	0	11½
Fore Topgallant-Yard	43	0	8⅝	40	6	8¼	40	0	8¼	40	6	8¼	34	6	7	31	6	6¼	32	0	6⅜	32	9	6½
Fore-Royal-Yard	27	0	6⅞	25	0	6⅜	24	0	6⅛	25	0	6½	23	0	6	23	0	5½	23	0	5⅜	23	0	5¾
Mizen-Yard	87	0	16	84	0	15	76	0	14½	84	0	15½	77	0	14	—		—	—		—	—		—
Mizen-Gaff	—		—	—		—	—		—	—		—	—		—	44	0	11½	36	8	11	38	0	11½
Mizen-Topsail-Yard	49	0	10⅛	47	0	9¾	43	0	8⅞	47	0	9¾	43	3	9	39	6	8¼	39	6	8	40	0	8¼
Mizen-Topgallant-Yard	32	9	6½	31	0	6¼	29	0	5¾	31	6	6⅜	29	0	5¾	26	0	5¼	25	9	5⅛	27	9	5½
Mizen-Royal-Yard	24	0	5	23	0	4⅞	21	0	4⅜	23	0	4⅞	21	0	4½	19	0	4⅛	19	0	4	20	0	4⅛
Spritsail-Yard	64	6	13¾	61	0	12½	58	0	12¼	62	0	13	57	6	12⅛	52	6	11	52	0	10	53	0	11½
Spritsail-Topsail-Yard	43	0	8⅝	40	6	8¼	40	0	8¼	40	6	8¼	34	6	7	31	6	6¼	32	0	6⅜	32	9	6½
Driver-Yard	—		—	—		—	—		—	—		—	—		—	31	6	6¼	32	0	6⅜	32	9	6½
Cross-Jack-Yard	64	6	13¾	61	0	12⅞	58	0	12¼	62	0	13	57	6	12⅛	52	6	11	52	0	10¾	53	0	11½
Lower-Studdingsail-Yard	33	0	6¾	31	0	6¼	29	6	6	31	0	6¼	29	0	5¾	26	0	5¼	25	6	5⅛	26	0	5¼
Main-Topmast ditto	29	6	6	28	0	3¾	26	6	5¼	27	3	5½	25	9	5⅛	23	6	4¾	23	0	4¾	23	6	4¾
Main-Topgallant ditto	21	0	4¼	20	0	4	19	0	3¾	20	0	4	18	9	3¾	17	6	3½	16	6	3¼	17	0	3½
Fore-Topmast ditto	25	6	5⅛	24	6	5	23	4	4¾	24	6	5	22	9	4½	20	6	4⅛	20	0	4	20	6	4⅛
Fore Topgallant ditto	18	6	3¾	17	6	3½	16	6	3¼	17	9	3½	16	6	3¼	15	0	3	15	0	3	15	0	3

Names of the Yards.	36 Guns. 871 Tons.			32 Guns. 677 Tons.			28 Guns. 594 Tons.			24 Guns. 513 Tons.			20 Guns. 429 Tons.			Sloops of 300 Tons.			Brigs of 200 Tons.			Cutters of 200 Tons.		
	Length. Ft.	In.	Dia. In.	Length. Ft.	In.	Dia. In.	Length. Ft.	In.	Dia. In.	Length. Ft.	In.	Dia. In.	Length. Ft.	In.	Dia. In.	Length. Ft.	In.	Dia. In.	Length. Ft.	In.	Dia. In.	Length. Ft.	In.	Dia. In.
Main-Yard	79	0	18⅝	74	4	17⅛	71	3	16½	65	6	15	63	0	14½	55	0	12¾	42	0	9¾			
Main-Topsail-Yard	57	0	12	75	0	11⅜	52	0	11	47	0	9⅜	45	6	9½	39	6	8¼	31	6	7	52	0	7½
Main-Topgallant-Yard	34	9	7	33	6	6⅝	32	6	6½	29	6	6	28	4	5¾	25	0	5	23	6	5	26	0	6
Main-Royal-Yard	25	0	6	25	0	5⅜	25	0	5½	23	6	4⅞	20	0	4¾	17	0	4⅛	15	9	3½	—		—
Fore-Yard	69	4	16	65	0	15	62	2	14½	57	8	13¼	55	0	12⅝	48	5	11¼	42	0	9¾	—		—
Fore-Topsail-Yard	51	9	11	48	0	10	46	0	9¾	41	0	8⅝	40	0	8½	35	0	7⅜	31	6	7	—		—
Fore-Topgallant-Yard	31	6	6⅜	29	6	5⅞	28	6	5⅝	25	10	5⅜	25	0	5	22	0	4⅜	23	6	5	—		—
Fore-Royal-Yard	23	0	5½	23	0	5	23	0	4⅞	20	6	4⅜	18	0	4¼	16	0	3¾	15	9	3½	—		—
Mizen-Gaff	36	0	11	35	0	10	32	6	9¾	29	4	8⅝	28	6	8½	24	9	7⅜	—		—	—		—
Main-Gaff	—		—	—		—	—		—	—		—	—		—	—		—	28	0	7¼	—		—
Mizen-Topsail-Yard	39	4	7⅞	36	9	7	35	0	7¼	31	5	6½	31	0	6¼	26	4	5½	—		—	—		—
Mizen-Topgallant-Yard	27	0	5½	24	0	5	22	0	4⅜	19	6	4	19	6	4	16	6	3¼	—		—	—		—
Mizen-Royal-Yard	19	0	3⅞	18	0	3½	17	6	3⅜	15	8	3¼	15	6	3⅛	13	2	2¾	—		—	—		—
Spritsail-Yard	51	9	11	48	0	10	46	0	9¾	41	0	8⅝	40	0	8½	35	0	7⅜	31	6	7	—		—
Spritsail-Topsail-Yard	31	6	6⅜	29	6	5⅞	28	6	5⅝	25	10	5⅜	25	0	5	22	0	4⅜	23	6	5	—		—
Driver-Yard	31	6	6⅜	29	6	5⅞	28	6	5⅝	25	10	5⅜	25	0	5	22	0	4⅜	—		—	—		—
Cross-Jack-Yard	51	9	11	48	0	10	46	0	9¾	41	0	8⅝	40	0	8½	35	0	7⅜	—		—	58	0	9½
Lower-Studdingsail-Yard	25	6	5⅛	23	6	4¾	23	0	4⅝	21	0	4¼	20	0	4	17	6	3½	13	6	2¾	—		—
Main-Topmast ditto	22	6	4½	21	3	4¼	20	6	4	19	0	3⅞	18	6	3¾	15	6	3⅛	12	0	2½	—		—
Main-Topgallant ditto	16	6	3¼	16	0	3¼	15	0	3	13	6	2¾	13	0	2⅝	11	6	2¼	8	9	2¼	—		—
Fore-Topmast ditto	20	0	4	18	6	3¾	17	9	3½	16	6	3¼	15	9	3⅛	14	0	2⅞	12	0	2½	—		—
Fore-Topgallant ditto	15	0	3	14	0	2⅞	13	3	2⅝	11	9	2⅜	11	6	2¼	10	0	2	8	9	2¼	—		—
Water-sail-Yard	—		—	—		—	—		—	—		—	—		—	—		—	—		—	12	0	2½
Spread-Yard	—		—	—		—	—		—	—		—	—		—	—		—	—		—	60	0	9½
Ring-tail-Yard	—		—	—		—	—		—	—		—	—		—	—		—	—		—	12	0	2½

TABLE

Of the Dimensions of Yards now used in the Merchant-Service.

Names of the Yards.	1300 Tons.			1200 Tons.			1100 Tons.			1000 Tons.		
	Length. Ft.	In.	Dia. In.	Length. Ft.	In.	Dia. In.	Length. Ft.	In.	Dia. In.	Length. Ft.	In.	Dia. In.
Main-Yard	86	0	21	84	0	20½	80	0	20	78	0	19
Main-Topsail-Yard	58	0	14	58	0	14	57	6	13½	54	0	13
Main-Topgallant-Yard ...	38	0	8½	37	6	8¼	37	0	8	36	0	8
Main-Royal-Yard	24	0	5	24	0	5	23	0	5	22	0	5
Fore-Yard	82	0	20	80	0	19½	76	0	18½	72	0	18
Fore-Topsail-Yard.......	56	0	13½	54	0	13¼	52	0	13	52	0	12½
Fore-Topgallant-Yard....	36	0	8	36	0	8	36	0	7¾	34	0	7½
Fore-Royal-Yard........	22	0	5	22	0	5	21	0	5	20	0	4½
Mizen-Yard	72	0	13	70	0	13	68	0	13	66	0	12½
Mizen-Topsail-Yard	40	0	9½	39	0	9¼	38	0	9	38	0	9
Mizen-Topgallant-Yard ..	26	0	5½	26	0	5½	26	0	5	26	0	5½
Mizen-Royal-Yard	16	0	4½	16	0	4½	16	0	4½	14	0	4
Spritsail-Yard..........	56	0	11½	55	0	11¼	54	0	11¼	54	0	11¼
Spritsail Topsail-Yard....	38	0	8	38	0	8	38	0	8	36	0	8
Cross-Jack Yard	56	0	11½	55	0	11¼	54	0	11¼	54	0	11¼
Lower Studdingsail-Yard .	30	0	7	29	0	7	28	0	7	28	0	7
Main Topmast ditto......	24	0	6	23	0	6	22	0	6	22	0	5¾
Main-Top-gallant ditto. ..	18	0	5	17	0	5	16	0	5	15	0	5
Fore-Topmast ditto......	24	0	6	23	0	6	22	0	6	22	0	5¾
Fore-Topgallant ditto	17	0	5	16	0	5	14	0	5	14	0	5

Names of the Yards.	900 Tons.			800 Tons.			700 Tons.			600 Tons.		
	Length. Ft.	In.	Dia. In.	Length. Ft.	In.	Dia. In.	Length. Ft.	In.	Dia. In.	Length. Ft.	In.	Dia. In.
Main-Yard	76	0	18½	74	0	18	70	0	16½	62	0	15½
Main-Topsail-Yard	52	0	12½	50	0	12	50	0	12	48	0	12
Main-Topgallant-Yard ...	34	0	8	33	6	8	32	0	7½	32	0	7½
Main-Royal-Yard	21	0	4½	20	0	4½	—		—	—		—
Fore-Yard	70	0	17	68	0	16	64	0	15	58	0	14½
Fore-Topsail-Yard.......	50	0	12	48	0	12	48	0	12	46	0	11½
Fore-Topgallant-Yard....	32	0	7½	30	0	7½	30	0	7½	30	0	7½
Fore-Royal-Yard........	19	0	4½	18	0	4½	—		—	—		—
Mizen-Yard	64	0	12	62	0	10½	62	0	10½	60	0	10½
Mizen-Topsail-Yard	35	0	8½	34	0	8½	34	0	8½	32	0	8
Mizen-Topgallant-Yard ..	23	0	5	23	0	5	23	0	5	22	0	4½
Mizen-Royal-Yard	—		—	—		—	—		—	—		—
Spritsail-Yard..........	50	0	10½	50	0	10½	48	0	9½	48	0	9½
Spritsail Topsail-Yard....	36	0	8	34	0	7½	32	0	7	32	0	7
Cross-Jack Yard	52	0	10½	50	0	10½	49	0	10	48	0	9½
Lower Studdingsail-Yard .	25	0	6½	23	0	6	22	0	5½	20	0	5
Main Topmast ditto......	20	0	5½	18	0	5½	17	0	5	16	0	5
Main-Top-gallant ditto. ..	14	0	5	14	0	4½	13	0	4	12	0	4
Fore-Topmast ditto......	19	0	6	17	0	5½	16	0	5	15	0	5
Fore-Topgallant ditto	14	0	5	14	0	4½	13	0	4	12	0	4

Names of the Yards.	500 Tons.			400 Tons.			360 to 350 Tons.			330 to 300 Tons.		
	Length. Ft.	In.	Dia. In.	Length. Ft.	In.	Dia. In.	Length. Ft.	In.	Dia. In.	Length. Ft.	In.	Dia. In.
Main-Yard..	52	0	13	50	0	13	48	0	12	47	0	12
Main-Topsail-Yard	41	0	10½	39	0	10	36	0	9	35	0	9
Main-Topgallant-Yard ...	31	0	7	29	0	6½	28	0	6½	26	0	6
Main-Royal-Yard	20	0	5	20	0	5	18	0	5	16	0	4
Fore-Yard..............	52	0	12½	49	0	12½	47	0	12	45	0	11½
Fore-Topsail-Yard.......	41	0	10½	39	0	10	35	0	9	32	0	8½
Fore-Topgallant-Yard....	28	0	6½	27	0	6	27	0	6¼	25	0	5¾
Fore-Royal-Yard........	19	0	5	19	0	4¾	17	0	4½	15	0	4
Mizen-Gaff.............	32	0	8	30	0	7½	28	7	7	26	0	6½
Mizen-Topsail-Yard	32	0	8	30	0	7½	30	0	7½	23	0	7
Mizen-Topgallant-Yard ..	21	0	5	20	0	5	18	0	5	16	0	4
Spritsail Yard	39	0	8½	36	0	7½	34	0	7½	30	0	7
Driver-Yard............	—		—	18	0	5½	22	0	5½	20	0	5
Cross-Jack-Yard	44	0	9	39	0	8	36	0	7½	35	0	7
Lower-Studdingsail-Yard	19	0	5	18	0	5	16	0	4½	15	0	4½
Main-Topmast ditto......	15	0	4½	15	0	4½	14	0	4	12	0	4
Main-Topgallant ditto....	12	0	4	12	0	4	10	0	3½	10	0	3½
Fore-Topmast ditto..	15	0	4½	15	0	4½	14	0	4	12	0	4
Fore-Topgallant ditto . ..	12	0	4	12	0	4	10	0	3½	10	0	3½
Trysail-Gaff	—		—	—		—	—		—	—		—
Wingsail-Sprit	—		—	—		—	—		—	—		—
Spread-Yard	—		—	—		—	—		—	—		—
Squaresail-Yard.........	—		—	—		—	—		—	—		—
Spread-Yard	—		—	—		—	—		—	—		—
Gaff..................	—		—	—		—	—		—	—		—

Names of the Yards.	280 to 250 Tons.			Brigs of 150 Tons.			Ketches of 170 Tons			Schooners of 110 Tons.		
	Length. Ft.	In.	Dia. In.	Length. Ft.	In.	Dia. In.	Length. Ft.	In.	Dia. In.	Length. Ft.	In.	Dia. In.
Main-Yard..	44	0	11	38	0	9½	51	0	12	—		—
Main-Topsail-Yard	32	0	8	30	0	7½	36	0	7½	38	0	6
Main-Topgallant-Yard ...	23	0	5½	21	0	4½	24	0	4	27	0	5½
Main-Royal-Yard	15	0	4	—		—	—		—	—		—
Fore-Yard..............	42	0	10½	36	0	9	—		—	—		—
Fore-Topsail-Yard.......	32	0	8	30	0	7½	—		—	30	0	6½
Fore-Topgallant-Yard....	23	0	5½	21	0	4½	—		—	25	0	5
Fore-Royal-Yard........	15	0	4	—		—	—		—	—		—
Mizen-Gaff.............	25	0	6	—		—	18	0	5½	22	0	6½
Mizen-Topsail-Yard	27	0	6	—		—	21	0	4	—		—
Mizen-Topgallant-Yard ..	—		—	—		—	—		—	—		—
Spritsail Yard	32	0	7	30	6	6	31	0	6	—		—
Driver-Yard............	18	0	5	—		—	—		—	—		—
Cross-Jack-Yard	32	0	7	—		—	29	0	5½	—		—
Lower-Studdingsail-Yard	—		—	—		—	—		—	—		—
Main-Topmast ditto......	12	0	4	10	0	3½	—		—	—		—
Main-Topgallant ditto....	—		—	—		—	—		—	—		—
Fore-Topmast ditto..	12	0	4	10	0	3½	—		—	—		—
Fore-Topgallant ditto . ..	—		—	—		—	—		—	—		—
Trysail-Gaff	—		—	—		—	16	0	5	—		—
Wingsail-Sprit	—		—	—		—	16	0	5	—		—
Spread-Yard	—		—	—		—	—		—	47	0	7
Squaresail-Yard.........	—		—	—		—	—		—	38	0	7½
Spread-Yard	—		—	—		—	—		—	40	0	7
Gaff..................	—		—	—		—	—		—	28	0	7

SLOOPS, of 70 Tons. { Square-Sail, or Cross-Jack-Yard, 40 feet long, and 8 inches in diameter,
Main-Topsail Yard............24 ditto, and 6 ditto.
Gaff.........................28 ditto, and 7 ditto.

See the ART OF MAKING MASTS, YARDS, &c. as practised in the Royal Navy &c

The following is a list of spars necessary for making a temporary lower-yard at sea:

One spare topsail-yard.
One spare top-gallant-yard,
One spare royal-yard. } To be placed in the centre.

Two topmast-studding-sail-booms, with the largest ends outward, to form the yard-arms.

Two top-gallant-studding-sail-bombs, their largest ends to butt against the small ends of the topmast-studding-sail-bombs.

When these spars are placed as above directed, and well lashed together, they will be found to answer every purpose of a lower-yard.

Yard-Arms, (*les bouts de vergues*, Fr.) the outer quarters or ends of a yard.

Yard-Arm and Yard-Arm, (*vergue à vergue*, Fr.) a phrase applied to two ships in time of action, when they are so near that their yard-arms nearly touch each other.

Yard-Arm *Pieces*, (*jumelles de rechange pour servir a jumeller un mât, ou une vergue*, Fr.) are pieces of timber kept in readiness to repair the yard-arms in the event of their being carried away or broken.

Quarters of a Yard. See the article Quarters.

Slings of a Yard. See the article Slings.

Sprit-Sail Yard, (*vergue de civadière*, Fr.) a name given to the yard which is placed immediately under the bowsprit: it is the same length as the fore-topsail yard, and is represented by *y*, fig. 3, plate XXXI.

Sprit-Sail-Topsail Yard, (*vergue de contre-civadière*, Fr.) the yard which is placed under the jib-boom, and is the same length as the fore-top-gallant-yard.

Yard-*Tackles*, (*palans des bouts des vergues*, Fr.) strong tackles suspended from the main and fore-yards of a ship of war, &c. whereby, with the assistance of the stay-tackles, the boats, and other weighty matters, are hoisted in and out.

To Brace the Yards, (*brasser les vergues*, Fr.) is to traverse them about the masts, so as to form greater or lesser angles with the ship's length. See the article Brace, &c.

To Square the Yards. See the articles Lift and Square.

Yards, (*arsenaux de marine*, Fr.) also denote places belonging to the royal navy, which are fitted with several wet and dry docks, wharfs, and slips, for building and repairing his Majesty's ships, and wherein are lodged great quantities of timber, masts, planks, anchors, cordage, blocks, and all other sorts of naval stores. There are, belonging to his Majesty's navy, six royal yards, viz. Portsmouth, Plymouth, Chatham, Deptford Woolwich, and Sheerness.

Portsmouth Yard, is situated in the town of Portsea, which is, properly, only a part of Portsmouth, on the island of Portsea, in the hundred of Portsdown and county of Southampton. Porchester, at the top of Portsmouth harbour, was the original sea-port; but, owing to the retiring of the sea from the upper parts of the harbour, which rendered Porchester less convenient, the inhabitants removed hither and built Portsmouth. The earliest historical notice of it is in the Saxon Chronicle, A.D. 501, where it is mentioned as the place at which a Saxon chief, called Porta, landed, in order to assist Cerdic in the subjugation of the Belgic provinces in England.

This yard was established in the reign of Henry VIII. King Charles II. after his restoration, made several docks and yards here, and raised several forts after the modern manner, which works were farther augmented under his brother's reign.

The wall which separates the yard from the town of Portsea, is of brick, about 14 feet high. The area within the wall is upwards of 100 acres; including the projecting parts of the jetties. The entrance to the dock-yard is from a street called the Common Hard, by a gateway for carriages, &c., and a small door for foot passengers. Immediately within the gates is the warden's residence. Turning to the right you pass the guard-house and the Navy Pay Office; a little farther stands the *Royal Naval* College, for an account of which, see that article. Proceeding in the same direction, you next come to the Commissioner's house and the chapel, both of which are handsome buildings; thence a visitor is generally conducted to the rope-house, a spacious pile 365 yards in length, 18 yards in width, and three stories high, the floors being lined underneath with plate iron, to prevent accidents by fire. On the second story twine is made, and yarn prepared for cables of the largest dimensions, which are layed below with immense labour.

At the south side of the rope-house is Anchor-lane, or row, where hundreds of anchors of almost every size and description are regularly stowed and painted, to preserve them from rust. At the top of this there is a row of houses, where the surgeon, master rope maker, clerk of the rope-yard, and boatswain, reside: and on the west side are the principal officers' offices. Proceeding from thence you pass various store-houses, the mould-loft, and vast piles of timber, for the service of the yard, when a sort of square presents itself to the view, and displays in its centre a statue of William III. in a Roman habit.

On the north-east side of this is a row of handsome houses, appropriated for the residence of the chief officers of the yard; in front of these are other store-houses, and the house-carpenters', wheelwrig ts', and joiners' shops.

You next come to the boat-houses, where boats of almost every description may be seen, from a canoe to a life-boat: then to the stables; and next to the stone-mason's yard near the north corner of the yard, which is thus far surrounded by a high wall. At the north-west end of the yard, are several piles of rough timber and saw-pits, built on some new made ground, which was lately taken from the harbour.

The next object worth notice is the blacksmith's shop, or anchor forge, on entering which, both the eye and ear are confounded by the smoke and noise, which spread throughout this vulcanic abode.

Many of the anchors made here, weigh from 40 to 90 cwt. each, and are said from experience to be the best anchors made in this country. Next to this stands the copper furnace, which is wrought by a steam-engine of 56 horse power; here all the old copper sheathing that is stripped off the bottom of the ships is melted, and afterwards rolled out into sheets for sheathing again.

As you proceed towards the side nearest to the harbour, you come to another small guard-house, and three large slips for building ships of the line in. You are then conducted over a bridge, erected on the caisson, across the camber-dock, to the steam-engine, or more properly the block machinery, which is peculiarly interesting, as there the whole process of block-making is exhibited: for an account of which see the article Block *Machinery*.

Water is conveyed by this engine to every part of the yard, and the pipes constantly kept full, in case of accidents by fire. On the north-east side of the steam-engine, are very extensive stores under-ground, for tar, pitch, and other commodities that are not reckoned of a perishable nature, under which is a reservoir for holding the water that may be pumped out of the docks, when ships are taken in to be repaired: and on the south side is the north-dock for repairing ships You then come to the new basin, which is of considerable extent, and communicates with the sea, by means of a caisson: here ships of war are taken in, with all their standing and running rigging, to be examined. From the basin are extended four dry-docks, three of which lie to the north-east, and one to the south-east, for repairing ships of the line, which, while under repair, are kept perfectly dry, by means of the gates which exclude the water.

The pitch-house stands between the basin and the harbour, and as you proceed toward the gates, there is another dock, called the south-dock, together with extensive wharfs, which bring you to the camber, or principal landing-place from the harbour, formerly known by the name of Hulk-Stairs, but now called the King's Stairs, where wherries and merchant-ships' boats are not allowed to land.

You next come to a long range of store-houses, running north and south, divided into three parts, called the north, the middle, and the south, stores, and between the middle store and the water side stands the rigging-loft; next to it, and opposite the south-store, stand the sail-loft and yard.

Between these store-houses and the sail and rigging-lofts is a commodious camber or slip, for transports and small vessels to load or land naval stores, &c.

The mast-pond, and a range of low buildings, called the mast-houses, are next in order, which reach round to the west side of the dock-gates, where you entered. There are also two large timber or mast-ponds without the yard, one on each side of the landing place on the Common Hard. These immense works are all peculiarly adapted for their respective purposes. The different employments, and manual activity, exhibited by the artificers, &c. in the various departments of this dock-yard, present a very interesting spectacle to those who have not been accustomed to appreciate the effects of human industry on so large a scale. The number of workmen employed here varies considerably: the peace establishment has seldom been less than two thousand; and, during war, frequently double that number has been employed. Here, as at the other yards, the shipwrights receive sixpence a day, as a commutation for their former perquisite of a bundle of chips.

Establishment of Officers at Portsmouth Yard, in 1814.

Resident Commissioner, Captain the Hon. Sir George Grey, Bart.
Three Master Attendants.
Master Shipwright.
Clerk of the Check.
Storekeeper.
Clerk of the Survey.
Clerk of the Rope-Yard.
Timber Master.
Master Shipwright's Assistants.
Surgeon.
Chaplain.
Masters of the different departments.
Warden.

Notwithstanding every precaution that can be devised is taken to guard against the destructive element of fire, three great conflagrations have occurred in this dock-yard since the year 1760. The first, which appears to have been accidental, broke out in the night of the 3d of July, 1761, and raged for a long time with dreadful fury.

The night had been extremely tempestuous; and the fire was attributed to the lightning striking upon the store-house containing the hemp, the windows of which had been left open to air it. By this conflagration, many hundred tons of tar, cordage, sails, and hemp, were totally consumed.

The second fire occurred on the morning of the 27th of July, 1770, when the damage done was still greater; and it was even for some time doubtful whether any part of the yard would escape destruction. From its bursting forth at different places at one time, and various other circumstances, great suspicions were entertained of its having been occasioned intentionally, but the officers were unable to discover the offenders.

The third fire happened on the 7th of December, 1776, and in this instance was undoubtedly the effect of design, as the incendiary was traced, tried, condemned, and executed, upon incontestible proof, afterwards confirmed by his own confession. The real name of this malefactor was John Aitken; but the appellation by which he is commonly known is that of "Jack the painter." He is supposed to have acted under foreign influence, and had previously attempted to destroy the docks at Plymouth, but failed in the attempt, though he excited very considerable alarm. His plans were laid with great sagacity and forethought; and in order the more effectually to ensure their success, and avoid suspicion, he had invented a very ingenious machine, which he contrived to lodge among the cordage, and setting fire to it, left it, and passed out at the gates unmolested. In the evening of the same day, the fire broke out, as it luckily happened, several hours before the incendiary had purposed; for had it not begun to display itself till after the fall of night, the destruction would probably have been much greater than it was. The immediate and effective assistance which was given to check the progress of the flames, and the favourable direction of the wind, confined the damage to the rope-house, and a few adjoining store-houses.

No strangers can be admitted to see the yard at any time, without an order from the Commissioner, and first subscribing their names and places of abode in a book kept at the gate for that purpose; a copy of which is daily transmitted to the Admiralty.

The ordnance department of this yard comprehends the gun-wharf and magazine.

The gun-wharf is included in the Portsea division of the town of Portsmouth. This arsenal contains several ranges of buildings for the reception of the naval and military artillery, ordnance stores, &c. Some of the store-houses are vast piles, particularly two, which, with their dependencies, are adapted to contain all kinds of necessaries for the sudden equipment both of a fleet and an army. On the wharf is the grand depôt for great guns, carronades, and mortars, with shot and shells of every size and weight all of which are arranged in immense piles. Here also in times of peace, the guns from the ships in ordinary at this port are lodged; each ship's guns being kept in a separate tier, while the carriages are stowed away in the same regular manner in proper store-houses; so that the whole may easily and readily be re-embarked.

The armoury is a building of late erection, capable of containing upwards of 25,000 stand of arms, and furnished with work-shops appropriated to the artificers employed to keep the arms in complete readiness for service. The houses of the store-keeper, and of other officers, who have the charge of the gun-wharf, are large and handsome structures.

On the west side of Portsmouth harbour, at Priddy's Hard, stands the magazine, where a sufficient supply of serviceable powder is always kept for the fleet frequenting this port: here are also handsome houses for the store keeper, and clerk of the check. Farther up the harbour is a floating-magazine, in which the damaged powder is deposited till it can be sent to the powder-kilns erected on a small island, in the harbour, called Horsea, to be dried in stoves by steam; when it is taken to buildings erected on the east side of the harbour, called Tipnor Point, to be mixed with other powder, in order to bring it to a proper strength; then stoved, and afterwards conveyed to the principal magazine for use.

The victualling-office, comprehends several extensive ranges of buildings, with a noble house for the agent-victualler, in St. Thomas's-street, and large store-houses, occupying a great part of St. Mary's-street, in Portsmouth, in which the provisions, wine, and liquors prepared for the supply of the navy are kept. Other houses are appropriated to the business of salting and baking.

The grain used in this office, is ground at the King's mills, a large building about a quarter of a mile from the Quay-gate, erected by government at an expense of near 7,000*l*. It is raised on immense piles, from 25 to 30 feet long, and shod with iron. This expensive foundation was necessary on account of the ground being remarkably marshy. This mill is worked by a stream of salt water, received from the harbour every tide, by means of a great sluice, into an extensive lake separating the two towns: this sluice is closed at high water, and on the ebbing of the tide the water is worked back again into the harbour.

At Weovil, on the west side of the harbour, opposite the dock-yard, are the royal brewery and cooperage, with an immense range of store-houses, for malt, hops, &c. Here are also good houses for the resident officers; and many small vessels belonging to this place are employed in conveying beer, water, &c. to ships in the harbour, Spithead, and St. Helen's.

On the same side of the harbour, nearly opposite its mouth, stands the Royal Naval Hospital; for an account of which, see Haslar Hospital.

Portsmouth harbour, for capaciousness, depth, and safety, decidedly excels every other in Great Britain. Secure from the most violent storms, the largest first rates may ride here at the lowest ebb, without touching ground.

The anchorage in every part is good; and the whole harbour, except at its mouth, is completely free from bars or impediments. It is, moreover, almost completely secured from assaults, by the strength and number of the batteries which defend it, particularly towards the sea, where the guns are nearly on a level with the edge of the water.

One material convenience with respect to this harbour, and which greatly adds to its importance is, the spacious reach or roadstead, and famous anchorage at Spithead, between Portsmouth and the Isle of Wight, where the greater part of his Majesty's ships and vessels rendezvous in time of war, so that it may justly be considered as the great national and central station of the navy.

At the Mother-bank, towards the Isle of Wight shore, the East India Company's ships anchor; and at Stoke's-bay, towards the opposite shore, the West India fleet, and other merchant ships ride safely at anchor.

Spithead is defended from the winds that blow from the west to the south-east, by the high lands of the Isle of Wight, and from most winds of the opposite quarter, by the main land of Hampshire, the town of Portsmouth fronting the middle of the road.

Plymouth Yard, is situated west of Plymouth, on the eastern bank of the Hamoaze, and contains four docks, one wet, the other three dry; two were built in the reign of William III.; and two in his present Majesty's reign, hewn out of a mine of slate, and lined with Portland stone. This yard, even in its present unfinished state, is acknowledged to be one of the finest in the world. The wall which separates the yard from the town, is of slate and lime-stone, and in some places 30 feet high. The area within the wall is upwards of 80 acres, including the projecting parts of the jetties.

The basin and dock, made in the reign of William III., is a large excavation, in which all the boats belonging to the yard, as well as the launches employed in moving ships, are kept; the water flows into it through an opening about 70 feet wide. Within the basin is a dock sufficiently capacious for a 74-gun ship; its length is 197 feet 3 inches; its width 65 feet 10 inches, and its depth 23 feet 1 inch

The rigging-loft, contiguous to the south jetty, is a handsome building, 480 feet long, and three stories high, forming one side of a quadrangle. The fabric is of lime-stone, with the quoins and cornices of Portland stone.

Over the rigging is the sail-loft, where all the sails are cut out and made, and old ones mended. The remaining three sides of the quadrangle are store-houses, in which the various articles necessary to equip the fleets are kept.

Advancing southward is a slip for hauling up and graving the bottoms of small vessels, such as sloops of war, cutters, &c.

Beyond this is the camber, a long canal, about 70 feet wide, terminating at the upper end in a basin, where boats lie; on the north side of which is the boat-house, where boats are built and repaired. Here, before the year 1768, were the bounds of the yard; all thence to the southward is still called the "New Ground."

The blacksmith's shop is situated south of the canal; it is a spacious building, about 210 feet square, and contains 48 forges. The anchor-wharf fronts the blacksmith's shop. Near this wharf are three slips, whereon large ships are built; and adjoining them is a boiling-house, in which the planks that are to receive a particular curve are boiled in water for a considerable time, and being afterwards applied hot to their places, are immediately fastened.

Northward of the slips are the mast-houses and pond; in the former the different masts and yards are made, and pressed together with iron hoops driven on red hot; and in the latter an immense number are kept to prevent their cracking from exposure to the sun. Near the south end of the mast-house is a small mount, generally called "Bunker's Hill," on the summit of which is a watch-house and a small battery. Under this hill is a small powder magazine and near it a slip for building cutters and small vessels.

Returning from the interior of the yard, the rope-houses first engage the attention. They are two stories high, and twelve hundred feet long, running parallel to each other. In the upper stories twine is made, and yarn prepared for the cables, which are *layed* below.

One of these buildings was partly consumed by an accidental conflagration in June, 1812. The whole of the machinery was destroyed, and only 400 feet of the rope-house saved. The loss was estimated at 15,000*l.* In 1776, considerable alarm was excited from John Aitken, who attempted to destroy the yard by fire; he was afterwards detected in a similar endeavour at Portsmouth, and hanged, as before-mentioned in the preceding article.

The mould-loft, where the different parts of ships to be built are laid down according to plans sent from the Navy Board, is in front of the store-house, and is the last building of importance in that part of the yard south of the basin.

Near the north jetty is a house, where pitch is kept continually boiling, to be applied to the bottoms and seams of ships

The double dock is the first of three, very near each other, for line of battle ships, and so denominated from its capability of containing two ships at the same time.

The second dock, called the Union, or North Dock, is 239 feet 4 inches long, 86 feet 7 inches wide, and 26 feet 10 inches deep, and is faced with Portland stone. The New Union, or North New Dock, 259 feet 9 inches long, 85 feet 3 inches wide, and 27 feet 8 inches deep, was made in the year 1789, and is on the same plan. Near the head of this latter dock is the burning place for old copper; and farther northward, are the plumbers, glaziers, and armourers shops, and bricklayers and stone-cutters yards.

Establishment of Officers at Plymouth Yard, in 1814.

Resident Commissioner, Capt. Robert Fanshawe.
Two Master Attendants.
Master Shipwright.
Clerk of the Check.
Storekeeper.
Clerk of the Survey.
Clerk of the Rope-Yard.
Timber Master.
Master Shipwright's Assistants.
Chaplain.
Surgeon.
Masters of the different departments.
Warden.

The gun-wharf is separated from the dock-yard by North-Corner-street; it was begun in 1718 or 1719, and completed about 1785. The quantity of ground within the walls is four acres and three quarters. Here are two principal store-houses, of three stories high, for 80,000 muskets, pistols, and other stores: a number of sheds for gun-carriages, &c.; a powder magazine, and a cooperage.

The bay or estuary of Hamoaze, on which the dock-yard is situated, is a commodious basin about four miles long, and half a mile wide, where in times of peace a very considerable part of the British navy is laid up in ordinary. The port admiral's hospital and prison ships are always stationed here, and those that have undergone, or are about to receive repair in the dock.

On the banks of this harbour are other establishments of government: the powder magazine, higher up, consists of several detached stone-buildings; the powder-kilns, erected for drying damaged powder; and on the bank of a creek of the harbour running to the S. W. is the South-Down brewery, where all the beer for the Plymouth division of the fleet is brewed.

The Victualling-Office, under the eastern walls of the citadel, is an extensive range of buildings, where the ovens for supplying the navy with bread and the ingenuity exercised in the baking and preparations, present interesting objects; though there are but two bake-houses, each containing four ovens, yet they are heated eight times a day, and in the course of that time bake a sufficient quantity of bread for 16,000 men. The granaries are large and well constructed; and the whole process of converting the grain into bread is simple and ingenious.

At Stonehouse, on the road leading from Plymouth

to Plymouth-Dock, is the Royal Naval Hospital; for an account of which, see PLYMOUTH HOSPITAL.

The port consists of three harbours, which are distinguished by different appellations. The division or inlet, Hamoaze, as before-mentioned, is the harbour for the British navy; being fitted out with moorings, viz. chains of iron, sixty fathoms long, consisting of 120 links, and having at the end a large anchor. Here is good anchorage for more than 100 ships.

Sutton-Pool is the harbour frequented by merchant-ships only, and is almost surrounded by the houses of the town. The entrance to the pool is between two large piers, erected by parliamentary grants in 1791—1799. It is encompassed with public and private quays, and all ships entering it are obliged to pay certain dues on landing their cargoes, as it is well calculated to afford protection to such ships as are moored in it.

Catwater harbour is formed by the estuary of the river Plym, and is capable of protecting from the south-west gales, which are prevalent, upwards of five hundred merchant-ships; but it is seldom entered by ships of war.

Proceeding from Catwater, we enter into that part of the harbour denominated the *Sound*, which is a large capacious bay peculiarly adapted for a naval rendezvous. It has, however, always been exposed to southerly and south-westerly winds, which occasion a tremendous swell. To obviate this objection, the lords of the Admiralty obtained the first parliamentary grant of 80,000*l*. in March, 1812, for erecting a pier or breakwater, which will afford an excellent shelter to a large fleet. See BREAK-WATER.

At the mouth of the Sound, five leagues south of Plymouth, is the Eddystone light-house, which is necessarily an appendage of the most important kind to the harbour, without which the entrance to it would be extremely dangerous. See LIGHT-HOUSE.

There is, besides the above, *Barn-pool*, which is a safe anchorage under Mount Edgecomb, within Drake's Island, where frigates and sloops of war lie while they are undergoing a temporary refit. There is also a good anchorage between Drake's Island and the main, which is frequented by gun-brigs, cutters, &c., as well as by merchant-vessels.

Chatham YARD, in Kent, was founded by Queen Elizabeth, and stood where the gun-wharf now is, there being only one small dock, and also too confined a spot, it was removed to its present situation, by James I., in 1622.

Here are two docks respectively denominated the *Old Dock*, and the *Royal Naval Dock*. The former with its connected ordnance wharf, store-houses, &c. are the places of deposit for the stores and ordnance. The guns belonging to his Majesty's ships are collected and ranged on this wharf in long tiers, and large piles of cannon balls, shells, &c. are accumulated here to be ready for service. In the range of store-houses are deposited the gun-carriages, and almost every other article of naval stores. In one apartment is a small armoury, which is deserving the inspection of the curious; being furnished with muskets, pistols, cutlasses, pikes, pole-axes, and other warlike instruments, classed and arranged in regular order.

The Royal Naval Dock-Yard is connected with the former, and ranges along the eastern bank of the river, for nearly a mile in length. The increase of the navy during the reign of Elizabeth occasioned these docks to be enlarged, and in the subsequent reigns of James and Charles I. they were considerably extended; within the last fifty years they have been greatly enlarged and improved; and within the fortified walls are now concentrated almost every requisite for the fitting out, repairing, &c. of a fleet. The storehouses, shops, and most of the buildings are extremely spacious and commodious; whilst those belonging to the commissioner and other principal officers are provided with every comfortable and elegant accommodation. One of the store-houses is 660 feet long, and the sail-loft is 209 feet in length. The rope-house measures 380 yards; in this cables of the largest dimensions are manufactured. Here are two spacious basins of water, where the timber for masts, &c. is kept continually floating till wanted for use. The smith's shop contains 21 forges, which are almost constantly in use, in making anchors, &c. In this yard are four spacious docks for docking and repairing vessels belonging to the royal navy; in these are six slips or launches, on which ships are constantly building. Ships of the first rate have been launched from this yard; among which are the Victory, the new Royal George, the Royal Charlotte, the Ville de Paris, the Howe, &c. The whole of the yard, except on the east side of the river, is guarded and environed by a high thick wall; it is entered through a large handsome gate-way, which is flanked by two towers. In time of war every stranger is precluded access within the walls, unless he has a satisfactory passport, and can obtain leave from the Commissioner.

Establishment of Officers at Chatham Yard, in 1814.

Resident Commissioner, Capt. Sir Robt. Barlow, Knt.
Two Master Attendants.
Master Shipwright.
Clerk of the Check.
Store-keeper.
Clerk of the Survey.
Clerk of the Rope-Yard.
Master Measurer.
Master Shipwright's Assistants.
Chaplain.
Surgeon.
Masters of the different Departments.
Warden.

At the entrance of Chatham from Rochester, is the Victualling-Office, containing a cooperage, pickling, baking, cutting, slaughtering, and other storehouses, from which the king's ships at Chatham and Sheerness are supplied with provisions. A new wharf has been added to it, and additional buildings have been erected for the convenience and service of this department of the navy.

Deptford YARD is seated on the Thames, in Kent. The Royal Dock, or King's Yard, as it is locally termed, appears to have been established early in the reign of King Henry VIII. but it has been greatly enlarged and improved since that period. The whole extent of the yard at present includes about 31 acres, which are occupied by two wet docks, a double and single one; three slips for men of war; a basin; two mast-ponds; a mould-loft; mast-houses; a large anchor-smith's shop, with about 20 forges; sheds for timber, &c. All the concerns of this yard are managed under the immediate inspection of the Navy Board. The old store-house is a quadrangular pile, and appears to have consisted originally only of one range of buildings on the north side; where, on what was formerly the front, is the date 1513, together with the initials H. R. in a cypher, and the letters A. X. for Anno Christi. The buildings on the east, west, and south sides of the quadrangle have been erected at different times; and a double front, towards the north, was added in 1721. Another store-house, parallel to the above and of the same length, having sail and rigging-lofts, was completed a few years ago; and there is also a long range of smaller store-houses, built under the direction of Sir Charles Middleton, about the year 1780.

The other buildings consist of various work-shops, and houses for the resident officers.

On the 4th of April, 1581, Queen Elizabeth visited the celebrated Blake here, and having dined on board his ship, conferred the honour of knighthood on him. The ship was afterwards laid up at this place by order of the queen till it went to decay, when, from some of the soundest timbers, an elbow-chair was made, and presented to the University of Oxford, where it is still preserved.

The number of artificers and labourers employed here in time of war, are about 1500, and in time of peace about 1000.

Establishment of Officers at Deptford Yard, in 1814.

Resident Commissioner, Capt. Charles Cunningham.
Master Attendant.
Master Shipwright.
Clerk of the Check.
Storekeeper.
Clerk of the Survey.
Master Measurer.
Master Shipwright's Assistants.
Surgeon.
Master of the different departments.
Warden.

Besides the King's Yard, here are two large private docks for ship-building, called Dudman's and Barnard's, where 74-gun ships are sometimes built.

At a short distance from the King's Yard, close to the river is the victualling-office, sometimes called the Red-house, from its standing on the site of a large range of store-houses, constructed with red bricks, which were burnt down in July 1639. In 1745, the victualling office was built upon this site, and burnt down in four years after, with great quantities of stores and provisions: it has since been rebuilt, and enlarged with store-houses of various descriptions, and dwelling-houses for the principal and inferior officers. Here is also a bakehouse for making biscuit, an extensive cooperage, spacious slaughter-houses, houses for curing beef, pork, &c. and other offices necessary for the important concern of victualling the navy.

Woolwich Yard, on the Thames, in Kent, was established early in the reign of Henry VIII. under the immediate inspection of the Navy-Board, and is called by the learned Camden, the Mother Dock. Bishop Gibson also supposes it to be the oldest dock-yard in England, from having discovered that the Harry Grâce de Dieu, of 1000 tons, was built here, in 1512. It is about five furlongs in length, and one in breadth; within this space are two dry-docks, several slips for building ships in, three mast-ponds, smith's shop, mould-loft, dwellings for the resident officers, storehouses of various descriptions, and other buildings. All its concerns, like that of Deptford yard, are managed under the immediate inspection of the Navy Board, and is generally visited weekly by the junior surveyor, and occasionally by the deputy comptroller.

Here also the old Elizabeth was built, when that queen was present at the launch, in 1559; and the Sovereign of the Seas, in Charles the First's time; the latter being launched in 1637.

The number of artificers and labourers employed here in war time, is about 4000, and about 1500 in time of peace.

Establishment of Officers at Woolwich Yard, in 1814.

Resident Commissioner, Capt. Charles Cunningham.
Master Attendant.
Acting ditto.
Master Shipwright.
Clerk of the Check.
Storekeeper.
Clerk of the Survey.
Clerk of the Rope-Yard.
Master Measurer.
Surgeon.
Warden.

Between this yard and the royal arsenal is an extensive building, 400 yards in length, including a rope-walk, where cables of all dimensions are made for the service of the navy, and where several hundred workmen are constantly employed.

The civil and military branches of the office of ordnance that have been established here, since the accession of George the First, have occasioned a very rapid increase, both in population and extent, particularly during the last war.

A royal academy is here established, under the Board of Ordnance, for the instruction and qualifying of young gentlemen, intended as candidates for the office of engineer in the military branch of that office: for farther particulars of this establishment, see the article *Royal Military* Academy *at Woolwich.*

At high-water, the Thames is about a mile broad at Woolwich, and the water brackish. As the channel lies due east and west for about three miles, the tide is strong, and the river being free from shoals and sands, and seven or eight fathoms deep, the largest ships may at all times ride here with safety.

Sheerness Yard, on the island of Sheppey, in Kent, was established in the reign of Charles II. with a dock intended chiefly for repairing ships that may meet with some sudden and unlooked-for accidents; but these are generally fifth and sixth rate, small frigates, yachts, &c. although sometimes large vessels are built here. The officers of the ordnance have here an office for the furnishing of every ship with military stores, and to check the officers of the navy in demand of those stores.

Establishment of Officers at Sheerness Yard, in 1814.

Resident Commissioner, Capt. the Hon. C. Boyle.
Master Attendant.
Master Shipwright.
Clerk of the Check.
Store-keeper.
Clerk of the Survey.
Chaplain.
Surgeon..
Master Measurer.
Master Shipwright's Assistants.
Warden.

There is at present such a line of heavy cannon, completely commanding the mouth of the river, that no fleet of men of war could attempt to pass by, without incurring the hazard of being totally destroyed.

There are also other minor dock-yards belonging to government, viz. at Milford, Deal, North-Yarmouth, Leith, &c.

YARN, (*fil de caret*, Fr.) in rope-making, one of the threads of which a rope is composed. See the articles CABLE, *Cable-laid* CORDAGE, HAWSER, ROPE-MAKING, ROPE-YARN, &c.

Yarn is ordered after the following maner:—after it has been spun upon spindles, spools, or the like, they wind it upon reels which are hardly two feet in length, and have but two contrary cross-bars, being the best and the least liable to ravel. In reeling of yarn, the better to keep it from fine ravelling, you must, as it is reeled, with a tye-band of big twist, divide the slipping or skein into several leys, allowing to every ley eighty threads, and twenty leys to every slipping, if the yarn is very fine; otherwise, less of both kinds. The yarn being spun, reeled, and in the slipping, the next thing to be done is to scour it. In order to fetch out the spots, it should be laid in luke-warm water for three or four days, each day shifting it once, wringing it out and laying it in another water of the same nature; then carry it to a well or brook, and rinse it till nothing comes from it but pure clean water: that being done, take a bucking-tub, and cover the bottom thereof with very fine ashen ashes; and then having opened and spread the slippings, lay them on those ashes, and put more ashes above, and lay in more slippings, covering them with ashes as before, and thus lay one upon another till all the yarn be put in; afterwards cover the uppermost yarn with a bucking-cloth, and in proportion to the bigness of the tub lay therein a peck or two more of ashes; then pour upon the uppermost cloth a great deal of warm water, till the tub can receive no more, and let it stand so all night. Next morning you are to set a kettle of clean water on the fire, and when it is warm, pull out the spiggot of the bucking-tub, to let the water run out of it into another clean vessel. As the bucking-tub wastes, fill it up again with the warm water on the fire; and as the water on the fire wastes, so likewise fill that up with the ley that comes from the bucking-tub, ever observing to make the ley hotter and hotter till it boils: then you must, as before, ply it with the boiling ley, at least four hours together, which is called driving a break of yarn. With yarn thus perpared the sails, cloths, &c. are made.

Rogues-YARN. See ROGUES-YARN.

Spun-YARN. See SPUN-YARN.

YAW, (*embardée*, Fr.) the movement by which a ship deviates from the line of her course towards the right or left in steering.

To YAW, (*lancer*, *embarder*, Fr.) is to steer wild, or out of the true line of the course.

YAWL, (*yole*, Fr.) a boat usually rowed with four or six oars. See the article BOAT.

YEAR, in astronomy and chronology, is the space of time occupied, apparently, by the sun in performing his course through the twelve signs of the zodiac, and in which is comprehended the several changes of the SEASONS; see that article.

The vicissitude of the seasons seems to have given occasion to the first institution of the year. Man, naturally curious to know the cause of that diversity, soon found it was the proximity and distance of the sun, and therefore gave the name year to the space of time in which that luminary performed his whole course, by returning to the same point of his orbit. According to the accuracy in their observations, the year of some nations was more perfect than that of others, but none of them quite exact, nor whose parts did not shift with regard to the parts of the sun's course.

The mean solar year, according to the observations of the best modern astronomers, contains 365 days, 5 hours, 48 minutes, and 48 seconds; the quantity assumed by the authors of the Gregorian calendar, is 365 days, 5 hours, and 49 minutes; but in the civil or popular account the year contains 365 days, 6 hours, or rather 365 days for three years in succession, and every fourth year 366 days.

The solar year is either astronomical or civil.

The Astronomical Solar YEAR, is that which is precisely determined by astronomical observations, and is of two kinds, *tropical* and *sidereal*, or *astral.*

Tropical or Natural YEAR, is the time which the sun, or rather the earth, employs in passing through the twelve signs of the zodiac, and which, as above-mentioned, contains 365 days, 5 hours, 48 minutes, and 48 seconds, which is the only natural year, because it always keeps the same seasons in the same months.

Sidereal or Astral YEAR, is the space of time the sun takes in passing from any fixed star, till his return to it again. This consists of 365 days, 6 hours, 9 minutes, and 11 seconds; being 20 minutes, 29 seconds longer than the true solar year.

Anomalistic YEAR, is the interval occupied by the sun in passing from apogee to apogee, or from perigee to perigee: it is greater than the sidereal year by the time required to describe the annual progression of the apogee. The length of the anomalistic year is 365 days, 6 hours, 14 minutes, and 1 second.

Lunar YEAR, is the space of 12 lunar months. Hence, from the two kinds of synodical lunar months, there arise two kinds of lunar years; the one astronomical, the other civil.

Lunar Astronomical YEAR, consists of 12 lunar synodical months; it contains 354 days, 8 hours, 48 minutes, 38 seconds, and is therefore 10 days, 21 hours, 10 seconds, shorter than the solar year, a difference which is the foundation of the epact.

Lunar Civil YEAR, is either common or embolismic

The Common Lunar YEAR, consists of 12 lunar civil months, and therefore contains 354 days: and

The Embolismic, or *Intercalary Lunar* YEAR, consists of 13 lunar civil months, and therefore contains 384 days.

Thus far we have considered years and months, with regard to astronomical principles, upon which the division is founded. By this, the various forms of civil years that have formerly obtained, or that do still obtain, in divers nations, are to be examined.

Civil YEAR, is that form of year which every nation has contrived or adopted for computing their time by. Or the civil is the tropical year, considered as only consisting of a certain number of whole days: the odd hours and minutes being set aside, to render the computation of time, in the common occasions of life, more easy. As the tropical year is 365 days, 5 hours and 49 minutes, or almost 365 days, 6 hours, which is 365 and a quarter; therefore, if the civil year be made 365 days, every fourth year it must be 366 days, to keep nearly to the course of the sun: and hence the civil year is either common or bissextile. The

Common Civil YEAR, is that consisting of 365 days; having seven months of 31 days each, four of 30 days, and one of 28 days, as indicated by the following well-known memorial verses:—

Thirty days hath September,
April, June, and November;
February twenty-eight alone,
And all the rest have thirty-one:
But leap-year coming once in four,
Gives February one day more.

Bissextile, or *Leap* YEAR, contains 366 days, having one day extraordinary, called the intercalary, or bissextile day, and takes place every fourth year. This additional day to every fourth year was first introduced by Julius Cæsar, who, to make the civil years keep pace with the tropical ones, contrived that the six hours which the latter exceeded the former, should make one day in four years, and be added between the 23d and 24th of February, which was the sixth of the calends of March; and as they then counted this day twice over, or had *bis sexto calendas,* hence the year itself came to be called *bis sextus bissextile.*

However, among us, the intercalary day is not introduced by counting the 23d of February twice over, but by adding a day at the end of that month, which, therefore, in that year, contains 29 days.

A further reformation was made in the civil year by Pope Gregory.

The civil or legal year, in England, formerly commenced on the day of the Annunciation, or 25th of March; though the historical year began on the day of the Circumcision, or 1st of January; on which day the German and Italian year also begins. The part of the year between these two terms was usually expressed both ways, as 1745-6, or 174⅚. But by the act for altering the style, the civil year now begins the 1st of January.

Ancient Roman YEAR. This was the lunar year, which, as first settled by Romulus, contained only ten months, of unequal number of days, in the following order, viz. March 31; April 30; May 31; June 30; Quintilis 31; Sextilis 30; September 30; October 31; November 30; December 30; in all 304 days, which came short of the true lunar year by 50 days; and of the solar by 61 days. Hence, the beginning of Romulus's year was vague, and unfixed to any precise season; to remove which inconvenience, that prince ordered so many days to be added yearly, as would make the state of the heavens correspond to the first month, without calling them by the name of any month. Numa Pompilius corrected this irregular constitution of the year, composing two new months, January and February, of the days that were used to be added to the former year. Thus, Numa's year consisted of 12 months, of different days, as follow; viz.:—January 29; February 28; March 31; April 29; May 31; June 29; Quintilis 31; Sextilis 29; September 29; October 31; November 29; December 29; in all 355 days; therefore, exceeding the quantity of a lunar civil year by one day; that of a lunar astronomical year by 15 hours, 11 minutes, 22 seconds; but falling short of the common solar year by 10 days; so that its beginning was still vague and unfixed.

Numa, however, desiring to have it begin at the winter solstice, ordered 22 days to be intercalated in February every second year, 23 every fourth, 22 every sixth, and 23 every eighth year. But this rule failing to keep matters even, recourse was had to a new way of intercalating; and instead of 23 days every eighth year, only 15 were to be added. The care of the whole was committed to the pontifex Maximus; who, however, neglecting the trust, let things run to great confusion. And thus the Roman year stood till Julius Cæsar reformed it. See CALENDAR, in the APPENDIX.

The Ancient Egyptian YEAR, called also the year of Nabonassar, on account of the epocha of Nabonassar, is the solar year of 365 days, divided into 12 months of 30 days each, besides five intercalary days added at the end. The names, &c. of these months are as follow:—

1. Thoth;	2. Paophi;	3. Athyr;
4. Chojac;	5. Tybi;	6. Mecheir;
7. Phamenoth;	8. Pharmuthi;	9. Pachon;
10. Pauni;	11. Epiphi;	12. Mesori.

The Ancient Greek YEAR, was lunar; consisting of 12 months, which at first had each 30 days, then alternately 29 and 30 days, computed from the first appearance of the new moon; with the addition of an embolismic month of 30 days every 3d, 5th, 8th, 11th, 14th, 16th, and 19th year of a cycle of 19 years; in order to keep the new and full moons to the same terms and seasons of the year. Their year commenced with that new moon, the full moon of which comes next after the summer solstice. The order, &c. of their months was thus: 1. Εκατομϐαιων, containing 29 days; 2. Μηταγειτνιων, 30; 3. Βοηδρομιων, 29; 4. Μαιμακτηριων, 30; 5. Πυανεψιων, 29; 6. Ποσειδεων, 30; 7. Γαμηλιων, 29; 8. Αιθεϛηριων, 30; 9. Ελαφηϐολιων, 30; 10. Μȣνυχιων, 30; 11. Θαργηλιων, 29; 12. Σκιροφοριων, 30.

The Ancient Jewish YEAR, is a lunar year, consisting commonly of 11 months, which alternately contain 30 and 29 days. It was made to agree with the solar year, either by adding 11, and sometimes 12 days, at the end of the year, or by an embolismic month. The names and quantities of the months stand thus: 1. Nisan or Abib, 30 days; 2. Jiar or Zius, 29; 3. Siban or Siwan, 30; 4. Thammuz or Tammuz, 29; 5. Ab, 30; 6. Elul, 29; 7. Tisri or Ethanim, 30; 8. Marchesvam or Bull, 29; 9 Cisleu, 30; 10. Tebeth, 29; 11. Sabat or Schebeth, 30; 12. Adar, in the embolismic year, 30. Adar, in the common year, was only 29 days.—Note, in the defective year, Cisleu was only 29 days; and in the redundant year, Marchesvam was 30.

The Modern Jewish YEAR, is likewise lunar, consisting of 12 months in common years, but of 13 in embolismic years; which, in a cycle of 19 years, are the 3d, 6th, 8th, 11th, 14th, 17th, and 19th. Its beginning is fixed to the new moon next after the autumnal equinox.

The Syrian YEAR, is a solar one, having its beginning fixed to the commencement of October in the Julian year, from which it only differs in the names of the months, the quantities being the same.

The Persian YEAR, is a solar year of 365 days, consisting of 12 months of 30 days each, with 5 intercalary days added at the end.

The Arabic, Mahometan, and Turkish YEAR, called also the year of the Hegira, is a lunar year, equal to 354 days, 8 hours, 48 minutes, and consists of 12 months, containing alternately 30 and 29 days.

The Ethiopic YEAR, is a solar year, perfectly agreeing with the Actiac, except in the names of the months.

The Hindoo YEAR differs from all these, and is indeed different in the different provinces of India. The best account that we have of it is by Mr. Cavendish, in the Philosophical Transactions of the Royal Society of London for the year 1792.

YEOMEN, (*mariniers des vaisseaux de guerre Anglois, qui sont chargés de l'arrangement et de la distribution des munitions et rechanges, du maître d'equipage, et du maître-canonnier, et d'en tenir compte*, Fr.) in a ship of war, are careful persons selected for the purpose of keeping the store-rooms clean, and the stores properly arranged They have the issuing of the stores and keeping an account of those issued for the uses of the ship, under the directions of the boatswain, gunner, and carpenter.

There are, besides the above-mentioned, the

YEOMAN *of the Powder-Room*, (*gardien de la Sainte-barbe*, Fr.) a careful person who is nominated by the gunner, having an optional rating on the ship's books, to attend the fore and after magazines, and to receive and deliver the powder when wanted for action, or otherwise.

YEOMAN *of the Sheets*, (*officier marinier de manœuvre chargé de veiller les écontes*, Fr.) a petty officer, whose duty it is to see that the fore or main sheets are properly belayed on working the ship, and that they are kept clear for running.

YOKE, (*palan mis à la barre du gouvernail*, Fr.) a small board which crosses the upper end of a boat's rudder at right angles, and having two lines extending from its opposite extremities to the stern-sheets of the boat, whereby she is steered as with a tiller.

YOKE, is also a name that was formerly given to the tiller when communicating with two blocks or sheaves affixed to the inner end of the tiller.

YOUNGSTER, (*mousse*, Fr.) a general name given to midshipmen on their first entering into the navy, and is used in contradistinction to *oldster*, which is generally applied to those who have been some years in the service.

Z.

ZENITH, (*zenith*, Fr.) in astronomy, the vertical point, or that point in the heavens directly over our heads. Or, the zenith is a point in the surface of the sphere, from which a right line drawn through the place of any spectator, passes through the centre of the earth.

The zenith of any place, is also the pole of the horizon, being 90 degrees distant from every point of it: and through the zenith pass all the azimuths, or vertical circles.

The point diametrically opposite to the zenith, is called the *nadir*, being the point in the sphere directly under our feet: and it is the zenith to our antipodes, as our zenith is their nadir.

ZENITH *Distance*, is the distance of the sun or star from our zenith; and is the complement of the altitude, or what it wants of 90 degrees.

ZENITH *Sector*. See the article SECTOR.

ZERO, a word used to express a cypher or nought.

ZODIAC, (*zodiaque*, Fr. *zodiacus*, Lat.) in astronomy, an imaginary ring or broad circle, in the heavens, in form of a girdle or belt, within which the planets all make their excursions. In the very middle of it runs the ecliptic, or path of the sun in his annual course; and its breadth, comprehending the deviations or latitudes of the planets, is by some authors accounted 16, or 18°, and by others 20 degrees.

The zodiac, cutting the equator obliquely, makes with it the same angle as the ecliptic, which is its middle line, continually varying, and is now 23° 28 2″, called the obliquity of the zodiac, or ecliptic; it is also the sun's greatest declination.

The zodiac is divided into twelve equal parts, of 30 degrees each, called the signs of the zodiac, being so named from the constellations which anciently passed them. But, the stars having a motion from west to east, those constellations do not now correspond to their proper signs; from whence arises what is called the precession of the equinoxes. Therefore, when a star is said to be in such a sign of the zodiac, it is not to be understood of that constellation, but only of that dodecatemory or twelfth part of it.

Cassini has also observed a tract in the heavens, within whose bounds most of the comets, though not all of them, are observed to keep, and which he therefore calls the *zodiac of the comets*. This he makes as broad as the other zodiac, and marks it with signs or constellations, like that; as Antinous, Pegasus, Andromeda, Taurus, Orion, the Lesser Dog, Hydra, the Centaur, Scorpion, and Sagittary.

ZODIACAL *Light*, a brightness observed sometimes in the zodiac, resembling that of the galaxy or milky way. It appears at certain seasons, viz. towards the end of winter and in spring, after sunset, or before his rising in autumn, and the beginning of winter, resembling the form of a pyramid, lying lengthways, with its axis along the zodiac, its base being placed obliquely with respect to the horizon. This phenomenon was first described and named by the elder Cassini, in 1683.

The zodiacal light, according to Mairan, is the solar atmosphere, a rare and subtile fluid, either luminous by itself or made so by the rays of the sun surrounding its globe, but in a greater quantity, and more extensively about his equator than any other part.

Mairan says, it may be proved from many observations, that the sun's atmosphere sometimes reaches as far as the earth's orbit, and there meeting with our atmosphere, produces the appearance of an aurora borealis.

The length of the zodiacal light varies sometimes in reality, and sometimes in appearance only, from various causes.

Cassini often mentions the great resemblance between the zodiacal light and the tails of comets. The same observation has been made by Fatio; and Euler endeavoured to prove that they were owing to similar causes.

This light seems to have no other motion than of the sun itself; and its extent from the sun to its point, is seldom less than 50 or 60 degrees in length, and more than 20 degrees in breadth; but it has been known to extend to 100 or 103 degrees, and from 8 to 9 degrees broad.

It is now generally acknowledged, that the electric fluid is the cause of the aurora borealis, ascribed by Mairan to the solar atmosphere, which produces the zodiacal light, and which is thrown off chiefly, and to the greatest distance, from the equatorial parts of the sun, by means of the rotation on his axis, and extending visibly as far as the orbit of the earth, where it falls into the upper regions of our atmosphere, and is collected principally towards the polar parts of the earth, in consequence of the diurnal revolution, where it forms the aurora borealis. And hence it has been suggested, as a probable conjecture, that the sun may be the fountain of the electric fluid, and that the zodiacal light, and the tails of comets, as well as the aurora borealis, the lightning and artificial electricity, are its various and not very dissimilar modifications. See HUTTON's *Philosophical Dictionary*: also VINCE's *Astronomy*.

ZONE, (*zone*, Fr. *zona*, Lat.) in geography and astronomy, a division of the earth's surface, by means of parallel circles, chiefly with respect to the degrees of heat in the different parts of that surface.

The ancient astronomers used the word zone, to explain the different appearances of the sun and other heavenly bodies, with the length of the days and nights: and the geographers. as they used the climates, to mark the situation of places; using the word climate when they were able to be more exact, and the term zone when less so.

The zones were commonly accounted five in number; one a broad belt round the middle of the earth, having the equator in the very middle of it, and bounded, towards the north and south, by parallel circles passing through the tropics of Cancer and Capricorn; this they call the *torrid zone*, which they supposed not habitable, on account of its extreme heat. Though this they sometimes divided into two equal torrid zones, by the equator, one to the north, and the other to the south; and then the whole number of zones was accounted six.

Next, from the tropics of Cancer and Capricorn, to the two polar circles, were two other spaces, called *temperate zones*, as being moderately warm; and these they considered to be the only habitable parts of the earth.

Lastly, the two spaces beyond the temperate zones, about either pole, bounded within the polar circles, and having the poles in the middle of them, are the two *frigid* or *frozen zones*, and which the ancients supposed not habitable, on account of the extreme cold.

Hence, the breadth of the torrid zone is equal to twice the greatest declination of the sun, or obliquity of the ecliptic, equal to 46° 56′, or twice 23° 28′. Each frigid zone is also of the same breadth, the distance from the pole to the polar circle being equal to the same obliquity, 23° 28. And the breadth of each temperate zone is equal to 43° 4′, the complement of twice the same obliquity.

The difference of zones is attended with a great diversity of phenomena: 1st. In the torrid zone, the sun passes through the zenith of every place in it twice a year, making as it were two summers in the year, and the inhabitants of this zone are called *amphiscians*, because they have their noon-day shadows projected different ways in different times of the year, northward at one season, and southward at the other. 2d. In the temperate and frigid zones, the sun rises and sets every natural day of 24 hours. Yet every where, but under the equator, the artificial days are of unequal lengths, and the inequality is the greater, as the place is farther from the equator. The inhabitants of the temperate zones are called *heterocians*, because their noon-day shadow is cast the same way all the year round, viz. those in the north zone towards the north pole, and those in the south zone towards the south pole. 3d. Within the frigid zones, the inhabitants have their artificial days and nights extended out to a great length; the sun sometimes skirting round a little above the horizon for many days together: and at another season never rising above the horizon at all, but making continual night for a considerable space of time. The inhabitants of these zones are called *periscians*, because sometimes they have their shadows going quite round them in the space of 24 hours.

ZOPISSA, (*zopissa*, Fr.) the best sort of pitch, or pitch which is scraped off from the sides of ships, and then tempered with wax and salt.

APPENDIX.

A.

ADVICE-Boat. Page 6, col. 1, after line 29, add,

It is said to have been employed, for the first time, in 1692, before the battle off La Hogue, in order to gain intelligence of what was passing at Brest. See Derrick's *Memoirs of the Royal Navy*.

ALTITUDE. Page 9, col. 2, after line 48, add,

Altitude *of a Shot or Shell*, in gunnery, is the perpendicular height of the vertex of the curve in which it moves above the horizon.

APPOINTMENT of Officers.—No commander in chief, or flag officer, or captain commanding a squadron of his Majesty's ships, is to appoint, by commission or warrant, any officer to a ship under his command, except when empowered to do so by the lords commissioners of the Admiralty; but may appoint them to act in vacancies till the pleasure of the Admiralty shall be known; and even when commanders in chief are invested with such authority, they are not to exercise that power whilst in the Channel Soundings, nor in the North or Baltic Seas, nor on any coast of the United Kingdom, without the sanction of the Admiralty. If any officer, except a flag officer, shall die, or be dismissed from the ship or service, by the sentence of a court-martial, the commander in chief, not being in any of the situations specified, may appoint a successor by commission or warrant. He must not, however, appoint an officer on half-pay to any commission or employment without express directions from the Admiralty; nor appoint any person to command a ship of 20 guns, or upwards, or have the rank of captain, until he has been one year a commander; neither shall any person command a sloop of war, nor have the rank of a commander, until he has been two years a lieutenant. If by necessity a lieutenant shall be appointed to command a ship of 20 guns or upwards, he is to have the permanent rank of commander only.

No person is to be appointed to the rank of lieutenant, or sub-lieutenant, until he has attained the full age of 19 years, and has actually served on board one or more of his Majesty's ships, six years, or when educated at the Royal Naval College at Portsmouth, four years; two of which he must have been rated as master's-mate or midshipman, and must produce such certificates of age, &c. as directed by the Admiralty: if any person shall produce false certificates, he shall be rendered incapable of receiving any commission; and even after promotion, if it be discovered, he shall immediately be discharged from the navy, whatever rank he may have attained to.

By a late order in council, no person will be allowed to pass an examination touching his qualifications to serve as master in the royal navy, before he shall have attained the age of twenty-one years, or after the age of thirty-five years, and have been at least seven years at sea, two of which must have been served in his Majesty's navy, in the quality of acting master by order, second master, master's mate, or midshipman; or being within the aforesaid ages, and having served seven years at sea, shall have actually been one year or more chief mate, and two years master; or two years chief mate, and one year master of a merchant ship; and shall produce certificates of diligence, sobriety, and good conduct, from the owners, for the time he served as master, and from the owners and master for the time he served as mate.

That no master shall be permitted to pass his examination for a higher rate, until he shall have served one complete year in a ship in commission, under his last qualification: and persons appointed to any station in which they are to have the charge of stores, must be able to read, write, and keep accounts. Every gunner, before he can be appointed by warrant, must serve one year as petty officer, when he is to produce certificates and undergo an examination: boatswains are also to serve one year as petty officers, and also produce certificates of their good conduct; nor are carpenters to be appointed by warrant, unless they have served an apprenticeship to a shipwright.

By a late order in council, a purser, before he can be appointed to any of his Majesty's ships, must not be less than twenty-one, or more than thirty-five years of age, and must have served two years as captain's clerk, or have been employed in the office of the secretary to a flag officer, for one other year; and shall, in every case, produce a certificate of his good conduct, pass the usual examinations, and give such security for an honest and faithful discharge of his duty, as may from time to time be required. That no purser shall be capable of being appointed to a 4th or 5th rate, who shall not have been two years a warranted purser on the list, nor to a 1st, 2d, or 3d rate, till he shall have been five years on the list of warranted pursers. An officer must be of the same rank, to be appointed to act in the absence of another. A captain or commander is not to be removed from the ship or vessel he commands without his own consent, except the captain of a ship in which a flag officer is embarked, or the senior captain commanding a frigate, whom the commander in chief may remove to the command of a ship of the line: neither are standing warrant officers to be removed without their consent.

If an officer, whose situation requires him to undergo an examination, before he can be promoted, shall have left England without having passed such examination, the commander in chief may order officers properly qualified to examine him. A candidate for a lieutenant or sub-lieutenant's commission must be examined by three of the senior captains of the squadron: a candidate for the appointment of master or second master, by one of the senior captains and three of the best qualified masters: a candidate for a gunner's warrant, by one master and three gunners, in the presence of a captain; and a candidate for the appointment of surgeon, or assistant surgeon, by the physician of the fleet, the physician and senior surgeon of the hospital, and three surgeons of the squadron; and to give such certificate as they shall find him really to deserve, which certificate is to be transmitted to the secretary of the Admiralty, or to the Navy Board or Sick and Hurt Board, as the appointment of such officer is invested, to be examined as the lords commissioners, or the respective boards, shall direct, before any appointment can be confirmed.

A commander in chief of one squadron may appoint proper officers to act in vacancies, in a ship belonging to another squadron, and he must send to the secretary of the Admiralty a return of all officers he shall have appointed to any ship under his command. When a ship is not in any part of the United Kingdom, nor in company with any other ship, if the captain of that ship should die, the senior lieutenant shall take the command, and must appoint a master's mate or midshipman to act as lieutenant; but if more ships than one are present, the senior captain shall make the necessary appointments, and information of such appointments are to be sent to the Admiralty; the senior captain present is also, in cases of death, to appoint commissioned or warrant officers to act till the pleasure of the Admiralty is known. If the captain and all the lieutenants and acting lieutenants are killed in battle, &c. the master is to succeed to the command, and is to appoint a sufficient number of master's mates and midshipmen to act as lieutenants. If the master also be killed, or taken prisoner, or die, the command will devolve on the second master; after him on the gunner, and so on. If a lieutenant or master succeed to the command of a ship, he shall receive the pay and allowances of a commander only: but if the second master, gunner, boatswain, or carpenter succeed to the command, he is to be paid as the Admiralty shall think his conduct deserves. For farther particulars, see the NAVAL INSTRUCTIONS.

ARTICLES OF WAR, are certain regulations for the better government of the navy, and may be altered at the pleasure of the king. They are printed and hung up in the most public place in every ship of the royal navy, and are ordered to be read once at least in every month to the ship's company.

The following is an Abstract of the Act of Parliament, commonly called the *Articles of War*.

After reciting several acts of parliament relating to the government and discipline of the navy, and declaring them to be repealed, it states—

That, for the regulating and better government of his Majesty's navies, ships of war, and forces by sea, whereon, under the good providence of God, the wealth, safety, and strength of this kingdom chiefly depend; be it enacted by the king's most excellent Majesty, by and with the advice and consent of the lords spiritual and temporal, and commons, in this present parliament assembled, and by the authority of the same, That from and after the twenty-fifth day of December, one thousand seven hundred and forty-nine, the articles and orders hereinafter following, as well in time of peace as in time of war, shall be duly observed and put in execution, in manner herein after mentioned.

I. All commanders, captains, and officers, in or belonging to any of his Majesty's ships or vessels of war, shall cause the public worship of Almighty God, according to the liturgy of the church of England established by law, to be solemnly, orderly, and reverently performed in their respective ships; and shall take care that prayers and preaching, by the chaplains in holy orders of the respective ships, be performed diligently; and that the Lord's Day be observed according to law.

II. All flag officers, and all persons in or belonging to his Majesty's ships or vessels of war, being guilty of profane oaths, cursings, execrations, drunkenness, uncleanness, or other scandalous actions, in derogation of God's honour, and corruption of good manners, shall incur such punishment as a court-martial shall think fit to impose, and as the nature and degree of their offence shall deserve.

III. If any officer, mariner, soldier, or other person of the fleet, shall give, hold, or entertain intelligence to or with any enemy or rebel, without leave from the king's Majesty, or the lord high admiral, or the commissioners for executing the office of lord high admiral, commander in chief, or his commanding officer, every such person so offending, and being thereof convicted by the sentence of a court-martial, shall be punished with death.

IV If any letter or message from any enemy or rebel be conveyed to any officer, mariner, or soldier, or other in the fleet, and the said officer, mariner, soldier, or other as aforesaid, shall not within twelve hours, having opportunity so to do, acquaint his superior officer, or the officer commanding in chief with it; or if any superior officer being acquainted therewith, shall not in convenient time reveal the same to the commander in chief of the squadron, every such person so offending, and being convicted thereof, by the sentence of the court-martial, shall be punished with death, or such other punishment as the nature and degree of the offence shall deserve, and the court-martial shall impose.

V. All spies, and all persons whatsoever who shall come, or be found, in the nature of spies, to bring or deliver any seducing letters or messages from any enemy or rebel, or endeavour to corrupt any captain, officer, mariner, or other in the fleet, to betray his trust, being convicted of any such offence by the sentence of the court-martial, shall be punished with death, or such other punishment as the nature and degree of the offence shall deserve, and the court-martial shall impose.

VI. No person in the fleet shall relieve an enemy or rebel with money, victuals, powder, shot, arms, ammunition, or any other supplies whatsoever, directly or indirectly, upon pain of death, or such other punishment as the court-martial shall think fit to impose, and as the nature and degree of the crime shall deserve.

VII. All the papers, charter-parties, bills of lading, passports, and other writings whatsoever, that shall be taken, seized, or found aboard any ship or ships which shall be surprised or taken as prize, shall be duly preserved, and the very originals shall, by the commanding officer of the ship who shall take such prize, be sent entirely, and without fraud, to the court of Admiralty, or such other court or commissioners as shall be authorized to determine whether such prize be lawful capture, there to be viewed, made use of, and proceeded upon according to law, upon pain that every person offending herein shall forfeit and lose his share of the capture, and shall suffer such further punishment as the nature and degree of his offence shall be found to deserve, and the court-martial shall impose.

VIII. No person in or belonging to the fleet shall take out of any prize, or ship seized for prize, any money, plate, or goods, unless it shall be necessary for the better securing thereof, or for the necessary use and service of any of his Majesty's ships or vessels of war, before the same be adjudged lawful prize in some Admiralty court; but the full and entire account of the whole, without embezzlement, shall be brought in, and judgment passed entirely upon the whole without fraud, upon pain that every person offending herein shall forfeit and lose his share of the capture, and suffer such further punishment as shall be imposed by a court-martial, or such court of Admiralty, according to the nature and degree of the offence.

IX. If any ship or vessel shall be taken as prize, none of the officers, mariners, or other persons on board her, shall be stripped of their clothes, or in any sort pillaged, beaten, or evil-intreated, upon pain that the person or persons so offending shall be liable to such punishment as a court-martial shall think fit to inflict.

X. Every flag officer, captain, and commander in the fleet, who, upon signal or order of fight, or sight of any ship or ships which it may be his duty to engage, or who, upon likelihood of engagement, shall not make the necessary preparations for fight, and shall not in his own person, and according to his place, encourage the inferior officers and men to fight courageously, shall suffer death, or such other punishment as from the nature and degree of the offence a court-martial shall deem him to deserve; and if any person in the fleet shall treacherously or cowardly yield or cry for quarter, every person so offending, and being convicted thereof by the sentence of a court-martial, shall suffer death.

XI. Every person in the fleet who shall not duly observe the orders of the admiral, flag officer, commander of any squadron or division, or other his superior officer, for assailing, joining battle with, or making defence against any fleet, squadron, or ship, or shall not obey the orders of his superior officer as aforesaid in time of action, to the best of his power, or shall not use all possible endeavours to put the same effectually into execution, every such person so offending, and being convicted thereof by the sentence of the court-martial, shall suffer death, or such other punishment, as from the nature and degree of the offence a court-martial shall deem him to deserve.

XII. Every person in the fleet, who through cowardice, negligence, or disaffection, shall in time of action withdraw or keep back, or not come into the fight or engagement, or shall not do his utmost to take or destroy every ship which it shall be his duty to engage, and to assist and relieve all and every of his Majesty's ships, or those of his allies which it shall be his duty to assist and relieve, every such person so offending, and being convicted thereof by the sentence of a court-martial, shall suffer death.

XIII. Every person in the fleet, who through cowardice, negligence, or disaffection, shall forbear to pursue the chace of any enemy, pirate, or rebel, beaten or flying; or shall not relieve and assist a known friend in view to the utmost of his power; being convicted of any such offence by the sentence of a court martial, shall suffer death.

XIV. If when action, or any service shall be commanded, any person in the fleet shall presume to delay or discourage the said action or service, upon pretence of arrears of wages, or upon any pretence whatsoever, every person so offending, being convicted thereof by the sentence of the court-martial, shall suffer death, or such other punishment as from the nature and degree of the offence a court-martial shall deem him to deserve.

XV Every person in or belonging to the fleet, who shall desert to the enemy, pirate, or rebel, or run away with any of his Majesty's ships or vessels of war, or any ordnance, ammunition, stores, or provision belonging thereto, to the weakening of the service, or yield up the same cowardly or treacherously to the enemy, pirate, or rebel, being convicted of any such offence by the sentence of the court-martial, shall suffer death.

XVI. Every person in or belonging to the fleet, who shall desert, or entice others so to do shall suffer death, or such other punishment as the circumstances of the offence shall deserve, and a court-martial shall judge fit: and if any commanding officer of any of his Majesty's ships or vassels of war shall receive or entertain a deserter from any other of his Majesty's ships or vessels, after discovering him to be such deserter, and shall not with all convenient speed give notice to the captain of the ship or vessel to which such deserter belongs, or if the said ships or vessels are at any considerable distance from each other, to the secretary of the Admiralty, or to the commander in chief; every person so offending, and being convicted thereof by the sentence of the court-martial, shall be cashiered.

XVII. The officers and seamen of all ships, appointed for convoy and guard of merchant ships, or of any other, shall diligently attend upon that charge, without delay, according to their instructions in that behalf; and whosoever shall be faulty therein, and shall not faithfully perform their duty, and defend the ships and goods in their convoy, without either diverting to other parts or occasions, or refusing or neglecting to fight in their defence, if they be assailed, or running away cowardly, and submitting the ships in their convoy to peril and hazard; or shall demand or exact any money or other reward from any merchant or master for convoying of any ships or vessels intrusted to their care, or shall misuse the masters or mariners thereof, shall be condemned to make reparation of the damage to the merchants, owners, and others, as the Court of Admiralty shall adjudge, and also be punished criminally according to the quality of their offences, be it by pains of death, or other punishment, according as shall be adjudged fit by the court-martial.

XVIII. If any captain, commander, or other officer of any of his Majesty's ships or vessels, shall receive on board, or permit to be received on board such ship or vessel, any goods or merchandise whatsoever, other than for the sole use of the ship or vessel, except gold, silver, or jewels, and except the goods and merchandizes belonging to any merchant, or other ship or vessel which may be shipwrecked, or in imminent danger of being shipwrecked, either on the high seas or in any port, creek, or harbour, in order to the preserving them for their proper owners, and except such goods or merchandizes as he shall at any time be ordered to take or receive on board, by order of the lord high admiral of Great Britain, or the commissioners for executing the office of lord high admiral for the time being; every person so offending, being convicted thereof by the sentence of a court-martial, shall be cashiered, and be for ever afterwards rendered incapable to serve in any place or office in the naval service of his Majesty, his heirs and successors.

XIX. If any person in or belonging to the fleet shall make, or endeavour to make, any mutinous assembly upon any pretence whatsoever, every person offending herein, and being convicted thereof by the sentence of the court-martial, shall suffer death: and if any person in or belonging to the fleet shall utter any words of sedition or mutiny, he shall suffer death, or such other punishment as a court-martial shall deem him to deserve: and if any officer, mariner, or soldier, in or belonging to the fleet, shall behave himself with contempt to his superior officer, such superior officer being in the execution of his office, he shall be punished according to the nature of his offence by the judgment of a court-martial.

XX. If any person in the fleet shall conceal any traitorous or mutinous practice or design, being convicted thereof by the sentence of a court-martial, he shall suffer death, or such other punishment as a court-martial shall think fit; and if any person in or belonging to the fleet, shall conceal any traitorous or mutinous words spoken by any, to the prejudice of his Majesty or Government, or any words, practice, or design, tending to the hindrance of the service, and shall not forthwith reveal the same to the commanding officer, or being present at any mutiny or sedition, shall not use his utmost endeavours to suppress the same, he shall be punished as a court-martial shall think he deserves.

XXI. If any person in the fleet shall find cause of complaint of the unwholesomeness of the victuals, or upon other just ground, he shall quietly make the same known to his superior, or captain, or commander in chief, as the occasion may deserve, that such present remedy may be had as the matter may require; and the said superior, captain, or commander in chief, shall, as far as he is able, cause the same to be presently remedied; and no person in the fleet, upon any such or other pretence, shall attempt to stir up any disturbance, upon pain of such punishment as a court-martial shall think fit to inflict, according to the degree of the offence.

XXII. If any officer, mariner, soldier, or other person in the fleet shall strike any of his superior officers, draw, or offer to draw, or lift up any weapon against him, being in the execution of his office, on any pretence whatsoever, every such person being convicted of any such offence by the sentence of a court-martial, shall suffer death; and if any officer, mariner, soldier, or other person in the fleet, shall presume to quarrel with any of his superior officers, being in the execution of his office, or shall disobey any lawful command of any of his superior officers; every such person being convicted of any such offence by the sentence of a court-martial, shall suffer death, or such other punishment as shall, according to the nature and degree of his offence, be inflicted upon him by the sentence of a court-martial.

XXIII. If any person in the fleet shall quarrel or fight with any other person in the fleet, or use reproachful or provoking speeches or gestures, tending to make any quarrel or disturbance, he shall, upon being convicted thereon, suffer such punishment as the offence shall deserve, and a court-martial shall impose.

XXIV. There shall be no wasteful expence of any powder, shot, ammunition, or other stores in the fleet, nor any embezzlement thereof, but the stores and provisions shall be carefully preserved, upon pain of such punishment to be inflicted upon the offenders, abettors, buyers, and receivers (being persons subject to naval discipline) as shall be by a court-martial found just in that behalf.

XXV. Every person in the fleet who shall unlawfully burn or set fire to any magazine or store of powder, or ship, boat, ketch, hoy, or vessel, tackle or furniture thereunto belonging, not then appertaining to an enemy, pirate, or rebel, being convicted of any such offence, by the sentence of a court-martial, shall suffer death.

XXVI. Care shall be taken in the conducting and steering of any of his Majesty's ships, that through wilfulness, negligence, or other defaults, no ships be stranded, or run upon any rocks or sands, or split or hazarded, upon pain that such as shall be found guilty therein be punished by death, or such other punishment as the offence, by a court-martial, shall be judged to deserve.

XXVII. No person in or belonging to the fleet shall sleep upon his watch, or negligently perform the duty imposed on him, or forsake his station, upon pain of death, or such other punishment as a court-martial shall think fit to impose, and as the circumstances of the case shall require.

XXVIII. All murders committed by any person in the fleet shall be punished with death, by the sentence of the court-martial.

XXIX. If any person in the fleet shall commit the unnatural and detestable sin of buggery or sodomy with man or beast, he shall be punished with death, by the sentence of a court-martial.

XXX. All robbery committed by any person in the fleet shall be punished with death, or otherwise, as a court-martial, upon consideration of circumstances, shall find meet.

XXXI. Every officer or other person in the fleet, who shall knowingly make or sign a false muster or muster-book, or who shall command, counsel, or procure the making or signing thereof, or who shall aid or abet any other person in the making or signing thereof, shall, upon proof of any such offence being made before a court martial, be cashiered, and rendered incapable of further employment in his Majesty's naval service.

XXXII. No provost-martial belonging to the fleet shall refuse to apprehend any criminal, whom he shall be authorized by legal warrant to apprehend, or to receive or keep any prisoner committed to his charge, or wilfully suffer him to escape, being once in his custody, or dismiss him without lawful order, upon pain of such punishment as a court-martial shall deem him to deserve; and all captains, officers, and others in the fleet, shall do their endeavour to detect, apprehend, and bring to punishment all offenders, and shall assist the officers appointed for that purpose therein, upon pain of being proceeded against, and punished by a court-martial, according to the nature and degree of the offence.

XXXIII. If any flag officer, captain, commander, or lieutenant belonging to the fleet, shall be convicted before a court-martial, of behaving in a scandalous, infamous, cruel, oppressive, or fraudulent manner, unbecoming the character of an officer, he shall be dismissed from his Majesty's service.

XXXIV. Every person being in actual service and full pay, and part of the crew in or belonging to any of his Majesty's ships or vessels of war, who shall be guilty of mutiny, desertion, or disobedience to any lawful command, in any part of his Majesty's dominions on shore, when in actual service, relative to the fleet, shall be liable to be tried by a court-martial, and suffer the like punishment, for every such offence, as if the same had been committed at sea, on board any of his Majesty's ships or vessels of war.

XXXV. If any person who shall be in actual service and full pay in his Majesty's ships and vessels of war, shall commit upon the shore, in any place or places out of his Majesty's dominions, any of the crimes punishable by these articles and orders, the persons so offending shall be liable to be tried and punished for the same, in like manner, to all intents and purposes, as if the same crimes had been committed at sea, on board any of his Majesty's ships or vessels of war.

XXXVI. All other crimes not capital, committed by any person or persons in the fleet, which are not mentioned in this act, or for which no punishment is hereby directed to be inflicted, shall be punished according to the laws and customs in such cases used at sea.

Provided always, That no person convicted of any offence shall, by the sentence of any court-martial to be held by virtue of this act be adjudged to be imprisoned for a longer term than the space of two years.

Provided always, That nothing in this act contained shall extend, or be construed to extend, to empower any court-martial to be constituted by virtue of this act, to proceed to the punishment or trial of any of the offences specified in the several articles contained in this act, or of any offence whatsoever (other than the offences specified in the fifth, thirty-fourth, and thirty-fifth of the foregoing articles and orders) which shall not be committed upon the main sea, or in great rivers only, beneath the bridges of the said rivers nigh to the sea, or in any haven, river, or creek, within the jurisdiction of the Admiralty, and which shall not be committed by such persons as at the time of the offence committed shall be in actual service and full pay in the fleet or ships of war of his Majesty, his heirs, or successors, such persons only excepted, and for such offences only, as are described in the fifth of the foregoing articles and orders.

Provided also, That nothing in this act contained, shall extend, or be construed to extend, to empower any court-martial to be constituted by virtue of this act, to proceed to the punishment or trial of any land officer or soldier on board any transport ship, for any of the offences specified in the several articles contained in this act.

And it is hereby further enacted, That from and after the twenty-fifth day of December, one thousand seven hundred and forty-nine, the lord high admiral of Great Britain, or the Commissioners for executing the office of lord high admiral of Great Britain, for the time being, shall have full power and authority to grant commissions to any officer commanding in chief any fleet or squadron of ships of war, to call and assemble courts martial, consisting of commanders and captains; and that in case any officer commanding in chief any fleet or squadron of ships of war (who shall be authorized by the lord high admiral, or the commissioners for executing the office of lord high admiral for the time being, to call and assemble courts martial in foreign parts) shall happen to die, or be recalled, or removed from his command, then the officer upon whom the command of the said fleet or squadron shall devolve, and so, from time to time the officer who shall have the command of the said fleet or squadron, shall have the same power to call and assemble courts martial, as the first commander in chief of the said fleet or squadron was invested with.

Provided always, and it is hereby enacted and declared, That no commander in chief of any fleet or squadron of his Majesty's ships, or detachments thereof, consisting of more than five ships, shall preside at any court martial in foreign parts, but that the officer next in command to such officer commanding in chief, shall hold such court martial, and preside thereat, any law, custom, or usage to the contrary notwithstanding.

And it is hereby further enacted, That from and after the twenty-fifth day of December, one thousand seven hundred and forty-nine, in case any commander in chief of any fleet or squadron of his Majesty's ships or vessels of war in foreign parts, shall detach any part of such fleet or squadron, every commander in chief shall, and he is hereby authorized and required by writing under his hand, to empower the chief commander of the squadron or detachment so ordered on such separate service (and in case of his death or removal, the officer to whom the command of such separate squadron or detachment shall belong) to hold courts martial, during the time of such separate service, or until the commander of the said detachment for the time being shall return to his commander in chief, or shall come under the command of any other his superior officer, or return to Great Britain or Ireland.

Provided always, and it is hereby further enacted, That if any five or more of his Majesty's ships or vessels of war shall happen to meet together in foreign parts, then, and in such case, it shall be lawful for the senior officer of the said ships or vessels to hold courts martial, and preside thereat, from time to time, as there shall be occasion, during so long time as the said ships or vessels of war, or any five or more of them, shall continue together.

Provided nevertheless, and be it also enacted, That where any material objection occurs, which may render it improper for the person who is next in command to the senior officer or commander in chief, of any fleet or squadron of his Majesty's ships of war in foreign parts to hold courts martial, or preside thereat, in such case it shall be lawful for the lord high admiral, or commissioners for executing the office of lord high admiral for the time being, as also the commander in chief of any such fleet or squadron of his majesty's ships in foreign parts respectively to appoint the third officer in command to preside at, or hold such court martial.

And it is hereby further enacted, That from and after the twenty-fifth day of December, one thousand seven hundred and forty-nine, it shall be lawful for the lord high admiral of Great Britain, or the commissioners for executing the office of lord high admiral for the time being, and they are hereby respectively authorized, from time to time, as there shall be occasion, to direct any flag officer, or captain of any of his Majesty's ships of war, who shall be in any port of Great Britain or Ireland, to hold courts martial in any such port, provided such flag officer or captain be the first, second, or third in command in such port, as shall be found most expedient, and for the good of his Majesty's service, and such flag officer or captain, so directed to hold courts martial, shall preside at such court martial, any thing herein contained to the contrary notwithstanding.

And it is hereby further enacted, That from and after the twenty-fifth day of December, one thousand seven hundred and forty-nine, no court martial to be held or appointed by virtue of this present act, shall consist of more than thirteen, or less than five persons, to be composed of such flag officers, captains, or commanders, then and there present, as are next in seniority to the officer who presides at the court martial.

Provided always, and be it enacted by the authority aforesaid, That nothing herein contained shall extend, or be construed to extend, to authorize or empower the lord high admiral, or the commissioners for executing the office of lord high admiral, or any officer empowered to order or hold courts martial, to direct or ascertain the particular number of persons of which any court martial, to be held or appointed by virtue of this present act, shall consist.

Provided always, and it is hereby enacted and declared, That in case any court martial shall by virtue of this act, be appointed to be held at any place where there are not less than three, nor yet so many as five officers of the degree and denomination of a post captain, or of a superior rank to be found, then it shall be lawful for the officer, at the place appointed for holding such court martial, who is to preside at the same, to call to his assistance as many of the commanders of his Majesty's vessels under the rank and degree of a post captain, as, together with the post captains then and there present will make up the number of five, to hold such court martial.

And it is hereby further enacted, That from and after the twenty-fifth day of December, one thousand seven hundred and forty-nine, no member of any court martial, after the trial is begun, shall go on shore till sentence be given, but remain on board the ship in which the court shall first assemble, except in case of sickness, to be judged of by the court, upon pain of being cashiered from his Majesty's service; nor shall the proceedings of the said court be delayed by the absence of any of its members, provided a sufficient number doth remain to compose the said court, which shall and is hereby required to sit from day to day (sunday always excepted) until the sentence be given.

And it is hereby further enacted, That from and after the twenty-fifth day of December, one thousand seven hundred and forty-nine, upon all trials of offenders by any court-martial, all the officers present, who are to constitute the said court martial, shall, before they proceed to such trial, take such oath as is herein after mentioned, upon the holy evangelists, before the court, which oath the judge advocate, or his deputy, or the person appointed to officiate as such, is hereby authorized and required to administer in the words following (that is to say)

I, A. B. *do swear, That I will duly administer justice, according to the articles and orders established by an Act passed in the twenty-second year of the reign of his Majesty King* George *the Second, for amending, explaining, and reducing into one Act of Parliament, the laws relating to the government of his Majesty's ships, vessels, and forces by sea, without partiality, favour, or affection; and if any case shall arise, which is not particularly mentioned in the said articles and orders, I will duly administer justice according to my conscience, the best of my understanding, and the custom of the navy in the like cases; and I do further swear, That I will not upon any account, at any time whatsoever, disclose or discover the vote or opinion of any particular member of this court martial, unless thereunto required by act of parliament.*

So help me God.

And as soon as the said oath shall have been administered to the respective members, the president of the court is hereby authorized and required to administer to the judge advocate, or the person officiating as such, an oath in the following words:

I, A. B. *do swear, That I will not upon any account, at any time whatsoever, disclose or discover the vote or opinion of any particular member of this court martial, unless thereunto required by act of parliament.*

So help me God.

And it is hereby further enacted, That from and after the twenty-fifth day of December, one thousand seven hundred and forty-nine, in case any person in the fleet, being called upon to give evidence at any court martial, shall refuse to give his evidence upon oath, or shall prevaricate in his evidence, or behave with contempt to the court, it shall and may be lawful for such court martial to punish every such offender by imprisonment, at the discretion of the court, such imprisonment not to continue longer than three months, in case of such refusal or prevarication, nor longer than one month in the case of such contempt; and that all and every person and persons who shall commit any wilful perjery, in any evidence or examination upon oath, at any such court martial, or who shall corruptly procure or suborn any person to commit such wilful perjury, shall and may be prosecuted in his Majesty's court of King's Bench, by indictment or information; and every issue joined in any such indictment or information, shall be tried by good and lawful men of the county of Middlesex, or such other county as the said court of King's Bench shall direct; and all and every person and persons being lawfully convicted upon any such indictment or information, shall be punished with such pains and penalties as are inflicted for the like offences respectively by two acts of parliament, the one made in the fifth year of the reign of Queen Elizabeth, intituled, "An Act for Punishment of such Persons as shall procure or commit any wilful perjury;" and the other made in the second year of the reign of his present Majesty, intituled, "An Act for the more effectual preventing any further Punishment of Forgery, Perjury, and Subornation of Perjury; and to make it Felony to steal Bonds, Notes, or other Securities for Payment of Money."

And be it further enacted by the authority afore said, That, in every information or indictment to be prosecuted by virtue of this act for any such offence, it shall be sufficient to set forth the offence charged upon the defendant, without setting forth the commission or authority for holding the court martial, and without setting forth the particular matter tried or to be tried, or directed or intended to be tried before such court.

And it is hereby further enacted, That from and after the twenty-fifth day of December, one thousand seven hundred and forty-nine, no sentence of death given by any court martial held within the narrow seas (except in cases of mutiny) shall be put in execution till after the report of the proceedings of the said court shall have been made to the lord high admiral, or the commissioners for executing the office of lord high admiral, and his or their directions shall have been given therein; and if the said court shall have been held beyond the narrow seas, then such sentence of death shall not be carried into execution but by order of the commander of the fleet or squadron wherein sentence was passed; and in cases where sentence of death shall be passed in any squadron, detached from any other fleet or squadron upon a separate service, then such sentence of death, (except in cases of mutiny) shall not be put in execution, but by order of the commander of the fleet or squadron from which such detachment shall have been made, or of the lord high admiral, or commissioners for executing the office of lord high admiral; and in cases where sentence of death shall be passed in any court martial held by the senior officer of five or more of his Majesty's ships, which shall happen to meet together in foreign parts pursuant to the power herein before given, then such sentence of death (except in cases of mutiny) shall not be carried into execution but by order of the lord high admiral, or commissioners for executing the office of lord high admiral.

And be it further enacted and declared, That from and after the twenty-fifth day of December, one thousand seven hundred and forty-nine, the judge advocate of any fleet for the time being, or his deputy, shall have full power and authority, and is hereby required to administer an oath to any witness at any trial by court martial: and in the absence of the judge advocate and his deputy, the court martial shall have full power and authority to appoint any person to execute the office of judge advocate.

And be it further enacted by the authority aforesaid, That from and after the twenty-fifth day of December, one thousand seven hundred and forty-nine, all the powers given by the several articles and orders established by this act, shall remain in full force with respect to the crews of such of his Majesty's ships as shall be wrecked, or be otherwise lost or destroyed: and all the command, power, and authority given to the officers of the said ship or ships, shall remain and be in full force as effectually as if such ship or ships to which they did belong were not so wrecked, lost, or destroyed, until they shall be regularly discharged from his Majesty's further service, or removed into some other of his Majesty's ships of war, or until a court-martial shall be held, pursuant to the custom of the navy in such cases, to enquire into the causes of the loss of the said ship or ships: And if upon such enquiry it shall appear by the sentence of the court martial, that all, or any of the officers or seamen of the said ship or ships, did their utmost to preserve, get off, or recover the said ship or ships, and since the loss thereof, have behaved themselves obediently to their superior officers, according to the discipline of the navy, and the said articles and orders herein before established, then all the pay and wages of the said officers and eamen, or of such of them as shall have done their duty as aforesaid, shall continue and go on, and be paid to the time of their discharge or death; or if they shall be then alive, to the time of the holding of such court-martial, or removal into some other of his Majesty's ships of war; and every such officer and seaman of any of his Majesty's ships of war, who, after the wreck or loss of his ship, shall act contrary to the discipline of the navy, and the several articles and orders herein before established, or any of them, shall be sentenced by the said court-martial, and punished as if the ship to which he did belong was not so wrecked, lost, or destroyed.

And be it further enacted, That from and after the said twenty-fifth day of December, one thousand seven hundred and forty-nine, all the pay and wages of such officers and seamen of any of his Majesty's ships as are taken by the enemy, and upon enquiry at a court-martial shall appear by the sentence of the said court, to have done their utmost to defend the said ship or ships, and since the taking thereof, to have behaved themselves obediently to their superior officers, according to the discipline of the navy, and the said articles and orders herein before established, shall continue and go on, and be paid, from the time of their being so taken, to the time of the holding of such court-martial, or until they shall be regularly discharged from his Majesty's service, or removed into some other of his Majesty's ships of war, or (if they shall die in captivity, or not live to the time of the holding of such court-martial to the time of their death, in such manner, and not otherwise, as if the said ship or ships to which they did belong respectively, was not, or were not so taken.

Provided always, and be it further enacted, That no person or persons, not flying from justice, shall be tried or punished by any court-martial for any offence to be committed against this act, unless the complaint of such offence be made in writing to the lord high admiral, or to the commissioners for executing the office of lord high admiral for the time being, or any commander in chief of his Majesty's squadrons or ships empowered to hold courts-martial, or unless a court-martial to try such offender be ordered by the said lord high admiral, or the said commisioners, or the said commander in chief, either within three years after such offence shall be committed, or within one year after the return of the ship, or of the squadron, to which such offender shall belong, into any of the ports of Great Britain or Ireland, or within one year after the return of such offender into Great Britain or Ireland.

And whereas by the said act, intituled, "An Act for the more effectual suppressing of Piracy," it is, amongst other things, enacted in the following words, That the said captain, commander, or other officer of the said ship or vessel of war, and all and every the owners and proprietors of such goods and merchandizes put on board such ship or vessel of war as aforesaid, shall lose, forfeit, and pay the value of all and every such goods and merchandizes so put on board as aforesaid; one moiety of such full value to such person or persons as shall make the first discovery, and give information of or concerning the said offence; the other moiety of such full value to and for the use of Greenwich Hospital: all which forfeitures shall and may be sued for and recovered in the high court of Admiralty: Now, for making the said in part recited act more useful and effectual, be it enacted by the authority aforesaid, That from and after the twenty-fifth day of December, one thousand seven hundred and forty-nine, if any captain, commander, or other officer of any of his Majesty's ships or vessels, shall receive on board, or permit or suffer to be received on board such ship or vessel, any goods or merchandizes, contrary to the true intent and meaning of the eighteenth article in this act before-mentioned and hereby enacted, every such captain, commander, or other officer, shall, for every such offence, over and above any punishment inflicted by this act, forfeit and pay the value of all and every such goods and merchandizes so received or permitted, or suffered to be received on board as aforesaid, or the sum of five hundred pounds of lawful money of Great Britain, at the election of the informer, or person who shall sue for the same, so that no more than one of these penalties or forfeitures shall be sued for and recovered by virtue of this and the said in part recited act, or either of them, against the same person, for one and the same offence; one moiety of which penalties or forfeitures shall be forfeited and paid to the person who shall inform and sue for the same; and the other moiety thereof to and for the use of the Royal Hospital at Greenwich; which forfeiture shall be sued for and recovered by action of debt, bill, plaint, or information, in any of his Majesty's courts of record, at Westminster, or in the high court of Admiralty, at the election of the informer, or person who shall sue for the same; and the court shall award such costs to the parties as shall be just; and in all cases where judgment or sentence shall be given against any such offender, the court where such judgment or sentence shall be given, shall, with all convenient speed, certify the same to the lord high admiral, or to the commissioners for executing the said office.

Provided always, That nothing in this act contained, shall extend, or be construed to extend, to take away from the lord high admiral of Great Britain, or the commissioners for executing the office of lord high admiral of Great Britain, or any vice-admiral, or any judge or judges of the admiralty, or his or their deputy or deputies, or any other officers or ministers of the Admiralty, or any others having or claiming any Admiral power, jurisdiction, or authority within this realm, or any other the king's dominions, or from any person or court whatsoever, any power, right, jurisdiction, pre-eminence, or authority, which he or they, or any of them, lawfully hath, have, or had, or ought to have and enjoy, before the making of this act, so as the same person shall not be punished twice for the same offence.

Provided nevertheless, and be it enacted, That the repeal of the said before-recited statutes, or any part thereof, or any thing herein contained, shall not extend, or be deemed to extend, to discharge or avoid, or prevent any prosecution or suit commenced, or at any time hereafter to be commenced, against any person or persons, for any offence committed on or before the said twenty-fifth day of December, one thousand seven hundred and forty-nine, or to be committed against the said statutes, or any part or parts thereof, but that all persons who have been, or shall, before the said twenty-fifth day of December, be guilty of any such offence, shall and may be prosecuted, sued, condemned, and punished for the same, as well after as before the said twenty-fifth day of December, as if the said statutes had not been repealed.

Anno Decimo Nono GEORGII III. Regis.

An Act to explain and amend an Act, made in the Twenty-second Year of the Reign of his late Majesty King George the Second, intituled, "An Act for amending, explaining, and reducing into One Act of Parliament, the Laws relating to the Government of His Majesty's Ships, Vessels, and Forces by Sea."

Whereas by an Act, made in the twenty-second year of the reign of his late Majesty King George the Second, intituled, "An Act for amending, explaining, and reducing into One Act of Parliament, the Laws relating to the Government of His Majesty's Ships, Vessels, and Forces by Sea;" it is, among other things, enacted, That from and after the twenty-fifth Day of December, one thousand seven hundred and forty-nine, no member of any court-martial, after the trial is begun, shall go on shore till sentence be given, but remain on board the ship in which the court shall first assemble, except in case of sickness, to be judged of by the court, upon pain of being cashiered from his Majesty's service; nor shall the proceedings of the said court be delayed by the absence of any of its members, provided a sufficient number doth remain to compose the said court, which shall, and is thereby required to sit from day to day (Sunday always excepted) until the sentence be given: and whereas it hath been found by experience, that the confining members of courts-martial to the ship in which such courts-martial shall first assemble, until sentence be given, hath been attended with great inconveniences and prejudice to the healths of officers summoned to attend as members of courts-martial; and it is highly necessary and expedient that such inconveniences should be prevented in future; may it therefore please your Majesty that it may be enacted; and be it enacted by the King's most Excellent Majesty, by and with the advice and consent of the lords spiritual and temporal, and commons, in this present parliament assembled, and by the authority of the same, That so much and such part of the said recited act as directs that no member of any court-martial, after the trial is begun, shall go on shore till sentence be given, but remain on board the ship in which the court shall first assemble, except in case of sickness, to be judged of by the court, upon pain of being cashiered from his Majesty's service; and that the proceedings of the said court shall not be delayed by the absence of any of its members, provided a sufficient number doth remain to compose the said court, which is thereby required to sit from day to day (Sunday always excepted) until the sentence be given, shall be, and the same is hereby repealed and made void, to all intents and purposes whatsoever.

Provided always, and be it enacted, That the proceedings of any court-martial shall not be delayed by the absence of any of its members, provided a sufficient number doth remain to compose such court, which shall, and is hereby required to sit from day to day (Sunday always excepted) until the sentence be given; any thing herein-before contained to the contrary thereof in anywise notwithstanding: and no member of the said court-martial shall absent himself from the said court during the whole course of the trial, upon pain of being cashiered from his Majesty's service, except in case of sickness, or other extraordinary and indispensable occasion, to be judged of by the said court.

And whereas, by two clauses in the said act, passed in the twenty-second year of the reign of his late Majesty King George the Second, it is enacted and declared, That every person in the fleet, who, through cowardice, negligence, or disaffection, shall, in time of action, withdraw or keep back, or not come into the fight or engagement, or shall not do his utmost to take or destroy every ship which it shall be his duty to engage, and to assist and relieve all and every of his Majesty's ships, or those of his allies, which it shall be his duty to assist and relieve, and being convicted thereof by the sentence of a court-martial, shall suffer death; and also that every person in the fleet, who, through cowardice, negligence, or disaffection, shall forbear to pursue the chase of any enemy, pirate, or rebel, beaten or flying, or shall not relieve or assist a known friend in view, to the utmost of his power, and being convicted of any such offence by the sentence of a court-martial, shall suffer death: and whereas the restraining of the power of the court-martial to the inflicting of the punishment of death, in the several cases recited in the said clauses, may be attended with great hardship and inconvenience; be it enacted, That, from and after the passing of this act, it shall and may be lawful, in the several cases recited in the said clauses, for the court-martial to pronounce sentence of death, or to inflict such other punishment as the nature and degree of the offence shall be found to deserve.

ASTROLABE, an instrument chiefly used for taking altitudes at sea, as the altitude of the pole, the sun, or the stars, and is represented by fig. 3, plate XXVI. It consists of a large brass ring, about 15 inches in diameter, whose limb, or a convenient part of it, is divided into degrees and minutes. It is fitted with a moveable label or index, which turns upon the centre, and carries two sights; and having a small ring A, to hang it by in time of observation.

To make use of the astrolabe in taking altitudes; suspend it by the ring A, and turn it to the sun, &c. so as that the rays may pass freely through both the sights F and G; then will the label cut or point out the altitude on the divided limb.

The astrolabe, though now grown into disuse, is by many esteemed equal to any other instrument for taking the altitude at sea; especially between the tropics, where the sun comes near the zenith.

AUTUMN, *(autumnus*, Lat.) the season of the year between summer and winter, beginning at the descending equinox, which, in the northern hemisphere, is when the sun enters the sign Libra, or about the 22d day of September, and ends about the same day in December.

AUTUMNAL *Equinox*, the time when the sun enters the descending point of the ecliptic, where it crosses the equinoctial, and is so called, because the nights and days are then equal.

AUTUMNAL *Point*, the point of the ecliptic answering to the autumnal equinox.

AUTUMNAL *Signs*, are Libra, Scorpio, and Sagittary, through which the sun passes during the autumn.

B.

To Back *an Anchor*. Page 28. col. 1, after line 14, add,

When a ship is obliged to anchor in a road where the water is deep, the ground bad for holding, and a lee shore, it has been often recommended to let go one of the bowers, as represented by *a*, fig. 10, plate XI. veering away a good scope, and stoppering the cable: then let go the other bower, and veer away nearly the whole of the cable, *a*, till the other checks her. A stout hawser *b*, must be next bent to the cable with a midshipman's hitch, and the end seized down to it: the other end of the hawser being taken into the hawse-hole, *c*, carried to the windlass or bits, and there made fast. The end of the starboard cable is then to be hauled out of the hawse-hole, and taken with a running clinch round the larboard cable *d*; after this, the stopper *b* must be cut, the clinch running by the ship's dragging; and when she has driven so as to bring it down to the ring of the larboard anchor, the larboard cable is veered away its proper scope, when she will ride with one anchor a-head of the other.

The difficulty of this method seems almost insurmountable; for when it blows so hard as to make something of this nature necessary, there is generally a heavy sea running, and the ship may pitch so violently, as to render it almost impracticable to clinch one cable round the other, as it must be done without board. The stopper *b*, from the great strain on it, may not hold. From the time which it would take to accomplish this, the ship might drive ashore before it could be effected.

A small check a-head of the anchor will prevent its coming home; therefore, when there is a lee-shore, and plenty of room to veer away a long scope, the stream cable, or even a stout hawser, should be got upon deck, and coiled upon a kind of platform made of spars on the forecastle, to keep it clear of timber-heads, &c. The under end may be taken round the bows, and bent to one of the bower an chors, and the upper one to the stream or kedge on the opposite side, letting go the latter when the ship is hove up in the wind, and the former when the last fake of the hawser is clear; she will thus ride as before, the small anchor backing the large one.

If the ship be off a lee-shore, and there is no room to veer away a length of cable, then recourse is had to letting go all the anchors, that she may be secured by their united power; for which purpose the square-sails are handed, and she keeps her way under the stay-sails: the anchors are let go, beginning with the weather one, then the next, and after it, the foremost one on the starboard side, &c. till the whole are gone. The stay-sails are hauled down, and the ship rides with all her anchors a-head. See Lever's *Practical Seamanship*, &c.

BARREL, a vessel or cask for holding liquids, &c. The English barrel, wine-measure, contains the eighth part of a tun, the fourth of a pipe, and one-half of a hogshead; viz. it contains $31\frac{1}{2}$ gallons: a barrel, beer-measure, containing 36 gallons. See the article Cask.

Barrel, in machinery, is also applied to any thing that is hollow and cylindrical, as the barrel of a pump, of a gun, &c.

To Bend *a Course in fair Weather*. Stretch the sail athwart the deck, the starboard side of the sail to the starboard side, the larboard to the larboard side; then bend yard-ropes to the earing cringles, and make fast the head earings a few feet up upon the yard-ropes, bunt-lines, leech-lines, clew-garnets, and all the jear bent; make fast a rope-band to each bunt-line and leech-line leg, that the men may be enabled to catch the head of the sail from the yard; then man well the yard-ropes, bunt-lines, leech-lines, and clew-garnets, and run the sail up to the yards. The sail aloft, send the hands up to bring it to, and let them haul out the weather earing first, then the lee, and, if it is a new sail, let them ride the head-rope to stretch it: the sail being hauled square out upon the yard, make fast the rope-bands, keeping the head of the sail well upon the yard.

To Bend *a Top Sail in fair Weather*. Overhaul the leeches of the sail, put in the earings, bend the bow-line legs, lay out the clews, and open them if necessary, and make the sail up snug again; then round down upon the lee topsail-halyards till the weather fly-block is high enough to bring the sail up over the guard-iron; then rack the tie over to the weather-rigging. Now pile the sail upon slings, with the lee side uppermost, hook on the topsail-halyards, and run the topsail up into the top, and stretch the sail round the fore part of the top, bend the jear, and make fast the head-earings a few feet up upon the reef-tackle pendants, with a rope-band or two to each bunt-line leg. The jear being bent, man the reef-tackles, bunt-lines, and clew-lines, and haul out the sail: this being done, let the hands lay out upon the yards, and haul out the weather earings first; then haul out to leeward, and ease off to windward till the sail is square, and make fast the rope-bands, keeping the head of the sail well up upon the yard. See the *Theory* and *Practice* of Seamanship; also Sauseuil's *Skilful Seaman*, translated from the French of Bourdé de Villehuet.

Bill-Boards, are pieces of elm plank about four inches thick by two feet wide, fixed on the fore-part of the fore-channels for the reception of the anchors: they have iron plates nailed to them, called *bill-board plates*, which enable the anchor to move freely, when an order is given to let it go.

Page 41, col. 1, after line 40, add

Bits, in block-making, are iron instruments resembling the shank of a gimlet, about six inches long, and from half an inch to one inch in diameter, having their ends a screw, a sharp point or edge, to cut or bore holes with. They are fixed in a stock or handle. There are various sorts, according to the purposes for which they are intended; as,

The Centre-Bit, which has in the middle of its end a small steel point, with a sharp edge on one side to cut horizontally, and a sharp tooth on the opposite side to cut vertically: holes bored with this instrument are not liable to split.

The large Centre-Bit, for boring holes through elm, does not differ from the former in the cutting part; but the plate from which the cutting part is made is let into a wooden shaft or handle, and rivetted, a ferrule or small hoop being first let over the end to prevent its splitting; at the other end of the shaft is a male screw cut to the size of the mandrel, as this instrument is used in the block-mill or horse engine. For an account of the other bits, see the articles Countersunk-*Bit*, Gouge-*Bit*, and Nose-Bit.

BOATS. Page 49, col. 1, after line 10, add,

Fig. 1, plate XXIV, represents the sheer-draught or elevation of a 14-oared double-banked pinnace; fig. 2, the half-breadth, and fig. 3, the body plan thereof.

Fig. 4, shews the sheer-draught of a 12-oared double-banked launch; fig. 5, the half-breadth, and fig. 6, the body plan of the same.

Fig. 7, the sheer-draught of a 6-oared gig; fig. 8, the half breadth, and fig. 9, the body plan.

This boat is adapted for expedition, on account of its lightness, &c. and is used by captains in the royal navy, being furnished at their own expense.

Fig. 10, the sheer-draught of a wherry; fig. 11, the half-breadth, and fig. 12, the body plan.

These boats are respectively drawn to the one-sixth of an inch to a foot.

Booms. Page 52, col. 1, after line 14, add

Proportions for the Lengths and Diameters of Booms.

Lengths of Booms.—Lower studdingsail-booms, five-ninths of the main-yard.

Top Studdingsail-Booms, one-half the length of the yards they go on.

Flying Jib-Boom, five-sevenths of the bowsprit.

Driver-Boom, the same length as the main-topsail-yard.

Diameters of Booms.—Studdingsail-booms, one inch to every five feet in the length.

Flying Jib-Booms, seven-eighths of an inch to every yard in the length.

Driver-Boom, the same as the fore-topsail-yard.

The following Tables contain the lengths and diameters of all the Booms now used in the different classes of ships and vessels, both in the Royal Navy and Merchant-Service:

Dimensions of Booms used in the Royal Navy.

Names of the Booms.	110 Guns. 2164 Tons. Length. Ft.	In.	Dia. In.	90 Guns. 1870 Tons. Length. Ft.	In.	Dia. In.	80 Guns. 1620 Tons. Length. Ft.	In.	Dia. In.	74 Guns. 1800 Tons. Length. Ft.	In.	Dia. In.	64 Guns. 1369 Tons. Length. Ft.	In.	Dia. In.	50 Guns. 1044 Tons. Length. Ft.	In.	Dia. In.	44 Guns. 879 Tons. Length. Ft.	In.	Dia. In.	38 Gnns. 951 Tons. Length. Ft.	In.	Dia. In.
Jib-Boom	53	0	15¼	50	9	14½	48	0	14	50	4	14½	43	9	13	39	8	11¾	39	0	11½	38	7	11½
Driver-Boom	—	—	—	—	—	—	—	—	—	—	—	—	—	—	—	60	9	11½	57	6	10¾	59	0	11½
Lower-Studdingsail-Boom.	56	9	11¼	54	6	11	51	6	10¼	53	9	10¾	50	0	10	45	6	9⅛	44	6	9	45	0	9
Main-Topmast ditto	51	2	10⅛	49	0	10	46	6	9¼	48	6	9¾	45	0	9	41	0	8⅛	40	0	8	41	0	8⅛
Main-Topgallant ditto	36	6	7¼	35	0	7	33	3	6¾	35	0	7	32	8	6½	30	5	6	28	9	5¾	29	6	6
Fore-Topmast ditto	44	6	9	42	10	8½	40	8	8⅛	42	6	8½	39	9	8	36	0	7¼	35	0	7	35	9	7⅛
Fore Topgallant ditto	32	3	6½	30	6	6⅛	29	0	5⅞	31	0	6¼	28	9	5¼	26	3	5¼	26	0	5¼	26	6	5¼
Fire-Boom	38	0	11¼	36	0	11	34	0	10¼	35	6	10¾	33	3	10	30	0	9⅛	29	6	9	30	0	9

Names of the Booms.	36 Guns. 871 Tons. Length. Ft.	In.	Dia. In.	32 Guns. 677 Tons. Length. Ft.	In.	Dia. In.	28 Guns. 594 Tons. Length. Ft.	In.	Dia. In.	24 Guns. 513 Tons. Length. Ft.	In.	Dia. In.	20 Guns. 429 Tons. Length. Ft.	In.	Dia. In.	Sloops of 300 Tons. Length. Ft.	In.	Dia. In.	Brigs of 200 Tons. Length. Ft.	In.	Dia. In.	Cutters of 200 Tons. Length. Ft.	In.	Dia. In.
Jib-Boom	38	0	11¼	36	10	10¾	35	0	10¼	32	6	9⅜	31	0	9	27	0	8	24	0	7	57	0	10
Driver-Boom	57	0	11	55	0	10	52	0	9¾	47	0	8⅝	45	6	8½	39	6	7⅞	—	—	—	42	0	8
Main-Boom	—	—	—	—	—	—	—	—	—	—	—	—	—	—	—	—	—	—	45	0	10½	66	0	14¾
Lower-Studdingsail-Boom.	44	0	9	41	0	8¾	40	0	8	36	9	7¼	35	0	7	30	6	6	23	6	4¾	—	—	—
Main-Topmast ditto	39	6	8	37	2	7½	36	0	7⅛	33	0	6½	32	0	6⅜	27	0	5⅜	21	0	4⅛	—	—	—
Main-Topgallant ditto	28	6	5¾	27	6	5½	26	0	5¼	23	6	4¾	22	6	4½	19	9	4	17	9	3	—	—	—
Fore-Topmast ditto	34	8	7	32	6	6½	31	1	6⅛	28	10	5¾	27	6	5½	24	2	5	21	0	4⅛	—	—	—
Fore-Topgallant ditto	25	10	5⅛	24	0	4⅞	23	0	4⅝	20	6	4⅛	20	0	4	17	6	3½	17	9	2¼	—	—	—
Fire-Boom	29	3	9	27	3	8¾	26	6	8	24	0	7¼	23	6	7	20	6	6	—	—	—	—	—	—

Dimensions of Booms used in the Merchant-Service.

Names of the Booms.	1300 Tons.			1200 Tons.			1100 Tons.			1000 Tons.		
	Length. Ft.	In.	Dia. In.	Length. Ft.	In.	Dia. In.	Length. Ft.	In.	Dia. In.	Length. Ft.	In.	Dia. In.
Jib-Boom	44	0	12½	42	0	12¼	40	0	12	40	0	12
Driver-Boom	62	0	12	60	0	11½	58	0	11	57	0	11
Lower-Studdingsail-Boom.	44	0	9	42	0	9	41	0	9	40	0	8¾
Main Topmast ditto	43	0	8½	41	0	8½	40	0	8½	39	0	8¼
Main-Top-gallant ditto	29	0	6	28	0	6	27	0	6	27	0	6
Fore-Topmast ditto	41	0	8½	39	0	8¼	37	0	8½	36	0	8½
Fore-Topgallant ditto	27	0	6	26	6	5¾	26	0	5½	26	0	5½

Names of the Booms.	900 Tons.			800 Tons.			700 Tons.			600 Tons.		
	Length. Ft.	In.	Dia. In.	Length. Ft.	In.	Dia. In.	Length. Ft.	In.	Dia. In.	Length. Ft.	In.	Dia. In.
Jib-Boom	40	0	12	38	0	12	36	0	10½	36	0	10½
Driver-Boom	50	0	10½	49	0	10½	50	0	10½	50	0	10½
Lower-Studdingsail-Boom.	37	0	8½	36	0	8	35	0	8	34	0	8
Main Topmast ditto	35	0	8	34	0	8	33	0	7½	32	0	7½
Main-Top-gallant ditto	25	0	6	25	0	5½	24	0	5½	24	0	5½
Fore-Topmast ditto	34	0	8½	33	0	8	31	0	7½	30	0	7½
Fore-Topgallant ditto	25	0	5½	24	0	5½	23	0	5½	22	0	5½

Names of the Booms.	500 Tons.			400 Tons.			360 to 350 Tons.			330 to 300 Tons.		
	Length. Ft.	In.	Dia. In.	Length. Ft.	In.	Dia. In.	Length. Ft.	In.	Dia. In.	Length. Ft.	In.	Dia. In.
Jib-Boom	35	0	10	34	0	10	34	0	9½	33	0	9
Driver-Boom	46	0	9	40	0	8½	38	0	8	38	0	7½
Lower-Studdingsail-Boom.	30	0	7½	28	0	7	26	0	6½	25	0	6
Main-Topmast ditto	28	0	7	26	0	6½	25	0	6	24	0	5¾
Main-Topgallant ditto	22	0	5	20	0	5	19	0	4½	18	0	4¼
Fore-Topmast ditto	28	0	7	25	0	6½	24	0	6	23	0	5¾
Fore-Topgallant ditto	20	0	5	19	0	5	18	0	4½	19	0	4¼
Main-Boom	—		—	—		—	—		—	—		—

Names of the Booms.	280 to 250 Tons.			Brigs of 150 Tons.			Ketches of 170 Tons.			Schooners of 110 Tons.		
	Length. Ft.	In.	Dia. In.	Length. Ft.	In.	Dia. In.	Length. Ft.	In.	Dia. In.	Length. Ft.	In.	Dia. In.
Jib-Boom	32	0	8	26	7	7	28	0	7¾	—		—
Driver-Boom	36	0	7½	—		—	—		—	—		—
Lower-Studdingsail-Boom.	—		—	—		—	—		—	—		—
Main-Topmast ditto	23	0	5	20	0	4½	—		—	—		—
Main-Topgallant ditto	—		—	—		—	—		—	—		—
Fore-Topmast ditto	22	0	5	19	0	4½	—		—	—		—
Fore-Topgallant ditto	—		—	—		—	—		—	—		—
Main-Boom	—		—	34	0	8½	—		—	47	0	10½

SLOOPS, of 70 Tons. { Main-Boom, 48 feet long, and 10 inches in diameter.
Jib-Boom .. 32 ditto, and 6½ ditto.

To BRAIL UP and Haul Down a *Main Top-mast Stay-Sail.* Man well the lee-brail and down-haul, having a few hands to gather in the slack of the weather-brail; then let go the halyards, ease off the sheet, haul down and brail up as briskly as possible. When the sail is down, let go the tack, and stop the sail over to the lee fore-rigging.

To Brail Up *a Mizen.* Man well the lee-brails, and in particular the throat-brails, which should have a whip-purchase; case off the mizen-sheet, and brail up briskly, taking in at the same time the slack of the weather-brail After the sail is hauled up, stop its foot by passing the gasket round to leeward, which will spill it.

Brake, in block-making, an instrument to confine blocks while boring the hole through the sheave: it is formed of a piece of elm plank, 3 inches thick, about 13 inches broad, and 3 feet long, in which, near one end, are 2 pegs, 2 inches and a half square, fixed in mortise-holes. Two inches from the side edge, towards the other end, is a double row of holes; to one of these holes a brake is attached by a wooden pin, which brake is 2 inches thick, 20 inches long, and 4 inches broad at one end, and tapering to a handle at the other. It is used as a lever, by pressing against it with the knee, and thereby confining the block.

BREAST of a Block, in block-making, the opposite end to that which reeves the fall.

BREECHING. Page 57. col. 2. after line 16, add,

The use of the breeching, as it checks the recoil of the cannon, is shewn in plate IX. fig. 9, where it is expressed by *e e*, passing through the ring-bolts *f f*, on the side of the carriage *g g*, being fastened to the cable *h*. It is also exhibited in the Midship-Frame, where it is employed to lash the cannon when housed during the course of a voyage. See the article Cannon.

BUOYS. Page 60, col. 2, after line 19, add,

Mr. Robert Dickinson, of 58, Great Queen Street, Lincoln's Inn Fields, London, has lately obtained his Majesty's letters patent, for his invention of Metallic Beacon, Transport, Coast, and Harbour Buoys, with buoyant iron chains, for mooring them: also ships Metallic Nun Buoys, with sustaining ball and chain.

Fig. 2, plate IV. represents a sustaining, or buoy-rope ball, which supports the slack of the buoy-rope, and prevents its falling on the anchor, and fouling; it is fitted with a certain length of chain next the anchor, and however it may cross the flook, it cannot cut asunder and be lost.

It is also applicable on a larger scale, to defend the ship's cable from being injured by corally or rocky ground.

Fig. 3, is a patent metallic life-buoy, which denotes its situation to the sufferer in the darkest night by means of a self-ringing bell; it tows to the ship like a boat, and besides being a life-boat, is a good single anchor buoy.

The *Patent Iron Nun Buoys* consist of four sizes; their prices are from two to seven guineas each, and designated by Nos. 1, 2, 3, and 4.

No. 1, for small craft, up to 40 tons; No. 2, to 200 tons; No. 3, to 700 tons; and No. 4, to 1,200 tons.

The iron buoy, which has for some time been used

in the royal navy, and now recommended to the commercial world, is perfectly smooth-sided, so that the current of the water slides away from it, without taking hold, and pulling it down, as is the case with corded wood; it never worm-eats, rots, nor wants tapping, and will always watch. It is also cheaper, in the first instance, than the wooden one, when the strapping is included. It can perform all the duty, and yet will last half a century; it never tires, and therefore is as well employed as not, is the same in all situations, remaining equally unaffected, whether in the sea or hung up against the mast, whether exposed to all the alternations of the sun, wind, and rain, or in the hold.

On the merits of the patent iron buoys, it seems sufficient to say, that the patentee states he has supplied his Majesty's navy with them for the last three years, for foreign service, and that the satisfaction they have given has induced the honourable commissioners of his Majesty's navy *now to appoint and order them for home as well as for foreign service.*

Many authentic reports from different captains of the royal navy, who have given the patent iron buoys a fair trial, both in strong tides and heavy blowing weather, are produced in corroboration of their superior advantage and durability over the wooden ones, and may be found in Mr. Dickinson's *Treatise on Iron Tanks, Casks, and Buoys,* lately published

C.

CALENDAR, or Kalendar, (from *calendæ,* Lat.) a distribution of time as accommodated to the uses of life; or an Almanac, or table, containing the order of days, weeks, months, feasts, &c. occurring in the course of the year: being so called from the word *Calendæ,* which among the Romans denoted the first days of every month, and anciently was written in large characters at the head of each month. See the articles Almanac, Month, Time, Year, &c.

Noah, as appears by the sacred writings, reckoned by months of 30 days each, and from him that mode of computing the year was adopted by the Chaldeans, Egyptians, &c.

The Roman Calendar, which, in point of equality of months, nearly resembled the sacred division of time, was first formed by Romulus, the founder and first king of Rome. He divided the year into 10 months, of 304 days; beginning on the 1st of March, and ending with December.

Numa Pompilius, the immediate successor to Romulus, made some alterations and amendments in that calendar; and although he did not disturb the names given by his predecessor to his ten months, yet he in some measure altered their rank, by adding the months of January and February, so that it consisted of 355 days. But as this was evidently deficient of the true year, he ordered an intercalation of 45 days to be made every 4 years, in this manner, viz. every two years an additional month of 22 days, between February and March; and at the end of each two years more, another month of 23 days; the month thus interposed, being called Marcedonius, or the intercalary February.

The Roman calendar was governed by Numa's amendment, until Julius Cæsar's third consulship, when the beginning of the year was found to have anticipated its real station 67 days, which that great man, with the aid of Sosigenes, a celebrated astronomer of those times, intercalated between the months of November and December, from whence arose the Julian calendar, and the Julian or old style.

Julius Cæsar, who was no less renowned for his general learning and acquirements than for his military talents, caused calculations to be made of the annual course of the sun, which he found to consist of 365 days and 6 hours; and agreeably thereto a new calendar was formed and established by public edicts. To bring to account the 6 hours in each year unprovided for by such regulation, he ordered that one day should be intercalated every fourth year, making it to consist of 366 days: this supplementary day was introduced between the 23d and 24th of February, and which the Romans called the 6th Kalendas Martii, or the 23d of February, reckoned twice over by accounting those two days as one. This practice of reckoning the two days as one was adopted and confirmed by an act of our own legislature in the reign of Henry III., which enacts, that the day increasing the leap-year, and the one before it should be accounted but as one day; whence the Roman term, yet retained, of *Bissextile,* or what we also call Leap-Year, which article see.

In the Common Prayer Books of King Edward, it was noted that the 25th of February, which in leap-years is counted for two days, shall neither alter the Psalms or Lessons, but those said on the first day shall also serve for the second day.

Upon the restoration of King Charles II. the Revisers of the Liturgy altered the Rubric, by causing the 29th day of February in the ecclesiastical computation to be regarded as the supplementary day, instead of the one between the 23d and 24th of that month, thereby making the church regulation conform to the civil mode of reckoning, which had adopted the 29th as the intercalary day. This regulation was long universally followed by the Reformed church, and in some countries is still observed at this day.

The Julian calendar was farther reformed by order of Pope Gregory XIII., from whence arose the term Gregorian calendar, or new style, as we call it, which is now observed by almost all European nations. Finding that a difference of 10 days had arisen between the calendar and actual time, and being desirous of celebrating Easter according to the original institution, he caused those 10 days, which had been over-run since the time of the council of Nice, in the year 325, by the surplus of 11 minutes each year, to be abated on the 4th of October 1582, which was done in all countries under the popish influence. And in order to prevent the seasons of the year from retrograding as they had done before, the pope, therefore, by the advice of Clavius and Ciaconius, ordained that 3 intercalary days should be omitted in every 400 years, by reckoning all those centurial years whose date consists of entire hundreds not divisible by 4, or into hundreds without a remainder, such as 1700, 1800, 1900, 2100, &c. to be only common years, and not bissextile years, as they would otherwise have been.

The reformation in this calendar was admitted in England in the year 1752, when the 10 days expunged by the pope, and another day which, since his time, had accrued, were taken out of our calendar, and the 3d of September reckoned the 14th that year; as the overplus minutes had then amounted to 11 days.

Though the Gregorian calendar be more accurate than the Julian, yet it is not without imperfections, as Scaliger and Calvisius have shewn; nor is it perhaps possible to devise any one that shall be quite perfect.

Yet the Reformed Calendar, and that which is ordered to be observed in England, by an act of parliament made the 24th George II., come very near to the point of accuracy; the register of time having been brought so near to the actual solar year as not to amount to a day in the period of 6000 years: and by that act it is ordered that "Easter-day, on which the rest depend, is always the first Sunday after the full-moon, which happens upon, or next after the 21st of March; and if the full-moon happens upon a Sunday, Easter-day is the Sunday after."

Julian Christian CALENDAR, is that in which the letters A, B, C, D, E, F, G, by means of the solar cycle; and the new and full-moons, particularly the paschal full-moon, with the feast of Easter, and the other moveable feasts depending upon it, by means of golden numbers or lunar cycles, are rightly disposed through the Julian year. See the article CYCLE *of the Moon.*

In this calendar, it is supposed that the vernal equinox is fixed to the 21st day of March; and that the golden numbers or cycles of 19 years, constantly indicate the places of the new and full moons; though both are erroneous; and from hence arose a great irregularity in the time of Easter.

Gregorian CALENDAR, is that which, by means of Epacts, rightly disposed through the several months, determines the new and full-moons, with the time of Easter, and the moveable feasts depending upon it in the Gregorian year. This differs therefore from the Julian calendar, both in the form of the year, and in as much as epacts are substituted instead of golden numbers. See the article EPACTS.

Reformed, or *Corrected*, CALENDAR, is that which, rejecting all the apparatus of golden numbers, epacts, and dominical letters, determines the equinox, and the paschal full-moon, with the moveable feasts depending upon it, by computation from astronomical tables. This calendar was introduced among the protestant states of Germany in the year 1700, when 11 days were omitted in the month of February, to make the corrected style agree with the Gregorian. This alteration in the form of the year, they admitted for a time, in expectation that, the true quantity of the tropical year being at length more accurately determined by observation, the Romanists would agree with them on some more convenient intercalation.

The Russians, who did not adopt the Christian æra until the year 1725, still adhere to the Julian style with strict pertinacity; and by another leap-year having occurred in the year 1800, they are now 12 days before us in the date of their time.

French New CALENDAR, a new form of calendar, framed by Fabre De Eglantine, and presented, on the 5th of October, 1793, to the convention then ruled despotically by Robespierre, when the rulers of that infatuated nation employed themselves alternately in deeds of death and minute attention to trifles. This calendar was formed on so republic a model as to destroy every allusion either to things before held sacred, as relating to the Deity, or respectable, as complimentary to human virtue.

The year commences at midnight the beginning of that day in which falls the true autumnal equinox for the observatory of Paris. The year is divided into 12 equal months, of 30 days each; after which 5 supplementary days are added, to complete the 365 days of the ordinary year: these 5 days do not belong to any month. Each month is divided into three decades of 10 days each; distinguished by 1st, 2d, and 3d decade. All these are named according to the order of the natural numbers, viz. 1st, 2d, and 3d, &c. month, or day of the decade, or of the supplementary days; the months and days of them having new names. The years which receive an intercalary day, when the position of the equinox requires it, which we call embolismic or bissextile, they call olympique; and the period of four years, ending with an olympic year, is called an olympiade; the intercalary day being placed after the ordinary five supplementary days, and making the last day of the olimpic year. Each day, from midnight to midnight, is divided into 10 parts, each part into 10 others, and so on to the last measurable portion of time.

This silly innovation commenced in France on the 22d of September, 1792, and was persisted in until some time after Buonaparte, the late ruler of that nation had destroyed all the benefits that might originally have been expected from the Revolution, and nearly overturned every regulation that had any reference to the Republic, for the establishment of which that unhappy country had been so lavish of her blood and treasure! See BRADY'S *Clavis Calendaria*; and DR. HUTTON'S *Mat. and Phil. Dictionary.*

CASK, a round wooden or iron vessel to keep liquors or provisions in; of these there are various sorts, which are generally denominated according to the articles they contain; as, a cask of water, a cask of rum or wine, a cask of sugar, a cask of raisins, a cask of flour, a cask of beef, a cask of pork, &c.

Mr. Robert Dickinson's patent *iron casks* seem now to be coming into general use, both inthe navy and merchant-service; and from the following testimony, from the first chemist of the age, to the patentee, there can be no doubt as to the propriety of using them for water and provisions.

Extract of a Letter from Sir Humphrey Davy.

"I do not conceive it possible that water can be in the slightest degree injured by being kept in air-tight casks of cast or wrought iron, and such casks would, I am fully inclined to believe, be much preferable to those of wood for preserving biscuits, flour, or any kind of stores from all attacks of worms and insects.

Royal Institution, "I am, &c.
Feb. 18, 1809. "H. DAVY."

"These iron casks," says the patentee, "are adapted to all purposes to which wooden barrels are now applied: they are tight, sweet, convenient, cheap, compact, light, and durable; and being lined with coatings suited to the nature of the contents—coatings, linings, or varnishes, for which we are indebted to the modern advances in chemistry, the contents, when this is desirable, are prevented from ever coming in contact with the iron.

"Numerous advantages resulting from the use of iron in place of wooden vessels for many purposes, must strike even a superficial reader; but they can be properly appreciated only, by taking also into consideration the improved system of stowage, which they offer to all who are interested in the economy of ship tonnage, as connected with the commercial prosperity of this maritime country.

"The proper stowage of liquids and other commodities destined for long voyages, foreign garrisons, and colonies, has always been an object of the first importance in the consideration of those who have the direction of naval and military equipments; or who are interested in the trans-marine commerce.

"Upon the whole, therefore, when the difference of expense between *wood* and *iron*, the superior durability of the latter, with the advantages to be gained in point of *stowage-room*, and the *preservation of cargoes and stores*, are taken into consideration, the patentee may be fairly allowed to estimate the superiority of this invention as equal to *cent. per cent.* in comparison over the common *wooden casks* now in use."

The iron barrels hitherto used for beer, ale, wine, spirits, &c. have been found to give such universal satisfaction, that the patentee is manufacturing them, of all the convenient sizes, at his factory, Elliot's Wharf, Bankside, London, where they may be seen, and information obtained respecting them.

CHASE. Page 84, col. 2, after line 2, add,

A vessel that chases another ought to have the advantage of sailing; because, were the ship chased as good a sailer as the chaser, she could never come up to her, if they manœuvred equally and at the same time. It is then useless to chase a ship over which you have not the superiority in sailing, unless it be found that she does not know how to take the benefit of her equality.

In order to know if your ship sails quicker than your adversary, you must get on the same tack, under the same sails, and keep the same course with the vessel you wish to chase, and set her exactly with a compass. If you sail best, the chase will soon be drawn a point more aft; but if she has the advantage, you will in short time bring her a point farther forward; but if you sail equally, she will remain on the same point you set her at first.

To CHASE *a Ship which is to Windward, and to join her in the shortest Method.*—When the chaser is to leeward of the vessel he means to pursue, he ought to run on the same tack as the enemy till he brings her to bear exactly perpendicular to his course; if he has not, however, already passed that point, then tack and continue the second board till he brings the chase again perpendicular to the direction on which he is standing by the wind, and he must then heave about again, always continuing the same manœuvre, by tacking every time he brings the chase perpendicular to his course on either board. In this manner, the chaser will, by the superiority only of his sailing, join the other in the shortest method.

When the ship A. fig. 9, plate XV. chases the ship B, which is three leagues to windward, A having one-fourth the advantage in sailing, the chaser is not to tack till he reaches the point C, because then the ship B will be right on his beam at the point D. He then runs on the tack C E, till he brings the chase again perpendicular to his course, at the point F. The ship A is to continue working thus every time she brings the vessel B right a-breast of her, whether the chase continues on the same tack or not; and thus the chaser will join the other at H, so that she will neither be able to change her course nor fly from him.

You continue on the same tack as the enemy, when first seen, in order to lose no time; because you will always bring the ship you are in chase of right on your beam, when you have a superiority of sailing, whatever may be the tack she is on, provided you are careful not to pass that point; but, if you should by chance, you must get on the other tack with all possible dispatch.

The chaser heaves about as soon as the vessel he is in pursuit of is on his beam; because she is, at this time, at the shortest possible distance, if he chases on the same tack and steers on the same course with the vessel chased. If the chaser runs on a different tack from the vessel chased, he is still to tack when the latter is upon his beam, because the distance is the least possible between them on the different boards they hold.

This mode is preferable to all others, being not only the shortest, but because you force the chase to fly from you close upon a wind, pressing her more and more from the leeward, by never passing the point at which the distance between the two vessels, in plying to windward, is the shortest possible.

Observations for the Ship to Windward which is Chased.

The weather ship, which flies, will always be joined by the chaser, since it is granted that she does not sail so well as the pursuing vessel. It is therefore her advantage constantly to keep one course without losing time to heave about, as tacking cannot be so favourable to her as to her adversary, whose sailing is superior.

If the chaser should mistakingly *stand on* a long

way, and tack in the wake of the chase, the best thing she can do is to heave-in stays, and pass to windward of him on the other tack (unless you suppose your vessel would have a superiority large): for if the chaser persists in tacking in the wake of the other ship, the chase will be very much prolonged.

To chase a Ship which is to Leeward.—When you sail to windward of a vessel that you wish to chase, keep the ship away, to cut her off; and, steering continually on that course, you come at last together at the point where the courses run by the two vessels intersecting each other. This will be exactly executed by the chasing ship, if, in the course she has chosen, she constantly keeps chase on the same degree of the compass as at the beginning of the pursuit. This principle applies equally to all the courses which the retreating ship steers; for overtaking can only be obtained by keeping on a straight line, which is the shortest possible that can be drawn between any two points.

If you take another course than that which keeps you in the same point of bearing you were in, with respect to the vessel pursued, at the beginning of the chase, you would fail, by being either too far a-head, or too far a-stern; that is, if the chaser keeps his wind too close, he will be too much a-head, and consequendy prolong the chase; and, if he keeps too much away, he will be too far a-stern. These are the only two considerations to be made for the performance of this manœuvre; considerations, which are easily observed, and corrected with an azimuth compass; for when you see that at the end of a certain time you bring the chase more aft than her first point of bearing, it is evident you keep your wind too much; if, on the contrary, you draw her forward, it is a proof you keep her too much away. These errors are easily corrected, by steering for the first case so as to see that the chase is always kept on the same degree of the compass; and, for the second, you keep your wind a little more, till you see that you rest on the same point of bearing with respect to one another. Then, it is evident, you chase by the shortest and most certain method, since you reach the chase, in running on a straight line.

Observations for the Ship to Leeward, which is Chased.

She ought to run on the course that will carry her most immediately from the chaser. Some vessels have more advantage in going large than others; some with the wind right aft, and others again go best close hauled; so that attention should be paid to the known qualities of a ship, in order to take the most advantageous and convenient directions capable to effect a retreat. It is, however, nearly certain, that, if the chase does not sail, at least at an equal rate with the chaser, whatever manœuvre she may practise, she will at length be overtaken by a skilful chaser adhering to principles.

If the chase be found right a-head, and so the chaser be put to a stern-chase, then the best sailer shall carry it, if there be sea-room and day-light. Being come up close with the chase, endeavour to cross her fore-foot, and by that means you will both hinder her way, and avoid the fury of her ordnance (except those in her chase), and use your own, if required, to more advantage; and that as well your chase-pieces, at your first getting up within reach, as your broadside and quarter-pieces, as you pass athwart her hawse, and scour her decks from stem to stern. If she makes away from you, ply your guns, as many as possible, at her sails, yards, &c. and, being near, spare not your chase-shot, or cross-bar shot, to make the greatest damage. See the *Theory* and *Practice* of SEAMANSHIP: also SAUSEUIL'S *Skilful Seaman*, translated from the French of BOURDÉ DE VILLEHUET.

CHOCKS, in block-making, are cylindrical pieces of wood or iron, screw-cut at one end, to screw into the end of a mandrel; at the other end, some have a sort of peg to be driven into the work, which is retained by the assistance of an opposite centre: some chocks have, instead of a wooden peg, three pieces of iron in the end; this sort is chiefly used for broad thin work, without an opposite centre.

Iron chocks are screw-cut, similar to wooden ones, and have a box at the end, into which the work is driven, and retained with or without an opposite centre; in the sides of the box are holes to admit a punch to punch out the work.

CHRONOLOGY, (from χρονος, time, and λογος, discourse) is the art of measuring time, distinguishing its several constituent parts, such as centuries, ages, years, months, weeks, &c. by appropriate marks and characters; and adjusting these parts in a methodical and orderly manner, to past transactions, by means of æras, epochs, cycles, &c. See the articles TIME, CALENDAR, GOLDEN *Number*, &c.

Nautical CHRONOLOGY, comprehends the method of finding the Golden Number, Leap-Year, Epact, the Moon's Age and Southing, the time of High-Water, &c. all of which are fully explained under their respective heads.

CIRCULAR *Saw*, in block-making, an implement to cut sheave-holes by the block-mill. The saw is fixed to the mandrel, and the block to a sliding puppet, by which it is conveyed to saw in the direction of the sheave-hole. The saw cuts it half way through in the middle, the block is then reversed, and cut as before, to meet the former incision: then, by boring a hole at each end of the sheave-hole to its size, the wood is completely removed by sawing a small portion left by the circular saw.

Page 88, col. 2, after line 59, add,

CISTERNS, are also small trunks lined inside with lead, and fixed on each side of a ship of war, near the barricade of the head, for the convenience of the people, and to prevent the decks from being wetted in other places. They empty themselves into the wooden shoots beneath, which are bolted to the ship's sides, and serve to convey the water into the sea; by these means the bows of the ship are always kept clean.

There are also cisterns in the light rooms of ships of war, to contain water for the safety of the lights.

CLASS, a term used to distinguish the description of boys entered into his Majesty's sea-service, of whom there are three sorts, borne on separate lists, after the ship's companies.

The 1st Class, is composed of young gentlemen intended for the sea-service; the sons of officers may be entered at 11 years, but all others must not be under 13 years of age: they are styled *volunteers*, and are allowed wages at the rate of 9*l.* per annum.

2d Class, are boys between 15 and 17, who are divided into watches with the seamen, in order to bring them forward as such, at 8*l.* per annum.

3d. Class, consists of boys between 13 and 15, from whom the lieutenants, and other officers who are allowed servants, are permitted to recommend to the captains, each of them one, to be the attendants on such officers, at 7*l.* per annum.

CLEATS. Page 89, col. 2, after *Thumb* CLEAT, add,

Belaying CLEAT, fig. 15, plate VII, is shaped like that of the range cleat, only smaller, and used for belaying the small ropes to.

Girt-Rope CLEAT, fig. 16, is nailed to the inside of a ship in ordinary, and used for passing a rope through, for the convenience of hauling a boat forward or abaft, as occasion requires.

Iron CLEAT, fig. 17, is used for belaying the port-tackles to, when the port-lid is hoisted up.

Shoulder CLEAT, fig. 18, is commonly used for the same purposes as the stop cleat, except on the yard-arms.

Stanchion CLEAT, fig. 19, is fastened to the head and keel of a stanchion in ships in ordinary.

CLERK OF THE CHECK. Page 84, col. 2, after line 23, add,

It is also his duty to muster the several persons belonging to the yard, upon their coming to, or quitting work; to keep an account of their wages; to make out, weekly, the pay-books of ordinary; to muster the companies of his Majesty's ships in commission, and the persons employed on board the ships in ordinary, and to check the absentees of their provisions; to muster at certain periods the marines at head-quarters; to view, in conjunction with the other officers, the quality of all stores received, and to make out certificates of the quantities, when required by the contractors, in order to their obtaining payment of the Navy Board; to attend in turn with the storekeeper and clerk of the survey, the receipt of all stores, and the delivery of lots of old stores to the purchasers: he likewise pays bounty to volunteer seamen, and the contingencies of the yard.

COAKING or BUSHING, in block-making, is the act of letting through the middle of a sheave a cylindrical piece of metal with a hole through its centre, agreeable to the size of the pin that is to be admitted as an axis for the sheave to turn on, &c.

Captain's COCKSWAIN, in a ship of war, a person chosen by the captain from among the crew, as a steady man, to sit at the stern of his barge, which he steers to and from the ship.

COMPASS. Page 100, col. 1, after line 21, add,

Description of Col. M. Beaufoy's Compass for accurate Observations on the Magnetic Variation.

The principal part of this instrument, viz. its needle, with the box and divided arc, to measure the angle of variation, is very similar to the instrument in the possession of the Royal Society; but it is greatly improved by the addition of a small transit telescope, which very readily and accurately determines the true meridian, or the zero from which the degree of the variation is to be measured. This addition renders it particularly well adapted for travellers who wish to make accurate observations in different parts of the globe, as it contains every thing necessary for the observation within itself; and as every part admits of adjustment, by reversing, it may be depended upon for accuracy.

Fig. 7, plate XXXV. contains a representation of the whole instrument in perspective, explaining all its parts at once. The box containing the needle, is not fixed, as in many compasses, but turns horizontally on a centre, and has an index fastened to it, pointing to a divided arc, on the brass frame on which it turns; and the method of observing is to move the box till a line drawn on it points exactly to the end of the needle, which being done, the angle of variation is shown by the divisions on the arc. F F, is a mahogany board, which is the support of the whole instrument; it stands on the points of three screws, G, H, I, by which it can be levelled. Above this is a flat plate of brass, E E, attached to the board by a centre pin, and resting upon three studs, projecting from the board, to ensure its having a perfect bearing, whilst it admits of a small horizontal motion round the centre, by means of the screw W X. The plate E E, has the divided arc, *m*, fixed upon one end of it; and a vernier, D, traverses against the divisions; it is divided upon a projecting part of a brass plate, C C, which moves on the common centre pin of the plate E, and also the needle A A B. The plate C C, has two segments of brass at each end of the needle, and these have the centre lines drawn upon them, which are brought to the points of the needle when the observation is made. A light brass box, *d d*, with a glass cover, is fitted over the plate C C, to preserve the needle from disturbance from the wind; it also supports a small double microscope, M, intended to assist in examining when the index line points exactly to the end of the needle; the stem of the microscope is fitted in a dovetail groove, and can be removed to the opposite end of the box at pleasure. Pl. XXXV. Fig. 7.

The centre pin of the plates E and C, terminates in an extremely fine point, on which the needle is suspended, having an agate cap, B, to diminish the friction as much as possible. The needle is of a cylindrical figure, about the five-hundredth of an inch in diameter, and ten inches long, and weighs $65\frac{1}{2}$ grains, and is terminated by two conical points; the circular enlargement of the centre has a hole through it, for the reception of the brass socket, B, which has the agate cap fixed in it; the needle is, of course, provided with the usual apparatus for lifting it off the point when not in use, to avoid wearing the point of suspension. *a*, *a*, are two arms, screwed down upon the plate, carrying the vernier, D; they support a brass frame, *b*, which has the usual clamp screw, L, and tangent screw, R, the former to fasten it to the arc *m*, and the latter to give a slow motion, and adjust the box for the observation.

The transit telescope, O P, is supported over the instrument by two pillars, N N, fixed on the brass plate, E E, and having small frames or boxes, *f*, at the top, for the reception of the Y's, in which the pivots of the axis, Q, of the telescope are supported; this axis is conical, and is fixed exactly at right angles to the tube, O P, in the same manner as other transit instruments.

On the extremity of one of the pivots of the axis, a small divided circle, R, is fixed, and has an arm at the opposite side, turning about on the centre, and provided with verniers to read the divisions on the circle; to this arm, a small level, S, is attached; the whole forming a flying index to set the telescope at any required altitude, by setting the index at the proper division, and then moving the telescope till the bubble of the level, S, shews the index to be horizontal; the eye-piece, *h*, of the telescope has a small dovetail in it to admit a dark glass for observation of the sun; *g*, are the screws for adjusting the wires in the eye-piece as usual; the separate figure, T V is a level to adjust the instrument.

Manner of adjusting the Instrument previous to Observation.

First place the feet, *k*, *l*, of the level, upon the brass plate, E E, in different directions, and bring it level by the screws, H, I, K; then apply the level upon the two pivots of the transit, the covers of the boxes, *f*, opening with hinges for this purpose, and the bottom of the feet, *k*, *l*, have notches to rest upon the pivots; if the axis does not prove level, one of the Y's must be elevated or depressed, by a screw in the farther pillar, N, until the level stands horizontally, and, reversing it end for end, proves every thing to be correct.

Now set the vernier, D, at zero, and put an additional object glass over the glass, P, of the telescope; and in this state the marks near A A, against which the needle reads, can be seen through it when directed to them; this proves the zero of the division to be exactly in the plane of the telescope motion; a small screw near Q will rectify it, if necessary, by moving the Y a small quantity in the box, *f*; the transit may be reversed in its Y's to rectify every thing.

The telescope is now to be adjusted to the meridian by the transit of the stars in the usual manner, the screws, W, X, being used to turn the whole instrument round when requisite; but after these observations have been once made in a satisfactory manner, distant marks should be set up, both north and south, and these will give the means of adjusting it at once, and in the day time.

When the instrument is properly placed, the taking of the observation is extremely simple. The needle being suffered to settle, the box is turned about on its centre till the mark comes very near the point of the needle; the clamp screw, L, is then fastened, and the screw *k*, is employed to make the coincidence exact, in which the microscope, M, greatly assists the eye. The vernier, D, now shows the angle of variation. It is proper, after the needle has been once observed, to attract it with a piece of iron, and cause it to make a slight vibration. When it settles again, it will not in all cases come to the same point, because of the friction of the suspending point; it is therefore advisable to make this trial three or four times, and take the mean of the whole.

The instrument is placed on a pillar, in the open air, free from any iron, and is adjusted before every observation; the vernier is divided to half minutes, and if the meridian be not at a considerable distance, an advantage will be gained by diminishing the aperture of the object-glass by a pasteboard cap. "I have found that the needles," says Col. Beaufoy, "should not exceed the weight mentioned, 65½ grains: and it may be worth while to remark, that this instrument will serve as a portable transit instrument for naval officers to examine the rate of their chronometers while on shore."

Page 100, col. 1, after line 46, add,

Binnacle Compass. The advantages of the binnacle now made and sold by Messrs. P. and G. Dollond, St. Paul's Church-yard, London, consist in the steadiness and equality of the light, which in the night is obtained by prismatic reflection. The facility and security with which the lantern may be removed to an enclosed place to be trimmed, for which there is seldom a necessity, it having been proved that the lamp will burn without trimming from twelve to fifteen hours.* The lamp being enclosed in a lantern prevents the possibility of the light being blown out, or the oil falling on the compass.

The lens in the front of the binnacle, which magnifies the compass, is also a very considerable advantage, as it enables the steersman to see the point more distinctly; and the whole of the apparatus being enclosed, the light is prevented from appearing in the night to any person except the steersman, and the weather from affecting the compass.

The improvements relating to the compass are as follow:—It is suspended in gimbals as usual, with the addition of a spiral spring to each axis; these springs relieve the compass when any sudden or violent motion of the ship takes place. The compass card is so constructed as to prevent its vibrating, and is suspended on a spiral spring, acting within two cylinders. By this suspension the point is preserved, and the card secured in its place, although the motion may be excessively violent.

Plate XXXV. represents the binnacle in three points of view: with a perpendicular view of the compass—a section of the lantern—and of the point on which the compass card is suspended.

References to the Plate.

A, fig. 1, is the lens through which the steersman is to look at the compass, and by which it is magnified. *B*, a flap attached to the slider, *C*. This flap opens to admit the compass, and when closed prevents the rain, &c. from splashing the compass, and the light from appearing on the deck. It is made fast by a long button at *d*. The pin at *e*, secures the slider *C*, at the proper place. This place is regulated by the height and distance of the steersman. Fig. 1.

Fig. 2, represents the back of the binnacle, with the lantern in its place opened, to shew the lamp *a*. *B*, is the slider that encloses the lantern, and is secured at its place by the pin *c*. Fig. 2.

Fig. 3, represents the binnacle with the side taken away, to shew the interior. *A*, the lantern. *B*, a section of the compass in its place. *C*, the opening to admit the light in the day-time, and for which there is a shutter. Fig. 3.

Fig. 4, a perpendicular view of the compass, shewing the spiral springs at each axis. Fig. 4.

* The best sperm oil is recommended to be used, and the cotton or wick should be cut about one-eighth of an inch above the holder.

Note. F, fig. 1 and 2, rings for lashing the binnacle to the deck.

Fig. 5. Fig. 5, a section of the lantern with the prism and lamp. *A*, the prism. *B*, the lamp. By the prism the light is reflected from the lamp, upon the face of the compass.

Fig. 6. Fig. 6, a section of the point on which the card of the compass is suspended.

COOK, or Ship's Cook. Page 103, col. 1, line 3, after the word boilers, add,

He is to have the charge of the steep-tub, and the meat put into it, and to be held responsible for any part of it which may be lost through his want of care.

He is to see that the salt meat is duly watered, that all the provisions are carefully and cleanly boiled; and that they are delivered to the men according to the established practice of the navy.

When fresh meat is served out, he is to see that the greens and all other vegetables boiled with them, are carefully washed before they are put into the coppers: and when he serves out soup, or burgoo, he is strictly charged to do it impartially, giving to each man an equal quantity. See the Naval Instructions.

Besides the above, there is the admiral's, captain's, and ward-room cook in a ship of war.

Cross-Trees. Page 112, col. 1, after line 31, add,

The topmast cross-trees are differently fitted from the lower ones. They are one-third longer than the trestle-trees; depth, seven-eighths the depth of the trestle-trees; and breadth, one-fourth more than the breadth of the Trestle-trees, which article see.

The cross-trees are trimmed (mostly sawed) straight on their upper sides, but moulded straight only one-third the length in the middle, and round aft from thence on each side, with a curve one-third the diameter of the top-mast, and one-half their breadth: they hold a parallel depth in the middle, three-sevenths the length, and taper from thence to half the depth at the ends, which are rounded, and a hole bored four inches within, for the topgallant-shrouds.

The topmast cross-trees are let into the trestle-trees with scores, as the lower ones, to half the depth of the cross-tree, and face on half an inch. The middle of the trestle-trees is kept well with the top-mast-head at the fore side, and the middle cross-tree let in between that and the topgallant-mast-heel; the fore side of the after cross-tree to the exact size of the topmast-head, and the aft side of the foremast cross-tree about one inch less in the clear, and both fastened to the trestle-trees with bolts seven-eighths to five-eighths of an inch in diameter, driven through the middle from the upper side with a saucer-head, and secured underneath with a forelock turned upon a ring. They are then got over the topmast-head, and fixed firm on the stop of the hounds.

CUTLASS, a sword about three feet long, rather broad, and straight; it has a japanned hilt, and is used by officers and seamen in the royal navy.

Naval Cutlass *Exercise*, is the art of attack and defence with the above-mentioned weapon in boarding and taking possession of an enemy's ship.

Mr. Angelo, jun. of Bolton-row, Berkeley-square, has reduced this exercise to a regular system, which is now very generally practised in the navy, and has, under the sanction of the lords commissioners of the Admiralty, lately published a large plate for the purpose of explaining the principles of this art.

The plate or target of instruction should be fixed flat to any place, the centre rather below the height of your shoulder; mark a chalk line along the deck in the direction of the centre of the target, and at about the distance of ten feet, take your station, with your body sideways, the feet at right angles, so that the left heel and right foot are upon the line, the right pointing to the figure, the arms behind, supporting each other, which forms the first position.

The figures in the upper part of the plate represent the three positions, the balance and extension motions, which, as they give the proper command of body, ought to be performed with ease and accuracy, before they are allowed to use the cutlass.

The circular figure shews the seven cuts, and the same number of guards:—The cuts are directed through the centre, distinguished by lines, and named according to that figure from which each cut is commenced: the guards are formed by holding the sword opposite those lines which have sword hilts added to them.

The *point* or *thrust* is made by placing the cutlass horizontally, just above the right hip, the edge to the right, the point to the front; the *parry*, by holding the sword opposite the centre perpendicular line, the edge to the right, and the back leading form a circle round the figure from left to right of your own front.

The *proving distance* is formed in the first position, by raising the cutlass, the edge to the left, perpendicularly above the head, to the full extent of the arm, and then lowering it horizontally to the right, the edge to the rear;—and the front prove distance by raising the cutlass as before, and stepping out to the 3d position, with the point to the front, the edge to the right: the salute is as usual with officers. The *proving distance* in the attack and defence should be in the first position, the sword arms extended, with the point of each cutlass just reaching the shell of the other.

By *guard*, is meant the most secure form to engage; it is in the second position, the hand raised as high as the head, the arm bent, so as to see your antagonist through the angle which it forms.

The *assault* consists in combining the first six cuts, which, in the regular exercise, is performed in the second position and at the word of command, *Head! cut* 7 is given with the third position, the *St. George*, which follows, with the first; so each cut and point is delivered with the third position, and each defence with the first.

The cuts given separately, conclude with the point opposed to the centre of the figure, and the defence follows each cut.

The *shift*, is slipping from the cut, the arm raised to St. George.

The *guards* and *parry* should be formed with the fort of the blade, gaining the foible of your opponent's; the *cuts* within six or eight inches of the point, that the cutlass may clear itself.

Mr. Angelo has kindly furnished the above article for this work, and means shortly to publish a more detailed account of it, which it is hoped, in future, will form a part of the education of our naval officers. The plate may be had at Egerton's Military Library, White-hall, and of the principal booksellers at the different sea-port towns.

D.

DISTANCE. Page 124, col. 1, after line 37, add, *To Regulate the* Distance *of Ships.* It is recommended to station them, by observing the angle formed between the main-topgallant-truck and water line of the ship next a-head, or a-breast, which, as there is no great dissimilarity between the height of one ship and another, must evidently produce a near resemblance to the truth.

The more readily to effect this purpose, a calculation has been made, on the supposition that the height of a large ship from the water line to the truck, is 168 feet. Such being the case, at half a cable off, the angle subtended between the truck and water line will be 21° 1′, at a cable off, 13° 8′, and so on.

The commodore having assigned, by signal, the distance of the lines, and the distance of the ships from each other; the officer of the watch is to set the index of his quadrant to the assigned angle, and so to regulate the distance of his ship from the next a-head or a-breast, that the reflected image of her main-top-gallant-truck, when viewed through his quadrant, shall appear in contact with the water-line.

Dog-Shore, among shipwrights, a piece of oak four feet long, and five or six inches square, placed at the foremost end of the bilgeways; its foremost end is fitted to a chock bolted to the bilgeways, and the after end rests against the ribband, which is securely fastened to the sliding-plank; by this contrivance it retains the ship in her place till she is ready to be launched. The dog-shore, represented by *a*, fig. 10, plate XIII, when knocked down, lies smoothly on the inclined plane, and the ship immediately slides into the water.

DOMINICAL Letter, in chronology, properly called the Sunday Letter, is one of the seven letters of the alphabet, A B C D E F G, used in almanacs, ephemerises, &c. to denote the Sundays throughout the year.

In our almanacs, the first seven letters of the alphabet are commonly placed to shew on what days of the week the days of the month fall throughout the year. And because one of those seven letters must necessarily stand against Sunday, it is printed in a capital form, and called the dominical letter, the other six being inserted in different characters, to denote the other six days of the week.

The dominical letter may be found universally, for any year of any century, thus,

Divide the centuries of the given year by 4, and take twice what remains from 6; then add the remainder to the odd years above the even centuries, and their 4th. Divide their sum by 7, and the remainder taken from 7 will leave the number answering to the dominical letter. Thus, for the year 1814,

$\frac{18}{4} = 4$ and 2 over $\left\{\right.$ and 2×2 subtracted from 6 leaves 2.

Last remainder 2
Odd years 14
$\frac{1}{4}$ odd years 3
$\overline{19}$

$\frac{19}{7} = 2$ and 5 over, and $7 - 5 = 2$.

Two, then, is the number of the dominical letter; and as 1 is the number of A, 2 must be the number of B, the dominical letter for 1814.

Dolphin-Striker, a boom or spar directed downwards from the cap of the bowsprit, through which is rove a stay or martingale, to support the jib-boom.

To DRIVE. Page 127, col. 2, after line 46, add, The art of driving consists in keeping the ship in a fair way, by the management of the rudder and sails.

When it happens that there is not room enough to work in a tide way, through a crowd of ships, or in a narrow channel, but that the ship must drive by the help of the tide, it may be done, provided the tide be sufficiently strong, in proportion to the wind.

To Drive *to Windward, when the Wind is against the Tide.*—If the channel is sufficiently broad, the ship should be drifted broad-side to the wind, as the tide will then have the greatest power on her; and, could the ship be backed a-stern, or shot a-head at pleasure, she might be kept drifting upon the same tack with safety; but ships in a tide-way can never be backed so far a-stern as they will shoot a-head. At the first of a stern-board a ship will go briskly a-stern, but will soon fall off, and drift with the wind abaft the beam, forging a-head; for this reason, she must be drifted with the helm a-lee. It follows, as a ship will shoot more a-head than she can be backed a-stern, that she will at length arrive at the opposite shore, when she must be stayed or veered, and drifted upon the other tack.

If she is to be stayed (which is preferable, because less drift will be lost by it), let the sails be filled in time, to give the ship sufficient head-way to bring her about; then put the helm a-lee. Should she come about, the sails and helm, having now a proper position for a stern-board upon the other tack, need not to be touched till her stern-way ceases, when the helm must be shifted a-lee; but should the ship refuse stays, then brace sharp round the head-yards, and box-haul her, by which method she will lose much less drift than by veering.

If the ship, now drifting broad-side, is approaching a narrow channel, where drifting in this position, she must be veered and dropped, stemming the tide, stern formost.

In this case, that the drift may be as much as possible, it will be necessary to take in sail, and reduce the ship's head-way, till she has only steerage-way left; thus a vessel may be dropped through a fleet of ships at anchor without danger.

To DRIVE, *when the Wind is across the Tide.*—Should the wind be a little across the tide, a ship may be easily drifted in the fair way, with her head towards the weather-shore; for thus it will be found that she can be backed and filled at pleasure, and generally be drifted with the sails shivering, in which position they oppose least power to prevent the drift.

It frequently happens in serpentine rivers, that the tide sets across; in this case, the ship must be drifted with her head to the side from which the tide sets. These sets are best discovered by observing the opening or shutting of two objects in the direction of the channel. See the *Theory* and *Practice* of SEAMANSHIP.

E.

To EDGE *a Plank*, in ship-building, is to hew it narrower, according to the shape required for use.

EPACTS (from επαγω, *induco*, *intercalo*) in chronology, the excesses of the solar month above the lunar or synodical month, and of the solar year above the lunar year of twelve synodical months; or of several solar months above as many synodical months, and several solar years above as many dozen of synodical months.

The epacts then are either annual or menstrual.

Menstrual EPACTS, are the excesses of the civil or calendar month above the lunar month. Suppose, for example, it were new moon on the first day of January; since the lunar month is $29^d\ 12^h\ 44'\ 3''$, and the month of January contains 31 days, the menstrual epact is $1^d\ 11^h\ 15'\ 57''$.

Annual-EPACTS, are the excesses of the solar year above the lunar. Hence, as the Julian solar year is $365^d\ 6^h$, and the Julian lunar year $354^d\ 8^h\ 48'\ 38''$, the annual epact will be $10^d\ 21^h\ 11\ 22''$; that is, nearly 11 days, consequently the epact of 2 years is 22 days; of 3 years, 33 days; or rather 3, since 30 days make an embolismic or intercalary month. Thus the epact of 4 years is 14 days, and so of the rest; and thus every 19th year the epact becomes 30 or 0; consequently the 20th year the epact is 11 again; and so the cycle of epacts expires with the golden number, or lunar cycle of 19 years begins with the same; these are Julian epacts, the Gregorian depend upon the same principles, accounting only for the difference of the respective years.

As the new moons are the same, that is, as they fall on the same day every 19 years, so the difference between the lunar and solar years is the same every 19 years: and because the said difference is always to be added to the lunar year, in order to adjust or make it equal to the solar year; therefore the said difference respectively belonging to each year of the moon's cycle is called the epact of the said year, that is, the number to be added to the same year, to make it equal to the solar year.

Rule to find the Gregorian Epact.—The difference between the Julian and Gregorian years being equal to the difference between the solar and lunar year, or 11 days; therefore the Gregorian Epact for any year, is the same with the Julian Epact for the preceding year; and hence the Gregorian Epact may be found by subtracting 1 from the golden number, multiplying the remainder by 11, and rejecting the 30's. This rule will serve till the year 1900; but after that year, the Gregorian Epact will be found by this rule:—Divide the centuries of the given number by 4, multiply the remainder by 17; then to this product add 43 times the quotient, and also the number 86, and divide the whole sum by 25, reserving the quotient: next multiply the golden number by 11, and from the product subtract the reserved quotient, so shall the remainder, after rejecting all the 30's contained in it, be the epact sought.

The following table contains the Golden Numbers, with their corresponding Epacts, till the year 1900.

TABLE *of Gregorian Epacts.*

Golden Number	Epacts.	Golden Number.	Epacts.	Golden Number.	Epacts.
I	0	VIII	17	XV	4
II	11	IX	28	XVI	15
III	22	X	9	XVII	26
IV	3	XI	20	XVIII	7
V	14	XII	1	XIX	18
VI	25	XIII	12	I	0
VII	6	XIV	23		

At present, if we add 1 to the number of the year, and divide by 19, the remainder will be the golden number; thus, for 1814, $\frac{1814+1}{19} = \frac{1815}{19} =$ 95 and 10 over as a remainder, which is the golden number.

If we subtract 1 from the golden number, then multiply by 11, and divide by 30, the remainder will be the epact, which was the Moon's age on the 1st of of January; thus, for 1814, we have 9 (10—1) × $\frac{11}{30} = \frac{99}{30} = 3$ and 9 over, which, as a remainder, is the epact.

EXPORTATION, the act of sending goods out of one country into another. In modern times it has been the principle object of commercial policy, in almost every country, to encourage exportation, except with respect to a few particular articles; the export of manufactured goods has been promoted with the view of encouraging the internal industry of the country, and the export of foreign produce, as a mean of drawing wealth from other countries by the profits of the carrying trade.

The principal regulations relative to exportations are as follow:—All goods shipped before the custom duties are paid, shall be forfeited. No person shall take on board any ship beyond sea any English goods or merchandise, until they are first entered in the books of the proper officer of the customs, outward; and shall bring to such officer the contents in writing nnder his hand, of the names of persons who shall have laden on board any goods, together with the mark or number thereof, under penalty of 100*l.*

Wharfingers, or keepers of cranes or quays, suffering uncustomed or prohibited goods to be landed upon their wharfs or quays, or shipped on board from thence, without the presence of some officer of the customs, shall forfeit 100*l.*

Goods exported and carried coastwise without a cocket or sufferance from the custom-house, are forfeited; and masters of vessels shall take out cockets where such goods were laden, and give bond and security that they shall be landed at the place from which they were entered, or some other within the United Kingdom, and shall return a certificate within six months (danger of the seas excepted), from the custom-house at the port where they were unladen, to the officer to whom the security was given, otherwise the bond and security will be forfeited.

No goods can be exported or imported in a packet-boat (unless allowed by the chief officer of the customs), under penalty of forfeiture, and 100*l.* upon the master of such vessel.

No woollen yarn or woollen cloth shall be exported unwrought, under penalty of forfeiture; nor English coin, nor bullion, unless first stamped at Goldsmith's-Hall, and oath made that no part of it was the current coin of this realm, or clippings thereof, or plate-wrought within this kingdom, under penalty of forfeiture of the bullion, and imprisonment of the owner.

No sheeps' wool, wool-fels, mortlings, shortlings, yarn made of wool, wool-flocks, fuller's earth, fulling clay, or tobacco-pipe-clay, shall be exported out of England, Wales, or Berwick upon Tweed, under penalty of forfeiture of the vessel, and treble costs; and persons assisting in exporting the same, shall suffer three years imprisonment.

Official value of exports from Great Britain, for the year ending 5th Jan. 1792,.....£.19,881,552 17 8
1804,.....£ 27,819,337 10 8
1813,.....£.37,647,874 12 9

F.

PAGE 148, col. 2, after line 38, add,

Fire-Buckets, (*seau à feu*, Fr.) are leather buckets, containing about three gallons; they have lanyards fastened to their handles, and used to draw the water out of the sea, in order to quench fire that may happen in any part of the ship during an engagement, &c.

Sometimes wooden buckets are used for the above purpose. See the article Buckets,

Fire-*Ship*. Page 149, col. 1, after line 59, add,

Immediately under the main and fore shrouds is fixed a wooden funnel, M, whose lower end communicates with a fire-barrel, by which the flame passing through the funnel is conducted to the shrouds. Between the funnels, which are likewise called fire-trunks, are two scuttles, or small holes, in the upper deck, serving also to let out the flames. Both funnels must be stopped with plugs, and have sail-cloth, or canvas, nailed close over them, to prevent any accident happening from above to the combustibles laid below.

FLAG. Page 152, col. 2, after line 49, add,

All flag-officers and captains are strictly forbidden to carry any other flag or pendant than that which belongs to their rank, except when his Majesty or any of the royal family are on board, or unless they shall be directed to do so by the Admiralty, or by their superior officer.

Where two flag-officers, carrying the same flag at the same mast-head, shall serve together, the commander in chief may order either of them to carry such other flag as he shall think fit, to prevent the confusion which might otherwise be occasioned.

When flag-officers shall think fit to carry their flags at the head of their boats, their ranks are to be distinguished as follow, viz:

The admiral of the fleet, and the admirals of the red, white, and blue, are to carry their proper flags; the vice-admirals of the red and blue, are to carry one white ball, and vice-admirals of the white one blue ball, in their respective flags. Rear-admirals of the red and blue, to carry two white balls, and rear-admirals of the white two blue balls, in their respective flags. The balls are to be large enough to be easily distinguished, and are to be in the upper part of the flag, and near the staff.

Packets employed by the post-office, and having commanders appointed by commission from the Admiralty, are permitted to carry a red ensign, a jack, and a pendant; but his Majesty strictly forbids all masters of merchant ships, and all ships and vessels employed in the service of any public office, or in raising seamen, or in transporting troops, except those on board of which a commissioned officer of the navy shall be embarked, to carry any pendants, or what may be taken for them; and if they shall offend herein, and presume to wear flags or pendants, or any other colour than those herein specified, it is his Majesty's pleasure that the captains, or any other officers of his Majesty's ships of war, who shall see them, do seize the said colours, flags, or pendants, and report the names of the master, and of the ship or vessel, the place to which she belongs, and the merchants who are owners of her, to the secretary of the Admiralty, together with affidavits of two witnesses to the fact; in order that the person so offending may be proceeded against in the high court of Admiralty.

Merchant ships are to carry a red ensign, with the union jack in a canton, at the upper corner next the staff, and a white jack with a red cross, commonly called St. George's cross, passing quite through it.

Ships having private commissioners, or letters of marque, or letters of reprisal, are to carry the same ensigns as merchant ships, and an union jack with a broad red border at the end and foot thereof.

Ships and vessels employed in the service of any public office, are to carry the same ensign and jack as ships having letters of marque; except that, in the fly of the ensign, there shall be described the seal of the office to which they belong.

The commanders of his Majesty's ships are not to suffer any foreign ship to ride in his ports or roads, with false colours; but if they persist in doing so, after being admonished, they are to put the ships under arrest, and to send an account thereof to the secretary of the Admiralty. See the NAVAL INSTRUCTIONS.

FRAME-SAW, in block-making, has an iron bow over the back, confined at the ends by rivets, and is used by two persons to saw with: some are set in motion by a mill, and have wooden frames.

G.

GAGE, a term used to denote various instruments for measuring the state of rarefaction in the air-pump, variations in the barometer, &c.

There are three sorts of marine gages which we shall notice in this work, namely, *Sea*-Gage, *Bucket Sea*-Gage, and *Tide*-Gage.

Sea-GAGE, is an instrument invented by Dr. Hales and Dr. Desaguliers, for finding the depth of the sea, the description of which is this: A B, fig. 13, plate XIV is the gage-bottle, in which is cemented the gage-tube E *f*, in the brass cap at G. The upper end of the tube E, is hermetically sealed, and the open lower end *f*, is immersed in mercury, marked C, on which swims a small quantity of treacle. On the top of the bottle is screwed a tube of brass H G, pierced with several holes, to admit the water into the bottle A B. The body K, is a weight hanging by its shank L, in a socket N, with a notch on one side at *m*, in which is fixed the catch *l* of the spring *s*, and passing through the hole L, in the shank of the weight K, prevents its falling out when once hung on. On the top, in the upper part of the brass tube at H, is fixed a large empty ball, or full-blown bladder I, which must not be so large, but that the weight K may be able to sink the whole under water.

The instrument thus constructed, is used in the following manner:—The weight K being hung on, the gage is let fall into deep water, and sinks to the bottom; the socket N is somewhat longer than the shank L, and therefore after the weight K comes to the bottom, the gage will continue to descend, till the lower part of the socket strikes against the weight; this gives liberty to the catch to fly off the hole whereby the weight K is disengaged; when this is done, the ball or bladder I, instantly buoys up the gage to the top of the water. While the gage is below, the water having free access to the treacle and mercury in the bottle, will by its pressure force it up into the tube; and *E f*, the height to which it has been forced by the greatest pressure, *viz.* that at the bottom, will be shewn by the mark in the tube which the treacle leaves behind it, and which is its only use. This shews into what space the whole air in the tube E *f* is compressed, and consequently the height or depth of the water, which by its weight produced that compression, which is the thing required.

If the gage-tube E *f* were of glass, a scale might be drawn on it with the point of a diamond, showing by inspection what height the water stands above the bottom. But the length of 10 inches is not sufficient for fathoming depths at sea; because, when all the air in such a length of tube is compressed into half an inch, the depth of water is not more than 634 feet, which is not half a quarter of a mile.

If to remedy this, we make use of a tube 50 inches long, which for strength may be a musket-barrel; and suppose the air compressed into an hundredth part of an inch; then, by saying, as 1 : 99 : : 400 : 39,600 inches, or 3,300 feet; even this is but little more than half a mile, or 2,640. But since it is reasonable to suppose the cavities of the sea bear some proportion to the mountainous parts of the land, some of which are more than three miles above the earth's surface; therefore, to explore such great depths, Dr. Hales contrived a new form for his sea-gage, or rather for the gage-tube in it, which is as follows: B C D F, fig. 14, plate XIV is a hollow metallic globe, communicating on the top with a long tube A B, whose capacity is a ninth part of that globe. On the lower part at D, it has also a short tube D E, to stand in the mercury and treacle. The air contained in the compound gage-tube is compressed by the water as before; but the degree of compression or height to which the treacle has been forced, cannot here be seen through the tube; therefore, to answer that end, a slender rod of metal or wood, with a knot on the top of the tube A B, will receive the mark of the treacle, and shew it when taken out.

If the tube A B be 50 inches long, and of such a bore that in every inch of length shall be a cubic inch of air, and the contents of the globe and tube together 500 cubic inches; then, when the air is compressed within an hundredth part of the whole, it is evident the treacle will not approach nearer than five inches of the top of the tube, which will agree to the depth of 3,300 feet of water, as above. Twice this depth will compress the air into half that space nearly, *viz.* $2\frac{1}{2}$ inches, corresponding to 6,600, which is a mile and a quarter. Again, half that space, $1\frac{1}{4}$ inch, will show double the former depth, *viz.* 13,200 feet, or $2\frac{1}{2}$ miles, which is probably very nearly the greatest depth of the sea.

Bucket Sea-GAGE, is an instrument contrived by Dr. Hales to find the different degrees of coolness and saltness of the sea, at different depths, consisting of a common household pail or bucket with two heads to it. These heads have each a round hole in the middle, near four inches in diameter, and covered with valves opening upwards; and that they might both open and shut together, there is a small iron rod fixed to the upper part of the lower valve, and at the other end to the under part of the upper valve: so that as the bucket descends with its sinking weight into the sea, both the valves open by the force of the water, which by that means has a free passage through the bucket. But when the bucket is drawn up, then both the valves shut by the force of the water at the upper part of the bucket; whereby it is brought up full of the lowest sea-water to which it has descended.

When the bucket is drawn up, the mercurial thermometer fixed in it is examined; but great care must be taken to observe the degree at which the mercury stands, before the lower part of the thermometer is taken out of the water in the bucket, as it would otherwise be altered by the different temperature of the air.

In order to keep the bucket in a right position, there are four cords fixed to it, reaching about four feet below it, to which the sinking weight is attached.

Tide GAGE, is an instrument used for determining the height of the tides, by Mr. Bayly, in the course of a voyage towards the south pole, &c. in the Resolution and Adventure, in the years 1772, 1773, 1774, and 1775. This instrument consists of a glass tube, whose internal diameter was seven-tenths of an inch, lashed fast to a ten-foot fir rod, divided into feet, inches, and parts; the rod being fastened to a strong post fixed firm and upright in the water. At the lower end of the tube was an exceedingly small aperture, through which the water was admitted. In consequence of this construction, the surface of the water in the tube was so little affected by the agitation of the sea, that its height was not altered the tenth part of an inch, when the swell of the sea was two feet; and Mr. Bayly was certain, that with this instrument he could discern a difference of the tenth of an inch in the height of the tide. See BARLOW's *Philosophical* and *Mathematical Dictionary*.

GAUGE, in block-making, a wooden instrument used to mark off distances; it is a square stick nine inches long, with an iron tooth through one end: this stick slides through a mortise in the middle of a semi-circular piece of wood that serves as a stop, the stick being moved at pleasure.

There is also another sort of gauge used by block-makers, to chop off rough wood from the circumference of sheaves. It consists of a cylindrical piece of wood, about six inches in diameter, hooped at each end: in the centre of one end is a pivot, about an inch in diameter; and through the centre of the cylinder, at right angles with the pivot, is a screw-cut handle, having a flat circular piece fixed at one end. By the screw handle it is set to any diameter wanted; and, the pivot at bottom being placed in the hole of the sheave, the flat piece at the end of the handle directs the axe to any nicety.

GEOMETRY, (from γη, earth, and μετρεω, I measure), signifying originally, to measure the earth, but since employed to denote one of the most general and important of the mathematical sciences.

Geometry, according to the present acceptation of the term, may be defined the science of extension, or of magnitudes, considered simply, generally, and abstractedly, or rather, it is that science which treats of the relative magnitude of extended bodies.

Geometry is distinguished into several denominations, as analytical, elementary, practical, &c.

Elementary GEOMETRY, is that which treats of the properties and proportions of right lines, and right-lined figures, as also of the circle and its several parts, and is either theoretical or practical.

Theoretical GEOMETRY, has for its object the demonstration of certain geometrical propositions.

Practical GEOMETRY, relates to the performance of certain geometrical operations, such as the construction of figures, and the drawing of lines into positions, as parallel, perpendicular, &c. to other given lines.

GLASSES. Page 166, col. 2, after line 34, add,

The half-minute glass, fig. 12, plate XIV. is composed of a glass vessel, having a narrow neck in the middle, and a wooden frame, both to preserve the glass from injury, and also to hold it by during the time of heaving the log. Such a quantity of sand is put into the glass, that it may run out at one end into the other, in the space of half a minute; cut steel is frequently substituted for sand; however, egg-shells dried in the sun, then beaten fine and sifted, have been proposed as the best material for this purpose.

It is often found that sand takes a longer time to run out at one end of the glass than it does at the other; the difference sometimes amounting to one or two seconds. That end of the glass, therefore, ought to be kept uppermost, out of which the sand runs in 30 seconds, or in the time nearest to that period. This distinction will, however, entirely be avoided in a glass of the following construction:—Provide a tube of glass in form of a barometer tube, having one end hermetically sealed, and the other applied to a bottle with an aperture of a less diameter than the bore of the tube, and let a sufficient quantity of sand be put into the bottle.

Now when the instrument is held in a vertical position, with the bottle uppermost, the sand will then run into the tube, and rise therein, and the time will be as the space occupied by the sand. To adopt a scale to this instrument, let the tube be sunk about half its thickness into a piece of wood of a convenient size, and by means of a clock or watch vibrating seconds, observe the height of the sand in the tube at the end of 30 seconds: hence, the bore of the tube being of the same diameter throughout, each single second may be easily marked on the scale.

GOLDEN NUMBER, in chronology, a number shewing what year of the Metonic, or lunar cycle, any given year is. To find the golden number, add 1 to the given year, and divide the sum by 19; what remains will be the golden number, unless 0 remains, for then 19 is the golden number.

The discovery of the Metonic cycle exhibited such extensive astronomical knowledge, that it obtained great success and reputation in Greece; insomuch that the order of the period was engraved in letters of gold, whence it acquired the name of *golden number*. See EPACTS.

GUN. Page 174, col. 2, after line 52, add,

General Congreve's New Pattern GUN. Analysis of the practice at Sutton Heath, in 1813, shewing the great general superiority of the ranges of General Congreve's New Pattern Gun of 7 feet 6 inches in length, and weighing 40 cwt. 2 qrs. 21 lb. compared with those of the long 24-pounder gun of 9 feet 6 inches in length, and weighing 50 cwt. 0 qr. 7 lb. and with those of General Blomefield's 24-pounders of 8 feet, and 7 feet 6 inches in length, the first weighing 43 cwt. 0 qr. 12lb. and the last 40 cwt. 0 qr. 24 lb. at point blank, $2\frac{1}{2}^\circ$, 5°, 8°, and 10°, elevation, single and double shotted, with *full charges* of powder to all the guns.

Observations on the foregoing Results.

That from the above points of comparison, single shotted, it appears, that with the *same windage*, and *charges of powder*, the first grazes of General Congreve's gun, at point blank, have in every case exceeded those of the three 24-pounders tried against it; *in some instances nearly double, and in none less than one-third more* than the point blank range of the other three; and that the ranges of its second grazes, at point blank, are greater than those of all the other three, except in one instance, where it was exceeded by General Blomefield's seven feet six inches gun, by 28 yards only. That at $2\frac{1}{2}°$ elevation, the ranges of the first graze from General Congreve's gun had the advantage in every instance but *one*, where the long gun beat it by 12 yards only; that at the same elevation, in *one* instance only, are its second grazes exceeded.

That at 5° elevation, the *first* grazes had very greatly the advantage in all the comparisons, except in two cases. That at 5° elevation, the *second* grazes also are very considerably in favour of General Congreve's gun.

It must be observed, however, that in the foregoing single-shotted practice, the extreme roll of the shot seems generally in favour of the long gun; but it is self-evident that this extreme roll is of very little importance, even on shore, and certainly means nothing in sea-practice, where, in fact, the first graze only is of any use whatever after $2\frac{1}{2}°$ of elevation. With regard to the double-shotted practice, it must be remarked, that the advantage also continues in favour of General Congreve's gun; and although the great excess of the weight of the long gun tells with the double-shotted charges, in point blank ranges, and in the point blank ranges *only*, still the advantage derived from it is not so great as might have been expected: the mean difference of the first grazes being only 74 yards; that of the second, 37; and that of the extreme, 94, in favour of the long gun: and here it must be observed, that General Congreve's gun in casting has turned out lighter than was intended; the calculation in his memoir giving 41 cwt. 0 qr. 22 lb.—and the gun actually weighing only 40 cwt. 2 qr. 21 lb.

In all other points, the advantage gained is evidently, without reference to weight, as the heavy and light guns are alike proved most decidedly inferior, on the average of all this practice: as, therefore, the charge and windage were the same in all the cases, it is clearly demonstrated, *as the grand result of this practice*, that the superiority of General Congreve's gun depends on *its form*, and on *its form alone*; that is, on the *additional quantity* of *metal applied to the breech*, by making the muzzle taper; a principle, the discovery of which appears to be one of the most important that has been made in the science of gunnery for many years, as realizing by *formation* only, even with a *diminution* of *weight*, not only a variety of points peculiarly desirable for sea-service, but an average increase of range the most valuable and extraordinary, sufficient, General Congreve trusts, to establish the importance of these guns, as a very superior arm for our frigates, more especially, as their recoil and general action are declared also to be most manageable, notwithstanding their great increase of powers.

"Another general result of the whole of this practice," says General Congreve, "seems to be, that 9 feet 6 inches is an useless length for any description of heavy ordnance; especially as, by my construction, an equal projection of muzzle may be obtained with a very reduced length. I conceive, therefore, that it would be extremely desirable, to make the experiment of the *heavy* 24-pounder upon this same principle of reduction of length, and increase of metal about the charge; and I trust also, that the investigation will be extended to the 32-pounder guns, by experiments, not only with a 32-pounder *medium* gun of my construction, weighing about 50 cwt. as proposed in my Memoir to the Admiralty, but with the heavy 32-pounder also, reducing its length and thickening its breech; as I cannot but anticipate, *that by a due modification of these principles, every class of line of battle ships may ultimately be armed with 32-pounder guns; and every class of frigates with 24-pounders, the lightest of which shall exceed, in general powers, the heaviest of their present respective natures; while the heavy natures themselves shall also receive, by the same system, a proportionate increase of power.* It is, perhaps, difficult to conceive any thing of more importance to our naval service!"

General Congreve concludes this account by a copy of the report of the naval officers of the committee to the lords commissioners of the Admiralty, date, Nov. 22d, 1813, and from Captain Phillimore, commanding the Eurotas frigate, which is armed with 24-pounder guns of his construction, dated October 11th, 1813.

GUNNER. Page 177, col. 1. after line 53, add,

The *gunner* is not to go into the magazine without being ordered; nor to allow the doors of it to be opened, until the proper officer is in the light-room, and he is to be careful that no person enters the magazine without wearing the leather slippers supplied by the ordnance. He is not to stave the empty powder barrels, but to return them, when the ship comes into port, to the store-keeper of the ordnance, who is desired to pay him one shilling for every empty powder-barrel he shall return in good condition. He is to turn the powder-barrels or see that they are turned once every three months; to examine the cartridges; and not to suffer the bare gun metal adzes to be struck against the copper hoops of the powder-barrels; but always to have the wooden setters applied to them, as strong sparks of fire have been frequently produced from the collision of a metal adze against a metal setter, or a copper hoop. See the NAVAL INSTRUCTIONS.

Lady of the GUN-*Room*, in line of battle ships, a name given to a trusty old sailor, who is not capable of doing the duty of a seaman, and who is generally appointed by the captain to keep the gun-room clean, &c.

GUNNERY. Page 179, col. 1, after line 52, add,

We shall conclude this article by transcribing from Dr. Hutton's Mathematical and Philosophical Tracts, in 3 vols. 8vo. his new approximate rule to find the elevation of a gun to hit an object at a given distance.

'Let D denote the given distance of the object in feet; *d*, the diameter of the ball in inches, obtained from the table of weights and diameters in problem 10: *b* the weight of the ball, and *c* that of the charge of powder, both in pounds; $V = 1600\sqrt{\frac{2c}{b}}$ the projectile velocity as given in problem 13; *v*, the last velocity with which the ball strikes the object; and *t* the time of the ball's flight. Then

'Divide D by 1338 *d*, considering the quotient $\frac{D}{1338^d}$ as a log.

'Take N the natural number of the log. $\frac{D}{1338^d}$.

'Take $v = \frac{V - q + q}{N}$ the final velocity: q being $= 231$.

'And $t = \frac{1338^d}{q} +$ log. of $\left(\frac{V-q}{v-q}\,\frac{v}{V}\right)$ by problem 11.

'Or $t = \frac{2D}{V+v}$, an approximation near enough.

'Then, $16\,t^2 = \frac{64\,D^2}{(V+v)^2}$ is the height above the object to be pointed at.

'Or $\frac{16\,t^2}{D} = \frac{64\,D}{(V+v)^2}$, is the tangent of the angle of elevation.

'So that, the height of the mark to be pointed at, above the object, is nearly as the square of the distance, and the angle of elevation simply as the distance, the projectile velocity being the same. But, in the case of different velocities, the height and the angle will be reciprocally as the square of the velocity nearly.'

It should, however, be remembered, that the above rule is given merely as an approximation. Much pains having been taken in applying it to the results of a great many accurate experiments by Dr. Hutton and others, it has been found, that if the angle of elevation obtained by these theorems be diminished by its *fifteenth* part, it will then agree very nearly with the actual practice of artillery.

H.

HARPINGS, *(la partie de l'avant des préceintes d'un vaisseau*, Fr.) in ship-building, are pieces of timber trimmed to the shape of the ship's body, and only used forward and aft in the weak of the cant-timbers; they are placed diagonally below the main-breadth, to secure the timbers until the planks are wrought. The upper harping being well shored, it remains till the main-wale is both worked and shored. When the planks are wrought, they become as harpings, and are accordingly shored before the harpings immediately below them are disturbed.

HATCHELLING HEMP, in rope-making, is the combing and preparing it ready for the spinner, in doing which, care and attention must be paid by the hatchellers, that they do not use too great a quantity of oil; as a small quantity of train oil sprinkled, or daubed on the hemp with a wad, facilitates the business of hatchelling, and is very necessary when the hemp is too dry; the spinners are better able to perform their business when the hemp has received such assistance; but when too much is used, it prevents the yarn from imbibing a proper proportion of tar, and thereby proves a serious injury.

Page 187, col. 1, after line 42, add,

To HAUL IN *a lower Studding-Sail.* When blowing fresh, lead one of the sheets clear aft, and man it well; then lower away briskly the outer halyards, to spill the sail; ease off the tack, run in upon the sheet, and lower away the inner halyards as required.

To HAUL DOWN *a topmast Studding-Sail.* Man well the deck-sheet and down-haul, ease off the yard-sheet, and haul the yard close out to the tack-block; then ease away the tack, and haul down both upon the deck-sheet and down-haul.

HEAD. Page 189, col. 1, after line 21, add,

In fig. 6, P represents the gammoning fish, which prevents the gammoning from wounding the bowsprit.

In fig. 7, M is the main head-rail, which is a greater support to the knee of the head than any other rail heretofore used.

N, the rail to which the birthing of the head is secured.

O, the lace-piece on the knee of the head.

HEAD-MONEY, an allowance granted by his Majesty, of 5*l.* per head, for all persons captured that appear to be borne on the ship's books previous to the commencement of action with an enemy's ship of war or privateer; and is paid by bill from the Navy Board, and generally distributed with the prize-money.

By act of parliament, the following documents are to be produced to the commissioners of the navy, before payment of head money can be made on any ship of war or privateer, belonging to the enemy, that may be captured or destroyed.

1st. An authentic copy of the sentence of condemnation under the seal of the court wherein she must appear to be a ship of war or privateer. This document, unless the enemy's ship shall have been sunk, burnt, or otherwise destroyed, will in no case be dispensed with.

2nd. A certificate of the number of men (not including passengers) living on board of her at the beginning of the engagement, to be granted by the mayor, or other chief magistrate of the first port within his Majesty's dominions, to which the prize is carried, or from the consul, or vice consul, if first carried to a neutral port. But if not from the mayor or chief magistrate of the first port, then from the mayor or chief magistrate of any other port; proof being first made upon oath, before such mayor or chief magistrate, of the inability of obtaining the same at the first port: and this certificate is to be founded upon the oaths of three or more of the chief officers of men taken in her.

If it shall not be practicable to procure the certificate or affidavit precisely in the manner above pointed out, the circumstances which prevent it, are to be particularly specified in the voucher that may be produced, and without the most satisfactory reasons they will not be dispensed with in any respect.

3d. A letter of attorney from the captain, officers, and company, or from a majority of them, of the captor ship, appointing an agent or agents to claim and recover the head-money.

4th. A list of the captor's ship's company on board at the time of the capture, attested by the captain and officers.

5th. If the enemy's ship is burnt, or otherwise

destroyed, the above vouchers are to be produced without the condemnation. See the NAVAL INSTRUCTIONS.

HELM. Page 192, col. 1, line 24, after tiller, add,

Five turns of the tiller-rope are usually wound about the barrel of the wheel; and the midship-spoke, which is generally painted black, is always perpendicular with the stanchion in which the wheel is shipped, when the helm is amidships; a mark from which the helmsman derives much assistance in the act of conning.

Page 195, col. 2, after line 61, add,

Slippery HITCH, (*amarrage coulant*, Fr.) a hitch that will not bear a stress or strain.

Page 196, col. 2, after line 5, add,

Lady's HOLD, is a small place in the after part of the ship, with a scuttle in the gun-room, which is appropriated to the use of the gunner.

Page 197, col. 2, after line 34, add,

Iron HOOKS, for tackles, have an eye at one end to receive the thimble, and taper at the point; they are made hollow on the back, that the seizing may lie on them without slipping, or a catspaw-hitch; and a hole is generally punched through the point for mousing. Iron hooks for the futtock-shroud have a square point, to prevent their unhooking.

Page 199, col. 1, after line 2, add,

HORSE, in mast-making, a wooden frame, used to support light work that wants rounding, &c. so that a man may stand easily to perform it. It consists of two sides, about four inches broad, two inches and a half thick, and four feet six inches long, scored or lapped together about the middle, like a cross, with two long legs and two short arms; these are farther supported by two legs, at right angles with these two sides, their upper ends being shaped and nailed thereto.

J.

JACOB'S LADDER, in a ship of war, a name given to two four inch ropes that have about seven or eight wooden rounds thrust through them, as steps, the upper part being fastened to the main and fore rigging, whilst the lower is supported by iron outriggers fixed in the side of the ship, with a score made at the ends for the better security of the ropes: and, in order to set up the legs or sides, there are two eye-bolts driven into the spirketting, to receive the lanyards which reeve into the eyes spliced at the ends. These ladders are of great use, both on the quarter-deck and forecastle, to facilitate the men's going aloft, and have a neat appearance.

JACOB'S *Ladder*, is also a rope-ladder leading up and down the masts of polacres, and small vessels used in the Mediterranean.

JACOB'S *Ladder*, among shipwrights, is also a name given to pieces of oak plank, fastened on each side of the pillars in ships' holds, with scores cut in them for the men to put their feet on, by which, and the help of a rope, they are enabled to go into and out of the holds. In some ships, scores are cut in the solid of the pillar for that purpose.

IMPORTATION. Page 203, col. 2, after line 49, add,

Official value of Imports into Great Britain, for the year ending 5th Jan. 1792..£16,958,807 3 2
1804..£24,728,191 18 2
1813..£24,362,124 13 7

L.

To LAUNCH. Page 216, col. 1, after line 49, add,

Fig. 10, plate XIII, represents a perspective view of one of the large class of frigates ready to be launched from the slip, all obstructions being previously cleared away for that purpose, excepting the dog-shore, exhibited by *a*, which, being knocked down, causes her immediately to advance down the inclined plane into the water.

When any of his Majesty's ships are going to be launched, the different flags, *viz.* the union jack, Admiralty flag, royal standard, navy flag, and ensign, are displayed in the above order from head to stern, on flag-staffs erected in the ship for that purpose.

To Stop a LEAK. Page 217, col. 1, after line 20, add,

Mr. Hill's invention for stopping leaks occasioned by gunshot-holes, consists of a circular piece of elm, from two to four inches thick, and of a circumference sufficient to fill the shot holes; the outside is convex and the inside somewhat concave; to be lined with several folds of flannel or kersey dipped in warm tallow. Through the centre of this piece of elm and flannel is a hole for a rope to be passed through, from two to four inches in circumference, and about six feet in length, with a double wall-knot on the outer end, tapering and served to an eye at the other end: the whole is thus used:—From the inside of the vessel is thrust through the shot-hole, spun-yarn or small line, of a sufficient length to float on the water, the inner end being retained on board, the outer or floating one is taken up, and fastened to the eye in the tapered rope; the spun-yarn is then hauled upon, until the stopper covers the shot-hole, and if a greater strain is wanted to press the flannel round the edges of the hole, a tackle must be used. Should the shot-hole not be sufficiently clear to admit the spun-yarn through it, an auger must be used.

LIFE-PRESERVER. Page 227, col. 1, after line 6, add,

Mr. Bosquet's life preserver is simply an improvement on the cork jacket, by being made more compact to the body, and obtained the silver medal from the Humane Society. See the article CORK JACKET.

Mr. Butter's life-preserver is the earliest for confining air; it is a copper girdle, divided into cells or partitions, and is fastened to the person much in the same manner as the *Marine* SPENCER (see that article); and may be had at No. 36, Wigmore-street, Cavendish-square, London.

Mr. George's life preserver is similar to Mr. Butter's, but made of tin lacquered over. It is preferable, not being so bulky, and considerably cheaper, and may be had at his residence, No. 422, Strand, London.

The patent air balls are two bladders filled with air, and covered with water-proof leather. They are fixed upon the arms, and by a strap prevented from falling off, similar to the mode in which school-boys use corks.

The Arabs dwelling on the banks of the Tigris and Euphrates, both men and women cross these rivers in perfect confidence, with the assistance of inflated goat-skins. They sew the whole skin closely together except one leg, through which, as a tube, it is blown up; then twisting the channel, they hold it firm while engaged in their watery excursions.

The inhabitants of Chili obtain still greater convenience, by uniting two very large inflated seal-skins, of which all the apertures are carefully secured. A few spars are laid across, whereon one or more can rest in safety, while occupied in fishing or passing rivers.

Lumber-Irons, a kind of iron stanchions, fitted on the gunwale of boats of burden, for the reception of the masts, yards, oars, &c. in order to keep them clear, when about to be loaded with casks, &c.

Lunar *Observations*. Page 249, col. 2, after line 62, add,

Mr. Stephen Leguin has also invented instruments for facilitating the knowledge of the longitude at sea; for which he obtained his Majesty's royal letters patent. These instruments are made and sold by Messrs. Watkins, Charing-Cross.

The first instrument is composed of a new-invented pair of compasses with four legs, which move on an axis.

The second instrument represents a semi-circle divided into 180 degrees, and carries a nonius to shew the true distance from the points of the legs of the compass.

The use of the instrument is, to obtain the true angular distance of the moon from the sun or a star, after having the apparent distance from the centres, the apparent altitude of the centres, and the correction of these altitudes. By this instrument the horary angle is also obtained, and thereby the problem of apparent or solar time is easily solved. The horary angle, found by an observed altitude of the sun, and reduced into time, shews the apparent time when the observation was made; and to the horary angle in time, found by a star, its right ascension must be applied, and it gives the right ascension of the mid-heaven; and the difference between the sun's right ascension for the place and time, and that of the mid-heaven, is the apparent time of the star's observation; and the watch to be regulated accordingly.

Page 250, col. 1, after line 18, add,

The Rev. Dr. Brinkley, Professor of Astronomy in the university of Dublin, has pointed out an easy and concise method for clearing the apparent distance of the moon from the sun or a star, from the effects of parallax and refraction, in the XI. Vol. of the Transactions of the Royal Irish Academy. To determine the longitude at sea, all possible means have been attempted for bringing the lunar method to perfection, and for accommodating the necessary computations to the abilities and habits of seamen in general, who are well known to be much averse to laborious calculations, such as naturally attend the problem of clearing the lunar distance from the effects of parallax and refraction. With this view, a variety of solutions has been proposed. They may, however, be divided into two principal classes; viz. that in which the correction of the observed distance is given, and not the corrected distance itself, and which therefore necessarily involves a distinction of cases; while the other, giving the corrected distance, is free from this source of embarrassment, but requires a longer and more tedious calculation.

The object of Dr. Brinkley, in this memoir is, to simplify the solution in the latter case; and we have no doubt that, if his method were generally known, it would be practised in preference to most others of the same kind—by many in preference to those of the first class, even when the purchase and incumbrance of the necessary tables are considered as no material objections.

Of the methods depending on the use of extensive tables, those which are proposed by Mr. Mendoza Rios are the most complete. This author, to whose exertion nautical astronomy is much indebted, has given, in a valuable paper in the Philosophical Transactions for 1797, forty different formulæ for computing directly the true distance. Of these, that which he selected as best adapted to practice affords a very concise and plain method of computation, requiring only the addition of five versed sines: but even here, says Dr. Brinkley, "It may be doubted whether it is preferable to Mr. Dunthorne's first method, improved by the substitution of versed sines for cosines, as was done by the late Dr. Mackay, in his treatise on the longitude." Mr. Mendoza's method requires an extensive table for an auxiliary angle, (equivalent to the auxiliary table in Mr. Dunthorne's method, and in those derived from it,) and the formation of five different arguments; and also, to practice it with convenience, a complete table of versed sines, for at least the semicircle. Mr. Mendoza, anxious to improve still farther the solution of the problem, published his very extensive tables, by which he reduced the method to be equivalent to taking out, and adding together, three numbers, and not requiring the formation of arguments. His ingenuity and perseverance in forming, computing, and publishing his tables, are deserving of the greatest praise.

But on several accounts those tables will not supersede the use of methods in which shorter tables are employed, although these methods should be somewhat longer in practice, provided they be equally plain. To many persons the necessary expense of the volume will be an objection, many will consider its bulk inconvenient: and many, disliking such extensive tables with double arguments, will even prefer the former methods, which those tables were intended to facilitate.

Having paid this tribute to the ingenuity and perseverance of Mr. Mendoza, and at the same time having fairly and correctly stated the inconveniences still attending his method, Dr. Brinkley gives the investigation of his own solution; which he candidly acknowledges is nearly allied to Mr. Dunthorne's method, and still nearer to that of Dr. Mackay. In point of practice and application, however, it seems to exceed either; and we question whether the solution by Mr. Mendoza's tables can be obtained much more readily than by Dr. Brinkley's method, although the calculation in the latter case will be somewhat longer. It involves no distinction of cases, requires only short tables with single arguments, and no proportional parts are necessary, except what may be taken out by inspection, and these for only one quantity besides the conclusion. We should be glad to see this rule introduced into some practical and popular work, in order to its being more generally known than it can be while it is confined to the Irish Transactions. Would it not be an improvement in the Nautical Almanac, to make a practice of devoting a few additional pages to the insertion of useful nautical problems and their investigations, after the manner of the *Connoissance des Temps?*

Page 250, col. 2, after line 57, add,

LYING-TO. In a more limited sense, it denotes the art of disposing the sails in such a manner, as that, counteracting each other, they render the ship as it were immoveable, with respect to the disposition she has, by her form, to divide the fluid, at either extremity, with facility.

Lying-to in fair weather is seldom practised but under three topsails; yet it is indifferent whether the fore or main-topsail be braced a-back or kept full; because they have nearly equal surfaces, and have nearly equal power, either to stop the ship's way, or to cause her to run a-head. When these two sails act together, or one against the other, one always tends to pay the ship's head off, whilst the other keeps her to the wind.

Lying-to *under a Mainsail.* In this situation, the ship does not so easily fall off as in the last-mentioned mode, because its effect passes abaft the centre of gravity of the ship; but it keeps her more to the wind, and consequently occasions less lee-way.

Lying-to *under the Mizen.* Ships keep better to the wind under the mizen, than under any other sail, because it is farther abaft the centre of gravity than any of the rest, and consequently ought to keep the vessel from drifting more than any of the others; but is inconvenient, should you have occasion to veer suddenly.

Lying-to *under the Main-Staysail.* A ship will not make so much lee-way under the main-staysail as under a fore-sail, because its effort passes very near the centre of gravity, but it will cause her to drift more than the main-sail: so that this mode of lying-to is a mean between the two others, and is preferable when it blows strong enough for that sail to support the rolling of the ship. It ought likewise to be preferred, because the ship will veer under that sail, the action of which passes at a small distance from the centre of gravity, and the power of which overcomes the resistance which all ships meet from the fluid under their lee; a resistance which always gives them a great inclination to fly up in the wind when it blows hard, or when under a heavy press of sail.

Lying-to *under a Foresail.* This is advantageous for veering when you are well to windward; but it augments the lee-way, and is subject to break the sea on-board, on account of the ship continually falling-off: because, in that movement, she gathers way by yielding to the impulse of the gale, and is afterwards recalled to the wind by the helm; so that, in springing the luff, she meets the wave which comes from windward.

Lying-To *in a Gale of Wind.* This is effected by keeping as close to the wind as possible, under some one sail well trimmed, with the helm lashed a-lee as much as may be requisite for the ship; and, as ships commonly bring-to from the stress of contrary winds, care should be taken to heave-to under such sail as will least strain the ship; because there are some ships which lie-to better under the fore-sail than the main-sail, others are more easy under the main-sail, some under a mizen, and many vessels lie-to best under a main-stay-sail.

Lying-to *under the Fore, Main, and Mizen Staysails.* All the preceding modes of lying-to have their peculiar faults; but the preferable way is under the fore-staysail, the main-staysail, and mizen-staysail; because, under these sails, the ship will steer, and is in a better situation for veering than under any other sail; for, only haul down the mizen-staysail, and put the helm a-weather; when the two other sails, being before the centre of gravity, will cause her to fall off; she will then soon gather way, and steer easily.

Should the gale continue very hard, and one of these staysails be blown away, the loss is not of much consequence, as the courses, in case of an emergency, are ready to set; whereas the courses are not so readily replaced when lost. This mode, therefore, appears preferable in every respect,* whether you wish to veer or keep your wind; because, if the ship does not sufficiently keep the wind, you may haul out the balanced mizen, or take in the fore-staysail, or even the main-staysail. One of these staysails, before the centre of gravity of a ship, is sufficient to make her veer as soon as the aft ones are suppressed. There are, besides, the following considerations for so doing: the ship will carry sail better; because, as the centre of effort of those on her is very low, she drifts less, holds a better wind, and goes faster through the water; and these three or four sails are so situated as to give the whole body of the ship play, which will strain her less than when under one single sail, which cannot by itself work it from aft forward. See Sauseuil's *Skilful Seaman,* translated from the French of Bourde de Villehuet.

M.

MASTER *Attendant.* Page 268. col. 2, after line 63, add,

; to superintend the works carrying on in the sail-loft and rigging-house; to attend the launching, docking, masting, and ballasting his Majesty's ships; to attend to the ship's coming into and going out of port, to see to the timely provision of the rigging and sails of the ships preparing for sea; to inspect the quality of all stores used in their department, either

* Should the sea run too high for the lower staysails to keep the ship steady, a close-reefed main-topsail (particularly if it has four reefs in it to come close down to the cap) will be found to answer very well.

purchased or served in upon contract, and to certify to the Navy-Board their fitness for the service.

MASTER OF A SHIP OF WAR. Page 267, col. 2, after line 25, add,

By an order in council of the 26th of July, 1814, it was resolved, "that masters of the royal navy, instead of being required, as at present, to pass an examination before the master and wardens of the Corporation of Trinity House for each rate of ships separately, shall be required only to pass an examination touching their qualifications to serve as masters of the following rates jointly.—First and second rates jointly; third and fourth rates jointly; fifth and sixth rates jointly; sloops and smaller vessels jointly.

"And in order to prevent any unnecessary, or improvident increase of the number of masters, and, at the same time to ensure, that all persons warranted as masters in his Majesty's navy, be men of experience and judgment, no person shall be warranted as master, until the number of masters on the list be reduced below the number of ships on the list of the navy (exclusive of such as are building, and have not been launched); and that no more masters shall be ever appointed, than the number of such ships at the time that any appointment of master is to be made."

MURAL ARCH (from *murus*, Lat.), a wall or arched wall, placed exactly in the plane of the meridian, for fixing a large quadrant, sextant, or circular instrument, to observe the meridian altitude, &c. of the heavenly bodies.

Tycho Brahe was the first who used a mural arch in his observations; after him Helvelius, Flamstead, De Lahire, &c. used the same means. See a description of the mural arch at Greenwich Observatory, constructed by Mr. Graham, in Smith's Optics, book 3. ch. 7, with the improvements of Mr. Bird, in the "Method of constructing Mural Quadrants," published by the Commissioners of Longitude in 1768.

MURAL *Circle*. — A grand mural circle of six feet radius, constructed by Troughton, under the direction of the late Dr. Maskelyne, was placed in the Royal Observatory in May, 1812, with which Mr. Pond, the present astronomer royal, made his observations on the summer and winter solstices of that year, which, with a catalogue of some of the principal fixed stars, he published in the first volume of the Philosophical Transactions for 1813. From the construction of this instrument, no use whatever is made either of the level or plumb-line; but the position of the polar point on the instrument is found by assiduously observing the pole-star, and is inferred, by comparing the observed with the real places of certain numbers of standard stars, previously determined by the instrument itself, by which means its position may always be ascertained to within a tenth of a second, a degree of precision to which no plumb-line or level can ever be expected to arrive.

Mr. Pond states that by sixty observations of γ draconis, of which only three differ so much as 1″ from the mean, it appears that the zenith point which he assumed is 2′ 18″ 64, south of the mean position of γ draconis for the beginning of the year 1812, which is the same quantity that is found by the observations with the zenith sector, in 1811.

We are happy to find that this beautiful instrument is likely to possess that extreme degree of accuracy which the celebrity of the artist, and the well-known talents and experience of its original projector, led us to anticipate. Dr. Maskelyne had long looked forwards with pleasure to the time at which he might commence a course of observations with this new circle, but unfortunately he did not live to see it completed; it is now, therefore, from his successor, Mr. Pond, that astronomers are expecting the confirmation or correction of former observations, and as Dr. Bradley's tables of refractions, are suspected, both by the French and the English astronomers, to require some modification; should the astronomer royal succeed with this new instrument in making the necessary corrections relative to this subject, he will doubtless render an essential service to the science of astronomy.

N.

NAIL and BOLT-DRAWER, a wrought iron tool used by shipwrights, &c. for extracting bolts and nails from the sides of ships, &c. It is represented by fig. 1, plate IV. in the act of drawing a spike-nail, B, out of a piece of wood, A, by the power applied to the lever, C.

NAVAL *Discipline*. Page 303. col. 2, after line 33, add,

British seamen, like the peasantry and mechanics of a state, are the most important class of all; being literally the sinews of our naval strength: like children, they may be moulded to almost any shape, by dexterous mananagement, yet like froward youths, they may also be spoiled or rendered useless, not to say dangerous to the state by an injudicious mode of training.

Mr. James Johnson, surgeon in the royal navy, and author of the Essay on the Influence of Tropical Climates on European Constitutions, has justly observed, "that the grand or fundamental principle of naval discipline, as promotive of health and comfort among seamen, consists in so artfully *employing both mind and body, that the one may not be affected by apathy or chagrin, nor the other by indolence or over-exertion.*

"In exact proportion as this principle is kept in view, and acted on, will the end and object of naval discipline be attained; and wherever it is disregarded, the inevitable consequences will be anarchy and disease! For this purpose, the code of interior regulations should be *mild* and *judicious*, in order that a rigid *performance* may be enforced; and instead of the many different orders that are suspended under the decks of ships throughout the navy; there should be one simple uniform system of interior economy, signed by the lords commissioners of the Admiralty, punctually adhered to.

"When seamen are convinced that nothing but an unforeseen exigency or indispensable necessity will cause a deviation from the routine of duty prescribed, they know exactly what they have to do: their minds are accordingly made up to the performance of it, and they go through it with alacrity, in order to have the *intervals* for their own amusement or private occupations.

"But where order and punctuality are not rigidly enacted and followed up, the equilibrium in the division of labour becomes unhinged, and the greatest share of toil often falls on the best men: never being certain of the exact periods of duty and relaxation, they frequently become listless, lazy, dissatisfied, and careless about their personal cleanliness; the consequences of which need not be pourtrayed.

"The great art then seems to consist in properly appreciating the capacities of the men, and so appointing their respective duties, in every department, as to get them into a kind of mechanical train, when the future superintendence will be easy and pleasant, and the health of the crew secured."

O.

OVER-HANGING, in ship-building, denotes the part projecting over; as, the stern is said to over-hang when it rakes much.

To Over-Set. Page 329, col. 1, after line 10, add,

A common, but not always a certain, method to recover a ship, when overset on her side, is to cut away the masts; however, as this expensive method may fail, stop-waters only on the lee quarter at sea, may cause the ship to veer; or, where there is ground, an anchor or anchors, dropped from the lee-bow, may bring the wind a-head, and take the sails a-back, so as to cast the ship on the other tack.

Out-Rigger. Page 329, col. 1, after line 63, add,

Mr. Hoskins, master of his Majesty's store-ship, Sir Francis Drake, has contrived an out-rigger for the support of the top-gallant-mast of a ship, to which is affixed a back stay, called by the sailors, "Hoskins's flying back-stay," which can be shifted to answer any point of sailing, in an instant of time, without sending a man off the deck. It has been proved to answer every requisite purpose; gives more support to the mast than all the other rigging; and the weight of all attached to the improvement does not exceed the common back-stay.

P.

PASSPORT, or Pass. Page 334, col. 2, after line 49, add,

The violation of safe-conducts, or passports, expressly granted by the king or his ambassadors to the subjects of a foreign power in the time of mutual war, or committing acts of hostility against such as are in amity, league, or truce with us, who are here under a general implied safe-conduct, are breaches of the public faith, without which there can be no intercourse or commerce between one nation and another; and such offences may, according to the writers on the law of nations, be a just ground of a national war. And it is enacted by the statute 31 Hen. VI. cap. 4, now in force, that if any of the king's subjects attempt or offend upon the sea or in any port within the king's obeysance, or against any stranger in amity, league, or truce, or under safe conduct, and especially by attacking his person, spoiling him, or robbing him of his goods; the lord chancellor, with any of the justices either of the King's Bench or Common Pleas, may cause full restitution, and amends to be made to the party injured.

Pasquier takes passport to have been introduced for *passe par-tout*. Balzac mentions a very honourable passport given by an emperor to a philosopher in these terms: "If there be any one on land or sea, hardy enough to molest Potamon, let him consider whether he be strong enough to wage war with Cæsar."

Page 388, col. 1, after the tables on Pay, add,

Half-Pay, a compensation, or retaining fee, given to commissioned and warrant officers, who have retired from the service, through age, inability, &c.

By late orders in council an increase of the half-pay of officers in the royal navy took place on the 1st of July, 1814, subject to the same regulations and restrictions as before, and is now as follows:

Flag Officers.	per day.		
Admiral of the fleet	£3	3	0
Admirals	2	2	0
Vice-Admirals	1	12	0
Rear-Admirals	1	5	0
Captains.			
The first 100 on the seniority list	0	14	9
The next 150	0	12	6
The remainder	0	10	6
Commanders.			
The first 150 on the seniority list	0	10	6
The remainder	0	8	6
Lieutenants.			
The first 300 on the seniority list	0	7	0
The next 700	0	6	0
The remainder	0	5	0
Masters.			
The first 100 on the seniority list, qualified for first and second rates	0	7	0
The next 200, qualified for 3d and 4th rates	0	6	0
The remainder	0	5	0
Surgeons.			
Not having served six years	0	5	0
After six years active service	0	6	0
After twenty years service, except those serving on board receiving ships, convalescent ships, and any other than hospital ships employed in harbour duty	0	10	0
After thirty years service	0	15	0
Pursers.			
The first 100 on the seniority list	0	5	0
The next 200	0	4	0
The remainder	0	3	0

The *half-pay* of the commissioned officers is, in future, to be paid *quarterly*, instead of *half-yearly*.

PAY. Page 339, col. 2, after line 1 from the bottom, add,

Half-PAY *of the Royal Marine Officers.*

By an order in council an increase of the half-pay of officers in the royal marines took place on the 1st of July, 1814, and is now as follows:

Colonels	£0	14	6
Lieutenant Colonels	0	11	0
Major	0	9	6
Captain	0	7	0
First Lieutenant	0	4	0
Ditto if com^d. 7 years as first lieut.	0	4	6
Second Lieutenant	0	3	0

Stay-PEAK, is when a cable is drawn in until it is almost perpendicular, and is called long or short, according to the angle which it makes from the perpendicular.

PITCH. Page 345, col. 2, after line 49, add,

Pitch is properly the juice of the wild pine, or pitch-tree; and is conceived to be no other than the oil thereof inspissated, and turned black, farther than in the balsam.

The best is that brought from Sweden and Norway. Its goodness consists in its being of a glossy black colour, dry, and brittle.

PLANK-*Sheers*, or *Plan-Sheers*, in ship-building, are the pieces of plank wrought horizontally over the heads of the timbers of the fore-castle, quarter-deck, and round-house, for the purpose of covering the top of the side; hence sometimes called *covering-boards*.

POLE, or POLAR-STAR. Page 349, col. 2, after the last line, add,

Mr. Pond, the present astronomer royal, has published a table of the north polar distances of 44 of the principal fixed stars; which he conceives to be much more accurate than any that has hitherto been offered to astronomers. The *maximum* of error, he thinks, seldom exceeds half a second, and only in four cases amounts to one second; *e. gr.* the polar star in summer is 1° 41′ 22″ 07, and in winter, 1° 41′ 21″ 47 from the north pole of the heavens.

Page 350, col. 1, after line 44, add,

PONTOON, or PONTON, a kind of flat-bottomed boat, whose carcass of wood is lined within and without with tin. The French pontoons, and those of most other powers, are covered on the outside with strong copper plates, without any inside lining, and these have the advantage of our's, because copper is much stronger than tin, and is not damaged by rust. Our pontoons are 21 feet 6 inches long, 4 feet 9 inches broad, and 3 feet deep within.

General Congreve's wooden pontoons are 26 feet long at top, 23 at bottom, 2 feet 8 inches deep, and 2 feet 3 inches wide.

PONTOON-*Bridge*, is made of pontoons, slipped into the water, and placed about five or six feet asunder, each fastened with an anchor, when the river has a strong current, or to a strong rope that goes across the river, and runs through the rings of the pontoons. Each boat has an anchor, cable, baulks, and chesses.

The baulks are about 5 or 6 inches square, and 22 feet 8 inches long. The chesses are boards joined together by wooden bars, about 3 feet broad and 12 feet long. The baulks are laid across the pontoons at some distance from one another, and the chesses upon them joined close. When several of these are thus connected, they form one firm, uniform bridge, by means of which a train of artillery, cavalry, infantry, &c. may pass over a river, arm of the sea, &c. If pontoons, &c. are wanting, bridges are formed of empty casks or barrels that support baulks and planks.

PONTOON-*Carriage* is made with two wheels only, and two long side-pieces, whose fore-ends are supported by a limber: it serves to carry the pontoon, boards, cross-timbers, anchors, and every thing else necessary for making a bridge.

PORTS. Page 351, col. 2, after line 15, add,

The lower and upper edges of the ports are always parallel to the deck, so that the guns, when levelled in their carriages, are all equally high above the lower extremity of the ports, which is called the *port-sils*.

PROVOST-MARTIAL. Page 359, col. 2, after line 39, add,

The office of provost-martial is frequently held by the master at arms of the flag-ship at the port, and he generally receives his warrant to act as such on the day appointed for the court-martial to take place, which warrant is signed by the president of the court.

PURSER. Page 365, col. 2, after line 11, add,

By an order in council of the 26th of July, 1814, it was resolved that, "in order to prevent any improvident or unnecessary increase of the number of pursers, and at the same time to ensure the appointment of such persons only as may be duly qualified for the duties of the situation, no person shall be warranted as purser, until the number of pursers be reduced below the number of ships on the list of the royal navy (exclusive of such as are building and have not been launched), and that hereafter the number of pursers shall never exceed the number of such ships.

"And in order to give a greater respectability to the situation of this class of officers, his Royal Highness the Prince Regent has been graciously pleased to confer on pursers of the royal navy the same rank as granted by his Majesty's order in council of the 25th of January, 1805, to the surgeons of the fleet."

R.

RATE *of Sailing*. Page 386, col. 2, after line 2, add,

To shew the velocity of a ship under sail, it may be proper to state under this article the result of three voyages, recently made by his Majesty's frigates Medusa and Caroline, and brig Actéon.

The Actéon, Captain Cator, in the course of 18 months, ending the 29th of February, 1812, sailed from England to the Cape of Good Hope, twice; to the Isle of France, twice; and to Bengal, once; comprising in the whole distance, per log, 53,152 miles, which is nearly 100 per day.

The Caroline, Captain Cole, made a remarkable quick passage home from India, in 94 days, seven of which she was becalmed. The Medusa, Sir John Gore, after taking out the late Marquis Cornwallis, as governor, came home from India in 84 days (averaging 160 miles per day), which perhaps was the quickest voyage ever made.

RATING. Page 386, col. 2, after line 14, add,

The captain is to observe that none are to be rated as ordinary, who have not been twelve months at sea, and able to perform some of the duties of a seaman; nor to be rated as able who have not been three years at sea, and capable of performing most of the duties of a seaman; and that none are to be rated as master's mates or midshipmen who have not been three years at sea, either in the navy or merchant-service.

RED-HOT SHOT. Page 388, col. 1, after line 51, add,

They were used on shore with great effect against the Spanish floating batteries, which were bomb-proof, at the late siege of Gibraltar; and at sea by the French in the action with Lord Howe, on the 1st of June, 1794; and again by the combined fleets of France and Spain in the battle of Trafalgar, with the celebrated Lord Nelson, on the 21st of October, 1805.

REFRACTION. Page 394, col. 2, after line 21, add,

If the result or the co-latitude be not always the same, some of the corrections must be faulty, supposing the observations to be accurately performed. According to Mr. Groombridge, who is one of our most accurate astronomers, the correction of refraction, as deduced by Bradley's formula, is erroneous; not, however, exactly as that great man gave it, but altered according to the suggestion of the late Dr. Maskelyne, in its numerical co-efficient.

Mr. Groombridge states, that, by his observations of different circumpolar stars, and of the same star in different states of the thermometer, there seems to require a small alteration in the numbers of Dr. Bradley's formula for refraction.

'The formula so altered is,

$$\text{Refraction} = 56'', 9 \times \text{tang.} \{\text{Zen. dis.} - 3, 2 \text{ ref.}\} \times \frac{\textit{height of barometer}}{29,6} + \frac{500}{450 + \text{ther.}}$$

By means of this formula, the observations of circumpolar stars considerably distant give the same co-latitude to a great degree of exactness.'

Congreve's ROCKETS. Page 410, col. 1, after line 32 add,

By means of Colonel (now General) Congreve's rockets, as much may be accomplished in a few minutes by a few small vessels, as, in the ordinary mode of naval bombardment, would be the work of hours, with the same number of the largest bomb-vessels.

Thus it is possible to equip a gun-brig so as to discharge twenty 32-pounder rockets in a broadside, firing them as quick as may be desired in succession. Ten gun-brigs so equipped, therefore, would in the same space of a few minutes, discharge 200 such rockets, each equal, or rather, as has been proved, superior to a 10-inch spherical carcass; whereas, 10 ships would not throw the same number of spherical carcasses in less than two hours; during which time, the gun-brigs would be able to repeat their broadsides at least five or six times, throwing in that time not less than 1,000 or 1,200 rockets, which convey *nearly four tons* of combustible matter.

For a naval attack, therefore, of any strongly fortified town or harbour, where the object is the conflagration of the town or shipping, it surely is not presuming too much, to say, if the reasoning on the powers of the rocket is admitted, that there is no weapon at present known which possesses the same powers.

But there is even a stronger case which may be put, where, for instance, it is desirable, for the sake of security, and on account of any particular formidable defence on the part of the enemy, to perform any such service, *absolutely* by a *coup de main*, and in the shortest possible time—20 gun-brigs would discharge, *in one broadside*, 400 rockets, and this number would make an impression on any place; nor is there any thing difficult in supposing the employment of 20 gun-brigs in such an exploit; whereas, to accomplish the same thing by mortars in the same time, would require no less than 200 mortar vessels, a number totally impracticable on all accounts. For services of this description, General Congreve, therefore, does not hesitate to affirm, that were this weapon properly put into the hands of the navy, its importance would be extreme. In short, every vessel, and every boat in the navy, becomes capable of the powers of the bomb-vessel, as far as throwing carcasses is concerned, by having only a few boxes of rockets and a frame put on board.

For Naval Purposes. A launch fitted with a frame to discharge two rocket carcasses of the largest natures in a volley, either rowing or under sail, shew, that by means of the rocket, the boats of the navy may then be made to possess the power of the largest bomb-vessel as to the throwing of carcasses.

It is asserted that General Congreve's rockets are by no means subject to be easily damaged in the sea-service. Rockets which were kept on board for three years have been fired, and are not found to have lost in the least, either as to range or any other power; nor is any danger to be apprehended from them when kept in store, as it is impossible they can be ignited except by positive design to do so.

By the report of an experiment tried at Malta, by order of Sir Sidney Smith, it is asserted, "That one of these rockets ranging 2350 yards, struck a stone wall in falling, penetrated it, and broke several large stones in pieces, and that the carcass burnt well." Also, by the report of Baron Ehen, after the capitulation of Copenhagen, "that a house was struck by one of the rockets, which went through the roof and three of the floors, and stuck into the side of the wall.

At the extreme range, the rocket falls from a height not much less than 1000 yards; and the 32-pounder descends after the rocket composition is expended, with a weight not less than 24 lbs. See General CONGREVE's *Account of the Origin and Progress of the Rocket System.*

ROLLER. Page 412, col. 2, after line 32, add,

There has been an improvement made in the patent rollers, by which they are simplified and rendered less liable to disorder: in this the rollers do not work on an axis supported at each end, but on a central axis, supported in the middle by a single plate of brass, upon which it works, and which has round its edge semi-circular cavities to receive the rollers.

Page 417, col. 1, after line 35, add,

Clerk of the ROPE-YARD, an officer in his Majesty's dock-yards, whose duty seems to combine the offices of the clerk of the check and store-keeper: he superintends the making of ropes, and musters the men in that department, upon their coming to and quitting work; and is charged with the hemp, tar, and other articles, used in rope-making, of the expenditure of which he keeps an account.

S.

SCUDDING or BEARING AWAY in a *Storm.* When the waves run high, and sudden necessity requires to bear away, it should be considered that the lower sails forward, which the ship may be veered under when she comes before the wind, may be becalmed by the height of the waves breaking violently against the stern; and that therefore a close-reefed main top-sail should be set to catch the wind because it is a loftier sail, and may always be kept drawing full above the waves. This increases the ship's headway so much, that the waves will not strike her abaft with so great a velocity as when her headway is less. Hence it follows, that, when going to scud before high waves, the close-reefed main-topsail should be the last square-sail taken in, in a laboursome ship.

SCURVY. Page 437, col. 1, after line 57, add,

Upon long cruises or voyages, when there is not a sufficiency of lemon-juice for the whole of the ship's company, the surgeon, in his inspection of the men from time to time, is to endeavour to ascertain whether any of them have obscure symptoms of sea scurvy, too slight for them to withdraw from duty, and to be put on the sick-list; and also to enquire what men have subsisted longest on salted provisions, and making out a list of such men, he is to present it to the captain, in order that he may give directions for their being supplied, in preference, with the usual allowance of lemon-juice and sugar, put into the purser's charge for that purpose. See the NAVAL INSTRUCTIONS.

To SEARCH, a term used among seamen when they overhaul any vessel in quest of contraband goods; also, when they examine any foreign ship or vessel to discover and take away such British seamen as may be found therein.

When a captain meets with any foreign ship or vessel, he is immediately to send a lieutenant to enquire whether there be on board of her any seamen who are subjects of his Majesty, and if there be, he is to demand them, provided it does not distress the ship, and to require the master to pay them the wages due to them to that day: but he is to do this without detaining the vessel longer than shall be necessary, or offering any violence to, or in any way ill treating, the master or his crew. See the NAVAL INSTRUCTIONS.

SOLSTICE. Page 489, col. 1, after line 31, add,

Mr. Pond, our present astronomer royal, has presented us with the first series of observations, made with the new Mural Circle at Greenwich, on the summer and winter solstices of the years 1812 and 1813, in the Philosophical Transactions of the last year. Mr. Pond observes, "That the distance of the sun from the pole at the moment of the solstice, may either be considered a simple arc, or as composed of two others, namely, the distance of the sun from the zenith, and the distance of the zenith from the pole. But in whatever way we consider the subject, it is evident that the sole object of the practical astronomer is to obtain by some mechanical means the measure of this total arc."

The Mural Circle at Greenwich has neither level nor plumb-line, nor does Mr. Pond think that the least advantage could be derived from the application of either, as its principle enables the observer to determine this total arc without any intermediate or zenith point; hence the co-latitude, the knowledge of which is so essential from the construction of other instruments, but in this case it becomes rather a question of curiosity than of absolute necessity. Mr. Pond finds it convenient, in observing polar distances, to assume an intermediate point, which he calls the zenith point, without considering whether it is so or not; as he finds he has the means of determining the position of this *imaginary* point of departure on the circle, to within the tenth of a second, and from this point of departure he measures the distance of the sun to the southward, and of the pole to the northward, and the sum of these two measures is evidently the north polar distance of the sun, which in every method is the ultimate object to be attained. He has annexed as an example, a computation of the same solstice obtained by direct measurement from the pole without the aid of the intermediate point above mentioned at 38° 31′ 21″5, and from 120 observations of Polaris with it at 38° 31′ 21″15. The results do not in this case differ above a quarter of a second from each other; and there appears indeed no other difference between these two methods but that in the former case, the part of the arc Z P is obtained rather more accurately by a great number of observations, both before and after the solstice, than could be done in the short interval of time in which the solstice itself is observed.

STAFF. Page 497, col. 1, after line 65, add,

Proportions for the Lengths and Diameters of Staffs.

Length of the *Ensign-Staff*, one-third of the main-mast, above the taffrail.

Diameter of ditto, half an inch to every yard in the length.

Length of the *Jack-Staff*, one-half the length of the ensign-staff above the taffrail.

Diameter of ditto, three-fourths of an inch to every yard in the length.

The following Tables contain the lengths and diameters of all the Ensign and Jack-Staffs now used in the different classes of ships and vessels, both in the Royal Navy and Merchant-Service:

Dimensions of those used in the Royal Navy.

Names of the Staffs.	110 Guns. 2164 Tons.			90 Guns. 1870 Tons.			80 Guns. 1620 Tons.			74 Guns. 1800 Tons.			64 Guns. 1369 Tons.			50 Guns. 1044 Tons.			44 Guns. 879 Tons.			38 Guns. 951 Tons.		
	Length. Ft.	In.	Dia. In.	Length. Ft.	In.	Dia. In.	Length Ft.	In.	Dia. In.	Length. Ft.	In.	Dia. In.	Length. Ft.	In.	Dia. In.	Length. Ft.	In.	Dia. In.	Length. Ft.	In.	Dia. In.	Length. Ft.	In.	Dia. In.
Ensign-Staff.........	45	0	7½	40	0	6½	39	0	6½	40	0	6½	38	0	6¼	36	0	6	35	0	5⅝	35	0	5⅝
Jack-Staff...........	19	6	4¾	18	0	4½	17	6	4¼	18	0	4½	16	6	4¼	15	6	3¾	15	6	3¾	15	6	3¾

Names of the Staffs.	36 Guns. 871 Tons.			32 Guns. 677 Tons.			28 Guns. 594 Tons.			24 Guns. 513 Tons.			20 Guns. 429 Tons.			Sloops of 300 Tons.			Brigs of 200 Tons.		
	Length. Ft.	In.	Dia. In.	Length. Ft.	In.	Dia. In.	Length. Ft.	In.	Dia. In.	Length. Ft.	In.	Dia. In.	Length. Ft.	In.	Dia. In.	Length. Ft.	In.	Dia. In.	Length. Ft.	In.	Dia. In.
Ensign-Staff	35	0	5⅝	34	0	5½	30	0	5¼	28	0	5	28	0	5	24	0	4½	20	0	3½
Jack-Staff...........	15	6	3¾	15	0	3¾	14	0	3⅜	13	0	3¼	13	0	3¼	12	0	3	10	0	2½

Dimensions of those used in the Merchant-Service.

Names of the Staffs.	1300 Tons.			1200 Tons.			1100 Tons.			1000 Tons.			900 Tons.			800 Tons.			700 Tons.			600 Tons.		
	Length. Ft.	In.	Dia. In.	Length. Ft.	In.	Dia. In.	Length. Ft.	In.	Dia. In.	Length. Ft.	In.	Dia. In.	Length. Ft.	In.	Dia. In.	Length. Ft.	In.	Dia. In.	Length. Ft.	In.	Dia. In.	Length. Ft.	In.	Dia. In.
Ensign-Staff.........	40	0	7	38	0	7	36	0	6½	36	0	6½	34	0	6½	34	0	6	32	0	6½	30	0	6½
Jack-Staff...........	26	0	5½	24	0	5	24	0	5	22	0	5	22	0	4½	22	0	4½	21	6	4½	21	0	4½

Names of the Staffs.	500 Tons.			400 Tons.			360 to 350 Tons.			330 to 300 Tons.			280 to 250 Tons.			Brigs of 150 Tons.			Ketches of 170 Tons.		
	Length. Ft.	In.	Dia. In.	Length. Ft.	In.	Dia. In.	Length. Ft.	In.	Dia. In.	Length. Ft.	In.	Dia. In.	Length. Ft.	In.	Dia. In.	Length. Ft.	In.	Dia. In.	Length. Ft.	In.	Dia. In.
Ensign-Staff.........	30	0	6	28	0	6	28	0	6	26	0	5½	25	0	5½	23	0	4½	22	0	4
Jack-Staff...........	20	0	5	20	0	5	14	0	4	13	0	4	10	0	3	9	0	3	8	0	3

T.

TABLES. Page 521, col. 1, after line 42, add,

Of all the tables of the principal planets, those of Jupiter and Saturn are the most difficult to establish with precision; the accuracy of them depending on the exact determination of the elliptic elements, which themselves depend on the precise knowledge of the masses of the planets: whence it is obvious that we can only arrive at the true results by determining at the same time, in a general manner, all the elliptic elements, and the masses on which the perturbations of the planets depend. The tables of Jupiter and Saturn, published by Halley, were for a long time the only tables employed by astronomers: but, though they were more exact than any which had been previously furnished, yet in 1760 they were erroneous to the amount of 11 minutes for Jupiter, and about 22 for Saturn. This circumstance induced Lalande to construct a new series: but even that celebrated astronomer, notwithstanding all the advantages which he derived from the labours and observations of his predecessors, succeeded little better than Halley had done so many years before. The fact is, that it was impossible to form correct tables of these planets, before the recent perfection in the theory of attractions, which was alone sufficient for explaining the true cause of the great inequalities experienced in the observations of them. Laplace was the first (see *Memoires de l'Académie des Sciences*, 1784) who gave the complete theory of the perturbations of these two planets. Delambre afterwards constructed his tables, in which the error seldom exceeded half a minute; and even this might have been rendered still less, had the author been in possession of a greater number of modern observations. Some uncertainty remained likewise with regard to the mass of Saturn, and consequently respecting the inequalities of Jupiter. M. Bouvard, having availed himself of several excellent observations made since 1789, when Delambre's tables were published, and having acquired also a more accurate idea of the mass of Saturn, has formed a new set of tables, which are extremely accurate; the greatest error being only 13″, of which a part may doubtless be attributed to defects of observation. They are calculated according to the new division of the day and the circumference of the circle.

TATOOING, a cant term used by seamen to denote drawing their names, and other devices, on their arms, or different parts of their bodies. The operation consists in first sketching out with Indian ink their names, &c. where they chuse to be marked; then with needles, or some pointed instrument, they prick the skin through the lines; this being done, they rub into the holes some fine-grained gunpowder, which can never be extracted.

This practice, borrowed from the Indians is generally encouraged, and has frequently led to the discovery of British seamen and soldiers who have deserted, and who, much to their disgrace, have sometimes been found in the service of hostile nations,

V.

VOYAGES *round the World.* Page 609, col. 2, after line 30, add,

After an elapse of three years, a very general solicitude was directed towards the uncertain fate of the two frigates La Boussole and L'Astrolabe, and it must ever remain as an honourable trait in the character of the National Assembly of France in 1791, that, in the heat and ferment of a great change then operating on men's minds, and amidst concerns of the highest importance, they found time and inclination to pay so much deference to the public feeling and private distress, as to adopt the only means which could afford the least chance of gratifying the one, and relieving the other. Accordingly, two frigates, La Recherche and L'Espérance, were ordered to be equipped at Brest, and Vice-Admiral Dentrecasteaux, from his distinguished talents, prudence, and long experience, was appointed to the command of an expedition, which, to use the words of the unfortunate Louis, " présentoit une occasion de perfectionner la description du globe, et d'accroître les connoissances humaines."

Vice-Admiral Dentrecasteaux sailed from Brest with these frigates on the 29th of September, 1791, and on the 13th of the following month, anchored in the bay of Santa Cruz, in the island of Teneriffe.

On the 17th of January, 1792, the expedition arrived at the Cape of Good Hope, where the Vice-Admiral received a dispatch that had been transmitted to him from the Isle of France, by M. de Saint-Felix, commander of the French squadron in the East Indies, to inform him that there had been some of the natives of the Admiralty Islands seen with articles of dress, &c. supposed to belong to the late expedition of La Pérouse.

On the 28th of March they saw the Island of St. Paul; and in April, arrived at the Bay of Storms, on the south-east point of Van Diemen's Land, on the north side of which the frigates anchored in a harbour called by them *Port du Nord*, and which is described as being safe and commodious, the water smooth in the worst of weather, and the surrounding shores beautifully covered with forest trees. While the frigates remained in this bay, a discovery was made of a passage between Van Diemen's Land and the island to the southward of it; which, though of little importance to the general interest of navigation, since the discovery of Bass's Straits, furnishes a secure and commodious anchorage for ships in need of such a retreat. They gave it the name of Canal de Dentrecasteaux.

Having been nearly a month at this place, they set sail, and made the Island of Pines on the southern extremity of New Caledonia: from this to the northern extremity they examined with minute attention every point along the western coast, standing in with both frigates close to the reef of coral rock, which seemed to bind the whole extent of the island, as it were, with an uninterrupted chain, whose edge rising to the surface of the sea, shews, even in the calmest weather, a silvery rippling line, as far as the eye can reach; thus warning the adventurous navigator of the dangers that surround him. The unbroken reef of New Caledonia would not allow even their boats to pass to the shore, and as they saw neither natives nor canoes, it was presumed that it might equally prevent their coming out, and thus contribute to that mildness of character, which Captain Cook had described the natives on the opposite side to possess. The Vice-Admiral, however, adds, that being in latitude 22° 4′ S., they did see an opening in the reef, but that on approaching close to it, they observed the sea breaking with violence within a creek, to which on that account they gave it the name of *Le Havre Trompeur.* This was rather an unfortunate appellation, for had they examined it with strict attention, they would have found the *Havre Trompeur* to be one of the most beautiful and extensive harbours perhaps in the world. for the discovery of which we are indebted to Captain Kent of the Buffalo, who in 1803 anchored there for some time. To this harbour Captain Kent has given the name of Port St. Vincent, and from the mean of several observations, has laid it down as follows:—Latitude 22° 0′ 10″ S. longitude 160° 55′ 20″ E. He describes the passage into it as three quarters of a mile in width, the anchorage well sheltered, the water perfectly smooth, the depth 12 fathoms down to 4, and the bottom muddy. Leaving New Caledonia and the line of breakers which extend fifty leagues to the N. W. far out of sight of any land, the Vice-Admiral passed, on the 9th of July, the Eddystone Rock, which was discovered by Lieutenant Shortland. The following day they fell in with the Treasury Islands, and passed close to Bougainville's Island on the north-western extremity of the group of Solomon's Islands, all of them surrounded by reefs of coral rock.

From the 17th to the 24th of July, the frigates remained at anchor in Carteret Bay, New Ireland, where incessant torrents of rain prevented the navigators from making a single observation, either on the sun or stars. Keeping the track of Captain Carteret, through St. George's Channel, on the 29th of July they reached Vendola, one of the Admiralty Islands, which was supposed to be near to the spot where the depositions in M. de Saint-Felix's dispatches stated the traces of the ships of La Pérouse to have been seen. Observing some natives, a boat was sent on shore, and a friendly intercourse held with the natives for some days.

Having satisfied themselves that not the least trace of the ships or crew of La Pérouse was discoverable among the Admiralty Islands, they stood away for the Negro Islands of Morelle, which they observed to be covered with cocoa-nut trees. Numbers of canoes appeared on every side among the reefs and small islands, all seemingly employed in fishing. Some of them came off to the frigates, and were anxious to procure knives, nails, and iron of any kind, for which they offered in exchange different sorts of ornaments, arms, and shells. They appeared to be, in all respects, the same happy, cheerful, and contented people as those of the Admiralty Islands.

Passing *Los Ermitanos*, the *Thousand Isles* of Morelle, and numberless rocks and islands which appeared to be bound together by reefs of coral; they arrived on the 17th of August, on the coast of New Guinea. In sailing through Dampier's Straits, they observed a large proa bearing a flag; but the appearance of the two frigates excited no curiosity in the people on board, nor in those of the numerous craft that were seen passing and repassing on every side; a certain proof of their near approach to some European settlement. On the 5th of September they accordingly arrived at Amboyna. Having remained some time at this place to refit the frigates and refresh the crews, M. Dentrecasteaux, again set sail; and keeping along the west coast of New Holland, and doubling Cape Leween, or S. W. Cape, anchored on the 9th of December, in a bay, to which he gave the name of Espérance, after one of the frigates. This bay is within a cluster of islands to which Vancouver two years before had given the name of Termination Islands. Standing off at this point to the southward, Dentrecasteaux continued along the coast about nine degrees farther to the eastward than Vancouver had done. Had he stood on a little farther, he would have effected the important discovery of Bass's Strait, which separates New Holland from Van Diemen's Land; the merit of which was reserved for a young man, surgeon of the Reliance, whose name it properly bears, and who in 1798, solicited a whale-boat of Governor Hunter at Port Jackson, to run down the coast and make observations. This discovery was made near six years after Dentrecasteaux conceived there might be such a passage, but had not the resolution to stand on and determine so important an object for the benefit of navigation in general.

In January, 1793, they again visited the Bay of Storms, and anchored in Port du Sud: they had now much intercourse with the natives, whom they found to be a mild, inoffensive, and good humoured people, without suspicion or distrust.

The frigates again came to anchor in Adventure Bay. Here they searched for the fruit-trees which had been planted by Captain Cook, and discovered some of them still in existence. They now steered a direct course for the Friendly Islands, where they arrived on the 23d of March, and brought the frigates to anchor at Tongataboo. The inhabitants of this island had a perfect recollection of the visits of Captain Cook, whose memory was held dear by many of them, especially by the family of Fatafé. Not the least vestige could be traced of any thing that had belonged to La Pérouse; no medals which he carried out with him, and of which they had exact copies; no trinkets, no French clothes, nor any thing belonging to their nation. They proceeded therefore to the north-westward, and on the 17th of April, came to anchor in the harbour of Balade, on the north-east extremity of New Caledonia, having passed the islands of Tanna, Annatom, and Erronun, three of the New Hebrides, without discovering any trace of the objects of their search. They took their leave of this 'effroyable peuple,' who had even forgotten Captain Cook, and gave no indications that Europeans had ever visited their coast, except by their avidity for iron; and on the 21st of May, the frigates came to anchor off the Isle of Santa Cruz: here they quarrelled with the natives, and shot one of them in an affray. On passing Solomon's Islands, their frigates were attacked by a number of light, well-constructed, and gaily-decorated proas. The men in them were naked, but ornamented with plumes, bracelets, and necklaces of shells and mother-of-pearl. From hence, stretching along the northern coast of Louisiade, and the south-eastern side of New Guinea, they passed the Straits of Dampier, along the north coast of New Britain.

M. Dentrecasteaux's health had for some time been on the decline, and the scurvy had found its way into the ship. M. Huon, the commander of the Espérance, died before they left New Caledonia; their provisions were now becoming short; their wine turning sour; and every thing considered, it was deemed advisable to make the best of their way to the Island of Java. On the 8th of July, they made the coast of New Britain, and on the 20th of the same month M. Dentrecasteaux departed this life. The command now devolved upon M. D'Auribeau, who determined to execute the intentions of the Vice-Admiral, and make the best of his way to Java. Passing the Admiralty Islands a second time, and coasting New Guinea, they anchored for a few days at Boni, in the Island of Waigiou. Hence they stood for Cayeli, in the Island of Bouro, where they continued several days; but this being a Dutch settlement, and inhabited by Portuguese, Chinese, and Malays, offered nothing worthy notice. Directing their course through the Strait of Bouton, they arrived at Sourabaya, in the Island of Java, on the 27th of October, 1793, where they had the mortification of learning the unfortunate situation of their country, the sanguinary proceedings of the revolutionists, the massacre of the king, and the war with England, in which the Dutch were likewise involved. In consequence of this intelligence, the voyage was here declared to be at an end; the frigates were dismantled, and the officers and crews separated. Voyez *Voyage de* DENTRECASTEAUX, *envoyé à la Recherche De la Pérouse:* redigé par M. de Rossel. 2 tom. avec un Atlas, à Paris, 1808.

In the year 1787, his Majesty's arm'd ship Bounty, under the command of Lieutenant William Bligh, sailed from England on a voyage to Otaheite. The object government had in view from this voyage was, to introduce the bread-fruit and other plants into the British colonies in the West Indies from Otaheite. The object of the voyage, so far as related to the collecting and getting a sufficient quantity of the said plants on board, being completed, the Bounty, after being on the most friendly terms with the natives of Otaheite, departed from that island with the fairest prospect of success; but in the prosecution of the voyage towards the West Indies, and in sight of the small high island of Toofoa, the westernmost of the Friendly Islands, a large portion of the crew, headed by a Mr. Christian, one of the master's mates, then acting as lieutenant, mutinied, and after gaining complete possession of the ship, without bloodshed, forced her commander and some others into the ship's launch, and put her adrift. Lieutenant Bligh and his companions, being in distress from the very scanty supply of provisions and water supplied them by the mutineers, shaped a course for the Island of Toofoa, where they arrived, and, with a hope of obtaining a supply of water and provisions, landed upon the island; but in this deplorable situation they were attacked by the natives and nearly overcome: they, however, made good their retreat to the launch, and proceeded to sea with a most dreadful prospect, steering towards the coast of New Holland, where they arrived after a passage of six weeks, having in the course of that time from hunger and thirst, wet and cold by night, and a hot sun during the day, in an open boat only 22 feet long, endured the most unparalleled hardships: they passed through Endeavour Strait, which separates New Guinea from New Holland, and arrived, nearly exhausted with fatigue and hunger, at the principal Dutch settlement upon the island of Timor: here they experienced the most hospitable treatment, and after being in some measure recruited, proceeded on board a small vessel to Batavia; from thence they sailed in a Dutch vessel to Europe, and finally arrived in England; having from the very untoward circumstance of the mutiny totally failed in that part of the object of the voyage, the introduction of the bread-fruit plants into the West Indian colonies. They, however, in their passage to the coast of New Holland discovered a fine cluster of islands, to which, in honour of their commander they gave the name of Bligh.

His Majesty's ship Pandora, under the command of Captain Edwards, was dispatched from England, in the year 1790, to the South Seas, particularly to the island of Otaheite, in quest of the seamen who had mutinied, as before-mentioned, in the armed ship Bounty. On the Pandora's arrival at Otaheite, and on their landing on that island, they fortunately found several of those concerned in the mutiny, who in consequence of a quarrel among themselves, had separated from the rest, deserted from the ship, and chosen to remain upon the island; they were taken on board the frigate, and secured.

The Pandora sailed from Otaheite, and in attempting the passage between New Holland and New Guinea, she was wrecked upon one of those dangerous reefs with which that passage abounds: some lives were lost, but the captain, the officers, and the remainder of the crew, with the mutineers, arrived in the boats that were fortunately saved at the principal settlement at Timor, and finally arrived in England in the year 1793, when several of the surviving mutineers, who were found to have been ringleaders, were executed at Portsmouth, agreeable to the sentence of a court-martial.

The following relation was stated to have been transmitted officially to the Admiralty from Rio de Janeiro, in 1808, by Sir Sidney Smith:—"Captain Folgar, of the American ship Topaz, of Boston, relates, that, upon landing on Pitcairn's Island, in latitude 25° 2′ S. longitude 130° 0′ W he there found an Englishman of the name of Alexander Smith, the only person remaining of nine that escaped in his Majesty's late armed ship Bounty, Captain William Bligh.

"Smith relates, that after putting Captain Bligh in the boat, Christian the leader of the mutiny, took command of the ship and went to Otaheite, where great part of the crew left her, except Christian, Smith, and seven others, who each took wives, and six Otaheitean men servants, and shortly after arrived at the said island, where *they ran the ship on shore*, and *broke her up*, in the year 1790.

"Smith gave to Captain Folgar a chronometer made by Kendall, which was taken from him by the governor of Juan Fernandez."

The two circumstances stated in the above narrative, and on which the story must stand or fall, are the name of the mutineer, and the maker of the time-piece; both of which have been ascertained to be facts from what follows. Alexander Smith appears on the books of the Bounty thus: 'Entered 7th Sept. 1787. Ab.—Born in London.—Aged 20. Run 28th April, 1789. One of the mutineers:' and it appears also that the Bounty was actually supplied with a time-piece made by Kendall.

In the year 1791, his Majesty's ship Providence, commanded by Captain William Bligh, and his Majesty's armed brig Assistant, commanded by Lieutenant Nathaniel Portlock, sailed from England for the South Seas; the object of the voyage, as well as discovery, was to make a second attempt towards the introduction of that valuable bread kind the bread-fruit into our West Indian colonies from Otaheite. In the prosecution of the voyage, they visited the Cape de Verd islands, where they replenished their stock of water and fresh provisions.

They next visited the Cape of Good Hope and Van Diemen's Land, and arrived at Otaheite, where they found a part of the crew of the merchant ship Matilda, who had been employed upon a whaling expedition. This ship had been wrecked upon a small island to the south-east of Otaheite, to which place the men had made their passage, after experiencing much hardship in the boats belonging to their ship that were fortunately saved from the wreck, and were very kindly received and treated by the natives; they were taken on board the Providence and Assistant for a passage to England: the residue of the crew, after arriving at Otaheite, and previous to the arrival of the Providence and Assistant, had sailed from that island under the direction of a Mr. Campbell, one of the mates of the Matilda, and an officer upon the half-pay list of his Majesty's marine forces, in one of the ship's whale-boats to the British settlement at New South Wales, as well to give information respecting the loss of the ship, as to apply for the means of removing their shipmates from the island; but there is sufficient reason for believing that, in the execution of this hazardous project, the boat and every one on board of her were lost, as they have not been heard of to this day. The king's ships, the Providence and Assistant, were received at Otaheite in the most friendly manner, the crews supplied with refreshments, and a large collection of the bread-fruit and other plants were received on board. Previous to their departure, they witnessed the barbarous ceremony of a human sacrifice being offered to the Otaheitean deities. Sailing from thence, they passed through a number of isles, particularly a cluster discovered by Lieutenant (now Captain) Bligh, when in the launch, on his passage from Toofoa to Timor, in a before-mentioned voyage.

They proceeded to the dangerous reefs in the Straits between New Holland and New Guinea, when the Assistant, in the act of leading a-head of the Providence, in this intricate navigation, was suddenly attacked by the natives in their canoes with bows and arrows, from a small island bordering upon the straits; a short but desperate conflict ensued, in which the canoes were beaten off with a serious loss on their part; and three, out of a complement of twenty-seven, being severely wounded on board the Assistant, one of whom, with a poisoned arrow, was wounded, and died soon afterwards under the most excruciating agonies. After much peril, they made their passage through the straits, and arrived at Timor, where they received from the governor and inhabitants of the principal Dutch settlement on that island, the most hospitable treatment.

Departing hence, they visited the Island of St. Helena, where, in the charge of Governor Brook, they deposited a portion of the bread-fruit and other plants. They then proceeded to the West Indies, visited the Island of St. Vincent, where, in his Majesty's botanical garden, and in the charge of the king's botanist, Mr. Anderson, they deposited a large portion of the bread-fruit and other plants. Proceeding thence to Jamaica, they there landed the remainder, which were distributed to the principal estates in the island. They sailed from Jamaica, and, after having in the most complete manner accomplished the object of the voyage, arrived in the River Thames in August, 1793, after being absent from England two years.

Captain George Vancouver and Lieutenant William Robert Broughton, the former commanding the Discovery sloop of war, and the latter the armed tender Chatham, sailed on the 1st of April, 1791, from Falmouth, upon a voyage of discovery, which was undertaken by his Majesty's command, principally with a view to ascertain the existence of any navigable communication between the North Pacific and North Atlantic Oceans; and also carefully to examine and survey the north-west coast of America.

After having made some judicious remarks on crossing the equator, Captain Vancouver, in compliance with the instructions he had received, proceeded to make an accurate survey of the coast of New Holland, comprehending an extent of 110 leagues, which gives us a more correct idea of it than we had before seen, and entirely overthrows an opinion of Dampier, who considered the whole of the western part of that country as consisting of a cluster of islands.

Having accomplished this object they proceeded to the northward, and soon after mid-day, on the 17th of April, 1792, they discovered the coast of New Albion, on which a surf broke with great violence. The shore appeared straight and unbroken, of a moderate height, with mountainous land behind, chiefly covered with stately forest trees. Having proceeded along the coast to the northward, ranging as near it as was possible, they continued to explore its various windings, until they arrived at those regions, where the imagination of closet philosophers, since the days of John de Fuca, the Greek pilot, in 1592, has expatiated with a luxuriancy worthy of the relations of the Arabian Sinbad.

Captain Vancouver, on leaving the solitary regions he had so minutely explored, by his own perseverance, assisted by the unwearied diligence of Messrs. Broughton, Mudge, Puget, Whidbey, Johnstone, and other officers, thus concludes his survey of the north-west continental shore of America, in 1792:—

"Having the greatest reason to be satisfied with the result of our summer's employment, as it had, by the concurrence of the most fortunate circumstances, enabled us finally to trace and determine the western continental shore of North America, with all its various turnings, windings, numerous arms, inlets, creeks, bays, &c. from the latitude of 39° 5′, longitude 236° 36′, to Point Menzies, in latitude 52° 18′, longitude 232° 55′; we took our leave of these northern solitary regions, whose broken appearance presented a prospect of abundant employment for the ensuing season, and directed our route through Calvert's Islands, in order to make the best of our way towards Nootka."

On Captain Vancouver's arrival at Nootka, he sent home Lieutenant Broughton, with dispatches of what had been done since their departure from England. Captain Vancouver again visited the two Spanish settlements in New Albion, and proceeded to examine Columbia river; thence steered for the Sandwich Islands, where he anxiously endeavoured to promote a humane and friendly intercourse between the monarchs of Owhyhee and Mowee, in order that the natives might live in peace with each other, return to their habitations, and to their former employments of cultivating the land, and other peaceful arts. This he did with much success during the winter months he remained there from his survey northward.

On the return of spring 1793, Captain Vancouver again proceeded to the north; touched at Nootka, and on the 26th of May arrived at Fitzhugh's Sound; and recommenced the survey of the preceding year with great diligence and incredible fatigue. In a channel only a mile in width, they traversed repeatedly from shore to shore without finding bottom with 185 fathoms of line, though within half a cable's length of the rocks. The survey of the branches of various inlets continued to be performed in boats detached in different directions. Captain Vancouver at one time was absent on this perilous service twenty-three days; during which, from their outset to their return, he traversed in an open boat above 700 geographical miles; but without having advanced his primary object, of tracing the continental boundary more than twenty leagues from the station of the vessels. The limits of this survey of the American coast appear to be from Fitzhugh's Sound to Cape Decision, and from Monterry to the southern extent of their intended investigation, which was finished on the 14th of December in the same year.

Captain Vancouver regrets, that he had not with him one or two vessels of 30 or 40 tons burden, calculated as well for rowing as for sailing, to assist him in the unremitting investigation he was obliged to pursue; as by these means much dispatch would have been given to the survey, and their labours would have been carried on with much less danger and hardship than they now constantly endured.

On the 22d of December, at a considerable distance from the tracks of former navigators, they discovered an island, named Oparo, in the latitude of 27° 36′, and, by their lunar observations of the two preceding days, reduced to its centre by the chronometer, in longitude 215° 58′ 28″, the mean of the variation was 5° 40′ eastwardly. Its principal character was a cluster of high craggy mountains, forming in several places most romantic pinnacles, with perpendicular cliffs, nearly from their summits to the sea.

Our navigators again visited the Sandwich Islands for the winter months, and at the Island of Owhyhee met with that hospitality and liberal treatment which would do honour to more polished nations; they saw the character of those islanders in a more favourable light than had generally been conceived since the death of Captain Cook.

They next proceeded up what had been termed Cook's River, and discovered its final termination, when Captain Vancouver found it to be only an inlet, the northern extent of which was ascertained to be in latitude 61° 29′, longitude 211° 17′: he adds, "Thus terminated this very extensive opening on the coast of North America; to which, had the great discoverer of it, whose name it bears, dedicated one day more to its further examination, he would have spared the theoretical navigators who have followed him in their closets, the task of ingeniously ascribing to this arm of the ocean a channel, through which a north-west passage, existing according to their doctrines, might ultimately be discovered."

Thence they sailed to Prince William's Sound, which they surveyed in boats. Mr. Whidbey. during his last boat excursions, particularly noticed the rapid encroachments of the sea upon the land. Captain Vancouver also noticed this phenomenon, and made some remarks thereon.

On quitting Prince William's Sound, in June 1794, Captain Vancouver makes some interesting remarks relative to some errors that had crept into Captain Cook's last voyage. The reader will be struck with that passage where he speaks with so much feeling of Captain Cook's not living to superintend the last publication of his labours; as Vancouver himself died before the third volume of his own voyage was printed off.

Captain Vancouver now proceeded to Cross Sound, where he anchored: thence sailed to the southward along the exterior coast of King George the Third's Archipelago, and completed the survey of the continental shores of North-west America, on the 22d of August, 1794.

After visiting St. Jago, the capital of Chili, they doubled Cape Horn, and searched in vain for Isla Grande, which had been mentioned by former navigators, and laid down in some charts as being in the South Atlantic Ocean, nearly parallel with Cape Blanco.

Captain Vancouver then proceeded to St. Helena; and, on his return to England, joined the convoy of Dutch prizes under the protection of the Sceptre, Captain Essington, who had sailed from St. Helena on the very day the Discovery and Chatham arrived there. They anchored in the Shannon in perfect safety on the 13th of September, 1795; and in the Thames on the 20th of October following, after an absence of 4 years and 6 months.

The late indefatigable Captain Cook, had already shewn that a southern continent did not exist, and had ascertained the important fact of the near approximation of the northern shores of Asia to those of America. To those great discoveries the exertions of Captain Vancouver have added a complete certainty, that within the limits of his researches on the continental shore of North-west America, no internal sea, or other navigable communication whatever exists, uniting the Pacific and Atlantic Oceans.

Captain James Colnett was nominated by the Board of Admiralty at the close of the year 1792, to undertake a voyage to the South Atlantic, and round Cape Horn into the Pacific Ocean, in consequence of a memorial from the merchants in the city of London, concerned in the South Sea Fisheries, for the purpose of extending the Spermaceti Whale Fisheries and other objects of commerce, by ascertaining the ports, bays, harbours, and anchoring places in certain islands and coasts in those seas, at which the ships of the British merchants might be refitted.

Captain Colnett had sailed with Captain Cook, as a midshipman on board the Resolution; and afterwards, while a lieutenant, had been engaged in various commercial undertakings on the north-west coast of America, where he had examined that coast from 36° to 60° north, surveyed several bays, harbours, and creeks, and discovered many inlets, particularly between 50° and 53° north. He had also made two voyages to China, and was deemed a very proper person by all parties for this undertaking.

Captain Colnett left Yarmouth Roads in the Rattler sloop of 374 tons burden, with a crew of only 17 officers and seamen, 3 landmen, and 5 boys, to work a ship, which in his Majesty's service had a complement of 130 men. Amid the snow and blowing weather he had to contend with at that season of the year, on leaving the Chops of the Channel, the Captain observes, the *Marine Barometer* was of the greatest service, as it warned him against making sail when there was an appearance only of moderate weather, and to shorten sail on the approach of foul tempestuous weather. They touched at Rio Janeiro on their way out, and at day light on the 13th of April, they saw the Isle of Diego Ramieres, bearing N. by E. three or four leagues. Captain Colnett makes them by observations corrected, in longitude 68° 58′ west, and in latitude 56° 30′ south; appearing to lie in an east and west direction; and makes the following remarks on the navigation round Cape Horn:—"That the beginning of winter, or even winter itself, with moon-light nights, is the best season; for then the winds begin to vary to the eastward, as has often been observed at the Falkland Isles." He recommends commanders of vessels to keep near the coast of Staten Island, and Terra del Fuego, because the winds are more variable in with the shore, than at a long offing. The common course is by keeping between Falkland Isles and the Main, and through the Straits Le Maire, which not only lengthens their distance, but subjects them to a heavy and irregular sea, occasioned by the current and tides in that channel, which may be avoided by passing to the eastward.

He also observes, that there is no necessity for going to 60° south, as most navigators do, only to 57° 30′, so as to give the Isle of Diego Ramieres a good birth, or if winds and weather would permit, make it, for a fresh departure, especially if one has not been taken at Cape St. John, Staten Land, or the east end of Falkland Isles. Staten Land is well situated as a place of rendezvous, both for men of war and merchant-ships, and contains plenty of wood and water; while the harbours on the north and south sides, which are divided by a small neck, would answer the purpose of ships bound out, or home. But the north side offers the best place for any establishment for carrying on the black whale fisheries towards the south pole; and it is one of the easiest land-falls a sailor can make.

The buccaniers of America have asserted, and Lord Anson seems to confirm their opinion, that there was no reason to apprehend danger on the coast of Mexico, from the middle of October till May; but Captain Colnett thinks the middle of January too early to expect good weather, as he found to the southward of Cape Corientes, and to the northward of Cape St. Lucas, that the thunder, lightning, and heavy rains, had not subsided at the beginning of November; and that the Spaniards themselves never leave the port of St. Blas for Acapulco till the latter end of that month, when the north winds set in, and blow steadily.

In the beginning of February, 1794, the Rattler came to anchor in the Gulf of Panama, and moored in the bay of Port de Dames in 19 fathoms, before the island of Quibo, latitude by observation 7° 27′ north, and longitude 82° 10′ west, and remained here till the 17th. The Captain represents Quibo as the most commodious place for cruisers, of any he had seen in these seas, as all parts of it furnish plenty of provisions, wood, and water, and a vessel may lie so near the shore as to take off her water; but the time of anchoring must be considered, as the flats run off a long way, and it is possible to be deceived in the distance. The time of high-water at full and change, by his observations, was at half-past three o'clock. The flood comes from the north, and returns the same way, flowing seven hours and ebbing five; and the perpendicular rise of the tide is about two fathoms. Captain Colnett thinks it would not be advisable for armed vessels to anchor far in, as the wind throughout the day blows fresh from the eastward, and right on shore; so that an enemy would have a very great advantage over ships in such a situation: he also states, that there is good anchorage throughout the bay, at five or six miles distance, in 33 and 35 fathoms, with a mud bottom and firm holding ground.

Captain Colnett quitted Quibo on the 18th, and cruised between the Isle of Quicara and Cape Mariatto, till the last day of February; during this time the winds were light, and mostly southerly in the day, and sometimes a stiff breeze from north-east in the night. As the sun drew near the equator, and long calms were expected, it became necessary to steer for the Gallipagos Isles, to procure salt, for the purpose of salting seal-skins, at the islands of St. Felix and St. Ambrose, in latitude 26° 15′ south, having failed in the whaling business. Accordingly, on the 1st of March, he determined to steer to the southward, in a direct line for the isles. As they sailed up the coast of Chili and Peru from latitude 38° south, during April, May, and June, they had no occasion to take in a reef from the strength of the wind, while the barometer stood mostly at 29.9, and the thermometer at 60, rising gradually till in latitude 1° 30′ south, until it reached 72, but in the evenings it was generally below summer heat in England. Captain Colnett tried the currents on this coast several times, and found them very irregular, sometimes setting one way, and sometimes another, generally from half a mile, to two miles an hour. Soundings were also tried for in many places, at the distance of five or six leagues from the shore, but they could not obtain any bottom with 150 fathoms of line. During the whole of their passage down the coast of Chili in the month of July, they had south-east and easterly winds, with variable, but in general pleasant weather, accompanied with occasional showers. In latitude 33° south, the wind came to the southward, then veered to the west, and continued mostly between that and north, till they got to 47° south: here it would blow for a few hours between the west and south-west, but never continued. In latitude 48° and 49°, the winds were light for 48 hours, in the south-east quarter, with a strong southerly current. On the 1st of August, they doubled Cape Horn, at the distance of fifteen leagues, amid frequent showers of rain, snow, and hail. At

10 o'clock in the morning of the 2d of September, they anchored in James Bay, St. Helena, and sailed thence on the 13th. On the 11th of October, the head of the mizen-mast was carried away, and on the 15th, the head of the main-mast sprung in a squall.

When they approached the Western islands, they housed their boats, knocked down their try-work, and fresh painted the ship, to make her appear like a man of war. They made the Eddystone on the 1st of November, and reached Portland in the course of the night, stood off and on till day-light, when they ran up and anchored in Cowes Road, at the S. W. end of the Isle of Wight.

This voyage occupied 22 months, and forms a most valuable addition to the labours of Cook and Vancouver; and is in fact, in many respects, connected with, and explanatory of the voyage of the latter. Excepting the loss of one man by accident, the whole of the crew, consisting of 25 men and boys, were preserved during this fatiguing and perilous voyage.

Captain William Robert Broughton, in his Majesty's sloop Providence, with her tender, sailed on a voyage of discovery in February, 1795, to the North Pacific Ocean: in order to examine and survey such parts of the north-west coast of America, as had not been completed by Captain Vancouver, and also the opposite coast of Asia, from the latitude of 35° north, to the latitude of 52° north, the island of Insu, (commonly known by the name of the land of Jesso), the North, South, and East Coasts of Japan, the Lieuchieux, and the adjacent isles, as well as the Coast of Corea.

It will be recollected that Captain King observes, in the third volume of Cook's last voyage, that the navigation of the sea between Japan and China afforded a large field for discovery; and the Honourable Daines Barrington also says, in the Preface to his Miscellanies, that, "the coast of Corea, the northern part of Japan, and the Lieuchieux Islands, should be explored." Captain Vancouver has also remarked in his 3d vol. "that the Asiatic coast, from about the latitude of 35° to 52° north, is at present very ill defined; and the American coast, from about the latitude of 44° south, to the southern extremity of Terra del Fuego, is very little known." These suggestions alone would appear sufficient for undertaking the present voyage; and it has been previously stated in Vancouver's voyage, that Captain (then Lieutenant) Broughton sailed with Captain Vancouver, in his voyage of Discovery, as commander of the Chatham tender, in 1791; and that he was sent home with the dispatches from Nootka.

Captain Broughton sailed from England on his mission in February, 1795. His orders were secret, with an additional one to put himself under the command of Captain Drury, of his Majesty's ship Trusty, and to proceed to sea with his convoy, then bound for the Mediterranean. Captain Broughton proceeded by way of Teneriffe and Rio Janeiro; on the 10th of June made Gough's Island; and, on the 18th of August, anchored in Port Stephens, where our voyagers remained a week. On the 27th of August, the Providence reached Port Jackson, where she remained till the 13th of October. Thence Captain Broughton directed his course to the north of New Zealand, touched at Otaheite; and on the 1st of January, 1796, reached the Sandwich Islands, where he learnt from an American vessel, that Captain Vancouver, with the Discovery and Chatham, had sailed for England. At Owhyhee, Captain Broughton obtained ample supplies from Ta-maah-maah; and had the satisfaction of seeing that the cattle which had been left on the island by Captain Vancouver had bred, and were in excellent order: the goats had multiplied prodigiously. Captain Broughton added a male and female to their number, also left some geese, ducks, and pigeons; and is of opinion that any vessel may now touch at Owhyhee in safety, and be amply supplied with refreshments.

On the 22d of February, the Providence sailed for Nootka Sound; and on the 25th, Captain Broughton says, "We altered our course to the west, intending to search for an island called Donna Maria Lajara, said to be discovered by the Spaniards in the ship Hercules, in 1781; and laid down in Arrowsmith's charts, from the authority of Mr. Dalrymple. The centre of it is situated in 28° 30′ north, and in longitude 202° 30′ east. By the chart it is of considerable extent, in a north and south direction. The afternoon sights for the watch made our longitude at noon 204° 1′ 30″ east; and we could see half a degree to the east. There was a large swell in that direction, but no indication that could induce us to suppose there was land in that quarter.

We had now run 5° of longitude nearly from 200° to 205° east, in the parallel of 28° 30′ north: the situation of this island must therefore be to the east or west of the above longitude; most probably to the east of 206°, as Captain Cook passed the parallel in longitude of 200° 15′ east, and many other navigators to the westward of that longitude."

On the 17th of March, the Providence anchored in Nootka Sound, which the Spaniards had then evacuated. The ship requiring much repair, she was here hove down, and did not quit the Sound till May. Captain Broughton then proceeded along the north-west coast of America, and stopped in the Bay of Monterry. "It was now necessary," says he, "I should come to some determination respecting my future proceedings. My orders from the Admiralty were, that I should survey the N. W. coast of America, upon the idea that Captain Vancouver, who had similar orders, would not be able to fulfil them. But as I now had certain intelligence, that he had left this port eighteen months before, and that both the ships, Discovery and Chatham, under his command, were in good condition, I had not the smallest doubt of his ability to comply with his instructions; particularly as I had information of his sailing from Valparaiso, for that purpose. This being the case, and wishing to employ his Majesty's sloop under my command, in such a manner as might be deemed most eligible for the improvement of geography and navigation; I therefore, after consulting the officers, whose opinions coincided with my own, determined to survey the coast of Asia, commencing at the Island of Sakhalin, situated in 50° north latitude in the southern part of the sea off Ochotz, and ending at the Nanking River, in 30° north latitude." But as this survey could not probably be completed before the middle of the year

1798, Captain Broughton proposed spending his time in that pursuit till Christmas, then to go to Canton, for stores and provisions, and to continue the survey early in the following year. He accordingly steered for the Sandwich Islands; and thence sailed across the North Pacific Ocean, to the Japanese and Kurile Islands, which he continued surveying during the months of September, October, and November; and in December he arrived at China. Here Captain Broughton purchased a very fine schooner of between 80 and 90 tons burden; and, in April, 1797, as soon as the monsoon permitted, he sailed to prosecute his voyage, with fifteen months' provisions on board of both the vessels. On the 17th of May, however, a dreadful and unexpected accident happened, which had nearly terminated their farther progress, viz. the Providence struck upon a reef of rocks, off some islands distinguished in the charts by the name of Typinsan, about 100 leagues from the east part of Formosa: every means to extricate her from this perilous situation were employed, but from adverse circumstances, she in the course of five hours became a perfect wreck, and they left her to the mercy of the sea, and embarked in the schooner.

The groupe of islands off which the Providence was lost, consists of seventeen. They are of different sizes, and many of them very small and uninhabited. They extend from 24° 10′ to 24° 52′ 30″ north latitude, and from 103° 2′ to 125° 37′ east longitude. The inhabitants distinguish them by the name of Madjicosemah; they are tributary to Great Lieu-chieux, or the Liquieux Islands.

It is but justice to the natives of Typinsan to remark, that they behaved with the most distinguished humanity and kindness to Captain Broughton and his crew, who, after having been liberally supplied with water and all kinds of refreshments, returned to China, where they arrived safe in the schooner, on the 4th of June.

The crew of the Providence, with the exception of thirty-five officers and men, whom Captain Broughton retained with him in the schooner, were here distributed on board the Swift sloop of war, and different East India ships, for the purpose of taking their passage to England. The Swift was afterwards lost, and the whole of her crew perished.

Captain Broughton, being determined to pursue and finish the survey which he had begun, sailed again on the 17th of the same month (June), with five months provisions, on board the schooner. Having accomplished his purpose, at least in its principal points, he returned to China in November. Captain Broughton proceeded thence through the Straights of Malacca, and from Madras to Trincomalé; a track of sea well known to navigators: thence he took the accustomed passage to England, where he arrived in February, 1799, after an absence of four years. It may be proper to observe, that Captain Broughton was tried at Trincomalé for the loss of the Providence, and honourably acquitted.

Mr. John Turnbull also made a voyage round the world, in the years 1800, 1, 2, 3, and 4, in which he visited the principal islands in the Pacific Ocean, and the English settlements at Port Jackson and Norfolk Islands.

As the object of this voyage was not to make discoveries, but solely undertaken to speculate and trade with the natives of the said islands, we shall not enter into any particulars, but refer the reader to the voyage itself, in 3 vol. 12mo. which was published in 1805.

W.

Stilling WAVES *by Means of Oil.* Page 625, col. 1, after line 15, add,

Captain Anthony Pool relates, that his ship, being one of a fleet, in 1761, whereof there was a ship laden with oil, which escaped through the seams of the casks containing it, and mixing with the water in the hold, both of which were pumped up together, while it was remarked, that the water in the wake of the ship was as smooth as a mirror: the longer the pumping continued, the more was the wake enlarged, and, notwithstanding the agitation of the sea continued, the waves did not break.

In regard to the method of employing oil, if only designed to smooth the surface of the sea, so as to expose the view of what is below, it is said to be enough to dip a feather in it, which is drawn through the water. If a more important purpose be designed in averting the presence of danger, a quantity must be allowed to escape slowly through a tube, the size of a goose quill, which will be sufficient to quell the turbulence of the waves; and as the effect is gradually lost, the effusion must be repeated.

We must not suppose, however, in here describing the properties of oil, that a calm and level plain is produced by its effusion on the sea: on the contrary, the swell remains unabated; but a vessel will safely mount the waves, and lie in the troughs between them, for the breakers, which are most of all to be dreaded, disappear. The lofty precipices, which would otherwise overhang the stern, threatening destruction in their fall, gradually decline when under the influence of the repelling fluid, and instead of washing the decks of the vessel from above, elevate the hull on their successive summits.

Y.

YARD. Page 652, col. 1, after line 56, add,

The yards of a 74-gun ship are represented in plate XX. where *d d* is the main-yard, *e e* the main-topsail-yard, *f f* the main-top-gallant yard, and *k k* the main royal yard.

ERRATA.

Page.	Col.	Line.	
2	2	40	Aburton. For a correct definition of this article, see the Appendix.
6	1	26	See Appendix, *read* See the articles Pay and Wages.
16	1	23	against, *read* again.
27	2	59	verticle, *read* vertical.
36	2	60	is being hoisted, *read* is hoisting.
37	2	34	coin, *read* quoin throughout the work. l. 41, ship-building, *read* mast-making.
40	1	62	Birth, *read* Berth throughout the work.
45	1	4	voyal, *read* voyol.
			The reference to the boats, from line 10 to line 41, in column 1 page 49, as in the original work, is cancelled; and the sheer-draughts, half-breadths, and body-plans of the different modern boats are substituted in plate XXIV.
50	2	7	from the bottom, granades, *read* grenades.
59	2	62	Plate IV. fig. 20, in the margin, *read* Plate V. fig. 11.
66	2	21	ship-building, *read* mast-making.—Plate V. fig. 3, in the margin, *read* Plate IV. fig. 3.
67	1	24	is, *read* are.
71	1	63	fig. 11, plate VI. *read* fig. 21, plate VII.
84	2	4	*Bow* Chasers, *read Bow* Chases.—l. 7, *Stern* Chasers, *read Stern* Chases.—l. 37, steps, *read* stops.
86	2	57	or the head, *read* and heel.
89	2	32	*Belaying* Cleat, *read Range* Cleat.—l. 50, *Single* Cleat, *read Sling* Cleat.
92	2	24	an officer, *read* the person.
95	1	32	Pendent, *read* Pendant.
98	1	62	Gillibrand, *read* Gellibrand.
110	2	30 and 31.	by plate VIII. fig. 7, *read* in fig. 10, plate XIII.
121	1	23	from the bottom, cargo, *read* cargoes.

Page.	Col.	Line.	
133	1	28	Eiking, *read* Ekeing.
144	2	21	hook, *read* hoop.
145	1	22	ship-building, *read* mast-making.
152	2	23	military, *read* naval.—l. 31, white, *read* red.
157	1	39	buntin, *read* buntine or bunting.
163	1	38	*Hard, read Stiff.*
176	2	24	from the bottom, longways, *read* lengthways; and for midship, *read* midships.
177	1	5	and, *read* an.
192	1	40	fig. 10, plate XI. *read* fig. 1, plate XXXIV.
214	1	14	Langrage, *read* Langrage-Shot.
219	2	56	fig. 1, plate XIV. *read* fig. 2, plate XXXIV.
242	2	7	fig. 9, } fig. 10, }
246	1	18	fig. 10, } plate XV. { fig. 9, } Plate
254	2	36	fig. 11, } *read* { fig. 7, } XIX.
261	2	16	fig. 12, } fig. 8, }
320	2	17	fig. 6, plate XVII, *read* fig. 7, plate VIII.
410	1	12	£1 11 0½, *read* £1 1 11
412	1	56	two, *read* eight.
423	1	4	fig. 1, plate XXII. *read* fig. 3, plate XXXI.
435	2	59	fig.3, } pl.XXIII. { figs. 5 & 6, pl. XIX.
441	2	35	fig.4, } *read* { fig. 1, plate XIV.
445	1	57	fig. 1, plate XXIV. *read* fig. 2, plate XXVI.
453	1	6 and 7,	fig. 2, plate XXIV. *read* fig. 3, plate XXXII.
460	2	52	first, *read* second; and for fig. 3, *read* fig. 13.
467	1	6	after the table, mold-loft, *read* mould-loft.
470	2	40	fig. 5, plate XXIII. *read* fig. 4, plate XIX.
488	2	30	from the bottom, fig. 2, *read* fig. 1.
503	1	44	plate XXIV. *read* plate XXX.
579	2	15	Treaddle, *read* Treadle throughout the work.
673	2	40	fig. 3, *read* fig. 5.

Explanation *of the Abbreviations used in the* Vocabulary, *viz.*

s. m. Substantive masculine.
s. f. Substantive feminine.
s. m. p. Substantive masculine plural.
s. f. p. Substantive feminine plural.
adj. Adjective.
adj. & s. m. Adjective and substantive masculine.
adj. f. Adjective feminine.
v. a. Verb active.
v. n. Verb neuter.
v. r. Verb reciprocal, or reflected.
v. a. & n. Verb active and neuter.
v. a. & r. Verb active and reciprocal, or reflected.
v. a. n. & r. Verb active, neuter, and reciprocal, or reflected.
part. act. Participle active.
part. pass. Participle passive.
adv. Adverb.
prep. Preposition.
int. Interjection.
Ex. Example.

A

VOCABULARY

OF

FRENCH SEA-PHRASES, AND TERMS OF ART.

A.

ABATTÉE, *s. f.* Casting, or falling off of a ship to leeward. *Ex.*

☞ Attention à L'ABATTÉE pour mettre la barre au vent! *Watch her falling off to put the helm up!*

ABATTRE, *v. n.* To fall off, or to cast. *Ex.*

Le vaisseau abat sur tribord. The ship casts to starboard.

☞ Laisse ABATTRE! *Let her fall off!*

ABATTRE *un vaisseau en carène, v. a.* To heave down, or careen a ship.

ABATTRE *un vaisseau en quille.* To heave down a ship so as to discover her keel.

ABORDAGE, *s. m.* The act of boarding; also the shock produced by two vessels falling accidentally a-board of each other, or otherwise.

Aller à L'ABORDAGE. To run down upon a ship with an intention to board her.

Sauter à L'ABORDAGE. To jump a-board.

Eviter L'ABORDAGE. To avoid an enemy's boarding.

ABORDAGE *de long en long.* Boarding fore and aft.

ABORDAGE *par la hanche.* Boarding on the quarter.

ABORDAGE *de bout-au-corps.* Boarding end-on.

ABORDAGE *de deux vaisseaux qui chassent l'un sur l'autre.* Falling, or driving a-bord of a ship, or running foul of one another, &c.

ABORDAGE *d'un vaisseau contre un quai, une balise, un écueil, &c.* Running foul of a bank, &c.

ABORDER, *v. a.* To board—To fall aboard, *or* to run foul of—To land. *Ex.*

Nous ABORDÂMES *le vaisseau ennemi.* We boarded the enemy's ship.

Un vaisseau de ligne nous ABORDA *en appareillant.* A line of battle ship fell aboard of us, in getting under way.

Nous ABORDÂMES *à Portsmouth.* Wel nded at Portsmouth.

ABOUGRI. *Voyez* RABOUGRI.

ABOUT, *s. m.* Butt-end of a ak: also, the place where the ends of two planks are joined on the ship's side, &c.

ABRAQUER, *v. a.* To haul taught upon any cask or weighty body that is slung or seized for that purpose. *Ex.*

☞ ABRAQUE le mou! *Haul taught!*

ABRI, *s. m.* Shelter. *Ex.*

Se mettre à L'ABRI *d'une côte.* To shelter a ship under the lee of a high shore.

ABRIER, *v. a.* To becalm. *Ex.*

Vaisseau ABRIÉ *par la lame.* A ship becalmed in the trough of the sea.

Ce vaisseau va nous ABRIER. That ship is going to becalm us.

La misaine est ABRIÉE. The fore-sail is becalmed.

ABRITER, *v. a.* To shelter. *Ex.*

ABRITÉ *de tous les vents.* Sheltered from all winds.

ACADÉMIE, ACADÉMIE *Royale, s. f.* A place in which young gentlemen are trained for the navy.

ACCALMIE. *Voyez* CALMIE.

ACCASTILLAGE, *s. m.* (from the Spanish, *castillo.*) Upper-works; a general name for the quarter-deck, poop, and forecastle; hence *haut accastillé* answers to deep-waisted. *Ex.*

Vaisseau qui a L'ACCASTILLAGE *ras.* A straight-sheered ship, or a ship whose upper-works are low-built.

Vaisseau qui a L'ACCASTILLAGE *élevé.* A round-sheered ship, whose upper-works rise very high.

ACCASTILLÉ, *adj.* *Vaisseau haut* ACCASTILLÉ. A deep-waisted ship.

ACCLAMPER. *Voyez* JUMELER.

ACCON, *s. m.* A flat-bottomed boat, for catching shell-fish.

ACCORD, *s. m.* Regularity in plying the oars, or hoisting sails, &c. *Ex.*

☞ Hisse D'ACCORD! *Voyez* ENSEMBLE.

☞ Nage D'ACCORD! *Voyez* ENSEMBLE.

ACCORER, *v. a.* To prop or sustain any weighty body. *Ex.*

ACCORER *un vaisseau sur le chantier.* To prop a ship on the stocks.

ACCORER *un tonneau, une malle, &c.* To wedge a cask, trunk, &c.

ACCORES, *s. f. p.* Shores fixed under a ship's wales, &c. to support her whilst building, &c.

Accores *d'un banc.* The edges of a sand-bank.
ACCOSTER, *v. a.* To near—To come along-side of a ship, speaking of a boat. *Ex.*
Accoster *un bâtiment la mèche à la main.* To come a long-side of a ship, being clear for action.
Accoster *la terre de près.* To make free with the land.
Le canot est il accosté*?* Does the boat lay fair along-side?
☞ Accoste! *Come along-side!*
ACCOURSIE, *s. f.* A passage formed in a ship's hold by a separation of her stores, cargo, or provisions, to go fore and aft, as occasion requires.
ACCROCHER, *v. a.* To grapple an enemy's ship.
ACCUL, *s. m.* The depth of a bay, or small road. (A term used in the French West India Islands.)
ACCULÉ, *adj.* (from *acculer.*) *Ex.*
Varangues acculées, *varangues demi-*acculées. *Voyez* Varangues.
ACCULEMENT, *s. m.* Rising of the timbers.
s'ACCULER, *v. r.* To be pooped; to receive a shock in pitching.
ACHEVER, *v. a.* To complete. *Ex.*
Ils auront bientôt achevé *leur cargaison.* They will soon have completed their cargo.
ACOTARS, *s. m. p.* Filling pieces. *Voyez* Clés.
ADENT, *s. m.* (In ship-building) A scoring in a piece of timber.
☞ ADIEU-VA! *int. Helm's a-lee! About ship!* the order for tacking.
ADONNER, *v. n.* To draw aft; speaking of the wind when it becomes favourable. *Ex.*
Le vent adonne. The wind draws aft.
AFFALER, *v. a.* To lower, to overhaul. *Ex.*
Affaler *les cargues fonds.* To overhaul the buntlines, &c.
Affaler *un palan.* To fleet a tackle.
s'Affaler. To get embayed. *Ex.*
Le navire étoit affalé *sur la côte.* The ship was embayed on a lee shore.
AFFINER, *v. n.* To clear up, speaking of the weather.
AFFLEURER, *v. a.* (In ship-building) To fay, as, a plank fays to the timbers.
AFFOLÉE, *adj.* An epithet given to a magnetic needle which has lost its virtue, and is become erroneous and defective.
AFFOURCHER un vaisseau, *&* s'Affourcher, *v. a.* To moor a ship by the head. *Ex.*
Affourcher *à la voile.* To moor across under sail.
AFFRANCHIR, *v. n.* To free the ship, or to clear her hold of water by the means of the pumps.
AFFRÉTEMENT, *s. m.* The freighting of a merchant ship.
AFFRÉTER, *v. a.* To hire.
AFFÛT d'un canon, *s. m.* Carriage of a gun.
Parties de l'Affut. The parts of a carriage.
Les flasques. The sides or cheeks.
Les essieux The axle-trees.
Les roues. The trucks.
L'entretoise. The transom.
La sole. The sole or bottom.
Le cousin. The bed.
Le coin de mire. The quoin.
Les coins pour mettre aux roues des canons dans le mauvais tems. The quoins for quoining guns in bad weather.
Les ferrures de l'Affût. The iron works of a carriage.
Les plattes bandes. The cap-squares, or clamps.
Chevilles à œillets. Eye-bolts.
Chevilles à goupilles. Joint-bolts.
Cheville qui lie l'entretoise avec les flasques. The transom-bolt.
Cheville qui traverse l'Affût *vers l'arrière sur laquelle porte la culasse du canon.* The bed-bolt.
Chevilles pour la brague. Breeching-bolts.
Chevilles qui lient les flasques avec l'essieu de derrière. Hind axle-tree bolts.
Œillets à accrocher les palans à canons. Loops, or eye-bolts, to which the gun-tackles are hooked.
Affût *de mortier, s. m.* Carriage of a mortar.
AFFÛTER un canon, *v. a.* To mount a gun upon its carriage.
AGENT de la marine, *s. m.* Navy Agent.
AGITATION, *s. f.* Violent motion, speaking of the sea. *Ex.*
l'Agitation *des flots.* The surges.
AGITER, *v. a.* To swell or run high, expressed of a turbulent sea.
La mer est agitée. The agitation of the sea.
Les lames agitoient *le vaisseau.* The ship was tossed by the waves.
AGRÉER. *Voyez* Gréer.
AGRÉEURS, *s. m. p.* Riggers.
AGRÈS, *s. m. p.* Rigging. The French term is of greater extent, and comprehends the rigging of the ground-tackle, yards, sails, blocks, cables, &c.
AIDE, *s. m.* Mate or assistant. *Ex.*
Aide *de port.* An officer under the commissioner of a French dock-yard.
Aide *Major des signaux.* The under officer belonging to the signals.
Aide-*Calfat.* Calker's mate.
Aide-*Canonnier.* Quarter-gunner.
Aide-*Pilote.* A petty officer on board French ships of war, whose duty is like that of a quarter-master in the English navy.
Aides-*Charpentiers.* Carpenter's crew.
Aides-*Chirurgiens.* Surgeon's mates.
Aides-*voiliers.* Sail-maker's crew.
AIGUADE, *s. f.* Watering-place for shipping.
AIGUILLE, *s. f.* Aiguille aimantée. Magnetic needle.
Aiguille, *ou flèche. Voyez* Flèche.
Aiguille *de fanal.* A crank used to support a poop lantern.
Aiguille *à œillet.* Eyelet-hole needle.
Aiguille *à ralingue.* Bolt-rope needle.
Aiguille *à voile.* Sail needle.
Aiguilles *de carène.* Spars employed to support a lower-mast in heaving down a ship.
AIGUILLETTES, *s. f. p.* Aiguillettes de porques. The upper futtock-riders.

AILE, *s. f.* AILES *d'une galère, félouque*, &c. Two planks set vertically to deck the stern of gallies, or other such vessels.
AILES *d'une armée navale.* The wings of a fleet.
AILES, *terme d'arrimage.* Wings of a ship's hold.
AILES *de dérive.* *Voyez* SEMELLES *de dérive.*
AIMANT, *s. m.* A loadstone; magnet.
AIMANTER, *v. a.* To rub or touch with a load-stone.
AIN, *s. m.* A shark hook.
AIR, *s. m.* Way of a ship. *Ex.*
Avoir de L'AIR. To have fresh way through the water.
Amortir L'AIR *d'un vaisseau.* To stop a ship's way.
☞ Donne de L'AIR au vaisseau pour virer de bord! *Give the ship way, that she may stay!*
AIRE DE VENT, *s. m.* A point of the compass.
AISSIEU. *Voyez* ESSIEU.
AJUSTE, *s. m.* A bend, or knot, by which the ends of two ropes are fastened together.
AJUSTER, *v. a.* To bend or tie two ends of ropes together.
ALARGUER. *Voyez* LARGUER.
ALIDADE, *s. f.* The index of a quadrant, sextant, nocturnal, astrolabe, or any other geometrical instrument used to measure angles or distances.
ALIGNER, *v. a.* To level. *Ex.*
☞ ALIGNEZ la batterie! *Level your guns!*
ALISÉS (vents.) *s. m. p.* Trade winds.
ALLÈGE, *s. f.* A lighter, pram, or craft.
ALLÉGER, *v. a.* To lighten. *Ex.*
ALLÉGER *un vaisseau.* To lighten a ship, by taking out a part of her lading.
☞ ALLÉGE le cable! *Light up the cable!*
ALLÉGER *un cable.* To buoy up a cable.
ALLÉGER *une écoute, ou une manœuvre.* To ease up any sheet or rope.
ALLER, *v. n.* To go, &c. *Ex.*
☞ Où ALLEZ vous? *Where are you bound to?*
ALLER *à la côte.* To run a-shore.
ALLER *à terre.* To go a-shore.
ALLER *à la découverte.* To go upon the look out.
ALLER *en course.* To cruise against, or in search of an enemy.
ALLER *à la sonde.* To run by the lead.
ALLER *à la touée.* To warp.
ALLER *à voiles & à rames.* To sail and row.
ALLER *à la bouline.* To stand close-hauled.
ALLER *au-plus-près du vent.* To sail close-hauled.
ALLER *de bout au vent.* To go head to wind.
ALLER *entre deux écoutes.* To sail right afore the wind, with both sheets aft.
ALLER *de bout à la lame.* To bow the sea.
ALLER *de conserve.* To keep company with a ship.
ALLER *de l'avant.* To go fast through the water.
ALLER *en dérive.* To go adrift.
ALLURE, *s. f.* The trim of a ship; point of sailing.
ALMADIE, *s. f.* An African canoe, usually formed of the bark of a tree.
ALONGE, *s. f.* A futtock, or futtock-timber. 1ere Alonge, 2de alonge, 3e alonge, 4e alonge. So they call the futtocks of any timber, beginning from the lower futtock, which is called GENOU; the next above it takes the name of 1ere alonge. *Ex.*
1ere ALONGE. Second futtock.
2e ALONGE. Third futtock, &c.
ALONGE *de revers.* Top-timber.
ALONGES *de porques.* Futtock-riders.
ALONGES *d'écubiers.* Hawse-pieces.
ALONGES *de cornière.* Top-timbers of the fashion-piece.
ALONGES *de poupe.* Stern-timbers.
ALONGES *de tableau.* Taffrail timbers.
ALONGER, *v. a.* To lengthen, &c. *Ex.*
ALONGER *un vaisseau.* To lengthen a ship.
ALONGER *le cable.* To haul up a range of the cable upon the deck ready for anchoring.
ALONGER *une ancre.* To carry out a small anchor.
ALONGER *un grelin.* To carry a warp out of the ship.
ALONGER *la terre.* To coast or sail along shore.
ALONGER *la vergue de civadière.* To get the sprit-sail-yard fore and aft under the bowsprit.
ALONGER *les écoutes des huniers.* To stretch along the topsail-sheets.
ALONGER *un vaisseau.* To run a ship along-side of another.
AMAN, *s. m.* ITAGUE; AMAN *de drisse.* Tye of a lateen yard.
AMARINER, *v. a.* To man a prize.
AMARQUE. *Voyez* BALISE.
AMARRAGE, *s. m.* Lashing, or seizing. *Ex.*
AMARRAGE *à plat.* Lashing of the shrouds and stay.
AMARRAGE *en étrive.* Throat-seizing of a dead-eye.
AMARRAGE *à fouet.* *Voyez* FOUET.
AMARRAGE *coulant.* Slip-knot.
AMARRE, *s. f.* A mooring, &c. *Ex.*
AMARRE *de bout.* A head-fast.
AMARRE *de travers.* A breast-fast.
AMARRE *de poupe.* A stern-fast.
AMARRER, *v. a.* To make fast, seize, or belay a rope. *Ex.*
☞ AMARRE! *Belay!*
☞ AMARREZ les canons! *Secure your guns!*
AMARRER *un vaisseau.* To make a ship fast.
AMARRER *de l'avant et de l'arrière.* To moor head and stern.
AMATELOTER, *v. a.* To mess together, to associate as comrades or messmates.
AME D'UN GRAND CORDAGE, *s. f.* The middle strand of a four-stranded rope.
AME *d'un canon*, *s. f.* The bore of a gun. *Voyez* CALIBRE.
AMENER, *v. a.* To lower, to strike, &c. *Ex.*
☞ AMÈNE! *Strike!*
AMENER *les perroquets sur le ton.* To lower down the topgallant-sails.
AMENER *une terre.* To make the land.
AMENER *deux amers l'un par l'autre.* To bring two sea-marks in one.
AMER, *s. m.* Sea-mark. *Ex.*
Conserver un AMER *au sud de la boussole.* To keep a sea-mark south of the compass.
AMIRAL, *s. m.* Admiral. *Ex.*
L'AMIRAL *d'Angleterre.* Admiral of the fleet.
Vice-AMIRAL. *Voyez* VICE.

Contre-AMIRAL. *Voyez* CONTRE.
Vaisseau AMIRAL *d'une armée navale.* The admiral-ship of a fleet.
Vaisseau AMIRAL *d'un port.* Guard-ship; an old ship kept at the entrance of a royal port.
Maître canonnier AMIRAL,—*pilote* AMIRAL. The senior gunner and pilot are thus styled in the French navy.

AMIRAUTÉ, *s. f.* Admiralty.
AMOLETTES, *s. f. p.* The bar-holes of any capstan or windlass.
AMORCER, *v. a.* To prime, to bait. *Ex.*
☞ AMORCEZ! *Prime!*
AMORCER *un hameçon.* To bait a fishing-hook.
AMORTI, *adj.* *Vaisseau* AMORTI *dans un port.* A vessel which is sewed in a harbour.
AMORTIR, *v. a.* To deaden, to sew. *Ex.*
AMORTIR *l'air d'un vaisseau.* To stop a ship's way.
Le navire est AMORTI. The ship has sewed.
AMPLITUDE, *s. f.* Amplitude.
AMPOULETTE, *s. f.* The watch-glass which is generally kept in the binnacle.
AMURE, *s. f.* Tack of a sail. *Ex.*
Ce vaisseau à les AMURES *à babord.* That ship is on the larboard tack.
AMURER, *v. a. & n.* To haul aboard the tack of the main, or fore-sail. *Ex.*
☞ AMURE la grande voile! *Aboard main-tack!*
ANCETTES. *Voyez* PATTE.
ANCRAGE, *s. m.* Anchoring place; also Anchorage.
Droit D'ANCRAGE. Duty of anchorage.
ANCRE, *s. f.* An anchor.
Parties de L'ANCRE. Parts of the anchor.
La verge. The shank.
L'œillet. The eye.
Les tenons. The nuts.
L'organeau. The ring.
Le collet. The crown.
Les pattes. The palms.
Les oreilles. The flooks.
Les bras. The arms.
Le bec. The bill.
Le jât, ou jouail. The stock.
La grand-ANCRE, ou ANCRE *de miséricorde.* The sheet-anchor.
Seconde ANCRE. The best bower anchor.
ANCRE *d'affourche ou de jêt.* The small bower anchor.
ANCRE *de flot.* The flood anchor.
ANCRE *de jusant.* The ebb anchor.
ANCRE *de touée.* The stream anchor.
ANCRE *à empenneller.* The kedge-anchor.
ANCRE *de terre.* The shore anchor; or, that which lies towards the shore.
ANCRE *du large.* The sea-anchor; or that which lies towards the offing.
ANCRE *de rechange.* The spare-anchor.
Les ANCRES *de bossoir.* The bower-anchors.
ANCRE *à demeure.* A large anchor sunk in a road,' or harbour, to warp ships in and out, or secure them for a short time.
ANCRE *de veille.* An anchor which is a cock-bill.

Être à L'ANCRE. To ride at anchor.
Empenneller une ANCRE. To back an anchor.
Mouiller L'ANCRE. To anchor.
Lever L'ANCRE. To hoist or weigh the anchor.
Chasser sur les ANCRES. To drag the anchors.
ANCRE *dont le cable a fait un tour.* Foul anchor.
L'ANCRE *est à-pic.* The anchor is a-peak.
Gouverner sur L'ANCRE. To sheer the ship to her anchor.
L'ANCRE *est dérapée, ou chasse sur son* ANCRE. The anchor is a-weigh.
L'ANCRE *est à vue.* The anchor is in sight.
Caponer une ANCRE. To cat an anchor.
Dégager L'ANCRE. To clear the anchor.
Traverser L'ANCRE. To fish the anchor.
Saisir L'ANCRE *contre le bord.* To secure the anchor.
Parer L'ANCRE. To see the anchor clear.
Mettre L'ANCRE *à poste.* To stow the anchor.
Draguer une ANCRE. To sweep an anchor.
ANCRER, *v. n.* To anchor. *Jetter* L'ANCRE, *mouiller, prendre fond, laisser tomber* L'ANCRE; all these terms are synonymous, and signify, To bring up, To come to an anchor, or, To let go the anchor.
ANGES. *Voyez* BOULETS *ramés.*

ANGUILLÈRES, *s. f. p.* Limbers. *Ex.*
Canal des ANGUILLÈRES. Limber-Holes.
Bordages des ANGUILLÈRES. Limber-boards.
Cordage des ANGUILLÈRES. Limber-ropes.
Chaines des ANGUILLÈRES. Limber-chains.
ANGUILLES, *s. f. p.* The ways, or bulge-ways, on which a ship's cradle is supported when launching into the water.
ANGUILLES *d'un canon de coursier.* The ways on which the great gun, which gallies carry a-head, slides fore and aft.
ANGUIS, *s. m.* Parral truss, in a galley, &c.
ANNEAU, *s. m.* A ring, of any kind. *Ex.*
ANNEAU *pour attacher les vaisseaux.* A mooring-ring on a wharf, buoy, &c.
ANNEAUX *de bois.* Hanks.
ANNEAUX *de cordes.* Grommets.
ANNEAUX *de fer.* Rings, or iron cringles.
ANNEAUX *d'écoutilles.* Hatch-way rings.
ANNEAUX *de sabords.* Port-rings.
ANSE. *s. f.* Creek, or cove.
ANSPEC, *s. m.* Hand-spike, or lever. *Ex.*
ANSPEC *de vindas.* Boatswain's handspike.
ANSPEC *pour le canonnage.* Gunner's handspike.
ANSPEC *de vingt-quartre.* Handspike for twenty-four-pounders.
ANTENNE, *s. f.* A lateen-yard.
ANTOIT, *s. m.* A wraining-bolt or staff.
A-PIC, *adv.* A-peak, or perpendicularly above the anchor, with an extended cable. *Ex.*
Le vaisseau est A-PIC. The ship is up and down.
Côte A-PIC. A bold-shore, or a bluff-shore.
L'ancre est A-PIC. The anchor is a-peak.
Virer A-PIC. To heave short.
APIQUER UNE VERGUE, *v. a.* To peak a yard, or top it up and down.
APOSTIS, *s. m. p.* The row-locks of a galley, or more properly the rails which support the row-locks, or tholes.

APÔTRES, *s. m. p.* The knight-heads, or bollard-timbers—the hawse-pieces which are nearest the stem.

APPARAUX, *s. m. p.* The whole furniture of a ship; as the masts, sails, yards, blocks, anchors, cables, helm, artillery, &c.—This term is more comprehensive than AGRÈS. *Ex.*

Agrès & APPARAUX. The rigging and furniture of a ship.

APPAREIL, *s. m.* Machinery, or purchase. *Ex.*

APPARIEL *de carène.* A general name for the machinery employed in careening a ship.

APPAREIL *de pompe.* The pump-jear, as the boxes, brake, spear, &c.

APPAREILLAGE, *s. m.* The act and manner of getting a ship under sail.

APPAREILLER, *v. n.* To get under sail, or underway, &c. *Ex.*

Ce vaisseau APPAREILLE. That ship is getting under-way.

Un vaisseau APPAREILLE *vent arrière.* A ship sailing before the wind.

APPAREILLER *une voile.* To set any sail.

APPAREILLER *une ancre.* To get ready, or to clear an anchor.

APPELER, *v. n.* To grow, &c. *Ex.*

Le cable APPELE *de tribord.* The cable grows on the starboard bow.

Aussière qui APPELE *de loin.* A rope which is fastened at a great distance.

APPOINTEMENS, *s. m. p.* The pay of officers.

APPRÊTER, *v. a.* APPRÊTER *un brûlot.* To prime a fire-ship.

APPROCHER, *v. n.* To near, &c. *Ex.*

S'APPROCHER *du vent.* To luff. To haul close to the wind.

APPUYER, *v. a.* APPUYER *la chasse.* To chase nimbly; to pursue a ship closely. *Ex.*

APPUYER *les bras du vent.* To haul taught the weather-braces of any yard.

ARAIGNÉE, *s. f.* Crow-foot of the tops. *Ex.*

Moque D'ARAIGNÉE. Crow-foot dead-eye.

ARBALÉTE ou ARBALESTRILLE, *s. f. (Instrument d'Astronomie.)* A cross-staff, or fore-staff, also called Jacob's-staff.

ARBALÉTRIERE, *s. f.* Planks nailed on the sides of a galley, between the row-locks and the thwarts, on which the soldiers stand to fire their musketry, &c.

ARBORER, *v. a.* ARBORER *un pavillon.* To hoist a flag, &c. *Ex.*

ARBORER *un mât*, in gallies, xebecs, &c. To step or set up a mast.

ARC, *s. m.* ARC *d un vaisseau.* The cambering of a ship's deck or keel.

ARC *d'une pièce de construction.* Compass of a piece of timber.

ARC, *instrument de construction.* A bow; an instrument employed by shipwrights, for drawing on paper the sheer of the wales, &c.

ARCASSE, *s. f.* Stern-frame of a ship; also the shell of a block.

Parties de L'ARCASSE. The parts of a stern-frame.

L'étambot. The stern-post.

Le contre-étambot intérieur. The inner post.

La lisse d'hourdi. The wing-transom.

La barre d'écusson. The counter-transom.

Le fourcal d'ouverture. The lower-transom.

La barre du premier pont. The deck-transom.

La barre de la soute du maître canonnier. The first-transom.

Les barres D'ARCASSE. The transoms; but more particularly the second, third, and fourth transoms.

Les estains, ou cornières. The fashion-pieces.

Les alonges de cornière. The top-timbers of the fashion-piece.

La contre-cornière. *Voyez* CORNIÈRE.

Courbe D'ARCASSE. *Voyez* COURBE, &c.

ARC-BOUTANT, *s. m.* A boom. *Ex.*

ARC-BOUTANS, *ou les boute-hors de grande voile.* Main-sail booms; or booms to extend the bottom of the main-sail, in square-rigged vessels.

ARC-BOUTANS *de misaine.* Fore-sail booms, in square-rigged vessels.

Les ARC-BOUTANS *de bonnettes.* Studding-sail booms.

ARCHIPEL, *s. m.* Archipelago.

ARCHIPOMPE, *s. f.* The pump-well.

ARCHITECTURE NAVALE, *s. f.* Naval architecture; or the art of ship-building.

ARDENT, *adj.* The quality of griping in the steerage, or carrying a weatherly helm. *Ex.*

Navire ARDENT. A ship which carries a weatherly helm.

ARÊTE, *s. f.* (In naval architecture) The edge or angle of a square piece.

ARGANEAU. *Voyez* ORGANEAU.

ARGILE, *s. f.* Clay.—*Fond* D'ARGILE. Clay ground.

ARGOUSIN, *s. m.* A petty officer belonging to a galley, whose duty it is to fix on, and to take off the shackles of the slaves, &c.

ARMADILLE, *s. f.* A small squadron of Spanish frigates of war, employed as cruisers to prevent illicit trade.

ARMATEUR, *s. m.* Owner of a merchant-ship, or privateer.

ARMÉE NAVALE, *s. f.* A large fleet of men of war.

Les ARMÉES *navales d'Angleterre.* The maritime forces of England.

ARMEMENT, *s. m.* The equipment, or fitting out of a ship, or squadron.

ARMER, *v. a.* ARMER *un navire.* To fit out a ship.

ARMER *les avirons.* To ship the oars.

ARMES À FEU, *s. f. p.* Fire-arms.

ARMURE, *s. f.* ARMURE *d'un bau.* (*Terme de construction.*) The middle piece of a beam, when made of three pieces.

ARMURES *d'un mât.* (*synonyme de* JUMELLES.) The cheeks or side pieces of a made-mast.

ARMURIER, *s. m.* The armourer of a ship of war.

ARONDELLES DE MER, *s. f. p.* A general name for all small vessels, as brigs, settees, &c.

ARQUER, S'ARQUER, *v. n. & r.* To camber; to become hogged, or broken-backed. *Ex.*

Vaisseau ARQUÉ. A broken-backed ship. *Quille* ARQUÉE A cambered keel.

La quille du Souverain est ARQUÉE *de* 20 *pouces.* The Sovereign's keel is cambered by 20 inches.

ARRÊT, *s. m.* ARRÊT *des vaisseaux dans un port* An embargo laid on vessels in a port or harbour.

ARRÊTER, S'ARRÊTER, *v. a.* & *r.* To detain, to stop, &c. *Ex.*

Le convoi pour Lisbonne est ARRÊTÉ *par des vents contraires.* The Lisbon convoy is detained by contrary winds.

ARRIÈRE, *s. m.* & *ad.* The stern, or after part of a vessel—a-stern, aft, &c. *Ex.*

Mettre un vaisseau de L'ARRIÈRE. To out sail a ship, or leave her a-stern.

Rester de L'ARRIÈRE. To remain a-stern or behind, by slow sailing. To drop a stern.

Se faire de L'ARRIÈRE. To be a-head of one's reckoning.

Passer de L'ARRIÈRE. To go under the stern of a ship.

Tirant-d'eau de L'ARRIÈRE. Draft of water abaft.

Les voiles de L'ARRIÈRE. The after sails.

ARRIÈRE-GARDE, *s. m.* The rear division of a squadron of men of war.

ARRIMAGE, *s. m.* The stowage of a vessel's cargo, or whatever is contained in her hold.

ARRIMER, *v. a.* To stow the hold. *Ex.*

Vaisseau mal ARRIMÉ. A ship badly stowed.

ARRIMEUR, *s. m.* A stower.

ARRIVÉE, *s. f.* The movement of bearing away, also of falling off when lying-to. *Ex.*

Ce vaisseau fait son ARRIVÉE. That ship is falling-off.

ARRIVER, *v. n.* To bear away, to bear up, or bear down; to arrive, &c. *Ex.*

ARRIVER *sur un vaisseau.* To bear down upon a ship.

ARRIVER *vent arrière.* To bear away right before the wind.

ARRIVER *tout plat.* To bear round up.

☞ARRIVE tout! *Hard a-weather! Hard up!*

☞N'ARRIVE pas! *Don't fall off! Luff! Keep her to!*

☞Sans ARRIVER! *Nothing off!*

☞Laisse ARRIVER d'un quart! *Keep her away a point!*

Le vaisseau est ARRIVÉ *aux Dunes.* The ship is arrived in the Downs.

ARRONDIR, *v. a.* ARRONDIR *un cap.* To weather, or sail round a cape.

ARSENAL, *s. m.* A royal dock-yard, with its gun-wharf.

ARTILLÉ, *adj.* Mounted with cannon.

Un vaisseau ARTILLÉ *de toutes ses pièces.* A ship mounting all her guns.

ARTILLERIE, *s. f.* Ordnance, artillery.

Parc D'ARTILLERIE. Artillery ground; or gun-wharf.

ARTILLERIE *d'un vaisseau.* The ordnance of a ship of war.

ARTIMON, *s. m.* Mizen.

Mât D'ARTIMON. Mizen-mast.

ASPECT. *Voyez* VUES.

ASSAILLIR, *v. a.* To assault, &c. *Ex.*

Nous avons été ASSAILLIS *d'un coup de vent.* We have been caught in a gale of wind.

ASSÉCHER, *v. n.* To appear dry, as a rock or shore, when the tide has left it.

ASSEMBLAGE, *s. m.* Rabbeting, scarphing, scoring, tenoning, or anywise uniting pieces of timber.

ASSEMBLAGE *à queue d'hironde.* A swallow-tail scarf.

ASSIETTE D'UN VAISSEAU, *s. f.* Trim of a ship.

ASSURANCE, *s. f.* Insurance. *Ex.*

Chambre D'ASSURANCE. Insurance office.

Prime D'ASSURANCE. Premium of insurance.

ASSURER UN NAVIRE, *v. a.* To insure a ship.

ASSUREURS, *s. m. p.* Underwriters.

ASTROLABE, *s. f.* Astrolabe; nocturnal.

ATTAQUER, *v. a.* To attack. *Ex.*

L'ennemi nous ATTAQUA *au mouillage.* The enemy attacked us at our anchorage.

ATTEINDRE, *v. a.* To join or come up with a ship at sea, either by accident or pursuit.

ATTELIERS, *s. m. p.* Sheds, workshops, or storehouses in a dock-yard. *Ex.*

ATTELIER *de charpentiers.* The carpenter's shop.

ATTELIER *de forgerons.* The smith's shop.

ATTELIER *de grêement.* The rigging-loft.

ATTELIER *de la mâture.* The mast-house.

ATTELIER *de menuisiers.* The joiner's-shop.

ATTELIER *de la peinture.* The painter's-shop.

ATTELIER *des poulis.* The block-house.

ATTENTION, *s. f.* Attention. *Ex.*

☞ATTENTION à la barre! *Mind the helm!*

ATTERAGE, *s. m.* Land-fall.

ATTERRER, *v. n.* To make the land. *Ex.*

Nous ATTERRÂMES *sur le Cap Lezard.* We made the land at Cape Lizard.

ATTOLES, *s. m.* A cluster of small islands, or chain of rocks.

ATTRAPER, *v. a.* To catch, &c. *Ex.*

☞ATTRAPE à garant de candelette! *Clap on the fish!*

ATTRAPER *la brise.* To catch the breeze.

ATTRAPES, *s. f. p.* Relieving-tackles, used in careening a ship.

AVANCES, *s. f. p.* Advance-money. A sum paid to the sailors before they sail, on account of their wages. *Ex.*

Trois mois D'AVANCE. Three months' advance-money.

AVANT, *s. m.* and *adj.* Head—a-head—a-fore, forward, or the forepart of the ship. *Ex.*

AVANT *d'un vaisseau.* The head, or bow of a ship.

AVANT *maigre.* A lean bow.

AVANT *élancé.* A flaring bow.

AVANT *joufflu, ou* AVANT *renflé.* A bluff bow.

Vogue AVANT. The strokesman in a boat or galley.

☞AVANT! *(Commandement aux rameurs!) Pull away!*

☞AVANT tribord! *Pull away starboard, or pull with the starboard oars!*

☞AVANT qui peut! *Pull with the oars that are shipped!*

Vaisseau sur L'AVANT. A ship which is too much by the head.

Le vaisseau est en AVANT *de mon point.* The ship is a-head of my reckoning.

De L'AVANT *à nous.* A-head of us.

☞AVANT babord, & scie tribord! *Pull the larboard oars, and hold the water with the starboard oars!*

De L'AVANT *à l'arrière*. Fore and aft.
☞AVANT par tout! *Give way fore and aft!*
AVANTAGE, *s. m.* AVANTAGE *du vent*. To wind-ward; the weather-gage.
Avoir L'AVANTAGE *du vent*. To have the weather-gage.
AVANT-CALE, *s. f.* The launch of a ship from the slip.
AVANT-GARDE, *s. f.* The van of a fleet.
AVARIE. *s. f.* The damage or loss which a ship sustains by accidents or bad weather, in her voyage, or in port.
AVARIÉ, *adj. Ex. Merchandises* AVARIÉES. Damaged goods.
AUBARÉTRIÈRES, *s. f. p.* A sort of balustrade erected on the sides of a row-galley, to support the rails of the gang-way, &c.
AUBINET, *ou Sainte* AUBINET, *s. m.* No Man's Land.
AUBOUR, *ou* AUBIER, *s. m.* The sap in timber.
AVEUGLER, *v. a. Voyez* BOUCHER.
AUGE À GOUDRON, *s. f.* A tar-bucket.
AVIRON, *ou* RAME, *s. m.* An oar. *Ex.*
AVIRON *de galère*. An oar of a galley.
AVIRONS *de vaisseaux*. Ship-oars; also sweeps.
Parties de L'AVIRON, The parts of an oar.
Le manche, ou le giron. The handle.
Le bras. The arm, or inner part.
Le plat, ou la pelle. The wash, or blade.
La menille, ou main tenante. An implement nailed to the oar, for a number of men to hold it by.
La galaverne. A small piece of wood nailed on each side the oar, to prevent its galling the thole, and to strengthen it in that part.
AVIRONS *à couple*. Double-banked oars, or scullers.
☞Armez les AVIRONS! *Get your oars to pass!*
Sabords des AVIRONS. *Voyez* SABORDS.
AVIRONNIER, *s. m.* Oar-maker.
AVIRONNIER *maître*. Master oar-maker.
AVIRONNIERS, *s. f. ou attelier des* AVIRONNIERS. The oar-makers shed.
AVISO, *s. m.* Advice boat.
AVITAILLEMENT, *s. m.* AVITAILLEMENT *d'un vaisseau*. The provision of a ship, or sea-victualling.
AVITAILLER, *v. a.* To victual. *Ex.*
AVITAILLER *une flotte*. To victual a fleet.
AVITAILLEUR, *s. m.* A commissioner of the victualling-office.
☞ AU-LOF! *Luff!* The order from the pilot to steer nearer the wind. *Voyez* LOF.
AULOFÉE, *s. f.* The act of luffing, or springing the luff.
AUMONIER, *s. m.* The chaplain of a ship.
AU-PLUS-PRÈS, *adj. ou* AU-PLUS-PRÈS *du vent*. Close to windward; close-hauled. *Ex.*
Gouverner AU-PLUS-PRÈS, To keep her to.
AURIQUE, *adj. Voiles* AURIQUES. Shoulder of mutton sails.
AVOIR, *v. a.* To have, &c. *Ex.*
AVOIR *affaire à des forces supérieures*. To have to contend with an enemy of superior force.
AVOIR *beaucoup de bau*. To have a great breadth of beam.
AVOIR *beaucoup de creux*. To have a roomy hold.
AVOIR *vent arrière*. To have the wind aft.
AVOIR *vent de bout*. To have the wind right a-head.
AVOIR *besoin de secours*. To be in want of assistance.
AVOIR *communication avec la terre*. To have free intercourse with any port after performing quarantine.
AVOIR *connoissance de terre*, To be in sight of the land.
AVOIR *le beaupré à terre*. To have the land close aboard.
AVOIR *le pied marin*. To have sea feet or legs; or to tread the deck like a sailor.
AVOIR *l'avantage de la marche*. To have the advantage in sailing.
AVOIR *de l'eau à courir*. To have good sea room.
AVOIR *de l'eau sous la quille*. To be in deep water.
AVOIR *son branle-bas fait*. To have the hammocks up and stowed, and the ship clear.
AVOIR *de l'entre-deux de mâts*. To have a great distance between the masts.
AVOIR *du largue dans les voiles*. To be going free.
AVOIR *du mou dans ses cables*. To be slack-moored.
AVOIR *la marée, ou le courant pour soi*. To have the tide or current in one's favour.
AVOIR *le fond*. To be in soundings.
AVOIR *le vent sur vn bâtiment*. To have the weather-gage of a ship.
AVOIR *ses ancres en mouillage*. To have the anchors clear for running.
AVOIR *de l'empâture*. To have a great spread for the rigging.
AVOIR *sa mâture sur l'arrière, ou sur l'avant*. To have the masts raking aft or forward.
AVOIR *ses couleurs*. To shew the colours.
AVOIR *les mâts de hune calés & les basses vergues amenées*. To have the yards and top-masts struck.
AVOIR *tous les ris pris*. To be close reefed.
AVOIR *toutes les voiles sur les cargues*. To have all the sails clewed up.
AUSSIÈRE, *s. f.* A hawser or small cable, and in general any rope made of three or four strands, and only once laid.
AUTAN, *s. m.* A gust or squall of the south wind.
AXIOMÈTRE, *s. m.* The tell-tale of the tiller.
AZIMUTH, *s. m.* Azimuth, vertical circle.

B.

BABORD, *s. m. & adv.* Larboard; port. *Ex.* BABORD, *Côté gauche d'un vaisseau*. The larboard side of any ship.
BABORD (partie de l'équipage) *ou les babordais*. The larboard watch.
☞ Avant BABORD! *Pull to larboard!*
☞ *Brasse* BABORD! *Brace to larboard!*
☞ Feu BABORD! (commandement aux canonniers). *Fire the guns on the larboard side!*
☞ BABORD, ou BABORD la barre! (commandement au timonnier). *Port the helm!*
☞ BABORD un peu! *Port a little!*

☞ BABORD tout! *Hard a-port!*
Passer à BABORD. To heel to port.
BAC, *s. m.* A ferry-boat.
BAC *à naviguer*, A punt, or small boat used by shipwrights to carry pitch, tar, &c.
BACALAS, *s. m. p.* A sort of standing-knees, on the deck of a galley, or xebec, projecting on each side above the row-locks.
BACASSAS, *s. m.* A sort of lighter, somewhat resembling an American periagua.
BACHOT, *s. m.* A yawl or wherry.
BACHOTEUR, *s. m.* The wherry-man.
BACLAGE, *s. m.* A tier of boats moored alongside of each other in a port or harbour.
BACLER UN PORT. *Voyez* CHAÎNE *de port.*
BADERNE, *s. f.* Mat, or paunch. *Ex.*
BADERNE *des mâts majeurs.* Dolphin of the masts.
BAGNE, *s. m.* A spacious building for the galley-slaves in a French dock-yard.
BAGUES, *s. f. p.* Hanks, grommets. *Ex.*
BAGUES *des voiles d'étai.* Grommets of a stay-sail.
BAGUETTE DE FUSIL, *s. f.* Ramrod of a musket.
BAIE, *s. f.* A bay, or road.
BAILLE, *s. f.* Half-tub. *Ex.*
BAILLE *de sonde.* Half-tub, or bucket, to contain the hand-lead and line on the deck, ready for sounding.
BAISSER. *Voyez* DESCENDRE.
BAISSER *les voiles.* *Voyez* AMENER.
BALAI, *s. m.* A broom.
BALAI *du ciel.* A name given by sailors to the north-west winds of America, which always produce fine weather.
BALANCEMENT, *s. m.* In naval architecture, *Couples de* BALANCEMENT. Balance-timbers or frames.
BALANCIERS, *s. m. p.* BALANCIERS *de boussole.* Gimbals of a sea-compass.
BALANCIERS *de lampe.* Rings of a lamp.
BALANCIERS *de pirogue.* Outriggers of a canoe.
BALANCINES, *s. f. p.* Lifts. *Ex.*
BALANCINES *de grande vergue.* The main-lifts.
BALANCINES *de misaine.* The fore-lifts.
BALANCINES *du grand hunier.* The main-top-lifts.
BALANCINES *du petit hunier.* The fore-top-lifts.
BALANCINES *du grand perroquet.* The main-top-gallant-lifts.
BALANCINES *du petit perroquet.* The fore-top-gallant lifts.
BALANCINES *de la vergue sèche.* The cross-jack-lifts.
BALANCINES *du perroquet de fougue.* The mizen-top-lifts.
BALANCINES *de la perruche.* The mizen-top-gallant-lifts.
BALANCINES *de la civadière.* The sprit-sail-lifts.
BALANCINES *de la contre-civadière.* The bowsprit top-sail lifts.
BALANCINE *de gui.* Topping-lift.
Fausses BALANCINES. Preventer-lifts.
BALANT, *s. m.* Bight, or slack of a rope. *Ex.*
☞ Abraque le BALANT des boulines d'avant! *Steady forward the head bowlines!*
BALAYER, *v. a.* To sweep.
BALESTON, *s. m.* Sprit of a shoulder-of-mutton-sail.
BALISE, *s. f.* The beacon, or buoy of a shoal.
BALISER, *v. a.* BALISER *une passe, &c.* To lay down buoys in a channel.
BALLOT, *s. m.* A pack or bale of goods.
BALON, *s. m.* A sort of galley or barge of Siam.
BANC, *s. m.* BANC *de sable.* A sand-bank.
BANC *de brume.* A fog bank.
BANC *de glace.* An island, or field of ice.
BANC *de rameurs.* The thwart or seat of rowers.
BANC *de quart.* A bench placed on the quarter-deck, in French ships, for the officer who commands the watch.
BANC *de poisson.* A shoal of fish.
BANCHE. *Voyez* BANC.
BANDE, *s. f.* The side of a ship; also of a coast.
Etre à la BANDE. To heel.
Donner la BANDE *sur tribord.* To heel to starboard.
BANDE *du nord, ou du sud.* The northern, or southern shore of an island
BANDES *de ris.* Reef-bands.
BANDES *de fer.* Iron plates, in general.
En BANDE! *adv.* Amain!
☞ EN BANDE! *Let go amain!*
BANDER, *v. a.* BANDER *une voile.* To line a sail at the edges, in order to strengthen it.
BANDER *les sartis.* To set up the shrouds in a galley,
BANDIÈRE, *s. f.* *Front de* BANDIÈRE. A line formed a-breast.
BANDINS, *s. m. p.* BANDINS *de l'espale.* Benches and balustrades placed on the sides of a row-galley, afore the canopy.
BANDINS *de la poupe.* The ballustrades which deck the sides of the canopy of a galley.
BANDINETS, *s. m. p.* The rails of the above-mentioned balustrades.
BANNE, *s. f.* Awning, or tilt.
BANNER, *v. a.* To cover with a tilt, or awning.
BANNIÈRE, *s. f.* A broad pendant cleft in two, and hoisted at the peak of lateen yards in a row-galley.
BANQUETTES, *s. f.* The stretchers of a galley, or row-boat.
BANQUIER, *s. m.* A banker; a vessel employed in the Newfoundland fishery.
BAPTÊME, *s. m.* Ducking, as practised in crossing the line or tropics.
BAPTISER, *v. a.* To duck, &c.
BAPTISER *un vaisseau.* To give a ship her name at the time of launching.
BARACHOIS, *s. m.* A snug anchorage formed by the land on one side, and a ledge that breaks off the sea on the other.
BARATERIE DE PATRON, *s. f.* A term of law used in the sea-trade, expressing the cheats and tricks which may be employed by the master of a merchant ship, for embezzling or altering any part of the cargo, or suffering it to be damaged by the master.
BARATTE, *s. f.* Mats, or paunches to strengthen the fore-sail in a storm.

ship's crew; for which he is punished by the laws, according to the nature of the case.

BARBE, *s. f.* BARBE *d'un bordage.* Hooding ends of a plank.

Mouiller en BARBE. To come to, with two anchors a-head.

BARBE (*Sainte*). *Voyez* SAINTE-*Barbé.*

BARBES *d'un vaisseau.* The entrance, or fore-foot of a ship.

BARBEYER. *Voyez* FASIER.

BARCO-LONGO, *s. m.* A Spanish coasting vessel.

BARDIS, *s. m.* Water-boards, or weather-boards, sometimes used when a ship is to be hove down.

BARGE, *s. f.* A river barge with a flat bottom.

BARIL, *s. m.* A small cask, also a rundlet. *Ex.*

BARIL *à poudre.* Powder-barrel.

BARIL *de galère.* Breaker

BARIL *à bourse.* A budge-barrel.

BARILLARD, *s. m.* The steward, or officer having charge of the wine and water in a row-galley.

BÂRIQUES À FEU *ou* FOUDROYANTES, *s. f. p.* Thundering-barrels, or casks containing the fire-pots, in a fire-ship.

BARQUE, *s. f.* A bark, or lighter; also, a kind of polacre, used in the Mediterranean.

☞ BARQUE droite! *Trim the boat!*

BARQUE *d'avis.* An advice-boat.

BARQUE *à eau.* A watering-boat.

BARQUE *de descente.* A flat-bottomed boat.

BARQUE *de vivandier.* A provision-boat; a bum-boat.

BARQUE *en fagot.* A boat in frame.

BARQUE *longue, ou double chaloupe.* A sort of pinnace, or large long-boat.

BARQUÉE, *s. f.* A boat full of any goods, stores, &c. *Ex.*

BARQUÉE *de lest.* A boat loaded with ballast, &c. reckoned to be of ten tons.

BARQUEROLE, *ou* BARQUETTE, *s. f.* A sort of passage-boat.

BARRE, *s. f.* Transom, the tiller, &c. *Ex.*

BARRES *d'arcasse.* Transoms.

BARRE *d'écusson.* The counter-transom.

BARRE *de pont.* The deck-transom.

BARRE *de la soute du maître canonnier.* The first or lower transom.

BARRE *du gouvernail.* The tiller of the helm.

BARRE *franche.* A hand-tiller without a wheel.

BARRE *de rechange.* A spare tiller.

BARRES *de cabestan.* Bars of the capstan.

BARRES *de hune.* The frame of the cross-trees and tressel-trees.

BARRES *Maîtresses de hune.* Trestle-trees

BARRES *traversières de hune.* Cross-trees.

BARRES *de perroquet, ou Croisettes.* Cross-trees of the top-masts.

BARRES *de panneaux d'écoutille.* The carlings, or ledges placed athwart under the covers of the hatchways.

BARRES *d'écoutilles.* Hatch-bars.

BARRES *de cuisine.* Cook-room bars.

BARRES *de sabords.* The gun-port bars.

BARRES *de trélingage.* A staff set across the main and fore shrouds, for the lower cat-harpings.

BARRE *de prisonniers.* Irons, or fetters.

BARRE *d'un port.* The bar of a harbour.

☞ Tribord la BARRE! *Starboard the helm!*

☞ Babord la BARRE! *Port the helm!*

☞ Change la BARRE! *Shift the helm!*

☞ Dresse la BARRE! *Right the helm!*

☞ Droit la BARRE! *Put the helm a-midships!*

☞ BARRE au vent! *Up with the helm!*

☞ BARRE dessous! *Down with the helm!*

☞ Mollis la BARRE! *Ease the helm!*

BARRER, *v. a.* BARRER *un bâtiment.* To give a ship too much helm, so as to yaw her about.

BARRER *un port.* To fix a boom across the mouth of a harbour to secure its entrance.

BARROT. *Voyez* LATTE.

BARROTÉ, *adj.* Full to the beams;—a vessel which is over-laden.

BARROTER, *v. a.* To fill a vessel quite up to the beams—to overload a ship.

BARROTINS, *s. m. p.* Ledges put across the deck beams, in French men of war, in order to strengthen the deck.

BARROTINS *d'écoutilles.* The spurs of the beams, which fortify the deck abreast of the hatchways.

BARROTINS *de caillebotis.* Ledges of the gratings.

BARROTS, *s. m. p.* Flat beams of the quarter deck, poop, and fore-castle; also the small beams placed between the larger ones.

BAS, *adj. Vaisseau de* BAS-BORD. A low-built vessel.

☞ BAS le pavillon! The order to haul down the colours.

BAS-BORD. *Voyez* BABORD.

BAS-FOND, *s. m.* Shallow water; shoal.

BASSE, *s. f.* A shoal, or flat.

BASSE-EAU, *ou* BASSE-MER, *adj.* Low water.

BASSES-VOILES, *s. f. p.* The courses, or principal lower sails.

BASSIN, *s. m.* Basin, or dock.

BASSIN *de radoub.* A dry dock.

BASSIN *de port.* A wet dock.

BASTET, *s. m. ou Quenouillette des Haubans.* Foot-hook staff.

BASTINGAGE, *s. m.* Barricado. Quarter, waist, and fore-castle nettings, and painted cloths.

Filets de BASTINGAGE. Nettings.

BASTINGUER UN NAVIRE, *v. a.* To barricade a ship.

BATAILLE, *s. f.* BATAILLE *navale.* A sea fight; an engagement.

BÂTARD, *s. m.* BÂTARD *de racage.* Parral-rope.

BÂTARDE, *s. f.* The largest sail of a galley; a large lateen sail, used in fair weather.

BÂTARDELLES, *s. f. p.* Square-sterned row-gallies.

BATAYOLES, *s. f. p.* Wooden stanchions of the netting. *Ex.*

Montans des BATAYOLES. The stanchions of the netting.

Lisses des BATAYOLES. The rails of the netting.

BATAYOLES *des hunes.* Stanchions of the nettting of the tops.

BATEAU, *s. m.* Boat—a general name for many kinds of small vessels.

BATEAU *d'office.* A small boat, or yawl employed by ships of war for bringing off fresh provisions from the shore.
BATEAU *Bermudien.* A sloop in the West Indies.
BATEAU *de loc.* The log.
BATEAU *lesteur.* A ballast lighter.
BATEAU *pêcheur.* A fishing boat.
BATELAGE, *s. m.* A waterman's fare.
BATELÉE, *s. f.* A boat's load; the number of passengers to be carried in a boat.
J'ai ma BATELÉE. I have got my fare.
BATELIER, *s. m.* A waterman, boatman, sculler, ferry-man, wherry-man, &c.
BÂTIMENT, *s. m.* A vessel, or ship of any kind. *Ex.*
BÂTIMENS *à rames.* Row-boats.
BÂTIMENT *à trois mâts.* A three-masted ship.
BÂTIMENT *latin.* A lateen vessel.
Mauvais BÂTIMENT. A bad vessel.
BÂTIMENT *condamné.* A condemned vessel.
BÂTIMENT *trompeur.* A deceitful looking vessel.
BÂTIMENT *très maniable.* A handy vessel.
BÂTIMENT *ramassé.* A snug vessel.
BÂTIMENT *lège.* A light ship.
BÂTIMENT *négrièr.* A slave ship, or Guinea-man.
BATIMENT *armé en flute.* Store-ship.
BÂTIMENT *de transport pour les troupes.* A troop-ship.
BÂTIMENT *qui a le côté droit.* A wall-sided ship.
BÂTIMENT *qui a le côté foible.* A crank ship.
BÂTIMENT *qui a le côté fort.* A stiff ship.
BÂTIMENT *qui a les fonds fins.* A sharp-bottomed ship.
BÂTIMENT *de peu d'apparence.* A ship that appears of much less force than she really is.
BÂTIMENT *qui capéye bien.* A ship that lies too well in a gale of wind.
BÂTIMENT *qui coule bas d'eau.* A ship that makes more water than the pumps can discharge.
BÂTIMENT *qui fatigue à la mer.* A ship that strains and labours much in a heavy sea.
BATIMENT *qui manque de liaisons.* A weak-built ship.
BÂTIMENT *qui se comporte bien à la mer.* A ship that behaves well at sea.
BÂTIMENT *à varangues plattes.* A flat-floored ship.
BÂTIMENT *armé en course.* A privateer, or letter of marque.
BÂTIMENT *armé en flûte.* A store-ship.
BÂTIMENT *armé en guerre.* A vessel equipped for war.
BÂTIMENT *armé en guerre & en marchandise.* An armed merchant-man.
BÂTIMENT *barbaresque.* A vessel of the coast of Barbary.
BÂTIMENT *bien soutenu de l'avant.* A ship with a full bow.
BÂTIMENT *bien taillé de l'avant.* A vessel very sharp forward.
BÂTIMENT *bien évide de l'arrière.* A ship with a clean run aft.
BÂTIMENT *charbonnier.* A collier.
BÂTIMENT *contrebandier.* A smuggler.
BÂTIMENT *croiseur.* A cruiser.
BÂTIMENT *de charge.* A vessel of burden.
BÂTIMENT *de commerce.* A trading vessel.
BÂTIMENT *de transport.* A transport.
BÂTIMENT *ennemi.* An enemy.
BÂTIMENT *flibustier.* A free-booter.
BÂTIMENT *pirate, ou forban.* A pirate.
BÂTIMENT *neutre.* A neutral vessel.
BÂTIMENT *parlementaire.* A cartel.
BÂTIMENT *pécheur.* A fishing vessel.
BÂTIMENT *marchand.* A merchant-man.
BÂTIMENT *de la compagnie des Indes.* An East India-man.
BÂTIMENT *taillé pour la marche.* A vessel built for sailing fast.
BÂTIMENT *servant de prison.* A prison-ship.
BÂTIMENT *servant d'hôpital.* An hospital-ship.
BÂTON, *s. m.* Staff, boom, &c. *Ex.*
BÂTON *de foc.* Jib-boom.
BÂTON *de flamme.* The stick of a pendant.
BÂTON *de gaffe.* The stick or staff of a boat-hook.
BÂTON *de girouette.* A spindle of a vane at the mast-head.
BÂTON *de commandement.* Ensign staff at the mast-head.
BÂTON *astronomique, ou de Jacob.* *Voyez* ARBALÉTE.
BÂTON *de guipon.* The handle of a long tar-brush, or pitch-mop.
BÂTONNÉE D'EAU, *s. f.* The quantity of water thrown out of a ship's pump at each stroke of the brake or handle.
BATTANT, *s. m.* BATTANT *d'un pavillon.* The fly of an ensign, &c.
Vaisseau bien BATTANT. A ship fit for battle.
BATTERIE, *s. f.* The whole range of the cannon placed on both sides of any one deck of a ship of war; also a battery. *Ex.*
Ce vaisseau a 5 pieds 8 pouces de BATTERIE. This ship's lowest gun-deck ports are 5 feet 8 inches above the surface of the water.
BATTERIE *& demie.* A deck and a half of cannon, said of a frigate which carries cannon on her upper-deck and quarter-deck only.
Vaisseau qui a une belle BATTERIE. A ship that carries her ports a proper height out of the water.
Vaisseau qui a sa BATTERIE *noyée.* A ship that carries her ports too near the surface of the water.
BATTERIE *qui s'adapte à la culasse d'un canon.* The lock applied to the breech of a cannon.
BATTERIE *flottante.* A floating battery.
☞ Mettez la BATTERIE de hors! *Run the guns out!*
☞ Mettez la BATTERIE de dans! *Run the guns in!*
BATTRE, *v. a.* To beat. *Ex.*
BATTRE *la diane & la retraite.* To beat the reveille and the retreat, as is customary on board men of war, when in port.
BATTRE *aux champs.* To sound a chase at sea.
BATTRE *la marche.* To give the signal for sailing.
BATTU PAR LE VENT, *adj.* Weather-beaten.
BATTURE, *s. f.* A flat, shallow, or reef.
BAU, *s. m.* Beam. *Ex.*
*Demi-*BAU. Half-beam.
*Maître-*BAU. Midship-beam.

BAU *du coltis.* Collar-beam.
BAU *de dale.* The hindmost, or aftmost beam.
BAU *de lof.* The foremost beam.
BAU *de pont.* Deck-beams.
BAUDET, *s. m.* A sawyer's frame, horse, block, or trestle.
BAUME, *s. f.* Spanker.

BAUQUIÈRES, *s. f. p.* In ship-building, clamps, or inner planks on which the beams rest upon the ship's sides. *Ex.*
BAUQUIÈRE *du premier pont.* Main-deck clamp.
BAUQUIÈRE *du second pont.* Upper-deck clamp.
BAUQUIÈRE *des gaillards.* Clamp of the quarter-deck and forecastle.
BAUQUIÈRE *de la dunette.* Clamp of the poop.
DEMI-BAUX, *s. m. p.* The spurs or crow-feet of the beams.
Faux BAUX. The orlop-beams.
BAYE. *Voyez* BAIE.
BAYONNETTE, *s. f.* A bayonet.

BEAUPRÉ, *s. m.* *Mât de* BEAUPRÉ. Bowsprit. *Ex.*
Bout de BEAUPRÉ. A small jib-boom, in merchant ships.
BEAUPRÉ *sur poupe.* Close behind, is said of a ship which is so near the stern of another, that the bowsprit of the former reaches almost over the stern of the latter.
BEC, *s. m.* BEC *d'une galère, felouque,* &c. The beak, or prow of a galley, felucca, &c.
BEC *d'une ancre.* The bill of an anchor.
BEC *de corbin.* A ripping-iron; an instrument used by calkers, to take the old oakum out of the seams.

BÉCASSE, *s. f.* A sort of Spanish boat used in the bay of Cadiz, and adjacent seas.

BÉLANDRE, *s. f.* *Bilander.* A merchant vessel with two masts.

BÉLIER, *s. m.* A wooden machine used for driving the wedges under a ship's bottom, &c.
BELLE, *s. m.* The main-deck or waist.
☞ Pointe en BELLE! *Point your guns horizontally!*

BÉQUILLES, *s f. p.* Shores used to support small sharp-built vessels, when laid a-ground.
BERCEAU, *s. m.* A cradle.
BERGE, *s. f.* High beach, or bluff-shore.
BERNE, *s. f.* Waft or weft. *Ex.*
Mettre le pavillon en BERNE. To hoist the ensign with a waft.
BERTELOT, *ou* BERTHELOT, *s. m.* The prow of a lateen vessel.
BESSON. *Voyez* TONTURE.
BETTE, *s. f. ou* MARIE-SALOPE. A small flat-bottomed boat belonging to a sort of lighter or barge, made use of to clean the ports of France; which little boat serves to carry away the mud.
BIDON, *s. m.* A can used by French sailors to fetch their drink.
BIDOT, *s. m.* À-BIDOT, *adv.* A-back, or to windward of the mast, speaking of a lateen sail.
BIGOTS DE RACAGE, *s. m. p.* Ribs of a parral.
BIGUES, *s. f. p.* Sheers.
BILLARD, *s. m.* A poker or iron bar used to drive on the iron hoops of the masts and yards.

BILLARDER, *v. a. & n.* To drive on the iron hoops of the masts and yards by means of the poker.
BILLER, *v. a.* To fasten a rope to a boom, in order to ride or tow a boat.
BILLOTS, *s. m. p.* Dead-wood, or short pieces of timber laid upon the keel, between the crotchets afore and abaft.
BISCAYENNE, *s. f.* A Biscayan long-boat, or Biscay *barca-longa.*
BISCUIT, *s. m.* BISCUIT, or sea bread.
BITORD, *s. m.* Spun-yarn. *Ex.*
BITORD *blanc.* White or untarred spun-yarn.
BITORD *goudronné.* Tarred spun-yarn.
BITORD *de trois fils.* Three-yarn spun-yarn.
BITTER LE CABLE. *Voyez* CABLE.
BITTES, *s. f. p.* The bits. *Ex.*
Grandes BITTES. The main bits.
Montans des BITTES. The bits.
Traversin des BITTES. The cross-piece of the bits.
Crochets des BITTES. Hasps, or hooks, fastened to the cross-piece.
Coussin des BITTES. Doubling of the bits.
Courbes des BITTES. The standards of the bits.
BITTONS, *s. m. p.* The topsail-sheet bits; also, the knight-heads, kevel-heads, &c.
BITTURE, *s. f.* A range of the cable.
BLANC, *adj.* White, or untarred, speaking of ropes or yarn.
BLIN. *Voyez* BÉLIER.
BLINDER, *v. a.* BLINDER *un vaisseau.* To cover a ship's sides with fenders of old cables, when she is moored near an enemy's battery, to preserve her from the effects of the shot.
BLOQUER, *v. a.* To block up. *Ex.*
La flotte françoise est BLOQUÉE *à Toulon.* The French fleet is blocked up in Toulon.
BOIS, *s. m.* Wood or timber. *Ex.*
BOIS *de chêne.* Oak.
BOIS *de cèdre.* Cedar.
BOIS *de chêne vert.* Live oak.
BOIS *de hêtre.* Beech.
BOIS *d'ormeau.* Elm.
BOIS *de sapin.* Fir.
BOIS *de charme.* Yoke-elm, or horn-beam.
BOIS *de gayac.* Lignum-vitæ.
BOIS *d'acajou.* Mahogany.
BOIS *de tec.* Teak wood.
BOIS *de construction.* Timber for ship-building.
BOIS *courbans.* Compass-timber.
BOIS *de haute-futaie.* Straight timber.
BOIS *d'arrimage.* Fathom wood, employed in stowing the hold.
BOIS *de chauffage.* Fuel, or fire-wood.
BOIS *de démolition.* Timber taken from an old ship when broken up.
BOIS *de rebut.* Timber unfit for service.
BOIS *de membrure.* Crooked timber for the floors or futtocks of a ship's frame.
BOIS *de recette.* Timber fit for service.
Faire du BOIS. To make a provision of wood for fuel, &c.
Tirer en plein BOIS. To hull a ship.

BOÎTE DU GOUVERNAIL, *s. f.* The rudder-case.
BOÎTE *à mitraille.* Case, or canister-shot.
BOMBARDE, *s. f.* A bomb-vessel.

BOMBARDER, *v. a.* To bombard.
BOMBE, *s. f.* Shell.
L'eclat d'une BOMBE. The bursting of a bomb.
BOMERIE, *s. f.* Bottomry.
BONACE, *s. f.* Calm weather, with a smooth sea.
BON-FRAIS, *s. m.* A steady breeze, or fresh gale.
BONNE DE NAGE. Swift of rowing; a fine rower.
BONNE TENUE. *Voyez* TENUE.
BONNE-VOGLIE, *s. m.* A volunteer-rower in a galley.
BONNEAU, *s. m.* The buoy of an anchor. *Voyez* BOUÉE.
BONNETTES, *s. f. p.* A general name for all studding-sails.
BONNETTES *basses.* The lower studding-sails.
BONNETTES *de hunes.* Top-mast studding-sails.
BONNETTES *de perroquet.* Top-gallant studding-sails.
BONNETTE *de brigantine.* Ring-tail.
BONNETTE *lardée.* A bag or basket charged with cinders, chopped oakum, &c. to be used in the act of fothering a ship.
Lacer la BONNETTE. To lace on the bonnet of a sail to its principal part.
BON-TOUR, *s. m.* A favourable turn or swing; expressed of a ship when she keeps her hawse clear by winding the right way.
BORD, *s. m.* Board, or the ship's side. *Ex.*
☞ Passe du monde sur le BORD! *Man the side!*
Aller à BORD. To go a-board.
Quitter le BORD. To leave the ship.
J'irai demain à mon BORD. I shall go to morrow on board my ship.
BORD, also signifies a board in tacking.
BORD *alongé, ou qui alonge.* A long board; said of a vessel plying to windward.
Faire un BORD. To make a tack.
BORD *sur* BORD. Tack for tack, hank for hank.
BORD *à terre.* Standing in.—Standing off shore.
BORD *au large.* Standing off the shore.
BORD-*à*-BORD. Along-side; speaking of two ships.
Bon BORD. A good board.
Mauvais BORD. A bad board.
BORD *de la mer.* The sea shore, or the strand.
BORDAGES, *s. m. p.* The planks of a ship's side. *Ex.*
Franc BORDAGE. The outside planks.
BORDAGE *de fond.* The planks of the bottom or floor.
BORDAGE *des ponts.* The planks of the decks.

BORDÉE, *s. f.* A board, tack, or stretch. *Ex.*
Nous gagnerons le port dans trois BORDÉES. We shall get into the harbour by making three tacks.
Courir à petites BORDÉES. To ply to windward by small boards.
BORDÉE. A broadside. *Ex.*
BORDÉE. The sea-watch.
Lâcher la BORDÉE. To fire a broadside.
Faire la grand BORDÉE. To set the sea-watch.
Faire la petite BORDÉE. To set the quarter-watch.
BORDER, *v. a.* BORDER *les côtés d'un vaisseau.* To plank a ship, or lay on her outside planks.
BORDER *les ponts.* To lay a ship's decks.
BORDER *à clin.* To plank a vessel with clincher-work.
BORDER *en louvelle.* To lay on the planks level.
BORDER *les avirons.* To ship the oars.
BORDER *une voile, ou une écoute.* To haul aft the sheet of any sail. *Ex.*
☞ BORDE la misaine! *Haul aft the fore sheet!*
☞ BORDE l'artimon! *Haul close aft the mizen sheet!*
☞ BORDE le perroquet de fougue! *Sheet home the mizen top-sail!*
BORDER *& brasser au vent.* To trim the sails by the wind.
BORDER *les écoutes tout plat.* To tally the sheets flat aft.
BORDIER, *adj. Bâtiment* BORDIER. A lap-sided ship.
BORDIGUE, *s. f.* A crawl for taking fish.
BORDURE, *s. f.* The foot of a sail.

BORÉAL, *adj.* Northern, northerly.

BORÉE, *s. m.* The north-wind.
BOSPHORE, *s. m.* A strait or narrow channel; as the Thracian Bosphorus.
BOSSAGE, *s. m.* Crooked timber for knees, &c.
SERRE BOSSE. *Voyez* SERRE.
BOSSEMAN, *s. m.* Quarter-master's mate.
N.B. This word must not be mistaken for boatswain; the charge of the boatswain in English ships answers to that of *maître d'équipage*; whereas the BOSSEMAN is a kind of inferior boatswain's mate in French ships of war.
BOSSER. *v. a.* To stopper. *Ex.*
BOSSER *le cable.* To stopper the cable.
BOSSER *l'ancre.* To stow the anchor upon the bow.
☞ BOSSE! *Belay!* or *stopper!*
BOSSES. *s. f. p.* Stoppers of the shrouds or stays. *Ex.*
BOSSES *à bouton.* Deck-stoppers.
BOSSES *à fouet.* Long-stoppers for the rigging.
BOSSES *à éguillette.* Stoppers with lanyards.
BOSSES *de bout.* Stoppers of the anchor.
BOSSES *des cables.* Ring-ropes, or stoppers of the cables.
Maîtresse BOSSE. Dog-stopper.
BOSSES *qui prennent des ailes de la fosse aux cables.* Wing-stoppers.
BOSSES *de chaloupe, ou de canot.* The boat's painter or mooring-rope
BOSSOIR, *s. m.* The cat-head. *Ex.*
☞ La misaine au BOSSOIR! *The fore-tack to the cat-head!*
BOT, *s. m.* A kind of Dutch boat.
BOTTE, *s. f. En* BOTTE, *adv.* In frame. *Ex.*
Chaloupe en BOTTE. A long boat in frame.
Bâtiment en BOTTE. The frame of a vessel in pieces, numbered for putting together.
Futaille en BOTTE. A cask in frame.
BOUCANIER, *adj.* Bucaneer. *Ex.*
Fusil BOUCANIER. A kind of long sea musket.
BOUCAUT, *s. m.* A dry cask. *Ex.*
Le navire la Marie est arrivé de la Martinique avec 500 BOUCAUTS *de café,* &c. The ship Mary is arrived from Martinico, with 500 casks of coffee, &c.
BOUCHE D'UN CANON, *s. f.* The mouth of a cannon.

BOUCHER, *v. a* Boucher *une voie d'eau.* To stop a leak.
BOUCHES. *Voyez* Embouchure.
BOUCLES de sabord. Port-rings.
BOUDIN, *s. m.* The middle rail of the head, in French ships, so denominated from its round form. This piece is not known in English ships, whose heads are differently adorned.—Also a general name for all rails of a round form.
Boudin *de trinquerin.* In a galley, a rail opposite the water-ways.
BOUDINURE, *s. f.* The puddening of the anchor.
BOUÉE, *s. f.* A buoy. *Ex.*
Bouée *de liége.* A cork-buoy.
Bouée *en forme de cône.* A can-buoy, such as are floated over shallows, banks, &c. *Voyez* Balise.
Bouée *en baril.* A nun-buoy.
Bouée *de sauvetage.* A life-buoy.
Mettre la bouée *à l'eau.* To stream the buoy.
☞ Jette la bouée! *Stream the buoy!*
☞ Rentre la bouée! *Haul in the buoy!*
BOUFFÉE, *s. f.* Bouffée *de vent.* A puff of wind.
BOUGE, *s. m.* The convexity of the beams. *Ex.*
Bouge *des baux.* Rounding of the beams.
Bouge *vertical de la lisse d'hourdi.* The rounding-up of the wing-transom.
Bouge *horizontal de la lisse d'hourdi.* The rounding-aft of the wing-transom.
BOUILLONNEMENT, *s. m.* The rippling of a river, as it falls into the ocean.
BOULETS, *s. m.* Shot. *Ex.*
Boulets *enchaînés.* Chain-shot.
Boulet *ramé.* Bar-shot, or double-headed shot.
Boulet *rond.* Round-shot.
BOULIER, *s. m.* A sort of fishing-net. *Voyez* Seine.
BOULINE, *s. f.* Bowline. *Ex.*
Bouline *de la grande voile.* Main bowline.
Boulines *du grand hunier.* Main-top bowlines.
Boulines *du grand perroquet.* Main-top-gallant bowlines.
Boulines *du grand perroquet volant.* Main-top-gallant-royal bowlines.
Boulines *de la misaine.* Fore bowlines.
Boulines *du petit hunier.* Fore-top bowlines.
Boulines *du petit perroquet.* Fore-top-gallant bowlines.
Boulines *du petit perroquet volant.* Fore-top-gallant-royal bowlines.
Boulines *du perroquet de fougue.* Mizen-top bowlines.
Boulines *de la perruche.* Mizen-top-gallant bowlines.
Boulines *du kakatois de perruche.* Mizen-top-gallant-royal bowlines.
Boulines *de revers.* The lee bowlines.
Pattes de bouline. Bridles of the bowline.
Aller à la bouline. To sail with a scant wind.
Vent de bouline. Scant wind.
BOULINER, *v. a.* Bouliner *une voile.* To haul the bowline of a sail.
Vaisseau qui bouline *bien.* A ship that sails well upon a wind.
BOULINETTE, *s. f.* Fore-top bowline. *Ex.*
☞ Choque la boulinette! *Check the fore-top bowline!*
BOULINIER, *s. m.* A plyer.
Bon boulinier. A good plyer, or a weatherly ship.
Mauvais boulinier. A bad plyer, or a leewardly ship.
BOULON, *s. m.* A square bolt.
BOURCER, *v. a.* Bourcer *une voile.* To carry a sail clewed up, or hauled up in the brails.
BOURCETS, *s. m. p.* A lugger's largest lugs.
BOURDE, *s. f.* A large sail in a galley.
BOURGEOIS, *s. m.* The proprietor or owner of a merchant-ship; also, a contractor or ship's husband.
BOURGUIGNON, *s. m.* An island of ice.
BOURRASQUE, *s. f.* A violent squall of wind, or sudden storm, but not of long duration.
BOURRELET, *s. m.* Puddening of a mast.
BOURRER, *v. a.* Bourrer *un canon.* To ram home the charge of a cannon.
BOUSSOLE, *s. f.* A Sea compass. *Voyez* Compass.
Boussole *affolée.* *Voyez* Affolée.
BOUT, *s. m.* The end. *Ex.*
Bout *de beaupré.* A bumkin.
Bout *de bordage.* Butt of a plank.
Bout *de lof.* Out-rigger.
Bout *de vergue.* Yard-arm.
Bout *pour* bout. *adv.* End-for-end. *Ex.*
Filer le cable par le bout. To veer away the cable end-for-end.
Bout *du cable.* The end of the cable.
Bout *de cordage.* A rope's end.
Bouts *de vieux cables.* Junk.
BOUTE-FEU, *s. m.* Lint-stock.
BOUTE-HORS, *s. m.* A general name for booms. *Ex.*
Boute-hors *de bonnettes.* The studding-sail booms.
BOUTEILLES, *s. f. p.* The quarter-galleries of a ship.
Fausse-bouteille. Quarter-badge.
BOUTON, *s. m.* Bouton *d'un canon.* The button, cascable, or pomelion of a cannon.
Bouton *de refouloir.* Knob at the end of a rammer.
Bouton *de bosse.* Knot of a stopper.
Bouton *d'étai.* Mouse of a stay.
BOYER, *s. m.* A kind of Dutch sloop.
BRAGUE, *s. f.* Span.
Brague *de canon.* Breeching of a cannon.
Brague *de gouvernail.* Rudder-stoppers, made use of in French ships to prevent the rudder from being unshipped.
Tirer à brague *sêche.* To fire with the breeching shortened, in order to prevent the gun's recoiling too much in a high sea.
Brague *pour lancer un vaisseau.* A large span used in dock-yards to help haul a ship down, in the operation of launching her.
BRAGUET, *s. m. ou* Braguette, *s. f.* A rope, serving to take hold of the heel of a top-mast, to help the top-rope, while the mast is hoisting up.
BRAI, *s. m.* Pitch. Brai *gras.* *Voyez* Goudron
BRANCHE de bouline. *Voyez* Patte *de bouline.*
Branche *supérieure d'une courbe.* The upper part of a knee or standard.

BRANCHE *d'en bas.* The lower arm of a knee, &c.
BRANCHE *de cyprès.* Beaconage.
BRANLE, *s. m.* Hammock. *Ex.*
BRANLE-BAS, *ou faire le* BRANLE-BAS. To lash up and take down all the hammocks from between decks, &c.
☞ BRANLE-BAS! *Up all hammocks!*
Faire BRANLE-BAS *général, ou* BRANLE-BAS *de combat.* To clear for action.
Faire BRANLE-BAS *de propreté.* To clear away for washing below.
BRAS, *s. m. p.* Braces of the yards. *Ex.*
Les grands BRAS, *ou les* BRAS *de la grand vergue.* The main-braces.
Les BRAS *de la misaine.* The fore-braces.
Les BRAS *du grand hunier.* The main-top braces.
Les BRAS *du petit hunier.* The fore-top braces.
Les BRAS *du grand perroquet.* The main-top-gallant braces.
Les BRAS *du petit perroquet.* The fore-top-gallant braces.
Les BRAS *du grand perroquet volant.* The main-top-gallant royal braces.
Les BRAS *du petit perroquet volant.* The fore-top-gallant royal braces.
Les BRAS *de la vergue sèche.* The cross-jack braces, or braces of the cross-jack yard.
Les BRAS *du perroquet de fougue.* The mizen-top braces.
Les BRAS *de la perruche.* The mizen-top-gallant braces.
BRAS *du kakatois de perruche.* Mizen-top-gallant royal braces.
Les BRAS *de la civadière.* The sprit-sail braces.
Les BRAS *de la contre-civadière.* The sprit-sail top-sail braces.
Faux BRAS. Preventer-braces.
BRAS *du vent.* Weather-braces.
BRAS *de dessous le vent.* Lee-braces.
Tenir un BRAS. To haul in and fasten the brace.
Bon BRAS. Braced in.
BRAS *d'une ancre.* The arms of an anchor.
BRAS *de mer, ou de rivière.* Arm of the sea, or of a river.
BRASILLER, *v. n.* To sparkle, speaking of the sea.
La mer BRASILLE. The sea sparkles.
BRASSE, *s. f.* A fathom, or measure of six feet.
BRASSER, *v. a.* To brace a yard. *Ex.*
☞ BRASSE tribord! *Brace to starboard!*
☞ BRASSE babord! *Brace to larboard!* or, *haul in the larboard braces!*
☞ BRASSE quarré! *Square the yards by the braces!*
☞ BRASSE à contre! *Brace the sails a-back!*
☞ BRASSE tout à culer! *Lay all flat a-back!*
☞ BRASSE sous le vent! *Haul in the lee braces!* or, *brace the yard!*
BRASSEYAGE, *ou* BRASSAGE, *s. m.* Tha act of bracing the yards; also, the inner quarters of a yard, between the shrouds.
BRASSIAGE, *s. m.* The depth of water, when known by means of the sounding lead.
BRAYE, *s. f.* The canvas coat of a mast.
BRAYE *du gouvernail.* The rudder canvas coat.
BRAYER, *v. a.* To pay the seams, planks, &c. with hot melted pitch.
BREDINDIN, *s. m.* A small tackle affixed to the main-stay in French ships.
BRESIN, *s. m.* A tackle-hook.
EN-BRETON, *adv.* Speaking of the stowage, across, or cross-wise.
BREVET, *s. m.* BREVET *d'officier.* The commission, or warrant of an officer.
BREUVAGE, *s. m.* A mixture of equal parts of wine and water, given to the crew on board French ships, to drink during an action.
BRICOLE, *s. f.* The uneasy rolling of a ship, frequently occasioned by great weights being placed above the centre of gravity.
BRIDER, *v. a.* To frap, or seize, &c. *Ex.*
BRIDER *des cordages.* To seize any complication of ropes, as the gammoning of the bowsprit, with a small rope or seizing.
BRIDER *l'ancre.* To stow the anchor to the bow, and seize it with the shank-painter; also to shoe the anchor.
BRIDER *les haubans.* To swift in the rigging.
BRIDER *les haubans pour en frapper le trélingage.* To span in the rigging for seizing on the cat-harping legs.
BRIG, *s. m. ou* BRIGANTIN. A brig or brigantine.
BRIGADIER, *s. m.* BRIGADIER *d'un canot.* The bowman of a row-boat.
BRIGANTINE, *s. f.* Spanker.
BRIN, *s. m.* *Le premier* BRIN. The best part of hemp when combed out and prepared.
Le second BRIN. The combings of the hemp, out of which they make a coarser sort of ropes.
BRINGUEBALE, *ou* BRIMBALE. *s. f.* The brake, or handle of a ship's pump.
BRION, *s. m.* The fore-foot, placed at the extremity of the keel forward.
BRISANS, *s. m.* A shelf or ridge of rocks; also, the breakers which beat against rocks or sand-banks, above the surface of the water.
BRISE, *s. f.* Breeze. *Ex.*
BRISE *de terre.* The land breeze.
BRISE *du large.* The sea breeze.
BRISE *réglée.* A steady breeze.
BRISE *carabinée.* A hard or stiff gale.
BRISES *folles.* Baffling winds.
BRISE *molle.* A light breeze.
BRISER, *v. n.* To break. *Ex.*
La mer BRISE *beaucoup.* The sea breaks much.
BROCHETER, *v. n.* The practice of shipwrights, which consists in throwing a small line over a plank, to which line small sticks are fastened at equal distances, in order to work it to a proper shape, so as to fay it to the next plank, &c.
BROUILLARD, *s. m.* Fog. *Voyez* BRUME.
BRUINE, *s. f.* Small drizzling rain.
BRULER, *v. a.* To burn. *Ex.*
BRULER *des amorces.* To let off powder in small quantities, as night signals, instead of false fires.
BRÛLOT, *s. m.* A fire-ship.
BRUME, *s. f.* Mist or fog. (This term is preferable to BRUILLARD, which is seldom used at sea.) *Ex.*
Il y a de la BRUME. It is foggy weather.
BRUMEUX, *adj. Temps* BRUMEUX. Foggy weather.

BRUT, *adj.* *Bois* BRUT. Timber in the rough.

BÛCHE, *s. f.* A herring buss, or small fly-boat.

BUGALET, *s. m.* A sort of small vessel, or bark, mostly used about Brest and the neighbouring ports.

BULLETIN, *s. m.* A certificate given to French seamen, when registered in a port, to testify their age, privileges, qualities, time of service, &c.

BUREAU DE SANTE, *s. m.* The office for regulating the quarantine of ships, &c.

BURIN, *s. m.* A sort of large toggel, or a wooden roller made use of in the rigging.

BUTIN. *Voyez* PILLAGE.

C.

CABANE, *s. f.* A small cabin.

CABANES also denote small cabins built close aft on the poop, in French ships of war, for the use of the master and pilot.

CABANER. *Voyez* CHAVIRER.

CABESTAN, *s. m.* A capstan. *Ex.*

Le grand CABESTAN. The after or main capstan.

Le CABESTAN *double.* A double capstan.

Le petit CABESTAN, *ou* CABESTAN *d'avant.* The jear capstan.

CABESTAN *volant.* A crab, which may be shifted from place to place, as occasion requires.

Parties du CABESTAN.	Parts of the capstan.
La tête.	The drum-head.
La tête de la cloche inférieure d'un CABESTAN *double.*	The trundle-head.
Les flasques.	The whelps.
La mèche.	The barrel.
Le pivot.	The spindle, or pivot.
L'écuelle.	The saucer.
Les barres.	The bars.
Les élinguets.	The pawls.
Les petites chevilles qui passent par la tête du CABESTAN.	The pins.
Le garde-corps.	The swifter.
La carlingue.	The step.
Armer le CABESTAN, *ou garnir le* CABESTAN.	To rig the capstan.
Mettre du monde au CABESTAN.	To man the capstan.
Choquer au CABESTAN.	To surge the capstan.
Virer au CABESTAN.	To heave in at the capstan.
Dévirer au CABESTAN.	To come up the capstan.
Mettre les linguets au CABESTAN.	To pawl the capstan.
Envoyer un homme au CABESTAN.	To send a man to the capstan to be flogged.

CABILLOT, *s. m.* A toggle; also a wooden pin for belaying ropes.

CABINE. *Voyez* CHAMBRE.

CABLE, *s. m.* A Cable.

Le maître ou le grand CABLE. The sheet cable.

Le second CABLE. The best bower cable.

CABLE *d'affourche.* The small bower cable.

CABLE *de touée.* The stream cable.

CABLE *de vingt pouces.* A twenty inch cable.

CABLE *de vingt-quatre pouces.* A twenty-four inch cable.

CABLE *de rechange.* A spare cable.

CABLES *de retenue.* Two thick cables fastened to the fore part of a ship on the stocks when about to be launched, and which are cut the moment she is to go off.

Faire bitture. To get up a range of the cable.

Bitter le CABLE. To bit the cable.

Débitter le CABLE. To unbit the cable.

Fourrer le CABLE. To serve the cable.

Filer le CABLE *par le bout.* *Voyez* FILER.

Bosser le CABLE. *Voyez* BOSSER.

Mettre une queue de rat à un CABLE. To point a cable.

☞ Abraque le mou du CABLE! *Heave in the slack of the cable!*

Le CABLE *est surjouailé.* The cable is foul of the anchor-stock.

☞ File du CABLE! *Pay out, or veer away the cable!*

Tout est-il paré au CABLE? Is the cable all clear?

☞ Gâre au CABLE! *Voyez* GÂRE.

CABLEAU, *ou* CABLOT, *s. m.* A cablet, or small cable; particularly the painter or moorings of a boat, &c.

CABOTAGE, *s. m.* The art of coasting; also the coasting trade.

CABOTER, *v. n.* To sail along the coast, or to trade from port to port.

CABOTEUR, *s. m.* *Bâtiment* CABOTEUR. A coasting vessel.

CABRIONS, *s. m. p.* Wedges or quoins for securing the guns in bad weather.

CADRE, *s. m.* A frame in general; but particularly a cot, or bed-frame, such as are used by the officers, or supplied to the sick. *Ex.*

Nous avons des hommes sur les CADRES. We have some men in the sick berth.

CAGE, *s. f.* CAGE *à drisse.* A sort of tub, open on the sides, used in French men of war, to coil in it the remaining parts of the main-topsail-halyards.

CAGE *à poule.* A hen coop.

CAGUE, *s. f.* A kind of Dutch sloop.

CAÏC, *s. m.* The long-boat, or pinnace of a galley.

CAILLEBOTIS, *s. m.* Gratings. *Ex.*

☞ En place les CAILLEBOTIS! *Lay over the gratings!*

☞ Oté les CAILLEBOTIS! *Unlay the gratings!*

CAISSE, *s. f.* CAISSE *flottante, pour amarrer les vaisseaux.* A mooring buoy.

CAISSE *à accorer les vaisseaux.* A contrivance used in the French navy, to prevent ships from being cambered while laying up in ordinary: it consists of two large empty chests, carefully calked up, which they place under the stern of a ship, the weight of which bears upon them, by means of several shores reaching therefrom to the ship's counter.

CAISSE *de poulie.* Shell of a block.

CAISSON, *s. m.* Locker. *Ex.*

CAISSONS *de la grande chambre.* Lockers of the ward-room.

CAISSONS *à gargousses.* Cartridge chests which contain the filled cartridges in a ship's magazine.

CAJOLER, *v. a. & n.* CAJOLER *la marée.* To drive with the tide against the wind.

CALANQUE, *s. f.* A cove, or small harbour, behind a rising ground on the sea coast, fit only to receive small vessels or boats.

CALCETS, *s. m. p.* The cheeks or hounds of the masts, which support the brazen blocks in galleys.

CALE, *s. f.* The hold of a ship. *Ex.*

Cale *à l'eau.* The fore-hold.

Cale *au vin.* The after-hold.

Fond de cale. The lower part of the hold, also the hold in general.

À fond de cale. Down into the hold.

Avant-cale. *Voyez* Avant-cale.

Cale, also signifies the punishment of ducking.

Cale *ordinaire.* Ducking.

Grande cale. Keel-hauling.

Cale *sèche.* A punishment used in some nations, which consists in letting a man drop from the yard-arm close to the surface of the water, but without plunging him into it.

Cale *de construction.* A slip for ship-building.

Cale *de magasin.* A slip of a store-house, or mast-house.

Cale *d'un quai.* A slip of a key or wharf.

Cale. A quoin, wedge, or chock.

CALER, *v. a.* & *n.* To bring down, or sink into the water. *Ex.*

Vaisseau qui n'est pas assez calé. A ship which has not sunk enough into the water.

Caler *une pièce d'arrimage.* To chock or wedge up a cask.

CALFAT, *s. m.* A calker.

CALFATAGE, *s. f.* Calking.

CALFATER, *v. a.* To calk.

CALFATIN, *s. m.* An oakum boy, who spins and twists oakum.

CALHAUBANS. *Voyez* Galhaubans.

CALIBRE d'un canon, *s. m.* The bore of a cannon.

CALIORNE, *s. f.* A winding tackle, or any large two-fold or three-fold tackle.

Caliorne *de misaine.* The fore-tackle.

Caliorne *du grand mât.* The main winding tackle.

CALME, *s. m.* & *adj.* Calm; cessation of wind. *Ex.*

Calme *tout plat.* A flat or dead calm.

Mer calme. A calm or smooth sea.

Tems calme. Calme or still weather.

CALMER, *v. n.* To fall calm, or to becalm.

Le vent a calmé. The wind has becalmed, or blows less strong.

CALMIE, *s. f.* Lull: the time during which the weather falls more calm. *Ex.*

☞ Vogue à la calmie! *Pull away now the wind lulls!*

CAMBUSE, *s. f.* The steward's room.

CAMPAGNE, *s. f.* A voyage, or a cruise at sea during a season or limited space of time. *Ex.*

Campagne *de côte.* A cruise along-shore from port to port.

Campagne *de croisière.* A cruise at sea.

Campagne *d'évolutions.* The time which a fleet spends at sea, for the purpose of exercising.

Campagne *de long cours.* A long voyage.

Campagne *de rade.* The time which men of war remain in a roadstead, and afterwards return into harbour without sailing.

Campagne *des Indes.* A voyage to the East Indies.

Vivres de campagne. The provisions allowed for a voyage or cruise.

CANAL, *s. m.* A channel, also a canal.

Faire canal. To cross over a channel from one shore to the opposite shore.

CANARD, *adj.* A term used by sailors to express the pitching of a ship. *Ex.*

Notre navire est grand canard. Our ship pitches extremely.

CANDELETTE, *s. f.* The fish, or the tackle which is used for fishing the anchor.

Garant de candelette. *Voyez* Garant.

CANON, *s. m.* A cannon or piece of ordnance.

Parties du canon.	Parts of a cannon.
La longueur du canon.	The length of a cannon.
La culasse.	The breech.
Le bouton.	The button, or pomelion.
Le cul-de-lampe.	The breech mouldings.
La plate-bande de la culasse.	The base ring and ogee.
Le champ de la lumière.	The vent-field.
La lumière.	The vent.
L'astragale de la lumière.	The vent astragal and fillets.
Le premier renfort.	The first reinforce.
La plate-bande du premier renfort.	The first reinforce ring and ogee.
Le second renfort.	The second reinforce.
La plate-bande du second renfort.	The second reinforce ring and ogee.
La volée.	The chace.
La ceinture de la volée.	The chace girdle.
La bouche.	The muzzle.
L'astragale de la volée.	The chace astragal and fillets.
L'astragale de la bouche, ou simplement L'astragale.	The muzzle astragal and fillets.
Le bourrelet.	The swelling of the muzzle.
La ceinture de la bouche.	The muzzle mouldings.
Les tourillons.	The trunnions.
L'âme ou le calibre.	The bore or calibre.

Canon *de dix-huit.* An eighteen pounder.

Canon *de vingt-quatre.* A twenty-four pounder.

Canon *de trente-six.* A thirty-six pounder.

Canons *de fer.* Iron guns.

Canons *de fonte.* Brass guns.

Canon *de rebut.* A gun unfit for service.

Canons *de chasse.* Bow-chases.

Canons *de coursiers.* The bow-chases of a galley.

Canons *de retraite.* Stern-chases.

Canons *en serre.* The guns housed athwart.

Canons *alongés contre le bord.* The guns housed fore-and aft.

Enclouer un canon. To nail up or spike a piece of ordnance.

Amarrer un homme sur un canon. To tie a man to a gun to be flogged with a rope's end.

Mettre les canons *a la serre.* *Voyez* Mettre.

CANONNADE, *s. f.* Cannonade, or the act of cannonading.

CANONNAGE, *s. m.* Gunnery.

Instrumens de cannonage. The implements necessary for charging guns.

Officiers mariniers de CANONNAGE. The gunner and his mates, &c.
CANONNER, *v. a.* To cannonade. *Ex.*
CANONNER *en plein bois.* To hull a ship.
CANONNER *un fort.* To cannonade à fort.
CANONNIER, *s. m.* CANONNIER *de vaisseau.* The gunner of a ship. *Ex.*
Maître-CANONNIER *d'un bâtiment de guerre.* Master-gunner of a man of war.
Second-Maître CANONNIER. Gunner's mate.
Aide-CANONNIER. Quarter-gunner.
CANOT, *s. m.* A boat. *Ex.*
Le grand CANOT *d'un vaisseau de guerre.* The barge of a ship of war.
Le second CANOT, *ou* CANOT *de l'état major.* The pinnace.
Le Petit CANOT. The jolly-boat.
CANOT *Indien. Voyez* PIROGUE.
CANOTIER, *s. m.* A rower of a boat, cutter, pinnace, &c.
CAP, *s. m.* The head of a ship, &c. *Ex.*
Le CAP *du vaisseau.* The head.
Où est le CAP? How is the head? &c.
Mettre le CAP *en route.* To stand on the course.
☞ Le CAP en route! *Steer the course!*
☞ Gouverne ou tu as le CAP! *Steer as you do!*
CAP *d'ouvriers.* A quarterman.
CAP. A cape, *Ex.*
Le CAP *de Bonne Espérance.* The Cape of Good Hope.
CAP *de mouton, s. m.* A dead-eye.
CAP *de mouton à corde.* A rope-bound dead-eye.
CAP *de mouton ferré.* An iron-bound dead-eye.
CAP *de mouton à croc.* An iron-bound dead-eye, with a hook.

CAPACITÉS, *s. f. p.* The burden or tonnage of a ship; or, in general, the space contained within her hold. *Ex.*
Vaisseau qui a de grandes CAPACITÉS. A full-built ship, or a ship of a great burden.
Vaisseau qui manque de CAPACITÉS. A ship that has too narrow a floor.
CAPE, *s. f.* Trying, or lying-to in a storm.
Etre à la CAPE *sous la grande voile.* To be lying-to under a main-sail.
Mettre à la CAPE. *Voyez* METTRE.
Ce bâtiment est à la CAPE *sous la pouilleuse et cet autre sous la misaine.* That ship is trying under her main stay-sail, and the other under her fore-sail.
CAPELAGE, *s. m.* The shrouds, rigging, &c. that go over the mast-head.
CAPELER, *v. a.* To fix the shrouds, &c. on the mast-head.
CAPELER *une hune.* To get a top over.

CAPÉYER, *v. n.* To lie to, or to try. *Ex.*
Bâtiment qui CAPÉYE *bien.* A ship that is easy when lying-to.
CAPION, *s. m.* The upper part of the stern-post and of the stem, in a galley or xebec. This term being applied to such vessels only. *Ex.*
De CAPION *à* CAPION. From stem to stern.
CAPION *de poupe.* The upper part of the stern-post.
CAPION *de proue.* The upper part of the stem.
CAPITAINE, *s. m.* A captain. *Ex.*
CAPITAINE *de vaisseau.* A post captain.
CAPITAINE *de pavillon.* The captain of a flag-ship.
CAPITAINE *de frégate.* An officer next in rank to a post captain, in the French navy, who is appointed to the command of frigates, large sloops of war; and also as second in command in a line-of-battle ship.
CAPITAINE *de brûlot.* The captain of a fire-ship.
CAPITAINE *de vaisseau marchand.* The captain of a merchant-man.
CAPITAINE *guarde-côte.* A captain of the militia guarding the coast.
CAPITAINE *de port.* Master-attendant; also, a harbour-master in a sea port.
CAPITAINE *d'armes.* Master at arms.
CAPLANIER, *s. m.* A cod-fisher; also her crew.
CAPON, *s. m.* The cat, or tackle employed for catting the anchor, *Ex.*
Poulie de CAPON. Cat-block.
Croc de CAPON. Cat-hook.
Garant de CAPON. *Voyez* GARANT.
☞ Croche le CAPON! *Hook the cat!*
☞ Hale sur le CAPON! *Haul away the cat!*
☞ Amarre le CAPON! *Belay the cat!*
CAPONNER L'ANCRE, *v. a.* To cat the anchor.
CAPOSER, *v. n.* To bring a ship to, with her helm a-lee.
CAPOT, *s. m.* CAPOT *d'échelle.* Companion.
Faire CAPOT. To overset; speaking of a ship.
CAPUCINE. *Voyez* COURBE *de Capucine.*
CARACORE, *s. f.* A light vessel used in the Indian seas by the Dutch and the natives, particularly those of the Island of Borneo.
CARAQUE, *s. f.* A name given by the Potuguese to large ships which they employed in the Brazil and East India trade.
CARAVELLE, *s. f.* A small square-sterned Portuguese vessel, with lateen sails.
CARCASSE, *s. f.* The carcass, or ribs of a ship, before the planks are laid on, or after they are ripped off.

CARÉNAGE, *s. m.* The act of careening ships; also a wharf, or careening place.

CARÈNE, *s. f.* Careen; also the outside of a ship's bottom.
Abattre un vaisseau en CARÈNE. *Voyez* ABATTRE.
CARÈNE *entière.* A thorough careen, by which a ship is hove keel out.
Demi-CARÈNE. A parliament-heel, or boot-topping.

CARÉNER, *v. a. & n.* To careen and boot-top, or to grave and pay a ship.
CARET, *s. m.* *Fil de* CARET. Rope-yarn.
CARGADOR, *s. m.* The person who procures a freight, or voyage for a merchant-ship.
CARGAISON, *s. f.* The cargo of a merchant-ship. *Ex.*
Quelle est votre CARGAISON? What are you loaded with?
CARGUE-BAS, *s. m.* The down-hall tackle of a topsail-yard.

CARGUER, *v. a.* To clew, or haul up a sail; to brail up. *Ex.*
☞ CARGUE la misaine! *Up fore-sail!*
☞ CARGUE le petit hunier. *Clew up the fore-top-sail!*
☞ CARGUE l'artimon! *Brail up the mizen!*
☞ CARGUE les perroquets! *Clew up the top-gallant-sails!*
CARGUES, *s. f. p.* A general name for the brails of a sail, comprehending the clew-lines, bunt-lines, leech-lines, &c. *Ex.*
CARGUES-*boulines.* Leech-lines.
CARGUES-*fonds.* Bunt-lines.
CARGUES-*points.* Clew-lines.
Mettre les huniers sur les CARGUES. To clew up the top-sails.
CARGUES *du vent.* Weather brails.
CARGUES *de dessous le vent.* Lee brails.
Fausses CARGUES. Spilling-lines.
CARLINGUE, *s. f.* The keelson, or kelson.
CARLINGUE *d'un mât.* Step of a mast.
CARLINGUE *du grand mât.* The step of the main-mast.
CARLINGUE *du mât de misaine.* The step of the fore-mast.
CARLINGUE *du mât d'artimon.* The step of the mizen-mast.
CARLINGUE *du beaupré.* The step of the bowsprit.
CARLINGUE *de cabestan.* The step of the capstan.
CARNAL ou CARNAU, *s. m.* The lower part of a lateen yard.
CARONADE, *s. f.* A carronade.

CARRÉ, *s. m.* A sledge, used in rope-yards.
CARREAU. *Voyez* PLAT-BORD.
CARROSSE, *s. m.* The canopy in the stern part of a galley, or xebec.
CARROSSE, *ou chariot de corderie.* A frame on four wheels, used in rope-yards to carry the laying-top.
CARTAHU, *s. m.* A girt-line.
Poulie de CARTAHU. A single block, through which a girt-line is reeved.
CARTE, *s. f.* A chart, or map of the sea. *Ex.*
CARTE *marine.* A sea-chart.
CARTE *plane.* A plane-chart.
CARTE *réduite.* A Mercator's chart.
CARTEL, *s. m.* Cartel, a ship employed to exchange prisoners; also to carry requests and proposals.
CARTOUCHE, *s. f.* A cartridge.
CASERNET, *s. m.* A sort of log-book kept by the pilots on board French ships of war.

CASSE-TÊTE, *s. f.* Over-head netting.
CASSER, *v. n.* To cashier, to break, &c. *Ex.*
CASSER *un officier.* To break or cashier an officer.
CATIMORAN, *s. m.* A catamaran.
CAYENNE, *s. f.* An old hulk or receiving ship, in which they lodge and victual the sailors that are ordered to a French dock-yard, for fitting out a squadron, &c.
CAYES, *s. f. p.* Keys or chains of rocks, nearly even with the surface of the sea.

CÉDER, *v. n.* To settle. *Ex.*
Ponts qui ont CÉDÉS. Decks which have settled.
CÉDER *au mauvais tems.* To bear up and put before it in a gale of wind.
CÉDER *l'avantage du vent.* *Voyez* AVANTAGE *du vent.*
CEINTURE, *s f.* CEINTURE *d'un bâtiment.* A swifter, or rope which encircles a vessel above the load-water-line.
CEINTURE *de canot.* Swifter of a boat.
CEINTURE *de la bouche d'un canon.* *Voyez* CANON.
CEINTRER, *v. a.* CEINTRER *un bâtiment.* To frap a ship.
CENTRE DE GRAVITÉ, *s. m.* The centre of gravity, or point of equilibrium.
CENTRE *d'une armée navale.* *Voyez* CORPS.
CERCLE, *s. m.* CERCLE *de réflexion.* A reflecting circle.
CERCLE *pour amurer le foc.* Jib-iron.
CERCLES *de boute-hors.* The studding-sail boom-irons.
CERCLE *de hune.* Hoop of a top.
CERCLES *de vergues.* Yard-hoops.
CERCLES *de pompe.* Pump-hoops.
CERCLES *de mât.* Mast-hoops
CERCLES *des roustures des mâts.* Wooden hoops nailed on the lower masts, above and below the wooldings.
CERCLES *de jât d'ancre.* Anchor-stock hoops.
CERCLES *d'épontilles.* Hoops of the stanchions between decks.
CERCLES *de cabestan.* Capstan-hoops.
CHABLEUR, *s. m.* A water-officer, who has the care of the wherries.

CHAÎNE, *s. f.* A chain. *Ex.*
CHAÎNE *de paratonnerre.* Chain of the conductor.
CHAÎNE *de port.* The boom, or chain of a harbour.
CHAÎNES *de haubans.* Chains of the dead-eyes of the lower rigging.
CHAÎNES *de vergues.* The top-chains.
CHAÎNES *de gouvernail.* Rudder-chains.
CHAÎNES *d'amarrage.* Mooring-chains.
CHAÎNES *de pompe.* Pump-chains.
CHAÎNES *des grands haubans.* The main-chains.
CHAÎNES *des haubans de misaine.* The fore-chains.
CHAÎNES *des haubans d'artimon.* The mizen-chains.
CHAÎNES *de rochers.* A ridge of rocks, or ledge.
CHALAN, *s. m.* A sort of lighter.
CHALAN *à cheval.* A lighter for shipping horses, &c.
CHALOUPE, *s. f.* A sort of large boat. *Ex.*
CHALOUPE *d'un vaisseau de guerre.* The launch or long-boat of a ship of war.
CHALOUPE *armée.* A launch manned and armed.
CHALOUPE *canonnière.* A gun-boat.
CHAMBRE, *s. f.* A cabin.
Grande CHAMBRE. The ward-room in line-of-battle ships, and the great cabin in smaller vessels.
CHAMBRE *de conseil.* The captain's cabin in a large man of war, and what is called the round-house, in a large East-Indiaman.
CHAMBRE *de poupe.* The aftermost cabin of a small vessel.
CHAMBRE *d'un canot.* The stern sheets of a boat.
CHAMBRE *de canon.* The chamber, or charged cylinder of a cannon.
CHAMBRE *de mortier.* The chamber of a mortar.

CHAMBRES, are also the flaws, cracks, or honeycombs in a piece of ordnance.
CHAMBRÉ, *adj. Pièce* CHAMBRÉE. A gun that is honeycombed.
CHAMBRIÈRES, *s. f. p.* Beckets for the tacks and sheets.
CHAMEAU, *s. m.* A camel.
CHAMPAN, *s. m,* A sampane, a kind of Chinese vessel.
CHANDELIERS, *s. m. p.* All sorts of stanchions or crochets that are fixed on the gunwale, &c.
CHANDELIERS *de lisses pour bastingage.* Stanchions for the netting.
CHANDELIER *de chaloupe.* The crotch of a boat which sustains the main-boom, or the mast and sail, when they are lowered.
CHANDELIERS *d'échelle hors le bord.* The stanchions which support the entering ropes at the gangway.
CHANDELIER *de pierrier.* The iron crotch of a swivel; also the wooden stock in which it is shipped.
CHANGER, *v. a. & n.* To alter, to turn, to change, to shift, &c. *Ex.*
CHANGER *la grande voile d'un cutter,* &c. To gybe.
CHANGER *la barre.* To shift the helm.
CHANGER *le tournevire.* To shift the messenger.
CHANGER *l'arrimage.* To alter the stowage.
CHANGER *d'amures.* To put about.
Le vent CHANGE. The wind shifts.
CHANGEMENT, *s. m.* CHANGEMENT *de la marée.* The turn of the tide.
CHANGEMENT *de lune.* The change of the moon.
CHANTIER, *s. m.* The stocks upon which a ship is built. *Ex.*
CHANTIER, *ou attelier de construction.* A shipwright's yard, or wharf.
CHANTIERS *de chaloupe.* Chocks for a boat.
CHANVRE, *s. m.* The hemp employed to make the sails and cordage of a ship.
CHAPE, *s. f.* CHAPE *de boussole.* The inner box of a sea compass.
CHAPEAU, *s. m. Droit de* CHAPEAU *du maître.* A captain's perquisite, in the merchant service, on freighted goods.
CHAPELLE, *s. f.* To chapel a ship; to broach-to.
CHARBONNIER, *s. m. Vaisseau* CHARBONNIER. A collier.
CHARGE, *s. f.* CHARGE *d'un bâtiment.* The cargo or lading of a vessel.
Morte CHARGE. Overlading, or the utmost burden that a ship can carry.
CHARGE *de canon.* Shot; also the quantity of powder used in loading a gun.
CHARGE *de combat.* The quantity of powder necessary for loading of a gun for service.
CHARGE *de salut.* The quantity of powder used in loading a gun for saluting.
CHARGE *à mitraille.* Case-shot,
CHARGE *à la suédoise, ou* CHARGE *en grappe.* Grape-shot.
CHARGEMENT, *s. m.* Cargo, or lading.
Etre en CHARGEMENT. To be loading.
Prendre un CHARGEMENT. To take in a cargo.
CHARGER, *v. a. & n.* CHARGER *un navire.* To load a ship.
CHARGER *à cueillette.* To load a ship with goods belonging to several owners.
CHARGER *à fret.* To take in freight.
Vaisseau CHARGÉ *à couler bas.* A ship overloaded, or loaded so as to be ready to sink.
Vaisseau CHARGÉ *par un grain.* A ship laid upon her side by a sudden gust of wind.
Vaisseau CHARGÉ *en côte.* A ship embayed on a lee-shore with a hard wind.
CHARGER *un canon.* To load a gun.
☞ CHARGEZ les canons à boulet! *Shot your guns!*
CHARGER *la pompe.* To fetch the pump.
CHARGER *en grenier.* To load a ship in bulk.
CHARGEUR, *s. m. Marchand* CHARGEUR The owner of a ship's cargo.
CHARNIER, *s. m.* Scuttle-butt; a cask placed on the quarter-deck, containing water for the use of the ship's crew.
CHARPENTIER, *s. m.* A carpenter, also a shipwright.
Maître-CHARPENTIER *d'un vaisseau.* Carpenter of a man of war.
Second Maître-CHARPENTIER. Carpenter's mate.
Aides-CHARPENTIERS, *ou gens du* CHARPENTIERS. Carpenter's crew.
CHARIER, *v. a.* CHARIER *de la voile.* To crowd sail; to carry all the sail the masts can bear.
CHARTE-PARTIE, *s. f.* Charter-party.
CHASSE, *s. f.* A chase at sea. *Ex.*
Donner la CHASSE *à un vaisseau.* To chase, or to give chase to a ship.
Prendre CHASSE. To sheer off; to stand away from a vessel in sight.
En CHASSE, *adv.* Speaking of guns. *Ex.*
☞ Pointez en CHASSE! *Point before the beam!*
CHASSÉ, *part. pass. & adj. Vaisseau* CHASSÉ. The chase, or a ship chased.
CHASSE-MARÉE, *s. m.* A sort of decked boat employed for the conveyance of fish, and also in the coasting trade.
CHASSER, *v. a. & n* To chase a ship. *Ex.*
CHASSER *en retraite.* To fight in the act of retreating.
Vaisseau qui CHASSE *au vent.* A ship chasing to windward.
CHASSER *la terre.* To look out for the land.
CHASSER *sur les ancres. Voyez* ANCRE.
CHASSER *sur un navire.* To drive on board of a ship by dragging the anchors.
CHASSEUR, *s. m.* A ship that chases another.
CHAT, *s. m.* Cat; a sort of vessel used in the Baltic, and on the north-east coast of England.
CHÂTEAU. *Voyez* GAILLARD.
CHÂTRER, *v. a.* CHÂTRER *une gargousse.* To reduce, as the piece gets warm, the quantity of powder in a cartridge.
CHATTE, *s. f.* A sort of lighter, so called at Rochefort.
CHAUDIÈRES, *s f. p.* The ship's coppers.
CHAUDRON DE POMPE, *s. m.* A plate of lead or copper, perforated with holes, to cover the bottom of a pump.

CHAUFFAGE, *s. m.* The act of breaming a ship, or burning the filth from off her bottom.

CHAUFFER, *v. a.* To bream, &c. *Ex.*

CHAUFFER *un navire pour la carène.* To bream a ship.

CHAUFFER *des bordages.* To bend planks, or to make them pliant by heating them.

CHAUFFER *la soute à pain.* To heat or dry the bread-room, in order the better to preserve the biscuit during a sea voyage.

CHAVIRER, *v. a. & n. ou* FAIRE CAPOT. To overset; to capsize.

Le canot est CHAVIRÉ. The boat is upset.

CHÉBEC, *s. m.* Xebec; a sort of vessel used mostly in the Mediterranean.

CHEF, *s. m.* CHEF *d'escadre.* A title formerly given to rear-admirals in the French navy, who are now called Contre-Amiraux. CHEF *d'escadre, ou* CHEF *de division*, is the present title given to a commodore.

CHEF *de gamelle.* The caterer of an officer's mess.

CHEF *de pièce.* Captain of a gun.

CHEF *de file.* The leading ship of a fleet or of a squadron, when drawn up in a line of battle.

CHEF, the stem or head of a boat; also the end of a cable used as a head-fast to a ship, &c.

CHEMIN, *s. m.* Way, or head-way. *Ex.*

CHEMIN *de l'avant.* Head-way.

CHEMIN also denotes a range of skeeds, whereon to roll full casks, either on shore or on board.

Vaisseau qui fait beaucoup de CHEMIN. A ship that has fresh way through the water.

Faire soixante lieus de CHEMIN. To run sixty leagues.

CHEMIN *fait en longitude. Voyez* LONGITUDE.

CHEMISE, *s. f. ou* CHEMISE *souffrée.* A curtain, or piece of old canvas, dipped in a composition of oil, petrol, camphire, and other combustible materials, to be nailed to the planks of an enemy's ship, when it is intended to set her on fire.

CHEMISE *de chargement.* A piece of canvas used to enclose the hold, pump-well, &c. to prevent the cargo from getting wet.

Serrer les huniers en CHEMISES. To furl or hand the sails; to take in the sails.

CHENAL, *s. m.* A narrow channel.

CHENALER, *v. n.* To sail through a channel.

CHENETS, *s. m. p.* A sort of iron claws used to bend the planks of a ship by fire.

CHEVALET, *s. f.* A roller for passing the cables from one place to another.

CHEVAUCHER, *v. n.* In ship-building, to ride, or be fayed upon.

CHEVILLE, *s. f.* An iron bolt, of which there are several sorts used in the construction of a ship, &c. *Ex.*

CHEVILLE *à goujon.* A common bolt.

CHEVILLE *à goupille.* Fore-lock bolt.

CHEVILLE *à œillet.* An eye-bolt.

CHEVILLE *à boucle.* Ring-bolt.

CHEVILLE *à croc.* Hook-bolt.

CHEVILLE *à boucle & à croc.* A bolt with a ring and a hook.

CHEVILLE *de hauban.* Chain-bolt.

CHEVILLE *à tête de diamant.* Square-headed bolt.

CHEVILLE *à pointe perdue.* A short drove-bolt, whose head is sunk into the timber in which it is driven.

CHEVILLE *de pompe.* A pump-bolt.

CHEVILLE *de potence de pompe.* A long pump-bolt, or bolt which fastens the brake to the cheeks or ears of the pump.

CHEVILLE *d'affût. Voyez* AFFÛT.

CHEVILLE *de fer à charger le canon.* Langrage-shot.

CHEVILLER, *v. a.* To bolt a ship, or to drive the bolts which fasten one part to another.

CHEVILLOT, *s. m.* Toggel; also a belaying-pin, fixed in the rail of a small vessel.

CHÈVRE, *s. f.* A gin, or triangle with pullies.

CHEVRON, *s. m.* Scantlings in general, or square timber, from five or six inches in thickness downwards; also a long wedge for securing guns.

Mettre les CHEVRONS *à la batterie.* To apply the wedges to the trucks of the gun-carriages.

CHICANER, *v. a.* CHICANER *le vent.* To hug the wind too close. *Ex.*

☞ Sans CHICANER le vent! *Don't hanker her in the wind!* or, *Don't hug the wind so close!*

CHIOURME, *s. f.* The gang, or crew of slaves belonging to a row-galley.

CHIRURGIEN, *s. m.* CHIRURGIEN-*major d'un vaisseau de guerre.* The surgeon of a man of war.

Second CHIRURGIEN. The surgeon's first mate.

Aides-CHIRURGIENS. The surgeon's mates.

CHOC, *s. m.* A surge, occasioned by suddenly slacking a taught rope.

CHOPINE DE POMPE, *s. f.* The lower pump-box.

CHOQUER, *v. a. & n.* To surge, to check, &c. *Ex.*

CHOQUER *le* tournevire. To surge the messenger.

CHOQUER *la bouline.* To check the bowline.

CHOSES DE LA MER. A wreck, or whatever is found floating at sea.

CHOUQUET, *s. m.* A cap of the mast-head.

CHOUQUET *de grand mât.* Cap of the main-mast.

CHOUQUET *du baton d'enseigne.* Cap of the ensign-staff.

CHOUQUET *en fer.* An iron cap.

CHÛTE, *s. f.* CHÛTE *d'une voile.* The depth, or drop of a sail, when speaking of the principal square-sails. *Ex.*

Le grand hunier a beaucoup de CHÛTE. The main-top-sail is very deep.

CHÛTE *des courans.* The settings of the tides, or currents.

CITERNE, *s. f.* A cistern.

CIVADIÉRE, *s. f.* The sprit-sail of a ship.

CLAMP, *s. m.* A sheave or small wheel, placed in a mortise, as in the foot of a topmast, to pass a rope through.

CLAN, *s. m.* A mortise or hole cut in a plank, mast, &c. lengthways, to admit a sheave; also a breast-hook in a large lighter.

CLAPET DE POMPE, *s. m.* Clapper of a pump-box.

CLAPETS *des dalots.* Scupper-leathers.
CLAPOTAGE, *s. m* Turbulent motion of the sea, or running in heaps.
CLAPOTEUSE, *adj.* *Mer* CLAPOTEUSE. A turbulent short sea, such as is seen upon a lee shore.
CLASSES, *s. f. p.* The divisions of pilots, gunners, and seamen, in the French service, who are registered in the several departments, so that they may be immediately called upon to serve whenever required.
CLAVETTE, *s. f.* The fore-lock of a bolt.
CLEF, *s. f.* A sort of knot or hitch, &c.
Demi-CLEF. A half-hitch.
CLEF *des mâts de perroquet & de hune.* Fid of the top-masts and top-gallant-masts.
Passer la CLEF *dans un mât de hune.* To fid a top-mast.
CLEF *de beaupré.* In ship-building, the collar-beam, which is raised a little above the second deck, to fortify the bowsprit.
CLEF *de pierrier.* The fore-lock of a swivel-gun.
CLEFS *de virevaut.* The cheeks of the windlass.
CLEFS *entre les varangues* In ship-building, filling-pieces put between the floor-timbers under the kelson, in French ships; also such filling-pieces thrust in the timber's room and space in two parallel ranges, running fore and aft.
CLIN—À CLIN, *adv.* *Ex.*
Border à CLIN. To plank a vessel with clincher-work.
Bordé à CLIN. Clincher-built.
CLOCHE, *s. f.* A bell.
☞ Pique quatre! *Strike the bell four!*
CLOCHE *de cabestan.* The part of a capstan round which the messenger, &c. is wound, comprehending the drum head; so called from resembling a bell.
CLOCHE *de plongeur.* A diver's bell.
CLOISON, *s. f.* Bulk-head, &c.
CLOISON *de la gatte.* Manger-board.
CLOUS, *s. m. p.* Nails or spikes.
CLOUS *au poids.* Weight-nails or spikes.
CLOUS *de* 20 *pouces, de* 12 *pouces, &c.* 20-inch spikes, 12-inch spikes, &c.
CLOUS *de double caravelle.* A sort of nails 5¼ French inches in length, much like the English deck-nails.
CLOUS *de caravelle.* A sort of nails 4¾ French inches in length, like some of the English single deck-nails.
CLOUS *de demi-caravelle.* Nails of 4¼ French inches in length, called, in English, single deck-nails.
CLOUS *de lisse.* Nails like the English twenty-four penny nails.
CLOUS *de double tillac.* Nails of about 2½ inches in length, French measure, much like English twenty-penny drawing nails.
CLOUS *de tillac.* A sort of nails about 2 French inches in length, like the English ten-penny nails.
CLOUS *de demi-tillac.* Six-penny nails.
CLOUS *à plomb.* Lead nails, commonly called two-penny nails.
CLOUS *à pompe.* Pump nails, ¾ of an inch in length.
CLOUS *à maugère.* Scupper nails, made with a broad flat head.
CLOUS *de ferrure de gouvernail.* Rudder nails, or nails used in the iron-work of the rudder.
CLOUS *à vis.* Clincher nails.
CLOUS *à tête ronde.* Round-headed nails.
CLOUS *à tête piquée.* Clasp-headed nails.
CLOUS *de cuivre.* Copper nails.
CLOUS *de sabords.* Doubling-nails, to line the gun-ports.
CO-BOURGEOIS, *s. m.* A co-partner in, or part-owner of a ship.
EN COCHE, *adv.* *Avoir les huniers en* COCHE. To have the top-sails a-trip, or hoisted up.
En COCHES *ou dents d'affût de bord.* The notches or steps of a gun-carriage.
COCOTIER, *s. m.* A cocoa-palm. *Ex.*
Bâtiment COCOTIER. A ship so rigged that her blocks are uncommonly conspicuous.
COFFRE, *s. m.* A chest, &c. *Ex.*
COFFRE, *d'arme.* A chest used to contain the small arms on the quarter-deck.
COFFRE *d'ammunition.* Ammunition chest.
COFFRE *à médicine.* Medicine chest.
COFFRE *de signaux.* Colour chest.
COFFRE *de bord.* A sea chest; a sailor's chest.
COIFFER, SE COIFFER, *v. a. & r,* To back, or to lay a-back.
Se COIFFER *en chicanant le vent.* To be taken a-back by hugging the wind too close.
COIFFER *un vaisseau.* To lay all flat a-back.
COIFFER *une voile.* To lay a sail a-back.
COIN, *s. m.* COIN *de mire.* The quoin of a gun.
COINS *de mât.* Quoins or wedges, by which a mast is confined in the partners.
COINS *de bois.* Small wooden quoins, or wedges.
COINS *à fendre.* Iron wedges.
COINS *à manche.* Horsing-irons, used by calkers, to open the seams.
COINS *de chantier.* A sort of wedges used in the launching of a ship.
COINS *d'arrimage.* Wedges, or chocks, used in trimming or stowing the hold, &c.
COINCER, *v. a.* To wedge. *Ex.*
COINCER *un mât.* To wedge up a mast.
COITTES. *Voyez* ANGUILLES.
COLLET, *s. m.* COLLET *d'étai.* The eye of a stay.
COLLET *d'une ancre.* The crown of an anchor.
COLLET *d'une courbe.* The throat of a knee.
COLLIER, *s. m.* COLLIER *d'étai.* The collar of a stay.
Faux COLLIER. Collar of a preventer stay.
COLLIER *de défense.* The puddening of a boat's stem.
COLLIER *de chouquet, ou* COLLIER *de mât.* An iron clamp to fasten the top to the mast or bowsprit in small vessels.
COLOMBIERS, *s. m. p.* Two shores employed to block up a ship's cradle.
COLONNE, *s. f.* In naval tactics, a line of ships. *Ex.*
L'escadre est rangée sur trois COLONNES. The fleet is drawn up in three lines.
COLTIS, *s. m.* The foremost frame of a ship.
Cloison du COLTIS. The beak-head bulk-head.
COMBAT NAVAL, *s. m.* An engagement, or sea-fight.
COMBATTRE, *v. a.* COMBATTRE *des deux bords.* To engage on both sides.
COMBLÉ, *adj.* *Port* COMBLÉ. A harbour choked or filled up with mud, &c.

COMBUGER, *v. a.* COMBUGER *des futailles.* To soak; to rince out casks.

COMITÉ DE MARINE, *s. m.* A secret council held by the king and his ministers, to deliberate about the affairs of naval war.

COMMANDANT, *s. m.* Commander, &c. *Ex.*

COMMANDANT *d'une escadre, d'une division.* The commander in chief of a fleet, or division; also the commodore or commanding officer of any detachment of ships.

COMMANDANT *de la marine dans un port.* A flag officer of the navy, who superintends the military business of the navy in a port, &c.

Vaisseau COMMANDANT. The admiral ship of a fleet or squadron.

COMMANDANT *d'une prise.* A prize-master.

COMMANDE, *s.f.* A gasket, rope-band, or knittle.

COMMANDEMENT, *s. m.* An order; also, the command of a ship, &c.

COMMANDER, *v. a.* To command.

☞ COMMANDE! *Holloa!* This is the answer made by the sailors to the boatswain and his mates, on board French ships of war, whenever they give a preparatory whistle, in order to command attention, and previous to their signifying any general orders verbally and by winding their call.

COMME CELA! ou COMME *ça! adv.* In conning, Thus! as you go! Steady as you go!

COMMANDEUR, *s. m.* The master or commander of any trading merchant ship.

COMMETTAGE, *s. m.* Laying of ropes and cables.

COMMETTRE, *v. a.* To lay; speaking of ropes, &c. *Ex.*

Cordage COMMIS *en trois.* A three-stranded rope.

Cordage COMMIS *en quatre.* A four-stranded rope, &c.

COMMIS DE LA MARINE, *s. m.* The clerks in a dock-yard; also captains' clerks.

COMMIS *de bâtiment marchand, s. m.* The super-cargo of a merchant-ship.

COMMIS *aux vivres, s. m.* Purser, or steward in a French ship of war, whose duties consist in surveying the delivery of provisions, and keeping an account of them; he being accountable to the contractor, and not to the state, as in the English service.

COMMIS *des bureaux des douanes.* The surveyors of the customs who visit shipping.

COMMIS *des vivres.* A clerk or steward to the commissary or purser of a ship of war.

COMMIS *général des convois.* The overseer or inspector of the customs, with respect to shipping.

COMMISSAIRE, *s. m.* COMMISSAIRE *de la marine.* A commissioner of the navy or marine.

COMMISSAIRE *général de la marine.* The principal officer under the *intendant de marine*, who inspects the equipping, manning, and victualling of the ships, the performances of the artificers, &c. the affairs of the dock-yards, the prizes, &c.

COMMISSAIRE *général de la marine ambulante.* An officer like the preceding, but who visits occasionally every port and harbour, having no fixed residence.

COMMISSAIRE *ayant inspection sur les vivres d'un port.* An agent victualler of any port.

COMMISSAIRE *des ports, ou* COMMISSAIRE *pour les constructions des vaisseaux.* A master-attendant.

COMMISSAIRE *ordinaire de la marine.* The officer who superintends the ordinary, &c.

COMMISSAIRE *de l'artillerie de la marine.* The officer who has the charge of the foundery, the proof of cannon and mortars, and all other arms, gunpowder, ammunition, instruments, and implements of war.

COMMISSAIRE *ordinaire de l'artillerie de la marine.* An officer who officiates for the COMMISSAIRE *de l'artillerie de la marine*, when he is absent.

COMMISSAIRE *des armemens.* An officer similar to our clerk of the check.

COMMISSION, *s. f.* The commission of a principal officer in the navy.

COMMISSION *en guerre.* A warrant or permission from the state, to entitle a privateer to seize on the enemy's vessels. *Voyez* LETTRES DE MARQUE, &c. which are more generally used.

COMMUNIQUER, *v. n.* To have a free intercourse or communication with the crew of a ship, or the people of any country, after having performed quarantine, or otherwise.

COMPAGNE, *s. f.* The cabin of the steward of a row-galley.

VAISSEAU DE COMPAGNIE, *s. f.* A good company keeper.

Aller de COMPAGNIE. To sail in company.

COMPAGNIE *des Indes.* The East India Company.

COMPAGNIE *de commerce.* A trading company.

COMPAGNONS, *s. f. p.* Ships bound to the same place.

COMPAS, *s. m.* A compass.

COMPAS *courbes.* Callipers, or calliper compasses.

COMPAS *azimuthal, ou de variation.* Azimuth compass.

COMPAS *renversé.* Hanging compass.

COMPAS (*ou* BOUSSOLE) *d'inclinaison.* Dipping compass.

COMPAS *de carte.* A pair of compasses used to prick a chart.

COMPAS *de route.* Steering compass.

Quarts du COMPAS Points of the compass.

NORD.	NORTH.
Nord quart-nord-est.	North by east.
Nord nord-est.	North north-east.
Nord-est quart de-nord.	North-east by north.
Nord-est.	North-east.
Nord-est quart-d'est.	North-east by east.
Est nord-est.	East north-east.
Est quart nord-est.	East by north.
EST.	EAST.
Est quart sud-est.	East by south.
Est sud-est.	East south-east.
Sud-est quart-d'est.	South-east by east.
Sud-est.	South-east.
Sud-est quart de sud.	South-east by south.
Sud sud-est.	South south-east.
Sud quart sud-est.	South by east.
SUD.	SOUTH.
Sud quart sud-ouest.	South by west.
Sud sud-ouest.	South south-west.
Sud-ouest quart-de sud.	South-west by south.
Sud-ouest.	South-west.
Sud-ouest quart-d'ouest.	South-west by west.
Ouest sud-ouest.	West south-west.
Ouest quart sud-ouest.	West by south.

Ouest.	West.
Ouest quart nord-ouest.	West by north.
Ouest nord-ouest.	West north-west.
Nord-ouest quart-d'ouest.	North-west by west.
Nord-ouest.	North-west.
Nord-ouest quart-de-nord.	North-west by north.
Nord nord-ouest.	North north-west.
Nord quart nord-ouest.	North by west.

Quarts du COMPAS. *Voyez* AIRE *de vent*, *&* RHUMB *de vent*.

AU COMPLET, *adv. Notre équipage est au* COMPLET. We have our full complement on board.

COMPORTER. *se* COMPORTER, *v. a. & r.* To behave, &c. *Ex.*

Cette frégate COMPORTE *une haute mâture.* That frigate requires to be taunt-masted.

Vaisseau qui se COMPORTE *bien à la mer.* A good sea-boat, or a ship that behaves well at sea.

COMPOST, *s. m.* Tide-duty, or the revenue arising from shipping.

CONDAMNÉ, *adj.* Unfit for service.

CONDAMNER, *v. a.* To condemn, &c. *Ex.*

CONDAMNER *un bâtiment.* To condemn a ship, as being unserviceable.

CONDAMNER *les sabords d'un vaisseau.* To calk in a ship's ports.

CONDUIRE, *v. a.* To carry, to lead. *Ex.*

Le pilote nous CONDUISIT *hors du port.* The pilot carried us out of harbour.

Notre navire CONDUIT *le convoi.* Our ship leads the convoy.

CONDUITE, *s. f.* Conduct-money, being money paid to the sailors to carry them to their several and respective homes, when discharged from a ship.

CONGÉ, *s. m.* CONGÉ *de bâtiment marchand.* A pass or passport, granted to the master of a merchant-ship, when he is ready to sail.

CONGÉDIER, *v. a.* CONGÉDIER *l'équipage d'un navire.* To pay off a ship's company.

CONGRÉAGE, *s. m.* The act of keckling or worming.

CONGRÉER, *v. a.* CONGRÉER *un cordage.* To worm a rope.

CONGRÉER *un cable.* To keckle a cable.

CONNOISSANCE, *s. f. Avoir* CONNOISSANCE *d'une terre, ou d'un bâtiment.* To get in sight of any land or ship. *Ex.*

A la hauteur du Cap de Bonne Espérance nous eûmes CONNOISSANCE *de deux bâtimens de guerre.* Off the Cape of Good Hope we saw two men of war.

Prendre CONNOISSANCE *de terre.* To make the land distinctly.

CONNOISSANCE *des tems.* Ephemeris, or Nautical Almanac.

CONNOISSEMENT, *s. m.* A ship's bill of lading.

CONSEIL DE GUERRE, *s. m.* Council of war; also a court-martial.

Tenir CONSEIL. To hold a council of war.

CONSEIL *de construction.* A council of the commissioner, principal officers, builders, &c. held in any of the ports.

CONSEIL *de l'amirauté.* A jurisdiction exercised under the name and authority of the lord admiral, who has certain claims called the dues of the Admiralty.

CONSENTIR, *v. n.* To spring, or give way, speaking of masts or yards. *Ex.*

Notre grande vergue a CONSENTI. Our main-yard is sprung.

CONSERVE, *s. f.* A consort. *Ex.*

Nous perdîmes de vue notre CONSERVE *après trois jours de mer.* We lost sight of our consort on the third day after our departure.

Aller de CONSERVE, *ou naviguer de* CONSERVE. To sail in company. *Ex.*

Nous fîmes voile de CONSERVE *avec l'Invincible mais nous le perdîmes bientôt de vue dans une brume.* We sailed in company with the Invincible, but soon lost sight of her in a fog.

CONSERVER, *v. a.* To keep. *Ex.*

CONSERVER *à vue un vaisseau.* To keep sight of a ship.

CONSERVER *deux marques, l'une par l'autre.* To keep two marks in one.

CONSOLE. *s. f.* In ship-building, a kind of bracket, or the part where two pieces of timber are united by a bracket.

CONSOMMATION, *s. f.* The expenditure of the stores of a ship, &c.

CONSTRUCTEUR, *s. m.* A ship-builder.

CONSTRUCTEUR *d'embarcations.* A boat-builder

CONSTRUCTION, *s. f.* Ship-building, or the practical part of Naval Architecture. *Ex.*

Bâtimens en CONSTRUCTION. New ships building.

La CONSTRUCTION *Française.* The French ship-building.

CONSTRUIRE, *v. a.* To build ships.

CONSTRUIT. Built. *Ex.*

CONSTRUIT *en chêne.* Oak-built.

CONSTRUIT *en cedre.* Cedar-built.

CONSTRUIT *en cutter.* Cutter-built.

CONSTRUIT *en goëlette.* Schooner-built.

CONSUL, *s. m.* A consul. An officer commissioned by government, to protect the commerce of the nation in foreign countries.

CONTRE-AMIRAL, *s. m.* A rear-admiral.

CONTRE-BITTES, *s. f. p.* The standards supporting the cable-bits.

CONTRE-BRASSER, *v. a.* To brace about the yards.

CONTRE-CIVADIÈRE, *s. f.* The sprit-sail top-sail.

CONTRE-CORNIÈRE, *s. f.* A piece framed and bolted with the fashion-piece afore it, in French ships.

CONTRE-COURANT, *s. m.* A counter-current.

CONTRE-ÉTAIS, *s. m. p.* The after back-stays.

CONTRE-ÉTAMBOT, *s. m.* CONTRE-ETAMBOT *intérieur.* The inner stern-post, in French ships.

CONTRE-ÉTAMBOT *extérieur.* The back of the inner-post.

CONTRE-ÉTRAVE, *s. f.* The apron which supports the scarph, in ship-building.

CONTRE-HILOIRES, *s. f. p.* Strakes of plank on the decks, close to the binding-strakes.

CONTRE-MAÎTRE, *s. m.* The boatswain's mate.

CONTRE-MAÎTRE *charpentier.* The foreman of the shipwrights in a dock-yard.

CONTRE-MAÎTRE *de la cale.* The captain of the hold.

CONTRE-MARCHE, *s. f.* In naval tactics, the general tacking, bearing up, or coming to the wind of a fleet or squadron drawn up in a line of battle, each ship successively performing the same manœuvre in the wake of another, so as to

preserve its former disposition on the opposite tack.
CONTRE-MARÉE, *s.f.* A counter or contrary-tide.
CONTRE-POINT *d'une voile*, *s. m.* The double rope and marling attached to the clew of a sail.
CONTRE-RODE *de poupe*. *Voyez* RODE.
CONTRE-QUILLE, *s.f.* The stemson.
CONTRE-SABORDS. *Voyez* MANTELETS.
CONTRE-SALUT, *s. m.* The return of a salute.

CONTRÔLEUR DE LA MARINE, *s. m.* A comptroller of the navy.
CONVERSO, *s. m.* The waist, or main deck of a ship.
CONVERTIR, *v. a.* To convert, to turn. *Ex.*
CONVERTIR *le tems en degrés de longitude.* To turn time into longitude.
CONVERTIR *du chanvre en cordages.* To convert hemp into ropes.
CONVOI, *s. m.* A convoy.
Mettre un CONVOI *en sureté.* To place a convoy in safety.
Donner dans un CONVOI. To run into the middle of a convoy.
Notre navire est la charette du CONVOI. Our ship is the dullest sailer of the convoy.
CONVOYER, *v. a.* To convoy.
COQ, *s. m.* The cook of a ship.
COQUE, *s.f.* COQUE *de cordage.* A kink in a rope.
COQUE *d'un navire.* The hull of a ship.
COQUERON, *s. m.* The cook-room, fore-castle, or cuddy, of a lighter or hoy.
COQUET, *s. m.* A cock-boat; a sort of small boat, which passes between Normandy and Paris.
COQUETER, *v. n.* To manage a boat, by paddling in her stern with an oar or paddle.
CORALINE, *s.f.* A light small long-boat, used for fishing coral in the Mediterranean.
CORBEAU. *Voyez* GRAPIN.
CORBILLON, *s. m.* A small kid, or tub, to contain the daily allowance of biscuit for the several messes.
CORDAGE, *s.f.* Cordage, a general name for all the ropes employed to rig or work a ship. *Ex.*
CORDAGE *du premier brin.* Cordage made of the best part of hemp.
CORDAGE *du second brin.* Cordage made of the combings of hemp.
CORDAGE *deux fois commis.* Cable-laid cordage.
CORDAGE *en trois.* A rope made with three strands.
CORDAGE *une fois commis.* Hawser-laid cordage.
CORDAGE *commis en quatre.* A four-strand rope.
CORDAGE *de brou de coco.* Cordage made of the bark of the cocoa-palm.
CORDAGE *blanc.* White, or untarred rope.
CORDAGE *goudronné.* Tarred cordage.
CORDAGE *refait.* Twice-laid cordage, or twice-laid stuff.
CORDAGE *de rechange.* Spare-ropes, spare-cordage.
CORDAGE *étuvé* Cordage which has passed through a stove, to discharge its moisture.
CORDAGE *raqué.* Cordage well rubbed.
CORDE, *s.f.* A rope of any kind.
CORDE *de retenue.* A guy, used to steady a cask, &c.
CORDES *de défense.* Fenders of junk or old cable.
CORDE. In the building of gallies, are two strong pieces of timber placed above the deck, on the starboard and larboard sides, lengthwise, serving to support the thwarts.
CORDELLE, *s.f.* A warp or tow-line.
wORDERIE, *s.f.* A rope-house, or rope-walk.
La CORDERIE. The art of rope-making.
CORDIER, *s. m.* A rope-maker.
Maître-CORDIER. Master rope-maker.
CORDON, *s. m.* A strand of rope-yarns.
CORDON. The lowest rail or ribband in a galley, and what in ships is called the main-channel.
CORNE, *s. f.* CORNE *d'amorce.* A powder-horn.
CORNE *de vergue.* The crutch or cheeks fixed on the inner end of a gaff or boom.
CORNE *de gui.* Crutch of a boom.
Vergue à CORNE. A gaff.
CORNE *d'artimon.* The peak of the mizen-yard.
CORNET, *s. m.* CORNET *de mât.* Case of a mast, peculiar to those vessels in which it is occasionally lowered.
CORNETTE, *s.f.* A broad pendant, displayed at the mast-head of a commodore's ship.

CORNIÈRES, *s. f. p.* Fashion-pieces.
CORPS, *s. m.* CORPS *de bataille.* The centre of a fleet of men of war.
CORPS *d'un vaisseau.* The hull of a ship.
CORPS *morts.* Transporting-buoys; also bollards in a dock-yard.
CORPS *de pompe.* The chamber of a pump.
CORPS *d'une poulie.* The shell of a block.
Les quatre CORPS *de voile.* The courses, and fore and main top-sails of a ship.
CORRECTION, *s.f.* A correction made in the calculations of navigation or astronomy.
CORRIGER, *v. a.* CORRIGER *l'estime.* To correct the dead-reckoning.
CORSAIRE, *s. m.* A privateer; also a pirate.
CORVETTE, *s.f.* A sloop of war; also a general name for all armed vessels under twenty guns.
COSSE, *s.f.* COSSE *de fer.* An iron thimble.
COSSE *de bois.* A bull's eye.

CÔTE, *s. f.* CÔTES *d'un vaisseau.* The ribs of a ship.
CÔTE *de la mer.* The sea-coast or shore.
CÔTE *saine.* A clear coast.
CÔTE *dangereuse.* A dangerous coast.
CÔTE *de fer, ou à pic.* A bold shore; a shore steep-to.
CÔTE *accore.* A bluff or bold shore.
CÔTE *basse.* A flat shore.
Faire CÔTE. To run a-shore.
Tenir sur une CÔTE. To remain upon a coast.

CÔTÉ, *s. m.* CÔTÉ *d'un navire.* The side of a ship.
CÔTÉ *du vent.* The weather side.
CÔTÉ *de dessous le vent.* The lee-side.
Prêter le CÔTÉ *à un vaisseau ennemi.* To bring the broadside to bear upon the enemy.
Se coucher sur le CÔTÉ. To tumble over; speaking of a vessel aground.
Mettre un vaisseau sur le CÔTÉ. To lay a ship on the careen. *Voyez* ABATTRE.
Vaisseau qui a un faux CÔTÉ. A lap-sided ship.

Vaisseau qui a le CÔTÉ *droit.* A wall-sided ship.

CÔTIER. *Pilote* CÔTIER, *s. m.* A coasting pilot.

Bâtiment CÔTIER. A coaster, or coasting vessel.

COTONNINE, *s. f.* *ou toile de* COTONNINE. A species of thick sail-cloth used in the Mediterranean sea: it is formed by a mixture of hemp and cotton, the warp being of the former, and the woof of the latter.

COTTIMO, *s. m.* A duty or exaction of so much per cent. demanded by consuls of the shipping of their nation, when a ship enters a port where such consuls are established.

COUBAIS, *s. m.* A Japanese barge or galley, ornamented and rowed with forty oars.

COUCHE, *s. f.* The pillow of a stay.

COUCHES *de mât.* The pieces of a made mast.

COUCHETTE, *s. f.* A sea bed-frame, or couch.

COUCHETTE *foncée.* A bad place with a boarded bottom.

COUILLARDS, *s. m. p.* Bunt slab-lines.

COULADEUX, *s. m. p.* The lanyards of the shrouds in lateen vessels.

COULAGE, *s. f.* Leakage.

COULER, *v. a. & n.* To leak; to sink. *Ex.*

COULER *bas, ou* COULER *à fond.* To sink or founder.

COULER *bas un vaisseau.* To sink a ship by cannonading.

Nous avons COULÉ *la frégate et pris la corvette.* We have sunk the frigate, and taken the sloop.

Nous futailles COULENT. Our casks are leaky.

COUP, *s. m.* COUP *de canon.* A gun fired, also a shot received, &c. *Ex.*

COUP *de canon d'assistance.* A gun fired as a signal for assistance.

COUP *de canon de partance.* A gun fired as a signal for sailing.

COUP *de canon d'assurance.* A gun fired by a ship when she shews her colours, to affirm the truth of her being really of the nation whose colours she displays.

COUP *de canon de diane.* The morning gun.

COUP *de canon de retraite.* The evening gun.

COUP *de canon à l'eau.* A shot received under water, or between wind and water.

COUP *de canon en plein bois.* A shot received in the upper-works of a ship.

COUP *de barre, ou de gouvernail.* Wild steering.

COUP *de mer.* The shock of a heavy sea, striking the vessel violently.

Nous resumes un COUP *de mer sur le gaillard derrière.* A heavy sea broke upon our quarter-deck.

COUP *de rame.* A stroke, in rowing.

COUP *de roulis.* A roll.

COUP *de sonde.* A cast of the lead.

COUP *de tangage.* A pitch.

COUP *de vent.* A gale of wind.

COUP *de vent forcé.* Stress of weather; a storm.

COUPE, *s. f.* COUPE *verticale.* A vertical section.

COUPE *horizontale.* A horizontal section.

COUPER, *v. a.* To cut, &c. *Ex.*

COUPER *le cable.* To cut away the cable.

COUPER *un mât.* To cut away a mast.

COUPER *la ligne.* In naval tactics, to break the line.

COUPER *la route à un vaisseau.* To fore-reach a ship in a chase, and cross her.

COUPER *la terre à un vaisseau.* To gain the land side of a ship when chasing her.

COUPER *la ligne équinoxiale.* To cross the equator.

COUPLE *s. f.* COUPLE *de haubans.* A pair of shrouds.

COUPLES *de levée.* The frames of a ship; understood only of those which are designed on the shipwright's plan, and moulded.

COUPLE *de coltis.* Knuckle-timbers.

COUPLE *de lof.* The loof-frame, or loof-timbers.

Maître-COUPLE. The midship-frame.

COUPLES *de l'avant.* The fore-body.

COUPLES *de l'arrière.* The after-body.

COUPLES *de remplissage.* Filling-timbers.

COUPLE, *s. m.* A frame. *Ex.*

COUPLES *de balancement.* Balance-timbers, in French ships.

COURADOUX. *Voyez* ENTRE-PONT.

Direction des COURANS. *Voyez* CHÛTE.

COURANT, *s. m.* A current.

Lit d'un COURANT. The main stream of a current.

Le COURANT *porte du nord.* The current sets to the north.

COURBANS, *adj.* *Bois* COURBANS. Compass-timber.

COURBATONS, *s. m. p.* Small knees.

COURBATONS *de l'éperon.* Brackets of the head.

COURBATONS *de hune.* Knees of the top.

COURBES, *s. f. p.* A general name for the larger knees of a ship.

COURBES *du premier pont.* The hanging knees of the lower-deck, in a two-decked ship.

COURBES *du second pont.* The hanging knees of the upper or main-deck.

COURBES *de gaillard.* The hanging knees of the quarter-deck, and fore-castle.

COURBES *horizontales.* Lodging knees.

COURBES *verticales, ou obliques des baux.* Hanging knees.

COURBES *d'arcasse.* Transom knees.

COURBE *d'étambot.* Knee of the stern-post.

COURBES *de fer.* Iron knees.

COURBES *de bittes.* The standards of the cable-bits.

COURBE *de bossoir.* The supporter of the cat-head.

COURBE *de capucine.* The standard which fastens the cut-water to the stem.

COURBES *de jottereaux, ou d'écubier.* Cheeks of the head.

COURBE *à équerre.* A square knee.

COUREAU, *s. m.* A small yawl of the Garonne.

COURIR, *v. a.* To sail; to stand, &c.

COURIR *à sec.* To scud under bare-poles.

COURIR *à toutes voiles.* To croud all sail.

COURIR *au large.* To stand off shore.

COURIR *au plus près.* To stand close upon a wind.

COURIR *au nord, au sud, à l'est, à l'ouest.* To stand to the northward, southward, eastward, or westward.

COURIR *bord à contre.* To stand upon a different tack to a vessel in sight.

COURIR *debout au corps sur un bâtiment.* To run end-on upon a ship.

COURIR *des bords.* To ply to windward.

Courir *en latitude, ou en longitude.* To run down latitude or longitude.
Courir *la bouline.* To run the gauntlet.
Courir *la grande bordée.* To keep watch and watch at sea.
Courir *le même bord.* To stand upon the same tack as some other ship in sight.
Courir *les écoutes largues.* To sail with a flowing sheet.
Courir *sur un parallèle.* To keep in the same latitude.
Courir *sur un danger.* To stand for a shoal.
Courir *sur sa panne.* To shoot, when lying-to, main or fore topsail to the mast.
Courir *sur son air.* To hold one's way.
Courir *sur son ancre.* To go over the anchor.
Courir *tribord, ou babord amure.* To stand upon the starboard or the larboard tack.
Courir *vent arrière.* To run before the wind.
Courir *vent largue.* To go large.
Faire courir *une manœuvre.* To pass along a rope.
COUROI, *s. m.* A composition, or coat of stuff, used for paying a vessel's bottom.
COUROIR. *Voyez* Coursive.
COURRONNEMENT, *s. m.* The after-part of the taffrail of a ship, which is usually ornamented with sculpture.
COURS, *s. m.* Cours, *file, ou virure de bordages.* Streaks of planks.
Cours *de vaigres.* A streak of the planks of a ship's ceiling, continued from stem to stern.
COURSE, *s. f.* A cruise, or cruising.
Aller en course. To cruise at sea, or to go a privateering.
Bâtiment armé en course. A privateer.
COURSIER, *s. m.* A fore-and-aft passage between the benches or rowers seats in a row-galley; also a bow-chase, or great brass gun, placed on the fore part of that passage.
COURSIVE, *s. f.* The waist, or more properly that part of the upper-deck which is open between the quarter-deck and fore-castle.
COURTIER, *s. m.* A broker.
COUSSIN, *s. m.* Coussin *de beaupré.* The pillow of the bowsprit.
Coussin *de bittes.* The fir-lining or doubling of the bits.
Coussin *d'écubiers.* The bolsters of the hawse-holes.
Coussin *de mire.* The bed of a cannon, which supports the breech.
Coussin *de fourrure.* The bolster.
Coussin *d'élongis.* Pillow of the mast-head.
Coussins *de vieux cordages.* Mats.

COUSTIERES, *s. f. p.* The shrouds of a galley.
COUTELAS, *s. m.* Cutlass; a broad cutting sword.
COUTURE, *s. f.* A seam. *Ex.*
Coutures *de voiles.* Seams of the sail.
Coutures *des bordages.* Seams between the planks of the deck or side of a ship.
COUVERTE, *s. f.* The deck of a ship, in the dialect of Provence.
Couverte *de l'isoscèle de proue.* The fore-castle or fore-deck of a galley, together with the space beneath it, where the cannon are planted.

CRACHER les étoupes, *v. a.* To expel the oakum out of the seams; speaking of an old ship in a heavy sea.

CRAÏER, *s. m.* A sort of small Swedish vessel, without top-masts or top-sails.
CRAMPE, *s. f.* A cramp-iron, or staple.
CRAPAUD, *s. m.* The goose-neck of the tiller.
CRAVAN, *s. m.* A barnacle, or small shell-fish, which fastens to a ship's bottom in a long voyage.
CRAVATE, *s. f.* Navel-line.
CREVER, *v. a.* To burst, &c. *Ex.*
Un de nos canons creva *dans le combat.* One of our guns burst in the action.
Crever *une gargousse.* To prick a cartridge.
Crever *une embarcation.* To stave a boat.
CREUSER, *v. a.* Creuser *un port.* To deepen a port or harbour, by taking out the mud or gravel from the bottom.
CREUX, *s. m.* Creux *d'un vaisseau.* The depth of a ship's hold, measured from the lower-deck beams to the keel.
Creux *d'une voile.* The belly or cavity of a sail, which retains the wind.
CRIBLER, *v. a.* To damage by shot, &c. *Ex.*
Vaisseau Criblé. A ship much damaged by worms or cannon shot.
La frégate eut ses voiles criblées *dans le combat.* The frigate had her sails much damaged in the action.
CRIQUE, *s. f.* A creek or cove, or small natural harbour.
CROC, *s. m.* A hook.
Croc *de capon.* A cat-hook.
Croc *de candelette.* The hook of the fore-tackle.
Croc *de palan.* Tackle-hook.
Croc *à trois branches.* A creeper or grapling with three hooks, used to heave up any thing from the bottom, especially an anchor-cable.
CROCHER, *v. a.* Crocher *un fardeau.* To seize any thing with a tackle-hook, &c.
CROCHET, *s. m.* Crochets *des bittes.* Hooks to fasten the cross-pieces to the bits.
Crochets *d'épontilles.* Hooks of the stanchions.
Crochets *d'armes.* Crutches used to support the small-arms in a ship of war.

CROISÉE, *s. f.* Croisée *d'une ancre.* The crown of an anchor.
CROISER, *v. a. & n.* To cruize. *Ex.*
La Belle Poule partira demain pour croiser *par le travers du Texel.* The Belle Poule will sail to-morrow to cruise off the Texel.
CROISETT, *s. f.* A pin or bolt used as a fid to any flag-staff.
CROISEUR, *adj. & s.* *Vaisseau* croiseur. A cruiser.

CROISIÈRE, *s. f.* Cruising latitude; also a cruiser. *Ex.*
Etre en croisière. To be cruising.
Tenir croisière. To keep on a particular cruising station.
Se tenir en croisière. To continue cruising.
CROIX, *s. f.* Croix *dans les cables.* An elbow or cross in the hawse.
En croix, *adv.* *Brasser en* croix. To square the yards by the braces.

L'escadre ennemie a ses perruquets en CROIX. The enemy's fleet have their top-gallant yards across.
CRONE, *s. m.* A wheel-crane erected on a wharf.
CROULER UN BÂTIMENT, *v. a.* To shake a ship by jumping on her decks, in order to launch her from the stocks.

CROUPIÈRE, *s. f.* A stern-fast.
Mouiller en CROUPIÈRE. To cast an anchor by the stern.
CUEILLE, *s. f.* One of the cloths of a sail.
CUEILLIR, *v. a.* CUEILLIR *une manœuvre.* To coil up a rope.
CUILLER, *s. f.* CUILLER *à brai.* A pitch ladle.
CUILLER *à canon.* A gunner's ladle.
CUILLER *de machine à creuser.* A sort of forked ladle used by the ballast lighters to heave the mud, &c. from the bottom of a harbour or river.
CUILLER *de pompe.* A pump-borer,
CUILLER *de tarrière.* An auger-bit.
CUIRS VERDS, *s. m. p.* Raw hides, such as are used on board ships to serve the cables, &c.
CUISINE, *s. f.* The galley, or cook-room of a ship.
CUL, *s. m.* The stern, or after part. *Ex.*
CUL *de poulie.* The arse of a block.
CUL *de lampe, s. m.* The lower finishing of the quarter-gallery.
CUL *de porc, s. m.* A wall-knot.
CUL *de porc simple.* A single wall-knot.
CUL *de porc double.* A double wall-knot.
CUL *de porc à tête de mort.* A crown-knot.
CUL *de porc à tête d'alouette.* A double crown-knot.
CUL *de sac, s. m.* A name given by the French, in the West India islands, to a small cove or harbour.
CULASSE, *s. f.* CULASSE *d'un canon.* The breech of a cannon.
CULER, *v. n.* To fall a-stern, or make stern-way.
Faire CULER. To make a stern-board.
Mettre les voiles à CULER. To back the sails; to lay all flat a-back.
CURATEUR DE LA MARINE, *s. m.* An officer formerly appointed over the Trierarques; purser.
CURETTE, *s. f.* A pump-scraper, used to clean the inside of a pump.
CUSE-FORNE, *s. m.* A small, long, and sharp rowing-boat of Japan, without decks, employed to fish whales.
CUTTER, *ou* COTTER, *s. m.* A cutter.

D.

DAILLOTS, *s. m. p.* The grommets of a stay-sail. *Voyez* BAGUES.
DALE, *s. f.* DALE *de pompe.* A pump-dale.
DALE *de brûlot, ou* DALE *à feu.* The trough or channel in which the train is laid in a fire-ship.
DALOTS, *s. m. p.* Scuppers, or scupper-holes.
DALOTS *du premier pont d'un vaisseau.* Scuppers of the lower-deck.
DALOTS *de la gatte.* Scuppers of the manger.
Manches de DALOTS. Scupper hoses.
DAME-JEANNE, *s. f.* A demijan, or large bottle of about four or five gallons, covered with basket-work, and much used in merchant-ships.
DAMELOPRE, *s. f.* A vessel navigated in the canals in Holland.
DANGER, *s. m.* A general name for all rock, banks, &c. that may bring a ship up.
DARCE, *s. f.* A wet dock; a small harbour within a large one, wherein ships are secured and moored.
DARD, ou HARPON, *s. m.* Harpoon.
DARD *à feu.* A fire-arrow, used to burn the sails of an enemy's ship.
DAVIED, ou DAVIER, *s. m.* The davit in a long-boat.

DÉ, *s. m.* Coak of a block-sheave.

DÉBÂCLE, ou DÉBÂCLAGE, *s. m.* The act of clearing or opening a harbour, by removing the lightened vessels to make room for such as are laden.

DÉBÂCLEUR, *s. m.* An officer whose duty it is to regulate the mooring of vessels, and to keep the passage, or fair-way open and clear.

DÉBARCADOUR, *s. m.* A place destined for the landing of a ship's cargo; a wharf, or store-house.

DÉBARDAGE *Voyez* DÉCHARGER.

DÉBARQUEMENT, *s. m.* Landing, or debarkation. *Ex.*
Lieu de DÉBARQUEMENT. Landing-place.
Des troupes de DÉBARQUEMENT. Landing troops.

DÉBARQUER, *v. a. & n.* To land, &c. *Ex.*
DÉBARQUER *des marchandises.* To unload or land goods.
DÉBARQUER *un homme d'un vaisseau.* To discharge a man from a ship.
Je m'embarquai à Cherbourg & je DÉBARQUAI *à Calais.* I took shipping at Cherburg, and landed at Calais.

DÉBILLER, *v. a. & n.* To loosen the horses that track a vessel up and down a canal or river.

DÉBITTER, *v. a.* DÉBITTER *le cable.* To unbit the cable.

☞ DÉBORDE! *Push off! Voyez* AU LARGE!
☞ DÉBORDE les huniers! *Let go the top-sail sheets!*

DÉBORDER, *v. a. & n.* DEBORDER *un navire.* To rip off the planks of a ship.
DÉBORDER *les avirons.* To unship the oars.
DÉBORDER *une voile.* To let go a sheet.
DÉBORDER. To sheer off from some other ship, &c.

DÉBOSSER, *v. a.* DÉBOSSER *le cable.* To take off the stoppers from the cable, &c.

DÉBOUQUEMENT, *s. m.* The act of sailing out into the open sea, from amongst islands or narrow channels.

DÉBOUQUER, *v. n.* To sail out from a narrow passage. *Ex.*
Nous avons DÉBOUQUÉ *par les îles Philippines.* We sailed out from between the Philippines.

DEBOUT, *adv.* DEBOUT *au vent.* Head to wind.
DEBOUT *à la lame.* Head to the sea.
☞ DEBOUT à terre! *Head in for the land!*
DEBOUT *au corps.* End-on.
Courir DEBOUT *au corps sur un vaisseau.* To run end-on upon a ship.

DÉBRIS, *s. m. p.* Pieces of a wreck.

DÉCAPELER, *v. a.* To take off the shrouds and other rigging from the mast-head.
DÉCAPELER *les haubans.* To strip the masts.
DÉCAPELER *une hune.* To get a top off.

DÉCAPER, *v. n.* To sail beyond a remarkable cape or head-land.

DÉCHARGEMENT, *s. m.* The act of unloading a merchant-ship.

DÉCHARGER, *v. a.* To unload. *Ex.*
DÉCHARGER *un navire.* To discharge a ship.
DÉCHARGER *une voile.* To fill a sail, after it has lain a-back.
☞ DÉCHARGE devant! *Haul off all!* or, *Let go and haul!*
☞ DÉCHARGE derrière! *Haul main-sail, haul!*
☞ DÉCHARGE le petit hunier! *Fill the fore-top-sail!*
DÉCHARGER *un canon.* To take the charge out of a gun by means of the worm.
DÉCHARGER *la chaloupe.* To clear the launch.

LA GRAND-VOILE FUT DECHIRÉE, *adj.* The main-sail was torn in the storm.

DÉCHOUER, *v. a.* DÉCHOUER *un bâtiment.* To get a ship afloat, or off the ground, into deep water.

DÉCLINAISON, *s. f.* DÉCLINAISON *de l'aiguille aimantée.* The variation of the magnetic needle.
DÉCLINAISON *d'un astre.* The declination of the sun, or a fixed star.

DÉCOINCER, *v. a.* DÉCOINCER *un mât.* To knock up the wedges of a mast.

DÉCOLLEMENT, *s. m.* The forming of a tenon upon the end of a piece of timber.

DÉCOMBRES, *s. m. p.* The chips and useless pieces of timber left on a shipwright's wharf.

DÉCOUDRE, *v. a.* DÉCOUDRE *des bordages.* To rip off the planks from a ship's side, in order to examine her timbers, &c.

DÉCOUVERTE, *s. f.* A look-out at the mast-head.

À la DÉCOUVERTE, *adv.* On the look out.

DÉCOUVRIR, *v. a.* To discover, &c. *Ex.*
DÉCOUVRIR *une voie d'eau.* To find out a leak.
Cette roche ne se DÉCOUVRE *qu'à la basse mer.* That rock appears only at low water.

DÉCROCHER, *v. a.* To unhook. *Ex.*
☞ DÉCROCHE la candelette! *Unhook the fish!*

DÉFAIRE, *v. a.* DÉFAIRE *les tours d'une manœuvre.* To take the turns out of a rope.
DÉFAIRE *la croix des cables.* To clear the hawse, when there is a cross in it.

DÉFENSES, *s. f. p.* Skids, or skeeds of a ship's side.
Cordes de DÉFENSES. Fenders of junk, or old cable.

DÉFERLER, *v. n.* To unfurl, or to loose; speaking of sails; also to break, speaking of the sea. *Ex.*
DÉFERLER *le petit hunier.* To unfurl the fore-top-sail.
La mer DÉFERLE *sur la côte.* The surf breaks on the coast.

DÉFIER, *v. a. & n.* To bear off, &c. *Ex.*
☞ DÉFIE du vent! *You are all in the wind! Keep her full! No-near!* or *No-nearer!*
☞ DÉFIE la chaloupe du bord! *Bear off! Fend the boat!*
☞ DÉFIE l'ancre du bord! *Bear off the anchor!*
☞ DÉFIE des embardées! *Don't yaw the ship about so!*

DÉFONCER, *v. a.* To stave, &c. *Ex.*
DÉFONCER *une futaille.* To stave a cask.

DÉGAGER, *v. a.* DÉGAGER *l'ancre.* To clear the anchor.
DÉGAGER *un cordage.* To clear a rope.
DÉGAGER *un palan.* To clear away a tackle-fall.
DÉGAGER *un bâtiment.* To rescue a ship from the possession, attack, or pursuit of an enemy.

DÉGARNIR, *v. a.* To unrig, &c. *Ex.*
DÉGARNIR *un mât, ou une vergue.* To uurig a mast, or yard.
DÉGARNIR *le cabestan.* To unrig the capstan, by taking off the voyol, and unshipping the bars.
DÉGARNIR *les canons.* To unrig the guns.
DÉGARNIR *les avirons.* To unship the oars.
DÉGARNIR *un vaisseau.* *Voyez* DÉGRÉER.

DÉGAUCHIR, *v. a.* To trim, or form a piece of timber, so as to fit exactly to the place for which it is designed.

DÉGORGEOIR, *s. m.* The priming-wire, or priming-iron.

DÉGORGER, *v. a.* DÉGORGER *la pompe.* To clear the pump when it is choked.
DÉGORGER *la lumière d'un canon.* To clear the vent of a cannon.

DÉGRADÉ, *adj.* *Matelot* DÉGRADÉ. A sailor escaped from shipwreck.
Bâtiment DÉGRADÉ. A ship blown off a coast which she wanted to keep aboard.

DÉGRAPINER, *v. a.* To warp a ship off from the ice by means of graplings, when she approaches too near it; also to lift the grapnel or small anchor.

DEGRÉ, *s. m.* The 360th part of a circle.

DÉGRÉER, *v. a.* To unrig, &c. *Ex.*
Notre vaisseau est DÉGRÉÉ. Our ship is unrigged.
DÉGRÉER *les perroquets.* To unbend the top-gallant gear.

DEHORS, *adv.* The offing; out or set, speaking of sails. *Ex.*
Toutes voiles DEHORS. Every sail set.
Il y a grosse mer DEHORS. The sea runs high in the offing.
Mettre DEHORS *un bâton de foc.* To get out a jib-boom.
Mettre DEHORS *un boute-hors.* To rig out a boom.

DÉJAUGER, *v. n.* *Vaisseau* DÉJAUGÉ *dans un port.* A ship sewed in a harbour.
Notre navire a DÉJAUGÉ *de quatre pieds.* Our ship has sewed four feet.

DÉJOUER, *v. n.* To flutter, or turn in the wind; said of flags, pendants, &c.

DÉLACER, *v. a.* To unlace. *Ex.*
DÉLACER *la bonnette.* To unlace or take off the bonnet from the foot of a sail.

DÉLAISSEMENT, *s. m.* An instrument or act by which the loss of a ship is announced by the master or owner to the insurer, summoning him to pay the stipulated insurance.

DÉLESTAGE, *s. m.* The discharging or unloading of ballast from a ship.

DÉLESTER, *v. a.* DÉLESTER *un navire.* To unballast a ship, or get the ballast out of her.

DÉLESTEUR, *s. m.* An officer appointed to receive the ballast of ships; also a ballast lighter.

DÉLIVRER, *v. a.* DÉLIVRER *des bordages.* To rip off planks.

À LA **DEMANDE**, *adv.* *Filer à la* DEMANDE *du cable, ou d'un cordage.* To ease off, or ease away the cable, or any rope.
Notre corvette DEMANDE *que sa mâture soit sur l'arrière.* Our sloop requires her masts to hang aft.

DEMANDER, *v. a.* To require, &c. *Ex.*
Le vaisseau DEMANDE *du cable.* The ship must have more cable.

DÉMARRAGE, *s. m.* The act of unmooring; also of unbending any rope.

DÉMARRER, *v. a. & n.* To unmoor; to cast loose. *Ex.*
Nous allons DÉMARRER. We are going to unmoor.
☞ DÉMARREZ les canons! *Cast loose the guns!*

DÉMÂTER, *v. a. & n.* DÉMÂTER *un vaisseau.* To take out the lower masts of a ship.
DÉMÂTER *un vaisseau à coups de canon.* To shoot away a ship's masts.
Nous fûmes DÉMÂTÉS *par la tempête.* We were dismasted in a storm.

À **DEMEURE**, *adv.* *Mettre les ancres à* DEMEURE. To stow the anchors for sea.

DEMI-BANDE, *s. f.* A parliament heel.
DEMI-CLÉ, *s. f.* A half-hitch.
Deux DEMI-CLÉS. A clove-hitch.
DEMI-PIQUE, *s. f.* A half-pike, which is sometimes used to oppose the boarders in a sea-fight.

DEMI-TOUR, *s. m.* DEMI-TOUR *dans les cables.* An elbow in the hawse.

DÉMOLIR, *v. a.* DÉMOLIR *un bâtiment.* To break up a ship.

DÉMONTER, *v. a.* To dismount, &c. *Ex.*
Nous eûmes plusieurs canons DÉMONTÉS *dans le combat.* We had several guns dismounted in the action.
DÉMONTER *le gouvernail.* To unhang the rudder.
DÉMONTER *un capitaine.* To supersede a captain.

DÉPAQUETER, *v. a.* DÉPAQUETER *une voile.* To open out a sail.

DÉPART, *s. m.* Departure. *Ex.*
Prendre son point de DÉPART. To take one's de-departure.

DÉPARTEMENT, *s. m.* A marine arsenal, or dock-yard with its gun-wharf, such as Brest, Toulon, &c.; also certain districts in France under the jurisdiction of a commissioner of the navy.

DÉPASSER, *v. a.* To out-sail, &c. *Ex.*
DÉPASSER *un bâtiment.* To run past a ship.
DÉPASSER *un port, ou une île.* To sail beyond a harbour, or an island.
DÉPASSER *la tournevire.* To take the messenger off the capstan.
DÉPASSER *une manœuvre, ou un cordage.* To unreeve a rope.
DÉPASSER *les tours des cables.* To clear the hawse.
DÉPASSER *un mât de perroquet.* To get a top-mast, or top-gallant-mast down upon deck.

DÉPÉCER, *v. a.* DÉPÉCER *un vaisseau.* To break up or take to pieces an old vessel.

EN **DÉPENDANT**, *adv.* *Ex.*
Arriver en DÉPENDANT. To sail edging away.
Aller en DÉPENDANT. To sail in company, &c.

DÉPLANTER, *v. a.* To start the anchor.

DÉPLOYER, *v. a.* DÉPLOYER *le pavillon.* To let fly, or display the ensign.
DÉPLOYER *une voile.* To heave out; to set a sail.

DÉPRÉDER, *v. a.* To plunder. *Ex.*
Effets DÉPRÉDÉS. Goods plundered or robbed from a wreck, contrary to law.

DÉPRESSION, *s. f.* DÉPRESSION *de l'horizon.* The dip of the horizon.

DÉRADER, *v. n.* To drive with the anchors a-head; to be driven from an anchorage, and forced out to sea.

DÉRALINGUER, *v. a.* To take off the bolt-rope of a sail.

DÉRAPER, *v. n.* To loosen the anchor from the ground, or to trip it.
L'ancre est DÉRAPÉE. *Voyez* ANCRE.

DÉRIVE, *s. f.* The angle of lee-way, or drift.
Avoir belle DÉRIVE. To have good sea-room.
En DÉRIVE, *adv.* Adrift. *Ex.*
Vaisseau qui est en DÉRIVE. A vessel which is adrift.

Etre en DÉRIVE. To be adrift.

DÉRIVER, *v. n.* To drive, or to sag to leeward.

DÉROBER, *v. a.* DÉROBER *le vent à un vaisseau.* To take the wind from a ship.

DÉSAFFOURCHER, *v. n.* To unmoor.

DÉSARMEMENT, *s. m.* The act of paying off the officers and crew, and laying up a ship.
Etre en DÉSARMEMENT. To be clearing a ship for laying her up.

DÉSARMER, *v. n.* DÉSARMER *un navire.* To lay up a ship.
Vaisseau DÉSARMÉ. A ship laid up.

DÉSARRIMER, *v. a.* To shift the stowage of the hold.

DESCENDRE, *v. a* DESCENDRE *une rivière.* To fall down a river with the tide.
DESCENDRE *à terre, v. a.* To go ashore.
DESCENTE, *s. f.* A descent. *Ex.*
Faire un DESCENTE. To make a landing in an enemy's country.

DÉSEMBARQUER. *Voyez* DÉBARQUEMENT.

DÉSEMPARÉ, *adj.* *Vaisseau* DÉSEMPARÉ. A disabled ship, or a ship crippled in battle, &c.

DÉSEMPARER, *v. a.* To disable, &c. *Ex.*
La frégate fut DÉSEMPARÉE *dès la première bordée du vaisseau.* The frigate was disabled by the first broadside of the ship of the line.

DÉSENVERGUER, *v. a.* To unbend; speaking of any sail.
Il faut DÉSENVERGUER *la misaine.* The foresail must be unbent.

DÉSERTER, *v. n.* To run away.

DÉSERTEUR, *s. m.* A deserter; a run-away.
DESTINATION, *s. f.* DESTINATION *d'un bâtiment.* A ship in commission bound to any port.

DÉTACHEMENT, *s. m.* A detachment.

DÉTACHER, *v. a.* DÉTACHER *un vaisseau d'une escadre.* To detach a ship from a fleet for a particular service.

DÉTALINGUER UN CABLE, *v. a.* To unbend a cable, or take it off from the anchor.

DÉTAPER, *v. a.* DÉTAPER *les canons.* To take out the tompions from the mouth of the guns.

DÉTERMINER, *v. a.* DÉTERMINER *la latitude & la longitude d'un lieu.* To find the latitude and longitude of any place.

DÉTRESSE, *s. f.* Distress.

DÉTROIT, *s. m.* A strait or narrow channel between two lands.
Donner dans un DÉTROIT. To enter a strait.
Se tenir à l'ouvert d'un DÉTROIT. To keep a strait open.
Le DÉTROIT *de Gibraltar.* The straits of Gibraltar.
DEVANT. *Voyez* AVANT.

DÉVENTER, *v. a.* DÉVENTER *une voile.* To spill a sail, or make it shiver in the wind.

DÉVERGUER. *Voyez* DÉSENVERGUER.

DÉVERS, *s. m.* The moulding of any piece of timber.
Marquer le bois suivant son DÉVERS. To mould the timber according to its compass or inclination.

DÉVIRER, *v. a.* To heave back. *Ex.*
☞ DÉVIRE au cabestan! *Come up the capstan!*
DÉVIRER *le cable.* To surge the cable about the capstan, in order to prevent it from riding one part over the other.

DÉVIS, *s. m.* A scheme containing the dimensions of a ship, from which the builder forms a draught for constructing her; also a statement made by a captain, containing his observations on the qualities of the ship under his command, during her navigation.

DÉVOIÉ, *adj.* *Couples* DÉVOIÉS. Cant-timbers.

DÉVOIEMENT, *s. m.* DÉVOIEMENT *de l'estain.* the flaring of the fashion-piece.
DIABLOTIN, *s. m.* The mizen top-mast stay-sail.
DIANE, *s. f.* The morning watch.
Coup de canon de DIANE. The morning watch-gun.
On bat la DIANE. The morning drum beats.
DIGON, ou DIJON, *s. m.* (In naval architecture) A triangular piece of timber, sometimes employed to fill up the angle of the knee of the head in French ships: also the stock or staff of a vane or pendant.
DIMINUER, *v. a. & n.* To lessen, &c. *Ex.*
Le vent a beaucoup DIMINUÉ. The wind is much abated.
DIMINUER *de voiles.* To shorten sail, to take in sail.
DIMINUER *de fond.* To shoal the water.
DIRECTION, *s. f.* *La* DIRECTION *d'un courant.* The setting in of a current.
DIRIGER, *v. a.* DIRIGER *la route vers une île,* &c. To shape a course towards an island, &c.
DISPUTER, *v. a.* DISPUTER *le vent.* To contend for the weather-gage, or to endeavour to get to windward of some ship or fleet in sight.
DIVISION, *s. f.* A division, or squadron.
DIXIÈME, *s. m.* An allowance made by an agent-victualler of one cask in every ten, to answer for waste or leakage.
DOGRE, *s. m.* A Dutch dogger; a dogger-boat.
DOGUES D'AMURE, *s. m. p.* The holes of the chess-trees.
DONNER, *v. a. & n.* To give, &c.
DONNER *de l'air au bâtiment.* To give the ship good way.
DONNER *à la bande.* To heel or incline to either side.
Le vaisseau DONNE *à la bande à babord.* The ship heels to port.
DONNER *à la côte.* To run a-ground.
DONNER *de dans.* To run right in for a port, &c.
DONNER *des vivres à un bâtiment.* To spare provisions to a vessel.

DONNER *du tour à une roche.* To give a good birth to a rock.
DONNER *un coup de sonde.* To get a cast of the lead.
DONNER *les perroquets à un navire.* To spare the top-gallant sails, &c. to a ship in company.
DONNER *sur la barre d'une rivière.* To take the bar of a river.
DONNER *vent devant.* To heave in stays.
DONNER *un suif, ou couroi à un bâtiment.* To pay a vessel's bottom.
DONNER *chasse.* *Voyez* CHASSE.
DONNER *la cale.* *Voyez* CALE.
DORER UN VAISSEAU. *Voyez* ESPALMER.
DORMANT, *s. m.* DORMANT *d'un cordage.* The standing part of a tackle, or other running rope.
EAU DORMANTE, *adj.* Standing water, or water where there is no tide or current.
DOSSIER, *s. m.* *Le* DOSSIER *de canot.* The back-board of a boat.
DOUANE, *s. f.* The custom-house.
DOUBLAGE, *s. m.* The sheathing applied to the bottom of a ship. *Ex.*
DOUBLAGE *de chêne.* Oak-sheathing.
DOUBLAGE *de sapin.* Fir-sheathing.
DOUBLAGE *de la quille.* The sheathing of the keel.
DOUBLAGE *de cuivre.* Copper-sheathing.
DOUBLAGES, *ou renfort des voiles.* Lining of the sails.
DOUBLE, *s. m.* DOUBLE *d'une manœuvre, ou d'un cordage.* The bite of a rope.
DOUBLER, *v. a. & n.* To sheath; to double upon, to weather, &c. *Ex.*
DOUBLER *les vaisseaux ennemis.* To double upon the enemy's ships, in a sea-fight.
DOUBLER *un cap.* To double or pass beyond a cape.
DOUBLER *un bâtiment en cuivre.* To sheath a ship with copper.
DOUBLER *le sillage d'un bâtiment.* To sail as fast again as another ship.
EN DOUCEUR, *adv.* Handsomely, gently. *Ex.*
☞ Amène en DOUCEUR! *Lower handsomely!*
☞ File en DOUCEUR! *Ease gently!*
☞ Vire en DOUCEUR! *Heave handsomely!*
DRAGAN, *s. m.* The extremity of the stern, or the wing-transom of a galley.
DRAGON. *Voyez* TOURBILLON.
DRAGON *de vent, s. m.* A sudden gust, or violent squall of wind.
DRAGUE, *s. f.* A drag.
DRAGUER, *v. a.* To drag.
DRAGUER *une ancre.* To drag or sweep the bottom for an anchor.
DRAILLE, *s. f.* DRAILLE *d'une voile d'étai.* A stay-sail's stay.
DRÈGE, *s. f.* A sort of net for catching soles and turbot.
DRESSER, *v. a.* To trim, &c. *Ex.*
DRESSER *la barre.* To right the helm.
DRESSER *les vergues.* To square the yards by the braces and lifts.
☞ DRESSE les vergues par-tout! *Square the yards fore and aft!*
DRESSER *un canot.* To trim the boat. *Voyez* BARQUE *droite.*
DRISSES, *s. f. p.* Jear-halyards. *Ex.*
☞ Alonge les DRISSES! *Man the halyards!*
DRISSES *des basses vergues.* Jears.
DRISSES *de la grande vergue.* Main-jears.
DRISSES *de la vergue de misaine.* Fore-jears.
DRISSES *de la vergue d'artimon.* Mizen-jears.
DRISSES *du grand hunier.* Main-top-halyards.
DRISSES *du petit hunier.* Fore-top halyards.
DRISSES *du perroquet de fougue.* Mizen-top halyards.
DRISSE *du grand perroquet.* Main-top-gallant halyard.
DRISSE *du petit perrequet.* Fore-top-gallant halyard.
DRISSE *du grand perroquet volant.* Main-top-gallant royal-sail halyard.
DRISSE *du petit perroquet volant.* Fore-top-gallant royal-sail halyard.
DRISSE *d'artimon.* Mizen-throat halyard.
DRISSE *de la grande voile d'étai.* Main stay-sail halyard.
DRISSE *de la voile d'étai de hune.* Main-top stay-sail halyard.
DRISSE *de la contre voile d'étai de hune.* Halyard of the upper main-top stay-sail.
DRISSE *de la voile d'étai du grand perroquet.* Main-top-gallant stay-sail halyard.
DRISSE *de la voile d'étai d'artimon.* Mizen stay-sail-halyard.
DRISSE *de la voile d'étai du perroquet de fougue.* Mizen-top stay-sail halyard.
DRISSE *du grand foc.* Jib halyard.
DRISSE *du petit foc.* Fore-top stay-sail halyard.
DRISSE *du contre foc.* Middle jib halyard.
DRISSE *de la trinquette.* Fore stay-sail halyard.
DRISSES *des bonnettes basses.* Lower studding-sail halyards.
DRISSES *des bonnettes des huniers.* Top-sail studding-sail halyards.
DRISSE *du paille en cul.* A ring-tail halyard.
DRISSE *de pavillon.* The ensign halyard.
DRISSE *flamme.* The halyard of a pendant.
DRISSE *du pic, ou corne d'artimon.* A throat halyard.
DROIT, *s. m.* Duty.
DROIT *d'ancrage.* *Voyez* ANCRAGE.
DROIT, *adv. & adj.* ☞ DROIT la barre! *Voyez* BARRE.
☞ DROIT! *Steady!*
☞ DROIT comme ça! *Right so! Thus!*
☞ Barque DROITE! *Voyez* BARQUE.
EN DROITURE, *adv.* *Faire sa route en* DROITURE. To make a strait course.
DROME, *s. f.* A raft of masts, spars, &c.
DROSSE, *s. f.* A truss, &c. *Ex.*
DROSSE *de basse vergue, ou de racage.* A parral-truss.
DROSSE *d'artimon.* Mizen-truss.
DROSSES *de la grande vergue.* Main-trusses.
DROSSES *de la vergue de misaine.* Fore-trusses.
DROSSE *de canon.* A gun-tackle.
DROSSE *du gouvernail.* A tiller-rope.
DROSSER, *v. a.* To drive. *Ex.*
Etre DROSSÉ *par les courans.* To drive with the currents.
Les courans nous DROSSENT *sur la terre.* We are driving a-shore by the currents.

DUNES, *s. f. p.* Downs, or heights on the sea-coasts.
DUNETTE, *s. f.* The poop of a ship of war.

Échelles de la DUNETTE. *Voyez* ÉCHELLES, &c.
Logemens sur la DUNETTE. The poop-royal.

E.

EAU, *s. f.* Water. *Ex.*
EAU *de mer, ou salée.* Sea, or salt-water.
EAU *douce.* Fresh-water.
EAU *saumâtre.* Brackish water.
EAU *corrompue.* Stinking water.
EAUX *qui se trouvent au dessous du vaigrage d'un bâtiment.* Bilge-water.
Entre deux EAUX. Between wind and water.
Etre dans les EAUX *d'un bâtiment.* To be in the wake of a ship.
Faire de L'EAU. To water a ship.
Faire EAU. To leak, or make water..
Coup de canon à L'EAU. A shot under the water.
Etre en grande EAU. To be in deep water.
Avoir du L'EAU *à courir.* To have good sea-room.
EAU *peu profonde.* Shoal water.
EAU *plate & courtoise.* Very smooth water; the state of water in a dead calm.
EAU *changée.* Foul water; or water whose colour is changed, which is always perceived in ships coming from abroad, when they approach the sound.
Lancer un navire à L'EAU. *Voyez* LANCER.
EAUX *mortes.* Neap-tides.
EAUX *vives.* Spring-tides.

ÉBAROUIE, *adj.* Abounding with shakes or rents; expressed of the planks of a ship when they are split, &c.

ÉBAUCHER, *v. a.* To make the first draught, to rough-draw.

ÉBAUCHER *un pièce de bois.* To rough-hew a piece of wood.

ÈBE, *s. f.* (JUSANT, REFLUX, *ou* MARÉE DESCENDANTE). The ebb-tide.
Il y a ÈBE. The tide ebbs or falls.

ÉBRANLEMENT, *s. m.* The cracking or straining of a ship in a tempestuous sea.

ÉCARLINGUE. *Voyez* CARLINGUE.

ÉCART, *s. m.* ÉCART *double.* A scarph, or two ends of timber laid over each other.

ÉCART *simple, ou quarré.* Butt and butt, or the junction of the butt-ends of two planks.

ÉCARTS *de la quille.* The scarphs of the keel.

ÉCARTS *de l'étrave.* The scarphs of the stem.

ÉCARVER, *v. a.* To scarph, or join two pieces of timber together.

ÉCHAFAUD, *s. m.* A light stage, used in Newfoundland to dry cod-fish, also a stage hung over a ship's side to calk or repair any breach.

ECHANDOLE. *Voyez* ESCANDOLE.

ÉCHANTILLON, *s. m.* The scantling, or thickness of the timber, used in ship-building. *Ex.*
Vaisseaux forts D'ÉCHANTILLON. Ships that have very thick timbers.

ÉCHARPE. *Voyez* ARCASSE.

ÉCHELLE, *s. f.* A ladder formed of ropes, to ascend to or descend from the ship's sides, &c.
La grande ÉCHELLE *d'un vaisseau.* The quarter-deck ladder.

L'ÉCHELLE *de la sainte-barbe.* The gun-room ladder.

ÉCHELLES *de la dunette.* Poop-ladders.

ÉCHELLE *de commandement.* Accommodation-ladder.

ÉCHELLES *de poupe.* The stern ladders.

ÉCHELLE *hors du bord.* The gangway-ladder.

ÉCHELLES *du Levant.* The sea-port towns of the Levant.

ÉCHELLES *de Barbarie.* The sea-port towns of Barbary.

ÉCHELONS, *s. m. p.* The steps of the ladders.

ÉCHILLON. *Voyez* SIPHON.

EN **ÉCHIQUIER**, *adv.* *Escadre formée en* ÉCHIQUIER (In naval tactics). A squadron standing on in a bow and quarter-line.

ÉCHÔMES. *Voyez* TOLLETS.

ÉCHOUEMENT, *s. m.* The act of getting a-ground, or of being stranded on a coast.

ÉCHOUER, *v. n. & a.* To run a-ground, or on shore. *Ex.*
Faire ÉCHOUER *un bâtiment.* To run a ship a-shore.
Le vaisseau ÉCHOUA *sur un banc de sable.* The ship ran a-ground, or struck upon a sand-bank.

ÉCLABOUSSURE D'EAU DE MER, *s. f.* The spray of the sea.

ÉCLAIRCIE, *s. f.* A clear spot in a cloudy sky.

ÉCLAIRCIR, *v. a. & r.* *Le tems s'est* ÉCLAIRCI. The wind has cleared up the weather.

ÉCLAT, *s. m.* ÉCLAT *de bois.* A splinter torn from a piece of timber, by the force of a shot, &c.

ÉCLUSE, *s. f.* A sluice, or drain.

ÉCOPE. *Voyez* ESCOP.

ÉCOUETS, *s. m. p.* The tacks of the main-sail and fore-sail.

ÉCOUTES, *s. m. p.* The sheets of a sail. *Ex.*

ÉCOUTES *de la grande voile.* Main sheets.

ÉCOUTES *du grand huniere.* Main-top sheets, &c.

ÉCOUTE *de bonnette en étui.* The tack or guy of a studding-sail boom.
Etre sous L'ÉCOUTE *d un bâtiment.* To be under the lee of a vessel, and within hail of her.
Larguer une ÉCOUTE. To ease off any sheet.
Border une ÉCOUTE. To haul aft any sheet.

Avoir les ÉCOUTES *largues.* To sail with flowing sheets.

ÉCOUTILLE, *s.f.* A hatchway. *Ex.*
La Grande ÉCOUTILLE. The main hatchway.

ECOUTILLE *de la fosse aux cables.* Fore hatchway.

L'ÉCOUTILLE *d'arrière.* The after hatchway.

ÉCOUTILLE *aux vivres.* The steward's hatchway.

ÉCOUTILLE *de la soute aux poudres.* The magazine hatchway.

ÉCOUTILLE *qui s'emboîte.* A hatchway with a scuttle covering its border.

ÉCOUTILLE *vitrée.* A sky-light.
Fermer les ÉCOUTILLES. To shut up the hatches.

ÉCOUTILLES *à huit pans,* ÉCOUTILLES *du mât.* The holes and partners of the mast.

ÉCOUTILLON, *s. m.* A scuttle, or small hatchway.

ÉCOUVILLON, *s. m.* The sponge of a gun.

ÉCOUVILLONNER, *v. a.* ÉCOUVILLONNER *un canon.* To sponge a gun; also to clean or cool it with a wet sponge.

☞ ÉCOUVILLONNEZ les canons! *Sponge your guns!*

ÉCRITURES, *s.f. p.* The papers of a ship, comprehending journals, registers, passports, &c.

ÉCRIVAIN. *Voyez* COMMIS *aux revues.*

ÉCUBIERS, *s. m. p.* Hawse-holes; also the hawse-pieces through which those holes are cut.

ÉCUEIL, *s. m.* A dangerous rock or shoal.

ÉCUELLE, *s.f.* ÉCUELLE *de cabestan.* The saucer of a capstan.

ÉCUME, *s. m.* The froth or foam of a breaking sea.

ÉCUMER, *v. a.* ÉCUMER *la mer.* To scour or infest the sea, as a pirate.

ÉCUMEUR DE MER, *s. m.* A pirate.

ÉCUSSON, *s. m.* The escutcheon or ornament on the middle of the upper counter, where the ship's name is written.

s'EFFACER, *v. r.* To present the head or stern to an enemy.

ÉGORGEOIR, *s. m.* A spilling-line.

ÉGORGER UN HUNIER, *v. a.* To haul up a top-sail lengthways.

ÉGOUT, *s. m.* The running down of water through the seams of a ship.

ÉGOUTTER, *v. a.* ÉGOUTTER *des fauberts.* To wring swabs.

ÉGOUTTOIR, *s. m.* A grating or drain wherein to lay cordage after it is tarred.

ÉGUILLETER, *v. a.* To seize, to rack, &c. *Ex.*

ÉGUILLETER *un cordage.* To seize any complication of ropes.

ÉGUILLETER *un palan.* To rack a tackle.

ÉGUILLETTES, *s. f. p.* Knittles, &c.

ÉGUILLETTES *de bosses.* Lanyards of the stoppers

ÉGUILLETTES *de carène.* Careening gear.

ÉGUILLOT, *s. m.* ÉGUILLOTS *de gouvernail.* The pintles of the rudder.

ÉLAN. *Voyez* LANS.

ÉLANCÉ, *adj. Avant fort* ÉLANCÉ. A flaring-bow.

ÉLANCEMENT, *s. m.* ÉLANCEMENT *de l'étrave.* The rake of the stem.

ÉLEVER, s'ÉLEVER, *v. a. & r.* To raise; to rise; to gain, &c. *Ex.*

ÉLEVER *l'arcasse.* To raise the stern-frame of a vessel.

s'ÉLEVER *au vent.* To get to windward.
Vaisseau qui s'ÉLEVE *bien sur la lame.* A vessel that rises easily, or tops the sea well.
Vaisseau qui s'ÉLEVE *en latitude.* A ship which gains fast latitudinally.

ÉLINGUER, *v. a.* To sling a cask, &c.

ÉLINGUES, *s. f. p.* Slings. *Ex.*

ÉLINGUES *à pattes.* Can-hooks.

ÉLINGUES *d'embarcation.* Boat-slings.

ÉLINGUES *pour les canons.* Gun-slings.

ÉLINGUES *à jarre.* Jar-slings.

ÉLINGUETS, *s. m. p.* The pawls of a capstan.
ELME. *Voyez* FEU *St. Elme.*

ÉLONGER, *v. a.* ÉLONGER *un vaisseau.* To come along-side of a ship, or to lay a-breast of her.
s'EMBANQUER, *v. n. & r.* To come upon a fishing-bank.

EMBARCADÈRE, *s. m.* A small port or bight along the sea coast, commodious for landing and shipping of goods.
EMBARCATION, *s.f.* A general name for boats and small craft.
☞ Les EMBARCATIONS à la mer! *Hoist the boats out!*

EMBARDÉE, *s.f.* A yaw; lee-lurch; wild steering.
Faire une EMBARDÉE. To give a yaw.
EMBARDER, *v. n.* To yaw, to steer wild.
EMBARGO, *s. m.* An Embargo.
Mettre un EMBARGO *sur tous les vaisseaux.* To lay an embargo on all ships.
EMBARQEMENT, *s. m.* Embarcation.
EMBARQUER, s'EMBARQUER, *v. a. & r.* To embark, to ship goods, &c. *Ex.*
EMBARQUER *un officier,* To enter an officer in a ship.
EMBARQUER *des troupes.* To ship troops.
Nous venons D'EMBARQUER *nos provisions.* We have just taken in our provisions.
EMBARQUER *un coup de mer* To ship a heavy sea.
s'EMBARQUER. To embark; to take shipping.
Je me suis EMBARQUÉ *à Malte & j'ai débarqué en*

Sicile. I embarked at Malta, and landed in Sicily.

☞ EMBARQUE les chaloupiers! *Man the launch.*

☞ EMBARQUE! *Come aboard!* EMBARQUE du Rivoli! *Rivolis away!* A general warning given by the cockswain, that the boat is going to put off.

EMBELLE. *Voyez* BELLE.

EMBOSSER, *v. a.* EMBOSSER *un bâtiment.* To bring the broadside of a ship to bear upon some fort, &c. by clapping a spring upon the cable.

EMBOSSURE, *s. f.* A spring.

Faire EMBOSSURE. To clap a spring upon the cable.

EMBOUCHURE, *s. f.* EMBOUCHURE *d'une rivière.* The mouth of a river, also the opening of a bay or gulf.

EMBOUDINURE. *Voyez* BOUDINURE.

EMBOUQUER, *v. n.* To enter a strait or narrow passage.

EMBRAQUER. *Voyez* ABRAQUER.

EMBRUMÉ, *adj.* *Tems* EMBRUMÉ. Foggy weather.

Ciel EMBRUMÉ. A cloudy sky.

Terre EMBRUMÉE. A land seen in a fog.

EMERILLON, *s. m.* A large fishing-hook.

EMMANCHER, *v. n.* To enter into the channel; said of a ship which is entering the British channel.

D'après la sonde nous devons être EMMANCHÉS. By our soundings we should be in the channel.

EMMÉNAGEMENS, *s. m. p.* The accommodations and conveniences of a ship; the contrivances of the hold. *Ex.*

Le paquebot a peu D'EMMÉNAGEMENS *pour les passagers.* The packet-boat has few accommodations for passengers.

EMMIELLER UN ÉTAI, *v. a.* To worm a stay.

EMMORTAISER, *v. a.* To drive a piece of wood into a mortise.

EMPÂTURE, *s. f.* A scarph; also the spread of the lower rigging. *Ex.*

La Ville de Paris a beaucoup D'EMPÂTURE. The Ville de Paris spreads her rigging well.

EMPENNELLE, *s. f.* A kedge-anchor.

EMPENNELLER, *v. a.* EMPENNELLER *une ancre.* To back an anchor.

Ancre à EMPENNELLER, *ou pennaut.* A small anchor, or kedge-anchor, used to back another large one, to prevent it from coming home to the ship.

EMPESER LA VOILE, *v. a.* To wet a sail.

EMPIRANCE, *s. f.* The deficiency of a ship's cargo at the time of delivery.

EMPORTER, *v. a.* To carry away, &c. *Ex.*

Notre gouvernail fut EMPORTÉ. Our rudder was knocked off.

Notre grand canot fut EMPORTÉ. Our long-boat was washed overboard.

Le grand mât fut EMPORTÉ. The main-mast was carried away, or broken by tempestuous weather, &c.

Un grain EMPORTA *notre perroquet de fougue.* A squall blew away our mizen top-sail.

La bouteille de tribord de ce vaisseau est EMPORTÉE. That ship's starboard quarter-gallery is carried away.

ENCABLURE, *s. f.* A cable's length, which contains 120 fathoms. *Ex.*

Nous étions à deux ENCABLURES *de terre quand nous échouâmes.* We were within two cable lengths of the shore when we grounded.

ENCAPÉ, *adj.* *Vaisseau* ENCAPÉ. A ship embayed, or entered between two capes.

ENCAPER. *v. n.* To embay, or to be embayed.

ENCASTILLAGE. *Voyez* ACCASTILLAGE.

ENCLAVER, *v. a.* To mortise, to let in a rabbet.

ENCOCURE, *s. f.* The situation of the eye of a pendant, or studding-sail boom-iron, &c. fixed on a yard-arm.

ENCOIGNURE, *s. f.* The elbow or angle of a knee or standard.

ENCOLURE, *s. f.* The thickness of timbers amidships.

Ligne D'ENCOLURE. The cutting-down line.

ENCOMBRÉ, *adj.* *Bâtiment* ENCOMBRÉ. A ship that is much lumbered.

Port ENCOMBRÉ A port crowded with shipping.

ENCOMBREMENT, *s. m.* Incumbrance; or the space which light and embarrassing goods occupy in the hold.

ENCOMBRER, *v. a.* To lumber. *Ex.*

Votre frégate doit être très ENCOMBRÉE. Your frigate must be very much lumbered.

ENCOQUER, *v. a.* To fasten a block-strap, or the eye of any rope, upon the yard-arm.

ENCORNAIL. *Voyez* CLAN.

ENDENTÉ, *adj.* Indented.

ENDENTER, *v. a.* To dove-tail; to indent.

ENDOMMAGER, *v. a.* To damage. *Ex.*

Ce bâtiment à sa quille endommagée. That ship has her keel damaged.

ENFILADE, *s. f.* Rake, or enfilade; in gunnery.

ENFILER, *v. a.* ENFILER *un vaisseau.* To rake a ship in action.

ENFLÉCHURES, *s. f. p.* The ratlings of the shrouds.

ENFONCEMENT, *s. m.* A bight in the land.

Dans L'ENFONCEMENT *de la baie il y a bon mouillage.* In the bight of the bay there is good anchorage.

ENGAGÉ, *adj. & part. past.* *Vaisseau* ENGAGÉ. A water-logged ship.

Manœuvre ENGAGÉE. A foul rope; or, a rope entangled.

ENGAGEMENT, *s. m.* The contract, or articles of agreement made between the seamen and the commander of a merchant-ship.

ENGAGER, *v. a.* ENGAGER *le combat.* To engage, to bring to an action.

ENGORGÉ, *adj. & part. past.* Choked. *Ex.*

La pompe est ENGORGÉE. The pump is choked.

ENGORGER, *v. a.* To choke up.

ENGRAVER, *v. a.* ENGRAVER *les pièces d'arrimage.* To bury the casks in the ballast.

ENGRENER, *v. a.* ENGRENER *la pompe.* To fetch the pump.

ENHUCHÉ, *adj.* *Navire* ENHUCHÉ. A moon-sheered ship; a ship whose upper-works rise very high fore and aft.
ENJALER, *v. a.* ENJALER *une ancre.* To stock an anchor; or to fix the stock upon it.
ENLEVER, *v. a.* ENLEVER *un bâtiment à l'abordage.* To carry a vessel by boarding her.
ENNEMI, *s. m.* An enemy. *Ex.*
Donner, ou faire porter sur L'ENNEMI. To run down upon an enemy.
Mettre L'ENNEMI *entre-deux feux.* To place the enemy between two fires.
Rester en présence de L'ENNEMI. To remain in sight of the enemy.
s'ENSABLER, *v. r.* To stick in the sand.
ENSEIGNE, *s. m.* ENSEIGNE *de vaisseau.* A commissioned officer in the French navy, next in rank to a lieutenant.
ENSEIGNE, *s. f.* An ensign.
Bâton D'ENSEIGNE. The ensign staff.
ENSEMBLE, *adv.* Together. *Ex.*
☞ Nage ENSEMBLE! *Pull together!*
☞ Hisse ENSEMBLE! *Hoist together!*
ENTENNE. *Voyez* ANTENNE.
ENTOURAGES. *s. m. p.* Screens.
ENTRAVERSER, *v. a.* To set across.
Le vaisseau S'ENTRAVERSE. That ship lies across.
ENTRE-DEUX, *s. m.* ENTRE-DEUX *des lames.* The hollow or trough of the sea.
ENTRE-DEUX *des mâts.* The distance between the masts.

ENTRÉE, *s. f.* L ENTRÉE *d'un port.* The entrance of a harbour or port.
Avoir L'ENTRÉE. To have communication with the shore (a term of quarantine).
ENTREMISES, *s. f. p.* (In ship-building) Straight and square pieces of timber, used by French ship-builders in the same place where the English put a horizontal or lodging-knee
ENTRE-PONT, *s. m.* Between decks.
ENTRE-PONT *de frégates.* The orlop-deck.
Hauteur de L'ENTRE-PONT. *Voyez* HAUTEUR.
ENTREPRENEUR, *s. m.* A contractor.
ENTRER, *v. a.* To enter, &c.
Faire ENTRER *la clé dans un mât de hune.* To enter the fid of a top-mast.
Faire ENTRER *un bâtiment dans le bassin.* To take a ship into dock.
ENTRE-SABORDS, *s. m. p.* The intervals between the ports of a ship's side.
s'ENTRETENIR, *v. r.* S'ENTRETENIR *sur le même bord.* To keep on the same tack.
ENTRETOISE, *s. f.* L'ENTRETOISE *d'affût.* The transom of a gun carriage.
ENTRETOISE *du mât de beaupré.* A kind of partners to the bowsprit, placed vertically at a small distance before its step, on the lower-deck, in French men of war.
s'ENVASER, *v. r.* To stick in the mud.
ENVERGUER. ENVERGUER *une voile, v. a.* To bend a sail to its yard.
ENVERGURE, *s. f.* The length of the yards of a ship, and the extent of the sails upon them.
Vaisseau qui a beaucoup D'ENVERGURE. A ship whose yards are very square.

ENVOYER, *v. a.* To send. *Ex.*
ENVOYER *un homme à la tête du mât.* To send a man to the mast-head.
ENVOYER *les embarcations remorquer un bâtiment.* To send the boats out to tow a ship.
☞ ENVOYE! *Helm a-lee!*

ÉPAVES, *s. f. p.* Whatever is cast by the sea upon shore, and has no owner.

ÉPAULES D'UN VAISSEAU, *s. f. p.* The bows of a ship.

ÉPAULETTES, *s. f. p.* Epaulets, or shoulder-straps, worn by naval and military officers, as a mark of distinction.

ÉPERON, *s. m.* The cut-water, or knee of the head of a ship; the parts of which are,
La gorgère. Part of the knee of the head.
La frise. The upper part of the knee of the head.
Les courbes de jottereaux. Cheeks of the head.
Mouchoir, ou remplissage sous les jottereaux. Filling-timber under the cheeks of the head.
Courbatons de herpes. The Timbers of the head.
Lisses de herpes. Rails of the head.
Boudin. Middle rail of the head.
N. B. The three last pieces form the ornament of the head of a French ship.
Bras de la poulaine, ou porte vergues. Wooden rails reaching from the cat-head to the hair of the figure or lion of the head, in French ships. *Voyez* GORGÈRE, &c.

ÉPI, *s. m.* ÉPI *du vent.* The wind's eye.

ÉPINEUX, *adj.* Rocky above water; full of rocks and breakers.

ÉPINGLETTE, ou DÉGORGEOIR, *s. f.* Priming-wire, or priming-iron.

ÉPISSER, *v. a.* ÉPISSER *un cordage.* To splice a rope.

ÉPISSOIRE, *s. f.* A marling spike; also a splicing-fid of hard wood.

ÉPISSURE, *s. f.* A splice of any kind.

ÉPISSURE *longue.* Long splice.

ÉPISSURE *courte.* Short splice.

ÉPISSURE *en portière de vache.* Cut-splice.

ÉPISSURE *de cable.* Cable splice.

ÉPISSURE *d'estrop, de gance, ou d'œillet.* Eye-splice.

ÉPONTILLE, *s. f.* A stanchion, or shore.

ÉPONTILLES *d'entre-pont.* The stanchions between decks.

ÉPONTILLES *de la cale.* The stanchions or pillars of the hold.

ÉPONTILLER, *v. a.* To prop, or shore up.

ÉPREUVE, *s. f.* Trial, experiment, proof. *Ex.*
Faire L'ÉPREUVE *du canon.* To try or prove a gun.
L'ÉPREUVE *de la poudre.* To prove the strength of gunpowder.

ÉQUARRISSAGE, *s. m* The square of a piece of timber. *Ex.*
Cette piece de bois a 18 *pouces* D'ÉQUARRISSAGE. That piece of timber is 18 inches square.

ÉQUATEUR, *s. m.* Equator; equinoctial line.

ÉQUERRAGE, *s.m.* The bevel of a ship's timbers.
Travailler une pièce suivant son ÉQUERRAGE. To bevel a piece of timber.

ÉQUERRE, *s. f.* A square, rule.
Pièce à L'ÉQUERRE. A square piece of timber.
Pièce à fausse ÉQUERRE. A piece out of square.
Courbe à ÉQUERRE. A square knee.

ÉQUIPAGE, *s. m.* The ship's company, exclusive of the captain and principal officers.

ÉQUIPAGE *d'attelier.* A general name for the machinery and furniture of a dock-yard, &c.

ÉQUIPEMENT, *s. m.* The fitting out, or out-fit of a ship.

ÉQUIPER, *v. a.* ÉQUIPER *un bâtiment.* To fit out; to man and provide a ship with whatever is necessary to prosecute war or commerce.

ÉQUIPET, *s. m.* A small open locker.

ÉRAILLÉ, *adj.* Chafed or galled, speaking of cables and ropes. *Ex.*
Cette manœuvre est beaucoup ÉRAILLÉE. That rope is very much chafed.

s'ÉRAILLER, *v. r.* To chafe, &c.
ERSE. *Voyez* ESTROPES.
ESCADRE, *s. f.* A fleet, or squadron of men of war.
ESCADRE *d'evolutions.* A squadron fitted out for the purpose of exercising and practising naval evolutions.
ESCADRE *legère.* A flying squadron.
ESCANDOLE, *s. f.* The cabin of the Argousin of a row-galley.
ESCARBIT ou ESCARBITE, *s. f.* A calker's oil-box.

ESCARPÉE, *adj.* Steep-to; expressed of a shore which may be approached without danger.
ESCOP, *s. m. Grand* ESCOP. A skeet or large scoop for wetting the sails of a ship, or a vessel's side.
ESCOP *à main.* A boat's scoop.
ESCORTE, *s. f.* Convoy. *Ex.*
Donner L'ESCORTE *à un bâtiment.* To convoy a ship.
Nous partîmes sous L'ESCORTE *d'une frégate.* We sailed under the convoy of a frigate.
ESPALE, *s. f.* The aftmost bank or thwart of a row-galley.
ESPALIER, *s. m.* The first oar in a galley; also the person who rows with it.
ESPALMER, *v. a.* To pay a vessel's or boat's bottom with soap, &c., after having breamed her.
ESPARRES, *s. f. p.* Spars.
ESPINGOLE, *s. f.* A Blunderbuss.
ESPONTON, *s. m.* A kind of half-pike, used in the act of boarding an enemy's ship.
ESPOULETTE, *s. f.* A tin canteen, or case, used to carry fine powder to the guns, in time of an engagement.

ESQUIF, *s. m.* A skiff, yawl, or small boat belonging to a ship.
ESSES, *s. f. p.* ESSES *d'affût.* The forelocks which traverse the axle-trees of a gun-carriage, to confine the wheels in their places.
ESSIEU, *s. m.* ESSIEU *d'affût.* The axle-tree of a gun-carriage.
ESSIEU *de pouliè.* The pin of a block.
EST, *s. m.*. The east.
ESTACADE, *s. f.* A boom, strong chain, &c. laid across the mouth of a harbour, to prevent an enemy's entering it.
ESTAIN, *s. m.* The lower part of the fashion-pieces of the stern.
Dévoiement de L'ESTAIN. Flaring of the fashion-pieces
ESTAIN *dévoié.* Cant fashion-pieces.
ESTERRE, *s. f.* A small haven or creek.
ESTIME, *s. f.* The dead-reckoning. *Ex.*
Erreurs dans L'ESTIME. The errors of a dead-reckoning.
Nous avons passé le tropique par notre ESTIME. By our reckoning we are past the tropic.

ESTIMÉE, *adj. & part. past.* *Route* ESTIMÉE. The ship's course as by dead-reckoning.
Latitude ESTIMÉE. Latitude as by dead-reckoning.
ESTIVE, *s. f.* The trim or disposition of the cargo in a ship's hold.
ESTIVER, *v. a.* ESTIVER *le chargement d'un bâtiment.* To screw down a vessel's cargo.
ESTRAN, *s. m.* A flat and sandy sea-coast.
ESTROPES, *s. m. p.* A general name for block-straps, &c. *Ex.*
ESTROPE *de poulie.* A block-strap.
ESTROPE *d'aviron.* The grommet of an oar.
ESTROPE *de culasse.* The breech-strap of a gun.
ESTROPE *d'affût.* The strap, or eye in the train of a gun-carriage.
ESTROPE *de gouvernail.* Rudder-straps.
ESTROPE *à rider les haubans.* A selvagee used in setting up shrouds.
ESTROPES *des marche-pieds.* The stirrups of the horses.
ESTROPER, *v. a.* To strap.
ESTROPER *une poulie.* To strap a block.

ÉTABLIR, *v. a.* ÉTABLIR *la marguerite.* To clap a purchase on the cable.

ÉTABLISSEMENT, *s. m.* A settlement, &c. *Ex.*

ÉTABLISSEMENT *de la marée.* The time of high-water at any place on the days of new and full moon.

ÉTABLISSEMENT *d'un port.* The situation of a harbour, and the time of high-water in it at full and change-days.

ÉTAI, *s. m.* The stay of a mast.

ÉTAI *du grand mât, ou grand* ÉTAI. Main-stay.

ÉTAI *du grand perroquet.* Main-top-gallant stay.

ÉTAI *de misaine.* Fore-stay.

ÉTAI *du petit mât de hune.* Fore-top-mast stay.

ÉTAI *du petit mât de perroquet.* Fore-top-gallant stay.

Draille D'ÉTAI *de voile de perroquet.* Main-top-gallant stay-sail stay.

ÉTAI *du grand perroquet.* Main-top-gallant stay.

ÉTAI *de misaine.* Fore stay.

ÉTAI *du petit mât de hune.* Fore-top-mast stay.

ÉTAI *du petit mât de perroquet.* Fore-top-gallant stay.

ÉTAI *du petit mât de perroquet volant.* Fore-top-gallant royal stay.

ÉTAI *d'artimon.* Mizen-stay.

ÉTAI *du mât de perroquet de fougue.* Mizen-top-mast stay.

ÉTAI *du mât de la perruche.* Mizen-top-gallant stay.

Faux grand ÉTAI. Main spring-stay.

Faux ÉTAI *du grand mât de hune.* Main-top preventer stay.

Faux ÉTAI *de misaine.* Fore-preventer stay.

Faux ÉTAI *du petit mât de hune.* Fore-top preventer stay.

Draille de voile D'ÉTAI *de misaine.* Fore-stay-sail stay.

Draille de la pouilleuse. Main-stay-sail stay.

Draille de la fausse voile D'ÉTAI. Middle stay-sail stay.

Collet D'ÉTAI. Eye of a stay.

Collier D'ÉTAI. Collar of a stay.

Pomme D'ÉTAI Mouse of a stay.

ÉTALE, *adj. Mer* ÉTALE. Slack water, at which time the tide apparently neither flows nor ebbs.

Etre en ÉTALE. To be at a stand.

Vent ÉTALE. A settled wind; a steady wind, without squalls.

ÉTALER, *v. a.* ÉTALER *la marée.* To pursue a ship's course by means of the tide, with a foul wind, and anchoring during the intervals of the contrary tide.

ÉTALER *un coup de vent.* To weather out a storm.

ÉTALINGUER, *v. a.* ÉTALINGUER *un cable.* To clinch, or bend a cable to the ring of an anchor, &c.

ÉTALINGURE, *s. f.* The clinch of a cable.

ÉTALINGURE *d'orin des grandes ancres.* The clinch of a buoy-rope to its anchor.

ÉTALINGURE *de bouée.* The clinch of a buoy-rope to its buoy.

ÉTALINGURE *de grapin.* The clinch of a grapling.

ÉTAMBOT, ou ÉTAMBORD, *s. m.* The stern-post.

ÉTAMBRAI, *s. m.* The partners of a mast, or of a capstan.

Pièce D'ÉTAMBRAI. A piece of timber fit for partners.

ÉTAMINE, *s. f.* Buntine or bunting.

ÉTANCE *à taquets, s. f.* Samson's post; a stanchion erected in a ship's hold.

ÉTANCE. *Voyez* ÉPONTILLE.

ÉTANCHE, *adj. Vaisseau* ÉTANCHE. A tight ship.

ÉTANCHER, *v. a.* ÉTANCHER *un bâtiment.* To free a ship of water by means of the pumps.

ÉTANCHER *une voie d'eau.* To stop a leak.

ÉTANÇONS, *s. m. p.* Stanchions or shores. *Voyez* ÉPONTILLE.

ÉTANÇONNER, *v. a.* To shore up; speaking of a ship.

ÉTARQUER, *v. a.* ÉTARQUER *une voile.* To hoist a sail taught up by the leech. *Ex.*

☞ ÉTARQUE! ÉTARQUE bien! *Hoist! and a taught leech!*

ÉTAT, *s. m.* ÉTAT *de situation d'un bâtiment.* The state and condition of a ship.

ÉTAT *de la marine.* The establishment of the navy.

ÉTAT *d'armement.* A list, or register, containing the number of ships and officers, destined for a naval armament; as also the quality and proportion of cordage, sails, and furniture of a ship, &c.

ÉTAT *des malades.* Sick-list.

ÉTAT-MAJOR, *s. m.* L'ÉTAT-MAJOR *d'un vaisseau de guerre.* The commissioned officers of a French man of war, including the surgeon and chaplain.

ÉTAT-MAJOR *d'une escadre.* The admiral, the first captain of the fleet, the signal-officers under him, the intendant and his deputy.

ÉTEINDRE, *v. a.* To extinguish.

☞ ÉTEINDRE les feux! *Put out the fire and lights!*

ÉTENDARD, *s. m.* The standard. *Ex.*

ÉTENDARD *d'Angleterre.* The standard of England.

ÉTOILES, *s. f. p.* Stars.

ÉTOILE *polaire.* The pole, or polar-star.

ÉTOILES *fixes.* The fixed stars.

ETOUPE, *s. f.* Oakum. *Ex.*

ÉTOUPE *blanche.* White or untarred oakum.

ÉTOUPE *noire, ou goudronnée.* Black, or tarred oakum.

ÉTOUPILLONS, *s. m. p.* Toppings.

ÉTRANGLER, *v. a.* ÉTRANGLER *l'artimon.* To balance the mizen.

ÉTRAVE, *s. f.* The stem of a ship.

ÉTRAVE *élancée.* A raking or flaring stem.

ÊTRE, *v. n.* To be, &c. *Ex.*

ÊTRE *à flot.* To be afloat.

Être *à l'accore d'un banc.* To be at the edge of a bank.

Être *à sec, ou à mâts & à cordes.* To be under bare-poles.

Être *dans les eaux d'un vaisseau.* To be in the wake of a ship.

Être *pratique de la mer.* To be accustomed, or inured to the sea.

Être *en armement.* To be fitting out.

ÉTRIERS, *s. m. p.* The stirrups of a horse.

Étriers *de porte-haubans.* Iron stirrups, or straps of the channels.

Étrier *de gouvernail.* The stirrup of the tiller.

Étriers *des chaines de haubans.* The lower links of the chains of the shrouds, which are bolted to the wales.

ÉTRIQUÉ, *adj. Bâtiment* étriqué. A narrow-floored vessel.

ÉTUVE, *s. f.* Étuve *de corderie.* A stove in a dock-yard, fitted with furnaces and cauldrons, for tarring cordage, &c.

Étuve *à bordage.* A place in which planks are immersed in hot water, in order to bend them to certain curves for ship-building.

ÉVENT, *s. m.* Évent *du boulet.* The windage of a cannon, or the difference between the diameter of the bore and the diameter of the shot.

☞ Évente le perroquet de fougue! *Fill the mizen top-sail!*

ÉVENTER, *v. a.* Éventer *les voiles.* To fill the sails.

Éventer *la quille d'un bâtiment.* To heave down a vessel, so as to discover her keel.

ÉVITÉE, *s. f.* The space wherein a ship may swing round; also, the act of a ship swinging round her anchor at the length of her cable. *Ex.*

Le vaisseau n'a pas son évitée *dans cet endroit.* The ship has not sufficient birth in that place.

Le vaisseau fait son évitée. The ship is swinging.

ÉVITER, *v. a. & n.* To avoid, to swing or tend. *Ex.*

Éviter *l'abordage.* To avoid being boarded.

Éviter *au vent.* To swing head to windward; to stem the wind.

Éviter *à la marée.* To swing at the change of the tide, or to stem the tide.

ÉVOLUER, *v. n.* To perform the several evolutions practised at sea.

ÉVOLUTIONS, *s. f p.* Evolutions, or the movements of a fleet in forming the line of battle, &c.

Art des évolutions *navales.* Naval tactics.

EXERCICE, *s. m.* Exercise in general.

Exercice *des voiles.* The exercise of the sails.

Faire l'exercice *du canon et de la mousqueterie.* To exercise the great guns and small arms.

EXPÉDITION, *s. f.* An expedition. *Ex.*

L'escadre de Rochefort est partié pour une expédition *particulière.* The Rochefort squadron is sailed upon a secret expedition.

F.

FABRIQUE, *s. f.* The particular structure of a ship.

FABRIQUER. *Voyez* Construction.

FAÇONS, *s. f. p.* The rising of a ship's floor. *Ex.*

Hauteur des façons *de l'arrière.* The rising of the floor abaft.

Hauteur des façons *de l'avant.* The rising of the floor a-fore.

Lisse des façons. The rising line, or floor-ribband.

FAIRE, se faire, *v. a. n. & r.* To do; to make; to cause, &c. *Ex.*

Faire *attention à la barre.* To mind the helm.

Faire *bitture.* To get up a range of the cable.

Faire *côte. Voyez* Côte.

Faire *courir.* To run with full sails; to keep her full. *Ex.*

☞ Fais courir! *Keep her full!*

Faire *des vivres.* To take in provisions.

Faire *du bitord.* To spin spun-yarn.

Faire *feu des deux bords.* To fire on both sides.

Faire *l'appel du quart.* To muster the watch.

Faire *la ronde.* To row guard.

Faire *le nord, le sud, &c.* To stand to the north, south, &c.

Faire *naufrage.* To be cast away.

Faire *pavois.* To dress a ship.

Faire *penneau.* To get every thing clear to come to anchor.

Faire *tête.* To stem the wind, or current.

Faire *voile.* To set sail; to get under weigh.

Faire *de la voiles.* To make sail.

Faire *force de voiles.* To crowd sail.

Faire *petites voiles.* To be under easy sail.

Faire *route.* To stand on the course.

Faire *une bordée.* To make a board, or tack.

Faire *vent arrière.* To run before the wind.

Faire *branle-bas. Voyez* Branle-bas.

Faire *de l'eau. Voyez* Eau.

Faire *gouverner.* To cun the ship, or to superintend and direct the helmsman.

Faire *quarantaine. Voyez* Quarantaine.

Faire *une descente. Voyez* Descente.

FAIT, *part. past. Tems* fait. Settled weather.

FAIX, *s. m.* Faix*-de-pont.* That part of the deck which supports a great weight, and which requires to be strengthened underneath by stanchions, &c.

FALAISE. *Voyez* Côte *accore.*

FALQUES, *s. f. p.* The wash-boards of a boat.

FANAL, *s. m.* A lantern.

Fanal *de poupe.* Poop-lantern.

Fanal *de hune.* Top-lantern.

Fanaux *de signaux,* Signal-lanterns.

Fanaux *de combat.* Fighting lanterns, or the lanterns used between the guns in time of battle.

FANAL *sourd, ou lanterne sourde.* Dark lantern.
FANAL *de soute.* Store-room lantern.
Aiguilles de FANAL. Lantern braces.
Cercles des FANAUX *de poupe.* Lantern girdles.
FANAL, ou *Tour à feu.* *Voyez* PHARE.
FANON, *s. m.* The balance of the mizen.
FARAILLON, *s. m.* A small sand-bank, separated from a larger one by a narrow channel.
FARDAGE, *s. m.* The dunnage laid in a ship's hold, when she is to be laden in-bulk; as with corn, salt, &c.
FARGUES, ou FARDES, *s. f. p.* Wash-boards of the ports.
FASIER, *v. n.* To shiver; speaking of the sails.
FATIGUER, *v. a. & n.* To strain. *Ex.*
Cette frégate a une mâture qui la FATIGUE *beaucoup.* That frigate's masts strain her all to pieces.
FAUBERT, *s. m.* A swab.
FAUBERTER, *v. a.* To swab a ship's decks, &c.
FAUX, FAUSSE, *adj.* False, &c.
FAUX *baux.* Orlop-beams, or platform.
FAUSSES *balancines.* *Voyez* BALANCINES.
FAUSSES *cargue-boulines.* Quarter slab-lines.
FAUX *collier.* *Voyez* COLLIER.
FAUSSES *drisses.* Slings for the yards in action.
FAUX *foc.* Middle jib.
FAUX *canons.* Wooden guns, placed on a ship's side to deceive the enemy.
FAUX *pont* *Voyez* PONT.
FAUSSE *quille.* False-keel.
FAUX *racage.* Preventer parral, used to confine the yard to the mast, in case the parral should be shot away in battle.
FAUX *sabords.* False-ports.

FÉLOUQUE. *s. f.* A felucca.
FEMELLES, *s. f. p.* The goodgeons used to hang the rudder on the stern-post.

FÉMELOTS. *Voyez* FEMELLES.
FER, *s. m.* FER *d'arc-boutant.* The fork of a fire-boom.
FER *de boute-hors.* Studding-sail boom-iron.
FER. A name given to the anchors of a row-galley.
FER *de calfat.* A calking-iron.
FER *de calfat double, ou bec de corbin.* A calker's making-iron.
FER *de chandelier de pierrier.* The iron socket in which the swivel of a pedrero rests and traverses.
FER *de girouette.* The iron spindle which supports the vane at the mast-head.
FERS *de prisonniers.* Fetters, or bilboes, to confine criminals.
Mettre un homme aux FERS. To put a man in irons.
FER *de gaffe.* *Voyez* GAFFE.
FERLER, *v. a.* FERLER *les voiles, ou les serrer.* To furl or hand the sails.
FERMER, *v. a. & n.* To shut, &c. *Ex.*
☞ FERME les sabords! *Lower the ports down!*
☞ FERME les hublots! *Shut the port scuttles!*
☞ FERME l'ecoutille de la soute aux poudres! *Cover the magazine scuttle!*

FERMÉ, *adj.* *Vaisseau* FERMÉ *par les glaces.* An ice-bound ship.
Navire FERMÉ *entre les terres.* A land-locked ship.

FERRÉ, *adj.* Iron-bound.
FERRURE, *s. f.* Iron-work.
FERRURES *d'un vaisseau.* The iron-work of a ship; as chains, bolts, spikes, nails, &c.
FERRURE *d'affût.* The iron-work of a gun-carriage.
FERRURE *de gui.* The goose-neck of a boom.
FERRURE *de sabords.* Port-hinges.
FERRURES *de gouvernail.* The pintles and braces of a ship's rudder.
FESSES, *s. f. p.* The buttocks of a ship.
FEU, *s. m.* Fire, &c. *Ex.*
☞ FEU! *Fire!*
☞ FEU babord! *Fire to larboard!*
☞ FEU tribord! *Fire to starboard!*
FEU *vif.* A brisk fire.
FEU *roulant.* A constant fire.
FEU *bien servi.* A fire well kept up.
☞ Bas le FEU par tout! *Leave off firing fore and aft!*
FEU *grégeois.* Wild-fire.
Mettre le FEU *à un bâtiment.* To set fire to a ship.
Voilà un bâtiment qui a le FEU *à bord.* There is a vessel on fire.
FEU *St. Elme, s. m.* Saint Elmo's fire, or Jack with a lantern.
FEUILLES BRETONNES, ou VAIGRES BRETONNES, *s. f. p* Spirketting.
FEUX, *s. m. p.* Lights, used as signals on board, or in light-houses. *Ex.*
FEUX *d'artifice.* Artificial fires used at sea.
Les FEUX *de la tour de Cordouan.* The fires of the tower of Cordouan.
L'ennemis ont leurs FEUX *éteints.* The enemy have their lights out.
FICHE, *s. f.* A rag-bolt.
Anneaux à FICHES. Ring-bolts ragged.
FICHURE. *Voyez* FOÊNE.
FIGALE, *s. f.* An Indian vessel with one mast; usually rowed with oars, or paddles.
FIGURE, *s. f.* The ornamental figure erected on the fore-part of a ship's stem. *Voyez* EPERON.
FIL, *s. m.* Twine, &c. *Ex.*
FIL *à voile.* Sail-twine.
FIL *de caret.* Rope-yarn.
FIL *de cable.* Cable-yarn.
FIL *de manœuvre.* Fine rope-yarn.
FIL *blanc.* White rope-yarn.
FIL *goudronné.* Tarred rope-yarn.
Tenir ses huniers sur les FILS *de caret.* To keep the topsails stopped with rope-yarns.
FIL *du bois.* The grain of wood.

FILADIÈRE, *s. f.* A small flat-bottomed boat used on the Garonne.
FILANDRES, *s. f. p.* Sea-weeds, which adhere to the bottom of a ship, in the course of a long voyage, and make her sail heavy.
FILARETS, *s. m. p.* Rails, or rough-tree rails, used to extend the nettings on a ship's quarter, waist, or fore-castle.
FILE. *Voyez* VIRURE.
☞ FILE le grand canot de l'arrière! *Veer the barge a-stern!*

☞ File un peu de la grande écoute! *Ease off a little of the main sheet!*

☞ Ne file plus, amarre! *Keep fast the cable! Stopper the cable!*

FILER, *v. a.* To veer; to spin, &c. *Ex.*

Filer *du cable.* To veer away, or pay out the cable to the anchors.

Filer *le cable par le bout.* To veer out the cable end for end, or to slip the cable.

Filer *une manœuvre, ou un cordage.* To ease off, or ease away any rope.

Filer *la grande écoute en bande.* To let fly the main sheet, as in a squall of wind, which endangers the ship.

Filer *du chanvre.* To spin hemp.

Filer. To run; speaking of a ship's way. *Ex.*

Nous filâmes *dix nœuds par heure.* We ran ten knots an hour.

FILERIE, *s. f.* A spinning-house, or the place where hemp or flax is spun.

FILETS, *s. m. p.* Filets *de bastingage.* The nettings of a ship's quarter, waist, or fore-castle.

Filets *de pêche.* Fishing-nets.

Filet *de merlin.* A small line, called marline.

FIN, *adj.* *Vaisseau* fin. A sharp-bottomed ship.

Fin *voilier.* A prime sailer; a ship that sails very fast.

FISCELLE, *s. f.* Whipping-twine; also a fox formed of a single rope-yarn.

FISOLÈRE, *s. f.* A Venetian sculler.

FLAMBER, *v. a.* Flamber *un vaisseau.* To hoist a particular ship's signal in a fleet, to denote that she has not executed her manœuvre according to orders, &c.; also, to light up.

FLAMME, *s. f.* A pendant; streamer.

Bâton de flamme. *Voyez* Bâton, &c.

Drisse de flamme. *Voyez* Drisse, &c.

Flamme *d'ordre.* A pendant displayed by the commander of a squadron to call officers of other ships on board for orders, &c.

FLANC, *s. m.* Flanc *d'un vaisseau.* The side of a ship.

Présenter le flanc *a l'ennemi.* To bring the ship's broadside to bear upon the enemy.

Etre flanc *à* flanc. To lie along-side of; to be broadside and broadside.

FLASQUES, *s. f. p.* Flasques *d'affût.* The sides or cheeks of a gun-carriage.

Flasques, ou *jottereaux de mât.* The cheeks or upper parts of the hounds of a mast.

Flasques *de cabestan.* Whelps of a capstan.

Flasques *de carlingue.* Pieces inserted in the step of a mast, in French ships, in order to give it greater strength.

FLÈCHE, *s. f.* The upper part of the knee of the head in ships of war.

Flèche. The pole of a mast.

Flèche, is also a long boom projecting from the prow of a lateen vessel.

Flèche *en cul, s. m.* A gaff top-sail.

À FLEUR-d'eau, *adv.* Level with the surface of the water. *Ex.*

Coup du canon à fleur-*d'eau.* A shot between wind and water. *Voyez* Coup, &c.

Rochers à fleur-*d'eau.* *Voyez* Rochers, &c.

FLEURS, *s. f. p.* Rung-heads, or floor-heads; also, the rising-line of the floor-timbers.

FLIBOT, *s. m.* A fly-boat; a vessel used by the Dutch, having a main-mast and fore-mast, without any top-mast.

FLIBUSTIERS, *s. m. p.* Free-booters; bucaneers.

FLOT, *s. m.* The flood-tide. *Ex.*

Il y a flot. The tide flows.

*Demi-*flot. Half-flood.

Quart de flot. The first quarter flood.

Troisième quarts de flot. The last quarter flood.

Flots *de la mer.* The surges or waves of the sea.

Abandonner un vaisseau à la merci des flots. To let a ship drive at the mercy of the wind and waves.

À-Flot, *adv.* A-float.

Etre à flot. To be a-float, on the surface of the water.

FLOTILLE, *s. f.* A flotilla.

FLOTTAISON, *s. f.* That part of the ship's side outward, which lies about the load-water line.

Ligne de flottaisson. The load-water-line.

FLOTTE, *s. f.* A fleet of merchantmen. *Ex.*

Flotte *des Indes.* The India fleet.

Flotte *d'Espagne.* The spanish fleet, or Armada.

Flottes *de cables.* Barrels, or pieces of wood fastened to a cable, to float it up from a rocky ground.

FLOTTER, *v. n.* To float, or swim upon the surface of the water.

Flotter *un cable.* To buoy up a cable.

FLÛTE, *s. f.* A store-ship. *Voyez* Bâtiment *armé en flute.*

FLUX, *s. m.* The flood, or flood-tide.

Flux *et reflux.* The flux and reflux of the tide.

FOC, *s. m.* A jib, &c. *Ex.*

Grand foc. Standing jib.

Second foc. Fore-top-mast stay-sail.

Petit foc, *ou trinquette.* Fore-stay-sail.

Clin foc. Flying jib.

FOÈNE, *s. f.* A fish-gig, or forked instrument used to strike fish withal.

FONCET, *s. m.* A long flat-bottomed barge, for carrying goods up or down a river, &c.

FOND, *s. m.* The ground or bottom of the sea.

Fond *de sable.* A sandy bottom.

Fond *de vase.* Oozy ground; muddy bottom.

Fond *d'argile.* Clay bottom.

Fond *de roche.* Rocky bottom.

Fond *de gravier.* Gravelly bottom.

Fond *de coquillage.* Shelly ground.

Fond *de corail.* A bottom of coral.

Fond *dur.* Hard bottom.

Fond *pierreux.* Stony bottom.

Fond *mou.* Soft ground

Bon fond. A good anchoring ground.

Fond *de bonne tenue.* Good holding ground.

*Bas-*fond, ou *Hauts* fond. *Voyez* Bas-Fond.

Il n'y a point de fond. Out of soundings.

Couler à fond. To sink; to go to the bottom.

Fond *de cale.* *Voyez* Cale.

Fond *d'une basse voile.* The foot of a lower sail.

Fond *d'une hune, ou sole.* The platform, or flooring of a top.

FONDS *plats.* Flat-bottomed.
Bâtiment à FOND *plat.* A flat-bottomed vessel.
FOND *d'un affût.* *Voyez Parties de* L'AFFÛT.
FORAINE, *adj.* *Rade* FORAINE. An open road-stead.
FORANTS, *s. m. p.* Balks.
FORBAN, *s. m.* A pirate, or rover.
FORÇAT, *s. m.* A convict, or galley-slave.
FORCE, *s. f.* Strength, might, power, &c. *Ex.*
FORCE *du courant, ou du vent.* The force of the current, or of the wind.
FORCES *navales.* Naval forces.
☞ Nage de FORCE! *Voyez* NAGE, &c.
FORCÉ, *adj.* *Coup de vent* FORCÉ. A stress of weather.
Faire vent arrière FORCÉ. *Voyez* FAIRE, &c.
FORCER, *v. n.* To force, &c. *Ex.*
FORCER *à travers les glaces.* To force through the ice.
FORCER *de voiles.* To crowd all sails; to carry a press of sail.
FORCER *un navire à faire côte.* To drive a vessel on shore.
FORCER *de rames.* To row strongly, so as to redouble the efforts of the oars.
FORCER *un mât.* To wring a mast, by setting up the rigging too taught.
FORME, *s. f.* A dry-dock.
FORT, *s. m.* *Vaisseau dont le côté est* FORT. A stiff ship.
FORT *d'un vaisseau.* The extreme breadth of a ship.
FORT, *adj.* FORT *du côté lorsqu'il porte bien la voile.* When the ship does not incline too much under the wind.
FORTUNE, *s. f.* FORTUNE *de mer.* Any accident or damage that may happen to a vessel when at sea, when not occasioned by inattention or neglect.
FORTUNE *de vent.* A tempest or violent storm, in the dialect of Provence.
Mât de FORTUNE. A jury-mast.
Vergue de FORTUNE. Cross-jack-sail-yard, or the square-sail yard in sloops, schooners, &c.
Voile de FORTUNE. Cross-jack-sail; or the square-sail of a sloop, &c.; also, the square or lug-sail of a galley or tartan.
FOSSE, *s. f.* FOSSE *aux cables.* The cable-tier, in the orlop, &c.
FOSSE *aux lions.* That part of the fore cock-pit containing the boatswain's store-room, &c.
FOSSE *aux mâts.* A mast-pond, where masts are kept afloat in salt-water, in or near a dock-yard.
FOSSE *sur un banc.* A pit or well upon a bank.
FOSSE, *ou ponton pour le carénage.* *Voyez* PONTON.
FOUET, *s. m.* *Amarrage à* FOUET. The lashing of a tail-block.
Poulie à FOUET. A tail-block.
FOUET, is also a cat or whip, used on board French ships to punish the youngsters for misbehaviour, &c.
Etre condamné au FOUET. To be sentenced to be whipped.
FOUETTER, *v. a.* To flap; speaking of the sails, also to lash a tail-block, &c.
FOUETTER. To whip, lash, scourge, &c.
FOUGON, *s. m.* The coboose, or fire-place of small vessels, in the dialect of Provence.
FOUGUE, *s. f.* *Mât de perroquet de* FOUGUE. The mizen top-mast.
Voile de perroquet de FOUGUE. The mizen top-sail.
Vergue de FOUGUE. The mizen top-sail yard.
FOUR, *s. m.* An oven.
Faire du pain FOUR. To bake.
FOURCATS, *s. m. p.* The crotches, or floor timbers placed in the fore and after hold of a ship.
*Demi-*FOURCATS. The cross-chocks of the timbers in the fore and after part of the hold in French ships.
FOURCATS *de porques.* The crotches of the riders, or floor-timbers of the riders in the fore and after hold.
FOURCAT *d'ouverture*, ou FOURCAT *horizontal.* The lower transom, which is generally forked.
FOURCHES DE CARÈNES, *s. f. p.* Breaming-hooks.
FOURRER, *v. a.* FOURRER *un cable, ou une manœuvre.* To serve a cable, or rope.
FOURRURE, *s. f.* Service, whether of plat, mat, hide, parsling, or spun-yarn.
Vieille toile à voile pour FOURRURE. Old canvas for service.
FOURRURE. (In naval architecture.) Furr, or furring in a piece of timber.
FOURRURE *de gouttière.* Water-ways.
FRAÎCHEUR, *s. f.* A light air of wind.
FRAÎCHIR, *v. n.* To freshen. *Ex.*
Le vent FRAÎCHIT. The wind freshens.
FRAIS, *adj.* *Vent* FRAIS. A fresh wind.
Petit FRAIS. A light breeze.
Bon FRAIS. A fresh breeze.
Grand FRAIS. A stiff breeze, or gale.
Il vente bon FRAIS, *grand* FRAIS, &c. It blows a fresh or stiff breeze, &c.
FRANC, *adj.* *Vaisseau* FRANC *d'eau.* A ship free of water.
FRANC-BORD, *d'un bâtiment, s. m.* The planks of the bottom of a ship.
FRANC-FUNIN, *s. m.* A white hawser, or large untarred rope. *Voyez* FUNIN.
Port-de-mer FRANC. *Voyez* PORT.
FRANCHIR, *v. a.* To forge over; to sew. *Ex.*
FRANCHIR *la lame.* To sail easily over a high sea; to head the sea.
Vaisseau FRANCHI *dans un port.* A vessel sewed in a harbour during the neap-tides.
Le navire FRANCHISSOIT *de six pieds à la basse marée.* The ship was sewed by six feet at low-water.
FRANCS, *s. m. p.* A name given to the Christians in the Levant trade, from which the term *Langue Franche*, or *Lingua Franca*, is derived; being the language made use of by the different nations in the Levant trade.
FRAPPER, *v. a.* To lash, or seize. *Ex.*
FRAPPER *une manœuvre, ou un cordage.* To seize, or lash a rope in its place.
FRAPPER *une poulie sur une vergue.* To lash a block upon a yard, &c.
☞ FRAPPE la tournevire sur le cable! *Bring to the messenger!*
☞ FRAPPE doubles garcettes sur la tournevire! *Clap on thick and dry!*

L'heure FRAPPE. The bell strikes.
☞ FRAPPE six! *Strike the bell six!*

FRÉGATE, *s.f.* A frigate.
FRÉGATE *d'avis.* A sloop of war, packet-boat, or tender. *Voyez* RANG.
FRÉGATÉ, *adj.* Frigate-built.
FREGATON, *s. m.* A sort of Venetian ketch.
FRET, *s. m.* The freight or hire of a ship.
Prendre un FRET. To take a freight.

FRÉTEMENT, *s. m.* The act of freighting a merchant-ship.

FRÉTER, *v. a.* To freight, hire, or take up a vessel. *Ex.*
Bâtimens qui sont FRÉTÉS *par l'état.* Transports, or vessels hired to carry troops, &c.

FRÉTEUR, *s. m.* The merchant or person who freights a ship.
FRISE, *s.f.* FRISE *de l'éperon.* The trailboard, or ornament of the head.
FRISE, is also baize, or kersey, used for lining the gun-ports, and covering the scarphs of the keel, &c. to prevent the water from entering at sea.
DE FRONT, *adv.* A-breast.
Aller de FRONT. To sail a-breast.
Ligne de FRONT. *Voyez* LIGNE.
FRONTEAU, *s. m.* The breast-work, consisting of a moulding, ornamented with sculpture, and a sort of ballustrade reaching athwart the ship.
FRONTEAU *du gaillard d'arrière.* The breast-work of the quarter-deck.
FUIR, *v. n.* To run, &c.
FUIR *devant le tems.* To bear up and scud.
FUNER LES MÂTS, *v. a.* To rig the masts. *Voyez* GRÉER.
FUNIN, *s. m.* Cordage of a certain size, used for running-ropes, and sometimes for the standing-rigging.

FUSÉE, *s.f.* FUSÉES *d'artifice, ou* FUSÉES *volantes.* Sky-rockets, or signal-rockets.
FUSÉE *d'aviron.* A mouse raised on the middle of an oar.
FUSÉE *de vindas.* The barrel or body of the windlass, in which the handspikes are lodged to turn it about.
FUSÉE *de brûlot.* The assemblage of troughs wherein the trains of a fire-ship are laid.
FUSIL, *s. m.* FUSIL *de bord.* A ship's musket.

FÛT, *s. m.* FÛT *de la girouette.* The stock of a vane.
FUTAILLES, *s. f. p.* The water and provision-casks of a ship.
Attelier des FUTAILLES. The cooper's shed.

G.

GABARE, *s. f.* A flat-bottomed lighter or barge, used in the river Loire, &c. to lade and unlade ships.
GABARER, *v. a.* GABARER *un canot.* To scull a boat.

GABARIER, *s. m.* A lighter-man, who conducts the *Gabare*; a lumper.
GABARIER, *v. a.* GABARIER *une pièce de bois.* To mould a piece of timber.
GABARIT, *s. m.* A draught, model, mould, &c.
Tracer les GABARITS *d'un vaisseau.* To lay down the moulds of a vessel
*Maître-*GABARIT. The mould of the midship-frame.
Faux GABARITS. Moulds for fitting timbers or slab-timbers.
Salle des GABARITS. *Voyez* SALLE.
GABET, *s. m.* The vane of a quadrant, &c.
GABIE, *s.f.* The top, in the dialect of Provence. *Voyez* HUNE.
GABIER, *s. m.* A top-man.
GABIERS *de la grande hune.* Main-top-men.
GABIERS *de la hune de misaine.* Fore-top-men.
GABIERS *de la hune d'artimon.* Mizen-top-men.
GABORD, *s. m.* The garboard-streak.
GABURONS. *Voyez* JUMELLES.
GAFFE, *s.f.* A boat-hook.
Fer de GAFFE. The hook of a boat-hook staff.
GAFFER, *v. a.* To hook and draw any thing with a boat-hook.
GAGNER, *v. a. & n.* To gain, &c. *Ex.*
GAGNER *la brise.* To catch the breeze.
GAGNER *le vent à un vaisseau.* To get the wind or weather-gage of a ship.
GAGNER *au vent.* To get to windward.
GAGNER *les vents alisés.* To get into the trade-winds.
GAGNER *sur un vaisseau.* To fore-reach, or gain ground of some ship in company, &c.
GAGNER *un port, ou un mouillage.* To reach a harbour, or anchorage; also, to secure a harbour in bad weather.
GAGNER *de l'avant.* To gain ground.
GAILLARD, *s. m.* GAILLARD *d'avant.* The forecastle.
GAILLARD *d'arrière.* The quarter-deck.
GAINE, *s. f.* GAINE *de flamme, ou de pavillon.* The canvas edging of a pendant, or ensign, &c.

GALÉASSE, *s. f.* A galeass, or a large Venetian galley, now little in use.

GALÈRE, *s. f.* A galley, or row-galley.
GALÈRE *réale.* A royal galley; a name given to the largest and finest galley belonging to any monarchical state.
GALÈRE *patrone.* The first-rate galley of Malta, &c.
☞ Avirons en GALÈRE! *Rest upon your oars!* A term used in the Levant.
*Demi-*GALÈRE. A half-galley.
Les GALÈRES. The gallies, understood as a punishment.
Criminel condamné aux GALÈRES. A convict condemned to serve on board a galley.
GALERIE, *s.f.* GALERIE *de poupe.* The stern-gallery.
GALERIES *du faux-pont, ou du fond de cale.* The passages formed in the wings of a ship of war, for the discovery of leaks in time of action.

GALÉRIEN, *s. m.* A galley-slave. *Voyez* FORÇAT.
GALETTE, *s. f.* A round and flat sea-biscuit.

GALHAUBANS, *s. m. p.* Breast back-stays.
GALHAUBANS *du grand mât de hune.* The main-top-mast back-stays.
GALHAUBANS *du petit mât de hune.* The fore-top-mast back-stays.
GALHAUBANS *du perroquet de fougue.* The mizen-top-mast back-stays.
GALHAUBANS *du grand perroquet.* The main-top-gallant back-stays.
GALHAUBANS *du petit perroquet.* The fore-top-gallant back-stays.
GALHAUBANS *de la perruche.* The mizen-top-gallant back-stays.
Rides des GALHAUBANS. *Voyez* RIDES, &c.
GALHAUBANS *volans.* Preventer back-stays.
GALION, *s. m.* A galleon, or Spanish ship of war of the Indian fleet.
GALIOTE, *s. f.* A galliot, also a Dutch fishing vessel.
GALIOTE *à bombes.* A bomb-ketch.
GALIOTE *servant de yacht d'avis.* A packet or advice-boat.
GALIOTE *barbaresque.* A quarter-galley.
GALOCHE, *s. f.* A belaying cleat; also a shoe-block.
GALOCHES *de fer.* Hanging-clamps.
GAMBES DE HUNE, *s. f. p.* Futtock, or foot-hook shrouds.
GAMELLE, *s. f.* A wooden bowl or platter, in which the victuals of the sailors are served up.
Faire GAMELLE *ensemble.* To mess together.
GARANT, *s. m.* GARANT *d'un palan.* The fall of a tackle, or a tackle-fall.
GARANT *de caliorne.* The fall of a winding-tackle.
GARANT *de candelette.* The fish-tackle-fall.
GARANT *de capon.* The cat-rope, or cat-fall.
GARANS *de palanquins de sabords.* The lanyards of the gun-ports, or port-tackle falls.
En GARANT, *adv.* Gently.
GARCETTES, *s. f. p.* Gaskets, &c. *Ex.*
Maîtresse-GARCETTE. The bunt-gasket, or middle-gasket of a yard.
GARCETTES *de bout de vergue.* Yard-arm gasket.
GARCETTES *de bonnettes.* Keys or buttons of the bonnets.
GARCETTES *de fourrures de cables.* Plat, for serving the cables.
GARCETTES *de ris.* The reefed points.
GARCETTES *de tournevire.* The nippers of the cable.
GARDE, *s. f.* The anchor-watch.
GARDE-CORPS, *s. m.* Any man-rope or netting, serving as a ballustrade to prevent the men from falling overboard, &c.
GARDE-CORPS *de beaupré.* Man-ropes of the bowsprit.
Bâtiment GARDE-CÔTE, *s. m.* A man of war employed to protect the coast, and to prevent illicit trade.
GARDES DES CÔTES. A sort of militia employed to defend the coast in time of war.
GARDE *de la marine, ou* ASPIRANT, *s. m.* A midshipman, or naval cadet.
GARDE-FEU, *s. m.* A cartridge-box.
GARDE-MAGAZIN, *s. m.* A store-keeper.
GARDER, *v. a.* GARDER *un vaisseau.* To keep sight of, or protect a ship.
GARDIENS, *s. m. p.* The men employed to take care of ships of war laid up in ordinary.
GARDIEN *de la Sainte Barbe.* The lady of the gun-room, in English ships; but the gunner's-yeoman, in French ships.
GARDIEN *de bureau.* The messenger.
GARDIEN *de la fosse aux cables.* The boatswain's-mate, who has the charge of the cables, &c.
GARDIEN *de la fosse aux lions.* The boatswain's-yeoman
GARDIEN *d'un quai.* A wharfinger.

☞ GÂRE au cable! *Stand clear of the cable!*
☞ GÂRE dessous! *Stand from under!*
GARGOUSSE, *s. f.* The cartridge of a cannon or other fire-arm.
Fil à GARGOUSSE. Thread for sewing cartridges.
☞ LA GARGOUSSE dans le canon! *Load with cartridges!*

GARGOUSSIÈRE, *s. f.* A cartridge-box; or a pouch for cartridges.
GARNIR, *v. a.* To rig, &c. *Ex.*
GARNIR *un mât.* To rig a mast.
GARNIR *une vergue.* To rig a yard.
GARNIR *le cabestan.* To rig the capstan.
GARNIR *les avirons.* To ship the oars.
GARNIR *une voile d'étai.* To bend a stay-sail.
GARNIR *le plomb.* To arm the lead.
GARNITURE, *s. f.* The standing and running rigging of a ship.
GARNITURE *de la bouée.* The slings of the buoy.
GARNITURE *de pompe.* Pump-gear.
GARNITURE *& rechange.* The rigging in use, and the spare rigging.
Attelier de la GARNITURE. *Voyez* ATTELIER *de grécment.*
GATTE, *s. f.* The manger of a ship.
Cloison de la GATTE. *Voyez* CLOISON.
GAVITEAU, *s. m.* A buoy, in the dialect of Provence.

GÉNÉRAL, *s. m.* GÉNÉRAL *d'une armée navale.* The commander in chief of a fleet of ships of war.

GÉNOPE, *s. f.* A temporary racking, seizing, or lashing.
GENOU, *s. m.* The lower futtock of any timber.
GENOUX *de fond*, The lower futtocks of the timbers amidships.
GENOUX *de revers.* The lower futtocks of the timbers fore-and-aft.
GENOUX *de porques.* The lower futtocks of the riders.
GENOUX *d'aviron.* The loom of an oar.

GENS DE L'EQUIPAGE. *Voyez* ÉQUIPAGE.
GERÇURE, *s. f.* A flaw.
GERSEAU. *Voyez* ESTROPES.

GÉSIR, *v. n.* To lie (employed only in the present of the indicative). *Ex.*
La côte GÎT *nord & sud.* The coast lies north and south.
GIBELOT. *Voyez* COURBE *de capucine.*
GIROUETTE, *s. f.* The vane of a mast-head.

Fer de GIROUETTE. *Voyez* FER.
Fût de la GIROUETTE. *Voyez* FÛT.
Pomme de la GIROUETTE. *Voyez* POMME.
GIROUETTES *carrées.* Very broad vanes, to shew the direction of the wind.
GISSEMENT, *s. m.* GISSEMENT *d'une côte.* The direction of a coast.
GLACE, ou GLAÇON, *s. f.* Ice.
Banc de GLACE. *Voyez* BANC.
GLÈNE, *s. f.* A coil.
Filer sa GLÈNE. To drop the coil from one's hands.
GLÉNER, *v. a.* GLÉNER *un cordage, ou une manœuvre.* To coil up a rope.
GOËLETTE, *s. f.* A schooner.
GOLFE, *s. m.* A gulf.
GOND, *s. m.* GONDS *de sabords.* Port-hooks. *Voyez* ANNEAU *de sabords.*
GONDS *de gouvernail.* *Voyez Ferrures de* GOUVERNAIL.
GONDOLE, *s. f.* A Venetian gondola.
GONDOLE, is also a name given in the Mediterranean to a long-boat or pinnace.
GONDOLÉ. *adj.* Round-sheered. *Ex.*
Un bâtiment GONDOLÉ. A round-sheered vessel.
GONDOLIER, *s. m.* A gondolier, or rower of a gondola.
GORET, *s. m.* A hog, or large brush for scrubbing a vessel's bottom under water.
GORETER UN BÂTIMENT, *v. a.* To hog a vessel; to apply the hog to her bottom.
GORGÈRE, *s. f.* The upper and inner part of the knee of the head.
GOUDRON, *s. m.* Tar, pitch and tar.
GOUDRONNÉ, *adj.* Tarred.
Cordage GOUDRONNÉ. Tarred cordage.
GOUDRONNER, *v. a.* To tar, or pay with tar.
GOUESMON, ou GOËMON, *s. m.* Sea-weed.
GOUFFRE, *s. m.* A gulf, race, or whirlpool.
GOUGE, *s. f.* A gouge, or hollow chisel.
GOUJURE, *s. f.* GOUJURE *d'une poulie, ou d'un cap-de-mouton.* The score of a block, or dead-eye.
GOUJURE *de chouquet.* The notch or channel cut on each side of a cap, through which the jears are reeved.
GOULET, *s. m.* A strait and narrow entrance of a harbour, as that of Brest, &c.
GOUPILLE, *s. f.* The fore-lock of a bolt. *Voyez* CLAVETTE.
GOUPILLER, *v. a.* To fore-lock a bolt.
GOURNABLES, *s. f. p.* Tree-nails.
GOURNABLER, *v. n.* GOURNABLER *un vaisseau.* To drive the tree-nails into a ship's sides, &c.
GOUTTIÈRES, *s. f. p.* The water-ways of a ship's decks.
GOUTTIÈRES *renversées.* Straight and long pieces of timber placed on the sides, in some French ships of war, to support the beams in the ward-room, instead of hanging knees.
GOUVERNAIL, *s. m.* The rudder, and, when speaking in general terms, the helm.

Parties du GOUVERNAIL.	Parts of the rudder.
La mèche.	The main-piece of the rudder.
Le safran.	The after-piece.
Le doublage.	The back of the rudder.
La tête.	The head of the rudder.
La mortaise.	The mortise.
Ferrures du GOUVERNAIL.	The rudder hinges.
Les éguillots.	The pintles.
Les fémelots.	The goodgeons, or braces.
La barre.	The tiller.
La roue.	The wheel.
Drosse du GOUVERNAIL.	The tiller-rope.

Monter le GOUVERNAIL. *Voyez* MONTER.
☞ Un homme au GOUVERNAIL! *A hand to the helm!*
Vaisseau sensible à son GOUVERNAIL. A ship that answers the helm readily.
GOUVERNEMENT, *s. m.* *Le* GOUVERNEMENT *d'une vaisseau.* The steerage of a ship.
GOUVERNER, *v. a. & n.* To steer a ship or boat. *Ex.*
GOUVERNER *à la lame.* To steer by the sea, in stormy weather.
GOUVERNER *sur son ancre.* To steer a ship to her anchor.
GOUVERNER *à la voile.* To steer with the sails, when the rudder is unshipped or lost, and no substitute provided.
GOUVERNER *à contre.* Inverse steering.
Le vaisseau GOUVERNE. The ship steers well.
Le vaisseau ne GOUVERNE *plus.* There is no steerage-way.
☞ GOUVERNE où tu as le cap! *Steer as you go!*
Faire GOUVERNER. *Voyez* FAIRE, &c.
GRAIN, *s m.* A squall. *Ex.*
GRAIN *pesant.* A heavy gust or squall of wind.
GRAIN *blanc.* A white squall, or a squall accompanied with white clouds.
GRAIN *noir.* A black squall.
Un GRAIN *nous vient de l'avant.* A squall is coming a-head.
Un bâtiment chargé par un GRAIN. A ship laid upon her side by a squall.
GRAPIN, *s. m.* A grapling, or grapnel.
GRAPIN *à main.* A hand grapnel.
GRAPIN *d'abordage.* A fire grapnel.
GRAPINS *de bout de vergue.* Grapnel-irons of the yard-arms, or sheer-hooks.
GRAPINER, *v. a.* To grapple; to warp a vessel towards a flake of ice by graplings and ropes.
GRAS, *adj. le tems est* GRAS. Hazy weather.
GRATTE, *s. f.* A scraper, used to clean a ship's sides, decks, &c.
GRATTE *double.* A double-headed scraper.
Grande GRATTE. A large scraper.
GRATTE *triangulaire.* A triangular scraper.
GRATTER, *v. a.* GRATTER *un vaisseau.* To scrape a ship.
GRATTER *les ponts.* To scrape the decks.
GRECQUES, *s. f. p.* Sky-scrapers.
GRÉE, *adj.* Rigged.
Vaisseau mal GRÉE. A ship not rigged ship-shape.

GRÉEMENT, *s. m.* A general name for the rigging.

GRÉER, *v. a.* To rig, &c. *Ex.*
GRÉER *un bâtiment, &c.* To rig a ship, &c.
GRÉER *un mât. Voyez* GARNIR.

GRÉLIN, *s. m.* A cablet; small cable, or stream-cable; and in general all cables from 12 to 15 inches in circumference.
GRÉLIN *en queue de rat.* A pointed stream-cable.
GRENADE *s. f.* A grenade.
EN GRENIER, *adv.* In bulk.
Charger en GRENIER. *Voyez* CHARGER.

GRÈVE, *s. f.* A sandy shore.
GRIBANE, *s. f.* A small vessel navigated on the coast of Normandy, carrying a main-mast and fore-mast, without any tops.
GRIGNON. *Voyez* MÂCHEMOURE.
GROS, *adj.* GROS *tems.* Blowing or stormy weather.
GROS *vent.* A high wind.
GROSSE *mer. Voyez* MER.
GROSSE, ou GROSSE *aventure, s. f.* Bottomry.
Donner à la GROSSE. To lend money on bottomry.
GROSSIR, *v. a.* To rise; speaking of the sea.
GRUE, *s. f.* A crane with a wheel, used on wharfs and quays.
EN GRUME, *adv.* *Bois en* GRUME. Rough timber.
GUEULE DE RAIE, *s. f.* A cat's paw made in a rope.
GUI, *s. m.* The main-boom of a sloop, brig, &c.
GUI *de baume.* A spanker-boom.
Balancine de GUI. A topping-lift.
Mettre le GUI *au repos.* To crutch the boom.

GUIBRE. *Voyez* ÉPERON.
GUIDE, *s. m.* GUIDE *de bassin.* A land-mark.
GUIDON, *s. m.* A triangular flag, or broad pendant.
GUINDAGE, *s. m.* The act of hoisting with tackles in lading or unlading ships; also the money paid for such labour.
GUINDANT, *s. m.* The hoist of a flag, &c.
GUINDER, *v. a.* To sway up; speaking of the top-masts, &c.
Les mâts de hune sont GUINDÉS. The top-masts are an-end.
GUINDERESSE, *s. f.* The top-rope, used to sway up, or strike the top-mast.
GUINDERESSE *double.* A double top-rope.
GUINDERESSE *simple, ou à l'Angloise.* A single top-rope.
Rouets de GUINDERESSE. The sheaves in the heels of the top-masts, for the top-rope to work in.
GUIPON, *s. m.* A mop.
GUIRLANDES, *s. f. p.* (In naval architecture). Breast-hooks, or fore-hooks.
GUIRLANDE *du premier pont.* The lower-deck breast-hook.
GUIRLANDE *des écubiers.* The breast-hook nearest the hawse-holes.
GUIRLANDE *du faux pont.* The orlop-deck-hook.
GUIRLANDE, is also the name for the snaking of a seizing.

GUITERNE, *s. m.* A shore to support the sheers which are employed in masting or dismasting ships.
GUITRAN, *s. m.* A sort of bitumen, or pitch, used to pay a ship's bottom.

H.

Note.—*When this mark* (ʻ) *is prefixed to* H, *in the following leading words, it denotes that the* H *must be aspirated.*

HABITACLE, *s. f.* The binnacle, formerly written Bittacle.
Lampe D'HABITACLE. *Voyez* LAMPE.
ʻ HACHE, *s. f.* A hatchet.
Grande HACHE. An axe.
Petite HACHE, ou HACHE *à main.* A small hatchet.
HACHE *d'arme.* A pole-axe, or battle-axe.
ʻ HALAGE, *s. m.* The tracking a vessel along the banks of a river, &c.
Chemin du HALAGE. The road used for tracking.
ʻ HALE-BAS, *s. m.* The down-haul of a stay-sail.
ʻ HALER, *v. a.* To haul, or pull upon any rope, &c. *Ex.*
☞ ʻ HALE de force sur la bouline de revers de misaine! *Haul well taught the fore bowline!*
☞ HALE-bas le grand foc! *Haul down the jib!*
☞ HALEZ ensemble! *Haul together!*
☞ HALE la chaloupe à bord! *Haul the boat a-board!*
HALER *un bâtiment.* To track a vessel from one place to another.
Se HALER *dans le vent.* To haul the wind, or come nearer to its direction.
HAMAC, *s. m.* A hammock.
HAMAC *à l'Angloise, ou cadre.* A cot.
ʻ HANCHE, *s. f.* *La* HANCHE *d'un vaisseau.* The quarter of a ship.
Aborder un vaisseau par la HANCHE. To board a ship upon the quarter.
Prendre un vaisseau par la HANCHE. To play upon a ship's quarter.
ʻ HANGAR, *s. m.* A boat-house.
ʻ HARPON, *s. m.* A harpoon, or harping-iron, used in striking whales, &c.
ʻ HARPONNER, *v. a.* To harpoon.
HARPONNER *une baleine.* To harpoon a whale.
ʻ HARPONNEUR, *s. m.* A harpooner.
HAUBANER, *v. a.* To fasten the stay of a gin, triangle, or such sort of machine, to a stay or peg.
ʻ HAUBANS, *s. m. p.* The shrouds of the masts, &c. *Ex.*
Les grands HAUBANS. The main shrouds.
HAUBANS *de misaine.* Fore shrouds.
HAUBANS *d'artimon.* Mizen shrouds.
HAUBANS *du grand mât de hune.* Main top-mast shrouds.
HAUBANS *du petit mât de hune.* Fore-top-mast shrouds.
HAUBANS *du mât de perroquet de fougue.* Mizen-top-mast shrouds.

Haubans *du mât de grand perroquet.* Main-top-gallant shrouds.
Haubans *du mât de petit perroquet.* Fore-top-gallant shrouds.
Haubans *de la perruche.* Mizen-top-gallant shrouds.
Haubans *de beaupré.* Bowsprit shrouds; also the standing lifts of the sprit-sail yard.
Haubans *de fortune, ou faux* HAUBANS. Preventer shrouds.
Haubans *du minois* Bumkin-shrouds.
Haubans *de bâton de foc.* Jib guys.
Haubans *à colonnes.* Shrouds that set up with a runner and tackle, used in lateen vessels.
Mollir les HAUBANS. To ease the shrouds.
Rider les HAUBANS. *Voyez* RIDER, &c.
Fourrer un HAUBAN. To serve a shroud.
Haubans *de chaloupe.* The lashings or gripes of a boat.
' HAUSSER, *v. a.* To raise, &c. *Ex.*
Hausser *un bâtiment.* To raise a ship; said of a vessel or any other object which appears more elevated as we approach it. *Ex.*
Nous HAUSSONS *le bâtiment en chasse à vue d'œil.* We raise the chase very fast.
HAUSSIERE. *Voyez* AUSSIÈRE.
' HAUT, *adj.* High, lofty, aloft, &c. *Ex.*
Vaisseau de HAUT-*bord.* A ship of the line.
Capitaine de HAUT-*bord.* A post captain.
Haute *marée.* *Voyez* MARÉE.
Hautes-*voiles.* The top and top-gallant sails.
☞ En HAUT, *adv.* En HAUT tout le monde! *All hands hoay!*
☞ En HAUT en vigie! *A hand to look out at the mast-head!*
' HAUTE-SOMME, *s. f.* Contingent money expended for extraordinary charges in the sea trade.
' HAUT-PENDU, *s. m.* A small cloud charged with a heavy squall.
' HAUTEUR, *s. f.* Height, &c. *Ex.*
Hauteur *de l'entre-pont.* The height between decks.
Hauteur *des sabords.* The depth of the gun-ports.
Hauteur *des façons.* The height of the floor.
Hauteur *de l'étambot.* The height of the stern-post.
Hauteur *de l'étrave.* The height of the stem.
La HAUTEURE *du pôle.* The elevation or height of the pole.
Hauteur also signifies an observation of the sun's, or of a star's meridian altitude.
Prendre HAUTEUR. To make an observation of the sun's altitude.

À la HAUTEUR, *adv.* Off. *Ex.*
Etre à la HAUTEUR *d'une île,* &c. To be off an island, &c.
Nous sommes par notre point à la HAUTEUR *du cap Finisterre.* We are by our reckoning off cape Finisterre.
HAUTS, *s. m. p.* **Hauts** *d'un bâtiment.* The upper-works of a ship.
☞ **Hauts** & bas! *Take long strokes!* the order to the men at the pump.
' HAUTURIER, *s. m.* A sea pilot, or such as directs the course of a ship by celestial observations. *Voyez* PILOTE.
' HAVRE, *s. m.* A harbour or haven. *Voyez* PORT.
Havre *de toutes marées.* A port accessible at any time of the tide.
HEAUME, *s. m.* The tiller or bar of the helm in a small vessel.
HELER, *v. a.* To hail. *Ex.*
Heler *un navire.* To hail a vessel at a distance.

HÉLINGUE. *Voyez* ÉLINGUE.

HÉMISPHÈRE, *s. m.* The hemisphere.
HERMINETTE, *s. f.* An adze.
Herminette *courbe.* A hollow adze.
HERPES, *s. f. p.* **Herpes** *de l'éperon.* The rails of the head. *Voyez* EPERON.
Herpes Marines. The natural treasures of the sea.
' HERSE. *Voyez* ESTROPES.
' HERSEAU, *s. m.* A cringle in the bolt-rope of a sail.
' HERSES, *s. m. p.* Iron cringles; also straps.
Herses *de fer.* Iron cringles, or thimbles of a large sort.
Herses *de vergues.* Straps to which the halyards are hooked.

' HERSILIÈRES, *s. f. p.* Certain knees placed horizontally on the quarters or bows of a ship, close to the gunwale.
' HEU, *s. m.* A sort of hoy, much used in Holland.
A LA BONNE HEURE, *adv.* Aye, aye; a reply given by the crew in French ships to the orders given by the boatswain.
' HEUSE, *s. f.* **Heuse** *de pompe.* The spear of a pump, together with its box.
HILOIRES, *s. f. p.* (In naval architecture). The binding strakes of the deck.
Hiloires *renversées..* The fore-and-aft earlings under the beams.
' HISSER, *v. a.* To hoist or pull upon any thing by a tackle. *Ex.*
Hisser *en douceur.* To hoist handsomely, or gradually.
Hisser *le pavillon.* To hoist the colours.
☞ **Hisse**! **Hisse** ensemble! *Hoist away! Pull together!*
☞ **Hisse** le petit hunier! *Hoist up the fore top-sail.*
HIVERNAGE, *s. m.* The winter season, also a wintering place.
HIVERNER. *v. a.* To winter; to lie up in a port during the winter season.

' HOLA! int. **Holà** *du bâtiment! Hoa the ship a hoay!* The manner of hailing a vessel.
Holà-ho! A cry which answers to *yoe-hoe!*
HOMME DE MER, *s. m.* A hardy sailor, or an experienced seaman.
☞ Encore un HOMME à la barre! *Another hand to the helm!*

HÔPITAL, *s. m.* An hospital *Ex.*
Vaisseau servant D'HÔPITAL. An hospital ship for the relief of the sick and hurt seamen.

HORIZON, *s. m.* The horizon.
L'HORIZON *rationel.* The rational, or true horizon.
L'HORIZON *sensible.* The sensible, or visible horizon.
L'HORIZON *est embrumé.* The horizon is foggy.
L'HORIZON *s'éclaire.* The horizon is getting fine.
HORLOGE, *s. m.* HORLOGE *de sable.* A sand-glass, or half-hour glass used to regulate the watch.
HORLOGE *marine.* A time-keeper, or chronometer, for determining the longitude at sea.
' HOUAICHE. *Voyez* OUAICHE *et* SILLAGE.
' HOUARI, *s. m.* A wherry; a two-masted small vessel, rigged with triangular sails.
' HOULE, *s. f.* The swell of the sea.
Il y a de la HOULE. There is a great swell.
' HOULEUX, *adj.* (used only in the feminine HOULEUSE.) *Ex.*
Mer HOULEUSE. A turbulent or swelling sea.
HOURCE, *s. f.* The vangs of a mizen-gaff, or yard.
HOURDI. *Voyez* LISSES.
' HOURQUE, *s. f.* A howker; a sort of hoy used mostly in Holland.
HUBLOTS, *s. m. p.* Scuttles in the ports.
' HUNE, *s. f.* The top.
La grande HUNE The main-top.
La HUNE *de misaine.* The fore-top.
La HUNE *d'artimon.* The mizen-top.
Mât de HUNE. The top-mast.
' HUNIERS, *s. m. p.* The top-sails; meaning only the fore and main.
Petit HUNIER. The fore-top-sail.
Grand HUNIER. The main-top-sail.
Amener les HUNIERS. To lower the top-sails.
Border un HUNIER. To haul aft the sheets of a top-sail.
Hisser un HUNIER. To hoist a top-sail.
Carguer les HUNIERS. To haul the top-sails up in the brails.
Larguer un HUNIER. To loose a top-sail.
Serrer un HUNIER. To take in a top-sail.
' HUTTER, *v. a.* HUTTER *les vergues.* To lower the lower yards down a port-last, and peak them up so as to hold less wind.
HYDROGRAPHE, *s. m.* An hydrographer; a teacher of navigation in the sea-ports.
HYDROGRAPHIE, *s. f.* Hydrography, or the science of navigation, &c.

I.

ILE, *s. f.* An island.
Les ÎLES *de dessous le vent.* The Leeward islands of the West Indies.
Les ÎLES *du vent.* The Windward islands of the West Indies.
ÎLE *de glace.* *Voyez.* BANC & GLACE.
ILOIRES. *Voyez* HILOIRES.
ÎLOT, *s. m.* A small island; an isle.
IMPÉRIALE, *s. f.* Roof.
L'IMPÉRIALE, *d'un tendelet.* The roof of a boat's awning.
L'IMPÉRIALE *de la cuisine d'un bâtiment.* The roof of a ship's galley.
L'IMPÉRIALE *du four.* The roof of the oven.
INCLINAISON, *s. f.* INCLINAISON *de l'aiguille aimantée.* The dip of the magnetic needle.
INCOMMODÉ, *adj.* *Bâtiment* INCOMMODÉ. A distressed ship.
INCOMMODITÉ, *s. f.* Distress.
Signal D'INCOMMODITÉ, *ou de détresse.* A signal of distress.
INGÉNIEUR, *s. m.* INGENIEUR *de la marine.* A naval engineer, or a naval architect.
INGENIEUR *constructeur.* A ship-builder.
INSPECTEUR, *s. m.* INSPECTEUR *des constructions.* A surveyor of the navy.
INSTRUCTIONS, *s. f. p.* Orders; sailing orders.
INTENDANT, *s. m.* INTENDANT *général de la marine.* A commissioner-general of all the royal dock-yards and ports of the kingdom.
INTENDANT *de la marine.* An officer, who, by his duty and authority, resembles the commissioner resident of one of our dock-yards.
INTENDANT *des armées navales.* An officer appointed to regulate the expenses, police, justice, &c. of a naval armament.
INTERLOPE, *s. m.* A smuggling vessel; contraband trader.
INTERLOPE, *adj.* *Commerce* INTERLOPE. Smuggling trade.
INVALIDES, *s. m. p.* INVALIDES *de la marine.* Sick, or wounded and superannuated seamen, who are no longer fit for sea-service.
INVENTAIRE, *s. m.* INVENTAIRE *d'armement.* An inventory of a ship's stores.
ISTHME, *s. m.* An Isthmus.
ITAGUE, *s. f.* A tie, or pendant. *Ex.*
ITAGUE *de drisse de hunier.* A top-sail tie.
Fausse ITAGUE. A preventer tie.
ITAGUE *de palan de ris.* A reef-tackle pendant
ITAGUE *de sabords.* *Voyez* SABORDS.

J.

JACK, *s. m.* (to be pronounced YAC.) The union flag, or jack.
JALOUX, *adj.* *Canot* JALOUX, *&c.* A crank boat, &c. in the dialect of Provence.
JAMBE-DE-CHIEN, *s. f.* A sheep-shank. *Ex.*
Faire une JAMBE-*de-chien.* To sheep-shank.
JARDIN, *s. m.* A name sometimes given to the gallery or balcony of a ship.
JARLOT, *s.* A channel cut in the stem afore, in the stern-post abaft, &c.; and into the keel, to receive the ends of the planks enveloping the sides and the bottom of a ship.
JARRE, *s. f.* A water-jar.
JARRON, *s. m.* A small jar.
JAT, *s. m.* JÂT *d'ancre, ou* JOUAIL. An anchor-stock, or the two pieces of which it is composed.
JAUGEAGE, *s. m.* The tonnage, or burden of a ship, also the act of measuring her for tonnage.

JAUGER, *v. a.* To measure a vessel in order to know her burden.

JAUMIÈRE, *s. f.* The helm-port, or the hole through which the rudder-head passes.

JET, *s. m.* Any part of a cargo thrown overboard in the time of danger, for the vessel's safety.

JETÉE, *s. f.* A mole, peer, or jetty.

JETER, *v. a.* To throw, to heave, &c. *Ex.*

JETER *l'artillerie à la mer.* To throw the guns overboard.

JETER *des mâts à bas.* To spend the ship's masts.

JETER *l'ancre.* To let go the anchor; to drop anchor.

JETER *la sonde, ou le plomb.* To sound or heave the lead.

JETER *la bouée.* *Voyez* BOUÉE.

JETER *dehors l'eau d'un canot.* To bale out a boat.

Navire JETÉ *sur la côte.* A vessel cast away, or wrecked on a lee-shore, bank, &c.

JEU, *s. m.* Play.

JEU *de gouvernail.* The play of the helm or rudder.

Donner du JEU *au mât.* To give the mast some play.

JEU *de voiles.* A complete suit of sails.

JONCTION, *s. f.* Joining or coming together. *Ex.*

JONCTION *de deux flotes, ou de deux armées navales.* The junction of two fleets of ships of war, or merchantmen.

JONQUE, *s. f.* A Chinese junk.

JOTTEREAU, *s.m.* JOTTEREAUX *ou flasques des mâts.* The cheeks of the masts.

JOTTEREAUX *de l'éperon, ou courbes de* JOTTEREAUX. The cheeks of the head.

JOUER, *v. n.* To play, &c. *Ex.*

Mât qui JOUE *dans son étambrai.* A mast that fetches way, or that plays in the partners.

Navire qui JOUE *beaucoup.* A ship that works much.

Faire JOUER *la pompe.* To pump ship.

JOUES, *s. f. p.* Cheeks. *Ex.*

JOUES *de vindas.* The cheeks of the windlass.

JOUES *d'un vaisseau.* *Voyez* ÉPAULES, &c.

JOUETS, *s. m. p.* Plates of iron, used to prevent the bolt-heads from cutting the timber into which they are driven.

JOUETS *de pompe.* The iron clamps nailed on the cheeks or ears of the pump.

JOUETS *de sep de drisse.* Plates of iron nailed on the knight-heads of the jears.

JOUFFLU, *adj.* *Avant* JOUFFLU. A bluff-bow.

JOURNAL, *s. m.* JOURNAL *de navigation, ou* JOURNAL *nautique.* A journal, or log-book.

JUMELLER, *v. a.* JUMELLER *un mât, ou une vergue.* To fish a mast, or a yard.

JUMELLES, *s. f. p.* The fishes of the masts and yards.

JUMELLES *de rechange.* Spare fishes.

JUMELLES, mean also the side-pieces of a made-mast.

JUMELLE *de bau.* The clamp of a beam.

JUSANT, *s. m.* The ebb-tide.

Le JUSANT *est à sa fin.* The ebb is almost spent.

K.

KAKATOIS,* *s. m.* A royal.

Le grand KAKATOIS. The main-top-gallant royal-sail.

Le petit KAKATOIS. The fore-top-gallant royal-sail.

Le KAKATOIS *de perruche.* The mizen-top-gallant royal-sail.

* This word being spelt in several different ways by the French writers, as CATAKOI, CATACOUA, &c. the author has thought proper to write it as it is always pronounced by the French sailors.

KETCH, *s. m.* A ketch.

KIOSQUE, *s. m.* Kiosk; a sort of turret elevated on the poop of a Turkish caravelle.

KOFF, *s. m.* A sort of Dutch vessel of burden.

L.

LABOURER, *v. a. & n.* To drag, speaking of the anchor.

L'ancre LABOURE. The anchor comes home.

LABOURER *le fond avec la quille.* To touch the bottom with the keel, when under sail, but so as not to hang.

LAC, *s. m.* A great lake of water, such as the Caspian Sea, &c.

LACER, *v. a.* LACER *une voile.* To lace a sail; to reef a course with a reef-line.

LÂCHE, *adj.* *Le bâtiment est* LÂCHE. The ship carries a slack helm.

LÂCHER, *v. a.* LÂCHER *une bordée à un bâtiment.* To pour a broadside into a ship.

LACUNES, *s. f. p.* (In naval tactics.) The too great openings or spaces between ships in a line-of-battle.

LAISSE, *s. f.* LAISSE *de basse mer.* The low-water-mark.

La LAISSE *de basse mer d'eau vive.* The mark of low-water in spring-tides.

LAISSE *de basse mer de morte eau.* The mark of low water in neap-tides.

LAISSER, SE LAISSER, *v. a. n. & r.* To leave, to loosen, &c. *Ex.*

LAISSER *tomber l'ancre.* To drop anchor.

L'ancre a LAISSÉ. The anchor is a-trip or a-weigh.

☞ LAISSE courir! *Keep her full!*

☞ LAISSE abattre! *Voyez* ABATTRE.

☞ LAISSE venir au vent! *Bring her to the wind!*

☞ LAISSE tomber la misaine! *Down foresail!*

LAISSER *une bouée sur un cable.* To leave a ship's buoy upon a cable.

LAISSER *tomber une basse voile.* To drop a course.

Se LAISSER *coiffer.* To be taken a-back.

LAISSES DE LA MER, *s. f. p.* Sea-wreck, or whatever the sea casts upon the beach.

LAISSES OU RELAIS. A sort of bank thrown up by the waves of the sea upon any coast.

LAMANAGE, *s. m.* Coasting pilotage, river or harbour pilotage; or the act of piloting a vessel in or out of a harbour or river.
LAMANEUR, *s. m.* *Pilote* LAMANEUR, ou *locman.* a harbour, or river pilot; loadsman, or lockman.
LAMES, *s. f. p.* The waves or billows of the sea.
La LAME *vient de l'avant.* There is a head sea.
La LAME *vient de l'arrière.* The sea comes a-stern.
L'entredeux des LAMES. The trough of the sea.
Vaisseau qui va debout à la LAME. A ship that heads the sea.
La LAME *prend le vaisseau par le travers.* The sea strikes the ship upon her broadside.
La LAME *nous prend par la hanche.* The sea strikes on the quarter.
Longue LAME. A long sea.
LAME *courte.* A short sea.
LAME *de fond.* A ground swell.
Les grosses LAMES. The high seas.
Donner à la LAME. To touch the log-line when a following sea sends the log home.
Prendre la LAME *de bout.* To bow the sea.
LAMPE, *s. f.* LAMPE *d'habitacle.* The binnacle lamp.
LANCE, *s. f.* A fire arrow.
LANCES. Wooden guns.
Fausses LANCES. False guns.
LANCER, *v. a. & n.* To launch; also to yaw.
LANCER *un vaisseau à l'eau.* To launch a ship from the stocks into the water.
☞ LANCE sur babord! *Yaw her to port!*
☞ LANCE sur tribord! *Yaw her to starboard!*
LANDES DE HUNE, *s. f. p.* Foot-hook, or futtock-plates.
LANGAR, *s. m.* The additional square-sail used in a brig.
LANGUETTES, *s. f. p.* A thin sort of wooden wedges, thrust under a ship's bottom in the act of launching; and which are used for other purposes.
LANS D'UN VAISSEAU, *s. m. p.* Yaws.
LANTERNE, *s. f.* LANTERNE *d'un mât de hune.* The hounds of a top-mast.
LANTERNE *à mitraille.* A mould for case-shot.
LARDER, *v. a.* LARDER *une bâderne, &c.* To thrum a mat.
LARDERASSE, *s. f.* A big and coarse rope, made of the toppings of hemp for common use.
LARGE, *s. m.* *Le* LARGE. The offing, or the main sea.
Prendre le LARGE To stand out to sea; to bear off; also to sheer off; to gain the offing.
Tener le LARGE. To keep the offing.
Vent du LARGE. A sea breeze.
Au LARGE, *adv.* In the offing.
Aller porter, ou courier au LARGE. To stand off to sea.
Passer au LARGE *d'un vaisseau.* To pass at some distance from a ship.
Pousser au LARGE. (speaking of a boat) To push off from a ship's side or wharf.
Porter un ancre au LARGE. To carry out an anchor.
Se tenir au LARGE *d'un banc, &c.* To keep without a bank, &c.
LARGEUR, *s. f.* LARGEUR *d'un vaisseau.* The extreme breadth of a ship.
LARGUE, *adj.* *Vent* LARGUE. A free or leading wind.
Vent grand LARGUE, *ou grand* LARGUE. A wind on the quarter.
Écoutes LARGUES. Flowing sheets.
LARGUER, *v. a. & n.* To ease, to let go, to loose, &c. *Ex.*
☞ LARGUE les galhaubans de dessous le vent! *Let go the jib down-haul!*
LARGUER *une manœuvre ou une amarre.* To let go, or cast off a belayed rope.
LARGUER *les écoutes.* To let go the sheets.
LARGUER *les bras du grand hunier.* To let go the main-top-sail braces.
LARGUER *la remorque.* To let go the tow-rope.
LARGUER *un ris.* To let go a reef.
LARGUER *une voile.* To loose a sail.
Un bordage qui a LARGUÉ. A butt that has sprung.
Navire qui LARGUE *de par-tout.* A vessel that strains violently in a high sea, so as to open in many places.
LASTE, *s. m.* A last; a weight that varies with several nations, but generally reckoned of 4000 lbs. weight.
LATIN, *adj.* *Bâtiment* LATIN. A lateen vessel.
Voile LATINE. A lateen sail.
LATIONE, *s. f.* A sort of row-galley, navigated on the coast of China.
LATITUDE, *s. f.* Latitude.
LATITUDE *estimée.* Latitude by account.
LATITUDE *corrigée*, Latitude corrected by observation.
LATITUDE *observée.* Latitude by observation.
LATITUDE *arrivée.* Latitude in.
LATITUDES *croissantes.* Meridional parts.
Etre en LATITUDE *Nord ou Sud.* To be in north or south latitude.
Se tenir en LATITUDE *d'une île, &c.* To keep in latitude of an island, &c.
LATTES, *s. f. p.* Broad, but thin beams, placed alternately between the larger deck-beams in French ships.
LATTES *de galère.* Broad and thin beams, to support the decks of a galley.
LATTES *de la dunette.* The beams of the poop.
LATTES *de caillebotis.* Battens of the hatches or gratings.
LAZARET, *s. m.* A Lazaretto, or building to receive persons and goods performing quarantine.
LÉBÈCHE. *s. m.* The south-west wind, in the dialect of Provence.
LÈGE, *adj.* Light; without a cargo of any kind.
Bâtiment LÈGE. A vessel in ballast, or a vessel which is not loaded enough.
LÉGER, *adj.* *Canot* LÉGER. A light boat.
Voiles LÉGÈRES. Small sails.
LEST, *s. m.* Ballast. *Ex.*
LEST *en fer.* Iron ballast.
LEST *en pierre.* Shingle ballast.
LEST *en galet.* Ballast consisting of small pebbles.
LEST *en gravier.* Gravel ballast.
LEST *volant.* Shifting ballast.

Aller en LEST. To go in ballast.
Bon LEST. Good ballast, or such as lies firmly in the hold, without incommoding the pumps.
Mauvais LEST. Bad ballast, as sand, salt, &c. which is apt to melt or penetrate through the ceiling, and choke the timbers and pumps.
Gros LEST. Heavy ballast, composed of large stones or pigs of iron.
Etre sur son LEST. To be in ballast.
Panier à LEST. A ballast basket.
LESTAGE, *s. m.* Lastage, or the furnishing a ship with ballast.
LESTER, *v. a.* LESTER *un bâtiment.* To ballast a ship.
Navire qui n'est pas assez LESTÉ. A vessel which has not ballast enough a-board.
LESTEUR, *adj.* *Bateau* LESTEUR. A ballast-lighter.
LETTRES DE MARQUE, ou LETTRES *de représailles*, *s. f. p.* Letters of mart, or marque; letters of reprisals.
LEVANT, *s. m.* The Levant, the East; also the east wind in the Mediterranean.
Greco LEVANT. East-north-east, in the language of the Mediterranean.
LEVANTINS, *s. m. p.* A name given to the inhabitants of the Levant, Turks, Greeks, &c.
LEVÉE, *s. f.* LEVÉE *de matelots.* A levy of seamen.
LEVÉE. The swell of the sea. *Ex.*
Il y a de la LEVÉE. There is a swell.
LEVER, *v. a.* To raise, to weigh, &c. *Ex.*
LEVER *l'ancre avec la chaloupe.* To weigh the anchor by the buoy-rope in the long-boat.
LEVER *les lofs.* To raise tacks and sheets.
LEVER *la fourrure du cable.* To take off the plat, or other service, from the cable.
LEVER *le lof de grande voile.* To haul up the weather clew-garnet of the main-sail.
LEVER *les callebotis.* To take off the gratings.
LEVER *la chasse.* To leave off chasing.
☞ LEVE rame! *Voyez* RAME.
LEVIER, *s. m.* A lever formed of a handspike or crow.
LEVIER *à croc.* A clawed handspike.
LIAISONS, *s. f. p.* Strengthening-pieces, such as knees, breast-hooks, &c.
Vaisseau manquant de LIAISONS. A vessel badly put together.
LIEUE, *s. f.* A league or measure of three miles.
LIEUTENANT, *s. m.* A lieutenant.
LIEUTENANT-*général des armées navales.* A rear-admiral in the French service.
LIEUTENANT *de vaisseau.* A lieutenant of a man of war; in the French service, officers obtain that rank after that of ENSEIGNE *de vaisseau.*
LIEUTENANT *en pied.* The first lieutenant in a ship of war.
LIGNE, *s. f.* A line. *Ex.*
LIGNES *d'eau* (in naval architecture). Water-lines.
LIGNE *d'eau en charge, ou* LIGNE *de flottaison.* Load-water-line
LIGNES. Small ropes, which are used on several occasions at sea.
LIGNE *d'amarrage.* A line for seizing or lashing.
LIGNE *d'axiomètre.* The spurling-line.
LIGNE *blanche.* White, or untarred line.
LIGNE *goudronnée.* Tarred line.
LIGNE *de six fils.* A six-thread line.
LIGNE *de pêche.* Fishing-line.
LIGNE *de loc.* The log-line.
LIGNE *de sonde.* Lead-line, or deep sea-line.
Petite LIGNE *de sonde.* A hand-line.
LIGNE *équinoxiale.* The equator.
LIGNE (In naval tactics). The line of battle.
LIGNE *de front.* The line-of-battle a-breast.
LIGNE *du plus près.* The line a-head close-hauled.
Couper la LIGNE. *Voyez* COUPER.
Se tenir hors de la LIGNE. To keep out of the line of battle.
Vaisseau de LIGNE. A ship of the line, or a line-of-battle ship.
LIGNEROLLE, *s. f.* (In rope-making). A small line.
LINGUETS, *s. m. p.* The pawls of the capstan.
LION, *s. m.* Lion; an ornament for the head of a ship.
LION, is also a prop made use of instead of a knee for the stanchions of the hold.
LISSE, *s. f.* (In naval architecture). LISSES *des couples.* Ribbands.
LISSE *du fort.* The extreme breadth-line.
LISSE *des façons.* Rising-line, or floor ribband.
LISSE *du fond.* A ribband between the floor ribband and the keel.
LISSES *des œuvres mortes.* Ribbands of the upper works, or top-timber ribbands.
LISSES *d'accastillage.* Rails.
LISSES *de plat-bord.* The drift-rails.
LISSE *d'hourdi.* *Voyez* ARCASSE.
LISSE *de vibord.* Waist-rail.
LISSES *de porte-haubans.* Sheer-rails.
LISSES *de herpes.* *Voyez* HERPES & ÉPERON.
LISSES *de batayoles.* Rough-tree rails.
LISSER, *v. a.* LISSER *un vaisseau en construction.* To place a ship's ribbands in their respective places.
LISTEAU, *s. m.* LISTEAU *de hune.* Top-brim, or top-rim.
LIT, *s. m.* LIT *d'une rivière.* The bed of a river.
LIT *du vent.* The direction of the wind, or wind's eye.
LIT *de la marée.* A tide-way.
LIURES, *s. f. p.* Seizings, or lashings.
Les LIURES *de beaupré.* The gammoning of the bowsprit.
LIVARDE, *s. f.* The sprit of a shoulder-of-mutton sail. *Voyez* BALESTON.
Voile à LIVARDE. A sprit-sail.
LIVARDER, *v. a.* LIVARDER *une voile.* To sprit a sail.
LOC, *s. m.* Log.
Tour du LOC. Reel of the log.
Table de LOC. Log-board.
Ligne de LOC. *Voyez* LIGNE.
Livre de LOC. A log-book.
Jeter le LOC. To heave the log.
LOCMAN. *Voyez.* LAMANEUR.

LOF, *s. m. & adv.* *Le* LOF. Luff, or the weather side.
Le LOF *du bâtiment.* The weather side of the ship.
LOF *de la grande voile, ou le grand* LOF. The weather tack or clew of the main-sail.
Aller au LOF. To go to windward.
☞ Lof, au LOF! *Luff boy, luff!* The order to steer the ship close to the wind.
☞ Lof tout! *Luff round! Luff all! Hard a-lee!*
☞ Lof un peu! *Luff a little!*
☞ Lof à la risée! *Ease the ship! Luff while it blows!*
☞ Largue le LOF, ou lève le LOF! *Up the tacks and sheets!*
☞ Largue le grand LOF! *Haul up the weather-clew of the main-sail!*
LOF *pour* LOF, *ou viver* LOF *pour* LOF. To veer the ship.
Couple de LOF. *Voyez* COUPLE, &c.
LOFER, *v. n.* To luff; to put the helm a-lee.
LONGER, *v. a.* LONGER *une côte, &c.* To coast along.
LONGIS, *s. m. p.* Trestle-trees. *Voyez* BARRE.
LONGITUDE *s. f.* Longitude.
LONGITUDE *observée.* Longitude by lunar observations.
LONGITUDE *par l'horloge marine.* Longitude by time-keeper.
LONGITUDE *estimée.* Longitude by account.
LONGITUDE *arrivée.* Longitude in.
Etre en LONGITUDE *est, ou ouest.* To be in east, or west longitude.
LONGUEUR, *s. f.* Length.
La LONGUEUR *d'un vaisseau.* The length of a ship from the stem to the stern-post, without board.
La LONGUEUR *de tête un tête.* The length of the gun-deck.
LONGUEUR *de la quille portant sur terre.* *Voyez* PORTANT.
LONGUEUR *de cable.* A cable's length, measuring 120 fathoms.
LOVER UN CABLE, *v. a.* To coil a cable.
LOUGRE, *s. m.* A lugger.
LOUP DE MER, *s. m.* A jack-tar; an epithet used to express a complete seaman.
LOUVOYER, *v. n.* To beat, or ply to windward.
LOXODROMIE, *s. f.* Loxodromic; an oblique course in navigation, or a course which crosses the meridians at equal and oblique angles.
TABLES LOXODROMIQUES. Loxodromic tables. *Voyez* TABLES, &c.
LUMIÈRE, *s. f.* *La* LUMIÈRE *d'une arme à feu.* The vent, or touch-hole of fire-arms.
La LUMIÈRE *d'une pompe.* The hole in the side of a pump, through which the water is discharged upon the deck, or into the pump-dale.
LUNE, *s. f.* The moon.
Clair de LUNE. Moon-light.
LUNETTE, ou LONGUE-VUE. A spy-glass.
LUNETTE *de nuit.* A night-glass,
LUZIN, *s. m.* Housing, or house-line.
LUZIN *blanc.* Untarred house-line.
LUZIN *goudronné.* Black house-line.

M.

MACHEMOURE, *s. f.* Bread-dust, or the remains of broken biscuit.
MACHINE, *s. f.* MACHINE *à curer.* A pontoon, with large forked ladles and machinery for working them, by means of which the mud or gravel is raised from the bottom of a harbour or the bed of a river, in order to deepen or cleanse it.
MACHINE *à purifier l'eau.* A machine for sweetening water.
MACHINE *à mâter.* *Voyez* MÂTURE.
MACRÉE, *s. f.* Tide-gate, or race.
MAGASIN, *s. m.* A storehouse, or warehouse.
MAGASIN *général.* A storehouse, containing naval stores in a dock-yard.
MAGASINS *particuliers*, are such as contain the rigging, cordage, &c. for the use of ships, &c.
*Garde-*MAGASIN. *Voyez* GARDE.
MAHON. *Voyez* GALÉASSE.
MAI ou MAIE, *s. f.* A sort of trough bored full of holes, wherein to drain cordage when it is newly tarred.
MAIGRE, *adj.* In ship-building, is said of a piece of timber which is too narrow in some places, thus;
Le coté MAIGRE. The thin side.
Le coté gras. The thick side.
MAILLES, *s. f. p.* (In naval architecture). The intervals left between a ship's timbers, commonly called room and space.
MAILLET, *s. m.* A maul, or mallet.
MAILLET *de calfat.* Calking mallet.
MAILLET *à épisser.* Driving mallet.
MAILLET *à fourrer.* Serving mallet.
MAILLET *de bois.* A wooden mallet.
MAILLETAGE, *s. f.* The sheathing or filling of a ship's bottom, with scupper-nails.
MAILLETER, *v. a.* To sheath or fill a ship's bottom with those nails.
MAIN-AVANT, ou MAIN SUR-MAIN, *adv.* Hand-over-hand.
☞ Hisse MAIN-SUR-MAIN! *Hoist away hand over hand!*
MAÎTRE, *s. m.* MAÎTRE *d'un bâtiment.* (Patron in the dialect of the Levant). A master or commander of a merchantman. The French having no masters in the navy, this term cannot be understood in that sense.
MAÎTRE *de port.* A harbour-master, or officer who takes care of a port, its booms and places of anchorage: also, an officer resembling our tide-surveyors of the customs in an out-port.
MAÎTRE *de quai.* A principal wharf-master, or master appointed to regulate the affairs of wharfs and quays, and the shipping moored along-side thereof.
MAÎTRE *d'équipage.* The boatswain of a man of war.

Second MAÎTRE, } Petty officers under the boat-
Contre-MAÎTRE, } swain.

MAÎTRE-*pilote.* A head pilot in the French navy, whose duty nearly resembles that of the master on board the English ships of war.

MAÎTRE-*canonnier.* *Voyez* CANONNIER.

MAÎTRE-*charpentier.* *Voyez* CHARPENTIER.

MAÎTRE-*calfat.* *Voyez* CALFAT.

MAÎTRE-*cordier.* *Voyez* CORDIER.

MAÎTRE-*voilier.* *Voyez* VOILIER.

MAÎTRE-*mâteur.* *Voyez* MÂTEUR.

MAÎTRE-*valet.* *Voyez* VALET.

MAÎTRE, *adj.* Midship. (speaking of the beams, &c.) *Ex.*

MAÎTRE *bau.* *Voyez* BAU.

MAÎTRE *couple.* *Voyez* COUPLE.

MAÎTRE *gabarit.* *Voyez* GABARIT.

MAÎTRESSE *varangue.* The floor-timber a midships.

MAJEUR, *adj.* *Mâts* MAJEURS. Lower masts, or standing masts.

MAJOR, *s. m.* *Major d'une escadre.* The captain of a fleet or squadron. *Voyez* OFFICIERS *Majors.*

Chirurgien MAJOR. The surgeon of a man of war.

MAJOR *de vaisseau.* An officer in the French navy, whose rank answers to that of a commander in the British service.

MAL DE MER, *s. m.* Sea-sickness.

MÂLE, *adj.* (speaking of the sea). *La mer est* MÂLE. The sea is rough.

MALINE, *s. f.* A spring-tide.

MANCHE, *s. f.* A great channel.

By *La* MANCHE simply, is understood the *British Channel*; for all others it is necessary to give its particular designation; as,

La MANCHE *de Bristol.* The Bristol Channel.

MANCHE *à eau.* A hose to convey water from the deck into the casks that are stowed in the hold.

MANCHE *de cuir.* A leather hose.

MANCHE *de toile.* A canvas hose.

MANCHE *de pompe.* The pump-hose.

MANCHE *de rame.* *Voyez* AVIRON.

MANCHE *à vent.* A wind-sail.

MANCHETTE, *s. f.* A sort of span, or bridle. *Ex.*

MANCHETTE *des bras de la grande vergue.* The span, fixed to the mizen rigging, at the extremities of which two bull's-eyes are spliced in, to keep up the height of the main-brace.

MANÉAGE, *s. m.* Hand-work; a name given to all sorts of employment for which the crew of merchantmen cannot demand additional pay.

MANÉGE DU NAVIRE, *s. m.* The general trim of a ship; also the disposition of the sails, and the efforts of the wind and sea.

MANGER, *v. a.* MANGER *du sable.* To flog or cheat the glass, by turning it before it has ran out.

MANGE *par la mer.* To lie in the trough of the sea, when it breaks on board.

La mer nous MANGE. The sea will swallow us.

Vaisseau MANG *par le terre.* A ship close under the land, and which disappears under it from other ships in the offing.

MANGER *le vent à un bâtiment.* To becalm a ship.

MANIABLE, *adj.* *Navire* MANIABLE. A ship that steers easily; a handy vessel.

Tems MANIABLE. Moderate weather and wind favourable for sea.

MANIER, *v. a.* To handle. *Ex.*

MANIER *bien un bâtiment.* To handle a ship well.

Vaisseau MANIÉ *par un courant.* A ship mastered, or overpowered by a stream.

MANIVELLE, *s. f.* A crank, winch, &c. *Ex.*

MANIVELLE *de corderie.* A laying-hook.

MANIVELLE *de meule.* An iron winch for a grindstone.

MANIVELLE *de rouet.* A crank.

MANIVELLE *de pompe.* A pump-winch.

MANNE, *s. f.* MANNE *d'osier.* A hand-basket.

MANŒUVRE, *s. f.* An evolution or manœuvre, speaking of a ship, or of a fleet; also the working of a ship by means of the helm, and the disposition of the sails to the wind.

MANŒUVRE *hardie.* A bold and difficult management of a ship in the act of working her.

MANŒUVRE *delicate.* A dexterous and fine manner of working a ship.

MANŒUVRE *prompte.* An expeditious management of a ship in the act of working her.

Commander la MANŒUVRE. To work a ship or to direct her movements.

Officiers mariniers de MANŒUVRE. The warrant and inferior officers in a ship of war, whose employment relates only to working the ship, and to her rigging, cables and anchors; the first of these is the maître d'équipage, and under him the contre-maître, quartier-maître, bossemans.

Grosse MANŒUVRE. The laborious work in a ship.

Mettre du monde sur une MANŒUVRE. To man a rope.

Mettre une MANŒUVRE *en queue de rat.* To point a rope, as the main and fore tacks.

Passer une MANŒUVRE *dans ses poulies*, &c. To reeve a rope, &c.

MANŒUVRER, *v. a. & n.* To manœuvre, to work a ship, or direct the movements of a fleet.

Bâtiment qui MANŒUVRE. A ship that works well.

Donner du mer dans les MANŒUVRES. To slack the running ropes.

MANŒUVRES *d'un vaisseau.* The rigging and cordage of a ship; and particularly all the standing and running ropes.

MANŒUVRES *courantes.* The running-rigging.

MANŒUVRES *dormantes.* The standing-rigging.

MANŒUVRES *basses.* The lower-rigging.

MANŒUVRES *hautes.* The upper rigging.

MANŒUVRES *de combat.* Preventer-rigging, which is only made use of in time of action.

MANŒUVRES *en bande.* Slack ropes which are unemployed.

MANŒUVRES *passées à contre.* Ropes leading forward; as those of the mizen-mast.

MANŒUVRES *passées à tours.* Ropes leading aft.

MANŒUVRIER, *s. m.* *Bon* MANŒUVRIER. An able and expert sea-officer, well skilled in working a ship by every method of sailing.

MANQUER, *v. a. & n.* To fail, &c. *Ex.*

MANQUER, *à virer.* To miss stays.

Le vent MANQUE. The wind fails.

MANQUER *un convoi.* To miss an enemy's convoy with which it was expected to fall in.

Le vent MANQUÉ. The wind fails.
MANQUER *de monde.* To be short of hands.
Vaisseau MANQUÉ. An ill-built ship; a vessel defective in her construction.
MANTELET, *s. m.* MANTELETS *de sabords.* Port-lids.
Faux-MANTELETS. False port-lids.
MAQUILLEUR, *s. m.* A decked boat, used in the mackerel fishery.
MARABOUT, *s. m.* A large main-sail, used in a galley when the weather is very tempestuous.
MARAIS SALANS, *s. m. p.* Salt-pits on the sea-coast; or reservoirs to contain sea-water, for making salt.
MARCHAND, *adj. Navire* MARCHAND; *bâtiment* MARCHAND. A merchantman.
Officier MARCHAND. The master or mate of a merchantman.
MARCHE, *s. f.* The progressive motion or the head-way of a ship through the water. *Ex.*
Vaisseau qui a la MARCHE *sur un autre.* A ship that sails swifter than another.
Navire construit, ou taillé pour la MARCHE. A vessel built for sailing fast.
MARCHE-PIED, *s. m.* The horse or foot-rope of any yard.
Etriers de MARCHE-PIED. *Voyez* ÉTRIERS.
Ride de MARCHE-PIED. The lanyard of a horse.
MARCHER, *v. n.* To sail; to have fresh way through the water.
MARCHER *dans les eaux d'un autre vaisseau.* To sail in the wake or track of another ship.
MARCHER *en ligne.* To sail in a line.
MARCHER *sur trois colonnes.* To sail in three lines.
Bâtiment qui ne MARCHE *pas.* A bad sailer.
MARCHEUR, *s. m. Bon* MARCHEUR. A good sailer (speaking of a ship).
Grand MARCHEUR. An extraordinary fine sailer.

MARÉE, *s. f.* Tide. *Voyez* FLUX *&reflux.*
Grandes MARÉES. Spring-tides.
Mortes MARÉES. Neap-tides.
MARÉE *haute.* High-water.
MARÉE *basse. Voyez* BASSE-EAU.
MARÉE *montante. Voyez* MONTANT.
MARÉE *descendante, ou jusant. Voyez* JUSANT.
Prendre la MARÉE *en travers.* To take the tide on the beam.
MARÉE *qui soutient.* A tide which counteracts the wind, with regard to a ship's course, enabling her to turn to windward better.
MARÉE *qui porte au vent.* A tide which runs to windward.
MARÉE *& contre-marée.* Tide and half-tide.
MARÉE, is also a name given to all sorts of sea-fish.
MARGOUILLET, *s. m.* A bull's-eye; also a seizing-truck..
MARGUERITE, *s. f.* A messenger.
Faire MARGUERITE *sur un cable.* To clap a messenger on the cable, when the anchor cannot be purchased by the voyol.
MARIAGE, *s. m.* MARIAGE *de tournevire.* The lashing together the eyes of the messenger.
MARIE-SALOPE, *s. f.* A mud-lighter; a sort of flat-bottomed boat, which is used to carry away the mud taken from the bottom of a harbour or river.
MARIN, *s. m.* A seaman, or sea-faring man.
Grand MARIN. An experienced sailor.
MARIN *d'eau douce*(by derision). A fresh-water Jack.
MARIN, *adj.* Any thing belonging to the sea.
Nœud MARIN. A reef-knot.
Sel MARIN. Sea-salt.
Plantes MARINES. Sea-weed.
MARINE, *s. f.* The navy; also the science of maritime affairs.
MARINE *marchande.* The merchant-service. *Ex.*
Avez vous servi dans la MARINE *marchande, ou dans celle de l'état?* Have you served in the merchant-service or in the navy?
MARINIER, *adj. Officiers* MARINIERS. *Voyez* OFFICIERS, &c.
MARITIME, *adj.* Maritime; naval; belonging to the sea.
Affaires MARITIMES. Naval, or marine affairs.
Côtes MARITIMES. Sea-coasts.
Puissances MARITIMES. Naval powers.
Service MARITIME. Naval service.
MARMOTTE, *s. f.* A sort of match-tub.
MARNOIS. *adj. Bateau* MARNOIS. A yacht, hoy, or smack, employed on the rivers Marne and Seine.
MARQUES, *s. f. p.* Sea-marks observed by the pilots upon any coast; as mountains, spires, wind-mills, &c.
MARSILIANE, *s. f.* A square-sterned ship of about 700 tons burden, navigated in the gulf of Venice, &c.
MARSOUIN, *s. m.* (In naval architecture). MARSOUIN *d'avant.* A name given to the stemson.
MARSOUIN *d'arrière.* A sort of false stern-post, used in the French construction.
MARTEAU, *s. m.* A hammer.
MARTEAU *à dents.* A claw-hammer used by ship-wrights.
MARTEGAU, ou JUMELLES *de brasseyage, s. m.* The fish of a yard.
MARTINET, *s. m.* MARTINET *du pic d'artimon.* Peak-halyards of the mizen.
MARTINET, is also a tracing-line of the crow-foot.
MASCARET, *s. m.* A violent eddy of the tide, &c.
MASQUE, *s. m.* A smoke-sail.
MASQUER, *v. a. & n.* To back; to take a-back, &c. *Ex.*
Etre MASQUÉ *par le vent.* To be taken a-back.
MASQUER *tout.* To throw all a-back.
MASQUER *une voile.* To lay a sail a-back.
MASSE, *s. f.* MASSE *de fer.* A pen-maul.
MASSE *de bois.* A commander, or large wooden mallet.
MASULIT, *s. m.* An Indian boat, made of the bark of trees and calked with moss.

MÂT, *s. m.* A mast.
Grand MÂT. Main-mast.
MÂT *de misaine.* Fore-mast.
MÂT *d'artimon.* Mizen-mast.
Grand MÂT *de hune.* Main-top-mast.
Petit MÂT *de hune.* Fore-top-mast.
MÂT *de grand perroquet.* Main-top-gallant-mast.

Mât *de petit perroquet.* Fore-top-gallant-mast.
Mât *de perroquet de fougue.* Mizen-top-mast.
Mât *de petit kakatois.* Fore-top-gallant royal-mast.
Mât *de grand kakatois.* Main-top-gallant royal-mast.
Mât *de kakatois de perruche.* Mizen-top-gallant royal-mast.
Mât *majeur.* *Voyez* Majeur.
Mât *de fortune.* *Voyez* Fortune.
Mâts *de hune de rechange.* Spare top-masts.
Mât *à pible.* A pole-mast.
Mât *de beaupré.* *Voyez* Beaupré.
Mât *d'assemblage.* A made-mast.
Mât *d'un brin.* A mast made of a single stick or tree; such are the bowspit and top-masts of all ships, and all the masts of a small vessel.
Mât *de plusieurs pièces.* A mast made of several pieces.
Mât *brut.* A rough mast.
Mât *forcé.* A mast which is sprung.
Mât *jumelé, reclampé, ou renforcé.* A mast fished in a weak place, or opposite to any spring.
Guinder les mâts *de hune.* To heave the top-masts an-end, or swayed up.
Bâtiment à trois mâts. Three-masted.
A mâts *& à cordes.* Under bare-poles.

MATAFIONS, *s. m. p.* (A term of the Levant). Gaskets of the sails of a galley.

MATAGE, ou matement, *s. m.* The placing or fixing of a ship's masts.

MÂTÉ, *part. past.* Masted.
Mâté *en brig.* Masted like a brig.
Mâté *en cutter.* Masted like a cutter.
Mâté *en heu.* Masted like a Dutch hoy.
Mâté *en galère.* Masted as a galley.
Haut mâté. Taunt-masted.
Mâté *à calcet.* Masted for lateen sails.

MATELOT, *s. m.* A sailor.
Il est un bon matelot. He is an able seaman.
Vaisseau matelot. Second to a ship in a fleet. This appellation is also given to a good company-keeper, or a ship that sails well and keeps her station in a fleet.
Matelots *du commandant.* Seconds to the admiral.
Matelot *d'avant.* Second a-head.
Matelot *d'arrière.* Second a-stern.

MÂTER, *v. a.* Mâter *un vaisseau, ou canot.* To mast, or place the masts of a ship, or boat.
Mâter *l'étambot d'un bâtiment.* To erect the stern-post of a ship.

MÂTEREAU, *s. m.* A small mast, pole, or spar.

MÂTEUR, *s. m.* A mast-maker.
Maître mâteur. Master mast-maker.

MÂTURE, *s. f.* The art of masting ships; also a general name for the maker.
Machine à mâter, ou mâture. A machine or sheers for masting ships, which in France are generally erected on shore.
Mâture *flottante.* A sheer-hulk.
La mâture. The mast-shed, or the place where masts are made.
Mâture *de lougre.* The way of masting a lugger.
Vaisseau qui a sa mâture *complete.* A ship which has all her masts and yards ready made.
Notre mâture *a beaucoup souffert dans le combat.* Our masts and yards have been much cut in the action.

MAUGÈRE, *s. f.* A cow-hide.
Maugères, ou mauges. Scupper-holes.

MÈCHE, *s. f.* Mèche *de gouvernail.* Main-piece of a rudder.
Mèche *de cabestan.* *Voyez Parties du* cabestan.
Mèche *d'un mât.* The middle piece of a made mast.
Mèche *d'un cordage.* The heart of a rope.
Mèche *à canon.* Match, or quick-match, by which a cannon is fired.

MÉLIS. *Voyez* Toile *de Mélis.*

MEMBRE, *s. m.* Membres *d'un vaisseau.* The ribs of a ship: as the floor-timbers, top-timbers, futtocks, &c.

MEMBRURE, *s. f.* All the ribs or frames of a ship.
Bois de membrure. Compass-timber; timber fit for the floors and futtocks of a ship's frames.

MÉNAGER, *v. a.* Ménager *les mâts d'un navire.* To favour the masts of a ship.

MENILLE, *s. f.* Menilles *de prisonniers.* Shackles upon the bilboe-bolts, to confine prisoners and criminals.

MENUISERIE, *s. f.* Menuiserie *des vaisseaux.* Joiner's work in a ship.

MER, *s. f.* The sea; also the tide.
Grosse mer. A great sea, or high sea.
Mer *calme.* A calm or smooth sea.
Il y a de la mer. The sea runs high.
Il y a grosse mer. There is a great sea.
La mer *est belle.* The sea is smooth.
La mer *est tombée.* The sea is abated.
La mer *est étale.* It is between flood and ebb-tide, or it is slack water at the highest of the flood.
La mer *monte, ou rapporte.* It is flowing water, or flood.
La mer *blanchit ou moutonne.* The sea foams or froths.
Tomber à la mer. *Voyez* Tomber.
Nous avions une grosse mer *de l'avant.* We had a great sea a-head.
Nous avions une mer *du N. O.* We had a great sea setting to the south-east, or from the north-west.
Coup de mer. *Voyez* Coup.
Loup de mer. *Voyez* Loup, &c.
Homme de mer. *Voyez* Homme.
La mer *sans fond.* An unfathomable sea.
La mer *est changée de couleur.* The colour of the water is changed.
Mer *basse.* *Voyez* Basse-eau.
Mer *haute.* A swelling sea.
Molle mer. *Voyez* Molle.
Etre en haute mer. To be at sea and out of sight of land.
Tenir la mer. *Voyez* Tenir.
La mer *a perdu.* The tide has fallen.
Mer *pleine.* *Voyez* Pleine.
La pleine mer. The main sea, or the offing.
Jeter à la mer. *Voyez* Jeter.

MERIDIEN, *s. m.* Meridian, or mid-day, when the sun is at his greatest altitude.

MÉRIDIONAL, *adj.* Southern, &c.

MERLIN, *s. m.* Marline.

Merlin *blanc.* White marline.

Merlin *noir.* Tarred marline.

MERLINER, *v. a.* Merliner *une voile.* To marl a sail to its foot-rope.

MESTRE, *s. m.* (A term of the Levant.) *Arbre de* mestre. The main-mast of a row-galley, &c.

Antenne de mestre. A main lateen-yard.

MÉTACENTRE, *s. m.* (In naval architecture). Metacenter.

METTRE, *v. a. & n.* To put, &c. This verb is much used in a naval sense, and applies to a variety of objects, of which the following are some instances.

Mettre *en mer.* To go to sea.

Mettre *à la voile.* To set sail.

Mettre *à la cape.* To bring-to, in a storm.

Mettre *à la bande.* To heel.

Mettre *en route.* To steer the course.

Mettre *à sec.* To hand all.

Mettre *en place le gouvernail.* To ship the rudder.

Mettre *la bouée à l'eau.* *Voyez* Bouée.

Mettre *les huniers sur les fils de caret.* To stop the top-sails with rope-yarn.

Mettre *les canons à la serre.* To house the guns.

Mettre *les canons aux sabords.* To run the guns out.

Mettre *l'ennemi entre deux feux.* To place the enemy between two fires.

Mettre *les voiles au sec.* To loose the sails to dry.

Mettre *ses ancres à demeure.* To stow the anchors for sea.

Mettre *un bâtiment dans le bassin.* To put a ship into dock.

Mettre *un officier aux arrêts.* To put an officer under arrest.

Mettre *les linguets au cabestan.* To pawl the capstan.

Mettre *un aviron de bout.* To toss up an oar.

MEURTRIÈRES, *s. f. p.* Loop-holes in a ship's sides or bulk-heads.

À MI-MÂT. Half-mast up.

MINOIS, ou boute-lof, *s. m.* Bumkin.

MIRAGE, ou mirement, *s. m.* The looming of a distant object.

MIRE, *s. f.* (In gunnery). Aim.

Prendre sa mire. To take aim with cannon or other fire-arms, to their objects.

MISAINE, *s. f.* The fore-sail.

Mât de misaine. Fore-mast.

Vergue de misaine. Fore-yard.

Étai de misaine. Fore-stay.

Haubans de misaine. Fore-shrouds.

Itague de la drisse de misaine. Fore-tie.

Drisses de la misaine. Fore-jears.

Trelingage des haubans de misaine. Fore cat-harpings.

Bouline de misaine. Fore bowline.

Hune de misaine. Fore-top.

MITRAILLE, *s. f.* Langrage-shot.

MODÈLE, *s. m.* Modèle *d'un vaisseau.* The model of a ship.

MOIS, *s. m.* Mois *de gages.* The monthly pay or wages of seamen.

MOLE, *s. m.* A mole, or mole-head.

MOLLE, *adj.* Molle *mer.* Slack water.

MOLLIR, *v. a. & n.* To ease off, &c. *Ex.*

Mollir *une corde.* To ease off a rope.

Mollir *les haubans.* To ease the shrouds.

☞ Mollis du bras du vent à la demande de la vergue! *Ease off the weather braces as the yard goes up!*

La marée a molli. The tide is slack.

Le vent mollit. The wind is going down.

MONDE, *s. m.* A general name for the crew of a ship, but often applied to a part, and best translated by the word hands.

Mettre du monde *sur les vergues.* To man the yards.

☞ Du monde à la pompe! *Man the pump!*

Nous manquons de monde *à bord.* We want hands on board.

☞ En haut toute le monde! *All hands hoay!*

Donner du monde *à un bâtiment.* To turn over men to another ship.

MONTANS, *s. m. p.* Stanchions of the bulk-heads, and also of all upright pieces.

Montans *des bittes.* *Voyez* Bittes.

Montans *de tente.* Stanchions of the awning.

Montans *de voûte.* Counter-timbers.

MONTANT, *s. m.* Montant *de la marée.* The flood.

MONTER, *v. a.* To mount, &c. *Ex.*

Monter *les canons.* To mount the guns upon their carriages.

Faire monter *le monde.* To turn the hands up.

Vaisseau bien monté *en hommes.* A ship well manned.

Vaisseau montant 80 *canons.* A ship carrying 80 guns.

MOQUE, *s. f.* A general name for dead-eyes, bull's-eyes, hearts, &c. except those used in setting up the shrouds.

Poulie à moque. A dead-eye.

Moque *à trois trous.* An oblong dead-eye with three holes.

Moque *d'étai.* A heart.

Moque *de trelingage.* *Voyez Moque* d'araignée.

MORDRE, *v. n.* To take hold, speaking of an anchor which is sunk into the ground.

MORNE, *s. m.* A high hill or mountain; a name applied by the French only when speaking of the West-India islands.

MORT. *adj.* *Corps* mort. Transporting buoys.

MORTAISE, *s. f.* A mortise, or hole to receive the end of a piece of timber, called the tenon.

Mortaise *du gouvernail.* The hole in the rudder-head which contains the tiller.

Mortaise *de poulie.* The channel, or vacant space in a block, which contains the sheave.

Mortaise *du mât de hune.* The fid-hole of a top-mast.

MORTIER, *s. m.* A mortar for throwing shell or carcasses from a ketch.

MOU, *adj.* Slack. *Ex.* *Vent* MOU. Slack wind.
Le cordage est MOU. That rope is slack.
Vaisseau MOU *à virer.* A ship slack in her stays.
Abraquer le MOU. *Voyez* ABRAQUER.
Donner du MOU. To slack out.
MOUCHE, *s. f.* MOUCHES *d'une escadre.* Small and fast-sailing vessels attached to a squadron or fleet of men of war, to carry the orders of the admiral, and to observe the motions of the enemy.
MOUFFLE DE POULIE. *Voyez* CAISSE *de poulie.*
MOUILLAGE, *s. m.* An anchoring-place, or anchorage.
Bon MOUILLAGE. Good anchoring ground.
Mauvais MOUILLAGE. Bad anchoring-ground, or foul bottom.
Aller au MOUILLAGE. To stand for an anchoring-place.
Prendre un MOUILLAGE. To take up an anchoring-ground.
Bâtiment au MOUILLAGE. A ship at anchor.
MOUILLER, *v. a. & n.* To anchor; to cast anchor; to come to an anchor, &c. *Ex.*
MOUILLER *une ancre.* To let go the anchor.
☞ MOUILLE! *Let go the anchor!*
MOUILLER *entre vent & marée.* Moored between wind and tide.
MOUILLER *une ancre en créance.* To moor with the long-boat, or to carry out an anchor.
MOUILLER *en croupière.* To moor with a spring upon the cable, in order to canonnade a fort, &c.
MOUILLER *en faisant embossure.* To bring up with a spring on the cable.
MOUILLER *en pagale* To come to all-standing.
MOUILLER *en barbe.* To come to, with two anchors a-head.
MOUILLER *en patte d'oie.* To moor with three anchors a-head, equally distant from each other.
MOUILLER *les voiles.* To wet the sails.
MOULE, *s. m.* MOULE *à gargousse.* A mould for cartridges.
MOULINET, *s. m.* MOULINET *à bitord.* A spun-yarn winch.
MOULINET, is also the name of a small windlass, as that of a launch or long-boat.
MOUSSE, *s.m.* A cabin-boy, younker, or youngster.
MOUSSON, *s. f.* Monsoon, or periodical trade-winds, prevalent in certain latitudes of the Indian ocean.
MOUSTACHES, *s. f. p.* Standing lifts.
MOUSTACHES *de la civadière.* Standing lifts of the sprit-sail yard.
MOUSTACHES *de la vergue sèche.* Standing lifts of the cross-jack-yard.
MOUTONNER, *v. n.* *La mer* MOUTONNE. The sea foams or froths.
MOUTONS, *s. m. p.* The foam, or froth of the sea, when it blows hard.
MOUVEMENS, *s. m. p.* *Avoir les* MOUVEMENS *durs à l'ancre.* To ride hard.
L'amiral a fait faire de beaux MOUVEMENS. The admiral made fine manœuvres.
MOYEN-PARALLÈLE (In navigation). The middle latitude, or the parallel that holds the middle place between the latitude departed from and the latitude arrived in.
MULET, *s. m.* A Portuguese vessel with three masts, and lateen sails.
MUNITIONNAIRE, *s. m.* An agent-victualler, or a contractor for sea provisions.
Commis du MUNITIONNAIRE. *Voyez* COMMIS, &c.
MUNITIONS, *s. f. p.* Stores, ammunition.
MUNITIONS *navales.* Naval stores.
MUNITIONS *de guerre.* Ammunition of war.
MUNITIONS *de bouche.* Sea provisions.

N.

NACELLE, *s. f.* A small row-boat or skiff, without masts or sails, employed to pass a river.
NAGE, *s. f.* The row-lock of a boat. *Voyez* TOLLETIÈRES.
Tente de NAGE. An awning. *Voyez* TENTE, &c.
La NAGE. The crew of rowers.
Donner un coup de NAGE. To pull the stroke oar.
Un homme à la NAGE, *adv.* A man swimming.
Se sauver à la NAGE. To swim away, or to escape by swimming.
NAGER, *v. n.* To row or pull with the oars, in a boat or small vessel. *Ex.*
NAGER *debout.* To row standing, with the face towards the boat's head.
NAGER *à sec.* To row ground, or touch the shore with the oars in rowing.
NAGER *de bout a la lame.* To head the sea in rowing.
NAGER *de bout au vent.* To pull head to wind.
NAGER *en arrière.* To back a-stern with the oars.
NAGER *de long, ou à l'Angloise.* To row a long stroke.
Bonne de NAGE. *Voyez* BONNE, &c.
☞ NAGE à bord! *Come aboard in the boat!*
☞ NAGE sec! *Row dry!* The order to row without wetting the passengers.
☞ NAGE tribord et scie babord! *Pull the starbord, and hold water with the larboard oars!*
☞ NAGE! *Pull away!*
☞ NAGE ensemble! *Pull together!*
☞ NAGE à faire abattre! *Pull to leeward!*
☞ NAGE à venir sur babord! *Pull to port!*
☞ NAGE au vent! *Pull to windward! or tow the ship to windward!*
☞ NAGE bien babord! *Pull away with the larboard oars!*
☞ NAGE bien tribord! *Pull away with the starboard oars!*
☞ NAGE qui est paré, ou avant qui peut! *Pull with the oars that are shipped!*
☞ NAGE de force! *Stretch out!*
NATION, *s. f.* Nation. When speaking of ships of different nations and places, both French and English express themselves as if speaking of a man. *Ex.*

Algérien,	Algerine.
Américain,	American.
Anglois,	Englishman.
Bremenois,	Bremener.
Chinois,	Chinese,

Danois,	Dane.
Espagnol,	Spaniard,
François,	Frenchman.
Génois,	Genoese
Hambourgois,	Hamburger.
Hollandois,	Dutchman.
Italien,	Italian.
Japanois,	Japanese.
Maltois,	Maltese.
Napolitain,	Neapolitan.
Norwégien,	Norwegian.
Portugais,	Portuguese.
Prussien,	Prussian,
Russe,	Russian.
Sicilien,	Sicilian.
Suédois,	Swede.
Turc,	Turk.
Venitien,	Venetian.

NATTE, *s. f.* A paunch, or mat.

NATTES, are also mats used to line the bread-room, &c.

NAVAL, *adj.* Of or belonging to the sea or ships; naval; maritime.

Armée NAVALE. A fleet of men of war.

L'armée NAVALE *d'Angleterre.* The English naval forces.

L'armée NAVALE *de France.* The French naval forces.

Combat NAVAL. Sea-fight.

Couronne NAVALE. Naval crown.

NAVETTE, *s. f.* A small Indian vessel.

NAUFRAGE, *s. m.* Shipwreck.

NAUFRAGÉ, *part. past. Notre bâtiment fut* NAUFRAGÉ *sur les côtes d'Espagne.* Our vessel was wrecked on the Spanish coast.

NAVIGABLE, *adj.* Navigable, or capable of navigation.

Rivière NAVIGABLE *des sa source.* A river navigable from its source.

NAVIGATEUR, *s. m.* A navigator, mariner, or seaman.

Habile NAVIGATEUR. A great navigator.

NAVIGATION, *s. f.* Navigation, or the knowledge of the theory and practice of that art.

NAVIGATION *hauturière.* Sailing out of sight of land.

NAVIGATION *côtière.* Coasting.

NAVIGUER, *v. n.* To navigate a ship, to direct her course at sea; to travel by sea; to lead a seafaring life.

NAVIGUER *en pleine mer.* To sail on the main.

Vaisseau qui NAVIGUE *bien.* A ship that sails well.

NAVIRE, *s. m.* A ship; or vessel; a sail; a general name for ships, but particularly for merchant-men. *Ex.*

☞ NAVIRE! *A sail!* The cry of the man who is on the look out, when he discovers a ship at sea.

Où va le NAVIRE? Where are you bound to?—hoay!

D'où vient le NAVIRE? Whence came you?—hoay!

NAVIRE *larguant de par tout.* A crazy vessel.

Un NAVIRE *marchand. Voyez* MARCHAND.

☞ NAVIRE au vent à nous! *A sail to windward!*

NAVIRE *étranger.* A foreign vessel.

Le NAVIRE *Argo.* The ship Argo.

NÉGRIER, *s. m.* A Guinea-man. *Voyez* BÂTIMENT, &c.

NEUVE. *Voyez* BÛCHE.

NEZ, *s. m.* The nose, beak, bow, or head of a ship.

Navire qui est trop sur le NEZ. A ship too much by the head.

NOCHER, *s. m.* (A term of the Levant). A boatswain; the ancient name for PILOT.

NOCTURLABE, *s. m.* A nocturnal.

NŒUD, *s. m.* A knot, or bend, &c. *Ex.*

NŒUD *plat, ou* NŒUD *marin.* A carrick-bend, or a proper sailor's knot.

NŒUD *de bouline.* A bowline-knot.

NŒUD *d'écoute.* A sheet-knot.

NŒUD *de vache.* A granny's bend.

NŒUD *d'anguille.* A timber-hitch.

NŒUD *de bois.* A jamming or countryman's knot.

NŒUD *d'agui à élingue.* A standing bowline-knot.

*Demi-*NŒUD. An overhand-knot not drawn close, with the seizings on.

NŒUD *de jambe de chien.* A sheep-shank.

NŒUD *à plein poing.* An over-hand knot.

NŒUDS *de la ligne de loc.* The knots of the log-line. This word becomes synonimous to a marine mile, or the 60th part of a degree. *Ex.*

Nous filons 12 NŒUDS *par heure.* We run twelve knots an hour.

NOLIS, *s. m.* (A term of the Levant). The freight or hire of a ship.

NOLISER, *v. a.* (A term in the Levant). To freight.

NOMMER, *v. a.* To appoint, speaking of an appointment in the navy. *Ex.*

Je viens d'être NOMMÉ *au commandement d'une frégate.* I have just been appointed (or promoted) to the command of a frigate.

NORD, *s. m. & adv.* North, or the north point.

Vent du NORD. Northerly wind.

NOVICE, *s. m. Matelot* NOVICE. A young sailor.

NOYALE. *Voyez* TOILE *de Noyale.*

NOYÉ, *adj.* Clouded or indistinct, said of the horizon when it cannot be easily distinguished by an observer, in taking an altitude of the sun, &c.

NOYER, *v. a.* To drown, &c. *Ex.*

Il y eut trois gabiers de NOYÉS *par la chûte de notre petit mât de hune.* There were three top-men drowned by our fore-top-mast going over the side.

NOYER *un vaisseau.* To drop a vessel, so as to lose sight of her by degrees.

NOYER *la terre.* To settle or lay the land.

Vaisseau qui a sa batterie NOYÉE. A ship that does not carry her ports to a proper height out of the water.

Se NOYER. To be drowned.

NUAGE, *s. m.* A cloud. *Ex.*

Ce NUAGE *amène un grain.* That cloud forebodes a squall.

NUAISON, *s. f.* A wind set in; a trade-wind.

O.

OBSERVATION, *s. f.* An observation. *Ex.* L'OBSERVATION *des astres.* The observation of the sun, moon, or stars.

Faire des OBSERVATIONS *de distance.* To take lunar observations.

Etre en OBSERVATION *devant un port.* To remain off a port, watching the enemy's motions.

OBSERVATOIRE, *s. f.* An observatory.

OBSERVER, *v. a.* To observe; to look out &c. *Ex.*

OBSERVER *la latitude.* To take an observation.

OBSERVER *des signaux, &c.* To look out for signals, &c.

OCCASE, *adj.* Westerly; speaking of the amplitude.

OCCIDENT, *s. m.* West. *Voyez* OUEST.

OCCIDENTAL, *adj.* Western; westerly.

OCÉAN, *s. m.* The ocean in general, but the French have given that name particularly to the Atlantic Ocean.

OCÉAN *Atlantique.* Atlantic Ocean.

OCÉAN *Pacifique, ou Mer du Sud.* The great South Sea, or Pacific Ocean.

OCÉAN *Indien, ou Mer des Indes.* The Indian Ocean.

Les ports de L'OCÉAN. The sea-port towns of France, which are situated on or near the ocean, in opposition to those of the Mediterranean.

OCTANT, *s. m.* A quadrant.

ŒIL, *s. m* ŒIL *d'ancre.* The eye of an anchor.

ŒIL-*de-bœuf.* *Voyez* YEUX-*de-bœuf.*

ŒIL-*de-bouc.* A water-gall, or weather-gall.

ŒIL *de pie.* *Voyez* YEUX *de pie.*

ŒIL *de roue.* The hole in the truck or wheel of a gun-carriage, through which the axle passes.

ŒILLET, *s. m.* An eye spliced on the end of any rope.

ŒILLETS *de voile.* Eye-let holes of a sail.

ŒILLET *d'étai.* The eye of a stay which goes over the mast-head.

ŒILLETS *de tournevire.* The eyes of the voyol.

ŒILLET *de fer.* The eye of an iron bolt.

ŒILLET *d'estrop de poulie.* The eye of a block-strap.

ŒUVRES, *s. f. p.* ŒUVRES-*vives.* The quick-work, or all that part of a ship which is under water.

ŒUVRES-*mortes.* The upper-works of a ship, or that part which is above water, comprehending the quarter-deck, poop, and forecastle.

ŒUVRES *de marée.* The graving, calking, or repairing a ship's bottom, when she is left dry a-ground during the recess of the tide.

OFFICIER, *s. m.* An officer.

OFFICIER *de la marine.* An officer in the navy.

OFFICIER *général.* A flag-officer.

OFFICIER *chargé du détail, ou lieutenant en pied.* The first lieutenant of a man of war.

OFFICIER *bleu.* An officer who performs the duty of another while absent or sick.

OFFICIER *de garde.* The officer of the deck in harbour.

OFFICIERS *de port.* Officers in a dock-yard, appointed to see that the shipping be properly moored, masted, rigged, ballasted, &c.

OFFICIER *de ronde.* The officer who rows guard in a port, or elsewhere.

OFFICIERS *de santé.* The officers who superintend the affairs of quarantine in a port.

OFFICIER *de quart.* The officer of the watch at sea.

OFFICIER *de manœuvre.* An officer selected by the captain of a French man of war to assist him in time of action on the quarter-deck.

OFFICIERS *mariniers.* Warrant and petty officers.

OFFICIERS *majors.* The principal officers in a fleet, and also the commissioned officers in a ship.

☞ OH! DU NAVIRE, hola! *Hoa, the ship ahoay!* The manner of hailing or calling to a ship whose name is not known.

☞ OH! d'en haut! *Yoa-hoa, aloft there! mast-head there!* &c. The call from the deck to those who are aloft, to attend to some order.

☞ OH! hisse! OH! hâle! OH! saille! OH! ride. The way of singing out, as a signal to hoist, haul, or rowse together, on a tackle or rope.

OING, *s. m.* Stuff, tallow, &c. used to pay the masts, &c.

ORAGE. *Voyez* TEMPÊTE.

ORDONNANCE, *s. f.* *Les* ORDONNANCES *de la marine.* The statutes of the navy.

ORDONNATEUR, *s. m.* An officer, whose employment is similar to that of a commissioner in the navy.

ORDRE, *s. m.* (In naval tactics). Order. *Ex.*

ORDRE *de battaile.* The order of battle in a naval engagement.

ORDRE *de croissant.* A line of battle in the form of a crescent.

ORDRE *de front de bandière.* A line of battle a-breast at half a cable's length distance.

ORDRE *de marche.* The order of sailing.

ORDRE *de marche en échiquier.* A bow and quarter line.

ORDRE *de retraite.* The order of retreat.

ORDRES *des vaisseaux.* *Voyez* RANG.

ORDRE *renversé.* The line inverted.

OREILLES, *s. f. p.* OREILLES *de lièvre.* Triangular sails in a galley, pink, &c.

OREILLES *d'une ancre.* *Voyez Parties de* L'ANCRE.

OREILLES *d'âne.* Kevels for belaying the tacks and sheets.

ORGANEAU, *s. m.* The ring of an anchor; also any iron ring.

ORIENTAL, *adj.* Eastern.

ORIENTER, *v. a.* To brace up; to trim the sails, or rigging. *Ex.*

Vaisseau bien ORIENTÉ. A well-trimmed ship.

Une voile mal ORIENTÉE. A sail badly trimmed.

Navire qui est ORIENTÉ *au plus près.* A sharp-trimmed ship.

ORIN, *s. m.* The buoy-rope of an anchor.

ORSE, *adv.* The larboard side; also the order to luff, in the language of the Levant.

ORSER, *v. n.* To row against the wind, in the same dialect.

ORTHODROMIE, *s. f.* A course which lies upon a meridian or parallel.

ORTIVE, *adv.* Easterly; speaking of the amplitude.
OSTE, *s. f.* A brace or vang used to steady a lateen yard.
Bragots des OSTE. The pendants of such a brace or vang.
OUAICHE. *Voyez* SILLAGE.
OUEST, *s. m.* West, or west point of the compass or horizon.
L'OUEST *du monde.* Due west.
L'OUEST *de la boussole.* West by the compass.
Brise D'OUEST. Westerly breeze.
OURAGAN, *s. m.* A hurricane.
OURSES, *s. f. p* OURSES *d'artimon.* Mizen bowlines.
OUVERT, *adj.* *Vaisseau* OUVERT *de partout.* A ship which is very leaky.
À L'OUVERT. *Etre à* L'OUVERT *d'un port d'une rade, d'un détroit, &c.* To be open with any port, roadstead, channel, &c.
OUVERTURE, *s. f.* An opening or valley between two hills, beheld from the sea, and serving frequently as a land-mark.
OUVRIERS, *s. m. p.* The artificers, &c. in a dock-yard; also the riggers of a ship.
OUVRIR, *v. a.* To open; to discover. OUVRIR *une baie, &c.* To open or discover a bay, &c. when sailing at some distance from it.
☞ OUVRE les hublots! *Open the scuttles!*

P.

PACFIS, *s. m. p.* The courses of a ship; as the main-sail and fore-sail; this word is now become nearly obsolete.
PACOTILLE, *s. f.* A certain quantity of goods which the owner of a vessel allows the master, or any other person, to ship on his own account.
PAGAYE, *s. f.* The paddle of a canoe.
PAGAYER, *v. a.* To paddle.
PAGE DE LA CHAMBRE DU CAPITAINE, *s. m.* A cabin-boy. *Voyez* MOUSSE.
PAILLE-EN-CUL, *s. m.* A gaff top-sail.
Boute-hors de PAILLE-EN-CUL. A ring-tail boom.
PAILLES *des bittes*, *s. f. p.* Long iron bolts thrust into the holes of the bits, to prevent the cable from starting off.
PAILLET, ou PAILLER, *s. m.* A mat.
PAILLOTT, *s. m.* The steward's room in a galley.
PALAMANTE, *s. m.* The oars of a row-galley; speaking of them collectively.
PALAN, *s m.* A tackle.
PALAN *simple.* A small single tackle.
PALAN *double.* A two-fold tackle.
PALAN *à caliorne.* *Voyez* CALIORNE.
PALANS *de canon.* Gun-tackles.
PALANS *de bout.* The sprit-sail halyards.
PALANS *de ris.* Reef-tackles.
PALAN *du serpente*, *Voyez* SERPENTE.
PALAN *à fouet.* A jigger.
PALAN *d'amure.* A tack-tackle.
PALAN *de drosse.* A truss-tackle.
PALAN *d'étai.* A stay tackle.
PALAN *d'écoute de brigantine.* A spanker sheet.
PALAN *de retenue.* *Voyez* RETENUE.
PALAN *de recul des canons.* A train tackle.
Grand PALAN. Main-tackle.
PALAN *de misaine.* Fore-tackle.
Poulie double de PALAN. *Voyez* POULIE.
Poulie simple de PALAN. *Voyez* POULIE.
Croc de PALAN. *Voyez* CROC.
Garant de PALAN. *Voyez* GARANT.
PALANQUER, *v. a. & n.* To bowse, or draw any thing by means of a tackle. *Ex.*
PALANQUER *les haubans.* To set up the shrouds.
PALANQUIN, *s. m.* A small tackle.
PALANQUINS *de ris.* *Voyez* PALANS *de ris.*
PALANQUINS *des mâts de hune.* Burtons.
PALANQUINS *de sabords.* Port-tackles.
PALANQUINS *simples de recage.* Nave-lines.
PALLE, *s. f.* A sort of East-Indian vessel.
PALME, *s. f.* A palm; a measure used to ascertain the thickness of masts.
EN **PANNE**, *adv.* Lying-to, or lying-by.
Etre en PANNE. Lying-to.
Mettre en PANNE. To lie-to, or bring-to.
PANNEAU, *s. m.* The cover or lid of a hatchway.
Le grand PANNEAU. The main-hatch.
PANNEAU *à boîte.* The cover of a hatchway, with a border round its edge.
PANNEAU *à vassale.* A great hatch without a border.
EN **PANTENNE**, *adv.* Disorderly; out of trim.
Amener les voiles en PANTENNE. To haul down the sails with the utmost expedition; as in a squall of wind.
Vaisseau en PANTENNE. A ship which rides a-peak, or with her rigging, sails, and yards disordered.
PANTOIRE. *Voyez* PENDEUR.
PAPIERS, *s. m. p.* PAPIERS *et enseignemens.* The papers of a ship, comprehending the bills of lading, manifests, cockets, &c.
PAQUE-BOT, *s. m.* A packet-boat.
EN **PAQUET**, *adv.* *Amener les voiles en* PAQUET. To strike the sails a-main.
PAR. *prep.* In, &c. *Ex.*
Mouiller PAR 20 *brasses.* To anchor in 20 fathoms of water.
Nous sommes PAR 5° *de latitude nord.* We are in 5° north latitude.
Etre PAR *le travers d'une île*, &c. To be off an island, &c.
PARADIS, *s. m.* A wet-dock, in the language of the Levant.
PARAGE, *s. m.* Any particular space of the sea, appointed as the station wherein to rendezvous or cruise; also, a part of the sea near the coast. *Ex.*
Nous ne rencontrerons pas l'ennemi dans ses PARAGES. We shall not meet the enemy in that latitude.
Les PARAGES *des pays chauds.* The warm latitudes.
Etre mouillé en bon PARAGE. To be moored in a good anchorage.
PARALÈLE, *s. m.* A parallel.
Moyen PARALLÉLE. *Voyez* MOYEN.

PARATONNERRE, *s. m.* A conductor to carry off lightning.
PARC, *s. m.* A term often used for *Arsenal de Marine*, which denotes a dock-yard and its several store-houses.
PARC *à boulets* The shot-locker.
PARC *à moutons*. A sheep-pen on a ship's deck.
PARC *d'artillerie*. *Voyez* ARTILLERIE.
PARCLOSES, *s. f. p.* Limber-boards. *Bordages des* ANGUILLÈRES is also used in the same sense.
PARCOURIR, *v. a.* PARCOURIR *les coutures*. To examine the seams of a ship's side, or deck, in order to calk them, if necessary.
PARE-PIEDS, *s. m.* A fender.
PARER, *v. a.* To clear or go clear off; to dub, &c.
PARER *un vaisseau*. To go clear of a ship.
PARER *un banc, &c.* To give a good birth to a bank, &c.
PARER *un Cap*. *Voyez* DOUBLER.
PARER *un cable, ou une ancre, &c.* To see a cable, or an anchor, &c. clear.
☞ PARE à mouiller! *See all clear for coming to!*
☞ PARE à virer! *Ready about! See all clear to go about!*
☞ PARE à virer lof pour lof! *See all clear for wearing ship!*
☞ PARE devant! *Fend off a-head!*
☞ Nage qui est PARÉ! *Voyez* NAGER.
☞ PARE les manœuvres! *Coil up the ropes!*
☞ PARE le canot! *Get the boat ready!*
PARFUMER, *v. a.* PARFUMER *un vaisseau*. To smoke or perfume a ship, and sluice her between decks with vinegar and gunpowder mixed together.
PARLEMENTAIRE, *s. m.* A cartel, or flag of truce.
PARLER, *v. a.* PARLER *à un bâtiment*. To speak to a vessel.
PARQUET, *s. m.* An inclosure.
PARQUETS *à boulets*. Shot-garlands.
PARQUETS *pour les bois*. Locks for keeping timber under water.
PARQUET (*ou caisson*) *du lest*. An inclosure where ballast is kept in a dock-yard.
PART, *s. f.* Share.
PARTS *de prises*. Prize-money.
PARTAGER, *v. a.* To share.
PARTAGER *le vent avec un vaisseau*. To hold way with a ship.
PARTANCE, *s. f.* The time and act of sailing, or departing from an anchorage.
Coup de canon de PARTANCE. A signal gun for setting sail.
Faire une belle PARTANCE. To sail with a fair and steady breeze.
Etre en PARTANCE. To be ready for sailing.
Point de PARTANCE. *Voyez* POINT, &c.
PAS, *s. m.* A narrow channel or strait.
Le PAS *de Calais*. The Straits of Dover.
PASSAGE, *s. m.* A passage.
Prendre PASSAGE *à bord d'un bâtiment*. To take a passage on board ship.
Bateau de PASSAGE. A passage-boat.
PASSAGE *de la lune au méridien*. The southing of the moon.
PASSAGER, *s. m.* A passenger in a vessel.
PASSE, *s. f.* The fairway, or track through a channel; the entrance of a harbour; also a narrow strait.
PASSE-AVANTS, *s. m. p.* The gangways leading from the quarter-deck to the forecastle.
PASSE-PORT, *s. m.* A ship's pass, or pass-port. *Voyez* CONGÉ.
PASSER, *v. a. & n.* To pass, &c. *Ex.*
PASSER *au vent d'un vaisseau*. To weather, or gain the wind of another ship.
PASSER *la revue*. To be mustered.
PASSE VOLANT, *s. m.* A false muster on the ship's books; also a wooden gun, which may terrify a ship at a distance.
PASSER *sous le beaupré d'un bâtiment*. To pass under the bowsprit; viz. to go a-head of, or before a ship, and run athwart her course.
PASSER *des canons d'un bord à l'autre*. To shift the guns over from side to side.
PASSER *les poudres*. To hand along powder.
PASSER *dessus un bâtiment*. To run down a ship.
PASSER *d'un ordre à un autre*. To change the disposition of a fleet or squadron from one order of sailing or of battle to another.
PASSER *un mât de hune, ou de perroquet*. To point a top-mast, or top-gallant-mast through the cap.
PASSER *une manœuvre dans ses poulies, &c.* To reeve a rope.
PATACHE, *s. f.* A guard-vessel; revenue cutter; an armed tender which attends a ship of war or a fleet.
PATACHE *d'avis*. *Voyez* FRÉGATE *d'avis*.
PATARAS, *s. m. p.* Preventer shrouds; also spare shrouds to be hooked on occasionally.
PATRON, *s. m.* The master or commander of a merchant-ship, in the dialect of the French sea-ports in the Mediterranean.
PATRON *de chaloupe, ou de canot*. The cockswain of a launch or long-boat of a man of war.
PATTE, *s. f.* *Les* PATTES *d'une ancre*. The flooks of an anchor.
PATTES *d'anspecs*. The claws of a gunner's hand-spike.
PATTES *de boulines*. *Voyez* BOWLINE.
PATTES *de voiles*. The tabling of the sails at the bolt-ropes.
Mouiller en PATTE *d'oie*. *Voyez* MOUILLER, &c.
PAVILLON, *s. m.* A flag; also a general name for colours. *Ex.*
PAVILLON *de beaupré*. A jack.
PAVILLON *de poupe*. An ensign.
PAVILLONS *de signaux*. The ensign and signal flags of different colours. *Ex.*
PAVILLONS *supérieures*. Superior flags.
PAVILLONS *inférieurs*. Inferior flags.
PAVILLONS *national, ou* PAVILLON *distinctif de chaque nation*. Colours or flags of any nation.
PAVILLON *François*. The French flag.
PAVILLON *Américain*. American colours.
PAVILLON *neutre*. The colours of a neutral nation.
PAVILLON *Hollandois*, Dutch colours.
PAVILLON *amiral*. Admiral-flag, displayed from the flag-staff at the head of the main-mast, attributed only to the high-admiral of France.
PAVILLON *vice-amiral*. A flag displayed from the fore-mast head.

PAVILLON *de lieutenant-général.* A flag displayed from the mizen-mast.
PAVILLON *en berne.* The ensign with a waft.
PAVILLON *de conseil.* The signal for a general council.
Etre sous un tel PAVILLON. To be under such a flag, or commanding officer.
Hisser le PAVILLON. *Voyez* HISSER.
Mât de PAVILLON. Ensign staff.
Drisse de PAVILLON. *Voyez* DRISSE, &c.
Porter le PAVILLON *& la flamme.* To wear the colours of a man of war.
PAUMELLE, *s. f.* A sail-maker's palm.
PAUMELLE *de cordier.* List.
PAUMOYER. *Voyez* POMOYER, &c.
PAVOIS, *s. m.* Waist and quarter-cloths of a ship.
PAVOIS *du gaillard derrière.* Quarter-cloths.
PAVOIS *de hune.* Top armours.
PAVOISER, *v. a.* PAVOISER *un vaisseau.* To dress a ship with a variety of flags, pendants, &c.
PAYE, *s. f.* Seamen's wages.

PÊCHE, *s. f.* A fishery. *Ex.*
La PÊCHE *de la baleine, ou de la morue, &c.* The whale, or cod fishery, &c.

PÊCHER, *v. a.* To fish.
PÊCHER *une ancre.* To sweep an anchor.

PÊCHEUR, *s. m.* A Fisherman.

PÉDAGNE, *s. f.* The stretchers of a row-galley.

PÉGOLIÈRE, *s. f.* A house or shed for melting pitch or tar.
PEIGNER, *v. a.* PEIGNER *du chanvre.* To hatchel hemp.
PEIGNER *un cordage.* To worm a rope.

PÉLICAN, *s. m.* An iron dog, used to confine pieces of wood.
PELLE, *s. f.* PELLE *d'aviron.* The blade or wash of an oar.
PENDEUR, *s. m.* A pendant.
PENDEURS *de grande caliorne.* Main runner pendants.
PENDEURS *de caliorne de misaine.* Fore runner pendants.
PENDEURS *des faux haubans.* Pendants of the preventer shrouds.
PENDEURS *de bras.* Brace-pendants.
PENDEURS *des grands palans.* Main-tackle pendants.
PENNAUT, *s. m.* *Faire* PENNAUT. To cock-bill the anchor, &c.
PENNE, *s. f.* PENNE *d'une antenne.* The peak of a lateen yard.
PENON, *s. m.* A dog-vane.
PENTURE, *s. f.* PENTURES *de sabords.* The port-hinges, or goodgeons of a port-lid.
PENTURES *à fer à cheval, ou de hublot.* Scuttle-hinges.

PÉOTTE, *s. f.* A light nimble Venetian wherry, used frequently as an advice-boat, to carry expresses.
PERCER, *v. a.* To pierce, &c. *Ex.*
Bâtiment PERCÉ *à* 20 *canons.* A ship pierced for 20 guns.

PERCEURS, *s. m. p.* The artificers that bore the holes for the tree-nails, bolts, &c. of a ship.
PERDRE, *v. a.* To lose, &c.
PERDRE *la terre de vue.* To lose sight of land.
La marée PERD. The tide falls.
PERDRE *son poste dans la ligue.* To lose one's station in the line of battle.
PERDRE *sur un bord et gagner sur l'autre.* To lose upon one tack and gain upon the other.
PERDRE *en virant vent devant.* To lose ground in stays.
Se PERDRE. To be lost, or cast away.
Nous avons manqué de nous PERDRE. We were about being cast away.

PÉRIR, *v. a.* PÉRIR *par le mauvais temps.* To be cast away through bad weather.
PERROQUETS, *s. m. p.* The fore and main-top-gallant sails. *Ex.*
Le grand PERROQUET. The main-top-gallant sail.
Le petit PERROQUET. The fore-top-gallant sail.
Le grand PERROQUET *volant.* The main-top-gallant royal sail.
Le petit PERROQUET *volant.* The fore-top-gallant royal-sail.
PERROQUET *de fougue.* The mizen topsail.
Temps à PERROQUET. *Voyez* TEMPS, &c.
PERRUCHE, *s. f.* The mizen-top-gallant-sails.
PERTUIS, *s. m.* An inlet, or narrow channel.
Le PERTUIS *d'Antioche.* The channel of Antioche.
PERTUISANNIERS, *s. m. p.* Guards employed to guard the galley-slaves in French dock-yards.
PESER, *v. a.* To hang upon any rope; to haul downwards on any rope over-head.
PESER *sur un levier.* To heave, or purchase with a handspike.
☞ PÈSE sur les balancines de misaine! *Bowse taught the fore lifts!*
☞ PÈSE sur la balancine de gui! *Top up the spanker boom!*
PHARE (ou TOUR *à feu*), *s. m.* A watch-tower, or light-house on the sea-coast.
PIC, *s. m.* A gaff.
PIC *d'artimon.* The mizen-peak.
À-PIC. *Voyez* À-PIC. (*Letter* A.)
Etre à-PIC *sur son ancre.* To be close a-peak upon the anchor.

PIÈCE, *s. f.* A piece, &c. *Voyez* CANON.
PIÈCE *de toile à voile.* A piece of canvas.
PIÈCE *de cordage.* A piece of cordage.
PIÈCE *de bois.* A piece of timber.
PIÈCE *de quille.* A piece of the keel.
PIÈCE *de charpente.* A general name for any piece of large timber, used in the construction of a ship.
PIÈCES *de liaison.* Sleepers.
PIÈCES *de quartier.* *Voyez* QUARTIER.
Petites PIÈCES *d'arrimage.* Small casks used for wingers.
PIÈCES *à vin.* Wine-casks.
PIÈCES *à l'eau, (ou futailles).* Water-casks.
PIÈCE *de quatre.* A leager, containing four hogsheads, or a tun.
PIÈCE *de trois.* A pipe.
PIÈCE *de deux.* A butt.
PIED, *s. m.* A foot, &c.

PIED *de mât.* The heel of a mast.
PIED *marin.* Sea legs; said of a man who treads firmly on the deck, when at sea.
PIED *droit.* A stanchion, or any piece of timber that is upright or an-end.
PIERRIER, *s. m.* A pedrero, or swivel-gun.
PIÉTER, *v. a.* PIÉTER *l'étrave et l'étambot.* To mark the stem and stern-post by feet, in order to ascertain the ship's draught of water afore and abaft.
PIGOU, *s. m.* A wooden candlestick.
PILIERS DES BITTES, *ou montans.* The main vertical pieces of the bits.
PILLAGE, *s. m.* The plunder taken from an enemy after an engagement.
PILOTAGE, *s. m.* The art and act of navigating, conducting, or steering a ship.
Droits de PILOTAGES. Pilots rates.
PILOTE, *s. m.* A pilot.
PILOTE *hauturier.* A sea pilot.
PILOTE *côtier, ou* PILOTE *lamaneur.* A harbour pilot, a river, or coasting pilot.
Maître-PILOTE. *Voyez* MAÎTRE, &c.
PILOTER, *v. a.* PILOTER *un vaisseau.* To pilot a ship into or out of a harbour or river, &c.
PINASSE, *s. f.* A sort of small vessel resembling a sloop.
PINCE, *s. f.* An iron crow.
PINCÉ, *adj.* Sharp. *Ex.*
Vaisseau PINCÉ *de l'avant.* A sharp-bottomed ship.
PINCEAU, *s. m.* PINCEAU *à goudronner.* A tar brush.
PINCER, *v. a.* To hug; speaking of the wind.
PINCER *le vent.* To hug the wind; to haul close; to trim all sharp.
PINNULE, *s. f.* The sight-vanes of an instrument, for observing or setting a distant object at sea.
PINQUE, *s. f.* A pink, or narrow-sterned ship, with a flat floor, used in the Mediterranean.
PIPE, *s. f.* A pipe.
PIQUER, *v. a.* PIQUER *l'horloge.* To strike the bell.
PIRATE, *s. m.* A pirate, or sea robber.
PIRATER, *v. a.* To commit piracy.
PIRATERIE, *s. f.* Piracy.
PIRIS, *s. m.* A canoe used by the negroes in Guinea, and the Cape de Verds.
PIROGUE, *s. f.* Periagua, or an American canoe.
PIROGUE *double.* A double canoe.
PIROGUE *à balanciers.* A canoe with out-riggers.
PISTOLET, *s. m.* A pistol, also a bumkin.
PISTON, *s. m.* PISTON *de pompe.* The upper box of a ship's pump.
PIVOT, *s. m.* PIVOT *de boussole.* The brass centre-pin of a compass.
PIVOT *de cabestan. Voyez Parties du* CABESTAN.
PLACEL, *s. m.* A rocky or sandy shore, of great extent near the surface of the water.
PLAFOND, *s. m. Le* PLAFOND *d'un bâtiment.* The floor or bottom of a ship.
PLAGE, *s. f.* A flat shore, without any capes or head-lands to form a road or bay for ships to anchor in; also a sandy beach.
PLAGE *de sable.* A sloping sandy coast.
Navire jeté sur la PLAGE. A ship stranded on a sandy beach.
PLAN, *s. m.* A plan, or draught.
PLAN *d'un port.* The plan of a harbour.
PLAN *d'un vaisseau.* The draught of a ship.
PLAN *d'élévation, ou de longueur.* Sheer-draught, or plane of elevation.
PLAN *horizontal.* Horizontal plane.
PLAN *de projection.* Vertical plane.
PLAN, also signifies a tier; speaking of the stowage of the hold. *Ex.*
Premier PLAN. The ground tier.
Second PLAN. The second tier.
Troisième PLAN. The third tier.
PLAN *supérieur.* The upper tier.
PLAN *du milieu.* The middle tier.
PLANCHE, *s. f.* PLANCHE *à débarquer.* The gang-board of a boat.
PLANCHES *à feu.* Boards put on a vessel's bottom, to prevent the flames rising beyond them, while the ship is breaming.
PLANCHES *de roulis.* Side-boards of a standing bed-place.
PLANCHER, *s. m.* A plat-form.
PLANCHERS *de la cale.* Platforms of the hold.
PLANCHER *de la soute aux poudres.* The platform of the magazine.
PLANCHER *de la fosse aux cables.* A platform for the cables.
PLANCHER *des malades.* The cockpit.
PLANCHER *de la poulaine.* The gratings of the head.
PLANÇON, *s. m.* Plank-timber.
PLANÉ, *adj. Carte* PLANE. A plane chart.
Navigation PLANE. Plane sailing.
PLAQUE, *s. f.* PLAQUE *de plomb.* Sheet-lead, which is used for several purposes on board ship.
PLAT, *s. m.* PLAT *de maîtres.* A warrant officer's mess.
PLAT *de matelots.* A sailor's mess.
PLAT *des malades.* The sick mess.
PLAT *de la varangue.* The flat or horizontal part of the floor-timbers.
PLAT *d'un aviron. Voyez* AVIRON & PELLE.
☞ Les avirons sur le PLAT! *Rest upon your oars!*
PLAT, *adj. Bâtiment à fond* PLAT. A flat-bottomed vessel.
Bateau PLAT. A flat-bottomed boat.
Bâtiment à varangues PLATES. *Voyez* BÂTIMENT.
PLAT-BORD, *s. m.* Gunwale, or gunnel. *Ex.*
Avoir le PLAT-BORD *à l'eau.* To have gunnel-in, or gunnel-to.
PLATE-BANDE, *s. f.* PLATE-BANDE *d'affût.* A cap-square, or clamp of a gun-carriage.
PLATE-FORME, *s. f.* A plat-form.
PLATINE, *s. f.* PLATINE *de canon.* The apron of a gun.
PLEIN, *adv. Porter* PLEIN. To keep the sails full.
☞ *Porte* PLEIN! *Voyez* PORTER.
PLEIN, *adj.* PLEINE *mer.* High-water.
En PLEINE *mer.* In the offing.
PLI, *s. m.* PLI *de cable, ou de manœuvre.* The fake of a cable, or rope.
PLIER, *v. n.* To heel, or lie along, &c. *Ex.*
Vaisseau qui PLIE. A ship that heels.

Faire PLIER *l'ennemi.* To make the enemy bear up.
PLOC, *s. m.* The hair and tar put between the bottom planks of a ship and the sheathing, in order to preserve the bottom from the worms.
PLOMB, *s. m.* The lead. *Ex.*
Le grand PLOMB *de sonde.* The deep-sea lead.
Petit PLOMB *de sonde.* The hand-lead.
PLONGER, *v. a. & n.* To immerse any thing into the water; to plunge, or dive into the water.
Le canon PLONGE. The gun is pointed too low.
PLONGEUR, *s. m.* A diver.
PLOQUER, *v. a.* To apply the sheathing-hair to a ship's bottom.
POGE! (In the language of the Mediterranean). The command to put the helm a-weather, in order to reef the sails.
POINT, *s. m.* Day's-work; also the ship's place pricked on a nautical chart.
Faire son POINT. To calculate the day's-work of a ship.
Etre à terre par son POINT. To be a-shore by one's reckoning.
POINT *de partance.* The place from which the departure is taken.
POINT *vélique, dans une route directe.* A point in a perpendicular, raised from the centre of gravity of the horizontal surface of the ship at the floating line, that is intersected by the direction of the absolute impulse of the sea on the vessel's head in a direct course.
POINT *d'une voile.* The clew of a sail.
POINT *du vent d'une voile.* The weather-clew of a sail; also, the tack of the main or fore-sail.
Contre-POINT. The double rope and marling on the clew of a sail.
POINTE, *s. f.* A point, or fore-land.
POINTE *de bouline.* Sharp-trimmed, or with sharp bowlines.
POINTE *de compas.* *Voyez Quarts du* COMPAS.
POINTE *de l'éperon.* The beak of a prow, or cutwater.
POINTER, *v. a.* To point, &c. *Ex.*
POINTER *les canons.* To take aim with the guns.
POINTER *à couler bas.* To aim at sinking, or between wind and water.
POINTER *en chasse.* To point before the beam.
POINTER *en belle.* To point quite strait.
POINTER *en retraite.* To point abaft the beam.
POINTER *à l'horizon* To point horizontally.
POINTER *à démâter.* To point at the masts.
POINTER *la carte.* To prick the chart.
POINTURE, *s. f.* The balance of a sail.
POINTURE *de ris.* *Voyez* RIS *des voiles.*
POIX, *s. f.* POIX *navale.* Tar; naval pitch. *Voyez* GOUDRON.
POLACRE, *s. f.* A polacre, or ship so called.
POLEGONE, *s. m.* A sort of sham battery, where young quarter-gunners are trained to the exercise of gunnery.
POLICE, *s. f.* POLICE *d'assurance.* Policy of insurance.
POLICE *de chargement.* A bill of lading.
POMME, *s. f.* A truck or acorn, also a mouse. *Ex.*
POMME *de pavillon.* The truck on the top of a flag-staff, or ensign-staff.
POMME *de flammes.* The acorn of a pendant-staff.
POMME *de girouette.* The trucks of the mast-heads.
POMME *d'étai.* The mouse of a stay.
POMMES *de tournevire.* The mouses of the messenger, but which are not raised on the messengers of English men of war.
POMMES *de racage.* *Voyez* RACAGE.
POMMES *gougées.* Seizing-trucks.
POMOYER, *v. n.* POMOYER *un cable.* To underrun a cable with the long-boat.
POMPE, *s. f.* A pump. *Ex.*
POMPE *ordinaire.* A common pump.
POMPE *à chapelet.* Chain-pump.
POMPE *de l'avant.* Head-pump.
POMPE *à main.* Hand-pump.
☞ A la POMPE! *Pump ship!*
La POMPE *est prise.* The pump is fetched.
Chopine de POMPE. *Voyez* CHOPINE.
Heuse de POMPE. *Voyez* HEUSE.
La POMPE *est franche.* The pump sucks.
La POMPE *se décharge.* The pump has lost water.
Verge de POMPE. Pump-spear.
Charger la POMPE. *Voyez* CHARGER.
POMPE *engorgée.* A choked pump.
POMPE *éventée.* A pump that blows, being split.
POMPE *d'incendie.* A fire-engine.
POMPER, *v. n.* To pump.
PONENT, *s. m.* The West (in the language of the Mediterranean); also the western coasts of France; and thus in the same dialect,
PONENTÉ-*lébèche.* The south-west wind.
PONENTÉ-*mistral.* The west north-west.
PONENTAIS, *s. m. p.* A name given to the seamen of the western coasts of France.
PONT, *s. m.* A deck.
Premier PONT. The lower gun-deck.
Second PONT. The middle gun-deck.
Troisième PONT. The upper gun-deck.
Faux-PONT. The orlop-deck.
PONT *arqué.* A cambered-deck.
PONT *coupé.* A deck open amidships.
PONT *de cordes.* Netting to cover a ship's waist, and prevent the impression of boarders.
PONT *entier.* A flush-deck.
PONT *sur gueule.* A spar-deck run a-cross from the quarter-deck to the forecastle of a deep-waisted vessel.
PONT *à caillebotis.* A grating deck.
PONT *volant.* A hanging stage.
PONTÉ, *adj.* Decked. *Ex.*
Bâtiment PONTÉ. A decked vessel.
PONTON, *s. m.* A sort of lighter; also a pontoon. *Ex.*
PONTON *de carénage.* A ponton for careening ships.
PONTON, is also a condemned ship without masts, &c.
Taquets de PONTON. *Voyez* TAQUETS, &c.
PONTONAGE, *s. m.* The hire of a pontoon.
PORQUES, *s. f. p.* Riders. *Ex.*
PORQUES *acculées.* The aftermost and foremost riders.
PORQUES *de fond.* The midships riders.
Varangues de PORQUES. *Voyez* VARANGUES, &c.
Genoux de PORQUES. *Voyez* GENOUX, &c.
Alonges de PORQUES. *Voyez* ALONGES, &c.

Aiguillettes de PORQUES. *Voyez* AIGUILLETTES, &c.
PORT, *s. m.* A port, or harbour.
PORT *brut.* A harbour made by nature.
PORT *de barre.* A harbour with a bar that can only be passed at or near high water.
PORT *de mer franc.* A free sea-port.
PORT *d'un bâtiment.* The tonnage of a vessel.
PORT *marchand.* A harbour for merchant ships.
PORT *de l'état.* A harbour for men of war.
PORT *de rivière.* A river harbour.
Donner dans un PORT. To get into a harbour.
PORTANT, *s. m. Le* PORTANT *de la quille sur terre.* The length of the keel on a right line.
PORTE, *s. f.* A door.
PORTES *de proue.* Head-doors of a ship.
PORTES *d'un bassin.* Gates of a dock.
PORTE-BOSSOIR, *s. m.* The supporter of the cat-head.
PORTE-COLLIERS, ou *Taquets de Mâts, s. m. p.* The cleats on the lower masts, serving to support the collars of the stays.
PORTE-HAUBANS, *s. m. p.* Chain-wales, or channels.
Grands PORTE-*haubans.* Main-chains.
PORTE-*haubans de misaine.* Fore-chains.
PORTE-*haubans d'artimon.* Mizen-chains.
PORTE-VOIX, *s. m.* A speaking-trumpet used at sea.

PORTÉE, *s. f.* (In gunnery). Range. *Ex.*
PORTÉE *de canon.* The range of a cannon shot.
PORTÉE *des mortiers.* Range of mortars.
PORTÉE *entière. Voyez* VOLÉE.
PORTÉE *de but en blanc.* Point-blank range.
Etre à PORTÉE. To be within range.

À demi-PORTÉE. Within half shot.

À PORTÉE *de fusil.* Within musket shot.

À PORTÉE *de pistolet.* Within pistol shot.

À PORTÉE *de la voix.* Within hearing.
Se tenir à PORTÉE *de la voix.* To keep within hail.
Se tenir à PORTÉE. To keep within gun-shot.
Se tenir hors de PORTÉE. To keep out of gun-shot.
PORTER, *v. a. & n.* To stand, to carry, &c. *Ex.*
PORTER *à l'est.* To stand to the east.
PORTER *en route. Voyez* ROUTE, &c.
PORTER *au vent, ou sous le vent d'un navire.* To look up for going to windward, or to leeward of a ship.
Faire PORTER. To keep the sails full.
Les voiles PORTENT. The sails draw.
Voile qui ne PORTE *pas.* A sail which does not draw.
☞ Fais PORTER! *Keep her full!*
Navire qui PORTE *bien la voile.* A ship that carries her sails well.
Bâtiment qui PORTE *mal la voile.* A crank ship.
Vaisseau qui PORTE *ses quatre corps de voile.* A ship that carries her courses, main and fore top-sails.
PORTER *les huniers hauts.* To carry the top-sails a-trip.
PORTER *les perroquets.* To carry top-gallant sails.
PORTER *sur l'ennemi. Voyez* ENNEMI, &c.
PORTER *des dépêches à un général.* To carry dispatches to an admiral.
PORTER *des troupes, des munitions, &c.* To carry troops, stores, &c.
PORTER *une amarre au large.* To run out a hawser.
PORTER *une ancre au large.* To carry out an anchor.
Le canon PORTE. The shot falls aboard.
Le canon ne PORTE *pas.* The shot falls short.
PORTER *du vingt-quatre.* To carry twenty-four pounders; speaking of the battery of a ship.
PORTUGAISE, *s. f.* The cross lashing of the head of a pair of sheers.
POSTE, *s. m.* Birth, station, &c. *Ex.*
POSTE *des chirurgiens.* The surgeon's mates birth.
POSTE *des malades.* The sick-birth.
POSTES *de combat.* Quarters.
☞ Chacun à son POSTE! *Every man to his quarters!*
POSTE *d'un vaisseau.* The station of a ship in a fleet or squadron.
Rôle de POSTE. *Voyez* RÔLE, &c.
Vaisseau qui a quitté son POSTE. A ship that has abandoned her station.
Vaisseau qui rejoint son POSTE. A ship crowding to her station.
Vaisseau à son POSTE. A ship in her station.
Mettre les ancres à POSTE. To stow the lower anchors to the bow.
Tenir son POSTE. To keep one's station.
Prendre son POSTE *dans la ligne.* To get into one's station in the line.
POT, *s. m.* POT *à brai.* A pitch-pot.
POT *à feu.* A fire-pot, or stink-pot.
POTENCE, *s. f.* Gallows-bits.
POTENCE *de pompe.* The fulcrum of a common pump.
Taquets de POTENCE. *Voyez* TAQUETS, &c.
POUDRE, *s. f.* POUDRE *à canon.* Gun-powder.
POUILLEUSE, *s. f.* The main stay-sail.
POULAINE, *s. f.* The head of a ship.
Ornemens de la POULAINE. The mouldings or carved-work of the head.
POULIE, *s. f.* A block.

Parties d'une POULIE.	The parts of a block.
Les corps de la POULIE.	The shell.
Les joues.	The cheeks.
Le cul	The arse.
La goujure.	The score.
Le canal, ou clan.	The sheave-hole.
Le rouet.	The sheave.
L'essieu.	The pin.

POULIE *simple.* A single block.
POULIE *double.* A double-block.
POULIE *à trois rouets.* A treble-block.
POULIE *d'étai à quatre rouets.* A block with four sheaves for the main-stay.
POULIE *simple à fouet.* A tail-block.
POULIE *simple à croc.* A single hook-block.
POULIE *double de palan.* A long tackle-block.
POULIE *coupée, ou à dent.* A snatch-block.
POULIE *de conduit, ou de retour.* A leading block.
POULIE *de guinderesse.* A top-block.
POULIE *à tourniquet.* An iron-bound block, with a swivel-hook.
POULIE *de bout de vergue.* A topsail-sheet block.
POULIE *de sous-vergue.* A clew-line block.
POULIE *de cargue-point.* A clew-garnet block.

Poulie *de balancine.* A lift-block.
Poulie *de capon.* *Voyez* Capon.
Poulie *de drisse à caliorne.* A jear-block.
Poulies *d'itague de drisse des huniers.* Bollock-blocks.
Poulie *plate pour itague des huniers et têtes de mâts.* A tie-block.
Poulie *simple estropée à œillet.* An eye block.
Poulie *double à canon.* A double-block for gun-tackles.
Poulie *de caliorne à trois rouets.* A Winding-tackle block with three sheaves.
Poulie *de caliorne à deux rouets.* A winding-tackle block with two sheaves.
Poulie *de retour pour caliorne.* The quarter-block for the main winding-tackle.
Poulie *de grande drisse à trois rouets.* The main jear-blocks with three sheaves.
Poulie *de drisse latine.* The jear-blocks in lateen vessels.
Râteau en poulies. *Voyez* Râteau, &c.
Poulie *à trois rouets pour apparaux de carène dans les ports.* Careening-blocks.
Poulie *à moque.* *Voyez* Moque.
POULIERIE, *s. f.* The shed where blocks are made.
POULIEURS, *s. m. p.* Block-makers.
POULIOTS, *s. m. p.* Pouliots *d'amure.* Chess-trees.
POUPE, *s. f.* The stern of a ship.
Bâtiment à poupe *quarrée.* A square-sterned ship.
Bâtiment à poupe *étroite.* A pink-sterned vessel.
Passer en poupe *d'un navire.* To pass a-stern of a ship.
Naviguer avec vent en poupe. To bear away before the wind.
POURSUIVRE, *v. a.* Poursuivre *l'ennemi.* To chase the enemy.
POUSSER, *v. a.* Pousser *au large.* To shove off.
PRAME, *s. f.* A pram, or sort of lighter; also a sloop of war, used formerly by the French, mounted with one tier of heavy guns, and, on account of its peculiar construction, used in shallow water, rivers, &c.
PRATIQUE, *adj.* *Etre* pratique *d'un lieu.* Is said of a pilot who is well acquainted with the bearings, soundings, marks, &c. of a particular place.
PRATIQUER, *v. a.* To have free intercourse with the inhabitants of a country where a contagious disease prevails, either before or after having performed quarantine.

PRÉCEINTES, *s. f. p.* The wales of a ship.
Préceinte-*basse.* The main wale
Troisième & quatrieme préceinte. The channel-wale, which is separated in two, in French line-of-battle ships.

PRÉLAT, *s. m.* A tar-pawling.
Prélats *de bastingage.* Hammock-cloths.
Prélat *à canot.* A boat's covering.
PRENDRE, *v. a.* To take, &c. *Ex.*
Prendre *hauteur.* To take the altitude of the sun, or a star.
Prendre *un mouillage.* *Voyez* Mouillage.
Prendre *du lest.* To take in ballast.
Prendre *les amures.* To get a-board the tacks.
☞ Prenez deux ris dans chaque hunier! *Take two reefs in the fore and main-top-sails!*
Prendre *vent devant.* To be taken with the wind a-head; or to chapel a ship.
Prendre *des prisonniers à son bord.* To take prisoners on board.
Prendre *des ris.* To reef, or take in reefs.
Prendre *possession d'une prise.* To take possession of a prize.
Prendre *son artillerie et ses poudres.* To take on board the guns and powder.

PRÈS, *adv.* Near; close. *Ex.*
*Vaisseau orienté au-plus-*près. A ship close hauled.
☞ Près & plein! *Full and by!* The order to the steersman to keep the ship close to the wind.
Ligne du plus près. *Voyez* Ligne, &c.
*Etre au-plus-*près. To be sharp trimmed.
☞ Gouverne au-plus-près! *Keep her close!*

PRÉSENTER, *v. n.* To lie up, &c. *Ex.*
Ce bâtiment présente *au vent deux quarts plus que nous.* That vessel lies up two points to windward of us.
Présenter *la grande bouline.* To snatch the main bowline, or to put it into the snatch-block.
Présenter *au vent.* To stem the wind, or to sail so as the ship stems, without making lee-way.
Présenter *au courant.* To stem the tide.
Présenter *à la lame.* To bow the sea.
Présenter *le travers.* To shew the broadside or length of a ship.

PRESQU'ÎLE, A peninsula.
PRESSE, *s. f.* Press, or impress.
Détachement de la presse, *ou gendarmes de la marine.* Press-gang.
Matelots provenant de la presse. Pressed men.

PRÊTER, *v. a.* Prêter *le côté à l'ennemi.* To range-abreast of a ship, ready for action.

PRÉVÔT, *s. m.* Prévôt *général de la marine.* A provost martial of the navy.
Prévôt *marinier.* The swabber of a ship, who also chastises the criminals: this part of his duty is performed in English ships by the boatswain's mate.
PRIME, *s. f.* Prime *d'assurance.* Premium of insurance.
PRISE, *s. f.* A prise, or a ship taken from the enemy at sea.
Prendre possession d'une Prise. *Voyez* Prendre, &c.
Amariner une prise. *Voyez* Amariner.
Part de prise. Prize-money.
PRISONNIER, *s. m.* A prisoner.
Fer de prisonniers. *Voyez* Fer, &c.
PROLONGER, *v. a.* Prolonger *une côte, &c.* To sail along a coast, &c.
Prolonger *un navire.* To lay a ship along-side of another.
Prolonger *la civadière.* To set the sprit-sail-yard along-side of the bowsprit.
PROMONTOIRE, *s. m.* A cape, head-land, or fore-land.
PROS, *s. m.* A proa, or boat peculiar to the La-

drones or Marianne Islands, and described in Lord Anson's voyage.

PROUE, *s. f.* The prow, or beak; also the head part of a ship.

PROVISIONS. *Voyez* VIVRES.

PRUD-HOMMES, *s. m. p.* A certain number of men chosen among the most ancient and reputable fishermen, to have an authority over the rest, and to settle the differences that may arise among them.

PURGER, *v. a.* PURGER *la côte.* To scour a coast.

PUITS, *s. m.* A well, or pit.

PUITS, *ou fosse sur un fond.* A pit, or well on a bank.

Q.

QUAI, *s. m.* A quay, or wharf, on the side of a harbour or river.

Amarré à QUAI, *rangé à* QUAI. Moored along-side of a quay or wharf.

Maître de QUAI. A wharfinger.

Droits de QUAI. Wharfage.

QUALITÉS D'UN BÂTIMENT, *s. f. p.* The qualities or properties of a ship.

QUARANTAINE, *s. f.* Quarantine.

Faire QUARANTAINE. To perform quarantine.

Prendre QUARANTAINE. To begin quarantine.

QUARANTENIER, *s. m.* Ratling stuff, or ratline, used as a lashing, &c.

QUARRÉ, *s. m.* A square.

QUARRÉ *naval.* The naval square, or a scheme formed on a ship's quarter-deck, to represent the division of a fleet into three columns; it is used to regulate the movements of each ship with regard to the rest, and preserve the whole fleet in uniformity.

QUARRÉ. A sledge. *Voyez* CARRÉ.

Voiles QUARRÉES. Square sails.

Vergues QUARRÉES. Square yards.

Brasser QUARRÉ. To square the yards.

QUART, *s. m.* The sea-watch.

QUART *de tribord.* The starboard watch.

QUART *de babord.* The larboard watch.

Officier de QUART. *Voyez* OFFICIER.

Faire le QUART. To keep the watch.

Commencer le QUART. To set the watch.

Relever le QUART. *Voyez* RELEVER.

☞ Bon QUART! *All's well!*

Faire bon QUART *sur la hune.* To keep a good look-out in the tops.

☞ Fais bon QUART d'avant! *Look well out afore!*

Le QUART *de la diane.* The day-break watch.

Le QUART *de quatre à huit heures du soir.* The dog-watch.

Rôle de QUART. *Voyez* RÔLE.

QUART *de vent.* A point of the compass.

QUARTIER, *s. m.* QUARTIER *Anglois, ou* QUARTIER *nonante.* A quadrant.

QUARTIER *de réduction.* A sinical quadrant, used by the French in working their day's works.

QUARTIER *sphérique.* A spherical quadrant.

Vent de QUARTIER. *Voyez* VENT, &c.

Pièces de QUARTIER. Harpings.

QUARTIER-*maîtres, s. m. p.* Quarter-masters.

QUENOUILLETTE, *s. f.* QUENOUILLETTE *de trélingage.* A foot-hook, or futtock-staff.

QUÉRATS, *s. m. p.* Shares. In the Levant trade, the property of a ship is divided into shares, according to which the master and crew of a vessel are paid by the owners.

QUÊTE, *s. f.* QUÊTE *de l'étambot.* The rake of the stern-post.

QUÊTE *de la poupe.* The length of the rake abaft.

QUEUE, *s. f.* QUEUE *de rat.* The pointing of a cable.

QUEUE *d'un cordage.* Rope tapering towards the end.

Grelin en QUEUE *de rat.* *Voyez* GRELIN.

Epissure en QUEUE *de rat.* A pointed splice.

QUEUE *d'hirondelle.* A dove-tail scarph.

QUEUE *d'une escadre.* The rear of a fleet of men of war.

Vaisseau de QUEUE. The aftermost ship.

QUEUE *d'une rade.* The innermost part of a road-stead.

QUILLE, *s. f.* *La quille d'un bâtiment.* The keel of a ship.

Fausse-QUILLE. The false-keel.

Contre-QUILLE. *Voyez* CONTRE-*Quille.*

Écarts de la QUILLE. *Voyez* ÉCARTS, &c.

Doublage de la QUILLE. *Voyez* DOUBLAGE.

Rablure de la QUILLE. *Voyez* RABLURE.

Hauteur de la QUILLE. The depth of the keel.

Largeur de la QUILLE. The thickness of the keel.

Largueur de la QUILLE *portant sur terre.* *Voyez* PORTANT.

QUINÇONNEAU, *s. m.* A toggle.

QUITTER, *v. a.* To leave; to give up, &c. *Ex.*

L'ancre à QUITTÉ. The anchor is a-trip.

QUITTER *un ancrage, &c.* To leave an anchorage, &c.

QUITTER *le commandement d'un bâtiment.* To give up the command of a ship.

R.

RABAN, *s. m.* A general name given to an earing, rope-band, knittle, &c. *Ex.*

RABANS *de pointure.* Earings.

RABANS *de pointure de ris.* Reef-earings of a sail.

RABANS *de têtière, ou d'envergure.* Rope-bands of the head of a sail.

RABANS *de ferlage.* Gaskets.

RABAN *de pavillon.* The rope-band of a flag.

RABANS *de sabords.* Port-ropes, or small ropes lashed to the inner rings of the port-lids.

RABANER, *v. a.* To fit a sail with rope-bands and earings, ready for bending to its yard.

RABATTUES, *s. f. p.* (In naval architecture) Hances, or falls of the rails, from the stern and forecastle to the gangways.

RABLURE, *s. f.* A rabbet, or channel. *Ex.*

RABLURE *de la quille.* The rabbet cut in the keel, stem, and stern-post, to receive the edges of the garboard-streaks, and the ends of the planks afore and abaft.

RABLURE *de l'étambot.* The rabbet of the stern-post.
RABLURE *de l'étrave.* The rabbet of the stem.
RABLURE *de la lisse d'hourdi.* The rabbet of the wing-transom.
RABOUGRI, *adj.* Cross-grained, knotty; a term applied by shipwrights to timber, which by this quality, is not fit for ship-building.
RACAGE, *s. m.* A parral with ribs and trucks.
Bâtard de RACAGE. *Voyez* BÂTARD.
Drosse de RACAGE. *Voyez* DROSSE.
RACAGE *simple, ou à l'Anglois.* Truss-parral.
Pommes de RACAGE. Trucks of a parral.
Bigots de RACAGE. *Voyez* BIGOTS, &c.
Palan de drosse de RACAGE. *Voyez* PALAN, &c.
Grand RACAGE. Main parral.
RACAGE *de misaine.* Fore parral.
RACAMBEAU, *s. m.* A traveller, or a slender iron ring, which sometimes encircles the mast of a long-boat, serving as a parral to the yard or gaff.
RACAMBEAU *du grand foc.* The jib traveller.
RACCOURCIR, *v. a.* To shorten. *Ex.*
RACCOURCIR *un mât.* To shorten a mast.
RACCOMMODER, *v. a.* To repair. *Ex.*
RACCOMMODER *le gréement.* To repair the rigging of a ship.
RÂCLE, *s. f.* A scraper. *Voyez* GRATTE.
RÂCLER, *v. a.* To scrape the sides, &c. of a ship.
RADE, *s. f.* A road, or roadstead.
Petite RADE. A little or inner road.
Grande RADE. A great or outer road.
RADE *foraine.* A free or open roadstead.
Etre en RADE. To be at anchor in a road.
Aller en RADE. To go into a roadstead.
RADEAU, *s. m.* A raft in general.
RADEAU *de bois de construction.* A raft of timber for ship-building.
RADEAU *de mâts.* A raft of masts.
RADEAU, *ou ras de carène.* *Voyez* RAS.
RADER, *v. a. & n.* To arrive in a road.
RADER *un vaisseau*; *le mettre en* RADE. To bring a ship into a road; to anchor in a road.
RADIER, *s. m.* An apron in a dock.
RADOUB, *s. m.* A repair; speaking of a ship's hull.
Bâtiment en RADOUB. A ship repairing.
RADOUBER, *v. a. & r.* To repair.
Se RADOUBER. To be in repair; speaking of a ship's timbers and planks.
Nous nous RADOUBONS. Our ship is repairing.
RAFALE, *s. f.* A sudden and violent squall of wind.
Le vent souffle par RAFALE. The wind blows by squalls.
RAFRAÎCHIR, *v. a.* To freshen, &c. *Ex.*
RAFRAÎCHIR *le cable.* To freshen the hawse by a renewal of service.
RAFRAÎCHIR *les canons.* To cool the guns in time of action, as with a wet sponge, sometimes dipped in vinegar.
RAFRAÎCHISSEMENS, *s. m. p.* A supply of fresh provisions; also the choice stores supplied for the use of the sick people on board.
RAGUÉ, *adj.* *Cable* RAGUÉ. A chafed cable.
RAISONNER, *v. n.* *Faire* RAISONNER *un navire.* To oblige a vessel to come near and speak.
RALINGUE, *s. f.* The bolt-rope of a sail.
RALINGUE *de têtière.* Head-rope.
RALINGUE *de chûte.* Leech-rope.
RALINGUE *de fond.* Foot-rope.
Aiguille à RALINGUE. *Voyez* AIGUILLE.
En RALINGUE, *adv.* *Mettre une voile en* RALINGUE. To shiver a sail in the wind.
☞ Mets en RALINGUE! *Luff, or shake her up in the wind! Let the sails touch!*
RALINGUER, *v. a. & n.* To lift and shiver; speaking of the sails.
RALINGUER *une voile.* To sew the bolt-ropes to a sail.
RALLIEMENT, *s. m.* *Signal de* RALLIEMENT. A signal given by the commander of a squadron, which calls the different ships of his squadron to resume their respective stations.
RALLIER, SE RALLIER, *v. a. & r.* To rally, &c.
RALLIER *la terre.* To stand in for the land.
RALLIER *un navire au vent.* To haul the wind again, or bring a ship to the wind after she has yawed to leeward.
RALLIER *un bâtiment.* To stand towards a ship.
Se RALLIER. To approach any object at sea.
RAMBADES, *s. f. p.* Two posts or platforms in the fore-part of a row-galley, whereon men are placed in time of action, to fi e with small-arms.
RAMBERGE, *s. f.* A sort of packet-boat, advice-boat, or tender.
RAME, *s. f.* An oar. *Voyez* AVIRON.
☞ Lève RAMES! *Unship the oars!*
RAMER, *v. n.* To pull. *Voyez* NAGER.
RAMEUR, *s. m.* A rower.
RANCÉ, *adj.* *Bâtiment* RANCÉ. An old ship, strengthened with timbers, fayed on from the sheer-rail to the top of the bends.
RANÇON, *s. f.* A ransom.
RANÇONNER, *v. a.* RANÇONNER *un bâtiment.* To ransom a ship or vessel.
RANG, *s. m.* Rank, or the rate of ships and vessels of war. *Ex.*
RANG *des vaisseaux et bâtimens de guerre François.* The rate of French ships and vessels of war.
Premier RANG. First rate, which comprehends all ships of three decks, from 110 guns upwards.
Second RANG. Second rate, ships of two decks of 80 and 84 guns.
Troisième RANG. Third rate, of 74 guns.
Quatrième RANG. Fourth rate, consists of frigates, from 38 to 44 guns.
Cinquième DANG. Fifth rate, from 32 to 36 guns.
Sixième RANG. Sixth rate, is composed of corvettes of 18 guns upwards.
Septième RANG. Seventh rate, includes advice-boats, brigs, cutters, luggers, schooners, &c.
Huitième RANG. Eighth rate, consists of armed store-ships and transports.
RANG *de canons.* A tier of guns.

RANG *d'un cable, ou d'autre cordage.* A tier of a cable, &c.
RANG *de rameurs.* A bank of rowers.
RANGER, *v. a. & r.* To range. &c. *Ex.*
RANGER *le terre, ou la côte.* To range along the coast; to coast.
RANGER *le vent.* To claw the wind, or haul close to the wind.
Le vent se RANGE *de l'avant.* The wind hauls forward; the wind heads us, or takes us a-head.
Le vent se RANGE *de l'arrière.* The wind draws aft.
Le vent se RANGE *au nord.* The wind veers to the north.
RANGER *à l'honneur d'un écueil.* To give a good birth to a shoal.
RANGER *du monde sur une manœuvre.* To man a rope.
Faire RANGER *du monde sur les vergues.* To man the yards.
☞ RANGE à carguer la grande voile! *Clap on the main clew-garnets briskly!*
Se RANGER *sur une vergue.* To lay out upon a yard.
RAPIDE, *adj. Le cours* RAPIDE *d'un fleuve.* The rapidity of a river; a fresh in a river.
RAPIQUER, *v. a.* RAPIQUER *au vent.* To bring a ship to the wind, after having bore up.
RAQUE. A general name for trucks, but particularly the trucks of a parral. *Voyez Pomme de* RACAGE.
RAQUE *de haubans.* A truck lashed to the shrouds.
RAQUE *encouchée.* A truck incircled with a notch, to receive the spun-yarn, &c.
RAQUE *gougée.* A truck hollowed on one side.
RAS, *s. m.* RAS *de carène.* A punt, or floating-stage.
RAS *de marée.* A tide-way, or tide-gate.
Le RAS *de Portland.* The race of Portland.
Le RAS *des Saints.* The race of the Saints, near Brest.
RAS, *adj. Vaisseau* RAS, *ou* RAS *à l'eau.* A low-built vessel; a strait-sheered ship.
Accastillage RAS. Upper-works strait-sheered and with low drifts.
RASÉ, *adj. Vaisseau* RASÉ. A razee, or a 74-gun ship which has been cut down, and thus become a frigate of the largest class.
RASER, *v. a.* RASER *un vaisseau.* To cut down a ship, or take off part of her upper-works, as the poop, quarter-deck, and forecastle.
RAT. *Voyez* QUEUE *de rat.*
RAT *d'eau.* A shipwright's floating stage.
RÂTEAU, ou RÂTELIER, *s. m.* A rack, &c. *Ex.*
RÂTEAU *en poulies.* A rack with running sheaves.
RÂTELIER *du beaupré.* The rack or range of blocks placed on each side of the gammoning of a ship's bowsprit.
RÂTEAU, ou RÂTELIERS *à chevillots.* The cross-piece of the windlass, or any ranges fixed to the shrouds, &c. in which belaying-pins are fixed.
RATION, *s. f.* The daily allowance of provisions distributed to the different messes in a ship.
*Demi-*RATION. Half-allowance.
RATION *et demie.* The allowance of a naval officer in the French fleet.
Double RATION. Double allowance, given on any particular occasion.
*Mettre l'équipage à la demi-*RATION. To put the crew to half-allowance.
RAVALEMENT, *s. m.* The fall of a ship's deck.
REÂLE, *adj. Galère* RÉALE. The royal galley; a name given to the principal galley of a kingdom.
REBANDER, *v. a. & n.* To carry over to the other side of the ship.
REBANDER *à l'autre bord.* To stand upon the other tack.
REBORDER, *v. a.* To board, or fall a-board of a ship a second time.
REBOUSSE. *Voyez* REPOUSSOIR.
DE REBUT, *adv. Bois de* REBUT. Refuse timber, or that which is unfit for service.
Canon de REBUT. A condemned cannon.
RECHANGE, *s. m.* A general name for the spare stores, rigging, sails, &c. of a ship, which are kept in reserve to supply what may be lost or disabled.
Les RECHANGES *du maître.* The boatswain's spare stores.
Les RECHANGES *du maître-canonnier.* The gunner's spare stores.
Voiles de RECHANGE. Spare sails.
Mâts et vergues de RECHANGE. Spare masts and yards.
RECLAMPER, *v. a.* To refish a mast or yard when it is sprung.
RECONNOISSANCE, *s. f.* A sea-mark; also the act of reconnoitring.
Signaux de RECONNOISSANCE. Reconnoitring signals.
RECONNOÎTRE, *v. a.* To reconnoitre, or take a view. *Ex.*
RECONNOÎTRE *un navire.* To approach a vessel, in order to discover her strength, and of what nation she is.
RECONNOÎTRE *une terre.* To survey the situation of a coast attentively.
RECOURIR, *v. a.* RECOURIR *les coutures d'un vaisseau.* To run over the seams of a ship, in calking; to calk them lightly and expeditiously.
RECOURIR *un cable, &c.* To under-run a cable, &c.
RECOUVRER, *v. a.* To rowse in, or haul any rope into the ship, when it hangs slack in the water, &c.
☞ RECOUVRE! *Rowse-in!* or *haul aboard!*
RECUL, *s. m.* RECUL *d'un canon.* The recoil of a gun.
Palan de RECUL. A relieving tackle for guns.
REDRESSE, *s. f. Cable de* REDRESSE. A relieving rope.
REDRESSER, *v. a.* REDRESSER *un vaisseau.* To right a ship that has been hove down.
REFLUX, *s. m.* Ebb, or ebb-tide. *Voyez* FLUX.
REFONDRE, *v. a.* REFONDRE *un vaisseau.* To give a ship a thorough repair.
REFONTE, *s. f.* A thorough repair.
Etre en REFONTE. To be undergoing a thorough repair.
REFOULER, *v. a.* REFOULER *la marée.* To stem the tide, or sail against its direction.

REFOULER *la charge d'un canon.* To ram home the charge of a gun.
REFOULOIR, *s. m.* The rammer of a cannon.
REFOULOIR *de corde.* A rope rammer, which is used in bad weather.

RÉFRACTION, *s. f.* Refraction.
REFUSER, *v. a.* To refuse, to scant or head; speaking of the wind. *Ex.*
Le vent REFUSE. The wind scants.
Le vaisseau a REFUSÉ. The ship would not come to the wind; or will not stay.
REGAGNER, *v. a.* To regain. *Ex.*
Après avoir croisé pendant quelques jours nous RÉGAGNÂMES *notre mouillage.* After having cruised a few days, we regained our anchorage.
REGISTRE, *s. m.* *Bâtiment de* REGISTRE. A Spanish register-ship.

RÉGLER, *v. a.* RÉGLER *un octant.* To regulate a quadrant.
Votre octant est-il bien RÉGLÉ? Is your quadrant well adjusted?

RÉGNANT, *part. pres.* *Vents* RÉGNANTS. The prevailing winds.

REGRÉER, *v. a.* To rig again.
REJOINDRE, *v. a.* To rejoin, &c. *Ex.*
REJOINDRE *une flotte, un convoi, &c.* To rejoin a fleet, convoy, &c.

RELÂCHE, *s. f.* The act of putting into a port; also the stay a ship makes in a port, when waiting for a fair wind, or for supplies.
Nous RELÂCHÂMES *à l'Ile de France en allant dans l'Inde.* We put in, or touched at the Mauritius on our passage to India.
Le Cape de Bonne Espérance est une bonne RELÂCHE. The Cape of Good Hope is a good place to touch at, or to put into to get refreshments.
Faire une RELÂCHE. To put into a port.

RELÂCHER, *v. a.* To touch at, or put into a port.
RELÂCHER *un bâtiment.* To permit a detained vessel to proceed on her voyage.

RELÈVEMENT, *s. m.* RELÈVEMENT *du pont d'un Vaisseau.* The sheer of a ship's deck, or the gradual rising of the deck afore and abaft.
RELÈVEMENT *d'une côte.* The survey of a coast.
Faire des RELÈVEMENS. To take bearings.
RELEVER, *Se* RELEVER, *v. a. & r.* To relieve; to set, &c. *Ex.*
RELEVER *une terre.* To take bearings.
RELEVER *un Vaisseau, ou tout autre objet avec la boussole.* To set a ship, or any other object, by the compass.
RELEVER *un vaisseau échoué.* To bring a ship a-float when lying a-ground.
RELEVER *les branles.* To lash up the hammocks, in order to make a clear passage between decks. *Voyez* BRANLE-BAS.
RELEVER *un bâtiment coulé.* To weigh a vessel that is sunk.
RELEVER *le quart.* To relieve the sea-watch.
RELEVER *le timonier.* To relieve the man at the helm.
RELEVER *une sentinelle.* To relieve a sentry.
RELEVER *une ancre.* To weigh an anchor again, in order to change its birth.
Se RELEVER *d'une côte.* To stretch from a shore where a ship has been embayed.
RELIGION, *s. f.* This term was understood for the Order of Malta, thus:
Vaisseau de la RELIGION. A Maltese ship.
Pavillon de la RELIGION. The Maltese colours.
REMBARQUER, SE REMBARQUER, *v. a. & r.* To re-embark. *Ex.*
Les troupes se REMBARQUÈRENT *dans la nuit.* The troops were re-embarked in the night.
REMBRAQUER, *v. a.* To haul taught; to haul in the slack of a cable or rope.
☞ REMBRAQUE le mou de l'aussière! *Haul in the slack of the hawser!*

REMÉDIER, *v. a.* REMÉDIER *à des voies d'eau.* To stop the leaks.
REMETTRE, *v. a.* REMETTRE *le quart.* To deliver up the charge of the watch at sea.
REMETTRE *à la voile.* To set sail again.
REMETTRE *un bâtiment à flot.* To bring a vessel a-float again.
REMISE, *s. f.* A shed.
REMONTER, *v. a. & n.* REMONTER *une rivière.* To sail up a river, or to tide it up.
REMONTER *au vent.* To sail to windward.
REMONTER *la côte.* To tide up the coast.
REMORQUE, *s. f.* A tow-rope.
Donner la REMORQUE. To tow.
Etre à la REMORQUE. To be in tow.
Tenir un bâtiment à la REMORQUE. To keep a vessel in tow.
Prendre à la REMORQUE. To take in tow.
REMORQUER, *v. a.* To tow a ship by a boat, or other small vessel with oars.
REMOUILLER, *v. a.* To anchor again.
REMOUILLER *une ancre.* To let an anchor go again.
REMOUX, *s. m.* The eddy, or dead-water, left behind a ship's stern when she is advancing under sail.
REMPLACEMENT, *s. m.* REMPLACEMENT *de vivres.* A supply of provisions.
REMPLISSAGE, *s. m.* Any piece of timber employed in filling up a vacancy, in ship-building.
Couples de REMPLISSAGE. Filling-timbers.
RENARD, *s. m.* A sort of handspike, or lever, with an iron claw, used to remove large pieces of timber, &c. in a dock-yard.
RENARD, is also a traverse-board.
RENCONTRER, SE RENCONTRER, *v. a. & r.* *Se* RENCONTRER *en mer avec un autre navire.* To fall in with another vessel at sea.
☞ RENCONTRE! *Shift the helm!*
☞ RENCONTRE à l'auloféc! *No nearer!*
☞ RENCONTRE à l'arrivée! *Veer no more!*
RENCONTRER. To shift the helm.
RENDEZ-VOUS, *s. m.* The rendezvous, or the place where ships are appointed to meet.
RENDRE, SE RENDRE, *v. a. & r.* To return; to surrender, &c. *Ex.*
RENDRE *un salut.* To return a salute.
Se RENDRE. To surrender. *Ex.*

La frégate la Chesapeake s'est RENDUE *au Shannon.* The Chesapeake surrendered to the Shannon.

RENFORCÉ, *adj.* *Canon* RENFORCÉ. A re-enforced gun.

RENFORCER, *v. a.* To re-enforce; to strengthen.

RENFORCER *une escadre.* To re-enforce a squadron.

RENFORCER *une vergue, &c.* To strengthen a yard, &c.

RENFORT, *s. m.* *Premier & second* RENFORT *d'un canon.* The first and second re-enforce of a gun.

SE RENGAGER, *v. r.* To engage again.

RENTRÉE, *s. f.* RENTRÉE *d'un vaisseau, ou rentrée des œuvres mortes d'un vaisseau.* The tumbling home of the top-timber, or housing in.

RENTRER, *v. a. & n.* RENTRÉR *dans un port.* To return to a port.

RENTRER *une bonnette.* To haul down a studding-sail.

RENTRER *les avirons.* To boat the oars.

RENTRER *les canons.* To run in the guns.

RENTRER *un cordage, &c.* To haul in a rope, &c.

RENTRER *un bâton de foc.* To rig in a jib-boom.

RÉPARER, *v. a.* RÉPARER *le gréement, &c.* To repair the rigging, &c.

REPASSER, *v. a.* REPASSER *des manœuvres.* To reeve running rigging, in lieu of any that has been cut in action.

RÉPÉTER, *v. a.* RÉPÉTER, *des signaux.* To repeat signals.

RÉPÉTITEUR, *s. m.* *Vaisseau* RÉPÉTITEUR, or simply RÉPÉTITEUR. The repeater, or ship that repeats the signals in a fleet or squadron.

RÉPONDRE, *v. a.* RÉPONDRE *aux signaux.* To answer the signals.

REPOUSSER, *v. a.* To repulse.

REPOUSSOIR, *s. m.* A driving-bolt, used by shipwrights to knock out some other bolt from its station.

REPRENDRE, *v. a.* REPRENDRE *une manœuvre.* To sheep-shank, or shorten a rope.

REPRENDRE *un palan.* To overhaul a tackle; to fleet a tackle.

REPRENDRE *les enfléchures des haubans.* To square the ratlings.

REPRENDRE *son poste.* To resume one's station.

REPRENDRE *une prise.* To recapture a prize.

REPRISE, *s. f.* A retaken ship.

RÉSERVOIR, *s. m.* RÉSERVOIR *d'un bateau pêcheur.* The well of a fishing-boat, wherein fish are kept alive.

RÉSINE, *s. f.* Rosin, used in paying a ship's sides or bottom.

RESSAC, *s. m.* Surf, or breaking of the waves upon the shores.

RESSIF, *s. m.* A reef or ridge of rocks lying under the water.

RESTAUR, *s. m.* The restoration, or loss made good by an insurer.

RESTER, *v. n.* To remain, to bear, &c.

RESTER *à l'ancre.* To remain at anchor.

RESTER *en station.* *Voyez* STATION.

RESTER *sous le feu d'un fort.* To remain exposed to the fire of a fort.

RESTER *en rade pour rallier un convoi.* To remain in a road-stead, in order to collect a convoy.

RESTER, implies also to bear upon any point of the compass, &c.

Le vaisseau nous RESTE *à l'est.* The ship bears east of us.

La terre nous RESTE *au sud.* The land bears south.

RESTER *à long pic.* To ride short.

RÉTABLIR, *v. a.* RÉTABLIR *l'ordre dans la ligne.* (In naval tactics). To form the line again, after it has been broken.

RETARDER, *v. a* To delay.

RETENIR, *v. a.* To detain.

Ce vent nous RETIENDRA *au mouillage.* This wind will keep us at anchor.

RETENUE, *s. f.* *Corde de* RETENUE. A guy, used to steady a heavy weight when hoisted into a ship.

Palan de RETENUE. A relieving tackle.

Palan de RETENUE *d'une voile aurique.* A sheet tackle in a lateen vessel.

RETORSOIR. *Voyez* MOULINET.

RETOUR, *s. m.* To return, &c. *Ex.*

Etre sur son RETOUR. To be homeward bound.

Vaisseau sur son RETOUR. A homeward-bound ship.

Les RETOURS Homeward-bound produce, or goods.

RETOUR *d'un cordage.* The hauling part of a rope that is lead through a leading block.

RETOUR *de la marée.* The turn of the tide, or the beginning of the ebb.

RETOUR *de courant.* The eddy of a stream.

Tenir à RETOUR *une manœuvre.* To hold on a rope with a turn.

Prendre un RETOUR *avec un cordage.* To take a turn with a rope, in order to hold fast.

Poulie de RETOUR. A quarter-block.

RETOURNER, *v. a.* To return, &c.

L'escadre vient de RETOURNER. The squadron has just put back.

RETRAITE, *s. f.* Retreat. *Ex.*

Canons de RETRAITE. Stern-chases, or those which are placed a-stern in men of war.

Sabords de RETRAITE. *Voyez* SABORDS, &c.

Coup de canon de RETRAITE. *Voyez* COUP, &c.

RETRAITE *de pirates.* A nest of pirates; a harbour of free-booters.

RETRANCHEMENT, *s. m.* RETRANCHEMENT *dans un vaisseau.* A temporary apartment formed in a ship on some particular occasion: also stopping part of the French sailors allowance of wine.

REVENTER, *v. a.* To fill the sails again; to brace about and fill.

REVERDIE, *s. f.* The increase of the tides immediately after the neaps.

REVERS, *s. m.* A general name for those pieces of timber, whose convexity lies inward in a ship's bottom or sides.

Alonges de REVERS. The top-timbers.

Genoux de REVERS. *Voyez* GENOUX, &c.

REVERS *d'arcasse.* The hollow of the stern timbers.

REVERS *du coltis.* The flaring of the knuckle-timber.
REVERS *de l'éperon.* The flaring of the cut-water.
Amures de REVERS. The lee-tacks.
Boulines de REVERS. *Voyez* BOULINE, &c.

Écoute de REVERS. The weather-sheet of a sail.
REVERSER, *v. a.* REVERSER *l'équipage d'un bâtiment sur un autre.* To turn over a ship's company from one vessel to another.
REVERSER *la cargaison d'un bâtiment dans un autre.* To unload a vessel, and put her cargo on board of another.
REVIREMENT, *s. m.* The act of going about again, either by tacking or veering.
REVIREMENT *par la tête, ou par la queue.* The tacking of a fleet or squadron by the van or rear, so that the foremost or aftmost ships go about first, to preserve the order of the line.
REVIRER, *v. n.* REVIRER *de bord.* To put about again, or bring the ship again on the tack she was before.
REVIRER *dans l'eau d'un navire.* To tack in a ship's wake, and stand on the same course astern of her.
REVOLIN, *s. m.* A sudden gust of wind, which blows off the shore, as by rebound, from the adjacent hills.
RHUMB, *s. m.* RHUMB *de vent.* A rhomb-line.
RIBORD, *s. m.* The second streak of planks on a ship's bottom, counting from the keel.
RIBORDAGE, *s. m.* The damage due from one ship to another, when the latter has sustained any hurt from the misconduct or neglect of the former.
RIDE, *s. f.* A lanyard.
RIDES *des haubans.* Lanyards of the shrouds.
RIDES *des galhaubans.* Lanyards of the back-stays.
RIDES *des marchepieds.* Lanyards of the horses of the yards.
RIDES *d'étai.* The lanyards of the stays.
RIDE *à palan.* A tackle, serving by way of a lanyard to a stay.
☞ RIDE bien les galhaubans! *Set the back-stays well up!*
RIDER, *v. a.* To set up, or tighten.
RIDER *un cordage.* To tighten a rope.
RIDER *les haubans.* To set up the shrouds.
RINGEOT. *Voyez* BRION.
RIPER, *v. n.* To slip through the nippers; speaking of the cable. *Ex.*
Le cable RIPE. The cable slips, or goes out as much as it comes in.
RIPOSTER, *v. a.* RIPOSTER *au feu de l'ennemi.* To return the fire of the enemy.
RIS, *s. m.* A reef. *Ex.*
RIS *des voiles.* Reefs of the sails.
Prendre un RIS. To take in a reef.
Etre aux bas RIS. To be close-reefed.
Larguer les RIS. To let the reefs out.
RIS *de chasse, ou premier* RIS. The first reef.
Bas RIS. Lower-reef.
Garcettes de RIS. *Voyez* GARCETTES, &c.
Palans de RIS. *Voyez* PALANS, &c.
Bande de RIS. *Voyez* BANDE, &c.
Œillets de RIS. Eye-let holes.
RISÉE, *s. f.* A gust or squall of wind.
RISÉE and RAFALE, are nearly synonymous, the latter implying a stronger gust of wind than the former.
☞ Lof à la RISÉE! *Luff while it blows!*
RISSER, *v. a.* To lash, seize, or frap, &c.
RISSER *la chaloupe.* To frap the long-boat.
RISSER *les vergues.* To strike the lower yards down a port-last.
RISSES, *s. f. p.* Gripes.
RISSES *de chaloupe.* Gripes, or ropes, by means of which they frap the boat on the ship's deck.
RISSONS, *s. m. p.* Grapnels with four claws, used as anchors in a galley.
RIVAGE, *s. m.* The beach, shore, or water-side.
RIVER UN CLOU, &c. *v. a.* To rivet a nail, &c.

RIVIÈRE, *s. f.* A river.
ROCHER, ROCHE, ROC. A rock; these three terms being synonymous in French.
ROCHER *sous l'eau.* A sunken rock.
ROCHERS *à fleur d'eau.* Rocks even with the water.
ROCHER *qui veille.* A rock above water.
ROCHER *qui découvre à mi-marée.* A half-tide rock.
RODE, *s. f.* The lower part of the stem and of the stern-post. This term is only used in the Mediterranean, and applied to lateen vessels.
*Contre-*RODE *de poupe.* The back of the stern-post of a galley.
*Contre-*RODE *de proue.* The apron of a galley.
ROIDIR, *v. a.* ROIDIR *les drisses.* To haul taught the halyards.
☞ ROIDIS les bras du petit hunier! *Haul taught the fore top-sail braces!*
ROINETTE, *s. f.* A marking-iron, to mark timber or casks which are shipped for a voyage, &c.

RÔLE, *s. m.* RÔLE *de combat.* A quarter bill.
RÔLE *d'équipage.* A muster-roll, or an exact muster-book of the ship's company.
RÔLE *de poste.* A station-bill.
RÔLE *de quart.* A watch-bill.
ROMBAILLET, *s. m.* (In naval architecture, synonymous with FOURRURE). A furr in a piece of timber.
ROMPRE, SE ROMPRE, *v. a. & r.* To break, &c.
ROMPRE *des bordages, &c.* To break up, or to rip off planks, &c.
La vergue d'artimon vient de se ROMPRE. The mizen-yard broke just now.
ROMPRE *la ligne.* To break the line of battle.
RONGÉ, *part. adj.* Worm-eaten; expressed of a ship's bottom, when it is much injured by the worms, as in a southern voyage.
ROSE, *s. f.* ROSE *de vents.* The card, or face of a sea-compass.
ROSE *de gouvernail.* *Voyez* FEMELOTS.
ROUANNE À MARQUER, *s. f.* An instrument to mark wine-vessels.
ROUANNE *de pompe.* A great pump-borer.
ROUANNER, *v. a.* To mark.
ROUANNER *une pompe.* To enlarge the bore or channel of a ship's pump.

ROUE, *s. f.* Roue *de gouvernail.* The steering-wheel, or wheel of the helm.
Roues *d'affût de canon.* *Voyez Parties de* l'Affût.
ROUER, *v. a.* To coil a rope.
Rouer *à tour.* To coil a rope with the sun; namely, according to the apparent course of the sun.
Rouer *à contre.* To coil against the sun.
ROUET, *s. m.* (In rope-making). A wheel in a rope-yard.
Rouet *de poulie.* The sheave of a block.
Rouet *de fonte.* A brass sheave.
Rouet *de gayac.* A lignum-vitæ sheave.
Rouets *de gayac à essieu dés de fonte.* Lignum-vitæ sheaves with brass coaks.
ROULEAU, *s. m.* A roller or cylindrical piece of wood, placed under any weighty body, in order to move it with greater facility by means of hand-spikes, &c.
ROULER, *v. n.* To roll.
Navire qui roule *beaucoup dans une mer houleuse.* A vessel that rolls much in a swelling sea.
ROULEUR, *adj.* *Bâtiment* rouleur. A ship which rolls much at sea.
ROULIS, *s. m.* Rolling, or the rolling motion of a ship.
ROUSTER, *v. a.* Rouster *un mât, ou une vergue.* To woold a mast or a yard.
ROUSTURES, *s. f. p.* The wooldings of a mast, &c.
Cercles des roustures *des mâts. Voyez* Cercles, &c.
ROUTE, *s. f.* The course, route, or track of a ship. *Ex.*
Etre en route. To be standing on the course required to be steered.
☞ Mets le cap en route! or merely *en* route! *Steer the course!* An order given to the helmsman to bring the ship in the appointed course.
Porter en route. To lie on the course.
Tenir une route. To steer a particular course.
Faire fausse route. To alter the course, in order to avoid an enemy, which is done in the night, or in a fog.
La Pérouse ne suivit aucune des routes *de Cook, voulant faire de nouvelles découvertes.* LaPérouse, wishing to make new discoveries, did not follow any of Cook's tracks.
Route *estimée.* The dead reckoning.
Route *corrigée de la variation et de la dérive, &c.* Corrected course.
Compas de route. *Voyez* Compas, &c.
ROUTIER, *s. m.* A book, or collection of charts, bearings, distances, soundings, and perspective views of the coasts of any country.
RUE, *s. f.* *La grande* rue. The waist, or that part of the main-deck comprised between the quarter deck and forecastle.

S.

SABATTE, *s. f.* Sabatte *d'une ancre.* The shoe of an anchor.
Sabatte, *patin, ou sole.* *Voyez* Sole *de bigues.*
SABLE, *s. m.* Sable *mouvant.* A quick-sand, or shifting sand.
Sable *dur.* Hard sand.
Sable *mou.* Soft sand.
SABLIER, *s. m.* A sand-glass, or hour-glass. *Ex.*
Sablier *de quatre heures.* Watch-glass.
Sablier *de deux heures.* Half-watch glass.
Sablier *d'une heure.* Hour-glass.
Sablier *d'une demi-heure.* Half-hour glass.
Sablier *d'une demi-minute.* Half-minute, or long glass.
Sablier *d'un quart de minute.* Quarter-minute, or short glass.
Le sablier *est-il paré?* Is the glass clear?
Il est paré. Clear glass.
☞ Vire le sablier! *Turn the glass!*
SABORD, *s. m.* A gun-port in a ship's side. *Ex.*
Sabords *de la première batterie.* The gun-ports of the lower tier or deck.
Sabords *de la seconde batterie.* The gun-ports of the middle tier or deck in a ship with three decks, and the upper-deck ports in a ship with two decks.
Sabords *de la troisième batterie.* The upper-deck ports in a three-decker.
Sabords *des gaillards.* The gun-ports of the quarter-deck and forecastle.
Sabords *de chasse.* Chase-ports of the forecastle.
Sabords *de retraite.* Stern-chase ports.
Sabords *de sainte-barbe.* Gun-room ports.
Sabords *de la batterie d'une frégate.* Main-deck ports of a frigate.
Sabords *des chambres des officiers, ou hublots.* Light-ports.
Sabords *des avirons.* Row-ports.
Sabord *à embarquer des bois, &c. dans la cale d'un vaisseau de charge.* Raft-ports.
Sabords *brisés, ou faux-sabords.* Half-ports; the upper part of which lets down for the admission of air.
Sabords *vitrés.* Sashes for the ports.
Sabords *de charge.* Raft-ports.
Fermer les sabords. To let fall or shut in the port-lids.
Anneaux de sabords. *Voyez* Anneaux, &c.
Seuillets de sabords. *Voyez* Seuillets.
Barres de sabords. *Voyez* Barres, &c.
Mantelets de sabords. *Voyez* Mantelets.
Palanquins de sabords. *Voyez* Palanquins, &c.
Itague de palanquin de sabords. Port-ropes.
SABRE, *s. m.* Sabre *d'abordage.* A ship's cutlass used in the act of boarding.
SACCADE, *s. f.* A violent jerk of the sails in light winds and in a heavy swell.
SACHETS de mitraille, *s. m. p.* Grape-shot, or partridge-shot.
SAFRAN, *s. m.* Safran *de gouvernail.* The after-piece of a rudder applied to increase its breadth.
Safran *de l'étrave.* An additional piece of timber fayed on the fore-part of the cut-water, to enlarge it immediately above the fore-foot, and enable the ship to hold the wind better.
SAILLER, *v. a.* To rowse, or haul upon a hawser. *Ex.*
☞ Saille! *Rowse!* or *Rowse together!*
☞ Saille hardi! *Rowse cheerly!*
SAIN, *adj.* Clear, safe, or clean; as

Côte SAINE. A coast free from rocks and shoals; also a clean bottom.]
Rocher SAIN. A rock steep-to.
SAINTE AUBINET. *Voyez* AUBINET.
SAINTE-BARBE, *s.f.* The gun-room.
Gardien de la SAINTE-BARBE. *Voyez* GARDIEN, &c.

SAÏQUE, *s.f.* A sort of Grecian or Turkish ketch.
SAISINE, *s.f.* A seizing or lashing of any kind. *Ex.*
SAISINES *des huniers.* Harbour leech-lines; hand leech-lines for the top-sails, used in French ships.
SAISINES *de beaupré.* *Voyez* LIURES, &c.
SAISIR, *v. a.* To seize or fasten any rope with a lashing; to nip, &c.
SAISON, *s. f.* Season. *Ex.*
Etre en SAISON. To be in time for making a passage at a proper season of the year.
*Etre à contre-*SAISON. To be too late for making a passage with the favourable periodical winds.
SALAISONS, *s. f. p.* Salt provisions, as salt fish and salt meat.
SALE, *adj.* Foul; said of a coast full of dangerous rocks, shallows, breakers, &c.
Vaisseau SALES. Ships with foul bottoms.
SALLE, *s.f.* SALLE *des gabarits.* The mould-loft in a dock-yard.
SALVAGE, *s. m.* Salvage.
SALUER, *v. a. & n.* To salute; to do homage at sea, by offering a salute. *Ex.*
SALUER *du canon.* To salute with guns.
SALUER *à boulet.* To salute by firing with shot; being an exclusive homage formerly paid to the king.
SALUER *de la mousquetterie.* To salute by firing a volley of small-arms.
SALUER *de onze coups de canons.* To fire a salute of eleven guns.
SALUER *des voiles.* To salute by lowering the top, or top-gallant-sails.
SALUER *du pavillon.* To strike the colours by way of salute.
SALUER *de la voix.* To salute with three cheers:—The English cry is, *Huzza!* and the French *Vive le Roi!*
SALUT, *s. m.* A salute of any kind offered at sea.
Faire un SALUT. To give a salute.
Rendre un SALUT. To return a salute.
SALVE, *s. f.* A platoon of a limited number of guns fired at once, by way of salute.
SAMEQUIN, *s. m.* A sort of Turkish merchant-man.
SANCIR, *v. a.* To founder, or sink. *Ex.*
SANCIR *à l'ancre, ou sur ses amarres.* To founder at anchor.
SANCIR *à la mer.* To founder at sea.
SANDALE, *s. f.* A sort of lighter or large boat used in the Levant.
SANGLES, *s. f. p.* Mats, or small paunches made of spun-yarn.

SANTÉ, *s. f. Bureaux de* SANTÉ. Board of health; an establishment in the Levant, to prevent the communication of epidemic disorders; they give certificates of health, or order ships into quarantine, &c.
Officiers de SANTÉ. Medical gentlemen; a general name for the physicians and surgeons belonging to the navy.
SAP, ou SAPIN, *s. m.* Fir or deal.
Planches de SAP *d'un pouce.* One-inch deals.
Bordages de SAP *de quatre pouces.* Four-inch deals.
SAPINETTES. *Voyez* CRAVAN.
SARAMBEAU, *s. m.* A fishing-raft of Manilla.
SARANGOUSTI, *s. m.* A peculiar gum, used to pay the seams of a ship in the East Indies, instead of pitch.
SART. *Voyez* GOËMON.
SARTIS, *s. m. p.* The shrouds of a galley, or other lateen vessel, in the language of the Levant.
SAUCISSON, *s. m.* The trough, or sausage filled with powder, which communicates the flame from the train to the fire-trunks, or powder-barrels in a fire-ship.
SAUGUE, *s. f.* A fishing-boat of Provence.
SAUMONS, *s. m. p.* SAUMONS *de fer.* A kentledge, or heavy weights of iron used as ballast.
SAUT, *s. m.* A waterfall in a river, which renders it unnavigable in that part; as those in St. Lawrence's river.
Donner un SAUT *à la bouline.* To check the bowline.
SAUT, also implies a jump, or leap.
SAUTE DE VENT, *s. f.* A sudden veering, or shifting of the wind.
SAUTER, *v. a. & n.* To jump; to blow up, &c.
SAUTER *en l'air.* To be blown up.
Faire SAUTER *un bâtiment.* To blow up a vessel.
SAUTER *à l'abordage.* *Voyez* ABORDAGE.
☞ SAUTE sur le beaupré! *Away out on the bowsprit!*
☞ SAUTE sur la vergue! *Go up to the yard! or out upon the yard!*
SAUTER. To veer suddenly; speaking of the wind. *Ex.*
Le vent a SAUTÉ. The wind has shifted.
SAUVE-GARDE, *s. f.* A man-rope, &c. *Ex.*
SAUVE-GARDES *de beaupré.* Man-ropes of the bow-sprit.
SAUVE-GARDE *de l'échelle hors le bord.* An entering-rope.
SAUVE-GARDES *de gouvernail.* The rudder-pendents, with their chains.
SAUVEMENT, *s. m.* Salvage; the payment of salvage.
SAUVER, *v. a.* To save. *Ex.*
Nous ne pûmes rien SAUVER *du navire naufragé.* We could not save any thing out of the wreck.
SAUVE-RABANS, *s. m. p.* The puddenings of the yards, used to preserve the rope-bands from being galled by the top-sail sheets.
SAUVETAGE, *s. m.* The act of saving, or recovering any thing that has been wrecked on the shore.
Bouée de SAUVETAGE. *Voyez* BOUÉE, &c.
SCALME, ou ESCAUME, *s. m.* (A term of the Mediterranean). Tholes. *Voyez* TOLLETS.
SCIER, *v. a.* To saw, &c. *Ex.*
SCIER *du bois.* To saw wood.
SCIER *avec les avirons.* To hold water.
☞ SCIE babord! *Hold water with the larboard oars!*
☞ SCIE tribord! *Hold water with the starboard oars!*
SCIER *à culer.* To back a-stern with the oars.

☞ Scie partout! *Back a-stern of all!*

Scier *sur le fer.* To support the cable of a galley by rowing with the oars, when she is at anchor in a storm, and in danger of running ashore.

SCITIE, ou setie, *s. f.* A settee, or particular kind of Italian bark with two masts.

SCORBUT, *s. m.* The scurvy; a well-known disease, incident to sailors on long voyages.

SCOUE, *s. f.* The extremity of a floor-timber, where it is joined to the lower futtock.

SCULPTEUR, *s. m.* A carver. *Ex.*

Maître-sculpteur. The master carver in a dock-yard.

SCULPTURE, *s. f.* Sculpture *des vaisseaux.* The mouldings of the head and stern of ships; carver's work.

SCUTE, *s. m.* A skiff or small boat, belonging to a ship.

SEAU, *s. m.* A bucket to draw water, for washing the decks, &c.

À SEC, *adv.* A-hull; high and dry, &c.

Vaisseau à sec, *ou à mâts et à cordes.* A ship a-hull, or under bare-poles.

Se tenir à sec. To keep under bare-poles.

Vaisseau à sec *sur le rivage.* A ship high and dry a-ground.

Etre à la cape à sec. To lie hull-to.

Mettre à sec. To hand all.

Mettre les voiles au sec. To loose the sails to dry.

Temps sec. *Voyez* Temps, &c.

SÈCHE, *adj. Vergue* sèche. *Voyez* Vergue, &c.

SÈCHES, *s. f. p.* Sands or rocks left dry in the sea, or near the coast.

SECOND, *ou vaisseau* second. *Voyez* Matelot.

Second *capitaine. Voyez* Lieutenant *en pied.*

SECOURS, *s. m.* Assistance. *Ex.*

Porter des secours *à un bâtiment.* To assist a ship.

SECRET, *s. m.* Secret *d'un brûlot.* That part of the train of a fire-ship where the match or fusee is laid by the captain, as ready for inflammation.

Secret *d'un canon.* The train of a piece of ordnance, which communicates with the touch-hole.

SEILLURE. *Voyez* Sillage.

SEIN *s. m.* A small bay or gulf with a narrow entrance. *Ex.*

Le sein *Persique.* The Persian gulf.

SEINE, *s. f.* A seine, or capacious fishing-net of a particular construction, used on the sea-coast.

Haler la seine. To haul the seine.

SEINER, *v. n. Donner un coup de* seine. To fish with a seine.

SÉJOUR, *s. m.* The time that a ship remains in any port whereat she touches in the course of a voyage.

SELLE, *s. f.* Selle *de calfat.* A calking-box, which contains the calker's implements and materials, and serves him as a seat.

SEMAQUE, ou semale, *s. f.* A smack, or fishing sloop.

SEMELLE, *s. f.* Semelle, *ou sole d'affut.* The bottom of a gun-carriage.

Semelles *de dérive.* Lee-boards.

SENAU, *s. m.* A snow; also a small Flemish vessel, rigged like a smack.

SENTINE, *s. f.* The well-room of a boat.

SENTINELLES, *s. f. p.* Sentinelles *d'un vaisseau.* Private marines who stand sentry in various parts of a ship of war, to prevent surprise, &c.

SENTIR, *v. a.* To feel, &c.

Bâtiment qui sent *bien son gouvernail.* A ship that answers or feels her helm well.

se SÉPARER, *v. r.* To separate, &c. *Ex.*

Les bâtimens du convoi se sont séparés *dans le coup de vent.* The vessels of the convoy have parted in the gale.

SEPS, *s. m. p.* Seps *de drisse, ou chaumard.* Knight-heads of the gears.

Seps *d'écoutes, ou bittons d'écoutes.* Topsail-sheet-bits.

SEPTENTRIONAL, *adj.* North; northern.

SERGENT, *s. m.* A wraining-bolt, used to bend a ship's planks into their places. *Voyez* Antoit.

SERINGUER, *v. a.* Seringuer *un vaisseau.* To rake a ship either fore or aft

SERPENTE, *s. f.* The act of snaking ropes.

SERPENTER, *v. a.* To snake.

Serpenter *deux cordages.* To snake two ropes together.

Serpenter *les étais.* To snake the stays down.

SERPER, *v. a.* Serper *l'ancre.* To weigh the anchor without the help of the capstan or windlass, as in xebecs, gallies, &c.

SERRAGE, *ou* serres *de vaisseau, s. m.* A general name for those planks of a ship, which are called *thick-stuff* by our shipwrights.

Faux serrage. Loose planks, laid occasionally as a platform for a ship's floor, when she has no ceiling.

en SERRE, *ou à la* serre, *adv.* Housed guns. *Ex.*

Canon en serre *a longé contre le bord.* A gun housed fore and aft.

Canon en serre *en travers du vaisseau, ou la bouche contre le haut du sabord.* A gun housed athwart.

Serre-banquières, *s. f. p.* Thick-stuff placed under the clamps in a ship's side.

Serre-bosse, *s. f.* The shank-painter.

Serre-file, *s. m.* (In naval tactics.) The rear ship of a fleet or squadron, when drawn up in a line of battle.

Serre-gouttières, *s. f. p.* The planks of the decks which are placed close to the water-ways.

SERRER, *v. a.* To close, &c. *Ex.*

Serrer *la ligne.* To close or contract the line of battle.

Serrer *la terre, ou la cote.* To keep close to the land; or to keep the land a-board.

Serrer *les voiles.* To take in, to furl, hand, or stow the sails.

☞ Serre les fonds de la misaine! *Bunt the fore-sail!*

Serrer *le grand hunier.* To take in the main top-sail.

Serrer *la brigantine.* To lower down the spanker.

Serrer *les bonnettes* To haul down the studding-sails.

SERRER *le vent*. To hug the wind; to ply or turn to windward.
SERVIR, *v. n.* *Faire* SERVIR. To fill the sails, after they had been shivering, or laid a-back for some time.
SERVIR, also implies to man, &c.; speaking of the guns.
Nos canons sont bien SERVIS. Our guns are well manned; also,
Nos canons furent bien SERVIS *dans le combat*. Our guns were well fought in the action.
SEUILLETS, *s. m. p.* SEUILLETS *de sabord*. The port-sills, or lower-part of the gun-ports.
Hauteur des SEUILLETS. The height of the port-sills from the deck immediately beneath them.
SEXTANT, *s. m.* A sextant.
SIAMPAN, *s. m.* A small expeditious coasting vessel of China, with one sail, and two, four, or six oars.
SIFFLEMENT, *s. m.* SIFFLEMENT *de balles et de boulets*. The whistling of balls and shot as they fly through the air, when discharged from a musket or piece of ordnance.
SIFFLET, *s. m.* A boatswain's call.
Donner un coup de SIFFLET. To wind a call.
SIGNAL, *s. m.* A signal.
SIGNAUX *de jour*. Day signals.
SIGNAUX *de nuit*. Night signals.
SIGNAUX *de brume*. Fog signals.
SIGNAL *d'incommodité*. *Voyez* INCOMMODITÉ.
SIGNAL *que la demande est accordée*. A signal of assent.
SIGNAL *de refus*. A signal of dissent.
SIGNAL *préparatoire*. A preparatory signal.
SIGNAL *de partance*. A sailing signal.
SIGNAUX *de combat*. Signals of battle.
SIGNAL *de ralliement*. A signal to leave off chase, or for all cruisers.
SIGNAL *de reconnoissance*. The private signal.
SILLAGE, *s. m.* The steerage-way; the track or wake of a ship; the trace which she leaves behind her on the surface of the sea.
Faire un bon SILLAGE, *ou un grand* SILLAGE. To sail swiftly, or to have fresh way through the water.
Doubler le SILLAGE *d'un vaisseau*. To sail with twice the velocity of another ship.
SILLER, *v. n.* To run a-head; to have head-way through the sea, &c.
SIMAISE, ou CYMAISE, *s. f.* A wave or ogee in the sculpture of a ship's mouldings.
SINGE, *s. m.* A sort of gin or machine, with a roller or winch in the middle, which is turned by handspikes; and used to discharge goods from a boat or small vessel.
SINGLER, *v. n.* To sail with a fair wind on a particular course.
SITUATION, *s. f.* SITUATION *d'une terre*. The bearings or distances of a coast.
SLOOP, *s. m.* (*Bateau Bermudien*). A sloop; a sort of vessel with one mast.
SŒUR, *s. f.* *Frégates, ou corvettes* SŒURS. Sister frigates or sloops.
SOLDAT, *s. m.* A soldier.
SOLDATS *de la marine*. Marines, or marine forces.
SOLE, *s. f.* SOLE *d'affût*. The sole or bottom of a gun-carriage.
SOLE *de sabord*. The sole or sill of a gun-port.
SOLE *de biguet*. A step for the heel of small shears that are used in a dock-yard.
SOLE *des taquets*. *Voyez* TAQUETS.
SOLIEL, *s. m.* The sun.
Lever du SOLIEL. Sun-rise.
Coucher du SOLEIL. Sun-set.
Le SOLEIL *a baissé*. The sun has fallen, or has passed the meridian; an expression used at the time of observing its altitude at noon.
Le SOLEIL *a passé le vent*. The sun has overtaken the wind; *i. e.* the wind being south, the sun, by passing from south to S. S. W. is said to have passed the wind: hence they say, in a contrary sense, *le vent a passé le* SOLEIL.
Le SOLEIL *chasse le vent*. The sun chases the wind; a phrase which implies the change of the wind from east to west, by the southern board, before sun-set.
Le SOLEIL *chasse avec le vent*. The wind keeps pace with the sun; an expression that denotes the change of the wind according to the course and progress of the sun.
Le SOLEIL *monte encore*. The sun continues to rise.
Le SOLIEL *ne fait rien*. The sun stands still.
Both of these latter phrases are peculiar to the operation of taking the meridian altitude.
SOMBRER, *v. n.* SOMBRER, *ou faire capot*. To overset in a squall, when under sail.
SOMME, *s. f.* A sort of Chinese vessel of burden.
Haute-SOMME. *Voyez* HAUTE-SOMME.
SOMMELIER, *s. m.* The captain's steward's mate.
SOMMER, *v. a.* To summon. *Ex.*
SOMMER *un fort, &c.* To summon a fort, &c. to surrender.
SONDE, *s. f.* Sounding; also the lead used to sound with, likewise the depth of water.
Plomb de SONDE. A plummet or sounding-lead.
Ligne de SONDE. *Voyez* LIGNE, &c.
Plomb pour les grandes SONDES. } *Voyez* PLUMB, &c.
Petit plomb de SONDE. ...... }
Aller à la SONDE. *Voyez* ALLER, &c.
Donner un coup de SONDE. *Voyez* DONNER, &c.
Etre sur les SONDES. To be in soundings.
Prendre la SONDE. To strike soundings.
Prendre les SONDES *d'une baie, &c.* To take soundings of a bay.
SONDE *de pompe*. The sounding rod of a pump.
SONDE *pour l'âme des canons, ou chat*. A searcher, or instrument used to ascertain the depth of the flaws or defects in a piece of ordnance.
SONDE *pour la lumière des canons*. *Voyez* ÉPINGLÉTTE, ou DÉGORGEOIR.
SONDER, *v. a. & n.* To sound, or heave the lead.
SONDER *l'archi-pompe*. To sound the pump-well.
SONNER, *v. a.* SONNER *la cloche*. To ring the bell.
SONNER *pour la pompe*. To strike the bell for pumping the ship.
SORTIR, *v. a. & n.* To go out, &c. *Ex.*
SORTIR *du port*. To depart from a harbour; to sail out, or put to sea.
SORTIR *le boute feu à la main*. To sail, prepared in every respect for immediate action; expressed of

a port whose entrance or opening is so conveniently situated, that a ship may sail from it with any wind, and be ready for engagement immediately after her departure.
SORTIR *un bâtiment du bassin.* To undock a ship.
SOUBERME, *s. f.* A fresh, or torrent, increased by the freshes in a river.
SOUFFLAGE, *s. m.* A thick sheathing put on a ship's bottom, in order to render her stiffer under sail, when she has proved to be crank. It is applied from about a foot above the load-water-line, and diminishes in thickness gradually, till it is reduced to nothing.
SOUFFLER, *v. a.* To double a ship with new planks and wales, so as to stiffen her when she is built too crank.
SOUFFLER *les canons.* To scale the guns; or cleanse them by blowing a little powder from them.
SOUILLE, *s. f.* The bed of a ship, or the impression made by her bottom on the mud, after having lain a-ground during the ebb-tide.
SOULIER, *s. m.* SOULIER *d'une ancre.* The shoe of an anchor.
SOUQUER, *v. a.* To stretch or taughten; to heave close any seizing or lashing.
☞ SOUQUE bien les garcettes de ris! *Tie your points close!*
☞ Attention à bien SOUQUER les garcettes de tournevire! *Mind how you clap your nippers on!*
☞ SOUQUE! *Hang, or swing upon!* A phrase used by the common sailors whilst they are pulling downwards on any rope or tackle.
SOURCE, *s. f.* SOURCE *du vent.* The point of the compass whence the wind blows.
SOUS, *prep.* Under.
Sous *le vent.* Under the lee.
Sous *la terre.* Under the shore.
Sous *voiles.* Under sail.
Sous *les basses-voiles.* Under the courses.
SOUS-ARGOUSIN. An officer in the gallies, who assists the Argousin in his duty. *Voyez* ARGOUSIN.
SOUS-BARBE, *s. f.* SOUS-BARBE *de beaupré.* A bob-stay.
SOUS-BARBE *d'un bâtiment sur le chantier.* A short shore placed under the stern-post of a ship, whilst on the stocks.
SOUS-BARQUE, *s. f.* The upper strake of a lighter, or the strake which lies close under the gunwale.
SOUS-FRÉTER *un bâtiment, v. a.* To under-freight a ship.
SOUS-INGÉNIEUR CONSTRUCTEUR. The builder's assistant in a dock-yard.
SOUTE, *s. f.* A store-room. *Ex.*
SOUTES *au pain.* The bread-rooms.
SOUTE *aux poudres.* The magazine, or powder-room.
SOUTE *vitrée.* The filling-room, or light-room.
SOUTE *aux voiles.* The sail-room.
SOUTE *aux hardes destinées à l'usage des matelots.* The slop-room.
SOUTE *aux vivres.* The hold for provisions.
SOUTE *aux rechanges du maître d'équipage.* The boatswain's store-room.
SOUTE *aux rechanges du maître-canonnier.* The gunner's store-room.
SOUTE *du chirurgien.* The surgeon's store-room.
SOUTE *aux provisions du capitaine.* The captain's store-room.
SOUTES *aux légumes.* Store-rooms for peas, &c.
SOUTES *à grain.* Store-rooms for corn.
SOUTES *des officiers.* Store-rooms for the officers' provisions.
SOUTENIR, SE SOUTENIR, *v. a. & r.* To support, &c.
SOUTENIR *la chasse.* To maintain a running fight.
Se SOUTENIR. To bear up against a scant wind or current, which acts upon the lee-side of the ship, and counter-balances the lee-way, when she is close-hauled, so as to keep her in the right course, without falling to leeward.
SPARTON, *s. m.* ou ESPARTE, *s. f.* A sort of junk or herb employed to make cordage in Spain, Italy, and other parts of the Mediterranean.
SQUELETTE, *s. m.* The carcass or skeleton of a ship; or the ribs, with the keel, stem, and stern-post, after the planks are ripped off.
STAMENAIS. *Voyez* GENOUX, &c.
STARIE, *s. f.* Demurrage.
STATION, *s. f.* A station, or cruizing-ground.
Etre en STATION. To be on a particular station.
Rester en STATION. To remain in a particular station.
SUBRÉCARGUE, *s. m.* A supercargo.
SUD, *s. m.* The south.
Etre au SUD *de la ligne.* To be in south latitude, or to the southward of the equinoctial line.
SUIF, *s. m.* A coat of soap, tallow, brimstone, &c. with which the bottom of a ship is paid.
Donner un SUIF *à un vaisseau.* To give her a coat.
SUPER, *v. n.* To stop or close accidentally; expressed of a leak which is choked or filled with sea-weed, &c. that may have entered with the water.
SURCHARGER, *v. a.* To overload. *Ex.*
SURCHARGER *un bâtiment des marchandises.* To overload a vessel with goods, &c.
SURCHARGER *de voiles.* To overpress with sails.
SURJALÉ, *adj.* Foul of the anchor-stock; expressed of the cable in that situation.
SURLIER, *v. a.* SURLIER *un cordage.* To whip a rope.
SURLIÛRE, *s. f.* The whipping of a rope.
SURPENTE, *s. f.* A guy, extending from the head of the main-mast to that of the fore-mast, to which a tackle is fixed for hoisting in or out weighty goods in merchant-ships.
SERPENTES *des basses vergues.* *Voyez* SUSPENTES, &c.
SURPRENDRE, *v. a.* SURPRENDRE *l'ennemi.* To surprise or take an enemy unawares.
Etre SURPRIS *par un coup de vent.* To be caught in a squall.
SURVENTER, *v. n.* To overblow, or to blow hard. *Ex.*
Il SURVENTE. It blows a storm.
SUSAIN, ou SUSIN. *Voyez* GAILLARD *d'arrière.*
SUSPENTES DES BASSES VERGUES, *s. f. p.* Slings of the lower yards.
SYRTES, *s. f. p.* Shifting-sands, quick-sands, or shelves.

T.

TABERNACLE, ou TENDELET, *s. m.* A place under the awning of a row-galley, where the captain sits to give his orders.

TABLE, *s. f.* *La* TABLE *de loc.* The log-board.

TABLES *loxodromiques.* Tables of the difference of latitude and departure.

TABLES *de la route d'un bâtiment.* Tables of the route of a ship.

TABLEAU, *s. m.* A flat part of the taffrail, between the cove and the upper moulding in French ships, whereon is placed some sculpture, bearing an allusion to the ship's name. *Voyez* ÉCUSSON.

TABLETTE, *s. f.* The rising-staff; a form or scale used by shipwrights when erecting the frames of the timbers.

TABLIER, *s. m.* TABLIER *des voiles.* Tabling in the bunt of the top-sails.

TABOURIN, *s. m.* The forecastle of a galley, with the space underneath it, where the guns are loaded and fired.

TACTIQUE NAVALE, *s. f.* Naval tactics, or the art of forming naval evolutions.

TAILLÉ, *adj.* (In naval architecture) TAILLÉ *pour la marche.* Built for sailing.

TAILLE-MER, *s. m.* The lowest and foremost part of a ship's cut-water, or of the prow in a galley.

TAILLER, *v. a.* *Tailler de l'avant.* To go fast through the water.

TALLARD, *s. m.* The space wherein the slaves of a row-galley are placed to manage their oars; it is situated between the *coursier*, or middle gangway, and the gunwale

TALON, *s. m.* TALON *de la quille.* The after-part of the keel into which the foot of the stern-post is tenoned.

TALON *de rode.* The heel of the stem, or stern-post of a row-galley.

TALONNER, *v. a.* To strike with the vessel's heel.

TAMBOUR, *s. m.* TAMBOUR *de l'éperon.* Wash-boards under the cheeks of the head, or doubling of the cut-water; or the planks nailed on the outside of it to defend it from the force of the waves.

TAMBOUR *de la roue du gouvernail.* The barrel of the steering-wheel.

TAMBOUR. A drum; also the person who beats it.

TAMISE, ou TAMISAILLE, *s. f.* The sweep of the tiller.

TAMISER, *v. a.* *La barre* TAMISE. The tiller shakes.

TAMISER also signifies, to let the wind through; expressed of sails that are worn very thin.

TAMPONS, *s. m. p.* Plugs, used on several occasions at sea. *Ex.*

TAMPONS *d'écubiers.* Hawse-plugs.

TAMPONS *de canon.* Tompions; wooden shot-plugs, employed to fill up the holes made in a ship's side by the balls received from an enemy; also plates of iron, copper, or lead, used for the same purpose.

TANGAGE, *s. m.* The act of pitching, or plunging with the fore and after parts of a ship.

TANGUER, *v. a.* To pitch or send; expressed of a vessel in a rough sea.

TANGUEURS, *s. m. p.* Lumpers. *Voyez* GABARIER.

TAPE-CUL, *s. m.* A driver.

TAPER, *v. a.* TAPER *les canons.* To lay the tompions on the guns.

Canon TAPÉ. A canon with its tompion on.

TAQUETS, *s. m. p.* Cleats kevels, wedges, &c. *Ex.*

TAQUETS *simples*, ou TAQUETS *à corne.* Common belaying cleats.

TAQUETS *à cœur.*
TAQUETS *à oreilles.* } Kevels.
TAQUETS *de tournage.*

Les soles des TAQUETS. The steps of the kevels.

Traversin des TAQUETS. *Voyez* TRAVERSIN, &c.

Cornes, ou oreilles des TAQUETS. The arms or horns.

TAQUETS *à gueule, ou à dent.* Snatch-cleats.

TAQUETS *de beaupré, ou violon.* The B's of the bowsprit.

TAQUETS *de fer, ou galoches de fer.* *Voyez* GALOCHES, &c.

TAQUETS *de cabestan, ou les flasques.* *Voyez Parties du* CABESTAN.

TAQUETS *de mâts.* *Voyez* PORTE-COLLIERS.

TAQUETS, *ou coins de mâts.* *Voyez* COINS, &c.

TAQUETS *de vergue, ou* TAQUETS *de bout de vergue.* Cleats of the yard-arms.

TAQUETS *d'échelle, ou échelons.* The steps which are nailed on the gang-way, to ascend or descend a ship's sides.

TAQUETS *de haubans.* Belaying cleats, seized on the shrouds.

TAQUETS *de hune.* *Voyez* COURBATONS, &c.

TAQUET *de la barre du gouvernail.* *Voyez* CRAPAUD.

TAQUETS *simples.* Cleats, quoins, or wedges, used to support or chock up any weighty body.

TAQUETS *d'amure.* *Voyez* POULIOTS, &c.

TAQUETS *d'écoutes.* The kevels or great cleats, whereon the tacks and sheets of the courses are belayed.

TAQUETS *de ponton.* Large hollow cleats fixed on the side of a pontoon or sheer-hulk, whereby to fasten the pendant of the relieving tackle.

TAQUETS *de potence.* The cheeks of a common sea-pump.

TARIÈRE, *s. f.* *Cuiller de la* TARIÈRE. An auger-bit.

Verge de la TARIÈRE. The shank of the auger.

Manche de la TARIÈRE. Trough of the auger.

TARIÈRE *de douze lignes.* A one-inch auger.

TARTANE, *s. f.* A tartan, or a sort of vessel with one mast and a lateen sail, navigated in the Mediterranean.

TAUREAU, *s. m.* A two-handled saw, or a cross-cut saw.

TEMPÊTE, *s. f.* A violent storm or tempest.

TEMPS, *s. m.* A general term for the weather.

Gros TEMPS, ou TEMPS *venteux.* Blowing weather.

Mauvais TEMPS. Bad weather.

TEMPS *orageux.* Stormy weather.
TEMPS *d'orages, ou de grains.* Squally weather.
TEMPS *couvert, ou* TEMPS *noir.* Thick weather.
TEMPS *pluvieux.* Rainy weather.
TEMPS *à giboulées.* Showery weather.
TEMPS *humide.* Damp weather.
TEMPS *de brouillard léger.* Misty weather.
TEMPS *embrumé. Voyez* EMBRUMÉ.
TEMPS *gras. Voyez* GRAS.
TEMPS *de brouillard. Voyez* BROUILLARD.
TEMPS *nébuleux.* Cloudy weather.
TEMPS *variable.* Changeable weather.
TEMPS *maniable. Voyez* MANIABLE.
TEMPS *calme. Voyez* CALME.
TEMPS *serein.* Clear weather
TEMPS *sec.* Dry weather.
Beau TEMPS. Fine weather.
TEMPS *fait. Voyez* FAIT.
Petit TEMPS. Light winds with fine weather.
TEMPS *à perroquet.* A top-gallant gale.
TENAILLE, *s. f.* A wooden engine formed like a pair of pincers, and employed to confine the planks of a ship in their places, till they can be nailed or bolted to the timbers.
TENDELET, *s. m.* The tilt of a boat; also the awning or canopy in the after-part of a galley.
TENDELET *de canot.* The tilt or awning of a boat.
L'impériale d'un TENDELET. *Voyez* IMPÉRIALE, &c.
TENIR, *v. a. n. & r.* To keep, to hold, to have, &c. *Ex.*
TENIR *la mer.* To keep the sea.
TENIR *le lit du vent.* To have the wind right on-end, or right in one's teeth.
TENIR *le vent, ou* TENIR *le plus près.* To keep the luff; to sail close to the direction of the wind.
TENIR *le largue.* To sail large, or with a large wind.
TENIR *le large. Voyez* LARGE.
TENIR *la bordée du large.* To stand off shore.
TENIR *la bordée de terre.* To stand in for the land.
TENIR *en travers, ou à la cape.* To hold it out.
Ce vent ne TIENDRA *pas.* This wind will not hold.
TENIR *bon sur un cordage.* To hold taught upon any rope, or to cease hauling on it.
☞ TIENS ferme la tournevire! *Hold on the messenger!*
☞ TIENS bon là! *Avast!*
☞ TIENS bon au cabestan! *Avast heaving!*
Se TENIR *à vue de terre.* To lie in sight of the land.
Se TENIR *sous les basses voiles.* To stand on the courses.
TENIR *sous voiles. Voyez* METTRE, &c.
TENIR *un bras. Voyez* BRAS.
Se TENIR *bord sur bord.* To make short tacks; to stand off and on.
Se TENIR *sur une seule amarre.* To lie at single anchor.
TENON, *s. m.* A tenon formed at the end of a piece of timber, to fix it in a mortise. *Ex.*
TENON *de l'étambot.* A tenon on the heel of the stern-post, which is let into the keel.
TENON *du mât de beaupré.* The tenon of the bowsprit.
TENONS *de l'ancre. Voyez Parties de* L'ANCRE.
TENTE, *s. f.* An awning. *Ex.*
Grande TENTE. Main awning.
TENTE *de gaillard.* Quarter-deck awning.
TENTE *de nage.* A boat's awning, that reaches from stem to stern.
TENTE *de dunette.* The poop-awning.
Montans de TENTE. *Voyez* MONTANS, &c.
TENTER, *v. a.* To spread an awning.
TENTER, also signifies to attempt. *Ex.*
TENTER *l'abordage.* To attempt boarding.
TENTER *un passage.* To attempt a passage.
TENUE, *s. f.* The hold that an anchor has of the ground in which it is sunk, also the quality of the ground as to its holding.
Bonne TENUE. Good holding.
Mauvaise TENUE. Bad anchoring ground.
TERMES, *s. m. p.* The quarter-pieces of a ship, by which the sides are terminated abaft.
TERRE, *s. f.* The land, or shore. *Ex.*
Etre à vue de TERRE. To be in sight of land.
Courir à TERRE. To stand in shore; to make for the shore.
TERRE *fine et claire.* Land which may be distinctly seen from the sea.
TERRE *défigurée.* Land which cannot be easily distinguished at sea, on account of the clouds.
TERRE *de beurre.* Cape fly-away, which implies an allusive appearance of land in the horizontal clouds, after sun-set, or before sun-rise.
Prendre TERRE. To touch at any port or place.
Dans les TERRES. In-land, or up the country.
Noyer la TERRE. *Voyez* NOYER.
TERRES *basses.* Low flat lands on the sea-coasts.
TERRE *hachée.* A coast with an opening between two mountains.
TERRES *hautes.* High lands on the sea-coasts; a bold, or iron-bound coast.
TERRE *maritime.* The sea-coast, or sea-shore.
Se tenir sous la TERRE. To keep under the land.
Se tenir tout à TERRE. To be quite close in shore.
Aller TERRE *à* TERRE. To coast, or sail along-shore.
Mangé par la TERRE. Shut in by the land; land-locked.
TERRE! TERRE! Land! Land! The cry of the men looking out at the mast-head, when they discover land.

À TERRE. A-shore, or on land.

À TERRE, *ou échoué, adv.* A-ground, or a-shore.
Le canot est allé à TERRE. The boat is gone a-shore.
TERRE-NEUVIER, *s. m.* A banker, or Newfoundland fishing-vessel.
TERRIR. *Voyez* ATTERRIR.
TERTRE, *s. m.* A hummock, or hillock, rising on a level shore, and seen from the sea.

TÊTE, *s. f.* Head, &c. *Ex.*
TÊTE *d'un mât.* The head or upper part of a mast.
TÊTE *de cabestan. Voyez Parties du* CABESTAN.
TÊTE *de gouvernail. Voyez Parties du* GOUVERNAIL.
TÊTE *de bordages.* A butt.
TÊTE *de potence des pompes.* The cheeks of the pump which support the brake.
TÊTES *de varangue.* Floor-heads.
TÊTES *d'alonge.* Futtock-heads.

TÊTE *de more.* A turk's head; a sort of knot.
Vaisseau de TÊTE. *Voyez* VAISSEAU, &c.
TÊTE (In naval architecture) *De* TÊTE *en* TÊTE. From stem to stern.
La longueur de TÊTE *en* TÊTE. *Voyez* LONGUEUR. &c.
Faire TÊTE. *Voyez* FAIRE, &c.

TÊTIÈRE, *s.f.* TÊTIÈRE *d'une voile.* The head of a sail.
Rabans de TÊTIÈRE. The head-lines of a sail.
Ralingue de TÊTIÈRE. *Voyez* RALINGUE, &c.
THON DE MÂT. *Voyez* TON.

TIERCÉ, *adj.* *Bâtiment* TIERCÉ. A vessel whose breadth is equal to one-third of her length.
TIERÇON, *s. m.* A tierce.
TIERS-POINT, *s. m.* A triangular sail, as a lateen-sail, stay-sail, &c. *Voyez Voile* LATINE.
TILLAC, *s. m.* A deck. *Ex.*
Franc-TILLAC. The main-deck.
Faux-TILLAC. The orlop-deck.
Clous de TILLAC. *Voyez* CLOUS, &c.
TILLE, *s. f.* The cuddy, or cabin, of a lighter or other undecked vessel; also the place where the helmsman stands, in a Dutch flight.
TIMON, *s. m.* TIMON, ou *Barre de gouvernail.* The tiller of the helm.
Etre au TIMON. To con, or cun, the ship.
TIMONNIER, *s. m.* The helmsman, or steersman.
TINS, *s. m. p.* The blocks upon which the keel and floor-timbers of a ship are laid while she is building.
TIRANT, *s. m.* TIRANT *d'eau.* The draught of water.
Différence du TIRANT *d'eau.* Difference in the draught of water.
Bâtiment qui flotte sans différence de TIRANT *d'eau.* A vessel that floats upon an even keel.
TIRE-BOURRE, *s. m.* The worm of a cannon.
TIRER. *v. a. & n.* To draw. *Ex.*
Un tel vaisseau TIRE *vingt pieds d'eau sur l'arrière.* Such a ship draws twenty feet of water abaft.
TIRER *un bâtiment à terre.* To ground a vessel.
TIRER *du canon.* To fire guns.
TIRER *le coup. de canon de retraite.* To fire the evening gun.
TIRER *des fusées.* To let off rockets.
TIRE-VEILLE, *s. f.* A man-rope, ladder, or side-rope. *Ex.*
Mettre les TIRE-VEILLES. To ship the side-ropes.
☞ Les TIRE-VEILLES à l'échelle! *Ship the ropes!*
TIRE-VEILLES *du mât de beaupré.* *Voyez* SAUVEGARDES, &c.
TIRE-VEILLES, ou *garde-corps d'un cabestan.* *Voyez Parties du* CABESTAN.
TOILE, *s. f.* Canvas in general.
TOILE *de dunette.* A canvas-covering for the poop.
TOILE *de tugue.* A canvas covering for the poop of a frigate.
TOILE *blanche pour les pavillons.* White linen cloth used for the French colours.
TOILE *à voile.* Sail-cloth.
TOILE *à prélat.* Thick canvas, or duck, used for tarpawlings.
TOILE *de Mélis.* A sort of cloth made at Mélis.
TOILE *à voile de Hollande.* Holland duck.
TOILE *de Noyale.* A sort of sail-cloth manufactured in the village of Noyale, in Britanny.
TOILE *à voile de Russie.* Russia duck.
TOILE *de cotonnine.* *Voyez* COTONNINE.
TOILE *écrue de* $\frac{7}{8}$. A sort of thin canvas, which is $\frac{7}{8}$ of an ell broad, and employed for top-gallant sails, stay-sails, and studding-sails.
TOISER, *v. a.* To measure by the fathom.

TOLLETIÈRES, *s. f. p.* Row-locks.
TOLLETS, *s. m. p.* Tholes.
TOMBER, *v. n.* To fall, &c. *Ex.*
TOMBER *sous le vent.* To fall or drive to leeward.
TOMBER *sur un vaisseau.* To fall aboard of a ship to leeward.
TOMBER *sur le côté.* To fall over; speaking of a ship a-ground.
TOMBER *à la mer.* To fall overboard.
Le mât TOMBE *en arrière.* The mast hangs, or rakes aft.
Le vent a TOMBÉ. The wind is spent, or has died away.
La mer est TOMBÉE, *Voyez* MER.
TON, *s. m.* TON *d'un mât.* The mast-head, or the upper part of the mast, above the trestle-trees.
Avoir les huniers sur le TON. To have the top-sails on the cap.
TONIES, *s. f. p.* A sort of Indian boats, which are usually lashed together in couples, in order to carry sail the better. The two thus paired are called *Catapanel.*
TONNE, *s. f.* A butt, or cask, containing two hogsheads.
TONNE, likewise signifies a can-buoy, placed over any shelf or rock in a channel; also the nun-buoy of a ship.
TONNEAU, *s. m.* A ton, employed in measuring the burden of shipping, containing 2000 *lb.* or 42 cubic feet, French measure.
Navire de cinq cens TONNEAUX. A ship of five hundred tons burden.
TONNEAU, *ou pièce de quatre.* A leager, or a cask containing four hogsheads.
TONNELIER, *s. m.* The cooper of a ship, who has the charge of all the provision casks, to keep them in proper repair.
Maître TONNELIER. The master cooper of a dock-yard.
TONNELLERIE, *s. f.* The shed in which the casks are made and repaired in a French dock-yard; also the magazine, or store-house, in which they are kept for use.
TONTURE, *s. f.* The sheer of the wales and decks, or the convexity of the transoms and beams of a ship.
TONTURE *des ponts.* The sheer of the decks.
TONTURE *des baux.* The round-up, or convexity of the beams.

TONTURÉ, *adj.* *Vaisseau fort* TONTURÉ. A ship built with a great sheer.
Vaisseau bien TONTURÉ. A ship built with a handsome sheer.
TORON, *s. m.* The strand of a rope.

Cordage a trois TORONS. A rope made of three strands.

TORS, *s. m.* TORS *d'un cordage.* The twisting, or lay of a rope.

Cordage qui a trop de TORS. A rope laid too hard.

Oter le TORS *a un cordage.* To untwist a rope, which is done immmediately after laying it, when it is too hard-twisted.

TORTUE, *s. f.* TORTUE *de mer.* A sort of transport-ship formed with a high deck, for the convenience of carrying troops, passengers, and their effects between decks.

TOUAGE, *s. m.* The act of towing, or of warping a ship from one place to another.

TOUCHER, *v. n.* To touch the ground; to strike against a rock or sand-bank.

TOUCHER *et parer.* To touch and go.

TOUCHER *terre.* Approaching the land.

TOUCHER *a un port.* To touch or call at any port.

Notre bâtiment TOUCHA *en entrant dans le port.* Our ship touched, or struck, in coming into harbour.

TOUCHER *un compas.* To touch the needle of a compass with a magnet.

TOUÉE, *s. f.* A warp, or tow-line.

Porter une TOUÉE *en déhors du vaisseau.* To carry a warp out of the ship.

La grande TOUÉE. A cable and a half.

TOUER, *v. a. & r.* To tow; to warp. *Ex.*

On est obligé de TOUER *les navires qui entrent ou sortent de ce port.* Vessels must be towed in and out of that harbour.

Se TOUER. To haul one's self a-head.

TOUPIN, *s. m.* (In rope-making) A top for laying ropes.

TOUR, *s. m.* A turn, &c. *Ex.*

TOUR *à bitord.* A reel for making spun-yarn.

TOUR *de bitte au cable.* A turn of the cable about the bits; the bitting of a cable.

TOUR-*et-choque.* A weather-bit of the cable, or a turn and a half-turn about the bits.

TOUR *de cable.* A foul hawse. *Voyez* CROIX.

Prendre des TOURS *dans ses cables.* To take turns in the hawse.

Cable qui a pris un TOUR. A cable which is foul in the hawse.

TOUR *de loc. Voyez* LOC.

TOUR *de la grande ligne de sonde.* Deep-sea reel.

TOUR *mort.* A clove-hitch.

TOUR, *s. f.* TOUR *à feu.* A light-house, or watch-tower on the sea-coast.

TOUR *de signaux.* A signal-tower.

TOURBILLON, *s. m.* A tornado, or whirlwind upon the sea.

TOURET, *s. m.* A winch, upon which the yarn is wound, in a rope-walk.

TOURILLONS, *s. m. p.* TOURILLONS *d'un canon.* The trunnions of a piece of ordnance.

TOURMENTE, *s. f.* A violent storm or tempest.

TOURMENTIN, *s. m.* A name sometimes given to the fore-top-mast stay-sail. *Voyez* PERROQUETS.

TOURNANT, *s. m.* TOURNANT *d'eau.* A whirlpool, or dangerous race in the sea.

TOURNER, *v. a.* To turn, &c. *Ex.*

TOURNER *le sablier.* To turn the glass.

TOURNER *sur son ancre.* To pass round the anchor; to make a foul anchor.

TOURNEVIRE, *s. f.* The messenger. *Ex.*

Garnir la TOURNEVIRE *au cabestan.* To get the messenger to pass.

Fusées de la TOURNEVIRE. Mouses of the voyol.

Œillets de TOURNEVIRE. *Voyez* ŒILLETS, &c.

Mariage de la TOURNEVIRE. *Voyez* MARIAGE.

TOURNIQUET, *s. m.* A roller.

TOUT, *adv.* All. *Ex.*

☞ En haut TOUT le monde! *Voyez* MONDE.

☞ TOUT le monde en bas! *Sit down close all hands!* The order to the ship's crew to lie snug upon deck or below, so as not to retard the ship's course by their motion upon deck, nor be discovered by the enemy, of whom they are in chase.

TRACER, *v. a.* To draw, &c. *Ex.*

TRACER *une route sur la carte.* To lay down a track upon the chart.

TRAIN DE BATEAUX, *s. m.* A train of boats in tow.

TRAIN *de bois.* A raft, or float of timber.

À LA TRAÎNE, *adv.* Towing overboard; expressed of any thing towed in the sea by a rope while the ship is advancing.

Canot à la TRAÎNE, A boat towing at the stern of a ship.

TRAÎNÉE, *s. f.* TRAÎNÉE *de poudre.* A train of gunpowder.

TRAIT, *s. m* *Voile à* TRAIT-*quarré.* A square sail; such are the courses, top-sails, top-gallant-sails, &c. of a ship.

TRAIT *du vent. Voyez* RHUMB, &c.

TRAITE, *s. f.* The trade or traffic carried on between shipping and the inhabitants of any country where they arrive; it is, however, seldom applied in that sense, expressing more particularly the barter carried on by the Europeans with the Indians and negroes.

TRAITE *des nègres.* The slave-trade. This inhuman traffic was abolished by the British Parliament in the year 1807.

TRAITEMENT, *s. m.* The full pay of officers. *Voyez* APPOINTEMENS.

TRAJET, *s. m.* TRAJET *de mer.* A passage over the sea.

TRAMONTANE, *s. f.* The north wind, and north point of the compass, in the dialect of the Mediterranean.

Greco-TRAMONTANE. North north-east.

Mestre-TRAMONTANE. North-north-west.

TRANSPORT, *s. m.* Conveyance. *Ex.*

Le TRANSPORT *des troupes a couté considérablement.* The conveyance of troops cost a great sum.

Bâtiment de TRANSPORT. *Voyez* BÂTIMENT, &c.

TRANSPORTER, *v. a.* To carry, &c. *Ex.*

TRANSPORTER *des troupes.* To transport troops.

Nous avons TRANSPORTÉ *deux régimens de Gibraltar à Minorque.* We carried two regiments from Gibraltar to Minorca.

TRAVADE, *s. f.* A tornado, or thunder-gust; prevalent on the coast of Africa.

TRAVAILLER, *v. a. & n.* To strain, &c.; when applied to a ship, it signifies to roll or pitch heavily, as in a high sea.

Le cable TRAVAILLE *beaucoup*. There is a great strain on the cable.

Le bâtiment TRAVAILLE *beaucoup par ce gros temps*. The ship strains much in this bad weather.

TRAVERS, *s. m.* A beam, &c.

Par le TRAVERS. On the beam.

Vent par le TRAVERS. A wind on the beam.

Par le TRAVERS *d'un vaisseau*. On the beam of a ship.

TRAVERS, in a naval sense, generally denotes athwart; abreast of, with sides parallel, and heads equally advanced; it is also applied to any piece of timber which is laid across others, and scored into them.

Etre par le TRAVERS *d'un port*. To be athwart a port or harbour.

Présenter le TRAVERS *à un vaisseau*. To bring the broadside of a ship to bear.

Tenir le TRAVERS *au vent*. To keep the broadside to the wind.

TRAVERS *par* TRAVERS. Hank for hank.

Passer par le TRAVERS *de Torbay, &c.* To cross or stand athwart Torbay, &c.

Etre mouillé par le TRAVERS *de l'Ile de Wight*. To be at anchor off the Isle of Wight.

La marée vient par le TRAVERS *du vaisseau*. The tide takes the ship athwart, or on the broadside.

En TRAVERS, *adv.* *Mettre en* TRAVERS. To bring to; which is done in the night when they do not wish to sail beyond a port, &c. or when they are obliged to wait for any thing.

Etre mouillé en TRAVERS *de la marée*. To ride athwart the tide.

TRAVERSÉE, *s. f.* A passage from one port to another; an outward, or homeward-bound voyage.

Faire une belle TRAVERSÉE. To have a good passage.

Notre TRAVERSÉE *a été de quinze jours*. We have had a passage of 15 days.

TRAVERSER, *v. a.* To cross, &c. *Ex.*

TRAVERSER *les voiles*. To flat in the sails.

☞ TRAVERSE la misaine! *Flat in the fore sheet!*

TRAVERSER *un vaisseau, ou s'entraverser devant un fort*. To bring the ship's broadside to bear upon a fort, or battery.

TRAVERSER *l'ancre*. To fish the anchor, or to get the anchor up along the bow, in order to stow it parallel to the gunwale.

TRAVERSER *une vergue*. To cross a yard, or get it across.

TRAVERSER *la lame*. To head the sea; to sail against the setting of the sea.

TRAVERSIER, *s. m.* A small fishing-vessel on the coast of Rochelle.

TRAVERSIER *de chaloupe*. The fore-beam, or fore-thwart of a long-boat.

Mettre la misaine au TRAVERSIER. To bring the fore-tack to the cat-head; as when the wind is large.

TRAVERSIER, *adj.* *Vent* TRAVERSIER *d'un port, d'une baie, ou d'une côte*. A wind that sets right into a harbour or bay, so as to prevent the departure of a ship from it; or a wind which sets directly to a shore, so as to render it difficult for a ship when embayed to claw off from the shore.

TRAVERSIN, *s. m.* (In naval architecture). TRAVERSINS *des baux, ou barrotins*. Small carlings placed between the beams in French ships of war.

TRAVERSIN *d'écoutille*. A gutter-ledge, or cross-bar laid in the middle of a hatchway, to support the covers.

TRAVERSIN *des bittes*. *Voyez* BITTES.

TRAVERSIN *d'élinguets*. The beam into which the pawls of the capstan are bolted.

TRAVERSIN *du gaillard d'avant*. The cross-piece of the forecastle, which contains the kevels and cleats for belaying ropes.

TRAVERSIN *de taquets*. The cross-piece of the main and fore-sheet kevels, also of other kevels.

TRELINGAGE, *s. m.* TRELINGAGE *des haubans sur les hunes*. Cat-harpings of the shrouds.

TRELINGAGE *des étais sous les hunes*. The crow-feet of the tops.

TRELINGAGE *de combat*. Lower catharpings, reeved below the former from the starboard to the larboard shrouds, forming a kind of netting overhead, the use of which is to receive any blocks or ropes which may fall from aloft in time of action, and thereby kill some of the ship's company.

TRELINGUER, *v. a.* To reeve a crow-foot, &c.

TRÉLUCHER, *v. a.* To gybe, or to shift lateen sails, in the language of the Mediterranean.

TRÉMUE, *s. f.* A sort of hood or companion to the hatchway of a fishing-boat.

TRÉMUE, is also a trunk, or sloping passage, formed in some merchant-ships, whereby the cables are conducted from the top of the fore-castle downward to the hawse; it is usually covered with a small grating.

TRÉOU, *s. m.* A lug-sail, or square-sail in lateen vessels, used in scudding; or the cross-jack-sail in some English boom-rigged vessels, galleys, tartans, &c.

Vergue de TRÉOU. The yard of such a square-sail, called the cross-jack-sail in some boom-rigged vessels among the English.

TRÉPORT, *s. m.* A stern-timber, whose lower end corresponds with the top of the stern.

TRÉSORIER, *s. m.* TRÉSORIERS *généraux de la marine*. Treasurers of the navy residing at Paris, each of them having one year's business to perform alternately.

TRESSES, *s. f. p.* Sennit; a sort of knittles frequently used as seizings.

TRESSES *de mèche à canon*. A large match formed of three matches twisted round each other, so as to fire a cannon with more certainty and expedition.

TRÉVIER. *Voyez* VOILIER.

TREVIRE, *s. f.* A Parbuckle.

TRIANGLE, *s. m.* A contrivance formed of three pieces of wood, for the purpose of making night signals.

TRIANGLE, is likewise a stage hung over a ship's side, to calk the seams, or pay the planks: also a stage composed of three capstan-bars or handspikes, whose ends being tied together form a triangle, to inclose any mast, along which this contrivance may be hoisted or lowered, to scrape the mast, or pay it with turpentine, resin, tallow, &c.

TRIBORD, *s. m.* The starboard side of a ship.

TRIBORD, *ou les tribordais.* A cant term for the starboard-watch.

☞ TRIBORD au quart! *Spell the starboard watch! Starboard hoay!*

☞ TRIBORD la barre! *Voyez* BARRE.

☞ TRIBORD un peu! *Starboard withal!*

☞ TRIBORD tout! *Hard a-starboard!*

☞ Ne viens pas sur TRIBORD! *Voyez* VENIR.

☞ Feu TRIBORD! *Voyez* FEU, &c.

☞ Nage bien TRIBORD! *Voyez* NAGE, &c.

TRIDENT, *s. m.* A fish-gig. *Voyez* FOÊNE.

TRINGLE, *s. f.* A thin lath, used occasionally to fill up the edges of a gun-port, dead-light, &c. so as to exclude the wind and water.

TRINGLE, is also a batten of wood about two feet long, nailed against the butts or joints of a boat's planks, to strengthen them in that place.

TRINGLER, *v. a.* To mark timber with a chalked or red line, in order to hew and bevel it to the exact form and scantling.

TRINQUERIN, *s. m.* The water-ways of a galley.

TRINQUET, *s. m.* The fore-mast in lateen vessels.

Voile de TRINQUET. The foresail of a galley.

TRINQUETTE, *s. f.* The fore-sail of a cutter or sloop-rigged vessel.

TRISSE DE BEAUPRÉ, *s. f.* The standing lifts of the sprit-sail yard.

TROMBE, *s. f.* A water-spout.

TROPIQUES, *s. f. p.* The tropics.

TROSSE, *s. f.* The parral of a lateen-yard.

TROU, *s. m.* A hole. *Ex.*

TROU *du chat, ou* TROU *du mât de hune.* Lubber's hole.

TROUS, *ou yeux de la civadiere.* The holes formed in the clews of a sprit-sail, to let out the water which falls into its cavity when the ship pitches.

TROUS *d'amure.* The holes in the chess-trees.

TROUS *d'écoutes.* The sheave-holes which are cut obliquely through the ship's side, wherein the main and fore-sheets are rove.

TROUS *de boulet.* Shot-holes.

TROUS *de cap-de-mouton.* The holes in a dead-eye.

TROUVER, *v. a.* To find. *Ex.*

TROUVER *fond.* To get bottom, or strike soundings.

TUGUE, *s. f.* The poop of a frigate, &c.

U.

UN, DEUX, TROIS! An exclamation, or song, used by the French seamen when hauling the bowlines; the greatest effort being made at the last word. English sailors, in the same manner, call out on this occasion; *Haul-in! Haul-to! Haul-belay!*

US, *s. m. p.* Us *et coutumes de la mer.* The uses and customs of the sea.

USANCE, *s. f.* The term of a month, among merchants, when relating to payments or bills of exchange; also an agreement, or contract, made between the master, owner, and freighters of a ship.

USTENSILES, *s. m. p.* The tools or implements used in and about a ship. *Ex.*

USTENSILES *de charpentier.* Carpenter's tools.

USTENSILES *de calfats.* Calker's tools.

USTENSILES *de tonnelier.* Cooper's tools.

USTENSILES *de canon.* The instruments used in charging and firing a cannon; as the rammer, the ladle, the sponge, the linstock, &c.

USTENSILES *de pêche.* Fishing gear.

V.

VADROUILLE. *Voyez* FAUBERT & GUIPON.

VAGANS, *s. m. p.* A name given to those who infest the sea-coast during a tempest, with intent to plunder the wrecked vessels.

VAGUES, *s. f. p.* Waves, or surges of the sea.

VAIGRAGE, *s. m.* The ceiling or foot-waling.

VAIGRER, *v. a.* To attach the planks and thick-stuff of a ship's ceiling to the timbers.

VAIGRES, *ou* SERRES, *s. f. p.* (In naval architecture). A general name for the clamps and thick-stuff used in a ship's ceiling.

VAIGRES *de fond.* The thick-stuff placed next to the keel.

VAIGRES *de fleurs.* The thick-stuff and ceiling placed next to those about the floor-heads.

VAIGRES *d'empâture.* The thick-stuff placed between the floor-heads and the clamps.

VAIGRES *de pont.* The clamps which support the ends of the beams.

VAISSEAU, *s. m.* A ship, but particularly understood of a man of war, or a ship of the line. *Ex.*

Votre frère commande-t-il une frégate, ou un vaisseau? Does your brother command a frigate, or a ship of the line?

VAISSEAU *de guerre.* A ship of war, or man of war.

VAISSEAU *du premier rang.* A first-rate man of war.

VAISSEAU *du second rang.* A second-rate man of war.

VAISSEAU *à trois ponts.* A three-decker.

VAISSEAU *de ligne.* A ship of the line from 50 guns upwards.

VAISSEAU *amiral.* *Voyez* AMIRAL.

VAISSEAU *rallongé.* A lengthened ship.

VAISSEAU *rasé.* *Voyez* RASÉ.

VAISSEAU *à l'ancre.* A ship at anchor.

VAISSEAU *servant de prison dans un port.* A prison-ship.

VAISSEAU *armé en guerre.* An armed ship, &c. *Voyez* BÂTIMENT, &c.

VAISSEAUX *frères.* Sister ships.

VAISSEAU *canard.* A ship that pitches deeply.

VAISSEAU *qui a les côtés ronds.* A round-sided ship.

VAISSEAU *de tête, ou* VAISSEAU *de queue d'une escadre.* The headmost or sternmost ship of a fleet.

VAISSEAU *beau de combat.* A roomy ship, advantageously built for battle, as carrying her lower tier high above the water, and having a good height between-decks.
VAISSEAU *de bas-bord.* *Voyez* BAS.
VAISSEAU *de haut-bord.* *Voyez* HAUT.
VAISSEAU *a la bande.* A ship lying along, or heeling gunnel-to, as under a press of sail in a fresh wind: this is frequently termed *lying down on her broad-side, or beam-ends.*

VAISSEAU *fort d'echantillon.* *Voyez* ECHANTILLON.
VAISSEAU *qui a beaucoup d'elancement.* A ship with a raking stem.
VAISSEAU *qui a beaucoup de quête.* A ship whose stern has a great rake.
VAISSEAU *qui a beaucoup de mailles.* A ship whose timbers stand very wide.
VAISSEAU *haut sur l'eau.* A ship that sits high upon the water
VAISSEAU *qui a l'air male.* A ship that has a very warlike appearance.
VAISSEAU *qui a un faux cote.* *Voyez* COTE.
VAISSEAU *qui a de la tonture.* *Voyez* TONTURE.
VAISSEAU *qui evolue bien.* A ship that works well.
VAISSEAU *qui gouverne bien.* A ship that answers her helm quickly.
VAISSEAU *qui rallie bien au vent.* A weatherly ship.
VAISSEAU *qui tire beaucoup d'eau.* A ship of a great draught of water.
VAISSEAU *bien suivi dans ses proportions.* A ship of good proportions.
VAISSEAU *de compagnie.* *Voyez* COMPAGNIE.
VAISSEAU *de la compagnie des Indes.* An East-India-man.
VAISSEAU *taillé pour la marche.* *Voyez* MARCHE.
VAISSEAU *servant de magazin.* A ship used as a repository for stores.
VALET, *s. m.* *Maître* VALET. The steward's man.
VALET *de la soute au pain.* Jack of the bread-room.
VALET *de menuisier.* A hold-fast.
VALET *a canon.* A wad.
VALETS *d'artillerie.* The boys who attend the great guns in a sea-fight, &c.
VALTURE, *s. f.* The lashing of the sheers; or a rope employed to lash two masts together in any particular place, when they are to be used as sheers.
VARANGUES, *s. f. p.* (In naval architecture) Floor-timbers.
Maîtresses VARANGUES, *ou* VARANGUES *de fond.* Midship floor timbers.
VARANGUE *plate, ou de fond.* A flat floor-timber.
VARANGUES *acculees.* The crotches or floor-timbers afore and abaft.
VARANGUES *demi-acculees.* The floor-timbers that are placed between the crotches and the midship floor-timbers.
VARANGUE *fort acculee.* A rising floor-timber amidships.
L'acculement de la maîtresse VARANGUE. The rising of the floor-timber amidships.
Les VARANGUES *de porques.* Floor-timbers of the riders, or floor-riders.

Demi-VARANGUE. A cross-chock. This piece is longer in French-built than in English ships, and is framed and bolted with the other timbers.
Tetes de VARANGUE. *Voyez* TETES, &c.
La longueur de la maîtresse VARANGUE. The length of the floor-timber amidships.
VARECH. *Voyez* CHOSES *de la mer* & GÖEMON.
VARIATION, *s. f.* VARIATION *du compas, ou de l'aiguille aimantée.* Variation of the compass, or of the magnetic needle.
VARIATION *nord-est.* East variation.
VARIATION *nord-ouest.* West variation.
Compas de VARIATION. *Voyez* COMPAS, &c.
La VARIATION *vaut la route.* The variation is on the weather-side, or opposite to the lee-way.
VASART, *ou* VASEUX, *adj.* Oozy, or slimy, expressed of a particular bottom, or soundings at sea.
Le fond de sable VASART. The bottom of oozy sand.
VASE, *s. f.* Mud, ooze, or slime.
VASE *dure.* Hard mud.
Fond de VASE. Muddy bottom.
Fond de VASE *molle.* Oozy ground.
VASSOLES, *s. f. p.* *Les* VASSOLES *des ecoutilles.* The coamings of the hatches.
VAT-ET-VIENT, *s. m.* A pass-rope, made fast to a ship and to a wharf, by means of which a boat may be hauled backwards and forwards by one man.
A LA VEILLE, *adv.* *Une bouee a la* VEILLE. A buoy floating in sight, as over the anchor.
Ancre a la VEILLE. An anchor a-cock-bill.
VEILLER, *v. a. & n.* To watch; to take care of any thing, &c. *Ex.*
☞ VEILLE aux drisses des huniers! *ou* Veille! Veille! *Stand by the top-sail halyards!* The order to have the top-sail-halyards ready to cast loose in case of a squall.
☞ VEILLE les ecoutes des huniers! *Stand by the top-sail sheets!*
Roche qui VEILLE. A rock always above water.
Il faut VEILLER *les mats, et non le côté.* We must look to the masts and not to the side; expressed of a ship, whose masts being good, will rather overset her, than be carried away.
☞ VEILLE! VEILLE! *Watch! Watch!* The word passed from the man who heaves the deep-sea lead to those who hold the line, all of whom in their turns repeat it, on dropping the coil from their hands.
VENIR, *v. n.* To come, &c. *Ex.*
VENIR *au vent.* To haul the wind; to come to windward; to luff.
VENIR *sur tribord, ou sur babord.* To haul to starboard, or to port.
VENIR *en travers.* To bring to, or heave to.
☞ Ne VIENS pas au vent! ou pas au vent! *No near! or No nearer!*
☞ Ne VIENS pas sur tribord! *Mind your starboard helm!*
☞ Sans VENIR sur babord! *mind your port helm!*
VENT, *s.m.* Wind.
Au VENT. To windward.

VENT *arriere, ou* VENT *en poupe.* A wind right aft, or a-stern.
Bon VENT. A fair wind.
VENT *contraire.* A foul wind.
VENT *debout, ou* VENT *de bouline.* A scant wind, with which a ship cannot lie on her course, even when close-hauled.
VENT *devant.* Head to wind.

VENT *etale.* *Voyez* ETALE.
VENT *frais.* *Voyez* FRAIS.
Gros VENT. *Voyez* GROS, &c.
VENT *maniable.* A moderate wind.
VENT *au-plus-pres.* A sharp wind, which makes it necessary to be trimmed sharp, in order to lie on the course.
VENT *de quartier.* A quarterly, or quartering wind.
VENT *traversier d'un port.* *Voyez* TRAVERSIER.
Sous le VENT. To leeward, or under the lee.
Tomber sous le VENT. *Voyez* TOMBER, &c.
Cote du VENT. *Voyez* CÔTE, &c.
Avantage du VENT. *Voyez* AVANTAGE.
VENT *dedans.* The situation of the sails which are filled, or inflated with the wind.
VENT *dessus.* The situation of the sails which are taken a-back.
Coup de VENT. *Voyez* COUP, &c.
Coup de VENT *force,* *Voyez* FORCE.
Gagner le VENT *a un vaisseau.* *Voyez* GAGNER, &c.
Derober, ou manger le VENT *a un vaisseau.* *Voyez* MANGER, &c.
Manche a VENT. *Voyez* MANCHE, &c.
Petit VENT. A small gale.
VENT *foible, ou* VENT *mou.* A slack wind.
Le VENT *est bon frais.* It blows a fresh gale.
VENT *largue.* *Voyez* LARGUE.
Les VENTS *alizés.* Trade winds.
Les VENTS *regnants.* *Voyez* REGNANTS.
Les VENTS *periodiques.* *Voyez* MOUSSON.
Le VENT *est rond.* The breeze is steady.
Le VENT *est inegal.* The wind comes in puffs.
VENTS *variables.* Variable winds, or such as are without the tropics.
Les VENTS *adonnent.* The wind draws aft.
Le VENT *est par rafales.* It blows hard in the squalls.
Le VENT *fait le tour du compas.* The wind flies all round the compass.
VENTS *de terre.* Land winds.
VENTS *de mer.* Sea breezes.
VENT *en poupe, largue la soute.* A large wind, large allowance; an expression used by seamen at the commencement of a fair wind, after they had been put to short allowance, in consequence of foul winds.
Le VENT *en poupe fait trouver la mer unie.* A stern wind brings an easy sea; expressed of a ship when sailing afore the wind, in which movement she is less strained by the agitation of the sea; than when she lies in the trough or hollow of it, side-ways.
Le VENT *est a-pic.* The wind is right down; a witticism amongst sailors, signifying a total cessation of wind, at which time the vanes hang right downward, instead of blowing directly out.

VENTER, *v. n.* To blow. *Ex.*
Il VENTE *grand frais du nord-est.* It blows very fresh from the north-east.
Il VENTE *fort.* It blows hard.
Il ne VENTE *pas.* There is not much wind.
VENTILATEUR, *s. m.* A ventilator used at sea.
VERBOQUET, *s. m.* A guy used by shipwrights to keep steady any piece of timber, so as to lodge it securely in a ship's frame.
VERGE, *s. f.* *La* VERGE *de girouette.* The spindle of a vane.
La VERGE *de l'ancre.* The shank of an anchor.
Une VERGE *de pompe.* A pump-spear.
VERGE, is also an English measure of three feet.
VERGUE, *s. f.* The yard of any principal sail, which traverses the mast at right angles. *Ex.*
Grande VERGUE. Main-yard.
VERGUE *de misaine.* Fore-yard.
La VERGUE *d'artimon.* Mizen yard.
VERGUE *de grand hunier.* Main-top-sail yard.
VERGUE *de grand perroquet.* Main-top-gallant yard.
VERGUE *de petit hunier.* Fore-top-sail yard.
VERGUE *de petit perroquet.* Fore-top-gallant yard.
VERGUE *de civadière.* Sprit-sail yard.
VERGUE *de contre-civadière.* Sprit-sail top-sail yard.
VERGUE *traversée.* The sprit which traverses a boat's sail diagonally.
VERGUE *sèche, ou* VERGUE *barree.* The cross-jack yard.
VERGUE *de perroquet de fougue.* Mizen-top-sail yard.
VERGUE *de la perruche.* Mizen-top-gallant yard.
VERGUE *de kakatois de perruche.* Mizen-top-gallant royal-yard.
VERGUE *de grand perroquet volant.* Main-top gallant royal yard.
VERGUE *de petit perroquet volant.* Fore-top-gallant royal yard.
VERGUE *de tape-cul.* Driver-yard.
VERGUES *de bonnettes de hunier.* Top-sail studding-sail yards.
VERGUE *en boute dehors.* *Voyez* GUI.
VERGUE *de tangon.* Fore-sail boom.
Bout de VERGUES. Yard-arms.
Taquets de bout de VERGUE. Cleats at the yard-arms.
VERGUES *quarrées.* Square-sail yards, or square-yards.
VERGUES *latines, ou antennes.* Lateen yards.
VERGUE *de tente,* A spread-yard for the awning.
VERGUE *à corne.* *Voyez* CORNE *de vergue.*
VERGUE *de tréou.* *Voyez* TREOU.
VERGUE *de fortune.* *Voyez* FORTUNE.
Mettre les voiles en VERGUE. To bring the sails to the yards.
VERGUE-A-VERGUE, *adv.* Yard-arm and yard-arm.
Se tenir VERGUE-À-VERGUE *d un batiment.* To continue along-side of a ship yard-arm and yard-arm.

VERIFIER, *v. a.* VERIFIER *la ligne de loc, &c.* To examine the log-line, &c.
VERIN, *s m.* An instrument nearly similar to a jack-screw, and used occasionally in launching a ship from a slip.

VESTES, *s.f.p.* The halyards of a lateen yard.
VIBORD, *s.m.* (In naval architecture). The quick-work, or that part of a ship's side which is comprehended between the drift-rails and the waist-rail.
Lisse de VIBORD. A waist-ribband; also the top timber line.
VICE-AMIRAL, *s. m.* A vice-admiral.
VICE-*consul*, *s. m.* A vice-consul; an officer under the consul, residing in a smaller part of the latter's district, to assist him in his duties.
VICTUAILLES. *Voyez* VIVRES.
VICTUALLEUR, *s. m.* A contractor, or agent-victualler.
VIGIES, *s.f. p.* VIGIES *ou écueils en pleine mer.* Rocks always above water, as those in the ocean, or main sea.
VIGIES, *ou poste d'un gardien des signaux.* Watch-places on the top of a hill, having a sight over the main sea, where watchmen are stationed, to make signals of what ships pass by. This term is peculiar to the French West-Indies, and is borrowed from the Spaniards.
Etre en VIGIE. To keep a look out.
VIGIER, *v a.* VIGIER *une escadre.* To dodge, or watch the motions of a fleet or squadron.
VINDAS, *s. m.* A windlass, or horizontal capstan, used in vessels of small burden.
VIOLON, *s. m.* VIOLONS *de beaupré.* The B's of the bowsprit.
VIRAGE *s. m.* The act of heaving up any weighty body by a crab or capstan.
VIRER, *v. a. & n.* To turn, to heave, &c. *Ex.*
VIRER *au cabestan.* To heave at the capstan.
VIRER *à-pic.* To heave short, or a-peak.
VIRER *de force au cabestan.* To heave a-strain at the capstan.
☞ VIRE! VIRE! ou VIRE hardi! *Heave cheerly!* The cry of the officers to encourage the men when they heave at the capstan.
☞ VIRE pour mettre les linguets! *Heave and pawl!*
VIRER *de bord.* To put about ship.
VIRER *de bord vent devant.* To tack, which is done by means of the following commands:
☞ Adieu-va! *About ship!*
☞ Pare a VIRER! *Ready about!*
☞ Leve les lofs! *Up tacks and sheets!*
☞ Decharge derriere! *Haul main-sail, haul!*
☞ Decharge devant! *Haul off haul! or, Let go and haul!*
VIRER *de bord vent arriere, ou* VIRER *lof pour lof.* To ware, wear, or veer; speaking of a ship.
☞ Pare a VIRER lof pour lof! *See all clear for waring ship!*
☞ En haut tout le monde pour VIRER lof pour lof! *All hands ware ship!*
VIRER *le beaupre a terre.* To tack close in shore.
VIRER *tous en semble.* To tack and ware together.
VIRER *un vaisseau en carene.* *Voyez* ABATTRE, &c.
Vaisseau VIRE *en quille.* A ship hove keel out of the water.
VIREVAUT, *s. m.* A small windlass, or winch.
VIROLE, *s. f.* A little iron ring placed on the small end of a bolt, which is driven through any part of a ship's decks or sides; it is used to prevent the fore-lock from cutting the wood.
VIROLET, *s. m.* A roller. *Voyez* MOULINET.
VIRURE, *s. f.* VIRURE *de bordages.* A streak of planks continued from the stem to the stern-post.
VIRURE, is also the sheer of any plank in a ship's side.
VISITER, *v. a.* To examine, &c. *Ex.*
VISITER *un bâtiment a la mer.* To examine, or search a ship at sea.
VIVIER, *s. m.* A fishing-boat, furnished with a well filled with water amidships, whereby the fish are kept alive till the vessel arrives in port.
VIVRES, *s. m. p.* Victuals; provisions for the subsistence of a ship's crew at sea, &c.
Faire des VIVRES. *Voyez* FAIRE. &c.
Directeur des VIVRES *de la marine.* An agent victualler of the navy.
Munitionnaire des VIVRES *de la marine.* *Voyez* MUNITIONNAIRE.
VOGUE, *s. f.* The act of rowing; the movement or course of a galley rowed with oars.
Mener la VOGUE. To give the stroke in rowing.
☞ VOGUE avant! *Pull away!*
VOGUE-AVANT, *s. m.* The strokesman in a boat or galley.
VOGUER, *v. n.* To row or pull.
VOIE, *s. f.* VOIE *d eau.* A leak.
Faire une VOIE *d eau.* To spring a leak.
Decouvrir une VOIE *d'eau.* To find out a leak.
Il s'est declare une VOIE *d'eau.* We sprung a leak.
Boucher une VOIE *d'eau.* To stop a leak.
VOILE, *s. f.* A sail; also a ship discovered at a distance.
La grande VOILE. The main-sail.
La VOILE *de misaine, ou la misaine* The fore-sail.
VOILE *d'artimon, ou l'artimon.* The mizen.
La VOILE *du grand hunier.* The main-top-sail.
La VOILE *du petit hunier.* The fore-top-sail.
La VOILE *du perroquet d'artimon, ou le perroquet de fougue.* The mizen-top-sail.
La VOILE *du grand perroquet.* The main-top-gallant sail.
La VOILE *du petit perroquet.* The fore-top-gallant-sail.
La VOILE *du grand perroquet volant.* The main-top-gallant royal sail.
La VOILE *du petit perroquet volant.* The fore-top-gallant royal-sail.
La VOILE *de la perruche d'artimon, ou simplement la perruche.* The mizen-top-gallant sail.
La grande VOILE *d'étai.* The main-stay-sail.
La VOILE *d etai d artimon.* The mizen-stay-sail.
La grande VOILE *d'etai de hune, ou du grand hunier.* The great main-top stay-sail.
La petite VOILE *d'etai de hune.* The small main-top stay-sail.
La VOILE *d'étai du grand perroquet.* The main-top-gallant stay-sail.
La VOILE *d'etai du perroquet de fougue.* The mizen-top stay-sail
La VOILE *d'etai de la perruche.* The mizen-top-gallant stay-sail.
La VOILE *de civadiere.* The sprit-sail.

La VOILE *de contre-civadière.* The sprit-top-sail.
La VOILE *de paille en cul.* The driver or ring-tail.

Parties et Accessoires d'une VOILE.	*The Parts and Appurtenances of a Sail.*
Pattes des VOILES.	The tablings of the sails, at the edges or bolt-ropes.
Bandes de ris.	The reef-bands.
Œillets.	The eyelet-holes.
Renfort ou doublage.	The tabling at the foot-ropes and leech-ropes.
Tablier.	The tabling in the bunt of a top-sail.
Ralingues.	The bolt-ropes.
Ralingue de fond.	The foot-rope.
Ralingues de chûte.	The leech-ropes.
Ralingue de têtière.	The head-rope.
Le point.	The tack or clew.
Les herseaux.	The cringles on the bolt-ropes.
Les garcettes de ris:	The reef-lines, or points.
Les rabans de ferlage.	The furling-lines, or rope-bands.
Les rabans de pointure	The earings.
Le fond d'une VOILE. *ou bordure.*	The foot of a sail.
Les fonds d'une VOILE.	The bunt of a sail.
Têtière d'une VOILE.	Head of a sail.
Chûte d'une VOILE.	Depth or drop of a sail.
Bordure d'une VOILE.	*Voyez* BORDURE.
L'envergure d'une VOILE.	The square of a sail.

VOILES *auriques. Voyez* AURIQUE.
VOILE *Angloise.* A boat's sail with a diagonal sprit.
Les VOILES *de l'arrière. Voyez* ARRIÈRE.
Les VOILES *de l'avant. Voyez* AVANT.
Les basses VOILES. *Voyez* BASSES-VOILES.
VOILE *latine. Voyez* LATIN.
VOILES *quarrées. Voyez* QUARRÉES.
VOILE *neuve.* A new sail.
VOILE *usée.* An old sail.
Toutes VOILES *dehors. Voyez* DEHORS.
VOILE *amenée.* A sail lowered.
VOILE *amurée.* A sail whose tack is hauled aboard.
VOILE *bordée.* A sail whose sheets are tallied or hauled aft.
VOILE *carguée, ou sur les cargues.* A sail clewed up, or hauled up in the brails.
VOILE *coiffée, ou sur le mât.* A sail taken or laid aback.
VOILE *criblée.* A sail shot through in several places.
VOILE *déferlée.* A sail loosened or unfurled.
VOILE *défoncée.* A sail split or rent asunder.
VOILE *dégarnie.* A sail unbent; speaking of jibs and stay-sails.
VOILE *déralinguée.* A sail whose bolt-rope is carried away.
VOILE *désenverguée.* A sail unbent from its yard.
VOILE *en bande.* A sail whose sheets are let fly.
VOILE *en ralingue.* A flowing sail, or a sail whose sheets being slackened, flies loose and flutters in the wind.
VOILE *enverguée.* A sail bent to its yard.
VOILE *qui faseye.* A sail shivering in the wind for want of being properly trimmed.
VOILE *garnie.* A sail bent; meaning a stay-sail or jib.
VOILE *hissée.* A sail hoisted.
VOILE *qui porte.* A sail which draws, or a taught sail.
VOILE *au sec.* A sail loosed to dry.
VOILE *serrée.* A sail taken in, furled, or stowed.
Faire de la VOILE. *Voyez* FAIRE, &c.
Faire force de VOILES. *Voyez* FAIRE, &c.
Porter bien la VOILE. To carry sail stiffly.
Se tenir sous les quatre corps de VOILES. To stand on under the courses, fore and main-top-sails.
VOILÉ, *adj. Bâtiment bien* VOILÉ. A vessel whose sails are well proportioned.
VOILERIE, *s. f.* A sail-loft, or the place where sails are made.
VOILIER, *s. m.* A sail-maker.
Maître VOILIER. A master sail-maker.
Bon VOILIER, *ou mauvais* VOILIER, when expressed of a ship, implies a good, or bad sailer, or one that sails swiftly, or heavily.
VOILURE, *s. f.* A general name for all sorts of sails belonging to a ship or vessel.
VOILURE *d'un vaisseau.* A complete suit of sails.
Ce vaisseau à sa VOILURE *complette.* This ship has her complete suit of sails.
VOILURE *à baleston.* Sprit-sails.
VOILURE *de lougre.* Lugger-sails.
VOILURE *de chébec.* Xebec-sails.
VOILURE *de galiottes.* Galliot-sails.
VOILURE *des cutters.* Cutter-sails.
Sous la même VOILURE. Under the same sail; expressed of two ships in company which carry the same quantity and number of sails.
Régler sa VOILURE. To regulate the quantity of sail to be carried, in order to keep company with some other ship or ships.
VOIX, *s. f.* The song used by sailors in heaving, hoisting, hauling, &c. *Ex.*
Donner la VOIX. To sing out.

☞ À la VOIX! *Mind the song!*
Porte-VOIX. *Voyez* PORTE-*voix.*
VOLAGE, *adj. Embarcation* VOLAGE. A crank boat.

VOLÉE, *s. f.* A platoon, or limited number of great guns, fired at once in a sea-fight; a broadside.

À toute VOLÉE. Random-shot.
Etre sous la VOLÉE *d'un bâtiment.* To be close under a ship's guns.
Au lieu de répondre à nos signaux, le corsaire tira sa VOLÉE. Instead of answering our signals, the privateer fired her broadside.
VOLET, *s. m.* A small sea-compass, used in a long-boat or cutter.
VOLONTAIRES, *s. m. p.* VOLONTAIRES *dans un vaisseau François de guerre.* Volunteers in a French ship of war. These are the same as our midshipmen, and are brought up to be officers in the navy, or masters of merchantmen.
VOLTE. *Voyez* REUTE.
VOLTIGLOLE, *s. f.* VOLTIGLOLE, *ou cordon de massane.* A term of the Levant, signifying the lowest rail or floor-ribband of a galley.

VOÛTE, *s. f.* The counter of a ship.

La grande voûte. The lower-counter.
Vaisseau qui a beaucoup de voûte. A ship that has a very hollow counter.
La petite voûte, *ou* voûte *de la galerie.* The Cove.
VOYAGE, *s. m.* A voyage.
Voyage *de long cours.* A long voyage, as to China or the Indies.
VRILLE, *s. f.* Vrille *à canon.* A drill, or long gimlet used to cleanse the touch-hole of a cannon.
Vrille *de charpentier.* A wimble, or drill, used by shipwrights, &c. to bore holes.
VUE, *s. f.* Sight, or eye-sight. *Ex.*
Vue *de la terre.* The looming or perspective view of the land from the sea.
Etre à vue, *avoir la* vue. To be in sight of; to make; to discover at sea some distant object.

Y.

YAC, *s. m.* A jack.
YACHT, *s. m.* A yacht.
YEUX, *s. m. p.* Yeux, *ou trous de la civadière.* The holes made in the clews of the sprit-sail, in order to let out the water.
Yeux-*de-bœuf.* Bulls-eye, or wooden travellers; also the trucks of a parral.
Yeux *de pie.* Eyelet-holes wrought in the reef of a sail, through which the points are reeved.
YOLE, *s. f.* A yawl, or very small boat.

Z.

ZÉNITH, *s. m.* The zenith.
ZÔNE, *s. f.* A zone.
La zône *torride.* The torrid zone.
Les zônes *tempérées.* The temperate zones.
Les zônes *froides, ou glaciales.* The frigid zones.
ZOPISSA. *Voyez* Goudron *&* Poix.

FIN.

ERRATA IN THE VOCABULARY.

Pages.	Col.	Lignes.		
5	1	11	*au lieu de*	Appariel, lisez Appareil.
5	2	4	———	*fourcal,* lisez *fourcat.*
6	2	25	———	*poulis,* lisez *poulies.*
7	1	54	———	*adj.* lisez *adv.*
9	2	1	———	Barre, lisez Barres.
10	1	46	———	*plattes,* lisez *plates.*
14	1	57	———	Tha, read The.
14	2	61	———	Bruillard, lisez Brouillard.
15	1	35	———	*élinguets,* lisez *linguets.*
20	1	33	———	*lieus,* lisez *lieues.*
20	2	38	———	*le,* lisez *la.*
25	1	31	———	*Nous,* lisez *Nos.*
25	1	51	———	*resumes,* lisez *reçûmes.*
29	1	20	———	his, read them.
32	1	30	———	ÉBAROUIE, lisez ÉBAROUI, E.
32	2	46	———	*huniere,* lisez *hunier.*
33	2	45	———	EMBARQEMENT, lisez EMBARQUEMENT.
36	1	38	———	ESCARPÉE, lisez ESCARPÉ, E.
36	2	24	———	ESTIMÉE, lisez ESTIMÉ, E.
37	2	32	———	Éteindre, lisez Éteignez.
38	2	37	———	*voiles,* lisez *voile.*
39	1	59	———	*hubolts,* lisez *hublots.*
39	2	34	*au lieu de*	*L'ennemis,* lisez *Les ennemis.*
40	2	12	———	*quarts,* lisez *quart.*
40	2	58	———	*Hauts,* lisez *Haut.*
41	2	6	———	*pain* Four, lisez *pain.*
41	2	16	———	Carènes, lisez Carene.
44	2	19	———	*d'une,* lisez *d'un.*
47	1	6	———	*s'éclairce,* lisez *s'éclaircit.*
47	2	41	———	JACK, lisez ÏAC.
49	1	54	———	*courier,* lisez *courir.*
49	1	60	———	*un,* lisez *une.*
49	1	51	———	*Tener,* lisez *Tenir.*
52	2	42	———	*mer,* lisez *mou.*
54	2	1	———	*complete,* lisez *complette.*
58	2	33	———	*battaile,* lisez *bataille.*
60	2	53	———	*supérieures,* lisez *supérieurs.*
61	2	6	———	*ligue,* lisez *ligne.*
66	2	31	———	*Largueur,* lisez *Longueur.*
68	1	5	———	*le terre,* lisez *la terre.*
69	1	34	———	*Cape,* lisez *Cap.*
75	2	4	———	SOLIEL, lisez SOLEIL.
75	2	5	———	Soliel, lisez Soleil.
75	2	24	———	Soliel, lisez Soleil.
76	2	39	———	*des,* lisez *de.*

Directions to the Binder.

The Binder is particularly requested to observe that pages 71, 72, 111, 112, 459, 460, 461, 462, 463, 464 have been cancelled; and that the reprinted pages must be carefully placed in their stead.

The Plates to be placed at the end of the Vocabulary, having regard to numbers; the double ones to be placed last, and pasted into papers, so as to have but one fold, and that in the middle.

Printed by Joyce Gold, 103, Shoe-lane, London.

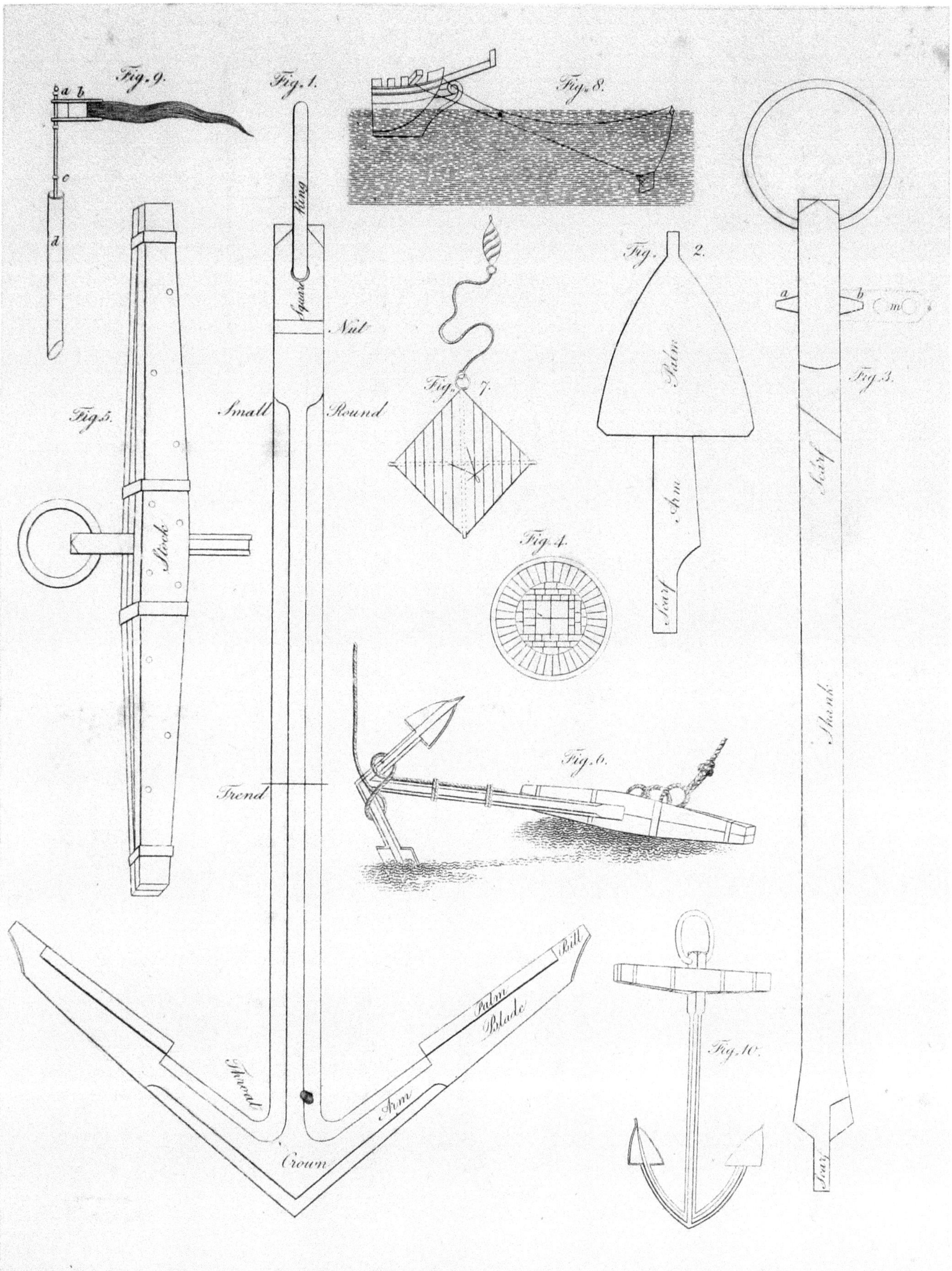

Mutlow Sc. Russell Ct.

Published March 1.1815, by T. Cadell and W. Davies, Strand London.

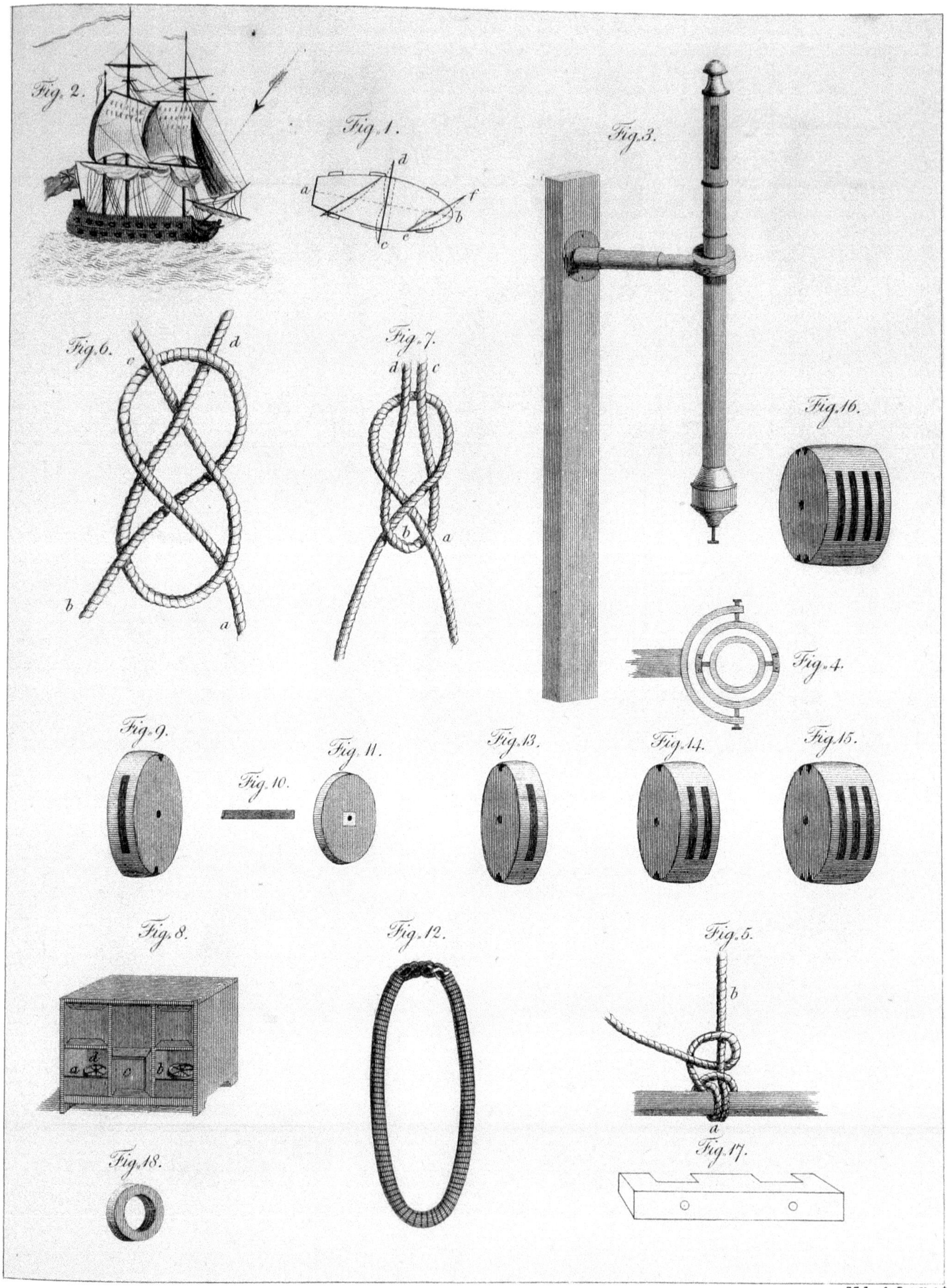

Mutlow Sc. Russell Ct.

Published March 1. 1815, by T. Cadell and W. Davies, Strand, London.

BLOCKS.

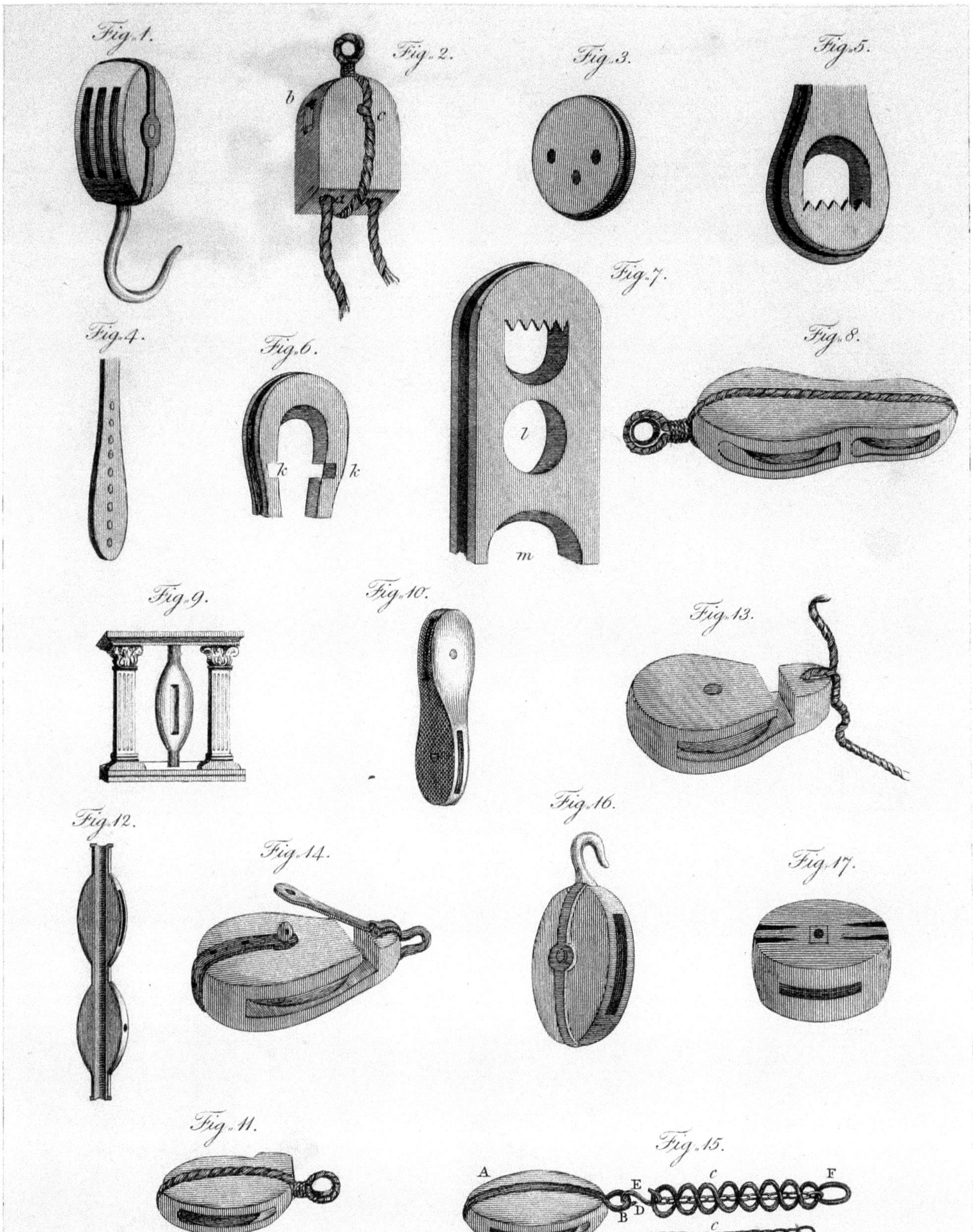

Mutlow Sc. Russell Co.

Published March 1.1815, by T. Cadell and W. Davies, Strand, London.

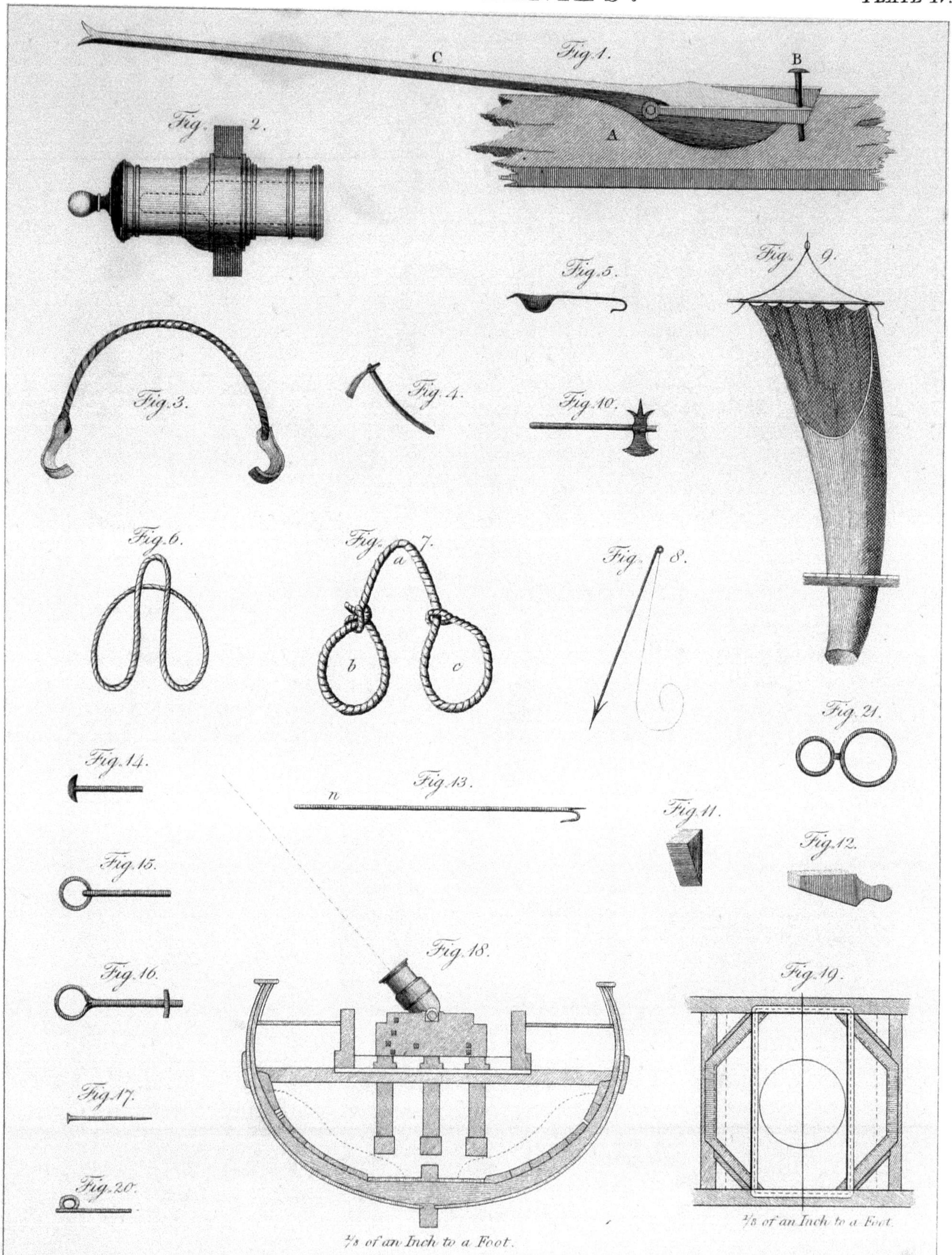

Mutlow So. Russell Ct.

Published March 1.1815. by T. Cadell & W. Davies, Strand, London.

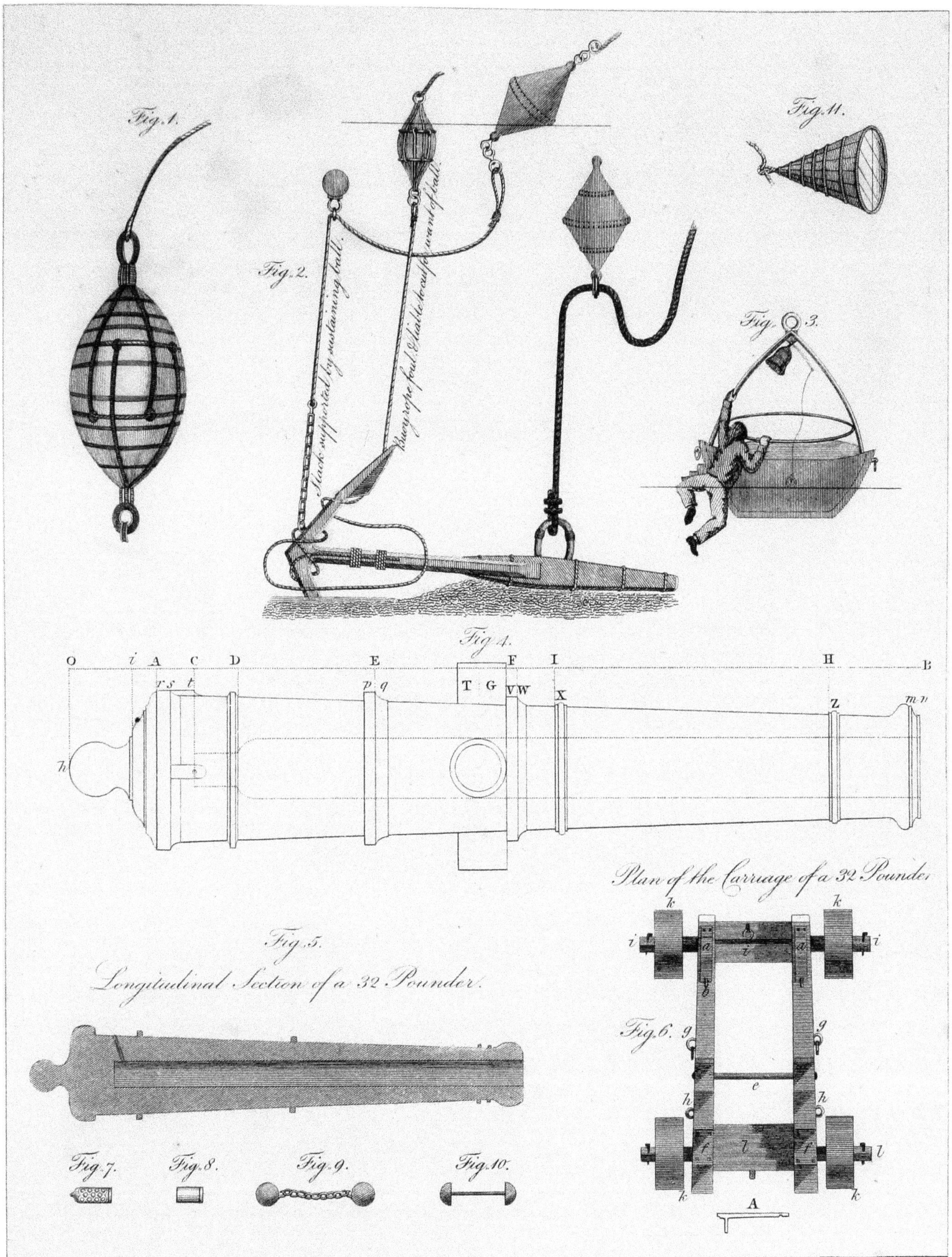

Mutlow Sc. Russell Cou.t

Published March 1. 1815, by T. Cadell and W. Davies, Strand, London.

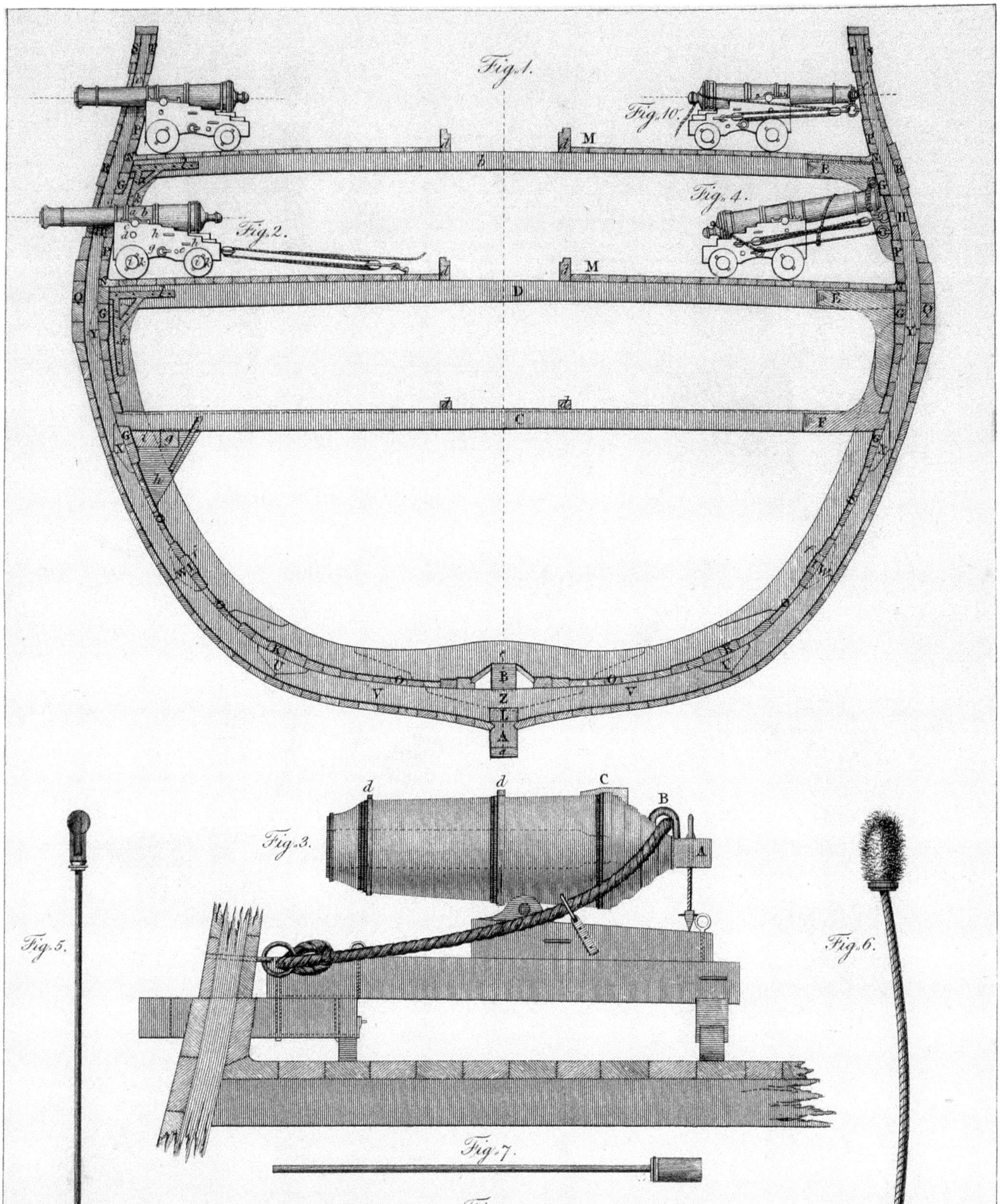

I. Taplin del. Mutlow Sc. Russell Co.

Published March 1. 1815. by T. Cadell and W. Davies. Strand, London.

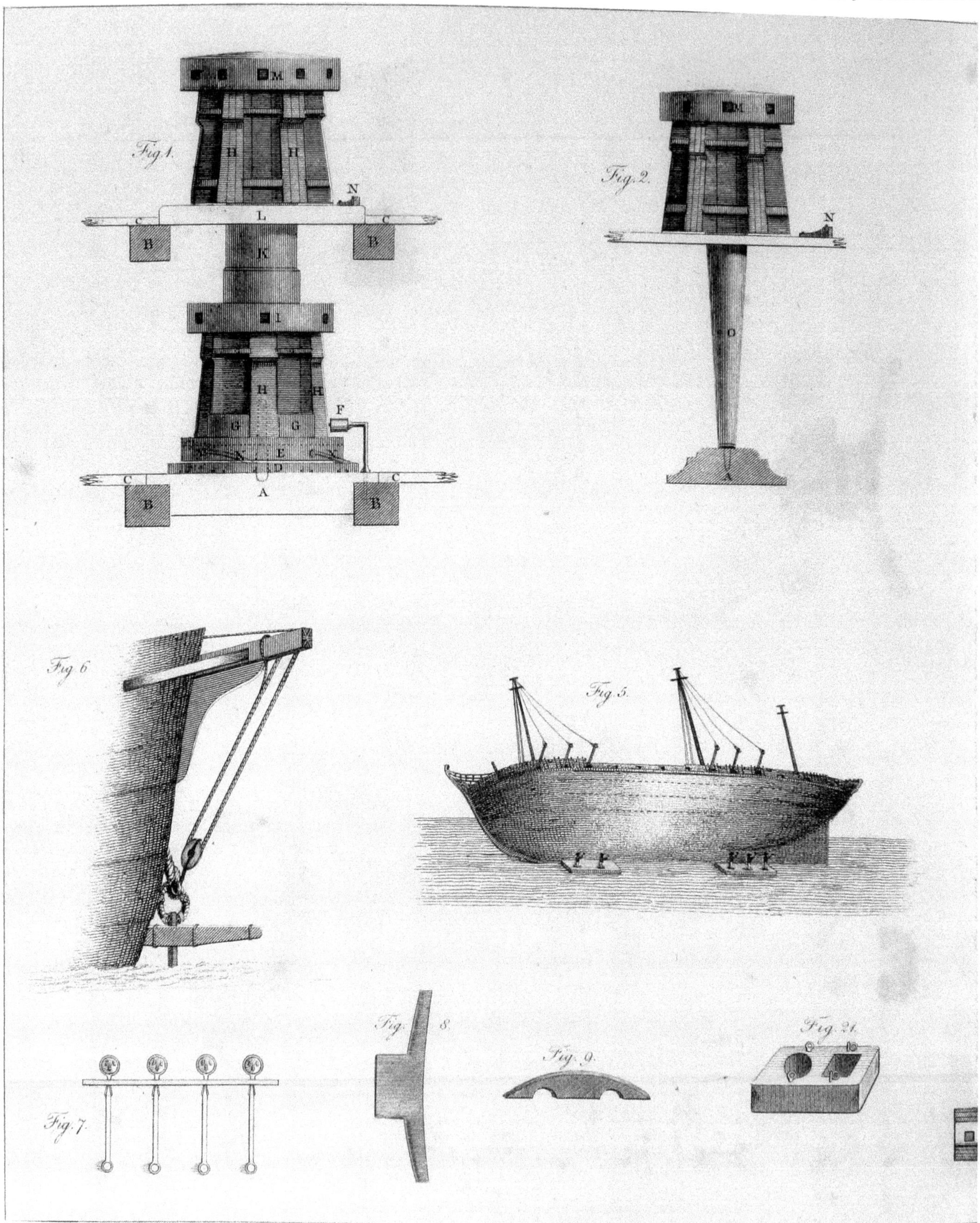

J Taplin, delin.

Published March 1.1815, by T. Cadell

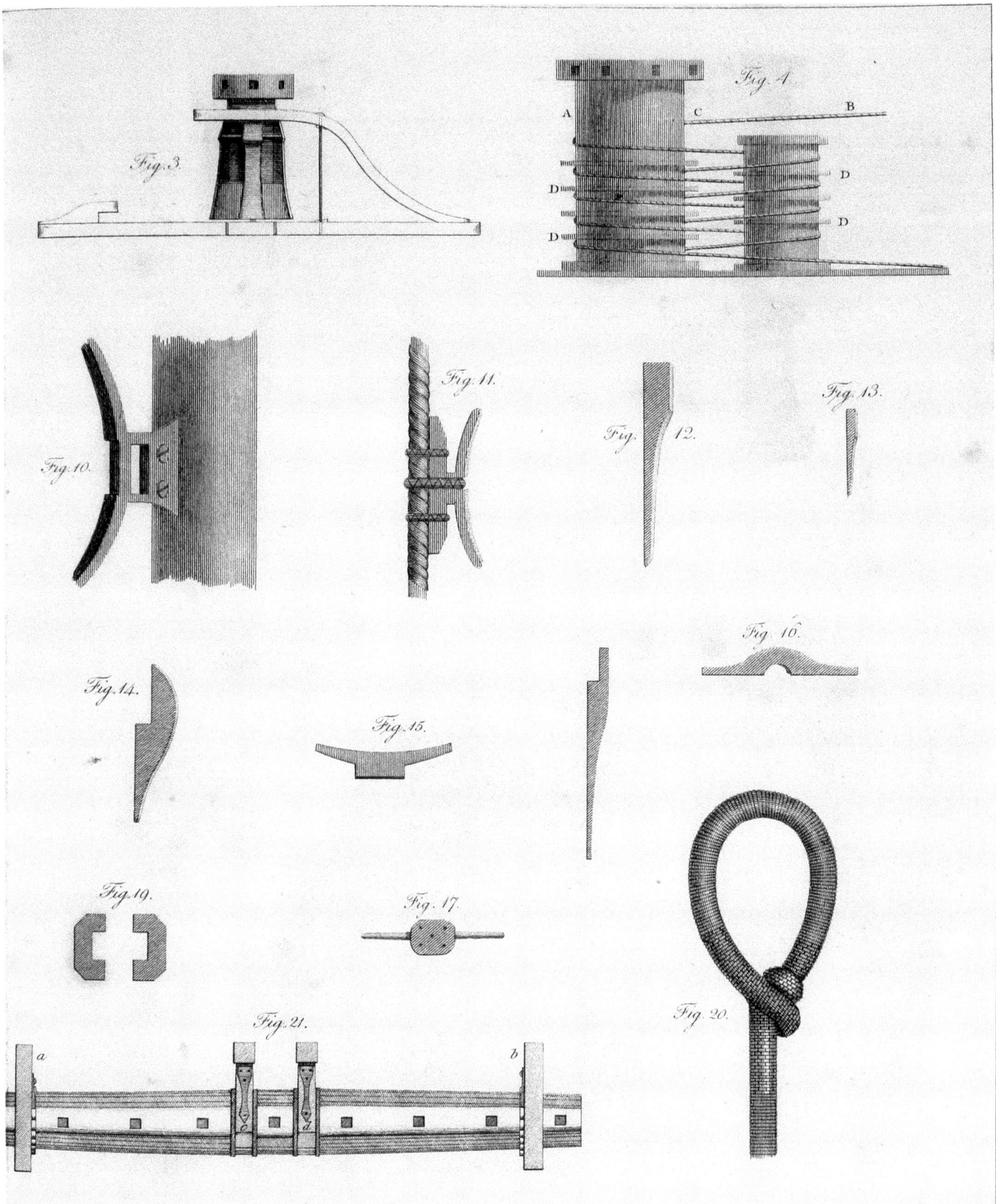

Mutlow Sc. Russell Co.
and W. Davies, Strand, London.

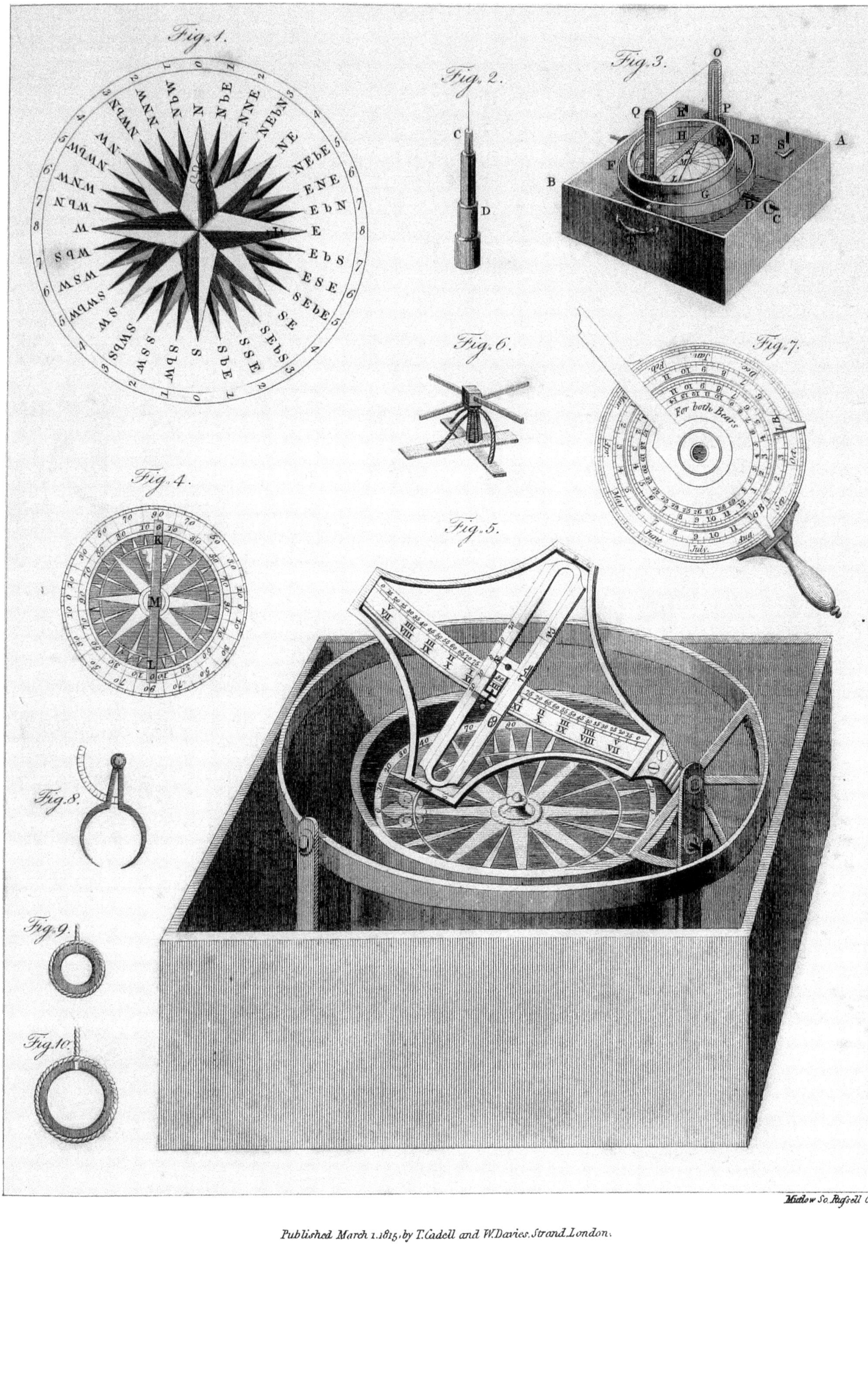

Midlow Sc. Russell Ct.

Published March 1.1815, by T. Cadell and W. Davies, Strand, London.

DECK, DIPPING-N

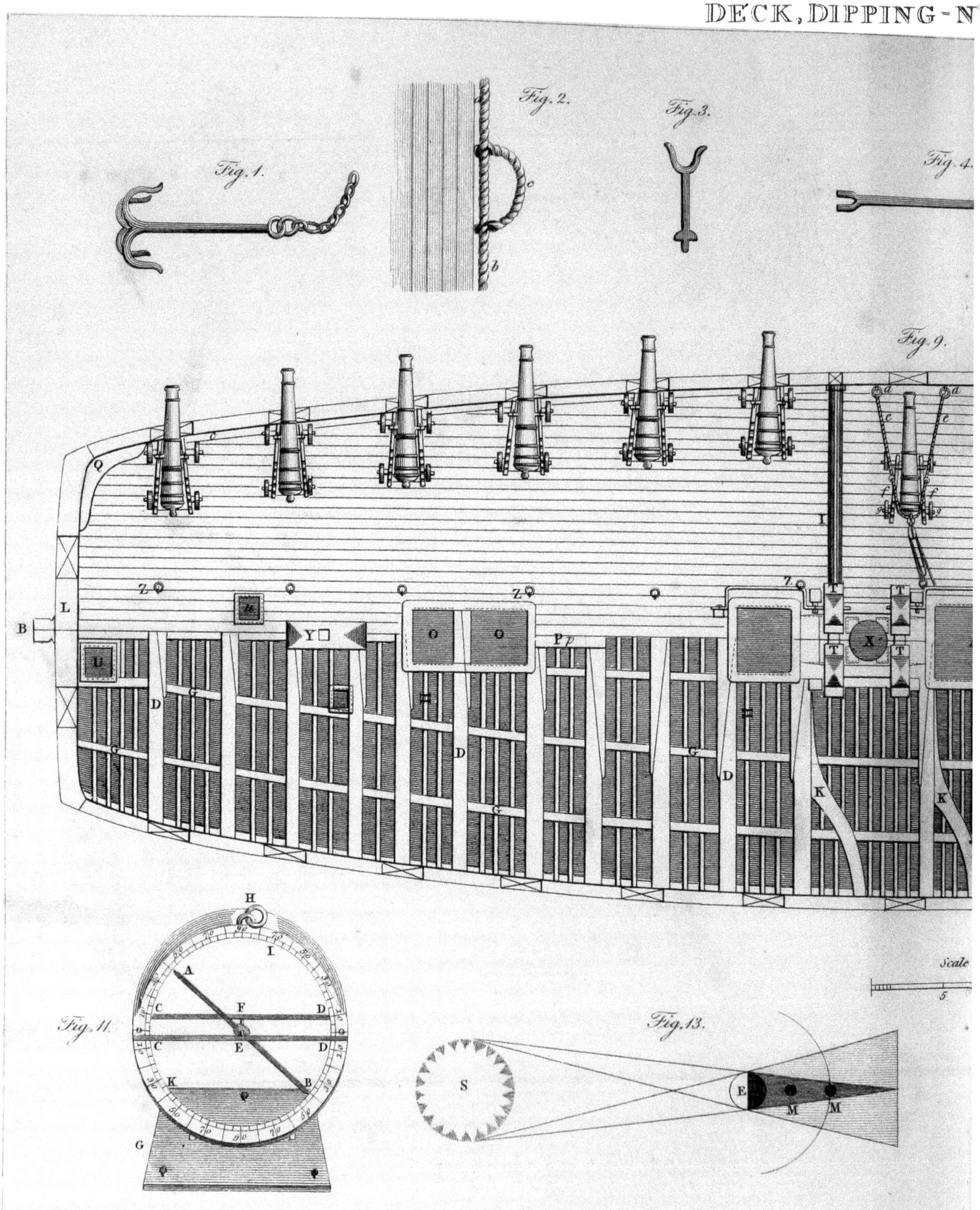

I. Taplin, Delin.

Published March 1.1815, by T. Cadell

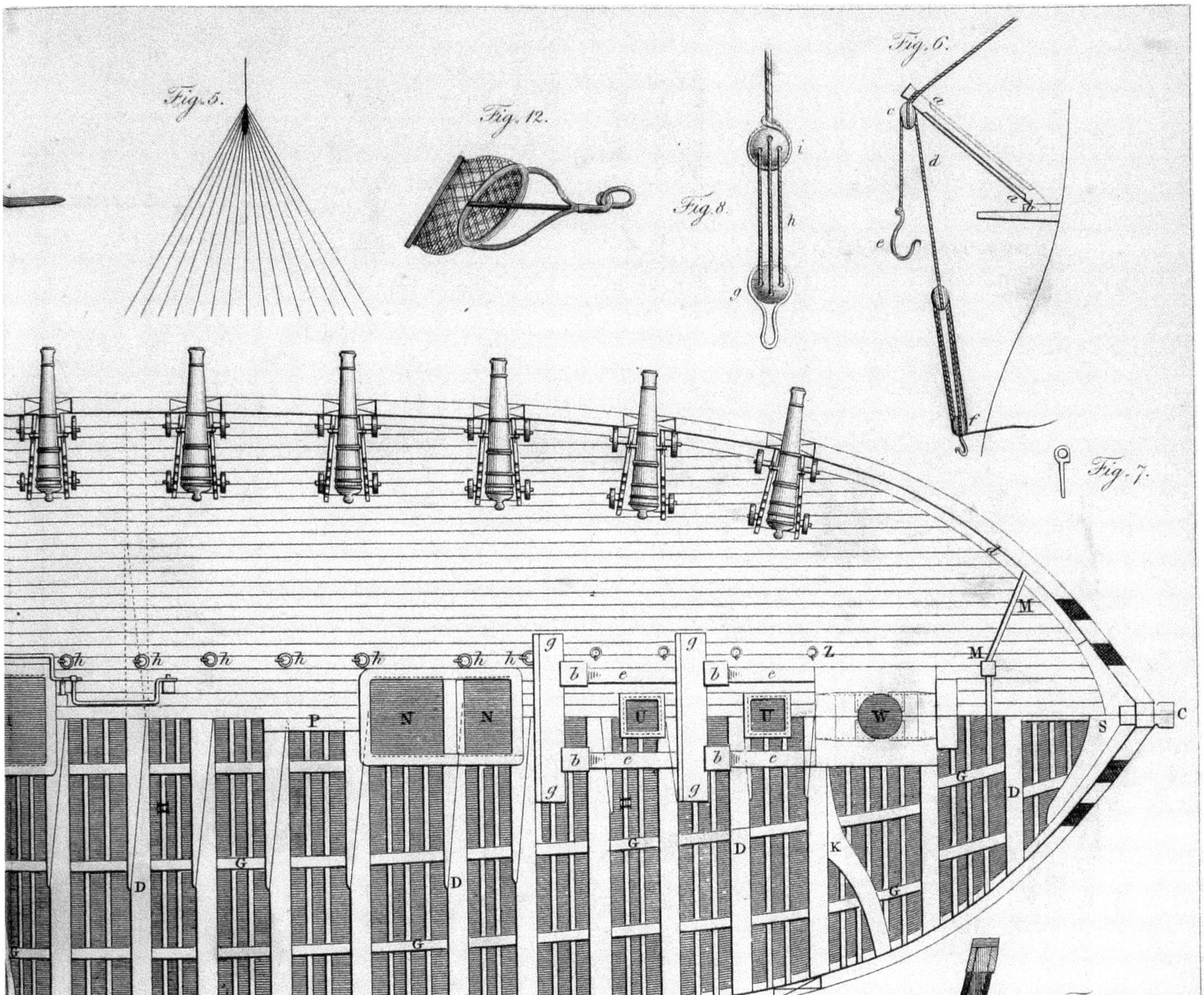

Mutlow So. Russell Co.

and W. Davies, Strand, London.

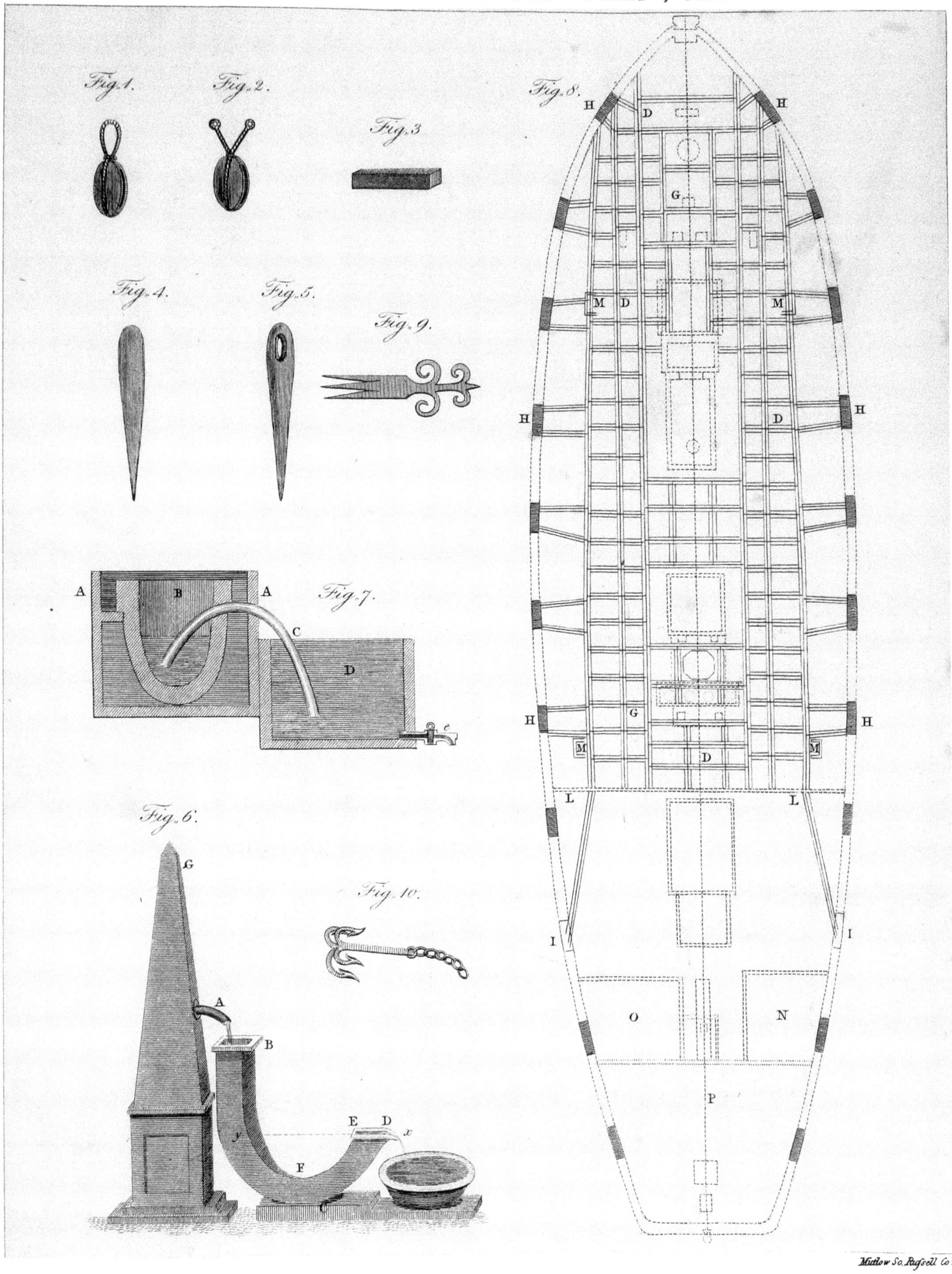

Midlow Sc. Russell Co.t

Published March 1. 1815, by T. Cadell and W. Davies, Strand, London.

HEAD AND QUARTER

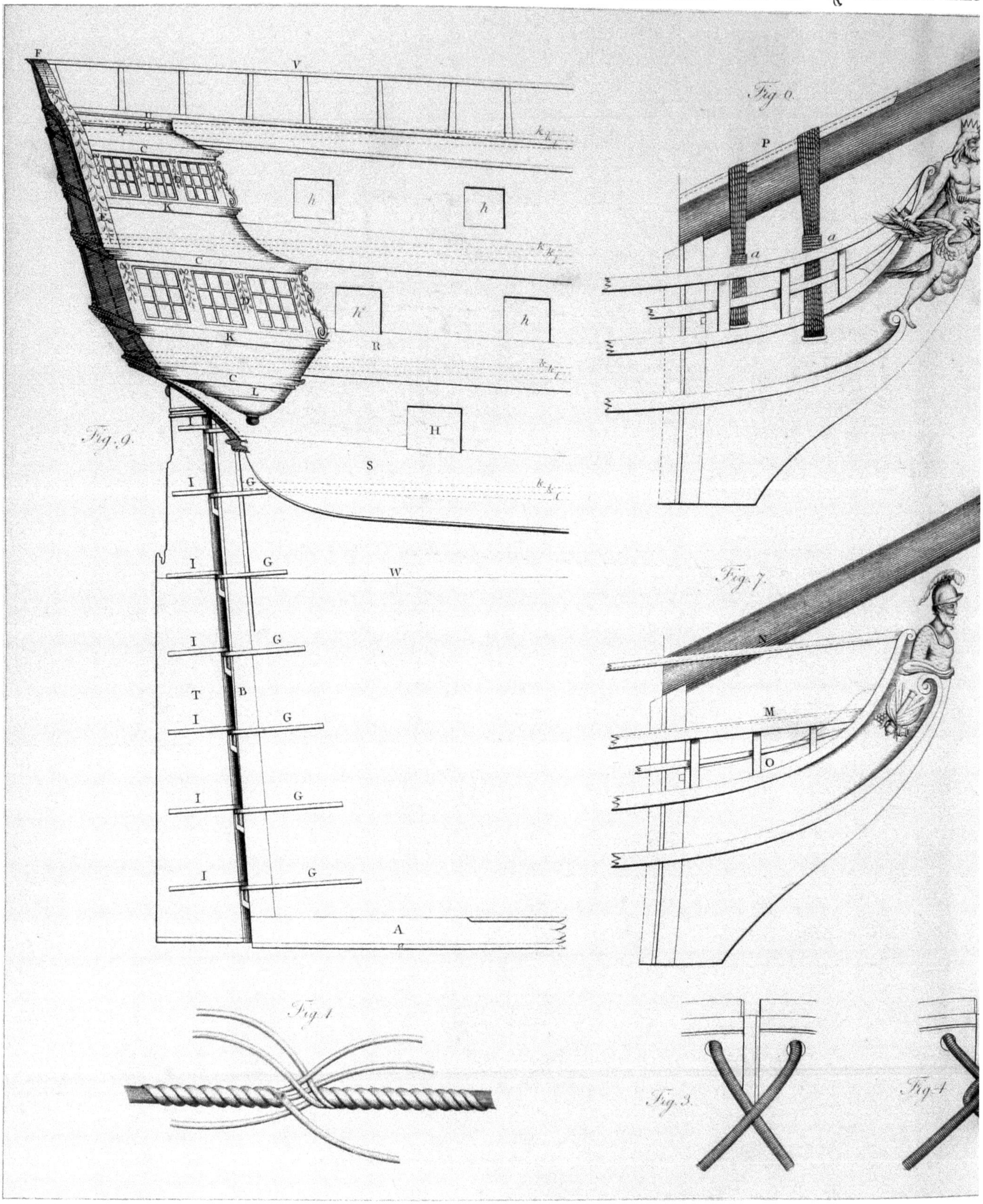

J. Taplin delin.

Published March 1.1815, by T. Cadell

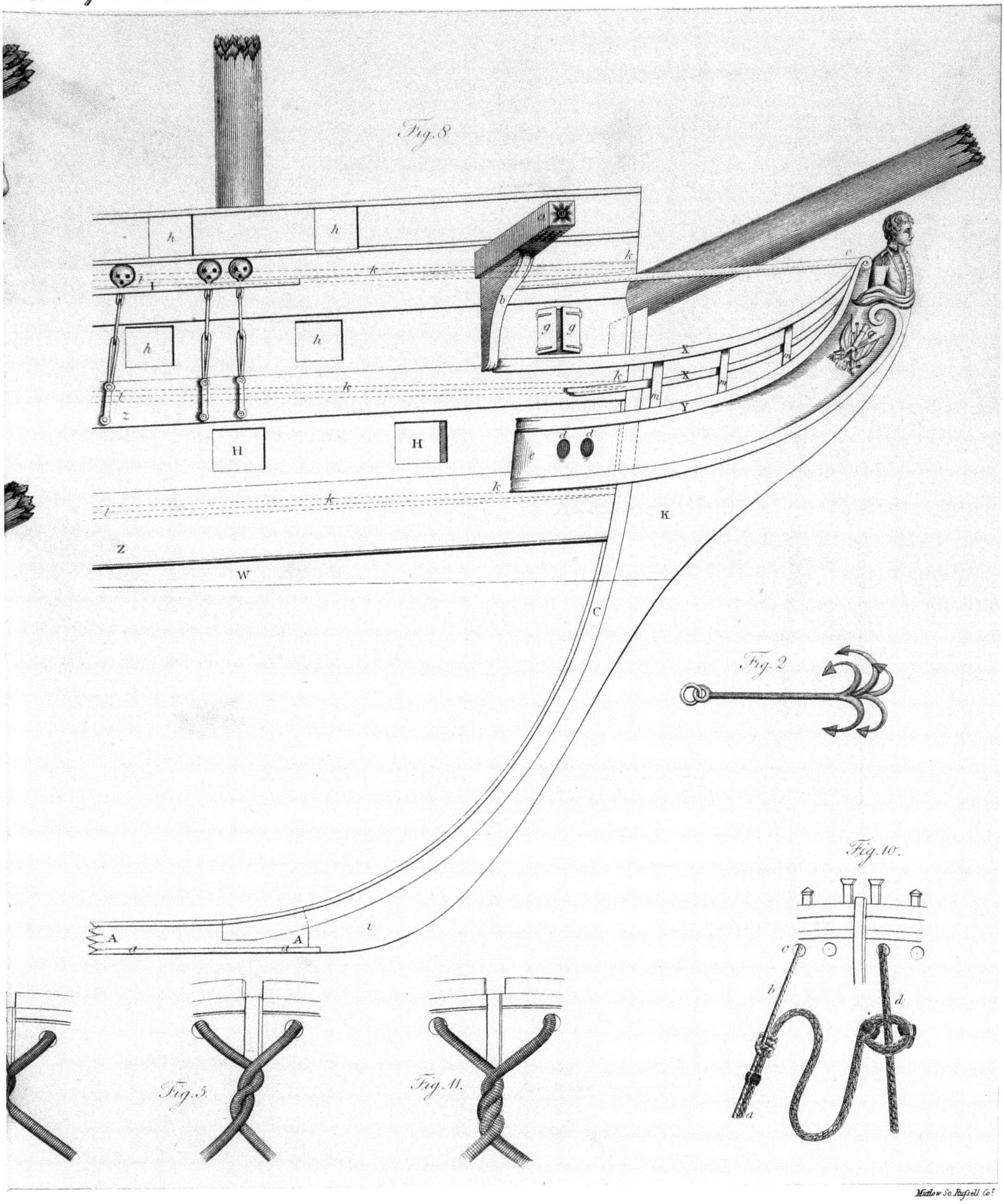
Fig. 8
Fig. 2
Fig. 10.
Fig. 5.
Fig. 11.
H
H
K
C
Z
W
X
X
Y
A
A

JURY-MAST, HITCHES, &c.

PLATE XII.

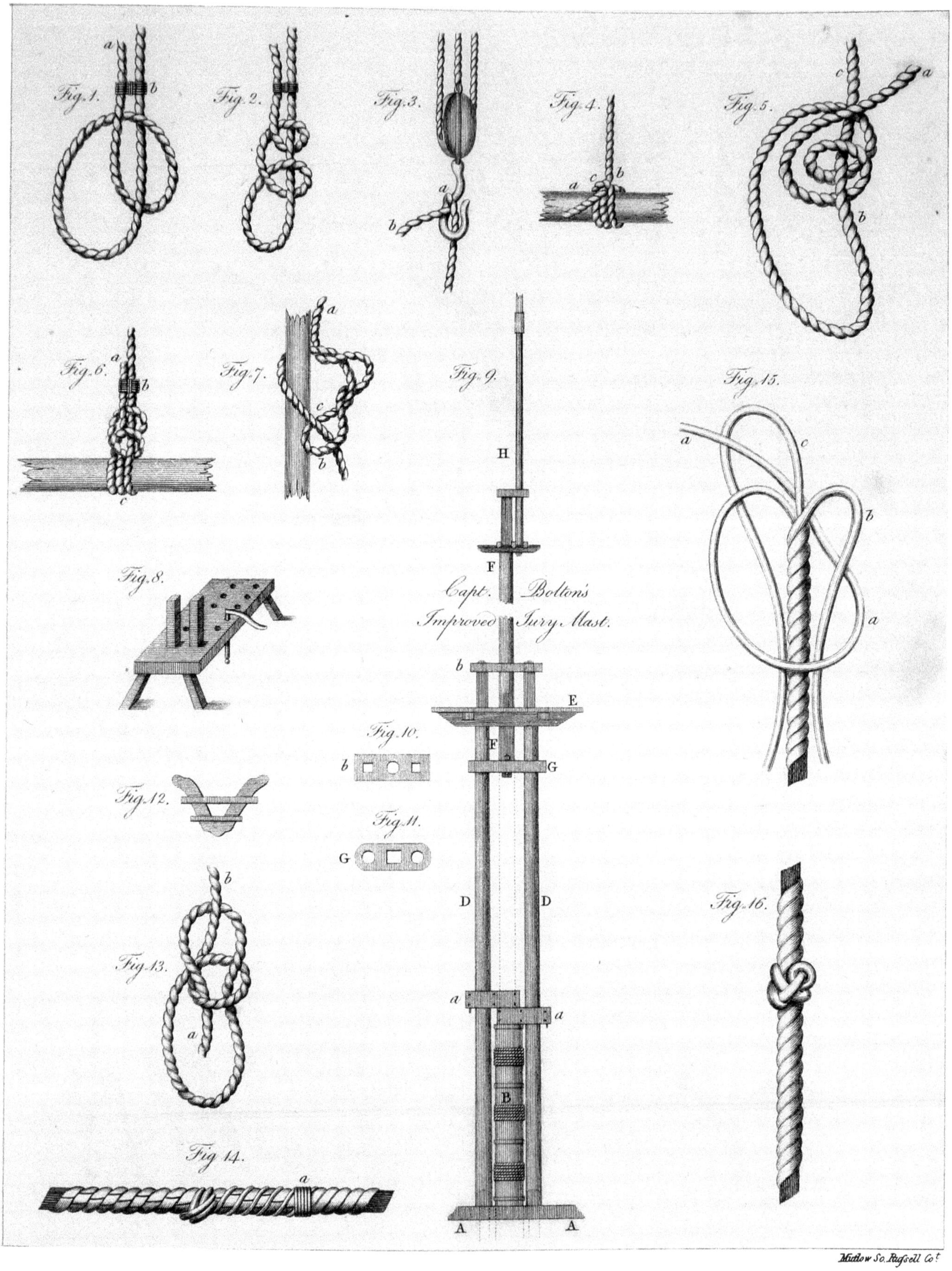

Mutlow So. Russell Ct.

Published March 1.1815, by T. Cadell and W. Davies, Strand, London.

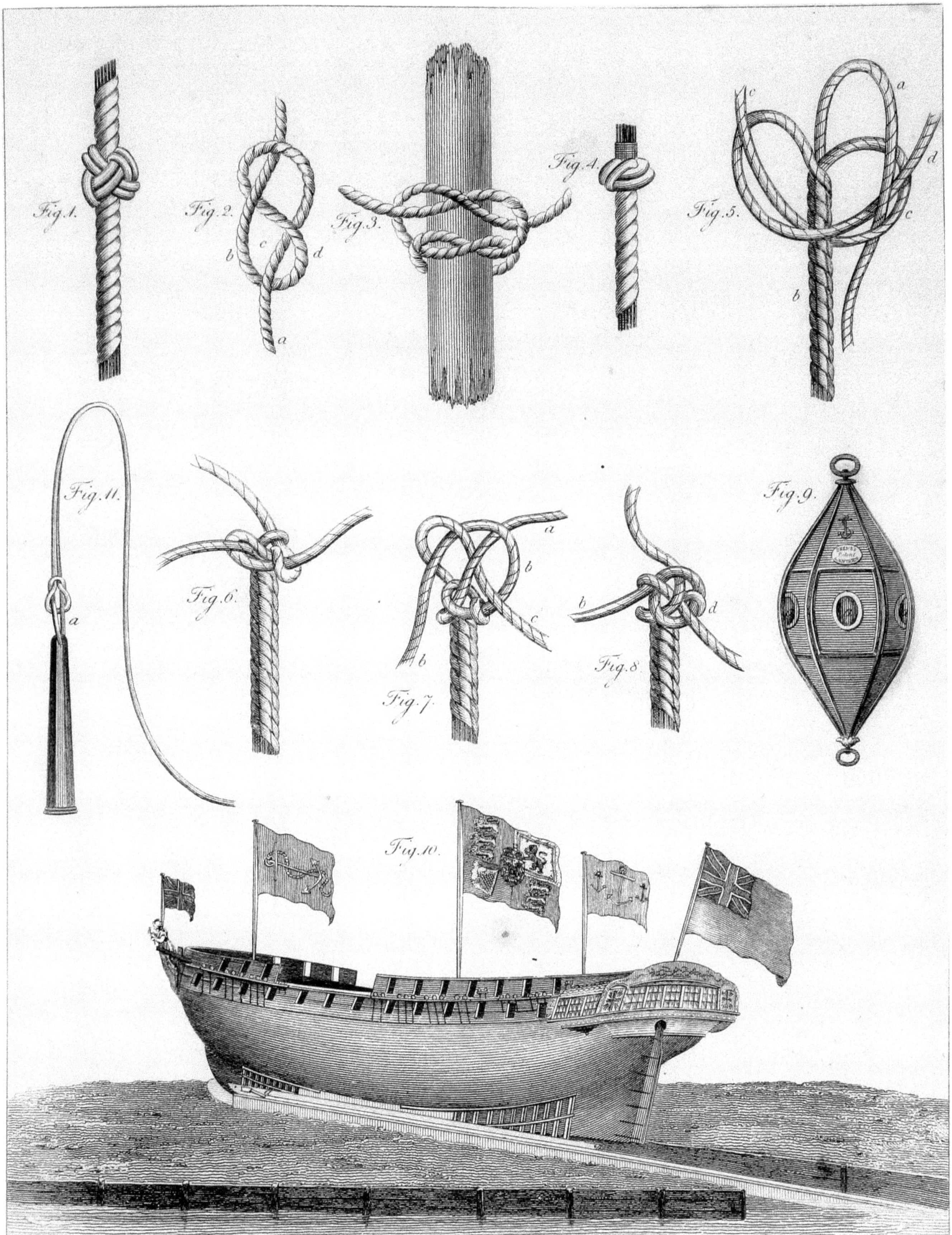

I. Taplin, delin. Mutlow So. Russell Co.

Published March 1st. 1815. by T. Cadell and W. Davies, Strand London.

LIFE-BOAT, LOG, &c.

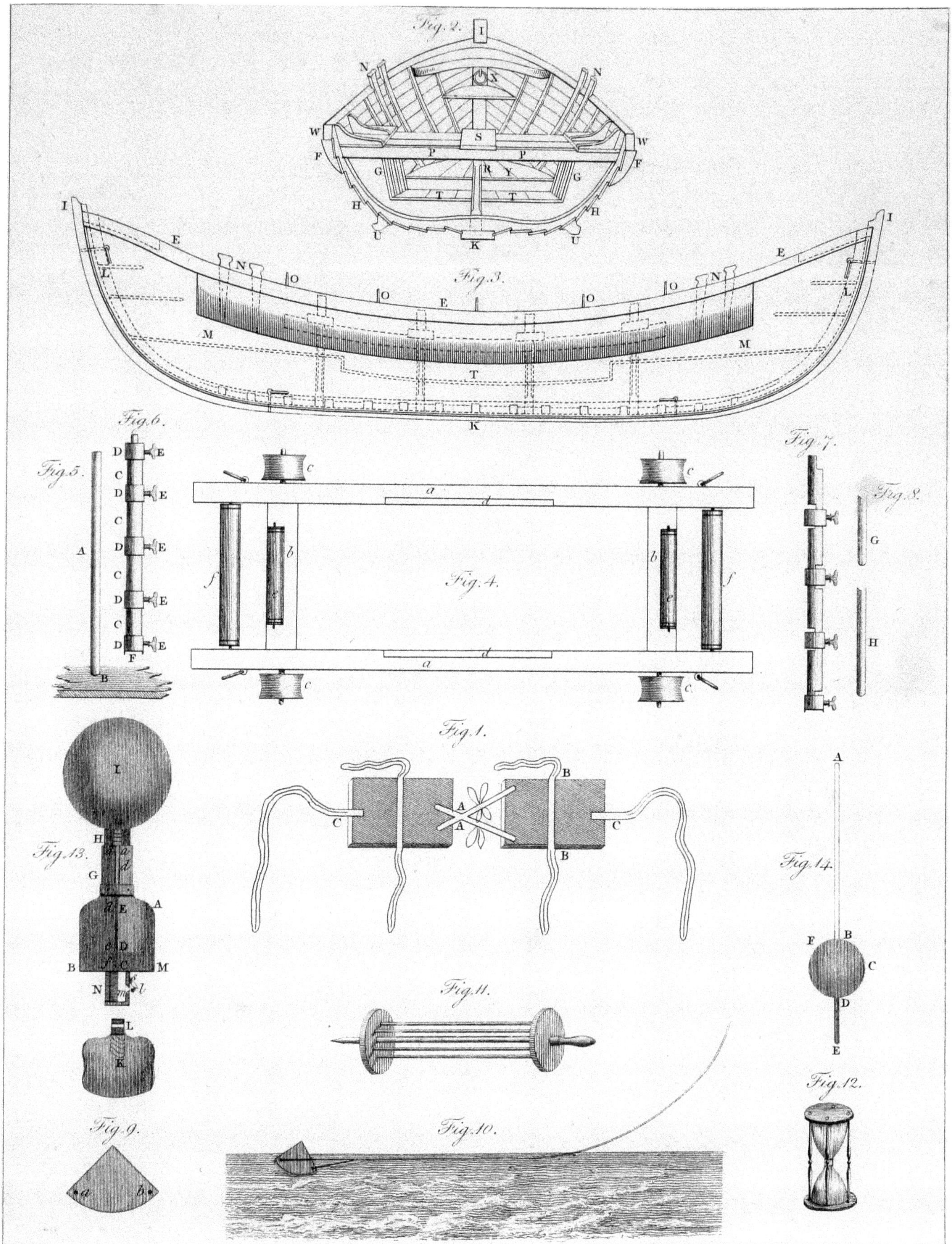

J. Taplin, delin.

Mutlow So. Russell Co.

Published March 1. 1815, by T. Cadell and W. Davies, Strand, London.

PERSPECTIVE VIEWS OF FLE

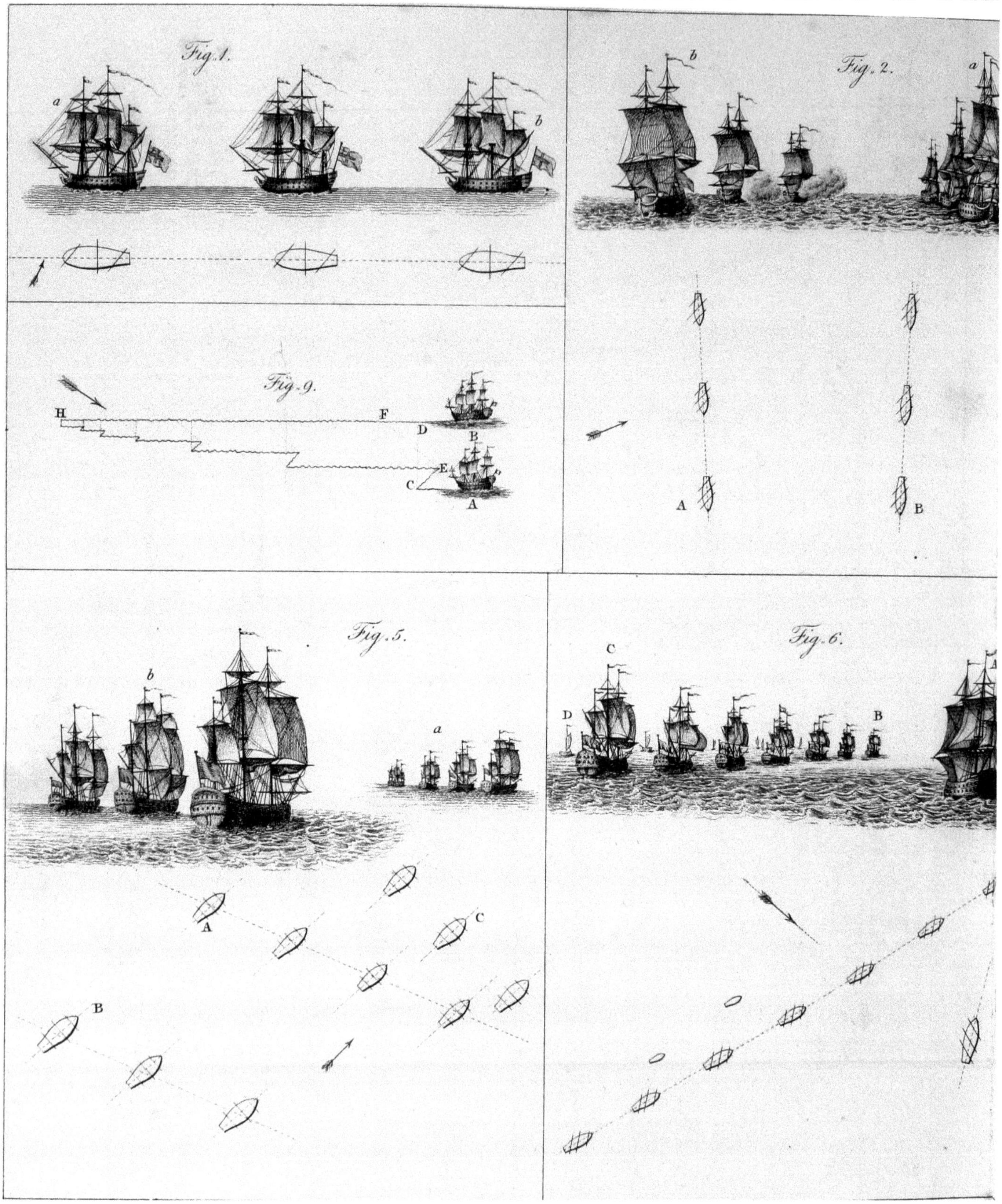

Published March 1.1815, by T. Cadell

ETS IN THE LINE OF BATTLE. PLATE XV.

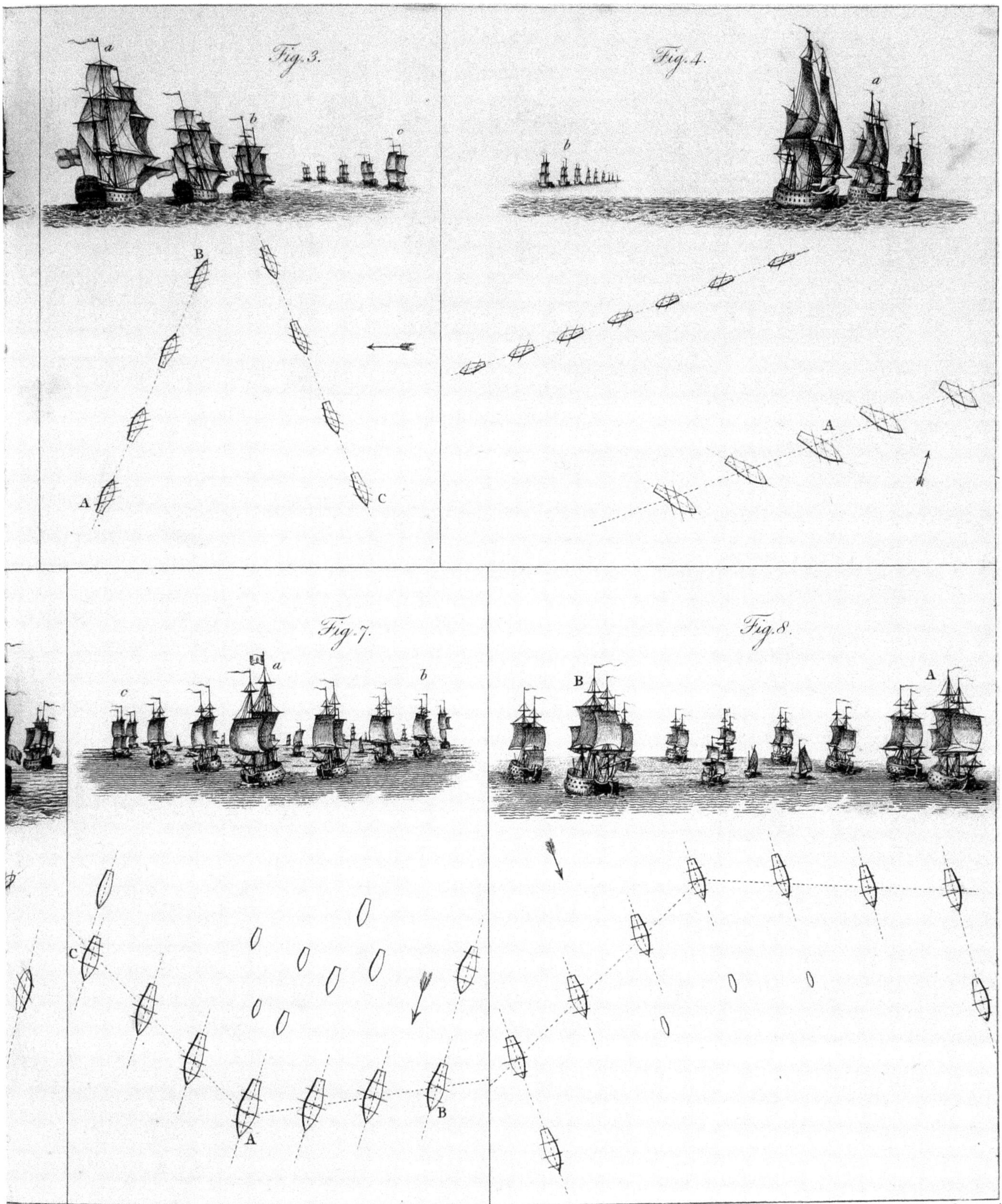

Mutlow Sc. Russell Ct.

and W. Davies, Strand, London.

MASTS, MORTARS, BED, BED-BOLSTER, SHELL, &c.

PLATE XVI.

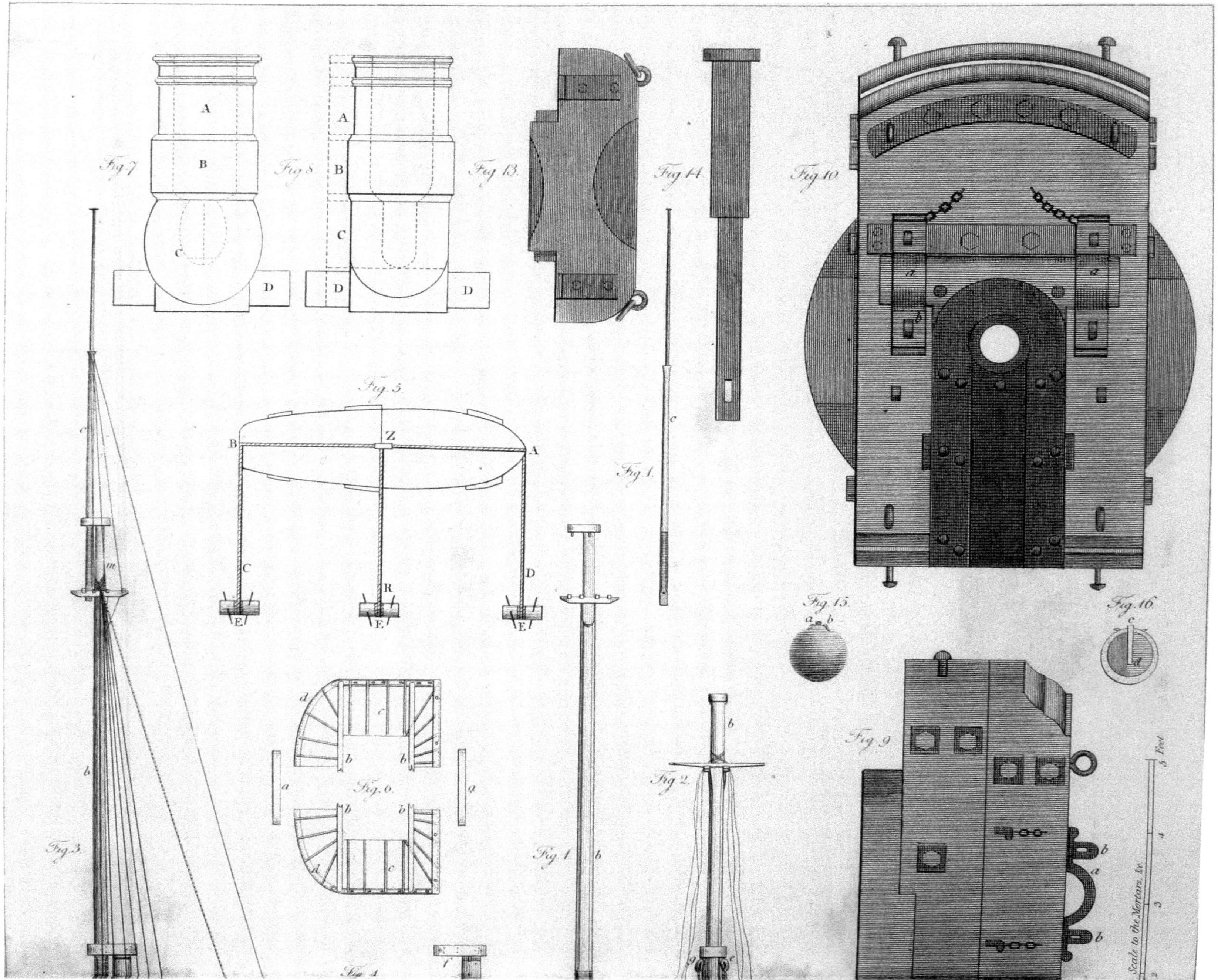

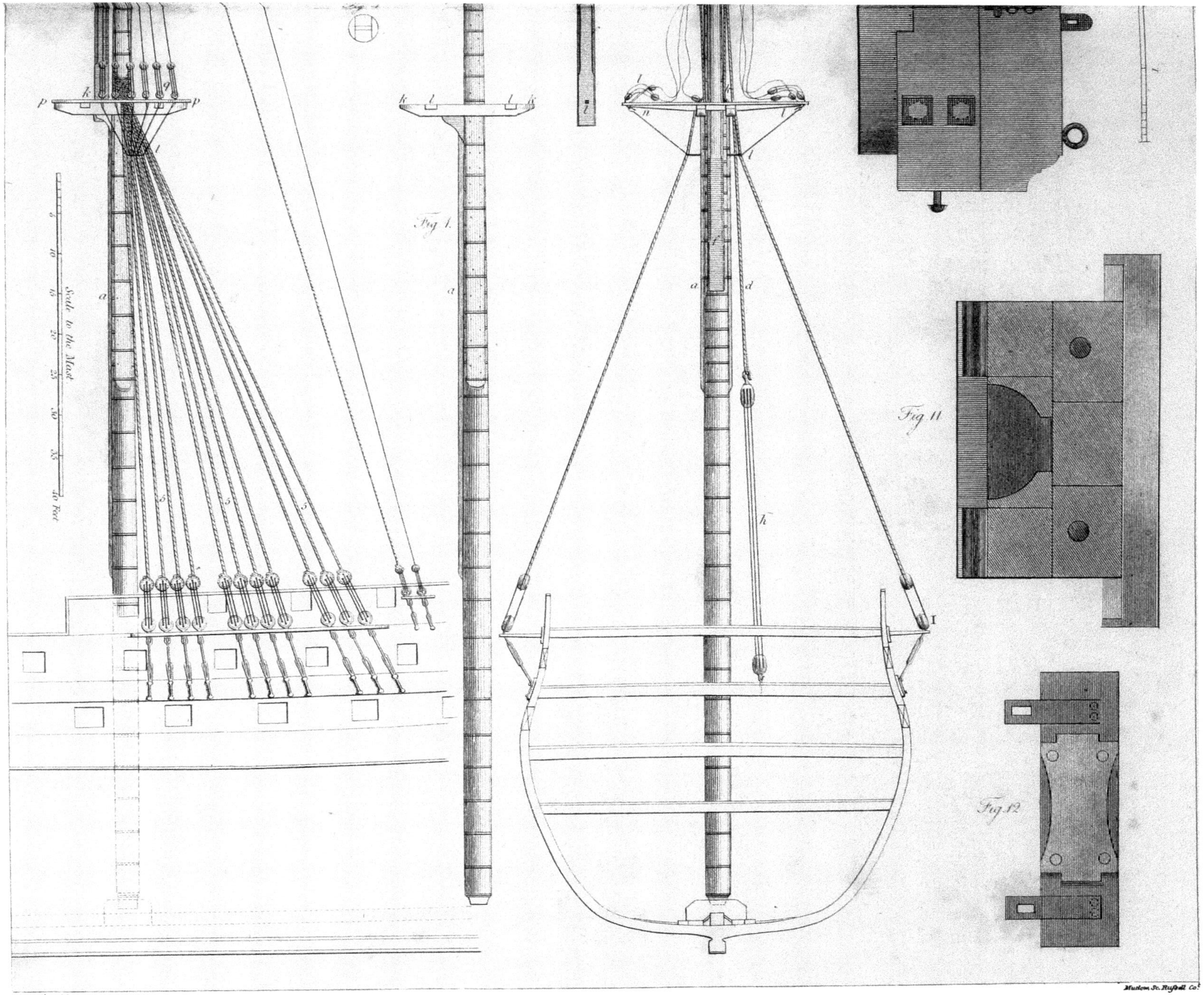

J. Taplin. del.

Mutlow Sc. Russell Co.

Published March 1.1815, by T. Cadell and W. Davies, Strand, London.

PLAN AND ELEVATION

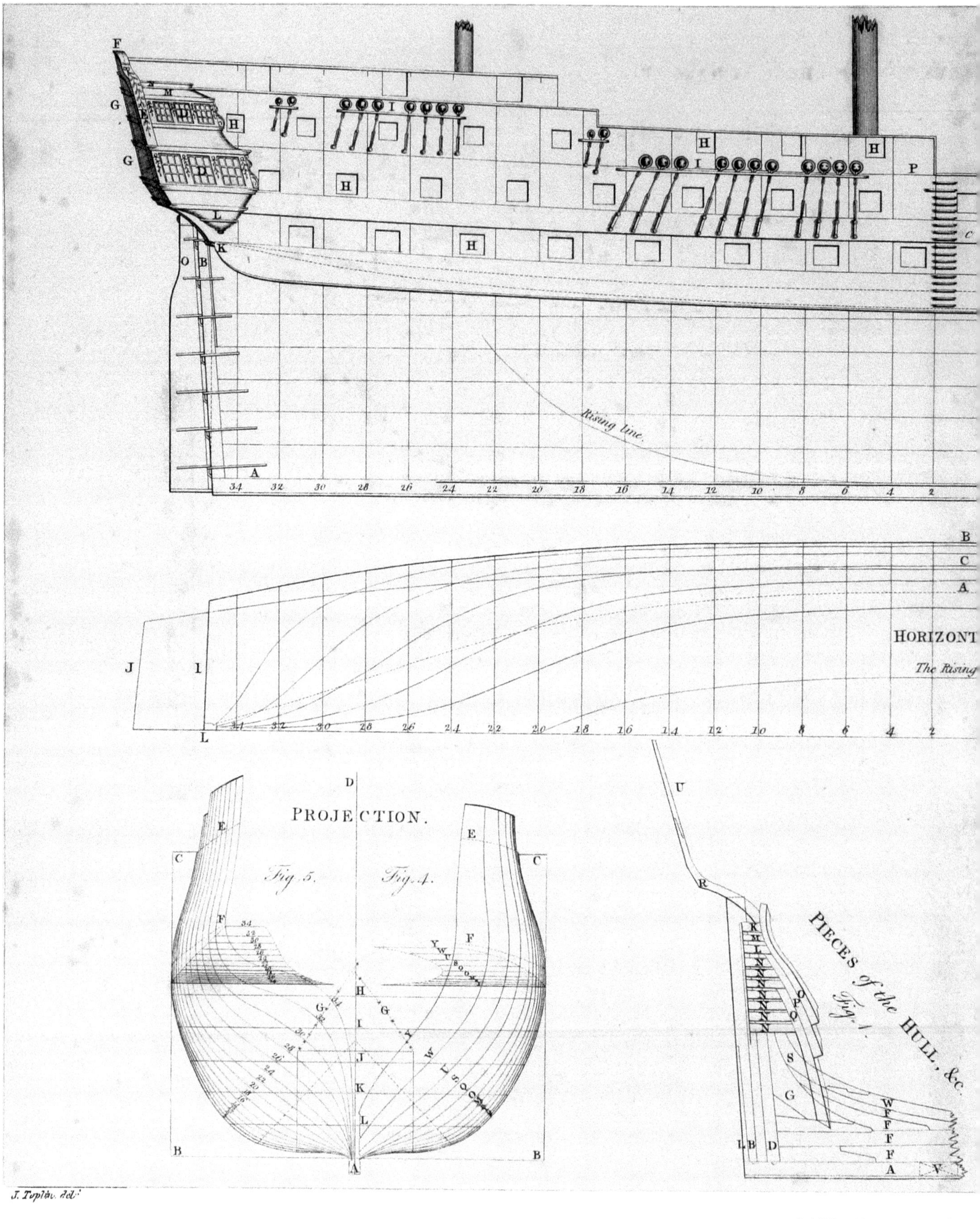

J. Topliss, del.

Published March 1.1815, by T. Cadell

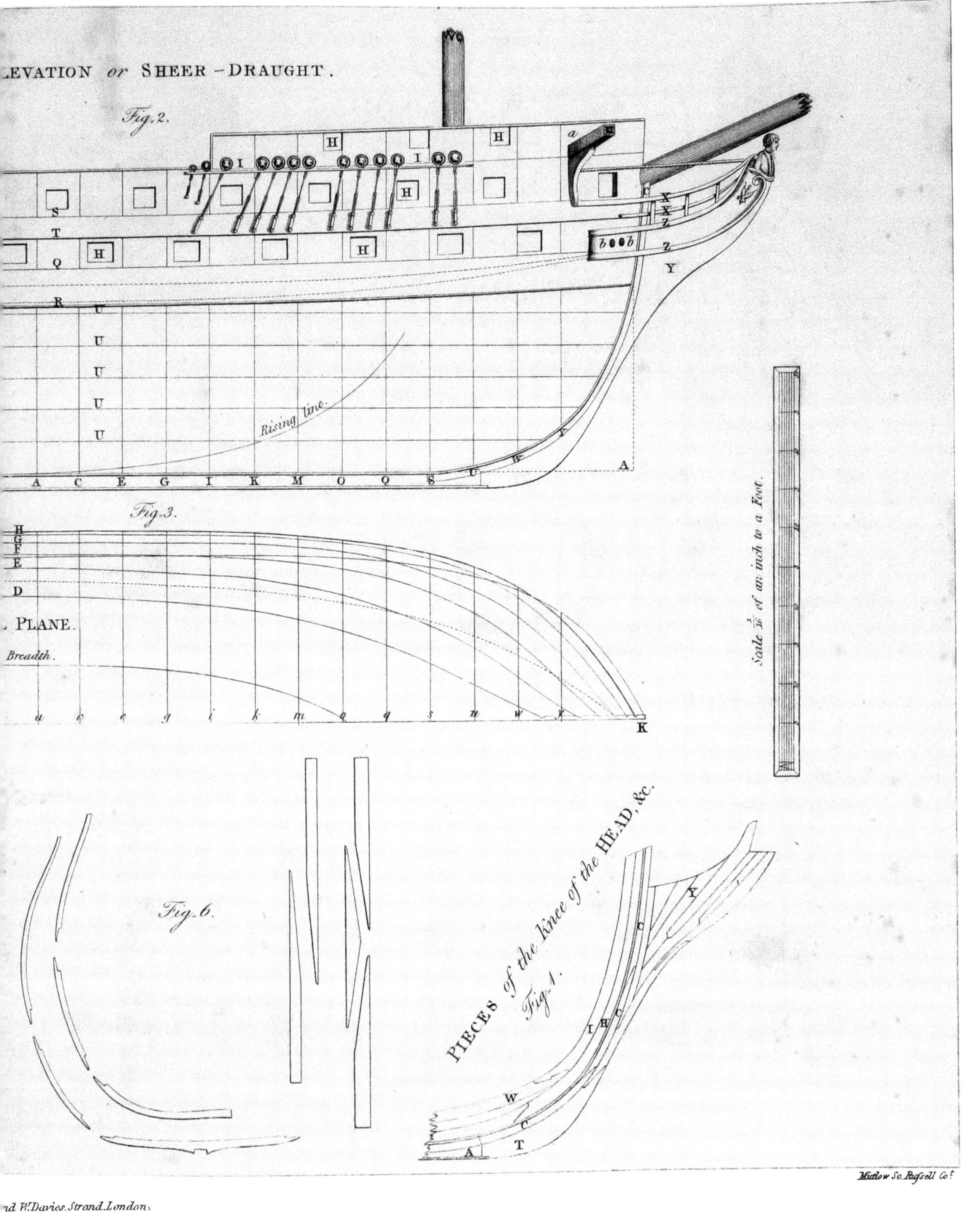

Mitlow So. Russell Co.

nd W. Davies. Strand. London.

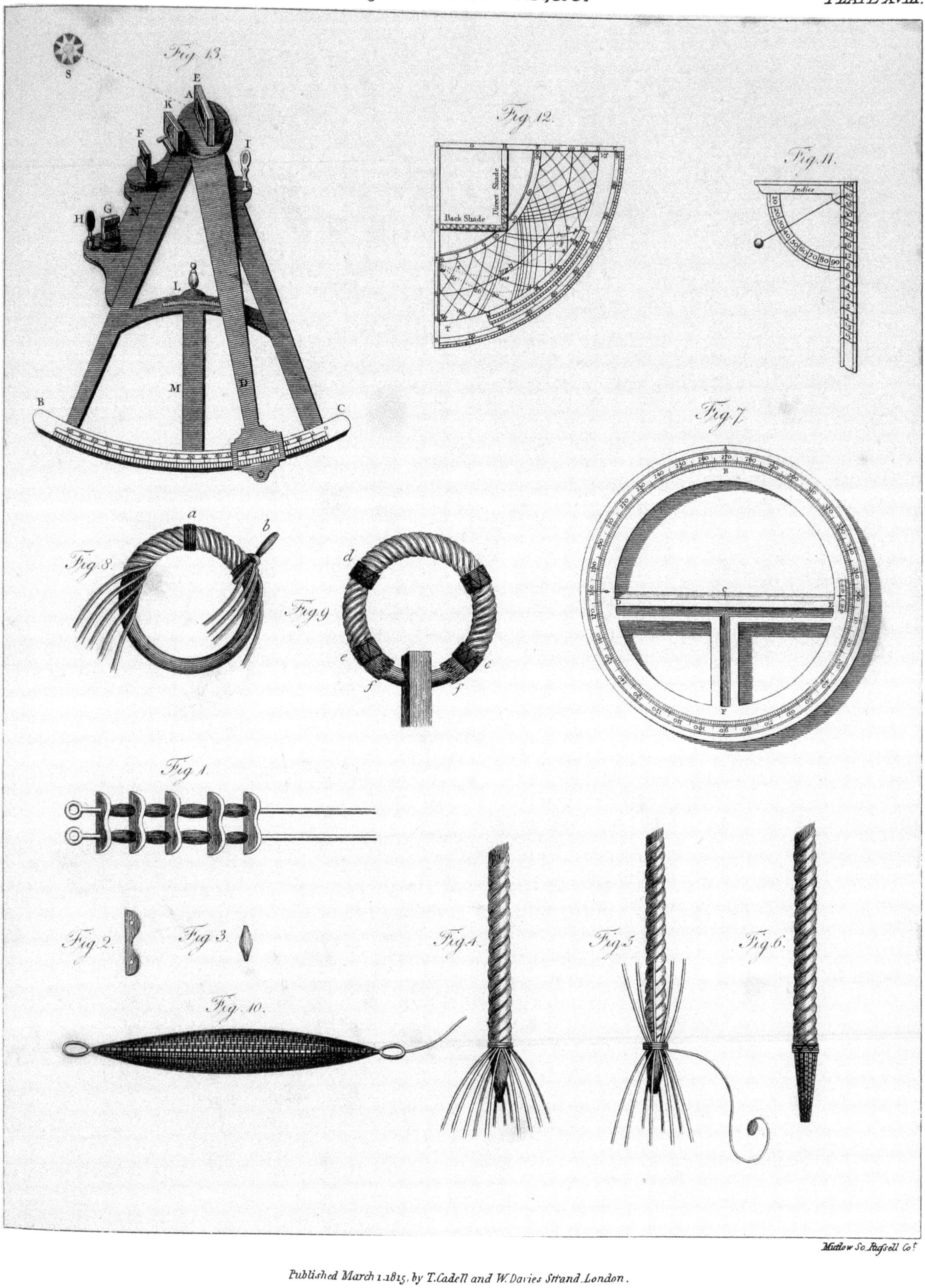

Mutlow So. Russell Co.

Published March 1.1815, by T. Cadell and W. Davies Strand. London.

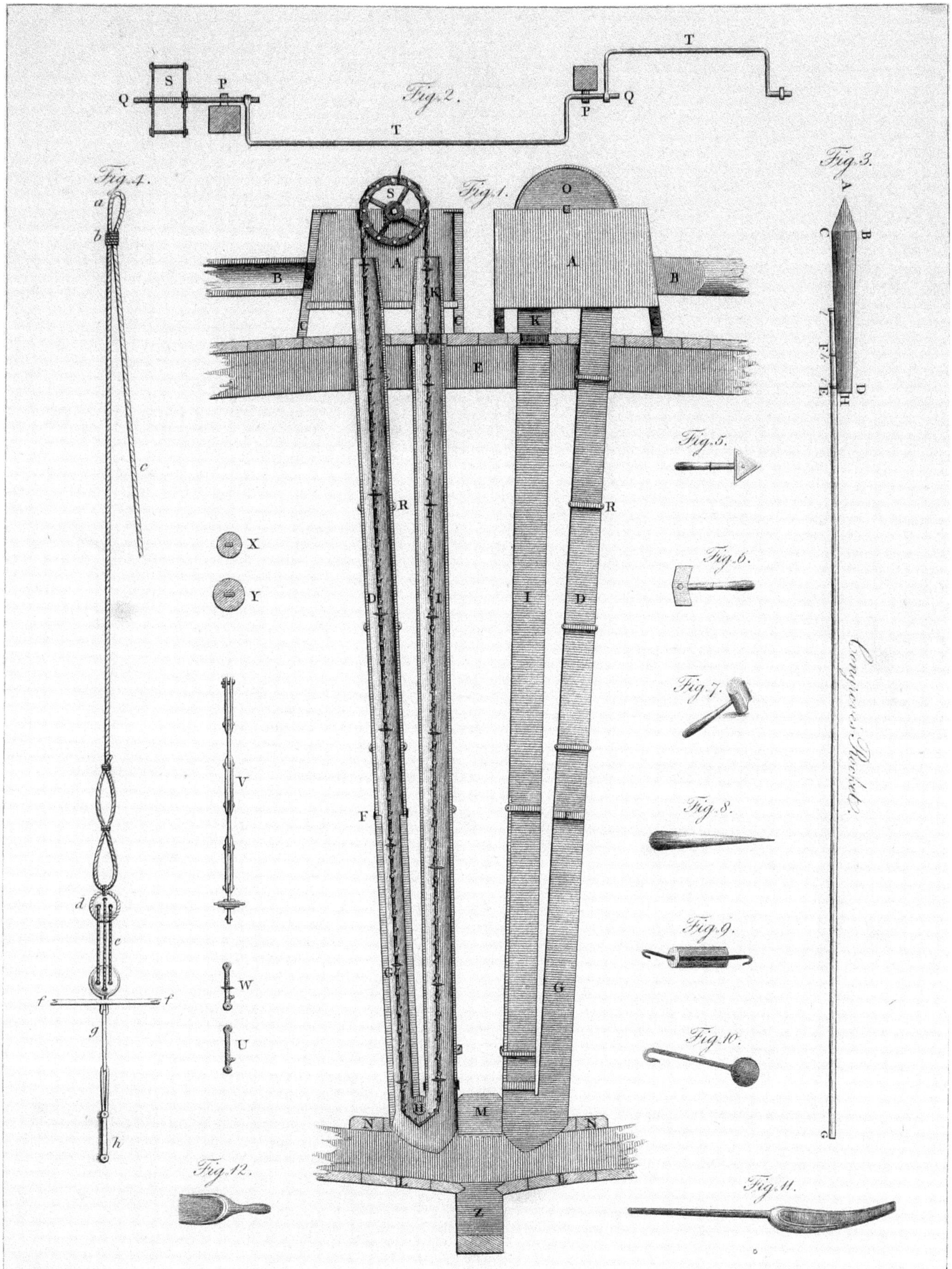

I. Taplin, Delin. Mutlow, Sc Russell Co.

Published March 1.1815, by T. Cadell and W. Davies, Strand, London.

MASTS, YARDS, SAILS, &c.

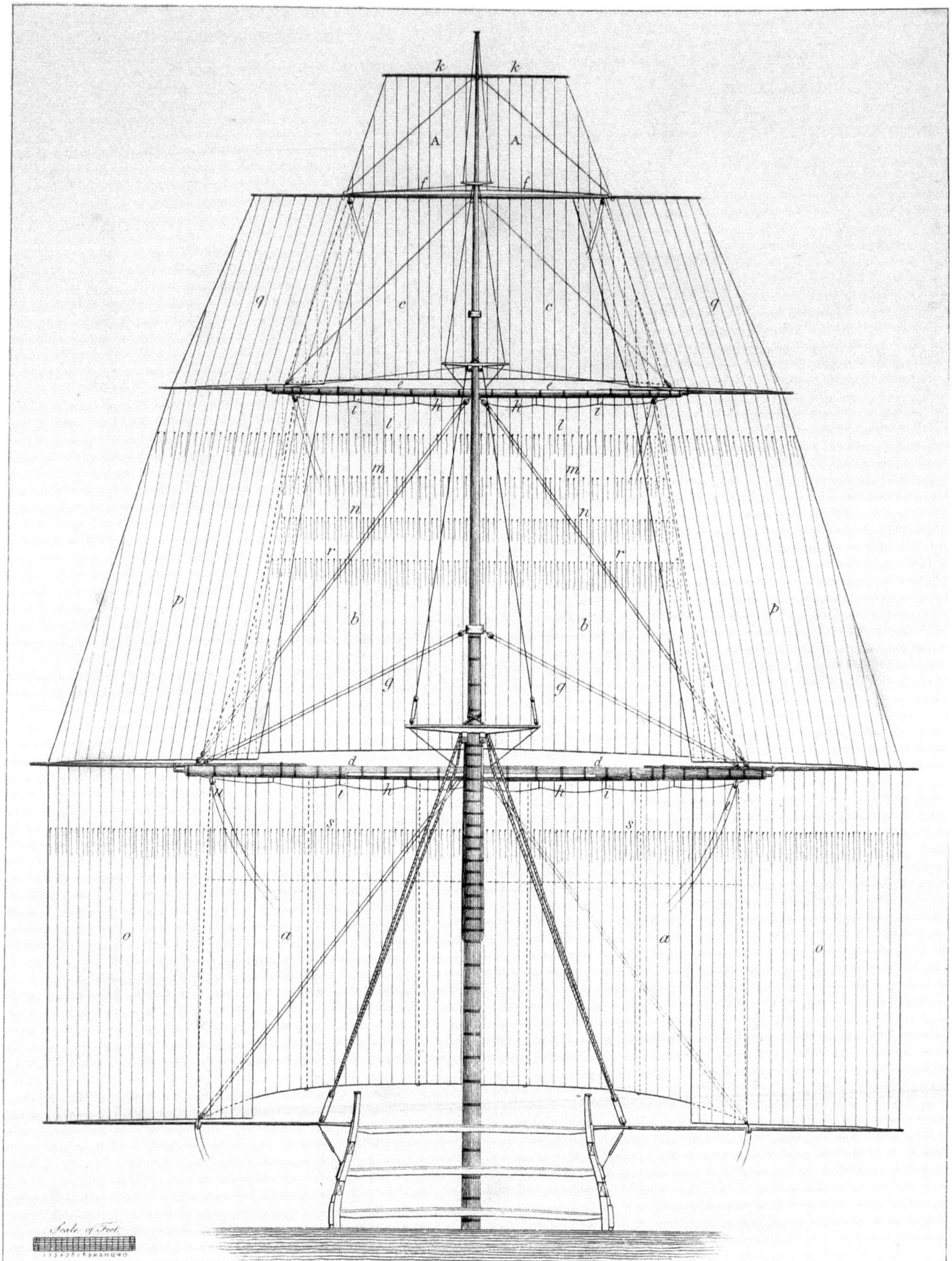

I. Taplin. Delin

H. Mutlow. Sc.

Published March 1.1815, by T. Cadell and W. Davies, Strand London.

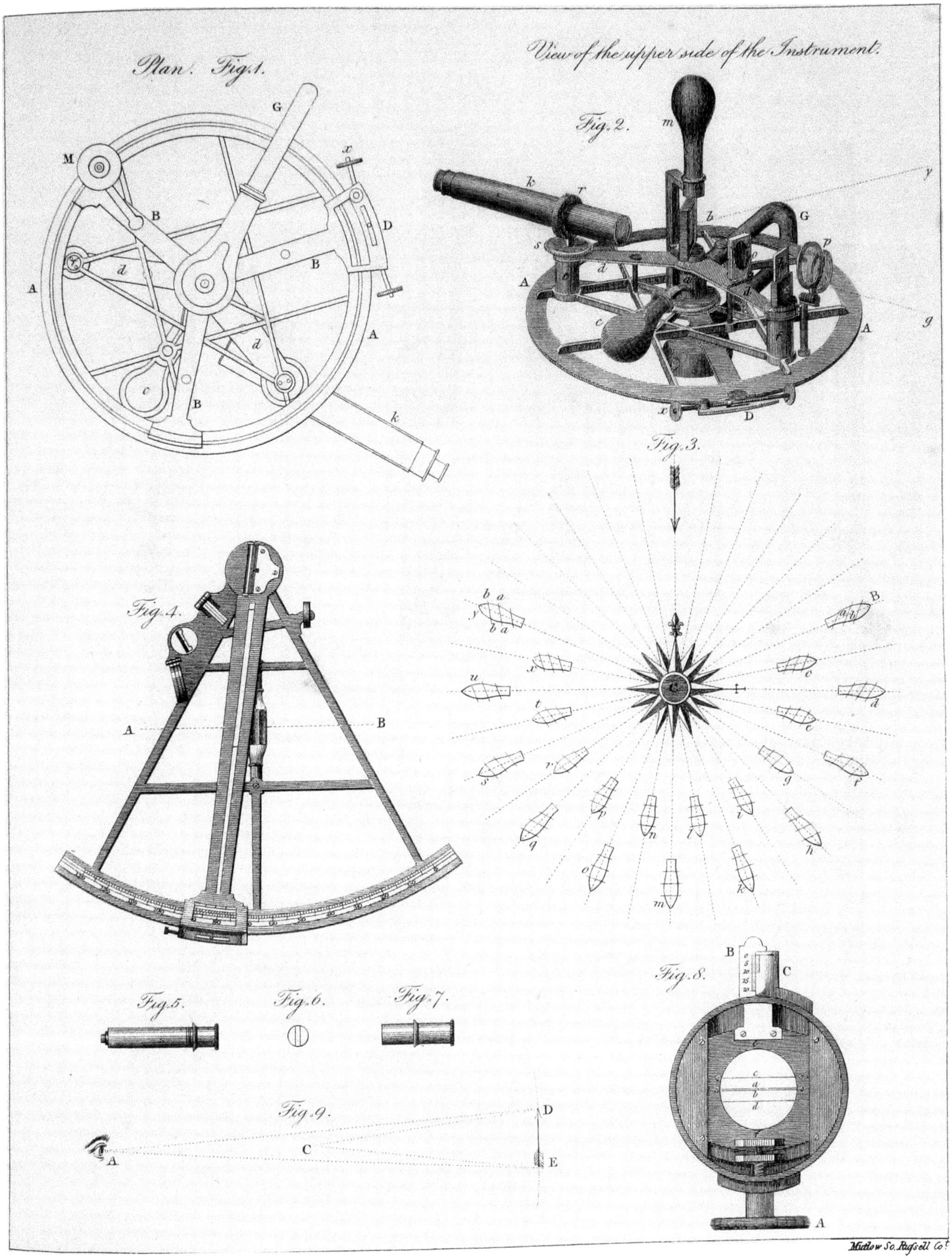

Mutlow Sc. Russell Ct.

Published March 1.1815, by T. Cadell and W. Davies, Strand London.

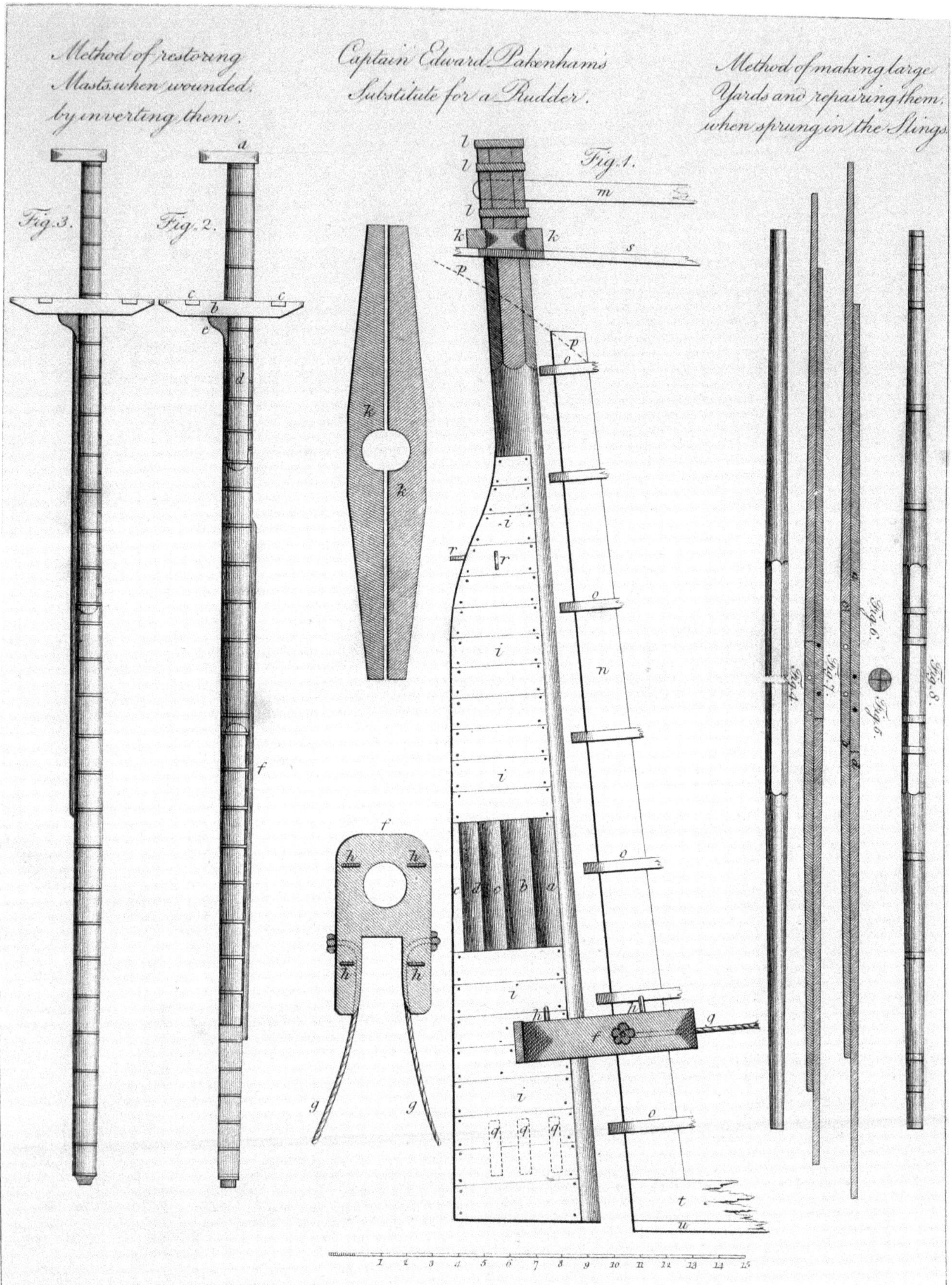

J. Taplin, del.t Mutlow So. Russell Co.t

Published March 1. 1815, by T. Cadell and W. Davies, Strand London.

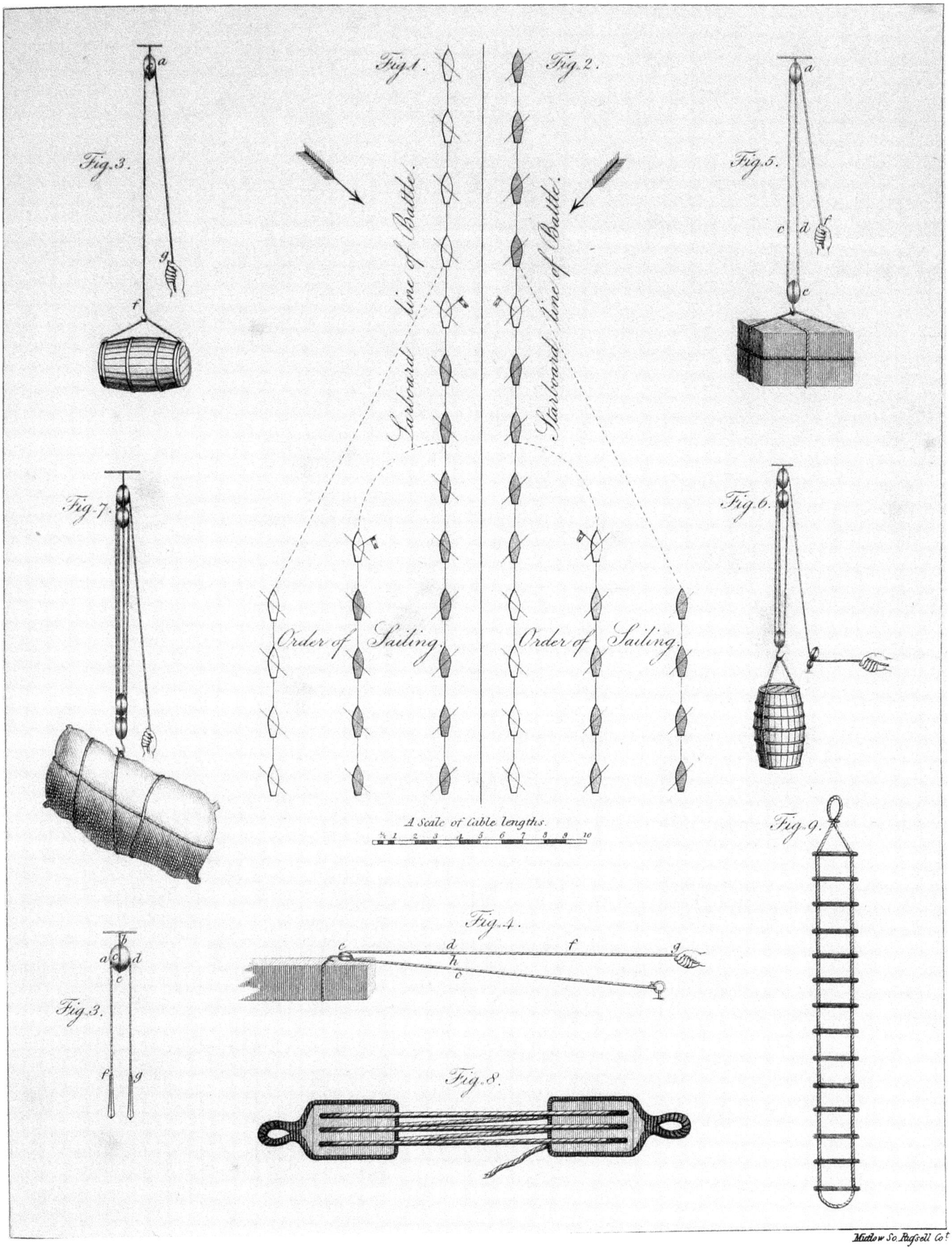

Published March 1.1815, by T. Cadell and W. Davies, Strand, London.

BOATS, AND THE FRAME

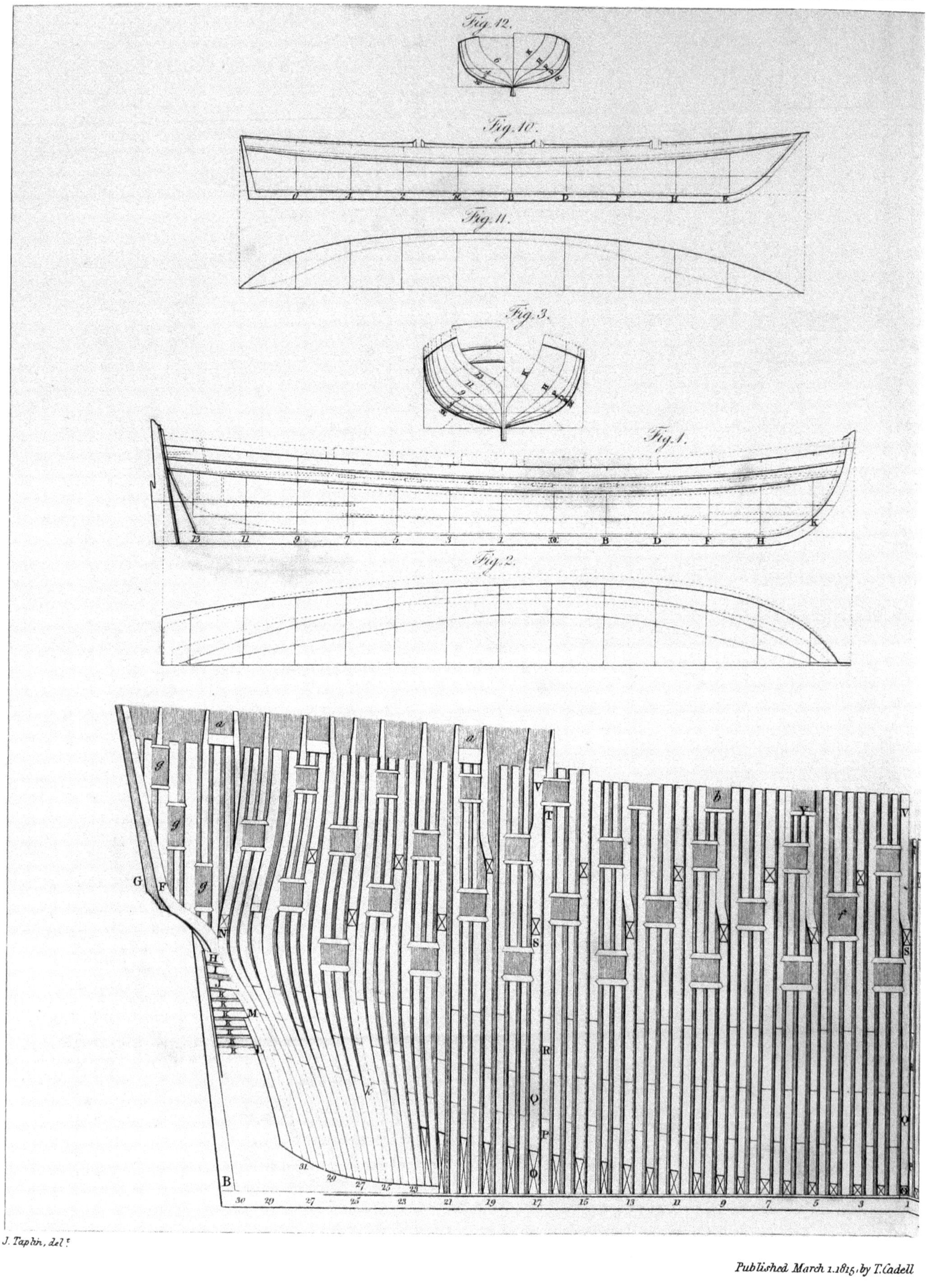

J. Taplin, delt.

Published March 1.1815, by T. Cadell

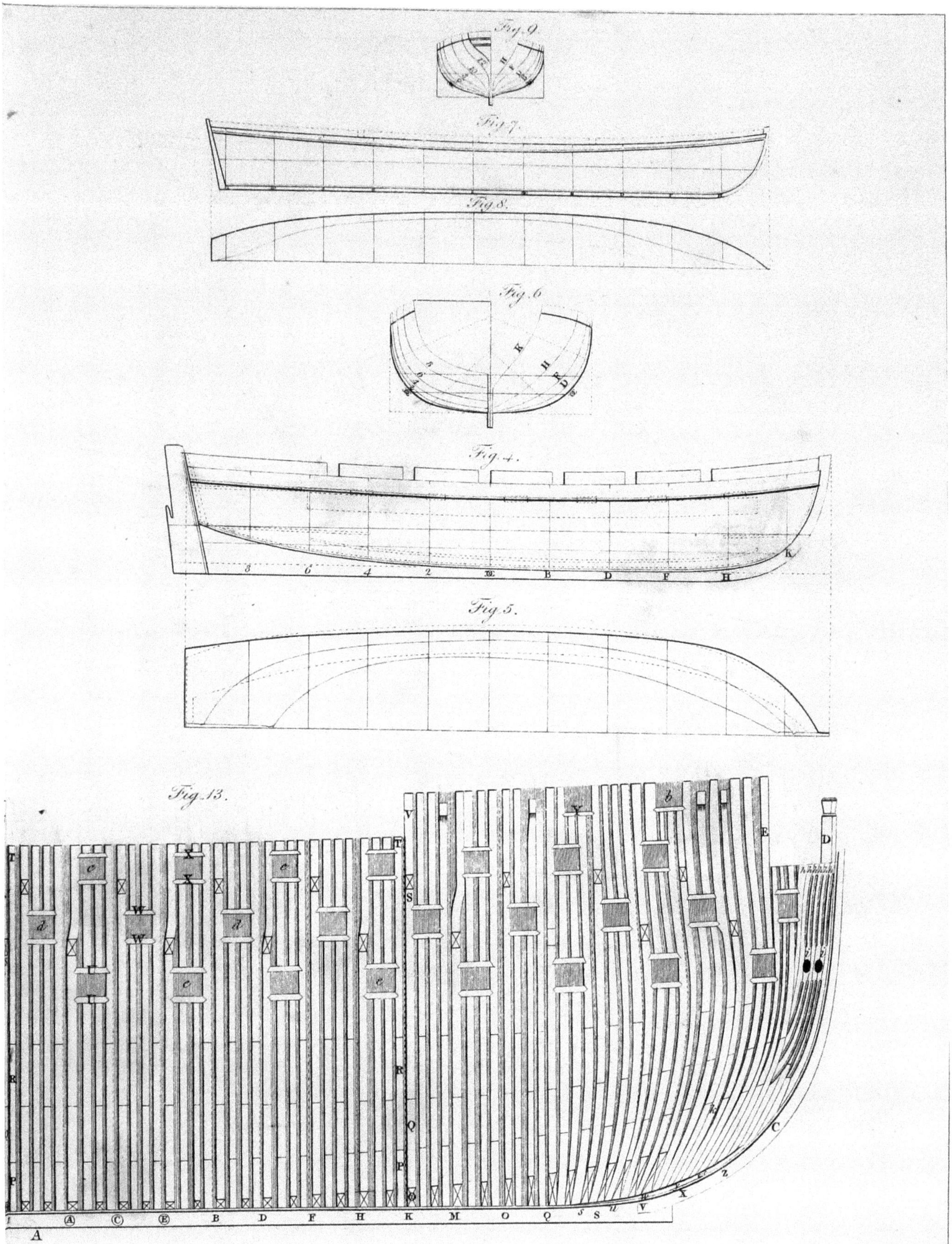

Mutlow Sc. Rufsell Co.t

and W. Davies, Strand, London.

A FIRST RATE SHIP AT ANCHOR.

PLATE XXV.

I. Taplin del. — Mutlow sc. Russell Co.

Published March 1. 1815, by T. Cadell and W. Davies Strand, London.

SOLAR SYSTEM & SEASONS.

PLATE XXVI.

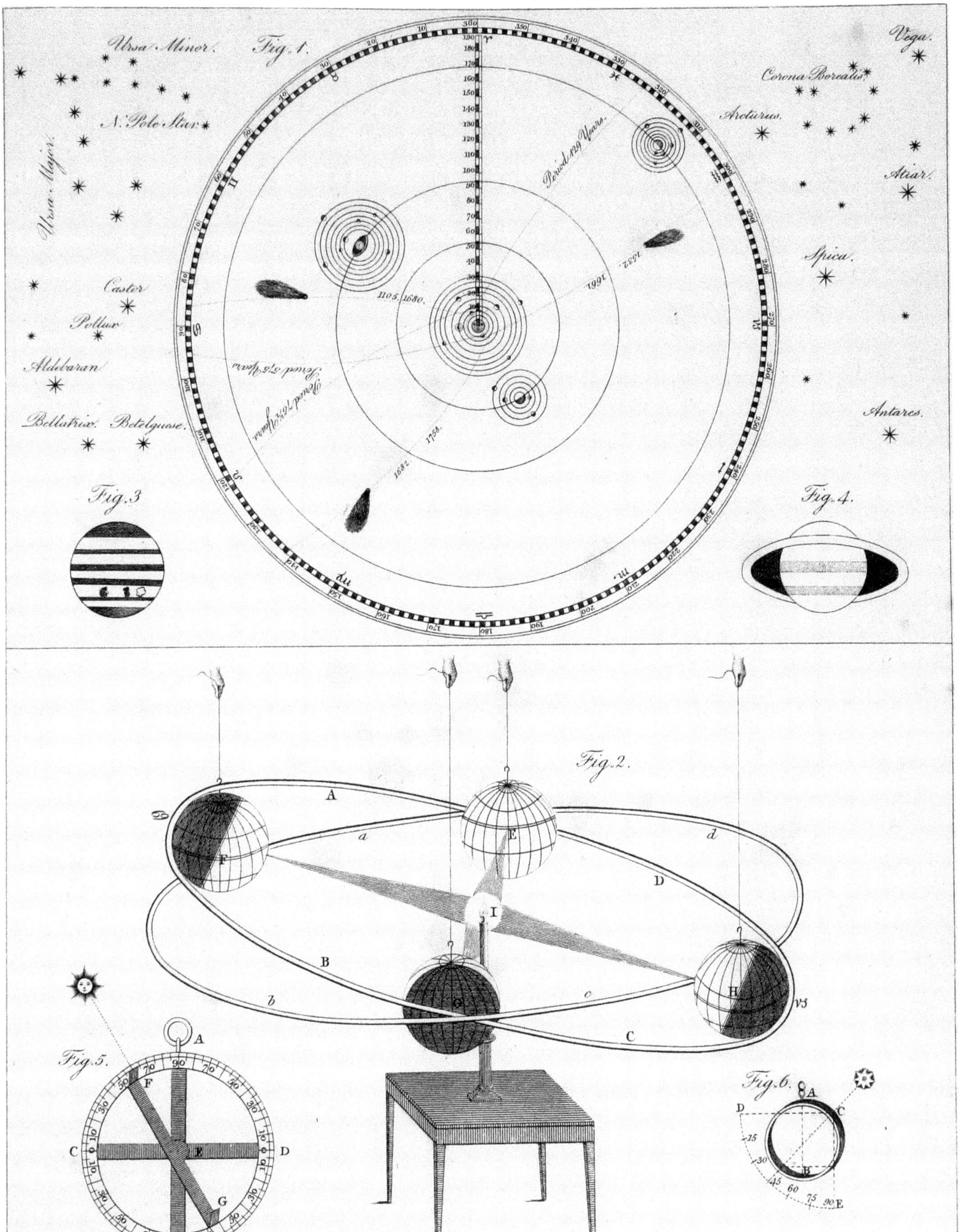

Mutlow Sc. Russell Co.^t

Published March 1. 1815, by T. Cadell and W. Davies, Strand London.

Signals.

PLAN OF TABULAR AND NUMERARY SIGNALS.

		A	B	C	D	E	F	G	H	I	J	K	L
	Superior Flags moveable.	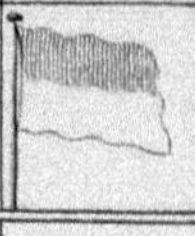		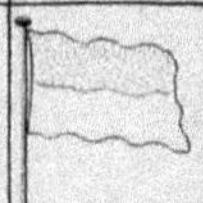	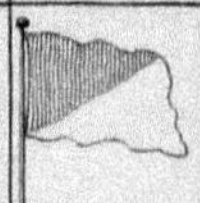	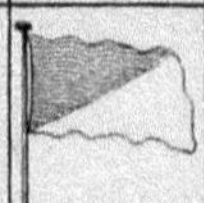			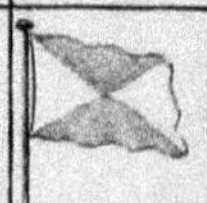		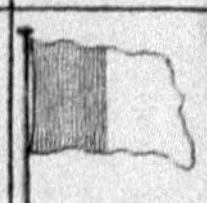		
	Index Numbers.	1	2	3	4	5	6	7	8	9	10	11	12
	General Heads.	Anchor-ing	Calling Officers.	Chasing	Convoys and Prizes.	Fighting Evolutions.			Sailing Evolutions.		Private Ship's Sig. to the Admiral.		Miscel-laneous.
M	 1	13	Y 25	37	49	61	73	85	97	109	121	133	145
N	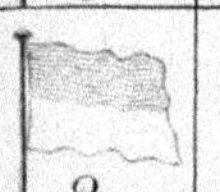 2	14	Z 26	38	50	62	74	86	98	110	122	134	146
O	 3	15	27	39	51	63	75	87	99	111	123	135	147
P	 4	16	28	40	52	64	76	88	100	112	124	136	148
Q	 5	17	29	41	53	65	77	89	101	113	125	137	149
R	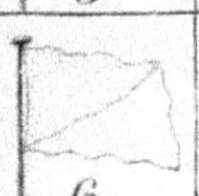 6	18	30	42	54	66	78	90	102	114	126	138	150
S	 7	19	31	43	55	67	79	91	103	115	127	139	151
T	 8	20	32	44	56	68	80	92	104	116	128	140	152
V	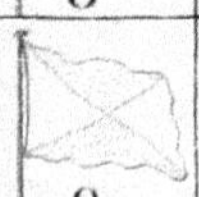 9	21	33	45	57	69	81	93	105	117	129	141	153
U	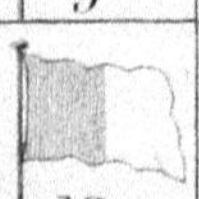 10	22	34	46	58	70	82	94	106	118	130	142	154
W	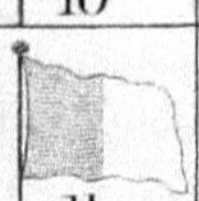 11	23	35	47	59	71	83	95	107	119	131	143	155
X	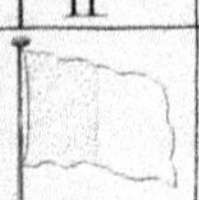 12	24	36	48	60	72	84	96	108	120	132	144	156

Substitute Flag. 

N.B. The Colours of the Flags are indicated thus, for RED, *vertical lines, for* BLUE, *horizontal, and for* YELLOW, *small dots.* —

Mutlow Sc. Russell Co.^t

Published March 1.1815, by T. Cadell and W. Davies, Strand London.

Fig. 1. POSITION 1st.

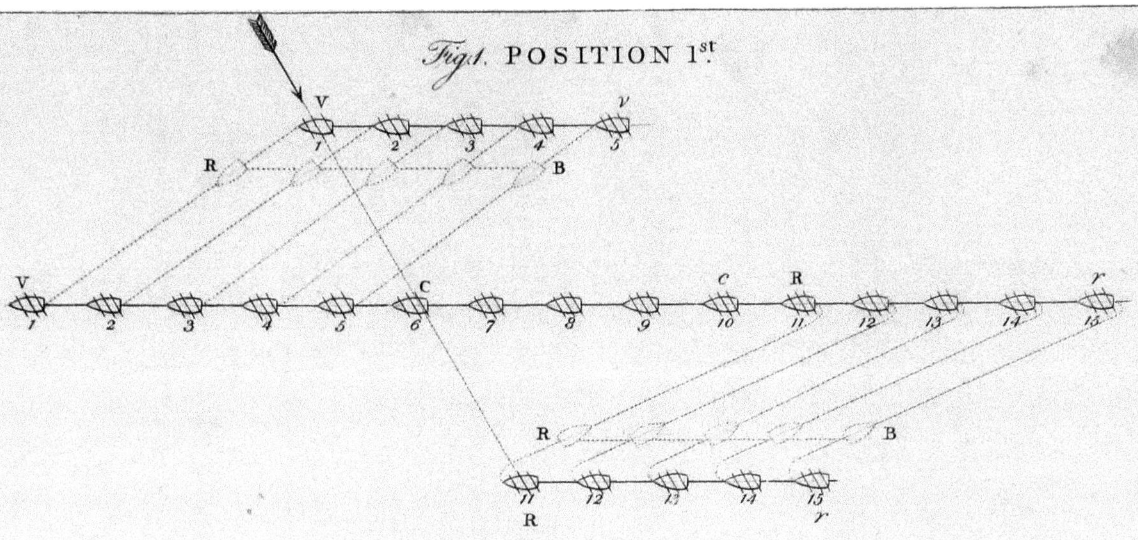

Fig. 2. POSITION 2nd.

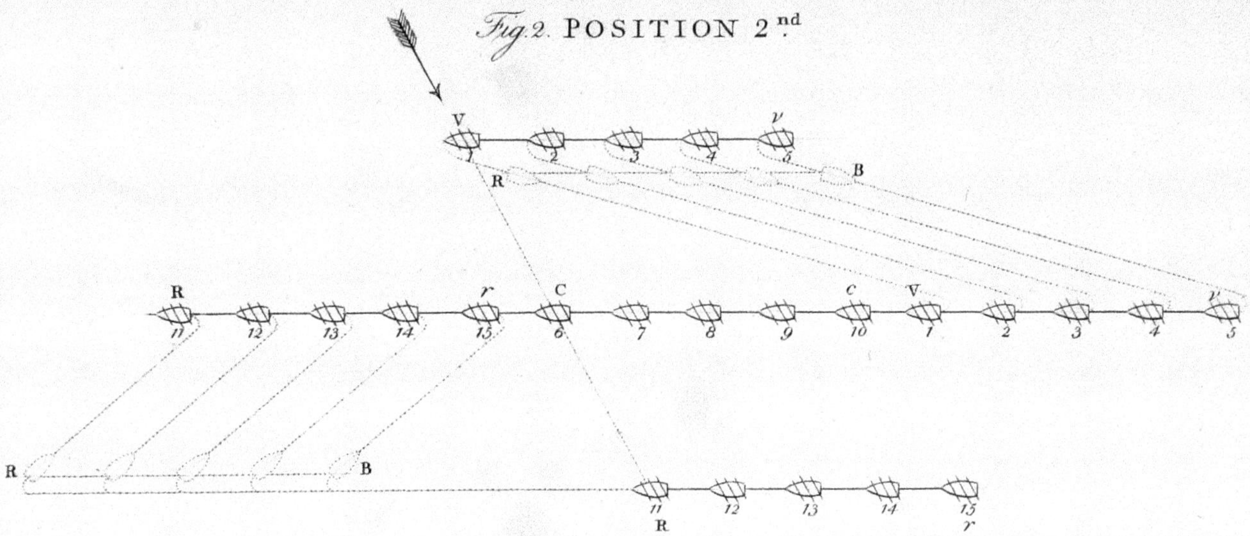

Fig. 3. POSITION 4th.

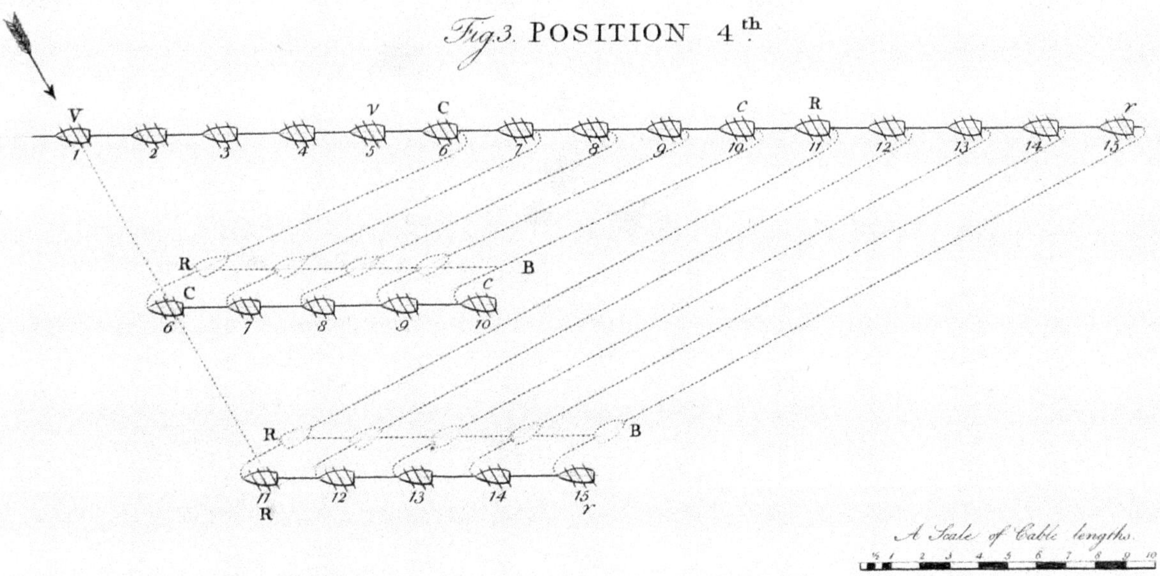

Mutlow Sc. Russell Co.

Published March 1. 1815, by T. Cadell and W. Davies, Strand, London.

Fig 4. POSITION 5th

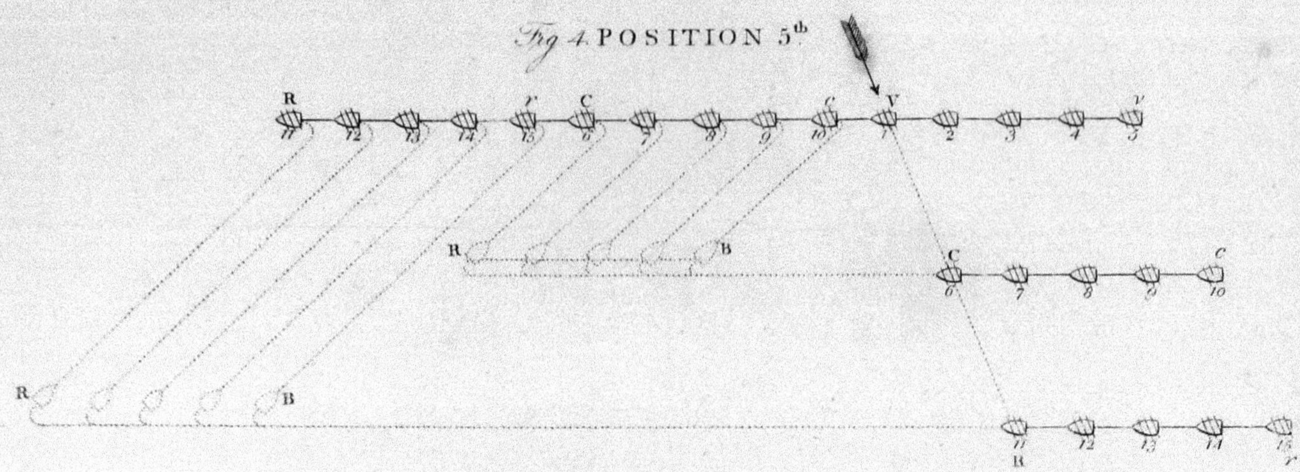

Fig. 5. POSITION 7th

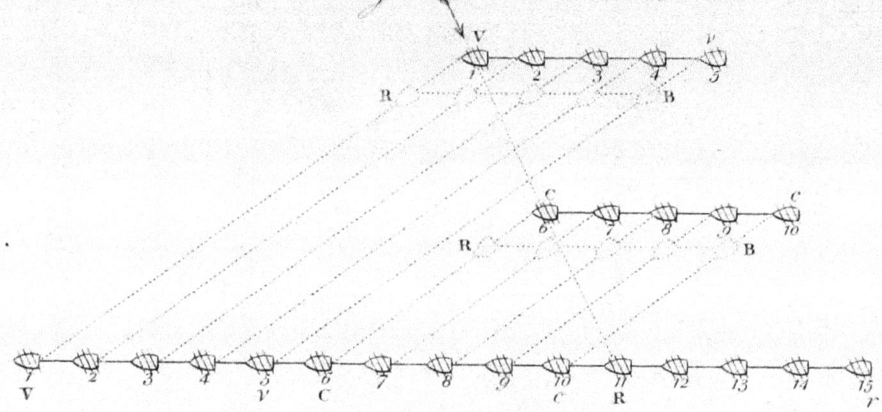

Fig 6. POSITION 8th

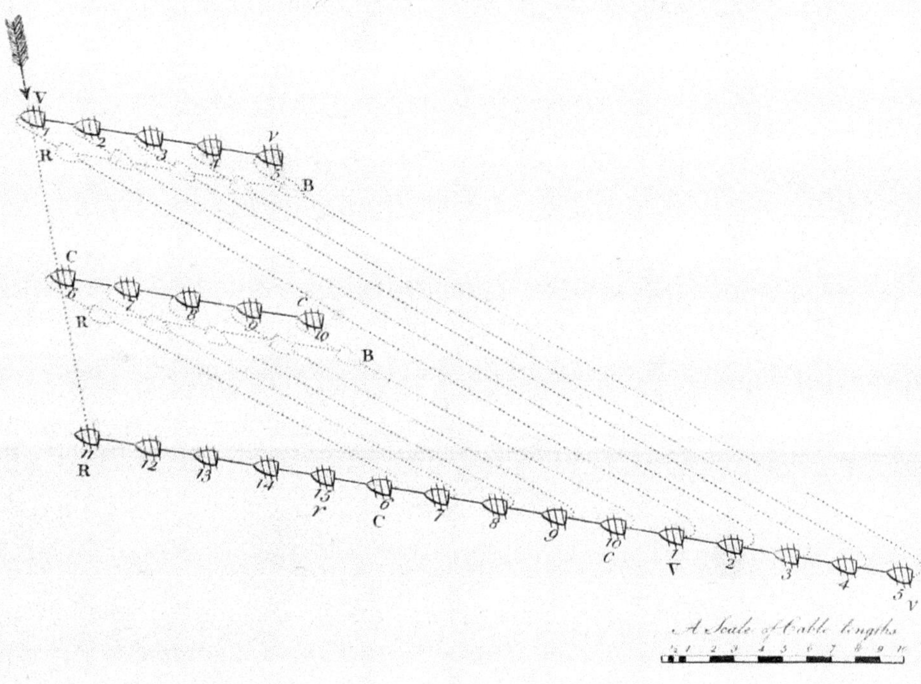

Mutlow Sc. Russell Co.

Published March 1. 1815, by T. Cadell and W. Davies, Strand, London..

STERN VIEWS OF SHIPS.

PLATE XXX.

Fig. 1.

Fig. 2.

J. Taplin, delt.

Mutlow Sc. Russell Cot.

Published March 1. 1815, by T. Cadell and W. Davies, Strand, London.

A 74 GUN-SHIP, &c.

I. Taplin, delin.

Mutlow Sc. Russell Co.

Published March 1. 1815, by T. Cadell and W. Davies, Strand, London.

J. Taplin delin. Mutlow Sc. Russell Co.

Published March 1. 1815, by T. Cadell and W. Davies, Strand, London.

VESSELS UNDER SAIL. PLATE XXXIII

J. Taplin. delin. Mutlow Sc. Russell Cot.

Published March 1. 1815, by T. Cadell and W. Davies, Strand, London.

THEORY OF WORKING SHIPS.

PLATE XXXIV.

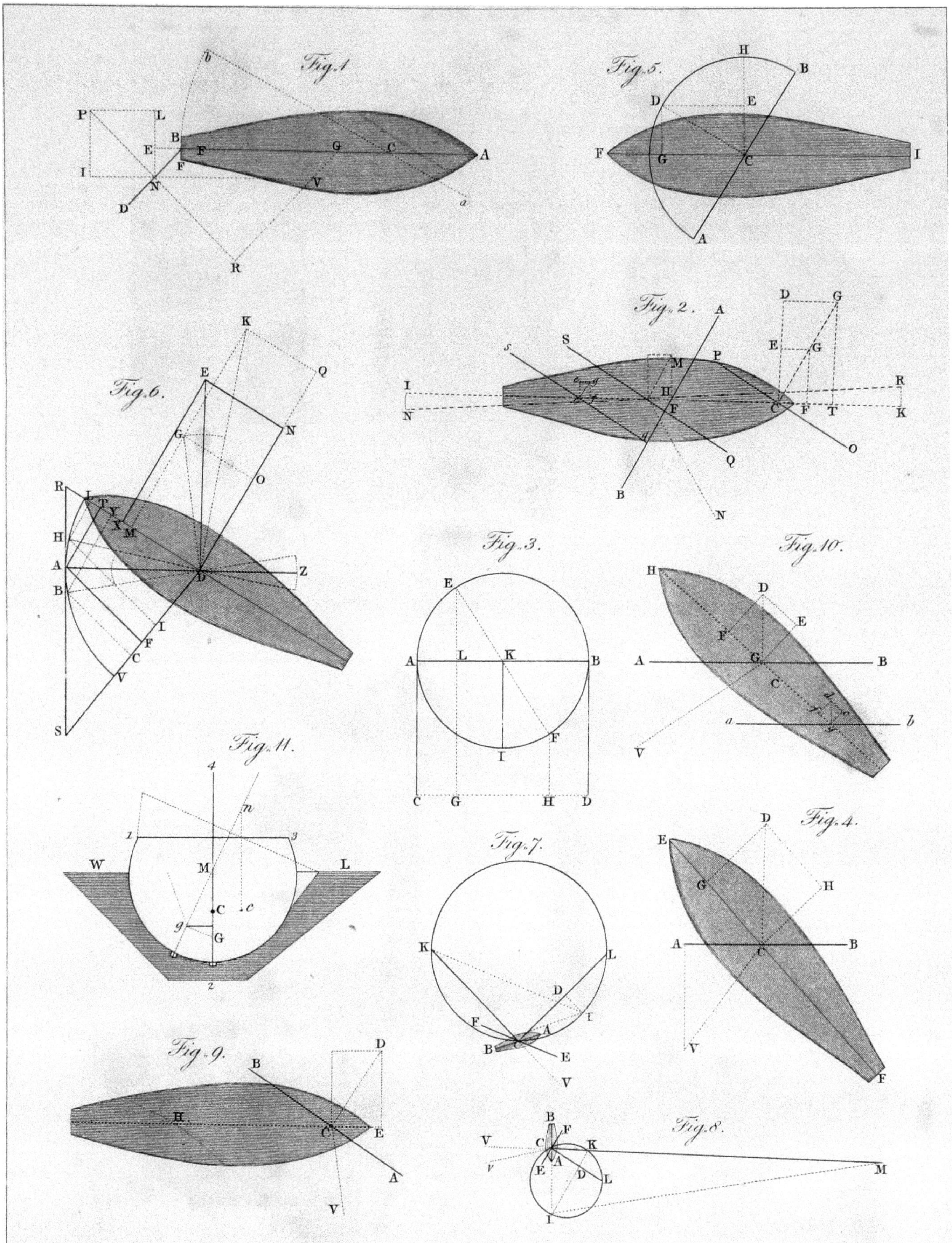

Mutlow Sc. Russell Co.t

Published March 1. 1815, by T. Cadell and W. Davies, Strand, London.

BINNACLE COMPASS, &c.

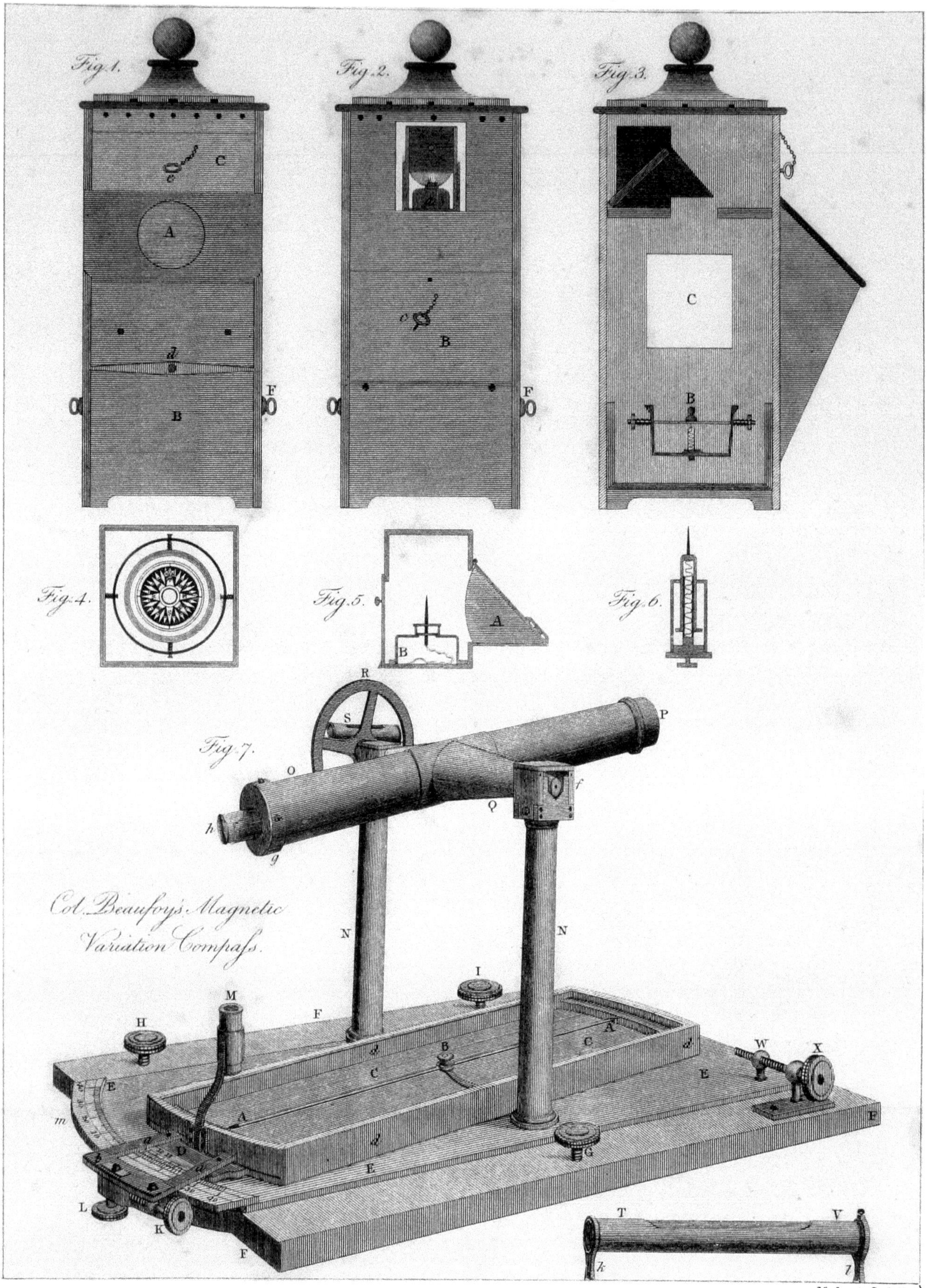

Mutlow Sc. Russell Co.

Published March 1. 1815, by T. Cadell and W. Davies, Strand, London.

For EU product safety concerns, contact us at Calle de José Abascal, 56–1°,
28003 Madrid, Spain or eugpsr@cambridge.org.

www.ingramcontent.com/pod-product-compliance
Ingram Content Group UK Ltd.
Pitfield, Milton Keynes, MK11 3LW, UK
UKHW052301280926
15471UKWH00006B/1038